AA004528

# XXVI R-S-P Seminar on Theoretical Foundation of Civil Engineering (RSP 2017)

MATEC Web of Conferences Volume 117 (2017)

Warsaw, Poland
21 - 25 August 2017

Part 1 of 2

**Editors:**

**S. Jemiolo**
**A. Zbiciak**
**M. Mitew-Czajewska**
**M. Krzeminski**
**M. Gajewski**

ISBN: 978-1-5108-4515-2

**Printed from e-media with permission by:**

Curran Associates, Inc.
57 Morehouse Lane
Red Hook, NY 12571

CURRAN ASSOCIATES INC. proceedings.com

**Some format issues inherent in the e-media version may also appear in this print version.**



Printed by Curran Associates, Inc. (2017)

For additional information, please contact EDP Sciences – Web of Conferences at the address below.

EDP Sciences – Web of Conferences
17, Avenue du Hoggar
Parc d'Activité de Courtabœuf
BP 112
F-91944 Les Ulis Cedex A
France

Phone: +33 (0) 1 69 18 75 75
Fax: +33 (0) 1 69 28 84 91

contact-edps@webofconferences.org

**Additional copies of this publication are available from:**

Curran Associates, Inc.
57 Morehouse Lane
Red Hook, NY 12571 USA
Phone: 845-758-0400
Fax: 845-758-2633
Email: curran@proceedings.com
Web: www.proceedings.com

# XXVI R-S-P Seminar on Theoretical Foundation of Civil Engineering (RSP 2017)

MATEC Web of Conferences Volume 117 (2017)

Warsaw, Poland
21 - 25 August 2017

Part 2 of 2

**Editors:**

**S. Jemiolo**
**A. Zbiciak**
**M. Mitew-Czajewska**
**M. Krzeminski**
**M. Gajewski**

ISBN: 978-1-5108-4515-2

**Printed from e-media with permission by:**

Curran Associates, Inc.
57 Morehouse Lane
Red Hook, NY 12571

CURRAN ASSOCIATES INC.
proceedings.com

**Some format issues inherent in the e-media version may also appear in this print version.**



Printed by Curran Associates, Inc. (2017)

For additional information, please contact EDP Sciences – Web of Conferences
at the address below.

EDP Sciences – Web of Conferences
17, Avenue du Hoggar
Parc d'Activité de Courtabœuf
BP 112
F-91944 Les Ulis Cedex A
France

Phone: +33 (0) 1 69 18 75 75
Fax: +33 (0) 1 69 28 84 91

contact-edps@webofconferences.org

**Additional copies of this publication are available from:**

Curran Associates, Inc.
57 Morehouse Lane
Red Hook, NY 12571 USA
Phone: 845-758-0400
Fax: 845-758-2633
Email: curran@proceedings.com
Web: www.proceedings.com

# TABLE OF CONTENTS

## PART 1

## PART 2

MATEC Web of Conferences 117, 00093 (2017) DOI: 10.1051/matecconf/201711700093
*XXVI R-S-P Seminar 2017, Theoretical Foundation of Civil Engineering*

# Proportions and static Analysis of a Historical Truss in a Rural House in Vápenná Village

*Peter* Krušinský[1,*], *Jozef* Gocál[1] , *Lucia* Augustínková[2] , *Eva* Capková[3] and *Renáta* Korenková[1]

[1]University of Žilina, Faculty of Civil Engineering, Žilina, Univerzitna 8215/1, Slovak Republic
[2]Technical University of Ostrava, Faculty of Civil Engineering, Ostrava, Ludvíka Podéště 1875/17, Czech Republic
[3]University of Žilina, Faculty of Humanities, Žilina, Univerzitna 8215/1, Slovak Republic

**Abstract.** We have analysed historical trusses based on previous building-historical researches, particularly focusing on sacral buildings, in chosen regions of Slovakia, with one of the primary goals to examine geometric concepts and proportional relationships used for their construction. The knowledge of proportional principles and relationships used in various historical sacral trusses, additionally supported by contemporary literature, was applied to a village house truss from 1774 in Vápenná, Jeseníky district of the Czech Republic. Spatial numerical model of the roof structure was created in order to perform a static analysis of the roof structure in accordance with present standards. With regard to the structural analysis results, it can be said that the original design of the structure based on geometrical principles satisfy the reliability conditions defined by current European standards for structural design.

## 1 Introduction

The analysis of proportional relationships in a rural house is based on historical-structural, archival and dendrochronological research of the object as a whole. The objective of this research is not only the analysis of truss's proportions, but also its relationship to the ground plan, whose concept is based on proportional relationships as well. The principle suggested by the contemporary literature as one of the possible means of designing the truss height has been clearly identified in case of historical trusses of sacral buildings from Slovakia, which we have previously analysed [3, 5], as well as in case of the researched secular house from Vápenná in the Czech Republic.

## 2 Building-historical development

House no. 98 in Vápenná, Jeseník region, Czech Republic (Fig. 1) was built in the 2nd half of the 18th century as a reflection of establishment's aims to raise the numbers of tax-

* Corresponding author: krusop@wp.eu

paying subjects [3]. The first written note originates in 1786, when Michal Hauke sold the house to his son, Michal [1]. Dendrochronological analysis of the ceiling - after 1775 is also an evidence for the house being built in the 2nd half of the 18th century 70s. The building is a typologically common rural house with a central room and a hood kitchen, but has been richly designed to contain decorations in form of painted red frames, which have been preserved in the western attic area. Despite the ceiling parts being largely preserved above the ground floor, the truss was newly built (1974/75 according to dendrochronological dating) with an inclusion near the southern gable.

**Fig 1.** The researched object – a rural house no. 98 in Vápenná

## 3 Truss construction

The building has a gable roof finishing with masonry gables. The main south facade has an ornate Baroque gable with windows and vents. Roof truss structure consists of two separate truss structures with a transition portion, which form a space with the same ridge height. The truss structures can be divided as follows: over the hall structure, south side of the building and the structure above the rooms in the northern part.

**Fig. 2.** The inner view of the truss

The truss over the southern part has a rafter – collar tie structure and consists of two collar tie levels and base struts located at the base of the rafters. Lateral stools, whose posts are placed on the threshold beams, were later incorporated into the design. Central binding rafters bearing the collar beams are tenoned to the posts. The posts are also stabilized using single-sided outer bracing structures clad together into binding beams. The posts in the marginal joints are longitudinally wind-braced by struts (three views of the truss are on the Fig. 2).
The truss over the northern part is also of a rafter - collar beam structure with two collar beam levels, positioned at the same height as in the southern part. Unlike the southern roof its central posts are located in the marginal joints, which are placed on the threshold beam. The posts are stabilized by base struts. The truss itself is longitudinally secured by inclined bracing structures passing through the anchor post into the columns and the anchor posts below the lower collar beams.

## 4 Geometric concepts and proportional relationships within the truss

After previous examinations of historical trusses of primarily sacral buildings it was found that one of the most commonly used principles for determining the essentials truss dimensions is a principle we dubbed "n plus one" [4]. It is a principle determining the height of a truss based on its width as follows: the width is divided into 2n units and the height is then (n + 1) units. The same principle was applied in case of the rural truss in Vápenná, with n = 5 and the ratio between the width and height of the truss being 10 : 6 (= 5 : 3). A basic square ABCD dimensions are 5x5 units was obtained (Fig. 3a, b). The unit used for determining the truss height based on its width is also five times the measure of length corresponding to the span (contemporary measure), in this case 19.3 cm [8]. The used measure (span) was found on the top side of the ceiling beam in the form of an engraved circle (Fig. 4b). (The cut of the beam with the engraved circle was included in the depository construction details Technical University of Ostrava under inventory number 176. A, [2].)

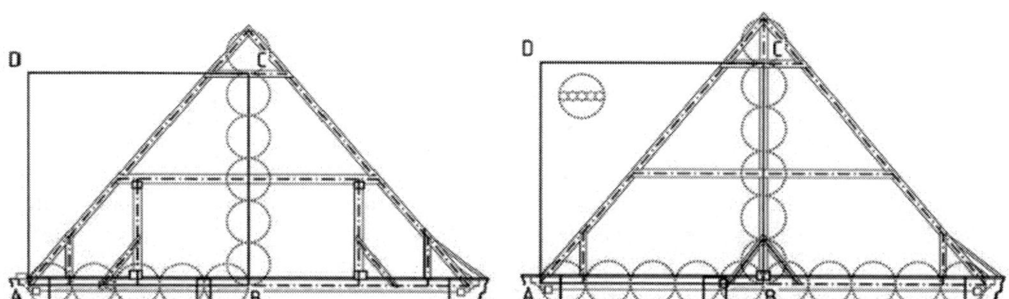


**Fig.3.** (a) Section of the south truss; (b) Section of the north truss.

For the design of the remaining important elements of buildings and trusses were formerly used a simple integer ratio, in the literature often named as "music" ratios [6, 7]. In the researched truss were used these ratios: the main collar tie is located at half of the height of the base square; upper collar tie is located at the height of the basic square; struts and rafters are placed on the tie beam in one tenth of the total width of the truss.The floor plan of the rural house (Fig. 4a) in relation to the truss is designed in such a way that the relationship between its length and width is 3:2, this relationship expressed by a span is 75:50. The original big front room on the South side of the house has internal dimensions in the same proportion as the entire floor plan of the house, hence the 3:2 (45:30).

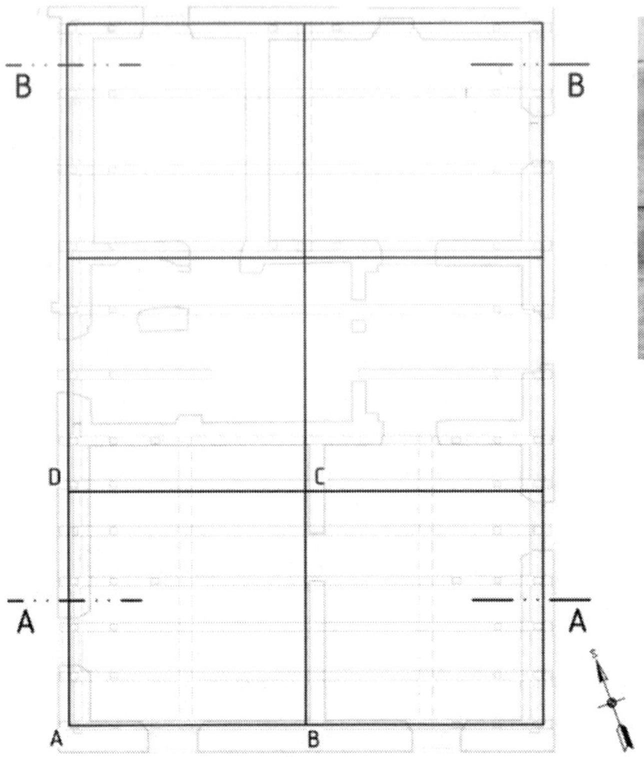


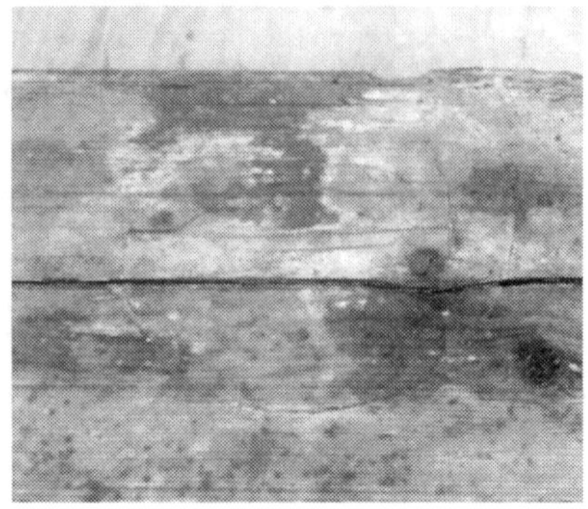

**Fig. 4.** (a) Floor plan; (b) Engraved circle

# 5 Static analysis of the roof structure

## 5.1 Description of the numerical model

Numerical model of the roof structure was developed in the finite element analysis (FEA) software SCIA Engineer [10]. The roof structure is modelled as three dimensional structure with beam elements. All member connections are modelled as hinge joints. Considering the stresses in the roof members and structural parameters being analysed, the isotropic material model with mechanical properties of C24 according to [11] is adopted, which corresponds to the applied type of wood. The geometry of numerical model is related to the roof structure's geometrical analysis presented hereinbefore. Visualization of the numerical models are presented in Fig. 5.

All the member connections are modelled as hinge joints with axial rigid connection and with capability of initial slip of 1 mm in the axial direction of member in order to consider an influence of gaps, cracks and geometry imperfections, occurring in historical carpentry joints. The tie beams of the secondary trusses are connected to the bottom chords of the longitudinal truss by joints, which are able to transfer only compression forces. The effect of local weakening of the cross sections by carpentry joints is neglected in the models.

The roof superstructures were loaded according to European standards [12, 13, 14] by permanent load (self-weight and weight of roofing) and variable loads (wind actions and snow loads). The self-weight load was generated by the FEA software. The combinations of load cases were generated according to the standard STN EN 1990 [12].

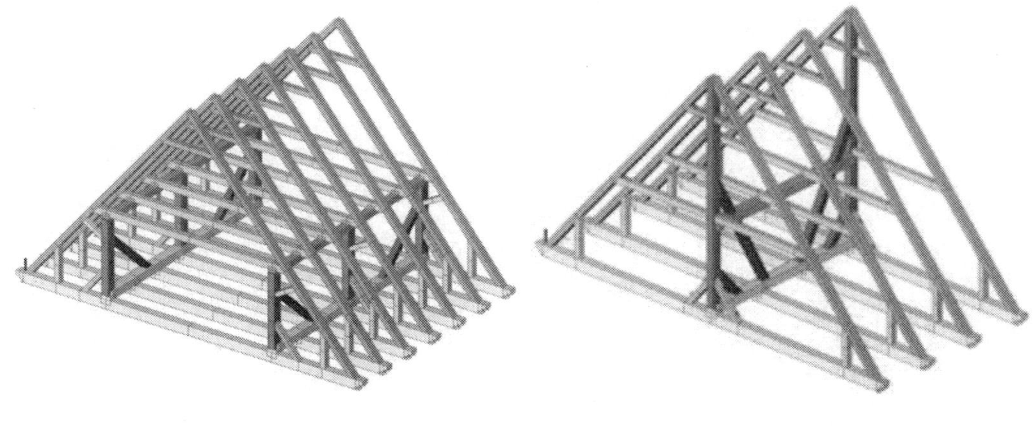

**Fig. 5.** Visualization of numerical models of the south truss (a) and the north truss (b).

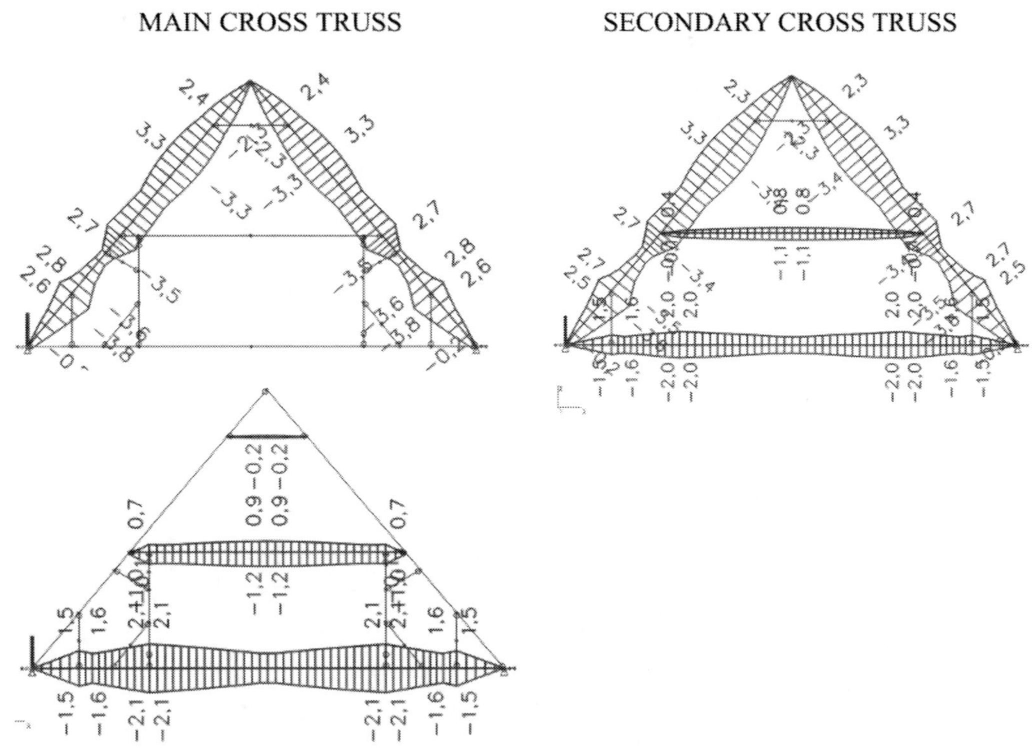


**Fig. 6.** Normal stresses (MPa) in the major cross truss members of the south truss.

### 5.2 Results of the numerical analysis

The results of numerical analysis are presented by the values of maximum tensile and compression stresses and displacements in the major structural members of the trusses, calculated for the decisive load combinations. The envelopes of maximum and minimum normal stresses in the major cross truss members are presented in Fig. 6 and Fig. 7. The presented values do not include possible buckling effects, which are negligible because of low share of compression forces (caused mainly by permanent loads). The envelopes of deformations caused by characteristic load combinations are presented in Fig. 8 and Fig. 9.
Based on the results of numerical analysis it can be stated out that the normal stresses in all truss members do not exceed the design values of bending and compression strength equal to 16.60 MPa and 14.53 MPa, respectively, determined according to STN EN 1995-1-1 [15], assuming the timber strength class C24, modification factor kmod = 0.9 and partial safety factor $\gamma M = 1.3$.
Normal stresses in the longitudinal truss members induced by the considered loads are negligible. The longitudinal truss represents mainly a longitudinal brace system ensuring the global stability of the roof structure. In addition, it ensures the load redistribution between the individual cross trusses in the case of their depression as a consequence of damaged joints or members during the life cycle of the roof structure.

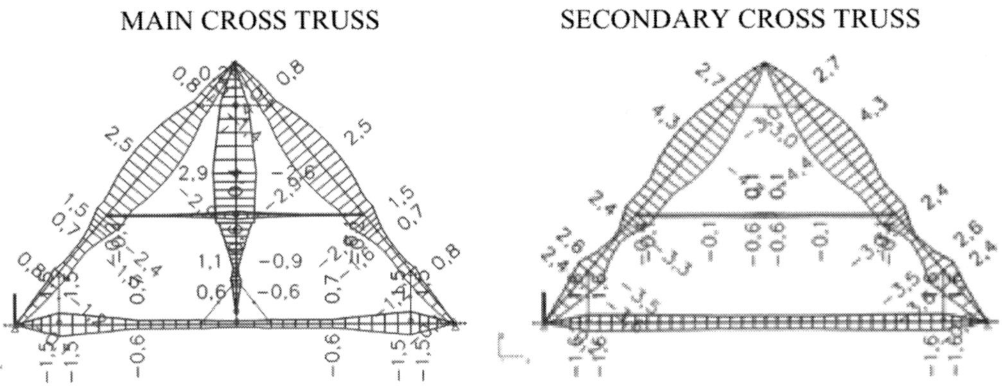


**Fig. 7.** Normal stresses (MPa) in the major cross truss members of the north truss.

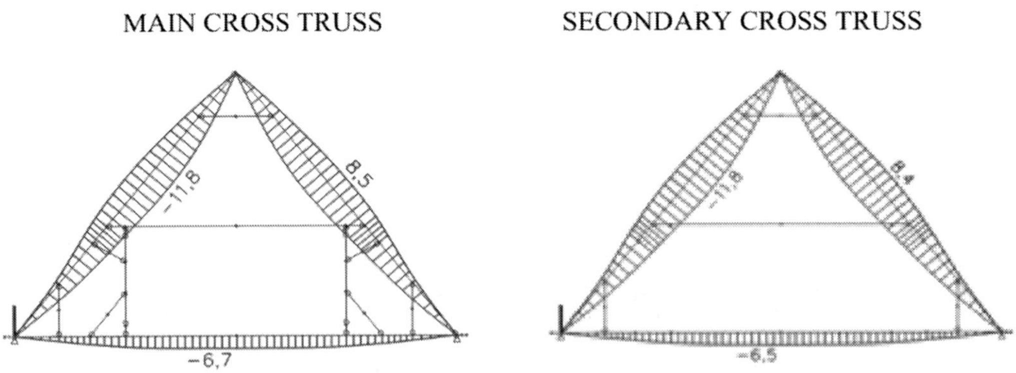


**Fig. 8.** Vertical deformations (mm) in the major cross truss members of the south truss.

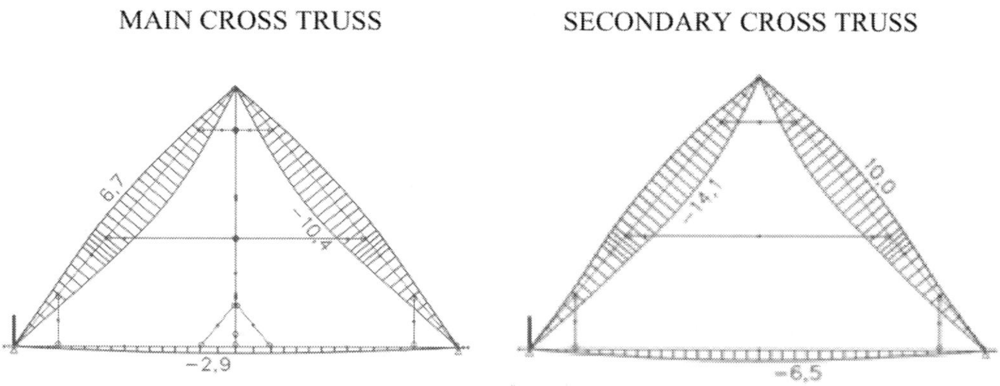


**Fig. 9.** Vertical deformations (mm) in the major cross truss members of the north truss.

## 6 Conclusion

There has been identified the geometrical and proportional concept of truss design on the researched object, in relation to the concept of the object floor plan, which is based on the use of contemporary system of measure. Geometric and proportional analysis of the original concept design of the truss and the object itself may be helpful in understanding the historical development of the object, used historical measures, proportions and structures. Knowledge of ratios and geometric concepts can highlight the impacts of cultural and socio-economic links with architecture in the region. Reached knowledge about proportional concept can be a guide for the methodology of recovery in relation to heritage preservation of historical objects.

To conclude, we can say that the observed Baroc truss is geometrically and proportionately very consistently designed with an emphasis on high level of reliability. The roof superstructure consists of quite massive solid elements, which results in quite high reliability of the roof structure. It can be assumed that with regard to great importance of the Rural House in Vápenná Village, there were high demands on the reliability and durability of the roof structure above the presbytery. Of course, it had to comply with adequate financial resources for realization of the building.

## References

1. L. Augustinková Venkovský dům čp. 98 ve Vápenné (k. ú. Vápenná, parc. č. 95). Stavebně historický průzkum. Ostrava 2013. Nepublikovaný materiál uložený na NPÚ, ú.o.p. v Olomouci.
2. L. Augustinková Předměty z venkovského domu čp. 98 ve Vápenné. In: Augustinková, L. – Krušinský, P. (edd.) Předměty z depozitářů stavebních detailů v Ostravě a v Žilině. Ostrava (2015)
3. P. Krušinský, E. Capková, J. Gogál, Advanced Materials Research **1122** (2015)
4. P. Krušinský, E. Capková, J. Gogál, M. Holešová, Civil and environmental engineering **11/2** (2015)
5. P. Krušinský, E. Capková, J. Gogál, M. Kekeliak, Advanced materials research **969** (2014)
6. A. Struhár, Geometric harmony of historical architecture in Slovakia. Bratislava. Pallas (1977)

7. R. Wittkover, Architectural Principles in the Age of Humanism. W. W. Norton & Company. (1971)
8. J. Heurich, Przewodnik dla cieśli obejmujący cały zakres ciesielstwa z 299 drzeworytami w tekście. Warszawa 1871. (Reprint 2010)
9. M. Holešová, Selected geometrical constructions, Acta mathematica 17 : conference proceedings of international conference 12th mathematical conference, Slovakia, Nitra, (2014)
10. SciaEngineer 2010. Software for structural analysis. http://nemetschek-scia.com/sk
11. STN EN 338 Structural timber. Strength classes. SÚTN, Bratislava (2004)
12. STN EN 1990 Eurocode. Basis of structural design. SÚTN, Bratislava (2009)
13. STN EN 1991-1-1 Eurocode 1. Actions on structures. Part 1-1: General actions. Densities, self-weight, imposed loads for buildings. SÚTN, Bratislava (2007)
14. STN EN 1991-1-3 Eurocode 1. Actions on structures. Part 1-3: General actions. Snow loads. SÚTN, Bratislava (2007)
15. STN EN 1991-1-4 Eurocode 1. Actions on structures. Part 1-4: General actions. Wind actions. SÚTN, Bratislava (2007)
16. STN EN 1995-1-1 + A1 Eurocode 5. Design of timber structures. Part 1-1: General. Common rules and rules for buildings. SÚTN, Bratislava (2008)

# The study of the calibration dependences used when testing the concrete strength by nondestructive methods.

*Azariy* Lapidus[1], *Tembot* Bidov[1], and *Alan* Khubaev[1]

[1] Moscow state university of civil engineering, Yaroslavskoye shosse, 26, Moscow, Russia, 129337

**Abstract.** The article presents the results of the investigation of universal calibration dependences used in testing of the concrete strength by shear, rebound and ultrasound test methods. Many of the instruments currently used do not provide the required measurement accuracy, especially when *using* universal calibration dependences. The introduction of new more sophisticated instruments for nondestructive testing of concrete strength requires an assessment of their accuracy, taking into account the composition of concrete and other parameters. The shear test method, called "direct", does not provide the required accuracy of measurements, and the use of this method for the calibration of sclerometers and ultrasonic devices is unacceptable.
The aim of the research is to clarify the dependences, taking into account the data on concrete composition, parameters of materials used, and the age of concrete. The studies will allow you to set the values of the coefficients to be used in the generalized calibration curve, with considering the concrete information obtained at a particular construction site (where concrete strength is tested).

**Key words**: Nondestructive strength testing, shear test method, rebound method, ultrasound method; universal calibration dependence, calibration based on direct measurements, direct and indirect measurements.

## Introduction

The control of the concrete strength in monolithic structures is carried out almost always by nondestructive testing methods. The most widely used instruments are ultrasonic devices and sclerometers (the rebound method) with the use of so-called universal[1] calibration dependences or their redefined variants.
To date, the only method of nondestructive testing that can be used without preliminary calibration (direct measurement) is the shear test method. [3]. Using this method, the calibration curve is refined by introducing coefficients that consider the aggregate size and the type of concrete – heavy or light. The method has become quite widespread [7]. However, according to the data of laboratory studies [8], at high reinforcement intensity of monolithic reinforced concrete structures, the result of the concrete strength measured by

[1] Corresponding author: alan_khubaev@mail.ru

the shear test method, is not exact because the scheme of concrete surface destruction often occurs along the reinforcement cell. Furthermore, this method requires considerable labour, damages concrete, and therefore by this method an insignificant number of measurements can be performed with significant variations in the readings. Practice shows that the use of the shear test method for the calibration of ultrasonic devices and sclerometers does not provide the required accuracy. The use of new improved instruments requires evaluating their accuracy and plotting calibration dependences for specific conditions with use of control samples (cubes) tested by destruction on a press.

## 1 Main part

For plotting a calibration curve for the **SilverSchmidt** device, **type N** (rebound method) for specific conditions of monolithic construction in Moscow, as well as for testing the possibility of using the universal calibration dependence stipulated in GOST 17624-2012 for the ultrasonic method and the reliability of determining the concrete strength by the shear test method (direct measurements), in the laboratory of the concrete plant PKF Stroybeton, LLC, Mytishchi, special researches were carried out. 4 batches of samples were made from one mix of concrete, to perform measurements in 4 time terms (at the age of 3, 7, 28 and 56 days). Each batch comprised 16 cubes of 100 × 100 × 100, 16 cubes of 150 × 150 × 150 (Fig. 1) and 8 blocks of 600 × 600 × 70 mm in size (for shear test) (Fig.2). All samples were stored under normal conditions (t = 20°; W ≥ 95%).

**Fig. 1.** – Samples placed in a normal hardening chamber.

**Fig. 2.** – Concrete block 600×600×70 m

In each time period, immediately before the determination of the concrete strength by means of the destruction of cubes on a press, in each cube 100 x 100 x 100, and in each of the two blocks, four measurements were performed with the **SilverSchmidt type N instrument**; in each cube of 150 x 150 x 150 and in blocks, 4 ultrasonic velocity measurements were performed with the device УК 1401; in each of the blocks, 4 measurements were taken by the shear test method (**ПОС-50МГ4**).

The studies were consistently performed on control concrete samples made from heavy concrete mixtures, ready for use, of the B22.5 and B30 compression strength class. When preparing concrete mixtures, the following raw materials were used:

- Portland cement CEM I 42.5 N, according to GOST 31108-2003, produced by Maltsovsky Portland Cement, CJSC;

- sand with a fineness modulus of 2.4 – 2.6 with the dust and rot particles content up to 2%, according to GOST 8736-93, produced by Yermolinsky Quarry, CJSC;
- crushed stone fr. 5 – 20 mm, according to GOST 8267-93, produced by Kamenogorsk Quarry Division;
- chemical additives – superplasticizers.

Component ratios for B22.5 heavy concrete mixtures — 1: 2.24: 3.28 (cement: sand: crushed stone), W/C = 0.5. Component ratios for B30 heavy concrete mixtures — 1: 1.7: 2.4 (cement: sand: crushed stone), W/C = 0.39. The graphs of the concrete strength set for tests of cubes (15 x 15 x 15) on a press are shown in Fig. 3.

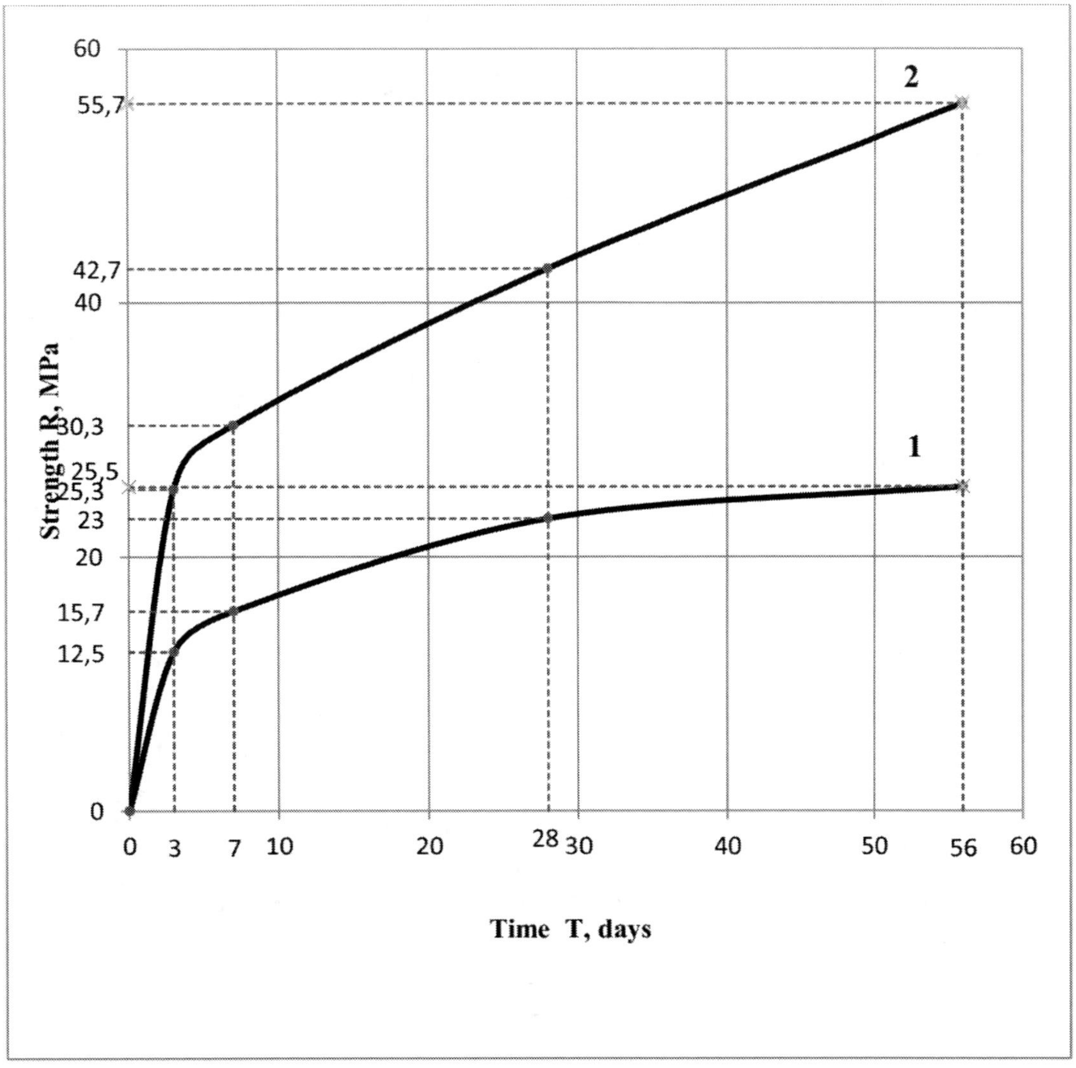


**Fig. 3.** – Graphs of the set of strength from the test of cubes
1 – Class of concrete B22.5;
2 – Class of concrete B30.

When determining the strength of concrete by the shear test method, the coefficient considering the crushed stone size $\boldsymbol{m_I} = \mathbf{1}$. (Crushed stone 5-10 mm). Figures 4 and 5 show the results of strength measurements by the shear test method in comparison with the actual values obtained by testing cubes on the press.

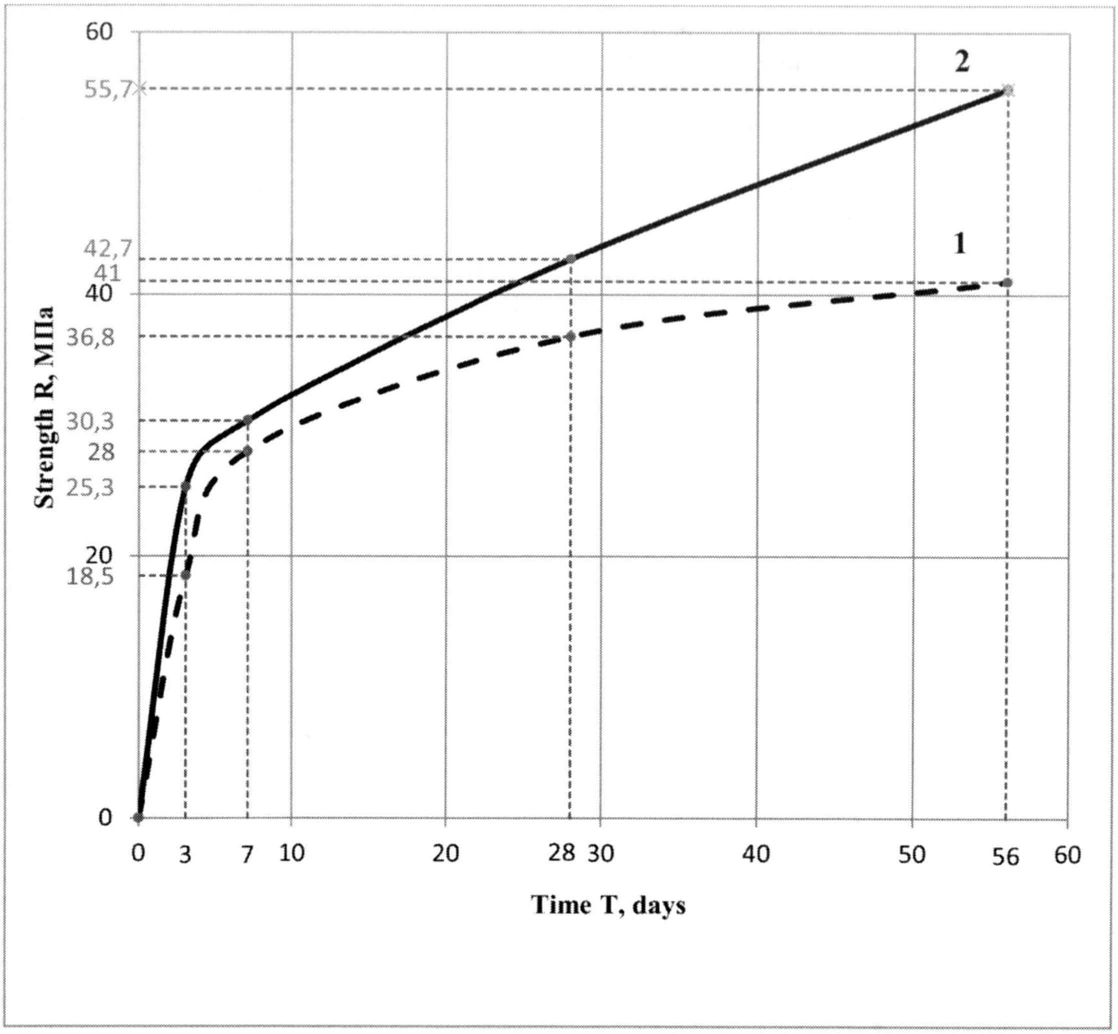


**Fig. 4.** – Determination of the strength of concrete B30 by the shear test method
1 – Curing schedule of measurements separation method;
2 – Curing schedule on the results of the test cubes.

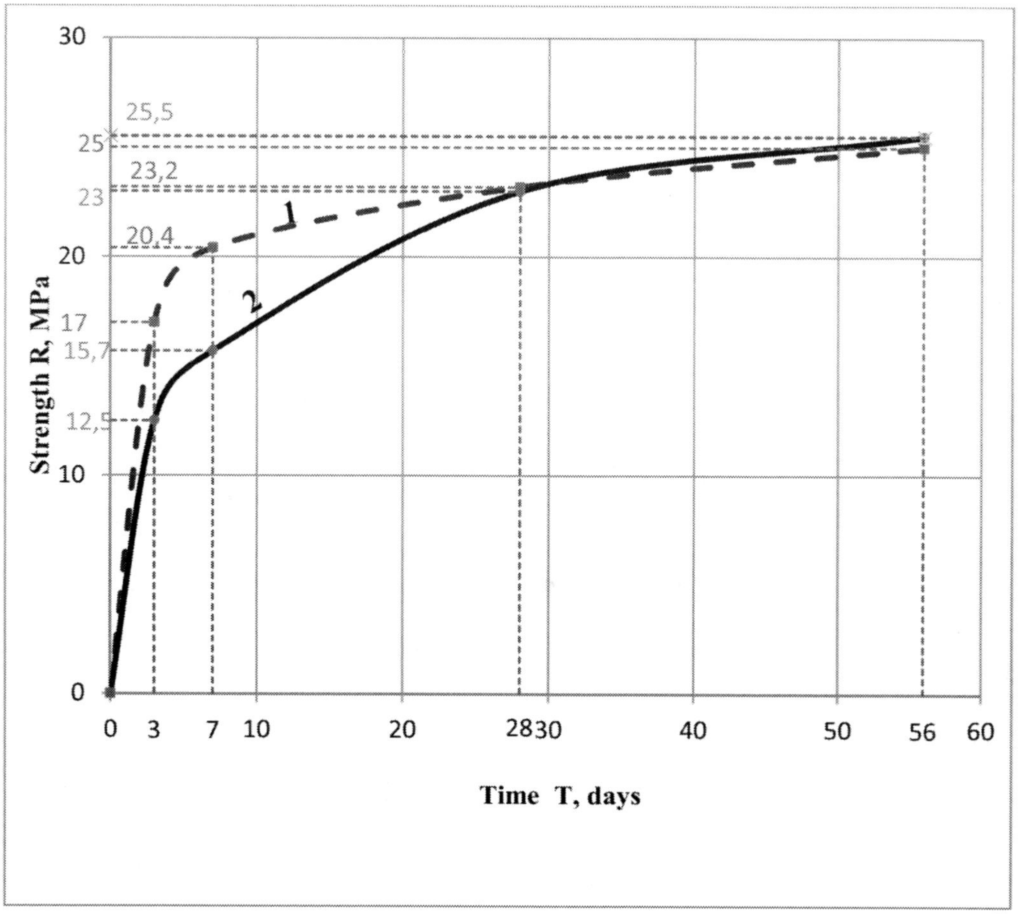


**Fig. 5.** – Determination of the strength of B22.5 concrete by the shear test method

1 – Curing schedule of measurements separation method;
2 – Curing schedule on the results of the test cubes.

It can be seen that when the strength is measured by the shear test method when the B30 concrete is hardened, the values obtained are underrated (the error is 14–27%), and when measuring the strength in B22.5 concrete, the values obtained are overrated by 30–36% when tested in early time terms (3 and 7 days), and a good match was found between the values at the age of 28 and 56 days.

Figure 6 shows the obtained calibration curves for the **УК 1401** device when determining the strength in B30 and B22.5 concretes.

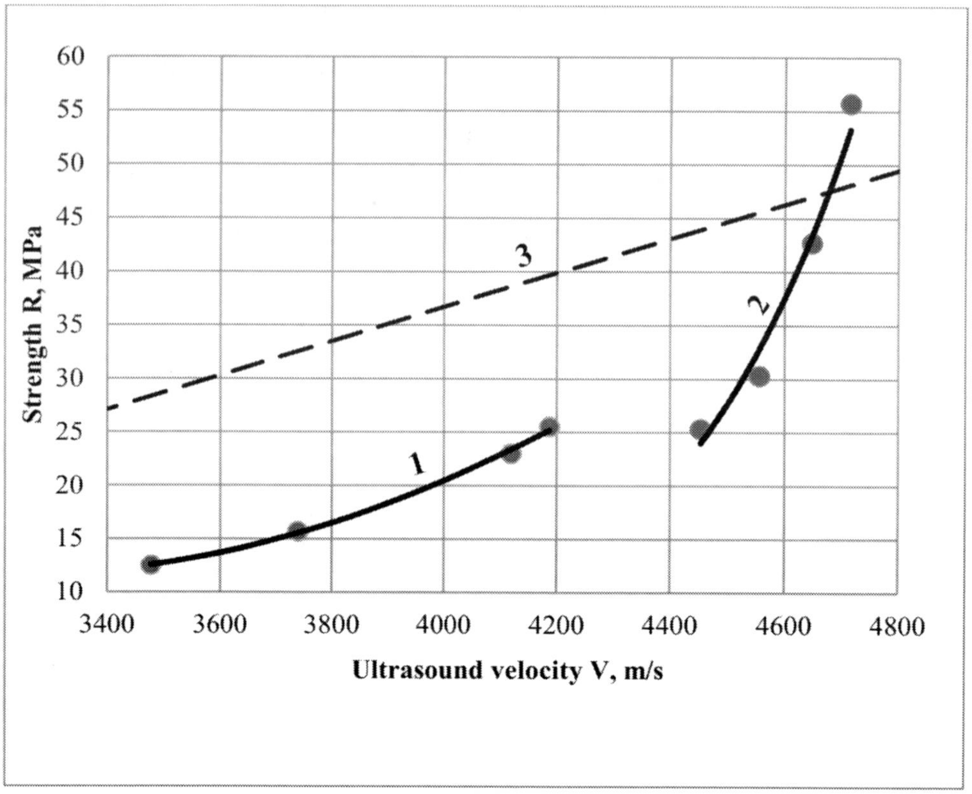


**Fig.6.** – Determination of strength by the ultrasonic method
1 – Graduating dependence "Ultrasound velocity – Strength" according to the results of tests of cubes of B22.5 class concrete;
2 – Grading dependence "Ultrasound velocity – Strength" according to the results of tests of cubes of B30 class concrete ;
3 – Universal calibration dependence according to GOST 17624-2012.

Comparing the obtained dependences with the universal calibration dependence in accordance with GOST 17624-2012, it is evident that the actual strength values differ from those specified in the universal one by about 2 times when tested at all-time terms for B22.5 concrete; and the actual calibration for B30 concrete is fundamentally different in form. Fig. 7 presents the obtained calibration dependences for the SilverSchmidt device and shows the universal dependence recommended by the device manufacturer.

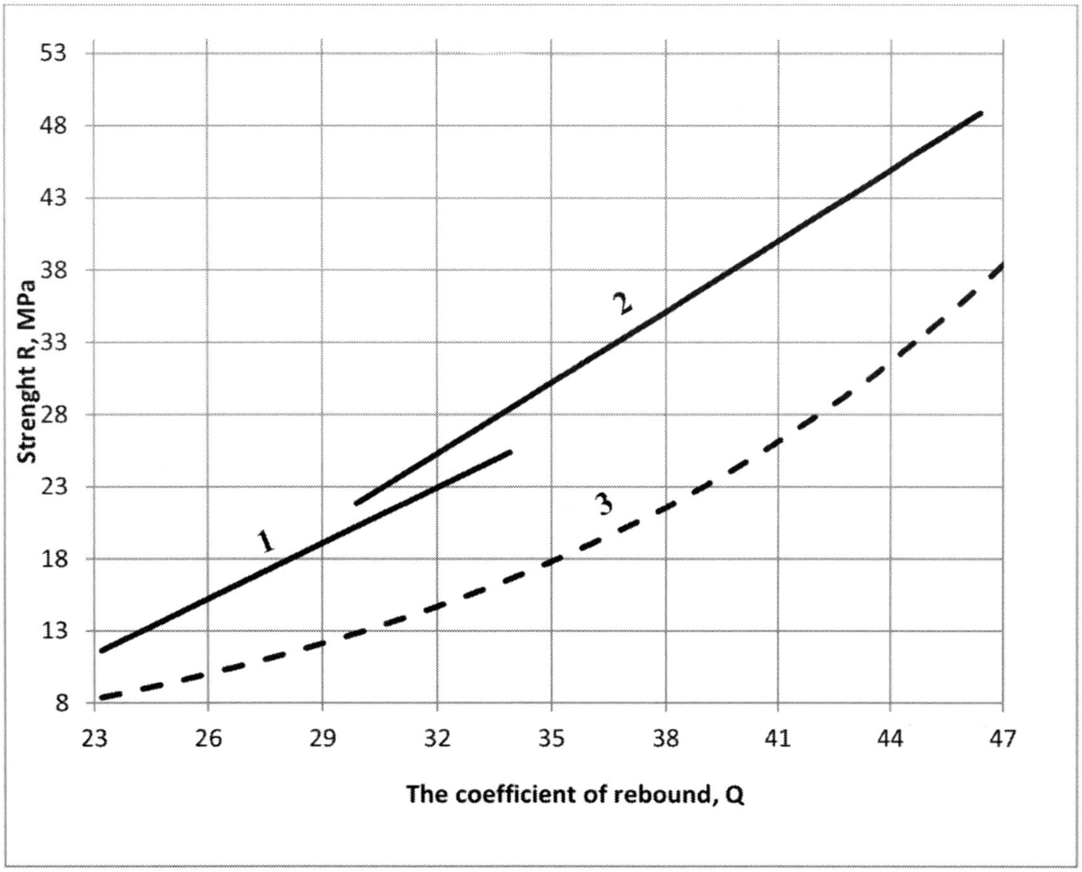


**Fig.7.** – Measurement method of strength rebound (SilverSchmidt)
1 – Gradual dependence on the results of tests for cubes for B22.5 concrete;
2 – Gradual dependence on the results of testing cubes for B30 concrete;
3 – Universal dependence for the device "SilverSchmidt type N" (from the manufacturer).

If we compare the obtained calibration dependence with the universal one, which is recommended by the manufacturer of this type of device (**SilverSchmidt, type N**), it is evident that with the universal dependence for the concrete compositions under study, the measurement errors range from 40% (when tested at early time terms of concrete hardening) and to 22% (at the age of 28 and 56 days).

## 2 Conclusion

1. As it was expected, the shear test method makes it possible to determine the concrete strength with the greatest accuracy without plotting real calibration curves. However, the significant errors obtained show that this method should not be considered as a direct one. When using it, universal calibration curves should be used, specified by the coefficients that take into account not only the size of crushed stone, but also other known data on concrete (component ratio, age of concrete, etc.).

2. Universal calibration dependence according to GOST 17624-2012 for ultrasonic method for determining concrete strength should be abandoned. It is known that the dependence "strength – the speed of ultrasound" has an exponential character $\mathbf{R} = \mathbf{k} \cdot \mathbf{e}^{nV}$. The coefficients k and n should be specified taking into account the data on concrete. It is

necessary to carry out additional studies to increase the reliability of the values of these coefficients. At present, the use of ultrasound provides the required accuracy in determining the strength only when plotting an individual calibration curve.

3. The application of the rebound method with the use of high-power devices (such as SilverSchmidt), in our opinion, is the most advanced for determining the concrete strength by the universal calibration dependence. Studies to clarify this dependence for the purpose of its use on construction sites in Moscow and the Moscow region and to determine statistical errors will be continued.

## References

1. S.S. Kovalev, A.A. Baranov, Yu.A. Artem'ev, Opredelenie prochnosti betona stroitel'nyh konstrukcij zdanij i sooruzhenij. (2007)
2. GOST 22690-88 Betony. Opredelenie prochnosti mekhanicheskimi metodami nerazrushayushchego kontrolya
3. GOST 17624-2012 Betony. Ul'trazvukovoj metod opredeleniya prochnosti
4. GOST 10180-2012 Betony. Metody opredeleniya prochnosti po kontrol'nym obrazcam.
5. GOST 18105-2010 Betony. Pravila kontrolya i ocenki prochnosti
6. GOST 22685-89 Formy dlya izgotovleniya kontrol'nyh obrazcov betona. Tekhnicheskie usloviya.
7. M.L. Berestevich, A.L. Osadovskij, Nerazrushayushchij kontrol' prochnosti betona v mostostroenii, **9**, 22-27 (2004)
8. L.M. Dobshic, A.L. Klibanov, A.V. Belov, O metodikah opredeleniya prochnosti betona konstrukcij, OAO NIIMosstroj, 63-73 (2015)

# The influence of description of fire curves on the thermal and reliability response of steel structure

*Katarzyna* Kubicka[1,*], *Urszula* Radoń[1] , and *Waldemar* Szaniec[1]

[1]Kielce University of Technology, Faculty of Civil Engineering and Architecture, Al. Tysiąclecia Państwa Polskiego 7, Kielce 25-314, Poland

**Abstract.** The paper presents a proposal for the assessment of the reliability of statically determinate steel truss in the persistent and accidental (fire) design situation. The steel truss was modelled as series system from the reliability standpoint. The global Hasofer-Lind reliability index was employed, computed in successive time steps for the whole structure, not for individual elements. Fire analysis was carried out according to three different time-temperature curves-standard, hydrocarbon and external. Monitoring the reliability index in the successive minutes of the fire makes it possible to estimate the probability of the structure failure, and to decide whether the required safety level is maintained.

## 1 Introduction

Building structures should be designed and executed in such a way so that, under fire conditions, the bearing capacity could be ensured for an assumed period of time, required to enable safe evacuation of the users from the building. The knowledge of the theoretical basis of the behaviour of steel elements subjected to fire is presented in the monographs of: Maślak, Biegus, Kosiorek, Franssen, Vila Real, Kodur, Zaharia, Purkiss and Li [2,6,7,8,12,16].

In recent years, some studies on the probabilistic evaluation of steel structures under fire conditions have been produced [1,4,10,11,17]. Those, however, concern the evaluation of simple elements, isolated from the whole of the structure. In the article, authors intend to analyse the structure as a whole.

* Corresponding author:ksuckert@tu.kielce.pl

## 2 Methods of analysis

### 2.1 Fire analysis of steel elements

In the paper, three types of fire curves (standard, hydrocarbon and external) were adopted. The temperature of fire gases ( $\theta_g$ ) is described as follows [14]:

-standard fire curve:

$$\theta_g = 20 + 345 \log_{10}(8t+1) \tag{1.1}$$

-hydrocarbon fire curve:

$$\theta_g = 1080(1 - 0.325e^{-0.167t} - 0.675e^{-2.5t}) + 20 \tag{1.2}$$

-external fire curve:

$$\theta_g = 660(1 - 0.687e^{-0.32t} - 0.313e^{-3.8t}) + 20 \tag{1.3}$$

in equations 1.1-1.3 t is the time of fire duration. All mentioned temperature-time curves are shown in figure 1.

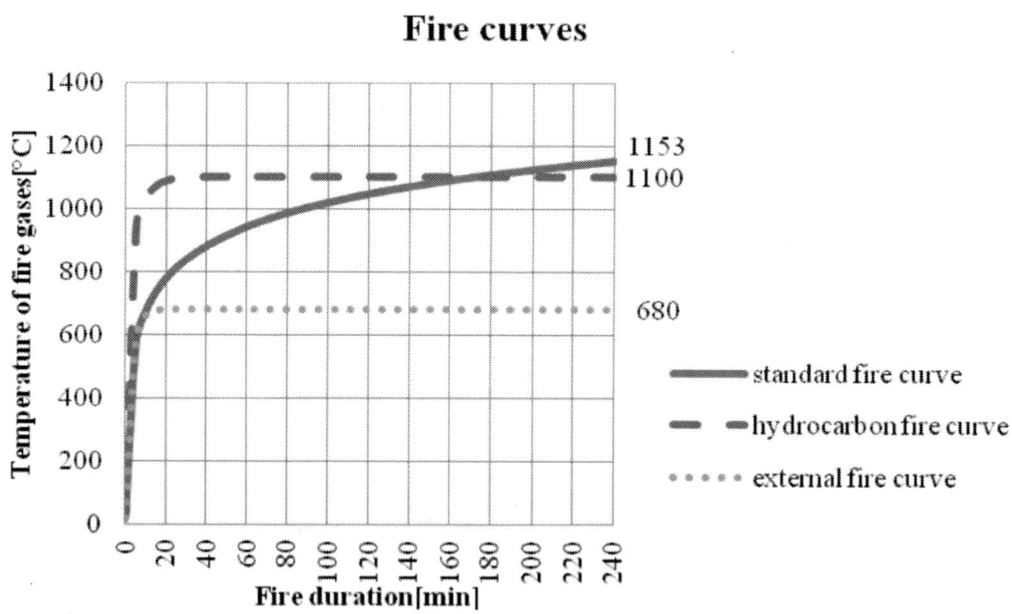


**Fig. 1.** Nominal temperature-time curves

When the temperature of fire gases is known, it is possible to determine the temperatures reached by individual bars and it is possible to calculate changes in steel mechanical properties that influence the bearing capacity.

When temperature distribution in the section is uniform, the temperature increment $\Delta\theta a,t$ of the insulated steel member in the time interval $\Delta t$ is given by the following formula [15]:

$$\Delta\theta_{a,t} = \frac{\lambda_p A_p / V(\theta_{g,t} - \theta_{a,t})}{d_p c_a \rho_a (1+\phi/3)} \Delta t - (e^{\phi/10} - 1)\Delta\theta_{g,t} \tag{2}$$

in addition:

$$\phi = \frac{c_p \rho_p}{c_a \rho_a} d_p A_p / V \tag{3}$$

where:

$A_p/V$- the section factor for steel members insulated by fire protection material, determined from Table 4.3 PN-EN 1993-1-2
$c_a$- temperature-dependent specific heat of steel
$c_p$- temperature-independent specific heat of the fire protection material
$d_p$ - the thickness of the fire protection material
$\Delta t$- the time interval (not greater than 30 s)
$\theta_{a,t}$- the steel temperature at time t
$\theta_{g,t}$- the ambient gas temperature at time t
$\Delta\theta_{g,t}$- the increase in the ambient gas temperature in the time interval $\Delta t$
$\lambda_p$- the thermal conductivity of the fire protection material
$\rho_a$- the unit mass of steel
$\rho_p$- the unit mass of the fire protection material

The design resistance at the member uniform temperature $\theta$ is given by the appropriate formulas in PN-EN 1993-1-2 [14].

## 2.2 Reliability analysis of the system

System-reliability analysis concerns the formulation of the limit state and calculation of the failure probability when the structure has more ways of failing. Mathematically we encounter system reliability analysis if the limit-state surface is composed by more pieces that generally intersect pairwise in sets of singular points, which are points at which the limit-state surface is not differentiable. Each of these pieces will usually be a part of a limit-state surface for a specific global failure mode or for a local failure. We can distinguish three main models of reliability system: series, parallel and mixed. The most common example of series system is a statically determinate structure. A series system is also denoted as a "weakest link" system. The reliability of structures, the static scheme of which is compliant with the series system, is computed according to the formula [5]:

$$R = \prod_{i=1}^{n} R_i = R_1 R_2 \ldots R_n \tag{4}$$

The parallel system is appropriate for some structures that are statically indeterminate. In the parallel system, the structure remains reliable as long as at least one element is reliable. The reliability of the parallel system is computed as follows [4]:

$$R = 1 - \prod_{j=1}^{m} (1 - R_j) \tag{5}$$

A majority of statically indeterminate structures is in line with mixed systems. The two basic systems are parallel-series and series-parallel. In real structures, however, mixed systems are usually more complicated.

## 3 Results and discussion

The presented method combines the fire and reliability analysis. The fire analysis was carried out according to the method described in [6,7,13,14]. The temperatures of fire gases and elements were calculated in the deterministic way. The analysis was carried out for the truss presented in figure 2. The profiles of elements and the effect of actions are shown in figure 2, where (-) means compression. All elements were assumed to be made from S235 steel. The structure was loaded by dead weight of layered roof plate (0.094kN/m2) and by the snow (0.72kN/m2-second snow zone).The whole load was converted to concentrated forces in the nodes.

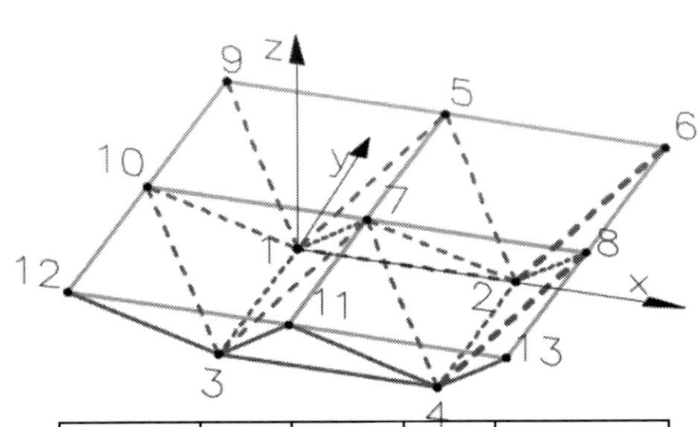


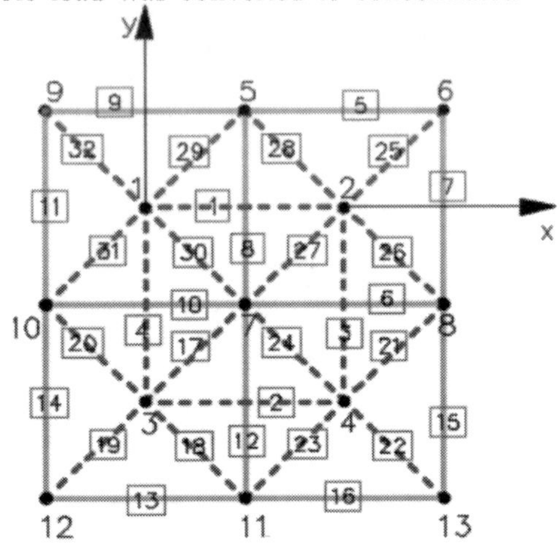


| Node's number | x [m] | y[m] | z [m] | blocked direction of translation |
|---|---|---|---|---|
| 1 | 0.00 | 0.00 | 0.00 | x,y,z |
| 2 | 3.00 | 0.00 | 0.15 | y,z |
| 3 | 0.00 | -3.00 | 0.00 | z |
| 4 | 3.00 | -3.00 | 0.15 | z |
| 5 | 1.50 | 1.50 | 1.50 | (-) |
| 6 | 4.50 | 1.50 | 1.65 | (-) |
| 7 | 1.50 | -1.50 | 1.50 | (-) |
| 8 | 4.50 | -1.50 | 1.65 | (-) |
| 9 | -1.50 | 1.50 | 1.35 | (-) |
| 10 | -1.50 | -1.50 | 1.35 | (-) |
| 11 | 1.50 | -4.50 | 1.50 | (-) |
| 12 | -1.50 | -4.50 | 1.35 | (-) |
| 13 | 4.50 | -4.50 | 1.65 | (-) |

| Element's number | Profile | Effect of actions [kN] |
|---|---|---|
| 1-4 | CHS 20x3.2 | -0.02 |
| 5,7,9,11,13-16 | CHS 20x3.2 | 2.00 |
| 6,8,10,12 | CHS 20x3.2 | 3.98 |
| 17,30 | CHS 30x3.2 | -3.42 |
| 18,21,26,29 | CHS 30x3.2 | -3.44 |
| 19,32 | CHS 30x3.2 | -3.35 |
| 20,23,28,31 | CHS 30x3.2 | -3.33 |
| 22,25 | CHS 30x3.2 | -3.46 |
| 24,27 | CHS 30x3.2 | -3.32 |

**Fig. 2.** Analyzed space steel truss

The thermal analysis was carried out for insulated structure. All the elements were assumed to be heated from each side. Three types of fire curves was analyzed: standard, hydrocarbon and external ones. Spray-applied vermiculite with cement with the thickness of 3 cm was assumed as an insulation. This material is characterized by the following parameters: density $\rho p$=550kg/m3. specific heat cp=1100J/(kgK). thermal conductivity $\lambda p$=0.2 W/(mK). The results of thermal analysis-temperature of fire gases and temperature of steel elements according to different fire curves are shown in the figure 3. As whole elements have the same thickness they heat up in the same degree. According to Eurocode section factor for circular hollow section can be calculated as 1/d, where d is thickness of section.

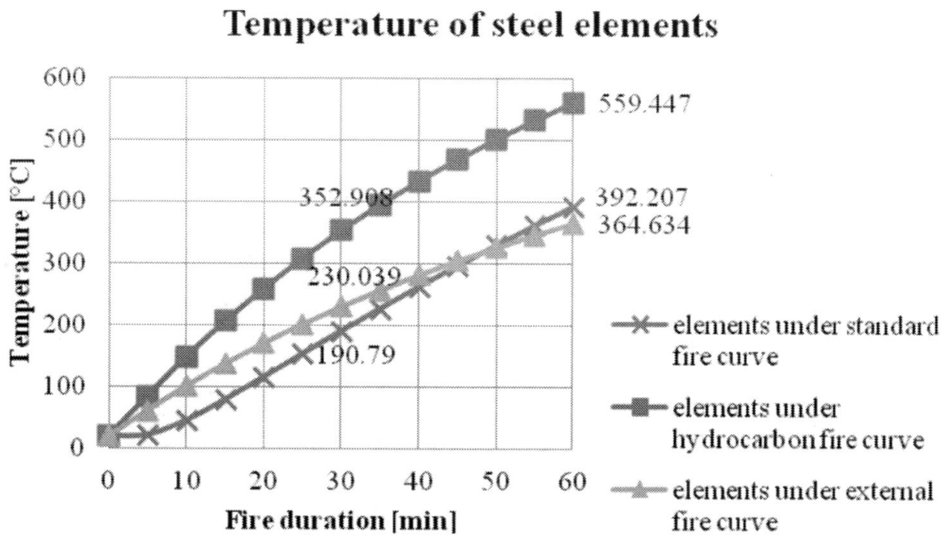


**Fig. 3.** Temperature of steel elements according to different types of fire curves

In the next part of analysis, for each step: 0, 5, 10, 15, 20, 25 and 30 minutes, the reliability analysis was carried out according to method presented in [9]. The influence of randomness was taken into consideration in the bearing capacity and the effect of actions. For both values, appropriate values of standard deviations, which are shown in the following table, was assumed.

**Table 1.** Random variables and their characteristics for the reliability analysis

| Random variable | Coefficient of variation [10,2] | Standard deviation |
|---|---|---|
| Effect of action for i$^{th}$ element ($E_i$) | $v_E = 0.06$ | $\sigma_E = 0.06 \cdot E_i$ |
| Yield strength ($f_y$) | $v_{f_y} = 0.08$ | $\sigma_{f_y} = 0.08 \cdot f_y$ |
| Area of cross-section (A) | $v_A = 0.06$ | $\sigma_A = 0.06 \cdot A$ |
| Bearing capacity (N) | $v_N = \sqrt{v_{f_y}^2 + v_{f_y}^2 \cdot v_A^2 + v_A^2} = 0.1$ | $\sigma_N = 0.1 \cdot N_i$ |

The entrance data were the mean values of the effect of actions and the bearing capacity, which was calculated according to [14]. Then, the reliability for each element of the structure was calculated. The structure has statically determinate scheme and corresponds to the series system, consequently, the failure of any element (1-32) results in the failure of the whole structure. The reliability of the structure (R) is calculated according to equation (5), particularly:

$$R = \prod_{i=1}^{32} R_i \qquad (6)$$

The calculation according to method described in chapter 2 was carry out in Mathematica program. The results of the reliability analysis are presented in Fig. 4. The required value of the reliability index (β) was assumed, according to [10], to equal 1.34. Dotted lines indicates the reliability indexes for the weakest elements (members 22 and 25), solid lines corresponds to the whole structure reliability.

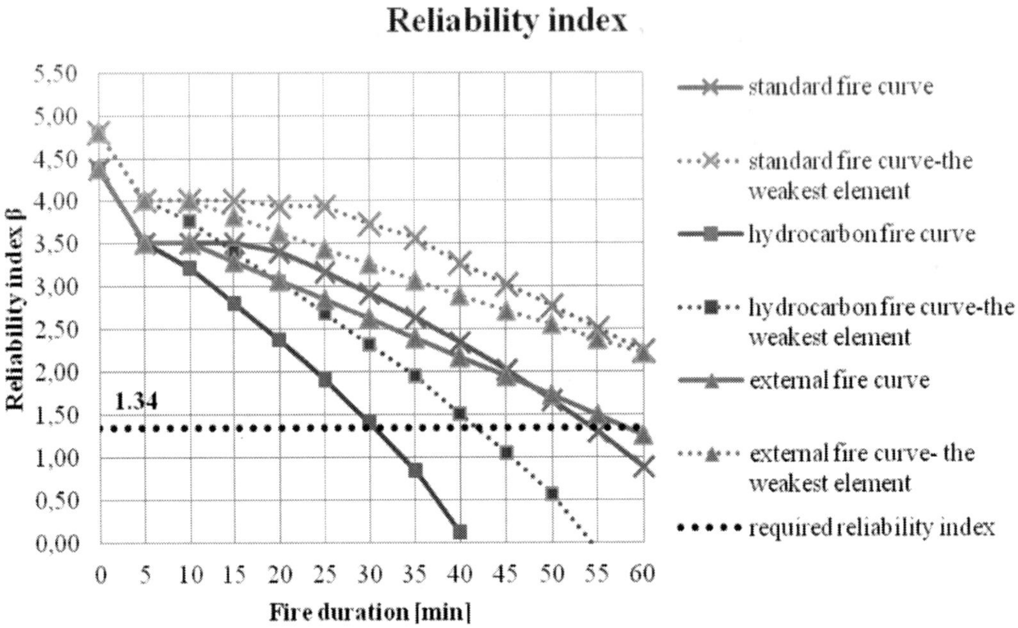


**Fig. 4.** Monitoring the reliability index in successive minutes of the fire

The application of the standard fire curve or the external fire curve made it possible to reach R45 fire resistance class. For the hydrocarbon curve, the reliability index falls below the required one after 30 minutes of fire duration. Additionally, in this case, the index drops below zero beyond the 40th minute.

## 4 Conclusions

This paper provided an analysis of a statically determinate spatial steel structure subjected to the action of fire. The first step involved the structure thermal response. The other step included the reliability analysis conducted using the Hasofer-Lind reliability index. Three types of nominal fire curves given in the Eurocode were taken into considerations. When the standard and external fire curves were applied, similar results of both the thermal and reliability analyses were obtained. However, with the hydrocarbon fire curve much more rapid increase in the temperature of the elements was found (Fig.3). That resulted in earlier decrease in the values of the mechanical properties of the materials, namely Young's modulus of elasticity and yield strength. Consequently, a very fast decrease in the reliability index was observed (Fig.4).

Using the standard fire curve or the external fire curve, R45 fire resistance class was achieved, whereas with the hydrocarbon fire curve, that was R30 fire resistance class. The fact illustrates the importance of appropriate curve selection which needs to account primarily for the use of a given object and the types of materials stored in it. If the ignition source are flammable substances, the hydrocarbon fire curve should be applied. In such a case, the use of the standard fire curve leads to the attainment of overestimated fire resistance class.

The reliability analysis conducted for the study shows a well-known and frequently reported effect of the static deterioration of the structure [3]. It is observed in statically determinate structures in which their reliability is computed in series system. In that case the reliability of a single element is much higher than reliability of the whole structure (Fig.4). That may raise doubts as to the correctness of Eurocode approach, in which individual elements are considered when dimensioning the structure in relation to the ultimate limit state. Structural design which utilizes the ultimate limit state method with partial safety factors, without taking into account the global reliability may lead to errors in evaluating the reliability of a given building structure. In the literature, the probabilistic measure of safety for the fire conditions is the minimum safety factor computed for the main structural elements at the ultimate limit state.

To sum up, it can be concluded that the system reliability analysis can be successfully combined with the fire analysis. For statically determinate structures, the determination of the reliability index with this method offers an option that is simple, and provides an alternative to traditional design methods.

## References

1. D. K. Banarjee, Fire Safety Journal, **61**, 65 (2013)
2. A. Biegus, *Design of steel structures according to Eurocode 3. Fire safety of steel structures* (Lectures in Polish) (Wrocław, 2013)
3. A. Biegus. *Probabilistic analysis of steel structures* (in Polish) (Polish Scientific Publishers PWN, Warszawa-Wrocław, 1999)
4. Q. Guo, K. Shi, Z. Jia., A. E. Jeffers, Fire Technology, **49**, 793 (2013)
5. M. Gwóźdź, A. Machowski, *The selected studies and calculations of building structures by probabilistic methods* (in Polish) (Politechnika Krakowska, Kraków, 2011)
6. J. M. Franssen, V. Kodur, R. Zaharia, *Designing steel structures for fire safety* (CRC Press, Taylor and Francis Group, Boca Raton, London, 2009)
7. J. M. Franssen, P. Vila Real, *Fire design of steel structures. Eurocode 1: Actions on structures. Part 1-2-General actions-Actions on structures exposed to fire. Eurocode 3: Design of steel structures. Part 1-2 General rules - Structural fire design* (Ernst&Sohn, 2012)
8. M. Kosiorek, *Fire resistance of building structures* (Arkady, Warszawa, 1988) (in Polish)
9. K. Kubicka, U. Radoń, ACE, **4**, 141 (2015)
10. M. Maślak, *Eight International Conference on Advances in Steel Structures Lisbon, Portugal, July 22-24* (2015)
11. M. Maślak, *Metal structures: ZK 2014: Scientific-Technical Conference: short papers: Kielce-Suchedniów, July 2014* (2014)
12. M. Maślak, *The fire durability of steel bar structures* (Politechnika Krakowska, Kraków, 2008) (in Polish)
13. PN-EN 1991-1-2. *Eurocode 1: Actions on structures - Part 1-2: General actions - Actions on structures expose to fire.*

14. PN-EN 1993-1-1. *Eurocode 3:Design of steel structures-Part 1-1: General rules and rules for buildings.*
15. PN-EN 1993-1-2. *Eurocode 3: Design of steel structures-Part 1-2: General rules-Structural fire design*
16. J. A. Purkiss, L. Y. Li, *Fire Safety Engineering Design of Structures* (CRC Press, Taylor and Francis Group, Boca Raton, London, New York, 2014)
17. Ch. Zhang, G.-Q. Li, Y.-Ch. Wang, Structural safety **50**, 16 (2014)

# Protective and Catching Safety Systems In Construction

*Marat* Kuzhin[1*]

[1]Moscow State University of Civil Engineering, 129337 Moscow, Russia

**Abstract.** In the article is described application of protective and catching systems in construction. Classification of similar systems, their types and purpose are listed. Dangerous zones on construction site and events to for limiting their influence or protection from the factors. Protective and catching systems is one of the most effective technical equipment, applied in recent time. Protective fences and catching systems are important part in the problem solution. Protective fences protect workers from falling from height. Protective and catching systems allows avoid injuries by workers, also catch debris, fallen from constructing buildings. In regard with continuing development in technical and technological solutions, protective and catching systems require adaptation to a new requirements of construction industry and requirements of normative documents. Technical regulations in the appliance sphere of protective and catching systems requires actualization and aligning with modern normatives. Important role should be given to developing organizational and technological documentation for application of the systems. Scientific studying of technical parameters of fences and protective catching nets also has great interest.

## 1 Introduction

Ensuring the required level of safety in construction industry is one of the major condition of project functioning and realization. Construction work have a lot of dangerous production factors, among which the most common are production work near an edge of buildings and work production in dangerous zones, appeared from objects falling from a building [14-17]. Modern conditions of construction assumes work production in tight urban conditions, what causes limits in construction production, which need to be considered in organizational and technological construction projecting [7,10,11,13]. In construction applied two kind of systems, protecting from falling objects: passive and active systems. Passive systems, while installed, protect workers without necessity of their participation in system activity. Active systems require each worker himself take action, from falling protection, such as wearing safety belt. In the article will be discussed only passive systems like safety nets and protective fences (fig.1).

* Corresponding author: kuzhinmf@mail.ru

**Fig. 1.** Protective and catching systems, installed on object.

Nowadays in sphere of application protective and catching systems exists following documents, regulating appliance of such systems:

1)GOST 12.4.059-89 Stroitelstvo. Ograzhdenia predohranitelnie inventarnie;

2)DIN EN 13374 Temporary edge protection systems. Product specification –Test methods;

3)DIN EN 1263-1 Part 1: Safety requirements, test methods.

4)DIN EN 1263-1 Part 2: DIN EN 1263-1 Part 2: Safety requirements for the positioning limits.

Review of the given documents shows that existing normative and technical documents regulates methods of testing and requirements for geometrical characteristics. And there is not enough information about technological and organizational issues [1-5].

## 2. Results and Discussion

Construction presents difficult system, which consists of working contract organizations, constantly moving materials and technical resources, machines and mechanisms, also temporary infrastructure, necessary for providing relevant conditions. Construction object is motionless element of such system, on which construction works are underway. Construction always causes appearance on territory dangerous production factors, related with work of installing machines and lifting mechanisms, with .works near border of high difference. Listed above factors can be divided into two groups: constantly-acting and temporary-acting (table 1).

**Table 1.** Dangerous areas on construction site and ways for their limitation

| № | Kind of factor | Work type | Relative dimensions | way for limitation |
|---|---|---|---|---|

| | | | | |
|---|---|---|---|---|
| 1 | Dangerous area from crane | Works, made with crane | Under crane boom with weight | Overhang limitation and rotation angle limitation |
| 2 | Dangerous area around building | Construction work | Repeats building's border | Protective and catching nets, protective screens, canopy above passes |
| 3 | Dangerous area from remote deck | At lifting and material unloading | On a contour of remote deck | Passive protection as a marking on the deck |
| 4 | Dangerous area of lifting mechanisms | Lifting construction materials on a floor | On a contour of lifting mechanism | Passive protection as a marking on the deck |
| 5 | Height difference | Work near an edge | Border of dangerous area | Protective fences |

Protective and catching system is one of the most effective technical equipment applied in recent years. Such systems according to type of work can have different purpose (table 2). Protecting fences and protective and catching systems are important part in the problem solution. Protecting fences prevent workers from falling from the height. Protective-catching nets allows avoid injuries, and also catch objects, fallen from constructing buildings [18-21,24].

**Table 2.** Comparison of fences according to existing norms

| № | Name | Purpose | Withstand loads |
|---|---|---|---|
| GOST 12.4.059-89 Stroitelstvo. Ograzhdenia predohranitelnie inventarnie | | | |
| 1 | Preventing fences | Fencing working places on the height and passes to them, construction of which placed in vertical position, used for preventing from falling | - |
| 1.1 | Protective fences | Protective fence, used for preventing unintentional accessto border of height difference | Protective fences calculated on strength and stability to alternately acting horizontal and vertical evenly spread loads 400 Н/м. |
| 1.2 | Insurance fences | Protecting fence providing retention when people loose stability near border of height difference | Calculated on strength and stability of acting horizontal fixed force, not lower than 700 Н, and insurance exterior calculated on strength for 100 kg, falling from 1 m height. |
| 1.3 | Signal fences | Preventing fences, used for marking dangerous zones | - |
| According to DIN EN 13374 Temp. edge protection systems. | | | |
| 2.1 | Fences class «A» | Class A protection provides | For posts and guardrails, |

| | | | |
|---|---|---|---|
| | | resistance to static loads only, based on the requirements to: support a person leaning on the protection or provide a handhold when walking beside it; and collectively stop a person who is walking or falling towards the protection. | *FT*1 = 300 N. For toeboards, *FT*2 = 200 N. |
| 2.2 | Fences class «B» | Class B protection provides resistance to static loads and low dynamic actions only, based on the requirements to: support a person leaning on the protection or provide a handhold when walking beside it; and collectively stop a person who is walking or falling towards the protection; collectively stop a person sliding/falling down a sloping surface. | Class B edge protection system shall fulfil the design requirements given in clause to class A Class B edge protection system shall be capable of absorbing a kinetic energy of 1100 J anywhere along the protection up to a height of 200 mm above the working surface and 500 J at all higher parts. |
| 2.3 | Fences class «C» | Class C protection provides resistance to high dynamic forces based on the safety requirements to prevent the fall of a person sliding down a steep sloping surface. Collectively stop a person sliding/falling down a steep sloping surface. | Class C has no static load requirement. Class C edge protection system shall be capable of absorbing 2200 J of kinetic energy anywhere along the protection up to a height of 200 mm above the working surface and 500 J at all higher parts. |

Catching systems also well known as safety nets appeared several years ago. Nowadays, there 4 main types of it. Each type is designed for a specific situation and construction phase (table 3).

**Table 3.** Comparison of fences according to existing norms

| № | Name | Purpose | Withstand loads |
|---|---|---|---|
| According to DIN EN 1263-1 Safety nets | | | |
| 1.1 | System type «T» | Holding construction garbage and workers, fell from the working horizon | Static tests for finding loads of tear. Dynamic tests with load 100 kg. |
| 1.2 | System type «S» | Holding construction garbage and workers, fell from the working horizon | Static tests for finding loads of tear. Dynamic tests with load 100 kg. |
| 1.3 | System type «V» | Holding construction garbage | Static tests for finding loads of tear. Dynamic tests with load 100 kg. |

| | | | |
|---|---|---|---|
| 1.4 | System type «U» | Preventing from falling workers | Static tests for finding loads of tear. Dynamic tests with load 75 kg rolling on inclined surface |

Classification of protective fences according to [1] and [5] have common requirements, differences only in character of applied loads and their type.

## Conclusions

Accounting of all above listed dangerous production factors is necessary condition for providing safety in construction [6,8,9,12]. In recent years technical solutions and equipment, needed for decision the problems, got a significant development. In accordance with their application causes additional expenses of time and work, which influences on construction period, technical decisions of such systems need constantly improvement. However, existing solutions allows provide minimal required level of protection from most of the factors [22,23].

According to constantly improving technical and technological solutions, protective-catching systems also need constant adaptation for modern requirements of construction production and requirements of normative documents. Technical regulations in applying protective-catching system sphere requires modernization and alignment with modern norms. Important role should be given to projecting organizational and technological documentation for using the systems. Scientific studying of technical parameters of fences and protective-catching nets also has great interest.

## References

1. GOST 12.4.059-89 Stroitelstvo. Ograzhdenia predohranitelnie inventarnie
2. SNiP 12-03-2001 Bezopasnost truda v stroitelstve
3. Prikaz Ministerstva truda i socialnoi zashiti RF ot 1 iuna № 336n "Ob utverzhdenii Pravil po ohrane truda v stroitelstve" (2015)
4. DIN EN 1263-1 Safety nets, English version (2002)
5. DIN EN 13374 Temporary edge protection systems. English version (2013)
6. Lyapin, A., Korolchenko, A., Meshalkin, E. Analysis of causes of combustible mixture explosions inside production floor areas. (2016) MATEC Web of Conferences, 86, art. no. 04030. DOI: 10.1051/matecconf/20168604030
7. B.V. Zhadanovskiy, M.F. Kuzhin Organizacionno-technologicheskie reshenia ustroistva navesnih fasadnih sistem pri reconstrukcii zhilih i obshestvennih zdaniy //Promishlennoe i grazhdanskoe stroitelstvo. № 1. P. 62-64 (2012)
8. Korolchenko A.Ya., Petrova E.A. Modern means for protection of wood against flame. (2003) Rossijskij Khimicheskij Zhurnal (Zhurnal Rossijskogo Khimicheskogo Obshchestva Im. D.I. Mendeleeva), 47 (4), pp. 49-55
9. Shebeko Yu.N., Tsarichenko S.G., Trunev A.V., Korolchenko A.Ya., Kaplin A.Yu. The influence of inert retardants on the combustion of hydrogen-oxygen mixtures under elevated temperatures and pressures. (1994) Combustion, Explosion, and Shock Waves, 30 (2), pp. 183-188. DOI: 10.1007.
10. B.V. Zhadanovskiy, S.A. Sinenko Pressure Method of Monolithic Concrete Structures of Buildings and Structures // International Journal of Applied Engineering, Research 11, Number 3, pp.1724-1727 (2012)

11. A.A. Lapidus, N.D. Cherednichenko The Impact of Technology Installation of Injection Mortar Systems for a Long Term Behavior of Anchor Connections // Procedia Engineering, 153, pp.371-377 (2014)
12. D.A. Korolchenko, A.F. Sharovarnikov Universalnost mechanizmov tusheniya plameni razlichnimi ognetushshimi veshestvami //Pozharovzrivobezopasnost. -T. 23, № 11. -P. 84-88 (2014)
13. A.A. Lapidus, A.A. Makarov Model for the Potential Manufacture of Roof Structures for Residential Multi-storey Buildings // Procedia Engineering, 153, pp.378-383 (2013)
14. P.P. Oleynik, V.I. Brodskiy Tehnicheskoe regulirovanie organizacii bezopasnosti truda v stroitelstve. Zhurnal TOSP, №1, p.27-32 (2014)
15. B.F. Shirshikov, S.A. Sinenko, A.M. Slavin, B.V. Zhadanovskiy, M.F. Kuzhin, V.I. Brodskiy Organizacionno-tehnologicheskie resheniya po bezopasnosti truda v proectah proizvodstva rabot. Moscow (2015)
16. S.A. Bocheev, M.F. Kuzhin Organizacionno-tehnologicheskie osobennosti proizvodstva fasadnih rabot. Sbornik: Integracii, partnerstvo i innovacii v stroitelnoy nauke i obrazovanii, p.229-234 (2017)
17. U.A. Vilman, M.F Kuzhin. Meropriyatiya po ogranicheniu opasnih zon na stroitelnoy ploshadke. Nauchnoye obozrenie, №17, p.341-344 (2015)
18. McCabe, B.Y., Alderman, E., Chen, Y., Hyatt, D.E., Shahi, A. Safety Performance in the Construction Industry: Quasi-Longitudinal Study Journal of Construction Engineering and Management Volume 143, Issue 4, 1 April 2017, Article number 04016113.
19. Fang, Y. ,Cho, Y.K. Effectiveness Analysis from a Cognitive Perspective for a Real-Time Safety Assistance System for Mobile Crane Lifting Operations Journal of Construction Engineering and Management Volume 143, Issue 4, 1 April 2017, Article number 05016025.
20. Kim, H. , Ahn, C.R. , Yang, K. Identifying Safety Hazards Using Collective Bodily Responses of Workers Journal of Construction Engineering and Management Volume 143, Issue 2, 1 February 2017, Article number 04016090.
21. Park, J., Marks, E. , Cho, Y.K.,Suryanto, W Performance test of wireless technologies for personnel and equipment proximity sensing in work zones Journal of Construction Engineering and Management Volume 142, Issue 1, 1 January 2016, Article number 04015049.
22. Korolchenko D. A.; Degaev E. N.; Sharovarnikov A. F. Dependence of Fire Extinguishing Efficacy of Low Expansion Foams Solutions Homology Sodium Sulfate on the Molecular Weight of the Surface-active Substances 2015 2nd International conference on material engineering and application (ICMEA 2015)
23. Korolchenko D., Tusnin A., Trushin S., Korolchenko A. Physical parameters of high expansion foam used for fire suppression in high-rise buildings, International Journal of Applied Engineering Research, Volume 10, Issue 21, 2015, Pages 42541-42548.
24. Pomares, J.C., Irles, R., Segovia, E. , Ferrer, B. Acceleration and deflection analysis for class C edge protection systems in construction work Journal of Construction Engineering and Management Volume 140, Issue 8, 1 August 2014, Article number 04014031.

# Filtration model of the unsteady suspension flow in a porous medium

*Ludmila* Kuzmina[1], and *Yuri* Osipov[2,*]

[1]National Research University Higher School of Economics, 101000 Moscow, Russia
[2]Moscow state university of civil engineering, Yaroslavskoye shosse, 26, Moscow, Russia, 129337

**Abstract.** The study of filtration of the suspension in a porous medium is a vital problem in the design and construction of tunnels and hydraulic structures. An exact solution is constructed for an unsteady flow of a monodisperse suspension in a homogeneous porous medium with size-exclusion mechanism for particle retention. The concentrations of suspended and precipitated particles are calculated in case of a linear blocking filtration coefficient.

## 1. Introduction

The study of the motion of the tiny solid particles in a porous soil is a vital problem in the design and construction of tunnels and hydraulic structures. During the filtration of the suspension in a porous medium some of the suspended particles pass through the pores along with the flow of the carrier fluid and others get stuck in the filter pores. Deep bed filtration, accompanied by the formation of deposit in the entire porous medium, and not only in its surface layer, significantly affects the properties of rocks and soil layers [1, 2]. In a variety of technical and biological systems for cleaning surface and waste water and liquid industrial waste, solid particles are retained when the suspension moves through the filter medium [3].

The mathematical model of the particle transport and retention in a filter is based on the mechanical and geometric interaction of particles with a porous medium [4–6]. A homogeneous porous medium with pores having the same cross section throughout its entire length is considered. It is assumed that the particles pass freely through large pores and get stuck at the entrance of pores if the particle diameter is larger than the pore diameter. One pore is blocked by one particle and vice versa. The retained particles can not be knocked out of the pores by the carrier fluid flow or another suspended particles, and remain permanently in the pores. As the retained particles concentration grows, the number of free small pores decreases, and the rate of deposit formation slows down. If all small pores are clogged by the retained particles, the retained particles concentration $S$ reaches the maximum possible limit value $S_M$. The coefficient of proportionality $\Lambda(S)$ between the growth rate of the deposit and the suspended particles concentration is called the

* Corresponding author: yuri-osipov@mail.ru

filtration coefficient. The blocking filtration coefficient tends to zero when all small pores are blocked $\Lambda(S_M)=0$.

For different filtration models of a stationary suspension flow in a porous medium, exact and asymptotic solutions are obtained [7–12]. In the absence of analytical solutions, numerical calculations are performed [13–15].

Standard filtration models for the suspension flow in a porous medium assume that a suspension with constant suspended particles concentration is injected into the filter inlet. In the paper the exact solution of the filtration problem for a suspension with variable concentration of suspended particles at the filter inlet is obtained. Numerical calculations of the suspended and retained particles concentrations are performed for a linear blocking filtration coefficient.

## 2. Governing equations

Consider two basic dimensionless equations for the suspended and retained particles of a suspension in a porous medium. Let $C(x,t)$ is the concentration of suspended particles, $S(x,t)$ is the retained particles concentration. The particles concentrations satisfy the equation of continuity, which, taking into account the immobility of the deposit, has the form

$$\frac{\partial C}{\partial t}+\frac{\partial C}{\partial x}+\frac{\partial S}{\partial t}=0. \tag{1}$$

The growth rate of the retained particles concentration is proportional to the concentration of suspended particles from which it is formed. It also depends on the current value of the retained particles concentration $S(x,t)$: the more pores of small diameter are clogged by particles, the slower the growth rate of the deposit. The equation for deposit growth is

$$\frac{\partial S}{\partial t}=\Lambda(S)C. \tag{2}$$

Here the filtration coefficient $\Lambda(S)>0$ for $S\geq 0$.

The system of equations (1), (2) is considered in the domain

$$\Omega=\{0<x<1,\ t>0\}.$$

The boundary conditions for the system (1), (2) are given at the filter inlet $x=0$ and at the initial time $t=0$:

$$C(x,t)\big|_{x=0}=p(t); \tag{3}$$

$$C(x,t)\big|_{t=0}=0; \quad S(x,t)\big|_{t=0}=0 \tag{4}$$

Conditions (4) indicate that at the initial moment $t=0$ the porous medium does not contain any suspended and retained particles. A suspension with variable suspended particles concentration (3) is injected into the porous medium and moves in the filter with the velocity $v=1$. The characteristic line $t=x$ of the equation (1) – the front of the suspended and retained particles concentrations is the mobile boundary of the suspension.

In the subdomain of an empty porous medium $\Omega_0 = \{0 < x < 1, 0 < t < x\}$ the solution is zero; in the subdomain of the suspension $\Omega_S = \{0 < x < 1, t > x\}$ the solution is positive.

## 3. The solution at the filter inlet

At the filter inlet $x = 0$ the equation (2) has the form

$$\frac{\partial S}{\partial t} = \Lambda(S)p(t). \tag{5}$$

Division of both sides of (5) by $\Lambda(S)$ and integration with respect to the variable $t$ gives

$$\int_0^t \frac{\partial S/\partial t}{\Lambda(S)} dt = P(t). \tag{6}$$

where $P(t) = \int_0^t p(t)dt$.

Using condition (4), the integral in the left part of (6) can be written as

$$\int_0^{S(0,t)} \frac{dS}{\Lambda(S)} = P(t). \tag{7}$$

Formula (7) specifies the retained particles concentration at the filter inlet.

## 4. The solution on the concentrations front

Denote the characteristic variables

$$\tau = t - x, \ x = x.$$

Equation (1) takes the form

$$\frac{\partial C}{\partial x} + \frac{\partial S}{\partial \tau} = 0. \tag{8}$$

On the suspended and retained particles concentrations front $t = x$ the deposit concentration $S = 0$. Substitution of (2) into (8) gives

$$\frac{\partial C}{\partial x} + \Lambda(0)C = 0. \tag{9}$$

The solution of the equation (9) with the condition $C|_{x=0} = p(0)$ is

$$C = p(0)e^{-\Lambda(0)x}. \tag{10}$$

## 5. The exact solution of the filtration problem

From (2)

$$C = \frac{\partial S/\partial t}{\Lambda(S)}. \tag{11}$$

Substitution of (11) into (1) gives

$$\frac{\partial}{\partial t}\left(\frac{\partial S/\partial t}{\Lambda(S)}\right) + \frac{\partial}{\partial x}\left(\frac{\partial S/\partial t}{\Lambda(S)}\right) + \frac{\partial S}{\partial t} = 0. \tag{12}$$

Using the formula

$$\frac{\partial}{\partial x}\left(\frac{\partial S/\partial t}{\Lambda(S)}\right) = \frac{\partial}{\partial t}\left(\frac{\partial S/\partial x}{\Lambda(S)}\right),$$

the equation (12) can be transformed to

$$\frac{\partial}{\partial t}\left(\frac{\partial S/\partial t}{\Lambda(S)}\right) + \frac{\partial}{\partial t}\left(\frac{\partial S/\partial x}{\Lambda(S)}\right) + \frac{\partial S}{\partial t} = 0. \tag{13}$$

Integration of (13) with respect to the variable $t$ gives

$$\frac{\partial S/\partial t}{\Lambda(S)} + \frac{\partial S/\partial x}{\Lambda(S)} + S = K(x). \tag{14}$$

The function $K(x)$ is determined by the initial condition (4): $K(x) = 0$.

The equation (14) in the characteristic variables takes the form

$$\frac{\partial S/\partial x}{\Lambda(S)} + S = 0. \tag{15}$$

Transfer of the term $S$ to the right side of (15), division by $S$, and integration with respect to $x$ gives

$$\int_0^x \frac{\partial S/\partial x}{S\Lambda(S)} dx = -x. \tag{16}$$

After the change of the variable in the integral equation (16) takes the form

$$\int_{S(0,\tau)}^{S(x,\tau)} \frac{dS}{S\Lambda(S)} = -x. \tag{17}$$

An inverse change of variables gives

$$\int_{S(0,t)}^{S(x,t-x)} \frac{dS}{S\Lambda(S)} = -x. \tag{18}$$

Formulas (7), (11), (18) determine the exact solution of the problem (1) – (4) in the domain $\Omega_S$.

## 6. The linear filtration coefficient

Consider a model with the linear blocking filtration coefficient

$$\Lambda(S)=\lambda(S_M-S),\ S_M>0\,. \tag{19}$$

In this case, the retained particles concentration at the filter inlet $x=0$

$$S(0,t)=S_M\left(1-e^{-\lambda P(t)}\right). \tag{20}$$

In the domain $\Omega_S$ the suspended and retained particles concentrations are given by explicit formulas

$$S(x,t)=\frac{S_M\left(e^{\lambda P(t-x)}-1\right)}{e^{\lambda S_M x}+e^{\lambda P(t-x)}-1};\qquad C(x,t)=\frac{e^{\lambda P(t-x)}p(t-x)}{e^{\lambda S_M x}+e^{\lambda P(t-x)}-1}\,. \tag{21}$$

## 7. Numerical calculation of the filtration periodic mode

Let the suspended particles concentration at the filter inlet vary periodically: $p(t)=1-\cos\pi t$. The calculation of the filtration problem (1) – (4) with the filtration coefficient (19) was performed for the parameters $\lambda=1$, $S_M=1$.

Fig. 1 shows the graphs of the suspended particles concentration: a) the time dependence at the filter inlet $x=0$ and the filter outlet $x=1$; b) the dependence on the coordinate $x$ for $t=1$ and $t=3$ (decreasing), $t=2$ and $t=4$ (increasing), and limit values for large even and odd integers $t$. The dashed line indicates the maximum value $C=2$ of the suspended particles concentration.

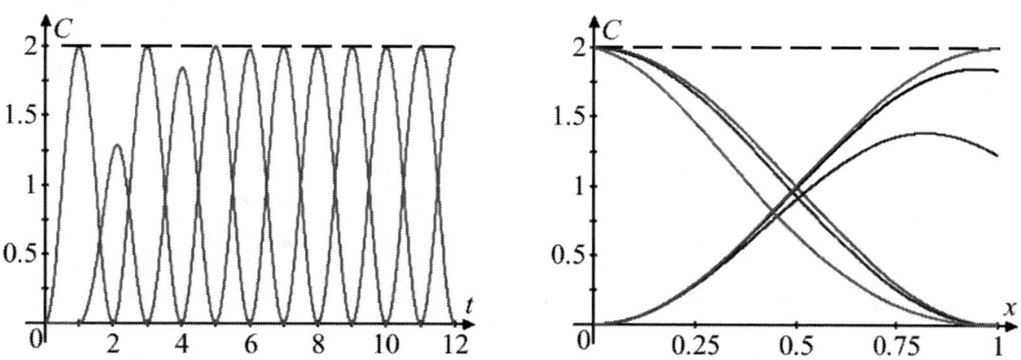


Fig. 1. a) The graphs of $C(x,t)|_{x=0}$ and $C(x,t)|_{x=1}$. b) The graphs of $C(x,t)|_{t=const}$.

Fig. 2 presents the graphs of the retained particles concentration: a) time dependence at the filter inlet (left) and filter outlet (right); b) dependence on the coordinate $x$ for $t=0.5$, 1, 2, 5 (from bottom to top). The dashed line indicates the maximum limit value $S=1$ of the retained particles concentration.

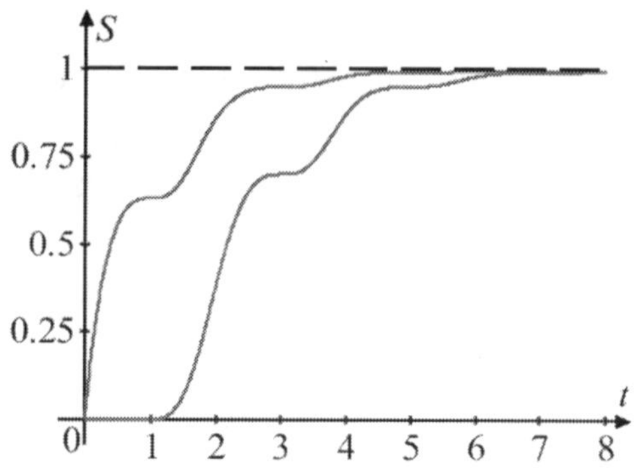


Fig. 2. a) The graphs of $S(x,t)|_{x=0}$ and $S(x,t)|_{x=1}$.

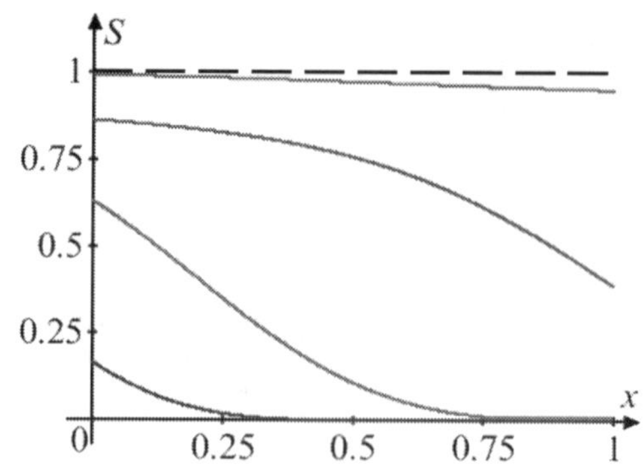


b) The graphs of $S(x,t)|_{t=const}$.

## 8. Conclusion

The exact solution of the one-dimensional filtration problem of a suspension in a homogeneous porous medium for any variable concentration of suspended particles at the filter inlet is constructed. In the case of a linear blocking filtration coefficient, explicit analytic formulas are obtained for the solution, generalizing the known solution for the flux of a suspension of constant concentration [7].

According to Fig. 1 a), b) the suspended particles concentration quickly goes to the periodic mode with the period of the injected suspension concentration at the filter inlet. With increasing time, the retained particles concentration tends to the limiting value (Figures 2a) and b)), which is determined by the blocking filtration coefficient (19).

The exact solution gives way to simplify the determination of parameters of the filtration process in laboratory and field conditions [16].

## References

1 P.G. Bedrikovetsky, *Mathematical Theory of Oil and Gas Recovery* (Kluwer 1994)
2 K.S. Basniev, N.M. Dmitriev, *Mechanics of fluid flow* (Wiley, 2012)
3 C. Noubactep, S. Caré, Chem. Eng. J. **163** (2010)
4 Z. You, A. Badalyan, P. Bedrikovetsky, SPE J. **18** (2013)
5 A. Badalyan, Z. You, K. Aji, P. Bedrikovetsky, T. Carageorgos, A. Zeinijahromi, Rev. Sci. Instrum **85**, ID 015111 (2014)
6 F. Bashtani, S. Ayatollahi, A. Habibi, M. Masihi, J. Memb. Sci. **435** (2013)
7 J. P. Herzig, D. M. Leclerc, P. Legoff, Ind. Eng. Chem. **62** (1970)
8 E.A. Vyazmina, P.G. Bedrikovetskii, A.D. Polyanin, Theor. Found. Chem. Eng. **41**, 5 (2007)
9 Z. You, P. Bedrikovetsky, L. Kuzmina, Abstr. Appl. Anal. **2013**, ID 680693 (2013)
10 Z. You, Y. Osipov, P. Bedrikovetsky, L. Kuzmina, Chem. Eng. J. **258** (2014)
11 L.I. Kuzmina, Yu.V. Osipov, Procedia Eng. **153** (2016)
12 M. Hayek, Appl. Math. Model. **38** (2014).
13 N. Sefrioui, A. Ahmadi, A. Omari, H. Bertin, Coll. Surf. A **427** (2013)
14 Yu.P. Galaguz, G.L. Safina, Procedia Eng. **153** (2016)
15 Y. Galaguz, G. Safina, Matec Web Conf. **86**, 03003 (2016)
16 P.G. Bedrikovetsky, D. Marchesin, F. Checaira, A.L. Serra, E. Resende, J. Petr. Sci. Eng. **32** 3 (2001)

# Comparative analysis of methods and results of numerical calculations of plywood panel

*Irene* Ladnykh *[1], *Alexander* Ibragimov[1]

[1]Moscow State University of Civil Engineering, Yaroslavskoye shosse 26, Moscow, 129337 Russian Federation

**Abstract.** The article discusses the new type of plywood panel for low-rise building. Calculating methods for the plywood panels are considered according with the codes of rules and the method for calculating sandwich panels with a honeycomb, proposed by Soviet scientists for aviation structures.

## 1. Introduction.

In the post-Soviet space, the plywood panels (figure 1) are widely used as wall and overlap-covering panels. They first appeared in the 40s of the XX century. The development of necessary computational formulas and reliable data on anisotropy was carried out by such scientists as E.K. Ashkenazi [1], E.V. Ganow, I.M. Linkov [2], V.M. Khrulev and others. The design of these panels is a wooden supporting frame made of timber, connected with plywood linings with glue. The main advantage of such panels is a high degree of factory readiness, but a significant disadvantage is the low degree of form stability of the main carrying ribs.

[1] Corresponding author: irenlad@yandex.ru

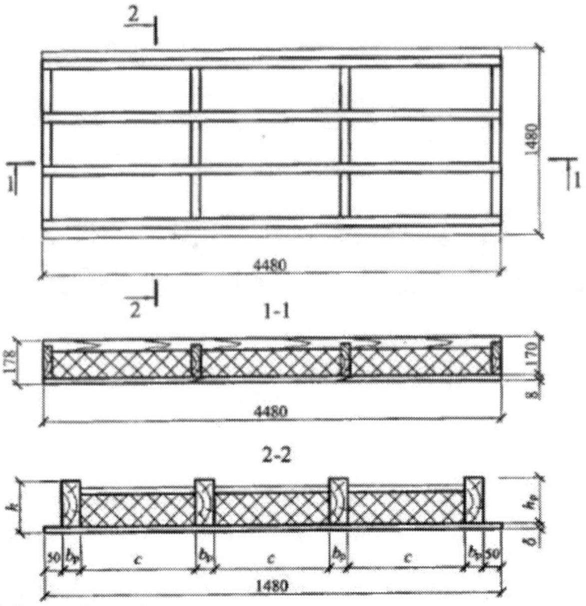


**Fig. 1** – The forms of plywood panel

In the late sixties and early seventies of the twentieth century another direction appeared in the wooden panel building based on layered structures. A huge contribution to its development was made by such scientists as V.A. Ivanov and B.E. Shungsky [3]. Laminated structures are a composition of two skins and a honeycomb core, connected with each other by glue, welding or soldering. The shape of the cells can be different, but the most common are hexagonal. Examples of honeycomb forms are shown in Figure 2.

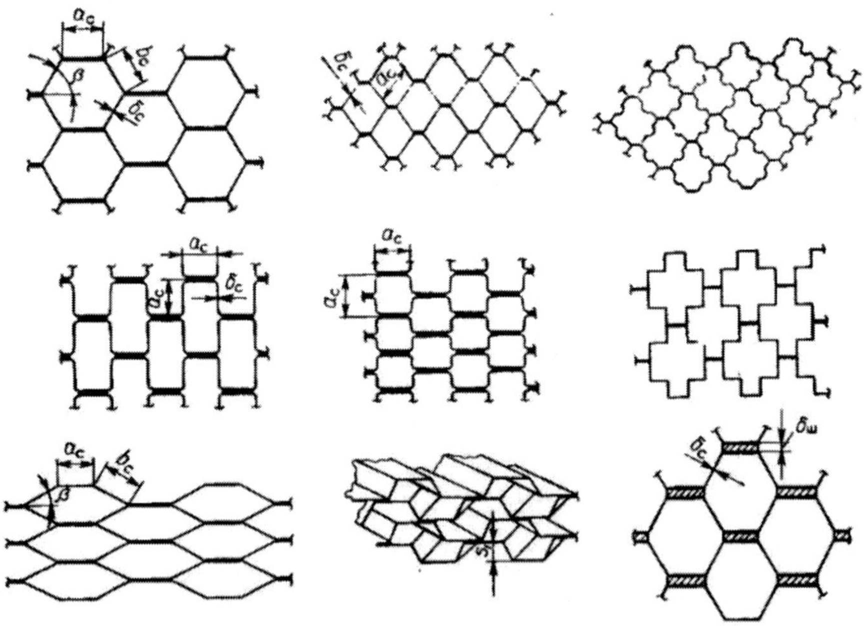

**Fig. 2** – The forms of honeycomb

The works of B.E. Shungsky and V.A. Ivanova considers a wood-paper honeycomb core with a hexagonal cell (Figure 3).

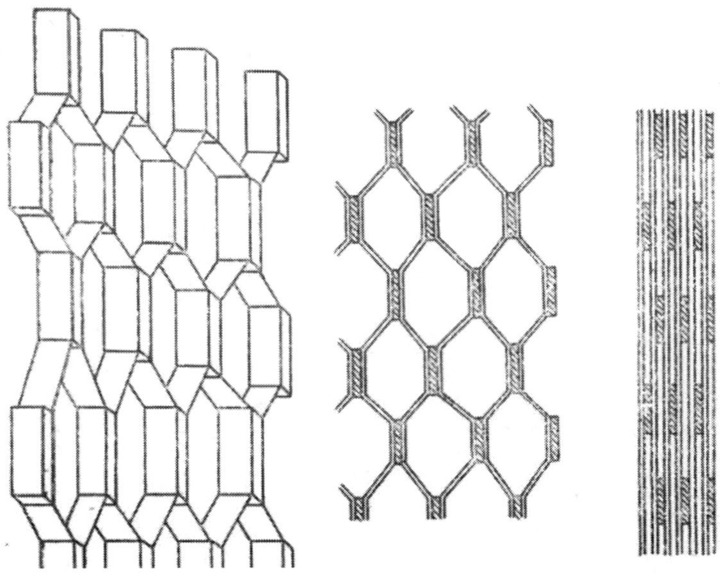

**Fig. 3** – Example of a wood-and-paper honeycomb

Based on the research data in 1968 at VDNH (Moscow) was presented the first garden house made of panels with a honeycomb core (Figure 4).

**Fig. 4** – Garden house made of panels with honeycomb - an exhibit of the USSR Exhibition of National Economic Achievements (1968)

The main disadvantages of the sandwich panels with honeycomb in the form of a hexagon is low form stability in a direction perpendicular to the position of the cells.

The authors of this article are invited to consider a new type of plywood panels with honeycomb for use in low-rise wooden housing construction, presented in Figure 6 [4]. This type of panel has several advantages in comparison with the panels presented above: a high degree of form stability. The cell is based on a triangle. However, at present, an obstacle to the widespread use of such panels is the lack of a methodology for calculating panels with a triangular cell.

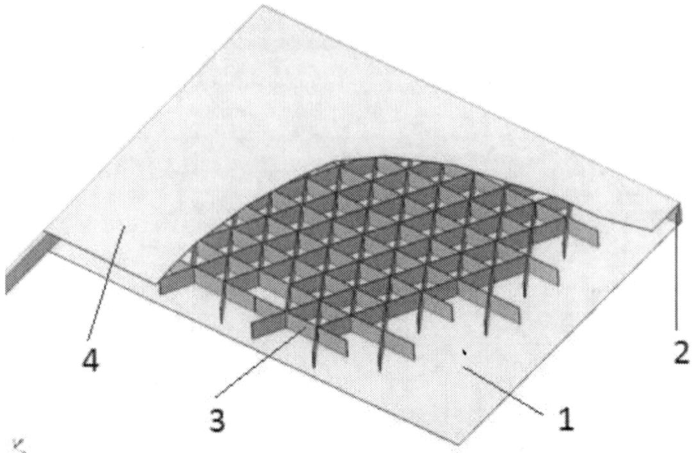


**Fig. 6** – The form of new type panel (1 – exterior covering; 2 – bundling beam; 3 – honeycomb; 4 – interior covering)

The height and width of the overlapping panels are determined by the architectural and construction requirements and the format of the sheathing sheets. The thickness of the edges is regulated by the thickness of the plywood, the thickness of the skin at the junction is equal to the thickness of the plywood except for one or two layers of veneer.

Due to the fact that the panels are intended to be used in the construction industry, the requirements, as well as structural loads, will therefore be accepted in accordance with the regulatory documents in force in the territory of the Russian Federation - a set of rules. Layered structures are calculated from two groups of limiting states: by their carrying capacity (strength or stability) and by deformation. When calculating the second limiting state, the ultimate deflections of the structures are taken in accordance with the values, given in Appendix E [5] and are equal to l/125.

Table 1 presents the numerical values of the panel's geometric characteristics, which will be used for further calculations with a view to performing a comparative analysis of the methods for calculating three-layered plywood panels with a honeycomb core.

**Table 1** - General parameters of the panel-overlapping.

| # | Symbol | Name | The numerical value for the example, unit. |
|---|---|---|---|
| 1 | $l$ | Panel length | 6100 mm |
| 2 | $b$ | Panel width | 1525 mm |
| 3 | $t$ | Panel thickness | 140 mm |
| 4 | $\delta f^{o}$ | Thickness of plywood skin | 7 mm |
| 5 | $\delta_f^{r}$ | Thickness of plywood ribs | 8 mm |

The value of the payload will be equal to 1.5 kPa, the load reliability factor is assumed to be equal to 1.3 according to [5]. When calculating the second limit state (in terms of deformations), the ultimate deflections of the structures are taken in accordance with the values given in Appendix E [5] and are 1/125.

**Methods.** In this article, we will consider two different methods for calculating sandwich plywood panels with a honeycomb.

1. Calculation using the methodology of code of rules 64.13330 "Wooden structures"

The normative documents [6, 7, 8, 9] give recommendations for the design and calculation of the glued-up panels. In [6], Section 6 describes the approach to the calculation of glued elements from plywood with wood, where the main method is the method of reduced cross-section. No matter how the panel is cut in the width direction, the same number of edges will fall into the section. To simplify further mathematical calculations and calculations, we introduce an additional value of n, which will be called the number of plywood ribs in the air gap over the width of the common b'.

$$n = 2 \cdot \frac{b' - \frac{a}{2} \cdot \tan\alpha}{\delta^{\mathrm{r}}_{\mathrm{f\phi}} + \frac{a}{2} \cdot \tan\alpha} + 1, \tag{1}$$

According to the technique, we calculate the geometric characteristics: the moment of inertia, the moment of resistance, the static moment with respect to the neutral axis. The moment of inertia of a given section will be:

$$I_{\mathrm{pr}} = I_f + n \cdot I^{\mathrm{r}}_{\mathrm{f}} + 2 \cdot I_{\mathrm{br}} \cdot \frac{E_{\mathrm{dr}}}{E_{\mathrm{f}}} \tag{2}$$

where $I_{\mathrm{f}}$ – moment of inertia of plywood panels skin
$I^{\mathrm{r}}_f$ – moment of intertia of plywood panels ribs
$I_{\mathrm{br}}$ – moment of intertia of beam
$E_{\mathrm{dr}}$ – elastic modulus of timber
$E_f$ – elastic modulus of plywood.

Table 2 shows the values of the geometrical characteristics of the cross section.

**Table 2** - Geometrical characteristics of a section

| # | Symbol | Name | The numerical value for the example, unit |
|---|---|---|---|
| 1 | $I_{pr}$ | Moment of inertia | 14010 $sm^4$ |
| 2 | $W_{pr}$ | Resistance moment | 2002 $sm^3$ |
| 3 | $S_{pr}$ | Static moment | 811 $sm^3$ |

We perform the basic verification calculations:

1. For the first group of limiting states:
1.1. Verification of the strength of the expanded plywood paneling of panels and panels, according to paragraph 6.25 [6].
1.2. Check for the stability of the compressed sheathing of slabs and panels in accordance with paragraph 6.28 [6].
1.3. Check for the chipping of the edges of the frame of panels and panels or sheathing along the seam at the junction with the ribs.
2. For the second group of limiting states:
2.1. Check the panel for the ultimate deflection.

The main results of the calculation are presented in Table 3.

2. Calculation by the method of calculating the honeycomb aggregate

The second method, which can be applied to the calculation, is the calculation of layered structures, based on the theory of A.Ya. Aleksandrov set forth in the works [10, 11, 12].

The main condition for calculating three-layer structures with a honeycomb core is the knowledge of the mechanical characteristics of the honeycomb core itself. Therefore, the reduced mechanical characteristics of the aggregate are first determined, which is a kind of conditional solid material endowed with all the properties of a real honeycomb core, and occupying the same volume as the honeycomb core itself.

In our case, let us single out the typical element of the honeycomb block, presented in Figure 7, for which we will determine the reduced mechanical characteristics.

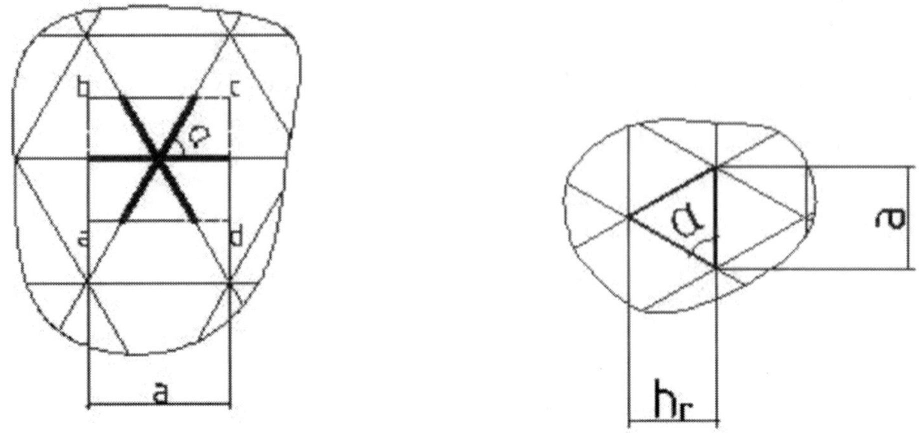

**Fig. 7** – A typical honeycomb element of panel

The main mechanical characteristics are: the module of normal elasticity $E_Z$; Shift modules $G_{XY}$, $G_{YZ}$; Allowable voltages $[\sigma_Z]$, $[\tau_{XZ}]$, $[\tau_{YZ}]$. It should be noted that the most difficult is to determine the shear moduli. Until now, there is not a single technique that could accurately reflect the actual value of the shear modulus, so an approximate value is taken in the calculations or the value is taken from the experimental data for the honeycomb.
Further:
1. Determine the critical stress loss of general and local stability for an ideally manufactured plate.
The deflection of the plate is determined as follows:

$$\omega = \sum_{n=1}^{n} \sum_{m=1}^{m} f_{mn} \sin\frac{m\pi x}{l} \sin\frac{n\pi y}{b} \tag{3}$$

$$f_{mn} = \frac{16\rho l^4}{\pi^6 mn\left\{D\left(m^4+2\frac{m^2n^2}{\psi_L^2}+\frac{n^4}{\psi_L^4}\right)-N_x\left[\frac{l^2}{\pi^2}+\frac{D}{hG_{xz}(m^2+\frac{n^2}{\psi_L^2})}\right]m^2\right\}} \tag{4}$$

where m=1,3,5,…, (2k-1); k=1,2,3,...; n=1,3,5,…, (2r-1); r=1,2,3,...; $\psi_L = \frac{b}{l}$
The maximum values of the bending moments (in the center of the plate):

$$M_{xmax} = \sum_{m=1}^{m} \sum_{n=1}^{m} f_{mn} \frac{\pi^2 D}{l^2}(m^2 + \mu^2 \frac{n^2}{\psi_L{}^2}) \sin\frac{m\pi}{2} \sin\frac{n\pi}{2} \tag{5}$$

$$M_{ymax} = \sum_{m=1}^{m} \sum_{n=1}^{m} f_{mn} \frac{\pi^2 D}{l^2}(\mu m^2 + \frac{n^2}{\psi_L{}^2}) \sin\frac{m\pi}{2} \sin\frac{n\pi}{2} \tag{6}$$

Maximum shearing forces (on the support contour in the middle of the support edges):

$$Q_{xmax} = \sum_{m=1}^{m} \sum_{n=1}^{m} f_{mn} \frac{\pi^3 D}{l^3}(m^2 + \frac{n^2}{\psi_L{}^2}) \sin\frac{n\pi}{2} \tag{7}$$

$$Q_{ymax} = \sum_{m=1}^{m} \sum_{n=1}^{m} f_{mn} \frac{\pi^3 D}{l^3}(m^2 + \frac{n^2}{\psi_L{}^2}) \sin\frac{m\pi}{2} \tag{8}$$

Maximum normal stresses in the bearing layers of the plate:

$$\sigma_{1x} = \frac{M_{xmax}}{H\delta_1} + \frac{N_x}{\delta_1(1+\psi_\delta)} \tag{9}$$

$$\sigma_{2x} = \frac{M_{xmax}}{H\delta_1\psi_\delta} - \frac{N_x}{\delta_1(1+\psi_\delta)} \tag{10}$$

$$\sigma_{1y} = \frac{M_{ymax}}{H\delta_1} \tag{11}$$

$$\sigma_{2y} = \frac{M_{ymax}}{H\delta_1\psi_\delta} \tag{12}$$

where $\psi_\delta = \frac{\delta_2}{\delta_1}$; δ1 – thickness of a carrier layer compressed under bending; δ2 – Thickness of the bending layer.
The maximum tangential stresses in the honeycomb:

$$\tau_{xzmax} = \frac{Q_{xmax}}{h} \quad (13)$$

$$\tau_{yzmax} = \frac{Q_{ymax}}{h} \quad (14)$$

Checking the strength of the honeycomb:

$$\frac{\sigma_z}{[\sigma_z]} + \left(\frac{\tau_{xz}}{[\tau_{xz}]}\right)^2 \leq 1 \quad (15)$$

2. Determine the forms of general and local loss of stability.
3. Given initial imperfections, a further calculation is carried out to determine the bearing capacity for wide plates.

**Results and Discussion.** The main results of the calculation are presented in Table 3.

**Table 3** – Calculation results

| # | Symbol | The method #1 | The method #2 |
|---|---|---|---|
| 1 | $\sigma$ | 6.796 MPa | 5.874 MPa |
| 2 | $f$ | 32.0 mm | 24.8 mm |

**Conclusions.** The analysis of methods and approaches to calculation, stated above, allows us to draw the following conclusions:

1. The procedure for calculating the glued panel according to [6] is simpler, and the calculation can be performed manually.
2. Calculation using the procedure [10,11,12] is more time-consuming, requires a large amount of experimental data or the introduction of approximate values, which must be further confirmed by experimental data.
3. Comparing the results of the calculations presented in Table 3, we note that the value of normal stresses and deflection is greater when calculated by the method proposed in the JV. This is explained by the fact that the second technique takes into account the greater number of parameters, both geometric and mechanical.

For the final choice of calculation methods, the authors suggest, compare the results obtained with the results of computer simulation and full-scale experiment.

## References

1. E. K. Ashkenazi, Anisotropy of wood and wood materials. - Moscow: Forest Industry, (1978)
1. I.M. Linkov, Development and research of structures of glued plywood reinforced panels. Tr. CNIISK them. V.A. Kucherenko. Moscow (1972)
2. B.E. Shungsky, Building structures with honeycomb aggregates. - M.: Stroiizdat, (1977)
3. Patent of the Russian Federation No. 160223. Panel / Tikhomirov LA, Zaitseva KV, Titunin AA, Ibragimov AM, Gnedina L.Yu .; Declared. 07/20/2015. Opubl. 03/10/2016.
4. SP 20.13330.2011 "Loads and impacts. Updated version of SNiP 2.01.07-85 * ». Moscow: FGUP TsPP, (2011)
5. SP 64.13330.2011 "Wooden structures. Updated version of SNiP II-25-80. Moscow: FGUP TsPP, (2011)
6. Manual on the design of wooden structures (to SNiP II-25-80) / TsNIISK im.Kucherenko // Moscow: Stroiizdat, (1986)

7. Recommendations on the design of panel structures using wood and wood materials for industrial buildings) / TsNIISK im.Kucherenko // M .: Stroiizdat, (1982)
8. Guide to the design of glued wooden structures. / TsNIISK im.Kucherenko // Moscow: Stroyizdat, (1977)
9. A. Ya. Alexandrov, E.P. Trofimova, Determination of the reduced elastic parameters of honeycomb aggregates: coll. Questions of calculating the elements of aviation structures. - Moscow: Oborongiz, (1959)
10. A.I. Endogur, M.V. Weiberg, Jerusalem KM Cellular constructions. Selection of design parameters. - M.:, Mechanical Engineering (1986)
11. M. Ya. Gofin, A.A. Ivanov, Mechanics of honeycomb structures: reference book in 2 volumes. - M .: TF "Mir" (2012)

# Investigation of aerodynamic instability of a thin plate

*Irina* Lantsova [1*] and *Irina* Afanasyeva [1,2]

[1]Moscow state university of civil engineering, Yaroslavskoye shosse, 26, Moscow, Russia, 129337
[2]Research & Educational Center StaDyO Ltd, Office 810, 18, 3-ya Ulitsa Yamskogo Polya, Moscow, 125040, Russia

**Abstract.** This paper is devoted to methods and principles of investigation of aerodynamic instability of structures. Object of interest is a thin plate. This is an important aspect of the design of large-span structures, bridges and other structures sensitive to wind loads. As Russian standards and Eurocode obliges to check the conditions for the occurrence of galloping, divergence and flutter for a certain class of structures. Modern computing facilities allow to calculate aerodynamic coefficients with high accuracy.

## 1 Introduction

The relevance of this work is determined by the need to take into account the sensitivity of structures to wind loads that produces aeroelastic vibrations, which can lead to collapse of the structure. This type of structures includes the large-span structures, large-span bridges, thin-walled cooling towers, high-rise buildings, roofs of long-span stadiums and so on.

Exciting analytical, experimental and numerical techniques that allow to assess the possibility of the occurrence of such aerodynamic instability effects as vortex shedding, galloping or divergence need to be improved [1, 2]. Nowadays the modern numerical methods and computational tools let to perform detailed simulations of the aeroelastic behavior of structures and fluid flow around it, but it is necessary to verify and validate them.

## 2 Problem statement

The object of interest is a thin plate (fig. 1) placed in a viscous incompressible turbulent flow ($Re = 7 \cdot 10^5$). The purpose of this study is an identification of the occurrence of such aerodynamic instability effects for the interesting object, such as vortex shedding, galloping or divergence. We consider two-dimensional problem (at the plane OXY). Therefore, in the third dimension (along the axis OZ) the domain has one element and size equal 0.1 m. For simulation of the turbulent flow RANS/URANS SST turbulence models are used (for steady state and unsteady simulations respectively). Simulation time for unsteady simulations is 40 s with Time step equal 0.005 s.

* Corresponding author: irina-lanzova@mail.ru

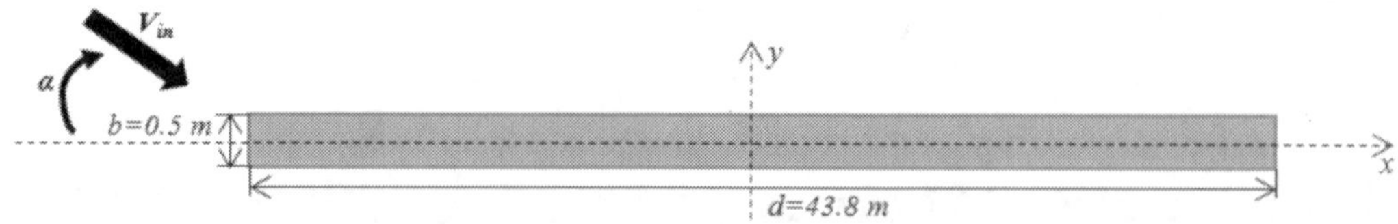


**Fig. 1.** Geometrical parameters of the model cross-section. $\alpha$ – angle of wind attack.

Following boundary conditions are set (Fig. 2):

- velocity at the domain *inlet* is constant and equal *21.7 m/s*, turbulence kinetic energy *42.9* $m^2/s^2$ and turbulence eddy dissipation *0.94* $m^2/s^3$;
- «Opening» boundary conditions are defined at domain *outlet* with zero average relevance pressure, turbulence kinetic energy equal *42.9* $m^2/s^2$ and turbulence eddy dissipation equal *0.94* $m^2/s^3$;
- «No slip wall» boundary conditions on the surface of the *plate* are set;
- «Symmetry» boundary conditions are set at the *left and right side* of the domain.

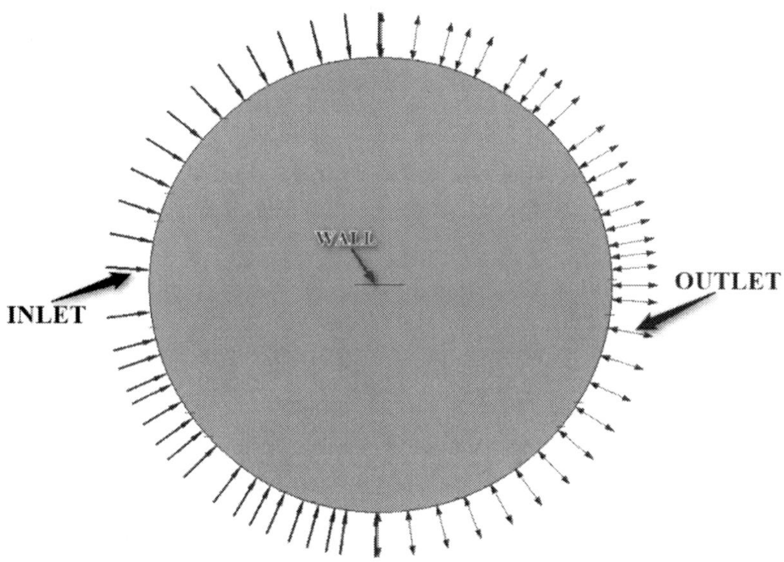


**Fig. 2.** Computational domain and boundary conditions.

## 3 Results

Computational domain was meshed by ANSYS Mechanical preprocessor. We considered 5 different computational grids (table 1). As controlled parameters following characteristics are chosen: aerodynamic coefficients of the drag (*Cd*) and the lift (*Cl*) forces. *Model 2* (fig. 3) was chosen as a basic model for steady state simulations, *Model4* (fig. 3) was chosen as a basic model for unsteady simulations.

**Table 1.** Variants of the computational grids and the corresponding values of the aerodynamic coefficients of the drag (*Cd*) and the lift (*Cl*) forces. Angle of the wind attack $\alpha=1°$.

| Name | Number of nodes | Y+ | Cd | Cl |
|---|---|---|---|---|
| **Model 1** | 745 294 | ~2 500 | 2.100 | -0.069 |
| **Model 2** | 1 014 764 | ~1 300 | 2.045 | -0.070 |
| **Model 3** | 1 676 424 | ~700 | 2.017 | -0.071 |
| **Model 4** | 1 827 780 | ~ 100 | 2.000 | -0.071 |
| **Model 5** | 338 092 | ~43 000 | 2.257 | -0.064 |
| **Ref. [3]:**<br>- plate with a square cross-section<br>- infinite thin plate with a rectangular cross-section | - | - | 2.03<br>2.12 | 0<br>- |

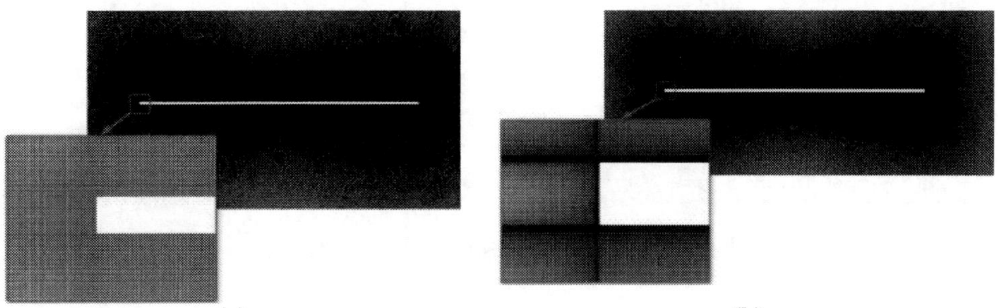

(a) (b)
**Fig. 3.** Computational grid. (a) *Model 2* (1 014 764 nodes) and (b) *Model 4* (1 827 780 nodes).

In Ref. [3] the author shows the dependence of the aerodynamic coefficients of the drag (*Cd*) and the lift (*Cl*) forces on angle of attack $\alpha$ for the infinite thin plate with a rectangular cross-section. Graphs of the dependence of the obtained aerodynamic coefficients on the angle of attack α in comparison with the coefficients from [3] are presented below.

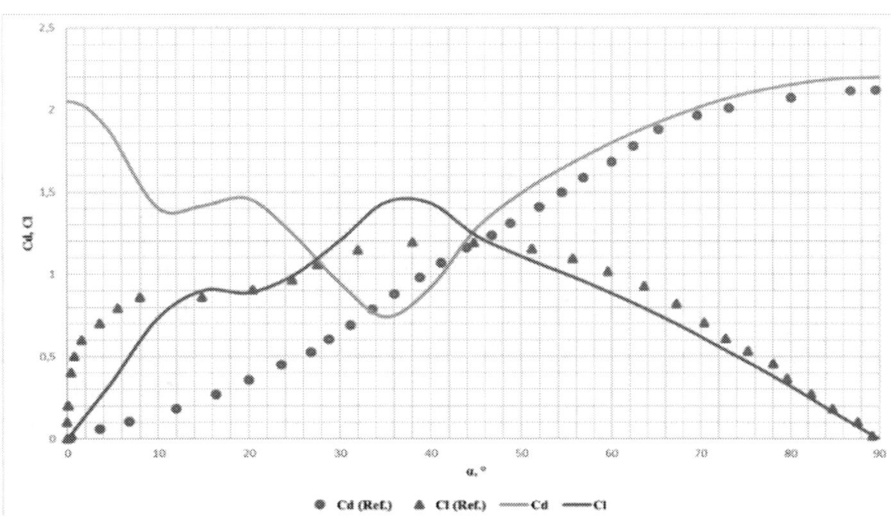


**Fig. 3.** Aerodynamic coefficients of the drag (*Cd*) and the lift (*Cl*) forces depending on the angle of attack $\alpha$. Comparison of the results of the simulations (Model 2) with the reference [3].

### 3.1 Estimation of galloping, vortex shedding and divergence

#### *3.1.1 Galloping*

Galloping is a self-excited oscillations of flexible structures in the form of bending vibrations along the normal to the direction of the wind

As a result of researches by scientists such as Den-Hartog and Glowrt, the necessary condition for the occurrence of an aerodynamic instability was obtained:

$$H = C_D + \frac{dC_L}{d\alpha} < 0 \tag{1}$$

According to the Eurocode [4], aerodynamic instability with galloping occurs when the wind speed $V_m$ reaches a value $V_{CG}$ at which the oscillating process begins with an increasing amplitude.

$$V_{CG} > 1.25 \cdot V_m \tag{2}$$

The critical wind velocity of galloping, $V_{CG}$, is given in Expression (3):

$$V_{CG} = \frac{2 \cdot Sc \cdot f_i \cdot b}{a_g} \tag{3}$$

where:

$f_i$ - cross-wind fundamental frequency of the structure;

b - the width;

$a_g$ - factor of galloping instability;

Sc - Scruton number:

$$Sc = \frac{2 \cdot \delta \cdot m}{\rho \cdot b^2} \tag{4}$$

where:

$\delta$ - structural damping expressed by the logarithmic decrement;

$\rho$ - air density under vortex shedding conditions;

m - equivalent mass;

b - reference width of the cross-section at which resonant vortex shedding occurs.

Values of the aerodynamic coefficients of the drag (*Cd*) and the lift (*Cl*) forces at various angles of attack were determined by performing steady state simulations using specialized software ANSYS CFX. Then, according to formula (1), the Glowrt – Den-Gartog criterion *H* was calculated, and the critical velocity of the beginning of the oscillatory process $V_{CG}$ was determined using formula (3) (at *fl* = 0.4 Hz, *Sc* = 551; *ag* = *ks*·*H*, where *ks* = 0.9). For all angles, the critical velocity exceeded the value of $1.25 \cdot V_m = 27.1$ m / s, despite the fact that for some angles of attack the Glowrt – Den-Gartog criterion *H* is fulfilled (i.e. there is a possibility of the occurrence of the galoping). In the table 2, the angels of attack for which the Glowrt – Den-Gartog criterion *H* is fulfilled are marked in red.

**Table 2.** Aerodynamic coefficients of the drag (*Cd*) and the lift (*Cl*) forces, the Glowrt – Den-Gartog criterion *H*, the critical wind velocity $V_{CG}$ on angle of the wind attack α.

| Angle α, ° | Cd | Cl | H | $V_{CG}$, m/s |
|---|---|---|---|---|
| *0* | 2.053 | 0.000 | -1.967 | 124 |
| *1* | 2.045 | -0.070 | -1.975 | 124 |
| *2* | 2.021 | -0.141 | -2.024 | 121 |
| *3* | 1.976 | -0.211 | -2.068 | 118 |
| *4* | 1.917 | -0.282 | -2.112 | 116 |
| *5* | 1.846 | -0.352 | -2.210 | 111 |
| *10* | 1.404 | -0.737 | -2.998 | 82 |
| *15* | 1.419 | -0.902 | -0.474 | 517 |
| ***20*** | 1.461 | -0.887 | 1.625 | 151 |
| ***25*** | 1.237 | -0.995 | 0.000 | 662657 |
| *30* | 0.947 | -1.203 | -1.431 | 171 |
| *35* | 0.742 | -1.436 | -1.927 | 127 |
| ***40*** | 0.918 | -1.433 | 0.955 | 256 |
| ***45*** | 1.274 | -1.235 | 3.540 | 69 |
| ***50*** | 1.493 | -1.108 | 2.944 | 83 |
| ***55*** | 1.661 | -0.997 | 2.940 | 83 |
| ***60*** | 1.801 | -0.884 | 3.094 | 79 |
| ***65*** | 1.920 | -0.759 | 3.354 | 73 |
| ***70*** | 2.021 | -0.620 | 3.606 | 68 |
| ***75*** | 2.099 | -0.474 | 3.780 | 65 |
| ***80*** | 2.156 | -0.319 | 3.931 | 62 |
| ***85*** | 2.191 | -0.160 | 4.012 | 61 |
| ***90*** | 2.202 | 0.000 | 4.034 | 61 |

### *3.1.2 Vortex shedding*

Vortex shedding occurs if the vortices periodic break from the opposite edges of the structure and, as the result, variable load perpendicular to the direction of the wind occur.

If the frequency of the vortex excitations coincides with the natural frequency of the structure, large amplitude oscillations may arise. This occurs at the so-called critical wind speed:

$$V_{cr,i} = \frac{b \cdot f_i}{Sh} \tag{5}$$

where:

b - reference width of the cross-section at which resonant vortex shedding occurs and where the modal deflection is maximum for the structure or structural part considered;

$f_i$ - natural frequency of the considered flexural mode i of cross-wind vibration;

*Sh* - Strouhal number.

According to [4], the effect of vortex shedding should be investigated when the ratio of the largest to the smallest crosswind dimension of the structure, both taken in the plane perpendicular to the wind, exceeds 6.

The effect of vortex shedding need not to be investigated when

$$V_{cr,i} > 1.25 \cdot V_m \tag{6}$$

where:

$V_{cr,i}$ - is the critical wind velocity for mode i;

$V_m$ - is the characteristic 10 minutes mean wind velocity.

Below there are the results of unsteady simulations, performed using the specialized software ANSYS CFX. The main frequencies of vortex shedding at various angles of attack were determined (Fig.4). Then, for two lowest (main) frequencies the Strouhal numbers *Sh* were calculated and the critical wind speed (at $f_1$ = 0.4 Hz) was determined using formula (5). For all angles of attack the critical wind speed did not exceed the value $1.25 \cdot V_m = 27.1$ m/s; therefore, according to the Eurocode, vortex shedding can occur at the considered angles of the wind attack (table 3).

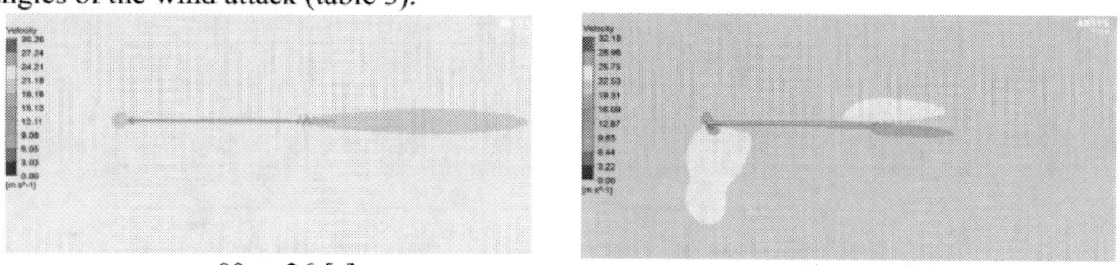

α=0°, t=26 [s] α=5°, t=26 [s]

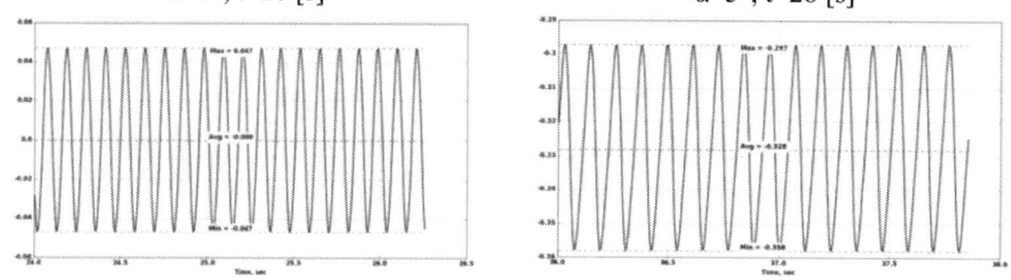

α=0°, Cl α=5°, Cl

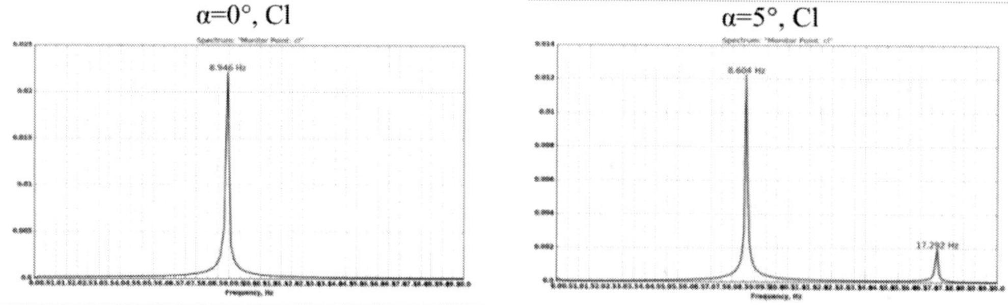

α=0°, f [Hz] α=5°, f [Hz]

**Fig.4.** Results of the unsteady simulations (*Model 4*) for angel of attack α=0° and α=5°– *from left to right*. Field of instantaneous velocity [m/s], aerodynamic coefficient of the lift force (*Cl*), PSD for *Cl* – *from top to bottom*.

**Table 3.** Strouhal numbers *Sh* and critical wind velocity $V_{cr,1}$ on angle of the wind attack α.

| Angle α, ° | Sh1 | Sh2 | Vcr,1(Sh1) | Vcr,1(Sh2) |
|---|---|---|---|---|
| *0* | 0.205 | - | 0.97 | - |
| *1* | 0.204 | 0.407 | 0.98 | 0.49 |
| *2* | 0.201 | 0.402 | 1.00 | 0.50 |
| *5* | 0.200 | 0.399 | 1.00 | 0.50 |

### *3.1.3 Divergence and Flutter*

According to [4], such types of dynamic instability as divergence and flutter, which occur if structural deformations cause an aerodynamic loads changing, should be excluded in principle. To be prone to either divergence or flutter, the structure satisfies all of the three criteria specified in the Eurocode. If all the conditions are met, it is necessary to check the structure under study for the possibility of a flutter or divergence arising from the condition:

$$V_{div} > 2 \cdot V_m \tag{7}$$

The critical wind velocity for divergence $V_{div}$ is given in Expression (8):

$$V_{div} = \left[ \frac{2 \cdot k_\Theta}{\rho \cdot d^2 \cdot \frac{dc_M}{d\Theta}} \right]^{\frac{1}{2}} \tag{8}$$

where:

$k_\Theta$ - torsional stiffness; $c_M$ - aerodynamic moment coefficient; $\frac{dc_M}{d\Theta}$ - rate of change of aerodynamic moment coefficient with respect to rotation about the torsional centre, $\Theta$ is expressed in radians; M - aerodynamic moment of a unit length of the structure;

$\rho$ - density of air; d - in wind depth (chord); b - width.

Results of the unsteady simulations are shown below. Values of the aerodynamic moment *Cmz* at different angles of attack were determined. Since all three conditions specified in the Eurocode are satisfied for the present structure, the critical wind speed for divergence was calculated (for $k_\Theta = 14.23 \cdot 10^4$ Nm$^2$, $\rho = 1.185$ kg / m$^3$) for all angels of attack according to formula (8). In the table 4, those angles for which the critical wind speed $V_{div}$ does not exceed $2 \cdot V_m = 43.4$ m/s are marked in red. For these angles, when a corresponding wind speed is reached, a divergence may occur (Table 4).

**Table 4.** Aerodynamic moment coefficients Cmz and critical wind velocity $V_{div}$ on angle of the wind attack α.

| Angle α, ° | Cmz | $V_{div}$, м/с |
|---|---|---|
| *0* | 0.000 | 10.39 |
| *1* | 0.020 | 10.39 |
| *2* | 0.040 | 10.43 |
| *3* | 0.060 | 10.64 |

| Angle α, ° | Cmz | $V_{div}$, м/с |
|---|---|---|
| 4 | 0.078 | 10.83 |
| 5 | 0.097 | 10.84 |
| 10 | 0.179 | 11.55 |
| 15 | 0.140 | 16.85 |
| 20 | 0.123 | 25.11 |
| 25 | 0.123 | 360.18 |
| 30 | 0.126 | 59.85 |
| 35 | 0.127 | 91.26 |
| 40 | 0.127 | -0.005 |
| 45 | 0.124 | -0.030 |
| 50 | 0.119 | -0.066 |
| 55 | 0.110 | -0.098 |
| 60 | 0.099 | -0.124 |
| 65 | 0.086 | -0.149 |
| 70 | 0.071 | -0.171 |
| 75 | 0.055 | -0.187 |
| 80 | 0.037 | -0.203 |
| 85 | 0.019 | -0.211 |
| 90 | 0.000 | -0.216 |

## 4 Conclusions

Present paper is devoted to the computational investigations of aerodynamic instability of thin flat plate placed in a viscous incompressible turbulent flow. For this structure a check of the conditions for the occurrence of galloping, divergence and flutter was conducted. Obtained results can be used as a reference for the investigation of the sensitivity of structures like thin roofs of the large-span stadiums or bridges to wind loads.

## Acknowledgments

The Reported study was funded by Government Program of the Russian Federation «Development of science and technology» (2013-2020) within Program of Fundamental Researches of Ministry of Construction, Housing and Utilities of the Russian Federation and Russian Academy of Architecture and Construction Sciences, the Research Project 7.1.2 and Ministry of Education and Science of the Russian Federation (RF President Grant, agreement #14.Z56.16.8493-MK).

## References

1. N. I. Afanasyeva, I. Yu. Lantsova, Numerical simulation of an elastic structure behavior under transient fluid flow excitation AIP Conference Proceedings 1800, Issue 1, (2017)
2. E. Simiu, R. Skanlan, Wind influence on buildings and structures. Moscow: Stroiizdat, (1984)
3. G.A. Savitsky Wind load on structures. Moscow: Publishing House of Literature on Construction (1972)
4. EN 199114:2005. Eurocode 1: Actions on structures – Part 14: General actions – Wind actions. Brussels: CEN TC 250, (2004)

# Statistical learning problem of artificial neural network to control roofing process

*Azariy* Lapidus[1], *Aleksandr* Makarov[1*]

[1]Moscow state university of civil engineering, Yaroslavskoye shosse, 26, Moscow, Russia, 129337

**Abstract.** Now software developed on the basis of artificial neural networks (ANN) has been actively implemented in construction companies to support decision-making in organization and management of construction processes. ANN learning is the main stage of its development. A key question for supervised learning is how many number of training examples we need to approximate the true relationship between network inputs and output with the desired accuracy. Also designing of ANN architecture is related to learning problem known as "curse of dimensionality". This problem is important for the study of construction process management because of the difficulty to get training data from construction sites. In previous studies the authors have designed a 4-layer feedforward ANN with a unit model of 12-5-4-1 to approximate estimation and prediction of roofing process. This paper presented the statistical learning side of created ANN with simple-error-minimization algorithm. The sample size to efficient training and the confidence interval of network outputs defined. In conclusion the authors predicted successful ANN learning in a large construction business company within a short space of time.

## 1 Introduction

In recent years, neural networks developed and tested for construction project organization and management can support decision making for construction process safety [1] and project management [2], they are also able to predict and evaluate specific construction processes such as: building underground structures [3], road building [4], earthwork [5], and other. The authors conduct a research in construction process organization and management to build roof structures at the planning and operation stage, assisted by the ANN methodology. Previously, the network architecture was designed (fig. 1) and functioning of each layer was determined. The main parameters of the system (*Qw* – qualification of workers, *Cs* – construction supervision, *It* – information technology, *Ts* – team structure, *Wc* – weather condition) are located on the second layer. These parameters are divided into basic organizational and technological factors (for example, $g_1$ - working experience and competencies, $g_4$ – foreman). Third layer includes four evaluation criteria of construction process (*T* – time, *S* – safety, *Q* – quality, *C* – cost) and last layer merges all information that passed through the network into an integrated indicator *P* (construction process potential). To measure ANN inputs, the authors used three linguistic variables represented by fuzzy sets [6] with arguments inside the interval [0,1]. The operator – an

* Corresponding author: anmakarof@yandex.ru

expert in the field – estimates each factors by determining a linguistic term and selecting a point of the membership function graph. The abscissa vector

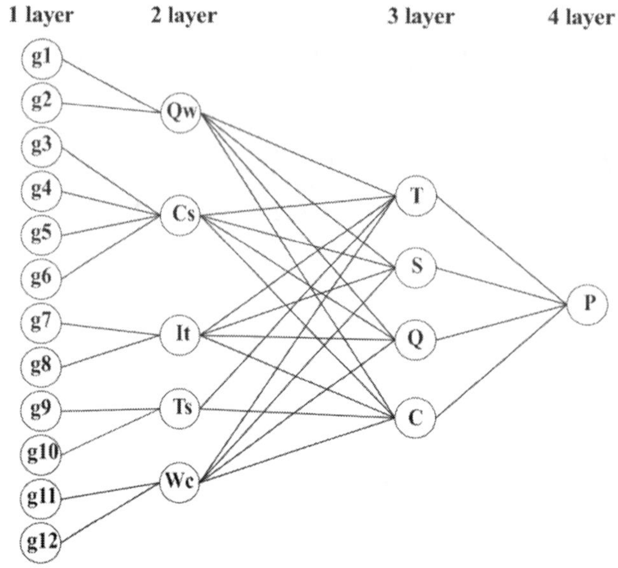


**Fig. 1.** ANN architecture.

of these points for all factors is input data of the network. ANN learning is a process whereby weight coefficients of synapses are tuned in a certain way so that the network deduces reliable results. A number of approaches to ANN learning exist. For a multilayer perceptron, supervised learning by mean the backward propagation of errors is one tried and trusted approach. This algorithm needs a training sample comprised of an input signal vector and a corresponding output signal vector. Formation this sample in real construction conditions is a laborious process that requires open access of researchers to all stages and participants of construction process. The organization of that access is impossible without the direct interest of the decision maker of construction project. Therefore, it is very important to estimate in advance a sufficient training sample size that must be obtained from construction site for the successful learning of the created ANN. Also providing enough training examples to make the learning algorithm statistically reliable is a key objective in the machine learning theory [7], and we are going to discuss a solution to it for the developed ANN.

## 2 Methods

ANN learning can be described as search for function $F(x,w)$ in the space of all possible function $\Omega$ , which approximates work of a real-life system with the permissible departure/error and sufficient reliability. According to the statistical learning theory [8], the objective is to minimize risk functional by parameter $w$:

$$R(w) = \int Q(x,y,w)P(x,y)dydy = \int [y - F(x,w)]^2 P(x,y)dxdy, \tag{1}$$

where $Q(x,y,w)$ - loss-function, $P(x,y)$ - unknown joint probability distribution of the network's ins and outs.

Because recovering density of probability distribution over the sampling is a labor-intensive task, all ANN learning algorithms based on precedents, particularly backpropagation, seek to minimize the empirical risk

$$R_{emp}(w) = \frac{1}{N}\sum_{N}[y - F_{emp}(x, w)]^2 \qquad (2)$$

Since a training sample size is always finite, the main issue about reaching the stopping criterion is about how function $F_{emp}(x, w)$ successfully obtained from experimental data can minimize the risk functional (1) – in other words, how the error after learning will differ from the network's error in real-life operation.

Solution to this objective is related to deciding the condition of convergence of empirical averages to expected values [9]; it appears as

$$P\{\sup | R_{emp}(w) - R(w) | \succ \varepsilon\} \prec p, \qquad (3)$$

and it means that with probability of $1 - p$, disparity $R_{emp}(w) - \varepsilon \le R(w) \le R_{emp}(w) + \varepsilon$ is true simultaneously for all input values; where $\varepsilon$ - bound of confidence interval or error boundary.

As they analyzed the conditions of convergence (3) for training samples of finite length $n = N$ and as they generalized the Glivenko theorem for classification of images, Soviet researchers V. Vapnik and A. Chervonenkis introduced the notion of capacity $h$ of function set $\Omega$, known internationally as the Vapnik-Chervonenkis dimension or *VC-dimension* [10]. We used this parameter to find permissible conditions for regular convergence over the sample size of $N \succ h$, from which one can isolate the expression for error boundary:

$$| R_{emp}(w) - R(w) | \le \frac{\varepsilon_0}{2}\left(1 + \sqrt{1 + \frac{4R_{emp}(w)}{\varepsilon_0}}\right), \text{ where} \qquad (4)$$

$$\varepsilon_0 = 4\frac{h(\ln\frac{2N}{h} + 1) - \ln p}{N}. \qquad (5)$$

Since $R_{emp}(w)$ □ $1$, this expression means that the sample size depends on $h$ alone. For neural network with sigmoid activation function, the established range of *VC-dimension* [11, 12] has the following orders: $\inf h = O(W^2)$, $\sup h = O(W^4)$, where W – number of free network parameters, i.e. synaptic weights.

Training of a created network stops at Layer 3, and the synaptic weights of neuron $P$ are beyond the training array; then $W = 32$ and *VC-dimension* will have the following orders:

$$\inf h = O(10^3), \; \sup h = O(10^6). \qquad (6)$$

Considering (4, 5) and the condition $N \succ h$ we find that training sample size is bounded by the values of (6).

The resulting theoretical training sample size sufficient to ensure statistical convergence of the ANN has a very large gaps between its boundaries; one of the reasons is that this is a conservative estimate for a scenario with random initial values of weights, zero *a priori* information about the system except the number of links, and no consideration for the specifics of different training algorithms. Subsequently, known results from practical runs of different neural networks demonstrated that to implement statistical convergence in real-life missions, a probably approximately correct learning (PAC-learning) [13, 14] is more preferable; it was also found that for the classification problem, the complexity of the training sample depends more on weights than on their number [15]. This is why, to evaluate the capacity of the function set $\Omega$ we proposed that the *VC-dimension* be substituted with its upscale version: *fat-shattering dimension* [16, 17], provided that the ANN learning algorithm ensures maximized share of accurate output values, i.e. with minimum error in learning data subjected to the restriction

$$\frac{1}{N}\Sigma[y - F_{emp}(x,w)]^2 \le \frac{1}{\sqrt{N}}. \tag{7}$$

Papers [18, 19] demonstrated that using the *fat-shattering dimension* $fat_F(\varepsilon)$ one can successfully determine complexity of the required training set for implementation of neural network convergence in problems of complex classification, and in recovery of regression. If we use a probably-approximately correct learning model with an SEM-algorithm, where the convergence condition (3) is described with the function of losses

$$P\{Q[y - F_{emp}(x,w)]^2 - \inf Q[y - F(x,w)]^2 \succ \varepsilon\} \prec p, \tag{8}$$

which means that with probability of $1-p$ the approximating function received from the learning sample will give a quadratic error not above $\varepsilon$, compared to the best function within the function set $\Omega$.

Thus, for a neural network with inputs from the interval of [0,1], and with outputs delimited by $A$, with $l$ of computing layers, whose each neuron has an activation function that satisfies the Lipschitz condition $|\varphi(v_1) - \varphi(v_2)| \le L|v_1 - v_2|$, and a limited sum of weights $\Sigma w \le V$, we are going to have

$$fat_F(\varepsilon) \le 16W(l\ln(LV) + 2\ln(32W) + \ln(\frac{1}{\varepsilon(LV-1)})). \tag{9}$$

We used $fat_F(\varepsilon)$ to find a sufficient training sample size to meet the condition of convergence (8)

$$N \ge \frac{c}{\varepsilon^2}(fat_F(\varepsilon)\log^2(\frac{1}{\varepsilon}) + \log(\frac{1}{\delta})), \tag{10}$$

being of the following order

$$N = O(\frac{1}{\varepsilon^2}(\frac{A^2V^{l(l+1)}}{\varepsilon^{2l}}\ln m + \ln(\frac{1}{\delta}))). \tag{11}$$

## 3 Results

The ANN built by these authors has three computing layers $l=3$, its activation function does meet the Lipschitz condition at $L=10$, $m=12$; therefore, considering condition $\Sigma w = 1$ (2) and assuming that $\varepsilon = 0,8$ and $p = 0,05$ quoted from above (9-11) we get $fat_F(\varepsilon) = 11874$, while the necessary training sample will have order $O(1486)$ for the size of $N \ge 929$ that is sufficient to meet condition (8). Our result does not contradict the lower boundary of our training set (6) that we found using the *VC-dimension.*

Assume the best function of the set yields a squared error $\inf Q[y - F(x,w)]^2 = 0.01$. Then based on the results and considering (8), the output value for the ANN $y_P$ from neuron $P$ will produce this confidence interval,

$$P(|y - y_p| \le 0,9) \ge 0,95. \tag{12}$$

and this means that following fuzzification the output from the ANN enters point B in the membership function graph, even as the actual value of the construction process potential is in section CD with probability of $p = 0,95$ (Fig. 2). Because neuron *P* unlike the rest is linear unit with activation function $y_p(v) = 10v$ and considering condition $\sum w = 1$ the argument interval of fuzzy sets will appear as [0,10].

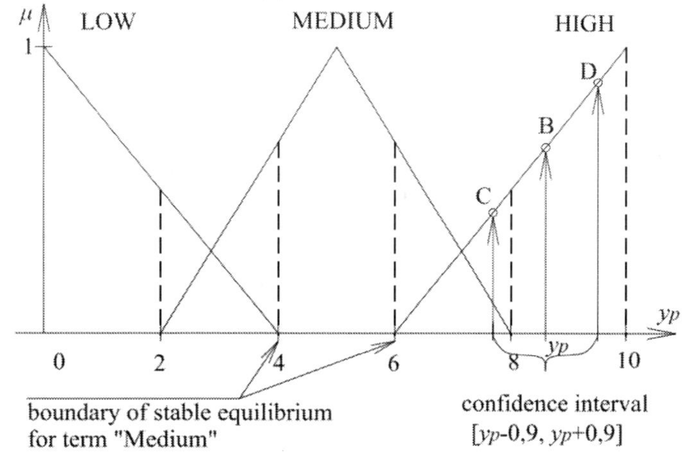


**Fig. 2.** Confidence interval of ANN outputs.

Because any high-rise construction process, including one related to roofing, is a complex system with a lasting life cycle, considerable human factor, and certain unpredictable random disturbances, the construction developer's main objective then is to keep the key process variables that all combined constitute potential *P*, within some specific preset boundaries that determine the system's sustainable equilibrium. The authors identified such boundaries, within which only one linguistic term can be assigned to the measured component (Fig. 2). The above confidence interval (12) of output values for the ANN is narrower than the stable equilibrium interval, and therefore, we can use the resulting interval to successfully evaluate and monitor the dynamics of both the integrated construction process potential *P*, and its values by one of criteria (*T*, *S*, *Q*, *C*). As an example, to evaluate *P* based on the quality criterion, we assume value $w_{QP} = 1$ to the quality criterion $Q - P$, the remaining weights of neuron *P* $w_p = 0$. Also the resulting confidence interval is narrower than in (12), considering the statistical aspects of learning discussed above.

## 4 Discussion

The final stage of learning for the created ANN understands its test run on real-life construction projects. If the construction company wants the ANN to be efficiently trained and if its project managers get actively involved, up to 2-3 training examples can be collected in a week of construction activities. Seeing that the average life cycle of a roofing process is 2 to 3 months, a project can quite realistically yield a sample size $N = 20 \div 30$. A large construction business, particularly one focused on roofing, or a profiled construction auditor may handle up to ten construction projects at the same time – so the average training sample size can be approximately $N = 800 \div 1000$ in a year. The academic results of sufficient training sample presented in this paper with the use of the SEM algorithm suggest that after a year or two of training in a major construction business the created ANN will be able to produce results with accuracy of (12). Meanwhile, our expressions (17, 22-23) lead us to believe that growth of sample size tends to make the confidence interval of ANN outputs narrower (12), and thus more accurate. Keeping the stability of learning algorithm under control, this will ensure continuous progress of the created ANN.

## References

1. D.A. Patel, K.N. Jha, SS, **89**, 240-248 (2016)
2. F. Costantino, G.D. Gravio, F. Nonino, IJPM, **33**, 1744-1754 (2015)
3. E. Siami-Irdemoosa, S.R. Dindarloo, M. Sharifzadeh, AiC, **58**, 85-94 (2015)
4. M. Li, W. Chen, PP, **25**, 1212-1217 (2012)
5. B. Hola, K. Schabowicz, AiC, **19**, 570-579 (2010)
6. A. Lapidus, A. Makarov, MATEC WoC, **86**, 05003 (2016)
7. M. Anthony, DAM, **156**, 883-902 (2008)
8. V. Vapnik, IEEE Trans. Neu. Net., **10**, 5, 988-999 (1999)
9. V. Vapnik, A. Chervonenkis, TP&A, **16**, 2, 264–280 (1971)
10. V. Vapnik, A. Chervonenkis, TP&A, **26**, 532–553 (1981)
11. P. Koiran, E.D. Sontag, JC&SS, **54**, 190-198 (1997)
12. M. Schmitt, NIPS 1999, 328-334, (1999)
13. D. Haussler, I&C, **100**, 78-150 (1992)
14. S. Floyd, M. Warmuth, ML, **21**, 269-304 (1995)
15. P.L. Bartlett, NIPS 1996, 134-140 (1996)
16. M.J. Kearns, R.E. Schapire, JC&SS, **48**, 3, 464-497 (1994)
17. N. Alon, S. Ben-David, N. Cesa-Bianchi, D. Haussler, JACM, **44**, 4, 615-631 (1997)
18. P.L. Bartlett, P.M. Long, R.C. Williamson, JC&SS, **52**, 434-452 (1996)
19. M. Anthony, P.L. Bartlett. *Neural network learning: theoretical foundations* (Cambridge university press, 1999)

# Stress relaxation of constructions elements

*Evgeniy* Larionov[1,*], *Evgeniy* Zveryaev[2]

[1]Moscow state university of civil engineering, Yaroslavskoye shosse, 26, Moscow, Russia, 129337
[2]Keldysh Institute of Applied Mathematics, Miusskaya sq. 4, Moscow, 125047, Russia

**Abstract.** A relaxation of stress in the elements of constructions is considered and an approach is proposed for solution of corresponding problems. It is notable that this approach is based on a modification Boltzmann's principle superposition of fraction creep deformations. This modification reduces the noted problems to solution of linear relative to so-called structural stress integral equations. Next a desired stress is defined by solution of algebraic equations. It should be underline that a material (concrete, steel, graph) of elements a considered as a union of its fractions with statistical disturbed strengths. This ascending to Weibull conception [1] permits to modify Boltzmann's principle superposition [2]. As a result this principle is applicable when a dependence on deformations from the stresses is nonlinear [3-7].

According to Berg [6] and Gvozdev [9] an increasing on cross-section loading $N(\tau)$ implies destruction a certain fractions of the element. Here $\tau$ is a current temp.
In consequence of this destruction an initial area $A$ of cross-section is decreased until $A(\tau)<A$. The area $A(\tau)$ represents a cross-section of all entire at instant $\tau$ fractions of the constructive element. $A$ connected with the structural damages of the element value

$$\sigma_c(\tau)=\frac{N(\tau)}{A(\tau)} \tag{1}$$

is called the structural stress. Since

$$\sigma_c(\tau)=\frac{N(\tau)}{A} \tag{2}$$

is the normal stress then according to the relations (1) and (2) we have

$$\sigma_c(\tau)=\frac{A}{A(\tau)}\sigma(\tau)=S^0(\tau)\sigma(t) \tag{3}$$

The function $S^0(\tau)$ describs the destruction of a certain part of element and is called the function of non-linearity of stress.

As $\sigma(\tau)=const$ ; $\tau\in[t_0,t]$ we obtain the creep deformation

* Corresponding author: larionov.ea@gmail.com

$$\varepsilon_c(t_0,t)=C_0(t,t)\sigma(t) \tag{4}$$

and the complete deformation is

$$\varepsilon(t,t_0)=\frac{\sigma(t)}{E(t)}+C_0(t,t_0)\sigma(t) \tag{5}$$

Here $\frac{\sigma(t)}{E(t)}$ is the instantaneous deformation $\varepsilon_m(t)$; $C_0(t,t_0)$ is the measure of the simple creep; $E(t)$ is the elasticity module.

The increment $\Delta\sigma(t)=\sigma(t)-\sigma(t_0)$ of stress implies the creep deformation $\Delta\varepsilon_c(t,t_0)$. In virtue of the Boltzmann's principle of superposition the fraction increment $\Delta\varepsilon_c(t,\tau_i)$ is defined only by the value and duration $\Delta\sigma(\tau_i)$, and is independent on the rest increments $\Delta\sigma(\tau_j)$ ; $j\neq i$.

Consequently

$$\Delta\varepsilon_c(t,\tau_i)=C_0(t,\tau_i)\Delta\sigma(\tau_i) \tag{6}$$

$$\Delta\varepsilon_c(t,t_0)=\sum_{i=1}^{n}C_0(t,\tau_i)\Delta\sigma(\tau_i) \tag{7}$$

$$\Delta\varepsilon_c(t,t_0)=\int_{t_0}^{t}C_0(t,\tau)d\sigma(\tau) \tag{8}$$

and integrating by parts, we have

$$\Delta\varepsilon_c(t,t_0)=C_0(t,t)\sigma(t)-C_0(t,t_0)\sigma(t_0)-\int_{t_0}^{t}\sigma(\tau)\frac{\partial C_0(t,\tau)}{\partial\tau}dt \tag{9}$$

Then, according to the equalities

$$\varepsilon(t,t_0)=\frac{\sigma(t)}{E(t)}+\varepsilon_c(t,t_0) \tag{10}$$

$$\varepsilon_c(t,t_0)=C_0(t,t_0)\sigma(t_0)+\Delta\varepsilon_c(t,t_0) \tag{11}$$

$$C_0(t,\tau)=\frac{1}{E(\tau)}-\frac{1}{E(t)}+C(t,\tau) \tag{12}$$

we obtain the linear rheological state equation

$$\varepsilon(t,t_0)=\frac{\sigma(t)}{E(t)}+C(t,t)\sigma(t)-\int_{t_0}^{t}\sigma(\tau)\frac{\partial}{\partial\tau}\frac{1}{E(\tau)}d\tau-\int_{t_0}^{t}\sigma(\tau)\frac{\partial C(t,\tau)}{\partial\tau}d\tau \tag{13}$$

In the equation (13) the value $C(t,t)\sigma(t)$ represents the so-called quick creep deformation [10].

In consequence of the destruction of a certain fractions of the element and the next redistribution of the loading $N(\tau)$ every increment $\Delta\sigma(\tau_k)$ intensifies the action of the previous increment $\Delta\sigma(\tau_i);(i<k)$. This means the dependence of fraction creep deformation $\Delta\varepsilon_c(t,\tau_i)$ also on all next increments of the stress $\sigma(\tau)$ and therefore the superposition of $\Delta\varepsilon_c(t,\tau_i)$ relatively $\Delta\sigma(\tau_i)$ is impossible.

Non we show that this superposition is realized relatively $\Delta\sigma_c(\tau_i)=S^0(\tau_i)\Delta\sigma(\tau_i)$, of the fraction increments of the structural stress $\sigma_c(\tau)$.

An interdependence on $\varepsilon_c(t,\tau_i)$ take place for an ideal element when all the fractions have the same strengths $R(\tau)$ and consequently the retribution of $N(\tau)$ is excluded. Therefore parallel with the given element we consider the geometrically identical to it ideal element with the same parameters $E(\tau)$ and $C(t,\tau)$. The ideal element under the axial loading $N_c(\tau)=\dfrac{A}{A(\tau)}N(\tau)=S^0(\tau)N(\tau)$ tests the stress $\sigma_c(\tau)$ and the increment $\Delta\sigma_c(\tau_i)$ generate $\Delta\varepsilon_c(t,\tau_i)$.

Thus we have

$$\Delta\varepsilon_c(t,\tau_i)=C_0(t,\tau_i)\Delta\sigma_c(\tau_i) \tag{14}$$

$$\Delta\varepsilon_c(t,t_0)=\sum_{i=1}^{n}C_0(t,\tau_i)\Delta\sigma_c(\tau_i) \tag{15}$$

$$\Delta\varepsilon_c(t,t_0)=\int_{t_0}^{t}C_0(t,\tau)d\sigma_c(\tau) \tag{16}$$

$$\Delta\varepsilon_c(t,t_0)=C(t,t)\sigma_c(t)-C(t,t_0)\sigma_c(t)-\int_{t_0}^{t}\sigma_c(\tau)\frac{\partial}{\partial\tau}\left[\frac{1}{E_{(\tau)}}+C(t,\tau)\right]d\tau \tag{17}$$

$$\varepsilon_c(t,t_0)=C(t,t)\sigma_c(t)-\int_{t_0}^{t}\sigma_c(\tau)\frac{\partial}{\partial\tau}\left[\frac{1}{E_{(\tau)}}+C(t,\tau)\right]d\tau \tag{18}$$

According to equation (18) and $\sigma_c(\tau)=S^0(\tau)\sigma(\tau)$ we obtain the nonlinear rheological state equation

$$\varepsilon(t,t_0)=S^0(t)\sigma(t)\left[\frac{1}{E(t)}+C(t,t)-\int_{t_0}^{t}\frac{S^0(\tau)\sigma(\tau)}{S^0(t)\sigma(t)}\frac{\partial}{\partial\tau}\left[\frac{1}{E(\tau)}+C(t,\tau)\right]d\tau\right. \tag{19}$$

We can present an alternate vision of the deduction for nonlinear rheological equation (19). Consider to this end the part $V_t$ of the constructive element consisting from the union of all the remaining entire in the interval $[t_0,t]$ fractions. The loading $N(\tau)$ generates on its cross-section (with the area $A(t)$) same structural stress $\sigma_c(\tau)$ that on cross-section of the considered element (with the area $A(\tau)$).

The element and its part $V_t$ have same fraction creep deformations and the absence of one in the interval $[t_0,t]$ of the redistribution of the loading $N_e(\tau)=\dfrac{A(t)}{A}N(\tau)$ implies their interindepency. Then, for $V_t$ we have the relations (14)-(18) and consequently again obtain the rheological equation (19).

Evidently linear equation (13) contradicts to the knowns non-linear diagrams «$\sigma-\varepsilon$».

Note that according to A.A. Gvozdev a non-linear part of «$\sigma-\varepsilon$» is generated by quick creep at the momentum of loading [19]. In addition he assumes the linearity of the dependence $\varepsilon_m(\tau)$ on $\sigma(\tau)$. By A.A. Gvozdev the creep deformation is the composition of the linear component $\varepsilon_{ce}(t,t_0)=\int_{t_0}^{t}C^*(t,\tau)d\sigma(\tau)$ and the non-linear component

$\varepsilon_{cn}(t,t_0)=\int_{t_0}^{t} L(t,\tau,\sigma)d\sigma(\tau)$, generated because of the structural damages. This conception implies a necessity of the choice of the functions $C^*$ and $L$ corresponding to the diagram «$\sigma-\varepsilon$». We note that this case reduce to the considerable inconvenience for the applications [12].

In connection with diagram «$\sigma-\varepsilon$» V.M. Bondarenko affirms that the dependence $\varepsilon_m(t)$ from $\sigma(t)$ also is non-linear and is described by the relation $\varepsilon_m(t)=S_m[\sigma(t)]/E(t)$ [13]. Since the instantaneous and creep deformations generated by the same stress $\sigma_c(\tau)$ their non-linear functions $S_m[\sigma(\tau)]$ and $S_c[\sigma(\tau)]$ coincide $S_m[\sigma(\tau)]=S_c[\sigma(\tau)]=A/A(\tau)$ [3, 4].

In [14–16] the non-linear rheological equation is presented in the following form

$$\varepsilon(t)=\varepsilon_m(t)-\int_{t_0}^{t}\sigma(\tau)\frac{\partial C(t,\tau)}{\partial\tau}d\tau \tag{20}$$

This means that it is assumed (along with $C(t,t)=0$ and $E(\tau)=const$) the possibility of the superposition $\Delta\varepsilon_c(t,\tau_i)$ relative to $\Delta\sigma(\tau_i)$. However (as is shown in section 2) such superposition takes place only in the linear creep theory.

**Remark 1.** The adequate to the physical essence of the creep phenomenon the non-linear rheological equations are deduced only on basis of the corresponding modification Boltzmann's principle superposition.

Papiers [14–16] are useful for an analysis of the applied in our actual norms creep theory. In addition they are connected with the problem of correspondence of these norms and Eurocod.

**Remark 2.** A creep phenomenon (increment of $\varepsilon(t)$ under $\sigma(\tau)=\sigma(t_0)$; $\tau\in[t_0,t]$) is the consequence of strength distribution on cross-section of the element. In fact a degradation of $R(\tau)$ implies an increment $\sigma_c(\tau)=\dfrac{A}{A(\tau)}\sigma(t_0)$ and therefore an increment of $\varepsilon(t)$.

**Remark 3.** In the course of taking down of the loading we have $S^0[\sigma(\tau)]=A/A(\tau)=const$, hence, the dependence $\varepsilon(\tau)$ on $\sigma(\tau)$ is linear, that is observed on diagram «$\sigma-\varepsilon$».

According to the equation (19) we have

$$\varepsilon(t,t_0)=\frac{\sigma_c(t)}{E(t)}+C(t,t)\sigma_c(t)-\int_{t_0}^{t}\sigma_c(\tau)\frac{\partial C_0}{\partial\tau}(t,\tau)d\tau \tag{21}$$

$$\sigma_c(t)=\hat{\sigma}_c(t)+\lambda(t)\int_{t_0}^{t}\sigma_c(\tau)\frac{\partial C_0(t,\tau)}{\partial\tau}d\tau \tag{22}$$

Here $\hat{\sigma}(t)=\dfrac{\varepsilon(t,t_0)E(t)}{1+E(t)C(t,t)}$; $\lambda(t)=\dfrac{E(t)}{1+E(t)C(t,t)}$ is the elastoplastic module of deformations. Solving the linear integral in equation (22) (par example by the iteration method) we come to an equation $S^0[\sigma(t)]=\sigma_c^*(t)$, where $\sigma_c^*(t)$ is the solution of equation (22).

In applications the function $S^0(\tau)$ is represented usually in form

$$S^0\left[\sigma(t)\right]=1+V\cdot\left[\frac{\sigma(t)}{R(t)}\right]^m \text{ or } S^0[\sigma(t)]=a[\sigma(t)]^b$$

where $V$, $m, a, b$ are the experimental parameters [17]. Then, we find $\sigma(t)$ as a solution of equation

$$\left[\sigma(t)\right]^{m+1}+\frac{\left[R(t)\right]^m}{V}\sigma(t)-\frac{\left[R(t)\right]^m}{V}\sigma_c^*(t)=0 \text{ or } a\left[\sigma(\tau)\right]^{b+1}=\sigma_c^*(t)$$

If $C(t,\tau)=C(0,\infty)\left[1-\beta e^{-\gamma(t-\tau)}\right]$ the stress $\sigma(t)$ is defined by solution of corresponding to (19) linear differential eqnation.

As the loading $N(\tau)=N(t_0)$ at $\tau\ge t_0$ we have $S^0\left[\sigma(t)\right]=S^0\sigma(t_0)$ and therefore $\sigma(t)=\sigma_c^*(t)/S^0\left[\sigma(t_0)\right]$. This case take place when $\varepsilon(\tau,t_0)=\varepsilon(t_0)$ at $\tau\ge t_0$.

Let us illustrate the simple example of the stress relaxation in the concrete. Suppose that a reinforced concrete rod tests with the axial loading

$$N(\tau)=\begin{cases}N;2kt\le\tau<(2k+1)T\\ N_0;(2k+1)T\le\tau<2(k+1)T\end{cases} \tag{23}$$

where $T>0$; $N_0<N$; $k=0,1,2,......,n$ .

The effort $N$ on the concrete part of the cross-section and $N_0$ generates the stresses

$$\sigma_1=\frac{N}{A_c(1+m\mu)} \text{ and } \sigma_2=\frac{N_0}{A_c(1+m\mu)} \tag{24}$$

at the moment $t_1=2kt$ and $t_2=(2k+1)T$ respectively. Here $m=E_s/E_c$; $\mu=A_s/A_c$; $E_c(E_s)$ is the elasticity module of the concrete (steel); $A_c(A_s)$ is the area of the cross-section of the concrete (steel).

Let $A_1$ be the corresponding to $\sigma_1$ area of the cross-section for all entire fractions. Taking into account that $A(\tau)=const$ on increasing loading and $\sigma_2<\sigma_1$ we obtain

$$S^0(\tau)=1+V\cdot\left[\frac{\sigma_1}{R}\right]^m;\tau\ge 0 \tag{25}$$

In given by relation (23) the intervals of temps we have

$$\sigma(t)=\sigma_i+\lambda\int_{t_i}^{t}\sigma(\tau)e^{-\gamma(t-\tau)}d\tau;i=1,2; \tag{26}$$

Here

$$\lambda=\frac{E_s\gamma\mu\beta C_o}{1+m\mu};t_1=2kt;\ t_2=(2k+1)T;k=0,1,2,......,n$$

The equations is reduced to the linear differential ones

$$\frac{d\sigma(t)}{dt}+\delta\sigma(t)=\sigma_i\gamma;\delta=\gamma+\lambda;\quad i=1,2 \tag{27}$$

The solution of the equation (27) is represented by the following function

$$\sigma(t)=\sigma_i\left[\frac{\gamma}{\delta}+\frac{\lambda e^{-\delta(t-nT)}}{\delta}\right] \tag{28}$$

where

$$\sigma_i = \begin{cases} \sigma_1; & n = 2p \\ \sigma_2; & n = 2p+1 \end{cases} \tag{29}$$

In [12] the noted problems are decided by the small parameter method. The stress $\sigma(t)$ is looked in the form

$$\sigma(t) = \sum_{i=0}^{\infty} \xi^i \sigma_i(t) \tag{30}$$

where $\xi$ is the small parameter and $\sigma(t)$ is the solution of corresponding linear integral equation. Evidently this approach reduces to a considerable calculs in applications.

## References

1. W. Weibull, Trans. Roy. Inst. Technol.(Stockholm), 27 (1949)
2. L. Boltzmann, [WA-1, 616] (Wien. Ber. 70. S. 275-306. (1874) und Pogg. Ann. Erg.-Bd. 7. S. 624 (1876))
3. V.M. Bondarenko, E.A. Larionov, Str. Mech. of Eng. Constr. and Build, 2 (2011)
4. E.A. Larionov, A.E. Larionov. Str. Mech. and Calculi build, 2 (2015)
5. N.T. Vasilkova, M.E. Bashcatova, E.A. Larionov, Str. Mech. of Eng. Constr. and Build,1 (2012)
6. E.A. Larionov, Vestnik MGSU, 4 (2016)
7. E.A. Larionov, Str. Mech. of Eng. Constr. and Build, 4 (2015)
8. O. Berg. *The physical bases of strength theorg of concrete and reinforced concrete.* (Stroyizdat, 1962)
9. A.A. Gvozdev. *A calcul of bearing capacity of constructions by limit equilibrium method.* (Stroyizdat, 1949)
10. Yu. N. Rabotnov. *A creep of constructive elements*, (Nauka, 1982)
11. A.A. Gvozdev, Mech. Solid, 5 (1972)
12. K.Z. Galustov. *Non-linear creep theory of Concrete and calculation of reinforced constructions*. (Fizmatlit, 2006)
13. V.M. Bondarenko, S.V. Bondarenko, *Ingeneering methods of non-linear theory of reinforced concrete*. (Stroyizdat, 1982)
14. R.S. Sanzharovskij. Str. Mech. of Eng. Constr. 1 (2014)
15. R.S. Sanzharovskij, M.M. Manchenko, Str. Mech. and Calculi build, 2 (2015)
16. R.S. Sanzharovskij, M.M. Manchenko, Str. Mech. of Eng. Constr. 3 (2016)
17. V.M. Bondarenko, Str. Mech. of Eng. Constr. 2 (2015)

# Application of type Timoshenko equations for shock action

*Evgeniy* Zveryaev[1,*], *Evgeniy* Larionov[2]

[1]Keldysh Institute of Applied Mathematics, Miusskaya sq. 4, Moscow, 125047, Russia
[2]Moscow state university of civil engineering, Yaroslavskoye shosse, 26, Moscow, Russia, 129337

**Abstract.** The generalized Timoshenko type equations are obtained from the plane problem equations of the theory elasticity. They might be applied for quick acting forces or displacements problems as the transversal shock on a beam or quick displacement of beam ends. The quick and slow solutions components are extracted. The quick ones describe the propagation of the shear wave due to the action of a concentrated force or an explosion; the slow solutions are generated by the common verification of the boundary conditions the slow and quick solutions.

## 1 Introduction

Under the notion of quick action we understand the quick varying action as a transversal shock on the beam with a force, an explosion near the height building considering as a beam of complex structure, or a moving of the ground resulting earthquake etc. It is believed the solution of such problems. It is believed, the solution of such problems should be made taking into account the wave character of the solution in the impacted body. In the theory of strength of materials it is proposed for these purposes to use the Timoshenko equation (1922) for a beam [1] and equations of the Timoshenko type, taking into account the inertia of the rotation and shear of the element, for a plate or shell. The procedure to find the wave solution of these equations reduces to apply the step load to the infinite body and the equations of state undergo a Laplace transformation with respect to the time and Fourier transformation with respect to coordinates. The inverse transformation is made numerically. It must be said Timoshenko's equation is extremely difficult to solve, and the author does not know the available understandings of either numerical or analytical solutions. The derivation and analysis of the equations of the theory of elasticity, given in this paper, showed that under rapid influences on the strip, two main types of motion are distinguished in it: longitudinal shear wave and transverse vibrational (not wave) low-frequency (classical).

In order to understand the behavior of the solution of the Timoshenko equation [2]

$$EI\frac{\partial^4 y^*}{\partial x^{*4}} - \rho I\left(1+\frac{E}{k'G}\right)\frac{\partial^4 y^*}{\partial x^{*2}\partial t^{*2}} + \frac{\rho^2 I}{k'G}\frac{\partial^4 y^*}{\partial t^{*4}} + \rho F\frac{\partial^2 y^*}{\partial t^{*2}} = 0 \qquad (1)$$

* Corresponding author: zveriaev@gmail.com

where $E$ is the Young's modulus, $G$ is the shear modulus, $k'$ is the coefficient depending on the shape of the cross section, $F$ is the area of the cross section, $\rho$ is the mass of the beam length unit, $y^*$ is the deflection, $x^*$ is the beam axis coordinate, $t^*$ is the time, $I$ is the moment of inertia of the cross section. The sign $^*$ marks the dimensional values.

There are various recommendations for choosing the value of the coefficient $k'$: for a rectangular cross-section the value 0.833 [2] is used, the value $2/3$ and value $3/4$ are for the circular section [3];it was pointed out $\pi^2/12$, but values $5/6$ and $5/(6-\nu)$ were also allowed [4].

If we put the time differentiation operator in equation (1) equal to zero, Timoshenko's equation turns into the classical bending equation of a beam. This means that the intuitive theory of Timoshenko exists only for dynamic problems. However, the constructive derivation of the equations on the basis of the mapping contraction principle from the equations of the theory of elasticity yields a static version of the Timoshenko type equations [5, 6]. The aim of this paper is to establish the additional equations to the classical ones taking into account the fast phenomena corresponding to the concentrated force in space and time.

## 2 Equations of static bending

We assign a long rectangular strip to a rectangular coordinate system $x^*, z^*$, so that $0 \le x^* \le l$, $-h \le z^* \le h$, where $2h$ is the thickness, $l$ is the longer of the strip. The strip bends under the action of the load $q^* = q^*(x)$ distributed on the top edge and directed downwards. We introduce dimensionless coordinates, $x = x^*/l$, $z = z^*/h$ dimensionless displacements $u = u^*/h$, $w = w^*/h$ along the axes $x, z$ respectively, and dimensionless stresses $\sigma_x = \sigma_x^*/E$, $\sigma_z = \sigma_z^*/E$, $\tau = \tau^*/E$. The equations of the plan problem of theory of elasticity

$$\frac{\partial \sigma_z^*}{\partial z^*} + \frac{\partial \tau^*}{\partial x^*} = 0, \quad \frac{\partial \sigma_x^*}{\partial x^*} + \frac{\partial \tau^*}{\partial z^*} = 0$$

$$\sigma_x^* = \frac{E}{1-\nu^2}(\varepsilon_x + \nu\varepsilon_z), \quad \tau^* = \frac{E}{2(1+\nu)}\gamma, \quad \sigma_z^* = \frac{E}{1-\nu^2}(\varepsilon_z + \nu\varepsilon_x)$$

$$\varepsilon_z = \frac{\partial w^*}{\partial z^*}, \quad \varepsilon_x = \frac{\partial u^*}{\partial x^*}, \quad \gamma = \frac{\partial u^*}{\partial z^*} + \frac{\partial w^*}{\partial x^*}$$

describing the stress-strain state of the strip we reduce with an identical transformations to the form allowing us to calculate all searching unknowns when the quantities $w, \tau$ are known

$$\frac{\partial u}{\partial z} = -\varepsilon\frac{\partial w}{\partial x} + 2(1+\nu)\tau, \quad \frac{\partial \sigma_z}{\partial z} = -\varepsilon\frac{\partial \tau}{\partial x}, \quad \varepsilon_x = \varepsilon\frac{\partial u}{\partial x}$$
$$\sigma_x = \varepsilon_x + \nu\sigma_z, \quad \varepsilon_z = (1-\nu^2)\sigma_z - \nu\varepsilon_x, \quad \frac{\partial w}{\partial z} = \varepsilon_z, \quad \frac{\partial \tau}{\partial z} = -\varepsilon\frac{\partial \sigma_x}{\partial x} \tag{2}$$

Here $\varepsilon = h/l$ is smoll parametr.

In fact from the first two equations when $w, \tau$ are known, we calcul $u, \sigma_z$ and via $u$ calcul $\varepsilon_x$. The relations of elasticity permit to calcul $\sigma_x, \varepsilon_z$. The last two equations again give the same values $w, \tau$. This fact in the functional analyses is treated as the the Banach fixed-point theorem (also known as the contraction mapping theorem or contraction mapping principle). Usually the exact values of the quantities $w, \tau$ are unknown. Therefore, we can choose the approximate quantities $w_0, \tau_0$ in the first two equations and calculate in the last two $w_1, \tau_1$ as the values of the first approximation. The process is treated as a method of simple iterations (the Banach theorem, 1922). The presence of the small parameter $\varepsilon = h/l$ makes it possible to compare the method of the simple iterations with the method of asymptotic integration and to estimate the conditions for asymptotic convergence.

We give the values of the initial approximation $w = w_0(x), \tau = \tau_0(x)$ as a function of only one coordinate $x$. This means that, in the zeroth approximation, there is no change in the thickness of the strip and a uniform distribution of the tangential stress that is usually used as a hypothesis in the derivation of the equations of the plates and shells theories. In the first approximation, a correction is calculated for the transverse compression and distribution of the tangential stress.

Calculating the integrals of the system (2) by the described method, we write the boundary conditions on the long sides of the strip

$$\tau = \sigma_z = 0 \text{ при } z = -1,\ \tau = 0,\ \sigma_z = -q \text{ при } z = 1; \qquad \left(q = q^*/E\right) \tag{3}$$

We shall use them as epy first-approximation quantities

$$\begin{aligned} \tau &= \left[\varepsilon^3 w_0''' - (2+\nu)\varepsilon^2 \tau_0''\right]\frac{z^2}{2} + \tau_0 - \left(\varepsilon^2 u_0'' + \nu\varepsilon\sigma_{z0}'\right)z \\ \sigma_z &= -\left[\varepsilon^4 w_0'''' - (2+\nu)\varepsilon^3 \tau_0'''\right]\frac{z^3}{6} - \varepsilon\tau_0' z + \left(\varepsilon^3 u_0''' + \nu\varepsilon^2\sigma_{z0}''\right)\frac{z^2}{2} + \sigma_{z0} \end{aligned} \tag{4}$$

Here and below, the prime denotes the differentiation with respect to $x$.

After substituting the stresses into the boundary conditions, we obtain the four equations for the determination of the functions $w_0, \tau_0, u_0, \sigma_{z0}$

$$\varepsilon^2 u_0'' + \nu\varepsilon\sigma_{z0}' = 0$$

$$\varepsilon^3 u_0''' + \nu\varepsilon^2\sigma_{z0}'' + 2\sigma_{z0} = -q$$

$$\varepsilon^3 w_0''' - (2+\nu)\varepsilon^2\tau_0'' + 2\tau_0 = 0,$$

$$\varepsilon^4 w_0'''' - (2+\nu)\varepsilon^3\tau_0''' + 6\varepsilon\tau_0' = 3q$$

The first two equations refer to the tension-compression, the second two to the bending, and we will be interested only in the last one. We present the searched unknowns and load in the form of the sum of rapidly varying components along $x$ and slowly varying components, i.e.

$$w_0 = w_0^s + w_0^q,\ \tau_0 = \tau_0^s + \tau_0^q,\ q = q^s + q^q$$

Substituting them into the first system and demanding that the slow and fast component go to zero separately in each equation, we obtain equations

$$\varepsilon^3 w_0^{s'''} + 2\tau_0^s = 0$$

$$\varepsilon^4 w_0^{s''''} + 6\varepsilon\tau_0^{s'} = 3q^s$$

$$\varepsilon^3 w_0^{q'''} - (2+\nu)\varepsilon^2\tau_0^{q''} + 2\tau_0^q = 0,$$

$$\varepsilon^4 w_0^{q''''} - (2+\nu)\varepsilon^3\tau_0^{q'''} + 6\varepsilon\tau_0^{q'} = 3q^q$$

In the first two equations, the terms with the second and third derivatives of $\tau_0^s$ are rejected as small compared with the principal $\tau_0^q$ and $\tau_0^{q'}$, respectively. Excluding $\tau_0^s$ from the second equation with the help of the first one we obtain the classical equation of bending and an additional equation for determining the tangential stress.

$$\frac{2}{3}\varepsilon^4 w_0^{s''''} = -q, \quad \varepsilon\tau_0^{s'} = \frac{3}{4}q$$

The second system of equations is transformed to the form

$$\begin{aligned} &\frac{2}{3}\varepsilon^4 w_0^{q''''} = -q^q + \frac{\varepsilon^2}{k^2}q^{q''} \\ &\frac{\varepsilon^2}{k^2}\tau_0^{q''} - \tau_0^q = \frac{1}{2}\varepsilon^3 k^2 w_0^{q'''} \end{aligned} \qquad \left(k^2 = \frac{2}{2+\nu}\right) \tag{5}$$

The two systems describe the action of the slowly and rapidly changing loads on a strip (beam) and can be declared in correspondence with equation (1) as a static version of the theory of Timoshenko's type. Having determined from this $w_0^s, w_0^q, \tau_0^s, \tau_0^q$ we obtain by means of the formulas (4) the stresses $\tau, \sigma_z$. Formulas for the remaining searched unknowns can be found in [4, 5].

If in the first equation of the system (5) the load is chosen so that it satisfies the equation $-q^q + \frac{\varepsilon^2}{k^2}q^{q''} = 0$, and this will be, for example, the function

$$q^q = \begin{cases} B_1 e^{k(x-b)/\varepsilon} & \text{при } x \le b \\ B_2 e^{-k(x-b)/\varepsilon} & \text{при } x \ge b \end{cases}$$

and choose the constants $B_1 = B_2 = Pk\varepsilon^{-1}/2$ ($P = const$), we see that

$$\lim_{\varepsilon\to 0} q^q = P\delta(x-b)$$

The value $P$ determines the intensity of the applied force corresponding to the local load $q^q$. In this case, the first equation of the system (5) becomes homogeneous and does not have any solution among the quickly varying functions, hence $w_0^q = 0$. Accordingly, the second equation also becomes homogeneous and has a boundary layer type solution.

## 3 Dynamic equations for strip

We apply the procedure described above for constructing the equations of motion for a long strip. Adding dimensionless inertial terms to the static bending equations (2), we obtain the following equations of motion

$$\begin{aligned} &\frac{\partial u}{\partial z} = -\varepsilon\frac{\partial w}{\partial x} + 2(1+\nu)\tau, \quad \frac{\partial \sigma_z}{\partial z} = -\varepsilon\frac{\partial \tau}{\partial x} + m^2\ddot{w} \\ &\sigma_x = \varepsilon\frac{\partial u}{\partial x} + \nu\sigma_z, \quad \varepsilon_z = -\nu\varepsilon\frac{\partial u}{\partial x} + (1-\nu^2)\sigma_z, \\ &\frac{\partial w}{\partial z} = \varepsilon_z, \quad \frac{\partial \tau}{\partial z} = -\varepsilon\frac{\partial \sigma_x}{\partial x} + m^2\ddot{u} \end{aligned} \qquad \left(m^2 = \frac{\rho h^2}{ET^2}\right) \tag{6}$$

where $\rho$ is the specific density of the strip material, $t = t^*/T$ is dimensionless time ($t^*$ is dimensional time), $T$ is some characteristic period of oscillation of a strip (beam), having length $l$ and height $2h$. The dot over a letter means the differentiation with respect to $t$.

Here, as well as in the static problem, it is possible consequentially calculate all the unknowns, assuming $w, \tau$ known. To find the approximate solution by the simple iteration method, we can choose the initial approximation and do the calculations according to the scheme

$$w_0, \tau_0 \Rightarrow u_0, \sigma_{z0} \Rightarrow \varepsilon_{x0}, \sigma_{x0}, \varepsilon_{z0} \Rightarrow w_1, \tau_1 \tag{7}$$

To show the applicability of the simple iterations method to the integration of the system (6), we transform it

$$\frac{\partial u}{\partial z} = -\varepsilon\frac{\partial w}{\partial x} + 2(1+\nu)\tau, \qquad \frac{\partial \sigma_z}{\partial z} = m^2\ddot{w} - \varepsilon\frac{\partial \tau}{\partial x}$$

$$\frac{\partial w}{\partial z} = -\nu\varepsilon\frac{\partial u}{\partial x} + (1-\nu^2)\sigma_z, \qquad \frac{\partial \tau}{\partial z} = -\varepsilon^2\frac{\partial^2 u}{\partial x^2} + m^2\ddot{u} - \nu\varepsilon\frac{\partial \sigma_z}{\partial x}$$

and rewrite it in vector form. We introduce vectors $\mathbf{u} = \{u, \sigma_z\}, \mathbf{w} = \{w, \tau\}$ and matrix operators

$$\mathbf{A}_1 = \begin{Vmatrix} -\varepsilon\dfrac{\partial}{\partial x} & 2(1+\nu) \\ m^2\dfrac{\partial^2}{\partial t^2} & -\varepsilon\dfrac{\partial}{\partial x} \end{Vmatrix}, \qquad \mathbf{A}_2 = \begin{Vmatrix} -\nu\varepsilon\dfrac{\partial}{\partial x} & 1-\nu^2 \\ -\varepsilon^2\dfrac{\partial^2}{\partial x^2} + m^2\dfrac{\partial^2}{\partial t^2} & -\nu\varepsilon\dfrac{\partial}{\partial x} \end{Vmatrix}$$

With them the system is written as follows

$$\frac{\partial \mathbf{u}}{\partial z} = \mathbf{A}_1\mathbf{w}, \quad \frac{\partial \mathbf{w}}{\partial z} = \mathbf{A}_2\mathbf{u}$$

Integrating these equations with respect to $z$, we reduce them to a single equation

$$\mathbf{w} = \int \mathbf{A}_2\left(\int \mathbf{A}_1\mathbf{w}dz + C\right)dz + D$$

where $C, D$ are the arbitrariness of integration, which are functions of the coordinate $x$. The solution of the equation can be searched by the simple iteration method

$$\mathbf{w}_{(n+1)} = \int \mathbf{A}_2\left(\int \mathbf{A}_1\mathbf{w}_{(n)}dz + C\right)dz + D$$

where $n$ is number of the iteration.

Thus, the formal feature of the applicability of the simple iteration method is fulfilled.

Performing the calculations according to scheme (7) and writing down the expressions for all the searched unknowns

$$\begin{aligned}
&u = -\varepsilon w_0' z + 2(1+\nu)\tau_0 z, \\
&w = w_0 + \left[\left(1-\nu^2\right)\varepsilon^2 m^2\ddot{w}_0 + \nu\varepsilon^2 w_0''\right]\frac{z^2}{2} - (1+\nu)^2\varepsilon\tau_0'\frac{z^2}{2}, \\
&\sigma_x = (-\varepsilon^2 w_0'' + \nu\varepsilon^2 m^2\ddot{w}_0)z + (2+\nu)\varepsilon\tau_0' z, \\
&\tau = \tau_0 + \left[\varepsilon^3 w_0''' - (1+\nu)\varepsilon m^2\ddot{w}_0' - (2+\nu)\varepsilon^2\tau_0'' + 2(1+\nu)m^2\ddot{\tau}_0\right]\frac{z^2}{2}, \\
&\sigma_z = -\left[\varepsilon^4 w_0'''' - (1+\nu)\varepsilon^2 m^2\ddot{w}_0'' - (2+\nu)\varepsilon^3\tau_0''' + 2(1+\nu)\varepsilon m^2\ddot{\tau}_0'\right]\frac{z^3}{6} - \varepsilon\tau_0' z + m^2\ddot{w}_0 z,
\end{aligned} \tag{8}$$

we use stresses $\tau, \sigma_z$ to satisfy the boundary conditions (3). We obtain equations for the definition of functions $w_0, \tau_0$.

$$\begin{aligned}
&\varepsilon^3 w_0''' - (1+\nu)\varepsilon m^2\ddot{w}_0' - (2+\nu)\varepsilon^2\tau_0'' + 2(1+\nu)m^2\ddot{\tau}_0 + 2\tau_0 = 0 \\
&\varepsilon^4 w_0'''' - (1+\nu)\varepsilon^2 m^2\ddot{w}_0'' - (2+\nu)\varepsilon^3\tau_0''' + 2(1+\nu)\varepsilon m^2\ddot{\tau}_0' + 6\varepsilon\tau_0' - 6m^2\ddot{w}_0 = 3q
\end{aligned} \tag{9}$$

From the second equation we subtract the first one, multiplied by $\varepsilon$ and differentiated with respect to $x$. Thus we obtain the relation

$$\varepsilon\tau_0' = 3\left(2m^2\ddot{w}_0 + q\right)/4 \tag{10}$$

Eliminating $\tau_0$ from the second equation with help of this equation, we obtain the equation

$$\frac{2}{3}\left[\varepsilon^4 w_0'''' - \left(4+\frac{5}{2}\nu\right)\varepsilon^2 m^2 \ddot{w}_0'' + 3(1+\nu)m^4 \ddddot{w}_0\right] + 2m^2\ddot{w}_0 = = -q + \frac{2+\nu}{2}\varepsilon^2 q'' - (1+\nu)m^2\ddot{q} \tag{11}$$

similar to the left-hand side of the equation of Tymoshenko (1) reduced to the dimensionless form,

$$\frac{2}{3}\left[\varepsilon^4 y'''' - \left(1+\frac{2(1+\nu)}{k'}\right)\varepsilon^2 m^2 \ddot{y}'' + \frac{2(1+\nu)}{k'} m^4 \ddddot{y}\right] + 2m^2\ddot{y} = 0$$

where, in addition to the previously introduced dimensionless quantities, the dimensional deflection was replaced by dimensionless $y^* = hy$. In the coefficients there is an empirical value $k'$ and substitutions $I = 2/3bh^3$, $F = 2bh$ ($b$ is the width of the beam section) are made. When $k' = 2/3$ this equation is close to the equation (10)

$$\frac{2}{3}\left[\varepsilon^4 y'''' - (4+3\nu)\varepsilon^2 m^2 \ddot{y}'' + 3(1+\nu)m^4 \ddddot{y}\right] + 2m^2\ddot{y} = 0$$

Thus, the Timoshenko equation (11) obtained from the system of two equations (9), we shall call it so, should be supplemented by some second equation from the same system or a combination of them. We differentiate the first equation of the system (9) with respect $x$, multiply it by $\varepsilon$ and add it with the second. It gives

$$2\varepsilon^4 w_0'''' - 2(1+\nu)\varepsilon^2 m^2 \ddot{w}_0'' - 2(2+\nu)\varepsilon^3 \tau_0''' + 4(1+\nu)\varepsilon m^2 \ddot{\tau}_0' + 8\varepsilon\tau_0' + 6m^2\ddot{w}_0 = 3q$$

Using the relation (10) in the form $2m^2\ddot{w}_0 = \frac{4}{3}\varepsilon\tau_0' - q$, we exclude $\ddot{w}_0$ from the second term. After reduction of such terms, we obtain the equation

$$\left[\frac{2}{3}\varepsilon^4 w_0'''' + 2m^2\ddot{w}_0\right] + \varepsilon\frac{\partial}{\partial x}\left[-\frac{16+10\nu}{9}\varepsilon^2\tau_0'' + \frac{4}{3}(1+\nu)m^2\ddot{\tau}_0 + \frac{8}{3}\tau_0\right] = q - \frac{1+\nu}{3}\varepsilon^2 q''$$

Here the first term in square brackets describes the transverse movement of the rod, the second describes longitudinal shear movement.

We add estimates of the inertial terms to these equations. To do this, we must select the unit of time in equations (6). Let us write out the known equations of longitudinal and transverse oscillations of a rod of rectangular cross section, having written them down in a dimensionless form

$$\varepsilon^2 u'' - m_l^2 \ddot{u} = 0, \quad \frac{2}{3}\varepsilon^4 y'''' - m_t^2 \ddot{y} = 0$$

where the notations are introduced $m_l^2 = \frac{\rho h^2}{ET_l^2}$, $m_t^2 = \frac{\rho h^2}{ET_t^2}$, $u$ is the longitudinal displacement of the points of the rod, $y$ is the transversal displacements of the neutral axis. Understanding under $T_l$ the characteristic period of the longitudinal vibrations of the rod and $T_t$ the characteristic period of the transverse vibrations, we conclude $m_l^2 \sim \varepsilon^2$ and $m_t^2 \sim \varepsilon^4$. In accordance with these estimates, we make the substitution $m_l = \varepsilon n_l$, $m_t = \varepsilon^2 n_t$ in equation (12), setting $n_l, n_t \sim \varepsilon^0$

$$\varepsilon^4\left[\frac{2}{3}w_0'''' + 2n_t^2\ddot{w}_0\right] + \varepsilon\frac{\partial}{\partial x}\left[-\frac{16+10\nu}{9}\varepsilon^2\tau_0'' + \frac{4}{3}(1+\nu)\varepsilon^2 n_l^2\ddot{\tau}_0 + \frac{8}{3}\tau_0\right] = q - \frac{1+\nu}{3}\varepsilon^2 q''$$

If the load on the right part is rapidly varying with respect to $x$, the particular solution is determined from the equation

$$\varepsilon\frac{\partial}{\partial x}\left[-\frac{16+10\nu}{9}\varepsilon^2\tau_0''+\frac{4}{3}(1+\nu)\varepsilon^2 n_t^2\ddot{\tau}_0+\frac{8}{3}\tau_0\right]=q-\frac{1+\nu}{3}\varepsilon^2 q''$$

or when $q$ is a slowly changing load it determined from the equation

$$\varepsilon^4\left[\frac{2}{3}w_0''''+2n_t^2\ddot{w}_0\right]=q$$

The general solution is determined from the equations

$$\frac{2}{3}w_0''''+2n_t^2\ddot{w}_0=0\,,\qquad \varepsilon\frac{\partial}{\partial x}\left[-\frac{8+5\nu}{3}\varepsilon^2\tau_0''+2(1+\nu)\varepsilon^2 n_t^2\ddot{\tau}_0+4\tau_0\right]=0$$

The first equation gives a slowly varying integral with respect to $x$, the second one gives the rapidly varying integral, depending on the argument $x/\varepsilon$, having wave properties.

## 4 Conclusion

So, the main conclusion that can be drawn from the present paper is the content in the well-known Timoshenko equation of two types of main motions, the high-speed shear wave and low-frequency transverse oscillations. They should be selected from the equations and integrated separately, regardless of the type of solution, analytically or numerical simulation.

## References

1. S. Timoshenko, Phil. Mag. **43**, 253 (1922)
2. S. Timoshenko, D.H. Young, W. Weaver, *Vibration Problems in Engineering* (N.Y.etc. : Wiley, 1974)
3. Ya. S. Uflyand, Prikl Mat Mekh, **12**, 3 (1948)
4. P. A. Zhilin, T.P. Il'icheva, J Appl Mech Tech Phy, **25**, **1**, (1984)
5. Ye. Zveryayev, G.I. Makarov, Prikl Mat Mekh, **72**, 2 (2008)
6. Ye. Zveryayev, Prikl Mat Mekh, **67**, 3 (2003)

# Piecewise-linear and bilinear approaches to nonlinear differential equations approximation problem of computational structural mechanics

*Roman* Leibov[1,*]

[1]Moscow state university of civil engineering, Yaroslavskoye shosse, 26, Moscow, Russia, 129337

**Abstract.** This paper presents a bilinear approach to nonlinear differential equations system approximation problem. Sometimes the nonlinear differential equations right-hand sides linearization is extremely difficult or even impossible. Then piecewise-linear approximation of nonlinear differential equations can be used. The bilinear differential equations allow to improve piecewise-linear differential equations behavior and reduce errors on the border of different linear differential equations systems areas. The matrices of bilinear differential equations system are estimated through nonlinear programming. The results of proposed approach application are presented.

## 1 Linear, piecewise-linear and bilinear approximation of nonlinear differential equations

Let's consider nonlinear differential equations [1-4].

$$\dot{\mathbf{x}} = \mathbf{f}(\mathbf{x},\mathbf{u}) \tag{1}$$

There are $\mathbf{x}$ — state vector, $\mathbf{u}$ — input vector.
The constant input vector ${}_s\mathbf{u}$ determines steady-state vector of state variables ${}_s\mathbf{x}$.

Nonlinear differential equations in a small neighborhood of $r$-th steady-state point are approximately described by linear differential equations [1-4].

$$\dot{\mathbf{x}}^n = {}^r\mathbf{A}\left(\mathbf{x}^n - {}^r_s\mathbf{x}^n\right) + {}^r\mathbf{B}\left(\mathbf{u}^n - {}^r_s\mathbf{u}^n\right) \tag{2}$$

or

$$\dot{x}_i^n = {}^r_i\mathbf{a}\left(\mathbf{x}^n - {}^r_s\mathbf{x}^n\right) + {}^r_i\mathbf{b}\left(\mathbf{u}^n - {}^r_s\mathbf{u}^n\right), \quad i = 1,\dots,n. \tag{3}$$

There are $\mathbf{x}^n$ — vector of normalized state variables, $\mathbf{u}^n$ — vector of normalized input variables,

$$x_i^n = \frac{x_i}{\max\left|{}_s x_i\right|}, \quad i = 1,\dots,n, \tag{4}$$

$$u_i^n = \frac{u_i}{\max\left|{}_s u_i\right|}, \quad i = 1,\dots,m. \tag{5}$$

* Corresponding author: r_leibov@mtu-net.ru

$$ {}^{r}\mathbf{A}=\begin{bmatrix} {}^{r}_{1}\mathbf{a} \\ \cdots \\ {}^{r}_{n}\mathbf{a} \end{bmatrix}, \tag{6} $$

$$ {}^{r}\mathbf{B}=\begin{bmatrix} {}^{r}_{1}\mathbf{b} \\ \cdots \\ {}^{r}_{n}\mathbf{b} \end{bmatrix} \tag{7} $$

are corresponding linear differential equations system matrices.

Also nonlinear differential equations in a small neighborhood of $r$-th steady-state point are approximately described by bilinear and quadratic differential equations [3], [4].

$$ \dot{x}_i^n=\left[{}^{r}_{i}\mathbf{a},{}^{r}_{i}\mathbf{b}\right]\begin{bmatrix} \mathbf{x}^n-{}^{r}_{s}\mathbf{x}^n \\ \mathbf{u}^n-{}^{r}_{s}\mathbf{u}^n \end{bmatrix}+\begin{bmatrix} \mathbf{x}^n-{}^{r}_{s}\mathbf{x}^n \\ \mathbf{u}^n-{}^{r}_{s}\mathbf{u}^n \end{bmatrix}^T {}^{r}_{i}\mathbf{\Gamma}\begin{bmatrix} \mathbf{x}^n-{}^{r}_{s}\mathbf{x}^n \\ \mathbf{u}^n-{}^{r}_{s}\mathbf{u}^n \end{bmatrix}, \quad i=1,\ldots,n, \tag{8} $$

$$ \dot{x}_i^n={}^{r}_{i}\mathbf{a}\left(\mathbf{x}^n-{}^{r}_{s}\mathbf{x}^n\right)+{}^{r}_{i}\mathbf{b}\left(\mathbf{u}^n-{}^{r}_{s}\mathbf{u}^n\right)+ $$

$$ \left(\mathbf{x}^n-{}^{r}_{s}\mathbf{x}^n\right)^T{}^{r}_{i}\mathbf{\Gamma}^{\mathbf{x}}\left(\mathbf{x}^n-{}^{r}_{s}\mathbf{x}^n\right)+\left(\mathbf{u}^n-{}^{r}_{s}\mathbf{u}^n\right)^T{}^{r}_{i}\mathbf{\Gamma}^{\mathbf{u}}\left(\mathbf{u}^n-{}^{r}_{s}\mathbf{u}^n\right), \quad i=1,\ldots,n, \tag{9} $$

where

$$ {}^{r}_{i}\mathbf{\Gamma}=\begin{bmatrix} {}^{r}_{i}\gamma_{11} & {}^{r}_{i}\gamma_{12} & \cdots & {}^{r}_{i}\gamma_{1,n+m} \\ 0 & {}^{r}_{i}\gamma_{22} & \cdots & {}^{r}_{i}\gamma_{2,n+m} \\ \cdots & \cdots & \cdots & \cdots \\ 0 & 0 & \cdots & {}^{r}_{i}\gamma_{n+m,n+m} \end{bmatrix}, \quad i=1,\ldots,n, \tag{10} $$

$$ {}^{r}_{i}\mathbf{\Gamma}^{\mathbf{x}}=\begin{bmatrix} {}^{r}_{i}\gamma^{\mathbf{x}}_{11} & {}^{r}_{i}\gamma^{\mathbf{x}}_{12} & \cdots & {}^{r}_{i}\gamma^{\mathbf{x}}_{1n} \\ 0 & {}^{r}_{i}\gamma^{\mathbf{x}}_{22} & \cdots & {}^{r}_{i}\gamma^{\mathbf{x}}_{2n} \\ \cdots & \cdots & \cdots & \cdots \\ 0 & 0 & \cdots & {}^{r}_{i}\gamma^{\mathbf{x}}_{nn} \end{bmatrix}, \quad i=1,\ldots,n, \tag{11} $$

$$ {}^{r}_{i}\mathbf{\Gamma}^{\mathbf{u}}=\begin{bmatrix} {}^{r}_{i}\gamma^{\mathbf{u}}_{11} & {}^{r}_{i}\gamma^{\mathbf{u}}_{12} & \cdots & {}^{r}_{i}\gamma^{\mathbf{u}}_{1m} \\ 0 & {}^{r}_{i}\gamma^{\mathbf{u}}_{22} & \cdots & {}^{r}_{i}\gamma^{\mathbf{u}}_{2m} \\ \cdots & \cdots & \cdots & \cdots \\ 0 & 0 & \cdots & {}^{r}_{i}\gamma^{\mathbf{u}}_{mm} \end{bmatrix}, \quad i=1,\ldots,n. \tag{12} $$

The derivatives of right-hand side of equation (11) are

$$ \frac{d\left[{}^{r}_{i}\mathbf{a}\left(\mathbf{x}^n-{}^{r}_{s}\mathbf{x}^n\right)\right]}{d\left(\mathbf{x}^n\right)^T}+\frac{d\left[{}^{r}_{i}\mathbf{b}\left(\mathbf{u}^n-{}^{r}_{s}\mathbf{u}^n\right)\right]}{d\left(\mathbf{x}^n\right)^T}+\frac{d\left[\left(\mathbf{x}^n-{}^{r}_{s}\mathbf{x}^n\right)^T{}^{r}_{i}\mathbf{\Gamma}^{\mathbf{x}}\left(\mathbf{x}^n-{}^{r}_{s}\mathbf{x}^n\right)\right]}{d\left(\mathbf{x}^n\right)^T}+\frac{d\left[\left(\mathbf{u}^n-{}^{r}_{s}\mathbf{u}^n\right)^T{}^{r}_{i}\mathbf{\Gamma}^{\mathbf{u}}\left(\mathbf{u}^n-{}^{r}_{s}\mathbf{u}^n\right)\right]}{d\left(\mathbf{x}^n\right)^T}= $$

$$ {}^{r}_{i}\mathbf{a}+\left(\mathbf{x}^n-{}^{r}_{s}\mathbf{x}^n\right)^T\left({}^{r}_{i}\mathbf{\Gamma}^{\mathbf{x}^T}+{}^{r}_{i}\mathbf{\Gamma}^{\mathbf{x}}\right), \quad i=1,\ldots,n, \tag{13} $$

$$ \frac{d\left[{}^{r}_{i}\mathbf{a}\left(\mathbf{x}^n-{}^{r}_{s}\mathbf{x}^n\right)\right]}{d\left(\mathbf{u}^n\right)^T}+\frac{d\left[{}^{r}_{i}\mathbf{b}\left(\mathbf{u}^n-{}^{r}_{s}\mathbf{u}^n\right)\right]}{d\left(\mathbf{u}^n\right)^T}+\frac{d\left[\left(\mathbf{x}^n-{}^{r}_{s}\mathbf{x}^n\right)^T{}^{r}_{i}\mathbf{\Gamma}^{\mathbf{x}}\left(\mathbf{x}^n-{}^{r}_{s}\mathbf{x}^n\right)\right]}{d\left(\mathbf{u}^n\right)^T}+\frac{d\left[\left(\mathbf{u}^n-{}^{r}_{s}\mathbf{u}^n\right)^T{}^{r}_{i}\mathbf{\Gamma}^{\mathbf{u}}\left(\mathbf{u}^n-{}^{r}_{s}\mathbf{u}^n\right)\right]}{d\left(\mathbf{u}^n\right)^T}= $$

$$ {}^{r}_{i}\mathbf{b}+\left(\mathbf{u}^n-{}^{r}_{s}\mathbf{u}^n\right)^T\left({}^{r}_{i}\mathbf{\Gamma}^{\mathbf{u}^T}+{}^{r}_{i}\mathbf{\Gamma}^{\mathbf{u}}\right), \quad i=1,\ldots,n. \tag{14} $$

The values of these derivatives at

$$ \begin{bmatrix} {}^{r}_{s}\mathbf{x}^n \\ {}^{r}_{s}\mathbf{u}^n \end{bmatrix} $$

are ${}^{r}_{i}\mathbf{a}$, ${}^{r}_{i}\mathbf{b}$, $i=1,\ldots,n$ and at

$$ \begin{bmatrix} {}^{r-1}_{s}\mathbf{x}^n \\ {}^{r-1}_{s}\mathbf{u}^n \end{bmatrix} $$

these derivatives are the rows of $(r-1)$-th steady-state point linear differential equations system matrices.

$$ {}^{r-1}_{i}\mathbf{a} = {}^{r}_{i}\mathbf{a} + \left({}^{r-1}_{s}\mathbf{x}^{n} - {}^{r}_{s}\mathbf{x}^{n}\right)^{T}\left({}^{r}_{i}\mathbf{\Gamma}^{\mathbf{x}T} + {}^{r}_{i}\mathbf{\Gamma}^{\mathbf{x}}\right), \quad i = 1,\ldots,n, \tag{15} $$

$$ {}^{r-1}_{i}\mathbf{b} = {}^{r}_{i}\mathbf{b} + \left({}^{r-1}_{s}\mathbf{u}^{n} - {}^{r}_{s}\mathbf{u}^{n}\right)^{T}\left({}^{r}_{i}\mathbf{\Gamma}^{\mathbf{u}T} + {}^{r}_{i}\mathbf{\Gamma}^{\mathbf{u}}\right), \quad i = 1,\ldots,n. \tag{16} $$

## 2 Quadratic differential equations system matrices estimation

Nonlinear differential equations time responses ${}^{\mathrm{NL}}\mathbf{u}^{n}(t_k)$, $k = 0,\ldots,N-1$, ${}^{\mathrm{NL}}\mathbf{x}^{n}(t_k)$, $k = 0,\ldots,N$, where $t_{k+1} = t_k + \Delta t$, $k = 0,\ldots,N-1$, are used for quadratic differential equations system matrices estimation. The ${}^{\mathrm{NL}}\mathbf{u}^{n}(t_k)$, $k = 0,\ldots,N-1$ stepwise time responses cross the border of $r$-th and $(r-1)$-th linear differential equations systems application areas.

Quadratic differential equations system matrices estimation can be reduced to nonlinear programming problem

$$ {}^{r}_{1}\mathbf{\Gamma}^{\mathbf{x}},\ldots,{}^{r}_{n}\mathbf{\Gamma}^{\mathbf{x}},{}^{r}_{1}\mathbf{\Gamma}^{\mathbf{u}},\ldots,{}^{r}_{n}\mathbf{\Gamma}^{\mathbf{u}} : \min \left\{ \sqrt{\frac{\sum_{k=1}^{N}\left[\mathbf{x}^{n}(t_k) - {}^{\mathrm{NL}}\mathbf{x}^{n}(t_k)\right]^{T} {}_{\mathbf{x}}\mathbf{W}(t_k)\left[\mathbf{x}^{n}(t_k) - {}^{\mathrm{NL}}\mathbf{x}^{n}(t_k)\right]}{\sum_{k=1}^{N}\left[{}^{\mathrm{NL}}\mathbf{x}^{n}(t_k) - {}^{\mathrm{NL}}\mathbf{x}^{n}(t_0)\right]^{T} {}_{\mathbf{x}}\mathbf{W}(t_k)\left[{}^{\mathrm{NL}}\mathbf{x}^{n}(t_k) - {}^{\mathrm{NL}}\mathbf{x}^{n}(t_0)\right]}} \right\} $$

$$ \left.\begin{array}{c} \dot{x}_i^{n}(t_k) = {}^{r}_{i}\mathbf{a}\left[\mathbf{x}^{n}(t_k) - {}^{r}_{s}\mathbf{x}^{n}\right] + {}^{r}_{i}\mathbf{b}\left[{}^{\mathrm{NL}}\mathbf{u}^{n}(t_k) - {}^{r}_{s}\mathbf{u}^{n}\right] + \\ \left[\mathbf{x}^{n}(t_k) - {}^{r}_{s}\mathbf{x}^{n}\right]^{T} {}^{r}_{i}\mathbf{\Gamma}^{\mathbf{x}}\left[\mathbf{x}^{n}(t_k) - {}^{r}_{s}\mathbf{x}^{n}\right] + \left[{}^{\mathrm{NL}}\mathbf{u}^{n}(t_k) - {}^{r}_{s}\mathbf{u}^{n}\right]^{T} {}^{r}_{i}\mathbf{\Gamma}^{\mathbf{u}}\left[{}^{\mathrm{NL}}\mathbf{u}^{n}(t_k) - {}^{r}_{s}\mathbf{u}^{n}\right], \\ x_i^{n}(t_{k+1}) = x_i^{n}(t_k) + \Delta t\, \dot{x}_i^{n}(t_k), \quad i = 1,\ldots,n, \quad k = 0,1,\ldots,N-1 \end{array}\right\}. \tag{17} $$

There are ${}_{\mathbf{x}}\mathbf{W}(t_k)$, $k = 1,\ldots,N$ positive definite diagonal weight matrices, ${}^{r}\mathbf{A}$, ${}^{r}\mathbf{B}$ are considered known, the initial values of quadratic differential equations system matrices ${}^{r}_{i}\mathbf{\Gamma}^{\mathbf{x}} = \mathbf{0}$, ${}^{r}_{i}\mathbf{\Gamma}^{\mathbf{u}} = \mathbf{0}$, $i = 1,\ldots,n$, and $\mathbf{x}^{n}(t_0) = {}^{\mathrm{NL}}\mathbf{x}^{n}(t_0)$.

## 3 Quadratic differential equations system matrices estimation example

In this example $m = 4$, $n = 5$, vectors of normalized steady-state values of input and state variables are

$$ {}^{r}_{s}\mathbf{u}^{n} = [1, 0.4515, 0, 0]^{T}, $$

$$ {}^{r}_{s}\mathbf{x}^{n} = [1, 1, 1, 1, 1]^{T}, $$

$$ {}^{r-1}_{s}\mathbf{u}^{n} = [0.9442, 0.4441, -0.0893, 0]^{T}, $$

$$ {}^{r-1}_{s}\mathbf{x}^{n} = [0.9741, 0.9834, 0.9656, 0.9783, 0.9790]^{T}, $$

linear differential equations system matrices

$$ {}^{r}\mathbf{A} = \begin{bmatrix} -1.8208 & 1.1770 & 0 & 0 & 0 \\ 0.0697 & -1.6507 & 0 & 0 & 0 \\ 9.7385 & 11.5191 & -20.1171 & -13.1868 & 0 \\ 9.6711 & 3.5115 & -5.5917 & -19.3568 & 0 \\ 2.9867 & -6.8163 & -3.4159 & -3.1171 & -30.0422 \end{bmatrix}, $$

corresponding eigenvalues

${}^{r}\lambda_1 = -1.4370$, ${}^{r}\lambda_2 = -2.0345$, ${}^{r}\lambda_3 = -11.1415$, ${}^{r}\lambda_4 = -28.3324$, ${}^{r}\lambda_5 = -30.0422$,

$$
{}^{r}\mathbf{B} = \begin{bmatrix} 0.6093 & 0.8201 & -0.1387 & 0.1788 \\ 0.4454 & 0.5736 & -0.0387 & -0.2204 \\ 10.8333 & -40.9063 & 2.6207 & 1.7204 \\ 7.1854 & -39.0036 & 2.1503 & 0.5060 \\ 16.0831 & 3.7726 & -0.8053 & -1.2318 \end{bmatrix}.
$$

At first quadratic differential equations system matrices for two differential equations are estimated.

$$
{}^{r}_{1}\mathbf{\Gamma}^{x} = \begin{bmatrix} -0.0162 & -0.0181 \\ 0 & 0.0278 \end{bmatrix},
$$

$$
{}^{r}_{2}\mathbf{\Gamma}^{x} = \begin{bmatrix} -0.1089 & 0.0133 \\ 0 & -0.0417 \end{bmatrix},
$$

$$
{}^{r}_{1}\mathbf{\Gamma}^{u} = \begin{bmatrix} -0.0319 & -0.1280 & 0.0823 & -0.2076 \\ 0 & 0.1117 & 0.0651 & 0.1743 \\ 0 & 0 & 0.0041 & 0.1553 \\ 0 & 0 & 0 & -0.4165 \end{bmatrix},
$$

$$
{}^{r}_{2}\mathbf{\Gamma}^{u} = \begin{bmatrix} -0.2445 & 0.0548 & 0.3299 & -0.1507 \\ 0 & -0.0267 & 0.0975 & -0.0024 \\ 0 & 0 & -0.0881 & -0.2008 \\ 0 & 0 & 0 & 0.0639 \end{bmatrix}.
$$

Then quadratic differential equations system matrices for 3-rd and 4-th differential equations are estimated considering the matrices ${}^{r}_{1}\mathbf{\Gamma}^{x}$, ${}^{r}_{2}\mathbf{\Gamma}^{x}$, ${}^{r}_{1}\mathbf{\Gamma}^{u}$, ${}^{r}_{2}\mathbf{\Gamma}^{u}$.

$$
{}^{r}_{3}\mathbf{\Gamma}^{x} = \begin{bmatrix} -1.2607 & 0.2927 & -1.6746 & -0.8985 \\ 0 & 0.0186 & -0.2216 & -0.3720 \\ 0 & 0 & 1.9339 & -0.5949 \\ 0 & 0 & 0 & 2.4800 \end{bmatrix},
$$

$$
{}^{r}_{4}\mathbf{\Gamma}^{x} = \begin{bmatrix} 0.3141 & 0.2177 & 0.0014 & -2.5816 \\ 0 & 0.7004 & 0.9653 & -0.9625 \\ 0 & 0 & -1.1875 & -0.6546 \\ 0 & 0 & 0 & -0.7394 \end{bmatrix},
$$

$$
{}^{r}_{3}\mathbf{\Gamma}^{u} = \begin{bmatrix} -2.6416 & 0.1334 & -2.2222 & -1.0052 \\ 0 & 0.4733 & 0.1910 & -2.4470 \\ 0 & 0 & -0.7613 & 0.1514 \\ 0 & 0 & 0 & -1.5057 \end{bmatrix},
$$

$$
{}^{r}_{4}\mathbf{\Gamma}^{u} = \begin{bmatrix} 0.6351 & 0.2250 & -0.5388 & 0.1228 \\ 0 & -3.3969 & -0.8555 & -0.2930 \\ 0 & 0 & -1.4348 & -1.3817 \\ 0 & 0 & 0 & 0.3208 \end{bmatrix}.
$$

At last quadratic differential equations system matrices for 5-th differential equations are estimated considering the matrices ${}^{r}_{1}\mathbf{\Gamma}^{x}$, ${}^{r}_{2}\mathbf{\Gamma}^{x}$, ${}^{r}_{1}\mathbf{\Gamma}^{u}$, ${}^{r}_{2}\mathbf{\Gamma}^{u}$, ${}^{r}_{3}\mathbf{\Gamma}^{x}$, ${}^{r}_{4}\mathbf{\Gamma}^{x}$, ${}^{r}_{3}\mathbf{\Gamma}^{u}$, ${}^{r}_{4}\mathbf{\Gamma}^{u}$.

$$
{}^{r}_{5}\mathbf{\Gamma}^{x} = \begin{bmatrix} 0.0764 & -3.4153 & -0.4633 & -0.6979 & 0.2337 \\ 0 & 0.4479 & -1.2369 & 0.8377 & 1.3469 \\ 0 & 0 & -0.9704 & -0.8798 & -0.5260 \\ 0 & 0 & 0 & -1.0099 & -1.5035 \\ 0 & 0 & 0 & 0 & -0.1620 \end{bmatrix},
$$

$$
{}^{r}_{5}\mathbf{\Gamma}^{u} = \begin{bmatrix} -1.5124 & 0.9865 & 0.3452 & -2.4686 \\ 0 & 1.2401 & 0.2039 & 0.5581 \\ 0 & 0 & 0.3440 & -1.4374 \\ 0 & 0 & 0 & 0.9207 \end{bmatrix}.
$$

There are $k = 1,\ldots,2001$, $\Delta t = 0.025$, and the weight matrices are identity matrices. Since matrix ${}^{r-1}\mathbf{A}$ is calculated through quadratic differential equations system matrices, additional constraints are used in order to provide matrix ${}^{r-1}\mathbf{A}$ real and different eigenvalues.

$$
{}^{r-1}\mathbf{A} = \begin{bmatrix} -1.8197 & 1.1765 & 0 & 0 & 0 \\ 0.0751 & -1.6497 & 0 & 0 & 0 \\ 9.8760 & 11.5266 & -20.1902 & -13.2445 & 0 \\ 9.7072 & 3.4703 & -5.5119 & -19.2194 & 0 \\ 3.0656 & -6.7466 & -3.2865 & -3.0073 & -30.0131 \end{bmatrix},
$$

corresponding eigenvalues

${}^{r-1}\lambda_1 = -1.4255$, ${}^{r-1}\lambda_2 = -2.0439$, ${}^{r-1}\lambda_3 = -11.1469$, ${}^{r-1}\lambda_4 = -28.2627$, ${}^{r-1}\lambda_5 = -30.0131$,

$$
{}^{r-1}\mathbf{B} = \begin{bmatrix} 0.6065 & 0.8198 & -0.1445 & 0.1752 \\ 0.4428 & 0.5622 & -0.0421 & -0.1941 \\ 11.3250 & -40.9378 & 2.8790 & 1.7846 \\ 7.1611 & -38.8895 & 2.4429 & 0.6247 \\ 16.2135 & 3.6811 & -0.8875 & -0.6247 \end{bmatrix}.
$$

Some results of proposed method application are presented at Fig. 1, 2. The piecewise-linear differential equations system state variable derivative ${}^{\mathrm{PL}}\dot{x}_5^n$ time responce surge on the border of two linear differential equations systems application areas is much less if matrix ${}^{r-1}\mathbf{A}$ is calculated through quadratic differential equations system matrices.

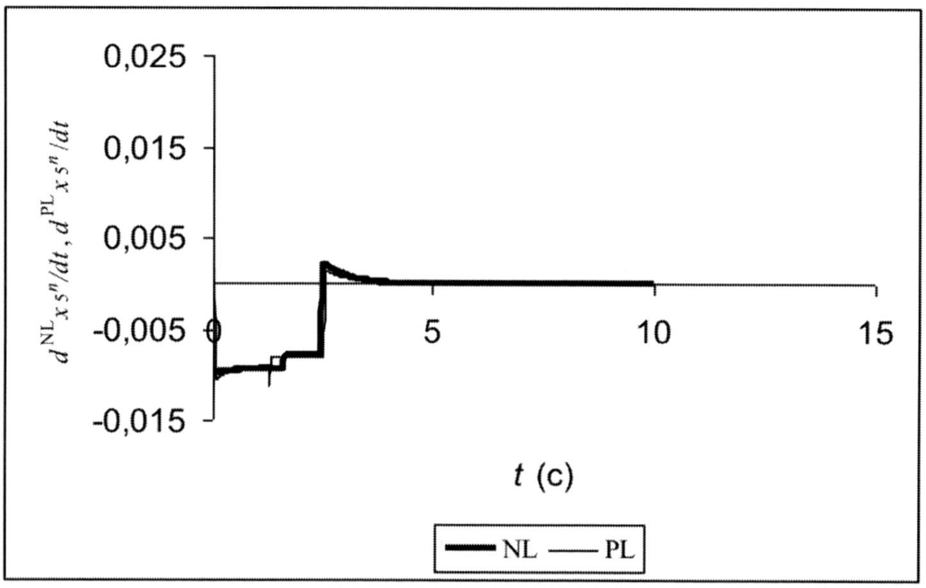


**Fig. 1.** Nonlinear and piecewise-linear differential equations systems state variables ${}^{\mathrm{NL}}x_5^n$ and ${}^{\mathrm{PL}}x_5^n$ derivatives ${}^{\mathrm{NL}}\dot{x}_5^n$ and ${}^{\mathrm{PL}}\dot{x}_5^n$ time responces (${}^{r-1}\mathbf{A}$ is calculated through quadratic differential equations system matrices).

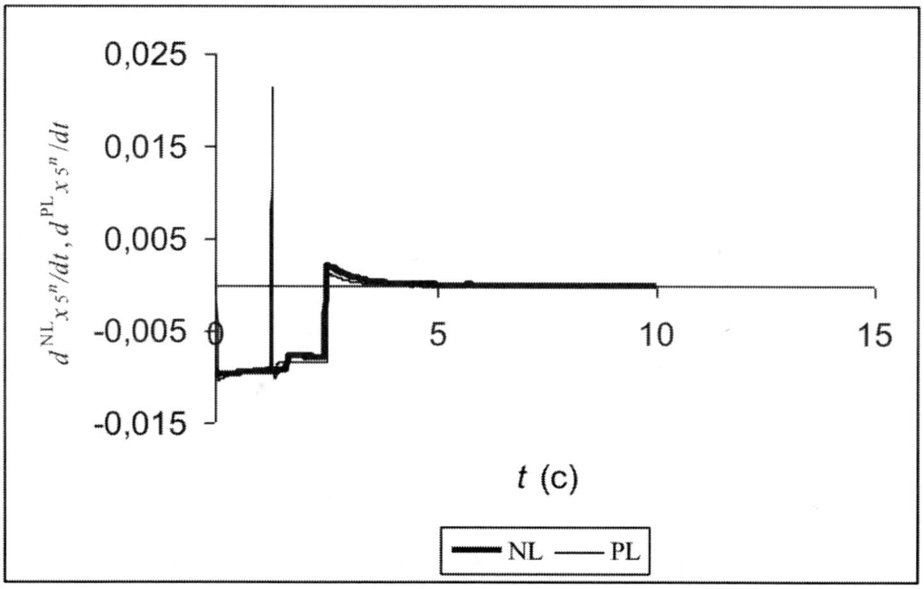


**Fig. 2.** Nonlinear and piecewise-linear differential equations systems state variables ${}^{NL}x_5''$ and ${}^{PL}x_5''$ derivatives ${}^{NL}\dot{x}_5''$ and ${}^{PL}\dot{x}_5''$ time responces.

## References

1. R.L. Leibov, IJCCSE **8**, 1, 96-107 (2012)
2. R.L. Leibov, IJCCSE **9**, 4, 186-198 (2013)
3. R.L. Leibov, IJCCSE **11**, 2, 107-130 (2015)
4. R.L. Leibov, IJCCSE **12**, 3, 40-57 (2016)

# Analysis of the concurrent operation of excavator actuating mechanisms as the determining factor in increasing its productivity

*Aleksandr* Lenivtsev[1], and *Ivan* Dudanov[1,*]

[1]Samara State Technical University, Institute of Architecture and Civil Engineering, 194, Molodogvardeyskaya St., 443001, Samara, Russia

**Abstract.** The most effective methods of organizing construction work are complex mechanization and automation of construction. Complex mechanization allows selecting the machines participating in the construction process according to their productivity. In this regard, it is necessary to take into account the structural features of construction machines that affect the overall pace of work. In order to shorten the cycle time, the design of a modern excavator's hydraulic system ensures the combined operation of two degrees of mobility in pairs: platform-boom, boom-arm, and arm-bucket. When designing and creating a system for automatic control of excavator actuating mechanisms, it is necessary to take into account the interrelation and mutual influence of hydraulic drives when working together from a source of limited power, the pump station of an excavator. Such a time combination of operations is possible when carrying out transportation to move the bucket to load and unload, while soil excavation is performed by turning the handle or bucket, and in these conditions, a large load of the feeding power station does not allow combining these operations. The paper presents a structural diagram of the concurrent operation of two hydraulic excavator actuators, which reflects the existence of connections between the hydraulic motors of the actuators.

## 1 Introduction

When organizing the construction process, special attention is paid to the timely performance of construction work. One of the main types of construction works is earthwork, where a large number of machines are used, mainly excavators and bulldozers. The most effective methods of organizing construction works are complex mechanization and automation of construction. Complex mechanization makes it possible to select the machines participating in the construction process rationally in terms of their productivity. In this case, it is necessary to take into account the structural features of construction machines that affect the overall pace of works.

The structural feature of the hydraulic excavator is the combination of two working operations [1]: turning the platform and lifting the boom, lifting the boom and turning the

* Corresponding author: iv_dudanov@mail.ru

handle, turning the handle and the bucket. These modes are used to improve the performance of an excavator and are only possible when carrying out transport operations (moving the bucket to load and unload), due to the large load on the feed pump during excavation.

In terms of working equipment management, the excavator is a 4-dimensional control object for actuators (platform turning, boom lifting, handle and bucket turnings), where the presence of interchannel connections is conditioned by the interaction of working equipment while changing the bucket spatial position and filling; it is also conditioned by the possibility of pairwise operation of hydraulic motors from one hydraulic pump [2]. When designing automatic control systems for excavators, it is extremely important to take into account the dynamic characteristics of the hydraulic drive when the excavator actuator mechanisms work together to ensure efficient operation modes.

Through the example of ЭО-4121 excavator, we have developed mathematical models (in the form of structures and computational models) of the concurrent operation of the three above-mentioned pairs of actuators using structural diagrams and computational models of actuator systems, hydraulic pump and algorithms for forming fluid flow rates [3].

## 2 Materials and methods

In our work, we used simulation modeling of the concurrent operation of platform drives turning and boom lifting, on computational models in the MatLab Simulink software environment. The generalized structure (Fig.1) presents channel operators in relation to control $A_{11}$, $A_{22}$ and disturbing influences - $C_{11}$, $C_{22}$; operators $B_{12}$, $B_{21}$ and $C_{12}$, $C_{21}$ show cross-channel links manifested when two working mechanisms operate together.

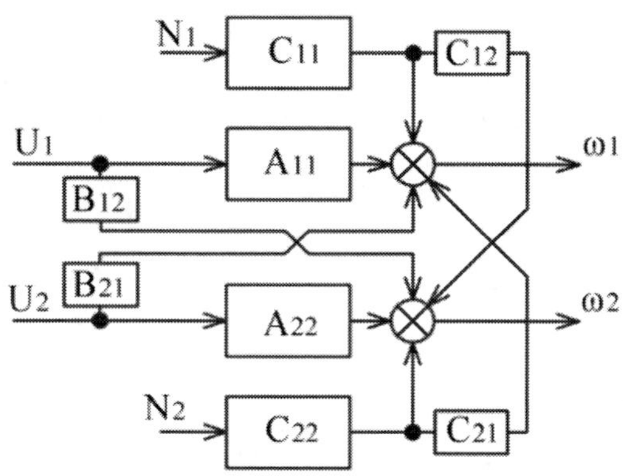


**Fig. 1.** The structure of the concurrent work of two simulation models.

It is analytically difficult to define the operators describing these connections. Therefore, we use the method of computational experiment [4-8]. Figure 2 shows the generalized structure of the computational model of the platform and the boom simultaneous operation dynamics, created in the MatLab Simulink software environment. Here, blocks *1,2* are models of the hydraulic drive of the platform and the boom, block *3* is the model of the hydraulic pump, block *4* is the model of the hydraulic distributor controlling the supply of fluid to the hydraulic motors of the actuators (PP-turning platform, S-boom), according to Table 1.

**Table 1.** Modes of hydraulic motors concurrent operation.

| Flow limitation for hydraulic motors | |
|---|---|
| PP | S |

| | |
|---|---|
| $Qpp \leq Q_H1$ | $Qs \leq Q_H2 + Q_H1 - Qpp$ |

Where $Q_{H1}$, $Q_{H2}$ are flows of working fluid from sections of the pumping unit; $Qpp$, $Qs$ represent supply of hydraulic fluid to hydraulic motors of the turning platform and the boom, respectively.

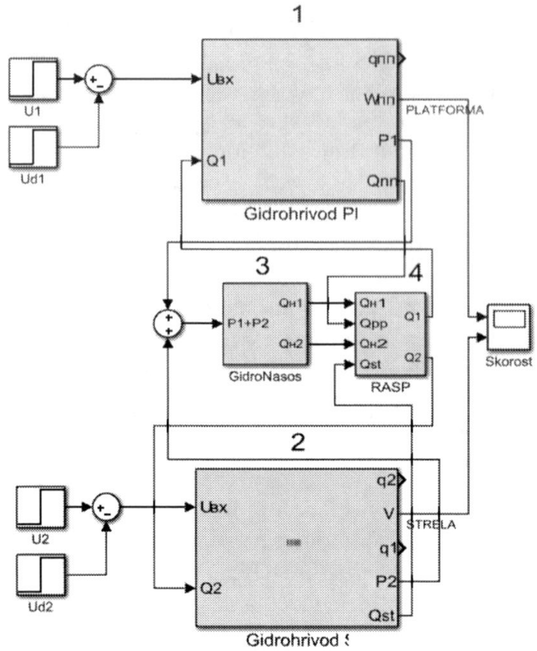


**Fig. 2.** Computational model for studying the dynamics of platform and boom simultaneous operation.

The presented model gives the opportunity to determine experimentally the operators shown in Fig. 1 from the analysis of transient characteristics in the two-dimensional system. When performing experiments, we use stepwise change in "small" and "large" control $U_1$, $U_2$ and perturbing $N_1$, $N_2$ influences.

## 3 Results

It has been experimentally established that during simultaneous operation of the hydraulic drives of the turning platform and the boom, the boom dynamics is significantly influenced by the degree of load of the hydraulic drive of the rotary platform [9]. This is explained by the fact that the amount of hydraulic fluid flow to the hydraulic motor of the boom is determined by the capacity of the two pump sections, except for the required flow rate of the hydraulic motor of the turning platform (Table 1). The hydraulic motor of the turning platform runs off one section of the pump unit only, so the boom does not affect the dynamics of its operation.

Using the computational model (Fig. 2), we obtained many transient characteristics reflecting the concurrent operation of the turning platform and the boom (Fig. 3.a, b).

When experiments were carried out, the same control influences $U_1 = U_2$ were simultaneously supplied to the inputs of both degrees of mobility. At the time $t = 3s$ (after the transient processes were completed), an additional stepwise control influence $U_{1D} = -0.3U_1$ was supplied to the input of the rotary platform model. This corresponds to the reduction in supply of hydraulic fluid to the platform hydraulic motor. In dynamic characteristics, this is reflected by 11% decrease in the speed of the rotary platform in statics (curve 1); the dynamic emission is 9%, while the boom speed in statics is increased by 29% (curve 2), and the transmission ratio of the boom as a separate control object increases by 39%. The performed experiment corresponds to the operation mode of the two-dimensional control object "in small". We determined interchannel dynamic connections of the *k-th* actuating mechanism influence on the *i-th* link; they can be represented as links of the second order:

$$W_{ik}(p) = \frac{k_{ik}}{T_{ik}^2 \cdot p^2 + 2 \cdot T_{ik} \cdot \xi_{ik} \cdot p + 1}, \qquad (1)$$

where the parameters vary depending on the magnitude of the jump $U_{ID}$. It was experimentally established that when jump $U_{ID}$ value varies from $-0.4U_1$ to $-0.1U_1$, the parameters of the transfer function change in accordance with the data in Table 2.

**Table 2.** Dynamic link $W_{ik}(p)$ parameters.

| $U_{ID}$ | $T_{ik}$, s | $\zeta_{ik}$ | $k_{ik}$, rad/V·s |
|---|---|---|---|
| $-0.4U_1 \ldots -0.1U_1$ | 0.04…0.15 | 0.59…0.35 | 0.005… 0.0039 |

If there is no supply of the power fluid to the hydraulic drive of the turning platform, the transmission ratio of the boom doubles, taking the value of $k_{OY2}$ = 0.068 rad / V · s, because in this case the entire flow from both sections of the pump unit is fed to the boom hydraulics. When the value of jump $U_{ID}$ of the control influence varies from level $U_{ID}$ = - 0.5U1, the object switches to the "large" mode, because the dynamics of its operation is influenced by the non-linearity "level limitation", manifested in the spool valve. This leads to a change in the form of the transition characteristic (Fig. 3b), which allows us to conclude that the control object is nonlinear. At the same time, it is difficult to assess the dynamics by the type of the transition process.

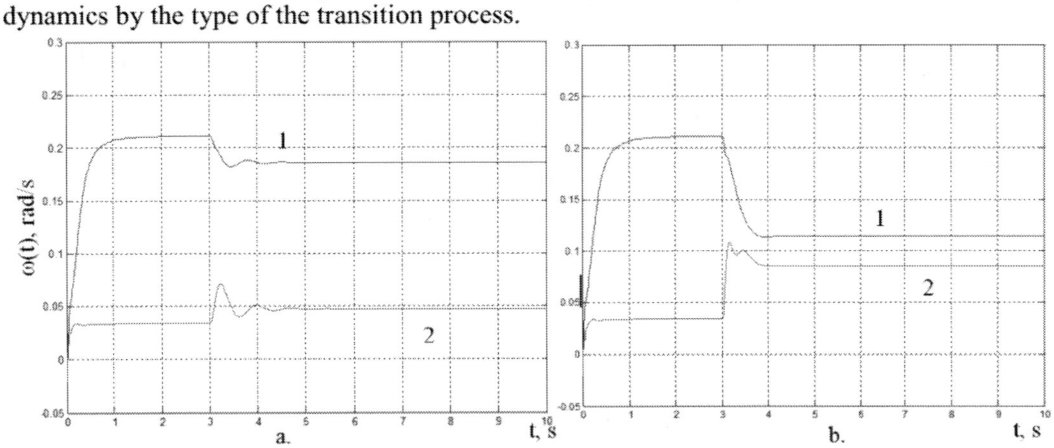


**Fig. 3.** Graphs of transient processes.

The study of the dynamics of the concurrent operation of the turning platform and the boom with variation of the hydraulic drive load [10] showed that in this case there is an interinfluence of these systems with each other.

Figure 4 shows the curves of transient processes in the turning platform drive (curve 1) and the boom drive (curve 2). By the time $t$ = *5s*, when transient processes in the boom drive were completed, stepwise control influence $U_1 = U_2$ was supplied to the input of the turning platform model. This causes reduction in the speed of the boom drive in static by

50%. At the time $t = 6s$, the load of the hydraulic drive of the turning platform $N_1 = 0.3N_{1nom}$ is set stepwise. This causes an additional decrease in speed by 13% in 0.4s.

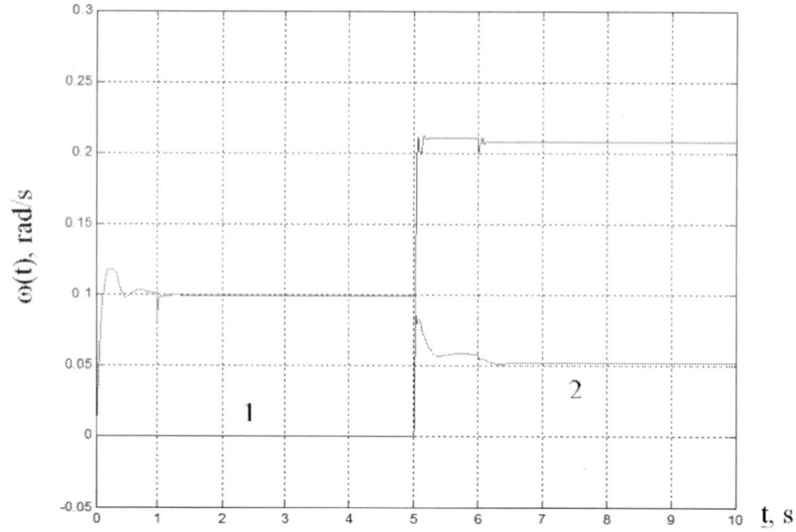


**Fig. 4.** Graphs of transient processes.

According to the form of transient processes, it was established that their dynamics of the *m-th* actuating mechanism influence on the *j-th* link can be represented, with a sufficient degree of accuracy, by a dynamic link of the second order with a transfer function: Interchannel dynamic links were defined, and they can be represented by links of the second order:

$$W_{jm}(p) = \frac{k_{jm}}{T_{jm}^2 \cdot p^2 + 2 \cdot T_{jm} \cdot \xi_{jm} \cdot p + 1}, \qquad (2)$$

where the parameters vary depending on the magnitude of the jump $N_1$. It has been experimentally established that when the magnitude of the jump varies from $0.1N_{1nom}$ to $0.3N_{1nom}$, the parameters of the transfer function change in accordance with the data in Table 2.

**Table 3.** Dynamic link $W_{ik}(p)$ parameters.

| $N_{1nom}$ | $T_{jm}$, s | $\zeta_{jm}$ | $k_{jm}$, rad/V·s |
|---|---|---|---|
| $0.1\ N_{1nom}...0.3\ N_{1nom}$ | 0.02…0.2 | 0.2…0.45 | 0.005… 0.0039 |

When the load is increased by more than $0.3_{Nom\_dist}$, it is impossible to estimate the dynamics by the type of the transient process due to the action of the control object nonlinearities.

## 4 Discussion

It has been experimentally established that the nature and level of dynamic connections between the hydraulic excavator actuating mechanisms in their simultaneous operation depend on the magnitude of the control influence, causing a decrease in the hydraulic boom speed up to 50% and the degree of the drives load, which causes a further decrease in the operating speed. This allows us to conclude that the dynamic links between the actuators are unsteady.

## 5 Conclusions

The computational experiments carried out and the results obtained make it possible to assess the nature and level of dynamic links between hydraulic excavator actuating mechanisms, which are extremely important to ensure effective modes of excavator operation, both of the leading machine and of the entire set of construction machines in the organization of construction.

## References

1. V.G. Krikun, *Calculation of the Main Parameters of Hydraulic Excavators with Working Equipment Backhoe*, 104, (2003)
2. S.Y. Galitckov, I.V. Dudanov, Mehanizatciia stroitelstva, **6**, 9 – 10 (2008)
3. S.Ya. Galitskov, I.V. Dudanov, *Control Systems and Computer Simulation of an Excavator Hydraulic Drive* [electronic resource] (Monograph) Samara, 2014. - Electronic Text and Graphic Data (48 Mb). - Scientific Electronic Edition of Combined Distribution: 1 CD. - Systemic Requirements: PC 486 DX-33; Microsoft Windows XP; 2-speed CD-ROM drive; Adobe Reader 6/0 - samgasu.ru Electronic text and graphic data (**48 Mb**)
4. S.Y. Galitskov, K.S. Galitskov, O.V. Samokhvalov, *Procedia Engineering,* (2016).
5. S.Y. Galitskov, K.S. Galitskov, M.A. Nazarov, *MATEC Web of Conferences* 86, **04010**, (2016)
6. S.Y. Galitskov, V.N. Mikhelkevich, A.S. Bolkhovetsky, *Materials of the 8th International Conference «Problems of Management and Control in Complex Systems»,* 347-350 (2016)
7. S.Y. Galitskov, K.S. Galitskov, O.V. Samokhvalov, A.S. Fadeev, MATEC Web of Conferences **86**, 04009 (2016)
8. M.A. Nazarov, V.I. Kichigin, A.S. Fadeev, Procedia Engineering **153** (2016)
9. V.N. Kuznetsova, V.V. Savinkin, Vestnik SibADI **6**, 26-33, (2014)
10. B.D. Kononikhin, Mechanization of Construction **4**, 15-20, (2005)

# Effects of Forces Applied in the Middle Plane on Bending of Medium-Thickness Band

*Andrey* Leontyev[1,*]

[1]Moscow state university of civil engineering, Yaroslavskoye shosse, 26, Moscow, Russia, 129337

**Abstract.** This paper examines the problem of calculating a plate of medium thickness uploaded by transverse loads and forces in the middle plane. We use variant equations from the theory of medium thickness plates as inputs, as proposed by B.F. Vlasov. The paper also considers the presence of tensile or compressive forces in the plane of the plate. The paper offers an example of calculation for a semi-infinite uploaded at the end of the transverse load and pivotally supported on the longitudinal edges. It is demonstrated that in the middle plane, the forces significantly influence the plate's deflections and bending forces. There are material differences between results obtained with the theory of plates of medium thickness and those of the classical the theory of bending of plates.

## 1 Introduction

A rectangular plate with various conditions on the contour, installed on an elastic base, is a widely used element in building structures. In certain cases, when calculating plates, particularly ones that rest on an elastic base, one needs to factor in some forces applied to the middle plane, in addition to the transverse loads. Such extra forces can be caused by seasonal and daily fluctuations of temperature, pre-tensed reinforcement steel, impacts from production equipment, and pressure of the enclosing walls on the foundation slab.

The stressed and deformed state of a medium thickness plate bending from the action of a transverse load, is described with these equations, as proposed by B.F. Vlasov [1]:

$$\nabla^2\nabla^2 t(x,y) = \frac{p(x,y)}{D}, \qquad \nabla^2\varphi(x,y) - \frac{12}{h^2}\varphi(x,y) = 0 . \tag{1}$$

Here, $D$ is the cylindrical rigidity of the plate, $h$ is its thickness. The deflection $w(x, y)$ and the rotation angles of the plate $t_x(x,y)$, $t_y(x,y)$ are associated with desired functions $t(x,y)$, $\varphi(x,y)$ as relations:

$$w = t - \frac{D}{Gh}\nabla^2 t, \qquad t_x = \frac{\partial t}{\partial x} - \frac{D(1-\nu)}{2Gh}\frac{\partial\varphi}{\partial y}, \qquad t_y = \frac{\partial t}{\partial y} + \frac{D(1-\nu)}{2Gh}\frac{\partial\varphi}{\partial x}, \tag{2}$$

where $G$ – the shear modulus, $\nu$ – Poisson's ratio.

We can write the following dependence:

* Corresponding author: LeontievAN@mgsu.ru

$$\nabla^2 t(x,y) = \frac{\partial t_x(x,y)}{\partial x} + \frac{\partial t_y(x,y)}{\partial y}, \qquad \varphi(x,y) = \frac{\partial t_y(x,y)}{\partial x} - \frac{\partial t_x(x,y)}{\partial y}. \tag{3}$$

$$M_x = -D\cdot\left(\frac{\partial^2 t}{\partial x^2} + \nu\frac{\partial^2 t}{\partial y^2} - \frac{D(1-\nu)^2}{2Gh}\frac{\partial^2 \varphi}{\partial x \partial y}\right),$$
$$M_y = -D\cdot\left(\frac{\partial^2 t}{\partial y^2} + \nu\frac{\partial^2 t}{\partial x^2} + \frac{D(1-\nu)^2}{2Gh}\frac{\partial^2 \varphi}{\partial x \partial y}\right), \tag{4}$$

$$M_{xy} = -D\frac{1-\nu}{2}\cdot\left[2\frac{\partial^2 t}{\partial x \partial y} + \frac{D(1-\nu)}{2Gh}\left(\frac{\partial^2 \varphi}{\partial x^2} - \frac{\partial^2 \varphi}{\partial y^2}\right)\right], \tag{5}$$

$$Q_x = -D\cdot\left(\frac{\partial}{\partial x}\nabla^2 t - \frac{1-\nu}{2}\frac{\partial \varphi}{\partial y}\right), \qquad Q_y = -D\cdot\left(\frac{\partial}{\partial y}\nabla^2 t + \frac{1-\nu}{2}\frac{\partial \varphi}{\partial x}\right). \tag{6}$$

## 2 State of the problem

If forces $N_x$, $N_y$, $N_{xy}$=$N_{yx}$ (Fig. 1) exist in the middle plane of the plate, the second equation (1) remains unchanged, and in the first equation must consider projections of such forces on axis $Oz$, created by the bending of the plate.

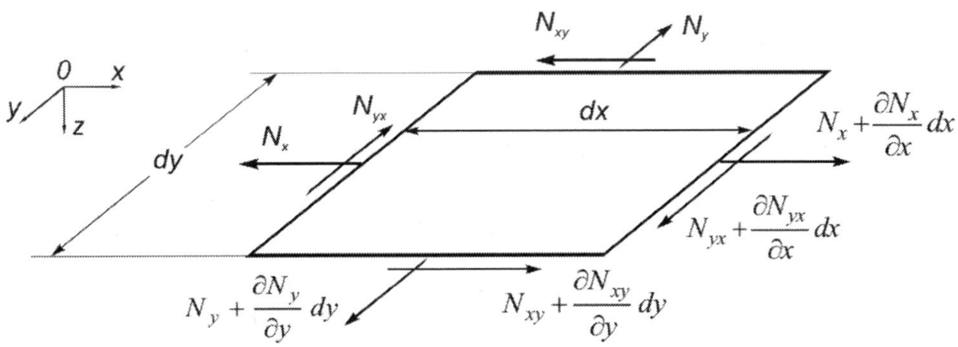


**Fig. 1.** Efforts in the middle plane of the plate

Notably, such forces are interrelated with equations of equilibrium:

$$\frac{\partial N_x}{\partial x} + \frac{\partial N_{xy}}{\partial y} = 0, \qquad \frac{\partial N_{yx}}{\partial x} + \frac{\partial N_y}{\partial y} = 0. \tag{7}$$

Project forces $N_x$, $N_y$, $N_{xy}$=$N_{yx}$ and their increments on the vertical axis. Then in the light of the expressions (2, 3, 7), the first equation (1) can be written as:

$$\nabla^2\nabla^2 t - \frac{N_x}{D}\frac{\partial^2 t}{\partial x^2} - \frac{N_y}{D}\frac{\partial^2 t}{\partial y^2} - 2\frac{N_{xy}}{D}\frac{\partial^2 t}{\partial x \partial y} = \frac{q}{D}. \tag{8}$$

## 3 Analytical method of solving

Let us assume that the longitudinal side edges of the plate have a hinged support (Fig. 2).

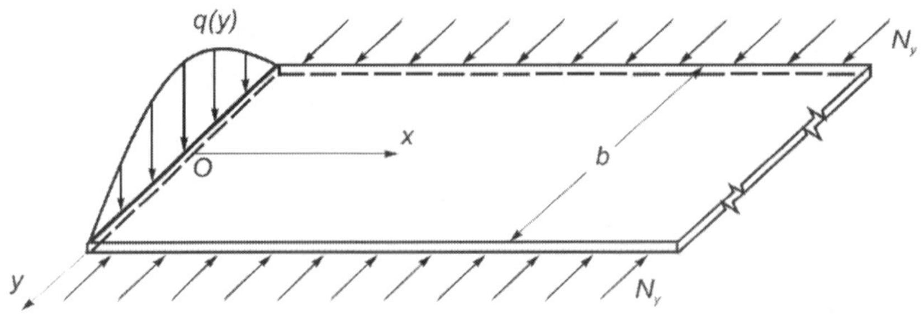


**Fig. 2.** Statement of the problem of calculation of the plate

Then our desired functions can be represented as:

$$t(x,y)=\sum_n W_n(x)\cos\beta_n y\,, \tag{9}$$

$$\varphi(x,y)=\sum_n F_n(x)\sin\beta_n y\,, \quad \text{where } \beta_n=\frac{n\pi}{b}. \tag{10}$$

We substitute decomposition (9) in equation (8), due to the orthogonality of trigonometric functions, for each of $n$ get an ordinary differential equation of the fourth order, which in the case of only longitudinal forces $N_x$ will appear as:

$$W_n^{IV}-2\left(\beta_n^2+\frac{N_x}{2D}\right)W_n^{II}+\beta_n^4 W_n=\frac{p_n}{D}, \tag{11}$$

and in the case of longitudinal forces $N_y$:

$$W_n^{IV}-2\beta_n^2 W_n^{II}+\beta_n^4\left(1+\frac{N_y}{\beta_n^2 D}\right)W_n=\frac{p_n}{D}, \quad \text{where } p_n=\frac{2}{b}\int_0^b q(x,y)\cos\beta_n y dy\,. \tag{12}$$

Equations (11) and (12) can be represented using the standard form, typical for a problem of the bending of the beam located on an elastic base:

$$W_n^{IV}-2r_n^2 W_n^{II}+s_n^4 W_n=\frac{p_n}{D}. \tag{13}$$

Depending on the relation of the coefficient values $r_n$ and $s_n$ we find the type of roots of the respective characteristic equations, and therefore the type of particular integrals for its solution.

First examine a case of the plate under the action of forces $N_y$, distributed evenly over its longitudinal edges (Fig. 2). By comparing equations (12) and (13) we can see that in this case, the coefficients of equation (13) will be values:

$$r_n^2=\beta_n^2, \qquad s_n^4=\beta_n^4\left(1+\frac{N_y}{\beta_n^2 D}\right), \tag{14}$$

relations between which depend on the sign of forces $N_y$: in the case of stretching plate $(N_y>0)$ $s_n>r_n$, in the case of compression $(N_y<0)$ $s_n<r_n$.

For $s_n>r_n$ the roots $k_n$ of the characteristic equation is determined by the formula:

$$k_n=\pm\alpha_n\pm i\gamma_n, \tag{15}$$

where

$$\alpha_n=\sqrt{\frac{s_n^2+r_n^2}{2}}, \qquad \gamma_n=\sqrt{\frac{s_n^2-r_n^2}{2}}, \tag{16}$$

while the general solution of differential equation (11) is written as:

$$W_n(x)=C_{1n}F_{1n}(x)+C_{2n}F_{2n}(x)+C_{3n}F_{3n}(x)+C_{4n}F_{4n}(x)+W_n^0(x). \tag{17}$$

Here, $F_{1n}(x),\ldots,F_{4n}(x)$ are hyperbolo-trigonometric functions, $W_n^0(x)$ is a particular solution that depends on the transverse load $p_n(x)$.

For $s_n < r_n$ the roots of the characteristic equation will appear as:

$$k_1 = -k_2 = \lambda_{1n} = \sqrt{r_n^2 + \sqrt{r_n^4 - s_n^4}}, \qquad k_3 = -k_4 = \lambda_{2n} = \sqrt{r_n^2 - \sqrt{r_n^4 - s_n^4}}, \tag{18}$$

and therefore the solution to equation (11) will have this form:

$$W_n(x) = C_{1n}\mathrm{sh}\lambda_{1n}x + C_{2n}\mathrm{ch}\lambda_{1n}x + C_{3n}\mathrm{sh}\lambda_{2n}(x) + C_{4n}\mathrm{ch}\lambda_{2n}(x) + W_n^0(x). \tag{19}$$

If the plate is under the action of forces $N_x$ coefficients of equation (13) appear:

$$r_n^2 = \beta_n^2\left(1 + \frac{N_x}{2\beta_n^2 D}\right), \qquad s_n^4 = \beta_n^4. \tag{20}$$

From the relations (20) we can see that this time, in the case of stretching plate $(N_x > 0)$ $s_n < r_n$, and in the case of compression plate $(N_x < 0)$ $s_n > r_n$, the solutions to the problem will be represented respectively as expressions (19) or (17).

Us now examine the problem of the bending of a semi-infinite plate under the action of transverse load $q(y)$ in its initial section and longitudinal compressing forces $N_y$, evenly distributed over its longitudinal edges (Fig. 2).

In this case, if the transverse load is represented as $q_y = q_0\cos\beta y$, where $\beta = \pi/b$, then even first term of decomposition (9) present an accurate solution, and so index $n$ can be omitted everywhere. Assuming the plate's infinite length, solution to equation (13) can be written as:

$$W(x) = C_1 e^{-\lambda_1 x} + C_2 e^{-\lambda_2 x}. \tag{21}$$

To completely solve the problem of the bending of a medium-thickness plate, we also need to add to expression (19) the integral of the second equation from (1), which in the light of decomposition of (10) and the infinite length of the plate, can be written as:

$$F(x) = D_1 e^{-gx}, \quad \text{where } g = \sqrt{\beta^2 + \frac{Gh}{D(1-\nu)}} = \frac{1}{h}\sqrt{12 + \beta^2 h^2}. \tag{22}$$

To find the three integration constants $C_1$, $C_2$, $D_1$ at the initial section ($x = 0$), one can set the following boundary conditions:

$$M_x = -D\left[W^{II} - \nu\beta^2 W - \frac{h^2(1-\nu)}{12}\beta F^I\right]\cos\beta y = 0,$$

$$M_{xy} = -D(1-\nu)\left[-\beta W^I + \frac{h^2}{24}(F^{II} + \beta^2 F)\right]\sin\beta y = 0, \tag{23}$$

$$Q_x = -D\left[W^{III} - \beta^2 W^I - \frac{(1-\nu)}{12}\beta F\right]\cos\beta y = -q\cos\beta y.$$

Revealing these conditions using equations (21) and (22), we obtain a system of three algebraic equations:

$$(\lambda_1^2 - \nu\beta^2)C_1 + (\lambda_2^2 - \nu\beta^2)C_2 + \frac{h^2(1-\nu)}{12}\beta g D_1 = 0,$$

$$\beta(\lambda_1 C_1 + \lambda_2 C_2) + \frac{h^2}{24}(\beta^2 + g^2)D_1 = 0, \tag{24}$$

$$\lambda_1(-\lambda_1^2 + \beta^2)C_1 + \lambda_2(-\lambda_2^2 + \beta^2)C_2 - \frac{1-\nu}{2}\beta D_1 = \frac{q_0}{D}.$$

Consider the same problem with existing tensile forces of $N_y$, the solution to (17) can be written as:

$$W(x) = C_1 e^{-\alpha x} \sin\gamma x + C_2 e^{-\alpha x} \cos\gamma x . \qquad (25)$$

To find the constants of integration, boundary conditions (23) and expressions (22) and (25) allow to obtain the following equation:

$$-2\alpha\gamma C_1 + (\alpha^2 - \gamma^2 - \nu\beta^2)C_2 + \frac{h^2(1-\nu)}{12}\beta g D_1 = 0,$$

$$-\beta\gamma C_1 + \beta\alpha C_2 + \frac{h^2}{24}(g^2 + \beta^2)D_1 = 0, \qquad (26)$$

$$\gamma(3\alpha^2 - \gamma^2 - \beta^2)C_1 + \alpha(3\gamma^2 - \alpha^2 + \beta^2)C_2 - \frac{1-\nu}{2}\beta D_1 = \frac{q_0}{D}.$$

Having solved the equation systems (24) and (26) and defining the constants $C_1$, $C_2$, $D_1$, we can determine our calculated values using the formulas of the theory of plate middle-thickness [1]. For example, the deflection $w(x,y)$ and the bending moment $M_y(x, y)$ are determine as:

$$w(x, y) = \left[ W(x) - \frac{12}{h^2}\left(W^{II}(x) - \beta^2 W(x)\right) \right] \cos\beta y , \qquad (27)$$

$$M_y(x, y) = -D\left[ -\beta^2 W(x) + \nu W^{II}(x) - \frac{h^2(1-\nu)}{12}\beta F^{I}(x) \right] \cos\beta y , \qquad (28)$$

where function $W(x)$ that is part of (27) and (28) appears as (21) when the plate is compressed, and as (25) when the plate is stretched.

## 4 Example of calculation

Fig. 3 has graphs representing the dependency of dimensionless deflection $\overline{w_0} = w(0,0)D / p_0 b^3$ in the middle of the plate's initial section on dimensionless longitudinal forces $\overline{N}_y = N_y b^2 / D$.

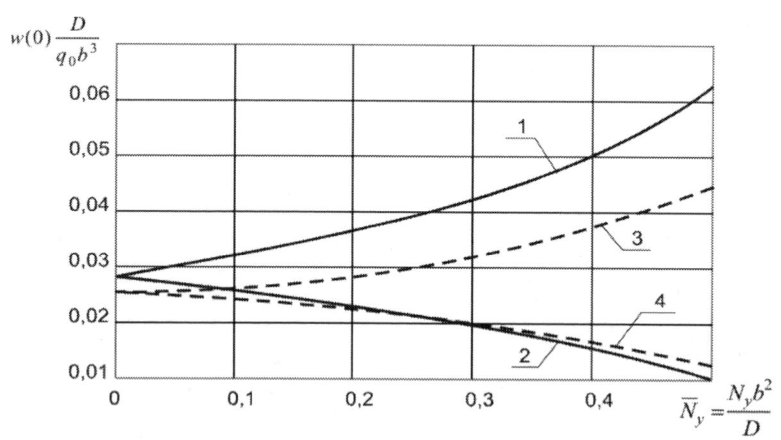


**Fig. 3.** Diagrams of dependence of deflection on longitudinal forces

In the diagram, solid lines 1 and 2 relate respectively to the cases compressed and stretched of the plate middle-thickness for $h/b = 0.2$, while dotted lines 3 and 4 describe the cases for the thin compressed and stretched plate.

Fig. 4 shows dependencies of dimensionless bending moments $\overline{M}_y^0$ at the middle of the initial section, on dimensionless longitudinal forces $\overline{N}_y$ .

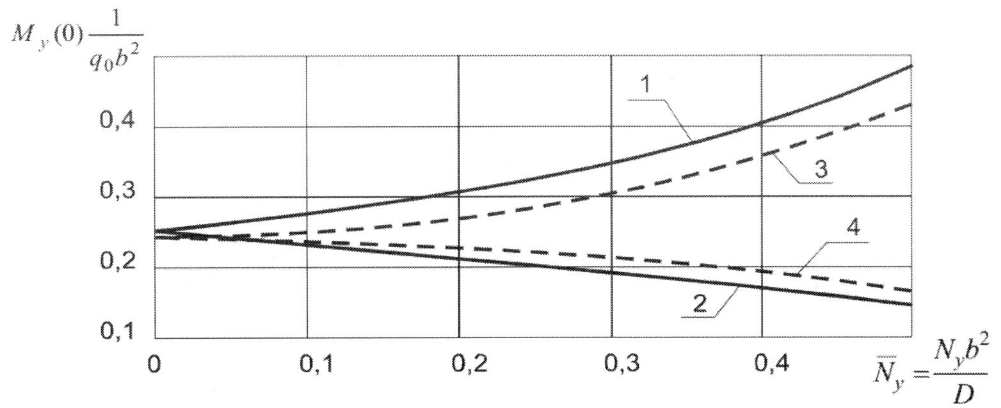


**Fig. 4.** Diagrams of dependence of the bending moments on longitudinal forces

## 5 Conclusions

Can see that an increase in longitudinal force $N_y$ causes material changes of deflection and bending moments: they are higher in the case of compression, and lower in the case of stretching. Expressed as percentage, deflections vary more than bending moments. It should also be noted that the effect of longitudinal forces on the plate middle-thickness is somewhat stronger than on a thin plate, and it is greater with compression than with stretching. Lines 1 and 3, nearly straight at low values of $N_y$, become curves tending to rise abruptly as $N_y$ increases, approaching their respective critical values.

## References

1. B.F. Vlasov, Equation of bending plates of average thickness. / Theoretical and experimental studies of strength and stiffness of structural elements, Moscow, MSUCE, (1989)
2. R.F. Gabbasov, V.V. Filatov, Calculation of compressed-bent plates in incomplete contact with an elastic Foundation, Moscow, MSUCE, (1999)
3. S.P. Timoshenko, S. Voynovskiy-Kriger. / Plates and shells, Moscow, Nauka (1966)
4. A.T. Turganbaev, Bending of a rectangular plate located on an elastic Winkler's Foundation with the influence of the longitudinal force. / Bases, foundations and soil mechanics, **3** (1993)

# Contact mechanic issue in non-typical connections in hybrid timber structures

*Arkadiusz* Leśko[1,*]

[1]Warsaw University of Technology, Faculty of Civil Engineering, Division of Wood Structures, Al. Armii Ludowej 16, Warsaw 00-637, Poland

**Abstract.** The paper relates to the issue of the contact mechanics in the innovative connections in hybrid timber structures. On the basis of the literature examples of the non-typical connections where the contact mechanic have to be investigated were discussed. In the paper are shown modern timber composites where connections are not systemized. Issue of the contact mechanic in finite element method related to the connections of the hybrid timber structures were elaborated. Spatial model of the "beam to beam" connection, based on the glued-in rods and advanced contact relations was shown. The necessity of the usage of advanced contact models was discussed and factors which affect the assumptions where presented.

## 1. Introduction

Connection in hybrid timber structures is an assembly of hybrid components, with usage of connectors and carpentry works to ensure mutual cooperation of the connected elements. It is the most responsive place in the structure of the object which has apart weakness inducted during montage and natural notches. Non-typical (experimental) connections are considered to be unique if they have at least one property listed below:

- compatibility with the atypical spatial positioning of the merged elements,
- compatibility with the atypical cross sections of joined elements,
- uniqueness of the solution is not subjected to the systemization,
- anility to transfer atypical load configuration,
- ability to carry loads under unusual service conditions.

Hybrid connections as the most loaded places in the constructions should be treated with distinctive precaution on each step of the creation of the structure. Created or designed connections ought to checked both: laboratory and in-situ tests. Examples of the hybrid connections are shown below.

* Corresponding author: a.lesko@il.pw.edu.pl

**Fig. 1.** Non-typical connections of timber structures.
(a) Join of the LVL structure [1]; (b) Screws connection without plates [2];

Connections shown above, based on non-typical structural design, allow for the merging specifically shaped elements with usage of "third" materials. However the internal structure of connected materials is constantly evolving what generate the need of creation new types of the joints based on sophisticated solutions and innovative materials. In particular, the development of the atypical architectural forms and the willing to create multistoreyed buildings in timber-composite technology, accelerates the evolution of hybrid materials.

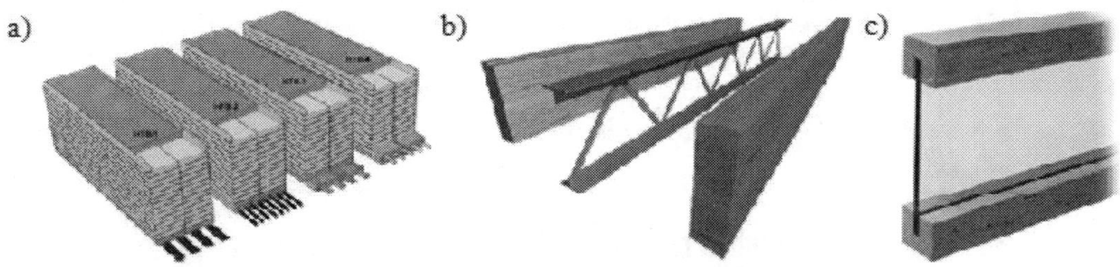

**Fig. 2.** Bar type composite elements (a) LVL based multi-composites [4];
(b) Hybrid beam Steel-LVL [5]; c) Innovative timber-glass beam composite [6];

Analysis of the connections of materials presented above requires the use of advanced scientific calculation methods and prioritization of the investigated phenomena. Contact mechanic is missed often enough and is replaced by "rigid conjunction" what is not always proper assumption. The issue of contact mechanics in term of timber structures connections is not systemized. Figure 3 depict unusual structural nodes what represents the complexity of the problem. Fig. 3 a) presents a geometrically complicated connection that does not require analysis of contact mechanic. Fig. 3 b) presents a geometrically simple connection that require analysis of contact mechanic.

**Fig. 3.** a) geometrically complicated column base with stiffening connectors [7];
b) connection based on friction forces [8];

## 2. Necessity of advanced contact models usage

Contact mechanic in modern, hybrid timber structures is an essential issue. Due to the multitude of possible steps of load transfer, it is not possible to establish a universal path that would allow quick application of the basic assumptions to the numerical models.

- direct load transfer:

- friction,
- pressure,

- indirect load transfer:

- glues,
- mechanical fasteners (wooden, steel, composite…).

- direct and indirect load transfer (depending on the step).

The most common case is the "c" variant – depending on the state of stress the load is transferred in a different way. By strictly following the assumptions it is not possible to accurately reproduce the work of the connection with usage of the finite element method. It is therefore necessary to use a number of simplifications.

Due to the development of sophisticated structures (in particular hybrid timber structures) it is indispensable to define "computational parameters" allowing to implement simplified dimensioning of folded structural elements. In addition the use of advanced contact models allows for analytical verification of experimental results such as:

- rigidity of the frame structures nodes $k_{\theta}$ [9]
- redistribution of a complex load system [10]
- delineation of material structure [11]
- fatigue degradation of materials [12]
- influence of temperature on fasteners behavior [13]

Application of new measurement methods [15], increasing the accuracy and precision of laboratory experiments [14] and necessity to reinforce existing structures also entails the development of analytical models.

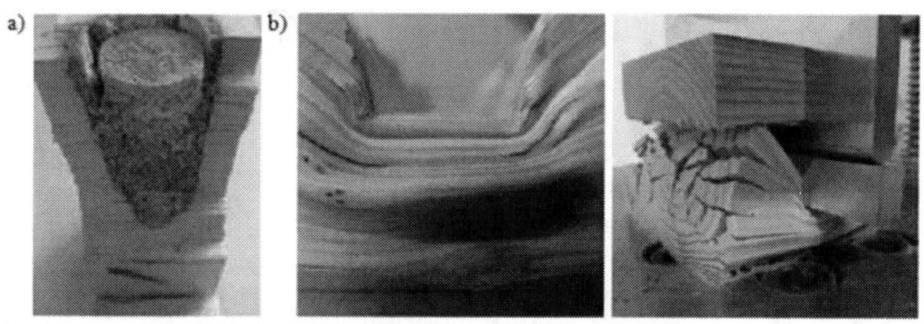


**Fig. 4.** Basic research based indirectly on the contact mechnics
(a) Junction based ona cement grout connector [17]; (b) Investigation of delamination [11]

## 3. FEM issues related to the contact mechanics in the hybrid timber structures connections

### 3.1. Detection and discretization of contact zone

The problem of frictionless contact for ideal bodies (smooth surface, lack of discontinuity, perfect adjust, etc.) comes down to analytical solution known as the "Hertz- Signorini-Moreau" problem [19,20,21]. Due to the complexity of hybrid connections in timber structures the analytical considerations are complicated. Wood (as a matrix) is a discrete fabric with numerous perturbations of a random nature. In addition typically hybrid connections are characterized by abrupt and high stiffness change in place of unmediated contact. Referring to the above it can be said that it is optimal to conduct detection analysis in hybrid timber structures by means of approximate methods – such as FEM.

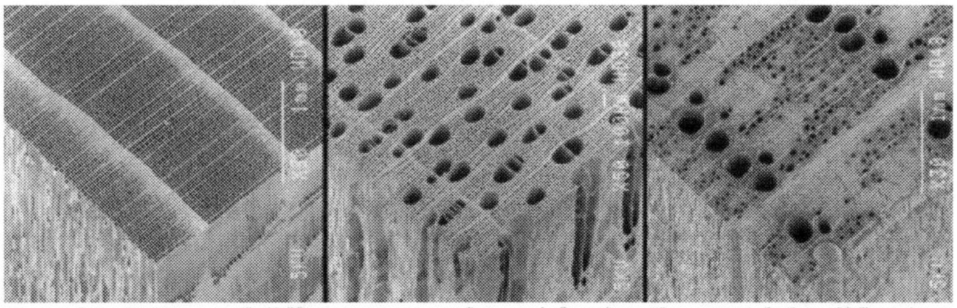

**Fig. 5.** Microscopic structure of different types of wood forcing the use of simplifications. [16]

### 3.2. Contact discretization

Discretization od the contact model is responsible for the proper propagation of stress between the combined materials. Publications such as [22,23,24] deal with the discretization of the continuum wherewith most of the analytical considerations referring to solutions with usage of FEM.

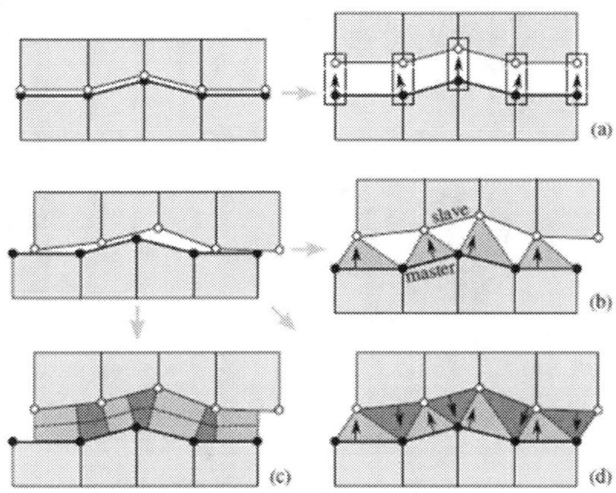


**Fig. 6.** Basic detection methods (related to discretization), method (a) for small displacement, methods (b,c, d) for large displacement [19] (a)"NTN" – node to node; (b)"NTS" – nodet to segment; c)"STS" – segment to segment of „contact domain method" type; d)"STS" – segment to segment of „mortar method" type;

The methods above have their limitations [21] all the more they are the most popular methods and still being developed. There are also other methods like "ALL to ALL" type or mixed techniques, but they do not have any ready-made formulas or modules in computational programs.

Use of these methods is strongly correlated with the geometry of the tested structural elements, methods of discretization of the contact and expected results. Depending on the combination of connected materials (wood, steel, glass, concrete, composite, bitumen, etc.) it is necessary to observe the results, adjust the parameters calculation parameters and methods used to detect the contact.

### 3.3. Issue of contact in FEM – methods of solving

Depending on the problem under investigation the method of solving the problem of contact should be selected. There are many relatively new methods such as: "Partitioning method", „SQP method", „Active set method" and many others. Due to simplicity and availability of software algorithms, despite their drawbacks, the most popular are:

- „Penalty Method" – it consist to select such a function of the "penalty" parameter (depending on whether or not the conditions of contact), which will not allow for interpenetration of connected elements. The advantages of this method are: easy physical interpretation of the phenomenon, lack of additional degrees of freedom and relatively good convergence. The disadvantages of penalty method are: the necessity of iterative matching penalty parameters and strong correlation between the results and the penalty function.
- „Lagrange Multiplier Method " – method taken from the optimization theory to obtain a convergent results at the expense of additional degrees of freedom.

These methods are used in one of the modules of "ABAQUS" program, such as „Contact pressure-overclosure relationship" or „Frictional behavior". These ready-made algorithms can be used to create models of interactions in hybrid timber structures based on built-in relations. There are already first studies to investigate the possibility of using contact modules in timber joints [18] but they only concern the effects of normal loads (without the analysis of tangential stresses).

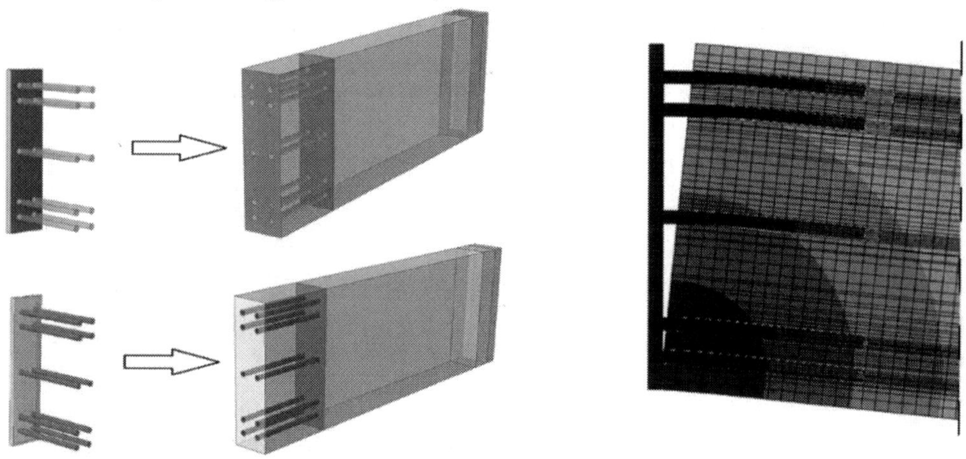

**Fig. 7.** Modeling attempt (maching of mechanical contact parameters) of the connection based on glued-in rods.[3]

## 4. Parameters affecting the issue of contact.

Apart from basic parameters being consequence of the implemented method of solving the contact problem, many factors affect both: the connection and the contact itself what should be considered in the investigation of the problem. These factors may have fundamental importance for the bearing capacity of the connected elements as a fasteners deadweight (or glue as a bonding medium) or the overall stability of the whole connection (SLS state). Two the most important parameters: the coefficient of friction and authentic (real) contact surface can be influenced by: duration of load, magnitude of load, degree of material effort, accuracy of adhesion of bonded surfaces, drop of roughness due to slipping, change in porosity due to the deformation, precision of the connection realization.

## 5. Conclusions

Creation of new solutions and examination of existing hybrid timber structure coupling technologies is not possible without analysis of the stress propagation using the contact mechanics and FEM. There are simulation systems (ABQUS type) that allow to analyze modern conjunctions. It is necessary to conduct a series of experiments, both: simulation and laboratory tests to correlate the assumed coefficients and parameters in numerical calculations.

### References

1. *Glued laminated timber structures. Part 2: construction and connection details., 9 Structural Timber Engineering Bulletin* (Structural Timber Association, 2014)
2. T. Gecys, A. Daniunas, T. K. Bader, L. Wagner, J. Eberhardsteiner, Engineering Structures **86**, (2015)
3. A. Leśko, Proc. Eng. **153** (2016)
4. M. Jahreis, M. Kastner, W. Hadicke, K. Rautenstrauch, Materials and Joints in Timber Structures **9** (2014)
5. K. Tavoussi, W. Winter, T. Pixner, M. Kist, WCTE (2010)
6. M. Kozłowski, *Badanie kompozytowych belek drewniano-szklanych* (Monografie Wydziału Inżynierii Lądowej, Warszawa, 2015)
7. J. Humbert, S. Lee, J. Park, M. Park, Materials and Joints in Timber Structures **9**, (2014)
8. K. Schober, T. Tannert, European Journal of Wood and Wood Products **74(3)** (2016)
9. K. Malo, H. Stamatopoulos, *World Conference on Timber Engineering e-book* (2016)
10. T. Laggner, G. Flatscher, G. Schickhofer, *World Conference on Timber Engineering e-book* (2016)
11. A. Tabuchi, *World Conference on Timber Engineering e-book* (2016)
12. Z. Christian, S. Aicher, *World Conference on Timber Engineering e-book* (2016)
13. M. Alam, G. Hadjisophocleous, J. Erochko, *World Conference on Timber Engineering e-book* (2016)
14. F. Wolfthaler, M. Augustin, *World Conference on Timber Engineering e-book* (2016)
15. F. Giudiceandrea, E. Ursella, E. Vicario, *World Conference on Timber Engineering e-book* (2016)
16. http://botanik.pl/tag/drewno/
17. M. Jahreis, K. Rautenstrauch, *World Conference on Timber Engineering e-book* (2016)
18. B. Iraola, J. Cabrero, B. Gil, *World Conference on Timber Engineering e-book* (2016)
19. P. Wriggers, *Computational contact mechanics, Second Edition* (Springer, New York, 2006)
20. A. Rashid, *Finite element modeling of contact problems, Linkoping Studies in Science and Technology, Dissertation No. 1736* (Linkoping, Sweden, 2016)

21. V.A. Yastrebov, *Numerical methods in contact mechanics* (Wiley, Great Britain, 2013)
22. P. Wriggers, *Nonlinear finite elements* (Springer, New York, 2006)
23. J. T. Oden, *Finite elements of nonlinear continua* (McGraw-Hill, New York, 1972)
24. O. C. Zienkiewicz, T. L. Taylor, *The finite element method – volume 2* (Oxford, United Kingdom, 2000)

# Nonlinear analysis of RC cylindrical tank and subsoil accounting for a low concrete strength

*Paweł M.* Lewiński[1,*] and *Sławomir* Dudziak[1]

[1] Building Research Institute (ITB), Filtrowa 1, 00-611 Warszawa, Poland

**Abstract.** The paper discusses deformational and incremental approaches to a nonlinear FE analysis of soil-structure interaction including the description of behaviour of the RC structure and the subsoil under short-term loading. Two kinds of constitutive models for ground and structure were adopted for a nonlinear interaction analysis of the RC cylindrical tank with subsoil. The constitutive laws for concrete and subsoil were developed in compliance with the deformational and flow theories of plasticity. Moreover, a non-linear elastic-brittle-plastic analysis of RC axi-symmetric structures using finite element iterative techniques is presented. The results of the two types of FE analysis of soil-structure interaction are compared taking into account a low concrete strength of tank structure.

## 1 Introduction

The subject of the paper is a new concept of a non-linear analysis of soil-structure interaction, including the description of the brittle and elastic-plastic behaviour of reinforced concrete structures, and the elastic-plastic behaviour of the subsoil. The analysis of the RC axisymmetric structure is performed with the use of the finite element iterative technique. The tensile cracking, the multiaxial compressive response of concrete and soil, and the yielding of steel reinforcement are the main non-linear effects studied. The present paper discusses two concepts of a nonlinear FE analysis of soil-structure interaction, including the description of the RC structure, one based on the deformation theory and the other – on the plastic flow theory of plasticity. This study analyses the impact of subsoil on stress resultants in the shell and ground slab of the RC cylindrical tank, as well as their redistribution which appears mainly due to cracks and, to some extent – uneven subsidence. Papers dealing with the finite element ultimate load analysis of thick-walled reinforced concrete structures subjected to radial pressure have been published since around 1968 in relation to its application for prestressed concrete reactor vessels. Likewise, widespread research were carried out on the subject of elastic foundations and soil-structure interaction. However, in the hitherto existing publications not all of the possible schemes of cracking patterns of RC axisymmetric structures have been exhausted and also with regard to RC tanks, it is rather difficult to refer to publications comprising numerical elastic-brittle-plastic analysis and soil-structure interaction. Nevertheless, many papers deal with axisymmetric RC structures, with the paper by Phillips and Zienkiewicz [1] and the book

* Corresponding author: p.lewinski@itb.pl

by Kotsovos and Pavlović [2] being among the first publications. In case of the total strain constitutive model, the analysis of soil-structure interaction was performed with the use of the FE direct iteration method, while in case of the plastic flow model, the *Newton-Raphson* method was applied. The results of the analysis indicate that in the case of structural concrete and subsoil of normal strength, the scope of redistribution is not very large, whereas in the case of weak concrete or subsoil the redistribution range is quite wide. This analysis shows an increase in radial displacements of the shell, as well as a rise in hoop steel stresses in relation to particularly weak structural concrete.

## 2 Material modelling

### 2.1 Failure criteria for concrete and cracking patterns

In case of the total strain constitutive model for concrete the ultimate surface proposed by J. Podgórski [3] was implemented. In the failure criterion assumed herein, the basic parameters can be written in the following manner:

$$r(\varphi) = \sqrt{2J_2} = [P(J)]^{-1},\ P(J) = \cos[(\arc\cos(\alpha J))/3 - \beta],\ J(\varphi) = \cos 3\varphi = \frac{3\sqrt{3}J_3}{2J_2^{\ 3/2}}, \tag{1}$$

where $\alpha$ and $\beta$ are constants fulfilling the following inequalities: $0 \le \alpha \le 1$ and $0 \le \beta \le \pi/6$. The parameters describing the proportion of radius $r$ for different $\varphi$ angles are as follows:

$$\lambda = r(\varphi = 0°)/r(\varphi = 60°) \qquad \vartheta = r(\varphi = 30°)/r(\varphi = 60°). \tag{2}$$

According to Podgórski [3], $\alpha$ and $\beta$ coefficients from equation (1) can be derived by means of iteration. However, the equations for the direct calculation of these parameters were proposed by Lewiński [4, 5]. From the set of equations (1) and (2) for $\varphi = 0°, 30°, 60°$ the trigonometric equations can be obtained which can be solved with regards to $\alpha$ and $\beta$:

$$\alpha = \sin 3\alpha_0,\ \beta = \frac{\pi}{6} - \arc\tan\left(\frac{\vartheta(1-\lambda)}{2\lambda \sin\alpha_0}\right),\ \text{where } \alpha_0 = \arccos\left(\frac{\vartheta}{2}\left(1 + \frac{1}{\lambda}\right)\right) \tag{3}$$

J. Podgórski confirmed the usefulness of the above expressions [6]. Assuming that in the octahedral set of coordinates the meridian of ultimate surface is described by a parabolic equation, the following equation for the failure surface for concrete was used (see Fig. 1a):

$$F(\sigma_{oct}, \tau_{oct}, \varphi) = \sigma_o - C_0 + C_1 P(J)\tau_o + C_2 \tau_o^{\ 2} = 0. \tag{4}$$

The failure surface (*JP*) described by the equations (1) and (4) is presented in Fig. 1a) (the inner blue surface). The values of $C_0$, $C_1$, $C_2$, $\alpha$ and $\beta$ coefficients can be derived on the basis of experimental data, utilizing the following strengths of concrete: uniaxial tensile strength $f_t$, uniaxial compressive strength $f_c$, biaxial compressive strength $f_{cc}$ ($\sigma_1/\sigma_2 = 1$), biaxial compressive strength $f_{oc}$ ($\sigma_1/\sigma_2 = 2$), triaxial tensile strength $f_t$. According to the tests performed by Kupfler *et al.* [7]: $f_{cc} = 1.1 f_c$ and $f_{oc} = 1.25 f_c$ were assumed. The values of $C_0$, $C_1$ and $C_2$ coefficients from the equation (4) can now be written in the following way:

$$C_0 = f_t,\ C_1 = \frac{\sqrt{2}}{P(\varphi = 0)}\left[1 - \frac{3}{2}\frac{(f_t / f_{cc})}{(f_{cc}/f_t - 1)}\right],\ C_2 = \frac{9}{2}\frac{f_t / f_{cc}}{f_{cc} - f_t} \tag{5}$$

Such defined 5-parameter failure surface for concrete, despite its apparent complexity, poses no problem with numerical implementation and complies well with experimental results. In the tensile zones of concrete, cracking is modelled by introducing five types of specific cracking patterns, including radial cracks and circumferential cracks, sheared both parallel and perpendicular to the cracks. In case of the incremental FE analysis, the "Smeared Cracking" (*SC*) material model [8] was adopted for concrete. The two aforementioned failure surfaces for concrete which are interpenetrating, are presented in Fig. 1a): *JP* (blue colour) and *SC* (green colour). In case of *SC* material model two intersecting conical surfaces are applied (see the axial cross-section of *SC* failure surface in Fig. 4a) – a thin blue line).

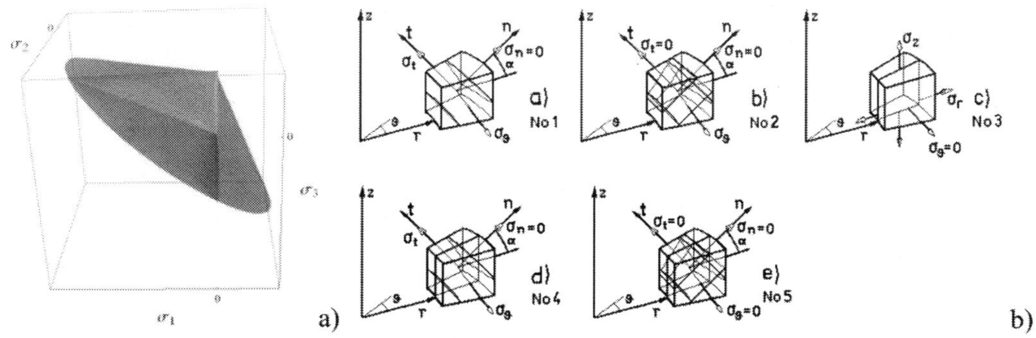


**Fig. 1.** Cracking criteria: a) failure surfaces: *JP* – blue colour, *SC* – green colour, b) cracking patterns of RC axi-symmetric structure.

## 2.2 Behaviour of concrete and soil

The total strain analysis is based on the deformation theory of plasticity. The plastic deformation model is based on deviatoric and hydrostatic components of stress and strain. This implies that the concrete remains isotropic under multiaxial stress. Thus the octahedral-based elastoplastic model is utilized by assuming the secant bulk modulus $K_S$ and the secant shear modulus $G_S$ to be functions of the volumetric and the octahedral strains. A model similar to this (but additionally sensitive to lateral pressure) is adopted for the description of the subsoil. In the present paper the original description of Kupfer and Gerstle [7] has been modified by using corrections with respect to $\gamma_{oct}$:

$$\frac{G_S}{G_0} = f_c \frac{\sqrt{f_c^2 + 4abG_0^2\gamma_{oct}^2} - f_c}{2abG_0^2\gamma_{oct}^2} - \gamma_{oct} \cdot \omega \cdot \left(\alpha_1 \cdot \gamma_{oct} + \alpha_2 \cdot \omega_{23}\right) \tag{6}$$

$$\frac{K_S}{K_0} = \frac{G_S}{G_0} - \gamma_{oct} \cdot \omega \cdot \left(\alpha_1 \cdot \gamma_{oct} + \alpha_2 \cdot \omega_{23}\right), \tag{7}$$

where $a$, $\alpha_1$, $\alpha_2$, $\omega$ and $\omega_{23}$ are the numeral coefficients dependent on $\alpha_m$, where $\alpha_m = 1000 \cdot f_c' / E_c$ for concrete and $100 \cdot R_c / E_0$ for soil. For concrete $b = 1$, while for soil $b$ means an empirical relationship (taking into account the value of lateral pressure). In the performed numerical analysis, the value of lateral stress $\sigma_r$ was taken into account instead of lateral pressure. $K_S$ and $G_S$ modules do not depend on the third strain tensor invariant due to the limitations imposed by the theory of the representations of tensor functions and *Noll* conditions (see [9]). For the incremental FE analysis, the "Smeared Cracking" elastic - plastic model [8] was adopted for concrete. Elastic - perfectly plastic material with the Coulomb-Mohr yield condition and the non-associated flow rule was applied for the soil.

### 2.3 Models for reinforcing steel and tension stiffening

Both the tension stiffening effect and the shear retention are taken into account, wherever the former is included, assuming that additional stress is carried by the reinforcement steel. In case of the deformational model of reinforcing steel the tension stiffening effect was adopted on the assumption that additional stress is carried by rebars. The stress-strain relationship for reinforcing steel is assumed as elastic-rigid plastic. The concept of modelling the *tension stiffening* by the control of reinforcing bars' stiffness is not quite new (see the paper [10]). The influence of the reinforcement bond in cracked concrete on shell or slab deformation was taken into account by the tension stiffening approach according to the provisions of Eurocode 2 [11]. The Expression (7.18) given in this standard can also relate to the strains in reinforcing steel in the arbitrary $i$ direction and take the form:

$$\varepsilon_{smi} = \zeta\, \varepsilon_{si} + (1-\zeta)\, \varepsilon_{ucri} \tag{8}$$

where:

$\varepsilon_{smi}$, $\varepsilon_{ucri}$, $\varepsilon_{si}$ = the mean strain in the reinforcement (according to EC2 [11]) between the cracks in $i$ direction (for intermediate bond of rebar) and the strains in the reinforcement calculated for the uncracked and fully cracked conditions, respectively,

$\zeta$ = a distribution coefficient allowing for tension stiffening (see Eqn. (7.19) in EC2 [11]).

Therefore, the *tension stiffening coefficient* can be introduced to control the effective elasticity modulus of the reinforcing bars ($\overline{E}_{si} = E_s / \psi_{si}$) as follows:

$$\psi_{si} = \varepsilon_{smi} / \varepsilon_{si} = \zeta + (1-\zeta)(\varepsilon_{ucri} / \varepsilon_{si}) \tag{9}$$

Substituting the coefficient $\zeta$ by the Expression (7.19) (see EC2 [11]), after some transformations, the following formula is obtained:

$$\psi_{si} = 1 - \beta(1 - \varepsilon_{ucri} / \varepsilon_{si})(\sigma_{sri} / \sigma_{si})^2, \tag{10}$$

where in turn the steel stress in cracked section $\sigma_{si}$ can be replaced by the ratio of its mean value and the tension stiffening coefficient: $\sigma_{smi}/\psi_{si}$, as the stress $\sigma_{smi}$ is directly available from the iterative numerical analysis. The ratio of the strains $\varepsilon_{ucri}/\varepsilon_{si}$ can also be determined in a numerical way and coefficient $\beta$ can be assumed according to EC2 [11]. After some rearrangements the coefficient $\psi_{si}$ can be formulated depending on mean stress $\sigma_{smi}$ [5].

For the incremental Abaqus model [9] the reinforcement was included with the "Rebar" option. An elastic-plastic analysis with hardening steel model was used.

## 3 Computational examples

In the paper, axisymmetric elements are used throughout. In the case of the first model, the analysis of soil-structure interaction was conducted by an own programme (namely an expanded FEAP version). That analysis was performed using the Lagrange quadrilateral isoparametric element. The second model was built using Abaqus software with 3-node axisymmetric shell elements (SAX2, for structure) and 8-node axisymmetric continuum elements (CAX8R, for soil). The analysis of the impact of low-strength concrete in the tank on the soil-structure interaction was performed taking into account the characteristics of the concrete given below. The purpose of this analysis is to show the impact of this significant, although an adverse factor on the redistribution of stress resultants. The computational example of a nonlinear FE analysis of the RC tank with cylindrical shell (with the inner radius of 6.90 m, height of 5 m, thickness of 20 cm) connected to the ground slab (with the depth of 20 cm) supported on elasto-plastic subsoil is given. For the deformational analysis the following properties of concrete were adopted: $f_c$ = 20 MPa, $f_{ct}$ = 0.4 MPa, $E_c$ = 20 GPa, $\nu_c$ = 0.167; subsoil strength: $R_c$ = 0.2 MPa, $E_0$ = 30 MPa, $\nu_0$ = 0.35; steel reinforcement was adopted as $f_y$ = 355 MPa, $E_s$ = 210 GPa. The vertical reinforcement of the shell: ∅ 16/20 cm and horizontal reinforcement: ∅ 20/20 cm were adopted on both sides, while in the ground slab – the radial and peripheral reinforcement: ∅ 18/13 cm. Near the junction of the shell with the slab, the hoop bars are concentrated: ∅ 20/12 cm. The following soil parameters were introduced for the incremental FE analysis: $E$ = 30 MPa, $\nu$ = 0.35, $c'$ = 25 kPa, $\varphi'$ = 20°, $\psi$ = 0°. The assumed soil parameters correspond to clay. For the incremental analysis the concrete properties were adopted as specified above. In regards to reinforcement, it was assumed that $f_u$ = 510 MPa.

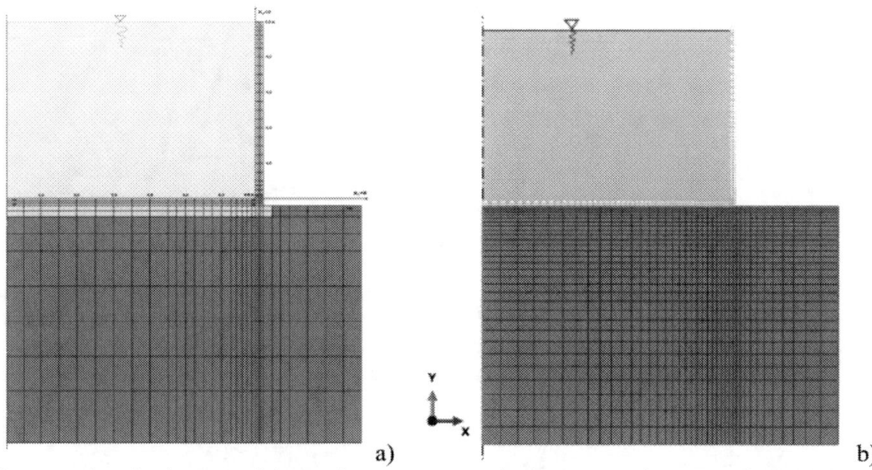


**Fig.2.** Computational schemes of the tank structure; vertical cross-sections showing division into the finite elements and the liquid level for both models: a) deformational, b) incremental.

## 4 Results and conclusions

The comparative analysis was performed for both incremental and small-strain stiffness constitutive models. Analytical results are presented in Figures 3 to 9.

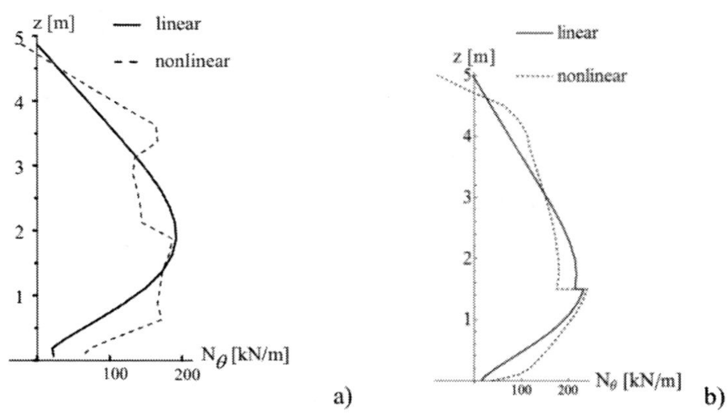


**Fig. 3.** Circumferential forces in tank wall due to hydrostatic pressure for both models: a) deformational, b) incremental.

The diagrams of internal forces, $N_\vartheta$ and $M_z$, in the shell in the elastic state and physically non-linear stage are shown in Fig. 3 and Fig. 5, respectively. The range of redistribution of internal forces obtained here is quite significant compared to the case of normal strength concrete. The redistribution of the forces $N_\vartheta$ and $M_z$ is a result of cracks in the shell when the concrete strength is lower than normal, as the shell is more susceptible to deformation. Moment at the connection of the shell and the slab is smaller than in case of normal strength concrete due to the cracking near the joint. The chart of circumferential forces $N_\vartheta$ shows the additional loading of non-cracked zones and unloading of cracked zones (due to decrease in stiffness), especially for the more restrictive *JP* criterion (Fig. 4a).

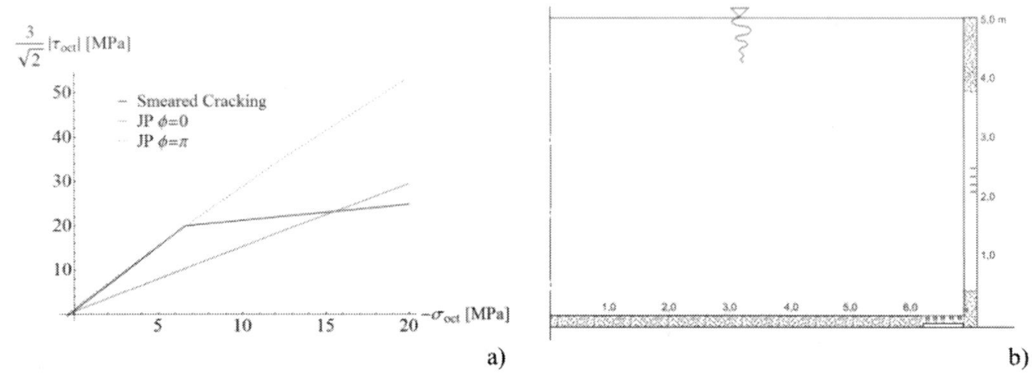


**Fig. 4.** Cracking phenomenon: a) failure criteria, b) cracks in tank structure for deformational model.

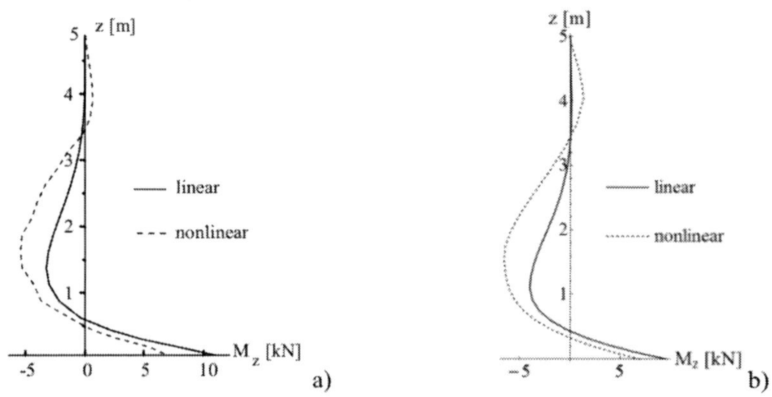


**Fig. 5.** Bending moment in the tank wall due to hydrostatic pressure for both models:
a) deformational, b) incremental.

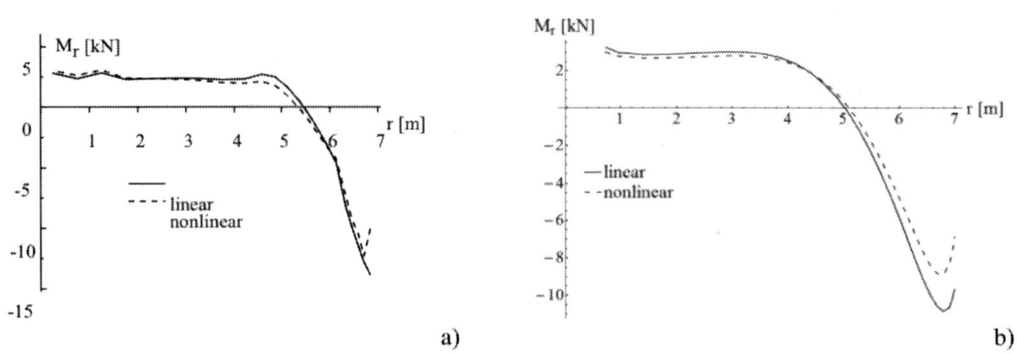


**Fig. 6.** Radial moments in the slab for the a) deformational and b) incremental model.

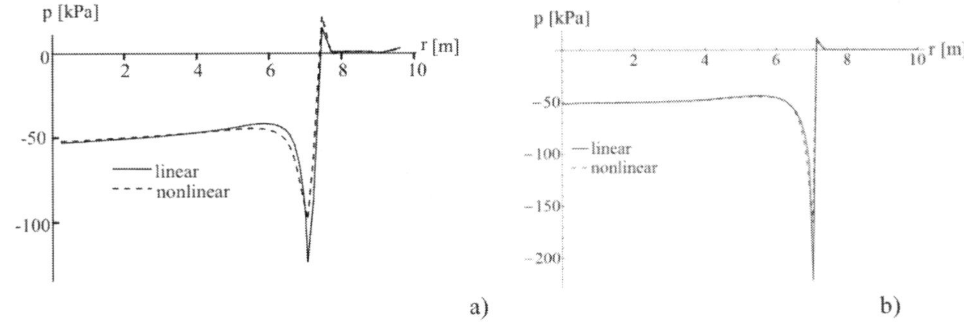


**Fig. 7.** Distributions of the ground reaction for both models: a) deformational, b) incremental.

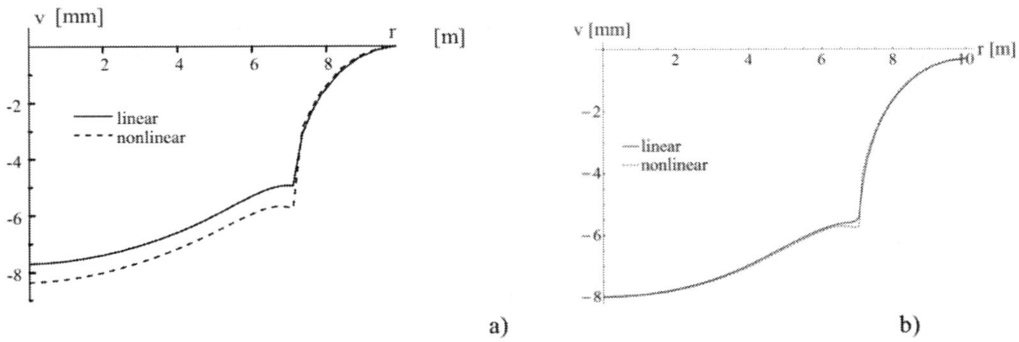


**Fig. 8.** Distributions of the ground subsidence for the: a) deformational and b) incremental model.

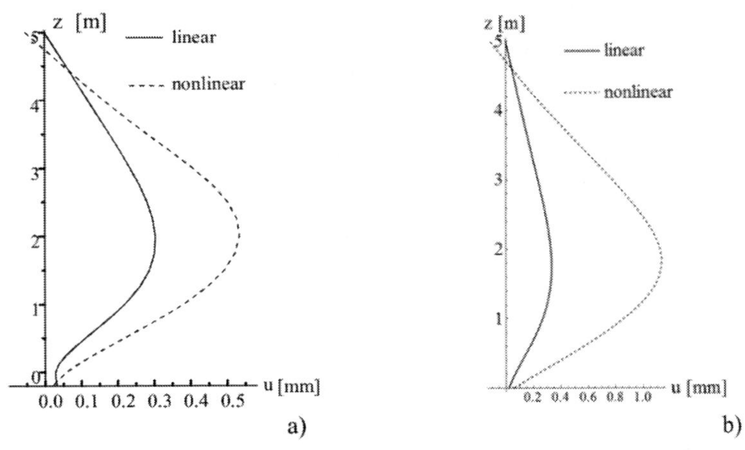


**Fig. 9.** Distributions of the radial displacements of the tank shell for the: a) deformational and b) incremental model.

The diagrams of bending moments, $M_r$, in the ground slab in the elastic state and physically non-linear stage are shown in Fig. 6. In this case, the slab moment $M_r$ at the connection of the slab and the shell is smaller than in the elastic stage. Different values of bending moments at the connection of the slab with the cylindrical shell also result from nonlinear behaviour of the ground under the slabs in both cases (see Fig. 8) as well as stress redistribution in the substrate soil. The subsoil reaction is also redistributed due to the development of plastic strains in the soil (see Fig. 7). Different values of subsoil reactions in elastic solutions for both models are mainly a result of different FE discretisiations. A significant increase in radial displacements of the tank shell is observed due to the cracks (from a maximum of 0.304 mm in the elastic stage to 0.534 mm in the inelastic stage for the deformational model and to more than 1 mm for the incremental model, see Fig. 9) and the increase of stress in the circumferential reinforcement of the shell. Different results obtained for both models (see Fig. 3 and Fig. 9) are linked with the differences of the shapes of both limiting surfaces (see Fig. 1 a) and Fig. 4a), which affects the different cracking patterns. During the deformational analysis there were registered cracking patterns № 3 and № 1 in the walls of the tank shell and, occasionally, cracking patterns № 1 and № 2 in the ground slab. Due to the weak concrete the scope of cracking is very extensive. Vertical cracks in the cylindrical tank shell designated by the iterative FE analysis, pass through (cracking pattern № 3) and occur in the area of the shell from 0.375 m to 3.75 m above the ground slab. The horizontal and skew cracks in the vertical cross-section occurred in the walls of the tank only in their lower part (near the connection with the ground slab), while the cracks in the ground slab resulting from the bending occurred in the area of the connection with the shell and they pass only through a part of the cross-section. As a result of the deformational analysis there were registered mainly the cracks extending from the top of the ground slab according to cracking pattern № 1 (occasionally № 2) over a distance of 0.9 m from the connection with the tank wall, and, furthermore, the cracks extending from the bottom of the slab according to the cracking pattern № 3, resulting from the bending in circumferential direction, occurring at the distance 0.4 m from the junction in the part of the ground slab along the edge.

The results of the analysis indicate that in the case of structural concrete and subsoil of normal strength, the scope of redistribution is not very large, whereas in the case of weak concrete (or subsoil) the redistribution range is quite wide, which is accompanied by the increase in the horizontal displacements of the shell and the width of the vertical cracks.

## References

1. D. V. Phillips, O. C. Zienkiewicz, *Proc. Inst. Civ. Engrs, Part 2*, **61 (3)**, 59 (1976)
2. M. D. Kotsovos, M. N. Pavlović, *Structural concrete. Finite-element analysis for limit-state design* (Thomas Telford Services Ltd, London, 1995)
3. J. Podgórski, Arch. Mech., **36 (3)**, 323 (1984)
4. P. M. Lewiński, *2nd Int. Sci. AMCM Conference, Łódź, Poland,* 405 (1996)
5. P. M. Lewiński, *Analysis of interaction of reinforced concrete cylindrical tanks with subsoil* (in Polish) (ITB, Warsaw, 2007)
6. J. Podgórski, Mechanics and Control, **30 (4)**, 229 (2011)
7. H. B. Kupfer, K. H. Gerstle, J. Engng Mech. Div., Proc. ASCE, **99 (EM4)**, 853 (1973)
8. *Abaqus Theory Manual, Ver. 6.12* (Dassault Systèmes, 2012)
9. S. Jemioło, A. Szwed, Logistyka, **6**, CD (2009)
10. H. Floegl, H. A. Mang, Journal of the Structural Division, Proc. ASCE, **108 (ST12)**, 2681 (1982)
11. EN 1992-1-1:2004 *Eurocode 2: Design of concrete structures – Part 1-1: General rules and rules for buildings* (2004)

# Modeling the dynamic behavior of railway track taking into account the occurrence of defects in the system wheel-rail

*Alexey* Loktev[1,*], *Vyacheslav* Sychev[2], *Boris Gluzberg*[3] and *Ekaterina* Gridasova[4]

[1]Moscow state university of civil engineering, Yaroslavskoye shosse, 26, Moscow, Russia, 129337
[2]MIIT, Department of Transport Construction, 125190 Moscow, Chasovaya st., 22/2, Russia
[3]VNIIZHT, The integrated office "Path and track facilities", 129626 Moscow, 3-ya Mytischenskaya, 10, Russia
[4]FEFU, Department of Welding Production, 690950 Vladivostok, st. Suhanova, 8, Russia

**Abstract.** This paper investigates the influence of wheel defects on the development of rail defects up to a state where rail prompt replacement becomes necessary taking into account different models of the dynamic contact between a wheel and a rail. In particular, the quasistatic Hertz model, the linear elastic model and the elastoplastic Aleksandrov-Kadomtsev model. Based on the model of the wheel-rail contact the maximum stresses are determined which take place in the rail in the presence of wheel defects (e.g. flat spot, weld-on deposit, etc.). In this paper, the solution of the inverse problem is presented, i.e., investigation of the influence of the strength of a wheel impact upon rails on wheel defects as well as evaluation of the stresses emerging in rails. During the motion of a railway vehicle, the wheel pair position in relation to rails changes significantly, which causes various combinations of wheel-rail contact areas. Even provided the constant axial load, the normal stresses will substantially change due to the differences in the radii of curvature of contact surfaces of these areas, as well as movement velocities of railway vehicles.

## 1 Introduction

Analysis of rail defects in railroad tracks reveals that the most substantial rail defects among detected are operational defects which occur due to the influence of a moving railway vehicle on rails and appear as hairline cracks, delamination and flaking of the rail head material (code 10.1-2, 11.1-2 according to [1]). It is known [1] that during railway operation defects of the codes 10.1-2 and 11.1-2 lead directly to defective rails of the codes 30G.1-2, 30B.1-2 and 21.1-2, which have to be eliminated immediately after their detection [1, 2]. Horizontal delaminations of the rail head appear due to the presence of nonmetallic impurities (defect of the code 30G.1-2) and transverse cracks, which are visible as light and dark spots in the rail head. Figure 1 shows the fractures formed due to the insufficient contact fatigue strength f the metal material (defect of the code 21.1-2) and welding joints (defect of the code 26.3).

* Corresponding author: aaloktev@yandex.ru

Substantial contact stresses between a wheel and a rail arise under large axial loading and high-speed movement of a railway vehicle [3, 4]. The contact area under extreme, alternating forces, which cause permanent plastic deformations of the microstructure, is only several millimeters large and rail hardening takes place. Therefore the main safety precaution is to avoid the transition of rails with defects of the codes 10.1-2 and 11.1-2 into fatal defects by carrying out so called rail improvement, which includes weld deposition and rail grinding provided that the defect sizes do not exceed the critical parameters and velocities of railway vehicles are not limited.

a) b)

**Fig. 1.** Rail defects: a) Defect of the code 30G.1 – Transverse cracks in the rail head; b) Defect of the code 26.3 – Transverse cracks in the rail head caused by violation of the joint welding procedure and processing of the welding joints, which led to the rail breakdown after passing of the warranted tonnage in the joining point.

## 2 Statement of the problem

According to the investigations of the Joint Stock Company Railway Research Institute (JSC "VNIIZhT") [3, 4], internal transverse or longitudinal cracks appear in the rail head of new railways at a depth of 3 to 14 mm from the tread surface (at axial load of rail cars of 150-250 kN) in the area of nonmetallic impurities, where a high concentration of sign-variable contact stresses is observed. The stress concentrators are located along rails and form microcracks under the influence of wheels of a moving railway vehicle and not necessarily result in formation of internal longitudinal cracks that demand the rail replacement. The following are the reasons for the similar formation and development of the microcracks: incorrect canting of rails and/or wheel defects caused by tread surface irregularities. This leads to the vertical wheel oscillations, as a result of which inertia forces appear in the wheel mass, which in their turn cause additional load on the rails. If this occurs, the impact of a moving wheel with isolated irregularities of the tread surface (i.e. flat spot) takes place, which has a very strong additional influence on the rail.

Figure 2 represents the scheme of a wheel with isolated roughness, i.e. wheel flat, where $l_n$, $h_n$ – the length and depth of a flat spot; $r$ – the radius of the rolling circle of the wheel; $z_k$ – the axis of ordinates of wheel pair trajectory; $l_{n3}$ – the length of the flat spot; $X$ – the coordinate along the railway path axis; $XABC$ – the trajectory of a wheel pair.

In the case the flat spot at a wheel surface is found, its depth has to be determined. The shape of the contact area and contact stresses, which can be represented as a distributed load applied along the contact surface, depend on the shape of contact bodies. Neglecting the conical nature of wheel-rail contact we assume for the analysis that the contact between a wheel and a rail can be represented as the contact between two cylinders – one is located perpendicular to the other one.

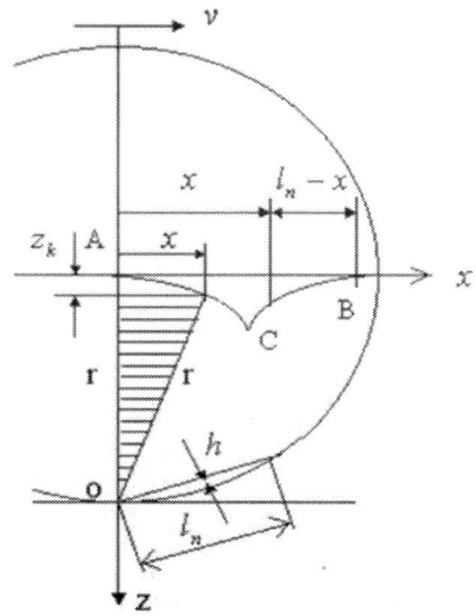


Figure 2. Schematic representation of a wheel with an isolated irregularity – flat spot.

## 3 Contact problem

In the general case the area of the wheel-rail contact has an ellipsoidal shape. Thus, the normal stresses at the wheel-rail boundary depend on the wheel load on the rail, radii of the tread surfaces between wheel and rail and properties of interacting materials.

In this case different contact models can be applied for the analysis, i.e., the quasistatic Hertz model, the linear elastic model, two elastoplastic contact models: the Aleksandrov-Kadomtsev model, the viscoelastic model with an exponential kernel of relaxation, the aggregated model based on the Hertz model and the unloading curve [5-10].

1) The quasistatic Hertz model [5, 7, 9] - $P = k a^{3/2}$, (1)

2) The linear elastic model [7, 8] - $P = E_1(a - w)$, (2)

3) The viscoelastic model [6,10] - $P(t) = E_1(\alpha - w) - \frac{E_1}{\tau_1}\int_0^t (\dot{\alpha} - \dot{w}) e^{-\frac{t-t'}{\tau_1}} dt'$, (3)

4) The elastoplastic Aleksandrov-Kadomtsev model [7, 10]

$$\alpha = \begin{cases} bP^{2/3}, & dP/dt > 0,\ P_{\max} < P_1, \\ (1+\beta)c_1 + (1-\beta)Pd, & dP/dt > 0,\ P_{\max} > P_1, \\ b_f P^{2/3} + \alpha_p(P_{\max}), & dP/dt < 0,\ P_{\max} > P_1, \end{cases} \quad (4)$$

5) The model where the force of interaction is divided into loading and unloading steps [11,12]. For a loading step – (1), for an unloading step –

$$P = P_m \left[ \frac{(\alpha - \alpha_0)}{(\alpha_m - \alpha_0)} \right]^q, \quad \alpha_0 = \begin{cases} \beta(\alpha_m - \alpha_{cr}); & \alpha_m > \alpha_{cr}, \\ 0; & \alpha_m \le \alpha_{cr}, \end{cases} \quad (5)$$

In the expressions (1) – (5), $\alpha$ is the local deformation of wheel and rail materials, $k$ – the coefficient depending on the geometry of contact bodies and elastic properties of their materials, $E_1$ – the elasticity modulus of the area of wheel-rail interaction, $w$ – the displacement of the lower edge of the rail, λ=5.7, $b=\left(\left(9\pi^2(k_1+k)^2\right)/16R\right)^{1/3}$, $k_1=\left(1-\sigma_1^2\right)/E_1$, $k=\left(1-\sigma^2\right)/E$, $\chi=\pi k_{pl}\lambda$, $P_1=\chi^3\left(3R(k_1+k)/4\right)^2$, $d=1/2\chi R$, $b_f=R_f^{-1/3}\left(3(k_1+k)/4\right)^{2/3}$, $R_p^{-1}=R^{-1}-R_f^{-1}$, $R_f=\left(4/3(k_1+k)\right)P_{\max}^{1/2}\chi^{-3/2}$, $\alpha_p(P_{\max})=(1-\beta)P_{\max}\left(2\chi R_p\right)^{-1}$, $\beta=0.33$, $c_1=3\chi^{1/2}(k_1+k)/8$, $k_{pl}$ – the lowest plastic constant of interacting bodies, $\sigma_1$, $E_1$ – the Poisson's ratio and modulus of elasticity of the wheel pair [2,8], $\tau_1=\eta_1/E_1$ , $\tau_1$ – the relaxation time in the case of the viscoelastic model, $\eta_1$ – the viscous resistance coefficient, $\gamma$ ($0<\gamma\leq 1$) – the parameter of fractionality, $P_m$ – the maximum value of the contact force before the unloading step, $\alpha_m$ – the maximum local deformation of the material in the contact area, $\alpha_0$ – the current local deformation value, $q$ and $\beta$ – the experimentally obtained constants, $\alpha_{cr}$ – the critical local deformation value [13].

For modelling of the dynamic interaction between a wheel and a rail under velocities of up to 90 km/h according to [14] one could apply expressions from the contact Hertz theory (1) in engineering analysis. The Hertz theory is valid at the following assumptions: the contact surfaces are homogeneous and isotropic; the frictional forces in the contact area can be neglected; the size of the contact area is negligible compared to the sizes of contact bodies and the characteristic radii of curvature of undeformed surfaces; for the contact problem the solution of a linear elastic half space is used; and the contact surfaces are smooth enough [15].

During the motion of a railway vehicle, the wheel pair position in relation to rails changes significantly, which causes various combinations of wheel-rail contact areas. Even provided the constant axial load, the normal stresses will substantially change due to the differences in the radii of curvature of contact surfaces of these areas, as well as movement velocities of railway vehicles.

## 4 The determination of the stresses in the interaction of wheel and rail

By taking into account defectiveness of wheel pairs and their tread surfaces it cannot be claimed that in the contact area there is only one radius of curvature. Therefore, the Hertz theory would not provide correct results. That is why in this case depending on the defect type, velocity of a railway vehicle and geometry of the upper railway construction (the tangent or the curve of a certain radius) expressions (2) - (5) are more reliable. For the determination of normal contact stresses for a non-conformal contact one could apply the approximation method of nonlinear geometry by using ellipses or carry out the modelling of contact bodies based on the elastic Winkler foundation, where the deformation of surfaces is proportional to the normal contact stresses. The maximum contact stress obtained in this case is 1.3 times higher than that obtained from the Hertz theory [5, 9, 11]. The size of the contact area at the normal wheel-rail geometry as well as the type of contact stresses depend on the normal load between a wheel and a rail, the profiles of a wheel and a rail, the transverse and angular position of the wheel pair on rails as well as the angle of the tread surfaces of rails [3, 7, 15]. According to the shape of the contact area and the type of deformation the intensity of pressure within contact surface is distributed by the semi-elliptic law.

If both cylinders are made from homogeneous material (e.g. carbon steel), the equation for the maximum stress in the contact area between a wheel and a rail ($Q_{max}$) is written as:

$$Qmax=(3/2)\cdot(P/a\cdot b\cdot\pi), \quad (6)$$

where $P$ – the contact load in kg, $a$ and $b$ – the semi-axes of the contact ellipse in cm, $(2/3)\cdot\pi\cdot a\cdot b\cdot Q_{max}$ – the volume of the semi-ellipsoid of stresses, which equals to the force $P$

in kg according to the law of equality between a force and its reaction force). The values of the semi-axes of ellipse of the contact area in the case of two cylinders are given by:

$$A=(1/2\cdot R_2) \text{ and } B=(1/2\cdot R_1), \quad (7)$$

where $R_1$ and $R_2$ – the radii of curvature of the wheel-rail surfaces ($R_2>R_1$). Hence, $A/B=2\cdot R_1/2\cdot R_2=R_1/R_2$.

Thus, the smaller radius is denoted as $R_1$ – the curvature of the rail head surface, and the bigger as $R_2$ – the wheel radius. In this case the maximum contact stress between a wheel and a rail is as follows:

$$Q_{max}=\alpha\cdot\sqrt[3]{P\cdot E^2/R^2}, \quad (8)$$

where α – the coefficient depending on the ratio $R_1/R_2$, taken for the calculation as 0.586; $P$ – the maximum dynamic vertical force of a wheel pressure on rail in kg; $E$ – the modulus of elasticity of rail's steel, which equals to $2.1\cdot10^6$ kg/cm$^2$. In the case of rail wear in the middle of the rail head, the radius becomes 320 mm, and at the distance of 20 mm from the longitudinal rail axis towards both sides – 100-200 mm, which is also corroborated by earlier investigations [3, 4, 6, 7, 14].

If both a wheel and a railway track possess defects – flat spot and rail head wear, respectively – the contact spot shifts to the edge of the rail head. In these cases [3, 5, 7, 9, 11, 12, 13] the maximum impact force of a wheel with a flat spot on rail is determined as:

$$P=V_{ud}\cdot\sqrt{m_{pr}\cdot G_{kt}}, \quad (9)$$

where $V_{ud}$ – the velocity of the striking point of the wheel, cm/s (Fig. 1); $m_{pr}$ – the reduced mass of the rail, kg·s$^2$/cm; $G_{kt}$ – the rigidity of the wheel-rail contact $0.5\cdot10^6$ kg/cm. The reduced mass of the rail is calculated according to the following formula [8]:

$$m_{pr}=(1,5\div2,3)\cdot m_r/k, \quad (10)$$

where $m_r$ – the mass per unit length of the rail 0.65 kg/cm; $k$ – the relative rigidity factor of the rail support on broken stone ballast and at rail type R-65 is assumed to be 0.01 cm$^{-1}$·981, since the sleeper density is 1840- 2000 pieces per 1 km and the rail wear is from 3 to 6 mm:

$$V_{ud}=V\cdot L_0/R_k, \quad (11)$$

where $V$ – the movement velocity, cm/s; $L_0$ – the length of the wheel flat spot, cm; $R_k$ – the radius of the wheel, cm.

## 5 Conclusions

The sorting out of defective wheels with flat spots of 1.5-2 mm in a locomotive depot makes 80% of all wheel defects. Therefore, one of the most important tasks in assuring the reliable functioning of the railway track is the diagnostics of causes and the determination of possibilities for reduction of dynamic forces from wheels of a moving railway car on the rails. Based on the impact force of a defect wheel upon rail (e.g. (3)) the value of the wheel defect can be determined, i.e. the inverse problem to the described above can be solved. This inverse problem – the determination of the value of a wheel defect (e.g. the size of a flat spot) from the impact force upon rail – can be solved by a system of automated diagnostics of a railway vehicle's technical condition. The system operation is based on measuring of vertical forces by determination of rail web deformations and sorting out of wheels by using force criteria of the evaluation of wheels roughness.

## References

1. "Normative and technical documentation/classification of rail defects/catalogue of rail defects/signs of a defective and fatal cropped rails" (NTD/TSP– 1, 2, 3 – 93). Moscow: TRANSPORT. 126 p. (1993)
2. Technical specifications for works on reconstruction (modernization) and repair of railway tracks. The order of JSC "RZD" from 18.01.2013 No. 75r.
3. L. G. Krysanov, A. Yu. Abdurashitov. Path and track facilities, **6**, 26-30 (1988)
4. Yu. Abdurashitov, M. N. Georgiev, L. G. Krysanov. Reliability of rails in various climatic conditions. Moscow: VNIIZHT, 138 p. (1987)
5. Yu. N. Mazov, A. A. Loktev, V. P. Sychev. Vestnik MGSU, **5**, 61-73 (2015)
6. A.A. Loktev. Acta Mech., 222(**1-2**), 17-25 (2011)
7. A.A. Loktev. I. J. of Eng. Science, 50, **1**, 46–55 (2012)
8. A.A. Loktev, A.V. Sycheva, V.V. Vershinin. *Proceeding of the 2014 International Conference on Theoretical Mechanics and Applied Mechanics, Venice, Italy, March 15-17*, 16-19 (2014)
9. A.Y. Kogan, A. A. Vergoten. Calculation of the impact on road wheelsets with the slider. Moscow: TRANSPORT, 224 p. (1984)
10. A.A. Loktev, G.N. Talashkin, K.D. Stepanov. Transport of the Rus. Fed. **2-3**(63-64), 40-44 (2016)
11. M. Agostinacchio, D. Ciampa, M. Diomedi, S. Olita. J. of Modern Transportation, **21**. **3**. 169-181. (2013)
12. R. Olsson, M.V.Donadon, B.G. Falzon. I. J. of Solids and Struct., **43**, 3124-3141 (2006)
13. S. Abrate. Compos. Struct., **51**, 129-138 (2001)
14. P. Chen, J. Xiong, Z. Shen. Mech. of Materials, **40**, 183-194 (2008)
15. A.P. Christoforou, A.A. Elsharkawy, L.H. Guedouar. Comp. and Struct., **79**, 2607-2619 (2001)

# About solution of multipoint boundary problem of static analysis of deep beam with the use of combined application of finite element method and discrete-continual finite element method. part 1: formulation of the problem and general principles of approximation

*Leonid* Lyakhovich[1], *Oleg* Negrozov[2,*]

[1]Tomsk State University of Architecture and Building, Department of Structural Mechanics, Solyanaya sq., Tomsk, 634003, Russia
[2]National Research Moscow State University of Civil Engineering, Department of Applied Mathematics; 26, Yaroslavskoe Shosse, Moscow, 129337, Russia

**Abstract.** This paper is devoted to formulation and general principles of approximation of multipoint boundary problem of static analysis of deep beam with the use of combined application of finite element method (FEM) discrete-continual finite element method (DCFEM). The field of application of DCFEM comprises structures with regular physical and geometrical parameters in some dimension ("basic" dimension). DCFEM presupposes finite element approximation for non-basic dimension while in the basic dimension problem remains continual. DCFEM is based on analytical solutions of resulting multipoint boundary problems for systems of ordinary differential equations with piecewise-constant coefficients.

## 1 Formulation of the problem

Let's consider multipoint boundary problem of static analysis of deep beam (Fig. 1) [7, 10-12,14,15,17-23,25,27-32]. Some elements of notation system is presented at Fig. 1. Corresponding design model is two-dimensional theory of elasticity [9].

Let's $\Omega$ be domain occupied by structure, $\Omega=\{(x_1,x_2):\ 0<x_1<l_1,\ 0<x_2<l_2\}$ or

$$\Omega=\bigcup_{k=1}^{n_b-1}\Omega_k\,,\quad \Omega_k=\{(x_1,x_2):\ 0<x_1<l_{1,k},\ x_{2,k}^b<x_2<x_{2,k+1}^b\}\,; \tag{1}$$

$x_1, x_2$ are coordinates ($x_2$ corresponds to basic dimension); $x_{2,k}^b$, $k=1,2,\ldots,n_b$ are coordinates of corresponding boundary points (cross-sections) along basic dimension (for mulipoint boudary peoblems $n_b>2$; $\Omega_k$, $k=1,2,\ldots,n_b-1$ are subdomains of $\Omega$.

* Corresponding author: genromgsu@gmail.com

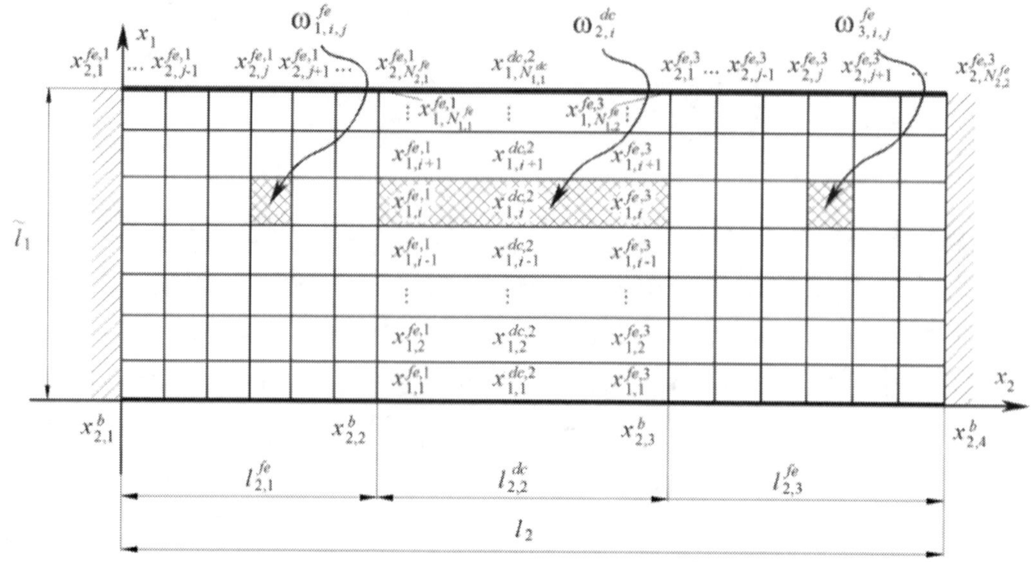

**Fig. 1.** Sample of three-point boundary problem.

Let $l_{1,k} = const$ for $x^b_{2,k} < x_2 < x^b_{2,k+1}$. However in general case $l_1 = l_1(x_2) \neq const$ (parameter $l_1$ can be piecewise-constant).

Let $\omega_k$, $k = 1,2,\ldots,n_b - 1$ are extended subdomains, embordering subdomains $\Omega_k \subset \omega_k$, $k = 1,2,\ldots,n_b - 1$,

$$\omega = \bigcup_{k=1}^{n_b-1} \omega_k\,, \quad \omega_k = \{(x_1,x_2):\ 0 < x_1 < \tilde{l}_1,\ x^b_{2,k} < x_2 < x^b_{2,k+1}\} \supset \Omega_k\,; \tag{2}$$

## 2 General principles of domain approximation

Let physical and geometrical parameters within one group of subdomains from (2) is constant in some dimension ("basic" dimension). Thus, it is recommended to use DCFEM [1-5,30,35] for approximation of these subdomains (discrete-continual design model is introduced). Let physical and geometrical parameters within other group of subdomains from (2) is arbitrary varying. FEM [6,8,9,13,24,26,33,34] can be effectively used for approximation here. Thus, combined application of DCFEM and FEM is advisable.

So-called approximation parameter $\rho_k$ can be introduced in accordance with the following rule : $\rho_k = 1$ for approximation with the use of FEM ; $\rho_k = 2$ for approximation with the use of DCFEM .

## 3 About numbering of subdomains

Various approaches can be used for numbering of subdomains.

The first approach provides separate numbering of subdomains with different types of approximation:

$$k_1 = k_1(k) = \sum_{s=1}^{k} |\,\rho_s - 2\,|\,; \quad k_2 = k_2(k) = \sum_{s=1}^{k} |\,\rho_s - 1\,|\,, \tag{3}$$

where $k$ is initial number of subdomain $\omega_k$; $k_1 = k_1(k)$ is the corresponding number of subdomain with approximation with the use of FEM; $k_2 = k_2(k)$ is the corresponding number of subdomain with approximation with the use of DCFEM.

Various approaches can be used for numbering of subdomains.

We can certainly construct inverse relationships by tabulating the results of calculations using formulas (3).

$$k = k(k_1)\;;\quad k = k(k_2)\,, \tag{4}$$

Thus we can rewrite (2) in the following form:

$$\omega = \bigcup_{k_1=1}^{N_{fe}} \omega_{k_1}^{fe} + \bigcup_{k_2=1}^{N_{dc}} \omega_{k_2}^{dc}\,. \tag{5}$$

Besides, the following relations are valid:

$$N_{fe} = \sum_{s=1}^{n_b-1} |\,\rho_s - 2\,|\;;\quad N_{dc} = \sum_{s=1}^{n_b-1} |\,\rho_s - 1\,|;\quad N_{fe} + N_{dc} = n_b - 1\,, \tag{6}$$

where $\omega_{k_1}^{fe}$, $k_1 = 1, 2, \dots, N_{fe}$ are subdomain with approximation with the use of FEM; $\omega_{k_2}^{dc}$, $k_2 = 1, 2, \dots, N_{dc}$ are subdomain with approximation with the use of DCFEM.

The second approach, on the contrary, is based on a linked numbering of subdomains with different types of approximation. Formula (2) can be used, where

$$\omega_k = \begin{cases} \omega_k^{fe}, & \text{if } \rho_k = 1 \\ \omega_k^{dc}, & \text{if } \rho_k = 2 \end{cases} \tag{7}$$

while formulas (3) and (4) are not required. The second approach is used in this paper.

## 4 About numbering of finite elements and discrete-continual finite elements

Let's consider arbitrary subdomain $\omega_k^{fe}$. We can introduce notation

$$l_{2,k}^{fe} = x_{2,k+1}^{b} - x_{2,k}^{b} \quad \text{if} \quad \rho_k = 1\,. \tag{8}$$

Let's $x_{1,i}^{fe,k}$, $i = 1, 2, \dots, N_{1,k}^{fe}$ and $x_{2,j}^{fe,k}$, $j = 1, 2, \dots, N_{2,k}^{fe}$ are coordinates coordinates (along $x_1$ and $x_2$) of nodes of finite elements, which are used for approximation of domain $\omega_k^{fe}$; $(N_{1,k}^{fe} - 1)$ and $(N_{2,k}^{fe} - 1)$ are numbers of finite elements along coordinates $x_1$ and $x_2$, which are used for approximation of $\omega_k^{fe}$. Three-index system is used for numbering of finite elements, which are used for approximation of $\omega_k^{fe}$. Typical number of has the form $(k, i, j)$, where $k$ is the number of subdomain with approximation with the use of FEM, $i$ and $j$ are numbers of elements (along $x_1$ and $x_2$); $\omega_{k,i,j}^{fe}$ is corresponding finite element.

Let's consider arbitrary subdomain $\omega_k^{dc}$. We can introduce notation

$$l_{2,k}^{dc} = x_{2,k+1}^{b} - x_{2,k}^{b} \quad \text{if} \quad \rho_k = 2 . \tag{8}$$

Let's $x_{1,i}^{dc,k}$, $i = 1, 2, \ldots, N_{1,k}^{dc}$ are coordinates (along $x_1$) of nodes (nodal lines) of discrete-continual finite elements, which are used for approximation of domain $\omega_k^{dc}$; $(N_{1,k}^{dc} - 1)$ is the number of discrete-continual finite elements, which are used for approximation of $\omega_k^{dc}$. Two-index notation system is used for numbering of discrete-continual finite elements, which are used for approximation of $\omega_k^{dc}$. Typical number of has the form $(k, i)$, where $k$ is the number of subdomain, $i$ is the number of element (along $x_1$).

We can introduce notation

$$N_{1,k} = \begin{cases} N_{1,k}^{fe} & \text{if} \quad \rho_k = 1 \\ N_{1,k}^{dc} & \text{if} \quad \rho_k = 2; \end{cases} \quad N_{2,k} = N_{2,k}^{fe} \quad \text{if} \quad \rho_k = 1 \; ; \tag{9}$$

$$x_{1,k,i} = \begin{cases} x_{1,i}^{fe,k} & \text{if} \quad \rho_k = 1 \\ x_{1,i}^{dc,k} & \text{if} \quad \rho_k = 2; \end{cases} \quad x_{2,k,j} = x_{2,j}^{fe,k} \quad \text{if} \quad \rho_k = 1 \; ; \tag{10}$$

$$l_{2,k} = \begin{cases} l_{2,k}^{fe} & \text{if} \quad \rho_k = 1 \\ l_{2,k}^{dc} & \text{if} \quad \rho_k = 2. \end{cases} \tag{11}$$

It should be noted that in the simples cases (such case in considered in the distinctive paper) discretization of structure is constant along $x_1$ throughout the domain (otherwise the mathematical constructions given below are substantially more complicated). We have

$$N_{1,k} = N_1, \quad k = 1, 2, \ldots, n_b - 1; \quad x_{1,k,i} = x_{1,i}, \quad k = 1, 2, \ldots, n_b - 1, \quad i = 1, 2, \ldots, N_1 . \tag{12}$$

In should be noted that notation from [3-5,30] is also used in this paper.

## 5 Discrete (finite element) approximation model for subdomain

Let's consider arbitrary subdomain $\omega_k^{fe}$. Discrete (finite element) approximation model for the considering two-dimensional problems presupposes finite element approximation along $x_1$ and $x_2$. Thus extended subdomain $\omega_k^{fe}$ is divided into finite elements,

$$\omega_k^{fe} = \bigcup_{i=1}^{N_1-1} \bigcup_{j=1}^{N_{2,k}-1} \omega_{k,i,j}^{fe}, \quad \omega_{k,i,j}^{fe} = \{(x_1, x_2): \; x_{1,i} < x_1 < x_{1,i+1}, \; x_{2,k,j} < x_2 < x_{2,k,j+1}\} . \tag{13}$$

Lame constants for finite element are defined by formulas:

$$\bar{\lambda}_{k,i,j} = \theta_{k,i,j} \lambda; \quad \bar{\mu}_{k,i,j} = \theta_{k,i,j} \mu, \quad \text{where} \quad \theta_{k,i,j} = \begin{cases} 1, & \omega_{k,i,j}^{fe} \subset \Omega_k; \\ 0, & \omega_{k,i,j}^{fe} \not\subset \Omega_k; \end{cases} \tag{14}$$

$\theta_{k,i,j}$ is the characteristic function of element $\omega_{k,i,j}^{fe}$.

Basic nodal unknowns are displacement components $u_1^{(k)}, u_2^{(k)}$ (superscript "$(k)$" hereinafter corresponds to the number of considered subdomain i.e. $\omega_k = \omega_k^k$ ). Thus for node $(k,i,j)$ we have the following unknowns: $u_1^{(k,i,j)}, u_2^{(k,i,j)}$.

Bilinear approximation of unknowns is used within finite element (conventional plane rectangular 4-node finite element of two-dimensional problem of elasticity theory (Fig. 2)).

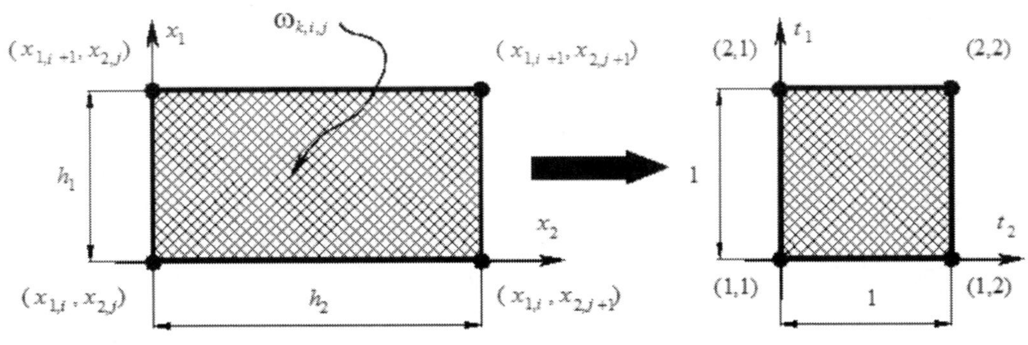


**Fig. 2.** Finite element and its local coordinate system.

Computing of partial derivatives of displacements, deformations and stresses within the finite element, nodal stresses and nodal deformations with allowance for averaging is described in [3-5,30].

As known, FEM is reduced to the solution of systems of $2N_1N_{2,k}$ linear algebraic equations:

$$K_k\overline{U}_k = \overline{R}_k, \tag{15}$$

where $\overline{U}_k$ is global vector of nodal unknowns (subscript "$(k)$" corresponds to the number of subdomain $\omega_k = \omega_k^k$ ),

$$\overline{U}_k = [(\overline{u}_n^{(k,1,1)})^T \;\; (\overline{u}_n^{(k,2,1)})^T \;\; \ldots \;\; (\overline{u}_n^{(k,N_1,1)})^T \;\; \ldots \;\; (\overline{u}_n^{(k,1,2)})^T \;\; (\overline{u}_n^{(k,2,2)})^T \;\; \ldots \;\; (\overline{u}_n^{(k,N_1,2)})^T \;\; \ldots \\ \ldots \;\; (\overline{u}_n^{(k,1,N_{2,k})})^T \;\; (\overline{u}_n^{(k,2,N_{2,k})})^T \;\; \ldots \;\; (\overline{u}_n^{(k,N_1,N_{2,k})})^T \;]^T; \tag{16}$$

$$\overline{u}_n^{(k,i,j)} = [\,u_1^{(k,i,j)} \;\; u_2^{(k,i,j)}\,]^T, \;\; i = 1,2,\ldots,N_1, \;\; j = 1,2,\ldots,N_{2,k}; \tag{17}$$

$K_k$ is global stiffness matrix of order $2N_1N_{2,k}$; $\overline{R}_k$ is global right-side vector of order $2N_1N_{2,k}$ (global load vector).

## 6 Discrete-continual approximation model for subdomain

Let's consider arbitrary subdomain $\omega_k^{dc}$. Discrete-continual approximation model is used for two-dimensional problems. It presupposes mesh approximation for non-basic dimension of extended domain (along $x_1$) while in the basic dimension (along $x_2$) problem remains continual. Thus subdomain $\omega_1$ is divided into discrete-continual finite elements

$$\omega_k^{dc} = \bigcup_{i=1}^{N_1-1}\omega_{k,i}^{dc}, \quad \omega_{k,i}^{dc} = \{(x_1,x_2):\; x_{1,i} < x_1 < x_{1,i+1}, \;\; x_{2,k}^b < x_2 < x_{2,k+1}^b\}. \tag{18}$$

Lame constants for finite element are defined by formulas:

$$\bar{\lambda}_{k,i}=\theta_{k,i}\lambda;\quad \bar{\mu}_{k,i}=\theta_{k,i}\mu,\ \text{where}\ \ \theta_{k,i}=\begin{cases}1, & \omega_{k,i}^{dc}\subset\Omega_k;\\ 0, & \omega_{k,i}^{dc}\not\subset\Omega_k;\end{cases} \tag{19}$$

$\theta_{k,i}$ is the characteristic function of element $\omega_{k,i}^{dc}$.

Basic nodal unknown functions are displacement components $u_1^{(k)},u_2^{(k)}$ and their derivatives $v_1^{(k)},v_2^{(k)}$ with respect to $x_2$ (superscript "$(k)$" hereinafter corresponds to the number of considered subdomain i.e. $\omega_k=\omega_k^{dc}$). Thus for node $(k,i)$ we have the following unknown functions: $u_1^{(k,i)},u_2^{(k,i)}$ and $v_1^{(k,i)},v_2^{(k,i)}$.

Linear approximation is used for unknown functions within discrete-continual finite element.

DCFEM is reduced at some stage to the solution of systems of $4N_1$ first-order ordinary differential equations:

$$\bar{U}_k'(x_2)=A_k\bar{U}_k(x_2)+\bar{\bar{R}}_k(x_2), \tag{20}$$

where $\bar{U}_k$ is global vector of nodal unknowns (subscript "$(k)$" corresponds to the number of subdomain $\omega_k=\omega_k^{dc}$),

$$\bar{U}_k=\bar{U}_k(x_2)=[\,(\bar{u}_k)^T\ \ (\bar{v}_k)^T\,]^T; \tag{21}$$

$$\bar{u}_k=\bar{u}_k(x_2)=[\,(\bar{u}_n^{(k,1)})^T\ \ (\bar{u}_n^{(k,2)})^T\ \ \ldots\ \ (\bar{u}_n^{(k,N_1)})^T\,]^T; \tag{22}$$

$$\bar{v}_k=\bar{v}_k(x_2)=[\,(\bar{v}_n^{(k,1)})^T\ \ (\bar{v}_n^{(k,2)})^T\ \ \ldots\ \ (\bar{v}_n^{(k,N_1)})^T\,]^T; \tag{23}$$

$$\bar{u}_n^{(k,i)}=\bar{u}_n^{(k,i)}(x_2)=[\,u_1^{(k,i)}\ \ u_2^{(k,i)}\,]^T;\ \ \bar{v}_n^{(k,i)}=\bar{v}_n^{(k,i)}(x_2)=[\,v_1^{(k,i)}\ \ v_2^{(k,i)}\,]^T; \tag{24}$$

$A_k$ is global matrix of coefficients of order $4N_1$; $\bar{\bar{R}}_k(x_2)$ is the right-side vector of order $4N_1$. Correct analytical solution of (21) is defined by formula

$$\bar{U}_k(x_2)=E_k(x_2)\bar{C}_k+\bar{S}_k(x_2); \tag{25}$$

$$E_k(x_2)=\varepsilon_k(x_2-x_{2,k}^b)-\varepsilon_k(x_2-x_{2,k+1}^b);\ \ \bar{S}_k(x_2)=\varepsilon_k(x_2)*\bar{\bar{R}}_k(x_2); \tag{26}$$

$\varepsilon_k(x_2)$ is the fundamental matrix-function of system (20), which is constructed in the special form convenient for problems of structural mechanics [1-5,35]; * is convolution notation; $\bar{C}_k$ is the vector of constants of order $4N_1$.

## 7 Multilevel approximation model for domain

Passing to multilevel approximation for domain is described in [3-5] for the simplest case of two-point boundary problem (Fig. 3). In a more general case, this transition is carried out in a similar way, although, of course, also with the use of more complicated formulas.

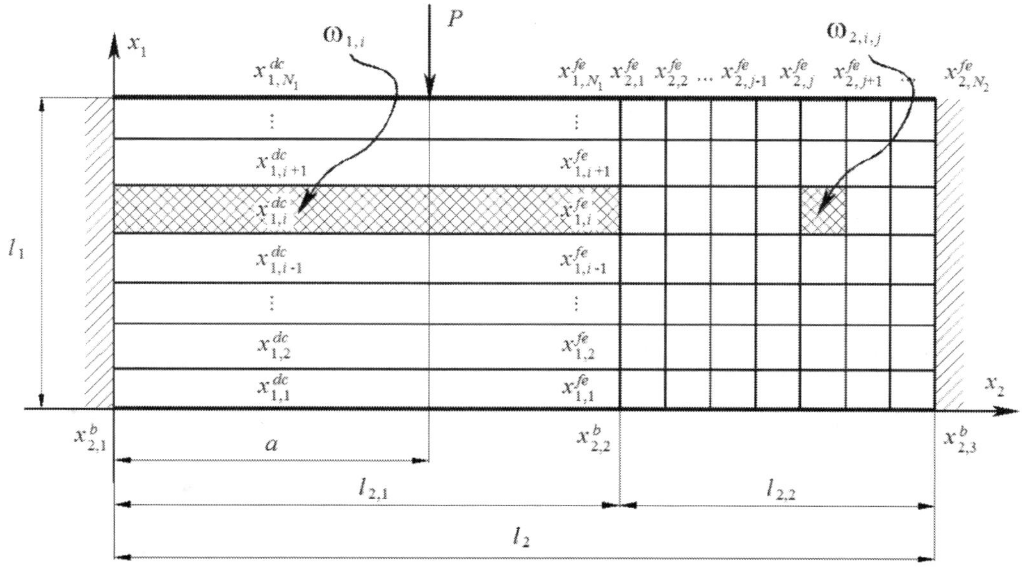

**Fig. 3.** Two-point boundary problem of static analysis of deep beam [3-5,30].

## 8 Software and verification samples

We should stress that all methods and algorithms considered in this paper have been realized in software. The main purpose of Analysis system CSASA2Dm (DCFEM + FEM) is semianalytical structural analysis (static structural analysis of deep beam within two-dimensional theory of elasticity), based on combined application of FEM and DCFEM. Programming environment is Microsoft Visual Studio 2013 Community and Intel Parallel Studio 2017XE (Fortran programming language [16]) with Intel MKL Library [8]. Software is designed for Microsoft Windows 8.1/10.

The Reported study was Funded by Government Program of the Russian Federation "Development of science and technology" (2013-2020) within Program of Fundamental Researches of Ministry of Construction, Housing and Utilities of the Russian Federation and Russian Academy of Architecture and Construction Sciences, the Research Project 7.1.1".

## References

1. P.A. Akimov, Applied Mechanics and Materials, **204-208**, 4502-4505 (2012).
2. P.A. Akimov, M.L. Mozgaleva, Applied Mechanics and Materials, **580-583**, 2898-2902 (2014).
3. P.A. Akimov, O.A. Negrozov, Procedia Engineering, **153**, 8-15 (2016).
4. P.A. Akimov, O.A. Negrozov, Procedia Engineering, **153**, 16-23 (2016).
5. P.A. Akimov, O.A. Negrozov, Procedia Engineering, **153**, 24-31 (2016).
6. E.M. Alawadhi, *Finite Element Simulations Using ANSYS* (CRC Press, 2009).
7. R. Azam, K. Soudki, Cement and Concrete Composites, **34(8)**, 946-957 (2012).
8. E.J. Barbero, *Finite Element Analysis of Composite Materials Using ANSYS* (CRC Press, 2013).
9. K.-J. Bathe, *Finite Element Procedures* (Prentice Hall Inc., 1996).

10. G. Campione, G. Minafo, Engineering Structures, **41**, 294-306 (2012).
11. M.-Y. Cheng, M.-T. Cao, Engineering Applications of Artificial Intelligence, **28**, 89-96 (2014).
12. Y.K. Cheung, F.T.K. Au, D.Y. Zheng, Thin-Walled Structures, **32(4)**, 289-303 (1998).
13. E.H. Dill, *The Finite Element Method for Mechanics of Solids with ANSYS Applitions* (CRC Press, 2011).
14. A.H. Gandomi, A.H. Alavi, D.M. Shadmehri, M.G. Sahab, Archives of Civil and Mechanical Engineering, **13(3)**, 354-369 (2013).
15. Y.H. Gedik, H. Nakamura, N. Ueda, M. Kunieda, Procedia Engineering, **14**, 2964-2971 (2011).
16. R.J. Hanson, T. Hopkins, *Numerical Computing with Modern Fortran (Applied Mathematics)* (SIAM-Society for Industrial and Applied Mathematics, 2013).
17. R.A. Hawileh, T.A. El-Maaddawy, M.Z. Naser, Materials & Design, **42**, 378-387 (2012).
18. M.A.T. Khatab, A.F. Ashour, T. Sheehan, D. Lam, Engineering Structures, **131**, 264-274 (2017).
19. H.S. Kim, M.S. Lee, Y.S. Shin, Procedia Engineering, **14**, 2212-2218 (2011).
20. Y.-B. Leng, X.-B. Song, H.-L. Wang, Journal of Constructional Steel Research, **106**, 89-98 (2015).
21. E. Lim, S.-J. Hwang, Engineering Structures, **108**, 104-112 (2016).
22. Z. Lin, M. Raoof, Engineering Structures, **17(10)**, 725-736 (1995).
23. J. Liu, B.I. Mihaylo, Engineering Structures, **112**, 81-89 (2016).
24. E. Madenci, I. Guven, *The Finite Element Method and Applications in Engineering Using ANSYS* (Springer, 2015).
25. I.M. Metwally, HBRS Journal, **13(1)**, 25-38 (2017).
26. S. Moaveni, *Finite Element Analysis: Theory and Application with ANSYS* (Prentice Hall, 2014).
27. A.R. Mohamed, M.S. Shoukry, J.M. Saeed, Alexandria Engineering Journal, **53(2)**, 329-339 (2014).
28. K. Mohamed, A.S. Farghaly, B. Benmokrane, K.W. Neale, Engineering Structures, **137**, 145-161 (2017).
29. M. Mohammadhassani, M.Z. Jumaat, M. Jameel, Construction and Building Materials, **30**, 265-273 (2012).
30. O.A. Negrozov, P.A. Akimov, I.Yu. Lantsova, Procedia Engineering, **153**, 926-932 (2016).
31. M. Pal, S. Deswal, Computers & Structures, **89(13-14)**, 1430-1439 (2011).
32. K. Uenaka, H. Tsunokake, Structures, **10**, 89-95 (2017).
33. O.C. Zienkiewicz, R.L. Taylor, D.D. Fox, *The Finite Element Method for Solid and Structural Mechanics* (Butterworth-Heinemann, 2013).
34. O.C. Zienkiewicz, R.L. Taylor, J.Z. Zhu, *The Finite Element Method: Its Basis and Fundamentals* (Butterworth-Heinemann, 2005).
35. A.B. Zolotov, P.A. Akimov, *Proceedings of the International Symposium LSCE 2002 organized by Polish Chapter of IASS* (Warsaw, Poland, 2002).

# About solution of multipoint boundary problem of static analysis of deep beam with the use of combined application of finite element method and discrete-continual finite element method. part 2: boundary conditions

*Leonid* Lyakhovich[1], *Oleg* Negrozov[2,*]

[1]Tomsk State University of Architecture and Building, Department of Structural Mechanics, Solyanaya sq., Tomsk, 634003, Russia
[2]National Research Moscow State University of Civil Engineering, Department of Applied Mathematics; 26, Yaroslavskoe Shosse, Moscow, 129337, Russia

**Abstract.** The formulation of considering multipoint boundary problem includes three main components: a description of the domain occupied by the structure and the corresponding subdomains; description of the conditions inside the domain and inside subdomains; description of boundary conditions (for boundaries of domain and boundaries between subdomains). These boundary conditions (interface conditions) in under consideration in the distinctive paper.

## 1 Analysis of options for boundary conditions

In practical applications the following variants of boundary conditions (interface conditions) between subdomains (2) from the first part of this paper are most often encountered (some of them are considered in this paper): interface “discrete-continual model – discrete model”, “internal” boundary condition of the type “perfect contact”; interface “discrete model – discrete-continual model”, “internal” boundary condition of the type “perfect contact”; interface “discrete-continual model – discrete-continual model”, “internal” boundary condition of the type “perfect contact”; interface “discrete model – discrete model”, “internal” boundary condition of the type “perfect contact”; interface “external boundary – discrete model”, boundary condition of the type “hinged support”; interface “external boundary – discrete model”, boundary condition of the type “free edge”; interface “external boundary – discrete-continual model”, boundary condition of the type “hinged support”; interface “external boundary – discrete-continual model”, boundary condition of the type “free edge”; interface “discrete model – external boundary”, boundary condition of the type “hinged support”; interface “discrete model – external boundary”, boundary condition of the type “free edge”; interface “discrete-continual model – external boundary”, boundary condition of the type “hinged support”; interface “discrete-continual model –

* Corresponding author: genromgsu@gmail.com

external boundary, boundary condition of the type "free edge". Of course, other variants of interface are possible, but similarly, somehow, in one way or another, as a rule, these variants are reduced to some combinations of the above-mentioned twelve [6-16].

## 2 Interface "discrete-continual model – discrete model", "internal" boundary condition of the type "perfect contact"

Let's consider arbitrary boundary point $x_{2,k}^b$, $1 < k < n_b$ and corresponding interface between subdomain $\omega_{k-1}^{dc}$ (subdomain is approximated with the use of DCFEM, $\rho_{k-1} = 2$) and subdomain $\omega_k^k$ (subdomain is approximated with the use of FEM, $\rho_k = 1$).

Boundary conditions at section $x_2 = x_{2,k}^b$ (perfect contact) have form ($4N_1$ equations):

$$u_1^{(k-1,i)}(x_{2,k}^b - 0) = u_1^{(k,i,j)}, \quad i = 1,2,...,N_1, \quad j = 1; \tag{1}$$

$$u_2^{(k-1,i)}(x_{2,k}^b - 0) = u_2^{(k,i,j)}, \quad i = 1,2,...,N_1, \quad j = 1; \tag{2}$$

$$\sigma_{1,2}^{(k-1,i)}(x_{2,k}^b - 0) = \sigma_{1,2}^{(k,i,j)}, \quad i = 1,2,...,N_1, \quad j = 1; \tag{3}$$

$$\sigma_{2,2}^{(k-1,i)}(x_{2,k}^b - 0) = \sigma_{2,2}^{(k,i,j)}, \quad i = 1,2,...,N_1, \quad j = 1, \tag{4}$$

where $\sigma_{1,2}^{(k-1,i)}(x_{2,k}^b - 0)$ and $\sigma_{2,2}^{(k-1,i)}(x_{2,k}^b - 0)$ are nodal (node $(k,i)$) values of functions of stress components $\sigma_{1,2}(x_2)$ and $\sigma_{2,2}(x_2)$ (after corresponding averaging) for discrete-continual finite element $\omega_{k-1,i}^{dc}$; $\sigma_{1,2}^{(k,i,j)}$ and $\sigma_{2,2}^{(k,i,j)}$ are nodal (node $(k,i,j)$) stress components $\sigma_{1,2}$ and $\sigma_{2,2}$ (after corresponding averaging).

Equations (1)-(4) can be rewritten in matrix form:

$$B_k^- \overline{U}_{k-1}(x_{2,k}^b - 0) = B_k^+ \overline{U}_k, \tag{5}$$

where $B_k^-$ is matrix of boundary conditions of size $4N_1 \times 4N_1$, which can be constructed in accordance with algorithm presented at Table 1; $B_k^+$ is matrix of boundary conditions of size $4N_1 \times 2N_1N_{2,k}$, which can be constructed in accordance with so-called method of basis variations [1-5,17].

**Table 1.** Algorithm of computing of nonzero elements of matrix $B_k^-$.

| **Numbers (indexes) of elements** | **Element value** | **Corresponding boundary condition** |
|---|---|---|
| 1 | 2 | 3 |
| $(i, 2i-1)$, $i = 1,2,...,N_1$ | 1 | (1) |
| $(N_1+i, 2i)$, $i = 1,2,...,N_1$ | 1 | (2) |
| $(2N_1+1, 2)$ | $\overline{\mu}_{k-1,1}\frac{1}{h_1}N_1'(0)$ | (3) $i = 1$ |
| $(2N_1+1, 4)$ | $\overline{\mu}_{k-1,1}\frac{1}{h_1}N_2'(0)$ | |

| 1 | 2 | 3 |
|---|---|---|
| $(2N_1+1, 2N_1+1)$ | $\overline{\mu}_{k-1,1}$ | (3) $i=1$ |
| $(2N_1+i, 2(i-1)),$ $i=2,3,\ldots,N_1-1$ | $\frac{1}{2}\overline{\mu}_{k-1,i-1}\frac{1}{h_{i-1}}N_1'(1)$ | (3) $i=2,3,\ldots,$ $N_1-1$ |
| $(2N_1+i, 2i),$ $i=2,3,\ldots,N_1-1$ | $\frac{1}{2}\left[\frac{\overline{\mu}_{k-1,i-1}}{h_{i-1}}N_2'(1)+\frac{\overline{\mu}_{k-1,i}}{h_i}N_1'(0)\right]$ | |
| $(2N_1+i, 2(i+1)),$ $i=2,3,\ldots,N_1-1$ | $\frac{1}{2}\overline{\mu}_{k-1,i}\frac{1}{h_i}N_2'(0)$ | |
| $(2N_1+i, 2(N_1+i)-1),$ $i=2,3,\ldots,N_1-1$ | $\frac{1}{2}(\overline{\mu}_{k-1,i-1}+\overline{\mu}_{k-1,i})$ | |
| $(3N_1, 2N_1-2)$ | $\overline{\mu}_{k-1,N_1-1}\frac{1}{h_{N_1-1}}N_1'(1)$ | (3) $i=N_1$ |
| $(3N_1, 2N_1)$ | $\overline{\mu}_{k-1,N_1-1}\frac{1}{h_{N_1-1}}N_2'(1)$ | |
| $(3N_1, 4N_1-1)$ | $\overline{\mu}_{k-1,N_1-1}$ | |
| $(3N_1+1, 1)$ | $\overline{\lambda}_{k-1,1}\frac{1}{h_1}N_1'(0)$ | (4) $i=1$ |
| $(3N_1+1, 3)$ | $\overline{\lambda}_{k-1,1}\frac{1}{h_1}N_2'(0)$ | |
| $(3N_1+1, 2N_1+1)$ | $\overline{\lambda}_{k-1,1}+2\overline{\mu}_{k-1,1}$ | |
| $(3N_1+i, 2i-3),$ $i=2,3,\ldots,N_1-1$ | $\frac{1}{2}\overline{\lambda}_{k-1,i-1}\frac{1}{h_{i-1}}N_1'(1)$ | (4) $i=2,3,\ldots,$ $N_1-1$ |
| $(3N_1+i, 2i-1),$ $i=2,3,\ldots,N_1-1$ | $\frac{1}{2}\left[\frac{\overline{\lambda}_{k-1,i-1}}{h_{i-1}}N_2'(1)+\frac{\overline{\lambda}_{k-1,i}}{h_i}N_1'(0)\right]$ | |
| $(3N_1+i, 2i+1),$ $i=2,3,\ldots,N_1-1$ | $\frac{1}{2}\overline{\lambda}_{k-1,i}\frac{1}{h_i}N_2'(0)$ | |
| $(3N_1+i, 2(N_1+i)),$ $i=2,3,\ldots,N_1-1$ | $\frac{1}{2}(\overline{\lambda}_{k-1,i-1}+\overline{\lambda}_{k-1,i})+$ $+\overline{\mu}_{k-1,i-1}-\overline{\mu}_{k-1,i}$ | |
| $(4N_1, 2N_1-3)$ | $\overline{\lambda}_{k-1,N_1-1}\frac{1}{h_{N_1-1}}N_1'(1)$ | (4) $i=N_1$ |
| $(4N_1, 2N_1-1)$ | $\overline{\lambda}_{k-1,N_1-1}\frac{1}{h_{N_1-1}}N_2'(1)$ | |
| $(4N_1, 4N_1)$ | $\overline{\lambda}_{k-1,N_1-1}+2\overline{\mu}_{k-1,N_1-1}$ | |

All other elements of matrix $B_k^-$ are equal to zero.

The algorithm for computing of matrix $B_k^+$ under the conditions (5) is described below.

1. Elements of the matrix $B_k^+$ are determined by the formula

$$(B_k^+)_{p,q} = \delta_{p,q}, \quad p = 1,2,\ldots,2N_1, \quad q = 1,2,\ldots,2N_1, \tag{6}$$

where $\delta_{p,q}$ is Kronecker delta.

It should be noted that computing of the elements of the matrix $B_k^+$ by formula (6) corresponds to the boundary conditions (1)-(2).

2. The following values are assigned sequentially: $i_g = 1, 2, \ldots, N_1$. For each fixed value of $i_g$, the actions listed below are performed.

2.1. The following values are assigned sequentially: $i = 1, 2, \ldots, N_1$. For each fixed value of $i$, the actions listed below are performed.

2.1.1. The following values are assigned sequentially: $j = 1, 2, \ldots, N_{2,k}$. For each fixed value of $j$, the actions listed below are performed.

2.1.1.1. The following values are assigned sequentially: $q = 1, 2$. For each fixed value of $q$, the actions listed below are performed.

2.1.1.1.1. Global index $j_g$ is computed in accordance with formula

$$j_g = q + 2(i-1) + 2N_1(j-1). \tag{7}$$

2.1.1.1.2. As a vector of unknowns (see the formula (21) from the first part of this paper), we set

$$\overline{U}_k = \overline{e}_{j_g}, \tag{8}$$

where $\overline{e}_{j_g}$ is the vector, $j_g$-th component of which is equal to 1, and the other components are equal to 0.

2.1.1.1.3. In accordance with the structure of vector of unknowns (8) with the use of corresponding formulas nodal components $[\sigma_{1,2}^{(k)}]^{(i_g,1)}$, $[\sigma_{2,2}^{(k)}]^{(i_g,1)}$ of stress tensor can be computed.

2.1.1.1.4. Elements of matrix $B_k^+$ are computed in accordance with formulas

$$(B_k^+)_{2N_1+i_g,j_g} = [\sigma_{1,2}^{(k)}]^{(i_g,1)}; \quad (B_k^+)_{3N_1+i_g,j_g} = [\sigma_{2,2}^{(k)}]^{(i_g,1)}. \tag{9}$$

It should be noted that computing of the elements of the matrix $B_k^+$ by formulas (9) corresponds to the boundary conditions (3)-(4).

## 3 Interface "discrete model – discrete-continual model", "internal" boundary condition of the type "perfect contact"

Let's consider arbitrary boundary point $x_{2,k}^b$, $1 < k < n_b$ and corresponding interface between subdomain $\omega_{k-1}^{fe}$ (subdomain is approximated with the use of FEM, $\rho_{k-1} = 1$) and subdomain $\omega_k^{dc}$ (subdomain is approximated with the use of DCFEM, $\rho_k = 2$).

Boundary conditions at section $x_2 = x^b_{2,k}$ (perfect contact) have form ( $4N_1$ equations):

$$u_1^{(k-1,i,j)} = u_1^{(k,i)}(x^b_{2,k}+0),\quad i=1,2,\ldots,N_1,\quad j=N_{2,k-1}; \tag{10}$$

$$u_2^{(k-1,i,j)} = u_2^{(k,i)}(x^b_{2,k}+0),\quad i=1,2,\ldots,N_1,\quad j=N_{2,k-1}; \tag{11}$$

$$\sigma_{1,2}^{(k-1,i,j)} = \sigma_{1,2}^{(k,i)}(x^b_{2,k}+0),\quad i=1,2,\ldots,N_1,\quad j=N_{2,k-1}; \tag{12}$$

$$\sigma_{2,2}^{(k-1,i,j)} = \sigma_{2,2}^{(k,i)}(x^b_{2,k}+0),\quad i=1,2,\ldots,N_1,\quad j=N_{2,k-1}; \tag{13}$$

where $\sigma_{1,2}^{(k,i)}(x^b_{2,k}+0)$ and $\sigma_{2,2}^{(k,i)}(x^b_{2,k}+0)$ are nodal (node $(k,i)$) values of functions of stress components $\sigma_{1,2}(x_2)$ and $\sigma_{2,2}(x_2)$ (after corresponding averaging) for discrete-continual finite element $\omega^{dc}_{k,i}$; $\sigma_{1,2}^{(k-1,i,j)}$ and $\sigma_{2,2}^{(k-1,i,j)}$ are nodal (node $(k,i,j)$) stress components $\sigma_{1,2}$ and $\sigma_{2,2}$ (after corresponding averaging).

Equations (1)-(4) can be rewritten in matrix form:

$$B_k^- \overline{U}_{k-1} = B_k^+ \overline{U}_k (x^b_{2,k}+0)\,, \tag{14}$$

where $B_k^-$ is matrix of boundary conditions of size $4N_1 \times 2N_1 N_{2,k-1}$, which can be constructed in accordance with so-called method of basis variations [1-5,17]; $B_k^+$ is matrix of boundary conditions of size $4N_1 \times 4N_1$, which can be constructed in accordance with algorithm presented at Table 2.

**Table 2.** Algorithm of computing of nonzero elements of matrix $B_k^+$.

| **Numbers (indexes) of elements** | **Element value** | **Corresponding boundary condition** |
|---|---|---|
| 1 | 2 | 3 |
| $(i, 2i-1),\ i=1,2,\ldots,N_1$ | 1 | (10) |
| $(N_1+i, 2i),\ i=1,2,\ldots,N_1$ | 1 | (11) |
| $(2N_1+1, 2)$ | $\overline{\mu}_{k,1}\dfrac{1}{h_1}N_1'(0)$ | (12) $i=1$ |
| $(2N_1+1, 4)$ | $\overline{\mu}_{k,1}\dfrac{1}{h_1}N_2'(0)$ | |
| $(2N_1+1, 2N_1+1)$ | $\overline{\mu}_{k,1}$ | |
| $(2N_1+i, 2(i-1)),\ i=2,3,\ldots,N_1-1$ | $\dfrac{1}{2}\overline{\mu}_{k,i-1}\dfrac{1}{h_{i-1}}N_1'(1)$ | (12) $i=2,3,\ldots,N_1-1$ |
| $(2N_1+i, 2i),\ i=2,3,\ldots,N_1-1$ | $\dfrac{1}{2}\left[\dfrac{\overline{\mu}_{k,i-1}}{h_{i-1}}N_2'(1)+\dfrac{\overline{\mu}_{k,i}}{h_i}N_1'(0)\right]$ | |
| $(2N_1+i, 2(i+1)),\ i=2,3,\ldots,N_1-1$ | $\dfrac{1}{2}\overline{\mu}_{k,i}\dfrac{1}{h_i}N_2'(0)$ | |
| $(2N_1+i, 2(N_1+i)-1),\ i=2,3,\ldots,N_1-1$ | $\dfrac{1}{2}(\overline{\mu}_{k,i-1}+\overline{\mu}_{k,i})$ | |

| 1 | 2 | 3 |
|---|---|---|
| $(3N_1, 2N_1-2)$ | $\bar{\mu}_{k,N_1-1}\dfrac{1}{h_{N_1-1}}N_1'(1)$ | (12) $i=N_1$ |
| $(3N_1, 2N_1)$ | $\bar{\mu}_{k,N_1-1}\dfrac{1}{h_{N_1-1}}N_2'(1)$ | |
| $(3N_1, 4N_1-1)$ | $\bar{\mu}_{k,N_1-1}$ | |
| $(3N_1+1, 1)$ | $\bar{\lambda}_{k,1}\dfrac{1}{h_1}N_1'(0)$ | (13) $i=1$ |
| $(3N_1+1, 3)$ | $\bar{\lambda}_{k,1}\dfrac{1}{h_1}N_2'(0)$ | |
| $(3N_1+1, 2N_1+1)$ | $\bar{\lambda}_{k,1}+2\bar{\mu}_{k,1}$ | |
| $(3N_1+i, 2i-3)$, $i=2,3,\ldots,N_1-1$ | $\dfrac{1}{2}\bar{\lambda}_{k,i-1}\dfrac{1}{h_{i-1}}N_1'(1)$ | (13) $i=2,3,\ldots,$ $N_1-1$ |
| $(3N_1+i, 2i-1)$, $i=2,3,\ldots,N_1-1$ | $\dfrac{1}{2}\left[\dfrac{\bar{\lambda}_{k,i-1}}{h_{i-1}}N_2'(1)+\dfrac{\bar{\lambda}_{k,i}}{h_i}N_1'(0)\right]$ | |
| $(3N_1+i, 2i+1)$, $i=2,3,\ldots,N_1-1$ | $\dfrac{1}{2}\bar{\lambda}_{k,i}\dfrac{1}{h_i}N_2'(0)$ | |
| $(3N_1+i, 2(N_1+i))$, $i=2,3,\ldots,N_1-1$ | $\dfrac{1}{2}(\bar{\lambda}_{k,i-1}+\bar{\lambda}_{k,i})+\bar{\mu}_{k,i-1}-\bar{\mu}_{k,i}$ | |
| $(4N_1, 2N_1-3)$ | $\bar{\lambda}_{k,N_1-1}\dfrac{1}{h_{N_1-1}}N_1'(1)$ | (13) $i=N_1$ |
| $(4N_1, 2N_1-1)$ | $\bar{\lambda}_{k,N_1-1}\dfrac{1}{h_{N_1-1}}N_2'(1)$ | |
| $(4N_1, 4N_1)$ | $\bar{\lambda}_{k,N_1-1}+2\bar{\mu}_{k,N_1-1}$ | |

All other elements of matrix $B_k^+$ are equal to zero.

The algorithm for computing of matrix $B_k^-$ under the conditions (14) is described below.

1. Elements of the matrix $B_k^-$ are determined by the formula

$$(B_k^-)_{p,q}=\delta_{p,q},\quad p=1,2,\ldots,2N_1,\quad q=1,2,\ldots,2N_1. \tag{15}$$

It should be noted that computing of the elements of the matrix $B_k^-$ by formula (15) corresponds to the boundary conditions (10)-(11).

2. The following values are assigned sequentially: $i_g=1,2,\ldots,N_1$. For each fixed value of $i_g$, the actions listed below are performed.

2.1. The following values are assigned sequentially: $i=1,2,\ldots,N_1$. For each fixed value of $i$, the actions listed below are performed.

2.1.1. The following values are assigned sequentially: $j = 1, 2, \ldots, N_{2,k-1}$. For each fixed value of $j$, the actions listed below are performed.

2.1.1.1. The following values are assigned sequentially: $q = 1, 2$. For each fixed value of $q$, the actions listed below are performed.

2.1.1.1.1. Global index $j_g$ is computed in accordance with formula

$$j_g = q + 2(i-1) + 2N_1(j-1) . \tag{16}$$

2.1.1.1.2. As a vector of unknowns (see the formula (21) from the first part of this paper), we set

$$\bar{U}_{k-1} = \bar{e}_{j_g} . \tag{17}$$

2.1.1.1.3. In accordance with the structure of vector of unknowns (8) with the use of corresponding formulas nodal components $[\sigma_{1,2}^{(k-1)}]^{(i_g,N_{2,k-1})}$, $[\sigma_{2,2}^{(k-1)}]^{(i_g,N_{2,k-1})}$ of stress tensor can be computed.

2.1.1.1.4. Elements of matrix $B_k^-$ are computed in accordance with formulas

$$(B_k^-)_{2N_1+i_g,j_g} = [\sigma_{1,2}^{(k-1)}]^{(i_g,N_{2,k-1})}; \quad (B_k^-)_{3N_1+i_g,j_g} = [\sigma_{2,2}^{(k-1)}]^{(i_g,N_{2,k-1})} . \tag{18}$$

It should be noted that computing of the elements of the matrix $B_k^-$ by formulas (18) corresponds to the boundary conditions (12)-(13).

## 4 Interface "discrete-continual model – discrete-continual model", "internal" boundary condition of the type "perfect contact"

Let's consider arbitrary boundary point $x_{2,k}^b$, $1 < k < n_b$ and corresponding interface between subdomain $\omega_{k-1}^{dc}$ (subdomain is approximated with the use of DCFEM, $\rho_{k-1} = 2$) and subdomain $\omega_k^{dc}$ (subdomain is approximated with the use of DCFEM, $\rho_k = 2$ ).

Boundary conditions at section $x_2 = x_{2,k}^b$ (perfect contact) have form ( $4N_1$ equations):

$$u_1^{(k-1,i)}(x_{2,k}^b - 0) = u_1^{(k,i)}(x_{2,k}^b + 0), \quad i = 1, 2, \ldots, N_1; \tag{19}$$

$$u_2^{(k-1,i)}(x_{2,k}^b - 0) = u_2^{(k,i)}(x_{2,k}^b + 0), \quad i = 1, 2, \ldots, N_1; \tag{20}$$

$$\sigma_{1,2}^{(k-1,i)}(x_{2,k}^b - 0) = \sigma_{1,2}^{(k,i)}(x_{2,k}^b + 0), \quad i = 1, 2, \ldots, N_1; \tag{21}$$

$$\sigma_{2,2}^{(k-1,i)}(x_{2,k}^b - 0) = \sigma_{2,2}^{(k,i)}(x_{2,k}^b + 0), \quad i = 1, 2, \ldots, N_1 . \tag{22}$$

Equations (19)-(22) can be rewritten in matrix form:

$$B_k^- \bar{U}_{k-1}(x_{2,k}^b - 0) = B_k^+ \bar{U}_k(x_{2,k}^b + 0) ; \tag{23}$$

where $B_k^-$ and $B_k^+$ are matrices of boundary conditions of size $4N_1 \times 4N_1$, which can be constructed in accordance with algorithms presented at Table 1 and Table 2 respectively.

## 5 Interface "discrete model – discrete model", "internal" boundary condition of the type "perfect contact"

Let's consider arbitrary boundary point $x_{2,k}^b$, $1 < k < n_b$ and corresponding interface between subdomain $\omega_{k-1}^{fe}$ (subdomain is approximated with the use of FEM, $\rho_{k-1} = 1$) and subdomain $\omega_k^{fe}$ (subdomain is approximated with the use of FEM, $\rho_k = 1$ ).

It should be noted that in this case special formulation of boundary conditions are not required. Subdomains $\omega_{k-1}^{fe}$ and $\omega_k^{fe}$ are normally considered jointly within FEM, boundary conditions of the type "perfect contact" are automatically met.

The Reported study was Funded by Government Program of the Russian Federation "Development of science and technology" (2013-2020) within Program of Fundamental Researches of Ministry of Construction, Housing and Utilities of the Russian Federation and Russian Academy of Architecture and Construction Sciences, the Research Project 7.1.1".

## References

1. P.A. Akimov, Applied Mechanics and Materials, **204-208**, 4502-4505 (2012).
2. P.A. Akimov, M.L. Mozgaleva, Applied Mechanics and Materials, **580-583**, 2898-2902 (2014).
3. P.A. Akimov, O.A. Negrozov, Procedia Engineering, **153**, 8-15 (2016).
4. P.A. Akimov, O.A. Negrozov, Procedia Engineering, **153**, 16-23 (2016).
5. P.A. Akimov, O.A. Negrozov, Procedia Engineering, **153**, 24-31 (2016).
6. K.-J. Bathe, *Finite Element Procedures* (Prentice Hall Inc., 1996).
7. G. Campione, G. Minafo, Engineering Structures, **41**, 294-306 (2012).
8. H.S. Kim, M.S. Lee, Y.S. Shin, Procedia Engineering, **14**, 2212-2218 (2011).
9. E. Lim, S.-J. Hwang, Engineering Structures, **108**, 104-112 (2016).
10. Z. Lin, M. Raoof, Engineering Structures, **17(10)**, 725-736 (1995).
11. J. Liu, B.I. Mihaylo, Engineering Structures, **112**, 81-89 (2016).
12. O.A. Negrozov, P.A. Akimov, I.Yu. Lantsova, Procedia Engineering, **153**, 926-932 (2016).
13. M. Pal, S. Deswal, Computers & Structures, **89(13-14)**, 1430-1439 (2011).
14. K. Uenaka, H. Tsunokake, Structures, **10**, 89-95 (2017).
15. O.C. Zienkiewicz, R.L. Taylor, D.D. Fox, *The Finite Element Method for Solid and Structural Mechanics* (Butterworth-Heinemann, 2013).
16. O.C. Zienkiewicz, R.L. Taylor, J.Z. Zhu, *The Finite Element Method: Its Basis and Fundamentals* (Butterworth-Heinemann, 2005).
17. A.B. Zolotov, P.A. Akimov, *Proceedings of the International Symposium LSCE 2002 organized by Polish Chapter of IASS* (Warsaw, Poland, 2002).

MATEC Web of Conferences **117**, 00111 (2017) DOI: 10.1051/matecconf/201711700111
*XXVI R-S-P Seminar 2017, Theoretical Foundation of Civil Engineering*

# Raising of geo ecological safety of engineering collectors: the new method of assessment of water tightness of reinforced concrete blocks with basalt plastic lining

*Olga* Lyapidevskaya[1,*], *Andrey* Sherstnev[1]

[1]Moscow state university of civil engineering, Yaroslavskoye shosse, 26, Moscow, Russia, 129337

**Abstract.** This work presents analytics ways to estimation and assessment of water tightness of reinforced concrete blocks with basalt plastic lining for increasing geo ecological safety of engineering collectors. It is proved the advisability of application basalt plastic lining instead secondary protection of concrete with hydraulic seal. The results of estimation filtration coefficient and assessment of water tightness of basalt plastic lining of engineering collectors are represented. It is carried out comparative analysis of filtration coefficient and water tightness data of basalt plastic lining and secondary protection. The geo ecological effect is proved at the expense of application of new materials and reducing breakdown events with collectors that may lead pollution of ground and subsoil water with sewage.

## 1 Assessment of water tightness of reinforced concrete blocks with basalt plastic lining and secondary protection

The exploration of city underground space with simultaneous reservation of environment is one of priority directions for building construction development.

In this connection the problem of increasing geo ecological safety and durability of underground construction (including tunneling and engineering communication) is especially sharp.

The insulating materials for secondary protection that use now (on bitumen or cement base) are not enough preserve structures from destructive action of aggressive waters. Besides the most of this materials have negative action to environment. For decision of this problem it is suggested to make lining of tunnels with prefabricated blocks with basalt plastic finishing.

Basalt plastic lining (fig.1) is the special multilayer finishing composed of polymer binder, cut basalt filament, basalt plastic anchors for connection with concrete and intended for securing protection of inner surfaces of reinforced block basalt plastic s from possible mechanical, thermal, physical and chemical damages [1, 2].

* Corresponding author : olga.lyapidevskaya@inbox.ru

**Fig. 1.** Prefabricated elements with basalt plastic lining

The existing determination methods of filtration coefficient and assessment of water tightness of concrete don't take into account presence of joints between the separate elements of construction. However, namely these joints are the most vulnerable places for sewage water penetration into ground and ground waters creating geo ecological danger for environment.

In this work it is presented the method of estimation of filtration coefficient and assessment of water tightness basalt plastic lining of reinforced block that takes into account the influence of joints between blocks on water tightness the whole engineering collector.

The estimation of water tightness of basalt plastic lining of reinforced blocks is carried out on the base of works of G. M. Lomize and V. M. Nasberg [3, 4].

Calculation of water leakages from engineering collector constructed with reinforced blocks with basalt plastic lining and self-swelling rubber laying is carried out in accordance with formula (1):

$$Q = \frac{2\pi LH}{\frac{1}{K_{nok}}\times ln\frac{R_{nok}}{R_0}+\frac{1}{K_{cr.0}}\times\frac{R}{R_{nok}}+\frac{1}{K_n}\times ln\frac{2t}{R}} \qquad (1)$$

Where $Q$ – leakage of water from engineering collector, l
$K_{nok}$ – filtration coefficient of lining, m/day;
$K_{cr.0}$ – filtration coefficient of reinforced concrete blocks, m/day;
$K_n$ – filtration coefficient of rocks around collector, m/day;
$R_{nok}$ – radius of outside zone of lining, m;
$R_0$ – radius of tunnel (inner), m;
$R$ – outside diameter of reinforced blocks, m;

2t – radius of influence, m; for tunnels with diameter 2–5 м it is recommended to take value 2t - 200 m;

L – length of engineering collector, m;

H – pressure of water inside collector, m

For the analogous engineering collector constructed with water tight lining made from cement mortar (secondary protection), the calculation of water leakages have made with the next formula (2)

$$Q = \frac{2\pi L(H_e - H_0)}{\frac{1}{K_b}\times ln\frac{R_b}{R_0}+\frac{1}{K_{cr.0}}\times\frac{R}{R_{b.n}}+\frac{1}{K_n}\times ln\frac{2t}{R}} \quad (2)$$

Where $K_b$ – filtration coefficient of secondary lining, m/day;

$R_{b.n}$– radius of outside secondary lining, m;

$H_e$ – max. pressure of ground waters, m of water;

$H_0$ – pressure of waters inside of engineering collector, m

Analyzing presented dependences (1), (2) it's received that in general water tightness depends on filtration coefficient of engineering collector's structures and rocks around them.

Proceeded from foregoing the main problem of development new type of lining is to work out the reliable method of calculation filtration coefficient taking into account joints between blocks.

## 2 Calculation of filtration coefficient of prefabricated reinforced blocks with basalt plastic lining

This time the trust worthy information about filtration coefficient of polymer lining is absent. To determine the filtration coefficient it's used indirect characteristics such as quantity of water that flow through the collector, the inner specific surface of pores, vapor permeability [6-8]. For production of lining it is possible to use polymer compositions on the base of thermo reactive (for example, Polyester) resin and also PVC and polystyrene.

Table 1 contains comparative characteristics of listed materials.

**Table 1.** Characteristics of polymer materials

| **Material** | **Inner specific pores surface, m$^2$/m$^3$** | **Vapor permeability mg/(m*h*Pa)** | **Filtration coefficient, m/day** |
|---|---|---|---|
| On the base of Polyester resins | - | $11{,}79\cdot10^{-5}$ <br> $26{,}63\cdot10^{-5}$ | $0{,}012\cdot10^{-7}$ |
| PVC | 250-300 | - | $<0{,}86\cdot10^{-7}$ |
| Polystyrene | 60-700 | - | $<0{,}86\cdot10^{-7}$ |

Data presented in Table 1 were treated with using dependence (3) [9],

$$K_f = (\alpha \times S - 5{,}125) \times 10^{-5} \quad (3)$$

Where K_f– filtration coefficient, m/day;

α – Coefficient, equal 0,0004142 (is taken by processing experimental data);

S – Specific pores surface, m$^2$/m$^3$;

The material is considered as water tight at filtration coefficient $K_f \leq 5\cdot10^{-6}$ m/day [6].

Values of the filtration coefficient from Table 1 show that polymer lining is essentially water proof. The binder on the base of polyester resins has been choice so it has the least filtration coefficient.

As it was noted the existing methods of calculation of filtration coefficient don't take into account presence joints between blocks of engineering collectors that cause different errors at calculation.

It is supposed that in engineering collectors of reinforced concrete blocks with basalt plastic lining conjunction would be made with self-swelling rubbers and joints would be filled by epoxy thermo reactive resin.

In this case calculation of the filtration coefficient of reinforced concrete blocks with regard to the filtration coefficient and width of joints would be according formula (4).

$$K_{cr.0} = \frac{B \times K_b \times l_0 + K_{st}[(B+l_0)\times\delta+\delta^2]}{(l_0+\delta)(B+\delta)} \qquad (4)$$

Where $K_{cr.0}$– filtration coefficient of reinforced concrete blocks with regard to width of joints between blocks, m/day;

$K_b$ – Filtration coefficient of reinforced concrete blocks, m/day;

$K_{st}$ – Filtration coefficient of joints between blocks, m/day;

$\delta$ – Width of joints, m;

B – Width of reinforced concrete block, m;

$l_0$ – size (thickness) of reinforced concrete block, m.

Filtration coefficient of reinforced concrete blocks take as $K_b$=0,069·10$^{-5}$ m/day [4].

Filtration coefficient of joints between blocks take as polymer lining $K_{st}$ = 0,012·10$^{-7}$ m/day.

Filtration coefficients of reinforced concrete blocks depending on the width of joint are calculated according formula 4 and represented in Table 3. The width of joints at construction of engineering collectors is from 0,005 up 0,03 m.

**Table 3.** Filtration coefficients of reinforced concrete blocks depending on joint width

| Width of joint, m | Filtration coefficients of reinforced concrete blocks Kф·10$^{-7}$, m/day |
|---|---|
| 0,005 | 1,94 |
| 0,010 | 2,01 |
| 0,015 | 2,04 |
| 0,020 | 2,09 |
| 0,025 | 2,18 |
| 0,030 | 2,29 |

Analyzing the data of Table 3 it is possible to come to conclusion that collector constructed with blocks with basalt plastic lining is practically water tight at any acceptable width of joint. During tests of reinforced concrete blocks with applied hydraulic seal on cement base the filtration coefficient was (0,81 –1,78)·10$^{-5}$ m/day that is much more than this index of concrete blocks with basalt polymer lining.

## 3 Calculation of water leakage from engineering collector

Calculation according formula 1 will be carried out with the use of the smallest filtration coefficient is equal 0,012·10$^{-7}$ m/day. Filtration coefficient of reinforced concrete blocks depending on the width of Diameter of engineering collector is 3,15m, inner diameter – 2,75 m, size (thickness) - 0, 20 m. The ring of engineering collector with width 1m consists of 6 blocks. Construction of engineering collectors mainly is conducted profundity and depth didn't exceed 20 m. At the depth of 20 m ground waters create pressure $H_e$ up to 14,9 m. To determine the resulting pressure H it would be considered the event of complete filling of engineering collector. In this case pressure of water inside collector will take $H_0$ = 2,6 m of water [5].

**Table 4.** Filtering parameters, taken as the basis in estimation according formulas (1), (2)

| Leakages of water from engineering collector | Pressure, m of water | | | Filtration coefficient, m/day | | |
|---|---|---|---|---|---|---|
| | H | $H_e$ | $H_0$ | Lining | Reinforced blocks | Secondary protection |
| Reinforced blocks with basalt plastic lining | 12,3 | 14,9 | 2,6 | $0{,}012 \cdot 10^{-7}$ | $0{,}069 \cdot 10^{-5}$ | - |
| Reinforced blocks with secondary protection | | | | - | $2{,}1 \cdot 10^{-5}$ | $0{,}82 \cdot 10^{-5}$ |

The results of estimation water leakages from engineering collector according formulas (1), (2) proceeding from filtering parameters presented in Table 4, are given in Table 5.

**Table 5.** Leakages of water from engineering collector

| Structures of engineering collector | Leakages of water (Q) at L=10 m during 0,5 hour | |
|---|---|---|
| | $m^3$ | l |
| Reinforced blocks with basalt plastic lining | 0,028 | 0,59 |
| Reinforced blocks with secondary protection | 0,094 | 1,97 |

## 4 Conclusion

The results of estimation (Table 5) have shown that hermetic nature of basalt plastic lining almost in 3,4 time exceeded secondary protection of engineering collectors (0,59<1,97l).

Thus, the application of basalt plastic lining for protection reinforced concrete blocks will increase geo ecological safety of city environment at the expense of contraction of failure at deterioration of water tightness of engineering collectors and contamination of soil and ground waters by sewage.

## References

1. O.B. Lyapidevskaya, A.K. Sherstnev, MATEC Web of Conferences, **86**, 03005 (2016)
2. O.B. Lyapidevskaya, A.K. Sherstnev, Procedia Engineering, **153**, 434-438 (2016)
3. G.M. Lomize. Filtration in cracked rocks (1951)
4. V. M. Nasberg Estimation of filtration of geotechnical tunnels («Izvestia VNIIG», T. 58, Gosenergoizdat, 1958 )
5. SP 129.13330.2011 External networks and constructions of water supply and Sewerag
6. Handbook for the processing of materials of engineering-geological surveys. - M: THE GIFT / VODGEO, (2005)
7. SP 74.13330.2011 Heat networks

8. A.A. Shilin, M.V. Zaytsev, I.A. Zolotarev, O.B. Lyapidevskaya, Waterproofing of underground and deepened constructions during building and repair (2003)
9. E.Y. Kulikova, Filtration reliability of city underground structures. (Peace of mountain book 2007)
10. GOST 12730.5-84 Concretes. Methods for determination of water tightness.

# Evaluation of the effectiveness of methods of compaction of sandy soil using physical modeling in the laboratory

*Andrey* Maltsev[1,*], *Alla* Ponomarenko[1], *Andrey* Karpov[1], and *Dmitrii* Popov[1]

[1] Samara State Technical University, Institute of Architecture and Civil Engineering, 194, Molodogvardeyskaya St., 443001, Samara, Russia

**Abstract.** Surface compacting of soils is one of the measures applied in the practice of construction for the improvement of deformation and strength indicators of soil characteristics. Compacting is used both in creation of artificial bases in the shape of sand blankets and in quality enhancement of sand matresses for constructions as an underlying and levelling layer within road surface. For creation of surface compacting at a site various methods dependent on machines and mechanisms being used are applied: tamping, rolling, vibrocompaction. The research has been executed for evaluation of the efficiency of the indicated means of compacting. The set task was solved in laboratory conditions on physical models in a laboratory tray. In the course of the research the special laboratory tray was made, sandy soil model was chosen, with the help of additional equipment and developed original methodology a modelling of three methods of surface compacting was done, experiments on compacting of physical models with different soil dampness were executed, quantitative and qualitative assessments of surface compacting of sandy soil were given. On the bases of the conducted experiments the most efficient means of surface compacting of sandy soil was found. Recommendations for builders were given.

## 1 Introduction

Soil compacting is one of the measures executed at sites with the aim of preparation of structure bases. The main aim of soil compacting is to improve their mechanical characteristics. With the help of compacting deformity can be reduced and strength can be increased of both artificial bases (for instance, a sand blanket) and sand mattress or as an underlying and levelling layer within a road surface [1-3]. It is known that security of the system «construction-base» is determined by security of every element within this system [4]. That is why the condition of the road surface or appearance of additional deformations of construction bases will depend on the quality of the sand mattress and first of all on the level of its compacting [5, 6].

* Corresponding author: geologof@yandex.ru

Analysis has shown that in the practice of construction various methods of surface compacting of sandy soil with the help of special machines and mechanisms are used: tamping, rolling, vibrocompaction (fig.1). Present research is conducted to answer the question which of the indicated means is more efficient?

The aim of this work is to value quality of various means of sandy soil compacting depending on its dampness.

The set task is solved in the laboratory conditions on physical models in the laboratory tray.

In the course of research the following actions were done:

- a laboratory tray for research conducting was issued;
- the necessary characteristics of the research object – sandy soil – were determined;
- in the laboratory tray with the help of additional equipment three means of surface compacting (tamping, rolling, vibrocompaction) were modeled;
- experiments on compacting of physical models with different soil dampness were conducted;
- quantitative and qualitative assessment of the considered means of surface compacting of sandy soil is given.

**Fig. 1**. Machines and mechanisms for surface compacting.

## 2 Materials and methods

The research was carried in several stages including preparation, experiment, processing and analysis of the experiment results.

At the stage of preparation to the physical experiment the laboratory flume especially for the research of surface compacting with different physical methods was designed and made (fig. 2,3). It represents a box with the dimensions of 24x38x50 cm. Vertical walls of the laboratory tray are made of three sheetings of MDF and plexi. The bottom of the laboratory tray is made of a polymer sheeting set on the wooden battens.

**Fig. 2.** Exterior of the laboratory tray.

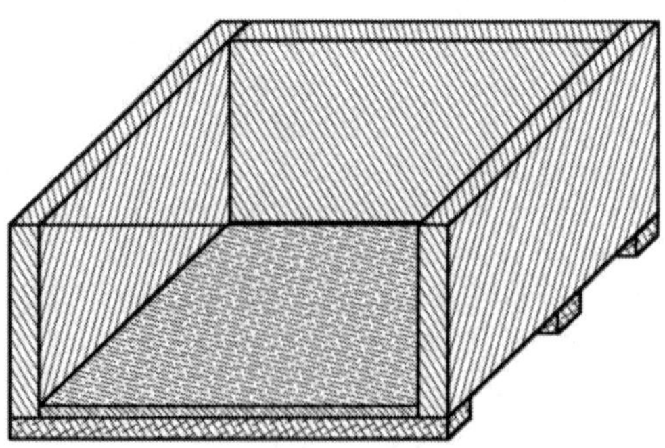

**Fig. 3.** Construction of the laboratory tray.

Homogeneous sand of medium size was chosen as a research object. Analysis of its mechanical composition was made with the help of standard sieves (fig. 4) in accordance with the indications [7].

Before the beginning of experiments the potential of the soil under investigation for compacting was defined.

This was evaluated with the following physical parametres:

- full density dry sand equal to $\rho_{dmax}$= 1.74 g/cm$^3$;
- sand optimum moister content which made $W_{opt}$ = 8.6 %.

Analysis of full density and optimum moister content was executed by the method of standard compacting in accordance with the requirements [8] with the help of special laboratory equipment (fig. 5).

**Fig. 4.** A set of standard sieves.

**Fig. 5.** Densometer.

An original methodology and special equipment permitting to model processes of surface soil compacting in the laboratory tray were developed to conduct the research conducting via the following means:

1. **Tamping** executed through the model of heavy compacting. In this case the level of sand (of 5 cm height) at the bottom of the laboratory tray was compacted by dynamic pressure done by a drop weight (total weight of 13 kg) from the height of 20 cm along the guide bar on the timber slab of 25x38 dimensions (fig. 6) set on the soil surface. Production experiments determined technical parameters of the sample under investigation and compacting weight as well as minimal number of strikes with the aim of gaining maximum possible compacting for this means – 5 strikes.

**Fig. 6.** Sand compacting in the laboratory tray by tamping.

2. **Rolling.** This type of compacting was executed by the model of a road roller – metal tubing with diameter of 70 mm and length of 50 cm filled inside with additional

counterweight (total weight of 14 kg). Sandy soil layer of 2.5 cm height was compacted under the influence of the tube weight which was relocated with the help of four preliminary rolled on it bundles of fishing line (fig. 7). Production experiments helped to define technical parameters of the sample under investigation and compacting weight as well as a minimal number of passages with the aim of gaining a full density for this means of compacting – 15 passes.

**Fig. 7.** Sand compacting in the laboratory tray by rolling.

3. **Vibrocompaction**. Was done in the laboratory tray with the help of a vibratory load created with the help of a special laboratory table vibrator (vibration frequency of 90 hz). The laboratory tray was filled with a sand layer of 2.5 cm height and was set on the table vibrator. Additional statictical weight was applied to soil surface in the shape of a beam (25x38 cm in dimension) and calibrated weight (fig. 8). Total weight was 27.7 kg. Production experiments determined that vibrocompaction for 30 seconds is enough to reach possible full density of a sandy soil sample.

**Fig. 8.** Vibrocompaction of sand in the laboratory tray.

While conducting the experiments apart from the means of external influence on the soil the influence of moister content of the sandy soil sample on the amount of its density was investigated as well. In the experiments took part samples having the following moister content: $W_1=W_{opt}=8.6$ %, $W_2=12.9$ %, $W_3=17.2$ %, $W_4=21.5$ %. Thus for every means of compacting with various moister content of samples were conducted series of 12 experiments (total number of experiments was more than 40).

Control over the compacting of sandy soil samples in the course of experiments was done visually via a transparent wall of a laboratory tray and with the help of instruments via measurement of height of a sample with a ruler. After execution of the next experiment from the layer of a compacted soil not less than three samples were taken to determine their density $\rho_i$ by method of cutting ring in accordance with the instructions [9]. Mass specific gravity of the compacted sample $\rho_{av}$ was calculated as an arithmetic mean of specials value of density of separate soil samples.

## 3 Results

For the evaluation of efficiency of the regarded means of compacting compaction factor $K_c$ which is applied in construction was chosen. This compaction factor is a relation of density of dry compacted sandy soil $\rho_d$ to its full density of dry soil $\rho_{dmax}$ determined at the stage of preparatory works.

Via experiments it was found out that the investigated sandy soil under optimum moister content $W_{opt}$=8.6 % had maximim compaction factor $K_c$=0.92. It should be marked that the given value is the least factor $K_c$ according to the Russian building regulations [10].

In the course of processing of experimental results on the basis of experimental data on the basis of recommended analytical decisions [7-10] for every experiment were determined:

- mass specific gravity of compacted soil sample $\rho_{av}$;
- density of dry compacted sample of sandy soil $\rho_d$;
- compaction factor of a sandy soil sample $K_{ci}$ .

The figures of the main indicators of density received as a result of the conducted experiments are given in table 1.

**Table 1.** Cumulative results of tests on sandy soil compacting.

| Soil dampness | Soil density | Different means of soil compacting | | |
|---|---|---|---|---|
| | | Tamping | Rolling | Vibrocompaction |
| $W_1$ = 8.6 % | $\rho_{av}$, г/см$^3$ | 1.743 | 1.747 | 1.745 |
| | $\rho_d$, г/см$^3$ | 1.605 | 1.609 | 1.607 |
| | $K_c$ | 0.92 | 0.92 | 0.92 |
| $W_2$ = 12.9 % | $\rho_{av}$, г/см$^3$ | 1.850 | 1.843 | 1.844 |
| | $\rho_d$, г/см$^3$ | 1.639 | 1.632 | 1.633 |
| | $K_c$ | 0.94 | 0.94 | 0.94 |
| $W_3$ = 17.2 % | $\rho_{av}$, г/см$^3$ | 1.992 | 1.845 | 1.870 |
| | $\rho_d$, г/см$^3$ | 1.700 | 1.574 | 1.596 |
| | $K_c$ | 0.98 | 0.90 | 0.92 |
| $W_4$ = 21.5 % | $\rho_{av}$, г/см$^3$ | 1.937 | 1.727 | - |
| | $\rho_d$, г/см$^3$ | 1.594 | 1.421 | - |
| | $K_c$ | 0.92 | 0.82 | - |

## 4 Discussion

For quantitative and qualitative assessment of efficiency of the regarded means of surface compacting of sandy soil the received experimental data were represented as a graph (fig. 9) [11].

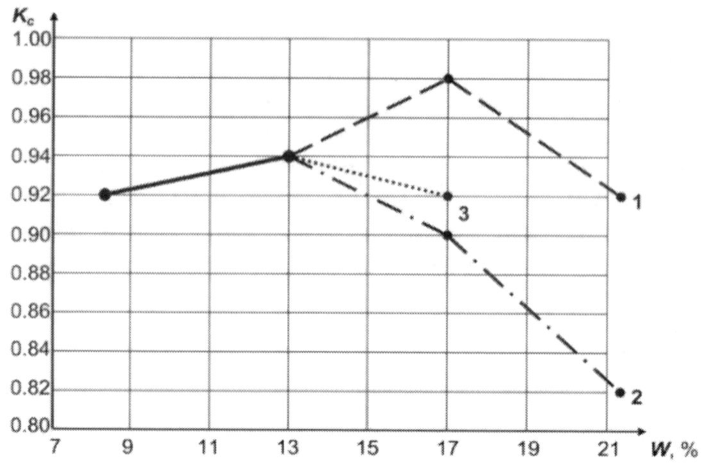


**Fig. 9.** Dependences between compaction factor and dampness with: **1** – tamping; **2** – rolling; **3** – vibrocompaction.

Comparison of special values of compaction factor $K_{ci}$ with maximum $K_c$ = 0.92 is given.

## 5 Conclusions

Conducted research of surface compacting of sandy soil with tamping, rolling and vibrocompaction showed the following:

1. When soil samples moister content is the limits from 9 % to 13 % all abovementioned means of sand compacting were characterised by approximately equal parametres.

2. All investigated means of compacting with sandy soil moister content from 13 % to 15 % showed compaction factor not less than the figure recommended by building regulations, i.e. of 0.92 [10].

3. When soil moister content was equal to 15 % (in the middle of the varied segment of soil samples dampness) compaction factor by tamping is more than the corresponding figures for vibrocompaction on 3 % and by rolling - on 4 %.

4. When sandy soil moister content increased on more than 15 % the most effective means of surface compacting is tamping, the least effective is vibrocompaction, and rolling has the amount of compaction factor below the figure recommended by building regulations [10].

5. Vibrocompaction as a means of surface compacting of sands is effective with dampness up to 17 %. To execute vibrocompaction of sandy soil with big moister content indicators is difficult as in damp and water saturated sands a vibratory load is transferred not only to sand elements but to pore water. Density effect of particles decreases significantly.

5. The most effective according to the results of the conducted experiment was acknowledged the means of compacting by tamping with soil moister content of 17 %. In this case the experiment showed the biggest sand compaction factor – 0.98.

6. Despite the received experimental data while choosing the means of surface compacting of sandy soil on site it should be considered that tamping is connected with the application of a significant in size dynamic pressure to the soil body. This pressure may have a negative dynamic influence on the surrounding soil semi-environment and worsen the parameters of soils. That is why tamping is recommended for application on sites free

from urban planning. In case of necessity in compacting of lower structures, foundations and soils of underground utility systems counterfill, construction and reconstruction of buildings in the constrained considtions on the territories with restrained urban conditions a means producing lower dynamic influence of the foundations of constructions placed nearby should be chosen. This may be a vibrocompaction, dynamic influence from which is more local and does not spread on considerable distances in a soil body.

## Acknowledgements

The authors of the article are grateful to the bachelors of 4 year from the Department of engineering geology, bases and foundations, Faculty of Industrial and Civil Construction, Architecture and Civil Engineering Institute of Samara State Technical University – Sergey Fedorov and Eduard Karimov – for participation in the carrying out of this research.

## References

1. L.G. Goverdovskaya, A.K. Yushantsev, Urban Construction and Architecture **1**, 72-80 (2015) DOI 10.17673/Vestnik.2015.01.12.
2. A.V. Filatova, T.V. Dormidontova, E.S. Wilderotter, V.S. Seliverstov, *Study of the processes and activities for major repairs of a road in the Samara region* (LLC "Consulting company Ucom", Tambov-Samara, 2016) DOI: 10.17117/mon.2016.08.01
3. M.N. Baranova, N.G. Chumachenko, V.V. Turnikov, Urban Construction and Architecture **1**, 80-85 (2014) DOI 10.17673/Vestnik.2014.01.14.
4. S.V. Evdokimov, T.V. Dormidontova, Urban Construction and Architecture **1**, 64-68 (2012) DOI 10.17673/Vestnik.2012.01.12.
5. T.V. Dormidontova, A.V. Filatova, Procedia Engineering **153**, 933-937 (2016) DOI: 10.1016/j.proeng.2016.08.256.
6. T. Dormidontova, S. Evdokimov, N. Shekhova, MATEC Web of Conferences **86**, 04008 (2016) DOI: 10.1051/matecconf/20168604008
7. GOST 12536-2014. *Soils. Laboratory determination of the grain (grain-size) and microaggregate composition* (2014)
8. GOST 22733-2016. *Soils. Laboratory method for determining maximum density* (2016)
9. GOST 5180-2015. *Soils. Methods of laboratory determination of physical characteristics* (2015)
10. SP 45.13330.2012. *Earthworks, bases and foundations. The updated edition of SNiP 3.02.01-87* (2012)
11. V.I. Isaev, A.V. Maltsev, A.A. Karpov, Procedia Engineering **153**, 223-227, (2016) DOI: 10.1016/j.proeng.2016.08.106

# The shear behaviour of pine wood in the Arcan test with the digital image correlation

*Leszek* Malyszko[1,*], *Andrzej* Rutkiewicz[1]

[1]University of Warmia and Mazury in Olsztyn, Faculty of Geodesy, Geospatial and Civil Engineering, Oczapowskiego 2, 10-719 Olsztyn, Poland

**Abstract.** The Arcan shear test is used together with the digital image correlation to study the shear stress-strain relationship for pine wood in the symmetry plane *LR*. The relationship of shearing perpendicular to the grain direction is shown by the straight line below the proportional limit and the nonlinear curve beyond it describing hardening up to the ultimate limit. Subsequent failure modes are shown during the load increase. Additionally, some results of the off-axis tension tests, as well as the uniaxial tension tests, with the mechanical and electric resistance strain gauges are presented to determine and compare the shear modulus from both methods of testing.

## 1 Introduction

Experimental determination of the shear behaviour has always been influenced by difficulties in obtaining a state of the pure and uniform shear in test specimens. Numerical simulations of the test indicate usually a combination of normal and shear stresses making difficult to interpret the pure shear behaviour. Deviations between input values of the shear modulus in numerical simulations and the resulting shear stiffness of the components as well as technical problems lead to development of many measuring techniques, such as e.g. the Arcan test and off-axis tests (Fig. 1). The Arcan test is characterized by a relatively uniform and pure state of shear over its critical section (Fig. 1a). Shear behaviour of clear wood obtained in the Arcan test were studied in [1-3] among others, however in our work the strains are measured by the digital image correlation system. The DIC technique is here introduced to determinate full-scale strain field produced by the test. The off-axis test enables to determine the shear modulus indirectly by measuring the vertical and horizontal strains on an inclined specimen at 45 degrees between the loading and grain direction (Fig. 1b). The determination of the shear modulus by the off-axis test in compression and tension was studied in [4-6] and, combined with the Iosipescu test, in [7,8], respectively.

For certain span widths, the asymmetric four-point bending proved to be applicable for testing of the shear modulus when combined with numerical analysis, although it was found that neither the three-point loading nor asymmetric four-point loading were suitable for determining shear strengths [9]. The ASTM [10] recommends the notched shear block test, while European standards [11] recommend usage of a specimen glued into two steel plates

* Corresponding author: leszek.malyszko@uwm.edu.pl

– both for obtaining the shear strengths. Torsion tests have also been used, although to a limited extent due to the complex underlying theory for obtaining the orthotropic shear characteristic. There are also other techniques, utilized not so often, like the bending ring test [12]. However, the nominal linear shear modulus ought to be usually modified for configuration and measurement effects to obtain more realistic values. Such discrepancies between pure material behaviour and experimental component behaviour are generally inherent in experiments and should be accounted for.

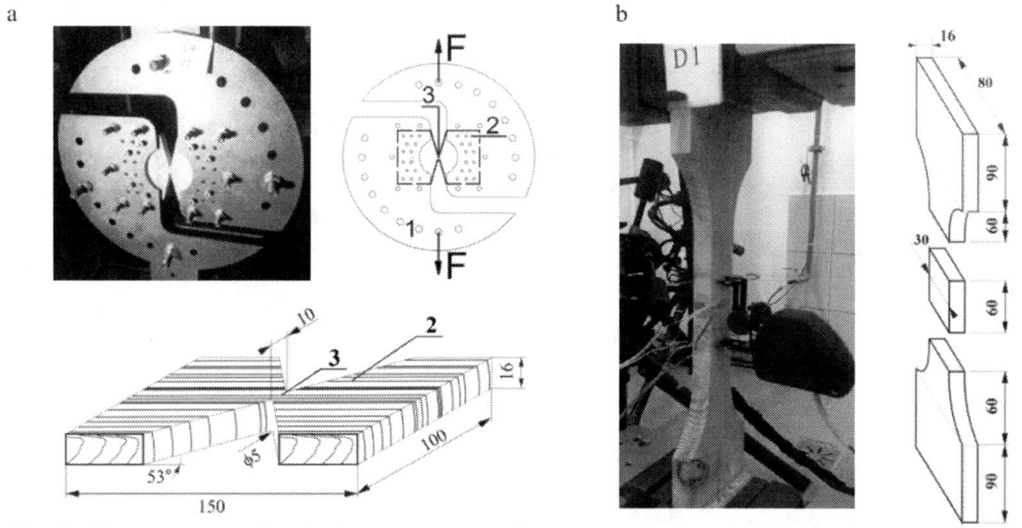


**Fig. 1.** Shear test setup [mm]: a) Arcan test: 1- fixture, 2- specimen, 3- measured area, b) off-axis tension tests.

In the paper the Arcan test is used together of the digital image correlation to study the shear stress – strain relationship of pine wood in the symmetry plane (*LR*). Additionally, some results of the off-axis tension tests are presented to determine the shear modulus of pine wood. Since both tests are used, it is possible to compare the some results and the mechanical behaviour from both methods.

## 2 Elasticity moduli of wood

Timber and many other wood products show an anisotropic mechanical behaviour that is commonly accepted as the technically orthotropic and linearly elastic behaviour for civil engineering purposes. The elastic symmetry planes of orthotropic material are distinguished by the three privileged directions that are perpendicular to each other: the longitudinal (L), tangential (T) and radial (R) direction (Fig. 2). The material orientation can be defined by

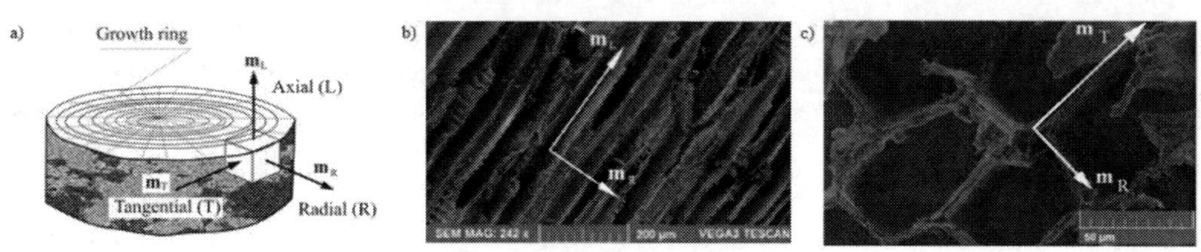


**Fig. 2.** a) Orthotropic material orientation of the tree, b) photos taken in the electron microscope at 242 times enlargement: cellular structure of wood in the L-R plane, c) elongated cells of wood microstructure for pine softwood in the R-T plane at 1340 times enlargement.

the unit normal vectors $\mathbf{m}_L$, $\mathbf{m}_T$ and $\mathbf{m}_R$ locally aligned with the privileged directions. The mechanical characteristics in radial and tangential directions are often assumed to be almost equal and two directions only are often distinguished for design purposes of structural members: longitudinal and perpendicular to the grain directions.

For technical purposes the moduli of elasticity are often used in the frame of reference $(x_L, x_R, x_T)$ that is aligned with the privileged directions, with so-called orthotropic axes. In that case these moduli are called the technical constants of the material and the relations between stresses and strains for the linear orthotropic material is defined by the twelve elastic parameters from which only nine are independent: three Young's moduli $E_L$, $E_R$ and $E_L$, three shear moduli $G_{LR}$, $G_{LT}$ and $G_{RT}$, six Poisson's ratios $\nu_{LR}, \nu_{RL}, \nu_{LT}$ and $\nu_{TL}, \nu_{RT}, \nu_{TR}$. However, the Young's modulus is a function dependent on the direction of the applied load as is a case for the orthotropic material. Moreover, Poisson's ratios and shear moduli can be specified in any plane that is determined by the normal vector or by the two mutually perpendicular vectors. Only the bulk modulus is a scalar. Young's modulus and Poisson's ratio ν are assumed to be constant only for the isotropic material, i.e. they are independent of the load orientation.

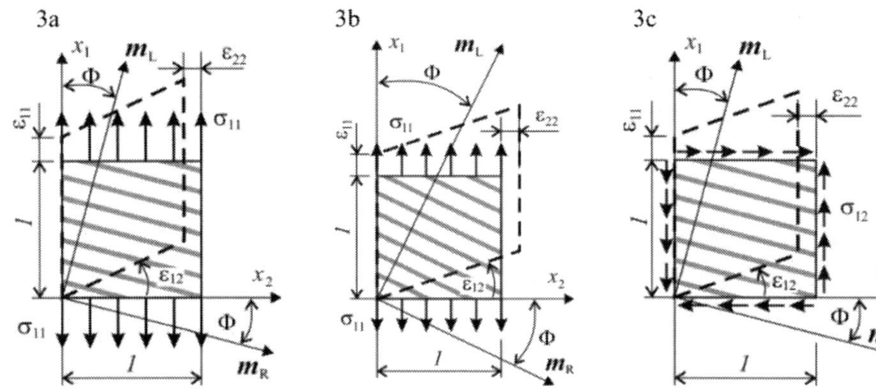


**Fig. 3.** Modes of the deformation for the angle $\Phi$ between the frame of reference $(x_1,x_2)$ and the privileged directions of the orthotropic material: a, b) uniaxial tension in $x_1$ - direction, c) pure shear.

In a state of plane stresses, e.g. in the plane $LR$ $(x_L x_R)$ and if $x_T$ – axis is coaxial with the third privileged direction of the material, i.e. with the third orthotropic axis, we have five technical in-plane moduli of elasticity: two longitudinal moduli of elasticity $E_L$, $E_R$ one shear modulus $G_{LR}$, two Poisson's ratios $\nu_{LR}$, $\nu_{RL}$. Four of them are independent due to the equality $\nu_{LR}/E_L = \nu_{RL}/E_R$. The tension of the orthotropic material in the direction that is not aligned with the orthotropic axes can cause shear strains apart from linear (normal) strains (Fig. 3). Depending on the values of the elastic moduli, the decrease or the increase of the specimen width may occur (Figs. 3a, 3b). In a state of the pure shear, apart from the shear strain component, the normal strain component may also occur (Fig. 3c).

Based on Fig. 3, the experimental moduli of elasticity can be determined in the frame of reference $(x_1, x_2)$ that is not aligned with the orthotropic axes as:

$$E_{11} = \frac{\sigma_{11}}{\varepsilon_{11}}, \quad E_{22} = \frac{\sigma_{22}}{\varepsilon_{22}}, \quad G_{12} = \frac{\sigma_{12}}{\varepsilon_{12}}, \quad \nu_{12} = -\frac{\varepsilon_{22}}{\varepsilon_{11}}, \quad \nu_{21} = -\frac{\varepsilon_{11}}{\varepsilon_{22}}, \tag{1}$$

where the values of stresses and strains are taken directly from the experiment.

We can also write the moduli of elasticity as functions of the angle $\Phi$ between the first axis of the reference frame and the first orthotropic axis of the material in the form:

$$\frac{1}{E(\Phi)} = \frac{1}{E_{11}} = \frac{1}{E_L}\cos^4\Phi + \left(\frac{1}{G_{LR}} - \frac{2\nu_{LR}}{E_L}\right)\sin^2\Phi\cos^2\Phi + \frac{1}{E_R}\sin^4\Phi,$$

$$\frac{1}{G(\Phi)} = \frac{1}{G_{12}} = 2\left(\frac{2}{E_L} + \frac{2}{E_R} + \frac{4\nu_{LR}}{E_L} - \frac{1}{G_{LR}}\right)\sin^2\Phi\cos^2\Phi + \frac{1}{G_{LR}}\left(\sin^4\Phi + \cos^4\Phi\right), \tag{2}$$

$$\nu(\Phi) = \nu_{12} = E(\Phi)\left[\frac{\nu_{LR}}{E_L}\left(sin^4\Phi + cos^4\Phi\right) - \left(\frac{1}{E_L} + \frac{1}{E_R} - \frac{1}{G_{LR}}\right)sin^2\Phi\, cos^2\Phi\right].$$

where the angle $\Phi(\mathbf{x})$ might be dependent on the point coordinates.

From the first and third equation (2) it is possible to obtain the formula for the shear modulus $G_{LR}$ in the following form:

$$G_{LR} = \frac{E(45^0)}{2[1+\nu(45^0)]} = \frac{E_{11}}{2(1+\nu_{12})}. \tag{3}$$

This equation means that the shear modulus can be calculated based on the experimental results of the uniaxial test of the inclined specimen for the $\Phi = 45^0$ if the stress and strain components in the loading direction $\sigma_{11}$ and $\varepsilon_{11}$ as well the strain in the direction perpendicular to the loading axis $\varepsilon_{22}$ are simultaneously measured.

## 3 The Arcan shear test

The specimens for the Arcan test were firstly cut from planks of pine wood with rectangular dimensions of 150 and 100 [mm], density $500\,[kg/m^3]$ and moisture content 9.95 $[\%]$ (Fig. 1a). Afterwards, the notches were cut utilizing a milling-machine with down spindle and a small bus saw. The specimens were strengthened in the grip section utilizing the glass fibre network on an epoxy resin. Tests were conducted on an experimental testing standpoint composed of the testing machine with a special fixture to mount specimens and the digital image correlation system [13]. The test was lead automatically, recording values of load from the testing machine and values of strains from the DIC, with a speed of 1.2 $[mm/m]$. If the direction of the applied load (*F*) is coincident with specimen transverse axis (*R*), a predominant shear stress $\sigma_{LR}$ is generated in the minimum cross-section of the specimen with longitudinal orthotropic axis (*L*) perpendicular to the direction of the applied load. In the test, the normal strain components $\varepsilon_{45}, \varepsilon_{-45}$ were measured by the digital image correlation at the centre of the specimen at an angle of $\pm 45^0$ with regard to its longitudinal axis.

As an example, the raw data measured directly in the Arcan test are illustrated in Fig. 4 for the specimen no 9. In Fig. 4 the load is shown as obtained by the load cell and the $\pm 45^0$ linear strains are measured by the DIC system. The letters *a-e* in Fig. 4 correspond to the different load level and subsequent failure modes. The subsequent failure modes are shown in Fig. 5 as maps of the engineering shear strain calculated by the system of the digital image correlation. In Fig. 5a the undamaged specimen is presented as the reference view for comparison purposes. No visible cracks appeared up to the proportional limit of engineering shear strain of the value *1%*. As the load increased, subsequent failure modes appeared. First, single horizontal crack occurred in the upper-right part of the minimum cross-section (the letter *b* in Figs. 4 and 5). Further increase of the load induced another crack in the lower-left part of the minimum cross-section (the letter *c* in Figs. 4 and 5) – both cracks are caused by tension in the radial direction. The map of the shear strain

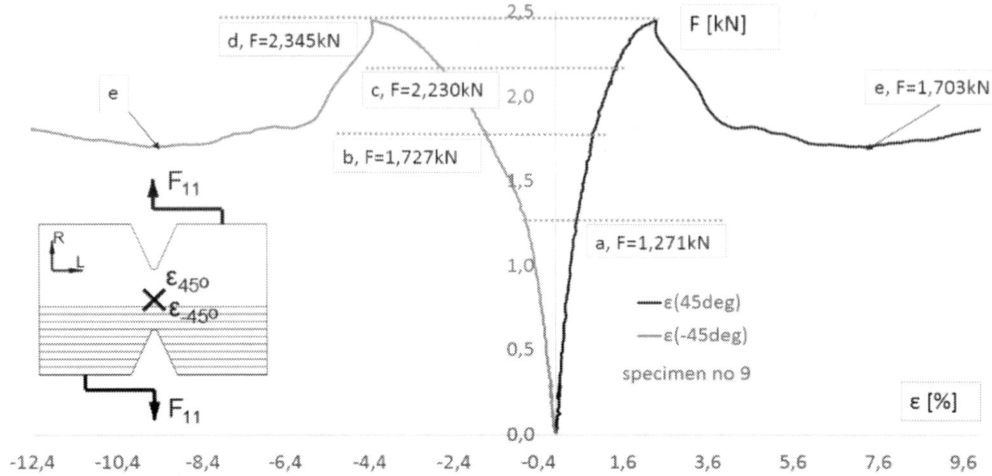


**Fig. 4**. Load as a function of the normal strain components $\varepsilon_{45}, \varepsilon_{-45}$

corresponding to the maximum load is presented in Fig. 5d. Further loading is associated with a visible rotation of the central part of the minimum cross-section, inducting not only the shear stresses but also normal stress components. Figure 5e presents softening of the material and an occurrence of the visible horizontal cracks caused by the compression in the radial direction. Finally, the specimen collapsed in failure modes shown in Fig. 5f and 5g. These views correspond to the modes shown just before and after the collapse of the specimen. The failure mode is visibly caused also by the normal, tensile components in the longitudinal direction of the specimen grain.

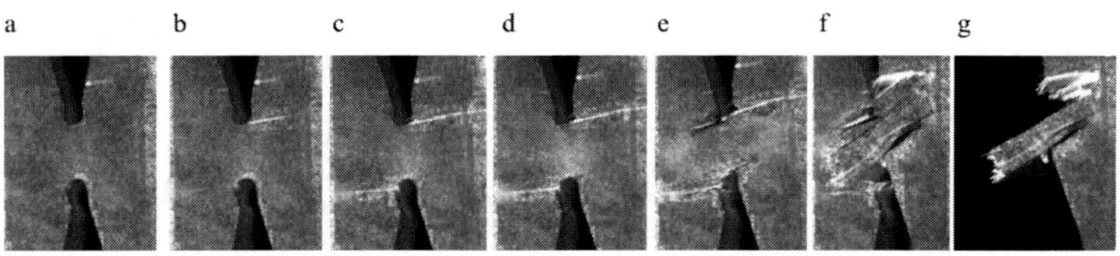


**Fig. 5.** Subsequent failure modes shown by maps of the shear strain from the DIC system.

From the raw data, both the average shear stresses $\sigma_{LR}$ at the cross-section with area $A$ between notches of the specimen and the average engineering shear strains $\gamma_{LR}$ can be determined as:

$$\sigma_{LR} = F_{11}/A, \qquad \gamma_{LR} = \varepsilon_{45} - \varepsilon_{-45} \tag{4}$$

where $F_{11}$ is measured by the load cell of the testing machine.

Figure 6 shows the shear stress-strain curves $\sigma_{LR}(\gamma_{LR})$ for seven specimens, where the average shear stresses $\sigma_{LR}$ and the average engineering shear strains $\gamma_{LR}$ are calculated according to Eq. 4. These non-linear curves are drawn up to the maximum stress obtained in the test. The stress-strain relationships of shearing perpendicular to the grain direction is

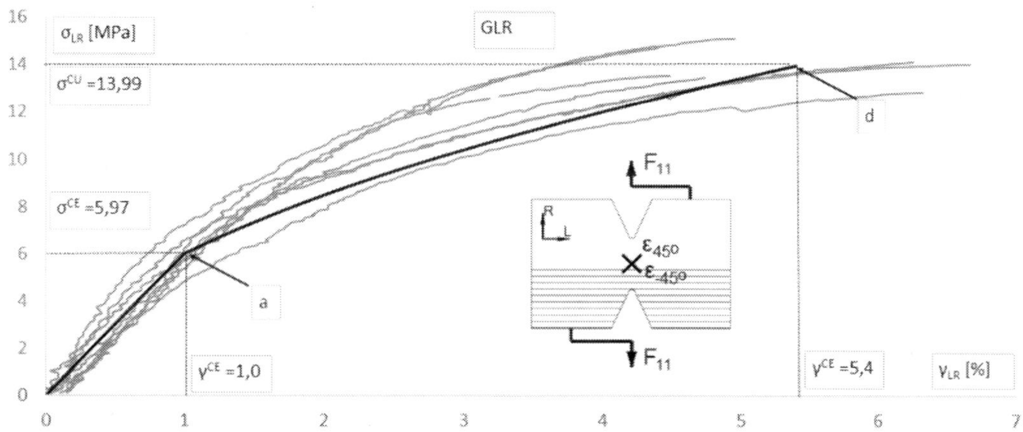


**Fig. 6.** The shear stress $\sigma_{LR}$ as a function of the shear strain $\gamma_{LR}$ for shearing-perpendicular-to-grain.

shown in Fig. 6 by the curve with the proportional limit $(\gamma_{LR}^{ce}, \sigma_{LR}^{ce})$ and the linear brunch below it and the nonlinear hardening brunch beyond it up to the ultimate limit $(\gamma_{LR}^{cu}, \sigma_{LR}^{cu})$. This curve can be expressed analytically for the purpose of the analysis of structural components as [2]:

$$\sigma_{LR}(\gamma_{LR}) = \left\{ \begin{matrix} G_{LR}\,\gamma_{LR} & \text{for } \gamma \le \gamma_{LR}^{ce}, \\ \mu(\gamma_{LR})^m & \text{for } \gamma_{LR}^{ce} \le \gamma_{LR} \le \gamma_{LR}^{cu} \end{matrix} \right\}, \qquad m = \frac{\ln(\sigma_{LR}^{ce}/\sigma_{LR}^{cu})}{\ln(\gamma_{LR}^{ce}/\gamma_{LR}^{cu})}, \qquad \mu = \frac{\sigma_{LR}^{ce}}{(\gamma_{LR}^{ce})^m}. \tag{5}$$

In the relationships (5) we can distinguish the linear brunch with the slope of the corresponding elastic shear modulus $G_{LR}$ and nonlinearity can be modelled by employing the power function of which the material constants $m$ and $\mu$ may be determined following the continuum conditions at the points of proportional and ultimate limits. For the pine wood from the experiment the following values are obtained: $G_{LR} = 600\,[MPa], m = 0.5$ and $\mu = 6.0\,[MPa]$. The modulus $G_{LR} = 660\,[MPa]$ for the strain levels $\gamma_{LR} = 0.05 \div 0.20\%$.

## 3 The off-axis test

The specimens for off-axis test were cut in a milling-machine with a down spindle as a full specimen using a stencil from pine planks. Afterwards the handlers were separated. Blocks of 45 degree angle were cut as slats 30 [mm] wide on a buzz saw and afterwards cut on 60 [mm] long blocks and glued into handlers using polyurethane glue used for timber constructions in civil engineering. The tests were conducted on an experimental testing standpoint composed of the testing machine with the mechanical strain gauge in the $x_1$ direction of tension and the electric resistance strain gauge bridge in the perpendicular $x_2$ direction (Fig. 1b). The test was conducted manually, step by step, recording values of load from the testing machine and values of strains from the mechanical and electric resistance strain gauges.

Figure 7 shows the shear stress-strain curve $\sigma_{12}(\gamma_{12})$ for the $\Phi = 45^0$ specimen obtained in the off-axis test based on the Eq. 3. This almost linear curve is drawn based on the formula $\sigma_{12} = G_{12}\,\gamma_{12}$, where the shear stress component $\sigma_{12}$ is calculated as $\sigma_{12} = 0.5\sigma_{11}$

and the engineering shear strain $\gamma_{12}$ is calculated as the difference between the normal strain components in the loading $x_1$-direction $\varepsilon_{11}$ and in the perpendicular $x_2$-direction $\varepsilon_{22}$,

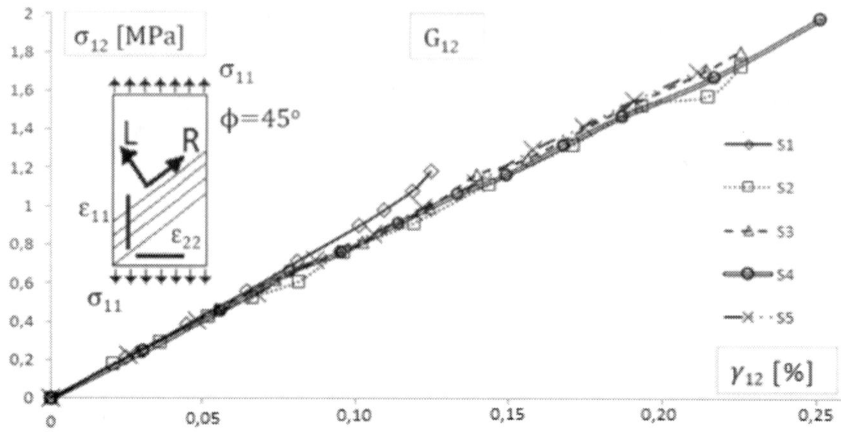


**Fig. 7.** The shear stress $\sigma_{LR}$ as a function of the shear strain $\gamma_{LR}$ in the off-axis tension test.

i.e. $\gamma_{12} = \varepsilon_{11} - \varepsilon_{22}$. To compare results with those from the digital image correlation, the values of the Young's modulus $E_{11}$ and Poisson's ratio $\nu_{12}$ were calculated based on Eq. 3 as a secant between the two strain levels of $\gamma_{12}$: 0.05% and 0.20%. Values obtained in the off-axis test and in the other tension tests, i.e. tension at longitudinal ($\Phi$=0$^0$) and radial ($\Phi$=90$^0$) directions, are presented in Tab.1.

**Table1.** Mean value of elastic moduli in off-axis tension test (five specimens)

| $E_{11}$ [MPa] | $\nu_{12}$ [ - ] | $G_{LR}$ [MPa] | $E_L$ [MPa] | $E_R$ [MPa] | $\nu_{LR}$ [ - ] | $\nu_{RL}$ [ - ] |
|---|---|---|---|---|---|---|
| 1850 | 0,154 | 800 | 14 240 | 1310 | 0,351 | 0,032 |

The value of the shear modulus calculated form the Eq. 3 $G_{12} = 800\,[MPa]$ is greater than value $G_{LR} = 660\,[MPa]$. The ratio is $G_{LR}/G_{12} = 660/800 = 0.825$.

## 5 Conclusions

In the work the application of the Arcan shear test and off-axis tension test, as well as other uniaxial tests, for measuring the shear module and other modules of pine wood were investigated in the longitudinal-radial (LR) plane of symmetry. For the identification of the LR moduli in the off-axis and uniaxial tension tests a two linear strain gauges were glued at $x_1$ and $x_2$ directions on each specimen – a total of fifteen specimens were investigated. Identification of the shear moduli in the Arcan test was determined using the digital image correlation system, where the ±45$^0$ linear values of strains were measured in the critical cross-section. Additionally, a full shear angle maps were presented for chosen stages of loading, referring to the characteristic points of the force-strain curves. The loading was conducted up to the moment of failure with full specimen share, which is unusually meet in the literature for the LR shear direction. A $\sigma_{LR}(\gamma_{LR})$ curve have been fit for mean values obtained in the test. A total of seven specimens were investigated.

The values of the LR moduli obtained in the tension tests (table 1) were compared with the literature - see [14] where comparable and close values of the moduli were obtained for the pine slash specie. However, wood is a variable material and many aspects are influencing its mechanical characteristics, such as growth conditions and climate, utilized measuring techniques and standards, and these must be taken into account when comparing the results.

The value of shear moduli obtained in the Arcan shear test is lower than obtained in the off-axis test (GLR/G12=0,825). The difference may occur, excluding the obvious differences in the testing methods, due to the fact that the height of the critical cross-section was relatively small, which affected the overall stiffness of the specimen and the central cross-section acted more or less closely to the joint. Manufacturing such small cross-section had the purpose of conducting the survey up to the moment of failure with full specimen share. The full process was recorded with the digital image correlation system, which proved to be efficient and usable for such purposes.

Further survey must consist, due to the relatively thick specimens, measurements on both surfaces and a larger number of specimens for proper statistical analysis and comparison of obtained results. In the further tests, the specimen shape and dimensions must be accurately analyzed.

## References

1. J.C. Xavier, N.M. Garrido, M. Oliveira, J.L. Morais, P.P. Camanho, F. Pierron, Compos. Part A-Appl. S., **35**, 827 (2004)
2. D. Huang, Y. Bian, A. Zhou, B. Sheng, Constr. Build. Mater., **77**, 130 (2015)
3. K.B. Dahl, K.A. Malo, Compos. Sci. Technol., **69**, 2144 (2009)
4. J.R. Aira, F. Arriaga, G. Iniguez-Gonzalez, Biosyst. Eng., **126**, 12 (2014)
5. N.T. Mascia, L. Vanalli, Constr. Build. Mater., **30**, 522 (2012)
6. G.Y. Jeong, M.J. Park, Constr. Build. Mater., **113**, 864 (2016)
7. G. Odegard, M. Kumosa, Compos. Sci. Technol., **60**, 2917 (2000)
8. J. Xavier, M. Oliveira, J. Morais, T. Pinto, Holzforschung, **63**, 217 (2009)
9. H. Yoshihara, Holzforschung, **63**, 211 (2009)
10. ASTM D143: Standard *Methods of Testing Small Clear Specimens of Timber, American Society for Testing and Materials,* 42 (1984)
11. EN 408: *Timber structures – Structural timber and glued laminated timber – Determination of some physical and mechanical properties* (2010)
12. J.J. Garcia, Ch. Rangel, K. Ghavami, Constr. Build. Mater., **31**, 52 (2012)
13. *Aramis v6 User Manual* (GOM GmbH, Mittelweg 7-8 D-38106 Braunschweig, Germany)
14. D.W. Green, J.E. Winandy, D.E. Kretschmann, *Forest Products Laboratory* (Department of Agriculture, Forest Service, Madison, USA, 1999)

# Soil plasticity with a different porosity

*Sergii* Klovanych[1], *Leszek* Malyszko[1,*]

[1]Faculty of Geodesy, Geospatial and Civil Engineering
University of Warmia and Mazury in Olsztyn,ul. Prawocheńskiego 15, 10-720 Olsztyn, Poland

**Abstract.** The model of soils with different porosity in the framework of the associated theory of plasticity is presented The single analytical function describes the loading surface in the stress space. The deformational hardening/softening and the phenomenon of dilatancy during plastic flow are incorporated in the model. The triaxial compression tests are simulated and compared with the experimental results for different values of the void ratio and initial hydrostatic stresses.

## 1 Introduction

In accordance with the classical plastic flow theory total strain rate $\dot{\boldsymbol{\varepsilon}}$ is decomposed into an elastic part $\dot{\boldsymbol{\varepsilon}}^e$ and a plastic part $\dot{\boldsymbol{\varepsilon}}^{pl}$, i.e. $\dot{\boldsymbol{\varepsilon}} = \dot{\boldsymbol{\varepsilon}}^e + \dot{\boldsymbol{\varepsilon}}^{pl}$. The plastic strains $\dot{\boldsymbol{\varepsilon}}^{pl}$ are determined by the flow rule $\dot{\boldsymbol{\varepsilon}}^{pl} = \dot{\lambda}\mathbf{g}$, where $\mathbf{g}$ is the second-order tensor of the plastic flow direction that is normal to a plastic potential surface in a stress space and the plastic multiplier $\dot{\lambda}$ determines the magnitude of the plastic strains. Adopting other governing equations, the stress rate $\dot{\boldsymbol{\sigma}}$ can be finally written in terms of the total strain rate $\dot{\boldsymbol{\varepsilon}}$ as

$$\dot{\boldsymbol{\sigma}} = \mathbf{D}^{ep} \cdot \dot{\boldsymbol{\varepsilon}} \equiv \left[ \mathbf{D} - \frac{(\mathbf{D} \cdot \mathbf{f}) \otimes (\mathbf{D} \cdot \mathbf{g})}{\mathbf{f} \cdot \mathbf{D} \cdot \mathbf{g} + H} \right] \cdot \dot{\boldsymbol{\varepsilon}}, \tag{1}$$

where $\mathbf{D}$ is the elastic stiffness tensor. The double contractions of tensors are denoted by one dot, the symbol $\otimes$ denotes a tensor product and the tensor of tangent elastoplastic modulus or the elastoplastic continuum fourth-order operator $\mathbf{D}^{ep}$ is generally no symmetric for arbitrary $\mathbf{g}$, except the case of an associative flow rule $\mathbf{f} = \partial f / \partial \boldsymbol{\sigma} = \mathbf{g}$ which is here discussed.

Depending on the hardening/softening modulus $H$ and the loading function $f$ or the gradient tensor **f**, the different material response can be obtained in the constitutive equation (1), see [1].

For isotropic materials a surface of the loading function $f$ is usually given in the stress space and can be convenient expressed in terms of scalar invariants of the stress tensor $\boldsymbol{\sigma}$. Using the deviatoric stress tensor $\mathbf{s}$ and the standard decomposition of the stress tensor into

* Corresponding author: leszek.malyszko@uwm.edu.pl

its spherical and deviatoric parts $\boldsymbol{\sigma}=\frac{1}{3}(tr\boldsymbol{\sigma})\,\mathbf{I}+\mathbf{s}$, we write the scalar invariants as

$$p=\frac{1}{3}tr\boldsymbol{\sigma}\,,\qquad q=\sqrt{\frac{3}{2}tr\mathbf{s}^2}\,,\qquad \zeta=\cos 3\theta=\frac{\sqrt{6}\,tr\mathbf{s}^3}{\sqrt{(tr\mathbf{s}^2)^3}}\quad, \tag{2}$$

where the invariant $p$ is equal to the mean effective stress or the octahedral normal hydrostatic stress, $q$ is the equivalent shear stress, $tr\boldsymbol{\sigma}$ is the first principal invariant of the stress tensor, $\frac{1}{2}tr\mathbf{s}^2$, $\frac{1}{3}tr\mathbf{s}^3$ are the second and third principal invariants of the stress deviator, $\theta$ is the Lode angle, $\zeta$ is the Lode parameter in the stress space. The equivalent shear stress reduces to $q=|\sigma_1-\sigma_3|$ for triaxial stress states with $\sigma_2=\sigma_3$, see [2].

## 2 The loading function

The shape of the loading function surface is illustrated in Fig.1, where its cross-sectional trace in the deviatoric plane and its meridians in the meridian planes are shown. The function is continuous, smooth and convex. Since the surface is closed at both ends, it entirely accounts for effect of the hydrostatic pressure. The apexes of the surface on the hydrostatic axis are assumed to pass through the point $p_c$ for the compressive hydrostatic stress and the point $p_s$ for the tensile uniform triaxial stress. Using the stress invariants, the loading function is expressed mathematically by the following simple analytical expression

$$f(p,q,\zeta)=q-A_o(\zeta)(p_s+p)\sqrt{p_c-p}\,, \tag{3}$$

where $A_o(\zeta)$ is a function of the Lode parameter. For the point $(p,q,\zeta)$ that corresponds to the current load level $f(p,q,\zeta)=0$ and $q=A_o(\zeta)(p_s+p)\sqrt{p_c-p}$.

The traces of the standard Drucker-Prager cones on the meridian planes are also shown in Fig. 1 as the straight lines in the $p-q$ plane, which can be described algebraically by linear equations with the two material constants. Using the most common approximations obtained by making the failure surfaces of the Drucker-Prager and Mohr-Coulomb criteria coincident either at the outer or inner edges of the Mohr-Coulomb surface, these constants can be matched to the cohesion $c$ and the friction angle $\varphi$ of the Mohr-Coulomb failure criterion. Since the ratio $q_{to}/q_{co}$ of the two characteristic lengths, the tensile length $q_t=q_{to}$ and compressive length $q_c=q_{co}$, for any deviatoric cross sections of the Mohr-Coulomb

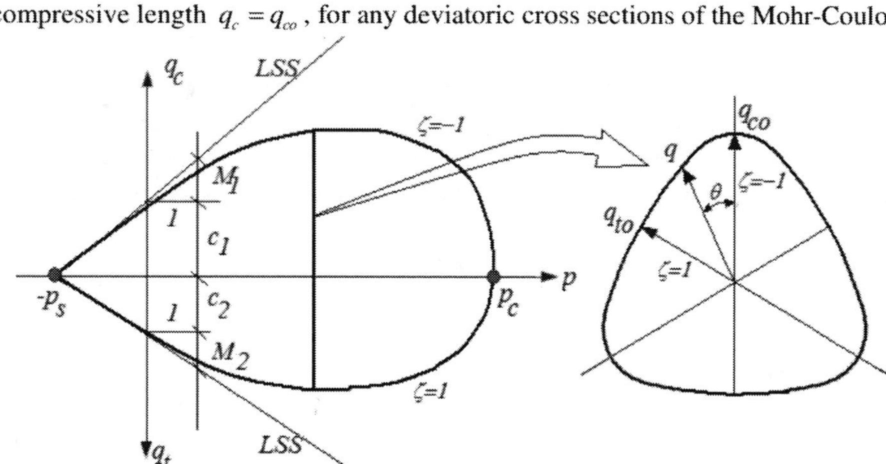


**Fig. 1.**Cross-sectional shape of the loading surface, LSS – Drucker-Prager limit state surface

surface is constant, i.e.

$$g=\frac{q_{to}}{q_{co}}=\frac{3-\sin\varphi}{3+\sin\varphi}, \tag{4}$$

the two surfaces of the Drucker-Prager and Mohr-Coulomb failure criteria can be made to agree along the compressive meridian or along the tensile meridian, if they are matched at their apex. For the outer compression cone, which passes through the compressive meridian where $\zeta=-1(\theta=0)$ , we have the linear equation $q_1=M_1p+c_1$ with the material constants

$$M_1=2\sqrt{2}\frac{\sin\varphi}{3-\sin\varphi},\quad c_1=2\sqrt{2}\frac{\cos\varphi}{3-\sin\varphi}c. \tag{5}$$

On the other hand, for the inner extension cone, which passes through the tensile meridian where $\zeta=1(\theta=\pi/3)$, we have the linear equation $q_2=M_2p+c_2$ with two sets of material constants $(M_2,c_2)$ and $(c,\varphi)$ related by

$$M_2=2\sqrt{2}\frac{\sin\varphi}{3+\sin\varphi},\quad c_2=2\sqrt{2}\frac{\cos\varphi}{3+\sin\varphi}c. \tag{6}$$

In the loading function (3) we have curved meridians. The trace of the failure surface on deviatoric sections is shown in Fig. 1. It is neither circular nor irregular hexagon and depends on the angle $\theta$ (or the Lode parameter $\zeta$ ). The cross-sectional shape of the loading surface on the deviatoric plane is a curvilinear triangle and it has the standard threefold symmetry with the maximal radius $q_c$ on the compressive meridian ( $\zeta=-1$ ) and the minimal radius $q_t$ on the tensile meridian with ( $\zeta=1$ ). Between these two limiting cases, for $-1<\zeta<1$ , the following simple approximation from [3] is used

$$\rho(\zeta)=\frac{2g}{1+g+(1-g)\zeta} \tag{7}$$

and the function $A_o(\zeta)$ in (3) can be written as

$$A_o(\zeta)=\frac{\rho(\zeta)M_1}{\sqrt{p_c+p_s}}. \tag{8}$$

The loading function (3) can be matched to the Drucker-Prager failure criterion which can serve as the limit state surface. Matching the coordinates of the apexes we will obtain $p_s\approx c_1/M_1$. From the condition $\partial q/\partial p=\rho(\zeta)M_1$ for $p=-p_s$ it is possible to obtain the coordinate of the point $p_c$ as

$$p_c=\frac{a^2p_s+p}{1-a^2},\qquad a=\frac{q}{\rho(\zeta)M_1(p+p_s)}. \tag{9}$$

It should be noted that the point $(p,q,\zeta)$ of the current loading corresponds to the softening state of the material if it lies outside the limit surface of the Drucker-Prager criterion.

## 3 Hardening and softening behaviour

The hardening/softening modulus $H$ in the constitutive equation (1) can be calculated in the coordinate system $(p,q,\zeta)$ by means of the matrices $[\mathbf{p}_\sigma]=[p,q,\zeta]^T$ and $[\mathbf{f}]$, see [1]

$$H=-\frac{\partial f}{\partial\kappa}[\mathbf{p}_\sigma]^T[\mathbf{f}], \tag{10}$$

where the scalar $\kappa$ is introduced as the internal measure for hardening/softening of the plastic flow. In the case of work hardening/softening, the scalar $\dot{\kappa}$ is the plastic work measure $\dot{\kappa}=\boldsymbol{\sigma}\cdot\dot{\boldsymbol{\varepsilon}}^{pl}$. Substituting $\kappa$ in the expression (10), we will obtain

$$H=-[\mathbf{f}]^T[\bar{\mathbf{D}}_p]^T[\mathbf{f}] \text{ and } [\bar{\mathbf{D}}_p]=\begin{bmatrix}3K_h & 0 & 0\\ 0 & G_s & 0\\ 0 & 0 & 1\end{bmatrix}, \tag{11}$$

where $3K_h=\partial p/\partial\varepsilon^{pl}$ and $G_s=\partial q/\partial\gamma^{pl}$ are plastic octahedral compressive and shear modules, respectively. We assume that the soil hardening is characterized by the modulus $K_h$ whereas the soil softening by the modulus $G_s$.

We also assume that the void ratio $e$ is a basic physical characteristic of soil behaviour and can be taken as a hardening internal variable. This variable is changing during the triaxial loading and is indirectly dependent on the octahedral normal stress $p$. The strength of soils increase as the void ratio decreases. Based on the experimental consolidation line shown in Fig. 2 in the $e$–$\ln p$ plane (the virgin compression line), we have the equation $e=e^0-\mu\ln p$ during the active loading and the equation $e=e^A-\delta\ln p$ during unloading, where $e^0$ is the initial void ratio, $e^A$ is the void ratio at the time of unloading, $\mu$ and $\delta$ are the experimental parameters, see [4,5]. The volumetric strain $\varepsilon_V=3\varepsilon$ caused by the change of the void ratio can be written in the form $\varepsilon_V=-(e-e_0)/(1+e)$. Considering the standard decomposition $\dot{\varepsilon}^{pl}=\dot{\varepsilon}-\dot{\varepsilon}^{el}$, where $\dot{\varepsilon}^{el}$ is the elastic strain, we can write the volumetric plastic strain as

$$\dot{\varepsilon}_V^{pl}=\frac{\mu-\delta}{1+e}\dot{p}. \tag{12}$$

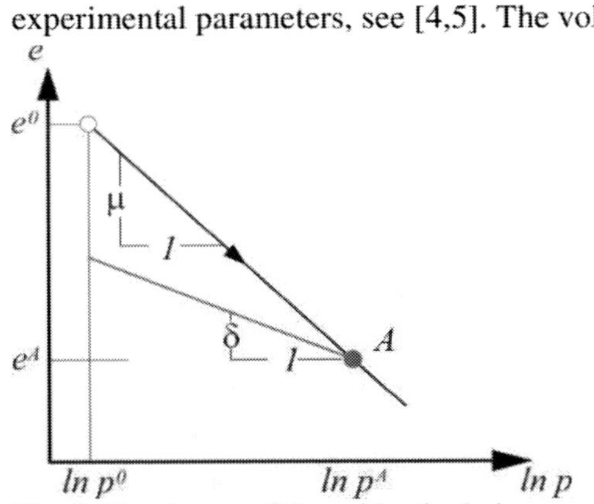

**Fig. 2.** The change of the void ratio during hydrostatic loading/unloading

Based on the relationship (12) the derivative in the hardening modulus (9) can be written as

$$\frac{\partial f}{\partial\kappa}=-\frac{3(1+e)}{\mu-\delta}\frac{\partial f}{\partial p} \text{ and } K_h=\frac{1+e}{\mu-\delta}. \tag{13}$$

If the stress state lies outside of the limit state surface (Fig. 1), it cannot be accepted as a solution because of violation of the classical Drucker's postulate about the positive dissipation of energy during the plastic deformation. This is a range of damaged behaviour of the soil that is caused by octahedral shear strains when octahedral shear stresses exceed limit values. However, we can observe in practice that the process of soil deformations continues as a softening of the material. In the $(\gamma-q)$-plane (Fig. 3b), the softening of the soil behaviour is represented by a descending branch of a curve. The derivative of $f$ in the expression (13) can be thus rearrange into a function of new variables as

$$\frac{\partial f}{\partial \kappa}=\frac{1}{q}\frac{\partial f}{\partial q}\frac{\partial q}{\partial \gamma}=\frac{G_s}{q}\frac{\partial f}{\partial q}=\frac{1}{q}\frac{G_o G_t}{G_o-G_t}\frac{\partial f}{\partial q}, \tag{15}$$

where $G_t$ is the tangent modulus of the octahedral shear strain and $G_s$ is the tangent plastic modulus. To determine the modulus $G_t$, the curve in the $(\gamma-q)$-plane can be again used, for example with the following analytical expression for the descending and ascending branch

$$\xi=\frac{q}{\hat{q}}=\frac{\kappa\eta}{1+(\kappa-2)\eta+\eta^2}, \tag{16}$$

where $\eta=\gamma/\hat{\gamma}$, $\hat{\gamma}=\kappa\hat{q}/(2G_o)$ and peak values of the curve are denoted by $\hat{\gamma}$ and $\hat{q}$.
As shown in Fig. 3a, if the current stress state with coordinates $(p, q)$ lies outside of the limit surface (the point *1* in Fig. 3a), the softening behaviour, associated with the descending branch of the curve in the $(\gamma-q)$-plane, should be determined by the strain state of soil, e.g. by the octahedral shear strain. The corresponding stress state, as defined by the curve, would be below the limit stress state. This formally means a return to the inside of the surface of the Drucker-Prager yield criterion (the point *2* in Fig. 3a).

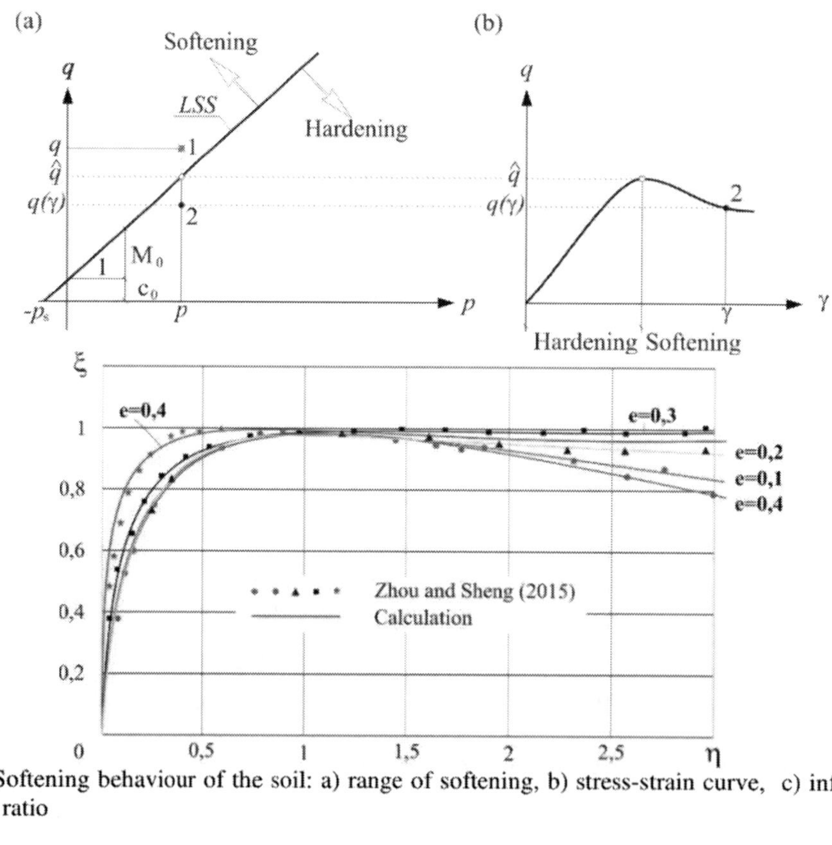


**Fig. 3.** Softening behaviour of the soil: a) range of softening, b) stress-strain curve, c) influence of the void ratio

The parameter $\kappa$ in the equation (15) characterizes the secant modulus of the shear strain. It depends on the void ratio of the soil. Based on an analysis of experimental data [6,7] the following expression can be recommended

$$\kappa = \kappa_p - (\kappa_p - 1)\exp(-0.1\, e) \tag{16}$$

where $\kappa_p$ is an experimental parameter ($\kappa_p = 2 \div 6$). Figure 3c shows curves drawn per the expression (15) with (16) for different values of $e$ and with a comparison to the experimental data from [7].

Taking the derivative of the function (15) with respect to $\gamma$, we will receive

$$G_t = \Omega G_0 \tag{17}$$

where the nonlinearity function $\Omega$ equals to

$$\Omega = \frac{1-\eta^2}{\left[1+(\kappa-2)\eta+\eta^2\right]^2}. \tag{18}$$

## 4 Dilatancy

The phenomenon of dilatancy during plastic flow is observed for many geomaterials. The associative plasticity usually employs as flow potential the classical yield function of the standard Drucker-Prager cone. However, the description of dilatancy in the framework of the theory of the plastic flow based on the relationship (1) is impossible because of violation of the classical Drucker's postulate about the positive dissipation of energy during the plastic deformation. To derive the appropriate flow rule one can write the plastic flow law by means of two equations of stress invariants, the one for the volumetric plastic strain and the second one for the octahedral shear plastic strain in the form [8]

$$\dot{\varepsilon}^p = \lambda_1 \frac{\partial f}{\partial p}, \quad \dot{\gamma}^p = \lambda_2 \frac{\partial f}{\partial q}. \tag{19}$$

Hence, considering the expression ($13_2$), we can write the following relationship

$$k_d = \frac{\dot{\varepsilon}_p}{\dot{\gamma}_p} = \frac{\lambda_1\, \partial f/\partial p}{\lambda_2\, \partial f/\partial q} = \lambda_o \frac{\partial f}{\partial p} = \lambda_o \frac{p_s - 2p_c + 3p}{2(p_s + p)(p_c - p)} q, \tag{20}$$

where for the materials without dilatancy we have $\lambda_0 = 0$ and for the dilatant materials we have $\lambda_0 > 0$. The expression of the parameter $\lambda_o$ as a function of the initial void ratio $e$ can be established by means of the experimentally determined results from [2] and [7].The following expression can be recommended

$$\lambda_o = \frac{0.0025}{1-0.95\, e} \tag{21}$$

## 6 Validation

The use of the model for the soil mechanics problems is validated next by a comparison with results available in the literature. In the first example, drained triaxial tests from [2] are investigated. The results of experimental data from [7] are compared with calculations of the own model for three types of soils with different density state: a loose state, a medium state and a dense state. The material parameters of the model used in the calculations are given in Table 1. The analyses are carried out with loading control and a step size of the value $\Delta\sigma_z = 0.02\,[MPa]$ until reaching the critical state of the soil. In the first step, the hydrostatic stress state is introduced with the value of $\sigma_h = 0.1\,[MPa]$.The comparison between numerical and experimental results is given in Fig. 4 in diagrams of the difference of principal stresses versus vertical axial strain (Fig. 4a) and the volumetric strain-vertical axial strain diagrams (Fig. 4b). Good agreement is found. The response of the model in the triaxial compression test is quite realistic.

**Table 1.** Material parameters for triaxial tests

| Density state | $E\,[MPa]$ | $\nu\,[-]$ | $\varphi\,[^0]$ | $c\,[MPa]$ | $\mu\,[-]$ | $\delta[-]$ | $e[-]$ |
|---|---|---|---|---|---|---|---|
| Dense | 40 | 0.2 | 40 | 0 | 0.09 | 0.05 | 0.3 |
| Medium | 30 | 0.2 | 35 | 0 | 0.09 | 0.05 | 0.8 |
| Loose | 20 | 0.2 | 30 | 0 | 0.09 | 0.05 | 0.9 |

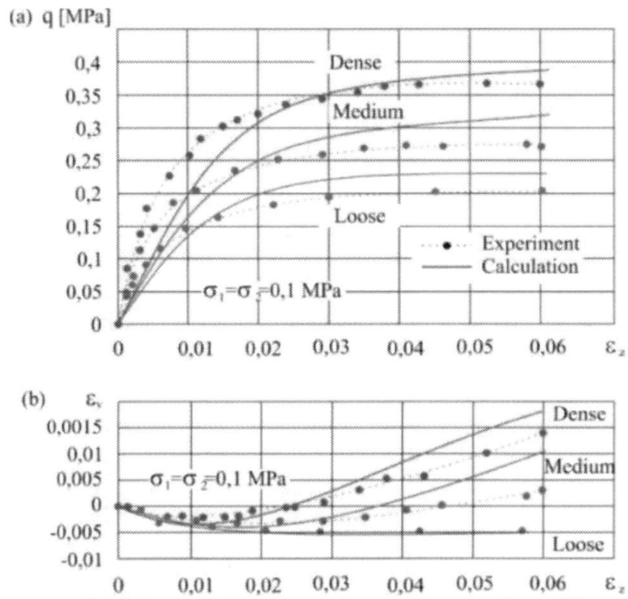


**Fig.4.** Comparison between plasticity model and experimental results from [7]: (a) equivalent shear stress versus vertical strain; (b) volumetric strain versus vertical strain.

In the second examples, deformations of the porous soil under the triaxial compression are investigated for various values of the initial hydrostatic stress. The comparison with the experimental results from [9] is shown in Fig. 5 in the similar diagrams as in the first example. The following material parameters are taken: $E = 9.75\,[MPa]$, $\nu = 0.25$, $\varphi = 34^o$, $c = 0\,[MPa]$, $\mu = 0.1$, $\delta = 0$, $e = 0.61$. The numerical analyses are conducted for the four values of the initial hydrostatic stress: $\sigma_h = 0.1; 0.3; 1.05; 2.0\,[MPa]$ in the first step of the loading. The calculations are carried out with loading control and a step size of the value $\Delta\sigma_z = 0.02\,[MPa]$ in the subsequent steps until reaching the critical state of the soil. Good agreement is again found. The response in the triaxial compression test confirmed utility of the model.

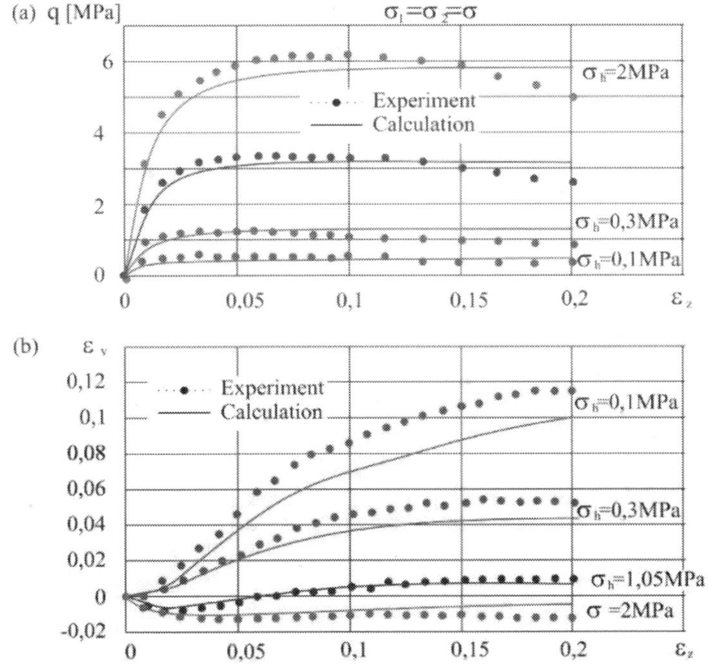


**Fig.5.** Comparison between plasticity model and experimental results from [9]: (a) equivalent shear stress versus vertical strain; (b) volumetric strain versus vertical strain.

## References

1. O. Zienkiewicz, A.H.C. Chan, M. Pastor, T. Shiomi, *Computational Geomechanics with special Reference to Earthquake Engineering* (London: John Wiley & Sons, 1999)
2. R.B.J. Brinkgreve, P.A. Vermeer, *Plaxis. Finite Element Code for Soil and Rock Analysis. Materials Models Manual. Part 3* (Rotterdam: Balkema, Brookfild, 1998)
3. J.H. Argiris, G. Faust, J. Szimmat, P. Warnke, K. Willam, Nuclear Engineering and Design, **28**, 42 (1974)
4. K. J. Roscoe, A. N. Schofield, *Proc. European Conf. on Soil Mechanics and Foundation Engineering,* **1**, 47 (1963)
5. K. Terzaghi, R.B. Peck. *Soil Mechanics in Engineering Practice*. 2nd ed. (New York: John Wiley &Sons, 1967)
6. D. Masin, *Int. J. Num. Anal. Meth. Geomech.*, **34**, 73 (2010)
7. A. Zhou, D. Sheng, *Comput. Geotech.*, **63**, 46 (2015)
8. R. de Borst, A. E. Groen. 2 *Computational Strategies for Standard Soil Plasticity Models. Modeling in Geomechanics* (Edited by Zaman M, Booker J, Gioda G., John Wiley & Sons, 2000).
9. K.L. Lee, H.B. Seed, *J. Soil. Mech. Found. Div.*, ASCE, **93(SM6)**, 117 (1967)

# A numerical solution for geometrically nonlinear bending plates problems subjected to local-loads

*Ali M.* Mansour[1,*], *Radek* Gabbasov [2], and *Vladimir* Filatov[3]

[1, 2, 3]Moscow state university of civil engineering, Yaroslavskoye shosse, 26, Moscow, Russia, 129337

**Abstract.** the following scientific content is demonstrating a new proposed numerical procedure for calculating deformations & stresses generated on geometrically nonlinear flexible (*thin*) plates under local loads. solving the predominating system of differential equations is based on generalized equations of the finite difference method. the proposed algorithm is verified by an example of solving a hinged-edges plate loaded with a point load centrally.

## Through manuscript

Nowadays, Flexible (*thin*) plates and shells are widely used in modern technology and by many various aspects such as aircraft construction, ship-building, engineering, energy, construction and much many others. The most common and universal numerical method for calculating such structures (*including nonlinear calculations*) is the extra-famous finite element method (*FEM*) [**1-13**]. At the same time, engineers are always in need to compare the results obtained by (*FEM*) (*especially for performed nonlinear calculations*), with the results of solutions obtained by other methods [**10, 13**]. Down-below, authors are suggesting a numerical technique for doing rectangular plates calculations with the aid of generalized equations of finite differences method [**15**]. As well-known [**14**], for geometrically nonlinear rectangular thin-plates, the next $4^{th}$ order differential equations system is considered to be the predominating one, which formulation can reduce to the solution to as follows:

$$\nabla^2\nabla^2 W = \frac{H}{D}\left(\frac{\partial^2\Phi}{\partial y^2}\frac{\partial^2 W}{\partial x^2} + \frac{\partial^2\Phi}{\partial x^2}\frac{\partial^2 W}{\partial y^2} - 2\frac{\partial^2\Phi}{\partial x\partial y}\frac{\partial^2 W}{\partial x\partial y}\right) + \frac{q}{D}; \quad (1)$$

$$\nabla^2\nabla^2\Phi = E\left[\left(\frac{\partial^2 W}{\partial x\partial y}\right)^2 - \frac{\partial^2 W}{\partial x^2}\frac{\partial^2 W}{\partial y^2}\right] \quad (2)$$

<u>Where</u> $W$- the deflection function; $\Phi$ - the stress function; $H$ - Plate thickness; $D$ - cylindrical rigidity (*stiffness*).

* Corresponding author: alaa_for_all@hotmail.com

Now, introducing dimensionless definitions for coordinates & parameters as:

$$\xi = \frac{x}{a};\ \eta = \frac{y}{a};\ w = \frac{W}{a};\ m = \frac{M}{qa^2};\ M = \frac{M_x + M_y}{1+\mu};\ p = \frac{q}{q_0} \tag{3}$$

Where $\mu$ - Poisson's ratio; $q_0$ - the value of $(q)$ at any calculation node.

And, The stress function in its dimensionless form:

$$f = \frac{\Phi}{Ea^2}$$

Where $E$ - modulus of elastic.

Rewriting EQ (1) with the terms of EQ (3) and EQ (4):

$$\left(\frac{\partial^2}{\partial \xi^2} + \frac{\partial^2}{\partial \eta^2}\right)\left(\frac{\partial^2 w}{\partial \xi^2} + \frac{\partial^2 w}{\partial \eta^2}\right) = \\ = 12\left(1-\mu^2\right)\frac{a^2}{H^2}\left(\frac{\partial^2 f}{\partial \eta^2}\frac{\partial^2 w}{\partial \xi^2} + \frac{\partial^2 f}{\partial \xi^2}\frac{\partial^2 w}{\partial \eta^2} - 2\frac{\partial^2 f}{\partial \xi \partial \eta}\frac{\partial^2 w}{\partial \xi \partial \eta}\right) + \frac{a^3}{D}q \tag{5}$$

Denoting that,
$$\frac{\partial^2 w}{\partial \xi^2} + \frac{\partial^2 w}{\partial \eta^2} = -m\,. \tag{6}$$

Substituting with EQ (6) back into EQ (5), obtaining a bi-equational system of 2nd order with terms of dimensionless unknowns

$$\frac{\partial^2 m}{\partial \xi^2} + \frac{\partial^2 m}{\partial \eta^2} = \\ = -12\left(1-\mu^2\right)\frac{a^2}{H^2}\left(\frac{\partial^2 f}{\partial \eta^2}\frac{\partial^2 w}{\partial \xi^2} + \frac{\partial^2 f}{\partial \xi^2}\frac{\partial^2 w}{\partial \eta^2} - 2\frac{\partial^2 f}{\partial \xi \partial \eta}\frac{\partial^2 w}{\partial \xi \partial \eta}\right) - \frac{a^3}{D}q\,. \tag{7}$$

Let the right-hand side of EQ (7) be defined as $(g)$ term:

$$g = \frac{a^3}{D}q + 12\left(1-\mu^2\right)\frac{a^2}{H^2}\left(\frac{\partial^2 f}{\partial \eta^2}\frac{\partial^2 w}{\partial \xi^2} + \frac{\partial^2 f}{\partial \xi^2}\frac{\partial^2 w}{\partial \eta^2} - 2\frac{\partial^2 f}{\partial \xi \partial \eta}\frac{\partial^2 w}{\partial \xi \partial \eta}\right). \tag{8}$$

Then EQ (7) could be reformed as :

$$\frac{\partial^2 m}{\partial \xi^2} + \frac{\partial^2 m}{\partial \eta^2} = -g\,. \tag{9}$$

From EQ (2), it's also possible to obtain a couple of 2nd order differential equations:

$$\frac{\partial^2 f}{\partial \xi^2} + \frac{\partial^2 f}{\partial \eta^2} = -s\,; \tag{10}$$

$$\frac{\partial^2 s}{\partial \xi^2} + \frac{\partial^2 s}{\partial \eta^2} = \frac{\partial^2 w}{\partial \xi^2}\frac{\partial^2 w}{\partial \eta^2} - \left(\frac{\partial^2 w}{\partial \xi \partial \eta}\right)^2. \tag{11}$$

Obtaining four 2nd order differential equations of the second order: EQ (9), EQ (6), EQ (11) and EQ (10); which to be solved in terms of EQ (8), considering the right-hand side of the differential equation EQ (11).

Let the right-hand side of EQ (11) be defined as ($l$) term & separately could be stated as:

$$l = \frac{\partial^2 w}{\partial \xi^2}\frac{\partial^2 w}{\partial \eta^2} - \left(\frac{\partial^2 w}{\partial \xi \partial \eta}\right)^2. \tag{12}$$

Then the equation takes the form:

$$\frac{\partial^2 s}{\partial \xi^2} + \frac{\partial^2 s}{\partial \eta^2} = l. \tag{13}$$

Solution algorithm consists of four main steps:

1. For $(g = \frac{a^3}{D}q)$ & with the aid of EQ (9), the value of ($m$) to be determined.
2. With EQ (6) & solving the system of equations, we can find the deflection value ($w$) at the calculated node.
3. The values of ($l$), ($s$) and then ($f$) to be determined using EQ (12) & EQ (13) and EQ (10) respectively.
4. The same wise, the new value of ($g$) to be calculated using EQ (8).

The previous steps to be repeated while moving to a new calculation cycle (*new calculation node*), considering that

$$g = \frac{1}{2}(g_{n-1} + g_n)$$

where $g_n$ - the value of ($g$) out of the last calculation cycle.

$g_{n-1}$ - the value of ($g$) out of the penultimate (*before-last*) calculation cycle.

The generalized equations of the finite difference method [**15**] are used to solve the above differential equations,which consider the discontinuity of the unknown function and its first derivative represented by the right-hand sides of the previous differential equations.

For the aim of illustrating the presented algorithm, Calculation of local load case is considered; Suppose that a load in the form of two symmetrically triangles acts on the center of a 4-sided hinge-supported square thin-plate as shown in (**Fig. 1**), noting that the load takes that triangular form affecting a small loaded area, thus; we can say that it acts as a point load relatively governed by the following formula:

$$P = 4\frac{1}{2}kh_a q_0, \tag{14}$$

Where $q_0$ – load intensity; $h$ – meshing step; $a$ - square plate side length; $K$ - coefficient that represents the contribution of the loaded part ($h$).

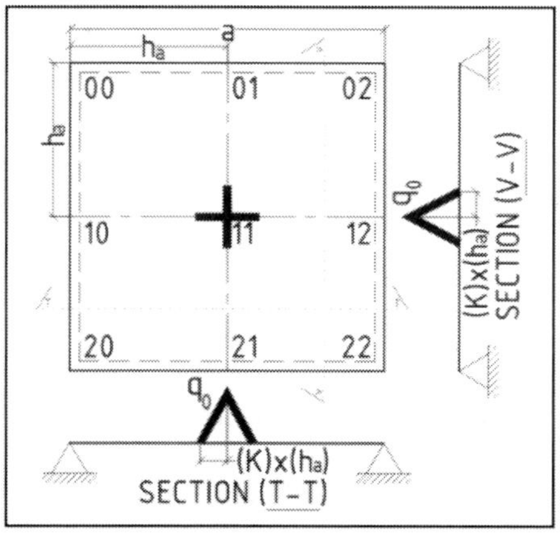


**Fig. 1.** A 4-sided hinge-supported square thin-plate loaded centrally.

Simplifying EQ (14) , separating ($q_0$) in a formula individual side getting that $q_0 = \dfrac{P}{2kha}$ which implies the load intensity in its dimensionless form ($q_0^*$) as:

$$q_0^* = \frac{q_0}{P/a} = \frac{1}{2kh}. \tag{15}$$

In the verifying example currently studied here, and in order to go through the solution steps; it is necessary to the considered the approximation of the differential EQ (9) taking into account the boundary conditions and the stripped pattern of the load centered onto the plate.

As per (3.1.24) (*Equation No.24, 1st Clause at the 3rd chapter*) reference [**15**] multiplied by ($h^2$), applied to node 11 (**Fig. 1**):

$$4m_{01} - 4m_{11} + h\left(\Delta^{\text{I-II}} m_{11}^{\xi} + \Delta^{I-III} m_{11}^{\eta}\right) = -h^2 g_{11}, \tag{16}$$

Where the shear force discontinuities are involved into calculations according to EQ (15) as $\Delta m^{\xi} = \Delta m^{\eta} = \dfrac{1}{2kh_a}$, $g_{11}$ - calculated by EQ (8) within the condition $\dfrac{a^3}{D} q = 0$.

As well as boundary conditions know cause the bending moments to be always equal to zero on its nodes, thus $m_{0j} = 0$; using EQ (16) we get:

$$m_{11} = \frac{h^2}{4} g_{11} + \frac{h}{4}\frac{1}{2kh} 2. \tag{17}$$

Concerning the non-linear formulation of our problem, besides $q$=0 – $g_{11}$=0, EQ (17) generates;

$$m_{11} = \frac{1}{4k}. \tag{18}$$

When ($k \to 0$); ($m_{11} \to \infty$)

For deflections determination, the difference syntax of EQ (6) should be applied to node 11. Once more using (3.1.25) of reference [**15**], multiplied by ($h^4$) getting:

$$4w_{01} - 4w_{11} = -h^2 m_{11}. \tag{19}$$

From the boundary conditions → $w_{01}$=0, then for $h$=1/2, $g_{11}$=0 and the value of $m_{11}$ (*calculated before via EQ (17)*), EQ (19) gives us:

$$w_{11} = \frac{h^2}{4} m_{11} = \frac{h^2}{4}\frac{1}{4k} = \frac{1}{4}\frac{1}{2^2}\frac{1}{4}\frac{1}{k} = \frac{1}{64k}. \tag{20}$$

In accordance to the solution algorithm stated above, difference syntax of EQ (9) to be written taking into calculations $g_{11}$=0, the algorithm is validated through EQ (18) that originally came from EQ (16). From EQ (6) ,associated with the resulted value of $m_{11}$; we can find the value of $w_{11}$. Also, with the aid of EQ (12); we can get the value of ($l$), the same fashion, solving EQ (13) generates the value of ($s$) noting that all these 2nd order differential operators are to be written in their difference forms.

On that, when ($g_{11}$ = 0) → $m_{11} = \frac{1}{4k}$; $w_{11} = \frac{1}{64k}$.

Returning to EQ (12) for expressing $l$ in terms of $w_{11}$: $l_{11} = 64w_{11}^2$ and to EQ (13) for determining the value of $s$ at node 11. Next to that; EQ (10) allows us to find the value of ($f_{11}$). By the end of this calculation cycle, confirming the value of $g_{11}$ (*on bases of EQ (8)*) when $q$ = 0 as follows:

$$g_{11} = 96\left(1-\mu^2\right)\frac{a^2}{H^2}\frac{1}{h^4} f_{11} w_{11}. \tag{21}$$

Finally, the previous cycle is repeated. Exact data for solving the problem numerically could be taken from (*84th page*) reference [**14**].

Progress of solution to be monitored in the light of the proceeded algorithm. For $g_{11}$=0 and depending on EQ (17); an initial value of $m_{11}$ to be determined, and once more substituting in EQ (17) when $k$=1; getting that ($m_{11}$ = 0.25). difference approximation for the differential equation EQ (6) gives us:

$$w_{11} = \frac{1}{4}\frac{1}{2^2} m_{11}.$$

Further proceeding through the previous mentioned algorithm results in the values; $s_{11} = -4w_{11}^2$ & $f_{11} = 0{,}0625 s_{11}$ and those values forming finally:

$$52{,}73m_{11}^3 + m_{11} = 0{,}25. \tag{22}$$

At the end & after determining the value of $m_{11} = 0{,}132$, following the same technique enables us to calculate the other parameters with the aid of the previous set of equations. Now, it became possible to solve that kind of problems, such that in reference [**14**], relatively;

$$\varphi = \frac{W}{H}$$

Where $W$- the maximum deflection; $H$ - Plate thickness

The numerical solution presented in this paper and other more similar numerical proposals become more acceptable in practical calculations, since all such solutions find their way to be performed on modern computers…

## References

1. T. Belytscko, W. Liu, B. Moran, Nonlinear finite elements for continua and structures. JWiley & Sons, 300 (2000)
2. J. Bonet, R. Wood, Nonlinear continuum mechanics for finite element analysis.Cambridge University Press, 248 (1997)
3. M.A. Crisfield, Non-linear finite elements analysis of solids and structures. J Wiley & Sons, **1**, 362 (1997)
4. P. Kyther, D. Wie, An introduction to linear and nonlinear finite element analysis. Birkhauer Verlag, 445 (2004)
5. J.N. Reddy, An introduction to nonlinear finite element analysis. Oxford University Press, 488 (2004)
6. A.N. Danilin, N.N. Zuev, D.V. Snegovsky, V.I. Shalashilin, CAD and graphics. **4**, 26-31 (2000)
7. V.V. Galishnikova, International Journal for Computational Civil and Structural Engineering. **5**, 35-44 (2009)
8. V.V. Galishnikova, Construction mechanics of engineering structures and structures, **1**, 42-50 (2010)
9. V.D. Kurguzov, Izvestiya Vysshikh Uchebnykh Zavedenii. Building. **3-4**, 14-22 (2009)
10. V.D. Kurguzov V.D. Computational technologies. **14**, 70-80 (2009)
11. S.V. Levyakov Applied Mechanics and Technical Physics. **53**, 128-136 (2012)
12. A.O. Scherbakova, Vestnik SUSU. Series Mathematics. Mechanics. Physics. **32**, 83-91 (2011)
13. V.A. Ignatiev, A. V. Ignatie, E.V.Onischenko, Vestnik MGSU. **2**, 20-33 (2016)
14. A.S. Volmir Flexible plates and shells. Gos. Ed. Technical and Theoretical Literature, Moscow, 419 (1956)
15. R.F. Gabbasov, A.R. Gabbasov, V.V. Filatov Numerical construction of discontinuous solutions to the problems of structural mechanics. Ed. ASV, Moscow, 280 (2008)

# Cost Benefit Analysis: Bypass of Prešov city

*Martina Margorínová*[1,*], *Mária Trojanová*[1]

[1] Department of Construction Management, Faculty of Civil Engineering, University of Žilina, Univerzitná 8215/1, 010 26 Žilina, Slovakia

**Abstract.** The paper describes decision making process based on economic evaluation, i.e. Cost Benefit Analysis for motorway bypass of the Prešov city. Three variants were evaluated by means of the Highway Development and Management Tool (HDM-4). HDM-4 is a software system for evaluating options for investing in road transport infrastructure. Vehicle operating costs and travel time costs were monetized with the use of the software. The investment opportunities were evaluated in terms of Cost Benefit Analysis results, i.e. economic indicators.

## 1 Introduction

Project of road infrastructure are concurrently assessed by means of cost benefit analysis (CBA). CBA is a tool which purpose is to determine all impacts (financial, economic, social, environmental and others) of the project and quantify all costs and benefits of the project. Cost benefit analysis consist of four types of analysis: financial analysis, economic analysis, sensitivity analysis and risk analysis. For the evaluation of road infrastructure, paramount is the economic analysis, which is not evaluated only through financial flows, but, in addition, identification and quantification of societal benefits. These social impacts are: travel time savings, reduction of accident rate, vehicle operating costs savings, reduction of noise levels and reduction of pollution. These effects are easily identifiable in practice, but their quantification is difficult. Various types of software might be utilized for quantification of these effects. For the bypass of Prešov city, was used toll HDM-4 (Highway Development and Management system). Three variants were evaluated. The best variant was chosen according to economic indicators calculated in the CBA.

## 2 CBA for road infrastructure projects

CBA is used in feasibility studies with the objective to ascertain most viable variant of road infrastructure solution. Feasibility analysis identifies and analyses the limits of proposed alternatives, which include legal, technical, economic, environmental and personal limits. Feasible project must achieve investor´s objectives, and, at the same time, accept these limits. Following the feasibility analysis, investment cost for each alternative

* Corresponding author: martina.margorinova@fstav.uniza.sk

need to be quantified. Concurrently, reference period of the project needs to be set, usually starting in the year of the analysis and ending in the final year of the project lifecycle. The reference period consists of the investment period – time of design and construction of the infrastructure, and operation period – time when the infrastructure is open for users, thus generating benefits but also operation costs for the administrator. To take the time factor into account, all costs and benefits are discounted. The discount rate for economic analysis is prescribed by the European Commission. Another important factor for calculating cost benefit analysis is traffic intensity on both existing and newly constructed road network. On this basis, program HDM-4 computes operating, user and societal costs. Project benefits are calculated as a difference between higher user costs in the alternative without the investments, and lower user costs in the alternative with investment. When all costs and benefits have been quantified the program computes economic indicators for proposed alternatives. [5] [6]

### 2.1 Formatting the title, authors and affiliations

Economic analysis is conducted from the perspective of whole society, in which financial gain is not the most important aspect, instead, so-called societal benefit is the main point of interest in the evaluation. Economic analysis doesn´t evaluate project only through financial flows, but, in addition, quantify various impacts entailed by the project. All costs and benefits of the project are computed from quantifying all these impacts. After all costs and benefits have been quantified, economic indicators are computed: economic net present value (ENPV), economic internal rate of return (EIRR), economic payback period (EPP) and economic cost benefit ratio (ECBR). Economic efficiency of the project is evaluated on the base of these economic indicators. [1]

#### *2.1.1 Economic net present value*

Economic net present value (ENPV) is difference between discounted costs and benefits of project.

$$ENPV = \sum_{t=0}^{n} \frac{B_t - IC_t - OC_t}{(1+0{,}01.\mathrm{u})^t} \tag{1}$$

Where: $ENPV$ – economic net present value, [€]
$B_t$ – benefits from investment in year t, [€]
$IC_t$ – investment cost in year t, [€]
$OC_t$ – operating costs in year t, [€]
$u$ – discount rate, [%]
$t$ – individual years, [years]
$n$ – reference period, [years]

Effective project has economic net present value more than zero. This economic indicator is together with economic internal rate of return the most decisive in assessing economic efficiency of the investment. [4]

### *2.1.2 Economic internal rate of return*

Economic internal rate of return (EIRR) represents a rate of return of the investment. EIRR is a discount rate, at which is the economic net present value equal zero.

$$\sum_{t=0}^{n} \frac{B_t - IC_t - OC_t}{(1+0{,}01.x)^t} = 0 \tag{2}$$

$$EIRR = x.100 \tag{3}$$

Where: $EIRR$ – economic internal rate of return, [%]
$B_t$ – benefits from investment in year t, [€]
$IC_t$ – investment cost in year t, [€]
$OC_t$ – operating costs in year t, [€]
$x$ – interest rate - searched value, [%]
$t$ – individual years, [years]
$n$ – reference period, [years]

The project is effective, if the internal rate of return exceeds the used discount rate. This indicator is usually the most decisive in economic efficiency evaluation of the investment. EIRR allows to compares projects of different sizes. [4]

### *2.1.3 Economic payback period*

Economic payback period (EPP) is the year, at which discounted benefits of project equal the discounted costs of a project. Project is effective, when its payback period is shorter than the lifecycle of project respectively shorter then reference period. This indicator is not decisive in project effectiveness assessment, but may be used as an additional criterion. [4]

$$0 = \sum_{t=1}^{EPP} \frac{B_t - IC_t - OC_t}{(1+0{,}01.u)^{(t-1)}} \tag{4}$$

Where: $EPP$ – economic payback period, [year]
$B_t$ – benefits from investment in year t, [€]
$IC_t$ – investment cost in year t, [€]
$OC_t$ – operating costs in year t, [€]
$x$ – interest rate - searched value, [%]
$t$ – individual years, [years]
$n$ – reference period, [years]

### *2.1.4 Economic cost benefit ratio*

Economic cost benefit ratio (ECBR) is a complementary economic indicator. ECBR is calculated by dividing the net present value of all benefits and the net present value of all project costs. Project is effective, when the EBCR exceeds one.

$$ECBR = \frac{\sum_{t=0}^{n} B_{t,DIS}}{\sum_{t=0}^{n} CC_{t,DIS}} \qquad (5)$$

Where:
- $ECBR$ – economic cost benefit ratio, [-]
- $B_{t,DIS}$ – discounted benefits from investment in year t, [€]
- $CC_{t,DIS}$ – discounted capital cost in year t, [€]
- $t$ – individual years, [years]
- $n$ – reference period, [years]

### 2.2 Software HDM-4

The software endorsed World Bank HDM-4 (Highway Development and Management System) is a versatile tool for the road management of existing or planned roads. HDM-4 has a defined system of relations and mathematical equations which describe traffic, pavement performance and road user effects. [2]

Based on input fed by the evaluator in the software and calibration file for particular region, HDM-4 monetizes travel times costs, vehicle operating costs and accident costs as well as administrator's cash flows. [2]

Travel time is considered in terms of passenger-hours during working and non-working time, and cargo holding hours. [6]

Vehicle operating costs include: fuel consumption, oil and lubricants consumption, tyre consumption, parts consumption, maintenance labour hours, depreciation, interest, crew hours and overheads. [6]

An accident is an event involving one or more road vehicles, which results in death, personal injury and /or damage to property. In HDM-4 are analysed three types of accident: fatal, injury and damage only. [6]

Administrator's cash flows include investment costs, maintenance and repair costs of the road. [2]

## 3 Road infrastructure of Prešov city

At present, Prešov is lacking motorway transport infrastructure directing traffic outside of built-up part of city. Full traffic volume is led throughout the communication system of the city with all the negative impacts on habitants and the environment of Prešov city. Traffic serviceability of Prešov is provided by municipal communications and trunk roads of first, second and third class. The brunt of traffic volume is distributet on first class roads I/18 and I/68. Daily intensity on critical sections on these roads exceeds 34 000 vehicles per day.

The main objective of the bypass is divert transit traffic outside of built-up area of Prešov, which significantly enhance the quality of life in the city. Second objective of the bypass is completion of highways and expressways connection of cities Bratislava and Kosice, i.e. fifth Pan-European transport corridor leading through Slovakia. The bypass, will provide safe, quick, and convenient connection in both North – South and East – West direction.

# 4 Bypass variants of Prešov city

Three variants were assessed: Red variant, Red modified variant and Blue variant. The following figure (Fig. 1) shows the assessed variants.

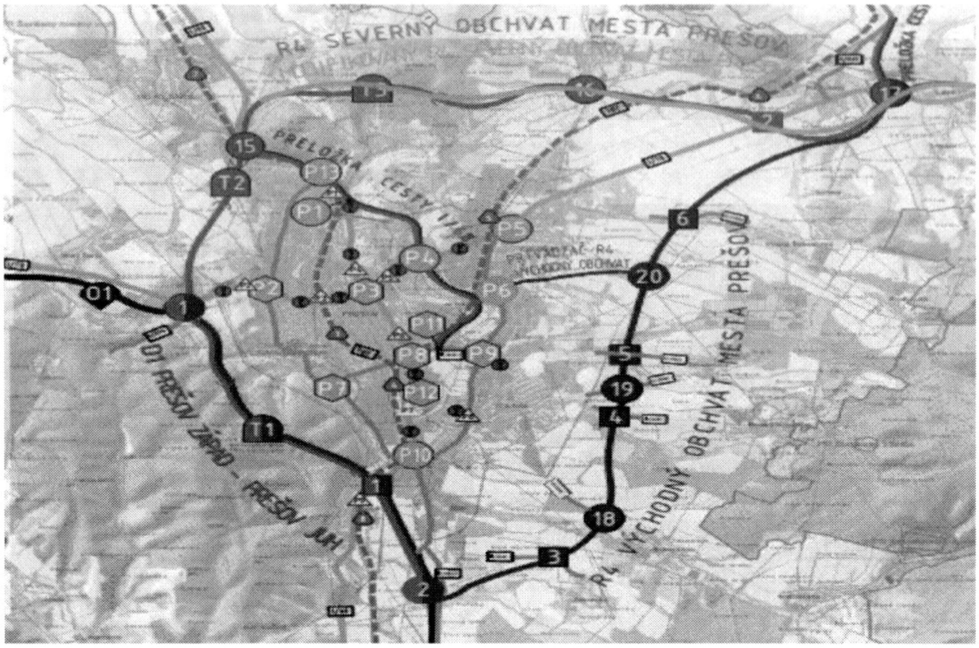


**Fig. 1.** Assessed variants of Prešov city [3]

## 4.1 Technical parameters of assessed variants

Assessed variants have different technical parameters such as: horizontal and vertical alignment, numbers of bridges, tunnels, elevated crossings, etc. The section of highway: D1 West Prešov – South Prešov, has the same technical parameters in all assessed variants. Red variant and Red modified variant are the same for elevated crossing – Dúbrava (ref.15). From this elevated crossing is the change in horizontal and vertical alignment, which is illustrated at Figure 1 by green colour. Technical parameters of assessed variants are listed in following table (Table 1). [3]

**Table 1.** Technical parameters of assessed variants

| VARIANT | Category of communication | Length | Number of elevated crossings | Number of bridges | Number of tunnels |
|---|---|---|---|---|---|
| **Red** | D 26,5/100<br>R 24,5/100 | 7,87 km<br>14,5 km | 2<br>3 | 18<br>21 | 1- Prešov<br>2 - Bikoš and Okrúhliak |
| **Red modified** | D 26,5/100<br>R 24,5/100 | 7,87 km<br>13,86 km | 2<br>3 | 18<br>18 | 1 – Prešov<br>1 - Bikoš |
| **Blue** | D26,5/100<br>R 24,5/130<br>R 22,5/80 | 7,87 km<br>14,8 km<br>3,73 km | 2<br>4<br>- | 18<br>21<br>5 | 1 – Prešov<br>-<br>- |

### 4.2 Traffic intensity

The aim of bypass is divert the transit traffic from Prešov on newly constructed motorway infrastructure. Traffic intensity is assumed the same in Red variant and Red modified variant, but different in Blue variant. For Prešov city, forty-three road sections, both new and existing, were evaluated. For each road section, HDM-4 requires input data regarding: construction parameters, pavement defects, actual traffic volume for Variant "Do nothing", traffic volume distribution for "Do something" variants. Design files as well as traffic study of the Bypass were supplied by design company DOPRAVOPROJEKT a.s. - division of Zvolen. Data regarding existing road network were obtained from the portal of road network database of Slovak Road Administration.

## 5 CBA for bypass of Prešov city

CBA was used for assessing economic efficiency of proposed variants of bypass Prešov city. CBA was performed in HDM-4 software.

### 5.1 Inputs into the software HDM-4

All necessary data about existing road network and proposed variants, which are mentioned higher, were inserted into the software HDM-4. Other data, which were inserted into HDM-4, were reference period and discount rate. Reference period of 35 years was set for evaluation, of which 5 years is the investment period and 30 years is the operation period, i.e. life expectancy. Discount rate for economic analysis is prescribed by the European Commission, and, currently it is 5 %.

### 5.2 Capital costs of assessed variants

Capital costs consist of investment and operating costs. Investment costs were quantified by project engineer and operating costs were quantified by program HDM-4. In the following table (Table 2), discounted and undiscounted capital costs are listed.

**Table 2.** Capital costs of assessed variants [mil. €]

| | UNDISCOUNTED | | | DISCOUNTED | | |
|---|---|---|---|---|---|---|
| **Variant** | **Red variant** | **Red modified variant** | **Blue variant** | **Red variant** | **Red modified variant** | **Blue variant** |
| **Capital costs** | 924.96 | 816.21 | 785.55 | 821.61 | 724.67 | 696.46 |

### 5.3 Benefits of assessed variants

Benefits for individual variants were calculated as a difference between higher user costs in the alternative without the investments, and lower user costs in the alternative with investment. Benefits were quantified in the software HDM-4. In the following table (Table 3), discounted and undiscounted benefits for assessed variants are listed.

**Table 3.** Benefits of assessed variants [mil. €]

| | UNDISCOUNTED | | | DISCOUNTED | | |
|---|---|---|---|---|---|---|
| **Variant** | **Red variant** | **Red modified variant** | **Blue variant** | **Red variant** | **Red modified variant** | **Blue variant** |
| **Benefits** | 1973.83 | 1975.46 | 1407.64 | 777.95 | 778.60 | 558.19 |

## 5.4 Economic indicators

After all costs (Table 2) and benefits (Table 3) were quantified, the HDM-4 software computed economic indicators of assessed alternatives. The most effective variant was chosen according to economic indicators. An effective variant must fulfil following rules: ENPV must be greater than zero, EIRR must be greater than discount rate 5%, EPP must be lower than lifecycle of project respectively lower than reference period, in this case it is 35 years and ECBR must be more than one.

### *5.4.1 Red variant*

| | Increase in Road Agency Costs | | | Savings in MT VOC | Savings in MT Travel Time Costs | Savings in NMT Travel & Operating Costs | Reduction in Accident Costs | Net Exogenous Benefits | Net Economic Benefits (NPV) |
|---|---|---|---|---|---|---|---|---|---|
| | Capital | Recurrent | Special | | | | | | |
| Undiscounted | 924.96 | 0.00 | 0.00 | 832.88 | 837.81 | 0.00 | 303.14 | 0.00 | 1,048.87 |
| Discounted | 821.61 | 0.00 | 0.00 | 330.13 | 329.56 | 0.00 | 118.27 | 0.00 | -43.66 |

Economic Internal Rate of Return (EIRR) = 4.7% (No. of solutions = 1)

**Fig. 2.** Output from software HDM-4 for Red variant

Red variant has reached following economic indicators: ENPV is -43.66 million €, EIRR is 4,7 %, EPP is longer than lifecycle of project and EBCR is 0,95. All economic indicators of Red variant show, that this variant is not effective.

### *5.4.2 Red modified variant*

| | Increase in Road Agency Costs | | | Savings in MT VOC | Savings in MT Travel Time Costs | Savings in NMT Travel & Operating Costs | Reduction in Accident Costs | Net Exogenous Benefits | Net Economic Benefits (NPV) |
|---|---|---|---|---|---|---|---|---|---|
| | Capital | Recurrent | Special | | | | | | |
| Undiscounted | 816.21 | 0.00 | 0.00 | 830.54 | 841.79 | 0.00 | 303.14 | 0.00 | 1,159.25 |
| Discounted | 724.67 | 0.00 | 0.00 | 329.20 | 331.13 | 0.00 | 118.27 | 0.00 | 53.93 |

Economic Internal Rate of Return (EIRR) = 5.6% (No. of solutions = 1)

**Fig. 3.** Output from software HDM-4 for Red modified variant

Economic indicators for Red modified variant are following: ENPV is + 53.93 million €, EIRR is 5,6 %, EPP is in lifecycle of project and ECBR is 1,07. These economic indicators fulfilled conditions of efficiency. Red modified variant is effective.

### *5.4.3 Blue variant*

| | Increase in Road Agency Costs | | | Savings in MT VOC | Savings in MT Travel Time Costs | Savings in NMT Travel & Operating Costs | Reduction in Accident Costs | Net Exogenous Benefits | Net Economic Benefits (NPV) |
|---|---|---|---|---|---|---|---|---|---|
| | Capital | Recurrent | Special | | | | | | |
| Undiscounted | 785.55 | 0.00 | 0.00 | 541.67 | 613.69 | 0.00 | 252.27 | 0.00 | 622.08 |
| Discounted | 696.46 | 0.00 | 0.00 | 217.51 | 242.51 | 0.00 | 98.17 | 0.00 | -138.27 |

Economic Internal Rate of Return (EIRR) = 3.5% (No. of solutions = 1)

**Fig. 4.** Output from software HDM-4 for Blue variant

Economic indicators for Blue variant are following: ENPV is – 138.27 million €, EIRR is 3,5 %, EPP is longer than lifecycle of project and EBCR is 0,80. Economic indicators of Blue variant show that this variant is not effective.

## 6 Conclusion

The dire current traffic situation in Prešov calls for the construction the bypass. The aim of the bypass is to divert the transit traffic outside the built-up area of Prešov city. Second objective of the bypass is completion of highways and expressways connection of cities Bratislava and Kosice, i.e. fifth Pan-European transport corridor leading through Slovakia. Three variants of bypass of Prešov city were assessed: Red variant, Red modified variant and Blue variant. The best variant was chosen according to economic indicators calculated in the cost benefit analysis.

Based on economic results presented in chapter 5, we conclude that the Red modified variant is the best choice from the economic point of view. Its economic indicators fulfilled a condition of efficiency. Red variant and Red modified variant have same traffic solution, but the Red modified variant has significant savings of investment costs 108.75 million €. This saving was achieved by removal of the tunnel Okúrhliak. The alternate route entailed to raise the motorway's longitudinal axis and requires significant earthworks as the motorway is positioned in the cut of hill Okrúhliak, the costs however justify this solution, i.e. capital costs of Red variant (Tab.2) has been higher than expected benefits of Red variant (Tab.3) and therefore this variant is not as effective.

The blue variant had the lowest capital costs (Tab.2), but, traffic wise, this variant is not as effective, thus scores poorly as shown in fig 4. Blue variant had scant benefits, because traffic study predicted less favourable relocation of freight transport compared to red variants. Traffic study expected that Blue variant takes over 19 % less of freight transport compared to red variants on the existing road network in Prešov. This fact reflected itself in the monetization of Blue variant's benefits which ultimately didn't exceed the capital costs.

## References

1. Ministry of transport and construction of the Slovak republic. Slovak guide to cost benefit analysis of investment projects in the transport sector, version 2.2., (2014).
2. Ministry of transport and construction of the Slovak republic. Technical conditions: TP 057 – Methodology for the use of HDM-4 in Slovak conditions, (2012).
3. National Motorway company a. s. Feasibility study – accompanying report D1/R2/R4 Prešov – Košice – Bidovce – Košice Šaca, (2015).
4. P.Sieber. Cost benefit analysis. Methodical guide, version 1.3. Ministry of regional development CZ, (2004).
5. J. Mikolaj, L. Remek. Life Cycle cost analysis-integral part of road network management system. In 23 Russian-Polish-Slovak Seminar on Theoretical Foundation of Civil Engineering. Poland: Wroclaw, Szklarska Poreba, (2014).
6. H.G.R. Kerali, J. B. Odoki. The Highway Development and Managment Series. Volume four: Analytical Framework and Model Descriptions. Paris: World Road Association (PIARC), (2000).

# Calculation of the viscosity of dispersions of nanoparticles with a polymer adsorption layer in a melt

*Tatyana* Matseevich[1*], *Andrey* Askadskii[1]

[1]Moscow state university of civil engineering, Yaroslavskoye shosse, 26, Moscow, Russia, 129337

**Abstract**. As a result of theoretical analysis of the influence of the chemical structure of the polymer and the chemically modified surface of nanoparticles, it was shown that it is possible to predict the dependence of the viscosity of the polymer melt on the concentration of nanoparticles, their size, and the molecular weight of the polymer. Two situations have been analyzed where a strong intermolecular interaction between polymer chains and polar groups located on the surface of nanoparticles is absent, and when a strong intermolecular interaction between polymer chains and polar groups located on the surface of nanoparticles takes place.

Currently, there are a variety of models and derived equations that relate the viscosity with volume fraction of dispersed particles. One of the equations describing the dependence of the viscosity of a dispersion containing spherical particles, is Einstein's equation:

$$\eta = \eta_0(1 + 2.5\varphi), \quad (1)$$

where η is the viscosity of the suspension, $\eta_0$ is the viscosity of initial fluid, $\varphi$ is the volume of the dispersed phase, 2.5 is the parameter applicable only for spherical particles.
Another most popular equation for the concentration dependence of the viscosity of the dispersions was proposed by Mooney [1]:

$$\eta = \eta_0 \exp\left[\frac{2.5\varphi}{1-(\varphi/\varphi^*)}\right], \quad (2)$$

where φ* is the so-called critical concentration (limit value of the filling degree of solid particles). This value depends on particle shape and method of packaging, i.e. coefficient packaging. For example, the value φ* = 0.74 in the case of hexagonal packing of solid spherical particles.

Other popular models and the corresponding equations, we note the model of Krieger and Dougherty [2]

$$\eta = \eta_0\left(1 - \frac{\varphi}{\varphi^*}\right)^{2.5\varphi^*} \quad (3)$$

and model of Quemada [3]

* MatseevichTA@mgsu.ru

$$\eta = \eta_0 \left(1 - \frac{1}{2} K\varphi\right)^{-2}, \tag{4}$$

where K takes values from 2.54 to 3.71.

Studies of concentrated suspensions [4, 5] belong to R. Pal. In the works cited, there is a large amount of experimental data that can be used to test the functionality of various models describing the dependence of the relative viscosity of the suspensions in a wide concentration range.

Look at some modern models. In the work [6] the role of normal stresses in curvilinear flows of suspensions is considered. The resulting equation for describing the relative viscosity dependence on volume fraction of the dispersed phase has the form:

$$\eta = \eta_0 \left[1 + 2.5\varphi\left(1 - \frac{\varphi}{\varphi^*}\right)^{-1} + 0.1\left(\frac{\varphi}{\varphi^*}\right)^2 \left(1 - \frac{\varphi}{\varphi^*}\right)^{-2}\right]. \tag{5}$$

In the work [7] the nature of the divergence in low shear viscosity of colloidal dispersions of hard spheres is discussed. In the reviews [8-9] the analysis of the viscosity of concentrated suspensions of hard nano-size particles and bimodal spheres is provided. Proposed ratio:

$$\eta = \eta_0 \left(\frac{1-\varphi}{1-\dfrac{\varphi}{\varphi^*}}\right)^{\frac{2.5\varphi^*}{1-\varphi^*}}. \tag{6}$$

In [10] considers the effective viscosity of concentrated suspensions of soft particles under conditions of static and high-frequency effects. The resulting ratio is:

$$\eta = \eta_0 \left(1 - \frac{1}{1 + \dfrac{1}{\varphi} - \dfrac{1}{\varphi^*}}\right)^{-2.5}. \tag{7}$$

In the last few years, considered numerous models to describe the relative viscosity of suspensions [11-12], and in [11] proposed and proved the generalized equation for suspensions subjected to steady flow at low Reynolds numbers. For suspensions containing solid particles, the equation is written like this:

$$\eta = \eta_0 \left[\frac{\varphi^* - \varphi}{\varphi^*(1-\varphi)}\right]^{-\frac{2.5\varphi^*}{1-\varphi^*}}. \tag{8}$$

All the considered models describe the dependence of the relative viscosity of the suspensions at different intervals of concentration of the dispersed particles depending on their volume fraction. However, no model allows one to calculate the relative viscosity of the suspension in which the dispersed particles stabilized by polymer coatings grafted on the surface of metal nanoparticles. This should be considered two options. One of them dispersed particles are inert and the other they have a strong intermolecular interaction with molecules of the liquid. In this regard, the aim of this paper is to develop a calculation scheme that allows prediction of the dependence of the volume fraction of the nanoparticles stabilized by the polymer. In the calculation scheme takes into account the chemical structure of grafted polymer. Two tasks are solved:

1) The direct problem, which allows calculating the value of relative viscosity based on chemical structure of polymer and its concentration. Such a calculation can be made

using any of the previously proposed equations, which include the volume fraction of dispersed particles.

2) Inverse problem, which allows on the basis of measurements of the relative viscosity to estimate the size of polymer stabilized nanoparticles. Here we consider single-layer and multi-coating polymer layers of metallic nanoparticles.

For the analysis the classical Einstein (1) and Mooney (2) equations have been used. Immediately, we note that equation (1), in the same way as equation (2), in pure form, becomes inapplicable, as a result of strong intermolecular interactions between the molecules of the liquid and solid particles formed stable adsorption layer.

Consider the calculation of the dependence of viscosity on the weight fraction of nanoparticles where there is a strong intermolecular interaction between the polar groups present on the surface of the nanoparticles and the molecules of the melt. In this case, the adsorption layer formed around the nanoparticles.

The $S_{np}$ surface of one nanoparticle is equal to

$$S_{np} = 4\pi R_{np}^2 , \tag{9}$$

where $R_{np}$ is the radius of the nanoparticle. The volume of one nanoparticle is equal to

$$v_{np} = \frac{4}{3}\pi R_{np}^3 . \tag{10}$$

The number of nanoparticles in the system is

$$n_{np} = \frac{c_{np}}{g_{np}} , \tag{11}$$

where $c_{np}$ is the weight of nanoparticles, $g_{np}$ is the weight of one nanoparticle. The weight of one nanoparticle is equal to

$$g_{np} = \rho_{np} v_{np} , \tag{12}$$

where $\rho_{np}$ is density of nanoparticle, $v_{np}$ is the volume of nanoparticle.

Consider 1g of the systems containing polymer melt and nanoparticles. In this composition $c_{np}$ is the weight of the nanoparticles and 1-$c_{np}$ is the weight of the solvent. Then the number of repeating units of the polymer in the composition of the $N_c$ is equal to

$$N_c = \frac{1 - c_{np}}{M_0} \cdot N_A , \tag{13}$$

where $M_0$ is the molecular weight of the repeating unit, $N_A$ is Avogadro's number.

The number of repeating units per macromolecule is equal to

$$N = \frac{M}{M_0} , \tag{14}$$

where $M$ is the molecular weight of the polymer. If the structure of the melt is globular, the volume of globule equal to

$$V_{gl} = \frac{\left(\sum_i \Delta V_i\right)_p \cdot N}{k} , \tag{15}$$

here $\left(\sum_i \Delta V_i\right)_p$ is the van der Waals volume of the polymer repeating unit, $k$ is the coefficient of molecular packing. Typically for polymer melts $k = 0.64$.

The value $\left(\sum_i \Delta V_i\right)_p$ is calculated using a computer program "Cascade" (INEOS RAS).

The radius of the globules can be calculated using the following formula:

$$R_{gl} = \sqrt[3]{\frac{3V_{gl}}{4\pi}} = \sqrt[3]{\frac{3\sum_i (\Delta V_i)_p N}{4\pi k}} . \quad (16)$$

The area occupied by 1 globule on the surface of the nanoparticle equal to

$$S_{gl} = \pi R_{gl}^2 = \pi \sqrt[3/2]{\frac{3\left(\sum_i \Delta V_i\right)_p N}{4\pi k}} . \quad (17)$$

The total surface of one nanoparticle is calculated using the formula (9). Limit the number of globules, which can be distributed on the surface of one nanoparticle is determined by the following ratio

$$n_{gl} = \frac{S_{np}}{S_{gl}} = \frac{4\pi R_{np}^2}{\pi \sqrt[2/3]{\frac{3\sum_i (\Delta V_i)_p N}{4\pi k}}} = \frac{4R_{np}^2}{0.518 \sqrt[2/3]{\sum_i (\Delta V_i)_p N}} \text{ (при } k = 0.64). \quad (18)$$

Taking into account formula (14) it can be written

$$n_{gl} = \frac{7.72R_{np}^2}{\sqrt[3/2]{\sum_i (\Delta V_i)_p \frac{M}{M_0}}} . \quad (19)$$

Let's consider an example. For polyamide-6 (PA-6) the van der waals volume of the repeating unit s equal to 116Å$^3$, $M_0$ = 113. If the radius of the nanoparticl0e is equal to 50 Å (5 nm) and the molecular weight of the polymer $M$ = 14000, the value of $n_{gl}$, calculated by the formula (19), equal to 32.6.

It is obvious that the radius of one nanoparticle together with the adsorption layer is equal to:

$$R_{np+ad.l.} = R_{np} + 2R_{gl} = R_{np} + 2\sqrt[3]{\frac{3\left(\sum_i \Delta V_i\right)_p N}{4\pi k}} = R_{np} + 1.44\sqrt[3]{\left(\sum_i \Delta V_i\right)_p \frac{M}{M_0}} . \quad (20)$$

The volume of single nanoparticle with adsorptive layer is

$$v_{np+ad.l.} = \frac{4}{3}\pi\left(R_{np} + 1.44\sqrt[3]{\left(\sum \Delta V_i\right)_p \frac{M}{M_0}}\right)^3 . \quad (21)$$

For nanoparticles that do not contain adsorption layers, their volume fraction can be calculated by the following formula

$$\varphi = \frac{v_{np} n_{np}}{v_{np} n_{np} + V_p} , \quad (22)$$

where $V_p$ is the volume of polymer in the composition.

For nanoparticles containing adsorption layers, their volume fraction can be calculated using the following formula

$$\varphi_{np+ad.l.} = \frac{v_{np+ad.l.} n_{np}}{v_{np+ad.l.} n_{np} + V_p - v_{ad.l.} n_{np}} , \tag{23}$$

where $v_{ad.l.}$ is the volume of the adsorption layer per one nanoparticle,

$$V_p = \frac{G_p}{\rho_p} , \tag{24}$$

where $G_p$ is the weight of the polymer in the composition, and $\rho_p$ is the density of the polymer.
If we analyze 1 g of the composition, $G_p = 1\text{-}c_{np}$. The density of the polymer is calculated by the following formula [13]:

$$\rho_p = \frac{kM_0}{N_A \left( \sum_i \Delta V_i \right)_p} . \tag{25}$$

Thus, we obtain

$$V_p = \frac{\left(1 - c_{np}\right) N_A \left( \sum_i \Delta V_i \right)_p}{kM_0} . \tag{26}$$

The volume of the adsorption layer of $V_{ad.l.}$ is:

$$V_{ad.l.} = n_{gl} \cdot V_{gl} . \tag{27}$$

Substituting relations (15) and (19) into formula (27), we get:

$$V_{ad.l.} = \frac{7.72 R_{np}^2 \sqrt[3]{\left( \sum_i \Delta V_i \right)_p}}{k} . \tag{28}$$

Substituting formula (11), (12), (26) and (28) into relation (23), we can write the relation for the volume fraction of nanoparticles containing adsorption layers:

$$\varphi_{np+ad.l.} = \frac{\left( R_{np} + 1.44 \sqrt[3]{\frac{M}{M_0} \left( \sum_i \Delta V_i \right)_p} \right)^3 \frac{c_{np}}{\rho_{np} R_{np}^3}}{\left( R_{np} + 1.44 \sqrt[3]{\frac{M}{M_0} \left( \sum_i \Delta V_i \right)_p} \right)^3 \frac{c_{np}}{\rho_{np} R_{np}^3} + \frac{\left(1 - c_{np}\right) N_A \left( \sum_i \Delta V_i \right)_p}{kM_0} - \frac{7.72 \sqrt[3]{\left( \sum_i \Delta V_i \right)_p} \cdot 3c_{np}}{4\pi \rho_{np} R_{np}}} \tag{29}$$

Now consider the examples of calculations.

The initial polymer is polyamide-6. The parameters of the polymer system are given in table 1.

**Table 1.** Parameters for the system "PA-6 + nanoparticles".

| | |
|---|---|
| Molecular weight of the repeating units, $M_0$ | 113 |
| Molecular weight of the polymer, $M$ | 14000 |
| Van-der-Waals volume of the repeating units of the polymer, $\left( \sum_i \Delta V_i \right)_p$, Å$^3$ | 116 |

| | |
|---|---|
| Coefficient of molecular packing for melt, $k$ | 0.64 |
| Weigh of the nanoparticles containing in 1 g of composition, g | 0.1 |
| Radius of nanoparticle $R_{np}$, Å | 50 |
| Density of nanoparticle $\rho_{np}$, g/cm$^3$ | 1.54 |
| Zero shear viscosity for PA-6 at 463K, N·sec/cm$^2$ | 15.2 |

If the adsorption layer is absent, the volume fraction of nanoparticles is calculated using the following formula

$$\varphi = \frac{\dfrac{c_{np}}{\rho_{np}}}{\dfrac{c_{np}}{\rho_{np}} + \dfrac{1-c_{np}}{\rho_p}} = \frac{\dfrac{c_{np}}{\rho_{np}}}{\dfrac{c_{np}}{\rho_{np}} + \dfrac{\left(1-c_{np}\right)N_A\left(\sum_i \Delta V_i\right)_p}{kM_0}} . \tag{30}$$

Substituting all the parameters available in the table 1, into formula (30), obtain $\varphi$ = 0.0695.

If the surface of the particles formed adsorption layer, using all parameters from table 1 and formula (29), have $\varphi_{np+ad.l} = 0.2226$.

Now determine the dependence of viscosity on the weight fraction of nanoparticles. The type of this dependence is determined by the possibility of the formation of a polymeric adsorption layer on the surface of the nanoparticles.

1 – There is no a strong intermolecular interaction between polymer chains and the polar groups on the surface of nanoparticles. Then, substituting all the system parameters from table 1 into equation (1), we obtain

$$\eta = 15.2\left(1 + \frac{1.6225c_{np}}{0.966 - 0.317c_{np}}\right). \tag{31}$$

2 – There is a strong intermolecular interaction between polymer chains and polar groups localized at the surface of nanoparticles. Then we get the following formula

$$\eta = 15.2\left(1 + \frac{7.9763c_{np}}{2.0419c_{np} + 0.9661}\right). \tag{32}$$

Next, let's calculate these dependencies using equation Mooney (2). When the strong intermolecular interaction between polymer chains and the polar groups on the surface of nanoparticles is absent, substituting all the parameters from table 1 into equation (1) with $\varphi^* = 0.74$ (hexagonal packaging), we get:

$$\eta = 15.2\exp\frac{1.6225c_{np}}{0.966 - 1.194c_{np}} . \tag{33}$$

When a strong intermolecular interaction is manifested, we obtain the following formula:

$$\eta = 15.2\exp\frac{7.9763c_{np}}{0.966 - 2.2695c_{np}} . \tag{34}$$

The dependence of viscosity on concentration of nanoparticles calculated by the Einstein equation (1) shown in figure 1. The dependencies of viscosity on concentration of nanoparticles calculated by Mooney equation (2) shown in figure 2.

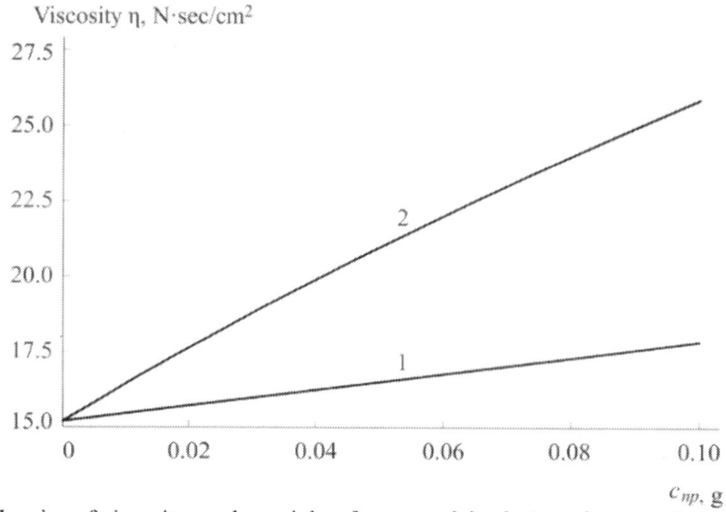


**Fig. 1.** Dependencies of viscosity on the weight of nanoparticles in 1 g of composite containing PA-6 and nanoparticles. 1 - There is no strong intermolecular interaction between polymer chains and polar groups located on the surface of nanoparticles. 2 - Strong intermolecular interaction between polymer chains and polar groups located on the surface of nanoparticles takes place. The calculations were performed using Einstein equation (1).

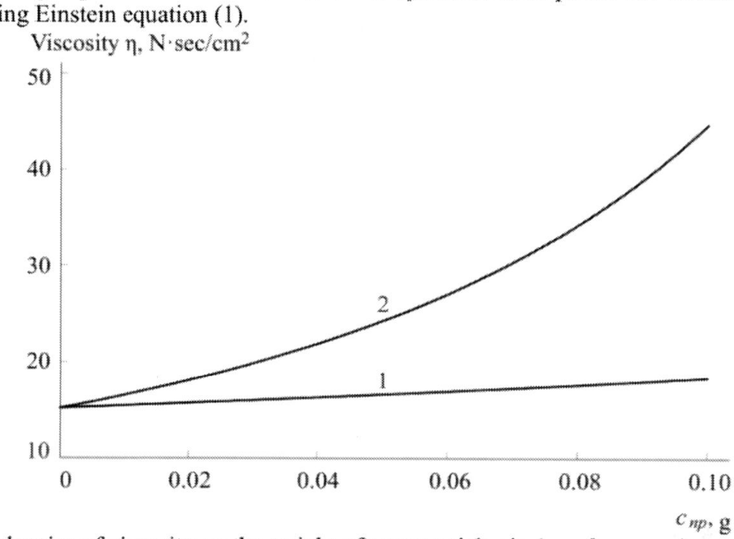


**Fig. 2.** Dependencies of viscosity on the weight of nanoparticles in 1 g of composite containing PA-6 and nanoparticles. 1 - There is no strong intermolecular interaction between polymer chains and polar groups located on the surface of nanoparticles. 2 - Strong intermolecular interaction takes place. The calculations were performed using Mooney equation (2).

Now let's calculate the dependencies of the viscosity on the radius of nanoparticles. When the adsorption layer is absent, the viscosity is independent of the size of nanoparticles. If the adsorption layer is formed, the relations for description of the dependencies of viscosity on the radius of nanoparticles are as follows (taking into account the values of all parameters are available in table 1). The dependencies of viscosity on radius of nanoparticles are described by relations (35) and (36).

The dependencies of the viscosity on the radius of nanoparticles are shown in figure 3. It can be seen that in agreement with equation Mooney the viscosity increases sharply when the size of nanoparticles decreases.

$$\eta = 15.2\left(1+\frac{0.16225\left(\frac{R_{np}+35}{R_{np}}\right)^3}{0.0649\left(\frac{R_{np}+35}{R_{np}}\right)^3+0.8685-\frac{0.912}{R_{np}}}\right), \text{(Einstein equation)} \quad (35)$$

$$\eta = 15.2\exp\left(\frac{0.16225\left(\frac{R_{np}+35)}{R_{np}}\right)^3}{0.8695-\frac{0.912}{R_{np}}-0.0228\left(\frac{R_{np}+35)}{R_{np}}\right)^3}\right). \text{(Mooney equation)} \quad (36)$$

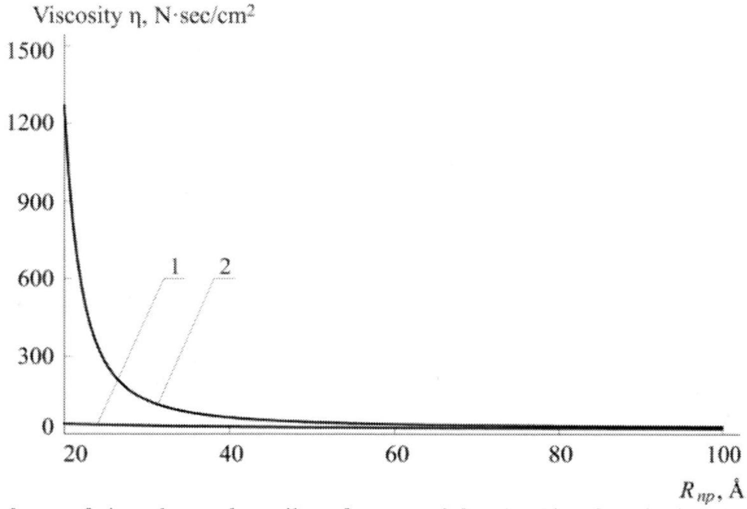


**Fig. 3.** Dependence of viscosity on the radius of nanoparticles. 1 – The viscosity is calculated by Einstein equation; 2 - The viscosity is calculated by Mooney equation. The adsorption layer takes place.

## References.

1. M. Moony, J. Colloid. Interface Sci. **6**, 2, 162 (1951)
2. I.M. Krieger, T.J. Dougherty, Trans. Soc. Rheol. **III**, 137 (1959)
3. D. Quemada, Rheol. Acta **16**, 82 (1977)
4. R. Pal, Ind. Eng. Chem. Res. **43**, 17, 5372 (2004)
5. R. Pal, J. Biomech. **36**, 7, 981 (2003)
6. J.F. Morris, F. Boulay, J. Rheol. **43**, 5, 1213 (1999)
7. Z. Cheng, J. Zhu, P.M. Chaikin, S.E. Phan, W.B. Russel, Phys. Rev. **65**, 4, 041405-1 (2002)
8. H.J.H. Brouwers, Phys. Rev. **5**, 051402-1 (2010)
9. H.J.H. Brouwers, Phys. Rev. **3**, 032202-1 (2013)
10. C.I. Mendoza, J. Chem. Phys. **135**, 054904-1 (2011)
11. S.A. Faroughi, C. Huber, Rheol. Acta. **54**, 85 (2015)
12. A. Dörr, A. Sadiki, A. Mehdizadeh, J. Rheol. **57**, 3, 743 (2013)
13. A.A. Askadskii, *Computational Materials Science of Polymers* (CISP, 2003)

# Experimental research of heat recuperators in ventilation systems on the basis of heat pipes

*Andrey* Matveev [1], *Danila* Zelentsov[1,*] and *Alexander* Louks[1]

[1]Samara State Technical University, Institute of Architecture and Civil Engineering, 194, Molodogvardeyskaya St., 443001, Samara, Russia

**Abstract.** The paper presents the results of experimental studies of heat pipes and their thermo-technical characteristics (heat power, conductivity, heat transfer resistance, heat-transfer coefficient, temperature level and differential, etc.). The theoretical foundations and the experimental methods of the research of ammonia heat pipes made of aluminum section AS – KRA 7.5 – R1 (made of the alloy AD - 31) are explained. The paper includes the analysis of the thermo-technical characteristics of heat pipes as promising highly efficient heat transfer devices, which may be used as the basic elements of heat exchangers - heat recuperators for exhaust ventilation air, capable of providing energy-saving technologies in ventilation systems for housing and public utilities and for various branches of industry. The thermo-technical characteristics of heat pipes (HP) as the basic elements of a decentralized supply-extract ventilation system (DSEVS) and energy-saving technologies are analyzed. As shown in the test report of the ammonia horizontal HP made of the section AS-KRA 7,5-R1-120, this pipe ensures safe operation under various loads.

## 1 Introduction

The purpose of this research is to create the scientific and technical basis for the design, manufacturing technology and tests of energy-saving decentralized supply-extract ventilation systems (DSEVS) based on new generation HP with tear-drop structure of the core and optimal constructive-processing and performance characteristics [1].

The objective of this research is to select optimal constructive-processing characteristics [2] for creating DSEVS based on high conductivity ammonia heat pipes (HP).

Special attention is paid to the rational and complex use of a flat HP with biphase heat-transfer fluid as a partition, and the organization of heat input in the case of flow-over of warm, exhausted air (the evaporation zone below) and heat removal (the condensation zone above), using fin fan computer heat exchangers (aluminum, aluminum-copper radiators ) [3, 4]. The objects of the research are HP and fans-heat recupretors based on high-conductivity heat pipes [5].

* Corresponding author: dvzelentsov@mail.ru

**Notation system**

| | |
|---|---|
| $\sigma$ | coefficient of surface tension |
| $\rho_l$ | liquid density |
| $\rho_v$ | steam density |
| $\mu_l$ | liquid dynamic viscosity |
| $\mu_v$ | steam dynamic viscosity |
| $d_{hp}$ | outside diameter of the pipe |
| $d_{in}$ | inside diameter of the pipe |
| $l_e$ | length of evaporation zone |
| $l_c$ | length of condensation zone |
| $l_{hp}$ | geometrical length (height) of the heat pipe |
| $l_{add,\,e}$ | length of evaporation zone under additional or bottom heater |
| $l_{add,\,c}$ | length of condensation zone under additional heater |
| $T_1$ | temperature of evaporator shell ($T_1$=$T_u$) |
| $T_2$ | temperature of condenser shell ($T_2$=$T_\kappa$) |
| $\lambda_l$ | heat conductivity of ammonia |
| $L$ | evaporation heat |
| $g$ | intensity of gravity |
| $Q$ | heat transfer ability of the thermal syphon or heat pipe |
| $\Delta P_s$ | capillary pressure of liquid |
| $\Delta P_f$ | capillary pressure of film |
| $D_v$ | diameter of steam passage channel |
| $G$ | amount of heat-transfer fluid put in the interior of the heat pipe |

## 2 Materials and methods

The paper presents the integrated development of methodological support for thermovacuum tests of honeycomb panels with built-in heat pipes, with simulators of heat-emitting equipment: physical analogues of vertically and horizontally located heat pipes for thermophysical tests; the formation of excess liquid in the HP and its influence on the thermotechnical characteristics of the HP are analyzed; the technique for performance calculation of the HP with heat input in the bottom part that operates in the biphase thermosyphon regime; the technique of performance calculation of the HP with several simulators of heat-emitting equipment that operates in the biphase thermal syphon regime under terrestrial conditions, and the main features of the thermal operating mode are revealed; the experimental values of heat transfer coefficients in the evaporation and condensation zones of the HP are given; the example of calculating the thermo-technical characteristics of the HP made of the section AS-KRA7.5-R2; the experimental results of the tests of the HP made of the section AS-KRA7.5-R1 with several heat sinks are analyzed; the experimental results of tests of the horizontal HP made of the section AS-KRA7.5-R1 with several simulators of heat-emitting equipment are analyzed [6].

**Formation of liquid droplets.** Due to the high speed of steam, it is impossible that free-floating liquid droplets exist for a long time. Such droplets, if created as a result of liquid film stripping from the wave crests by countercurrent flow of liquid and steam, will be carried away by steam and will settle at the end of the condensation zone, forming an inactive leg.

**Formation of a liquid slug.** Under overload conditions (in 0-g medium), excess liquid may accumulate at the end of the condensation zone of the heat pipe and form a liquid slug, that completely fills the cross section of the steam passage channel of the pipe. This so-called slug reduces the active effect of high temperature of the steam flow on the remaining active leg of the heat pipe.

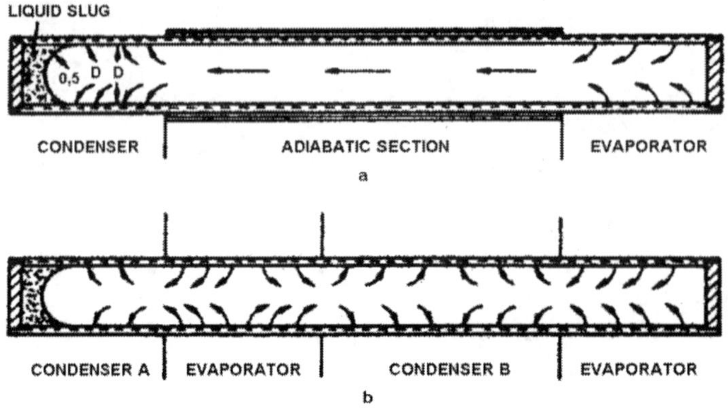


**Fig.1. a,b.** The configuration of a liquid slug in the heat pipe with different location of condenser and evaporator.

**Formation of a liquid film.** The liquid film layer on the condenser's wall may have a stable configuration if the condenser is not located close to at least one end of the manufactured heat pipe. Due to the low heat conductivity of liquid, the thermal characteristics (transfer ability, power) of the heat pipe can be significantly reduced.

However, combined with the slug, the liquid film is not a stable configuration. Let us consider a simple configuration of the HP with an evaporator at one end and a condenser at the other (Figure 1,a).

In the case of a liquid slug formation, the stable surface of the steam passage channel's interface would be a hemisphere (although there will be a form somewhat different from this ideal contour, due to the influence of the steam flow). If the excess liquid forms a film layer in the condensation zone adjacent to the slug, then the two separate and communicated (having hydraulic connection) liquid formations exist as a system [7, 8]. However, the capillary pressure of these two surfaces is absolutely different. The capillary pressure for the liquid $\Delta P_s$ and the cylindrical layer of film $\Delta P_f$ is expressed respectively as

$$\Delta P_S = \frac{4\sigma}{D_v} \tag{1}$$

$$\Delta P_f = \frac{2\sigma}{D_v} \tag{2}$$

That is, the capillary pressure of the formed liquid slug is twice more than that of the liquid film layer. So, we can make the conclusion that the liquid film layer in the condensation zone located close to the end of the pipe is not stable. That excess fluid will be involved in the more stable configuration of the liquid slug.

In Fig. 1,b we can see another, though typical situation - the condenser placed in the middle of the pipe is between two evaporation zones. The only liquid bond between this area and the area of the slug in the end of the pipe is through the capillary grooved structure. However, the capillary pressure of these grooves in the evaporation zone is higher than on the slug's surface, so the supply of the possible mass of liquid from the film layer to the slug can not take place. For this configuration it is necessary to consider completely different positions of various areas of the condenser and evaporator.

To provide the given heat power $Q$, besides the temperature drop $T_1$-$T_2$ along the interior wall of the thermal syphon (heat pipe), the thermo-physical characteristics $L$, $\rho_l$, $\lambda_l$, $\mu_l$, the geometry of the pipe, it is necessary to provide the geometry of the evaporation and condensation zones and the correlation between them. Moreover, this dependence is non-linear and not arbitrary. The condensation zone $l_c$ is located above the evaporator (heating area), and the condensate film moves only downward under the action of gravity.

In the horizontal position of HP ($\alpha = 0^0$), the number of evaporation zones along the pipe's length may be arbitrary, when the mass inflow of liquid condensate into the evaporation zone (simulator of heat-emitting equipment) takes place under the influence of capillary forces both to the left and to the right of the evaporator.

The summary temperature di $\Delta T$ between the evaporation and condensation zones along the exterior surface of the heat pipe consists of the temperature drops through the thickness of the shell's wall in evaporation $\Delta T_{w,e}$ and condensation $\Delta T_{w,c}$ zones, through the thickness of the condensate film in evaporation $\Delta T_{f,e}$ and condensation $\Delta T_{f,c}$ zones:

$$\Delta T=\Delta T_{w,e}+\Delta T_{w,c}+\Delta T_{f,e}+\Delta T_{f,c} \tag{3}$$

The sum $\Delta T_{f,e}+\Delta T_{f,c}$ is the temperature different along the interior surface of the heat pipe lengthwise, i.e. $\Delta T_{f,e}+\Delta T_{f,c}= T_1-T_2$.

The temperature different s $\Delta T_{w,e}$ and $\Delta T_{w,\kappa}$ are defined from the equations of heat transfer via heat conductivity through cylindrical wall:

$$\Delta T_{w,e} = \frac{Q}{\pi l_e}\frac{1}{2\lambda} ln\frac{d_{in}}{d_{hp}} \tag{4}$$

$$\Delta T_{w,c} = \frac{Q}{\pi l_c}\frac{1}{2\lambda} ln\frac{d_{in}}{d_{hp}} \tag{5}$$

The heat pipe has got a uniform cross-section and a thin wall. Having substituted the values of temperature drop components in equation (3), we will obtain

$$\Delta T = \frac{Q}{2\pi\pi} ln\frac{d_{in}}{d_{hp}}\left(\frac{1}{l_e}+\frac{1}{l_c}\right)+(T_1-T_2) \tag{6}$$

When making calculations for a smooth-wall heat pipe, the geometry, material of the shell, type of heat-transfer fluid, operating conditions of the heat pipe in the evaporation zone or in the condensation zone are presupposed.

The dependence $Q=f_1(T_e, \Delta T)$ by $T_c$ = const or $Q=f_1(T_e, \Delta T)$ is established, here by $T_e$ = const $T_c$, $T_e$ are the temperatures on the exterior surface of the heat pipe in the evaporation and condensation zones. Having made the calculations for a number of values of $Q$, the dependencies $Q=f_2(T_c, \Delta T)$ are established.

As shown by the experimental research [9, 10], the temperature drop in the steam passage channel depends on the amount of heat-transfer fluid filling the pipe, and on the tilt angle of the axis of the heat pipe to the horizon. The three operation modes of the smooth-wall heat pipe (Fig.2) differ in the amount of filling heat-transfer fluid.

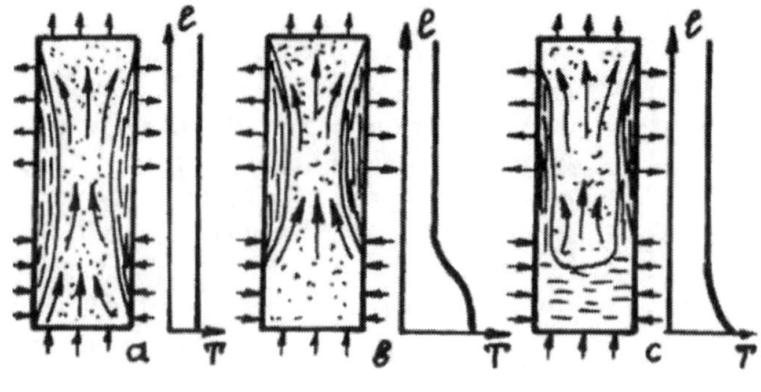


**Fig.2.** Influence of the heat pipe's filling degree on its operation mode.

1. Isothermal mode of the heat pipe's operation (Fig. 2,a). In this mode the amount of the filling heat-transfer fluid fully corresponds to the value of the transferred heat flow.

2. The mode of insufficient filling (Fig. 2,b). In this case there is not enough liquid for the film to cover the interior surface of the walls of the heat pipe's shell completely. In bare areas there is localized overheating of the wall.

3. The mode of excessive filling (Fig. 2,c). By excessive filling, a "puddle" forms on the bottom of the heat pipe. Excessive filling with heat-transfer fluid is less dangerous for the operation of the heat pipe, but there is significant temperature drop along the height of the "puddle".

For the HP made of section AS-KRA 7.5-R2, built into the thermopanel, the third mode (c) is typical, with excess heat-transfer fluid (up to about 10%) accumulating on the bottom of the thermal syphon, as the wall temperature in the condensation zone practically doesn't change, and in the evaporation zone it falls, starting from the lower limit.

For optimal filling of the heat pipe, we get

$$G = \left( \frac{4}{5} l_c + l_{hp} + \frac{4}{5} l_e \right) \sqrt[3]{\frac{3Q\mu_l \rho_l \pi^2 d_{hp}^2}{Lg}} \tag{7}$$

The calculation of the characteristics of the heat pipe with several simulators of heat-emitting equipment that operate in the biphase thermosyphon regime under terrestrial conditions is carried out with the help of the method described below.

When the heat pipe (HP) operates with several simulators of heat-emitting equipment in the thermal syphon regime, the heat flows differ in size and density

$$Q_i = \frac{4}{3} \pi d_{hp} \sqrt[4]{\frac{L_i \rho_{l,i}^2 g \lambda_{жс,i}^3 \left(T_{1,i} - T_{2,i}\right)^3 l_{e,i}^3 l_{c,i}^3}{4\mu_{l,i} \left(l_{e,i} + l_{c,i}\right)^3}} \tag{8}$$

I.e. the biphase thermosyphon as if consists of several component thermosyphons, each with its own evaporation $l_{u,i}$ and condensation $l_{\kappa,I}$ zones, joined together in the same shell.

$$\sum_{i=1}^{n} \left(l_{e,i} + l_{c,i}\right) + \left(l_{addi} + l_{addi}\right) = l_{hp} \tag{9}$$

By vertical arrangement of the honeycomb panel, the role of the additional heater may be played by the simulator of heat-emitting equipment, shifted to the lower limit position. Then the total heat load from the simulators of heat-emitting equipment (heaters), including the load on the additional heater, will be:

$$Q_\Sigma = \sum_{i=1}^{n} Q_i + Q_{add.h} = \sum_{i=1}^{n}\left[\frac{4}{3}\pi d_{hp}\sqrt[4]{\frac{L_i\rho_{l,i}^2 g\lambda_{l,i}^3\left(T_{1,i}-T_{2,i}\right)^3 l_{e,i}^3 l_{c,i}^3}{4\mu_{l,i}\left(l_{e,i}+l_{c,i}\right)}}\right]+Q_{add.h} \tag{10}$$

In this case, the total heat load may be determined from the heat release cyclogram of the heat-emitting equipment. The optimal filling value is calculated at the total heat load $Q_\Sigma$ with the help of the formula:

$$G=\left(\frac{4}{5}l_{c,i}+l_{hp,i}+\frac{4}{5}l_{e,i}\right)\sqrt[3]{\frac{3Q_\Sigma \mu_{l,i}\rho_{l,i}\pi^2 d_{hp}^2}{L_i g}} \tag{11}$$

During tests the value $Q_{add.h}$ is a variable, and it's the part $\xi$ of $\Sigma^n{}_{i=1}Q_i$ . Then the total heat load for the thermal syphon with simulators of heat-emitting equipment will be:

$$Q_\Sigma = \sum_{i=1}^{n} Q_i + \xi\sum_{i=1}^{n} Q_i = (1+\xi)\sum_{i=1}^{n}\left[\frac{4}{3}\pi d_{hp}\sqrt[4]{\frac{L_i\rho_{l,i}^2 g\lambda_{l,i}^3\left(T_{1,i}-T_{2,i}\right)^3 l_{e,i}^3 l_{c,i}^3}{4\mu_{l,i}\left(l_{e,i}+l_{c,i}\right)}}\right] \tag{12}$$

The complex ratio $\xi$ depends on the thermal conditions of the heat-emitting equipment (from the cyclograms of internal heat release), on the conditions of the launch modes of the honeycomb panel with built-in HP, on the location of the controls on the honeycomb panel, and on other factors, and it's obligatory to determine it experimentally during testing.

During the operation of the *i*-th heat-emitting device, the liquid condensate film on the top of the condenser flows to the *i*-th evaporator zone, and the liquid condensate film on the bottom part of the condenser summarizes, after the condensation of saturated steam, with the condensate film from the $(i + 1)$-th device, and flows under the action of gravity into the evaporation zone of the $(i + 1)$-th device. It can not move up to the evaporation zone of the top device. Thus, when arranging simulators of heat-emitting equipment, it is necessary to implement the principle: the increase in the heat power of devices from top to bottom when they are positioned along the height of the biphase thermosyphons.

## 3 Results

As the results of testing the HP made of the section AS-KRA7.5-R1-120 or R2 show, the symmetrical distribution of thermal power between two simulators of heat-emitting equipment lengthwise in horizontal position and heightwise in vertical position, when $Q_1$=$Q_2$=5; 10; 15; 20 W contributes to the increase in the degree of isothermality of the honeycomb panel's temperature pattern. The surface temperature does not exceed 40 °C, and the longitudinal and transverse temperature drops in the evaporation and condensation zones are quite acceptable.

**Table 1.** Development of the input control method of the thermo-physical characteristics of the HP with grooved structure of the core, in case of several heat sinks by way of computer fans in DSEVS

| № | τ, hrs | $T_{air.}$ $^0$C | Q, W | $T_1$ | | $T_3$ | | $T_2$ | | $\Delta T$ | Note |
|---|---|---|---|---|---|---|---|---|---|---|---|
| | | | | мВ | $^0$C | мВ | $^0$C | мВ | $^0$C | $^0$C | |
| 1 | $16^{15}$ | 21 | 20 | 0.58 | 29.857 | 0.57 | 29.714 | 0.52 | 29.0 | 0.857 | With exit fan № 1 |
| 2 | $16^{30}$ | 21 | 20 | 0.57 | 29.714 | 0.58 | 29.857 | 0.51 | 28.833 | 0.881 | With exit fan № 1 |
| 3 | $16^{45}$ | 21 | 20 | 0.55 | 29.429 | 0.56 | 29.429 | 0.5 | 28.66 | 0.769 | With two fans № 1, 2 |
| 4 | $17^{00}$ | 21 | 20 | 0.44 | 27.714 | 0.46 | 28.0 | 0.41 | 27.286 | 0.428 | With three fans № 1, 2, 3 |

The temperature sensor $T_1$ is installed in the heating (evaporation) zone in the middle of the flange in the place where the heater is fixed; the temperature sensor $T_2$ is in the zone of heat removal (condensation) in the middle of the flange along the same generatrix in the places where the cooler is fixed; the temperature sensor $T_3$ is in the middle of the transport zone.

**Table 2.** Development of the input control method of the thermo-physical characteristics of the HP with tear-drop grooved structure of the core and several computer fans

| № | τ, hrs | $T_{air}$, $^0$C | Q, W | $T_1$ | | $T_2$ | | ΔT | Note |
|---|---|---|---|---|---|---|---|---|---|
| | | | | мВ | $^0$C | мВ | $^0$C | $^0$C | |
| 1 | $11^{45}$ | 19 | 15 | 0.33 | 24.0 | 0.28 | 23.286 | 0.714 | With all the four fans |
| 2 | $12^{00}$ | 19 | 15 | 0.31 | 23.714 | 0.27 | 23.143 | 0.571 | With all the four fans |
| 3 | $12^{15}$ | 19 | 15 | 0.30 | 23.571 | 0.27 | 23.143 | 0.428 | With three fans № 1, 2, 3 |
| 4 | $12^{30}$ | 19 | 15 | 0.34 | 24.167 | 0.31 | 23.714 | 0.429 | With three fans № 1, 2, 3 |
| 5 | $13^{00}$ | 19 | 15 | 0.31 | 23.714 | 0.28 | 23.286 | 0.428 | With three fans № 1, 2, 3 |
| 6 | $13^{45}$ | 20 | 15 | 0.24 | 23.667 | 0.21 | 23.167 | 0.50 | With three fans № 1, 2, 3 |
| 7 | $14^{00}$ | 20 | 15 | 0.25 | 23.833 | 0.22 | 23.333 | 0.50 | With three fans № 1, 2, 3 |

## 4 Discussion

Compared to the regenerative heat recuperator of ventillation air with a massive copper heat exchanger ($\lambda_c \sim 308$ W/(m·K)), the use of the ammonia HP intensifies the transfer of heat.

The heat transfer coefficient for the ammonia aluminum pipe with a grooved core related to the cross-sectional area and to the surface area of the evaporator reaches $1.06 \times 10^5$ W/(m$^2$·K) and 2660 W/(m$^2$·K).

To obtain the best thermo-technical characteristics of the HP, it is necessary to use such constructions of capillary systems (CS), which will at the same time provide high mass velocity and high level of heat exchange in the heat input and removal zones. At the same time, there are conflicting requirements to the CS: large capillary potential, low hydraulic resistance, high effective heat conductivity.

These conditions are met by the ammonia aluminum HP with a tear-drop grooved core, that combines a round artery with a narrow split of the type AS-KRA7.5-R1 or R2. The heat transfer in the areas of input and removal of heat of the HP is intensified by the use of computer finned aluminum radiators with fans on top, providing the air capacity of up to 125 m$^3$/h in DSEVS (w ~ 0.5 m/s).

## Conclusions

Basing on the mathematical model for the heat pipe operating in biphase thermosyphon mode, the research of heat pipes with different Ω-shaped grooves was carried out [11]. When using various technical means of heat input and removal in Keldysh Center and in the laboratory of Samara State University, the thermo-physical characteristics on trials of the HP made of AS-KRA 7.5-R1-30 section 1762 mm long correlate well with each other. The maximum heat transfer capacity of the HP made of the section AS-KRA7.5-R2 is slightly different from the value ($Q_{max}$) for the HP made of the section AS-KRA7.0-R2.

This is the effect of the unequally spaced mechanical roughness on the upper surface of a large number of grooves in the core, which affects the formation of steam bubbles, thus intensifying heat exchange. This fact is confirmed by a sharp increase of heat transfer coefficient during evaporation ($\alpha_e$ = 15800 W/(m$^2$·K) for the HP made of the section AS-KRA7.0-R2.

When testing the HP, large values of heat transfer coefficients during evaporation and condensation were obtained. The revealed self-similarity of the heat transfer coefficient values in the evaporation and condensation zones of the HP with high heat conductivity from the density of heat flow opens up the possibility for the more reliable calculation of effective heat conductivity for structural cores with tear-drop grooved structure, which is significantly higher than for rectangular and trapezoidal grooved cores.

The value of heat transfer resistance of the HP is 0.01-0.02 K/W by maximum transmitted heat flow. The heat transfer coefficient values for evaporation are much higher than the same values for the known analogues of the HP. The HP qualification test program was carried out in strict compliance with the European standard [12]. Mechanical, thermal, radiation and service life tests of the pipes were performed.

## References

1. D.V. Zelentsov, A.L. Louks, Traditions and innovations in civil engineering. Construction technologies. Articles, 315-317 (2015)
2. E.A. Tsynaeva, A.A. Tsynaeva, Journal of Engineering Physics and Thermophysics. 1-8 (2016). doi: 10.1007/s10891-016-1505-6
3. W. Harwell, W.B. Kaufman, L. Tower, Re-Entrant groove heat pipe. AIAA Paper. **(773)** 1-8 (1977)
4. R. Schlitt, Performance Characteristics of Recently Developed High Performance Heat Pipes. 8th International Heat Pipe Conference. Beijing (1992)
5. L.L. Vasiliev, V.G. Kiselev, Yu.N. Matveev, F.F. Molodkin, Heat Exchangers-Heat Recovery Units on Heat Pipes (Science and Technology, Minsk, 1987)
6. A.I. Streltsov, Theoretical and Experimental Research of Heat Pipes Filling. Bulletin of Higher Educational Institutions. Power Engineering, 12 (1973)
7. S.S. Kutateladze, M.A. Styrikovich, Hydrodynamics of Liquid-Gas Systems (Energia, Moscow, 1976)
8. V.P. Isachenko, V.A. Osipova, A.S. Sukolin, Heat Transfer (Energia, Moscow, 1975)
9. M. Dubois, B.Mullender, W. Supper, Space Qualification of High Capacity Grooved Heat Pipes. 10lh International Heat Pipe Conference, September 21-25, Stuttgart, Germany. Prepr. of Sess. B, C, D, 1-7 (1997)
10. C. Brand, P. Stephan, M. Dubois, B. Mullender, Theoretical Investigation of Advanced Capillary Structures in Grooved Heat Pipe Evaporators for Space Applications. 30th International Conference on Environmental System SAE (2000)
11. A.L. Louks, E.A. Krestin, A.G. Matveev, V.I. Vesnin, Urban Construction and Architecture. 3 (24), 17-24 (2016). doi: 10.17673/Vestnik.2016.03.3
12. Heat Pipes Qualification Requirements. ESA PSS-49, Issue 2, March 1983

# Research on the technology of constructing a protection with joint use of no-fines concrete and bitumen

*Andrey* Mikhasek [1,*]

[1]Samara State Technical University, Institute of Architecture and Civil Engineering, 194, Molodogvardeyskaya St., 443001, Samara, Russia.

**Abstract.** To build a bank slope protection of a hydraulic structure, a technology is proposed that involves building a framework of no-fines concrete and pouring it with hot bituminous materials. Implementation of this technology requires conducting a research on bitumen penetration into the protection plate body. The research involved developing models of non-stationary filtration under varied bitumen viscosity. The research involved experimental observation in laboratory environment. Models of movement of a variable viscosity liquid in a porous medium were proposed, and their study in laboratory environment allowed for identifying the depth of penetration, which is an important result to solve engineering objectives.

## 1 Introduction

A search for solutions to the problem of strengthening the waterline to reduce the construction costs is an important and relevant objective [1-3]. A new technology [4] that involves pouring a no-fines concrete plate with hot bitumen is proposed to protect the bank slope of a hydraulic structure.

While moving, the bitumen gets colder and its viscosity changes; this results in velocity reduction and, consequently, the bitumen sticks in the body of no-fines concrete.

Thus, when choosing a model, the following objective (goal) has been set: to identify dependence of the depth of bitumen penetration (in general, a variable viscosity liquid) on its characteristics and the characteristics of no-fines concrete (porous medium).

The depth of viscous liquid penetration into a porous medium will depend on the following parameters:

- the liquid density $\rho$;
- viscosity $\mu$;
- the ratio $\Delta K$ of the initial liquid temperature to the temperature at the moment of movement termination - the temperature of "sticking";
- effective diameter d of the medium particles;
- porosity coefficient m;

* Corresponding author: andremixas@mail.ru

- free fall acceleration g.

## 2 Materials and Methods

To reach the first objective, we used a symmetric tube of flow that can be represented as a vertically positioned straight unit tube, centered relative to the axis of symmetry.

Taking into account that the issue of variable viscosity bitumen penetration into no-fines concrete has been studied very little, we use the dimension theory method, and then the correlation between the values can be written as follows:

$$h = f(\mu, \rho, d, g, \Delta K, m), \tag{1}$$

where h – depth of variable viscosity liquid penetration into a porous medium, Π – according to the theorem, any correlation between physical dimensions can be presented as a dependence between $n+1-k$ (k – the number of physical values with independent dimensions) and dimensionless values Π.

Let us take $\mu$, $\rho$, $d$ as parameters with independent dimensions. To check their independence, we write down their dimensions and calculate the determiner comprised of the indicators of base dimension exponent values:

$$[\mu] = L^{-1}MT^{-1} = \left[\frac{kg}{m \cdot s}\right]; [\rho] = L^{-3}M = \left[\frac{kg}{m^3}\right]; [d] = L = [m]. \tag{2}$$

The determiner can be represented as follows:

$$\begin{vmatrix} & L & M & T \\ \rho & -3 & 1 & 0 \\ \mu & -1 & 1 & -1 \\ d & 1 & 0 & 0 \end{vmatrix} \neq 0.$$

The determiner is other than zero, so the parameters have independent dimensions.

On the basis of the values in brackets, using Buckingham's method, we take m, ΔK for dimensionless values. We obtain dimensionless complex

$$P = \frac{h}{d}.$$

The dimensionless complex

$$P_1 = \frac{g}{\rho^{\alpha} \cdot \mu^{\beta} \cdot d^{\gamma}}$$

is obtained.

Then we find coefficients $\alpha$, $\beta$, $\gamma$ of the dimensionless complex P1.

The numerator and denominator have equal value, so

$$\frac{g}{\rho^{\alpha} \cdot \mu^{\beta} \cdot d^{\gamma}} = \frac{L \cdot T^{-2}}{L^{-3\alpha} \cdot M^{\alpha} \cdot L^{-\beta} \cdot M^{\beta} \cdot T^{-\beta} \cdot L^{\gamma}}. \tag{3}$$

We equate exponent values with equal dimensions and obtain the system of equations:

$$\begin{cases} -3\alpha - \beta + \gamma = 1 \\ \alpha + \beta = 0 \\ \beta = 2 \end{cases}. \tag{4}$$

We solve this system of equations (5) and obtain:

$$\begin{cases} \alpha = -2 \\ \beta = 2 \\ \gamma = -3 \end{cases}.$$

We substitute the obtained values into (3) and obtain (5):

$$P_1 = \frac{g}{\rho^{-2} \cdot \mu^2 \cdot d^{-3}} = \frac{\rho^2 \cdot g \cdot d^3}{\mu^2}. \tag{5}$$

The obtained dimensionless complex is a Galileo number [2] that can be otherwise written as:

$$Ga = \frac{\text{Re}^2}{Fr} = \frac{g \cdot d^3}{\nu^2}, \tag{7}$$

or we can express kinematic viscosity in terms of dynamic viscosity

$$\nu = \frac{\mu}{\rho}$$

and obtain:

$$Ga = \frac{g \cdot d^3 \cdot \rho^2}{\mu^2} = P_1. \tag{6}$$

According to V.V. Gukhman [5], Galileo number describes free flow of liquid, which we believe to most fully reflect the process of viscous liquid penetration into a porous medium due to gravity. Thus, the formula (2) can be represented as:

$$h = d \cdot f(Ga, \Delta K, m). \tag{7}$$

The type of dependence (7) can be obtained through processing experimental data that we plan to obtain using models.

## 3 Results

Preparatory work in experiments with no-fines concrete also included concrete preparation directly in the metal tube. To make concrete, we used crushed stone with fraction of 20-40; concrete mix composition and concrete preparation method in laboratory environment are described earlier. The metal tube was greased with oil for further separation of concrete from its surface, and then concrete was laid and compacted, in the same way as crushed stone. The no-fines concrete was cured for 24 hours.

Penetration results were evaluated using a measurement method and presented in Table 1. Then we conducted one-factor dispersion analysis of the obtained results, which is presented in Tables 2 and 3.

**Table 1.** Results of experiments with the bitumen model using no-fines concrete.

| Experiment No. | Penetration depth in m at the temperature of °C | | |
|---|---|---|---|
| | **120** | **160** | **200** |
| 1 | 0.16 | 0.4 | 0.75 |
| 2 | 0.13 | 0.36 | 0.64 |
| 3 | 0.14 | 0.37 | 0.67 |
| Average value | 0.14 | 0.38 | 0.69 |

**Table 2.** Dispersion calculation.

| Groups | Count | Sum | Average | Dispersion |
|---|---|---|---|---|
| Column 1 | 3 | 0.43 | 0.14 | 0.0002 |
| Column 2 | 3 | 1.13 | 0.38 | 0.0004 |
| Column 3 | 3 | 2.06 | 0.69 | 0.003 |

**Table 3.** Variation source.

| Variation source | SS | df | MS | F | P-Value | F critical |
|---|---|---|---|---|---|---|
| Between the groups | 0.45 | 2 | 0.223 | 171.4 | 5.08619E-06 | 5.14 |
| In the groups | 0.0078 | 6 | 0.0013 | | | |
| Total | 0.45 | 8 | | | | |

Thus

$$\hat{F} = 171.4 > F_{(k=1,n-k,\alpha)} = 5.14,$$

the penetration depth data with the bitumen model using no-fines concrete and crushed stone with fraction of 20-40 statistically belong to a single functional dependence.

The conducted research on physical models indicates that bitumen movement is affected by the following factors: bitumen density$\rho$, the ratio $\Delta K$ of its initial temperature to the temperature of sticking, viscosity coefficient $\mu$, effective diameter d of no-fines concrete particles, porosity coefficient m and free fall acceleration g. Using the dimension theory method, we identified functional dependence of the depth of bitumen penetration into no-fines concrete on the parameters indicated in Formula 7. We need to identify the type of this function.

Exponential, logarithmic and direct functions were considered as possible types of functional dependence. Logarithmic function was chosen by trial and error.

Thus, by data fitting, the following type of dependence was identified:

$$h = d \cdot \left( \ln \frac{Ga}{B \cdot \Delta K} \right) \cdot A \cdot m, \quad (8)$$

where d - effective diameter of the porous medium particles, Ga – Galileo number is identified by the dependence, $\Delta K = \frac{K_{in}}{K_{st}}$ – the ratio of the temperature at the start of injection Kinitial to the temperature of liquid sticking Ksticking (for water, the ratio is 273 K, for bitumen – 373K), m – porosity coefficient, A and B – correction coefficients obtained through the least squares method based on the results of experimental data processing.

## 4 Discussion

On the basis of the conducted research, we obtained the dependence that can be used for engineering calculations, namely: [1] process temperature of 170-180 °C is established and recommended for no-fines concrete used with the aggregate of 80-120 mm, with the layer thickness of 60 cm.

The dependence obtained is valid for different mediums. Yet comparison of penetration depths on different models is possible only with equal values of Ga numbers and equal volumes of the injected liquid.

The proposed method allows to predetermine the layer thickness depending on sealing compounds and penetration depth for no-fines concrete made with different aggregates, though this requires correction of A and B coefficients.

The theoretical data obtained were proved by a full-scale experiment on an experimental block of no-fines concrete: bitumen fully penetrated into the whole layer, the penetration depth and the shape of spreading were close to the estimated values.

The results of the research were used to identify the main process parameters [6-10].

## 5 Conclusions

The research conducted showed:

1. The proposed method of building bank slope protection that involves pouring a no-fines concrete plate with hot bitumen can be implemented in practice, main process parameters, such as the temperature of bitumen and the size of the aggregate, can be predetermined, and penetration depth can be predicted.

2. The experiment conducted proved that liquid of a certain variable viscosity sticks in porous medium at a certain depth. The experiment conducted on physical models provided us with data on the depth of hot bitumen penetration into no-fines concrete made of crushed stone with fraction of 20-40. It is established that statistical penetration depth depends on temperature. In case of non-stationary processes, the shape of spreading is described as a circle.

## References

1. I.S. Sobol, The Volga Region Scientific Magazine **2 (26),** 65-67 (2013)
2. A.V. Mishakova, A.V. Vakhrushkina, V.A. Murgul, T.V. Sazonova, *Procedia Engineering 15th. Cep. "15th International Scientific Conference 'Underground Urbanisation as a Prerequisite for Sustainable Development', ACUUS 2016"*, 1812-1817 (2016)
3. M.I. Balzannikov, I.S. Kholopov, M.V. Kholopov, M.G. Litvinova, Scientific Survey **22,** 65-71 (2015)
4. Andrey Mikhasek, Boris Ivanov, MATEC Web of Conferences **105,** 03022 (2017)
5. V.V. Gukhman, *Introduction to the Theory of Similitude* (Higher School, Moscow, 1963)
6. A.A. Mikhasek, A.A. Smyvalov, Scientific Survey **14,** 102-109 (2015)
7. A.A. Mikhasek, M.V. Rodionov, Procedia Engineering **91,** 82-88 (2015)
8. A.A. Mikhasek, Scientific Survey **14,** 67-72 (2016)
9. A.A. Mikhasek, M.V. Rodionov, V.A. Seliverstov, Y.E. Senitsky, MATEC Web of Conferences **86** (2016)
10. M.I. Balzannikov, M.V. Rodionov, V.A. Seliverstov, Urban Construction and Architecture **1,** 100-105 (2011)

# Influence of air tightness of the building on its energy-efficiency in single-family buildings in Poland

*Artur Miszczuk*[1,*]

[1]Warsaw University of Technology, Faculty of Civil Engineering, Institute of Civil Engineering, Team Building and Sustainable Development Al. Armii Ludowej 16, 00-637 Warsaw, Poland

**Abstract.** This publication focuses on assessing the impact of the tightness of single-family houses with a higher energy standard on their energy demand for heating. To formulate conclusions quantitative and qualitative research, including tightness test (blower door) has been conducted in energy-efficient and passive houses. In the next step, energy demand for heating has been estimated. Based on the observation and results, the simulation of the impact of reducing the flow of infiltrated air through leaks in the buildings for energy demand for heating is calculated. The simulation results confirm the dependence between the tightness of the building and energy demand.

## 1. Introduction

In the revised Regulation of the Minister of Transport, Construction and Maritime Economy in 2013 (on the technical conditions to be met by buildings and their location) [1], a program of gradual reduction of energy consumption by new buildings was introduced. Achievement of the energy standard of the building at the design stage is in many cases impossible to meet because of the low air-tightness of the external partitions. This problem occurs in energy-efficient buildings where mechanical ventilation with heat recovery is used. The use of recuperation is a solution that largely limits the heat loss to the heating of the ventilation air. In this situation, the heat loss caused by the infiltration of the outside air is acquired.

## 2. Examples of conventional tests and analyses on European market

Croatian [2] and Lithuanian [3] researchers analysed the possibility of a relationship between the airtightness of a building and its demand for energy. They came up with the assumption that the building's airtightness is a factor influencing the speed of air infiltration. Consequently, it also affects the air quality in the building and ventilation,

* Corresponding author: a.miszczuk@il.pw.edu.pl

but also the energy required for heating and/or cooling. The authors in them analyses prove that as the building's airtightness increases, heat losses decrease [4].

Analogical to Croatian and Lithuanian studies were conducted under Polish conditions [5, 6]. The obtained results confirm the effect of leakage on the value of the demand for utility energy, and consequently, on the value of the demand for non-renewable primary energy. This impact is greater for buildings with lower energy requirements and for buildings with significant leaks.

In the conclusion of other studies in Poland worth recalling [7], the researcher has shown that low leakage of partitions may increase the demand for utility energy for heating and ventilation of energy-efficient buildings by up to 40 %.

## 3. Required level of tightness for new buildings

The Polish Technical Conditions (WT) [1] define the level of recommended airtightness of the building, which is respectively for the building:

- with gravitational ventilation - $n_{50} < 3.0$ h$^{-1}$ (where $n_{50}$ is the number of air exchanges in a building, within 1 h, by infiltration, at a pressure difference of 50 Pa)
- equipped with mechanical ventilation or air conditioning - $n_{50} < 1.5$ h$^{-1}$.

In the provisions the legislator recommends conducting a leak test after completion of construction. Replacing in WT the word "required" with the word "recommended" causes the record to become dead and the clear majority of investors do not conduct the examination.

In Poland, there is a provision requiring a leakage test, but this regulation was only applied to buildings that were submitted to the energy and passive buildings subsidy program (National Fund for Environmental Protection and Water Management - NFOŚiGW [9]). In the case of applying for a grant, the building must meet several requirements specified by NFOŚiGW. One of the most important is to achieve the airtightness of the outer partitions at a relatively low level. Limit values depend on the building standard: for NF40 the airtightness should be no worse than $n_{50} \leq 1.0$ h$^{-1}$ and for NF15 $n_{50} \leq 0.6$ h$^{-1}$ (where NF is the demand for utility energy for heating and ventilation respectively: at 40 kWh/(m$^2$·a) and 15 kWh/(m$^2$·a)) [9].

## 4. Research

To conclude whether there is a relationship between the building's airtightness level and its energy demand, energy-efficient and passive buildings located in different regions of Poland were tested.

### 4.1. Assumptions for research

17 single-family buildings located in the most diverse climates zone in Poland [10], which were characterized by potentially low energy requirements, were qualified for the study. The selection of buildings for energy consumption was based on:

- Energy Performance Certificate (if the building has one),
- grants application send by the building owner to NFOŚiGW for subsidies to the energy efficient building (NF40) or passive (NF15),
- preliminary analysis of the structure of the building.

All the buildings qualified for the test were equipped with mechanical ventilation with heat recovery from the air removed.

### 4.2. Research procedure

Determination of the leakage level of individual buildings has been carried out in accordance with the applicable European Standard PN-EN ISO 9972:2015 [7], per method B. During the tests, all deliberately made openings in the building's enclosure (such as exterior doors, windows, chimney openings) and diffusers and ventilators in the ventilation system are closed or blinded [4]. Preparation of the building for the test also included the possibility of equalizing the air pressure by leaving all internal doors open in the heated zone of the building [5].

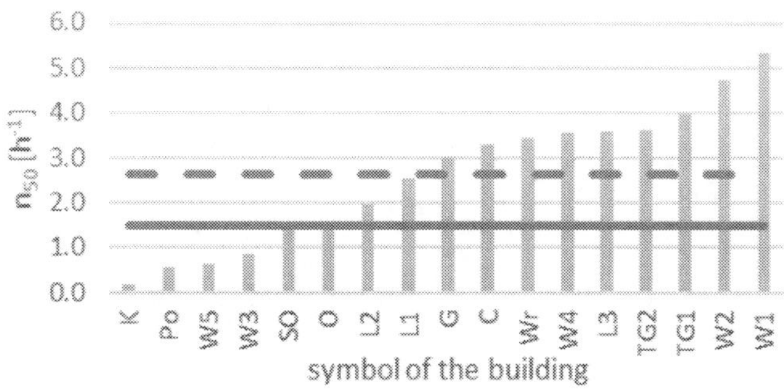


**Fig. 1.** Level of building's airtightness

The buildings in which the leakage test was carried out had a varying level of $n_{50}$ (from 0.17 $h^{-1}$ for building K to 5.33 $h^{-1}$ for W1) (fig.1). Most of them exceeded the ventilation recommended for buildings with air infiltration value ($n_{50} < 1.5\ h^{-1}$) [1] (continuous line in Fig. 1). The average value for all buildings was $n_{50} = 2.6\ h^{-1}$ (dotted line in Fig. 1), and it exceeds (less than 2 times) the value of the infiltration.

### 4.3. Calculation results

The calculation of the energy demand for heating and ventilation (EU) was carried out using the monthly method in accordance with PN-EN ISO 13790 [11].

The most important data on which the calculation of the EU index is based are:

- surface, construction and materials used in the building's external walls,
- the type and characteristics of window and door joinery and how it is assembled,
- level of recuperation (recovery of heat from removed air).

The above listed parameters were obtained from projects, inventories and from designers, manufacturers of materials/equipment and/or owners/developers of buildings.

The necessary meteorological data were read out for the nearest metrological station located in the building. The most important weather data obtained from the Ministry of Infrastructure and Development [12] include:

- average monthly outdoor air temperature,
- the sum of the total solar radiation intensity on the N, E, S, W orientation.

Shading elements (such as neighbouring buildings, arcades) have been included in the calculations to reduce the amount of solar heat gain.

Based on the research, calculations and analyses, the most important parameters characterizing each of the examined buildings were determined. The most important properties include:

- heated surface of the building (table 1),
- outer baffle surface,
- the area of windows and glazing,

- volume of heated part of the building,
- building shape factor (A/V, where A- is the surface of the building's external partition, V- building volume) (Table 1),
- heat transfer coefficients of the external partitions - walls, roofs, flooring and exterior windows and doors, the weighted average value of the heat transfer coefficient for all external partitions (walls, floors, roofs and windows and doors) - $U_{avg.}$ (Table 1),
- the orientation of glazing to the orientation of world's sides and to the gains from solar radiation,
- $n_{50}$ - airtightness of the building casing (table 1),
- EU - index of energy demand for heating and ventilation (Table 1).

Due to the confidentiality clause with the owners, the article has given only an indicative location of the buildings.

**Table 1.** Summary of results and calculations.

| **Location** | **Symbol of the building** | **Year of construction** | **Heated surface of the building** | **A/V** | $U_{avg.}$ | $n_{50}$ | **EU** |
|---|---|---|---|---|---|---|---|
| | | [-] | [m$^2$] | [m$^{-1}$] | [W/(m$^2$·K)] | [h$^{-1}$] | [kWh/(m$^2$·a)] |
| Czestochowa | C | 2013 | 101 | 0,8 | 0,23 | 3,29 | 75 |
| Gdansk | G | 2015 | 109 | 0,9 | 0,10 | 3,04 | 28 |
| Kluczbork | K | 2013 | 140 | 0,7 | 0,15 | 0,17 | 11 |
| Lublin | L1 | 2011 | 135 | 0,9 | 0,23 | 2,53 | 61 |
| | L2 | 2012 | 95 | 1,0 | 0,15 | 1,95 | 47 |
| | L3 | 2010 | 125 | 0,9 | 0,20 | 3,58 | 55 |
| Olsztyn | O | 2012 | 198 | 0,7 | 0,23 | 1,58 | 45 |
| Poznan | Po | 2015 | 125 | 0,7 | 0,16 | 0,55 | 15 |
| Strzelce Opolskie | SO | 2011 | 130 | 0,9 | 0,14 | 1,51 | 23 |
| Tarnowskie Gory | TG1 | 2012 | 145 | 0,7 | 0,19 | 4,02 | 49 |
| | TG2 | 2012 | 145 | 0,7 | 0,20 | 3,62 | 40 |
| Warszawa | W1 | 1997 | 113 | 0,7 | 0,42 | 5,33 | 82 |
| | W2 | 2012 | 168 | 1,0 | 0,23 | 4,74 | 73 |
| | W3 | 2010 | 141 | 0,9 | 0,17 | 0,86 | 41 |
| | W4 | 2013 | 133 | 0,8 | 0,29 | 3,55 | 41 |
| | W5 | 2015 | 102 | 0,9 | 0,16 | 0,63 | 34 |
| Wroclaw | Wr | 2011 | 138 | 0,7 | 0,28 | 3,46 | 70 |

One of the assumptions of the research was to determine if there is a relationship between the $n_{50}$ and the building characteristics.

At first, the relationship between the year of commissioning and the airtightness was sought (fig.2). The trend is visible (on the chart marked with a dashed line) between $n_{50}$ and the building's age. Despite the noticeable upward tendency of the $n_{50}$ level (relative to the age of the building), it cannot be considered as reliable. Considering that most of the buildings qualified for the study have been erected over the past 5 years, and the divergence of the $n_{50}$ values for individual facilities in those years is significant, only the trend and not the dependency can be indicated.

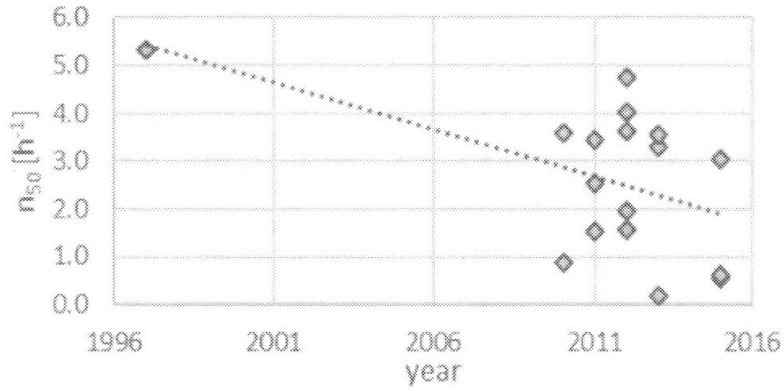


**Fig. 2.** The dependence of airtightness from buildings construction.

The usable area of buildings designated per the regulations in force [13] is in the range of 95 m$^2$ (for L2 building) and 198 m$^2$ (for building O). The relationship between the airtightness of the building and its usable surface and the surface of the external partitions (Fig.3, Fig.4) is not noticeable. This means that the usable area and the surface of the partitions do not affect the building's leakproof (if the airtightness is expressed with $n_{50}$).

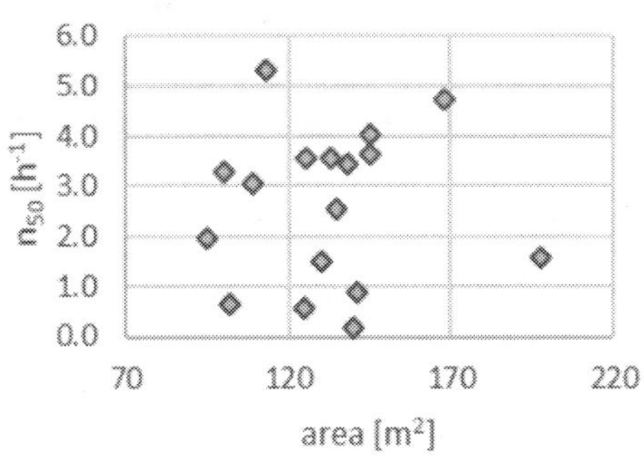


**Fig. 3.** The dependence of airtightness from floor area of the building.

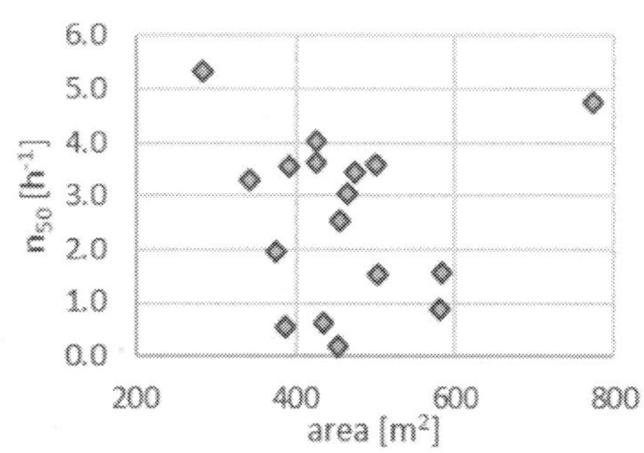


**Fig. 4.** The dependence of airtightness from the external surfaces of the building.

The comparison of the shape factor (A/V) of the building with its airtightness does not indicate the existence of a relationship between these two variables (Fig. 5). The chart shows that there were, among others, very compact buildings (low A/V) that were both having low (building W1) and high airtight (building K).

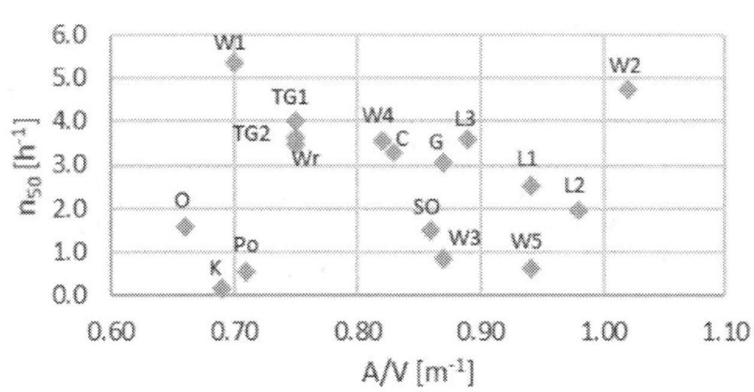


**Fig. 5.** The dependence of airtightness from A/V.

The weighted average heat transfer coefficient of all external barrier ($U_{avg}$.) was chosen for the next set. The graph shows the relationship between the $n_{50}$ and $U_{avg}$. This means that the high thermal insulation of partitions implies a better airtightness of the building.
In energy-efficient construction, the outer partition is usually made as multilayer (due to the requirement of achieving low heat transfer coefficient). For the main premise, for the presence of leaks in energy-saving buildings in Poland, the human factor can be identified. The lack of experienced staff results in sloppy and often improperly airtight coatings [14], for example, in the locations where systems are installed through external barrier [15]. Therefore, the use of laminated materials (when making multi-layered walls) in the bulkhead may lead to involuntary sealing of the building.

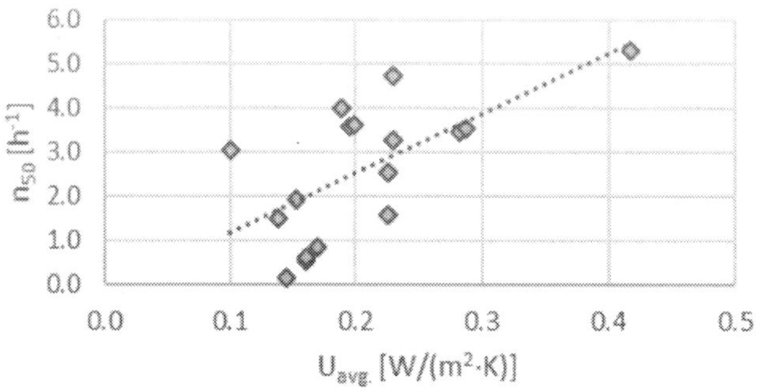


**Fig. 6.** The dependence of air-tightness from $U_{avg.}$

Problems arising from the negligence of the performers, i.e. the mentioned human factor, are difficult to eliminate. For example, if window and door joinery is to be installed, high-precision window-to-wall joints are required. When comparing the window joinery with $n_{50}$, the trend line (dotted line in Fig. 7) indicates that with the falling of the window surfaces, the value of $n_{50}$ also decreases. This tendency is due to the frequent occurrence of leaks in the place where the joinery connects with the wall. However, there are exceptions from the analysed dependencies (the W1 building with a high value of $n_{50}$, despite a small window area of 24 m$^2$, is the extreme case). The reason for this situation is the extremely poorly executed installation of window frames.

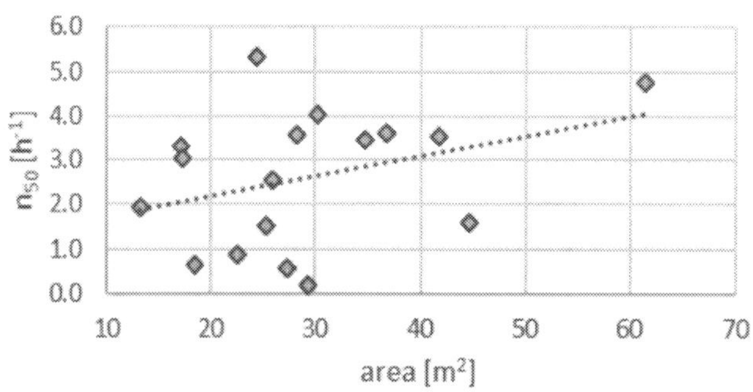


**Fig. 7.** The dependence of air-tightness from the surface of the windows.

The last parameter analysed was the need for heating and ventilation (EU) for leakage (Fig. 8). The lowest (i.e. the best) value of energy demand was 11 kWh/(m$^2$·a), for a building located in Kluczbork (K). The building was also marked by the best airtightness $n_{50} = 0.17$ h$^{-1}$. The highest EU value was determined for a building in Warsaw (W1) and was 82 kWh/(m$^2$·a), the building also has the highest value of $n_{50} = 5.33$ h$^{-1}$.

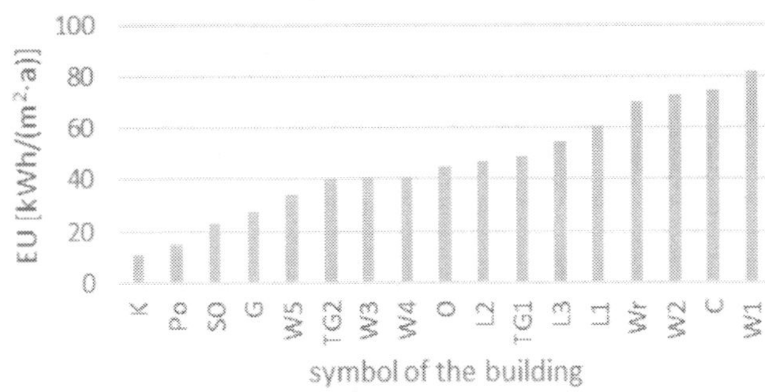


**Fig. 8.** Building's energy requirements.

After compiling the EU with the $n_{50}$, the level of infiltration directly translated into the heat demand for heating and ventilation of the tested buildings (Fig. 9).
As the building's airtightness increases (value $n_{50}$ decrease), the demand for the EU is decreasing (Fig. 9 shows the trend in terms of energy demand for heating and the airtightness of the building's casing).

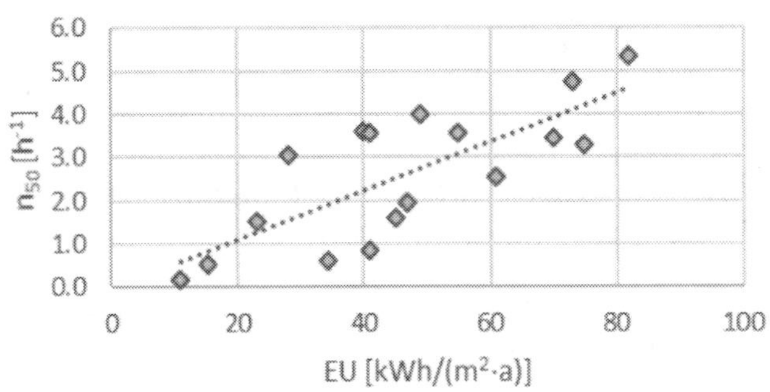


**Fig. 9.** The dependence of the energy demand of air tightness.

The next stage of the tests, the leakproof of the buildings was simulated. For this purpose, the energy demand for individual buildings has been calculated for the average leakproof of buildings ($n_{50} = 2.6\ h^{-1}$). EU demand calculations for $n_{50} = 1.5\ h^{-1}$ (recommended value for buildings with mechanical ventilation [1]), $n_{50} = 1.0\ h^{-1}$ (maximum permissible value for building NF40) and $n_{50} = 0.6\ h^{-1}$ (maximum permissible value for building NF15 [9]) are also used to check the change in demand.
The average decrease in energy demand for heating and ventilation losses, with a change in leakage from $n_{50} = 2.6\ h^{-1}$, is respectively:

- for $n_{50} = 1.5\ h^{-1}$ - 16%,
- for $n_{50} = 1.0\ h^{-1}$ - 22%, and
- for $n_{50} = 0{,}6\ h^{-1}$ is 28%

the decrease in the energy required for heating and ventilation.

## 5. Conclusion

Studies and analyses show that the level of airtightness of a building is influenced by:

a) the length of joinery windows with the wall (and hence the surface of the transparent partitions),
b) heat transfer coefficient through the outer partition,
c) insufficient knowledge of many designers and contractors, as well as inaccuracy and ignorance of the latest trends and solutions.

Disregarding the leakage test limits the ability to control the work carried out. In addition, the low leakage of the building envelope significantly translates into an increased level of energy demand for heating and ventilation (EU).

In summary, the impact of building tightness on the EU is greater when total energy consumption is lower, so it is important to raise investor, designer and contractor awareness of the relationship between building airtightness and the EU, and strive to eliminate any leaks.

## References

1. Rozporządzenie Ministra Transportu, Budownictwa i Gospodarki Morskiej z dnia 5.VII.2013 zmieniające rozporządzenie w sprawie warunków technicznych, jakim powinny odpowiadać budynki i ich usytuowanie
2. B. Milovanović, I. Banjad Pečur, N. Štirmer, *Quality assurance of an energy performance of buildings - airtightness testing* (Podgorica Univerzitet Crne Gore, 2012)
3. J. Šadauskienė, V. Paukštys, L. Šeduikytė, K. Banionis, Energies **7(8)** (2014)
4. *Passive house seminar for professionals from the building sector*, Intelligent Energy Europe Project Passnet.
5. K. Nowak, K. Nowak-Dzieszko, *Wpływ badań szczelności budynków na ich charakterystykę energetyczną* (Wydawnictwa Uczelniane Uniwersytetu Technologiczno-Przyrodniczego w Bydgoszczy, Bydgoszcz, rok???) ISBN 978–83–61314-67-7.
6. K. Nowak, K. Nowak-Dzieszko, Materiały Budowlane, **1**, 24 (2014)
7. S. Firląg, Czasopismo Techniczne, **109(3–2/B)**, 105 (2012)
8. PN-EN ISO 9972:2015 Cieplne właściwości użytkowe budynków -- Określanie przepuszczalności powietrznej budynków -- Metoda pomiaru ciśnieniowego z użyciem wentylatora
9. *Wytyczne do weryfikacji projektów budynków mieszkalnych, zgodnych ze standardem NFOŚiGW* (Warszawa, 2012)
10. A. Miszczuk, Proc. Eng. **153**, 461 (2016)
11. PN-EN ISO 13790; „Energetyczne właściwości użytkowe budynków - Obliczanie zużycia energii na potrzeby ogrzewania i chłodzenia"
12. https://www.mir.gov.pl/
13. Rozporządzenie Ministra Transportu, Budownictwa i Gospodarki Morskiej z dnia 25 kwietnia 2012 r. w sprawie szczegółowego zakresu i formy projektu budowlanego.
14. A. Miszczuk, K. H. Żmijewski, Mat. Bud. **1**, 24 (2015)
15. S. Firląg, A. Miszczuk, Rynek Instalacyjny **4**, 56 (2015)

# FEM modelling of deep excavation – parametric study, Hypoplastic Clay model verification

*Monika* Mitew-Czajewska[1,*]

[1]Warsaw University of Technology, Faculty of Civil Engineering, Institute of Roads and Bridges, Al. Armii Ludowej 16, 00-637 Warsaw, Poland

**Abstract.** A parametric case study of a 25,5 m deep excavation is presented. Finite Element analysis was performed using Modified Mohr-Coulomb and Hypoplastic Clay constitutive soil models. The results of the FE analysis were compared to in-situ displacements measurements taken during construction as well as to the results of calculations made using dependent pressures method. Final conclusions concerning the applicability of the Hypoplastic Clay constitutive soil model for modelling of deep excavations made in soft clay in terms of accurate determination of horizontal displacements of the excavation wall are presented.

## 1 Introduction

The very first challenge of the numerical modelling of geotechnical or underground structures by means of Finite Element Method (FEM) is the choice of appropriate constitutive soil model. In the engineering practice the most commonly used model is Mohr-Coulomb or Modified Mohr-Coulomb model due to the simplicity and availability of its parameters. This practice is acceptable in case of simple structures and simple geotechnical conditions. While for demanding projects, e.g. very deep excavations built in complex geotechnical conditions it leads to wrong estimation of structure displacements and internal forces. In Warsaw, in locations where Mio-Pliocene clays dominate in the subsoil, the choice of constitutive soil model for FE analysis is a great challenge. Several attempts of the use of other soil models were made, such as Hardening Soil model [1], Modified Cam-Clay [2] or Hypoplastic Clay model [3] - the latter showing very promising results.

The case study presented in the paper, describing 25,5 m deep excavation executed in a very complex geotechnical conditions, mostly in Mio-Pliocene clays, is aimed to assess the applicability of the Hypoplastic Clay model (HC model) implemented in GEO5 FEM software [4].

* Corresponding author: m.mitew@il.pw.edu.pl

## 2 Description of the case

The 25,5 m deep excavation for the construction of metro station was executed in a very complex geotechnical conditions (dominating Mio-Pliocene clays, and high water pressure). Cut and cover method was used for the construction of the station; the stability of 36 m deep diaphragm walls during excavation was ensured by 3 levels of underground slabs. Detailed description of the geometry and construction stages can be found in [5].

### 2.1 Geotechnical conditions

In general, the subsoil of Warsaw is composed of Tertiary deposits covered with a complex of Quaternary formations from the Pleistocene and Holocene periods. In the analysed case, there is a 3-5 m thick layer of fills directly under the surface, below which 6-7 m of fluvioglacial sands and 5-6 m glacial sandy clays occur. These layers are covering the Mio-Pliocene deposits (clays, silty clays or stiff silty clays) expanding to great depths. In order to take into account the change in parameters with depth, clay layer is divided into two layers in the model. Detailed description of soil conditions in the analysed case is included in [5].

The main groundwater table is located 8.5 m below the ground surface.

Geotechnical parameters of all soil layers specified above are given in Table 1 and 2 in the numerical analysis description chapter.

### 2.2 Construction stages

The following construction stages were considered:

Stage 1 – Construction of peripheral diaphragm walls
Stage 2 – Excavation - 4.9 m below ground surface
Stage 3 - Construction of the roof slab of the upper station (1.2m thick)
Stage 4 - Excavation - 9.8 m below ground surface
Stage 5 - Construction of the first/top intermediate slab (0.6m thick)
Stage 6 - Excavation - 17.2 m below ground surface
Stage 7 – Construction of the third/bottom intermediate (0.5m thick)
Stage 8 - Excavation -25.5 m below ground surface
Stage 9 - Construction of the foundation slab (1.2m thick).

## 3 Numerical analysis

### 3.1 Basic assumptions

The numerical FE model was built in GEO5 FEM software. Model dimensions are 70x120m. FE model mesh, generated automatically, was built of 11933 nodes, 7016 elements (4620 surface elements, 599 beam elements, and 1797 contact elements).

The following basic assumptions were adopted in the numerical FE model:

- Final excavation depth – 25.5 m,
- Diaphragm wall height – 36 m, and thickness - 1.2 m,
- 3 slabs supporting the diaphragm wall in the temporary and permanent state,
- Construction stages as specified in chapter 2.2.

The geometry of the model and the finite elements mesh are presented in Figure 1.

In addition following assumptions concerning underground water modelling were adopted in accordance with construction methods:

- Peripheral diaphragm walls embedded in clays cut the inflow of water into the excavation,
- The water level inside the excavation was lowered in accordance with the excavation depth (in each construction stage - 1 m below the current excavation level).

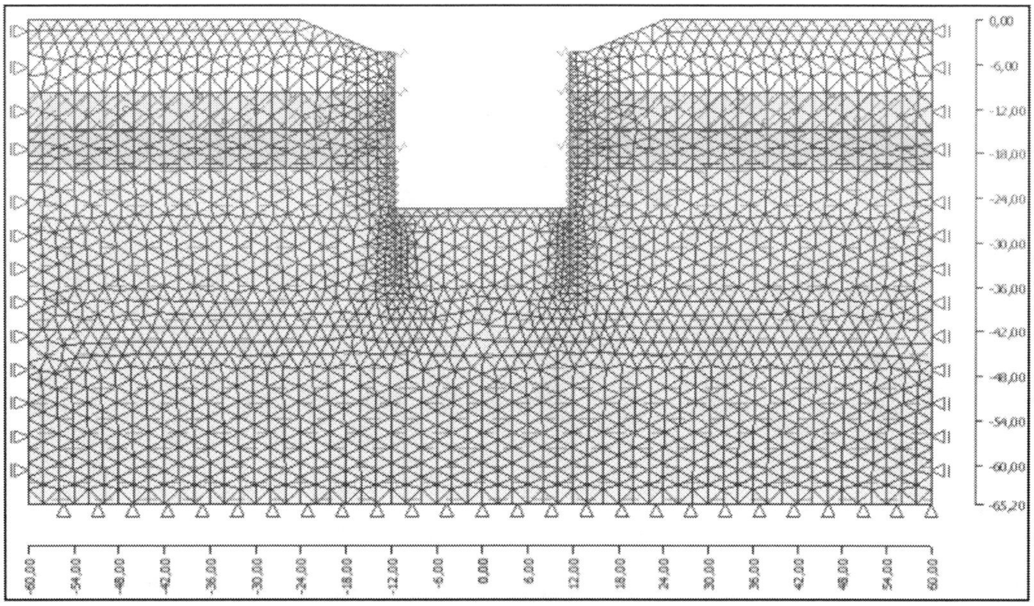


**Fig. 1.** The geometry of the model, FE mesh

## 3.2 Constitutive soil models

Two calculation series of the case were made with the use of following constitutive soil models: Modified Mohr-Coulomb, the formulation of which is presented in [4] and Hypoplastic Clay model. The Hypoplastic Clay model is a critical state model applicable for modelling of fine grain soft soils [4]. The model allows for reflecting a different stiffness in loading and unloading, and softening or hardening, in dependence with the soil compaction and the change of volume in shearing (dilation, compression). The state boundary of the Hypoplastic Clay model in meridian and deviatoric plane is shown on Figure 2 (left and right respectively). The mathematical formulation of the HC model is given in [6].

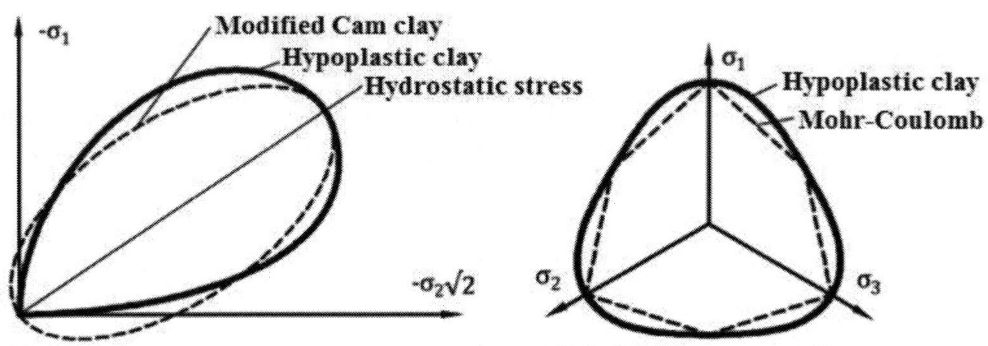


**Fig. 2.** State boundary of Hypoplastic Clay model: in the meridial plane (left), in the deviatoric plane (right)

### 3.3 Model parameters

Parameters of all soil layers occurring in the numerical model applied in the analysis are compiled in Table 1 (for Modified Mohr-Coulomb model) and Table 2 (for Hypoplastic Clay model). Please note also that the Hypoplastic Clay constitutive model was applied only to the Mio-Pliocene clay layer (divided into 2 sub-layers as explained before). Fill, Sand and Sandy Clay layers were in both calculation series modelled by means of Modified Mohr-Coulomb model.

**Table 1.** Soil parameters – Modified Mohr-Coulomb model

| | γ | φ' | c' | E | Ko | υ |
|---|---|---|---|---|---|---|
| | kN/m³ | ° | kPa | MPa | - | - |
| **Fill** | **18** | **25** | **0** | **25** | **0.577** | **0,30** |
| **Sand** | **19** | **34** | **0** | **50** | **0.441** | **0,30** |
| **Sandy Clay** | **21** | **31** | **3** | **50** | **0.485** | **0,29** |
| **Clay 1** | **20** | **18** | **10** | **80** | **0.917** | **0,35** |
| **Clay 2** | **21** | **18** | **15** | **100** | **0.783** | **20** |

**Table 2.** Soil parameters – Hypoplastic Clay model

| | γ | Ko | κ | λ | $e_0$ | $e_{max}$ | $\varphi_{cv}$ | r |
|---|---|---|---|---|---|---|---|---|
| | kN/m³ | - | - | - | - | - | ° | - |
| **Clay 1** | **20** | **0.917** | **0.019** | **0.071** | **0.57** | **2.5** | **13** | **0,5** |
| **Clay 2** | **21** | **0.783** | **0.019** | **0.071** | **0.57** | **2.5** | **13** | **0,5** |

All parameters except $e_{max}$, $\varphi_{cv}$ and r for the HC model where established based on laboratory and field tests carried out at the Department of Geotechnics and Underground Structures of the Warsaw University of Technology [2, 7, 8]. Parameters $e_{max}$, $\varphi_{cv}$ and r were taken as average for similar soil types [4]. Precise specification of all parameters of the Hypoplastic Clay model for Mio-Pliocene clays is a part of ongoing study.

### 3.4 Results of FE analysis

The results of calculation series in form of plots of the horizontal displacements fields in the final excavation stage, including graphs of horizontal diaphragm walls displacements are presented on Figure 3 and Figure 4 for Modified Mohr-Coulomb and Hypoplastic Clay models respectively. The maximum theoretical value of horizontal wall displacement obtained using MMC model is 30.6mm, and HC model – 13.6mm, while the maximum displacement measured during construction was 15.6 mm. This indicates that the Hypoplastic Clay model can be found valuable for modelling of Mio-Pliocene soft clays.

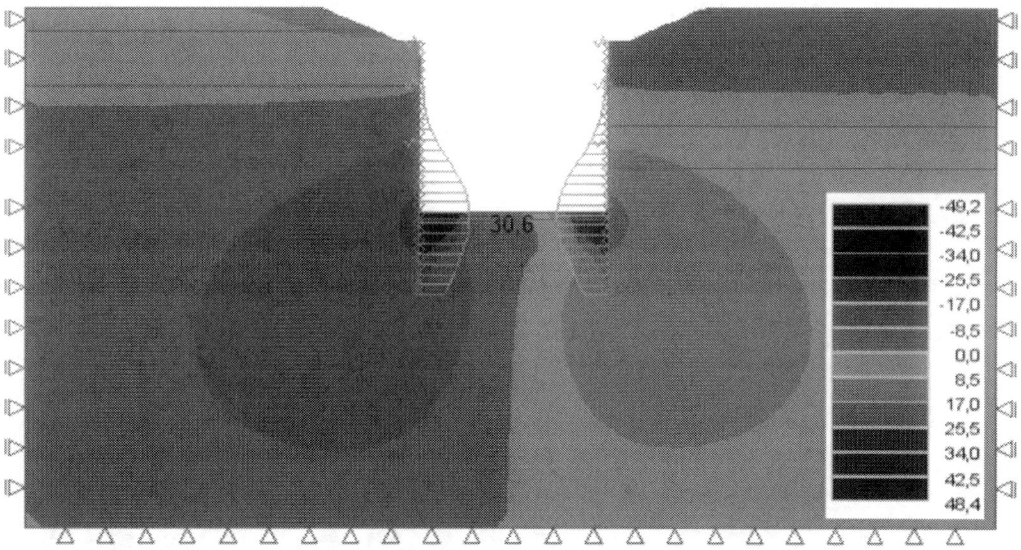


**Fig. 3.** Results of FE analysis, Modified Mohr-Coulomb model, horizontal displacements plot in final excavation stage

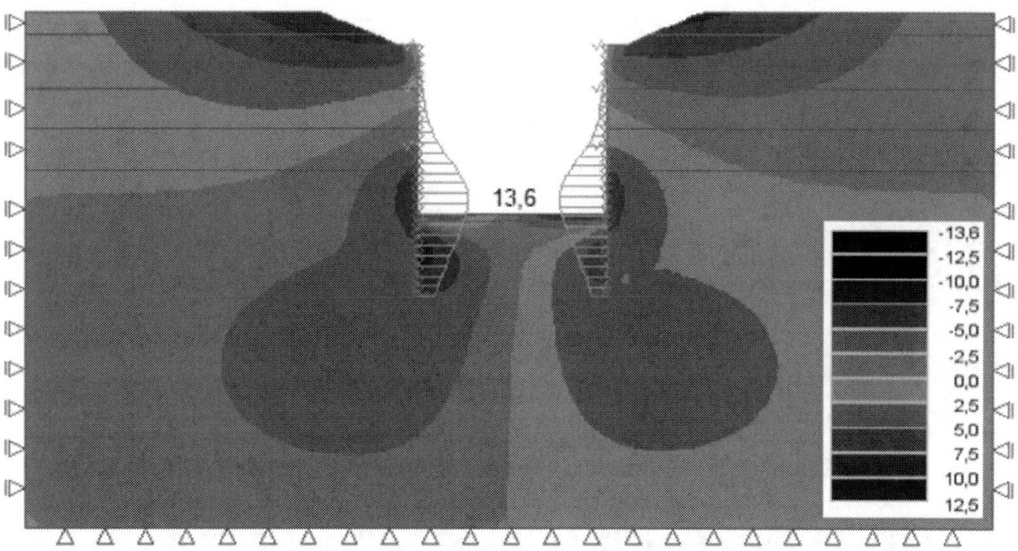


**Fig. 4.** Results of FE analysis, Hypoplastic Clay model, horizontal displacements plot in final excavation stage

## 4 Summary and conclusions

The Mio-Pliocene clay occurring on the large area of central Poland is a very difficult material to model in FE analysis. The use of Modified Mohr-Coulomb constitutive model, with simple and easily available model parameters, for this material usually leads to miss-estimation of displacements of the excavation wall (Fig. 5). In the paper, the attempt of application of Hypoplastic Clay constitutive model for modelling of Mio-Pliocene clay is presented. The comparison of maximum theoretical horizontal displacements of the excavation wall calculated using Modified Mohr-Coulomb (MMC) and Hypoplastic Clay

(HC) models with real wall displacements measured during construction is presented on Figure 5. In addition, reference result of the calculation of wall displacement using dependent pressures method is shown, based on [5]. It may be observed that Hypoplastic Clay model is suitable for modelling of deep excavations in Mio-Pliocene clays. The key issue is the proper calibration of model parameters. Further tests and studies will be made in this field.

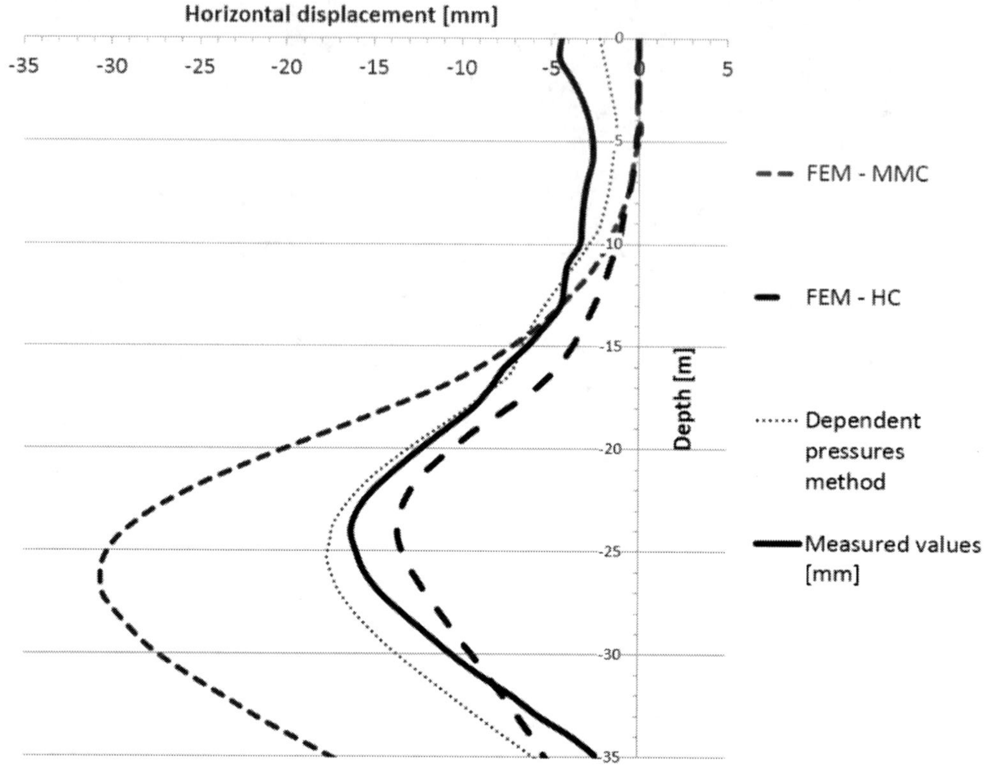


**Fig. 5.** Comparison of diaphragm wall maximum horizontal displacements

## References

1. A. Truty, R. Obrzud, Studia Geotechnica et Mechanica, **37**, 61 (2015)
2. M. Superczyńska, K. Józefiak, A. Zbiciak, ACE **62(3)**, 207 (2016)
3. M. Mitew-Czajewska, ACE **62(4)**, 73 (2016)
4. Fine Ltd, *GEO5 User Guide* (Prague, 2017)
5. M. Mitew-Czajewska, *Geotechnical Aspects of Underground Construction in Soft Ground* (CRC Press, 2014)
6. D. Mašín, International Journal for Numerical and Analytical Methods in Geomechanics **29(4)**, 311 (2005)
7. R. Kuszyk, M. Superczyńska, A. Lejzerowicz, Przegląd Komunikacyjny **9**, 59 (2012) In polish
8. M. Superczyńska, *Geologiczno-inżynierska ocena iłów formacji poznańskiej jako podłoża budowlanego* (PHD thesis, Warsaw University of Technology, 2015) In polish
9. A.Z. Ter-Martirosyan, Z.G. Ter-Martirosyan, Procedia Engineering, **111**, 756 (2015)
10. Z.G. Ter-Martirosyan, A.Z. Ter-Martirosyan, Soil Mech Found Eng **51**, 273 (2015)

# Study into the feasibility of manufacturing liquid glass using resource-saving technology

*Sergey* Mizyuryaev [1*], *Alexandr* Chiknovorian[1], and *Kirill* Kuriatov[1]

[1]Samara State Technical University, Institute of Architecture and Civil Engineering, 194, Molodogvardeyskaya St., 443001 Samara, Russia

**Abstract.** The authors' views on the problem of resource-saving in the production of building materials are outlined, with three main modes of resource-saving indicated: the use of cheap raw materials, a reduction in the production costs, and an increase in the efficiency of the produced materials and products. The research provides information on the production and use of liquid glass in industry, including the construction industry. The theoretical substantiation of the possibility of developing a resource-saving technology for the production of liquid glass for construction purposes is given. The work provides information on promising alternative raw material components - diatomite, natural rock and black ash, industrial waste. Their properties are given as well as the justification of their effective use as raw materials. The method of preparation of the components and their mixtures, the preparation of sodium silicate through roasting, and the identification of the suitability of the obtained product for the manufacture of efficient building materials are described. Conclusions are made in regards to the feasibility of producing liquid glass using resource-saving technology.

## 1 Introduction

The issue of resource-saving is a constant one. However, in the opinion of designers, production engineers, researchers and consumers, different things can be meant by this term. The same difference can be observed when the term is applied to construction, manufacture, and use of building materials and products.

Production engineers try to use the cheapest and most readily-available raw materials as well as to decrease production costs in order to reduce the cost price. [1-4]. Sometimes such policies can unfortunately lead to a decrease in product quality. Resource-saving for consumers of construction materials and products implies a level of reliability, durability and comfort of the buildings made from such materials.

In evaluating these processes, we found that the primary role in the provision of resource-saving as regards this matter should be defined by the producers of construction materials and by research aiming to develop and create new methods and technologies that ensure the production of economical but efficient building materials and products. Nonetheless, one of the main conditions for the use of relatively cheap industrial waste and

* Corresponding author: mizuriaev@gmail.com

so-called "local materials" is that such materials should in no case lead to a degeneration in the performance characteristics of building materials and products.

Therefore, scientific developments in this field should not aim at simply finding a replacement for a raw material with a cheaper one, but at developing technologies to produce highly effective materials and products that can reduce the consumption of resources in the long-run.

## 2 Materials and Methods

It is known that a material called "liquid glass" can be used as the foundation for producing highly effective building insulation materials, the distinguishing characteristic of which is high porosity and fire resistance. Such materials include foam-silicate plate products, highly porous granular aggregates, and others [10-14].

Usually, the term "liquid" or "soluble" glass refers to an aqueous solution of alkali metal silicates (usually sodium). Conventionally, the chemical composition of liquid glass can also be expressed by the formula $R_2O$ * $nSiO_2$ * $mH_2O$.

The technical production of liquid glass began only in the first half of the XIX century, although the technology was already well known in the Middle Ages.

Primary materials used in the production of liquid glass include silica components appearing in the form of quartz sand, natural and artificial silica, and alkaline components in the form of soda, potassium carbonate, sodium sulphate, and the like.

Liquid glass can be obtained by fusing together the raw components at a high temperature (about 1300 ° C), followed by cooling the resulting molten mass and then dissolving it in water.

Liquid glass has many diverse uses. About 30% of its production is used in mechanical engineering, about 31% in the chemical industry, about 15% in paper and cardboard production, and about 10% in construction. However, its high-level performance characteristics and high fire resistance of building materials which are made of it allow us to assert that the use of liquid glass in construction can be greatly enhanced [15-18].

The main factors limiting the use of liquid glass in construction include special requirements for its raw materials components and a relatively high final cost. This high cost is due to the complexities of its production process, including the extreme temperature at which it is produced, the difficulty in obtaining scarce raw materials in "pure" form, and high standards of the composition and characteristics of the finished liquid glass due to its widespread use in different applications. Therefore, the development of low-cost technology for the production of liquid glass for building applications is a very relevant topic.

The authors of this work propose a technology for the production of liquid glass, based on the following postulates:

1. Standard requirements for homogeneity, composition, etc. should not be applied to liquid glass used for construction purposes. In this case, it should only meet the requirements for thermal intumescence, and the products obtained on its basis should meet specified operational characteristics (porosity, thermal conductivity, strength, durability, fire resistance, etc.).

2. Primary materials should originate from accessible local sources, preferably industrial wastes.

3. Crystalline materials such as silica, being that they are more refractory and have a higher resistance to heat, should be replaced with materials of a more amorphous structure having a lower melting point and greater chemical activity.

In our opinion, the successful application of this method can greatly reduce the cost of production of liquid glass for construction, which will eventually result in an expansion of the nomenclature of effective thermal insulation materials.

On the basis of the stance put forward by the authors, sodium liquid glass was obtained at a temperature of 950-980 degrees Celsius. In this case, naturally-occurring diatomite was used as the main silica component and black ash ball derived from industrial waste was used as the sodium component.

Diatomite is a sedimentary, fine-porous siliceous rock. Deposits of diatomite are very significant. For example, in the Ulyanovsk region, the total explored reserves are more than 720 million cubic meters. Usually diatomite is composed of 65% to 97% amorphous silica. The authors used the diatomite of the Inzen deposit of the Ulyanovsk region, which was composed of 86.5% $SiO_2$ in its amorphous form. Its density was 3620 kg/cubic meter.

Black ash (molten soda), a by-product of industrial caprolactam production, was obtained from the enterprise "KuibyshevAzot" (Togliatti). It appeared in the form of broken plates 10-15 mm thick and light yellow colour. Its composition is represented by the following oxides, wt. %: $Na_2CO_3$-53.4; NaOH - 18.6, $Na_2SO_4$ - 13.0; NaCl - 2.5 and others.

Studies on the production of liquid glass were carried out as follows:

- the components (diatomite and black ash) were finely ground (until particles were able to pass through a sieve with an opening of 0.15 mm), mixed in various proportions and fired at temperatures which ensured the formation of a powdered mixture of sintered material;
- the obtained sintered samples were ground and kept in boiling water for 1 hour;
- the resulting compositions were evaluated for their ability to distend.

## 3 Results

It is known that chemical processes are significantly accelerated when the reacting components are in liquid form, for example in molten form. Therefore, our task at the first stage is to establish a minimum temperature at which test samples reach molten state, which in our opinion will allow for an intensive interaction of the raw components.

Tests showed that at a firing temperature of up to 950 degrees Celsius, samples remained in a free-flowing state, which indicated that the components had not melted or fused together. The samples were observed to sinter at temperatures above 950 degrees Celsius.

The second task was to determine the depth of interaction and the intensity of formation of sodium silicates. Since the resulting material is X-ray amorphous, the use of X-ray diffraction methods was not practical. Therefore, we used an indirect method based on the fact that aqueous sodium silicate solutions have the ability to expand when introduced to a heat source. Moreover, this method allowed for the immediate characterization of new porous structures in the resulting compounds.

It was found that the optimal temperature for roasting the mixture of diatomite and black ash is one equal to 950-980 degrees Celsius.

At these temperatures and at mass ratios of 5/5 to 3/7 of diatomite to black ash, sufficient distension of the resulting compositions is observed. The intensity of distension (coefficient of distension), constituting the increase in volume compared to the initial volume, was 5-7, which, in our opinion, is quite sufficient for liquid glass used for manufacturing insulation materials and construction products.

## 4 Discussion

Comparing the standard technologies of liquid glass production to the ones proposed by the authors, one can immediately note the pros and cons of the latter.

The obvious pros are as follows:

- The new technology allows for the use of more affordable components;

- The new technology also allows for lower firing temperatures, significantly reducing levels of energy consumption.

Among disadvantages of the technology can be the inconsistency of the characteristics of the resulting sodium silicate with the standard requirements for liquid glass. But is this consistency really necessary? Strict requirements for ordinary liquid glass production are necessary because of its widespread application (in metallurgy, in the chemical and paper industry, etc.). In other words, ordinary liquid glass must satisfy a spectrum of requirements, which are sometimes conflicting to each other. Is this necessary? In our opinion, no! It is similar to applying the same requirements and design for engines used both for wide fuselage aircrafts and smaller ones, such as 2-4 seaters.

Therefore, we believe that one of the conditions for implementing resource-saving methods of liquid glass production should be that only those requirements that will ensure the production of specific products of a specified quality must be adhered to, not that the same requirements must be fulfilled for each and every type of liquid glass being produced. For example, for the production of liquid glass for construction purposes specifically, it should fulfil the requirement to intensively distend.

## 5 Conclusions

Thus, the authors have proven that it is possible to use resource-saving and low-cost production technology for liquid glass intended for construction purposes, i.e. it is potentially suitable for manufacturing effective thermal insulation materials and products.

## References

1. I.A. Katkova, M.V. Rodionov, A.V. Shabanov, Urban Construction and Architecture **2**, 114-119 (2011) DOI: 10.17673 / Vestnik.2011.02.25
2. A.I. Khlystov, MV Konnov, A.V. Vlasov, Urban Construction and Architecture **4**, 87-92 (2011) DOI: 10.17673 / Vestnik.2011.04.17
3. S.F. Korenkova, Y.V. Sidorenko, *Traditions and innovations in construction and architecture. Construction technology*, 27-30 (2016)
4. I.Y. Doladov, I.P. Doladova, Urban Costruction and Architecture **2**, 132-134 (2011) DOI: 10.17673 / Vestnik.2011.02.29
5. Y.S. Vitchikov, *Increasing the energy efficiency of buildings and structures. Interdepartmental Scientific Workshop 7* (Samara, 2012)
6. A.Y. Zhigulina, Town planning 1, 80-81 (2012)
7. V.V. Saburov, S.Y. Galitskov, A.N. Alyoshin, Urban Construction and Architecture **4(17)**, 107-110 (2014) DOI: 10.17673 / Vestnik.2014.04.16
8. O.G. Orlov, S.Y. Galitskov, O.S. Velmyaykina, Urban Construction and Architecture **2(23)**, 56-62 (2016) DOI: 10.17673 / Vestnik.2016.02.11
9. S.A. Mizyuryaev, A.Y. Zhigulina, K.V. Ganechkina, Urban Construction and Architecture **2(23)**, 10-13 (2016) DOI: 10.17673 / Vestnik.2016.02.2
10. V.A. Lotov, V.A. Kutugin, Glass and ceramics **1**, 6-10 (2008)
11. V.A. Lotov, V.A. Kutugin, Glass and ceramics **1**, 6 -8 (2008)

12. S.A. Mizyuryaev, A.Y. Zhigulina, A.N. Mamonov, N.V. Ivanova, Building Materials **7**, 12-13 (2011)
13. Y.G. Ivashchenko, S.A. Evstigneev, A.V. Strakhov, Scientific review **8**, 177-180 (2015)
14. A.I. Kudyakov, TN. Radina, M.Y. Ivanov, Construction Materials **11**, 12-13 (2004)
15. O.Y. Suprun, M.S. Zolotov, Municipal economy of cities: Scientific and technical collection **72**, 44-52 (2006)
16. S.A. Mizyuryaev, A.Y. ZhigulinA, Urban Planning **1**, 82-84 (2012)
17. S.A. Korotaev, V.T. Erofeev, Bulletin of the University of Mordovia **4**, 54-59 (2008)
18. A.M. Guryanov, S.A. Miziuryaev, V.M. Lebedev, V.T. Lebedev, *Collection of Works of the 5th Eurasian scientific and practical conference "Strength of heterogeneous structures"*. Moscow, April 20-22, 136 (2010)

# Comparative analysis of results from experimental and numerical studies on concrete strength

*Oleg* Mkrtychev [1], *Dmitriy* Sidorov [1] and *Sergey* Bulushev [1*]

[1]Moscow state university of civil engineering, Yaroslavskoye shosse, 26, Moscow, Russia, 129337

**Abstract.** Some results of numerical experiments of testing concrete cubes and prisms on unconfined compression, and the comparison of results obtained with experimental and specified data, are presented in the article. When performing calculations of structures in a nonlinear setting, it is very important to choose adequate deformation diagrams or material models. Because of the fact that there are no instructions how to use the diagrams of concrete and armature deformation in collaboration of steel and concrete, the simulation of reinforced concrete structures by finite elements of the same type without any assumptions is impossible. Numerical experiments have been performed in the LS-DYNA software package. This software package allows simulating the collaboration of concrete and steeling with the help of three-dimensional (for concrete) and rod (for the reinforcement) finite elements. As samples, a cube and a prism with dimensions of 150x150x150 mm and 150x150x600 mm, respectively, have been taken. The samples are simulated by solid finite elements. For the simulation of concrete, the non-linear CSCM (Continuous Surface Cap Model) material is used. The tests were carried out with samples of the following classes of concrete as for cylinder compressive strength: C12, C16, C20, C25, C30, C35. This corresponds to the following classes of cube compression strength: B15, B20, B25, B30, B37, B45. The tests have been carried out considering the friction coefficients between the plates of a testing machine, and a sample. The performed researches have shown that the destruction nature of the samples in a numerical experiment corresponds to the failure nature in real tests. The investigated model of CSCM concrete can be used in the calculation of concrete and reinforced concrete structures with acceptable accuracy for main classes of concrete.

## 1 Introduction

When performing calculations of structures in a nonlinear setting, it is very important to choose adequate deformation diagrams or material models. The behaviour of steel structures can be precisely and simply modeled by finite elements using the diagrams described in [1]. The diagrams of deformation of concrete and reinforcement are given in [2]. However, there are no instructions how to use these diagrams in collaboration of steel

* Corresponding author: sergey.bulushev@gmail.com

and concrete. Thus, the simulation of reinforced concrete structures by finite elements of the same type without any assumptions is impossible. Modern software packages, such as LS-DYNA, ANSYS, and others, allow simulating the collaboration of concrete and steel with the use of solid (for concrete) and rod (for reinforcement) finite elements. In this setting, for rod-like finite elements, we can use the diagrams presented in [2]. The behaviour of concrete in the case of triaxial stress state cannot be fully described by a concrete deformation diagram obtained under uniaxial compression. It is necessary to use more complex models of concrete.

## 2 Research Objective

The purpose of this study is to perform numerical experiments to test concrete cubes and prisms on uniaxial compression and compare the results with experimental and normative data.

Numerical experiments are implemented in the LS-DYNA software package.

As samples, in accordance with [3], a cube and a prism with dimensions of 150x150x150 mm and 150x150x600 mm, respectively, were taken. The samples are modeled by solid finite elements. The lower and upper plates of the testing machine are also made of solid finite elements (Figure 1). The lower plate is fixed to avoid all possible displacements, the upper one can move only in vertical direction.

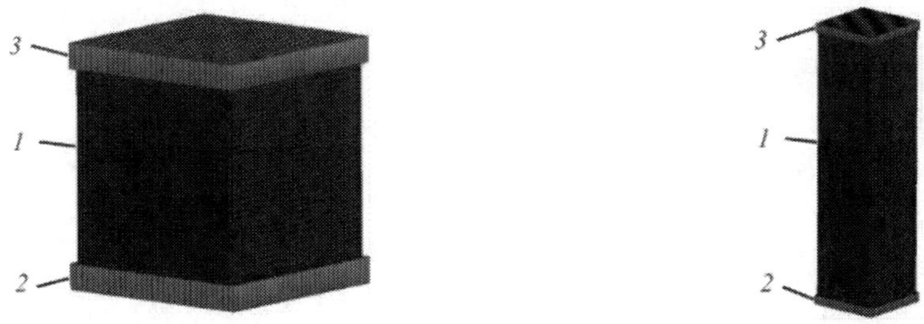


**Fig. 1.** Testing model. *1* – test sample, *2* – lower support plate of a testing machine, *3* – upper support plate of a testing machine (press).

Material of plates is steel. For the simulation of concrete, CSCM non-linear material [4-7] is used.

In the course of the experiment, a gradually increasing load is applied to the upper plate, and the sample is brought to destruction.

The tests were performed with samples of the following classes of concrete as for cylinder compressive strength: C12, C16, C20, C25, C30, C35 [8-10]. This corresponds to the following classes of cube compression strength: B15, B20, B25, B30, B37, B45 [8-10].

The tests were also carried out with different friction coefficients between the plates of the test machine and the sample.

## 3 Results of Investigations

The researches carried out have shown that the destruction nature of the samples in a numerical experiment corresponds to the nature of failure in real tests. The destruction of the cube with allowance for friction is shown in Figure 2.

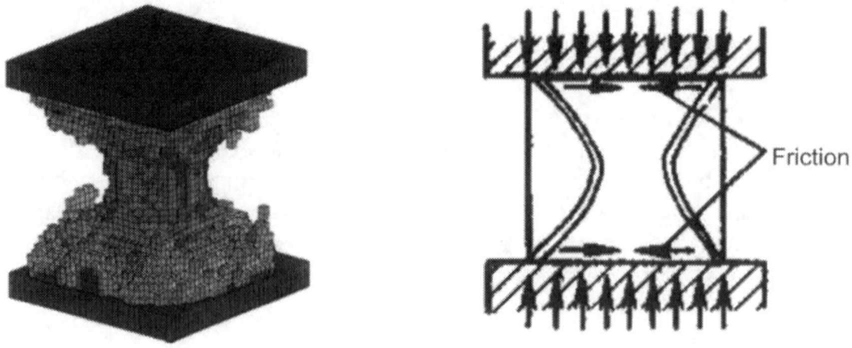


**Fig. 2.** Pattern of the concrete cube destruction with allowance for friction. Numerical experiment (left), experimental data (right).

When the friction decreases, the pattern of destruction changes (Figure 3). And herewith, the strength of the cube is reduced by 1.5-2 times. For a prism with specified dimensions, the friction variation hardly affects. These results also correspond to the experimental data [11]. The pattern of the destruction of the prism is shown in Fig 4.

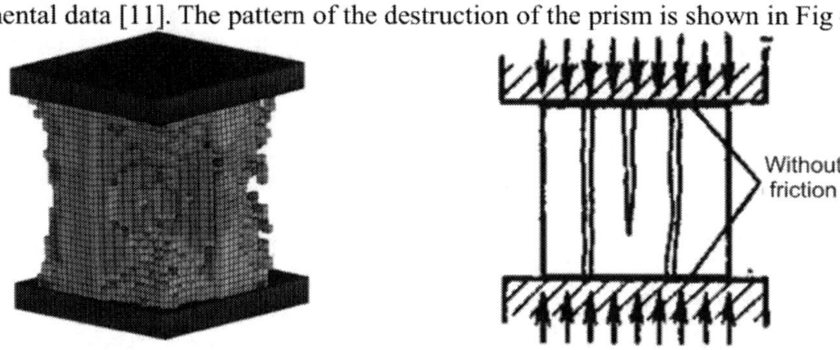


**Fig. 3.** Pattern of the destruction of the concrete cube without considering the friction. Numerical experiment (left), experimental data (right).

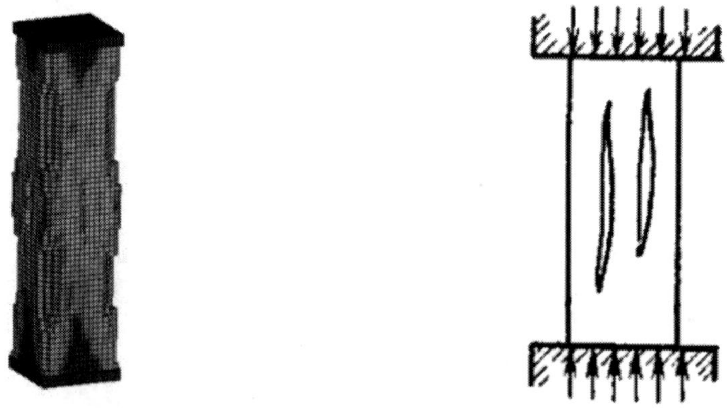

**Fig. 4.** Pattern of the destruction of the concrete prism without considering the friction. Numerical experiment (left), experimental data (right).

In the course of numerical experiments, the values of the destructive load for cubes and prisms were recorded with allowance for friction. These values correspond to the average cube strength and average prismatic strength. The data obtained were compared with the normative values. The results for different classes of concrete are given in Tables 1-6.

**Table 1.** Concrete B15.

| **Characteristics** | **Concrete B15 (C12)** | | |
|---|---|---|---|
| | **Specified value** | **Numerical experiment** | **Δ** |
| $f_{cm}$(mean cyl.) MPa | 20 | - | - |
| $f_{cm,cube}$(mean cube) MPa | 19.3 | 22.5 | 16.58% |
| $f_{cm,prism}$(mean prism) MPa | 14.5 | 15.3 | 5.52% |

**Table 2.** Concrete B20.

| **Characteristics** | **Concrete B20 (C16)** | | |
|---|---|---|---|
| | **Specified value** | **Numerical experiment** | **Δ** |
| $f_{cm}$(mean cyl.) MPa | 24 | - | - |
| $f_{cm,cube}$(mean cube) MPa | 25.7 | 26.1 | 1.56% |
| $f_{cm,prism}$(mean prism) MPa | 19.3 | 20.3 | 5.18% |

**Table 3.** Concrete B25.

| **Characteristics** | **Concrete B25 (C20)** | | |
|---|---|---|---|
| | **Specified value** | **Numerical experiment** | **Δ** |
| $f_{cm}$(mean cyl.) MPa | 28 | - | - |
| $f_{cm,cube}$(mean cube) MPa | 32.1 | 30.9 | -3.74% |
| $f_{cm,prism}$(mean prism) MPa | 24.1 | 25.5 | 5.81% |

**Table 4.** Concrete B30.

| **Characteristics** | **Concrete B30 (C25)** | | |
|---|---|---|---|
| | **Specified value** | **Numerical experiment** | **Δ** |
| $f_{cm}$(mean cyl.) MPa | 33 | - | - |
| $f_{cm,cube}$(mean cube) MPa | 38.5 | 35.1 | -8.83% |
| $f_{cm,prism}$(mean prism) MPa | 28.9 | 30.3 | 4.84% |

**Table 5.** Concrete B37.

| **Characteristics** | **Concrete B37 (C30)** | | |
|---|---|---|---|
| | **Specified value** | **Numerical experiment** | **Δ** |
| $f_{cm}$(mean cyl.) MPa | 38 | - | - |
| $f_{cm,cube}$(mean cube) MPa | 47.5 | 39.0 | -17.89% |
| $f_{cm,prism}$(mean prism) MPa | 35.6 | 33.9 | -4.78% |

**Table 6.** Concrete B45.

| **Characteristics** | **Concrete B45 (C35)** | | |
|---|---|---|---|
| | **Specified value** | **Numerical experiment** | **Δ** |
| $f_{cm}$(mean cyl.) MPa | 43 | - | - |
| $f_{cm,cube}$(mean cube) MPa | 57.8 | 42.5 | -26.47% |
| $f_{cm,prism}$(mean prism) MPa | 43.4 | 37.9 | -12.67% |

The specified values of the average cube strength are taken from [10]. The average prismatic strength is accepted as 75% of the cube [9].

In Figures 5 and 6, the results obtained are presented in the form of graphs.

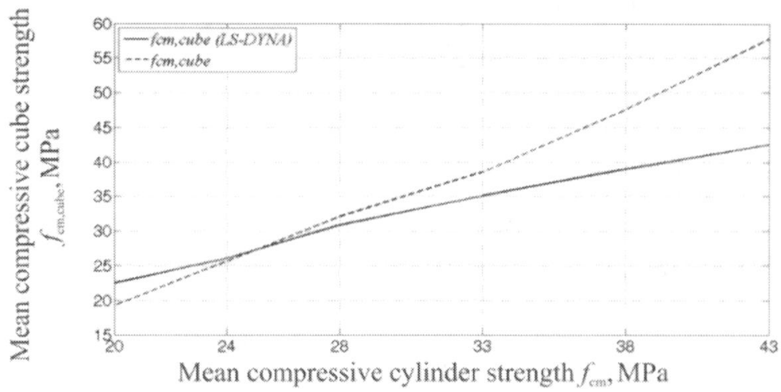


**Fig. 5.** Average cube strength depending upon the average cylinder strength.

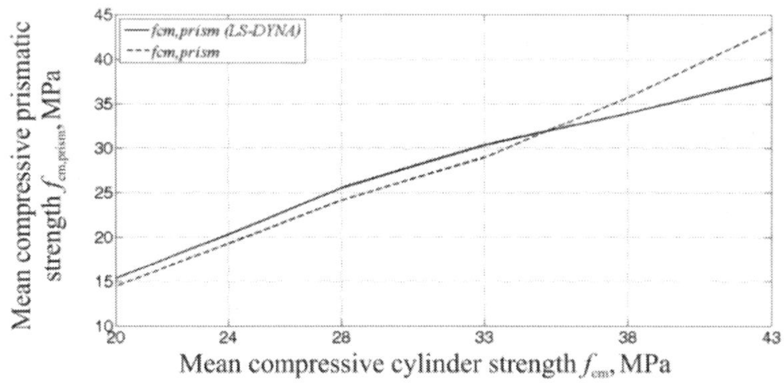


**Fig. 6.** Average prismatic strength depending upon the average cylinder strength.

## 4 Conclusions

The conducted researches showed:

1. The character of destruction of concrete cubes and prisms obtained in numerical studies corresponds to experimental data.
2. The divergence of cube strength for concrete classes B20-B30 is not more than 9% in comparison with the standard (experimental) values.
3. The obtained values of the average prismatic strength for all concrete classes studied, with the exception of concrete B45, show good convergence with normative values. The discrepancies are not more than 6%.

Thus, it can be concluded that the investigated CSCM concrete model can be used in the calculation of concrete and reinforced concrete structures with acceptable accuracy for the main classes of concrete used in mass construction. For concretes of class more than B35, additional studies are necessary.

## Acknowledgements

This study was performed with the financial support of the RF Ministry of Education and Science, grant #7.1524.2017/Project Part.

## References

1. *Proprietary standard ADSC 11251254.001-2015 «Composite structures. Design rules»* (Association of Business Development Associates for the Development of Steel Construction, Moscow, 2015)
2. *SP 63.13330.2012 Concrete and Reinforced Concrete Structures. Summary* (2012)
3. *State standard 10180-2012 Concrete. Methods for determining the strength of control samples* (2012)
4. *US Department of Transportation. Federal Highway Administration. Evaluation of LS-DYNA Concrete. Material Model 159.* **FHWA-HRT-05-063**, 190 (2007)
5. Y.D. Murray, *Users Manual for LS-DYNA Concrete Material Model 159*, **FHWA-HRT-05-062** (2007)
6. O. V. Mkrtychev, M. S. Busalova, Proc.Eng., **153,** 475 (2016)
7. O. V. Mkrtychev, M. S. Busalova, Adv. in Eng, Res., **104**, 160 (2016)
8. *EN 1992-1-1 Eurocode 2: Design of concrete structures - Part 1-1 : General rules and rules for buildings* (2004)
9. T. Telford *Designers' guides to the Eurocodes. Designers' guides to Eurocode 2 : Design of concrete structures designers' guide to EN1992-1-1 and EN1992-1-2 Eurocode 2 : Design of concrete structures general rules and rules for buildings and structural fire design* (2005)
10. T. Telford *Comite Euro-Internetional du beton CEB-FIP model code 1990. Design code* (1993)
11. V. N. Baikov, E. E. Sigalov *Reinforced concrete structures. General course* (Stroyizdat, Moscow, 1984)
12. *Heavy and fine-grained concretes. Specifications. State standard 26633-91* (State Construction Committee of the USSR, Moscow, 1991)

MATEC Web of Conferences 117, 00124 (2017) DOI: 10.1051/matecconf/201711700124
*XXVI R-S-P Seminar 2017, Theoretical Foundation of Civil Engineering*

# Verification of the spar model of a reinforced concrete beam

*Oleg* Mkrtychev [1], *Marina* Busalova [1*], and *Vladimir* Dorozhinskiy [2]

[1]Moscow state university of civil engineering, Yaroslavskoye shosse, 26, Moscow, Russia, 129337

**Abstract.** The paper presents the outcomes of the research aimed at the verification of the computational model simulating a reinforced concrete beam employing spar and 3D finite elements, with account taken of physical nonlinearity. Structural calculations involved in R&D and design effort must factor in non-linear structural behavior both of the material and of structural elements. In particular, this is valid for the analysis of special types of impact (seismic, accidental etc.). However, design models under development may incorporate significantly varying models of nonlinear materials as well as different kinds of finite elements, and both have to be verified. This article considers two models of a hinged (pinned) beam. In the first case, the modeling was based on 3D elements for concrete and on spar elements for reinforcement using Euler-Lagrange coupling of finite elements belonging to concrete and reinforcement (three-dimensional model). The second case the simulation based on spar finite elements (spar model). Both the spar model and 3D model allow for non-linear nature of concrete and reinforcement. In case of concrete the material was set using the Continuous Surface Cap Model, while the reinforcement was modeled involving a bilinear diagram of material behavior.

## 1 Problem setting

Design calculations of buildings and structures performed to take account of a particular case or type of impact require that the non-linear nature of structures' and materials' behavior be factored in. The present research presents a verification of a reinforced concrete beam. Two models of a hinged (pinned) beam were devised during the research: a three-dimensional model consisting of 3D elements with concrete and reinforcement coupled, and a spar model (Figure 1).

* Corresponding author: marina8busalova@gmail.com

**Fig. 1.** Model of a beam.

Beam material is B25, span 6 m, section 400x600(h) mm. The 3D model made use of the concrete non-linear material CSCM (Continuous Surface Cap Model), developed at the U.S. Department of Transportation, Federal Highway Administration [1]. Figure 2 presents the yield surface of primary stresses in space in accordance with this model of concrete [2].

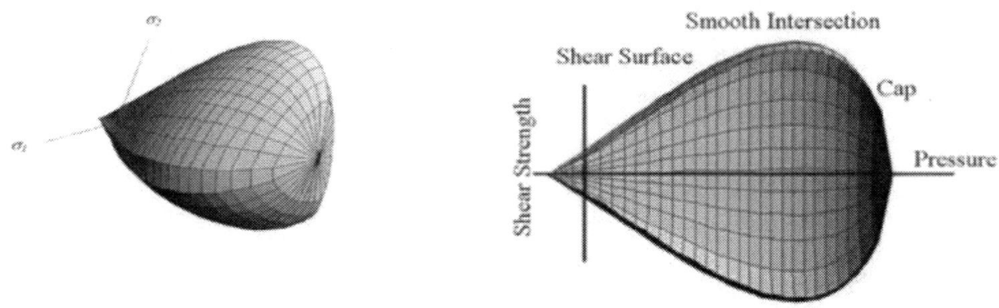


**Fig. 2.** The mathematical model of concrete (CSCM – Continuous Surface Cap Model).

The yield surface for this model is expressed in terms of stress invariants:

$$J_1 = 3P,\ J_2' = \frac{S_{ij}S_{ij}}{2},\ J_3' = \frac{S_{ij}S_{ik}S_{ki}}{3}, \tag{1}$$

where $J_1$ stands for the first invariant of the stress tensor, $J_2'$ - second invariant of the stress deviator, $J_3'$ - third invariant of the stress deviator, $S_{ij}$ - stress deviator, $P$ – pressure.

The CSCM concrete model allows modeling structural elements via 3D finite elements and taking account of the actual reinforcement of a structure (Figure 3), which constitutes an important factor when investigating the behavior of buildings and structures exposed to accidental and seismic impact [3,4]. Reinforcement for this A500C class beam was selected based on a static calculation of load encompassing the own gravity load and a 30 kN/m load distributed along the length of the beam. Based on the static calculation presuming crack resistance, the beam was reinforced in the lower area with four 20 mm rods, in the upper area with two 18 mm rods and 10 mm stirrups with 100 mm pitch at the bearing and 300 mm in the span.

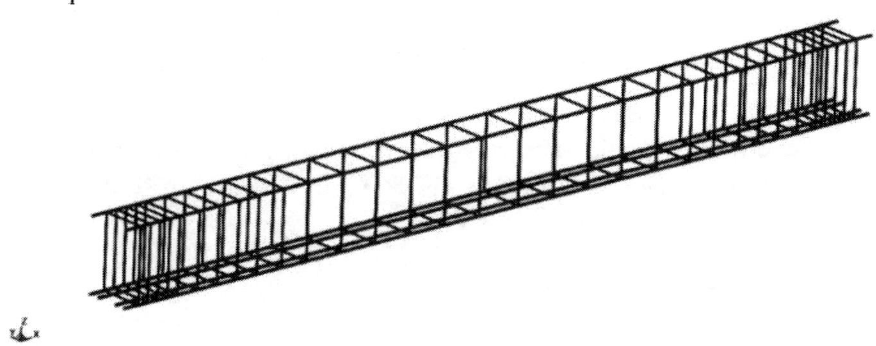

**Fig. 3.** 3D model of the beam's reinforcement cage.

The beam modeled with spar finite elements had the behavior of its material set via a bilinear diagram (Figure 4).

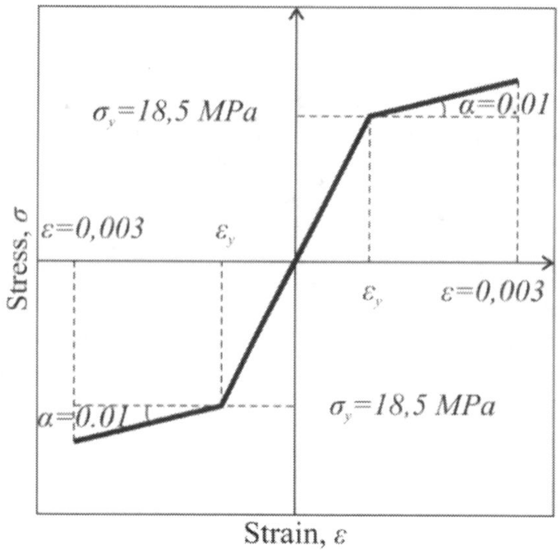


**Fig. 4.** Elastoplastic model of concrete.

The yield function for the diagram presented in Figure 4 is determined from the expression:

$$\phi = \frac{s_{ij}s_{ij}}{2} - \frac{\sigma_y^2}{3} \leq 0, \tag{2}$$

where

$$\sigma_y = \beta\left[\sigma_0 + f_k(\varepsilon_{eff}^p)\right]. \tag{3}$$

$f_k(\varepsilon_{eff}^p)$ denotes the hardening function, $\beta$ - speed of deformation.

Assuming the beam to be an isotropic rod is the most widely used approach utilized when developing computational models. However, this particular case considers the transformed cross section of the beam, with the reinforcement taken as integral and seen as evenly distributed across the section. Besides, the so-called reduced modulus of elasticity is introduced.

The calculations were performed in a nonlinear static setting. In case the finite element of the spar model (beam) reached the value of 0,003, the element was considered collapsed and was eliminated from the structural behavior. The failure of the three-dimensional beam is represented by the plastic deformation in the longitudinal effective reinforcement hitting the value of 0,05.

The numerical study has the following progression: the beams are exposed to their own gravity load and to a gradually increasing load distributed along the length. The load of failure is registered. The stress limits for concrete and reinforcement are considered equal to guideline values.

## 3 Findings of the research

Figures 5-7 demonstrate the isofields of intensity of plastic deformations in the concrete of the three-dimensional beam exposed to various loads.

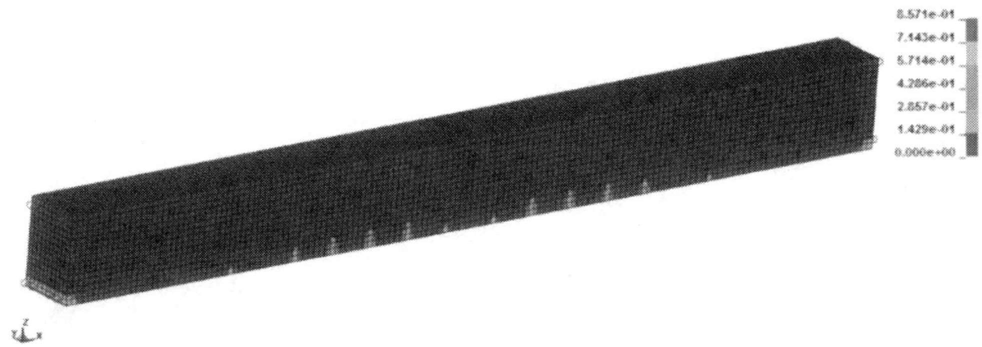


**Fig. 5.** Intensity isofields of plastic deformations in the beam. Commencement of crack opening. Load 32 kN/m.

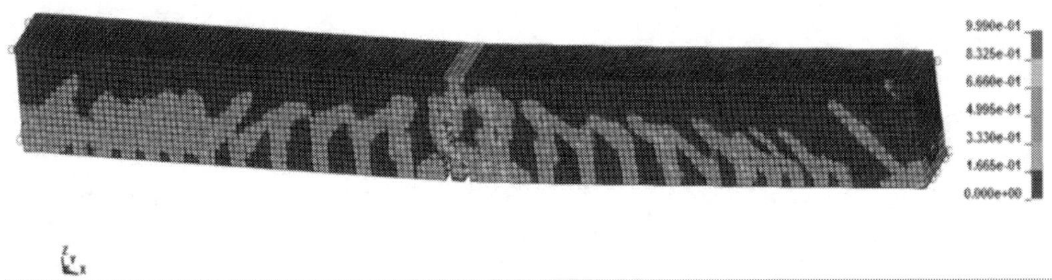


**Fig. 6.** Intensity isofields of plastic deformations in the three-dimensional beam. Plastic deformations intensifying in the upper layers of the section in the middle of the span, cracks opening in the lower layers of the section. Load 171 kN/m.

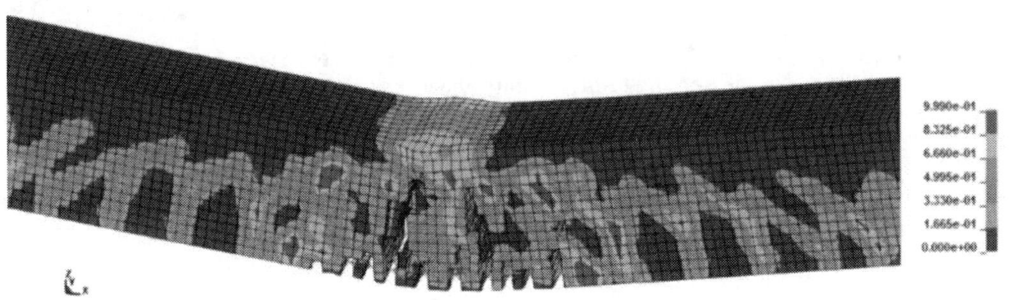


**Fig. 7.** Intensity isofields of plastic deformations in the concrete of the beam. The beam collapsing. Load 191 kN/m.

Figures 8-10 display the intensity isofields of plastic deformations in the reinforcement cage of the three-dimensional beam, taking place at different times under various loads.

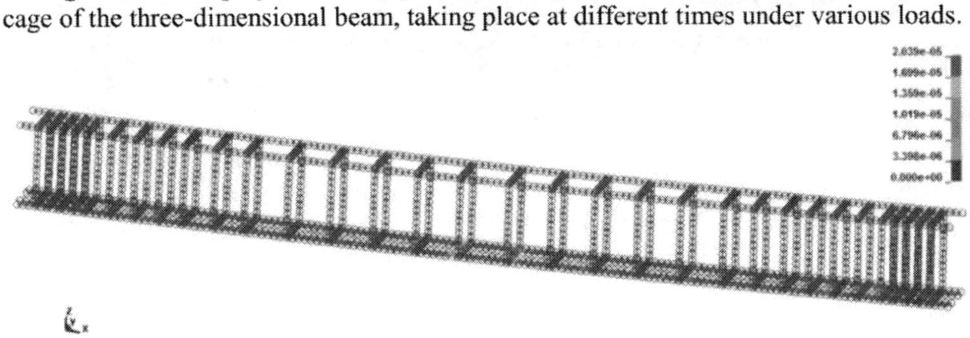

**Fig. 8.** Intensity isofields of plastic deformations in the reinforcement cage. Plastic deformations emerging in the lower working reinforcement. Load being 141 kN/m.

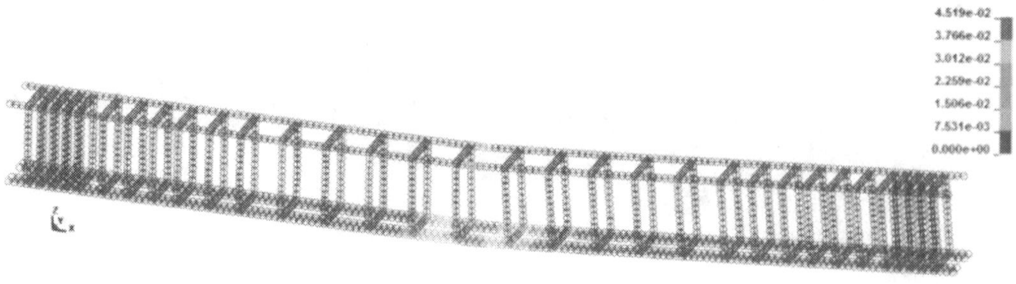


**Fig. 9.** Intensity isofields of plastic deformations in the reinforcement cage. Plastic deformations emerging in the lower working reinforcement. Load being 177 kN/m.

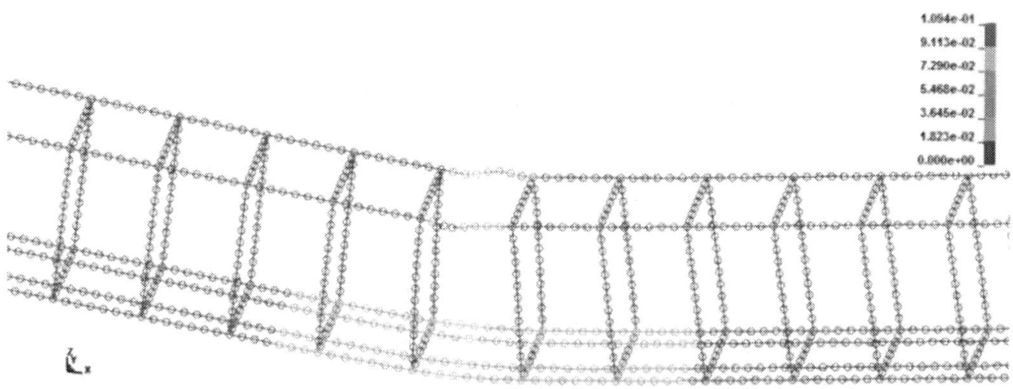


**Fig. 10.** Intensity isofields of plastic deformations in the reinforcement cage. Upper working reinforcement being compressed and losing stability, the beam collapsing. Load being 191 kN/m.

Figures 11-12 display isofields of intensity of plastic deformations in the beam modeled via spar elements, at different times under various loads.

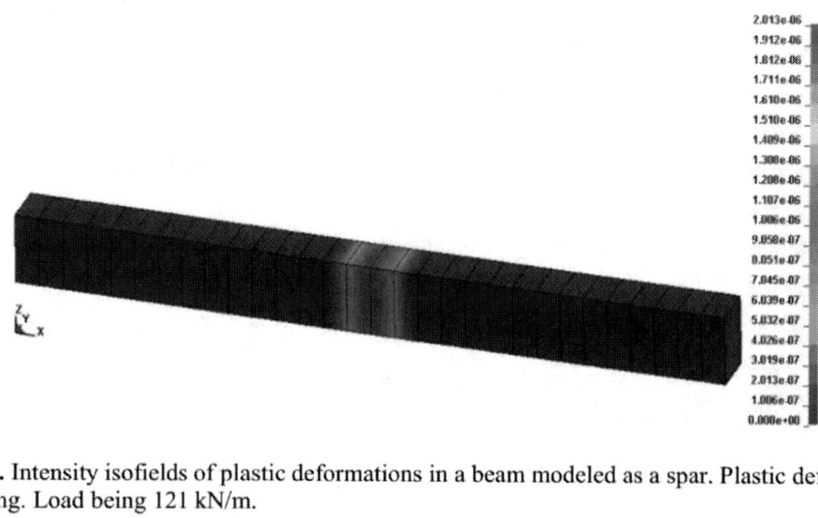


**Fig. 11.** Intensity isofields of plastic deformations in a beam modeled as a spar. Plastic deformations emerging. Load being 121 kN/m.

**Fig. 12.** Intensity isofields of plastic deformations in a rod model of a beam. Beam collapsing in the middle of the span. Load being 165 kN/m.

Table 1 presents the values of destructive load causing beams modeled in different ways to collapse.

**Table 1.** Destructive load values and collapse time for beams.

| | **Load distributed along the length, kN/m** |
|---|---|
| Beam modeled with spar elements | 165 |
| Beam modeled with 3D elements and rod-type reinforcement. Diameter of lower working reinforcement 20 mm | 191 |
| Beam modeled via 3D elements and rod-type reinforcement. Diameter of the lower working reinforcement 18 mm | 165 |

Thus, the verification of the reinforced concrete beam enables us to draw the following conclusions:

1. Mathematical models of beams comprised of spar elements are acceptabe for use in calculations in case the loads are nearly destructive and non-linear behavior of materials is accounted for;
2. If used, a mathematical model of beams comprised of spar elements made of a homogenous isotropic material reqiures that an assumption about 'optimal reinforcement' be introduced, meaning that the compressed area of concrete and the stretched working reinforcement collapse in their section simultaneously.

## Acknowledgements.

This study was performed with the financial support of the RF Ministry of Education and Science, grant #7.1524.2017/Project Part.

## References

1. *US Department of Transportation. Federal Highway Administration. Evaluation of LS-DYNA Concrete. Material Model 159.* **FHWA-HRT-05-063**, 190 (2007)

**2.** Y.D. Murray, *Users Manual for LS-DYNA Concrete Material Model 159*, **FHWA-HRT-05-062** (2007)
3. O. V. Mkrtychev, M. S. Busalova, Proc.Eng., **153,** 475 (2016)
4. O. V. Mkrtychev, M. S. Busalova, Adv. in Eng, Res., **104**, 160 (2016)

# Distribution of waves in the buildings and constructions having regular structure

*Vladimir* Mondrus[1,*], and *Stanislav* Shutovskiy[1]

[1]NRU MGSU, Editorial Department, 129337, 26, Yaroslavskoye Shosse, Moscow, Russia

**Abstract.** In article settlement models of the buildings and constructions having periodic (regular) structure with a masses concentrated in nods are considered. Passing of waves in such systems is considered. For a special case the picture of lines of equal frequency allowing to determine limit frequencies of waves which can pass through this model is constructed.

Article is devoted to research of settlement models of the buildings and constructions representing themselves periodic (regular) systems with a masses concentrated in nods. Further we assume that interaction takes place only between the next masses. In the most general view the three-dimensional settlement model of the building is shown in Fig. 1.

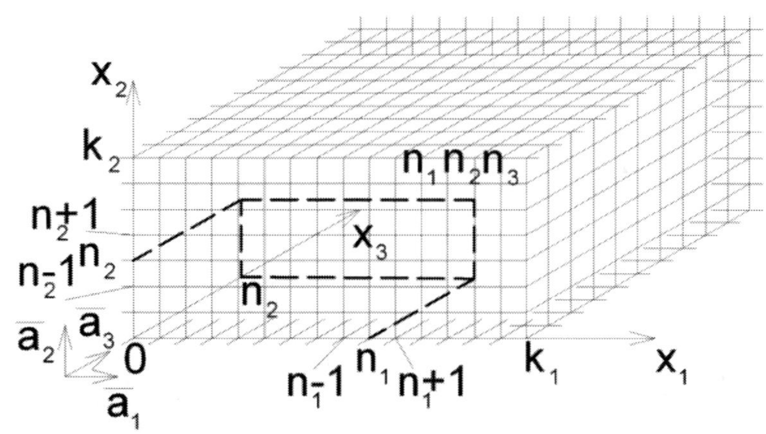


**Fig. 1.** Three-dimensional general model of the building

Let's consider the three-dimensional periodic orthogonal settlement model presented in Fig. 1. We believe that elements communications between masses, parallel to various axes of coordinates, possess various stiffness characteristics, and elements of one direction have various stiffness characteristics depending on the direction of mutual forward shift of masses[1].

* Corresponding author: mondrus@mail.ru

From all $\alpha(\alpha = x_1, x_2, x_3)$ the values of size, possible for communication of any one direction, $\alpha(\alpha = x_1, x_2, x_3)$ $E_{\Pi\alpha\beta}(nn')$ determined by a matrix

$$\begin{Vmatrix} \overset{\alpha}{E}_{\Pi 11}(nn') & \overset{\alpha}{E}_{\Pi 12}(nn') & \overset{\alpha}{E}_{\Pi 13}(nn') \\ \overset{\alpha}{E}_{\Pi 21}(nn') & \overset{\alpha}{E}_{\Pi 22}(nn') & \overset{\alpha}{E}_{\Pi 23}(nn') \\ \overset{\alpha}{E}_{\Pi 31}(nn') & \overset{\alpha}{E}_{\Pi 32}(nn') & \overset{\alpha}{E}_{\Pi 33}(nn') \end{Vmatrix} \tag{1}$$

let's consider only three, standing on the main diagonal and making real physical sense for settlement models of buildings. Characteristics of connections,

which axis is parallel to an axis $x_1$:

$$\overset{\alpha}{E}_{\Pi 11}(nn') = \gamma_{11};\ \ \overset{\alpha}{E}_{\Pi 12}(nn') = \gamma_{12};\ \ \overset{\alpha}{E}_{\Pi 13}(nn') = \gamma_{13}; \tag{2}$$

which axis is parallel to an axis $x_2$ :

$$\overset{\alpha}{E}_{\Pi 21}(nn') = \gamma_{21};\ \ \overset{\alpha}{E}_{\Pi 22}(nn') = \gamma_{22};\ \ \overset{\alpha}{E}_{\Pi 23}(nn') = \gamma_{23}; \tag{3}$$

which axis is parallel to an axis $x_3$ :

$$\overset{\alpha}{E}_{\Pi 31}(nn') = \gamma_{31};\ \ \overset{\alpha}{E}_{\Pi 32}(nn') = \gamma_{32};\ \ \overset{\alpha}{E}_{\Pi 33}(nn') = \gamma_{33}. \tag{4}$$

Other of possible values of size $\overset{\alpha}{E}_{\Pi\alpha\beta}(nn')$ it is believed equal to zero, as not making physical sense.

Thus sizes $\gamma_{\alpha\beta}$ characterize rigidityof connections which axis is parallel to an axis of coordinates $\alpha$ at mutual shift of the connected masses along the axis parallel to an axis of coordinates $\beta$ (Fig. 2)

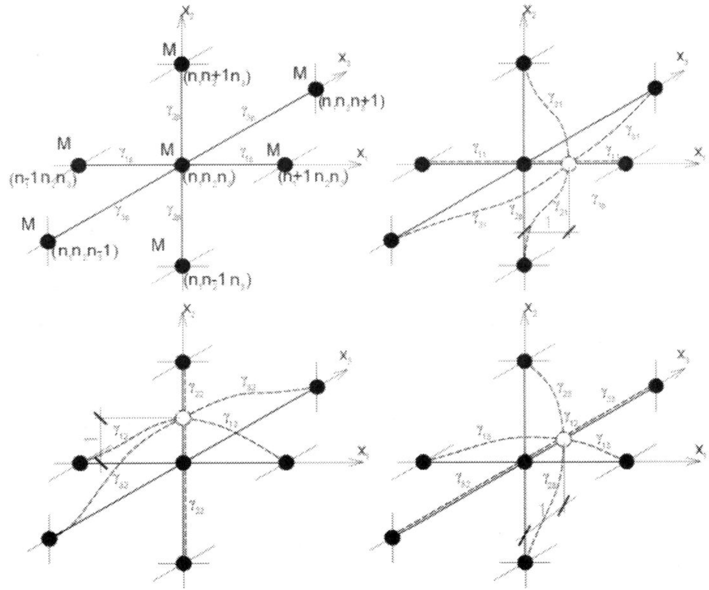

**Fig. 2.** Definition of stiffness of connections

Besides forward movements of masses also their turns round the axes parallel to axes of coordinates are possible. Such connections are designated as $\gamma^{*}_{\alpha\beta}$.

It is possible to write down the movement equation for $\alpha$ - any mass of settlement model which is working out shifts[2]. At $\alpha = x_1$ the equation will register in a look

$$M\ddot{\overset{x_1}{u}}(n_1n_2n_3) = \gamma_{11}\Delta_{n1}\overset{x_1}{u}(n_1n_2n_3) + \gamma_{21}\Delta_{n2}\overset{x_1}{u}(n_1n_2n_3) + \gamma_{31}\Delta_{n3}\overset{x_1}{u}(n_1n_2n_3) \tag{5}$$

where

$$\Delta_{n1}\overset{x_1}{u}(n_1n_2n_3) = \overset{x_1}{u}(n_1+1, n_2, n_3) - 2\overset{x_1}{u}(n_1n_2n_3) + \overset{x_1}{u}(n_1-1, n_2, n_3)$$

$$\Delta_{n2}\overset{x_1}{u}(n_1n_2n_3) = \overset{x_1}{u}(n_1, n_2+1, n_3) - 2\overset{x_1}{u}(n_1n_2n_3) + \overset{x_1}{u}(n_1, n_2-1, n_3)$$

$$\Delta_{n3}\overset{x_1}{u}(n_1n_2n_3) = \overset{x_1}{u}(n_1, n_2, n_3+1) - 2\overset{x_1}{u}(n_1n_2n_3) + \overset{x_1}{u}(n_1-1, n_2, n_3-1) \tag{6}$$

Solving this equation we find:

$$\frac{M\omega^2}{4} = \gamma_{11}\sin^2(\pi\bar{c}\bar{a}_1) + \gamma_{21}\sin^2(\pi\bar{c}\bar{a}_2) + \gamma_{31}\sin^2(\pi\bar{c}\bar{a}_3) \tag{7}$$

Having divided both parts of equality (7) on $\gamma_{11}$, we will receive expression in dimensionless coordinates

$$\Gamma_1 = \sin^2(\pi \bar{c}\,\bar{a}_1) + \chi_1 \sin^2(\pi \bar{c}\,\bar{a}_2) + \chi_2 \sin^2(\pi \bar{c}\,\bar{a}_3), \tag{8}$$

where

$$\Gamma_1 = \frac{M\omega^2}{4\gamma_{11}} \quad \chi_1 = \frac{\gamma_{21}}{\gamma_{11}}, \quad \chi_2 = \frac{\gamma_{31}}{\gamma_{11}} \tag{9}$$

- dimensionless parameters.

Equality (8) is a dispersive ratio for one of three branches of a spectrum of three-dimensional settlement model, the corresponding component of progress of masses along an axis X1.

Similarly dispersive ratios for other branches of a spectrum (turn out at forward and the movements of mass of settlement model and their turn round axes).

Let's consider rectangular two-dimensional settlement model - see Fig. 3.

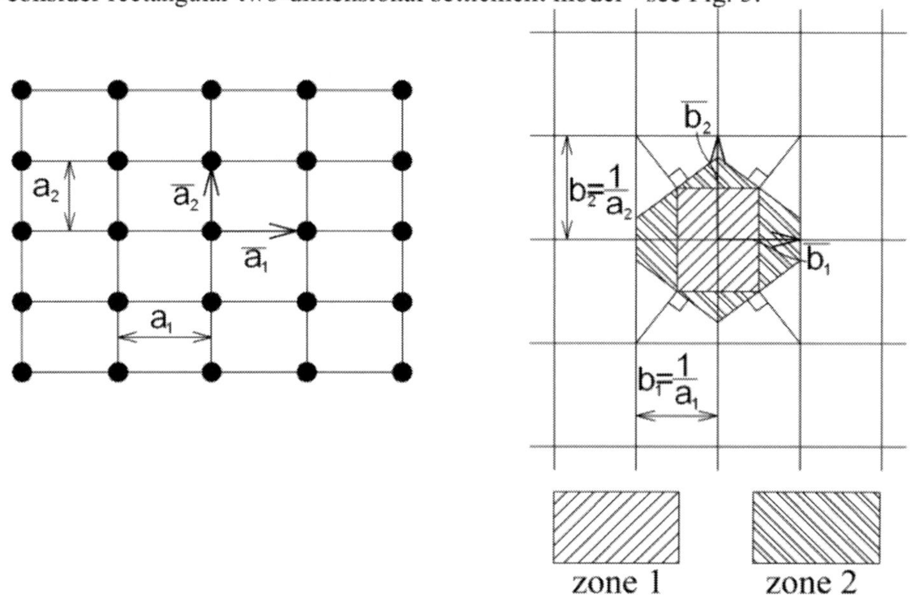


**Fig. 3.** Two-dimensional lattice of flat settlement model

For creation of the return lattice in a direct lattice the distance $\bar{a}_\alpha = |\bar{a}_\alpha|$ between parallel ranks of masses is defined $\bar{a}_\alpha = |\bar{a}_\alpha|$. Having carried out a piece $|b_\alpha| = \dfrac{1}{\bar{a}_\alpha} = \dfrac{1}{|\bar{a}_\alpha|}$ of the direction, perpendicular to the corresponding ranks, we receive a vector of the return lattice. For the considered lattice straight line the return lattice will also be rectangular. The structure of the two first zones of the return lattice is represented in Fig. 3. For two-dimensional settlement model function $\nu(\bar{c})$ or $\omega(\bar{c})$ depends on variables $c_1$ and $c_2$ (a vector component $\bar{c}$) and on border of the first zone its normal derivative has to address in zero. Using dispersive ratios for two-dimensional settlement models, in the plane $c_1c_2$ it is possible to construct family of curves, along each of which frequency is constant. For such lines of equal frequency B. P. Wolfson entered the name "izosikhnotit". All family of curves characterizing change of frequency of dependence from $\bar{c}$ is called as a picture izosikhnotit[3]. The example of a picture izosikhnotit for the two-dimensional settlement model represented in Fig. 3, it is provided on Fig. 4.

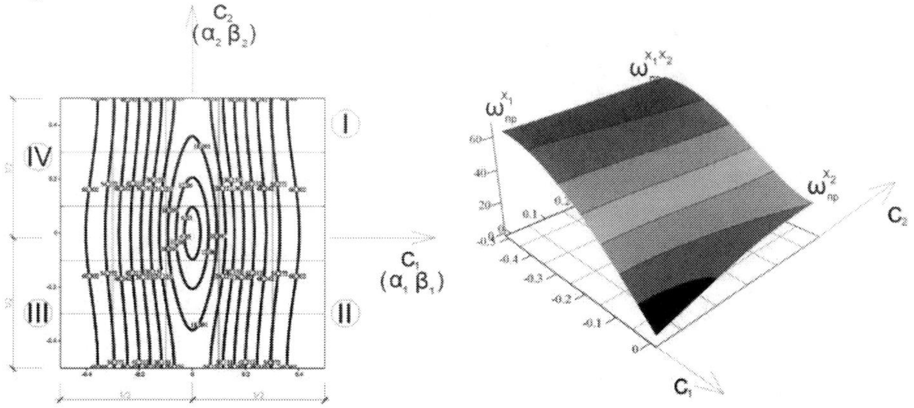


**Fig. 4.** The picture "izosikhnotit" (lines of equal frequency)

It is possible to consider a concrete dispersive ratio for two-dimensional settlement model. For example,

$$\Gamma_1 = \sin^2(\pi \bar{c}\,\bar{a}_1) + \chi_1 \sin^2(\pi \bar{c}\,\bar{a}_2), \tag{10}$$

where

$$\Gamma_1 = \frac{M\omega^2}{4\gamma_{11}}, \quad \chi_1 = \frac{\gamma_{21}}{\gamma_{11}} \tag{11}$$

This dispersive ratio corresponds to one of branches of a spectrum of two-dimensional model. We will be set by concrete sizes of parameters:

$$M = \frac{20000kg}{980cm/s^2} \approx 20\frac{kg \cdot s^2}{cm}; c_1 = 360cm; c_2 = 270cm; \gamma_{11} = 2 \cdot 10^4 \frac{kg}{cm^2}; \gamma_{21} = 2 \cdot 10^3 \frac{kg}{cm^2}$$

At such values of parameters the picture izosikhnotit $\omega(\bar{c})$, constructed in the first zone of the return lattice, has the appearance shown in Fig. 4. The ratio for determination of frequency thus registers in a look:

$$\omega^2 = 400\left(10\sin^2 \pi c_1 + \sin^2 \pi c_2\right) \tag{12}$$

Follows from the ratios given above that frequency $\omega$ - size significantly positive and is the periodic function of coordinates $c_\alpha$ of a wave vector $\bar{c}$ symmetric relatively, axes of coordinates, having an absolute maximum in points $c_1 = \pm 1/2$, $c_2 = \pm 1/2$ and stationary saddle points with coordinates $c_1 = \pm 1/2$, $c_2 = 0$ and $c_1 = 0$ $c_2 = \pm 1/2$. The period, functions in the direction of both axes of coordinates is equal 1, as

$$\Gamma_1 = \sin^2 \pi(c_1 + 1) + \chi_1 \sin^2 \pi(c_2 + 1) = \sin^2 \pi c_1 + \chi_1 \sin^2 \pi c_2 \tag{13}$$

Existence of special points specifies that for any direction of distribution of a wave for discrete periodic settlement models there are limit frequencies which we will call further characteristic. They are defined from a ratio (7):

$$\omega_{np}^{x_1} = 2\sqrt{\frac{\gamma_{11}}{M}} \quad c_1 = \frac{1}{2} \ () \ c_2 = 0 \tag{14}$$

- local limit frequency for the direction of the wave vector coinciding with an axis $Ox_1$;

$$\omega_{np}^{x_2} = 2\sqrt{\frac{\gamma_{11}}{M}} \quad c_1 = 0 \ () \ c_2 = \frac{1}{2} \tag{15}$$

- local limit frequency for the direction of the wave vector coinciding with an axis $Ox_2$;

$$\omega_{np}^{x_1 x_2} = 2\sqrt{\frac{\gamma_{11} + \gamma_{21}}{M}}, \left( c_1 = c_2 = \frac{1}{2} \right) \tag{16}$$

- absolute limit frequency of settlement model for the direction of the wave vector coinciding with a diagonal of any of zone quadrants. If the frequency of the extending wave is more or is equal $\omega_{np}^{x_1 x_2}$, irrespective of the direction of distribution amplitude of this wave at its movement in settlement model will decrease from weight to weight.

The spatial discrete settlement model of the building or construction generally can have 6 branches of a spectrum of own frequencies corresponding to the 3rd to components of progress and the 3rd possible turns of masses. The dispersive ratio testifies to the certain restrictions imposed by various factors on possibility of distribution in settlement model of the corresponding wave mode (fluctuation)[4]. These restrictions, are generally connected with that for any direction of a wave vector of the wave extending in the building there is quite certain pass-band limited to limit frequencies. Fluctuations with the frequencies which are not getting to a pass-band practically in the building cannot extend. Thus, if to the building (construction) the wave of the general type capable to cause all six possible types of fluctuations comes, that, depending on characteristics of the falling wave and design of the building in the last waves of one types can actively extend and practically waves of other types cannot extend.

In real buildings and constructions the same seismic wave in standing nearby on a surface of the earth and buildings and constructions, various on the constructive decision, can cause fluctuations of absolutely different types (vertical, horizontal, tortional). The same wave in the constructions, identical on the constructive decision located under different corners to the front of a wave can or cause active fluctuations, or not cause almost any fluctuations. Similar results obviously have to take place at impact on buildings and constructions of a blast wave, a wind, etc.

Depending on parameters of a wave and characteristics of a construction it can appear that the wave from a source in a construction mainly extends only in any chosen directions lying in certain space angles - for three-dimensional settlement models or certain flat corners - for two-dimensional settlement models. In special cases the wave can extend only in the direction of some beams proceeding from a construction point in which the source is located. Thus in certain directions mainly extends waves of one type. Along with the specified special cases perhaps also that the wave extends in all directions from a source without attenuation, or, on the contrary, in all directions actively fades.

## References

1. L.Brilluen, M.Parody, *Distribution of waves in periodic structures*. (SILT, Moscow, 1959)
2. B. P. Wolfson, Researches on the theory of constructions, under. edition of the prof. B. G. Korenev, member correspondent of Academy of Sciences of the USSR I. M. Rabinovich, prof. A. F. Smirnov, **XYII** (1969).
3. B. P. Wolfson, Researches on the theory of constructions, under. edition of the prof. B. G. Korenev, member correspondent of Academy of Sciences of the USSR I. M. Rabinovich, prof. A. F. Smirnov, **XIX** (1972).
4. M. A. Dashevsky, Works TsNIISK of V. A. Kucherenko, **43** (1975).

# Features of the solution of a problem of dynamics of a plate with use of the Python programming language

*Vladimir* Mondrus [1,*], and *Dmitrii* Sizov[1]

[1]Moscow State University of Civil Engineering, 129337, Moscow, Yaroslavskoe shosse, 26, Russia

**Abstract.** The article describes the features of the application of the modern general-purpose language Python for solving the dynamic problem of oscillations of a rectangular plate. The focus is often on libraries used for numerical computation and visualization of results.

## 1 Python as the language of numerical calculations

One of the modern languages of programming of general purpose is the Python language. Feature of writing of programs in this language is limit clarity of a program code, the maximum orientation to increase of labor productivity of the developer [1],[2]. Language is freely extended and has implementations of the interpreter practically for all modern computer platforms.

Considering that fact that the speed of implementation of programs by the interpreter is obviously much lower, than the executed files of the programs written in the compiled programming languages such as Fortran, C, C++, it is necessary to pay at implementation of computing procedures the main attention to perhaps bigger use of numerical libraries which can be freely used in the Python language. In particular, the free software environment of programming of Anaconda already contains a set of the numerical libraries allowing to possess convenience of manipulation with arrays and array-like objects at the level of the Matlab language, or Fortran in a standard distribution kit, thus the speed of calculations substantially is defined by algorithms of numerical libraries. Using the Python language as the connecting language (language glue) at the appeal to the libraries written in the different compiled languages it is possible to receive the program which is expedited also as well as using the compiled programming languages. It is known that now the community of developers of the Python language at the same time keeps two branches developments of language: Python2x and Python3x, as reference implementation is most often considered the version of the CPython interpreter (www.python.org/dev/implementations/) though there is also some other interpreters [1],[2], in particular, of Jython, IPython based or on Java-classes, or oriented for work in Microsoft.NET Framework. The special attention is deserved by the experimental version of the PyPy interpreter implemented in the Python language. In scientific calculations it is

* Corresponding author: mondrus@mail.ru

most often applied Python2x versions that is explained, first of all by a little more prime object model of language and higher speed of work of Python2x in comparison with Python3x.

One of the most often used libraries applied in Python NumPy and SciPy [1] which allow to expand standard objects of Python with arrays, functions for work with arrays, matrix objects, functions of work with polynoms, functions of linear algebra.. Use of these libraries allows to create fast numerical programs so far the majority of operations is made not over scalars, and over arrays and matrixes - data objects of libraries. For display of scientific graphics - matrixes, cuts of arrays the Matplotlib package including the main instruments of display of function graphs of one and many variables is often applied. Use and such means of visualization of results of calculations as Gnuplot possessing the built-in own language and the system of teams is possible.

## 2 Modelling of oscillations of a rectangular plate

The solution of a problem of dynamic calculation of a plate which separate elements have the set external shift is provided in article, sources of vibration influence are on a contour of a plate and in five internal areas in it. Change of coordinates of the points located on a contour of a plate and in it happens under the harmonious law to different frequencies. In Fig. 1 the scheme of a plate taking into account the enclosed shifts is shown.

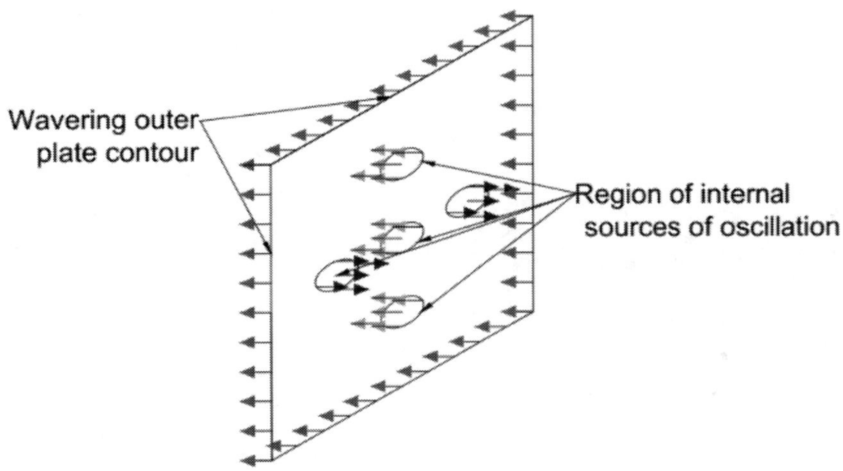


**Fig. 1** The arrangement of the sources of oscillations on a square plate.

As appears from [3] at constant flexural rigidity of a plate the equation of fluctuations has an appearance:

$$D\Delta\Delta w + \rho h \frac{\partial^2 w}{\partial t^2} = q(x,t), \tag{1}$$

where:

$D = \dfrac{Eh^3}{12 \cdot (1-\nu^2)}$ - flexural rigidity of a plate

$E, \rho, \nu, h, w$ – respectively plate material elastic modulus, density, Poisson's ratio, thickness of a plate and deflection of a median surface.

$q(x,t)$ – external load of a plate, depends on coordinates of a point of application and time $t$.

The equation of fluctuations given in article is guided by the classical theory of Kirchhoff-Lyava. Briefly we will remind its main hypotheses:

1) the rectilinear element, normal to a median surface before deformation remains direct and normal to a median surface, without changing the length;

2) normal stresses between separate layers of a plate $\sigma_z = 0$

It is convenient to apply the finite difference method based on introduction of a differential grid, splitting area of a plate into separate sites and the further decision of the turned-out algebraic system of equations to the solution of dynamic equation.

As it is known [4],[5] solution of initial boundary value problems for wave equation is possible with use explicit, explicit and implicit and implicit schemes. Use of the explicit scheme allows, being guided by the known solution of a task on the previous time layers at once to determine the decision on the current step by time. This approach does not demand any decision of systems of equations, but imposes rigid restriction on step size on time, at violation of this condition gradual reduction of reliability of the received decision (Courant's condition) is observed. Use of the implicit scheme demands on each of steps of the decision of the turning-out system of equations, as in the sampled equation and unknown, and required sizes on the current temporary step. Explicit and implicit schemes, combine advantages of an explicit method and the guaranteed stability of the implicit scheme. In article for the solution of a task the implicit scheme demanding on each step of the decision of systems of equations, however possessing the proved guaranteed stability of the account is used. In drawings the illustrations characterizing change of a deflection of a plate in different instants are given in Table 1.

**Table 1** Field plate deflections at various times

| | |
|---|---|
| 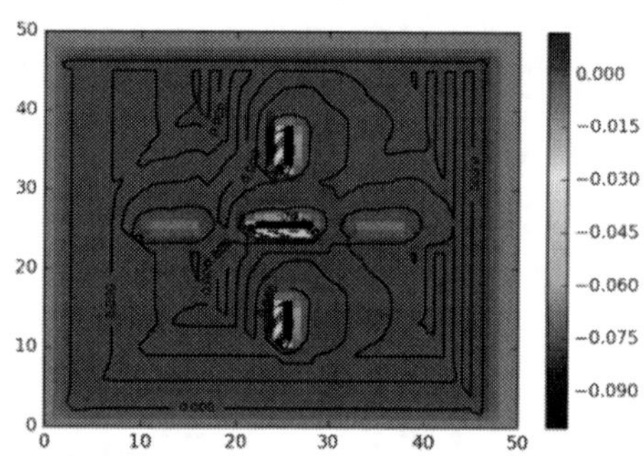 | 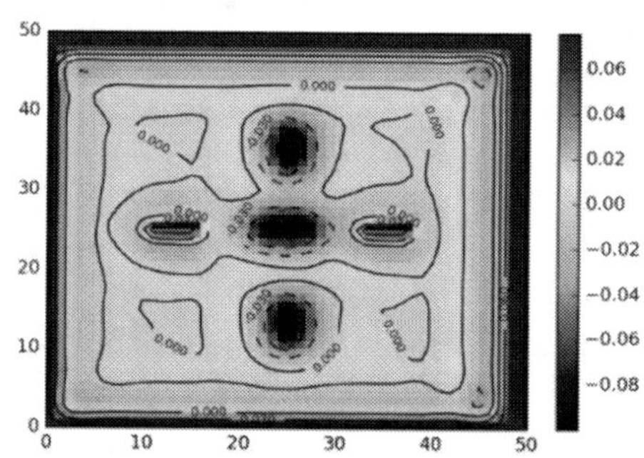 |
| t = 0.0 sec. | t = 40,00 sec. |

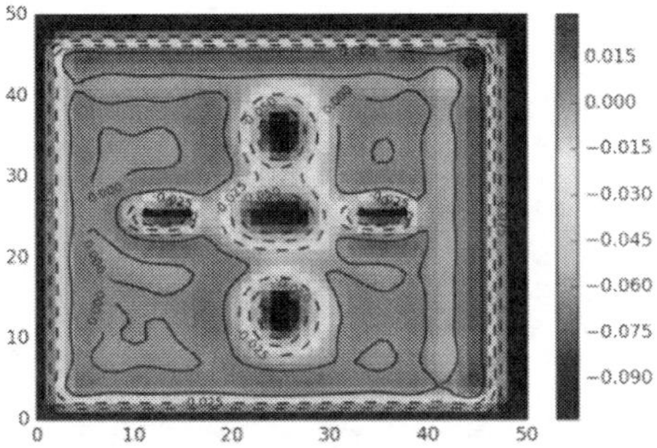


t = 80.0 sec.

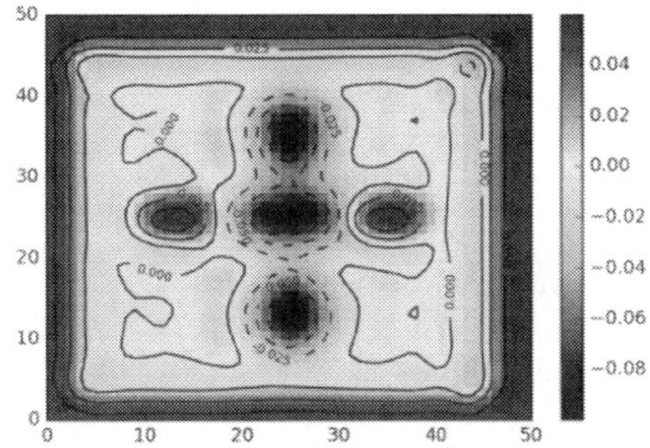


t =120.0 sec.

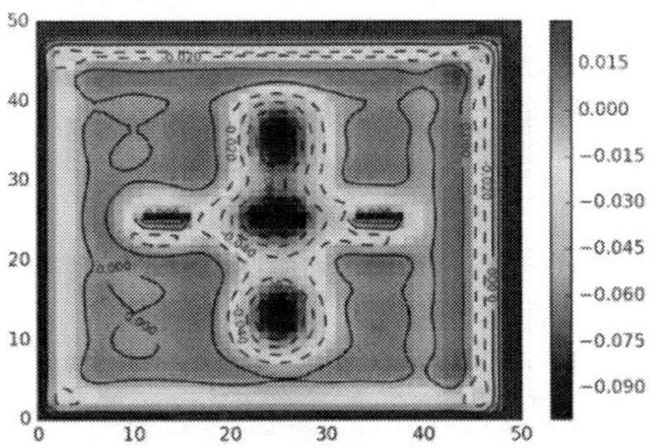


t = 160.0 sec.

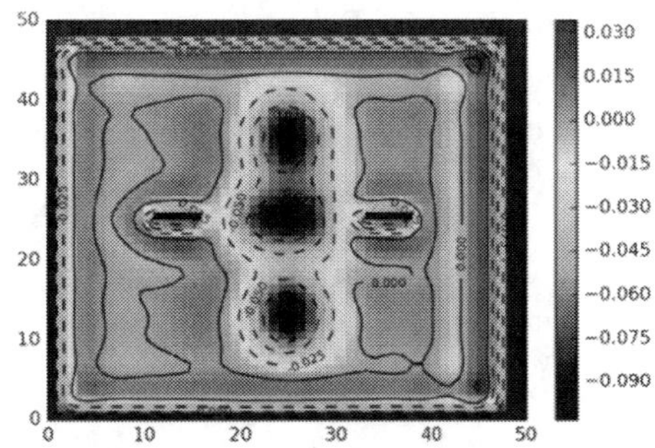


t = 200.0 sec.

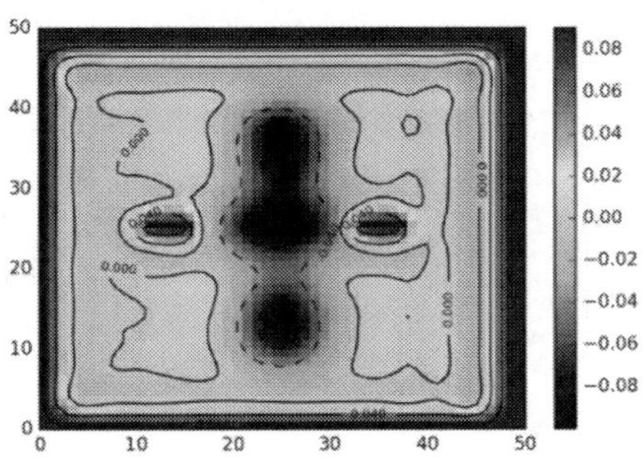


t = 240.0 sec.

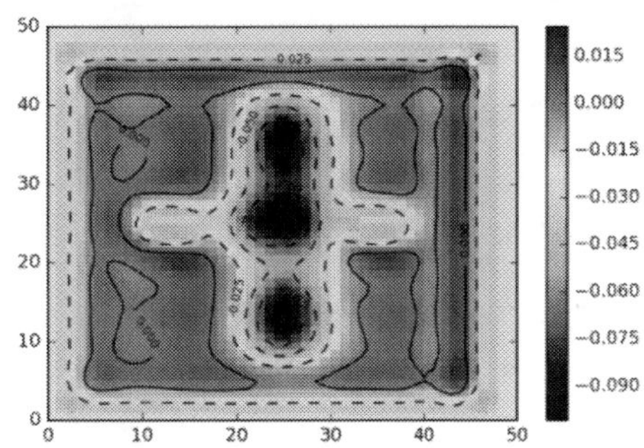


t = 280.0 sec.

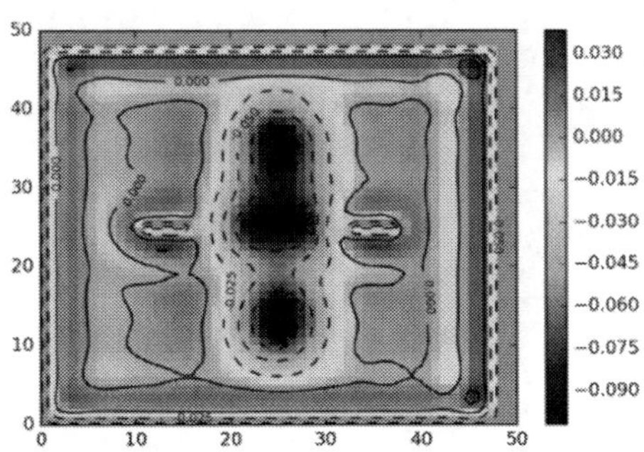


t = 320.0 sec.

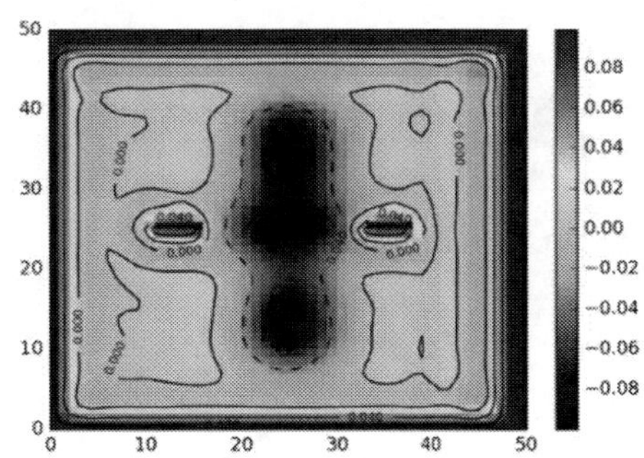


t = 360.0 sec.

As the conclusion it would be desirable to note that the developed Python-script allows not only to receive the numerical solution of a task for acceptable time, but also in the dynamic mode to watch change of the numerical decision on each step of calculation that is reached by use of the connected library of scientific graphics of Matplotlib.

## References

1. Luciano Ramaglio *Python To skill tops* (M: DMK publishing house, 767 pages ISBN 978-5-97060-384-0, 2016)
2. Wes McKinney *Python and the analyses of data* ( M: DMK publishing house, 767 pages. ISBN 978-5-97060-315-4, 2015)
3. K. Vasidzu *Variation methods in the theory of elasticity and plasticity* - (M: World, 542 pages, 1987).
4. E.E. Tyrtyshnikov *Numerical analyses: manual for the student. higher education institutions* ( M: The publishing center "Akademiya",.320 page ISBN 978-5-7695-3925-1, 2007)
5. N.P. Savenkova, O. G. Provorova, A.U. Mokin *Numerical methods in a mathematical simulation* (M: INFA-M ARGAMAK-MEDIA, 176 p, ISBN 978-5-00024-019-9 (ARGAMAK-MEDIA), ISBN 978-5-16-009705-3 (INFA-M, print), 2014)

# Meshless simulation of dam break using MLPG-RBF and shallow water equations

*Juraj* Mužík[1,*], *Martina* Holičková[1]

[1]University of Zilina, Faculty of Civil Engineering - Department of geotechnics, Univerzitná 8215/1, Slovakia

**Abstract.** This article focuses on the application of the meshless local Petrov-Galerkin (MLPG) method to solve the shallow water equations (SWE). This localized approach is based on the meshless weak formulation with the use of radial-basis functions (RBF) as the trial functions. Comparing with mesh-based methods, the present method is more efficient for large-scale problems with complex geometries. In this work, the numerical model is applied to simulate a dam-break problem as one of most descriptive benchmark problems for SWE. As a result, the adopted meshless method not only shows its algorithm applicability for class of problems described by SWE, but also brings more efficiency than several conventional mesh-based methods.

## 1 Introduction

The numerical models based on the shallow water equations (SWE) are commonly applied in hydraulic problems such as dam break analysis, open channel flows and sloshing wave problem.

Many numerical methods have been used to discretize numerical model described by the shallow water equations. For example, finite difference method (FDM) is very popular method that is simple to understand and code[1], finite volume method (FVM) has significant progress in simulation of flow phenomena[2] and also finite element method (FEM) has been successfully applied to the hydraulic simulation problems[3].

Due to the advantage of no mesh generation, the applications of meshless methods is current essential topics. In generally meshless methods have been classified into two main categories: boundary type and domain type. Under the premise that the fundamental or general solution of the governing partial differential equation (PDE) exists, boundary-type meshless methods have advantages from requiring relatively fewer computing nodes only around the boundary of the domain, such as the method of fundamental solutions (MFS) [4] and boundary knot method [5]. On the other hand, domain-type mesh-free methods have wider applications to the PDEs without fundamental solutions but need more nodes distributed in the whole domain. The numerical approximations of these methods are mainly based on the radial-basis functions (RBFs). For the SWE simulations, RBF strong form methods apply the interpolation of RBF to approximate the governing equations for

* Corresponding author: muzik@fstav.uniza.sk

examples multiquadrics (MQ) [6] or compactly supported radial-basis function (CSRBF) [7]).

The MLPG-RBF scheme has been developed by combining the RBF-based concept and the MLPG method. Atluri and his co-workers [8] initially introduced the MLPG method which followed the concept of weak form generalization. The conventional MLPG method has advantages of the high accurate solution, however the use of moving least-squares approximation (lack of delta property), brings some problems to the implementation of the boundary conditions. To overcome these difficulties the RBF interpolation method is adopted. Moreover, the band-width of the characteristic matrix can be adjusted using the concept of locally supporting nodes.

In this paper, we examine the applications of the MLPG-RBF to solve the SWE. First, the formulations will be introduced before the application of numerical procedures. The MLPG-RBF is introduced in chapter 3. In this chapter, the governing equations are discretized by the MLPG-RBF method in spatial domain and by Euler scheme in temporal direction. Chapter 4 demonstrates numerical example of the dam break. The model is verified by comparing the results with the benchmark solution in the literature.

Finally, conclusions and suggestions will be given based on the numerical model results.

## 2 Two-dimensional shallow water equations

The shallow water equations (SWEs, which are also called the de Saint-Venant equations) are used to describe flow behaviors in bodies of water where the horizontal length scales are much greater than the flow depth (i.e., the long wavelength phenomena). Here we briefly review the 2D SWEs.

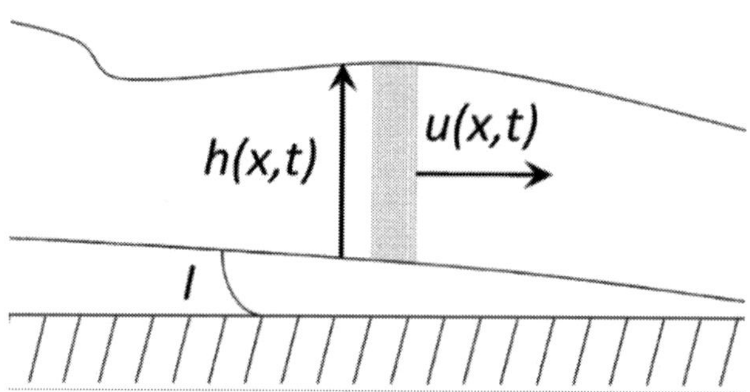


**Fig. 1.** Definition of the shallow-water flow region.

A shallow-water flow region is illustrated in Fig. 1. By the assumption of the hydrostatic pressure and uniform velocity profile along the vertical direction, the 3D incompressible Navier-Stokes equations can be simplified to the 2D SWEs.

In the primitive form, the governing equations including a continuity equation and a momentum equation in each of the $x$- and $y$-directions are shown as follows:

$$\frac{\partial h}{\partial t}+\frac{\partial (uh)}{\partial x}+\frac{\partial (vh)}{\partial y}=0 \tag{1}$$

$$\frac{\partial(uh)}{\partial t}+\frac{\partial(u^2h)}{\partial x}+\frac{\partial(uvh)}{\partial y}+\frac{1}{2}g\frac{\partial h^2}{\partial x}=\nu_e\left(\frac{\partial(hu_x)}{\partial x}+\frac{\partial(hu_y)}{\partial y}\right)+F_x \qquad (2)$$

$$\frac{\partial(vh)}{\partial t}+\frac{\partial(uvh)}{\partial x}+\frac{\partial(v^2h)}{\partial y}+\frac{1}{2}g\frac{\partial h^2}{\partial y}=\nu_e\left(\frac{\partial(hv_x)}{\partial x}+\frac{\partial(hv_y)}{\partial y}\right)+F_y \qquad (3)$$

where $t$ is time, $h$ is the total water depth, $g=9.81\ m/s^2$ is the acceleration of gravity, $\nu_e$ is the kinematic eddy viscosity. $u$ and $v$ are depth-averaged velocity components in the horizontal $x$- and $y$-directions, respectively. $u_x,u_y$ and $v_x,v_y$ are the derivatives of depth-averaged velocity components to $x$- or $y$-direction according to its subscript. The forcing term used in our simulation contains only the bed-shear stress, which is given as

$$F_x=\frac{gu\sqrt{u^2+v^2}}{C_Z^2};\ F_y=\frac{gv\sqrt{u^2+v^2}}{C_Z^2} \qquad (4)$$

where $C_Z = h^{(1/6)}/n$ is the Chezy friction coefficient and $n$ is the Manning coefficient. It is noted that the kinematic eddy viscosity $\nu_e$ of water is much smaller than the bed shear stress. Hence the viscous effect of kinematic eddy viscosity can be neglected in many hydrodynamics problems. By substituting continuity equation (1) into Eqs. (2) and (3) the momentum equations can be simplified as

$$\frac{\partial u}{\partial t}+u\frac{\partial u}{\partial x}+v\frac{\partial u}{\partial y}+g\frac{\partial h}{\partial x}=\frac{F_x}{h} \qquad (5)$$

$$\frac{\partial v}{\partial t}+u\frac{\partial v}{\partial x}+v\frac{\partial v}{\partial y}+g\frac{\partial h}{\partial y}=\frac{F_y}{h} \qquad (6)$$

the former form (5) and (6) of governing equations has been used for the numerical model.

## 3 The meshless local Petrov-Galerkin method with radial basis functions

The MLPG method has been introduced by Atluri et al. [8]. It is characterized as meshless since distributed nodal points, covering the domain, are employed. All needed integrals (weak form) are carried out on the local subdomain centred at every point. All unknown variables are approximated by some interpolation method to obtain a system of non-linear equations. Solving this system of equations leads to a numerical solution of the problem. Atluri et al. [8] used the moving least squares (MLS) approximation scheme but nowadays the radial basis functions (RBFs) interpolation can be used instead (see e.g. [9]). An important advantage of RBF interpolation is an existence of the delta property and therefore the boundary conditions of the first kind can be easily defined.

### 3.1 Methodology

The entire domain $\Omega_s$ is covered by nodes located inside the area and also on the global boundary $\Gamma$ (see Fig. 2).

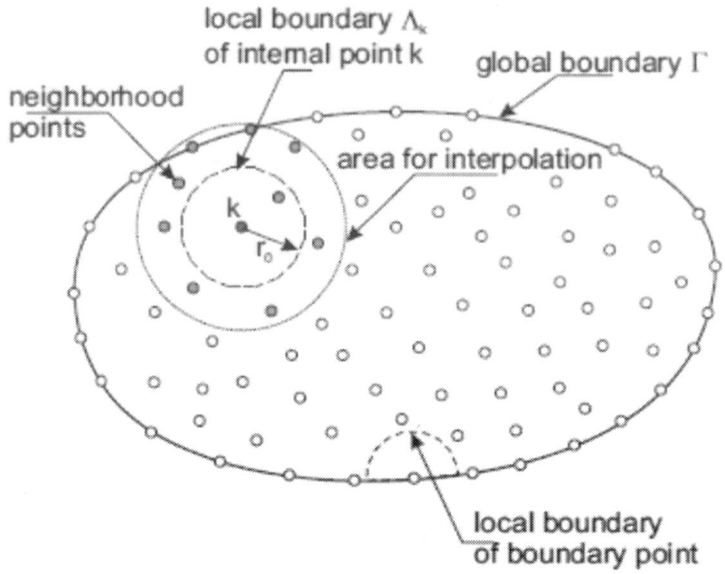


**Fig. 2.** Geometrical definition of the meshless model.

The mutual relationship of particular nodes is based on some interpolation algorithm. The local radial basis functions (RBFs) are used to approximate unknown height and velocities in the neighbourhood or support of a reference point *i*. Multiquadric functions are one of the most popular radial functions used for this purpose and they have been used in our paper. They can be defined as

$$R(r_{ij}) = \sqrt{r_{ij}^2 + c^2} \tag{7}$$

where $r_{ij}$ is a distance between points *i* and *j* and *c* is the so-called shape factor of multiquadric function. The formula of Hardy [10] with a slight modification is applied to the local RBFs (see [9]) to find the optimal value of the shape factor, which can be computed in point *i* as

$$c = \frac{0.815}{N} \sum_{j=1}^{N} r_{ij} \tag{8}$$

The interpolation of the unknown variables can be written using the basis functions $\varphi_{ij}$ in the form (more details can be found in [9])

$$h_i = \sum_{j=1}^{N} \varphi_{ij} h_j, \quad u_i = \sum_{j=1}^{N} \varphi_{ij} u_j, \quad v_i = \sum_{j=1}^{N} \varphi_{ij} v_j \tag{9}$$

where *N* is the number of neighbourhood points in the support domain.

### 3.2 The discrete form

The five supporting points for point $i$ inside of computational domain $\Omega$ (including the point $i$ itself) were used for the unknown values interpolation at node $i$.

To obtain a local weak form of the shalow water flow in the reference point $i$, (1), (5) and (6) is weighted by test functions $w_i$ and integrated over the local sub-domain $\Omega_s$ and the following equations are obtained

$$\int_{\Omega_{S_i}} w_i \left[ \frac{\partial h}{\partial t} + h\frac{\partial u}{\partial x} + u\frac{\partial h}{\partial x} + h\frac{\partial v}{\partial y} + v\frac{\partial h}{\partial y} \right] d\Omega_{S_i} = 0 \tag{10}$$

$$\int_{\Omega_{S_i}} w_i \left[ \frac{\partial u}{\partial t} + u\frac{\partial u}{\partial x} + v\frac{\partial u}{\partial y} + g\frac{\partial h}{\partial x} \right] d\Omega_{S_i} = \int_{\Omega_{S_i}} w_i \left[ \frac{F_x}{h} \right] d\Omega_{S_i} \tag{11}$$

$$\int_{\Omega_{S_i}} w_i \left[ \frac{\partial v}{\partial t} + u\frac{\partial v}{\partial x} + v\frac{\partial v}{\partial y} + g\frac{\partial h}{\partial y} \right] d\Omega_{S_i} = \int_{\Omega_{S_i}} w_i \left[ \frac{F_y}{h} \right] d\Omega_{S_i} \tag{12}$$

Spline weight functions $w_i$ with compact support have been considered in this paper

To compute integrals in (10), (11) and (12) we used a quadrilateral shape of the sub-domain. The non-linear 2D SWEs consisting of Eqs. (10), (11) and (12) are discretized by the MLPG-RBF method in spatial domain and by full-implicit scheme (including all the external forcing terms) in temporal direction, and the discrete form, obtained using principles described in Chapter 3.1, at $i$th node can be written as follows:

$$\begin{aligned} &\int_{\Omega_{S_i}} w_i \left[ \frac{1}{\Delta t}\sum_{j=1}^{N} \varphi_{ij} h_j^{n+1} + \sum_{j=1}^{N} \varphi_{ij}\tilde{v}_j \sum_{j=1}^{N} \frac{\partial \varphi_{ij}}{\partial y} h_j^{n+1} + \sum_{j=1}^{N} \varphi_{ij}\tilde{u}_j \sum_{j=1}^{N} \frac{\partial \varphi_{ij}}{\partial x} h_j^{n+1} \right] d\Omega_{S_i} + \\ &\int_{\Omega_{S_i}} w_i \left[ \sum_{j=1}^{N} \varphi_{ij}\tilde{h}_j \sum_{j=1}^{N} \frac{\partial \varphi_{ij}}{\partial x} u_j^{n+1} \right] d\Omega_{S_i} + \int_{\Omega_{S_i}} w_i \left[ \sum_{j=1}^{N} \varphi_{ij}\tilde{h}_j \sum_{j=1}^{N} \frac{\partial \varphi_{ij}}{\partial y} v_j^{n+1} \right] d\Omega_{S_i} = \\ &\int_{\Omega_{S_i}} w_i \left[ \frac{1}{\Delta t}\sum_{j=1}^{N} \varphi_{ij} h_j^{n} \right] d\Omega_{S_i} \end{aligned} \tag{13}$$

$$\begin{aligned} &\int_{\Omega_{S_i}} w_i \left[ \frac{1}{\Delta t}\sum_{j=1}^{N} \varphi_{ij} u_j^{n+1} + \sum_{j=1}^{N} \varphi_{ij}\tilde{v}_j \sum_{j=1}^{N} \frac{\partial \varphi_{ij}}{\partial y} u_j^{n+1} + \sum_{j=1}^{N} \varphi_{ij}\tilde{u}_j \sum_{j=1}^{N} \frac{\partial \varphi_{ij}}{\partial x} u_j^{n+1} \right] d\Omega_{S_i} + \\ &\int_{\Omega_{S_i}} w_i \left[ g\sum_{j=1}^{N} \frac{\partial \varphi_{ij}}{\partial x} h_j^{n+1} \right] d\Omega_{S_i} + \int_{\Omega_{S_i}} w_i \left[ \frac{F_x}{h} \right] d\Omega_{S_i} = \int_{\Omega_{S_i}} w_i \left[ \frac{1}{\Delta t}\sum_{j=1}^{N} \varphi_{ij} u_j^{n} \right] d\Omega_{S_i} \end{aligned} \tag{14}$$

$$\int_{\Omega_{S_i}} w_i \left[ \frac{1}{\Delta t} \sum_{j=1}^{N} \varphi_{ij} v_j^{n+1} + \sum_{j=1}^{N} \varphi_{ij} \widetilde{v}_j \sum_{j=1}^{N} \frac{\partial \varphi_{ij}}{\partial y} v_j^{n+1} + \sum_{j=1}^{N} \varphi_{ij} \widetilde{u}_j \sum_{j=1}^{N} \frac{\partial \varphi_{ij}}{\partial x} v_j^{n+1} \right] d\Omega_{S_i} +$$
$$\int_{\Omega_{S_i}} w_i \left[ g \sum_{j=1}^{N} \frac{\partial \varphi_{ij}}{\partial y} h_j^{n+1} \right] d\Omega_{S_i} + \int_{\Omega_{S_i}} w_i \left[ \frac{F_y}{h} \right] d\Omega_{S_i} = \int_{\Omega_{S_i}} w_i \left[ \frac{1}{\Delta t} \sum_{j=1}^{N} \varphi_{ij} v_j^{n} \right] d\Omega_{S_i} \tag{15}$$

to solve non-linear system of equations the fixed value iteration procedure is adopted. The upper index in the equations (13), (14) and (15) represents the time step, the values marked with tilde are the values obtained from the last iteration (fixed values). The weight function used in the equations (10), (11) and (12) resp. (13), (14) and (15) is 4th order spline function, more details can be found in [9].

## 4 Numerical results - 2D dam-break problem by non-linear SWE model

This problem models a partial dam-break for a rapid opening of a sluice gate with a non-symmetric breach and its ability to simulate discontinuous flows. The computational domain is a *200 m* by *200 m* region. A dam is located in the middle of the domain with *10 m* thickness. The initial water depth is *10 m* on one side and *5 m* on the other side of the dividing wall. At time *t=0*, the dam fails, and the water is released through the 75 m wide non-symmetric breach, as shown in Fig. 3.

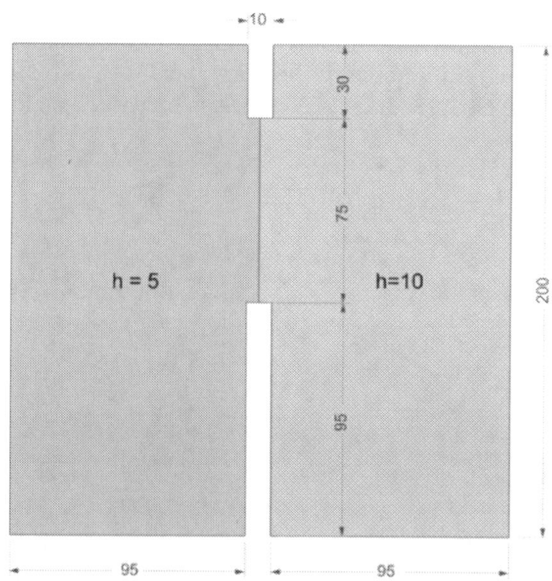


**Fig. 3.** Geometrical definition of the Dam-break model.

When the downstream water depth is *5 m*, the flow is subcritical everywhere. The boundary conditions at *x=0* and *x=200 m* are assumed to be transmissive and all other boundaries are considered as reflective. At the instant of the dam break, water is released through the breach, forming a positive wave propagating downstream and a negative wave spreading upstream.

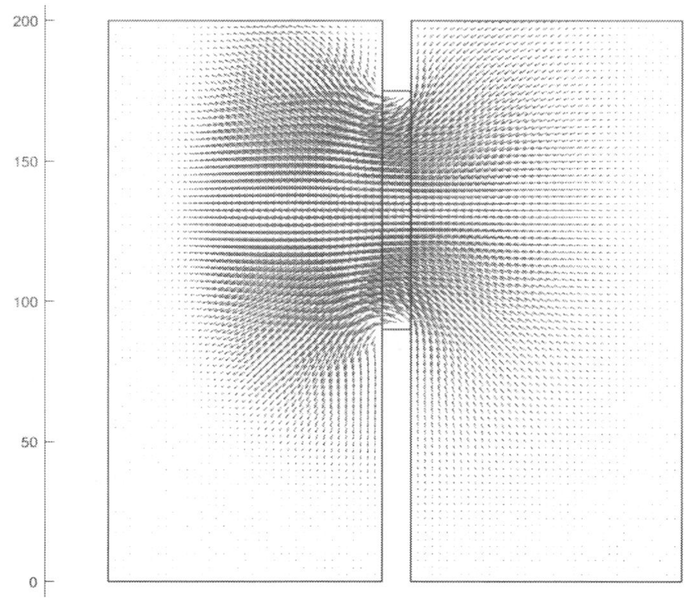


**Fig. 4.** Dam-break flow field vectors at *t=7.2 s*.

The scale of time step is given by the CFL condition as follows. At the beginning of the dam break caused by the gravity, we take the value of $u_{max}$ by the velocity caused from the difference of the initial water elevation. By the initial water elevation, the main difference causing the dam break is on the $x$-direction and hence we also focus on the rule of $x$-direction. For $N=81\times 81$, $\Delta x=200/80=2.5\ m$ and by CFL condition $\Delta t$ needs to satisfy

$$\Delta t \le \frac{\Delta x}{u_{\max}} = \frac{2.5}{7} \approx 0.357s\,.$$

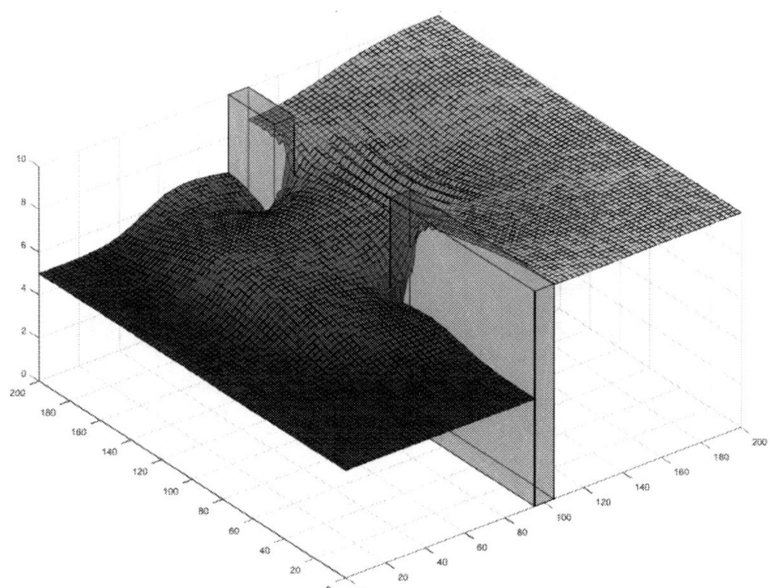

**Fig. 5.** Water surface profile of the 2-D dam-break problem at *t=7.2 s*.

We compare our results by $N = 81 \times 81$ at $t=7.2s$ (see Fig.4, Fig.5), when the waves have not yet reached all the boundaries, with least-squares finite-element method (LSFEM) [11]. Other results of the form and propagating speed of the wave can be referenced in [3,11]. The left moving positive wave and right moving negative wave are both well resolved. The results were confirmed more stable to capture the fine details of the flow. The behaviour of the numerical scheme is in satisfactory agreement with computed results of these researches.

## 5 Conclusions

In this study, the applied local meshless method - the local meshless Petrov-Galerkin radial-basis-function (RBF) method is successfully applied to establish 2D shallow-water models for real engineering problems. The MLPG-RBF based 2D shallow-water models are verified by solving the dam-break problem compared with results of LSFEM. Results show good agreement with the results reported by other researches.

It is convinced that the proposed model not only provides accurate result but also has applicability to do inverse numerical analysis in real engineering project problem. In the further work, we plan to apply our SWEs to simulate more challenging flow problems, especially debris flow phenomena.

This contribution is the result of the project implementation: "Formulation of the new progressive numerical approaches for debris flow simulation" No. 1/0716/17 supported by the Scientific Grant Agency of Slovak Republic (VEGA).

## References

1. J.C. Galland, N. Goutal, J.M. Hervouet, Advances in Water Resources, **14**, 138, (1991)
2. E.F. Toro, *Shock-capturing methods for free-surface shallow flows*. (John Wiley and Sons, NewYork, 2001)
3. D.L. Young, Journal of the Chinese Institute of Engineers, **14**, 143, (1991)
4. J. Xin-rong, W. Chen, Chinese Journal of Computational Mechanics, **28**, 338, (2011)
5. W. Chen, M. Tanaka, Computers and Mathematics with Applications, **43**, 379–391, (2002)
6. E.J. Kansa, Computers and Mathematics with Application, **19**, 127–145, (1990)
7. X. Zhou, Y.C. Hon, K.F. Cheung, Engineering Analysis with Boundary Elements, **28**, 967–973, (2004)
8. S.N. Atluri, T. Zhu, Computational Mechanics, **22**, 117-127, (1998)
9. K. Kovářík, J. Mužík, Engineering Analysis with Boundary Elements, **37**, 187-196, (2013)
10. R.L. Hardy, Journal of Geophysical Research. **76**, 1905–1915, (1971)
11. S.J. Liang, J.H. Tang, M.S. Wu, Acta Mechanica Sinica, **224**, 523, (2008)

# Two-dimensional convection–diffusion problem solved using method of localized particular solutions

*Juraj* Mužík[1,*] and *Martina* Holičková[1]

[1]University of Zilina, Faculty of Civil Engineering - Department of geotechnics, Univerzitná 8215/1, Slovakia

**Abstract.** A meshless local method of approximated particular solutions (LMAPS) is used to analyze problem described by the convection-diffusion equation. The method solves the steady convection-diffusion equation with reaction term. The discretized system of equations is derived via interpolation procedure using integrated multiquadrics (MQ) radial basis functions (RBF). The solution of the equation is performed over simple geometry with non-uniform velocity field and results are presented in the article. The LMAPS method is capable to produce stable solutions with results comparable to the analytical solutions.

## 1 Introduction

Meshless approaches of approximated particular solution are receiving increased attention due to their accuracy associated with the special type of RBF function used and the flexibility they offer as the meshing requirements are either eliminated or reduced.

The main drawback of mesh-based numerical methods is the requirement of the mesh generation, which may be difficult to process during the numerical implementations, especially for multi-dimensional problems. In order to avoid mesh generation process many meshless numerical methods, such as the local boundary integral element method (LBIEM) [1], smoothed particle hydrodynamics (SPH) method [2], meshless strong (collocation) method [3] and meshless local Petrov-Galerkin method (MLPG) [4] have been developed.

The local method of approximated particular solution (LMAPS) was proposed by Cheng et al. and was applied to elliptic problems [5] and non-linear problems [6]. In LMAPS the domain is covered by cloud of scattered nodes. In the work on LMAPS reported so far, the support of the any computational node is taken to be a simple sub-domain in a shape of a circle though in theory the domain can be of any shape, with the computational node in the centre of the circle. The most often used interpolation for field variables were the moving least-squares, though Sellountos and Sequeira [7] used augmented thin plate spline (ATPS) radial basis functions (RBFs) for interpolation of the field variable and gradients over the circular boundaries.

The LMAPS formulation used in this article represents a stable, accurate solution tool. Numerical experiment of the quasi two-dimensional linear convection-diffusion-reaction

* Corresponding author: muzik@fstav.uniza.sk

equation solution is presented in the article with the comparison against the analytical solution, which verifies the performance of the LMAPS.

## 2 Localized method of approximated particular solutions

In order to apply the LMAPS, the following procedure may be adopted to solve physical variable u with any given differential operator. Even though the implementation details of the LMAPS are quite similar to the meshless collocation methods, its origin follows the globally defined method of approximated particular solution (MAPS) (see [8]). In this study the LMAPS is treated as special case of the meshless collocation method with the support domain concept already included in the formulation.

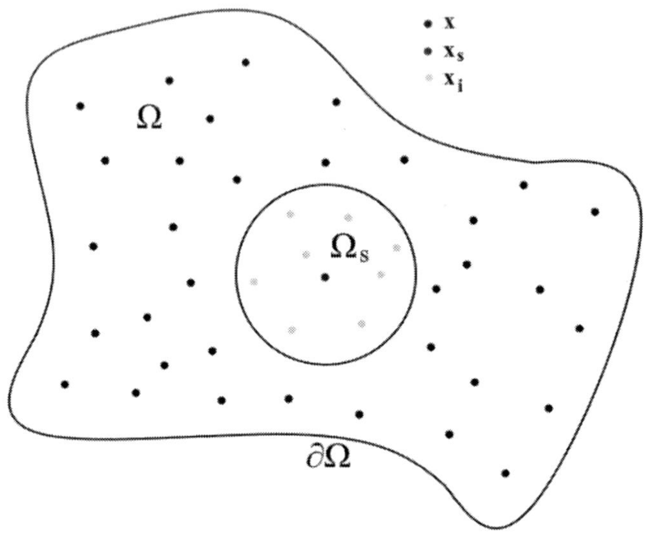


**Fig. 1.** The diagram of global domain $\Omega$, local support domain $\Omega_s$ of point $x_s$, global points $x$ and local point $x_i$.

The area of interest $\Omega$ with the boundary $\partial\Omega$ is covered by points within the area and also on the global boundary (see Fig.1). Consider a local circular (or any simple shape e.g. rectangle) sub-domain $\Omega_s$ centered at every point $s$. This sub-domain is called support domain and using the points in a particular support domain any function can be expressed using just nodal values as

$$u(\mathbf{x})=\sum_{i=1}^{NS}\alpha_i F(\|\mathbf{x}-\mathbf{x}_i\|),\quad x\in\Omega_s \tag{1}$$

where $NS$ is the total number of computational nodes in the support domain, $F$ is the radial basis function, $\mathbf{x}$ is coordinate vector and $\alpha_i$ are the local weighting coefficients to be determined. Because the values of $u$ expressed using (1) at the computational nodes should be the same as $u$ specified in these nodes, we can write following matrix expression

$$\mathbf{u}=\mathbf{F}\boldsymbol{\alpha} \tag{2}$$

where

$$\mathbf{u} = \left[u(\mathbf{x}_1) \quad \ldots \quad u\left(\mathbf{x}_{N_s}\right)\right]^T \tag{3}$$

$$\mathbf{F} = \begin{bmatrix} F\left(\|\mathbf{x}_1 - \mathbf{x}_1\|\right) & \cdots & F\left(\|\mathbf{x}_1 - \mathbf{x}_{NS}\|\right) \\ \vdots & \ddots & \vdots \\ F\left(\|\mathbf{x}_{NS} - \mathbf{x}_1\|\right) & \cdots & F\left(\|\mathbf{x}_{NS} - \mathbf{x}_{NS}\|\right) \end{bmatrix} \tag{4}$$

$$\boldsymbol{\alpha} = \left[\alpha_1 \quad \cdots \quad \alpha_{NS}\right]^T \tag{5}$$

Since $\mathbf{F}$ is invertible, we will obtain the unknown weighting coefficients as

$$\boldsymbol{\alpha} = \mathbf{F}^{-1}\mathbf{u} \tag{6}$$

Now, for operators concerned in solving convection-diffusion-reaction equations, we firstly consider the operator with highest order, the Laplace operator $\nabla^2$, and let

$$\nabla^2 u(\mathbf{x}_i) = \sum_{i=1}^{NS} \alpha_i f\left(\|\mathbf{x} - \mathbf{x}_i\|\right) = \mathbf{f}\boldsymbol{\alpha} \tag{7}$$

where $f$ is the radial basis function with the essential condition as $f=\nabla^2 F$. Replacing $\boldsymbol{\alpha}$ with Eq.(6) we get

$$\nabla^2 u(\mathbf{x}_i) = \mathbf{f}\mathbf{F}^{-1}\mathbf{u} = \boldsymbol{\varphi}\mathbf{u} \tag{8}$$

where $\boldsymbol{\varphi}$ is the vector of shape functions Laplacian. We apply similar procedures to the gradient operator $\nabla$ and we get

$$\nabla_{,j} u(\mathbf{x}_i) = \sum_{i=1}^{NS} \alpha_i \frac{\partial F\left(\|\mathbf{x} - \mathbf{x}_i\|\right)}{\partial x_j} = \mathbf{G}_{,j}\boldsymbol{\alpha} = \mathbf{G}_{,j}\mathbf{F}^{-1}\mathbf{u} = \boldsymbol{\gamma}_{,j}\mathbf{u} \tag{9}$$

where $\gamma_j$ is the vector of shape function gradients and subscript $j$ represents the coordinate of the gradient. The processed operators may now be applied directly into numerical implementations, which discretize the governing equations and the essential conditions into formulation of linear systems. All boundary conditions in this paper have also been processed using the LMAPS.

## 3 LMAPS implementation of convection-diffusion equations

### 3.1 Integrated multiquadrics radial basis functions

The trial function basis used in the LMAPS incorporates multiquadrics RBF, but different approach than the standard meshless collocation methods is adopted. In the case of LMAPS the MQ-RBF [8] represents the form of basis function after differentiation using highest order operator. To obtain the basis function, MQ-RBF needs to be integrated.
In this article the two-dimensional integrated multiquadrics radial basis function (integrated MQ-RBF) is used as the basis function $F$. The MQ-RBF

$$f(r) = \sqrt{r^2 + c^2} \tag{10}$$

and the integrated MQ-RBFs [8]

$$F(r) = \frac{1}{9}\left(r^2 + 4c^2\right)\sqrt{r^2 + c^2} - \frac{1}{3}c^3 \ln\left(\sqrt{r^2 + c^2} + c\right) \tag{11}$$

where $r$ depicts the distance between the collocation points and $c$ is a shape parameter.

### 3.2 Handling derivative boundary conditions

Strong-form methods can produce accurate results for partial differential equations, when the boundary conditions are all of Dirichlet type. If there is any derivative boundary condition, the accuracy of the solution deteriorates drastically, and the solution can be unstable; small changes in the setup of the problem can lead to a large change in the solution.

The discretized system equation behaves like an ill-posed problem in which errors introduced into the system are magnified in the output. A number of strategies can be used to impose the derivative boundary conditions in the strong form methods. The method using fictitious points is used in this article to enforce the pressure gradient on the boundary. Along the derivative boundaries, a set of fictitious points is added outside the problem domain. In this case, two sets of equations are established at each derivative boundary node, one for the derivative boundary condition, and the other for the governing equation.

### 3.3 Handling derivative boundary conditions

In this section, the numerical procedures for the approximation of the convection-diffusion-reaction equations via the LMAPS will be explained. The solved equation is assumed in following form:

$$D\frac{d^2u}{dx^2} - V_x\frac{du}{dx} - ku = 0 \tag{12}$$

Using the LMAPS principles the equation (12) is discretized as follows

$$\begin{gathered} D\boldsymbol{\varphi}\mathbf{u} - V_x\boldsymbol{\gamma}_x\mathbf{u} - \boldsymbol{\psi}\mathbf{u} = 0 \\ \left[D\boldsymbol{\varphi} - V_x\boldsymbol{\gamma}_x - \boldsymbol{\psi}\right]\mathbf{u} = 0 \end{gathered} \tag{13}$$

Which results in classical linear system of equations **Ax=b**.

## 4 Numerical example

The example used is a convection-diffusion-reaction equation with variable velocity field and reaction term as given in subchapter 3.3. A rectangular domain with length $L$=1 *m* and width $W$=0.2 *m* is considered. The following BCs were applied:

$$u(0,y)=U_0=300;\ u(L,y)=U_1=10;\ \left.\frac{\partial u}{\partial y}\right|_{y=W/2}=\left.\frac{\partial u}{\partial y}\right|_{y=-W/2}=0; \tag{14}$$

The velocity field is defined as

$$V_x=\ln\frac{U_1}{U_0}+k\left(x-\frac{1}{2}\right);\ V_y=0 \tag{15}$$

The analytical solution of the above problem [9] for $L=1$ m and $D=1\ m^2s^{-1}$ is given by

$$u(x)=U_0\exp\left[\frac{k}{2}x^2+\left(n\frac{U_1}{U_0}-\frac{k}{2}\right)x\right] \tag{16}$$

Though the analytical solution is 1D, the solution is a 2D one since the code needs to accurately predict that in the $y$-direction the potential is constant for a given $x$, i.e. $u(x,y)=u(x)$, and also that the flux is zero in the $y$-direction. This example is particularly challenging to solve because the velocity changes direction inside the domain. For example, for $k=10$ and $40$ it is in opposite direction to the $x$-axis for $x$ smaller than 0.15988 and 0.41497, respectively, and is in the direction of the $x$-axis for $x$ larger than 0.15988 and 0.41497, respectively. The conservation of mass forces the solution to have high gradients in the part of the domain close to $x=0$, in order for the diffusion to counterbalance the convection in the opposite direction to the $x$-axis. The maximum Peclet numbers for the examples are $Pe_{max}$=8.4, for $k$=10, and $Pe_{max}$=23.4 for $k$=40. This example has been used as a benchmark in a number of previous studies [9,10,11].

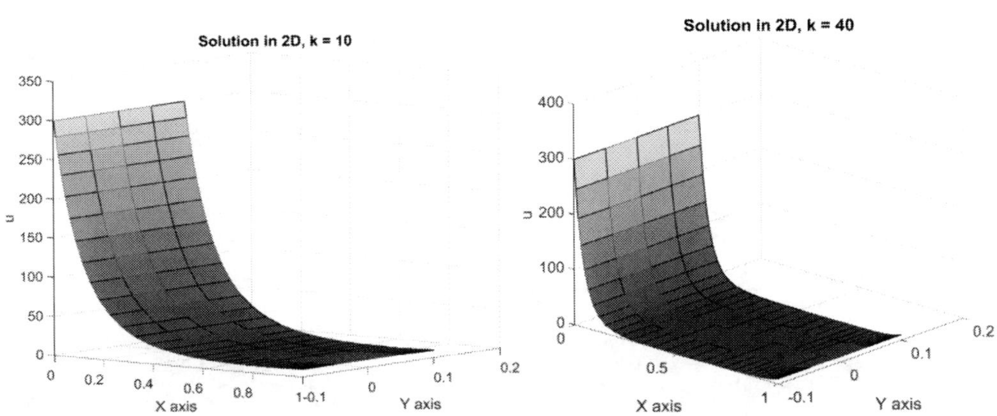


**Fig. 2.** Numerical results in visualised in 3D, obtained with the LMAPS ($k$=10 and $k$=40)

Two values of $k$ were considered: 10 and 40. The analytical results and the numerical results obtained by using the LMAPS for the case of $k=10$ and $k$=40 are shown in Fig. 2. Fig. 3 shows that the results obtained from the meshless approach agree well with the analytical results.

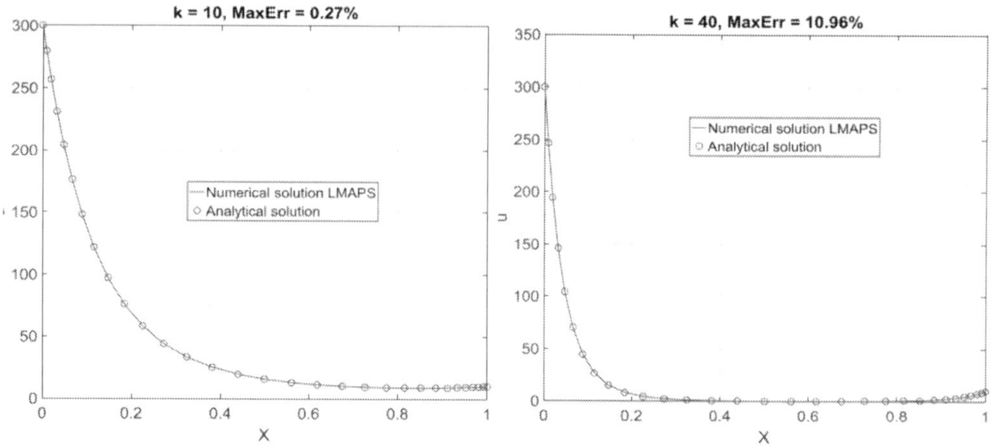


**Fig. 3.** Comparison of the results obtained with the LMAPS with the analytical solution ($k$=10 and $k$=40)

In Table 1 the value of potentials obtained using the LMAPS are compared with the analytical solution (16), the results obtained using the DRM-MD (dual reciprocity-multiple domain) [9], RBIEM (radial boundary integral equation method) [9] and the results obtained using Galerkin FEM (finite element method). In this case the main comparison of the accuracy of the method is considered to be with the RBIEM approach. The relative errors for the potential obtained using the four numerical methods are given in the same table. The relative error is calculated by the following formula:

$$e_r = \frac{\left|u_i^n - u_i\right|}{u_i} 100\% \tag{17}$$

where $u_i^n$ is the analytical solution for $u$ and $u_i$ are potential values at node $i$ obtained by the numerical method. It can be seen that the LMAPS produced results with accuracy comparable with RBIEM.

**Table 1.** Comparison of the results for potential obtained using the MLPG compared with the analytical solution and results of RBIEM, DRM-MD and Galerkin FEM for a regular scheme of 185 computational nodes (case k=40).

| x | Analytical solution | LMAPS | RBIEM | DRM-MD | Galerkin FEM | Error (%) LMAPS | Error (%) RBIEM | Error (%) DRM-MD | Error (%) Galerkin FEM |
|---|---|---|---|---|---|---|---|---|---|
| 0.00 | 300.00000 | 300.00000 | 300.00000 | 300.00000 | 300.00000 | 0.00000 | 0.00000 | 0.00000 | 0.00000 |
| 0.02 | 189.38055 | 190.53640 | 179.04666 | 187.61755 | 187.68159 | 0.60662 | 5.45668 | 0.93093 | 0.90054 |
| 0.04 | 121.47817 | 124.03792 | 118.35698 | 118.98971 | 119.41530 | 2.06368 | 2.56934 | 2.04848 | 1.69814 |
| 0.07 | 64.30894 | 66.02516 | 63.18108 | 62.42024 | 61.66200 | 2.59934 | 1.75382 | 2.93691 | 4.11598 |
| 0.10 | 35.29222 | 37.19934 | 35.16106 | 34.32921 | 33.14000 | 5.12675 | 0.37164 | 2.72869 | 6.09828 |
| 0.14 | 16.77039 | 17.86164 | 17.05505 | 16.68902 | 15.10380 | 6.10947 | 1.69740 | 0.48519 | 9.93769 |
| 0.19 | 7.23952 | 7.75684 | 7.47060 | 7.60905 | 6.12290 | 6.66924 | 3.19192 | 5.10446 | 15.42395 |
| 0.25 | 3.01465 | 3.38576 | 3.14630 | 3.44925 | 2.36960 | 10.96102 | 4.36701 | 14.41637 | 21.39718 |
| 0.32 | 1.30135 | 1.41994 | 1.34937 | 1.63377 | 0.96120 | 8.35223 | 3.69001 | 25.54530 | 26.13824 |
| 0.40 | 0.63337 | 0.69743 | 0.66253 | 0.85084 | 0.45580 | 9.18422 | 4.60394 | 34.33515 | 28.03575 |
| 0.56 | 0.32339 | 0.33747 | 0.34728 | 0.43739 | 0.25190 | 4.17010 | 7.38737 | 35.25292 | 22.10643 |
| 0.75 | 0.55040 | 0.57254 | 0.59498 | 0.69063 | 0.48900 | 3.86825 | 8.09956 | 25.47853 | 11.15552 |
| 0.85 | 1.30052 | 1.31763 | 1.32975 | 1.56886 | 1.21380 | 1.29873 | 2.24756 | 20.63368 | 6.66810 |
| 0.93 | 3.45102 | 3.47837 | 3.28041 | 3.92384 | 3.33370 | 0.78630 | 4.94376 | 13.70113 | 3.39957 |
| 1.00 | 10.00000 | 10.00000 | 10.00000 | 10.00000 | 10.00000 | 0.00000 | 0.00000 | 0.00000 | 0.00000 |

## 5 Conclusions

A meshless method based on the strong formulation and combined with the local support domain approach has been proposed for solving the convection-diffusion equation. The strong form of convection-diffusion equations is solved at each node. Integrated radial basis (RBF) function interpolation is applied in order to obtain the values of the field variable.

The current approach does not require any integration over the domain or boundary of the model but the application of Neuman type boundary condition requires use of "ghost" points in order to apply the governing equation together with boundary operator of the boundary condition. The accuracy of the method has been compared with the accuracy of the RBIEM, DRM-MD with overlapping sub-domains and the FEM. In all cases the current approach has shown equal to superior accuracy to the formulation which uses a mesh. Though the method has been applied to 2D problems, extension of the approach to 3D problems is straightforward.

This contribution is the result of the project implementation: "Formulation of the new progressive numerical approaches for debris flow simulation" No. 1/0716/17 supported by the Scientific Grant Agency of Slovak Republic (VEGA)..

## References

1. K. Kovářík, J. Mužík, M.S. Mahmood, Engineering Analysis with Boundary Elements, **36**, 738-743, (2012)
2. G.R. Liu, M.B. Liu, *Smoothed Particle Hydrodynamics: a meshfree particle method* (World Scientific, Singapor, 2003)
3. E.J. Kansa, Computers and Mathematics with Application, **19**, 127–145, (1990)
4. S.N. Atluri, T. Zhu, Computational Mechanics, **22**, 117-127, (1998)
5. C.S. Chen, C.M. Fan, P.H. Wen, Numerical Methods for Partial Differential Equations, **28**, 506–522, (2012)
6. C.S. Chen, M.A. Golberg, M. Ganesh, A.H.-D. Computers and Mathematics with Application, **43**, 359–378, (2002)
7. E.J. Sellountos, A. Sequeira, Comput Mech, **41**, 617, (2008)
8. J. Mužík, L. Kais, R. Bulko, Communications - Scientific Letters of the University of Zilina, **18**, 77-82, (2016)
9. V. Popov, T.T. Bui, Engineering Analysis with Boundary Elements, **34**, 680-689, (2010)
10. V. Popov, H. Power, International Journal for Numerical Methods in Engineering, 44, 327–353, (1999)
11. V. Popov, H. Power, Engineering Analysis with Boundary Elements, 23, 175–88, (1999)

# Modelling the Reduction of Project Making Duration

*Pavel* Oleinik[1], *Tatyana* Kuzmina [1,*]

[1] Moscow state university of civil engineering, Yaroslavskoye shosse, 26, Moscow, Russia, 129337

**Abstract.** The article points out why earlier patterns of investment process were ineffective in developing the construction projects and shows sources for reducing of its total duration. It describes the procedure of statistical modeling and obtaining medium-term time parameters required for modern pattern of project-making; offers design formulas for assessment of total time required for project-making as well as for its main stages; reveals advantage of modern system of project-making against traditional one by comparing indicators of their duration.

## 1 Introduction

Shorter periods of constructing buildings and facilities are a challenging problem of capital construction. Investment process in project-making includes many stages, most important of them are: feasibility study, engineering, preparation for construction, construction, commissioning [1, 2, 3, 4].

Pre-2000 pattern of traditional project-making was ineffective due to variety of reasons. Firstly, all types of resources were scattered among numerous projects being built at the same time. E.g. year 1980 saw simultaneous construction of 600,000 industrial design projects, and, with residential, public, preschool, administrative, medical, sport and trade projects added, total amount of the projects under construction was not less than 800,000. Consequently, each project employed 7.5-8 workers on average. That explained why ingrained notions of 'start-up', 'extremely important' project laid almost legal foundation under a rushed work, as it was not earlier than the final year when such projects received the largest part of the funds allocated. Secondly, the first year of construction spent usually 3-12 % of construction-and-assembling operations cost, the sum sufficient just for some preliminary works at the site, with many similar works and main construction-and-assembling operations remained not fulfilled. The result was a severe shortage of resources and performance of the most profitable resource-demanding operations as a key factor. These and some other reasons made de-facto duration of construction 1.8 – 2.5 times longer than the normative one [2, 4, 5]. Years of economic reforms managed to agree intricate multi-layer construction procedure and capacities of economic mechanism. On average, duration of buildings and facilities construction takes 1.3-1.4 less time thanks to management-organizing factors, that does not need attraction of any extra funds [6, 7, 8].

* Corresponding author: KuzminaTK@mgsu.ru

## 2 Methods

The findings of Russian and foreign researchers give a reason to conclude that further reducing of facility making investment process duration is reached mainly via revealing and considering of factors that harmonize various steps of construction and reducing of their duration [4, 9, 10, 11].

Economic reforms contributed a lot to combining and shortening of investment process steps [1, 12, 13], e.g.: uniform requirements for structure and contents of project documentation sections for construction of industrial and non-industrial projects, cancellation of design staging, and (for some projects) wider content of feasibility study in replace of traditional construction management project, etc. Of equal importance are implementation results of innovative methods, ways and forms of construction management: long-tern-flows-based construction, component assembly and complete containerized construction methods, advance design preparation of territories, conveyer block building system, mobile forms of work (rotation, expedition, rotation and expedition, etc.) [2, 6, 14].

To create a modern topology pattern for facility making investment process and obtaining its mean time parameters, we scrutinized more than 60 projects in various sectors of industry that were built and commissioned in different years and required significant capital investments. Statistical parameters for empiric distribution of duration included absolute and relative frequencies, mathematical expectation, statistical dispersion, root-mean-square deviation with plotting of bar graphs. Statistical parameters were calculated separately for actual and normative duration of each facility making step.

The research proved that approximately 65.6 % of all failures fall on imperfect planning, procurement and financing. At the same time, inadequate organizational and procedure decisions make 29.8 %. Functioning system of making project has typical drawbacks like segregation of participants, huge lags between planned actual dates of obligations fulfilment, time-consuming procedures of making key decisions. Additional factors are wrong choice of the construction site, inappropriate correction of documents and processing of orders, separate construction of substructure blocks of the buildings and infrastructure, many provisional buildings and facilities, unprepared field of operations, etc.

Calculations proved the hypotheses for normal distribution of durations in each step of project making. From graph bars follows that the mean total duration of feasibility study, designing and preparation for construction (before construction proper) takes 13 months or 22 %, and mean duration of the construction, 75 % of total duration. Mean actual duration of facility making ($\overline{\mathrm{T}}$) was obtained as a weighted mean value

$$\bar{T} = \frac{\sum_{i=1}^{n} V_i\,(t_i^1 + t_i^2)}{\sum_{i=1}^{n} V_i} \tag{1}$$

where $V_i$ is: construction costs for the $i^{\text{th}}$ facility, ths. RUB;

$t_i^1$ is: duration of the phases of: feasibility study, design, preparation of the facility for construction prior to the start of construction at the $i^{\text{th}}$ facility, months;

$t_i^2$ is: duration of the construction phase at the $i^{\text{th}}$ facility, months.

$j = \overline{1, n}$, $n = 56$

## 3 Main part

Mean duration of facility making makes 58 months for the sample mentioned. Feasibility study, required approvals inclusive, takes about 6 % of a total duration of project making, and duration of designing, including time for the peer review of project documentation, takes 54 %. It is implied that after the project approval detailed documents are to be compiled for complex of operations: substructure block, superstructure block, infrastructure, finishing, etc, that allows the construction company to support high rate of operations. Preparation for construction comes at the same time with designing and includes: concluding a construction contract between the general contractor and the technical customer, delivery of project documents and notarized copies of a building permit and a land allocation certificate, developing of construction projects, etc.

Construction-and-assembling operations begin in month 14 and finish in 43 months. Commissioning takes approximately 2 months. Time parameters of facility making are shown in Table 1.

**Table 1.** Time parameters of the facility making

| # | Main phases | Temporal phases of indicator | | |
|---|---|---|---|---|
| | | **start** $t^s$ | **finish** $t^f$ | **duration** $t^f - t^s$ |
| 1 | Feasibility study | $t_1^s = 0$ | $t_1^f = 0.1 \cdot T$ | $t_1 = t_1^f - t_1^s$ |
| 2 | Design | $t_2^s = 0.06 \cdot T$ | $t_2^f = 0.6 \cdot T$ | $t_2 = t_2^f - t_2^s$ |
| 3 | Preparation of the facility for construction | $t_3^s = 0{,}1 \cdot T$ | $t_3^f = 0.81 \cdot T$ | $t_3 = t_3^f - t_3^s$ |
| 4 | Construction | $t_4^{н} = 0.22 \cdot T$ | $t_4^{о} = 0.97 \cdot T$ | $t_4 = t_4^f - t_4^s$ |
| 5 | Comissioning phase | $t_5^s = 0.97 \cdot T$ | $t_5^f = T$ | $t_5 = t_5^f - t_5^s$ |

Total duration of investment process, as a part of facility making, is obtained from its parameters as follows:

$$T = t_1 + t_3 + (1 - u_4) \cdot t_4 + t_5 \tag{2}$$

where $u_4 = 0{,}79$ is: a coefficient of combining the construction phase with the phase of preparation of the facility for construction.

Indications of advance lags make 58.3% for feasibility study, 7.9% for designing, 17.1% for preparation for construction.

Implementing of modern patterns for investment process of facility making while planning, designing and construction revealed their advantages, in particular:

- validation of priority and duration of construction, dates of developing the project documents and implementation preparation for construction measures;
- expediency of advanced development of work performance projects for the part of working documents to be brought;
- reality of negotiated construction price agreed between the customer and the contractor (with regard to methods of construction, job management, etc.).

Comparison studies of mean durations of traditional and modern patterns facility making (Table 2) strongly support significant advantage of the modern pattern in almost every stage of the investment process. The gap is largest in the designing step (more than 2 times) and steps before the start of construction.

**Table 2.** Time indicators for the traditional and modern patterns of the facility making

| # | Name of indicator | average values in days | |
|---|---|---|---|
| | | according to traditional scheme | according to modern scheme |
| 1 | Total duration | 112 | 58 |
| 2 | Duration of feasibility study taking | 4 | 3.5 |
| 3 | Duration of design phase | 77 | 31.5 |
| 4 | Duration of preparation of the facility for construction phase | 53 | 41 |
| 5 | Duration of construction phase | 58 | 43 |
| 6 | Duration of commissioning phase | 2 | 2 |
| 7 | Duration before construction phase | 54 | 13 |

## 4 Findings

The necessity for modelling the investment process in facility making allows establishing firm organizational-procedure interrelations between its main steps presented compactly and rationally. Organizational-procedure interrelations are described via time indicators (start, finish, and duration of the stage), coefficients of coincidence and advance lags of the steps.

Modelling the investment process in facility making is attractive for various sectors of industrial construction for, firstly, finding the sources for reduction of total duration, and, secondly, for harmonization of facilities in terms of strategic and current planning and creating the work program for the construction company.

## References

1. P. Oleinik, T.K. Kuzmina, *Modeling the activities of a technical customer* / Civil and industrial engineering, **11**. pp. 42-43 (2012)
2. P.P. Oleinik, *Modeling the shortening of the duration of the investment process* / Natural and technical sciences, **10**, pp. 412-414 (2015)
3. T. Kuzmina and N. Cherednichenko, *Systematization of the major stages of the client in certain branches of construction production* / MATEC Web Conf. **86**, **05012** (2016)
4. P. Oleinik, T. Kuzmina, V. Zenov, *Intensification of the investment process of construction* / MATEC Web Conf. **86**, **05019** (2016)
5. L.V. Kievsky, *From organizing the construction process - to organizing the investment processes in construction* / included: Urban development. Scientific digest edited by L.V. Kievsky, Moscow, pp. 205-221 (2014)
6. P. Oleinik, V.I. Brodsky, *Method of rating the performance indicators for preparatory construction works* / Technology and organization of construction process, **1(2)**, pp. 27-31 (2013)
7. V. Telichenko, *Modelling of processes of the life cycle of facilitits built on the basis of CALS-technologies* / Russian Academy of Architecture and Construction Sciences (RAASN), Newsletter of civil engineering sciences. Moscow, **3**, pp. 178-183 (2000)
8. Clifford f. Gray, Erik W Larson, *Project management: a practical guide.* / translation from English, Moscow: Business and Service Publishing house (2003)
9. B.F. Shirshikov, A.M. Slavin, *Shortening of the investment cycle period, based on smoothing out contradictions among its participants* / Civil and industrial engineering. **8**, pp. 92-95 (2016)

10. Frame J. Davidson, *The New Project Management: tools for an age of rapid change, complexity, and other business realities* / by J. Davidson Frame. 2nd ed. San Francisco: Jossey-bass, A Wiley Company. p. 360 (2002)
11. Jonsons, J.C., *Contemporary Logistics* / J.C. Jonsons, D.F.Wood. – New York : MacMillan, p. 325 (2007)
12. S.A.Sinenko, T.K. Kuzmina, *Modern information technologies in customer service (technical customer)* / Scientific review, **18**, pp. 156-159 (2015)
13. Charvat Jason Project Management Methodologies–Selecting, Implementing, and Supporting Methodologies and Processes for Projects. New Jersey: John Wiley & Sons inc. p. 264 (2003)
14. L.V. Kievskiy, I.L. Kievskiy, *Multiplier effects of the Moscow construction complex* / International Journal of Applied Engineering Research. **11. 1**. pp. 304-311. (2016)

# Optimization of the annual construction program solutions

*Pavel* Oleinik* and *Alexey* Yurgaytis

Moscow State University of Civil Engineering, 129337, 26, Yaroslavskoye Shosse, Moscow, Russia

**Abstract.** The article considers potentially possible optimization solutions in scheduling while forming the annual production programs of the construction complex organizations. The optimization instrument is represented as a two-component system. As a fundamentally new approach in the first block of the annual program solutions, the authors propose to use a scientifically grounded methodology for determining the scope of work permissible for the transfer to a subcontractor without risk of General Contractor's management control losing over the construction site. For this purpose, a special indicator is introduced that characterizes the activity of the general construction organization - the coefficient of construction production management. In the second block, the principal methods for the formation of calendar plans for the fulfillment of the critical work effort by the leading stream are proposed, depending on the intensity characteristic.

**Keywords.** Annual production program of the construction company, scheduling, annual (current) planning, optimization solutions, optimization, the coefficient of construction production management, intensity of work performance, human resource, labor

## 1. Introduction

Production planning plays a decisive role in the development of a construction organization, and as a system, it consists of separate planning complexes (subsystems), which include general planning (3-5 years, long-term planning), strategic planning (1-3 years, medium-term planning ), Current (annual) planning and operational planning (up to 1 year, short-term planning) (Fig. 1). In the conditions of the modern economy, it is the current (annual) planning of the production program of a construction organization that is a particularly complex multifactor task, often solved intuitively and chaotically in oranizations, without an established system and methodology. Such an extensive production and economic program envisages the implementation of the concluded construction contracts during the reporting period, compliance with the requirements of the codes, mobilization of production resources and obtaining the maximum possible profit from the activities (Fig. 2). Within the framework of such planning, we can isolate the

* Corresponding authors: aljurgaitis@gmail.com

planning of using the labor resources with the aim of ensuring the output of finished construction products of high competitive quality in a timely manner.

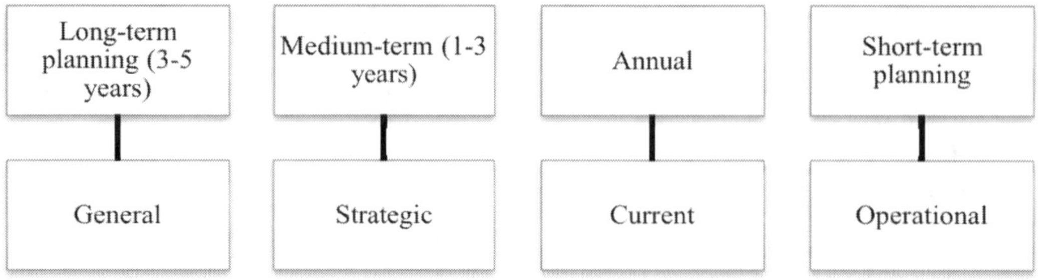


**Fig. 1.** Scheme of Production Planning

The optimization instrument of the annual planning can be represented in the form of a two-component complex, shown in Fig. 3.

The principal task of planning, on the one hand, is to withstand the timing of the commissioning of construction projects; on the other hand, - to ensure a constant, uniform and, as far as possible, maximum full mobilization of the resources of all production units. In addition to the available algorithms for constructing and optimizing the calendar plans, mainly considered in the works of foreign authors [1-10], this article proposes a two-stage instrument that allows the General Contractor Organization to first isolate the permissible amount of work for transfer to the subcontractor balance, and then proceed to the formation of a calendar plan and resource schedules (Fig. 3). On the basis of heuristic methods that have become widespread in planning practice, it is possible to develop a clear scientific methodology for each stage for production implementation.

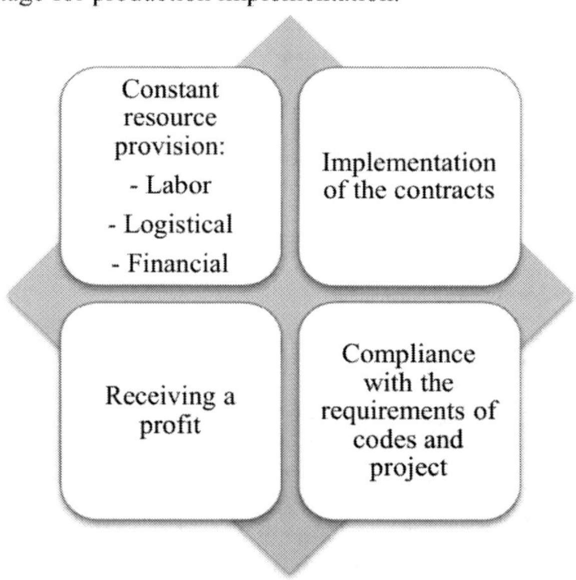


**Fig. 2.** Matrix of the current (annual) planning

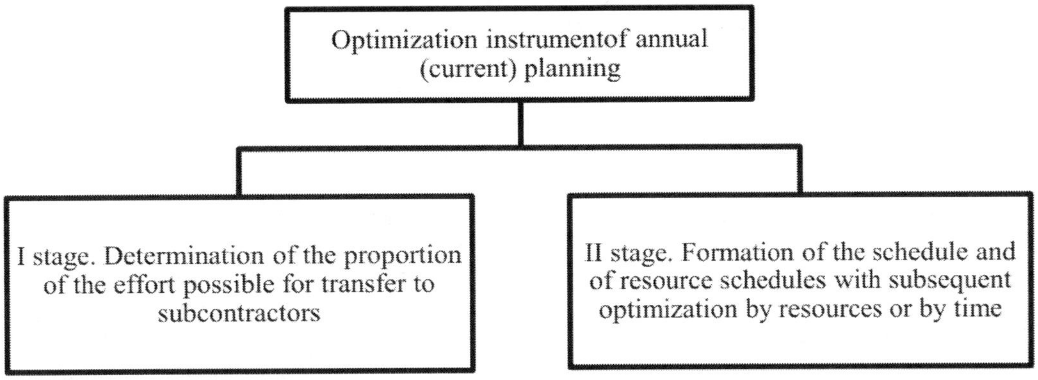


**Fig. 3.** Optimization instrument

## 2. Metods, Results and discussion

### 2.1 Stage I

The construction project effort or production program (a collection of objects) is generally represented as a sum of two components that form an integral binary system (Figure 4). We call such system a schematic diagram of the construction works binary separation (SDSW) and mathematically represented as a sum of two terms (formula 1).

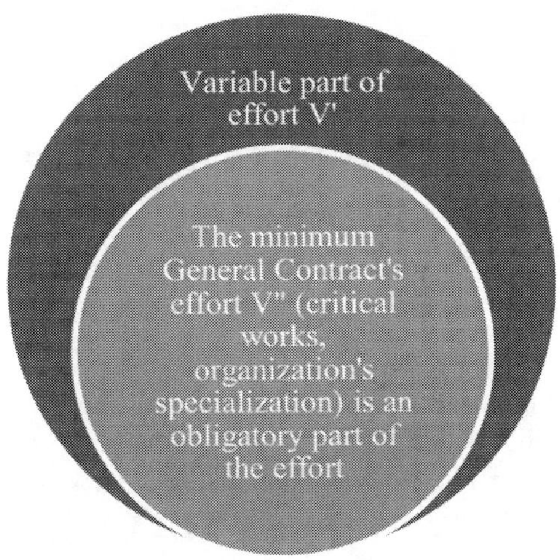


**Fig. 4.** A schematic diagram of the construction works binary separation (SDSW)

$$V = V' + V'', \tag{1}$$

где $V$ - the total effort of the construction object or the annual production program of the construction company (a set of efforts);

$V'$ - a certain proportion of the effort, realized at the discretion of the General Contractor with the involvement of another resource - the so-called Variable Part of SDSW;

$V''$ - the minimum proportion of the effort, which is necessarily realized directly with the involvement of the General Contractor's resource (includes the effort performed by the leading flows) - the so-called obligatory part of the SDSW.

As a result of the analysis of the production activity of a significant number of objects-representatives of a certain general population (construction organizations of various capacities), it is possible to obtain empirical dependences of the effort performed directly by the General Contractor and the amount of work that, for various reasons, is transferred to a subcontractor, from the General Contractor's capacity and total effort of the production program. The limiting factor in this distribution will be the need to maintain production control over construction sites for general contracting facilities and to avoid such distribution of volumes within the binary system, in which production management is transferred to subcontractors in fact absolutely. For a numerical description of this factor, we introduce a relative indicator - the coefficient of construction production management ($K_{csm}$), describing the nature of the efforts distribution between the General Contractor and the subcontractor and establishing the minimum allowable value of the General Contractor's effort to provide management control over the production process.

We define the term "capacity of a construction organization" as the amount of work on the objects of the production program performed by the labor resource (all production teams of the organization) per unit of time - index CF (company facilities).

So, let's select the first (formulas 2, 3) and the second (formulas 4, 5) principal cases of distribution of effort inside a binary system:

$$\begin{cases} V'' \to max(V'') \to 1 \\ V' \to min(V') \to 0 \end{cases} \Rightarrow V \approx V'' \quad (2)$$

$$K_{csm} = \frac{V}{max(V'')+min(V')} \approx 1'' \quad (3)$$

$$\begin{cases} V'' \to min(V'') \to 0 \\ V' \to max(V') \to 1 \end{cases} \Rightarrow V \approx V' \quad (4)$$

$$K_{csm} = \frac{V}{min(V'')+max(V')} \approx 1' \quad (5)$$

Case 1. The case of absolute production control. In the first case, $K_{csm}$, equal to one, indicates that the production process is under absolute control of the General Contractor ($\approx$ 100% of $V$).

Case 2. The case of the absence of production control. In the second case, $K_{csm}$ is also equal to one (execution$\approx$ 100% of $V$), but the performer (or a group of performers) will be the subcontractor, which ensures that there is no actual production control by the General Contractor.

The empirically established range of permissible values of the coefficient $K_{csm}$ is illustrated on Fig. 5.

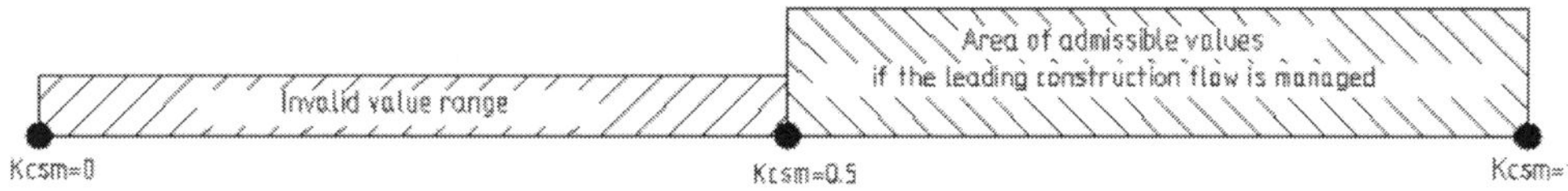


**Fig 5.** The range of admissible values of the coefficient Kcsm

**Table 1. SDSW components description**

| *SDSW component* | *Obligatory part of the General contractor's effort, V''* | *Variable part of the General contractor's effort, V'* | |
|---|---|---|---|
| ***How to determine the share of the SDSW component from the total effort V?*** | It is determined depending on the capacity of the General Contractor Organization ($CF$) and the total amount of contracts effort ($V$) according to empirical dependencies: $$V'' = f(CF; V) \quad (6)$$ | $$V' = V - V'' \quad (7)$$ | |
| ***How to evaluate the admissibility of calculations?*** | Characterized by the input indicator - the coefficient of construction production (construction site) management $K_{csm}$: $$K_{csm} \in [0.5; 1] \quad (8)$$ | | |
| ***Resource availability of the component*** | Own labor resource | Own labor resource due to the implementation of the effort $V''$ | Labor resource of the subcontractor |
| ***Additional burdensome actions*** | Are absent under condition of provision by a constant resource on the given volume of works | -Hiring additional workforce;<br>-Training of workers;<br>- Provision with snap;<br>- Creation of domestic production conditions;<br>- The maintenance of an additional labor resource after performance of effort $V''$ | None |

## 2.2 Stage II

The main conditions for the "correctness" of the calendar plan are: the fulfillment of the contractual terms of the technical customer (developer) and stable continuous loading of labor.

The achievement of these conditions is provided in various ways. At the same time, the process of formation and updating of the calendar plan is carried out continuously and not only because of the inclusion of new objects in the plan, but also for a number of other reasons (shortage of workers, disruption of financing, etc.). Nevertheless, several basic approaches to the formation of calendar plans can be identified. But at the same time for all approaches, as a rule, the principle of internal coordination of the efforts parameters is observed. For example, object construction flows can include both combined and non-combined specialized flows. But even from the number of non-combined specialized flows that make up the object flow, the leading stream, which determines the total duration of the entire object flow, is allocated, as a rule, according to the labor input, and the leading private stream is already allocated in it. Therefore, the task of isolating and linking the leading flows among themselves is primary, and the task of subordinating the found intensity of other flows is secondary. At a minimum, such flows have reserves (temporary, resource), which provide the possibility of optimization actions. The main approaches to the formation of calendar plans can be grouped into the following principal schemes (Table 2).

**Table 2**

| Number of the scheme in order | Attribute | Initial condition | Algorithm |
|---|---|---|---|
| Scheme 1 | Constant intensity of work performed<br>$J_i^B \rightarrow const,$<br>where $J_i^B$ - work intensity of $i$-leading flow. | The intensities of the leading flows have a constant value throughout the calculation period | 1. The contractual terms are checked.<br>2. Determine the need for labor $R_i^B$.<br>3. Linked $R_i^B$.<br>4. Schedules of labor's movement.<br>5. The schedules are corrected.<br>6. Determine $R_i^B$ for another efforts |
| Scheme 2 | The varying intensity of the work performed | The intensity of the leading flows is set by the variable throughout the calculation period. | 1. The contractual terms are checked.<br>2. The intensities for the time periods and the technological stages are calculated.<br>3. Determine $R_i^B$ of leading processes.<br>4. Adjustment is made in the workforce<br>5. Determine the size of the need for labor for other, non-leading processes. |
| Scheme 3 | With the prioritization of | 1. The intensity of the execution of the | 1. All objects are allocated by priority |

| Number of the scheme in order | Attribute | Initial condition | Algorithm |
|---|---|---|---|
| | objects | leading streams can be either constant or variable.<br>2. The lead time should not exceed the prescriptive (normative) duration | 2. Contractual terms are checked for all objects lying on the critical path<br>3. The intensity of the leading streams is calculated<br>4. Determine $R_i^B$ with the necessary adjustment<br>5. Determine other $R_i$ |

## 3. Conclusion

3.1 Authors proposed a two-component instrument for the formation and optimization of calendar plans and resource support for annual production programs of construction organization, capable of promptly responding to fluctuations in the economic and construction systems under the influence of destabilizing factors (addition of new contract, etc.).

3.2 In the practically widespread intuitive scheme of transferring the share of effort to a subcontractor, a scientific methodology is developed that allows, depending on the organization's capacity and the total amount of effort, to determine the required minimum amount of General Contractor's effort that provides management and control over the construction sites.

3.3 The indicator $K_{csm} \in [0.5; 1]$ is entered with the range of admissible values, which establishes the boundary conditions in determining the Variable part of the SDSW.

3.4 The minimum value of the indicator $K_{csm}$, necessary and sufficient for the General Contractor's management instrument, is proposed.

3.5 Three principal schemes for the formation of calendar plans are proposed depending on the parameter of the intensity of the work performed by the leading stream on critical works.

## References

1. Said M. Easa Resource Leveling in Construction by Optimization // ASCE. Journal of Con-struction Engineering and Management. (1989)
2. S. Farzad Moosavi and Osama Mosehli Review of Detailed Schedules in Building Construction // ASCE. Journal of Legal Affairs and Dispute Resolution in Engineering and Construction. (2014)
3. Yuanjie Tang, Rengkui Liu and Quanxin Sun Two-Stage Scheduling Model for Resource Leveling of Linear Projects // ASCE. Journal of Construction Engineering and Management. (2014)
4. Min-Yuan Cheng, Doddy Prayogo, Duc-Hoc Tran Optimizing Multiple-Resources Leveling in Multiple Projects Using Discrete Symbiotic Organisms Search // ASCE. Journal of Computing in Civil Engineering. (2016)
5. Selen Mubarak Construction Project Scheduling and Control. Canada: John Wiley & Sons, (2010)

6. Alcaraz, J., Maroto, C., Ruiz, R. Solving the multi-mode resource-constrained project scheduling problem with genetic algorithms // Journal of Operational Research Society, **54**, 614-626 (2003)
7. Bouleimen, K. and Lecocq, H. A new efficient simulated annealing algorithm for the resource-constrained project scheduling problem and its multiple mode version // European Journal of Operational Research, **149**, 268-281 (2003)
8. Hsiang-Hsi Huang, Chian-Ying Huang, Yu-Tsung Chen // Proceedings of the 4th International Conference on Engineering, Project, and Production Management (EPPM 2013)
9. Hong-Hai Tran and Nhat-Duc Hoang A Novel Resource-Leveling Approach for Construction Project Based on Differential Evolution // Journal of Construction Engineering (2014)
10. Ying XIONG, Ya Ping KUANG Ant colony optimization algorithm for resource leveling problem of construction project // The CRIOCM 2006 International Symposium on "Advancement of Construction Management and Real Estate" – (2006)

# Calculation of the filtration of polydisperse suspension with a small rate

*Yuri* Osipov[1,*]

[1]Moscow state university of civil engineering, Yaroslavskoye shosse, 26, Moscow, Russia, 129337

**Abstract.** The study of filtration problem is important when designing and constructing structures on water saturated ground. The filtration of a suspension with particles of different sizes moving at the same speed in a homogeneous porous medium is considered. A global asymptotic solution is constructed for a small filtration coefficient. The numerical calculation is based on experimental data.

## 1. Introduction

The study of filtration of suspended particulate matter - suspension in a porous medium is an important problem of underground hydromechanics [1]. For a long-term evaluation of the foundations strength it is necessary to take into account the composition of the soil and the penetrating capacity of groundwater.

A model of deep bed filtration of a suspension with suspended particles of various sizes moving with the same velocities in a homogeneous porous medium is considered. It is assumed that the particle and pore size distributions overlap, and the dominant cause of particle capture is the size-exclusion [2]. The suspended particles pass freely through large pores and are stuck at the inlet of pores, smaller than the diameter of the particles. One particle completely blocks the small pore and can not be knocked out of its throat by a flow of carrier fluid or other particles.

The mathematical model of filtration includes the equation of mass balance of suspended and retained particles, which is an analog of the continuity equation, and also the kinetic equation describing the rate of deposit formation proportional to the concentration of suspended particles [3]. The coefficient of proportionality, called the filtration coefficient, depends on the value of the deposit. As the deposit forms, the amount of free small pores decreases, and the growth rate of the deposit slows down.

Two quasilinear differential equations form a hyperbolic system. The boundary conditions that uniquely determine the solution of the system are set at the filter inlet and at the initial time. It is assumed that at the initial moment the porous medium does not contain any particles. A suspension containing $n$ different types of suspended particles of constant concentration is injected at the filter inlet.

A number of filtration problems have an exact analytical solution [4-8]. In the absence of an exact solution, asymptotics are constructed [9-12], or the problem is solved

* Corresponding author: yuri-osipov@mail.ru

numerically [13-15]. In the paper an asymptotic solution of the filtration problem of a polydisperse suspension in a porous medium is constructed assuming that the filtration coefficient is small. Numerical calculation of the asymptotics at the filter outlet is performed for 3-size particles on the basis of experimental data [10].

## 2. The mathematical model

In the domain

$$\Omega = \{0 < x < 1,\ t > 0\}$$

consider a system of equations with respect to the concentrations of suspended $C_i$ and retained $S_i$ particles

$$\frac{\partial C_i}{\partial t} + \frac{\partial C_i}{\partial x} + \frac{\partial S_i}{\partial t} = 0\,; \tag{1}$$

$$\frac{\partial S_i}{\partial t} = \varepsilon \Lambda_i(\vec{S}) C_i;\ \ i = 1,\ldots,n \tag{2}$$

with the boundary condition

$$x = 0 : C_i = p_i \tag{3}$$

and the initial conditions

$$t = 0 : C_i = 0,\ S_i = 0\,. \tag{4}$$

Here $\varepsilon > 0$ is a small parameter, $\vec{S} = (S_1,\ldots,S_n)$; the filtration coefficients $\Lambda_i(\vec{S})$ are continuous and positive; the constants $p_i > 0$.

The concentrations front of the suspended and retained particles moves from the filter inlet $x = 0$ to the outlet $x = 1$ at a constant velocity $v = 1$ and divides the domain $\Omega$ into two subdomains $\Omega_0 = \{0 < x < 1,\ 0 < t < x\}$ and $\Omega_S = \{0 < x < 1,\ t > x\}$. In $\Omega_0$ the system (1) – (4) has a zero solution; in $\Omega_S$ the solution is positive. Since conditions (3) and (4) are not matched at the origin, the solution has a strong discontinuity on the concentration front -- the characteristic line $t = x$. The solution $S_i(x,t)$ is continuous in $\Omega$ and has a weak discontinuity on the concentration front.

With increasing time *t*, the concentration of retained particles increases, and the number of free small pores is reduced. The growth rate of the deposit slows down and the suspended particles concentration increases:

$$\frac{\partial S_i}{\partial t} > 0;\ \frac{\partial C_i}{\partial t} > 0\,. \tag{5}$$

The retained particles concentration decreases with increasing distance $x$ from the porous medium entrance, since the filtration process of the suspension starts at the inlet of the filter. With a smaller deposit, the particle retention occurs more intensively, and the concentration of suspended particles also decreases with increasing $x$.

$$\frac{\partial S_i}{\partial x} < 0; \ \frac{\partial C_i}{\partial x} < 0. \tag{6}$$

Consider the condition on the characteristic line

$$S_i\big|_{t=x} = 0. \tag{7}$$

In the domain $\Omega_S$ the problem (1) – (4) is equivalent to the Goursat problem (1) – (3), (7).

In the characteristic variables (Riemann variables)

$$\tau = t - x, \ x = x$$

in the domain $\tilde{\Omega}_S = \{0 < x < 1, \tau > 0\}$ the system (1) – (3), (7) takes the form

$$\frac{\partial C_i}{\partial x} + \frac{\partial S_i}{\partial \tau} = 0; \tag{8}$$

$$\frac{\partial S_i}{\partial \tau} = \varepsilon \Lambda_i(\vec{S}) C_i \tag{9}$$

with the conditions

$$x = 0: C_i = p_i; \tag{10}$$

$$\tau = 0: S_i = 0; i = 1,...,n. \tag{11}$$

## 3. The asymptotic solution of the filtration problem

Assume that the filtration coefficients $\Lambda_i(\vec{S})$ have the form

$$\begin{gathered}\Lambda_i(\vec{S}) = \lambda_i + \sum_{m=1}^{n} a_i^m S_m + \frac{1}{2}\sum_{k,l=1}^{n} b_i^{k,l} S_k S_l + O\left(\left|\vec{S}\right|^3\right); \\ \lambda_i = \Lambda_i(\vec{0}), \ a_i^m = \frac{\partial \Lambda_i}{\partial S_m}(\vec{0}) > 0, \ b_i^{k,l} = \frac{\partial^2 \Lambda_i}{\partial S_k \partial S_l}(\vec{0}), \ \vec{0} = (0,0,...,0), \ i = 1,...,n.\end{gathered} \tag{12}$$

In the domain $\tilde{\Omega}_S$, the asymptotics of the problem (8) – (11) is constructed in the form

$$S_i = \varepsilon s_i^1 + \varepsilon^2 s_i^2 + \varepsilon^3 s_i^3 + O(\varepsilon^4); \ C_i = p_i + \varepsilon c_i^1 + \varepsilon^2 c_i^2 + \varepsilon^3 c_i^3 + O(\varepsilon^4); \ i = 1,...,n. \tag{13}$$

Substituting the expansions (12), (13) into equations (8), (9) and equating terms for identical powers of the small parameter $\varepsilon$, we obtain a recurrent system of differential equations

$$\frac{\partial c_i^1}{\partial x} + \lambda_i p_i = 0; \tag{14}$$

$$\frac{\partial s_i^1}{\partial \tau} = \lambda_i p_i; \tag{15}$$

$$\frac{\partial c_i^2}{\partial x} + p_i \sum_{m=1}^{n} a_i^m s_m^1 + \lambda_i c_i^1 = 0 ; \qquad (16)$$

$$\frac{\partial s_i^2}{\partial \tau} = p_i \sum_{m=1}^{n} a_i^m s_m^1 + \lambda_i c_i^1 ; \qquad (17)$$

$$\frac{\partial c_i^3}{\partial x} + p_i \sum_{m=1}^{n} a_i^m s_m^2 + \frac{p_i}{2} \sum_{k,l=1}^{n} b_i^{k,l} s_k^1 s_l^1 + \sum_{m=1}^{n} a_i^m s_m^1 c_i^1 + \lambda_i c_i^2 = 0 ; \qquad (18)$$

$$\frac{\partial s_i^3}{\partial \tau} = p_i \sum_{m=1}^{n} a_i^m s_m^2 + \frac{p_i}{2} \sum_{k,l=1}^{n} b_i^{k,l} s_k^1 s_l^1 + \sum_{m=1}^{n} a_i^m s_m^1 c_i^1 + \lambda_i c_i^2 . \qquad (19)$$

The initial conditions for the asymptotic terms follow from (3), (7):

$$c_i^j \Big|_{x=0} = 0; \;\; s_i^j \Big|_{\tau=0} = 0; \;\; i = 1,...,n; \;\; j = 1,2,3 . \qquad (20)$$

The terms of the asymptotic expansions are obtained by successive solution of the equation (14) – (19) with conditions (20)

$$c_i^1 = -\lambda_i p_i x ; \qquad (21)$$

$$s_i^1 = \lambda_i p_i \tau ; \qquad (22)$$

$$c_i^2 = \lambda_i^2 p_i \frac{x^2}{2} - p_i A_i^1 \tau x ; \qquad (23)$$

$$s_i^2 = -\lambda_i p_i A_i^1 \frac{\tau^2}{2} - \lambda_i^2 p_i \tau x . \qquad (24)$$

$$c_i^3 = -p_i (A_i^3 + B_i) x \frac{\tau^2}{2} + p_i (A_i^2 + 2\lambda_i A_i^1) \frac{x^2}{2} \tau - \lambda_i^3 p_i \frac{x^3}{6} ; \qquad (25)$$

$$s_i^3 = p_i (A_i^3 + B_i) \frac{\tau^3}{6} - p_i (A_i^2 + 2\lambda_i A_i^1) x \frac{\tau^2}{2} + \lambda_i^3 p_i \frac{x^2}{2} \tau ; \qquad (26)$$

where

$$A_i^1 = \sum_{m=1}^{n} a_i^m \lambda_m p_m ; \;\; A_i^2 = \sum_{m=1}^{n} a_i^m \lambda_m^2 p_m ; \;\; A_i^3 = \sum_{m=1}^{n} a_i^m A_m^1 p_m ; \;\; B_i = \sum_{k,l=1}^{n} b_i^{k,l} \lambda_k \lambda_l p_k p_l ; \;\; i = 1,...,n .$$

Substituting the solutions (19) - (26) into the expansions (13) and performing an inverse change of variables, we find the asymptotic solution of the problem (8) – (11) in $\Omega_S$ :

$$\begin{aligned} C_i(x,t) = {} & p_i - \lambda_i p_i x \varepsilon + p_i x \left( \lambda_i^2 \frac{x}{2} - A_i^1 (t-x) \right) \varepsilon^2 - \\ & - p_i x \left( (A_i^3 + B_i) \frac{(t-x)^2}{2} - (A_i^2 + 2\lambda_i A_i^1) \frac{x}{2} (t-x) + \lambda_i^3 \frac{x^2}{6} \right) \varepsilon^3 + O(\varepsilon^4); \end{aligned} \qquad (27)$$

$$S_i(x,t)=\lambda_i p_i(t-x)\varepsilon - p_i\left(-A_i^1\frac{(t-x)^2}{2}+\lambda_i^2(t-x)x\right)\varepsilon^2 +$$
$$+p_i(t-x)\left((A_i^3+B_i)\frac{(t-x)^2}{6}-(A_i^2+2\lambda_i A_i^1)x\frac{(t-x)}{2}+\lambda_i^3\frac{x^2}{2}\right)\varepsilon^3+O(\varepsilon^4);\ i=1,\dots,n. \quad (28)$$

The applicability of the asymptotics (27), (28) is determined by the inequalities (5), (6).

## 4. Numerical simulation

The calculation is performed for spherical particles of 3 radii $r_1 = 1.5675$ μm, $r_2 = 2.179$ μm and $r_3 = 3.168$ μm, for the small parameter $\varepsilon = 0.25$. The concentrations of suspended particles at the filter inlet $p_1 = 1$, $p_2 = 0.2$, $p_3 = 0.07$ are chosen so that the masses of the 3 types of injected suspended particles are the same. The filtration coefficients were chosen in accordance with experimental data [10]

$$\Lambda_1(S) = 0.110 - 0.014S + 4,49\cdot 10^{-5}S^2,$$

$$\Lambda_2(S) = 0.510 - 0.00596S + 2,29\cdot 10^{-6}S^2,$$

$$\Lambda_3(S) = 1.551 - 0.00347S - 1,16\cdot 10^{-6}S^2.$$

Here $S = S_1 + S_2 + S_3$ is the total retained particles concentration.

Fig. 1 shows the graphs of suspended particles concentrations for 3-size particles a) at the filter outlet and b) at $t = 0.5$ (the smaller the particle size, the higher the graph). The dashed line indicates the maximum value $C = 1$.

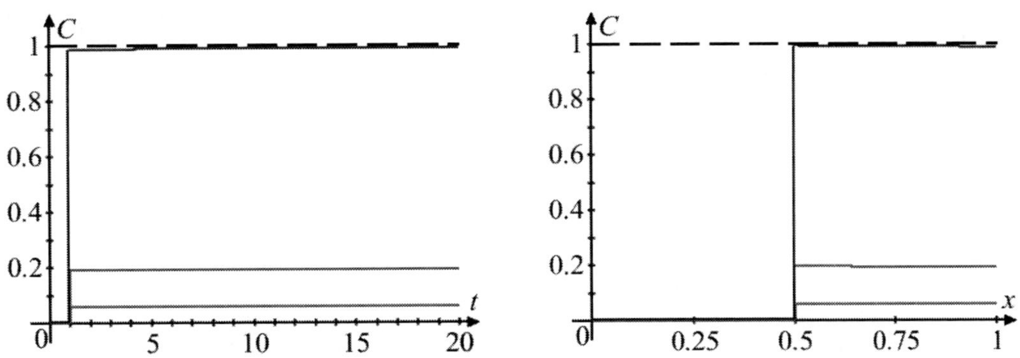


Fig. 1. a) The graphs of $C_i(x,t)|_{x=1}$. b) The graphs of $C_i(x,t)|_{t=0.5}$.

Fig. 2 presents the graphs of the retained particles concentrations for the particles of 3 sizes a) at time $t = 0.5$ and b) at time $t = 5$ (at the filter inlet $x = 0$ the graphs are higher for larger particles).

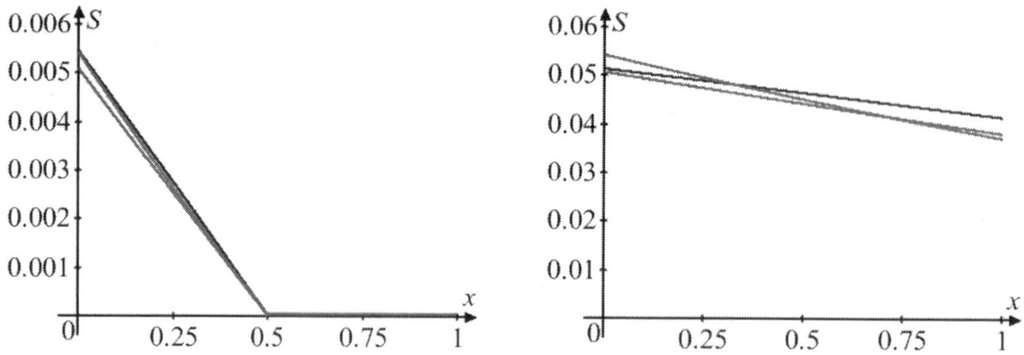


Fig. 2. a) The graphs of $S_i(x,t)\big|_{t=0.5}$. b) The graphs of $S_i(x,t)\big|_{t=5}$.

## 5. Conclusion

The global asymptotic solution of the one-dimensional filtration problem of a polydisperse suspension in a homogeneous porous medium for small filtration coefficients is constructed. In the paper the methods of [16] are extended to the case of $n$-size particles. The solution has a break on the characteristic line. Fig. 1 a) and b) show a gap in the suspended particles concentration on the concentration front $t = x$. On the graph of the retained particles concentration there is a kink at the point of intersection with the concentration front $x = 0.5$ (Fig. 2 a)).

The rate of decrease in the retained particles concentration with increasing distance to the filter inlet depends on the particle size (Fig. 2 b)). The larger the particle, the faster the rate of decrease The graphs confirm the experimental fact that larger particles mostly get stuck near the filter inlet, and fine particles are retained uniformly throughout the entire porous medium.

The obtained asymptotic solution improves the qualitative tuning of the mathematical model and reduces the number of laboratory experiments [17].

Mathematical methods used to construct the asymptotics can be generalized to the case of variable porosity and permeability, depending on the concentration of retained particles [18].

## References

1 K. Khilar, S. Fogler, *Migration of fines in porous media* (Kluwer, 1998)
2 A. Santos, P. Bedrikovetsky, S. Fontoura, J. Memb. Sci. **308** (2008)
3 M. Elimelech, J. Gregory, X. Jia, R.A.F. Williams, *Particle Deposition and Aggregation: Measurement, Modelling and Simulation*, (Butterworth-Hein., 2013)
4 J. P. Herzig, D. M. Leclerc, P. Legoff, Ind. Eng. Chem. **62** (1970)
5 E.A. Vyazmina, P.G. Bedrikovetskii, A.D. Polyanin, Theor. Found. Chem. Eng. **41**, 5 (2007)
6 Z. You, P. Bedrikovetsky, L. Kuzmina, Abstr. Appl. Anal. **2013**, ID 680693 (2013)
7 A.D. Polyanin, S.A. Lychev, App. Math. Mod. **40** (2016)
8 N.E. Leont`ev, D.A. Tatarenkova, Mos. Univ. Mech. Bull. **70**, 3 (2015)
9 A. Brillard, M. El Jarroudi, M. El Merzguioui, Math. Comp. Mod. **51** (2010)
10 Z. You, Y. Osipov, P. Bedrikovetsky, L. Kuzmina, Chem. Eng. J. **258** (2014)
11 L.I. Kuzmina, Yu.V. Osipov, Procedia Eng. **111** (2015)
12 L.I. Kuzmina, Yu.V. Osipov, Matec Web Conf. **86**, 01005 (2016)

13 V.I. Golubev, D.N. Mikhailov, Proc. MIPT **3**, 2 (2011) [in Russian].
14 Yu.P. Galaguz, G.L. Safina, Procedia Eng. **153** (2016)
15 Y. Galaguz, G. Safina, Matec Web Conf. **86**, 03003 (2016)
16 L.I. Kuzmina, Yu.V. Osipov, Vestnik MGSU, **1** (2015)
17 A. Vaz, D. Maffra, T. Carageorgos, P. Bedrikovetsky, J. Nat. Gas Sci. Eng. **34** (2016)
18 Z. You, A. Badalyan, P. Bedrikovetsky, SPE J. **18** (2013)

# The use of prestressed fiberglass to reinforce concrete bending elements

*Denis* Panfilov[1,*], *Alexander* Pischulev[1], *and Yuriy* Zhiltsov[1]

[1]Samara State Technical University, Institute of Architecture and Civil Engineering, 194Molodogvardeyskaya St., 443001, Samara, Russia

**Abstract.** The article presents the research results of mathematical models of bending concrete elements reinforced with prestressed fiberglass reinforcement. The investigations are built on the basic positions of the nonlinear deformation model, taking into account the nonlinear work of concrete. The samples were calculated with consideration for the deformation diagrams of materials obtained experimentally using the Lira CAD 2014 software package. The article also covers the current practice of using fiberglass to reinforce concrete bearing structures without prestressing, as well as the current normative documents regulating the use of fiberglass reinforcement in the Russian Federation, regulatory and state standards containing information on the strength and deformation properties of fiberglass reinforcement, and the results of physical tests of fiberglass reinforcement for tension. Based on the results of the study, we performed the analysis of fiberglass reinforcement use with and without prestressing. The studies presented in the article show that the use of pre-stressed fiberglass reinforcement makes it possible to increase the rigidity and fracture toughness of bending concrete elements.

## 1 Introduction

The use of fiberglass as a reinforcing element of concrete structures is hampered by the following factors:

- low adhesion of fiberglass reinforcement with concrete;
- low deformation modulus of fiberglass reinforcement.

The first drawback can be eliminated by using special anchoring devices to ensure the concurrent work of concrete and fiberglass reinforcement. The second disadvantage can be eliminated by changing the mechanical properties of fiberglass fibers, which will lead to the creation of a new material and the rejection of existing samples of fiberglass reinforcement manufactured in accordance with State Standard 31938-2012 [2], or by creating a pre-stressed state.

* Corresponding author: panda-w800i@yandex.ru

## 2 Materials and methods

The purpose of this work was to study the strength and deformation properties of fiberglass reinforcement, manufactured at a specialized plant SVSK Production company in Togliatti (Russia) in accordance with Specification No. 5769-001-21121765-2013 [1] and State Standard 31938-2012 [2]; the work also carries out experimental studies of fiberglass reinforcement samples, in continuation of the studies undertaken [3-9], to assess the feasibility of using such reinforcement as prestressed one in bending concrete elements. The prestress allows creating a reverse sign force in the unloaded structure that will increase the rigidity and fracture toughness of reinforced concrete elements.

## 3 Results

To obtain experimental data on the actual relationship between stresses and deformations, we carried out tests of fiberglass reinforcement for tension. For this reason, we made samples of fiberglass reinforcements with diameters of 4, 6, 8, 10, 12 mm. The length of the samples was assumed to be 1.2 m. The prototype designs are shown in Figure 1 and include pieces of fiberglass reinforcement, which are equipped with gripping devices at their ends. The construction of the gripping devices is a steel tube with a length of 350 mm (35 diameters of the test rod) and a diameter of 20 mm. The prototype of the rod is pasted into the gripping device using epoxy resin. The adopted design of the gripping device makes it possible to fix the test specimen in the tensile machine securely and transfer the breaking force to the sample, not allowing its local damage in the press clamps. The general view of samples of fiberglass reinforcement is shown in Figure 1.

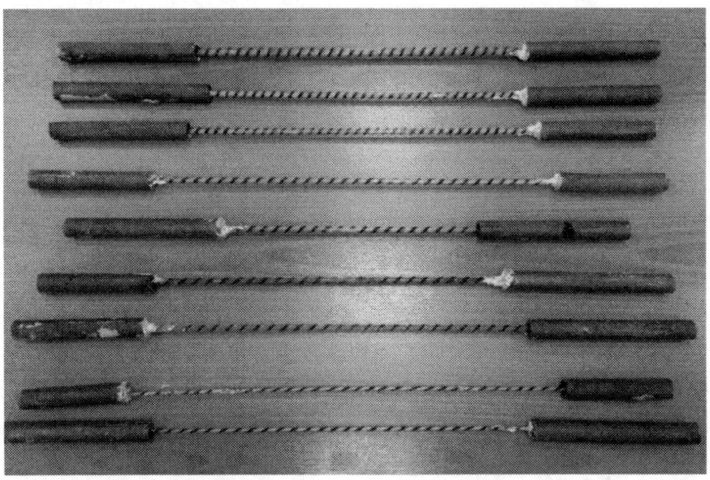

**Fig. 1.** Samples of fiberglass reinforcement.

To measure the controlled parameters, we developed a clamp design to fix the hour-type indicators on the sample during the research, with a measurement accuracy of 0.01 mm. The general view of the test set for the investigation of fiberglass reinforcement is shown in Figure 2.

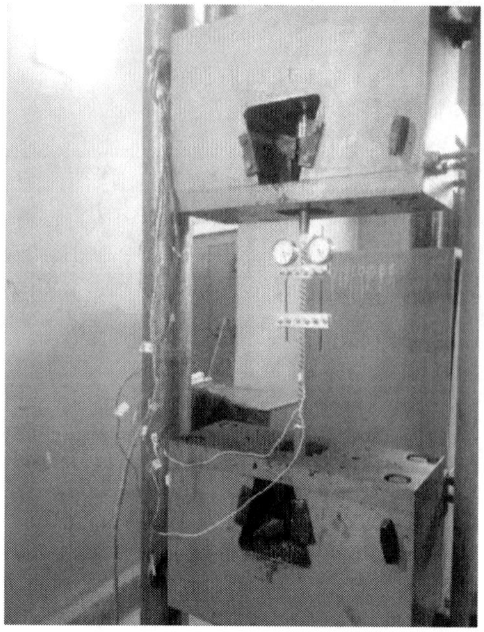

**Fig. 2.** General view of a test set for the research of fiberglass reinforcement.

To measure the relative deformations of the order of $1*10^{-6}$ Units of Relative Deformations, tensometric sensors were attached to the test sample. The tensometric station IDZ–1 was used as the recording equipment. During the tests, the error due to temperature deformation was eliminated by using tensometric compensation.

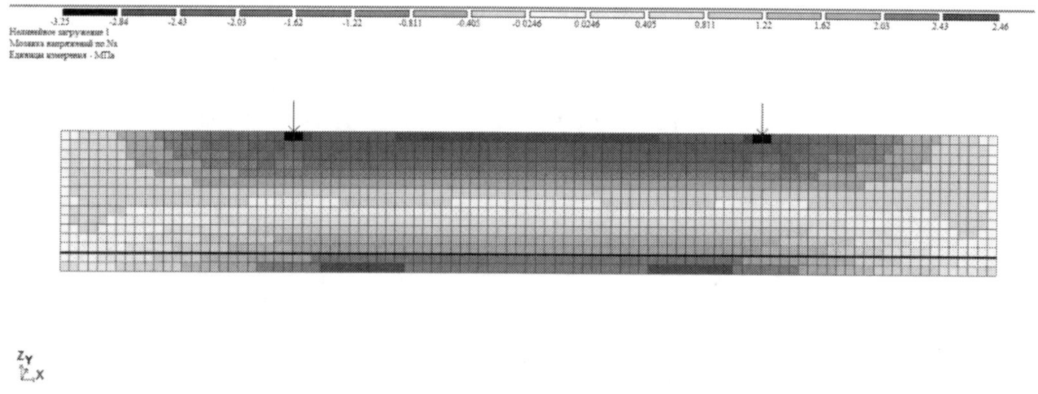

**Fig. 3.** The diagram of normal stresses distribution of the finite element model.

Tests were carried out on the basis of the research and industrial laboratory of the Building Structures Department of Samara State Technical University, Institute of Architecture and Civil Engineering on a tensile machine R-50. The load was set in steps of 1 kN.

According to the testing results of the reinforcement for the break, it was established:

1. The relationship between stresses and deformations is practically linear, which characterizes the work of fiberglass reinforcement as a conditionally elastic material.

2. The elasticity modulus along the fibers is about 55,000 MPa and remains conditionally constant at all loading stages.

3. Destruction of the sample with a diameter of 10 mm occurred at a load of 39 kN. The character of the sample destruction is brittle, as a result of a consistent instantaneous breakage of fiberglass reinforcement fibers.

4. Temporary resistance to tearing of fiberglass reinforcement was 39kN.

As prototypes, it is proposed to manufacture single-span hinged-supported beams loaded with a uniformly distributed load along the length. The beam structures are shown in Figure 4.

The sample is made of concrete B25 with the following geometric characteristics: height –120 mm, width – 120 mm, length – 1000 mm.

The analysis of the tension-deformation state of a bending concrete element reinforced with prestressed fiberglass reinforcement was performed with the help of a spatial finite element model created in the Lira-CAD 2014 software package with stepwise loading of the structure and taking into account the nonlinear deformation of concrete.

The design of the sample included a rod of prestressed fiberglass reinforcement installed in the stretched zone of the element as well as a spatial metal frame consisting of longitudinal working reinforcement and transverse reinforcement installed in the bearing zones of the beam. Longitudinal reinforcement installed along the upper edge of the beam at the stage of prestressing the sample (the tension of the reinforcement to the concrete) takes the tensile forces from the moment of the opposite sign. As the sample is loaded with an external load, the longitudinal steel reinforcement begins to perceive compressive forces, i.e. it is in the compressed zone of concrete. Transverse reinforcement provides the strength of oblique sections in the bearing zones of the beam. When manufacturing a sample in a stretched zone a channel is provided for installing prestressed fiberglass reinforcement. The tension of the reinforcement is carried out on concrete, after it takes up 70% of strength. After the rod is tensioned, it is inserted into the previously prepared channel, and the space between the gripping devices of the rod (tubes) and the thrust plate of the beam is selected by calibrated metal washers.

The samples were tested on a universal test machine GRM–1. The general view of the test machine and the nature of the destruction of the experimental sample are shown in Figures 4 and 5. The graphs of dependence of the deflection and width of the structure crack opening from the load are shown in Fig. 6 and 7.

**Fig. 4.** General view of the test machine with a sample.

**Fig. 5.** Brittle destruction of the sample at a load of 660 kN.

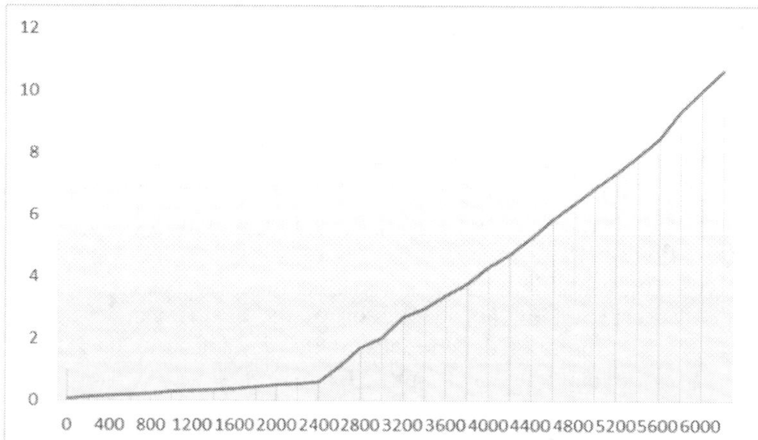


**Fig. 6.** Graph of dependence of the structure deflection on the load.

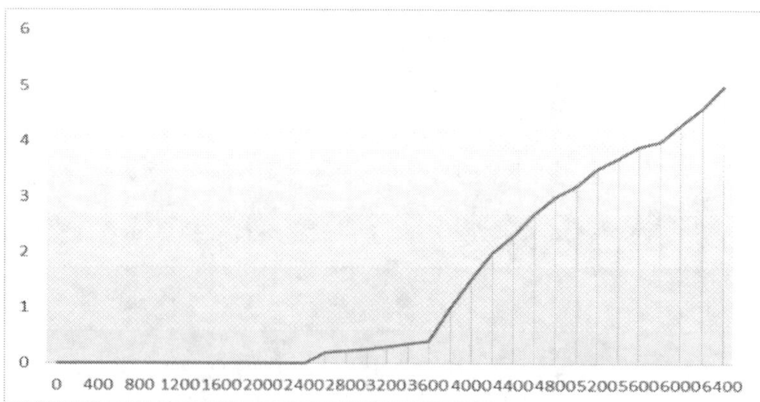


**Fig. 7.** Graph of the dependence of the structure crack width on the load.

## 4 Conclusion

Based on the results of research and comparison of calculated and experimental data, we can draw the following conclusions:

1. Fiberglass reinforcement has higher strength characteristics, but has a lower deformation modulus compared to steel. A low elasticity modulus limits the use of fiberglass reinforcement as a reinforcing element of concrete. To reduce the negative effect of a low elasticity modulus in work, it is suggested to use prestressed reinforcement.

2. Based on the results of non-linear calculations in Lira-CAD 2014 software and full-scale experiments, it is proved that the use of pre-stress fiberglass reinforcement makes it possible to increase the rigidity and fracture toughness of bending elements.

3. Based on the results of experimental studies of bending concrete beams with prestressed fiberglass reinforcement, the following was established: after reaching the limiting value of the cracking formation moment and the formation of the first crack, due to the loss of adhesion of the prestressed reinforcement to the concrete of the stretched zone, following the increased load, there is a sharp increase in deflections and width of crack opening, and the nature of the destruction of the sample is brittle, with the delamination of the compressed concrete zone.

## References

1. Specification № 5769-001-21121765-2013 *Fiberglass Plastic Reinforcement (Composite)* (2013)
2. The State Standard 31938-2012 *Composite Polymer Reinforcement for the Reinforcement of Concrete Structures. General Specifications.* 34 (2013)
3. D.A. Panfilov, Yu.V. Zhiltsov, K.I. Gimadetdinov, *Traditions and Innovations in Construction and Architecture. Construction*, 111-115 (2015)
4. A.A. Budko, A.A. Potekhin, A.A. Akopyan, *Technical Sciences: Tendencies, Prospects and Technologies of Development. Collection of scientific works on the results of the international scientific and practical conference*, 148-152 (2016)
5. T.A. Mukhamediev, D.V. Kuzevanov, Construction Mechanics and Calculation of Structures **4**, 18-22 (2016)
6. A.R. Wolik, A.A. Pochebyt, Modelling and Mechanics of Structures **4**, 18 (2016)
7. Valery B. Filatov, Alexander A. Suvorov, Procedia Engineering **153**, 144-150 (2016)
8. S.S. Mordovsky, Urban Construction and Architecture **3**, 26-28 (2013)
9. M.A. Ivlev, I.V. Nedoseko, Urban Construction and Architecture **4**, 86-91 (2012)

# Principles of formation of mobile dwelling of cluster type with application of additive technologies

*Alexander* Panfilov[1], *Vladimir* Panfilov[1,*]

[1]Institute of Architecture and Design, Industrial University of Tyumen, 625001 Tyumen, Russian Federation

**Abstract.** One of the most acute problems of modern society are issues related to the globalization of mankind caused by the next small scientific and technological revolution that is taking place and the emerging transition to a post-industrial society. Some of their characteristic features of this society are: deterritoriality, weakening of functional-spatial ties between places of residence, the application of labor and recreation rights. The peculiarity of the emerging new architecture with the arrival of the postindustrial information society is becoming completely new approaches to the formation of cities and the human dwelling itself. The article deals with the application of additive technologies in the production of mobile residential objects of high factory readiness. The article describes the concept of the formation of a mobile dwelling cluster type based on the principles of the organization of intelligent self-adapting systems (Smart Development), gives examples of author's developments in the field of cluster housing, with a description of the fundamental principles necessary for the possibility of further development and technical implementation.

One of the most burning issues of the modern society, along with ecology, social inequality and military conflicts, are issues of the globalization of mankind caused by the ongoing small scientific and technological revolution taking place and the emerging transition to a postindustrial society. Some of their characteristic features are: deterritoriality, weakening of functional-spatial connections between places of residence, and the application of labor and recreation [1-4]. It should be noted that with the change of society, there will also be a gradual change in human psychology in relation to the above-mentioned problems. In the stage of changing human society, there is also a rethinking of approaches to the formation of the world order that surrounds man and, in particular, the surrounding architecture. This happened in Ancient Rome, in Medieval Europe, in each of the occurred scientific and technological revolutions.

The feature of the emerging new architecture with in the postindustrial information society is completely new approaches to the formation of cities, a new tectonics of the human dwelling itself [5-7]:

1. The residential cell is ceasing to be an integral structural part of the residential building, gaining an increasing degree of autonomy in providing energy and other resources.

* Corresponding author: archi-zoom@yandex.ru

2. The residential cell requires to refuse of geographic attachment to the place of permanent deployment and the granting of it as freedom of movement, and universal systems of connection to engineering and other communications.
3. The outer contour of the residential cell, as well as its internal filling, should no longer be capital objects - they are light, informationally saturated, dynamically changing shells with a certain degree of adaptability.
4. Modern telecommunication and information networks allow a person to be in those places where his physical presence is not required, reducing the value of space, distance and attachment to one or several close places almost to zero. The idea of "remoteness" disappears, which is confirmed also by the words of the Director of the Paris Architectural School, Paul Verilio: "There are no more spaces!" [8], prophetically told them long before the present time.
5. Based on the first four features, a new - the fifth feature - follows: structural mobility. "Structural mobility" of a residential cell, unlike the usual "spatial mobility", means not only the possibility of spatial displacement, but also the saturation of a residential cell with such properties as:
   - **Interactivity of Interactions** at the System "Home - nature - man";
   - **Versatility and Multiplicity** of solutions that enable reflect infinitely of people activities in real mode Foot time;
   - **Dynamics and Variability** - when it is possible to reconfigure residential cells as a function of changing the initial settings, external and internal conditions (it is significant that with the development of control systems based on artificial intelligence becomes possible transition to neural network Spatial modeling of objects);
   - **Multifunctionality** as the possibility to change tasks Carried out Object or Separate him Parts without Making Cardinal Changes at him volumetric - planning decision.

The answer to these challenges may be came the system model of a mobile transformed residential cell of the cluster type. This article deals with two issues related to its creation:

1. General description and main characteristics of the system.
2. Principal decisions related with its creation on the basis of additive technologies.

**General description and main characteristics of the system**

Considering the growing human need in global mobility, as well as the acceleration of dynamic life processes, it is proposed to consider the possibility of creating a model of a mobile transformable cluster-type residential cell (hereinafter: cluster mobile housing).

Mobile housing is a product of high factory readiness and relates to the housing construction industry and to machine building, and also to such its part that deals with transport systems and other means of transport. This related state of two independent areas of industrial production has been repeatedly noted in earlier author's publications and it is based on a study of the historical evolution of the human dwelling, mobile home and vehicles [9-11]. The difference between these system begins to gradually fade at the present stage of development of them. We can live in the car with equal comfort, as well as move in our own house. I will give some examples from the real practice of our time:

1. Volkner Mobil Performance [12]

**Fig. 1.** Volkner Mobil Performance, general view

**Fig. 2.** Volkner Mobil Performance, option of interior solutions

This line of products of the company, belonging to the category of house-liners is not just a full-fledged three-room apartment, but a whole house with a garage with a degree of autonomy, reaching up to 1 week.

2. Greenmoxie Tiny House [13]

You can give another example of this solution: Greenmoxie Tiny House, this is not a machine, adapted for living - it's a house, put on wheels.

**Fig. 3.** Greenmoxie Tiny House, general view

3. Anderson Mobile Estates [14]

It should also be noted that the need for such homes is often fueled by many famous personalities whose lives are associated with a large number of movements and who have chosen this path for themselves - always and everywhere to stay in their home. For example, such actors as Ashton Kutcher, Sharon Stone and Vin Diesel, who ordered their homes from Anderson Mobile Estates

**Fig. 4.** Example of Anderson Mobile Estates, general view

An additional feature of projects implemented by Anderson Mobile Estates is their multicomponent volumetric transformation, which allows to significantly increase the volume of internal premises in operation mode and not to go beyond the dimensions of a standard vehicle in the mode of transportation.

4. «Manned Cloud » [15]

In this line, a separate line is based on promising developments based not on ground but air vehicles. An example of such solutions is the Manned Cloud project jointly developed by Jean-Marie Massaud and the French Aerospace Research Center ONERA (Office National d'Etudes et Recherches Aérospatiales). This hotel-airship is designed for 40 passengers and 15 crew members, has on board 20 cabins, a restaurant, salon, library, gym, spa, bar and terraces.

**Fig. 5**. Manned Cloud, general view

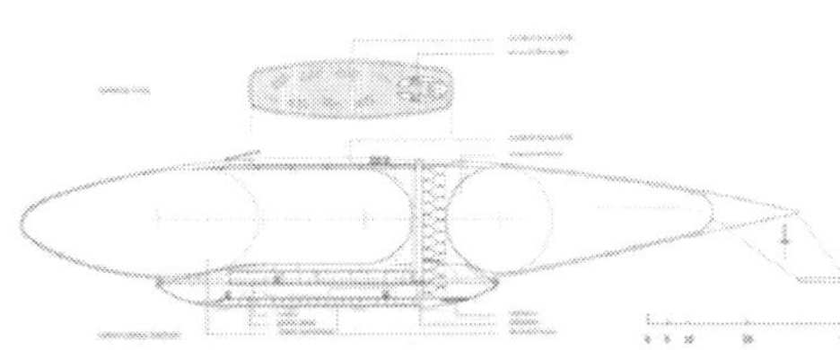

**Fig. 6.** Manned Cloud, basic technical solution

It should be given a brief description returning to the described project of a cluster mobile home. A mobile transformed residential cell of a cluster type consists of a system of transforming volume-spatial elements (blocks) performed in an open, closed or partially closed loop. In addition, there is a separate set of transverse elements (membranes) for forming a closed loop or separating of the internal space in the process of installation between individual open-type units or their groups.

Blocks of an open type are designed for forming the base of the mobile home. Blocks of a partially closed type - for inclusion of individual functional blocks of increased factory readiness (for example, sanitary units, cabins) in the structure of the house, Closed-type blocks are mainly technological elements (blocks of autonomy, energy supply and other necessary resources) full factory readiness.

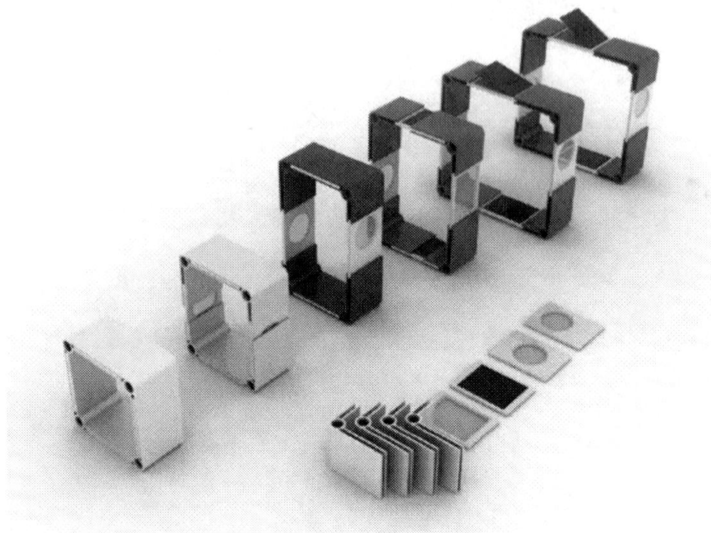

**Fig. 6**. Transformation of an element by the example of an open type block (A.V. Panfilov ©)

It is noted that each unit can be manufactured in monofunctional and in polyfunctional design. In the first variant the unit is delivered with assembled (or partially assembled) functional set (furniture, equipment, etc.). In the second case, it is possible to independently fill the internal space at the request of the owner.

Each block has outside measurements that correspond to the possibilities of high-quality exploitation and movement. So in the transported state they have dimensions 1.6x1.6m, which allows them to be moved element-by-element by any kind of trailers for passenger transport and moved on location without using of heavy equipment. In the operational position, the size increases to 3.2x3.2 m, which corresponds to the spatial human ergonomics. The width of the unit of reception is equal to 0.8 m, Which also corresponds to modern requirements for ensuring the comfort of space and ease of movement. The system also includes blocks of increased width (1.6 m) and a reduced width (0.4 m), for accommodation of larger and smaller equipment, respectively.

The construction of the cell allows for hot-swapping, adding or removing of individual blocks during exploitation.

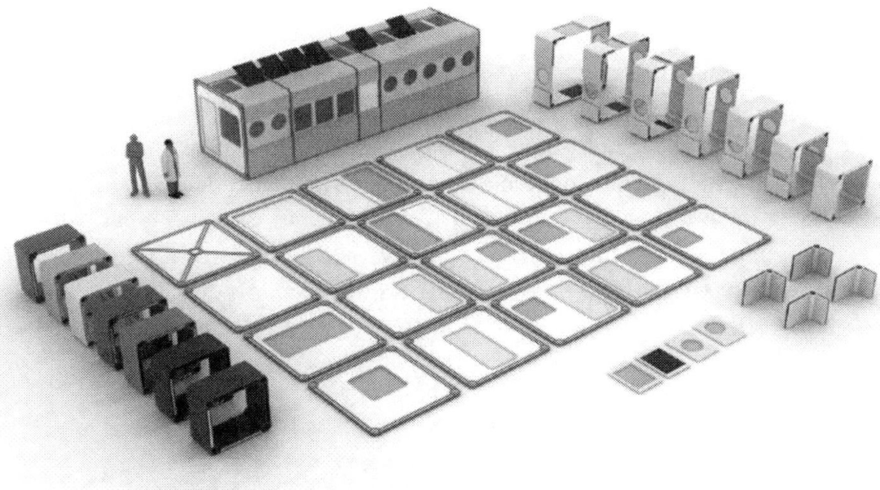

**Fig. 7.** An example of a set (home-kit) of a residential cell of a cluster type (A.V. Panfilov ©)

The project provides that there are sections of communications and engineering networks in each block, which in the process of linking blocks into a single building are connected to each other and they fasten the blocks into a single structural circuit by means of a locking element (electromagnetic, vacuum, etc.). In this case, the option of using an electromagnetic lock seems to be more optimal due to the possibility of switching off individual circuits for inserting or removing individual blocks. However, it is inferior to vacuum locks in terms of the degree of energy dependence and the density of contiguity to the elements to each other.

**Principal decisions related to its creation on the basis of additive technologies**

As an alternative to this project, it is possible to apply additive technologies (3D printing) [16-18] and individual elements of the supporting frame of blocks (in the solving the application of transformable blocks) or the whole block element in the event of a failure of the transformable configuration.

Using of additive production technologies in combination with parametric projection allows to solve a set of problems allowing to achieve the necessary qualitative characteristics that ensure the possibility of combining and recombining blocks:

- Accuracy of reproduction of elements and their separate parts;
- Accuracy and the possibility of introducing into the structure of a one unit of engineering network and communications elements;
- Shell formation of the one unit block taking into account the specified parameters of environmental extremes in operation mode;

- The possibility of a constructive combination of different materials, both in terms of technical characteristics and in terms of their typology (metal, polymers, foam and fiber aggregates, etc.) in a single system;
- The possibility of structural integration of elements of internal design (furniture, equipment) into a single design with a bearing contour of the block;
- Thus, the above listed positions allow us to talk about the beginning of the formation of fundamentally new approaches to the creation of residential cells, the possibility of using them under the condition of the emergence of a sufficient level of demand is quite broad: recreation, alternative or second dwelling of a citizen, expeditionary and research activities, Oil companies a comfortable home that meets all modern requirements.

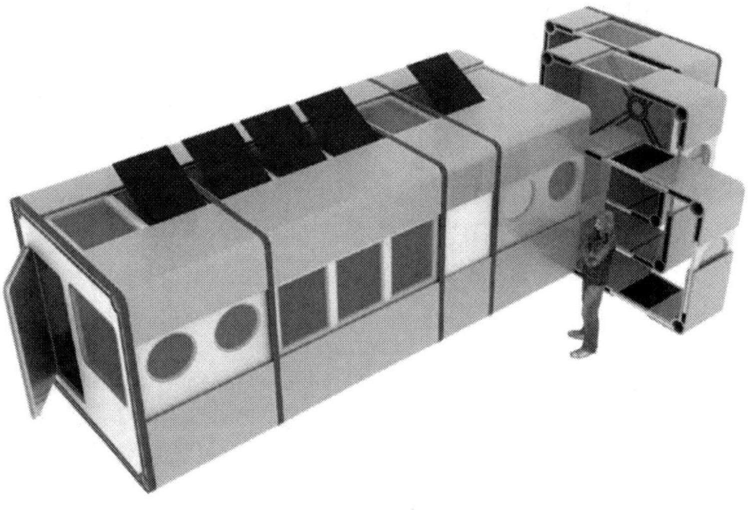

**Fig. 8.** Mobile residential cell. General view during assembly and reconfiguration (A.V. Panfilov ©)

At the same time, the present and perspective possibilities of using additive technologies allow to talk about the potential for using various kinds of local materials (earth, sands, etc.) and production waste (sawdust, slag, foam concrete) as the basic material of the supporting framework provided that an environmentally safe and weatherproof binder . In addition to this postulate, you can add the possibility of forming on this basis of expedition modules for the creation of bases and settlements in places where the possibility of delivery of finished equipment, buildings and structures is difficult or impossible. These can be represented by the Polar, Arctic and mountain territories, territories remote from traditional human habitats over considerable distances, including the exploration of outer space.

**References**

1. Schmitt G.N.A., Duarte J.P. Future Cities : proceedings of the 28th Conference on Education in Computer Aided Architectural Design in Europe, September 15-18, 2010, Zurich, Switzerland, ETH Zurich / G.N.A. Schmitt, J.P. Duarte, vdf Hochschulverlag AG and der ETH Zürich, 2010. 257-264 c.
2. Jacque Fresco, Roxanne Meadows. Designing The Future. – The Venus Project, Inc, 2007.
3. Vitrenko Andrii. The Essence of Intangible Service as a Special Theoretical Category in the Modern Post-industrial Economics System. Procedia Economics and Finance **27** (2015). Elsevier. 2015. 267-273c.
4. Feldman, V.P., Audretsch D.B. Innovation in Cities: Science based Diversity, Specialization and Localized Competition-European Economic Review. – 1999. - **№ 43**. – P. 409-429.
5. Landry C. Creative City. - Moscow: Publishing house "Classics-XXI, 2011. - 399p.
6. Saprykina, N.A. Utopian design of the 20th century. Futurological concepts of forecasting // "FUNDAMENTAL RESEARCHES of RAASN on the scientific provision of development of architecture, town-planning and construction industry of the Russian Federation in 2011" // Scientific works of RAASN, **vol. 1**, Moscow, 2012. - P. 262-267.
7. Hunt J. Architecture in the "cybernetic era" // Architectural Design. - **№ 11-12** / 1998 Retrieved from: http://www.i-home.ru/site.xp/049055048050124051056049055124.html
8. Virilio P. Information bomb. Strategy of deception. Translation from Fr. I. Okuneva. - M.: "Gnosis", "Pragmatics of Culture", 2002. -192 p.
9. Panfilov A. The Cluster Model of Development of the City - MATEC Web Conf. **73** 06008 (2016). DOI: 10.1051/matecconf/20167306008
10. Panfilov A.V. Mobile home: on the road from mobile to self-adaptive // Collection of scientific works SWorld. Materials of the International Scientific and Practical Conference "Modern Directions of Theoretical and Applied Research" 2013 ". - **Return 1. Volume 43**-Odessa: KUPRIENKO, 2013 C. 81-93
11. Panfilov, AV Evolution, features of development and classification bases for the formation of mobile homes for temporary stay / A. V. Panfilov // Electronic Journal AMIT **№ 4 (17),** 2011. (No. of state registration 0421100089): Retrieved from http://marhi.ru/AMIT/2011/4kvart11/panfilov/panfilov. Pdf, free. - Heading. From the screen..
12. Volkner Mobil Retrieved from http://www.volkner-mobil.com/perfgalgarage .html
13. Greenmoxie Retrieved from https://www.greenmoxie. com/tiny-house
14. Anderson Mobile Estates Retrieved from http://andersonmobileestates .com
15. Manned Cloud Retrieved from http://www.massaud.com/node/386
16. Gibson, I., Rosen, D. W., & Stucker, B. Additive Manufacturing Technologies. Additive Manufacturing Technologies 2010 (pp. 17–40). https://doi.org/10.1007/978-1-4419-1120-9
17. Hopkinson N., Hague R.J.M., Dickens P.M. Additive Manufacturing Technologies for the Construction Industry. Rapid Manufacturing An Industrial Revolution for the Digital Age, 2006. (pp. 249–273). https://doi.org/10.1002/047 0033991.ch17
18. Fusero, P., Massimiano, L., Tedeschi, A., & Lepidi, S. (2013). Parametric Urbanism: A New Frontier for Smart Cities. Planum The Journal of Urbanism, **2(27),** 1–13. Retrieved from http://www.planum.net/planum-magazine/ themes-online/parametric-urbanism-a-new-frontier

# Numerical study of the applicability of the Saint-Venant principle

*Alexander* Paushkin[1*], *Sergey* Saiyan[1]

[1]Moscow state university of civil engineering, Yaroslavskoye shosse, 26, Moscow, Russia, 129337

**Abstract.** The limits of the applicability of the Saint-Venant principle are investigated while using the LIRA-CAD app. A large number of bars loaded at the end with different loads are considered. A graph of the distribution of the zones of influence of the Saint-Venant principle is plotted, depending on the size of the bars. It can be concluded that the stress-strain state near the perturbation zones can be represented by three components, two of which are linear, and the third is self-balanced in force and moment.

## 1 Introduction

One formulation of the Saint-Venant principle allows us to assume that the stress-strain state of the bar depends on the mode of load applications only near the load placement location [1]. In the resistance of materials, it is usually believed that the "perturbations" of the stress-strain state decay approximately at a distance of one transverse dimension of the bar. Calculations were carried out with various grids on the LIRA-CAD software complex.

## 2 Research

In calculation No.1 (fig. 1, a), a rigidly clamped bar with dimensions of 800x100x10 cm, loaded with a central tensile force $F$ = 100 kN, was considered. In calculation No. 2 (fig.1, b) two forces $F/2$ = 50 kN were applied along the edges of the free end of the bar.

In the calculation of No. 3, a central compressive force $F$ = 1000 kN (fig.3) was applied to the end of the bar.

The criterion for determining the zone where the results of the numerical calculation coincide with the results obtained from the material resistance formulas was the area where the discrepancy reached no more than 5%.

Fig. 3 demonstrates the diagrams of normal stresses over the five sections indicated in fig. 2. Irregular distribution of stresses decays with distance from the place where the load is applied (from section 1 to section 4 in 20 cm spacing). An error of 5% with the results obtained from the material resistance formula $\sigma = F/A$ is indicated for the section 5 at a distance of 92 cm from the application point. In this case, cross sectional dimension was 100 cm.

* Corresponding author: alexglebov@bk.ru

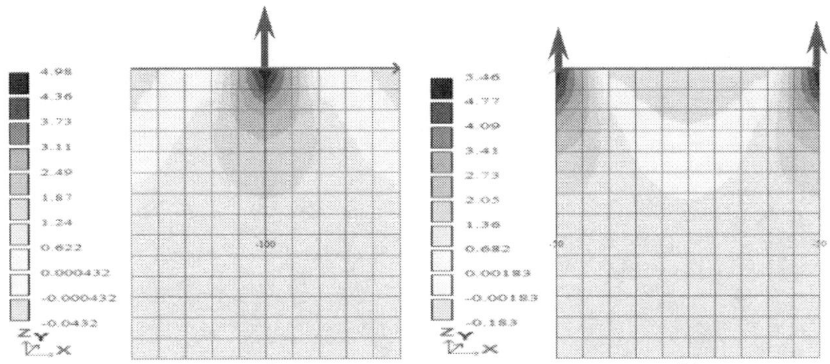


a) b)

**Fig. 1.** Results of calculations No.1 (a) and No.2. (b). Distribution of normal stresses.

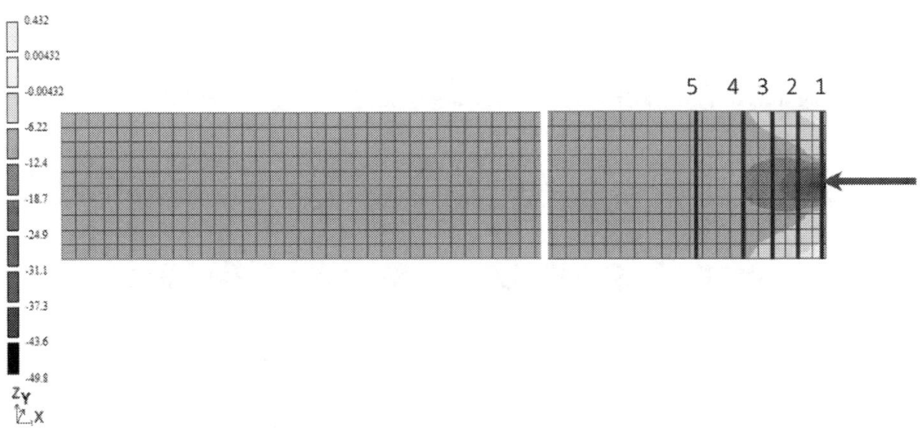


**Fig. 2.** To calculation No.3.

In the calculation No. 4 (Fig. 4), the same bar was tested for bending to determine the zone on which the formula $\sigma = (M/J)y$ is observed. The bending moment was formed by a pair of forces $F = 1000$ kN of the opposite direction.

From fig. 5 it follows that the form of the normal stress diagram tends to a linear function when the load is displaced from the place of application of the load at a distance of half the cross-section of the beam (section 4).

About a hundred numerical experiments on these loads were held, and the ratio of the transverse dimension of the beam to its length and the dimensions of the grid changed. The criterion for the matching of the results was the 5% discrepancy deviation barrier. For such types of loading, the perturbation of the stressed state did cease at a distance not exceeding one transverse dimension.

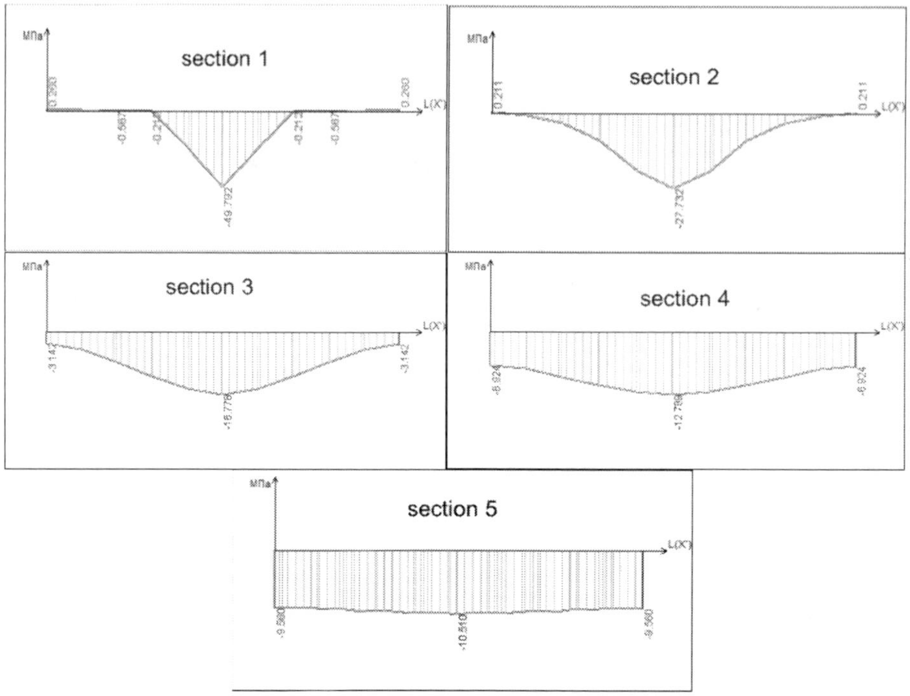


**Fig. 3.** Diagrams of normal stresses (to calculation No. 3)

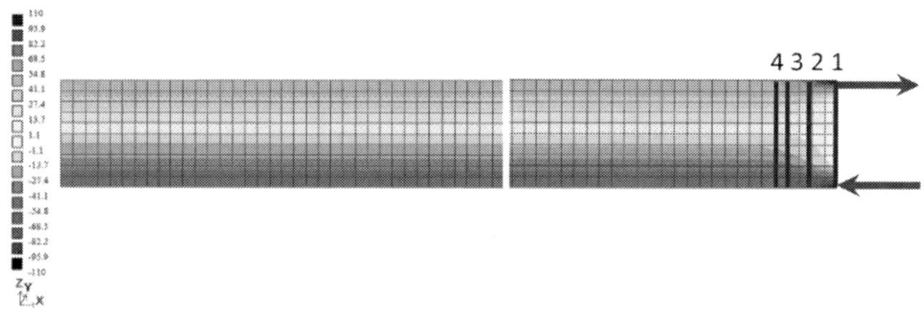


**Fig. 4.** To calculation No. 4

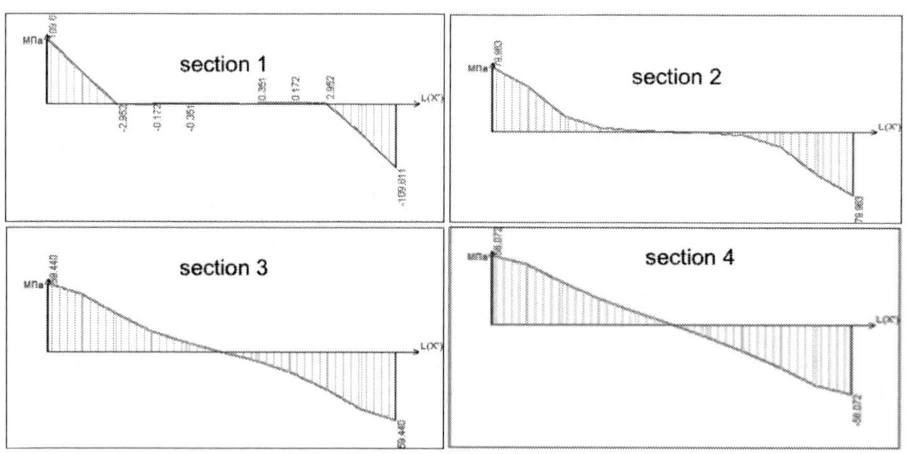


**Fig. 5**. Diagrams of normal stresses along sections 1 - 4 to calculation No. 4.

Additionally, the bars were calculated for the following loads:
a). Two opposing horizontal forces applied at a distance 20 cm at the free end (fig. 6, a).
b). One vertical force at the free end (fig. 6, b).
c). Two vertical forces at the free end (fig. 6, c).
d). Eccentric force at the free end (fig. 6, d).

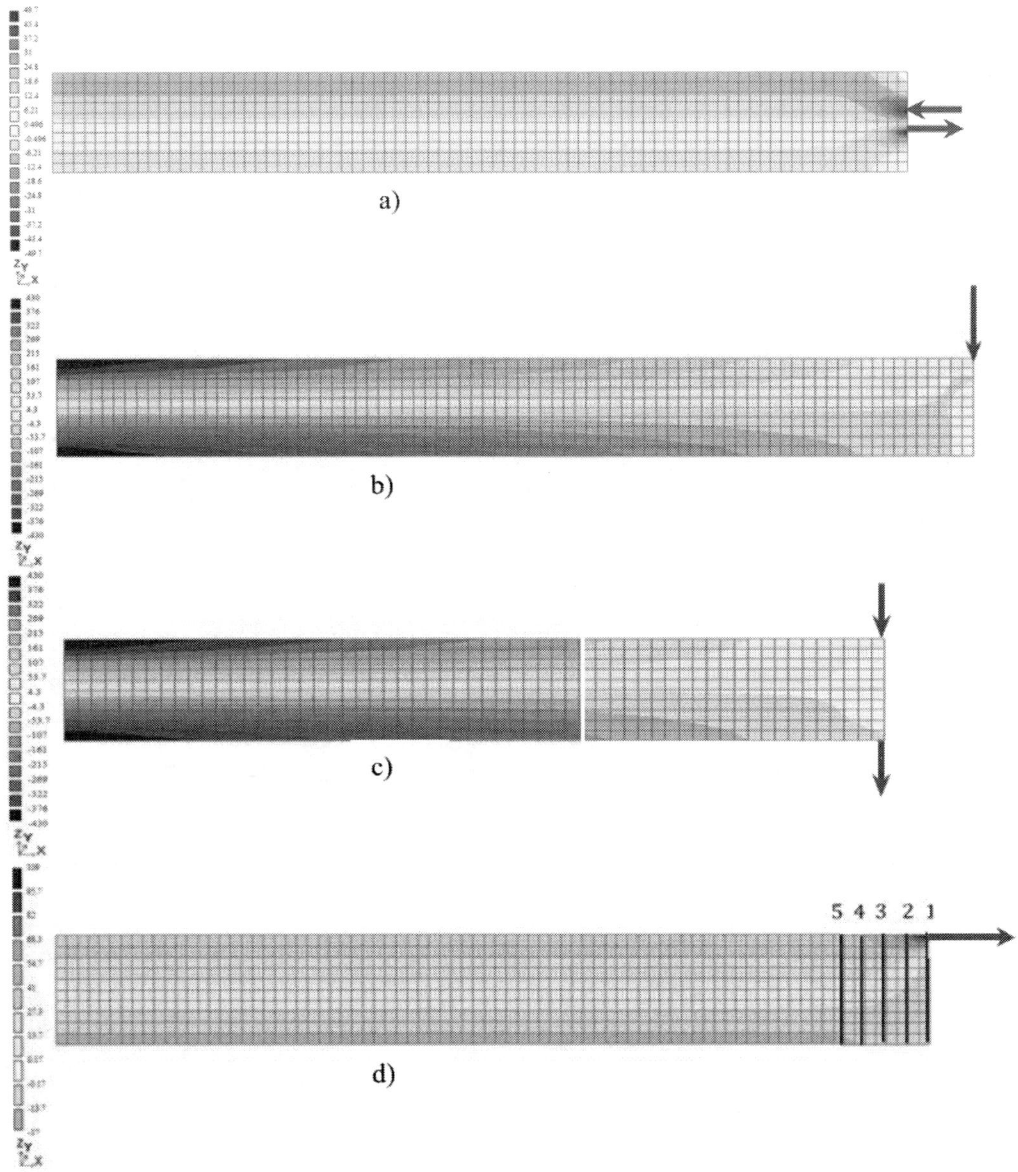


**Fig 6.** Calculation schemes and the results of experiments a, b, c, d.

Fig. 7 shows normal-stress diagrams in five sections, obtained from the calculation for eccentric tension (fig. 6, d).

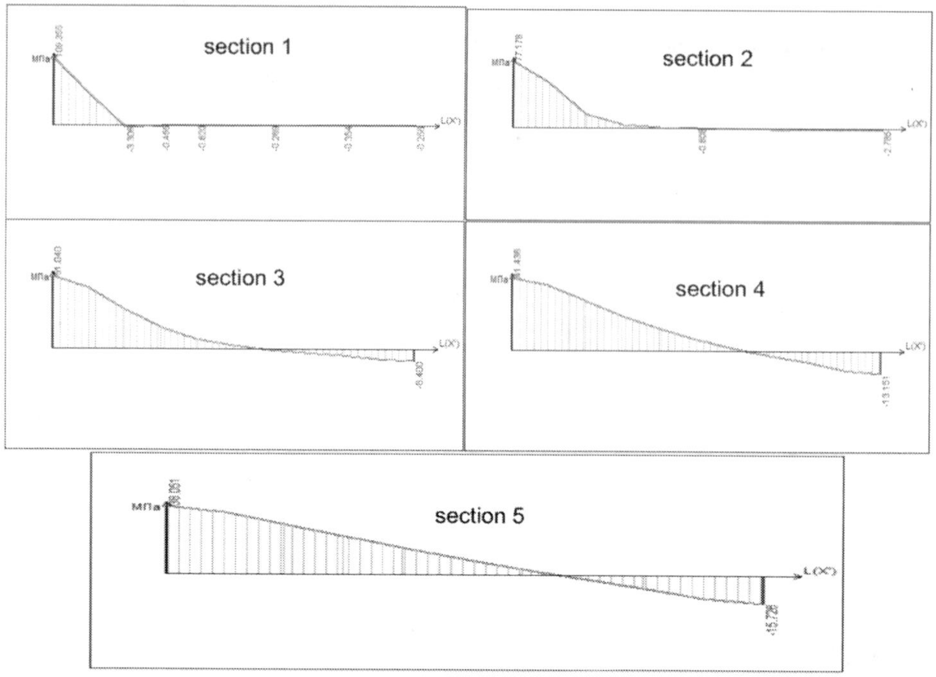


**Fig. 7.** Diagrams of normal stresses in five sections for calculation of eccentric tension

## 3 Conclusions

Based on the results of the research, the following schedule was constructed.

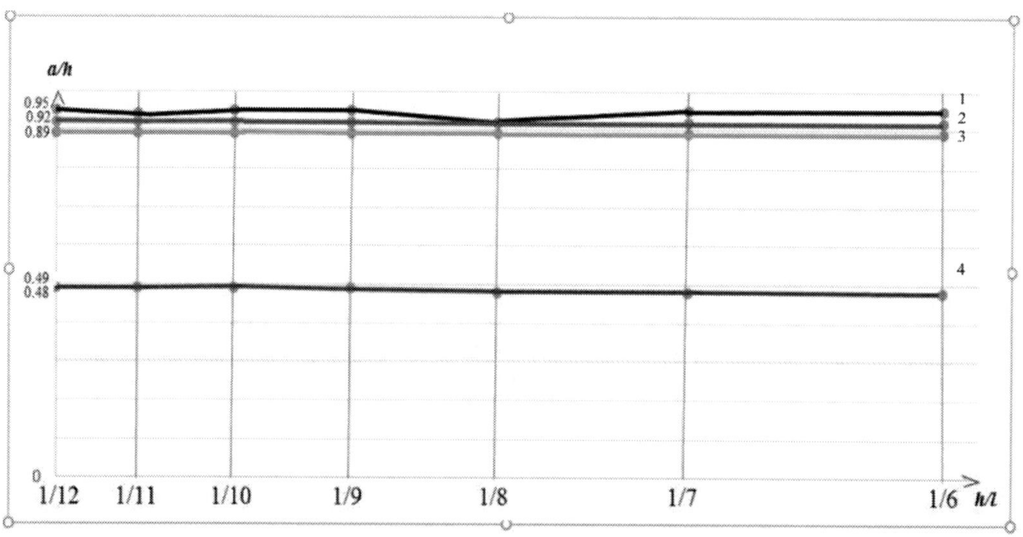


**Fig. 8.** Dependency graph of the implementation of the principle of Saint-Venant with a 5% discrepancy to the ratio of linear dimensions.

The graph (fig. 8) indicates: *a* – length of the disturbance zone, *h* - transverse dimension, *l* – length of the bar.

Loading:
1 - Central extension by force $F$ (fig.1, a).
2 - Tension by two forces $F/2$ (fig. 1, b).
3 - Bending by a pair of forces $F$ of the opposite direction (fig. 4).
4 - Bending by two forces $F/2$ (fig.6, c).

In all numerical experiments, the problem of studying the nature of the change in the stress-strain diagram along different sections near the end of the bar was also posed. The stress state near the places of application of loads is usually ignored in engineering calculations. However, it is possible to refine the state of stress. Near the place where the load is applied, the diagram of normal stresses can be represented as a sum of three components:

1. $\sigma$ = const, the area of which is equal to the normal force:

$$N = \int_A \sigma dA;$$

2. $\sigma = ay$ – linear function corresponding to the condition:

$$M = \int_A \sigma y dA;$$

3. A self-balanced function for which the following conditions are valid:

$$N = \int_A \sigma dA = 0;$$
$$M = \int_A \sigma y dA = 0.$$

Studies in this direction can be continued to create a library of self-balanced functions corresponding to different loads, which will allow them to be superimposed on a linear diagram of normal stresses in order to clarify the stress state in places close to the points of application of loads.

## References

1. V. Andreev, A. Paushkin, A. Leontiev, *Technical mechanics*, 18 (2013)

MATEC Web of Conferences 117, 00135 (2017) DOI: 10.1051/matecconf/201711700135
*XXVI R-S-P Seminar 2017, Theoretical Foundation of Civil Engineering*

# Revisiting the convergence of successive approximations in the problem of disturbed body motion

*Vera* Petelina[1*]

[1]Moscow state university of civil engineering, Yaroslavskoye shosse, 26, Moscow, Russia, 129337

**Abstract:** This work discusses the problem of the convergence of Picard's successive approximations, solving the differential equation system of disturbed body motion through assembling a majorizing linear differential equation. The problem is to determine in which region of variables used these successive approximations will converge and to estimate the error, if a finite number of approximations is used.

## 1 Introduction

One of the most important mechanics problems is the approximation of orthogonal coordinates, body velocity components and time by algebraic polynomials of the lowest degree in relation to an auxiliary variable with a prescribed accuracy degree. Sundman realized the Poincare's idea of series generation by degrees of an independent variable convergent for any real time value, but the Sundman series converge very slowly and therefore have not practical application potential. Another way for solution of the three-body problem was specified by G.A. Merman in [1]: in his solution, orthogonal coordinates and time are represented as Mittag-Leffler series, members of which are polynomials in an independent regularizing variable. These polynomial series converge faster than the Sundman series, since the convergence region is broader. New methods for determination of disturbances in orthogonal coordinates preserve standard features of conventional methods. For disturbance calculation a small parameter method is used; it enables asymptotic solution expansion and applies the Picard method of motion equations integration. This method leads to a convergent process of successive approximations, providing a solution of the disturbed motion differential equation system. The solution error depends on the accuracy of the initial approximation of the disturbing function.

The work [3] provides the idea of orthogonal coordinates, regularized velocity and time components for disturbed body motion in the second approximation as polynomials by degrees of a certain regularizing variable. The same can be performed to find the third, the fourth and higher approximations, since analytical features of these functions are known. Thus, it provides an iterative process for finding a solution of the disturbed body motion differential equation system.

* Corresponding author: verapetelina51@gmail.com

## 2 Mathematical model

Let us prove the convergence of successive approximations and determined conditions, when successive approximations converge to an exact solution. Let us consider a differential equation system of the following type:

$$\frac{dC_i}{du} = \varepsilon\left[f_{i0}(u) + f_{i1}(u, C_1, C_2, ..., C_7)\right] \ (i = 1,2,...,7) \tag{1}$$

where: $\varepsilon$ - a small parameter,

$u$ - an independent variable, where $-1 \le u \le 1$

$C_i$ ,$(i = 1,2,...,7)$ - in the problem under review these are variations of osculating initial values of orthogonal coordinates $\xi$, $\eta$, $\zeta$, regularized velocity components $\xi'$, $\eta'$, $\zeta'$, time point $\tau$.

$f_{i0}(u)$ - right-hand members of the differential equation system [2, formulas (30)-(33)] at:

$$\xi = \xi_0 ; \eta = \eta_0 ; \zeta = \zeta_0 ; \xi' = \xi'_0 ; \eta' = \eta'_0 ; \zeta' = \zeta'_0 ; \ \tau = \tau_0 . \tag{2}$$

$$f_{i1}(u, C_1, C_2, ..., C_7) = f_i(u, C_1 + \xi_0, C_2 + \eta_0, ..., C_7 + \tau_0) - f_i(u, \xi_0, \eta_0, ..., \tau_0) \tag{3}$$

$f(u, \xi, \eta, ..., \tau)$- right-hand members of the differential equation system under review [2, formulas (30)-(33)]. The small parameter $\varepsilon$ is introduced for further representation of the solution of the differential equation system (1) as series by degrees $\varepsilon$ .

It is possible to select such $C^*$ that in the hyperparallelepiped $D$: $-1 \le u \le 1; |C_i| \le C^*$, the right-hand member of the system (1) will be continuous in $u$ and will satisfy generalized Lipschitz conditions (4):

$$\left|f_{i1}(u, C_1^*, C_2^*, ..., C_7^*) - f_{i1}(u, C_1^{**}, C_2^{**}, ..., C_7^{**})\right| \le \sum_{j=1}^{7} L_{ij}(u) \cdot \left|C_j^* - C_j^{**}\right| \tag{4}$$

where: $L_{ij}(u) = \sup\left|\frac{\partial f_{i1}}{\partial C_j}\right|$ in the hyperparallelepiped $D_u$ , which is a cross-section of the hyperparallelepiped $D$ by the plane $u = u_0$.

Let us find a solution of the differential equation system (1) by the successive approximations method. Assuming $C_i = 0$,$(i = 1,2,...,7)$, in the first approximation it is:

$$C_i^{(1)}(u) = \varepsilon \cdot \int_{-1}^{u} f_{i0}(u) du, \tag{5}$$

in the second approximation it is:

$$C_i^{(2)}(u) = \varepsilon \cdot \int_{-1}^{u} \left(f_{i0}(u) + f_{i1}\left(u, C_1^{(1)}(u), ..., C_7^{(1)}(u)\right)\right) \cdot du, \tag{6}$$

the k-th approximation it is:

$$\begin{aligned} C_i^{(k)}(u) &= C_i^{(1)}(u) + \varepsilon \cdot \int_{-1}^{u} f_{i1}\left(u, C_1^{(k-1)}(u), ..., C_7^{(k-1)}(u)\right) \cdot du = \\ &= \varepsilon \cdot \int_{-1}^{u} \left(f_{i0}(u) + f_{i1}\left(u, C_1^{(k-1)}(u), ..., C_7^{(k-1)}(u)\right)\right) \cdot du \end{aligned} \tag{7}$$

Let us review the auxiliary functions $P(u)$ and $Q(u)$, for which:

$$\left|f_{i0}(u)\right| \leq P(u)\ ;\ \sup\left|\frac{\partial f_{i1}}{\partial C_j}\right| = L_{ij}(u) \leq L(u)\ ;\left|Q(u)\right| = 7 \cdot L(u) \tag{8}$$

$(i, j = 1,2,...,7)$

Let us generate a majorizing linear differential equation with variable coefficients:

$$\frac{dC}{du} = \varepsilon\left[P(u) + Q(u)\cdot C\right]\ , \tag{9}$$

which is reviewed within a range of $-1 \leq u \leq 1$.

Let us analyze the solution convergence for the equation (9), which is as follows:

$$C(u) = e^{\varepsilon\cdot\int_{-1}^{u} Q(u)\cdot du}\cdot\varepsilon\cdot\int_{-1}^{u} P(u)\cdot e^{-\varepsilon\cdot\int_{-1}^{u} Q(u)\cdot du}\cdot du, \tag{10}$$

where $C(-1) = 0$. It can be proved that $C'(u) \succ 0$, $C(u)$ is a steadily increasing function and is represented as an absolutely convergent series:

$$C(u) = \sum_{k=1}^{\infty}\varepsilon^k \cdot \tilde{C}^k(u)\ , \tag{11}$$

If:

$$C(1) = e^{\varepsilon\cdot\int_{-1}^{l} Q(u)\cdot du}\cdot\varepsilon\cdot\int_{-1}^{l} P(u)\cdot e^{-\varepsilon\cdot\int_{-1}^{u} Q(u)\cdot du}\cdot du \leq C^{*}. \tag{12}$$

Let us find a solution for the equation (9) by the successive approximations method. After $l$ – iterations we have:

$$C^{(l)}(u) = \sum_{1}^{(l)}\varepsilon^k \cdot \tilde{C}^k(u), \tag{13}$$

where: $\varepsilon^k \cdot \tilde{C}^k(u) = C^{(k)}(u) - C^{(k-1)}(u)$, $C^{(k)}(u)$ - solution of the majorizing equation (9) in k-th approximation.

Let us now prove that if the condition (12) is satisfied, successive approximations, representing the solution of the system (1), converge to functions that represent the solution of the equation (9). Considering the conditions (4) and the formulas (7), we have:

$$\begin{aligned}\left|C_i^{(k)} - C_i^{(k-1)}\right| &\leq \varepsilon\cdot\int_{-1}^{u}\left[f_{i1}\left(u, C_1^{(k-1)}, C_2^{(k-1)}, ..., C_7^{(k-1)}\right) - f_{i1}\left(u, C_1^{(k-2)}, C_2^{(k-2)}, ..., C_7^{(k-2)}\right)\right]\cdot du \leq \\ &\leq \varepsilon\sum_{j=1}^{7}\int_{-1}^{u} L_{ij}(u)\cdot\left|C_j^{(k-1)} - C_j^{(k-2)}\right|\cdot du\end{aligned} \tag{14}$$

Supposing that the inequality:

$$\left|C_i^{(m)} - C_i^{(m-1)}\right| \leq \varepsilon^m \cdot \tilde{C}^{(m)}(u) \tag{15}$$

is true for $(m = 1,2,...,k-1)$, we will know that is true for $(m = k)$. It follows from the inequality (14). Thus, if the condition (12) is satisfied, the inequality (15) is true with any $m$, i.e. successive approximations, representing the solution of the system (1), converge to functions, representing the solution of the majorizing equation (9). If the solution of the equation (9) exits the hyperparallelepiped at $-1 < \alpha < 1$, i.e. at $C(\alpha) = C*$, the successive approximations $C_i^{(m)}(u)$ absolutely converged within $[-1;\alpha]$. If we estimate $C*$, where

$|C_i| \le C^*$, $(i = 1,2,...,7)$, we can find a time interval, where the approximations under review converge.

## 3 Assembling a majorizing linear differential equation

Let us consider the simplest variant of the problem, when the right-hand equation member of the system [2 of the formulas (30) - (33)] and the functions $L_{ij}(u)$ are amplified in the hyperparallelepiped $D$ $-1 \le u \le 1$, $|C_i| \le C^*$ by their maximum values.

$$\max P(u) = A, \qquad \max Q(u) = B \tag{16}$$

The usage of regularized differential equations [2 of the formulas (30) - (33)] will automatically remove singularities due to collision of the body with the Earth. The convergence of successive approximations for regularized equations depends on the error in selection of initial approximations of unknown boundary conditions to a much lesser degree.

Let us assume that $\varepsilon = 1$, and the equation (9) will appear as follows:

$$\frac{dC}{du} = A + B \cdot C \tag{17}$$

and its solution will be:

$$C(u) = \frac{A}{B}\left(e^{Bu} - 1\right) \tag{18}$$

Let us set the right-hand part (18) equal to the value *C**, and we will receive the following transcendental equation:

$$\frac{A}{B}\left(e^{Bu} - 1\right) = C^* \tag{19}$$

from which we can determine the value of the independent regularizing variable $u = u_1$, corresponding to $C(u) = C^*$.

$$u_1 = \frac{1}{B} \cdot \ln\left(\frac{B \cdot C^*}{A} + 1\right) \tag{20}$$

For estimation of the value *C** it is necessary to estimate values A and B. For estimation of the value A it is necessary to evaluate the right-hand parts of the differential equation system for disturbed body motion, where the desired variables are osculating initial values of orthogonal coordinates $\xi, \eta, \zeta$, components of regularized velocity $\xi', \eta', \zeta'$, regularized acceleration $\xi'', \eta'', \zeta''$, the time point $\tau$, the radius-vector $\rho$ and its derivatives $\rho'$ and $\rho''$ by the independent regularizing variable $u$ considering the assumptions made. For this purpose, the osculating initial values of orthogonal coordinates, components of regularized velocity, regularized acceleration and time of geocentric motion of the body are supplemented with members dependent of the parameter *C**, assumed with their scaling factors, i.e. the members $C_i = \gamma_i C^*$ $(i = 1,2,...,7)$, where the scaling factors $\gamma_1, \gamma_2, \gamma_3$ are dimensionless, $\gamma_4, \gamma_5, \gamma_6$ are dimension factors (km/s), $\gamma_7$ is a dimension factor (s/km). For each function in the right-hand parts of the differential equation system [2 of the formulas (30) - (33)] the right and left boundaries were estimated. They depend on the value *C**, and to find the left boundary the minimum value was sought after, and the components $\gamma_i C^*$, $(i = 1,2,...,7)$ were assumed with the minus sign; to find the right boundary the maximum value was sought after. The maximum-modulus estimate was selected as the desired estimate. Thus, rough function estimates were produced, using mainly geometrical considerations. More precise estimates of these functions are built considering behavior of functions $x, y, z, x', y', z', \xi'', \eta'', \zeta'', \sigma$ and their derivatives with respect to the variable $\rho$. Note that the inequality (12) includes only integral characteristics of the functions $P(u)$ $Q(u)$.

This ensures a more precise generation of the majorizing equation (9), and the behavior of its solution does not differ much from the behavior of solution of the initial equation (1).

## 4 An example of successive approximation convergence domain evaluation

To illustrate the developed research methodology for the convergence of successive approximations, calculations were carried out with specified initial conditions, parabolic motion conditions of the body, considering assumptions made, and this resulted in the numerical estimate of the convergence domain. Several variants were calculated with different values of the variable $u$; the dependence of the variable $u$ on $C^*$ was established, and it was proved that this dependence has the extremum.

Dependence of the parameter C* on the quantity of the scale factors $\gamma_i$ (i=1,2,3,4,5,7) was analyzed. In the case of degeneracy the dependence of the quantity C* on the variable $u$ has no extremum with $\gamma_i = 0$(i=1,2,3,4,5,7), since the quantities A and B do not depend on the C*. When increasing the factors $\gamma_i$, this dependence has an extremum, which indicates an extension of the convergence domain. The quantity C* is limited, so it can be assumed that the scale factors fall within a certain range of variation.

When giving the value of the parameter C*, it was defined more precisely using the developed technique. For the given initial conditions, we found the extremal value of the quantity C*=4,987 and the corresponding value of the independent variable $u_1$=0.24. Thus, if the mismatches in the rectangular coordinates of a body perturbed motion at the same instants of time at different stages of approximation do not exceed the quantity C*=4,987 the successive approximations which represent the solution of the perturbed motion differential equations system converge absolutely.

Figure 1 shows the dependence of the parameter C* on the standardized regularizing variable $u$ (the dotted line shows the dependence in the case of degeneracy with $\gamma_i = 0$, the solid line shows the same with $\gamma_1 = \gamma_2 = 0.08$ $\gamma_4 = \gamma_5 = 0.001$.

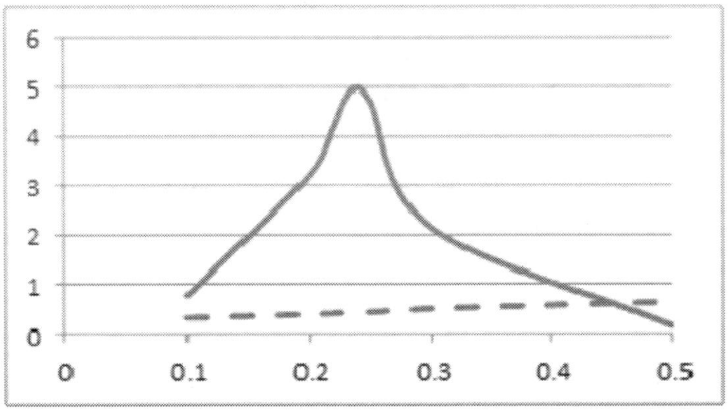


**Fig. 1.** The dependence of the parameter «C*» on the normalized regularizing variable «$u$ »

## 5 Conclusions

A research method for the convergence of Picard successive approximations, representing a solution of the differential equation system for disturbed body motion, was developed. A method for finding majorizing function estimates, required for determination of the of successive approximations convergence, was developed. These estimates were determined and analytical formulas for determination of the convergence domain boundaries were produced.

The developed technique for study of the Picard's successive approximations convergence which represents the solution of the body perturbed motion differential equations system can be used in underground construction.

## References

1. G.A. Merman. On representation of the general solution of the three-body problem using convergent series. Bulletin of the USSR AS Theoretical Astronomy Institute, **6**, **10**, 713-731 (1958)
2. V.D. Petelina. Disturbances of the second and higher orders for geocentric body motion. In the compilation "Universe study problems. Development of astronomy research methods", the USSR AS Theoretical Astronomy Institute, L. **8**, 247-258. (1979)
3. M.S. Yarov-Yarovoy. Analytic theory of body motion toward the Moon. Materials of the P.K. Sternberg's State Astronomy Institute, **36**, 28-160 ( 1967)

# Evaluating the load-carrying ability of compressed reinforced concrete elements with heterogeneous strength properties of concrete

*Alexander* Pischulev[1], *Denis Panfilov*[1,*], and *Andrey* Aleshin[1]

[1]Samara State Technical University, Institute of Architecture and Civil Engineering, 194, Molodogvardeyskaya St, 443001, Samara, Russia

**Abstract.** This article presents a methodology of determining the load-carrying ability of compressed reinforced concrete elements with layers of various concrete strength and describes the stress-strain behaviour of these elements. The methodology is based on the use of a deformation model and diagrams of materials deformation [1, 2]. The authors used this methodology to develop an algorithm. The algorithm makes it possible to evaluate stress and strain state of normal sections at all stages of loading up to destruction. The solution of the problem is shown as a data table. Analysis of this data table allows to estimate an expected pattern of the samples destruction. The paper also gives an example of calculation of a compressed reinforced concrete element with heterogeneous strength properties of a compressed area with symmetrical reinforcement (4 bars). The calculation methodology in this article has been successfully used to evaluate the load bearing capacity of the described samples and those of other authors.

## 1 Introduction

Inspections and examinations of buildings and other structures revealed that reinforced concrete elements with different strength of concrete layers are quite frequent within these constructions. The shift in the strength of concrete layers may be caused by technological factors or by mechanical or corrosive damage. The methods currently available to determine the depth of corrosion damage and résidual mechanical strength characteristics of concrete are generally based on a review of corrosion of materials over time, with the effect of a certain reagent based on laboratory research. However, the operating conditions of real structures are often different from laboratory conditions, since the structure may be susceptible to the aggressive effects of several reagents, and the actual temperature-humidity conditions may differ from those based on calculations. Errors in corrosion processes prediction in reinforced concrete structures may reduce the accuracy of calculations and, in some cases, lead to reassessing their load bearing capacity. The question of calculation of structures with disrupted strength of concrete is very actual and it has been the center of analysis in in recent years by many Russian and foreign researchers.

* Corresponding author: panda-w800i@yandex.ru

Theoretical and experimental research in this area (6-9) is currently being carried out in the Department of Building Structures of Samara State Technical University. The authors of the research suggest that load bearing capacity capacity of disrupted structures should be assessed by means of calculation algorithms using non-linear diagrams of concrete and reinforcement deformation.

## 2 Materials and Methods

Modern methods for estimating mechanical strength characteristics of concrete [3-5] allow to estimate mechanical strength characteristics of concrete in the section with high accuracy. By analyzing strength properties of concrete, relatively homogeneous concrete areas can be identified in the section under investigation. Once data on strength characteristics of the relatively homogeneous concrete areas are obtained, coefficients $a_c$ and $b_c$ are defined for the exponential function 1 proposed by professor G.V. Murashkin [1]

$$\sigma_b(\varepsilon) = a_c \cdot \varepsilon^{b_c} \cdot \exp\left(\frac{-b_c \cdot \varepsilon}{p}\right) , \tag{1}$$

where $\sigma_b$ is the stress in the concrete; $\varepsilon$ –relative deformation of concrete; ac и bc – $a_c$ and $b_c$ – coefficients of exponential function $f(\varepsilon) = \varepsilon \cdot E$, which describes shift in the elasticity module from the magnitude $\varepsilon$; $p$ – relative deformation of concrete in the chart of the maximum value of the stresses.

The deformation model of the elements shown in Figure 1 is accepted for calculation.

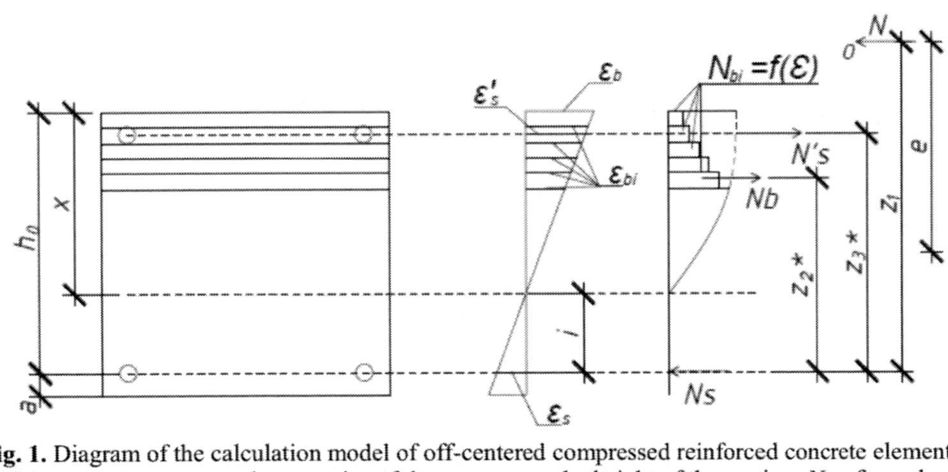


**Fig. 1.** Diagram of the calculation model of off-centered compressed reinforced concrete element with heterogeneous strength properties of the concrete at the height of the section. $N_b$ – force by the concrete of the compressed section area; $N_{bi}$ – force taken by the concrete of the relatively chosen layer; $\varepsilon_{bi}$ – relative deformation of concrete in the most compressed part of the section; $\varepsilon_s$ – relative deformation of reinforcement; $N_s$ – is the force of reinforcement.

## 3 Results and Discussion

Deformation of concrete in the most compressed concrete zone can be calculated by equation of equilibrium of internal and external forces is formed 2).

$$N = N'_s + N_b - N_s , \tag{2}$$

where $N_s$ ,$N_s'$– a force taken, respectively, by stretched and compressed reinforcement; $N_b$ – a force taken by compressed concrete.

Force $N_s$ is determined according to the level of deformation in reinforcement. If reinforcement deformations have not reached the yield limit, force $N_s$ is determined by the formula (3), otherwise by the formula (4).

$$N_s = A_s \cdot \sigma_s(\varepsilon_s)\ , \tag{3}$$

where $A_s$ the cross-sectional area of reinforcement; $\sigma_s(\varepsilon_s)$ – a function of stresses in reinforcement from deformation; $\varepsilon_s$ –relative deformation at the center of gravity level of the section of tensile reinforcement.

$$N_s = A_s \cdot R_s\ , \tag{4}$$

where $A_s$ – the cross-sectional area of reinforcement; $R_s$ – designed resistance of reinforcement to axial stretch;

Force$N_s'$ is defined in a similar way By formula (5), (6).

$$N_s' = A_s' \cdot \sigma_s(\varepsilon_s')\ , \tag{5}$$

where $A_s'$ – the cross-sectional area of compressed reinforcement; $\sigma_s(\varepsilon_s')$ – a function of stresses in reinforcement from deformation; $\varepsilon_s$ – relative deformations at the center of gravity level of the section of compressed reinforcement.

$$N_s' = A_s' \cdot R_{sc}\ , \tag{6}$$

where $A_s'$ – the cross-sectional area of compressed reinforcement; $R_{sc}$ – designed resistance of reinforcement to compression;

Force of $N_b$ is determined as a sum of the forces (7) taken by concrete of relatively homogeneous layers:

$$N_b = \sum_{i=1}^{n} N_{bi}\ , \tag{7}$$

where $N_{bi}$ – a force taken by compressed concrete of layer $i$ by formula (8).

$$N_{bi} = b \cdot \int_{mi}^{ni} \sigma_{bi}(x)dx\ , \tag{8}$$

where $b$ – width of the rectangular section of the bending element; $\sigma_b(x)$ – function of shift in stresses in concrete (9); $m_i$ *and* $n_i$ – are the lower and upper limits of integration corresponding to the lower and upper boundaries of the corresponding selected layers in the compressed section area.

$$\sigma_b(x) = a_c \cdot \left(\frac{\varepsilon_b \cdot x}{k}\right)^{b_c} \cdot \exp\left(\frac{-b_c \cdot \left(\varepsilon_b \cdot \frac{x}{k}\right)}{p}\right)\ , \tag{9}$$

where $\sigma_b$ – the stress in the concrete; $\varepsilon_b$ – relative deformation of the most compressed face of the element resulting from the reparsing; $x$ –the coordinate of the compressed concrete zone point in question; ; $k$ – the full height of the compressed concrete area resulting from the reparsing; $a_c$ and $b_c$ – coefficients of exponential function $f(\varepsilon)= \varepsilon\cdot E$, which describes the shift of the elasticity module from magnitude $\varepsilon$; $p$ – relative deformation of concrete in the chart f the maximum value of the stresses.

Next we define the eccentricity $e$ from equation (10):

$$N_s'(e-h/2+a)+N_b(e+h/2-a-S_b/A_b-i)-N_s(e+h/2-a)=0 \tag{10}$$

The equation of moments of external and internal forces is drawn relative to the center of gravity of the section of tensile reinforcement (11):

$$\sum M=0:\ N_b\left(\frac{S_b}{A_b}+i\right)+N`_s(h_0-a`)-N(e+h/2-a)=0 \quad , \tag{11}$$

where $Z_2^*$ – a shoulder of $N_b$ force as regard to the center of gravity of stretched reinforcement; $Z_3^*$ – a shoulder of effort $N_s'$ as regard to the center of gravity of tensile reinforcement;$Z_1$ – a shoulder of $N$ force as regard to the center of gravity of stretched reinforcement.

To determine the distance between the center of gravity of stretched reinforcement and the application point of the resultant force$N_b$ we first found a static moment of concrete stress diagram and its area.

The area of the diagram of a compressed area of concrete is determined according to formula (12):

$$A_b=\sum_{i=1}^{n}A_{bi}\,, \tag{12}$$

where $A_{bi}$ – the area of a diagram of the i-th layer determined by formula (13).

$$A_{bi}=\int_{mi}^{ni}\sigma_{bi}(x)dx \quad , \tag{13}$$

where $\sigma_b(x)$ is the function of changing stresses in concrete (7);$m_i$ *and* $n_i$ – are the lower and upper limits of integration corresponding to the lower and upper bounds of the corresponding selected layers in the compressed section area.

The static moment of the diagram of a compressed area of concrete is determined according to formula (14):

$$S_b=\sum_{i=1}^{n}S_{bi}\,, \tag{14}$$

where $A_{bi}$ – the area of a diagram of the i-th layer determined by formula (1).

$$S_{bi}=\int_{mi}^{ni}\sigma_{bi}(x)xdx\,, \tag{15}$$

Then the shoulder of force $N_b$ is to be calculated by formula (16):

$$Z_2' = i + \frac{S_b}{A_b}. \tag{16}$$

The calculation algorithm is given below.

The basic data are: $b$, $h$ – an actual width and height of the calculation section; $h_0$ – working height of the section; $a_{cj}$, $b_{cj}$ – coefficients of analytical dependence describing stress and strain state of concrete with relatively homogeneous layer; $P_j$ – relative deformation of concrete corresponding to the stress in concrete of equal to $R_b$ in a separately chosen layer; $E_{bj}$ – a starting module of concrete elasticity; $A_s$, $R_s$, $E_s$ – area, design resistance, and elasticity module of working reinforcement; shoulders $Z_1$, $Z_2$*, $Z_3$*, external force moment $M_i$, force in concrete $N_b$ and reinforcement $N_s$ and $N_s$\`.

Cycle 1. During this cycle, the values are set sequentially from 0 to 0.0055, to the value of $\varepsilon_{b,max}$, where ($\varepsilon_{b,max}$ – deformation of concrete at the most compressed part.)

Cycle 2 (a cycle in Cycle 1). During this cycle, we successively assign to the variable i values from -100 to $h_0$, где i, where $i$ is the distance from the center of gravity of the section of stretched reinforcement to the zero line.

Altitude of compressive zone of concrete is determined by the formula: $X_i=h_0-i$

By assigning soma value to the values $I$ and $\varepsilon_{b,\ max}$ determine deformations:

- in compressed reinforcement: $\varepsilon`_s = \varepsilon_{b,\max}(h_0 - i - a`)/(h_0 - i)$
- in stretched (least compressed) reinforcement: $\varepsilon_s = (\varepsilon_{b,\max} - i)/(h_0 - i)$.

Determination of $N_s$ force: if $0 \le \varepsilon_s \le R_s/E_s$ , then $N_s=E_s \cdot \varepsilon_s \cdot A_s$, otherwise $N_s=A_s \cdot R_s$.

Determination of $N_s'$ force: if $0 \le \varepsilon_s \le R_{sc}/E_s$, then $N_s'=E_s \cdot \varepsilon_s' \cdot A_s'$, otherwise $N_s'=A_s' \cdot R_{sc}$.

Determination of the number of layers$i$, that are in the compressed section area.

Determination of $N_b$ force by formula (8).

Determination of $N$ force $N=N_s'+N_b-N_s$

Eccentricity $e$ can be defined from equation (10).

The condition of breaking cycle 2 is calculated by an equation (11).

Then intermediate values are calculated, which are the solution of internal forces equation with some value of $\varepsilon_{b,max}$, which goes as follows:

1. The area of the diagram of a compressed area of concrete is determined according to equation (12):
2. The static moment of the diagram of a compressed area of concrete is determined according to equation (14):
3. $N$-force eccentricity is : $e$
4. Force $N$
5. Shoulders $Z_1$; $Z_2$*;$Z_3$*
6. Moment of external force $M_i=N \cdot e$.
7. Force in concrete $N_b$ and reinforcement$N_s$ and $N_s$'.
8. The values of relative deformations in the most compressed face of concrete $\varepsilon_{bi}$ and at the center of gravity of the cross-section of stretched and compressed reinforcement $\varepsilon_{si}$ and $\varepsilon_{si}$'.

Interruption of Cycle 1 occurs when the following inequality is performed: $M_i > M_i+1$.

The results of the calculations are presented as a table of values for all intermediate solutions.

## 4 Conclusions

The paper provides a methodology for determining the capacity of compressed reinforced concrete elements, including off-centered compressed elements, with varying strengths of concrete layers based on the deformation model and non-linear material deformation diagrams (concrete and steel). The paper also introduces a step-by-step algorithm for estimating stress and strain state of compressed reinforced concrete elements with symmetrical reinforcement, with the possibility of taking into account the heterogeneous strength properties of concrete in a compressed zone. The solution of the problem is shown as a data table. Analysis of this data table allows to estimate an expected pattern of the samples destruction.

The results of this scientific research are recommended for practical application in design organizations and can be used to improve existing regulatory documents for calculating reinforced concrete elements.

## References

1. G.V. Murashkin, V.G. Murashkin, News of Higher Educational Institutions. Construction **10**, 4-6 (1997)
2. D.A. Panfilov, A.A. Pishchulev, K.I. Gimadetdinov, Industrial and civil engineering **3**, 80-84 (2014)
3. V.O. Krichke, G.V. Murashkin, Yu.V. Volkov, D.V. Goryainov, RF Patent 2179722. Means of analysing structure and strength of concrete in the process of sampling from concrete structures by drilling, Ann. 6 March 2000 (Published 20.02.2002)
4. V.O. Krichke, G.V. Murashkin, Yu.V. Volkov, RF Patent 2198399. Information system for analysing structure and strength of concrete in the process of sampling from concrete structures by drilling, Ann.10 March 2000 (Published 10.02.2003)
5. V.B. Gavrilov, *Assessment of mechanical and structural characteristics of concrete in building constructions by a method of local destruction* (2004).
6. Valery B. Filatov, Alexander A. Suvorov, Procedia Engineering **153**, 144-150 (2016)
7. E.M. Rudakov, A.E. Frolov, Urban Construction and Architecture **3**, 98-102 (2013).
8. A.N. Ilyin, D.A. Panfilov, D.V. Litvinov, P.N. Slavkin, Urban Construction and Architecture **1**, 82-89 (2015).

# Methodology of numerical and experimental research of wind effects on high-rise complex

*Olga* Poddaeva[1] , *Julia* Gribach[1,*]

[1]Moscow State University Of Civil Engineering 129337, 26, Yaroslavskoe shosse, Moscow, Russian Federation

**Abstract.** Nowadays there is a tendency for the requirement toughening when the construction projects are designed. The existing standard and rule control in terms of the identification of the external load on buildings and constructions is directly connected with the reliability and safety control of building constructions, and one of the most important parts of the design is to take into account the wind effect. Air flows have a different impact on buildings due to their height; it is connected with the surface air existence. Consequently, the wind effect on the high-rise should be well-researched. The acting normative documents (SP 20.13330.2011 "Loads and impacts") formalize the inclusion of a special evaluation study of the wind conditions and aerodynamic property in the list of project documentation for the reliability and safety control of the high-rise building constructions. The identification of the wind conditions and aerodynamic property is available only in accordance with the theoretical and experimental analysis, of which a procedure is described in the present article.

## 1 Introduction

The development and progression of engineering of the construction provide a means of the new off-standard space planning decision creation and improve the activity of designers and architects. The high-rise building is not exception: today in most major cities of the world construction of skyscrapers is first ranks in the entire construction complex [1]. The buildings surpassing the height of 100 m can save the territory and also give opportunities to create the unique architectural masterpieces (fig. 1).

**Fig. 1.** High-rise buildings in Moscow (Moscow City)

* Corresponding author: js-995@mail.ru

However, despite the technological progress and the opportunity to create the unique structures, the main problem of architects and engineers, while developing the design documentation, is the procuring control of reliability and safety of constructions. First of all, this requirement applies to the high-rise, because high-rise buildings belong to the class of increased danger.

Based on experience, which gained of the monitoring of high-rise buildings, it was made a conclusion of the necessity of wind effect [2]. Firstly, construction of the skyscraper changes the surface air circulation conditions [3], what produces changes of climate near the building. Secondly, the building itself is adversely affected by the alternate wind load, which cause deformation of the high-rise building and destruction.

The hard research of the building design features and the evaluation of the area climate references encourage the improvement of reliability and safety of the skyscrapers while being affected by the wind load. The aerodynamics of high-rise buildings has own specificity. The coexisting of several factors of the wind effect impact on the structural analysis and calculations takes an important problem for architects and engineers: the lack of recommendations on the shape factor designation [4].

The acting normative documents (SP 20.13330.2011 "Loads and impacts") formalize the inclusion of a special evaluation study of the wind conditions and aerodynamic property in the list of project documentation for the reliability and safety control of high-rise building constructions. However, the all determination is available only by the test of the high-rise building layout in the architectural wind tunnels.

The aim of the present work is to describe the procedure of theoretical and experimental analysis of the wind effect on the high-rise buildings and constructions for their reliability and safety, which was produced in the laboratory of MGSU, and to approbate it on the example of the high-rise complex, which is situated in Moscow.

## 2 Methodology

The present work has the procedure of theoretical and experimental analysis of the wind effect, which was produced and approbated at the Educational research and production laboratory of aerodynamic tests of building construction of MGSU.

To realize the theoretical and experimental analysis of the wind effect on the high-rise complex it is necessary to implement sequentially the following steps:

- The climate terrain analysis, including the topographical and meteorological data analysis: the nature of the average wind speed change in level, the atmospheric turbulence structure, the average speed and turbulence dependence on the underlying surface roughness of the site are determined to evaluate the wind load.
- The designing 3D-model of the construction, the elaboration and layout with scale down, including the assessment of the construction project geometrical adjectives, the installation of the measuring equipment to determine the wind loads.
- The experimental research in the specialized wind tunnel and the processing of the obtained results.
- A series of numerical testing on the analytic model in the certified software (MGSU use software package ANSYS).
- The verification results of the theoretical and experimental analysis. In this case the allowable discrepancy mustn't exceed 15-20%, otherwise the numerical modeling is repeated with the modified parameters of the grid and bounding actions until the acceptable convergence is achieved.
- The data analysis and the formulation of the conclusions and recommendations for ensure the structural reliability and safety.

The experimental research realize at the Educational research and production laboratory of aerodynamic and aero acoustic tests of building construction of MGSU. The large gradient wind tunnel (fig. 2) of architectural type with 18,9 m long work space, use as a test rig. It allowed to simulate the surface atmosphere layer.

**Fig. 2.** Large gradient wind tunnel of laboratory of MGSU.

Moreover, for the aerodynamic researches the following equipment is being used:

• Integrated differential pressure sensors - MPXV7002DP. POP - from -2 to 2 kPa. The error after calibration is less than 0.5%.

• Pitot pressure modification tube, designed to measure the volumetric flow rate of liquid and gas at one point of the cross section of the channel with an equivalent diameter of at least 300 mm. It's used in determining the speed and volume flow in gas flues and ventilation systems complete with differential manometers.

• Differential digital differential pressure gauge DMC-01M with data processing is a device for measuring pressure, rarefaction and gas pressure difference, as well as for determining the speed and flow rate of gas and dust and air streams using Pitot pressure modification tubes.

• Six-component sensors for measuring the forces and moments of Schunk FTD. Main technical characteristics: measuring range Fx, Fy: ± 1000 N, measurement range Fz: ± 2500 N, measuring range Mx, My: ± 120 Nm, measuring range Mz: ± 120 Nm, accuracy for measuring Fx , Fy, Fz: ± 1/16 N, accuracy for measuring Mx, My: ± 1/160 N*m, accuracy for measuring Mz: ± 1/320 N*m.

## 3 Approbation

The approbation of methodology is made using an example of a projected residential complex, located in the south of Moscow (fig. 4,a). The project includes 4 various-storey residential towers (height is 173 and 300 m) and two-storey non-residential premises built into the stylobate. The towers are located along the perimeter of the site, forming a geometric arrangement with a individual well-maintained precinct.

As previously noted, primarily carried out a climate terrain analysis. A type of terrain is chosen in virtue of climate analysis under the requisition of SP 20.13330.2011 "Loads and impacts", and it's rated a relevant frontal velocity profile [5]. In this case, the type of development site corresponds to the type "B", a fig. 3 shows the velocity profile obtained in a wind tunnel.

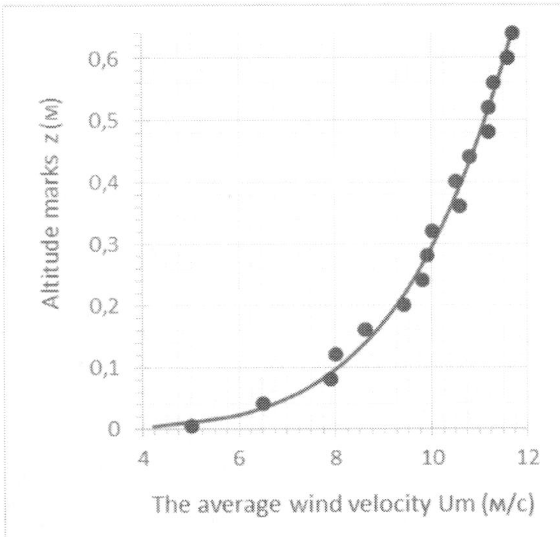


**Fig.3**. Dependency graph of average wind speed from altitude (for the "B" terrain type)

There were designed 3D-models of high building for the accomplishment of a number of computational tests (fig. 4, b) and made layout for experimental studies. The scale of the layout chosen from the conditions of adherence to the established limitations for this pipe (in this case, It is 1:250) (fig. 4, c) (GOST-R "Russian National Standard" 56728-2015. Buildings and constructions. Methods of determining the wind loads on enclosing structures). The drainage system installed to measure the velocity of flow and pressure at distinguished points (fig. 4, d). The pressure transferred through the holes in copper and silicone tubes to differential pressure sensors.

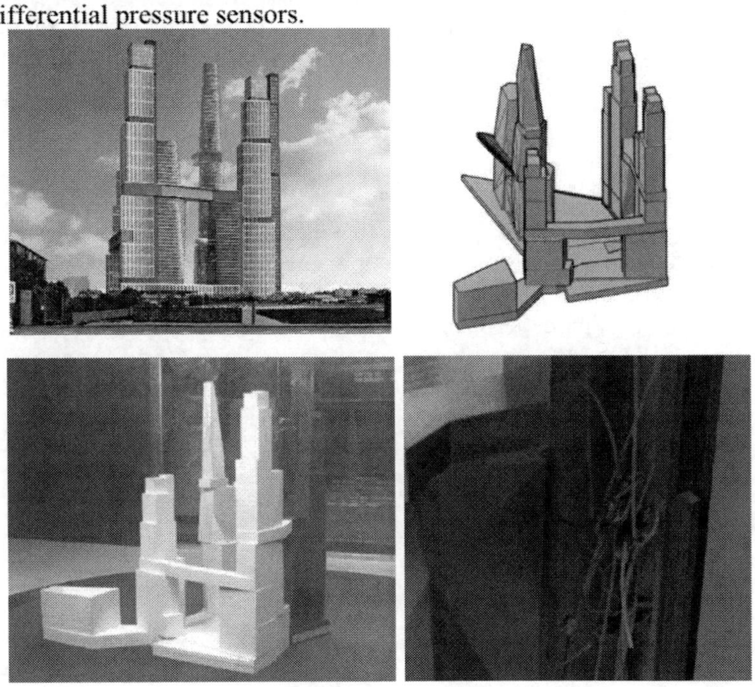

**Fig. 4.** Projected residential compound (a), 3D-models (b), a layout (c) mounting of a tube transfer system (d)

The model of the explored complex installed on a specialized turntable in the wind tunnel workspace (fig. 5). After that, the tunnel is started up and the air flow is supplied to the model. Specific airflow velocity in the pipe is chosen to be sufficient for providing self-similarity according to the Reynolds number [6], when further velocity increase does not lead to a significant change of averaged values of shape factor in control points on the layout (GOST-R "Russian National Standard" 56728-2015). In this test, the wind speed in the work space was 15 m/s.

**Fig. 5.** Model of the explored complex in the wind tunnel workspace

## 4 Results and discussion

The result of the tests is the pattern of pressure distribution over the surface of the layout. According to the received data, it is possible to calculate the average values of the pressure shape factor for different angles of wind attack, from 0° to 360°, using formula (1).

$$C_p=(2*p)/(\rho*v^2), \quad (1)$$

where $C_p$ is a coefficient of pressure;

p is a pressure (obtained by the experimental method);

ρ is a density of air (at a temperature of +20° ρ = 1,2041 kg/m³);

υ is a speed of air (15 m/s).

In the same way it is possible to calculate peak values of pressure shape factor, necessary for calculating the peak component of the wind load on enclosing structures and facade systems.

Also during the experimental studies using 6-component strain-gage weigher and software ATI DAQ F/T can identify total forces $F_x$, $F_y$ and $M_z$ and a torque along the axes X, Y and Z, can also calculate total shape factors of resistance to axes X, Y and Z, $C_x$, $C_y$ and $CM_z$ on which can make a shape factor dependency graph $C_x$, $C_y$, $CM_z$ with a change in the angle of attack from 0° to 345° (fig. 6).

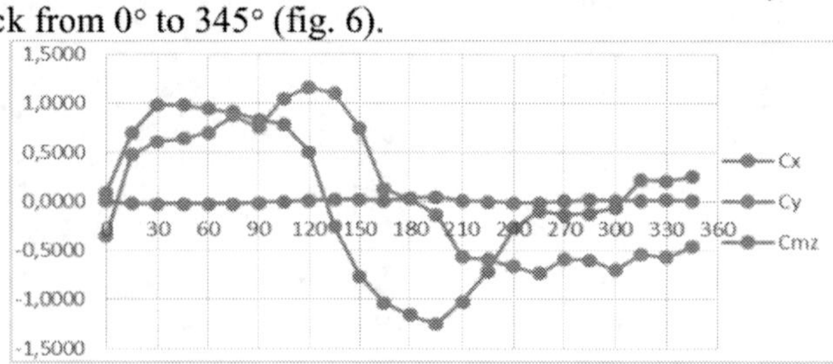


**Fig. 6.** Shape factor dependency graph Cx, Cy, Cmz with a change in the angle of attack from 0° to 345°

Numerical simulation of the flow around the building is carried out in parallel with the implementation of experimental studies to clarify wind load data. Performing computer calculations it is computed the unsteady flow and the distribution of the dimensionless pressure coefficient Cp by the facades along the perimeter of the horizontal section of the building and height along the building [7]. For the numeral studies realization it is necessary to install computational grid with the necessary grinding (fig. 7) and establish the boundary conditions.

**Fig. 7**. General view of computational grid

Reliability of the results of numerical simulation is confirmed by the use of a certified software bundle ANSYS and by the results of the verification tests. In this project it showed good consistency of the calculated data with the experimental data (the discrepancy of the data wasn't more than 20%).

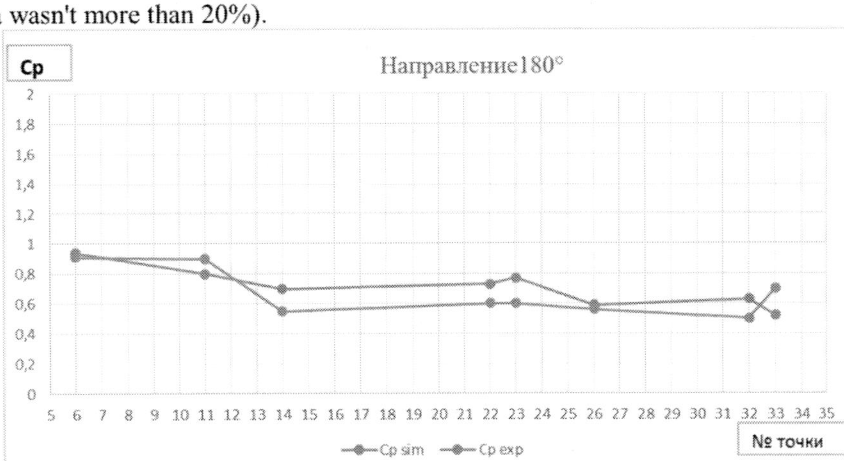


**Fig 8.** Verification test results

Fig. 9 shows the distribution patterns of the average pressure shape factor on the facades of high-rise buildings. Numerical modeling provides more complete information, which is necessary for the wind load calculation (the results of experimental studies are distinct states at points on the layout surface) [8], it also identifies all problem areas.

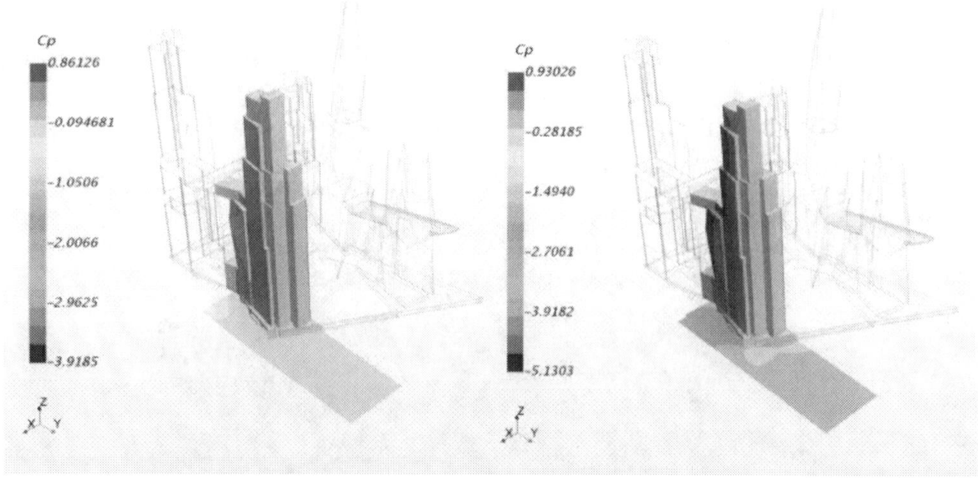


**Fig. 9.** Distribution of the average pressure shape factor on the facades of high-rise buildings of the explored residential complex for 0° and 45°

## 5 Conclusions

A methodology of numerical and experimental research of wind effects on high-rise complex has been developed.

In the framework of approbation of the developed methodology, the following results were obtained: patterns of wind pressure distribution along building facades, average and peak values of aerodynamic coefficients, and dimensionless aerodynamic coefficients.

Due to the obtained results, it can be asserted that the average values of the pressure shape factor are approximately in the range ± 1,5. At the same time, there are zones of increased pressure at the corner areas of facade structures with the peak values: 3 – 3,5.

The values of the dimensionless shape factor Cx, Cy, Cmz are defined relative to the center of the base of the explored object, the maximum values of the coefficient Cx arise at the blowing angles 30°and 195° (Cx=0,98 and Cx=-1.24). In general, the obtained values are well correlated with the shape factor values, which are considered in SP 20.13330.2011 "Loads and impacts", and do not exceed them, so the need to develop specialized recommendations for the redesign is excluded [9, 10].

Using the values of the shape factor obtained during the processing of calculated and experimental data values, the designer can independently determine the average and peak values of the wind load per unit area of the building facade in accordance with the requirements of SP 20.13330.2011 "Loads and impacts".

It should be noted that a multifaceted approach is important to the solution of project tasks that ensure safety and reliability of high-rise buildings and constructions. It should include the work of both engineers and scientists. Currently, there is no other system that simplifies the implementation of this procedure, because the design of unique buildings with a height more than 100 m is quite a challenge [11].

## References

1. O.I. Poddaeva, Y.S. Buslaeva, D.S. Gribach, Advanced Materials and Engineering Materials IV, 246-249 (2015)

2. O.I. Poddaeva, Y.S. Buslaeva, D.S. Gribach, Newsletter of the BSTU (Belarusian State Technological University) V.G. Shukhova, **6**, 58-62 (2014)
3. O.O. Egorychev, P.S. Churin,Housing construction, **6**, 20-22 (2015)
4. O.I. Poddaeva, I.V. Dunichkin,S.A. Andreeva, Materials of the international practical conference. Effective organizational and technological solutions and energy-saving technologies. Kharkov, 64-66 (2014)
5. O.I. Poddaeva, Industrial and civil construction. **9,** 35-38 (2013)
6. D. Čehel'ová, M. Janák, B. Bielek, Applied Mechanics & Materials. **820** (2016)
7. O.O. Egorychev, S.I. Dubinsky, A.N. Fedosova, Applied Mechanics and Materials, **713-715**, 1729 (2015)
8. S.V. Gouvernyuk, V.G. Gagarin, Journal "ABOK" (Russian Association of Engineers for Heating, Ventilation, Air-Conditioning, Heat Supply and Building Thermal Physics). 16-22 (2007)
9. Q. Yong, G. Ming, T. Yukio, Wind and Structures, **8, 5**, 309-324 (2005)
10. H. Dongmei, Z. Ledong, D. Quanshun, Z. Xue, C. Wei, Journal of Fluids and Structures, **69**, 355-381 (2017)
*11.* A. Stepek, I.L. Wijnant, *Interpolating wind speed normals from the sparse Dutch network to a high resolution grid using local roughness from land use maps // KoninklijkNederlandsMeteorologischInstituut, Technical Report TR-321.* – (2011).
12. A. Stoffelen, J. Vogelzang,. Wind bias correction guide // KNMI, Version. – (2011)
13. P. Irwin, R. Denoon, D. Scott D. *Wind Tunnel Testing of High-Rise Buildings. //* Routledge, (2013)
14. F. Fan, et al., PowerTech, 2015 IEEE Eindhoven, IEEE, 1-6 (2015)
15. A. Allik, et al., Agronomy Research, **12, 2,** 433-444 (2014)

# The use of thermodynamic characteristics when selecting fire retardants

*Elena* Pokrovskaya and *Feodor* Portnov*

Moscow State University of Civil Engineering, Jaroslavskoe Shosse 26, Moscow, Russian Federation

**Abstract.** In the present study we consider the modification of wood by phosphites to reduce flammable properties. We define thermodynamic parameters of modified wood surface. The study explains high values of free surface energy while studying the surface structure and the surface layer of modified wood. Based on the findings, the possibility of thermodynamic optimization of modifiers upon development of protective compositions is shown.

**Keywords.**
Chemical modification, thermodynamic characteristics, phosphites, thermodynamic optimization, fire resistance

## Introduction

Modification is an effective way to improve performance of building materials, including wood [1-4]. There are mechanical, thermal and chemical modification. Technologically the simplest and most available is surface chemical modification, which allows to reduce fire risk of structures, hydrophilic properties, to create a high biocorrosion resistance. The effectiveness of surface chemical modification is determined by the selection of a modifier, which reacting with wood components causes the durability of the acquired performance properties [5-6].

For effective modification the selection of modifier is the most important and difficult task. It was natural, that with surface modification, surface properties affect the efficiency of the modification process. Energy properties are characterized by changes in the following thermodynamic characteristics: the contact angle, the surface tension, and the change in Gibbs thermodynamic potential of the surface during the modification.

The surface energy intensity factor is surface tension [7]. The dynamic analysis of Gibbs thermodynamic potential of the modified surface is based on the combined I and II principles of thermodynamics [8-12].

$$\Delta G = \sigma dS_{sp} + S_{sp} d\sigma$$

The study of surface modification of wood makes it possible to estimate in advance the effectiveness of surface layer modifiers using thermodynamic quantities ($\Delta G, \sigma$).

* Corresponding author: wastingtimefilmart@gmail.com

Selecting the optimum characteristics allowing to estimate the efficiency of the substrate-modifier system is most rational for the selection of modifiers reducing the fire-hazardous properties of materials. That is the **aim** of the study.

To achieve the aim we set an objective to study the changes of the energy characteristics of the wood surface by the example of wood modification with phosphites, as well as the study of the thermodynamic optimization of modifiers reducing the fire-hazardous properties of materials.

## Research methods

The following experimental methods are used to estimate the characteristics of modified wood.

The chemical analysis is performed using a Quanta 200 scanning microscope with an Apollo 40 attachment for element analysis using energy dispersive spectroscopy [13]. During the tests the phosphorus percentage in the samples of modified wood was estimated. The Quanta 200 microscope makes it possible to obtain high-definition images with magnification exceeding 100,000 times [14,15].

Data on the contact angle $\theta$ of modified wood samples are obtained by the method of a neutral drop on the Easy Drop DSA25S installation [16]. The Easy Drop installation consists of a sample holder, a liquid feeding gun mechanism and a high-definition camera. During the tests, the mechanism ensures the supply of a liquid drop and the camera provides a fixation of the moment of liquid drop contact with the sample.

Quantachrome NOVA 4200e device [17] and the Brewer-Emmett-Teller (BET) method, which measures the mass of the gas adsorbed by the sample after degassing [18], are used to determine the specific surface of the initial and modified wood.

The valuation of the protective action of the composition is performed by the method determining the group of fire-rating effectiveness of fire retardant and by the method of experimental determination of the smoke-developed index of solids and materials [19].

## Results and discussion

The determination of thermodynamic characteristics is carried out on samples of pine sapwood in size of 40x40x10 mm, the moisture content of the samples is 8%. The 20% solutions of phosphites (dimethylphosphite (DMP), diethylphosphite (DEP), dipropylphosphite (DPP), dibutylphosphite (DBP), diphenylphosphite (DFP)) are used as modifiers. The literature notes an increase in the fire resistance of wood when modifying it with the indicated modifiers [3-4].

The objective of the study is to determine the change in Gibbs thermodynamic potential of the modified wood surface depending on the modifiers of the surface layer, and also to determine the relationship of the energy characteristics to the physico-chemical properties of the modified samples.

The modifiers are applied on the sample surface in two layers at temperatures of 20-25 ° C, the amount of modifiers used is 300 g / m$^2$. To remove the excess amount of a modifier not reacting with the surface, the samples are extracted in Soxhlet apparatus (solvent - acetone, water). The samples are dried in a drying oven at 40 ° C for 6 hours, after which the samples are placed in a desiccator over anhydrous $CaCl_2$ and dried at 25 ° C for 7 days.

The degree of chemical interaction of the modifiers in the wood surface layer is determined by the percentage of phosphorus in the samples after extraction. The % P data is shown in Table 1 for various modifiers.

**Table 1.** The content of phosphorus in the wood surface layer.

| | Surface layer modifier | | | | |
|---|---|---|---|---|---|
| | **DMP** | **DEP** | **DPP** | **DBP** | **DFP** |
| **%P of thermal destruction** | **2.87** | **3.57** | **2.6** | **2.65** | **2.45** |

The surface tension σ is determined by the neutral drop method [9, 10] at the EasyDrop installation and the corresponding software. The values $S_{sp}$ of wood samples are obtained using NOVA Chrome 2200 installation by the nitrogen sorption method [16, 17, 18] (Table 2).

We obtain the dependence cosθ = f (σ) of the contact angle on the standard surface tension of a drop of the aqueous ethanol solutions of various concentrations in the form of an averaged straight line using the EasyDrop installation.

Extending the line equation to cosθ = 1, it is possible to determine the critical value of the surface tension σ, which is a characteristic of the surface energy of a surface unit (Table 2)

**Table 2.** The values σ and $S_{sp}$ for different modifiers.

| | | Surface layer modifier | | | | |
|---|---|---|---|---|---|---|
| | **Initial wood** | **DMP** | **DEF** | **DPP** | **DBP** | **DFP** |
| $\sigma \cdot 10^3$,N/m | **25.33** | **19.91** | **25.03** | **21.97** | **24.19** | **23.03** |
| $S_{sp}$ | **42.52** | **58.56** | **17.081** | **71.49** | **70.37** | **18.289** |

The surface of modified wood samples is also studied by electronic microscopy using a scanning electron microscope JSM-840 [14, 15]. The images of modified wood surface with an increase of 300 times are shown in Figure 3.

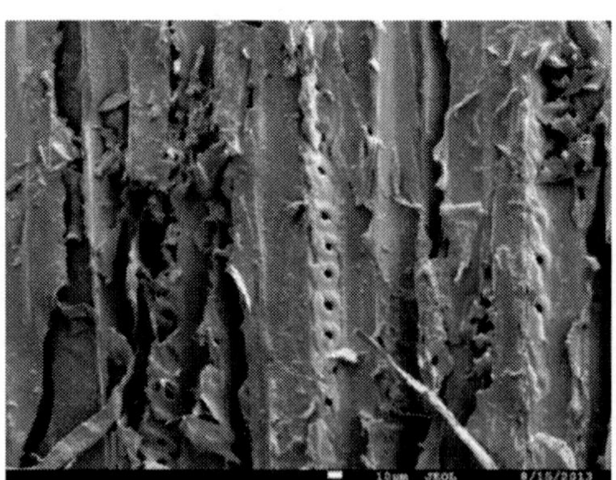

a

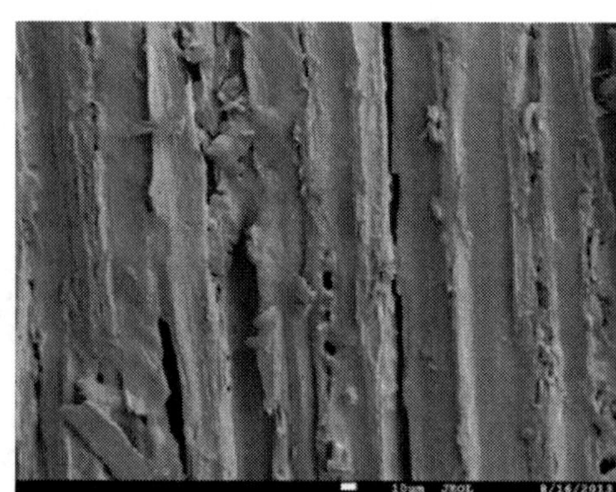

b

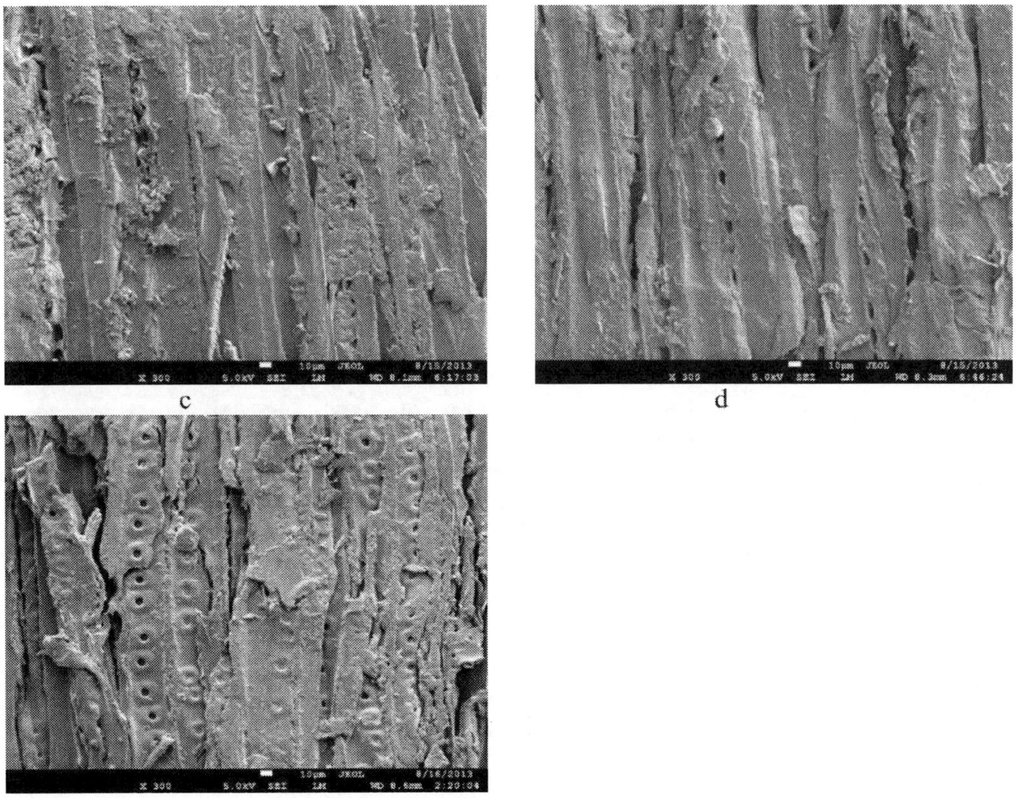


**Figure 3.** The electron microscope images of the wood surface, modified by dimethylphosphite (a), diethyl phosphite (b), dipropyl phosphite (c), dibutyl phosphite (d), diisopropylfluorophosphat (e), are obtained by electronic microscopy.

As can be seen from the obtained images, the wood surface varies depending on the value of the OR-group alkyl in phosphites. Using modifiers with the radicals $-C_3H_7$, $-C_4H_9$, $-C_6H_5$, a complete covering of the wood surface takes place (dipropyl phosphite, dibutyl phosphite, diisopropylfluorophosphate).

Granting smaller radicals $-CH_3$, $-C_2H_5$, the oriented structure covering occurs and these modifiers (dimethylphosphite, diethyl phosphite) are characterized by the decrease of the specific surface, which is possible when covering the inner surface of the capillaries. By effective chemical modification, the walls of the capillaries are strengthened with a modifier, which results in pore size decrease and adsorption decrease.

Data on the smoke-forming capacity of wood, fire resistance [19,20], and water vapor sorption [21-23] are obtained to assess the operational characteristics of the initial and modified wood (Table 3).

**Table 3.** Fire-hazardous properties and water vapor sorption of the initial and modified wood.

| | **Initial wood** | **Surface layer modifier** | | | | |
|---|---|---|---|---|---|---|
| | | **DMP** | **DEF** | **DPP** | **DBP** | **DFP** |
| **$\Delta G$, J** | **-** | **-12.68** | **-14.5** | **-4.94** | **-6.97** | **-10.16** |
| **$\Delta m$%** | **79.0** | **6.9** | **8.5** | **12.3** | **11.5** | **9.5** |
| **$a_m$, %** | **421.1** | **215.8** | **155.5** | **221.9** | **193.4** | **163.52** |

| | | | | | | |
|---|---|---|---|---|---|---|
| **Dm, m²/kg** | **1010-1050** | **450-470** | **130-140** | **640-680** | **600-640** | **740-780** |

Based on the obtained data of $S_{sp}$ and σ (Table 2), Gibbs thermodynamic potential of the modified sample surface is calculated (Table 3)

Table 3 shows that all values ΔG < 0, this characterizes the inadvertent process of surface modification in these conditions. The most negative values of Gibbs thermodynamic potential variation of the modified wood surface are peculiar in case of using diethyl phosphite as a modifier, which largely interacts chemically with the lignin-carbohydrate complex. The thermodynamic characteristics of the modification depend on the degree of chemical interaction of the substrate with the modifier.

## Conclusion

With a surface modification of wood by 20% solutions of phosphites, an inadvertent chemical interaction of phosphorus-containing compounds with the wood substrate occurs. The greatest degree of chemical interaction is peculiar for diethyl phosphite, the smallest for diisopropylfluorophosphate. However, the wood surface is covering with modifier by diethyl phosphite modification. In addition, based on $S_{sp}$ data, we can conclude on the chemical modification of the internal part of the capillaries. Strengthening of the surface and capillary structure during chemical modification with phosphorus-containing compounds leads to the reduction of fire risk and smoke-forming ability of wood.

All the data obtained are correlated with the change in Gibbs thermodynamic potential (ΔG). The most negative values ΔG characterize the surface of wood materials with lower values of fire risk.

Thus, we defined the method of thermodynamic optimization of wood with the aim to improve its operational characteristics, which allows to select the most effective modifiers by value ΔG. Based on the availability of the present method, we can talk about its great practical importance. The prospect of further study is the evaluation of the effectiveness of other classes of modifiers based on the present method.

## References

1. E.N. Pokrovskaya, F.A. Portnov, A.A. Kobelev, D.A. Korolchenko, Fire and Explosion Safety, **10**, 40-46 (2013)
2. A.A Stenin, *Construction - the formation of the environment of life*, 553-559 (2013)
3. V.A. Shamaev, *International Scientific and Technical Conference "Modern Problems of Mechanical Wood Technology"*, 11-17 (2010)
4. E.U. Tariva, *VI International Scientific and Technical Conference "Actual problems of construction and construction industry"*, 55-56 (2005)
5. E. N. Pokrovskaya, *Preservation of monuments of wooden architecture with the help of organoelement compounds* (2009)
6. E. N. Pokrovskaya, *Chemical and physical basis for increasing the longevity of wood* (2003)
7. V.I. Roldugin, *Physicochemistry of the surface* (2011)
8. A.S. Tutygin, A.A. Shinkaruk, A.M. Aisenstadt, V.S. Lesovik, *J. of International* Scientific Publications: Ecology & Safety, **8**, 54-61 (2014)
9. A.S. Tutygin, A.A. Shinkaruk, A.M. Aisenstadt, M.F. Frolova, T.A. Pospelova., J. of International Scientific Publications: Ecology & Safety, **7**, 37-45 (2013)
10. A.M. Aisenstadt, *Innovative materials and technologies for construction in extreme climatic conditions*, 244 (2014)

11. B.V. Deryagin, I.I. Abrikosova, E.M. Lifshits, Successes in physical sciences, **3**, 494-526 (1958)
12. B.V. Deryagin, N.V. Curaev, *Wetting membranes* (1984)
13. P. Dzh. Gudkh'yu, *Practical Methods in Electron Microscopy* (1980)
14. Dzh. Gouldsteyn *Practical raster microscopy* (1978)
15. B.A. Kalin, *Raster electron microscopy. Laboratory work* (2008)
16. A.A. Abramzon, L.Ye. Bobrova, L.P. Zaychenko, *Surface phenomena and surface-active substances: handbook* (1984)
17. A.S. Vyacheslavov, Ye.A. Pomerantseva, *Measurement of surface area and porosity by the method of capillary nitrogen condensation (preparation of masters - operators of modern scientific equipment): Methodological development* (2006)
18. YU.A. Podkamennyy, A.A. Nosenko *Innovative method for determining the specific surface area of disperse systems and porous materials.* Available at: http://conf.sfu-kras.ru/sites/mn2013/section060.html (2013)
19. Interstate Standard 53292-2009. Fire protection compounds and substances for wood and materials based on it. General requirements. Test methods (2007)
20. Interstate Standard 12.1.044–89*. Occupational safety standards system. Fire and explosion hazard of substances and materials. Nomenclature of indices and methods of their determination (1989)
21. N.V. Kel'tsev *Basics of adsorption technique* (1984)
22. S. Greg, K. Sing *Adsorption. Specific surface area. Porosity* (1984)
23. D. Ruthven *Principles of Adsorption and Adsorption Processes* (1984)

# The results of the first periodical testing vent-panels (VP)

*Yuri* Polandov, *Denis* Kukin*

Moscow State University of Civil Engineering, Jaroslavskoe Shosse 26, Moscow, Russian Federation

**Abstract.** The article is devoted to analysis of obtained during testing of sandwich panels and insulating glass as VP results. Here are the details on the first periodic test VP method GOST R 56289-2014 «Design translucent legkousvaivaemyh for buildings» using measuring instruments and testing equipment according to this standard, and, accordingly, the results of tests with additional observations obtained during the research.

## 1. Introduction

In recent years in construction are widely used lightweight wall materials, such as sandwich panels and glass units (hereinafter the Panel). Faced with a clear need for use as VP. However, we know the effect of material density VP on its effectiveness as a means of protection in gas explosions at the premises: the greater the density, the higher the pressure the explosion. Obviously, that is confirmed by the Standard certification is necessary in such cases such as VP remedies. In civil engineering at the Institute for integrated security for the first time in Russia there are all conditions for production acceptance and periodic tests such VP. In the article the results of Russia's first periodic tests such VP.

It should be noted that the Russian demands to VP does not deal with sandwich panels. For example, in the rules [1] considered requirements for Windows made of glass, so they specify minimum window size and requirement of single of their covers. But it should be noted that these Regulations apply only to the boiler. Another famous document, such as a standard [2], contains requirements for only one VP class, namely to devices of translucent materials (DTM), i.e. to Windows and modern so called restful. Despite the fact that it discussed in detail various aspects of the technical requirements to this class of VP, wonderfully issue limiting their density does not pay attention. Even in one document Standard [3], describes this type of VP test conditions (i.e., from DTM) for triggering at a given pressure, the conditions according to requirement should be close to reality. The dignity of this standard include what he, in spite of its name, and formally against only the VP from DTM, methodology and tools described in it have substantial community and can be applied to all known types of VP.

It follows from this that, at this stage, the question of the applicability of the sandwich panels as VP should be determined in each case, the designers. It is clear that, as a result of obtaining evidence of their effectiveness, obtained by direct testing, for example, according to [3].

The number of research papers in Russia regarding the topics raised is very limited. You can mark only 2 work, this dissertation (PhD) [4-7], which dealt with mathematical models and experimental results of the development process of gas explosion at operation VP. Unfortunately, the first job you can note the lack of skilled works on VP, close in size to the actual, and in the second-it can be seen that the results of an experiment on SP real size seem unconvincing.

In Europe and the United States panels have been used earlier than we have. To date, there are enough serious conducted studies on their use as VP. For the most part, this issue is considered solved, and its solution is reflected in the relevant normative instruments [8, 9] in the sections dealing with the security device in rooms with gas explosions. Incidentally, they also recommend the use of specific gravity with VP, specified in our materials. But these documents are summarized and the rationale for this restriction, which is not in our.

In addition to standards, there are also work [10-15] which are developing as mathematical models, sound empirical data that describe the behavior of the explosion, and a description of the physical processes occurring in the process of exploding.

Naturally, in addition to the formal aspects of the tests we accompanied them with additional research. In our opinion is of interest in practical and scientifically obtained experimental results.

Test task: practicing methods of testing; verification of the claimed efforts autopsy VP in the dynamics of explosion; Evaluation of the effectiveness of the VP as a remedy when the gas blast.

## 2. Method of testing

The tests were carried out on the test bench vent-panels (VP). The installation consists of two chambers. The main explosion Chamber, standards compliant, has dimensions (2000x2000x2000) mm. Inside the installed small explosion

* Corresponding author: dekuk07@mail.ru

chamber size (1000x1000x1000 mm), which supplied propane gas in volume sufficient to produce stoichiometry mixtures with air. There, in a small Chamber positioned fan for gas mixing and ignition device. To ensure the isolation of small volume Chambers in the cooking process blend side, open and facing to the escape hole, blast Chamber closed the curtain is a reusable action. When the screen opens and the gases from the small camera coming into explosive, increasing the pressure there. Increase pressure ensures destruction sites fastening of panels and reset them.

When testing was carried out two major measurement parameters characterizing the trigger nodes bracing of vent-panels: pressure inside the blast Chamber and the beginning of motion design. For measurements were used the following measuring tools: pressure sensor, power supply and analog-to-digital converter la USB 20, production of company "Rudnev-Shilyaev"; the device, move the sensor type gap "lever" and video recording; personal computer.

On a computer that has produced a record of the following signals: from a pressure sensor, the signal "explosion" combined with a signal lamp, signal "open circuit" and displacement sensor. When processing and conversion of analogue signals used guaranteed manufacturers specifications and codes for them, and when you build charts-Excel 2010.

Of course, for spending explosion used tools such as stopwatches, measuring rulers, sensors, computer codes, ADC and programs for data processing, which have confirmed with certificates and manufacturers because organizations. But it is difficult to list other parameters and conditions that, in one way or another, affect the outcome of the explosion. Consider in advance the result of sharing of devices and means of attorneys, as well as the impact of the entire set of factors is almost impossible, therefore, to assess the impact of an uncontrolled experiment on the outcome of the total conducted the installation tests. We conduct them in quantities of 10 experiments during one period of time. The results are given in Figure 1. Visible to their acceptable reproducibility. The results obtained after installation panels, obviously, are the result of it is the installation of the panels, rather than by some other factor. Therefore, the difference in received data can be interpreted.

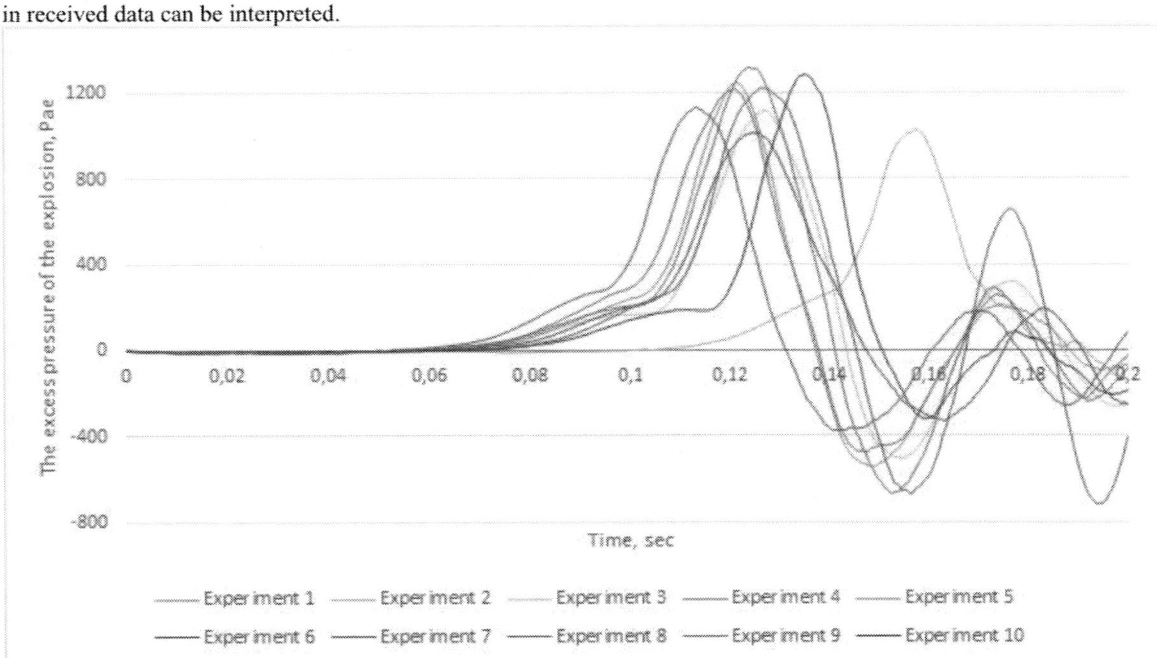


**Fig.1** Schedule installation tests

## 3. The results of the tests

### 3.1 The results of the tests of sandwich panels as VP

Tested characteristics of sandwich panels. The Panel consists of steel plates (facing) thickness of 0.8 mm, inside of which is thermal insulation (insulation) on the basis of mineral wool. Outside sheets coated with polymeric material in white color. Weight of panels with dimensions (1000x1640) mm thick 100 mm was 30.4 kg. The total number of nodes, retention, collapsing in the explosion-4, they placed at a distance of 60 mm from the edges of the Panel. Each node consists of a screw with a diameter of 10 mm, inserted in a hole with a diameter of 10.5 mm Panel head outdoors, indoors on it attached to the sleeve, which is on the housing of the camera. In the sleeve and its axis perpendicular to the bolt drilled a hole with a diameter of 1.0 mm, which are inserted steel wire with a diameter of 0.75 mm flanged brand. In the case of explosion inside Panel begins with it moves relative to the bolt sleeve, cutting off the wire. Force cut wire in field 8 is an obstacle that must overcome the pressure of the blast.

There are two unknown factors when gathering with fixation points in case of explosion may cause skewing sandwich panels, that can affect the outcome of the test. Firstly, when a large number of destructible items (8 slices) is quite reason, the question of their simultaneous firing. And, secondly, it is obvious that the weight of the load is distributed between bolts unevenly, which can lead to different values of the forces of friction of rest between bolts and bushings that can also affect the results of the experiment.

Tests of sandwich panels made using the fan unit. Objective test was to evaluate the pressure at which collapse fixture equal to 5000 Units. The footage shows how the moves ploskoparallel'no installation after the explosion. Data processing showed that it moves at an average speed of 18 m/s. Other test data are shown in Figure 2. You can see that the destruction of the attachment node occurs at the same pressure of 4500 Units in both experiments. On the face of a good repeatability. Concluded that the fixture are destroyed at a pressure closer to the declared.

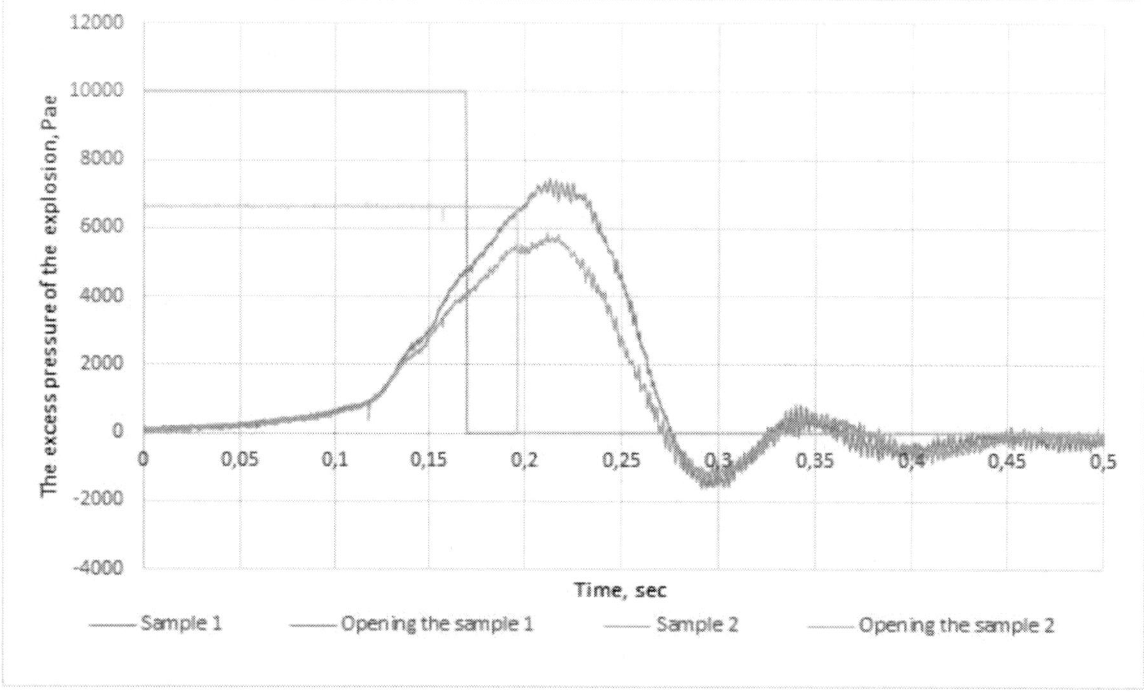


**Fig.2** Test schedule of sandwich panels

Noteworthy that the pressure in the Chamber continues to grow slightly after the destruction of the attachment node, although according to estimations it must fall with the beginning reset panels.

Storyboard videos tests of sandwich panels is shown in Figure 3.

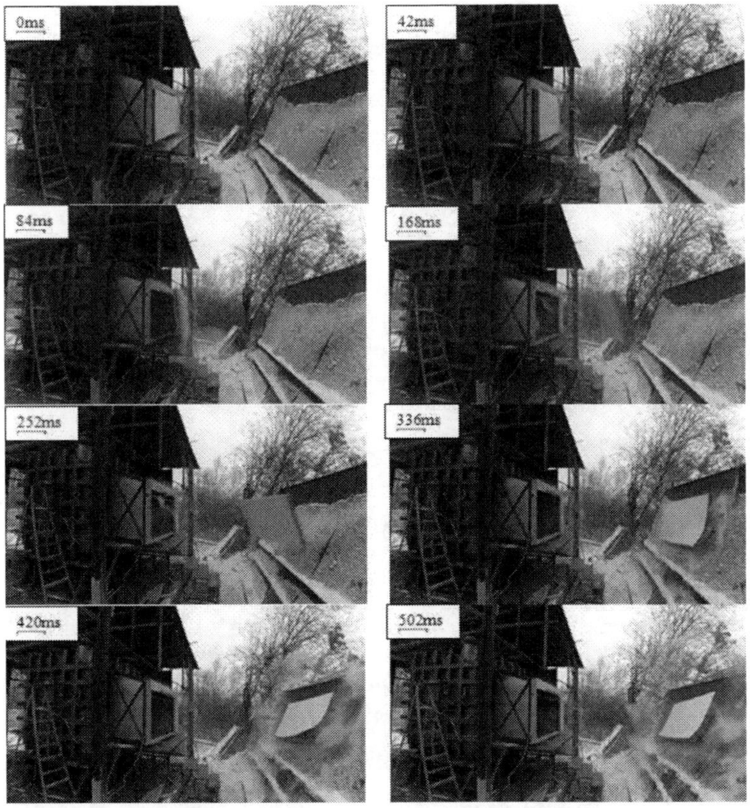


**Fig.3** Video release of sandwich-panels with the explosion

### 3.2 The results of the tests of insulating glass as VP

Characteristics of the tested insulating glass. The specimen is a vent-panels (VP) in the form of a window frame with two double-glazed Windows with dimensions (1650hh1450) mm. Frame made of aluminum profiles post and beam system. F50:

-Reinforced stand 80 mm ALT. (F) 50.0103 + ALT. (F) 50.0303

-Reinforced bolt 85 mm ALT. (F) 50.0206 + ALT. (F) 50.0303

-fill-single chamber energy-saving glazing thickness of 24 mm (6/12/6).

Weight of a double-glazed window in the collection stood at 50.0 kg. Panel sealed metal waste hole blast Chamber in six points. In the opening holes top and bottom patch installed lateral corners (75x75x5) mm using wedge anchors "HILTI" length 160 mm with step of 600 mm. Through the special holes in the corner of the 300 mm increments the frame inserted self-tapping screw-Stud, then for the centric fitting guide Bush. After putting all six landing points on studs were wrapped fluorine plastic washers F4. After mounting the frame inserted double glazing using the clamping plates. The principle of operation of the VP was founded at offset frames with glazing (destruction of the attachment node VP)-roll it out when exposed to excessive pressure.

Held only 3 tests. Objective test was to evaluate the pressure at which collapse fixture, equal to 700 Units. The first test of insulating glass units conducted using fan console showed a strong influence on the development process of the explosion and the test results. The remaining 2 explosion taken place without diffuser and their results are given in Figure 4. It can be seen that glazing all 6 mounting nodes are destroyed at the same time, and the package moves almost ploskoparallel'no at an average speed of 16 m/s shows that the destruction of the attachment node occurs at the same pressure of 2500 Units in both experiments. On the face of a good repeatability. Concluded that the fixture are destroyed at a pressure that varies more than three times with a declared, but is below the acceptable for categories a and b class premises on fire safety.

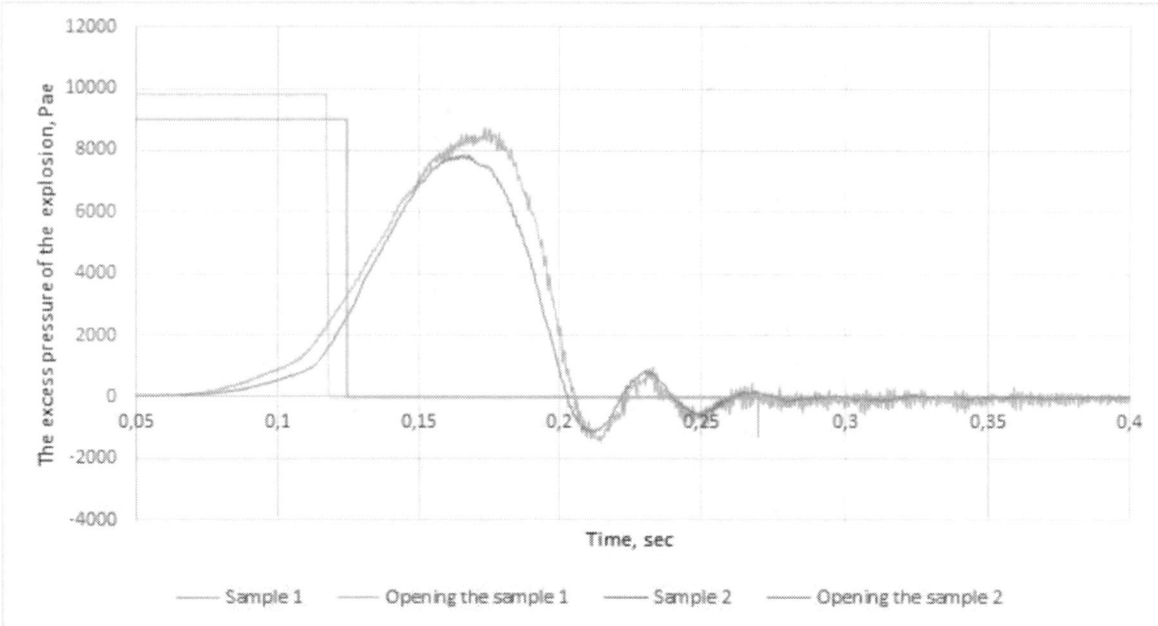


**Fig. 4** Test schedule of double-glazed Windows

Noteworthy, that in this case the Chamber pressure continues to grow after the destruction of the attachment node, although according to estimations it must fall with the beginning reset panels.

Storyboard videos tests of sandwich panels is shown in Figure 5.

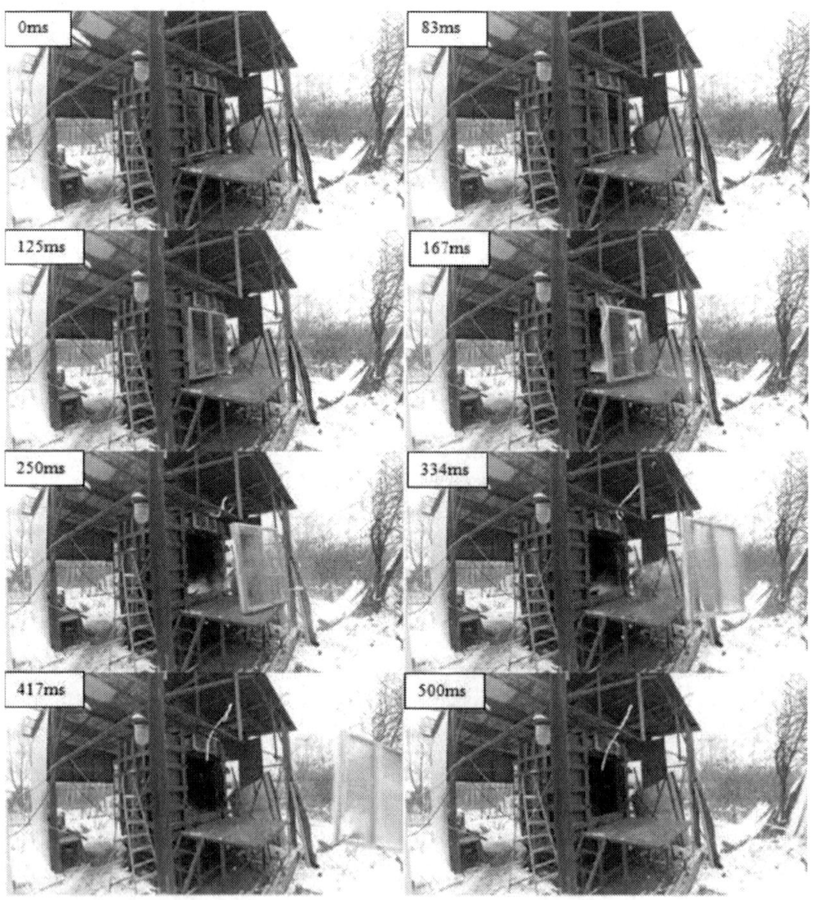


**Fig. 5** Video release of sandwich-panels with the explosion

The move on the gauge shows that the destruction of sites of fastening takes place in several stages, among which, in our view, there are several stages: sampling clearances on the node are attached; the existence of a plot by the elastic deformation of the specimen and fluidity and destroyed material attachment sites. The first 10 cm panel movement is accompanied by a build-up of pressure in the Chamber. The sensor in the future due to the fact that the Panel has gone.

## 4. Conclusion

Based on the results of the works you can make the following conclusions:

- Reproducibility shows the adequacy of instrumentation and test management system-wide;
- The results of the test panels in one case showed convergence with data stated in another case-non-conformity;
- Revealed an interesting effect is to increase pressure on a certain period of time after the initiation of the dumping of panels. In our opinion, should evaluate the effectiveness of panels as VP incl. plot increases in pressure after the destruction of sites of attachment. The reason for the growth of pressure after the destruction sites should install more fasteners.

## References

1. GOST R 56288-2014. Designs windowing fiberglass legkosbrasyvaemye for buildings. Technical conditions. 07.01.2015 entered. Moscow, Standartinform, (2014)
2. Snip II-35-76 *. Boiler installation. Section 3. Objectively-planning and design solutions. -Enter. 01.01.1978.- m.: Stroiizdat, (1977)
3. GOST R 56289-2014 "Designs for buildings" legkosbrasyvaemye translucent. Test methods for impact internal emergency Moscow blast Standartinform (2015)
4. A.A. Komarov, A.Y. Mishuev, The dissertation on competition of a scientific degree of doctor of technical sciences-"prediction of loads from emergency deflagracionnyh blasts, and assessment of the impact of their effects on buildings and structures; Moscow (2001)
5. E.Y. Salymova, V.A. Gorev v.a.-Dissertation on competition of a scientific degree of candidate of technical sciences-«Dynamics hazards in buildings with protecting structures from the three-layer sandwich panels in fires and explosions; Moscow (2015)
6. Yu.N. Shebeko, A.Ya. Korolchenko, S.G. Tsarichenko,V.Yu Navtsenya, V.L. Malkin, Combustion, Explosion, and Shock Waves, **25 (3)**, 289-292 (1989)
7. Yu. N. Shebeko, A. Ya. Korolchenko, E.D. Zamishevski, A.V. Trunev, V.Yu. Navzenya, A.A. Zaitzev, Combustion and Flame, **117 (1-2)**, 438-440 (1999)
8. EN 14994:2007. Gas explosion venting protective systems.
9. NFPA 68. Standard on explosion protection by deflagration venting. 2007 Edition. – Quincy, MA: National Fire Protection Association (2007)
10. R. Siwek, R, O. Scov, VDI Berichte, **701**, 569–616 (1988).
11. H.S. John Lee The gas dynamics of explosions 216 (2016)
12. A.F. Sharovarnikov, D.A. Korol'chenko, Applied Mechanics and Materials, **475-476,** 1344-1350 (2014)
13. D. Korolchenko, S. Voevoda, MATEC Web of Conferences **86** (2016)
14. Y. Polandov, A. Korolchenko, S. Dobrikov, MATEC Web of Conferences, **86** (2016)

# Impact of external wall insulation thickness on internal surface temperature behaviour

*Radoslav* Ponechal[1,*], *Daniela* Staffenova[1]

[1]University of Zilina, Faculty of Civil Engineering, Department of Building Engineering and Urban Planning, Univerzitna 1, 01026 Zilina, Slovakia

**Abstract.** During the last years, the concept of low-energy buildings based on high insulation levels becomes the reality. The aim of this paper is to assess some alternatives of insulated and uninsulated external walls with respect of thermal inertia. The thermal damping factor, phase shift, together with the daily courses of indoor surface temperature of the external wall have been analysed. Analysed surface temperatures show the ability of constructions to accumulate heat gains, which can arise during the day.

## 1 Introduction

By design of the material for the perimeter wall structures, the static and thermo-insulating aspect is the most important. If the values of thermal resistance are approximately at the same level, the builders would come to a number of other criteria such as the construction technology or internal environment, created by these construction materials. The role of thermal inertia in creating a high-quality indoor environment is indisputable. Its effects on structures and buildings are manifold and they are expressed by a number of indicators. Many indicators of the thermal inertia impact assessment have been developed and published till today. Most of them are described in the standard EN ISO 13786:2007 [1]. In addition, a dynamic simulation methods are increasingly used for the evaluation. The paper from Italy is to assess the deviation arising by the use of different approaches for the calculation of the dynamic thermal characteristics of an opaque envelope element. The EN ISO 13786 procedure has been firstly applied by decomposing the external forcing temperature by means of the Fourier Transform analysis and the comparison with different approaches, such as Finite Difference Methods and Transfer Function Methods, has been carried out [2]. It concludes that there is a good accord among the Transfer Functions Methods and the Finite Difference Methods as regard periodic thermal transmittance. The time shift calculations show deviations of the order of magnitude of one hour, mainly due to the time discretization of the results.

The college from Eindhoven University of Technology considered the 700 mm depth ultra-lightweight concrete wall in terms of energy efficiency and thermal comfort [3]. At first, they use the periodic thermal transmittance, which combines the properties of thermal transmittance, time shift and decrement factor in one metric. Analytical validation study

* Corresponding author: radoslav.ponechal@fstav.uniza.sk

showed that the conduction finite difference scheme in EnergyPlus is able to predict the transient heat transfer phenomena in ultra-lightweight concrete structures with sufficient accuracy, but users of simulation should carefully select an appropriate simulation tool, because this study has also demonstrated that not all transient heat conduction methods that are used in the widely-used software tools are able to predict the heat transfer in ultra-lightweight slabs without errors. The results of the dynamic simulation of the house operation were shown on discomfort hours.

Nemecek solved similar problems in the passive house [4]. In his contribution based on simulation calculations of summer overheating, according to the importance he ranks an energy storage capacity of construction in a $4^{th}$ place, behind heat gains from solar radiation, internal heat gains and ventilation regime. The difference between air temperatures by light-weight or heavy-weight construction is about 1-1.5 K.

## 2 Dynamic thermal properties

Because of the fact that the overheated room temperature is influenced by many factors, it is advisable to look at the comparison of basic dynamic characteristics of building materials separately. The International Standard ISO 13786 describes dynamic thermal parameters of multi-homogeneous-layered walls/roofs based on sinusoidal variations of temperature or heat flow rate on the one hand and constant air temperature on the other hand. The parameters relate cyclic heat flow rate to cyclic temperature variations. These parameters are expressed as complex numbers.

Three groups of parameters for the thermal evaluation of building envelope variants in not air-conditioned buildings are analysed in this work: thermal damping factor (DFs), phase shift (LTs) and behaviour of indoor surface temperature. The thermal damping factor is calculated by

$$DFs = \frac{(\theta_{si,\max} - \theta_{si,\min})}{(\theta_{se,\max} - \theta_{se,\min})} \ (\text{-}) \qquad (1)$$

where $\theta_{si,max}$ and $\theta_{si,min}$ are the maximum and minimum values of the indoor surface temperature during a day, respectively, and $\theta_{se,max}$ and $\theta_{se,min}$ are the maximum and minimum values of the outdoor surface temperature.

The phase shift is defined as:

$$LTs = t(\theta_{si,\max}) - t(\theta_{se,\max}) \ (\text{h}), \qquad (2)$$

where $t(\theta_{si,max})$ and $t(\theta_{se,max})$ are the time, when the indoor surface and outdoor surface temperatures reach their maximums.

## 3 Structural variants

Three wall configurations with four different insulation thicknesses are studied and presented: perforated brick wall (BR), heavy weight concrete wall (HC), porous concrete wall (PC) with thermal insulation made of expanded polystyrene (EPS) or mineral wool (MW). These three wall variants have very similar U-values, but differ from each other in the dimension, material density and thermal conductivity. Each of the wall structures (according to Fig. 1) has some selected different insulation thicknesses. The first variant fulfils the requirement for maximal U-values $U_{max}$, the second variant fulfils desired value $U_N$, the third one fulfils recommended value $U_{r1}$ (considered for the year 2016) and the last variant fulfils the final thermal resistance $U_{r2}$ (considered for the year 2021) by the STN 73 0540:2012 standard [5]. The wall layers compositions are described in Fig. 1.

## 4 Thermal conductivity coefficient and DFs

The experiment analyses several of the above mentioned external wall compositions with the calculation of thermal damping factor and phase shift of the temperature oscillation, which was conducted on the basis of EN ISO 13786. The results from all calculations are shown in Table 1 and Fig. 4.

Constructions designed for new buildings (with the thermal resistances $U_{r1}$ and $U_{r2}$) reach significantly higher values of the thermal damping factor than the constructions with minimum requirements for thermal resistance. There are strongly significant differences among the porous concrete wall and the other two alternatives. While by the other insulated walls, the thermal damping factor is as high as it is meaningless, this does not go for the porous concrete wall. Comparing the differences by phase shift, it is necessary to take into account the values of temperature setback, too.

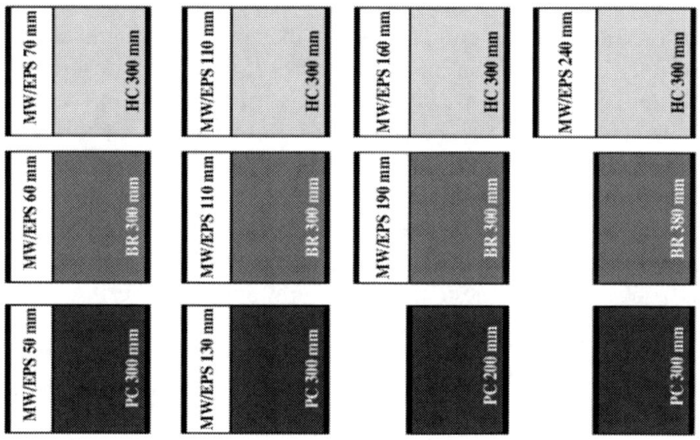


**Fig. 1.** Evaluated variants of external wall, perforated brick (BR), Porous concrete (PC), heavy-weight concrete (HC), expanded polystyrene (EPS) or mineral wool (MW).

**Table 1.** Results of thermal performance variables reached for evaluated wall structure variants

| | *Structure A* | *U (W/(m².K))* | *DFs* | *LTs* |
|---|---|---|---|---|
| 1 | *HC 300 + MW 70* | *0.454* | *160.8* | *11.7* |
| 2 | *HC 300 + EPS 70* | *0.454* | *156.1* | *10.5* |
| 5 | *HC 300 + MW 110* | *0.307* | *294.8* | *13.4* |
| 6 | *HC 300 + EPS 110* | *0.307* | *255.0* | *11.5* |
| 7 | *HC 300 + MW 160* | *0.219* | *575.1* | *15.6* |
| 8 | *HC 300 + EPS 160* | *0.219* | *417.8* | *12.9* |
| 9 | *HC 300 + MW 240* | *0.150* | *1555.0* | *19.1* |
| 10 | *HC 300 + EPS 240* | *0.150* | *827.6* | *15.0* |

| | *Structure B* | *U (W/(m².K))* | *DFs* | *LTs* |
|---|---|---|---|---|
| 11 | *BR 380* | *0.438* | *261.4* | *18.4* |
| 12 | *BR 300 + MW 60* | *0.318* | *333.3* | *17.2* |
| 13 | *BR 300 + EPS 60* | *0.318* | *314.9* | *16.1* |
| 14 | *BR 300 + MW 110* | *0.211* | *726.3* | *19.3* |
| 15 | *BR 300 + EPS 110* | *0.211* | *597.3* | *17.5* |
| 16 | *BR 300 + MW 190* | *0.146* | *2071.6* | *22.8* |
| 17 | *BR 300 + EPS 190* | *0.146* | *1277.4* | *19.6* |

| | *Structure C* | *U (W/(m².K))* | *DFs* | *LTs* |
|---|---|---|---|---|
| 18 | *PC 200* | *0.444* | *28.2* | *8.1* |
| 19 | *PC 300* | *0.303* | *87.0* | *12.2* |
| 20 | *PC 300 + MW 50* | *0.217* | *119.3* | *14.4* |
| 21 | *PC 300 + EPS 50* | *0.217* | *95.0* | *13.5* |
| 22 | *PC 300 + MW 130* | *0.149* | *413.1* | *17.9* |
| 23 | *PC 300 + EPS 130* | *0.149* | *262.6* | *15.7* |

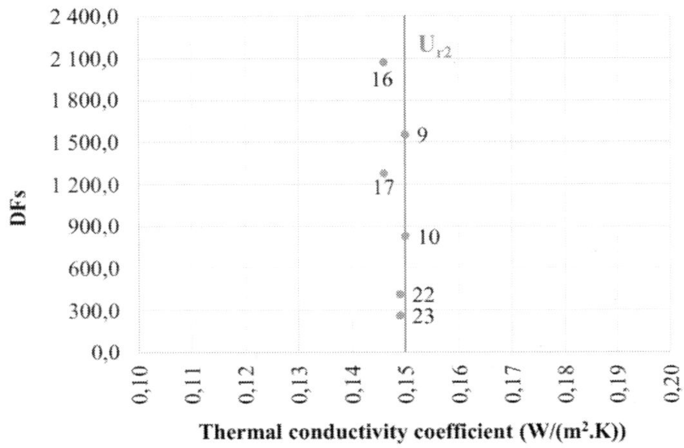


**Fig. 2.** Thermal damping factor (DFs) for the first group of walls.

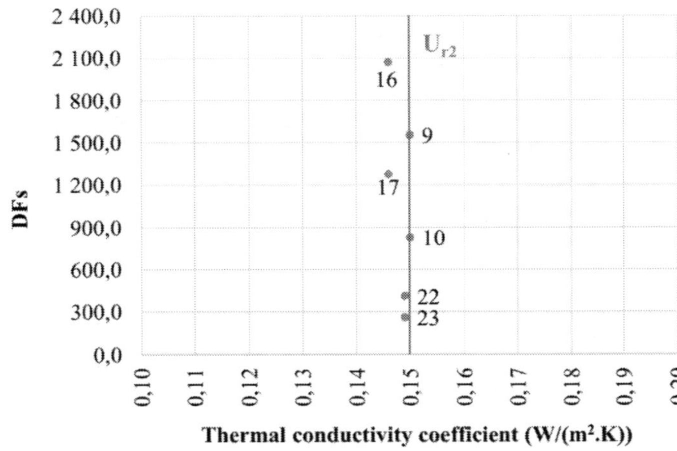


**Fig. 3.** Thermal damping factor (DFs) for the second group of walls.

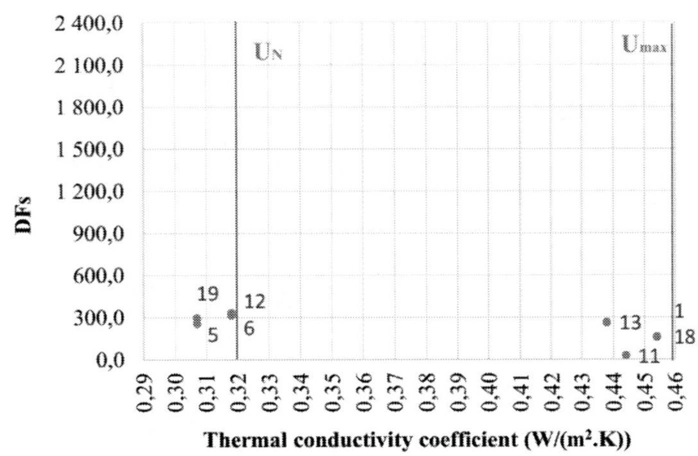


**Fig. 4.** Thermal damping factor (DFs) for the third group of walls.

If the value of temperature damping is high (e.g. 200 and more), the phase shift does not play such a role and the inner surface temperature will increase up to 0.5 degree only. It is not so crucial according to the fact at what time this occurs.

Calculations of DF and LT are meaningful only in terms of heat transfer fragment of building envelope constructions (walls, roofs, etc.). The complete image of the building or a room in the building, in terms of summer overheating, will give us only complex calculations of thermal stability or computer simulations.

## 5 Simulations of internal surface temperature

Insulated wall can resist temperatures from the inside and from the outside with a difference. Graphs on Fig. 5 and Fig. 6 show a behaviour of internal surface temperature with 24 hour period change of internal air temperature. The internal air temperature has risen since the morning and it was falling late in the evening. It is a heat conduction behind a non-stationary state. The temperature varies both inside and outside. Lower internal surface temperature means better thermal comfort. The lower internal surface temperature at times of high heat gains means the possibility of accumulating a portion of the heat from the air into the walls.

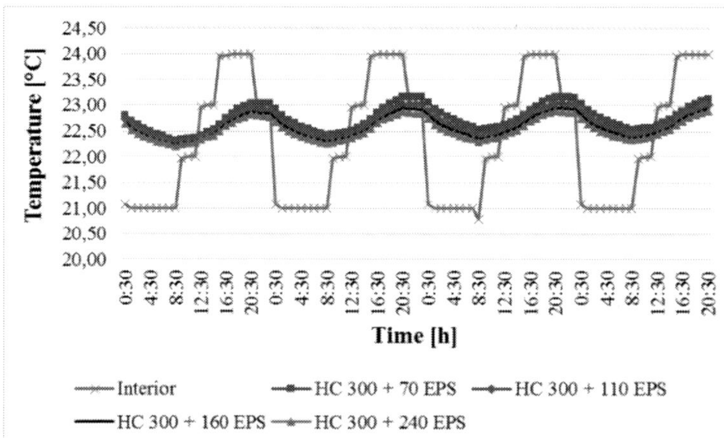


**Fig. 5.** Internal surface temperature for a heavy weight concrete wall with thermal insulation

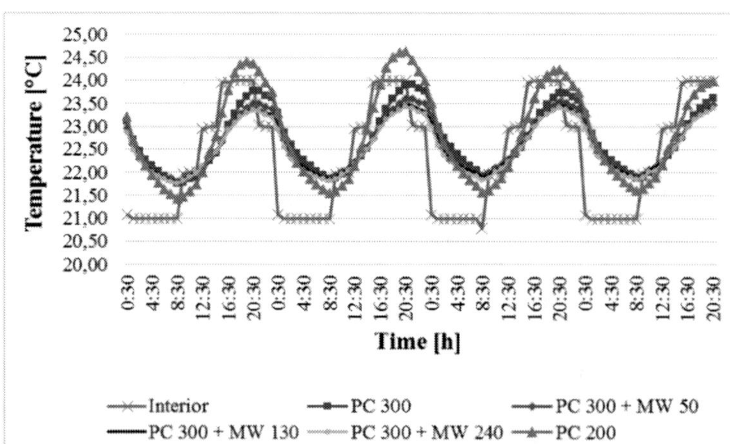


**Fig. 6.** Internal surface temperature for a porous concrete wall with thermal insulation

## 6 Conclusion

The impact of external wall on overheating could be defined in two ways: as the structure, through which the heat is transferred from the exterior and it warms the interior or (less usual) as the structure, which contributes to the thermal stability of interior by its accumulative ability.

In the first case, the thermal damping factor, which is very high for the walls in new buildings, is a crucial criterion. Thermal insulation protects the wall from rapid temperature changes. High values of thermal damping factor in insulated walls indicate that its assessment in terms of the daily cycle of temperature rise and fall is an insignificant phenomenon.

A much more substantial contribution can the external wall have on heat loss in the summer season. Heat loss in the summer helps cool the interior at night. However, there is still a conclusion from the contributions [3, 4] that the effect on overheating is affected by the control of building ventilation and heat gains.

Another case, when the external walls could affect the interior overheating is its accumulative ability. Integrating of simulation approach could eliminate this imperfection. It appears that the walls insulated in compliance with the new requirements for thermal conductivity coefficient can contribute to thermal stability of the room, because their internal surface temperature remains low even in the evening. The external wall, which holds more heat, has a higher potential to stabilize the interior temperature. However, the thickness of the insulation does not play here a significant role.

Presented results were obtained with the support of the grant project VEGA no. 1/0945/16.

**References**

1. STN EN ISO 13786 (2007)
2. A. Gasparella, G. Pernigotto, M. Baratieri, P. Baggio, Energy and Buildings, **43**, 2509-2571 (2011)
3. F. Roberz, R.C.G.M. Loonen, P. Hoes, J.L.M. Hensen, Energy and Buildings, **138**, 432-442 (2017)
4. M. Nemecek, M. Kalousek, www.tzb-info.cz
5. STN 73 0540:2012 (2012)

# Exceeding the permissible noise levels from rail transport in the Kysuce region

*Alžbeta* Pultznerová[1*] *and Martin* Mečár[1]

[1]University of Žilina, Faculty of Civil Engineering, Univerzitná 8215/1, 010 26 Žilina, Slovak Republic

**Abstract.** Rail transport is a source of noise that has disturbing effects on the environment. Adverse noise emissions must be observed and at the same time keep the permissible noise levels that depend on the type of protected area given by the Slovak legislation. The paper deals with compliance and analysis of these permitted noise values in some selected locations of the railway line in the Kysuce region.

## 1 Introduction

Noise emissions represent a significant environmental burden on the environment. The largest polluters include transport.

Although the railway noise is considered to be less burdensome it is the second most dominant source of the environmental noise in Europe with nearly 7 million people exposed to levels above 55 dB $L_{den}$ in 2012 considering people exposed in urban areas. [1]

The topography of the terrain, its absorption properties and obstacles on the way of spreading from the source, for spreading noise in the external environment is very important. The range and intensity of noise is also influenced by the substance through which it spreads. In the case of the spread of noise from railway transport in outside, it is air and its climatic conditions. However, the climatic factors are not considered in the paper and it will only deal with the spread of noise from rail traffic in the free space in the Kysuce region and its reduction, depending on the distance from the source.

## 2 permissible noise limits in transport

Unlike to road traffic, where permissible noise limits at the source have existed in the EU since the 1970s, noise standards for trains only came into force at the beginning of the twenty-first century [2]

In 2007, Regulation 549/2007 of the Ministry of Health of the Slovak Republic, in which the permissible values of noise $L_{Aeq,p}$ for rail transport compared to the values defined in Government Regulation 339/2006 have not changed. This decree is still valid and in full wording the permissible limits are given in Tab. 1.

* Corresponding author: alzbeta.pultznerova@fstav.uniza.sk

The noise levels are evaluated for the purposes of spatial planning and control at 1.5 m ± 0.2 m above the ground outside the building or in a front of peripheral wall of a building, the noise are assessed at the distance of 1.5 m ± 0.5 m of the wall and at 1.5 m ± 0.2 m above the floor of the relevant floor levels. The Surroundings means the area within 100 m away from an adjacent axis of driving lane of road or from the axis of the adjacent railroad tracks [3].

**Tab. 1** The permissible noise limits defining the parameters of environmental noise [3]

| Category of the area | Description of the protected area or outer space | The reference. time interval | Permissible noise limits (dB) | | | | |
|---|---|---|---|---|---|---|---|
| | | | Traffic noise | | | | The noise from other sources $L_{Aeq,p}$ |
| | | | Land a) and water transport b),c) $L_{Aeq,p}$ | Railway transport c) $L_{Aeq,p}$ | Air transport | | |
| | | | | | $L_{Aeq,p}$ | | |
| I. | Territory with special protection against noise (e.g. spa towns, spa and wellness resorts). | day<br>evening<br>night | 45<br>45<br>40 | 45<br>45<br>40 | 50<br>50<br>40 | -<br>-<br>60 | 45<br>45<br>40 |
| II. | Space in front of the windows of habitable rooms of the block of flats and residential houses, the space in front of windows protected rooms of school buildings, health care facilities and other protected objects, d) recreation area. | day<br>evening<br>night | 50<br>50<br>45 | 50<br>50<br>45 | 55<br>55<br>45 | -<br>-<br>65 | 50<br>50<br>45 |
| III. | Territory as in the category II around the highways, roads I and II. classes, local roads with public transport, railways and airports, city centres. | day<br>evening<br>night | 60<br>60<br>50 | 60<br>60<br>55 | 60<br>60<br>50 | -<br>-<br>75 | 50<br>50<br>45 |
| IV. | Territory without housing function and without protected outdoor space, production zones, industrial parks, resorts plants | day<br>evening<br>night | 70<br>70<br>70 | 70<br>70<br>70 | 70<br>70<br>70 | -<br>-<br>95 | 70<br>70<br>70 |

(a) The permissible values apply to the dry road surface and uninterrupted terrain, in the case of seasonal installations, noise shall be assessed under conditions which may be assumed in operation.
(b) Land transport is the transport of roads, including tramway.
(c) Stops for local public transport, bus, rail, waterway and taxi stations for the entry and exit of persons are assessed as part of the land and waterways.
d) The permissible values in front of the non-residential facade are applied at the time of their use, for example schools during classes, etc.

# 3 Noise study in the Kysuce region

Measurements were carried out during the night-time on railway line No. 127 in various places in the village Čadca and Svrčinovec in 10 measuring points according to the valid Regulation.

## 3.1 Evaluation of the noise limits compliance

To evaluate the assessed values for comparison with the limit value under current legislation, the equivalent noise pressure level A - weighted sound pressure level is used.

Equivalent A - weighted sound pressure level of the continuous sound $L_{pAeq}$ - time-averaged mean A level, which is used to describe the sound with varying levels of sound. It is calculated as the sound pressure that is energetically equivalent to the long-term effects of intermittent noise or noise with varying sound level [4].

The sound pressure level was counted by the formula 1:

$$L_{pAeq,T} = 10.log\left(\frac{1}{T}\cdot\int_0^T \frac{p_A^2(t)}{p_0^2}dt\right) \quad (dB) \quad (1)$$

where:

$T$ – measurement duration, (s),
$p_0$ – reference value of sound pressure, ($2.10^{-5}$ Pa),
$p_{A(t)}$ – instantaneous acoustic pressure of the sound signal, (Pa).

The measured $L_{R,Aeq,n}$ value is measured value $L_{pAeq,\ T}$ increased by uncertainty of measurement U.

Uncertainty of measurement and sound prediction determined according the guidance No: NRÚ / 3116/2005 dated 2.5.2005. Classification of the measured noise, depending on the frequency composition and its directional properties, results in the resulting expanded measurement uncertainty U = 1.8 dB [5].

From the particular hour measurements of LAeq,1h noise events, which were not caused by railway operation were "cut out". The particular hourly noise levels were energy calculated by the formula [6]:

$$L_{Aeq,p} = 10.log\left(\frac{1}{T}\sum_{j=1}^{8} 10^{0,1L_{Aeq,j}}\right) \quad (dB) \quad (3)$$

where:

$L_{Aeq,p}$ – equivalent noise pressure level [dB];
$L_{Aeq,j}$ – equivalent noise pressure level measured in one hour intervals [dB];
j – number of equivalent noise pressure levels measured during night time;
T – measurement duration, in our case it is 8 hour (from 22,00 to 6,00). [hour]

### 3.2 Assessment of noise study in the Kysuce region

In-situ measurements of the non- modernised track were used to assess the value compared to the permissible value according to valid legislation, with the measurement points being chosen in the residential area beside of the railway track under consideration

- in category of the area II. - at the distance over 100m from the adjacent railway track No. 127,
- in category of the area III. – at the distance to 100m from the adjacent railway track No. 127.

**Tab. 2** Sample of the one hour equivalent noise pressure level measured in point M4 – 2 m in front of the window of residential room on 2nd floor of house No. 370, by Špindel Street, Čadca town, approximately 39 m from track No. 127 and 30 m from the road I/11[7]

| Time (h) | 22-23 | 23-00 | 00-01 | 01-02 | 02-03 | 03-04 | 04-05 | 05-06 |
|---|---|---|---|---|---|---|---|---|
| $L_{Aeq,1h}$ (dB) | 63.3 | 63.03 | 62.4 | 66.4 | 66.6 | 68.7 | 67.6 | 67.0 |

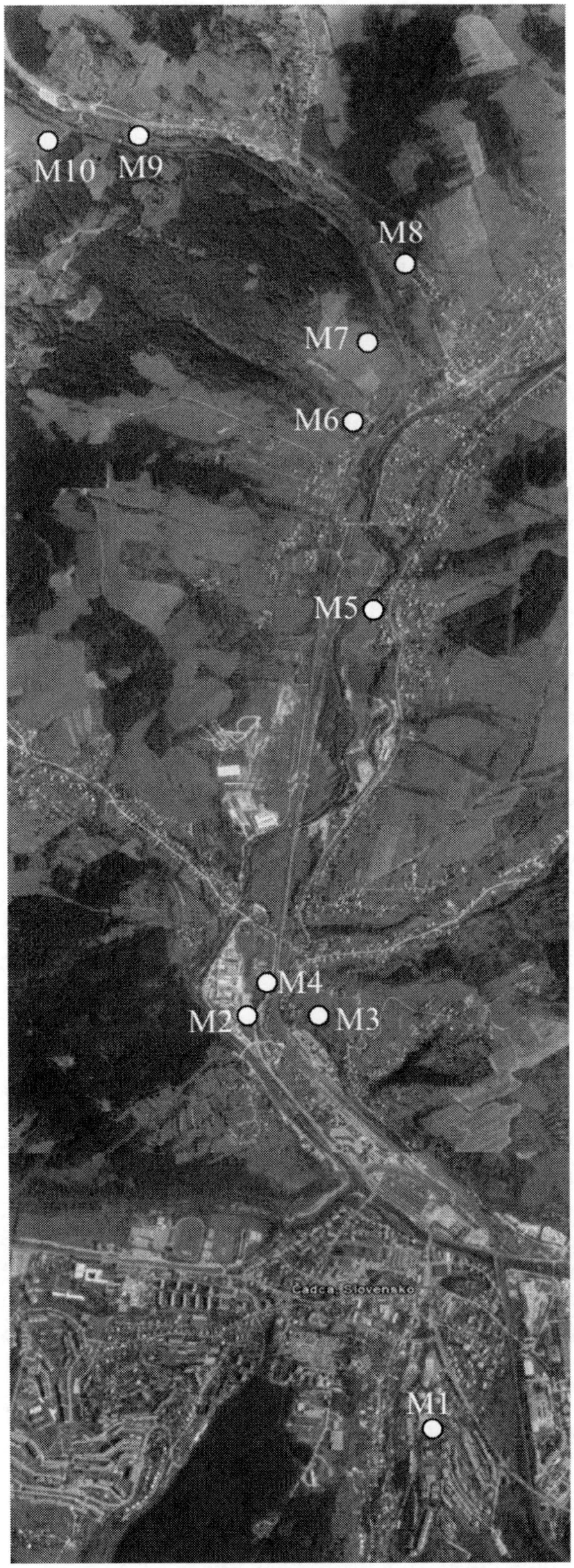


**Fig. 1** Measuring points in the Čadca town and Svrčinovec settlement [7]

List of measuring points - see Fig. 1 [7]:

M1 - 2m in front of the window of residential room on 7 th of residential house No. 2589 J. Kráľ street, Čadca town, approximately 145 m from railway track No. 127, 118 m from the road I/11 and approximately 15 m from J. Kráľ street,
Total equivalent sound pressure level A measured during night time 22:00 - 06:00:
$L_{Aeq,8\ h\ night}$ = 61.5 dB... overall noise - environment noise,
$L_{Aeq,8\ h\ night}$ = 60.1 dB... only noise from the railway line No. 127,

M2 - 2m in front of the window of residential room on 2nd floor of house No. 297 Slovanská cesta street, Čadca town, approximately 60 m from railway track No. 127,
Total equivalent sound pressure level A measured during night time 22:00 - 06:00:
$L_{Aeq,8\ h\ night}$ = 60.3 dB... overall noise - environment noise,
$L_{Aeq,8\ h\ night}$ = 59.6 dB... only noise from the railway line No. 127,

M3 – on the fence of the garden of house No.219, Čadca - Drahošanka, altitude from the ground 1,5 m and approximately 300 m from railway track No. 127,
Total equivalent sound pressure level A measured during night time 22:00 - 06:00:
$L_{Aeq,8\ h\ night}$ = 51.2 dB... overall noise - environment noise,
$L_{Aeq,8\ h\ night}$ = 48.6 dB... only noise from the railway line No. 127,

M4 - 2 m in front of the window of residential room on 2nd floor of house No. 370, by Špindel street, Čadca town, approximately 39 m from railway track No. 127 and 30 m from the road I/11,
Total equivalent sound pressure level A measured during night time 22:00 - 06:00:
$L_{Aeq,8\ h\ night}$ = 66.2 dB... overall noise - environment noise,
$L_{Aeq,8\ h\ night}$ = 65.3 dB... only noise from the railway line No. 127,

M5 - 2m in front of the window of residential room on 2nd floor of house No. 141, Svrčinovec settlement, approximately 240 m from railway track No. 127 and approximately 50 m from the road I/11,
Total equivalent sound pressure level A measured during night time 22:00 - 06:00:
$L_{Aeq,8\ h\ night}$ = 56.4 dB... overall noise - environment noise,
$L_{Aeq,8\ h\ night}$ = 55.4 dB... only noise from the railway line No. 127,

M6 - 2m in front of the window of residential room on 2nd floor of house No.835, Svrčinovec settlement, approximately 43 m from railway track No. 127,
Total equivalent sound pressure level A measured during night time 22:00 - 06:00:
$L_{Aeq,8\ h\ night}$ = 68.4 dB... overall noise - environment noise,
$L_{Aeq,8\ h\ night}$ = 68.2 dB... only noise from the railway line No. 127,

M7 - 2m in front of the window of residential room on 2nd floor of house No. 354, Svrčinovec settlement, approximately 115 m from railway track No. 127 and approximately 300 m from the road I/11,
Total equivalent sound pressure level A measured during night time 22:00 - 06:00:
$L_{Aeq,8\ h\ night}$ = 46.7 dB... overall noise - environment noise,
$L_{Aeq,8\ h\ night}$ = 44.6 dB... only noise from the railway line No. 127,

M8 - 2m in front of the window of residential room on 2nd floor of house No. 335, Svrčinovec settlement, approximately 160 m from railway track No. 127 and approximately 15 m from the road I/11,
Total equivalent sound pressure level A measured during night time 22:00 - 06:00:
$L_{Aeq,8\ h\ night}$ = 67.7 dB... overall noise - environment noise,
$L_{Aeq,8\ h\ night}$ = 61.9 dB... only noise from the railway line No. 127,

M9 - 2m in front of the window of residential room on 1st floor of house No. 490, Svrčinovec settlement, approximately 100 m from railway track No. 127,
Total equivalent sound pressure level A measured during night time 22:00 - 06:00:
$L_{Aeq,8\ h\ night}$ = 56.2 dB... overall noise - environment noise,
$L_{Aeq,8\ h\ night}$ = 55.5 dB... only noise from the railway line No. 127,

M10 - 2m in front of the window of residential room on 3[th] floor of house No. 882, Svrčinovec settlement, approximately 145 m from railway track No. 127,
Total equivalent sound pressure level A measured during night time 22:00 - 06:00:
$L_{Aeq,8\ h\ night}$ = 47.7 dB... overall noise - environment noise,
$L_{Aeq,8\ h\ night}$ = 47.2 dB... only noise from the railway line No. 127.

**Tab.3** Evaluation of the measured and the calculated values [own research]

| Assessment point (altitude from the ground/ distance from the rail. track) (m) | | Land category | PV (dB) | Measured values (dB) | Assessed measured value $L_{R,Aeq,n}$ (dB) | $L_{R,Aeq,n}$ - PH (dB) |
|---|---|---|---|---|---|---|
| M1 | 7[th] floor /145 | II. | 45 | 60.1 | 61.9 > 45 | 16.9 |
| M2 | 2[nd] floor /60 | III. | 55 | 59.6 | 61.4 > 55 | 6.4 |
| M3 | 1,5/300 | II. | 45 | 48.6 | 50.4 > 45 | 5.4 |
| M4 | 2[nd] floor /39 | III. | 55 | 65.3 | 67.1 > 55 | 12.1 |
| M5 | 2[nd] floor /240 | II. | 45 | 55.4 | 57.2 > 45 | 12.2 |
| M6 | 2[nd] floor /43 | III. | 55 | 68.2 | 70 > 55 | 15.0 |
| M7 | 2[nd] floor /115 | II. | 45 | 44.6 | 46.4 > 45 | 1.4 |
| M8 | 2[nd] floor /160 | II. | 45 | 61.9 | 63.7 > 45 | 18.7 |
| M9 | 1[st] floor /100 | III. | 55 | 55.5 | 57.3 > 55 | 2.3 |
| M10 | 3[th] floor /145 | II. | 45 | 47.2 | 49 > 45 | 4.0 |

## 4 Conclusions

In Tab. 3 all the measurement points are evaluated from the point of view of the territory category, it means the distance from the noise source and their compliance with the permissible values. The measurements were done before the modernization of the railway line. From the noise study it is clear that the following modernization of the railway line will require effective noise reduction measures. At all measuring points, the permissible noise values were exceeded for the night period. There is no visible effect of the distance from the noise source to the noise exposure level, because noise of the road and geographically rugged terrain also affected to the noise level.

Observance of nighttime noise level value for residential areas of category II. (45 dB) at the distance of 100 m to 200 m from the railroad is quite demanding. Compared with other countries (Fig. 2) it is clear that the Slovak republic belongs to the countries that have some of the highest permissible values in this area.

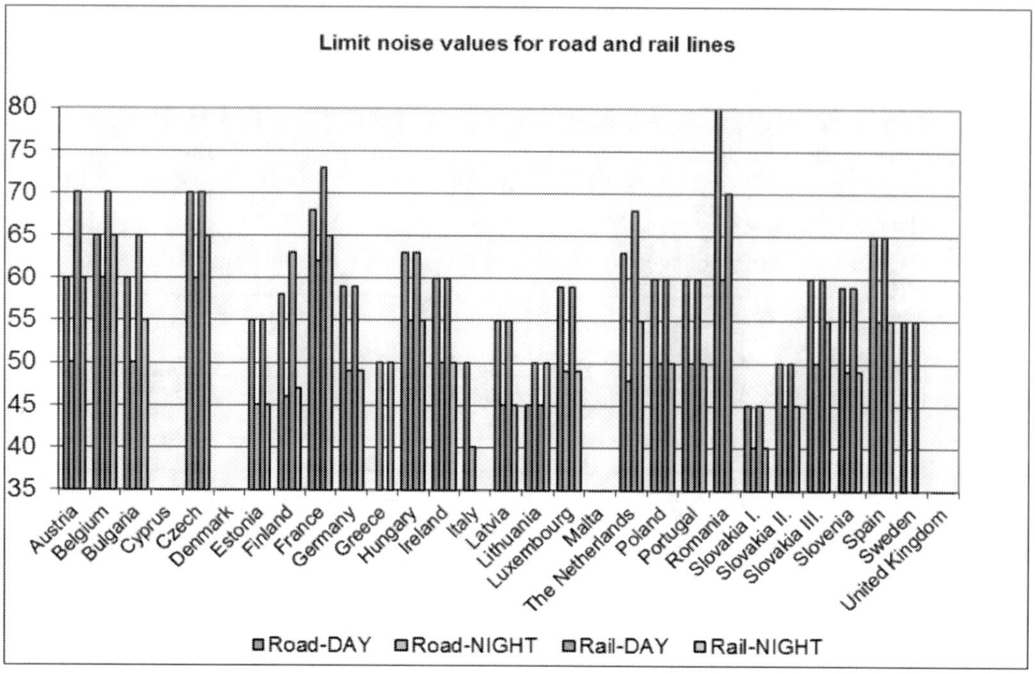


**Fig. 2** Permissible noise levels for road and railway communications in EU countries from 2010 [8]

There are partial results of the grant VEGA 1/0766/15" Research of noise emissions sources in the railway transport and methods of their effective reduction" in the paper.

## References

1. European Environment Agency, Noise in Europe 2014, EEA report 10. Luxembourg (Publications Office of the European Union 2014)
2. Murphy and King, (2014)
3. Decree of MZ No. 549/2007 Z.z. stating the details on the permissible levels of noise, infrasound and vibration and the requirements on objectification of noise, infrasound and vibration in the environment of 16th August 2007 (2007)
4. STN EN ISO 3095 Railway applications. Acoustics. Measurement of noise emitted by railbound vehicles (2005)
5. Professional guidance: NRÚ/3116/2005 (2005)
6. J. Veverka, V. Kozel, L. Ládyš, M. Liberko, J. Chybík, *Stavební fyzika*, (University of Technology Brno, 1998)
7. Archive of the ZPS Club in vibroacoustics, s.r.o. (2010)
8. M, Decký, A. Krokker, Hluk od cestnej dopravy ako limitujúci faktor kvality vonkajšieho prostredia. Fyzikálne faktory prostredia, 1, 205 (2011)
9. Grenčík, J., Skrúcaný, T., Volna, P.: Energy consumption as an important factor for decision on proper transport mode, In: TTS Technika transportu szynowego: koleje - tramwaje - metro. - ISSN 1232-3829. - Vol 22, no. 12 (2015).

# Design features of vortex dust collectors

*Svetlana* Puring[1], *Denis* Vatusov[1,*], and *Nicolay* Turin[1]

[1]Samara State Technical University, Institute of Architecture and Civil Engineering, 194, Molodogvardeyskaya St., 443001, Samara, Russia

**Abstract.** This article states that during the construction of industrial buildings it is necessary to provide engineering communications (including systems of supply and exhaust ventilation) which ensure uninterrupted production, its technological and environmental safety. It is necessary to use cleaning devices to remove pollutants from ventilation emissions in local exhaust systems. As for devices for removing highly disperse non-sticky dust, it is possible to offer vortex dust collectors, the efficiency of which depends on the design and operating mode. A method is proposed to increase the efficiency of a vortex dust collector, to improve the quality of the purified air, and to reduce energy consumption during air purification by installing watering nozzles in primary and secondary air pipes made in the form of a Venturi tube. In consequence of multidirectional movement of dust particles and sprayed water, active coagulation of dust particles takes place, as a result of which the separation increases. Apart from that, watering dusty air flows leads to the formation of a liquid film on the inner surface of the separation chamber, which prevents dust particles from rebounding off the separation chamber and facilitates their capture and flushing into the collecting hopper

## 1 Introduction

Effective functioning of industrial buildings is ensured by the installation of necessary engineering services. The systems of supply and exhaust ventilation are one of the basic systems in the construction of buildings. Among many components that ensure the productivity of local exhaust ventilation systems, it is necessary to distinguish air purification devices [1, 2].

At present, one of the most effective devices for removing dry fine dust from air is vortex dust collector (VDC). VDCs were for the first time patented in Germany in the 50s of the previous century [3]. As in cyclones, the operating principle of VDCs is based on the action of centrifugal forces. However, in a cyclone the polluted air is supplied through a single inlet, whereas air supply in a VDC is provided through two channels, the lower one and the upper one. The polluted air coming through the lower channel whirls and moves along the axis of the separator chamber upwards, where, under the action of centrifugal forces, dust particles are thrown to the periphery. At the same time, through the upper channel, air (contaminated or clean, depending on the design of the device) is fed, which,

* Corresponding author: vatuzov74@email.ru

whirls and moves down along the walls of the separator chamber, strengthening the effect of centrifugal forces affecting the dust particles that having reached the wall of the chamber fall down into the receiving hopper. Moreover, unlike cyclones, the use of a VDC makes it possible to achieve effective purification even for highly dispersed particles ($d_p$ <1 mkm) [4-8].

The main parameters influencing the efficiency of dust collectors, including VDCs, are efficiency of dust collection, hydraulic resistance of the device, its dimensions and cost of cleaning.

Some of these parameters are determined by the internal design features of the device. These include hydraulic resistance, dimensions of the device and efficiency of dust collection.

In addition to internal factors, the cost is also influenced by external ones, coming from the cost of materials and electricity, features of the technological process, etc., the forecasting of which is either difficult or specified by a particular production.

Therefore, the main task of designing an effective VDC is to optimize internal factors [9,10].

## 2 Materials and methods

### 2.1 Main theoretical dependencies used in the design of the VDC

The dimensions of the device are quite a significant factor that determines not only the volume required for the device installation, but also the convenience of installation, operation and maintenance. The required diameter of the device $D_1$ is determined by formula [11]

$$D_1 = \sqrt{4V/(\pi u_g)}, \tag{1}$$

where: $V$ – volumetric flow of dusty gas, m$^3$/s; $u_g$ – gas velocity in the working area of the dust collector, m/s (recommended to take within 5…12 m/s).

The efficiency of cleaning is the main criteria by which the work of the vortex dust collector is evaluated. The degree of air purification (efficiency) is characterized by the ratio of the mass of dust trapped in the device to the mass of dust introduced into it. Since dusty air can be supplied into the dust collector by two streams - through the upper and lower supply pipes, the overall cleaning efficiency will be determined depending on the distribution of the flows along the inputs:

$$\eta = (V_1\eta_1 + V_2\eta_2)V \tag{2}$$

where: $\eta_1$ ,$\eta_2$ – efficiency of dust collection in primary and secondary gas flows; $V_1$,$V_2$ – efficiency of dust collection in primary and secondary gas flows, m$^3$/s.

In addition to the degree of purification, the minimum particle size fully captured by the vortex dust collector is also of great significance for the device efficiency [4]:

$$d_{p\min} = \sqrt{\left(18\mu_g u_g \ln[D_1/D_t]\right)/\left(H\left[\rho_p - \rho_g\right]\omega^2\right)} \tag{3}$$

where: $H$ – height of dust collecting chamber, m; $D_t$ – diameter of the branch pipe for the supply of dusty gas, m; $\omega$ – angular velocity of the gas flow in the device, s$^{-1}$; $D_1$ – device diameter, m; $\mu_g$ – dynamic viscosity of gas, Pa·s; $\rho_p$ – particle density, kg/m$^3$; $\rho_g$ – gas density, kg/m$^3$.

Hydraulic resistance of the device, $\Delta p$, Pa, determines the energy efficiency of the device. The hydraulic resistance of vortex dust collectors is calculated using a ratio similar to that used to calculate cyclones:

$$P=\xi u_g^{\ 2}\rho_g/2 \tag{4}$$

where: $u_g$ – gas velocity in the device working zone, m/s; $\xi$ – coefficient of hydraulic resistance, calculated by speed.

Optimization of any of the considered factors leads to an increase in the VDC efficiency.

### 2.2 Improving VDC design

Research on the trajectories of particles, their velocity fields and distribution of static pressures in different areas of the dust collector, revealed the presence of turbulent vortex formations that enable the entrainment of the separated dust into the near-axis zone of the purified gas removed from the VDC. This phenomenon has a significant effect on the degree of cleaning of the dust collector, decreasing it from "theoretically possible" to "practically observed".

Therefore in process of designing such devices, one of the main tasks is to reduce entrainment, and to achieve this various design modifications are applied [5, 7, 9].

However, despite the design improvements, dry cleaning of dust and gas stream in vortex dust collectors, does not always allow to achieve the required degree of purification, especially from dust of submicron sizes, for a number of certain reasons, associated with the entrainment of this fraction [9]. An additional effect of reducing the entrainment of highly dispersed dust particles can be achieved during wet cleaning.

Therefore, in order to increase the efficiency of the dust collection device, to improve the quality of the cleaned air and to reduce energy consumption during air purification in a vortex dust collector containing a cylindrical separation chamber with upper and lower air supply channels, it is proposed to additionally supply each channel with an injector making it possible to wet the supplied airflow in a device having the form of a Venturi tube [12, 13].

Preliminary coagulation of fine dust particles takes place in wet low-velocity Venturi tubes at flow rates of the cleaned air in the Venturi tube neck $V_r$ to 40 m/s. Besides, the coefficients of the local resistance of the confuser and diffuser of the Venturi tube are assumed to be minimal. The device in the form of a Venturi tube is characterized by the following parameters: $V_n \leq 2V_t$; $l_{dif} \approx 2l_{con}$; $d_n \approx 0{,}7d_t$ ($V_t$ and $V_n$ – air velocity in the branch pipe and in the neck, m/s; $l_{dif}$ – diffuser length, m; $l_{con}$ – confuser length, m; $d_n$, $d_t$ – diameter of the neck and diameter of the branch pipe, m).

## 3 Results

The vortex dust collector (Fig. 1) contains a separation chamber, the upper part of which has an axial branch pipe for discharging purified gas and a branch pipe for introducing secondary air, equipped with a Venturi-type device fitted with a nozzle wetting the secondary airflow, which are installed in front of the Venturi tube and meet the air flow [14].

In the lower part of the housing, there is a branch pipe for the input of the primary dusty air, a blade vortex, a cylinder cowl and a hopper with a lower lateral outlet of the pulp. Similar to the branch pipe for the secondary dusty air, the branch pipe for the input of the primary dusty air is equipped with a Venturi-shaped device with a nozzle. Due to the multidirectional movement of dust particles and sprayed water, active coagulation of dust particles and water particles occurs. Rotation of two counter flows (primary and secondary air) inside the separation chamber has one direction.

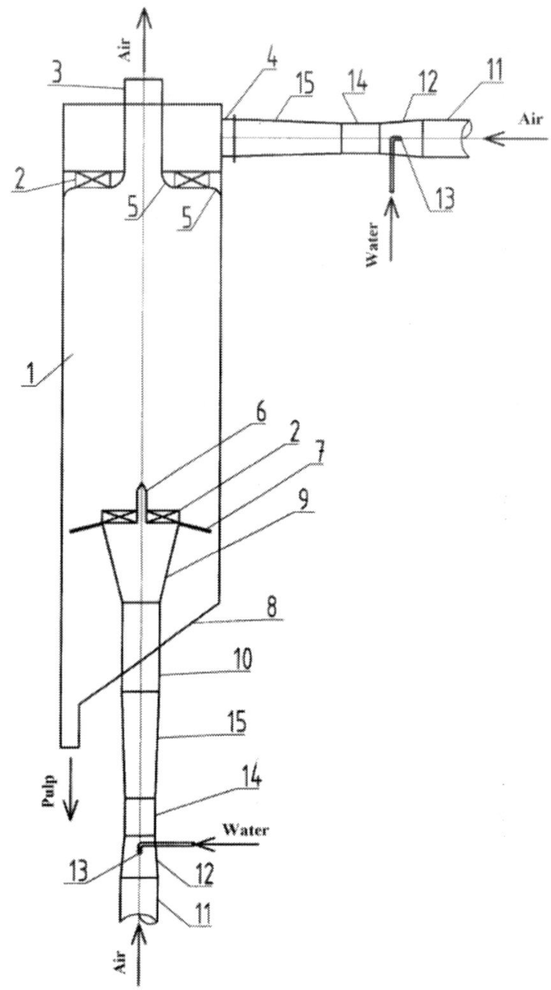


**Fig. 1.** The longitudinal section of the vortex dust collector: 1 – separation chamber, 2 – blade vortex, 3 – outlet branch pipe, 4 – branch pipe for introducing secondary air, 5 – stabilizing cowl, 6 – cylinder cowl, 7 – bumper washer, 8 – hopper compartment, 9 – stabilizing device, 10 – branch pipe for introducing primary air, 11 – supply air duct, 12 – confusor, 13 – injector, 14 – neck 15 – diffuser.

Enlarged due to coagulation agglomerates are separated efficiently in the separation chamber. Since during the separation the inner surface of the separation chamber gets covered by a film of running water, the fine dust particles moving in the laminar sublayer do not bounce off the inner surface of the separation chamber, but settle on it under the influence of gradient coagulation.

To prevent "overgrowth" of internal surfaces of the vortex dust collector and to improve flushing of the trapped dust, the amount of water sprayed by the injectors into the primary dusty air flow is 2-3 times bigger than into the secondary air flow.

The dispersed composition of the particles sprayed with water jets into the primary air flow is 10 ÷ 70 µm, which enables their better separation in the lower part of the separation chamber and better flushing of the trapped dust. The dispersed composition of particles sprayed with water jets into the secondary air flow is 2 ÷ 10 μm, which provides better mutual coagulation of dust and liquid particles throughout the whole separation chamber.

## 4 Discussion

Thus, the work resulted in creating a VDC of improved design that provides a higher cleaning efficiency.

The use of the device in the form of a low-velocity Venturi tube in the primary and secondary air supply canals reduces energy costs in the process of cleaning dusty air.

This VDC is recommended for removing non-splashing and water-soluble dusts from air during air purification.

## 5 Conclusions

The proposed invention allows to significantly improve the efficiency of the dust collection device and the degree of purification of dusty gases by pre-enlargement of dust particles, the separation of which considerably increases. Furthermore, watering dusty air flows leads to the formation of a liquid film on the inner surface of the separation chamber, which prevents the dust particles from rebounding off the separation chamber and facilitates their capture and flushing into the collecting hopper.

The considered design modifications enable an increase in the degree of purification of the VDC. However, the proposed options do not cover the whole range of possible improvements in vortex dust-cleaning devices and they are only a small link in the chain of possible design innovations.

## References

1. M. B. Generalov [and oth.], *Machinery and equipment of chemical and petrochemical industries, Mechanical engineering, P. IV–12* (Mechanical engineering, Moscow, 2004)
2. N.E.Pospelova, E.C.Timirbulatova, Urban Construction and Architecture **3 (16)**, 84-88 (2014) doi: 10.17673/Vestnik.2014.03.15
3. V.N. Azarov, *Dust collectors with counter swirling flow* (RPK «Politechnik», Volgograd, 2003)
4. M.V. Vasilievsky, *Dedusting of gases by inertial devices* (Tomsk Technical University Publishing, Tomsk, 2008)
5. I.A. Khurin, N.P. Turin, Regional architecture and construction **2**, 149-154 (2011)
6. V.V. Kuzmin, M.A. Komarov, Modern trends in the development of science and technology **V. 8, I. 1,** 118-120 (2015)
7. D.V. Azarov, D.P. Borovkov, Biosphere compatibility: people, region, technology **3**, 55-60 (2013)
8. U. V. Knyshenko, A. E. Deshko, Technical mechanics, **1**, 64-72 (2008)
9. S.M. Puring, D.N. Vatuzov, Modern science: actual problems of theory and practice. Series: Natural and technical sciences **1-2**, 10-12 (2013)
10. S.M. Puring, D.N. Vatuzov, N.P. Tyurin, Procedia Engineering **153**, 563-568 (2016)
11. V.N. Uzhov, A.U. Waldberg, B.I. Myagkov, I.K. Reshidov, *Purification of industrial gases from a dust* (Chemistry, Moscow, 1981)
12. M. Balzannikov, S.Piyavsky, A. Kurmanayev, MATEC Web of Conferences **86**, 02006 (2016)

13. E.A.Krestin, Urban Construction and Architecture **2 (23),** 48-55 (2016) doi: 10.17673/Vestnik.2016.02.10
14. N.P. Tyurin, D.N. Vatuzov, S.M. Puring, D.N. Tyurin, *Vortex dust collector, utility model patent* RUS 167446, (2016)

# Inhibitive Protection of Low-Carbon Steel in Citric Acid Solutions

*Yaroslav* Avdeev[1], *Maria* Tyurina[1], *Vladimir* Rabinkov[2,*], *Andrei* Luchkin[1]

[1] A.N. Frumkin Institute of Physical Chemistry and Electrochemistry, Russian Academy of Sciences, Leninsky pr. 31, Moscow, 119071 Russian Federation
[2]Moscow state university of civil engineering, Yaroslavskoye shosse, 26, Moscow, Russia, 129337

**Abstract:** An effective mixture has been developed on the basis of the inhibitor IFKhAN-92 and KCNS to protect low-carbon steel in citric acid hot solutions. The proposed mixture inhibits the corrosion of steel in these solutions in a wide range of their concentrations (0.25÷2.0 M) and temperatures (20÷95°C). The effective inhibition of the corrosion of steel by the mixture under study is due to the strong retardation of the metal's both electrode reactions.

## 1 Introduction

Thermal power equipment is chemically cleaned with citric acid solutions ($C_6H_8O_7$) before the launch (pre-operational chemical cleaning) to remove rust, scaling and other sediments and in the process of operation (operational cleanings) to remove iron oxide and carbonate (crust) formations [1-3]. These solutions demonstrate low aggressiveness towards steel equipment and remove roll scale, magnetite sediments and crust formations well, including through their complexing properties. Treatment with citrate solutions is quite frequently carried out at temperatures ($t$) close to 100°C. The solutions of other organic acids (adipic, maleic, oxalic, succinic) are less effective in thermal power equipment cleaning than citric acid solutions. Mixtures containing pyridine toxic derivatives – Catapin K and the inhibitor I-1-A [1], benzotriazol [4], cinnamic aldehyde [5] and various vegetal extracts [6-8] are recommended as additives to inhibit the corrosion of steels in citric acid solutions. In this case, benzotriazol, cinnamic aldehyde and vegetal extracts do not allow obtaining the metal protection degree of $Z > 90\%$. Lately, 1,2,4-triazole derivatives [9-11] have been gaining importance for protecting metals and alloys in mineral and organic acid solutions. The inhibitor of this class, IFKhAN-92, which we have developed, allows protecting low-carbon steel in mineral acids [12-14], and also in acetic and formic acid environments [15,16]. As its essential advantage, IFKhAN-92 is stable in hot acidic solutions.

It appeared expedient to develop an inhibitor based on IFKhAN-92, which is a substituted triazol, to protect low-carbon steel in citric acid hot solutions (up to 95°C).

* Corresponding author: rabinkov@yandex.ru

## 2 Materials and methods

### 2.1 Gravimetric measurements

The corrosion rate of low-carbon steel St3 in $C_6H_8O_7$ solutions was measured by the mass loss of specimens (no less than three specimens per experimental point) sized 50 mm × 20 mm × 1.5 mm, using 30 ml of the acid solution per specimen. Before the experiment, the specimens surface was conditioned by grinding on an abrasive disc (ISO 9001, 60 grit) and degreased with acetone. The duration of experiments was 2 hours.

The efficiency of inhibitors was estimated from the inhibition coefficient, $\gamma = k_0/k_{in}$, and the degree of protection $Z = [(k_0 - k_{in})/k_0]\cdot 100\%$, where $k_0$ and $k_{in}$ are the corrosion rates in the non-inhibited solution and in the solution with the additive being studied, respectively. Coefficients of mutual influence were calculated to quantify the effect of inhibiting mixture components on the efficiency of the protection of steel:

$$K_m = \frac{\gamma_{mix}}{\prod_{i=1}^{m} \gamma_i},$$

where $\gamma_{mix}$ is the corrosion inhibition coefficient for the inhibitor mixture and $\prod_{i=1}^{m} \gamma_i$ is the product of the corrosion inhibition coefficients for individual mixture components. If $K_m < 1$, the protection coefficient of the inhibitor components is mutually reduced; at $K_m = 1$, it shows an additive behavior; and only at $K_m > 1$ it should be concluded that the components of the inhibitor mixture mutually enhance protection.

Citric acid monohydrate (chemically clean) and distilled water were used to prepare solutions. KCNS (chemically clean) was studied as an additive to IFKhAN-92. Due to its low solubility, IFHAN-92 was introduced into $C_6H_8O_7$ solutions in the form of an ethanol solution, with the ethanol concentration in the pickling solution equaling 0.24 mol/ l.

### 2.2 Electrochemical measurements

Electrochemical measurements were carried out on a cylindrical electrode made of steel St3 (0.72 $cm^2$) in the argon-deaerated 2.0 M $C_6H_8O_7$ solutions at $t = 95°C$. The steel electrode potentials were measured with reference to a silver chloride electrode and recalculated to the standard hydrogen electrode scale. The electrode that was preliminarily roughed up and degreased by acetone was held in the test solution for 30 minutes, after which the anodic and cathodic polarization curves were sequentially recorded with the help of the potentiostat IPC-PRO FM at a dynamic scan rate of 0.0005 V/s.

## 3 Results and their discussion

### 3.1 Gravimetric measurements

The corrosion of steel St3 in 0.25÷2.0 M $C_6H_8O_7$ solutions at a temperature of $t = 20÷95°C$ occurs at a relatively low rate of no more than 106 g/($m^2$·h) (Table 1). The corrosion of steel accelerates with the increase of the concentration $C_{C6H8O7}$ and $t$ in the solution, which is consistent qualitatively with the data of the authors [3] on the study of the dissolution of low-carbon steels' kinetics in citric solutions.

**Table 1**. Corrosion rates ($k$, g/(m$^2$·h)) and corrosion inhibition coefficients ($\gamma$) of steel St3 in citric acid solutions

| Inhibitor | Acid concentration | | | | | | | |
|---|---|---|---|---|---|---|---|---|
| | 0.25 M | | 0.50 M | | 1.0 M | | 2.0 M | |
| | $k$ | $\gamma$ | $k$ | $\gamma$ | $k$ | $\gamma$ | $k$ | $\gamma$ |
| 20°C | | | | | | | | |
| - | 1.93 | - | 1.95 | - | 1.98 | - | 2.01 | - |
| 5.0 mM IFKhAN-92 | 1.32 | 1.5 | 1.35 | 1.4 | 1.37 | 1.4 | 1.41 | 1.4 |
| 5.0 mM KCNS | 1.96 | 0.98 | 2.01 | 0.97 | 2.15 | 0.92 | 2.21 | 0.91 |
| 4.5 mM IFKhAN-92 + 0.5 mM KCNS | 0.049 | 39.4 | 0.054 | 36.1 | 0.068 | 29.1 | 0.071 | 28.3 |
| 40°C | | | | | | | | |
| - | 4.82 | - | 5.22 | - | 6.14 | - | 6.26 | - |
| 5.0 mM IFKhAN-92 | 3.51 | 1.4 | 3.56 | 1.5 | 3.64 | 1.7 | 3.70 | 1.7 |
| 5.0 mM KCNS | 8.83 | 0.55 | 9.20 | 0.57 | 10.7 | 0.57 | 11.0 | 0.57 |
| 4.5 mM IFKhAN-92 + 0.5 mM KCNS | 0.21 | 23.0 | 0.23 | 22.7 | 0.27 | 22.7 | 0.29 | 21.6 |
| 60°C | | | | | | | | |
| - | 15.1 | - | 17.6 | - | 19.5 | - | 22.5 | - |
| 5.0 mM IFKhAN-92 | 8.09 | 1.9 | 8.43 | 2.1 | 9.02 | 2.2 | 9.53 | 2.4 |
| 5.0 mM KCNS | 17.3 | 0.87 | 18.3 | 0.96 | 19.2 | 1.0 | 20.3 | 1.1 |
| 4.5 mM IFKhAN-92 + 0.5 mM KCNS | 0.29 | 52.1 | 0.30 | 58.7 | 0.33 | 59.1 | 0.35 | 64.3 |
| 80°C | | | | | | | | |
| - | 30.3 | - | 39.7 | - | 45.2 | - | 52.5 | - |
| 5.0 mM IFKhAN-92 | 15.7 | 1.9 | 16.7 | 2.4 | 17.3 | 2.6 | 16.1 | 3.3 |
| 5.0 mM KCNS | 34.9 | 0.87 | 40.5 | 0.98 | 44.5 | 1.0 | 44.4 | 1.2 |
| 4.5 mM IFKhAN-92 + 0.5 mM KCNS | 0.34 | 89.1 | 0.35 | 113 | 0.38 | 119 | 0.40 | 131 |
| 95°C | | | | | | | | |
| - | 57.4 | - | 66.6 | - | 90.6 | - | 106 | - |
| 5.0 mM IFKhAN-92 | 24.1 | 2.4 | 26.9 | 2.5 | 30.8 | 2.9 | 34.6 | 3.1 |
| 5.0 mM KCNS | 62.1 | 0.92 | 70.4 | 0.95 | 78.8 | 1.1 | 85.0 | 1.2 |
| 4.5 mM IFKhAN-92 + 0.5 mM KCNS | 0.43 | 133 | 0.50 | 133 | 0.55 | 165 | 0.57 | 185 |

The addition of 5.0 mM IFKhAN-92 to 0.25÷2.0 M $C_6H_8O_7$ solutions weakly reduces $k$ of steel: $Z_{IFKhAN-92}$ = 29÷70%. One of the methods to enhance the inhibitory effect of IFKhAN-92 in mineral acid solutions is to add hydrophobic compounds, among which KCNS is distinguished, which we successfully used for the protection of steel in phosphoric acid media [17,18]. Such additive itself either stimulates the corrosion of steel or slightly inhibits it with the value of $Z \leq 17\%$. The additive's maximum stimulating effect is observed at $t = 40$°C.

In the presence of 4.5 mM IFKhAN-92 + 0.5 mM KCNS in $C_6H_8O_7$ solutions, the corrosion of steel was significantly slowed down: $Z$ = 96÷99%. In this case, the corrosion rate of steel in acid solutions inhibited by this mixture did not exceed 0.57 g/(m$^2$·h). With $t$ > 40°C, the $\gamma$ values of the mixture of inhibitors grow systematically with the increase of $t$. It is important to note that the temperature peak of the inhibition efficiency of the IFKhAN-92 and KCNS mixture was not reached in the tested range of $t$. The increase of $t$ by 75°C increases the value of $\gamma$ by 3.4÷6.5 times. The mixture's protective effect at $t$ = 20÷40°C decreases with the growth of $C_{C6H8O7}$ and, on the contrary, rises at higher $t$.

The mixture of IFKhAN-92 and KCNS displays the synergism of components, with $K_m$ demonstrating quite high values (Table 2). It intensifies with the increase of $t$ and declines with the increase of $C_{C6H8O7}$.

**Table 2.** Coefficients of mutual influence of 4.5 mM IFKhAN-92 and 0.5 mM KCNS components in corrosion of steel St3 in citric acid solutions

| $t$, °C | Acid concentration | | | |
|---|---|---|---|---|
| | 0.25 M | 0.5 M | 1.0 M | 2.0 M |
| 20 | 27 | 27 | 23 | 22 |
| 40 | 30 | 27 | 23 | 22 |
| 60 | 32 | 29 | 27 | 24 |
| 80 | 54 | 48 | 46 | 33 |
| 95 | 60 | 56 | 52 | 50 |

The inhibitive mixture under study retains its protective effect in citric acid solutions during at least 6 days (Table 3). Over this time, the rate of corrosion in the base solution decreases by 14.4 times and in the presence of 4.5 mM IFKhAN-92 + 0.5 mM KCNS it declines slightly less – by 7.7 times. A marked decrease in the $\gamma$ value of the inhibitive mixture is observed only on the 6th day of observations.

**Table 3.** Corrosion rates of steel St3 ($k$, g/(m$^2$·h)) and coefficients of corrosion inhibition ($\gamma$) in 2.0 M citric acid inhibited by 4.5 mM IFKhAN-92 and 0.5 mM KCNS mixture at specimens' different exposure time. $t$ = 20°C

| Inhibitor | $k$, $\gamma$ | Specimen exposure time, h | | | | | | | | |
|---|---|---|---|---|---|---|---|---|---|---|
| | | 0.5 | 1.0 | 2 | 4 | 10 | 24 | 48 | 96 | 144 |
| - | $k$ | 3.32 | 2.96 | 2.01 | 1.42 | 0.79 | 0.48 | 0.34 | 0.29 | 0.23 |
| IFKhAN-92 + KCNS | $k$ | 0.10 | 0.09 | 0.071 | 0.051 | 0.029 | 0.018 | 0.013 | 0.012 | 0.013 |
| | $\gamma$ | 33.2 | 32.9 | 28.3 | 27.8 | 27.2 | 26.7 | 26.2 | 24.2 | 17.7 |

The IFKhAN-92 + KCNS mixture with fixed $C_{KCNS}$ = 0.5 mM demonstrates an $S$-shaped dependence of $Z_{mix}$ on IFKhAN-92 concentration (Fig. 1). The values of Z ≥ 90% are reached at $C_{IFKhAN-92} \geq 0.1$ mM, which, based on the shape of experimental $Z_{mix}$ – $C_{IFKhAN-92}$ dependencies, corresponds to the maximum coverage of the metal's surface with the inhibitor. With $C_{IFKhAN-92} < 0.01$ mM, the mixture loses its protective effect and can stimulate corrosion. We discussed the nature of the stimulating effect of low $C_{IFKhAN-92}$ values on the corrosion of metal in acids earlier [19] and it is most probably related to the solution's surface tension weakened by this inhibitor, which facilitates the removal of hydrogen bubbles from the metal's surface.

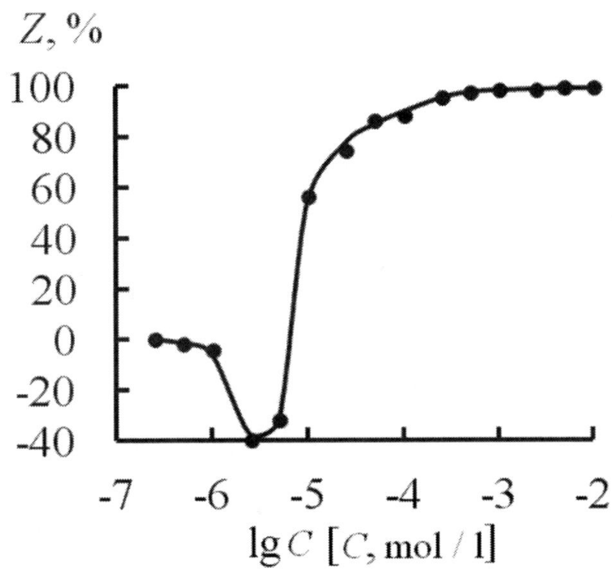


**Fig. 1.** Dependence of the degree of protection by IFKhAN-92 + KCNS (0.5 mM) mixture on IFKhAN-92 concentration for St3 steel corrosion in 2.0 M $C_6H_8O_7$ at 60°C

The same mixture with $C_{IFKhAN-92}$ = 5.0 mM also manifests an *S*-shaped dependence of $Z_{mix}$ on $C_{KCNS}$ (Fig. 2). This composition loses its efficiency at $C_{KCNS}$ < 0.05 mM in comparison with 5.0 mM IFKhAN-92. The maximum protective effect of the 5.0 mM IFKhAN-92 + KCNS mixture is observed at $C_{KCNS}$ = 0.25 mM ($\gamma$ = 98). A further increase in $C_{KCNS}$ decreases $\gamma$. For example, $\gamma_{mix}$ for the mixture of 5.0 mM IFKhAN-92 + 10.0 mM KCNS is 1.6 times lower.

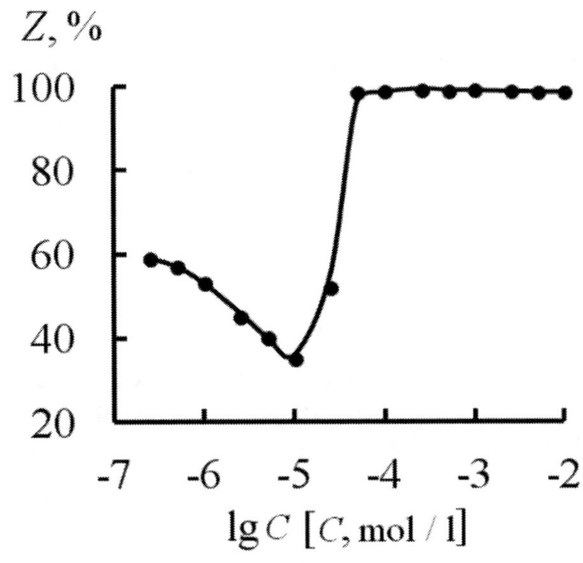


**Fig. 2.** Dependence of the degree of protection by IFKhAN-92 (5.0 mM) and KCNS mixture on KCNS concentration for St3 steel corrosion in 2.0 M $C_6H_8O_7$ at 60°C

### 3.2 Electrochemical measurements

The electromechanical measurements performed on steel in 2.0 M $C_6H_8O_7$ ($t$ = 95°C) have shown that the use of the 5.0 mM KCNS additive in the solution somewhat slows down the cathodic reaction of steel and disinhibits the anodic reaction probably due to the participation of rhodanide anion in this process as a complexing reagent (Fig. 3). Addition of 5.0 mM IFKhAN-92 refines the corrosion potential ($E_{cor}$), inhibiting mainly the anodic process. The use of the 4.5 mM IFKhAN-92 + 0.5 mM KCNS mixture causes a considerably stronger increase in $E_{cor}$, and also the inhibition of the cathodic and especially the anodic reaction. Therefore, the corrosion of steel in $C_6H_8O_7$ solutions is inhibited more effectively by the IFKhAN-92 + KCNS mixture due to the stronger inhibition of the metal's electrode reactions by this mixture compared to its components.

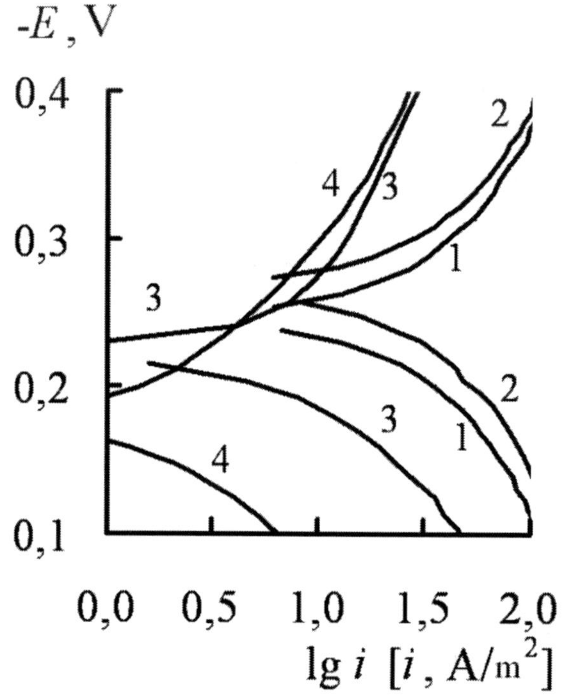


**Fig. 3.** Steel St3 polarization curves in 2.0 M C6H8O7 (1) in the presence of: 2 – 5.0 mM KCNS, 3 – 5.0 mM IFKhAN-92, 4 – 4.5 mM IFKhAN-92 + 0.5 mM KCNS. t = 95°C

## 4 Conclusions

4.1 The IFKhAN-92 metal acid corrosion inhibitor can be successfully used in a mixture with the KCNS additive for protecting low-carbon steel in citric acid solutions in a wide range of its concentrations (0.25÷2.0 M) and temperatures (20÷95°C) during a long period of time.

4.2 The effective inhibition of the corrosion of steel in citric acid solutions by the IFKhAN-92 + KCNS mixture occurs due to the strong inhibition of the metal's both electrode reactions by this mixture.

## References

1. Ivanov E.S., Ingibitory korrozii metallov v kislyh sredah (Ye.S. Ivanov, Inhibitors of Metals Corrosion in Acid Media), in Reference Book, Metallurgiya, Moscow, (1986) pp. 112-118 (in Russian).
2. Rozenfel'd I.L., Ingibitory korrozii (I.L. Rozenfeld, Corrosion Inhibitors), in Khimiya, Moscow (1977) pp. 235-242 (in Russian).
3. Fateev V.D., Klyuchnikov N.G., Izuchenie rastvoreniya kotel'nyh stalej i okislov zheleza v rastvorah limonnoj kisloty (V.D. Fateyev, N.G. Klyuchnikov, Study of the Dissolution of Boiler Steel and Iron Oxides in Citric Acid Solutions), in Ingibitori Korrozii Metallov, Sudostroyeniye, Moscow (1965) pp. 192-209 (in Russian).
4. Matheswaran P., Ramasamy A.K., Influence of Benzotriazole on Corrosion Inhibition of Mild Steel in Citric Acid Medium, in E-Journal of Chemistry (2010) **V. 7 (3)**, pp. 1090-1094.
5. Sathiya Priya A.R., Muralidharan S., Velmurugan S., Venkatachari G., Corrosion inhibitor for the chemical decontamination of primary coolant systems of nuclear power plants, in Materials Chemistry and Physics (2008) **V. 110**, pp. 269-275.
6. Matheswaran P., Ramasamy A.K., Corrosion Inhibition of Mild Steel in Citric Acid by Aqueous Extract of Piper Nigrum L., in E-Journal of Chemistry (2012) **V. 9 (1)**, pp. 75-78.
7. Alka Singh, Kalpana S., Inhibition of the Corrosion of Iron in Citric acid solutions by Aqueous extract of Fenugreek seeds, in Ultra Chemistry, India, Bhopal, 2012, V. 8 (2), pp. 175-179.
8. Anand B., Balasubramanian V., Corrosion Behaviour of Mild Steel in Acidic Medium in Presence of Aqueous Extract of Allamanda Blanchetii, in E-Journal of Chemistry (2011) **V. 8 (1)**, pp. 226-230.
9. Ya.G. Avdeev, Yu.I. Kuznetsov, Physicochemical aspects of inhibition of acid corrosion of metals by unsaturated organic compounds, Russ. Chem. Rev., 81 (2012) 1133-1145.
10. Korolchenko, D., Tusnin, A., Trushin, S., Korolchenko, A. Physical parameters of high expansion foam used for fire suppression in high-rise buildings. (2015) International Journal of Applied Engineering Research, **10 (21)**, pp. 42541-42548.
11. Sharovarnikov A.F., Korol'chenko D.A., Fighting fires of carbon dioxide in the closed buildings // Applied Mechanics and Materials (2014) **Vol. 475-476**. – P.1344-1350.
12. Avdeev YA.G., Luchkin A.YU., Kuznecov YU.I., Gorichev I.G., Tyurina M.V., Zashchita nizkouglerodistoj stali v serno-kislyh rastvorah ot vysokotemperaturnoj korrozii (do 200C) (Ya.G. Avdeev, A.Yu. Luchkin, A.Yu. Kuznetsov, I.G. Gorichev, M.V. Tyurina, Protection of Low-Carbon Steel in Sulphuric Acid Solutions from High-Temperature Corrosion (up to 200□C)), in Korroz.: mater., zashch., Moscow (2011) **No. 8**, pp. 20-26 (in Russian).
13. Avdeev YA.G., Luchkin A.YU., Kuznecov YU.I., Gorichev I.G., Tyurina M.V., Zashchita nizkouglerodistoj stali v solyanokislyh rastvorah v usloviyah vysokotemperaturnoj korrozii (do 160C) (Ya.G. Avdeev, A.Yu. Luchkin, Yu.I. Kuznetsov, I.G. Gorichev, M.V. Tyurina, Protection of Low-Carbon Steel in Hydrochloric Acid Solutions in the Conditions of High-Temperature Corrosion (up to 160□C)), in Korroz.: mater., zashch., Moscow (2011) **No. 10**, pp. 26-31 (in Russian).
14. Shebeko, Yu.N., Smolin, I.M., Korolchenko, A.Ya., Shevchuk, A.P., Borodkin, A.N., Malkin, V.L., Simonov, O.A., Gurinovich, L.V., Popov, S.A., Kolosov, V.A., Smirnov, E.V. Some aspects of fire and explosion hazards of large LPG storage vessels. (1995) Journal of Loss Prevention in the Process Industries, **8 (3)**, pp. 163-168. DOI: 10.1016/0950-4230(95)00011-O.

15. Avdeev YA.G., Kuznecov YU.I., Tyurina M.V., Ob ingibirovanii korrozii nizkouglerodistoj stali v goryachih rastvorah organicheskih kislot (Ya.G. Avdeeev, Yu.I. Kuznetsov, M.V. Tyurina, Inhibition of Low-Carbon Steel Corrosion in Hot Solutions of Organic Acids) in Korroz.: mater., zashch., Moscow (2012) **No. 3**, pp. 24-28 (in Russian).
16. Shebeko, Yu.N., Tsarichenko, S.G., Trunev, A.V., Korolchenko, A.Ya., Kaplin, A.Yu. The influence of inert retardants on the combustion of hydrogen-oxygen mixtures under elevated temperatures and pressures. (1994) Combustion, Explosion, and Shock Waves, **30 (2)**, pp. 183-188. DOI: 10.1007/BF00786125
17. Avdeev YA.G., Kuznecov YU.I., Tyurina M.V., Zashchita nizkouglerodistoj stali v rastvorah fosfornoj kisloty ingibitorom IFHAN-92. CH. 1 (Ya.G. Avdeev, Yu.I. Kuznetsov, M.V. Tyurina, Low-Carbon Steel Protection in Phosphoric Acid Solutions by the IFKhAN-92 Inhibitor. Part 1) in Korroz.: mater., zashch., Moscow (2012) **No. 5**, pp. 22-26 (in Russian).
18. Shebeko, Yu.N., Tsariichenko, S.G., Korolchenko, A.Ya., Erofeev, A.N. Characteristics of combustion of hydrogen - methane - air mixtures in a closed vessel. (1991) Fizika Goreniya i Vzryva, **(6)**, pp. 52-56.
19. Avdeev YA.G., Belinskij P.A., Kuznecov YU.I., Zel' O.O., Zashchita stali ot serno-kislotnoj korrozii ingibitorom IFHAN-92 (Ya.G. Avdeev, P.A. Belinsky, Yu.I. Kuznetsov, O.O. Zel', Protection of Steel from Sulfuric Acid Corrosion by the IFKhAN-92 Inhibitor), in Korroz.: mater., zashch., Moscow (2008) **No. 8**, pp. 16-21 (in Russian).

MATEC Web of Conferences **117**, 00144 (2017) DOI: 10.1051/matecconf/201711700144
*XXVI R-S-P Seminar 2017, Theoretical Foundation of Civil Engineering*

# Problems in Carrying Out Construction Projects in Large Urban Agglomerations on the Example of the Construction of the Axis and High5ive Office Buildings in Krakow

*Elżbieta* Radziszewska-Zielina[1*], *Ewelina* Kania[2]
[1, 2]Department of Construction Technology and Organisation, Institute of Management in Construction and Transport, Faculty of Civil Engineering, Cracow University of Technology, Poland

**Abstract.** The paper discusses problems in the carrying out of construction projects in large urban agglomerations on the example of two structures located in Krakow. The location and the specifics of the work that had to be performed generated numerous technological and organisational problems. The manner of solving them, in addition to propositions of guidelines for similar projects are discussed in the paper.
*Keywords: technology, organisation, construction project, urban agglomeration*

## 1. Introduction

The term construction project is used to describe an action that is undertaken in order to meet the needs of a property developer, as well of the future users of a structure. A construction project includes work focused on the preparation, organisation and carrying out of construction work, settling the accounts for construction and providing a finished structure that is ready for operation [1,2]. The carrying out of a construction project is a complicated process associated with unique external factors and the individual qualities of a given project.

Urban agglomerations are areas with the highest density of construction projects. In such a case, a property developer is often a public entity and adhering to public procurement laws is required [3]. An urban agglomeration is an area with a very dense built environment and a high density of population - both permanent and that which temporarily stays in its area. This area is typified by a rapid spatial development, well-developed infrastructure and widely understood socio-economic and cultural conditions [4]. The rapid development of urban agglomerations is associated with numerous construction problems. Table 1 contains examples of the causes of problems and a selection of problems that occur in construction projects carried out in large urban agglomerations.

The goal of the paper is to discuss problems related to construction projects that are carried out in large urban agglomerations in Polish conditions on the example of two office building projects located in Krakow.

* Corresponding author: eradzisz@izwbit.pk.edu.pl

**Table 1** Examples problems in the carrying out of construction projects on the territory of large urban agglomerations, as well as their causes. Original work based on [5,6,7,8,9,10,11,12,13]

| | | |
|---|---|---|
| CONSTRUCTION PROBLEMS THAT ARE INDEPENDENT OF EITHER PARTY | | • A dense built environment (the need to perform deep excavation, height limits for buildings)<br>• Limitations in available space for performing work and setting up storage areas<br>• Difficulties in logistics (traffic jams)<br>• Dynamic (e.g. vibrations) and mechanical loads influencing the structure and the ground<br>• Negative influence on humans due to noise, traffic difficulties, dirt<br>• Difficult soil and water, as well as atmospheric conditions (wind, snow, rain, temperature)<br>• Natural disasters (flooding, hurricane, landslides)<br>• Equipment failure<br>• Intermissions in power and water supply<br>• Sudden price increases, inflation<br>• The requirement of conservation supervision in the case of historical buildings or work that is being carried out in the vicinity of areas of historical importance (specific construction and material requirements, additional work that is difficult to foresee) |
| DEPENDENT CONSTRUCTION PROBLEMS | ON THE PROPERTY DEVELOPER | • Financial difficulties<br>• Requirements of constructing more representative structures first (for instance from the side of the street with increased traffic in the case of a multiple-building project)<br>• Delays in the decision-making process<br>• Problems with coordination and communication between project participants |
| | ON THE DESIGNER | • Changes in structural and material solutions in the design of a structure already under construction<br>• Administrative and legal problems<br>• Mistakes in the technical documentation<br>• An insufficient level of detail in the design<br>• Delays in the decision-making process<br>• Problems with coordination and communication between project participants |
| | ON THE CONTRACTOR | • Changes in the timetable resulting in delays in performing work and material deliveries<br>• Picking insufficient scheduling tools<br>• Inappropriate construction technology choices<br>• Lack of appropriate equipment for given construction work<br>• Lack of specialist management and construction staff<br>• Mistakes in material and service commissions<br>• Delays in material and service commissions<br>• Lack of financial resources for the procurement of materials and equipment rental<br>• Not adhering to health and safety regulations, resulting in work accidents, intermissions in work and financial penalties<br>• Improperly organised logistics (transports on the construction site and the work of tower cranes)<br>• Delays in the decision-making process<br>• Problems with coordination and communication between project participants |
| | ON THE SUPPLIERS | • Problems with timely order delivery<br>• Materials out of stock<br>• Difficulties in the delivery of non-typical materials and construction products<br>• Providing products and services of inferior quality<br>• Insufficient transport pool<br>• Production plant failure (e.g. concrete mixing facility) |

## 2. Characteristics of the analysed structures

**Characteristic of the first structure - Axis Krakow**

The Axis building (fig.1) is an office building with services on the ground floor. The construction of a part of the structure of the Axis office building began in 2014 and took 22 months to complete. The office surface area of the building amounts to 19 298 m$^2$, while retail and service areas amount to 1183 m$^2$. The building has 9 storeys above ground and three underground levels with parking lots. The building is a class A office building [14]. The load-bearing structure is composed of monolithic and prefabricated reinforced concrete elements. The subterranean levels were constructed as entirely monolithic, while those above ground used prefabricated reinforced concrete pillars, beams, staircases and FILIGRAN-type floor slabs, in addition to monolithic shafts and landings [15].

**Fig. 1** View of the finished Axis building. Original photograph.

**Characteristics of the second structure - the High5ive complex (buildings D and E)**

The High5ive office building complex (fig.2) is being built in Krakow near Pawia street. The surface area of two of the buildings from among the five-building complex amounts to around 23 500 m$^2$. Both of the buildings have six storeys and have subterranean levels with parking lots (building D – one underground level, building E – two underground levels). They are class A buildings. The load-bearing structure of both buildings is a spatial post and slab structural system with monolithic shafts that provide rigidity.

**Fig.2** Visualisation of the High5ive buildings [16]

## 3. Problems in constructing the analysed buildings in Krakow

**Problems associated with a dense built environment**

The Axis office and service building is located in the central part of the city of Krakow in the area of Rondo Mogilskie. The building's surroundings include a hotel, a green square and some of Krakow's largest and most important streets. The area designated for the construction of the second building - the High5ive office and service building complex is located in the strict city centre of Krakow. The immediate vicinity includes a shopping gallery and hotels. The construction site is adjacent to the main train station and a street with a tram line that is partially on the surface and partially in a tunnel. A part of the site is located over an existing road tunnel. Such a site (fig. 3) generates serious difficulties and limitations in construction work and influences their duration. In the case of a construction project carried out in a dense built environment, the preparatory and design phases are particularly crucial, including: a comprehensive identification of soil and water conditions both underneath the designed structure, as well as in its zone of influence on surrounding buildings, determining the functional and spatial solutions of the underground section, a selection of the support and construction method of the underground part of the designed building and the monitoring of the condition of adjacent structures [8,18].

**Fig. 3** The location of the AXIS office building [15] and the buildings of the High5ive complex [17]

## Problems associated with deep excavation

Performing work associated with deep excavation is often problematic, especially in the case of large cities. The authors of [10] while discussing projects in New York, tackle problems that have also been observed by the author of [19] who discusses projects with deep excavations in Krakow. The height of the structures is limited by urban zoning and determined in the spatial development plan of the city. Oftentimes, this results in the necessity of building several subterranean levels that are to serve as parking lots and the logistics hub of a building. Due to the proximity of neighbouring buildings and the varied nature of subsurface conditions (groundwater, ground type), specialised support of the excavation walls is necessary. Important threats demand consideration, such as: the relaxation of the ground after performing excavation, a lowering of the level of ground water, the relocation, cracking and lowering of the load-bearing capacity of existing foundations and damage to the structure and installations of existing structures [8].

The projects that are being discussed are linked by the need to perform deep excavations in a dense built environment, whilst they differ in terms of their surroundings, soil and water conditions and the type of excavation wall support. The construction site of the Axis office building is located at a distance of around 1 km from the Białucha river. This is the cause of the stabilisation of the ground water level at around 4,40 m below the surface, while the third subterranean level of the office building has been placed at a depth of 13,28 m. The depth of the required excavation and the proximity of a densely built-up area, including that of tall buildings are the cause of the employment of a specific technology of the subterranean elements of the building that has been built. The three-level underground section has been constructed using the top-down method (fig. 4). The sequence of actions in the case of this method must be performed in stages without the possibility of work overlap, which lengthens the construction time and generates additional costs. In the case of the multiple-building project of the High5ive complex, a different type of deep excavation has been used. The circumference of building D has been secured using a Berliner retaining wall. In the case of this technology, the task that posed a problem was the construction of the walls of the subterranean levels in contact with the Berliner walls while including the rigors of preserving the technology of waterproof concrete. Building E, with two subterranean levels, required the use of a concrete diaphragm wall. In this case, temporary steel trench strut supports have been installed between the diaphragm walls.

When comparing the time of the construction of subterranean levels in the event of using deep excavations with the time of the construction of levels above ground, we can observe just how a demanding and time-consuming process it is. The structure of the High5ive complex has been made entirely of monolithic concrete. In building D, the first subterranean level took 4 months to build, while 6 above-ground levels took 6. In building E, 2 subterranean levels took 6 months to build while 6 above-ground levels took 7.

**Fig. 4** The construction of the underground levels of the AXIS office building using the top and down method. Original photograph.

### Problems associated with logistics and limited storage area

A large project scope with a limited construction site area, short completion time and a specific location add up to a complex logistical problem. The partnering relations between the participants of the project and effective logistics are the groundwork for success and are responsible for the delivery of the required resources, such as: materials, equipment, people, money and information [20, 21].

In urban agglomerations, internal on-site circulation is often a problem. In both of the analysed cases, expensive parcels in the city centre have been completely developed with the constructed buildings in mind, at the same time limiting storage area to a minimum (fig. 5). The newly built levels, the work of tower cranes and the quick pace of work, in addition to the simultaneous presence of several hundred people at the construction site required the establishment of a separate logistics department. Delivery deadlines and locations in the case of such a location and the difficulties that had been encountered had to be met in accordance with a strict schedule that were given by construction supervisors to the logistics department. A common practice in the event of having limited space at a site is transport and assembly without storage. This method has been used in the projects that are discussed. This was a very important matter in the case of the construction of the Axis building. The delivered elements needed to be assembled and incorporated into the structure in an established sequence, which is why transport delays, alterations in the sequence of the arrival of trucks with material or the wrong orientation of an element on a truck would cause delays. The small storage area resulted in limited capabilities of ordering elements that were supposed to be installed ahead of time (for instance steel rebar was delivered slightly ahead of the assembly time). Due to the lack of space on the site, the site's backrooms had to be located in a place some distance away from it.

Traffic jams are a serious problem in city centres. Both construction projects are located near large, busy streets, which is a plus for future users as it provides good transport accessibility to the rest of the city, but has proven to be difficult during construction. The greatest logistical difficulties in this regard occur during the excavation stage and the transport of large masses of earth, as well as during the erection of the concrete structure. Concrete transport trucks would often stand in traffic jams long enough that the mix would not be fit for construction due to changing its consistency. After performing concrete slump tests by an accredited laboratory, the mix had to be sent back to the production plant for processing, which generated costs for the contractor and supplier and negatively impacted the environment due to air pollution from the concrete transport trucks' exhaust.

Another additional cost affecting a project's budget in a large city in the case of limited construction site area is the need to temporarily or permanently take over walkways or streets that belong to a city or a private owner (e.g. that of Galeria Krakowska in the case of High5ive). Sometimes heavy transports can damage the pavements or streets that are used, which requires unscheduled repairs. Such areas also need to be kept appropriately clean, which requires additional work.

**Fig. 5** A view of the limitations associated with the location of buildings E and D of the High5ive complex (a dense built environment, lack of a storage area, the train station platforms, the parking lot of the shopping gallery). Original photograph.

One of the most important criteria of carrying out work in any location is the safety of the workers on the site. A dense built environment and the site's location near busy streets causes a project's influence on the safety of people that are not associated with the construction to be quite large. This has been a particularly important problem in the case of High Five. It is located near a busy walkway. The problem of protecting people was solved by constructing an additional overhang above the walkway and the fitting of safety nets on each of the levels of the buildings under construction.

**Problems with the influence of vibrations on both existing buildings and those under construction**

In the case of carrying out a construction project in an urban agglomeration, we can divide the problem of vibrations into two types: the vibrations of the structure under construction under the influence of the project's surroundings and the vibrations of the structures adjacent to the site under the influence of construction work. Performing an analysis of the influence of the space that surrounds the planned project and taking it into account already during the design process is an important problem. Both of the projects that are being discussed are located near busy streets with heavy traffic from cars and trams. Besides, the buildings of the High5ive complex are placed in difficult conditions in terms of vibrations. They are located near the main train station, which services around 240 trains of various types per day. On the other side of the site, it is influenced by the presence of a tunnel for high-speed municipal trams. These surroundings required the adoption of an appropriate excavation technology, the ability to provide the buildings with a safe foundation depth and forced the contractor to constantly monitor the tension of the excavation wall supports, the successively constructed levels, the ground and the walls of the tunnels near the site. The problem of the influence of vibration from traffic has been widely discussed in numerous works, including [22, 23, 24, 25]. Construction work carried out in a dense built environment exerts influence on the surroundings of a site. The work of heavy construction machinery induces vibrations in the ground which affect the buildings that surround a site. This can lead to foundation damage and cause cracking of an entire building's structure. The state of the structures near the site needs to be monitored and taken into account during the technology selection process.

**Problems with noise**

Noise is a natural problem in an urban environment. Most studies in the case of urban agglomerations are concentrated on noise from traffic. We can also find important studies [26, 27] regarding persons who are directly tied with the process that is being carried out - engineers and workers. Due to the dense built environment of an urban agglomeration, noise from construction sites affecting neighbouring buildings becomes a factor. Although it is a lesser inconvenience in the case of places that are temporarily occupied (in the case of the discussed projects: a court building, an office building, a train station), this becomes a considerable problem in the case of residential buildings and high class hotels. Studies on the effect of noise from construction in an agglomeration and the means of dealing with it are discussed by the author of [28]. In the case of the analysed projects, the schedule featured work only between the hours of 6-22 in order to limit the negative impact of noise and in order not to significantly affect the business of neighbouring hotels.

**Problems associated with archaeologically important sites**

The site of the Axis building was located in the vicinity of an archaeological dig (Fort Lubicz), which caused excavation work to fall under the supervision of the Voivodship Conservator of Historical Buildings. No elements of historical value had been discovered during construction, but, as the authors of [12] state, this can cause numerous difficulties and would probably lengthen the unplanned duration of the project. In the case of a project under archaeological oversight, the discovery of an element can lead to problems such as [29]: halting the work for the duration of research, the need to change the planned technology and the design, the possibility of the occurrence of so-called additional work.

## 4. Conclusions

Table 2 contains the most common causes of problems in the carrying out of a construction project in a large city based on literature and the analysed structures. Guidelines were proposed in order to avoid them or limit their occurrence in similar future projects.

**Table 2** Causes of technological and organisational problems in the carrying out of construction projects in urban agglomerations and proposals of limiting or avoiding them. Original work

| FACTORS | RECOMMENDATIONS |
|---|---|
| Dense built environment | A thorough development of a project during the preparatory and design phase by a comprehensive identification of soil and water conditions both beneath the designed structure and in the zone of its influence on neighbouring structures. Determining the zones of influence of the designed building on neighbouring structures and the monitoring of the condition of existing structures. |
| The need to perform deep excavation | Determining the requirements of the functional and spatial layout of the subterranean section, selecting the most appropriate form of the support and construction of the subterranean section of the designed building, constant monitoring of the tension of the supports and the ground. |
| Logistical difficulties | Appointing a person to be responsible for effective logistics. Strict adherence to a previously established delivery timetable. |
| Vibrations | Performing appropriate analyses regarding the propagation of vibrations and the monitoring of the structure under construction, as well as of existing structures. |
| Noise | Limiting the work to daytime hours, the use of temporary acoustic screens |
| Archaeological digs | Thoroughly analysing the historical context of the construction site and performing available research as needed. |

The carrying out of construction projects in a large urban agglomeration is a difficult undertaking. The space in which construction work is performed, the requirements of the property developer and a long construction time generate organisational and technological problems. Despite difficulties, the analysed projects was completed in accordance with the planned schedule, which means that the implemented actions had been appropriate.

## References

1. T. Kasprowicz, Technical Transactions **2,** 177 (2010)
2. K. Kähkönen, Automation in construction, **8(6)**, 625 (1999)
3. J. Korytarova, T. Hanak, R. Kozik, E. Radziszewska-Zielina, Procedia Engineering, **123**, 276 (2015)

4. Gawron H., Journal of Management and Finance **12(4),** 135 (2014)
5. E. Radziszewska-Zielina, B. Szweczyk, Procedia Engineering **161**, 1096 (2016)
6. A. Sobotka Logistyka przedsiębiorstw i przedsięwzięć budowlanych (AGH, Kraków, 2010)
7. M. Książek, P. Nowak, S.Kivrak, J.Rosłon, L.Ustinovichius, Journal of Civil Engineering and Management **21(2),** 248 (2015)
8. L. Runkiewicz, J.Sieczkowski, Przegląd Budowlany **9**, 18 (2015)
9. S. A. Assaf, S. Al-Hejji, Causes of delay in large construction projects International Journal of Project Management **24(4)** 349 (2006)
10. A.J. Alzamora, L.E. Ford, A.J. Ciancia, IFCEE 2015, **256**, 1522 (2015)
11. G. Maas, F. Van Gassel, Automation in Construction **14**, 435 (2005)
12. G. Śladowski, E. Radziszewska-Zielina, Technical Transactions **6**, 117 (2014)
13. M. Krzemiński, ACE, **62(2),** 89 (2016)
14. E. Radziszewska-Zielina, R.Rumin, Web Of Conferences, **10**, 00079 (2016)
15. http://axis.byskanska.com/ 07.05.2017
16. http://high5ive.pl/ 07.05.2017
17. http://investmap.pl/watek/krakow-kompleks-biurowy-high5ive 07.05.2017
18. N. Ibadov, J.Rosłon, **ACE 61(3),** 105 (2015)
19. M. Pająk, Górnictwo i Geoinżynieria **31/3**, 319 (2007)
20. A. Sobotka, A. Czarnigowska, Journal of Civil Engineering and Management **11(1),** 73 (2005)
21. E. Radziszewska-Zielina, ACE **62(2)**, 167 (2016)
22. J. Hong, W. Liu, Journal of Vibration and Shock, 2006-04
23. B. Kożuch, T. Tatara, E3S Web of Conferences **10**, 00047 (2016)
24. D. Connolly, G. Kouroussis, O. Laghrouche , C. Ho, M. Forde, Construction and Building Materials **92**, 64 (2015)
25. K. Stypuła, T. Tatara, Transport Miejski i Regionalny **7** (2015)
26. S.L. Lusk, M.J. Kerr, S.A. Kauffman, American Industrial Hygiene Association Journal **59(7),** 466 (2010)
27. M. D. Fernandez, S. Quintana, N. Chavarria, J.A. Ballesteros, Applied Acoustics **70(5),** 753 (2009)
28. NG Cheuk Fan, Journal of Environmental Psychology **20**, 375 (2000)
29. E. Radziszewska-Zielina, G. Śladowski, Journal of Cultural Heritage (2017)
30. M. Krzemiński, Acta Physica Polonica A, **130(6)**, 1439-1442 (2016)
31. M. Kaftanowicz, M. Krzemiński, Procedia Engineering, **111**, 364-370 (2015)
32. B. Jankowski, J. Prokocki, M. Krzemiński, Procedia Engineering **111**, 351-355 (2015)

# Assessment of the Conditions that Determine the Reclaiming of Stone Materials and their Use in the Revalorisation of Historical Buildings

*Elżbieta* Radziszewska-Zielina[1,*], *Grzegorz* Śladowski[2]

[1,2] Cracow University of Technology, Faculty of Civil Engineering, Institute of Management in Construction and Transport, Chair of the Technology and Organization of Construction

**Abstract.** The paper focuses on the problem of the effective management of materials obtained from demolition in the scope of the revalorisation of historical structures. The conditions that influence the reclamation and reuse of materials were analysed. The entirety of the discussion was based on the practical example of the revalorisation of the Outer Courtyard of Wawel Castle, for which the importance of these conditions has been determined, as well as the relationships between them.

## 1 Introduction

Renovation work and demolition work belong to the group of construction works that generate larger amounts of waste than in the case of erecting new structures. In such cases, appropriate waste management, which needs to be taken into account already at the planning and design stage, and needs to be implemented during the carrying out of a project, is of particular importance [1]. Examples of good practice of such an approach can be found in literature [2-5]. Within the scope of these processes, physical flows of waste occur, which need to be carried out in accordance with the principles of reverse logistics [6,7]. The implementation of the principles of reverse logistics should be preceded by an analysis of the conditions related to a given project.

An interpretation of the fundamental conditions associated with the recycling of materials during renovation projects can be found in literature [8].

The goal of this article is the broadening of this set of conditions for the purposes of analysing the problem of recycling in projects associated with the revalorisation of historical buildings. In addition, an analysis of the impact of these conditions and the relations between them - in relation to the project of the revalorisation of the Outer Courtyard of Wawel Castle - has been performed.

## 2 The conditions that determine material reclamation and reuse in the scope of the revalorisation of historical buildings

***Environmental protection conditions*** require us to respect the regulations contained in the Waste Act of the 14th of December 2012 [9]. In it are "defined the means that serve the protection of the environment, life and health of people, the prevention and lessening of negative influence on the environment and human health resulting from the production of waste and its management, as well as the limitation of the overall effects of the use of resources and the improvement of the effectiveness of said use". The manner of planning,

* Corresponding author: eradzisz@ztob.pk.edu.pl

design and performing of work, including construction work, should include first and foremost the prevention of the production of waste and, in the case of its occurrence, the preparation of materials for reuse. The recycling of materials or other reclamation processes should be considered next, with neutralisation being the final option. In accordance with the "principle of proximity" provided in the Act, we should first strive to process waste on the site of its production (on the construction site).

**Technical conditions** associated with the evaluation of the physical, chemical, mechanical and aesthetic parameters of the materials obtained through demolition in terms of their reuse within a structure. Many tools and diagnostic methods are currently available: ones based on, for instance, electromagnetic or acoustic procedures, which can constitute an alternative to the typical destructive methods. Whether a piece of material is going to be cleared for reuse depends on the meeting of quality criteria outlined in the technical documentation.

**Architectural conservation conditions** associated with conservational requirements, determined individually for each project involving the performing of work on a historical building or within the surroundings of one. The basic goal of conservation authorities is the securing and preservation of original historical substance, which is why there is a tendency to minimise demolition work on elements of historical structures to a minimum and to reclaim and reuse as much material that is of historical value as possible.

**Financial conditions** apply to the technology and organisation of work in the context of the reclamation of construction materials and their reuse through reincorporation. These conditions apply to the financial costs of the performance of: construction work, the transport of materials obtained through demolition, the necessary quality analysis, as well as work that prepares an element for reincorporation or its recycling and neutralisation.

**Temporal conditions** associated with the time needed for performing the abovementioned technological and organisational processes in the context of the reclamation of construction materials and their reincorporation, or recycling, as well as neutralisation.

## 3 The proposed approach to the evaluation of the importance of conditions and the relationships between them

The analysis of the conditions described above indicates that their descriptions mutually overlap, which justifies the necessity to include interdependency in their analysis, including so-called feedback loops between them. As a result, this train of thought leads us to establish a network structure of the relations between the conditions. The ideological scheme of the proposed network structure is presented on (figure 1).

The notion of a system as an object composed of various elements, between which there are certain relations (dependencies) [18] lies at the foundation of numerous methods used in the analysis of various problems of an inter-dimensional character, for instance [10-17]. In order to synthetically describe and analyse the dependencies above, it is necessary to select an appropriate tool which will allow us to properly model and analyse the dependencies between the elements of the system in question.

One of the effective concepts of the replication of such a system is structural modelling, understood as a set of diverse techniques which make it possible to understand the properties of complicated systems [19].

Many structural modeling methods can be found in the literaturę [20,21]. Recently a method has also been developed WINGS method (Weighted Influence Non-linear Gauge System) [22].

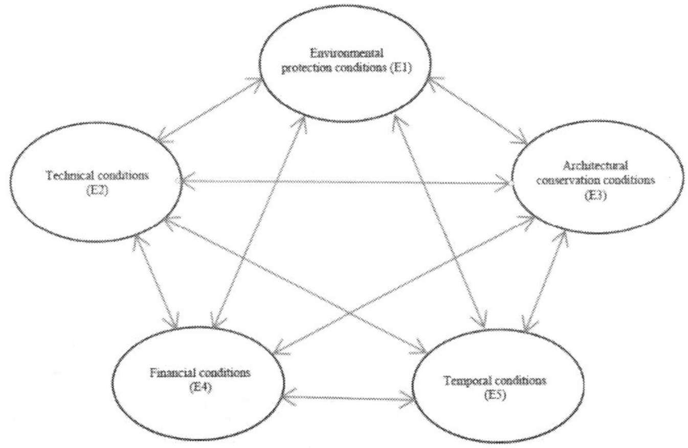


**Fig. 1.** Ideological scheme of the network structure of the relationships between the elements of the analysed system. Source: original work

In these methods, the tool used to model the dependencies between the elements of the system is a directed graph, the tips of which symbolise the elements of the system and the arches determine the relations (mutual influence between the elements of the system).

The evaluation of the impact of the elements and the intensity of the influence between these elements is performed by an expert with the help of a four-point verbal scale, to which numerical values have been assigned. The input values of evaluations are introduced into a matrix, the sum of all the powers of which, in the liminal sense, returns the output values in the analysed model.

The effect of the use of the WINGS method is the determining of the value of total impact $( r_i )$ of the system's $i$ element, as well as its total susceptibility $( c_i )$.

The sum of the values of $( r_i )$ and $( c_i )$ determine the total involvement of element $i$, while their difference informs us of the role of the element in the system. A positive value of the difference $( r_i - c_i )$ states that element $i$ constitutes a cause, while a negative value labels it an effect within the analysed system. The absolute value of the difference $( r_i - c_i )$ determines the strength of the abovementioned causal (effectual) character of the element under discussion.

In order to make the analysis of the relations between the elements of the analysed system easier, the values that are obtained can be depicted in a graphical form on a two-dimensional set of coordinates, forming a so-called impact-relation map - IRM. The values of the difference $( r_i - c_i )$are placed on the axis of ordinates, while the values of the sum $(r_i+c_i)$ are placed along the abscissa.

# 4 Case study

## 4.1 Basic information on the project

In order to present the proposed systemic approach to the evaluation of the impact of and the relations between the conditions that determine the reuse of materials obtained through dismantlement, the authors used the example of the revalorisation of the Outer Courtyard of the Wawel Castle, see fig. 2.

The main element of the project in question was the comprehensive replacement of paved surfaces (around 10 000 m$^2$) along with the redevelopment of external stairs and the construction of a driveway and a ramp for disabled persons. The revalorisation work was an activity that was sorely needed due to the worsening technical condition of the stone surfaces of the Outer Courtyard, the driveways, as well as the poor condition of the underground infrastructure (water pipes, sewerage, power cables, fire safety infrastructure and telecommunications).

The recreation and improvement of the condition of immobile historical artefacts, as well as increasing the quality level of services for domestic and foreign tourists as a part of the popularisation of global cultural heritage allowed the project to be qualified for the "Priority XI Operational Programme Infrastructure and the Environment, action 11.1 "The protection and preservation of cultural heritage of supralocal importance".

**Fig. 2**. a) The Outer Courtyard during revalorisation, b) Wawel Hill parking lot surfaces

## 4.2 Structural analysis of the conditions

The description of the conditions for the project in question is as follows:

***Environment protection conditions (E1)***

The specifics of the project have, to a large degree, made it possible to achieve the goals outlined in the act of law. The strategy of achieving the maximum possible reclamation and reincorporation of the materials obtained through dismantlement and recycling, as well as the neutralisation of produced waste, constituted the achievement of the basic goals of the project. However, due to the specific nature of the site and the difficulties associated with, among other things, the lack of sufficient space for storing the reclaimed materials, the "principle of proximity" could not be followed during their processing. All of the processes associated with the evaluation and adaptation of the materials for reuse or recycling and processing took place in an area located a couple of kilometres away from the construction site, see fig. 3.

**Fig. 3**. a) Loading the stone elements on a lorry, b) The storage area away from the construction site, where the stone elements were kept

***Technical conditions (E2)***

The evaluation of the technical condition of the stone elements obtained through dismantlement was performed primarily on the basis of an initial inspection of the elements in order to eliminate those which showed signs of cracking or delamination, see fig. 4a. Samples were taken for each batch of elements in order to perform laboratory analyses and determine the compatibility of the physical, chemical and mechanical properties of the analysed materials with the requirements outlined in the design. Over the course of the analysis it turned out that the existing sandstone tiles did not meet the requirements outlined in the design regarding compressive strength, despite having high parameter values. However, taking into account the fact that the remaining qualities that were analysed did meet the requirements of the design, including tensile strength (in the case of large tiles this parameter is more important than compressive strength) and, in the face of a high reclamation rate approaching 50%, the designer approved the sandstone tiles in question to be reincorporated by request of the property owner and the conservator, as well as on the basis of an expertise, see fig. 4b. Rejecting such materials would have been a case of far-reaching wastefulness and mismanagement.

**Fig. 4.** a) Initial selection of sandstone tiles for reuse, b) The reclaimed tiles composed with new tiles

### *Architectural conservation conditions (E3)*

The Wawel Hill is a listed site, which is why, at the behest of the conservator, the condition of the maximum possible reclamation of stone materials obtained through dismantlement was introduced. The "arched" layout of the sandstone tiles suggested by the conservator, including the mixing of original tiles with new ones, as well as the running pattern of porphyry cobblestone, determined additional procedures associated with the cleaning, proper processing and cutting of the reclaimed tiles before their reincorporation. In addition, a part of the existing sandstone tiles was cut and used as cobblestone, see fig. 5a, or as tiles with irregular shapes and sizes, see fig. 5b.

**Fig. 5.** a) The laying of cobblestone obtained from the cutting up of existing reclaimed tiles, b) Laying leftover elements from cutting sandstone tiles.

### *Financial conditions (4)*

Due to the specifics of the location and the performing of the work on a so-called active property (unlimited access by tourists and the employees of Wawel Hill), all technological and organisational processes generated high costs. These costs were associated with: dismantlement work, the initial stocktaking of the elements and their transport to the storage area, the performance of costly quality analyses and processing, transport to the construction site and, in the case of producing waste - its processing. Due to the multi-stage structure of the project, all logistics activities were based on cyclically developed designs for each stage: the organisation of the site and traffic management, which was not without impact on the costs.

***Temporal conditions (E5)***

The costs of the duration of the completion of the project, similarly to the general costs of the project, were associated with the aforementioned technological and organisational processes. A lot of time needed to be dedicated to the quality analyses of the reclaimed stone elements, and, in particular, the recreation of 48 cycles of freezing and thawing during the process of assessing the frost resistance of these elements. Under the conditions of limited required completion time for each stage of the work, the above-mentioned technological and organisational process, including the time-consuming material analysis, required the development of detailed project schedules.

Using the aforementioned systemic approach and the WINGS procedure (Weighted Influence Non-linear Gauge System), an expert evaluated the significance of each condition (the elements of the system) in a linguistic manner (using an evaluation scale, presented in table1, that had been developed for the purposes of this method), as well as their direct influence and its intensity, see table 2.

**Table 1.** Linguistic evaluations of the importance and influence of an element and their representation in on a four point scale

| Assessing the significance of an element in the system | | Assessing the influence of an element in the system | |
|---|---|---|---|
| Linguistic assessment | Point scale | Linguistic assessment | Point scale |
| No significance (NS) | (0) | No influence (NI) | (0) |
| Low significance (LS) | (1) | Low influence (LI) | (1) |
| Average significance (AS) | (2) | Average influence (AI) | (2) |
| High significance (HS) | (3) | High influence (HI) | (3) |
| Very high significance (VHS) | (4) | Very high influence (VHI) | (4) |

**Table 2.** Linguistic assessments of the significance and the mutual, direct influence of the system's elements and its intensity

| Elements of the system | E1 | E2 | E3 | E4 | E5 |
|---|---|---|---|---|---|
| E1 | **HS** | HI | AI | VHI | VHI |
| E2 | LI | **VHS** | AI | HI | HI |
| E3 | HI | VHI | **VHS** | VHI | VHI |
| E4 | AI | AI | LI | **HS** | LI |
| E5 | AI | AI | LI | AI | **AS** |

As a consequence, the process of algebraic conversions, while taking into account indirect relations, has produced the values of the total significance and the influence between the elements of the system in question. As a result, the total impact values have been determined for element *i*, as well as its total susceptibility. Figure 6 illustrates the impact-relation map for the analysed elements of the system.

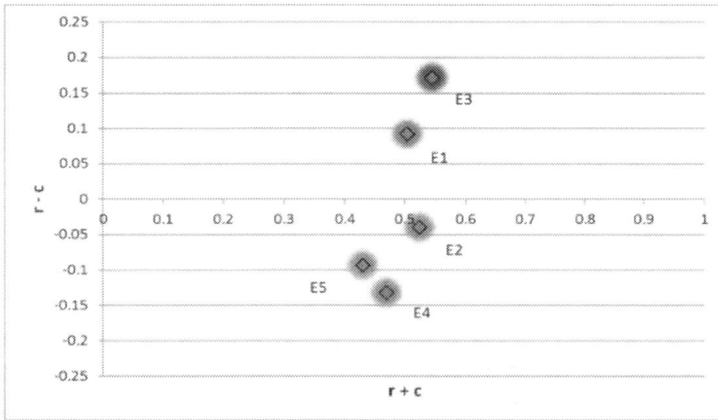


**Fig. 6.** The impact-relation map of the elements of the system

### 3.3 Analysis of the results

The total involvement values $( r_i+c_i )$ determine the ranking of the elements within the system. For the case under analysis, the value $( r_i+c_i )$ is the highest for architectural conservation conditions (E3) (fig.2). In addition, the highest positive value of the relations parameter $( r_i - c_i )$ that has been obtained for this parameter is evidence of its strong impact on the remaining elements of the system. Technical conditions (E2) placed second in terms of total involvement, however, the negative value of $( r_i - c_i )$ indicates that this element is susceptible to influence exerted by architectural conservation conditions (E3), as well as the conditions of environmental protection (E1). This is because the role of condition (E1), similarly to (E3), due to their positive $( r_i - c_i )$ values, has a character that impacts other elements within the system. The subsequent places within the ranking of the conditions in terms of total involvement, are, respectively, occupied by the conditions associated with environmental protection (E1), as well as cthe onditions associated with costs (E4) and temporal conditions (E5). However, despite the fact that temporal conditions (E5) placed last, they are less susceptible to influence within the system than element (E4).

## 5 Conclusions

The implementation of the principles of reverse logistics within a given construction project should be preceded by an analysis of its conditions, the type and character of which is going to depend on its specifics. Expanding the set of basic conditions that are known from literature [8] by including additional ones will allow a comprehensive analysis of the problem in question to be performed in the scope of projects associated with the revalorisation of historical buildings. The paper contains the proposal of a systemic approach to the evaluation of these conditions, while taking into account the interdependencies between them. The entirety of the discussion has been backed by a practical example of the evaluation of the conditions that determined the reclamation process and the reuse of materials during the revalorisation of the Outer Courtyard at Wawel Hill. In the future, the degree of subjectivity can be reduced in the proposed approach to the assessment of conditions by introducing the formulation of evaluations by a group of experts, as well as elements of fuzzy logic due to the uncertainty of the opinions that will be formulated [23]. The evaluation of the impact and relations of these conditions can be helpful in defining the acceptable variants of construction waste management within the scope of construction projects in the future, particularly concerning the revalorisation of historical buildings.

## References

1. A. Sobotka, J. Czaja, Procedia Engineering **122 (Orsdce)** (2015)
2. A. Sobotka, J. Sagan, A. Sikora, Materiały Budowlane **526(6)** (2016)
3. J. Czaja, M. Szczygielski, Ł Stopa, M. Blajer, Logistyka **4** (2015)
4. A. Sobotka, J. Sagan, *The World Multidisciplinary Civil Engineering-Architecture-Urban Planning Symposium, June, Prague* (2016)
5. J. Czaja, J. Jaskowska-Lemańska, D. Wałach, Logistyka **4** (2015)
6. J. Czaja, A. Sobotka, Logistyka **2** (2015)
7. J. Sagan, A. Sobotka, Builder (2016)
8. J. Sagan, A. Sobotka, Inżynier Budownictwa **12** (2016)
9. *Ustawa z dnia 14.12.2012 o odpadach. Dz.U. 2013 p. 21*
10. N. Ibadov, J. Rosłon, ACE **61(3)** (2015)
11. MV Ksiażek, P. Nowak, S. Kivrak, JH, Roslon, L. Ustinovichius, Journal of Civil Engineering and Management **21(2)** (2015)
12. E. Radziszewska-Zielina, B. Szewczyk, ACE **1(3)** (2015)
13. Radziszewska-Zielina E., The Application of Multi-Criteria Analysis in the Evaluation of Partnering Relations and the Selection of a Construction Company for the Purposes of Cooperation, „Archives of Civil Engineering", **2(62)** (2016)
14. Radziszewska-Zielina E., Szewczyk B., *WMCAUS 2016*, Procedia Engineering **2(62)** (2016)
15. Krzemiński, M., ACE **2(62)** (2016)
16. Radziszewska-Zielina E., Rumin R., *SEED 2016*, E3S Web Of Conferences **10** (2016)
17. Korytarova J., Hanak T., Kozik R., Radziszewska-Zielina E., Procedia Engineering **123** (2015)
18. J. Michnik, *Wielokryterialne metody wspomagania decyzji w procesie innowacji* (Uniwersytet Ekonomiczny w Katowicach, Katowice, 2013)
19. F.S. Roberts, Journal of Mathematical Psychology **3 (29)** (1985)
20. T. L. Saaty, *Decision Making with Dependence and Feedback, The Analytic Network Process* (University of Pittsburgh, RWS Publications, 1996)
21. A. Gabus, E. Fontela, *World Problems an Invitation to Further Thought within the Framework of DEMATEL* (Battelle Geneva Research Centre, Switzerland Geneva, 1972)
22. J. Michnik, European Journal of Operational Research **228** (2013)
23. E. Radziszewska-Zielina, G. Śladowski, Journal of Cultural Heritage (2017), https://doi.org/10.1016/j.culher.2017.02.008
24. B. Jankowski, J. Prokocki, M. Krzemiński, Procedia Engineering **111**, 351-355 (2015)
25. M. Krzemiński, Acta Physica Polonica A, **130(6),** 1439-1442 (2016)
26. M. Kaftanowicz, M. Krzemiński, Procedia Engineering, **111**, 364-370 (2015)

# Multilayered dynamic strain gauge using of piezoelectric PVDF films

*Nelly* Rogacheva*

Moscow state university of civil engineering, Yaroslavskoye shosse, 26, Moscow, Russia, 129337

**Abstract.** The strain gauge made of a piezoelectric film is proposed. The operation of the gauge is based on the piezoelectric effect: the deformation of the gauge causes an electric potential proportional to the deformation. The gauge consists of piezoelectric layers, covered by metal electrodes and adhesive layers that connect the film layers to a multilayer rectangular bar. Since the piezoelectric film has a small modulus of elasticity, the strain gauge can be used to measure the deformation of structures from soft materials.

## 1 Introduction

Analysis of stress-strain state of structures, as a rule, is performed on the basis of measuring deformations on the surface of the body under study. The strain gauge must satisfy the following requirements:

– The gauge measurements must have temporal and temperature stability;

– The gauge measurements must have good accuracy;

– The dimensions of the gauge should be small enough to correctly measure the deformation at the point of the body;

– The influence of the gauge on the stress-strain state of the body in the vicinity of the point under study should be negligible;

– The gauge should be of low cost and easy to use;

Among the experimental methods of measuring deformations, the most widely used strain gauges based on the idea of Lord Kelvin [1, 2]. Kelvin found, the conductor subjected to deformation, slightly changes resistance. This small change in resistance can be measured. The strain resistance gauge measures a secondary small value - a change in resistance and requires special equipment to measure it.

The dynamic strain gauge from piezoceramics was first proposed in [3] and then in [4].The gauge is a piezoceramic rectangular thin plate with a thickness pre-polarization. The plate faces are covered with electrodes. The plate is glued to the surface of the body being examined and deformed together with it. Due to the piezoelectric effect, when the plate is deformed, an electric charge appears on its electrodes, which is easy to measure. If the piezoceramics possesses a strong piezoelectric effect, then a significant part of the mechanical deformation energy transforms into electrical energy. For example, for

* Corresponding author: nrogache@yandex.ru

piezoceramics PZT-4, PZT-5, more than 30% of mechanical energy is converted into electrical energy.
The elastic compliance of piezoceramics is a quantity of the same order as the elastic compliance of aluminum. Such a gauge will distort the stress-strain state at the studied point of the deformable body from a soft material with a small modulus of elasticity.

## 2 Basic equations and boundary conditions

Consider a piezoelectric rectangular bar. Its length is $2l$, the width is $2m$, and the thickness is $2h$. We assume that the length of the bar is much larger than its width and the thickness: $l>>m$ and $l>>h$. The bar consists of $n$ layers of piezoelectric film. Each layer is pre-polarized to the thickness direction and it faces covered by electrodes. The layers of the piezoelectric film are glued together. One of external faces is glued to the investigated body with an insulating adhesive.
We place the piezoelectric bar in a Cartesian coordinate system: the $x_1$ axis is directed along the length of the bar, the $x_2$ axis is directed along the width of the bar, the $x_3$ axis is perpendicular to the plane $x_1x_2$ and is directed along the thickness of the bar: $-l\le x_1\le +l$, $-m\le x_2\le +m$, $-h\le x_3\le +h$.The gauge consists of repeating sets of three layers: an adhesive layer, an metal layer (electrode), a PVDF film layer. Since any set of three layers has the same stress-strain state, we will consider only one set to be called the composite layer. The thicknesses of the PVDF layer, the electrode, and the adhesive layer are $2h_f$, $2h_m$, and $2h_a$ respectively.
We write out only those equations which we will need later:
The constitutive relations for elastic layers (Hooke's law)

$$\sigma_{i,c} = \frac{E_c}{1-\nu_c^2}(e_i + \nu_c e_j) + \frac{\nu_c}{1-\nu_c}\sigma_{3,c} \tag{1}$$

The index $c$ should be replaced by $a$ for the adhesive layer and by $m$ for the electrode.
Piezoeffect equations

$$\sigma_i = \frac{1}{s_{11}^E(1-\nu^2)}(e_i + \nu e_j) - \frac{s_{13}^E}{s_{11}^E(1-\nu)}\sigma_3 - \frac{d_{3i}+\nu d_{3j}}{s_{11}^E(1-\nu^2)}E_3 \tag{2}$$

$$D_3 = \varepsilon_{33}^T E_3 + d_{31}\sigma_1 + d_{32}\sigma_2 + d_{33}\sigma_3 \tag{3}$$

Strain-displacement formula

$$e_i = \frac{\partial u_i}{\partial x_i} \tag{4}$$

Electrostatic equation

$$E_3 = -\frac{\partial \psi}{\partial x_3} \tag{5}$$

Here $u_1$ and $e_1$ are the displacement $u_1$ and strain $e_1$ in the $x_1$ direction, $E_3$ and $D_3$ are the components of the electric field vector and electric induction vector in the $x_3$ direction, $s_{11}^E$ is the elastic compliance at constant electric field, $d_{31}$ and $d_{32}$ are the piezoelectric constants, $\varepsilon_{33}^{T}$ is the permittivity for zero stresses, $\psi$ is the electric potential. The notation in use coincides with taken previously [5]. Every equations with subscripts $i$ and $j$ contains two equations: one for $i=1$ and $j=2$ and the other for $i=2$ and $j=1$.

We assume that on the contact area $x_3=-h$ of the gauge and the body the displacements of the bar are $u_i$, $u_3$ equal to the displacement of the studied body $v_i$, $v_3$:

$$x_3 = -h : v_i = u_i , \quad v_3 = u_3 \tag{6}$$

On the remaining faces of the bar we have the conditions:

$$\begin{aligned} x_3 &= +h : \sigma_{i3} = \sigma_3 = 0 \\ x_1 &= \pm l : D_1 = 0, \quad \sigma_{11} = \sigma_{13} = \sigma_{12} \\ x_2 &= \pm m : D_2 = 0, \ \sigma_{21} = \sigma_{23} = \sigma_{22} \end{aligned} \tag{7}$$

The electrical potential on the electrodes of each layer is known (measured):

$$\psi\big|_{x_3=h_a} = -V, \qquad \psi\big|_{x_3=h_a+2h_f} = V \tag{8}$$

Since the electrodes are connected by an external contour, the integral condition should hold

$$\int_s \dot{D}_3 \, ds = 2VY$$

Where the integral is taken over the surface $s$ of one of the electrodes and the point denotes the time derivative.

For simplicity we will assume that the body and the gauge vibrate by the law $e^{-i\omega t}$

$$-i\omega \int_s D_3 ds = 2VY \tag{9}$$

where $\omega$ is the angular frequency of vibrations and $t$ is the time. Since the vibrations are harmonic, we will describe them by equations written for the amplitude values of the needed quantities

## 3 One-dimensional strain gauge

Piezoelectric film with greatly different piezoelectric constants is an ideal material for a one-dimensional gauge. For example, for Kureha KF piezoelectric film PVDM MDMs the piezoelectric constants $d_{31}$ is much larger than $d_{32}$ [6]:

$$d_{31} = 2.17 \cdot 10^{-11} m/V , \ d_{32} == 1.87 \cdot 10^{-12} m/V$$

The measured voltage is proportional to the value $d_{31}$.

The gauge theory is based on the hypotheses similar to those proposed for piezoceramic gauge [5].

First hypothesis

We assume that the dimensions of the gauge are small enough to treat deformations in the contact area of the gauge and the body under investigation as constant.
Second hypothesis

In the equations (1) – (3) the stresses $\sigma_2$ and $\sigma_3$ can be neglected as compared to the principal stress $\sigma_1$. Using this hypothesis, we rewrite the equations in a simplified form: for the elastic layers

$$\sigma_{1c} = \frac{E_c}{1-\nu_c^2}(e_1 + \nu_c e_2), \qquad e_2 + \nu_c e_1 = 0 \tag{10}$$

for the piezoelectric layers

$$\sigma_1 = \frac{1}{s_{11}^E(1-\nu^2)}(e_1 + \nu e_2) - \frac{d_{31} + \nu d_{32}}{s_{11}^E(1-\nu^2)}E_3$$
$$\frac{1}{s_{11}^E(1-\nu^2)}(e_2 + \nu e_1) - \frac{d_{32} + \nu d_{31}}{s_{11}^E(1-\nu^2)}E_3 = 0 \tag{11}$$
$$D_3 = \varepsilon_{33}^T E_3 + d_{31}\sigma_1$$

We find the strain $e_1$ from the second equation (10) and substitute it in the first equation, as a result we obtain

$$\sigma_{1,c} = E_c e_1 \tag{12}$$

Similarly we transform the equations (11) for f the piezoelectric layers

$$\sigma_1 = \frac{e_1}{s_{11}^E} - \frac{d_{31}}{s_{11}^E}E_3 \tag{13}$$

$$D_3 = \varepsilon_{33}^T(1-k_{31}^2)E_3 + \frac{d_{31}}{s_{11}^E}e_1 \tag{14}$$

Third hypothesis
We assume that the electrical potential varies with the thickness by the linear law

$$\psi = \psi_{,0} + x_3\psi_{,1}$$

Hence, using the relation (5) and the conditions (8), we get

$$E_3 = -\frac{V}{h} \tag{15}$$

Since the components of the strain $e_1$ and the strength of the electric field $E_3$ are constant over the whole area of the gauge, from Eq. (13) follows that the component $D_3^{(k)}$ (14) of the electrical induction vector is constant too. Then the equation for a closed electric circuit (9) after integration over the surface of one electrode will take the form

$$-i\omega s D_3 = 2VY$$

We solve this equations according to the deformation $e_1$ taking into account the relations (14), (15)

$$e_1 = i\frac{2VY}{\omega s}\frac{s_{11}^E}{d_{31}} + d_{31}\frac{1-k_{31}^2}{k_{31}^2}\frac{V}{h_f} \quad (16)$$

where s equals $2l \times 2m$.

From the formula (16) the strain $e_1$ for harmonic vibrations is calculated using the measured electrical potential difference.

To increase the measured electrical potential difference, we should connect the electrodes on the surfaces of the piezoelectric film layers in series. If the gauge consists of *n* composite layers, the electrical potential will increase *n* times.

In the constructed theory of the 1D gauge the boundary conditions (7), (9) – (11) are satisfied approximately. The complete error is defined by the formula [5] $\varepsilon = O(m/l, h/l)$.

## 4 Electromechanical coupling coefficient

An important characteristic of the performance of piezoelectric elements is the electromechanical coupling coefficient (EMCC). The EMCC is commonly used as a measure of a piezoelectric element's efficiency. It gives the ratio of the electrical (mechanical) energy stored in the volume of the piezoelectric body and capable of conversion to the total mechanical (electrical) energy supplied to the body. In [5, 7] it was shown that the energy formula the EMCC is the universal formula. The energy formula for computing the EMCC $k_e$ has the form

$$k_e^2 = \frac{U^{(d)} - U^{(sh)}}{U^{(d)}} \quad (17)$$

where $U^{(d)}$ is the internal energy of the body when the electrodes are disconnected and $U^{(sh)}$ is the internal energy for short-circuited electrodes.

In order to find $U^{(d)}$ and $U^{(sh)}$ we should solve first our initial problem and then to solve two additional problems. In solving the additional problems, we suppose that

$$e_1^{(d)} = e_1^{(sh)} = e_1$$

where $e_1$ is the component of strain found from the solution of the initial problems.

We find the quantity $U^{(d)}$ using the condition for disconnected electrodes

$$D_3^{(d)} = \varepsilon_{33}^T(1-k_{31}^2)E_3^{(d)} + \frac{d_{31}}{s_{11}^E}e_1 = 0 \quad (18)$$

From this equation we obtain

$$E_3^{(d)} = -\frac{k_{31}^2}{1-k_{31}^2}\frac{e_1}{d_{31}}, \quad \sigma_1^{(d)} = \frac{1}{1-k_{31}^2}\frac{e_1}{s_{11}^E} \quad (19)$$

To determine $U^{(sh)}$ we use the conditions for short-circuited electrodes

$$\psi^{(sh)}\Big|_s = 0 \quad (20)$$

Hence it follows

$$E_3^{(sh)} = 0, \qquad \sigma_1^{(sh)} = \frac{e_1}{s_{11}^E} \tag{21}$$

Taking into account the energy of the elastic layers we write the formulas for the quantities $U^{(d)}$ and $U^{(sh)}$

$$U^{(d)} = \int_v (\sigma_1^{(d)} e_1 + E_3^{(d)} D_3^{(d)} + \sigma_{1m} e_1 + \sigma_{1a} e_1) dv$$

$$U^{(sh)} = \int_v (\sigma_1^{(sh)} e_1 + E_3^{(sh)} D_3^{(sh)} + \sigma_{1m} e_1 + \sigma_{1a} e_1) dv$$

After integrating by the volume $v$ and simple transformations we get

$$U^{(d)} = 2se_1^2 \left( \frac{h_f}{s_{11}^E (1 - k_{31}^2)} + h_m s_{11}^E E_m + h_a s_{11}^E E_a \right)$$

$$U^{(sh)} = 2se_1^2 \left( \frac{h_f}{s_{11}^E} + h_m s_{11}^E E_m + h_a s_{11}^E E_a \right)$$

We substitute the quantities $U^{(d)}$ and $U^{(sh)}$ into the formula (17) and find $k_e^2$

$$k_e^2 = \frac{k_{31}^2}{(1 - k_{31}^2)(1 + s_{11}^E E_m h_m / h_f + s_{11}^E E_a h_a / h_f)} \tag{22}$$

This formula can be simplified: the value $k_{31}^2$ is a negligibly small quantity as compared with 1. The $k_{31}$ is of the order 0.02.

$$k_e^2 = \frac{k_{31}^2}{(1 + s_{11}^E E_m h_m / h_f + s_{11}^E E_a h_a / h_f)} \tag{23}$$

## 5 Conclusions

The advantage of the PVDF film gauge are the following:

- the gauge is simple in structure and easy to use;
- the gauge can be made of any size;
- the gauge is suitable for soft structures.

## References

1. W. Thomson (Lord Kelvin), *Proc. R. Soc.,*(London, 1856)
2. A. S. Kobayashi, ed., *Handbook on experimental mechanics* (Prentice-Hall, New Jersey, 1987)
3. A. A. Infimovskaya, G. N. Chernyshev, N. N. Rogacheva, Mech. of Solids, **24**, 2 (1989)
4. H. S. Tzou, *Piezoelectric Shells* (Kluver, Dordrecht, 1993)

5. N. N. Rogacheva, *The theory of piezoelectric shells and plates* (CRC Press, Roca Raton, 1994)
6. T. Sato, H. Ishikawa, O. Ikeda, Appl. Optics, **21,** 20 (1982)
7. C. C. Chou, N. N. Rogacheva, S. H. Chang, IEEE Trans. Ultr., Ferr. and Freq. contr., **42**, 4 (1995)

# The selecting of roofing material for pitched roofs by the analytic hierarchy process

*Irina* Romanova[1,*], and *Nataliya* Skanavi[1]

[1]Moscow state university of civil engineering, Yaroslavskoye shosse, 26, Moscow, Russia, 129337

**Abstract.** Today's in the market of building materials there are a large number of roofing materials for pitched roofs. All of these materials have their advantages and disadvantages, so the task of choosing the optimal option is complex and requires special methods to eliminate the subjectivity of evaluation. The article describes the approach to selecting the best option of roofing material from the available options by the set of its quantitative and qualitative characteristics. It is proposed to carry out the evaluation of options and selection of the best option based on the analytic hierarchy process. The article presents the results of the selection of the most appropriate roofing material for pitched roofs using the specified method.

## 1. Introduction

The roof of the building is one of the most important structural elements that performs the protective and insulating functions (heat preservation and overheat protection, protection from external adverse factors – all types of precipitation, wind, melt water). There are two parts to a roof, its supporting structure and its outer skin, or uppermost weatherproof layer, which is directly exposed to atmospheric influences. This part of the roof shows great variation dependent upon availability of material [1].

The basic requirements to the roofing materials are [2-4]:

1. Weather resistance.
2. Watertightness- the ability of a material to pass water under a certain pressure for a certain time.
3. Water and frost resistance.
4. The strength of roofing material is the ability to carry dynamic (wind, atmospheric precipitation) and static (the weight of the snow masses, etc.) loads.
5. Biostability and corrosion resistance.
6. Good sound insulation properties – the ability to create reliable and highly efficient protection against external noise.
7. Durability.
8. Environmentally friendly.
9. Fire safety - an advantage have a non-combustible or slow-burning materials.
10. Efficiency.

*Correspondingauthor: nirs-isa@mail.ru

11. Adaptability - it is possible to lower the complexity of works on roofing.

12. Aesthetic requirements - in the case of pitched roofs the roof is an important architectural element of the building.

In total, the concept of "aesthetic requirements" includes the so-called prestige of the material. Considered among the most prestigious materials such as ceramic tiles and natural slate. There are many materials that mimic ceramic tile, and even their name includes the word "tile": shingles, metal tile, cement-sand tile [5].

## 2. Experimental section

The materials for pitched roofs on the basis of its composition can be divided into mineral (cement concrete and clay roof tiles, asbestos cement sheets and tiles), metal (metal roof tile, profiled steel sheet, clicklock, composite roof shingles) and organic (asphalt shingle, onduline).

Clay roof tile is made of fired clay, it is durable (lifespan to 300 years), eco-friendly, non-combustible, has a high frost resistance and corrosion resistance. Because of their inherent thermal mass, clay roof tiles are able to absorb heat during the day and release it during the night, which makes them an energy efficient option in cooler climates. The disadvantages are the large weight cover – 50-65 kg/m$^2$, the brittleness, the complexity of installation and high cost [6]. Cement concrete tiles made of cement-sand mixture with the addition of mineral pigments by stamping or pressing. It is lighter than ceramic, but its lifespan is not so long (about 50 years). Slate (asbestos sheet) is obtained by molding mixture consisting of portland cement, asbestos fiber and water. It's cheap, durable, non-combustible, technological. The disadvantages are brittleness, high water absorption (material absorbs moisture, which leads bleaching and the growth of moss), the contents of asbestos. The latter is associated with unsatisfactory environmental friendliness of the material, although it is believed that the products asbestos is bound and not released into the environment [7].

Metal roofing materials consist of a galvanized steel sheet (0.4 mm to 0.5 mm thick), coated on two sides by anti-corrosion (zinc, aluzinc or galfan) coating. Outer side is coated with a special polymer protective structure (metal roof tile) or colored granules of natural stone (composite roof shingles). The advantages of these materials are the low weight of the coating (4-5 kg/m$^2$), durability (lifespan is about 30-50 years), weather resistance, simplicity and speed of installation, aesthetics. The disadvantages of metal roofing materials (except composite tile) include poor sound insulation, high thermal conductivity, low stiffness of the coating, the possibility of corrosion damage of the protective coating, for example, during the installation and uneconomical - a large amount of waste (up to 40%) when assembling a complex roof [8].

Asphalt shingles are made of a fiberglass mat coated with bitumen-polymer binder to both sides and on the front surface covered with colored mineral granules. Onduline is made of corrugated cardboard (or fiberglass), impregnated with bitumen-polymer binder and painted on the front side with weatherproof vinyl-acrylic paint. Advantages of organic roofing materials are low weight of the coating (less than 10-12 kg/m2), simplicity and speed of installation, low cost, flexibility, good sound and thermal insulation. The disadvantages include combustibility (such materials are prohibited for use in most buildings of social importance (hospitals, schools, kindergartens, sports facilities)), low temperature brittleness, low stiffness of the coating, and for asphalt shingles , additionally, the need of continuous sheathing. A significant disadvantage of ondulin is bleaching in the sun [9-11].

Table 1 presents the most popular roofing materials for pitched roofs and their characteristics. The price includes the cost of the roofing material, the cost of installation (calculated for simple gable roof) and the cost of additional materials (truss, insulation, antiseptic, vapor barrier, consumables) [12, 13].

**Table 1.** Roofing materials for pitched roofs

| Roofing materials | Fire safety class | Weight of the coating, kg/m$^2$ | Lifespan / warranty, years | Cost (material + installation + additional materials), RUB/m$^2$ | Corrosion resistance | Sound insulation | Thermal conductivity |
|---|---|---|---|---|---|---|---|
| Metal roof tile Metal Profile Monterrey PURETAN | KM1 | 4.5 | 50/ 20 | 405+1200+ 1900 = 3505 | good | low | high |
| ClickLock Grand Line Satin | KM1 | 4.81 | 20/ 10 | 404+1150+ 1900 = 3454 | satisfacto ry | low | high |
| Cement concrete tiles Baltic tile Sea Wave | KM0 | 43 | 50/ 30 | 356+1700+ 2200 = 4256 | great | high | low |
| Clay roof tile Roben FLANDERN | KM0 | 39 | 100/ 30 | 2185+1800 +2300 = 6285 | great | high | low |
| Asphalt shingles Tegola Nordland, Classic | KM5 | 9.5 | 30/ 15 | 240+1500+ 2200 = 3940 | good | average | average |
| ONDULINE roofing sheets Onduline Tile | KM5 | 3.5 | 25/ 15 | 245+1200+ 2000 = 3445 | good | average | average |
| Composite roof shingles Decra Decra classic | KM1 | 6.7 | 50/ 30 | 930+1700+ 2000 = 4630 | great | high | average |

All of these materials have their advantages and disadvantages, so the task of choosing the optimal option is complex and requires special methods to eliminate the subjectivity of evaluation. One of such methods is the analytic hierarchy process, which allows to solve the problem of choice on set of qualitative and quantitative characteristics [14-17]. The decision problem is decomposed into a hierarchy of more easily comprehended sub-problems, each of which can be analyzed independently. The method is based on constructing a hierarchy starting from the top (goal), through intermediate levels (a group of factors or criteria that relate the alternatives to the goal) to the very bottom level (the list of alternatives for reaching the goal). Each element of the system, except for the top one, is subordinate to one or more other elements. The criteria can be further broken down into subcriteria, sub-subcriteria, and so on, in as many levels as the problem requires. Then the hierarchy is analyzed through a series of pairwise comparisons: the criteria are pairwise compared against the goal for importance, the alternatives are pairwise compared against each of the criteria for preference. The comparisons are processed mathematically: at each level the set of matrix of pairwise comparisons is built (table 2) [18].

**Table 2.** Matrix of pairwise comparisons

| | $A_1$ | $A_2$ | $A_3$ | ... | $A_n$ |
|---|---|---|---|---|---|

| | | | | | |
|---|---|---|---|---|---|
| $A_1$ | $w_1/w_1$ | $w_1/w_2$ | $w_1/w_3$ | ... | $w_1/w_n$ |
| $A_2$ | $w_2/w_1$ | $w_2/w_2$ | $w_2/w_3$ | ... | $w_2/w_n$ |
| $A_3$ | $w_3/w_1$ | $w_3/w_2$ | $w_3/w_3$ | ... | $w_3/w_n$ |
| ... | ... | ... | ... | ... | ... |
| $A_n$ | $w_n/w_1$ | $w_n/w_2$ | $w_n/w_3$ | ... | $w_n/w_n$ |

$A_1$, $A_2$, ..., $A_n$ is the set of n elements and $w_1$, $w_2$,..., $w_n$ respectively, their weight, or intensity. Weight, or intensity, of each element is compared with the weight or intensity of any other element of the set in relation to the common property or the goal (i.e. $w_1/w$ means comparison, not dividing the weights of these elements). The comparison of weights can be represented as follows: the elements of any level are compared with each other regarding their effects on guided element on 9-point scale (from 1 – equal importance to 9 – very strong superiority). If the element $A_1$ is dominant over the element $A_2$, the cell corresponding to the row $A_1$ and column $A_2$ is filled with an integer, and the cell corresponding to the row $A_2$ and column $A_1$ is filled with the integer reciprocal (fraction). Then at each level the synthesis of priorities is made, i.e. for each row the geometric mean is calculated. The vector of priorities is obtained by dividing each geometric mean by the sum of all geometric means (table 3) [19].

**Table 3.** Calculating of the vector of priorities

| | $A_1$ | ... | $A_n$ | The rating of the eigenvector components by row | Vector of priorities |
|---|---|---|---|---|---|
| $A_1$ | $w_1/w_1$ | ... | $w_1/w_n$ | $\sqrt[n]{\frac{w_1}{w_1}\times\ldots\times\frac{w_1}{w_n}}=a_1$ | $\frac{a_1}{\sum_{i=1}^{n} a_i}=x_1$ |
| ... | ... | ... | ... | ... | ... |
| $A_n$ | $w_n/w_1$ | ... | $w_n/w_n$ | $\sqrt[n]{\frac{w_n}{w_1}\times\ldots\times\frac{w_n}{w_n}}=a_n$ | $\frac{a_n}{\sum_{i=1}^{n} a_i}=x_n$ |

Multiplication of the matrix by the vector of priorities is as follows:

$$\begin{pmatrix} \frac{w_1}{w_1} & \frac{w_1}{w_2} & \cdots & \frac{w_1}{w_n} \\ \frac{w_2}{w_1} & \frac{w_2}{w_2} & \cdots & \frac{w_2}{w_n} \\ \cdots & \cdots & \cdots & \cdots \\ \frac{w_n}{w_1} & \frac{w_n}{w_2} & \cdots & \frac{w_n}{w_n} \end{pmatrix} * \begin{pmatrix} x_1 \\ x_2 \\ \cdots \\ x_n \end{pmatrix} = \begin{matrix} \frac{w_1}{w_1}x_1 + \frac{w_1}{w_2}x_2 \ldots + \frac{w_1}{w_n}x_n = Y_1 \\ \frac{w_2}{w_1}x_1 + \frac{w_2}{w_2}x_2 \ldots + \frac{w_2}{w_n}x_n = Y_2 \\ \cdots \\ \frac{w_n}{w_1}x_1 + \frac{w_n}{w_2}x_2 + \ldots + \frac{w_n}{w_n}x_n = Y_n \end{matrix} \quad (1)$$

It is important to note that in the matrix of pairwise comparisons there is no ratio $w_i/w_j$, there are only integers or integer reciprocals from a scale. This matrix in the general case is inconsistent. Algebraically the problem of consistency is the solution of the equation $Aw=nw$, $A=(w_i/w_j)$, and the total task is the solution of the equation $A'w'=\lambda_{max}\cdot w'$, $A'=(a_{ij})$, where $\lambda_{max}$ is the largest eigenvalue of the matrix of pairwise comparisons $A$.

To check the consistency of each matrix the eigenvalues of the matrix are calculated (as the sum of the vector components obtained by multiplying the matrix of pairwise comparisons by the vector of priorities):

$$\lambda_{max}=\sum_{i=1}^{n} Y_i \tag{2}$$

Next, the index of consistency (IC) and consistency ratio (CR) are calculated:

$$IC = (\lambda_{max} - n)/(n-1), \tag{3}$$

where $\lambda_{max}$ – eigenvalue of the matrix, n – the number of compared elements.

$$CR=IC/RI, \tag{4}$$

where RI – random index.

CR and IC should not exceed 10%. Otherwise, the quality of the judgments should be improved, perhaps by revising the way in which questions are asked when conducting pairwise comparisons.

When conducting assessments it is important to keep in mind all compare items to comparison was relevant. To conduct a reasonable numerical comparisons should not compare more than 7±2 elements.

Then hierarchical synthesis is conducted, i.e. the sum of all the weighted components of the corresponding eigenvectors of the hierarchy level lying before is calculated [20].

## 3. Results section

The described method was used for selecting the best option of roofing material for the device pitched roofs. The results are presented in table 4.

**Table 4.** The results of the selecting the best option of roofing material

| Roofing materials | Global priorities |
|---|---|
| Metal roof tile<br>Metal Profile Monterrey PURETAN | 0,144 |
| ClickLock<br>Grand Line Satin | 0,116 |
| Cement concrete tiles Baltic tile Sea Wave | 0,149 |
| Clay roof tile Roben FLANDERN | 0,236 |
| Asphalt shingles Tegola Nordland, Classic | 0,071 |
| ONDULINE roofing sheets<br>Onduline Tile | 0,144 |
| Composite roof shingles Decra Decra classic | 0,142 |

## 4. Discussion section

Thus, the best option by the set of its characteristics is clay roof tiles Roben FLANDERN, its disadvantages are high cost and a large mass of 1 $m^2$, however, durability, fire safety, environmental friendliness, corrosion resistance, sound insulation and thermal conductivity are significantly ahead of other options of roofing materials. Metal roof tile Metal Profile PURETAN Monterrey, ONDULINE roofing sheets Ondulin Tile, Cement concrete tiles Baltic Sea Wave tile and composite roof shingles Dekra Dekra classic have approximately equal values of global priorities. The least successful of the above options can be considered the Asphalt shingles Tegola Nordland, Classic – its disadvantages are high cost with a relatively small lifespan and high combustibility.

## 5. Conclusions

The proposed technique can be applied for selection of the most appropriate building material or technological equipment from the available options based on its qualitative and quantitative characteristics.

## References

1. J. Testa, M. Krarti, Sust. Cities and Soc., **31**, 62-73 (2017)
2. J. Y. Lee, G. Bak, M. Han, Envir. Poll., **162**, 422-429 (2012)
3. B. Vale, Ecol.and Sust. Built Environ., **2**, 67-112 (2017)
4. P. Berdahl, H. Akbari, R. Levinson, W. A. Miller, Constr. and Build. Mater., **22(4)**, 423-433 (2008)
5. S. Pişkin, A. K. Figen, E. özkan, ü. Özçay, IFAC Proc. Vol., **46(16)**, 484-487 (2013)
6. V. Ducman, A. S. Škapin, M. Radeka, J. Ranogajec, Ceram. Int., **37(1)**, 85-91 (2011)
7. S. Emmitt, C. A. Gorse, Barry's Intr. to Constr. of Build, **1**, 208 (2013)
8. M.I. Sánchez de Rojas, F.P. Marín, M. Frías, E. Valenzuela, O. Rodríguez, Constr. and Buildi. Mater., **25(6)**, 2888-2897 (2011)
9. A. Arnold, Proc. Eng., **145**, 760-765 (2016)
10. C.R. Dixon, F.J. Masters, D.O. Prevatt, K.R. Gurley, T.M. Brown, J.A. Peterka, M.E. Kubena, J.of Wind Eng. and Industrial Aerodyn., **130**, 30-40 (2014)
11. B. Nam, H. Maherinia, A. H. Behzadan, Constr. and Buildi. Mater., **50**, 308-316 (2014)
12. Available at: http://krovlja.ru/
13. Available at: http://smol-krovlya.ru/
14. S. Rahman, H. Odeyinka, S. Perera, Y. Bi, Exp.Syst.with Appli., **39(8)**, 6857-6871 (2012)
15. I. Kafiev, P. Romanov, Bull. Bashkir Univ., **2(26)**, 83-86 (2013)
16. I. Kafiev, P. Romanov, I. Romanova, Bull. Bashkir Univ., **3(35)**, 48-52 (2015)
17. I. Kafiev, P. Romanov, I. Romanova, Bull. Bashkir Univ., **1(33)**, 82-90 (2015)
18. T. L. Saaty, RACSAM, Series A: Mathematics, **102(2)**, 251-318 (2008)
19. B.O. Saracoglu, Eur. J. of Ind. Eng. Ind. Enter., **7(4)**, 416-441 (2013)
20. Z. Q. Geng, S. S. Zhao, G. C. Tao, Y. M. Han, Appl. to food safety, **78**, 33-42 (2017)

# Types of Cadastral Maps in Slovak Republic and Accuracy of the Land Area

*Anna* Seidlová[1,*], *Jakub* Chromčák[1]

[1]Department of Geodesy, Faculty of Civil Engineering, University of Žilina, Slovak Republic

**Abstract.** The technical progress has affected in all parts of science. This paper is reflecting the changes of creation and accuracy of the maps, used for tax purposes from Austria-Hungary to present. From the precision of coordinates of the break points, which are defining the running of boundary, depends the precision of the parcel area as well. This information is more important because of the prices of land in present.

## 1 Introduction

The total area of Slovak republic is round 49000 km$^2$. This area is divided onto 7.3 million of existing lands evidenced in Cadastre of real Estates as the parcels of register C. From this number, 5.7 million have owner´s folio, which means that 78% of all parcels have defined proprietary relation. The rest 1.6 million of areas that are registered in register E does not exist in terrain; they were merged because of political or historical reasons into bigger agricultural or wood units or their configuration changes trough usage in history. In the register E, there are over 8.2 million of lands and the proprietary relations were restored and are evidenced on owner´s folio for register E.

## 2 History

The area of Slovakia was till 1918 part of Austria-Hungary. In 1647, there was the first registration of land for the tax purposes. The very first ruler, who wanted to correct tax on scientific base and create stable cadastre of 19$^{th}$ Century in Austrian empire, was Emperor Franz I. The maps of stable cadastre are dated from 1849. Slovak region, also named Upper Hungary was part of Budapest reference system with beginning in observatory of Budin on hill Gellért Hegy, the name of system was Gellért Hegy datum.

The original cadastre maps were created manually. Most of them are colour, in the build-up area, there are differences between wooden and brick buildings, in non-build-up area are colour differences between main roads, side roads, watercourses and areas, which helps in global clarity. The information on these maps are written in Slovak, Hungarian and German, in the east side of Slovakia, there are occasions, when there are information in Russian.

* Corresponding author: anna.seidlova@fstav.uniza.sk

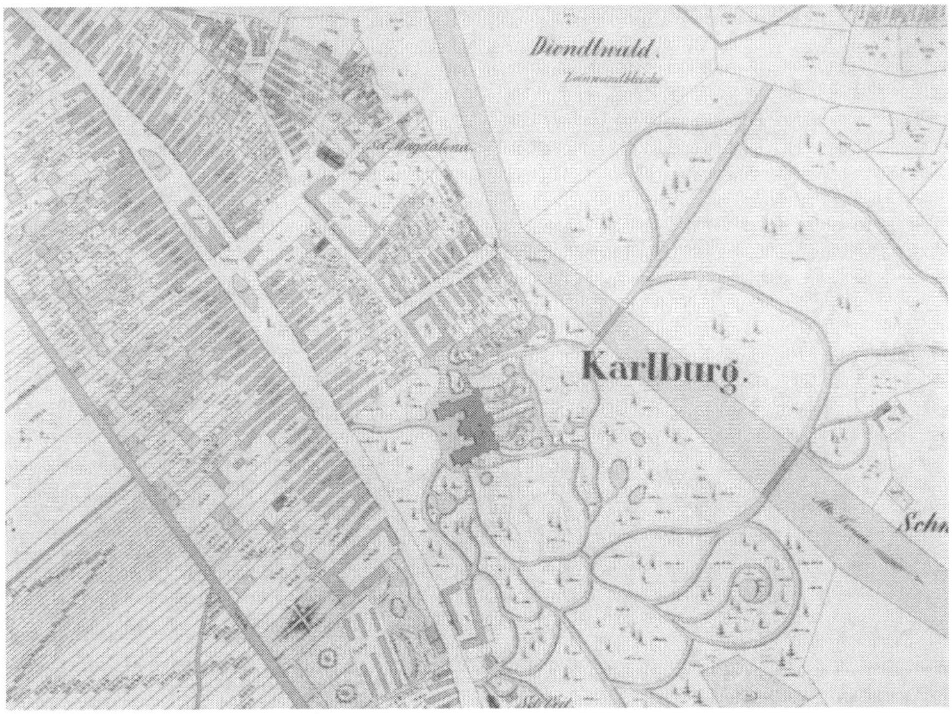


**Fig. 1.** The original map of village Rusovce from 1856

The Czech land surveyor František Horský was working on alignment of trigonometrical network by least squares method in 1863 and suggested to replace Gellért Hegy datum by stereographical projection and to align and recalculate the trigonometrical network in this new projection. The maps in stereographical geodetic datum were in Slovakia created in fathom scale. These maps were created in 1864-1875 (on the most on east part of Slovakia) and in 1886-1913 (on the most on south part of Slovakia). There is one problem, that maps in this datum have too big distortion, so they failed to satisfy the necessities of cadastre, so they were stopped to be created as well as used.

In the area of past Hungary was from 1908 established new datum called Fasching projection in fathom scale (conform universal cylindrical projection) The mapping in this projection was in Slovakia carried out until 1924, but only in less quantity.

In the Czechoslovakia was the first aim to create the united cadastre of lands. The conditions of creation were established in Cadastral law n. 177/1927 Zb. z. The maps created in the wording of this law were called the maps of Czechoslovakian cadastre of lands, these were the first maps created in metric scale.

### 1.1 Sort composition of the cadastral maps

From the perspective of original cadastral maps′ division onto fathom and metric scale, 33.9% of whole Slovakia was measured in fathom scale (on the most 1:2880) in Hungarian coordinate datum. In the effect of law n. 177/1927 Zb. z. about cadastre of lands was measuring done just in metric scale, in this scale are maps of rest 66.1% of area, mapped after 1927.

The maps in metric scale can be divided onto: 12.1% of area measured between World War One (1914-1918) and World War Two (1939-1945), in very good standard. 16.4% of area were mapped in the wording of methodises for the technic-economical map in 1971-

1986 and the coordinates of lands´ break points were calculated from the measurement. After 1986 were mapped the rest of area which presents 23% of area, the lands´ break points were documented as well with high accuracy and these maps presents the best cadastral map with the highest precision and the most sufficient content. These maps were mapped in the class of required precision of survey n. 3 (0.14 m) sometimes n. 4 (0.26 m) (forests with use of photogrammetry method).

## 3 Current situation

Nowadays, according to the UGKK SR (The Geodesy, Cartography and Cadastre Authority of Slovak Republic) legislation in force, there can be used the measuring with GNNS. This technology allows first time in history of cadastre of real estates processing of high precise data of lands´ break points measurement processed just in terms of European Terrestrial Reference System 1989 (ETRS89). [5] Fixed steps of transformation from ETRS89 to S-JTSK03 and to S-JTSK makes using of GNSS so there is possible to reconstruct each and every new measured break point back into terrain without any undesirable deformation in the vectorial cadastral map. To ensure a functionality of whole system, UGKK SR unelaborated an amendment of generally legally binding legislative regulations and a whole series of related technical regulations. From 30.1.2013 there is available the sectoral transformation service that allows the transformation between ETRS89, S-JTSK03 and S-JTSK. This transformation is allowed to be built in the GNSS rover. In the same time there is a possibility to free download of an appropriate conversion interpolation table to each computer.

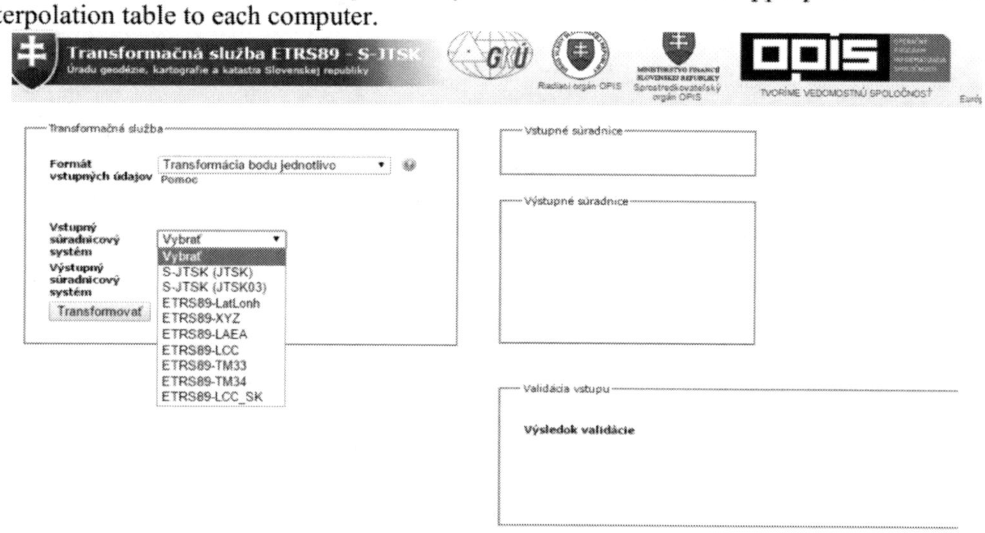


**Fig. 2.** The conversion interpolation table

On the base of these changes was established new category of accuracy. It is characterized by quality of every cadastral map break point of detailed planimetric survey, the way of its´ creation, the coefficients for calculation of areas quality and the values of coefficient for checking the precision of points´ displaying.

**Table 1.** The definition of points′ quality code. [1]

| Code | The point of detailed planimetric surveys′ quality |
|---|---|
| 1 | The point defined numerical in system S-JTSK from § 55 par. 4 (mxy = 0, 08 m) |
| 2 | The point defined numerical in system S-JTSK without real placing; this code cannot be used in the occasion, if it is necessary to signalize the point in terrain or if the point is signalized yet (mxy = 0, 08m) |
| 3 | The point defined numerical in system S-JTSK defined via geodetic terrestrial method without connecting to used geodetic base (mxy = 0, 14 m). |
| 4 | The point defined non-numerical in system S-JTSK defined via geodetic terrestrial method without connecting to used geodetic base (mxy = 0, 26 m). |
| 5 | The point defined non-numerical, the coordinates are defined by using cartometry (mxy is non-defined). |

On the base of points of detailed planimetric survey are vectorial cadastral maps divided to:

- VKMč- numerical vectorial cadastral maps, where the accuracy of break point quality code is from 1 to 3,
- VKMn- non-numerical vectorial cadastral maps, where the accuracy of break point quality code is from 4 to 5.

**Fig. 3.** The current vectorial cadastral map of village Rusovce

## 4 The lands´ area

The lands´ area is presented by the area of land projection into reference plane in square meters. The area is defined by using Euler formula from coordinates of break points of boundaries. The value of area is rounded to whole square meters.

The difference between the area identified in the file of survey data and the area defined in the file of descriptive information is judged by the value of permissible deviation calculated from formula:

$$m_p = a.\sqrt{P} - b \ ; \qquad (1)$$

where a and b are the coefficients, which are taking into account the scale of map and P is the area of parcel evidenced in the file of descriptive information. [1] The values are defined for numerical and non-numerical cadastral map and are possible to see in table number 2 and 3.

**Table 2.** Coefficients a, b used in different scales for numerical cadastral map. [2]

| VKMč | | |
|---|---|---|
| **Map scale** | **a** | **b** |
| 1:1000 | 0,42 | 0,4 |
| 1:1250 | 0,53 | 0,5 |
| 1:2000 | 0,84 | 0,8 |
| 1:2500 | 1,05 | 1,0 |
| 1:5000 | 2,10 | 2,0 |

**Table 2.** Coefficients a, b used in different scales for non-numerical cadastral map of old analogue map. [2]

| VKMč | | |
|---|---|---|
| **Map scale** | **a** | **b** |
| 1:1000 | 0,84 | 0,8 |
| 1:1250 | 1,05 | 1,0 |
| 1:1440 | 1,21 | 1,2 |
| 1:2000 | 1,68 | 1,6 |
| 1:2500 | 2,10 | 2,0 |
| 1:2880 | 2,42 | 2,4 |
| 1:3600 | 3,02 | 3,0 |
| 1:5000 | 4,20 | 4,0 |
| 1:7200 | 6,05 | 6,0 |

## 5 Conclusions

In present age are all break points of detailed planimetric survey defined in coordinates intended by terrestrial method. The position criteria are defined in relation with point intended by method GNSS and transformed from geocentric coordinate datum ETRS89 into coordinate system S-JTSK, which is used in cadastre of real estates. This acceptable variation is defined for terrestrial measurement on 0.24m and after its´ exceeding it is necessary these coordinates to adjust.

## Acknowledgements

This article is the result of the implementation of the project VEGA 1/0275/17 "Application of numerical methods to define the changes of geometrical track position" supported by the Scientific Grant Agency of the Ministry of Education, science, research and sport of the Slovak Republic and the Slovak Academy of Sciences. This article is the result of the implementation of the project ITMS 26220220156 "Broker centre of air transport for transfer of technology and knowledge into transport and transport infrastructure" supported by the Research & Development Operational Programme funded by the ERDF.

## References

1. Regulation of The Geodesy, Cartography and Cadastre Authority of the Slovak Republic, n. USM_UGKK SR 3/2014 from 07.04.2014 (Correction of the detailed survey points and areas evidenced as parcel of C and E register)
2. Edict 461/2009 Z.z. of The Geodesy, Cartography and Cadastre Authority of the Slovak Republic from 28.10.2009
3. Šafář V., Ždímal V., Identification of land cover in the past using infrared images at present, International Archives of the Photogrammetry, Remote Sensing and Spatial Information Sciences, Volume XXXIX-B7, **229-234**, 2012, Melbourne, Australia
4. Pravda J., The language of maps, Comenius University, pp. **39-48,** 1977, Bratislava, Slovakia
5. Ižvoltová J., Pisca P., Černota P., Mancovič M., Adjustment of code ranging of GNSS observations. Communications, 18 (4), pp. 15-18, 2016, Žilina, Slovak Republic

# Properties of dry masonry mixtures based on hollow aluminosilicate microspheres

*Vyacheslav* Semenov[1] *

[1]Moscow state university of civil engineering, Yaroslavskoye shosse, 26, Moscow, Russia, 129337

**Abstract.** At present, there is a steady increase in the volume of housing construction in the Russian Federation. The modern trends in the field of energy and resource saving determine the need of the use of efficient building materials that ensure the safety, comfort and minimum cost of housing construction. Among the materials, often used for erecting of fencing structures, it is possible to note effective small-piece elements (ceramic and light-weight concrete units, etc.). To ensure the solidity of such structures, it is necessary to use the masonry mortars whose properties correspond to those of the main wall material. The existing dry mixes for obtaining of such mortars are expensive and often do not meet the minimum physical-and-mechanical and exploitation requirements. The solution of this problem is the usage of the hollow ceramics (aluminosilicate) microspheres as a filler for such mixes. The article presents the results of studies of the main physical-and-mechanical and exploitation characteristics of dry masonry mixes with hollow ceramics microspheres modified with various chemical additives. The effect of the compounding factors on the average density and strength of dry masonry mixes was studied. The compositions have been optimized by the methods of mathematical planning.

## 1 Introduction

The housing construction is one of the largest parts of the construction industry in our country and since the 2000s there has being a significant increase in the construction of residential areas. According to the data published by the Research Institute of Building Physics, in the Russian Federation the buildings consume up to 45% of the total volume of produced energy, 18 ... 45% of which are the heat losses through external walls [1]. A consequence of this is the development of rather high normative requirements for providing of thermal protection to fencing structures for the performance of which it is expedient to use the efficient building materials. An important factor in assessing the heat-shielding properties of the fencing structure is the coefficient of thermal uniformity. This coefficient takes into account the presence of thermally heterogeneous fragments in the fencing structure – the "cold bridges" [2]. Their presence leads to an increase in heat losses through the fencing structures, as well as to the risk of condensate formation in the zone of heat conducting inclusions.

* Corresponding author: science-isa@yandex.ru

A significant part of residential buildings are erected with the use of two or three-layer fencing structures made of effective small-piece elements - ceramic units, aerated concrete blocks, polystyrene concrete blocks, etc. For the construction of fencing structures made of similar materials, it is necessary to use the light-weight masonry mortars whose average density should not exceed the average density of the wall material, and the compressive strength of these mortars should be 20...25% higher than the strength of the wall material, which is the condition for ensuring the design strength of the fencing structure. For example, for masonry made of ceramic units, the grade strength of the mortar should be 5.0...15.0 MPa, for masonry made of aerated concrete blocks this value should be 5.0...7.5 MPa. At the same time, the existing dry mixes for preparation of such mortars are expensive and do not provide the required strength. Such mixes are produced by introducing into their composition the light-weight fillers: expanded perlite or vermiculite sands, foam glass or expanded polystyrene granules, etc. The mortars with expanded vermiculite and perlite sands have low strength. At an average density of 400...600 kg/m$^3$, the compressive strength of such mortars varies between 1...2.5 MPa, while their freeze-thaw durability is not more than 35 cycles [3-4]. The expanded pearlite and vermiculite sands have high water absorption. For perlite sand the water absorption can reach 900% by mass, for vermiculite sand it value can be up to 400%, which significantly reduces the strength, freeze-thaw durability and durability of mortars. To solve this problem, the fillers are hydrophobized by creating a thin mineral film on the surface of grains, which leads to a significant increase in the cost of aggregates (by 1.5...2.2 times). In addition, pearlite sand with a low bulk density has a strength value up to 0.1 MPa, and, due to its high brittleness, it can break down during transportation as well as during the process of mortar mixing. Vermiculite sand has a higher strength, but its cost is 3...5 times higher than the cost of perlite sand.

Also, expanded polystyrene and foamed glass granules are used as the light-weight aggregates for mortars and concretes [5]. The granules of aggregate float up at stirring and are unevenly distributed in the mortar. Such aggregates are difficult to use in masonry mortars, because the size of the granules does not allow evenly distributing the mortar with a thin and cohesive layer over the surface, which leads to the denudation of the granules and to the low strength of the mortar joint.

At present, the cement compositions using the hollow microspheres (glass, ceramic (aluminosilicate), polymeric) as a light-weight aggregate are widely developed [6-9]. The spherical shape of this aggregate granules provides for greater plasticity of the mortars and its higher compressive strength. The introduction of hollow microspheres into the dry mixes as a filler will make it possible to obtain an effective composite material with high strength and thermophysical parameters [10-12]. However, the analysis of earlier studies demonstrates a need in an integrated approach to the formation of the structure of such composites.

The aim of the research was the development and study of the properties of modified dry masonry mixes with hollow ceramics microspheres (CMS).

## 2 Materials and methods

In this study, the Portland cement CEM I 42.5 N (Holcim), in accordance with GOST 31108-2003 and GOST 30515-2013, was used as the binder. The most promising is the application of the hollow aluminosilicate microspheres as a filler for light-weight dry masonry mixes [15]. The aluminosilicate microspheres are a component of fly ash (mineral waste generated by flaring coal at thermal power plants) [14-16]. In this work, the hollow ceramics (aluminosilicate) microspheres with the fraction size from 1 to 500 μm (INOTEK, Siberia Kuznetskaya) were used as a filler. The hollow ceramics microspheres are a gray powder consisting of thin-walled balls. The thickness of the microsphere's wall is 0.1 to 10 μm. The coefficient of thermal conductivity of the microspheres is 0.08 W/(m·°C) at the temperature of 20°C, the softening point is over 1000°C. The composition of the gas phase inside the spheres is a mixture of $CO_2$ (70%) and $N_2$ (30%). The main properties of aluminosilicate thin-walled microspheres are given in Table 1.

**Table 1.** The main properties of aluminosilicate thin-walled microspheres (INOTEK, Siberia Kuznetskaya)

| Parameters | Units | Parameter value |
|---|---|---|
| Density of the shell material | kg/m$^3$ | 2450 |
| Bulk density | kg/m$^3$ | 370…390 |
| Average particle size | μm | 300 |
| Minimum strength in hydrostatic compression (10% fracture) | MPa | 15…28 |
| Flotation | % of volume, type result | 96 |

Since the subject-matter of the study is a dry mix, all chemical additives must be rapidly dispersible (the dissolution time at a water temperature of 15...30°C should not exceed 2 to 10 min), should have low hygroscopicity and a sufficient period of activity, and also be well distributed in the mixture in a process of dry mixing. To reduce the water-cement ratio of the mortar mixture, the superplasticizer PERAMIN SMF 10 was used at a dosage of 0.4% of the cement mass, the chemical composition is a sulfonate melamine powder. The air-entraining additive (AEA) ASCO 93 which is an anionic surfactant based on a high molecular weight olefin sulfonate, and the redispersible polymer powders (RPP) Vinnapas 4023 N and Vinnapas 8034 H were used as the structure modifiers. The introduction of the air-entraining additive improves the workability of the mortar, increases the freeze-thaw durability of the mortar and reduces its average density. The redispersible polymer powders increase the physical-and-mechanical properties and improve the technological properties of the mortar.

The experimental studies were carried out according to standard testing methods in accordance with GOST. The mobility of the mortar was determined from the depth of immersion of a standard cone with a mass of 300 g. For all compositions, the mobility which corresponds to the immersion value of a cone of 4 to 8 cm was adopted. The average density of mortar was determined by a steel cylindrical vessel with a capacity of 1000 cm$^3$, the determination of the water-retention capacity was performed with the help of the DW-RC device. The retention of the mortar was determined according to EN 1015-9: 2007 (procedure B). This parameter determines the usefulness of the masonry mortar for use (min), after which the mobility of the mortar in the standard test changes by 30 mm as compared to the initial mobility. The physical-and-mechanical characteristics of the masonry mortar were determined at the age of 28 days on samples with the dimensions of 4×4×16 cm. The moisture and water absorption of the masonry mortar were determined by standard methods. The determination of the freeze-thaw durability of the masonry mortar was carried out according to the basic method with repeated freezing and thawing on samples with the dimensions of 4×4×16 cm. The sorption humidity was determined by the desiccator method at the relative humidity of 40%, 60%, 80%, 90% and 97% (an aqueous solution of sulfuric acid was used). The coefficient of thermal conductivity of the mortar was determined in the dried state.

## 3 Results and discussion

Earlier, the authors in the paper [17] used the computer modeling of the structure of cement mortar with hollow ceramics microspheres for the calculation of the optimum content of aggregate (CMS) of a given granulometric composition, in which the most dense packing of particles is achieved (60% of the binder mass). In this case, theoretically, the highest specific strength of the composite should be achieved. The average density of the obtained mortar in dry state was 950 kg/m$^3$, the compressive strength was 17.1 MPa, and a bending strength was 2.8 MPa. The further increase of the content of CMS in the mortar does not provide an effective reduction in its average density, increasing the water requirement of the mortar, reducing its strength and causing an unnecessary increase in the cost of the dry mix. Meanwhile, for the construction of fencing structures with effective small units, it is necessary to obtain a masonry mortar with a lower average density. One of the ways to solve the aforesaid problem is to modify the structure of the mortar with an air-entraining additive at compensating the decrease in physical-and-mechanical characteristics by the introduction of the RPP.

All studies were carried out on the previously obtained optimal composition of the dry mix containing 60% of the CMS from the mass of the binder and a superplasticizer at a dosage of 0.4% of the binder mass. The consumption of AEA and RPP was varied in the range of 0.01...0.03% and of 1...5% of the mass of the binder, respectively. Two types of RPP were used. The mobility of all masonry mortars corresponds to the immersion value of a standard cone of 4 to 8 cm. The water-retention capacity for all compositions was higher than the minimum normalized value (90%), therefore the introduction of a water-retaining additive into the mortar is not required. This is due to the fact that hollow ceramics microspheres actively adsorb water, and, due to the surface forces, keep it on its surface. The retentivity of the initial mobility of the mortar was 4.5 to 5 hours, which is sufficient for the production of masonry work. The compositions and properties of dry mixes in the solidified state are given in Table 2.

**Table 2.** Compositions and properties of light-weight dry mixes with hollow ceramics microspheres, redispersible polymer powders and air-entraining additive

| No | Composition, mass, % | | | | | | W/C | Average density of mortar in dry state, kg/m$^3$ | Strength at the age of 28 days, MPa | |
|---|---|---|---|---|---|---|---|---|---|---|
| | PC | CMS | SP | AEA | V 4023 N | V 8034 H | | | compression | bending |
| 1 | 100 | 60 | 0,4 | – | – | – | 0,65 | 951 | 17,14 | 2,76 |
| 2 | 100 | 60 | 0,4 | 0,02 | – | – | 0,71 | 576 | 5,01 | 1,67 |
| 3 | 100 | 60 | 0,4 | 0,02 | 1 | – | 0,71 | 645 | 7,96 | 1,88 |
| 4 | 100 | 60 | 0,4 | 0,02 | – | 1 | 0,73 | 583 | 6,84 | 1,73 |
| 5 | 100 | 60 | 0,4 | 0,02 | 3 | – | 0,73 | 811 | 10,68 | 2,03 |
| 6 | 100 | 60 | 0,4 | 0,02 | – | 3 | 0,76 | 598 | 8,38 | 1,89 |
| 7 | 100 | 60 | 0,4 | 0,02 | 5 | – | 0,75 | 887 | 11,37 | 2,19 |
| 8 | 100 | 60 | 0,4 | 0,02 | – | 5 | 0,78 | 701 | 8,92 | 1,99 |

Notes: PC - Portland cement; CMS - hollow ceramics microspheres; SP - superplasticizer; AEA - air-entraining additive; V 4023 N and V 8034 H -redispersible polymer powders Vinnapas 4023 N and Vinnapas 8034 H, respectively.

From Table 2, it can be seen that the introduction of different amount of RPP into the dry mix neutralizes the effect of air entrainment. With the help of mathematical planning of the experiment and the processing of its results, two-factor mathematical models of the properties of light-weight masonry mortar with hollow ceramics microspheres, AEA and RPP were obtained, their compositions were optimized. For the production of mortar with an average density of 500 kg/m$^3$, the introduction of a RPP is not required, and the AEA flow rate is 0.04% of the binder mass (composition No. 2). To obtain a mortar with an average density of 600 kg/m$^3$, the optimum consumption of the modifying additives was 0.02% for AEA and 3% for RPP of the mass of the binder, respectively (composition No. 6).

An important exploitation parameter of the masonry mortar is water resistance characterized by a softening factor. The coefficient of softening for mortar of all compositions was in the range 0.8... 0.89, which makes it possible to classify the developed dry mixes as water resistant ones.

The coefficient of thermal conductivity of the mortar obtained from the developed light-weight dry mortar mixture depends on the content of microspheres and the porosity of the cement stone. The results of measuring of the coefficients of thermal conductivity for the control and modified compositions of mortars are presented in Table 3.

**Table 3.** Main exploitation properties of mortar obtained from a light-weight dry masonry mix with hollow ceramics microspheres, redispersible polymer powders and air-entraining additive

| No | Average density of mortar in dry state, kg/m$^3$ | Coefficient of thermal conductivity of the mortar in dry state, W/(m·°C) | Water absorption, % | | Water absorption at capillary suction, kg/(m$^2$·hours$^{0,5}$) | Freeze-thaw durability, cycles |
|---|---|---|---|---|---|---|
| | | | mass | volume | | |
| 1 | 951 | 0,244 | 8,1 | 8,1 | 0,3 | 100 |
| 2 | 576 | 0,146 | 13,9 | 13,9 | 0,4 | 50 |
| 6 | 598 | 0,152 | 9,2 | 9,2 | 0,2 | 75 |

From Table 3 we can see that the composition with AEA (No. 2) has the smallest coefficient of thermal conductivity. Compared with the control composition (No. 1), this parameter decreased by 40% due to the formation of a highly porous structure and a decrease in the average density. The coefficient of thermal conductivity of the mortar with AEA and RPP (No. 6) decreased by 34% compared to the control composition.

The water absorption of the mortar is provided a significant effect on its performance characteristics. From Table 3 it can be seen that the water mass absorption of the mortar with AEA increased by 72% in comparison with the control composition. Since Vinnapas 8034 H has a hydrophobizing effect, the water absorption of the composition with this RPP decreased by 51% as compared with the composition with the AEA. The sorption humidity of the developed dry mixes is in the range of 5...7% at a relative air humidity of 97% and in the range of 0.7...1% at the relative humidity of 40%.

The freeze-thaw durability is the most important parameter determining the durability of the building material. The results of the determination of the freeze-thaw durability of a mortar obtained from a light-weight dry masonry mix are shown in Table 3. According to the GOST 530-2012, the freeze-thaw durability of the face brick should not be lower than 50 cycles, therefore, taking into account that, the water absorption of mortar (by mass) did not exceed the normative value of 15%, the softening factor is not less than 0.8, the developed dry mixes can be used for facade masonry.

The composition and structure of the light-weight dry masonry mixes in the hardened state were studied. The results of microstructural and X-ray phase analysis of samples of dry masonry mixes in the solidified state with CMS, modified by the AEA and RPP are shown in Figures 1, 2.

It is established that the microstructure of the samples of the modified mortar is represented by uniformly distributed in volume microspheres and cement matrix over the solid partially crystallized calcium hydrosilicates with inclusions of low-strength needle-like calcium hydrosilicates.

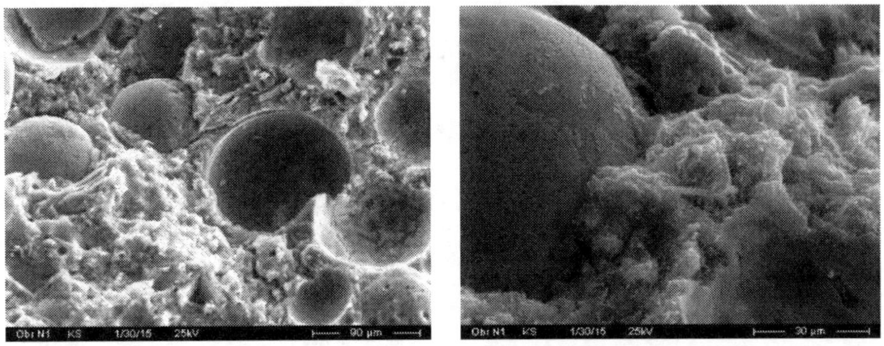

**Fig. 1.** Microstructure of the light-weight masonry mortar with hollow ceramics microspheres, redispersible polymer powders and air-entraining additive

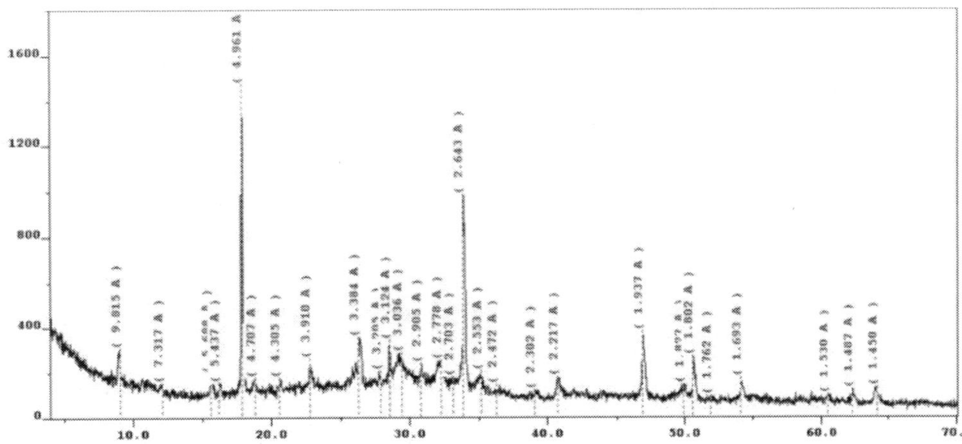


**Fig. 2.** X-ray diagram of the light-weight masonry mortar with hollow ceramics microspheres, redispersible polymer powders and air-entraining additive

The surface of the microspheres is covered by the products of the chemical interaction of their walls with the cement matrix, which leads to an increase in the strength of the contact zone "microsphere – cement matrix". On the X-ray diagram the clinker minerals, the portlandite, partially crystallized calcium hydrosilicates, hydroaluminate and hydroferrite of calcium, calcite, ettringite are identified. The degree of cement hydration is 66.5% (at the age of 28 days).

## 5 Conclusions

As a result of the studies, the compositions of modified light-weight dry mixes for masonry work with hollow aluminosilicate microspheres with an average density of 576...608 kg/m$^3$, compressive strength of 5...8.38 MPa, bending tensile strength of 1.67 ... 1.89 MPa, water-retaining capacity of the mortar not less than 93%, the retainability of the initial mobility of the mortar mixture not less than 4.5 hours were obtained. The coefficient of thermal conductivity of the developed light-weight mortar is 0,146...0,162 W/(m·°C), water absorption by mass is 13,9...9,2%, softening factor is 0,8 ... 0,89, sorption humidity at relative humidity of air 97 % is 6.9...5.7%, the freeze-thaw durability is 50...75 cycles.

## References

1. V.G. Gagarin, V.V. Kozlov, Vestnik MGSU, **3(1)**, 192 (2011)

2. A.S. Gorshkov, A.A. Gladkikh, Magazine of Civil Engineering, **3**, 39 (2010)
3. R. Demirboğa, İ. Örüng, R Gül, Cement and Concrete Research, **31 (11)**, 1627 (2001)
4. Yu.M. Tikhonov, V.I. Kolominetc, Bulletin of Civil Engineers, **3**, 83 (2006)
5. V. Ferrándiz-Mas, T. Bond, E. García-Alcocel, C.R. Cheeseman, Construction and Building Materials, **61**, 285 (2014)
6. V.S. Semenov, T.A. Rozovskaya, D.V. Oreshkin, Advanced Materials Research, **860–863**, 1244 (2014)
7. V.S. Semenov, T.A. Rozovskaya, IOP Conference Series: Materials Science and Engineering, **71**, 012042 (2015)
8. A.K. Suryavanshi, R.N. Swamy, Cement and Concrete Research, **32**, 1783 (2002)
9. S.P. McBride, A. Shukla, A. Bose, Journal of Materials Science, **37 (19)**, 4217 (2002)
10. Y. Wu, J.Y. Wang, P.J.M. Monteiro, M.H. Zhang, Construction and Building Materials, **87**, 100 (2015)
11. D.P. Bentz, M.A. Peltz, A. Durán-Herrera, P. Valdez, C.A. Juárez, Journal Of Building Physics, **34 (3)**, 263 (2011)
12. A.S. Inozemtcev, E.V. Korolev, V.A. Smirnov, Structural Concrete, **18(1)**, 67 (2017)
13. A.S. Inozemtcev, E.V. Korolev, Industrial and civil Engineering, **10**, 80 (2013)
14. L.Ya. Kizilshtein, Science and Life, **5**, 42 (2008)
15. A.A. Sagradyan, G.A. Zimakova, News of higher educational institutions. Construction, **2**, 43 (2012)
16. M. Żyrkowski, R. C. Neto, L. F. Santos, K. Witkowski, Fuel, **174**, 49 (2016)
17. V.A. Smirnov, T.A. Rozovskaya, V.S. Semenov, Science Review, **10-2**, 78–83 (2015)

MATEC Web of Conferences 117, 00150 (2017) DOI: 10.1051/matecconf/201711700150
*XXVI R-S-P Seminar 2017, Theoretical Foundation of Civil Engineering*

# In the consideration of internal friction forces in nonstationary dynamics problems

*Yuriy E.* Senitsky [1], and *Elena N.* Elekina[1*]

[1] Samara State Technical University, Institute of Architecture and Civil Engineering, 194, Molodogvardeyskaya St., 443001, Samara, Russia

**Abstract.** In this paper, the original methodology for taking into consideration internal friction forces is given in the example of nonstationary oscillations of a beam with elastically clamped edges. Bearing in mind the experimentally confirmed fact that the forces of internal friction practically do not affect the forms of structural vibrations, they are introduced into the equation of motion after separation of the spatial variable. This decomposition approach of forming a mathematical model in conjunction with the frequency independent Voigt hypothesis, with a known loss factor, made it possible to represent the solution in the form of spectral decomposition. For this purpose, we used the structural algorithm of the finite integral transform (FIT) method with the definition of the transformation kernel in the solution process. In fact, the proposed method is a method of quasinormal coordinates and represents an effective method of solving dynamic problems for mechanical systems in the presence of internal friction forces.

## 1 Introduction

Spectral expansions obtained as a result of the application of the method of integral transformations with finite limits (FIT) represent the effective solution of linear nonstationary problems of mechanics. One of the main conditions that determines the possibility of such an approach is the self-adjointness of the initial-boundary value problem under consideration. At the same time, as a result of taking into consideration the resistance forces, including internal friction, this important property of the mathematical model is lost, and a direct way of constructing an analytical solution is impossible [1-8]. However, the structural algorithm of FIT method developed by the author [9-10] in combination with the frequency independent Voigt hypothesis [11] by introducing viscous drag forces in the solution process made it possible to obtain a new qualitative solution of the corresponding dynamic problems. In this paper, this approach is demonstrated by the example of a non-stationary problem for a beam under sufficiently general conditions for its fixing and loading.

* Corresponding author: Elekina-E1@yandex.ru

## 2 Materials and methods

We consider an elastic beam (prismatic rod EJ = const) loaded with a transverse dynamic load q (x, t) distributed along the span x ∈ (o, l) and in time t> 0 in an arbitrary law. A rather general case of elastic pinching of the end sections of a rod with a known coefficient of elasticity α is investigated. The material of a beam is homogeneous, isotropic. The mathematical model takes into account the forces of internal friction, with a given loss factor for each mode of oscillation γ_i (i = (1, ∞)̅). The solution algorithm proposed below is based on the method of quasinormal coordinates [11] and corresponds to an experimentally confirmed fact about the significant effect of internal friction forces on the behavior of structures in time and practically does not affect the shape of the oscillations. For this purpose, the structural algorithm of the method of finite integral transformations (FIT) is very convenient, in the procedure of which all the components of the resulting solution are determined. It is essential that the results obtained in this case reflect the nature of the damped oscillation process.

Differential equations of motion and boundary conditions in the study of transverse oscillations of a beam are formulated as follows [9]:

$$EJ\frac{\partial^4 y(x,t)}{\partial x^4} + m\frac{\partial^2 y(x,t)}{\partial t^2} = q(x,t), \tag{1}$$

$$y(x,0)|_{t=0} = y_0(x), \qquad \left.\frac{\partial y(x,t)}{\partial t}\right|_{t=0} = v_0(x), \tag{2}$$

$$y(x,t)|_{x=0,l} = 0, \quad EJ\frac{\partial^2 y(x,t)}{\partial x^2} = \pm\alpha\left.\frac{\partial y(x,t)}{\partial x}\right|_{x=0,l}, \tag{3}$$

Here y (x, t) is the required function of transverse displacements, J and m are respectively the moment of inertia and the distributed mass of the beam, y_0, v_0 are the initial displacements and the velocity of displacements. Relations (2) and (3) represent the initial and boundary conditions for the boundary value problem under consideration. It should be noted that the forces of internal friction are introduced later in the process of solving the problem (1) - (3).

We introduce a finite integral transformation (FIT) with respect to the variable x, defined by such a pair of formulas [9]:

$$\bar{y}(\lambda_i,t) = \int_0^l y(x,t)G(\lambda_i,x)dx;$$

$$y(x,t) = \sum_{i=1}^{\infty}\bar{y}(\lambda_i,t)G(\lambda_i,x)\|G_i\|^{-2}. \tag{4}$$

Where y¯(λ_i, t) is the transform of the original y (x, t), G (λ_i, x) is the kernel of the integral transformations defined below, λ_i are the positive parameters i = ((1, ∞)̅).

$$\|G_i\|^2 = \int_0^l G^2(\lambda_i,x)dx, \tag{5}$$

– square of the norm of a function $(\lambda_i, x)$

Relations (4) are valid in the case when the functions G (λ_i, x) represent an orthogonal system.

We apply FIT (4) to the equation (1) and the initial conditions (2). We have

$$EJ\int_0^l \frac{\partial^4 y}{\partial x^4}Gdx + m\frac{\partial^2}{\partial t^2}\int_0^l yGdx = \int_0^l qGdx; \tag{6}$$

$$\int_0^1 yGdx\Big|_{t=0} = \int_0^1 y_0 Gdx, \qquad \frac{\partial}{\partial t}\int_0^1 yGdx\Big|_{t=0} = \int_0^1 V_0 Gdx. \tag{7}$$

Here, in accordance with (4), the quadratures in equalities (6), (7) represent the transformants of the functions q (x, t), y_0 (x), V_0 (x), i.e.

$$\{\bar{q}(\lambda_i,t), \overline{y_0}(\lambda_i), \overline{V_0}(\lambda_i)\} = \int_0^1 \{q(x,t), y_0(x), V_0(x)\} G(\lambda_i, x) dx. \tag{8}$$

Using the algorithmic procedure of the structural algorithm of FIT method [1-4], we integrate the first quadrature (6) by parts. As a result, we obtain

$$EJ\{\Phi(y,G)|_0^1 + \int_0^1 yG^{IV}dx\} + m\frac{\partial^2 y(\lambda_i,t)}{\partial t^2} = \bar{q}(\lambda_i,t), \tag{9}$$

$$\Phi(y,G)|_0^1 = \left[\frac{\partial^3 y}{\partial x^3}G - \frac{\partial^2 y}{\partial x^2}G' + \frac{\partial y}{\partial x}G'' - yG'''\right]|_0^1, \tag{10}$$

$$1)\ \Phi(y,G)|_0^1 = 0, \qquad 2)\ \int_0^1 yG^{IV}dx = \lambda_i^4 \int_0^1 yGdx. \tag{11}$$

Taking into account (8), (10), (11), equations (9) and conditions (7) take the form:

$$\frac{\partial^2 \bar{y}(\lambda_i,t)}{\partial t^2} + \omega_i^2 \bar{y}(\lambda_i,t) = \frac{1}{m}\bar{q}(\lambda_i,t); \tag{12}$$

$$\bar{y}(\lambda_i,0)|_{t=0} = \overline{y_0}(\lambda_i); \qquad \frac{d}{dt}(\bar{y}\lambda_i,t)|_{t=0} = \bar{v}_0(\lambda_i) \tag{13}$$

where $\omega_i = \lambda_i^2\sqrt{\frac{EJ}{m}}$ circular oscillation frequencies of the beam.

Relations (11) allow us to formulate a differential equation and boundary (boundary) conditions for the kernel of the integral transformation G (λ_i, x). We have

$$G^{IV}(\lambda_i,x) - \lambda_i^4 G(\lambda_i,x) = 0; \tag{14}$$

$$G(\lambda_i,x)|_{x=0,1} = 0, \quad G''(\lambda_i,x) \mp \mu G'(\lambda_i,x)|_{x=0,1} = 0, \quad \mu = \alpha(EJ)^{-1} \tag{15}$$

## 3 Results

Thus, as a result of the application of the QIB (4), the original boundary value problem (1) - (3) has split into two - (12), (13) and (14), (15), i.e. there was a separation of variables. In accordance with the method of quasinormal coordinates and the frequency-independent Voigt hypothesis, we introduce into the differential equation (12) the forces of internal friction at each mode of oscillation $\gamma_i\omega_i\frac{d\bar{y}(\lambda_i,t)}{dt}$. We have

$$\frac{\partial^2 \bar{y}(\lambda_i,t)}{\partial t^2} + \gamma_i\omega_i\frac{d\bar{y}(\lambda_i,t)}{dt} + \omega_i^2\bar{y}(\lambda_i,t) = m^{-1}\bar{q}(\lambda_i,t) \tag{16}$$

The solutions of equations (16) and (14), taking into account conditions (13), are written in the form:

$$\bar{y}(\lambda_i,t) = e^{-\beta_i t}[\bar{y}_0(\lambda_i)\cos\alpha_i t + (\beta_i\bar{y}_0(\lambda_i) + \bar{v}_0(\lambda_i)\alpha_i^{-1}\sin\alpha_i t] + $$
$$+(\alpha_i m)^{-1}\int_0^t \bar{q}(\lambda_i,\tau)e^{-\beta_i(t-\tau)}\sin\alpha_i(t-\tau)d\tau; \tag{17}$$

$$G(\lambda_i,x) = C_{1i}\mathrm{ch}\lambda_i x + C_{2i}\mathrm{sh}\lambda_i x + C_{3i}\cos\lambda_i x + C_{4i}\sin\lambda_i x. \tag{18}$$

$$\alpha_i = \sqrt{\omega_i^2 - (0{,}5\gamma_i\omega_i)}, \quad \beta_i = 0.5\gamma_i\omega_i. \tag{19}$$

For various dynamic effects q (x, t) on the beam, in expression (17) it is only necessary to calculate the Duhamel integral.

The arbitrary constants C_1i, C_2i, C_3i, C_4i are determined as a result of substitution of the relation (20) into the boundary conditions (16). In this case C_3i = -C_1i, and then a non-trivial solution of the resulting system of the remaining three algebraic equations is sought. Equating its principal determinant to zero, we find

$$\begin{vmatrix} 2\lambda_i & -\mu & -\mu \\ (ch\lambda_i l - cos\lambda_i l) & sh\lambda_i l & sin\lambda_i l \\ [\lambda_i(ch\lambda_i l + cos\lambda_i l)] + \mu(sh\lambda_i l + sin\lambda_i l) & (\lambda_i sh\lambda_i l + \mu ch\lambda_i l) & (\mu cos\lambda_i l - \lambda_i sin\lambda_i l) \end{vmatrix} = 0_i \tag{20}$$

The roots of the transcendental equation (20) are the parameters λ_i and, in accordance with (14), the oscillation frequencies ω_i can be determined. Substituting expressions (17) into the inversion formula (4), and taking into account the C_1i, C_2i, C_3i, C_4i and (18) found, we obtain the required displacements of the beam y (x, t).

## 4 Discussion

Using the method of quasinormal coordinates, the structural algorithm for the finite integral transforms, and the frequency independent Voigt hypothesis, a new analytic solution of the dynamical problem for the rod is constructed in the presence of internal friction forces. The results are obtained for arbitrary conditions of loading and fixing the rod. Spectral decompositions are constructed by decomposition of a mathematical model, and all components of the reduced structure (load integral, integral transformation core, oscillation frequencies) are determined directly in the process of solving.

## 5 Conclusions

In conclusion, it should be noted that the numerical analysis of the calculation of the frequencies and modes of oscillations of an anisotropic cylinder and a sphere in the presence of internal friction forces is given in the author's papers. The proposed methodology of consideration of viscous drag forces can also be useful in studying the oscillations of the bottoms of navigable locks.

## References

1. D.A. Shlyakhin, Procedia Engineering, **150**, pp. 68-73 (2016)
2. D.A. Shlyakhin, O.V. Kazakova, Procedia Engineering **153**, 662-666 (2014)
3. V.S. Shirokov, I.S. Kholopov, A.V. Solovejv, Procedia Engineering **153**, 655-661 (2016)
4. A.A. Abashkin, Russian Math. (Iz. VUZ) **60**:2, 1–6 (2016)
5. A.O. Lukin, V.Y.Alpatov, D.D. Chernyshev, Urban Construction and Architecture **2**, 4 - 9 (2016) DOI 10.17673/Vestnik.2016.02.1
6. A.A.Romanov, O.V. Leonov, S.V. Evdokimov, Urban Construction and Architecture **4**, 66-70 (2011) DOI 10.17673/Vestnik.2011.04.13

7. K.V. Vorob'eva, K.P. Zainulabidova, M.V. Frize, Vestnik of Dagestan Satate technical University. Technical Sciences **1(40)**, 108 – 119 (2016)
8. A.M. Uzdin, D.V. Schelkunov, Seismic resistant construction. Safety of structures **2**, 47-50 (2011)
9. Y.E. Senitsky, Vestnik Samara State Technical University. Series «Physics and mathematics» **2**(19), 78 – 89 (2009)
10. Y.E. Senitsky, Procedia Engineering **153**, 649 – 654 (2016)
11. A.I. Tseitlin, A.A. Kusainov, *Methods of accounting for internal friction in dynamic design calculations* (1987)

# Technique of performing construction works by machines with hybrid: manual and remote control

*Nadezhda* Sevryugina[1,*]

[1] Moscow state university of civil engineering, Yaroslavskoye shosse, 26, Moscow, Russia, 129337

**Abstract.** The article discusses issues dealing with efficiency of construction work mechanization. It offers a mathematical model for assessment of mutual influence between the members of the 'construction site-machine-operator' system triad, that can give a quantitative assessment of how the efficiency of a technological task varies with more comprehensive use of operational capacities of the machine, while lower effect that limiting parameters of production environment and technical condition of the machine have on the operator. The article contains a constructive remote control solution for upgrade of the base machine. It describes the conditions for using the machines with hybrid: manual and remote control at construction sites. There is also an imitation model of operator's scanning pattern and data experimental research that prove the efficiency of remotely controlled technological operations. The article proves that lower psychological load on the operator and better comfort contribute to positive economic effect and higher quality of the construction process.

## Introduction

Economic stability of any country relies not only on the activity of the people, but, mainly, on the efficiency of machinery and technologies used. Progressiveness of the construction machinery development is supported by implementation of innovations obtained by scientific research of experts in many countries.

The review of developments existing has shown that the research, though systematic, is still narrow-scale. Acceleration of technical progress needs innovative updates.

Scientific foundations for managing the enhanced mechanization of construction, building the systems for remote control of the construction machinery and automatization of the construction technological processes were laid in the researches of the most prominent scientists in the sphere: P. M. Astafurov, V. I. Balovnev, N. G. Dombrovsky, A. A. Kononov, B. D. Kononykhin, M. N. Kotrovsky, Ye. M. Kudryavtsev, E. N. Kuzin, M. I. Malenkov, V. P. Pavlov, P. S. Sologub, A. N. Tarakanov, Yu. F. Ustinov and many others [1-5, 10, 12, 13, 18].

Mechanization tools, in particular, construction machinery are one of the elements that affect the efficiency of the construction plan implementation. The quality of construction

* Corresponding author: SevryuginaNS@mgsu.ru

works depend on technological perfection of machinery, experience of the operator and environment conditions [7, 9, 11, 14-17].
Studies of how the elements interact in the 'construction site-machine-operator' triad are still urgent for search of innovative technologies and management methods in construction.
*Aim of Research:* theoretical-practical justification of how the construction site can make effective use of machines with hybrid: manual and remote control that increase the quality of technological operations performed.
The aim can be achieved through the solving of the following tasks:
- creating the mathematical model for the system triad 'construction site-machine-operator' adequate for comprehensive assessing the parameters of factors influence on the quality of the construction works;
- computer modelling of the operator's scanning area, when an upgraded machine with remote control performs technological operations at the construction site;
- experimental justification for the efficiency of the remote control the operator has over the construction machine;
- economic justification for expediency of construction mechanization with machines having hybrid: manual and remote control.

## Materials and Methods

### *Mathematic modelling*

Mathematic model was created after the review of the existing methods offered by best scientists in the sphere [2, 6, 10, 18].
The model suggests a hypothesis that technology operations performed at the construction sites create interconnections of element influence in the system triad "Construction site-machine-operator" (CsMO).
Static data prove that the person is the main component of the CsMO system, and, with regard to the nature of its interaction with the other components, safety during construction operations at the site.
Each element of the system can be described by a package of characteristics presented in a certain range. For CsMO elements, functioning limits are:

$$F_{CsMO} = \sum (f_C; f_M; f_O) \tag{1}$$

Integral nature of dependence is explained by the fact that the range of system elements functioning varies with constant change of base parameters.
Function of element 'construction site' $f_C$, with consideration to base parameters, becomes:

$$f_C = f(x_{C+1}) = f(x_C) + e_C, \tag{2}$$

$$\int_{C_{\min}}^{C_{\max}} (f_C) = \int_{C_{\min}}^{C_{\max}} (x_{C+1}) = \int_{C_{\min}}^{C_{\max}} (f(x_C) + e_C), \tag{3}$$

where $f(x_C)$ means a set of parameters characterized by physical and climatic characteristics conditions of environment; $x_{C+1}$ means quantitative changes of parameters when the whole system is in function;
$e_C$ means a nonnegative discrete random value that describes variability and randomness of parameters change in an element of the 'construction site' system;
$C_{\min}$; $C_{\max}$ – means limits of the 'construction site' function definition; these limits are base for any parameters of the CsMO system and their probable changes.

The function parameters of 'Machine' and 'Operator' elements of the system are described in similar way.
With regard to the base parameters, the $f_M$ function of the 'machine' element becomes:

$$f_M = f(x_{M+1}) = f'_C + f(x_M) + e_M, \tag{4}$$

$$\int_{M_{\min}}^{M_{\max}} (f_M) = \int_{M_{\min}}^{M_{\max}} (x_{M+1}) = \int_{M_{\min}}^{M_{\max}} (f'_C + f(x_M) + e_M), \tag{5}$$

where $f_M$ means a function of the machine runnability;

$f(x_M)$ means a set of technical and physical parameters of the machine;

$x_{M+1}$ means quantitative changes of parameters when the whole system is in function;

$e_M$ means a nonnegative discrete random value that describes variability and randomness of the machine functioning parameters;
$M_{\min}$; $M_{\max}$ means limits of the 'machine' function definition; they limit technical functioning of the machine and depend on current parameters of $f(x_C)$ and its increment.

With regard to the base parameters, the $f_O$ function of the 'operator' element becomes:

$$f_O = f(x_{O+1}) = f'_C + f'_M + f(x_O) + e_O, \tag{6}$$

$$\int_{O_{\min}}^{O_{\max}} (f_O) = \int_{O_{\min}}^{O_{\max}} (x_{O+1}) = \int_{O_{\min}}^{O_{\max}} (f'_C + f'_M + f(x_O) + e_O) \tag{7}$$

where $f_O$ means a function of the 'operator' condition;

$f(x_O)$ means a set of parameters that describe physical and psychological condition of the construction machine operator;

$x_{O+1}$ means quantitative changes of parameters when the whole system is in function;

$e_O$ means a nonnegative discrete random value that describes variations in physical and psychological condition of the construction machine operator;
$O_{min}$; $O_{\max}$ means limits of the 'operator' function definition; they are limited by psychophysical condition, depend on current parameters of the function of the CsMO system elements and their probable changes.

The work of the construction machines operator can only be viewed in its connection with the construction site, as the influence on the latter forms a subject of the technological operation. The condition of the environment is the main factor that contributes to safe and comfortable operation of the operator and operational limitations of the construction machine.
For design reequipment of the base unit and using a remote control, the article studies the efficiency of CsMO functioning from the point of view of the remote control ($CsM^{MC}O$) (Fig. 1). The model $CsM^{MC}O$ offered describes information and energy processes of the system in terms of technical and psychology interaction for some elements: 'Operator', 'Machine$^{ДУ}$', 'object of production operation', that are placed into the 'Environment', the one of more importance, without which their objective existence is impossible. The 'environment' element is presented as a macroelement, as well as a varying border of the 'environment' local territory defined, with certain set of parameters and a border of their action. This means that production process can be managed both at the local territory, under the current conditions, and outside it.
It should be noted that the influence of parameters gives a general description for the operable state of the system. The interrelations in the CsMO system allow conclusions on

the varying efficiency of the technological task implementation at the expense of more comprehensive use of operational capacity of the machine when the operator is exposed to varying limiting parameters that come from the environment and the machine.
To summarize, the quality of any technological operation the operator performs at the construction site $f'_O$ in a designated period of time is a function of factors of environment influence $f'$, condition of machine $f'_M$ and components of operator's physical condition and professional skills:

$$f'_O = \Delta f'_C + \Delta f'_M + f(x_O) + e_O, \qquad (8)$$

where we offer to treat $\Delta$ as a *nonnegative* variation of parameters set and their values for the appropriate function.
This mathematical model for establishing the relationship in the CsMO system affords a quantitative estimation of changing the efficiency of the technology task implementation at the expense of more comprehensive use of operational capacity of the machine when the operator is exposed to varying limiting parameters that come from the environment and the machine.

### *Imitation Modelling*

Further research required development a computer model of the remotely controlled machine. A sample excavator was taken as a prototype for a 3D model with visualization of the construction site area the operator can scan from a work place in manual or remote control while doing technological operation of buildings disassembly. Then, we studied fields of view for *bucket working position* of the excavator that operates under various conditions and various positions of its *track* (Fig. 1).
1. *Scanning from the operator's cabin*
The quality of scanning and visibility from the operator's place in the cabin depends on the cabin position about the boom and the body of the excavator, position of the seat in the cabin, configuration of windows, quality of glass, their proper cleaning and heating, eyes protection from the direct sunlight blinding, avoiding the glaring.
Scanning patterns obtained provide more refined data for the quality of scanning from the operator's work place. For visual control of the processes that take place outside the objects of primary observation (chassis, ground under the excavator etc.) these values are very low and limit time and quality of technological operations.
2. *Scanning from a Remote Control Point*
The position of a remote control point should consider certain parameters, like distance, safety, availability, comfort etc.

**Fig. 1.** Modelling the construction site space visualization for buliding disassembly: a0 manual control b) remote control.

The scanning picture obtained for remote control demonstrates comprehensive visual control for the excavator bucket and working equipment. It was found that the operator's perception distorts the distance varying with moving away, as moving at 6.9 – 201.6 m gives 1:5.7 proportion for visual change of the excavator size.

## Results

### *Experimental research*

Experimental research was aimed at estimating the time required for performing a technological operation with a predetermined sequence for moving the elements of the excavator working equipment to preset marks (Fig. 2).

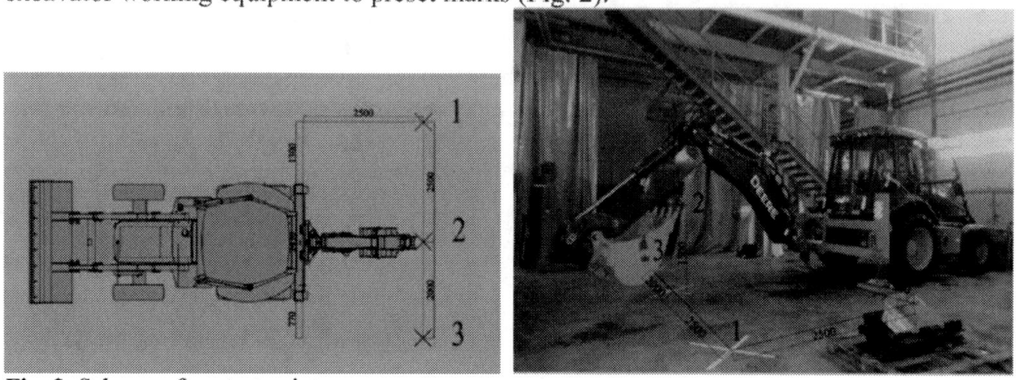


**Fig. 2.** Scheme of contact points

The experiment noted the following parameters:
- time the operator required for making one complete cycle that includes a sequence of touching preset points with the bucket, sec;
- vibration level at the operator's working place, Hz;
- area of visual perception from the operator's working place.

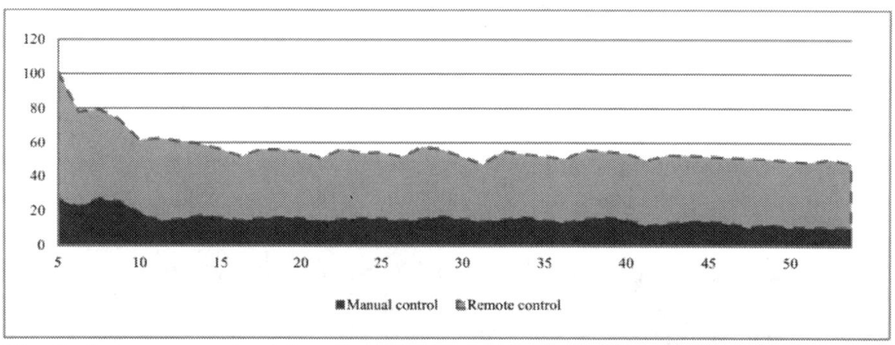


**Fig. 3.** Summary graph of time the operator needs for making predetermined operations in manual and remote control modes

**Table 1.** Extent of machine operator's visual perception of the territory involved in construction technological process

| Position of scanning evaluation | Object scanning while manual control | Object scanning while remote control |
|---|---|---|
| Condition of the base surface | Scanning of object is insufficient. For better view, the operator has to move his body | The objects are in full view |

| | | |
|---|---|---|
| | forward | |
| Scanning of boom | The object overlaps the field of vision at the construction site.<br>The boom is in full view | No influence to perception of the construction site<br>The boom is in full view |
| Scanning of handle | The object overlaps the field of vision at the construction site.<br>The handle is in limited view. | No influence to perception of the construction site.<br>The handle is in full view. |
| Scanning of bucket | The object overlaps the field of vision at the construction site.<br>The bucket is in sufficient view. | No influence to perception of the construction site.<br>The bucket is in full view. |
| Information display medium. Interval for scanning of information display media. | Available for scanning. The operator has to look at the control panel on his left. | A part of information block is placed at the remote control panel.<br>Partial information on the condition of the machine. |
| Positions 1,2,3 (zones of primary scanning) | Visibility of object varies when elements of working equipment gets into the operator's field of view. | Fully available for scanning |
| Additional comments | Cabin stands and glazing reduce scanning, as they overlap the vision and form glare. | Changing of remote control position will find the safest point of control. |

The survey of time required for the operator to make operations in manual and remote control of the excavator working equipment were processed and are shown in the graph of Fig. 3.

Experimentally, we found that the time the operator needs for making the same operations with the excavator working tool are objectively lower in case of the remote control. This is shown in the Table 1 that contains qualitative estimation for the extent of the operator's visual perception of the territory involved in the construction technology.

Switching from manual to remote control has proved the operator is psychologically ready for changing conditions of technology operations performance. The operator who had 7 years of experience needed 20–25 minutes for practicing the remote control. The control position was opposite to that of working equipment, the inversion explained by a time delay the operator had to take decision while controlling. We noted the possibility for changing the control position, body position (sitting/standing); free visual access for the main parts of the base machine to be observed; absence of vibration and other hazardous factors; higher quality of the technological task to be performed by the operator.

### *Economic Assessment of the Results Obtained*

The reasearch proved the convergence of practical and theoretical results. Then, we had to confirm economic expediency for the remote control the machine operator performs at the construction sites.

We calculated and presented at the diagram (Fig. 4) aggegated annual operating costs for 10 working years of the machine with hybrid manual and remote control.

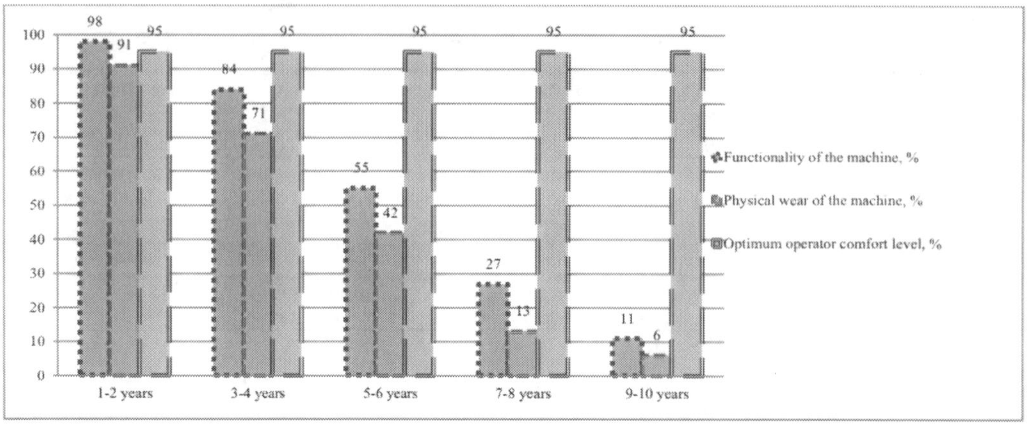


**Fig. 4.** Correlation of changing annual costs of a construction company that operates a remotely controlled machine

The diagram shows that the provision of operator's comfort peseet by the manufacturer gets more costly with years of operation. Physical wear of the machine manifests itself in increase of psychphysical load of the operator, the key ones being nose level and vibration in the cabin, dust content and air temperature in the cabin. The diagram shows that the remote control technology makes operator's comfort an invariant value throughout the whole period of the machine service.

## Discussion

It is well known that the manufacturer procures the functionality of the machine that operator implements according to his skills and experience, that is, his psychophyical capacities and proficiency.

When the operator performs technological operatons at the construction sites, he is to disperse his attention between the factors of outer (technological operation object – environment) and inner (comfortable control of the machine – stability of technical condition) perception.

Logical conclusion is that the maintenanace of the operator's comfort depends directly on the constructive perfection and and functionality of the machine, as well as on the quality of percpation of the space involved in the techonological process at the construction site. The remote control by the machine operator rreduces the effect of negative factors. Lower psychological load contributes to the feeling of security, beacuase the operator stays away from the hazards at functional zone of the machine working tool and better control over the construction object.

The methods offered is distinguished from earlier methods by the fact that its results consider for the factors of mutual influence the CsMO system elements have, thus raising the level of the machine operator's comfor to the range of constant values.

## Conclusions

The article offered the solution for the task of machine remote control technology application at the technology sites that affords higher quality of the construction management.

Making the scanning pictures affords assessment of changes in the 'operator-machine-construction site' system: security of operator and persons at the construction object, quality of the construction management and management of machine operation efficiency.

The possibility for the operator to be outside probably hazardous area while using the remote control is an effective way of creating comfortable and safe conditions of work.

We revealed that psychophysical condition of the operator increases dynamically when he performs the remote machine control.
The comparison between the results obtained from research of technological operations performance by the construction machine operator who works in manual or remote control mode has proven the developments offered are promising.
The article has proven the economical expedience for adding remote control modules to the machines of base constructions.

**References**

1. R.V. Abaimov, Y.S. Bahracheva, A.V. Vasilyev, A.A. Konev, V.I. Kychkin, E. I. Lezhneva, N.S. Sevryugina, K.M. Selivanov, V.S. Yushkov, R. M. Ahmednabiev (Eds.), *Road transport*, Novosibirsk, 2013.
2. A.A. Bogomolov, N.S. Sevryugina, *Variational interpretation of the life cycle of technical systems*, Construction and road machines, (2010), No. **10**, pp. 48-52. (In Russian).
3. S.N. Glagolev, N.S. Sevryugina, *Efficiency of functioning of the owner-car-service system as result of the choice its accented components* , Automotive industry, (2012), No. **6**, pp. 10-11. (In Russian).
4. S.N. Glagolev, N.S. Sevryugina, A.A. Konev, *Mathematical model of an estimation of efficiency of development of the Territory of the region with the formation of clusters of roadside service,* World of transport and technological machines, (2013), No. **3(42)**, pp. 121-125. (In Russian).
5. O.N. Didmanidze, A.M. Karev, G.E. Mityaev, *On the development prospects of road transport in the agricultural sector*, International scientific journal, (2016), No. **1**, pp. 53-65.
6. V.A. Zorin, *Bases of operability of technical systems*, M.: LLC Magistr-Press, (2005), 536 p. (In Russian).
7. P.D. Kapyrin, E.S. Romanova, *Analysis of current state of the building materials industry and the factors contributing to the development production*, Vestnik MGSU, (2010), No. **4-1**, pp. 165-170. (In Russian).
8. P.D. Kapyrin, E.S. Romanova, *Modern technological line for production of three-layer wall panels*, Vestnik MGSU, (2011), No. **4**, pp. 490.
9. S. M. Pulyaev, M. A. Stepanov, B.A. Kaitukov, N.A. Lukyanov, N.A. Deacon, P.D. Kapyrin, *Mechanical equipment and technological complexes*. Textbook, Moscow: MGSU, (2015). (In Russian).
10. A.N. Novikov, O.A. Ivashchuk, V.V. Vasilieva, *Ecological monitoring of the impact of transport on the acoustic environment of the city*, The Repair. Recovery. Modernization, (2006). No. **6**, pp. 33-34. (In Russian).
11. E.S. Romanova, P.D. Kapyrin, *Modern technological lines for production of plates of overlappings*, The MGSU Bulletin, (2011), No. **4**, pp. 499. (In Russian).
12. N.S. Sevryugina, *Modernization of building and road machines as a factor of realization of the targets*, Construction and road machine us, (2007), No. **7**, pp. 28-29. (In Russian).
13. N.S. Sevryugina, V.M. Babin, *Resource modernization of self-propelled machines*, Building and road machines, (2007), No. **9**, pp. 49-53. (In Russian).
14. V.I. Telichenko, E.A. Korol', P. B. Kagan, E.N. Kulikova, *Systemic administration of the trust building programs*. Procing manual, Moscow: ASV, (2010). (In Russian).
15. A.D. Ishkov, N.G. Werstine, V.I. Telichenko, N.G. Miloradov, The patent for the invention. *The diagnostic method of the level of claims person*, RUS 2444979 from 18.08.2010. (In Russian).

16. V.I. Telichenko, E.A. Korol', M.S. Khlistunov, S.I. Zavalishin, *Global risks and new security threats responsible governmental construction projects in the metropolis*, The Urban complex and the problems ispovedition of citizens, the Collection of reports of scientific-technical conference, M.: (2005), pp. 211-218. (In Russian).
17. R.R. Sharapov, A.A. Mamedov, E.V. Kharlamov, I.V. Bragin, V.V. Klochkov, *Performance of illicit small self-propelled crane cargo lifting capacity of over 100 tons of improved mobility and stability*, Bulletin of Belgorod state technological University. V. G. Shukhov, (2016), No. **3**, pp. 112-115. (In Russian).
18. N.S. Sevryugina, S.B. Melikhova, E.A. Volkov, *The solution of applied problems of optimization of stability of system "environment-man-technics"*, Modern Applied Science, (2015) Vol.9 No. 3, pp. 200-207. DOI: 10.5539/mas.v9n3p200 (Publisher: Canadian Center of Science and Education)

# Analysis of the spectrum distribution of oscillation amplitudes of the concrete mix at shock vibration molding

*Rashid* Sharapov[1*] and *Vitaly* Vasiliev[1]

[1]Moscow state university of civil engineering, Yaroslavskoye shosse, 26, Moscow, Russia, 129337

**Abstract.** In the production of concrete structures widespread shaking tables of various designs. The effectiveness of vibroforming concrete items largely depends on the choice of rational modes of vibroeffect to the compacting mixture. The article discusses the propagation of a wave packet in the concrete mixture under shock and vibration molding. Studies have shown that the spectrum of a wave packet contains a large number of harmonics. The main parameter influencing the amplitude-frequency spectrum is the stiffness of elastic gaskets between mold and forming machine vibrating table. By varying the stiffness of the elastic gaskets can widely change the spectrum of the oscillations propagating in the concrete mix. Thus, it is possible to adjust the intensity of the process of vibroforming.

The theory and practice of vibratory compaction of concrete mixtures are still relevant in the production of concrete items and structures [1-3]. The main tasks in easel forming concrete items are the choice of effective modes of vibration effects and rational parameters of the shaking tables.

Unfortunately, when using standard molding conditions of existing construction vibrating equipment does not meet modern requirements. This is due to their large resource consumption over the low quality of manufactured concrete items, low operational reliability, as well as the excess of admissible norms of noise and vibration [4, 5].

To compact firm concrete mix B. V. Gusev [6] developed the theoretical foundations of shock and vibration technology. Staff members of his school were created and installed equipment for implementing low-frequency modes forming concrete items [1, 7]. In other works [8, 9], as effective modes of compaction of firm mixes it is proposed to use the so-called asymmetrical shock-free oscillations. The advantage of applying these oscillations is the absence of oscillating vibrotable with the balancing frame.

In articles [10, 11] described optimization of parameters of shock and vibration shaking tables for efficient modes of forming concrete items of firm concrete.

It is interest to consider the process of propagation of a wave packet in concrete mix under shock and vibration effects on the mold and explore a range of pulses, which propagate in the mixture.

* Corresponding author: ptdm_zavkaf@mail.ru

The mold with concrete mixture is placed on the shaking table through the elastic element – the gasket. Under the influence of the vibrotable to the mold, they continue to move towards each other. With deformed gasket, the area of contact increases and the repulsive force from the mold of the vibrotable increases. Under the action of this force the relative speed between the mold and the working body decreases to zero. Then the elastic force to return mold and a working body of the shaking table in the original dynamic position.

In the absence of the gasket, the role of the elastic element performs the material from which made the working body.

Consider the effects that occur when oncoming traffic the mold and working body of the shaking table. Thus we investigate the pulse shape, its spectrum and duration of the process of collision.

It is known [12] that the magnitude of deformation of the elastic element (gasket or material of vibrotable) depends on the compressive strength and stiffness of the gasket.

The duration of impact can be determined by the formula:

$$\tau = \Delta T = 2{,}94 \cdot \sqrt[10]{\frac{1{,}22}{P_0 \cdot \delta \cdot c^4}} \cdot \sqrt{M}, \tag{1}$$

where $P_0$ – compressive force that occurs between the mold and the working body with elastic contact, $\delta$ – the maximum distance (amplitude of oscillations) the mold from of in the process of their mutual motion, m; $c$ – the stiffness of the elastic element between the mold and the vibrotable, N·m$^{-1}$; $M$ – the reduced mass of the system "working body – mold with a mixture" determined by the formula:

$$M = \frac{M_1 \cdot M_2}{M_1 + M_2}, \tag{2}$$

where $M_1$ is the mass of the working body, kg; $M_2$ – mass of the mold with concrete mix, kg.

From (1) shows that $\tau$ little depends by gap size $\delta$ and strength $P_0$. Basically, this value is determined by the weight of the mold with concrete mix and the working body and the stiffness of the intermediate elastic element.

The force exerted at the moment of contact mold with a work body of the shaking table increases according to the following law [10]:

$$P_k(t) = 4{,}69 \frac{\sqrt{MP_0\delta}}{\Delta T} \sin \frac{\pi}{\Delta T} \cdot t, \tag{3}$$

where $\Delta T$– the moment of maximum convergence of the vibrotable and the mold with concrete mix:

$$P_k(t) = 0 \text{ if } t \rangle \Delta T.$$

The convergence $y(t)$ of the mold with the working body undur elastic contact:

$$y(t) = \left(1{,}25 \frac{M}{c}\right)^{0{,}4} \cdot \upsilon_0^{\,0{,}8} \sin \frac{\pi}{\Delta T} t, \tag{4}$$

where $\Delta T$– the moment of maximum convergence of the vibrotable and the mold with concrete mix:

$$P_k(t) = 0 \text{ if } t \rangle \Delta T.$$

It is known [13] that the spectrum of some almost periodic functions $y(t)$ is determined by the formula:

$$S(\omega)=\int_{-\infty}^{\infty} y(t)\cdot e^{-j\omega t}\,dt\,. \tag{5}$$

Substituting (4) into (5) and made the necessary computing, we obtain the dependence for calculating the amplitude-frequency spectrum:

$$S(\omega)=\frac{1,5\left(MP_0\delta\right)^{1/2}\cdot\cos\dfrac{\omega\cdot 2,94\sqrt[10]{\dfrac{1,22}{P_0\cdot\delta\cdot c^4}}\sqrt{M}}{2}}{\left[1-\left(\dfrac{\omega\cdot 2,94\sqrt[10]{\dfrac{1,22}{P_0\cdot\delta\cdot c^4}}\sqrt{M}}{\pi}\right)^2\right]}, \tag{6}$$

where $y$ – a deflection value of the element for elastic collision the mold with vibrotable, m; $\omega_i$ – the $i$ –th harmonic component of the spectrum, $s^{-1}$.

According to the formula (6) carried out calculations of the spectral distribution of the vibration amplitudes of the mold for forming concrete items with a length of 5 m, width 2 m and height 0.5 m of moderately firm concrete mix [10]. The calculated density (ρ) of the concrete mix is 2400 kg/m3, the scope of working body is 1.5 mm, the main oscillation frequency of 12.5 Hz. The calculation results are presented in table. 1.

As can be seen from table 1, the spectrum of a wave packet under shock and vibration effect to the mold with the concrete mixture contains a large number of harmonics. The scatter of amplitudes of individual harmonics is large enough.

When forming concrete items on the shaking table is often occur uneven distribution of oscillation amplitudes along the surface of the vibrotable and impacts the mold on the frame of the table.

Studies of the movement of the mold under its free installation on the vibrotable of shaking table was carried out in [10]. It was shown that the spectrum propagating in a concrete mix of frequencies is wider than the periodic fluctuations the mold, which is rigidly attached to the shaking table. Multicomponent oscillations of the mold under its free installation on vibrotable generate in concrete mixture wide and almost continuous range of frequencies.

To determine the optimal parameters of the effects from working body of the forming machine on the system "mold –concrete mixture", let us analyze the formula (6). After appropriate transformations, it can be represented as follows:

$$S(\omega_i)=\frac{A_1\cdot \mathrm{Cos}\omega_i\cdot A_2}{\left(1-A_3\cdot\omega_i^2\right)}, \tag{7}$$

**Table 1.** The results of the calculation of the spectrum distribution of oscillation amplitudes of the mold.

| No of harmonic | The frequency of the working body | | $A_i$, mm |
|---|---|---|---|
| | $\omega_i$, $s^{-1}$ | $f_i$, Hz | |
| 1 | 78,5 | 12,5 | 7,64 |
| 2 | 157 | 25 | 0,01 |
| 3 | 235,5 | 37,5 | 0,8 |

| | | | |
|---|---|---|---|
| 4 | 314 | 50 | 0,6 |
| 5 | 392 | 62,5 | 0,2 |
| 6 | 471 | 75 | 0,005 |
| 7 | 549,5 | 87,5 | 0,15 |
| 8 | 628 | 100 | 0,16 |
| 9 | 706,5 | 112,5 | 0,08 |
| 10 | 785 | 125 | 0,003 |
| 11 | 863,5 | 137,5 | 0,06 |
| 12 | 942 | 150 | 0,08 |
| 13 | 1020 | 162,5 | 0,04 |
| 14 | 1099 | 175 | 0,002 |
| 15 | 1177,5 | 187,5 | 0,03 |
| 16 | 1256 | 200 | 0,0005 |

where $$A_1 = 1,5\left(c\cdot y^{3/2}\cdot\delta\right)^{1/2}, \tag{8}$$

$$A_2 = 2,94\cdot\sqrt[10]{\frac{1,22}{\delta\cdot y^{1,5}}}\cdot\sqrt{\frac{M}{c}}; \tag{9}$$

$$A_3 = 2,94\cdot\sqrt[10]{\frac{1,22}{\delta\cdot y^{1,5}\cdot c^{5}}M}. \tag{10}$$

Assessing the contribution of individual components and after some simplifications, the expression (6) can be written as:

$$|S(\omega_i)| = A_1\cdot|\mathrm{Cos}\,\omega_i\cdot A_2|. \tag{11}$$

It is obvious that the amplitude of the spectral components will be minimal when the value $|\mathrm{Cos}\omega_i\, A_2| = 1$, where

$$A\cdot\omega_i = n\pi \tag{12}$$

where $n = 1, 2, 3,\ldots$

$$A_2 = \frac{n\pi}{\omega_i}. \tag{13}$$

In the transition from one spectral component to another ($\omega_{i..} - \omega_{i+1} = \omega_t$) amplitudes of them will be maximum if:

$$A_2 = \frac{\pi}{\omega_t}. \tag{14}$$

After transformation the expression (9) $A_2$ can be represented as follows:

$$A_2 = 3\left(\delta\cdot y^{3/2}\right)^{-0,1}\cdot\left(\omega_0\right)^{-1}, \tag{15}$$

where $\omega_0 = \sqrt{\frac{c}{M}}$ - natural frequency oscillation on an elastic element.

Substituting the expression (15) into (14) we get:

$$\omega_t = \frac{\pi}{A_2} \cong \frac{\pi}{3}\omega_0\left(\delta\cdot y^{3/2}\right)^{0,1} = 1,05\omega_0\left(\delta\cdot y^{3/2}\right)^{0,1}. \tag{16}$$

In table 2 are some results of calculations of values of $\omega_t$ with fixed values of $\delta = 7\cdot10^{-3}$ m, $y = 0,1\cdot10^{-3}$ m and different values of the stiffness of the elastic gasket $c$ and the mass $M$.

**Table 2.** Calculated values of the maximum amplitudes of the spectral components.

| **The stiffness of the elastic gasket *c*, N/m** | **Reduced mass *M*, kg** | **Freguency $\omega_t$, Hz** |
|---|---|---|
| $8\cdot10^7$ | 88,5 | 27,9 |
| $6\cdot10^7$ | 76,7 | 24,5 |
| $5\cdot10^7$ | 49,5 | 14,1 |
| $4\cdot10^7$ | 36,1 | 11,6 |
| $3\cdot10^7$ | 27,1 | 9,3 |
| $2\cdot10^7$ | 19,8 | 5,2 |
| $1\cdot10^7$ | 12,8 | 0,5 |

From table 2. it follows that the stiffness of elastic contact between the mold with concrete and vibrotable significantly affects the range of vibrations which are transmitted to the concrete mixture.

For confirmation of theoretical results were carried out experimental studies of the spectrum of a wave packet in a concrete mixture using the method of holographic interferometry [10, 14]. In the experiments we changed the elasticity of the contact, the amplitude of oscillations of the working body, the ratio of the mold mass with the concrete mixture and of the vibrotable and the frequency of the vibrodrive, which was varied in the range of 12.5...350 Hz. As elastic elements were used materials with a hardness in the range of $0.7\cdot10^7$ to $1.66\cdot10^{11}$ N/m.

The research program included the determination of the frequency and amplitude of the harmonic components of a wave packet spectrum and the effective acceleration of the particles of the mixture, calculated according to the formula:

$$a=\sum_{i=1}^{n}a_i\cdot\omega^2_i, \tag{17}$$

where $a$ and $n$ are the amplitude and number of harmonics is taken.

Some results of experiments are presented in Fig.

Analysis of the research results showed that the amplitude-frequency spectrum of oscillations propagating in the concrete mix is changed according to a complicated law. For large stiffnesses of the contact between the mold and the working body of the forming machine, the distribution of amplitudes of harmonics can be seen clearly. The analysis of expression (7) shows that at large stiffnesses, high frequency harmonics, defined $\mathrm{Cos}\omega_i$, change little

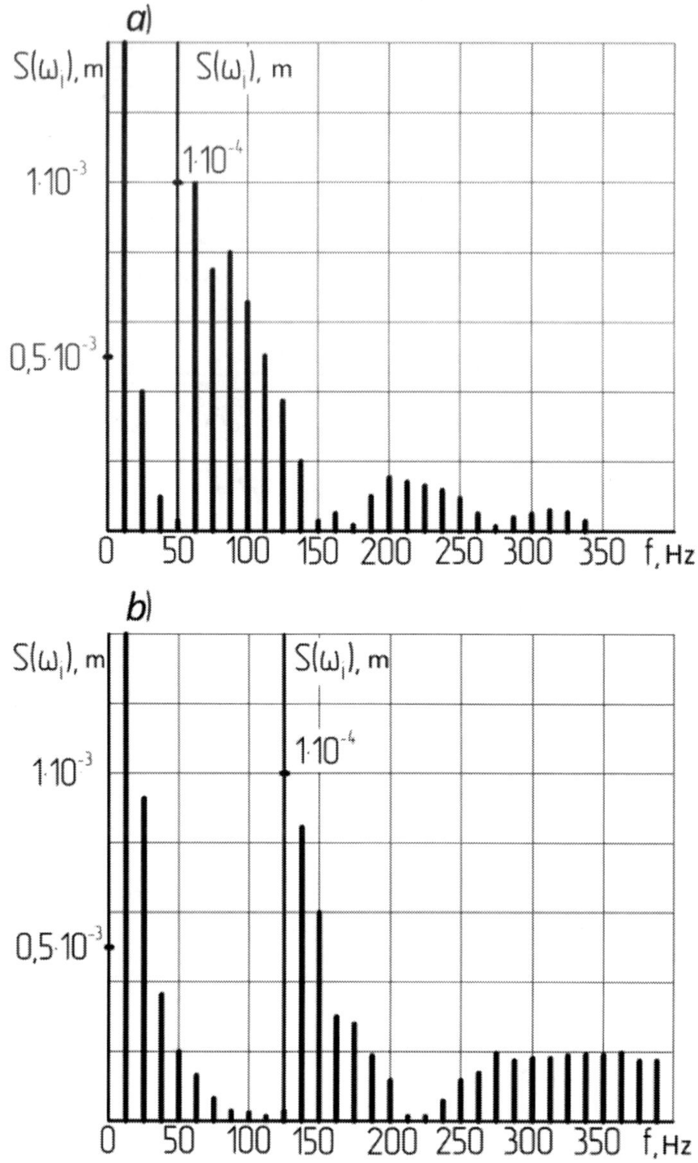


Figure. Spectra of pulses in concrete mixtures when the collision mold with vibrotable through the rubber gaskets of different hardness *c*: *a*) $c = 3.53 \cdot 10^7$ N/m; *b*) $c = 2.08 \cdot 10^7$ N/m.

Under elastic contact the mold with vibrjtable through the rubber gasket with less hardness, spectra of the pulse at impact have a more complex distribution close to the law (Cos*X*)/*X*. It has been the practical extinction of some harmonic components with increase in the amplitude of the others. Thus arise opportunities to manage spectrum by replacing one elastic gasket to another, or to use elements with adjustable stiffness [10, 15].

Thus, when we use the elastic gaskets with nonlinear characteristics in the system "working body - the elastic element - the mold with concrete mix" are prerequisites for substantial expansion of the spectrum generated frequencies in a mixture. Look at it from an energy point of view. The processes of vibration compaction of the concrete mix will flow more intensely when exposed to a formable mixture of a wide range of pulses.

Moreover, the spectrum of the pulses should contain a large number of components, which carry high energy (above threshold).

As shown in [3, 10], for estimating the spectrum of a wave packet in the concrete mix can be successfully use such a parameter as the acceleration.

Theoretical and experimental studies have a good agreement with the results of works on creation of shaking table with controlled elastic elements [10, 15] and shaking table with a freely mounted the mold on the vibrotable [16, 17].

## References

1. B.V. Gusev, V.G. Zazimko, *Vibration technology of concrete*. Kiev: Budivelnik, 158 p. (1991)
2. O.A.Savinov, E. V.Lavrinovich. *Vibration machinery for compacting and molding concrete mixtures*. L.: Stroiizdat, 278 p, (1986)
3. B.V. Gusev, E.W. Axelrod, A.I. Zvezdov, *Manual on technology of forming concrete products*, (to SNiP 3.09.01-85)/ M.: Stroyizdat, 111 p, (1988)
4. K.A. Olekhnovich, V.G. Zazimko, Main trends in the development of easel compacting concrete mixes, New technologies in precast concrete. M.: Znanie, pp. 90-96, (1985)
5. L.A.Volkov, S.A. Genkin, V.F. Lapeev, Universal vibro-impact pad with multi-component oscillations, Building materials, **5**, pp 6-7, (1996)
6. B.V. Gusev, The theory and practice of compaction of concrete mixtures at the low frequency vibration modes, Authoref. doctor. dis. of doctor technical sciences, M.: 25 p, (1977)
7. R.R. Sharapov, A.M. Agarkov, Rinat R. Sharapov, Matrix Modeling of Technological Systems Grinding with Closed Circuit Ball Mill, World Applied Sciences Journal, **24**, No. 10, pp. 1399-1403, (2013)
8. V.G. Vasiliev, F.N. Anatolly, T.A. Rakishev, Effective regimes for creation of the equipment for compaction of concrete mixtures: Chelyabinsk, pp. 18-20, (1991)
9. V.G. Vasiliev, E.G. Zinoviev, *Development and investigation of asymmetric unstressed shaking table*, MGAKHIS, M., 15 p, (1999)
10. E.G. Zinoviev, *Shaping large size concrete products with the use of guided modes of vibration compaction*: Authoref. Cand. Dis. of tech. Sciences, M., 25 p, (1989)
11. V.G. Vasiliev, The choice of optimal parameters of shock-vibrating tables, Mechanization of construction, **11**, pp. 30-33, (2015)
12. V.E. Karamzin, Some features of the behavior of the concrete mix at the easel vibration, Vibration technology of concrete products, M:, Stroyizdat, pp. 67-84, (1970)
13. I.V. Shrubchenko, A.V. Hurtasenko, R.R. Sharapov, T.A. Duyun, N.A. Shchetinin, Investigation of Characteristics of Contact Bandages and Support Rollers of Rotating Technological Drums, Modern Applied Science, **9**, No 1, pp. 195–203, (2015)
14. E.G. Zinoviev, V.G.Vasiliev, E.P. Parfenov, *Development of methods of investigating the system "machine – environment" with the implementation of the various oscillations*, MGAKHIS, M., 12 p, (1999)
15. E.G. Zinoviev, V.G.Vasiliev, V.A. Serov, *Shaking table with adjustable parameters*, MGAKHIS, M., 10 p. (2001)
16. S.A. Genkin, V.F. Lipeev, E.V. Parkhaev, Patent No 2002615, Russian Federation, Shaking table for compaction of the concrete mix in the mold
17. A.V. Ovchinnikov, V.N. Baryshnikov, L.A. Volkov, A.G. Pyzhikov, Patent No. 2252137, Russian Federation, Shaking table

# Energy-saving processes simulation in reconstruction of educational institutions edifices

*S.G.* Sheina[1], *I. Y.* Zilberova[1], *A.O.* Vongay[1,*]

[1]Department of urban development and economy, Don State Technical University, Academy of Construction and Architecture, 1 Gagarina square, Rostov-on-Don, 344000, the Russian Federation

**Abstract.** One of the high-priority directions in the development of the Russian and global economies is energy saving and reduction of exhaustible energy resources consumption. It is possible to make the existing urban development pool correspondent with the required energy efficiency class by way of reconstruction. Despite the fact that the said law entered into force in 2010, there is still a lack of methods of systematic development of energy-saving policies for educational institutions edifices within the framework of reconstruction. Methods applied separately for reconstruction and introduction of energy saving measures do not allow a simulation of acceptance of effective solutions for level enhancement of technical operation of the edifices. Methods of efficiency enhancement for the technological processes of edifices reconstruction are represented in this article for the first time; evaluation criteria for efficiency of technological processes in reconstruction of higher educational edifices have been developed; a method of criterion analysis has been applied taking the binding conditions of technologies introduction into consideration. The practical significance of the method developed has been proved by way of introduction thereof into the framework of reconstruction of the edifices of the higher educational institution.

## 1 Foreword

Under the current economic situation, the civil engineering market of Russia demonstrates a trend to reduce the construction of new objects. Nevertheless, the problems of energy efficiency of public edifices are becoming more and more pressing, especially after the enactment of the federal law FZ-261 "Of Energy Saving and increase of Energy Efficiency …".

According to the said law, the operated edifices shall fall under the energy efficiency class no below C "Normal", but almost 80 of the deficits in Rostov Region belong to the D and E classes. Energy efficiency class enhancement for the existing edifices can be carried out within the framework of reconstruction. Despite the fact the issue of energy efficiency have since long enough become the issues of state importance there is still a lack of adequate methods of systematic development of energy-saving policies for public edifices

* Corresponding author: vao1090@gmail.com

within the framework of reconstruction.

Methods applied separately for reconstruction and introduction of energy saving measures do not allow a simulation of acceptance of effective solutions for level enhancement of technical operation of the edifices.

Urgency of the issue is stipulated by the fact that the enhanced arrangement of technological processes in the course of reconstruction of edifices is one of the ways to provide for efficient solutions for improvement of the level of technical operation of the edifices granted the application of energy-saving measures. The solution of this problem is offered by means of an innovative method of arrangement efficiency enhancement for technical processes of reconstruction of public edifices with the use of energy-saving measures.

# 2 Main problems of technological processes arrangement in reconstruction of educational institutions edifices

## 2.1 Peculiarities of energy-saving programs implementation in budget-funded institutions

Analysis of rules and regulations in the sphere of energy saving and energy efficiency enhancement has shown that the normative framework in the Russian Federation is built by the federal laws, orders and resolutions of the Russian Federation Government, orders of the federal ministries, and other legal acts and documents. As the main regulatory act guiding the energy saving policy in the Russian Federation, the Federal Law No. 261 "Of Energy Saving and increase of Energy Efficiency …" has been currently adopted.

The categories of entities for whom the compliance with the requirements to energy efficiency is mandatory have been determined by the law 261-FZ. Among such entities, there are budget-funded institutions.

According to the Russian Federation Government Decree No. 20 dd. January 25, 2011, the budget-funded institutions are included into the state information system for energy saving and enhancement of energy efficiency (EE)[1].

Now let us highlight the main peculiarities of the budget-funded organizations in the sphere of energy saving:

- Determination of indicators for calculation of key indicators values for the period till the year 2020 and adjustment of energy saving programs with consideration thereof;
- Conduction of works aimed at installation of instrument gauges at the budget-funded objects;
- Conduction of systematic energy inspections at the budget-funded objects;
- Periodic control over compliance with the legislative requirements in the sphere of energy saving, implementation of programs for energy saving.

Based on the analysis of normative documents, a conclusion can be made that in order to comply with the requirements of modern legislation to energy efficiency (EE) of the budget-funded institutions, a complex (program) method of measure-taking is necessary.

Funding of energy saving programs is yet another important aspect of budget-funded institutions operation.

Following are the peculiarities of funding of the budget institutions:

1) Sources of funding – measures taken within the federal state (municipal) institutions$ these are implemented within the framework of current financing by the federal executive power authorities;
2) Utilization of saved funds – grant given for fulfillment of the state (municipal) task proceed to the current account of the budget-funded institutions. These are used by the

budget-funded institutions within the limits of balances reflected on the individual accounts of such institutions without representation of documents confirming the occurrence of cash commitments. This allows to use the saved funds in the direction selected by the institution itself.

The funding of the energy efficiency programs in the budget-funded institutions is thus determined by the state for the period until the year 2020. The funding is effected within the framework of the state program for energy efficiency and assumes the use of both budget and non-budget sources of funding.

## 2.2 Analysis of technological processes arrangement in the course of reconstruction of educational institutions edifices

Analysis of theoretical researches in the sphere of technological processes arrangement has shown that these issues are covered with sufficient fullness in the works of such scientists as V.A. Afanasyev, L.G. Dikman, A.A. Lapidus, Y.B. Monfred, V.I. Telichenko, T.N. Tsay, A.K. Schreiber, et al.

Reconstruction of educational institutions edifices has a number of specific peculiarities of conduction of works: limited time of works implementation, while the reconstruction is carried out without interruption of educational process in the period of minimal edifice utilization rate; limited field of operation due to implementation of works within the existing development pool; increased requirements to safety techniques, due to implementation of works in the operated building; limited funding.

The common drawback of all the mentioned works is the lack of theoretical research and practical advice as of assessment and choice of rational variants of technological processes arrangement in the course of reconstruction of edifices and buildings based on the certain terms of works implementation. In particular, such issues as possibility to simulate technological processes of building and installation works under preservation of educational process have not been discussed for the educational institutions.

Pursuant to the requirements of the current Russian legislation, energy-efficient technologies must be utilized in the course of reconstruction of public edifices and buildings. Sometimes this can become one of the reasons for reconstruction if, for instance, it is necessary to provide for the necessary value of thermal transmission resistance of the walls in connection with the alternation of normative values. [1].

Such researchers as V.G. Gagarin, L.A. Danilov, A.N. Dmitryev, V.I. ledenev, V.V. Malyavina, Y.A. Matrosov and others have been dealing with the developments in the sphere of energy efficiency for various branches of economy.

The works of the mentioned researchers reflect the methods of implementation of energy-saving and energy-efficient measures (ESEM) within the framework of the "State Enerfy Efficiency Program till 2020", and technical and economical efficiency evaluation methods for implementation results of such measures.

On the basis of analysis presented in the works by I.A. Bashmakov and V.I. Bashmakov, Sh. G. Zinganshin, S.B. Sborschikov et al., it has to be noticed that the energy saving policy in the Russian Federation still does not cover all the main spheres of development economy. ESEM fulfillment is not taken into account in the course of fulfillment of works for reconstruction of buildings as the ESEM are the part of other programs for development of Russian economy and entail sufficient expenditures in the course of reconstruction.

Besides, in order to comply with the requirements made by the current legislation in the sphere of energy saving and enhancement of energy efficiency for the budget-funded institutions, a complex (program) method of measures implementation is necessary.

The main evaluation criteria for the technological processes in the course of

reconstruction of buildings under consideration of ESEM are:
- Duration of works;
- Labor input;
- Cost of works;
- Energy efficiency of the building;
- Efficiency of energy saving measure.

# 3 Method development of the resource and technological model formation

## 3.1 Principal technological processes of public buildings reconstruction

### *3.1.1 Technological processes of public buildings reconstruction*

Depending on the scope and nature of the measures implemented, the reconstruction of the building or edifice can be partial or complete. The possibility and expediency of either complete or partial reconstruction of the building are determined by the two main reasons:
1) Depreciation rate of the main bearing constructions (walls and floor slabs);
2) Compliance of the building plan and amenities with the modern requirements and a possibility to use the external sanitary utilities without substantial rearrangement.
3) In case of complete (complex) reconstruction of the building, durability and amenities of the building are also enhanced at the same time (including, thermal protection improvement). The complete (complex) reconstruction is mostly expedient provided the good state of walls and foundation (depreciation rate below 40%).

Following the complete reconstruction, the building shall comply with the modern operating requirements, sanitary and household requirements, design requirements, energy efficiency requirements, and other norms to the fullest possible extent, and in some cases, even exceed those.

The main groups of reconstruction works include:
1) reinforcement of constructions
2) replacement of constructions
3) perforation and boring of holes and apertures in the constructions. Filling of holes, gains and grooves.
4) topping-out and extension of buildings
5) reconstruction of buildings
6) dismantling of structures
7) demolition of buildings

All reconstruction works including topping-out may be carried out as an independent reconstruction method or be a part of the accompanying measures for complex reconstruction of the entire building.

### *3.1.2 Principal energy saving measures in the course of reconstruction of buildings and edifices of the educational institutions*

The requirements contained in the Federal Law No. 261-FZ dd. November 23, 2009 and the SRO standards are made to the energy saving measures; the main requirements are:
1. Recommended measures including the proposed results of implementation thereof must be easily comprehendible for the head, the engineering and technical team, and managerial staff charged with the implementation of the measures;

2. Measures must be specific and targeted;
3. Measures must be oriented on the existing and actually available methods and possibilities of implementation thereof;
4. Implementation of the measures must allow for the assessment of the deliverables by relatively simple methods;
5. Measures must obligatorily take the specific social and economical situation at the enterprise into account;
6. Offering several alternative variants of measures is desirable.

Compliance with these requirements requires the knowledge of, and expertise in classification of measures, and a guidance under the method of classification thereof.

At IV Energy Forum "Energy Efficiency Standards: Educational and Scientific Organizations", the Russian Federation Ministry of Education and Science presented a list of recommended ESEM for educational organizations including the higher educational institutions.

Based on the considered technological processes (TP's) of educational institutions edifices reconstructions, and the ESEM list recommended by the Russian Federation Ministry of Education and Science, a list of technological processes carried out in the course of reconstruction with regard to the ESEM is being formed. Technological processes carried out in the course of educational institutions edifices reconstructions must comply with the modern energy saving requirements.

## 3.2 Resource and technological models of the principal energy saving measures in the course of reconstruction of the educational institutions

The buildings reconstruction works comprise complex technological processes and a multitude of variants of implementation thereof. A technological process is chosen on the basis of the existing technologies for each type of works. Each technology is assigned a sequential number $i=\{1, 2, ..., n\}$.

Rational choice of technological process for carrying out one type of works or another on the pre-design stages of civil engineering production from the economic, material and resource expediency viewpoints is made on the basis of the resource and technological model (RTM) which essentially is a specially prepared (aggregated) set of material and labor resources invariable within a long period of time which is formed on the basis of data about representative objects [2].

The data on the representative object allow to simulate the most rational arrangement and technological solutions of civil engineering system retrieved by means of trial design and detection of alternative solutions that comply the most with the specific production conditions.

The resource and technological model is comprised by two blocks:

1. Resources block, which contains specially elaborated rated volumes of materials and products in physical terms, and the normative value of labor costs of the employees engaged in building and installation works.
2. Cost block, which includes the cost assessment of the volume unit of a resource (resource price) and the complete resource volume cost. The cost assessment block contains the values of resource costs for the reference period.

The basis for forming of resources block is the definition of material resources application volume in the course of reconstruction works implementation at the representative objects depending on the technological process.

The total volume of applied resources is determined by means of fulfilled works listing ($\sum a_i$) for the object, and determination of volume of materials, machinery and mechanisms applied as per type of works or as per object as a whole. According to the valid state cost estimate standards, each work $a_i$ is separated into the list of resources: materials, machinery, mechanisms, labor costs; also, the duration of the works $a_i$ and technical process *TPi* as a whole are established.

For the homonymous objects, the resource indicators undergo a procession, and then used for the RTM formation for the purpose of work implementation costs determination with the use of statistical istrumentarium

The cost block is formed by means of determination of the average weighted value for the group of materials as per basic price level for the materials in the region under consideration.

### 3.3 Assessment of arrangement efficiency of the technological processes in the course of reconstruction of the buildings and edifices as per economic and energy criteria

There is a significant number of methods and models used in order to solve the problems of efficiency assessment for the technological processes arrangement (TPA) in the course of reconstruction [3]. Unfortunately, the modern researches lack recommendations as of the choice of top-priority technical process for specific kind of work. For this purpose, the evaluation of efficiency research was made for TPA, and criterion analysis and the weight coefficient (μ) were chosen as the scientific research methods.

Quite naturally, a question arises what criteria are the defining ones for the choice of the most effective technological process. To solve this problem, the analysis of peculiarities of educational institutions in the course of arrangement of TP reconstruction, and the analysis of requirements to energy saving and energy efficiency enhancement were carried out. As a result, a set of criteria taken into account in determination of the TP priority was compiled, which comprises 2 groups:

1) economic criteria of TP efficiency evaluation;

2) energy-related criteria of the TP efficiency evaluation.

The range of economic criteria included:

1) expenditures of technological process implementation within the range of basic process as of January 01, 2001;

2) expenditures of technological process implementation at the current price level;

3) duration of works;

4) labor input into works.

The following criteria are energy-related:

5) energy efficiency of TP in natural volumes;

6) energy efficiency of TP in monetary terms;

7)payback period.

The energy efficiency of the technological process supposes the use of simplified calculation scheme without consideration of time factor (discounting) as the evaluation of each measure with discounting will not be credible enough, and the process itself is quire durable and labor intensive.

The TPA efficiency enhancement procedure is essentially a multi-criterion task in which several factors characteristic of the technological process under specific limitations have to be considered at the same time.

In order to solve this problem, it is possible to use the criterion analysis method to determine the weight coefficient μ and to carry out an expert evaluation.

On the first stage of criterion analysis, a criterion matrix Q(i; j) is prepared, where i={1,2…n}, j={1, 2, … m} or

$$Q=\begin{pmatrix} q_{11} & q_{12} & \cdots & q_{1m} \\ q_{21} & q_{22} & \cdots & q_{2m} \\ \cdots & \cdots & \cdots & \cdots \\ q_{n1} & q_{n2} & \cdots & q_{nm} \end{pmatrix} \quad Q=\begin{pmatrix} q_{11} & q_{12} & \cdots & q_{1m} \\ q_{21} & q_{22} & \cdots & q_{2m} \\ \cdots & \cdots & \cdots & \cdots \\ q_{n1} & q_{n2} & \cdots & q_{nm} \end{pmatrix} \tag{1}$$

Where each line i corresponds to a technological process $TP_i$, each column j corresponds to a specific criterion, and $q_{ij}$ is a value of the criterion j of the $TP_i$. Thus, we obtain the table 1.

**Table 1.** Overview of the criterion assessment table of the technological process.

| | Criterion 1 | Criterion 2 | ... | Criterion m |
|---|---|---|---|---|
| $TP_1$ | | | | |
| $TP_2$ | | | | |
| ... | | | | |
| $TP_n$ | | | | |

The next stage assumes a verification of compliance of the criteria ($q_{ij}$) with the limitation conditions and exclusion conditions for the process TPi in case of failure to fulfill the condition (2)

$$q_{ij} \le Q_{jmax}, \tag{2}$$

where $Q_{jmax}$ is the limit value of the criterion $q_j$.

Thus, the number of the excluded TP (N) can be N=[0; n]. in this case, the dimension of the matric will be n-N × m.

Further, an expert evaluation of the technological processes TPi is carried out.

One person acts as an expert, or one ECM (computer) is used for scoring. The score is attributed by the criterion value $q_{j\ within}$ the range from 1 to (n-N) in order of priority increase of the criterion.

The final stage assumes a calculation of the weight coefficient μ calculation for the TPi and verification of the calculation.

The weight coefficient μ is calculated under the formula (3):

$$\mu_i = \frac{\Sigma g_{ij}}{\Sigma\Sigma g_{ij} g_{ij}} \tag{3}$$

where $\mu_i$ is the weight coefficient, $g_{ij}$ is the individual indicator value represented in scoring as per each j-th characteristic.

Verification: $\sum\mu_i = 1$; if the condition is not fulfilled, calculation verification is necessary.

Based on the criterion analysis and calculation of the weight coefficient, the expert draws a conclusion on the most appropriate technological process (TPi) proposed for introduction.

Economic assessment of the TP arrangement efficiency in course of the public buildings reconstruction is made as per the following efficiency indicators: Net discounted income (NDI), Profitability index (PI), Payback period ($T_{OK}$ / PbP).

On the basis of the above calculations, a simulation algorithm for TP arrangement in the course of reconstruction of public buildings under consideration of the ESEM was developed (see fig. 1).

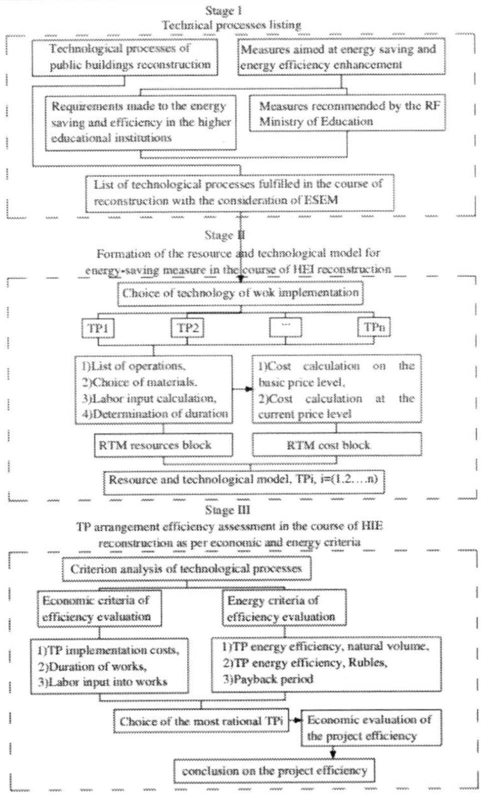


**Fig. 1**. RTM forming algorithm.

The resource and technological model allows for an optimization of the civil engineering production, maximal level of adaptation of the object to the specific conditions of the region, and analysis of investment efficiency on the preliminary design stages. Nevertheless, the RTM forming method has its advantages and disadvantages, too. Having a high degree of assessment probability, it does at the same time require a processing of a vast database on the resources for the representative objects built in the specific region. Economic assessment of the technological process efficiency on the basis of which a conclusion on the expediency of reconstruction is made, is an important stage of technological processes optimization.

## 4 Conclusions

On the basis the analysis of regulatory documentation, it can be concluded that in order to comply with the requirements of modern legislation in the field of energy saving and increase of energy efficiency of educational institutions, a comprehensive (program) method of implementation of measures is needed.

Funding of the energy saving programs and energy efficiency enhancement is defined by the state for the period until the year 2020.The funding is carried out within the framework of the state program and supposes the utilization of both budget and non-budget sources of funding. Such issues as the possibility of technological processes simulation for building and installation works without interruption of the educational process have been considered herein.

The main assessment criteria for technological processes in the course of buildings reconstruction with consideration of the ESEM are:

- Duration of works;
- Labor input into works;
- Cost of works;

- Energy efficiency of the building;
- Efficiency of energy–saving measure taken.

It was proposed to increase the efficiency of technological processes arrangement in the course of reconstruction of buildings by means of use of the resource and technological modeling and to evaluate the efficiency of technological processes by economic and energy parameters using the criterion analysis.

On the basis of the above calculations, an algorithm of technological processes simulation in the course of public buildings reconstruction with the consideration of ESEM was elaborated. The resource and technological model allows for an optimization of the civil engineering production, maximal level of adaptation of the object to the specific conditions of the region, and analysis of investment efficiency on the preliminary design stages. Nevertheless, the RTM forming method has its advantages and disadvantages, too. Having a high degree of assessment probability, it does at the same time require a processing of a vast database on the resources for the representative objects built in the specific region.

Economic assessment of the technological process efficiency on the basis of which a conclusion on the expediency of reconstruction is made, is an important stage of technological processes optimization.

Practical implementation of the elaborated method of TP arrangement efficiency enhancement has taken place in the course of reconstruction of the public buildings of Academy of Construction and Architecture under the Don State Technical University. By using the method of criterion analysis, the most effective technological process of thermal protection improvement was chosen for the buildings. Economic assessment of the project efficiency was carried out.

This article has been prepared with the support of the Russian Foundation for Fundamental Research. Copyright certificate for registration of software on the ECM was duly obtained.

## References

1. B.N. Volynsky, Constructive solutions for energy-saving building, Energy-saving No. 3, **67** (2001)
2. V.G. Gagarin, Regarding the rationale for increasing the thermal protection of the enclosing structures of buildings, *Stroyprofil*, No. 1, **21-23** (2010).
3. Grabovy, P.G., Kharitonov, V.A., Reconstruction and renewal of the existing urban development: A textbook. 2nd ed., revised. Moscow, Prospekt (2013)
4. I.Y. Zilberova, System-flow arrangement of civil engineering production in the course of reconstruction of residential houses, Author's abstract form PhD Tech. thesis, Rostov-on-Don, (2000)
5. S.G. Sheina, N.D. Cherednichenko, A.O. Vongay, The choice of the most effective option for increasing the thermal protection of buildings of higher educational institutions by the use criterion analysis method, *Bulletin of Construction Machinery* No. 9, **55-58** (2009).
6. S. Sheina, A. Khamavova *Technique for the Russian Federation regional territories assessment used to create industrial parks network* J. Procedia Engineering **150**, 1960-1965 (2016)

# Information Modelling of Engineering Systems as an Improvement Factor for the Energy Efficiency of Building

*Leonid* Shilov[1*] and *Liubov* Shilova[1]
[1]Moscow state university of civil engineering, Yaroslavskoye shosse, 26, Moscow, Russia, 129337

**Abstract.** The article presents the analysis of the heat power of a heating system in accordance with valid normative documents. It also reveals some other factors influencing both the general heat balance of a building and its microclimate parameters but not taken into account in the analysis. The recommendations have been developed on the improvement of energy efficiency parameters of the buildings. The authors suggest some possible ways of solution of the problem of air-exchange disturbances due to the introduction of automation systems responsible for a continuous monitoring of climatic and micro-climatic conditions as well as for a combined regulation of heat-transfer agent and of fresh air inflow.

## 1 Introduction

Recently, both the home building industry and the foreign one have paid top priority attention to the improvement of energy efficiency of residential buildings. It was caused by high prices for power resources as well as by the deterioration of environment "[1]-[5]".

The improvement of energy efficiency of residential buildings reduces the heat losses by 15-25% for foundation and substructures, by 15-25% for walls, by 20-40% for windows, by 10-20% for roofing structures and by up to 30-35% for the ventilation equipment, respectively.

The reduction in heat losses may be achieved by the construction of energy-efficient buildings and structures as well as by the use of special technologies of control over the heat balance of buildings.

The article considers one of the key factors influencing the choice of ventilation and heating equipment – the climatic conditions of the building region - as the consumption of heat energy depends on the ambient air temperature. A proper control over the heat supply system is impossible without consideration of the dynamics of temperature changes.

## 2 Methodology of analysis

The climatic conditions in the regions of our country differ much, but every region has some common features, e.g. temperature conditions, average wind velocity, pressure values.

* Corresponding author: Leonid.A.Shilov@gmail.com

In this connection, special regional maps for the Russian Federation territory have been developed in accordance with the climatic characteristics concerned (SP 20.13330.2011. Loads and effects (SNIP 2.01-07.85*).

The heat energy necessary for residential buildings is established through the Rules of normative energy consumption for communal services stated by the RF Government Decree №306 dated 23.05.2006, and it depends on the average daily ambient air temperature during the heating period. As a rule, the temperature values are taken from the building norms and rules or the climatic regional maps for the Russian Federation territory.

The main criteria of a proper operation of heat supply and ventilation systems for residential buildings are the following: heat balance, air balance and humidity balance. The general case of a heating system analysis is reduced to the solution of the heat balance equation.

The heat consumption of residential buildings is determined through the analysis methods which consider the factual heat-engineering characteristics of external enclosure structures and the normative air exchange values.

Neglecting the heat input due to the solar radiation, we can determine the heat power of the heating system for a typical room through the formula (1):

$$Q_{HS} = Q_{HL} + Q_{inf/vent} - Q_{dom} \quad (1)$$

$Q_{HL}$ - heat losses through the enclosure structures, W, which are determined by the formula (2):

$$Q_{HL} = Q_{WALL} + Q_{WIND} \quad (2)$$

The calculation of the optimum thermal resistance, the thickness of warmth-keeping material and the heat-transfer coefficient for enclosure structures is performed through SP 50.13330.2012. Here the required value of the thermal resistance $R_0^{rv}$, $m^2 \cdot K / W$, of enclosure structures is calculated through the formula (3) with consideration of energy-saving conditions and depends on the degree-days of the heating period (DDHP):

$$R_0^{rv} \rightarrow DDHP = (t_{in} - t_{hp}) \cdot z_{hp} \quad (3)$$

The formula (4) gives the required value of the heat-transfer resistance:

$$R_0^{rv} = a \cdot DDHP + b, \quad (4)$$

where ***a*** and ***b*** – coefficients taken from the Appendix to SP 50.13330.2012 Heat insulation of buildings. or obtained through the interpolation.

The total heat losses through the structural elements $\sum Q_{TП}$ include main and additional losses which consider the additions $\beta_i$ and are calculated as the sum of heat losses through all the enclosure rooms by the formula (5):

$$\sum Q_{WALL/WIND} = k_{0,i} \cdot F_{0,i} \cdot (t_{in} - t_{ex5}) \cdot n_i \cdot (1 + \sum \beta_i), \text{ W} \quad (5)$$

where $k_{0,i} = \frac{1}{R_0^{rv}}$ - the coefficient of the heat transfer through the enclosure element, W/($m^2 \cdot K$);

$F_{0,i}$ - the design enclosure surface, $m^2$;

$t_{in}, t_{ex5}$ - the design temperature values inside and outside the room, °C;

$n_i$ - the coefficient depending on the location of the external surface of the enclosure structure with respect to the ambient air (for windows and walls it is taken equal to 1);

$\beta_i$ - the coefficient considering additional heat losses as the parts of the main heat losses, and it depends on the orientation of building (it will be taken equal to 0.1 for the northern building side);

$Q_{\text{inf/vent}}$ – the greatest value of heat losses for the heating of the infiltration air or necessary for the compensation of the air outlet from the building, W. For the further calculations we take the maximum value of $Q_{vent}$ as there are practically no leakages in modern buildings.

For residential rooms, the heat expenditures for the heating of the infiltration air are taken not less than the values obtained by the formula (6):

$$Q_{vent} = L_{vent} \cdot c_a \cdot \rho_a \cdot (t_{in} - t_{ex5}) / 3.6 \text{, W;} \tag{6}$$

where $t_{in}$ – the internal temperature in the room;

$t_{ex5}$ – the temperature of the coldest 5 day-long period;

$c_a = 1{,}005$ kJ/(kg·K) – the specific heat of the air;

$\rho_a = 1{,}2041$ kg/ m$^3$ – the density of the air;

$L_{vent} = l_n \cdot F_{fl}$, where $F_{fl}$ - the surface of the floor in the heated room.

$l_n$ - the specific normative consumption of the incoming air taken as $l_n = 3m^3/h$ for 1 $m^2$ of residential area (the air circulation factor for the flat area of not more than 20 $m^2/person$). The aforesaid parameters are established in the normative documents SP 54.13330.2011 Residential buildings (blocks of flats), the actual edition SNIP 31-01-2003.

For the calculation of the heat power of heating equipment we should consider the constant heat emissions in the rooms:

$$Q_{dom} = q_{dom} \cdot F_{fl} \text{, W} \tag{7}$$

where $q_{dom}$ - the specific domestic heat emission taken equal to 10 W/m$^2$,

$F_{fl}$ - the floor area in heated rooms.

This value is a constant; it does not depend on the ambient air temperature.

## 3 An example of the analysis of the heat power of a heating system

Let us consider a typical residential room with dimensions of 3x4x2.5m$^3$ and the window of 1.47x1.55 (the room looks north). In accordance with the formula (3), we obtain

$$R_0^{rv} \rightarrow DDHP = (20 + 2.2) \cdot 205 = 4551,$$

where $t_{in}$ = 20 °C – (GOST 30494-2011 Residential and public buildings. Microclimate parameters in rooms);

$t_{hp}$ = -2.2 °C; $z_{hp}$ = 205 days. – (SP 131.13330.2012 Constriction climatology) .

we obtain the required heat-transfer resistance through the formula (4):

$$R^{rv}_{DDHP} = 2.8 + \frac{3.5 - 2.8}{6000 - 4000} \cdot (4551 - 4000) = 2.99$$ (m$^2$K/W) – for the wall;

$$R^{rv}_{DDHP} = 0.45 + \frac{0.6 - 0.45}{6000 - 4000} \cdot (4551 - 4000) = 0.49$$ (m$^2$ K/W) – for the windows and balcony doors.

In accordance with DDHP, the external wall should have the minimum required heat-transfer resistance equal to $R_0 = 2.99$ m$^2$K/W.

Calculate the total heat losses through the walls and windows in accordance with (5)

$\sum Q_{WALL} = 0.334 \cdot (2.6 \cdot 3.1 - 1.47 \cdot 1.55) \cdot (20 - (-25)) \cdot 1 \cdot (1 + 0,1) = 95.6$ W

$\sum Q_{WIND} = 2.04 \cdot (1.47 \cdot 1.55) \cdot (20 - (-25)) \cdot 1 \cdot (1 + 0.1) = 230.1$ W

And the total heat losses in accordance with (2)

$Q_{HL} = Q_{WALL} + Q_{WIND} = 95.6 + 230.1 = 325.7$ W

The heat expenditures for the heating of ventilation air for this example is calculated through (6):

$Q_{vent} = 36 \cdot 1.005 \cdot 1.2041 \cdot (20 - (-25)) / 3.6 = 544.6W$, with $L_{vent} = 3 \cdot 12 = 36 m^3 / h$

The domestic heat emissions in this example are $Q_{dom} = 10 \cdot 12 = 120W$

As a result, we obtain $Q_{HS} = 325.7 + 544.6 - 120 = 750.2W$

To provide for the required microclimate (+20$^0$C) with the average temperature of the coldest 5day-long period (-25$^0$C) it is necessary to have the heat power of 750.2 W. However, a wrong operation of the ventilation system causes a surplus heat power which upsets the heat balance.

Let us consider the dynamics characteristics of the required heat power, that is the operation of the heating system depending on the ambient air temperature. The results of the calculations through the aforesaid formulas are presented in Table 1; Fig.1 presents the heating system power output depending on the ambient air temperature.

**Table 1**. Heat Expenditures

| $t_{ex}$, °C | $L_{vent}$, m$^3$/h | $Q_{vent}$, W | $Q_{WALL}$, W | $Q_{WIND}$, W | $Q_{HL}$, W | $Q_{dom}$, W | $Q_{cons}$ with $L_{vent}$, W |
|---|---|---|---|---|---|---|---|
| -30 | 36 | 605,1 | 106,2 | 255,6 | 361,9 | 120,0 | 846,9 |
| -25 | 36 | 544,6 | 95,6 | 230,1 | 325,7 | 120,0 | 750,2 |
| -20 | 36 | 484,0 | 85,0 | 204,5 | 289,5 | 120,0 | 653,5 |
| -15 | 36 | 423,5 | 74,3 | 179,0 | 253,3 | 120,0 | 556,8 |
| -10 | 36 | 363,0 | 63,7 | 153,4 | 217,1 | 120,0 | 460,1 |
| -5 | 36 | 302,5 | 53,1 | 127,8 | 180,9 | 120,0 | 363,5 |
| 0 | 36 | 242,0 | 42,5 | 102,3 | 144,7 | 120,0 | 266,8 |
| 5 | 36 | 181,5 | 31,9 | 76,7 | 108,6 | 120,0 | 170,1 |
| 10 | 36 | 121,0 | 21,2 | 51,1 | 72,4 | 120,0 | 73,4 |

The analysis of the aforesaid data proves that the heating system does not reach its balance state because of permanent not-controlled and unpredicted changes in the ambient air temperature.

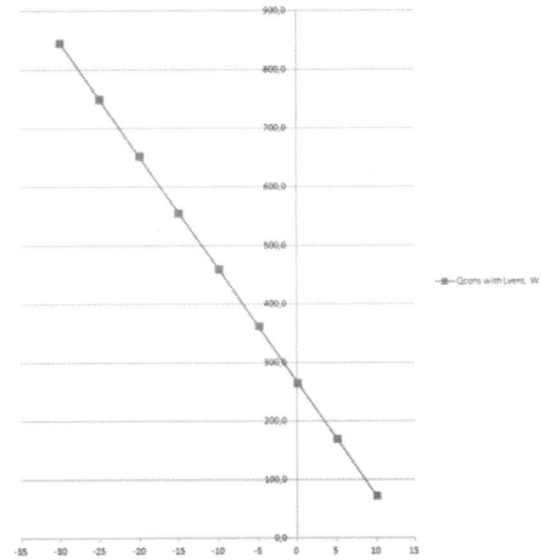

Qcons with Lvent, W - the real heat power consumption in case when the ventilation system operates properly

**Fig.1.** The heating system power depending on the ambient air temperature

Nowadays, the main kind of operation mode is the centralized control over the heat power output as it makes possible to use the minimum number of simple regulators for the normative temperature of hot water in the hot water supply systems.

It should be noted here that the heat power output in heat supply systems considers only one meteorology factor – the ambient air temperature – which is taken equal for all buildings under heat supply.

The problem of surplus heat emissions could be partly solved through the use of heating equipment with lock accessories (in accordance with SNIP 41-01-2003 and MGSN 3.01.01, the automated thermostatic regulation of the heat flows should be provided for). However, such regulation is possible only within a narrow temperature range.

When the thermoregulator is out of its operation range (as a rule, it happens when the windows are open), the delivery of heat-transfer agent is stopped. Thus, a wrong operation of the valve causes the decrease in the internal air temperature and the shut-down of heating system.

## 4 Conclusion

The aforesaid studies revealed some problems connected with the disturbances in the operation of heating supply and ventilation systems.

On the one hand, there is a problem of the air-exchange disturbance because of no infiltration through the windows; on the other hand, the surplus heat power causes the increase in the room temperature.

The required air-exchange may be provided for through the use of inlet valves. But it is necessary to consider the fact that any changes in the ambient air temperature require dynamic changes in the valve flow rate which is practically impossible nowadays.

At the same time, the factors influencing the heat supply requirements of buildings and their particular rooms are quite different; so it is practically impossible to consider them in the process of centralized regulation of the heat power output. In this connection, a combination of various regulation methods seems to be expedient.

Here the measurement of internal microclimate parameters is very labour-consuming; it requires both the equipment and the qualification concerned.

In this connection, the next logical stage in the development of heating and ventilation systems should be the development and introduction of software considering the dynamics of changes in external and internal temperatures and including all-the-year-round operation cycle of the heating and ventilation systems. The further stage will be the creation of fully automated computer systems for a continuous monitoring of climate and microclimate parameters which perform a combined regulation of the heat-transfer agent and the ventilation.

## Acknowledgements

This study was performed with the financial support of the RF Ministry of Education and Science (RF President Grant, agreement #14.Z56.3456-MK).

## References

1. S. Eleftheriadis, D. Mumovic, P. Greening, RSER, **67**,811–825 (2017)
2. F.H. Abanda, L. Byers, E,**97**,517–527,(2016)
3. J.B. Kim, W. Jeong, M.J. Clayton, J.S. Haberl, W. Yan, AC, **50**, 16–28 (2015)
4. G. Martinopoulosa, K.T. Papakostasb, A.M. Papadopoulos, EB,**124**, 79–87 (2016)
5. J. Vanus, P. Vojcinak, R. Martinek, M. Kelnar, Z. Machacek, P. Bilik, J. Koziorek,J. Zidek, PS,**7**, 114–121 (2016)

# Calculation of the particles transfer in a porous medium

*Olga* Shirokova[1,*]

[1]Moscow state university of civil engineering, Yaroslavskoye shosse, 26, Moscow, Russia, 129337

**Abstract.** The transfer and deposition of solid particles in porous rocks are encountered in many fields of science and technology, including the design and construction of waterproof foundations and hydraulic structures. The paper considers polydisperse suspesion with suspended particles of various types moving in the porous medium at the same velocity. The asymptotics of the solution of the filtration problem with various linear filtration coefficients is constructed. A numerical calculation of the asymptotics is performed for the case of two different types of particles.

## 1 Introduction

The transfer and retention of solid suspended particles of suspension and colloids in the porous medium is one of the main problems of underground fluid mechanics [1]. Such problems are encountered in many fields of science and technology, in many fields of science and technology, including the design and construction of waterproof structures and hydraulic structures.

Classical filtration models of a suspension in the porous medium consider a flow of a carrier fluid with the identical suspended particles [2, 3]. Porous medium has a complex structure, and various factors can influence the retention of particles in the filter pores: electrical forces, gravity, dispersion, viscosity, etc. If the particle and pore sizes are close, then the determining mechanism of pore capture is size-exclusion. The mathematical model of deep bed filtration includes the equation of mass balance and the kinetic equation of deposit growth rate, forming a quasilinear hyperbolic system. The kinetic equation determines the growth rate of the deposit in proportion to the concentration of suspended particles. The coefficient of proportionality is called the filtration coefficient. In the simplest cases, the filtration coefficient is assumed to be constant. A linear relationship describing the change in the growth rate deposit with increasing the concentration deposit is used in more complex models. Analytic and numerical solutions of the classical filtration problem are known [3-9]. Models with a complicated equation of mass balance take into account the change in porosity and permeability porous medium when forming a deposit [10-14].

Consider a suspension with suspended particles of two types. Particles, differing in size or shape, move in the porous medium at the same velocity. It is assumed that particles of

* Corresponding author: shirokovaol@mgsu.ru

different types interact differently with pores and have different filtration coefficients. In the paper the asymptotic solution of the filtration problem is constructed near the concentrations front of suspended and retained particles and a numerical calculation of the asymptotics is performed.

## 2 Formulation of the problem

In the domain

$$\Omega = \{0 < x < 1,\ t > 0\}$$

a mathematical model is given by a system of equations

$$\frac{\partial C_i}{\partial t} + \frac{\partial C_i}{\partial x} + \frac{\partial S_i}{\partial t} = 0 \tag{1}$$

$$\frac{\partial S_i}{\partial t} = \Lambda_i(\mathbf{S})C_i;\ i = 1,\dots,n \tag{2}$$

with the boundary condition

$$x = 0 : C_i = p_i\ . \tag{3}$$

and the initial conditions

$$t = 0 : C_i = 0,\ S_i = 0\ . \tag{4}$$

Here $\mathbf{S} = (S_1,\dots,S_n)$; the filtration coefficients $\Lambda_i(\mathbf{S})$ are continuous and positive; the constants $p_i > 0$ .

At the initial moment the porous medium does not contain any particles. A suspension with a constant concentration $p_i$ of suspended particles begins to flow into the filter inlet. The concentrations front of the suspended and retained particles moves from the filter inlet $x = 0$ to the outlet $x = 1$ at a constant velocity $v = 1$ and divides the domain $\Omega$ into two subdomains $\Omega_0 = \{0 < x < 1,\ 0 < t < x\}$ and $\Omega_1 = \{0 < x < 1,\ t > x\}$. In $\Omega_0$ the system (1) – (4) has a zero solution; in $\Omega_1$ the solution is positive. Since conditions (3) and (4) are not matched at the origin, the solution $C_i(x,t)$ has a strong discontinuity at the concentrations front, the characteristic straight line $t = x$.

The solution $S_i(x,t)$ is continuous in $\Omega$ and has a weak discontinuity on the concentrations front.

We define a condition on the characteristic

$$S_i\big|_{t=x} = 0\ . \tag{5}$$

In the domain $\Omega_1$ the problem (1) – (4) is equivalent to the Goursat problem (1) – (3), (5).

In the characteristic variables (Riemann variables) $\tau = t - x,\ x = x$ in the domain

$$\tilde{\Omega}_1 = \{0 < x < 1, \tau > 0\}$$

the system (1) – (3), (5) takes the form

$$\frac{\partial C_i}{\partial x}+\frac{\partial S_i}{\partial \tau}=0\,; \tag{6}$$

$$\frac{\partial S_i}{\partial \tau}=\Lambda_i(\mathbf{S})C_i \tag{7}$$

with initial and boundary conditions

$$x=0: C_i=p_i\,; \tag{8}$$

$$\tau=0: S_i=0;\ i=1,...,n\,. \tag{9}$$

## 3 The exact solution on the concentrations front

Let us determine the concentrations of suspended particles on the concentrations front. Substituting (7) into (6)

$$\frac{\partial C_i}{\partial x}+\Lambda_i(\mathbf{S})C_i=0;\ i=1,...,n\,. \tag{10}$$

The concentrations front is given by formula $\tau=0$. According to condition (5), on the concentrations front equation (10) has the form

$$\frac{\partial C_i}{\partial x}+\Lambda_i(0)C_i=0,\ C_i\big|_{x=0}=p_i\,. \tag{11}$$

The solution of (11)

$$C_i=p_i e^{-\Lambda_i(0)x},\ i=1,...,n\,. \tag{12}$$

The formula (12) gives the exact solution on the concentrations front.

## 4 Asymptotics near the concentrations front

Assume that the filtration coefficients linearly depend on the concentrations of $n$ types of retained particles

$$\Lambda_i(\mathbf{S})=\lambda_i+\sum_{k=1}^{n}a_k S_k,\ i=1,...,n\,, \tag{13}$$

where $a_k$ are constants.

In the neighborhood of the concentrations front $\tau=0$ the solution of problem (6) – (9) can be obtained in the form

$$S_i=S_i^1\tau+\frac{1}{2}S_i^2\tau^2+O(\tau^3);\ \ C_i=C_i^0+C_i^1\tau+O(\tau^2);\ \ i=1,...,n\,. \tag{14}$$

Here $C_i^0$ are given by the formulas (12), the expansions coefficients $S_i^1$, $S_i^2$, $C_i^1$ depend on the variable $x$.

Substituting the expansions (14) into equations (6) and (7) and equating the terms of the same powers of $\tau$, we obtain a system of recurrent algebraic and differential equations for the leading terms of the asymptotics

$$S_i^1 = \lambda_i C_i^0 ; \tag{15}$$

$$\left(C_i^1\right)' + \lambda_i C_i^1 + C_i^0 \sum_{k=1}^{n} a_k S_k^1 = 0 ; \tag{16}$$

$$S_i^2 = \lambda_i C_i^1 + C_i^0 \sum_{k=1}^{n} a_k S_k^1 ; \quad i = 1,...,n . \tag{17}$$

The initial condition for the differential equation (16) is obtained from (3)

$$C_i^1 = 0; \quad i = 1,...,n . \tag{18}$$

The solution of the system (15) – (17)

$$S_i^1 = \lambda_i p_i e^{-\lambda_i x} \quad ; \tag{19}$$

$$C_i^1 = p_i e^{-\lambda_i x} \sum_{k=1}^{n} p_k a_k \left(e^{-\lambda_k x} - 1\right) ; \tag{20}$$

$$S_i^2 = p_i e^{-\lambda_i x} \sum_{k=1}^{n} p_k a_k \left((\lambda_i + \lambda_k) e^{-\lambda_k x} - \lambda_i\right) . \tag{21}$$

Substituting the terms (12), (19) – (21) into the expansions (14) and performing an inverse change of variables, we obtain the asymptotics of problem (1) – (4) in the domain $\Omega_1$ near the concentrations front

$$S_i(x,t) = \lambda_i p_i e^{-\lambda_i x}(t-x) + \frac{1}{2} p_i e^{-\lambda_i x} \sum_{k=1}^{n} p_k a_k \left((\lambda_i + \lambda_k) e^{-\lambda_k x} - \lambda_i\right)(t-x)^2 + O(t-x)^3 ; \tag{22}$$

$$C_i = p_i e^{-\lambda_i x} + p_i e^{-\lambda_i x} \sum_{k=1}^{n} p_k a_k \left(e^{-\lambda_k x} - 1\right)(t-x) + O(t-x)^2 ; \quad i = 1,...,n . \tag{23}$$

## 5 Numerical simulation

The numerical calculation of the asymptotic solution is performed for 2-size particles ($n = 2$) for the linear filtration coefficients

$$\Lambda_1 = 1 - S_1 - S_2 ; \quad \Lambda_2 = 2 - S_1 - S_2 ,$$

and the initial suspended particles concentrations $p_1 = 1, \ p_2 = 1$.

Fig. 1 shows 3-D graphs of the suspended and retained particles concentrations separately for each type of particles.

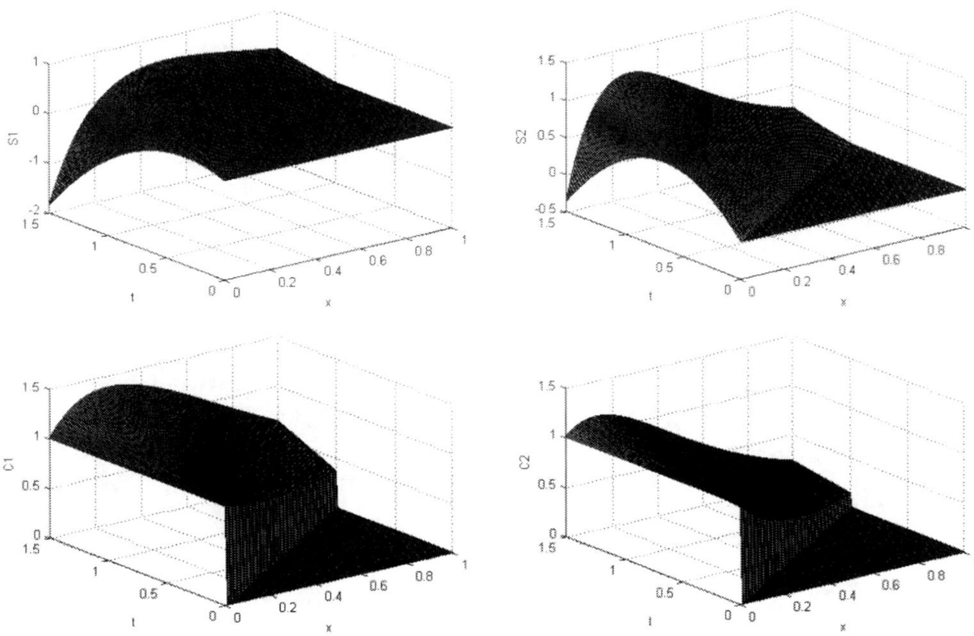


Fig. 1. Suspended and retained particles concentrations $S_1(x,t)$; $S_2(x,t)$; $C_1(x,t)$; $C_2(x,t)$.

The graphs of suspended and retained particles concentrations for different fixed $x$ are presented in Fig. 2.

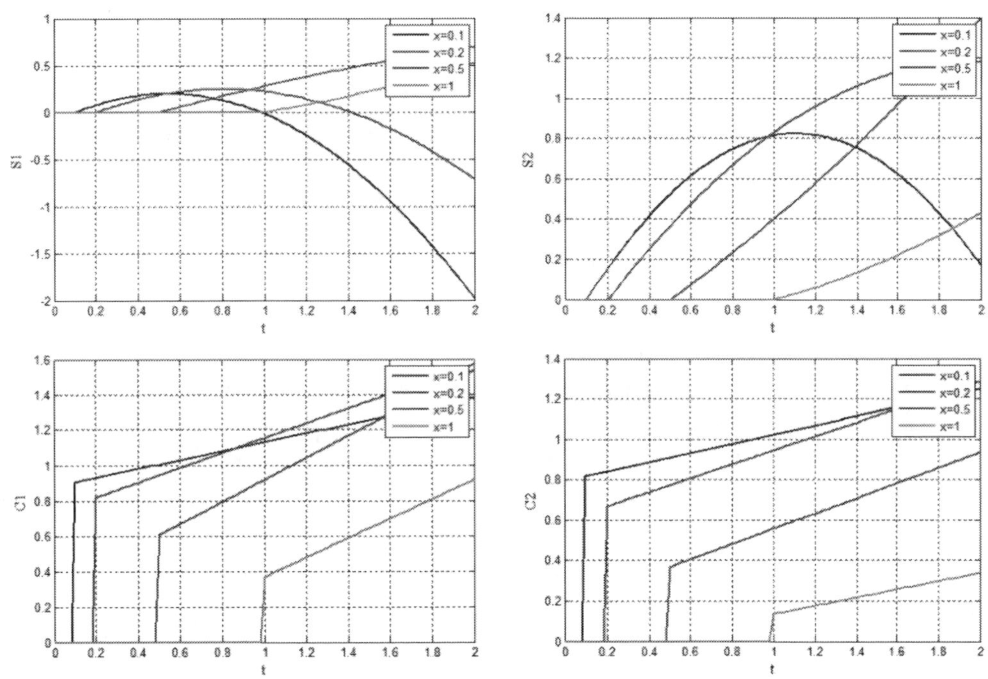


Fig. 2. Particles concentrations $S_1(x,t)\big|_{x=const}$; $S_2(x,t)\big|_{x=const}$; $C_1(x,t)\big|_{x=const}$; $C_2(x,t)\big|_{x=const}$.

## 6 Conclusion

The asymptotic solution of the problem of particles transfer is constructed near the concentrations front. Particles of $n$ different types move in a porous medium with the same velocity. It is assumed that the filtration coefficients of all types of particles are linearly dependent on the retained particles concentrations.

A numerical calculation of the asymptotic solution for the particles of two different types is carried out. The calculation indicates that the applicability of the asymptotics is limited by the small parameter $\tau = t - x$. Figures 1 – 2 show that the asymptotic solution is valid for $\tau < 0.4$. For small values of $\tau$, the asymptotics determines the analytic dependence of the solution on the parameters, reducing the amount of laboratory and field research [15].

With increasing $\tau$, the retained particles concentration of the first type begins to decrease and at $\tau \sim 1$ becomes negative. Respectively the suspended particles concentration of the first type exceeds the initial concentration at the filter inlet, which contradicts the physical meaning of the problem.

A further development of the method involves the construction of a global asymptotic solution of the filtration problem on a small limiting deposit.

## References

1 G.I. Barenblatt, V.M. Entov, V.M. Ryzhik, *Theory of fluid flows through natural rocks*, (Kluwer Acad. Publ. 1989)
2 M. Sharma, Y. Yortsos, AIChE J. **33** (1987) 1636 – 1643
3 J. P. Herzig, D. M. Leclerc, P. Legoff, Ind. Eng. Chem. **62** (1970) 9 – 35
4 E.A. Vyazmina, P.G. Bedrikovetskii, A.D. Polyanin, Theor. Found. Chem. Eng. **41**, 5 (2007) 556 – 564
5 A.D. Polyanin, V.F. Zaitsev, *Handbook of nonlinear partial differential equations*, (CRC pr. 2003)
6 A. Vaz, D. Maffra, T. Carageorgos, P. Bedrikovetsky, J. Nat. Gas Sci. Eng. **34** (2016) 1422 – 1433
7 L.I. Kuzmina, Yu.V. Osipov, Int. J. Comp. Civil Struct. Eng.**10** (3) (2014) 17 – 22
8 L.I. Kuzmina, Yu.V. Osipov, Vestnik MGSU **2** (2016) 49 – 61
9 Yu.P. Galaguz, G.L. Safina, Procedia Eng. **153** (2016) 157 – 161
10 Z. You, P. Bedrikovetsky, L. Kuzmina, Abstr. Appl. Anal. **2013**, ID 680693 (2013)
11 Z. You, Y. Osipov, P. Bedrikovetsky, L. Kuzmina, Chem. Eng. J. **258** (2014) 374–385
12 L.I. Kuzmina, Yu.V. Osipov, Procedia Eng. **111** (2015) 491 – 494
13 L.I. Kuzmina, Yu.V. Osipov, Matec Web Conf. **86**, 01005 (2016)
14 Y. Galaguz, G. Safina, Matec Web Conf. **86**, 03003 (2016)
15 P.G. Bedrikovetsky, D. Marchesin, F. Checaira, A.L. Serra, E. Resende, J. Petrol. Sci. Eng. 32, 3 (2001) 167 – 177

# Non-stationary axially symmetric goal of thermo-elasticity for a fixed rigid circular multi-layer plate

*Dmitry* Shlyakhin[1*]

[1]Samara State Technical University, Institute of Architecture and Civil Engineering, 194, Molodogvardeyskaya St., 443001, Samara, Russia

**Abstract.** This paper considers axially symmetric goal of thermos–elasticity for a fixed rigid circular multi-layer isotropic plate with non–stationary heat flows on its front faces. The mathematical model is presented in the form of a not self–adjoint system that includes differential motion equations and a linear equation of the thermal-elastic state in a three-dimensional model. A new closed solution in an unconnected setting is built by the method of finite integral transformation. At the same time, a standardization procedure is conducted at each stage of the solution, which allows to implement an appropriate conversion algorithm. The calculated ratios provide an opportunity to perform a qualitative and quantitative analysis of the associated temperature and mechanical fields and also to examine stress–strain behaviour of the multi–layer systems both under the local thermal shock, which is observed, for example, in the interaction with the laser beam and in the case of uneven unsteady surface heating, which is frequent during the operation of protective construction structures.

## 1 Introduction

There are structures working under uneven heating which are used in construction, aviation and missile equipment as well as in a number of branches of mechanical engineering [1–6]. To describe their work, it is necessary to perform an in-depth analysis of non-stationary processes, without which it is impossible to understand the effect of temperature and mechanical stress fields. Typically, the design structure is presented as an infinitely long plate or an infinitely long cylinder. In other cases the task is investigated on the basis of the solution of the thermoelasticity equation only, without regard to the deformation of the building [3.4].

The task is greatly complicated when multi–layer finite dimension systems with a number of properties that qualitatively distinguish them from traditional structures are investigated [7–13]. In this case, the mathematical model of the initially boundary problem with account of the connection between movement and temperature, consists of a not self-adjoint system of differential equations of thermoelestic movement [14]. Their solution and fulfillment of boundary conditions is only possible for some elastic systems with some private

* Corresponding thor: d-612-mit2009@yandex.ru

docking options [14–16]. In addition, the paper aims to try using accurate and approximative methods of calculation in their combination [4].

In this paper, in order to solve the non-stationary task of thermo-elasticity for a multilayer round rigid plate, the authors use methods of finite integral transformations (Hankel, Fourier, generic transformation) to obtain a closed solution in an unbound setting.

## 2 Materials and methods

Let a fixed rigid circular multi-layer plate in the cylindrical coordinate system occupies region $(r_*,\theta,z_*)\ \Omega:\ \{0\le r_*\le b,\ \ 0\le\theta\le 2\pi,\ 0\le z_*\le h^*\}$. For definiteness we assune that the construction consists of three layers. Besides, exterior and interior layers are in hard contact and have thickness $h_1^*$ and $h_2^*$ ( $h^*=2h_1^*+h_2^*$ ) respectively. Heat flows affect the end surfaces of the plate, $\omega_1^*(r_*,t_*),\omega_2^*(r_*,t_*)$ which are arbitrary time and radial coordinate functions $r_*$ and $t_*$ .

Differential equations of motion and thermal conductivity in a linear position in the absence of internal heat sources, as well as the initially-boundary conditions in the dimensionless shape and in cylindrical coordinate system, are as follows [14]:

$$\frac{\partial}{\partial r}\nabla U+\frac{(1-2v_j)}{2(1-v_j)}\frac{\partial^2 U}{\partial z^2}+\frac{1}{2(1-v_j)}\frac{\partial^2 W}{\partial r\partial z}-B_j\frac{\partial T}{\partial r}-F_j\frac{\partial^2 U}{\partial t^2}=0\,, \tag{1}$$

$$\frac{(1-2v_j)}{2(1-v_j)}\nabla\frac{\partial W}{\partial r}+\frac{\partial^2 W}{\partial z^2}+\frac{1}{(1-v_j)}\frac{\partial}{\partial z}\nabla U-B_j\frac{\partial T}{\partial z}-F_j\frac{\partial^2 W}{\partial t^2}=0\,,$$

$$\nabla\frac{\partial T}{\partial r}+\frac{\partial^2 T}{\partial z^2}-\frac{1}{k_j}\frac{\partial T}{\partial t}-\eta_j\frac{\partial}{\partial t}\left(\nabla U+\frac{\partial W}{\partial z}\right)=0\,;\ r=0,1$$

$$W(0,z,t)<\infty,\ \ U(0,z,t)<\infty,\ \ T(0,z,t)<\infty\ , \tag{2}$$

$$U(1,z,t)=0\ ,\ W(1,z,t)=0,\quad \frac{\partial T}{\partial r}_{|r=1}=0;$$

$$z=0,h\quad \frac{\partial W}{\partial r}+\frac{\partial U}{\partial z}=0\ ,\quad \nabla U+\frac{(1-v_1)}{v_1}\left(\frac{\partial W}{\partial z}+T\right)=0\,,\quad T=\omega_{1(2)}; \tag{3}$$

$$z=h_1,h_1+h_2\quad \left(\frac{\partial W}{\partial r}+\frac{\partial U}{\partial z}\right)_{|j=1}=b_1\left(\frac{\partial W}{\partial r}+\frac{\partial U}{\partial z}\right)_{|j=2}, \tag{4}$$

$$\left[\nabla U+\frac{(1-v_1)}{v_1}\left(\frac{\partial W}{\partial z}+T\right)\right]_{|j=1}=b_2\left[\nabla U+\frac{(1-v_2)}{v_2}\left(\frac{\partial W}{\partial z}+T\right)\right]_{|j=2},$$

$$U(z+0)=U(z-0),\quad W(z+0)=W(z-0),\quad T(z+0)=T(z-0)\ ;$$

$$t=0 \quad U(r,z,0)=U_0(r,z),\ \dot{U}(r,z,0)=\dot{U}_0(r,z), \tag{5}$$

$$W(r,z,0)=W_0(r,z),\ \dot{W}(r,z,0)=\dot{W}_0(r,z),\ T(r,z,0)=T_0(r,z);$$

where $\{U,W,r,z,h,h_1,h_2\}=\{U^*,W^*,r_*,z_*,h^*,h_1^*,h_2^*\}/b$, $\quad t=\dfrac{A}{b}t_*$

$$A=\sqrt{\frac{E_1(1-\nu_1)}{(1+\nu_1)(1-2\nu_1)\rho_1}},\ \{T,\omega_1,\omega_2\}=\{T^*,\omega_1^*,\omega_2^*\}\frac{(1-\nu_1)}{(1+\nu_1)\alpha_t^{(1)}},\ B_1=1,$$

$$B_2=b_3\frac{(1+\nu_2)}{(1+\nu_1)}\frac{\alpha_t^{(2)}}{\alpha_t^{(1)}},\quad F_2=\frac{b_3}{b_1}\frac{\rho_2}{\rho_1}\frac{(1-2\nu_2)}{(1-2\nu_1)},$$

$$F_1=1,\quad \eta_j=AE_jb\frac{(1-\nu_1)}{(1+\nu_1)(1-2\nu_j)}\frac{\alpha_t^{(j)}}{\alpha_t^{(1)}}T_0,\quad k_j=\frac{k_j^*}{Ab},\nabla=\frac{\partial}{\partial r}+\frac{1}{r},$$

$$b_1=\frac{E_2(1+\nu_1)}{E_1(1+\nu_2)},\quad b_2=b_1\frac{(1-2\nu_1)\nu_2}{(1-2\nu_2)\nu_1},\quad b_3=\frac{(1-\nu_1)}{(1-\nu_2)},$$

$U^*(r_*,z_*,t_*)$, $W^*(r_*,z_*,t_*)$ – components of movement vector; $T^*$ – temperature change; $T_0$ – absolute temperature of the initial state of the body; $E_j,\nu_j,\alpha_t^{(j)},k_j^*$ – elastic modulus, Poisson ratios, linear thermal expansion and thermal diffusivity of layer $j$.

The first three inequalities (2) are the regularity conditions for the solution of a continuous plate, and the latter equality takes into account lack of heat exchange between the plate and rigid fixing environment. Equations (4) are the conditions of joint stress, displacement, and temperature.

Formulae (1) – (5) represent a mathematical statement for the theory of thermo-elasticity boundary problem.

Here, equalities (1) present a not self–adjoint system of equations for which a common integral cannot be obtained. It is therefore necessary to consider the taske (1) to (5) in an unbounded setting. New closed solutions of non–stationary tasks of thermoelasticity and elasticity theory are built by consistently using the method of partial separation of variables as finite integral transformations. The authors in consecutive stages use Hankel's transformations with finite limits along the axial coordinate and a generalized final transformation (kip) along the radial variable At the same time, a standardization procedure is conducted at each stage of the solution, which allows to implement an appropriate conversion algorithm.

## 3 Results

The first step is to solve the equation of thermoelasticity (1) relative to the function $T(r,z,t)$ with the corresponding initially boundary conditions:

$$\nabla\frac{\partial T}{\partial r}+\frac{\partial^2 T}{\partial z^2}-\frac{1}{k_j}\frac{\partial T}{\partial t}=\eta_j\frac{\partial}{\partial t}\left(\nabla U+\frac{\partial W}{\partial z}\right), \tag{6}$$

$$r=0,1 \qquad T(0,z,t)<\infty,\ \frac{\partial T}{\partial r}_{|r=1}=0; \tag{7}$$

$$z=0,h \qquad T(r,0,t)=\omega_1(r,t) \quad T(r,h,t)=\omega_2(r,t); \tag{8}$$

$$z=h_1,h_1+h_2 \qquad T(z+0)=T(z-0), \tag{9}$$

$$\left[\frac{(1-v_1)}{v_1}T\right]_{|j=1}-b_2\left[\frac{(1-v_2)}{v_2}T\right]_{|j=2}=b_2\left[\nabla U+\frac{(1-v_2)}{v_2}\frac{\partial W}{\partial z}\right]_{|j=2}-\left[\nabla U+\frac{(1-v_1)}{v_1}\frac{\partial W}{\partial z}\right]_{|j=1};$$

$$t=0 \qquad T(r,z,0)=T_0(r,z). \tag{10}$$

It is initially assumed that functions $U,W$ are equal to zero. In this case we notice that the last condition (9) will not run.

Taking into account the theorems of solutions superposition and the linear nature of the original calculation ratios, we enter the new function $R(r,z,t)$ associated with the function $T(r,z,t)$, as follows:

$$T=(r^2-2r)(\omega_1+\omega_2)+R. \tag{11}$$

This view allows the functions to be deferred to a $\omega_1,\omega_2$ series in the future if they do not change at the radial coordinate.

As a result of substitution (11) in (6) – (10), a new task of thermoelasticity is obtained in relation to function $R$. This task is solved by the method of Hankel's final integral transformations along radial coordinate using the following transformant and the reciprocation formula:

$$R_H(n,z,t)=\int_0^1 R(r,z,t)rJ_0(j_n r)dr, \quad R(r,z,t)=2\sum_{n=0}^{\infty}\frac{R_H(z,t)J_0(j_n r)dr}{J_0(j_n)^2}. \tag{12}$$

Where $j_n$ – is positive zeros of function $J_1(j_n)$ $(n=0, j_0=0)$, $J_0(...), J_1(...)$ – the Bessel function of the first type of zero and first order.

When resolving a task relative to transformant $R_H$, we use a generalized final integral transformation (kip) [18.19] is used along the radial variable:

$$L_H(n,m,t)=\int_0^h R_H(n,z,t)N(\lambda_m,z)dz, \tag{13}$$

$$R_H(n,z,t)=\sum_{m=0}^{\infty}L_H(n,m,t)N_1(\lambda_{im},z)\|N_{im}\|^2, \quad \|N_{im}\|^2=\int_0^h N_1(\lambda_{im},z)^2\,dz.$$

Where $N_1(\lambda_{im},z)$ is a function of transformation edge along function $z$, $\lambda_{im}$ – - its own values makeing up the counting set.

Applying reciprocation formula (12) ,(13) in consecutive stages to the transformant $L_H(n,m,t)$, with account of (11), we obtain the expression for $T(r,z,t)$:

$$T=(r^2-2r)(\omega_1+\omega_2)+\frac{4}{h}\sum_{j=0}^{\infty}\frac{J_0(j_n r)}{J_0(j_n)^2}\sum_{m=1}^{\infty}L_H(n,m,t)N(\lambda_m,z), \tag{14}$$

The second step is to solve a tsk of the theory of elasticity with regard to function $U,W$ based on a given (well-known) function $T$:

$$\frac{\partial}{\partial r}\nabla U+\frac{(1-2v_j)}{2(1-v_j)}\frac{\partial^2 U}{\partial z^2}+\frac{1}{2(1-v_j)}\frac{\partial^2 W}{\partial r\partial z}-F_j\frac{\partial^2 U}{\partial t^2}=B_j\frac{\partial T}{\partial r}, \tag{15}$$

$$\frac{(1-2v_j)}{2(1-v_j)}\nabla\frac{\partial W}{\partial r}+\frac{\partial^2 W}{\partial z^2}++\frac{1}{(1-v_j)}\frac{\partial}{\partial z}\nabla U-F_j\frac{\partial^2 W}{\partial t^2}=B_j\frac{\partial T}{\partial z};$$

$$r=0,1\quad W(0,z,t)<\infty,\quad U(0,z,t)<\infty,\quad U(1,z,t)=0\ ,\quad W(1,z,t)=0; \tag{16}$$

$$z=0,h\qquad \frac{\partial W}{\partial r}+\frac{\partial U}{\partial z}=0\ ,\qquad \nabla U+\frac{(1-v_1)}{v_1}\frac{\partial W}{\partial z}=-\frac{(1-v_1)}{v_1}T; \tag{17}$$

$$z=h_1,h_1+h_2\qquad \left(\frac{\partial W}{\partial r}+\frac{\partial U}{\partial z}\right)_{|j=1}=b_1\left(\frac{\partial W}{\partial r}+\frac{\partial U}{\partial z}\right)_{|j=2}, \tag{18}$$

$$\left[\nabla U+\frac{(1-v_1)}{v_1}\frac{\partial W}{\partial z}\right]_{|j=1}-b_2\left[\nabla U+\frac{(1-v_2)}{v_2}\frac{\partial W}{\partial z}\right]_{|j=2}=b_2\left[\frac{(1-v_2)}{v_2}T\right]_{|j=2}-\left[\frac{(1-v_1)}{v_1}T\right]_{|j=1},$$

$$U(z+0)=U(z-0),\quad W(z+0)=W(z-0)\ ;$$

$$t=0\qquad U(r,z,0)=U_0(r,z),\ \dot{U}(r,z,0)=\dot{U}_0(r,z), \tag{19}$$

$$W(r,z,0)=W_0(r,z),\ \dot{W}(r,z,0)=\dot{W}_0(r,z).$$

To accomplish tasks (15) – (19), Hankel's finite integral transformations along the radial coordinate and the generalized transformation (kip) [18.19] are used on the axial variable. Firstly, formulae (15) – (19) are reduced to a standard form making it possibly to carry put a procedure of separation of variables along radial coordinate. For that purpose, Formula 12 is changed for that of absence of shearing stress:

$$\sigma_{rz|r=1}=\frac{C_{55}^{(s)}}{C_{11}^{(s)}}\left(\frac{\partial W}{\partial r}+\frac{\partial U}{\partial z}\right)=0, \tag{20}$$

and on the basis of the superposition of solution theorem, a new function $w(r,z,t)$ related to ratio $W(r,z,t)$ is introduced:

$$W(r,z,t)=W_1(t)+w(r,z,t), \tag{21}$$

where $W_1(t)-$ is twice-differentiated function that is defined in the process of resolving the task from the condition that there are no vertical movements of the median surface of the plate under $r=1$.

When we put (21) in (15) – (19), (20), we obtain a new boundary problem as regard to functions $U, w$. Here, the boundary conditions of the elasticity theory (16) take the form:

$$r=0,1 \quad U(1,z,t)=0, \quad \frac{\partial w}{\partial r}_{|r=1}=0, \quad U(0,z,t)<\infty, \; w(0,z,t)<\infty. \tag{22}$$

We a apply Hankel transformation with the finite bound along variable $r$ to the boundary problem $U,w$ using the following transformants and reciprocation formula:

$$u_H(j_n,z,t)=\int_0^1 U(r,z,t)rJ_1(j_nr)dr, \quad w_H(j_n,z,t)=\int_0^1 w(r,z,t)rJ_0(j_nr)dr, \tag{23}$$

$$U(r,z,t)=2\sum_{n=1}^{\infty}\frac{u_H(n,z,t)}{J_0(j_n)^2}J_1(j_nr), \quad w(r,z,t)=2\sum_{n=0}^{\infty}\frac{w_H(n,z,t)}{J_0(j_n)^2}J_0(j_nr).$$

Using transformations (23) allows us to obtain the boundary problem relative to the functions $u_H$, $w_H$. To further implement the solution algorithm, it is necessary to adopt heterogeneous boundary conditions for the variable $z$ to homogeneou conditions. To do this, we introduce a new function $W_H$ related to $w_H$ by the ratio:

$$w_H(n,z,t)=H_1(n,z,t)+W_H(n,z,t). \tag{24}$$

Where $H_1(n,z,t)-$ is the function that allows the conversion of boundary conditions.

We solve the obtained initially boundary task for $u_H, W_H$ by using a structural algorithm of the KIP method, entering [*0, h*] instrument with unknown components on the segment $K_1(\lambda_{in}r), K_2(\lambda_{in}r)$:

$$G(\lambda_{in},n,t)=\int_0^h\left[u_H(n,z,t)K_1(\lambda_{in},z)+W_H(n,z,t)K_2(\lambda_{in},z)\right]dz, \tag{25}$$

$$u_H(n,z,t)=\sum_{i=1}^{\infty}G(\lambda_{in},n,t)K_1(\lambda_{in},z)\|K_{in}\|^{-2},$$

$$W_H(n,z,t)=\sum_{i=1}^{\infty}G(\lambda_{in},n,t)K_2(\lambda_{in},z)\|K_{in}\|^{-2},$$

$$\|K_{in}\|^2=\int_0^h\left[K_1^2(\lambda_{in},z)+K_2^2(\lambda_{in},z)\right]dz.$$

The use of the KIP method results in expressions for the component $K_1(\lambda_{in}r), K_2(\lambda_{in}r)$ and transformant $G(\lambda_{in},n,t)$. Entering $K_1(\lambda_{in}r), K_2(\lambda_{in}r)$ in a variable $z$ boundary conditions creates a system of algebraic equations that allow to define permanent integrations $D_{1in}..D_{12in}$ and own values $\lambda_{in}$.

The final expression of functions $U(r,z,t)$, $W(r,z,t)$ are obtained by applying to transformant $G(\lambda_{in},n,t)$ reciprocation formula KIP (25) and Hankel's final transformations (23) in consecutive stages. As a result, taking into account (21), (24), we have:

$$U(r,z,t)=2\sum_{n=1}^{\infty}\frac{J_1(j_n r)}{J_0(j_n)^2}\left[\sum_{i=1}^{\infty}G(\lambda_{in},n,t)K_1(\lambda_{in},z)\|K_{in}\|^{-2}\right], \quad (26)$$

$$W(r,z,t)=W_1(t)+2\sum_{n=0}^{\infty}\frac{J_0(j_n r)}{J_0(j_n)^2}\left[H_1(n,z,t)+\sum_{i=1}^{\infty}G(\lambda_{in},n,t)K_2(\lambda_{in},z)\|K_{in}\|^{-2}\right].$$

Function $W_1(t)$ can be found on condition when the middle surface is stationary with account of $r=1$.

$$W_1(t)=-2\sum_{n=0}^{\infty}J_0(j_n)^{-1}\sum_{i=1}^{\infty}G(\lambda_{in},n,t)K_2(\lambda_{in},h/2)\|K_{in}\|^{-2}. \quad (27)$$

The resulting calculation ratio (26) satisfies the differential equations (15) and boundary conditions (16) – (19) (the last boundary condition (16) is only performed for the medial surface of the plate), i.e. represent a closed solution of the elasticity theory.

## 4 Discussion

The calculated functions $U,W$ in the next phase of the research allow to improve value $T$. This iterative process is performed until the numerical results of the boundary tasks in question are matched with the specified precision.

For the purpose of the elasticity theory, the primary assumption is to replace the last boundary condition (16) by (20). As a result, the cylindrical surface of the structure is docked with the additional function $W_1(t)$ used only at one point at the height of the section. Paper [13] shows that in case of a bi-morph plate, this assumption results in a slight change in the tense and deformed state and can be used in the study of thin multi-layer structures.

It is also important that in the present paper, when considering an axisymmetric initially-boundary problem of thermoelasticity has theory, the authors took into account the effect of the effect of reinforcement of a multi-layer structure on its dynamic reaction in the case of operation under nonstationary heat flows and to build a closed solution as a 3d production.

## 5 Conclusions

On the basis of our algorithm the following conclusions can be formulated:

1) A new closed solution has been built within the framework of the thermoelasticity theory for a three-layer rigidly fixed circular plate affected by temperature effect on its front faces. This approach makes it possible to abandon various kinematic hypotheses used in applied theories.

2) The introduced algorithm also works for composite plates with an arbitrary number of layers and with different physical and elastic properties, which are placed symmetrical and asymmetrical in relation to the middle surface.
3) The temperature asymmetrical load, which is applied to the end surfaces of the structure as defined in general terms, allows the study of broad range of tasks of thermoelasticity.
4) The achieved ratio design makes it possible to determine not only the stress-strain behaviour of the system in question, but also the thickness of the layers and physical characteristics of the material necesary for the effective work of the construction.
5) The calculated ratios allow to determine the frequency spectrum of multi-layer structure own variations without regard of dissipative forces.

## References

1. Yu.S. Vytchikov, I.G. Belikov, E.N. Mokhrina, Urban Construction and Architecture **3**, 103-110 (2014) DOI:10.17673/Vestnik. 2014.03.18.
2. Yu.S. Vytchikov, M.E. Saparev, Urban Construction and Architecture **4**, 108-114 (2015) DOI:10.17673/Vestnik.2015.04.15.
3. V.A. Kudinov, A.V. Eremin, A.E. Kuznetsova, E.V. Stefanuk, Higher education establishments news. Aviation Equipment **1**, 30-35 (2014)
3. V.A. Kudinov, A.E. Kuznetsova, A.V. Eremin, E.V. Kotova, Vestnik of Samara State Technical University Series: Physico-Mathematical Sciences **1** (30), 215-221 (2013)
5. T.A. Golova, A.P. Denisova, Civil Engineering magazine **8**, 9-19 (2014)
6. M.E. Frantsev, Shipbuilding **1**, 38-42 (2005)
7. I.S. Makarova, Vestnik of Samara State Aerospace University **5** (36), 108-111 (2012)
8. D.C.D. Oguamanam, J.S. Hansen, G.R. Heppler, Journal Applied Mechanics **1**(71), 49-56 (2004)
9. N.V. Smetankina, E.V. Svet, A. Shupikov, Problems of Mechanical Engineering **3** (7), 25-31 (2004)
10. E.I. Grigoluk, G.M. Kulikov, Vestnik TGU **2a**(11), 439-448 (2005)
11. G. I. Rastorguev, S.I. Snisarenko, Applied Mechanics and Technical Physics **1** (50), 187-196 (2009)
12. Yu.E. Senitskiy, I.E. Kozma, Vestnik of Samara State Technical University Series: Physico-Mathematical Sciences **43**, 52-67 (2006)
13. D. Shlyakhin, Procedia Engineering 150, 68-73(2016). DOI: 10.1016/j. proeng. 2016.06.717
14. S.A. Lychev, Vestnik of Samara State Technical University Natural Studies Series **4** (30), 118-124 (2003)
15. S.A. Lychev, A.V. Manzhirov, S.V. Uber, Bullutin of Russian Academy of Sciences Solid Mechanics **4**, 138-154 (2010)
16. S.S. Kurennov, A. G. Nikolaev, Applied Mechanics and Technical Physics **1** (45), 92-98 (2004)
17. P.N. Knyazev, *Integral transformations* (Moscow, Editorial URSS, 2004)
18. Yu.E. Senitskiy, News of Saratov University. New series. Series: Mathematics. Mechanics. Informatics **3-1** (11), 61-89 (2011)
19. Yu. E. Senitskiy, Procedia Engineering **153**, 649-654 (2016)
20. D.A. Shlyahin, Industrial and civil engineering **6**, 11-15 (2015)

# Continuous structural health monitoring of selected geotechnical quantities within Kościuszko Mound in Cracow

*Rafał* Sieńko[1,*], *Łukasz* Bednarski[2] , and *Tomasz* Howiacki[1]

[1] Cracow University of Technology, Warszawska 24, 31-155 Kraków, Poland
[2] AGH University of Science and Technology, Al. Mickiewicza 30, 30-059, Poland

**Abstract.** Kościuszko Mound in Cracow is earthen structure particularly susceptible to deformations, among other things, due to the unfavourable physical and mechanical properties of the ground and a significant tilt of slopes. Since the beginning of its existence to the present day, structural failures were repeatedly occurred, and the biggest took place after 1997, when heavy rains led to the greatest destruction of the entire mound solid ever. Overall renovation, however, did not produced the expected results and did not stop negative atmospheric influences on the monument behaviour. Therefore, in 2012, structural health mo nitoring was installed on-site, providing remote, continuous and automatic measurements of selected physical quantities, which are important from mound technical condition point of view. System is measuring, among other things, changes in vertical displacements, circumferential and radial deformations, ground pore pressure of water within a mound body and humidity of its surface layers. Correlation of results from geodetic, inclinometer and meteorological (derived from the weather station which is part of the monitoring system) measurements will enable a comprehensive assessment of ground layer work mechanisms within the whole mound solid and will be one of the bases for effective strengthening design in the future.

## 1 Introduction

Kosciuszko Mound in Cracow is a unique geotechnical structure, because of its geometry and material, but also for historical and tourist aspects. To this day it is a symbol of free and independent Poland, the National Monument, which annually attracts thousands of tourists from all over the world. The mound was raised on the hill of Blessed Bronisława in Cracow between 1820 and 1823 [1] in order to commemorate the outstanding Polish and American hero, general and engineer - Tadeusz Kościuszko, whose name is now worn by the Cracow University of Technology.

* Corresponding author: rsienko@pk.edu.pl

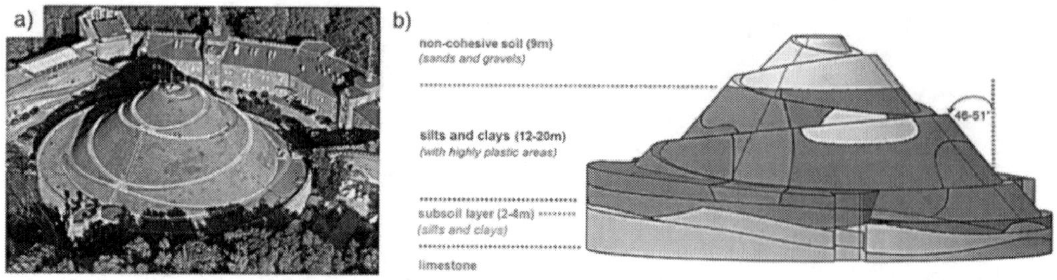


**Fig. 1.** Kościuszko mound: a) view [5]; b) visualization of ground layers [own elaboration].

The object under the assumption of builders was meant to be a symbol of durability, which cannot be destroyed neither by enemies nor time. Unfortunately, both the construction method and the building material used for this purpose have caused the structure to be susceptible to atmospheric influences. The mound was raised mainly using silt and clays, which are sensitive to water actions. Under the influence of hydrostatic pressure, the particles of soil move and organize, which has resulted in the past and result today in significant deformations of the mound body. Another important thing is steep slope, which exceeds (in some places more than twice) the angle of internal friction of such kind of grounds. This fact makes the Kościuszko Mound in Krakow a unique geotechnical structure worldwide. The shape of the limestone ceiling (inclination to the east) located in the immediate subsoil of the building, and the geometry of the surrounding brick wall (tholobate), which is indeed lower at the eastern side, are also significant for main directions of ground displacements.

## 2 Damages and failures of Kościuszko Mound

In the past, the solid of Kościuszko Mound has been destroyed many times, mainly in the form of landslides, cracks, ruptures and path damages through which rain water could freely enter the structure interior. Prior to 1979, no significant geological research was carried out, as well as engineering works that would improve the technical condition of the facility. Activities which were done were limited to determining the type ground in the superficial layers and describing damages resulting from the external loads. In the later period, until 1997, a number of minor repairs were performed. Among others, ruptures were sealed with clay, irrigation system was built and mechanical slope protection was applied by micropile palisade and ground anchor system. Unfortunately, all repairs proved to be ineffective, primarily because they were designed with incomplete recognition of the geological profile of the mound and did not constitute a comprehensive repair.

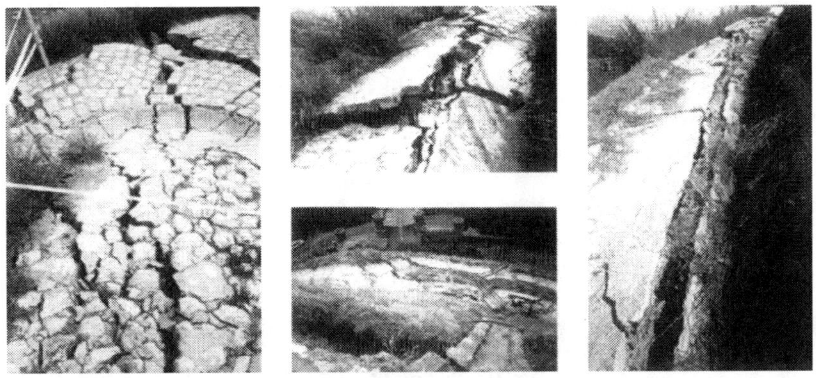

**Fig. 2.** Damages of Kościuszko Mound after flood in 1997 [1].

Large-scale repairs were carried out only after the 1997 flood, when the technical condition of the mound has been recognized as structural disaster. After its completion, about 50 geodetic benchmarks were installed on the mound peripheries as well as 4 inclinometer tubes. On the basis of vertical and horizontal displacement measurements it was found that for the next 10 years, the mound top settled about 0.50 m and moved in the east about 0.20 m. It means that repairs did not stop the mass movements. Similarly to previous strengthening, designers did not have a full geotechnical documentation of deeper layers of the mound body.

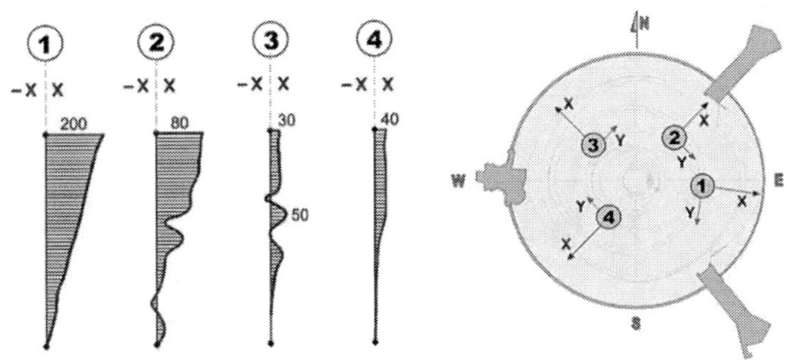


**Fig. 3.** Selected displacement profiles [mm] from inclinometer measurements between 2012-2014 [own elaboration].

## 3 Structural health monitoring system

Currently the condition of the monument allows for its normal operation, although surface layers sliding and slope erosion are locally observed. The underperformance of earlier repairs and the progressive degradation of the mound as a result of environmental impacts (the biggest during the floods in 2010) necessitate the development of solutions which aim is to protect the monument from another catastrophe and ensure its safe operation for next decades. For this reason, in 2012, while performing geological-engineering studies within the whole solid of mound and hill of Blessed Bronisława, it was decided to install an automatic measuring system, which allows to observe and control the work of whole volume of the mound in real-time. Continuous measurement of selected physical quantities is intended to indicate the mechanisms of monument destruction.

One of the basic assumptions of structural health monitoring system design was to recognize the work of ground layers inside the body to perform a full three-dimensional analysis of its structural behavior. [2] It was decided to measure vertical displacements including changes in the thickness of selected soil layers (vertical extensometers), horizontal deformations (circumferential and radial), as well as changes in ground water pressure, soil humidity and temperature. Applied sensors are based on vibrating wire technology, which, thanks to its advantages (accuracy, resistance to external conditions, time stability) is widely used in long-term measurement systems [3, 4, 5, 6].

**Fig. 4.** Installation of SHM within Kościuszko Mound in Cracow [courtesy of SHM System].

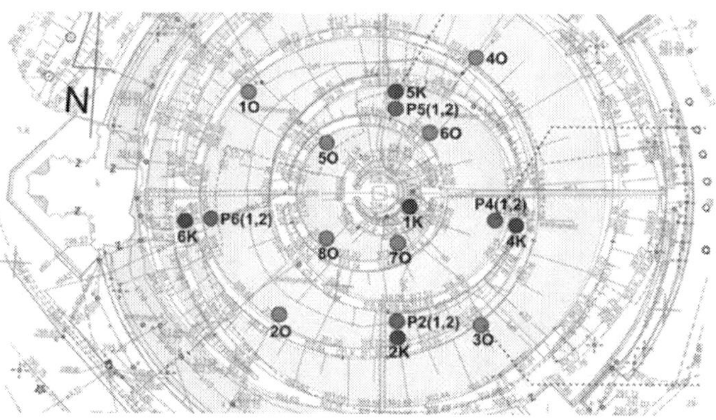


**Fig. 5.** Localization of measuring points („K" – vertical displacements, „O" – horizontal displacements (circumferential and radial) and ground humidity; „P" – ground water pressure and temperature) [courtesy of SHM System].

Measurements of vertical displacement are performed by means of tendon extensometers (Fig. 6a) in five locations. Details of its operation principle, sample data and results interpretation are presented in Chapter 4. Measurements of horizontal deformation are carried out by extensometers installed circumferentially along the peripheral surface and radially inside the body (Figure 6b). Located in eight places, sensors have measuring bases of 2 000 mm (radial extensometers) and 1 000 mm (peripheral extensometers). As the main active cause of Kościuszko Mound deformation is water infiltration and the associated change in soil moisture, within SHM system at four measuring points and three different depths, vibrating wire piezometers were installed (Figure 6c).

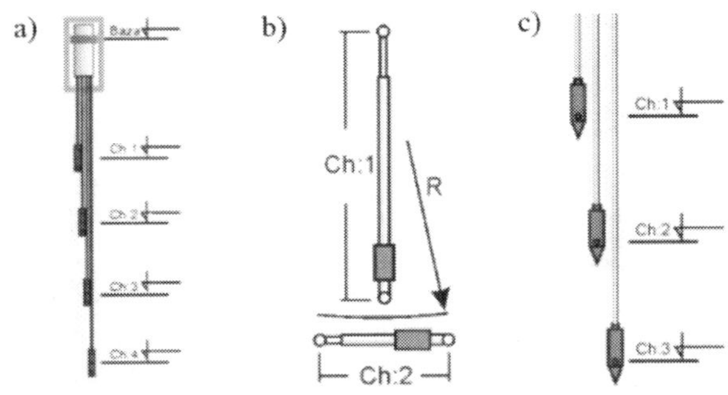


**Fig. 6.** Measuring sets: a) vertical extensometers, b) horizontal extensometers, c) piezometers [courtesy of SHM System].

All vibrating wire sensors are equipped with thermistors for parallel temperature measurements. This allows for thermal compensation of measuring data and for evaluation of temperature impact on the global response of geotechnical structure [7, 8].

The above-described elements of the SHM system should be assigned to structural (technical) monitoring subsystem. In 2014 the installation was supplemented with meteorological monitoring. A weather station consisting of a rain gauge, anemometer, air relative humidity and temperature sensor and pyranometer were installed. Obtained information about atmospheric conditions within the mound (precipitation, sunshine, wind force and direction) can be correlated with data obtaining from the building interior, which will allow to construct dependence between the external environment influences and the behavior of the whole body.

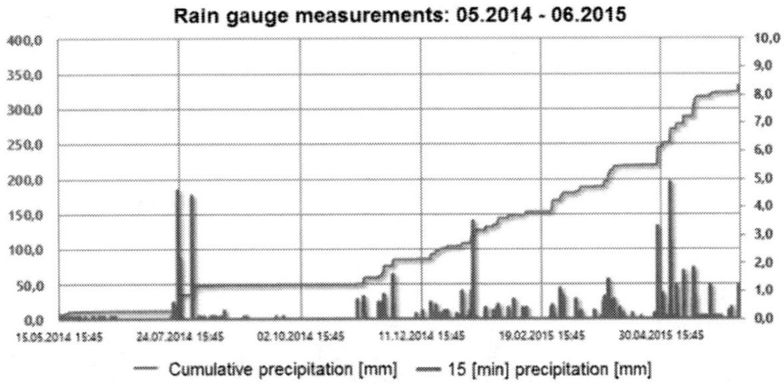


**Fig. 7.** Exemplary plot of the amount of precipitation [courtesy of SHM System].

The SHM system of the automatically performs measurements and archives data every 15 minutes from 2012 and eventually next over several decades. All data are converted according to the developed procedures and the results in tabular and graphical form (graphs) are displayed on the screen of any computer with Internet access.

## 4 Vertical extensometers

Extensometric sensors allow for determination of defined segments lengths and they are often used in geotechnical measurements [9]. Within the monitoring system, five extensometers were installed, each consisting of four sensors. One end of a particular displacement sensor is attached to the surface of the mound perimeter (base) and the second, through a stainless steel tendon, is anchored inside the body at various depths - Fig. 7. Changes in the displacement of the base are determined by geodetic measurements, so it is possible to determine absolute changes in thickness of selected ground layers within the analyzed vertical measuring line. The rock ceiling was assumed as a fixed, unmovable reference, so the deformations of five layers are analyzed in a given vertical. The change in length recorded by sensor 1 will be included in the measurement of the length change recorded by sensor 2. Analogously, measurements are made by other sensors. By subtracting the individual displacement values, we obtain changes in the thickness of the ground layers. The sum of the thickness changes of the four layers is equal to the direct value of displacement resulting from measurement carried out by the sensor with the longest tendon (no. 4). If this value is subtracted from the base displacement, we will get information about the displacement of the extensometer anchor.

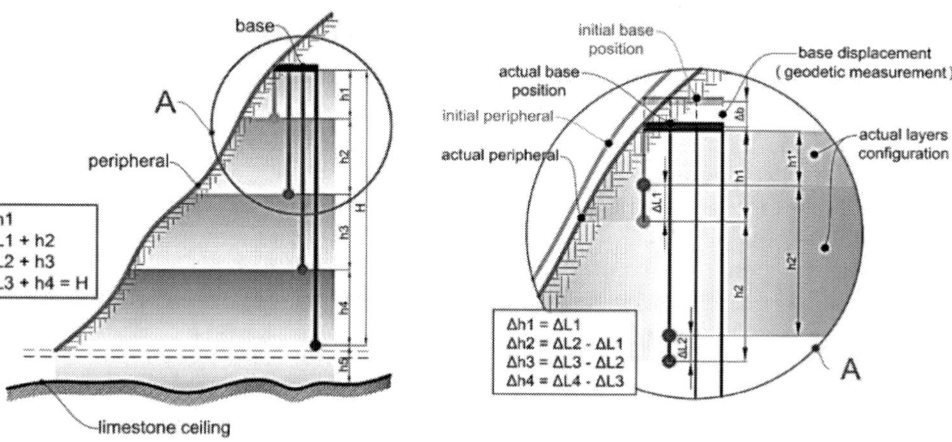


**Fig. 8.** Construction and operation principle of vertical extensometers [own elaboration].

The sum of the thicknesses $h_i$* of all layers (from the rock to surface) in the current configuration will be equal to the sum of the initial thicknesses $h_i$ of the layers minus the base offset $\Delta b$.

$$\Sigma h_i^* = \Sigma h_i^* - \Delta b \tag{1}$$

Knowing the change in length $\Delta H$ measured by sensor 4 and base displacement $\Delta b$, we can determine the change in the thickness of the lowest ground layer $\Delta X$ lying directly above the limestone layer.

$$\Delta X = \Delta b - \Delta H \tag{2}$$

**Table 1.** The changes of ground layers [mm] between 06.2013 - 09.2014

| Point | Δh1 | Δh2 | Δh3 | Δh4 | Σ = ΔH |
|---|---|---|---|---|---|
| 1K | 4,84 | 0,96 | 12,82 | -12,77 | **5,85** |
| 2K | 4,96 | -0,81 | 2,47 | -1,94 | **4,67** |
| 4K | 6,91 | 1,46 | 0,65 | -0,49 | **8,53** |
| 5K | 1,32 | 0,52 | -2,96 | 0,07 | **-1,05** |
| 6K | 3,06 | 0,70 | 0,44 | -0,90 | **3,31** |

It is worth to notice that the behavior of particular ground layers from which the mound was built is dependent on the depth. It is interesting that the thickness of certain layers is increased (red color and the minus sign in Table 1), which may indicate the presence of bulging soils. Nevertheless, during the entire considered period, there was a systematic decrease in the height of the extensometers bases. A qualitative graphical visualization of changes in the thickness of particular ground layers over a given period of time is shown in Figure 8.

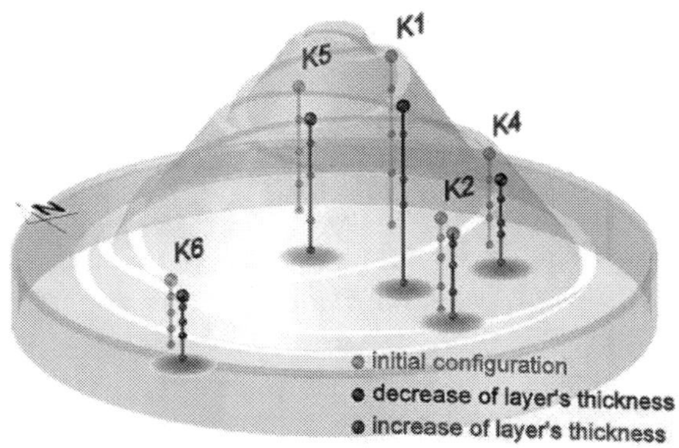


**Fig. 9.** Visualization of ground layers thicknesses changes between 06.2013 and 09.2014 [own elaboration].

Geodetic measurements showed the largest subsidence of K1 base during analyzed period. It was 26 mm. Base K2 and K6 moved vertically by 7 mm and base K4 and K5 by 13 mm. The largest displacement measured by the K1 extensometer results, among others, from the location closest to the vertical central axis of the mound body. Concentration of the largest ground masses in this area generates the largest subsidence - Fig. 9.

The analyzed section at K5 (northern slope) was the only one with a slight elongation, but it was accompanied by a noticeable decrease in the base position by 13 mm as well as by the longest tendon anchor by 14 mm. The elongation of the section results in the reduction of the thickness of the lowest layer (located directly above the rock). This deformation can be caused by washing away silt and clays from mound solid and moving them into limestone rock crevices.

## 5 Conclusions

The structural health monitoring system installed on the Kościuszko Mound in Cracow automatically provides information regarding selected physical quantities associated with the mound behavior. The qualitative and quantitative analysis of SHM data, geodetic, inclinometer and meteorological measurements is intended to help answer the question how to permanently protect the Monument against progressive degradation.

The monitoring system described in this paper is an example of how geotechnical design can be supported by in-situ measurements. Certain geoscientific problems require such kind of measurements, since classical geological recognition is insufficient. It is worth to remember that data acquisition about mechanical properties of the ground from the test takes place at any given time, whereas, due to different influences, these parameters may change over the time. SHM systems can actually enrich this information.

SHM systems allow to observe the structural work during construction process, reconstruction or during operation. The system described in this paper is also intented to perform such kind of function. Thanks to the results obtained from automatic sensors, it will be possible to verify the design assumptions and theoretical analyzes with in-situ measurements. This approach to geotechnical design is recommended by EN 1997 [10].

Currently, increased application of structural health monitoring systems in geotechnical issues is being observed in Poland. This phenomenon is in line with global trends, where most of the large and complex structures made from the ground, whose destruction consequences would be very significant, are equipped with measuring devices that allow for continuous observation of their behavior over time and quick reaction in case of unpredictable phenomenon.

## References

1. Rokosz M., *Katastrofa Kopca Kościuszki w 1997 roku i pierwsze próby ratowania zabytku* (Kraków, 2007)
2. Dunnicliff J., *Geotechnical instrumentation for monitoring field performance* (John Wiley & Sons, 1988)
3. Bednarski Ł., Sieńko R., Inżynieria i Budownictwo **11**, 615 (2013)
4. Bednarski Ł., Sieńko R., Howiacki T., Procedia Engineering **108**, 536 (2015)
5. Bednarski Ł., Sieńko R., Howiacki T., Cement-Lime-Concrete **6**, 396 (2014)
6. Bordes J.L., Debreuille P.J., T.R.B. Record **No. 1004**, 20 (1985)
7. Lucas D. R., Fankhauser K., Springman S. M., Engineering Geology **219**, 32 (2017)
8. Florin V. A., Industrial Services **47 (2)**, 12 (2013)
9. Sieńko R., *I Konferencja Naukowo-Techniczna: Problematyka Projektowania i Wykonawstwa w Aspekcie Nowych Technologii, Materiałów i Nowoczesnej Techniki w Budownictwie "TECH-BUD 2013"* (Kraków 2013)
10. EN 1997, *Eurocode 7, Geotechnical Design*

MATEC Web of Conferences **117**, 00158 (2017) DOI: 10.1051/matecconf/201711700158
*XXVI R-S-P Seminar 2017, Theoretical Foundation of Civil Engineering*

# Studying the effects of coarse pore expanded clay concrete on a permanent formwork made of cement bonded particle boards

*Aleksandr Sigalov*[1,*], *Sergey Sinenko, Tatiana Poznahirko*

[1] Moscow State University Of Civil Engineering 129337, 26, Yaroslavskoe shosse, Moscow, Russian Federation

**Abstract**:This paper suggests using a formwork system based on cement bonded particle boards (CBPB) filled with coarse pore expanded clay concrete in order to improve the technical efficiency of the building envelope. Such a structure is made of lightweight or honeycomb concrete units, has a thickness equal to that of heavyweight concrete blocks and provides increased resistance to heat, better fire resistance and longer durability. The relatively modest price of the suggested option is seen as a notable advantage. The effects of coarse pore expanded clay concrete have been shown as being similar to the effects of heavyweight concrete, with the nature of the effects demonstrated on a hydrostatic pressure curve. The author has substantiated the geometric parameters for the suggested heat-efficient envelope design: a 800x800x370 mm block of CBPB filled with coarse-pore expanded clay concrete $\gamma$= 500kg/m3.

## Introduction

This problem is urgent, because it is always necessary to reduce construction costs. The DSP has proved itself well on the part of the economy, but few have considered the issue of reliability. In connection with this, a number of experiments were in order to improve the technical efficiency of a building envelope, the author suggests using a formwork system based on cement bonded particle boards (CBPB) filled with coarse pore expanded clay concrete. Such a structure is made of lightweight or honeycomb concrete units, has a thickness equal to that of heavyweight concrete blocks and provides increased resistance to heat, better fire resistance and longer durability. The relatively modest price of the suggested option is seen as a notable advantage.

The issues in question have been studied by Ye.M. Pugach (Ph.D. in Technical Science), S.A. Sinenko (Doctor of Technical Science), I.Ye. Sigalov (Doctor of Technical Science), A.A. Afanasiev Doctor of Technical Science and many others. Some of the scholars suggested using either a temporary formwork or the types of formwork that

had poorer properties versus CBPB; others spoke in favor of CBPB filled with heavyweight concrete types. The combined use of CBPB and coarse pore expanded clay

* Corresponding author: sigalov2012@inbox.ru

concrete is a relatively new approach in construction science and, therefore, it needs further review. Hence the study presented in this paper.

Construction scientists have determined the elasticity modulus of lightweight concretes to be dependent on the elastic characteristics of an aggregate and a mortar part and their relative amount in the concrete mass. Lightweight concrete has a relatively low modulus of elasticity due to increased deformation of porous aggregates. This is one of the most significant features of lightweight concrete, which, depending on the area of application and workability in structures, can play both a negative and a positive role.The elasticity modulus of a lightweight concrete aggregate is much lower than that of cement stone, but given the high degree of cement stone's intergrowth with the aggregate, the mortar part reduces stresses that occur in porous aggregate elements and causes a homogeneous distribution of such stresses. A large number of formulas are used to calculate the strength of lightweight concrete; however, these do not always take into account the effects of the aggregate contacts in coarse-pore structures and the adhesion of grains to the mortar, which results in these formulas being inaccurate.

Simulating the structure formation process for coarse-pore expanded clay concrete allows us to substantiate with more precision which processes occur in the material while it is mixed and cast in terms of the strength properties of the model.

The effects made by the deformative properties of the mixture allows the following conclusions to be drawn:

1. The deformability of a porous aggregate creates more favorable conditions for shrinkage in cement stone without forming cracks in it.

2. The full shrinkage value for densely structured lightweight concretes exceeds that for heavy concretes by 15-20% and can reach 1.5 mm/m.

3. The post-mold shrinkage value for most concretes using artificial porous aggregates ranges between 0.3 and 0.8 mm/m. The larger the volume and porosity of cement stone, the higher the post-mold shrinkage value.

4. After being cast, a lightweight concrete mixture differs from a regular heavyweight concrete mixture not only by a lower weight but by a significant air content some of which is compressed in the pores of lightweight concrete and causes the cast mixture to swell.

## Results

Let us calculate the geometric parameters for a formwork made of CBPB.

For the purpose of this paper, an experimental unit shown in Figure 1 was used to carry out an experimental study of the lateral pressure of coarse pore expanded clay concrete on the walls of a formwork. This is a unit for experimental determination of deflections.

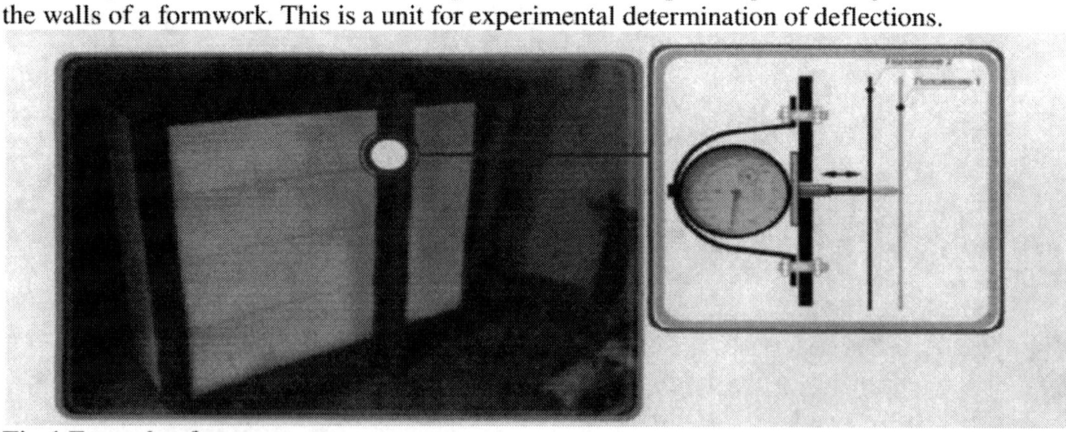

**Fig.1** Example of mesurement

Based on the computational model (Figure 2), we will calculate steel strips to determine the deflections in the side beam:

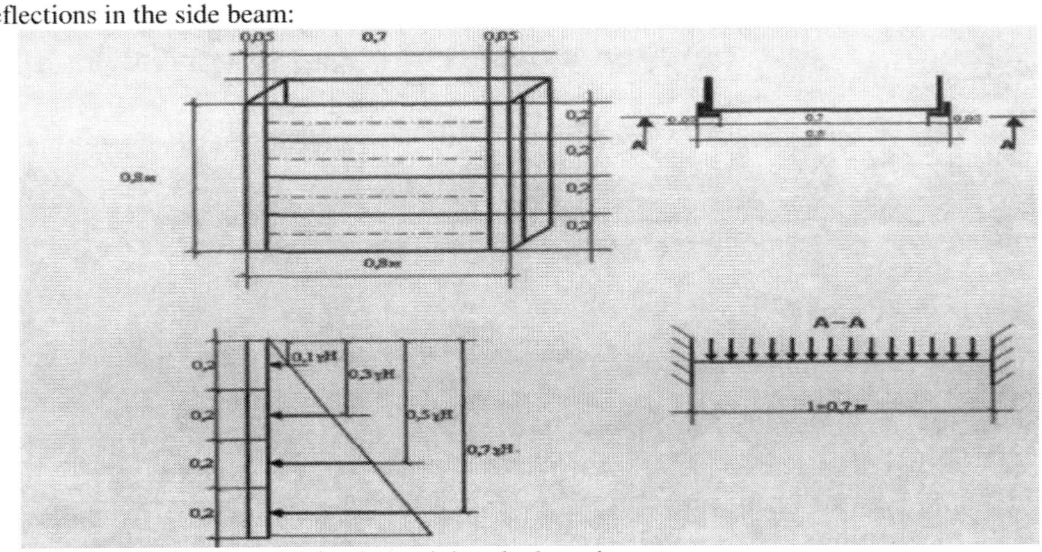

**Fig. 2**. The computational model for determining the lateral pressure

According to the Formula I, the load on the strip will be:

$$P_i = \gamma H_i \bullet F \quad \text{(I)}$$

$\gamma$ - the bulk weight of coarse-pore expanded clay concrete, $\gamma$ = 500 kg/m3;

Hi – the arm of the force acting on the strip, H;

F – strip area, m2;

$$F = 0.7m \bullet 0.2 = 0,14m^2$$

$$P_1 = 500 \bullet 0.1 \bullet 0.14 = 7kg = 70N$$

$$P_2 = 500 \bullet 0.3 \bullet 0.14 = 21kg = 210N$$

$$P_3 = 500 \bullet 0.5 \bullet 0.14 = 35kg = 350N$$

$$P_4 = 500 \bullet 0.7 \bullet 0.14 = 49kg = 490N$$

Using Formula II, we will determine the deflection in the strip along the central axis

$$\int_i = \frac{P_i \bullet l_i}{E \bullet J} \bullet \frac{1}{384} \quad \text{(II)}$$

Pi – concentrated load, N;

$l_i$ - span, m;

E – elasticity modulus, Pa;

J – inertia moment, m4.

Steel's elasticity modulus E = 206 • 109 Pa.

The cross-section of the strip has the following geometric characteristics: h=0.003m, b=0.2,

$$J = \frac{b \bullet h^3}{12} = \frac{0.2 \bullet 0.003^3}{12} = 4.5 \bullet 10^{-10} m^4$$

h – thickness (height) of the strip, m; B-strip width, m

Using Formula III we will find the deflection values for strips:

$$f_4 = \frac{P_4 l^3}{EJ} \bullet \frac{1}{384} = \frac{490 \bullet 0.7^3}{206 \bullet 10^9 \bullet 4.5 \bullet 10^{-10}} \bullet \frac{1}{384} = 4.721mm,$$

$$f_2 = \frac{P_2 l^3}{EJ} \bullet \frac{1}{384} = \frac{210 \bullet 0.7^3}{206 \bullet 10^9 \bullet 4.5 \bullet 10^{-10}} \bullet \frac{1}{384} = 2.023mm,$$

$$f_3 = \frac{P_3 l^3}{EJ} \bullet \frac{1}{384} = \frac{350 \bullet 0.7^3}{206 \bullet 10^9 \bullet 4.5 \bullet 10^{-10}} \bullet \frac{1}{384} = 3.372mm,$$

$$f_4 = \frac{P_4 l^3}{EJ} \bullet \frac{1}{384} = \frac{490 \bullet 0.7^3}{206 \bullet 10^9 \bullet 4.5 \bullet 10^{-10}} \bullet \frac{1}{384} = 4.721mm.$$

We will construct dependence (IV) from the results of the calculation (f  H):

$$\frac{\Delta f}{l} \le \frac{1}{400}, \Delta f \le \frac{1}{400} = \frac{0.7}{400} \le 0.0017m = 1,7mm \prec 2,5mm. \qquad (IV)$$

The calculation results as well as the experiment's results are summarized in Table 1. Proceeding from the results of our calculations, for the purposes of further experimental study let us take a heat-efficiency envelope design characterized as follows: a 800x800x370mm permanent formwork, wall thickness of 32 mm filled with coarse-pore expanded clay concrete, $\gamma$= 500 kg/m3 bulk density, R6 strength = 0.5 MPa, coarse-pore expanded clay concrete with gravel fraction of 20-40 mm, γbulk = 400kg/m3 and in-cylinder crushing strength σ3 = 1,2Mpa - Portland cement 400 and C-3 water reducing agent (0.2% of cement mass).

**Table 1.** The experiment's results

| Test No | Strip No | Deflection value, mm | | | Deflection value, theoretical, mm | Deviation, % |
|---|---|---|---|---|---|---|
| | | reference | reference | deflection | | |
| 1 | 1 | 0.00 | 0.60 | 0.80 | 0.76 | 5 |
| | 2 | 0.00 | 2.13 | 2.13 | 2.02 | 6 |
| | 3 | 0.00 | 3.94 | 3.94 | 3.72 | 6 |
| | 4 | 0.00 | 5.03 | 5.03 | 4.72 | 7 |
| 2 | 1 | 1.00 | 1.82 | 0.82 | 0.76 | 8 |
| | 2 | 1.00 | 3.16 | 2.16 | 2.02 | 7 |
| | 3 | 1.00 | 4.99 | 3.99 | 3.72 | 7 |
| | 4 | 1.00 | 5.98 | 4.98 | 4.72 | 6 |
| 3 | 1 | 2.00 | 2.81 | 0.81 | 0.76 | 7 |
| | 2 | 2.00 | 4.15 | 2.15 | 2.02 | 6 |
| | 3 | 2.00 | 6.10 | 4.10 | 3.72 | 10 |
| | 4 | 2.00 | 7.03 | 5.03 | 4.72 | 7 |
| 4 | 1 | 0.00 | 0.79 | 0.79 | 0.76 | 4 |
| | 2 | 0.00 | 2.17 | 2.17 | 2.02 | 7 |
| | 3 | 0.00 | 3.96 | 3.96 | 3.72 | 6 |
| | 4 | 0.00 | 5.09 | 5.09 | 4.72 | 8 |
| 5 | 1 | 1.00 | 1.83 | 0.83 | 0.76 | 9 |
| | 2 | 1.00 | 3.18 | 2.18 | 2.02 | 8 |
| | 3 | 1.00 | 5.05 | 4.05 | 3.72 | 9 |
| | 4 | 1.00 | 6.15 | 5.15 | 4.72 | 9 |
| 6 | 1 | 1.00 | 1.85 | 0.87 | 0.80 | 9 |

| | | | | | | |
|---|---|---|---|---|---|---|
| | 2 | 1.00 | 3.24 | 2.18 | 2.10 | 8 |
| | 3 | 1.00 | 5.09 | 4.10 | 3.74 | 9 |
| | 4 | 1.00 | 6.17 | 5.22 | 4.80 | 9 |
| 7 | 1 | 1.00 | 1.93 | 0.90 | 0.80 | 9 |
| | 2 | 1.00 | 3.30 | 2.20 | 2.12 | 8 |
| | 3 | 1.00 | 5.12 | 4.20 | 3.60 | 9 |
| | 4 | 1.00 | 6.19 | 5.21 | 4.90 | 9 |

At the next step we make calculations for the CBPB boards according to the diagram in Figure 3.

a) Formwork beam loading diagram

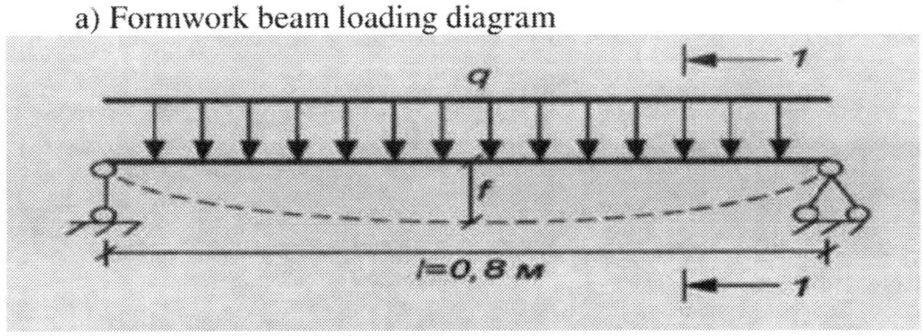


b) Side beam's cross-section

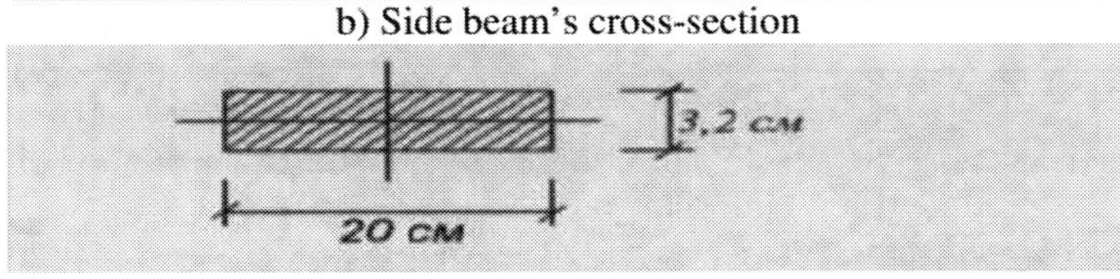


**Fig. 3**. a) Formwork beam loading diagram, b) Side beam's cross-section

Using this algorithm the lateral pressure exerted by coarse-pore expanded clay concrete will be calculated under the same loading conditions on the formwork beam.

Initial loads: q=70,210,350,490 N/m; f=5ql4/384Elx , E=3 hPa

Next, we will define the deflections, the CBPB material with the E = 3,000-3,500 mPa modulus and construct the graph (f H) as shown in Fig. 4.

fmax=1,6mm<fn=2,5mm.

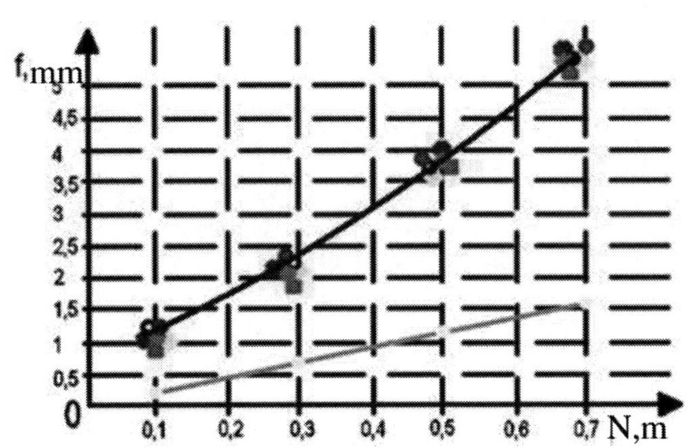


**Fig. 4.** Design and experimentally derived values of f-H dependencies

Thus, the formwork's geometrical parameters are 800x800x370 mm and it can be loaded up to its full height at the time of casting coarse-pore expanded clay concrete; however, for technical reasons and taking into account the need to create an optimum structure while casting as well as the effect of external impulse compaction on the formwork, the layer to be cast should not exceed 400 mm.

With experimental deflection values not exceeding the values obtained by calculation by an average of 6%, the following conclusion can be made: operational effects of coarse-pore expanded clay concrete are similar to those of heavyweight concrete and, therefore, it would be correct to assume that the accepted method for calculating lateral pressure has been chosen correctly.

Further calculations and studies have shown that the span of the formwork board can be doubled without compromising its design characteristics, but the increased dimensions and weight of the formwork itself (up to 130 kg) would make it difficult to manually assemble and transport. Since a maximum use of manual labor is one of the conditions for undertaking low-rise individual construction projects, we used a 800x800x370mm formwork to develop further technological solutions. It stands to reason that the use of mechanized tools will allow for a 1,600x800x370mm formwork to be constructed with relative ease.

## Conclusions

The experimental part of our study has resulted in the following conclusions:

1. Experimental findings on determining the lateral pressure of the mixture on the walls of the formwork (operational effects) have confirmed that the effects of coarse pore expanded clay concrete are similar to those of heavyweight concrete with the nature of the effects demonstrated on a hydrostatic pressure curve.

2. The geometric parameters of the suggested heat-efficient building envelope have been substantiated: a 800x800x370 mm block of CBPB filled with coarse-pore expanded clay concrete $\gamma$= 500kg/m3.

## References

1. Anpilov S. M.: Shuttering systems for monolithic construction Anpilov S. of AVS M / publishing house - M.: Educational edition. (2005)
2. A.A. Afanasyev, Construction of buildings and constructions from monolithic reinforced concrete / A.A. Afanasyev - Stroyizdat - M (2012)
3. M.Z. Weinstein, Deformativnye of property of light concrete on porous fillers at compression / M. Z. Weinstein//Works NIISM - Alma-Ata. - Collection 5(7)-(2008)
4. V.N. Vernigorova, Modern methods of studying the properties of building materials. Vernigorova, P.I. Makridin, Yu.A. Sokolova // Publishing house ABC - M., (2003)
5. E.M. Pugach, Technology of production of three-layer blocks for the construction of energy-efficient enclosing structures // Thesis for the competition uch. Degrees Ph.D. - M.: - MSSU (2005)
6. SNiP 2.01.02-85 *. Fire regulations / / Stroyizdat - M., (1995)
7. A.A. Afanasyev, Technology of construction processes / A.Afanasyev, N.N. Danilov, V.D. Kopylov // Proc. - Higher School - M., (2006)
8. N.I. Makridin, Structure, Deformability, Strength and Criteria for the Destruction of Cement Composites / N.I. Makridin, I.N. Maksimova / / Ed. Saratov University – (2005)
9. Modern walls and facades. Overview of technical capabilities and materials // Know-how - 2 – (2010)

10. Typical technological map for concrete and reinforced concrete works 31032K. Concreting of typical floors in the block-panel formwork // SNIIMTOP-M., (2005)

# Justification of application of new types of fastening of slopes of hydraulic engineering constructions

*Anatoly* Smyvalov[1], *Maxim* Rodionov [1,*], and *Yulia* Galitskova [1]

[1]Samara State Technical University, Institute of Architecture and Civil Engineering, 194, Molodogvardeyskaya St., 443001, Samara, Russia.

**Abstract.** The article represents the analysis of various types of fastenings of the soil slopes which have found broad application in the hydrotechnical construction practice for the last 10-15 years. It is noted that, generally, new types of fastenings represent composite materials made of concrete, metal and polymers. Experience shows that producers of the construction materials used in bank protection designs seek to recommend them for the solution of a wide range of tasks, at the same time it isn't always founded. The authors, on the basis of the analysis of experience with the types of fastenings and also modern domestic and foreign recommendations about their design and construction, offer a technique of justification of new types of fastenings of slopes of hydraulic engineering constructions. The use of concrete filled geotextile mats in bank protection is brought out as a practical realization of the represented technique.

## 1 Introduction

The construction industry dynamically develops that is connected with the new building constructions, technologies and materials. In the hydrotechnical construction it is possible to observe similar tendencies, in particular in the field of protection of soil slopes against the destroying external influence. The reliable protection of soil slopes is extremely important as their destruction can lead to serious accidents in the water-engineering system or at bank protection constructions and as a result on adjacent territories. Therefore the slope fastening design has to be quite reliable for a long period of time.

For the last 10-15 years in Russia and the Samara region, in particular, excepting traditional types of fastenings made of stone, concrete and reinforced concrete, "new" constructive decisions [1-6] are being made to use universal flexible protective concrete mats (UFPCM), concrete filled geotextile mats (CFGM), the polymeric and composite tongue, gabions, etc.

Structurally new types of fastenings are almost all composite materials made of concrete, metal and polymers. So, UFPCM consist of the concrete blocks which are tied around by a metal or polymeric cable, CFGM is a woven polymeric cover filled with concrete; the composite tongue consists of a fabric basis and a special hardener.

* Corresponding author: rodionov_max@mail.ru

Bank protection fastenings function in very severe conditions: waves, current, ice, vessels, etc. Therefore it is essential for them to withstand all possible influences.

Design features of new types of fixings don't allow to fully use the existing normative and technical basis for their application. The area and conditions of application of new types of fixings are, as a rule, are specified in the technical documentation provided by the vendor. Often vendors tend to recommend the produce to solve a wide range of problems (fixing in a zone higher than a water boundary and variable level, berthing facilities, in a zone of ice and wave actions), which isn't always justified. As a rule, reasoning comes down to the theoretical calculations which aren't confirmed with full-scale data. So, vendors of UFPC represent theoretical researches of their produce functioning in wave and ice conditions. But they do not consider ice piling or berth impact. Also there are no full-scale confirmations of reliability of the received theoretical researches.

## 2 Materials and Methods

To justify the application of new types of fastening of slopes of hydraulic engineering constructions the following methods were used:

- the statistical analysis of the data obtained as a result of the natural survey and observation of coastal slopes, elements of fastening of soil dams, building the fastening of earth constructions of bridge crossings;
- studying of materials properties, with the subsequent calculation of economic efficiency to choose the optimal constructive and technology solution.

The authors have analysed the experience with new types of fastenings of slopes at a number of sites in the Samara region as well. The authors were personally involved in developing and justification of constructive decisions [1].

CFGM INKOMAT have been applied on the site of the Kirovsky bridge crossing. Laying of the first square meters was executed in 2010. Fastening is executed from CFGM on a slope of a road embankment to protect it from washout in the spring high water. The average thickness of fastening is 18 cm (figure 1).

Similar CFGM (figure 2) were applied on the Mikhaylo Ovsyansky water-engineering system in 2015. The Mikhaylo-Ovsyansky water-engineering system is Class IV and works of laying CFGM were among urgent antiflood actions connected with a considerable destruction of fastening of a pressure head slope of the soil dam and a possible danger of its destruction by the next high water. The works were executed in the winter period and allowed to gather water in a reservoir during the first high water practically to mark NBL without danger of destruction of the soil dam.

UFPCM were applied when performing fastening of the river Pestravochka bank (Pestravka, the Samara region). The design of UFPCM provides reliable «sticking» of a protective cloth from mats to the protected surface. Moreover, the effect of immersion of extreme ranks of the UFPCM concrete blocks in soil is noted. When laying UFPCM on the soil of clay type (sandy loam, loam, clay) it should be made with a rough party of a mat down which allows to increase coupling with soil. The edge of a mat of UFPCM-202 contacts the mattress of Renault laid on the horizontal surface.

The composite and PVC tongue have been applied on bank protection sites in the village of Rozhdestveno and Lunacharsky of the Samara region (figure 4).

**Fig. 1.** Construction of "Kirovsky Bridge" (author's photo).

**Fig. 2** A reconstruction stage of the water-engineering system in the village of Mikhaylo-Ovsyanka (the author's photo).

**Fig. 3**. Fastening of a slope of the river Pestravochka (author's photo).

**Fig. 4.** Building of fastening with the use of tongues: 1 – PVC, 2 – composite.

At justification of the fastening design standard calculations of durability, stability were used. However, the peculiarities of the considered designs are the following: the producers gave no recommendations how to use these materials in the conditions of ice influences as they were used only in "soft" climatic conditions. In this regard, the research of "new" types and sorts of materials in hydrotechnical construction is urgent.

Also on the basis of the technical and economic calculations [7], it is found out that the most economical designs are those of a better type with the use of UFPCM and CFGM. Vertical designs are applied in case of impossibility to use batter designs. The PVC tongue is recommended to be used with a vertical clearance of about two meters and sandy soil of the basis, the composite one – with a vertical clearance of about four meters.

## 3 Results

At designing bank protection constructions we have developed the following research methodology.

During researches the following types of works were carried out:

- theoretical researches and calculations;
- production on the building site;
- laboratory researches of the bearing ability;
- production of the pilot block;
- organization of a hydrometeorological office;
- field researches of ice influences.

The vertical profile wasn't further considered owing to being less economical. On the batter fastening the ice cover has various impacts, the most important of them are the following loadings:

- of the moving ice floe on the batter profile construction;
- of the influence of the stopped even ice floe leaning on the construction under the influence of a water current and wind;
- on constructions from a continuous ice cover at its temperature expansion;
- of the ice cover frozen to the construction at a water level change.

In engineering calculations in the conditions of ice influences it is necessary to consider the following loadings: of the moving ice floe on the batter profile construction; of the influence of the stopped even ice floe leaning on the construction under the influence of a water current and wind; on constructions from a continuous ice cover at its temperature expansion; of the ice cover frozen to the construction at a water level change.

The reliability of a bank protection design in the conditions of ice influences has to be defined with ensuring its being resistant to the shift along a slope and capsizing relative to a top or lower edge.

On the basis of methods of calculation of ice influences and the accounting of design features the conditions of its being resistant to the shift and capsizing are determined by formulas 1 and 2. The providing condition of concrete mats being resistant to the shift within the ice foe is defined as follows:

$$K = \frac{\sum T_1}{\sum T_2} \geq \gamma_n, \tag{1}$$

where K – stability coefficient; $\Sigma T_1$ – the sum of all holding forces, kN; $\Sigma T_2$ – the sum of all shifting forces, kN; $\gamma_n$ – reliability coefficient accepted depending on a class of hydraulic engineering constructions.

$$K = \frac{\sum M_1}{\sum M_2} \geq \gamma_n, \tag{2}$$

where $\Sigma M_1$ – the sum of the moments of all holding forces; $\Sigma M_2$ – the sum of the moments of all shifting forces.

For the majority of "new" materials similar researches weren't conducted therefore, they were used in the zone of a variable level at the considered objects construction sites. For justification of using CFGM we have conducted special researches, including the pilot block setting-up where observations are currently carried out. Similar researches were conducted for UFPCM [8] which have shown that they are allowed to be used on flat laying (more than 4), the minimum thickness of ice (approximately about 0,7 m) and/or in an underwater part of the construction. In this regard all further results of researches are devoted to studying of CFGM.

On the basis of field observations the basic data have been established for calculation. The ice influences observed: the maximum thickness of ice is 0,60 m; the maximum fluctuations of water level in a reservoir at a freeze-up are one meter; the maximum fluctuations of temperature of external air within a day are seven degrees Celsius, the maximum speed of the water current under ice in the ice drift period is V=3,0 m/s, the maximum wind speed in the period of the ice drift is V = 10,0 m/s, the average length of the ice foe in the downstream is L=30.0 m, a bias of a stream surface, i=0,07. On the basis of the actual data obtained as a result of the observation, the calculations for three types of mats are carried out and represented in Table 1.

The results of the observations made by us at the pilot block are represented in the following scientific works [9-10]. As a result of the conducted researches at the pilot block the possibility of using CFGM in the zone of the ice drift and influence of ice foes on the rivers and reservoirs has been proved.

**Table 1.** Calculation results of mats stability.

| Mat thickness, м | Of ice drift | | | Of leaning | | | Of temperature expansion | | | Of level change | | | Capsizing | | |
|---|---|---|---|---|---|---|---|---|---|---|---|---|---|---|---|
| | $\Sigma T_1$ | $\Sigma T_2$ | K | $\Sigma T_1$ | $\Sigma T_2$ | K | $\Sigma T_1$ | $\Sigma T_2$ | K | $\Sigma T_1$ | $\Sigma T_2$ | K | $\Sigma M_1$ | $\Sigma M_2$ | K |
| 0,1 | 47,3 | 117,2 | 2,5 | 37,1 | 53,2 | 1,4 | 46,4 | 53,3 | 1,2 | 48,3 | 69,2 | 1,4 | 70,2 | 187,4 | 2,7 |
| 0,2 | 54,4 | 127,4 | 2,3 | 44,2 | 63,4 | 1,4 | 46,4 | 70,7 | 1,5 | 55,4 | 79,5 | 1,4 | 70,2 | 258,0 | 3,7 |
| 0,3 | 61,6 | 137,6 | 2,2 | 51,4 | 73,7 | 1,4 | 46,4 | 88,1 | 1,9 | 62,54 | 89,7 | 1,4 | 70,2 | 329,0 | 4,7 |

## 4 Discussion

Theoretical researches have shown that owing to numerous factors of the ice influences and conditions of interaction of the sheeting and it isn't possible to develop universal recommendations to use a certain new material type and design. When calculating it is necessary to take into account all ice influences, four types possible, on the basis of these long-term observations of ice foes formation around the alleged construction site. Theoretical researches have shown that the most adverse are influences from the thermal ice expansion and the frozen ice cover at a water level change.

Theoretical researches of mats standard types have shown that the most adverse are influences from the thermal expansion of ice and the frozen ice cover at a water level change. In this regard for mats to be reliable, an important factor of stability is ensuring collaboration of separate "pillows" of a mat. The collaboration is ensured by preserving woven covers of a mat, first of all the lower part and connecting elements between the top and lower fabric for the entire period of operation of the construction, and also by connecting contacting rolls according to the laying instruction. Stability calculations have to be carried out as for all massif of the mat laid on a slope.

CFGM can be used in ice conditions at observance of certain requirements.

## 5 Conclusions

The conducted researches have shown:

1. Modern «new» materials allow considerably to lower costs of bank protection constructions.

2. The most economical are designs of a batter type with use of UFPCM and CFGM. Vertical designs are applied in case of impossibility to use the batter ones.

The PVC tongue is recommended to be used with a vertical clearance of about two meters and sandy soil of the basis, the composite one – with a vertical clearance of about four meters.

3. For the majority of «new» materials researches of ice foes influence weren't conducted therefore, at the considered objects construction sites they are recommended to be used in the zone of a variable level higher than the ice level and ice drift.

4. For UFPCM and CFGM special researches are conducted, for BNTM we have created the pilot block where observations are currently carried out. The research has shown a possibility to use widely UFPCM in the conditions of ice foes influence. The conducted researches have shown a possibility to use them in conditions shift stability, capsizing stability and durability.

## References

1. M.I. Balzannikov, A.A. Mikhasek, Procedia Engineering **91,** 183-187 (2014)
2. M.G. Litvinova, Scientific Survey **13**, 70-74 (2016)
3. Andrey Mikhasek, Boris Ivanov MATEC Web of Conferences **105**, 03022 (2017)
4. A.A. Mikhasek, A.A. Smyvalov, Scientific Survey **14,** 102-109. (2015)
5. Y.E. Senitskij, A.A. Mikhasek, A.V. Kozlov, Scientific Survey **23,** 64-70 (2015)
6. A.A. Mikhasek Scientific Survey **14,** 67-72 (2016)
7. A.A. Mikhasek, M.V. Rodionov, Procedia Engineering **111,** 82-88 (2015)
8. Scientific Report for Contract № 72/12-N (Voronezh State University of Architecture and Civil Engineering, Voronezh, 2012)
9. A.A. Mikhasek, M.V. Rodionov, M.G. Litvinova, Scientific Survey **4,** 125-134 (2015)
10. Andrey Mikhasek, Maxim Rodionov , Vladimir Seliverstov, and Yury Senitsky, MATEC Web of Conferences **86** DOI: 10.1051/ matecconf/2016 86 8603002 (2016).

# Modification of Concrete Damaged Plasticity model. Part I: Modified plastic potential

*Aleksander* Szwed[1,*], and *Inez* Kamińska[1]

[1]Warsaw University of Technology, Al. Armii Ludowej 16, 00-637 Warszawa, Poland

**Abstract.** In order to extend the Concrete Damaged Plasticity model [1], a modification of flow rule is considered. A new plastic potential is proposed. It takes into account the difference in behavior of concrete under compressive and tensile loading. In the present paper the proposed function is examined and compared to the original potential. Influence of parameters of the potential choice on the shape of its surface is analysed. Formulation and preliminary validation of the modified model is shown in the subsequent paper (Part II).

## 1 Introduction

Experiments show that prediction of the response of concrete structures subjected to the loading exceeding plastic limit should take into account its elastic and plastic behaviour as well as degradation of elastic stiffness due to plastic deformation and initiation of cracks. Basic models for concrete and soil use Drucker-Prager type functions as both the yield condition and the plastic potential [1-7]. The experiments showing deviation of plastic strain rate tensor from the gradient of yield function are scarce, compare [8] for soils, however it is frequently assumed that the flow rule should be non-associative [5,6,8]. Isotropic damage described by one scalar parameter is often accepted as a rough approximation but also models taking into account anisotropic damage of material are developed [9].

The basic model for numerical computation implemented in Abaqus [1] was proposed by Lubliner [6] and further developed by Lee and Fenves [5]. It is formulated in the effective stress space, contains Lubliner's yield condition and the non-associative flow rule, as well as a scalar damage parameter. Material damage evolves along with the plastic strain growth; however the coupling between damage and plasticity is weak. The model has been continuously modified to better fit experimental data [7,10,11]. The main features taken into consideration by the researchers are the hardening rule and the failure criterion [10,11]. In the present study other refinement of this elastoplastic damage model for concrete is presented - an improved plastic potential is proposed.

The plastic potential introduced in this paper takes into account different behavior of concrete under the compressive and tensile loading. Two conic surfaces, according to the Lubliner's yield condition or the two-surface Drucker-Prager condition, devoted to the tension and compression zones respectively, are used as the envelope for definition of

* Corresponding author: a.szwed@il.pw.edu.pl

a new plastic potential. The proposed potential is smooth, so its gradient used in the flow rule is uniquely defined. Basic four material parameters involved in the definition of function can be calibrated on the basis of results of four experiments: the uniaxial tension and compression, the equal biaxial compression and the triaxial compression tests. Additional four free parameters used in the definition of potential can be calibrated to fit observed behavior of material. Including plasticity condition and other equations, the model requires input of twelve parameters and four material functions. Regarded modification of the existing CDP Abaqus model can improve performance of the model to fit a wider range of loading regimes observed in concrete structures. Plastic flow defined by the proposed function can be calibrated independently for the tension and compression zone of loading. Attempts of calibration of the CDP model have been made in the literature e.g. [12-14].

In Part I, the difference between the original Abaqus potential based on two free parameters and the proposed potential is shown, as well as typical angle of deviation of direction of the plastic flow from the normal to the yield surface. The influence of change of model parameters on the shape of potential's surface is investigated. Subsequent paper (Part II) deals with the model formulation and computations for the case of uniaxial loading.

## 2 Plastic potential

The Drucker-Prager type potentials are commonly used to cover observed performance of concrete structures, compare [1-7] among others. They characterize properly compressive behavior of the material though they fail to model its tensile properties. As an extension, the two-surface Drucker-Prager condition should be used to properly portray the compressive and tensile behavior of material. However, in this case discontinuity of the gradient of plastic potential at the joint of two surfaces is observed. The proposed potential is intended to overcome this inconvenience and to extend flexibility of calibration of the model. The function is based on the selected meridional sections of Lubliner's yield condition [6] determined separately for the tension ($\sigma_{\max} = \sigma_1 > 0$) and compression ($\sigma_1 \le 0$). It is smooth thus the flow rule is properly defined for all points of the stress space.

The envelope for the plastic potential (meridians of the Lubliner's condition) is defined as follows:

$$f(r,\xi) = \begin{cases} \xi + \alpha_1 r - \beta_1 = 0 & for \quad \xi_W \ge \xi \ge \xi_P, \\ \xi + \alpha_2 r - \beta_2 = 0 & for \quad \xi < \xi_P, \end{cases} \tag{1}$$

where:

$$\xi = \frac{1}{\sqrt{3}} \mathrm{tr}\boldsymbol{\sigma}, \quad r = \sqrt{\mathrm{tr}\mathbf{s}^2}, \tag{2}$$

$$\alpha_1 = \frac{3 + 2m_1\beta}{\sqrt{2}(3\alpha + \beta)}, \quad \beta_1 = \frac{\sqrt{3}(1-\alpha)}{3\alpha + \beta}\sigma_C, \quad \alpha_2 = \frac{3 + 2m_2\gamma}{\sqrt{2}(3\alpha + \gamma)}, \quad \beta_2 = \frac{\sqrt{3}(1-\alpha)}{3\alpha + \gamma}\sigma_C,$$

$$\alpha = \frac{\sigma_{BC} - \sigma_{C0}}{2\sigma_{BC} - \sigma_{C0}}, \quad \beta = (1-\alpha)\frac{\sigma_C}{\sigma_T} - (1+\alpha), \quad \gamma = -(1-\alpha)\frac{\sigma_{C0}}{\sigma_{TC}} - (2+\eta)\alpha + \eta - 1,$$

$$\xi_P = \frac{2\sqrt{3}(1-\alpha)(m_1\beta - m_2\gamma)}{3(1-2m_2\alpha)\gamma - 3(1-2m_1\alpha)\beta + 2(m_1 - m_2)\gamma\beta}\sigma_C, \quad \xi_W = \frac{\sqrt{3}(1-\alpha)}{3\alpha + \beta}\sigma_C = \beta_1 .$$

Notations $\sigma_{C0}$, $\sigma_{T0}$ and $\sigma_{BC}$ stand for the initial compressive yield stress, the tensile yield limit and the biaxial compressive yield stress respectively ( $\sigma_C$ and $\sigma_T$ are the current values), and $\eta\sigma_{TC}$ is the triaxial compressive yield limit in a test where the stress tensor has representation $\boldsymbol{\sigma} \to -\sigma_{CT}\, diag[1,1,\eta]$, compare [15]. Both $m_1 = \cos\theta_1$ and $m_2 = \cos\theta_2$ are parameters describing selected meridians of the Lubliner's condition. Angles $\theta_1$ and $\theta_2$ belong to the interval $\langle 0, \pi/3 \rangle$, so the parameters $m_1, m_2 \in \langle 0.5, 1.0 \rangle$. If both asymptotes are located on the tensile meridian then $m_1 = m_2 = 1.0$, and on the compressive meridian - $m_1 = m_2 = 0.5$. The proposed range for $m_1$ and $m_2$ parameters can be widened when the calibration of the model requires it.

The first formula in (1) is applicable for the tension zone while the second applies to the compressive zone.

To avoid discontinuity of the gradient and to properly portray both the tensile and compressive behavior of concrete the following plastic potential is proposed:

$$\Phi(\boldsymbol{\sigma}) = \xi - f_s(r), \tag{3}$$

where $f_s$ is a function of the stress deviator invariant $r$:

$$f_s(r) = \frac{1}{2}\left( f_{1n} + f_{2n} - \sqrt{(m_4 A)^2 + (f_{1n} - f_{2n})^2} \right), \tag{4}$$

$$f_{1n}(r) = \beta_1 - \sqrt{A^2 + (\alpha_1 r)^2}, \quad f_{2n}(r) = \beta_2 - \sqrt{A^2 + (\alpha_2 r)^2},$$

$$A = \frac{2(1-m_3)\beta_1(\beta_2 - m_3\beta_1)}{\beta_1 + \beta_2 - 2m_3\beta_1 + \sqrt{(\beta_2 - \beta_1)^2 + m_4^2 \beta_1 (\beta_2 - m_3\beta_1)(1-m_3)}}.$$

$m_3 \in (0,1)$ is a parameter characterizing location of the vertex of the plastic potential surface. It describes the closeness of the apex to the point where envelope $(1)_1$ intersects axis $\xi$, compare Fig. 1(a) and Fig. 2(c). Parameter $m_4 > 0$ is chosen arbitrarily to properly model the vertex region and the transition zone from meridian $\theta_1$ to $\theta_2$ associated with the branches of envelope (1). It influences strongly the curvature of surface (3) in the region near the potential apex - the derivative changes rapidly for great values of $m_4$, while for smaller values the change is gradual, see Fig. 1(b) and Fig. 2(d).

The asymptote for meridional section of (3) is the compressive cone $(1)_2$. As it is depicted on Fig.1(b), the potential's derivative with respect to $r$ is positive for whole domain, which results in the curve showing its meridional section being a growing function of $r$. In the vertex point, $\xi = m_3\xi_W$, the derivative equals 0, so then the potential is smooth for $r = 0$.

It is worth mentioning that during plastic hardening (or softening) the values of $\beta$, $A$, $\alpha_1$, $\beta_1$ and $\beta_2$ change as the development of plastic limits occurs, compare equations (2). The values of $\alpha$, $\gamma$ and $\alpha_2$ remain constant.

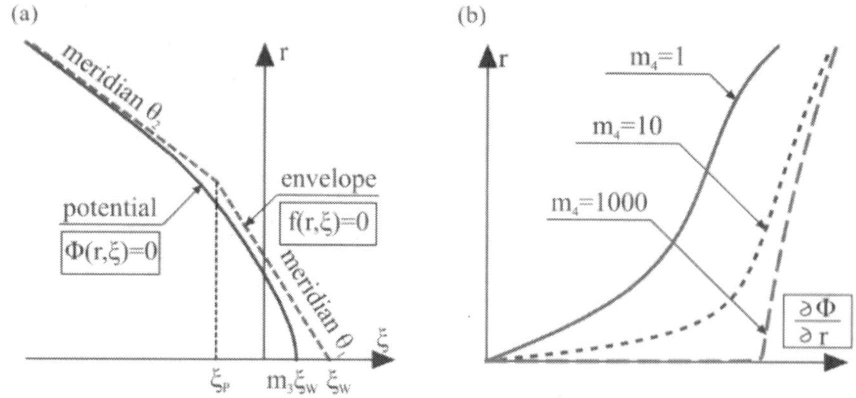


**Fig. 1.** The proposed potential: (a) meridional section and envelope, (b) derivative of the potential due to variable $r$ for various values of parameter $m_4$.

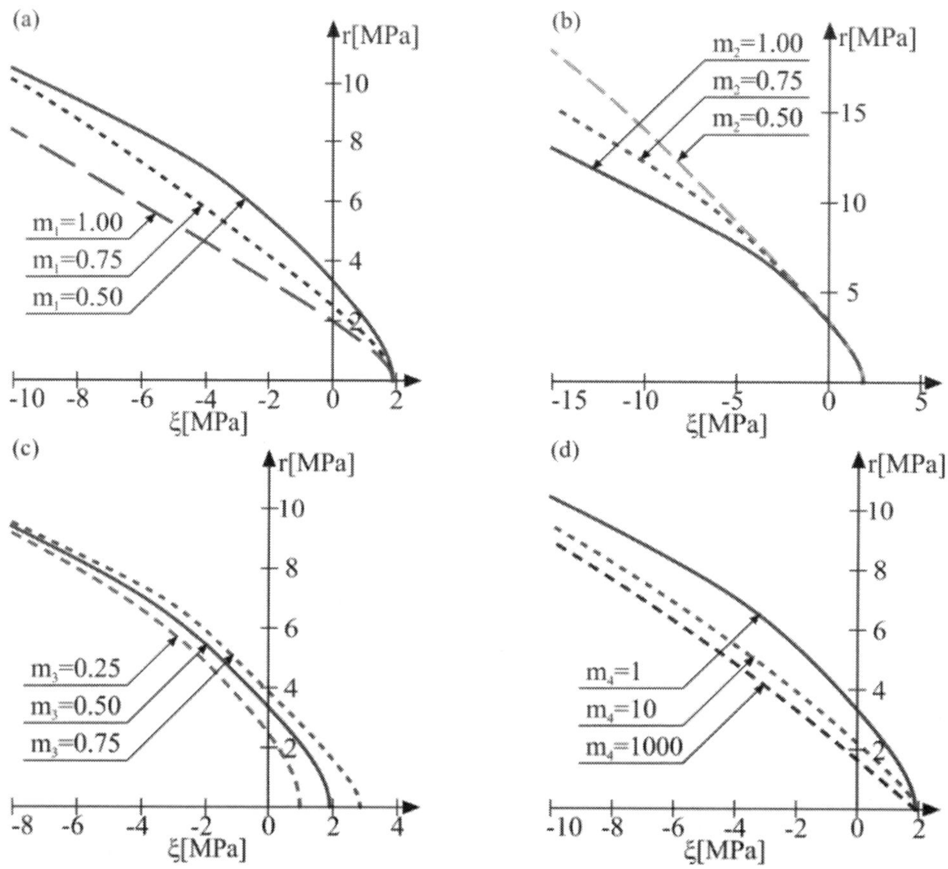


**Fig. 2.** Meridional section of the plastic potential for several values of: (a) $m_1$, (b) $m_2$, (c) $m_3$, (d) $m_4$; $\sigma_{C0} = 20MPa$, $\sigma_{T0} = 2MPa$, $\sigma_{BC} = 23MPa$, $\eta\sigma_{TC} = 123MPa$.

In comparison to the original plastic potential [1], the proposed surface (3) offers possibility of distinguishing the compressive and tensile stress states, see Fig. 3(a). Gradient of the primary function has approximately the same direction for almost all $r$, while the modified plastic potential changes the slope when passing from the compressive to the tensile states, as it is shown in Fig. 3(b). Potential (3) allows to adjust the direction of plastic flow through change of the parameters $m_1$ and $m_2$ responsible for the slope of meridians, see Fig. 2(a),(b) and Fig. 3(b). For the plastic potential to be convex one needs to ensure $\alpha_1 \le \alpha_2$, which imposes a condition on values of $m_1$ and $m_2$, see equations (2).

The proposed potential (3) imitates the yield condition, however its normal differs from the normal of the yield surface. The deviation of gradient of the plastic potential from the normal to the yield surface is approximately constant along chosen meridians of the Lubliner's condition, see Fig. 4. Sudden changes appear in the zone of transition between the compressive and tensile states (i.e. passing through $\sigma_{\max} = \sigma_1 = 0$ plane), and close to the vertex zone.

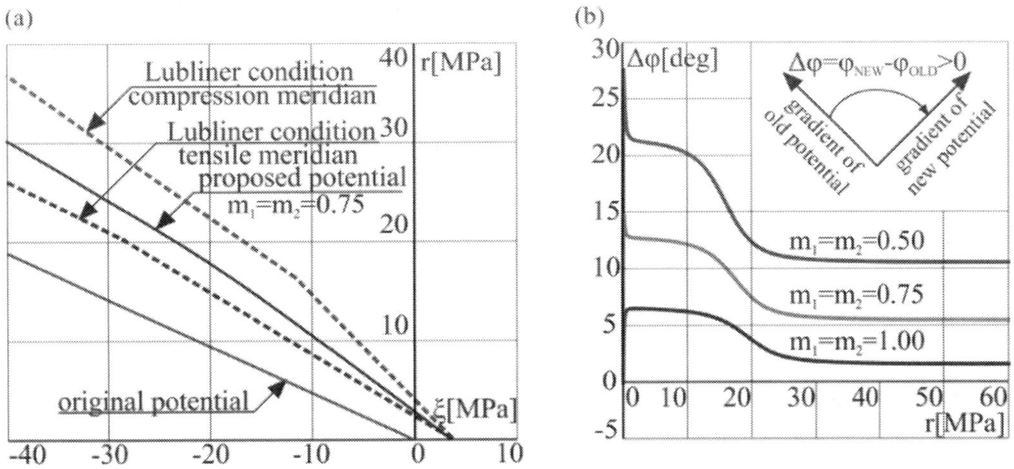


**Fig. 3** The proposed potential versus original potential for: $\sigma_{C0} = 20MPa$, $\sigma_{T0} = 2MPa$, $\sigma_{BC} = 23MPa$, $\eta\sigma_{TC} = 123MPa$, $m_3 = 0.9$, and $\phi = 45^0$, $\beta_M = 0.1$ [1]; (a) meridional section, (b) the difference of directions (angles) of the potential's gradients in relation to $r$ axis.

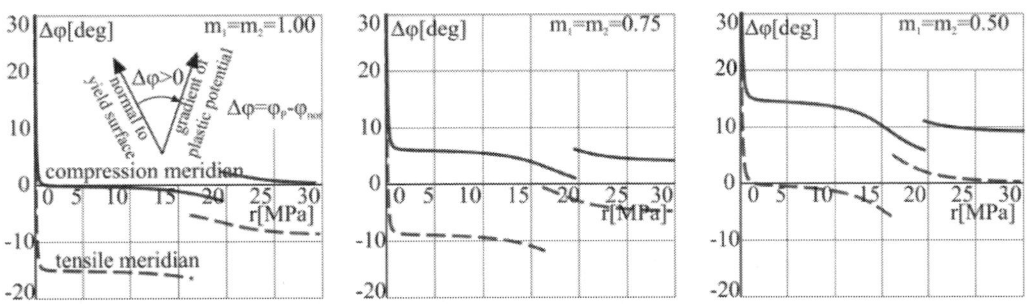


**Fig. 4** Angle between normal to the plastic surface and gradient of the proposed potential for various parameters $m_1$ and $m_2$ - meridional section for: $\sigma_{C0} = 20MPa$, $\sigma_{T0} = 2MPa$, $\sigma_{BC} = 23MPa$, $\eta\sigma_{TC} = 123MPa$, $m_3 = 0.9$, $m_4 = 10$.

## 3 Conclusions

A new plastic potential for the refined elastoplastic model for damaged material allows for distinguishing the tensile and the compressive stress states which is of great importance when describing concrete, rock and soil behavior. Free parameters $m_1$ and $m_2$ permit of adjusting the potential to the chosen slopes of meridians of the Lubliner's yield condition. The deviation of direction of the plastic flow gradient from the yield surface gradient can be easily handled by selecting adequate coefficients. Parameter $m_3$ governs location of the vertex of plastic potential, while parameter $m_4$ models smoothness of the potential in a zone surrounding the apex, as well as the smoothness of zone of transition from the compressive to the tensile states. In original Abaqus CDP model only two free parameters for the slope (dilation angle) and smoothness at the vertex (eccentricity) are available.

Parameters of plastic potential are to be calibrated on basis of four material tests: the uniaxial tension and compression, biaxial uniform compression and triaxial compression. Omitting this assumption the new potential can be reduced to the original one introducing:

$$\alpha_1 = \alpha_2 = \frac{3}{\sqrt{2}\, tg\, \psi}, \quad A = \sqrt{3}\, \beta_M\, \sigma_{T0}, \quad \beta_1 = \beta_2 = \frac{1}{2} m_4 A, \tag{5}$$

where $\beta_M$ is eccentricity and $\psi$ - dilation angle defined for the original potential [1].

A possible improvement of the function concerns the deviatoric section of potential surface - it is reasonable to introduce the dependency of plastic potential of the third stress invariant [15-17]. Dependence of plastic potential on the third invariant can enhance flexibility of the model and effectively control the non-associative flow in deviatoric cross-section. Extensive study on defining the shape of deviatoric cross-section in plasticity surfaces is presented in [17] and other literature.

## References

1. *Abaqus Theory Manual*
2. R. Eid, P. Pautre, Eng. Struct. **29**, 3301 (2007)
3. J. Jiang, Y. Wu, Int. J. Solids Struct. **49**, 445 (2012)
4. W. B. Kratzig, R. Polling, Comput. Struct. **82**, 1201 (2004)
5. J. Lee, G. Fenves, Earthquake Eng. Struct. Dyn. **27**, 937 (1998)
6. J. Lubliner, J. Oliver, S. Oller, E. Onate, Int. J. Solids Struct. **25(3)**, 299 (1989)
7. O. Omidi, V. Lofti, Earthquake Eng. Struct. Dyn. **42**, 2129 (2013)
8. J. Lee, G. Fenves, Earthquake Eng. Struct. Dyn. **27**, 937 (1998)
9. S. Murakami, *Continuum Damage Mechanics* (Springer, 2012)
10. N. J. Hany, E. G. Hantouche, M. H. Harajli, Eng. Struct. **125**, 1 (2016)
11. Z. Tao, Z. Wang, Q. Yu, J. Constr. Steel Res. **89**, 121 (2013)
12. I. Kamińska, A. Szwed, *Concrete and concrete structure – experiments*, E. Szmigiera, P. Łukowski, S. Jemioło (eds.), 93 (OWPW, Warsaw, 2015) (in Polish)
13. P. Kmiecik, M. Kamiński, Archives of Civil and Mechanical Engineering **XI(3)**, 623 (2011)
14. M. Szczecina, A. Winnicki, ASEM15 (Korea, 2015)
15. N. S. Ottosen, ASCE J. Eng. Mech. Div. **103**, 527 (1977)
16. A. Szwed, *Failure criteria and constitutive equations of material with degradation*, Ph. D. Dissertation (Warsaw, 2000) (in Polish)
17. A. Szwed, *Deformation and Stregth of Materials and Structures*, S. Jemioło, A. Szwed (eds.) **3**, 51 (Warsaw, 2013) (in Polish)

MATEC Web of Conferences 117, 00161 (2017) DOI: 10.1051/matecconf/201711700161
*XXVI R-S-P Seminar 2017, Theoretical Foundation of Civil Engineering*

# Modification of Concrete Damaged Plasticity model. Part II: Formulation and numerical tests

*Inez* Kamińska[1], and *Aleksander* Szwed[1,*]

[1]Warsaw University of Technology, Al. Armii Ludowej 16, 00-637 Warszawa, Poland

**Abstract.** A refined model for elastoplastic damaged material is formulated based on the plastic potential introduced in Part I [1]. Considered model is an extension of Concrete Damaged Plasticity material implemented in Abaqus [2]. In the paper the stiffness tensor for elastoplastic damaged behaviour is derived. In order to validate the model, computations for the uniaxial tests are performed. Response of the model for various cases of parameter's choice is shown and compared to the response of the CDP model.

## 1 Introduction

An alteration to the Abaqus' Concrete Damaged Plasticity model is made in attempt to improve its flexibility. The new potential shown in Part I [1] is introduced to the equations of elasticity and plasticity theory in order to derive the stiffness tensor for elastoplastic damaged behavior of material. As mentioned in Part I, the extended model requires input of twelve parameters and four material functions.

To confirm validity of this material model, its consistency with basic experiments has to be proven. The objective of the second part of the present paper is to show and analyze numerical results of the uniaxial loading test for the proposed model concerning improved potential. In order to achieve this goal, a simple forward-Euler algorithm was created using Mathematica [3]. The influence of free parameters $m_1$, $m_2$, $m_3$ and $m_4$ on the uniaxial test results is considered. The first two parameters indicate the meridians of Lubliner's condition used as asymptotes of the potential, the third one - location of the potential's apex and the last one - the curvature in the vertex region and the zone of transition between potential's asymptotes (near $\sigma_1 = 0$). The results of basic tests are compared for various sets of parameters' values.

As it has been noted in Part I, the difference between the new and original Abaqus potential [2] allows more flexibility when it comes to reflecting material properties. Examples of numerical results are shown in comparison to those acquired using the CDP model.

* Corresponding author: a.szwed@il.pw.edu.pl

## 2 Formulation of the model

Considered elastoplastic damage model for concrete is based on the Lubliner's two-surface failure criterion with isotropic damage described by one scalar damage variable. Plasticity and damage surfaces are the same. The non-associative flow rule is proposed, which is based on the plastic potential described in Part I [1]. Constitutive equations are formulated in the effective stress space [4]. The basic equations and assumptions of the model are as follows.

- The relation between nominal stress tensor and effective stress tensor is:

$$\hat{\boldsymbol{\sigma}} = \frac{\boldsymbol{\sigma}}{1-d}, \tag{1}$$

where $d$ is a damage parameter.

- Additive decomposition of the tensor of strain rate into elastic ($\dot{\boldsymbol{\varepsilon}}_E$) and plastic ($\dot{\boldsymbol{\varepsilon}}_P$) parts and the Hooke's constitutive law for elastic behavior hold:

$$\dot{\boldsymbol{\varepsilon}} = \dot{\boldsymbol{\varepsilon}}_E + \dot{\boldsymbol{\varepsilon}}_P, \qquad \hat{\boldsymbol{\sigma}} = \mathbf{C} \cdot \boldsymbol{\varepsilon}_E, \tag{2}$$

where $\mathbf{C} = K\,\mathbf{I}\otimes\mathbf{I} + 2G\left(\mathbf{1} - \tfrac{1}{3}\mathbf{I}\otimes\mathbf{I}\right)$ is the stiffness tensor for isotropic elastic undamaged material.

- The dissipation during plastic flow is positive:

$$D = \hat{\boldsymbol{\sigma}} \cdot \dot{\boldsymbol{\varepsilon}}_P > 0 \text{ and } D = \boldsymbol{\sigma} \cdot \dot{\boldsymbol{\varepsilon}}_P > 0 \text{ when } \dot{\boldsymbol{\varepsilon}}_P \neq \mathbf{0}. \tag{3}$$

- The yield criterion is Lubliner's condition [5,6]:

$$f\left(\hat{\boldsymbol{\sigma}}, \varepsilon_{PL,C}\right) = \begin{cases} \dfrac{1}{1-\alpha}\left(\sqrt{3}\,\alpha\,\hat{\xi} + \sqrt{\dfrac{3}{2}}\,\hat{r} + \gamma\,\hat{\sigma}_1\right) - \hat{\sigma}_C\left(\varepsilon_{PL,C}\right) = 0 & \textit{when } \hat{\sigma}_1 \le 0 \\ \dfrac{1}{1-\alpha}\left(\sqrt{3}\,\alpha\,\hat{\xi} + \sqrt{\dfrac{3}{2}}\,\hat{r} + \beta\left(\varepsilon_{PL,C}, \varepsilon_{PL,T}\right)\hat{\sigma}_1\right) - \hat{\sigma}_C\left(\varepsilon_{PL,C}\right) = 0 & \textit{when } \hat{\sigma}_1 > 0 \end{cases} \tag{4}$$

where:

$$\hat{\sigma}_1 = \frac{1}{\sqrt{3}}\hat{\xi} + \sqrt{\frac{2}{3}}\,\hat{r}\cos\hat{\theta}, \quad \hat{\theta} = \frac{1}{3}\arccos\left(\frac{\sqrt{6}\,\mathrm{tr}\,\hat{\mathbf{s}}^3}{\sqrt{\mathrm{tr}^3\hat{\mathbf{s}}^2}}\right). \tag{5}$$

- The plastic potential $\Phi$ is described in Part I. The non-associative flow rule and the consistency condition state:

$$\dot{\boldsymbol{\varepsilon}}_P = \lambda \frac{\partial \Phi(\hat{\boldsymbol{\sigma}})}{\partial \hat{\boldsymbol{\sigma}}} = \lambda \mathbf{p}, \qquad \dot{f}\left(\hat{\boldsymbol{\sigma}}, \varepsilon_{PL,C}\right) = 0. \tag{6}$$

- Plastic hardening is prescribed by the following formulae:

$$\hat{\sigma}_C\left(\varepsilon_{C,PL}\right) = \sigma_{C0} + E_P\,\varepsilon_{C,IN}\left(\varepsilon_{C,PL}\right), \quad \hat{\sigma}_T\left(\varepsilon_{T,PL}\right) = \sigma_{T0} + E_P\,\varepsilon_{T,IN}\left(\varepsilon_{T,PL}\right), \tag{7}$$

where $E_P$ is a hardening modulus and $\varepsilon_{C,IN}$, $\varepsilon_{C,PL}$, $\varepsilon_{T,IN}$ and $\varepsilon_{T,PL}$ are equivalent inelastic and plastic compressive and tensile strains respectively.

- Relationships considering equivalent strains are:

$$\dot{\varepsilon}_{C,PL} = (r^*-1)\,\dot{\varepsilon}_{P,MIN}\,,\quad \dot{\varepsilon}_{T,PL} = r^*\,\dot{\varepsilon}_{P,MAX}\,,\quad r^* = \frac{1}{2}\frac{\hat{\sigma}_1+\hat{\sigma}_2+\hat{\sigma}_3}{|\hat{\sigma}_1|+|\hat{\sigma}_2|+|\hat{\sigma}_3|}\,, \tag{8}$$

$$\varepsilon_{T,IN} = \varepsilon_{T,PL} + \frac{d_T}{1-d_T}\frac{\sigma_T}{E_0}\,,\quad \varepsilon_{C,IN} = \varepsilon_{C,PL} + \frac{d_C}{1-d_C}\frac{\sigma_C}{E_0}\,,$$

where $\varepsilon_{T,IN}$ and $\varepsilon_{T,PL}$ are equivalent inelastic and plastic tensile strains, $\varepsilon_{C,IN}$ and $\varepsilon_{C,PL}$ - equivalent inelastic and plastic compressive strains respectively, $\varepsilon_{P,MAX}$ and $\varepsilon_{P,MIN}$ are the maximum and minimum principal plastic strains, $d_T$ and $d_C$ are tensile and compressive damage parameters and $E_0$ is an initial (undamaged) Young modulus.

- The formulas for damage parameters are:

$$d = 1-(1-s_T d_C)(1-s_C d_T)\,,\quad d_T = H_T\varepsilon_{T,IN}\,,\quad d_C = H_C\varepsilon_{C,IN}\,, \tag{9}$$

where $H_T$ and $H_C$ are material parameters, and $s_T$ and $s_C$ reflect both material properties and the current effective stress state. See reference [2] for more details.

The Lubliner's yield criterion (4) contains four material constants: values of the compressive yield limit, the compressive biaxial yield limit, the compressive triaxial yield limit and the tensile yield limit of material. To define the plastic potential one needs additionally to determine constants $m_1$, $m_2$, $m_3$ and $m_4$. The other equations contain four more material constants and four material functions concerning hardening and evolution of damage parameters. Full description of the proposed model of elastoplastic material with isotropic damage requires determining twelve parameters (four of which are chosen arbitrarily) and four material functions, compare Tab. 1 and [7,8,9,10].

**Table 1.** Material constants and functions used in the proposed model.

| Equations | Material parameters/functions | Number of const. | Number of fun. | Test |
|---|---|---|---|---|
| Hooke's law | $E$ and $\nu$ (or $K$ and $G$) | 2 | 0 | uniaxial compression |
| Lubliner's yield condition | $\sigma_{T0}$, $\sigma_{C0}$, $\sigma_{BC}$, $\eta\sigma_{TC}$ | 4 | (2) | uniaxial tension and compression, biaxial compression, triaxial compression |
| Hardening | $\sigma_C(\varepsilon_{C,IN})$, $\sigma_T(\varepsilon_{T,IN})$ | 0 | 2 | uniaxial tension and compression |
| Gradient of plastic potential | $\sigma_{T0}$, $\sigma_{C0}$, $\sigma_{BC}$, $\eta\sigma_{TC}$ | 4 | 0 | uniaxial tension and compression, biaxial compression, triaxial compression |
| | $m_1$, $m_2$, $m_3$, $m_4$ | 4 | 0 | slopes and smoothness - calibration |
| Damage | $d_C(\varepsilon_{C,IN})$, $d_T(\varepsilon_{T,IN})$, $\omega_T$, $\omega_C$ | 2 | 2 | uniaxial cyclic compression-tension |

## 3 Stiffness tensor

The stiffness tensor for the introduced material model ( $\mathbf{C}^{EP}$ ) defines the relation between effective stress rate tensor and strain rate tensor:

$$\dot{\hat{\boldsymbol{\sigma}}} = \mathbf{C}^{EP} : \dot{\boldsymbol{\varepsilon}}\,, \tag{10}$$

To derive a formula for the stiffness tensor one has to investigate several cases. It is essential to consider separately two Lubliner's surfaces (for compression $\hat{\sigma}_1 \le 0$, and tension $\hat{\sigma}_1 > 0$ when $0 < \theta < \pi/3$), as well as special cases for the tensile and compression meridians only.

Taking into consideration Hooke's law with the additive split of strain rate tensor (2), the flow rule and consistency condition (6) for $0 < \theta < \pi/3$, one arrives at:

$$\mathbf{C}^{EP} = 2G\mathbf{1} + \left\{\left[1 - \frac{\sqrt{3}\,u(\hat{\theta})}{a_0(1-\alpha)}\right]K - \frac{2}{3}G\right\}\mathbf{I}\otimes\mathbf{I} - \frac{1}{a_0(1-\alpha)}\left[2\sqrt{3}\,KG\left(\frac{w(\hat{\theta})}{\hat{r}}\mathbf{I}\otimes\hat{\mathbf{s}} + \frac{2v(\hat{\theta})}{\hat{r}^2}\mathbf{I}\otimes\hat{\mathbf{s}}^2\right)\right.$$
$$\left. + 2Gz(\hat{r})\left(\frac{u(\hat{\theta})}{\hat{r}}\hat{\mathbf{s}}\otimes\mathbf{I} + \frac{2G\,w(\hat{\theta})}{\hat{r}^2}\hat{\mathbf{s}}\otimes\hat{\mathbf{s}} + \frac{4G\,v(\hat{\theta})}{\hat{r}^3}\hat{\mathbf{s}}\otimes\hat{\mathbf{s}}^2\right)\right], \tag{11}$$

where:

$$\mu = \begin{cases} \beta & when\ \hat{\sigma}_1 > 0 \\ \gamma & when\ \hat{\sigma}_1 \le 0 \end{cases},$$

$$v(\hat{\theta}) = \frac{\mu \sin\hat{\theta}}{\sin 3\hat{\theta}},\quad u(\hat{\theta}) = (3\alpha+\mu)K - \frac{4}{3}G\,v(\hat{\theta}),\quad w(\hat{\theta}) = \sqrt{\frac{3}{2}} + \sqrt{\frac{2}{3}}\,\frac{\mu\sin 2\hat{\theta}}{\sin 3\hat{\theta}},$$

$$z(\hat{r}) = -\frac{\partial\Phi}{\partial\hat{r}} = \hat{r}\left(\frac{\alpha_1^2(f_s - f_{2n})}{f_{1n} - \beta_1} + \frac{\alpha_2^2(f_s - f_{1n})}{f_{2n} - \beta_2}\right)\left(f_{1n} + f_{2n} - 2f_s\right)^{-1},$$

$$a_0 = \begin{cases} y(\hat{r},\hat{\theta}) - (1-r^*)E_P p_{\min} & when\ \ \hat{\sigma}_1 \le 0 \\ y(\hat{r},\hat{\theta}) - (1-r^*)\left(1 - \dfrac{\hat{\sigma}_1}{\hat{\sigma}_T}\right)E_P p_{\min} + r^* E_P p_{\max}\dfrac{\hat{\sigma}_1\hat{\sigma}_C}{\hat{\sigma}_T^2} & when\ \ \hat{\sigma}_1 > 0 \end{cases},$$

$$y(\hat{r},\hat{\theta}) = \frac{1}{1-\alpha}\left[\sqrt{3}(3\alpha+\mu)K + 2G z(\hat{r})\left(w(\hat{\theta}) + \sqrt{\frac{2}{3}}\,v(\hat{\theta})\cos\hat{\theta}\right)\right].$$

$f_{1n}$, $f_{2n}$ and $f_s$ are functions of $\hat{r}$ [1] and $p_{\min}$ is the minimum eigenvalue of gradient $\mathbf{p}$. Plastic multiplier is given by the equation:

$$\lambda = \frac{1}{a_0(1-\alpha)}\left\{\left[(3\alpha+\mu)K - \frac{4}{3}G\,v(\hat{\theta})\right]tr\,\dot{\boldsymbol{\varepsilon}} + \frac{2G\,w(\hat{\theta})}{\hat{r}}\hat{\mathbf{s}} : \dot{\boldsymbol{\varepsilon}} + \frac{4G\,v(\hat{\theta})}{\hat{r}^2}\dot{\boldsymbol{\varepsilon}} : \hat{\mathbf{s}}^2\right\}. \tag{12}$$

The stiffness tensor (11) and the plastic multiplier (12) for the tensile meridian can be determined by finding limits of expressions (11) and (12) for $\theta \to 0$. Situation is more complex for the compression meridian, where the Lubliner's yield condition does not have uniquely defined gradient and one has to assume its form, for example as the Drucker-Prager potential gradient or alternatively as the average of gradients of two adjacent patches of the yield surface.

Supposing weak coupling between plasticity and damage:

$$\dot{\boldsymbol{\sigma}} = (1-d)\dot{\bar{\boldsymbol{\sigma}}}, \qquad \text{which entails} \qquad \dot{\boldsymbol{\sigma}} = (1-d)\mathbf{C}^{EP}[\dot{\boldsymbol{\varepsilon}} = \mathbf{C}^{EPD}[\dot{\boldsymbol{\varepsilon}}. \tag{13}$$

## 4 Uniaxial test

Uniaxial test is a basic experiment determining properties of concrete. It is frequently used to adjust values of the material models' parameters. During the primary stage of loading material is elastic then plastic deformation emerges. Along with plasticity, damage occurs which is observed as the degradation of initial elastic moduli. As the loading reaches a certain point, decrease of volume appears.

In the computations a cyclic change of normal component of stress tensor $\sigma_{XX}$ is assumed. Applied load program is shown on Fig. 1(a). The parameters of the model are as follows: $E = 20GPa$ and $\nu = 0.2$, which yields $K = 11GPa$ and $G = 8.3GPa$ ; $\sigma_{C0} = 20MPa$, $\sigma_{T0} = 2MPa$, $\sigma_{BC} = 23.2MPa$ and $\sigma_{TC} = 25MPa$ with $\eta = 4.91$, which result in $\alpha = 0.121$ and $\gamma = 2.370$. $\beta$ is subjected to changes during the loading process with initial value of $\beta = 7.667$. Parameters $m_1 = 1.0$ and $m_2 = 0.5$, so the asymptotes for the plastic potential are associated to the tensile and compression meridian, accordingly. $m_3 = 0.5$ and $m_4 = 1000$, which yields the initial values of $\alpha_1 = 1.614$, $\beta_1 = 3.791MPa$ and $\beta_2 = 11.139MPa$, as well as $\alpha_2 = 1.389$ and $A = 8.349 \cdot 10^{-3} MPa$. The modulus of the assumed linear hardening is $E_P = 5GPa$ and the values of parameters describing damage are $H_C = 120$, $H_T = 120$, $\omega_C = 0.1$ and $\omega_T = 1$ [2]. For the original CDP model $\beta_M = 0.1$, $\phi = 30^0$ is assumed.

Obtained results of the numerical computations show consistency with the test results. For decreasing $\sigma_{XX}$, $\varepsilon_{XX}$ declines and the increase of $\varepsilon_{YY} = \varepsilon_{ZZ}$ is observed (Fig. 1(b),(c)) which is the outcome of the assumed constitutive law. The development of irreversible deformation begins in the first cycle of loading. Plastic components of the strain tensor become to dominate over elastic parts as the number of cycles grows, see Fig. 1(b). The relative volume (volumetric strain) grows for the initial stages of analysis but then it decreases (Fig. 1(d)).

The damage parameter $d$ grows during the plastic flow, see Fig. 1(e). Its development is responsible for the change of initial Young modulus, which is depicted on Fig. 1(c). $d$ reaches the value of 0.25, which is usually near the point of complete destruction of material.

Plastic dissipation during the process is controlled to ensure thermodynamic consistency of the model. In the presented analysis dissipation is positive and increases as the yielding advances, see Fig. 1(f).

The difference between the original and the new potential and influence of parameters $m_1$, $m_2$, $m_3$ and $m_4$ of the proposed potential are analyzed for monotonic uniaxial compression and tension tests, where $\sigma_{XX}$ is imposed as shown on Fig. 2.

For both tests, regardless of $m_i$ selection, the $\varepsilon_{XX}$ component of strain tensor remains unchanged. It also matches the result for the CDP model. Furthermore, the modification of potential does not seem to influence damage parameter, which is only slightly different between models and does not vary for distinct values of $m_i$.

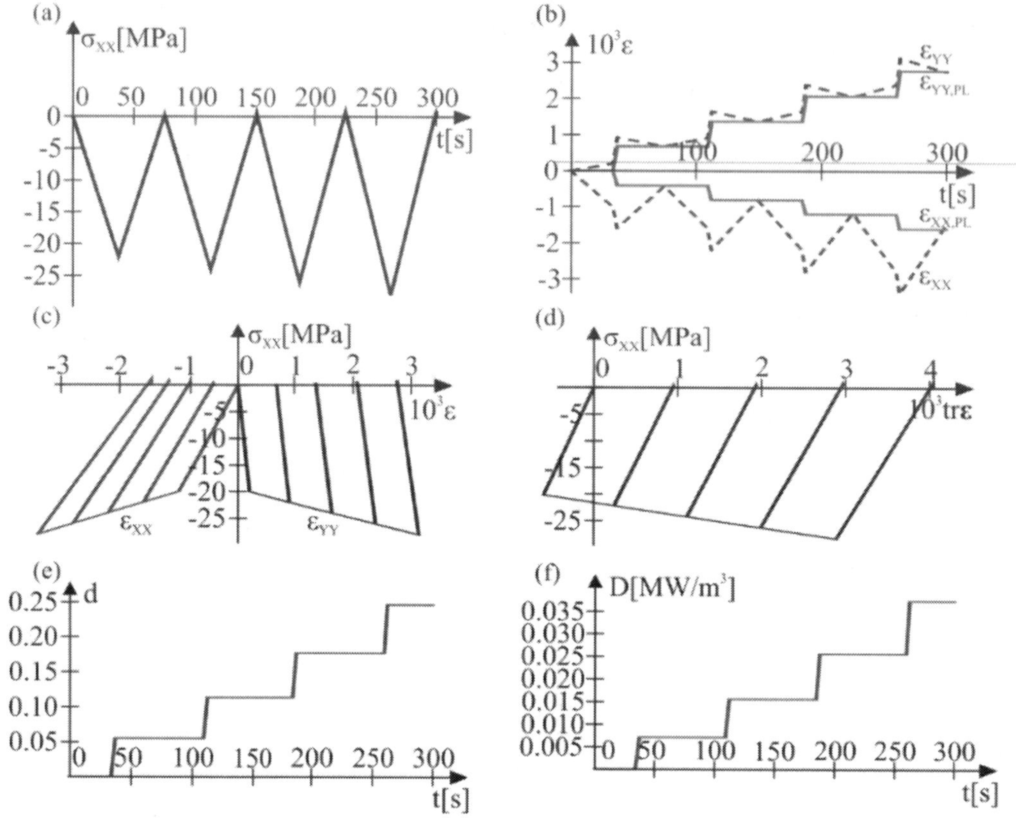


**Fig. 1.** Uniaxial cyclic test: (a) loading, (b) components of plastic and total strain tensor versus time, (c) components of strain tensor versus $\sigma_{XX}$, (d) volumetric strain versus $\sigma_{XX}$, (e) evolution of damage parameter, (f) specific dissipation.

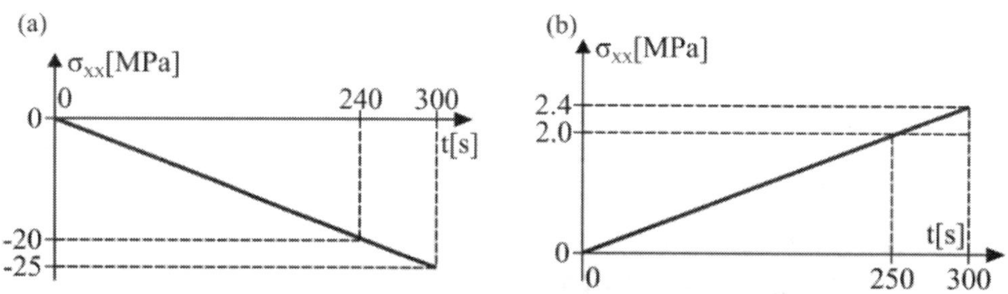


**Fig. 2.** Loading program in monotonic uniaxial tests: (a) compression, (b) tension.

For uniaxial compression test, the difference between original potential and new potential for variuos values of $m_1$ and $m_2$ is depicted on Fig. 3. An essential dissimilarity is observed in values of the lateral strain $\varepsilon_{YY} = \varepsilon_{ZZ}$, which implicates the difference of relative volume ($\varepsilon_V = \mathrm{tr}\boldsymbol{\varepsilon}$). As it is shown on Fig. 2(a), incerease of $m_1$ implicates growth of $\varepsilon_V$, while increase of $m_2$ causes $\varepsilon_V$ to decrease (Fig. 2(b)). Parameters $m_1$ and $m_2$ govern the slope of the plastic potential's asymptotes, so their great impact on computation results is apparent. However, $m_2$ affects plastic strain level stronger than $m_1$, because it controlls the compressive surface, where the considered test is located. Plastic lateral strain (or $\varepsilon_V$) for the CDP model is appreciably smaller than for the modified model.

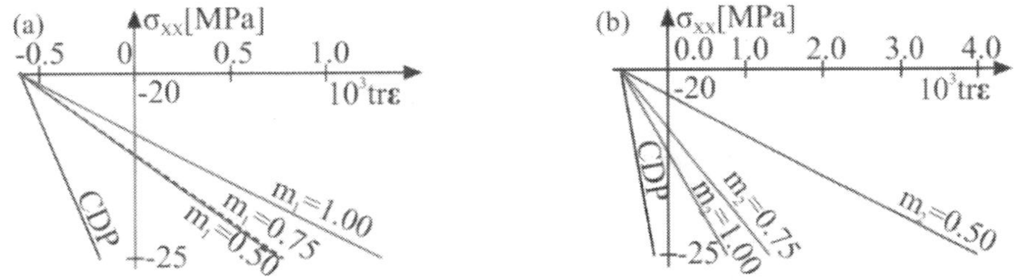


**Fig. 3.** Influence of values of parameters: (a) $m_1$ and (b) $m_2$ on relative volume ($\varepsilon_V = \mathrm{tr}\boldsymbol{\varepsilon}$) in the uniaxial compression test after crossing yield limit for: $\sigma_{C0} = 20MPa$, $\sigma_{T0} = 2MPa$, $\sigma_{BC} = 23MPa$, $\eta\sigma_{TC} = 123MPa$, $m_1 = 0.5$, $m_2 = 1.0$, $m_3 = 0.5$, $m_4 = 1$.

For the uniaxial compression test parameters $m_3$ and $m_4$ do not have any noticeable impact on the obtained results. The lack of influence is easily explicable, for the parameters control mainly the region near the potential's apex, whereas the test is located on compression cone.

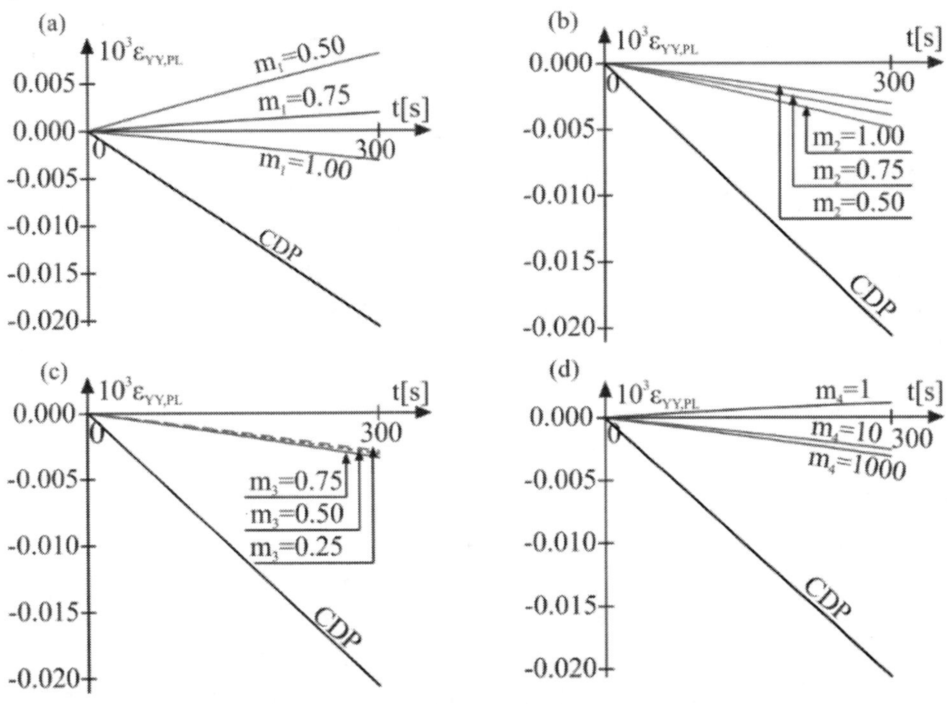


**Fig. 4.** Influence of selection of: (a) $m_1$, (b) $m_2$, (c) $m_3$ and (d) $m_4$ on the lateral components of plastic strain ($\varepsilon_{YY,PL} = \varepsilon_{ZZ,PL}$) in the uniaxial tension test; $\sigma_{C0} = 20MPa$, $\sigma_{T0} = 2MPa$, $\sigma_{BC} = 23MPa$, $\eta\sigma_{TC} = 123MPa$, $m_1 = 1.0$, $m_2 = 0.5$, $m_3 = 0.5$, $m_4 = 1000$.

The results of uniaxial tension test are depicted on Fig. 4. It is shown that they depend on $m_1$, $m_2$ and $m_4$, while the influence of $m_3$ is negligible, compare Fig. 4(c). For lower values of $m_1$ (Fig. 4(a)) and low values of $m_4$ (Fig. 4(d)) negative lateral component of plastic stress is obtained which implicates that one should choose the tensile meridian to calibrate tension surface of potential ( $m_1 = 1$ ) and to select the potential meridians to be as close as possible to asymptotes (large values of $m_4$ ), compare Part I [1]. With the increase of $m_2$ lateral plastic strain becomes smaller, but the influence of $m_2$ on the result is appreaciably smaller than for $m_1$ and $m_4$ . $\varepsilon_V$ grows during the loading process. The difference between the new and original potential is noticeable, regardless of allowable change of $m_i$ parameters.

The considerable differences between curves for variuos values of parameters enable to calibrate $m_1$ and $m_4$ on basis of the uniaxial tension test and $m_2$ on grounds of the uniaxial compression experiment.

Assumed values of the material parameters are chosen arbitrarily to depict typical properties of concrete. To reflect properties of particular type of concrete and prove quantitative compatibility of the model it is necessary to propose a proper calibration procedure.

Strain-stress curves for concrete subjected to uniaxial load include decreasing branch not shown in the simulation. It is possible to model material behavior in this area assuming an appropriate law of hardening (softening) instead of (7).

## 5 Conclusions

The results of a simple simulation of the uniaxial cyclic test prove that the proposed model properly reflects basic features of material behavior, although it is necessary to perform more complex numerical computations to ensure its qualitative and quantitative compatibility with the observed results of experiments.

Another important issue to be addressed is the calibration of the free parameters included in the definition of potential. A preliminary analysis shows that the uniaxial tests results are affected mainly by $m_1$ , $m_2$ and $m_4$ . As for $m_3$ , another test should be performed to determine its advisable value, preferebly located in the triaxial tension zone.

Extended flexibility of the presented potential allows for more accurate calibration of the material response. This feature is desirable when considering structures where material failure can occur both by tension and compression in some regions. Original Abaqus CDP model cannot cover such cases correctly.

## References

1. A. Szwed, I. Kamińska, Part I – accompanying paper
2. *Abaqus Theory Manual*
3. www.wolfram.com
4. S. Murakami, *Continuum Damage Mechanics* (Springer, 2012)
5. J. Lubliner, J. Oliver, S. Oller, E. Onate, Int. J. Solids Struct. **25(3)**, 299 (1989)
6. J. Lee, G. Fenves, Earthquake Eng. Struct. Dyn. **27**, 937 (1998)
7. I. Kamińska, A. Szwed, *Concrete and concrete structure – experiments*, E. Szmigiera, P. Łukowski, S. Jemioło (eds.), **93** (OWPW, Warszawa, 2015) (in Polish)
8. P. Kmiecik, M. Kamiński, Archives of Civil and Mechanical Engineering **XI(3)**, 623 (2011)
9. M. Szczecina, A. Winnicki, *ASEM15* (Korea, 2015)
10. A. Truty, T. Zimmermann, *Numerics in Geotechnics and Structures* **1** (2015)

# Reduce of bearing strength of the bent reinforce-concrete elements on a sloping section with the corrosive damage of transversal armature

*Ashot* Tamrazyan[1], and *Dmitriy* Popov[1*]
[1] Moscow State University of Civil Engineering , Moscow,Russia

**Abstract**: the technique of calculation of the reinforced concrete element exposed to the aggressive environment is considered, the quantitative reduction estimation of power resistance of concrete and armature is given.

## 1 Introduction

Corrosive influences entail considerable degradation of bearing strength of reinforce-concrete elements of building and constructions in the period of their exploitation. [1]. The protracted exploitation without realization of events on major repairs leads to a problem about structural safety and vitality of building. Tasks on reduction of bearing strength and as a result prognostication of risks of bringing down of building, which is inalienable part of modern science [2-4].

Numerous authors engaged in the account of decline of power resistance, construction of models of empiric dependences and depth-finding of corrosive damage of reinforce-concrete element [5-7].

The probability of breaking a bent element along an inclined section is equally probable, as in the normal section, so it becomes necessary to analyze this case

As is generally known, there are three types of calculation and accordingly destruction of the bent reinforce-concrete element on sloping sections: fracture from the action of flexion moment M; breaking up of the compressed concrete on a stripe between sloping cracks; change on a sloping section from the action of transversal force of $Q$.

The strength of the element along the inclined cross section to the action of shear force under the influence of an aggressive environment cannot be determined by means of normative literature within the framework of the actual evaluation of the bearing capacity of the element.

This complexity is stipulated by variation of types of damage (pitting, continuous, uneven), kinetics of advancement (colmatation, filtration, avalanche), presence and thicknesses of zones of corrosive damage (fig 1). Under the influence of the aggressor, the physical and mechanical characteristics of the reinforcement and concrete change.

* Corresponding author: popovds89@mail.ru

This leads to the need of solving new problems with variable cross-section of transverse reinforcement, heterogeneous strength of concrete, the length of the projection of the inclined section to determine the actual bearing capacity of the element at different stages of operation.

## 2 Methods

In this article the sloping section of reinforce-concrete element is examined at the one-sided contact of aggressive environment with the external fibre of the compressed zone. It is used two and three-zone layout of power resistance chart [8] (fig. 1). Reduction of cross-sectional reinforcement cross-section is achieved by entering a correction factor $\omega_{sw}$. On an image 1 reduction of area of transversal armature is shown in a section depending on character of corrosive destruction. So at existence of the destroyed fibre layer ($z^{*}$>0), power resistance of transversal armature and concrete on a height $z^{*}$ is not taken into account in calculations (fig .1B)

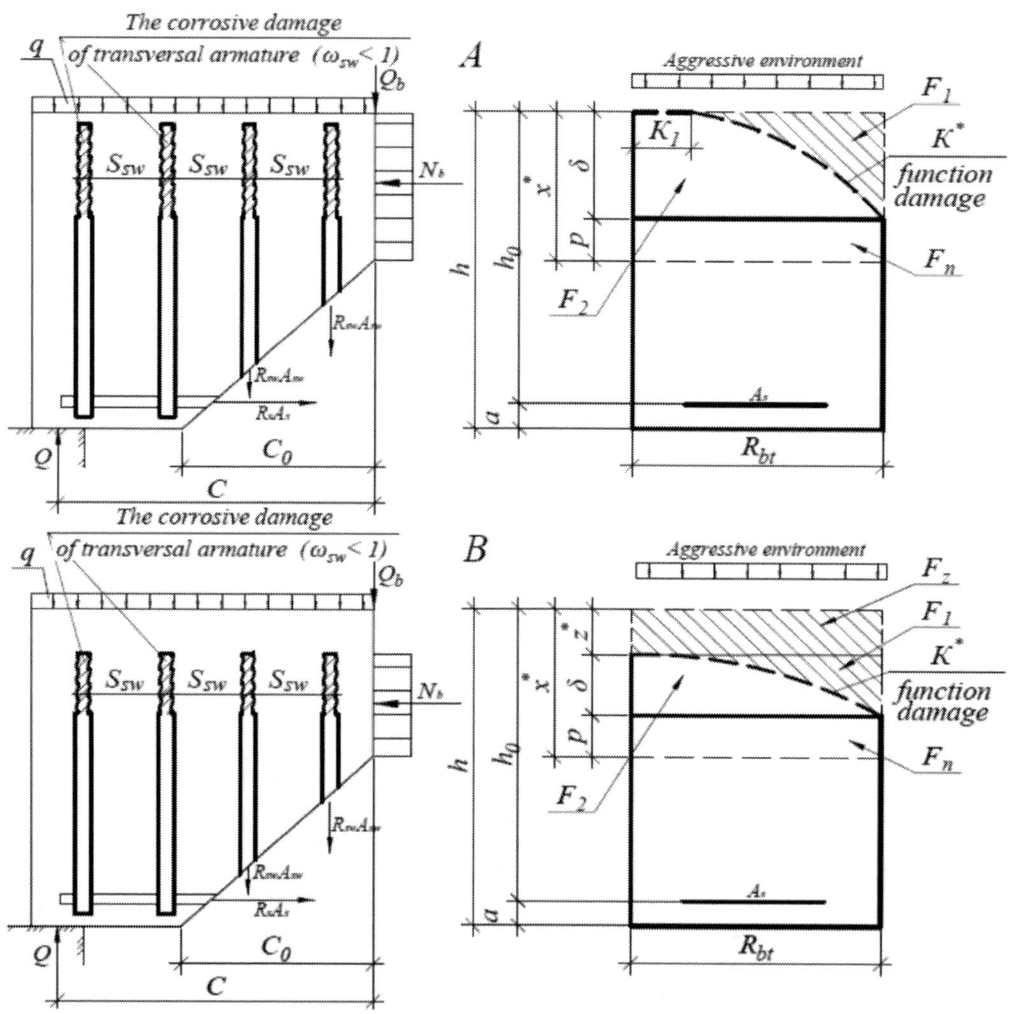


**Fig. 1.**Calculation charts of the bent reinforce-concrete element damaged by corrosion : A - in default of the fully lost resistance of fibre layer ($z^{*}$=0); B - at presence of the destroyed fibre layer ($z^{*}$>0)

Here $z^*$ - is a thickness of complete corrosive destruction of concrete; $\delta$ - it is a thickness of incomplete corrosive damage of concrete; p is a thickness of uncrippled concrete; $a, x^*, h, h_0$ - are geometrical sizes of concrete body (standard); $K^*$ - is a curve of function of damages (coefficient of maintenance of initial mechanical descriptions); $K_1$ - is the fixed value of stake of maintenance of durability of concrete on an external fibre; $Fz^*$ - is a size of the lost part of power resistance of concrete; $F_1$ - is a size of the lost part of power resistance of concrete; $F_2$ - is a size of the saved part of power resistance of concrete; $Fs$ - is a size of power resistance of longitudinal armature; $S_{sw}$ - is a step of transversal armature; $\omega_{sw}$ - is a coefficient of damage of transversal armature; $C$ - is distance from the top of calculation sloping section to support; $C_0$ - is a projection of sloping calculation section; $N_b$ - resultant normal tensions in a concrete.

Classification of a large number of corrosive processes of concrete is combined into three main types. In the first form of corrosion, the composition of concrete $Ca(OH)_2$ dissolves under the influence of atmospheric precipitation, the second type is characterized by the interaction with acids contained in water and the third type of corrosion of concrete occurs as a result of crystallization of hardly soluble substances. Under natural conditions, corrosion of one or another species is rarely found apart from the others, but usually one species predominates. The author does not consider any kind of corrosion damage, taking into account the reduction in the mechanical characteristics of materials exposed to the corrosive environment

Researches of the bent corrosive damaged reinforce-concrete elements on a sloping section in home literature on the real stage small enough [9, 10].

In foreign literature it is possible to distinguish some works [11,12] which give the results of researches of durability and character of destructions of the reinforce-concrete beams on a sloping section, exposed to influence of aggressive environment.

Durability of sloping section from сдвиговых destructions (caused by transversal force) without the "unbended" re-enforcement it is obliged to be provided concordantly [13]. In this work [10] the calculation of sloping section is executed taking into account the corrosive damages of reinforce-concrete element.

$$Q^* = Q^*_{sw} + Q^*_b \tag{1}$$

where, $Q^*$- remaining resource of power resistance of the damaged reinforce-concrete element on a vertical cut; $Q^*_{sw}$ - is a stake of this resource, being transversal armature; $Q_b{}^*$- stake of this resource, being on the cut away part of the damaged concrete.

Pursuant to [9], it is possible to write down :

$$\begin{gathered} Q^*_{sw} = \sum \omega_{sw} A_{sw} R_{sw} = q^*_{sw} c \\ q^*_{sw} = \frac{\omega_{sw} A_{sw} R_{sw}}{S_{sw}}; Q^*_{sw} = \omega_{sw} A_{sw} R_{sw} \frac{c}{S_{sw}} \end{gathered} \tag{2}$$

$$\begin{gathered} Q^*_b = \frac{M^*}{c}; \;\; M^* = \varphi(h_0 - z^*) N^*_b \\ N^*_b = b[(h_0 - z^*) R_{bt} - F_1] \;\; F_1 = \frac{1 + k_1}{3} \delta b R_{bt} \end{gathered} \tag{3}$$

where, $S_{sw}$ - is distance between transversal reinforcing bars; $c$ - is an ichnography of sloping crack; $b$ - is a width of element; $\varphi$ = 1.5 integral empiric correcting coefficient; $R_{bt}$ - ultimate strength of concrete by tension; $q_{sw}$ - is effort in a transversal armature by length of element unit; ωsw is a coefficient of damage of transversal armature; $A_{sw}$ - is an area of section of transversal armature; $R_{sw}$ - a calculated resistance of transversal armature; $F_l$ - is a size of the lost part of power resistance if $z$*=0 and $K1$>0.

Now the sought after resistance of the damaged section to transversal force will get a record:

$$Q^* = q_{sw}^* c + \frac{M^*}{c} \quad (4)$$

for that a value of C is from the condition of minimization (4) at $c$:

$$\frac{dQ^*}{dc} = 0, \text{ т.е. } c = \sqrt{\frac{M^*}{q_{sw}^*}} \quad (5)$$

Developing a method [10] for next corrosive damages we will get:

- on condition of damage: $z^*>0$, $K_l=0$, $0<K^*<1$. i.e. at existence of the fully destroyed fibre layer of $z^*>0$ and absence of the saved durability on the fibre layer of $K_l=0$, in equalization (6) from the working height of section subtracted the fully destroyed layer of $z^*$ and 1/3 part of the not fully destroyed (transitional) layer, and from the shoulder of force the size of the fully destroyed layer of $z^*$ is subtracted.

$$M^* = 1{,}5bh_0^2 R_{bt} = 1{,}5b(h_0 - z^*)R_{bt}[h_0 - (z^* + \frac{1}{3}\delta)] \quad (6)$$

- on condition of damage: $z^*=0$, $K_l=0$, $0<K^*<1$. i.e. when the fully destroyed fibre layer of $z^*=0$ and saved durability is absent on the fibre layer of $K_l=0$ in equalization (7) from the working height of section subtracted only 1/3 part of the not fully destroyed (transitional) layer.

$$M^* = 1{,}5bh_0^2 R_{bt} = 1{,}5bh_0 R_{bt}(h_0 - \frac{1}{3}\delta) \quad (7)$$

Losses of power resistance of transversal effort from influence of aggressive environment defining in a percentage ratio is possible on a formula:

$$P = \frac{Q - Q^*}{Q} 100 \quad (8)$$

where, $Q = Q_{sw}+Q_b$ is primary (project) power resistance of element on a vertical cut; $P$ - are losses of power resistance in percents.

Being base on an offer methodology (4), (8) the calculation of reinforce-concrete beam is executed on a sloping section at two different terms of damages.

Condition of damage 1: $z^*=0$, $\omega_{sw}=1$ and $0 < \delta_{max} < x$, id est, when an area of complete destruction is not, a transversal armature is not damaged, and the area of incomplete corrosive damage of concrete is varied from 0 to $x$.

Condition of damage 2: $z^*=0$, $\omega_{sw}=0{,}5$ and $0 < \delta_{max} < x$, id est, when an area of complete destruction is not, the area of cross-sectional is diminished on 50%, and the area of incomplete corrosive damage of concrete is varied from 0 to $x$.

## 3 Results

We will accept some conditional descriptions of durability and corrosive damages : the bent element of rectangular section of $h$=600mm, $b$=300mm, $a$=30mm, concrete of B15 ($R_b$=8,5MPa, $R_{bt}$=0,75MPa), with re-enforcement in the stretched zone by an armature by the class of A400 ($Rs$=350MPa), border relative height of the compressed zone $\xi_R$ =0,533, transversal armature of class of A240 ($R_{sw}$=170MPa), $A_{sw}$=157mm$^2$, with the step of $S_{sw}$=100mm. The corrosive damage of longitudinal working armature is $\omega_s$=1. The maximal depth of damages $\delta_{max}$ is accepted to the equal border height of the compressed zone of section $\delta_{max} = x$, where $x = \xi_R \cdot h_0$=30,38cm.

The results of calculation are presented on the fig. 2, that illustrates the relation of depth of damage and to the decline of bearing strength of element in percents, so at the first set condition the decline of power resistance arrives at no more than 10%, but if to accept the terms of development of corrosion in a transversal armature, on the example of 50% reduction of area of cross-sectional, general losses develop to 36%.

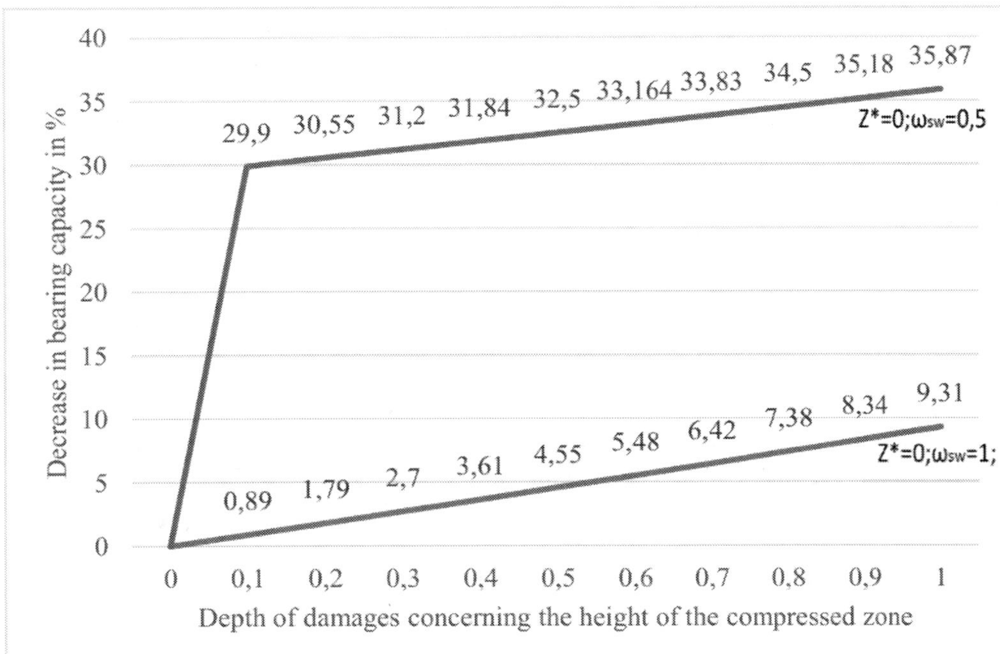


**Fig.2** Chart of loss of power resistance of reinforce-concrete element on a sloping section at the increase of depth of corrosive damage δ.

As is generally known bearing strength of reinforce-concrete element is provided by a concrete and armature, coming from principle of calculation on [13] the general losses of power resistance of element on a sloping section from influence of aggressive environment will consist of 57% due to a concrete and 43% armature.

According to [13], the strength condition for the oblique section must be observed:

$$Q < Q^* \quad (9)$$

In the event that condition (9) is not observed, the element must be strengthened. One of the possible methods of strengthening is the installation of additional vertical clamps, the necessary cross-sectional area in the corresponding zone.

$$Q < Q_b^* + Q_{sw}^* + Q_{fw} \tag{10}$$

Where, $Q_{fw}$ - is the transverse force perceived by the additionally mounted vertical clamps.

$$Q_{fw} = \frac{A_{fw} R_{fw} c}{S_f} \tag{11}$$

Where, $S_f$ - is the distance between the additional transverse clamps; C is the horizontal projection of the inclined crack; $A_{fw}$ - cross-sectional area of additional transverse clamps; $R_{fw}$ - is the design resistance of the additional clamps.

## 4 Conclusion

Thus, the method of calculation is considered on the sloping section of the reinforce-concrete corrosive damaged element. It is set that the decline of power resistance of the corrosive damaged reinforce-concrete element arrives at to 36% at reduction of area of transversal armature on 50%. The decline of bearing strength on 57% takes place due to a concrete, and 43% due to an armature.

## References

1. E. Larionova MSSU Announcer.**6**, 26 (2016).
2. A.Tamarzyan.M.Orlova Construction in housing. **6**, 32 (2015).
3. A.Tamarzyan.M.Orlova MSSU Announcer. **4**, 181 (2014)
4. A.Tamarzyan,V. Falikman Construction and reconstruction. **3**(65), 71 (2016)
5. A. Popesku *Capacity of reinforce-concrete constructions subject to corrosion. SASU,* **182** (1996).
6. G. Smolyago, A.Krychkov, S. Drokin, A. Dronov. Announcer of Belgorod State technological university the name of. Shukhov. **2**, 22, (2014).
7. O. Morozova, S, Markov, I. Stavskaya Construction and reconstruction. **3**(65), 71 (2016)
8. V Bondarenko Concrete and reinforced concrete. **2**, 25 (2008).
9. N. Klyueva, D. Karpenko, A. Kashavtsev Structural mechanics of engineering constructions and buildings. **5**, 77 (2015)
10. B. Yagupov., R. Migal Concrete and reinforced concrete. **3**, 28 (2007)
11. I . Khan., R. Francois,. A. Castel. Materials and Structures. V **47**, I 9, 1467 (2014)
12. B.Almassri, J. Barros, F. Mahmoud, R. Francois. Composite Structures. V **131**, 731 (2015)
13. СП 63.13330.2012. Constructions from concrete and reinforced concrete.

# Optimal design of reinforced concrete structures taking into account the particular calculation for progressive destruction

*Ashot* Tampazyan [*1] and *Svetlana* Zubareva [**1]

[1] Moscow state university of Civil Engineering, 129337, 26, Yaroslavskoe shosse, Moscow, Russian Federation.

**Abstract**. The problems of optimization of reinforced concrete structures are considered taking into account their properties and features under fire impacts and dynamic loading.

Experimental data on the dynamic strength of concrete and reinforcement under fire impacts and their approximation in analytical form are presented. It is shown that the dynamic strength of materials at temperatures above 3500C leads to their sharp decrease, which adversely affects the resistance of buildings to progressive destruction.

To resist the progressive destruction of buildings and structures, they require a high cost of providing them, which in most cases is inefficient. The search method is proposed, which allows taking into account a set of constraints, and the objective function, including the risk of collapse, to find the optimal solution.

On the example of a reinforced concrete bezel plate of overlap, its optimum parameters are found.

Keywords. Optimization, objective function, constraint system, risk analysis, search method, progressive destruction, fire effects, dynamic impact, dynamic coefficient, beamless slab

## Introduction

Nowadays, special attention is paid to the problem of ensuring the safety of buildings and structures from progressive and other types of destruction, drawing back the issues of their optimal design [1]

The traditional design method in these conditions does not allow to reveal significant reserves of saving materials and structures. Economic efficiency can be significantly increased by optimizing cross sections and reinforcing structural elements [2]. The economic approach to optimization problems is reduced to constructing the objective function of the cost of the construction and finding its minimum.

## Results and discussion

When calculating for progressive destruction, the objective function F2 is found as the sum of the costs, taking into account its manufacture, operation, necessary to maintain a certain level of reliability throughout the service life, the volume of restoration work, the risk of failure, the risk of failure, and the cost of elimination of consequences of possible emergency situations.

*tamrazian@mail.ru
**zubareva494@mail.ru

$$F_2 = k_c k_z \left[ C_1 V_b + C_2 V_b^2 + \sum_r \left( C_3 V_s + C_4 V_s^2 \right) + \sum_i C_{ad} + (1 - P_t) \cdot C_{ser} + R(t) \right], \quad (1)$$

where $k_c$ – coefficient of conversion;
$k_z$ – coefficient of winter appreciation;
$C_1$ - production cost of concrete for 1m$^3$;
$V_b$ - volume of concrete;
$C_2$ - production cost of reinforcing steel for 1m$^3$;
$V_s$ – volume of reinforcing steel of the given type;
$\sum C_{ad}$- additional costs, including the consumption of reinforcement on the framing of technological holes, overlapping and anchoring;
$(1-P_t)\cdot C_{ser}$ - operational costs in view of the probability of failure-free operation of the structure;
$R(t)$ — loss risk arising from a design failure.

$$R(t) = C_y \cdot Q, \quad (2)$$

$Q$ — probability of design failure;
$C_y$ — potential damage in case of failure of structures.

Compared to the function

$$F_1 = k_c k_z \left[ C_1 V_b + C_2 V_b^2 + \sum_r \left( C_3 V_s + C_4 V_s^2 \right) + \sum_i C_{ad} \right], \quad (3)$$

taking into account only the initial constraints corresponding to the limiting states of СП 63.13330.2012 (limiting bending moments, forces, deflections, crack opening width, etc.), it is clear that a fundamentally new approach is required to solve it.

With optimal design, the system of constraints will complement the following positions:

- restrictions on the expenditure of resources that must be established, based on the level of funding, or by analogy with previously known solutions. This includes restrictions on the amount of concrete, reinforcement, labor, energy costs, etc. The restrictions of this group have the form

$$V_b \le [V_b], G_s \le [G_s], T \le [T], \ldots \quad (4)$$

- Constructive, architectural, technological limitations that significantly affect both the appearance of the structure and the shape or size of its elements

$$\varphi_k \le [\varphi]_k, \; k = 1, 2, \ldots \quad (5)$$

- Restriction on the coefficient of renovation during repair work:

$$k_n = \frac{C_{ser}}{F_K} \le [k_n], \quad (6)$$

$C_{ser}$ - value of newly input funds;

$F_K$ - total cost of capital investments after reconstruction.

- Grade of initial reliability:

$$P_0 \ge [P_0]. \quad (7)$$

- Allowable risk of collapse of structures $R$:

$$R(t) \le [R]. \quad (8)$$

Moreover, the more input restrictions, the higher the actual cost of the construction becomes.

To minimize the alleged damage, it is best to consider the risk of failure. The quantitative risk assessment allows to solve tasks for determining the risk of an accident, and the expected expected material losses, determined by the area of destruction. Approximate risk levels and the corresponding volumes of damage are determined by the exponential dependence [3]:

1. Collapse of the 1st rank ($R = 10^{-5}$) - the area of destruction of $100\ m^2 \le S \le 240\ m^2$. Collapse to 6 floors - leads to a disproportionate collapse along the perimeter or adjacent spans in all directions;

2. Collapse of the 2nd rank ($R = 5 \cdot 10^{-6}$) - the area of destruction is $240\ m^2 \le S \le 360\ m^2$. Collapse to 9 floors;

3. Collapse of the 3rd rank ($R = 10^{-6}$) - the area of destruction $S \ge 360 m^2$. Collapse more than 9 floors.

This approach is more justifiable from an economic point of view and does not lead to an understatement of the bearing capacity [2].

To find the value of the function, the range of admissible values is first chosen and the variable parameters are set. The values of the function are obtained by the method of search optimization using the criterion of minimum cost.

Consider a searching technique of the most favorable decision of optimization problem. We choose a parameter that has a large effect on the values of the function. Increasing its value, we fall into the permissible region of the function. After this parameter is released to the most acceptable value, the search continues with a change in the percentage of reinforcement on which the cost parameters depend. By reducing the discrepancy by constraints, we reach the upper limit of the allowed values, which is determined to a greater extent by the possible consumption of resources. The optimal value will be at the minimum point of the function at the boundary of the admissible region. The position of this boundary is determined by groups of constraints, including taking into account the risk of collapse of the structure.

Consider an example of optimization of a non-beam slab with the following initial parameters. The size of the plate cell $l_1 x l_2$= 600x400cm, column cross-section $b_C$=40cm, class of concrete B30, armature A500, protective layers of concrete $a_1$=2 cm, $a_2$=1,7cm, load q=20 kN/m$^2$. Cost of concrete $C_1$=4000 rub./m$^3$, armature $C_4$ =1.775.000 rub./m$^3$.

Denote the boundary conditions. To ensure the load-carrying capacity of the assembly of columns with a flat slab, check the bearing capacity of the slab in the normal section for the action of the negative bending moment:

$$M \le M_{ult}. \tag{9}$$

Check for the effect of transverse forces:

$$Q \le Q_{ult}. \tag{10}$$

Check for forcing:

$$F \le F_{b,ult} + F_{sw,ult}, \tag{11}$$

For units, the load-carrying capacity test must be followed:

$$\frac{P^{разр}}{A_b + A_s n} \le [R_{be}]. \tag{12}$$

where $R_{be}$ - strength of concrete.

For each zone, the condition for limiting the width of the opening of normal and diagonal cracks:

$$a_{Ti} \le [a_T], \tag{13}$$

as well as the condition of limiting the deflection:

$$W_i \leq [W]. \tag{14}$$

When calculating the slab in the operational stage, the consumption of concrete varies within the limits 0,4-0,7m$^3$/m$^2$, consumption of armature 35-70kg/m$^2$, therefore, the constraints on the material will have the form:

$$V_b \leq 0{,}7m^3/m^2,$$
$$G_s \leq 70kg/m^2. \tag{15}$$

It is recommended to use a concrete class not higher B50.

$$R_b \leq 27{,}5MPa \tag{16}$$

The coefficient of renewal is assumed equal to

$$[k_n] = 0{,}7...0{,}9 \tag{17}$$

The restriction on the stability of the building against the progressive collapse, which is ensured if for any element the condition:

$$F' \leq [S], \tag{18}$$

where $F'$ and $S$ respectively, the force in the structural element, found from the performed static calculation, and its design load-carrying capacity.

Determination of an acceptable level of risk

$$P(F) < P_{adm}, \tag{19}$$

where $P_{adm}$ – allowable risk.

In conditions of man-caused hazards (technical risk), individual risk is considered acceptable if its value does not exceed $10^{-6}$.

Consider the zone of maximum moments. In the results obtained, we fix the value of the plate height and in the future we will only vary the reinforcement coefficient.

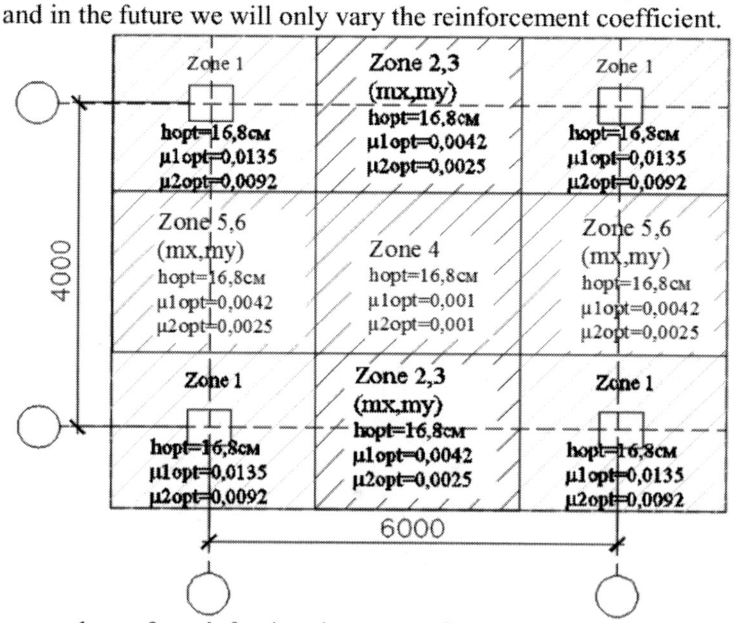


**Fig. 1**. Optimum scheme for reinforcing the zones of a non-beam slab

The calculation results with optimal parameters for different zones are schematically shown on the reinforcement plan (fig.1)

We describe the method of searching for the optimal solution. Depending on the initial data and the adopted variable parameters, the stiffness characteristics for all zones are determined, the forces and their limit values are given for the given stiffnesses, the objective function F for any combination of plate parameters is calculated. Further, a small standard increment $\Delta x_i$ is introduced, and an additional factor that is the sensitivity factor of the value of the objective function at the optimum point to the resources:

$$\Delta C = \frac{\Delta P}{\Delta F}. \tag{20}$$

To reach the optimum point, ΔP should be as large as possible and ΔF smaller. The objective function F is computed and a generalized discrepancy P is found for the constraints. If the discrepancy is more than zero, it is eliminated and the final value of the cost function F.

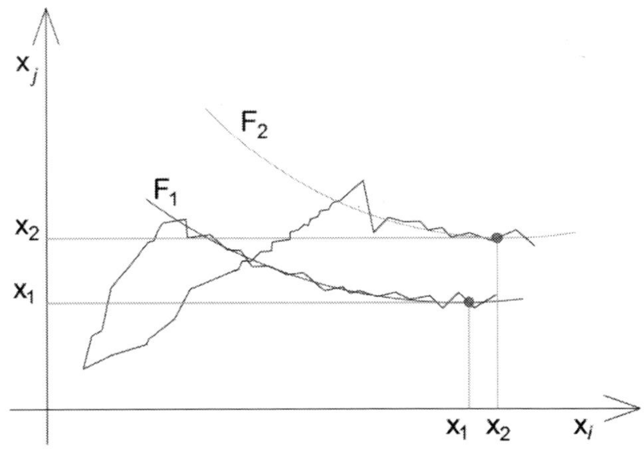


**Fig.2**. Scheme of searching for the optimum of the final element function.
$x_1$ and $x_2$ – variable parameters of a reinforced concrete element.

In the calculations for progressive destruction, an instantaneous failure of one supporting element is assumed [4,5]. However, this is practically possible with explosive and shock effects accompanied by fires [6,7]. In case of fire in a loaded condition, the structures are exposed to high-temperature fire action, which changes the properties of concrete and reinforcement [8].

At the same time, dynamic loads begin to act on some structural elements, which, in combination with short-term fire effects, can be greater than the bearing capacity. This inevitably leads to destruction [9,10]. This kind of impact should be taken into account in a special combination of loads. In this case, it will be sufficient to satisfy only the safety requirements of the element or the structure as a whole [11].

As features of the calculation for progressive destruction, it is necessary to take into account the properties of materials and the design work under fire and dynamic influences [12]. The effect of a dynamic load application is characterized by a dynamic factor, that is, the ratio of concrete strength under dynamic loading to static strength [13-15]. Its value at normal temperatures is 1.8, but at temperatures from 300°C and more it drops to 0,8-0,6, which is 2,25-3 times smaller (fig. 3) [16].

With dynamic actions, the moments must be calculated:

$$M_{u,d} = \alpha_m R_{bd} b h_0^2, \text{ when } \alpha_m = \xi(1-0,5\xi); \tag{21}$$

$$\xi = \frac{R_{sd} A_s}{R_{bd} b h_0}, \tag{22}$$

$$R_{sd} = K_{ds} \cdot R_s \tag{23}$$

$$R_{bd} = K_{db} \cdot R_b \tag{24}$$

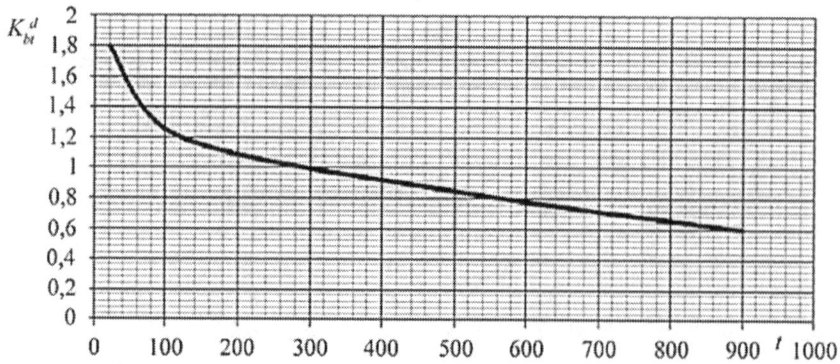


**Fig.3** Graph of dependence of the coefficient of dynamism of concrete $K_{bt}^d$ on temperature

The dynamic coefficient depends on several indicators, including temperature, time and loading speed. And with an increase in the loading speed, its unevenness also increases, that is, certain areas of concrete prove to be highly overstressed. The dynamic load application changes the design design scheme. Based on the tests carried out, the analytical dependence of the dynamic strength of concrete on temperature at impact speed 0,4 m/s:

$$K_{db} = 1.27 + 9.97 \cdot 10^{-12} T^4 - 1.87 \cdot 10^{-8} T^3 + 1.10 \cdot 10^{-5} T^2 - 2.93 \cdot 10^{-3} T \tag{25}$$

## Conclusion

At the moment, in the calculations for progressive destruction in complexes and methodological instructions, the dynamic coefficient is taken to be greater than one [17-20], however, the experiments show a real decrease in its value when heated, which leads to a significant decrease in the bearing capacity of the element and an increase in the values of the first three terms of the cost function (1).

Comparing the functions $F_1$ and $F_2$ (fig. 2), we can say that the cost of manufacturing the element of the function $F_2$ will be higher than that found by the function $F_1$. However, the safety function $F_2$ will be the most profitable solution, because we reduce the risk of failure of the design. Verification calculations showed that $F_2$ is more economical by 20-30% compared to $F_1$.

Using the cost function (1) when designing the cost function, taking into account the risk of loss, leads to an increase in the level of safety. As a rule, a decrease in the magnitude of the risk leads to an increase in the cost of constructing. On the other hand, increasing the risk may lead to the failure of structures in a shorter period. Therefore, the definition of the accepted value of risk is a very important task. The proposed objective function provides the most objective assessment of the costs for the projected facilities, taking into account the emergence of emergency situations. This approach will significantly improve the quality of design and avoid the occurrence of progressive destruction.

## References

1. DoD UFC: *Design of buildings to resist progressive collapse*, UFC 4-023-03, U.S. DoD. (2010).
2. A.G. Tamrazyan, E.A.Filimonova, *Vestnic MSUCE* **10** 68—74 (2013).
3. A.G. Tamrazyan Izvestiya OSTU **4-16** 107-108 (2007).
4. D.T. Pham, P. Buhan, C. Florence, J.-V. Heck, H.H. Nguyen ES **90** 38-47 (2015).
5. H. Wanga,*,Y.Sub, Q.Zengc SEP **1** 48–54 (2011).
6. A.G. Tamrazyan / A.G. Tamrazyan, L.A. Avetisyan TTP, Switzerland 638-640 (2014).
7. A.G. Tamrazyan ICSMIM 16-17, (2013).
8. MDV 21-2.2000.
9. A. Colombo and P. A Negro ES **27(8)** 1164-1174 (2005).
10. C.T.Davie. H.LZhang, A.Gibson Comput Struct., 94–95 (3) 54–69 (2012)..
11. V.O.Almazov, A.I.Plotnikov, B.S.Rastorguev *Vestnic MSUCE* **2** 15-20 (2011)
12. D.D. Capua, A.R. Fire Saf. **42(2)** 139–49 (2007).
13. A.G. Tamrazyan, L.A. Avetisyan *Vestnic MSUCE* **10** 14—23 (2013).
14. M. Bikhiet, F.Nasser, El-Shafey,Hany M. El-Hashimy *Alexandria EJ* **53** 643–653 (2014)..
15. Y.S.Sin, J.E Park, J.Y. Kim *Journal of the Korea Concrete Institute* **23** 37 (2011).
16. A.G. Tamrazyan, L.A. Avetisyan *ICC* **4** 24-28 (2014).
17. S.Kokot, A.Anthoine, P.Negro and G.Solomos ES **40** 205-217 (2012).
18. Kwasniewski Leslaw **32** 1223–1235 (2010).
19. Y.J. Kwon, N.H. Kim., J.Y. Lee, D.J. Kim, K.Harada, *Proceedings of Fire Science and Engineering in Japan Association* 282-283 (2010).
20. J.C. Mindeguia, H.Carré, P. Pimienta, *La Borderie, In Rencontres Universitaires de Génie Civil. La Grande Motte*. **1-2**, (2006).

# Landscape Architecture and green spaces in Russia

*Valery* Telichenko[1], *Andrey* Benuzh[1*] and *Ilya* Mochalov[2]

[1]Moscow State University of Civil Engineering (National Research University), 26, Yaroslavskoe Sh., Moscow, 129337, Russian Federation
[2]Chairman of the Board of ROO "Moscow Association Landscape architects ", 129594, Moscow, Sheremetyevskaya St., 34

**Abstract.** The article considers the challenges of designing sustainable landscape architectural projects in Russian cities such as high-raised linear parks and green roofs. Sustainable landscape projects create great and green cities where the built and the natural environments are interwoven. Sometimes creating green spaces is considered as a great challenge, particularly in the big and dense cities, where there is pressure for space and green development is very limited. In this case high-raised parks and green roofs could be the only solution.

## 1 Introduction

Sustainable landscape architectural projects create ecological designs for the outdoor and urban environment. It begins with appropriate systems which address function, cost, energy efficiency, beauty and the environment. Broadly speaking, sustainable landscape architecture is the integration of ecological, social, cultural and economic factors in designing landscapes to help protect natural habitats, contribute to stormwater management, conserve water, among the other objectives.

Sustainability and conservation landscaping both strive to work with nature to reduce air pollution, increase water quality, lower water consumption, utilize native plants, and reduce usage of pest control. However, sustainability and conservation differ in the emphasis sustainability places on addressing social and economic factors in addition to environmental factors. In other words, conservation can be seen as the environmental part of the sustainability concept.

The Sustainable Sites Initiative grew out of a conference hosted in 2005 by the American Society of Landscape Architects (ASLA) and the Lady Bird Johnson Wildflower Center. The initiative's definition of sustainability derives from the well-known Brundtland report. To be sustainable, a site has to have "design, construction, operations, and maintenance practices that meet the needs of the present without compromising the ability of future generations to meet their own needs." Sustainable sites do not only mitigate negative impacts on the environment, but are a mutual benefit to the site itself and the people who use it. In addition, a sustainable site must address social, environmental, and

* Corresponding author: ABenuzh@gmail.com

economic concerns. The economic aspects addressed must take into account the value of the natural systems of the site [1]. This is the root of the program's ecosystems services framework. In other words, the guidelines have been built around the concept that people receive benefits, in the form of goods and services from healthy ecosystems.

## 2 Sustainable cities and green spaces

170 Green open spaces are an important part of sustainable cities. Cities are comprised of more than just buildings and people. The most "liveable" cities – and some of the world's most famous cities – are as known for their open spaces as they are for their culture. Hyde Park in London, Central Park in New York, the Bukit Timah Nature Preserve in Singapore, Phoenix Park in Dublin - all are attractions in their own right for inhabitants and visitors alike and even landmarks.

Open space in urban environments provides many advantages: formal and informal sport and recreation, preservation of natural environments, provision of green space and even urban storm water management. Thus green space must be a key consideration in urban planning if the health of a city and its people are both considered important [2]. A new, broader view of parks has also recently been emerging. This new view focuses on how policymakers, practitioners, and the public can begin to think about parks as valuable contributors to larger urban policy objectives, such as job opportunities, youth development, public health, and community building.

As the world's cities continue to grow, continuing to value green space in cities is vital: but is also a challenge, particularly in developing countries where there is pressure for space, resources and development.

Urban green spaces are considered as important contributor and can be a significant part of sustainable development. Developments of urban green spaces need to consider interdisciplinary and integrative approaches such as economic, social, cultural, management and planning aspects to improve existing urban spaces' facilities and services and to optimize urban green space policies. The quality of cities depends on how the urban green spaces are designed, managed and protected. The social aspects of urban green spaces include diversity of land uses, contribution to health and active life styles in cities, social justice by incorporating all groups and ages of people into green spaces, opportunities to interact and expand social network, enhancement of cultural life for different communities living in the city by providing a platform to share views, feelings and to celebrate different groups occasions and, a venue for environmental education for the schoolchildren and a play ground for children for the social, mental and physical development [3].

From the planning aspects, urban green spaces include business, retail, leisure development, tourism development; employment centers besides residential areas and the good planning of urban green spaces can play a role as a visual screen, a function of noise protection and a place for commuting and recreation by providing well-designed networks within the park and with the other areas.

The economic aspects of urban green spaces incorporate- as a place for production and supply of fruits, wood to green business centers, and as a place for new jobs creation and increasing economic value of the area by integrating the environment friendly behavior and attracting tourists provided with convenient atmosphere, security and facilities for the tourists. Most importantly the ecological perspective considers urban green spaces as a facilitator to reduce the impact of human activities through absorbing pollutants and re-leasing oxygen, contributing to the maintenance of a healthy urban environment with clean air, water and soil and preserving the local natural and cultural heritage with a diversity of urban wildlife and urban resources [4].

## 3 High-raised linear parks as a part of modern urban green city system

Traditional urban parks are public spaces designed for either passive recreation - users sit to read, to eat, to watch other people or active facilities such as sports. These parks are destinations in itself, places to go to. In contrast, urban linear parks take you somewhere, in some cases from one transportation hub to another. They are mostly facilities for active recreation, for walking, jogging, or biking. They are "transit spaces, as they combine elements of transportation together with recreation.

Linear parks in cities and towns are popular partly because they contribute to today's interest in physical fitness. They offer urban populations changing vistas, including riversides, ports and cityscapes. They draw consumers to commercial places they might otherwise not visit. They increase residential real estate values, attracting new tenants and owners to homes and apartments with these nearby amenities.

They are remarkably diverse. Some pass through old railroad tunnels; one is being constructed on a disused elevated railway; many follow old canal berms. There are oceanside linear parks as well as riverside examples. Many transform deteriorated industrial and residential areas into popular features of the urban renaissance. Some are at least a hundred years old, while some impressive ones have been created in the past decade. They are a very active part of any city green infrastructure, connecting not only the hubs, but being a city "green veins", with a very active "green blood" flow [5].

Several recent examples illustrate quite well how these convenient and poplar public spaces have been created and the benefits they bring to urban populations.

### 3.1 High Line Park, New York City

A pair of Urban Linear Parks in Lower Manhattan combine a remarkable feature—reuse of an old elevated freight line 30 feet above the street—parallel to a second urban trail passing along the Hudson from the Battery to the Javits Center in Mid-town. The elevated High Line offers an elevated garden as well as a trail. The views and the variety of recreational offerings on the adjacent Hudson River piers—everything from a full soccer field to boating—already makes this partially completed public space among the city's most popular. A rare feature among Urban Linear Parks, this pair will provide a round trip experience for city hikers (Figure 1).

The High Line design is a collaboration between James Corner Field Operations (Project Lead), Diller Scofidio + Renfro, and Piet Oudolf. Converting each section of the High Line from an out-of-use railroad trestle to a public landscape entailed not only years of planning, community input, and work by some of the city's most inventive designers, but also more than two years of construction per section.

The High Line's planting design is inspired by the self-seeded landscape that grew on the out-of-use elevated rail tracks during the 25 years after trains stopped running. The species of perennials, grasses, shrubs and trees were chosen for their hardiness, sustainability, and textural and color variation, with a focus on native species. Many of the species that originally grew on the High Line's rail bed are incorporated into the park's landscape. The conditions of the plants in 2017 are perfect - planted species are very well developed and established and the new species are appearing by being seeded by birds.

**Fig. 1.** High Line Park, New York City.

### 3.2 Jardins de la rambla de sants', Barcelona, Spain

Jardins de la rambla de sants' is an elevated 20,000 square meter park that opened earlier this year in Barcelona. Rather than similar designs in New York, Barcelona's elevated 'rambla' does not take the place of the rail line, but instead hides it with an extruded box that contains over 160 trees and 85,000 plants and shrubs organized into different gardens.

A structural system of beams infilled with glazing supports the park, while allowing pedestrians at ground level views of the passing trains. At certain points, the structure soars 12 meters above street level. Over 650 meters of climbing plants have already started to ascend the concrete structure, blending the natural aspects of the scheme with the built environment.

Visitors enter the park at plaza de sants, where the scheme starts to slope upwards, offering elevated views across the city. The promenade, which measures a total of 760 meters, has been designed by architects Sergi Godia and Ana Molino and hosts a range of different public amenities. Completed after a decade of development, the park is situated on a raised structure that covers existing railway lines below. Plants are recently planted, it is essential to use an automatic water irrigation due to a very harsh weather conditions.

### 3.3 Skygarden in Seoul

The Skygarden aims to regenerate and connect places near the main railway station that have been fragmented by roads and railway tracks.

**Fig. 2.** Skygarden in Seoul.

The Skygarden, which will be open to all 24 hours a day, re-uses an existing structure – like the High Line – in the form of a 1970 motorway flyover that was no longer deemed safe for its original purpose.

It is also part of a bigger set of ideas about taking a big, dense – sometimes ugly – city, one which was created without a great deal of concern for public space and pedestrian movement, and giving it qualities of walkability, neighbourliness, human scale and shared enjoyment of its places (Figure 2). The Skygarden is one of the more eye-catching examples of several initiatives promoted by the first holder of this job, Seung H-Sang, and his successor

and ally Young Joon Kim.

The ambitious mayor of a big city backs a project to put a garden on a bridge. The mayor's works programme aims to redress the balance. His successive chief architects have aimed to make the city more pedestrian-friendly and where possible to adapt the existing fabric rather than erase it in favour of grand new structures. "Revitalisation" rather than "reconstruction" is how Seung H-sang puts it. The city government is also considering the introduction of a congestion charge for the historic centre of the city.

### 3.3 "Moscow sky" project

This project was designed by Moscow based landscape architectural practice "Ilya Mochalov and Partners" in 2016 as a design concept proposed to the city department of Nature uses and environmental protection. The idea was to use the upper level of the shops (stylobate), which were built in New Arbat Avenue (former prospect Kaliningrad) in early 1960. There was a special technical survey carried out to find out the structural possibilities for the buildings to carry the additional weight of the newly designed green space. The survey shoved that the buildings are perfectly all right and the design was proposed. The idea of the design was to use the former grids and foot-prints of buildings and roads which were in this place before the great demolition in 1963. The liner park on the roof was called "Moscow sky" to reflect the idea that the gardens are raised in the sky. The park offers all the recreational facilities such as kids play areas, sports, boardwalks and of course night time lightning. Several escalators take people from the ground level on the top of the roof and back. Hopefully this design is realised at some near future. This will bring another dimension to the city landscape and provide another sustainable landscape in the sky.

## 4 Conclusion

Typically full of people, Urban Linear high-raised Parks rarely evidence public safety concerns. An exception emerged from Wilmington, DE's linear park planning where transformation from a largely abandoned riverfront to residential, retail and entertainment produced a Business Improvement District (BID) that organized and financed uniformed security personnel and cleaning up after the crowds.

Urban Linear Parks appeal to people as public facilities with an abundance of engaging things to do and see. They produce urban places where people want to be active.

To be successful, Urban Linear high-raised parks must:

1. Exist in high density places with ready access on foot and via transit—real urban places;
2. Be both attractive and interesting, offering changing views.
3. Be wide enough to safely accommodate bikers and pedestrians, where both are expected.
4. Provide opportunities to purchase food and pleasant places to take occasional breaks.
5. Be at least a half-kilometre long.

## References

1. V.I. Telichenko, V.M. Roitman, A.A. Benuzh, *Integrated safety in construction* (MISI-MGSU, 2015)
2. V.I. Telichenko, A.A. Benuzh, Ind. and Civ. Eng. **10**, 40 (2014)
3. V.I. Telichenko, A.A. Benuzh, Adv. Mat. Res., **1065-1069**, 2169 (2015)
4. A.A. Benuzh, E.N. Orenburova, Housing construction **2**, 14 (2015)
5. B. Cayce, Y. Chia-Hui, UTSOA (2009)

MATEC Web of Conferences 117, 00165 (2017) DOI: 10.1051/matecconf/201711700165
XXVI R-S-P Seminar 2017, Theoretical Foundation of Civil Engineering

# Interdisciplinary diagnostics in the process of revitalization of historic buildings, in terms of changing their function

*Wojciech* Terlikowski[1,*]

[1]Warsaw University of Technology, The Institute of Building Engineering, al. Armii Ludowej 16, 00-637 Warsaw, Poland

**Abstract.** Revitalization of historic buildings in crisis areas is a complex, interdisciplinary process that addresses not only the technical problems of buildings, but also the social, environmental, economic and cultural issues of the local community. Proper diagnosis of the dangers and social problems in the crisis area should be the basis for identifying new functionalities for this area. It should help to resolve the diagnosed problems, as well as to stimulate growth and revival in all spheres related to the life of the local community. The construction process, which results in the change of the function of the building and its adaptation to modern demands - technical, utility and formal, requires innovative interdisciplinary diagnostic methods. Each building, its structure and its components, has a certain adaptability that can be presented in a measurable and comparable manner. Evaluation of adaptability is a part of the expanded, interdisciplinary diagnostics of the historic building, which should be applied in the process of its revitalization. Determination of adaptability is an element of wider diagnostic procedures, which results in assessment of the revitalization capacity of a building.

## 1 Change of functions of historic buildings in the process of revitalization

Revitalization is an interdisciplinary process which main purpose is to revitalize a specific area that, in accordance with objective evaluation and diagnostic tests, is in a state of crisis [1]. Crisis is a set of undesirable, destructive processes that take place in the spatial, social and economic spheres, causing permanent degradation of the area of occurrence. The complexity of the revitalization process consists in simultaneously conducting spatial, building-architectural, urban, social and economic changes. They are usually undertaken in the public interest and include, among others, restoration of the revitalized area to its former functions, or to create new ones and to create conditions for its further development.

* Corresponding author: w.terlikowski@il.pw.edu.pl

The change of function is particularly important for historic buildings in a crisis area [2]. Historical buildings are usually central objects of these areas and the process of degradation starts from them. Degradation of buildings - in the material, functional and social spheres, directly affects the degradation of the area. The poor technical condition of the structures of historic buildings is often a result of the age of the buildings, their improper use and the wear of structural elements that constitute the building's supporting structure. This usually jeopardize the safety of construction and use. An abandoned and unused building usually falls into a state of ruin because of the acceleration of destructive activities. Lack of management of the building and its surroundings results in pathological behavior, which increases the social risk by deepening the crisis.

Correct diagnosis of the dangers and social problems in the crisis area should be the basis for identifying the new functionalities for this area. They should help to resolve the diagnosed problems, as well as to stimulate the development and revival in all spheres related to the life of the local community. It is clear that the key issue is the proper selection of utility functions of buildings located in the revitalized area. In the process of revitalization of a historic building as an adaptation to the needs of a given area by the application of specific technical solution, the effect could be obtained in beyond-technical issues - social, economic, cultural, educational and environmental. It derives from the definition of revitalization as well as the principles of sustainable development [3] in the civil engineering.

## 2 Technical and formal conditions of changes in the function of historic buildings

Changing the function of a historic building is often a difficult task and requires the conservator's permission. For formal reasons, changing the function of historic buildings is often not possible. According to the Act on the Protection of Monuments and the Care of Monuments [4] (Article 3, point 1), a monument is a property or a movable property, parts or assemblies that are made by human or related to its activity and constitute a testimony of a past age or event, whose protection is in the public interest because of their historical, artistic or scientific value. According to the Act [4] (Article 7) there are four forms of protection of monuments in Poland:
- entry in the register of monuments,
- recognition as a monument of history,
- creating a cultural park,
- define protection in a local land use plan or in a decision to establish a public purpose location, a decision to authorize a road investment, a decision to set a railway line, or a decision to authorize an airport investment project.

Historical buildings [4] (Article 9 point 1) are included in the "register of monuments". The basis of the entry is the decision of the voivodeship conservator of monuments, issued ex officio, at the request of the owner of immovable monument or perpetual lessee of land on which a permanent monument is located. According to the Act, other elements such as the surroundings of a registered monument and the geographical, historical or traditional names of this monument may be entered in the register. As it appears from the above records, the utility function of buildings included in the register of monuments may also be protected. This means that it may not be possible to change this feature.

Changing the function of a historic building may often involve the need to change the architectural and structural layout, which may even imply a change in the entire footing of a building or its part. The preservation of the architectural form of the monument and its building elements is directly related to the rules governing its protection. It applies to: the dimensions of the roof, the type of roof structure and the type of roofing, the facade design and composition, architectural details and construction joinery, the type of building materials used, the traditional architectural and functional arrangement of the building. All these requirements, and often - restrictions, are placed in the local spatial development plan of each municipality. They are justified because a change of the architectural and structural system obviously reduces the historic value of the monument. The preservation of the original architectural and functional layout is particularly important when it concerns historic buildings representing specific, often unique representatives of architectural styles that characterize historical epochs. In the local spatial development plan, arrangements for the protection of monuments of architecture and construction should be made, including :

- defining the protected values of existing buildings (historical line of construction, form and size of buildings, form of roofs, facade design, colors, materials, etc.) and establishing a ban on its alteration or restoration order,
- rules of land development - building plots (location of buildings);
- definition of acceptable changes and conditions of their implementation;
- definition of dimensions and aesthetic, technical and utility parameters for the new building (course of the building line, form and size of the objects, form of the roofs, development of the facade, colors, materials, etc.);
- maintenance, restoration or redistribution of the parcel division (basic element of organization and space physiognomy);
- identification of other elements of the landscape context (including greenery, pavement and accompanying small architectural objects) to maintain or restore protected values.

Among the historic buildings subjected to revitalization processes three types of buildings can be distinguished:

- buildings used with a function unaltered from the moment of implementation (designed function = implemented function),
- buildings used with the function changed during the use of the building (designed function ≠ implemented function), with the adaptation of the building to the new function,
- buildings used with the function changed during the use of the building (designed function ≠ implemented function), without the adaptation of the building to the new function

The use of a building with a function incompatible with the originally designed function negatively affects its technical condition. This can even contribute to the deterioration of the structural safety and usability, as well as to the reduced durability of the structure, the finishing materials and the décor of the building. Adaptation to new functions often results in changes to the original architectural-architectural system that can radically change the original layout and be irreversible. In such a case analysis of these changes are necessary in order to properly determine the functional capabilities of a revitalized historic building.

Changing function of a historic buildings implies the necessity of assessing the basic requirements of buildings, and in particular [2] the safety of construction and safety of use. Adaptation of a historic building to a new function usually involves the adaptation of the architectural-structural system, installations and equipment of the building. All this, and especially the interference with the existing structure of the historic building, often destroyed or worn, in addition to the usual revitalization activities, may cause additional effects resulting from adaptation:

- force the change of the structural system or its part,
- force the usage of additional stiffening elements and structural solutions resulting from the weakening of the stiffness of the building,
- stiffening or replacing of basic structural elements such as walls, ceilings, columns, resulting from their weaknesses, by making new communication holes and new architectural arrangements in them, as well as by increasing or changing of loads,
- adaptation, strengthening, repair, replacement of ceilings and other structural elements (Fig.1);

**Fig. 1** Ceilings stiffened with steel beams with reinforced concrete strips in the thickness of the floor in the building at ul. Hoża 42 in Warsaw (photo by W. Terlikowski)

- change of existing architectural layout, adaptation of courtyards (Fig.2), extension, superstructure;

**Fig. 2.** The roofing of historic building in the old town of Leipzig (Photo: W. Terlikowski)

- force the adaptation, stiffening, repair, replacement of ceilings and other structural elements,

- force the implementation of new installations, including such disturbing the existing structure,
- installation of new equipment and fittings and new finishing elements,
- force the repair, expansion, raise, strengthen of foundations, due to the increment of stresses under them.

In many historic buildings, the usability of premises and buildings has been changed during their use, for example the original residential function (in palaces, mansions, residences, villas) has been transformed into a public function (offices, hotels, commercial functions). This trend is current in modern times. All structural changes in historic buildings leave a trace in the form of genetic transformations. Under the concept of genetic transformation of the structure of historic buildings, it is necessary to understand all the changes that took place in the history of the object, in relation to the initial pattern of construction technology and materials (for example changing brick bonding in the structure of the wall, the type of mortar or masonry units) [5]. According to [5] the following types of structural systems can be distinguished:
- not transformed structural systems, characterized by original, period-specific construction technology, construction materials that were used, architectural and structural design, composition, with the primary function of the whole object and individual rooms preserved;
- construction systems transformed in terms of structure, technology, form, function in the past that partially or to a large extent meet the requirements and characteristics of a given historical age, although they are actually reconstructed by anastylosis or materials and elements analogous to primary ones in primary technology or reproducing primary technology;
- structural systems transformed in such a way that the structure, technology, form, function, decor lost the features of historical authenticity, differing substantially from the original.

In all types of structural systems, remembering the principles of preservation and rehabilitation of historic buildings, it is necessary to properly assess their historical and scientific value using the criteria of authenticity, age, aesthetic, decorative, emotional and social values. The reliable assessment of these values is necessary when a proper assessment regarding the possibility of changing the function is made that results in an interference with the architectural and structural system. Sometimes the restoration of the primary function may involve the removal of transformations made during the period of early use of the building.

## 3 Adaptability of building and structural elements

Adaptability of a building [6] and its components is a set of features and properties that determine its adaptation to new functional and structural conditions (work in old or new structural systems, implying specific static schemes, the possibility of changes, repairs, exchanges, changes of loads resulting from new imposed loads, etc.). Specific load-bearing systems, in defined architectural systems, imply specific functional solutions. This relationship is reversible - the specific usable function determines the appropriate architectural-structural solution. The different structural systems of buildings have different adaptive capabilities, which in consequence allows for functional changes in various ways. Different types of wall systems (longitudinal, transverse, cross and mixed) provide high spatial rigidity of the building, but give no freedom in shaping the interior space of the building. They are designed for functions where multiple rooms can be used, divided by walls- residential, offices. Skeletal systems give the possibility of arbitrary shaping of space, without interference with the architectural and structural system. However, these are structures that have less spatial stiffness and are therefore more susceptible to all structural changes.

In order to objectify and regulate the assessment of the adaptability of a building and its components, specific points may be assigned to certain features. That provide the ability to perform a SMART analysis that can show the strengths and weaknesses of an adaptive project and identify potential advantages and problems to solve. Table 1 summarizes the characteristics of the selected types of ceilings found in historic buildings in Poland. Table 1 uses the symbols: 0,1,2,3,4 - rating scale, w- most often necessary replacement, + applied, - not applied, +/- sometimes used certain types.

**Table 1.** Evaluation of selected properties of the historical ceiling to determine the adaptability [6].

| Type of historic ceiling | | | Nowadays used ceilings | Adaptab ility | Load bearing capacity | Ceiling spread | Ceiling thickness (cm) |
|---|---|---|---|---|---|---|---|
| Wooden ceilings | | | +/- | 0,w | 2 | 2 | 25-50 |
| Ceilings on steel beams | Unreinforced with a vault or ceramic plate | Segmental ceiling | +/- | 3 | 2 | 2 | 10, 12, 15, 25 |
| | | Ekscelsior and Foerstera type ceiling | - | 0,w | 1 | 2 | 33 |
| | | Hourdis type ceiling | - | 1 | 1 | 2 | >20 |
| | | K.B. type ceiling | - | 1 | 1 | 2 | 24 |
| | Reinforced ceramic plate | Klein ceiling | + | 4 | 3 | 3 | >25 |
| | | Pomorze ceiling | - | 0, w | 1 | 2 | >20 |
| | Concrete | Segmental ceiling | - | 1 | 1 | 2 | >14 |
| | | With unreinforced plate | - | 1 | 1 | 2 | >14 |
| | | With reinforced plate | + | 4 | 4 | 3 | >6 |
| | | Plate ceiling | - Another types of plates are used | 1 | 1 | 2 | 8-10 |
| | With plates from another materials | Agri ceiling | - | 0, w | 1 | 2 | >14 |
| | | M.W. type ceiling | - | 0, w | 1 | 2 | 22 |

By analogy, all structural elements of the building's supporting system must be analyzed. Only such an extended analysis can provide a full representation of the adaptability of an existing structure. While assessing the adaptability of a historic building, some various aspects of this issue should be taken into account, including the adaptability of the entire building's load bearing system, the adaptability of structural components, the foundation, walls, columns, ceilings, beams, etc., possibility of adaptation to new functions (possibility of changing the load values, static schemes, architectural and constructional systems), possibility of adaptation of the building to the needs of the disabled people, possibility of improving the energy efficiency of the building partitions and the whole building.

## 3 Interdisciplinary diagnostics - assessment of the revitalization capacity of historic buildings

The assessment of the adaptability of a historic building, beyond the area of standard technical diagnostics, is part of extended, interdisciplinary diagnostics (Figure 3).

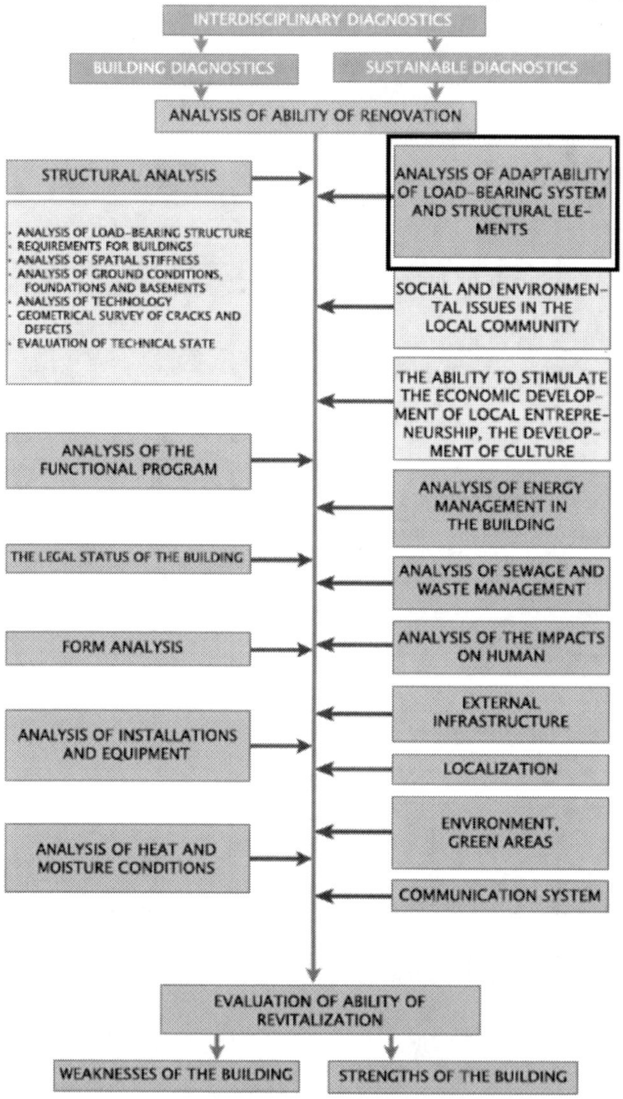


**Fig. 3.** Diagram of interdisciplinary diagnostics with adaptability analysis [7].

Such diagnostics should be used in the process of revitalization of historic buildings. It consists of examination of all technical and non-technical aspects of the state of the revitalized building, taking into account those resulting from the definition of revitalization and the principles of sustainable development. Such a diagnosis is the assessment of the revitalization capacity of a historic building, in which the assessment of adaptability is an element. The revitalization capacity [7] of a building (including the historic one) is a set of features, properties and states of the building, its construction, form, function, location, environmental, social and health values, and the economic viability of the planned revitalization including all aspects of revitalization and the principle of sustainable development [3].

## 4 Conclusion

Changing the function of a historic building is often a crucial part of the process of revitalization of the area where the building is located. It is an important tool for revitalization, through which not only technical and functional changes to the building are made, but above all, the problems of the local community are being solved. The construction process, which results in the change of the function of the building and its adaptation to modern demands - technical, utility and formal - requires innovative interdisciplinary diagnostic methods. Each building, its structure and its components have a certain adaptability that can be measured in a measurable and comparable manner. Evaluation of adaptability is part of the expanded, interdisciplinary diagnostics of the historic building, which should be applied in the process of its revitalization. Evaluation of adaptability is an element of wider diagnostic procedures, which results in assessment of the revitalization capacity of a building

## References

1. N. Doratli, S. Hoskara, M. Fasli, Cities **21**, 329 (2004)
2. D. Dvornik Perhavec, N. Šuman., J. Civ. Eng. Arch. **7**, 712 (2013)
3. L. Czarnecki, M. Kaproń, Int. J. Soc. Mat. Eng. Res. **17**, 99 (2010)
4. *Act on the Protection of Monuments and the Care of Monuments* (Dz. U. **80**., Warsaw, 2003)
5. W. Borusewic, *Konserwacja zabytków budownictwa murowanego* (Arkady, Warsaw, 1985)
6. W. Terlikowski, Mat. Bud. **9**, 82 (2013)
7. W. Terlikowski, JCEEA **33**, 531 (2016)

# Settlement of a foundation slab, non-uniform in depth

*Zaven* Ter-Martirosyan[1,*], *Armen* Ter-Martirosyan[1], and *Ivan* Luzin[1]

[1]Moscow state university of civil engineering, Yaroslavskoye shosse, 26, Moscow, Russia, 129337

**Abstract.** The paper describes the formulation and solution of problems for the quantitative evaluation of settlements and bearing capacity of rectangular and circular foundation slabs constructed on the base with deformation ($K$, $G$ or $E$, $\nu$) properties of continuously varying heterogeneity in the depth due to the conditions of its formation. It is shown that the inhomogeneity of the deformation properties of the foundation soils over depths has a significant effect on the formation of additional stress and strain states under the influence of a uniformly distributed external load over the area of a rectangle and circle, where, with growth of loading area ($A$) relative settlement $(S/\sqrt{A})$ depends nonlinearly on $\sqrt{A}$ and decreases with growth of $\sqrt{A}$ up to two or more times. As a computational model for the soil base, the paper considers the nonlinear geomechanical Klein model, according to which the stress-strain modulus of soils increases with depth according to the law of a power function of the form: $E(z) = E_1 z^n$, where $n \leq 1$. The solution of the SSC problems for an inhomogeneous ground half-space under the influence of a local load was obtained by an analytical method using the Mathcad software complex on the basis of the Boussinesq-Frohlich concentrated force problem. The results of the solution are presented in tables and distribution diagrams $\sigma_z$ and $S(z)$, as well as $S = f_2(\sqrt{A}, n)$.

## 1. Introduction

Experience in the design, construction and operation of high-rise buildings and heavy energy facilities on large-scale foundation slabs with an area of more than 1000 m$^2$ has shown that the design settlement calculated from the homogeneous half-space model significantly exceeds the actual settlements (more than 2 times) and that the scale factor of the slab foundation requires consideration.

The heterogeneity of the foundation ($E$ ($z$)) has a significant effect on the nature of the formation of an additional stress-strain condition (SSC) under the influence of an external load transmitted through a large-scale slab foundation.

* Corresponding author: gic-mgsu@mail.ru

The present paper describes the formulation and solution of problems on the quantitative estimation of the SSC of an inhomogeneous foundation under the influence of an external load uniformly distributed over the area of a rectangle and a circle, and for different proportions of length and width $(l/b,\ b=2a)$, as well as ($l{\cdot}b$ or $\pi R^2$), including dependencies $S(P)=f(\sqrt{A},n)$.

As a computational geomechanical model, a hypothetical Klein foundation model was adopted, in which the stress-stain modulus increases continuously with depth in accordance with the power law

$$E(z)=E_1\cdot z^n,\ (n\le 1) \tag{1}$$

where $E_1$ - stress-strain modulus at a depth of 1 m, $n$- heterogeneity parameter.

To determine additional SSC of an inhomogeneous foundation from an external load, the well-known solution of the Boussinesq-Frohlich problem of force applied to an inhomogeneous half-space is considered as the initial one:

$$\sigma_z=\frac{\nu\cdot P}{2\pi}\cdot\frac{z^{\nu}}{R^{\nu+2}} \tag{2}$$

where $\nu$ - stress concentration factor ($\nu\ge 3$). At $\nu=3$, the solution corresponds to a homogeneous half-space.

## 2. The SSC of a reconsolidated ground half-space under the influence of concentrated force (the Boussinesq and Frohlich-Klein problem)

It is known [1-2] that under the action of a concentrated force on a ground-half-space, homogeneous in depth and along strike, the stress components and surface settlements are determined by the Boussinesq formulas:

$$\begin{gathered}
\sigma_z=\frac{3P}{2\pi}\cdot\frac{z^3}{R^5},\\
\sigma_x=\frac{3P}{\pi}\left\{\frac{zx^2}{R^5}+\frac{1-2\nu}{3}\left[\frac{R^2-R\cdot z-z^2}{R^3(R+z)}-\frac{x^2(2R+z)}{R^3(R+z)^2}\right]\right\},\\
\sigma_y=\frac{3P}{\pi}\left\{\frac{zy^2}{R^5}+\frac{1-2\nu}{3}\left[\frac{R^2-R\cdot z-z^2}{R^3(R+z)}-\frac{y^2(2R+z)}{R^3(R+z)^2}\right]\right\},\\
\sigma_m=\frac{P}{\pi}(1+\nu)\frac{z}{R^3},\\
S=\frac{P(1+\nu)}{2\pi E}\left[\frac{z^2}{R^3}+\frac{2(1-\nu)}{R}\right],
\end{gathered} \tag{3}$$

where $R=x^2+y^2+z^2$, $\nu$ - Poisson's ratio, $P$ - concentrated force.

These dependences do not take into account the intrinsic weight of the soils and the initial stress state, which can cause the inhomogeneity of the initial SSC of the array in depth, and, consequently, the inhomogeneity in deformation properties. Given this circumstance, Frohlich [3] proposed the formula for $\sigma_z$ of Boussinesq (3) to introduce the stress concentration parameter $\nu$ to better match the results of experimental studies, i.e. suggested to present the dependency $\sigma_z(P,x,y,z)$ if the form of (2). Klein has showed [4] that dependence (2) satisfies the equilibrium conditions for any value, and the conditions of deformation continuity only at the following values:

$$\nu = n + 3 \text{ at } \mu_0 = \frac{1}{2+n}$$
$$\nu = n + 2 \text{ at } \mu_0 = \frac{1}{1+n} \qquad (3)$$

where $\mu_0$ - coefficient of lateral expansion.

In addition, Klein has considered the problem of stress distributions $\sigma_z$ under the center of a rectangular and circular shape of the areas and compiled the table of corresponding coefficients $K_0(b/z)$ and $K_0(a/z)$. For the case $\nu = 3$, Love obtained formulas for determining the stresses along the vertical lines passing under the center and at the angle of the rectangle [5]. According to this decision, the tables were compiled listed in SP 22.13330.2011. Below we consider the problem of determining stresses $\sigma_z$ on any vertical within and outside the rectangle at $\nu > 3$, i.e. at $\nu = 5$.

## 3. SSC of an inhomogeneous ground half-space under the influence of a uniformly distributed load over the area of a rectangle for the case of an inhomogeneous half-space

For the reliable In the case of a uniformly distributed load with intensity $p = P/A$ over the area of the rectangle $(2b \geq 2a)$, stresses $\sigma_{zp}(z)$ on any vertical passing through the points *M*(*x*,*y*) on the surface, on the basis of (1) can be determined by the formula:

$$\sigma_{zp} = \frac{\nu p z^{\nu}}{2\pi} \int_{-a}^{+a} \int_{-b}^{+b} \frac{d\xi d\eta}{\left[(x-\xi)^2 + (y-\eta)^2) + z^2\right]^{\frac{\nu+2}{2}}} \qquad (5)$$

Comparative evaluation $\sigma_{zp}$ at $\nu = 3$ and $\nu = 5$ is shown in Fig. 1 with an indication of the lower boundary of the compressible thickness, determined from the condition $\sigma_{zp} = 0{,}5 \cdot \sigma_{zg}$.

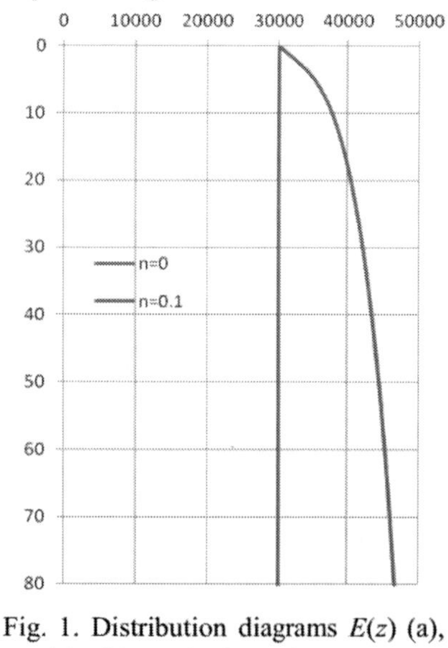


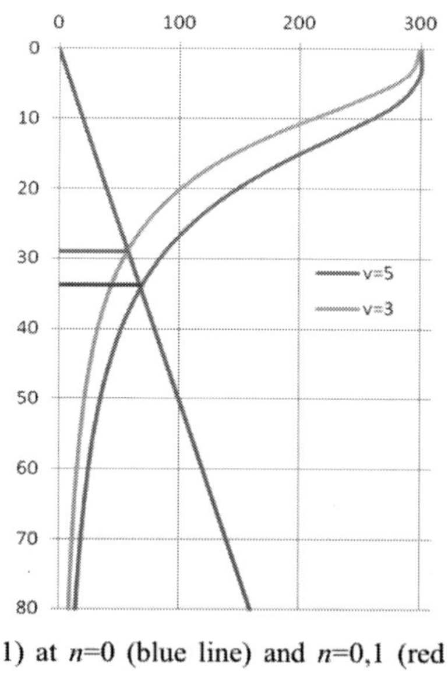


Fig. 1. Distribution diagrams *E*(*z*) (a), made from (1) at *n*=0 (blue line) and *n*=0,1 (red line), and $\sigma_{zp}(z)$ (b), made from (5) at $\nu = 3$ (green line) and $\nu = 5$ (blue line) with an indication of the corresponding lower limits of the compressible thickness at $\sigma_{zg} = 0{,}5 \cdot \sigma_{zp}$

Settlements at any point of the surface of an inhomogeneous half-space, $M(x,y)$ on the basis of (8) can be determined by integrating the relative deformation $\varepsilon_z(x,y)$, i.e.

$$S(x,y)=\int_0^h \varepsilon(z)\cong 0{,}8\int \frac{\sigma_z(z)}{E(z)}dz \qquad (6)$$

where $E(z)$ is determined by (1), $\sigma_z(z)$ - by (8), $h$- power of compressible thickness. Putting (7) into (6), we obtain

$$S(x,y)=\frac{\nu p}{2\pi E_1}\int_0^h z^{\nu}\int_{-a}^{+a}\int_{-b}^{+b}\frac{\dfrac{d\xi dz}{\left[(x-\xi)^2+(y-\eta)^2+z^2\right]}}{z^n}dz \qquad (8)$$

where $h$- lower boundary of the compressible thickness. Integration of this expression with the help of the Mathcad PC showed that the settlement in the center of the loaded area of the rectangle is 4 times larger than at the corner point, which coincides with the solution of a homogeneous half-space. The advantage of this method of solving the Love problem lies in the fact that it allows determining the settlement of any point of the surface both of as heterogeneous $(\nu>3)$ and homogeneous $(\nu=3)$ ground half-space within and outside loading area and to build settlement isolines of the surface. This allows determining the bedding value under a rectangular foundation $C(x,y)=\sigma_z(x,y)/S(x,y)$.

## 4. Influence of the loading area of a rectangular foundation on the settlement value of an inhomogeneous half-space

To determine the deformation characteristics Expression (10) makes it possible to quantify the effect of the loading area $A=l\cdot b$ on the value of relative average settlement $\bar{S}$, where $\bar{S}=(S_0+S_{yr})/2$, at any values of the inhomogeneity index $n$ (1), at any aspect ratio of a rectangular loading area $(l:b)$ and for a given distributed load $p$ for $\nu=3;4$ and 5. Given that $S_0=4S_c$, we can write $\bar{S}=5S_0/8$. Calculations performed for different values $\sqrt{A}=10,25,50,75,100$ and at the same load values $p$ have shown a nonlinear dependence between the average settlement $\bar{S}$ and value of $\sqrt{A}$ in the form

$$\bar{S}_i=S_{10}-\alpha(n)\ln(A_i/A_{10}) \qquad (9)$$

where $S_{10}$ - settlement of a foundation with area of 10 m$^2$, $\alpha(n)$ - coefficient depending on the degree of heterogeneity of the foundation $n$, $A_i$ - area of $i$-th foundation, wherein $A_i>A_{10}$. By constructing dependencies $\bar{S}-\sqrt{A}$ for values $n=0;0{,}25;0{,}5$ and 1,0, we obtain a family of lines (Figure 2). Dependence of average settlement $\bar{S}$ on the degree of heterogeneity of the foundation $n$ can also be obtained by comparing the foundation settlements with fixed values of the foundation area ( $A=a\cdot b=const$ ), stress concentration factor ν and the intensity of the applied load $p$ given (8) for different values of $n$. Analysis of these dependences at $\nu=5$ have shown that under identical loads $p_1=p_2=p$ at $\nu=5$ we can obtain a relation of the form

$$\frac{\bar{S}_1}{\bar{S}_2}=\left(\frac{\sqrt{A_1}}{\sqrt{A_2}}\right)^{1{,}3-n} \qquad (10)$$

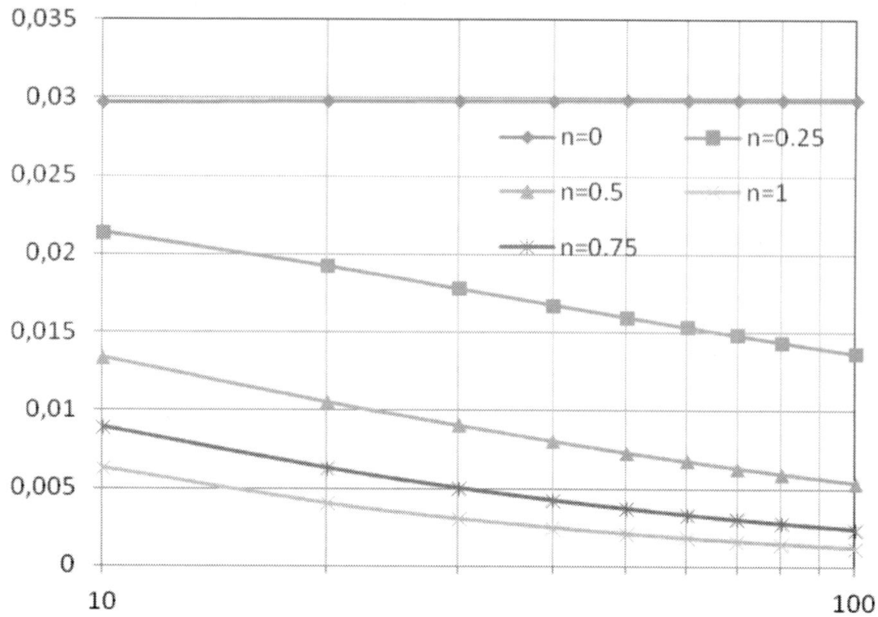


Fig. 2. Dependence of relative settlement ( $S/\sqrt{A}$ ) on $\sqrt{A}$ on a foundation, homogeneous ($n$=0) and a heterogeneous ($n$=0,25; 0,5; 0,75 and 1) in depth, on a semilogarithmic scale

## 5. Determination of the settlement of a reconsolidated heterogeneous foundation under a distributed load along the area of the circle

Dependence of average settlement $\bar{S}$ on the degree of heterogeneity of the foundation $n$ can also be obtained by comparing the foundation settlements with fixed It is known that the SSC of a homogeneous foundation $(n=0, \nu=3)$ under the influence of the distributed load with intensity $p[\kappa\Pi a]$ over the area of a circle of radius $r=a$ is determined by the formulas

$$\sigma_{zp}=p\left[1-\frac{z^3}{(a^2+z^2)^{3/2}}\right] \tag{11}$$

$$S(r=0)=\frac{2(1-\nu^2)ap}{E} \tag{12}$$

For the general case, when $n\le 1$, $\nu>0{,}3$, the stress distribution is $\sigma_z$ on any vertical within and outside the loaded area of the circle (Figure 3) can be determined given (6) by the formula:

$$\sigma_z=\frac{\nu\cdot z^\nu p}{\pi}\int_0^a\int_0^{2\pi}\frac{\rho d\rho d\theta}{\left(\rho^2+b^2+z^2-2bp\cdot\cos\theta\right)^{\frac{\nu+2}{2}}} \tag{13}$$

where at $0<b<a$, the point is within the area of the circle, and at $b>a$ - outside. At $\nu=5$ and $b=0$ we obtain $\sigma_z$ for the center point

$$\sigma_z=5\cdot z^5 p\int_0^a\frac{\rho ds}{(\rho^2+z^2)^{3,5}} \tag{14}$$

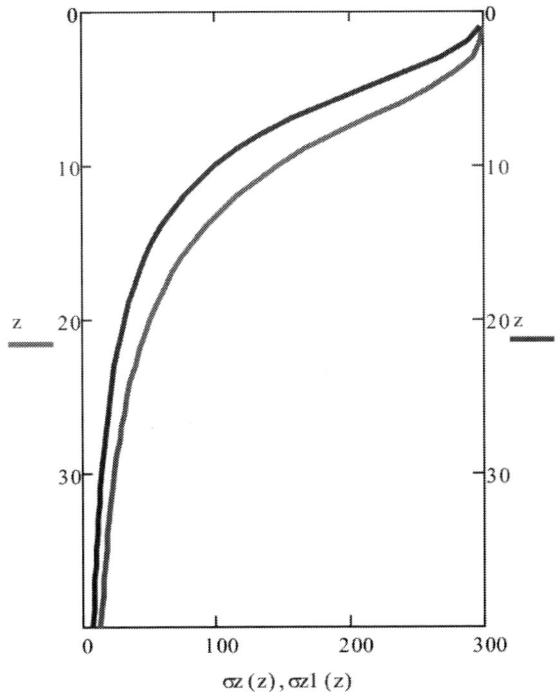


Fig. 3. Distribution of vertical stresses $\sigma_z$ at $\nu = 3$ (blue line) and $\nu = 5$ (red line) by depth calculated by the formula (13)

The integral over (14) is known, i.e.

$$\sigma_z = \frac{5 \cdot z^5 p}{3}\left[\frac{1}{z^3} - \frac{1}{(\sqrt{a^2 + z^2})^3}\right] \tag{15}$$

To determine the settlement at the point $r = 0$, we can use the expression (8); then, taking (15) into account, we obtain

$$S(0) = \frac{4p}{3E_1}\int_0^z \frac{z^5}{z^n}\left[\frac{1}{z^3} - \frac{1}{(a^2 + z^2)^{3/2}}\right]dz \tag{16}$$

Calculations from (16) show that $S(0)$ depends nonlinearly on the radius of the circle $r = a$, and with increasing radius, the settlement $S(0)$ decreases; for the case of a homogeneous $n = 0$ and heterogeneous $n = 0{,}5$ of the foundation, the settlements at $a = 5$ m differ by a factor of 2.

To determine the settlements of other points on the surface of an inhomogeneous foundation, it is necessary to integrate (13) at $b > a$. Calculations show that the deflection of the surface of an inhomogeneous foundation under uniform load along the area of the circle has the form (Fig.). Comparing this curve with Zaretsky influence curve [6], we see that they coincide qualitatively. Comparing the settlements of the surface of two circular foundations with area of $A_1$ and $A_2$ for an inhomogeneous foundation, all other conditions being equal, we see that the dependence $S_1 / S_2$ has the form:

$$\frac{S_1}{S_2} = \frac{\sqrt{A_1}}{\sqrt{A_2}} \tag{17}$$

It should be noted that Yu.K. Zaretsky hasn't considered when solving the problem of drafting a circular foundation on a surface, non-homogeneous by Klein. Only the influence coefficient was used $K(r)$, which to some extent is an approximation. The results of this solution in the form of a function $S_1/_2 = f(A_1/A_2, n)$ were included in the SNiP 2.02.02 - 85. However, stress distribution $\sigma_z$ by depth in calculation of the settlement by the method of layerwise summation was determined for a homogeneous foundation, which is also a certain approximation.

The results of long-term observations of settlements of the foundations of buildings and structures show that the design settlement often exceeds the actual one by a factor of 2-5 [7]. Consequently, based on the above solution, it is possible to introduce an averaged correction coefficient that increases the stress-stain modulus in calculating the sediment of the foundations of buildings and structures.

## 5. Conclusion

1. The SSC of a foundation, heterogeneous in depth, ( $E(z) = E_1 z^n, n < 1$ ) is substantially different from the SSC of a homogeneous foundation (*n*=0), including foundations with an area of more than 1000 m$^2$.
2. When designing large-scale foundations of high-rise buildings and heavy power plants, it is necessary to take into account the variability of the soil deformation modulus of the foundation according to Klein.
3. The heterogeneity of deformation properties of soils in depth *E*(*z*) may be due to the history of formation of the earth cover, as well as the variability of the initial stress state $E(z) = f(\sigma_{zg})$ .

## Acknowledgements

This study was performed with the financial support of the RF Ministry of Education and Science, grant №7.3225.2017/ Project Part.
All tests were carried out using research equipment of Head Regional Collective Research Centre of Moscow State University of Civil Engineering.

## References

1. Y. Boussinesq. Application des potentiels à l'étude de l'équilibre et du mouvement des solides élastiques, Paris, 1885
2. V.A. Florin. *Osnovy mekhaniki gruntov*. Moscow, Gosstroyizdat, **T.I**, (1961).
3. O.K. Frohlich. *Raspredeleniye davleniy v grunte*. Moscow, Stroyizdat, (1933).
4. G. K. Klein. Uchet neodnorodnosti razryvnykh deformatsiy i drugikh mekhanicheskikh svoystv gruntov pri raschete sooruzheniy na sploshnom osnovanii. MCEI collected papers. **14**, pp. 85-88, (1956).
5. A. Love. *Matematicheskaya teoriya uprugosti*. ONTI, (1935).
6. Yu.K. Zaretsky, M.Yu Garitselov. *Glubinnoye uplotneniye gruntov udarnymi nagruzkami*. Moscow, Energoizdat, (1989).
7. Z.G. Ter-Martirosyan. *Mekhanika gruntov*. Moscow. ASV, (2009).
8. M. Mitew-Czajewska, ACE **62(4)**, 73 (2016)

# Theoretical computation background for transformation of foundations using pile drains

*Zaven* Ter-Martirosyan[1], *Armen* Ter-Martirosyan[1], and *Vitalii* Sidorov[1]

[1]Moscow state university of civil engineering, Yaroslavskoye shosse, 26, Moscow, Russia, 129337

**Abstract.** In the design of foundations for buildings and structures of various purposes, including improved risk, weak water-saturated clay soils with low mechanical characteristics are often found on a construction site. One of the possible ways of using them as a foundation is to seal them in various ways, including using pile drains of sand or rock stone material that are capable of both absorbing the load at the base and accelerating the process of filtration consolidation. This paper describes an analytical solution to the problem of interaction between the pile and the mattress with the surrounding soil of the foundation, taking into account the possibility of expanding the pile shaft. Solutions are obtained for determining the stresses in the shaft of the pile drain and in the soil under the mattress. The solution takes into account the influence of the pre-stressed state of the foundation after compaction on the formation of a stress-strain state during the erection and operation of structures. The solutions are relevant for consolidating pile drains made of rubble or for jet grouting piles, the rigidity of which is comparable to the rigidity of the surrounding soil. The paper describes the technique for determining the characteristics of the strength and deformability of the converted foundation and the results of large-scale tests at the experimental site for the construction of a large energy facility in Russia.

## 1. Design technique and tools for consolidating of weak bases

In the presence of a high capacity of weak soils in the base, it is not economically expedient to replace it completely with sand or ground bed. In addition, it is quite difficult to provide a quality compaction (ensuring the design quality parameters of the embankment - the density of dry soil or the compaction factor over the entire thickness of the bed) [1]. At present, two methods are used for compaction of weak water-saturated bases: surface compaction using sand draines and embankments, and deep compaction with the use of deep ramming, etc., creating considerable radial stresses in the weak base.

The proposed method is based on the consolidation of cohesive soils under a constantly acting distributed load in the presence of drains, which are rock stone columns. After the construction of the piles, a crushed stone or sand mound is made designed to collect and drain water from the pile drains. It makes it possible to reduce the unevenness of the deformations of the foundation in the process of filtration consolidation.

[1] Corresponding author: gic-mgsu@mail.ru

In this technique, it is proposed to realize the compaction of a weak foundation of high capacity with the help of a device in it piles of crushed stone.

1. Construction of piles of rock stone is made by batch indenting of an inert material (sand, rock stone, etc.) into the borehole with a large vertical force of the submerged projectile, which results in the occurrence of significant radial stresses and expansion of the walls of the borehole. In this case, excessive pore pressure appears in the water-saturated soil mass around the pile.

2. Water permeability of a pile from rock stone is higher than that of the surrounding weak ground, by several orders of magnitude, which allows them to fulfill the role of drains significantly accelerating the process of filtration consolidation, increasing the rate of the process of water squeezing out of the ground and also shortening the horizontal filtration path (see Fig.1).

3. After manufacturing rock stone piles and finishing the filtration consolidation process, the converted composite (consisting of the consolidating columns and surrounding compacted ground) foundation acts as a single transformed array for which it is possible to determine the specified mechanical characteristics.

The device of ground piles requires equipment for pushing the pilot borehole and pushing loose pile material into it.

For drilling, any standard equipment used in the construction area may be applied.

The pilot borehole is formed by pressing the working tool to the required depth. Thus, the first compaction takes place. The tool is not removed yet. Then, a bunker is placed on it, which is filled with rock stone. After that, the slide is opened, and the rock stone reaches the lower end of the tool, then the tool is lifted to the specified height, and the rock stone that has reached the bottom of the borehole remains at the same depth and is compacted by pressing the tool (tool immersion) for a certain distance. Thus, the pilot borehole expands significantly to form a rock stone column, and the soil around the column is sealed radially. Compaction of the ground surrounding the column causes activation of the consolidation process due to the appearance of excessive pore pressure.

The simplest pressing tool of a machine can be a pipe of appropriate rigidity (capable of perceiving a vertical load of up to 300 tons) with a closed lower end.

## 2. Analytical solutions

1. Design of pile drains. For the design of the transformation of a weak foundation with pile drains, we must consider two successive processes - the interaction of the surrounding ground with the expanding pile shaft (in the process of its arrangement) and the interaction of the pile shaft and surrounding soil under load as part of the foundation. In this case, the initial and final radius of the piles are usually pre-determined, that is, the radial movement during the construction of the compaction unit is a predetermined value. The external boundary of the calculated cell is determined by the step of the arrangement of the pile drains, on which the degree of transformation of the weak foundation depends.

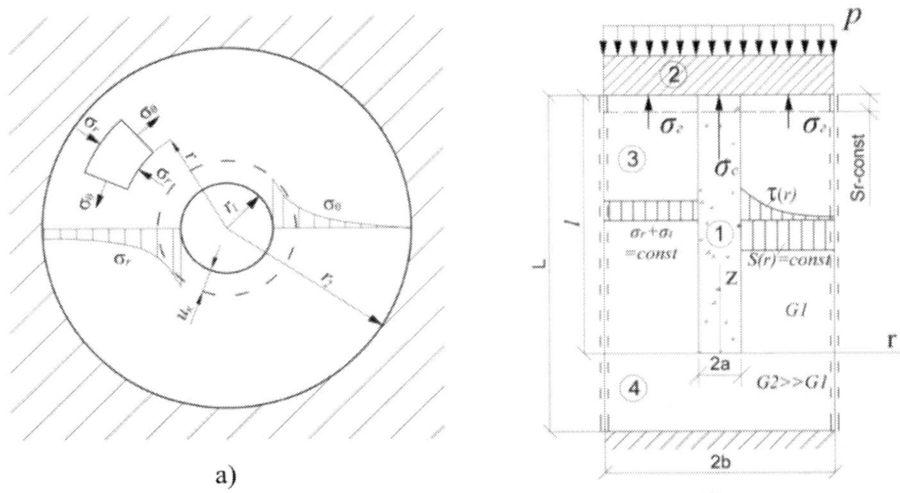


a)

б)

**Fig.1** – a) scheme for expanding the diameter of the pilot borehole during the pile manufacturing process; b) the design scheme of the pile's interaction with the surrounding soil and the foundation slab in the foundation structure according to the compression pile scheme). 1 – pile drain, 2 - foundation slab, 3 - compacted layer, 4 - underlying dense ground

On the basis of a joint examination of geometric, physical and equilibrium equations, the SSS problem of a ground shaft can be reduced to solving a differential equation in displacements [2] of the form:

$$\frac{d^2u}{dr^2} + \frac{1}{r} \cdot \frac{du}{dr} = \frac{u}{r^2} \tag{1}$$

Next, we can write the resulting expressions to find the radial and tangential stresses on the pile-ground contact.

$$\sigma_r = \frac{E_2 A}{(1+\nu_2)\cdot \nu_2'}\left(1 + \frac{r_2^2}{r^2} + \left(1 - \frac{r_2^2}{r^2}\right)\xi_2\right) + \frac{\sigma_{z2}\cdot \xi_2}{(1+\nu_2)\cdot \nu_2'} \tag{2}$$

$$\sigma_\theta = \frac{E_2 \cdot A}{(1+\nu_2)\cdot \nu_2'}\left(\left(1 - \frac{r_2^2}{r^2}\right) + \xi_2\left(1 + \frac{r_2^2}{r^2}\right)\right) + \frac{\sigma_{z2}\cdot \xi_2}{(1+\nu_2)\cdot \nu_2'} \tag{3}$$

In order to determine the vertical load, which must be transferred through the tamping equipment to the ground, it is necessary to use the limiting equilibrium equation, where the values of vertical and radial stresses will be present as the main stresses. At the same time, the vertical stress should be extreme, so that the layer of the trawl material may exhaust its strength, collapse and take its design position in the already extended well.

The limiting equilibrium equation takes the form:

$$\frac{\sigma_z^* - \sigma_r}{\sigma_z^* + \sigma_r + 2c \cdot ctg\varphi} = \sin\varphi \tag{4}$$

Then the value of the limiting vertical stress is expressed as follows:

$$\sigma_z^* = \frac{\sigma_r(1+\sin\varphi) + 2c\cdot\cos\varphi}{(1-\sin\varphi)} \tag{5}$$

Based of (5), it is possible to express the force N needed to expand the pilot borehole to the required size:

$$N^{*} = \pi \cdot r_1^2 \cdot \sigma_z^{*} \tag{6}$$

2. Interaction of the ground pile of extreme rigidity with the surrounding soil in the composition of the foundation, taking into account the expansion of its diameter in the loading process. Considering the equality of radial displacements on the radius of the pile (r = r1), and also taking into account the equality of deformations [3], we can obtain an expression relating the load to the base slab p and the resulting lateral extensions of the pile under the load u1.

The expression has a complicated form and is represented in a simplified way with the help of special factors, which are presented in [4] the following form:

$$u_1 = p\left(\frac{G}{F} + \frac{J}{K}\right) : \left(\frac{I}{K} + \frac{H}{F}\right) \tag{7},$$

Factors A,B,C,D,E,F,G,H,J,K depend on the geometric and mechanical parameters of the pile and surrounding ground both around it and under its toes.

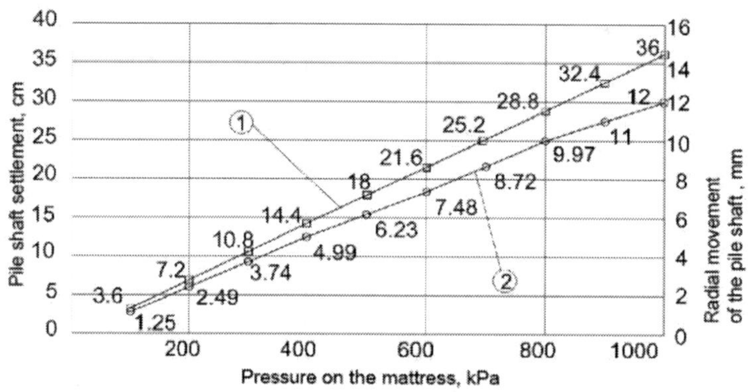


**Fig.2** – Dependency graph of the draft of the pile shaft (cm) on the pressure on the mattress slab of the design cell, kPa (1), radial displacement of the pile shaft (mm) on the pressure on the mattress slab, kPa (2)

3. Determination of the given values of the foundation deformation modulus, packed with rock stone pile drains. After the arrangement of pile drains on the site and consolidation of the weak foundation, it is necessary to consider the work of the composite substrate under load. The main issue at this stage is to find the bottom of the foundation of the structure. Since the bearing elements of such a system are the piles themselves and the surrounding soil, it is very difficult to consider them separately. Therefore, dependencies are required to determine the reduced strain modulus [5].

Dependences are obtained for determining the reduced strain modulus in cases where dense soils lie under the sole of the pile drains and the ground is not extruded under the sole, and also for the case when the soils under the sole are deformable and the punching is possible. In the first case, the expression looks like:

$$\overline{m} = \frac{m_c \cdot m_2}{m_2 \cdot \lambda + m_c(1-\lambda)} \tag{8},$$

where $\overline{m}$ - relative compressibility coefficient of the calculated cell as a whole;

$m_c$ - relative compressibility coefficient of the pile drain material;

$m_2$ - coefficient of relative compressibility of the surrounding compacted soil of the foundation.

If a punching of the foundation ground is possible, a more complex relationship the lower end of the pile drain can be obtained, which is more conveniently written in a reduced form:

$$\overline{m} = \frac{D \cdot E}{l} \tag{9}$$

where D,E – values that depend on the geometric and mechanical characteristics, and are described in detail in [4].

Strength characteristics of the transformed foundation are determined by performing a virtual experiment by the finite element method. A triaxial test of the transformed cell is simulated according to Fig. 1b at various hydrostatic pressures in order to obtain the Coulomb limit line and determine the angle of internal friction and the specific adhesion for the cell as a whole. In this case, in the design scheme, each of their constituent parts (pile material, surrounding soil, mound, etc.) is assigned its own parameters.

On the axial-symmetrical design problem in the form of two shafts (the shaft of the ground pile in the shaft of compacted weak ground), the following loads are applied:

- radial, simulating the sample compression;
- vertical, simulating the load on the sample.

The virtual experiment is carried out as close as possible to the triaxial test procedure in accordance with GOST 12248-2010.

Several calculations (at least three) are performed with different compressive stresses to construct stress circles and subsequent laying them on the general graph in order to construct the limit line.

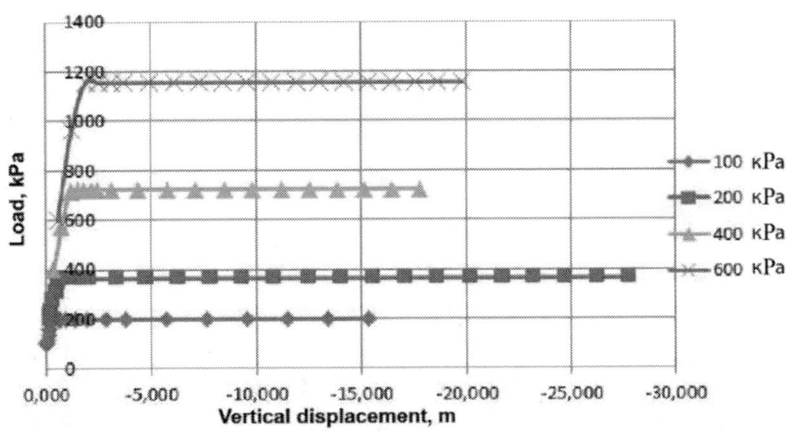


**Fig.3** – Dependency graph of vertical displacements (m) on the load (kPa) for different values of the compressive load (100 kPa, 200 kPa, 400 kPa, 600 kPa)

## 3. Field testing of soil substrates and the pile drain material

For the reliable design of a large nuclear power plant, a pilot site 50x15 m was created, on which three sections of a densified foundation were made with pile steps of 1 m, 1.2 m and 1.5 m with initial diameters of 325 mm, 426 mm and 530 mm, respectively.

After the installation of the consolidating piles and consolidation of the foundation, field tests were carried out with dies of the material of the resulting pile drains.

**Table 1.** Summary results of field tests of the pile drain material

| Depth of experiment h, m | Soil | Value of the strain modulus E, MPa | Average value of the strain modulus E, MPa |
|---|---|---|---|
| 7.0 | Rock stone of the ground pile | 120.5 | 140.52 |
| 7.0 | | 106.92 | |
| 7.0 | | 104.3 | |
| 7.0 | | 146.7 | |
| 7.0 | | 158.15 | |
| 7.0 | | 206.57 | |

In addition, tests of rock stone of pile drains were carried out in laboratory conditions using triaxial compression for specimens of 600 mm height and 300 mm in diameter. According to the results of triaxial tests, the deformation modulus of crushed stone at a density equal to the density of the material in the consolidating column is at least 103.57 MPa.

The tests showed that the rock stone material is sufficiently compacted during the process of piling using the technology in question. Obtaining high deformation characteristics of column material will lead to a significant increase in the overall specified deformation characteristics of the compacted foundation as a whole, since both columns and compacted soil around them will interact [6-8].

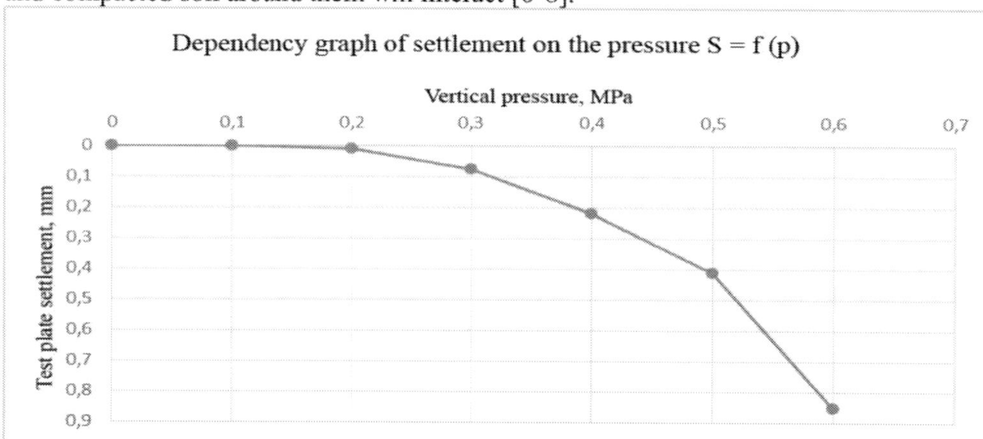


**Fig.4** – Curve of plate tests of rock stone material of a pile drain

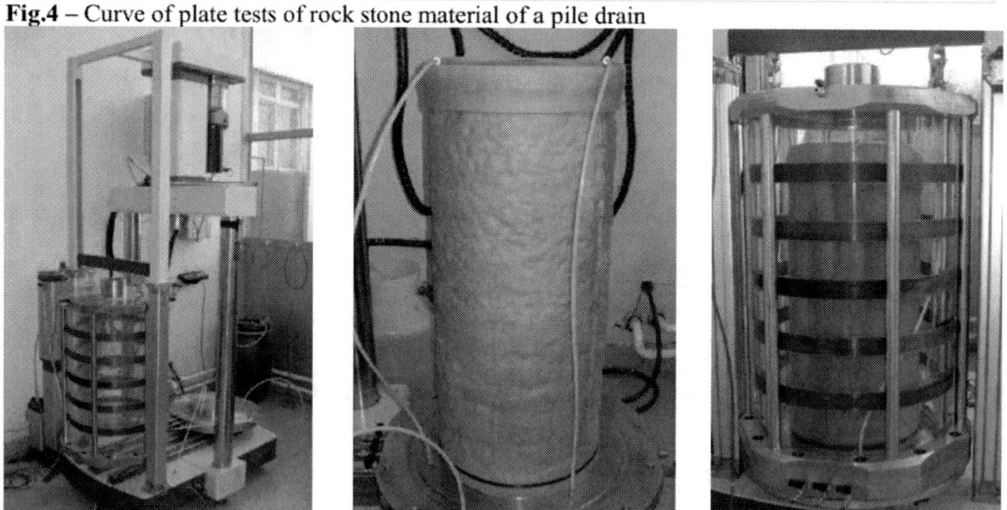

**Fig.5** - Triaxial tool for testing macrofragmental and heterogeneous soils with inclusions of large size

## 4. Nonstandard field tests of the transformed composite foundation

To determine the deformation characteristics of a consolidated foundation as a whole (a composite foundation consisting of a rock stone pile and compacted soil around it), the specialists of the MSUCE Research Institute performed tests of the design cell using flat plates over the area of 7854 $cm^2$ (1000 mm in diameter) in the well and at the bottom of the pit at a depth of 6.5 m, at the absolute mark of 145.00. A total of three tests were performed. The tests were carried out in accordance with GOST 20276-2012 *Soils. Field methods for determining strength and deformability characteristics*. The resulting strain moduli were 45.07 MPa, 42.88 MPa and 56.64 MPa, which is more than the design value of 40 MPa. The tests confirmed the assumptions about the joint operation of the elements of the composite foundation arranged according to the presented technology, as well as the correctness of the expressions themselves for determining the reduced strain modulus.

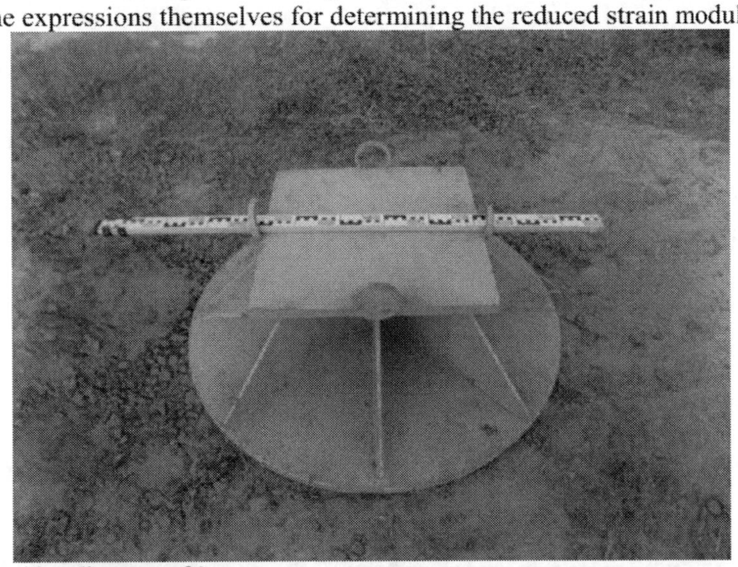

**Fig.6** – Test plate with a diameter of 1 m

**Fig.7** – General view of the test bench with a test plate of a non-standard size

## 5. Conclusion

1. The technology under consideration for transforming weak foundations is economically viable and effective. Low cost is ensured by the simplicity of the technological cycle, which requires fairly simple machines and mechanisms for its implementation.
2. Analytical solutions are proposed that are sufficient to determine the technological parameters of compaction - pile spacing, final diameter, working stroke of the consolidating equipment, and also to determine the specified deformation characteristics of the transformed foundation.
3. The tests of the design cell of the transformed base and the inter-sealed compacted soil carried out at the test site showed the efficiency of the applied technology and sufficient convergence of the mechanical parameters of the soils to the design ones.
4. Consequently, the developed technique of compaction of the foundation with rock stone pile drains can compete with traditional compaction methods and with more expensive deep foundations.

## Acknowledgements

This study was performed with the financial support of the RF Ministry of Education and Science, grant №7.3225.2017/ Project Part.
All tests were carried out using research equipment of Head Regional Collective Research Centre of Moscow State University of Civil Engineering.

## References

1. Z.G. Ter-Martirosyan, A.S. Abdulmalek. The stress-strain state of the compacted base. Soil Mechanics and Foundation Engineering. **6** pp. 8-11. (2007).
2. S.P. Timoshenko, J. Gudier. *Theory of elasticity.* Moscow, Nauka (Science), (1975).
3. N.I. Bezukhov. *Bases of theory of elasticity, plasticity and creep.* Moscow, Visshaya shkola (High school), (1968).
4. A.Z.Ter-Martirosyan, Z.G. Ter-Martirosyan, V.V. Sidorov. Interaction of compaction piles with surrounding soil with due regard for pile diameter expansion. Soil Mechanics and Foundation Engineering. pp. 1-8.(2016)
5. V.A. Barvashov. Experimental and theoretical studies of computational models of slab-pile foundations. Soil Mechanics and Foundation Engineering. **5** pp. 32-39. (2009).
6. A.A. Grevcev. The theory of expansion of the cavity and limiting resistance of the soil under the toe of driven piles in sandy soils. Zhilishnoe stroitel'stvo (Civil Engineering). **9** pp. 2-5. (2012).
7. Z.G. Ter-Martirosyan, V.V. Sidorov, P.V. Strunin. Calculation of the stress-strain state of a single compressible barrets and piles interactied with the soil. Zhilishnoe stroitel'stvo (Civil Engineering). **9** pp. 18-22. (2013).
8. S.G. Bezvolev. Problems of design and analysis of foundations when large groups of piles and other vertical elements to transform the soil massif. Geotechnics. **3** pp. 30-67. (2011).
9. M. Mitew-Czajewska, ACE **62(4)**, 73 (2016)

# Influence of parameters of mixing of the mortar mixtures on the performance of ornamental composites for facade coating

*Evgeniay* Tkach [1*], and *Kanapiya* Nurbaturov [2]

[1]Moscow state university of civil engineering, Yaroslavskoye shosse, 26, Moscow, Russia, 129337
[2]Limited Liability Partnership «INNOBUILD», Kazakhstan, Almaty, Auezov str., 84

**Abstract.** Studies have shown that the main physical-mechanical properties of decorative coatings based on colloidal cement systems greatly depend on the homogeneity of the structure of hardened stone, therefore, in the preparation of mortar mixtures were set the task of achieving this target. It is shown that vibrational mix of materials helps to ensure the preparation of mortar mixes with cement-based colloidal systems with a more homogeneous distribution of the components. The efficiency of vibrational mixing was determined by comparing the strength of the mortar mixes based on colloidal cement glue, cooked in vibromaster when the vibration acceleration with the strength of samples prepared in a standard mortar mixer of forced action without vibration. The results of the research confirmed some influence of the mineralogical composition of clinker the cement component of the colloidal material on the effect of vibration treatment solutions. Parameters preparation of the mortar mixtures based on cement colloidal material in a vibratory mixer. Optimum resonant operating frequency of the vibrations, at which is achieved the positive effect of mixing of the mixture is ensured with amplitude 5mm while accelerating 214,8 m/S$^2$ and duration of mixing 60-90s. It is established that vibropressure contributes to the intensification hydration processes to temperature is minus 5 $^0$C

## 1 Introduction.

The construction of modern buildings requires the use of a huge variety of finishes, the main purpose of which is not only the architectural appearance but also protection of structures against aggressive external environments [1]. The development of decorative coatings based on technologies of deep processing of raw materials using construction waste, the regulation of surface phenomena in disperse systems, selection and optimization of compositions of composite materials based on colloidal cement materials.

Currently acute problem of disposal of construction waste, including concrete waste that is generated in the result of the destruction of residential buildings and concrete structures [2-4]. When designing cement-colloid composites with the use of concrete scrap special

* Corresponding author: ev_tkach@mail.ru

attention is paid to the study of the influence of external factors on the processes of mixing and provision of operational properties of composites. Analysis of the research showed that improvement in processes of mixing of the mixture becomes possible only with the use of new effects, which are the vibration of the mixing device [5-7]. Therefore, to determine optimal conditions for a homogeneous distribution of the components of the mixture and obtain a homogeneous mixture it was investigated the influence of mixing on the basic properties of mortar mix.

## 2 Problems.

The goal of the presented research is set depending on the technological parameters of mixing particulate cement systems with the use of screenings from the concrete crushing to a uniform distribution of the components in the main mix volume, which ultimately determines the operational characteristics of decorative coatings for exterior designs. Studies have shown that the main physical-mechanical properties of decorative coatings based on colloidal cement systems greatly depend on the homogeneity of the structure of hardened stone, therefore, in the preparation of mortar mixtures were set the task of achieving this target. Studies have shown that when using conventional mixers for mortar mixes is dominated by processes that are not able fully to destroy agglomerates and do not provide the high homogeneity of the mixture. The analysis of previously performed studies have shown that a significant intensification of the processes of mixing can be achieved at the expense of vibration exposure [8-10]. The mobility characterizing properties such as viscosity, plasticity and uniformity of mortar mixture was determined by conventional stirring and mixing in a vibrating mixer. Water-cement ratio and mixing time of the mixtures was identical. The mobility of colloid-cement mortar was determined using the cylinder of Suttard the diameter of the spot spreading. Mixing of mortar mix produced with different amplitude (A) from 1; 3 to 5 mm and vibration frequency ($\omega$) of 100; 150 and 260 $s^{-1}$. The mixing time was changed from 60, 90, and 120s. The determination of optimal parameters of mixing will allow us to achieve the ultimate uniformity of distribution of finely dispersed solid phase with optimal content in the dispersion medium in the initial stages of technology and stabilization of the achieved homogeneity and dispersion after completion of phase transformation, i.e. after the formation of the solid structure. The deceleration or acceleration of the processes of formation and hardening of an artificial stone largely dependent on temperature of curing and the method of preparation of the mixture. When evaluating the efficiency of mixing the colloid-cement mortar mix it is important to determine how the method of mixing intensificare the process of setting and hardening. It is important to assess the impact of vibration effects on structure formation of hardening of the stone and on the process of curing the colloid-cement mortars in terms of positive and negative temperatures. Hardening at lowered positive and small negative temperatures is accompanied by insufficient development of exothermic processes due to the fact that the hardening of the decorative coatings occurs in a thin layer. The authors found that lowering the temperature slows down the hydration process and, consequently, increase the strength [11-12], as in the period of setting of the binder of the concrete freezing of the most dangerous. To determine the effect of low temperatures on the strength of the samples obtained by vibropressure was prepared mortar mixture in the ratio 1:3, in water the ratio W/R=0,35. Samples were collected with no external defects, the average density of which did not differ from the minimum values by more than 50 $kg/m^3$. Samples after 4 hours storage in air was placed in the freezer and a drying Cabinet, where they maintain the required temperature.

## 3 Results

These studies showed that the mobility of colloid-cement mortar mix that has been cooked in a conventional mortar mixer is much less than the mobility of the mixture is made in a vibrating mixer. Increase mobility in the vibrational mixing of the mixture was about 70%. It is because vibropressure eliminates negative phenomena by reducing the viscosity of mortar mix, whereby the cement paste becomes more homogeneous in composition, plastic and adhesive than conventional mixing. The obtained results of the influence of the content of the clinker components of the cement ($C_3S$, $C_2S$, $C_3A$, $C_4AF$) included in the cement composition of the colloidal material, the effect of vibropressure colloidal cement mortar mix is shown in figure 1. On the radiograph of the sample of Portland cement LLP "Bukhtarma cement company" are clearly visible the peaks of the major minerals: $C_3S$ (3,8660, 3,0320, 2,7754, 2,1826, 2,0519, 1,6225), $C_2S$ (2,6062, 1,7623, 1,4870), $C_3A$ (2,8819, 2,0519) and $C_4AF$ (7,3222, 2,6463). In the sample of Portland cement LLP "Standardalert" clearly visible changes in the composition alyumoferritnoy phase $C_4AF$ (7,6094, 2,6454) and the decrease in peak intensities of $C_3S$ (3.0318,2,7747, 2,6070), $C_2S$ (1,7642, 1,4868), $C_3A$ (2,8819, 2,0519).

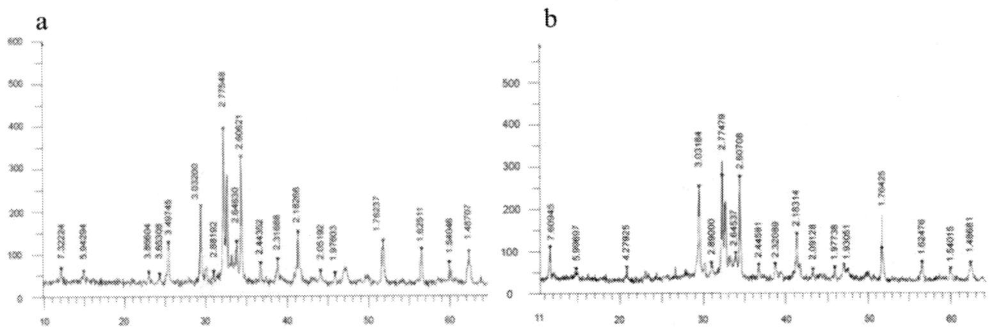

a – too "Bukhtarma cement company, b – too "Standardalert"
**Fig. 1.** Radiographs of ceramic components used in studies of Portland cement

The results are shown in table 1 (in the numerator of the filler of sand, in the denominator of screenings from the concrete crushing). Comparison of curing and mixing times allows to conclude the following: tensile strength of specimens in compression, cooked in vibromaster, in all periods of hardening higher than the corresponding values of the control samples prepared in the mortar mixer of forced action without vibration.

**Table 1.** The Influence of method and time of mixing of mortar on the basis of colloidal cement material and sand on the kinetics of curing

| Processing options | The limit of compressive strength at the age, MPa | | | |
|---|---|---|---|---|
| | 7 day | 14 day | 21 day | 28 day |
| Mixing without vibrations | 33,8/32,4 | 42,3/40,8 | 45,6/43,6 | 51,9/51,7 |
| Mixing in vibromaster for 60 s | 47,4/46,8 | 50,2/49,6 | 53,8/53,1 | 58,7/58,2 |
| Mixing in vibromaster for 90 s | 50,0/49,6 | 52,2/52,4 | 54,7/54,3 | 60,1/59,9 |

Laboratory tests showed that vibropressure shows a positive result, the process of hardening and the rate of strength development at different temperatures (table 2, in the numerator of the filler of sand, in the denominator of screenings from the concrete crushing). The results show that the effect of low temperatures, even for a short time negatively affects the process of curing

**Table 2.** The Effect of mixing and curing on durability of mortar mixtures

| The mixing of the mixture in mortar mixer | The temperature of hardening, $^{0}C$ | The tensile strength of the samples at the age, MPa | |
|---|---|---|---|
| | | 1 day | 3 day |
| Laboratory | +30 | 18,8/18,3 | 26,1/25,6 |
| Vibrating | | 27,6/28,0 | 34,1/32,6 |
| Laboratory | +20 | 22,6/23,3 | 32,5/31,4 |
| Vibrating | | 33,9/35,4 | 47,7/47,9 |
| Laboratory | 0 | 1,6/1,5 | 3,9/3,6 |
| Vibrating | | 2,4/2,2 | 5,7/5,3 |
| Laboratory | -5 | 1,2/1,1 | 2,7/2,8 |
| Vibrating | | 1,6/1,5 | 3,7/3,8 |

Hardening at low temperatures under the influence of early freezing due to the expansion of the mixing water slows down the hydration process, which negatively affects the formation of the structure of the stone, as freezing it will create a harmful amount of pores and channels that affect the strength characteristics of the material. However, the strength of samples obtained with the vibration stirring, it has an increase of strength compared to samples prepared in the laboratory mortar mixer to the age of 1 and 3 days and the curing temperature of 30 0C, from 30 to 50 %. At a temperature of 20 0C curing – from 77 to 52 %. At a temperature of 0 0C hardening from 36 to 39 %. At the temperature of hardening minus 5 0C – 33 to 37 %. Small increase in strength during curing at low temperatures can be explained by the fact that after thawing tumors retain the porous structure of the stone, and the increase in strength does not occur. However, vibrational mixing is more effective in this case. Increased strength of the samples can be explained by the microstructure of the hardened stone, studied at 28 days age. The microstructure of the contact zone (figure 2) binder and fine aggregate in the samples, prepared in different ways.

a)
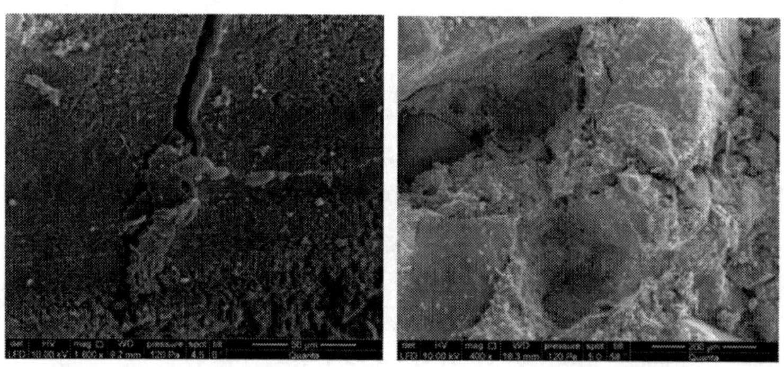

b) 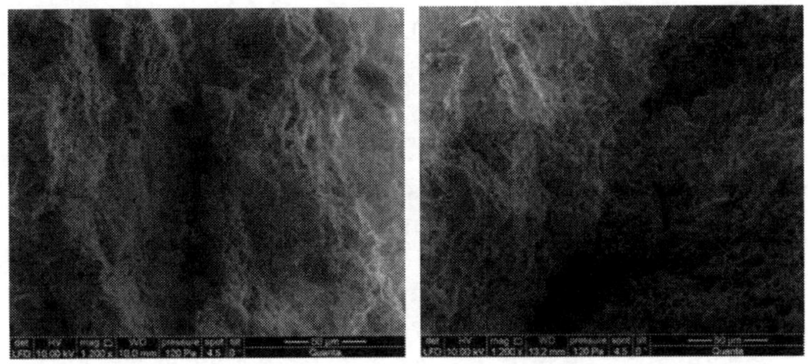

a) mixing without vibration; b) vibrational mixing

**Fig. 2.** The microstructure of the contact zone of binder and fine aggregate in the samples, prepared in different ways.

It is established that the strength characteristics of samples prepared in a vibratory mixer that harden at low and negative temperatures, comparable to the strength characteristics of the samples hardening at positive temperatures, which can be explained not only by high colloidal characteristics of cement composites modified with additives, but also mode mixing in vibratory mixer. The influence of vibratory impulses to the mixture improves the condition of the contact zone of sand with an astringent and also promotes the growth of tumors with improved composition and structure.

Table 3 presents comparative results of two methods of mixing, showing the dependence of the strength of mortar mix in different stages of hardening from the mixing equipment, the duration of the mixing ratio of the components of mortar mix.

**Table 3.** the dependence of the strength of mortar mix in different stages of hardening from the mixing equipment, the duration of the mixing ratio of the components of mortar mix

| The ratio of mixture components | Terms of hardening, day | Strength specimens in compression, prepared in laboratory mixer, MPa | The duration of mixing in vibromaster, s | | | |
|---|---|---|---|---|---|---|
| | | | 30 | 60 | 90 | 120 |
| | | | $R_{com}$, MPa | $R_{com}$, MPa | $R_{com}$, MPa | $R_{com}$, MPa |
| 1:1 | 1 | 28,1 | 37,92 | 39,50 | 39,93 | 38,72 |
| | 7 | 37,6 | 43,76 | 44,00 | 44,43 | 44,03 |
| | 14 | 46,4 | 57,86 | 56,92 | 60,36 | 59,62 |
| | 28 | 55,2 | 64,60 | 63,48 | 64,86 | 63,87 |
| 1:2 | 1 | 27,3 | 35,60 | 35,12 | 32,52 | 33,0 |
| | 7 | 36,1 | 41,64 | 42,56 | 40,83 | 42,0 |
| | 14 | 44,7 | 55,90 | 54,70 | 55,0 | 54,90 |
| | 28 | 54,9 | 64,28 | 62,90 | 62,60 | 62,60 |
| 1:3 | 1 | 23,7 | 29,93 | 30,15 | 31,57 | 31,22 |
| | 7 | 34,4 | 40,86 | 41,27 | 40,00 | 41,68 |
| | 14 | 42,6 | 51,14 | 47,70 | 51,20 | 52,0 |
| | 28 | 52,3 | 58,10 | 57,39 | 59,42 | 59,24 |

The data obtained show that the preferred ratio of components (colloidal cement material: screenings from the concrete crushing) provided given a certain "viability" of a solution it is necessary to choose proceeding from conditions of the production technology of works and conditions of use decorative coatings. This is the most effective method in the implementation of technological schemes that reduce excessive activity or, on the contrary, increase it, depending on the working conditions of coverage by increasing or reducing the ratio of the components. With optimal quantities and optimal mixing ratio (1:3) the greatest increase of strength is observed at a mixing time of 60-90s.

The effect of the duration of vibropressure manifests itself not linear – it rises up in the first 60s, then stabiliziruemost 90s, and with the increase of time of vibropressure to 120s, a decrease in the resulting effect, which is manifested in lowering of strength of the samples in all periods of hardening. This is confirmed by data of other authors, who observed a pronounced effect [5]. In the study based on the efficiency of mixing sand and mortar mixes from exposure time, shown that prolonged exposure to vibration mixing leads to a decrease of the results achieved technological effect and the hardening effect of stone (increasing the strength), and the achievement of the effect depends on the vibration parameters and the ratio of the components.

Given sufficient experimental data, it is important to establish the most rational ratio of the components of the colloidal cement glue and sand it the best combined grading. Experimental data showed an increase in the proportion of colloidal material cement strength mortar per unit volume of solution is increased in all terms of hardening. The greatest increase of strength is observed in the first 7 day hardening components at a ratio of 1:1 (from 42 to 70.5 %) – in the duration of vibropressure from 30 to 120 C. The most effective is a ratio of 1:3 (cement colloidal material : screenings from the concrete crushing), in which due to a sufficient concentration of particles of colloidal cement glue in the matrix created conditions favorable for the formation of the stone structure of low voltage intergrowths of crystals. It should take into account the fact that the increase in the content of screenings from the concrete crushing instead of natural sand in the mortar mix helps to reduce the development of deformation shrinkage, reduce the risk of spreading of shrinkage cracks and failure of the coating, and to reduce the share of the binder and therefore of cement, which is a priority for improving economic efficiency.

## 4 Conclusions and recommendations

Optimal vibration intensity and duration of mixing components of the mortar mixture based on cement colloidal materials contributes to obtaining a high uniformity of distribution of different phases. Determined that the optimum resonant operating frequency of the vibrations, at which is achieved the positive effect mixing of the mixture is provided with a vibration amplitude of 5 mm, when accelerating 214,8 $m/S^2$. The duration of stirring the mixture, necessary for obtaining adequate technological effect depends on the amount of kneading and 60-90 C. It is shown that vibrational mix of materials helps to ensure the preparation of colloidal mortar of higher quality by the degree of homogeneity. Vibratory mixing of mortar mixes with different particle size and density of components achieved greater homogenization, unattainable in the traditional way of mixing.

Confirmed a significant influence of the mineralogical composition of the clinker component of the colloidal cement glue on the effect of vibration treatment solutions. The difference in mineralogical composition affected the quantitative and qualitative effect. With the increase of tricalcium silicate and tricalcium aluminate in the cement clinker mixing efficiency is significantly increased.

It is established that vibropressure provides intense hydration, faster structure formation and, correspondingly, the growing strength of the mortar mixtures to a temperature of minus 50 °C. This can be explained by the dense structure of the hardened stone and more homogeneous structure of the tumors. Thus, the executed researches have allowed to establish the technological parameters of mixing particulate cement systems with the use of screenings from the concrete crushing with uniform distribution of components in the main volume with improved performance decorative composites for facade coatings.

## References

1. Oreshkin D. V. Problems of building materials and production of building materials // Building materials. 2010. **No. 11**. p. 6-9.
2. Pakhratdinov A. A., Tkach E. V., Oreshkin D. V. Environmental and geo-Environmental rationale for recycling concrete waste as crushed stone in the production of concrete products// Ecology of urbanized areas. – 2016. - **No. 3**. – p. 84-88.
3. Bezgodov I. M., Fahrutdinov A. A., Tkach E. V. Physical and mechanical characteristics of concrete as crushed stone recycled crushed concrete // Vestnik MGSU. – Moscow, 2016. **No. 10** – p. 24-35.
4. Bazhenov Yu. M., S. Murtazaev, A. Y. Efficient concretes and mortars for construction and rehabilitation of roads using concrete scrap and dump ashes in thermal power plants // Vestnik MGSU.- 2008.- **No. 3**. - p. 124-128.
5. Barantseva, E. A., Mizonov V. E., Khokhlova Yu. V. Processes of mixing of granular materials: modeling, optimization, design // GOUVPO "the Ivanovo state power University. V. I. Lenin". – Ivanovo, 2008. -p.116 .
6. Bazhenov, Y. M., Chernyshov E. M., Korotkikh D. N. Design of modern concrete structures: defining principles and technological platforms //Building materials. 2014. **No. 3**. p. 6-14.
7. Boldyrev A. A., Romanov I. V., Voronin V. V., Alimov L. A study of the formation of structure and properties of multi-component concrete // Scientific review. 2013. **No. 9**. -p. 177-181.
8. Uriev N. B. Modelling of the structural and mechanical characteristics of disperse systems under dynamic effects//Colloid journal. 2013. T. 75. **No. 5**. -p. 596.
9. Ruzhyn S. I. Influence of chemical characteristics of cement on the effectiveness of vibroactivity. Site access: /http://www.ibeton.ru/a158.php.
10. Duval R., Kadri E. H. Influence of silica fume on the workability and the compressive strength of high-performance concretes //Cem. and concr stock. Res. – 1998. – Vol.28. – **No. 4**. – p. 533-547.
11. Stark. The hydration of cement and microstructure of concrete // Cement and its application. - 2011. - **No. 2**.- p. 90-94.
12. Serdyukov A. A. Rakhimbaev S. M. Influence of negative temperatures on the hardening of the concrete // concrete and reinforced Concrete. -2013. - **№1(18)**. -p. 52-54.

# Advantages of modern collapsible systems for exterior finishing of buildings in urban areas in Russia

*Kristina* Tolstova [1,*], *Vitaly* Chulkov [1,**]

[1] Moscow State University Of Civil Engineering 129337, 26, Yaroslavskoe shosse, Moscow, Russian Federation

**Abstract.** Traditionally, the method of finishing interior spaces, as well as exterior surfaces and architectural and structural elements of buildings and structures is widely used, when layers of building materials or finishing structures are applied in a strict sequence, allowing them to eventually form an unbroken whole - a "monolith". This method of finishing, its technology and results not only depend significantly on the quality of the materials used and the mechanization of processes, but also require high qualification and specialization of construction workers. The execution of such a finish is more an art than a reproduced with a given level of quality of results by a technological process. An alternative to "monolith" is a variety of collapsible (they are also called "hinged") systems for finishing urban buildings, the elements of which are manufactured in the construction industry, requiring, as a rule, only installation and assembly in internal premises (the so-called "evroremont"), Or on the external surfaces of a building or structure.

## Introduction

Means of protection for worn out and lost to a certain extent their physical and aesthetic characteristics of the surfaces of premises and facades of building objects, people used for centuries. In the overwhelming majority, these were different versions of the known method, when solutions and paints were applied to the prepared substrate, allowing the previous layer to dry before applying the subsequent layer. Approximately from the 70s of the last century, materials and technologies of construction and installation works abroad allowed to use another method of finishing the surface of buildings - "hinged systems". The need for preliminary preparation of the base ("continental surface") has disappeared, since it is "hung" the metal frame to be mounted, to which then the facing products are fastened. The number of varieties of the results of this method of finishing is very great.

Publications of foreign authors (Carlsson, P., 2003, [1], 2005, [2], Deneyer, A., Moenssens, N., 2004, [3]; Lang W., Cremers J., Beck A., Manara J. et al., 2013, [4, 5] Hamidi S., Ewing R., 2014, [6] Hennig E., Schwick C., Soukup T., Orlitova E., Kienast F., Jaeger J ., 2015, [7], Zeković S., Maričić T., Vujošević M., 2016, [8] and a number of them preceded) are devoted to the promotion of certain types of hanging facades and used finishing materials (metal, plastic, glass and others) ).

Recently publications of domestic authors appeared in foreign professional publications (for example, Vatin N., Nemova D., Kazimirova A., Gureev K., 2014, [9], Aristov D., Glotova Yu., Sazonova Yu., Tyulenev M., 2015, [11] and a number of others), as well as dissertation research (for example, - Olshevsky V., 2014, [10]), who also consider local cases of using variants of hinged facades and finishing materials.

System studies of the class of collapsible hinged systems in Russia began in 2006 Kuzina O.N. She called the process of forming such systems "retreat" (from English Retrieve - to restore, find, attach) [12].

This article is devoted to continuing research on this topic. In Russia, in the conditions of a market economy, different kinds of market construction products, related to the class of collapsible hinged interior finishing systems and external surfaces of buildings and structures for different special purposes, have emerged and are competing with each other. Such a variety of names and manufacturers of components for different finishing systems is

* Corresponding author: kristi.tolstova@gmail.com
** Corresponding author: vitolch@gmail.com

encouraging in the competitive market in the construction products market. But, from the scientific point of view, all this diversity refers to one class of retrieval systems, methods and processes of their application in construction.

## Methods

The basis of scientific study of the class of retracement systems, methods and processes of their application in construction is the basic infographic model of the reorganization cycle, which consists of four stages (Fig. 1):

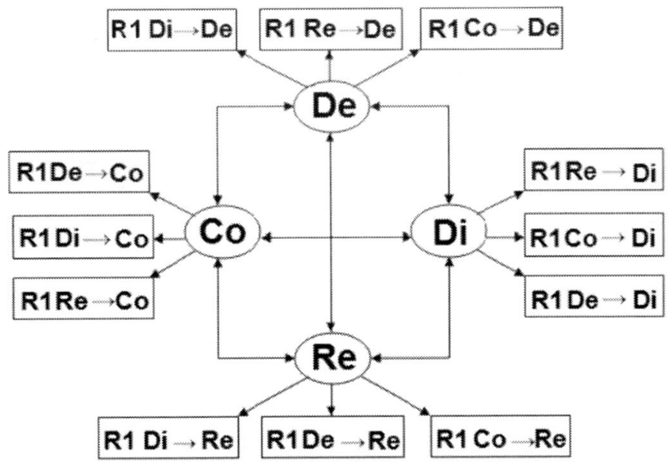


**Fig.1.** Interaction of the stages of the basic reorganization cycle in the construction

of ***De***, ***Di***, ***Re*** and ***Co*** and its results ***R1***

- ***The device***, ***De***, as a certain state of stability with predefined (projected and implemented) characteristics and parameters that meet the requirements of a certain position in the activities of society; In the construction industry, this may be a building or a structure ***erected anew*** on the basis of previously designed structural and organizational and technological projects;

- ***Disorganization***, ***Di***, as formed by a well-known array of personalities from a certain point of view and with known goals, an idea of the building or structure erected at the stage of "device", as inconsistent with the point of view and goals of the disorganization figures; It is assumed that these claims of disorganizers are set out in substantive texts ***reflected in documents available for review***;

- ***Reorganization***, ***Re***, as the process of ***reconstruction*** of a building or structure in accordance with proposals of disorganizers, discussed with specialists of the construction industry and interested public of the society; The results of such a discussion should form the basis for the developed constructive and organizational and technological projects for such reconstruction of these buildings or structures;

- ***Co-organization***, ***Co***, as the process of correct (approved, perceived positively) perception of necessity and realization of the construction reorganization formed by the disorganizers and developers of constructive and organizational-technological projects for the reconstruction of these buildings or structures; If the co-organization is executed qualitatively, then the reconstructed building or structure takes its place at the stage of "device" of the basic reorganization cycle.

The processes of retracement are realized at the interconnected stages ***De*** and ***Re*** of the basic reorganization cycle (Fig. 1), in which the industrial labor activity is directly realized. At the interrelated stages, ***Di*** and ***Co*** carry out a cogitative activity in the modeling and implementation of stages ***De*** and ***Re***.

## Results

The compounds obtained in non-collapsible and collapsible methods are referred to as "composites" - inhomogeneous continuous entities in which two types of components are functionally distinguished (Fig. 2):

**Fig.2.** Examples of the structure of the functional components of the "composite"

- reinforcing (provide the necessary mechanical characteristics of the composite);

- binders or "matrix" (ensure the joint work of reinforcing components).

The increase in the static strength of the separately considered components of the composite does not lead to a decrease, but to an increase in the fracture toughness characteristics of the composite as a whole.

Two states are distinguished in the formation and monitoring of composites:

• ***statics*** - fixed at a certain moment of time by the instrument-technical or model-analytical diagnostics, the unchanged state and the aggregate values of the parameters of the composite, their compliance with the norm at the time of diagnosis;

• ***dynamics*** - a sequence of static states of the composite in time and space, fixed by statistical computer monitoring to reveal the evolution of the performance of the composite.

Modern building collapsible systems are used both for interior planning and decoration (improving the comfort of living in the process of its vital activity or labor productive activity) and for protecting (prolonging the "life") of external enclosing surfaces and structures (the facade of a building or structure) from the harmful effects of the building environment, giving the facade an attractive appearance.

Collapsible finishing systems ensure the implementation of one of the new modern varieties of building reconstruction - ***retrevatsia***.

The study of regularities, properties and terminology in the field of retrieval is just beginning and, as is often the case, the practice of business processes of building reconstruction outstrips scientific research. Each business sector in the construction or a group of companies of a similar orientation uses or suggests the most appropriate combination for the advancement of its version of the collapsible system for the finishing of buildings and structures and the technology that implements it (that is, the business product and business services):

Facade systems;

Translucent finishing systems;

Planar glazing;

Glass facades;

Hinged facing systems;

Ventilated facades;

Wet facades; Hinged facades; Systems of "evroremont", etc.

The listed and many other business conjuncture names represent special cases of implementation of ***hinged retrieval finishing systems*** - composite structures from different products and materials, depending on the purpose of the erected or rearranged building, its design, the chosen architectural solution and the customer's wishes.

## Discussion

The most common representative of the retracement systems for exterior decoration of buildings is a **hinged ventilated facade** (HVF) - a cladding composite (protective decorative shield of plates or sheet protective and decorative materials) and a structure that is fixed to the wall so that between the protective and decorative coating and the wall there is air gap. External insulation of enclosing structures can improve the indoor climate.

Due to the pressure difference, the air gap acts as a "exhaust pipe" for the upward flow, with which excess heat, atmospheric and internal moisture leave the building's enclosing structure. A ventilated air gap is "a temperature buffer that reduces heat loss (air in it is about 3°C higher than outside). The wall of the building remains warm and dry, and in the summer the HVF works like a sun shield (it reflects a part of the heat flux incident on it).

Specially developed design and technology of installation of illegal armed units to the building provide the absorption of deformation from diurnal and seasonal temperature differences, reduction of internal stresses in the protective decorative screen and skeleton, reduction of the probability of cracking and destruction of illegal armed formations.

Among the advantages of HVF include:

- soundproofing;
- thermal insulation;
- heat insulation;
- protection against precipitation;
- compensation of thermal deformations;
- natural ventilation of water vapor;
- fire safety.

Let's consider in more detail each of the listed advantages of illegal armed formations of HVF.

***Soundproofing.*** Joint application of the hinged facade and soundproof materials and products provides the required level of sound absorption over a wide frequency range.

***Thermal insulation.*** The excessive accumulation of heat inside the building is hampered by the joint application of a special profile system for hinged facades and a heat insulator providing natural ventilation of the facade. The comfort of the microclimate in the room is achieved without the use of additional expensive air conditioning systems, which, in addition to the positive effect, can harm human health in the emergency mode of their application.

***Heat insulation.*** The effects of precipitation and the occurrence of condensation significantly reduce the heat-saving properties of the insulation. The presence of a heater, protected from the effects of precipitation and condensation and the profile system of the ventilated facade, allows reducing energy consumption for heating, reducing the thickness of the bearing walls, thereby reducing the load on the foundation of the building or structure.

***Protection against precipitation.*** The design of the main bearing profile of the HVF is designed in such a way that all the moisture that gets to the surface of the facade is removed into the drainage and has no direct contact with the heater and the building wall.

***Compensation for thermal deformations.*** The specially designed scheme of mounting and fixing to the wall of the NVF profile system provides the possibility of absorbing the temperatures of the so-called "thermal deformations" arising during the diurnal or seasonal changes. Thermal deformations cause internal stresses in the materials of the bearing structure of an HVF and in facing materials, which are compensated for by the movable or rigid fixation of the elements of the HVF fastening system. In this case it is necessary:

- ensure freedom of thermal deformation of HVF panels to avoid their destruction;

- to withstand the size and location of the rusts, so that disturbances in the geometry and location of the rust do not reduce the architectural expressiveness of the facade.

***Natural ventilation of water vapor.*** The water vapor arising in the walls of buildings and structures during their operation is removed due to natural ventilation provided for in the HVF system. Such ventilation significantly improves the thermal insulation properties of the building walls, while ensuring a comfortable temperature regime in the interior. The air layer in the HVF distinguishes it from other types of facades. Differential pressure transforms this air "jumper" into a kind of "exhaust pipe", through which internal and atmospheric moisture is removed. This ventilation duct also reduces heat loss, and the air temperature inside the facade is several degrees higher than outside.

***Fire safety.*** Hinged facades include materials and products that are classified as difficult to burn or not combustible (that is, preventing the spread of fire). The gap between the surface of the building and the decorative screen in the HVF constantly moves the air. This air flow with the use of materials with a high degree of flammability can spread fire throughout the building in a few minutes. The main material that is recommended in such cases for thermal insulation is stone wool, which is not combustible, so it has no restrictions on the number of storeys of buildings and is an excellent material that makes the facades an order of magnitude safer. High resistance to high temperatures, exceeding several hundred degrees, makes stone wool an excellent thermal insulation material.

Refusal from "wet" (for example - glutinous) technologies, their replacement by mechanical connections allows to install and reconstruct HVF at any time of the year. The absence of special requirements to the surface of the bearing wall of the building and the mobility of the junction points in order to compensate for dynamic loads (wind, temperature changes, etc.) ensure the simplicity and high speed of installation of all components of the HVF, their long overhaul life (from 25 to 50 years depending from the materials used), Fig.3.

In the Russian market there are many different domestic and foreign illegal armed groups:

Producer Russia: Alcon Trade (U-kon), Diat, Kaptechnostroy and MOSMEK, Plant Metalware (CTS), Technocom, ARKADA and others;

Manufacturer Austria: SLAVONIA (SPIDI), EUROFOX (in Russia it is represented by Duvils and Mirage companies);

Manufacturer Germany: BWM, WAGNER-SISTEM (represented by TECHNOCOM); Manufacturer Holland: HUNTER DUGLAS and many others.

Each of the listed systems HVF has a number of design features that ensure its optimality in solving specific problems (for example, when: smoothing the unevenness of the bearing wall of the building, reducing the negative effect of the "cold bridges", ensuring the possibility of fastening the small-sized cladding without significantly increasing the guide frame; Heat-insulating plates, etc.).

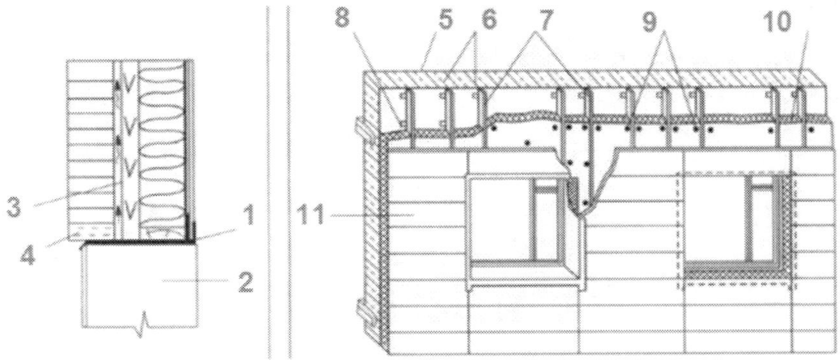


1 - waterproofing (membrane, roofing material, hydroglass, bokroelast); 2 - socle; 3 - ventilated gap; 4 - ventilated air; 5 - insulated wall; 6 - nodes of fastening of guides; 7 - metal guides; 8 - insulation, covered with windproof material; 9 - dowels; 10 - air layer; 11 - outer lining.

**Fig.3.** Structure diagrams (left) and installation (right) HVF

The practice of using foreign HVF in Russian conditions allowed to work out ways of adapting these hinged systems:

• reduction of the contact area of the metal with the wall ("point contact");

• use of different heat-insulating pads (foreign companies use different plastics, domestic companies often use an effective heat-insulating device "paronite").

## Conclusions

The introduction of HVF in Russia is accompanied by a number of problems not known to foreign manufacturers of these systems (significant wall irregularities, very significant differences in winter and summer temperatures, etc.). They force the builders of any foreign HVF system, even a very high level, to adapt to Russian conditions (in particular - solving the important problem of the possible formation of "cold bridges".

There are two approaches to solving this problem:

• reduction of the contact area of the metal with the wall ("point contact");

• use of different insulating gaskets (foreign companies use different plastics as such gaskets, Russian campaigns often use an effective heat-insulator "paronite").

Both approaches have advantages and disadvantages.

With point fixing, the thermal conductivity is lower, but at the same time the load-carrying capacity of the brackets to the wall decreases.

With increasing contact area, the problem of forming "cold bridges" becomes more complicated.

Alternative HVF - a wet facade, it is erected, as a rule, with the use of thin-layer plaster for repair and decoration works on historical sites, where it is important to preserve the architectural appearance of the building. The facade is decorated in a warm, dry season.

In European countries, "wet" facades are erected by attaching thermal insulation material to the wall surface with glue or mechanical fasteners, and then plaster, fiberglass and synthetic materials are applied.

Such technology provides an individual textured solution of facades, warmth of slopes, less laboriousness of work execution and the possibility of further renovation of the facade during repair or reconstruction.

While in the systems of "wet" facades import components prevail and it is difficult to single out an undoubted leader in the production of high-quality domestic materials and components for the "wet" facade technology.

## References


1. Carlsson, P., 2003. Glazed facades - double skin facades: requirements and methods. Arkus (The forum of research and development of the architects), Stockholm, Sweden (in Swedish).
2. Carlsson, P., 2005. Construction with glass. Swedish glass trade union, Stockholm, Sweden (in Swedish).
3. Deneyer, A., Moenssens, N., 2004. Ventilated double skin facades - Aspects regarding natural lighting and visual comfort. Belgian Building Research Institute, Dept. of Building Physics, Indoor Climate and Building Services.
4. New Envelopes for old Buildings - the Potential of using Membrane Systems for the thermal Retrofitting of existing Buildings. / Lang W., Cremers J., Beck A., Manara J. // Life-Cycle and Sustainability of Civil Infrastructure Systems Proceedings of the Third International Symposium on Life-Cycle Civil Engineering (IALCCE'12), Vienna, Austria, October 3-6, 2012 / Edited by A. Strauss, D. Frangopol, K. Bergmeister.- London, 2013.- P. 1737-1744
5. The system of external insulation of building facades. Encyclopedia of safety. Surviving in the city (http://survincity.com/2013/12. Decamber 6th, 2013)
6. S. Hamidi, R. Ewing, A longitudinal study of changes in urban sprawl between 2000 and 2010 in the United States, Landscape and Urban Planning, 128, pp. 72-82 (2014)
7. E. Hennig, C. Schwick, T. Soukup, E. Orlitova, F. Kienast, J. Jaeger, Multi-scale analysis of urban sprawl in Europe: Towards a European desprawling strategy, Land Use Policy, 49, pp. 483-498 (2015)
8. S. Zeković, T. Maričić, M. Vujošević: Megaprojects as an Instrument of Urban Planning and Development: Example of Belgrade Waterfront Project, in Tech4Dev 2016 UNESCO conference “From Innovation to Social Impact.” Lausanne: EPFL (2016)
9. N. Vatin, D. Nemova, A. Kazimirova, K. Gureev: Increase of energy efficiency of the building of kindergarten, Advanced Materials Research, 953-954, pp. 1537-1544 (2014)
10. V. Olshevskiy: Rezultaty eksperimentalnogo opredeleniya skorostey i raskhodov v ventiliruyemykh prosloykakh NVF [The results of the experimental determination of the velocity and expenditure in ventilated interlayers of ventilated suspended facades]: Master's thesis. St.Petersburg State Polytechnical University. Saint-Petersburg, (2014)
11. D. Aristov, Yu. Glotova, Yu. Sazonova, M. Tyulenev: Fasady Rossii: sostoyanie rynka i tekhnologii [Facades of Russia: The State of the Market and Technologies]. Stroitel'stvo: nauka i obrazovanie [Construction: Science and Education], N4, paper 5. Available at http://nso-journal.ru (2015)
12. V. Chulkov, O. Kuzina, Reorganization and reconstruction of building objects, P.232-280 // In .: Manufacture and use of building materials, products and systems. - Volume 1 / Under total. Ed. V.O.Chulkov (Moscow: SvR-ARGUS, 2009)

# Forecasting Parameters of Earth Surface Subsidence to Assess Safe Operation of Engineering Structures

*Vitaliy A.* Trofimov[1] , *Tamara G.* Makeeva[2,*]

[1]Research Institute of Comprehensive Exploitation of Mineral Resources, Russian Academy of Sciences, Moscow, Russia
[2]Moscow state university of civil engineering, Yaroslavskoye shosse, 26, Moscow, Russia, 129337

**Abstract.** Mining activities, particularly at shallow depths, create heavy environmental impact that causes irreversible changes in the environment's properties, structure and so on. Such unnatural interference frequently makes the environment unusable to an extent in future. It is known that exhaustion of stratified deposits causes notable subsidence of earth surface that may destroy structures, and interrupt the mode of water flows both on surface and underground. To describe the parameters of earth surface subsidence trough under the impact of underground mining of stratified deposits, this paper uses a three-parameter function expressed with the Gauss probability integral. We also propose a numerical algorithm to find the function's parameters based on the values of vertical shift of reference points, measured in situ with benchmark stations. The algorithm is based on the least-squares method, while the minimum value of the respective functionality discrepancies are found with random search. Analysis of the specifics of the function proper helped to propose an approximated method to find its parameters. The paper compares approximating curves produced by both methods, with field measurements.

## 1. Preamble

Forecast of earth surface deformations caused by underground mining is today a highly relevant problem yet to be resolved. Correctly evaluated nature of future surface shifts depending on the planned development of underground mining can greatly reduce the damage to above-ground infrastructure, including buildings, roads, pipelines, and other assets that happen to be shifting trough.

Since the middle of the 19$^{th}$ century when the issue was added to the order of business, researchers have proposed and designed a plethora of methods and ways to address the task, based both on the purely empirical approach and on the academic approach related to various mechanical and mathematical models of the rock mass [1]. Each such method is usable to an extent within the existing constraints, which are not however clearly defined. Obviously, the most preferable approach would be one that is simple enough to use, oper-

* Corresponding author: makeeva13new@yandex.ru

ates parameters with simple physical meanings that are easily definable and directly related to a specific mining situation; and has a broad enough scope of application.

Let us now assume that the mining situation at the current time is described by a certain existing system of excavations typical for working a stratified deposit and generally with a rather complex configuration both in plane and in depth as multiple strata are mined simultaneously at different depths. Here, earth surface subsidence is controlled by one or more profile lines of reference points on the surface, possibly created at different times and in different areas of the mine field. The objective is to build an algorithm by using data about shifts of reference points (including vertical shifts) and about the state of mining, for some point in time; the algorithm should allow computing earth surface shift parameters, both at a given time and for secondary activities developing in future.

Below, we are going to examine a simplest case of earth surface trough subsidence caused by underground mining. We assume that the following constraints are available and true:

- the mined deposit is a horizontal tabular deposit;
- the overlying rock mass features transversal isotropy with respect to mechanical properties, in particular, resistance to deformation and strength; as a rule, this is related to the stratified nature of the rock mass;
- the rock mass, in the rock shifting zone, has no fracturing anomalies, along which mutual shifts of fronts may emerge;
- subsidence is caused by only one headway, rectangular in plane, with one side is much larger than the other; in other words, we assume that conditions a flat deformation exists

The situation is illustrated in Figure 1, which shows a cut that runs perpendicular to the long side of the mined space and passes through its middle.

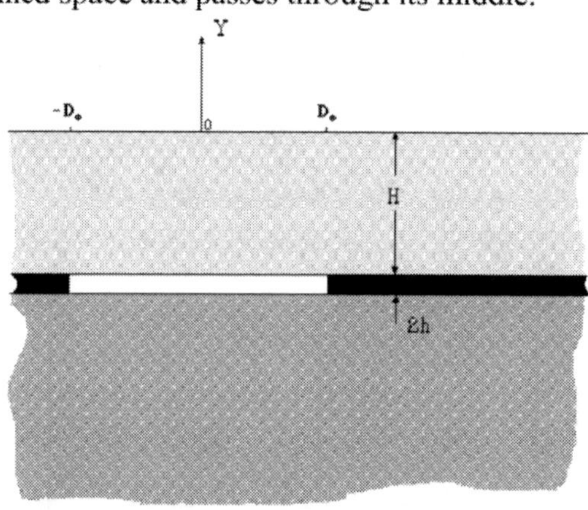


Fig. 1. Schematic representation of a model situation for stratified deposit mining

We need to examine deformation processes in this model setup is necessary to establish that the quantitative description approach represented below for earth surface subsidence is fundamentally valid and applicable. Using the results, in future we can apply the studied principles to more complex situations caused by rejection of certain problem-setting constraints.

Therefore, let us assume that we know the following values found experimentally during field measurements:

- location of observatory reference points, described by distance $x_i$ along the preset distance from a certain zero point that is frequently but necessarily congruent with the trough center ($i$ =1, 2, ..., $N$, $N$ – number of reference points on the profile line);

- values of earth surface subsidence $\eta_i$ at locations of reference points;
- mining parameters, namely:
  1. Depth of secondary activities $H$;
  2. Half-length of headway $D_\Phi$.

## 2. Universal Function of Subsidence

Within the limits of the accepted constraints, it is considered that vertical subsidence of earth surface can be described with function [2]

$$\eta(x) = -\frac{\eta_o}{2}\left[\Phi\left(\frac{D_p + x}{C_p H}\right) + \Phi\left(\frac{D_p - x}{C_p H}\right)\right], \tag{1}$$

which belongs under a three-parameter family, where $\eta_0$, $D_p$, $C_p$ – constants typical for specific conditions, and $\Phi(t) = \sqrt{\frac{2}{\pi}}\int_0^t e^{-\frac{z^2}{2}} dz$ - Gauss probability integral.

Here: $\eta_0$ – value of greatest subsidence of headway roof for complete undermining, typically related to formation thickness $m$; $D_p$ – parameter that correlates with value $D_\Phi$, equal to ~0.7÷0.9 thereof; $C_p$ – parameter related to mechanical properties of overlying rock mass.

The main objective is to demonstrate that a type (1) function can be used to approximate with sufficient accuracy experimentally measured values of reference point subsidence. Apparently, the listed numerical parameters are generally unknown in advance in any specific mining situation, and the objective can be resolved by constructing the algorithm proper to define their values.

Real shift values of $\eta_i$, differ from the above values $\eta(x_i)$ found under (1) at points $x_i$, due to both instrumental errors and local irregularities of the rock mass, and because function (1) corresponds to real-life processes of rock mass deformation. This gives rise to the problem of interpreting all $N$ values of $\eta_i$ within model (1).

Therefore, the objective now becomes largely specific and boils down to selecting a type (1) function – in other words, values of $D_p$, $C_p$, $\eta_o$ – in a way that all $\eta_i$ values lie close to the respective curve (1).

As a measure of proximity, we can use a regulated value brought into the space of point functions, selected partly due to its relative simplicity in the light of analytical research. These requirements are fully met by a quadratic norm

$$F = \frac{1}{N}\sum_{1}^{N}[\eta_i - \eta(x_i)]^2 \tag{3}$$

This is related to the so-called least-squares method. Parameters $D_p$, $C_p$, and $\eta_o$ should be selected with the least possible functionality $F$. Ultimately, the objective boils down to finding the minimum of function (3) by variables $D_p$, $C_p$, and $\eta_o$.

A highly efficient and fairly unsophisticated approach to finding the functionality minimum is based on the philosophy of random search methods. This encompasses a multitude of essentially multistep methods, their main feature being that the testing point for the function is selected in a more or less random manner. The core of such methods is that a point is random-selected within a specific area in the space of potential changes of control parameters. The functionality value is computed at this point, and if it is less than a certain input value, the point and its related functionality value are stored in memory as the minimum and as the input for the next probe. Otherwise, the point is rejected and the next one is selected. Despite its apparent uncertainty, the method is highly efficient for computer-based analysis, and it presents the least requirements to regularity of the function to be examined. By consistently going through the points, we will ultimately arrive at into any small area of the desired solution.

To reduce the required computations, the area of required calculations in the area of possible test values of $D_p$, $C_p$, $\eta_o$, from where they are selected with a randomizer, is limited by certain parameters $\Delta_D, \Delta_C, \Delta_\eta$, which set a moving *rectangular* window of size $2\Delta_D * 2\Delta_C * 2\Delta_\eta$ with its center at the selected test point. The parameter values can be randomized, and their optimal values are the result of a numerical experiment based on the condition of highest convergence rate of the multistep process.

The specifics of the selection window are that it always covers the neighborhood of the current "minimum" point. Random sample here comes only from this fairly small neighborhood, and this significantly narrows down the area of possible search of points, which is practically unlimited from the start. Besides, if the current window fails to yield a point of a smaller value after a certain number of steps (say 1,000), then the window is downsized and the search proceeds in a narrower set.

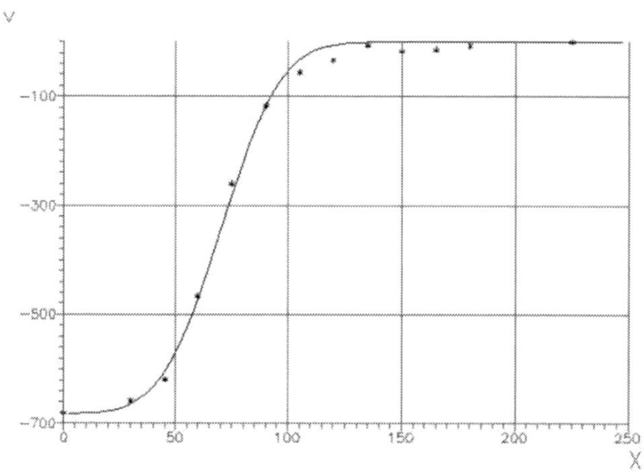


Fig. 2 – Example of approximated field measurements of earth surface subsidence for type (1) functions in incompletely undermined conditions

As a typical example, Figure 2 presents a simulation curve of type (1) and respective experimental data, represented with points, for a real-life case borrowed from the practices of the Donetsk Basin mines, with mining parameters: $H$=145 m and $D_\Phi$=95 m. Calculations produce the following values of target parameters for function (1): $\eta_o$ =682 mm, $D_P$=70 m, $C_P$=0.144. Proximity of the corresponding calculated curve to the experimental data is described by magnitude $\sqrt{F}$, which equals ~12 mm.

## 3. Approximat Method to Find Subsidence Function Parameters

Proposed below is a different approach to finding the parameters of function (1) based on experimental data, which involves research of certain shared properties of the family of function (1).

Let us introduce the dimensionless coordinate $\bar{x}$ and dimensionless parameter $\alpha$ into our case

$$\bar{x} = \frac{x}{D}, \quad \alpha = \frac{CH}{D} \tag{4}$$

In addition, let us substitute $\eta_o$ with the greatest subsidence $V_{max}$, which for the tabular deposit of the type in question apparently exists at the center of the trough, i.e. for $\bar{x}=0$. It is clearly demonstrated that

$$\eta_o = \frac{V_{max}}{\Phi\left(\frac{1}{\alpha}\right)} \tag{5}$$

In our notation, relation (1) should be represented as follows:

$$v = \frac{V_{max}}{2\Phi\left(\frac{1}{\alpha}\right)}\left[\Phi\left(\frac{1-\bar{x}}{\alpha}\right)+\Phi\left(\frac{1+\bar{x}}{\alpha}\right)\right] \tag{6}$$

By normalizing (6) by $V_{max}$ and by introducing $\bar{v} = v/V_{max}$, we ultimately arrive at the relationship for dimensionless value of subsidence in the trough

$$\bar{v} = \frac{1}{\beta}F(\bar{x},\alpha) \tag{7}$$

where

$$\beta = 2\Phi\left(\frac{1}{\alpha}\right) \tag{8}$$

$$F(\bar{x},\alpha) = \Phi\left(\frac{1-\bar{x}}{\alpha}\right)+\Phi\left(\frac{1+\bar{x}}{\alpha}\right) \tag{9}$$

Therefore the initial relationship (1) is represented as dimensionless, thus it is demonstrated that all vertical subsidence movements of earth surface, since they are described with equation (1), regardless the values of real-life mining parameters, they form a single-parameter family derivative from parameter $\alpha$. The appearance of the family (7) is illustrated in Fig. 3, where the values of parameter $\alpha$ are shown with numbers on the respective curve lines.

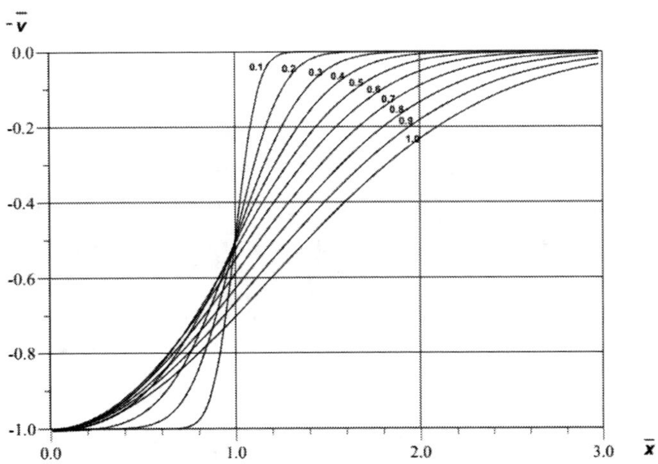


Fig. 3. Dimensionless curves of earth surface subsidence for various values of $\alpha$

Now we examine the dependence of $\bar{v}$ on $\alpha$ for $\bar{x}=1$. Let us note that $\bar{x}=1$ corresponds to parameter value $x=D$. Considering (8), (9), equation (7) gives

$$\bar{v}_1 = \bar{v}\,|_{\bar{x}=1} = \frac{\Phi\left(\frac{2}{\alpha}\right)}{2\Phi\left(\frac{1}{\alpha}\right)} \tag{10}$$

The graph represented in Fig. 4 shows that within the change limits of $0 \le \alpha \le 0.5$, the value of $\bar{v}_1$ is found to be a constant equal to 0.5, with accuracy of up to 5%. This is because for $\alpha$ <0.5 we have $\Phi\left(\frac{1}{\alpha}\right) \cong \Phi\left(\frac{2}{\alpha}\right) \cong 1$. This in turn means that within preset limits all graphs with $\alpha$ virtually intersect at the same point, if $\bar{x}$ =1, while $\bar{v} = 0.5$. In other words, if $x=D$, earth surface subsidence will be half of the maximum reached at the center of the trough.

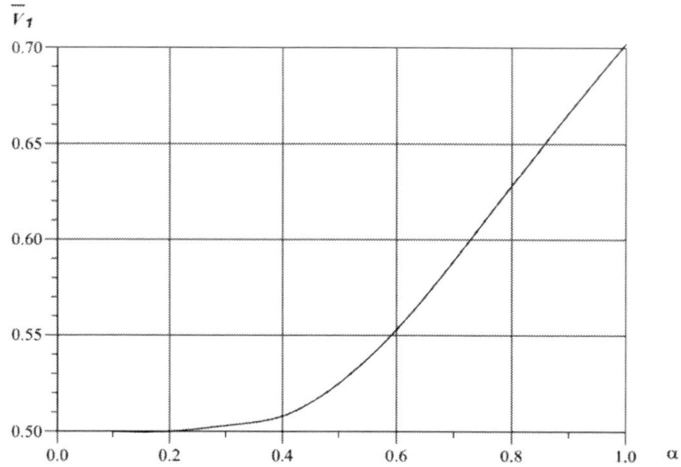

Fig. 4. Dimensionless subsidence of earth surface for $\bar{x}$ =1.

Let us now examine the inclination of subsidence curves of (7) for $\bar{x}$ =1. We differentiate the relation (7) by $\bar{x}$ and introduce this

$$tg\theta = \frac{\partial \bar{v}}{\partial \bar{x}}\,|_{\bar{x}=1} \tag{11}$$

finally arriving at

$$tg\theta = \frac{1}{\sqrt{2\pi}} \frac{1}{\alpha\Phi\left(\frac{1}{\alpha}\right)}\left(e^{-\frac{2}{\alpha^2}} - 1\right) \tag{12}$$

Therefore, our analysis has established biunique functionality dependencies $\bar{v}_1(\alpha)$ and $tg\theta(\alpha)$, i.e. dimensionless value of earth surface subsidence and inclination for $x=D$ are unambiguously determined by parameter $\alpha$ and reversely (also considering the constraints of $\alpha$).

Now we use our dependencies to find the parameters of $\eta_o$, $D$, and $C$ in (1) based on data produced during field measurements of earth surface subsidence at the shifting trough. To this end, we can use the following sequence of steps:

Figures 7 and 8 illustrate how the proposed methodology is used to process test data harvested from a Donbass mine. Measured values of earth surface subsidence are points on the graphs, while $V_{max}$=380 mm. The formation depth is $H$=200 m. As is obvious from Fig. 3 and as explained above, $D$=84.3 m. Fig. 7 represents actual data of the benchmark station line; Fig. 8 shows normalized curves.

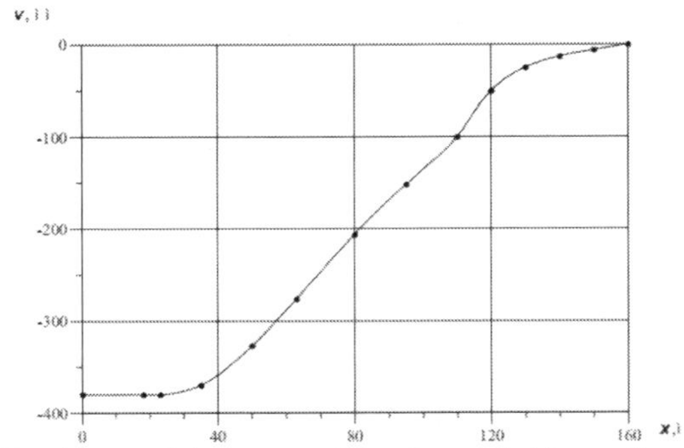


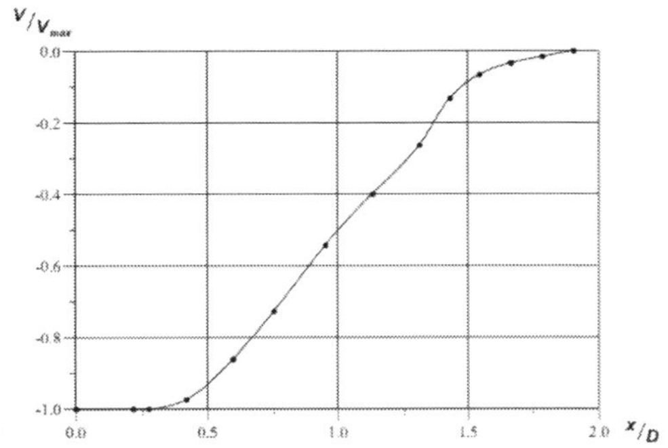


Fig. 5. Actual data on earth surface subsidence from benchmark station line
Fig. 6. Normalized subsidence curve

Directly measured on the graph (Fig.6), to the scale of the axis, we can find that $tg\theta = 0.885$. From this, we find parameter $\alpha = 0.465$ and then correspondingly $C$=0.196. After this, we can easily find the greatest value of subsidence for complete undermining case as $\eta_0$ =393 mm.

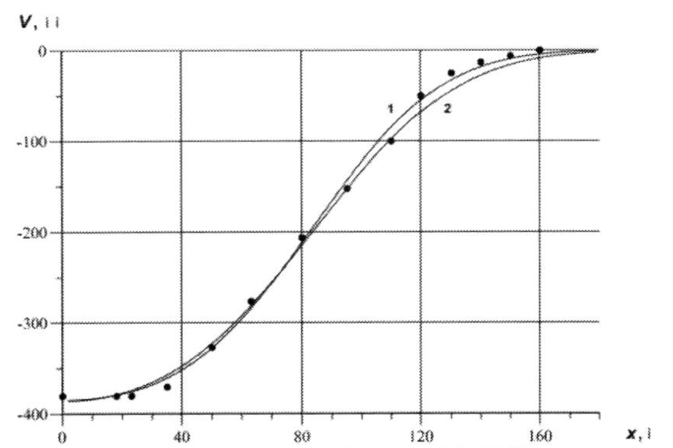


Fig. 7. Comparing earth surface subsidence curves found with different methods, and experimental data

Figure 7 contains comparable experimental data (marked as points), the subsidence curve of type (1) with parameters found through the methodology described above (curve No. 2) and the subsidence curve of type (1), that best approximates the experimental data (curve No. 1, for which $D$=82.3 m, $C$=0.18, and $\eta_0$ =395 mm), build using the random search method described above.

Therefore, it has been demonstrated that relation (1) can be used to describe subsidence troughs in the event of flat deformation scenario, achieved by stated algorithms used to find the parameters of function (1).

By differentiating relation (1) we can find angles of inclination of the surface and its longitudinal forces of stretching-compression; this is highly important for safe operation of structures built above ground.

## References

1. G.Cratch, *Sdvijenie gornih porod i zaschita podrabatyvaemih soorujeniy* (Movement of rocks and protection of underworked structures). Moscow: Nedra, (1978)
2. R.A. Muller, *Vliyanie gornih virabotok na deformatsiu zemnoy poverhnosty* (Influence of mine workings on deformation of the earth's surface). Moscow: Ugletehizdat, (1958)

# Nonlinear stability analysis of shallow cylindrical shells with initial imperfections

*Sergey* Trushin[1,*]

[1]Moscow state university of civil engineering, Yaroslavskoye shosse, 26, Moscow, Russia, 129337

**Abstract.** The present article describes numerical procedures for geometrically nonlinear stability analysis of shallow flexible cylindrical shells. Computing models are based on the theory of plates with account of transverse shear deformations. The finite difference energy method of discretization is used for reducing the continuum problem to finite dimensional problem. Solution procedures for discrete nonlinear problem are based on initial value formulation. The solution of Cauchy problem is performed using the self-correcting Runge-Kutta method with the constrained equation. For stability analysis of shells with initial imperfections small asymmetric perturbations are introduced in the initial conditions of the Cauchy problem for finding asymmetric shapes of shell equilibrium.

## 1 Introduction

In the general case the solution of the stability problem is reduced to the identification of qualitative changes in a behavior of the system at changes in the structure of the system. The system is structurally stable if sufficiently small perturbations lead to correspondingly small changes in its behavior. A measure of the deviation from the unperturbed states is internal state parameters (generalized coordinates).

The stability analysis of a shell structure studies its behavior under small perturbation of parameters including in the energy functional, as well as the initial shape, boundaries and boundary conditions. Such a generalization of the concept of stability is important for practical calculations as in real structures it is possible not only imperfections in the form or a small unbalance in the load but also a heterogeneity of the physical properties of material and a deviation from set boundaries and boundary conditions. In the analysis of shells the definition of such initial imperfections in the form that lead to the largest drop of the critical load compared to the same load for the perfect system becomes important. It allows to obtain the lower estimation of the actual critical load of a real shell.

## 2 Methods

In the analysis of cylindrical shells the model that takes into account the transverse shear strains is used. The shells are considered in Cartesian coordinate system. Geometric

* Corresponding author: trushin2006@yandex.ru

relationships between the strains and the displacements have the form [1-3]:

$$e_{11}=\frac{\partial u}{\partial x}+\frac{w}{R}+\frac{1}{2}\theta_1^2;\ e_{22}=\frac{\partial v}{\partial y}+\frac{1}{2}\theta_2^2;\ e_{12}=\frac{\partial v}{\partial x}+\frac{\partial u}{\partial y}+\theta_1\theta_2;$$

$$\kappa_{11}=\frac{\partial \theta_1}{\partial x};\ \kappa_{22}=\frac{\partial \theta_2}{\partial y};\ \kappa_{12}=\frac{\partial \theta_2}{\partial x}+\frac{\partial \theta_1}{\partial y};\ e_{13}=\frac{\partial w}{\partial x}+\theta_1;\ e_{23}=\frac{\partial w}{\partial y}+\theta_2\ , \tag{1}$$

where $e_{11}$, $e_{22}$, $e_{12}$ – linear and angular deformations; $\kappa_{11}$, $\kappa_{22}$, $\kappa_{12}$ – the changes of curvatures and curvature of torsion; $e_{13}$, $e_{23}$ – the transverse shear deformations; $u$, $v$, $w$ – displacements in the direction of coordinate axes $x$, $y$ and $z$; $\theta_1$, $\theta_2$ – angles of rotation of shell cross-sections.

The relationships (1) constructed with consideration of transverse shear deformations which allows to analyze the shells with medium thickness and made from the materials with low shear stiffness. If we assume that $e_{13}=e_{23}=0$, from the expressions (1) the geometric relations of Kirchhoff-Love theory of shells and plates are followed [4, 5].

For analyze the stress-strain state and stability of cylindrical shells the finite difference energy method is used [6, 7]. Lagrange functional of the theory of shells has the form:

$$\Pi=\frac{1}{2}\iint_{\Omega}\boldsymbol{\varepsilon}^{\mathrm{T}}\mathbf{D}\boldsymbol{\varepsilon}\,d\Omega-\iint_{\Omega}\mathbf{q}^{\mathrm{T}}\mathbf{u}\,d\Omega, \tag{2}$$

where $\boldsymbol{\varepsilon}$ – strains vector; $\mathbf{D}$ – the elasticity matrix linking stresses and strains; $\mathbf{q}$ – the external load vector whose components correspond to the five components of the displacement vector; $\mathbf{u}$ – the vector of nodal displacements; $\Omega$ – the area occupied by the shell.

In this method a grid is applied on the region $\Omega$ of the changing of independent variables. The desired function $\mathbf{u}$, delivers the stationary value to the functional (2), approximately determined by its values at the nodes, and the derivatives of $\mathbf{u}$ are replaced by finite differences. Using finite-difference formulas for displacements and its derivatives and replace integration by summation over the area $\Omega$ occupied by the plate, the functional (2) reduces to its discrete analog, which is a scalar function of a vector argument. Euler equations for the Lagrange functional (2) are the equations of equilibrium in displacements.

The equilibrium equations of the shell in the finite difference energy method can be represented as:

$$\mathbf{F}(\mathbf{u},p)=\nabla W(\mathbf{u})-p\mathbf{Q}=0, \tag{3}$$

where $\nabla W(\mathbf{u})$ – the gradient of the potential energy of deformation; $\mathbf{Q}$ – the normalized vector of nodal loads; $p$ – the parameter of the load.

All the components of the displacement vector $\mathbf{u}$ and the load parameter $p$ is assumed to dependent on some continuation parameter $s$. Each value $s$ will uniquely correspond to certain equilibrium $\mathbf{u}(s)$, $p(s)$. A continuous sequence of equilibrium states of the system is the equilibrium curve. For its determination method of differentiating with respect to parameter is used [8, 9]. It brings the system of nonlinear equations (3) to Cauchy problem for systems of ordinary differential equations with the constrained equation:

$$\frac{d}{ds}\left[\mathbf{F}(\mathbf{u}(s),p(s))\right]=0;\ \mathbf{u}(s_0)=\mathbf{u}_0,\ p(s_0)=p_0;$$
$$\left\|\frac{d\mathbf{u}}{ds}\right\|^2+\left(\frac{dp}{ds}\right)^2=1. \tag{4}$$

The solution of Cauchy problem is performed using the self-correcting Runge-Kutta method [1, 10]:

$$\nabla^2 W(\mathbf{u}(s))\frac{d\mathbf{u}}{ds}=\mathbf{Q}\frac{dp}{ds}+Z\left[\mathbf{Q}p(s)-\nabla W(\mathbf{u}(s))\right], \tag{5}$$

where $\nabla^2 W(\mathbf{u}(s))$ – Hessian matrix (matrix of second derivatives) of the function $W(\mathbf{u}(s))$; $Z=1/\Delta s$ – a positive scalar parameter.

The components of the gradient vector and Hessian matrix elements are calculated using the formulas [1, 10] based on expressions (1) and (2):

$$\frac{\partial W(\mathbf{u})}{\partial u_i}=\iint_{\Omega}\boldsymbol{\varepsilon}^T\mathbf{D}\frac{\partial\boldsymbol{\varepsilon}}{\partial u_i}d\Omega;\quad \frac{\partial^2 W(\mathbf{u})}{\partial u_i\partial u_j}=\iint_{\Omega}\left(\left(\frac{\partial\boldsymbol{\varepsilon}}{\partial u_i}\right)^T\mathbf{D}\frac{\partial\boldsymbol{\varepsilon}}{\partial u_j}+\boldsymbol{\varepsilon}^T\mathbf{D}\frac{\partial^2\boldsymbol{\varepsilon}}{\partial u_i\partial u_j}\right)d\Omega. \tag{6}$$

The application of procedure of the self-correcting Runge-Kutta method for solution of the system (3) provides an opportunity to build the efficient numerical algorithm with high precision and the absence of accumulation of the calculations errors. Self-correcting scheme of the Runge-Kutta method on the step $m$ includes [1, 10]:

1. Determination of the vector $\mathbf{u}(s_m)$ and parameter $p(s_m)$ according to the formulas:

$$\mathbf{u}(s_m)=\mathbf{u}(s_{m-1})+\Delta\mathbf{u};\quad p(s_m)=p(s_{m-1})+\Delta p;$$
$$\Delta\mathbf{u}=\frac{\Delta\mathbf{u}_1+2\Delta\mathbf{u}_2+2\Delta\mathbf{u}_3+\Delta\mathbf{u}_4}{6};\quad \Delta p=\frac{\Delta p_1+2\Delta p_2+2\Delta p_3+\Delta p_4}{6}. \tag{7}$$

2. Search values $\Delta\mathbf{u}_k$, $\Delta p_k$ through the serial solutions of the equation systems of the form:

$$\nabla^2 W(\mathbf{u}_k)\Delta\mathbf{u}_k=\mathbf{Q}\Delta p_k+\mathbf{Q}p_k-\nabla W(\mathbf{u}_k);\ \|\Delta\mathbf{u}_k\|^2+(\Delta p_k)^2=\Delta s^2. \tag{8}$$

The solution of equations (8) is performed by Newton-Raphson method. The calculations are performed for one iteration.

3. The adjustment values $\mathbf{u}_k$, $p_k$ for each current substep, using the increments computed in the previous substep:

$$\mathbf{u}_1=\mathbf{u}(s_{m-1});\quad p_1=p(s_{m-1});$$
$$\mathbf{u}_2=\mathbf{u}(s_{m-1})+\Delta\mathbf{u}_1/2;\quad p_2=p(s_{m-1})+\Delta p_1/2;$$
$$\mathbf{u}_3=\mathbf{u}(s_{m-1})+\Delta\mathbf{u}_2/2;\quad p_3=p(s_{m-1})+\Delta p_2/2; \tag{9}$$
$$\mathbf{u}_4=\mathbf{u}(s_{m-1})+\Delta\mathbf{u}_3;\quad p_4=p(s_{m-1})+\Delta p_3.$$

The expression (7), (8) are the fourth order accurate, so step $\Delta s$ can be chosen significantly larger than for Euler method which has the first order accuracy. Optionally, the solution obtained using the method of differentiating with respect to parameter is specified at a certain step using Newton-Raphson procedure.

## 3 Results

The developed method for studying the stability of the equilibrium forms of the shell structures were used in the analysis of shallow cylindrical shells under transverse uniformly distributed load with intensity $q_z$ (Fig.1).

Boundary conditions are hinged support along the contour. The load is uniformly distributed load with intensity $q_3 = q_z(2a/h)^4 p/E$. The construction material is linear elastic. The calculations are carried out at the following values of the dimensionless parameters: $g_{12} = G(1-\nu^2)/E = 0{,}35$; $g_{13} = g_{23} = G(1-\nu^2)/(2a/h)^2/E = 3500$; $q_z(2a/h)^4/E = 61{,}4$; $k_1 = (2a)^2/(Rh) = 60$; $k_2 = 0$.

Investigation of stability of cylindrical panels with curvatures $k_1 = 60$, $k_2 = 0$ is made by the method of differentiating with respect to parameter (4). The arc length of the equilibrium curve is taken as a parameter of continuation. The numerical solution was obtained using a self-correcting Runge – Kutta method (5-9) with increment $\Delta s = 0{,}5$. In Fig. 2 equilibrium curves of the considered panels with symmetric and asymmetric deformation are shown ($p$ is load parameter, $W_c$ is the normal displacement in the center of the shall). Curve 1 illustrates the symmetric deformation of the shell.

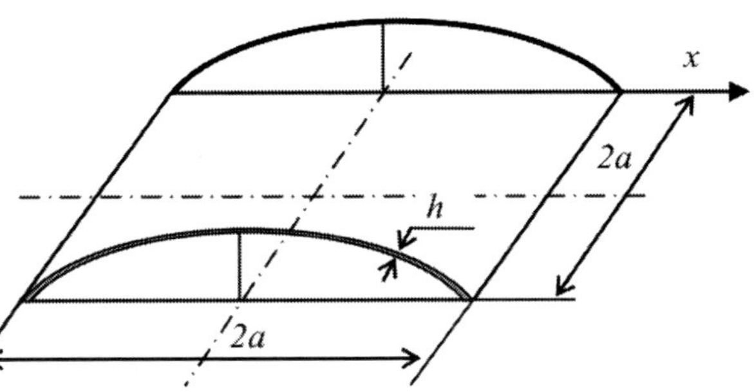


**Fig. 1.** The general form of a shallow cylindrical shell

When solving this problem using the Crisfield scheme (4) it is necessary to separate limit points from the bifurcation points, and thus to determine the sign of the increment of continuation parameter. On each step of leading parameter the value of stiffness $S_p$=$\mathbf{U}^{\mathrm{T}}\mathbf{Q}$ is calculated. Vector $\mathbf{U}$ is determined from the solution of the equation $\nabla^2 W(\mathbf{u}_k(s_m))\mathbf{U} = \mathbf{Q}$. The moment of the simultaneous change of the signs of the matrix determinant $\nabla^2 W(\mathbf{u}_k(s_m))$ and the parameter $S_p$ indicates the transition trough the limit point.

In the analysis of shallow shells the definition of such initial imperfections that cause the largest drop of the critical load compared to the same load for the perfect system becomes important. It allows us to obtain the lower estimation of the actual critical load of a real shell. Numerical studies have shown that the most dangerous are the initial imperfections, which are determined by its eigenform, obtained from the solution of the eigenvalue problem.

Small asymmetric perturbations corresponding to the first asymmetric eigenforms are introduced in the initial conditions of Cauchy problem (4) for finding asymmetric shapes of shell equilibrium:

$$w_0 = \beta \sin\frac{\pi x}{ia} \sin\frac{\pi y}{ja}, \qquad (10)$$

where β – small number, about $10^{-2}h$; $i, j$ – the number of half waves in the direction of the $x$ and $y$ axes respectively. Given small initial asymmetric imperfections (10) in the initial stages does not affect the stress-strain state practically but when approaching to the bifurcation point it leads to the transition to the asymmetric curve. In this case, the critical points corresponding to the asymmetric shape of the stability loss, defined not as the bifurcation points but as limit points.

For the transition to the asymmetric branch of the equilibrium curve 2 (Fig. 2) it is given the values $i = 1$, $j = 2$ in the formula (10), which corresponds a small asymmetry with respect to the line $x$=$a$. For the transition to the asymmetric branch of the equilibrium curve 3 (Fig. 2) it is given the values $i = 1$, $j = 1$ in the formula (10), which corresponds a small asymmetry with respect to the two straight lines $x$=$a$ and $y$=$a$.

The correction of the first asymmetric branch of equilibrium curve in the vicinity of the bifurcation point is performed as the following: after entering on the asymmetric branch the initial imperfection is deleted and the reverse motion along the equilibrium curve by setting the negative increment parameter is implemented. This method let specify the limit load in asymmetric deformation of shells.

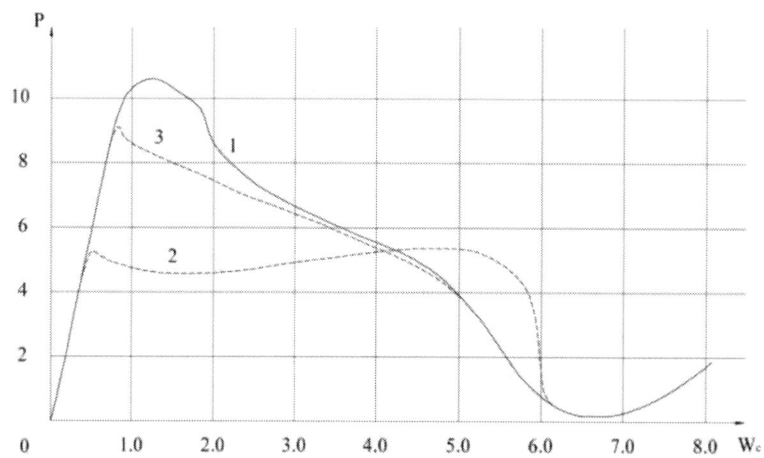


**Fig. 2.** Equilibrium curves of a cylindrical shell with curvatures $k_1$=60, $k_2$=0 when symmetric and asymmetric deformation

## 4 Conclusions

The analysis of the obtained results allows to draw some conclusions about the applied computational schemes:

1. Rapid restructuring of the equilibrium forms in some levels of loading makes a significant error in the solution of the problem using the standard procedures of Runge-Kutta method. Self-correcting form of Runge-Kutta method (5-9) compensates the error in the solution from the previous step which allows to determine the entire equilibrium curve

with a sufficient degree of accuracy.

2. Numerical integration of Cauchy problem (4) with the constraint equation, describing the spherical surface, requires a much smaller integration step than applying the constrained equation, describing the plane. If the calculations are performed with a large step, then in the second case there are oscillations of the obtained solutions with respect to the true equilibrium position.

3. Comparison of techniques based on the self-correcting method of increments and Runge-Kutta method shows that to obtain results with the accuracy within 1% is required to set the step for self-correcting method of increments four times less than for Runge-Kutta method. It leads to the same cost of computer time for the numerical problem solution of both methods.

4. Euler method has a slow convergence. Using of this method in numerical computations of nonlinear deformable shells is associated with high computer time due to the necessity of choosing a small step continuation parameter, especially in the stability problems.

5. The most effective methods for solving problems of this class are Cholesky decomposition and the $LDL^T$-factorization. Cholesky decomposition, sometimes called the square root method is possible only for symmetric, positive definite matrix of the algebraic equations system. In the study of stability of thin-walled structures in the supercritical region Hessian matrix of potential energy of deformation becomes negative definite, therefore the use of Cholesky decomposition becomes impossible. When using $LDL^T$-factorization the condition of positive definiteness of the matrix is not imposed, so this procedure can be successfully applied for construction of the equilibrium curves.

## References

1. I.E. Mileikowskii, S.I. Trushin, *Analysis of Thin-Walled Structures* (A.A. Balkema Publishers, Rotterdam, 1994).
2. S. Trushin, D. Goryachkin, Procedia Engineering **153**, 781 (2016)
3. S.I. Trushin, S.I. Zhavoronok, Space Structures 5, 1527 (2002)
4. G.S. Vardanyan, V.I. Andreev, N.M. Atarov, A.A. Gorshkov, *Mechanics of materials with the basics of the theory of elasticity and plasticity* (Infra-M, Moscow, 2014)
5. S. P. Timoshenko, S. Woinowsky-Krieger, *Plates and Shells* (Librokom, Moscow, 2009)
6. A.B. Zolotov, V.N. Sidorov, Structural Mechanics and Analysis of Constructions **5**, 36 (1975)
7. A.B. Zolotov, P.A. Akimov, V.N. Sidorov, M. L. Mozgaleva, *Mathematical methods in structural mechanics* (ASV, Moscow, 2008)
8. S.I. Trushin, *Structural Mechanics: Finite Element Method* (INFRA-M, Moscow, 2016)
9. S.I. Trushin, Structural Mechanics and Analysis of Constructions **4**, 27 (2007)
10. A.S. Ivanov, S.I. Trushin, Proceedings of the SEIKEN-IASS Symposium, 89 (1993)

MATEC Web of Conferences 117, 00172 (2017) DOI: 10.1051/matecconf/201711700172
*XXVI R-S-P Seminar 2017, Theoretical Foundation of Civil Engineering*

# Method to Increase the Coupling Force in a Construction Machine

*Il'ja* Tsipurskij[1,*]

[1]Moscow state university of civil engineering, Yaroslavskoye shosse, 26, Moscow, Russia, 129337

**Abstract.** This paper discusses a possible method to increase the coupling tractive force track-wheel locomotion of construction machines. Sufficient tractive coupling force allows organizing translational displacement of the machine under above-medium load modes during operation of overburden chain excavators, tower cranes and gantry cranes in outdoors environments. A mechanism is examined to convert rotary motion into rectilinear motion using the example of a gear and rail, with kinematic calculations quoted. Analysis of the "force couple" system is proposed to identify free traction forces. Factors are established that influence the machine's working movements. Equations to calculate tractive forces in track-wheel locomotion are described. A laboratory complex is presented where students of mechanical engineering gain practical skills in mastering the production process of soil excavation and the influence of the coupling tractive force during the machine's operation. As practical recommendation, the paper describes a device made of a balancing lever, drive cogwheel and tractive chain to implement the required tractive force of the trolley in coupling; this solution's efficiency is demonstrated for experimental works on hard soils with high coefficient of difficulty.

Improvements in mechanization of construction go centuries back. It's amazing how designers come up with new solutions that boost productivity and improve performance of construction machines. According to natural laws, sooner or later the quantity of such novelties must transform into a quantity leap, resulting (theoretically) in a fundamentally new machine or in a dramatically improved mechanization technology for civil construction and assembly. For some 40 years now, complexity of construction equipment has grown many times more compared to its efficiency. And yet, one of the ways to innovations lies through mechanical improvements.

The list of issues related to mechanical improvements in the drive of the track-wheel locomotion mechanism of construction machines includes a scant (by its contribution) solution for one option of boosting the tractive force for coupling of construction machines that run on rail tracks.

The purpose of construction machine undercarriage is to transfer loads to the base and to move the machine proper. The undercarriage includes the drive that interacts with the support base, a suspension and a support frame or axes; self-propelled machines also have a travel mechanism. One of the drive types is the track-wheel on machines used in a limited

* Corresponding author: cil@bk.ru

zone, such as tower cranes, bridge and gantry cranes. Wheel track locomotion is also used with excavators for a number of production manipulations such as digging pits along railroad tracks or leveling the track slope. Some types of open-mine continuous-operation cross-digging excavators are also track-mounted when used to strip-mine coal and clay or to level stony slopes around water bodies. The rail track ensures high-precision movements of the excavator and regular thickness of stripped soil, because the rail track is a sort of a guideway [1].

The positive motion drive on a construction machine along the route with a wheel-track enables use of designed modes for permissible tractive forces.

Progressive forward movement of any machine (crane, excavator) is possible if the tractive force generated by the carrier trolleys $T_{max}$ is at least equal to total resistance to movement $\sum W$ of the machine proper with the load (on a crane) or the digging equipment (on an excavator):

$$T_{max} \geq \sum W \quad (1)$$

Force $T_{max}$ is constrained by two factors: the torque *moment* $M_к$ of the drive for the lead metal wheels on the trolleys, and their *coupling* $T_c$ with the supporting rail. The proportion of torque moment $M_д$ of the motor (in a single-motor drive) applied to the lead wheels of the propulsion mechanism can be represented as tangential tractive force T on the rim of the metal wheel, multiplied by its radius R. If force $F_1$ that equals the tangential tractive force T is applied to the center of the driving wheel, and force $F_2$ equal and opposed to force T, is applied to the wheel rim, then forces $F_1$ and $F_2$ become a *pair of forces with arm* R and moment $M_к$. The discipline of "Classical Mechanics" offers this definition: "The moment of a joined pair as a result of reducing force A applied to point O, equals the moment of such force A relative to point O". The force couple has no equivalent force, and at the same time the combined forces of the couple are not in equilibrium. This means that a couple of forces is an example of a set of forces not in equilibrium and without any equivalent force" [2]. The force couple produces a rotating action that can be evaluated by the pair's moment with arm R:

$$M(F_1,F_2) = F_1R = F_2R, \quad (2)$$

The couple's rotary effect is characterized by a vector value known as the *couple's moment*. The moment of the force couple relative to point O:

$$M_O(F_1, F_2) = M_O(F_1) + M_O(F_2) \quad (3)$$

is independent of the selected point O and equal to the moment of one of the pair of forces relative to the other force's application point A at point B:

$$M(F_1,F_2) = M_A(F_1) = M_B(F_2). \quad (4)$$

This circumstance conveniently explains the origin of "free" tractive force T, applied at the center of the driving wheel (or two wheels), equal by magnitude to force $F_1$ and headed in the same direction. This is the force that drives the propelling wheeled trolley to/from its working pass if it is equal to the grip forces of driving and driven wheels on the rail.

The active traction force is created by the motor's moment $M_д$, applied to the driving wheels of the propulsion mechanism of the trolley. It is known as the tractive *force* by output, and is defined by this dependence:

$$T_п = M_д \cdot \eta_x \cdot \eta_к \cdot i_x / R, \text{ kN}, \quad (5)$$

where: $\eta_x$ – efficiency of the transmission of the propulsion drive;
$\eta_к$ - efficiency of the metal running wheels;
$i_x$ – gear ratio of the transmission of the propulsion drive;
R – radius of the rim of the обода trolley's running wheels

The numerator in the formula stands for torque $M_к$ on the trolley's driving wheels. This torque is none other but the couple of forces numerically equal to traction force $T_п$ with arm R. One force $T_п$ of the couple heads opposite to the moving direction and applies to the circumference of the driving wheel; the other is $T_п$ heading towards its axis and along the moving direction. This is exactly the force that moves the trolley, but only if the first force $T_п$ can be absorbed by the coupling of the driving and driven wheels with the rail – that is, only if coupling force $T_c$ exists.

Coupling force $T_c$ between the support wheels and the rail is realized by the friction of wheels on the rail. This force is expressed as part of the construction machine weight, and to be more precise, as a reaction of the rail to the drivingи ($R_1$) and driven ($R_2$) running wheels. For coupling coefficient $f_1$ of the metal wheel and the rail, total traction force *for coupling* is expressed as:

$$\sum T_c = f_1 \cdot (R_1 + R_2)\text{, kN} \qquad (6)$$

The equation of a construction machine moving on full-drive track-wheels in the general case [3] envisages that certain resistance forces must be dealt with: resistance inside the running mechanism assemblies running gear $W_{вн}$, resistance of inertial forces when starting from rest $W_и$, resistance to ascent $W_п$, resistance of wheel on rail $W_к$, resistance to rotation when passing the curve $W_{кр}$. During operation, chain-and-bucket cross-digging excavators have to overcome extra resistance on their working tool $W_{коп}$, and thus total resistance to working movement will be:

$$\sum W = W_{вн} + W_и + W_п + W_к + W_{кр} + W_{коп}\text{, kN} \qquad (7)$$

In the digging operation mode, one needs to factor in different combinations of resistance forces, and working tool loads $W_{коп}$ are made up by resistances to:

$W_к$ – ground digging by the excavator's chained bucket;

$W_г$ – gravity of the ground carried to the stacking conveyor;

$W_р$ – gravity of the chained bucket and its supporting frame

These extra resistances on the excavator's working tool tend to vary most because ground can differ greatly over different sections of the pit face. Even though calculations do envisage different operation modes of the excavator, insufficient coupling of the tractive force compromises the machine's performance. The project envisages that the coupling coefficient for track-wheel machines used outdoors should be $f_1 = 0.12 \div 0.15$, this theoretically ensures tractive force for coupling with the excavator's preset weight. The conclusion sounds commonplace: the excavator's weight (known as coupling weight) should be increased, but this is of course unrealistic.

Another circumstance is relevant for the coupling tractive force $T_c$: wet or iced up rail or the running wheel of the excavator working outdoors. The coupling coefficient declines, and so does coupling tractive force.

The condition that fully enables the movement of any self-propelled machine will be:

$$T_c \geq T_п \geq \sum W. \qquad (8)$$

This expression demonstrates that peripheral force $T_п$, received by driving wheels from the motor, is able to overcome sum total of resistance $\sum W$, and coupling force $T_c$ is sufficient to realize force $T_п$.

As the excavator is in use, the following typical condition may arise:

$$T_c \geq T_п < \sum W\,. \qquad (9)$$

This disparity demonstrates that efforts on the wheels are insufficient to overcome resistances and coupling forces, the wheels cannot turn, and the motor will stall.

If coupling coefficient $f_1$ is lower, accordingly coupling force $T_c$ will fade resulting in this disparity:

$$T_c < T_п > \sum W. \qquad (10)$$

The wheels skid, and the excavator remains without motion.

To increase the trolley's tractive force in coupling, sanding gear is used that raises the coupling coefficient to $f_1 = 0.2$, and in theory enables the excavator operation mode without changing its weight.

An option of increasing tractive force in coupling is of interest, as it is used at the laboratory test bench "Wheel-Type Rotary Trencher", built at the "Mechanization for Construction" Chair at MGSU (Fig. 1).

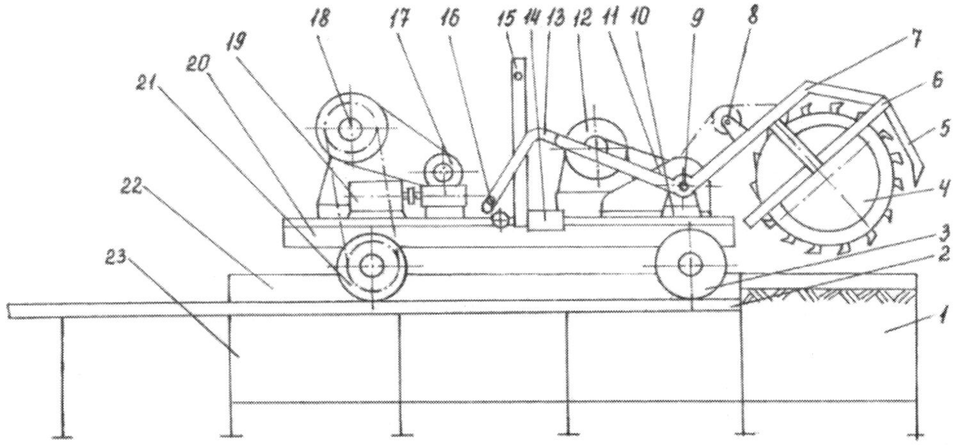


**Fig. 1** - Test bench with the wheel-type rotary trenching excavator ( transport position)

The test bench [4] is a physical model of a continuous-action rotary mill, mounted on a trolley (20) that runs on metal rail tracks (2) above a trench (22) filled with ground (1). The trench rests on posts (23). The rotor frame (6) with a shoe plate (5) is secured fast on the frame (7) made as a double-arm lever. The frame lever (13) is connected with a rope to an angle pulley (16) with a drum of worm-gear manual winch (14) used to lift and lower the swing frame and therefore the rotor. As the trench is being excavated while the rotor is fully sun, the lever is rigidly secured on the support post (15). A hinge (9), around which the double-arm lever turns as the rotor moves up and down, stands on the cross beam (10) of the running trolley frame on the test bench, the one nearest to the rotor. The frame rests on the axes of the driven (3) and driving (21) metal wheels that run along the rail track.

The running trolley carries the rotor drive mechanism and the propulsion mechanism. The rotor drive motor (12) acts by means of a parallel-shaft reduction gear (11) to engage a belt transmission and chain transmission and rotate the shaft (8) of the rotor drive sprocket, from which the torque moment is transferred directly to the rotor. The propulsion drive motor (19) acts by means of a worm reduction gear (17) and belt transmission to rotate the countershaft (18) of the propulsion drive sprocket, from which the moment is transferred directly to the two (left and right side of the trolley) driving wheels (21). Adhesion of the wheels and the rail ensures the coupling tractive effort of the trencher.

The operation equipment – the Rotor (4) has two disks: one fixed and one rotating. A sprocket ring is riveted to the inner side of the rotary disk; the ring is engaged with the drive gear that is overhung on the shaft of the driven sprocket of the rotor drive. Over the perimeter of the rotating disk, two removable half-sectors are secured with countersunk screws, to which cog-holding pockets for are welded in staggered rows.

Equipment installed on the running trolley allows no-load operation, rotor in/out movements, and continuous trenching, thus enabling complete operating simulation of a rotary trenching mill.

During trenching at the test bench, the excavator has to overcome the same sum total of reactive resistance $\sum W$, with breakdown as listed above. Such resistances are overcome with traction forces of power T and coupling $T_c$. (Fig. 2)

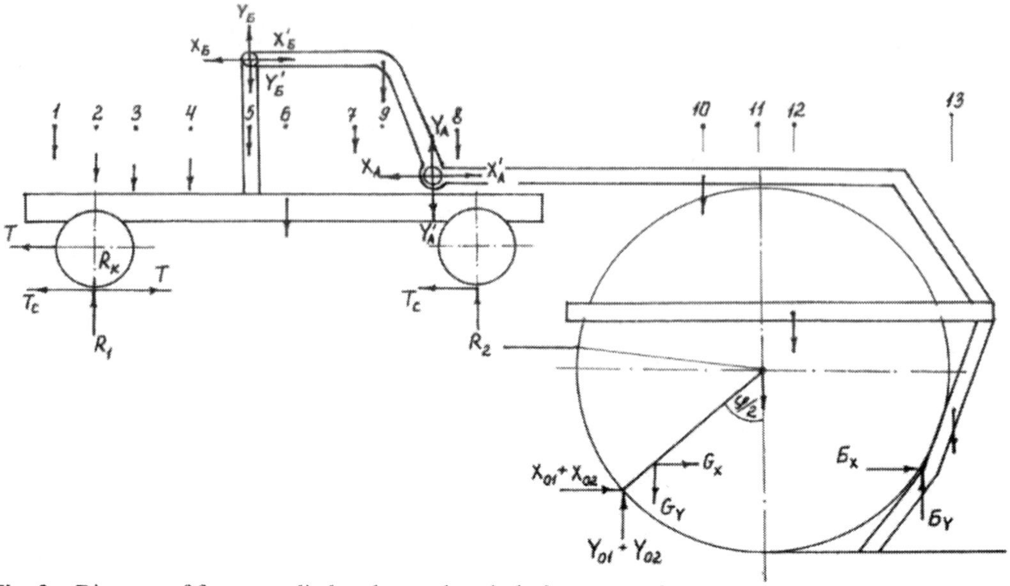


**Fig. 2** – Diagram of forces applied to the test bench during excavation

Active forces T, and $T_c$ and reactive forces $\sum W$ have their respective methodology to find their numeric values when the laboratory test bench trencher is used with different ground handling categories. In particular, resistance to movement by excavation forces $W_{коп}$ is created by existing two components (normal $P_{01}$, and tangential $P_{02}$), that create resistance of ground to excavation, distributed over all cogs currently on the pit face.

Experimental estimated values from the test bench demonstrate that normal operation with stripping thickness up to 10 cm is only possible with grounds of up to category 3 by excavation difficulty. Work in heavier modes is limited by the coupling force between the rail and the wheel in motion.

To increase the tractive force of the running trolley in coupling, the bench uses a mechanism that converts rotary motion to straight-line motion, such as the cogwheel-rail pair [5]. As the rail, the bench uses the sagging part of the chain supported by the rollers of the balancing lever. The chain's one end is secured on the metal structure opposite to the working pass of the trencher while excavating; the other one, partly going round the sprocket on a $120^0$ arch, lies in the guides along the trench in progress. The sprocket sits rigidly on the driving wheel shaft and a balancing lever is mounted centrally with two rollers (Fig. 3).

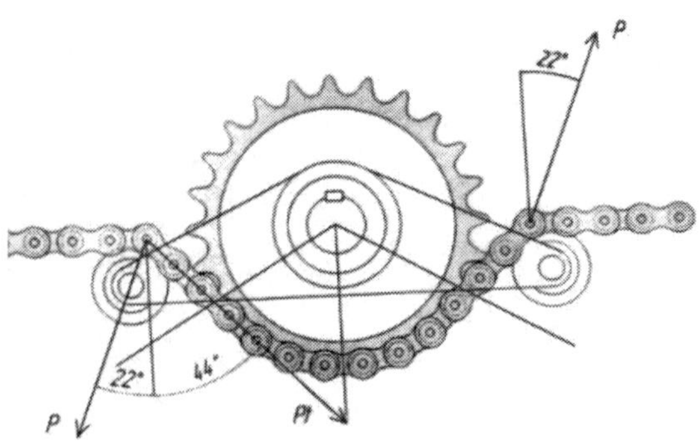


**Fig. 3** - Balancing lever and drive cogwheel engaged with chain

The diameter of the reference circle of the sprocket is equal to that of the outer diameter of the driving metal wheels (21) of radius $R_к$. The arch of sprocket engagement on the chain is adjusted with a dedicated balancing lever that has deflecting rollers. Locations of the rollers are assumed structurally such that the chain can engage the sprocket at an angle that ensures normal distribution of loads over the chain joints in contact. Chain tension $P_t$ adequately makes up for the tractive force in coupling in the event of disparity (10).

The progressive movement speed (mm/min) of the sprocket is found using this formula:

$$V = n \cdot \pi \cdot m \cdot z, \qquad (11)$$

where:

n – rotation frequency of sprocket (driving wheel, $min^{-1}$);
m – engagement modulus, mm;
z – number of sprocket cogs

The above method of increasing the tractive force in coupling by improving the mechanics of the rail track drive helps to realize heavier modes of excavator operation, because tractive force by power is usually sufficient to meet the requirement of necessary and sufficient movement of the excavator during ground excavation:

$$T_п \geq T_c \geq \sum W . \qquad (12)$$

Considering the problems with the national economy at present, it is best to invest in mechanical retrofitting or upgrade of the existing construction machines, ensuring rapid payback, rather than investing heavily to create fundamentally new samples and solutions, complete with expensive electronics that is also unreasonably difficult to service.

## References

1. N.G.Dombrovsky, M.I.Galperin, Construction. Machines, Ch.P., M. "Higher School", M.. – 224 (1989)
2. Internet, "Pair of forces", Section "Statics", "Theoretical mechanics", ISopromat.ru
3. I.L Tsipursky Settings digging and recruitment earthmoving machines: textbook/Publisher ASV 223 (2016)
4. I.L. Tsipursky Determination of the parameters of soil excavating, Manual, / MGSU, - 107 (1998)
5. Internet Toothed rack and pinion transmission www.promtechnoservice.ru
6. I.L. Tsipursky Excavator with working equipment dragline and grapple: Proc. Pos. / M . MGSU, 56 (2011)
7. I.L. Tsipursky Increase traction of the rail track model excavator // Mechanization of construction. **11 (857)**, 7-8 (2015)
8. Road-building machines and complexes: Textbook for high schools / V.I. Balovnev et al .; Under the Society. Ed. IN AND. Balovneva. - 2 nd ed., Supplement. And pererab. - Moscow-Omsk: Publishing House of SibADI, (2001)
9. I.L. Tsipursky The influence of teeth on the remote ability of the milling rotor // Interstroymech, International Conference M.- (2016)
10. A.A. Bogomolov Machines for earthmoving: Textbook for high schools / Belgorod: Izd. BSTU. VG Shukhov. - 314
11. Y.A., Verigin, N.S. Sevryugina. Synergetic foundations of processes and technologies - a new course in master's preparation in the direction of "Construction" // "Guarantees of the quality of vocational education". Abstracts of the reports of the ISPM, (2012)

12. S.N., Glagolev T.A., Duyun, N.S., Sevryugina Problems of engineering education in the field of engineering and technology: a training manual. - Belgorod: Belgorod State Technological University. V.G. Shukhova, (2013)
13. A.N. Drozdov Building machines and equipment: a textbook for the stud. Institutions of
14. E.M. Kudryavcev Construction machines and equipment: Textbook. -M.Discovery of the ASV, 2012.- 328 p higher education / M .: Izd. Center "Academy", (2012)
15. V.I. Balovnev Machines for excavation: construction, calculation, consumer properties:. Book 1. Excavators and earth-moving machines: Ucheb. Allowance. - Belgorod: BSTU Publishing House, (2011)

# Analysis of dynamic coefficients for damage to the middle support of two-span and three-span continuous beams

*Alexander* Tusnin[1*]

[1]Moscow state university of civil engineering, Yaroslavskoye shosse, 26, Moscow, Russia, 129337

**Annotation.** The paper deals with the operation of continuous two- and three-span beams with damage to the middle support. For a two-span continuous beam, the dynamic coefficients for the action of concentrated forces over the middle support and span of the beam are theoretically substantiated. The results of numerical studies of double-span beams at different loads and time intervals of damage to the middle support are presented. The method of dynamic calculation, which can be used to calculate the survivability of steel structures, has been worked out. A good correspondence of the dynamic coefficients obtained numerically with theoretical values is established. Depending on the order of the formation of the design, the dynamic coefficients vary from 1 to 3. For survivability of the design, the pre-bending of the beams above the middle support is most dangerous. With the use of a well-grounded numerical technique, the dynamic coefficients for a continuous three-beam beam are investigated. For continuous beams, if the middle supports are damaged, it is recommended to use a dynamic coefficient of 2 for the load in the spans adjacent to the damaged support

## 1 Introduction

In Russia and other countries, a scientific direction is developing, which studies the behavior of damaged structural structures. In the design practice, the analysis of the load-carrying capacity of such structures is called the calculation for progressive destruction or the calculation of survivability. The problem of studying the bearing capacity of damaged structures is very relevant because of the adverse consequences of the destruction of buildings [1, 2]. In Russia, when designing unique buildings and structures, the calculation of load-bearing structures is performed with damage to their individual elements [3-18]. One of the ways to calculate survivability is to apply a modified long-term load with a dynamic coefficient to the changed design. The magnitude of the dynamic coefficient with a quick turn-off of the element can be taken as equal to 2. For spatial systems, local damage, even with an instantaneous failure of the element, does not always lead to dynamic effects corresponding to the dynamic coefficient 2. Determination of the dynamic coefficients for common designs for various damage schemes is of great practical importance .

* Corresponding author: valeksol@mail.ru

## 2 Theoretical part

For a theoretical justification of the dynamic coefficient, simple beam structures can be used [19, 20]. Consider a beam on two supports with hinged support at the ends and an intermediate support in the middle of the span. In the middle of the span the beam is loaded with concentrated force. Damage to the structure consists in the destruction of the middle support ( Fig.1).

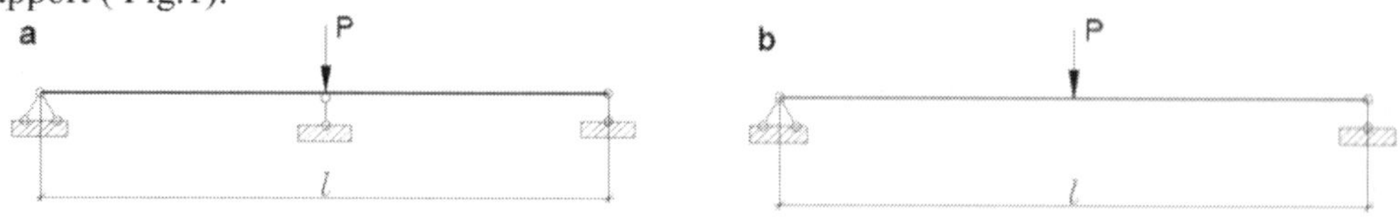


**Fig.1.** The beam is loaded with a concentrated force in the middle of the span: a - the original structure; B- damaged construction.

Perhaps several options for the interaction of the middle support with a beam:

1 variant. With the static action of the force of the beam the span of the span l curved to the deflection, then the middle support was fed. The reaction of the support is zero.

2 variant. The support is fed at a load equal to the force fraction, at the moment of contact of the beam with the middle support, the deflection of the beam.

3 variant. The beam is supported by an intermediate support before the load is applied, the initial deflection is zero

4 variant. The beam is supported on the intermediate support, while the beam is given the initial bend up equal to the deflection.

To determine the dynamic coefficient for each of the variants, one can use the well-known dependence [21]:

$$k_d = 1 + \sqrt{1 + 2H / w_c}\,, \tag{1}$$

where $H$ - the height from which the load $P$ falls on the beam, $w_c$ - the deflection of the beam by a span $l$ of a hinged-at-end force from the static action.

For the connection between force and deflection, we use the following relationships [21]:

$$P = cw_c \text{ и } P_d = cw_d\,, \tag{2}$$

where $c$ - system stiffness (the same for static and dynamic load action), $P_d$ - dynamic load, $w_d$ - dynamic deflection.

1 variant

In the first version, the beam initially has a deflection equal to the full static while supporting it at the ends and in the middle of the force $P$. The absence of an intermediate support during its damage does not affect the operation of the beam, $k_d = 1$.

2 variant

With the 2nd variant of the damage, the part of the full load $P_o$ is perceived due to deformations of the beam, the rest of the load is transferred to the middle support. With the initial deflection $w_o$, the fraction of force perceived by the deformation of the beam is:

$$P_o = P\, w_o / w_c \tag{3}$$

After removing the support on the beam, it will act as an instantly applied, unbalanced fraction of the force, and the total load on the beam will be:

$$P_d = P_o + 2\Delta P = P\left(w_o / w_c + 2 - 2w_o / w_c\right) = P\left(2 - w_o / w_c\right) \tag{4}$$

The dynamic coefficient in this case is:

$$k_d = P_d / P = 2 - w_o / w_c \quad (5)$$

3 variant

With the 3rd variant of the damage, the full load is transferred to the middle support. After the instant removal of this support, the entire load is immediately applied to the beam, and the dynamic coefficient is $k_d = 2$. This result is obtained both from formula (1) for, and from formula (5).

4 variant

With the 4th variant of structural damage, to determine the dynamic coefficient, let us consider the energy of the system deformation $U_d$, the deformation energy of the pre-curved beam $U_o$ and the work $A$ done by the load $P$ after removal of the support. Energy balance of the system:

$$U_o + A = U_d, \quad (7)$$

where

$$U_o = Pw_c / 2 \quad (8)$$

$$A = P(w_c + w_d) \quad (9)$$

$$U_d = P_d w_d / 2 = cw_d^2 / 2 = Pw_d^2 / (2w_c) \quad (10)$$

$w_d$ - the dynamic deflection of the beam, measured from the rectilinear axis of the beam.

After the simplest transformations, we get the quadratic equation:

$$w_d{}^2 - 2w_c w_d - 3w_c^2 = 0 \quad (11)$$

The solution of the quadratic equation has the form:

$$w_{d_{1,2}} = w_c \pm \sqrt{w_c^2 + 3w_c^2} \quad (12)$$

The roots of the quadratic equation:

$$w_{d_1} = 3w_c, \quad w_{d_2} = -2w_c \quad (13)$$

The second root contradicts the physical meaning of the problem, therefore $k_d = 3$. The same value $k_d$ can also be obtained from expression (5), if we take: $w_o = -w_c$.

Set the initial bend over the middle support $0.5w_c$ And carry out similar transformations. Then:

$$w_d = 2.581w_c, \; k_d = w_d / w_c = 2.581 \quad (14)$$

The dynamic coefficient obtained from expression (5) at: $w_o = -0.5w_c$, is 2.5, which differs from $k_d$ from the expression (14) 3%.

Thus, the coefficient of dynamism depends on how the structure was formed. It is necessary to take into account the presence of gaps between the supports and the structure, as well as the presence of preliminary bends during the construction of supports. More dangerous for the survivability of the design is its preliminary bending.

Consider the construction in the form of a continuous two-span beam, loaded in the middle of each of the spans by concentrated force $P$ (Fig.2).

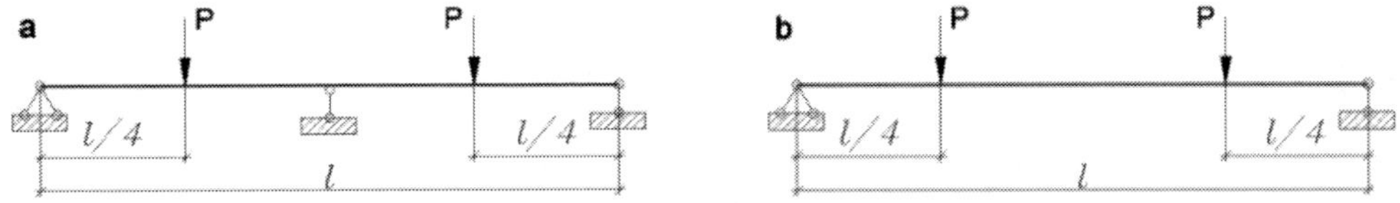


**Fig.2.** A double-span beam loaded with two concentrated forces in the middle of the span: a - the original structure; b - damaged construction

The reference reactions for the original construction are:

– on extreme supports:

$$F_1 = 5P/16; \tag{15}$$

– on medium support:

$$F_2 = 22P/16. \tag{16}$$

Deflection of the beam under each of the concentrated forces in the original structure:

$$w_o = 0.001139Pl^3/(EI) \tag{17}$$

The deflection of the beam under each of the concentrated forces with their static action in the damaged structure, measured from the rectilinear beam axis, is equal to:

$$w_c = 0.02083Pl^3/(EI) \tag{18}$$

After examining the energy balance of the system (7) and performing transformations similar to those carried out for the 4th variant, we obtain a dynamic deflection of the beam in the places of application of forces, measured from the rectilinear axis of the beam:

$$w_d = 1.972w_c. \tag{19}$$

The dynamic coefficient in this case:

$$k_d = w_d / w_c = 1.972 \tag{20}$$

## 3 Numerical studies

Numerical studies are performed using finite element models of a damaged design in a static and dynamic setting [22-25]. With dynamic calculations, the damaged element is removed, and the internal forces arising in the removed element are applied to the structure, which decrease to zero during the element's damage time [22-24].

Numerical studies were carried out using the Nastran computer complex. The beam was modeled by the beam elements "beam". Each beam span was broken into 6 elements. To take into account the dynamic mass calculations in places where concentrated forces are applied, an element of the "mass" type is used. Calculations are made for a beam from steel I-beam 20B1.

The study of the dynamic coefficients under the action of concentrated force over the middle support is carried out for several design schemes:

1 scheme: force 48.94 kN is applied over the middle support to the undeformed beam;

2 scheme: the force is applied to the beam by a span of 12 m, after reaching a half of the beam deflection by a span of 12 m from the force of 48.94 kN, the middle support is brought under the beam and the load is brought to 48.94 kN;

3 scheme: the beam is bent by displacing the middle support by an amount equal to half the beam deflection by a span of 12 m from the concentrated force of 48.94 kN, after which a concentrated force of 48.9 kN is applied to the beam to the middle support, the reference reaction is 73.41 kN.

Variants of the investigated structures are presented on Fig.3.

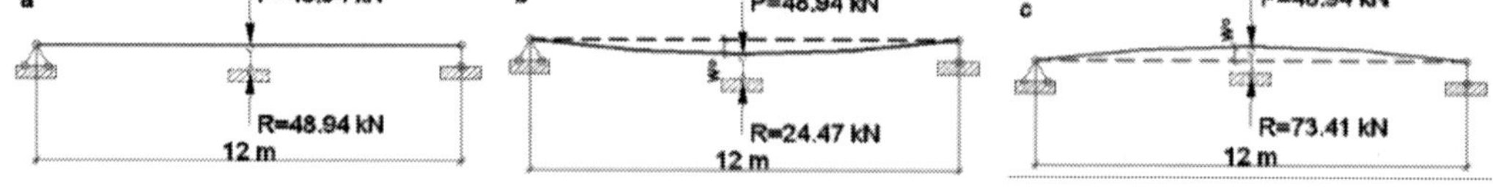


**Fig.3.** Schemes of the structures under investigation: a-1 scheme; b- 2 scheme; c- 3 scheme

With dynamic calculations, the weight of the beam (21 kg / m) and the concentrated mass of 4894 kg (equivalent to 48.94 kN force) are taken into account as the distributed masses. The frequency and period of the first form of natural oscillations were: 0.712 Hz and 1.404 seconds. The following time intervals for the reduction of the reference reaction are considered, sec .: 0.05, 0.1, 0.14, 0.4, 0.7, 1.0, 1.404, 2.0, 3.0. When calculating the 1 circuit, the external load is constant, and the reference reaction decreases for a specified period of time to zero. In the calculation of schemes 2 and 3, the external load and the reference reaction increase for 20 seconds. From zero to the full value, the external load then does not change, and the reference reaction decreases to zero within a specified time interval.

Table 1 shows the forces in a two-span beam, depending on the time of reduction of the support reaction in the event of damage to the middle support. In addition to the efforts in Table 1, the dynamical coefficients are equal to the ratio of the maximum dynamic forces to the corresponding forces calculated for the static loading of the damaged structure. Static load forces are equal: maximum bending moment *M*=48.94*12/4=146.82 кН·м, Maximum lateral force *Q*=48.94/2=24.47 кН.

**Table 1.** Efforts and coefficients of dynamism in two span beam

| Parameter | Efforts and coefficients of dynamism at the time of reaction decrease | | | | | | | | |
|---|---|---|---|---|---|---|---|---|---|
| | 0.05 | 0.10 | 0.14 | 0.40 | 0.70 | 1.00 | 1.401 | 2.00 | 3.00 |
| 1 scheme | | | | | | | | | |
| M, kN·m | 273.88 | 273.08 | 272.05 | 257.78 | 228.17 | 191.77 | 152.04 | 171.38 | 154.26 |
| $k_d$ | 1.87 | 1.86 | 1.85 | 1.76 | 1.55 | 1.31 | 1.04 | 1.17 | 1.05 |
| Q, kN | 45.95 | 45.82 | 45.65 | 43.24 | 38.23 | 32.07 | 25.35 | 28.62 | 25.35 |
| $k_d$ | 1.88 | 1.87 | 1.86 | 1.76 | 1.56 | 1.31 | 1.03 | 1.17 | 1.03 |
| 2 scheme | | | | | | | | | |
| M, kN·m | 210.25 | 209.95 | 209.43 | 202.14 | 187.49 | 169.30 | 149.43 | 159.10 | 150.54 |
| $k_d$ | 1.43 | 1.43 | 1.43 | 1.38 | 1.28 | 1.15 | 1.02 | 1.08 | 1.03 |
| Q, kN | 35.22 | 35.15 | 35.06 | 33.86 | 31.37 | 28.28 | 24.92 | 26.55 | 25.11 |
| $k_d$ | 1.44 | 1.43 | 1.43 | 1.38 | 1.28 | 1.15 | 1.02 | 1.08 | 1.02 |
| 3 scheme | | | | | | | | | |
| M, kN·m | 333.79 | 332.67 | 331.24 | 310.63 | 267.17 | 213.51 | 154.95 | 182.92 | 157.87 |
| $k_d$ | 2.27 | 2.27 | 2.26 | 2.12 | 1.82 | 1.45 | 1.06 | 1.25 | 1.08 |
| Q, kN | 56.12 | 55.94 | 55.70 | 52.20 | 44.85 | 35.76 | 25.85 | 30.58 | 26.34 |
| $k_d$ | 2.29 | 2.28 | 2.27 | 2.13 | 1.83 | 1.46 | 1.06 | 1.25 | 1.08 |

It is established, the faster the support reaction decreases, the greater the dynamism coefficient. The maximum design dynamic coefficients differ from the theoretical ones by 4-12% (the theoretical coefficients of dynamism are somewhat larger). For the considered constructive schemes, if the time of reduction of the reference reaction coincides with the period of the first form of natural oscillations $k_d$ is close to unity.

The work of beams with load in the flights of a continuous beam (Fig.4) is numerically investigated: a beam with concentrated forces in the middle of each span and a beam with a uniformly distributed load.

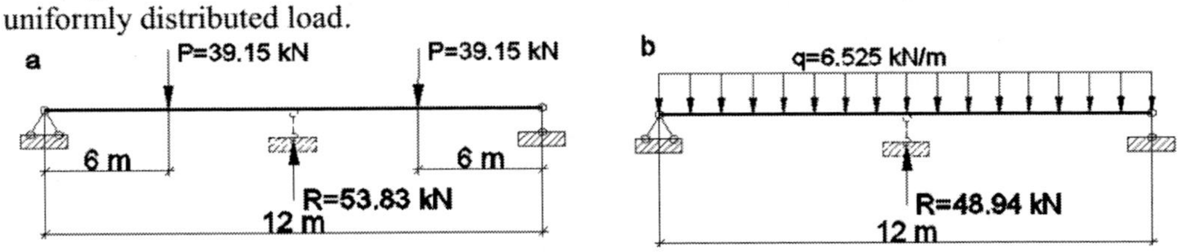

**Fig.4.** Schemes of the structures under study with a load in the span: a - concentrated forces; b- distributed load

As the dynamic masses, the weight of the beam (21 kg / m) is taken into account, in the places of action of the concentrated forces of mass 3915 kg, for the distributed load the distributed masses 652.5-21 = 631.5 kg / m are taken into account. For circuits with load in the spans, the frequency and period of the first form of natural oscillations were: with a concentrated load of 0.794 Hz and 1.260 seconds, with a distributed load of 0.812 Hz and 1.232 seconds. The following time intervals are considered for which the reference reactions decrease, sec .: 0.05, 0.1, 0.14, 0.4, 0.7, 1.0, 1.260 (concentrated forces), 1.232 (distributed load), 2.0, 3.0. When calculating beams with load in the span, the external load does not change with time, the reference reaction at the removal of the middle support decreases for the above-mentioned time intervals to zero.

Table 2 shows the forces and coefficients of dynamism in a double-span beam loaded in spans, with damage to the middle support. Efforts from the static load with concentrated forces in the span are equal: the maximum bending moment $M$=39.15*3=117.45 kN·m, the maximum lateral force $Q$=39.15 kN. With a uniformly distributed load, static forces: $M$=6.525*12^2/8=117.45 kN·m, $Q$=6.525*12/2=39.15 kN.

**Table 2.** Efforts and coefficients of dynamism in a double-span beam under the action of load in the span

| Parameter | Efforts and coefficients of dynamism at the time of reaction decrease | | | | | | | | |
|---|---|---|---|---|---|---|---|---|---|
| | 0.05 | 0.10 | 0.14 | 0.40 | 0.70 | 1.00 | 1.260 (1.232) | 2.00 | 3.00 |
| Concentrated forces in flight | | | | | | | | | |
| M, kN·m | 217.97 | 217.38 | 216.07 | 200.93 | 170.39 | 137.05 | 123.39 | 132.90 | 128.57 |
| $k_d$ | 1.86 | 1.85 | 1.84 | 1.71 | 1.45 | 1.17 | 1.05 | 1.13 | 1.09 |
| Q, kN | 72.57 | 72.30 | 71.94 | 66.91 | 56.84 | 45.57 | 41.13 | 44.28 | 42.85 |
| $k_d$ | 1.85 | 1.85 | 1.84 | 1.71 | 1.45 | 1.16 | 1.05 | 1.13 | 1.09 |
| Evenly distributed load | | | | | | | | | |
| M, kN·m | 221.81 | 219.99 | 219.80 | 204.40 | 173.69 | 139.26 | 122.87 | 133.86 | 128.62 |
| $k_d$ | 1.89 | 1.87 | 1.87 | 1.74 | 1.48 | 1.19 | 1.05 | 1.14 | 1.10 |
| Q, kN | 66.14 | 65.92 | 65.64 | 61.58 | 53.63 | 44.70 | 40.42 | 43.28 | 41.91 |
| $k_d$ | 1.69 | 1.68 | 1.68 | 1.57 | 1.37 | 1.14 | 1.03 | 1.11 | 1.07 |

With a decrease in the time of damage to the middle support, the dynamic coefficient increases. The maximum coefficient of dynamism under the action of concentrated forces in the middle of the spans, obtained numerically, differs from the theoretical one by 6% (theoretical value $k_d = 1.972$ more numerical).

The operation of continuous three-span beams is considered in case of damage to one of the middle supports (Fig.5). The removed support is marked in the figures with a dotted line. The value of the support reaction of the removed support is 1.1*6.525*6=43.07 kN.

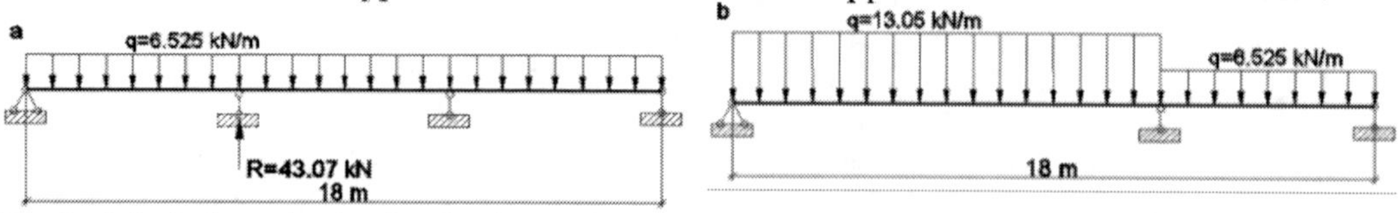


**Fig.5.** Calculation schemes for the destruction of the internal support of the three-span beam: a- for the calculation in a dynamic setting; b- for calculation in a static setting with $k_d$=2

In a three-span continuous beam, the maximum bending moment occurs in the region of the remote support. The dynamic moment is 141.92 kN·m, on and 159.05 kN·m According to static calculation with a dynamic coefficient 2. The maximum reference reaction occurs in an undamaged middle support and is 113.5 kN (dynamic calculation) and 139.5 kN (static calculation). Bending moments and support reactions in the three span beam, obtained by dynamic calculation are less, obtained by static calculation for 11-22%.

## 4 Conclusions

Depending on the initial state of a continuous two-span beam, the theoretical dynamic coefficients vary in the range from 1 to 3. The possibility of numerical calculation of damaged structures in a dynamic setting using the Nastran computer complex has been developed, and a numerical calculation method for the action of concentrated and distributed load has been worked out. The discrepancy between the theoretical coefficients of dynamism and the coefficients determined by the numerical calculation is 4-12%. For the constructions considered, the shorter the time of destruction of the middle support, the greater the coefficient of dynamism. If the average support is damaged during a time equal to the period of the first waveform, there is no significant increase in effort in the structure, while the dynamic coefficients are 1.02 - 1.06.

Taking into account the carried out researches for structures in the form of cut and continuous beams, in the absence of initial gaps or bends, it is recommended to use in the calculations in the static setting the load with the dynamic coefficient equal to 2 in the spans adjacent to the damaged support.

For simple beam structures (in the form of individual beams, girders, consisting of main and secondary beams), in the load capacity reserve it is allowed to perform a static calculation in the design using loads with the corresponding dynamic coefficients.

## References

1. *Report of the Inquiry into the Collapse of Flats at Ronan Point, Caning Town*. – MSO, (TsINIS, perevod 18736), (1968)
2. K. A. Marchand, F. Alfawakhive. *Blast and Progressive Collapse*. – AISC, (2005)
3. Yu.M. Strugatskiy, G.I. Shapiro. *Bezopasnost moskovskih zhilyih zdaniy massovyih seriy pri chrezvyichaynyih situatsiyah*.– Prom.i grazhd. stroit., **8**, 37-41. (1998)
4. P.G. Eremeev. *Predotvraschenie lavinoobraznogo (progressiruyuschego) obrusheniya nesuschih konstruktsiy unikalnyih bolsheproletnyih sooruzheniy pri avariynyih vozdeystviyah*. – Stroit. meh. i rasch. Sooruzh, **2**, 65-72 (2006)
5. A.I.Danilov. *Kontseptsiya upravleniya protsessom razrusheniya stroitelnogo ob'ekta* - Prom.i grazhd. Stroit **8**, 74-77 (2014)
6. Yu.M. Strugatskiy. *Obespechenie prochnosti panelnyih zdaniy pri lokalnyih razrusheniyah ih nesuschih konstruktsiy*. - Sb.: Issled. nesuschih bet. i zhelezobet. konstr. sbornyih mnogoet. zd. – M.: MNIITEP, 3-19 (1980)
7. Yu.P.Nazarov, A.S.Gorodetskiy, V.N.Simbirkin. *K probleme obespecheniya zhivuchesti stroitelnyih konstruktsiy pri avariynyih vozdeystviyah* - Stroit. meh. i rasch. sooruzh., **4**, 5-9 (2009)
8. S.A.Zenin, R.Sh.Sharipov, O.V.Kudinov, G.I.Shapiro, A.A.Gasanov. *Raschyotyi krupnopanelnyih zdaniy na ustoychivost protiv progressiruyuschego obrusheniya metodami predelnogo ravnovesiya i konechnogo elementa* – Stroit. Nauki **4**, 109-113 (2016)
9. A.M. Belostotskiy i dr. *Raschetyi zdaniy na ustoychivost protiv progressiruyuschego obrusheniya s uchetom fizicheskoy i geometricheskoy nelineynostey*. – Teor. i prakt. rasch.zd., sooruzh. i elem. konstr. Analit i chisl. metodyi: Sb. trudov mezhd.nauchno-prakt..konf., M.: MGSU, 183-193 (2008)

10. F. B. Xavier , L. Macorini , and B. A. Izzuddin. *Contribution of masonry cladding for robustness enhancement of multi-storey buildings under sudden column loss.* Proc. of the 16th Int. Brick and Block Masonry Conf., Padova, Italy, 26-30 June (2016)
11. D. A. Nethercot. *Design of Building Structures to Improve their Resistance to Progressive Collaps.* The 20-th East Asia-Pacific Conf. on Struct. Eng. and Const.. Proced. Eng. **14,** 1–13 (2011)
12. Hang Yu, Bassam A.Izzuddin, and Xiao-Xiong Zha., Adv. Steel Const. **6, 4,** 932-948 (2010)
13. P.M. Stylianidis¨ D.A. Nethercot, B.A. Izzuddin, A.Y. Elghazouli. Vol. **115**, 78-95 (2016)
14. B.R.Ellingwood, R.Smilowitz, D.O.Dusenberry, D.Duthinh, H.S.Lew, and N.J. Carino. (2007), *Best Practices for Reducing the Potential for Progressive Collapse in Buildings,* Nat. Inst.of Stand. and Tech., NISTIR 7396, USA.
15. Kirk A. Marchand, David J. Stevens. *Progressive Collapse Criteria and Design Approaches Improvement.* J. of Perf. of Const. Fac. 29(5):B4015004 · June (2015).
16. P.M. Stylianidis', D.A. Nethercot. *Modelling of connection behaviour for progressive collapse analysis.* J. of Const. Steel Res. Vol. **113**, 169–184 (2015)
17. A.Astaneh-Asl. *Progressive Collapse of steel truss brides, the case of I-35W collapse.* Proc., 7$^{th}$ Inter.l Conf. on Steel Brid., Guimaraes, Portugal, June 4-6, 1-10 (2008)
18. Kamel Sayed Kandil, Ehab Abd El Fattah Ellobody, Hanady Eldehemy. *Progressive Collapse of Steel Frames.* W. J. of Eng. and Tech.,**1**, 39-48 (2013)
19. Griengsak Kaewkulchai, Eric B. Williamson. Comp. & Struct.vol. **82**, **7-8**, 639-651 (2004)
20. P.M.Styliandis, D.A.Nethercot, B.A.Izzuddin, A.Y.Elghazouli. *Modelling of beam response for progressive collapse analysis.* Struct. **3**, 137-152 (2015)
21. N.M. Belyaev. *Soprotivlenie materialov.* M., (1965)
22. Yu.Kudishin, D.Drobot. *Zhivuchest konstruktsiy v avariynyih situatsiyah.* Met. zd., **4[8]**, 20-22 (2008)
23. D.Yu. Drobot. *Otsenka zhivuchesti Kryitogo Konkobezhnogo tsentra v Kryilatskom.* Vestnik MGSU. **2**, 116-119 (2009)
24. Yu.I.Kudishin. *Kontseptualnyie problemyi zhivuchesti stroitelnyih konstruktsiy.* Vestnik MGSU. **2**, 28-36 (2009)
25. R.A.K. Sanches, H.B. Coda. Lat.Am. J. of Solids and Struct. **14,** 52-70 (2017)

# The investigation of evolution of the harmonic perturbation the stationary solution of the boundary value problem for a system of the Carleman equations

*Olga* Vasil'eva[1, *]

[1]Moscow state university of civil engineering, Yaroslavskoye shosse, 26, Moscow, Russia, 129337

**Abstract.** The evolution of the harmonic perturbation of one component of the stationary solution of the Carleman system is studied. The system can be used for mathematical modelling of technical problems in such fields: as the gas dynamics, the chemistry, acoustics etc. In particular, for the case of one dimensional space variable the system can be used to describe autokatalys problems for research of building materials. In this case the important property of the mathematical model is the stability of its stationary solutions. Numerical results of investigation of evolution perturbated stationary problem solution for special values of parameters are presented and analyzed

## 1 Introduction

We consider the evolution of the harmonic perturbation of the stationary solution of initial boundary problem for the kinetic Carleman system of equations. This system is a system of two nonlinear differential equation of first order. It can be used for mathematical modelling different problems of the kinetic theory of gasses, chemistry and in particular, for research of building materials. The Carleman system can be used to modelling autokatalys problem in the case one space variable. However the system has the main properties more complete system. It explains the fact that the investigation of the Carleman system is very interested now [1-6]. One of the important properties of the Carleman system is the stability of the stationary solutions. We consider the stationary solutions of the initial boundary problem for the Carleman system equation which components have opposite signs. The positive component of stationary solution is perturbated. The response of the solution to periodic perturbation is investigated. The investigation of the stationary solution properties is based on using the finite-difference methods and the application of quadrature formulas of the second order of accuracy

* Corresponding author: vasilijeva.ovas@yandex.ru

## 2 Formulation of the problem

We consider the initial boundary problem for the Carleman system equations

$$u_t + u_x = \varepsilon^{-1} (w^2 - u^2), \quad t > 0,\ 0 < x < 1$$
$$w_t - w_x = \varepsilon^{-1} (u^2 - w^2), \qquad (1)$$

with initial and boundary conditions

$$u|_{t=0} = u^0(x)\,,\ w|_{t=0} = w^0(x)$$
$$u(t,0) = u_0(t),\ w(t,1) = w_1(t),\ u_0(0) = u^0(0),\ w_1(0) = w^0(1).$$

The functions $u^s(t,x) = u^s(x) = c$ and $w^s(t,x) = w^s(x) = -c$ are two components of the stationary solution (1) with following initial and boundary conditions $u^0(x) = c$, $w^0(x) = -c$, $u_0(t) = c$, $w_1(t) = -c$ for any constants $\varepsilon > 0$ and $c > 0$.

For the investigation properties of the stationary solutions we consider problem (1) with pertrubated initial condition (witn perturbated of the first pozitive component)

$$u^0(x) = c - \alpha\sin(2\pi\omega x)\,,\ w^0(x) = -c \qquad (2)$$

and the following boundary conditions

$$u_0(t) = c,\ w_0(t) = -c. \qquad (3)$$

The problem have parameters $\varepsilon$, $\alpha$, $c$, $\omega$.

Two components of energy of stationary solution perturbation $u^s(x)$, $w^s(x)$ denote by

$$e_1(t) = \left(\int_0^1 (u(t,x) - u^s(x))^2 dx\right)^{\frac{1}{2}},$$

$$e_2(t) = \left(\int_0^1 (w(t,x) - w^s(x))^2 dx\right)^{\frac{1}{2}}$$

and total enargy of perturbation denote by

$$E(t) = (e_1^2(t) + e_2^2(t))^{0.5}$$

The maximum deviation of the first component solution from stationary solution $u^s(x) = c$ denote by

$$Du(t) = \max_{0 \le x \le 1} |\, u(t,x) - c\,|\,..$$

The maximum deviation of the second component solution from stationary solution $w^s(x) = -c$ denote by

$$Du(t) = \max_{0 \le x \le 1} |\, w(t,x) + c\,|\,.$$

## 3 Results of numerical investigation

We consider the problem (1) – (3) for the $c = 1.8$, $\varepsilon = 0.57$, $\alpha = 0.01$, $\omega = 1$. Figures 1, 2 present initial condition of problem $u^0(x)$, $w^0(x)$. Figure 3, 4 show numerical solution of problem $u(t,x)$ and $w(t,x)$ the values of the time variable $t_1 = 1.5$, $t_2 = 3$, $t_3 = 4.5$, $t_4 = 6$, $t_5 = 7.5$..Figures 5, 6 demostrate the dependences of the components energy. Figure 7 demonstrates the total energy E($t$) which tends to zero. The dependences have the oscillating character.

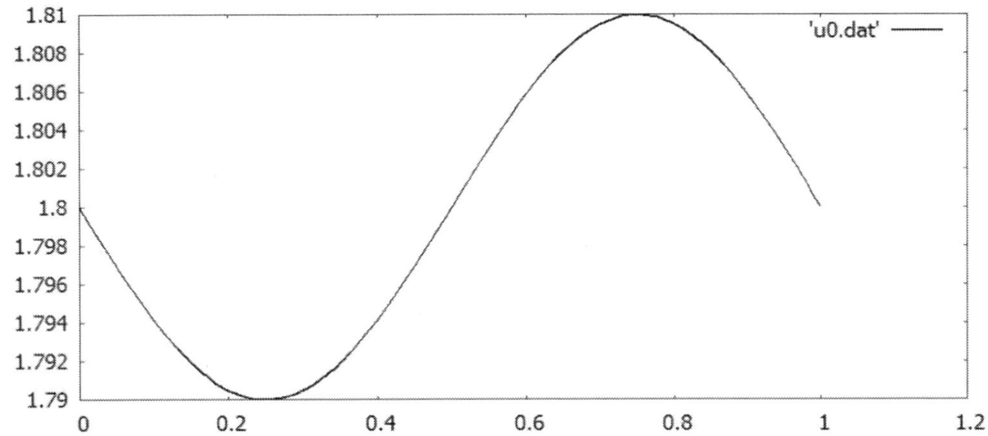


**Fig. 1.** Initial condition for the first solution component $u^0(x)$.

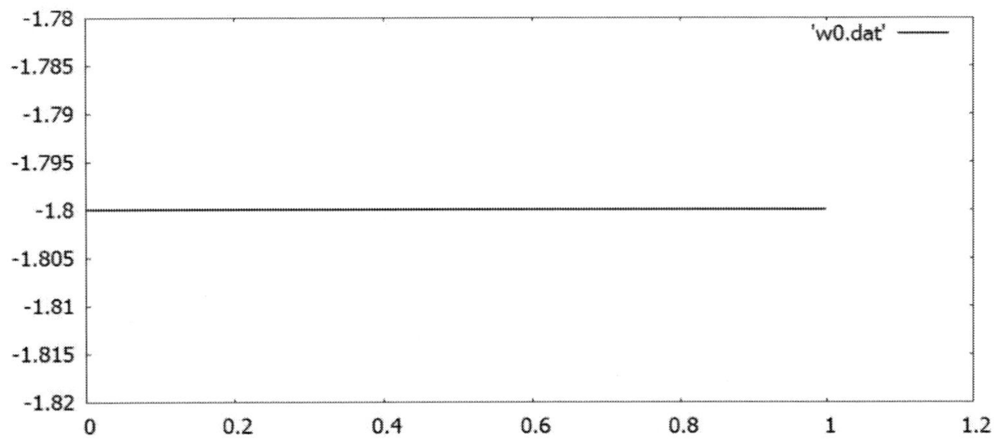


**Fig.2.** Initial condition for the second solution component $w^0(x)$.

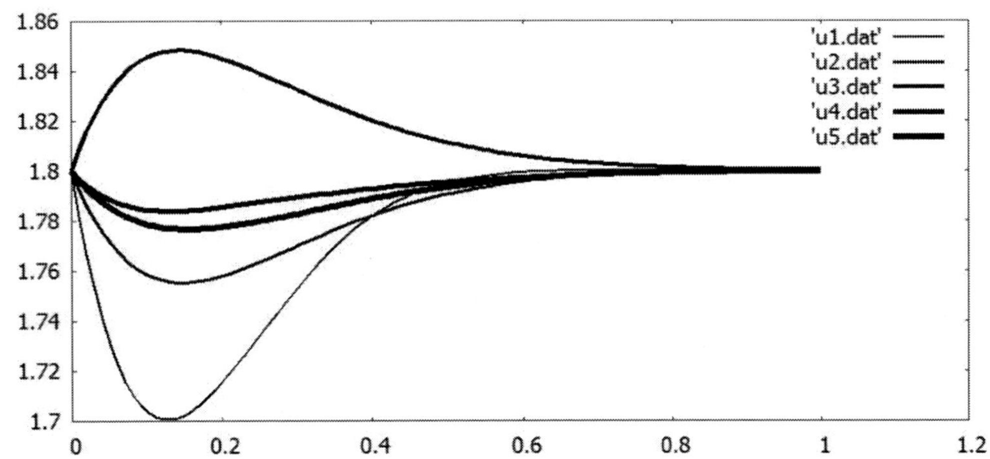


**Fig. 3.** The first component (the perturbated component) of numerical solution of problem $u(t,x)$.

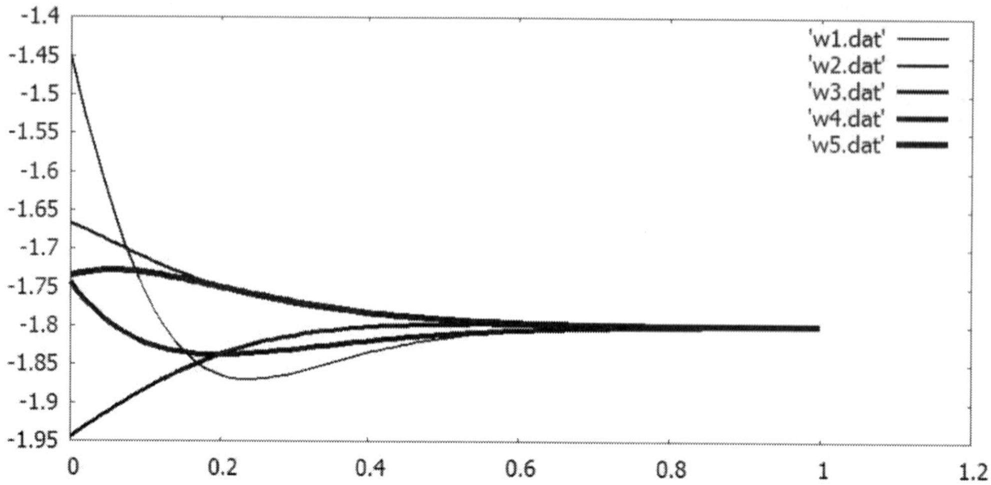


**Fig. 4.** . The second component of numerical solution of problem $w(t,x)$.

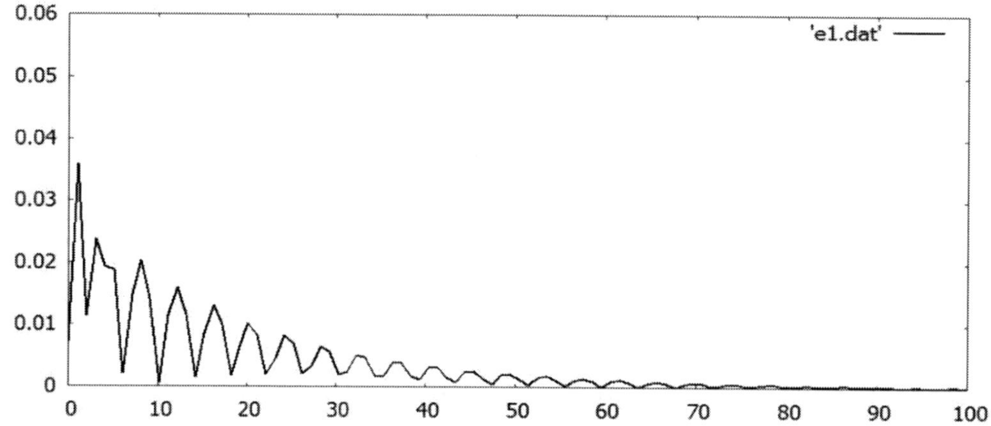


**Fig. 5.** Energy of perturbations $e_1(t)$.

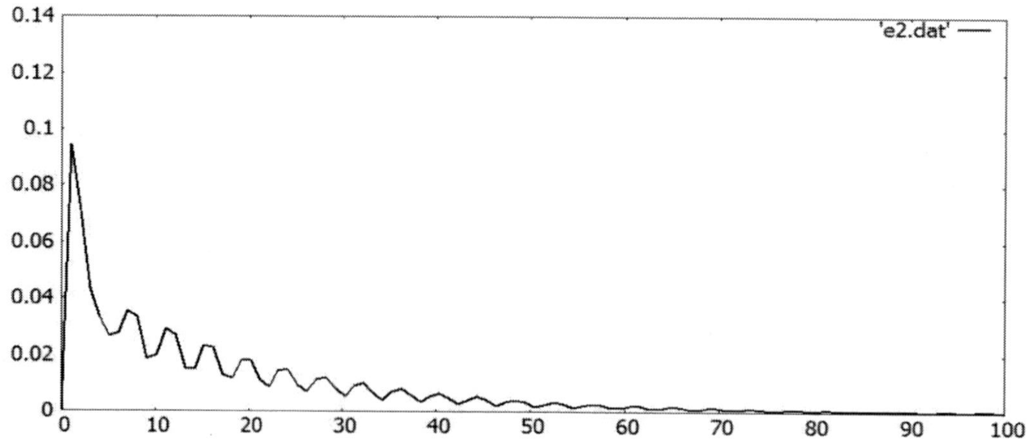


**Fig. 6.** Energy of perturbations $e_2(t)$.

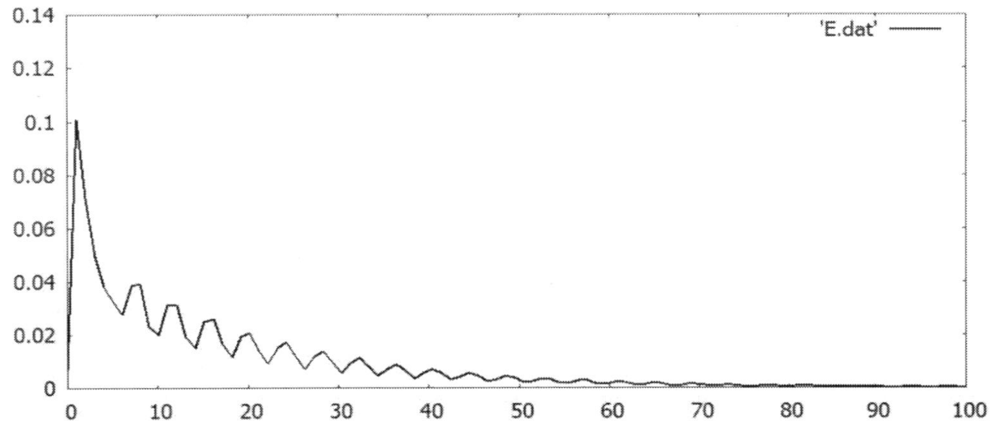


**Fig. 7.** Total energy E($t$).

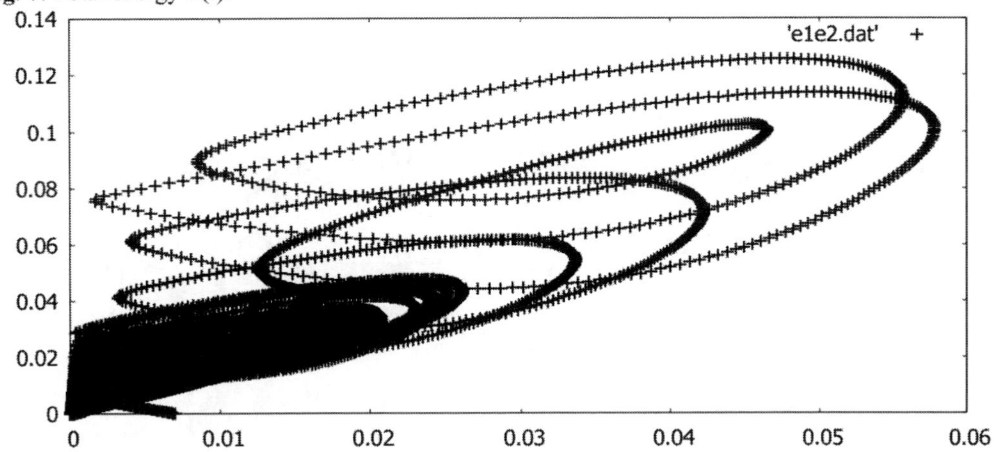


**Fig. 8.** Phase plane ($e_1(t)$, $e_2(t)$).

Figures 8 present the phase plane ($e_1(t)$, $e_2(t)$). Figure 9 demonstrates the dependences of maximum deviations of two component of the solution.

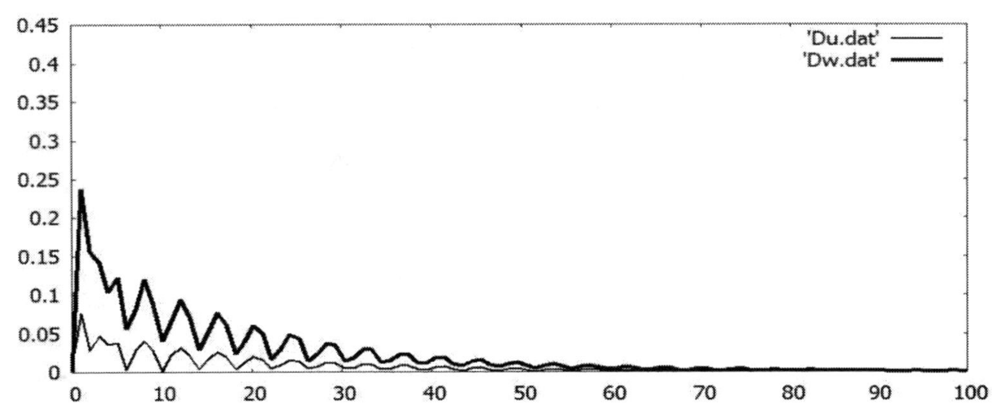


**Fig. 9.** Maximum deviations from stationary solution.

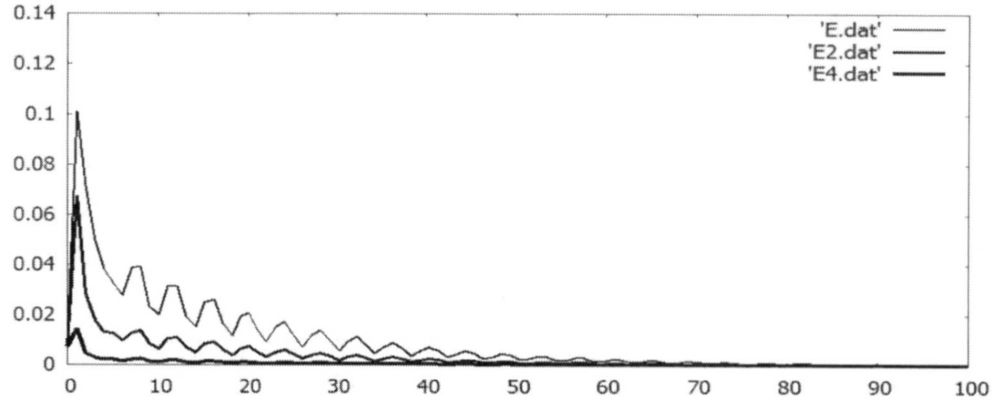


**Fig. 10.** Total energies for different $\omega$.

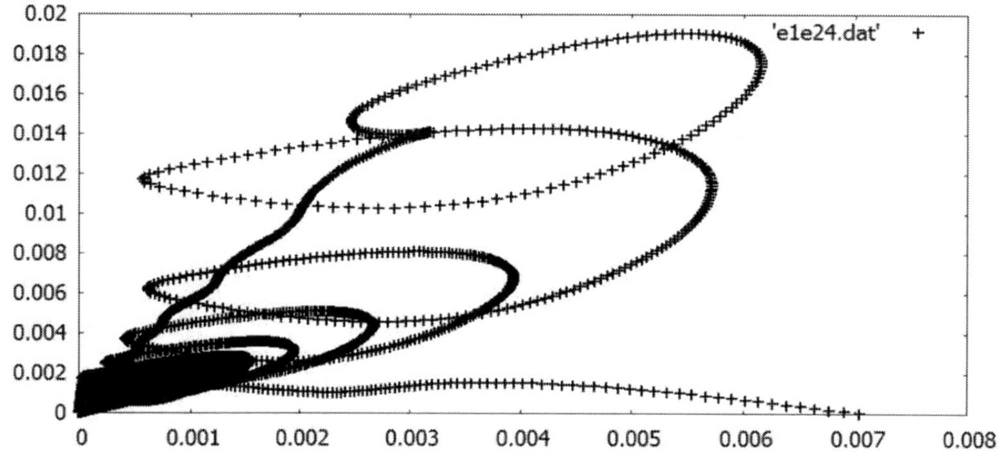


**Fig. 11.** Phase plane ($e_1(t)$, $e_2(t)$).

Figures 10 presents results of numerical investigation of the problem (1) - (3) for $c = -1.8$, $\varepsilon = 0.57$, $\alpha = 0.01$ and different $\omega$. Curves E, E2, E4 correspond the following $\omega$: 1, 2, 3 Here we can see that the stabilisation time essentially increasing with growth $\omega$. But there is not the dependence between $\omega$ and the frequency of oscillation of the total energy $E(t)$. Figure 11 demonstrate the phase plane ($e_1$,$e_2$) for $c = -1.8$, $\varepsilon = 0.57$, $\alpha = 0.01$ and $\omega$=4.

## 4 Conclusion

The results of numerical investigation of the perturbed stationary solution of the Carleman system for the case different sings of solution component are presented and analyzed. So, we can see that there is solution stabilization for considered values of parameters. Dependence of total energy and of maximal deviation solution from stationary solution has oscillating character with constant frequency. The stabilization time essentially depends from frequency perturbation of the first component initial condition.

The author would like to express gratitude to Professor E.V.Radkevich for formulation of the problem and useful discussions and comments.

## References

1. S. K. Godunov, U. M. Sultangazin. Math. Surv. **26**, 3 (1971)
2. E. V. Radkevich. J. Math. Sci. **181**, 5 (2012)
3. E. V. Radkevich. J. Math. Sci. **184**, 4 (2012)
4. V. V. Aristov, O.V. Ilyin. Phys. Let. A. **374,** 4381-4384 (2010)
5. E. V. Radkevich, O.A.Vasil'eva, S.A.Duhnovskij. J. Math. Sci. **201,** 32 (2015)
6. E. V. Radkevich, O.A.Vasil'eva, J. Math. Sci. **224,** 5 (2017)

# Effect of Environmental Actions on Engineering Constructions

*Josef* Vičan[1*], *Anna* Spiewak,[2] *Jaroslav* Odrobiňak[1] and *Malgorzata* Ulewicz[2]

[1]University of Žilina, Faculty of Civil Engineering, Univerzitná 8215/1 10 26 Zilina, Slovakia
[2]Czestochova University of Technology, Faculty of Civil Engineering, Akademicka 3, 42-200 Częstochowa, Poland

**Abstract.** Durability is a very important parameter for design of the engineering constructions because of influence due to degradation processes in the form of structural steel and reinforcement corrosion caused by the environmental actions within structure lifetime. Those degradation processes are affecting the cross-sectional and structural member resistance and deformation parameters of structure. The paper is focused on the study of the corrosion characteristics of the steel bearing structures of engineering constructions to know corrosion effects on the structural steel and to develop the methodology of design for durability.

## 1 Introduction

Increased activities of the environmental actions become the worldwide problem significantly influencing serviceability of the engineering constructions whose maintenance and repairs or reconstructions require large funds [1, 2]. Whereas the environmental actions cause the massive degradation of engineering constructions, it is needed to predict in advance their influence on reliability and mainly durability of structures. Standard [3] specifies two approaches to classification of the corrosion aggressiveness of the atmospheric environment. The first one is based on the measurements of the corrosion velocity using standard test samples. The second approach processes information related to the environment obtained from measurements of air pollution. Both methods have some uncertainness and limitations. The degree of corrosion aggressiveness determined using corrosion losses measured within one year reflects the specific environmental situation of this year of exposition. On the other hand, determination of degree of corrosion aggressiveness by means of dose-response function reflects statistical uncertainness of the given function. But the determination based on informative approach using comparison of local conditions describing typical atmospheric environment could give incorrect interpretation.

Measurements of environment aggressiveness run already a long-time in the World and also in Slovakia. The information obtained is processed into so-called corrosion maps that indicate environmental actions on building objects in terms of increase of corrosion losses. Similarly, in Slovakia, the map was gradually developed in frame of projects of the

* Corresponding author: vican@fstav.uniza.sk

Research Centre UNIZA [4]. Now the research activities continue in the frame of SK-PL project [5] in cooperation with the Polish partner - University of Czestochova, Faculty of Civil Engineering. The main goal of the project is to obtain information related to performing environmental actions in the term of corrosion effects on bridge structural members using in-situ measurements and also those performed in laboratory conditions. Based on the obtained information, the time-dependent corrosion models for structural steel could be developed to define cross-sectional and member time-dependent resistance. The obtained information should be implemented in the concept of the bridge members design for durability respecting the limit state principles using form of partial safety factor method.

## 2 Experimental research in Poland

Firstly, the analysis of the impact of selected factors ($SO_2$, $NO_{xx}$, atmospheric particulate matter PM10 and relative humidity) affecting the rate of corrosion process of steel bridge structural members was investigated. The concentration of air pollutants in Silesia region was evaluated as the highest in comparison to other polish counties due to high rate of urbanization. Therefore the Silesia region was selected for the observation of the corrosion effects on structural steel caused by the air pollution. As the reference samples, 36 test samples in vertical and horizontal position were placed in three independent measurement stations constructed to directly investigate the effect of air pollution on the degradation of structural steel due to corrosion.

**Fig. 1.** Location of observed bridges in the Silesian province.

The obtained information related to the corrosion losses should be used to accurate the formulas for dose-response function [6] determined by means of air pollution data from IMWM - Institute of Meteorology and Water Management.

Secondly, the influence of chloride ions, which are a product of the Winter Road Maintenance (WRM), on the corrosion process was analyzed. In Poland, depending on traffic, all roads are classified into one of the five standards of the WRM designated by means of the Roman numerals from I to V. Based on the analyzed factors of air pollution and type of WRM standards, the Silesia region was divided into three zones in accordance with the Road Administrative divisions and type of WRM. In total, 11 steel bridge structures were selected for placing the measurement samples (see Fig. 1).

Test samples were installed in November 2015 at all sites. For each bridge object, 12 samples on the surface of the vertical web and horizontal lower flange of the main girder were discretely glued (see Fig. 2) to simulate the effect of web and flange corrosion.

**Fig. 2.** Test samples laid out on the steel girder bridge in Marcinków.

The test samples were made of structural steel S355J2+N having nominal geometrical parameters of 100x150x2 mm. Before the installation, each sample was weighed with accuracy of 0.01 g and the dimensions of length, width and thickness were measured. The chemical composition of test samples was determined too. In July 2016 after one year of exposition to aggressive environment due to WRM, six samples were taken off (three samples placed on lower flange and three ones placed on the web) from each of the 11 bridge structures. The samples were then weighed and mechanically cleaned from corrosion products and secondly weighed. Triple weighing allowed ascertaining the weight of the corrosion product and the overall mass loss of test samples due the corrosion after the first observation period. Concurrently, six reference test samples were taken off (three samples located in horizontal positions and three ones from vertical positions) from each of the 3 measurement stations.

Table 1. Results of statistical evaluation of corrosion weight losses.

| Type of WRM | I (g/m$^2$) | II (g/m$^2$) | III (g/m$^2$) | All types of WRM (g/m$^2$) |
|---|---|---|---|---|
| Horizontally located samples | m = 97.06<br>s = 18.86 | m = 187.50<br>s = 192.82 | m = 114.59<br>s = 48.03 | m = 67.38<br>s = 29.76 |
| Vertically located samples | m = 63.50<br>s = 9.72 | m = 67.17<br>s = 31.59 | m = 72.85<br>s = 41.73 | m = 134.73<br>s = 126.16 |
| Both locations | m = 80.28<br>s = 22.51 | m = 127.33<br>s = 150.69 | m = 93.72<br>s = 49.60 | m = 101.06<br>s = 97.65 |

The same process of investigation of corrosion losses was realized with the reference test samples as in the case of bridge samples to compare obtained results and to determine effect of WRM. The chemical composition of all investigated samples was also determined after exposition. The obtained results of corrosion weight losses ($ML_{steel}$) were statistically evaluated with characteristics (m - mean value, s - standard deviation) showed in Table 1. Reference test samples corroding in normal atmospheric environment showed the following statistical characteristics of corrosion weight losses after one year of exposition: m = 80.81 g/m$^2$, s = 36.41g/m$^2$ for horizontally located samples and m = 87.44 g/m$^2$, s = 20.87 g/m$^2$ in the case of vertically located samples.

The corrosion losses of samples thicknesses (μm/year) were statistically evaluated too. Results of evaluation are presented in Table 2.

Table 2. Results of statistical evaluation of corrosion losses of thickness.

| Type of WRM | I μm/year) | II (μm/year) | III (μm/year) | All types of WRM |
|---|---|---|---|---|
| Horizontally located samples | m = 12.27<br>s = 2.37 | m = 23.61<br>s = 24.00 | m = 14.43<br>s = 6.05 | m = 16.98<br>s = 15.88 |
| Vertically located samples | m = 8.05<br>s = 1.24 | m = 8.48<br>s = 3.98 | m = 9.19<br>s = 5.26 | m = 8.51<br>s = 3.75 |
| Both locations | m = 10.16<br>s = 2.83 | m = 16.04<br>s = 18.97 | m = 11.81<br>s = 6.25 | m = 12.75<br>s = 12.29 |

Reference samples corroding in normal atmospheric environment showed the following statistical characteristics of thickness losses after one year of exposition: m = 9.24 μm/year, s = 3.58μm/year for horizontally located samples, while m = 11.44 μm/year, s = 2.58μm/year in the case of vertically located samples. Result comparison indicates the higher corrosion thickness losses of samples placed on the bridges. Also the standard deviations are higher in the case of bridge samples as reference ones. The significant differences of standard deviations of the observed values means considerable dispersion of corrosion weight losses measured on bridges. Both type of tested samples proved the more significant corrosion effect of samples located on bridge in horizontal position.

## 3 Experimental research in Slovakia

The results of measurements of concentrations of air pollution carried out by the Slovak Hydro meteorological Institution (SHMI) from 1971 are the basic background for evaluation of the air quality in Slovakia. Using this information, the dose-response functions were presented for structural steel [7] and for zinc [8]. The in-situ measurements were carried out using standard test samples having nominal dimensions 150x100 mm in accordance with [9]. The thicknesses of samples were specified of 3 mm with respect of long-term measurements. In total, 360 test samples were made of steel S355J2. Each of them was designated, weighed and the real sample dimensions were measured. The location of test samples was oriented on the Žilina self-governing region (Fig.3). The 24 sets of 15 test samples were placed on the following locations:

- 1 set was placed in own corrosion station in the area of University of Žilina (UNIZA) in cooperation with the Research Centre of UNIZA (Fig. 4 -left);
- 11 sets were placed (Fig. 4-rigth) on chosen bridges in frame of Žilina self-governing region;
- 8 sets were placed on roofs of buildings in the cities of Žilina self-governing region;
- 4 sets of samples will be used for corrosion chamber measurements.

**Fig. 3.** Location of observed bridges and building roofs in the Žilina self-governing region.

**Fig. 4.** Location of test sample in the area of UNIZA (left) and on the chosen bridge (right)

Covering heterogonous environment of the Žilina self-governing region was the basic criterion for sample location on bridges. Test samples were installed in modified plastic holders in slope of 45° (see Fig. 4 - right). The measurements of the corrosion losses will be realized in June 2017. The obtained results in-situ measurements will be completed by rapid testing in corrosion chamber. The pilot analyses were carried out to define amount of NaCl causing the same effects as the results recorded by in-situ measurements. The course of dependence of NaCl on the value of corrosion loss obtained within one year of sample exposition on bridge was derived. The calibration should also continue after two years of sample exposition on bridges.

## 4 Conclusions

The contribution described the preparation and particular results of the research related to the investigation of air pollution and its effect on degradation of structural steel due to corrosion. The outputs of measurements of corrosion losses of test samples located on bridges in Silesian region after one year of exposition were presented. Future research activities in this field of study were also introduced. Based on the obtained information, the time-dependent corrosion models for structural steel should be developed [10] to define cross-sectional and member time-dependent resistance. The results obtained in frame of long-term observations should be implemented in the concept of the bridge members design for durability [11] respecting the limit state principles using form of partial safety factor method.

The paper presents results of the research project SK-PL-2015-0004 and the research project APVV-14-0772 supported by the Slovak Research and Development Agency and the project 012ŽU-4/2016 supported by the Cultural and Education Agency of Slovak Republic.

## References

1. Z. Agocs, J. Brodnianský, J. Vičan, and J. Ziolko, *Assessment and refurbishment of steel structures,* Taylor & Francis (2005)
2. J. Bujňák, R. Hlinka, J. Odrobiňák and J. Vičan, *Procedia Engineering,* **40** (2012)
3. STN EN ISO 9223, *Corrosion of metals and alloys – Corrosivity of atmospheres – Classification, determination and estimation,* SOSMT Bratislava (2012)
4. J. Vičan, B. Hadzima at all, *Research of progressive materials for construction of transportation path and production of transportation means.* Project of Research Centre, University of Žilina (2013-2015)
5. J. Vičan at all, *Effect of environmental actions on engineering,* Project of the SRDA SK-PL-2015-0004, University of Žilina (2016-2017)
6. V. Kucera, J. Tidblad, K. Kreislova, K. et al. *Water Air Soil Pollut: Focus* 7 (2007)
7. M. Ivašková, P. Koteš, M. Brodňan, *Procedia Engineering*, **108** (2015)
8. M. Ivašková, P. Koteš, B. Hadzima, Konstrukce, **6** (2015)
9. STN EN ISO 9226, *Corrosion of metals and alloys – Corrosivity of atmospheres – Determination of corrosion rate of standard specimens for the evaluation of corrosivity,* SOSMT Bratislava (2012)
10. V. Křivý, V. Urban, K. Vavrušová, *14th International Conference on Civil, Structural and Environmental Engineering Computing, Civil-Comp Proceedings*, **102**, (2013)
11. M. Krejsa, R. Čajka, *Insights and Innovations in Structural Engineering, Mechanics and Computation, Proceedings of the 6th international conference, SEMC* (2016)

MATEC Web of Conferences **117**, 00176 (2017) DOI: 10.1051/matecconf/201711700176
*XXVI R-S-P Seminar 2017, Theoretical Foundation of Civil Engineering*

# Application of expert systems in the construction sector

*Arkadiusz* Węglarz[1*] and *Paweł Grzegorz* Gilewski[2]
[1]Warsaw University of Technology, Faculty of Civil Engineering, al. Armii Ludowej 16, 00-637 Warsaw
[2]Warsaw University of Technology, Faculty of Building Services, Hydro and Environmental Engineering, ul. Nowowiejska 20, 00-653 Warsaw

**Abstract.** Nowadays in the construction sector, there are many challenges regarding technical and economic problems as well as those associated with the impact on human health and environment. Solving these problems is not straightforward and requires the involvement of experts from different fields. Expert systems are interesting and complex solutions for the above-mentioned aspects. The article presents characteristics of Polish and world experts system in which the authors participated.

## 1 Expert systems

An expert system is a single or multiple sets of software which intends to support the use of knowledge and facilitate decision-making process [1, 2]. Expert systems can assist or even replace human experts in a certain field. They can provide the user with advice, recommendations or diagnosis in the problems related to the study field.

The following characteristics distinguish expert systems from conventional systems:
- explicit representation of knowledge,
- application to solving specific problems according to the reasoning procedures (inference)
- ability to explaining the solutions found by the system,
- knowledge processing mainly concerns symbol processing rather than numerical computations.

Unlike classical software knowledge in the knowledge, database describes the problem domain without providing an exact algorithm for solving a given problem. Knowledge is written according to a specific knowledge representation language, most commonly consisting of:
- description of the facts,
- description of the rules used in the inference process,
- meta description of rules that show a strategy for solving the problem,

A typical expert system is shown in Fig. 1.

* Corresponding author: a.weglarz@il.pw.edu.pl

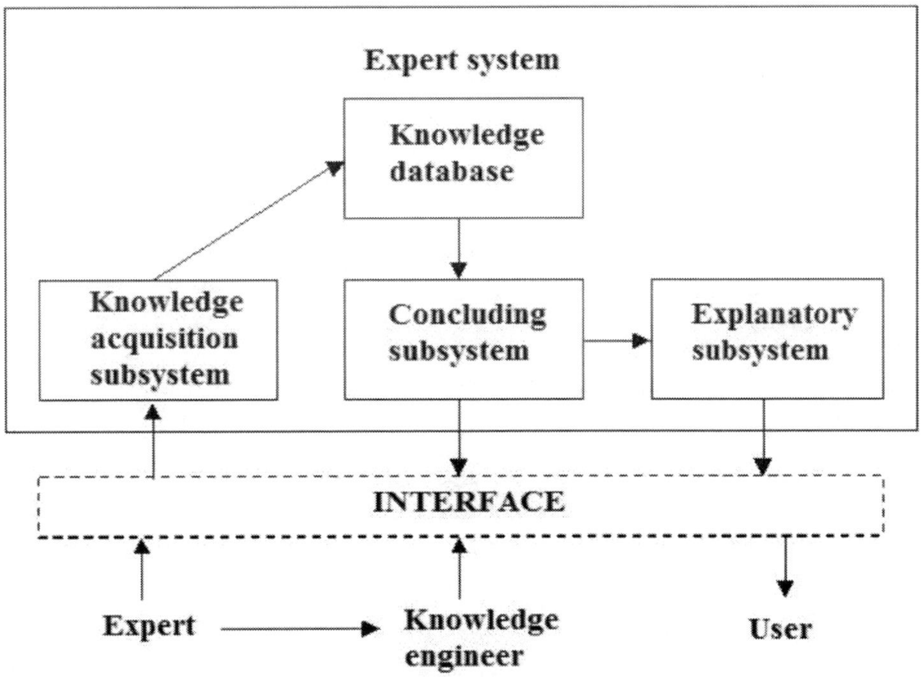


**Fig. 1.** Diagram of expert system (own elaboration based on: http://www.rosczak.com/mlp/ai.html)

## 2 Existing expert systems in the construction sector

One of the first expert systems in the construction sector was HI-RISE program which was elaborated to help with the design of tall buildings [3]. Another interesting program is called ScaRC and is used to create construction schedules during the construction of repeatable storey floors [3]. The expert system called RUMBAu uses the 3D technique to design wall formations in the PERI TRIO system. Also, the ELPSON program is used for the support in decors design in the PERI system. It is characterized by such a level of service simplicity that it can be used by people with very little professional experience. An interesting diagnostic system called REPCON is used during the assessment of buildings structure damage. It also provides hints and proposes methods for repairing them.

The system SPERIL [4] is being developed at Purdue University. Its aim is to analyze the building damage which results from hazardous events such as an earthquake or roof overload due to precipitation. It proposes a mathematical model of analysis, using rules of the IF-THAT'S-DIFFERENT type. The inaccuracy of reasoning is modeled by fuzzy logic (between truth and falsity there are many indirect values), and the full program is implemented in C-language.

For the evaluation of a construction investment, the HICOST program [3] was implemented as a rule-based program. It consists of a series of subsystems to find the cost of the building's components.

The application called COMVOC proposes support for project planning while operating on estimates of the final value of the construction project as a function of decisions affecting the project. It allows exporting the resulting database to the well-recognized MS environment.

The ACPROM system [5] provides support for the investment implementation phase, estimating the progress of building works based on project and photographic documentation, allowing to export the resulting databases to MS Project. It allows to integrate:
• executive drawings,
• digital pictures of the construction site,
• digital progress photography,
for project monitoring and decision support by the building process managers. As a result of the system automatically interprets the drawings in [*.dwg] format and extracts construction element data so that, comparing the above with digital images, generate information about the percentage progress of the planned work.

The "Value Management for Construction Projects" program supports the process of identifying financial parameters for construction projects. The process of acquiring knowledge into the system consists of an aggregation of knowledge available in the literature on value management in the construction industry by means of a checklist and knowledge available from experts using the submitted forms. It was implemented in C ++ and later verified by experts in the field of value management.

The COMIX program [6] has an advisory role in the design of concrete mix based on New Zealand standards. It is based on rules and framework and helps to choose the type of mix and the type of construction, calculates the correct ratio for the appropriate load and sand mass, aggregate and all concrete components.

The BIDEX program offers support of the process of preparation of offers for a given building objects. It is used by construction engineers when making bid decisions. Built using the expert system shell "Exsys", the program makes a decision in two stages. The first decision concerns the cost of works, materials, and equipment. The second decision concerns the size of the overheads. This last decision includes the type of owner, type of work, the size of work, place of work, market power, the degree of danger and confidence level of subcontractors.

The platform "Advisory System for Managers" [6] platform is an operational prototype of decision support for planning and management of building projects that can be used by building managers in their day-to-day duties and during the pre-design phase. So far the program has been used to analyze small projects. The main purpose of the application is to systematize the decision-making process. It includes models of cost, raw materials, administration. The system has mainly an advisory function, among others like:
• information management,
• financial management,
• current and future technical problems on the construction site.

The CONFAULT system is intended to support diagnostic processes related to defects in reinforced concrete construction by identifying subgroups of defects. The knowledge database of the program is divided into modules corresponding to the six basic types of defects. Metadata are used to control and restrict the research. The original approach to confidence allows uncertainty modeling.

## 3 Expert system for the building energy source selection

At the University of Zielona Góra the “Expert system for the selection of energy source for the building” [7] has been created in the framework of the research task No. 6 "Analysis of technical and operational requirements for buildings with power from centralized heat sources", implemented as a part of the strategic project "Integrated system for reducing the energy consumption of buildings". The concept of the system was based on the integration of many computing tools to create a unified computer system supported by knowledge databases.

The main purpose of the expert system is to analyze the energy of a given building or group of buildings and to make recommendations on energy sources based on $CO_2$ emissions and investment costs based on the indicators contained in knowledge databases.

The expert system consists of nine major elements:
• module for entering and editing data of research buildings,
• the input generation module for the simulation application (EnergyPlus) calculating the energy characteristics of the researched buildings,
• simulation application based on ready-made calculation engine (EnergyPlus),
• input file generation module for calculation and selection of energy source,
• an independent calculation module prepared in the MATLAB environment and designed to calculate and select a source of energy for a research building (including a centralized source) based on $CO_2$ emission,
• investment analysis module for the construction of district heating and local energy sources for the needs of a group of buildings,
• investment analysis module related to the necessary repairs of the district heating network when the recommendation concerns a centralized energy source for a group of buildings,
• investment analysis module for an embedded energy source,
• the module of recommendations for energy sources based on $CO_2$ emission and investment costs of the heat source.
• executive drawings,

The expert system for both pathways conducts all analyses considering the following possibilities of the recommendation of energy sources.

Analysis of a single building:
• energy source built into the facility (gas or oil boiler room),
• built-in microgeneration unit,
• inserting the building into an existing heating system with a centralized heat source.

Analysis of a group of buildings:
• energy source built into each building separately (gas or oil boiler room),
• heating micro-grid with group energy source in the form of gas or oil boiler room,
• heating micro-grid with a group of energy sources in the form of a micro-cogeneration unit,
• inserting a group of buildings into an existing district heating system with a centralized heat source.

As a result, the expert system provides a list of energy recommendations for a single building or group of buildings that can be ranked on both criteria, i.e. $CO_2$ emissions and estimated investment costs, or only one criterion.

The software is ideal for the decision support, investment planning related to changes in the way that heat is supplied to buildings or small settlements. Due to its expert nature, it can also be regarded as an advisory tool for planning changes in the energy policy of a settlement unit.

## 4 Description of the expert system supporting the energy management process in a smart single-family house

In a properly newly designed and constructed an energy-saving building or in a deeply thermo-modernized existing building the theoretically calculated energy-saving effects are often not achieved. There are many reasons for that. One of them is poor management of energy in the process of building operation. The solution to this problem can be the use of smart building systems such as the KNX system. Such issues have already been described in the literature [8, 9]. However, the problem is complicated when dealing with energy management in buildings with hybrid systems of heat and cold production such as gas boiler and wood burning fireplace. At this point, it is advisable to use expert systems to control the KNX installation.

In order to get the database input for the expert system controlling the KNX installation in a single-family house, the calculation of the demand for heat and cold (power and heat/cold consumption) was made using the hourly method using the Design Bilder program. The subject of the study was a multi-generation single-family house located in Warsaw. The analyses were conducted for the standard year (365 days) and the year with extreme weather conditions (frosty winter and hot summer).

The simulation results are shown in Fig. 2.

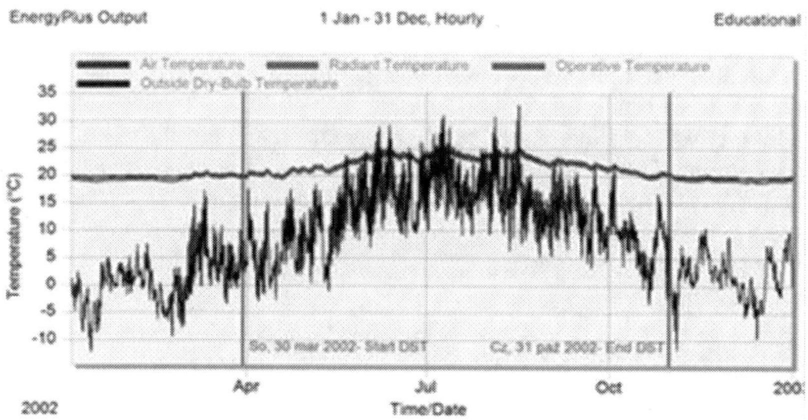


**Fig. 2.** Design Bilder simulation results.

The purpose of the expert system is to determine when to run the selected device so that the final energy consumption of the indoor climate control system (energy management) is minimal and the costs as small as possible.

The expert system must decide whether to start heating or cooling. The KNX system will do it.

In the case of the heating mode the following options are available:
- turn on the ground source heat pump,
- activate the electric heater for supply air (1 kW),
- start the electric heater for supply air (3 kW) and turbine with the productivity of 540 $m^3/h$,
- start the fireplace (15 kW) and turbine with the productivity of 700 $m^3/h$.

For cooling mode, the following options are available:
- turn on the ground source heat pump,
- active the cooler for supply air (1 kW),
- active the cooler for supply air (3 kW) and turbine with the productivity of 540 $m^3/h$.

Each room has the option of setting the required internal temperature with a KNX thermostat that controls the throttle located before the diffusers in the DGP Hot Air Distribution System. If all the throttles are closed then the turbine is shut down. If all are fully open then the turbines are running at a maximum efficiency. In intermediate states, the productivity of turbines and the opening of throttles is proportionally controlled.

Considering the results of the simulation of Design Bilder, the following rules of the expert system are proposed:

**Heating season = (Yes/No)**

$t_z$ – temperature on the external sensor,

$t_w^{max}$ – maximum value from temperature measurements on sensors inside the building [°C].

**Rule 1:**

If $t_z \leq -5°C$ or $t_z \geq 25°C$ then turn on the ground source heat pump, otherwise follow the Rule 2.

**Rule 2:**

If $t_z > -5°C$ and $t_z < 25°C$ then turn off the ground source heat pump from the ventilation system.

**Rule 3:**

If the temperature sensor over the fireplace indicates a temperature of > 30°C and at least one throttle is open, then turn on the turbine with a capacity of 700 $m^3/h$.

**Rule 4:**

If $t_w^{max} > 26°C$ then start up the supply air cooler with 1 KW and turn off the heater.

**Rule 5:**

If $t_w^{max} > 26°C$ and a 1 KW ventilated air cooler was switched on and the heating was switched off then start up a 3 KW cooling radiator and the turbine of a capacity of 540 $m^3/h$.

**Rule 6:**

If Heating season=Yes and the temperature sensor over the fireplace indicates a temperature of < 20°C and at least one throttle is open then start up 3 kW heater and the turbine of a capacity of 540 $m^3/h$.

The system was launched in September 2016 and the first results show the correctness of the assumed thesis and the convergence of the results of the digital simulation with measurements of the outside and inside temperatures.

## 5 Summary and conclusions

In the world construction industry, for many years, the expert systems have been used to support decision-making at every stage of the investment process from design to building use. Unfortunately, in Poland, the use of these artificial intelligence tools is still very rare. This is due to the building tradition and the awareness of the participants in the building process: investors, designers, contractors and property managers. In recent years, the concept of native expert systems has emerged as an example of the systems described above. It seems necessary to develop further research in this field towards hybrid artificial intelligence systems with the key role of expert systems.

## References

1. J. Mulawka, *Expert systems* [in Polish] (Wydawnictwo Naukowo-Techniczne, 1997)
2. R. Tadeusiewicz, *XVI Conference of Automation Rytro* (CD), (2012)
3. D.R. Rehak, S.J. Fenves, *Expert systems in civil engineering, construction and construction robotics* (Research Showcase CMU, 1984)
4. S.J. Fenves, M.L. Mahler, D. Siriam, IABSE PER. **4**, 63 (1985)

5. Z.A. Memon, *First International Conference on Construction in Developing Countries* (ICCIDC–I, 2008)
6. M. Akram, I.A. Rahman, I. Memon, Int. Jour. of Civ. Engin. and Bui. Env., **1**, 24 (2014)
7. P. Ziembicki, A. Węglarz, Ciepłownictwo, Ogrzewnictwo, Wentylacja, **5,??** (2013)
8. R. Radajewski, A. Kamińska, Elektro.info **11,??** (2008)
9. M. Horyński, *Energy-efficient automated energy management systems in residential buildings* (Wyd. Pol. Lub., 2015)

MATEC Web of Conferences 117, 00177 (2017) DOI: 10.1051/matecconf/201711700177
*XXVI R-S-P Seminar 2017, Theoretical Foundation of Civil Engineering*

# OMA Research of Sky Tower in Wrocław, Poland

*Zbigniew* Wójcicki[1,*], *Jacek* Grosel[1], *Alexander* Belostotsky[2,3,7], *Pavel* Akimov[2,3,4], *Vladimir* Sidorov[5,6,7]

[1]Civil Engineering Faculty, Wrocław University of Technology, Wyb. Wyspiańskiego 27, 50-370 Wrocław, Poland
[2]Scientific Research Center "StaDyO"; office 810, ul. 3ya Yamskogo Polya, Moscow, 125040, Russia
[3]National Research Moscow State University of Civil Engineering, Department of Applied Mathematics; 26, Yaroslavskoe Shosse, Moscow, 129337, Russia
[4]Russian Academy of Architecture and Construction Sciences; 24-1, ul. Bolshaya Dmitrovka, Moscow, 107031, Russia
[5]Kielce University of Technology; Al. 1000-lecia Państwa Polskiego 5, 25-541, Kielce, Poland
[6]Moscow State University of Railway Engineering; 9b9 Obrazcova Street, Moscow, 127994, Russia
[7]Perm National Research Polytechnic University; 29 Komsomolsky prospekt, Perm, Perm Krai, 614990, Russia

**Abstract.** Sky Tower is a skyscraper in Wrocław, Poland. Construction began in December 2007 with the demolition of the 24-story Poltegor structure, until then the tallest building in the city. Sky Tower is the tallest residential building in Poland in the category of height to roof and category of highest floor but is shorter than the Palace of Culture and Science. On the 49$^{th}$ floor is a publicly accessible viewpoint. Operational Modal Analysis (OMA) of Sky Tower was done in 2012 when the tower was almost completed. No excitations with the exception of ambient excitations were used 8 eigenfrequencies and eigenforms were detected.

## 1 Subject of study

The Sky Tower is a skyscraper in Wrocław, Poland, Figure 1.

The Sky Tower complex consists of 3 connected buildings:

- B1 – 3-storey platform, in which the shopping gallery is located.
- B2 – 50-storey tower, levels from 28 to 48 contain 184 apartments and the remaining levels provide office space.
- B3 – 19-storey "sail" containing 52 apartments on levels from 11 to 18, as well as offices.

The highest tower with spire was originally planned to be 258 m tall (roof height 221 m). The construction of a reinforced concrete structure was completed in September 2011, and its final height is 205.82 m. Total building height is 212 m.

* Corresponding author: zbigniew.wojcicki@pwr.edu.pl

**Fig. 1.** Sky Tower complex.

## 2 Measurements

Measurements were made at the end of the construction works. Measurements were carried out using practically all the available measuring equipment. Fifteen seismic accelerometers were connected to two system PULSE cassettes.

Two variants of dynamic research were prepared:

- the first variant included the connection of two system PULSE cassettes and their full synchronization via 2 optical fibre junction stations. However, due to their weight, it proved impossible to carry the suitcases containing the optical fibre stations so high and the availability of lifts was very limited.
- the second variant included connecting two PULSE system cassettes and synchronizing them directly without the optical fibre station using a long LAN cable.

The second variant was finally realized. Signals from fifteen accelerometers mounted in the horizontal direction were recorded. Three DeltaTron 8430 accelerometers were installed on each of the selected floors - three measurements on each level identify the movement of the floor as a rigid plane body.

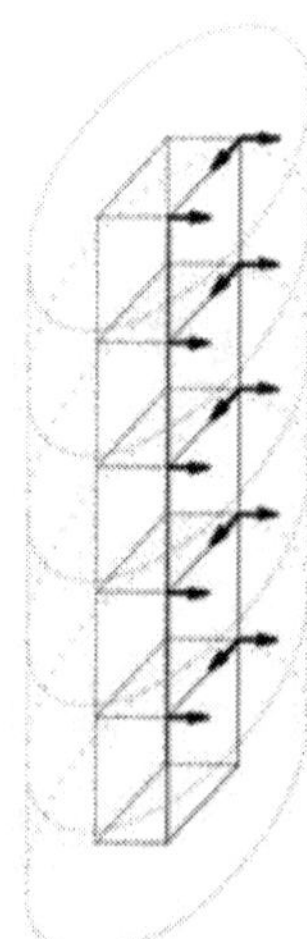

**Fig. 2.** The placement of accelerometers in the Sky Tower in Wrocław (OMA scheme).

Accelerometers were mounted on the five storeys of the B2 (Figure 2):

- level + 87.04 m (floor 20),
- level + 116.96 m (floor 28),
- level + 139.40 m (floor 34)
- level + 161.84 m (floor 40),
- level + 184.28 m (floor 46).

The schematic location of accelerometers is shown in Figure 2, where the dotted line indicates the outline of the B2 building (tower), the dashed line schematically shows the elevator-stair shaft. The solid line shows the shaft that forms the geometric model in the OMA analysis.

Because the vibration were measured mainly for OMA, the measurements were long – 60 minutes. The distribution and assembly of the equipment took much longer than the measurements itself. Nevertheless, the entire measuring system was mounted, applied to measurements and disassembled during one day.

## 3 Results

Table 1 summarizes OMA analysis results obtained using the stochastic methods SSI-PC.

**Table 1.** The results of Operational Modal Analysis (SSI-PC methods).

| Mode N° | Eigenfrequency [Hz] | Damping ratio [%] |
|---|---|---|
| 1 | 0,3107 | 0,6473 |
| 2 | 0,3889 | 0,6730 |
| 3 | 0,6426 | 0,8550 |
| 4 | 1,176 | 1,345 |
| 5 | 1,177 | 1,119 |
| 6 | 1,261 | 2,536 |
| 7 | 1,538 | 1,869 |
| 8 | 1,539 | 1,860 |
| 9 | 1,615 | 1,250 |
| 10 | 2,051 | 1,677 |

In order to compare the eigenforms and check for their linear dependence on each other, the value of the Modal Assurance Criterion (MAC) was calculated. The results of this analysis are shown in Fig. 3 and in Table 2.

**Table 2.** Modal Assurance Criterion (MAC) matrix values.

| | 0.3107 Hz | 0.3899 Hz | 0.6426 Hz | 1.175 Hz | 1.177 Hz | 1.261 Hz | 1.538 Hz | 1.539 Hz | 1.615 Hz | 2.051 Hz |
|---|---|---|---|---|---|---|---|---|---|---|
| 0.3107 Hz | 1 | 0.0001178 | 0.008871 | 0.007059 | 0.007735 | 0.002152 | 0.001324 | 0.002525 | 0.003118 | 0.04495 |
| 0.3899 Hz | 0.0001178 | 1 | 0.1898 | 0.002171 | 0.0003759 | 0.02441 | 0.005253 | 0.006795 | 0.002806 | 0.0002133 |
| 0.6426 Hz | 0.008871 | 0.1898 | 1 | 0.001516 | 0.0005745 | 0.01694 | 0.009476 | 0.01161 | 0.02018 | 0.006749 |
| 1.175 Hz | 0.007059 | 0.002171 | 0.001516 | 1 | 0.9148 | 0.6632 | 0.02229 | 0.01336 | 0.2172 | 0.1668 |
| 1.177 Hz | 0.007735 | 0.0003759 | 0.0005745 | 0.9148 | 1 | 0.6292 | 0.01089 | 0.0272 | 0.2451 | 0.1352 |
| 1.261 Hz | 0.002152 | 0.02441 | 0.01694 | 0.6632 | 0.6292 | 1 | 0.1438 | 0.114 | 0.004757 | 0.28 |
| 1.538 Hz | 0.001324 | 0.005253 | 0.009476 | 0.02229 | 0.01089 | 0.1438 | 1 | 0.9647 | 0.004005 | 0.04622 |
| 1.539 Hz | 0.002525 | 0.006795 | 0.01161 | 0.01336 | 0.0272 | 0.114 | 0.9647 | 1 | 0.04324 | 0.07119 |
| 1.615 Hz | 0.003118 | 0.002806 | 0.02018 | 0.2172 | 0.2451 | 0.004757 | 0.004005 | 0.04324 | 1 | 0.1537 |
| 2.051 Hz | 0.04495 | 0.0002133 | 0.006749 | 0.1668 | 0.1352 | 0.28 | 0.04622 | 0.07119 | 0.1537 | 1 |

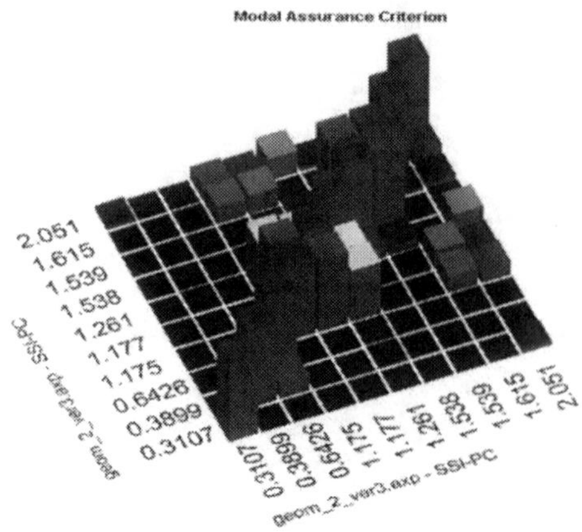


**Fig. 3.** Modal Assurance Criterion visualise.

Because of the very large value of the MAC (0.9), it was considered that the forms N° 4 and N° 5 are one eigenform of (MAC = 0.91). The result of the comparison of eigenforms N° 7 and N° 8 (MAC = 0.96) were interpreted similarly – compare Table 3.

It should be noted that the above eigenforms have been identified using stochastic methods SSI-PC. The use of identification methods in frequency domain allowed the identification of only one eigenfrequency in the region up to 3 Hz – 1.545 Hz. Deeper analysis showed that the incidence of 1.545 Hz corresponds to the same form that has been detected by the SSI-PC method, i.e. form N° 7 and N° 8, which was previously considered to be one form that corresponds to the frequency (1,538-1,539) Hz.

The final results of the OMA analysis within the scope of the detected eigenfrequencies and the corresponding damping ratios are shown in Table 3.

**Table 3.** The final results of Operational Modal Analysis.

| Mode N° | Mode description | Eigenfrequency [Hz] | Damping ratio [%] |
|---|---|---|---|
| 1 | 1st bending eigenform in the direction of a lesser rigidity (in a direction perpendicular to the plane of the model) | 0,3107 | 0,6473 |
| 2 | 1st bending eigenform in the direction of a greater rigidity (in the plane of the model) | 0,3889 | 0,6730 |
| 3 | 1st torsional eigenform | 0,6426 | 0,8550 |
| 4 | 2nd bending eigenform in the direction of a lesser rigidity (in a direction perpendicular to the plane of the model) | 1,177 | 1,232 |
| 5 | 2nd bending eigenform in the direction of a greater rigidity (in the plane of the model) | 1,261 | 2,536 |
| 6 | 2nd torsional eigenform | 1,538 | 1,865 |
| 7 | Complex bending-torsional eigenform | 1,615 | 1,250 |
| 8 | Complex bending-torsional eigenform | 2,051 | 1,677 |

The eigenforms corresponding to the detected eigenfrequecies are shown in the Fig. 4, Fig. 5 and Fig. 6.

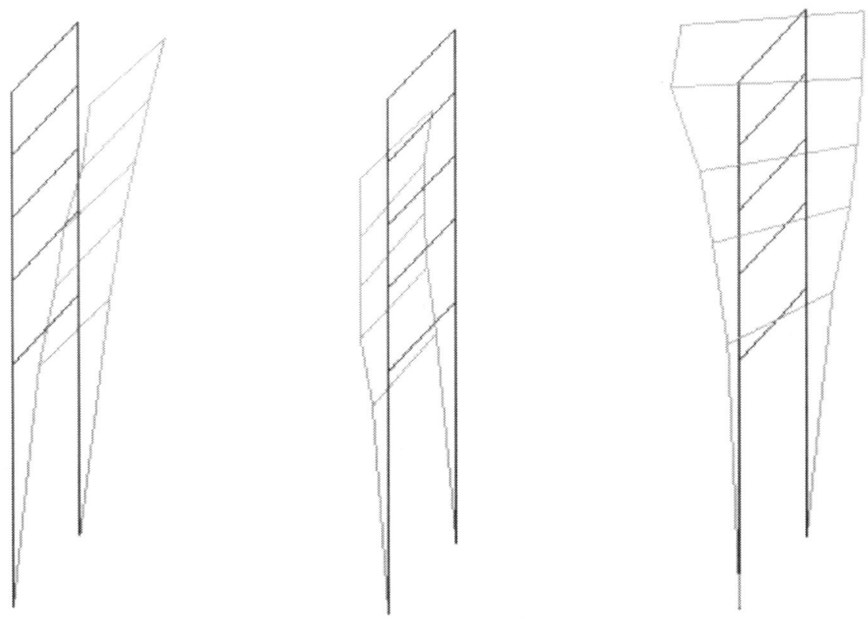

**Fig. 4.** The first three eigenforms (bending – 0,3107 Hz, bending – 0,3899 Hz, torsional – 0,6426 Hz).

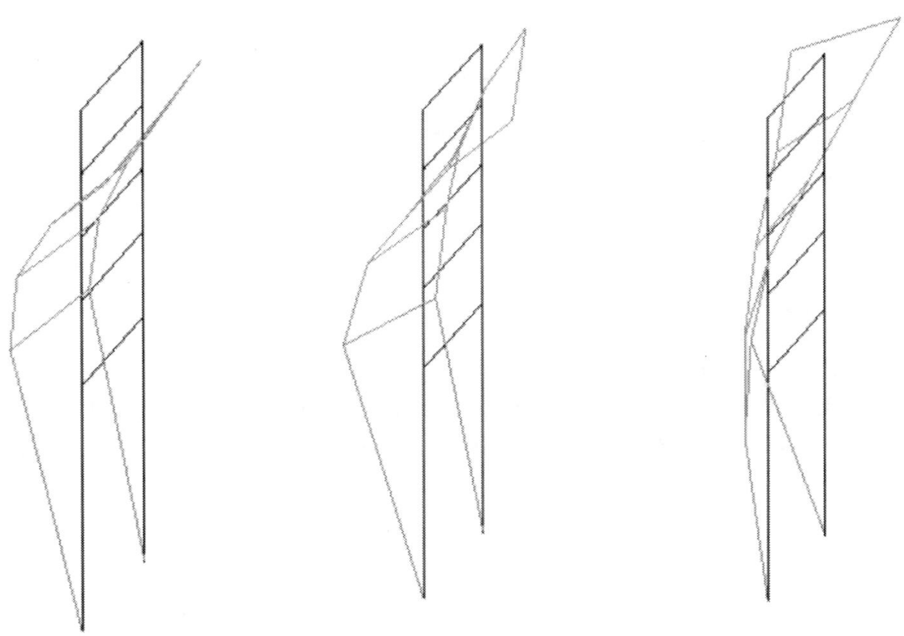

**Fig. 5.** The next three eigenforms (bending – 1,176 Hz, bending – 1,261 Hz, torsional – 1,538 Hz).

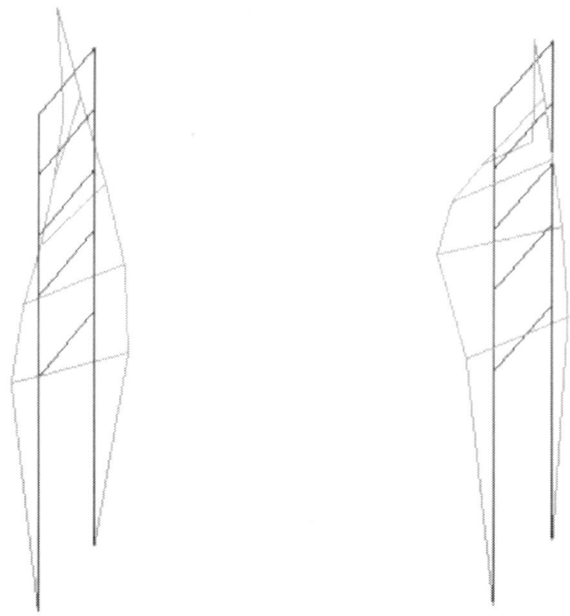

**Fig. 6.** The last two complex eigenforms (1,615 Hz, 2,051 Hz).

## 3 Conclusions

Based on the analysis of the results obtained, it is clear that stochastic methods are the more efficient methods of OMA analysis of a tall buildings. Within the analyzed range up to 3 Hz, 8 eigenfrequencies were detected. The use of frequency domain identification methods allowed for the detection of only one eigenfrequency in the range up to 3 Hz, which was consistent with one of eigenfequencies detected by the stochastic method. This confirms the correctness of the operation of the stochastic algorithms, but at the same time demonstrates the essential advantage of using stochastic methods in the detection of relatively low eigenfrequencies of large construction and engineering structures.

According to the authors, OMA modal analysis methods based on stochastic algorithms should be the standard for modal analysis of large structures, especially high and very high buildings.

## References

1. R. Brincker, L. Zhang, P. Andersen: Proceedings of IMAC, 625-630 (2000)
2. R. Brincker, C. Ventura, *Introduction to operational modal analysis* (WILEY 2015)
3. Z. Wójcicki, J. Grosel, W. Sawicki, *Experimental dynamic research of civil engineering structures* (in Polish), (DWE 2014)
4. Z. Wójcicki, S. Kostecki, J. Grosel, Procedia Engineering, **153**, 874-881 (2016)
5. J. Grosel, W. Sawicki, J. Marcinowski, Z. Wójcicki, IOMAC'15, 39-45 (2015)

# Shrinkage characteristics of cement stabilized rammed earth

*Piotr* Woyciechowski[1,*], *Piotr Leon* Narloch[1], and *Damian* Cichocki[1]

[1]Warsaw University of Technology, Faculty of Civil Engineering, Institute of Building Engineering Al. Armii Ludowej 16, 00-637 Warsaw, Poland

**Abstract.** Among the most commonly studied properties of building materials, their strength and physical properties should be mentioned. There is, however, one more feature that exist at the border of these two areas and not taking it into account during the design process of a given building can lead to negative results. This is the susceptibility of the material to the phenomenon of shrinkage. The purpose of the research was to get to know the phenomenon of shrinkage of cement stabilized rammed earth in the context of its composition and changing environmental conditions, by examining such factors as: aggregate content, water/binding materials ratio, relative humidity of the environment RH [%]. For the purpose of the research four groups of samples were made that differed in mixture composition and curing conditions. Based on the results obtained, it was found that the shrinkage of rammed earth is larger than commonly used mineral composites. In addition, it is very sensitive to changes in composition, so it is recommended to continue research on composition optimization, taking into account not only strength and durability but also the phenomenon of shrinkage.

## 1 Introduction

Earth-based construction is one of the oldest types of technology used to construct buildings. Its beginnings stretch out to prehistory and its development, depending on the region and the climate, lasted until the 20$^{th}$ century. It is estimated that about one third of the human population live in buildings built from earth [1]. There are many methods of using earth in constructing buildings. Technologies changed over time and under the influence of local conditions in which they were used. During this time raw earth was used in masonry, plastering, furnace construction, constructing bearing and partition walls, constructing ceilings, making bricks or even waterproof roof tiles. Today attempts are being made to find an use for raw earth in construction of vertical support elements and multi-storey residential buildings.

For this purpose, two techniques of building made of raw earth are used. The first of these assumes the erection of walls from pre-formed (manually or mechanically) blocks. This technology may resemble making bricks but it differs when it comes to heat treatment. This type of product does not undergo the firing process. The second technique involves the

* Corresponding author: p.woyciechowski@il.pw.edu.pl

construction of walls by compacting the raw earth in the formwork. As in the moulding of blocks, in the same way, the elements can be made in a non-mechanized manner as well as using modern construction equipment. In this method, called rammed earth, the mixture is compacted in layers. After compacting one layer the next one is laid on top and this done until the desired height of the element is achieved.

## 2 Rammed earth shrinkage

Although there are numerous publications dedicated to the usage of earth as a construction material, the results of shrinkage tests of rammed earth are rarely presented. The acceptable values of linear shrinkage according to different documents are featured in Table 1.

**Table 1.** Acceptable values of rammed earth shrinkage

| Country / Author | Document / Publication | Source | Acceptable value of linear shrinkage [%] |
|---|---|---|---|
| Germany | Lehmbau Regeln | [2] | ≤ 2 |
| New Zealand | NZS 4298:1998 | [3] | ≤ 0.5 |
| Poland | BN-62/6738-02 | [4] | 1.5 - 2.0 |
| Australia/Walker P. | HB 195-2002 | [5] | < 2.5 |
| United Kingdom /Keable J. | Rammed Earth Structures. A code of Practice (1996) | [6] | < 2 |

Unfortunately the analysis of publications does not give a satisfactory answer concerning the method of testing of rammed earth shrinkage. The New Zealand standard NZS 4298: 1998 [3] assumes testing rammed earth shrinkage on samples rammed in steel or wooden moulds with a square cross-section 5cm x 5cm and length 60cm. According to the specification the moulds should be covered with paper so that the samples do not stick to mould. After ramming the samples remain in their moulds. Together they undergo 7 days of conditioning under a plastic covering and after this they are dried for 21 days in the sun, without any cover. The measurement of the deformation is therefore done at least 28 days after forming the samples. It is done using the feeler gauge that is inserted between the smaller wall of the mould and the sample itself. Unfortunately the method of moulding and studying the samples presented in the NZS 4298:1998 standard cannot be accepted. The accuracy of the measurement itself is also doubtful.

Rammed earth stabilized by cement is a material technologically similar to low cement concrete that has a dry consistency, therefore it seems reasonable to adopt the Amsler method during the research. This method is used successfully in testing concretes [7], that is the standard procedure according to Polish standard PN-84/B-06714/23 [8], further supported by the recommendations contained in the ITB instruction no. 194 [9].

## 3 Materials used in the research

The purpose of the research was to investigate the phenomenon of shrinkage of cement stabilized rammed earth in the context of its composition and changes of environmental conditions. On the basis of the analysis of publications regarding the researched material [10,11,12] and the phenomenon of shrinkage in technologically similar materials [13,14,15] the most important factors determining the shrinkage characteristics of the material were selected as follows:

- aggregate content,
- mineralogical and chemical composition of the used loam,
- water/binding materials ratio,
- relative humidity of the environment [%].

The research was performed using two soil mixtures. Each mixture contained 70% of aggregate and 30% of loam. The mixtures granular composition of the aggregate differed from each other as far as the 703 mixture contained only 70% of sand fraction and the 433 mixture contained 40% of the sand fraction and 30% of the 2-4mm gravel fraction (Fig. 1). The mineralogical and chemical compositions of the loam used in the research is in Table 2 and Table 3, respectively.

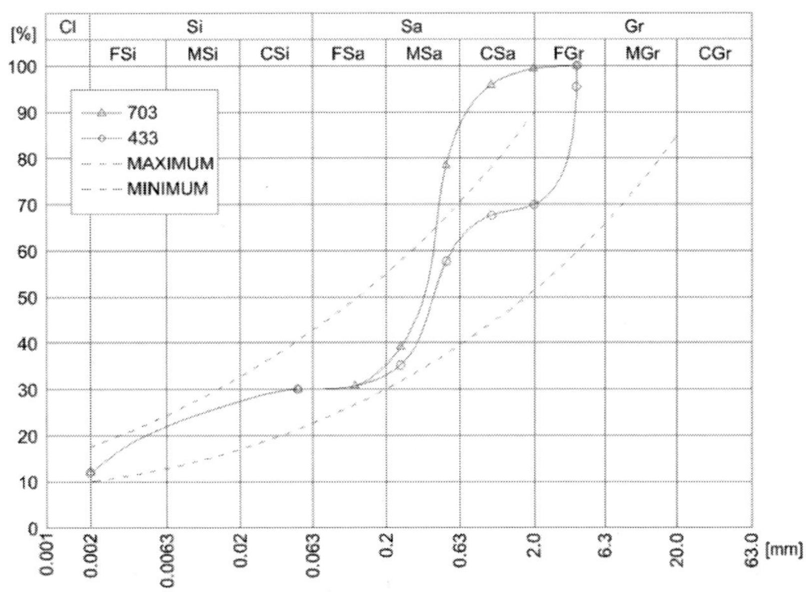


**Fig. 1.** The granular composition curves of soil mixtures used in the research. The dotted lines mark the recommended range of soil granular composition used in the rammed earth technique [16].

**Table 1.** The mineralogical composition of the loam [%]

| Clay minerals | | | | goethite | siderite | Calcareous minerals | Organic substance | quartz and others |
|---|---|---|---|---|---|---|---|---|
| total | beidellite | kaolinite | illite | | | | | |
| 43,7 | 8,9 | 8,6 | 26,2 | - | 6,0 | - | - | 50,3 |

**Table 2.** The chemical composition of the loam [%]

| $SiO_2$ | $Al_2O_3$ | $Fe_2O_3$ | $K_2O$ | CaO | $TiO_2$ | Other oxides |
|---|---|---|---|---|---|---|
| 61,78 | 19,63 | 10,65 | 3,18 | 0,66 | 0,89 | 0,33 |

Before moulding the samples the optimum moisture content was determined for both of the soil mixtures that had the addition of the stabilizing agent in the form of Portland cement CEM I 42,5R in amount of the 9% of dry mass of the soil mixture. The optimum moisture content was determined using the 6,5 kg rammer that was then used to mould the samples for shrinkage testing. For the soil-cement mixture marked as 703, the optimum moisture content of 10% of mass was determined and for the soil-cement mixture 433 it was 9% of mass (a mixture with 1% lower moisture was also examined, ie. 8% of moisture). The shrinkage was tested from the moment of demoulding (after 24 hours) till about 1 year. In order to investigate the effect of the given factors on shrinkage, the study included four groups of samples, differing in composition and curing conditions (Table 3).

**Table 3.** Group of rammed earth samples tested in the research

| Symbol of the group of rammed earth samples | Composition of the soil mixtures [% of mass] | | | Amount of Portland cement ([%] of mass of soil mixture) | Amount of water ([%] of mass of soil mixture with cement) | Curing conditions [%] RH |
|---|---|---|---|---|---|---|
| | Sand | Gravel | Loam | | | |
| 703-9C-10W 35%RH | 70 | 0 | 30 | 9 | 10 | 35 |
| 433-9C-8W 35% RH | 40 | 30 | 30 | | 8 | 35 |
| 433-9C-9W 35%RH | 40 | 30 | 30 | | 9 | 35 |
| 433-9C-9W 95%RH | 40 | 30 | 30 | | 9 | 95 |

Samples used in the research had the dimensions of 100x100x500mm. The samples were rammed in steel moulds superimposed with a profile, that was used as a guide for the 6,5 kg rammer that ensures vertical movement trajectory. The samples were rammed in three equal thickness layers that had 3.33 cm in height. The detailed ramming process of samples was described in [17].

## 4 Research results

The shrinkage test results are shown in Fig. 2, which also includes the shrinkage curve for C25/30 concrete with low proportion of CEM I cement - that is slightly below 300kg/m3 and gravel aggregate to 8mm.

The shrinkage of rammed earth is much larger than the shrinkage of concrete, which is confirmed even by the most rigorous NZS 4298: 1998 [3] standard which allows shrinkage after 28 days at 5 ‰. The increase in shrinkage in concrete is also different from that of cement stabilized rammed earth, where it is much more dynamic in the initial phase than in the case of concrete. The highest shrinkage - at the 3.5‰ level - was recorded in samples made from soil without any gravel aggregate. The shrinkage value in this case is approximately 2 times that of samples made of a mixture of similar composition and stored under the same conditions but containing 2/4 mm gravel fractions.

Samples stored in 95% relative humidity are characterized by a total shrinkage of nearly 20% less than samples from the same mixture and the same moisture, but stored in an environment with a relative humidity of 35%.

After 28 days the mixture tested at 35% humidity reached a similar percentage of maximum shrinkage (about 85% of its value), showing that shrinkage growth is similar regardless of differences in soil composition, if it is under the same environmental conditions.

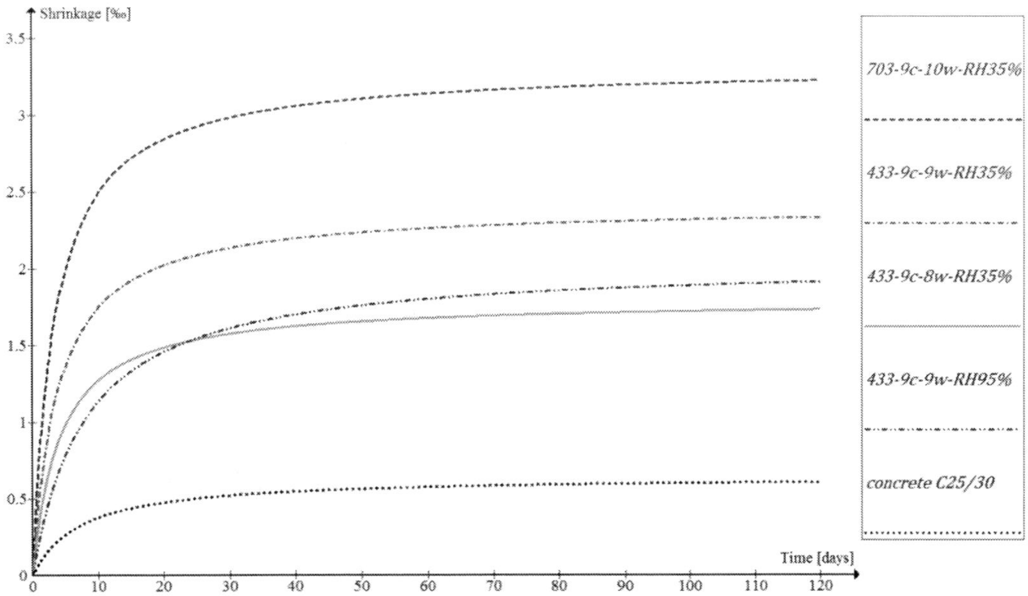


**Fig. 2.** Overview of the process of shrinkage of rammed earth and concrete

Using the principles of regression analysis, the empirical mathematical model of the curve was determined using the "curve fitting" method, statistically best suited to the results obtained. Parabolic, logarithmic, hyperbolic models were tested, but the best compatibility, measured by the $R^2$ coefficient of determination was achieved in the case of the curve known as "saturation growth rate" in the form of a hyperbole:

$$S = \frac{a \cdot t}{b+t} \tag{1}$$

where:
S - linear shrinkage [‰]
t - time [days]
a,b - coefficients of the model

The shrinkage model (Fig. 3, Fig. 4, Fig. 5, Fig. 6) in the form of a hyperbolic curve that has a horizontal asymptote means that the ordinate of this asymptote is the hypothetical maximum shrinkage value of the material being tested under accepted environmental conditions. Table 4 provides these values for 4 tested cases.

**Table 4.** Maximum shrinkage values according to mathematical models

| Symbol of the group of rammed earth samples | Mathematical model | Coefficients of the model | | Maximum shrinkage value [‰] | Coefficient of determination $R^2$ |
|---|---|---|---|---|---|
| | | a | b | | |
| 433-9C-9W 35%RH | 2.4068·x/(3.8169+x) | 2.4068 | 3.8169 | 2.4 | 0.9822 |
| 433-9C-8W 35% RH | 1.7924·x/(4.1279+x) | 1.7924 | 4.1279 | 1.8 | 0.9962 |
| 703-9C-10W 35%RH | 3.3126·x/(3.3242+x) | 3.3126 | 3.3242 | 3.3 | 0.9940 |
| 433-9C-9W 95%RH | 2.0427·x/(8.0376+x) | 2.0427 | 8.0376 | 2.0 | 0.9951 |

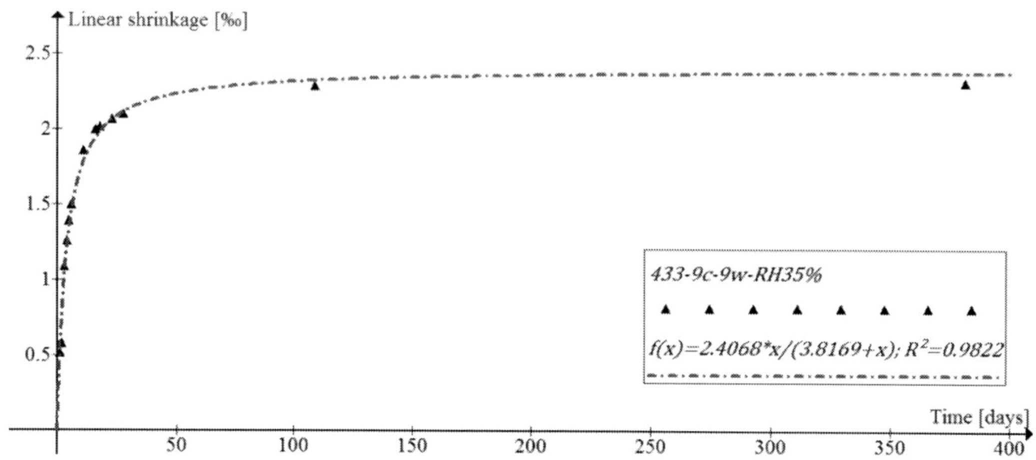


**Fig. 3.** Shrinkage model of the 433-9C-9W-RH35% mixture

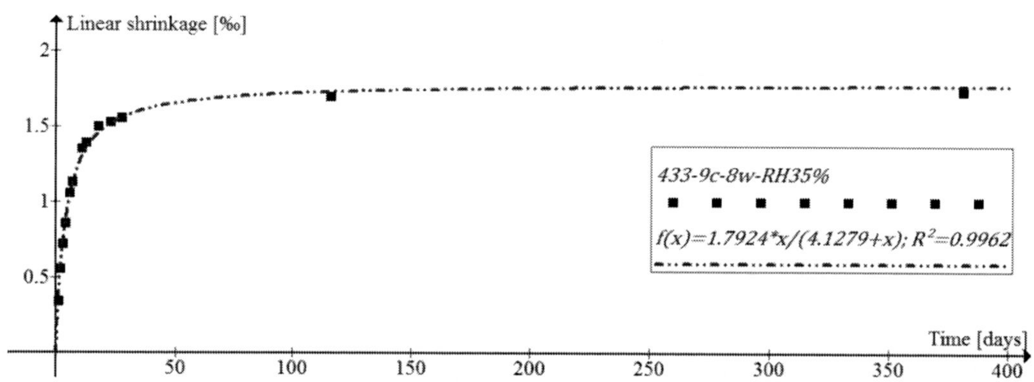


**Fig. 4.** Shrinkage model of the 433-9C-8W-RH35% mixture

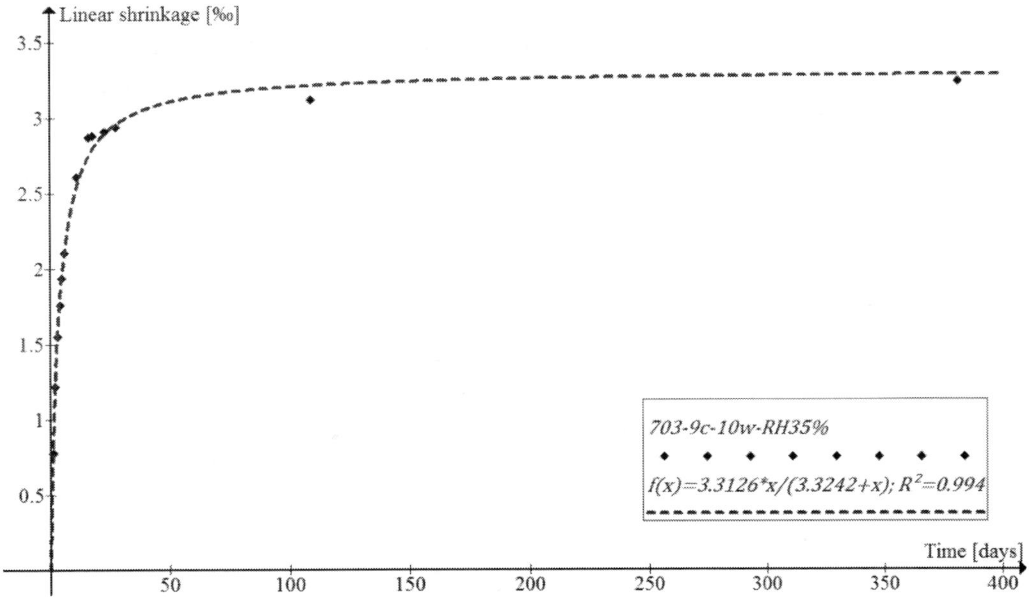


**Fig. 5.** Shrinkage model of the 703-9C-10W-RH35% mixture

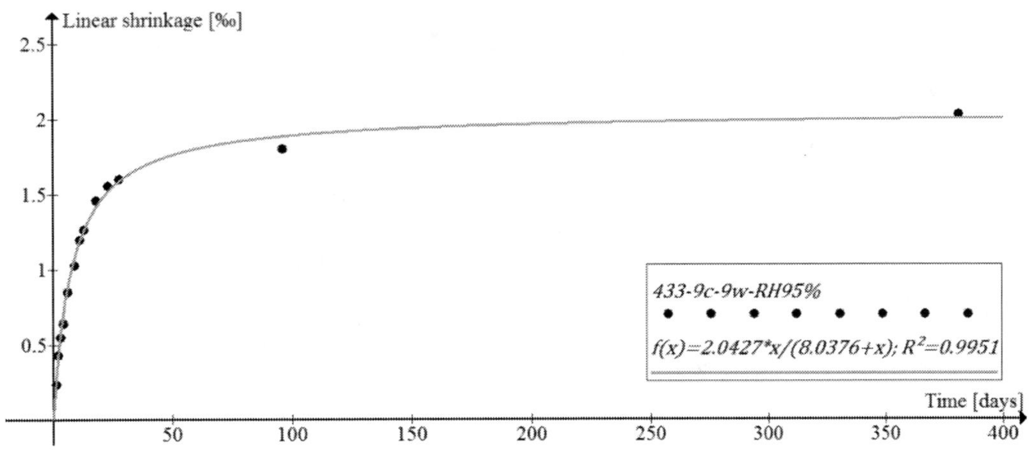


**Fig. 6.** Shrinkage model of the 433-9C-9W-RH95% mixture

## 5 Conclusions

All of the tested earth mixtures met the shrinkage requirements but forward in the New Zealand standard NZS 4298:1998 [2]. The maximum total shrinkage value that was the result of testing is significantly lower than the 28 days shrinkage that is given as acceptable in the standard.

As predicted, the shrinkage of "rammed earth" is significantly greater than the shrinkage of commonly used mineral composites. The maximum shrinkage value determined on the basis of the tests carried out is nearly 7 times higher than for medium strength concretes. This demonstrates additional restrictions of the use of "rammed earth" technology in larger engineering constructions.

The characteristics of the cement stabilized rammed earth shrinkage are also different from the popular mineral composite that is the C25/30 concrete. The shrinkage characteristics of the material under investigation are characterized by extremely dynamic incremental shrinkage in the initial phase. Within the first 28 days of hardening in low humidity, 85% of total shrinkage was recorded. Such a dynamic course of the phenomenon causes a significant part of the shrinkage to occur during the erection of the object, and thus deformation in the use stage will not be so large, despite the high total shrinkage.

Research has shown that the greatest impact on the effective shrinkage reduction of rammed earth has the presence of the gravel aggregate. The samples made of the mixtures that did not contain this type of aggregate were characterized by even two times higher shrinkage than mixtures having 30% of 2-4 mm aggregate. This means that when using local earth mixture it is advisable to enrich them with gravel fraction aggregate.

Differences in the total shrinkage value of rammed earth resulting from changes in the composition of the mixtures are large (even twice the value) and greater than for other mineral composites used in construction. This demonstrates the high sensitivity of the material to changes in composition and confirms the need to maintain a technological regime on the construction site to obtain the assumed parameters. At the same time, this confirms the great potential for the development of this technology and confirms the validity of further research aimed at optimization.

## References

1. G. Minke, *Building with Earth. Design and Technology of a Sustainable Architecture* (Birkhäuser – Publishers for Architecture, Kassel, Germany, 2006)
2. *Lehmbau Regeln – Begriffe, Baustoffe* (Wiesbaden, Germany, 2009)
3. *NZS 4298:1998: Materials and Workmanship for Earth Buildings* (Standards New Zealand, Wellington, New Zealand, 1998)
4. *BN-62/6738-02: Budownictwo z gliny – masy gliniane,* (1962)
5. P. Walker, *HB 195-2002 The Australian Earth Building Handbook* (Standards Australia, Sydney, Australia, 2002)
6. J. Keable, R. Keable, *Rammed Earth Structures: A Code of practice*, 2nd edition, Practical Action (2011)
7. P. Woyciechowski, K. Chilmon, J. J. Sokołowska, MATEC Web of Conferences, EDP Sciences **86**, 1 (2016)
8. *PN-84/B-06714/23 – Kruszywa mineralne. Oznaczanie zmian objętościowych metodą Amslera* (1984)
9. Instrukcja ITB nr 194: *Badania cech mechanicznych betonu na próbkach wykonanych w formach* (ITB, Warsaw, Poland, 1998)
10. P. L. Narloch, *Ziemia ubijana stabilizowana cementem jako konstrukcyjny materiał budowlany w klimacie umiarkowanym,* PhD thesis, (Warsaw University of Technology, Faculty of Civil Engineering, Warsaw, Poland, 2017)
11. P. L. Narloch, P. Woyciechwski, E. Dmowska, K. Halemba, ACE **61 (2)**, 73 (2015)
12. P. L. Narloch, P. Woyciechowski, P. Jęda, ACE **61(1)**, 73 (2015)
13. K. Chilmon, *Analiza wpływu charakterystyki kruszywa grubego na całkowity skurcz betonu*, MSc thesis (Warsaw University of Technology, Faculty of Civil Engineering, Warsaw, Poland, 2016)
14. D. Wilk, Ł. Bratasz, R. Kozłowski, Cement and Concrete Research **53**, 168 (2013)

15. D. Wilk, *Structural response of Roman cement mortars on drying*, PhD thesis (Polish Academy of Sciencies, Jerzy Haber Institute of Catalysis and Surface Chemistry, Cracow, Poland, 2012)
16. H. Houben, H. Guillaud, *Earth construction – a comprehensive guide*, 2nd edition (Intermediate Technology Publications, London, United Kingdom, 1996)
17. P. L. Narloch., M. Lidner, E. Kunicka, M. Bielecki, *XXIV Russian-Polish-Slovak Seminar Theoretical Foundation of Civil Engineering*, Procedia Engineering, **111**, 589 (2015)

# Reasons for the failure of the reactor unit lining at a petrochemical plant

*Margarita* Yakovleva[1], *Alexey* Khlystov[1], *Irina* Frolova[1], *Olga* Kotkova[1]*

[1]Samara State Technical University, Institute of Architecture and Civil Engineering, 194, Molodogvardeyskaya St., 443001, Samara, Russia

**Abstract.** In Russia most of the thermal units in petrochemistry are still being built with the use of piece ceramic refractories. Analysis of causes of frequent destruction of thermal units lining operated in contact with a carbon-containing medium indicates a corrosive process leading to the failure of the liner material due to the formation of soot carbon by the Bell reaction. Even high-alumina cements used for synthesizing heat-resistant dry mixtures intended to work in contact with a carbon-containing medium should not contain more than 0.1% each of $Fe_2O_3$ and $SiO_2$ oxides.

## 1 Introduction

Petrochemical plants are classified as hazardous production facilities and the resulting damage in the elements of technological installations may lead to a violation of the facility safety.

In this article the issue of establishing the quality of refractory lining and meeting the operating conditions by carrying out full-scale surveys was raised.

## 2 Materials and methods

Violations of the internal devices of the unit were manifested in the detachment and destruction of the internal concrete refractory lining of the reactor after reconstructive work. Visual surveys made it possible to establish the nature of the destruction (Figures 1 and 2) detected after the first start-up of the reactor, as well as to conduct instrumental and laboratory studies of selected samples of the protective lining material.

* Corresponding author: kotkovaon@yandex.ru

**Fig. 1**. Second cyclone lining concrete destruction, separation of metal anchors.

**Fig. 2**. Destruction of the transport line in zone 2 lining layer, concrete peeling of the horizontal grid lining.

For the analysis of possible violations of the lining installation technology, the customer presented work logs.

The nature of the visually detected damages of the liner allows us to establish the following. The inspection was carried out in accessible places of the reactor at the level of the hatches, including the transport zone in the upper, middle and lower zones of the system.

Typical for zone 1 is the vortex flow formed by the supplied catalyst stream under a pressure of 2.5 atm and a temperature of up to 600°C.

Due to the dynamic effect of the vortex flow, separated at the exit by the guiding ribs-plates, there may be unequal damage to the elements of the cyclone lining (Figures 3 and 4). In this area, concrete is completely destroyed in the local zones and anchor elements are partially torn off, the concrete of the lining is partially coked.

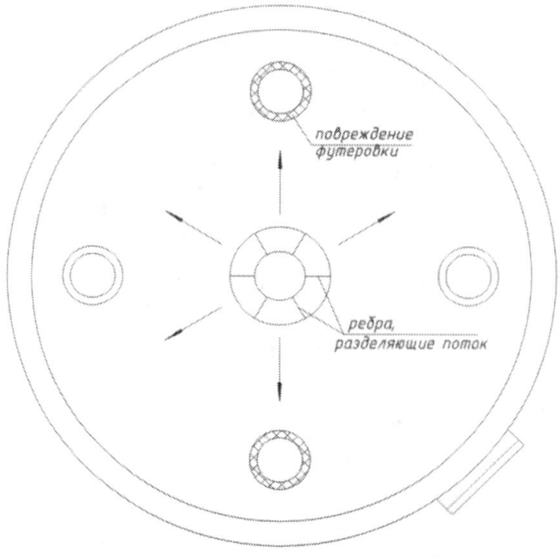


**Fig. 3**. Distribution of the vortex flow inside the reactor at elevations of ~ + 21,000 ÷ + 22,000 m.

**Fig. 4.** Destruction of one of the cyclones concrete lining, separation of metal anchors.

In Figure 5 of the general view of the reactor zones 1 and 2 are seen where the lining is located outside the pipeline systems under the influence of hydrocarbon vapors, catalysts and a temperature of approximately 600°C.

In Figure 6 in the lower part of the transport line zone 3 is seen. Its lining is located on the inner surface of the pipeline system and is under the influence of the regenerated catalyst.

Figures 7 and 8 show the state of the lining in the conditional zone 1, where, in addition to the medium indicated in the passport, there occurs an intensive dynamic effect of the vortex flow of the catalyst delivered under pressure from the transport system.

One of the reasons for the damage is the difference in the coefficients of thermal expansion of steel and concrete at a temperature of 600 °C, which leads to disruption of the tight contact between concrete and metal [1]. The detachment of concrete from reinforcing elements and surface of metal elements reduces the continuity and rigidity of the lining system. The consequence was a fracture of free anchoring elements under the influence of oscillations from a mechanical vortex action.

**Fig. 5.** Reactor unit No. 1, reactor P-5 and regenerator P-14, general view.

**Fig. 6.** General view of transport lines near the mark of 0.000 m.

To determine the possibility of an "explosive" character of the fracture due to vaporization in the structure of concrete, the analysis of materials was carried out according to the regime of temperature rise during the start-up period of the reactor unit. The temperature regime according to the documents did not violate the regime of technological requirements that contribute to the destruction of the lining.

**Fig. 7.** Destruction of cyclone lining concrete, detachment of metal anchors.

**Fig. 8.** Occurrence of soot carbon in the pores of concrete lining.

The carried out research revealed physical and chemical parameters of the concrete sample, formation of soot carbon on the surface and inside the lining layer; erosion of refractory materials and the disruption of the continuity of concrete lining, as well as the violation of the mating of concrete with metal equipment. The composition of the ramming paste used in the lining [2,3] was analyzed.

## 3 Results

Cyclones, transport lines and other thermal installations with the operating temperature ranging from 600 to 800 °C are one of the main units of plants (reactors) for the production of hydrocarbon monomers and polymers.

Under the terms of service, the lined metal structures of these plants must be resistant to the action of a gaseous carbon-containing medium, in particular to a reducing medium (CO and $H_2$ gases) and to the particle flux of a pulverized corundum-containing catalyst. Lining constructions for thermal petrochemical units are made mainly by applying a plastic concrete mix to the galvanized surface of metal structures. S-shaped spikes are made of heat-resistant steel 12x18N10T and are welded to metal structures. Modern dry heat-resistant mixtures are used as a refractory material for the formation of erosion-resistant lining, which makes it possible to obtain composites with a strength of up to 90-120 MPa.

Many petrochemical enterprises (ZAO Togliattisintez, OAO Novokuibyshevsk Petrochemical Company) supply such mixtures to Aliter-Aksi LLC established with the help of foreign partners in the Leningrad Region.

Heat-resistant mixtures, which are characterized by their high strength characteristics, contain high-alumina cement, chamotte, bauxite and corundum-containing aggregates. The durability of refractory materials in technological structures of petrochemical industry depends to a certain extent, and in some cases to a greater extent, on the composition of the gaseous medium.

When processing natural gas, as well as its combustion in thermal petrochemical plants, the content of reducing gases increases: CO, $H_2$, $CH_4$, $C_2H_6$.

Hydrocarbons - methane, ethane, and, consequently, natural gas emit soot carbon in the presence of catalysts. Soot carbon is also formed according to the reaction

$$CO+H_2 \rightarrow C_{soot}+H_2O \tag{1}$$

Of the oxygen materials corundum refractories with a low porosity are the most erosion-resistant to the action of the supportive catalyst carriers under oxidizing conditions. Corundum ceramic products possess high-erosion resistance. The wear of refractory materials under the influence of a pulverized catalyst carrier is mainly due to permanent wear and tear changes in place and during the processes of wear and tear as a result of friction of solid particles with the surface of the lining.

When choosing a refractory properties such as density, porosity, compressive strength, heat resistance are also taken into account. In the case when non-flammable, heat-resistant concrete is applied, dry mixtures on aluminous cement are developed as a refractory lining material, which allows obtaining high strength results in the early periods of hardening.

Carbon monoxide CO is one of the most aggressive substances with respect to refractories. Its corrosive effect consists in the reduction of refractory oxides and in the action of soot carbon.

Soot carbon emission reaction flows in the volume of the porous body (refractory) slowly and can be accelerated by the action of catalysts. The catalyst is, first of all, metallic iron. The formation of soot carbon can occur in a reactor with a reducing medium according to the reaction

$$CO+H_2 \rightarrow C_{soot}+H_2O \tag{2}$$

## 4 Discussion

The deposition of soot carbon in the pores of refractories leads to reactions of reducing of iron oxides to metal and the interaction of metallic iron with carbon monoxide with the formation of carbides and oxides ($Fe_3C$; FeO; $Fe_3O_4$). These latter reactions come with a significant increase in volume.

The increase in volume leads to colmatization of the concrete structure and partial hardening, but as the carbides and iron oxides accumulate, internal stresses and cracking of the concrete are created.

The increase in volume causes stresses in the body of the lining material, exceeding the strength, which leads to destruction (Figures 1, 2). Studying the wear of refractory materials under the influence of gaseous and dust-like components, German engineers established that rapid breakdown of furnace lining is caused by erosion associated with high velocity of gas and dust particles contained in it [4]. However, the erosion process only rarely causes the wear of the refractory masonry.

## 5 Conclusions

Thus, it is necessary to pay attention to the chemical composition of the mixture, especially to the presence of oxides of elements with variable valency (iron oxides) [5] for lining of thermal petrochemical units in the selection of heat-resistant mixtures, in addition to high strength that provides erosion and abrasion resistance.

In this regard, for the lining of thermal petrochemical units with a possibility of a hydrocarbon medium, it is recommended to use dry heat-resistant mixtures with a minimum content of iron oxide, which has a prevailing influence in the processes of destruction of practically any refractory lining materials [6,7].

## References

1. A.I. Khlystov, L.S. Konnova, V.A. Shirokov, A.O. Levinkov, Scientific Survey **23**, 22-26 (2015)
2. M.M. Buslovich, *Lining materials for electric furnaces with controlled atmospheres* (Moscow, 1975)
3. I.V. Kudinov, L.S. Abisheva, A.N. Branfilyova, Urban Construction and Architecture **3**, 6-10 (2014)
4. A.I. Khlystov, *Increasing the efficiency and improving the quality of refractory lining materials* ( Samara State University of Architecture and Civil Engineering, Samara, 2004)
5. A.M. Cherepanov, S.G. Tresvyatskiy, *High-refractory materials and products made of oxides* (Moscow, 1964)
6. V.I. Krayukhin, O.P. Khomyakova, N.V. Khayuk, Urban Construction and Architecture **1**, 78-84 (2012)
7. A. Frolov, I. Frolova, O. Kotkova, MATEC Web Conf. **86,** 02007 (2016)

# Influence of changes in hydrogeological conditions on industrial safety of buildings and structures

*Margarita* Yakovleva[1*], *Evgeniy* Frolov[1], *Veniamin* Isayev[1], and *Sergey* Zubanov[1]

[1]Samara State Technical University Institute of Architecture and Civil Engineering, 194, Molodogvardeyskaya St., 443001, Samara, Russia

**Abstract.** Technical condition surveys of a number of multi-purpose industrial facilities have shown that the hydrological conditions of the occupied sites often change in the course of operation. This is especially often observed in oil refining and petrochemical industries, the distinguishing feature of which is high water consumption for cleaning the machines and for general cleaning; with a low culture of the facility operation, water often flows out of the main water conduits, and when the drainage systems are disturbed, basement soils are steeped and plant territories are watered. This was observed even at sites where, at the time of the construction of the facility, the groundwater was buried deep enough and did not affect the surveyed building. Monitoring one of the processing plants made it possible to identify the source of flooding and investigate the nature of the change in the groundwater surface after long-term operation of the facility. The results of this work are given in this article.

## 1 Introduction

The main feature of oil refining and petrochemical industry is the presence of hydrocarbons in the technology, which have an increased explosion and fire hazard as well as a high combustion temperature of their products. In addition, the production technology is characterized by high water consumption, more than 10 $m^3$ of water is necessary per 1 $m^3$ of production.

Water and aqueous solutions are widely used for washing and testing technological apparatuses on external installations, in pedestal buildings and production plants, which leads to soaking the basements and foundations of the buildings and structures, and causes their premature aging [1], [2], [3] .

Trays are usually provided for water drainage, but eventually they fail, become clogged with rubbish, and streams of water and water solutions enter the basement grounds. Flooding of the industrial enterprise territory occurs even at sites where during construction the groundwater was deep and did not represent any threats to nature. After several years of operation, the groundwater level rose sharply, flooding the foundations of industrial buildings and structures. This phenomenon was observed by the staff of the branch research

* Corresponding author: md-63@yandex.ru

laboratory "Reconstruction" in numerous surveys of such objects in the cities of Volgograd, Volzhsky, Solikamsk, Togliatti, Kursk, Samara and others [1], [4] - [6].

Water leaks when mixed with toxic materials lead to flooding of the territory and contamination of groundwater, causing damage to the environment and leading to economic losses.

Ground watering and the ways in which moisture may affect the foundation ground impact the industrial safety of buildings and structures.

It is known that the properties of ground largely depend on the relative content of water in them.

Water content in the ground does not remain constant, but depends on the inflow of capillary moisture that fills the pores of the ground partially or completely at the required height from the level of flowing groundwater.

Capillary moistening of ground primarily affects its mechanical strength.

## 2 Materials and methods

Groundwater is divided into subground water, groundwater, spring water and water of technological origin.

The site under consideration is characteristic of ground and subground water. Subground water is seasonal in nature and is often short-lived.

Ground water is a supported gravity moisture, located entirely in the ground thickness; it is capable of flowing out of the sections, but does not have a hydraulic connection with the underlying groundwater. Excessive water can, however, flow down into the underlying layers of rocks. A variety of ground water is open groundwater, i.e. the accumulation of water caused by the location of slightly water-permeable layers of clay, alluvial horizons in the ground.

Groundwater lies below the ground layer on the first waterproof layer of the ground from the day surface; it is not blocked by water permeable rock and is the source of capillary waters.

Capillary waters become one of the main sources of moistening and destruction of underground structures of buildings.

Capillary phenomena arise in structures depending on the kind of material, temperature conditions, physical, chemical and biological destructive processes.

Groundwater also includes waters of technical origin, which can appear in the course of leaks from technological production water lines and sewage systems, as well as from water vapour condensate in the ground.

The phenomenon of water rising through the capillaries is especially often manifested with negative temperatures, when ice crystals are formed in large pores of the soil. The structure of frozen grounds changes with the degree of freezing.

The main causes of moisture migration are as follows:

- osmotic forces that arise in the form of the suction force of the solution in the presence of a temperature gradient (the higher the gradient, the greater the forces);
- stresses caused by uneven freezing;
- capillary forces.

Migration of moisture at negative temperatures is also observed in materials of structures, foundations and walls [7], [8].

Underground waters always contain dissolved gases and minerals. The waters can be soft or hard depending on the content of dissolved mineral particles.

The hardest water is found in carbonate rocks.

Soft waters are characterized by a large dissolving capacity (aggressiveness) when concrete, mortars and many other silicate materials are exposed to them.

The most common of acidic corrosive waters is carbon dioxide containing oxygen dioxide (CO2), which produces carbon dioxide as compound with water.

$$CO_2 + H_2O = H_2CO_3 \quad (1)$$

The more free carbon dioxide there is in the water, the stronger is its aggressiveness.

Among aggressive waters are those containing organic acids, for example, humic acids of marsh waters, destroying concrete.

Water containing sulphate acids and chlorine is also aggressive.

Humid-aggressive effects of groundwater influence both the foundations and the base [9], [10].

The most soluble bases are those composed of limestones, dolomites and gypsum; the least soluble are igneous and clayey rocks.

The choice of a solution to protect the foundations from the effects of groundwater is based on studying the factors that cause a change in the groundwater regime.

These factors are divided into natural and artificial.

Natural factors are related to the features of the geological structure, climate, terrain and the development of the hydrographic network. The nature of rocks bedding and their lithological composition affect the groundwater regime to a greater extent. When slightly permeable rocks are located close to the day surface, underflooding is very likely. The same rocks contribute to the formation and retention of capillary water.

Seasonal and annual fluctuations in the groundwater level are associated with climatic factors. When percolating through layers containing soluble salts, waters that are not chemically harmful often turn into aggressive ones.

Without a proper drainage system and with a waterproof layer at a shallow depth, water accumulations can cause waterlogging of the terrain.

Changes in the atmospheric pressure also affect the fluctuations in the groundwater level. The change in the level occurs simultaneously with the change in atmospheric pressure and consists in the partial or complete transformation of capillary water into gravitational and, vice versa, (with decrease in pressure) in the increase of the capillary level due to the partial transfer of gravitational water to capillary water.

The filtration action of a groundwater stream directed downwards sometimes causes large ground settlement and considerable deformation in the buildings.

Upward movement of moisture can occur due to evaporation process as well as a rise in the groundwater level.

It should be noted from the foregoing that at the surveyed site the following phenomena can occur in the presence of groundwater level:

- fluctuations in the groundwater level;
- compaction of base grounds with lowering of groundwater level;
- soaking and swelling of the ground layer with increasing groundwater level;
-freeze heaving and freezing of wet ground with foundations;
- destructive changes in foundation materials under the influence of aggressive agents.

## 3 Results

We carried out an examination at one of the enterprises of the city of Samara to study the nature of flooding of the plant site where a high groundwater level was found during the piercing of the object foundation. To study the dynamics of changes in the level of the groundwater surface, we carried out monitoring, designed for a long service life. After studying the documentation, a working program was drawn up, according to which we planned the sites for drilling boreholes in the ground and for placing casing pipes.

Drilling of boreholes was carried out with a manual drill with a diameter of 140 mm. The depth of drilling depended on the depth of the waterproof layer and the level of groundwater. Soil samples with broken structure were taken in the process of drilling boreholes from a depth of up to four meters and delivered to a ground laboratory, where the characteristics of soil plasticity were determined.

It has been established that the soils of the base are clay loams from soft-plastic to flowing ones, as well as flowing sandy loams (Table 1).

**Table 1.** List of soil properties

| Borehole number | Sampling depth, m | Spec. Weight, kN/m$^3$ | Moist, Units | Flowing limit | Plasticity limit | Plasticity number | Flowing index | Set water level, m | Ground type |
|---|---|---|---|---|---|---|---|---|---|
| 1 | 4,0 | 27,1 | 0,22 | 0,26 | 0,16 | 0,1 | 0,60 | 3,71 | soft-plastic clay loams |
| 2 | 4,0 | 27,0 | 0,19 | 0,18 | 0,14 | 0,04 | 1,25 | 2,59 | flowing sandy loams |
| 3 | 2,0 | 27,1 | 0,30 | 0,35 | 0,18 | 0,17 | 0,71 | 0,64 | soft-plastic clay loams |
| 4 | 4,0 | 27,1 | 0,22 | 0,24 | 0,16 | 0,08 | 0,75 | 1,03 | soft-plastic clay loams |
| 7 | 4,0 | 27,1 | 0,22 | 0,24 | 0,15 | 0,09 | 0,77 | 2,40 | flowing-plastic clay loams |
| 8 | 4,0 | 27,1 | 0,25 | 0,31 | 0,16 | 0,15 | 0,60 | 1,52 | soft-plastic clay loams |
| 11 | 4,0 | 27,1 | 0,25 | 0,24 | 0,17 | 0,07 | 1,14 | 2,00 | flowing clay loams |
| 12 | 4,0 | 27,1 | 0,24 | 0,25 | 0,16 | 0,09 | 0,89 | 2,03 | flowing-plastic clay loams |
| 13 | 4,0 | 27,1 | 0,31 | 0,31 | 0,17 | 0,14 | 1,00 | - | flowing-plastic clay loams |
| 14 | 4,0 | 27,1 | 0,30 | 0,30 | 0,16 | 0,14 | 0,93 | 2,05 | flowing-plastic clay loams |
| 15 | 4,0 | 27,0 | 0,22 | 0,22 | 0,16 | 0,06 | 1,00 | 1,78 | flowing sandy loams |
| 16 | 4,0 | 27,1 | 0,24 | 0,29 | 0,17 | 0,12 | 0,58 | 2,25 | soft-plastic clay loams |
| 17 | 4,0 | 27,1 | 0,23 | 0,24 | 0,15 | 0,09 | 0,89 | 2,17 | flowing-plastic clay loams |
| 18 | 4,0 | 27,1 | 0,25 | 0,28 | 0,16 | 0,12 | 0,75 | 2,56 | soft-plastic clay loams |
| 19 | 4,0 | 27,0 | 0,20 | 0,20 | 0,16 | 0,09 | 1,00 | 2,18 | flowing sandy loams |

The layout of the observation boreholes is shown in Figure 1. The depth of the borehole was 4 m, and the casing pipes are made of two separate elements 2 m each. Holes with a pitch of 250 mm were drilled in staggered order in the wall of the lower pipe. The upper pipe is without holes. The design of borehole completion is given in Figure 2.

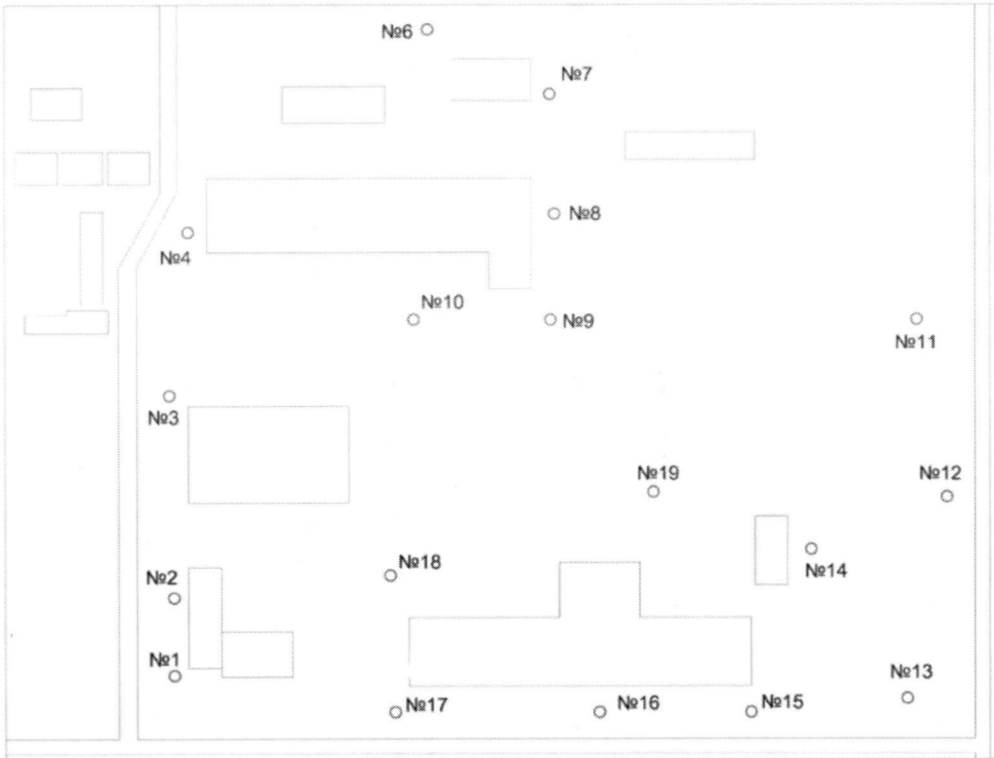


**Fig.1.** The layout of control boreholes at the industrial site

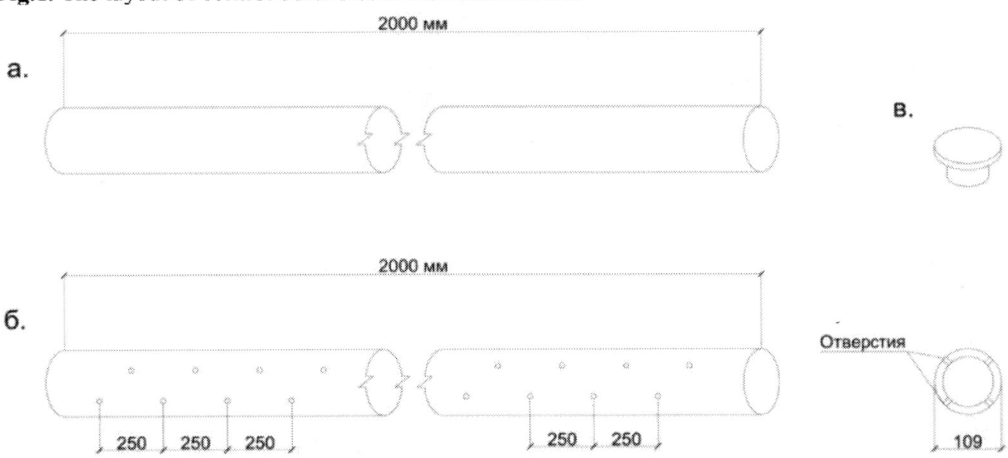


**Fig.2.** Casing design for monitoring groundwater level. а) - the upper part of the pipe; б) the lower part of the pipe with holes; в) the lid.

Both pipes (Fig. 3b) were lowered into the borehole drilled in the ground (Figure 3a). A special cover is provided from above to prevent clogging of the pipe. The groundwater level was measured by measuring dipstick, the results of measurements were recorded in a special register. Figure 4 shows a pipe with a lid installed in place.

a) b)

**Fig. 3**. a) Drilling the borehole b) Installing the lower casing pipe into the borehole.

Monitoring of the groundwater level was carried out on a quarterly basis. Data showed the following results during the two years period (Table 2).

**Fig. 4.** The borehole is equipped. The pipe is closed with a special lid.

**Table 2.** Measurement of the groundwater level from the ground surface, m.

| Period | Borehole number | | | | | | | | | |
|---|---|---|---|---|---|---|---|---|---|---|
| | 1 | 2 | 3 | 4 | 5 | 6 | 7 | 8 | 9 | 10 |
| Year 1 | 3,71 | 2,59 | 0,64 | 1,03 | - | 2,48 | 2,40 | 1,52 | 2,47 | 2,58 |
| Year 2 | 3,58 | 3,45 | 1,30 | 2,0 | - | 2,42 | 2,44 | 1,92 | 2,58 | 2,72 |
| | 11 | 12 | 13 | 14 | 15 | 16 | 17 | 18 | 19 | |
| Year 1 | 2,00 | 2,03 | - | 2,05 | 1,78 | 2,25 | 2,17 | 2,56 | 2,18 | |
| Year 2 | 1,85 | 1,76 | - | 1,93 | 1,51 | 1,36 | - | 2,48 | 1,63 | |

## 4 Discussion

Analysing the results of the work performed on the production site, which are partially given in Tables 1 and 2, we make the following conclusions:

- water-bearing materials (grounds) on the entire survey site are soft-plastic clay loams;
- the established level of groundwater within the object is located at a depth of more than 2 m;
- boreholes №3, 4 and 8 have a dome-shaped rise in the groundwater level;
- a seasonal change in the water level within 0.5-1.0 m was noted based on the results of long-term tests,;
- groundwater did not rise above the ground freezing level;
- sources of flooding were identified;
- water surface measurement data from the industrial site is taken into account when designing new facilities with a deep foundation level.

## 5 Conclusions

Installing a system of hydrological observations in the form of a field of boreholes with special casing pipes at operating industrial sites enabled us to control the dynamics of fluctuations in the rise of groundwater. This makes it possible to prevent flooding of individual units of large enterprises through timely protective measures.

These observations are especially important during reconstruction, repair work or modernization.

## References

1. M.V. Yakovleva, O.N. Kotkova, *Estimation of the Change in the Load- Bearing Capacity of Building Structures during Operation: Guidelines* (Samara State University of Architecture and Civil Engineering, Samara, 2009)
2. *Rules for Inspection of Load-Bearing Building Constructions of Buildings and Structures* (State Committee for Construction of Russia, Moscow, 2004)
3. *Buildings and Structures. Rules for Inspection and Monitoring Technical Condition*. (State Committee for Construction of Russia, Moscow, 2011)
4. F. Parkins, *Reinforced Concrete Structures. Waterproofing Repair and Protection* (Moscow, 1980)
5. M.V. Yakovleva, E.A. Frolov, A.E. Frolov, K.I. Gimadetdinov, *Inspection of the Technical Condition of Buildings and Structures: Training Manual* (Samara State University of Architecture and Civil Engineering, Samara, 2011)
6. M.V. Yakovleva, E.A. Frolov, A.E. Frolov, *Building Construction. Preparation, Reinforcement, Protection against Corrosion* (Forum, Moscow, 2015)
7. Dobromyslov A.N. *Diagnosis of Damage to Buildings of Engineering Structures*. (ASV Publishing House, Moscow, 2006)
8. Ushakov I.I., Bondarev B.A. *Fundamentals of Diagnostics of Building Structures* (Phoenix Publishing House, Rostov-on-Don, 2008)
9. *Manual for Monitoring the State of Metal Constructions of Buildings and Structures in Aggressive Environment. Carrying out Surveys and Designing Corrosion Protection* (Stroiizdat, Moscow, 1989)

10. *Recommendations for the Inspection and Monitoring the Technical Condition of Existing Buildings Located near New Construction or Reconstruction* (Moscow Commission of Construction and Architecture, Moscow, 1998) .

# Vibration isolation systems, considered as systems with single degree of freedom

*Mohammed* Zebilila[1] and *Jury* Chernov[2]
[1,2] Moscow State University of Civil Engineering, 129337, Yaroslavskoe shosse 26, Moscow, Russia

**Abstract**: The research considers and analyzes vibration isolation systems, whose design schemes are single degree of freedom systems, including nonlinear elements - displacement limiter and viscous damper. Presented are calculation formulas in closed form for linear systems in operational modes (for harmonic and impulse loads), algorithms and examples of calculation of linear and nonlinear systems in operational and transient modes. The calculation method and the above dependences are written using the transfer (TF) and impulse response functions (IRF) of linear dynamical systems and dependencies that determine the relationship between these functions. The effectiveness of 2 options of vibration isolation systems in transient modes is analyzed. There is significant reduction of load from the equipment to the supporting structures in the starting-stopping modes by the use of displacement limiter.
**Keywords:** vibration isolation, displacement limiter, viscous damper, nonlinear system, Duhamel integral.

## 1. Introduction

Vibration-active equipment stationed close to buildings or installed on structures cause such structures to undergo oscillation, which often go above the limits set in building standards and design manuals [1]. High displacements can also damage springs and flexible elements (for example, pipelines) which are connected to such equipment. Ways of reducing such dynamic loads interest a lot of scientists and engineers when putting up structures with vibration-active equipment. Assessment of vibration isolation system using auxiliary mass damper to reduce structural vibration was investigated in [2].

There has been growing interest in variable stiffness isolation systems[3-9]. These systems work by changing the structural stiffness thereby altering the natural period of the structure and thus escaping resonance. The changing of the structural stiffness is done by switching devices which are controlled by control laws [10-12]. Some of these devices incorporate dampers such that there is either stiffness or damping added to the structure, depending on the feedback response[11,13]. In [5] smart rubber material with capability of changing its stiffness was studied to be used in variable stiffness base isolation system.
In order to reduce vibrations in transient modes in many fields of technology, including the operation of vibration isolated equipment, systems with non-linear elements - displacement limiters and viscous dampers are widely used. In a significant part of the design process, dampers are constantly associated with an oscillating body.

[1] Corresponding author: halisghx@mail.ru

Such dampers, including a stator and a rotor, are tuned to the frequency of oscillations of the object and are effective in damping vibrations in transient modes. The disadvantage of such dampers is that their natural frequencies depend on the viscosity of the liquid, which varies with the ambient temperature. Such dampers, including a stator and a rotor, are tuned to the frequency of oscillations of the object and are effective in damping vibrations in transient modes. The disadvantage of such dampers is that their natural frequencies depend on the viscosity of the liquid, which varies with the ambient temperature. More stable in the operation of the damper are the options in which the damper is activated in operation directly in areas close to resonance, in particular, during start-up and shutdown modes. Such a scheme is shown in Fig. 1 [14].

This paper outlines the algorithm and calculation method to be used in analyzing vibration isolation system with nonlinear elements for a single degree of freedom system. Using the relationship between transfer function and impulse response function, the method could be used for 2 or more degrees of freedom systems [15,16] .

Two kinds of nonlinear elements are studied - nonlinear stiffness element (displacement limiters), and non-linear viscous damper.

In this study the control information (feedback structural information) is the reference structural displacement (y0). The structure and the standby stiffness engage together as soon as the displacement peaks to the reference point (y0). It maintains in this state until the displacement goes below the reference point. Below the reference point, the standby stiffness and the structure disengage separately.

These types of vibration isolation could be employed in reducing the levels of oscillation of equipment with rotating parts (Fans, compressors) and on screen machines in transition modes - starting and stopping modes[16].

## 2. Algorithm of calculation

The calculation schemes of single degree of system with displacement limiter and viscous damper are shown in figure 1. The algorithms of solutions of such systems in operational and transient modes are considered.

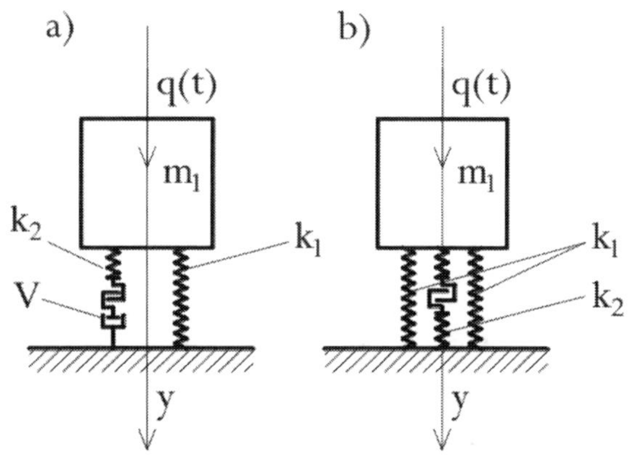


**Fig. 1**: a) system with displacement limiter ;b) system with viscous damper

### 2.1Algorithm of solving vibration isolation system with displacement limiter (fig. 1a):

The equation of motion of vibration isolation in fig. 1a is of the form:

$$m\ddot{y}+\left(1+2v\frac{d}{dt}\right)c(y)y=q(t) \tag{1}$$

The characteristics of the non-linearity considered in this systems is:

$$c(y)y = k_1 y \text{ при } y \le y_0;$$
$$c(y)y = k_1 y_0 + (k_1 + k_2)(y - y_0) \text{ при } y > y_0 \quad (2)$$

Taking the nonlinear components of equation (1) to the right side, we have:

$$\ddot{y} + \left(1 + 2v\frac{d}{dt}\right)p_1^{\,2} y = \frac{q(t)}{m} + \left(1 + 2v\frac{d}{dt}\right)p_1^{\,2} y - \left(1 + 2v\frac{d}{dt}\right)\frac{c(y)y}{m}; \quad (3)$$

After steps of simplification of equation (3), we have:

$$\ddot{y} + \left(1 + 2v\frac{d}{dt}\right)p_1^{\,2} y = \frac{q(t)}{m} - \left(1 + 2v\frac{d}{dt}\right)\frac{k_2(y - y_0)}{m}; \quad (4)$$

The solution of equation (4) is of two parts; from the linear aspect from the applied dynamic load ( $y_{lin}$ ) and from fictitious load which depends on the type of nonlinearity ( $y_{nonlin}$ ).

$$y = y_{lin} - y_{nonlin} \quad (5)$$

The solution from the applied dynamic load is expressed in terms of Duhamel integral [16,17]:

$$y_{lin} = \frac{1}{p_1^* m}\int_0^t q(\tau) e^{-n_1(t-\tau)} \sin p_1^*(t-\tau) d\tau;$$

$$= \frac{1}{p_1^* m}\int_0^t q(\tau) e^{-n_1(t-\tau)} (\sin p_1^* t \cos p_1^* \tau - \cos p_1^* t \sin p_1^* \tau) d\tau$$

$$= \frac{1}{mp_1}\left[d_1(t)F_2(t) - d_2(t)F_1(t)\right]; \quad (6)$$

$$\text{где } 2n_1 = 2vp_1^2,\ p_1^* = \sqrt{p_1^2 - n_1^2},$$

$$d_1 = e^{-n_1 t}\sin p_1 t;\ d_2 = e^{-n_1 t}\cos p_1 t \quad (7)$$

$$F_1 = \int_0^t q(\tau)\cdot e^{n_1\tau}\sin p_1^*\tau d\tau;\ F_2 = \int_0^t q(\tau)\cdot e^{n_1\tau}\cos p_1^*\tau d\tau; \quad (8)$$

The solution from the fictitious load is also expressed in terms of Duhamel integral. For the reason that the integrand contains the expression of displacement (y), the second-order integral is nonlinear.

$$y_{nonlin} = \frac{1}{mp_1^*}\int_{t_0}^t \left(1 + 2v\frac{d}{dt}\right) k_2 (y - y_0) e^{-n_1(t-\tau)} \sin p_1^*(t-\tau) d\tau \quad (9)$$

where $t_0$ – time of switching on the additional stiffness

Following equations (6) – (8), we have:

$$y_{nonlin} = \frac{k_2}{mp_1^*}\int_{t_0}^t (y - y_0) e^{-n_1(t-\tau)} (\sin p_1^* t \cos p_1 \tau - \cos p_1^* t \sin p_1^* \tau) d\tau$$

$$= \frac{k_2}{mp^*_1}\left[d_1(t)F_2(t_0,t) - d_2(t)F_1(t_0,t)\right]. \quad (10)$$

where $F_2(t_0,t) = \int_{t_0}^{t} (y - y_0)\cdot e^{n_1\tau}\cos p^*_1\tau d\tau$;

$$F_1(t_0,t) = \int_{t_0}^{t} (y - y_0)\cdot e^{n_1\tau}\sin p^*_1\tau d\tau . \quad (11)$$

The total displacement is found from equation (5), which is solved for each time step by iterations at each step [18,19].

The force of reaction to the supporting structure is calculated using:

$$Q_{sup} = Ak_i \quad (12)$$

where $A$ - amplitude of displacement, $k_i$ - system stiffness

## 2.2 Algorithm of calculation of system with viscous damper (fig. 1b):

The dissipative force arising when the piston moves in a working medium with a velocity in the vertical direction is determined by the formula:

$P_{gk} = h_k\upsilon_k = \mu l_p\Psi$, where $h_k -$ the coefficient of resistance, in which $\mu(Pa.s) -$ the dynamic viscosity of the medium; $l_p -$ Working height of a layer of viscous liquid; $\Psi(t) -$ Coefficient determined from Fig. 35 [14].

In the presence of a viscous damper according to the scheme in Fig. 1b, the equation of motion of a system with one degree of freedom takes the form (in particular, during start-up and in operating mode)

$$m\ddot{y} + \left(1 + 2\mu\frac{d}{dt}\right)\kappa y + h_k\left|\frac{dy}{dt}\right| = q(t). \quad (13)$$

At $y_0 < y < y_1$, where $y_0(t_0)$ and $y_1(t_1) -$ the boundary of the zones of inclusion in the work of damping and shutdown. In the rest of the zone $h_k$ equals to zero.

Taking into account the above conditions, using the Duhamel integral, the solution of equation (13) can also be represented as the sum of two solutions: a linear system for the applied load $q(t)$ and a fictitious load, which takes into account the nonlinear dependence of the dissipative forces. The solution of the linear systems for the load $q(t)$ is given above and can be rewritten as follows:

$$y_{lin} = \frac{1}{p^*_1 m}\int_0^t q(t)V_1(p^*_1, t-\tau)d\tau ; \quad (14)$$

where $V_1(p^*_1,t) = e^{-n_1 t}\sin p_1 t -$ Impulse response function.

$$y_{nonlin} = \frac{1}{p^*_1 m}\int_{t_0}^{t} h_k\frac{dy}{d\tau}V_1(p^*_1, t-\tau)d\tau - \text{when } t \le t_1 \quad (15)$$

Integrating (15) by parts, we write (in the interval $t_0 \le t \le t_1$)

$$y_{nonlin} = \frac{1}{p^*_1 m}\left\{h_k y V_1(p^*_1,t)\Big|_{t_0}^{t} - \int_{t_0}^{t} h_k y V_2(p^*_1, t-\tau)d\tau\right\}; \quad (16)$$

$V_2 = \frac{d}{d\tau}V_1$, after similar simplification (see expressions 6-11) takes the form:

$$y_{nonlin}(t) = \frac{h_k}{p^*_1 m}\{\left[n_1 d_1(t) - p_1 d_2(t)\right]F_2(\tau) + \left[n_1 d_1(t) - p_1 d_2(t)\right]F_1(\tau)\}. \quad (17)$$

The nonintegral terms are 0 for $\tau = t$ ; and

$$y_{non\,int} = \frac{h_k}{mp_1^*} y(t)V_1(p_1^*, t - t_0) \text{ при for } \tau = t_0 . \tag{18}$$

$d_i, F_i$; $i = 1,2$ see formula (8), (9)

Without giving numerical solutions, we note that this algorithm corresponds to the start solution, and the numerical algorithm is similar to the algorithm used in the first problem. In the operating mode, the upper limit of the integral is to be put $t = t_1$. The displacements of the systems in the stop mode are determined by the algorithm given above in the interval $t_3 - t$, where $t_3$ – the time for switching on the damper.

## 3. Numerical example

A single degree of freedom of mass 10 ton with displacement limiter is investigated, amplitude of excitation force (Q) –350kN, frequency of excitation force (ω) – 78rad/s, damping coefficient - 0.1, reference structural displacement (y0) = 0.015m.

For the linear displacement, four values of stiffness k1 were used as input to get the displacement for the start-up and shut-down modes. Additional stiffness (k2) with values 500, 1000 and 1500 kN/m were used for each value of k1 to get the nonlinear displacement. Results are shown in table 1.

The support reactions were computed and are shown in table 2.

The effect of values of the reference structural displacement(y0) was investigated and results are in table 3.

The input dynamic load is taken from [2]. The time taken is 12seconds for start-up, 60 seconds for operational and 45 seconds for shut-down mode.

### 3.1 Results and discussion

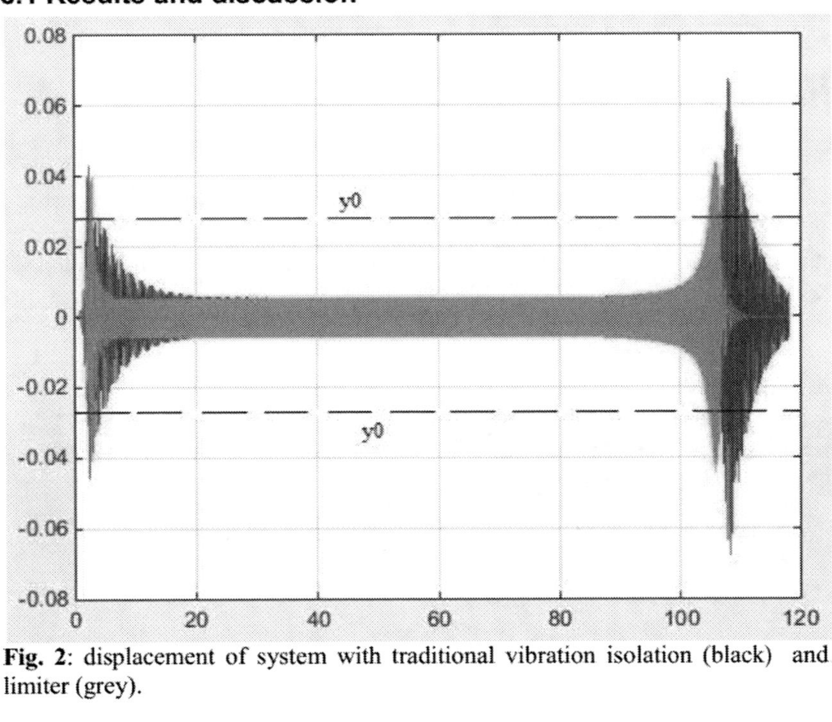


**Fig. 2**: displacement of system with traditional vibration isolation (black) and with displacement limiter (grey).

**Table 1**: displacement of system with traditional vibration isolation and with displacement limiter

| | displacement of system with traditional vibration isolation (m) | | | displacement of system with displacement limiter (m) | |
|---|---|---|---|---|---|
| k1 (kN/m) | in start-up | in shut-down | k2 (kN/m) | in start-up | in shut-down |
| 2500 | 0.0497 | 0.0772 | 500 | 0.0535 | 0.0700 |
| | | | 1000 | 0.0530 | 0.0537 |
| | | | 1500 | 0.0717 | 0.0328 |
| 3000 | 0.054 | 0.0818 | 500 | 0.0566 | 0.0760 |
| | | | 1000 | 0.0569 | 0.0636 |
| | | | 1500 | 0.0565 | 0.0458 |
| 3500 | 0.0561 | 0.0853 | 500 | 0.0596 | 0.0804 |
| | | | 1000 | 0.0604 | 0.0704 |
| | | | 1500 | 0.0548 | 0.0545 |
| 4000 | 0.0592 | 0.0881 | 500 | 0.0624 | 0.0841 |
| | | | 1000 | 0.0629 | 0.0771 |
| | | | 1500 | 0.0575 | 0.0652 |
| 4500 | 0.0608 | 0.0905 | 500 | 0.0642 | 0.0872 |
| | | | 1000 | 0.0654 | 0.0817 |
| | | | 1500 | 0.0602 | 0.0739 |

For the k1 values of 2500kN/m and 3000kN/m there a decrease in displacement in the start- of up to 57% and 44% in the shut down mode. There is however an increase in displacement in the startup mode. Therefore the optimum parameters of k1 and k2 are 3500 kN/m and 1500kN/m, where the reduction in displacement is 36% in the shut-down mode and marginal decrease in start-up mode (table 1).

There is a 32% reduction in the support reaction in the shut-down mode where k1 and k2 are 2500kN/m and 1500kN/m. In the start-up mode there is increase in the support reaction for all values of k1 and k2. (table 2).

The optimum value of y0 is 0.025m for k2 value of 1500kN/m which results in the most reduction of shut-down support reaction (table 3).

**Table 2**: support reaction of system with traditional vibration isolation and with displacement limiter

| | support reaction of system with traditional vibration isolation(kN) | | | support reaction of system with displacement limiter (kN) | |
|---|---|---|---|---|---|
| k1 (kN/m) | in start-up | in shut-down | k2 (kN/m) | in start-up | in shut-down |
| 2500 | 124.25 | 193.00 | 500 | 160.50 | 210.00 |
| | | | 1000 | 185.50 | 187.95 |
| | | | 1500 | 286.80 | 131.20 |
| 3000 | 162.00 | 245.40 | 500 | 198.10 | 266.00 |
| | | | 1000 | 227.60 | 254.40 |
| | | | 1500 | 254.25 | 206.10 |
| 3500 | 196.35 | 298.55 | 500 | 238.40 | 321.60 |
| | | | 1000 | 271.80 | 316.80 |
| | | | 1500 | 274.00 | 272.50 |
| 4000 | 236.80 | 352.40 | 500 | 280.80 | 378.45 |
| | | | 1000 | 314.50 | 385.50 |
| | | | 1500 | 316.25 | 358.60 |
| 4500 | 273.60 | 407.25 | 500 | 321.00 | 436.00 |
| | | | 1000 | 359.70 | 449.35 |
| | | | 1500 | 361.20 | 443.40 |

**Table 3**: Variation of values of reference structural displacement(y0) on displacement

| k1(kN) | k2 (kN) | **y0**(m) | displacement(m) | | support reaction (kN) | |
|---|---|---|---|---|---|---|
| | | | in start-up | in shut-down | in start-up | in start-up |
| 3500 | 500 | 0.015 | 0.0596 | 0.0804 | 238.4 | 321.6 |
| | | 0.025 | 0.0579 | 0.0745 | 231.6 | 298 |
| | | 0.035 | 0.0563 | 0.0723 | 225.2 | 289.2 |
| | 1000 | 0.015 | 0.0604 | 0.0704 | 271.8 | 316.8 |
| | | 0.025 | 0.0583 | 0.0609 | 262.35 | 274.05 |
| | | 0.035 | 0.0561 | 0.0676 | 252.45 | 304.2 |
| | 1500 | 0.015 | 0.0548 | 0.0545 | 274 | 272.5 |
| | | 0.025 | 0.0573 | 0.0512 | 286.5 | 256 |
| | | 0.035 | 0.0558 | 0.093 | 279 | 465 |

## 4. Conclusion

The paper looked at variants of nonlinear systems of vibration isolation (with displacement limiter and viscous damper) and algorithms for their calculation as systems with single degree of freedom under the action of harmonic load (in operating mode) and in transient modes. From the numerical example it was observed that the system with the displacement limiter could reduce the maximum values of displacements in transient modes by 30-35%. The algorithm for calculating the system with a viscous friction dampener is illustrated by the example of calculating the system in the starting mode.

### References

1. TSNIISK im. V. A. Kucherenko Nits construction, (2015).
1. M.V Osipov, Seismic construction. Seismic resistant construction. Safety of buildings, **1**, 31 - 34 (2014)
2. Ya. M. Eisenberg, Stroiizdat, (1976)
3. S.A. Alotaibi, M.A. Al-Ajmi, International Journal of Mechanical, Aerospace, Industrial, Mechatronic and Manufacturing Engineering, **8, 3,** 565-567 (2014).
4. Y. Li, J. Li, Australasian Conference on the Mechanics of Structures and Materials (ACMSM23). Byron Bay, Australia. **II,** 913-918 (2014).
5. Y. Xinghua, Model and analysis of variable stiffness semi-active control system. *12WCEE*. (2000)
6. D.Y. Kim, J. Joung, Experimental and Numerical Studies of a Newly Developed Semi-Active Outrigger Damper System. *CTBUH 2011 World Conference*. Seoul. Korea. (2011)
7. S. Nagarajaiah, S. Ertan, Journal of Structural Engineering, **33, 1**, 67-77 (2007).
8. S. Nagarajaiah, S. Sahasrabudhe, Earthquake Engineering and Structural Dynamics, **35, 2**, 177-197 (2006)
9. S. Kamagata, T. Kobori, Proceedings of the 1st World Conference on Structural Control. Los Angeles, U.S.A. (1994).
10. J.N. Yang, J.H. Kim, A.K. Agrawal, Journal of Structural Engineering, 1427-1433 (2000)
11. J.W. Joung, A.W. Smyth, L. Chung, Smart Materials and Structures, **19** (2010)
12. K. Kawashima, Proc.of International Workshop on Structural Control. (1993).
13. TSNIISK im. V. A. Kucherenko. Rekomendatsii po vibrozashchite nesushchikh konstruktsiy proizvodstvennykh zdaniy. [Recommendations on vibration isolation of load-bearing structures of industrial buildings]. (1986).
14. V.V. Solodovnikov, Staticheskaya dinamika lineynykh sistem avtomaticheskogo upravleniya [Static dynamics of linear automatic control systems]. Moscow: Fizmatgiz, (1960).
15. J.T. Chernov **2**nd ed. Moscow: Publisher ASV, 384 (2011).
16. A.K. Chopra, Pearson Education Asia Ltd and Tsinghua University Press, 846 (2004)
17. I.A. Petrov, M.V. Osipova, Internet-Bulletin VolgGASU. Ser.: Polit., 3 (23), (2012).
18. A.S. Gorodetsky, I.D. Evzerov, Kompyuternyye modeli konstruktsiy [Computer models of constructions] Moscow: Publishing Association of Construction Universities, 360 (2009).

# Assessment of impact of construction materials on the ecological safety of home

*Anna Zhigulina* [1,*], *Natalia Chumachenko*[1]

[1] Samara State Technical University, Institute of Architecture and Civil Engineering, 194 Molodogvardeyskaya St., 443001, Samara, Russia

**Abstract.** The article deals with the problems of creating environmentally friendly aerial environment within residential premises. The main sources causing air pollution of urban housing are determined and classified. The origins of air pollution sources of residential premises are adopted as the classifying criterion. The sources of contamination are defined and assessed. Particular attention is paid to the choice of environmentally friendly building materials. The methodology for assessing toxicity of industrial waste used in the production of housing materials is developed to assess the comfort and environmental safety of home. The idea of creating "Residential buldings ID" containing information on the construction materials used is introduced.

## 1 Introduction

Collaborative studies of hygienists, architects and builders in the course of design and construction of housing in the USSR rendered it possible to realistically assess the quality of the residential premises environment, to introduce necessary adjustments and updates to the construction regulation norms and thus produce healthy housing [1-7]. Such cooperation is just as essential now, when residential construction is done not only within government contracts, but on a commercial basis as well. Today, of course, environmental, sanitary and health departments, including the Academy of Medical Sciences, are actively working to create healthy living environment, they control the quality of building materials, are involved in the assessment of environmental conditions of residential premises, monitor air quality, water bodies, soil [8 -10]. Some of the authors [11-17] have focused on the issues of environmentally friendly housing design.

However, there are a number of topical issues concerning ecology and hygiene of residential buildings that deserve special attention from both the sanitary, engineering and construction companies, as well as the manufacturers of building materials.

## 2 Materials and Methods

According to the World Health Organization experts, city dwellers spend about 80% of their time indoors. The results of various studies of air quality in an urban apartment have

* Corresponding author: auzhigulina @ mail.ru

shown that the indoor air is 4-6 times as polluted and 6-8 times as toxic compared to outdoors. Hygienists believe that indoor air pollution can cause health problems of various levels of severity, from simple discomfort and headaches to severe allergic reactions, asthmatic symptoms or even cancer.

Sources of indoor air pollution can roughly be divided into five main groups (Table 1)

Sources of indoor air pollution

Table 1

| Pollution sources groups | Pollution source | Location |
|---|---|---|
| 1 | substances entering premises with polluted urban air | external sources |
| 2 | anthropotoxins (toxic human waste) | internal sources |
| 3 | substances related to household activity (household chemicals, cooking gas combustion products) | internal sources |
| 4 | microbial air pollution (biodeterioration of construction units caused by fungus, mold) | internal sources |
| 5 | toxins released from building materials. | internal sources |

We give below a more detailed description of these major groups of pollutants.

**Group 1**.The main sources of urban air pollution and, as a consequence, of air in urban apartments, are motor vehicles, industrial plants and heating systems. Contamination occurs due to tailpipe and aerosol pollution. According to *Health Care in Russian Federation* magazine, in 2005 only 15 % of the urban population in Russia lived in the areas with pollution levels within hygienic standards. With every year of increase in the number of cars, the level of tailpipe pollution grows only higher. Improvement in quality of urban air can be achieved by the town planning activities, such as landscape gardening and watering areas, i.e. arrangement of parks, gardens, artificial lakes, ensuring sufficient aeration. Elimination of harmful effects of industrial enterprises on residential areas should be achieved by reasonable functional zoning of urban areas. This involves the remoteness of industrial enterprises from residential areas, the establishment of sanitary buffer zones which size shall be appointed in accordance with the enterprise hazard level and adjusted in line with the wind rose. It is quite difficult to introduce dramatic changes to current urban planning. New construction areas offer more opportunities for air pollution reduction. In this case, the contamination of air entering the apartment from the outside may be reduced by city planning decisions at the blueprint stage.

**Groups 2,3,4.** Sources of pollution from the second, third and fourth groups may controlled by the residing people to a greater or lesser extent.

**Group 5**.Building and finishing materials make for one of the most significant internal sources of air pollution in residential premises. They are the source of 80% of chemicals detected in the air of urban residences [18]. Harmful substances may originate from raw materials used for the production of building supplies, or be the result of destructive processes occurring during building operation.

Ensuring environmental and hygienic safety of the dwelling, in terms of choosing building materials, should provide for the application of construction and decorative materials with minimal or no emission of toxic substances throughout the "product life cycle". The concept of "product life cycle", introduced in a series of ISO 14000 standards, suggests the cycle of "Production - Maintenance - Disposal" for building materials.

At present, the mineral resource base of the construction industry includes two types of raw materials: natural and man-made. If construction or decoration procedures utilize natural building materials such as wood or stone, or natural raw materials (clay, sand, limestone, chalk, etc.) are used for manufacturing building supplies, their environmental safety is almost out of question. But even in this case, monitoring radiation background of mineral stock may be worth the time.

However, synthetic and composite materials are the most commonly used in modern construction. The current state of natural resource base is such that, on the one hand, the depletion of natural conditioned raw materials takes place, and, on the other hand, the amount of man-made by-products while extracting and processing raw materials is constantly growing. Very often, industrial wastes are used as additives for various purposes, or substitutes of some natural raw materials in production of construction supplies. The development of industry in our country has led to formation of large amounts of waste, which harmful effect on the biosphere is beyond doubt. The establishment of cost-effective methods for their disposal, one of which is the use of industrial waste in production of building materials, is, by all means, a very important issue. The results obtained in the course of research, including those lead by scientists of Institute of Architecture and Civil Engineering of Samara State Technical University [19], confirm the prospects of this approach.

Although, waste recycling can be considered effective if both economic criteria and environmental properties of materials produced from waste are taken into account, which should eliminate harmful effects of wastes on human beings.

The toxicity of industrial waste needs to be assessed while determining the way of its application. The **methodology for evaluating toxicity of waste** by comparing the toxicity of industrial waste emission with MPC of carcinogenic (toxic) substances and elements has been developed. According to the **methodology**, the waste is divided into three types (см табл.2)

Evaluation of industrial waste types in aspect of toxic ingredients content and applicability to building materials production

Table 2

| Industrial waste type | Toxic ingredients content | Applicability to building materials production |
|---|---|---|
| 1 | waste containing significant quantities of toxic substances, the concentration of which exceeds MPC | cannot be used without special cleaning measures |
| 2 | waste composed of a small amount of heavy metals | are recommended to be used for materials production only at baking technologies |
| 3 | concentration of emission does not exceed MPC of carcinogenic (toxic) substances and elements | are recommended for next evaluation level |

Wastes from the first group cannot be used in manufacturing of building materials without any special decontamination and should be sent for disposal. If waste contains elements of heavy metals (second group), it might be recommend for usage in roasting technologies provided that raw mass has sufficient molten alloy for conservation (encapsulation) of heavy metals. The technology not only ensures reliable waste preservation, but also provides considerable increase in strength of ceramic material by heavy metal elements performing the role of heterogeneous catalysts promoting crystallization of melt and consequently increasing strength. Wastes from the third group are assessed on a finer scale. After a multi-level evaluation, waste is assigned a certain status [20-22].

Another source of indoor air pollution is polymers. If building materials are composed of polymers, degradation processes, inevitable in the course of operation, can lead to poor indoor climate. The list of polymeric materials used in construction contains more than 100 items. Polymers are used in manufacturing of wall finishing materials, heat-insulation materials, flooring, waterproofing materials, sealants, manufacturing of windows and door units, cladding panels, etc. The intensity of volatile toxin emission from them depends on the microclimate conditions of their application, i.e. temperature, humidity, air change rate, operation time. With increasing amount of polymeric materials in the air environment of residential premises and public buildings, the concentration of formaldehyde, phenol, benzene, ethyl benzene, xylene, ethyl acetate, toluene, butyl acrylate and other toxic substances expectedly increases as well.

Unfortunately, there are practically no regulations concerning environmental safety of home in terms of application of building materials. An exception is the State Standard GOST 30108-94 *Building Materials and Products. Determination of Specific Effective Activity of Natural Radionuclides (with Amendments 1, 2),* controlling the radioactivity of technogenic waste and radon emission from building materials. State examination at the designing stage only controls structural strength. Even if building inspection controls the quality of construction materials, they only consider conformity certificates. Monitoring indoor atmosphere at the commissioning stage is not obligatory. This problem needs to be addressed at the state level.

In the course of research, the authors classified information about emissions from basic groups of construction materials and maximum permissible concentrations (MPC) of harmful substances in indoor air. The method of evaluation of ecological safety and comfort of living accommodations in accordance with used construction and finish materials was proposed. Materials were grouped by sections "Construction Materials", "Construction-Heat-Insulation Materials", "Heat-Insulation Materials", "Finishing Materials", "Window and Door Materials." The study evaluated chemical composition of materials, a generic list of possible additives and their potential chemical emission into surrounding environment. Distribution of materials by their category of comfort was done according to five-point evaluation system: 5 – excellent (no toxic emission); 4 – very good (little emission, significantly below MPC); 3 – permissible (below MPC) 2 – permissible for areas of infrequent occupation; 1 – unacceptable for habitation.

The elaborated method of building materials evaluation gives an opportunity to put in perspective housing ecological parameters and to choice the least toxic ones. This information could be useful for private developers which control construction process more than housing buyers.

In order to enable construction companies to confirm the quality of the housing built and to provide citizens with complete information about their homes, the authors consider it necessary to create "Residential buldings ID", which, along with the assessment of planning concepts, design and engineering solutions, and the availability of infrastructural elements, should also evaluate the parameter of ecological safety of housing.

## 3 Results

The research yielded the following results:

- air pollution sources of residential premises were identified;
- building and finishing materials were recognized as the main source of pollution among them;
- regulatory framework on environmental security in the scope of construction materials was reviewed;

- information about basic construction materials, emissions from them, and maximum permissible concentrations of harmful substances in indoor air was collected and classified;
- the obtained data were grouped by sections "Construction Materials", "Construction-Heat-Insulation Materials", "Heat-Insulation Materials", "Finishing Materials", "Window and Door Materials."
- evaluation methodology of ecological safety of home was introduced.

## 4 Discussion

Construction of a modern apartment suggests using large amounts of materials. In the classical evaluation on the famous triad of Vitruvius – "Usefulness – strength – beauty» – environmental component (usefulness) is at the first place for a certain reason. Aesthetic beauty of home should go hand in hand with its ecological safety. Therefore, it is vital to control building and finishing materials for creating healthy environment. The idea of saying 'no' to unsustainable building materials is very difficult to implement in practice. The authors ' methodology of evaluating the quality of materials for residential housing will let consumers objectively evaluate environmental performance of housing. The methodology is being constantly improved. This is due to the fact that it often happens so, that even without exceeding the maximum permissible concentrations of selected construction materials, indoor air quality leaves much to be desired. The thing is that the existing MPC regulations are calculated for each substance separately, without taking into account a combination of pollution sources. Therefore, in the course of further work the authors plan to develop a methodology for integrated assessment of housing contamination by toxic emissions from construction and finishing materials, as well as to investigate the variations of toxic emission volume and its intensity over time resulting from material degradation and aging.

## 5 Conclusions

Creating environmentally friendly living environment is a complex task that requires the interaction of hygienists, architects, builders and manufacturers of construction materials.

Improvements to the ecological safety of urban dwellings are possible if we:

- control the quality of air environment of premises put into operation at the state level;

- using the methods developed, carry out a toxicity assessment of the composite material components used in construction of residential buildings and the toxins released from synthetic materials;

- create "Residential buldings ID" containing information on the construction materials used;

- make an integral estimation of toxic emission from building materials.

## References

1. V. K. Litskevich, Residential Construction **4**, 24-26 (2011)
2. SANPIN 2.1.5.980-00 (Date of updating 12.02.2016) Hygienic Requirements for Surface Water Protection
3. SANPIN 2.1.7.1287-03 (Date of updating 12.02.2016) Sanitation and Disease Control Regulations for Soil Quality

4. GOST 17.2.3.01-86 (Date of updating 06.04.2015) Nature protection. Atmosphere. Air quality control regulations for populated areas
5. GOST 17.1.2.04-77 (Date of updating 12.02.2016) Nature protection. Hydrosphere. Idices of state and regulations for valuation survey of fishery waters
6. SANPIN 2.1.2.1002-00 Sanitation and Disease Control Requirements for Residential Occupancies and Spaces (Date of updating: 01.02.2017)
7. MU 2.1.2.1829-04 Sanitary and Hygienic Evaluation of Polymer and Polymer-cotaining Building Materials and Structures for Residential, Public and Industrial Construction. Methodology Guidelines (Russian Ministry of Health, Russia, 2004)
8. SANPIN 2.1.2.729-99 Higienist requirements for safety polymeres building materials (Date of updating: 12.02.2016)
9. GN 2.1.6.1338-03 Maximum Permissible Concentration (MPC) of Pollutants in the Ambient Air of Populated Areas (as amended on 01.02.2017) (Russian Ministry of Health, Russia, 2003)
10. State Standard GOST 30494-2011 Residential and public buildings. Microclimate Parameters in Buildings (amended)
11. N. G. Chumachenko, A. Y. Zhigulina, Residential Construction **4**, 21-23 (2011)
12. A. Y. Zhigulina, *Traditions and Innovations in Construction and Architecture: Materials of the 69th All-Russian Scientific and Technical Conference on the results of R & D, 2011,* 150-151 (Samara State University of Architecture and Civil Engineering, Samara, 2012)
13. A. Y. Zhigulina, *Traditions and Innovations in Construction and Architecture: Materials of the 71st All-Russian Scientific and Technical Conference on the results of R & D, 2013*, 384-385 (Samara State University of Architecture and Civil Engineering, Samara, 2014)
14. A. Y. Zhigulina, *Traditions and Innovations in Construction and Architecture: Materials of the 69th All-Russian Scientific and Technical Conference on the results of R & D, 2011,* 363-365 (Samara State University of Architecture and Civil Engineering, Samara, 2012)
15. A. Y. Zhigulina, *Traditions and Innovations in Construction and Architecture*: Materials of the 70th Anniversary All-Russian Scientific and Technical Conference on the results of R & D, 2012, **1**, 337-338 (Samara State University of Architecture and Civil Engineering, Samara, 2013)
16. A. Y. Zhigulina, Town Planning **6**, 46-48 (2013)
17. A. Y. Zhigulina, N. G. Chumachenko, Urban Construction and Architecture **3**, 94 – 100 (2015). DOI: 10.17673/Vestnik. 2015.04.12
18. *Accommodation: Comprehensive View,* 32 (Moscow, 2001)
19. Scientific works of Samara School of Materials Scientists (Samara State University of Architecture and Civil Engineering, Samara, 2006)
20. N. G. Chumachenko, *Ecology and Human Health. Proceedings of the VII All-Russian Congress,* 201-203 (Samara, 2006)
21. N. G. Chumachenko, Urban Construction and Architecture **3**, 112-116 (2011). DOI: 10.17673/Vestnik.2011.01.22

# Technology of obtaining thermal insulation material on the basis of liquid glass by a method of low temperature processing

*Anna* Zhigulina[1,*], and *Sergey* Mizuriaev[1]

[1]Samara State Technical University, Institute of Architecture and Civil Engineering, 194, Molodogvardeyskaya St., 443001, Samara, Russia.

**Absract.** The paper deal with the development of energy saving technology for obtaining thermal insulation material on the basis of liquid glass. The aim of the research is to study processes of chemical and physical modification of soluble sodium silicate and change its rheology. The paper also considers development of technology for the production of thermal insulation material with porosity uniformly distributed throughout the volume and with the specified pores size. In this research the authors used both traditional methods (chemical analysis, physical-mechanical tests) and "thin" methods, i.e. differential thermal analysis, electron microscopy. For research in the field of rheology changes the authors developed a methodology for determination of viscoelastic properties of materials with variable rheology. For the development of resource-saving technologies the researchers used the method of laboratory simulation which is as close as possible to real production conditions. The research identifies regularities of processes of modification and swelling of liquid-glass compositions and suggests methods for increasing viscosity of liquid-glass compositions by introducing various modifying additives. It also introduces an energy-saving technology of porous granular material production, and proposes options for the production of effective thermal insulation material based on sodium silicate composition.

## 1 Introduction

Sodium silicate, and its aqua solution in particular, is widely used in a variety of industries. In Russia, an sodium silicate aqua solution is better known under its trade name "liquid glass" or "soluble glass." Liquid glass is obtained from sodium silicate (silicate blocks) and water during autoclave curing. Liquid glass is a transparent sticky liquid with a density of 1200-1500 kg/m$^3$, characterized by high adhesion properties.

It is widely known that liquid glass swells when heated to temperatures of about 150-250 $^O$ C and increases 50-70 times in its volume formating high porous foam mass [1-2]. This property makes it possible to use liquid glass in the production of heat-insulating materials and products. The main advantages of these thermal insulation products, which distinguish them from traditional insulants are low density and low thermal conductivity.

* Corresponding author: auzhigulina@mail.ru

Besides, they do not evolve toxic substances when being used or even burnt; are resistant to chemical and biological impacts, characterized by low water absorption (not more than 5 % (thorough the volume) and, as a consequence, by durability (at least 100 years); by high strength (their compressive strength is not less than 0.7 MPa); and are easy to process.

That's why obtaining highly porous structures based on liquid glass by thermal processing holds much promise. This problem is studies by many research teams [3-16]. Research in this direction, both in Russia and abroad, is focused mainly on improving traditional technology of liquid glass and foam glass production. Various components are put into the raw mixture as additives, adjusting the specific properties of the product. These additives, being of natural or man-made origin, are local materials, typical for the region in which the specific raw material composition was developed. Analysis of existing technologies, literature and patent information suggests that the development of energy-saving production technology of modified liquid glass production with the use widespread materials as additives is an actual scientific and technological challenge.

## 2 Materials and Methods

To achieve the aim of the research it was necessary to study processes which take place during thermal synthesis of sodium and sodium-based components and lead to the formation of sodium-silicate compounds. The authors analyzed state diagram of the system $Na_2O$-$SiO_2$. This system served as the basis for the equilibrium system $SiO_2$-$Na_2O$-aq, widely used in insulation materials technology based liquid glass. According the theoretical data [1, 2] melting in a triple system $SiO_2$-$Na_2O$-aq begins at temperatures above 48 $^{O}$C. This is due to the fact that the cation and anion $Na_2$ and $SiO_3$ bind only electrostatic interaction. They are able to make only weak hydrogen bonds with hydroxyl groups. It means that even at a temperature of 48 $^{O}$C crystal structure collapses and crystallohydrate melts in its own water.

The volume of hydrated water affects the melting temperature of alkaline silicate. If this silicate contains 20% of hydrated water, it melts at a temperature of 100 $^{O}$C. With rapid heating up to 200 $^{O}$S, hydrated silicate deliquates and hydrated water turns into steam. Due to the viscosity of liquated silicate, water vapors are trapped in it, forming bubbles with thin walls.

The research also explored swelling processes in liquid glass compositions. The study confirmed published data on the high degree of pore formation in liquid-glass masses during heat treatment and found that the increase of liquid glass density leads to an increase in its swelling ability. Experimental results showed that even from ordinary liquid glass with high density it is possible to obtain only porous pellets with arbitrary shape (not spherical), and without a crust (indicating a high water absorption and low strength). When ordinary liquid glass is heated, the obtained pellets stick together, forming an irregularly shaped material with arbitrary sizes.

Thus, it was proved that the basic technical and engineering difficulty of producing insulation materials on the basis of liquid glass lies in the fact that porosity which is formed when heated is very uneven (both in pores size and their distribution throughout the volume). For obtaining granular material with uniform porosity the most difficult thing was to obtain spherical micro-porous pellets with compacted outer crust (similar to ceramsite granulars) from sodium silicate composition. Therefore, in the course of further research we examined regularities of processes of modification and swelling of liquid-glass compositions. Tests showed that the form of obtained foamed pellets became more spherical with the simultaneous appearance of the crust when viscosity of liquid glass was increased.

Experimental research of methods to increase the viscosity of sodium silicate composition showed that the most effective way was introducing the acids and salts of alkali metals. Interaction of liquid-glass compositions with solutions of alkali metal salts is a complicated physical-chemical process, the details of which are not yet clarified. It is assumed that chemical interaction between alkali metal salts and components of liquid glass able of forming new compounds is hardly possible. We believe that it is rather a process of silica gel coagulation by electrolytes of alkali metal salts present in solution [1, 16]. According to the data, the process of coagulation of liquid glass during the introduction of different salts of acids is influenced by both cations and anions of these salts. The ability to cause a coagulation uniform throughout the volume that results in obtaining the mixture capable of being formed into raw pellets of specified size and later form spherical pellets after heat treatment is considered to be the main criterion in determining the effectiveness of additives. The following elements were tested as additives: hydrochloric acid and phosphoric acid, sodium chloride, calcium and ammonium, sodium nitrate, sodium and potassium sulphates, sodium hydroxide.

Table 1 shows the results of experiments of influence of liquid glass some chemical composites on its reology characteristics and type of thermal distensibility of liquid glass with additives

**Table** 1. Additives influence of reology characteristics and distensibility of liquid glass

| Additive type | % | Results | |
|---|---|---|---|
| | | **Raw mass coagulation** | **Formed pellets distention** |
| **1** | **2** | **3** | **4** |
| Hydrochloric acid | 2 | Rapid coagulation | No distention |
| | 5 | Rapid coagulation | |
| Orthophosphoric acid | 2 | Rapid coagulation | No distention |
| | 5 | Rapid coagulation | |
| Sodium chloride | 5 | Coagulation with unformable mass | Distention with pellets of irregular form |
| | 10 | Coagulation with formable mass | Distention with pellets of sphericity form |
| Calcium chloride | 10 | White precipitation | No distention |
| | 50 | White precipitation, progressive curing of mass | |
| хлорид аммония | 5 | | No distention |
| | 10 | Rapid curing of mass | |
| Sodium nitrate | 10 | White precipitation | No distention |
| | 50 | White precipitation, progressive curing of mass | |
| Sodium sulfate | 10 | White precipitation | No distention |
| | 50 | White precipitation, mass surface crusting | |

The experiments demonstrated that the most effective modifying additive was sodium chloride. The introduction of this substance resulted in gradual coagulation of liquid glass. Viscosity of the mass gradually increased and acquired plastic-elastic properties, which made possible to form pellets. Heat treatment of these pellets resulted in their considerable swelling, they squired spherical shape, then they solidified and preserved the acquired form. Foamed spherical pellets had homogeneous high-porous inner structure with external compacted crust. In addition, it was taken into account that among all tested substances sodium chloride was the most affordable and cheap material.

In the next phase of works the authors studied processes of swelling of liquid glass compositions which were modified with sodium chloride, and determined the impact of the type and quantity of the components on the swelling ability and formability of liquid glass compositions. These experiments demonstrated that the minimum density of pellets (0.22 g/cm$^3$) is achieved when using liquid glass of 1.51 g/cm$^3$ density (Fig. 1) and while adding modifying additives in the amount of 3-10 % of the mass of liquid glass (см. табл. 1).

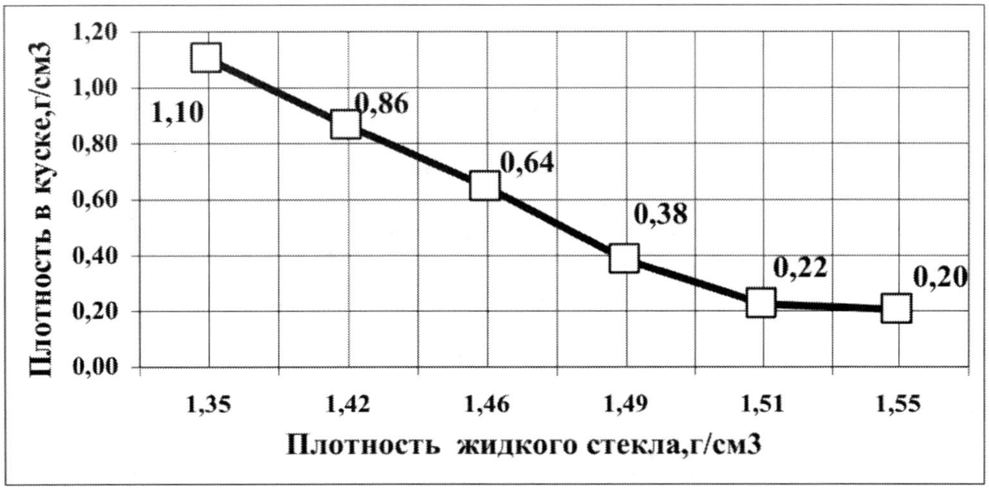


Fig.1. Influence of liquid glass density on pellet surface of aggregates

Further experiments allowed to define optimal swelling modes for the developed materials. The material of developed mixtures was pelletized and subjected to heat treatment at 250-300 $^{\circ}$C. Foamed pellets which were heated on the bottom of stationary kilns were of an irregular shape. Th improve the form of pellets the researchers used a rotating disk pelletizer, integrated in a laboratory kiln. After heat treatment at this installation the form of the pellets became close to spherical. The obtained pellets had a dense solid crust and internal high-porous structure. Then the researches applied an energy-saving technology of low-temperature swelling at 300 $^{\circ}$C for 10 minutes. In a kiln pelletizer spherical porous pellets with the following characteristics derived from physico-mechanical tests were obtained: density in the piece – 0.22 g/cm$^3$; bulk density – 150 kg/m$^3$; splitting strength – 0.07 MPa; water resistance factor – 0.3.

Thus, it was possible to create composite materials with specified properties on the basis of the developed highly porous material by introducing various additives. The authors received heat granulated material and granulates with increased strength, water resistance and biostability [17-19]. The characteristic feature of these granular materials was their low density and their constant "ceramsite" structure, i.e. the presence of external and internal shell of compacted high porous foam structure. Further results showed sufficient stability of these materials properties. Results of microscopic studies of materials samples showed the presence of nanoscale pores inside these materials [20]. It makes possible to consider the obtained material suitable for creating effective innovative insulating material on its basis.

## 3 Results

The researchers explored possibilities of modifying liquid glass in order to obtain granulated material with homogeneous porous structure. These experiments yielded the following results:

- Processes taking place during thermal synthesis of sodium-and sodium-based components and leading to the formation of sodium-silicate compounds were studied and described.
- The influence of different technological modes on the nature of liquid glass swelling masses was analized and the optimal mode was found.
- Regularities of processes of modifying liquid glass compositions modification were investigated and a unique component (sodium chloride) was pointed out. Sodium chloride has the ability to cause a coagulation uniform throughout the volume that results in obtaining the mixture capable of being formed into raw pellets of specified size and later form spherical pellets after heat treatment.
- The authors studied processes of swelling of liquid glass compositions which were modified with sodium chloride, and determined the impact of the type and quantity of the components on the swelling ability and formability of liquid glass compositions.
- Technology of obtaining granulated porous material on the basis of liquid glass by a method of low temperature processing was worked out.
- The possibility of introducing additives capable of obtaining material with specified properties into the developed composition was investigated.

## 4 Discussion

Research results show that liquid glass can be modified by introducing a certain amount of sodium chloride. As the result, we get uniformly coagulated viscous mass which can be formed in different ways. It allows to solve the problem of obtaining granulated material from liquid glass. At low temperature swelling of raw granules we obtain high-porous material of "ceramsite" structure with an external compacted shell and an internal high-porous structure. This material is unique because it is possible to attach different properties to it by entering various components in the raw mixture composition.

## 5 Conclusions

The developed energy-efficient technology makes it possible to obtain granular high-porous material on the basis of liquid glass by a method of low temperature processing. The obtained granulated high-porous material might be used as a basis for innovative insulating material which properties may be adjusted by the introduction of additives. It is possible to use widespread available materials, including the industrial wastes, as additives.

## References

1. P.N. Grigorev, M.A. Matveev, *Fluid glass* (Promstrojizdat, Moscow, 1956)
2. V.I. Korneev, V.V. Danilov, *Production and application of fluid glass* (Stroiizdat, Leningrad, 1991)
3. Y.P. Karnaukhov, V.V. Sharov, Construction materials **11,** 14-16 (1994)
4. A.I. Kudyakov, T.N. Radina, M.Yu. Ivanov, Building materials **11**, 12-13 (2004)
5. Yu.G. Ivashchenko, A.A.Surnin, N.V. Zobkova, Liquid glass composition (Patent № 2158717 RU, MPK C 04 B 28/26 req. Saratov State Technical University 16/02/99; pub. 10.11.2000. Bul. № 31)

6. V. A. Reshetov, V. T. Pavlov, A. T. Pavlov, Means to obtain porous material on the basis of liquid glass. (Patent № 2177922 RF, MPK C 04 B 28/26, C 04 B111:40 req. Rechetov Vyacheslav Aleksandrovich, Pavlov Vladimir Tikhonovich, Pavlov Anatoly Tikhonovich 24/08/00; pub. 10.01.02. Bul. № 1.)
7. A.I. Kudyakov, T.N. Radina, M.Yu. Ivanov, Raw materials mixture and method of obtaining foam silicate insulating material (Patent № 2267468 RU, MPK C 04 B 28/26 req. State educational institution of higher professional education "Bratsk State Technical University 30/03/04; pub. 20.02.2012. Bul. № 5.)
8. Yu.G. Ivashchenko, R.V Fomin, Liquid glass composition (№ 2245861 RF, MPK C 04 B 28/26. req. Saratov State Technical University 15/11/02; pub. 10.02.2005. Bul. № 4)
9. I.B.Bessonov, A.N.Salepin, N.P.Kordyukov, Method of obtaining construction-insulation building material based on aluminosilicate microspheres (Patent № 2455253 RU. MPK C 04 B 28/26 req. Federal State budgetary institution "Scientific Research Institute of building physics. Russian Academy of architecture and building science 01/03/11; pub.: 10.07.2012. Bul. № 12)
10. Yu.G. Ivashchenko, R.V Fomin, Liquid glass composition (Patent № 223597 RF, MPK C 04 B 28/26 req. Saratov State Technical University 19/11/2002; pub. 10.09.2004. Bul. № 25)
11. A.B. Fashchevskiy, A.A.Fashchevskiy, M.A.Fashchevskiy, Method of producing a building material (Patent № 2333176 RU MPK C 04 B 28/26, C 04 B 111/20 ; req. A.B. Fashchevskiy, A.A.Fashchevskiy, M.A.Fashchevskiy 05/03/2007; pub. 10.09.2008. Bul. № 25)
12. A.P. Korsakov, A.A.Korsakov, P.A.Korsakov, Method of obtaining construction-insulation foam glass (Patent 2451644 RF, MPK C 03 C 11/0000, C 03 B 19/08; req. KONAK INDUSTRY Inc. (CONAC INDUSTRY INC) (VG) 10/22/2010; pub. 5.27.2012, Bul. № 15)
13. V.Z. Leonidov, Production of foam glass (Patent 2455245 RU, MPK C 03 C 11/0000, C 03 B 19/08; req. Leonidov V.Z. 25/02/2011; pub. 10.07.2012, Bul. № 19)
14. V.A. Smoliy, E.A.Yatsenko, V.A. Guziy, Foamed slag glass (Patent 2448919 RU, MPK C 03 C 11/00 req. State educational institution of higher professional education "South-Russian State Technical University (Novocherkassk Polytechnical Institute. 09.09.2010; pub. 27.04.2012. Bul. № 12)
15. Yu.A. Shchepochkina, Production of glazed foam glass (Patent 2431619 RF, MPK C 03 C 11/00, C 03 B 19/08; req. Shchepochkina Yu.A. 15/03/2010; pub. 20.10.11, Bul. № 29)
16. Yu.A. Shchepochkina. Production of foam glass (variants) (Patent 2452701 RF, MPK C 03 C 11/00, C 03 B 19/08. Applicant and patentee:Shchepochkina Yu.A. №2010149109/03; req. Shchepochkina Yu.A. 30/11/2010; pub. 10.06.12, Bul. № 16)
17. O.l. Figovskiy, P.G. Kudryavtsev, Don Engineering bulletin **29**, 117 (2014).
18. S.A. Mizuriaev, A.Yu. Zhigulina, A.N. Mamonov, N.V. Ivanova, Construction materials **7**, 12-13 (2011)
19. S.A. Mizuriaev, A.Yu. Zhigulina. Urban planning **1,** 82-84 (2012)
20. S.A. Mizuriaev, A.Yu. Zhigulina, G.S. Solopova, Procedia Engineering **111,** 540-544 (2015)
21. A.M. Guryanov, S.A. Mizuriaev, V.M. Lebedev, V.T. Lebedev, *Proceedings of the V Eurasian scientific-practical Conference "Strength of non-homogeneous structures",* 136 (2010)

# The investigation of expanded polystyrene creep behaviour

*Aleksey* Zhukov[1], *Vyacheslav* Semenov[1 *], *Ivan* Gnip[2], and *Saulus* Vaitkus[2]

[1] Moscow state university of civil engineering, Yaroslavskoye shosse, 26, Moscow, Russia, 129337
[2]Scientific Institute of Thermal Insulation Vilnius Gediminas Technical University (VGTU), 10223 Saulėtekio al. 11, Vilnius, Lithuania

**Abstract.** The results obtained in long-term testing under constant compressive stress of the cut from the Slabs EPS 50/100 and EPS 150 with the density ranging from 15 to 24 kg/m$^3$, which were manufactured by the same manufacturer by foaming EPS solid granules (beads) in closed volume. The creep strain of the above described specimens was used as a criterion for estimating the deformability of the EPS slabs under long-term compressive stress. It was measured using special stands EN 1606, maintaining constant stress during the fixed time interval $t_n$=122 days. Creep strains were determined by the methods described in EN 1606 for constant stress $\sigma_c$=0.35$\sigma_{10\%}$ (compressive stress $\sigma_{10\%}$ was determined in accordance with EN 826:2013. The long-term compressive stress measurement error did not exceed 1 %, while the creep strain measurement error was not larger than 0,005 mm. The tests were conducted at the ambient temperature of (23±2)°C and relative humidity of (50±5) %.The long-term constant compressive load $\sigma_c$=0.35$\sigma_{10\%}$. The method of mathematical and statistical experimental design optimization models taking into account the thickness of specimens is proposed to determine the creep compliance $I_c$ ($t_n$) the creep strain $\varepsilon_c$ ($t_n$) and predictive point estimate of creep strain $\varepsilon_c$ ($T$). Graphical interpretation of the abstained models is also presented. It should be noted that the abstained equations may be used in practice for estimating the creep strains at time $t_n$=122 days and predictive estimates of $\varepsilon_c$ ($T$) for the load time of 10 years.

## 1 Introduction

The criterion for the effectiveness of any insulation, in addition, the thermal characteristics of the material are its resistance operational and manufacturing costs. Foamed plastics are used in transport systems as an insulating bases pavement [1, 2]. Therefore, it is important to assess their strength properties and creep. Foamed plastics are used in transport systems as an insulating bases pavement. Therefore, it is important to assess their strength properties and creep [3, 4].

According to EN 1606:1996/A1:2006 (GOST EN 1606-2011), EPS creep behaviour should be studied by considering square specimens cut from EPS slabs with the side of 50,

* Corresponding author: science-isa@yandex.ru

100, 150, 200, 300 mm and thickness equal to slab thickness. The creep level of expanded polystyrene is declared now based on the investigation of the specimens (50×50×50) mm (EN 13163:2013. (GOST R 56148-2014)).

The results obtained in standard tests often do not correlate with building material behaviour in real conditions. The investigations described in the paper presents were aimed at optimizing the thickness of EPS specimens of various densities for determining their creep strains under constant compressive stress [5, 6].

In this paper presents, the results obtained in long-term testing under constant compressive stress of the specimens cut from the slabs EPS 80, EPS 100 and EPS 150 with the density ranging from 15 to 24 kg/m$^3$, which were manufactured by the same manufacturer by foaming EPS solid granules (beads) in a closed volume. Thus, the slabs EPS 150 were made of the raw material (EPS granules) F 295 available from BASF (Ø 1,0-2,3 mm), the slabs EPS 100 were made of the granules F 215N available from BASF (Ø 1,0-2,0 mm) and the slabs EPS 80 were made of the granules NF-514 available from STYROCHEM (Ø 1,4-2,5 mm).

## 2 Experimental

The research methods and statistical design of experiments. Square specimens with the side of $b$=100 mm and thickness $d$, equal to the thickness of the tested EPS slabs (50, 75 and 100 mm) (accepted in accordance with EN 1606), have been tested [7, 8].

The creep strain of the above described specimens was used as a criterion for estimating the deformability of the EPS slabs under long-term compressive stress. It was measured using special stands EN 1606, maintaining constant stress during the fixed time interval $t_n$=122 days. Creep strains were determined by the methods described in EN 1606 for constant stress $\sigma_c$[1]$=0.35\sigma_{10\%}$ (compressive stress $\sigma_{10\%}$ was determined in accordance with EN 826:2013 (GOST EN 826-2011)). The long-term compressive stress measurement error did not exceed 1%, while the creep strain measurement error was not larger than 0,005 mm. The tests were conducted at the ambient temperature of (23±2)°C and relative humidity of (50±5) %.

Based on the data obtained in testing EPS specimens under short-term compression, the regression line for compressive stress $\bar{\sigma}_{10\%}$ is given (Fig. 1), depending on the results obtained in determining their density EN 13163.

Statistical processing of the experimental data for specimens under long-term compressive stress was made for the experiment of $t_n$=122 days. Relative creep strains $\varepsilon_c(t)$, %, were determined by the formulas given in EN 1606 and approximated by the Findley equation [9, 10]:

$$\bar{\varepsilon}_c(t) = b_0 t^{b_1}, \tag{1}$$

where $\bar{\varepsilon}_c(t)$ is the mean creep strain value at time $t$, %; $b_0$, $b_1$ are the constants, depending on the material properties; $t$ is time, h.

Creep strains may be quantitatively measured by creep compliance, which is only the time function:

$$I_c(t) = \frac{\varepsilon_c(t)}{\sigma_c}, \tag{2}$$

where $I_c(t)$ is a relative unit creep strain at the fixed time $t$, caused by unit stress applied at time $t \leq 15$ s; $\sigma_c$ is constant compressive stress at time t; t is readout time.

[1] According to EN 1606, the levels of $\sigma_{10\%}$ corresponding to 0.15; 0.20; 0.25; 0.30 and 0.35 are recommended for stress $\sigma_c$.

Thus, creep strains may be obtained by multiplying the creep compliance caused by the unit stress by a certain stress $\sigma_c$:

$$\varepsilon_c(t) = \sigma_c \cdot F\left[I_c(t)\right], \quad (3)$$

where $F[I_c(t)]$ represents some time function, expressing the increase in the creep strain caused by the application of unit stress equal, for example, to 1 MPa.

Then,

$$I_c(t) = \frac{\varepsilon_c(t)}{\sigma_c} = F\left[I_c(t)\right]. \quad (4)$$

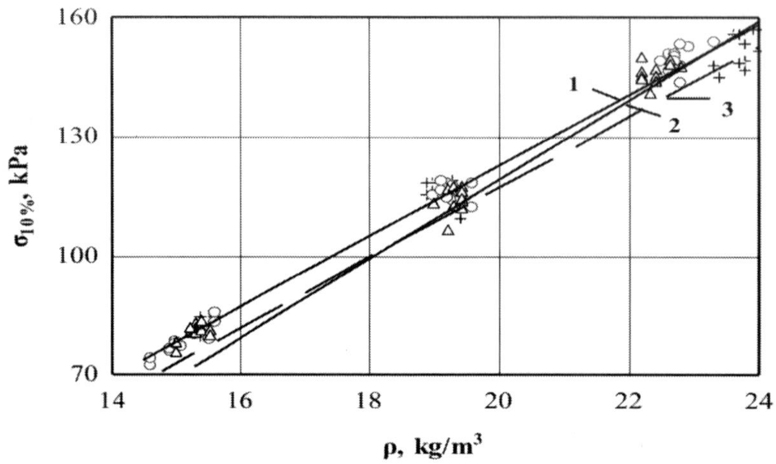


**Fig. 1.** The empirical regression lines for EPS specimens under short-term compressive stress $\bar{\sigma}_{10\%}$ based on the results obtained in determining their density, ρ; 1, 2 are the regression lines for the mean ($\sigma_{10\%}$) value plotted in accordance with EN 13163, respectively; 3 same for the lower (minimum) bound of stresses $\sigma_{10\%}$, according to the experimental data. Points are the experimental data for slab specimens of the thickness, mm: + means 50; ○ means 75; Δ means 100.

In processing the experimental data, three output parameters ($I_c(t_n)$, $\varepsilon_c(t_n)$, $\varepsilon_c(T)$) and two controlled factors (expanded polystyrene density ρ, and the specimen thickness $d$) were used. The slabs with the density ranging from 15 to 24 kg/m$^3$ were tested according to a full factorial experiment of the type $N=n^k$ (where $n$ is the number of levels and $k$ is the number of factors). The number of tests was $N=3^2=9$ (there were three specimens of the same density in a set to ensure the reliable results under compressive stress) were tested [11, 12].

In testing, the values of the factors were maintained at the fixed level according to the experiment design. For simplifying the record of the experimental conditions and experimental data processing, the factors were encoded by the formula:

$$x_j = \frac{\bar{x}_j - \bar{x}_{j(i)}}{I_j}, \quad (5)$$

where $x_j$ is the encoded value of a factor; $\bar{x}_j$ is the natural value of a factor; $\bar{x}_{j(i)}$ is the natural value of the basic level; $I_j$ is the variation interval; $j$ is the number of the factor.

The basic level and the variation interval of the controlled factors are given in Table 1, while the matrix of the experimental design is presented in Table 2 (the last 5 columns present the experimental results).

The following characteristics of expanded polystyrene were determined under the conditions of constant (long-term) compression of the specimens, which were cut from EPS slabs and selected according to the experiment design matrix: $y_1$ is creep compliance $I_c$,

MPa$^{-1}$, at time $t_n$=122 days; $y_2$ is creep strain $\varepsilon_c$, %, developed under constant compressive stress $\sigma_c$=0.35$\sigma_{10\%}$ during the experiment, $t_n$=122 days; $y_3$ is the predictive point estimate of creep strain, $\varepsilon_c(T)$, %, for the lead time of 10 years.

For describing the relationship between the factors $x_i$ and the output $y_i$ in the investigated domain, a quadratic model was used.

**Table 1.** Basic characteristics of experimental design

| Characteristic | Density of EPS specimens, $x_1$, kg/m$^3$ | Thickness of EPS specimens, $x_2$, mm |
|---|---|---|
| Basic level | 19.3 | 75 |
| Interval of variation | 4.0 | 25 |
| Upper level | 23.3 | 100 |
| Lower level | 15.3 | 50 |

**Table 2.** Matrix of full factorial design of the type 3$^2$ and the experimental results

| Number of experiments | $x_1$ | $x_2$ | $y_1$, MPa$^{-1}$ | | $y_2$, % | | $y_3$, % |
|---|---|---|---|---|---|---|---|
| | | | $\bar{y}_1^{э}$ *) | $S_{y_1}$ | $\bar{y}_2^{э}$ *) | $S_{y_2}$ | $\bar{y}_3^{pred.}$ |
| 1 | -1 | -1 | 0.3564 | 0.0077 | 1.0239 | 0.0220 | 2.163 |
| 2 | 0 | -1 | 0.2019 | 0.0153 | 0.8206 | 0.0622 | 2.203 |
| 3 | +1 | -1 | 0.0842 | 0.0117 | 0.4426 | 0.0618 | 1.074 |
| 4 | -1 | 0 | 0.3117 | 0.0248 | 0.8652 | 0.0772 | 2.420 |
| 5 | 0 | 0 | 0.1792 | 0.0053 | 0.7263 | 0.0214 | 2.055 |
| 6 | +1 | 0 | 0.0760 | 0.0072 | 0.3980 | 0.0374 | 0.927 |
| 7 | -1 | +1 | 0.3138 | 0.0180 | 0.8920 | 0.0524 | 2.065 |
| 8 | 0 | +1 | 0.1847 | 0.0095 | 0.7421 | 0.0324 | 2.262 |
| 9 | +1 | +1 | 0.0719 | 0.0019 | 0.3689 | 0.0102 | 0.873 |

*) Mean values of output parameters $y_1$, $y_2$ for three parallel determinations.

Output settings:

$y_1$: $\sum_{i=1}^{N} S_{y_1}^2 = 0.001543$, $G = \frac{S_{y_1(\max)}^2}{\sum_{i=1}^{N} S_{y_1}^2} = 0.399$, $S_{reprod.}^2 = \frac{S_{y_1}^2}{N} = 0.000171$, $S_{bj} = \sqrt{\frac{S_{reprod.}^2}{Nm}} = 0.00252$.

$y_2$: $\sum_{i=1}^{N} S_{y_2}^2 = 0.020314$, $G = 0.293$, $S_{reprod.}^2 = 0.00226$, $S_{bj} = 0.00914$.

$y_3$: $S_{reprod.}^2 = 0.00226$, $S_{bj} = 0.00914$.

The table values: $G_{1-\alpha(2;9)}$=0.478, $t_{1-\alpha(18)}$=2.10 (if $\alpha$=0.05), $G$ – Cochran's test, $t$ – Student's t test.

In general, a quadratic model may be expressed by the equation:

$$y = b_0 + b_1x_1 + b_2x_2 + b_{12}x_1x_2 + b_{11}x_1^2 + b_{22}x_2^2 \quad (6)$$

The significance of constant coefficients was validated by the Student's $t$-statistic at the significance level $\alpha$=5%. Non-significant coefficients were not included in the equation. The adequacy of Eq. (6) was tested by Fisher's criterion [13, 14].

## 3 Results

In Fig. 1, the experimental values and the regression lines of the compressive stress $\sigma_{10\%}$ based on the results obtained in determining density $\rho$ of the specimens, as well as the regression dependence from EN 13163 (Fig. 1, line 2), are presented. In accordance with EN 13163, a linear dependence, approximated based on the experimental data by the regression equation (line 1), was used: $\bar{\sigma}_{10\%} = 8.9\rho - 55.5$, with the standard deviation $S_r$=3.64 kPa ($n$=107).

The minimum predictive estimates, $\bar{\sigma}^{\min}_{10\%}$, were approximated by the equation (Fig. 1, line 3): $\bar{\sigma}^{\min}_{10\%} = 8.9 - 59.7$.

**Table 3.** Experimental test data and results of the regression analysis of creep deformation of polystyrene specimens under constant compressive stress $\sigma_c$=0.35$\sigma_{10\%}$

| Number of test series | The test data for specimens under compressive stress | | | | | | | Regression analysis data [a] | | | The predicted point values of creep strain $\bar{\varepsilon}_c(T=10 years)^e$, % |
|---|---|---|---|---|---|---|---|---|---|---|---|
| | Shor-term[b] | | Long-term[c] | | | | | Constant coefficients in Eq. (5) | | $\sum_{i=1}^{n}(y_{3(i)}-\bar{y}_{3(i)})^2$, %[d] | |
| | ρ, kg/m³ | $\sigma_{10\%}$, kPa | ρ, kg/m³ | $d_s$, mm | $\sigma_c$, kPa | $\varepsilon_0$, % | $\varepsilon_c(t_n)$, % | $b_0$ | $b_1$ | | |
| 1 | 15.4 | 82.1 | 15.3 | 48.0 | 28.7 | 1.374 | 1.0239 | 0.1520 | 0.2333 | 0.01329 | 2.163 |
| 2 | 19.3 | 116 | 19.2 | 49.4 | 40.6 | 1.155 | 0.8206 | 0.0738 | 0.2984 | 0.00426 | 2.203 |
| 3 | 23.8 | 152 | 23.6 | 48.6 | 52.6 | 1.277 | 0.4426 | 0.0333 | 0.3222 | 0.00071 | 1.074 |
| 4 | 15.1 | 78.5 | 15.3 | 72.8 | 27.7 | 1.145 | 0.8652 | 0.0989 | 0.2809 | 0.01100 | 2.420 |
| 5 | 19.3 | 116 | 19.3 | 74.4 | 40.5 | 0.951 | 0.7263 | 0.0582 | 0.3132 | 0.00253 | 2.055 |
| 6 | 23.0 | 150 | 23.0 | 73.3 | 52.4 | 1,005 | 0.3980 | 0.0533 | 0.2511 | 0.00022 | 0.927 |
| 7 | 15.3 | 81.1 | 15.3 | 98.2 | 28.4 | 0.985 | 0.8920 | 0.0814 | 0.2841 | 0.00833 | 2.065 |
| 8 | 19.3 | 115 | 19.3 | 99.5 | 40.2 | 0.767 | 0.7421 | 0.0490 | 0.3367 | 0.00496 | 2.262 |
| 9 | 23.0 | 147 | 23.0 | 98.0 | 51.3 | 0.932 | 0.3689 | 0.0427 | 0.2652 | 0.00137 | 0.873 |

[a] Linear regression analysis was performed starting with the values of creep strain $\varepsilon_c(t=24\ h)$; [b] The results of testing 12 specimens for selecting long-term compressive stress parameters; [c] ρ, $d_s$, $\sigma_c$, $\varepsilon_0$, $\varepsilon_c(t_n)$ are the average values for 3 specimens; [d] The sums of standard deviations, $y_{3(i)}$, from their estimated values, $\bar{y}_{3(i)}$, for the number of retrospective observations n=18; [e] Calculated by time extrapolation EN 1606.

Thus, and the data obtained in creep tests may be used for describing the slabs of the types EPS 80[2], EPS 100, EPS 120 and EPS 150.

In Table 3, the experimental data and the regression equation of the relative creep strains of expanded polystyrene specimens, calculated by Eq. (5) based on the experiments with the duration $t_n$=122 days, are given. The data obtained in their mathematical-statistical analysis are given in Table 4.

The coefficients of Eq. (6), calculated based on the data (encoded variables) found in Tables 2 are presented in Table 4.

In general, we can see from the coefficient values in the equations that not all the factors equally affect the considered output parameters (Table 4). The regression coefficient is a quantitative measure of the effect exerted by a particular factor on an output parameter, when going from a zero level to the lower or higher levels. The larger the coefficient, the stronger the factor. The signs of the coefficients indicate the strength of the factor's influence.

**Table 4.** Coefficient values of regression equations based on coded variables

| Coefficient | Coefficient |
|---|---|

[2] According to EN 13163, these are the EPS slabs with the compression stress at 10% deformation $\sigma_{10\%}$ equal to 80, 100, 120 and 150 kPa, respectively.

| value for equations | $b_0$ | $b_1$ | $b_2$ | $b_{12}$ | $b_{11}$ | $b_{22}$ |
|---|---|---|---|---|---|---|
| $y_1$ | 0.1798 | -0.1250 | -0.0120 | 0.00758 | 0.0137 | 0.0132 |
| $y_2$ | 0.7284 | -0.2619 | -0.0474 | - | -0.0979 | 0.0518 |
| $y_3$ | 2.1916 | -0.6290 | -0.0400 | -0.0257 | -0.5863 | -0.0273 |

## 4 Discussion

Considering the equation for the parameter $y_1$, we have found that the extremum point (which is a 'nearly stationary area') is in the interval of (79÷94) mm, corresponding to the variation of $x_1$ from 23.3 to 15.3 kg/m$^3$. The investigation results for $y_2$ show that the extremum point, $y_2$ , at which the parameter $y_2$ is minimum with respect to the specimens, is the point $x_2$=86 mm. Considering the equation for the parameter $y_3$, we have found that, when $x_1$ varies from 15.3 to 23.3 kg/m$^3$, the extremum point $x_2$ ('almost stationary area') is in the interval of (68÷45) mm.

When we considered natural (real) variables instead of the coded variables and averaged the regression coefficients, the equations for the output parameters took the form of with the standard deviation $S_r$=0.0052 MPa$^{-1}$ ($N{\cdot}m$=27 test results).

$$I_c(t_n=122)=1.3664-0.0700\cdot\rho-0.00511\cdot d+0.0000758\cdot\rho\cdot d+0.000856\cdot\rho^2+0.0000211\cdot d^2, \quad (7)$$

$$\varepsilon_c\left(t_n=122\right)=0.3213+0.1707\cdot\rho-0.0143\cdot d-0.00612\cdot\rho^2+0.000083\cdot d^2, \quad (8)$$

[$S_r$=0.0212 %, $N{\cdot}m$=27].

$$\varepsilon_c\left(T=10\ years\right)=-8.9209+1.2765\cdot\rho+0.0099\cdot d- \\ -0.0002575\cdot\rho\cdot d-0.03664\cdot\rho^2-0.0000437\cdot d^2, \quad (9)$$

[$S_r$=0.180 %, $N{\cdot}m$=9].

When the value of the factor $x_1$ (EPS density) is fixed at three levels, the figures present the relationships between the output parameters $I_c$ and $\varepsilon_c$ and the factor $x_2$ (specimen thickness), proving the existence of a particular optimum, i.e. 'nearly stationary area', corresponding to this factor (dotted lines in Figures 2a and 3a).

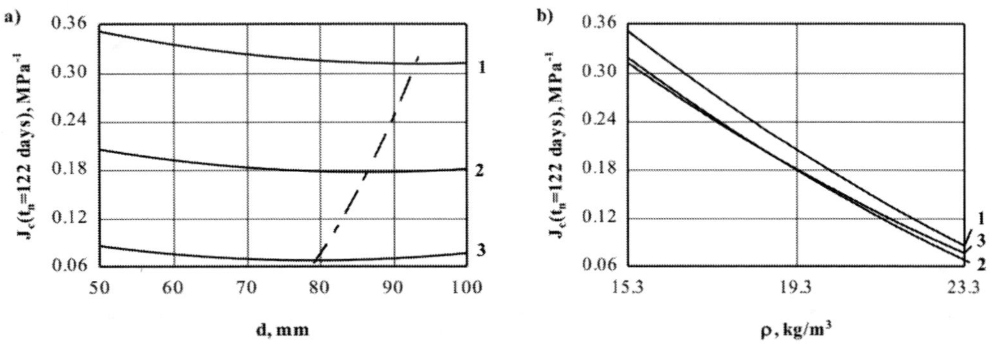


**Fig. 2.** Graphs of Eq. (7): a) when ρ=15.3 (1); 19.3 (2); 23.3 kg/m$^3$ (3); b) when $d$=50 (1); 75 (2); 100 mm (3). A dotted line represents a 'nearly stationary area'.

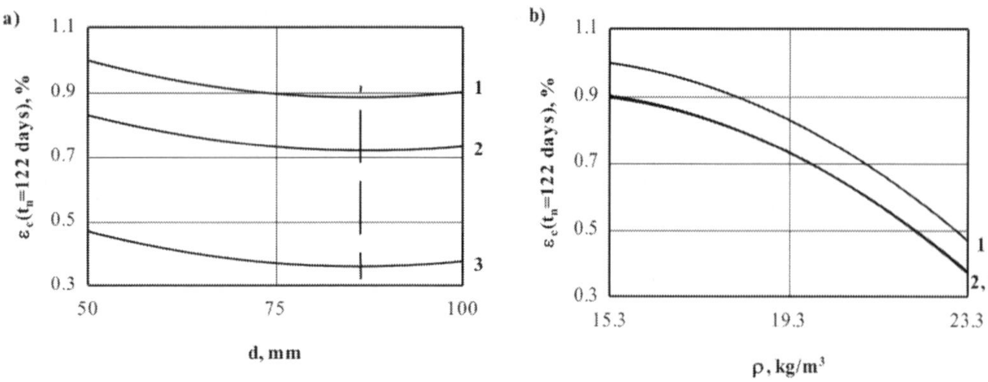


**Fig. 3.** Graphs of Eq. (8). The notation is the same as given in Fig. 2.

Considering the response generators of the output parameters $I_c$ and $\varepsilon_c$, with the value of factor $x_2$ (specimen thickness) fixed at three levels, we presented (Figures 2b and 3b) the dependences of the output parameters on factor $x_1$ (specimen density). When the density of the EPS specimens is in the range of 15.3–23.3 kg/m$^3$, creep compliance, ($I_c$), values determined for 50 mm-thick specimens is 15% higher than that for 75 and 100 mm-thick specimens (when $d$ is equal to 75 and 100 mm, the values of $I_c(t_n=122$ days) are actually equal).

The value of the creep strain $\varepsilon_c(t_n=122$ days) in EPS density range of 15.3–23.3 kg/m$^3$, when specimen thickness varies from 50 to (75 and 100) mm, decreases by about 11 and 26% (rel.) [3] with respect to its density (Fig. 3b). The creep strains $\varepsilon_c(t_n=122$ days) of 75 and 100 mm-thick specimens are actually equal. It is worth noting that, in the considered EPS density range, the absolute values of $\varepsilon_c(t_n=122$ days) obtained for 50 mm-thick specimens and 75 and 100 mm-thick specimens differ by 0,1% [15].

In Fig. 4*b*, the dependence of the output parameter of the creep strains $\varepsilon_c(T=10$ years) on the factor $x_1$ (EPS density) are presented for the factor $x_2$ fixed at three levels. When EPS density is 15.3–23.3 kg/m$^3$, and the specimens are 50 and 75 mm-thick, the predictive point estimates of $\varepsilon_c(T=10$ years), calculated according to their density, are by (2–13)% (rel.) larger than those found for 100 mm-thick specimens.

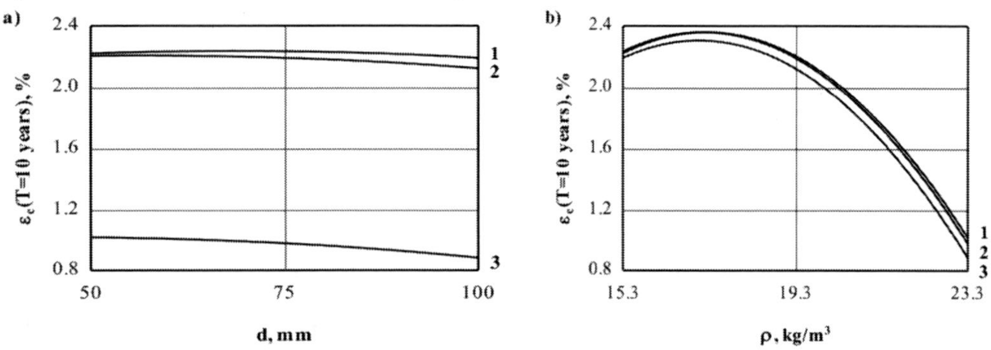


**Fig. 4.** Graphs of Eq. (9). The notation is the same as given in Fig. 2.

It should be noted that the obtained equations (3-5) may be used in practice for estimating the creep strains at time $t=122$ days and the predictive estimates of $\varepsilon_c(T=10$ years) for the lead time of 10 years.

[3] The abbreviation (rel.) indicates a relative difference in values, when creep strain $\varepsilon_c$ is measured in percent.

## 5 Conclusions

In conclusion, it is necessary to mention that the mathematical models for determining $I_c(t_n$=122 days), $\varepsilon_c(t_n$=122 days) and $\varepsilon_c(T$=10 years) of expanded polystyrene specimens under constant compressive stress $\sigma_c=0.35\sigma_{10\%}$ have been constructed, using a method of experimental design.

The empirical equations (3–5) for determining the predictive estimates of creep compliance $I_c(t$=122 days), creep strains $\varepsilon_c(t$=122 days) and $\varepsilon_c(T$=10 years) for the lead time of 10 years have been obtained for expanded polystyrene slabs with the density of 15.3–23.3 kg/m$^3$.

In EPS density range of 15.3–23.3 kg/m$^3$ and under the variation of specimen thickness from 50 to (75 and 100) mm, the experimental value of the creep strain $\varepsilon_c(t$=122 days) decreases by about 11 and 26 % (rel.) with respect to its density. The influence of specimen thickness on the predictive estimates of the creep strains for the lead time of 10 years $\varepsilon_c(T$=10 years) is insignificant. In EPS density range of 15.3–23.3 kg/m$^3$, the values of $\varepsilon_c(T$=10 years), calculated from the experimental data obtained for specimens with the thickness of 50 and 75 mm, are by 2–13 % (rel.) larger compared to 100 mm thick specimens.

The work carried out in accordance with the local project "Conducted in the field of energy saving and energy efficiency of buildings and structures" in accordance with the plan of research works of MSUCE. Work was carried out on the basis of the initiative.

## References

1. U. Knaack, Sh. Chung-Klatte, R. Hasselbach, *Prefabricated systems: Principles of construction* (De Gruyter, 2012)
2. I.J. Gnip, V.J. Keršulis, S.J. Vaitkus, Mechanics of Composite materials, **41(4)**, 357 (2005)
3. A.D. Zhukov, N.V. Naumova, R.M. Mustafayev, Industrial and civil building, **7**, 39 (2014)
4. I. Gimenez, M.K.Faroog, A. El Mahi, A. Kondrotas, M. Assarar, Materials Science (Med iagotyra), **10 (1)**, 34 (2004)
5. V.I. Telichenko, D.V. Oreshkin, Ecology of urbanized areas, **2**, 31 (2015)
6. B.M. Rumiantcev, A.D. Zhukov, D.B. Zelenshikov, A.S. Chkunin, K.K. Ivanov, Yu.V. Sazonova, MATEC Web of Conferences, **86** (2016)
7. A.D. Zhukov, E.Yu. Bobrova, D.B. Zelenshchikov, R.M., A.O. Khimich, Advanced Materials, Structures and Mechanical Engineering, **1025-1026**, 1031 (2014)
8. J.S. Horvath, *Manhattan College Research report,* 39 (No CE/GE-98-3, New-York, USA, 1998)
9. B.M. Rumiantcev, A.D. Zhukov, E.Yu. Bobrova, I.P. Romanova, D.B. Zelenshikov, T.V. Smirnova, MATEC Web of Conferences, **86** (2016)
10. A.D Zhukov, I.V. Bessonov, A.N. Sapelin, N.V. Naumova, A.S. Chkunin, Italian Science Review, **2**, 155 (2014)
11. P.R. Head, *Proceedings of the ICE – Civil Engineering*, **144(3)**, 113 (2001)
12. A.D .Zhukov, A.B. Chugunkov, Vestnik MGSU, **1-2**, 273 (2011)
13. I.J. Gnip, V.J. Keršulis, Construction Materials, **4**, 22. (2003)
14. I.J. Gnip, V.J. Keršulis, S.J. Vaitkus, Mechanics of Composite materials, **41(5)**, 407 (2005)
15. I.J .Gnip, S.Vaitkus, V. Keršulis, S. Vejelis, Polymer Testing, **29**, 693 (2010)

# Calculation of the hydraulic characteristics of the protective coating used in trenchless technologies for the construction and renovation of pipelines to extend their service life

*Vladimir* Orlov[1,*], *Sergey* Zotkin[1], *Irina* Dezhina[1] and *Irina* Zotkina[1]

[1] Moscow state university of civil engineering, Yaroslavskoye shosse, 26, Moscow, Russia, 129337

**Abstract.** The lifetime of the pipeline, which undergoes trenchless repair by pulling and then fixing various types of internal protective coatings inside, is determined by the strength characteristics of the two-layer structure "old pipeline + internal protective coating". Hydraulic parameters of internal protective coatings, such as surface roughness and degree of hydrophobicity (water repellency), can play a role in prolonging the life of the repaired pipeline. With a high degree of hydrophobicity of the protective coating, the contact of the transported liquid with the walls of the pipeline during its operation in the non-pressure regime decreases, which leads to a decrease in the negative impact on the walls, in particular, their attrition, interaction with the specific structure of the coating material, which in turn reduces the energy costs when transporting water. Given article deals with a mathematical description of the interaction of the surface structure (with its inherent wetting ability) and the transported liquid in a mini-stream created on an inclined open chute imitating the inner surface of a pressureless pipeline. The theoretical positions and methodology for calculating the degree of hydrophobicity are based on research conducted on the basis of frontal and coaxial (along the flow) digital photo and video. T

## 1. Introduction

The wide application of new polymer repair materials in the form of hoses applied on the inner surface of pipes with bearing capacity has opened a new page in the practice of efficient and operative renovation of water supply and sewage pipelines [1].

The modern construction market is replete with materials that can potentially be reliable internal protective coatings for the pipeline transport of liquids and gases. Polymeric materials have not only high strength characteristics, but also allow to achieve the energy saving effect in the transportation of water due to the corresponding hydraulic characteristics of the materials used with low coefficients of hydraulic friction [2]. In

* Corresponding author: <orlov950@yandex.ru>

addition, polymeric materials are not subject to gas, chemical and biocorrosion, such as steel, cast iron and reinforced concrete, which is another additional advantage over other materials used for the production of pipes [3].

In recent years, there has been some interest in the study of such materials, especially in terms of studying their hydrophobicity and its relationship to the service life of pipelines used in the field of water supply, wastewater and gas supply [4, 5]. In the construction sector, the use of unique nanomaterials is also actual, which can be used as internal coating for pipeline networks.

It should be noted that the capabilities of modern polymer and nanomaterials used as protective coatings of pipelines are unlimited thanks to their unique properties. The problem of studying the smoothest surfaces has been addressed by modern technology for a long time. Obviously, the smaller the roughness, the smoother the surface is and the less friction and wear of the surface are caused by the impact of the transported environment (natural water, sewage, etc.), as well as the higher energy efficiency of the pipelines in operation is provided. During the choosing repair materials, first of all it's necessary to rely on established criteria. Unevenness for which the ratio of the height of the protrusions $S_c$ to the diameter of the pipeline $D_c$ is $S_c$ / $D_c$ <40, is generally estimated as roughness [6]. As a rule, the height of unevennesses is measured by optical instruments (light and shadow cross-sections), by microinterference methods, using rasters (special grids with a small step and a scanning microscope).

Mechanical methods for determining roughness are widely used, for example, probe electromechanical devices like profilometers are used, and if necessary, the character of the irregularity along the length is recorded with profilographs. When choosing a construction material to use, in particular, from nanomaterials, it is necessary to take into account their toxicity, since it has to do with the pipelines of transport for both sewage and natural waters [7].

Investigations of the structure of the fluid near hydrophobic surfaces and, in particular, in the wall zone [8] have special interest.

The analysis of research in the field of static and dynamic wettability of various surfaces of building materials used in trenchless laying and renovation of pipeline networks contributed to focus on finding an effective protective coating of pipes with the identification of the relationship between the degree of hydrophobicity of the coatings and their hydraulic parameters, in particular, with roughness [9, 10 ]. In connection with this, a compact test bench (open tray) was created and patented in the Moscow State University of Civil Engineering in order to determine the degree of hydrophobicity of materials for manufacturing pipes and repairing pipelines [11]. As a mechanism for calculating geometric and hydraulic parameters functioning according to the corresponding algorithm, a computer program was developed and patented [12].

The geometric and hydraulic characteristics of the mini-stream formed as a result of the flow of a certain mass of liquid through the open tray at various values of its slope, created by a special mechanical jack and controlled by a system of communicating tubes (Fig. 1) were studied at the stand.

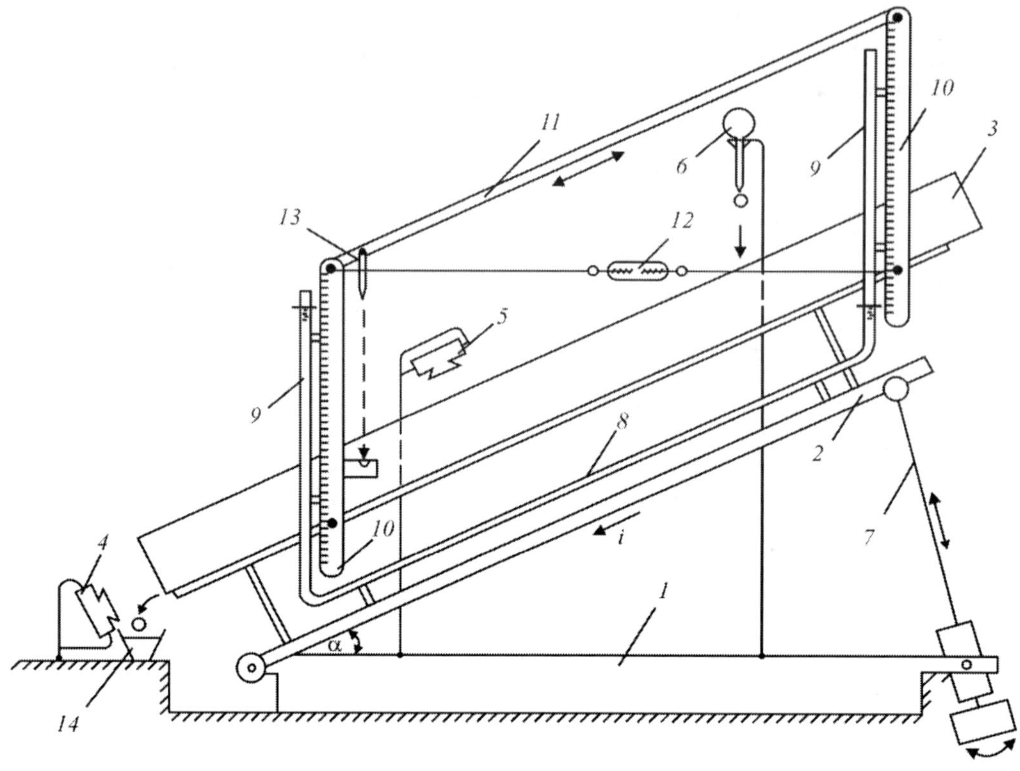


**Fig. 1.** Sketch diagram of the device of the test bench to determine the degree of hydrophobicity of materials

1- metal support frame; 2-ramp with a variable gradient; 3- open trench located on the trestle (tray) of the investigated pipe (about 1 m long); 4- the camera rigidly connected to the frame for frontal shooting; 5- the camera rigidly connected to the frame for co-axial shooting (fixing the length of the moving critical mass); 6 - a dispenser of liquid in the form of a pipette, 7 - a rod of a mechanical jack lifting an overpass; 8- transparent polyethylene line of the system of communicating vessels; 9 - transparent polyethylene tubes of the system of communicating vessels; 10- movable measuring rulers for fixing the water level in two tubes; 11- control bar for the position of the tubes and measuring rulers when the gradient changes; 12- screw cable ties; 13 - laser plummet; 14 - collection of drops (liquid flow); i - slope of the surface; α- the angle of the tray

During the experiments on various types of protective coatings used in trenchless construction and repair of pipelines, the parameters of the mini-stream were recorded using photo and film techniques. When conducting the experiments, a multi-frame serial shooting was carried out with the Sony SLR 550 (DT 1.8 / 50 SAM lens in conjunction with the Kenko Extension Tube system) and the Sony HDR-CX250 digital video camera.

The method of operation in the mentioned above installation is to compare the behavior of a mini-stream of a liquid (drinking water) on different working planes with different slopes of the sliding surface and to fix with measuring rulers and a stopwatch both the way and the rates of flow of the mini-stream through the tray. On the basis of the results of the experiment, visualized multimedia results are obtained in the form of digital photographs of the frontal and coaxial surfaces of the mini-stream. For the subsequent processing and interpretation of the data, the relationship between the volume and geometric parameters of the flow with the structure of a given surface is estimated, and, as a consequence, this relationship is extended to the parameters of the wetting regime in the form of a corner angle, referring the coating to hydrophilic or hydrophobic type, respectively [13].

Fig. 2 demonstrates an axonometric diagram of a mini-stream formed as a result of the flow of a critical volume of liquid on an inclined surface in the form of a tray. The flow is formed as a result of dosing a certain number of drops from the pipette onto the tray, which, when collected under the appropriate conditions (slope of the tray, the nature of the slip surface, etc.), form a critical mass starting to move along the inclined surface of the tray.

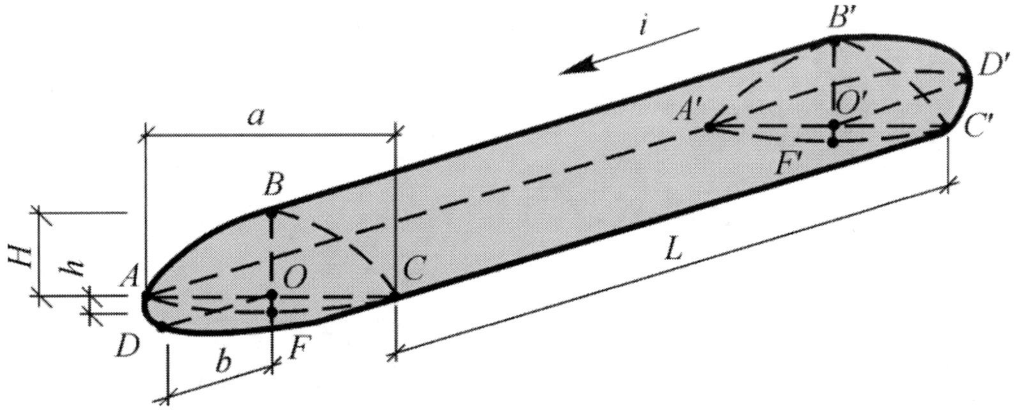


**Fig. 2.** Axonometric scheme of a mini-stream with geometric dimensions
*a, L*-respectively, are the average values of the width and length of the compact part of the mini-stream; *H, h-* - respectively the height of the top and bottom lenses; *b-* the height of the half of the ball segment of the head part of the mini-stream; AFC, AA`C`C - respectively wetted perimeter and wetted surface of the compact part of the mini-stream

Considering the cumbersome calculation of all intermediate geometric and hydraulic parameters, including the degree of hydrophobicity, an automated program is used, which dialog box is shown in Fig. 3.

According to the calculation algorithm, the parameters of fixation and subsequent automated processing are the following:

-The average length of the compact part of the mini-stream *L, mm*;
-The average width of the mini-stream *a, mm*;
- Top lens height (distance from the horizontal surface to the top of the lens) *H, mm*;
- Bottom lens height (distance from the tray to the horizontal surface) *h, mm*
- Number of drops forming a critical mass (mini-stream) *N, pcs.*
-Height of the head ball segment of the mini-stream $H_g$, *mm*
-The height of the tail ball segment of the mini-stream $H_k$, *mm*
- The path traversed by the head of the mini-stream *U, mm*
- Travel time *T, s*

The most significant practical effect obtained from the calculation results is the determination of a number of parameters, in particular the relative degree of hydrophobicity of the protective coating, the coefficients of Manning's and Chezy's relative roughness for the type of protective coating in question [14]. The presence of an automated program for calculating the above-described coefficients substantially accelerates and increases the accuracy of obtaining the required quantities.

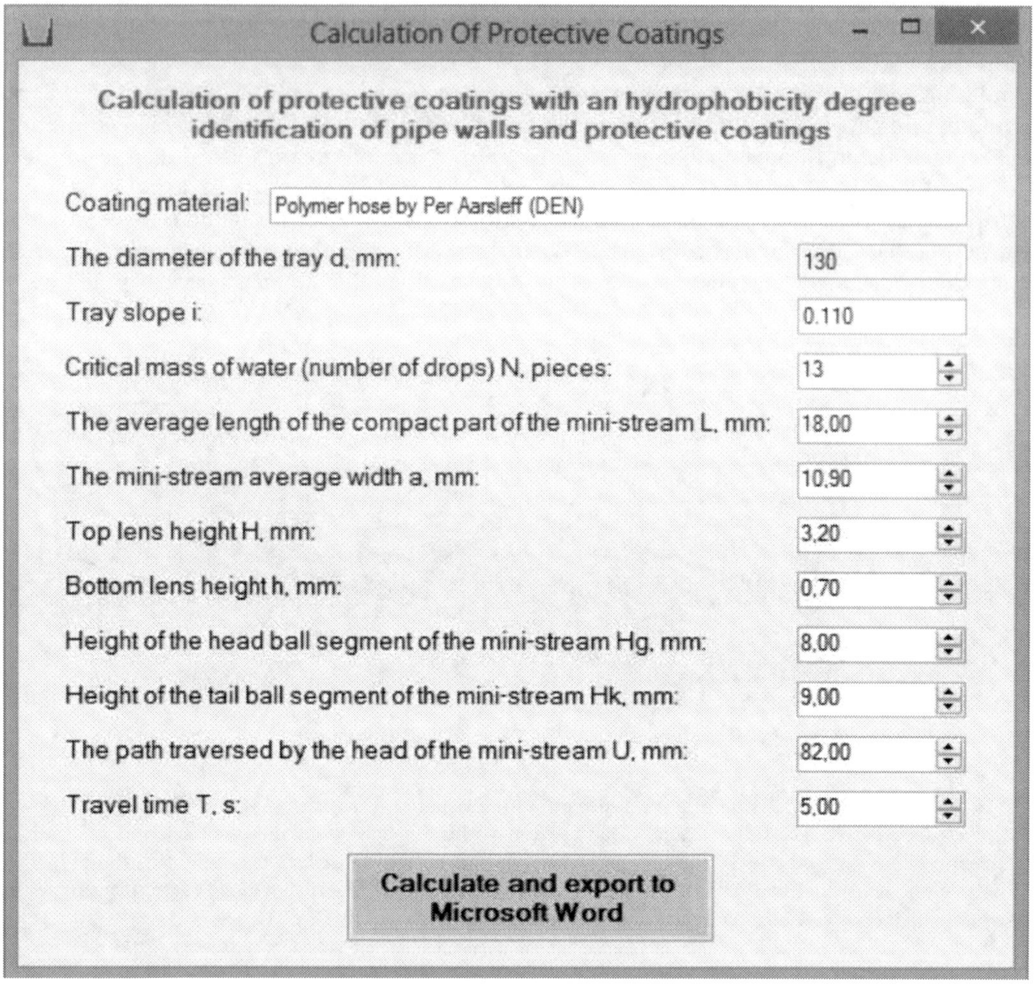


**Fig. 3.** The working interface of the automated program for the calculation of protective coatings with an hydrophobicity degree identification of pipe walls and protective coatings

Below there is a shortened final calculation of the geometric, hydraulic and hydrophobic characteristics of protective coating material made by Per Aarsleff (Denmark) as a result of the automated complex work. This type of protective coating is chosen as a model, which has found wide application in the field of renovation of water supply and drainage pipelines in many countries of the world [15].

Calculated data of output information after input according to Fig. 3:

1. The total volume of the mini-flow$W_1$, ml:

$$W_1 = 0{,}0568N + 0{,}4943 = 1.232$$

2. The mini-stream total volume $W$, mm$^3$:

$$W = 10^3 W_1 = 1232.700$$

3. Top lens area $S_v$, mm$^2$:

$$S_v = 2aH/3 = 23.253$$

4. Bottom lens area $S_n$, mm$^2$:

$$S_n = 2ah/3 = 5.086$$

5. Small chord length m, mm:

$$m = [(a/2)^2 + h^2]^{1/2} = 5.494$$

6. Wetted perimeter p, mm:

$$p = 2m + (2m - a)/3 = 11.019$$

7. Total area of the top and bottom lenses $S_s$, mm$^2$:

$$S_s = S_n + S_v = 28.340$$

8. Theoretical volume of the compact part of the mini-stream $W_t$, mm$^3$:

$$W_t = S_s L = 510.120$$

9. The wetted surface of the compact part of the mini-stream $S$, mm$^2$:

$$S = pL = 198.349$$

10. Relative index of hydrophobicity $K_o$:

$$K_o = S_v/(Si) = 1.0658$$

11. Edge angle (angle of wetting) α, deg.:

$$\alpha = 49.6$$

12. The estimated volume of the mini-stream $W_r$, mm$^3$:

$$W_r = W - (1/12)[H_g(H_g^2 + 0.75a^2) + H_k(H_k^2 + 0.75a^2)] = 510.120$$

13. Estimated length of the mini-stream compact part $L_r$, мм:

$$L_r = W_r / S_s = 18.051$$

14. The approximation between the projected length $L_r$ and the average length L of the mini-stream $P_L$, %:

$$P_L = |100 - L \cdot 100/L_r| = 0.287$$

15. Tray filling $(h/d)$:

$$(h/d) = h/d = 0.00538$$

16. Hydraulic radius $R_1$ (through filling), mm:

$$R_1 = 62.2(h/d)^{0.8833} = 0.616$$

17. Hydraulic radius $R_2$ (through the wetted perimeter and live section), mm:

$$R_2 = S_n / p = 0.461$$

18. The absolute approximation value $P_R$ between $R_1$ and $R_2$, %:

$$P_R = |100 - R_1 \cdot 100/R_2| = 33.488$$

19. The adjusted exponent "x" in the formula to determine the hydraulic radius:

$$R_2 = 62.2(h/d)^X; \quad x = 0.938$$

20. The general corrected formula for determining the hydraulic radius R and its value, mm:

$$R = 62.2(h/d)^X = 62.2(h/d)^{0.93859} = 0.461$$

21. Average mini-stream speed V, mm / s:

$$V = U/T = 16.40$$

22. The Chezie's Coefficient $C$:

$$C = V/(R \cdot i)^{1/2} = 72.779$$

23. Manning's relative roughness coefficient $n$:

$$n = (1/C)R^{1/6} = 0.0121$$

24. Hydraulic friction coefficient λ:

$$\lambda = 8g/C^2 = 0.0148$$

The results of the automated calculation are output directly in Microsoft Word, which must be preinstalled on the user's computer (the version does not matter). The technology developed by Microsoft OLE (Object Linking and Embedding) is applied [16].

Making a conclusion of conducted researches and obtained results it is possible to note that thanks to the use of the developed technique, including a mathematical model describing the interaction of the structure of the investigated surfaces with the fluid, the algorithm and the automated program for calculating geometric, hydraulic and hydrophobic parameters, the researcher can get an overall picture of such indicators for any type of potential protective coating of pipelines subjected to trenchless repair or reconstruction.

## Conclusions

1. On the basis of experimental studies, an automated calculation of the hydraulic parameters of one of the types of internal protective coatings used in technologies of trenchless construction and renovation of pressure and non-pressure pipeline systems was carried out.

2. According to an automated calculation, the Manning's roughness coefficient "n" for the protective coating made by Per Aarsleff enterprise was equal to n = 0.0121, the hydraulic friction coefficient $\lambda$ = 0.0148, and the relative coefficient of hydrophobicity Ko=1.0658. This coating should be attributed to the type of hydrophilic (the contact angle at the front section of the mini-stream was equal to 49.6 degrees).

3. The relatively low value of the degree of hydrophobicity (practically the lower limit in the range 1.06-2.0) of the studied material can indicate that, despite its wide use in practice of trenchless construction and reconstruction technologies, which should provide high strength characteristics of pipelines and guarantee the resource-saving effect, the material is surpassed by other types of protective coatings, which are hydrophobic type and provide a greater energy saving effect.

4. The proposed algorithm and automated program for calculating the hydraulic characteristics of protective coatings enable to make an operative evaluation of any types of materials that can be used in construction technologies.

## References

1. A Kuliczkowski, *Rury kanalizacyjne.* Wydawnictwo Politechniki Swietokrzyskiej, p. 507 p., (2004)
2. S.V. Khramenkov, O.G. Primin, V.A. Orlov, *Rekonstruktsiya truboprovodnyh system*, ASV., 214 p., (2008)
3. L. Zhang, P. De Schryver, B. De Gusseme *Chemical and biological technologies for hydrogen sulphide emission control in sewer systems*, Water Research. 42. p. 1-12, (2008)
4. P. Roach, N.J. Shikcliffe, V.I. Newton, *Progress in superhydrophobic surface development*, The Roual Society of Chemistry, **4**, 224-240 (2008)
5. T. Min, J. Kim., *Effect of hydrophobic surface on stability and transition*. Physics of fluids, **17**, 106-108 (2005)
6. V.I. Margolin, V.A. Zhabriaev, G.N. Lukyanov, V.A. Tupik, *Vvedenie v nanotekhnlogiyu, Lan'*, 457 p. (2012)
7. A.V. Kilesnitchenko, M.A. Timofeev, M.V. Protopopova, *Toksichnost' nanomaterialov*, Rossiyskie nanotekhnologii, 3, 3-4, 54-61 (2008)

8. C. Choi, U.Ulmanella, J. Kim and C. Kim, *Effect slip and friction reduction in nanograted superhydrophobic microchannels,* Physics of fluids, **18**, 087105 (2006)
9. S. Grossmann, D. Lohse, *Curvature effects on the velocity profile in turbulent pipe flow*, Eur. Phys. J. E., **40**, 16 (2017)
10. C. Dorrer, J. Ruhe, Langmuir J., *Condensation and Wetting Transitions on Microstructured Ultrahydrophobic Surfaces,* **23**, 3820-3824 (2007)
11. V.A. Orlov, I.S. Dezhina, E.V. Orlov, I.A. Averkeev, *Ispytatelniy stend po opredeleniyu stepeni hydrofobnosti materialov diya izgotovleniyq trub i remonta truboprovodov,* Patent na poleznuyu model RF № 157695 (12015)
12. V.A. Orlov, S.P. Zotkin, I.S. Dezhina, A.A. Pelipenko, *Programma rascheta stepeni hydrofobnosti i hydravlitcheskih parametrov trub i zaschitnyh pokrytiy,* Svidetelstvo o gosudarstvennoy registracii programmy dlya EVM № 2017612281 (2017)
13. L.B. Boynovich, A.M. Emelyanenko, *Hydrofobnye materialy i pokrytia: principy sozdania, svoystva i primenenie,* Uspekhi khimii Journal, № 77 (7), 621-638 (2008)
14. M.Rameil, *Handbook of pipe bursting practice.* Vulkan verlag, 351 p. (2007)
15. A. Kuliczkowski, E. Kuliczkowska, A. Zwierzchowska *Technologie beswykopowe w inzeynierii srodowiska,* Wydawnictwo Seidel-Przywecki Sp. 735 p. (2010)
16. S.P. Zotkin, B.S. Sadovskiy, I.A. Zotkina, *K voprosu o sopryazhenii prikladnyh programm s Microsoft Word. Voprosy prikladnoy matematiki i vychislitelnoy mekhaniki,* Sbornik nauchnyh trudov MGSU, **17**, p. 260-263 (2014)

# AUTHOR INDEX

# AUTHOR INDEX

# AUTHOR INDEX

# AUTHOR INDEX

**EDP Sciences – Web of Conferences**
17, Avenue du Hoggar
Parc d'Activité de Courtabouf
BP 112
F-91944 Les Ulis Cedex A
France

ISBN 978-1-5108-4515-2

# XXVI R-S-P Seminar on Theoretical Foundation of Civil Engineering (RSP 2017)

MATEC Web of Conferences Volume 117 (2017)

Warsaw, Poland
21 - 25 August 2017

Part 1 of 2

edp sciences

# TABLE OF CONTENTS

## PART 1

## PART 2

ISSN (electronic edition): 2261-236X

edp sciences

MATEC Web of Conferences

# Statement of Peer review

In submitting conference proceedings to MATEC Web of Conferences, I certify to the Publisher that I adhere to the **Policy on Publishing Integrity** of the journal in order to safeguard good scientific practice in publishing.

1. All articles have been subjected to peer review administered by the proceedings editors.
2. Reviews have been conducted by expert referees, who have been requested to provide unbiased and constructive comments aimed, whenever possible, at improving the work.
3. Proceedings editors have taken all reasonable steps to ensure the quality of the materials they publish and their decision to accept or reject a paper for publication has been based only on the merits of the work and the relevance to the journal.

Title, date and place of the conference

RSP 2017
XXVI R-S-P Seminar 2017 Theoretical Foundation of Civil Engineering 21-25 August 2017
WARSAW POLAND

Proceedings editor(s):

Prof. Stanisław Jemioło
Prof. Artur Zbiciak
Dr Monika Mitew
Dr Michał Kneminski
Dr Marcin Gajewski

Date and editor's signature

28.06.2017

17, avenue du Hoggar - PA de Courtabœuf – BP 112 - 91944 Les Ulis Cedex A (France)
Tél. : 33 (0)1 69 18 75 75 - Fax : 33(0)1 69 07 45 17 – www.edpsciences.org

# Advantages of pre-fabricated reinforced concrete construction in Iraq

*Ivan* Abramov[1], *Aleksander* Stepanov[1],[2.1] *and Ibrahim F. Ibrahim* [1]
[1]Moscow State University of Civil Engineering, 26 Yaroslavskoye Shosse, Moscow, Russian Federation

**Abstract:** With its building infrastructure almost completely destroyed during the military conflict, modern Iraq is facing an urgent need for a number of measures to increase the quality of installation, construction and reconstruction jobs and to reduce timeframes taking into account local and climatic features. To achieve this, the priority task is to develop methods and theoretical recommendations based on the analysis of existing methods for calculating construction timeframes, ones that would most accurately reflect the influence of such factors as cement class, type and an average daily air temperature. The climatic conditions in Iraq impact negatively on the properties of hardened concrete. Specifically, concrete mixture shrinks more rapidly, which entails a change in its physico-mechanical properties and further corrosion of reinforcement bars at joints in monolithic prefabricated housing construction. This paper aims to determine factors that affect construction timeframes and quality for prefabricated reinforced concrete buildings, and to develop a model on the basis of statistical research to determine the concrete hardening time at joints in pre-fabricated frame housing construction.

## Keywords:

Pre-fabricated frame housing construction, grouting of joints, pre-fabricated reinforced concrete structures

## Introduction

Reconstruction and renovation of housing stock is one of the most important tasks on the agenda in a country's housing policy [1]. This article looks into the trends that are currently taking shape as regards to the choice of effective construction technology solutions suitable for house building in Iraq.

Construction technology solutions should be economical, technically feasible and functional.

When studying the available construction solutions or individual structural designs one should choose by comparing their technical parameters to those provided for in local detailed designs.

When comparing the available design options for building structures, the necessary criteria would be: identical purpose, same payload values and project's compliance with current construction design standards. When carrying out a feasibility study of structures, the following array of parameters should be evaluated: cost of manufacture, the amount of capital investment, operating costs, weight, volumes, consumption of basic materials (steel, sawn timber, cement, etc.), installation and labor input. As best practice suggests, a

2.1 Contact the author at: inferno1020@mail.ru

prerequisite for finding the right design solution for buildings and related structures is a decent feasibility study with due account of all factors affecting the efficiency of solutions [2, 3].

One of the main directions is monolithic reinforced concrete construction. Reinforced concrete is, and in the near future will remain, the main type of material used in the construction of buildings and structures of various purposes in Iraq. However, the actual technical and economic parameters are affected by a number of factors considered in this article. The Iraqi climate is diverse with very hot summers and cold winters; temperatures may vary during the day + (50-53 ° C) and at night to - (5-8 ° C), especially in the north of the country, in the mountainous Iraqi Kurdistan. The large temperature differences increase labor intensity and costs in the process of monolithic housing construction in Iraq.

Performance of construction and installation jobs on site is associated with a number of negative factors [4] that normally affect construction timeframes, the use of construction machinery on the construction site and consumption of construction materials. Consequently, the factors in question drive up significantly the costs of performing the entire range of construction and installation jobs beyond initial projections. Thus, all these factors affecting the quality of the final result during construction can be divided into two groups [5]:

1. negative factors typical of the entire construction and installation period

- construction of facilities on the slopes of mountains with sharply intersected relief with a severe continental climate;

2. negative factors only typical of certain types of construction work at various stages of construction;

- construction of new facilities in cramped conditions: on the premises of the existing enterprises, which entails cramped material storage conditions;

- reconstruction, restoration of facilities after man-caused impacts.

When erecting monolithic reinforced concrete buildings the process of casting a concrete mixture is exposed to one of the above factors, specifically the severe continental climate. As is known, concrete's binding property is the key indicator in monolithic housing construction and, in many respects, it determines its physical and mechanical properties. The rapid gaining of strength at the early stages of concrete hardening leads to the formation of dense shells around new tiny bodies that impede water access to non-hydrated cement grains. At the same time, hydration is a fast process, which limits concrete's strength from growing further. The structure of concrete can also be broken as a result of periodic daily and seasonal changes in air temperature and humidity, which cause compressive and tensile strength as well as the elasticity modulus to grow. Reinforced concrete structures (prefabricated reinforced concrete) after gaining strength (of 85% and higher) upon being manufactured are less affected by the hot climate and are widely used as an efficient construction technology solution for erecting buildings and structures in Iraq [6,7].

Prefabricated reinforced concrete products are structures or products manufactured at a plant and transported to a construction site for subsequent assembly or installation.

One of the main processes in the construction and assembly work from prefabricated reinforced concrete is filling joints with concrete.

When constructing such joints, a formwork needs to be used for casting concrete and mortar and it should be properly treated to ensure it gains all the properties necessary. The erection technology used in prefabricated reinforced concrete construction, depending on the number and type of elements connected together, connections are divided into joints, fittings and seams. The joining of two structural elements in one place is called a joint; three or more elements are called a fitting (for example, joining a column, crossbar and floor slabs in multi-story frame buildings). A seam is the point of connection (contact) between individual structural elements, for example, slabs of coatings and wall panels. Grouting allows not only fixing, but also stabilizing the joints of prefabricated elements, as well as obtaining a spatial rigidity of the entire structure. Stability, design position, reliability and durability of prefabricated structures largely depend on how well the joints of prefabricated elements and structures are grouted [8].

When erecting prefabricated reinforced concrete buildings, grouting of joints is considered an important stage and is recorded in a special joint grouting log.

To make sure that concrete at joints has gained due strength, it needs proper treatment. When work is performed in hot and dry weather, manufacture of concretes should generally use plasticizing and complex additives accelerating the hardening process. One of the important treatment techniques is reducing the amount of water evaporating from concrete's surface, a measure especially relevant in the dry and hot climate of Iraq.

In order to reduce the amount of water evaporating from the surface (the surface of the concrete mixture) the recommendation is to apply additives that are less dense than water and do not dissolve in it forming a thin protective layer on the surface. The use of such additives greatly simplifies treatment of the concrete mixture after casting, reduces labor input and lowers the costs of water so scarcely found in Iraq [9].

Prefabrication companies are rare in this country despite the fact that in 2013 the Iraqi Ministry of Housing and Construction declared that prefabricated system will be the main construction system. The first precast concrete plant was opened recently on April 2014 at Bismayah, Baghdad which mainly produces pre-fabricated reinforced concrete structures[10].

The air temperature in Iraq in summer is + 44 ° C on average [11]. Table 1 shows the study of concrete hardening times for different temperatures and cement classes.

Table No. 1. Concrete hardening times for different temperatures and cement classes.

| *Type and class of cement used in the manufacture of concrete* | *Concrete strength, % of the design value* | *Concrete hardening times in days at an average hardening temperature* | | | |
|---|---|---|---|---|---|
| | | *25* | *30* | *35* | *44* |
| *Portland Cement 22.5* | *50* | *4* | *3* | *2.5* | *2* |
| | *70* | *7* | *6* | *5* | *4.5* |
| | *100* | *16* | *14* | *12* | *11* |
| *Portland Cement B 32.5* | *50* | *3* | *2.5* | *2* | *1.5* |
| | *70* | *6* | *5* | *4* | *3.5* |
| | *100* | *14* | *12* | *10* | *9* |
| *Portland Cement B 42.5* | *50* | *2* | *1.8* | *1.5* | *1.3* |
| | *70* | *3* | *2.5* | *2* | *1.8* |
| | *100* | *10* | *8* | *6* | *5* |
| *High Early Strength Portland cement* | *50* | *1.5* | *1.2* | *1* | *0.8* |
| | *70* | *2.5* | *2* | *1.5* | *1* |
| | *100* | *8* | *6* | *4* | *3* |

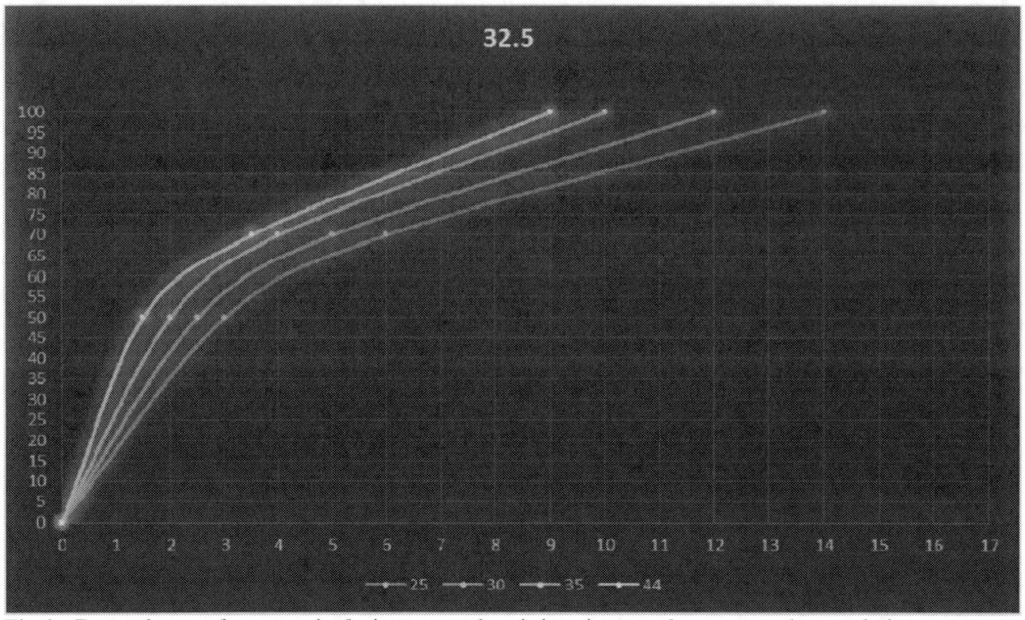


Fig.1 . Dependence of concrete's design strength gaining time on the cement class and air temperature

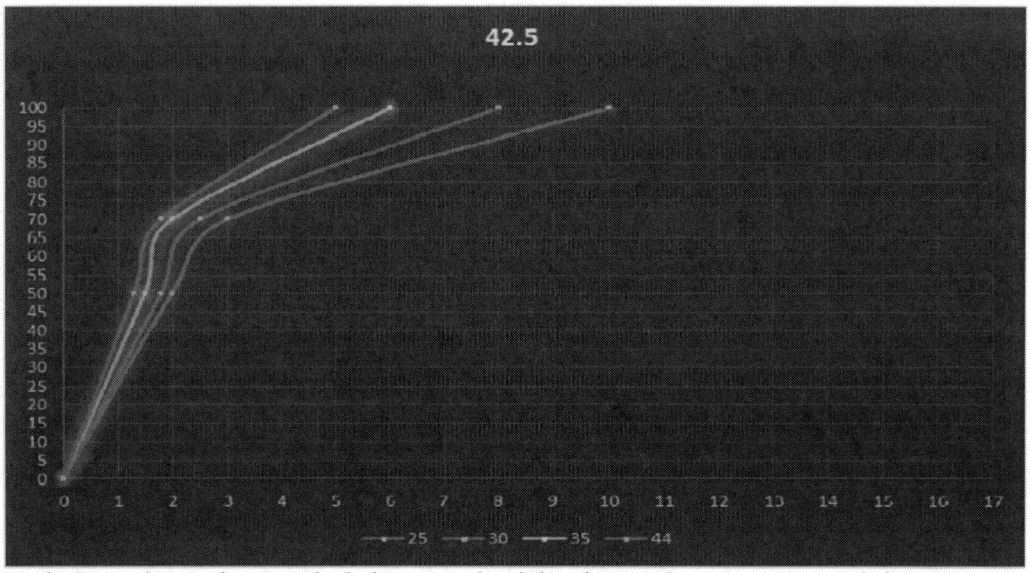


Fig.2 . Dependence of concrete's design strength gaining time on the cement class and air temperature

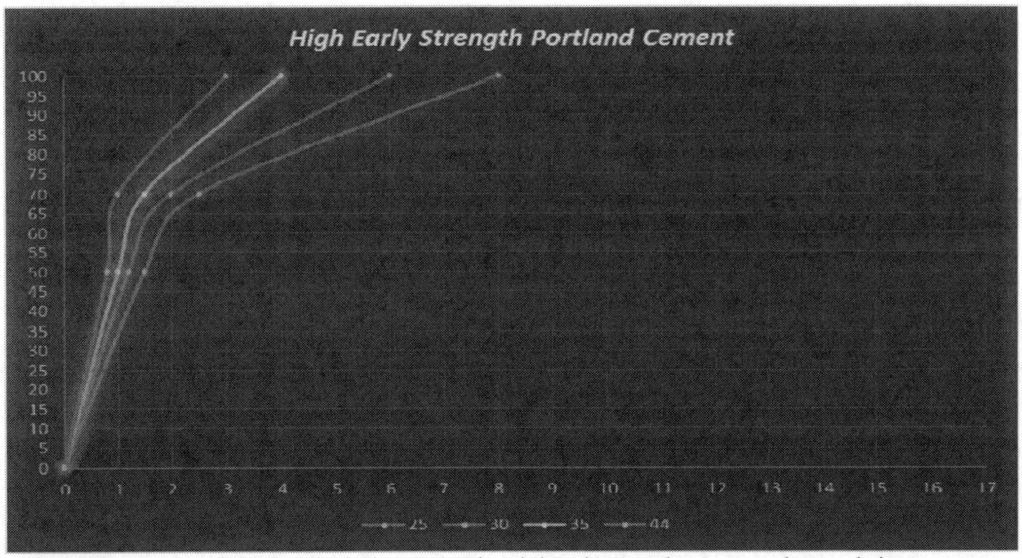


Fig.3 . Dependence of concrete's design strength gaining time on the cement class and air temperature.

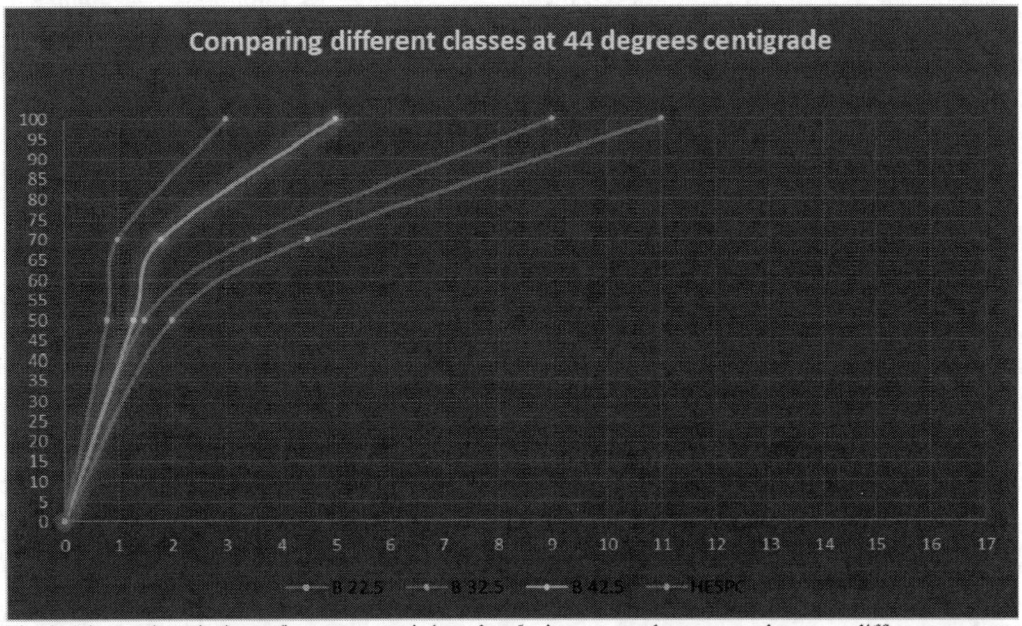


Fig.4. Linear description of concrete gaining the design strength compared across different cement classes

## Conclusion

The comparative data suggests that, at a temperature of +44 C °, the type of concrete that gains strength fastest (three days) is the one using High Early Strength Portland Cement. Concrete using Grade 42.5 Portland Cement gains the design strength of 100% on the 5$^{th}$ day.

Concrete with the use of Class 32.5 Portland Cement gains maximum design strength on the 9$^{th}$ day and concrete with the use of Class 22.5 Portland Cement - on the 11$^{th}$ day.

The greatest 4-day difference in the gaining of strength is most noticeable when comparing the classes 42.5 and 32.5, which can significantly affect the timing of construction and restoration of buildings.

It is worth a special note that the quality of prefabricated structures and products is preserved by manufacturing them at a plant. Also, due to the exclusion of labor-intensive and highly skilled work in monolithic construction on the construction site, construction timeframes are shortened.

The use of prefabricated monolithic technology in the construction, restoration of buildings and structures and the use of the method reviewed in this article to calculate construction timeframes, taking into account the influence of various factors such as the use of different cement grades, plasticizers and air temperature will allow building and carrying out reconstruction work in the time required for the speedy restoration and provision of housing for those in need.

## References

1. O.M. Akunina, *Foreign experience of state regulation of the reproduction of housing stock* // Politics, state and law No. 3. URL: http://politika.snauka.ru/2015/03/2521 (2016).
2. A. A. Lapidus, *Efficiency Potential of Organizational and Process-Related Solutions of the Construction Project.* Bulletin of Moscow State University of Civil Engineering – No. 1. **175-180** (2014).
3. The Government of Western Australia. *The Impact of New Building Techniques and Technologies on the Residential Housing Sector of the Construction Industry.* Construction Training Fund (2015)
4. I.L. Abramov, T.Y. Poznakhirko, A. Sergeev, *The analysis of the functionality of modern systems, methods and scheduling tools*, MATEC Web of Conferences 86, 04063 (2016) IPICSE-2016
5. http://www.1c-esk.ru/about/spr/03.html#osnfakt
6. A.A. Afanasyev, *Technology of erection of fully pre-fabricated buildings.* M: Publishing house of the Association of Construction Universities, **361** (2012).
7. A.A. Afanasyev, E.P. Matveev, *Technologies of reconstruction of residential buildings and buildings.* Part I and II. // M., **367** and **363** (2009)
8. V.I. Snarsky, M.M. Aigumov Technology of construction processes. Volume 2. The processes of erection of load-bearing structures (2007)
9. I.G. Sovalov, Ya. G. Mogilevsky, *Concrete and reinforced concrete works* (1988)
10. Department of the Army. U.S. Army Corps of Engineers. Manual No. 1110-2-3506 *Grouting Technology.* Washington D.C. (2017).
11. https://world-weather.ru/archive/iraq/summer/
12. SNiP 48.13330.2011 Construction Rules, *Construction Organization,* (2011)
13. P. P. Oleynik, V. A. Grigoryev, *Current Methods of Modeling Standards for Residential Construction Timelines*, Constr. Techn. and Org., **42-44** (2014)
14. P. P. Oleynik, V. I. Brodsky, *Construction Planning Arrangements.* Constr. Techn. and Org., **40-43** (2013)
15. Ammar W. Abbood, Karam M. Al-Obaidi, *Achieving energy efficiency through industrialized building system for residential buildings in Iraq.* Inter. J. of Sust. Built Envir.. Volume 4, Issue 1, **78–90** (2015).
16. A.A.Lapidus, I.L. Abramov. Calculating a construction company's resource potential as a bidder – Science Bulletin 9-2017-p.6-9.

# Industrial technologies of the residential buildings reconstruction of the first mass-produced series

*Aleksandr* Afanas'ev[1,*] and *Aleksandr* Kurochkin[2]

[1]Grand Ph.D in Engineering, Prof., A.M. of RAACS, Moscow State University of Civil Engineering, Jaroslavskoe Shosse 26, Moscow, Russian Federation
[2]Ph.D in Engineering, 105318, LLC "InTechProject", Mironovskaya ul., 25, Moscow

**Abstract.** The article dwells upon the reconstruction technologies of the residential buildings of the series that are not subjected to demolishing by way of superstructing of the attic floors and standard floors made of unitized folding units, adding of lifts, unitized elements of kitchens, living rooms and loggias. Their application makes it possible to increase the areas of kitchens by 6.0…8.2 $m^2$, of bedrooms and other premises - by 3.5-4.2 $m^2$. The technology of the attic units manufacture under plant conditions has been worked out. It has made it possible to optimize the design concept of the articulated joints, ensuring the transport adaptability due to flatwork elements folding. The technologies of the high-speed superstructing and building up of the buildings, using line production of works have been investigated.

The moral and aesthetic depravity of the five-storey residential development gets worse by the fact that the only type of the sectional house has been realized without regard to the demographical and architectural traditions of the districts and climatic conditions.

It is worth noting that in spite of enough high structural reliability of five-storey buildings, they have small reconstructive capability. It relates to residential panel houses the interior wall of which are bearing ones and are located with a narrow pitch of 2.6 and 3.2 m. [1].

The RF Government and the Moscow Government has taken a decision to demolish the five-storey housing stock.

The first program of the five-storey buildings demolishing is being completed. 160 thousand families have been rehoused in the course of its implementation.

According to the new program it is presupposed to demolish more than 20 mln $m^2$ housing accommodations, it is meant for 15-20 years. 1 mln. and 800 thousand dwellers will get the new modern housing accommodations.

* Corresponding author: 7450688@gmail.com

The new flats' area will be bigger by 20% than the area of the residential premises, what will make it possible to increase to the regulatory values the kitchens, lavatories, corridors, loggias or balconies [3].

At the same time, it should be noted that a part of the five-storey buildings of the series that are not subjected to demolishing can be reconstructed. Due to lack of funds in the RF regions, the primary way of the obsolescence and physical deterioration liquidation is reconstruction without the dwellers being rehoused.

Possessing of the high obsolescence, low performance characteristics and low-level architectural and planning standard, such housing stock needs full-fledged rehabilitation through comprehensive reconstruction, with due regard to the modern requirements lodged to the housing accommodations [4-6]. The annual physical deterioration deepens the obsolescence process of the buildings and approximates to the emergency condition.

The characteristic traits of the obsolescence are the following ones: small dimensions of the kitchens (4.0-6.0 $m^2$), entranceways, utility cores; absence of lifts and rubbish ducts; discomfort of the residential premises due to disturbance of humidity conditions and sound insulation; big physical depreciation of the utility networks; high heat losses through the building envelopes resulting in multiple excess of the energy carriers consumption [7].

The technical and economic study, the foreign and domestic reconstruction experience of the residential houses having few floors shows that the most effective is the superstructing with few floors (1…2 floors) and adding of small volumes to the elements of the buildings. Such concept excludes a cycle of works for shoring of foundations and other elements of the buildings, what reduces significantly the prime cost of the additional areas and makes it possible to perform the reconstructive works without or with partial rehousing of the dwellers [2].

The conceptual basis of the buildings reconstruction of the first mass-produced series is the development and creation of highly industrial flexible technologies based on use of the conceptually new unitized elements of factory production. Such concept makes it possible to turn the technological reconstruction process into the assembly and integration cycle and to reduce many times the works duration.

The design concept provides for the use of several types of the unitized units: one-/double-level – for superstructing of the buildings (fig.1), of bay window type – for adding, aimed at obtaining of additional areas for kitchens and living rooms; of special units – for adding of lifts (fig.2). The main requirements to the unitized units for superstructing and building up of the buildings is a high level of the factory readiness, reducing to a minimum the works volumes at a construction site, ensuring the transport adaptability and serviceability. An additional requirement to the units being superstructed is an ability to the dimensional instability owing to the articulated joints use, what ensures the transport adaptability and an ability to obtain the different geometry of the floors being superstructed.

The project designs for the residential houses reconstruction of series 1-464, 1-447, 1-510, 1-515 and others, have shown that the adopted technology use provides for the total areas increment in the following range:

| | |
|---|---|
| - adding of unitized bay windows | - 19.4…20.6% |
| - adding of unitized bay windows and superstructing of the attic floor | - 42.6…50.3% |
| - adding of the unitized units along the perimeter of the buildings with superstructing of a full and attic floors | -70.5…72.8% |

One- and double-storey unitized units have been developed, ensuring the superstructing of the residential buildings of different series with the attic floors, with one full and attic floor. Thereby, the overall dimensions of the units are adopted according to a pitch of the interior bearing walls for large-panel buildings and which is multiple of the window areas placement for the buildings with brick walls. The outstanding feature of such systems is an opportunity to obtain a space of hall type, what ensures a flexible planning of the floors being superstructed.

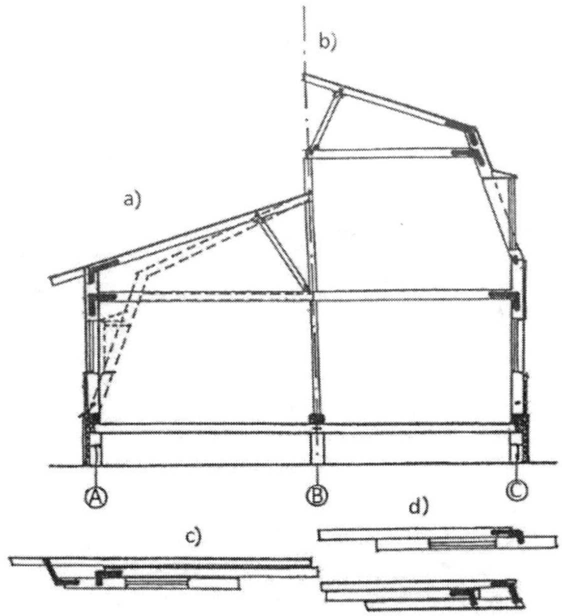


**Fig.1.** The basic diagrams of one-storey (a) and double-storey (b) of the folding unitized units for superstructing of the buildings; (c, d) - the transport position of the units.

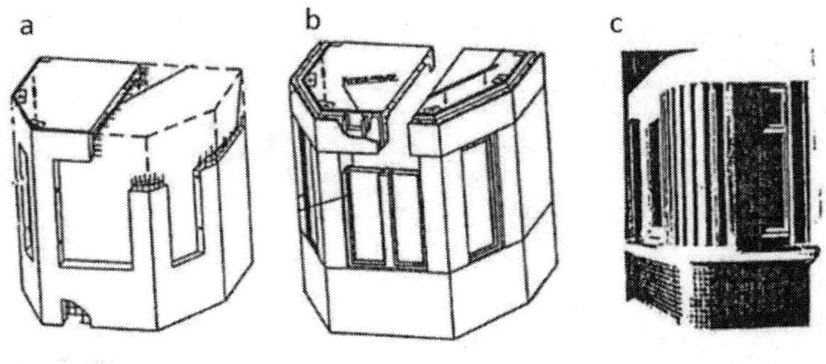


**Fig.2.** The structural scheme of the added-on unitized units.
a – non-heat insulated loggia unit; b – unit with 3-layered exterior walls; c – overall view of a ready-to-operate unitized unit.

Use of the composite structures of metal, slabby and casting heat insulator, heavyweight concrete and lightweight concrete has resulted, beside the total mass reduction – in increase of the heat and sound insulation indicators. For the unitized units of the floors being superstructed metal consumption amounts to 24 … 26 kg/m$^2$ of the area.

The added-on unitized elements of different geometrical shape have been worked out for the buildings of different framing schemes: the kitchen units with the area of 4.5…6.08 m$^2$; bay window blocks to increase the rooms' area 3.5…4.6 m$^2$; the loggias' blocks – 3.2 … 4.3 m$^2$, what ensures a diversity of the architectural solutions and an opportunity to avoid attribution to a certain classification type [2].

The architectural and planning solutions for reconstruction of a three-section residential house with brick walls is specified in fig.3, where a standard block with the area of 8.2 m$^2$ with three-layered wall structures, complying with the thermotechnical requirements as well as the blocks of the lift shafts are used as building extensions along the perimeter of a building.

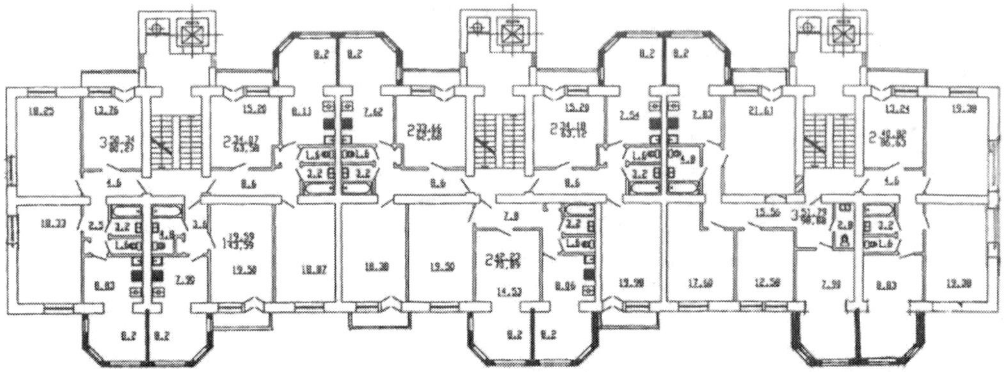

**Fig.3.** Plan of a standard storey with addition of the unitized units.

In fig.4 there is a layout of an attic block after opening and assembly of half blocks, installation of a translucent window block, envelopes. Pre-assembly can be carried out in immediate proximity to a facility under reconstruction, as well as at an individual site.

**Fig.4**. Overall view of an attic unit after pre-assembly.

The comparative data for reconstruction with superstructing of attic floors according to the traditional technology and the unitized units have shown the unit labor costs reduction by 6-8 times under building conditions and prime cost reduction for 1 m$^2$ of the area - by 26-30%.

A special site is used for pre-assembly, where opened half blocks are installed successively, are united in the central zone by means of the bolted-type connections and welding. A special team of the installation fitters takes part in the work.

The technological stages of the works production include several construction flows that are consistent in time.

Prior to perform the installation works of the unitized units, the zero works cycle is performed that consists in basement work for added-on volumes in form of short bored and cast-in-place piles and grillage beams. A cycle for the monolithic girt creation in a parapet part is implemented in parallel to these works for a subsequent placement on it of the units of the attic floor.

The erection technologies of the floors being superstructed using the methods of the half blocks installation and unitized elements enlarged to a span, as well as their launching, have been worked out.

When building up the buildings with the unitized elements and superstructing with an attic floor, the most reasonable is the works performance when two construction flows are formed, one of which produces the cycle of superstructing, and the other one – installs the added-on volumes from the transportation facilities.

When reconstructing large-panel buildings to perform the building up works, dismantling of the external wall panels is required that is fulfilled by a separate technological flow.

To install the blocks of the attic floor, jib-type crane Liebherr LTM 1160 with the telescopic jib is used, with the lifting capacity at a working radius of operation of 24 m – 14.6 t. Mass of the blocks being installed with the width of 3.2 and 2.6 m amounts to 7.2 and 8.4 t.

For the pre-assembly of the attic units, truck crane with the telescopic jib with the lifting capacity of 10 t (fig.5) is used.

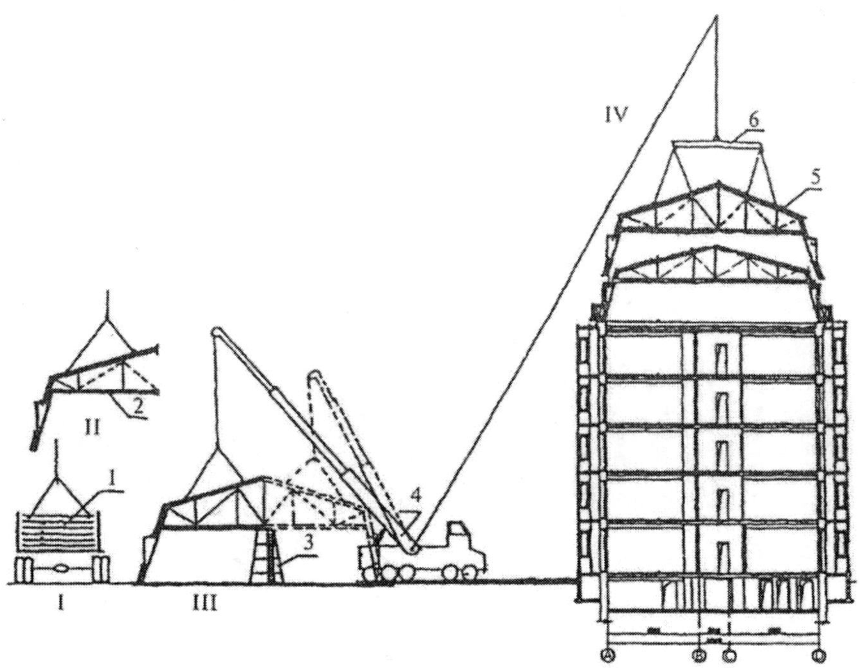


**Fig.5.** Technological schemes of the units enlargement and their installation.

I – transport facility with volumetric sides; II – opening of units; III – pre-assembly at a guide; IV – installation of a unitized unit to a bay of a building; 1 – units in a transport position; 2 – opened unit; 3 – supporting platform; 4 – motor crane; 5 – unit within an installation period; 6 – cross arm.

The works performance technology includes three cycles: transportation, pre-assembly and installation of the unitized units.

Taking into account a construction site's lack of space, two zones of the pre-assembly are created, what ensures an estimated consistency and duration of the installation works and associated works.

While installing the attic units, the works cycle for installation of end walls, stairways, lift shafts, interior planning works is fulfilled.

The detailed study is given in the technological documentation in form of the works production plan, flow diagrams and special calculations.

The exploratory tests have been carried out through superstructing with the attic units of 80-flat 4-section residential house. In standard labor intensity of one unit installation 6.8 people/hour and membership of the installation fitters' team 5 people, the works duration at one grip (6 units) amounts to 16.8 people/hour. The total installation duration of the attic units has amounted to 8.5 machine-shifts of the crane work. The total erection duration of an attic floor and the associated works duration has amounted to 70 working days. Thereby, an additional area of about 870 $m^2$ has been obtained.

When using 2-storey unitized units (fig.6), they apply more powerful cranes Lakomo A395NR, special systems of slinging operations with the remote removal of slings, and means ensuring stability of geometrical shape of a unit in the process of hoisting.

To reduce transport costs, pre-assembly is done in the zone of the main crane operation with the use of motor crane KS-3577. Special supporting platforms with the hydraulic jacks are used to align the mating elements of the lower and upper storeys to the permanent location and to fix them in the permanent location. In the course of the pre-assembly, the works quality control is exercised and compliance with the design concept is made.

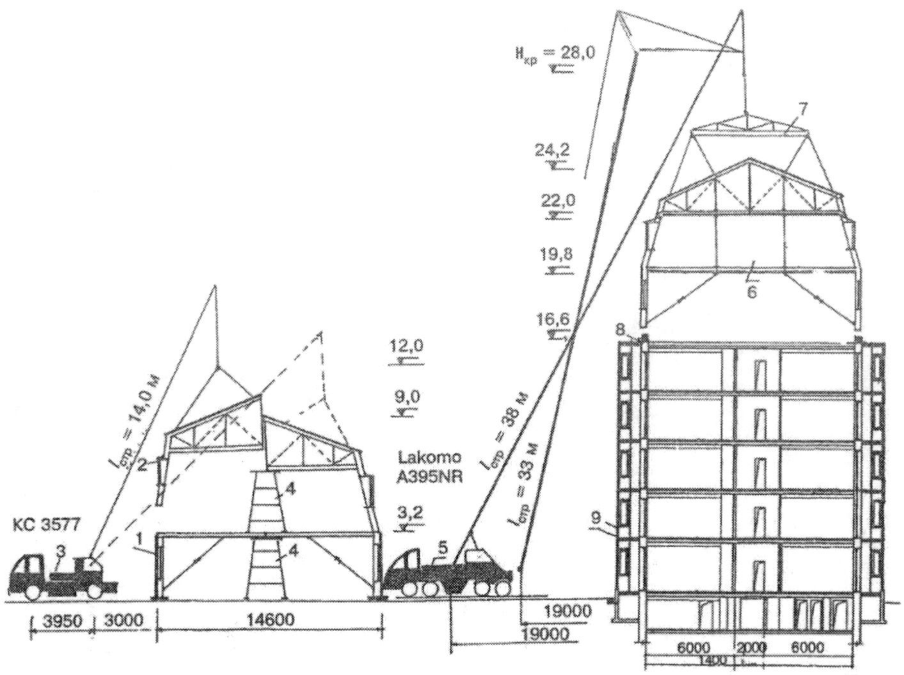


**Fig.6.** Technological schemes of pre-assembly and installation of 2-storey unitized units to superstruct the buildings.
1 – a unitized unit of a storey being superstructed; 2 – the same of the attic floor; 3 – jib-type crane with the telescopic jib for pre-assembly; 4 – supporting platforms; 5 – jib-type crane Lakomo to install 2-storey unitized unit; 6 – a unitized unit in an installation position; 7 – cross arm; 8 – a monolithic girt; 9 – installed bay windows.

Erection of 2-storey superstructures needs the use of not only more powerful cranes but also actions for ensuring safe production of works, spatial rigidity and stability, fencing of working places etc.

Optimization of the installation processes is made through the cost estimation of the works performance with due regard to the factors of erection cranes' down time or queue formation consisting of the transport facilities for the unitized units delivery, their time of maintenance for pre-assembly and following up. The variants of the models have been dwelt upon in which assembly components are supplied directly below an erection crane, and a model with an interim site – stock for pre-assembly.

The results of the numerical calculation methods have shown that in installation intensity of the unitized units $\mu_1$=2,0, $\mu_2$=1,5, $\mu_3$=1.0 unit/hour, the continuous work of two erection cranes will be achieved in terms of the supplying intensity equaling to 3.5; 2.7 and 2.2 units per hour accordingly.

In these values the minimum costs are achieved and the continuous cycle of the installation works performance is ensured and it has the essential value when designing the total term of the building reconstruction.

The building up and superstructing method of the buildings made of unitized elements is universal and can be used for the majority of the mass-produced series of the large-panel and brick buildings.

## References

1. STO 000433363-01-2008, *Reconstruction and modernization of deterioration housing*, RAASN (2008).
2. A. Afanas'ev, E. Matveev, *Reconstruction of residential buildings, Part I, Part II* (2008).
3. S. Bulgakov, *Reconstruction of apartment houses of the first mass series*, (1998).
4. P. Oleinik, V. Brodskii, *Features of the organization of construction in the reconstruction of buildings and structures*, TOS, **4(5)** (2013).
5. A. Ishin, D. Tumanov, M. Ershov, A. Lapidus, *Technological and organizational aspects of building reconstruction*, TOS, **3(8)** (2014).
6. M. Ershov, G. Gamrekilidze, *Reconstruction buildings of the series that are not subjected to demolishing*, TOS, **2(3)** (2014).
7. M. Ershov, G. Gamrekilidze, A Manin, *On the issue of the relevance of the reconstruction and overhaul of residential buildings in the mass construction of 1953-1975*, TOS, **4(9)** (2014).

# Physical modeling of stresses caused by volume centrifugal forces in a compound rotation body

*Elephan* Agakhanov[1], *Murad* Agakhanov[2,*]

[1]Dagestan State Engineering University, Imam Shamil Avenue, 70, 367015, city of Makhachkala, Russia
[2]Moscow state university of civil engineering, Yaroslavskoye shosse, 26, Moscow, Russia, 129337

**Abstract.** The article considers the question of physical modeling of stresses in a compound rotation body of a complex form with a complex loading distribution. According to the similarity parameters, the stresses, deformations and displacements caused by the volume forces are reduced proportionally to the similarity scale for geometrical dimensions which complicates their direct modeling with the use of the models made from conventional epoxy materials. The general methods of study of stresses and deformations usually register them with an inadequate sensitivity. On the basis of the independence principle, the initial problem is presented as a superposition of two problems. In the first (uniform) problem, the stresses in a rotation body caused by centrifugal forces are simulated through the standard "freezing" method. For the solution of the second (non-uniform) problem, the "freezing" of stresses is carried out in the model domain corresponding to the centrifugal forces acting there, and, at the room temperature, the models in their natural state are glued to the model, and the "annealing" of the compound model is performed. The components of radial, tangential and axial stresses at the outlines as well as the cross-sections of models have been obtained through the methods of normal translucence and of numerical integration of the equilibrium equation.

## 1. Introduction

The modeling of the problems in mechanics of deformable solid bodies is carried out with the help of the similarity parameters, on the basis of which the model is performed, the loading conditions are determined, and the transition from the measured model values to the corresponding values for the full-scale structure is put into practice. The similarity factors for stresses $K_\sigma(t)$, deformations $K_\varepsilon(t)$, displacements $K_u(t)$, geometrical dimensions $K_L$, volume forces $K_F(t)$ and elasticity modulus $K_E$ in the process of modeling of problems in the creep theory and the elasticity theory (in the problems of the elasticity, the similarity factors become constant values, i.e. similarity scales) are connected through the following formules [1].

* Corresponding author: muradak@mail.ru

$$\frac{K_L K_F(t)}{K_\sigma(t)}=1,\quad \frac{K_L K_F(t)}{K_\varepsilon(t)K_E}=1,\quad \frac{K_L^2 K_F(t)}{K_u(t)K_E}=1. \tag{1}$$

It follows from (1) that the stresses, deformations and displacements caused by volume forces are reduced proportionally to the scale of similarity of geometrical dimensions. This phenomenon complicates a direct modeling even for the most simple case of volume forces (mass forces) through the photo-elasticity method with the use of models made from conventional epoxy materials. The general methods of study of stresses and deformations usually register them with an inadequate sensitivity. The modeling procedures for volume forces are considered in works [2, 3, 4].

When modeling the problems with different characteristics $\nu, E, \gamma, \alpha$ in different body domains (in contrast to the problems when these characteristics have constant values in the whole body), we have to observe some additional relationships between the similarity factors (scales) [5]:

$$\begin{gathered} K_{\nu 1}=K_{\nu 2}=\ldots=K_{\nu n}=1, \\ K_{E1}=K_{E2}=\ldots=K_{En}, \\ K_{F1}(t)=K_{F2}(t)=\ldots=K_{Fn}(t), \\ K_{\alpha\Delta T}=1\ K_{\alpha 1}=K_{\alpha 2}=\ldots=K_{\alpha n}, \end{gathered} \tag{2}$$

where n is the number of domains with different characteristics.

These conditions create some restrictions and difficulties for the use of traditional techniques. Some new potentialities for the solution of such problems may appear with the use of the properties of polymers displayed during the process of polymerization (methods of polymerization, straitened shrinkage and fixation of temperature stresses) [6].

In addition to that, the methods are being developed especially for the determination of stresses caused by mechanical loading and temperature on the models made from a standard optical sensitive material in the problems with different characteristics in different body domains [7].

Taking into consideration the aforesaid methods allowing us to obtain effective solutions, we take the characteristics $\nu, E, \alpha$ to be constant in the whole body in the process of solution of the problem of effect of volume forces.

## 2. Methodology

The engineering practice, in particular, the machine building, more and more uses compound (composite) structural elements made from materials with different volume masses. In their earlier works, the authors considered only uniform rotation bodies [8]. The analysis of stress-and-strain state of compound structural elements is reduced to the solution of complicated systems of partial differential equations, the direct solution of which is quite difficult. Thus, many important practical problems for compound bodies cannot be solved theoretically yet. The determination of stress-and-strain state of compound structural elements with complicated forms and with a complicated distribution of loading is one of such problems.

Let us consider a volume rotation body composed of two materials with the density values $\rho_1$ and $\rho_2$ (Fig. 1). This compound structural element rotates with the angular velocity $\omega_{fsc}$, and it is under the influence of volume centrifugal forces

$$F_x = \rho_1 \omega_f^2 x, \quad F_y = \rho_1 \omega_f^2 y, \quad F_z = 0 \ \text{в} \ V_1, \tag{3}$$

$$F_x = \rho_2 \omega_f^2 x, \quad F_y = \rho_2 \omega_f^2 y, \quad F_z = 0 \ \text{в} \ V_2. \tag{4}$$

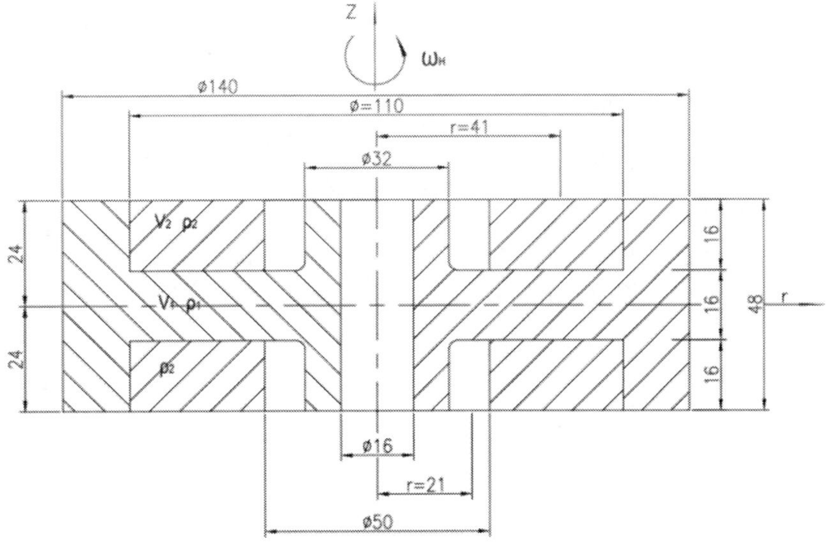


Fig. 1. Compound rotation body under the effect of centrifugal forces.

## 3. Main part

We use the principle of independence of action of forces and interpret the solution of the initial problem as the sum of two problems.

1. In the whole compound body ($V = V_1 + V_2$) the following volume forces act:

$$F_x = \rho_2 \omega_{fsc}^2 x, \quad F_y = \rho_2 \omega_{fsc}^2 y, \quad F_z = 0 \ \text{в} \ V. \tag{5}$$

2. The compound body is under the action of the volume forces

$$F_x = (\rho_1 - \rho_2)\omega_{fsc}^2 x, \ F_y = (\rho_1 - \rho_2)\omega_{fsc}^2 y, \ F_z = 0 \ \text{в} \ V_1, \tag{6}$$

$$F_x = F_y = F_z = 0 \ \text{в} \ V_2. \tag{7}$$

In the process of modeling of these problems, a standard optical sensitive material is used with the following characteristics

$$T_{fr} = 130^0\text{C}, \ E_{fr} = 20{,}0 \text{ MPa}, \ \nu_{fr} = 0.5, \ \sigma_{0,fr}^{(1,0)} = 0{,}046 \text{ MPa}.$$

The stresses in a uniform rotation body caused by centrifugal forces in the first problem are simulated as usual: in the model of the V domain with the angular rotation velocity $\omega_{mod}$, the "freezing" of stresses corresponding to the unknown quantities is carried out.

For the solution of the second problem, we perform in the model of the $V_1$ domain the "freezing" of stresses corresponding to the centrifugal forces acting in the $V_1$ domain. Further, we glue to the model of the $V_1$ domain at room temperature the models of the $V_2$ domain which are in their natural state, and then we carry out the "annealing" of the composed model.

The methods of normal translucence of sections and of numerical integration of the equilibrium equation along the z-axis gave the components of radial $\sigma_r$, tangential $\sigma_\theta$ and axial $\sigma_z$ stresses at the outlines of the models and at two cross-sections.

The stress diagrams in the models are shown in Figures 2-4.

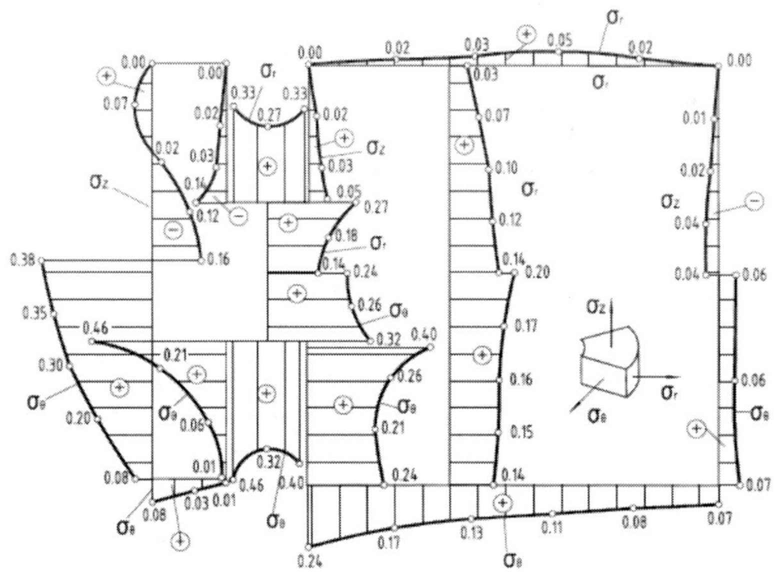

Fig. 2. Stress diagrams $\sigma_r$, $\sigma_\theta$ и $\sigma_z$ [MPa] in the model of the V domain.

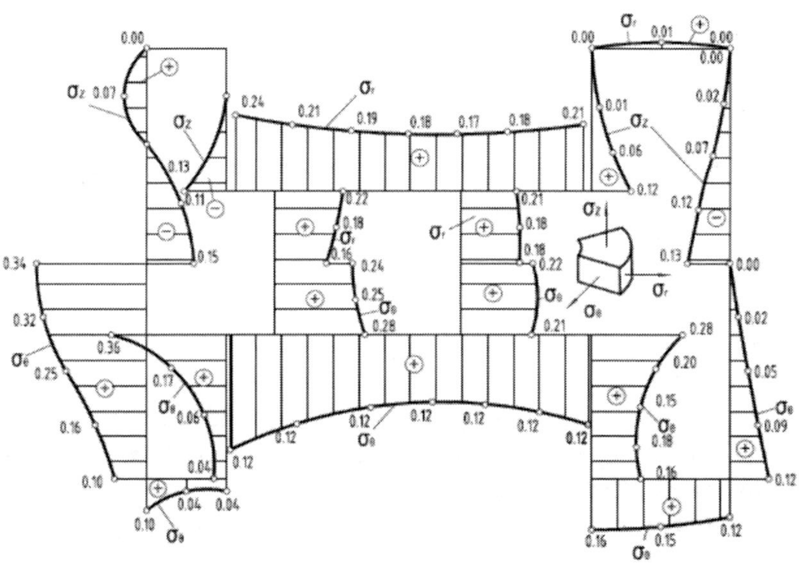

Fig. 3. Stress diagrams $\sigma_r$, $\sigma_\theta$ и $\sigma_z$ [MPa] in the model of the $V_1$ domain.

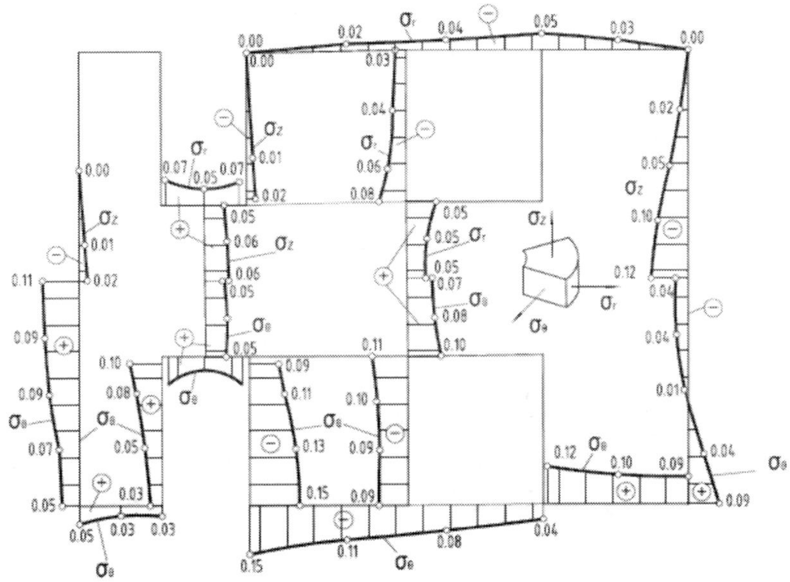

Fig. 4. Stress diagrams $\sigma_r$, $\sigma_\theta$ и $\sigma_z$ [MPa] in the model, composed of areas $V_1$ and $V_2$, after annealing.

In accordance with the parameters of modeling of a volume static elastic problem, we obtain the following formula of transition from the stresses in the models to the stresses in a full-scale structural element:

$$\sigma_{ij}^{(fsc)} = K_\omega^2 K_\rho K_l \left[ K\sigma_{ij}^{(mod)} + (1-K)(\overline{\sigma}_{ij}^{(mod)} - \overline{\overline{\sigma}}_{ij}^{(mod)}) \right], \tag{8}$$

where: $K_\omega = \dfrac{\omega_{fsc}}{\omega_{mod}}$, $K_\rho = \dfrac{\rho_1}{\rho_{mod}}$, $K_l = \dfrac{l_{fsc}}{l_{mod}}$, $K = \dfrac{\rho_2}{\rho_1}$;

$\sigma_{ij}^{(mod)}$ - stress in the model of area V;

$\overline{\sigma}_{ij}^{(mod)}$ - stress in the model of area $V_1$;

$\overline{\overline{\sigma}}_{ij}^{(mod)}$ - stresses in models made up of areas $V_1$ and $V_2$ after annealing.

The presented experimental studies allow us to analyze the influence of the ratio of the densities of materials from which the body is composed on the stressed state of the whole structural element.

The results of the analysis are shown in Figures 5 and 6.

For the structural element under consideration, the studies showed:

- the stresses $\sigma_z$ are negligible as compared with $\sigma_r$ and $\sigma_\theta$;

- the ratio of densities in a compound body effects (both quantitatively and qualitatively) on the stressed state of the structure.

Therefore, changing the ratio of densities of the materials in a compound body, i.e. choosing the materials with different volume masses, we can change the value and the distribution of stresses at either cross-section of the structural element, which is of great practical interest.

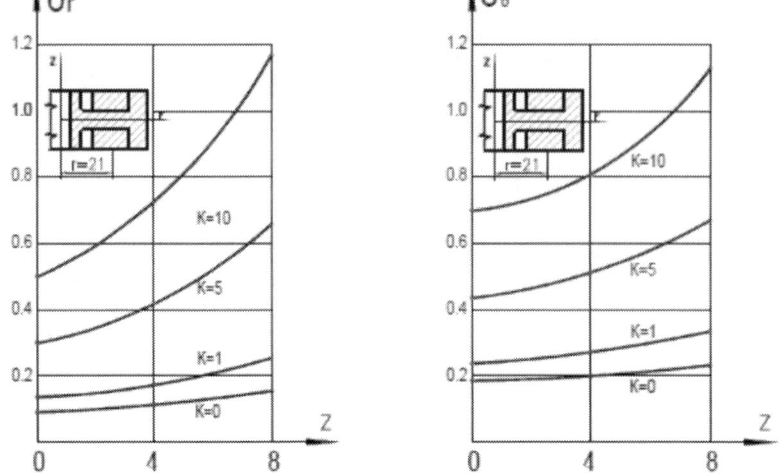


Fig. 5. The effect of the parameter $K = \rho_2 / \rho_1$ on the stresses $\sigma_r$ and $\sigma_\theta$ [MPa] in the cross-section $r = 21$ in the fractions of $K_\omega^2\ K_\rho\ K_\ell$ .

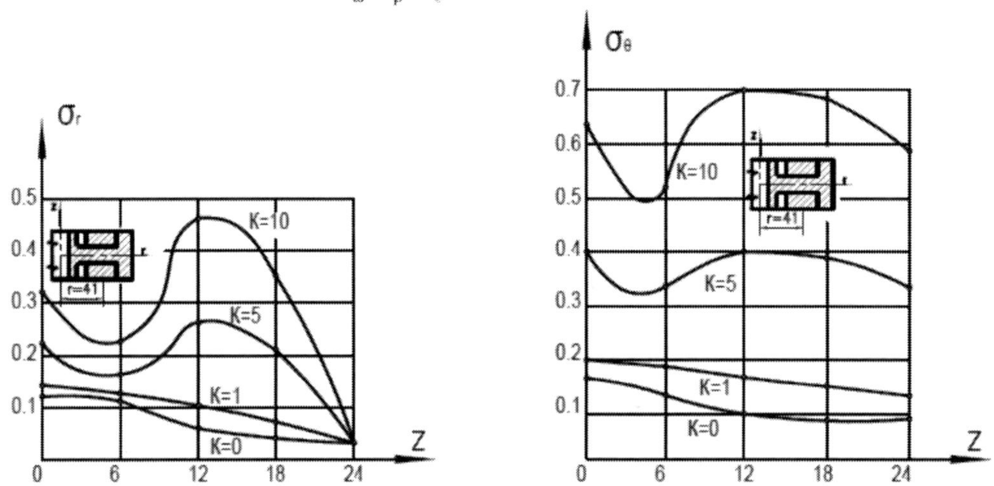


Fig. 6. The effect of the parameter $K = \rho_2 / \rho_1$ on the stresses $\sigma_r$ and $\sigma_\theta$ [MPa] in the cross-section $r = 41$ in the fractions of $K_\omega^2\ K_\rho\ K_\ell$ .

Suppose that in the considered compound structural element it is required that there were no radial expanding stresses at the outline points located at the joint of two materials. The radial expanding stresses may be eliminated through an appropriate choice of the ratio K of densities.

Supposing in (8) $\sigma_r^{(fsc)} = 0$ , we obtain the equation

$$K_\omega^2 K\rho K_l \left[ K\sigma_r^{(mod)} + (1-K)(\overline{\sigma}_r^{(mod)} - \overline{\overline{\sigma}}_r^{(mod)}) \right] = 0 , \qquad (9)$$

from which with known experimental values $\sigma_r^{(mod)}$ , $\overline{\sigma}_r^{(mod)}$ , $\overline{\overline{\sigma}}_r^{(mod)}$ and $K_\omega^2 K\rho K_l \neq 0$ we can find

$$0{,}4K + 0{,}5(1-K) = 0$$

$$K=5$$

Thus, with $K \geq 5$ there will be no radial expanding stresses at the outline points located at the joint of two materials.

## 4. Conclusions

Since the changes in the ratio of densities of materials in a compound body cause considerable changes in both the values of stresses and their distribution, it is necessary to consider the inequality of volume masses of the compound body when studying such structural elements.

## References

1. G. S. Vardanyan, Applied mechanics: the use of the similarity theory methods and the methods of the analysis of dimensions for the simulation of problems in mechanics of deformable rigid bodies. Textbook, - Moscow: NITs INFRA-M, 2016, -174 p.
2. E. K. Agakhanov, M. K. Agakhanov, *About the possibility of the use of the effect equivalence in analytical solutions to the elasticity theory problems* // MGSU Bulletin, 4/2010, **No. 3**, p.p. 140-143.
3. E. K. Agakhanov, M. K. Agakhanov, *About the simulation of the space-forces effect in a creep elastic body* // Higher education proceedings. North-Caucasus Region. Engineering sciences. -2005. **No. 1**, p. 25.
4. E. K. Agakhanov, M. K. Agakhanov, *Modeling of weighing and filtration forces in the "dam-foundation" system* / MATEC Web Conf. **86, 01012** (2016)
5. B. N. Ushakov, I. P. Frolov, Stresses in composite structural elements. Moscow, Mashinostroyenie, 1979, 134 p.
6. B. N. Ushakov, *Study of stresses in composite structural elements. VIII All-Union conference devoted to the photo-elasticity method.* Tallinn, 1979, p.p. 283-290.
7. Z. E. Abdulaliev, *Determination of temperature stresses in multi-material work-pieces*. Plant laboratory, 1970, **No. 3**, p.p. 347-349.
8. A. K. Preiss, F. A. Granenko, *Application of the "freezing" method to the determination of stresses on rotating models.* The polarization-optics method of study of stresses, edited by Prigorovsky N. I., Moscow, 1956, p.p. 271-279.

# Numerical method and calculation of two-phase swirling flows with rigid particles for technical applications

*Vadim* Akhmetov*, *Valentina* Akhmetova

Moscow state university of civil engineering, Yaroslavskoye shosse, 26, Moscow, Russia, 129337

**Abstract.** Swirling flow with particle deposition effects at the lateral surface is numerically investigated. The flow field calculation results have been obtained as the solutions of the Navier-Stokes equations. Various flow regimes with the formation of axial recirculation zones are presented. The convection-diffusion model is used for the determination of the flow particle concentration and the formation of typical sedimentation zones.

## 1 Introduction

Swirling flows have many applications in industrial engineering, in particular, in hydrotechical construction, in gas turbine combustion chambers and in vortex chambers, for the intensification of heat and mass transfer processes [1-5]. The stability of the swirling axisymmetric flows is considered in [6-7]. In this work, a vortex atomizer model which generates aerosol particle flows is discussed. The investigated device is modeled as a vertical pipe with a tangential swirler at the bottom. It is placed at some distance from the horizontal plane with the input powder. The airflow is moved through the tangential swirler into the pipe by creating a pressure gradient at the top. The powder is inducted into the pipe through the bottom section. After the intermixing of the two-phase blend, an aerosol flow at the top of the pipe is formed.

## 2 Formulation of the problem and numerical procedure

The present analysis is based upon the numerical solution of the full Navier-Stokes equations for laminar axisymmetric viscous flow. In the cylindrical coordinate system $r, \varphi, z$ the Navier-Stokes equation can be represented in terms of the stream function $\psi$, the vorticity $\Omega$ and azimuthal velocity $V_z$ in form

$$\frac{1}{r}\frac{\partial^2\psi}{\partial z^2}+\frac{\partial}{\partial r}\left(\frac{1}{r}\frac{\partial\psi}{\partial r}\right)=-\Omega \qquad (1)$$

* Corresponding author: AhmetovVK@mgsu.ru

$$\frac{\partial\Omega}{\partial t}+\frac{\partial}{\partial z}(V_z\Omega)+\frac{\partial}{\partial r}(V_r\Omega)=\frac{1}{\mathrm{Re}}\left[\frac{\partial^2\Omega}{\partial z^2}+\frac{\partial^2\Omega}{\partial r^2}+\frac{\partial}{\partial r}\left(\frac{\Omega}{r}\right)\right]+G^2\frac{1}{r}\frac{\partial(V_\varphi)^2}{\partial z} \quad (2)$$

$$\frac{\partial V_\varphi}{\partial t}+\frac{\partial}{\partial z}(V_zV_\varphi)+\frac{1}{r}\frac{\partial}{\partial r}(rV_rV_\varphi)+\frac{V_rV_\varphi}{r}=\frac{1}{\mathrm{Re}}\left[\frac{\partial^2 V_\varphi}{\partial z^2}+\frac{1}{r}\frac{\partial}{\partial r}\left(r\frac{\partial V_\varphi}{\partial r}\right)-\frac{V_\varphi}{r^2}\right] \quad (3)$$

$$V_r=-\frac{1}{r}\frac{\partial\psi}{\partial z},\quad V_z=\frac{1}{r}\frac{\partial\psi}{\partial r},\quad \Omega=\frac{\partial V_r}{\partial z}-\frac{\partial V_z}{\partial r} \quad (4)$$

The system of equation (1)-(4) is written in the conservative form and contains two dimensionless parameters: the Reynolds number $\mathrm{Re}=UR/\nu$, where $\nu$ is the kinematic viscosity, and the swirl ratio $G=W_0/U$.

Let $z_1$ be the distance from the pipe to the plane with input powder. The flow enters the pipe throughout the opened lateral part, $0\le z\le z_1$, with the radial velocity $V_0$ and the uniformly perforated pipe surface, $z_1\le z\le z_0$, with the radial velocity $kV_0$ and the azimuthal velocity $W_0$. The pore coefficient $k$ is the ratio of open to total area for the lateral surface $z_1\le z\le z_0$. The value $V_0$ is defined as

$$V_0=\frac{RU}{2}\frac{1}{(z_0-z_1)k+z_1} \quad (5)$$

where $U$ is the given mean flow velocity at the outlet section $z_k$.

The flow is considered in the cylindrical domain $D$ $(0\le z\le z_k,\ 0\le r\le 1)$. The boundary conditions involve specifying the velocity profiles at the inlet section $r=1$, $0\le z\le z_0$, no-slip conditions on the rigid surfaces, and symmetry conditions on the axis $r=0$. Soft boundary conditions are imposed in the outlet section $z=z_k$. The set of boundary conditions can be written in form

$$\psi=0,\ V_\varphi=0,\ \frac{\partial\psi}{\partial z}=0,\quad 0\le r\le 1,\ z=0 \quad (6)$$

$$\psi=f_1(z),\ V_\varphi=0,\ \frac{\partial\psi}{\partial r}=0,\quad 0\le z\le z_1,\ r=1 \quad (7)$$

$$\psi=f_2(z),\ V_\varphi=1,\ \frac{\partial\psi}{\partial r}=0,\quad z_1\le z\le z_0,\ r=1, \quad (8)$$

$$\psi=const,\ V_\varphi=0,\ \frac{\partial\psi}{\partial r}=0,\quad z_0\le z\le z_k,\ r=1 \quad (9)$$

$$\psi=0,\ V_\varphi=0,\ \Omega=0,\quad 0\le z\le z_k,\ r=0 \quad (10)$$

$$\frac{\partial\psi}{\partial z}=\frac{\partial\Omega}{\partial z}=\frac{\partial V_\varphi}{\partial z}=0,\quad 0\le r\le 1,\ z=z_k \quad (11)$$

To solve numerically the boundary value problem (1)-(4), (6)-(11) the finite difference method was used. The solution of the Poisson equation (1) was obtained using the incomplete reduction method, and the transport equations (2)-(3) were solved using the implicit block iteration method. The diffuse terms were approximated by means of central differences. For approximating the convective terms we used a modified Leonard scheme with quadratic upstream terms of third-order accuracy. Initially, for each time step equation (1) was solved with respect to $\psi$, then the values of $V_z$ and $V_r$ were calculated from formulas (4) and equation (3) was solved to determine $V_\varphi$.

The pressure distribution $p$ was found by taking the calculated velocity flow field into account. The Poisson equation with respect to the pressure gradient has the form

$$\frac{\partial^2 p}{\partial z^2}+\frac{\partial^2 p}{\partial r^2}+\frac{1}{r}\frac{\partial p}{\partial z}=2\left(-\frac{\partial V_z}{\partial r}\frac{\partial V_r}{\partial z}+\frac{\partial V_r}{\partial r}\frac{\partial V_z}{\partial z}+G^2\frac{V_\varphi}{r}\frac{\partial V_\varphi}{\partial r}\right) \qquad (12)$$

The boundary conditions for equation (12) are obtained from the equations of motion and have the form

$$\frac{\partial p}{\partial z}=-\frac{1}{\mathrm{Re}}\left(\frac{\partial \Omega}{\partial r}+\frac{\Omega}{r}\right),\quad 0\le r\le 1,\quad z=0,\quad z=z_k \qquad (13)$$

$$\frac{\partial p}{\partial r}=-\frac{1}{\mathrm{Re}}\frac{\partial \Omega}{\partial z}+\frac{V_\varphi^2}{r}+\frac{V_r^2}{r},\quad 0\le z\le z_k,\quad r=1 \qquad (14)$$

$$\frac{\partial p}{\partial r}=0,\quad 0\le z\le z_k,\quad r=0 \qquad (15)$$

For solving the boundary-value problem (12)-(15) the implicit block iteration method was used.

## 3 Flow field calculation results

The flow fields corresponding to the boundary-value problem (1)-(11) were calculated for $\mathrm{Re}=100$–$2000$, $G=0$–$8$, $k=0.4$–$0.9$, $z_0=1.4$–$2.4$, and $z_1=0.4$–$1.4$. The results show that when a critical value of swirl is reached, a recirculation zone is formed.

The typical streamline patterns for the calculated flows are presented in Fig. 1. Consider the flow field structure for an increase of swirl ratio $G$ and fixed parameters $\mathrm{Re}=100$, $k=0.5$, $z_0=2$, $z_1=1$. When the swirl ratio is small $G=1$, the flow pattern is qualitatively similar to that produced by radial blowing in a pipe through the lateral surface. In this case the swirl does not significantly influence the flow and at the outlet section, $(z=z_k)$, a Poiseuille profile is established. With an increase in the swirl, $G=3$ (Fig. 1,a), a reverse flow zone is generated near the wall and the typical looping bend of the streamline is formed. With a further increase in the swirl to $G=6$, the center of this recirculation zone is moved to the plane $z=0$ and the length of the reverse flow is extended both in the axial and radial directions. The maximum value of the azimuthal velocity reaches $-0.5U$ and $-0.8U$ for $G=3$, 6 respectively and occurs near the wall.

As the Reynolds number increases, the flow structure changes. For small swirl numbers $(G=2)$ the flow structure and recirculation zone size do not change as Re is increase to 1000. However with an increase in swirl to $G=4$ and $\mathrm{Re}=500$ (Fig. 1,b), the shape of the reverse flow region is changed: the recirculation zone is U-shaped and the

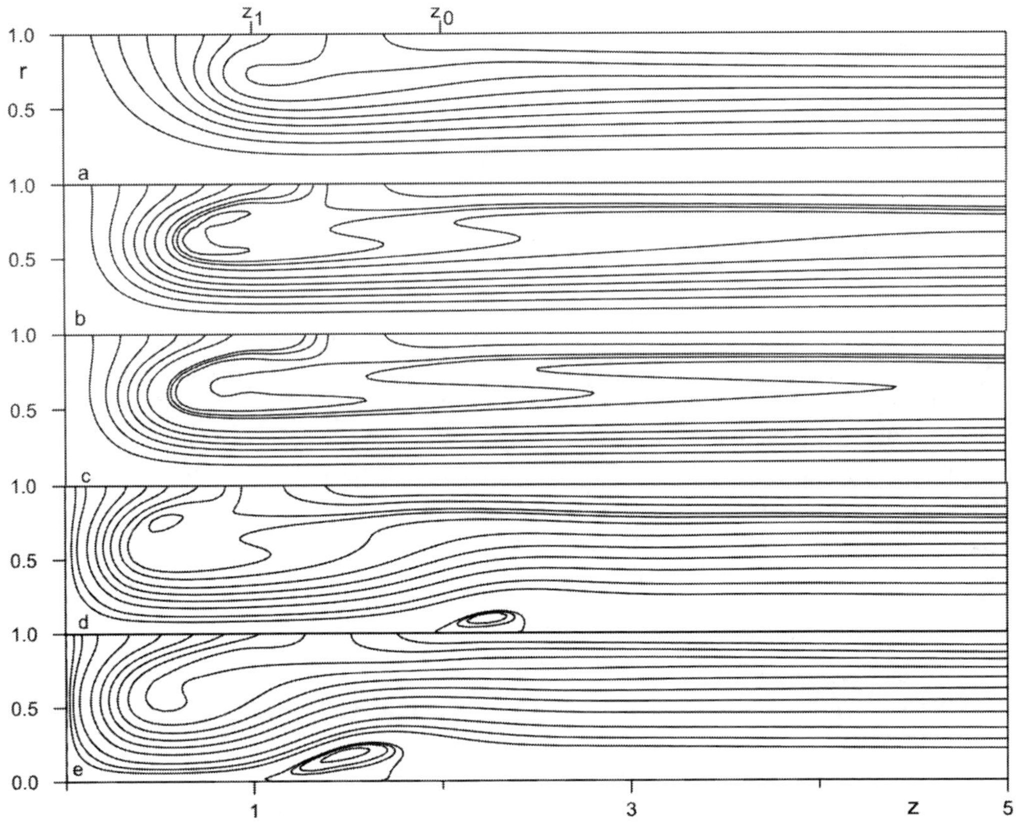


**Fig. 1.** Streamline patterns: (a) – Re=100, $G$=3, $k$=0.5, $z_1=1$; (b) – Re=500, $G$=4, $k$=0.5, $z_1=1$; (c) – Re=750, $G$=4, $k$=0.5, $z_1=1$; (d) – Re=250, $G$=4, $k$=0.5, $z_1=0.6$; (e) – Re=250, $G$=6, $k$=0.7, $z_1=1$.

streamlines behind it make a bend. At higher Reynolds numbers the length of the reverse flow zone increases and extends to the outlet section, $z=z_k$, for $\mathrm{Re}=750$ (Fig. 1,c).

When the pipe is moved to the plane $z=0$, the flow rate entering the pipe through the opened lateral part without the swirl is decreased. Subsequent changes in flow field structure are similar to those expected for an increase in swirl ratio, $G$. Thus for small Reynolds number and swirl ( $\mathrm{Re}=100$, $G=3$) with decrease $z_1=0.8$, 0.6 the recirculation zone is moved to the plane $z=0$, but the flow patterns are not changed qualitatively.

At higher Reynolds number and swirl ratio ( $\mathrm{Re}$=250, $G=4$) a decrease of the distance $z_1$ leads to an appearance of zigzag streamlines at $z_1=0.8$ and then at $z_1=0.6$ (Fig. 1,d). The recirculation zone is formed near the axis. There is analogous to the vortex breakdown phenomenon [8-10].

Increasing the pore coefficient $k$ for other fixed parameters leads to an increase of the flow rate entering the pipe with the swirl. As in the previous case there is no significant change to flow field structure for $k$ from 0.5 to 0.9, small Reynolds number $\mathrm{Re}=100$, 250, and medium values of $G=2$ at $z_0=2$, $z_1=1$. With further increase in the swirl to $G=6$ at $\mathrm{Re}=250$, and $k=0.6$ the typical looping bend of the streamline is formed. With an increase in $k$ to 0.7 a reverse flow zone with a toroidal structure appears near the axis which moves upstream to $k=0.8$ (Fig. 1,e).

## 4 Convection-diffusion model

The numerical investigations of two-phase flow with rigid particles are based upon the diffusion model. We assume that the influence of the particle motion on the basic flow is negligible. The flow field is defined as a sum of the particle velocity and the liquid phase velocity. In this case the equation of volume global continuity transforms to the diffusion equation for the particle concentration

$$\frac{\partial c}{\partial t}+\frac{\partial}{\partial z}(V_z c)+\frac{1}{r}\frac{\partial}{\partial r}(r V_s c)=\frac{1}{\mathrm{Re}\,Sc}\left[\frac{\partial^2 c}{\partial z^2}+\frac{1}{r}\frac{\partial}{\partial r}\left(r\frac{\partial c}{\partial r}\right)\right] \quad (16)$$

in which $c$ is the particle concentration, $Sc = \nu / D$ is the Schmidt number, and $D$ is the diffusion coefficient. The convection-diffusion equation (16) has been used for investigations of particle deposition in [11-13].

For the determination of $V_s$ in this case experimental results [14] have been used. It was found, that for the particle motion in a vortex chamber only two forces are important: the centrifugal force $F_c$ and the Stokes force $F_{st}$:

$$F_c = \frac{4}{3}\pi r_p^3 \rho_p \frac{V_\varphi'^2}{r}, \quad F_{st} = 6\pi\mu r_p (V_s' - V_r') \quad (17)$$

where the prime denotes dimensional velocities.

We require that the Taylor number $Ta$ meet the following condition:

$$Ta = \frac{W_0 r}{\nu} \ll 1 \quad (18)$$

Then it is permissible to assume that the azimuthal velocity of the particle is equal to the azimuthal velocity of the liquid phase. Hence, the deposition velocity $V_s$ can be found from the particle motion equation in the radial direction:

$$\frac{dV_s}{dt} = G^2 \frac{V_\varphi^2}{r} - \frac{1}{2\,St}(V_s - V_r), \quad St = \frac{\rho_p r_p^2 U}{9\nu\rho R} \quad (19)$$

The velocity $V_s(z)$ is defined from (19) as an average value with respect to $r$, taking into consideration that $St \ll 1$ and $G^2 \cdot St = O(1)$:

$$V_s(z) = 2StG^2\langle V_\varphi^2 / r\rangle + \langle V_r\rangle \quad (20)$$

The obtained distribution $V_s(z)$ is substituted into equation (16).

Let us assume that the input powder location is the domain $P(0 \le r \le r_1, z = 0)$. Therefore the boundary conditions for equation (16) can be written:

$$z = 0: \quad c = 1, \ 0 \le r \le r_1; \quad \frac{\partial c}{\partial z} = 0, \ r_1 \le r \le 1 \quad (21)$$

$$z = z_k: \quad \frac{\partial c}{\partial z} = 0, \quad 0 \le r \le 1 \quad (22)$$

$$r = 0, r = 1: \quad \frac{\partial c}{\partial r} = 0, \quad 0 \le z \le z_k \tag{23}$$

In this case reflection effects for particle deposition at the lateral pipe surface are not considered. We assume that the particle flux defined as $j = D\partial c/\partial r + V_s c$ has only a convection part and the influence of diffusion is negligible $(\partial c/\partial r = 0)$.

For solving the boundary value problem (16), (21)-(23) we used a finite difference relaxation method.

## 5 Calculation results for flow fields concentration

Numerical solutions of the boundary value problem (16), (21)-(23) were obtained for $Sc = 1-10$, $St = 10^{-5} - 10^{-1}$.

First consider the particle deposition process on the lateral pipe surface. The particle concentration flux through the wall is given as $Q_w = V_s c$. Fig. 2 shows the variation $Q_w$ with $z$ for $\mathrm{Re} = 100$, $G = 4, 4.5, 5, 6$ (curves 1-4). For $G = 4.5$ (curve 2) the distribution has a minimum value at $z \approx 1.95$. The more strongly expressed maximum is noted for $G = 5, 6$ (curves 3,4). At the same time with decreases in the swirl, $G = 4$ (curve 1), the deposition profile is practically uniform. Thus a zone with maximum particle deposition is formed. The presence of such a zone is connected with the swirl ratio.

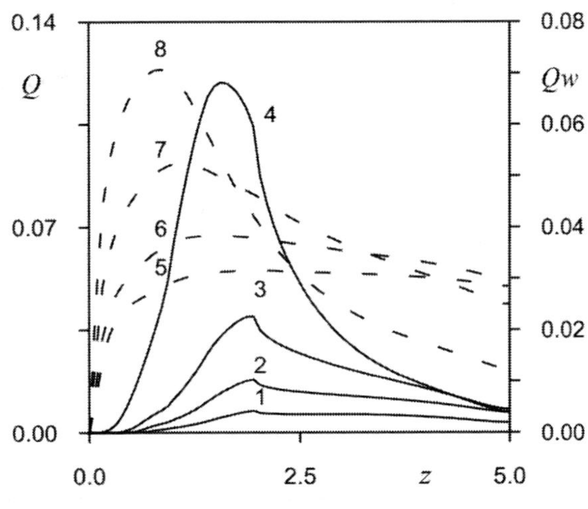


**Fig. 2.** Flux concentration $Q_w$ (1-4 : G=4, 4.5, 5, 6) and flow particle rate $Q$ (5-8: G=3, 4, 5, 6) for Re=100.

Significant particle deposition on the lateral pipe surface is an undesirable effect. The purpose of the considered device is to form an aerosol flow at the outlet section. Excessive swirl can lead to undesirable effects. This fact is illustrated in Fig. 2. Here the variation of the flow rate $Q(z)$ is given by

$$Q = 2\pi \int_0^1 V_z rc dr \tag{24}$$

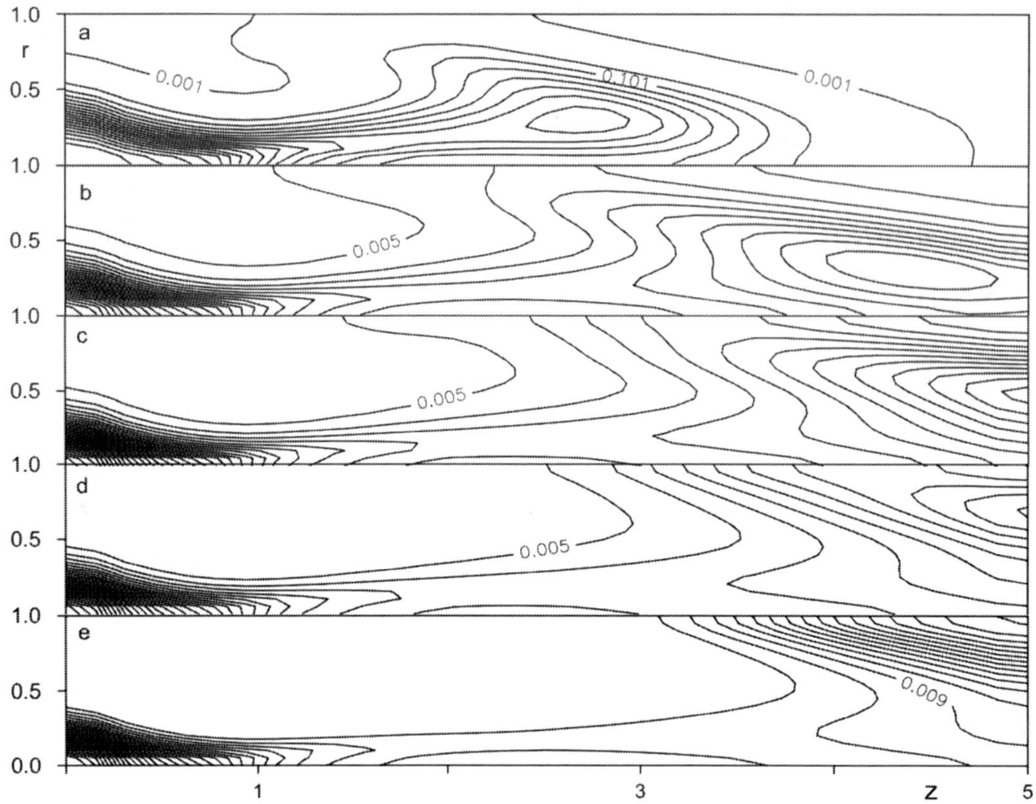


**Fig. 3.** Concentration fields, Re=100, *G*=3, (a) – (e): t=1, 2, 3, 4, 5, $\Delta c$ =0.05, 0.025, 0.05, 0.01, 0.002.

for $\mathrm{Re}=100$, $Sc=1$, $St=0.022$, $k=0.5$, $z_1=1$, $G=3, 4, 5, 6$ (curves 5-8). Fig. 2 shows that small swirl ($G=3$) is insufficient for effective mixing. At the same time, strong swirl ($G=6$) leads to significant particle deposition on the wall. The optimal swirl for the given parameters is $G=5$.

Consider the nonstationary problem for particle transport and deposition. We assume that initially the particles occupy the volume $0 \le r \le r_1$, $0 \le z \le z_1$. So we specify the initial particle concentration as follows:

$$c=1,\ 0 \le r \le r_1, 0 \le z \le z_1;\quad c=0,\ r > r_1,\ z > z_1 \tag{25}$$

The set of boundary conditions can be written:

$$\frac{\partial c}{\partial z}=0,\quad 0 \le r \le 1,\quad z=0,\quad z=z_k \tag{26}$$

$$\frac{\partial c}{\partial r}=0,\quad 0 \le z \le z_k,\quad r=0,\quad r=1 \tag{27}$$

The solution of the value boundary problem (16), (25)-(27) was carried out for the following parameters: $Sc = 1$, $St = 0.022$, $k = 0.5$, $z_1 = 1$, $z_0 = 2$, $r_1 = 0.5$. The typical concentration field for the different time with the steps ( $\Delta c$ ) is presented in Fig. 3.

## Conclusions

The above examples give a fairly detailed picture of the atomization of a powder to obtain an aerosol using an open cylindrical channel in which a swirling flow is formed. The mathematical model used makes it possible qualitatively to describe the basic properties of the flow, namely, the formation of recirculation zones, the onset of rarefaction in the neighborhood of the flow axis under the action of the swirl, and the entrainment of powder by the flow with partial deposition on the lateral surface of the dispersing apparatus.

## References

1. A. Gupta, G. Lilley, *Swirl Flows* (Abacus Press, 1984)
2. S.V. Alekseenko, P.A. Kuibin, V.L. Okulov, *Introduction to theory of concentrated vortices* (Institute of Computer Science, Moscow, 2005)
3. V.K. Akhmetov , V.Ya. Shkadov, *Numerical simulation of viscous vortex flows for applications to technics* (ASV Publishing House, Moscow, 2009)
4. V.K. Akhmetov, *Structure and hydrodynamic stability of swirling flows with recirculation zones* (MGSU, Moscow, 2009)
5. V.K. Akhmetov, V.Ya. Shkadov, V.P. Shkadova, Fluid Dynamics, **41** (4), 504 (2006)
6. V.K. Akhmetov, V.Ya. Shkadov, Vestn. Mosk. Univ., Ser. I (5), 54 (2005)
7. V.K. Akhmetov, V.Ya. Shkadov, Vestn. Mosk. Univ., Ser. I (1), 66 (2003)
8. T. Sarpkaya, Trans. ASME, J. Fluids Eng., **111**, 5 (1989)
9. M. Escudier, Progr. Aerosp. Sci., **25**, 189 (1988)
10. M.A. Herrada, V.N. Shtern, J.M. López-Herrera, Physics of Fluids, **25**, 093604 (2013)
11. R. Tsai, Y.P. Chang, T.Y. Lin, J. Aerosol Sci., **29,** 811 (1998)
12. J. Mang, E. Minkov, U. Schaflinger, M. Ungarish, Trans. ASME. J. Fluids Eng., **120**, 676 (1998)
13. B.T. Tan, Y.S. Liow, L. Mununga, M.C. Thompson, K. Hourigan, Physics of Fluids, **21**, 024104 (2009)
14. S.S. Kutateladze, E.P. Volchkov, V.I. Terekhov, *Aerodynamics and heat and mass exchange in bounded vortex flows* (Institute of Thermophysics, Novosibirsk, 1987)

# Operational and Variational Formulations of Boundary Problems of Anisotropic Plate Analysis, Adapted for Application Within Discrete-Continual Methods

*Pavel* Akimov[1,2,*], *Vladimir* Sidorov[3,4,5], *Zbigniew* Wójcicki[6]

[1]Russian Academy of Architecture and Construction Sciences; 24-1, ul. Bolshaya Dmitrovka, Moscow, 107031, Russia
[2]Scientific Research Center "StaDyO"; office 810, ul. 3ya Yamskogo Polya, Moscow, 125040, Russia
[3]Kielce University of Technology; Al. 1000-lecia Państwa Polskiego 5, 25-541, Kielce, Poland
[4]Moscow State University of Railway Engineering; 9b9 Obrazcova Street, Moscow, 127994, Russia
[5]Perm National Research Polytechnic University; 29 Komsomolsky prospekt, Perm, Perm Krai, 614990, Russia
[6]Civil Engineering Faculty, Wrocław University of Technology, Wyb. Wyspiańskiego 27, 50-370 Wrocław, Poland

**Abstract.** This paper is devoted to operational and variational formulations [1] of boundary problems of anisotropic plate analysis (with the use of so-called method of extended domain [2]), adapted for applications within discrete-continual methods (discrete-continual finite element method (DCFEM), discrete-continual variation-difference method (DCVDM)) [1]. Generally, the field of application of these methods, which are now becoming available for computer realization, comprises structures with regular (in particular, constant or piecewise constant) physical and geometrical parameters in some dimension. Considering problems remain continual along "basic" direction while along other directions discrete-continual methods presuppose mesh approximation.

## 1 Introduction

As is known, structures from anisotropic materials are widely used in modern construction. Anisotropic materials normally have sharp difference in the elastic properties for different directions. A classic vital sample here is natural wood. Its elastic modulus in tension parallel to the grain is well above corresponding modulus in tension across the grain. Besides, elastic constants of natural wood depend on the direction with respect to the wood grains. Anisotropic (and inhomogeneous) materials are also include composite materials (particularly rather popular in aircraft engineering), crystals, some rocks, concrete, etc. Here we indicated materials with "natural" anisotropy. However, materials with so-called "artificial" anisotropy are applied in construction as well. We should mention here as a samples corrugated plates and shells from isotropic material or rib-

* Corresponding author: pavel.akimov@gmail.com

reinforced plates and shell. Thus we have to consider boundary problems of anisotropic plate analysis [3-6].

## 2 Standard operational and variational formulations of boundary problem within method of extended domain

Operational formulation of the problem in the domain $\omega$ has the form:

$$\sum_{i=1}^{2}\sum_{j=1}^{2}\partial_j^2\overline{B}_{i,j}\partial_i^2 w+4\partial_1\partial_2\overline{B}_{3,3}\partial_1\partial_2 w+$$
$$+2\sum_{i=1}^{2}(\partial_i^2\overline{B}_{i,3}\partial_1\partial_2 w_k+\partial_1\partial_2\overline{B}_{i,3}\partial_i^2 w_k)+\overline{c}w=\theta q+\delta_\Gamma' M_\Gamma+\delta_\Gamma Q_\Gamma, \quad (1)$$

where $\Omega$ is the domain, occupied by structure with boundary $\Gamma=\partial\Omega$; $\omega$ is extended domain, embordering $\Omega$; $x_1,x_2$ are coordinates in plan; $x_3$ is coordinate of plate thickness; $\theta=\theta(x_1,x_2)$ is the characteristic function of domain $\Omega$; $\delta_\Gamma=\delta_\Gamma(x_1,x_2)$ is the delta-function of border $\Gamma=\partial\Omega$; $\overline{n}=[\,n_1\;\;n_2\,]^T$ is unit normal vector of domain boundary $\Gamma=\partial\Omega$; $\delta_\Gamma'=\delta_\Gamma'(x_1,x_2)$ is corresponding normal derivative; $w$ is plate deflection in domain $\Omega$; $q$ is the load density in domain $\Omega$; $Q_\Gamma$, $M_\Gamma$ are shear force and torque moment at the border $\Gamma=\partial\Omega$; $c$ is modulus of foundation;

$$\partial_k=\partial/\partial x_k,\; \partial_k^*=-\partial/\partial x_k,\; k=1,2; \quad (2)$$

$$\overline{c}=\theta c; \quad \delta_\Gamma'=\delta_\Gamma'(x_1,x_2)=\partial\delta_\Gamma(x_1,x_2)/\partial\overline{n}; \quad (3)$$

$B_{1,1}, B_{2,2}, B_{3,3}, B_{1,2}=B_{2,1}, B_{2,3}=B_{3,2}, B_{1,3}=B_{3,1}$ are constants (parameters) that characterize the elastic properties of structure;

$$\overline{B}_{1,1}=\theta B_{1,1},\; \overline{B}_{1,3}=\theta B_{1,3},\; \overline{B}_{2,3}=\theta B_{2,3},\; \overline{B}_{1,2}=\theta B_{1,2},\; \overline{B}_{3,3}=\theta B_{3,3};$$
$$\overline{B}_{1,2}=\theta B_{1,2},\; \overline{B}_{3,3}=\theta B_{3,3},\; \overline{B}_{2,2}=\theta B_{2,2}; \quad (4)$$

and in the particular case of an isotropic plate, we have

$$B_{1,1}=B_{2,2}=D;\quad B_{1,2}=B_{2,1}=D\nu;\quad B_{3,3}=0.5\cdot D(1-\nu);\quad B_{1,3}=B_{2,3}=B_{3,1}=B_{3,2}=0. \quad (5)$$

Internal moments and plate forces are determined by formulas

$$M_1=B_{1,1}\chi_1+B_{1,2}\chi_2+B_{1,3}\chi_{12};\quad M_2=B_{1,2}\chi_1+B_{2,2}\chi_2+B_{2,3}\chi_{12}; \quad (6)$$

$$M_{12}=M_{21}=B_{1,3}\chi_1+B_{2,3}\chi_2+B_{3,3}\chi_{12}; \quad (7)$$

$$N_1=B_{1,1}\partial_1\chi_1+3B_{1,3}\partial_2\chi_1+(B_{1,2}+2B_{3,3})\partial_1\chi_2+B_{2,3}\partial_2\chi_2; \quad (8)$$

$$N_2=B_{1,3}\partial_1\chi_1+(B_{1,2}+2B_{3,3})\partial_2\chi_1+3B_{2,3}\partial_1\chi_2+B_{2,2}\partial_2\chi_2; \quad (9)$$

where $\chi_1, \chi_2, \chi_{12}=\chi_{21}$ are strains for $x_3=1$,

$$\chi_1=-\partial_1^2 w;\quad \chi_2=-\partial_2^2 w;\quad \chi_{12}=\chi_{21}=-2\partial_1\partial_2 w \quad (10)$$

Variational formulation of the problem is formulated in the form of the corresponding energy functional (this formulation is convenient in the algorithmic sense):

$$\Phi(w)=\frac{1}{2}\iint_{\omega}[\theta(M_1\chi_1+M_2\chi_2+2M_{1,2}\chi_{1,2})+\bar{c}w^2]dx_1dx_2-\iint_{\omega}qwdx_1dx_2\,. \tag{11}$$

## 3 Operational and variational displacement formulations of boundary problem within method of extended domain

If we combine (6)-(10) we can get formulation (1) in the following form

$$Lw=F\,, \tag{12}$$

where $L$ is so-called stiffness operator,

$$\begin{aligned}\mathrm{L}&=\partial_1^2\bar{B}_{1,1}\partial_1^2+\partial_2^2\bar{B}_{1,2}\partial_1^2+\partial_1^2\bar{B}_{2,1}\partial_2^2+\partial_2^2\bar{B}_{2,2}\partial_2^2+4\partial_1\partial_2\bar{B}_{3,3}\partial_1\partial_2+\\&+2\partial_1^2\bar{B}_{1,3}\partial_1\partial_2+2\partial_2^2\bar{B}_{2,3}\partial_1\partial_2+2\partial_1\partial_2\bar{B}_{1,3}\partial_1^2+2\partial_1\partial_2\bar{B}_{2,3}\partial_2^2+\bar{c};\end{aligned} \tag{13}$$

$$\mathrm{F}=\theta q+\delta'_\Gamma M_\Gamma+\delta_\Gamma Q. \tag{14}$$

We can rewrite (11) in matrix form:

$$\Phi(w)=\frac{1}{2}\iint_{\omega}[(\theta\bar{N},\bar{\chi})+\bar{c}w^2]dx_1dx_2-\iint_{\omega}qwdx_1dx_2\,, \tag{15}$$

where

$$\bar{N}=[M_1\ \ M_2\ \ M_{1,2}\ \ M_{2,1}]^T;\quad \bar{\chi}=[\chi_1\ \ \chi_2\ \ \chi_{1,2}\ \ \chi_{2,1}]^T\,. \tag{16}$$

Using (6)-(9), we get

$$\bar{N}=A\bar{\chi}\,,\qquad \bar{\chi}=Bw\,, \tag{17}$$

where

$$A=\begin{bmatrix}B_{1,1}&B_{1,2}&0.5\cdot B_{1,3}&0.5\cdot B_{1,3}\\B_{1,2}&B_{2,2}&0.5\cdot B_{2,3}&0.5\cdot B_{2,3}\\B_{1,3}&B_{2,3}&0.5\cdot B_{3,3}&0.5\cdot B_{3,3}\\B_{1,3}&B_{2,3}&0.5\cdot B_{3,3}&0.5\cdot B_{3,3}\end{bmatrix};\quad B=-\begin{bmatrix}\partial_1^2\\\partial_2^2\\2\partial_1\partial_2\\2\partial_1\partial_2\end{bmatrix}; \tag{18}$$

Thus we have:

$$\Phi(w)=\frac{1}{2}\iint_{\omega}[(B^*\bar{A}Bw,w)+\bar{c}w^2]dx_1dx_2-\iint_{\omega}qwdx_1dx_2\,, \tag{19}$$

where

$$\bar{A}=\theta A=\begin{bmatrix}\bar{B}_{1,1}&\bar{B}_{1,2}&0.5\cdot\bar{B}_{1,3}&0.5\cdot\bar{B}_{1,3}\\\bar{B}_{1,2}&\bar{B}_{2,2}&0.5\cdot\bar{B}_{2,3}&0.5\cdot\bar{B}_{2,3}\\\bar{B}_{1,3}&\bar{B}_{2,3}&0.5\cdot\bar{B}_{3,3}&0.5\cdot\bar{B}_{3,3}\\\bar{B}_{1,3}&\bar{B}_{2,3}&0.5\cdot\bar{B}_{3,3}&0.5\cdot\bar{B}_{3,3}\end{bmatrix};\quad B^*=-[\partial_1^2\ \ \partial_2^2\ \ 2\partial_2\partial_1\ \ 2\partial_2\partial_1\,]. \tag{20}$$

## 4 Operational and variational formulations of boundary problem in displacements with extraction of basic direction

Without loss of generality we suppose constancy of physical and geometrical parameters of plate along coordinate $x_2$ (it is so-called "basic direction").

We can rewrite (12) in the following form:

$$L = -L_4\partial_2^4 + L_3\partial_2^3 + L_2\partial_2^2 + L_1\partial_2 + L_0 + \bar{c} , \tag{21}$$

where

$$L_4 = -\overline{B}_{2,2} ; \quad L_3 = 2\overline{B}_{2,3}\partial_1 + 2\partial_1\overline{B}_{2,3} ; \quad L_2 = \overline{B}_{1,2}\partial_1^2 + \partial_1^2\overline{B}_{2,1} + 4\partial_1\overline{B}_{3,3}\partial_1 ;$$

$$L_1 = 2\partial_1^2\overline{B}_{1,3}\partial_1 + 2\partial_1\overline{B}_{1,3}\partial_1^2 ; \quad L_0 = \partial_1^2\overline{B}_{1,1}\partial_1^2 . \tag{22}$$

We can rewrite (21):

$$-L_4\partial_2^4 w + L_3\partial_2^3 w + L_2\partial_2^2 w + L_1\partial_2 w + (L_0 + \bar{c})w = F . \tag{23}$$

Let us introduce the following notation

$$y_1 = y_1(x_1, x_2) = w(x_1, x_2) ; \quad y_i = y_i(x_1, x_2) = \partial_2^{i-1} w(x_1, x_2), \quad i = 2, 3, 4 . \tag{24}$$

If we combine (23) and (24) we get:

$$-L_4\partial_2 y_4 + L_3 y_4 + L_2 y_3 + L_1 y_2 + (L_0 + \bar{c}) y_1 = F . \tag{25}$$

Uniting (24) and (25) we obtain

$$\begin{bmatrix} 1 & 0 & 0 & 0 \\ 0 & 1 & 0 & 0 \\ 0 & 0 & 1 & 0 \\ 0 & 0 & 0 & L_4 \end{bmatrix} \begin{bmatrix} y_1' \\ y_2' \\ y_3' \\ y_4' \end{bmatrix} = \begin{bmatrix} 0 & 1 & 0 & 0 \\ 0 & 0 & 1 & 0 \\ 0 & 0 & 0 & 1 \\ L_0 + \bar{c} & L_1 & L_2 & L_3 \end{bmatrix} \begin{bmatrix} y_1 \\ y_2 \\ y_3 \\ y_4 \end{bmatrix} - \begin{bmatrix} 0 \\ 0 \\ 0 \\ F \end{bmatrix} , \tag{26}$$

where

$$y_i'(x_1, x_2) = \partial_2 y_i(x_1, x_2), \quad i = 2, 3, 4 . \tag{27}$$

Thus we have the following system of the first-order ordinary differential (with respect to $x_2$) equations with operator coefficients

$$\begin{bmatrix} y_1' \\ y_2' \\ y_3' \\ y_4' \end{bmatrix} = \begin{bmatrix} 0 & 1 & 0 & 0 \\ 0 & 0 & 1 & 0 \\ 0 & 0 & 0 & 1 \\ \mathrm{L}_4^{-1}(\mathrm{L}_0 + \bar{c}) & \mathrm{L}_4^{-1}\mathrm{L}_1 & \mathrm{L}_4^{-1}\mathrm{L}_2 & \mathrm{L}_4^{-1}\mathrm{L}_3 \end{bmatrix} \begin{bmatrix} y_1 \\ y_2 \\ y_3 \\ y_4 \end{bmatrix} - \begin{bmatrix} 0 \\ 0 \\ 0 \\ \mathrm{L}_4^{-1}\mathrm{F} \end{bmatrix} \tag{28}$$

or

$$\overline{U}' = \tilde{L}\overline{U} + \overline{\overline{F}} , \tag{29}$$

where

$$\widetilde{L}=\begin{bmatrix}0 & 1 & 0 & 0\\ 0 & 0 & 1 & 0\\ 0 & 0 & 0 & 1\\ L_4^{-1}(L_0+\bar{c}) & L_4^{-1}L_1 & L_4^{-1}L_2 & L_4^{-1}L_3\end{bmatrix};\quad \overline{\overline{F}}=-\begin{bmatrix}0\\0\\0\\L_4^{-1}F\end{bmatrix};\tag{30}$$

$$\overline{U}=\begin{bmatrix}y_1\\y_2\\y_3\\y_4\end{bmatrix};\quad \overline{U}'=\partial_2\overline{U}=\begin{bmatrix}\partial_2 y_1\\\partial_2 y_2\\\partial_2 y_3\\\partial_2 y_4\end{bmatrix}=\begin{bmatrix}y_1'\\y_2'\\y_3'\\y_4'\end{bmatrix}.\tag{31}$$

Equations (29), of course, should be supplemented with boundary conditions given in cross-sections with coordinates $x_{2,k}^b$, $k=1,\ldots,n_k$ This boundary conditions have form

$$B_k^-\overline{U}_{k-1}(x_{2,k}^b-0)+B_k^+\overline{U}_k(x_{2,k}^b+0)=\overline{g}_k^-+\overline{g}_k^+,\quad k=2,\ldots,n_k-1;\tag{32}$$

$$B_1^+\overline{U}_1(x_{2,1}^b+0)+B_{n_k}^-\overline{U}_{n_k-1}(x_{2,n_k}^b-0)=\overline{g}_1^+ +\overline{g}_{n_k}^-,\tag{33}$$

where $B_k^-, B_k^+$, $k=2,\ldots,n_k-1$, $B_1^+$, $B_{n_k}^-$ are matrices of boundary conditions of the fourth order; $\overline{g}_k^-, \overline{g}_k^+$, $k=2,\ldots,n_k-1$, $\overline{g}_1^+$, $\overline{g}_{n_k}^-$ are right-side vectors of boundary conditions.

Operators (18) and (21) can be rewritten in the following form:

$$B=\partial_1^2 B_{2,0}+\partial_1\partial_2 B_{1,1}+\partial_2^2 B_{0,2};\quad B^*=\partial_1^2 B_{2,0}^*+\partial_1\partial_2 B_{1,1}^*+\partial_2^2 B_{0,2}^*,\tag{34}$$

where

$$B_{2,0}=-[1\ \ 0\ \ 0\ \ 0]^T;\ B_{1,1}=-[0\ \ 0\ \ 2\ \ 2]^T;\ B_{0,2}=-[0\ \ 1\ \ 0\ \ 0]^T;\tag{35}$$

$$B_{2,0}^*=-[1\ \ 0\ \ 0\ \ 0];\ \ B_{1,1}^*=-[0\ \ 0\ \ 2\ \ 2];\ \ B_{0,2}^*=-[0\ \ 1\ \ 0\ \ 0].\tag{36}$$

Thus, we have:

$$\begin{aligned}(B^*\overline{A}Bw,w)&=((\partial_1^2 B_{2,0}^*+\partial_2^*\partial_1^* B_{1,1}^*+\partial_2^2 B_{0,2}^*)\overline{A}(\partial_1^2 B_{2,0}+\partial_1\partial_2 B_{1,1}+\partial_2^2 B_{0,2})w,w)=\\&=(\partial_1^2 B_{2,0}^*\overline{A}\partial_1^2 B_{2,0}w,w)+(\partial_1^2 B_{2,0}^*\overline{A}\partial_1\partial_2 B_{1,1}w,w)+\\&+(\partial_1^2 B_{2,0}^*\overline{A}\partial_2^2 B_{0,2}w,w)+(\partial_2^*\partial_1^* B_{1,1}^*\overline{A}\partial_1^2 B_{2,0}w,w)+\\&+(\partial_2^*\partial_1^* B_{1,1}^*\overline{A}\partial_1\partial_2 B_{1,1}w,w)+(\partial_2^*\partial_1^* B_{1,1}^*\overline{A}\partial_2^2 B_{0,2}w,w)+\\&+(\partial_2^2 B_{0,2}^*\overline{A}\partial_1^2 B_{2,0}w,w)+(\partial_2^2 B_{0,2}^*\overline{A}\partial_1\partial_2 B_{1,1}w,w)+\\&+(\partial_2^2 B_{0,2}^*\overline{A}\partial_2^2 B_{0,2}w,w)=(B_{0,2}^*\overline{A}B_{0,2}y_3,y_3)+\\&+(B_{0,2}^*\overline{A}\partial_1 B_{1,1}y_2,y_3)+(\partial_1^* B_{1,1}^*\overline{A}B_{0,2}y_3,y_2)+\\&+(B_{0,2}^*\overline{A}B_{2,0}\partial_1^2 y_1,y_3)+(\partial_1^2 B_{2,0}^*\overline{A}B_{0,2}y_3,y_1)+\\&+(\partial_1^* B_{1,1}^*\overline{A}B_{2,0}\partial_1^2 y_1,y_2)+(\partial_1^2 B_{2,0}^*\overline{A}\partial_1 B_{1,1}y_2,y_1)+\\&+(\partial_1^* B_{1,1}^*\overline{A}B_{1,1}\partial_1 y_2,y_2)+(\partial_1^2 B_{2,0}^*\overline{A}B_{2,0}\partial_1^2 y_1,y_1),\end{aligned}$$

and finally

$$\begin{aligned}(B^*\overline{A}Bw,w)&=(B_{0,2}^*\overline{A}B_{0,2}y_3,y_3)+(B_{0,2}^*\overline{A}\widetilde{B}_{1,1}y_2,y_3)+(\widetilde{B}_{1,1}^*\overline{A}_k B_{0,2}y_3,y_2)+\\&+(B_{0,2}^*\overline{A}\widetilde{\widetilde{B}}_{2,0}y_1,y_3)+(\widetilde{\widetilde{B}}_{2,0}^*\overline{A}B_{0,2}y_3,y_1)+(\widetilde{B}_{1,1}^*\overline{A}\widetilde{\widetilde{B}}_{2,0}y_1,y_2)+\\&+(\widetilde{\widetilde{B}}_{2,0}^*\overline{A}\widetilde{B}_{1,1}y_2,y_1)+(\widetilde{B}_{1,1}^*\overline{A}\widetilde{B}_{1,1}y_2,y_2)+(\widetilde{\widetilde{B}}_{2,0}^*\overline{A}\widetilde{\widetilde{B}}_{2,0}y_1,y_1),\end{aligned}\tag{37}$$

where

$$B_{2,0} = -[\partial_1^2 \ \ 0 \ \ 0 \ \ 0]^T; \ B_{1,1} = -[0 \ \ 0 \ \ 2\partial_1 \ \ 2\partial_1]^T; \tag{38}$$

$$B_{2,0}^* = -[\partial_1^2 \ \ 0 \ \ 0 \ \ 0]; \ \ B_{1,1}^* = -[0 \ \ 0 \ \ 2\partial_1^* \ \ 2\partial_1^*]. \tag{39}$$

We have

$$(B^*\overline{A}Bw, w) = (K\overline{U}^r, \overline{U}^r), \tag{40}$$

where

$$K = \begin{bmatrix} \widetilde{B}_{2,0}^*\overline{A}\widetilde{B}_{2,0} & \widetilde{B}_{2,0}^*\overline{A}\widetilde{B}_{1,1} & \widetilde{B}_{2,0}^*\overline{A}B_{0,2} \\ \widetilde{B}_{1,1}^*\overline{A}\widetilde{B}_{2,0} & \widetilde{B}_{1,1}^*\overline{A}\widetilde{B}_{1,1} & \widetilde{B}_{1,1}^*\overline{A}_k B_{0,2} \\ B_{0,2}^*\overline{A}\widetilde{B}_{2,0} & B_{0,2}^*\overline{A}\widetilde{B}_{1,1} & B_{0,2}^*\overline{A}B_{0,2} \end{bmatrix}; \ \ \overline{U}^r = \begin{bmatrix} y_1 \\ y_2 \\ y_3 \end{bmatrix}. \tag{41}$$

Thus, functional (19) can be rewritten in the following form:

$$\Phi(w) = \Phi(\overline{U}^r) = \frac{1}{2}\iint_\omega (\widetilde{K}\overline{U}^r, \overline{U}^r)dx_1dx_2 - \iint_\omega (\overline{F}, \overline{U}^r)wdx_1dx_2, \tag{42}$$

where

$$\widetilde{K} = \begin{bmatrix} \widetilde{B}_{2,0}^*\overline{A}\widetilde{B}_{2,0} + \overline{c} & \widetilde{B}_{2,0}^*\overline{A}\widetilde{B}_{1,1} & \widetilde{B}_{2,0}^*\overline{A}B_{0,2} \\ \widetilde{B}_{1,1}^*\overline{A}\widetilde{B}_{2,0} & \widetilde{B}_{1,1}^*\overline{A}\widetilde{B}_{1,1} & \widetilde{B}_{1,1}^*\overline{A}_k B_{0,2} \\ B_{0,2}^*\overline{A}\widetilde{B}_{2,0} & B_{0,2}^*\overline{A}\widetilde{B}_{1,1} & B_{0,2}^*\overline{A}B_{0,2} \end{bmatrix}; \ \ \overline{F} = \begin{bmatrix} F \\ 0 \\ 0 \end{bmatrix}. \tag{43}$$

The solution of this problem is the point (function) of the constrained extremum of this functional with the condition (27), taking into account (32), (33).

These formulations are the initial ones for the realization of the discrete-continual methods for the problems of anisotropic plate analysis.

The Reported study was Funded by Government Program of the Russian Federation "Development of science and technology" (2013-2020) within Program of Fundamental Researches of Ministry of Construction, Housing and Utilities of the Russian Federation and Russian Academy of Architecture and Construction Sciences, the Research Project 7.1.1".

## References

1. P.A. Akimov, Applied Mechanics and Materials, **204-208**, 4502-4505 (2012).
2. P.A. Akimov, M.L. Mozgaleva, Applied Mechanics and Materials, **580-583**, 2898-2902 (2014).
3. K.-J. Bathe, *Finite Element Procedures* (Prentice Hall Inc., 1996).
4. O.C. Zienkiewicz, R.L. Taylor, D.D. Fox, *The Finite Element Method for Solid and Structural Mechanics* (Butterworth-Heinemann, 2013).
5. O.C. Zienkiewicz, R.L. Taylor, J.Z. Zhu, *The Finite Element Method: Its Basis and Fundamentals* (Butterworth-Heinemann, 2005).
6. A.B. Zolotov, P.A. Akimov, *Proceedings of the International Symposium LSCE 2002 organized by Polish Chapter of IASS* (Warsaw, Poland, 2002).

# Self-compacting fine-grained concretes with compensated shrinkage

*Lev* Alimov[1] , *Igor* Kharchenko[2*] , and *Viktor* Voronin[1]

[1] Moscow state university of civil engineering, Yaroslavskoye shosse, 26, Moscow, Russia, 129337

[2] Moscow state university of civil engineering, Yaroslavskoye shosse, 26, Moscow, Russia, 129337

**Abstract.** This paper substantiates the efficiency of application of fine-grained concrete for erection of cast-in-place concrete and reinforced concrete structures of different purpose. On the basis of analysis of experimental research results it was established that the introduction of microfillers with expansion effect to composite binder allows not only improving the rheological properties of fine-grained concrete, but also decreasing of value of shrinkage strain and improving of concrete crack resistance and durability. The analysis of the results of industrial use of fine-grained concretes with compensated shrinkage is given.

## 1 Introduction

Almost in all regions of Russia where 80% of construction capacities are concentrated, there is an acute deficiency of broken stone for production of qualitative concrete and reinforced concrete structures. Herewith, the fraction of transportation costs in final costs of commercial concrete in different regions reaches 50%. In this connection the large-scale implementation in construction practice of non-broken-stone or fine-grained concrete is the crucial task of modern cast-in-place construction. However, there exist a number of factors restraining the application of fine-grained concretes for erection of cast-in-place concrete and reinforced concrete structures. These factors include excessive cement consumption, excessive shrinkage, low crack resistance, excessive porosity and deformability, etc. Besides, fine-grained concrete mixes prepared with no account of their rheological properties possess low processibility, which affects the quality of structures.

Improvement of quality and efficiency of application of fine-grained concrete mixes may be provided by means of implementation of principles underlying the technology of self-compacting concretes (SCC) that find wide application in foreign construction practice [1-5].

Nowadays, SCC didn't come into common use yet in Russia, but there are lots of examples of its effective application.

It is known that self-compacting concrete mixes may be obtained only if its separate components, particularly fine-ground mineral additives, are agreed in grain-size composition. Introduction of fine-grained and finely dispersed microfillers into

* Corresponding author: iharcenko@mail.ru

composition of binders allow substantial increase in strength characteristics and durability of concretes [6, 7]. One of effective mineral fillers essentially increasing quality of concrete mix and concrete is the microsilica dust, as well as fine-grained carbonate flour. Introduction of these fillers into concrete composition allows both filling of intergranular openings in cement matrix and substantial improvement of contact area between the cement stone and the filler. [8, 9, 10]

Optimum utilization of microfillers along with the application of highly effective superplasticizers is the primary prerequisite for obtaining of high-quality fine-grained self-compacting concretes [11 - 15].

At production of self-compacting concretes (SCC) composite multicomponent binders are used, which microstructure includes microparticles of cement clinker, mineral-based microfillers with different fineness and hydraulicity, water, common pore structure.

## 2 Materials and mixture composition

Experimental research of the effect of mineralogical composition of Portland cement and composite binders on its basis on rheological properties of fine-grained concrete of different strength grades were performed with the application of polycarboxylate superplasticizers (PCE) of different type ("Sika Visko Crete"). On the basis of analysis of obtained results the defining effect of $C_3A$ on the efficiency of PCE application was established. Also the research of properties of binder solutions at different content of $C_3A$ in cement component was conducted. According to $C_3A$ content the cements of three types were applied:

- ordinary Portland Cement type CEM I 42.5 N with $C_3A <4\%$;
- ordinary Portland Cement type CEM I 42.5 N with $4\%<C_3A <6\%$;
- ordinary Portland Cement type CEM I 42.5 N with $C_3A >7\%$.

For optimization of binder and fine-grained concrete properties the concrete mix was added with microfillers with different degree of dispersion and mineral composition. Different grades of fine-grained microfiller on the basis of microcement with expansive additive (MEA) possessing the ability to independent hydraulic hardening were used as hydraulically active microfillers. MEA has constant and smoothly changing grain-size composition and stable chemical and mineralogical composition.

By grain-size composition MCГТ is divided into 3 basic grades restricting the maximum size of grains not exceeding weight percentage of d95:

- MEA-Extra (E) – d95 <5,5 μm
- MEA-Ultra (U) – d95 <8,5 μm
- MEA-Standard (S) – d95 <12,5 μm

The parameters characterizing different MEA grades by the degree of independent hydraulicity and mineral composition are presented in Table 1.

**Table 1.** Different grades of fine-grained microfiller applied in composite binders with expansive additive (MEA).

| Type | Ability to independent hydraulic hardening and strength gain, % |
|---|---|
| Ordinary Portland Cement type CEM I 42.5 N | 100 |
| Microcement with expansive additive - Standard (MEA-S) | 42 |
| Microcement with expansive additive – Ultra (MEA-U) | 64 |

| | |
|---|---|
| Microcement with expansive additive - Extra (MEA-E) | 102 |

The results of parallel tests on electronic viscosimeter for determination of cement paste viscosity with different content of microfillers were considered.

## 3 Results

The results of research of indices of viscosity and shear resistance of binder solution with adding of different microfillers are presented in fig. 1

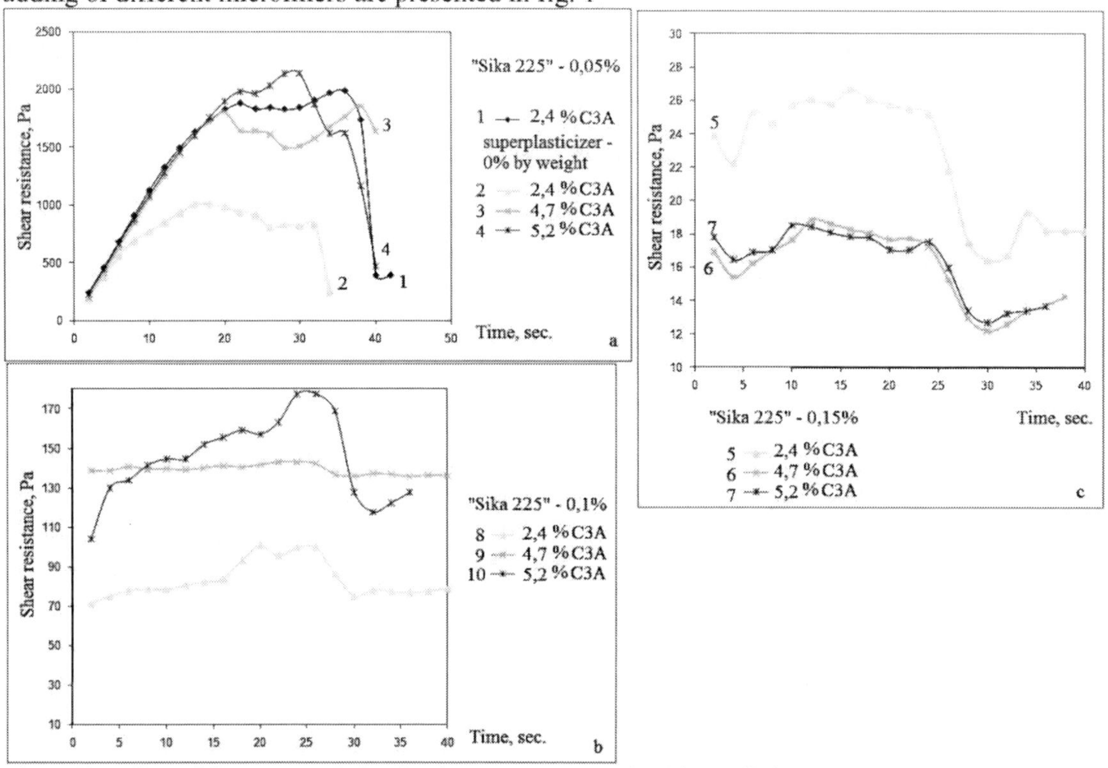


**Fig.1.** The effect of $C_3A$ content in Portland cement on superplasticizer efficiency: *a*,*b* and *c* – with superplasticizer content of 0,05; 0,1 and 0,15%, respectively.

As it seen from fig.1, with the increase of tricalcium aluminate ($C_3A$) content in binder the efficiency of superplasticizer decreases (Table 2).

**Table 2.** The dependence of shear resistance from the content $C_3A$ and Sika225.

| Sample No. | $C_3A$, % | Superplasticizer PCE Sika 225, % | Shear resistance, Pa |
|---|---|---|---|
| 1 | 2,4 | 0 | 1884 (standard) |
| 2 | 2,4 | 0,05 | 903,64 |
| 3 | 4,2 | 0,05 | 1660,091 |
| 4 | 7,5 | 0,05 | 2002,714 |
| 8 | 2,4 | 0,1 | 91,186 |

| | | | |
|---|---|---|---|
| 9 | 4,2 | 0,1 | 140,889 |
| 10 | 7,5 | 0,1 | 158,236 |
| 5 | 2,4 | 0,15 | 25,315 |
| 6 | 4,2 | 0,15 | 18,010 |
| 7 | 7,5 | 0,15 | 17794 |

It was found out that at exceeding of certain limit of superplasticizer dosage, the influence of aluminate constituent on consistency of "binder-water" system becomes negligible (mixtures 5, 6, 7). Test results demonstrate that the maximum degree of efficiency belongs to hyperplasticizer Sika ViscoCrete – 225, which was further used.

The distinctive feature of MEA is the reproducible smooth grain-size composition at high dispersion degree and stable mineralogical composition.

During experimental research as particulate filler was used MEA on the basis of Portland cement clinker, as well as the blast-furnace slag, which possesses high resistance under conditions of sulfate corrosion and is widely used for impregnation injection of soils, injection of foundation structures, brickwork, remediation of waterworks and subsurface structures, stabilizing additive for concretes. In particular, the meaning of MEA microfillers in composite binder shows up in the reduction of composition water demand at the expense of substitution of free water in intergranular space (Fig. 2).

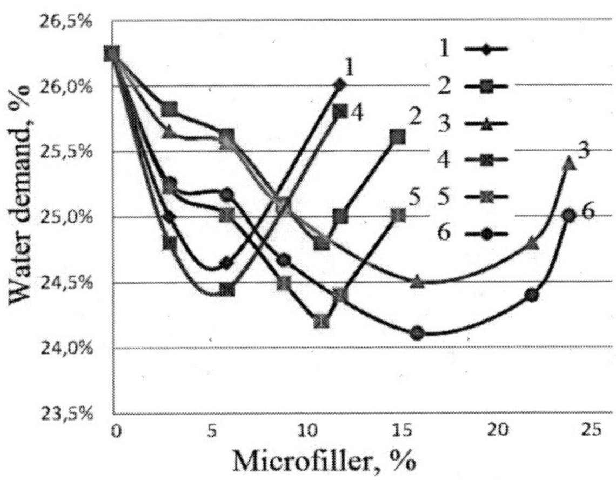


**Fig. 2.** The effect of microfiller consumption on composite filler water demand (1 – Standard, 2 – Ultra, 3 – Extra with expansive additive; 4, 5, 6 – the same, but without expansive additive).

Fig.2 shows the effect of microfiller consumption in composite binder on its water demand. Depending on microfiller fineness its content in binder varies from 7% to 19%: 7% - the most coarse with d95 < 12,5 μm (Standard), and 19% - the finest with d95 < 6 μm (Extra).

At introduction to composite binder of microfiller with expansive additive the minimum water demand is provided at its content in composite binder of not more than 10%. At application of microfiller without expansive additive its optimum consumption increases up to 15...17%.

Strength gain for molded samples with composite binder varies in the range from 7 to 27% relative to use of initial cement CEM I 42,5 N at the age of 28 days.

Similarly, water demand is reduced at application of composite binder in fine-grained concrete composition. (Fig. 3)

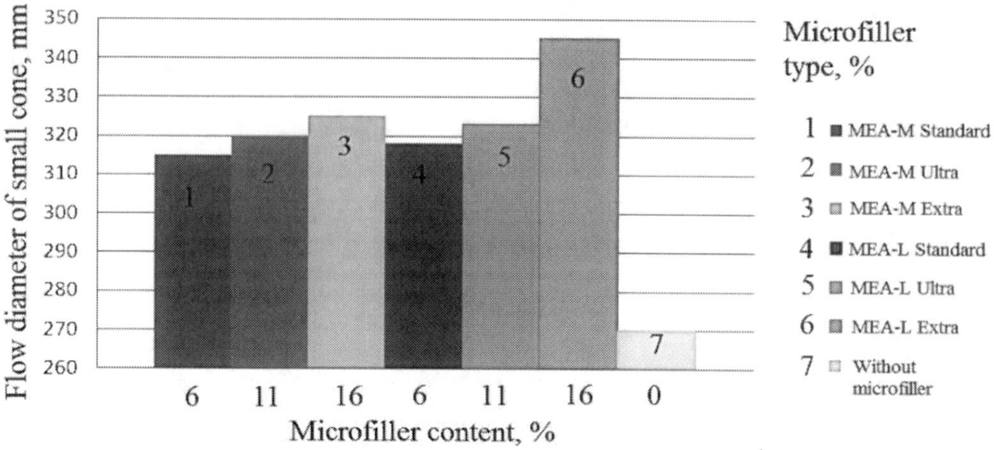


**Fig.3.** The effect of microfiller on mix flowability at W/B=0,37

Fig.3 represents the change of flowability of fine-grained concrete with microfiller as compared to composition without additives. Fine-grained concrete with composite binder has substantially higher flowability at the expense of free water release from intergranular space. Herewith, fine-grained concrete with Extra microfiller possesses maximum flowability, which is explained with its maximum ability to fill intergranular space with displacement of maximum quantity of free water from intergranular space.

As the result of conducted experimental research was extablished the dependence W/C ratio of fine-grained concrete on the type and amount of microfiller in composite binder composition. Fig.4.

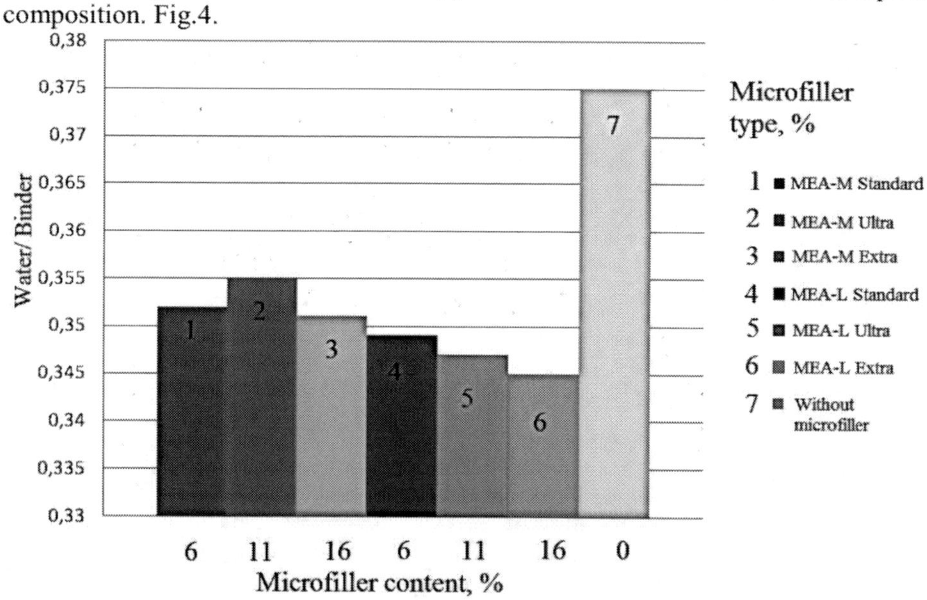


**Fig.4.** The effect of microfiller on W/C ratio of fine-grained concrete at mix fixed flowability.

As is seen from fig. 4 and 5, fine-grained concrete mixtures based on composite binders with different composition have substantially lower water content as compared to mixtures with basic Portland cement. Considering the presence in composite binders of mineral microfiller possessing high hydraulicity, fine-grained concretes on its basis are characterized with the intense strength gain that reaches 70% of grade strength within the period of up to 3 days.

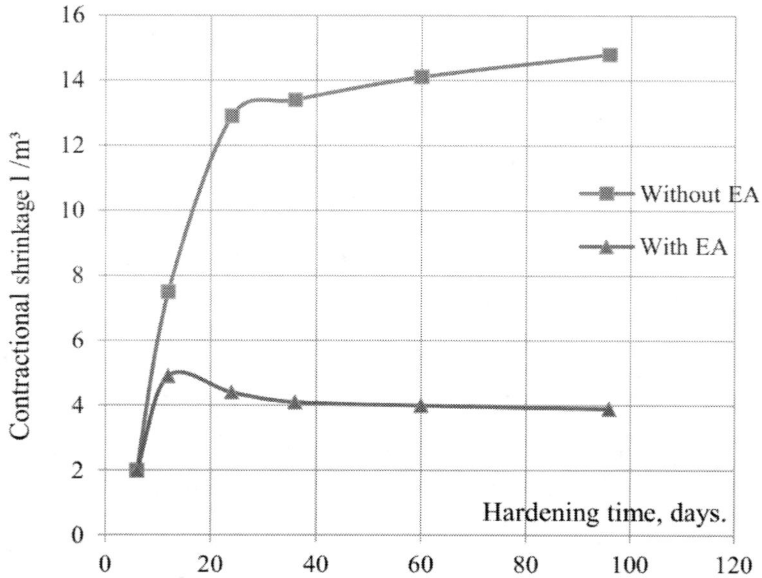


**Fig.5.** Deformations of fine-grained concrete with expansive component and without it in composition of modified additive.

Fine-grained concretes are intended for erection of cast-in-place concrete and reinforced concrete structures with a wide range of properties with concrete strength class from B30 to B60. Pore structure of modified fine-grained concrete of different production compositions was investigated with the application of mercury porosimetry method. Herewith, it was established that the total pore volume of fine-grained concrete of B50 class amounts 8,91% with average radius of 0,0501 μm, for concrete of B40 class total pore volume is 14,12% with average radius of 0,0646 μm, and for concrete of B30 class total pore volume amounts 15,13% with average radius of 0,1205 μm. Relatively fine-grained structure of fine-grained concrete of different strength classes assumes correspondingly its high durability. In compliance with the accepted procedure the durability of nanomodified fine-grained concrete with production composition was evaluated by weight loss and reduction of dynamic modulus of elasticity after freeze-thawing cycles in water-saturated state at the temperature of -50 °C. Analysis of the results of freeze-thaw resistance tests demonstrated that the weight loss of fine-grained concrete of all strength classes after 56 freeze-thawing cycles did not exceed 200 g/m$^2$. Herewith, the value of dynamic modulus of elasticity remained almost unchanged or even increased for 3-8%. This can be explained with the high content of clincker fund, which is used for recovery of microdestructions of structure at freezing.

Due to elevated content of cement in fine-grained concrete composition in order to eliminate shrinkage impact on its structure and properties, the composition of nanomodified complex additive was added with expansive component on sulfoaluminate basis, which allowed shrinkage reduction for more than 5 times and considerably increase crack resistance of cast-in-place reinforced concrete structures. (Fig.6) The analysis of practical application of expansive cements on the basis of ettringite formation revealed their high efficiency at erection of cast-in-place reinforced concrete structures made of fine-grained concrete.

Thus, at construction of sewage facility complex in Varyogan of Nizhnevartovsk district of the Tyumen Region the main objective of application of expansive cement was the secured achieving of design impermeability to water and durability of erected structures. Expansive cement that was used for manufacturing of cast-in-place structures was made by means of mixing of commercial Portland cement and expansive additive at the ratio of 85:15. Expansive additive was made by means of mixing of 70% of aluminate cement and 30% of calcium sulfate dihydrate. Concrete with expansive additive after 28 days had the following characteristics: expansion value of 0,4%, compression strength of 48 MPa, self-stressing or expansion pressure at single-axis restraint of expansion process amounted 1,8 MPa. Concrete was placed at air temperature of up to -20°C. Sodium nitrate in the amount of 7% of mixing water was used as anti-freezing admixture. Herewith, compression strength after 7 days was about 13 N/mm$^2$, and after 28 days it was 25 MPa. The analysis of concrete quality after 15 years of operation of sewage facilities demonstrated that all reinforced concrete elements and joints of expansive concrete do not require any repairs and are in operating condition.

The advantages of expansive concrete (high impermeability to water, durability, crack resistance) were used at construction of the number of sports facilities.

For instance, in 2000…2002 was build a sports complex in Gubkinsky, the Yamal-Nenets Autonomous District, with a slab with the area of 1 800 m$^2$ made of self-stressing reinforced concrete. Herewith, the requirements to quality of concrete works at erection of reinforced concrete slab were rather high: roughness of < 1,0mm; high crack resistance under thermal stresses; minimum number of expansion joints on surface etc. Due to high percentage of slab reinforcement – more than 10%, at application of common concrete there was a risk of cracking at stress concentration due to development of shrinkage and thermal strain. Herewith, an effective measure against cracking within the reinforced concrete structures is the concrete prestressing. However, conventional methods of prestressing of reinforced concrete structures were inapplicable for relatively thin slab (14 cm) with large area. In this connection it was decided to apply concrete on the basis of expansive cement providing plain stress condition, i.e. self-stressing of reinforced concrete slab.

Prepared expansive cement had self-stressing index of 1,45 MPa and 28-day compression strength at free hardening of 48 MPa, as well as bending tensile strength of 6,0 MPa. These indices corresponded to design requirements. Concrete composition and technological activities for concrete placing were defined during preliminary studies of slab models with the size of 0,14x 2x6 m. Concrete (containing 560 kg/m$^3$ of expansive concrete and quartz sand with the humidity of 5% in the quantity of 1 360 kg/m$^3$, water: 280 l/m$^3$, plasticizer C-3 in the amount of 0,5% of weight oa expansive cement). Placed concrete was cast into the formwork without additional compacting, manually floated and after placing covered with polyethylene film. Concrete works were performed continuously on a three-shift basis. Placing of fresh concrete was performed by staggered strip method. After placing of a strip 2m wide, the next strip was left empty and was concreted in 10 ÷ 20 hours depending on temperature. Due to unfavorable climatic conditions (air temperature during night was up to + 5 °C, and during day in sunny weather under the film it was up to +50°C) the concrete surface after 24 hours was covered with a high-absorbent material. As a result of this treatment it was possible to avoid serious temperature gradient across the cross-section of reinforced concrete slab and to provide wet storage necessary for expansive concrete. Measurements performed during 28 days showed the expansion of 4,1 mm for reinforced concrete slab 35 m long. Herewith, tendon stressing amounted 16 MPa, stress from concrete prestressing in longitudinal direction amounted 0,84 MPa, and in transverse direction - 0,93 MPa. After 28 days the compression strength was determined, which amounted 52 MPa. Reinforced concrete slab made of expansive cement after 15 years of operation and up to the present day is in good condition.

Due to severe operating conditions related to high static and dynamic loads, high hydrostatic pressure, agressivity of ground water, very high demands are placed on the quality control of concrete and reinforced concrete structures and tunneling and metro engineering. Ten-year experience demonstrates that expansive cements and concretes on their basis may also be used for this purpose with high efficiency.

## Conclusions

It was established that the introduction of microfiller with independent hydraulicity and optimized grain-size and mineral composition to composite binder facilitates the reduction of water demand of composite binder, and respectively, of fine-grained concrete on its basis. Introduction to composite binder of microdisperse expansive additive allows substantial decrease of shrinkage value with simultaneous increase of crack resistance, impermeability to water, freeze-thaw resistance and strength of self-compacting fine-grained concretes. The analysis of the results of production implementation of self-compacting fine-grained concretes with expansive mineral microfiller demonstrated high technical-and-economic efficiency of their application.

## References

1. B. Craeye, G. De Schutter, B. Desmet, J. Vantomme, G. Heirman, L. Vandewalle, Ö. Cizer, S. Aggoun, E.H. Kadri, Cem. and Conc. Res., 40, 6 (2010)
2. M. Bouasker, P. Mounanga, P. Turcry, A. Loukili, A. Khelidj, Cem. and Conc. Comp., 30, 1 (2008)
3. B. Craeye, G. De Schutter, B. Desmet, J. Vantomme, G. Heirman, L. Vandewalle, Ö. Cizer, S. Aggoun, E.H. Kadri, Cem. and Conc. Res., 40, 6 (2010)
4. M. Bouasker, P. Mounanga, P. Turcry, A. Loukili, A. Khelidj, Cem. and Conc. Comp., 30, 1 (2008)
5. F. Han, L. Li, Sh. Song, J. Liu, Powd. Techn., 315 (2017)
6. R. Martinez-Lopez, J. Escalante-Garcia, Const. and Build. Mat., 119 (2016)
7. F. Han, R. Liu, D. Wang, P. Yan, Therm. Act., **586** ( 2014)
8. K. Lee, H. Yoon, K. Yang, Con. and Buil. Mat., **146** (2017)
9. Mehdipour, K. H. Khayat, Cem. and Conc. Comp., **78** (2017)
10. F. Lollini , E. Redaelli , Luca Bertolini, Cem. and Conc. Comp., **46** (2014)
11. O. Larsen, V. Naruts, Mag. of Civ. Eng.(in , 8 (2016)
12. A. Adamtsevich, S. Pashkevich, A. Pustovgar, Magazine of Civil Engineering (in Russian - Inzhenerno-stroitelnyj zhurnal), 3 (2016)
13. Q. Yu , H. Brouwers, Cem. and Conc. Comp., **34**, 9 (2012)
14. R. Magallanes-Rivera, C. Juarez-Alvarado, P. Valdez, J. Mendoza-Rangel, Cons. and Build. Mat, 37 (2012)
15. V. Soloviev, A. Eremin, D. Eliseev, A. Buryanov, Building materials (in Russian - Stroitelnye materialy), 1-2 (2017)

# The task of multi-criteria optimization of metal frame structures

*Vadim* Alpatov[1,*]

[1]Samara State Technical University, Institute of Architecture and Civil Engineering, 194, Molodogvardeyskaya St., 443001, Samara, Russia

**Abstract.** Optimal design of a frame structure with a specified geometric scheme consists in finding control parameters that provide the highest or lowest value of composite functions which present some quality criteria. Searching for optimal parameters is related to a number of design and calculation constraints. When it is necessary to vary a geometrical scheme, node coordinates are also considered as unknown varied parameters that affect the quality criteria. When designing frames with a specified scheme, the volume of material is typically the primary criterion for solving an optimization task and is written as a function of control parameters and state settings. In problem specification it is also important to reduce the deformation of the system. This is accomplished by introducing an additional criterion -maximum moments of inertia of the sections of the system. There is a two-phase design and calculation model existing in design practice now. In the first stage, the work is based on the experience or existing prototype. On their basis stiffness of the bars is assigned, and then a load vector is calculated. In the second stage, the sections are chosen according to known forces.

## 1 Introduction

Optimal design of a frame structure with a specified geometric scheme consists in finding control parameters that provide the highest or lowest value of composite functions which present some quality criteria. Searching for optimal parameters is related to a number of design and calculation constraints [1-3].

The dimensions of cross sections of metal elements (width, thickness, and surface of rolling parts) are the basic parameters from which the system properties change. These settings are called controlling variables.

When it is necessary to vary a geometrical scheme, node coordinates are also considered as unknown varied parameters that affect the quality criteria. The search for an optimum scheme of a structure or a construction should be based on a algorithmic comparison of quality criteria for individual cases [4, 5].

[*] Corresponding author: avu75@mail.ru

## 2 Materials and methods

As any frame structure consists of many elements, the vector of the system control variables itself consists of vectors of parameters of each of its constituent bars, that is:

$$Z = (Z_1, Z_2, \ldots, Z_m), \tag{1}$$

where $m$ – is the number of bars in the system.

Each embedded vector $Z_i (i = 1,2,\ldots,m)$ looks like this:

$$Z_i = (Z_1, Z_2, \ldots, Z_{di}), \tag{2}$$

where $d_i$ is the number of control variables in the bar number $i$.

When choosing $Z_k$, the following condition must be observed

$$Z_k \in D_k \tag{3}$$

Here $D_k$ is a limited set, where $k = 1,2,...,m_1$; $m_1$ – the total amount of typycal sizes of elements.

Let us specify control parameters by using double indexing, as $Z_{jk} (j = 1,2,\ldots,d)$, then the vector of the control variables is written as:

$$Z = Z(Z_{jk}), Z_{jk} \in D_k, \; Z_{jk} \geq 0. \tag{4}$$

To develop a mathematical model for finding an optimal solution for frame structures, we have to consider parameters defining geometry of the system and impact of external environment on the system in question (topology of the construction, node coordinates, loads, and other). At the first stage, we accept these parameters as fixed. Let's call them status parameters. A vector of status parameters of the frame structure includes state vectors of individual elements:

$$B = (B_1, B_2, \ldots, B_p, \ldots, B_{m_3}). \tag{5}$$

Here $B_p = (b_{1p}, b_{2p}, \ldots b_{zp})$ is the state vector corresponding to the load $p$; $p = 1,2,...,m_3$.

When designing frame structures with a specified geometric scheme, the volume of material is typically the primary criterion for solving an optimization task and is written as a function of control parameters and status parameters.

$$V = V(Z, B) = \min. \tag{6}$$

When specifying an optimization task, it is important to take into account the requirement to limits of the system deformation . It can be achieved, for example, by introducing the second criterion = maximum moments of inertia of sections of the system:

$$I = [I_1(Z_1), I_2(Z_2), \ldots, I_z(Z_d)] = \max. \tag{7}$$

Some properties connected with energy condition of the system can be shown as a vector of minimized composite functions:

$$\phi = [\phi(Z,B_1),\phi(Z,B_2),\ldots\phi_{m3}(Z,B_3)] = \min. \quad (8)$$

The same properties can be written in the form of constraints-equalities, such as an equation of the force method or the method of movement.

Restrictions should be placed on construction and design characteristics of elements of the system. Such restrictions are presented in the form of approximate equalities:

$$\bar{G}(Z,\bar{B}) \geq 0, \quad (9)$$

where: $\bar{G}(Z,\bar{B}) = [\bar{G}_1(Z,\bar{B}),\bar{G}_2(Z,\bar{B}),\ldots,\bar{G}_{m2}(Z,\bar{B})]$; $m_2$ – the number of constraint vectors.

There is a two-phase design and calculation model existing in design practice of building structures now. In the first stage, the work is based on the experience or existing prototype. On their basis stiffness of the bars is assigned, and then a load vector is calculated.

$$X = (X_1, X_2, \ldots, X_d). \quad (10)$$

Here $X_i = X_{i1}, X_{i2}, \ldots, X_{im3}$ - the forces in bar *i*.

In the second stage, the sections (their parametres) $Z$ are chosen according to known forces $X$.

This two-stage design model can be used for optimal design of frame structures with specified geometry. If it is necessary to find the system's geometric shape, the task becomes a three-stage one with an additional set of varied parameters: geometric parametres of the system. In this case, the solution of the generalized optimization task should be organized in stages, based on its decomposition. It is proved that the convergence of this process to the global minimum will be ensured if the main criterion and the limitations expressed in $X_i$ are bumped [6].

At the level of one bar of the system, the search for its optimum cross-section (see Figure 1) can be reduced to a variation and selection from the gauge material of discrete characteristics of the compound section (flange, wall, edge, etc.) satisfying several quality criteria, design standards requirements, and manufacturing and mounting conditions. The algorithm described can be applied both to the private task of finding optimal section parameters of a certain type, and of solving the problem of finding the optimal section shape, steel grade, or section dimention.

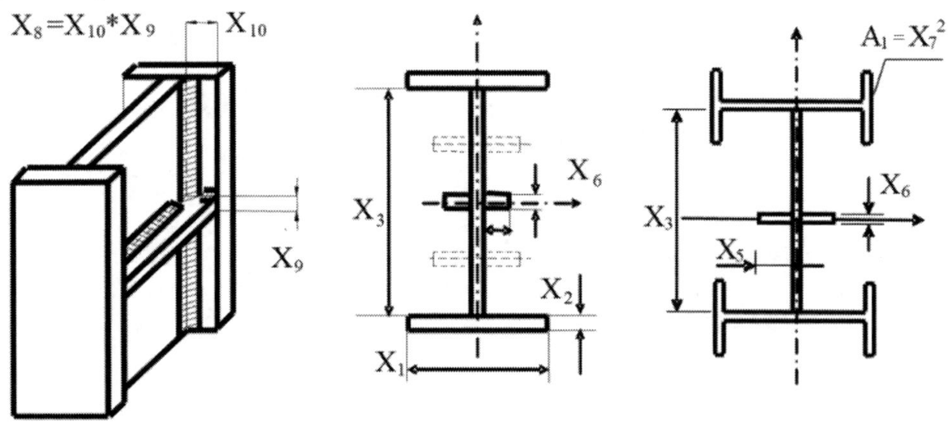


**Fig. 1.** Cross-sections diagram.

It is known that there are many best and closest to the best solutions to the optimum. Finding multitudes that are called pareto-optimal in the tasks of multi-criteria analysis is the purpose of the this work.

Volume, labour input, and rigidity criteria can be presented in general terms by expressing them through the system's accepted parameters, that is, by formulating three target functions:

$$V = \sum_{i=1}^{m_1} \sum_{j=1}^{m_2} a_{ij} \cdot X_i \cdot X_j \,; \; |X_i \in B_1(B_3); \; X_j \in B_2(B_3) \tag{11}$$

$$T = K \cdot \sqrt{\left(\sum_{i=1}^{m_1} \sum_{j=1}^{m_2} a_{ij} \cdot X_i \cdot X_j\right) \cdot \gamma \cdot n} \tag{12}$$

$$I = \sum_{i=1}^{m_1} \sum_{j=1}^{m_2} a_{ij} \cdot X_i \cdot X_j \cdot \left(k_{ij} \cdot X_j^2 + b_{ij}\right), \tag{13}$$

where $k_{ij} = n_i / 12, b_{ij} = n_i \delta_i^2$ ( $n_i$ – - the number of elements in the section).

Equations (11)-(13) are hypersurfaces in the sphere of accepted parameters $X_i$ and $X_j$. Hypersurfaces *V*, *T*, *J* have a number of important features:

1. they are are non-convex, so there is no limit to the minimum (except for a trivial one) without additional restrictions;
2. they ascend in the increase $X_{i(j)}$ (when $X_{i(j)} > 0$ );
3. their minimum lies in the area adjacent to the boundary of the restrictions;
4. their optimum can be multi-valued when there are several combinations of pairs $X_i, X_j$ or the same volume or there are slightly different (from the optimum) solutions (quasi optimal).

The view of surfaces *V*, *T*, *I* when $X_i > 0$; $X_j > 0$ is shown on Fig 2.

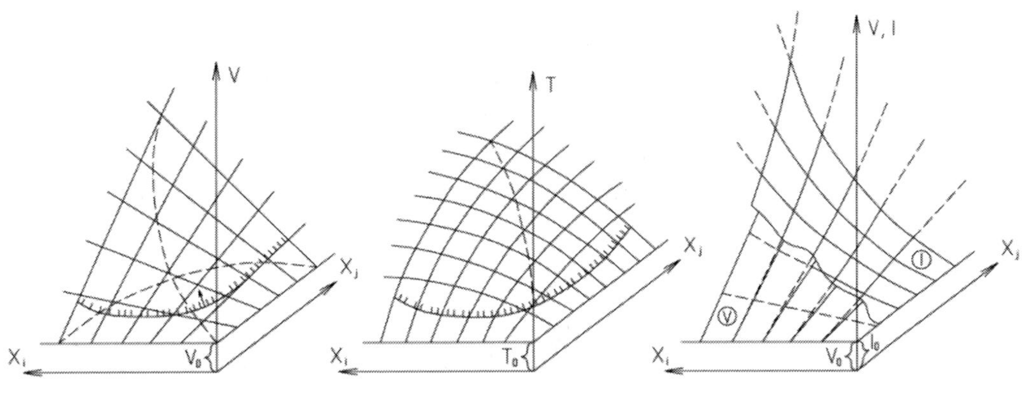


**Fig. 2.** Hypersurfaces of criteria *V*, *T*, *I*.

The limitations placed on target functions and system parameters can be divided into two types: constructive limitations and design limitations.

Constructive limitations are related to the limits of parameter ratios introduced by construction regulations. They can be presented as:

$$f_k(X_i, X_j) - C_k \geq 0 \tag{14}$$

Here $k$ is the restriction number $C_k = const$ , $D_1 \leq X_i \leq D_2$; $D_3 \leq X_j \leq D_4$; $X_i > 0$; $X_j > 0$.

Limitations on strength, general and local sustainability are as fpllows:

$$f_k(X_i, X_j) - g_k(X_i, X_j) \geq 0 \tag{15}$$

When optimizing according to multiple criteria, a method of succesive concessions can be applied [7]. In this case, the first step is to find a minimum value of the leading criterion $V(X_i, X_j) = \min$. Besides, $X_i \in G_1, X_j \in G_2$, $G_1 \subset B_1; G_2 \subset B_2$. Multiples $G_1$ and $G_2$ are formed from $B_1$ and $B_2$ under limitations. In the second stage, criteria limitations are formulated.

$$\min V \leq V \leq V^* \big| V^* = \left[\min V + \varepsilon_1(\min V)\right], \tag{16}$$

$$\min T \leq T \leq T^* \big| T^* = \left[\min T + \varepsilon_2(\min T)\right], \tag{17}$$

$$\max I \geq I \geq I^* \big| I^* = \left[\max I - \varepsilon_3(\max I)\right]. \tag{18}$$

Here, $\varepsilon_1(\min V)$, $\varepsilon_2(T\min)$, $\varepsilon_3(\max I)$ – the values of concession", by volume, labour input and rigidity, respectively.

The selection of optimum sets is based on the solution of two minor tasks:

1. Defining the standard ( $V^{ont}; T^{ont}; I^{ont}$ ).
2. Formation of many permissible (pareto-optimal) solutions.

The search for the optimum consists of the following operations:

1. Fixation $X_1, X_2, \ldots, X_{n-1}, X_n \in B_1, B_2$;
2. Verifying the ability to exhaust the set step of $B_1$, $B_2$;
3. Variation $X_i, X_j$ in a manner that provides $I \rightarrow \max$ and defines $\min V$, $\min T$;
4. Finding rigidity appropriate to its critical limitation.

If the solution is found for the first time, it is taken for a standard $V_1^0 = V_1, T_1^0 = T_1, I_1^0 = I_1$. The subsequent variation analyzes the optimum by comparing $V_n$ with $V_{n-1}^0$, where $n$ is the step number. There are three possible options:

1. If $V_n \leq V_{n-1}^0(1+\varepsilon_1)$ and $I_n > I_{n-1}^0$, then the decision $n$ is optimum and accepted as a standard;
2. If $V_{n-1}^0 \leq V_n \leq V_{n-1}^0(1+\varepsilon_1)$, a $I_{n-1}(1-\varepsilon_2) \leq I_n \leq I_{n-1}^0$, then the standard for comparison is still $V_{n-1}^o$ and $I_{n-1}^o$ , though $V_n$ and $I_n$ are remembered as valid (multitude $D$);
3. If $V_n > V_{n-1}^0(1+\varepsilon_1)$, the decision is discarded even before limitations are checked.

Criteria limitations in the optimization process are specifies when the standard $V^o, T^o, I^o$ is specified. Therefore, after $D$ is found, a final sorting is performed and a multitude of effective (Pareto) points is allocated. Then the search for optimal sections for the specified parameters is over.

In a discrete task, there are many equivalent options for optimal variants and variants closest to optimal. Therefore, the optimum scope will be fully defined only after all parameters have been traversed. At the same time, there are areas where it is meaningless to traverse parameter combinations because they do not provide any structural, technological or computational limitations. That is why it makes sense to use a method direct search of the values of discrete parameters when performing a series of "cutting-off multitudes" techniques.

## 3 Results

The above method of solving a multi-criteria task through the sequential concessions is implemented in a computing complex developed in Samara state university of architecture and civil engineering. The complex is designed to optimize the selection of steel structure sections with different section topologies, steel marks and dimensions.

For space-framed constructions, questions of optimal design are most actual in relation to their weights or labour input indicators. For such structures, because of their constructive features, the contradiction between materials consumption and labour input is most evident. Traditionally for such constructs (in the case of serial and generic structural slabs), the design priority was to reduce labour intensity by simplifying the design, including joint connection. This was often done at the expense of their materials consumption. Therefore, because of the large number of elements, firstly, and because of the strict unification of elements of structural constructions, secondly, they are characterized by increased material production discharge. Much of the cost of these designs is offset by a decrease in labour input in manufacturing, transport and assembling. However, today there is no convincing justification for the advantages of such a strict unification for any known spatial-frame structures. An economic justification of this type can only be given only when there is a task of optimizing with a clear mathematical formulation of the problem is solved. It is obvious, however, that this problem should be raised either as a multi-criteriatask of parametric optimization, or in an even broader setting, as a function of synthesizing a structure. It follows from the above that at least two criteria, mass and labour input, should be introduced in the formulation of such tasks. When designing real-world objects, there is typically a need to take into account many additional factors (such as rigidity, etc.) that can be entered as limitations or as quality criteria in a structural optimization task. In the latter case, there is an optimization task with more than two criteria.

The author of this paper has solved a number of tasks for optimizing space-framed constructions [8-10], [20]. These tasks were formulated as a search for optimal geometry of structural constructions. Because of the complexity of the mathematical model of the task, it was devided into two subtasks, that is, the decomposition was performed. Decomposition of the tasks was accomplished by using the hierarchy of parameters, limitations, and target functions for each subtask. Each subtask there was only a stage of optimization for the main task, but at the same time was an independent complete optimization task.

Splitting into two subtasks is due to the fact that an iterative recalculation method was adopted to solve the main task. The search for a solution requires two stages of optimization: finding optimal geometry of the structural construction and finding optimum distribution of the material in the structural construction with the unchanged geometry. The two-stage devision is natural in terms of definition with variable and non-variable parameters at each stage of the design, the form of a mathematical record of the target functions and the limitations expressed through the accepted parameters. Dividing the main task into two phases is natural and in terms of classifying optimization tasks [7], [8], [14-16].

## 4 Discussion

In general, the task of a high hierarchy is to find optimal geometry of a structural construction. It is formulated as follows:

Let us assume, that there is a structural construction with topology and geometry in a certain plan (a span in one direction, a span in the other direction). Its support conditions, materials of the elements, and the structure load diagram are known. We need to define an optimal geometric shape of the structural construction.

A low-hierarchy task is to find the optimum distribution of material in a structural construction with unchangeable geometry. It is formulated as follows:

Let us assume there is a structural construction with a specific topology and geometry. It means we know a span in one direction, a span in the other direction, coordinates of all structure nodes. The conditions of the support, material of the elements and the structure load diagram (see Figure 3) are also defined. We need to define optimum distribution of the material in a structural construction when the mass of the structure is minimal.

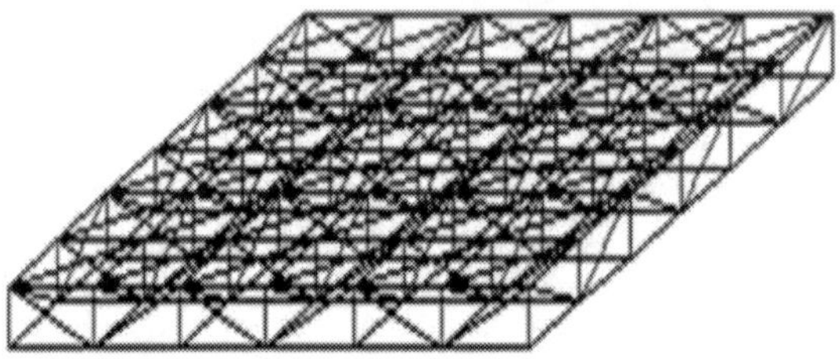

**Fig. 3.** Structural plate.

For the first task, a solution algorithm was implemented. It presented an iterative process to find the minimum mass based on the random search method. The task was formulated and solved as a single-criterion (with mass as a a criterion) task.

For the second task, the methodology described above, the algorithm and the programme complex on their basis [15], [17] were applied. This task was solved as a two-criteria task, with criteria bing volume and rigidity.

The implementation of the mathematical model described above resulted in the creation of a software complex for calculation and optimization of structural constructions. The complex is called "Search for the optimal form of structural constructions." The complex operates in a computer-to-people dialogue mode: static analysis of structural constructions; recalculation of bar sections; optimization calculations of proportioning elements sections based on the discrete size of the rolling profiles according to minimum volume and maximum rigidity criteria; search for optimum structural structure geometry on the minimum mass criterion [18], [19].

## 5 Conclusions

Figure 4 shows some of spatial constructions that resulted from finding the optimal geometric shape of these structures when using the introduced software complex.

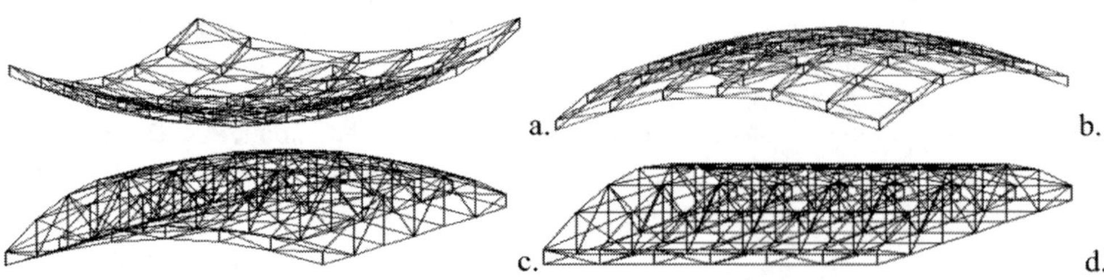


**Fig. 4.** Spatial structures obtained when searching for optimal geometry.

## References

1. I.S. Kholopov, Vestnik of Research Science Center "Building Construction" **2**, 114-125 (2010)
2. I.S. Kholopov, Industrial and civil engineering **8**, 13-17 (2010)
3. I.S. Kholopov, Metal structures **4**, 285-297 (2008)
4. A.V. Ginzburg, A.A. Vasilkin, Vestnik MGSU **6,** 52-62 (2014)
5. M.R. Garifullin, S.A. Semenov, S.V. Belyaeva and others, Construction of unique buildings and structures **2 (17)**, 107-124 (2014)
6. L.S. Lasdon, *Optimizing large systems* (Moscow, Science, 1975)
7. I.M. Rabinovich, *On the theory of static undefined trusses* (Moscow, Transzheldorizdat, 1933)
8. V.Yu. Alpatov, *Thesis for PhD in Technical Sciences* (2002)
9. V.Yu. Alpatov, Efficient building structures: Theory and practice, 199-206 (2002)
10. V.Yu. Alpatov, I.S. Kholopov, Actual problems of improvement and development of metallic, wooden and plastic structures in construction and transport, 193–198 (2002)
11. V.Yu. Alpatov, Actual problems in construction and architecture. Education. Science. Practice, 476-477 (2007)
12. A.O. Lukin, V.Yu. Alpatov, D.D. Chernyshev, Urban Construction and Architecture **2(23)**, 4-9 (2016)
13. A.I. Vinogradov, Papers of Kharkov Institute of Railway Engineering **25**, 30-36 (1955)
14. A.I. Vinogradov, *Study on construction theory* (Moscow, Gosstrojizdat, 1965)
15. I.S. Kholopov, *Thesis for Ph.D. in Technical Sciences* (1992)
16. I.B. Lazarev, *Basics of optimal design. Tasks and methods* (Novosibirsk, SGAPS, 1995)
17. I.S. Kholopov, A.V. Solovyev, A.V. Simbirev, *The 59th Scientific and Technical Conference abstracts*, 44 (2002)
18. V.Yu. Alpatov, *The 56th scientific and Technical Conference abstracts,* 277-278 (1999)
19. V.Yu. Alpatov, Postgraduate Bullitin of Samara region **2**, 83-86 (2001)
20. V.Yu. Alpatov, A.O, Lukin, A.A. Sakharov, D.I. Zhuchenko, Urban Construction and Architecture **4(25)**, 19-22 (2016)

# The contact layer method in calculating of the shear compounds.

*Vladimir* Andreev[1,*], *Robert* Turusov[1] , and *Nikita* Tsybin[1]

[1] Moscow state university of civil engineering, Yaroslavskoye shosse, 26, Moscow, Russia, 129337

**Abstract.** The paper presents a solution of the problem of the stress-strain state of a three-layer model under a shear load. The interaction between layers is accomplished by the contact layer in which the substances of adhesive and substrate are mixed. We will consider the contact layer as the transversal anisotropic medium with such parameters that it can be represented as a set of short elastic rods, which are not connected to each other. Three variants of the resolving system of equations are considered. For each of the variant, a calculation is made for the three load cases. The curves of distribution of the normal forces and tangential stresses in the contact layers were obtained.

## 1 Statement of problem

Consider the model of a multilayer beam (Fig. 1)

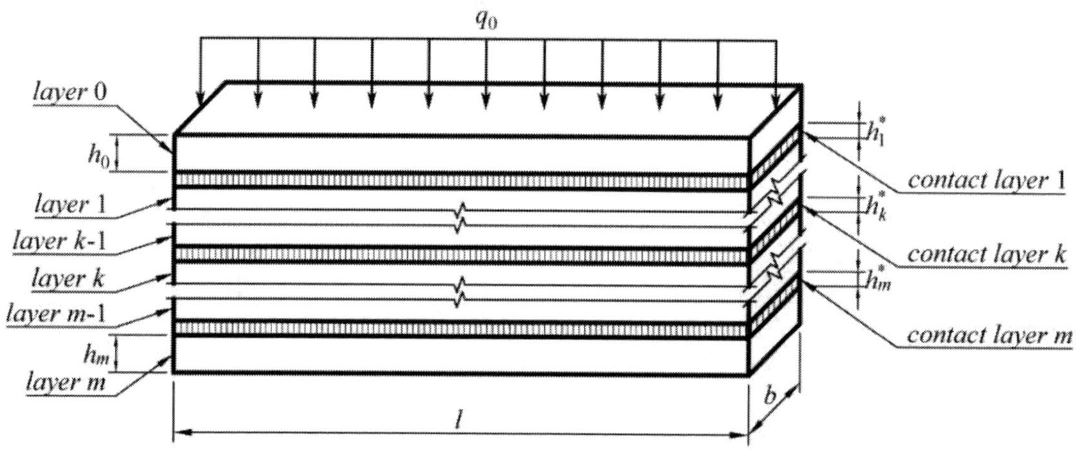


**Fig. 1**. The model of a multilayer beam.

The resolving equations for the external layers with taking into account the bending obtained in [1] are written below

* Corresponding author: asv@mgsu.ru

$$\left.\begin{aligned} &\frac{d^2N_k}{dx^2}=\frac{dP^t_{x,k}}{dx}-\frac{dP^b_{x,k}}{dx};\\ &-D_k\frac{d^4\upsilon_k}{dx^4}+\frac{h_k}{2}\left(\frac{dP^t_{x,k}}{dx}+\frac{dP^b_{x,k}}{dx}\right)=P^t_{z,k}-P^b_{z,k}.\end{aligned}\right\} \tag{1}$$

Disregarding of the bending will give

$$\frac{d^2N_k}{dx^2}=\frac{dP^t_{x,k}}{dx}-\frac{dP^b_{x,k}}{dx}. \tag{2}$$

The values $P^t_{x,k}$, $P^b_{x,k}$, $P^t_{z,k}$, $P^b_{z,k}$ represent the external forces applied to the top and bottom faces of the external layers.

Consider the contact layer.

Because the rods are not connected, in Hooke's law for an orthotropic body Young's modulus $E^*_{x,k}$ and Poisson ratio $\mu^*_k$ must be set equal to zero. The symbol * will then be used to mark all values related to the contact layer.

The system of two differential equations with respect to the stresses in the contact layers can be obtained as a result of integration of the differential equations of equilibrium, with taking into account the continuity of the displacements of the external layers and contact layers.

$$\left.\begin{aligned} &\frac{\tau^*_k}{G^*_k}=\frac{\left(h^*_k\right)^2}{12E^*_k}\frac{d^2\tau^*_k}{dx^2}+\frac{1}{2}\frac{d}{dx}\left(\upsilon^*_{t,k}+\upsilon^*_{b,k}\right)-\frac{1}{h^*_k}\left(u^*_{t,k}-u^*_{b,k}\right);\\ &\sigma^*_k=-y^*_k\frac{d\tau^*_k}{dx}-\frac{E^*_k}{h^*_k}\left(\upsilon^*_{t,k}-\upsilon^*_{b,k}\right).\end{aligned}\right\} \tag{3}$$

In these equations: $E^*_{y,k}=E^*_k$, $G^*_{xy,k}=G^*_k$. Values $\upsilon^*_{t,k}$, $\upsilon^*_{b,k}$, $u^*_{t,k}$, $u^*_{b,k}$ represent the displacements of the top and bottom external layers and are determined from expressions

$$u^*_{t,k}=u_{0,k-1}-\frac{h_{k-1}}{2}\frac{d\upsilon_{k-1}}{dx};\; u^*_{b,k}=u_{0,k}+\frac{h_k}{2}\frac{d\upsilon_k}{dx};$$

$$\upsilon^*_{t,k}=\upsilon_{k-1};\; \upsilon^*_{b,k}=\upsilon_k.$$

The displacementы in the contact layer are a nonlinear function of the argument $y^*_k$ in this case, and the tangential stresses in the contact layer are a constant value.

In the case of a one-dimensional problem of the contact layer, the displacements are a linear function. As result, equations (3) taking form

$$\left.\begin{aligned} &\frac{\tau^*_k}{G^*_k}=-\frac{1}{h^*_k}\left(u^*_{t,k}-u^*_{b,k}\right);\\ &\sigma^*_k=-\frac{E^*_k}{h^*_k}\left(\upsilon^*_{t,k}-\upsilon^*_{b,k}\right).\end{aligned}\right\} \tag{4}$$

Simplifications used in (2) and (4) essentially simplify the solution of the problem; due to the fact that the number of unknown functions in the resolving system of differential equations is reduced. However, that simplifications lead to failure of the boundary conditions with respect to tangential stresses in the contact layer.

A three-layer model is considered for the calculations (Fig. 2)

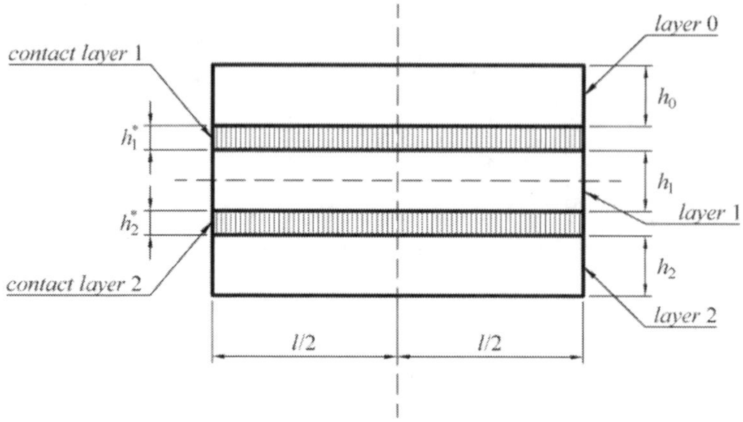


**Fig. 2.** Three-layer model.

This model is the most common in the calculation of adhesive compounds.

Three versions of the system of resolving equations are considered for comparison.

- Variant 1 represent a model that takes into account the bending of the external layers and the nonlinearity of the distribution of displacements in the contact layer. The resolving system of equations represented by relations (1) and (3) in this case.
- Variant 2 represent a model that takes into account the bending of the external layers and the linearity of the distribution of displacements in the contact layer. The resolving system of equations represented by relations (1) and (4) in this case.
- Variant 3 does not take into account the bending in the external and the nonlinearity of the distribution of displacements in the contact layer (2), (4).The resolving system of equations is represented by (2), (4). Some examples of calculations are given in [2, 3].

Below are the load cases of calculating for shear (Fig. 3).

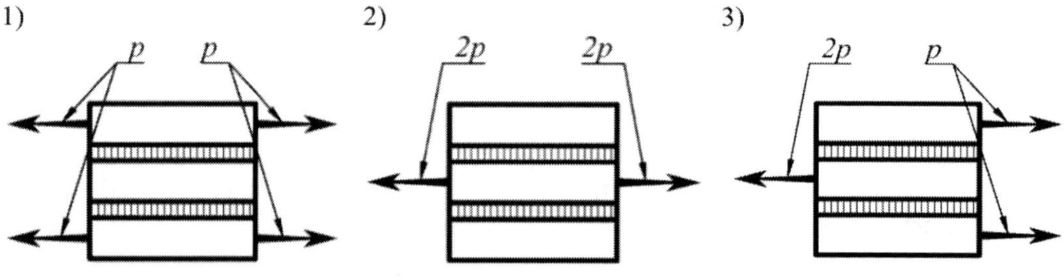


**Fig. 3.** Load cases.

Boundary conditions for axial forces are represented below

$$\left.\begin{aligned} &N_0\left(-\frac{l}{2}\right)=[p,\ 0,\ p];\ N_0\left(\frac{l}{2}\right)=[p,\ 0,\ 0];\\ &N_1\left(-\frac{l}{2}\right)=[0,\ p,\ 0];\ N_1\left(\frac{l}{2}\right)=[0,\ p,\ 2p];\\ &N_2\left(-\frac{l}{2}\right)=[p,\ 0,\ p];\ N_2\left(\frac{l}{2}\right)=[p,\ 0,\ 0]. \end{aligned}\right\} \tag{5}$$

Boundary conditions for bending moment and shear force

$$M\left(-\frac{l}{2}\right)=M\left(\frac{l}{2}\right)=0;\ Q\left(-\frac{l}{2}\right)=Q\left(\frac{l}{2}\right)=0. \tag{6}$$

Boundary conditions for shear stresses at the contact layer

$$\tau_1^*\left(-\frac{l}{2}\right)=\tau_1^*\left(\frac{l}{2}\right)=0;\ \tau_2^*\left(-\frac{l}{2}\right)=\tau_2^*\left(\frac{l}{2}\right)=0. \tag{7}$$

In this case, taking into account the symmetry with respect to layer 1, can obtain the following relations

$$\begin{gathered} N_2=N_0;\ N_1=2(p-N_0);\ Q_1=M_1=0; \\ \upsilon_2=-\upsilon_0;\ \upsilon_1=0;\ u_2=u_0;\ D_2=D_0;\ h_2=h_0; \\ P^t_{x,0}=P^b_{x,2}=P^t_{z,0}=P^b_{z,2}=0. \end{gathered} \tag{8}$$

Below are written the expressions that determine the stress state of the model.

$$\left.\begin{gathered} \tau_1^*=-\frac{dN_0}{dx};\ \tau_2^*=-\tau_1^*; \\ M_0=-D_0\frac{d^2\upsilon_0}{dx^2};\ M_2=-M_0; \\ Q_0=-D_0\frac{d^3\upsilon_0}{dx^3}-\frac{h_0}{2}\tau_1^*;\ Q_2=-Q_0. \end{gathered}\right\}$$

Analytical solutions of systems of equations 1-3 it is not possible to write down because they are cumbersome. We give only the calculation results for the following parameters.

$$h_0=2\text{mm};\ h_1=0.2\text{mm};\ b=2\text{mm};\ l=20\text{mm};\ h^*=0.1\text{mm};$$

$$E_0=2400\text{MPa};\ E_1=72000\text{MPa};\ E^*=1200\text{MPa};\ G^*=1200\text{MPa}.$$

Those parameters correspond to the [4] Kargin model – the simplest model of unidirectional reinforced plastic consisting of two polymer plates (layers 0 and 2) and a fiber reinforcement layer (layer 1).

Below are curves represented calculation results for load case 1. Variants of resolving equations are marked with the numbers 1, 2, and 3.

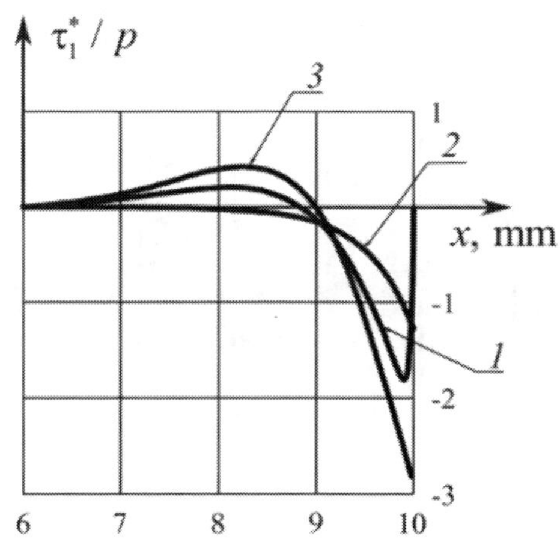


**Fig.** 5. Curves of tangential stresses in the contact layer 1 for the load case 1.

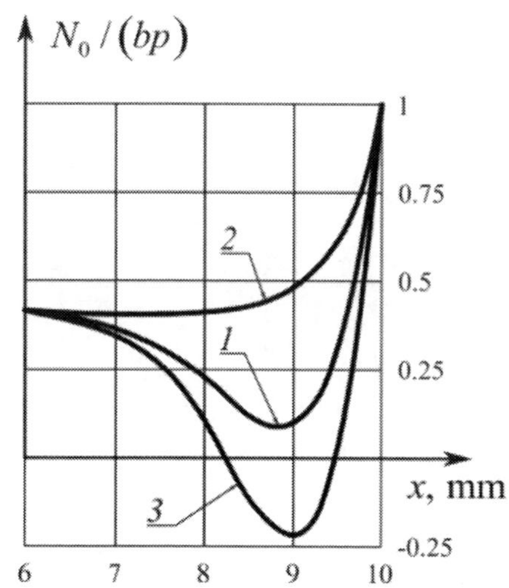


**Fig.** 6. Curves of axial force in the external layer 0 for the load case 1.

Below are curves represented calculation results for load case 2.

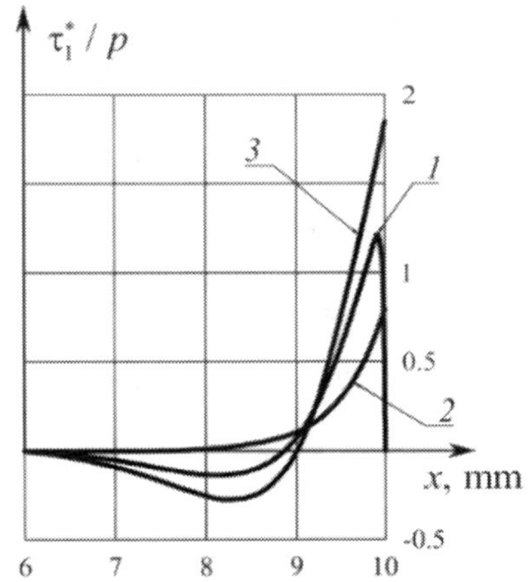


**Fig**. 7. Curves of tangential stresses in the contact layer 1 for the load case 2.

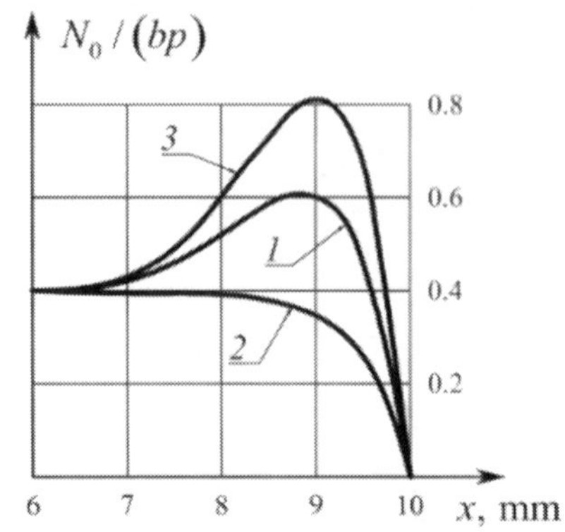


**Fig**. 8. Curves of axial force in the external layer 0 for the load case 2.

For the load cases 1 and 2, the calculation results are given only for the right-hand side in view of the symmetry.

Below are curves represented calculation results for load case 3.

The results of the calculation for load case 3 are given for the left and right side of the model, since there is no symmetry.

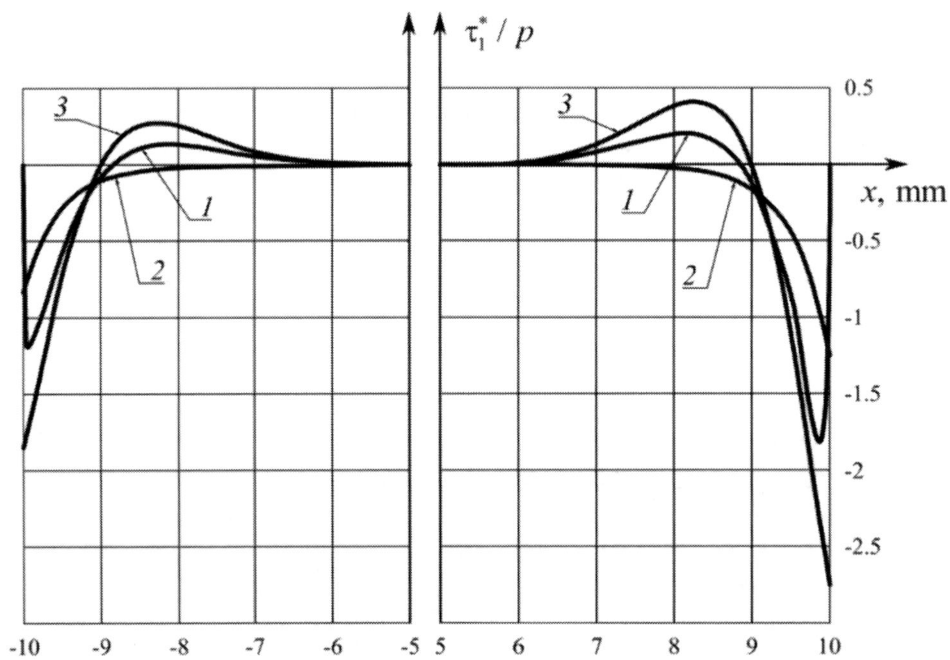


**Fig**. 9. Curves of tangential stresses in the contact layer 1 for the load case 3.

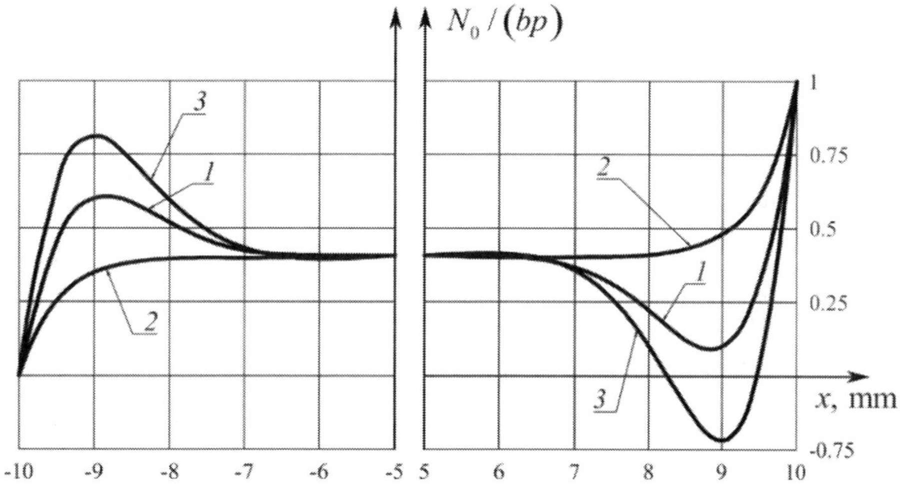


**Fig**. 10. Curves of axial force in the external layer 0 for the load case 3.

## 2 Conclusions

The results of the calculations show that variant 1 of resolving system of equations fulfills all the boundary conditions. However, it is not applicable for practical calculations, as result of its complexity.

Variant 2 can be used to calculate bending moments and shear forces, in models where these values are more important that axial forces and tangential stresses in the contact layer.

Variant 3 can be used to determine the zone of tangential stress concentration. Despite the fact that the value maximum of tangential stresses in this case is overestimated, it is the most suitable for practical calculations. Some results with using that model are given in paper [5].

Additional calculation models are given in paper [6].

## Acknowledgements

This study was performed with the financial support of the RF Ministry of Education and Science (the basic part of the state task for 2017-2019).

## References

1. Andreev V.I., Turusov R.A., Tsybin N.Yu. Application of the contact layer in the solution of the problem of bending the multilayer beam. Procedia Engineering. 2016. Vol. 153. P. 59-65.
2. Tsybin N.Yu., Turusov R.A., Andreev V.I. Comparison of creep in free polymer rod and creep in polymer layer of the layered composite. Procedia Engineering. 2016. Vol. 153. P. 51-58.
3. Andreev V.I., Turusov R.A., Tsybin N.Yu. Long strength of layered composite under normal fracture. Advances in Engineering Research, 2015. Vol. 39. P. 1703-1708.

4. Turusov R.A. Elastic and thermal behavior of a layered structure. Part II. Calculation results and their analysis. Mechanics of Composite Materials. 2015. Vol. 51(1). P. 127-134.
5. Turusov R.A., Kuperman A.M. Elastic properties of thin adhesive interlayers. Polymer Science. Series D. 2014. Vol. 7(1). P. 1-8.
6. Andreev V.I., Barmenkova E.V. The modeling of the real building object by using the model of a two layer beam of variable rigidity on an elastic basis. Applied Mechanics and Materials. 2012Vol. 204-208. P. 3596-3599.

# The profit as in-company evaluation of the construction site effectiveness

*Hubert* Anysz[1,*]

[1]Warsaw University of Technology, Faculty of Civil Engineering, Institute of Building Engineering, Division of Production Engineering and Construction Management, ul. Armii Ludowej 16, 00-637 Warsaw, Poland

**Abstract.** The paper discuss the profit made by construction sites operated by one company as a measure of their effectiveness. The different types of cost, that should have been covered by income, were described. The distinction between indirect cost made on building site and indirect cost made by headquarter of a company were shown, as well as necessity of including both on them in calculating building site effectiveness. Three different method of assigning company overhead cost to the building site were shown and discussed. Their significant influence on building site profitability was emphasised and calculated for exemplary values. The necessity of calculating full cost for each building site operating by one company was discussed in the article too, by showing which information is necessary for a company management, a building site management and for company shareholders.

## 1 The profit as a measure of effectiveness

It can be said, that a given project was successful if planned time schedule was kept and cost was not overrun [1], [6]. The time schedule is easy accessible document. It clearly shows the start and the end of a construction project, so it is easy to check if milestones were kept by a contractor and if the completion date of a project was not later than planned its end. The cost of a project is different for a client and a contractor. When the contract value is a lump sum, a client expects that the total cost of erecting a building object will not exceed the value stated in the contract signed with contractor. In the case of unit prices based contract, the client's expectations allows for some changes in the contract value according to the differences between the quantity take off and real number of works executed. A contractor's point of view on cost of a project is much different. The value stated in the contract signed with a client is an upper limit of the contractor's income that can be achieved on a given project. A cost of the project can be managed and it should be lower than income. It should cover direct cost of construction, indirect cost arisen on site and at the headquarter of a company. Remaining part of income – if appears – constitute profit of a building site.

* Corresponding author: h.anysz@il.pw.edu.pl

### 1.1 The necessity of making profit

Profit worked out by a company allows for development. New machinery can be acquired, contracts of bigger values can be won and financed. As a result of that higher profit can appears in a company financial statements. A tendency to making profit higher and higher can be explained by safety of a company existence on the market. The significant part of profit usually is not spent, but left in a company as a reserve for crisis time. This is one of the most important reasons for making a profit – willingness to survive [7], [8].

### 1.2 Calculating a profit for a company operating only one building site

Assuming that the company operates only one building site, income achieved on that should cover cost (direct and indirect) and remaining part constitute profit. Fig. 1 shows how income is divided.

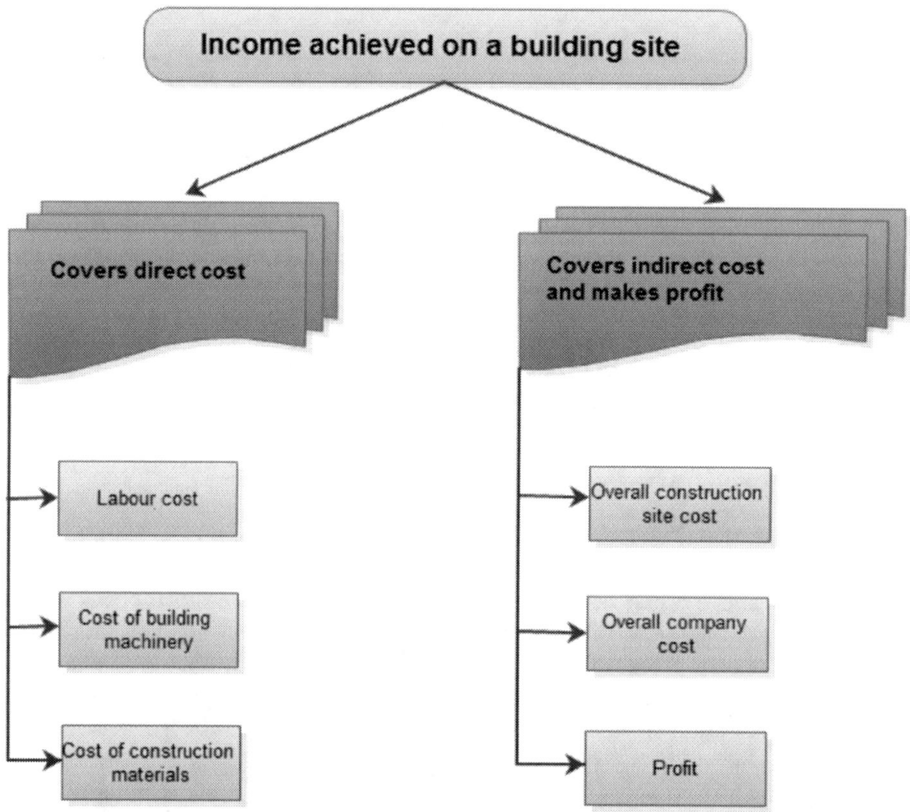


**Fig. 1.** Division of income achieved on construction site.

Direct cost can be easily assigned to the specific works being carried out. The cost of site management, office and social containers, a fencing, guard service, lightening the site serves to all activities. As the building site is settled as one unit, there is no need to divide these indirect cost to specific activities. Achieved income should cover them all. The company has its headquarter, accounting – department , human resources department etc.

Their cost should be covered by income too [3], [5]. To calculate the profit of the building site the following formula can be applied

$$Z = P - TKW - Kb \quad (1)$$

where:

$Z$ - profit of a building site,

$P$ - income achieved on a building site,

$Kb$ - overall construction site cost,

$TKW$ - direct cost of construction works,

Having $Z$ calculated, the company profit $Zp$ can be calculated too as

$$Zp = Z - Kp \quad (2)$$

where:

$Zp$ - profit of the comapany (operating only one building site),

$Kp$ - overall company cost.

### 1.3 Calculating a profit for a company operating a few building sites

Profit of the company that operates a few building sites can be calculated on the basis on formula (2) where profits from all building sites should be summarized. Then

$$Zp = (\Sigma Z_i) - Kp \quad (3)$$

where:

$Z_i$ - profit from $i$ building site being operated by the company

In order to calculate profit achieved on each individual building site, it is necessary to assign parts of overall company cost (Kp) to each building site.

## 2 Methods of assigning overall company cost to building sites

### 2.1 Methods description

There is no obligatory method of assigning overall construction cost to building sites in Poland. The decision is left to the CEO (Chief Executive Officer) of the company. He/she can be advised by financials but the decision is carried by CEO [2]. There are some methods widely applied [3], [4]:

- in equal portions,

$$Kp_i = Kp/n \quad (4)$$

where:
$Kp_i$ - the part of overall company cost assign to $i$ construction site,
$n$ - number of building sites operated by the company.

- proportionally to site income,

$$Kp_i=(P_i/(\Sigma P_i))*Kp \quad (5)$$

where:
$P_i$ - income achieved by *i* construction site

- proportionally to site profit,

$$Kp_i=(Z_i/(\Sigma Z_i))*Kp \quad (6)$$

Having Kpi defined, it is possible to write a formula for profit of the i building site after overall company cost assigned, marked as $Z_i'$.

$$Z_i'=Z_i - Kp_i \quad (7)$$

CEO can chose method, even propose his/her own one e.g. the ratio 3:7:6:8 for four building sites the company operates. Nevertheless, each method will have different consequences for separate building site.

### 2.2 Examples of company overhead cost assignments

Let's assume that the company operates three building sites A, B and C. The results achieved by each building site are shown in tab. 1, as well as the company profit is calculated. It is to emphasise, that the construction site profit is calculated before *Kp* assignment.

**Table 1.** Results achieved by building sites and the company

| Results achieved | Site A | Site B | Site C | Total |
|---|---|---|---|---|
| Income | 20000 | 5000 | 7000 | 32000 |
| Direct cost | 13400 | 3000 | 4550 | 20950 |
| Overall construction site cost | 6000 | 1500 | 2100 | 9600 |
| Construction site profit ** | 600 | 500 | 350 | 1450 |
| Overall company cost | * | * | * | 1200 |
| Company profit | - | - | - | 250 |

* not assigned yet ** before *Kp* assignement

Subtraction $Kp_i$ from the construction site profit, using three described above methods (as shown in tab 2) can give profit or loss for a given building site. The method of assignment does not affect the total profit of a company, of course.

Site B only shows the profit every time, independently from the *Kp* assignment method. Sites A and C can be evaluated as profitable or loss making in case on choosing different assignment methods. Which method is the right one, which method is fair. It can not be said. The contract A (the biggest one) was won due to low profit assumed. It has allowed for low bid price. Applying the method where the company's overhead cost are assigned proportionally to income, it puts this building on loss side. The contract C is middle profitable i.e. 5 % (350/7000), but its income is much lower that for contract A. Dividing company overhead cost on equal parts makes loss for Site C.

**Table 2.** Building sites profit after assigning company overhead cost

| Methods of overall company cost assigning | Profit after *Kp* assigning | | | |
|---|---|---|---|---|
| | Site A | Site B | Site C | Total |
| Equal portion | 200,00 | 100,00 | -50,00 | 250 |
| Proportionally to income | -150.00 | 312,50 | 87,50 | 250 |
| Proportionally to site profit | 103,45 | 86,21 | 60,34 | 250 |

The contract B is very profitable i.e. 10 % (500/5000) and for each method of company overhead cost assigning (chosen from aforementioned) Site B shows profit (varying from 86,21 to 312,5). On the other hand assignments are done proportionally to achieved already values. A project manager of contract A can be confused when according to his/her extraordinary engagement, planned profit (before assignment) has risen twice, but according to the assignment method applied, after deducting the part of company overhead cost, the contract he/she managed become much less profitable or even loss making.

Taking a decision about the method of company overhead cost assignment should take into account the assumptions made on the stage of preparing an offer for a given contract – the assumed profit level and the total value of an offer. Otherwise assignment of overheads can create false evaluation of building sites profitability.

## 3 The reasons of assigning company overhead cost to a building sites

There are three groups of people interested in proper evaluation of construction sites operated by a given company [6]:

- management of the company,
- building site management,
- shareholders of a company.

All of them in order to carry the right decision, require precise information about profitability of a company and its building sites. In a fact, it is rare case when a construction company operates one building site only. But even then, accounting, human resources, marketing and other functions of a company have to be provided and financed. There is no reason to omit these cost in calculating building site effectiveness. It refers to a company operating more than one building site too.

Owners of a company i.e. shareholders need to be sure that the size of administrative part of a company is matched to the size of a company operations. The total profit of the construction company is not sufficient information to evaluate it. Building site effectiveness is needed. The total profit can be on the satisfying level e.g. according to one, big, very profitable building site only, whereas other sites are poorly managed. In this case the most of headquarter effort is directed to this not profitable sites. Wrongly chosen assignment method can create false, misleading information.

The management of a company should be provided with accurate, precise information concerning building site effectiveness. Directors are responsible for selection of contracts for execution by a company. Proper financial information can help in finding the rules, and to determine which contracts are valuable for a company and which should be avoided. The profit of a building site based on full cost calculation is a measure of project manager abilities and effort too. False evaluation of a building site leads to wrong decisions.

A project manager have to know the system of overhead cost settlements too. He/she mustn't be surprised at the end of the project, that the positive financial effect was turned to negative by the administrative decision as choosing company overhead cost assignment method is. It is especially important for project managers who are not a company employees, hired for a certain project only.

## 4 Conclusions

Construction company overhead cost are generated by its headquarter. The functions of accounting, human resources, marketing departments etc. as a managerial functions are necessary for operations and have to be financed. The effort of exact assignment of aforementioned cost (of each headquarter department) to the building site would be too high. CEO makes a decision about the method of assigning these overhead cost to a building sites (projects). In order to create proper, precise and reflecting reality information about building site effectiveness, the method should be chosen very carefully, taking into account initial assumptions concerning each project being executed. Wrongly chosen assignment method can radically change the project evaluation. Profitable building site can be shown as a loss making one. Shareholders of a company, top company management and project managers require precise, reflecting reality information about construction project financial effectiveness in order to carry correct, future decisions aimed to making a profit by a company.

## References

1. D. Skorupka, D. Kuchta, *Zarządzanie ryzykiem w projekcie*, WSOWL, Wrocław (2012)
2. *Ustawa z dn. 29.09.1994 o rachunkowości* (2016) – Polish accounting law
3. C. Drury, *Management and Cost Accounting*, Cengage Learning (2015)
4. ACCA Texbooks. Management Information, AT Foulks Lynch Ltd (1998)
5. B. Kacprzyk, Kosztorysowanie obiektów i robót budowlanych, Polcen Sp. z o.o. (2010)
6. *A Guide to the Project Management Body fo Knowledge*, Project Management Institute (2000)
7. E. Głodziński, *Efektywność ekonomiczna – dylematy definiowania i pomiaru,* Zeszyty Naukowe Politechniki Śląskiej, Organizacja i zarządzanie Nr kol. 1919 (2014)
8. R. Borowiecki, *Pomiar i ocena procesów kreowania wartości w badaniu efektywności przedsiębiorstwa*. Wyd. UE w Krakowie, Kraków (2009)

# Vibrations of non-homogenous thick plate on two-parameter elastic foundation subjected to moving oscillator

*Magdalena* Ataman[1,*]

[1]Warsaw University of Technology, Faculty of Civil Engineering, Al. Armii Ludowej 16, 00-637 Warsaw, Poland

**Abstract.** In the paper the thick plate on the two-parameter Winkler foundation subjected to the one-mass oscillator is analysed. The oscillator is moving at uniform speed, parallel to the edge of the rectangular plate. The foundation is defined by two elastic constants $k_z$ and $k_t$. Equations of motion of the analysed system are submitted and solved. The paper includes numerical example.

## 1 Introduction

Equations of motion of thick plate can be obtained by a few means [1-2]. The momentum conservation law and the angular momentum conservation law in the case of an infinitesimal element of plate are determined by a resultant vector and a resultant moment. Simultaneous equation them to zero provides to equations of motion. The rule of virtual work (d'Alembert's rule), when forces and moments of inertia are taken into consideration, as well as the Hamilton rule are the most frequent used during derivation of equations of motion of plate. Particular attention should be paid to the Hamilton's rule, which allows to obtain, apart from equations of motion, natural kinematic, static and mixed boundary conditions. The variation rule is very important as a basis for computer methods like FEM used to static and dynamics computations. Dynamic equations of motion of thick plate were derived by Reissner [3], Vlasov [4], Mindlin [5], Ambarcumian [6] in 50's of last century and later by many others, e.g. [7-10]. Because of vertical and horizontal reactions of soil there are a shield-plate coupled effects in thick plate supported by three-parameter Winkler foundation. In general, these effects provide to tenth order partial differential equation of motion of analysed plate. In this paper an assumption concerning separation of shield and plate effects will be made. In spite of this assumption the obtained equation of motion is complicated when we want to solve it using analytical method.

* Corresponding author: m.ataman@il.pw.edu.pl

## 2 Governing equations of the problem

Assuming model of elastic Winkler foundation described by two parameters (Fig. 1), components of soil reaction can be written:

$$p_z = k_z w, \quad p_t = k_t u_i, \quad i = x, y. \tag{1}$$

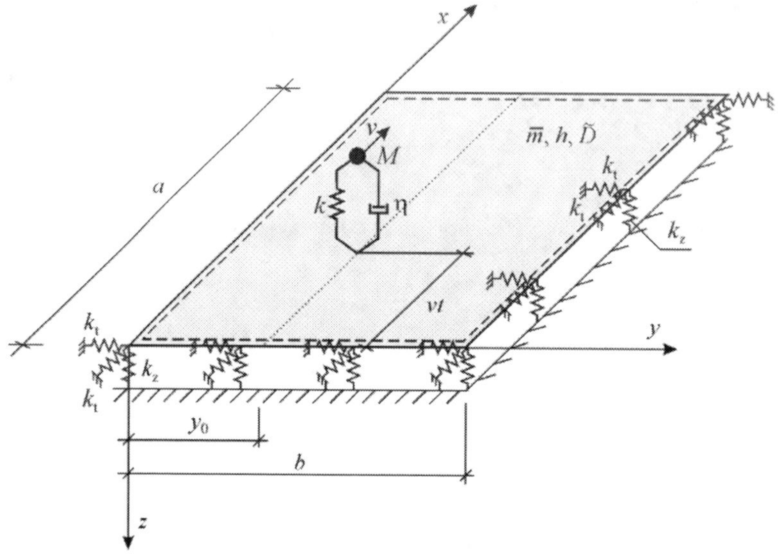


**Fig. 1.** Thick plate resting on generalised three-directional Winkler foundation subject to moving one-mass oscillator.

Tangential and normal forces of soil reaction at the bottom of plate are expressed:

$$X_1 = \frac{k_t h}{4}\frac{\partial w}{\partial x} + \frac{k_t h^3}{24(1-\nu)}\chi_x, \qquad Y_1 = \frac{k_t h}{4}\frac{\partial w}{\partial y} + \frac{k_t h^3}{24(1-\nu)}\chi_y, \qquad Z_2 = p - k_z w. \tag{2}$$

Basic equation of motion of the thick plate is following [2]:

$$\begin{aligned}
&\nabla^4 w + \frac{\rho h}{D}\frac{\partial^2 w}{\partial t^2} - \frac{17-6\nu}{60D(1-\nu)}\rho h^3 \nabla^2\left(\frac{\partial^2 w}{\partial t^2}\right) + \frac{\rho^2 h^6}{60D^2(1-\nu)}\frac{\partial^4 w}{\partial t^4} + \\
&+ k_z\left[\frac{w}{D} - \frac{(2-\nu)h^2}{10D(1-\nu)}\nabla^2 w + \frac{\rho h^5}{60D^2(1-\nu)}\frac{\partial^2 w}{\partial t^2}\right] + \\
&- \frac{1}{1-\mu}\left\{\left[\frac{k_t h^2}{4}\nabla^2 - \mu\left(\rho h\frac{\partial^2}{\partial t^2} + k_z\right)\right]\left[\frac{w}{D} - \frac{(2-\nu)h^2}{60D(1-\nu)}\nabla^2 w + \frac{\rho h^5}{360D^2(1-\nu)}\frac{\partial^2 w}{\partial t^2}\right]\right\} = \\
&= \frac{p}{D} - \frac{(2-\nu)h^2}{10D(1-\nu)}\nabla^2 p + \frac{\rho h^5}{60D^2(1-\nu)}\frac{\partial^2 p}{\partial t^2} + \\
&+ \frac{\mu}{1+\mu}\left[\frac{p}{D} + - \frac{(2-\nu)h^2}{60D(1-\nu)}\nabla^2 p + \frac{\rho h^5}{360D^2(1-\nu)}\frac{\partial^2 p}{\partial t^2}\right],
\end{aligned} \tag{3}$$

where $\mu = \frac{k_t h^4}{24D(1-\nu)}$ is the dimensionless factor describing influence of unknown functions $\chi_x$ i $\chi_y$ to the vertical forces at the bottom of the plate, $D = \frac{Eh^3}{12(1-\nu^2)}$ is the flexural rigidity of the plate, $w$ is the plate deflection, $\rho$ is the density of the plate, $h$ is the plate thickness, and $p$ is the external load per unit area normal to the top surface. It should be mentioned that for factors $k_t$ describing actual subgrades $\mu$ factor is considerably smaller than 1. When $\mu = 0$, the form of equation of motion (3) is simpler and can be written as follows:

$$\begin{aligned} &\nabla^4 w + \frac{\rho h}{D}\frac{\partial^2 w}{\partial t^2} - \frac{17-6\nu}{60D(1-\nu)}\rho h^3 \nabla^2\left(\frac{\partial^2 w}{\partial t^2}\right) + \frac{\rho^2 h^6}{60D^2(1-\nu)}\frac{\partial^4 w}{\partial t^4} + \\ &+ k_z\left[\frac{w}{D} - \frac{(2-\nu)h^2}{10D(1-\nu)}\nabla^2 w + \frac{\rho h^5}{60D^2(1-\nu)}\frac{\partial^2 w}{\partial t^2}\right] + \\ &- \left\{\left[\frac{k_t h^2}{4}\nabla^2\right]\left[\frac{w}{D} - \frac{(2-\nu)h^2}{60D(1-\nu)}\nabla^2 w + \frac{\rho h^5}{360D^2(1-\nu)}\frac{\partial^2 w}{\partial t^2}\right]\right\} \\ &= \frac{p}{D} - \frac{(2-\nu)h^2}{10D(1-\nu)}\nabla^2 p + \frac{\rho h^5}{60D^2(1-\nu)}\frac{\partial^2 p}{\partial t^2}. \end{aligned} \tag{4}$$

The equation of motion of one-mass oscillator should be connected to the equation of motion of the plate (4) [7]

$$M\frac{\mathrm{d}^2 w_1}{\mathrm{d}t^2} + \eta\left(\frac{\mathrm{d}w_1}{\mathrm{d}t} - \frac{\mathrm{d}w}{\mathrm{d}t}\right) + k(w_1 - w) = 0, \qquad \frac{\mathrm{d}w(vt,t)}{\mathrm{d}t} = \left.\frac{\partial w}{\partial t}\right|_{x=vt} + v\left.\frac{\partial w}{\partial x}\right|_{x=vt}. \tag{5}$$

where: $M$ is the mass of the oscillator,
$k$ is the elastic constant of the oscillator,
$\eta$ is the damping coefficient of the oscillator,
$w_1$ is the vertical displacement of the oscillator mass.

Dynamic pressure of oscillator acting to the plate describes formula:

$$N(t) = Mg - M\frac{\mathrm{d}^2 w_1}{\mathrm{d}t^2}. \tag{6}$$

Moving load can be written as follows:

$$p = p(x,y,t) = N(t)\delta(x - vt)\delta(y - y_0) = \left(Mg - M\frac{\mathrm{d}^2 w_1}{\mathrm{d}t^2}\right)\delta(x - vt)\delta(y - y_0). \tag{7}$$

When $P = Mg$ is introduced into equation (7), load and its derivatives can be expressed:

$$
\begin{gathered}
p = p(x,y,t) = p_1(x,y,t) + p_2(x,y,t) \\
p(x,y,t) = \left( P - M\frac{\mathrm{d}^2 \mathrm{w}_1}{\mathrm{d}t^2} \right)\delta(x - vt)\delta(y - y_0) = \sum_n \sum_m f_{n,m}(t)\sin\alpha_m x \sin\beta_n y, \\
f_{n,m}(t) = \frac{4}{ab}\left( P - M\frac{\mathrm{d}^2 \mathrm{w}_1}{\mathrm{d}t^2} \right)\sin\alpha_m vt \sin\beta_n y_0\,, \\
\nabla^2 p(x,y,t) = \\
= -\frac{4}{ab}\left( P - M\frac{\mathrm{d}^2 \mathrm{w}_1}{\mathrm{d}t^2} \right)\sum_n \sum_m \left(\alpha_m{}^2 + \beta_n{}^2\right)\sin\alpha_m vt \sin\beta_n y_0 \sin\alpha_m x \sin\beta_n y, \\
\frac{\partial^2 p(x,y,t)}{\partial t^2} = \\
= -\frac{4v^2}{ab}\left( P - M\frac{\mathrm{d}^2 \mathrm{w}_1}{\mathrm{d}t^2} \right)\sum_n \sum_m \alpha_m{}^2 \sin\alpha_m vt \sin\beta_n y_0 \sin\alpha_m x \sin\beta_n y + \\
-\frac{4}{ab}M\frac{\mathrm{d}^4 \mathrm{w}_1}{\mathrm{d}t^4}\sum_n \sum_m \sin\alpha_m vt \sin\beta_n y_0 \sin\alpha_m x \sin\beta_n y.
\end{gathered} \tag{8}
$$

To obtain solution of the problem a system of two differential equations (4) and (5) should be solved. We assume the elastic plate is rectangular of $a$ and $b$ sides and $h$ thickness. All sides of the plate are simply supported therefore the deflected form is represented by the double series

$$
w = w(x,y,t) = \sum_{n=1}^{\infty}\sum_{m=1}^{\infty} q_{n,m}(t)\sin\alpha_n x \sin\beta_m y, \qquad \alpha_n = \frac{n\pi}{a},\ \beta_m = \frac{m\pi}{b}. \tag{9}
$$

Substitution (9) in (4) and (5) for $n = m = 1$ Bubnov-Galerkin method in (4) yields the system of two coupled differential equations of sixth order depending on time:

$$
\begin{gathered}
M\frac{\mathrm{d}^2 w_1}{\mathrm{d}t^2} + \eta\frac{\mathrm{d}w_1}{\mathrm{d}t} + kw_1 = \\
= (\eta\dot{q} + kq)\sin\frac{\pi vt}{a}\sin\frac{\pi y_0}{b} + \eta q\frac{\pi v}{a}\cos\frac{\pi vt}{a}\sin\frac{\pi y_0}{b}, \\
q^{IV} + \frac{1}{12}\left(\frac{c_p}{h}\right)^2 A\ddot{q} + \frac{1}{576}\left(\frac{c_p}{h}\right)^4 Q^2 q = \left\{ \frac{2c_p{}^2}{\rho h^3 ab}\left( P - M\frac{\mathrm{d}^2 w_1}{\mathrm{d}t^2} \right)\left[ 10(1-v) + \right.\right. \\
\left.\left. + \tilde{\gamma}^2(2-v) - \frac{\rho h^5 \alpha^2 v^2}{12D} \right] - \frac{4M}{\rho hab}\frac{\mathrm{d}^4 w_1}{\mathrm{d}t^4} \right\}\sin\frac{\pi vt}{a}\sin\frac{\pi y_0}{b},
\end{gathered} \tag{10}
$$

where: $\theta = \dfrac{k_z}{k_t}$, $\mu = \dfrac{k_t h^4}{24D(1-v)}$, $\mu_z = \dfrac{k_z h^4}{D}$, $\tilde{\gamma} = h^2(\alpha^2 + \beta^2)$, $\lambda = \dfrac{a}{b}$,

$\xi = \dfrac{h}{a}$, $\gamma = \pi\xi^2(1 + \lambda^2)$, $c_p{}^2 = \dfrac{E}{\rho(1 - v^2)}$,

$$
A = 60(1-v) + (17 - 6v)\tilde{\gamma}^2 + \mu_z\left(1 + \frac{\theta\tilde{\gamma}^2}{24}\right),
$$

$$\hat{\gamma}^4 = \tilde{\gamma}^4 + \mu_z \left[ 1 + \frac{2-\nu}{10(1-\nu)} \tilde{\gamma}^2 + \frac{\theta \tilde{\gamma}^2}{4(1-\nu)} \left( 1 + \frac{2-\nu}{60(1-\nu)} \tilde{\gamma}^2 \right) \right],$$
$$Q = 4\hat{\gamma}^2 \left[ 15(1-\nu) \right].$$

The above system of linear equations is solved numerically using Mathematica code. Equations (10) are valid for $t \in \left\langle 0; \frac{a}{v} \right\rangle$. When $t \geq \frac{a}{v}$, the free vibrations of plate and oscillator are analysed. Assuming the oscillator is moving on the rigid surface, equations of motion of oscillator and plate are given by two uncoupled equations:

$$M \frac{\mathrm{d}^2 \tilde{w}_1}{\mathrm{d}t^2} + \eta \frac{\mathrm{d}\tilde{w}_1}{\mathrm{d}t} + k\tilde{w}_1 = 0, \quad \tilde{q}^{IV} + \frac{1}{12}\left(\frac{c_p}{h}\right)^2 A \ddot{\tilde{q}} + \frac{1}{576}\left(\frac{c_p}{h}\right)^4 Q^2 \tilde{q} = 0. \tag{11}$$

Substitution of dimensionless eigenvalues of plate yields:

$$\omega^4 - \frac{1}{12}\left(\frac{c_p}{h}\right)^2 A\omega^2 + \frac{1}{576}\left(\frac{c_p}{h}\right)^4 Q^2 = 0 \;\rightarrow\; \Omega_{1/3} = \frac{1}{\tilde{\gamma}^2}\left\{\frac{1}{2}\left[P \mp \left(P^2 - Q^2\right)^{\frac{1}{2}}\right]\right\}^{1/2},$$
$$\Omega_1 = \frac{\omega_1}{\omega_{kl}}, \quad \Omega_3 = \frac{\omega_3}{\omega_{kl}}, \quad \omega_{kl} = \pi^2\left(\frac{m^2}{a^2} + \frac{n^2}{b^2}\right)\left(\frac{D}{\rho h}\right)^{1/2} = \frac{\tilde{\gamma}_{m,n}{}^2 c_p}{h\sqrt{12}}, \quad m = n = 1. \tag{12}$$

Natural frequency of oscillator is given by expression $\omega_{osc} = \sqrt{\frac{k}{M}}$.
Solution of homogenous differential equations of motion of plate and oscillator is given by:

$$\tilde{w}_1 = A_1 \cos\omega_{osc} t + A_2 \sin\omega_{osc} t,$$
$$\tilde{q} = C_1 \cos\omega_1 t + C_2 \sin\omega_1 t + C_3 \cos\omega_3 t + C_4 \sin\omega_3 t. \tag{13}$$

The constants $A_1$, $A_2$ and $C_1 \div C_4$ are to be determined from the initial conditions for $t = \frac{a}{v}$. Introduction of notations $\tilde{w}_1$ and $\tilde{w}$ instead of $w_1$ and $w$ for $t \geq \frac{a}{v}$ gives following expressions:

$$t = \frac{a}{v} \;\rightarrow\; w_1 = \tilde{w}_1, \quad \frac{\mathrm{d}w_1}{\mathrm{d}t} = \frac{\mathrm{d}\tilde{w}_1}{\mathrm{d}t}, \quad \frac{\mathrm{d}^2 w_1}{\mathrm{d}t^2} = \frac{\mathrm{d}^2\tilde{w}_1}{\mathrm{d}t^2}, \quad \frac{\mathrm{d}^3 w_1}{\mathrm{d}t^3} = \frac{\mathrm{d}^3\tilde{w}_1}{\mathrm{d}t^3},$$
$$w = \tilde{w}, \quad \frac{\mathrm{d}w}{\mathrm{d}t} = \frac{\mathrm{d}\tilde{w}}{\mathrm{d}t}, \quad \frac{\mathrm{d}^2 w}{\mathrm{d}t^2} = \frac{\mathrm{d}^2\tilde{w}}{\mathrm{d}t^2}, \quad \frac{\mathrm{d}^3 w}{\mathrm{d}t^3} = \frac{\mathrm{d}^3\tilde{w}}{\mathrm{d}t^3}. \tag{14}$$

## 3 Transversally non-homogenous plate

When consider plate of modulus of elasticity $E(z)$, and uniform Poisson's ratio $\nu$, the flexural rigidity of the plate in Eq. (3) and (4) is given by formula:

$$D=\tilde{D}=\int_{-\frac{h}{2}-e}^{\frac{h}{2}-e}\frac{E(\tilde{z})\tilde{z}^2}{1-\nu^2}\,d\tilde{z},\qquad e=\frac{\int_{-\frac{h}{2}}^{\frac{h}{2}}E(z)z\,dz}{\int_{-\frac{h}{2}}^{\frac{h}{2}}E(z)\,dz}. \tag{15}$$

where $e$ is a distance for which shield and plate effects are separated.

For the three-layered plate $e$ can be found using expression:

$$e=\frac{E_1\int_{-\frac{h}{2}}^{-\frac{h}{2}+h_1}z\,dz+E_2\int_{-\frac{h}{2}+h_1}^{-\frac{h}{2}+h_1+h_2}z\,dz+E_3\int_{-\frac{h}{2}+h_1+h_2}^{-\frac{h}{2}+h}z\,dz}{E_1\int_{-\frac{h}{2}}^{-\frac{h}{2}+h_1}dz+E_2\int_{-\frac{h}{2}+h_1}^{-\frac{h}{2}+h_1+h_2}dz+E_3\int_{-\frac{h}{2}+h_1+h_2}^{-\frac{h}{2}+h}dz}, \tag{16}$$

where $E_i$ and $h_i$ are modulus of elasticity and thickness of $i$-th layer respectively.

Displacements and internal forces are referred to $Oxy$ - the equivalent neutral plane of the plate.

## 4 Numerical example

The system of two differential equations of motion of oscillator and plate presented in the paper is solved using MATHEMATICA code. Dimensionless coordinates are introduced into calculations. Two cases of vibrations are investigated in the paper. The first case when the oscillator is moving on the plate $\zeta=\frac{vt}{a}\le 1$, and the second one after the oscillator descending the plate $\zeta\ge 1$. According to the order of the system of differential equations of motion eight initial conditions equal to zero are assumed. The power of double series describing plate deflections is one.

In numerical example influence of linear viscous damping in the oscillator and in the plate was taken into account. Proportionality of the damping coefficient $c_e$ to the mass of the plate $\bar{m}$, and proportionality of the damping coefficient $c_i$ to Young's modulus $E$ were assumed and the substitutional damping number $\xi_n$ was defined:

$$c_e=\alpha_e\bar{m},\quad c_i=\alpha_i E,\quad \xi_n=\frac{1}{2}\left(\frac{\alpha_e}{\omega_1}+\alpha_i\,\omega_1\right). \tag{17}$$

Above assumptions change the equation of motion of the plate and were taken into account in numerical example.

We consider a square plate of sides $a=b=1\,\mathrm{m}$ and thickness $h=0.2\,\mathrm{m}$. Following parameters for plate material are assumed: Young's modulus is $E=20\cdot10^9\,\mathrm{Pa}$, Poisson's ratio is $\nu=0.167$, density is $\rho=2200\,\mathrm{kg/m^3}$, viscosity coefficient is $\eta_1=90000\,\mathrm{Ns/m}$, modulus of elastic subgrade reaction in vertical direction is $k_z = 50 \cdot 10^6$ N/m$^3$ and in horizontal directions $k_t = 0,5k_z$. Oscillator's mass is $M = 6000$ kg, its elastic constant is $k = 16 \cdot 10^5$ N/m and damping coefficient is $\eta = 14000$ kg/s. The oscillator is moving parallel to $x$ axis at uniform speed. Using Mathematica code dynamic deflection of the plate midpoint and deflection of the mass of the oscillator were determined and analysed for various values of oscillator velocity. In the figures dynamic case $v = 100$ m/s and static case $v = 0.1$ m/s, for verification purpose, are presented. Figures present deflection of the plate and the oscillator for forced and free vibrations.

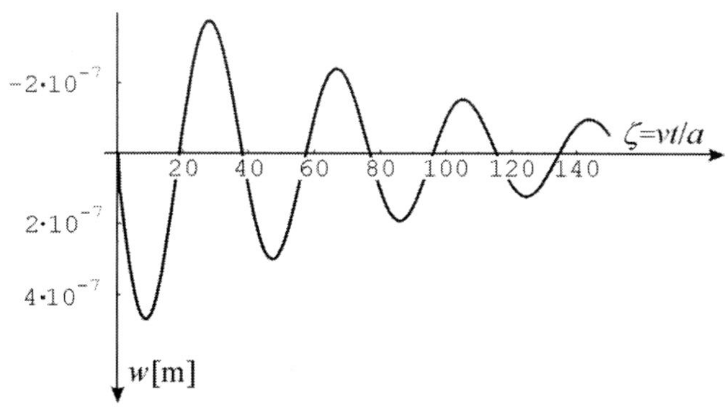


**Fig. 2.** Deflection of viscoelastic oscillator, velocity $v = 100$ m/s.

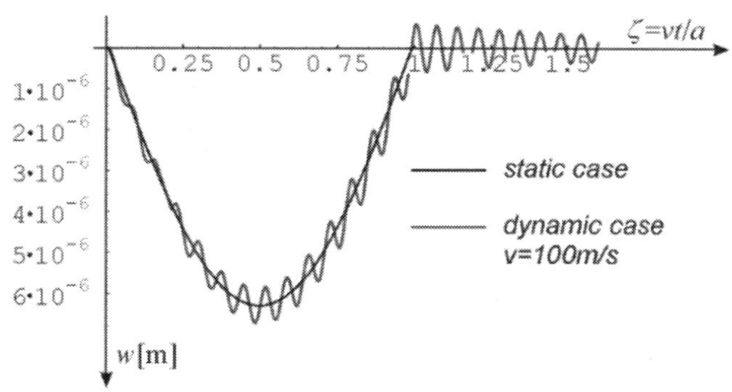


**Fig. 3.** Deflection of thick plate midpoint. Dynamic case $v = 100$ m/s.

## 4 Conclusions

In the paper theoretical basis of the thick plate on the generalised Winkler foundation under one-mass oscillator are submitted. The equations of motion of the analysed oscillator and the plate are of second and fourth order in respect to time. The instant when the oscillator is descending the plate requires further investigations because the oscillator impact in the plate support causes disturbance in the movement.

## References

1. W. Szcześniak, *Wybrane zagadnienia z dynamiki płyt*, OWPW, Warszawa (2000)

2. W. Szcześniak, Eng. Trans., IPPT PAN, **35**, 3, 347-374 (1987)
3. E. Reissner, J. Appl. Mech., **25**, 1 (1958)
4. B.F. Vlasov, Wiestnik MGSU, **2** (1957)
5. R.D. Mindlin, J. Appl. Mech. **18**, 1, 31-38 (1951)
6. S.A. Ambarcumian, *Theory of anisotropic plates*, Nauka, Moskwa (1987)
7. W. Szcześniak, Eng. Trans., IPPT PAN, **20**, 3, 455-478 (1972)
8. W. Szcześniak, ACE, **22**, 1, 107-127 (1976)
9. W. Szcześniak, Eng. Trans., IPPT PAN, **3**, 1-2, 37-53 (1985)
10. W. Szcześniak, Eng. Trans., IPPT PAN, **37**, 1, 87-115 (1989)

# Lean construction as an effective organization model in Arctic

*Elena S.* Balashova[1], and *Elizaveta A.* Gromova[2*]

[1]Peter the Great St. Petersburg Polytechnic University, Saint-Petersburg, Russia
[2]Peter the Great St. Petersburg Polytechnic University, Saint-Petersburg, Russia

**Abstract.** In recent time, due to the sharp climatic changes, the Arctic attracts an increased interest of the world powers as a strategically important object. In 2013, the development strategy of the Arctic zone of the Russian Federation and national security for the period up to 2020 was approved by the President. In this strategy, the socio-economic development of the region in terms of improving the quality of life, expressed in the implementation of housing and civil engineering is very important. The goal of the study is to identify effective organization model of construction in the Arctic zone of the Russian Federation. Lean construction as a dynamically developing methodology abroad is analyzed. Characteristics of this organization model of construction meet the necessary requirements for the construction of various infrastructure objects in the Arctic. Therefore, the concept of lean construction can be an effective strategy of development of the Arctic regions of Russia as well as other Arctic countries.

## 1 Introduction

In recent years, the world community is characterized by a turbulent development of economy. In conditions of extremely high global competition a well-built state program of development and ability to respond quickly to the changes in the external environment come to the fore. So, recently, the Arctic is increasingly attracting the interests of the world powers. Global warming, which triggered a rapid decline of the ice area in the Arctic ocean, is able to release water from the Arctic ice in summer, according to the latest modeling software, by 2020. According to data released by the U.S. geological service in 2008, up to a quarter of the world's undiscovered oil and gas reserves are concentrated in the Arctic. And 80% of these reserves are hidden in the Arctic shelf. Taking into account the trend of melting ice and high mineral potential, the region is very promising.

The Arctic is becoming a strategic vector of Russia's development, both from the point of view of strengthening positions on the international arena and ensure the national interests. In the future, this region will be able to contribute to the economic growth of the country. In February 2013, the development strategy of the Arctic zone of the Russian Federation and national security for the period up to 2020 was approved by the President. Plans of expanding Russia's military presence in the Arctic is an essential part of this

* Corresponding author: lizaveta-90@yandex.ru

document. But also the issues of ensuring sustainable socio-economic development of the Arctic zone of the Russian Federation occupy a significant place. The construction industry is a driver of economic growth. Therefore, the object of the research is the housing and civil engineering in the Arctic part of Russia.

The region has a number of features:

— extreme climatic conditions, including low temperatures, strong winds and ice cover in the waters of the Arctic seas;

— remoteness from basic industrial centers, high resource intensity and dependence of economic activities and livelihoods of supplies from other regions of Russia;

— low sustainability of ecological systems.

At the same time, the domestic construction industry is characterized by uncertainty of construction time, the project cost and scope of work. That's why, the choice of technology and construction organization are crucial. Lean construction represents a promising model of the systematic management of all stages of the life cycle of the construction project, which is focused on reducing the time and cost expenses.

The theoretical and practical issues of development of the effective models of construction organization from the point of lean manufacturing are explored by many scientists. Among them the following scholars are highlighted: L. Koskela (1992, 2000, 2002), G. Ballard (1994, 2000, 2003), J. Howell (1999, 2003, 2014), A. Mossman (2005, 2009). It is noteworthy that this concept has received wide recognition abroad, both theoretical and practical levels, possible methods of implementation are developed, the tools of lean construction are analyzed. In Russia, the spread of the concept is poorly expressed. The problems of applying the model of lean construction in the Arctic zone of the Russian Federation are quite undeveloped.

Thus, the objective of the study is to analyze the housing and civil engineering in the Arctic from a position of implementing the methodology of lean construction.

## 2 Methods

Lean construction is a kind of derivative of the lean production concept [1-7]. In the beginning of 1990-ies L. Koskela was one of the first who analyzed the application of the principles of lean production in construction industry [8-11]. As a result, he laid the foundation of lean construction. This concept is a management model that is similar to lean production in part of the pursued objectives, which are based on constant cost reduction, continuous quality improvement and optimization of the speed of material flow according to the logic of just in time. But it has several features in connection with the specifics of the construction field. One of the main distinguishing features is the predominant percentage of losses in construction compared with production [12] and a slightly different classification [13]. Losses in construction are along with the classical losses which were allocated by [14]: overproduction; waiting; unnecessary transportation; processing; inventory; unnecessary movement; production of defective products, the replacement is a monetary loss caused by the substitution of more expensive materials (with an unnecessary better performance), performing simple tasks by highly qualified staff or using high-tech equipment instead of the less technological; and the other related to the openness of construction sites (theft, vandalism, inclement weather, accidents). Another distinctive feature is the nature of the product movement: in industrial production, the products move through the stationary working stations, and in construction, the work areas move through stationary products [15]. Also, differential moment is a high degree of uncertainty of the construction project because of the scale of activity.

The Deming cycle (PDCA) [16] is the basis of the algorithm logics of improvement the processes of the lean construction.

Lean construction uses the entire set of tools of lean production and in addition has a number of specific such as:

— the last planner system (the system of operational planning based on the principle "down-up" and covering a period of one week, i.e. weekly and daily schedules are built for two weeks, where the graphs of the first week are working (fully secured by the material resources), and graphs of the second week provide solutions to the logistics works of the second week. The indicator (PPC, percent plan complete) that shows what share of weekly work plan completed on time, is introduced to describe the quality of the development and implementation of schedules. This system is one of the most effective ways to improve construction efficiency through improving the processes of the planning and controlling.) [17-21];

— concurrent engineering (parallel execution of various tasks using multidisciplinary teams with the aim of obtaining the best products from the point of functionality, quality and performance);

— the planned conditions and working environment in the construction (this tool is necessary to match the conditions for execution of a task with capabilities of workers with the aim of preserving their health and safety).

Unlike traditional approaches to organization of construction, which are based on project management techniques, the concept considers the construction process as a value stream creation. Stages of construction constitute the stream logically replacing each other. At that time, in the framework of the established approach: the production process consists of the several subprocesses that transform a variety of resources at the entrance to the finished product at the output and that can be implemented and analyzed separately from each other.

Thus, the main advantages of the lean construction include:

— reduction of project turnaround time;

— reduction of construction costs;

— high quality of the completed construction objects;

— high degree of controlling of construction project.

## 3 Results and discussion

According to the development strategy of the Arctic zone of the Russian Federation and national security for the period up to 2020 one of the priorities is an integrated socio-economic development of the region, which is provided, in particular, improving the quality of life of indigenous peoples and social conditions of economic activity in the Arctic. Accordingly, in order to improve the quality of life of people who are living and working in the Arctic zone of the Russian Federation, including indigenous small peoples, increase their level of social and cultural services and ensure positive demographic developments and necessary social conditions of economic activity are considered, among others, the following activities:

— housing development and upgrading of social infrastructure objects, including educational institutions, healthcare organizations and culture organizations;

— housing renewal and housing modernization, upgrade and modernization of the fixed assets of housing and communal services on the basis of modern energy saving technologies;

— active formation in the cities, small villages and towns new available for all segments of the population and mobile multifunctional cultural institutions.

The state program of the Russian Federation "Socio-economic development of the Arctic zone of the Russian Federation for the period till 2020" is the main mechanism for the implementation of the aforementioned strategy, in which the aim is the creation and development of the local livelihood systems in the region, where housing and civil engineering plays an essential role. Thus, this problem must be realized through the implementation of projects directly in the field of housing construction along with the construction of socially significant objects equipped with network engineering and technical support. Basis for the formation of the positive trends in the implementation of such projects is the financing on the required scale, focused on the mechanisms of the public-private partnership. By 2020 in the context of this task, the achievement of the following values of some targets is provided:

— average population of the Arctic zone will be 2341.2 thousand people;

— rate of the natural increase of the population of the Arctic zone (per 1000 persons) will grow by more than threefold and will amount to 9.5 persons per 1000 persons of the population;

— total area of living spaces will reach 25.9 sq.m in average per one resident.

### 3.1 Features of housing and civil engineering in the Arctic zone

The combination of extreme environmental conditions with low socio-economic development of the region predetermined the difficult conditions for construction. The process of housing and civil engineering has many engineering features. Objects have to be built on unstable terrain under which the soil is mixed with ice. In the permafrost the surface is constantly changing its structure. Built structures heat the soil, and it loses solidity, begins to melt and shift. For example, in Canada, which also represents the Arctic, the residents of two small towns were forced to leave their homes for this reason. Therefore, it is possible to build the foundation of buildings on permafrost, only by taking special measures to maintain constant temperature of the soil. The construction of the foundations must necessarily be accompanied by an engineering-geocryological investigations in order to adequately predict the behavior of soil during the construction and subsequent operating of buildings. Calculations of the thermal regime of soils are performed and the type of foundation is determined after the engineering studies. There are two principles on which the use of permafrost as the foundation of buildings is based. The first principle is the aspiration to maintain the permafrost in its natural state (the most popular and affordable solution). Under the second principle, a building is designed with the expectation that the foundation of the house will be in a thawing condition [22]:

1) in accordance with the first principle, the permafrost Foundation remains in its original state not only in the process of building construction, but during its subsequent operation. This principle is applied in situations where preservation of the frozen soil in its original state is cost-effective. The easiest way is to build a foundation on a sandy ground, which is not classified as plastic-frozen. For the latter cases, activities to reduce the temperature of the base to the calculated values are additionally provided, and possible plastic deformations of the base under the loading are considered in the calculations of the foundation. Following this principle mainly pile or pier foundations are satisfied. But, for example, strip foundation may also be the solution. The only one condition is to not give the top layer of soil to change its properties under the influence of heat from operated object. For this purpose, the underground is made cold, ventilated through the produhi in zabirka or the basement of the house. Also, a solid insulation layer with high insulation properties under the building can be performed. This will keep the soil in its natural state. Also, it is important to determine the depth of the soil. For different types of structures, its value is assigned separately:

— for pile foundations, the depth of laying should not be less than 2 m greater than the thickness of the layer of soil that freezes and thaws seasonally. The calculation is done on that layer of the permafrost soil will provide the required resistance to compression;
— for all other types of foundations, the depth of laying is set greater than the thickness of the seasonal thawing soil at 1 m;
— if the construction of the building on an artificial material with prescribed characteristics is designed, the depth of laying the soles is not standardized and it is based on the conditions of construction;

2) the second way is used less frequently and under conditions when the soil at the construction site is not heaving or subsidence and when in terms of temperature fluctuations, the soil deformations do not exceed the maximum permissible values. In this case, it is necessary either to thaw out the soil before construction of the foundation, or make all important calculations, and to assume that the foundation will thaw during the operating of the building. Construction of the building's basis involves the calculation of the depth of laying of the foundation sole in the comprehensive assessment of the thickness of the seasonally freezing soil and level of ground waters, focusing on the zone of thawing. And it will be formed in the subsequent operating of the building.

A specific choice is made by comparing the technical and economic calculations and the effectiveness of the considered solutions.

The necessity of application of pile foundations and thermo-insulating and innovative energy-intensive materials for construction of buildings walls in severe climatic conditions creates significant costs. For many areas in the Arctic region, such construction becomes uneconomical. For example, in Norilsk, the price of 1 sq. m of real estate is about 22 - 26 thousand rubles on the secondary market and the cost of construction of 1 sq. m in low-rise apartment building is about 75 thousand rubles. The example is significant from the point of view that Norilsk is allocated from the set of settlements included in the Arctic zone of the Russian Federation, by positive trend in population (in 2012 – 177273 people; in 2013 – 177738 people; in 2014 – 176559; in 2015 – 176251 people; in 2016 – 177428 people, according to the data of the Federal state statistics service of the Russian Federation).

In the Arctic, the accumulation of damages in the protective and decorative coverings and construction materials happens many times faster than the Average strip of Russia, because of the rapidly flowing vapor condensation processes under changing temperature and humidity conditions. Intense destruction is caused by the volume expansion of water in construction materials when it is freezing. Errors of the design, construction and operating of real estate quickly manifest. Correction of defects in the Arctic zone costs many times more expensive than in other areas [23]. In turn, the physical deterioration of buildings quite early reaches critical values, which mean the assignment to the category of dilapidated housing. The overhaul of such objects is difficult and often economically inefficient.

Therefore, the construction in the Arctic can be described as a costly project that requires the highest quality and rapid pace of work and, consequently, reduced construction time compared to the objects located in more favourable conditions.

The significant role in the implementation of this organization model of construction is given to the last planner system. The project of the low-rise residential building in Norilsk, which takes into account the characteristic features of the Arctic in relation to construction and installation works is presented in figure 1.

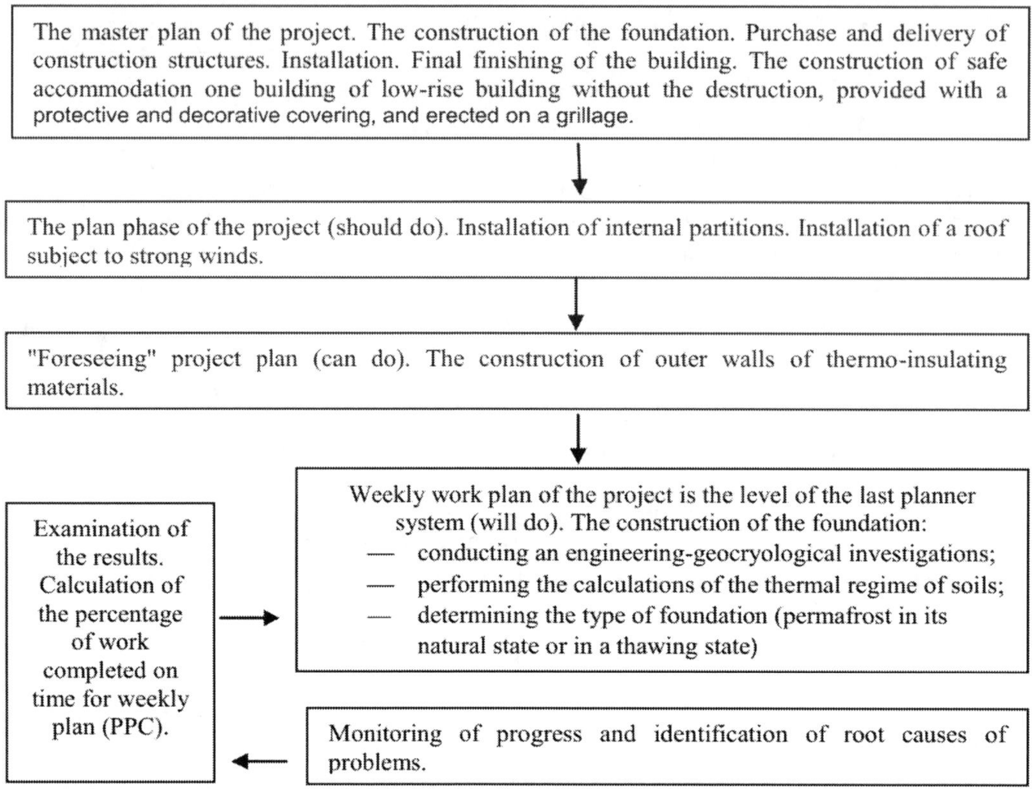


**Fig. 1.** The scheme of operational planning of construction, developed by the authors

In summary, it is possible to highlight the key features of housing and civil engineering in the Arctic zone of the Russian Federation:
— high cost of construction, which pushes private developers, along with insufficient state financing;
— required high professionalism of employees in terms of the design and construction of objects related to the effects of harsh climatic conditions;
— execution of typical projects;
— limited construction time.

Obviously, the main features of the concept of lean construction meet adequately to requirements, that apply to models of the housing and civil engineering in the Arctic regions of the Russian Federation and they are able to reach the target results. Due to the state support of development of the Arctic at the highest level, the issue of progress of infrastructure of life support for the population of the region, consisting for the most part in the development of residential objects and socially significant objects, is reduced to the rational methodology of construction organization. Certain features of lean construction contribute to a new, large-scale development and accelerated infrastructural development of the region.

## 4 Conclusions

Summing up the results of the study, the following conclusions should be allocated:
— high degree of importance of the Arctic as a strategic object is underlined;

— issues of socio-economic development of the Arctic zone of the Russian Federation in part of life-support systems creating are actualized;
— features of the housing and civil engineering in the Arctic zone of the Russian Federation are systematized;
— lean construction, which is aimed at reducing costs and construction time, along with high quality construction, has all the prerequisites to become a breakthrough development strategy;
— lean construction in the context of the Arctic regions may become the first large-scale example of implementation of this concept in Russia.

In summary, lean construction is a promising way of organizing housing and civil engineering both for the Russian part of the Arctic and its foreign parts.

## References

1. G. Ballard, G. Howell. Implementing lean construction: Improving downstream performance. Lean construction. Balkema, Rotterdam, Netherlands. 111–125 (1997)
2. J. Howell. Lean construction, Public Infastructure Bulletin. 1(9) (2014)
3. V.A. González, R. Sacks, I. Pavez, M. Poshdar, L. Ben Alon, V. Priven. Interplay of Lean Thinking and Social Dynamics in Construction. In: Proc. 23rd Ann. Conf. of the International Group for Lean Construction. Perth, Australia, July 29-31, 2015. 681-690. (2015)
4. O. Salem, J. Solomon, A. Genaidy, I. Minkarah. Lean Construction: From Theory to Implementation. Journal of Management in Engineering. 22(4). 168-175 (2006)
5. G. Howell. What is Lean Construction. Proceedings of the Seventh Conference of the International Group for Lean Construction (1999)
6. A. Mossman. Why Isn't The UK Construction Industry Going Lean With Gusto?. Lean Construction Journal. 5(1). 24-36 (2009)
7. S. Belayutham, V.A. Gonzalez. Process Complexity at the Pre-construction Stage: A Lean Based Solution. Proceedings of the 4th New Zealand Built Environment Research Symposium (NZBERS). Auckland, New Zealand (2014)
8. L. Koskela. An Exploration Towards a Production Theory and Its Application To Construction. Espoo, Finland, VTT Publications, 296 (2000)
9. L. Koskela. Application of the New Production Philosophy to Construction, Technical Report No. 72, CIFE, Stanford University, CA (1992)
10. A. Alsehaimi, L. Koskela. Critical evaluation of previous delay studies in construction. Proceedings of the 8th International Postgraduate Conference, Prague (2008)
11. S. Bertelsen, L. Koskela. Construction beyond lean: a new understanding of construction management. Proceedings of the 12th annual conference in the International Group for Lean Construction, Elsinore, Denmark (2004)
12. R.F. Aziz, S.M. Hafez. Applying lean thinking in construction and performance improvement. Alexandria Engineering Journal. 52. 679–695 (2013)
13. C. Formoso, L. Soibelman, C. De Cesare, E. Isatto, Method of Waste Control in the Building Industry, Proceedings ICLG-7, University of California, Berkeley, CA, USA. 325–334 (1999)
14. T. Ohno. Toyota Production System. Productivity Press (1988)

15. Ye.A. Chernykh. Primeneniye printsipa potoka v berezhlivom stroitel'stve [Application of the principle of flow in lean construction]. Quality management. 2. 102-121. (rus) (2010)
16. E. Deming. Out of the crisis. Cambridge MA: MIT Press (1982)
17. G. Ballard. The last Planner System of Production Control. University of Birmingham, Uk (2000)
18. A. Mossman, Last Planner Overview: Collaborative Production Planning, Collaborative Programme Coordination, Lean Construction Institute, UK (2005)
19. G. Ballard. The Last Planner. Northern California Construction Institute, Monterey, CA (1994)
20. D. Bhargav, H. Juho-Pekka, L. Koskela. Exploring the Recurrent Problems in the Last Planner Implementation on Construction Projects. Proceedings of the Indian Lean Construction Conference (ILCC 2015) (2015)
21. G. Ballard, G.A. Howell. An update on last planner. In Proc., 11th Annual Conf., International Group for Lean Construction, Blacksburg, VA (2003)
22. Smirnov K. Stroitel'stvo v vysokikh shirotakh. Printsipy, vozmozhnosti i perspektivy [Building in high latitudes. The principles, opportunities and prospects]. Proceedings of the I Russian architectural competition with international participation «Green project» (2015)
23. Y.A. Varfolomeev Osobennosti proyektirovaniya i stroitel'stva maloetazhnykh domov v Arktike [Features design and construction of low-rise buildings in the Arctic]. Arctic and Nord. 17. 28-43. (rus) (2014)

# Feasibility of the sizes of water-conveyance system in run-of-river power plants

*Mikhail* Balzannikov[1,*]

[1]Samara State Technical University, Institute of Architecture and Civil Engineering, 194, Molodogvardeyskaya St., 443001, Samara, Russia

**Abstract.** The paper deals with run-of-river power stations. The author notes the importance of the implementation of the technical-economic calculations in large-sized elements of the water-conveyance system in run-of-river power plant: turbine pits and intake pipes. The article shows that the volume of construction and the total cost of power plants construction depend on the size of water-conveyance system elements. The task of the research is the economic feasibility analysis of intake pipes sizes in run-of-river power plants, built in the European part of Russia. The author presents the methodology of economic analysis for justification of the length of the intake pipe, based on the method of the total cumulative effect. The methodology is used for calculations of low-pressure run-of-river power plants. The paper gives the results of feasibility of the optimal sizes of intake pipe. The article analyses the influence of operating conditions of power plants on the choice the optimal length of an intake pipe. The analysis showed that it is economically feasible to use intake pipes with a greater length. The length of the intake pipe will increase the efficiency of power plants at the expense of pressure losses of the output stream. The results are important for a correct choice of the main geometrical dimensions of the designed power plants.

## 1 Introduction

Hydroelectric power plants (HPP) play an important role in electricity supply of various consumers during the stress periods of the daily production schedule of power supply systems as they have very high maneuverable qualities [1, 2]. At the same time, stations of this type demand considerable costs of construction, especially construction of run-of-river power plants on the low-land rivers [3-7], besides, these stations produce significant negative impact on the environment [8, 9].

It is very important to conduct careful research and technical and economic calculations for reasonable feasibility of the geometrical sizes of large-size parts of the building of hydroelectric power plant, including, elements of the water-conveyance system for reduction of unfairly big expenses on construction of hydroelectric power plants [10-12].

* Corresponding author: balzannikov@samgasu. ru

At the same time, among different types of research, pilot studies play important role for such feasibility, as they allow choosing favorable forms and outlines of walls of elements providing the minor losses of a stream pressure [13-16].

The basic large-size elements of the water-conveyance system in run-of-river power plants are the turbine pits and intake pipe [17, 18]. Their turbine pits, as a rule, are spiral, produce defining impact on the overall construction dimensions in the plan. Intake pipes have the greatest value for formation of the general height of a power plant object.

Intake pipes of the run-of-river constructions of power plants with vertical hydraulic aggregates are most often built in the form of elbow devices (Fig. 1). It enables reduction of foundation embedding of an object, lowering work amounts of works and financial costs of its constructing. However, overall dimensions of an intake pipe influence not only the cost of construction works for all power water-conducting elements, but also losses of a pressure of water flow in them. Therefore, they define the general performance coefficient of the hydraulic turbine and overall performance of hydroelectric power plant in general.

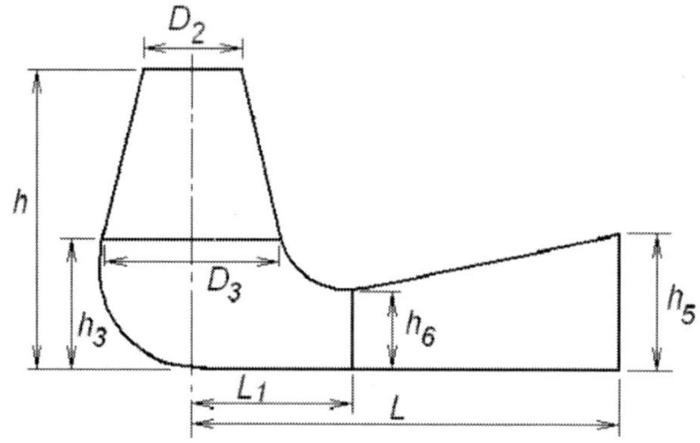


**Fig. 1.** Forms and parameters of an intake pipe in the cross-section of run-of-river power plant construction.

The task of this research was the feasibility of the optimal main sizes of the intake pipes in the run-of-river hydroelectric power plant and identification of influence of conditions in which the hydroelectric power plant operates on the choice of their key geometrical parameters.

## 2 Materials and Methods

The research was conducted, using the method of the total cumulative effect. The following equation was chosen as a criterion [6]:

$$\sum_{t=1}^{T}\left(\sum_{n=1}^{m}\Delta D_n\right)_t(1+i)^{\tau-t} - \sum_{t=1}^{T}\left(\sum_{n=1}^{m}\Delta P_n\right)_t(1+i)^{\tau-t} \geq 0\ , \qquad (1)$$

where: $\Delta D_n$ – additional revenue amount, , $\Delta P_n$ – amount of additional expenses (costs), $T$ – considered time interval, $t$ – current annual period, $\tau$ – the year of expenses, $i$ – discount rate.

The length of intake pipe $L$ greatly influences the general loss of water flow pressure in it. On the one hand, increase in length of the intake $\Delta L$ will cause expenses growth on construction works because of volume increase of the ground excavation and volume of the poured concrete:

$$\Delta P = c_1 W_1 + c_2 W_2 , \quad (2)$$

where $c_1$ and $c_2$ – the cost of unit volume of ground and concrete works, , $W_1$ and $W_2$ – are additional volumes of ground and concrete works.

Length increase of the intake pipe can also lead to growth of other expenses, for example, costs of the valves established at the stream exit or the cost of the lifting equipment.

On the other hand, the area of output section will increase with the length increase of a diffuser part of the intake pipe on $\Delta L$. At the same time the average speed of a water stream in output section will decrease that will cause reduction of losses of a high-speed pressure in output section at $\Delta h$:

$$\Delta h = \left( \frac{\alpha V^2}{2g} - \frac{\alpha V_1^2}{2g} \right), \quad (3)$$

where $V$ and $V_1$ – average speed of water flow in output section according to initial option and option with the extended diffuser area, $\alpha$ – irregularity coefficient of water stream speed in the output section.

Loss enhancement of a pressure will cause the generation of additional power $\Delta N$, additional power $\Delta E$ and the additional income $\Delta D$:

$$\Delta N = 9{,}81 \eta Q \Delta h , \quad (4)$$

$$\Delta E = \Delta N T_N . \quad (5)$$

$$\Delta D = \Delta Э b , \quad (6)$$

where $\eta$ – efficiency factor of hydraulic unit, $Q$ – water flow, $T_N$ – conditional annual quantity of hours of power usage of power plants, $b$ – charge rate per 1 kWh of electric power.

The analysis of constructive solutions of intake pipes in the run-of-river power plants constructed in Russia on the Volga Rivers and Kama – Lower Kama, Cheboksary, Zhigulyovsk, Saratov and Volgograd hydroelectric power plants. The analysis has shown that the size characterizing the length of intake pipe – its relative length $L'=L/d$ where $d$ – diameter of the turbine wheel, is close in the size equal to 3,5. This size has also been accepted as initial parameter during further research. The intake pipe of the Zhigulyovsk hydroelectric power plant has served as a compatible of technical and economic calculations. The length varied from initial size towards its increase during the calculation process. For different values of $L$ the quantities were counted as (2) – (6) and used for criterion (1).

Change of geometrical parameter $L$ towards its increase $\Delta L$ was considered as economically justified decision if the condition (1) was satisfied, i.e. if the sum of the provided additional income connected with increase in the sizes of the intake pipe for the considered period of time was not less than the sum of the year of additional expenses caused by change of this geometrical parameter, or: $R \geq 0$ .

## 3 Results

The use of the mentioned method helps to calculate the optimal main sizes of the intake pipes of run-of-river hydroelectric power plant and identification of influence of service conditions of hydroelectric power station on the choice of the best parameters of the intake pipes in hydraulic units. Calculations are executed for the conditions close to parameters and service conditions of the Zhigulyovsk hydroelectric power station on the Volga River. Ranges of the varied parameters are specified in tab. 1.

**Table 1.** Range of varied parameters of power plants.

| № | Parameter | Value |
|---|---|---|
| 1 | Hydro turbine pressure, m | 24 |
| 2 | Turbine parameter, m | 9,0 |
| 3 | Water discharge trough turbine, $m^3/s$ | 620 |
| 4 | Original relative length of intake pipe | 3,5 |
| 5 | Charge rate for the electric energy in the first year, rub/kWt | 2,4 |
| 6 | The use of installed capacity per year, hour. | from 1000 up to 7000 |
| 7 | Length increase of the intake pipe, m | from 0,5 up to 8,0 |
| 8 | Additional expenses on the construction works, thousands of rubles/$m^3$ | from 10 up to 15 |
| 9 | Period of calculations, years | From 5 up to 20 |
| 10 | Discount rate, % | from 2 up to 10 |
| 11 | Dynamics of charge rate of electric energy, % per hour | from 0 up to 20 |

Results of calculations were presented for descriptive reasons in the graphic form. In Fig. 2 the generalized results of these technical and economic calculations at $T_N$ = 3000 hours are presented for the ten year period $T$.

All constructed curve dependences $E = f(\Delta L, b)$ have a maximum. They demonstrate the presence of optimal size of the varied parameter – length of the intake pipe.

The initial size of the intake pipe with a relative length of 3,5 corresponded to the absolute value of 31,5 m. According to the obtained data the optimal size of increase in relative length of the intake pipe was 0,25-0,33 (absolute – 2,5-3,0 m). Data demonstrates that the maximum economic effect for the considered conditions can be reached in case the application of the intake pipe having the total relative length equal 3,75-3,83 (absolute – 34,0-34,5 m). At the same time all range of sizes $\Delta L$ for which R is a positive value, are related to economically justified decisions.

Results of calculations for identification of influence of external service conditions of run-of-river power station on the greatest possible increase in length of the intake pipe at which this increase will be economically justified i.e. at which the condition (1) will be satisfied are given in Fig. 3.

The calculations show that the greatest lengthening of the intake pipe: the relative size $\Delta L'$ = 0,90-1,01 (absolute – 8,1-9,1 m)if the number of hours of use of rated power $T_N$ = 3000 and the discount rate equal to 2% , and the relative size $\Delta L'$ = 0,72-0,80 (absolute – 6,5-7,2 m) is gained in case the discount rate is 10%. At the same time the total length of the intake pipe will be the following: relative size – 4,40-4,51 and 4,22-4,30; an absolute value – 39,6-40,6 m and 38,0-38,7 m.

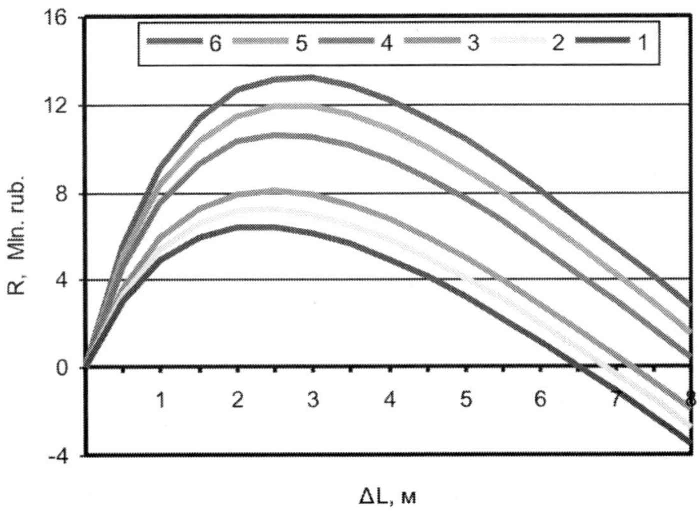


**Fig. 2.** Dependence diagram $R = f(\Delta L, b)$ where $b$ = 2,4 rub/kWh and $T$ = 10 years (bottom upwards): 1 – where $i$ = 10% and non-change rate for electric power, 2 – where $i$ = 10% and annual rate increase by 10 %, 3 – where $i$ = 10% and annual rate increase by 20 %, 4 – where $i$ = 2% and non-change rate for electric power, 5 – where $i$ = 2% and annual rate increase by 10 %, 6 – where $i$ = 2% and annual rate increase by 20 %.

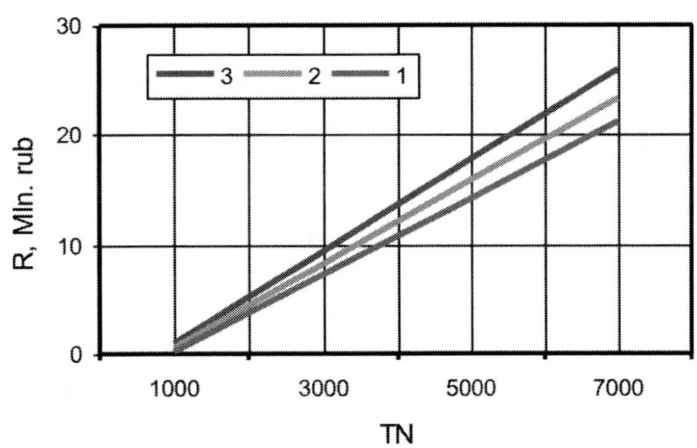


**Fig. 3.** Dependence diagram $\Delta L_{MAX} = f(i, b)$ where $T$ = 10 years (bottom upwards): 1 – non-change rat e for electric power, 2 – annual rate increase by 10 %, 3 – annual rate increase by 20 %.

The authors conducted research about the influence of an important indicator of regime service conditions of hydroelectric power station – number of hours installed capacity use $T_N$ on integrated economic effect. Fig. 4 shows dependence diagram of integrated economic effect on this parameter.

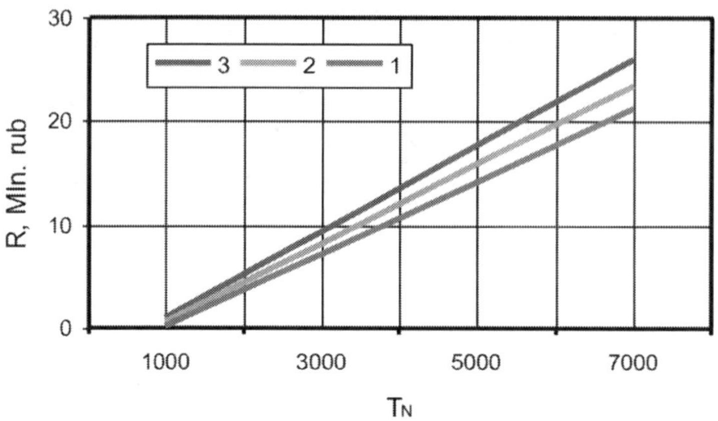


**Fig. 4.** Dependence diagram $R = f(T_N, i)$ for $\Delta L$=2,5 where $T$ = 10 years and non-change rate for electric power (bottom upwards): 1 – when the discount rate is equal to 4%, 2 – 6%, 3 – 8%.

The diagrams show that economic effect will considerably improve if number of hours of integrated capacity use increase for run-of-river power plants in case the optimal sizes of the intake pipe are identified. In the process of calculations the measure of an integrated economic effect where $T_N$ = 7000 hours was defined as the sum of 21-26 million of rubbles.

## 4 Discussion

The received results have allowed defining optimal geometric parameters of the most important element of the water-conveyance system of run-of-river hydroelectric power plant – the intake pipe under modern economic conditions. The general view of the set of the constructed curves of dependence shows that initial increase of the intake pipe length causes firstly economic effect increase for the considered ranges of conditions change of hydro units operation. However, in case of further increase in dimensions of pipe costs of construction works begin to increase more intensively and economic effect decreases. Existence of an optimum of the varied parameter – length of the intake pipe proves the necessity of amending recommendations for the determination of the intake pipe length that tend to increase. The reason for dimensional underrating of water-conveyance system of run-of-river power plant is the determination to reduce the initial costs for power plant construction.

The results given in Fig. 4 mean that special attention should be paid to justification of dimensions of the intake pipes for the run-of-river hydroelectric power plants working in a semi-peak and peak zone of the diurnal behavior of energy load. A considerable economic effect of the intake pipe length increase can be gained under these conditions

It is also necessary to note that the calculated data is provided for quite short base period – 10 years. It takes significantly long time to operate power plants. It is obvious that the amount of economic effect significantly depends on the duration of the considered base period. So, increase during the base period from 10 to 20 years will lead to increase in economic integrated effect by 45-60%.

The obtained results will be of great importance for the correct choice of the basic geometric parameters of the elements of the water-conveyance system and can be used for the determination of the sizes of the intake pipes at the designed power plants.

## 5 Conclusions

The main conclusions of the research can be formulated as follows:

1. Feasibility of the main sizes of elements of water-conveyance system, including the elbow intake pipe in hydroelectric run-of-river power plant, and the choice of their optimal parameters is of great importance for ensuring high overall performance of hydroelectric power station.

2. Length of the intake pipe regarded as recommended for the hydroelectric power stations which have been earlier constructed in Russian run-of-river power plant, can be considered as minimum admissible under the terms of normal operation of the hydraulic turbine now. Nowadays it is necessary to carry out technical and economic calculations for detection of viability of increase of this parameter and feasibility of its optimal value in each case in the design process of hydroelectric power station.

## References

1. V.V. Elistratov, *Renewable energy* (Nauka, Saint Petersburg, 2013
2. V.V. Elistratov, Research and technical bulletin of Saint-Petersburg State Politechnical University 3-1, 77-83 (2012)
3. Yu.S. Vasil'ev, L.I. Kubyshkin, Hydraulic engineering 7, 2-8 (2014)
4. F. Svitala, Yu.M. Galitskova, Scientific survey 10, 450-456 (2014)
5. F. Svitala, Yu.M. Galitskova, S.V. Evdokimov, Industrial and civil engineering 12, 87-90 (2014)
6. M.I. Balzannikov, Urban Construction and Architecture 4, 86-92 (2015)
7. M. Balzannikov, V. Alpatov, I. Kholopov, A. Saharov, A. Lukin, MATEC Web of Conferences 73, 01012 (2016)
8. S.V. Evdokimov, Urban Construction and Architecture 2, 68-74 (2012)
9. Yu.S. Vasilyev, I. G. Kudryashova, V.I. Maslikov, Works of the St. Petersburg polytechnical university of Peter the Great 502, 317-322 (2007)
10. V.M Ivanov., I.A. Bakhtina, T.Yu. Ivanova, S.V. Iliynykh, Urban Construction and Aarchitecture 2, 88-93 (2015)
11. M.I. Balzannikov, S.V. Evdokimov, N.V. Shekhova, Economy and managements of property 1, 68-72 (2015)
12. B.U. Urishev, M.M. Mukhammadiyev, F. Nosirov, S.R. Zhurayev, Urban Construction and Architecture 4, 49-53 (2013)
13. I.A. Bakhtina, V.M. Ivanov, S.V. Ilyinikh, P.V. Stepanova, Ye.S. Elizarov, Polzunovsky Bulletin 4-2, 12-19 (2013)
14. S.A. Piyavskiy, S.V. Evdokimov, Bulletin of higher educational institutions. Construction 6, 36-43 (2012)
15. M.I. Bal'zannikov, V.A. Seliverstov, Power Technology and Eng. 49-1, 22-26 (2015)
16. V.P. Popov, A.A. Mikhasek, V.A. Seliverstov, Bulleting of Kyrgyz state university of construction, transportation and architecture 1, 73-79 (2014)
17. A.A. Romanov, Zhigulyovsk hydroelectric power station. Operation of hydraulic engineering constructions (Agni, Samara, 2010)
18. M.I. Balzannikov, S.A. Piyavskiy, Bulletin of Moscow state construction university 10, 127-137 (2015)

# Research on flow in water intake of a run-of-river hydropower plant

*Mikhail* Balzannikov[1,*]

[1]Samara State Technical University, Institute of Architecture and Civil Engineering, 194, Molodogvardeyskaya St., 443001, Samara, Russia

**Abstract.** The work shows the importance of flow research in parts of canal systems at hydropower plants. The authors researched a mathematical model of water intake at run-of-river hydropower plant. ANSYS software was used. The 3D model created by the authors described the upper part of the water intake extended to the approach area in front of it. The work was aimed at identifying flow parameters in the flow-narrowing zone of the water intake. The work describes the research method and presents graphical calculation results. Whirling areas were discovered in the upper part of water intake and in gate grooves. It was noted that the flow in gate grooves is spiral-shaped and runs downwards. The work highlights the danger of such flow because it increases the amount of small rubbish gravitating onto the lower part of the groove, which can cause gate hang up (blocking) when it is shut in emergency. The results are of high importance for increasing safety of intakes at run-of-river hydropower plants.

## 1 Introduction

Hydropower plants are objects of complex construction that is why designing them involves solving safety issues. A lot of authors published their research on increasing safety of hydropower plants, for example, [1-4]. In these works, attention is drawn to researching strength properties of certain constructive elements under the influence of static and dynamic loads. As factors of such loads, they often consider weight of the construction itself, weight of equipment and its operation, automobile and railway transport [5-8].

Also, hydropower plants have a water supply canal which directs water flow to hydraulic power units and then to tailrace. Elements of the water supply canal also have quite complex construction [9-11]. They are also influenced by water flow hydraulic pressure, which should be considered while designing and justifying their geometrical properties [12-14]. Attention to hydrostatic and hydrodynamic influence on water supply canal units increased dramatically after the catastrophic accident at the Sayano-Shushenskaya Dam [15]. Researching physical and mathematical models of water supply canal units in order to study water flow parameters and characteristics in these units is crucial for justifying parameters of the water supply canal [16-18].

* Corresponding author: balzannikov@samgasu. ru

Water intakes at run-of-river hydropower plants have significant size and carry large amounts of water to the hydraulic power unit. Water is always supplied with significant flow narrowing (reduction in cross-section area). Most often, such narrowing in vertical plane is provided only in the upper part (Fig. 1). The constructive solution of such narrowing decreases the volume of earthwork while arranging the base plate. However, the flow narrowing at combined hydropower plants happens mostly in the lower part. (Fig. 2). The reason for such construction solution is that the entrance of bottom water collector is located lower than the intake of water supplied to the turbine chamber.

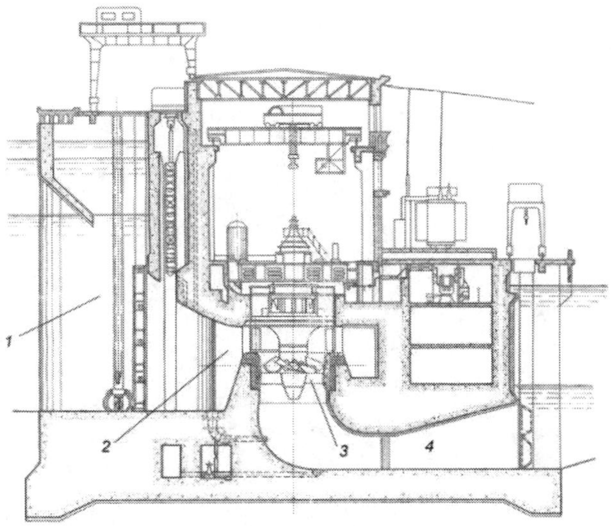


**Fig. 1.** Water supply canal at run-of-river hydropower plant [19]: 1 – water intake, 2 – turbine chamber, 3 – runner chamber, 4 – draft tube.

Additional constructive elements like gate grooves, curtain and divider walls are put exactly in the area of flow narrowing in the water intake. These elements significantly influence the flow character and the loss of pressure. If the sizes of constructive elements have been chosen incorrectly, this can lead to unfavourable hydraulic conditions of the flow [20].

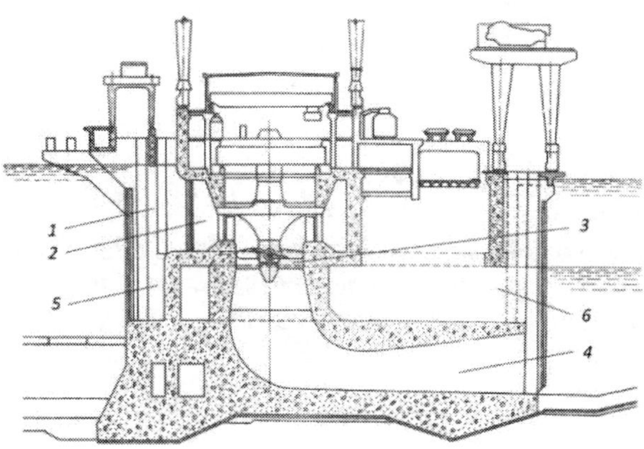


**Fig. 2.** Water supply canal at run-of-river hydropower plant [19]: 1 – water intake, 2 – turbine chamber, 3 – runner chamber, 4 – draft tube, 5 – tailwater entrance, 6 – tailwater exit.

The work was aimed at identifying flow parameters in the flow-narrowing zone of the water intake at the run-of-river hydropower plant and studying the influence of constructive elements on flow character and properties.

## 2 Materials and Methods

Research was done using ANSYS software with CFX solver [21]. In order to reach the research objective, a 3D model of water intake entrance was used, extended to the approach area in front of the water intake (Fig. 3). The modelling area was divided into multiple tetrahedron-shaped elements whose size was set by the programme depending on the surface curvature and proximity to other flow-limiting surfaces. In addition, a zone of boundary layer was separated in the net, which is important for getting a broad picture of water flow in gate grooves. The modelling net included about 390 000 elements.

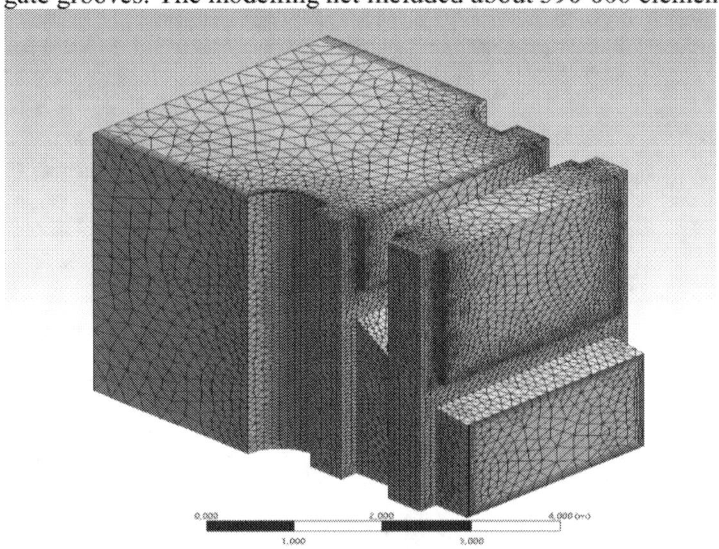

**Fig. 3.** 3D model of water intake with final elements net.

Gravity was taken into consideration and k-ε turbulence model was chosen while performing the task in the modelling flow. The turbulence model influences the correctness, but in the given situation, these effects are of minor importance as most of open flows work in quadratic zone of resistance. Density and dynamic viscosity were set for the flow. After initializing the definition area, the following boundary conditions were set:

- Inlet water: mass flow 10000 kg/s;
- Outlet water: mass flow 10000 kg/s;
- Opening: pressure ratio 0;
- Other walls, rate of flux 0 (according to Boundary layer theory).

When setting the boundary conditions, the following admission was introduced: open water surface is not deformed in the contact zone of water entrance and intake at hydropower plant. In order to reach the stated objective, the maximum iteration number of 500 was set, but the solver could stop calculation pattern at error value $10^{-4}$. To reach this error value, 200 iterations were enough, which proves good solution convergence. The convergent graphs are shown in Fig. 4.

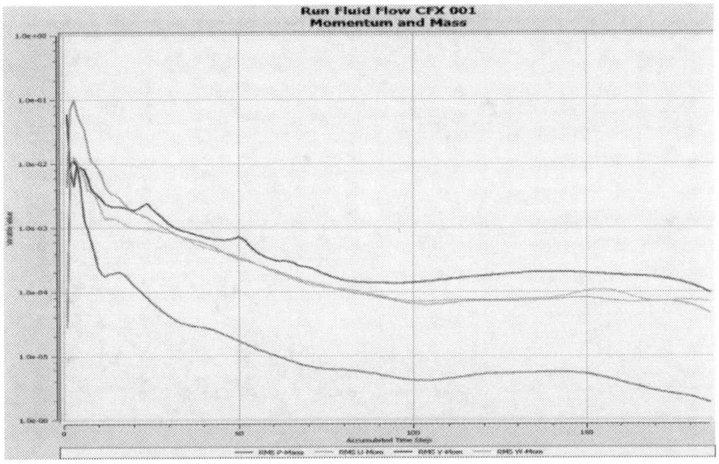


**Fig. 4.** Solution convergence graphs.

## 3 Results

The above mentioned methodology was used to calculate flow parameters in water intake with narrowing only in the upper part of the water supply canal at hydropower plant. Results of such calculation were visualized for presentation and further analysis.

Flow velocity parameters (direction and value) in specific vertical section are presented in Fig. 5. The received data shows that flow velocity increases when water supply canal cross-section area decreases in such a way that the maximum values are indicated in the upper part. In addition, a constant whirling zone and upward water flow appear between divider and intake walls.

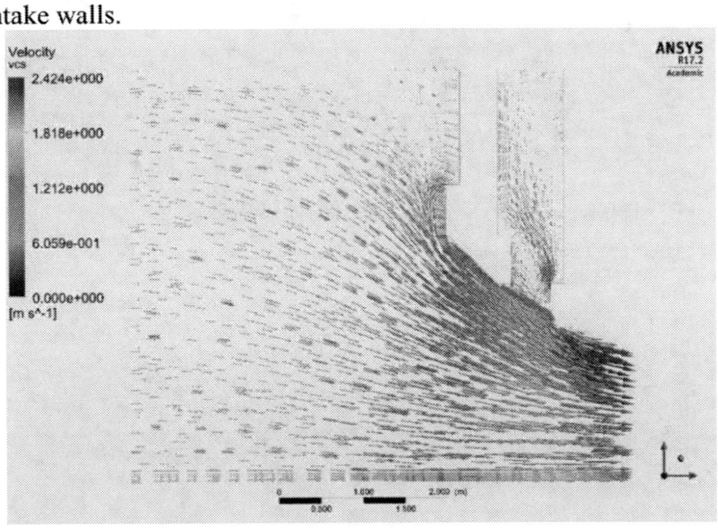


**Fig. 5.** Flow velocity in vertical plane.

Fig. 6 gives data of flow velocity in horizontal plane of water intake. The area where the water supply canal narrows and the flow velocity increases also has certain construction elements which influence the water flow. They are gate grooves.

Results of flow research in these grooves are presented in detail in Fig. 7 and 8. Fig. 7 shows distinctly that constant whirling areas are formed in the grooves. In addition, downward flow is detected in vertical section going through gate grooves (Fig. 8). This proves that the flow in gate grooves is spiral-shaped, curled and directed downwards.

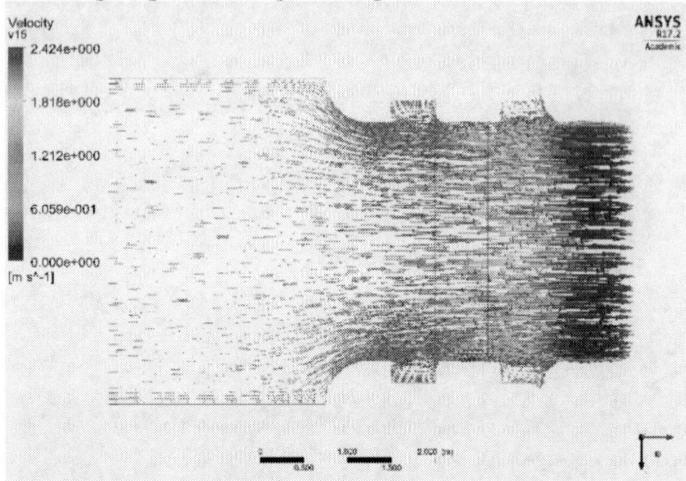


**Fig. 6.** Flow velocity parameters in horizontal plane.

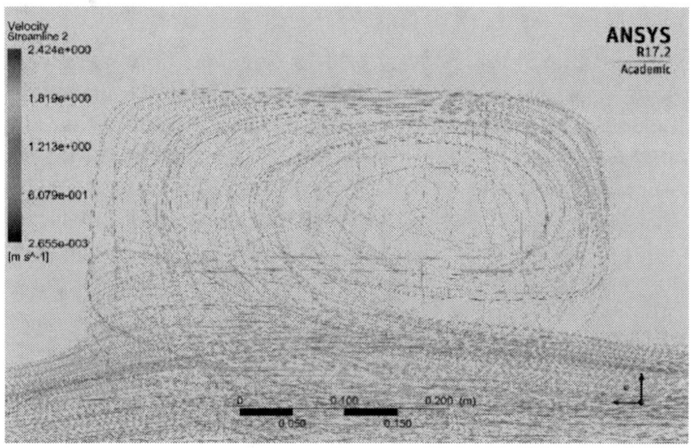


**Fig. 7.** Flow velocity parameters in water intake grove in horizontal plane.

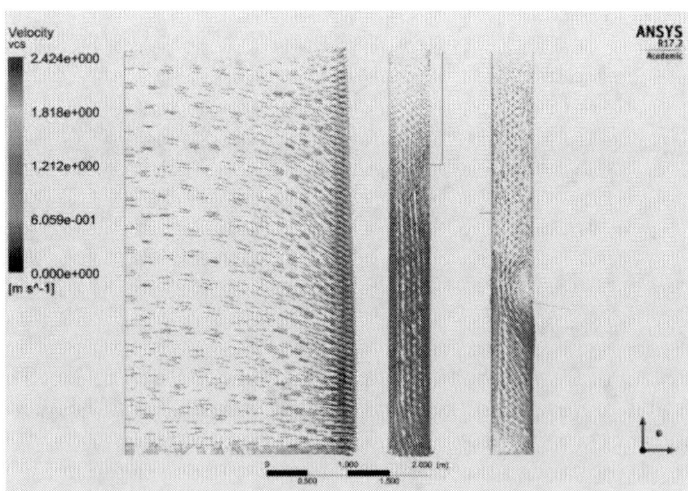


**Fig. 8.** Flow velocity parameters in water intake grove in vertical plane.

## 4 Discussion

Research of the flow in water intake discovered constant whirling areas in the zone of flow narrowing in cross section area at run-of-river water plant. In addition, upward and downward flows appear in separate areas. Whirling zones cause pressure loss and weaken the hydropower plant effectiveness. Besides, water spirals breakdowns can happen in whirling areas, which will inevitably increase the dynamic load on the intake and decrease safety of the whole construction.

In our opinion, the biggest danger is downward spiral-shaped flow in the intake gate grooves. The flow carries small rubbish, which gravitates onto the lower parts of the groove. Therefore, large amounts of such rubbish can cause hang up (blocking) of gate when it is shut in emergency. This is inadmissible under high safety conditions of a hydropower plant.

The achieved results are of high importance for correctly choosing the form and the geometrical parameters of the intake constructive elements. Besides, the results highlight the necessity of regular inspections of the lower part of the grooves in order to find and remove any gravitated rubbish. Such actions will provide high safety at hydropower plant.

## Conclusions

The main conclusions of the research can be formulated as follows:

1. Research on flow parameters in complex constructive elements of water supply canal are of high importance for correctly choosing the form and the geometrical parameters of these elements. Justified forms and sizes of water intake elements allow to decrease the dynamic load on the water intake and, as a result, to increase the water supply canal safety.
2. Research showed that the flow in gate grooves of the water intake at run-of-river hydropower plant that is characteristic of narrowing the flow section only at the upper part is spiral-shaped and runs downwards. Such flow direction is dangerous as it can carry small rubbish, which gravitates onto the lower parts of the groove. Large amounts of such rubbish can cause hang up (blocking) of gate when it is shut in emergency.
3. In order to maintain high safety of the hydropower plant and its water intake it is recommended to perform regular inspections of the lower part of the grooves in order to find and remove any gravitated rubbish.

## References

1. S.V. Evdokimov, Urban Construction and Architcture **2(5)**, 68-74 (2012)
2. Yu.S. Vasiljev, L.I. Kubyshkin, Hydraulic Engineering **7**, 2-8 (2014)
3. M. Balzannikov, V. Alpatov, I. Kholopov, A. Saharov, A. Lukin, MATEC Web of Conferences **73,** 01012 (2016)
4. A.A. Romanov, *Hydropower constructions maintenance* (Agni Publishing House, Samara, 2010)
5. A.A. Romanov, S.V. Evdokimov, V.A. Seliverstov, Industrial and Civil Engineering **6,** 44-48 (2015)
6. F. Svitala, S.V. Evdokimov, Yu.M. Galitskova, Industrial and Civil Engineering **12, 87-90** (2014)
7. F. Svitala, Yu.M. Galitskova, Scientific Srvey **10,** 450-456 (2014)
8. M.I. Balzannikov, Procedia Engineering **111,** 65-71 (2015)

9. S.A. Pijavskij, S.V. Evdokimov, Bulletin of higher education establishments. Construction **6**, 36-43 (2012)
10. V.P. Popov, A.A. Mihasek, V.A. Seliverstov, Vestnik KGUSTA **1**, 73-79 (2014)
11. L.G. Kozinets, Alternative Energetics and Ecology **13-14**, 26-30 (2015)
12. L.G. Kozinets, L.P. Potehin, Engineering Construction Journal **6**, 29-35 (2012)
13. Y.E. Senitsky, M.I. Balzannikov, A.A. Mikhasek, Urban Construction and Architecture **V.7, 1**, 64-69 (2017)
14. M.I. Bal'zannikov, V.A. Seliverstov, Power Technology and Engineering **49**, 22-26 (2015)
15. V.I. Fortov, M.P. Fiodorov, V.V. Yelistratov, Vestnik of Russian Academy of Sciences **7**, 579-586 (2011)
16. I.A. Bahtina, B.M. Ivanov, S.V. Iljinih, P.V. Stepanova, E.S. Yelizarov, Polzunovsky Vestnik **4-2**, 12-19 (2013)
17. B.U. Urishev, M.M. Muhammadiev, F. Nosirov, S.R. Zhuraev. Urban Construction and Architecture **4**, 49-53 (2013)
18. M. Balzannikov, S. Piyavsky, A. Kurmanayev, MATEC Web of Conferences **86**, 02006 (2016)
19. Yu.S. Vasiljev, V.V. Elistratov, *Hydroelectric power plants* (St. Petersburg Polytechnic University, Saint Petersburg, 2011)
20. Yu.S. Vasiljev, V.V. Elistratov, L.N. Kubishkin, Hydraulic Engineering **5**, 11-18 (2007)
21. V.A. Seliverstov, Scientific technical bulletin of St. Petersurgh Polytechnic University **4-2**, 149-153 (2009)

# The conservation of historical architectural heritage in Russia

*Ekaterina* Balzannikova[1,*], *Vitaliy* Samogorov[1]

[1] Samara State Technical University, Institute of Architecture and Civil Engineering, 194, Molodogvardeyskaya St, 194, 443001, Samara, Russia

**Abstract.** Much of the historically valuable architectural fabric of the city of Samara is currently in an unsatisfactory condition. Therefore it is advisable to develop and implement effective measures for the conservation and renewal of architectural heritage. This article describes a comprehensive method for the preservation of urban architectural properties of historical and cultural value. The proposed systematic approach enables the scope of renewal works to be justified. Furthermore, its implementation should result in a raised profile and increased tourist visitor numbers for heritage properties and their adjacent areas. The proposed method facilitates both the creation and the regeneration of spaces for recreation and consequently an increase in foot-fall within the adjacent social infrastructure, as part of a general broadening of the cultural-aesthetic life of the city.

## 1 Introduction

In the central regions of Russia, large cities founded on the banks of major rivers have a rich history. Such factors as natural resources and transport routes, facilitating industrial production and trade, have played an important role in the development and improvement of cities. In particular, the construction of industrial facilities which were large in their day, became vital for the formation of the historical character of Russian cities [1-5].

The authors' research on historically valuable architectural constructions in Samara has shown that many buildings are collapsing and may well be lost forever [6-11]. As an example, Figure 1 shows the current condition of one of the Samara Maslennikov factory workshops. The building is unoccupied, and the factory isn't functioning. At the same time, there is a wealth of international experience in the renewal and conservation of historically valuable buildings and constructions; for example, in the Polish city of Krakow [12].

## 2 Materials and methods

On the basis of the above-mentioned facts, it may be concluded that the urgent focus of town planning policy should be the development and realization of effective measures for the conservation and renewal of historical architectural heritage, with consideration of the specifics of style and form of each building and its details. Such measures should be

* Corresponding author: balzannikova@mail.ru

systematic and based on rigorous research; in particular, the current condition of the building and its existing immediate urban environment are to be taken into consideration.

**Fig.1.** Current condition of one of the factory buildings in Samara.

## 3 Results

The authors suggest that the development of a systematic method for the conservation of historical architectural heritage should consist in the following stages [12]:

1. the research stage, including the following procedures: investigation of the town planning features of the immediate urban environment, assessment of the historical and architectural value of the property, assessment of the dilapidation of the property and its suitability for further exploitation, assessment of the relationship between the property under investigation and the architectural character of the surrounding properties, and the development of a conservation programme;
2. the preparatory stage, involving investigations of the extent of wear and dilapidation of the property, the development of technologies for structural reinforcement, the delivery of building materials and the preparation of the construction site for renovation works;
3. the construction stage, including the removal of defunct structural elements, assembly and installation of new constructional elements, interior finishing works and renovation of the façades;
4. the completion stage, characterized by: historical and architectural evaluation of the restored property in relation to the aesthetic and architectural character of the adjacent territory.

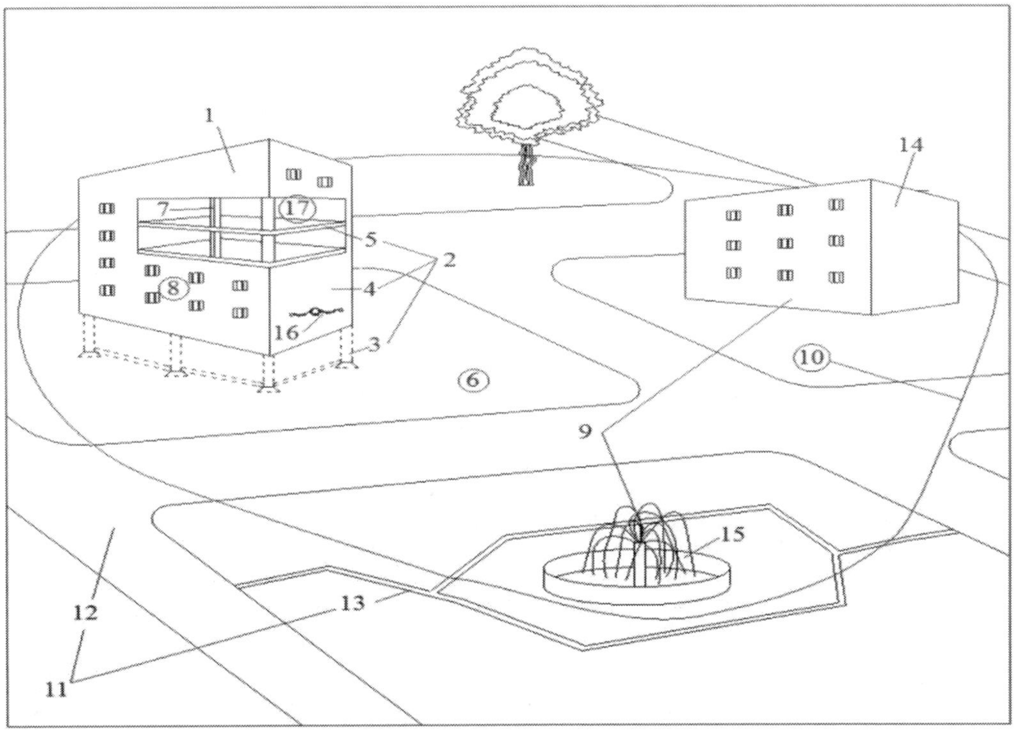


**Fig.2.** Diagram showing the historic monument in its urban setting.

## 4 Discussion

The authors' systematic approach is explained by the diagram in Figure 2. It illustrates a building of historical architectural heritage within its adjacent territory. The numeric annotations indicate the following: 1 – the building of historical architectural heritage, 2 – structural elements of the building: 3 – foundations, 4 – structural walls, 5 - structural floors. Further: 6 – building site, 7 – new structural elements, 8 – façade, 9 - surrounding buildings, 10 – adjacent territory, 11 – transport and pedestrian infrastructure: 12 – roads, 13 – pavements; 14 – social infrastructure facilities, 15 – historical and cultural landmark, 16 – historical architectural-constructional decorative feature, 17 – interior space.

While investigating the town planning features of the immediate urban environment (stage 1) the extent and condition of the transport and pedestrian infrastructure (11) of the district, the extent of development of social infrastructural facilities (14), and the number and function of historical and cultural landmarks (15) in the area adjacent (10) to the building of historical and cultural heritage should be identified. When assessing the historical and architectural value of the property the historical period to which it belongs should be confirmed, its architectural style identified and the condition of the historic architectural-structural decorative features (16) should be assessed.

Assessment of the dilapidation of the property and its suitability for further exploitation should include examination of its load-bearing elements: foundations (3), structural walls (4), structural floors (5) etc. When analysing the relationship between the property under investigation and the architectural character of the surrounding properties (9) it is necessary to consider the following: the scale (dimensions) of the building in relation to neighbouring structures, the stylistic particularities of the building in the context of the architectural fabric of the district in question, the visual attractiveness of the site as a whole and also the dialogue with other nearby historical and cultural landmarks (15).

A programme of conservation should define both the methodologies of building preservation adopted and detailed constructional and organisational proposals. The selection of preservation methodologies depends on the condition of the building. If the building is in satisfactory condition, works may include the restoration of the façade (8), and if necessary, reinstatement of the original interior layout. Where structural problems are present, works may involve reinforcement of load-bearing structural elements (2) in addition to façade restoration (8). If the building is in a poor condition, works will include the replacement of failed structural elements (2), façade restoration '(8) and re-planning of the interior layout (17) to suit the proposed new building function.

In Stage 4, the historical and architectural evaluation of the restored property in relation to the aesthetic and architectural character of the adjacent territory, the building in question must be understood within the context (10) as a whole. This final stage should include identification of required improvements of the adjacent territory (10) including, for example, construction of fountain ensembles and minor architectural interventions, separation of parking and recreational zones, and so on.

The proposed systematic approach allows the scope of work for the conservation and renovation of historical architectural heritage to be justified. Additionally, the realisation of the described approach will make historical and cultural heritage properties and their adjacent areas more attractive, resulting in increased visitor numbers. The proposed method facilitates both the creation and the regeneration of spaces for recreation and consequently an increase in foot-fall within the adjacent social infrastructure, resulting from the broadened possibilities for cultural-aesthetic recreation, particularly for children and elderly citizens.

## Conclusions

The main conclusions can be formulated as follows:

1. The construction of large industrial facilities and associated installations has played a formative role in the historical development of the character of Central Russian cities founded on the banks of major rivers. In the present day, much of the historically valuable architectural fabric is in decay. Considering this fact, a focal point of town planning policy should be the development and realisation of effective measures for the preservation and renovation of historical and architectural heritage.

2. The article details the authors' systematic approach for improving the methodology of restoration of historical architectural heritage. This approach includes implementation of research, preparatory, construction and completion stages. The characteristic feature of the authors' approach is the historical and architectural evaluation of the restored building in relation to the aesthetic and architectural qualities of the adjacent context.

3. The developed systematic approach allows the scope of work for the conservation and renovation of historical architectural heritage to be justified, making the property and its adjacent context more attractive and subject to increased visitor numbers.

## References

1. .A. Sysoeva, *Natural-science higher education: problems and outcomes*, 42-44 (2013)
2. T.V. Vavilonskaya, Vestnik MGSU 2, 14-17 (2009)

3. V.A. Samogorov, O.S. Rybacheva, Vestnik of Volgograd State University of Architecture and Civil Engineering, Series: Construction and Architecture 31-1 (50), 300-304 (2013)
4. E.A. Sysoeva, Collection of abstracts of the 1st International Scientific and Practical Conference «Cultural heritage in XXIst century: «conservation, use and promotion», 65-68 (2012)
5. T.V. Vavilonskaya, News of Orel State Technical University, Series: Construction and Transport 3-23, 68-72 (2009)
6. T.V. Vavilonskaya, V.V. Alekseev, Traditions and Innovations in Construction and Architecture, 537-538 (2014)
7. T.V. Vavilonskaya, Architecture and Construction in Russia 12, 2-9 (2014)
8. E.A. Grintsevich, Economics and Enterprise 5 (22), 122-124 (2011)
9. E.M. Balzannikova, Traditions and Innovations in Construction and Architecture, 401-404 (2012)
10. T.V. Baranova, N.A. Kosenkova, Synthetic image of orthodox architecture in the Middle Volga, Sacred architecture in shaping the identity of place, 149-157 (Politechnika Lubelska, Lublin, 2006)
11. M.I. Balzannikov, D.S. Privorotskiy, Architektura sakralna w ksztaltovaniu tozsamosci kulturowej miejsca, 345-357 (Wydawnictwo Politechniki Lubelskiej, Lublin, 2006).
12. E.M. Balzannikova, Vestnik MGSU 1, 15-24 (2014)

# Concrete composites of double structure formation

*Yuri* Bazhenov[1], *Lev* Alimov[1,*] and *Viktor* Voronin[1]

[1] Moscow state university of civil engineering, Yaroslavskoye shosse, 26, Moscow, Russia, 129337

**Abstract.** Modification of concrete composites includes the manufacture of a concrete product, followed by its impregnation with monomers such as styrene and methyl methacrylate, capable of curing in concrete pores. However, despite the high efficiency of concrete modification with polymers, their production is hampered by the complexity of the technology associated with the toxicity and explosiveness of these impregnating materials. The article proposes to use technical vegetable oil as an impregnating material. The study showed that the treatment of concretes with technical vegetable oil (TVO) reduces their water absorption to less than 1%, strength by 40%, and frost resistance by an order of magnitude.

## 1 Introduction

Modified concrete composites are materials pores of which are filled with polymers such as styrene, methyl methacrylate with their subsequent polymerization directly into concrete pores [1, 2, 3]. Because of sealing the material structure with the polymer, and the effect of a number of physicochemical factors, its strength increases by several times, and frost resistance and resistance significantly increases in aggressive media [4-7].

However, despite the high efficiency of concrete composite modification by polymers, their production is hindered by the complexity of the technology associated with toxicity, explosion and fire hazard of impregnating materials [8, 9].

In this connection, studies were carried out on the use of technical vegetable oils as technical impregnation compositions, belonging to drying oils, which can significantly simplify the technology [10]. The studies conducted showed that the simplest, most effective, and cheapest impregnating composition is technical vegetable oil (TVO) without any additives, heated to 130-140 °C and capable of polymerization in the body of concrete [11, 12].

The method of concrete modification with technical vegetable oil was developed on the concrete of grade 400, with the following composition: at the consumption rate of cement, water, filler 409, 193, 1,873 kg/m$^3$, respectively, having an average density of 2,169 kg/m$^3$ with hardness
of 55s.

* Corresponding author: AlimovLA@mgsu.ru

## 2 Materials and methods

Portland cement CEM I 42.5 B GOST 31108-2016, granite crushed stone - with fraction contents of 10-20-10%, 5-10-30%, 2.5-5-45%, 1.25-2 , 5-9%, 0.63-1.25 - 6%.

Samples were prepared, which after 28 days of normal hardening, were impregnated with TVO. Vacuum treatment was used to accelerate the impregnation process. The oil was heated to 130-140 °C, concrete samples to 200 °C.

When the heated concrete sample is immersed into oil, intensive impregnation occurs. Then concrete gradually cools, the air in the sample is compressed, vacuum is created and oil is drawn into concrete pores. The amount of absorbed monomer is increased by 20-25% compared to studies without self-vacuuming.

The polymerization of saturated samples, carried out with infrared radiation at a temperature of 150-160 °C for 2.5.8 and 24 hours, showed that about 70% of oil passes into a three-dimensional polymer. At the same time, the strength of the impregnated concrete increased in comparison with the control one, is equal to 42.3 MPa, after 2 hours - 19.2; 5 hours - 24; 8 hours - 42.3 24 hours - 59.2 MPa.

Thus, the strength of the modified sample after this treatment was 40% higher than the control sample after 24 hours of the polymerization.

## 3 Results

Based on the studies carried out, the following method for modifying concrete with heat treatment was adopted: the manufactured concrete products are dried at temperature of 180-200 °C, oil is heated to 130-140 °C. Then the product is immersed in oil. The impregnation is carried out for as long as it is necessary to penetrate oil to a predetermined depth. The impregnated articles are then polymerized at the temperature of about 200 °C for 6-8 hours.

Thus, concretes modified by technical vegetable oils are materials with the double structure formation. They consist of a concrete matrix and are distributed over the volume of the polymeric material network. Both components have continuity and a common interaction surface. Oil, getting polymerized in pores and capillaries of cement stone, forms a three-dimensional network, which reinforces the silicate base. Not all pores can be filled with polymer. Electron-microscopic studies of modified concrete showed that in pores larger than 200-600 μm there is a central zone not filled with oil. Processing of the experimental data showed that the maximum amount of absorbed oil (O), depending on the water absorption of concrete (W) was:

$$O = 0.8\ W - 21.28,\ \%\ \text{in mass}$$

The dependence of maximum oil absorption upon the concrete porosity (*P*) was:

$$O = 0.43\ P,\ \%\ \text{in volume}$$

The coefficient of saturation of pores equals 0.4 – 0.45, and that of cement stone is 0.5 – 0.55.

To accelerate the assessment of concrete frost resistance, a dilatometric method of structure analysis was used [13, 14, 15]. The measurement procedure for temperature-humidity deformations is as follows: prior the test, samples of tested concrete, 7x7x21 cm in size, are saturated with water, insulated with frost-resistant rubber, placed in a dial gauge, and then into the freezer; temperature control is carried out in the chamber and in the sample middle; strain readings are recorded every 5° C at the time when the temperature in the chamber and the sample middle is equalized. On the base of obtained data, a graph is plotted (Fig. 1).

The data in Fig. 1 show that the dilatometric curve of a water-saturated sample of a concrete based upon freezing is characterized by presence of anomalous expansion deformations that occur at temperatures of -5 ... -10 ° C and -40 ... -50° C. However, there are no such deformations in modified concretes. This indicates a high frost resistance of modified concrete.

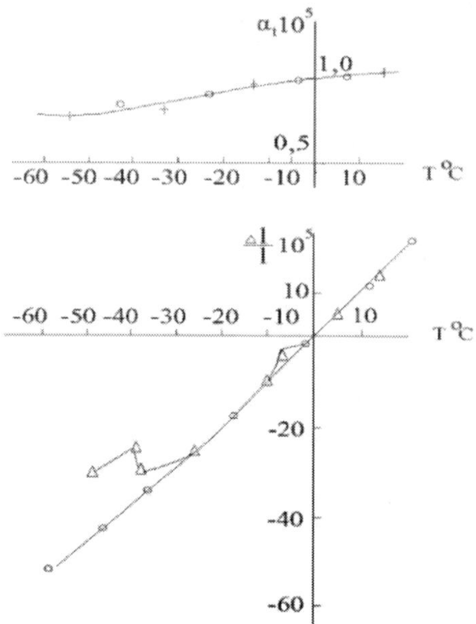


**Fig. 1.** Dilatometric curve and coefficient of thermal expansion of concrete in temperature range from +20 °*C* up to -60 °*C*: Δ-----Δ - thermal deformation of concrete foundation; o-----o - temperature deformation of modified concrete.

## 4 Conclusions

To predict properties of fair-faced concrete modified with heat treatment, the mathematical design of the experiment was carried out and polynomial mathematical models of the basic properties were calculated: strength, porosity and the amount of absorbed oil.
For the mathematical design of the experiment, the *D*-optimal plan for the dimension K-3 of the range B was used. At that, the main independent parameters were adopted:
- Cement consumption (C) - X1;
- binder activity (R) - X2;
- water flow rate (W) - X3

**Table 1.** Levels and intervals varying factors.

| Description | | Levels varying factors | | | Intervals varying factors |
|---|---|---|---|---|---|
| Natural | Code | +1 | 0 | -1 | |
| Cement | X1 | 740 | 490 | 240 | 250 |
| Strength | X2 | 600 | 500 | 400 | 100 |
| Water | X3 | 310 | 225 | 140 | 85 |

The following polynomial models were obtained:
- strength of modified concrete
R = -6.9 + 0.005 C + 0.29 R - 0.39 B - 0.000053 C 2 - 0.0002 R2 + 0.00035B2 + 0.000098 C R + 0.00024 CI - 0.00016 RV
- porosity of modified concrete
P = -0.85 + 0.014 C + 0.0031 R -0.1 B + 0.00000035 C 2 -0.00000022 R2-0.0000038 B2 + 0.0000015 C R + 0.00000059 C-0.00000147 RB

- amount of absorbed oil

O = -0.0057 – 0.0055 C + 0.000028 R + 0.043 B – 0.0000003 C2 – 0.000004 R2 – 0.0000047 B2 + 0.0000011C R + 0.00000012CB – 0.0000038 RB

The adequacy of all the models was checked by the Fisher's ratio test and showed satisfactory convergence of calculated and experimental data.

Thus, the treatment of concrete with heat treatment reduces their water absorption to less than 1%, increases strength by 40%, and increases frost resistance by an order of magnitude.

## References

1. V. Voronin, *PhD Thesis,* Moscow State University of Civil Engineering, Russian Federation, (1985)
2. A. Ouskern, *A Review of Properties of Polymer impregnated Concrete.* «New materials in Concrete Construction.», University of Illinoys Conference. Chicago, 1971
3. E Manuilova, *PhD Thesis,* Moscow State University of Civil Engineering, Russian Federation, (1995)
4. U. Magdeev, A Davidyuk, B. Sadov, Industry of construction materials of Moscow(in Russian - Promyshlennost stroitelnykh materialov Moskvy), 7(1986)
5. T. Sebök, , O. Stráněl, Cem. and Conc. Res., 34, 10 (2004)
6. O. Larsen, V. Naruts, Mag. of Civ. Eng., 8 (2016)
7. J Chen, D Zampini, , A Walliser, Cem. and Conc. Res., **32**, 1(2002)
8. D. Gemert, L. Czarnecki, M. Maultzsch, H. Schorn, A. Beeldens, P. Lukowski, E. Knapen, Cem. and Conc. Comp., 27, 9-10 (2005)
9. Marcelo Amianti, Vagner Roberto Botaro, Cem. and Conc. Comp., **30**, 1 (2008)
10. S. Yehia, M. Fawzy, Int. Jour. of Curr. Eng. and Tech., **7**, 2 (2017)
11. M. Bhuttaa, T. Maruyab, K Tsurutac, Con. and Build. Mat., 43 (2013)
12. C.- H. Chen , R. Huang, J.K. Wu, Journ. of the Ch. Inst. of Eng., 1 (2007)
13. G. Gondrashov, Eur. J. of Nat. Hist., **5** (2008)
14. M. Amianti, V. Botaro, Cem. and Conc. Comp., **30,** 1 (2008)
15. F. Gilabert, K. Van Tittelboom, J. Van Stappen, V. Cnudde, N. De Belieb, W. Van Paepegem, 77 (2017)

MATEC Web of Conferences 117, 00016 (2017) DOI: 10.1051/matecconf/201711700016
*XXVI R-S-P Seminar 2017, Theoretical Foundation of Civil Engineering*

# About Contemporary Problems of Numerical Modelling of Unique Structures, Buildings and Facilities

*Alexander* Belostotsky[1,2,3], *Taymuraz* Kaytukov [3,4,*] , *Irina* Afanasyeva [1,3], *Jacek* Grosel[5]

[1]Scientific Research Center “StaDyO”; office 810, ul. 3ya Yamskogo Polya, Moscow, 125040, Russia
[2]Perm National Research Polytechnic University; 29 Komsomolsky prospekt, Perm, Perm Krai, 614990, Russia
[3]National Research Moscow State University of Civil Engineering, Department of Applied Mathematics; 26, Yaroslavskoe Shosse, Moscow, 129337, Russia
[4]Russian Academy of Architecture and Construction Sciences; 24-1, ul. Bolshaya Dmitrovka, Moscow, 107031, Russia
[5]Civil Engineering Faculty, Wrocław University of Technology, Wyb. Wyspiańskiego 27, 50-370 Wrocław, Poland

**Abstract.** This paper is devoted to contemporary problems of numerical modeling of unique structures, buildings and facilities and corresponding directions of activity of Scientific Research Center “StaDyO” including development, verification and advanced of corresponding software, development and refinement of methods of structural analysis, solution of scientific and technical problems, scientific and educational activity. Experience in theoretical and practical computational analysis is under consideration as well (26-year history).

## 1 Introduction

International and Russian experience, reflected in the numerous papers and monographs, well-known events in the recent years in various cities all over the world indicate that problem of technological safety of metropolis is rather actual, science-intensive and, unfortunately, still far from practical solution. An effective and economically justified solution of these problems is normally based on predictive mathematical modeling of state (static and dynamic stress-strain state, gas- and hydrodynamic state, thermal state, etc.) of structures, buildings and facilities with the use of corresponding universal and specialized software systems, realizing numerical methods of mechanics (fluid mechanics, gas dynamics, solid mechanics, structural mechanics). In accordance with the state-of-the-art concept mathematical models must accompany objects at all stages of their life cycle (design and construction, operation, repair and reconstruction), providing an adequate and complete analysis and forecast of the state as part of computer-based structural health monitoring systems [1, 2]. Scientific Research Center “StaDyO” (SRC “StaDyO”) was founded in Moscow in 1991 as one of the Russian pioneers offering and performing

* Corresponding author: pavel.akimov@gmail.com

computer-oriented computation services for industrial purposes. Problems dealing with effects of static, temperature and dynamic loads on complex buildings, equipment and pipelines of nuclear and hydro power plants, wind structures, offshore structures etc. are under consideration. SRC "StaDyO" develops and distributes computer-aided design systems as well as the respective specialized software. Besides, SRC "StaDyO" distributes technical software in the area of structural mechanics and other related areas. During twenty six years SRC "StaDyO" worked in the field of design and numerical analysis of structures, equipment and pipe systems of nuclear power plants, hydro power stations and dam constructions, offshore structures and other complicated systems.

## 2 Directions of Activity of Scientific Research Center "StaDyO"

### 2.1 Contemporary problems of mathematical modelling of unique structures, buildings and facilities

Directions of activity of SRC "StaDyO" include the following contemporary problems of mathematical modelling of unique structures, buildings and facilities:

– modelling of interaction of buildings and structures with a foundation with allowance for real properties, stage-by-stage construction and actual operation history;

– structural analysis with allowance for physical, geometrical and other nonlinearities (plasticity of metal, creep and crack formation of reinforced concrete, nonlinear rheology of the foundation, large displacements, loss of stability, postbuckling behaviour, contact problems (friction separation, etc.));

– structural analysis with allowance for structural and technological specificity (structural (constructive) nonlinearity, genetic nonlinearity) of buildings and facilities (construction sequence, stage-by-stage construction, sensitivity of buildings and structures, assessment of the quality of the constructive solution from the position of the sensitivity of the stress-strain state to corresponding design deviations);

– numerical modelling of wind flows and loads (average and pulsation components, loads on facade structures, pedestrian comfort, vortex resonance oscillations), experimental validations of wind load analysis;

– seismic analysis (with allowance for acceleration spectra (versions of the linear-spectral theory), accelerograms, platform models, wave effects);

– progressive collapse analysis of buildings and facilities with allowance for real dynamic highly nonlinear effects of elastoviscoplasticity and large displacements;

– development and refinement of methods and algorithms for solution of large-scale computational problems (direct and iterative solvers, superelement technology, adaptive schemes, parallelizing, etc.);

– development of calibratable mathematical and computer models as part of structural health monitoring systems at the stages of erection and operation of buildings and facilities;

– application of algorithms of aerodynamics for modelling of snow sedimentations, explosion loads and distribution of hazardous emissions;

– numerical modelling of three-dimensional non-stationary problems of fire resistance;

– solution of coupled problems of aerohydroelasticity.

### 2.2 Development and verification of software

During twenty six years specialists from SRC "StaDyO" successfully worked in the field of design, development and verification of proprietary software systems. STADYO is the universal software system providing temperature fields, static, stability and dynamic analysis (including response spectra and accelerations definition) as well as fracture mechanics and strength analysis and optimization of arbitrary combined 2-D and 3-D solid, shell, plate and beam mechanical systems by the finite elements, superelement and other modern numerical methods. STADYO is verified and certificated in Gosatomnadzor RF (State Nuclear Safety Commission of the Russian Federation), it is applied in the leading design and research institutes and plants (for instance, Atomenergoproject (Russia), Hydroporoject (Russia), Teploenergoproject (Russia), Energoproject (Bulgaria), Siemens AG (Germany), etc.)). ASTRA-NOVA is the software system for nuclear power plants, fuel power plants, which provides petroleum pipelines stress-strain analysis in accordance with existing Russian and international design codes. SyMonEx is informational & diagnostic computer system of monitoring and expert safety evaluation of complicated power energetic structures. These software systems have powerful pre- and postprocessors, which allow to prepare input data and analyse results, correspond to world standards and can work together with various modern CAD/CAM systems and data bases.

Besides, it is vital to specify the following activities:

– review of results of structural analysis with the use of verified software;

– training of specialists-users of software systems which provide mathematical modelling of structures, buildings and facilities;

– qualification tests and certification of specialists dealing with computational structural analysis and corresponding experts reports;

– verification of numerical methods and software used in design and structural analysis;

– construction and technical expertise of buildings and facilities (including cases of local destruction).

### 2.3 Development and refinement of methods of structural analysis

Research and development of numerical (finite element method, boundary element method, variation-difference method, meshless method, finite volume method etc. [3]) and semianalytical (discrete-continual) methods of structural analysis are the most important aspects of ensuring safety of structures and buildings.

Particularly the field of application of discrete-continual methods (discrete-continual finite element method, discrete-continual boundary element method, discrete-continual variation-difference method), which are now becoming available for computer realization, comprises structures with regular (in particular, constant or piecewise constant) physical and geometrical parameters in some dimension (so-called "basic" direction (dimension)). Considering problems remain continual along "basic" direction while along other directions discrete-continual methods presuppose corresponding mesh approximation. Analytical solution along basic direction is apparently preferable in all aspects for qualitative analysis of calculation data. It allows investigator to consider boundary effects when some components of solution are rapidly varying functions. Due to the abrupt decrease inside of mesh elements in many cases their rate of change can't be adequately considered by conventional numerical methods while analytics enables study. Another feature of discrete-continual methods is the absence of limitations on lengths of structures. Hence it appears that in this context discrete-continual methods are peculiarly relevant.

Solution of corresponding resultant systems of equations with immense number of unknowns is the most time-consuming stage of the computing, especially if we take into account the limitation in performance of personal computers, contemporary software and necessity to obtain correct solution in a reasonable time. However, high-accuracy solution at all points of the model is not required normally, it is necessary to find only the most accurate solution in some pre-known domains. Generally the choice of these domains is a priori data with respect to the structure being modeled. Designers usually choose domains with the so-called edge effect (with the risk of significant stresses that could potentially lead to the destruction of structures, etc.) and regions which are subject to specific operational requirements. It is obvious that the stress-strain state in such domains is of paramount importance. Specified factors along with the obvious needs of the designer or researcher to reduce computational costs by application of sophisticated correct discrete and discrete-continual methods cause considerable urgency of constructing of special algorithms for obtaining local solutions (in some domains known in advance) of boundary problems. Wavelet analysis provides effective and popular tool for such researches. Solution of the considering problem within multilevel wavelet analysis is represented as a composition of local and global components.

### 2.4 Educational activities

Leading specialists of SRC "StaDyO" are closely involved in development, updating and realization of Master of Science programs (in the fields of "Applied Mathematics" and "Construction") and PhD programs (in the fields of "Informatics and Computer Engineering" and "Engineering and technology of construction"). Besides, we should also mention here the following activities:

– "open lectures", specialized training courses, seminars, counseling (including counseling in the preparation of graduate qualification works);

– training of specialists within the programs of supplementary vocational education in the fields of large-span and high-rise buildings, underground structures, nuclear, thermal and hydroelectric power stations, pipeline systems for various purposes etc);

– organization and development of scientific research & educational centers and laboratories in leading Russian universities;

– organization and participation in Russian and international scientific events (conferences, symposiums, seminars, etc.);

– preparation of textbooks, tutorials and monographs on topical problems of mathematical and computer simulation of the condition of buildings, structures and complexes.

## 3 Experience in theoretical and practical computational analysis

The reliability of research results, methods, algorithms, software systems and results of analysis is provided by multilevel verification and great (more than 40-years) experience in theoretical and practical computational analysis including the following problems, specified below (Figure 1, Figure 2).

1. Three-dimensional thermal and stress-strain state, stability, strength and reliability of pipelines, technological, electrotechnical and lifting-transport equipment, machines and mechanisms, structures, coupled systems "equipment – pipelines", "foundation – overground structure", "foundation – subground structure" of buildings, facilities and complexes, with allowance for design and actual load combinations (including temperature loads, static loads, wind loads, operational load (vibrations), special dynamic loads (seismic, shock-wave, emergency and other)):

– reactor compartments, engine rooms, generator halls, turbine buildings, reserve diesel power plants and cooling towers of nuclear power plants (NPP) and waste storages (Armenian, Kursk, Smolensk, Chernobyl, Leningrad, Ignalina, Bilibino, Novovoronezh, Kola, Balakovo, Volgodonsk, Kalinin, Zaporozhye, Beloyarsk, Lovisa, Kozloduy, Belene, Paksh, Temelin, Stendal, Kudankulam, Tianwan, Bushehr, new generation NPP (AS-NP 500, AS-NP 1000, NP 2006, NPP with WWER-TOI, etc.));

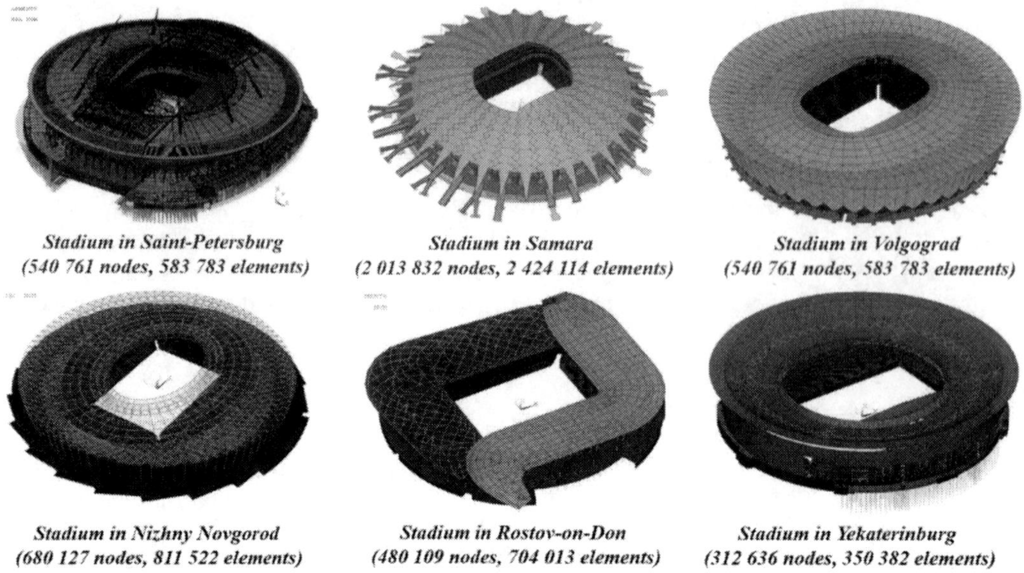


**Fig. 1.** Analysis of stress-strain state, strength and stability of load-bearing structures of football stadiums of the World Cup 2018.

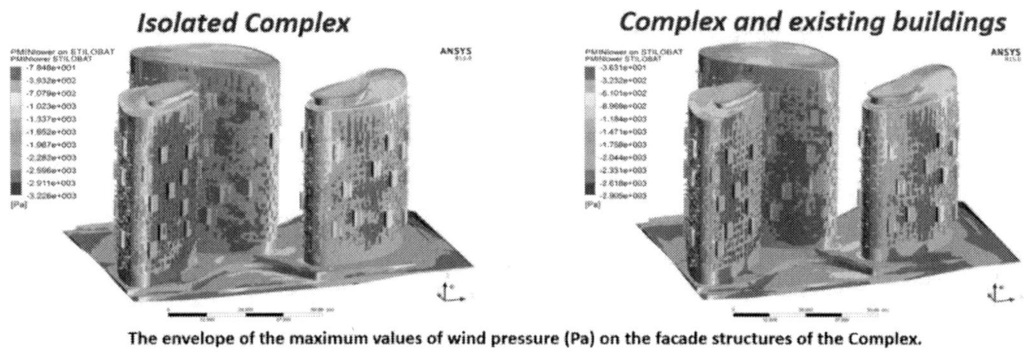


**Fig. 2.** Computing of wind loads on the load-bearing and facade structures of a multi-storey complex with an apart-hotel in Moscow.

– arched, gravity and earth dams, underground structures and buildings of hydroelectric power stations (Sayano-Shushenskaya, Krasnoyarskaya, Bratskaya, Boguchanskaya, Zeyskaya, Bureyskaya, Vilyuyskaya, Katun, Chirkeiskaya, Volzhskaya, Kamskaya, Inguri, Khudoni, Namakhvani, Kurpsai, Nurek, Rogun, Plyavinskaya , Gekhi, Hoabin, Kapanda, Teri, Tang-E-Duk, etc.), hydroelectric pumped storage power stations (Zagorskaya), tidal hydroelectric stations, coast protection structures, other hydraulic strctures;

– unique and typical buildings and structures of civil engineering (roof of Grand Sports Arena of Luzhniki Olympic Complex (Moscow), the monument to the 300th anniversary of the Russian fleet (Moscow), underground parking of shopping and entertainment mall "Manezhnaya Square" (Moscow), sports and fitness complex "Aquadrom", ice stadium "Megasport" located on Khodynka (Moscow), sports complex "Moskvich" (Moscow), indoor swimming pool of the "Iskra" sanatorium (Sochi), ice palace "Bolshoi" (Sochi), bobsleigh track "Sanki" and the ski complex "Gorki" (Sochi, Winter Olympic Games – 2014), football stadiums of the World Cup 2018 ("Zenit" (Saint-Petersburg), "Spartak" (Moscow), stadiums located at Volgograd, Samara, Nizhny Novgorod and Rostov-on-Don); multifunctional high-rise complexes (including Moscow International Business Center "Moscow-City", Poklonnaya (Moscow), Profsoyuznaya (Moscow), Leningradskaya

(Moscow), building located at Volgograd, Omsk, Vladivostok, Krasnodar, Kiev, Astana), multi-block shopping and entertainment centers, multi-storey panel block sections and monolithic buildings, schools and polyclinics, banners with allowance for initial tension of ropes, wind loads and other factors);

– overground and underground pipelines of heating systems, main oil and gas pipelines, petrochemical and gas pipelines and equipment (analysis of "Sakhalin – Khabarovsk – Vladivostok" gas pipeline system is among the last comprehensive and "breakthrough" researches);

– wind power plants of various types and capacities;

– floating structures and platforms for the offshore extraction of oil and gas;

– various bio- and nanostructures (for example, double linear and closed DNA (deoxyribonucleic acid) helices).

– complex mechanical engineering structures, machines and mechanisms including aerospace systems, transport, shipbuilding, power engineering, ferrous and non-ferrous metallurgy, consumer electronics, etc. (analysis of stress-strain state, dynamic response and strength of the coupled system "Science power platform – drive unit – solar batteries" and subsystems of the International Space Station (ISS) "Alpha" at the stages of launch and orbital activity.

2. Complex, including record-sized (up to 200 million unknowns) three-dimensional stationary and non-stationary problems of building aerodynamics dealing with computing of average and pulsating components of wind loads, wind loads on facade and enclosing structures, pedestrian comfort of numerous unique objects:

– high-rise buildings, structures and complexes (Moscow International Business Center "Moscow-City", "Gazoil City" (Moscow), "Zodiac" (Moscow), "Sky Fort" (Moscow), "Dirigible" (Moscow), "Rublevsky Lights" (Moscow), "Aquamarine" (Vladivostok), group of buildings of National Research Moscow State University of Civil Engineering, Ostankino TV Tower in Moscow, buildings located at Saint-Petersburg, Kiev, Astana, etc.);

– large-span buildings and structures (stadiums "Moskvich" (Moscow), "Zenit" (Saint-Petersburg), railway station in Adler, a ski jumping complex of Winter Olympic Games – 2014, etc.);

– monuments (monument on Poklonnaya Hill (Moscow), monuments "Conquerors of the Cosmos" and "Worker and Collective Farm Girl" at the Exhibition of Economic Achievements (VDNKh, Moscow), etc.);

– complex of basic structures of nuclear power plant with WWER (reactor compartments, engine rooms, evaporative cooling towers, etc.) with allowance for extreme (hurricane) wind and tornado.

Besides, contemporary problems of refined numerical modeling and aircraft crashes are considered for the basic structures of nuclear power plants of various types (WWER, RBMK and BN).

The Reported study was Funded by Government Program of the Russian Federation "Development of science and technology" (2013-2020) within Program of Fundamental Researches of Ministry of Construction, Housing and Utilities of the Russian Federation and Russian Academy of Architecture and Construction Sciences; Research Projects 7.1.1, 7.1.2".

## References

1. A.M. Belostosky, P.A. Akimov, Procedia Engineering, **153**, 83-88 (2016).
2. A.M. Belostosky, P.A. Akimov, Procedia Engineering, **153**, 89-94 (2016).
3. O.C. Zienkiewicz, R.L. Taylor, D.D. Fox, *The Finite Element Method for Solid and Structural Mechanics* (Butterworth-Heinemann, 2013).

# About optimal geometric parameters of a developed I-beam with a perforated web

*Viktor* Belov[1,*], *Oleg* Goryachevskij[2]

[1] Moscow state university of civil engineering, Yaroslavskoye shosse, 26, Moscow, Russia, 129337
[2] Moscow state university of civil engineering, Yaroslavskoye shosse, 26, Moscow, Russia, 129337

**Abstract.** For the first time, questions of optimization of geometric parameters of developed I-beam with a perforated web are considered with the use of the finite element method in PC ANSYS Mechanical. For the linear model of the material, strength and stiffness calculations are performed depending on the shape of the cut: the height of the I-beam development and the distance between the holes. For the I-beam with the highest height, the stability of the web was checked, the first two critical loads and forms of stability loss were obtained. The optimum degree of development is shown on the example of the 12-meter developed I-beam from the initial rolled I-beam with a height of 60 cm. The strength and rigidity of the beam is determined when the length of the distance between the holes is changed.

## 1 Status of the question

Developed I-beams with a perforated web in their design features occupy an intermediate position between a welded beam from rolled T-beams and Vierendeel trusses.

The exact method of their calculation is connected with certain difficulties, in view of the complicated configuration of the holes.

In TsNIISK together with VNIIMontazhspetsstroi [1] and at the experimental plant VNIIGAZA [2] in the 1970s, experimental studies of a series beam structures from developed I-beams with a perforated web made it possible to develop a calculation method based on the approximate calculation of Vierendeel trusses.

The purpose of the research was to assess the actual bearing capacity, deformability and the nature of the destruction of full beam structures from developed I-beams with a perforated web, working on transverse bending under the action of two concentrated loads located at a distance of 1 / 4-1 / 3 span from the supports.

The initial profile was the I-beams № 36, № 40 and № 50 in accordance with GOST 8239-59* from steel VSt.3ps.

The ratio of the height of the cross section of the perforated I-beam to the height of the original varied from 1.39 to 1.5.

The ratio of the height of the cross-section of the tested beams to the span was 1/12 - 1/18 and covered the characteristic range of bending beam structures.

* Corresponding author: belov@mgsu.ru

The transverse bending was provided by a system of links preventing the exit of the upper flange from the plane.

The entire sections of the web of the tested structures did not lose stability, including in places of application of concentrated loads, despite the absence of vertical ribs there.

The exhaustion of the load-carrying capacity of all tested structures occurred as a result of the appearance of yield in the extreme fibers of the T-section.

The appearance of relatively early yield in the sections of the web, where the holes begin to change their height, was of a local character and was practically no effect on the bearing capacity of the structure.

Taking into account the results of the tests, the calculation bearing strength of perforated I-beams working on transverse bending should be carried out in a weakened section opening at the maximum normal stresses arising in the T-sections and consisting of normal stresses from the bending moment $M$ and normal stresses from the action of the moment $M$ $\sigma_Q$ of conditionally applied shear forces, i.e. $\sigma = \sigma_M + \sigma_Q \le R$, where $R$ is the design resistance of steel to tension and compression.

The normal stresses from the bending moment in the cross sections near the holes are distributed linearly, in proportion to the distance from the longitudinal axis of the beam (Fig. 1).

at points 1, 4: $$\sigma_M = \pm \frac{M_{I-I} \cdot H}{2J_x}, \quad (1)$$

at points: 2, 3: $$\sigma_M = \pm \frac{M_{I-I} \cdot (H-h)}{J_x}, \quad (2)$$

where $H$ – height of developed I-beam, cm;

$h$ – height of the original I-beam, cm;

$J_x$ – moment of inertia of the perforated beam in section *I-I* (net section) with respect to the axis x-x, $cm^4$;

$M_{I-I}$ – bending moment in section *I-I*.

In the area of simultaneous action of the bending moment and the transverse force, the total transverse force in the sections along the hole is assumed to be distributed proportionally to the moments of inertia of the T-sections and concentrated in the cross section passing through the middle of the hole.

With a symmetrical section of the developed profile, the magnitude of the transverse force is distributed equally between the T-sections.

At a certain distance from the conventional point of application of the transverse force to the normal stresses from the bending moment in the sections along the apertures, stresses caused by the action of the applied lateral force are added.

Under these conditions, the maximum stresses from the transverse force arise in sections *I-I*, *III-III* (fig. 1) and for points 1-8 can be determined by the formulas:

at points: 1, 5, 4, 8: $$\sigma_Q = \pm \frac{Q_b}{2W_{\perp\, max}}, \quad (3)$$

at points: 2, 6, 3, 7: $$\sigma_Q = \pm \frac{Q_b}{2W_{\perp\, min}}, \quad (4)$$

where $b$ – T-section length, cm;

$Q$ – transverse force in section, N;

$W_{\perp\max}$, $W_{\perp\min}$ – the greatest and the least moments of resistance of a T-section of constant height, cm$^3$.

It is not difficult to verify that the maximum total normal stresses $\sigma = \sigma_M + \sigma_Q$ arise at points 2, 3, 5, 8, and their values can be determined by the following formulas:

at points 2 и 3: $$\sigma = \pm\left(\frac{M_{I-I}\cdot(H-h)}{J_x} + \frac{Q_{II-II}\cdot b}{2W_{\perp\min}}\right) \quad (5)$$

at points 5 и 8: $$\sigma = \pm\left(\frac{M_{III-III}\cdot H}{2J_x} + \frac{Q_{II-II}\cdot b}{2W_{\perp\max}}\right) \quad (6)$$

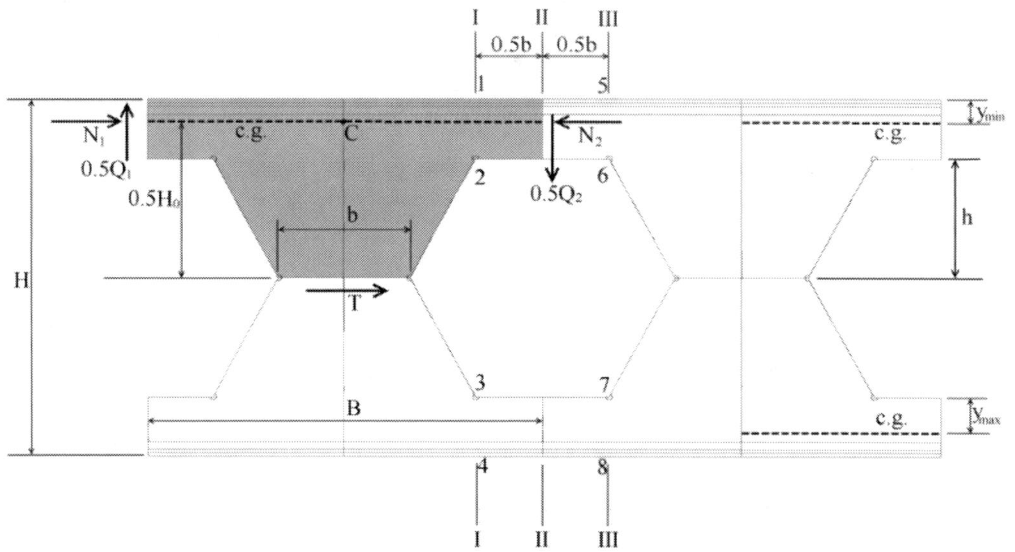


**Fig. 1.** Geometric parameters and the scheme of application of internal forces in the developed I-beam.

Thus, the strength of bent elements from perforated I-beams should be checked by formulas (5) and (6), and the values of normal stresses at points 2 and 3 should not exceed $\sigma \le mR$, at points 5 and 8 $\sigma \le R$, where $m = 1{,}2$ – the coefficient of conditions work, taking into account the local nature of the extreme stresses in the corners of the holes at points 2 and 3.

The magnitude of the tangential stresses in the weld region can be determined by taking into account the assumptions made for the normal stresses.

For this purpose, cut out a part of the beam, shown in Fig. 1, and apply the forces acting on it.

Forces $N_1$ и $N_2$ – the components of the bending moment are applied along the center of gravity of the T-section; $0.5Q_1$ и $0.5Q_2$ – values of the conditionally applied lateral force with symmetrical sections; $T$ – shear force in the web.

Having the sum of the moments of forces relative to the point $C$, the shearing force $T$ and the magnitude of the shearing stresses acting in the weld seam are determined from the equilibrium condition of the cut beam part:

$$\tau = \frac{QB}{H_0 t_w l_w}, \quad (7)$$

where $Q = \frac{Q_1 - Q_2}{2}$ – transverse force, N;

$B$ – step of holes in a perforated beam, cm.

$H_0$ – distance between centers of gravity of T-sections, cm;

$t_w$ – web thickness, cm;

$l_w = b - 2$ – the length of the welded joint, taking into account bad welding at the ends, cm.

The overall stability of perforated beams, as well as the stability of the web, is recommended to be checked in accordance with the requirements of SNiP for bending beams with entire web.

Particular attention should be paid to checking the stability of the web in the area of support and large concentrated loads.

The deformability of the bent perforated I-beam should be determined taking into account the effect of lateral force.

The total deflection $f$ is assumed to consist of a deflection $f_M$ from the action of the bending moment, defined as for a beam with entire web, and a deflection $f_Q$, caused by the action of shear forces.

$$b : f = f_M + f_Q$$

The values $f_M$ и $f_Q$, are determined from the general formulas for the resistance of materials.

For bending beam structures with a ratio $\frac{H}{L} \leq \frac{1}{15}$ value of deflection from the action of the transverse force $f_Q$ can be taken approximately equal to 6% of the deflection caused by the action of the bending moment: $f_Q = 0{,}06 f_M$

The basic calculation provisions were taken as a basis for the writing of Chapter 19 of SNiP II-23-81 *.

All the calculation provisions above relate to engineering methods of calculation using obsolete regulatory documents created in the 1970s of the last century.

In 70-ies of the last century the author carried out experimental studies of the shape of the cut and bearing capacity of developed I-beams at the experimental plant VNIIGAZA, on the basis of which for the first time in the USSR the developed I-beams were used in the frame of the compressor station in Bukhara, and then in the frames of other compressor stations [2].

The normative method for calculating the developed I-beams with a perforated web, set out in SNiP II-23-81* [3] and without actualization, transferred to the "Updated" edition (SP 16.13330.2011) [4].

In this direction, a number of scientists were studied both by engineering methods [5-10] and using finite element modeling [11-16].

Currently on the basis works of the author on the improvement of building metal structures was research using the finite element method for the optimal cut of the I-beam. [17, 18]

## 2 Finite Element Simulation

Calculation of developed I-beams was carried out in PC ANSYS Mechanical.

In the programming language APDL, a program was written that implements the calculation of a set of constructive forms of developed I-beams with the ability to select and vary physical and geometric parameters, calculate and output the quantities of interest.

Finite elements of the *Solid186* type with a quadratic form function were used, which are much more accurate in the problems of calculating structures with thin wall than linear *Solid185*. [19, 20]

In static analysis to reduce the calculation time, the symmetry condition was used.

The results of calculation of developed I-beams with a perforated web made of the initial profile №60 (GOST 8239-89) with a length of 12 meters are given.

Beams have hinged supports at the ends and loaded with a uniformly distributed load.

The material was modeled as linearly elastic with a modulus of elasticity $E = 2.06 \cdot 10^5$ MPa, Poisson's ratio $v = 0.3$.

Steel class – *C245* (GOST 27772-2015) with resistance to bending with ultimate strength $R_u = 360$ MPa and yield strength $R_y = 240$ MPa.

To verify the calculated finite element model, an additional verification model with a smaller grid was built. The main results of the calculation on the basis of both models are presented in Table. 1, and the geometric model and fragments of finite-element models in Fig. 2.

To compare the results by different criteria, the maximum load was used. The maximum load was calculated using the following formulas:

$$F_{\max A} = \frac{R_u}{\sigma_{\max A}} q,\ F_{\max B} = \frac{R_y}{\sigma_{\max B}} q,\ F_{\max C} = \frac{L}{200 \cdot u_{\max}} q,$$

where $F_{\max A}$ – the maximum load at which the stress of Mises at the points at which the hole begins to change its height $\sigma_{\max A}$ does not exceed $R_u$;

where $F_{\max B}$ – the maximum load at which the stress of Mises in flange $\sigma_{\max B}$ does not exceed $R_y$;

where $F_{\max B}$ – the maximum load at which the deflection $u_{\max}$ does not exceed *1/200* of span *L*.

The results of the calculations are shown in fig. 3-5. For the developed I-beams with the highest height 108 cm, the stability of the web was checked (fig. 6).

**Table 1.** Comparison of finite element networks

| **Characteristics of the finite element grid** | $\sigma_{\max A}$ **MPa** | $\sigma_{\max B}$ **MPa** | $u_{\max}$, **mm** |
|---|---|---|---|
| Calculated<br>• 20 elements in height of I-beam between perforations<br>• 8 elements along the cut line of the cut at the extreme perforations<br>Total nodes – 275 595 | 58,21 | 4,76 | 0,847 |
| Test<br>• 30 elements in height of I-beam between perforations<br>• 15 elements along the cut line of the cut at the extreme perforations<br>Total nodes – 393 979 | 58,71 | 4,76 | 0,847 |

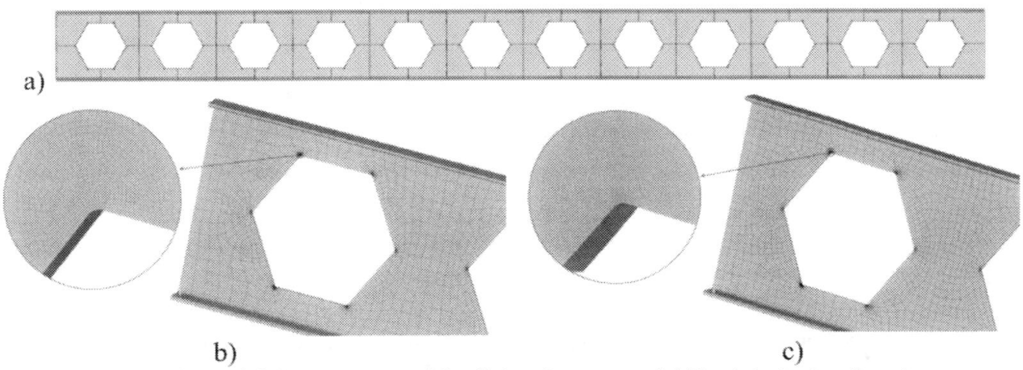


**Fig. 2. a)** geometric model; **b)** Fragment of the finite element model No. 1 (calculated) at the outermost hole; **c)** Fragment of the finite-element model No. 2 (test) at the outermost hole.

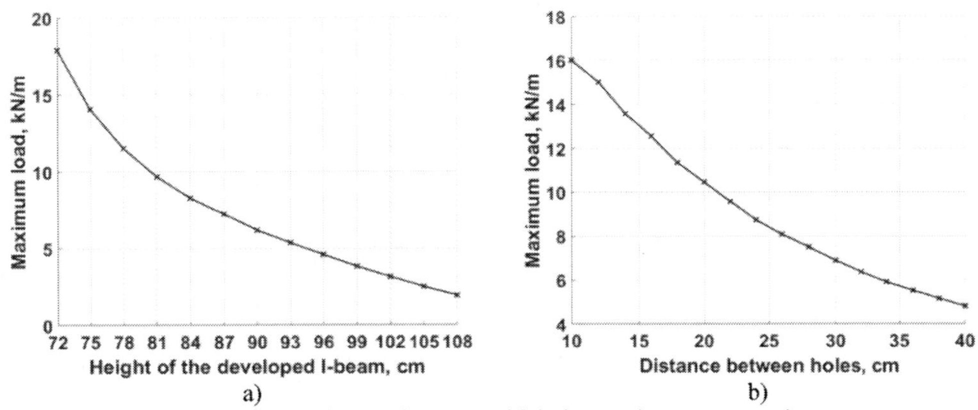


**Fig. 3.** The choice of the shape of the web cut, at which the maximum stress values in the nodes the change in the cross sections does not exceed $\boldsymbol{R_u}$.
a) depending on the height of the developed I-beam;
b) depending on the distance between the holes.

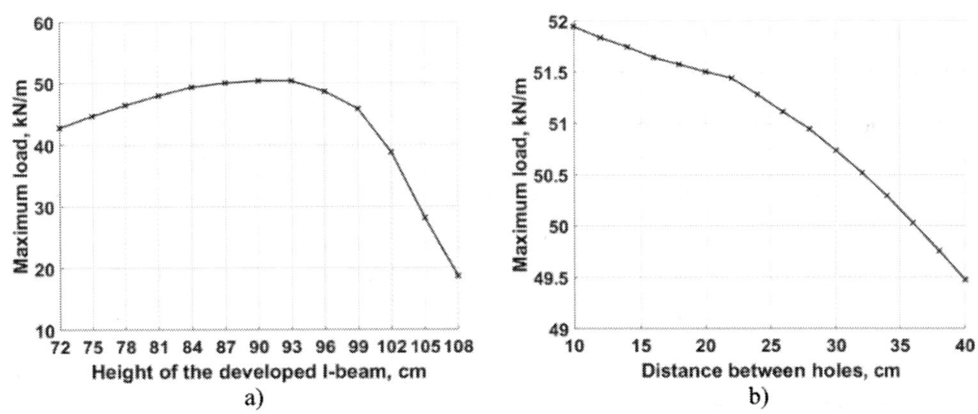


**Fig. 4.** The choice of the shape of the web cut, at which the maximum stress values in the flanges do not exceed $\boldsymbol{R_y}$.
a) depending on the height of the developed I-beam;
b) depending on the distance between the holes.

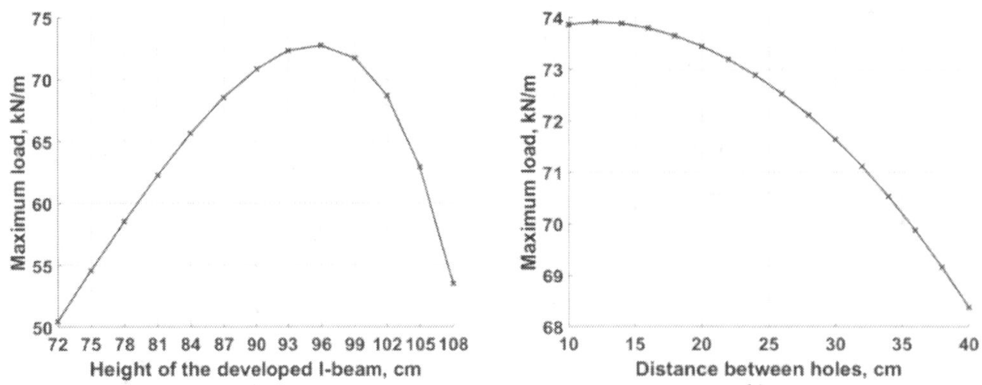


a) b)

**Fig. 5.** The choice of the shape of the web cut, in which the deflection does not exceed the 1/200 span.
a) depending on the height of the developed I-beam;
b) depending on the distance between the holes.

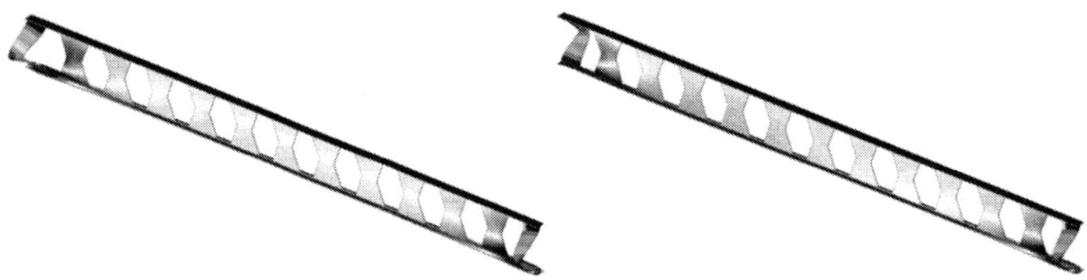

a) b)

**Fig. 6. a)** First (70,270 kN/m) and **b)** second (70,279 kN/m) buckling modes.

## 3 Conclusions

At this stage, dependencies are obtained for the ultimate load by three criteria: strength - at the points where the hole cross-sections change, strength - in the shelves and the maximum allowable deflections.

Based on the finite element method, the following

a) the optimum height of 90 cm. Of the developed I-beam from the original I-beam No. 60.

b) it is established that a decrease in the length of the bridge increases the strength and rigidity of the developed I-beam.

c) the stability of a developed I-beam is ensured even with the ratio $H / h = 1{,}8$.

## References

1. *Novye formy legkih metallicheskih konstrukcij*: monografiya pod redakciej V.I. Trophimova/CNIISK im. V.A. Kucherenko, akcionernoe ob"edinenie "Korporaciya Montazhspecstroj", firma "Stal'stroj". 200-203. (1993)
2. V.A. Belov. O nekotoryh voprosah tekhnologii izgotovleniya skvoznyh razvityh balok iz prokatnogo dvutavrovogo profilya. Tezisy dokladov vsesoyuznogo soveshchaniya: "Puti povysheniya proizvoditel'nosti truda i ehkonomii metalla pri izgotovlenii svarnyh stroitel'nyh konstrukcij" (g. CHelyabinsk, 20-22 iyunya 1973 g.). 85-88. (1973)

3. SNiP II-23-81*. *Stal'nye konstrukcii* (1981)
4. SP 16.13330.2011. *Stal'nye konstrukcii. Aktualizirovannaya redakciya SNiP II-23-81*.* (2011)
5. V.V. Suprunyuk. Stal'nye poperechno nagruzhennye perforirovannye arki. avtoref. dis. ... kand. tech. nauk. 23. (2007)
6. A.A. Yurchenko. Napryazhenno-deformirovannoe sostoyanie balok zamknutogo secheniya s perforirovannymi stenkami. avtoref. dis. ... kand. tech. nauk. 27. (2008)
7. V.M. Daripasko. Prochnost' i ustojchivost' dvutavrovyh ehlementov s perforirovannoj stenkoj pri obshchem sluchae zagruzheniya. avtoref. dis. ... kand. tech. nauk. 23. (2000)
8. V.M. Dobrachev. Puti povysheniya ehffektivnosti stal'nyh balok s perforirovannoj stenkoj.. avtoref. dis. ... kand. tech. nauk. 21. (1982)
9. V.V. Egorov. Povyshenie ehffektivnosti shprengeshl'nyh balok s perforirovannoj stenkoj kombinirovannym sposobom regulirovaniya napryazhenij. avtoref. dis. ... kand. tech. nauk. 24. (1986)
10. A.I. Sklyadnev. Konstruktivnye formy i metody rascheta balok s perforirovannoj stenkoj. avtoref. dis. ... kand. tech. nauk. 19. (1978)
11. A.G. Kozhikov. Sovershenstvovanie metodiki opredeleniya silovogo soprotivleniya i konstruktivnoj formy perforirovannyh stoek. avtoref. dis. ... kand. tech. nauk. 20. (2011)
12. E.V. Litvinov. Prochnost' i ustojchivost' stenki v linejno-perforirovannyh ehlementah stal'nyh konstrukcij s regulyarnymi otverstiyami. avtoref. dis. ... kand. tech. nauk. 24. (2006)
13. A.I. Pritykin. Razrabotka metodov rascheta i konstruktivnyh reshenij balok s odnoryadnoj i dvuhryadnoj perforaciej stenki. avtoref. dis. ... d-ra tech. nauk. 44. (2011)
14. T.M. Rogotovskih. Prochnosti stal'nyh szhato-izognutyh perforirovannyh ehlementov v uprugo-plasticheskoj stadii. avtoref. dis. ... kand. tech. nauk. 26. (2009)
15. E.Yu. Fomenko. Izgibno-krutil'naya forma poteri ustojchivosti vnecentrenno-szhatyh stal'nyh dvutavrovyh stoek s perforirovannoj stenkoj. avtoref. dis. ... kand. tech. nauk. p. 26. (2011)
16. E.V. Balashov. Sovershenstvovanie konstrukcij stalezhelezobetonnyh proletnyh stroenij mostov so skvoznymi balkami. avtoref. dis. ... kand. tech. nauk. 24. (2010)
17. V.A. Belov, K. Krul'. *Modelirovanie i raschyot metallicheskih konstrukcij zdanij i sooruzhenij*. Monografiya. (2012)
18. K. Krul', V.A. Belov, K. Oleyarchik, *Modelirovanie i raschet konstrukcij stroitel'nyh mashin i sooruzhenij*. Uchebnik. p. 232 (2008).
19. O.S. Goryachevskij, Issledovanie kachestva konechno-ehlementnoj modeli razvitogo dvutavra s perforirovannoj stenkoj // Pod"yomno-transportnye, stroitel'nye, dorozhnye, putevye mashiny i robototekhnicheskie kompleksy: Materialy nauch.konf. 14-15 aprelya 2016g. 138-139. (2016).
20. O.S. Goryachevskij, V.A. Belov. Ob aktual'nosti konechno-ehlementnogo modelirovaniya razvityh dvutavrov s perforirovannoj stenkoj. //Sbornik dokladov nauchno-tekhnicheskoj konferencii po itogam nauchno-issledovatel'skih rabot studentov instituta inzhenerno-ehkologicheskogo stroitel'stva i mekhanizacii, 14-18 marta 2016 g. 114-120. (2016)

MATEC Web of Conferences 117, 00018 (2017) DOI: 10.1051/matecconf/201711700018
*XXVI R-S-P Seminar 2017, Theoretical Foundation of Civil Engineering*

# Layered structures mechanical properties assessment by dynamic tests

*Alexey* Beskopylny[1,*], *Alexander* Lyapin[1], *Vladimir* Andreev[2]

[1]Don State Technical University, pl. Gagarina, 1, Rostov-on-Don, 344010, Russia
[2]Moscow state university of civil engineering, Yaroslavskoye shosse, 26, Moscow, Russia, 129337

**Abstract.** An effective method of the layered structures mechanical properties determination is presented. The method is based on the non-stationary analysis of the structure response that is acted by impact load on the surface of multi-layered structure. The assessment of the design parameters using a genetic algorithm is performed. During the solving procedure there are several methods for breeding and selection processes are used, that demonstrated different results. The most appropriate parameters for described algorithms are selected and analysed.

## 1 Introduction

Determination of the mechanical properties of multi-layered structures for various purposes is one of the most important tasks at any stage of their design and operation. In the construction field there are many examples of such structures include coated materials, sandwich panels, ground base construction of roads and others. Depending on the object the methods to determine the structures properties are different. These are indentation methods that used for non-destructive tests metal structures [1,2], determination of the rheological parameters for the creep models [3,4,5], identification of the structures properties [6]. There are many engineering applications in road construction field; methods of analysis of the dispersion properties of surface waves (MASW, SASW), used for soil and foundations.

A semi-analytical approach to the solution of the axisymmetric indentation problem for a multi-layered elastic half-space is presented in [7]. The solution is based on the Hankel transform and numerical computations of the solution are implemented for multi-layered composites.

Assessment of material characteristics using spherical indentation test is described in [8]. Three parameters of the LUDWIG's equation using the spherical indentation test and Neural Networks is determined. A Neural Networks is trained following the spherical indentation test using two parameters that are obtained from the *P-h* curve.

The indentation of a layer by a rigid circular cylinder is considered in [9]. The Green's functions for the indentation of elastic layer lying on or bonded to a rigid base by a line load are found by analytical solution. The peculiarities of this solution under various boundary conditions are considered.

A steady-state dynamic analysis of a response multi-layered transversely isotropic (TI)

*Corresponding author: besk-an@ya.ru

half-space generated by a point load moving along a horizontal straight line with constant speed is investigated in [10]. The solution is obtained using the double inverse Fourier transform with respect to the frequency and one horizontal wavenumber.

There are several ways to determine the mechanical characteristics of layered materials by pressing indenters of various forms. The choice of the shape of the indenter in this case is an important task [11,12]. Most common in modern practice are indenters in the form of a sphere [13,14], cone [15,16], pyramid, but they have one significant drawback – the complexity of the registration of the beginning of plastic flow

In all cases, the problem of the physical parameters determination of the structure uses the solution of the problem of dynamic impact on the area layer type or multi-layered half-space.

## 2 Problem statement

We consider the action of the impact load that specified subjected in the limited time interval on the surface of a layered viscoelastic half-space. The properties of the medium will determine a set of parameters: - elastic modulus $E_i$, - Poisson ratio $\nu_i$, - density $\rho_i$, $tg\alpha_i$, $tg\beta_i$ - the tangents of the angles of the losses for longitudinal and transverse waves.

Because of a sufficiently fast decay time signal due to the spatial scattering of the energy and physical dissipation we use the method of discrete harmonic analysis to find the displacement field of the structure points in the form of a finite set of solutions to the problems of stationary oscillations

$$\mathbf{u}(\mathbf{r},t)=\sum_{k=1}^{N} p_k\left[-\cos\eta_k \operatorname{Im}(\mathbf{U}(\mathbf{r},\omega_k)\exp(-i\omega_k t))+\sin\eta_k \operatorname{Re}(\mathbf{U}(\mathbf{r},\omega_k)\exp(-i\omega_k t))\right], \tag{1}$$

where $T$- limit time observation, $t^*$ - the decay time of the signal (it is possible to take $t^* = T$);

$p_k$ - the expansion coefficients of the impulse impact in a Fourier series on an interval $\left[-T/2, T/2+t^*\right]$ with a specified degree of accuracy $\varepsilon$

$$p^{(N)}(t)=\sum_{k=1}^{N} p_k\chi_k(t), \quad \left\|p(t)-p^{(N)}(t)\right\|_{L_2} < \varepsilon$$

$$\chi_k(t) = \sin(\omega_k t + \eta_k);$$

$$\omega_k = k\pi / (T + t^*);$$

$$\eta_k = 0.5 \cdot k\pi \cdot T / (T + t^*).$$

In accordance with this, the displacement field in a layered viscoelastic medium $\mathbf{U}(\mathbf{r},\omega)$ is sought separately for each component of the layered structure under stationary steady state with the frequency of the oscillations $\omega$. In particular for a half-space using the integral transform of the Hankel function $\mathbf{U}(\mathbf{r},\omega)$ can be represented in the form

$$\mathbf{U}(\mathbf{r},\omega) = \int_0^\infty \mathbf{P}(u,z)\cdot \bar{\mathbf{X}}(u) J_k(uR)u\,du\,, \tag{2}$$

where in cylindrical coordinates $\mathbf{r} = (R,z)$, $R \geq 0$, $z \leq 0$, $\mathbf{X}(R)$ - load distribution on the boundary of half-space and the upper layers,

$$\bar{\mathbf{X}}(u) = \int_0^{+\infty} \mathbf{X}(R)\, J_k(uR) R\, dR\,, \qquad \mathbf{P}(u,z) = \mathbf{U}(u,z)\cdot \mathbf{B}^{-1}(u,0)$$

$$\mathbf{U} = \begin{pmatrix} -u\exp(\sigma_{11}z) & -\sigma_{12}\exp(\sigma_{12}z) \\ \sigma_{11}\exp(\sigma_{11}z) & u\exp(\sigma_{12}z) \end{pmatrix};$$

$$\mathbf{B} = \begin{pmatrix} \varsigma_2^2\exp(\sigma_{11}z) & 2u\sigma_{12}\exp(\sigma_{12}z) \\ -2u\sigma_{11}\exp(\sigma_{11}z) & -\varsigma_2^2\exp(\sigma_{12}z) \end{pmatrix},\ \varsigma_2^2 = u^2 + \sigma_{12}^2,\ \sigma_{1k} = \sqrt{u^2 - \theta_{1k}^2}\,,\ k = 1,2\,; \tag{3}$$

$$\mathbf{B}^{-1}(u,0) = \begin{pmatrix} \zeta_2^2 & 2u\sigma_{21} \\ -2u\sigma_{11} & -\zeta_2^2 \end{pmatrix} \Delta_R^{-1} \equiv \mathbf{B}(u,0)\,\Delta_R^{-1},\ \Delta_R = \zeta_2^4 - 4u^2\sigma_{11}\sigma_{12};$$

$$\theta_{j1}^2 = \omega^2 a^2 / V_{Pj}^2, \qquad \theta_{j2}^2 = \omega^2 a^2 / V_{Sj}^2,$$

- the relative frequency of oscillations $j$-th component of the medium from the velocities of propagation of longitudinal and transverse waves $V_{Pj}, V_{Sj}$ (for half-space $j = 1$);
$a$ - the radius of the area distribution of surface impact efforts.

It's important that the expression (3) have only exponentially decreasing components (if $z < 0$).

Considering further the $j$-th layer , assume that on its sides is set to the vectors of stress

$$\mathbf{t}^{(j)}(R,0) = \mathbf{Y}^{(j,1)}(R),\quad \mathbf{t}^{(j)}(R,h_j) = \mathbf{Y}^{(j,2)}(R),\quad R \in (0,+\infty)) \tag{4}$$

The function $\mathbf{U}(\mathbf{r},\omega)$ that satisfies the equations of motion, according to the proposed method, we will seek as

$$\mathbf{U}^{(j,1)}(\mathbf{r},\omega) + \mathbf{U}^{(j,2)}(\mathbf{r},\omega)$$

Here are the components $\mathbf{U}^{(j,n)}(\mathbf{r},\omega)$, $n = 1,2$ of this representation are solutions of the Lame equations for a homogeneous half-space with satisfaction of the boundary conditions in stresses

$$\mathbf{t}^{(j,1)}(R,0) = \mu_j\, \mathbf{X}^{(j,1)}(R) \text{ for the first half-space } z > 0,$$

$$\mathbf{t}^{(j,2)}(R,h_j) = \mu_j\, \mathbf{X}^{(j,2)}(R) \text{ for second } z < h_j\,.$$

The vectors of displacements $\mathbf{U}^{(j,n)}(\mathbf{r},\omega)$, $n = 1,2$ and stresses $\mathbf{t}^{(j,n)}(\mathbf{r})$ can be obtained from their submission in the transform of Fourier-Bessel

$$\mathbf{U}^{(j,n)}(\mathbf{r},\omega) = \int_{\Gamma_+} du u J_k(uR)\, \mathbf{P}^{(j,n)}(u,z)\cdot \bar{\mathbf{X}}^{(j,n)}(u)$$
$$\mathbf{t}^{(j,n)}(\mathbf{r}) = \int_{\Gamma_+} du u J_k(uR)\, \mathbf{B}^{(j,n)}(u,z)\cdot \bar{\mathbf{X}}^{(j,n)}(u) \tag{5}$$

For matrix $\mathbf{P},\mathbf{Q}$ the relations are

$$\mathbf{P}^{(j,2)}(u,z) = \mathbf{P}(u,z-h_j)\,,\ \mathbf{B}^{(j,2)}(u,z) = \mathbf{B}(u,z-h_j)$$

$$P_{lm}^{(j,1)}(u,z) = P_{lm}^{(1)}(u,-z)(-1)^{\delta_{lm}}\,,\ Q_{lm}^{(j,1)}(u,z) = Q_{lm}^{(1)}(u,-z)(-1)^{\delta_{lm}+1}$$

$l,m = 1,2$, $\delta_{lm}$ -Kronecker symbol,

with the replacement of the half-space parameters by the parameters of the corresponding $j$-th layer.

Functions $\bar{\mathbf{X}}^{(j,n)}(u)$ in representation (5) are unknown. Their definition is carried out from the boundary conditions of joining the layers, as a rule, requiring the continuity of displacement and stress vectors when passing through interfaces, and the conditions on the surface of the structure. As a result, in the space of Hankel transformations we obtain a system of linear algebraic equations. The matrix of the obtained system has a block diagonal form that is well conditioned. Its determinant $\Delta(u)$ for $u \to +\infty$, $\Delta(u) = \det \mathbf{A}(u) \sim C(E_j, \nu_j) = const$, which ensures the numerical stability of the inversion of the system of equations for arbitrary $u$. The monotonicity of matrix representations $\mathbf{P}$ and $\mathbf{B}$ also makes it possible to effectively apply the proposed algorithm for constructing wave fields using parallel computations.

## 3 Solving method

As an example, the implementation of the model we consider the problem of determining the elasticity moduli of the layers of the road structure, consisting of a layer of asphalt, layers of foundation and soil subgrade. The remaining design parameters are assumed to be specified and correspond to the design values.

To assess the moduli of elasticity of the structural elements of the pavement required the comparison of characteristics in time or frequency region obtained from the model and similar experimental characteristics. This task refers to the number of inverse coefficient problems. The realization of the numerical solution of the direct problem in the form (1)-(2) significantly restricts the possibility of using basic methods of solving inverse problems based on analytical approaches.

The result of calculating frequency characteristics of the surface of the multilayer structure (with a short punch small binding energy) shows that the change in the variation of the elastic moduli of the system under study in their real change is relatively small for structures with decreasing elastic modulus with depth and significantly only in the presence of propagating surface waves. In our case, the most sensitive characteristic is the change of the maximum amplitude of vertical displacement at removal of surface design from the point of impact – "the Cup of the maximum dynamic deflections", obtained numerically. This feature can also be obtained on the basis of field measurements, the installation on the surface of the structure group of sensors at different distances from the point of impact.

In the general case the solution of the resulting inverse problem is reduced to a functional equation in the implicit form to determine the elastic moduli of the pavement layers and subgrade

$$W_z(R, E_j) = \varphi(R), \tag{6}$$

where $W_z(R, E_j) = \max_t u_z((R,z),t)_{z=H}$ - the estimated bowl of the maximum dynamic deflections (obtained numerically for given values of the parameters) on the surface $z = H$; $\varphi(R)$ - given the shape of the bowl (in the General case is obtained by processing experimental data).

It should be noted that to solve the system of equations (6) at fixed observation points $R = R_k$ the different methods were used: methods of the coordinate and the gradient descent method of conjugate gradients. However, in many cases due to the monotonicity of the influence of parameters investigated $E_j$ on the value of the maximum dynamic deflection (any increase $E_j$ leads to a drop deflection) the solution of system (6) is unstable. As a result, along with the possibility of regularization of the solution of the system in this study used a classical genetic algorithm.

The sequence of actions for solving the inverse problem of identifying properties of pavement structures is the following.

1. Filling of chromosomes the initial population with random values which correspond to values of moduli of elasticity for each layer

$$\mathbf{E}^p = \left\{E_1^p, \dots, E_n^p\right\} \quad , \quad E_j^p \in \mathbf{X}, \tag{7}$$

where $p$ is the number of population, $n$ is the population size, $\mathbf{X}$ is physically valid set of design parameters.

2. Calculation of fitness of a population

The adjustment of the population is determined by analysing the values obtained with the help of the fitness function (fitness). The function of fitness you need to pass the values of maximum vertical deflections at the points of observation that chromosome is determined by sequential calculation of amplitude-frequency characteristics amplitude-time characteristics, the magnitude of the dynamic deflection bowls. After these steps will the calculated values of displacements of layers of pavement structures that are passed as arguments to the function adaptation based on the method of least squares

$$S = \sum_{k=1}^{K} \left(W(R_k, E_j^p) - \varphi(R_k)\right)^2, \tag{8}$$

where $K$ – the number of control sensors;

$\varphi(R_k)$ – maximum time of observation of the vertical deflection of the surface at the $k$ point obtained in the course of the experiment;

$W(R_k, E_j^p)$ - values of deflections of the surface of the structure at the $k$ point obtained in the course of the genetic algorithm.

b. If the adjustment of one chromosome satisfies the accuracy of the solution, the genetic algorithm ends and output are the values of the elastic moduli of each layer of pavement structures.

If the fitness of the population does not satisfy the given accuracy of the solution, in this case, you should move on to creating a new generation.

b. The choice of the parents for the next generation.

Selection of parents is based on the tournament selection method or roulette method. The selected parents are then crossed to produce a new generation of chromosomes.

5. Crossing parents

Crossing parents is one of the methods of recombination.

b) Discrete recombination

During the discrete recombination is the exchange of genes (modules of elasticity) between the chromosomes of the parents.

b) Intermediate recombination

During the interim of recombination is determined by the numeric values of the genes of the descendants, which must contain the gene values of the parents. The descendants are generated by the following rule

$$E_1^2 = E_1^1 + \alpha\left(E_2^1 - E_1^1\right), \tag{9}$$

where multiplier $\alpha$ is a random number in the interval $[-d,1+d]$, $d>0$,the number $d$ is chosen for each task separately, but it was found experimentally that the optimum value of $d=0{,}25$.

c) Linear recombination

In the linear recombination, the descendants are created by the same rule used in the intermediate recombination, but the difference is that the multiplier $\alpha$ is chosen at random, and are pre-defined and fixed during the course of the algorithm

6. In the process of life of the population can occur in mutation a random gene (modulus of elasticity) in the chromosome. The mutation probability is specified as a parameter of genetic algorithm. The formula by which the mutation of the chromosome

$$E_j^p = E_j^p \pm \Delta , \tag{10}$$

where the amount of change of a gene chromosome , $\Delta \in R$ .

After that, there is the repetition of the algorithm starting from step 2.

On the basis of the presented algorithm was implemented to select optimal parameters of the genetic algorithm.

During the field experiment used 5 sensors – accelerometers at the maximum distance from the point of application of the blow, equal to 2,1 m.

During a series of experimental calculations the comparison was subject to two selection methods - tournament method and the roulette method. The minimum error value in the method of roulette is made up of 0.03. The method of tournament selection showed stable, adequate result. Achieved the desired accuracy of 0,00005.

## 4 Results and discussion

During a series of experimental calculations when selecting method parameters crossing comparison was subject to three methods – discrete, intermediate and line recombination. In the method of crossing the "intermediate recombination" is used elitist selection, which is that in the new generation of selected chromosome with the best fitness from the previous generation. Further, this chromosome can participate in the crossing and to give their descendants. In the other two methods of crossing an elite selection led to inadequate results of the calculation, therefore, was not used in the calculations. To select the most appropriate method there were 30 test calculations by 10 calculations in each method. As a method of selection the "method of tournament selection" was used.

Best results were recorded for the method of intermediate recombination as accuracy and execution time of calculations. Discrete recombination was better than the linear recombination time calculations.

On the basis of experimental calculations, the following optimal parameters of the genetic algorithm:

- probability of mutation = 0.02,
- standard deviation of the normal distribution - 10,
- number of chromosomes - 30,
- the order of error of 0.0001,
- step enter random chromosome – 3

As an example of the application of the genetic algorithm for calculating the dynamic moduli of the highway layers elasticity, the results of experimental determination using the FWD dynamic deflection unit in the Rostov-on-Don-Taganrog section are considered. The measurements were carried out sequentially at 10 points of dynamic action. As a model, a three-layered half-space model with the specified design properties of the layers was selected (Table 1).

**Table 1** Design value of the modulus of elasticity

| Layer Number | Layer Material | Design value of the modulus of elasticity, MPa |
|---|---|---|
| 1 | Light loam | 80 |
| 2 | Crushed gravel | 120 |
| 3 | Asphalt-concrete porous coarse-grained | 11030 |

Fig. 1-2 shows the results of restoration of the modulus of elasticity of the asphalt-concrete layer and the ground foundation using the software "RoSy DESIGN NG" and based on the genetic algorithm (GA). In both cases, fairly close results were obtained with the exception of point 3, which is special for both methods and removed from the figures. It can be seen that the genetic algorithm is more stable when restoring the elastic moduli of both the surface layer and the substrate. In this case, the calculation results using the software "RoSy DESIGN NG" for the ground are uniformly overestimated, which is explained by the difference in the mathematical models used in the calculations.

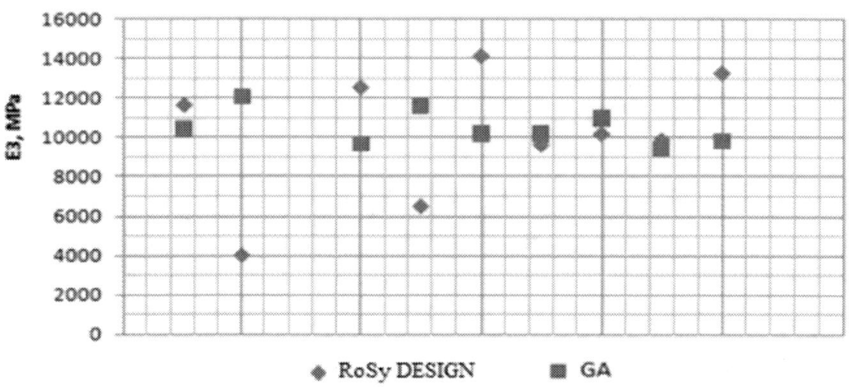


**Fig. 1.** Modulus E3 recovering

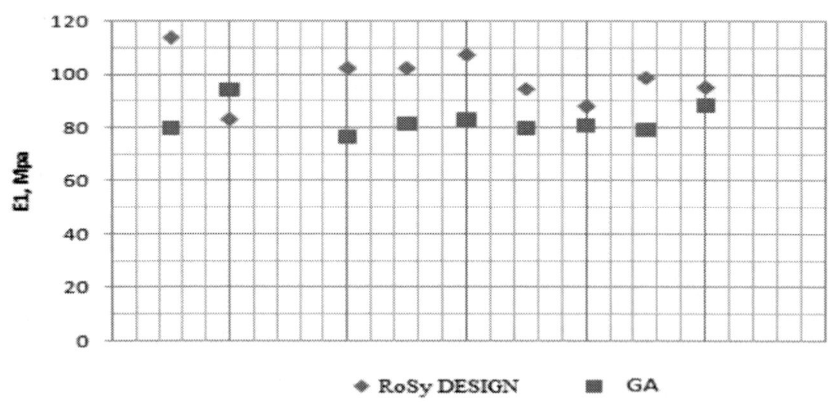


**Fig. 2.** Modulus E1 recovering

## 5 Conclusions

The developed approach with the use of the proposed mathematical model of the analysis of the dynamic behaviour of layered structures and the genetic algorithm for determining the properties of the medium on the basis of experimental data has shown good stability of restoring the design parameters.

Genetic algorithms can be used as refining (correcting) methods for solving inverse problems due to significant time costs for the calculation. We should note that the recovery time of the elasticity modules of a three-layer structure at a single measurement point using a genetic algorithm on a dual-core processor with a frequency of 2 GHz is from 2 to 5 minutes, which does not allow it to be used when operating FWD installations in real time.

A significant advantage of these algorithms is the increase in the number of parameters restored from the experiment, including not only the moduli of elasticity of the medium, but also the thickness of the layers.

## References

1. D.M. Belenkyi,A.N. Beskopylnyi, N.L. Vernesi, News of higher educational institutions. Construction**1**, 99-102 (2003)
2. A.N. Beskopylny, I.G. Kadomtsev, N.I. Beskopylnaya, Theoretical foundations of civil engineering,**83-84** (2005)
3. A.S. Chepurnenko, V.I. Andreev, B.M. Yazyev, A.N. Beskopylny, MATEC Web of Conferences**67**,06059 (2016)
4. N.Yu. Tsybin, R.A.Turusov, V.I. Andreev, Procedia Engineering, **153,** 51-58, (2016)
5. V.I. Andreev, E.V. Barmenkova, Applied Mechanics and Materials, **204-208**, 3596-3599 (2012)
6. A.N. Beskopylnyi, A.A. Lyapin, Recent trend in Science and Technology management **23-29**, 36-45 (2016)
7. A. Constantinescu, A.M. Korsunsky, O. Pison, A. Oueslati, International Journal of Solids and Structures **50**, Issue 18, 2798–2807 (2013)
8. A.H. Mahmoudi, S.H. Nourbakhsh,Procedia Engineering **10**, 3062-3067 (2011)
9. J.A. Greenwood, J.R.Barber, International Journal of Solids and Structures **49**, 2962–2977 (2012)
10. B.A. Zhenning, Jianwen Liang, Vincent W. Lee, Hang Ji, International Journal of Solids and Structures **100-101**, 427-445 (2016)
11. A.S. Vasiliev, S.S. Volkov, A.A. Belov, S.Yu. Litvinchuk, S.M. Aizikovich, International Journal of Engineering Science **112**, 63-75 (2017)
12. A.R.H. Midawi, C.H.M. Simha, M.A. Gesing, A.P. Gerlich, International Journal of Solids and Structures **104–105**, 81-91(2017)
13. J. Dean, T.W. Clyne, Mechanics of Materials **105**, 112-122 (2017)
14. A.S. Useinov, K.S. Kravchuk, A.A. Rusakov, I.V. Krasnogorov, A.P. Kuznetsov, T.V. Kazieva, Physics Procedia**72**, 194-198 (2015)
15. I. Mohagheghian, W.J. Stronge, G.J. McShane, European Journal of Mechanics**61**, 134-150 (2017)
16. I.I. Argatov, International Journal of Solids and Structures **48**, 3444-3452 (2011)

# The application of ANSYS software package in limit load analysis of structures made from anisotropic nonlinear elastic materials.

*Nina Blokhina*[1,*], and *Andrey Galkin*[2]

[1] Moscow state university of civil engineering, Yaroslavskoye shosse, 26, Moscow, Russia, 129337
[2] Acronis International GmbH, 127566, Moscow, Russia, Altufievskoye Shosse 44.

**Abstract.** The article deals with the analysis of structures made from anisotropic nonlinear elastic materials. A variation of the deformation theory of plasticity by G.A. Geniev was chosen as theoretical basis for the analysis. The article introduces the method of finding the limit load of a structure made of an anisotropic nonlinear elastic material in the state of plane stress. The analysis was carried out with the help of ANSYS software package. A special subroutine UserMat developed by the authors enables the user to specify a model in accordance with the basic theory of material behavior.

## 1. Introduction

In last two decades rapid software development has made it possible to generate computer models of quite complex processes. In structural engineering the application field of theoretical research on structural performance of materials carried out by Russian scientists towards the end of the 20th century is being expanded by new technologies.

Most modern software packages use numerical methods for structural modeling and analysis, and one of the most popular of them is the finite-element method that enables the user to create models of any complexity level. The computing power of an average PC can now suffice for the analysis of structures consisting of thousands of elements. Nevertheless, structural analysis has mostly been carried out with the use of simplified physical models. As a result, anisotropy and nonlinear properties of materials were rarely taken into account, and there was a strong need for integrate models adjusting to a larger variety of material properties.

Nowadays structural analysis is mostly performed using predetermined load analysis. Nevertheless, ultimate load analysis backed up by a certain strength theory is also an important tool in construction process. The problem is getting more complicated if physical nonlinearity and material anisotropy also have to be taken into consideration. In this way the problem can only be solved with the help of a powerful software solution with a modern

* Corresponding author: nsb_sapr@mail.ru

user interface [1]. This article gives an example of a limit load analysis of an orthotropic flat plate performed with the help of ANSYS software package.

## 2. Basic physical formulas used in the research

The analysis is based on a variation of deformation theory of plasticity by G.A. Geniev for an orthotropic nonlinear elastic material with the same compressive strength limit parallel and across the grain for the general and special case of plane stress. [1]. According to Geniev's theory, the relations between stress and deformation are as follows:

$$\sigma_x = \frac{E_0 \Psi}{1-\upsilon^2}\left(\varepsilon_x + \upsilon \varepsilon_y\right); \tag{1}$$

$$\tau_{xy} = G_0 \Psi \gamma_{xy},$$

$$\text{where } \Psi = 1 - \frac{\Gamma}{2\Gamma_s}$$

In this expression Γs is the ultimate shift deformation intensity defined by the following formula:

$$\Gamma_s = \Gamma_c \theta \sqrt{1-\sin\omega\cos\omega}, \tag{2}$$

where Γc is the ultimate shift deformation intensity in uniaxial compression;

θ – the value of the stated position vector of the material resistance curve in biaxial state of stress. It can be calculated through the yield criterion of an orthotropic nonlinear elastic material shown as follows:

$$\frac{1}{\theta} = \sqrt{\frac{1}{\mu^2} + \frac{1}{\eta}} - \frac{1}{\mu},$$

$$\text{where } \frac{1}{\mu} = \frac{1}{2}[(\frac{1}{k_{12}} - 1)\sin\omega + \frac{1}{k_{21}}\cos\omega];$$

$$\frac{1}{\eta} = [\frac{\sin^2\omega}{k_{12}} - \sin\omega\cos\omega + \frac{\cos^2\omega}{k_{21}}];$$

$$\tan\omega = \frac{\sigma_2}{\sigma_1};$$

$$\frac{1}{k_{12}} = \frac{\sin^2\beta}{\rho_1} + \frac{\cos^2\beta}{\rho_2};$$

$$\frac{1}{k_{21}} = \frac{\sin^2\beta}{\rho_2} + \frac{\cos^2\beta}{\rho_1};$$

$$\frac{1}{\rho_1}=\frac{R_c}{R_{1s}};\ \frac{1}{\rho_2}=\frac{R_c}{R_{2s}};$$

$\sigma_1$ и $\sigma_2$ - main stress vectors , $\beta$ - angle between stress vector $\sigma_1$ and grain direction; $R_{1t}$ - ultimate strength of the material in uniaxial tension parallel to the grain; $R_{2t}$ - in uniaxial tension across the grain; $R_c$ - in compression. Methods of construction analysis based on aforementioned relations are shown in articles [2,3].

According to relation (1), Ψ equals 0,5 when the structure reaches its limit state. The strength criterion (1) was developed for an orthotropic nonlinear-elastic material of uniform ultimate compressive strength, such as laminated plastic or asbestos cement. The articles describe the analysis of an asbestos cement flat plate under plane stress. The plate of dimensions 2000 x 2000 mm is fixed at the bottom edge and has a square orifice 400 x 400 mm. The central part of the plate, ¼ of its surface, is subject to a uniformly distributed vertical load. Physical constants required in the analysis are as follows: $R_{1t}$ = 15 MPa, $R_{2t}$ =11 MPa, $R_c$ =46 MPa . The Poisson coefficient equals 0,2; initial elastic modulus $E_0$ = 20000 MPa. In order to find the limit load in accordance with the strength criterion of an orthotropic nonlinear-elastic material we need to introduce a relation of minimal Ψ of the whole structure to the load applied to the structure. The relation can be displayed in a form of function

$$\Psi_{min}=f(P) \tag{3}$$

where P coefficient is used for varying the load. The limit load problem is solved by finding the value of P (with precision rate ε) with which Ψmin is maximally close to the limit value (0,5) for the element given. Numerical solutions have proved that function (3) is monotonically decreasing.

## 3. Computational analysis methods

The article [4] gives an example of applying the aforementioned theory to structural analysis with the help of ANSYS. The analysis is carried out via UserMat, a user subroutine which enables the user to specify a certain material. Thus an implementation model of Geniev's theory for a specific material is created. ANSYS also provides built-in interface for calculating 'userfc', user-defined failure criterion. Nevertheless it has a number of disadvantages such as the absence of parallel computing applications or doubling the number of computations required. So the authors have decided to develop a software solution for automating limit load analysis which has no disadvantages of this kind. First of all UserMat subroutine was modified so that the obtained Ψ can be exported into a file. The easiest way to support parallel computation was to export Ψ into numerous files at each iteration in order to avoid different process flows competing for files.

The software solution developed by the authors interacts with ANSYS in the following way:

- Adjusts the macro for ANSYS by changing P coefficient with the help of SetNewCoef subroutine;

- Initiates the ANSYS analysis using ExecuteAnalysis subroutine;
- Reads the generated files that contain Ψ and finds its minimal value Ψmin using ReadMinPsi.

The basic element of the solution is TryFindFsi, a recursive subroutine that initiates computation cycle for calculating Ψ value. If Ψ turns out to be maximally close to 0,5, the subroutine shuts down. If not, it reduces increment and calls itself again from the current point. An important part of the analysis is the comparison of linear and nonlinear analysis results. Fig.1 shows contour plots of vertical stresses for a linear elastic solution. Fig.2 shows the results of nonlinear analysis. In case of nonlinear solution peak stresses in their concentration areas are being smoothed out. Further comparison also demonstrates that peak stress tend to smooth out more as the load approaches the limit value.

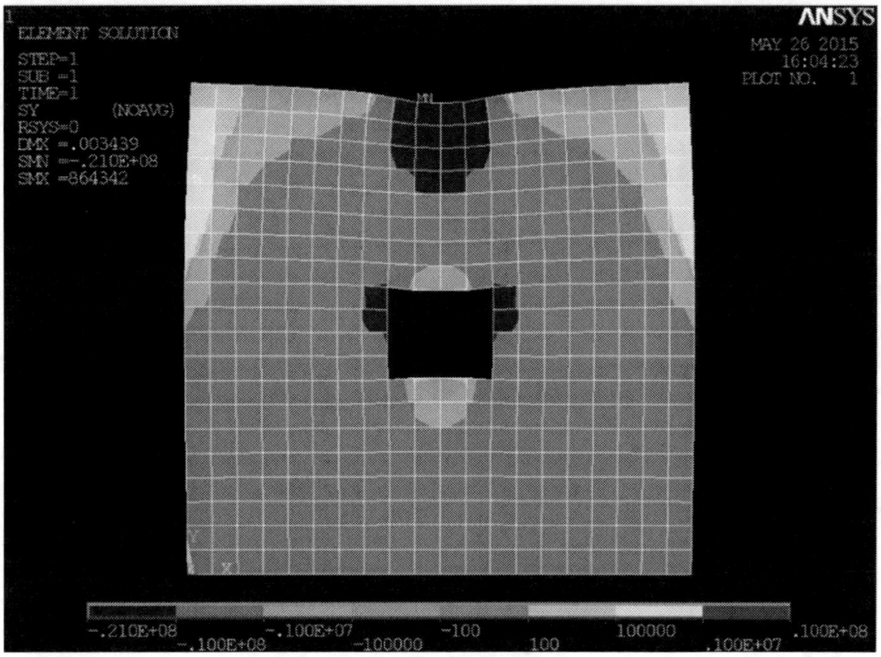


**Fig.1** Contour plots of vertical stresses in linear elasic solution

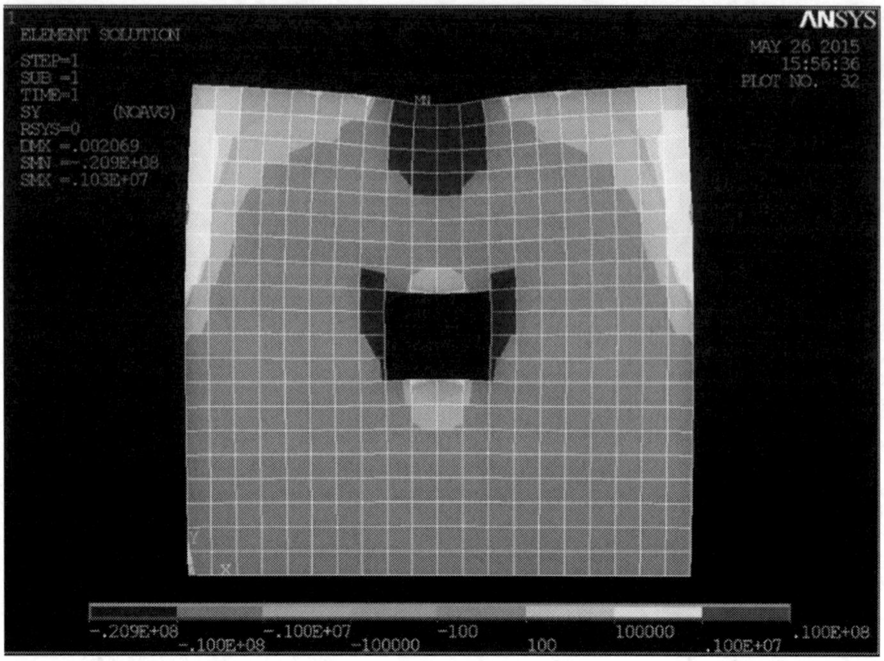


**Fig.2** Contour plots of vertical stresses in nonlinear solution

## References

1. G.A.Geniev, A.S. Kurbatov, F.A. Samedov. Voprosy prochnosti I plastichnosti anizotropnykh materialov [Anisotropic Materials Strength and Plasticity]. Moscow. Interbook Publ., (1993)

2. N.S. Blokhina, Applied Mechanics and Materials, **405-408**, 2686 (2013)

3. N.S Blokhina, ASV Publ. **6**, 384 (2011)

4. N.S. Blokhina, A. G. Galkin, S.V. Scherbina, Structural Mechanics and Analysis of Constructions, **2**, 3 (2015)

# Stress Distribution in Layered Elastic Creeping Array with a Vertical Cylindrical Shaft

*Tatiana* Bobyleva[1,*] , *Alexei* Shamaev[2]

[1]Moscow state university of civil engineering, Yaroslavskoye shosse, 26, Moscow, Russia, 129337
[2]Institute for Problems in Mechanics of the Russian Academy of Sciences, 101-1, Pr. Vernadskogo, Moscow, 119526, Russian Federation

**Abstract.** Construction should be taking into account the influence of time factor on the stability of the structures. In the paper hereditary creep and homogenization theories are used to determine stresses in the layered elastic creeping array with a vertical shaft. Volterra correspondence principle was applied. As a result, the reduction of a time-dependent elastic creeping problem to a corresponding elastic problem became possible. The method proposes a way to determine average (effective) elastic creeping properties and homogenized stress field from known properties of the layers' components. Creep kernels are of a convolution type and are taken in the exponential form. The problem of heterogeneous elastic creeping environment is reduced to a problem of homogeneous transversely isotropic medium. Different boundary conditions on the cylindrical shaft's surface were considered. An analytical solution was obtained. These explicit expressions can be useful for the necessary calculations in the construction practice.

## 1 Introduction

Building constructions on the grounds consisting of rocks or inside rocks is an actual problem. Most of the rocks have a heterogeneous structure. Partial differential equations which describing the behavior of such arrays contain rapidly changing coefficients. The gomogenization method is used to solve such problems. This method is developed in the works [1-3]. Examples of its application are given in [4-6]. The problem of a heterogeneous elastic medium is replaced by a problem of a homogeneous medium for which equations have constant coefficients. These coefficients are called effective. Problems for heterogeneous materials with continuously changing characteristics have been solved in [7-9].

In addition, almost all rocks have the property of creeping, which is the ability to deform with time at constant mechanical stresses [10-11]. Consideration of creeping is of great importance for the stability assessment of the rock massif during construction and exploitation of surface and underground structures. Examples of solving these problems are in the following works [12-16]. A characteristic feature of rheological processes is the dependence of deformation observed at given moment on the character of the whole

* Corresponding author: tatyana2211@outlook.com

previous load process. This property of the materials is called hereditary. Hereditary theory is widely used in rock mechanics because it agrees satisfactorily with experience [17-18].

One of the most powerful approaches in solving linear elastic creeping problems consists of the reduction to a corresponding pure elastic problem. Volterra correspondence principle can be applied to the solution of such elastic creeping problems. In the paper at first the Laplace transform was applied to all governing equations describing an elastic creeping array in equilibrium. As a result, all coefficients depend on the transform parameter. Because the array has layered structure, the theory of averaging is applied to the obtained equations. Finally, the equilibrium problem for a transversely isotropic homogeneous elastic array is obtained. The solution of this problem for an array with a vertical cylindrical cavity already exists [19]. We substitute previously obtained averaged (effective) coeficients in this solution and apply an inverse Laplace transform to obtain final solution.

## 2 Problem specification and decision

Let a layered array has pairs of alternating homogeneous layers of two elastic-creeping isotropic materials. These layers are parallel to the horizontal boundary plane. In addition, a vertical shaft extends from this boundary plane. The array and the shaft are assumed semi-infinite. The shaft is bounded by a cylindrical surface of circular cross section of radius R. The problem is to determine stresses in this array in case when the only body force is its weight.

The original constitutive relations are [13]:

$$-\frac{\partial}{\partial x_k}\left(a_{ijkh}\delta(t)+d_{ijkh}\right)*\frac{\partial u_j}{\partial x_h}=f_i(x,t),\ (i,j,k,h=1\div 3). \tag{1}$$

In (1) the symbol * denotes the convolution operator:

$$\left(d_{ijkh}(x,t)*\frac{\partial u_j}{\partial x_k}\right)=\int_0^t d_{ijkh}(x,t-\tau)\frac{\partial u_j(x,\tau)}{\partial x_k}d\tau\,. \tag{2}$$

$u_j\ (j=1,2,3)$ are components of the displacement, $\delta(t)$ is Dirac-delta function, operation $(a_{ijkh}\delta(t))*$ is equivalent to multiplication by $a_{ijkh}$. (Einstein convention for repeated indices is used.) We are considering isotropic materials therefore the tensors in (1) can be expressed in the form [1 - 3, 13]:

$$a_{ijkh}=\lambda\delta_{ij}\delta_{kh}+\mu(\delta_{ik}\delta_{jh}+\delta_{ih}\delta_{jk}), \tag{3}$$

$$d_{ijkh}=-(D_v(t)-\frac{1}{3}D_s(t))\delta_{ij}\delta_{kh}-\frac{1}{2}D_s(t)(\delta_{ik}\delta_{jh}+\delta_{ih}\delta_{jk}). \tag{4}$$

We denote here by $D_s$ and $D_v$ the regular part of the shear and the bulk relaxation respectively, by $\delta_{ij}$ Kronecker symbol. Let us assume that the amplitude of the bulk creeping kernel is proportional to the shear amplitude with a coefficient of proportionality $k_i$ for each layer that is: $(D_v)_i=k_i(D_s)_i$, $k_i=\text{const}$, $k>0$, $(i=1,2)$. Further, $D_s$ is denoted by $D$.

A cylindrical coordinate system $(r,\varphi,z)$ with the axis $oz$ directed vertically downward is chosen. Its origin is at the center of the upper circular cross section of the shaft. The thickness ratio of one layer to the cell period $\varepsilon$ is denoted by $h$. Consequently, the ratio of the thickness of the other layer to the cell period will be $1-h$. $(0<h<1)$. As usual, we denote by $\sigma_r,\sigma_\theta,\sigma_z$ stress components, by $e_r,e_\theta,e_z$ strain components, by $u_r,u_\theta,u_z$ elastic displacements, by $\lambda_i,\mu_i$ Lame parameters, by $\gamma_i$ specific weight, by $D_i$ creeping kernel for each layer $(i=1,2)$.

Suppose a surface of the shaft is load-free, namely $\sigma_r=0,\sigma_{rz}=0$ at $r=R$. Conditions on a horizontal boundary surface of the rock are: $\sigma_z=0,\sigma_{rz}=0$ at $z=0$.

The conditions of ideal contact are taken on the horizontal boundaries of the layers which are: two components of the displacement $u_r,u_z$ and the normal component of stress parallel the $z$-axis $\sigma_z$ are continuous on all contact planes of the layers, i.e. $[u_r]=0,[u_z]=0,[\sigma_z]=0$.

In this problem, all elastic modulus, specific weight and creeping kernels are periodic functions of the coordinate $\xi=\frac{z}{\varepsilon}$ ($\varepsilon$ is the cell period) and are piecewise constant functions of this variable, i.e., elastic modulus, specific weight and creep kernels have the form [16]:

$$\lambda_i=\begin{cases}\lambda_1,\xi\in[0;h],\\ \lambda_2,\xi\in[1-h;1],\end{cases}\quad \mu_i=\begin{cases}\mu_1,\xi\in[0;h],\\ \mu_2,\xi\in[1-h;1],\end{cases}\quad \gamma_i=\begin{cases}\gamma_1,\xi\in[0;h],\\ \gamma_2,\xi\in[1-h;1],\end{cases}$$

$$D_i=\begin{cases}D_1,\xi\in[0;h],\\ D_2,\xi\in[1-h;1],\end{cases}\quad 0<h<1,\ i=1,2. \tag{5}$$

Let us choose creep kernel of exponential type for each layer: $D_i=d_ie^{-\alpha_it}$, where $\alpha_i,d_i$ are constants, $\alpha_i>0$, $d_i>0$, $(i=1,2)$, $t$ is the variable that specifies time. For a long time period such kernel well describes the behaviour of rock massif as shown by the experiments.

We apply the Laplace transform in the time domain to the equations (1)

$$\tilde{f}(p)=\int_0^\infty f(t)e^{-pt}dt\ . \tag{6}$$

The result is the system of elasticity theory with a complex parameter p. We apply homogenization method described in [1-3] to this system. As a result, we have got a transversely isotropic homogeneous medium. The symmetric matrix of its effective coefficients is [16]:

$$\begin{pmatrix} <\tilde{c}_{11}> & <\tilde{c}_{12}> & <\tilde{c}_{13}> & 0 & 0 & 0 \\ & <\tilde{c}_{22}> & <\tilde{c}_{23}> & 0 & 0 & 0 \\ & & <\tilde{c}_{33}> & 0 & 0 & 0 \\ & & & <\tilde{c}_{44}> & 0 & 0 \\ & & & & <\tilde{c}_{55}> & 0 \\ & & & & & <\tilde{c}_{66}> \end{pmatrix} \quad (7)$$

$$<\tilde{c}_{11}>=<\tilde{c}_{22}>=L-\frac{G_1h}{p+\alpha_1}-\frac{G_2(1-h)}{p+\alpha_2}-$$

$$-\frac{h(1-h)\{[\lambda_2(p+\alpha_2)-j_2](p+\alpha_1)-[\lambda_1(p+\alpha_1)-j_1](p+\alpha_2)\}^2}{\{[L_2(p+\alpha_2)-G_2]h(p+\alpha_1)+[L_1(p+\alpha_1)-G_1](1-h)(p+\alpha_2)\}(p+\alpha_1)(p+\alpha_2)},$$

$$<\tilde{c}_{12}>=l-\frac{j_1h}{p+\alpha_1}-\frac{j_2(1-h)}{p+\alpha_2}-$$

$$-\frac{h(1-h)\{[\lambda_2(p+\alpha_2)-j_2](p+\alpha_1)-[\lambda_1(p+\alpha_1)-j_1](p+\alpha_2)\}^2}{\{[L_2(p+\alpha_2)-G_2]h(p+\alpha_1)+[L_1(p+\alpha_1)-G_1](1-h)(p+\alpha_2)\}(p+\alpha_1)(p+\alpha_2)},$$

$$<\tilde{c}_{13}>=<\tilde{c}_{23}>=$$

$$=\frac{[\lambda_1(p+\alpha_1)-j_1]h[L_2(p+\alpha_2)-G_2]+[\lambda_2(p+\alpha_2)-j_2](1-h)[L_1(p+\alpha_1)-G_1]}{[L_2(p+\alpha_2)-G_2]h(p+\alpha_1)+[L_1(p+\alpha_1)-G_1](1-h)(p+\alpha_2)},$$

$$<\tilde{c}_{33}>=\frac{[L_1(p+\alpha_1)-G_1][L_2(p+\alpha_2)-G_2]}{[L_2(p+\alpha_2)-G_2]h(p+\alpha_1)+[L_1(p+\alpha_1)-G_1](1-h)(p+\alpha_2)}, \quad (8)$$

$$<\tilde{c}_{44}>=<\tilde{c}_{55}>=\frac{[\mu_1(p+\alpha_1)-g_1][\mu_2(p+\alpha_2)-g_2]}{[\mu_2(p+\alpha_2)-g_2]h(p+\alpha_1)+[\mu_1(p+\alpha_1)-g_1](1-h)(p+\alpha_2)},$$

$$<\tilde{c}_{66}>=M-\frac{g_1h}{p+\alpha_1}-\frac{g_2(1-h)}{p+\alpha_2}.$$

In these formulas we have denoted ($i$ is the layer number):

$$L_i=\lambda_i+2\mu_i,\ G_i=d_i\left(k_i+\frac{2}{3}\right),\ j_i=d_i\left(k_i-\frac{1}{3}\right),\ g_i=\frac{d_i}{2},(i=1,2),$$

$$L=L_1h+L_2(1-h),\ l=\lambda_1h+\lambda_2(1-h),\ M=\mu_1h+\mu_2(1-h).$$

Constitutive stress-strain relations of received transversely isotropic medium look as:

$$\tilde{\sigma}_r=<\tilde{c}_{11}>\tilde{e}_r+<\tilde{c}_{12}>\tilde{e}_\theta+<\tilde{c}_{13}>\tilde{e}_z,$$

$$\tilde{\sigma}_\theta=<\tilde{c}_{12}>\tilde{e}_r+<\tilde{c}_{11}>\tilde{e}_\theta+<\tilde{c}_{13}>\tilde{e}_z,\ \tilde{\sigma}_{rz}=<\tilde{c}_{44}>\tilde{e}_{rz}. \quad (9)$$

Stress distribution problem for an array of transversely isotropic elastic material with vertical cylindrical cavity was solved in [19]. Tensor of elastic compliances was used in this solution. We reverse (9), so we define the components of deformation via stress components with help of the effective compliance tensor $<\tilde{s}_{ij}>$:

$$\tilde{\varepsilon}_r = <\tilde{s}_{11}>\tilde{\sigma}_r + <\tilde{s}_{12}>\tilde{\sigma}_\theta + <\tilde{s}_{13}>\tilde{\sigma}_z,$$

$$\tilde{\varepsilon}_\theta = <\tilde{s}_{12}>\tilde{\sigma}_r + <\tilde{s}_{11}>\tilde{\sigma}_\theta + <\tilde{s}_{13}>\tilde{\sigma}_z,\ \tilde{\varepsilon}_{rz} = <\tilde{s}_{44}>\tilde{\sigma}_{rz}. \tag{10}$$

Then we find components of this tensor by using (8).

For our problem, stress-strain state of the elastic creeping array can be found by using the aforementioned solution [19] and the received tensor of effective compliances. This is possible because the average environment is transversely isotropic one. The result is:

$$\tilde{\sigma}_r = \frac{<\tilde{s}_{13}>}{<\tilde{s}_{11}>+<\tilde{s}_{12}>}<\gamma>z\left(1-\frac{R^2}{r^2}\right),\ \tilde{\sigma}_\theta = \frac{<\tilde{s}_{13}>}{<\tilde{s}_{11}>+<\tilde{s}_{12}>}<\gamma>z\left(1+\frac{R^2}{r^2}\right),$$

$$\tilde{\sigma}_z = -<\gamma>z,\ \tilde{\sigma}_{rz} = 0. \tag{11}$$

After substitution of the effective compliances we have:

$$\tilde{\sigma}_r = -\left\{\frac{[\lambda_1(p+\alpha_1)-j_1]h}{L_1(p+\alpha_1)-G_1} + \frac{[\lambda_2(p+\alpha_2)-j_2](1-h)}{L_2(p+\alpha_2)-G_2}\right\}[\gamma_1 h+\gamma_2(1-h)]z\left(1-\frac{R^2}{r^2}\right),$$

$$\tilde{\sigma}_\theta = -\left\{\frac{[\lambda_1(p+\alpha_1)-j_1]h}{L_1(p+\alpha_1)-G_1} + \frac{[\lambda_2(p+\alpha_2)-j_2](1-h)}{L_2(p+\alpha_2)-G_2}\right\}[\gamma_1 h+\gamma_2(1-h)]z\left(1+\frac{R^2}{r^2}\right),$$

$$\tilde{\sigma}_z = -[\gamma_1 h+\gamma_2(1-h)]z,\ \tilde{\sigma}_{rz} = 0. \tag{12}$$

Now we use the inverse Laplace transform. As a result, we have stress state of the elastic creeping array with the vertical cylindrical shaft:

$$\sigma_r = -\left[e^{-\frac{L_1\alpha_1-G_1}{L_1}t}\frac{(G_1\lambda_1-j_1L_1)h}{L_1^2} + e^{-\frac{L_2\alpha_2-G_2}{L_2}t}\frac{(G_2\lambda_2-j_2L_2)(1-h)}{L_2^2} + \right.$$

$$\left. + \frac{\lambda_1L_2h+\lambda_2L_1(1-h)}{L_1L_2}\delta(t)\right][\gamma_1 h+\gamma_2(1-h)]z\left(1-\frac{R^2}{r^2}\right),$$

$$\sigma_\theta = -\left[e^{-\frac{L_1\alpha_1-G_1}{L_1}t}\frac{(G_1\lambda_1-j_1L_1)h}{L_1^2} + e^{-\frac{L_2\alpha_2-G_2}{L_2}t}\frac{(G_2\lambda_2-j_2L_2)(1-h)}{L_2^2} + \right.$$

$$\left. + \frac{\lambda_1L_2h+\lambda_2L_1(1-h)}{L_1L_2}\delta(t)\right][\gamma_1 h+\gamma_2(1-h)]z\left(1+\frac{R^2}{r^2}\right), \tag{13}$$

$$\sigma_z = -[\gamma_1 h+\gamma_2(1-h)]z,\ \sigma_{rz} = 0.$$

Laplace transforms of radial and axial **displacements** $u_r, u_z$ in this case can be found by the formulas:

$$\tilde{u}_r = \frac{(<\tilde{s}_{11}> - <\tilde{s}_{12}>)<\tilde{s}_{13}>}{<\tilde{s}_{11}> + <\tilde{s}_{12}>} <\gamma> z\frac{R^2}{r},$$

$$\tilde{u}_z = \frac{2<\tilde{s}_{13}>^2 - <\tilde{s}_{33}>(<\tilde{s}_{11}> + <\tilde{s}_{12}>)}{2(<\tilde{s}_{11}> + <\tilde{s}_{12}>)} <\gamma> z^2 -$$

$$-\frac{(<\tilde{s}_{11}> - <\tilde{s}_{12}>)<\tilde{s}_{13}>}{<\tilde{s}_{11}> + <\tilde{s}_{12}>} <\gamma> R^2 \ln r + C, \tag{14}$$

where C is arbitrary constant.
After substitution of the effective compliances these formulas take the form of:

$$\tilde{u}_r = \left\{\sum_{i=1}^{2} \frac{[j_i - \lambda_i(p+\alpha_i)]H_i}{[L_i(p+\alpha_i)-G_i]A}\right\}(p+\alpha_1)(p+\alpha_2)[h\gamma_1 + (1-h)\gamma_2]z\frac{R^2}{r},$$

$$\tilde{u}_z = -\left\{\sum_{i=1}^{n} \frac{(p+\alpha_i)H_i}{2[L_i(p+\alpha_i)-G_i]}\right\}[h\gamma_1 + (1-h)\gamma_2]z^2 + \tag{15}$$

$$+\left\{\sum_{i=1}^{2} \frac{[\lambda_i(p+\alpha_i)-j_i]H_i}{[L_i(p+\alpha_i)-G_i]A}\right\}(p+\alpha_1)(p+\alpha_2)[h\gamma_1 + (1-h)\gamma_2]R^2 \ln r + C,$$

We denoted by $A = 2M(p+\alpha_1)(p+\alpha_2) - d_1 h(p+\alpha_2) - d_2(1-h)(p+\alpha_1)$ and by

$$H_i = \begin{cases} h, \text{ if } i = 1 \\ 1-h, \text{ if } i = 2. \end{cases}$$

After applying the inverse Laplace transform we have :

$$u_r = \frac{1}{2M}\sum_{i=1}^{2} H_i \left\{ \frac{e^{-\frac{L_i\alpha_i - G_i}{L_i}t} G_i(j_i L_i - G_i\lambda_i)[G_i + (-1)^i L_i(\alpha_1 - \alpha_2)]}{L_i^2[L_i(m_1+\alpha_i)-G_i][L_i(m_2+\alpha_i)-G_i]} + \right.$$

$$\left. + \sum_{n=1}^{2} \frac{(-1)^n e^{m_n t}(m_n+\alpha_1)(m_n+\alpha_2)[\lambda_i(m_n+\alpha_i)-j_i]}{(m_1-m_2)[L_i(m_n+\alpha_i)-G_i]} - \frac{\lambda_i}{L_i}\delta(t)\right\}[h\gamma_1 + (1-h)\gamma_2]z\frac{R^2}{r},$$

$$u_z = \frac{[h\gamma_1 + (1-h)\gamma_2]}{2}\left\{-\left[\sum_{i=1}^{2} \frac{e^{-\frac{L_i\alpha_i - G_i}{L_i}t} H_i G_i}{L_i^2} + \frac{L_1(1-h) + L_2 h}{L_1 L_2}\delta(t)\right]z^2 + \right. \tag{16}$$

$$+\frac{1}{M}\sum_{i=1}^{2}H_i\left[e^{-\frac{L_i\alpha_i-G_i}{L_i}t}\frac{G_i(G_i\lambda_i-j_iL_i)[G_i+(-1)^iL_i(\alpha_1-\alpha_2)]}{L_i^2[L_i(m_1+\alpha_i)-G_i][L_i(m_2+\alpha_i)-G_i]}+\right.$$

$$\left.+\sum_{n=1}^{2}\frac{(-1)^{n+1}e^{m_nt}(m_n+\alpha_1)(m_n+\alpha_2)[\lambda_i(m_n+\alpha_i)-j_i]}{(m_1-m_2)[L_i(m_n+\alpha_i)-G_i]}+\frac{\lambda_i}{L_i}\delta(t)\right]R^2\ln r\Bigg\}+C.$$

In formulas (16) $m_1, m_2$ are the roots of the following quadratic equation:

$$2M(p+\alpha_1)(p+\alpha_2)-d_1h(p+\alpha_2)-d_2(1-h)(p+\alpha_1)=0,$$

that has a positive discriminant $D=[d_1h+d_2(1-h)+2M(\alpha_2-\alpha_1)]^2+4d_1d_2h(1-h)$, because $d_1>0, d_2>0, 0<h<1$. Therefore, $m_1, m_2$ there will always be real numbers.

Let's consider also the case when the shaft is filled with liquid to the brim. Pressure of this liquid on to the surface of the shaft is equal to $qz$. Stress components in this case are the following:

$$\sigma_r=-\left[e^{-\frac{L_1\alpha_1-G_1}{L_1}t}\frac{(G_1\lambda_1-j_1L_1)h}{L_1^2}+e^{-\frac{L_2\alpha_2-G_2}{L_2}t}\frac{(G_2\lambda_2-j_2L_2)(1-h)}{L_2^2}+\right.$$

$$\left.+\frac{\lambda_1L_1h+\lambda_2L_2(1-h)}{L_1L_2}\delta(t)\right][\gamma_1h+\gamma_2(1-h)]z\left(1-\frac{R^2}{r^2}\right)-\frac{qR^2}{r^2}z, \tag{17}$$

$$\sigma_\theta=-\left[e^{-\frac{L_1\alpha_1-G_1}{L_1}t}\frac{(G_1\lambda_1-j_1L_1)h}{L_1^2}+e^{-\frac{L_2\alpha_2-G_2}{L_2}t}\frac{(G_2\lambda_2-j_2L_2)(1-h)}{L_2^2}+\right.$$

$$\left.+\frac{\lambda_1L_1h+\lambda_2L_2(1-h)}{L_1L_2}\delta(t)\right][\gamma_1h+\gamma_2(1-h)]z\left(1+\frac{R^2}{r^2}\right)+\frac{qR^2}{r^2}z,$$

Now let's assume that the surface of the shaft is rigidly fixed, for example, with a pipe. In this case, boundary conditions on this surface are the following: $u_r=0, \sigma_{rz}=0$ if $r=R$. In this case, shaft has no effect on stress components they will be the same as in the rock without shaft. Array will exert the following pressure on the pipe's wall

$$\sigma_r=-\left[\sum_{i=1}^{2}e^{-\frac{L_i\alpha_i-G_i}{L_i}t}\frac{(G_i\lambda_i-j_iL_i)H_i}{L_i^2}+\frac{\lambda_1L_2h+\lambda_2L_1(1-h)}{L_1L_2}\delta(t)\right][\gamma_1h+\gamma_2(1-h)]z. \tag{18}$$

Stress distribution will be the same as in the array without shaft.

## 3 Conclusion

The article presents a method of calculating stress-strain state of layered elastic creeping array with a vertically shaft. The theory of homogenization and the creeping theory were applied to solve this problem. Prediction of subsidence of the earth's surface is possible on the basis of the obtained solution. A function of creep has a form of a spectral function of two parameters. Method, proposed in this paper, allows to reduce the solution of the elastic creep problem of the heterogeneous array to the problem of pure elasticity for a transversely isotropic medium.

As a result, obtained formulas give an exact analytical solution for determination the stress-strain state of elastic creeping layered array with a vertical shaft. This exact solution can be used in construction.

## References

1. O.A. Oleynik, A.S. Shamaev, G.A. Yosifian, *Mathematical Problems in Elasticity and Homogenization* (Elsevier, North-Holland, 1992)
2. D.I. Bardzokas, A.I. Zobnin, *Mathematical Modelling of Physical Processes in Composite Materials of Periodical Structures* (URSS, Moscow, 2003)
3. E. Sanchez-Palencia, *Non-Homogeneous Media and Vibration Theory* (Springer, Berlin, 1980)
4. A.S. Shamaev, V.V. Shumilova, Doklady Physics **56** (1), 43 (2011)
5. T. Bobyleva, Procedia Engineering **153**, 103 (2016)
6. T.N. Bobyleva, Scientific review **24**, 18 (2016)
7. V.I. Andreev, *Some problems and methods of mechanics inhomogeneous bodies* (ASV Publ. House, Moscow, 2002)
8. V.I. Andreev, D.A. Kapliy, Procedia Engineering **111**, 36 (2015)
9. N.Yu.Tsybina, R.A.Turusova, V.I. Andreev, Procedia Engineering **153**, 51 (2016)
10. Yu.N. Rabotnov, *Creep of Structural Elements* (Nauka, Moscow, 1966)
11. A.A. Ilyushin, B.E. Pobedrya, *Foundations of the mathematical theory of thermovisco-elasticity* (Nauka, Moscow, 1970)
12. A.S. Shamaev, V.V. Shumilova, Proceedings of the Steklov Institute of Mathematics **295**, 218 (2016)
13. A.S. Shamaev, V.V. Shumilova, Proceedings of the Steklov Institute of Mathematics **295**, 229 (2016)
14. Z.G. Ter-Martirosyan, A.Z. Ter-Martirosyan, Soil Mechanics and Foundation Engineering **49** (6), 219 (2013)
15. Z.G. Ter-Martirosyan, M.V. Proshin, Soil Mechanics and Foundation Engineering **43** (2), 56 (2006)
16. T.N. Bobyleva, MATEC Web of Conferences **86**, 01024 (2016)
17. S.A. Konstantinova, V.M. Pestrenin, I.V. Pestrenina, *Proceedings of higher educational institutions.* Mining journal **4,** 92 (2007)
18. M. H. Gabdrahimov, M.N. Dedyukin, *Research on mining. The proceedings of the conference on the results of scientific research* (Perm Polytechnic Institute, Perm, 1968)
19. S.G. Lehnickii, *Theory of the Elasticity of the Anisotropic Body* (Nauka, Moscow, 1977)

# Capture of Air by a Water Stream at Coupling of Sites of a Non-Pressure Collector with a Large Level Difference

*Valery* Borovkov[1,*], *Yulia* Bryanskaya[1], *Yuri* Fetisov[2], and *Aleksandra* Ostyakova[1,3]

[1]Moscow state university of civil engineering, Yaroslavskoye shosse, 26, Moscow, Russia, 129337
[2]Volgograd State Technical University, 28, Prospect imeni V.I. Lenina, Volgograd, 400005, Russia
[3]Water Problem Institute RAS, 3, Gubkina str., Moscow, 119333, Russia

**Abstract.** The article considers the modeling of aeration of water masses by water jets falling on the surface in order to improve the quality of water from surface water sources. The mechanism of the aeration process is described when a turbulent non-crushed jet falls onto the water surface. Calculated analytical and experimental studies of air entrapment by cylindrical jets falling into the water made at the Department of Hydraulics of MSUCE made it possible to obtain a universal dimensionless dependence, suitable for describing the process of air entrapment in various conditions. The results of experimental verification of this dependence are presented, as well as the results of simulation of air entrainment in specific conditions of the collector chamber when its sections are mated to a large level difference. The results of mathematical simulation of the release of air-gas inclusions from the water flow at the lower part of the collector are presented. The volumetric flow rate of the air-gas mixture in the collector below the collector chamber is identified. Estimated calculation of the parameters of the deaeration of the water flow: the length, time and the rate of ascent of air bubbles, is executed.

## 1 The justification of the calculated dependences on the identification of air entrapment at large differences and mathematical modeling of the capture of the air-gas mixture in the collector chamber

In the development of engineering measures for improving of the quality of water from surface water sources, improving of the water quality in urban water bodies, improving of the efficiency of the use of settling ponds for cleaning up of contaminated surface runoff and industrial wastewater, an artificial aeration of water masses is used [1], [2]. Aerators, perturbing the surface of the stream, serve as an artificial turbulence promoter of the water mass, therefore, for the known perturbation parameters initiated by the aerator on the free surface, physical modeling of this process must be considered taking into account the hydrodynamic features. The dimensions and power of the aeration devices are different, but the accuracy of their hydro-aerodynamic calculations based mostly on laboratory experimental data is currently insufficient

* Corresponding author: mgsu-hydraulic@yandex.ru

in connection with the lack of valid criteria for the similarity of the aeration phenomenon in installations of different scales and the lack of a developed physical modeling technique. Physical modeling of the aeration process is characterized by a complex interaction of forces of different nature and has not been sufficiently studied to the present time, although achievements in the field of mathematical modeling of aeration structures are known [3]. The problems of modeling of the aeration of water by jets falling on its surface have been studied little or not at all, in spite of the high efficiency of this aeration method.

Let us consider in more detail the physical essence of the process of aeration of the water mass by turbulent jets falling on its surface. The performed studies [4] allow illustrating jet aeration as follows. When a turbulent jet flows from the nozzle, perturbations develop on its surface due to the interaction of the jet with the surrounding air under the conditions of the stabilizing role of the surface tension. At some distance from the nozzle, perturbations in the jet develop so much that the forces of surface tension cannot resist them. There is a violation of the compactness of the jet and its disintegration into separate water masses, which are further broken into drops, when they interact with the surface of the water mass, it is aeration in the so-called "drop" mode. When the fractured jet falls on the surface of the liquid, due to the impact of the drops, caverns are formed, and the free surface of the water is locally curved. In these zones, Laplace forces of surface tension and hydrostatic pressure arise, which "slam" the cavern, forming an air bubble. When the following drops fall, the bubbles are fractured and dragged under the surface of the water. The actual process is much more complicated than the described scheme, since turbulent pulsations in the jet spread drops over a certain area, but the above description of the aeration of the water mass when a fragmented jet falls on it, is qualitatively correct and may be taken into account when developing the frameworks of its modeling.

Mathematical modeling of hydro-aerodynamic phenomena becomes possible in the presence of analytical dependencies describing the simulated phenomenon. Until recently, it was believed that the process of capturing air by water flows does not obey the laws of similarity and is an unmodified process [5]. This is explained by the simultaneous action of a sophisticated force complex, which includes the force of gravity, the force of viscosity, the buoyancy force, the inertial forces and the surface tension force. For this reason, many attempts to obtain dependencies describing the process of air capture were unsuccessful. The dependencies irregularly reflected the nature and extent of the influence of these factors, their comparison showed very significant quantitative discrepancies.

The physical picture of air entrainment during the aeration of water masses by jets incident on its surface makes it possible for us to identify the determining parameters and establish on their basis the similarity criteria. It seems obvious that the flow rate of the attracted air will depend on the velocity of the jet $V$, the diameter of the nozzle $d_0$, the length of the jet $L$, the angle of its incidence $\alpha$, the physical properties of the liquid (viscosity $\upsilon$, density $\rho$, surface tension $\sigma$), gravity force acceleration $g$ and the nozzle undulation $k_э$. In this case, the functional dependence for the flow rate of the attracted air $Q_a$ may be written in the form of the following functional connection:

$$Q_a = f(V, d_0, L, \upsilon, \rho, \sigma, \alpha, g, k_э) \quad (1)$$

Using the $\pi$-theorem this functional dependence may be represented in the form:

$$f(\pi_1, \pi_2, ..., \pi_7) = 0, \quad (2)$$

where $\pi_i$ are dimensionless complexes.

Using the principles of dimensionality, we represent the dependence for the air flow involved in the water by a falling jet stream with a flow rate $Q_j$ in the form:

$$\frac{Q_a}{Q_j} = f\left(\mathrm{Re}, \mathrm{Fr}, \mathrm{We}, \frac{L}{d_0}, \frac{k_э}{d_0}, \sin\alpha\right), \tag{3}$$

where $\mathrm{Re} = \frac{Vd_0}{\nu}$, $\mathrm{Fr} = \frac{V^2}{gd_0}$, $\mathrm{We} = \frac{\rho V^2 d_0}{\sigma}$.

The results of calculated analytical and experimental studies of air entrapment by cylindrical jets falling into the water performed at the Department of Hydraulics of Moscow State University of Civil Engineering are presented in Fig. 1 [6].

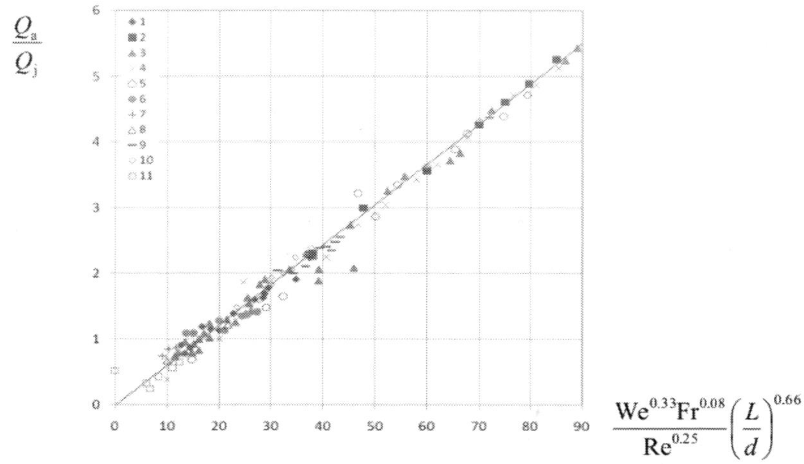


**Fig. 1.** Generalization of the results of experimental studies on the measurement of volumetric absorption of atmospheric air, 1-11 – groups of experiments with different values of numbers Fr, Re, We, *L/d*.

Experimental data made it possible to obtain a universal dimensionless dependence, suitable for describing the process of air entrapment under various conditions, which has the form:

$$\frac{Q_a}{Q_j} = 0.06\frac{\mathrm{We}^{0.33}\cdot\mathrm{Fr}^{0.08}}{\mathrm{Re}^{0.25}}\left(\frac{L}{d}\right)^{0.66}, \tag{4}$$

where $Q_a$ – volumetric air flow at atmospheric pressure; $Q_j$ – secondary water flow; $L$ – drop height; $d$ – the diameter of the jet in the plane of its contact with the mass of the liquid; $\mathrm{Re} = \frac{Vd}{\nu}$ - the Reynolds criterion, which takes into account the influence of fluid viscosity and inertia forces; $V$ – the velocity of the jet in the plane of its contact with the mass of the liquid; $\nu$ - the kinematic viscosity of the liquid; $\mathrm{Fr} = \frac{V^2}{gd}$ - the Froude criterion, taking into account the influence of gravity; $\mathrm{We} = \frac{\rho V^2 d}{\sigma}$ - the Weber criterion, taking into account the influence of surface tension forces, here $\sigma$ - coefficient of surface tension at the water-air interface.

It should be noted, however, that the experimental verification of this relationship is performed under laboratory conditions with limited water flow and with jets diameters up to 50 mm. For these reasons, the quantitative results of mathematical modeling of air entrapment when the water flow falls into the vertical shaft of the collector chamber should be treated with caution and considered only as estimates. In order to use the dependency (4) for simulating the air intake under specific conditions of the collector chamber, a conversion was performed, taking into account the fact that the height of the drop depends on the water flow rate, and the velocity and transverse dimension of the incident jet in the plane of contact with the level in the vertical shaft of the collector depend on its turn from the height of the drop.

Assuming that, at a significant height of the *L*-drop, the jet in the plane of its entry under the water level in the collector shaft moves vertically and, assuming the energy losses when falling insignificant, the velocity of the jet in the plane of the water level in the shaft is equal to:

$$V = \sqrt{2gL} \tag{5}$$

Dependence (5) reflects the apparent increase in velocity with an increase in the fall height of the flow. As the velocity of the incident jet increases, the area of its cross section $\omega_э$ decreases:

$$\omega_э = \frac{Q_j}{\sqrt{2gL}} \tag{6}$$

Assuming in the plane of the jet to the water level in the vertical shaft the circular shape of its cross section, we find the equivalent diameter of the jet in this plane:

$$d_э = \sqrt{\frac{4\omega_э}{\pi}} = \frac{2Q^{0.5}}{\pi^{0.5}(2gL)^{0.25}} \tag{7}$$

Using the relations obtained, we perform the transformations of the above criteria:

$$\mathrm{Re} = \frac{2}{\pi^{0.5}\nu} Q_j^{0.5}(2gL)^{0.25}, \quad \mathrm{Fr} = \pi^{0.5}(2g)^{0.25}\frac{L^{1.25}}{Q_j^{0.5}}, \quad \mathrm{We} = \frac{2\rho}{\sigma\pi^{0.5}} Q_j^{0.5}(2gL)^{0.75} \tag{8}$$

Taking into account the expressions obtained, the complex of dimensionless quantities entering into the dependence (4) is transformed as follows:

$$\frac{\mathrm{We}^{0.33}\mathrm{Fr}^{0.08}}{\mathrm{Re}^{0.25}}\left(\frac{L}{d}\right)^{0.66} = \frac{52.47Q_j^{0.165}L^{0.248}\cdot 1.11L^{0.1}}{39.41Q_j^{0.125}L^{0.0625}Q_j^{0.04}}\frac{L^{0.66}L^{0.165}}{0.66Q_j^{0.33}} = 2.24\frac{L^{1.11}}{Q_j^{0.33}} \tag{9}$$

The performed transformations of formula (1) to the form (9) made it possible for us to establish the nature and degree of dependence of the captured air intake $Q_a$ on the water flow $Q_j$ and the drop height $L$ in the following form:

$$Q_a = 0.134L^{1.11}Q_j^{0.67} \tag{10}$$

Dependence (10) shows that the consumption of the air captured by the flow increases with the flow rate of water and depends to a large extent on the height of the fall. The dependence also shows that in the real conditions of the collector chamber, the reduction in air intake can be achieved by reducing the height of the fall of the flow.

The obtained dependence (10) was further used in mathematical modeling of the air-capture process when the water flow fell into the vertical shaft of the collector chamber. Since the consumption of the air-gas mixture according to the dependence (10) depends on the water flow and the height of the fall of the water flow into the vertical shaft of the collector chamber, mathematical modeling of this process uses the results of mathematical modeling of the hydraulic regimes and the position of the piezometric line in the collector section. Mathematical modeling data show that the ratio of the average air-gas mixture consumption to the water discharge decreases with an increase in water flow from 1.2 (at $Q_j = 0.1\ Q_p$) to 0.46 (at $Q_j = Q_p$). However, the absolute value of the air flow $Q_a$ increases with increasing water flow (Fig. 2). For calculated conditions, the air flow rate is 4.1 $m^3/s$ at atmospheric pressure. The entrained air currents as they immerse in the liquid reduce their volume under the influence of excessive hydrostatic pressure.

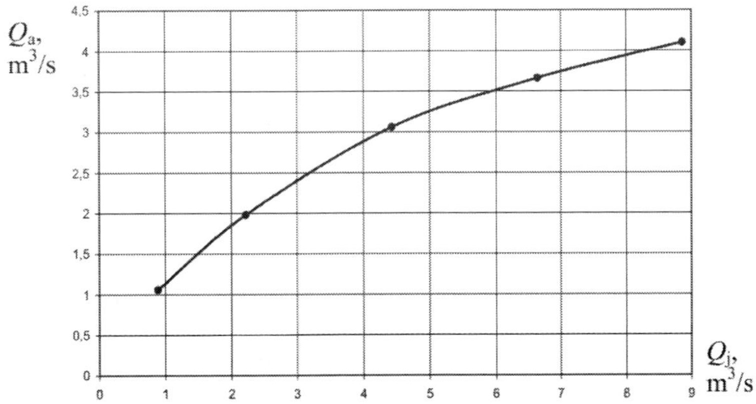


**Fig. 2.** Schedule of air-gas mixture flow rate change at different water flow rates $Q_a = f(Q_j)$.

Due to the partial decomposition of the water flow during its fall at the drop, the amount of air drawn in can vary (pulsate) with respect to the established average value.

## 2 Mathematical modeling of the release of air-gas inclusions from the water flow in the lower part of the collector

To substantiate the engineering measures for intercepting the air-gas mixture in the collector section adjacent directly to the collector chamber, it is necessary to establish a distance from the collector chamber, within which air-gas inclusions completely leave the water stream. To determine this distance, it is necessary to have data on the speed of movement of the water and air flow, the proportion of the cross-sectional area occupied by the water and air flow at various water flows, as well as information on the hydraulic size of the air-gas inclusions that characterizes the rate of their ascent.

According to the results of experimental studies [7], large air-gas inclusions are unstable and under the influence of the turbulence of the water flow they are split into spherical bubbles of 4-5 mm in size, which are stable. According to Wallis [8], the ascent rate of such bubbles in the water flow is close to $w$ = 0.25 m/s (Fig. 3). It should be noted that the asymmetry of the turbulent pulsations in the velocity of the water flow affects the bubble uptake. However, at the present time there are no reliable data characterizing the asymmetry of turbulence, therefore, in the mathematical modeling of the ascent process, the effect of vertical turbulent pulsations was not taken into account.

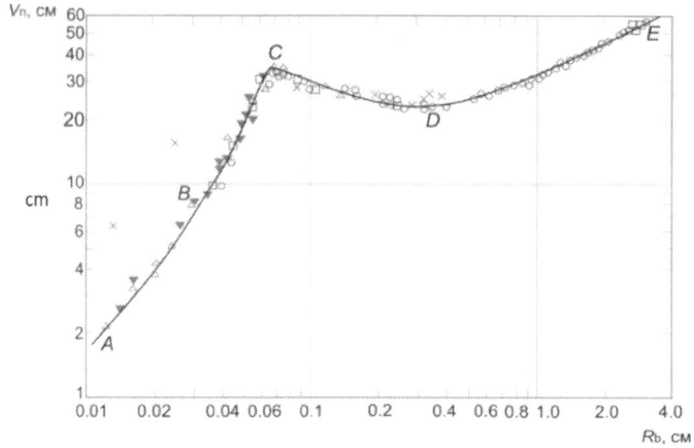


**Fig. 3.** Limit the rate of rise of air bubbles in water depending on their size, $R_b$ - the equivalent radius.

When the flow from the vertical shaft of the chamber to the collector changes, the air inclusions are uniformly distributed over the cross section of the water flow due to intensive turbulence. The greatest distance corresponding to the exit of air bubbles will be determined by the distance of the bubbles from the lower layers of the water flow. Air inclusions that leave the water stream are collected in the underwater space of the collector. The resulting air mass moves as a result of interaction with the water flow along the collector, with a speed close to the speed of the water flow. The motion of air masses in the underroof space can have a different character depending on the hydraulic characteristics of the flow, as illustrated by the Lockhart-Martinelli diagram [9], [10], [11] obtained on the basis of laboratory experiments. Using it for the conditions of a given collector requires additional research and full-scale research. Analysis of the data of the Research Institute of Power Structures on the motion of air masses in hydraulic tunnels suggests that in the conditions of the collector in the underroof scm e, the air masses will move in the form of separate volumes occupying a significant portion of the cross section.

To determine the proportion of the cross-section occupied by the air mass, it must be taken into account that the trapped air is in the collector under excessive hydrostatic pressure determined by the level mark in the collector chamber. Excessive hydrostatic pressure compresses the air-gas mixture, which leads to a decrease in its volumetric flow rate. To correct the volume associated with the action of hydrostatic pressure, the equation of state for the isothermal process has been used:

$$pW = const\,, \tag{11}$$

where $p$ – absolute pressure; $W$ – air volume at pressure $p$.

Taking into account expression (11), we can write in the form:

$$\frac{Q_p}{Q_a} = \frac{p_a}{p}\,, \tag{12}$$

where $p_a$ – atmosphere pressure.

This ratio is used to determine the volumetric flow rate of the air-gas mixture in the collector below the collector chamber. Estimated calculation of the deaeration length of the water flow can be performed taking into account the following relationships. The ascent time of the air bubble, the flow rate, the flow time of the deaeration flow are:

$$t_в = \frac{h}{w},\ V = \frac{q}{h},\ t_д = \frac{L_д h}{q} \qquad (13)$$

where $h$ - depth, $w$ – hydraulic size, $q$ – specific water discharge per unit of collector width, m$^2$/s, $L_д$ – deaeration length, m.

Taking into account (13) we find the condition for complete deaeration of the flow in the form:

$$L_д > \frac{q}{w} \qquad (14)$$

For the design conditions, the deaeration length $L_д$ = 16 m. With decreasing water flow, the extent of the deaeration section of the flow decreases. The results of mathematical modeling of the deaeration of the flow at flowrates less than the calculated ones are shown in Fig. 4. This estimated calculation does not reflect the effect of compression of the air-gas mixture under the influence of excessive hydrostatic pressure in the collector.

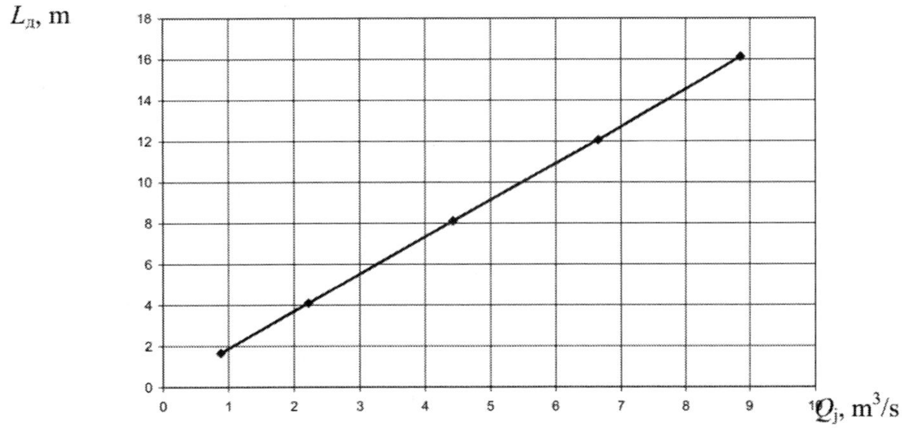


**Fig. 4.** Graph of changes in the length of the de-aeration flow in relation to the flow rate of water.

Calculation data show that the average volumetric flow rate of the air-gas mixture in the collector under consideration with allowance for compression under the influence of excess hydrostatic pressure at the design flow is close to 2 m$^3$/s. Thus, through the cross-section of the collector, the total flow of water and air-gas mixture passes. The filling of the collector with an air flow was determined by calculation, taking into account the geometric and hydraulic characteristics of the collector [12]. It should be noted that the maximum length of the deaeration of the flow, set taking into account the compression of the air-gas mixture, is close to that found above on the basis of estimated calculations of the maximum deaeration length.

The obtained data on the length of the deaeration of the water flow determine the location of the formation of the underroof air current and were taken into account in determining the location of the chamber for intercepting and diverting the air-gas mixture in the region directly adjacent to the collector chamber.

## References

1. Recommendations for calculating pressure losses along the length of the hydroelectric power lines (VNIIG imeni B. Ye. Vedeneyeva, Leningrad, 1981)

2. *Hydraulic calculations of spillway hydraulic structures. Reference Manual* (Energoatomizdat, Moscow, 1988)
3. G. Schlichting, *The theory of the boundary layer* (Science, Moscow, 1969)
4. *Handbook of hydraulic calculations*. Ed. P. G. Kiselev. (Energia, Moscow, 1972)
5. V. M. Lyatkher, A. M. Prudovsky, *Investigations of open flows on pressure models* (Energia, Moscow, 1971)
6. V. S. Borovkov, Yu. M. Fetisov, *Capture air when the jet interacts with a fluid at rest. Questions of hydraulics and water supply* (Proceedings of the MISI imeni V. V. Kuibyshev, 1980)
7. C. Sow, *Hydrodynamics of multiphase systems* (Mir, Moscow, 1971)
8. G. Wallis, *One-dimensional two-phase flows* (Mir, Moscow, 1972)
9. A. D. Altshul, P.G. Kiselev, *Hydraulics and aerodynamics* (Stroyizdat, Moscow, 1975)
10. Yu. V. Bryanskaya, *Refinement of velocity profile parameters for flows in smooth and rough pipes,* International Journal of Applied Engineering Research, Volume 11, 3 (2016)
11. Yu. M. Fetisov, V. S. Borovkov, O. V. Pavlova, *Methodical bases of physical modeling of artificial aeration of water by falling jets*, Ecology of urbanized territories, 3 (2013)
12. A. M. Kurganov, N. F. Fedorov, *Hydraulic calculations of water supply and sanitation systems. Directory*, (Stroyizdat, Leningrad, 1986)

# The flow of air in a moving and stationary permeable layer during drying of bulk building materials

*Valery* Borovkov[1,*], *Yulia* Bryanskaya[1], *Aleksandra* Ostyakova[1,2] , and *Manana* Medzveliya[1]

[1]Moscow state university of civil engineering, Yaroslavskoye shosse, 26, Moscow, Russia, 129337
[2]Water Problem Institute RAS, 3, Gubkina str., Moscow, 119333, Russia

**Abstract.** With the use of the equation of motion in a one-dimensional formulation, the flow in a stationary and moving permeable layer during its blowing is investigated. Resistance to the flow within the layer is assumed to be linear in Darcy. The thickness of the boundary layers was determined and the velocity distribution was obtained for various patterns of motion encountered in problems of drying and extraction.

In a number of technological processes of firing and drying of granulated granular materials, for example, in the production of building materials, gas or air is blown through a permeable layer of material moving in the apparatus on the belt of the conveyor, or sliding by gravity. To calculate the processes of mass transfer and heat exchange between the gas and the material, as well as with the walls of the apparatus, it is necessary to establish the kinematic characteristics of the gas flow blown through the permeable layer, depending on the characteristics of the flow that flows around the upper boundary of the layer (Fig. 1). A similar problem of the flow of a heavy permeable layer moving along an inclined wall under the influence of gravity can be encountered in the technology of washing of the finished product, extracting impurities, and the like.

When investigating flow characteristics in a permeable layer, we consider the motion of a homogeneoufs model gas (or liquid) occupying the entire volume of the permeable layer. A similar model is usually used in the solution of filtration problems [1]. In this paper, we consider cases when the motion of all particles of a heavy permeable layer occurs at the same rate, which does not depend on the velocity of the flowing gas or liquid flow. Therefore, we consider blowing (or flow) modes with moderate velocities that do not lead to erosion of the permeable layer and do not cause weighing and entrainment of its particles.

* Corresponding author: mgsu-hydraulic@yandex.ru

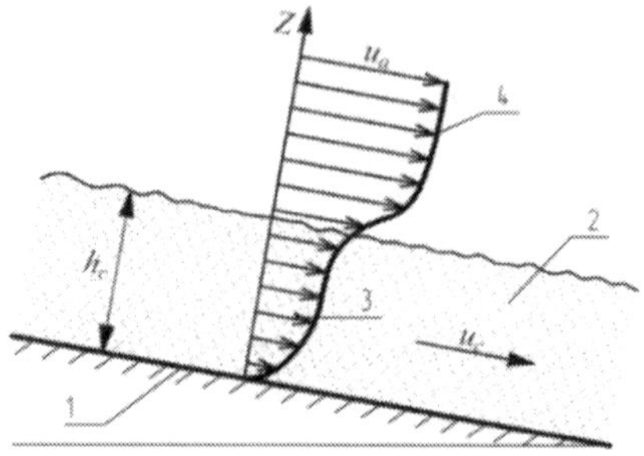


**Fig. 1.**

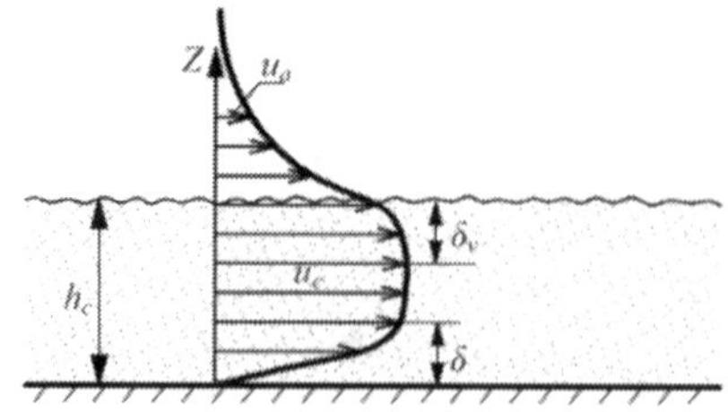


**Fig. 2.**

**Fig. 1.** The calculation scheme for blowing a permeable layer with a gas or a liquid. 1 - permeable bottom, 2 - layer of bulk material, 3 - velocity profile $u_c$, 4 - gas or liquid velocity profile.

**Fig. 2.** The calculation scheme when airflow is absent.

The equation of uniform motion of the model environment (model fluid) within the permeable layer is written as:

$$gi - \frac{dp}{\rho dx} + \nu_e \frac{d^2u}{dz^2} + f_g = 0\,, \tag{1}$$

where $gi$ – component of gravity; $\frac{dp}{dx}$ - longitudinal pressure gradient; $f_g$ – specific force of filtration interaction between a model fluid and a solid permeable environment; $u$ – longitudinal velocity of motion of a model environment in a fixed coordinate system associated with the wall; $\nu_e$ – the effective viscosity of the model fluid.

In the formulation of this problem, great difficulties arise with the formulation of the hypothesis on the effective viscosity; for the Stokes flow regime, one of the problems was considered by Saffman [2]. The controversial nature of such hypotheses usually casts doubt on the acceptability of the approaches developed. Therefore, in this article an attempt is made to obtain solutions in parametric form with a parameter that depends on the effective viscosity, without a detailed deciphering of the factors determining it. The formulated problem, in the general case, is quite complicated, it allows us to consider a number of important special cases for practice.

Under conditions where the component of the weight force and the pressure gradient are absent, the movement of the model fluid is initiated by the intrinsic displacement of the permeable layer (Fig. 2). The equation of motion for this case takes the form:

$$\nu_e = \frac{d^2u}{dz^2} + f_g = 0 \tag{2}$$

Note that the force of interaction between a liquid and a permeable environment in this case is an active driving force. Despite the fact that the velocity of the motion of all the particles in the layer is the same according to the condition of the problem, the rate of the filtration motion of the liquid will change, since tangential stresses play an appreciable role at the wall, and interaction with the layer of the liquid at rest (or gas) at the outer boundary of the layer. The assumed linearity of the filtration flow excited in the permeable layer as it moves and, taking into account that the interfacial interaction force is proportional to the velocity of the gas or liquid relative to the permeable layer ($u_c$ - $u$), force $f_g$ we write as follows:

$$f_g = g\frac{u_c - u}{k_\varphi},\tag{3}$$

where $k_\varphi$ - filtration coefficient.

Since the friction between the layers of the model fluid is related to the presence of the wall and is determined in the chosen coordinate system by the velocity gradient $u$ relative to the wall, the equation of motion takes the form:

$$\nu_e\frac{d^2u}{dz^2} + \frac{g(u_c - u)}{k_\varphi} = 0\tag{4}$$

By introducing the effective permeability of a layer $k_e = k_\varphi \nu_e / g$ it reduces to the equation:

$$\frac{d^2u}{dz^2} - \frac{u}{k_e} = -F,\tag{5}$$

where $F = u_c / k_e$ - constant for the given case under consideration. This differential equation has a particular solution:

$$u = k_e F,\tag{6}$$

and the general solution:

$$u = C_1 e^{z/\sqrt{k_e}} + C_2 e^{-z/\sqrt{k_e}}\tag{7}$$

Equation (5) thus has the form:

$$u = C_1 e^{z/\sqrt{k_e}} + C_2 e^{-z/\sqrt{k_e}} + k_e F\tag{8}$$

To find the unknown coefficients $C_1$ and $C_2$, we can use the boundary conditions on the lower and upper boundary of the liquid layer δ (Fig. 2). From the boundary conditions in the form: at $z=0$, $u=0$; at $z=\delta$, $u=u_c$, we get:

$$C_1 = \frac{u_c - k_e F + k_e F e^{-\delta/\sqrt{k_e}}}{e^{\delta/\sqrt{k_e}} - e^{-\delta/\sqrt{k_e}}},\quad C_2 = \frac{u_c - k_e F + k_e F e^{\delta/\sqrt{k_e}}}{e^{\delta/\sqrt{k_e}} - e^{-\delta/\sqrt{k_e}}}\tag{9}$$

Substituting the coefficients found in the solution of (8), taking into account that $e^{\delta/\sqrt{k_e}} - e^{-\delta/\sqrt{k_e}} = 2Sh\,\delta/\sqrt{k_e}$, since above the liquid layer under consideration ($z > \delta$) derivatives $\frac{du}{dz} = 0$ и $\frac{d^2u}{dz^2} = 0$ we find that $u_c = k_e F$. With this in mind, the solution looks like with the use of hyperbolic functions:

$$\frac{u}{u_c} = 1 - \frac{Sh\left(z/\sqrt{k_e} - \delta/\sqrt{k_e}\right)}{Sh\,\delta/\sqrt{k_e}}\tag{10}$$

The analysis shows that the profile obtained corresponds to a physically obvious condition $z = \delta$, $\frac{du}{dz} = 0$ only at $\delta \to \infty$, which agrees with the concepts of boundary-layer theory [3], [4], [5], [6], [7].

Approximate estimate of $\delta/\sqrt{k_e}$, which is a parameter of the velocity profile, can be obtained by specifying a certain finite value of the velocity gradient at the outer boundary of the layer δ. Considering the achievable accuracy of velocity measurements and taking the value of the velocity gradient at the upper boundary of the layer equal to 5% of the averaged gradient $u_c/\delta$, we find an approximate estimate $\delta/\sqrt{k_e} \approx 5$.

To determine the coefficients $C_1$ and $C_2$, being part of solution (8), we can also use the boundary conditions in the form: at $z = 0$, $u = 0$; at $z = \delta$, $\frac{du}{dz} = 0$. The resulting values of the coefficients have the form:

$$C_1 = -\frac{u_c e^{-\delta/\sqrt{k_e}}}{e^{\delta/\sqrt{k_e}} + e^{-\delta/\sqrt{k_e}}},\quad C_2 = -\frac{u_c e^{\delta/\sqrt{k_e}}}{e^{\delta/\sqrt{k_e}} + e^{-\delta/\sqrt{k_e}}} \tag{11}$$

Values $C_1$ and $C_2$ differ from the previously found expressions (9) and the velocity profile in the layer δ takes the form:

$$\frac{u}{u_c} = 1 - \frac{Ch\left[\delta/\sqrt{k_e}\,(1 - z/\delta)\right]}{Ch\,\delta/\sqrt{k_e}} \tag{12}$$

The resulting profile (12) provides a zero velocity gradient at $z = \delta$, but condition $u = u_c$ is provided only with $\delta \to \infty$.

Approximate estimate δ in this case may be found from condition $u = 0.97u_c$, as is usually done in boundary layer theory [3]. As is usually done in boundary layer theory (12) results in $\delta/\sqrt{k_e} \approx 4$, which is close to the previous estimate $\delta/\sqrt{k_e}$, obtained using the profile (10).

Comparison of the values of the velocity calculated from the profiles (10) and (12) shows that at $\delta/\sqrt{k_e} = 4$ they are sufficiently close to each other, which is connected with the smallness of the summand $e^{-\delta/\sqrt{k_e}}$ in the expression for the hyperbolic sine and cosine. Given this, the velocity profile in the layer δ may be written in the form of the following simple expression:

$$\frac{u}{u_c} = 1 - e^{-z/\sqrt{k_e}}, \tag{13}$$

fair in the above estimate $\delta/\sqrt{k_e}$. Layer thickness δ in this case it may most simply be determined under viscous flow conditions in the wall layer. For other cases of flow, one or another hypothesis is required regarding the effective viscosity, the analysis and verification of which is the subject of a separate study. The approximate estimates obtained above for $\delta/\sqrt{k_e}$ make it possible for us to establish that when the granular layer moves and the mean values of the filtration coefficient [8] for this case, the value δ turns out to be small and equal to several grain diameters. The value δ may be commensurable with the layer thickness $h_c$ only in special cases with very high layer permeability and a small thickness $h_c$.

At a layer thickness $h_c >> \delta$, friction at its upper boundary will be manifested only in the layer $\delta_v$, adjacent to the upper boundary of the permeable layer being transported. The frictional force arising at the upper boundary will either slow the course of the layer (Fig. 3, a) or accelerate it (Fig. 3, b). The latter case is considered by V.N. Spiridonov [9].

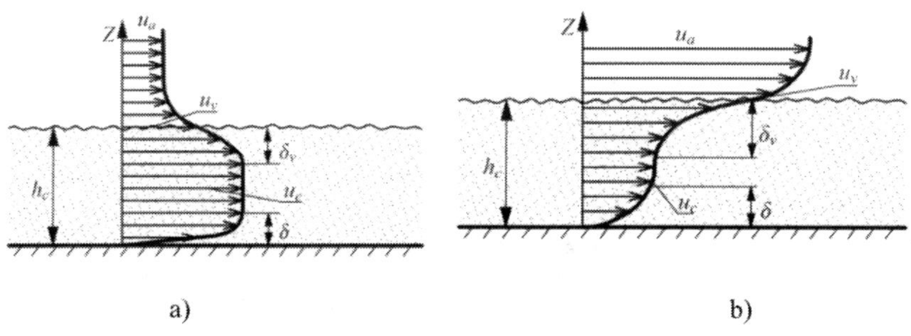


**Fig. 3.** Calculation scheme, a) $u_a > u_c$, b) $u_a < u_c$.

The above estimates for the layer thickness $\delta$ are apparently valid for the $\delta_v$ layer and allow us to assume that in most practically important cases for $u_a > u_c$ the velocity of the model fluid in the bulk layer will be equal to $u_c$. Thus, the solutions obtained above for layer $\delta$ are retained for the given design case.

The flow in the layer $\delta_v$ for $u_a < u_c$ will be similar to the considered flow in the layer $\delta$, therefore the solution (8) obtained in this case is also acceptable. The coefficients $C_1$ and $C_2$ can be determined from the conditions on the boundaries of the upper layer: $z = h_c - \delta_v = z_v$, $u = u_c$; $z = h_c$, $u = u_v$. Considering that $u_c = k_e F$, under these conditions, $C_1$ and $C_2$ may be determined for a known value of $u_v$. The velocity $u_v$ can be found from the consideration of the flow of a gas or liquid by the outer boundary of a layer. Thus, for example, if a turbulent boundary layer is formed in the external flow past a rough surface, a known logarithmic velocity profile can be used to determine $u_v$ [3], which for $u_a < u_c$ in the fixed coordinate system (see Figure 3) is written as follows:

$$\frac{u_c - u}{u_*} = \frac{1}{\kappa}\ln\frac{z'}{k_s} + 8.48 \tag{14}$$

The coordinate $z'$ is measured from the plane passing through the centers of the upper row of particles of the permeable layer; $k_s$ is the equivalent sand roughness of the permeable layer; $u_* = \sqrt{\dfrac{\tau_v}{\rho}}$ is the dynamic velocity determined by the tangential stress $\tau_v$ at the upper boundary of the layer.

In the case when the blowing velocity $u_a$ is greater than the velocity of the displacement of the layer $u_c$ (Fig. 3, b), the velocity at the upper boundary of the layer is determined similarly from the profile (14). The dynamic velocity $u_*$ is determined from the calculation of the flow of gas or liquid in the upper part of the apparatus above the permeable layer. With known dimensions of this part of the apparatus and a given flow rate of gas or liquid, the value of $u_*$ is found from the well-known relationship:

$$\frac{u_*}{V} = \sqrt{\frac{\lambda}{8}}, \tag{15}$$

where $V$ is the average velocity of gas or liquid flow over the permeable layer; $\lambda_v$ – coefficient of hydraulic resistance, determined by the known methods of hydraulics [10].

It should be noted that in the case under consideration ($u_a < u_c$) the positive increment of the shear stress $\nu_e \frac{d^2u}{dz^2}$ is balanced by the interfacial interaction force $f_g = g\frac{u-u_c}{k_\varphi}$, and the equation of motion of the model fluid in the fixed coordinate system for this layer is written as:

$$\nu_e \frac{d^2u}{dz^2} - \frac{g(u-u_c)}{k_\varphi} = 0, \tag{16}$$

which coincides with the already considered equation (4). Therefore: the solution (8) will also be suitable in this case also, taking into account the obvious peculiarities in determining the coefficients $C_1$ and $C_2$ on the basis of the corresponding boundary conditions. After finding the velocity profile in this layer, the estimate for the value $\delta_v / \sqrt{k_e}$ that is a parameter of the velocity profile can be obtained, as before, on the basis of setting the velocity value or its gradient at the lower boundary of the layer in fractions of the average velocity gradient equal in this case $\frac{u_v - u_c}{\delta_v}$.

In a number of practically important cases the motion of a model fluid can depend on the forces of gravity and the longitudinal pressure gradient $\frac{dp}{dx}$. The equation of motion is written as:

$$gi - \frac{1}{\rho}\frac{dp}{dx} + \nu_e \frac{d^2u}{dz^2} + f_g = 0 \tag{17}$$

where $gi$ - longitudinal component of gravity; $i$ – slope of the wall of the apparatus.

Depending on the part of the permeable layer in which motion is considered, the interfacial interaction force $f_g$, as well as the shear stress gradient, can change the sign, which should be taken into account in the calculations. Assuming, as before, the linear coupling of the force $f_g$ with the flow velocity of particles of the granular layer, we write in the form:

$$f_g = \frac{g(u_c - u)}{k_\varphi}$$

For a wall layer ($u < u_c$), the force $f_g > 0$ is attractive and must enter the equation (17) with a positive sign.

$$gi - \frac{1}{\rho}\frac{dp}{dx} + g\frac{u_c}{k_\varphi} + \nu_e \frac{d^2u}{dz^2} - g\frac{u}{k_\varphi} = 0$$

Determining $\frac{1}{\nu_e}\left(gi-\frac{1}{\rho}\frac{dp}{dx}+g\frac{u_c}{k_\varphi}\right)=F$, we write the equation of motion (17) in the form:

$$\frac{d^2u}{dz^2}-\frac{1}{k_e}u=-F \tag{18}$$

The form of equation (18) coincides with the form of equation (5) considered earlier, and the solution of equation (5) is also acceptable in this case.

As before, the value $\delta/\sqrt{k_e}$ may be found approximately for a given F from the condition of a definite value of the velocity gradient at the upper boundary of the layer. Using the boundary conditions $z=0$, $u=0$; $z=\delta$, $\frac{du}{dz}=0$ the velocity profile is written similarly to (12) as:

$$\frac{u}{k_eF}=1-\frac{Ch\frac{\delta-z}{\sqrt{k_e}}}{Ch\delta/\sqrt{k_e}} \tag{19}$$

It should be noted that in the case where the liquid or gas moves in the apparatus under the action of a pressure gradient as well as gravity, the flow velocity $u_c$ within the permeable layer will generally be less than the velocity $u_a$ above the permeable layer due to the considerable resistance of the layer (see Figure 3). However, if the blowing speed $u_a$ is small and the velocity of the layer $u_c$ in the apparatus is considerable, the flow scheme shown in Fig. 3, a.

If the permeability of the layer is large at a small thickness, the flow cases shown in Fig. 4. When calculating the flow pattern (Fig. 4, a), the use of boundary conditions $z=0$, $u=0$; $z=\delta$, $u=u_c$; $z=h_c-\delta_v=z_v$, $u=u_c$ perhaps, however, the condition: at z = δ, $\frac{du}{dz}=0$ should not be executed. Equality of derivatives $\left(\frac{du}{dz}\right)_{z=\delta}$ For the velocity profiles in the layer δ and in the layer $\delta_v$ it ensures the smoothness of the function $u=f(z)$ at $z=\delta$. In the case when the velocity $u_c$ is high at high layer permeability and its small thickness, it is possible that the maximum velocity will be less than $u_c$ (Fig. 4, b). Then, as the boundary conditions, the following should be used:

$$u = idem \text{ at } z=\delta;\ \left(\frac{du}{dz}\right)_{z=\delta}=0$$

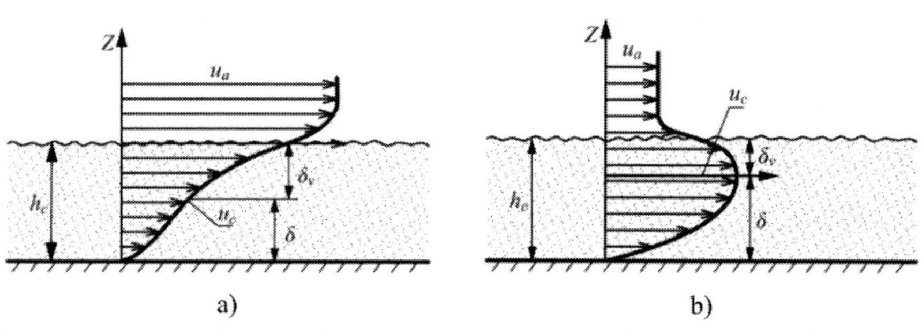


**Fig. 4.** Calculation scheme for a small thickness of the permeable layer, a) $u_a > u_c$, b) $u_a < u_c$.

We note that the solutions and estimates obtained above are also valid for the special case when the permeable layer is fixed relative to the wall of the apparatus (at $u_c = 0$). Thus, the analysis carried out made it possible to establish flow characteristics in the permeable layer of bulk material during its blowing and to obtain data that can be useful in solving drying and extraction problems in the production of building materials.

## References

1. P. Ya. Polubarinova-Kochina, *Theory of groundwater movement* (Nauka, Moscow, 1977)

2. P.G. Saffman, *Stadies in Applied Mathematics*, Vol. 50, **2**, p. 93 (1971)

3. G. Schlichting, *The theory of the boundary layer* (Science, Moscow, 1969)

4. *Handbook of hydraulic calculations*. Ed. P. G. Kiselev. (Energia, Moscow, 1972)

5. V. N. Spiridonov, Interaction between channel flow and filtration flow in disconnected soils (WPI RAS, Moscow, 1983)

6. A. D. Altshul, *Hydraulic resistance* (Nedra, Moscow, 1982)

7. V. V. Kafarov, *Osnovy massoperedachi* (Vysshaya shkola, Moscow, 1972)

8. G. Wallis, *One-dimensional two-phase flows* (Mir, Moscow, 1972)

9. Yu. V. Bryanskaya, I. M. Markova, A. V. Ostyakova, *Hydraulics of the water flows and flows with a suspension in a rigid and a deformable boundaries,* Ed. V. S. Borovkov (ASV, Moscow, 2009)

10. Bryanskaya Yu.V. *Refinement of velocity profile parameters for flows in smooth and rough pipes*, International Journal of Applied Engineering Research, Volume 11, 3, pp. 1694-1698 (2016)

# Data bases of aerological wind observations for industrial and civil construction

*Fedor* Bryukhan[1,*]

[1] Moscow state university of civil engineering, Yaroslavskoye shosse, 26, Moscow, Russia, 129337

**Abstract.** Due to extensive construction of high-rise buildings and structures, wind load calculations are becoming more and more important. These calculations require availability of credible wind data at the altitude of the atmospheric boundary layer and associated databases. This work is devoted to development of a framework for such databases, aimed at solution of various wind engineering tasks. Main principles for wind aerological observation database development for industrial and civil construction were established. Examples of wind characteristics and atmospheric dispersion characteristics are provided on the basis of radiosonde and SODAR/RASS observations. Prospects for development of similar databases for atmosphere pollution monitoring, wind power engineering, aviation are reported.

## 1 Introduction

Many buildings and structures 300-500 m high and more have been built or designed. These include Burj Khalifa building (Dubai, 828 m), Tokyo Skytree television tower (Tokyo, 634 m), Shanghai Tower (Shanghai, 632 m) and others. Besides this, many heat power stations use smoke stacks 200-400 m high. Taking into account that the wind speed progressively increases with the altitude, the wind exerts considerable loads on these buildings and structures [1, 2].

Design of high-rise buildings and structures requires calculation of static and dynamic load loads for development of engineering solutions for safe operation assurance [3]. Besides credible calculation models of wind loads, reliable wind observation data and related databases adapted to calculation models are required.

Most wind load calculations used ground meteorological observation data, extrapolated to the altitudes for which loads are determined [4]. However, due to diversity of atmosphere vertical stability types and the effect of wind vector rotation in the atmospheric boundary layer (ABL), actual wind characteristics and, therefore, wind loads differ from model loads very much. Due to the above circumstances, the most preferable wind data are observations at ABL altitudes. Such observations are carried out by using either radiosonde or acoustic sounding systems.

As applied to specific wind engineering tasks, one of the above observation methods is used, and sometimes – a combination of the two.

* Corresponding author: pniiis-gip@mail.ru

## 2 Measurement systems and databases of wind observations

Radiosonde systems for meteorological parameters measuring a hydrogen or helium-filled latex balloon with an attached temperature, pressure and humidity sensor is launched into the atmosphere. Measurement results and current coordinates of the balloon are received by a radio locator. Based on the balloon flight coordinates the wind speed and direction are determined. First of all, radiosonde observations are aimed at meteorological data acquisition for weather forecasting. However, ABL wind measurements can be used in wind engineering, for studying pollutant propagation in the atmosphere (atmospheric dispersion) and for other tasks.

Radiosonde data are stored in special archives for a 40-50 year period of observations at more than 100 observation points across the ex-USSR territory. Such data support calculations of various climatic atmosphere characteristics for a wide range of applications [1, 5]. Unfortunately, since 2015 the network of Russian aerological stations started to perform only 1-2 soundings per day instead of 2-4. In such situation, radiosonde data cannot fully support reliable load calculations since at low altitudes (up to 300-500 m) ABL meteorological parameters greatly vary within 24 hours.

The most suitable method for wind engineering is the ABL acoustic sounding using wind radio acoustic sodars. The operating principle of sodars is similar with that of radars. The difference is that sodars use acoustic impulses instead of electromagnetic impulses sent by radars. Based on such acoustic signal, reflected from turbulent atmospheric irregularities, the sodar integrated software determines atmospheric turbulence characteristics, including the wind vector. Unlike radiosonde systems, sodars enable virtually continuous acquisition of wind speed and direction data at the time resolution of about 10 minutes [6, 7]. The upper ABL sounding limit for sodars is about 1 km and is sufficient for wind load calculations. Another important fact is that sodars enable acquisition of wind data averaged to 10-minute intervals, normally used for wind load calculations.

For atmospheric dispersion studies the SODAR/RASS system (sodar coupled with the radio acoustic sounding system – RASS) is used [7, 8]. The operating principle of the temperature profile meter is that it sends electromagnetic waves, which are reflected from atmosphere irregularities, produced by acoustic waves from sodars. Electromagnetic waves reflected from such structural irregularities produce the diffraction effect and are received by the RASS antenna. Based on the received signal the sound speeds and consequently the air temperatures at different altitudes are determined [7].

The main disadvantage of sodars is that they have been widely used only for a relatively short period and no long-term data have been collected.

First automated archives of ABL radiosonde observation data were developed in the 1970's and their network covered 146 aerological stations in the USSR [1]. As the data were accumulated and their processing methods improved, the database structure and content changed considerably [5]. Since January of 1964 till present the All-Russian Research Institute for Hydrometeorological Information – World Data Centre has accumulated an immense data array for the Russian aerological network [9]. This array includes standard atmosphere sounding data, as well as information on specific points of vertical temperature and wind profiles, where the linearity of their vertical profile variations is disrupted. Additionally, the array includes ABL altitude data. Respective databases are presented by long-term months, but can be easily transformed in the chronological sequence. Data control, their vertical interpolation and statistical processing are performed by specialized software for calculation of various characteristics.

As the remote ABL sounding is carried out using the SODAR/RASS complex, current month readings are recorded and databases are formed. It is followed by data control and statistical processing, usually by current seasons, years and for a long-term period. As the remote ABL sounding is carried out using the SODAR/RASS complex, current month readings are recorded and databases are formed. It is followed by data control and statistical processing, usually by current seasons, years and for a long-term period.

# 3 Examples of database application

Below are examples of calculations of different wind and atmospheric dispersion characteristics using radiosonde and SODAR/RASS observation databases.

## 3.1 Radiosonde observation data

Tables 1 and 2 show the basic average annual wind characteristics according to the Nizhny Novgorod aerological station after long-term averaging. Table 1 shows rapid increase of the wind speed with the altitude. The mean value modulus of the wind vector is considerably lower than the average speed. This circumstance indicates frequent changes in the wind direction. The prevailing wind direction is western.

**Table 1.** Average annual wind characteristics at heights. Averaging period 1964-2012. Nizhny Novgorod.

| Height, m | Average speed, m/s | Average value of the wind vector | |
|---|---|---|---|
| | | Module, m/s | Direction, degrees |
| 0 | 2.6 | 0.5 | 240 |
| 100 | 4.0 | 1.2 | 244 |
| 200 | 5.0 | 1.7 | 248 |
| 300 | 5.8 | 2.1 | 251 |
| 500 | 7.2 | 2.9 | 256 |
| 1000 | 8.9 | 4.0 | 261 |

**Table 2.** Average annual combined repeatability of wind speed and direction in 8 compass points at 300 m. Averaging period 1964-2012. Nizhny Novgorod.

| Speed, m/s | Compass points | | | | | | | | Calm | Sum |
|---|---|---|---|---|---|---|---|---|---|---|
| | N | NE | E | SE | S | SW | W | NW | | |
| Calm | – | – | – | – | – | – | – | – | 0.97 | 0.97 |
| 1 | 0.72 | 0.56 | 0.60 | 0.58 | 0.69 | 0.75 | 0.90 | 0.88 | – | 5.68 |
| 2 | 1.21 | 1.01 | 0.93 | 0.91 | 1.12 | 1.30 | 1.45 | 1.77 | – | 9.69 |
| 3 | 1.60 | 1.20 | 1.04 | 1.07 | 1.23 | 1.71 | 2.38 | 2.27 | – | 12.51 |
| 4 | 1.94 | 1.16 | 0.93 | 0.89 | 1.31 | 1.81 | 2.73 | 2.44 | – | 13.21 |
| 5 | 1.84 | 0.96 | 0.81 | 0.88 | 1.16 | 1.85 | 2.57 | 2.54 | – | 12.62 |
| 6-10 | 4.82 | 1.93 | 1.63 | 2.25 | 3.03 | 5.73 | 7.86 | 7.06 | – | 34.31 |
| 11-15 | 0.85 | 0.29 | 0.35 | 0.54 | 1.01 | 1.81 | 2.49 | 1.61 | – | 8.93 |

**Table 2** – continuation.

| Speed, m/s | Compass points | | | | | | | | Calm | Sum |
|---|---|---|---|---|---|---|---|---|---|---|
| | N | NE | E | SE | S | SW | W | NW | | |
| 21-25 | 0.02 | 0.00 | 0.00 | 0.00 | 0.04 | 0.11 | 0.18 | 0.04 | – | 0.40 |
| 26-30 | 0.00 | 0.00 | 0.00 | 0.00 | 0.00 | 0.02 | 0.01 | 0.00 | – | 0.03 |
| > 30 | 0.00 | 0.00 | 0.00 | 0.00 | 0.00 | 0.00 | 0.00 | 0.00 | – | 0.00 |
| **Sum** | 13.07 | 7.14 | 6.35 | 7.18 | 9.85 | 15.48 | 21.12 | 18.83 | 0.97 | 100.00 |

Table 2 shows aggregate average annual combined repeatability of wind speed and direction in 8 compass points (complex wind rose) at 300 m for the same averaging period. Data from Table 2 confirm the fact that westerly winds prevail. Zero wind repeatability is low, less than 1%.

In the provided examples and below the wind direction is meteorological and means the azimuth of the direction from where the wind blows.

### 3.2 SODAR/RASS measurement results

As an example of practical application of SODAR/RASS measurement data, Figure 1 shows the horizontal distribution of the average annual coefficient of meteorological radionuclide dilution (ratio of radionuclide concentration to emission rate) near the Belorussian nuclear power plant. The averaging period is January-December of 2015. Calculations are made using the Gaussian dispersion model [10].

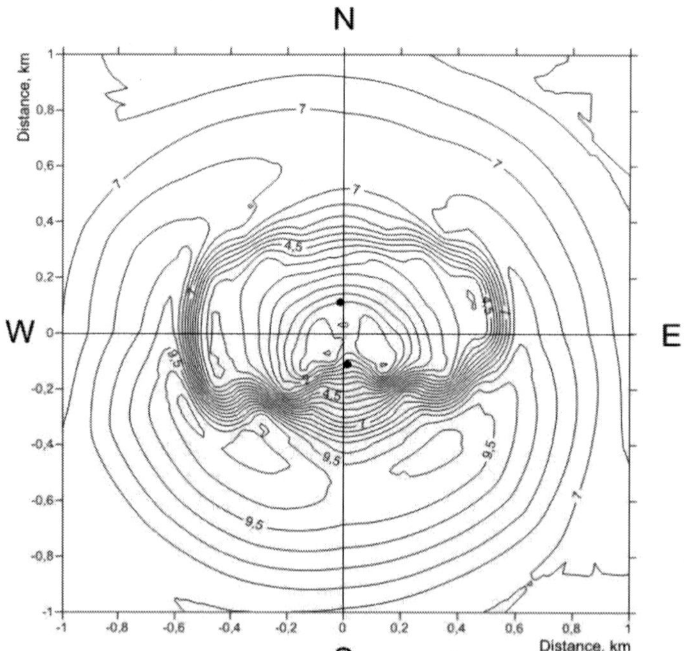


**Fig. 1.** Horizontal distribution of the average annual coefficient of meteorological dilution ($10^6 \times s/m^3$) at the ground level in normal operating conditions of the nuclear power plant. Emission source height is 100 m. Dots near the center indicate vent pipe locations.

## 4 Other fields of application for aerological databases

Fields of practical application of aerological observations, especially SODAR/RASS measurement results, are not limited to the above. Such measurements are widely used in air pollution monitoring in big cities, in wind engineering, in aviation.

Monitoring of meteorological parameters, determining pollutant dispersion conditions, becomes more and more important due to dependence of the big-city population health on the atmospheric air quality. Statistical analysis of atmospheric dispersion characteristics, measured by SODAR/RASS, enables development of air protection measures [11].

Radiosonde systems and sodars used in wind engineering support wind observations for assessment of wind-power resources. Such assessment are necessary for selection of locations for wind power stations and for generated power forecasting [12, 13].

Atmospheric vortexes, vertical wind shears, speed of winds transverse to runways are important factors that affect the safety of plane takeoff and landing [14]. Relevant operational information is provided by sodars.

The mentioned areas of application for sodar or SODAR/RASS measurements required development of relevant specific databases.

## 5 Conclusions

• Basic principles for development of aerological wind observation databases are established for solution of wind engineering tasks in industrial and civil construction.

• Examples of calculations of various wind characteristics, as well as atmospheric dispersion characteristics using radiosonde and SODAR/RASS observation databases are provided.

• Reference is made to potential development of similar databases in other areas: in air pollution monitoring, wind power engineering, aviation.

## Acknowledgments

The author thanks Dr O.A. Alduchov, the Leading Researcher of the All-Russian Research Institute for Hydrometeorological Information – World Data Centre for useful discussions.

## References

1. F.F. Bryukhan, *Methods of climatologic processing and analysis of aerologic information* (Gidrometeoizdat Publishing House, Moscow, 1983)
2. N. Perlin, S.P. De Szoeke, D.B. Chelton et al., Mon. Weather Rev., **142**, 4284-4307 (2014)
3. E. Simiu, R. Scanlan, *Wind effects on structures: an introduction to wind engineering* (J. Wiley & Sons, New York, 2000)
4. A. Peña Diaz, S.-E. Gryning, C.B. Hasager, Bound.-Lay. Meteorol., **129**, 479-495 (2008)
5. O.A. Alduchov, I.V. Chernykh, *Methods of analysis and interpretation of the atmosphere radio-sound observations data: quality control and data processing* (All-Russian Research Institute for Hydrometeorological Information – World Data Centre, Obninsk, Russia, 2013)
6. M.A. Lokoshchenko, E.A. Yavlyaeva, H.-J. Kirtzel, Meteorol. Z., **18**, 321-330 (2009)
7. *The METEK radio acoustic sounding system* (METEK GmbH, Elmshorn, Germany, 2013)

8. F.F. Bryukhan, A.Yu. Vinogradov, A.A. Lavrusevich, At. Energ., **118**, 365-368 (2015)
9. O.A. Alduchov, A.F. Bryukhan, At. Energ., **115**, 53-57 (2013)
10. *Generic models for use in assessing the impact of discharges of radioactive substances to the environment: IAEA safety report no. 19* (IAEA, Vienna, 2001)
11. B. Hennemuth, G. Peters, H.-J. Kirtzel, Atmos. Meas. Tech., **5**, 1399-1408 (2012)
12. F.F. Bryukhan, R.D. Diab, J. Appl. Meteorol., **34**, 2565-2571 (1995)
13. S. Lang, E. McKeogh, Remote Sens., **3**, 1871-1901 (2011)
14. S. Boluriaan, P. Morris, AIAA J., **39**, 1097-1105 (2001)

# Preliminary validation of the dynamic probing methods used in estimation of the relative density of cohesionless soils

*Karol* Brzeziński[1,*], *Maciej* Maślakowski[1] , and *Marta* Sokołowska[2]

[1]Warsaw University of Technology, Faculty of Civil Engineering, 16 Armii Ludowej Ave., 00-637 Warsaw, Poland,
[2]Polish Geological Institute, National Research Institute, 4 Rakowiecka Street, 00-975 Warsaw, Poland,

**Abstract.** The dynamic probing methods are commonly used for the subsoil identification. For example, the relative density/density index ( $I_D$ ) of cohesionless soils can be determined by using appropriate correlation relationships. The results obtained by various probing methods (DPL, DPH, DPM) may vary from one to another. Thus, the field tests were conducted in order to determine how important these discrepancies are. Each of the above mentioned methods were performed 10 times at one location. The effect of the influence of the penetrometer type on the relative density estimation results was determined. Based on the analysis, it can be concluded that the results obtained by the medium and heavy penetrometers are very close to each other. On the other hand, the discrepancies between the results obtained with the light penetrometer and the other two penetrometers are greater, both in terms of mean difference and scatter.

## 1 Introduction

Many methods are used to determine the parameters of a soil medium in the direct tests (in accordance with the definition of a given parameter) or in the approximate way using correlations. Such methods are often developed under specific soil conditions, and it is not clear how accurate they are when the conditions are different. Therefore, within the framework of a research project entitled "Modern Methods of Ground Recognition in Roads", research is being conducted to validate various methods of soil identification in Polish conditions. On this basis the new guidelines for the diagnosis of ground subsoil for road construction will be developed. The subject of this study is the validation of the methods using the results of the dynamic probe test to determine the relative density ( $I_D$ ) of non-cohesive soils. According to the assumptions presented in [1], validation of methods consists of analysing the difference of the results obtained by the reference method and the validated method

* Corresponding author: k.brzezinski@il.pw.edu.pl

$$d = I_D(r) - I_D(v) \tag{1}$$

where $d$ is discrepancy between $I_D$ obtained by using reference ($r$) and validated ($v$) method.

Within the methodology adopted for the grant, the reference method is the test that defines the parameter. In the case of relative density, the reference method consists of the laboratory test. However, in practice, the $I_D$ is usually estimated by the results of probing (static or dynamic). Subsequently, on this basis further conclusions about the other parameters of the soil can be drawn. Many studies available in the literature concern a study comparing results obtained from static (CPT) and dynamic (DP) surveys [2-5]. The purpose of this analysis is to compare the discrepancies in results from different dynamic penetrometers. These methods consist of determining the relative density by using the correlation dependence, based on the value of the resistance imposed by the soil. The measure of soil resistance is the number of strokes ($N_{10}$) needed to deepen the penetrometer by the assumed depth (10 cm). Dynamic probing are described in details in standard document [6]. The relative density interpretation was also performed on the basis of this document. For the purpose of the present study, the results of three methods of probing were used:

- light – DPL (10 kg),
- medium – DPM (30 kg),
- heavy – DPH (50 kg).

Due to the diminishing importance of direct determination of soil reference density by laboratory methods (due to their labor intensity) [7], the reference method was abandoned and the pairing method was compared to assess how the results obtained by the different methods are similar and whether they can be used interchangeably.

## 2 The site characteristics and the research methodology

Depending on the genesis of the soil, its density can be very diverse, even in a small area [8]. Therefore, the test was conducted on a compact area (probing was done on a grid of 2.0 m). This approach helped to minimize the impact of local factors on the obtained results [9]. In the middle of the area a borehole was drilled to a depth of 8.0 m. On this basis the soil layer depths was determined:

- 0. 0-0.15 – humus,
- 0.15-1.25 – FSa,
- 1.25-3.3 – MSa+Gr,
- 3.3-4.5 – FSa/MSa,
- 4.5-5.2 – MSa/FSa+Gr,
- 5.2-8.0 – MSa+Gr,
- water at a depth of 1.6 m.

As it can be seen in Figure 1, the results of the probes approximately reflect the lithology established on the basis of the borehole. According to the cumulative difference approach, uniformly spaced layers are separated along the sections with a constant slope of the cumulative sum.

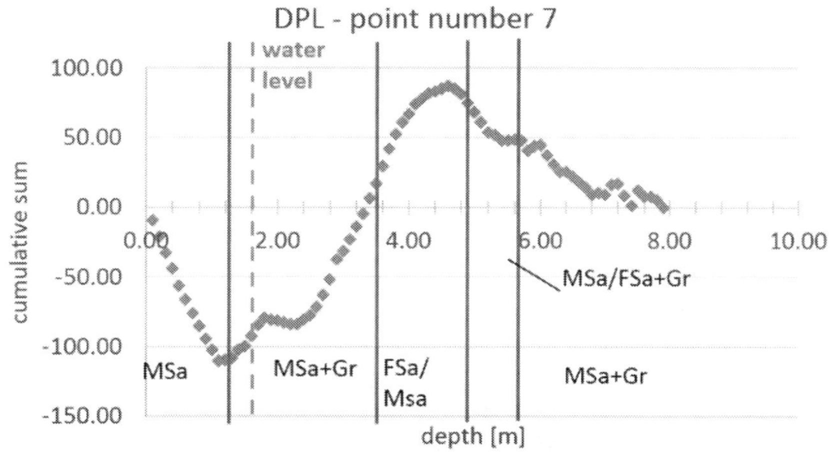


**Fig. 1.** Comparison of layer depths obtained on the basis of borehole and slope changes of cumulative sum of $N_{10}$ results.

It should be emphasized that the results of the probing conducted in different locations (even by the same method) are different. The homogeneity of the results consists of the soil medium homogeneity and the repeatability of the method. It is not possible to separate these two factors, so their combined effect should be taken into account.

## 3 Analysis of the results

The analysis was based on the distribution of differences in the $I_D$ results obtained between the methods (DPL-DMP, DPL-DPH, DPM-DPH). The results obtained in the nearest locations were paired (corresponding numbers and depth). About 700 results were obtained for the validation of the method in such a way. Results from a depth above 1.00 m were rejected due to the lack of the reliability requirements fulfilment ( $N_{10}$> 3).

The confidence interval depends on the value of the standard deviation that is the measure of the scatter of results. The first proposal was to use the T-Student distribution in order to approximate results' distribution. This distribution is used for populations with unknown standard deviation values. In order to be able to apply the T-Student distribution, it is necessary to meet the condition of normality of the discrepancy between the values of the reference method and the validated method [10]

The unimodal and almost symmetrical histograms of the results were obtained, but the Shapiro-Wilk test showed a deviation from the normality assumption. Due to the slight skewness, log-normal distribution was sought, but the transformation of the variable to normal distribution failed. Frequency histograms are shown in Figures 2-4. Because of the very large sample, confidence intervals were estimated by using quantiles of the sample. It was assumed that confidence interval should be 95%. It means 95% of the results (discrepancy between $I_D$ obtained by using different method in "the same" location) should be included in the interval. The confidence intervals for each method were determined on the basis of the sample quartiles (2.5% and 97.5%).

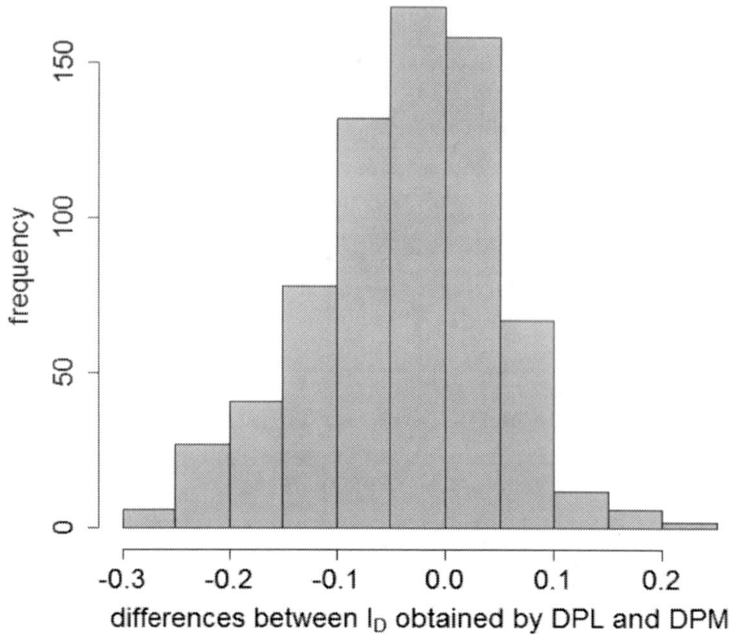


**Fig. 2.** Histogram of differences of ID results obtained using DPL and DPM.

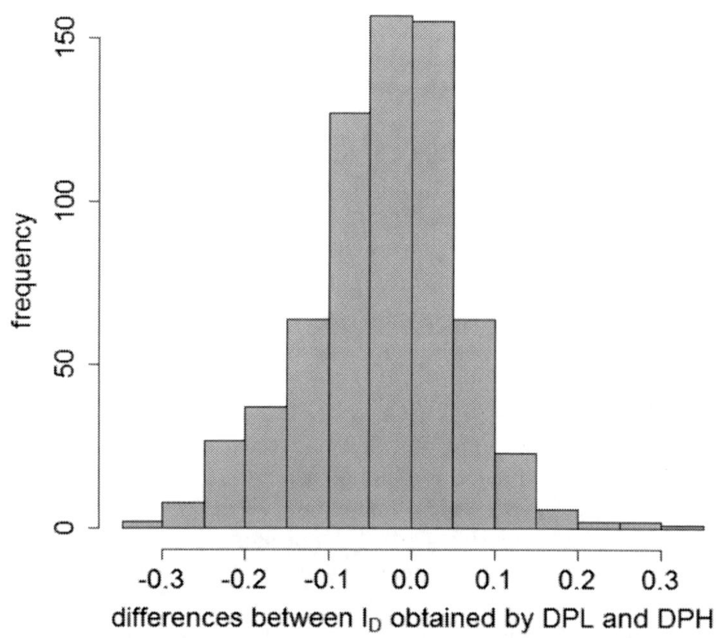


**Fig. 3.** Histogram of differences of ID results obtained using DPL and DPH.

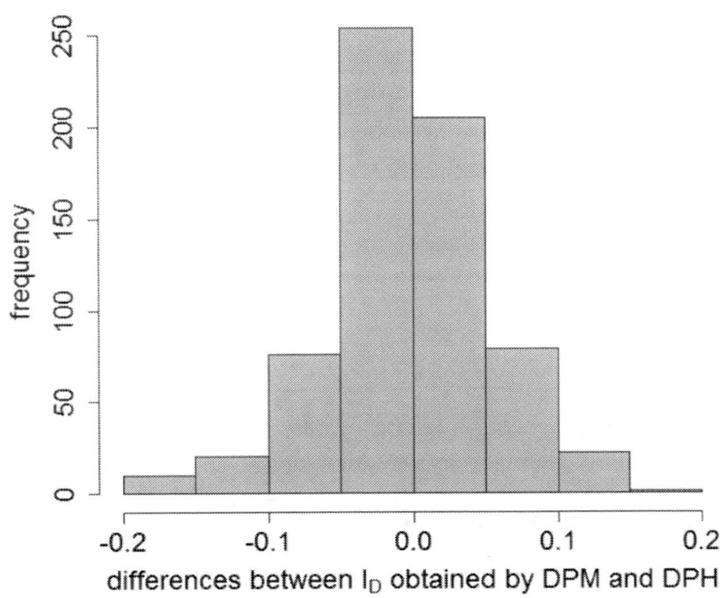


**Fig. 4.** Histogram of differences of ID results obtained using DPM and DPH.

Analysing the histograms one can see some dependencies, which are summarized in Table 1. The differences between the light penetrometer and the other penetrometers (medium and heavy) are similar in terms of both the average and the scatter. In contrast, the differences between the medium and heavy penetrometers results are much smaller.

**Table 1.** Basic statistics of discrepancies between $I_D$ obtained by using different methods of probing.

| | $d = I_D(DPL) - I_D(DPM)$ | $d = I_D(DPL) - I_D(DPH)$ | $d = I_D(DPM) - I_D(DPH)$ |
|---|---|---|---|
| mean, $\mu$ | -0.033 | -0.029 | 0.003 |
| std. dev., $sd$ | 0.084 | 0.092 | 0.056 |
| 95% conf. interv. | -0.220÷0.116 | -0.232÷0.132 | -0.114÷0.120 |

## 4 Conclusions

An attempt was made to analyse differences in the density results of $I_D$ obtained by various dynamic probing methods (DPL, DPM, DPH). Shapiro-Wilk's test showed failure to meet the normality of distribution. The attempts to find transformations of results that would make use of normality assumptions (e.g. log-normal distribution) also failed. The confidence intervals were directly derived from the distribution, based on the sample quantiles, because of the large number of results. The 95% confidence interval was 0.34 width for the DPL-DPM comparison and 0.36 for the DPL-DPH method. In contrast, the DPM-DPH method comparison resulted in significantly smaller differences,(the confidence interval width of 0.23). The average difference between DPL and the other validated methods was about -0.03. Thus, medium and heavy penetrometers give similar results to each other, but they slightly overestimate the degree of compaction of the $I_D$ relative to the results obtained with a light probe.

The $I_D$ parameter itself is rarely used in design process. By knowing its value, one can indirectly draw conclusions about other parameters of the tested soil. Therefore, the presented results will be used to further validation of the methods to determine other parameters of the soil and to optimization of the recommended methods of field testing.

## Acknowledgments

The "Modern Methods of Ground Recognition in Roads" research project is funded by the National Center for Research and Development and the General Directorate for National Roads and Highways.

The project is being implemented by a consortium:
• Polish Geological Institute, National Research Institute,
• Warsaw University of Technology,
• AGH University of Science and Technology.

## References

1. K. Brzeziński, M. Maślakowski, A. Zbiciak, K. Józefiak, Drogownictwo **LXXII (3)**, 85 (2017)
2. B. Czado, J.S. Pietras, CzT **109(z. 3-B)**, 21 (2012)
3. A. Makler, J. Szendefy, Period. Polytech. Civ. Eng. **53(2)**, 101 (2009)
4. G. Žaržojus, K. Dundulis, BALT J ROAD BRIDGE E **V(2)**, 69 (2010), doi: 10.3846/bjrbe.2010.10
5. M. Ura, M. Tarnawski, PG **62(10/2)**, 715 (2014)
6. PN-B-04452:2002 *Geotechnics. Field tests* (in Polish)
7. M. Tarnawski, Inż. Mors. i Geotechn. **3**, 441, (2010)
8. M. Sokołowska, Pr. ITB **R. 39(4)**, 3 (2010)
9. M. Borowczyk, Z. Frankowski, Inż. i Bud. **7-8**, 277, (1985)
10. J. Mazerski, *Statistical analysis of experimental results* (in Polish), (Malamut, Warsaw 2009).

# Deformation parameters of soils determined by the cone penetration test

*Roman* Bulko[1*], *Filip* Gago[2]

[1] University of Žilina, Faculty of Civil Engineering, Department of Geotechnics, Univerzitná 8215/1, 010 26, Slovakia

[2] University of Žilina, Faculty of Civil Engineering, Department of Geotechnics, Univerzitná 8215/1, 010 26, Slovakia

**Abstract.** The article deals with the derivation of correlation relations from the cone penetration test, which are applied as input parameters in the calculations of advanced constitutive soil models. Numerical modelling has become important part of geotechnical calculation. Hand in hand there is a big demand for accurately determined input parameters for constitutional models, that are not usually determined by IG surveys. On the model area we applied fieldworks as cone penetration tests and boreholes to obtain detailed parameters. Soil samples were examined in laboratory to create own database of values. Derived correlations include compression index $C_c$ and recompression index $C_r$.

## 1 The principle of a cone penetration test

The cone penetration test (*Fig.1*) is one of the geotechnical field methods used to determine the geotechnical properties of soils. The measuring cone is statically pushed through a set of steel rods into the soil by a constant speed of 20 mm.s$^{-1}$. During penetration, number of variables are recorded at the cone head or along the sleeve. Values as cone resistance $q_c$, friction on sleeve $f_s$ and pore pressure $u_1/u_2$ are continuously recorded. At the cone the cone resistance is recorded, which expresses the resistance of the soil to penetration. The total resistance of the penetrated system $q_t$ is then summation of the cone resistance $q_c$ and the friction on sleeve $f_s$. Changing the penetration probe from the mechanical tip to the piezocone, enables to measure the pore pressures $u_1$ (measurement at the cone) or $u_2$ (measurement just behind the cone).

* Corresponding author: roman.bulko@fstav.uniza.sk

**Fig. 1** Demonstration of cone penetration test

Cone Penetration Tests are divided according to used type of the measuring cone on:

- *CPTm* – Cone Penetration Test with mechanical Begemann cone,
- *CPTu* – Cone Penetration Test with piezocone,
- *SCPTu* – Seismic Cone Penetration Test.

## 2 The fieldwork on the model area

Various programs using finite element methods have been used as Plaxis, Midas, Adina and others to analyse some geotechnical tasks. Constitutional models of soils and their input parameters are base data to be entered into the different programs. In our research we have focused on the development of new correlations for the input parameters of advanced constitutional models derived from the cone penetration test.

Input data for derivation of correlation were obtained by Engineering Geological (EG) survey on model area (*Fig.2*). The *EG* survey was performed by six core boreholes 5 m long. Four cone penetration probes (*CPTm*) each of the length of 5.6 m were allocated to each core borehole. Representative soil samples were taken from the core boreholes from the depths of 1.0 m to 5.0 m at intervals of 0.2 m. Soil samples were tested in the Geotechnical Laboratory of the University of Žilina. Laboratory tests were done as follows:

- Granulometric analysis, humidity test, evaluation of plasticity and fluid limits;
- Shear box test;
- Determining the apparent density of solid particles;
- Oedometric test.

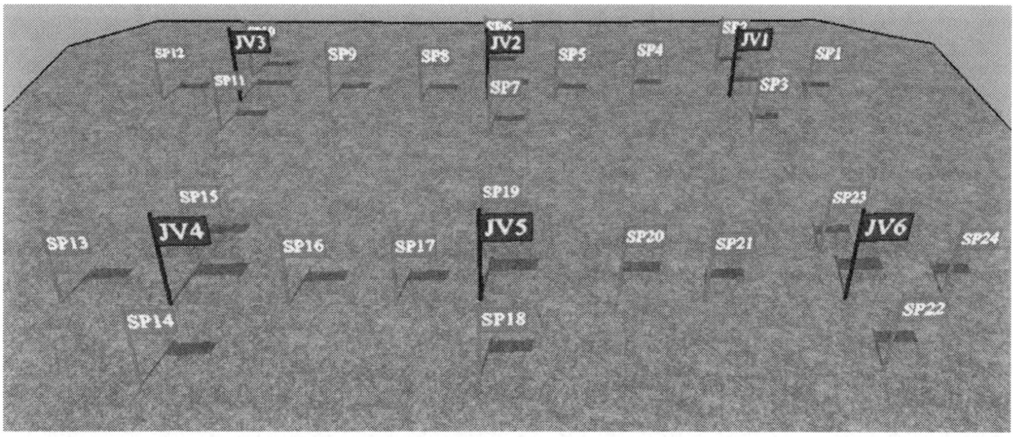


**Fig. 2** Deployment of CPT probes and core boreholes on model area

The soils were classified in the graph (*Fig.3*) depending on the normalized penetration resistance $q_{CN}$ (1) and the friction ratio $F_r$ (2) based on the continual recording of cone penetration tests. This classification was introduced by Robertson, who described each area of the graph.

$$q_{CN} = q_c . C_q \quad ak \quad C_q = \frac{1.7}{\left(\frac{\sigma_v}{p_a} + 0.7\right)} \tag{1}$$

where:

$q_c$ – cone resistance [MPa],
$p_a$ – reference stress [0.1MPa],
$\sigma_v$ – geostatic vertical stress [MPa],
$C_q$ – stress normalization factor [MPa].

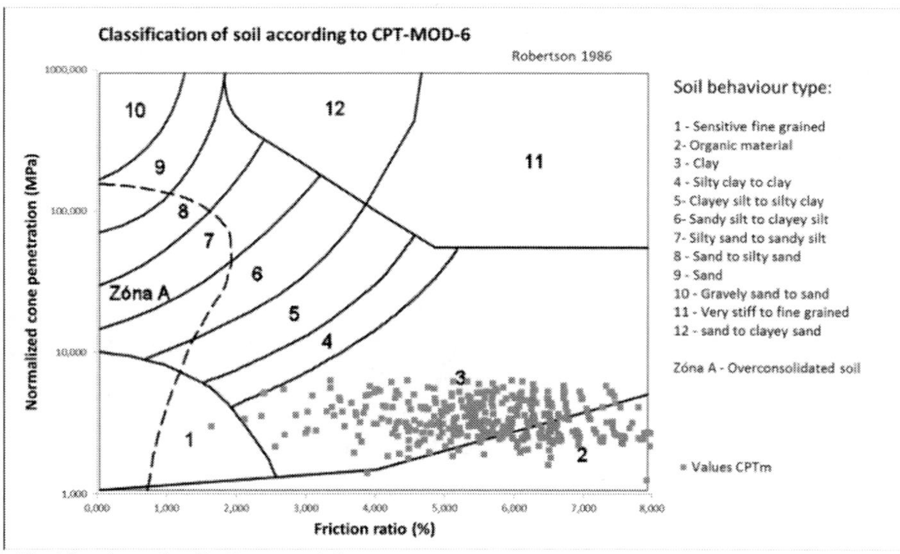


**Fig. 3** Clasification of soils from model area

$$F_r = \frac{100.fs}{q_c - \sigma_v} \tag{2}$$

where:

$f_s$ – sleeve friction [MPa],
$q_c$ – cone resistance [MPa],
$\sigma_v$ – geostatic vertical stress [MPa].

## 3 Constitutive models

As it has been already mentioned, various FEM programs are used to analyse complex geotechnical tasks. The more complex the geotechnical tasks are, the better understanding of the situation is required. This is the case of more advanced constitutive soil models. An important question in numerical modelling in geotechnics is therefore the question, how to use the constitutive model for proper calculations and how to determine model input parameters. Another question is the degree of precision of the description of soil behaviour using individual models versus widely known facts.

Soil models - also called constitutional models - can be divided into two basic groups, linear and nonlinear. Linear models give a fairly fast, but none too accurate estimation of actual soil behaviour. They can be used in cases where we are only interested in stress or deformation of the massif. They are not well applicable for the method of plane of possible failure. However, if we try to do a credible description of the behaviour of the earth construction, or when we are interested in the distribution of possible areas of failure, it is necessary to use non-linear models. We chose an advanced Soft soil model for our work.

One of the input parameters is the modified compression index $\lambda*$ (3). This parameter is obtained from a triaxial test and entered into the Soft soil model. If the modified compressibility index $\lambda*$ is not known from a triaxial test, it can be roughly determined from the compression index $C_c$, determined by the oedometric test:

$$\lambda* = \frac{C_c}{2,3.(1+e)} \tag{3}$$

where:

$C_c$ – compression index,
$e$ – void ratio.

Another input parameter is the modified swelling index $\kappa*$ (4). This parameter is entered into the Soft soil model and retrieved again from the triaxial test. If the modified swelling index $\kappa*$ is not known from the triaxial measurement, it can be approximated by the $C_r$ recompression index, which is determined by the oedometric test:

$$\kappa* = \frac{C_r}{2,3.(1+e)} \tag{4}$$

where:

$C_r$ – recompression index.

If results from laboratory tests or in-situ tests are not available, we can use the following relationship to determine the $C_r$ recompression index:

$$C_r = \frac{1}{5} \sim \frac{1}{10} C_c \tag{5}$$

## 4 New correlations depending on cone penetration test

Thirty one oedometric tests were carried out in the geotechnical laboratory, with a load of 50-100-150-200-150-100-50-100-150-200 kPa. Such load patterns also provided a branch of unloading and re-strain, from which it was later possible to determine the compression index $C_c$ and the recompression index $C_r$.

The compression index $C_c$ was obtained from the oedometric test, which describes the dependence of the porosity change $e$ on the change of the effective stress $\sigma_{ef}$ plotted in the logarithmic scale. It is therefore the deformation characteristic of the overconsolidated soil, which is given by the relation (6):

$$C_c = \frac{\Delta e}{\Delta \log \sigma_{ef}} \tag{6}$$

where:

$\Delta e$ – void ratio,

$\Delta \log\sigma_{ef}$ – logarithmic effective stress.

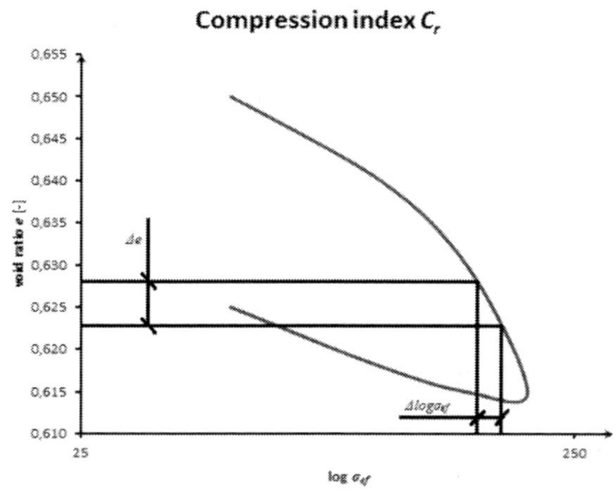


**Fig. 4** The oedometric test and the dependance of the change of the void ratio on the change of the effective stress

The recompression index $C_r$ is obtained from the oedometric test, the unloading/reloading curve and the dependence of the porosity change $e$ on the change in the effective stress $\sigma_{ef}$ plotted on the logarithmic scale. It is a deformation characteristic of overconsolidated soil. The determination of the recompression index $C_r$ is given by the relation (7):

$$C_r = \frac{\Delta e}{\Delta \log \sigma_{ef}} \tag{7}$$

where:

$\Delta e$ – changing void ratio for curve unloaded – reloaded,

$\Delta \log\sigma_{ef}$ – changing effective stress for curve unloaded - reloaded.

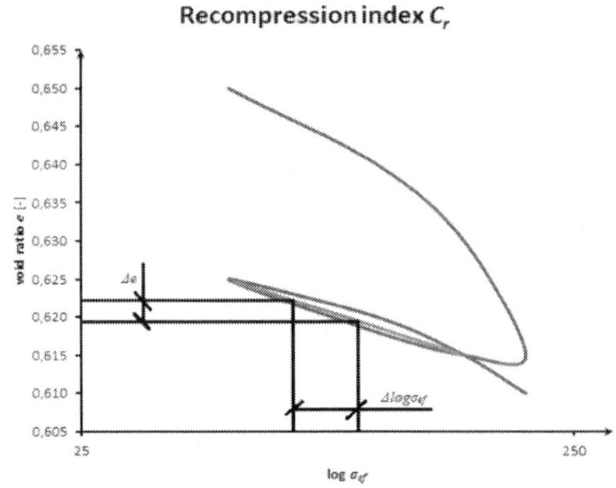


**Fig. 5** The oedometric test and the dependance of the change of void ratio on the change of the effective stress for curve unloaded - reloaded

On the base of the tests, having a sufficiently large database of values of compression index $C_c$ and the $C_r$ recompression index, we deduced the dependence between these values and the measured values of the normalized penetration resistance $q_{CN}$. The correlation relation refers to the medium plasticity clays *F6 (CI)* classified according to *STN 73 1001*. On the figure (*Fig. 8*) we can see the distribution of 31 measured values of the compression index. We have to point out that the measured compression index values ranged from 0.07 to 0.24. The compression index $C_c$ is given by the relation (8):

$$C_c = 0.0077.q_{CN}^{\ 3} - 0.0776.q_{CN}^{\ 2} + 0.2591.q_{CN} - 0.1863 \tag{8}$$

Based on the determination of the standard deviation, no extreme values were found. To *100 %* of the results, we can derive the compression index $C_c$ with *77%* accuracy based on the relationship (8)

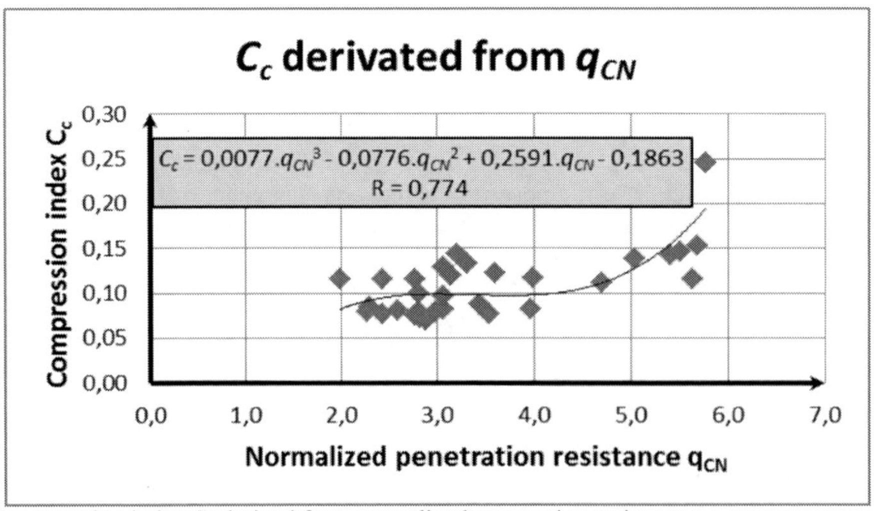


**Fig. 6** Compression index $C_c$ derived from normalized penetration resistance $q_{CN}$

Recompression index $C_r$ (9) is derived from normalized cone resistance and is given by the relation:

$$C_r = 0.0008.q_{CN}^{\ 3} - 0.01.q_{CN}^{\ 2} + 0.0425.q_{CN} - 0.0385 \tag{9}$$

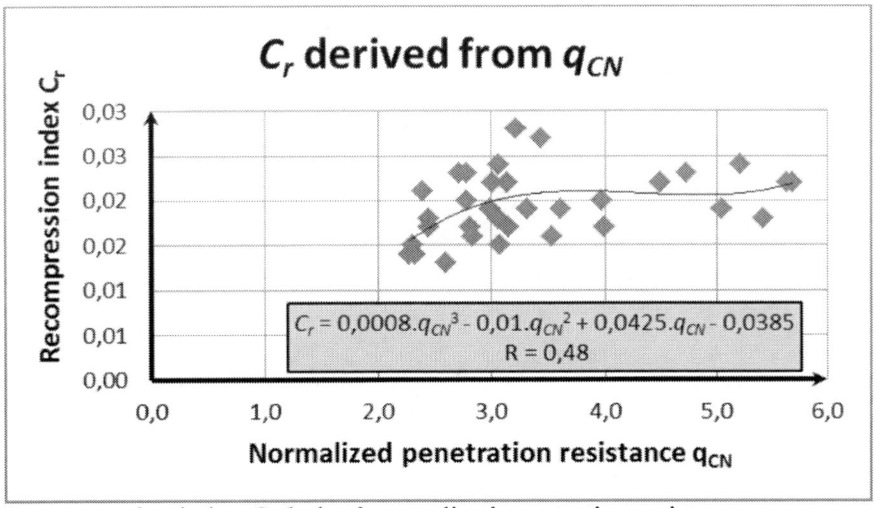


**Fig. 7** Recompression index $C_r$ derived normalized penetration resistance $q_{CN}$

## Conclusion

The ability to determine the advanced geotechnical parameters from Cone Penetration Tests is an indisputable asset. It is promising and providing possibilities for their extended usage. New correlations need to be verified in practice and, of course, they need to be refined or improved by feedback. In the future, the authors plan to deal with the correlation of the presented parameters in relation to the assessment of geotechnical structures, or for their direct usage for numerical analyses. From the engineering and geological point of view, it is evident that the knowledge of the local properties of fine-grained soils obtained by penetration probing is unique. But their extend usage should be examined for certain conditions. They can be used for similar problems to contribute to the discovery of hazards and pitfalls in the rock environment, as potential liquefaction during seismic shaking, overconsolidation, etc.

## References

1 M. Decký, M. Drusa, Navrhovanie a kontrola kvality zemných konštrukcií inžinierskych stavieb. Krupa Print ISBN 978-80-970139-1-2.

2 M. Drusa, Proceedings of SGEM, **2**, 151-158 (2012)

3 P. W. Mayne, Cone Penetration Testing State of Practice, Final Report NCHRP Project 20-05, Transportation Research Board, Washington (2007)

4 Research project VEGA No. 1/4215/07 Evaluation of penetrating methods for determining the deformation properties of subsoil of transport constructions and their verification in geotechnical practice (2009)

5 P. K. Robertson, Canadian Geotechnical Journal, **27/1**, 151-158 (1990)

6 J. Vlček, R. Bulko, WMESS world multidisciplinary earth sciences symposium CD-ROM, s. 104 (2015)

7 R. Bulko, L. Kais, J. Vlček,: Použitie statickej penetračnej skúšky pre určovanie geotechnických parametrov. Nové poznatky v geotechnickom inžinierstve – Bratislava (2014)
8 R. Bulko, Drusa, J. Vlček, M. Mečár, Civil and environmental **11/2**, 152-157 (2015)
9 STN 72 1001 : 2010 : Klasifikácia zemín a skalných hornín.
10 STN 72 1027 : 1983 : Laboratorní stanovení stlačitel'nosti zemin v edometru

# Simulating the structure of gypsum composites using pulverized basalt waste

*Aleksandr* Buryanov[1], *Victoria* Petropavlovskaya[2,*], *Tatiana* Novichenkova[3], *Kirill* Petropavlovskii[4]

[1]Moscow state university of civil engineering, Yaroslavskoye shosse, 26, Moscow, Russia, 129337
[2]Associate Professor, Building products and constructions Department, TvSTU, 22 Af. Nikitin, Tver, 170026, Russia
[3]Associate Professor, Building products and constructions Department, TvSTU, 22 Af. Nikitin, Tver, 170026, Russia
[4]Graduate student, Higher mathematics Department, TvSTU, 22 Af. Nikitin, Tver, 170026, Russia

**Abstract.** This paper examines the possibility of simulating the structure of gypsum composite modified with basalt dust waste to make materials and products based on it. Structural simulating of the topological space in gypsum modified composite by optimizing its grain-size composition highly improves its physical and mechanical properties. Strength and density tests have confirmed the results of the simulation. The properties of modified gypsum materials are improved by obtaining of denser particle packing in the presence of hemihydrate of finely dispersed basalt and plasticizer particles in the system, and by engaging basalt waste in the structuring process of modified gypsum stone.

## 1 Introduction

The issue of creating a health-saving environment is directly related to the construction material manufacturing, because microclimate created indoors will be determined by chemical composition, properties and quality of the materials to be used for construction [1, 2, 3, 4]. Polymer compositions are frequently used as an effort to cut the costs of construction projects [5, 6]. However, many materials are a health risk, some even trigger chronic disease. For example, formaldehyde found in the resin that is used to make many finishing materials (such as chipboard, fiberboard or putty) can provoking allergic reactions, even diseases of the eyes and respiratory organs.

Environmentally friendly traditional mineral materials and products based on gypsum, clay, lime and cement compounds, cements and ceramics can serve as great substitutes for polymer finishing materials [7, 8, 9].

Gypsum glue possesses important advantages compared to the above, because it enables implementing most daring and diverse art design solutions, while creating most favorable indoor microclimate. This makes it more viable in the construction industry.

* Corresponding author: victoriapetrop@gmail.com

Production of gypsum glue is power-efficient, that is it does not require high firing temperatures; therefore, gypsum panels are safe not only as finished product but also during production [10].

Technology-wise, gypsum panel manufacturing can recycle waste from other sectors [11, 12, 13, 14, 15]. For instance, basalt dust (waste of production of basalt fiber) can be used as a reinforcing component; currently, it is not recycled in real production [16, 17, 18].

Use of basalt dust as modifier in gypsum products helps to improve the structure and properties of gypsum stone [19, 20, 21]. Therefore, tests in this domain are needed to determine optimized composition of raw mix in production of facing panels of gypsum with better physical and mechanical characteristics and performance.

In addition to better strength characteristics, use of basalt dust can cut production costs, being waste from basalt fiber manufacturing. Pulverized particles possess the same properties as fibers proper: fire resistance, high thermal stability, and vapor permeability [22, 23, 24].

## 2 Methods

The raw material used was high-strength gypsum of grade GVVS-16 – α-modification gypsum glue made by Samara Gypsum Plant (SGK) based in Samara, Russia. High-strength gypsum of α-modification, brand GVVS-16 made by Samara Gypsum Plant, is described by low water consumption rate of 35-40 %, it begins setting after 4 – 5 minutes, and complete sets within 20 minutes. Its grain structure is compliant under GOST 125-79 with 0.5 % of 0.2 mm oversize sieve residue. Water absorption rate is 18 %. Compressive strength is 16 MPa; tensile strength in bending is 6-7 MPa. 1 kg contains no more than 10 mg of metal impurities.

For chemical composition of glue GVVS-16 made by the Samara Plant, see Table 1.

**Table 1.** Chemical composition of Samara high-strength glue, modification α

| $Si0_2$ | $A1_20_3$ | $Ti0_2$ | $Fe_20_3$ | $Ca0$ | $Mg0$ | $S0_3$ | $Na_20$ | $K_20$ | $P_20_5$ | F |
|---|---|---|---|---|---|---|---|---|---|---|
| 0.8 | traces | traces | - | 37.52 | 0.10 | 53.78 | 0.05 | 0.007 | - | - |

Figure 1 presents a registered diffractometric scanned image of Samara high-strength gypsum, modification α.

Key performance characteristics of Samara high-strength gypsum, modification α, are presented in Table 2.

Figure 2 presents integral and differential distribution of particles by size in the structure of high-strength gypsum GVVS-16.

The modifier additive was basalt dust that is waste from basalt fiber production. Methods of differential thermal analysis, X-ray diffraction analysis and chemical analysis were used to determine the aggregative and chemical composition of powder waste (Table 3).

Figure 3 illustrates integral and differential particle distribution by sizes in basalt powder.

According to the grain-size analysis, average particle size of basalt dust was $d_{50}$=24.866 mcm, largest particle size was $d_{99}$ = 111.743 mcm, with 9.66 % by mass of particles smaller than 2 mcm.

The plasticizer added was Melflux 1461f hyper-plasticizing agent, made by BASF Construction Polymers (Trostberg, Germany).

Studies of the properties of the gypsum modified composite were researched on standard sample bars, sized 40x40x160 mm, tested to meet the provisions of GOST 23789.

Particles distribution in dust samples was measured with the laser diffraction method under ISO 13320-1:2009 "Size Particle Analysis. Methods of Laser Diffraction", by Analysette 22 laser particle size analyzer. Diffractometric images of gypsum glue and the modified composite on its basis were registered using diffractometer ARL X'tra (Switzerland).

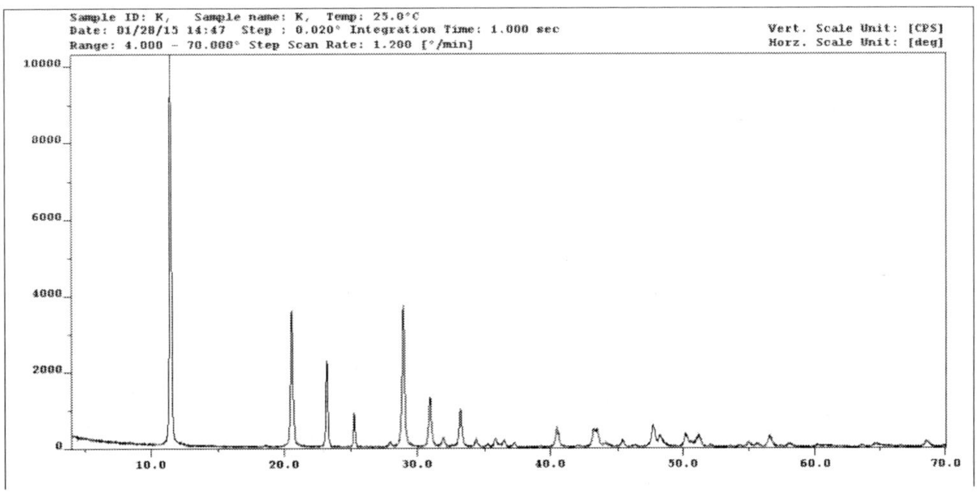


**Fig.1.**Registered diffractometric scanned image of GVVS-16

**Table 2.** Key performance characteristics of high-strength gypsum, modification α

| Brand of glue | Specific area, $m^2/kg$; | Expansion by volume, %; | Insoluble impurities in HC1, %. |
|---|---|---|---|
| Г-16 | 6.893 | 0.13 | 0.17 |

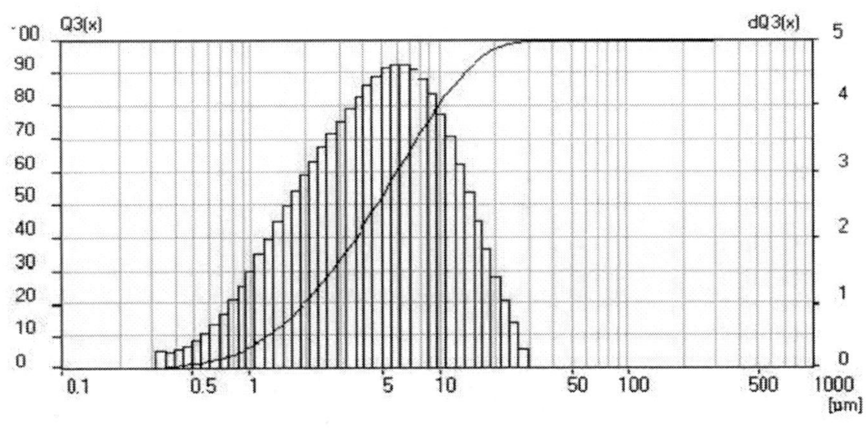


**Fig. 2.** Integral and differential distribution of particlesin GVVS-16

**Table 3.** Chemical composition of waste basalt

| $CO_2$ | $Na_2O$ | $MgO$ | $Al_2O_3$ | $SiO_2$ | $SO_3$ | $K_2O$ | $CaO$ | $TiO_2$ | $Fe_2O_3$ |
|---|---|---|---|---|---|---|---|---|---|
| 1.45 | 1.58 | 9.06 | 12.14 | 52.67 | 0.65 | 0.57 | 13.92 | 1.10 | 6.87 |

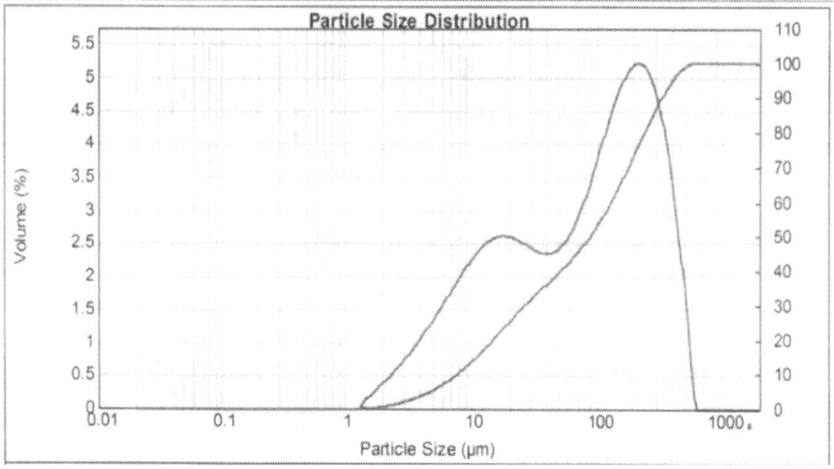


**Fig.3.** Results of grain size analysis of basalt powder

## 3 Results and Discussion

This paper examines the possibility of simulating the structure of gypsum composite modified with basalt dust waste to make materials and products based on it. The topological structure of the composite was selected with TvSTU software solution [25] by ensuring possibly tight packing of particles of each component. The respective grain size composition of each of the above was used as basis to find the optimal grain composition (Figure 4). Strength and density tests have confirmed the results of the simulation.

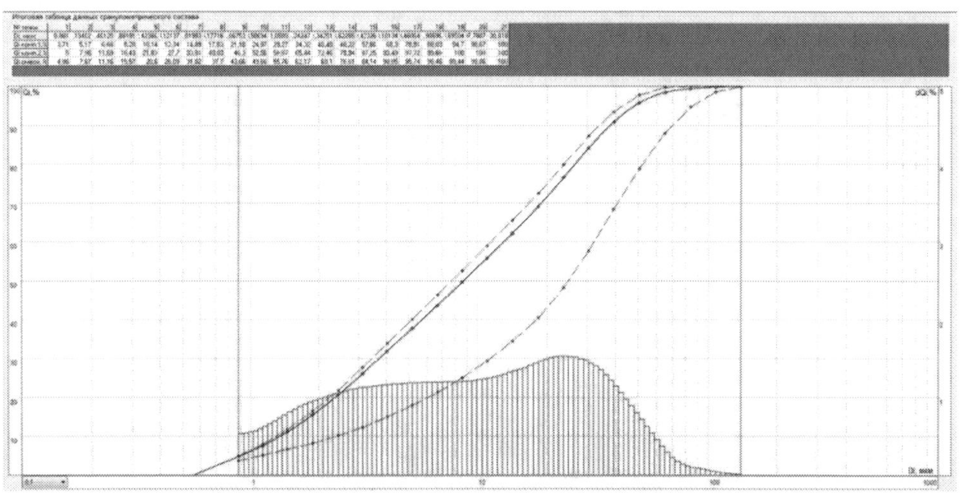

**Fig. 4.** Results of grain-size structure simulation of gypsum composite by TvSTU software solution

Highest strength of gypsum composite was 7.96 MPa, achieved by adding the modifier in 10 % proportion. Average density of composite was 1.687 kg/m$^3$. The results have experimentally confirmed the data from mathematical calculation. Therefore, optimal content of added modifier based on basalt powder waste was 10 % for tested of gypsum composite.

The next stage conducted a two-factor experiment to research the properties of gypsum composite modified with added basalt powder and Melflux hyper-plasticizer. The adopted variable parameters were: water-to-solid ratio ranging between 0.33 and 0.37, and percentage of plasticizer ranging between 0 and 0.2 %. The results are represented in Figures 5, 6, and 7.

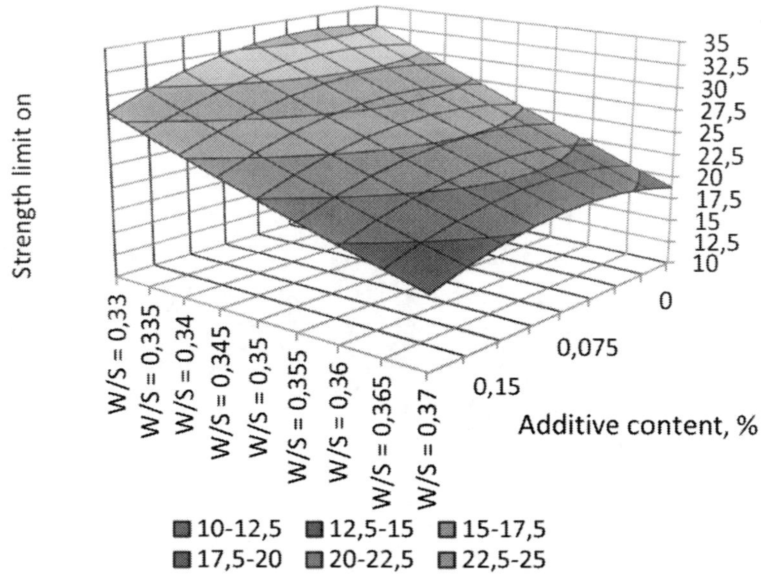


**Fig. 5.** Dependency of gypsum composite strength on content of Melflux 1461f hyper-plasticizer

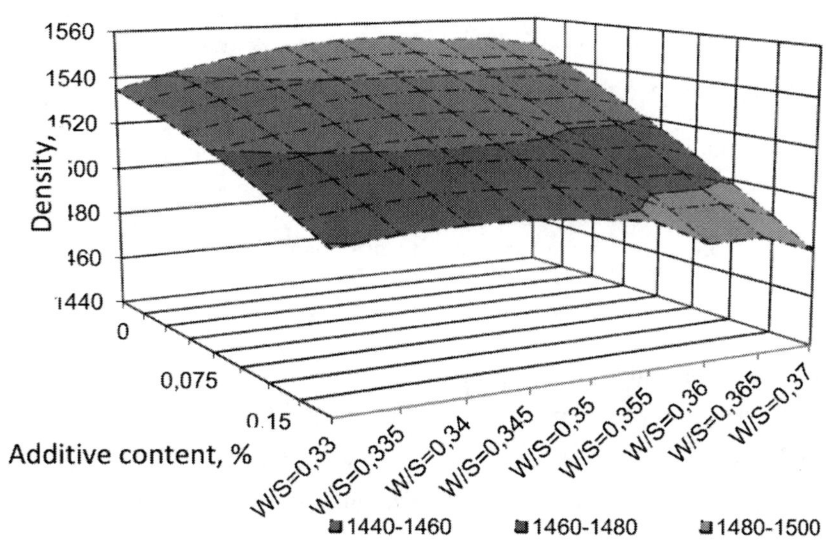


**Fig. 6.** Dependency of gypsum composite density on content of Melflux 1461f hyper-plasticizer

It has been found that strength of samples aged 7 days for all values of the water-to-solid ratio tends to rise with additive content rate of 0 to 0.075, as illustrated in Fig. 5. Further increase of additive content rate works to decrease strength properties.

Highest strength of 33.3 MPa is achieved at the water-to-solid ratio of 0.33; the low of 17.5 MPa occurs at the water-to-solid ratio of 0.37.

Increase of plasticizer content slightly decreases density of samples, while porosity rises; this is also due to excessive plasticizing effect (Fig. 2 and 3). Added plasticizer has virtually no effect on the material's density and porosity; this is different from the water-to-solid ratio – this effect can be explained by pores emerging when surplus water evaporates and changing the density of the samples. The lowest density was registered at the water-to-solid ratio of 0.37; the highest was registered at 0.33.

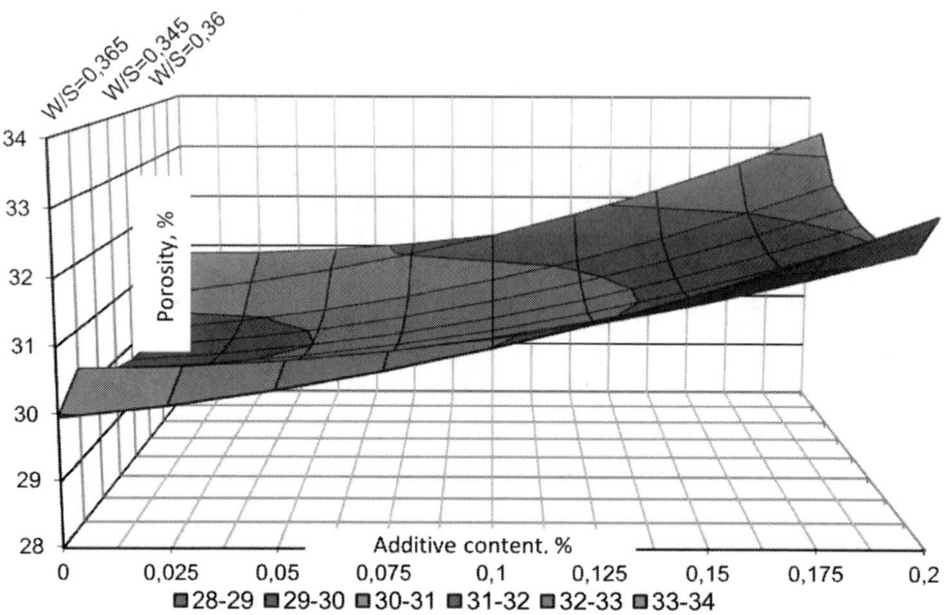


**Fig. 7.** Dependency of gypsum composite porosity on content of Melflux 1461f hyper-plasticizer

The properties of modified gypsum materials are improved by obtaining of denser particle packing in the presence of hemihydrate of finely dispersed basalt and plasticizer particles in the system, and by engaging basalt waste in the structuring process of modified gypsum stone.

Alkaline components found in basalt raise the environment's alkalinity, thus boosting the process of structuring in the system, because sulfates precipitate best in alkaline environments, as confirmed by the results illustrated in the X-ray photo (Figure 8) and microphotography of the structure of modified composite (Figure 9). Active fusion of calcium sulfate dehydrate particles is observed. The registered diffractometric image of gypsum composite with added basalt fiber demonstrates a rise of background activity under d = 2.7 - 3.1 Å, indicating the presence of an amorphous component.

## 4 Conclusions

Therefore, seeing the outcome of the above experiments, we can conclude that using basalt dust as additive is a positive effect on the structure of gypsum modified stone. Structural simulating of the topological space in gypsum modified composite by optimizing its grain-size composition highly improves its physical and mechanical properties. The morphology of the crystals depends on the environment's alkalinity. Use of Melflux hyper-plasticizer with modified gypsum composites has borne no significant effect in the researched material.

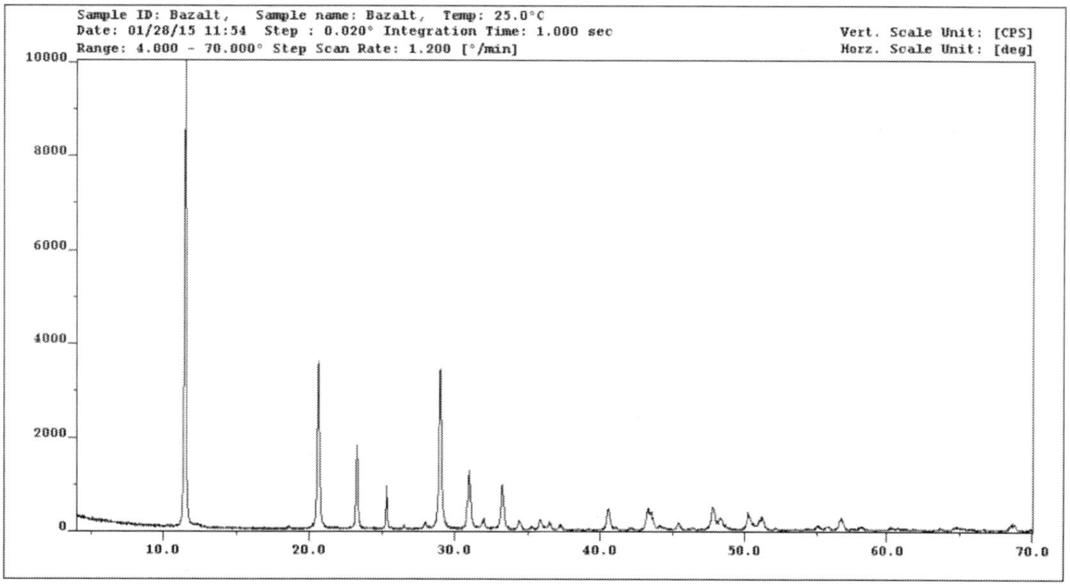


**Fig. 8.** Registered diffractometric image of gypsum composite with industrial basalt fiber

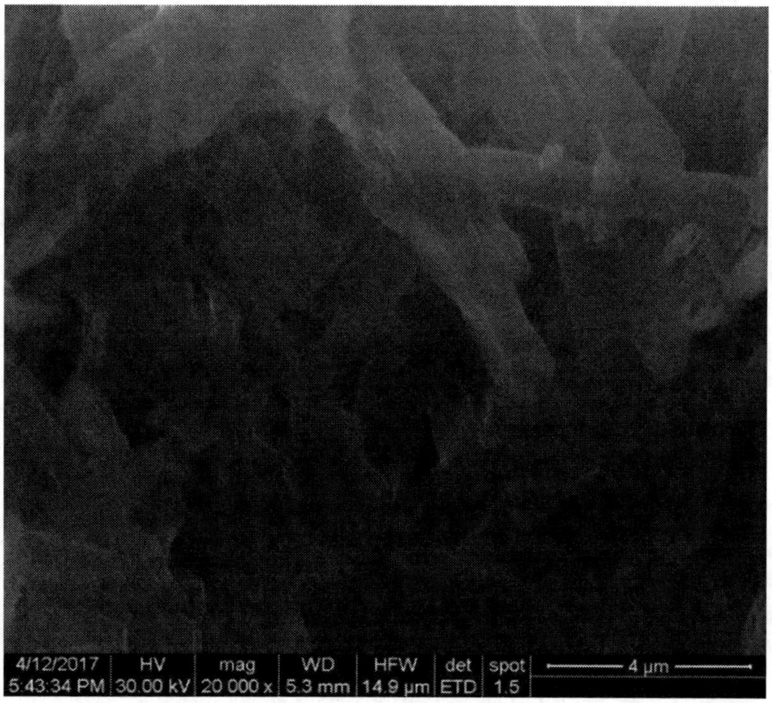


**Fig.9.** Microphotography of the structure of gypsum composite with industrial basalt fiber

## Acknowledgement

These authors are profoundly grateful to Professor M.S. Garkavi for his assistance with the research.

## References

1. A. V. Ferronskaya, ets. Gypsum materials and products. ASV (2004).
2. V.B. Petropavlovskaya, T. B. Novichenkova, A.F. Bur'yanov, Yu. Yu. Poleonova, *Procedia Engineering* **57**, pp. 906 -913 (2013)
3. V. V. Belov, A. F. Bur'yanov, G. I. Yakovlev, V.B. Petropavlovskaya, T. B. Novichenkova, etc. M.; De Nova (2012)
4. V.B. Petropavlovskaya, A. F. Bur'yanov, T. B. Novichenkova. *Stroitel'nye Materialy*, **7**, pp. 8-9 (2006)
5. Yu.V. Gontar. *Sukhie stroitelnye smesi*, **1** (2012)
6. S.P. Sivkov, E.N. Potapova, ets. *Sukhie stroitelnye smesi*, **3** (2014)
7. N.N. Morozova, N.V. Maysuradze, T.F. Galiev, L.I. Potapova. Vestnik Kazanskogo tekhnologicheskogo universiteta. T. 19. **2** (2016)
8. N.A. Galtseva Innovations and modeling in building materials science. Collection of scientific papers. Under a general edition of V.B. Petropavlovskoy. Tver (2016)
9. V.B. Petropavlovskaya, T. B. Novichenkova, A. F. Bur'yanov, K.S. Petropavlovskiy, H.-B. Fisher. Stroitel'nye Materialy, **7** (2014)
10. R.V. Chizhov, N.I. Kozhukhova, ets. Stroitel'nye Materialy. **3** (2015)
11. L. V. Zavadskaya, G.I. Berdov, ets. Izvestiya vysshikh uchebnykh zavedeniy. Stroitelstvo. **10** (658). (2013)
12. L. V. Zavadskaya, G.I. Berdov, Ya. S. Agalakova, etc. Izvestiya vysshikh uchebnykh zavedeniy. Stroitelstvo. **8** (2013)
13. D. O. Bondarenko, V. V. Strokova, A. M. Rykunov, V. V. Nelyubova. Modern construction materials, technologies and designs: materials of the International scientific and practical conference devoted to FGBOU VPO'S 95 anniversary of "GGNTU of the academician M. D. Millionshchikov" (2015)
14. G.I. Berdov, E.V. Parikova, V.F. Khritankov. Izvestiya vysshikh uchebnykh zavedeniy. Stroitelstvo, **8** (2006)
15. R. R. Ryazapov, R. H. Mukhametrakhimov, ets. Izvestiya KGASU. **3** (2011)
16. H. A. Khezhev, Yu. V. Pukharenko. Vestnik grazhdanskikh inzhenerov. **2** (2013)
17. M. S. Osnos, S. P. Osnos. Composite World. **5**(26). (2009)
18. H. A. Hezhev, Yu. V. Pukharenko. Vestnik grazhdanskikh inzhenerov. **2** (2013)
19. D. Mognonov, [et al.] Stroitel'nye Materialy. **10** (2012)
20. L.V. Zavadskaya, L.V. Ilina, G.I. Berdov. Izvestiya vuzov. Stroitelstvo. **7** (2011)
21. M.S Garkavi. Magnitogorsk: MGTU (2005)
22. M. S. Garkavi, third ed. Suhie stroitel'nye smesi. **2**, pp. 38 (2013)
23. V. G. Babayev, V. V. Strokova, V. V. Nelyubova. Vestnik Belgorodskogo gosudarstvennogo universiteta imeni V. G. Shukhova: BGTU. **4** (2012)
24. V. Petropavlovskaya, A. Buryanov, T. Novichenkova. *Advanced Materials Research*, **T.467**, pp. 91-96 (2014)
25. V.V. Belov, I.V. Obraztsov. Stroitel'nye materialy, oborudovanie, tekhnologii XXI veka. **7** (2012)

# Application of composite reinforcement for modernization of buildings and structures

*Alina* Cherkas[1*], *Vladimir* Rimshin[2]

[1]Moscow state university of civil engineering, Yaroslavskoye shosse, 26, Moscow, Russia, 129337
[2]Moscow state university of civil engineering, Yaroslavskoye shosse, 26, Moscow, Russia, 129337

**Abstract.** Properties of new reinforcing element — composite reinforcement, are considered. Feasibility study of using this material for modernization of buildings, namely for vertical extension with additional storey, is carried out. Comparative analysis of performance is conducted for metal and composite reinforcement, as well as calculations are performed which results will be used for determination of advantages and disadvantages of this innovative reinforcing method. An important factor for development of projects of building vertical extension after quite a long operation period is a proper selection of construction materials and strength analysis considering the structures to be erected. Calculation of a floor slab erected with using composite fiberglass reinforcement is represented. Results obtained in the calculation and the analysis of the data prove high efficiency of this reinforcing method through the decrease of loads on lower storeys of the building, economic feasibility and other factors.

## 1 Introduction

During its long history construction industry has experienced a great number of changes. Today, there is a variety of scientific developments and innovations introduced which can manifold improve construction quality and reliability of the construction process. This paper describes a new reinforcing element — high-strength non-metallic reinforcement made of composite materials. Fiberglass reinforcement is represented by extra strong rods with diameter from 4 to 20 mm with spiral-shaped ribbed surface available in any construction length. Composite fiberglass reinforcement is designed for use in concrete structures with pretensioned and non-tensioned reinforcement instead of conventional steel rebars. Non-metallic reinforcement is manufactured in the form of rods with spirally ribbed surface produced from glass or basalt fibers soaked with chemically resistant polymer. Composite materials are a group of rebars which differ by type of raw material and consist of two or more components. The first element — fibers produced from various types of raw materials, the second — thermosetting or thermoplastic polymer (resin) [1-4].

* Corresponding author: alina.tcherkas@yandex.ru

## 2 Materials and Methods

While exploring the topic of structure strengthening, it is worth mentioning that during a significant period of time steel rebars have been used for reinforcing concrete structures. Main disadvantages of this type of reinforcement are corrosion and electrical conductivity. These facts encouraged researchers to improve quality of reinforcing elements, and among other consequences was creation of non-metallic composite reinforcement. Composite reinforcement was developed as a consequence of the fact that steel rebars not always comply with the service standards on structures intended for operation in corrosive environment due to corrosion and lack of antimagnetic and dielectric properties required for many articles and structures, and also because they cannot provide for light weight of the reinforced structures. However, in case of building vertical extension even with only one storey, strengthening of floor structure raises many questions concerning reliability and light weight of the new structure, as well as stability of the existing one. Provision of solutions for these questions will substantially help to strengthen the structure and extend its service life without need for regular repairs [5-7].

## 3 The main part

The simplest and most efficient technology for vertical extension is erection of usable roof levels which provide up to 20-25% of additional living space with cost not exceeding 45-50% of new construction. In this case, it's essential to obtain data which confirm efficiency of the selected method of floor structure strengthening. Above all, attention should be paid to performance comparison of floor reinforcing materials.

Let's consider the advantages of composite reinforcement which include absence of electrical conductivity, high resistance to corrosive environment, high tensile strength, light weight, and the same for the finished structure.

However, it also has disadvantages which are worth mentioning. Among them: Composite reinforcement is not suitable for bending on site, as the shape is determined by drawings and configured by reinforcement rod during manufacturing. Heating up to 550°C leads to softening of binder contained in the composite material, which results in total loss of thermal insulating properties and require further measures to provide thermal insulation. Due to this fact, use of electric welding is excluded. Besides, elastic modulus of composite rebars is 3-5 times lower than that of steel structures. It means that for one and the same diameter fiberglass reinforcement will show much heavier deflection than their steel analogues. These characteristics allow successful using this material for construction of basements and foundations. However, use of this type of reinforcement in floor structures requires further calculations (i.e. quantity, surface area, material, diameter, etc.) [8, 9].

Further on, we will represent the comparative characteristics of the next material — steel rebars. Their main advantages include very high strength as compared to non-metallic analogues, high resistance to various impacts, including chemical attack, sufficient flexibility for performing construction operations, as well as possibility to use welding for connection.

However, in spite of all the advantages, there are also some disadvantages: high density results in considerable weight of such system and the fact that steel rebars are highly susceptible to corrosion, which jeopardizes service life of the materials.

The main factor for vertical extension with one storey is compliance with design loads associated with such modernization of the building. In this case, use of composite reinforcement for strengthening the floor being constructed is the most cost effective technical and economic solution. Further operation of the building shall be taken into account, as well as the fact that use of non-metallic reinforcement will extend service life of structures by 2-3 times as compared to metal reinforcement, especially when operated in corrosive environment including chlorine salts, alkalies and acids. It should be emphasized that metal rebars incorporated into floor structures and other reinforced concrete structures are the most critical, but also most weak and vulnerable part. Influenced by moisture and alkali contained in concrete metal rebars suffer from corrosion, that leads to limited service life, need for repair and eventually to destruction of reinforced concrete articles and structures. In order to provide for proper service life of reinforced concrete structures fiberglass reinforcement is utilized, that leads to long-term preservation of the buildings and structures being erected.

Table 1 below shows performance comparison of metallic and fiberglass reinforcement.

**Table 1.** Comparative characteristics of metallic and fiberglass reinforcement

| *Characteristics* | *Metallic reinforcement A400C Class according to GOST 5781-82* | *Non-metallic composite fiberglass reinforcement (АСП), according to Tech. Spec. 2296-001-12655746-2012* |
|---|---|---|
| Material | Steel 35ГС, 25Г2С and other | АСП — glass fibers with diameter 13-16 microns with polymer binder. |
| Tensile strength, MPa | 360 | 1,200 |
| Elastic modulus, MPa | 14 | 2.2 |
| Percentage elongation, % | Curved line with yield plateau under load | Straight line of linear elastic dependency under load below failure |
| Behavior under load (stress-strain curve) | 13-15 | 9-12 |
| Density, t/m3 | 7 | 1.9 |
| Resistance to corrosive environment | Corrosion with rust formation | Corrosion-proof material of the first chemical resistance group including resistance to alkali environment of the concrete. |
| Thermal conductivity | Yes | No |
| Electrical conductivity | Yes | No, dielectric |
| Profiles produced | 6-80 | 4-20 |
| Length | Rods with length 6-12 m | Various length on customer's request |
| Environmental friendliness | Environmental friendly | Safety and health certificate available, no hazardous or poisoning emissions |
| Durability | According to construction standards | Expected lifetime not less than 80 years. |
| Replacement for rebars in terms of physical and mechanical properties | 6 A400 C<br>10 A400 C<br>12 A400 C<br>14 A400 C<br>16 A400 C<br>18 A400 C | АСП-4<br>АСП-6<br>АСП-8<br>АСП-10<br>АСП-12<br>АСП-14 |
| Parameters of reinforcement cage equal in strength under load of 25 kg | When using rebars 8A400C, mesh size 14x14 cm, weight 5.5 kg/m2 | When using rebars 8АСП, mesh size 23x23 cm, weight 0.61 kg/m2 Less weight by 9 times |

| | | |
|---|---|---|
| Cost saving | Today, metal cost tends to rise by 67% on average during last 6 months | Money saving from replacement of metallic rebars with composite reinforcement of equal strength is 10-30%. Price growth dynamics is 2-4% per year |

Use of fiberglass reinforcement for strengthening concrete structures is possible due to specific properties of this material: fiberglass has high tensile strength; temperature deformation coefficient of glass insignificantly differs from that of concrete. One of the methods of strengthening concrete with fiberglass provides for using it as part of high-strength fiberglass reinforcement [14]. Calculation technique for brief analysis of floor slab shown below is based on strength characteristics of metal and fiberglass. Below we introduce grounds for calculation method of replacement of metal reinforcement with fiberglass reinforcement.

Input data:

Tensile strength: for metal $\sigma_{мет}$ =390 MPa, $\sigma_{асп}$ = 1000 MPa.

Ratio $\sigma_{мет}$ / $\sigma_{асп}$ – determines conversion coefficient:

$$K = \sigma_{асп} / \sigma_{мет} = 1000/390 = 2{,}56 \quad (1)$$

as $\sigma_{мет}$ and $\sigma_{асп}$ - are loads per unit square, so conversion coefficient can be applied to cross section ratio:

$$A_{асп} = \frac{A_{мет}}{K} \quad (2)$$

where

$A_{асп}$ – cross-section area of fiberglass reinforcement,

$A_{мет}$ – cross-section area of metal rebars, or

$$\pi D_{2асп} = \pi D_{2мет} / K \quad (3)$$

where

$D_{мет}$ and $D_{асп}$ - diameters of metal and fiberglass reinforcement respectively, therefore

$$D_{асп} = \sqrt{D_{2мет}} / K \quad (4)$$

with $K = 2{,}56$, we obtain

$$D_{асп} = 0{,}625\ D_{мет} \quad (5)$$

For example, 0,625*12 ($D_{мет}$) = 7,5 => $D_{асп}$ = 8

Based on these grounds for replacement of steel rebars with fiberglass reinforcement the following can be concluded [15]:

- Reinforcement of floor slab between storeys with span width not exceeding 3 meters in case of steel rebars can be assumed as Ø8 A-500C with pitch 200
- For reinforcement of floor slab between storeys with the same span width when using fiberglass reinforcement, several options can be assumed: Ø14 АКП-СП with pitch 200 Ø10 АКП-СП with pitch 100

It's worth mentioning that composite reinforcement have the advantage of lighter weight as compared to steel rebars, which is an important factor for the floor slab being reinforced. The reinforcement is produced in the form of rods with length of 6 meters or 12 meters. With diameter up to 12 mm (included) can be produced in the form of coils (reels), with length 100 meters. With diameter 10 mm can be produced in the form of coils (reels), with length 50-100 meters. 1 meter of fiberglass reinforcement АСП Ø 10 mm weighs 0.75 N. Correspondingly, total weight of 2,000 meters of composite reinforcement will be 0.75 N * 2000= 1,500 N. Therefore, 2,000 m of reinforcement АСП Ø 10 mm will be equal to 20 coils per 100 meters each. One 100-meter coil weighs 75 N. All 20 coils will weigh 1,500 N. All this can be transported in a jeep, car trunk, minibus or light truck. Regarding metallic reinforcement: 1 meter of metallic reinforcement A-500C, Ø 12 mm weighs 8.9 N. Correspondingly, total weight of 2000 meters of steel reinforcement will be 8.9 N * 2,000=

17,800 N. Transportation of this amount of material requires long truck or small hopper barge with a cargo body of proper length.

It can be concluded that steel rebars are heavier than composite reinforcement by 11.87 times. For transportation of this reinforcement large size vehicles must be involved, which is not economically feasible. At the same time, composite reinforcement can be transported in private cars without engaging a carrier company.

While analyzing possible risks of using both types of reinforcement for erection of floor structures for vertical extension of buildings, the economic factor shall also be considered. Real economic effect form using fiberglass reinforcement will be much greater than just the difference in the cost of one running meter [16-18]. Let's consider cost comparison diagram (**Fig. 1.**).

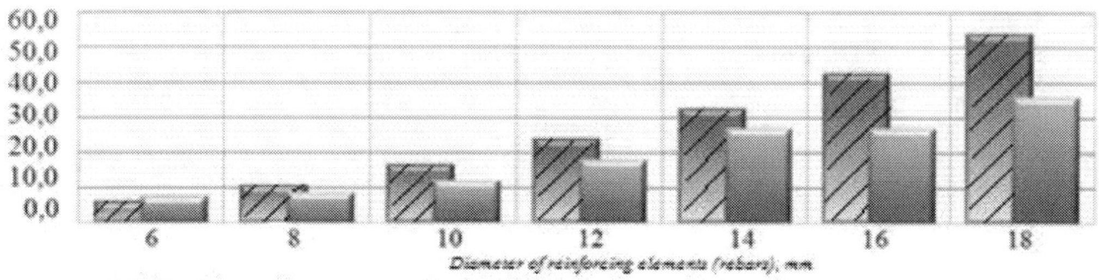


**Fig. 1.** Comparison of cost per one meter of metallic rebars A500C and fiberglass reinforcement

## 4 Conclusions

As it appears from the data above which are based on the assessment of technical and economic feasibility of using composite reinforcement for strengthening of floor slabs, the following conclusions can be made:

1. Based on comprehensive analysis of bearing capacity of bowing reinforced concrete elements which involves integrated practice of manufacturing, reinforcing and testing of structures, new data was obtained using the method of direct comparison.
2. Today, increasing efficiency of repair and reconstruction works is impossible without improving design solutions to strengthen structures which could provide their reliability, durability and economic feasibility. These indicators are equally dependent on the reduction of costs and labor intensity of production processes of strengthening, efficient utilization of materials and energy, as well as use of new materials.
3. Reinforcement of building structures using composite materials (composite reinforcement) for vertical extension of buildings which due to such properties as high tensile strength, light weight, adaptability to manufacture and resistance to corrosive environment become more and more popular in practice.
4. In the future we will develop and expand practical recommendations for designing reinforcing elements for building structures from composite materials intended for vertical extension. These recommendations will include: effect of concrete elements reinforced with composite materials on bearing capacity; behavior of structure under loading with total percentage of steel and composite reinforcement higher than limiting level; role and efficiency of anchor devices application for overreinforced sections when changing the type of composite material.
5. Introduction of newly developed recommendations will allow increasing reliability and efficiency of strengthening bowing concrete elements using unconventional reinforcing method.

## References

1. S. Korotaev, V. Kalashnikov, V. Rimshin, I. Erofeeva, V. Kurbatov *The impact of mineral aggregates on the thermal conductivity of cement composites*. EEC **22**, **3**, 1159-1164 (2016)
2. V.T. Erofeev, E.V. Zavalishin, V.I. Rimshin, V.L. Kurbatov, M.B. Stepanovich *Frame composites based on soluble glass*. RJPBC Sciences **7**, **3**, 2506-2517 (2016)
3. A.L.Krishan, E.A.Troshkina, V.I.Rimshin, V.A.Rahmanov, V.L. Kurbatov *Load-bearing capacity of short concrete-filled steel tube columns of circular cross section* . RJPBC Sciences **7**, **3**, 2518-2529 (2016)
4. V.Erofeev, S.Karpushin, A. Rodin, I.Tretiakov, V.Kalashnikov, M.Moroz, O.Smirnova, V.Rimshin, A. Matvievskiy *Physical and mechanical properties of the cement stone based on biocidal portland cement with active mineral additiv*. SSP **871**, 28-32 (2016)
5. Y.M.Bazhenov, V.T.Erofeev, V.I.Rimshin, S.V. Markov, V.L. Kurbatov *Changes in the topology of a concrete porous space in interactions with the external medium*. ESM **4**,**4**, 219-225 (2016)
6. V.D. Antoshkin, V.T.Erofeev, V.I.Travush, V.I.Rimshin, V.L.Kurbatov *The problem optimization triangular geometric line field*. MA Science **9**,**3**, 46-50 (2015)
7. V.T. Erofeev, A.D.Bogatov, S.N.Bogatova, V.F.Smirnov, V.I. Rimshin, V.L. Kurbarov *Bioresistant building composites on the basis of glass wastes*. BBRA **12**, **1**, 661-669 (2015)
8. A.Krishan, V.Rimshin, S.Markov, V.Erofeev, V.Kurbatov *The energy integrity resistance to the destruction of the long-term strength concrete* PE **117**, 211-217 (2015)
9. V.I.Rimshins, E.A.Larionov, V.T.Erofeyev, V.L. Kurbatov *Vibrocreep of concrete with a nonuniform stress state*. LSJ **11**,**11**,278-280 (2014)
10. В.И.Римшин, С.И.Меркулов *Элементы теории развития бетонных конструкций с неметаллической копозитной арматурой* ПГС **5**, 338-342 (2015)
11. D.Duriatti , A.Beakou , R. Levillain *Optimisation of the crimping process of a metal end-fitting onto a composite rod* CS **73**, **3**, 278-289 (2006)
12. S.Wen, W.Cai, H.Deng, R.Ke, S.Du, P.Xiong *Fault simulation experiment on composite insulators with fiber bragg grating implanted*. GJ **39**, **1**, 81-87 (2013)
13. M.Kumosa, Y.Han, I.Kumosa *Analyses of composite insulators with crimped end-fittings: part i - non linear finite element computations*. CST **62**, **9**, 1191-1207 (2002)
14. M.Kumosa, D.Armentrout, I.Kumosa, Y.Han, S.Carpenter *Analyses of composite insulators with crimped end-fittings: part ii - suitable crimping conditions*. CST **62**, **9**, 1209-1221 (2002)
15. C.Pinto, E.Andrad, L.Silva *Study of ionizing radiation on the properties of polyamide 6 with fiberglass reinforcement*. RPC **76**, **11-12**, 1708-1710 (2007)
16. A.Biddah *Structural reinforcement of bridge decks using pultruded gfrp grating*. CS **74**, **1**, 80-88 (2006)
17. F.Vilaseca, J.A.Mendez, J.P.Lopez, M.E.Vallejos, I.Barbera, M.A.Pelach, X.Turo, P.M utje *Recovered and recycled kraft fibers as reinforcement of pp composites*. CEJ **138**, **1-3**, 586-595 (2008)
18. V.F. Stepanova, A.V. Buchkin, D.A. Ilyin *Investigation of the features of the operation of concrete structures with combined reinforcement*. AAC **1**, 124-128 (2017)

# Research of the flow-over of the mechanical core of four pipes by a wind current, using Control Volume Method

*Dmitry* Chernyshev [1], and *Vadim* Alpatov[1,*]

[1]Samara State Technical University, Institute of Architecture and Civil Engineering, 194, Molodogvardeyskaya St., 443001, Samara, Russia

**Abstract.** One of the most important stages of designing the framework of high-rise stacks is the calculation of loads on the stack and on the mechanical core of the gas outlet pipes located inside the stack. In current normative documents there is no data on the aerodynamics of the mechanical core. At the same time the aerodynamic coefficients of pipes in the core differ from the aerodynamic coefficients of single cylinders. When dealing with gas dynamics problems, program complexes in which control volume method is applied are getting widespread use. When problems are solved in such complexes, the speed of wind current and the size of the structure are selected, and the required Reynolds numbers similarity is achieved. This paper presents the results of the research which deals with the flow-over of the mechanical core of four pipes in the gas-dynamic complex STAR-CD, the work of which is based on control volume method. In the research two options of the mechanical core model are considered, depending on the cylinders' relative position. The first option is the ratio of the distance between the cylinders' centers 1.5d, the second option is the ratio 2. For each design option the flow-over by the wind current at three angles of attack $0^0$, $30^0$ and $45^0$ is simulated.

## 1 Introduction

One of the most important stages of designing the framework of high-rise stacks is the calculation of loads on the stack and on the mechanical core of the gas outlet pipes located inside the stack. In current normative documents there is no data on the aerodynamics of the mechanical core. At the same time the aerodynamic coefficients of pipes in the core differ from the aerodynamic coefficients of single cylinders [1-3].

## 2 Materials and methods

To determine the real values of the aerodynamic parameters of the pipes in the mechanical core, it is necessary to carry out field observations of their flow-over in a wind tunnel. During flow-over the model, it is necessary to observe the Reynolds numbers similarity of

* Corresponding author: avu75@mail.ru

the model and the real structure. Models of high-rise structures turn out to be quite sizeable. That's why in some cases such similarity can not be achieved. It requires the use of large wind tunnels, which are not always accessible.

Recently, when dealing with gas dynamics problems, program complexes in which control volume method is applied have been getting widespread use. When problems are solved in such complexes, the speed of wind current and the size of the structure are selected, and the required Reynolds numbers similarity is achieved. This paper presents the results of the research which deals with the flow-over of the mechanical core of four pipes in the gas-dynamic complex STAR-CD, the work of which is based on control volume method.

The main idea of the method is that the domain of computation is divided into a number of control volumes, so that one nodal point is contained in one control volume. Differential equations are integrated over each volume. Using patch profiles, a discrete analogue of the differential equation is found, which includes the values of the variable of interest for several nodal points.

One of the important features of control volume method is that it implies the exact integral preservation of such values as mass, momentum and energy for any group of control volumes and, therefore, for the entire domain of computation. This is true for any number of nodal points, not only for the borderline case of their very large number. So, even the solution based on a very rough mesh satisfies accurate integral balances.

## 3 Results

Purposes of the research:

1) To determine the values of the pressure caused by the wind current at certain points on the surface of the pipes;

2) To obtain the integral parameters of the current (longitudinal and transverse aerodynamic coefficients) for each pipe and for the whole mechanical core;

3) To research the patterns of direction and the values of speeds, as well as the nature of the current in the areas near the cylinders.

Research objective. Stacks with a mechanical core of flue gas pipes are designed 100 m - 250 m high [4]. The diameter of the gas-outlet pipes in the mechanical core varies from 1 to 3.5 m. The distance between the centers of the flue gas pipes with respect to the diameter of the pipe is usually in the ratio of $(L/D)=1{,}5-2$, where L is the distance between the pipes, and D is the diameter of the flue gas pipe.

The characteristic value of wind pressure $W_0$ (Pa) is determined from the formula:

$$W_0 = 0{,}5 \cdot \rho \cdot V^2 , \qquad (1)$$

where $V$ is wind current speed, and $\rho = 1{,}22\,kg/m^3$ is atmospheric density. In current normative documents the speed is the speed of wind at the height of 10 m above ground level for the type of locality A by ten-minute averaging and the frequency of 5 years.

The characteristic value of the central component of wind load $W_н$ at the height $z$ above ground level, according to [5], is:

$$W_н = W_0 \cdot k_z \cdot c , \qquad (2)$$

where $k_z$ is the factor taking into account the height variation of wind pressure, and $c$ is the aerodynamic coefficient.

The movement of air mass in the boundary layer of up to 200 m is turbulent. The speed of air movement consists of the average component $\overline{V}$ and the pulsation component $V'$: $V = \overline{V} + V'$. The pulsation component is positive at some time points, at other time points it is negative, and its average time value is zero. As shown in paper [6], turbulent flow is considered to be a stationary random process with the average value equal to zero.

Solving the problems of turbulent fluid flows allows obtaining satisfactory data and thus simulating the behavior of real turbulent flows. The models of turbulence are described by a system of additional equations (as well as associated algebraic relations and a set of constants), that are solved together with the Navier-Stokes equation.

There is a wide variety of models of turbulence. In this paper calculations using STAR-CD program are made basing on the so-called two-parameter "high-Reynolds $k-\varepsilon$ model". Unlike other two-parameter models, $k-\varepsilon$ model is the most universal one from the point of view of forecasting turbulent effects [7]. Here $k$ is the specific kinetic energy of turbulent pulsations, $k = (3 \cdot \gamma^2)/2$, where $\gamma$ is turbulence intensity, $\varepsilon$ is dissipation rate of turbulent energy, connected with the scale of turbulence $L$ (the characteristic size of turbulent vortices in the direction of the midstream).

Modeling the flow-over of the mechanical core of pipes using STAR-CD program.

The requirements to make calculations for different angles of attack by the wind current lead to the necessity to rotate the researched core of cylinders. For this purpose, an intermediate cylindrical element [5], [8] is built in the mesh, which allows obtaining a new mesh with the given angle of attack of ram air current, by means of circular shift of the calculation model nodes in the required direction. The program has modeled the cross sections of the cylinders of the core, and the plane problem has been dealt with.

The cell dimensions of the mesh of voxels are described in detail in paper [8].

The initial conditions of the problem are: a resting medium.

Boundary conditions of the problem:

1. The input for the domain of computation is the speed of the current $V_n = 25\ m/s$ and the turbulence intensity $\gamma = 1\%$.
2. The output is the condition for pressure constancy $p = p_\infty$.
3. On the remote external lateral boundaries it is the condition of an aerodynamically smooth wall (the speed vector is parallel to the lateral boundary of the domain of computation). The dimensions of the workspace are defined in paper [1].
4. On the internal boundary (on the perimeter of the cylinders) the adhesion condition (zero speed) is set.
5. The time step is 0.001 (s).

## 4 Discussion

The results of the research of the flow-over of the mechanical core of four pipes. In the research two options of the mechanical core model are considered, depending on the cylinders' relative position. The first option is the ratio of the distance between the cylinders' centers $L/D = 1{,}5$, the second option is $L/D = 2$. For each design option flow-over by the wind current at three angles of attack $0^0$, $30^0$ and $45^0$ is simulated.

The results of the longitudinal and transverse aerodynamic coefficients of separate cylinders and of the whole mechanical core are presented in Table 1. The patterns of flow-over of the mechanical core by the wind current at different angles of attack for the option $L/D = 2$ are shown in Figures 1-3.

**Table 1**. The results of the research of the flow-over of the mechanical core of four pipes.

| № | Distance between the pipes' diameters and the angle of attack by the wind current | Longitudinal $C_X$ and transverse $C_Y$ aerodynamic coefficients | |
|---|---|---|---|
| | | $C_X$ | $C_Y$ |
| 1 | 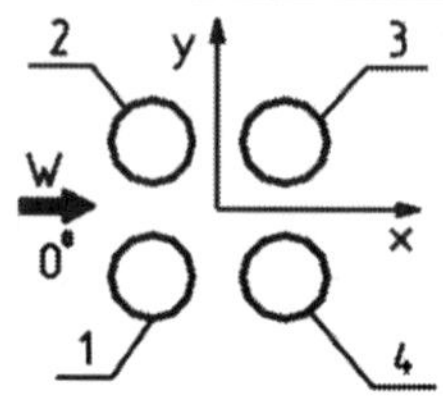  $L/D = 1{,}5$ ; angle of attack $0^0$ | $C_{X1}$= 0,29<br>$C_{X2}$= 0,29<br>$C_{X3}$= 0,32<br>$C_{X4}$= 0,32<br>$\Sigma$ = 1,22 | $C_{Y1}$= -0,12<br>$C_{Y2}$= 0,12<br>$C_{Y3}$= -0,28<br>$C_{Y4}$= 0,29<br>$\Sigma$ = 0,01 |
| 2 | 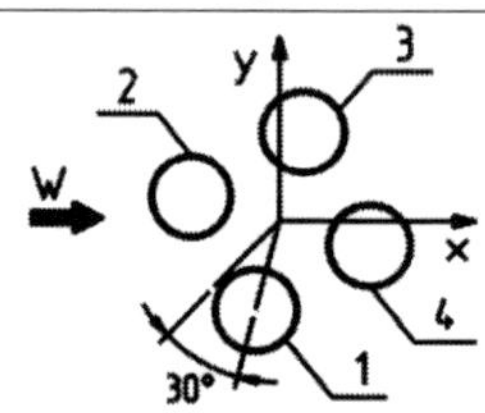  $L/D = 1{,}5$ ; angle of attack $30^0$ | $C_{X1}$= 0,41<br>$C_{X2}$= 0,62<br>$C_{X3}$= 0,47<br>$C_{X4}$= -0,07<br>$\Sigma$ = 1,43 | $C_{Y1}$= 0,04<br>$C_{Y2}$= 0,42<br>$C_{Y3}$= 0,28<br>$C_{Y4}$= -0,2<br>$\Sigma$ = -0,54 |
| 3 | 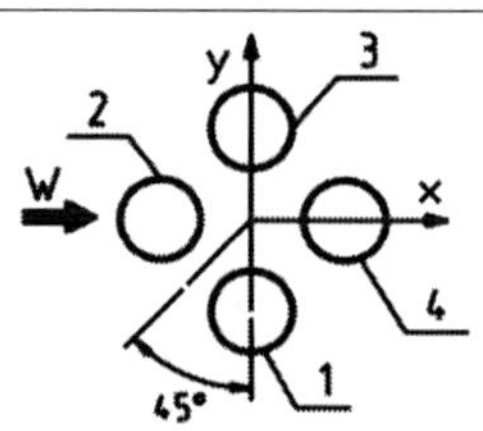  $L/D = 1{,}5$ ; angle of attack $45^0$ | $C_{X1}$= 0,43<br>$C_{X2}$= 0,75<br>$C_{X3}$= 0,43<br>$C_{X4}$= -0,17<br>$\Sigma$ = 1,44 | $C_{Y1}$= -0,07<br>$C_{Y2}$= 0<br>$C_{Y3}$= 0,07<br>$C_{Y4}$= 0<br>$\Sigma$ = 0 |
| 4 | 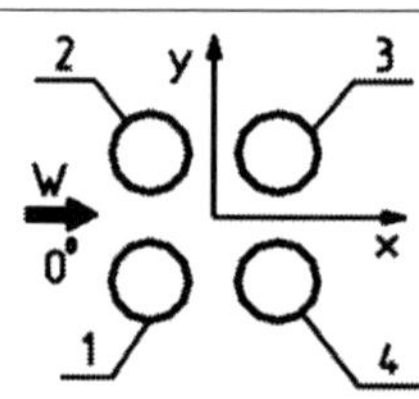  $L/D = 2$ ; angle of attack $0^0$ | $C_{X1}$= 0,26<br>$C_{X2}$= 0,26<br>$C_{X3}$= 0,37<br>$C_{X4}$= 0,37<br>$\Sigma$ = 1,26 | $C_{Y1}$= -0,05<br>$C_{Y2}$= 0,05<br>$C_{Y3}$= -0,21<br>$C_{Y4}$= 0,21<br>$\Sigma$ = 0 |
| 5 | 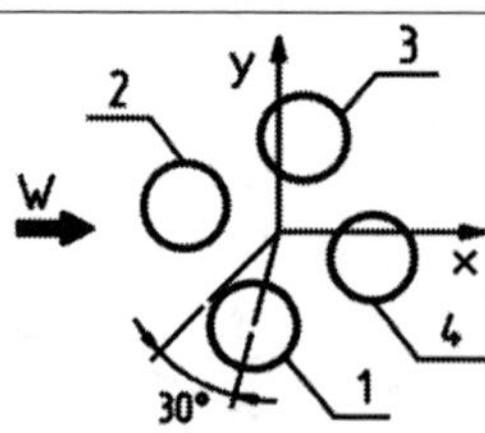  $L/D = 2$ ; angle of attack $30^0$ | $C_{X1}$= 0,33<br>$C_{X2}$= 0,46<br>$C_{X3}$= 0,44<br>$C_{X4}$= 0,16<br>$\Sigma$ = 1,39 | $C_{Y1}$= 0,09<br>$C_{Y2}$= 0,18<br>$C_{Y3}$= 0,04<br>$C_{Y4}$= -0,04<br>$\Sigma$ = 0,27 |

<table>
<tr>
<td>6</td>
<td>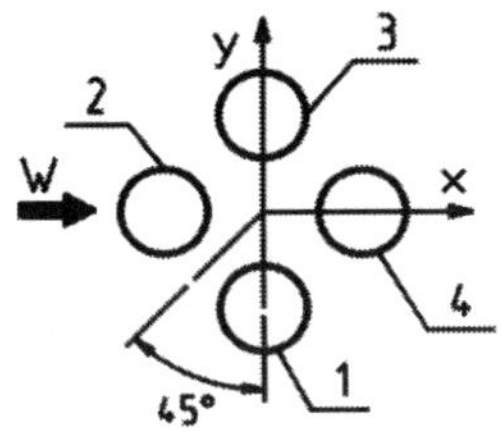

$L/D = 2$ ; angle of attack $45^0$</td>
<td>$C_{X1} = 0{,}39$<br>$C_{X2} = 0{,}5$<br>$C_{X3} = 0{,}39$<br>$C_{X4} = 0{,}11$<br>$\Sigma = 1{,}39$</td>
<td>$C_{Y1} = 0{,}02$<br>$C_{Y2} = 0$<br>$C_{Y3} = -0{,}02$<br>$C_{Y4} = 0$<br>$\Sigma = 0$</td>
</tr>
</table>

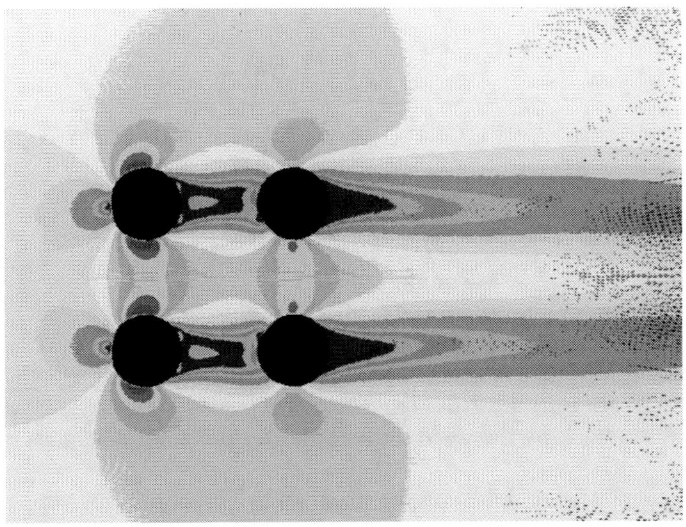

**Fig. 1.** Wind current speed field, by $L/D = 2$ , angle of attack $0^0$.

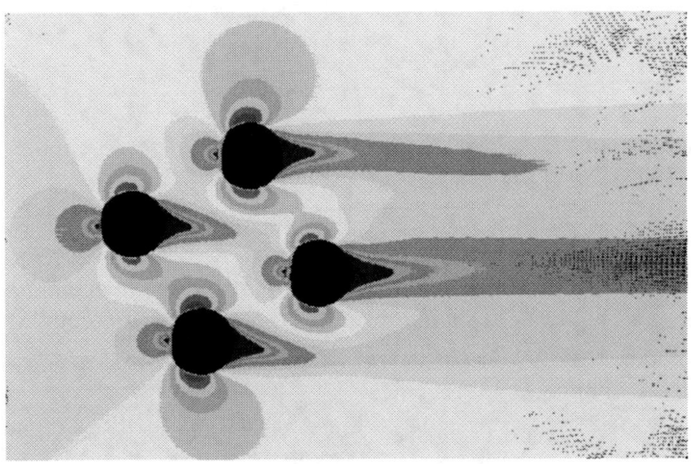

**Fig. 2.** Wind current speed field, by $L/D = 2$ , angle of attack $30^0$.

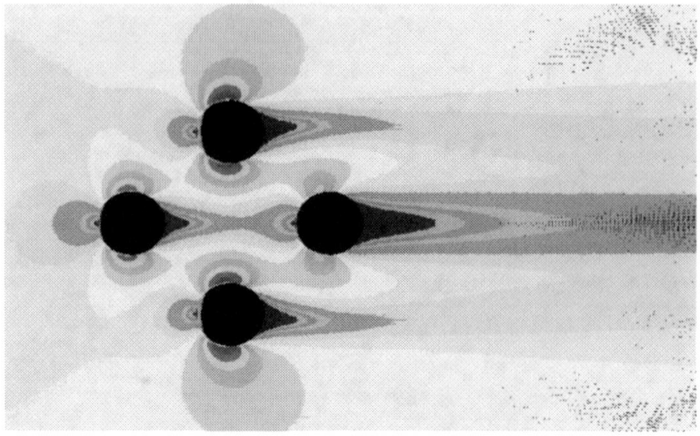

**Fig. 3.** Wind current speed field, by $L/D = 2$, angle of attack $45^0$.

## Conclusions

Conclusions and results of the research:

1. On the basis of control volume method, the problem of the flow-over of the mechanical core of four flue gas pipes' cylinders by a wind current is simulated.
2. The longitudinal and transverse aerodynamic coefficients of separate pipes and of the whole mechanical core are determined. The aerodynamic coefficients are determined for different angles of attack by the wind current and for different distances between the pipe's centers.
3. Wind current speed fields of the mechanical core of four pipes are obtained for different angles of attack and for different distances between the pipe's centers.
4. The results obtained may be useful for calculating the wind load when designing stack structures with the mechanical core of four similar gas-outlet pipes.

## References

1. A.V. Atamanchuk, *Extended abstract of candidate of technical sciences dissertation* (Samara, 2005)
2. I.S. Kholopov, *Analysis of structures and buildings under dynamic loading* (Samara, 2008)
3. I.S. Kholopov, *Analysis of structures and buildings under dynamic loading* (Association of higher educational institutions of civil engineering, Moscow, 2008)
4. M.B. Solodar, M. V. Kuznetsova, Yu.S. Plishkin, *Metal structures of stacks* (Stroyizdat, Leningrad, 1975)
5. D.D. Chernyshev, K.V. Yashukov, A.V. Atamanchuk, News of higher educational institutions. Civil engineering **9**, 107-112 (2010)
6. E. Simiu, R. Skanlan, *Impact of wind on buildings and structures* (Stroyizdat, Moscow, 1984)
7. L.G. Loitsansky, *Mechanics of liquid and gas* (Drofa, Moscow, 2003)
8. D.D. Chernyshev, I.S. Kholopov, A.V. Atamanchuk, Industrial and Civil Engineering **11**, 40-42 (2009)

# The estimation of transition curves geometry in railway engineering from measured data

*Jakub* Chromčák[1,*]

[1]Department of Geodesy, Faculty of Civil Engineering, University of Žilina, Slovakia

**Abstract.** The position of the track is defined as a course of the track in its ground plan. [1] In the simplest way the course of track can be defined as an axis, which is constructed not only from straight parts, but from the arcs and transit curves as well. In consideration of the normative changes it is necessary to change the point of view on measuring of track geometry parameters and hand in hand to adjust the way of processing and the results interpretation.

## 1 The current status of the geodesy in the railway engineering

Through track geometry measurements, which are defining the course of the track, some normative changes enter into force. From 1st July 2015 there is in force norm STN 73 6360-1: *Railway applications. Track. Part 1: Geometrical position arrangement of 1 435mm gauge railways* and norm 73 6360-2 *Railway applications. Track. Part 2: Acceptance of construction works, maintenance works and assessment of service condition track gauges 1 435mm*, which replaced norm STN 73 6360. These norms are interested in the requirements necessary for the designing, the construction, the reconstruction and the modernization of the normal track gauge with speed to 300km/h, than the technical parameters of the structural and the geometrical setting of the track, the rail branches and their spatial position, the taking over of the construction and the maintenance works. Through the track condition assessing is needed to check the geometrical position of track by the geodetic instrument with the continual recording. There are measured the position coordinates of track axis and the elevation of the non-overridden tracks roof. In special, reasonable occasions it is possible to control the position by conventional geodetic method. In other reasonable occasions it is possible to check track gauge and overriding by manual measuring method without continual recording. [2] The usage of these suppositions needs the technology of measurement and calculations to be adjusted depending on the design, the distance and the location of the object. [3] The possibility of the movements' estimation is the analysis of measurement as the parametrical defined geometrical parts. These can be after the measurement of displacement and deformation stages compared between stages. The biggest benefit of this method is closely total elimination of errors depending from centration and signalisation of points. In railway engineering, there are not expected discreet deformations, on the other hand deformations of whole parts are.

* Corresponding author: jakub.chromcak@fstav.uniza.sk

## 2 The characteristic of measured area

Hand in hand with the trans-European corridors modernisation, Railways of Slovak republic realized the reconstruction of track Nové Mesto nad Váhom- Púchov, where in one part- Turecký vrch, first time ever in Slovakia was designed ballastless track system RHEDA 2000®. [3] The whole length of Turecký vrch tunnel in the axis is 1775 m, the length of tunnels tube is 1738.5 m plus 25m continuing part from south side and 10m continuing part from the North. The double rail track in the tunnel is designed for speed 200km/h with opposite arcs with radius 2000 m. [4] The measurement was done not only in the part of ballastless track system but on the north side of the tunnel as well. In the transition part was first time used new type of construction in total length of 20 m that is using standard components of the rail superstructure without its´ stabilisation. [5]

## 3 The estimation of the track positioning from the measured data

The easiest is to estimate the parameters of the straight part. For the most precise interpretation of rails´ course is sufficient to fold the regression line through measured data (it is better to estimate the rail axis but it is possible to estimate one of the track as well). This regression line can be described algebraically by linear equation [6]

$$y = ax + b. \quad (1)$$

Through the estimation of arc parameters are estimated coordinates of the arc centre and its´ radius. In Cartesian coordinate system with the arc centre S [$Y_0$, $X_0$] and radius R is the set of points defined by linear equation

$$(Y - Y_0)^2 + (X - X_0)^2 = R^2. \quad (2)$$

### 3.1 The comparison of measured data with the different types of transition curves

The troubles through estimation rise because of its parametrical definition. The basic functions of transition curves are:

- a smooth transition between the straight section and the circular arc, the curvature gradually changes from zero to the final value;
- provide space for smooth change when creating or changing an override;
- ensure a smooth curvature change in complex arcs or between opposite arcs.

The transition curve can be only the curves that meet these conditions:

- one end of the curve has a cross-section with a downstream straight line;
- the other end of the curve has a common cross-section with a downstream arc;
- the curvature of the transition point at the point following the circular arc have to be the same as the curvature of the circular arc;
- the curvature of the transition point at the point following the straight section have to be equal to zero;
- the course of the change of the curvature over the entire length of the transition period corresponds to the course of the override change.

### *3.1.1 Parabola*

A parabola is a two-dimensional mirror-symmetrical curve. The parabola is described as a lotus of points that are equidistant from both a directrix (line) and a focus (point). The focus does not lie on the directrix.

The parabola can be algebraically described by equation

$$y = ax^2 + bx + c\,. \tag{4}$$

The comparing calculation was done by the least squares method it is necessary to create design matrix, which contains from partial derivatives of the *a, b, c* parameters with total length n x r (n- number of measured points, r- number of unknown parameters), the weight matrix could be neglected, every point was measured with same precision. From the results of the calculation, parameter a=0.000508 b= -0.866551and c= 0.102152. The mean error $m_0$ calculated from equation

$$m_0 = \sqrt{\frac{v^T v}{n-k}}\,, \tag{5}$$

where k means the number of needed measures (3) and v means the vector of residuals. The value of $m_0$ is the best comparative criterion. The size of parabolas′ mean error is 0,147m.

### *3.1.1 Klothoide*

Whereas a Klothoide is forbidden for Railway Engineering in Slovakia, it is used in every other country in European Union but it is often used worldwide. The klothoide is also known as the curve with linear increase of curvature. From the Figure 1 it is possible to see that the growing in axis x and y are marked by dx and dy:

$$dx = dl.\cos\tau\,, \tag{6}$$

$$dy = dl.\sin\tau\,. \tag{7}$$

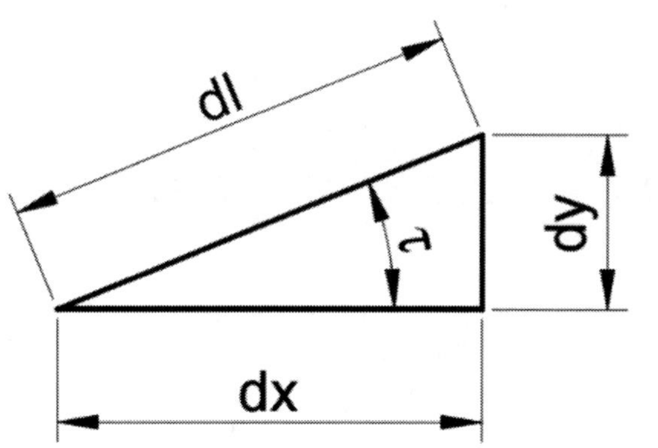


**Fig. 1.** The growing of x and y coordinates in klothoide

$$\tau = \frac{l^2}{2.L_K R} \tag{8}$$

The l means the distance from the start of transition curve, LK is the distance of whole curve and R is the radius of klothoide curvature with is equal to radius of arc, which transition curve is connected to.

To possible use of the least square method it is necessary to derivate equations (6) and (7) so these functions must be linearized by Taylor series.

$$\begin{aligned} dx &= dl.\cos\tau = dl.(1-\frac{\tau^2}{2!}+\frac{\tau^4}{4!}-\frac{\tau^6}{6!}+...) = \\ &= dl.(1-\frac{l^4}{8.L_K^2 R^2}+\frac{l^8}{384.L_K^4 R^4}-\frac{l^{12}}{46080.L_K^6 R^6}+...) \end{aligned} \tag{9}$$

$$\begin{aligned} dy &= dl.\sin\tau = dl.(\tau-\frac{\tau^3}{3!}+\frac{\tau^5}{5!}-\frac{\tau^7}{7!}+...) = \\ &= dl.(\frac{l^2}{2.L_K R}-\frac{l^6}{48.L_K^3 R^3}+\frac{l^{10}}{3840.L_K^5 R^5}-\frac{l^{14}}{645120.L_K^7 R^7}+...) \end{aligned} \tag{10}$$

After the integration by the distance l we become equations:

$$\begin{aligned} dx &= \int(1-\frac{\tau^2}{2!}+\frac{\tau^4}{4!}-\frac{\tau^6}{6!}+...)dl = \\ &= (l-\frac{l^5}{8.L_K^2 R^2}+\frac{l^9}{384.L_K^4 R^4}-\frac{l^{13}}{46080.L_K^6 R^6}+...) \end{aligned} \tag{11}$$

$$\begin{aligned} dy &= \int(\tau-\frac{\tau^3}{3!}+\frac{\tau^5}{5!}-\frac{\tau^7}{7!}+...)dl = \\ &= (\frac{l^3}{2.L_K R}-\frac{l^7}{48.L_K^3 R^3}+\frac{l^{11}}{3840.L_K^5 R^5}-\frac{l^{15}}{645120.L_K^7 R^7}+...) \end{aligned} \tag{12}$$

After the adjustment by the least square method it is necessary to astimate the values, $l_i$ is distance of every point from the beginning, other valuea ase $L_K$- the lenght of whole curvature and the radius of curvature R. The number of unknown is equal to number of meausred points n+2. The number of equation that are used in method is 2.n, every coordinate has it own equation.

The shape of desing matrix is

$$A_{2n,n+2} = \begin{vmatrix} \frac{\partial f(x_1)}{\partial R} & \frac{\partial f(x_1)}{\partial L_K} & \frac{\partial f(x_1)}{\partial l_1} & \frac{\partial f(x_1)}{\partial l_2} & \cdots & \frac{\partial f(x_1)}{\partial l_n} \\ \frac{\partial f(y_1)}{\partial R} & \frac{\partial f(y_1)}{\partial L_K} & \frac{\partial f(y_1)}{\partial l_1} & \frac{\partial f(y_1)}{\partial l_2} & \cdots & \frac{\partial f(y_1)}{\partial l_n} \\ \vdots & \vdots & \vdots & \vdots & & \vdots \\ \frac{\partial f(x_n)}{\partial R} & \cdots & & & & \frac{\partial f(x_n)}{\partial l_n} \\ \frac{\partial f(y_n)}{\partial R} & \cdots & & & & \frac{\partial f(y_n)}{\partial l_n} \end{vmatrix} =$$

$$= \begin{vmatrix} A_{1,1} & A_{1,2} & A_{1,3} & \cdots & A_{1,n+2} \\ A_{2,1} & A_{2,2} & A_{2,3} & \cdots & A_{2,n+2} \\ \vdots & \vdots & \vdots & & \vdots \\ A_{2n-1,1} & \cdots & & & A_{2n-1,n+2} \\ A_{2n,1} & \cdots & & & A_{2n,n+2} \end{vmatrix}, \qquad (13)$$

To simplify the calculation it was not needed to calculate with more than first three units of Taylor series because of their low impact to calculations presented by thousandths of millimetre. Through calculations were the values of R, $L_K$ and $l_i$ estimated and there was only calculation of the supplements. The values were $R_0$=1997m, $L_{KO}$=450m and $l_{i0}$ was calculated from equation $l_{i_0} = \sqrt{x_i^2 + y_i^2}$ .

The final values of R=1997.362m and $L_K$=432.233m. The equations of coordinate calculations look like:

$$y = \frac{l^3}{5{,}179954.10^6} - \frac{l^7}{3{,}088626.10^{19}} + \frac{l^{11}}{2{,}302050.10^{31}}, \qquad (20)$$

$$x = l - \frac{l^5}{2{,}9981326.10^{13}} + \frac{l^9}{1{,}919873.10^{27}}. \qquad (21)$$

The size of klothoides´ mean error is 0,220m.

## Conclusion

After the legislative changes in present age, it is necessary to change the access of measurement adjustment. It is needed not only to change the methods of measurement from discrete to continual but to change the method of adjustment as well. One possibility is to divide the track into the parts with same directional guidance (line, transition curve and arc) and calculate their parameters. These can be compared between epochs and show better conclusion without impact of pointing error etc.

## Acknowledgements

This article is the result of the implementation of the project VEGA 1/0275/17 "Application of numerical methods to define the changes of geometrical track position" supported by the Scientific Grant Agency of the Ministry of Education, science, research and sport of the Slovak Republic and the Slovak Academy of Sciences. This article is the result of the implementation of the project ITMS 26220220156 "Broker centre of air transport for transfer of technology and knowledge into transport and transport infrastructure" supported by the Research & Development Operational Programme funded by the ERDF.

## References

1. J. Ižvoltová, A. Villim, P. Kozák, Determination of Geometrical Track Position by Robotic Total Station. Theoretical Foundation of Civil Engineering, Wroclaw, XXIII R-S-P seminar, (2014)
2. STN 73 0275 Accuracy of geometric parameters in building industry. Check measurement of line structures
3. STN 73 6360-2 Railway applications. Track. Part 2 : Acceptance of construction works, maintenance works and assessment of service condition track guages 1 435mm
4. J. Ižvoltová, P. Pisca, Vyhodnotenie 3D laserového skenovania pevnej jazdnej dráhy. Faculty of civil Engineering STU, Bratislava, (2015)
5. J. Ižvoltová, P. Pisca, V. Koťka, M. Mancovič, 3D laser scanning of railway line. Communications. University of Žilina. (2015)
6. L. Ižvolt, Súhrnná správa o vstupných skúškach pre monitoring časti úseku nekonvenčnej konštrukcie železničného zvršku a jej prechodových oblasí. University of Žilina. Žilina. (2012)

# To the question of peculiarities of thermal activation of natural siliceous raw material

*Natalya* Chumachenko[1,*], *Diana* Galiullina[1] , and *Euphym* Mironenko[2]

[1] Samara State Technical University, Institute of Architecture and Civil Engineering, 194, Molodogvardeyskaya St., 443001, Samara, Russia
[2] CJSC «TechStroyProekt», 146, Samarskaya St., 443010, Samara, Russia

**Abstract.** The results of research of activity enhancement of natural siliceous raw material are given in the article. Fossil meal of Khotynetsky deposit, diatomite of Sharlovsky deposit, silica clay of Balasheika deposit were used as natural active mineral admixtures. The influence of heat-treating temperature and dispersion on activity of different types of siliceous raw material is studied. The increase of activity of fixation of $Ca(OH)_2$ in several times is traced after heat-treating at a certain temperature in the range from 100 to 800 $^0$C. The type of activity change is discovered. Explanation is given connected with the change of silica structure in the surface layer. Parameters of the highest activity are defined for every type of siliceous raw material.

## 1 Introduction

In the structure of many cementing, first of all portland pozzolan cement and limestone pozzolan cementing concretes are present active mineral admixtures (AMA) [1 - 3].
When maturing AMA adsorb at first, and then chemically fix Ca(OH)2, received from interaction of alite with water into solventless hydrated calcium silicate according to reaction:

$$mCa(OH)_2 + SiO_2(act.) + nH_2O \rightarrow (0,8\text{-}1,5)CaO{\cdot}SiO_2{\cdot}pH_2O \quad (1)$$

This process happens under usual conditions of cementing: moist environment with positive temperatures. The amount of AMA in the structure of cementing depends on mineral composition of cementing and activity of the admixture. Activity of AMA is determined via fix of $Ca(OH)_2$.

The activity amounts of different admixtures of natural and artificial origin which are given in table 1 are known.

As it is seen from table 1 the highest activity is with silica wastes, and from natural ones with natural siliceous raw material.

The siliceous raw material is used widely in the structure of airbinding [4] concretes [5, 6], in the materials of construction-related use [7].

* Corresponding author: uvarovang@mail.ru

**Table 1.** Activity of admixtures.

| Group of admixtures | Name of admixtures | Activity of admixture on soaking of lime, lime-stone mortar during 30 days, in mm CaO on 1 g of admixture, not less |
|---|---|---|
| Natural admixtures of sedimentary origin | Diatomites, fossil meals, silica clays | 150 |
| | Burnt clays | 30 |
| Natural volcanogenic admixtures | Ashes, pumice-stones, pozzolans | 50 |
| | Trasses | 60 |
| Artificial admixtures | Silica waste | 200 |
| | Burnt clay, furnace bottom ashes and slags | 50 |

The efficiency of admixtures of natural and industrial raw material can increase due to mechanochemical activation [8] and thermal activation [9 -11].

Taking into consideration the spread of natural siliceous raw material and resource-saving approach to the raw-material base of construction industry [12] a goal was set: to study the influence of heat-treating temperature on activity of natural siliceous raw material of different dispersion.

## 2 Research material

As natural active mineral admixtures (AMA) were applied: fossil meal of Khotynetsky deposit, diatomite of Sharlovsky deposit, silica clay of Balasheika deposit. Chemical structures of silica raw material are presented in table 2.

**Table 2.** Chemical structure of silica raw material.

| Type of silica material | Oxide content, mass. % | | | | | | |
|---|---|---|---|---|---|---|---|
| | $SiO_2$ | $Al_2O_3$ | $Fe_2O_3$ | CaO | MgO | $R_2O$ | Polypropylene |
| Silica clay | 82,0 | 4,6 | 2,8 | 0,6 | 1,5 | 4,3 | 4,2 |
| Diatomite | 76,0 | 6,0 | 3,3 | 0,4 | 2,7 | 3,2 | 8,4 |
| Fossil meal | 56,0 | 14,8 | 8,6 | 1,6 | 2,8 | 4,5 | 11,7 |

Characteristics of silica raw material are studied well enough [13 -14].

## 3 Methods

In the research two methods were used: method of thermal activation and method of activity determination.

Method of thermal activation lies in change of activity due to heat-treating. Activity of AMA is determined first of all by the origin of admixture (table 1). At the same time initial activity can be changed by heat-treating. This is the so called thermal activation effect [15].

With the aim of activity enhancement of natural siliceous raw material was studied the effect of thermal activation which is mentioned in literature. The research was held with natural siliceous raw material.

Method of activity determination: after heat-treating activity was determined on the basis of limestone soaking from limestone mortar in mg CaO on 1 g of admixture.

## 4 Experiment

To determine the temperature of regional activation of silica containing AMA they were subject to heat-treating under temperatures from 100 to 800 $^0$C with the interval in 100 $^0$C.

The experiment was conducted parallel on silica raw material of three types of dispersion:
1 – with particle size of 0,3 - 1,0 mm;
2 – with particle size less than 0,3 mm (100 % undersize 0,3 mm);
3 – with particle size less than 80 micron (100 % undersize 0,08 mm).

After heat-treating of silica raw material of a certain dispersion their fixation activity of $Ca(OH)_2$ was determined.

## 5 Experiment results

The results are given in figure 1. The initial activity of raw material is different and changes with dispersion. For large particles of 0,3-1,0 mm, when natural dispersion is saved the highest activity was with diatomite and fossil meal as more permeable rocks. When regrounding the raw material till 100% undersize 0,3 mm the activity falls which is explained with the erosion of initial permeable structure, decrease of specific surface. Further grounding (till 100 % undersize 0,08 mm) enables activity enhancement proportionally to the initial rock density.

The most essential activity enhancement due to heat-treating was reached for diatomite with dispersion < 80 micron at t = 200 $^0$C .

## 6 Discussion

Activity dependence on temperature for investigated types and dispersion of silica raw material has a peculiar sinusoidal character and according to Snider's data [15] is explained by the change of structure of silica in the surface layer. On the surface of silica in different proportions can be five types of groups:

- silanol (combination water - free, separate OH-groups);
- physically-combined water – water molecules having hydrogen bonds with silanol groups;
- dehydrated oxide - siloxane groups;
- bigeminal OH group fixed with one silicon atom;
- reactive vicinal OH groups prevailing in fine-pored silica – neighbouring, situated nearby OH-groups fixed by hydrogen bond.

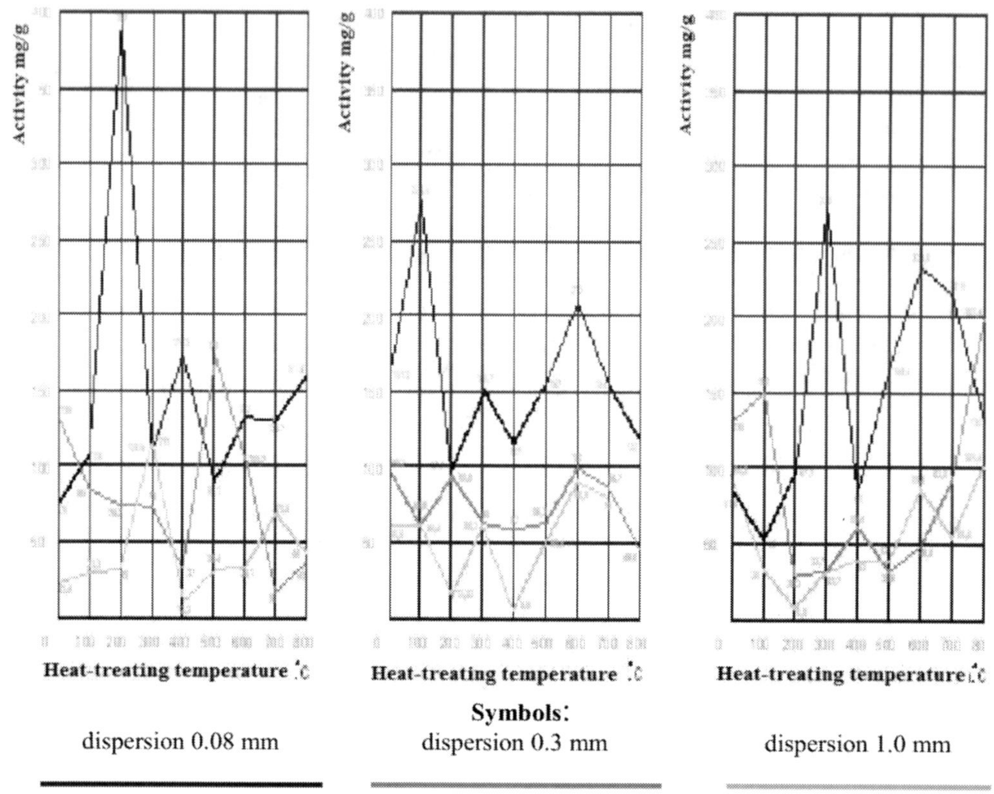


**Fig. 1.** Thermal activation of natural natural siliceous AMA.

Under heating silica first of all physically-combined water is removed. If temperature increases silanol groups are able to condensate making thus siloxane groups. As a result the material gets water-repellent property and consequently the lowest chemical activity. The highest chemical activity have reactive vicinal groups.

Though the research [15] has found that to differentiate these processes is not possible as they can occur simultaneously, but explain the received results. Maximum activity at low temperatures (100 - 350 $^0$C) is connected with removal of physically-combined water, at high temperatures – with the creation of reactive groups. Minimal activity is explained by creation of siloxane groups.

## 7 Use of the experiment results

Diatomite of the highest activity received with heat-treating of finely dispersed diatomite (< 80 micron) at t = 200 $^0$C was used in the structure of cementing of masonry mortar which guarantees absence of efflorescence from $Ca(OH)_2$ [16].

## Conclusions

1. Original and artificial active mineral admixtures generally made of amorphous silica ($SiO_2$ more than 50 %) is widely used in the structures of cementing, mortars and concretes. AMA adds to the reduce cement consumption and enhances corrosive resistance of hardened cement paste.
2. Is defined thermal activation after heat-treating within temperature range from 100 up to 800 $^0$C of natural siliceous raw material: diatomite, silica clays, fossil meal.

3. It is found out that the activity amount is influenced by the type of raw material, heat-heating temperature and dispersion.
4. Dependence activity on temperature has a sinusoidal character and is explained by change of silica structure.

## References

1. GOST P 56592 Mineral admixtures for concretes and water mortar. General technical conditions (2015)
2. U.G. Ivashenko, N.A. Kozlov, D.K. Timokhin, Vestnik of Saratov State Technical University 3, 25-29 (2010)
3. U.M. Bazhenov, V.S. Demyanova, V.I. Kalashnickov. *Modified high-quality concretes*, 368 (Publishing house of Association of Universities of Building and Construction, Moscow, 2006)
4. T.U. Saphonova, Izvestia of St.Petersburg Railroad University **2(31)**, 93-98 (2012)
5. E.G. Matveeva, *Concrete efficiency enhancement by admixture of nanodispersed silica: dissertation ... candidate degree in technology* : 05.23.05 ( Belgorod, 2011)
6. V.I. Cherkashin, B.D. Toturbiev, V.Y. Matsapulin, Concrete and armed concrete **6**, 2-5 (2013)
7. S.F. Koren'kova, U.V. Sidorenko, Achievements of modern natural sciences **3**, 172 (2014)
8. V.P. Suchkov, Urban Construction and Architecture **4**, 82-86 (2011) DOI: 10.17673/Vestnik.2011.04.16
9. E.I. Evtushenko, *Management of structure and formation of building materials with the use of raw material thermal activation : dissertation ... candidate degree in technology: 05.23.05* (Belgorod, 2005)
10. I.N. Anisina, A.G. Chetverikova, O.N. Kanugina, Vestnik of OSU **4(140)**, 170-174 (2012)
11. O.N. Kanugina, A.G. Chetverikova, Study of materials **6**, 44-49 (2011)
12. N.G. Chumachenko, Urban Construction and Architecture **1**, 112-116 (2011) DOI: 10.17673/Vestnik.2011.01.22
13. N.G. Chumachenko, M.N. Baranova, S.F. Koren'kova, Construction materials **8**, 1-4 (2011)
14. N.G. Chumachenko, M.N. Baranova, S.F. Koren'kova, Construction materials **12**, 72-75 (2012)
15. *Modified silica in sorption, catalyse and chromatography,* 24-26 (Chemistry, Moscow, 1986)
16. E.V. Mironenko, *Physical and chemical processes of efflorescenceformation in brickwork and methods of its elimination : dissertation ... candidate degree in technology: 05.23.05.* (Samara, 2006)

# Choice of design solutions for BREEAM international certificate

*Katarzyna* Ćwik[1], *Paweł* Nowak[1*]

[1] Warsaw University of Technology, Civil Engineering Faculty, Armii Ludowej 16, 00-637 Warsaw, Poland

**Abstract.** The aim of this paper is to describe and explain the influence of BREEAM International system on the choice of design solutions. Authors present the algorithm for implementation during the process of applying for BREEAM certificate. The algorithm was prepared to create a clear scheme that would enable any party of the investment process, especially designers, to understand and effectively go through all the fundamental stages of assessing the performance of the building and to become successfully certified.

## 1 Introduction

BREEAM (Building Research Establishment's Environmental Assessment Method) is the world's most popular, multi-criteria environmental assessment method for buildings. At the moment (end of year 2016), 554 733 buildings are certified and 2 256 752 are registered in this system in 78 countries [1]. First standard concerning assessment of commercial buildings in Europe was launched in 2008 by BRE. Then, in 2009 and 2010 the schemes were updated. It was decided to launch a new improvement of BREEAM International in 2011, expanding the scope of the certification to residential buildings and buildings outside of Europe. The main aim of BREEAM certification is to measure and describe the environmental performance of the building during the Design Stage (DS) and Post-Construction Stage (PCS). The system aims mainly at reducing the impact of buildings on the environment and broadening the concept of sustainable buildings. The aforementioned is applied by ensuring that best environmental practice is implemented during the Design Stage and Post-Construction Stage of the building by setting the criteria and standards to be followed [5, 6, 7, 9] - all concerning lowering the environmental impact of the building. BREEAM enables to raise the public awareness and allows for demonstrating progress towards environmental objectives. All is performed according to BRE Global's guides, publications, standards and certification schemes.

There are two stages distinguished as mentioned before – Design Stage and Post-Construction Stage. The Design Stage (DS) assessment is the intermediate measurement of building's performance, mostly completed at the stage of design that allows for checking and reporting the accordance to the technical guidance. This stage leads to the interim BREEAM Certificate. The Post-Construction Stage leads to final BREEAM certificate and

* Corresponding author: p.nowak@il.pw.edu.pl

represents the final performance of the building. It is completed and certified when all building works are completed. Here two approaches can be distinguished: post-construction review confirming the rating achieved at the design stage or the full review in case the interim assessment has not been carried out.

The crucial definitions need to be enclosed in order to perform the further analysis of influence of BREEAM on the choice of design solutions [2, 9, 10]:
- Shell and Core – Typically Shell and Core building covers base building elements such as structure, envelope and fit out of common areas. A core HVAC system may be provided to allow for tenant connections
- Green Building Guide – a formal document that provides detailed advice to the proposed/actual tenant(s) of the building on how to minimize the environmental impacts of the building. The guide will have a particular focus on those impacts that can be influenced by the tenant(s) as a result of their building fit out.

BREEAM rating is carried out by the assessor – licensed and monitored specialist in BREEAM issues. The specialist must determine the rating using Assessor's Spreadsheet Tool and associated calculators. Firstly, according to the detailed assessment criteria presented in the manual, a number of credits awarded is determined in ten categories. Then, the percentage of credits achieved is calculated for each category and it is multiplied by corresponding weighting (listed in the table 1), to give the section score.

**Table 1.** BREEAM sections weighting, source: [2]

| BREEAM section | Weighting [%] |
|---|---|
| | New buildings, extensions and major refurbishments |
| Management | 12 |
| Health & Wellbeing | 15 |
| Energy | 19 |
| Transport | 8 |
| Water | 6 |
| Materials | 12,5 |
| Waste | 7,5 |
| Land Use & Ecology | 10 |
| Pollution | 10 |
| Innovation | 10 |

Finally, provided that all minimum standards have been met, calculated scores are added together to give the overall BREEAM score. Comparing to the table below, the BREEAM rating is achieved. Rates are: Unclassified $< 30$; Pass $\geq 30$; Good $\geq 45$; Very Good $\geq 55$; Excellent $\geq 70$; Outstanding $\geq 85$. There is also an additional (up to 10%) score for Innovation, available to projects that demonstrate that they have gone above and beyond the best practice levels of BREEAM.

## 2 Algorithm Proposal for Office Building BREEAM Assessment

Authors present the algorithm for implementation during the design process when it is foreseen to apply for the BREEAM certificate. The aim of this algorithm is to show the scheme that would enable any party of the investment process to effectively go through all the fundamental stages and become successfully certified. The certification process begins with choosing an appropriate BREEAM scheme. This and all the following steps should be guided by a competent person- licensed assessor. To become an assessor, one should had been trained and examined by BRE Global before. Typically, the knowledge of a person is assessed via undertaking a multiple choice examinations. Once the tests are passed, the assessor can apply for a license and is able to carry out the projects as a professional.

Firstly, the assessor carries out a pre-assessment, which helps to predict a likely score by using a Pre-Assessment Estimator [10]. Once this part is finished, the assessor registers the project for the assessment – and it is done only once. BRE Global evaluates and makes certification decision only on registered assessments. The project can be assessed both at the Design Stage and Post-Construction Stage. The first part of the algorithm is presented on Fig. 1.

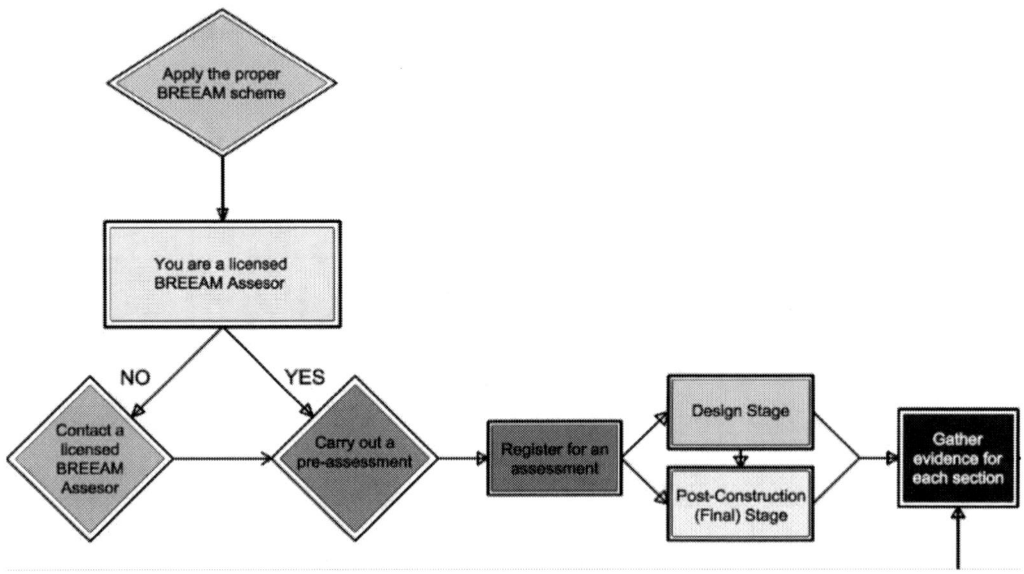


**Fig. 1.** Initial part of the BREEAM assessment algorithm [own source]

The next step is to collect all the evidence for each of ten sections in order to demonstrate the degree of compliance of a client's project with the performance criteria defined in the technical manual. The aforementioned is done by assessor. The client's responsibility lies in ensuring that all the relevant parties (consultants, designers, contractors, building owners, etc.) will supply the information needed to the assessor at the appropriate time. Once all the evidence is gathered, the Internal Quality Assurance takes place. At this stage, all the documents are checked by the appointed person before they are issued to the assessor.

The assessor is responsible for maintaining a record of this information and ensuring that it is correctly interpreted against the assessment criteria. That is why close cooperation of assessors, their clients and the project team is strongly advised. If the evidence is missing or not clearly prepared, BRE Global will not be able to make a positive certification decision.

As a consequence, the certification process may be delayed. The elements to be considered during the design process in terms of BREEAM certification should be as follows [3, 4]: management and maintenance (starting with Commissioning, taking under consideration Life Cycle Cost Analysis), Health and Wellbeing, Energy use, Internal transport and routes, Water use and management, Materials, Waste management, Land use and Ecology, Pollutions, Innovative aspects of the building.
Figures 2 and 3 shows the most important steps of the algorithm (examples) – gathering and assessment of the data.

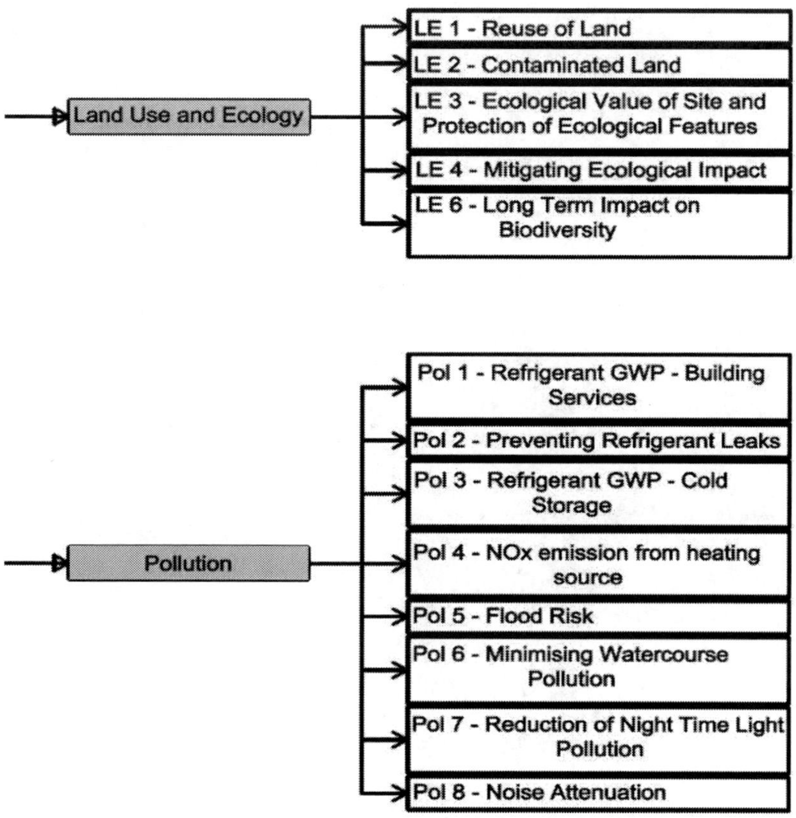


**Fig. 2.** Crucial part of the BREEAM assessment algorithm Land Use and Ecology elements example [own source]

Based on the evidence supplied from the third parties, the assessor determines the rating of the building and prepares a report containing the details of the building's performance as measured against the BRE Environmental and Sustainability Standard.

If the assessor has got any remarks, the third parties in charge of supplying the evidence should make all the corrections needed and submit the evidence to the assessor for repeated check. Then the assessor prepares the report which contains Executive Summary, BREEAM Certificate Information, Project Team and Building Details, Summary of Building's Assessment Performance, Detailed Assessment of Building Performance and Appendices. Once the registered assessment is completed and verified, it is submitted to BRE Global for the Quality Assurance (QA) and a certification decision. The records can be submitted on paper or in electronic form via BREEAM online system or file transfer protocol. All should be submitted in coherent, structured format. Figure 4 shows Final steps of the BREEM assessment algorithm.

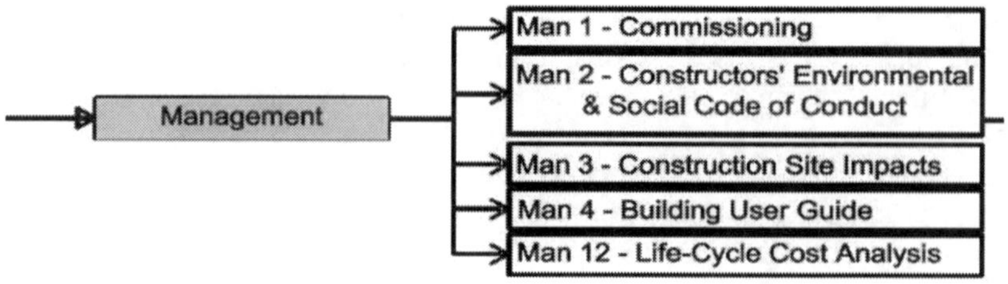


**Fig. 3.** Crucial part of the BREEAM assessment algorithm - Management elements example [own source]

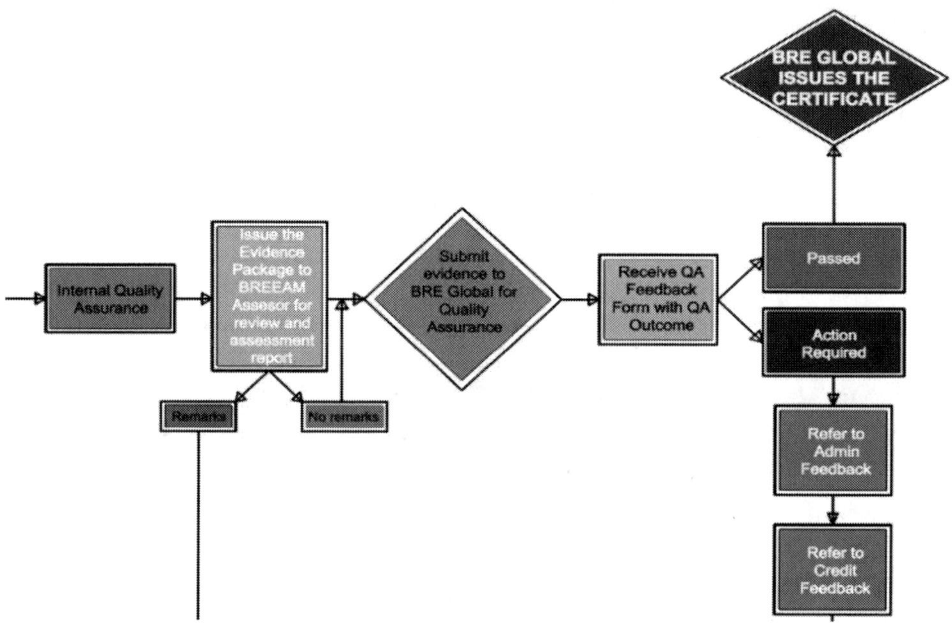


**Fig. 4.** Final steps of the BREEAM assessment algorithm [own source]

The Quality Assurance process is based on a 'risk-based' approach.
Within the QA analysis, only random criteria are checked, considering the key risk factors likely to affect the quality of the certification, to make sure that the assessor has produced an assessment and rating in accordance with the procedures and processes defined within the Technical Manual. Depending on the scheme type, three or four QA audit levels are distinguished and can be assigned to an assessor after submission an assessment for the certification:

- Administrative audits (carried out on all assessment reports),
- Credit check audit (carried out on a selection of assessments determined by the audit status of a licensed assessor submitting the assessment),
- Partial audits (carried out on a selection of assessments determined by the audit status of a licensed assessor submitting the assessment),
- Full audits (carried out on a selection of assessments determined by the audit status of a licensed assessor submitting the assessment).

However, when the licensed assessors submit their first assessment, it is subjected to a full audit.

## Summary

After submission of the assessment, BRE Global informs the assessor about the latest date of receiving the QA feedback (depending on the type of audit chosen for the project) and about the possible re-submission required as well. After specific time, Quality Assurance feedback is received. Number of issues, criteria and supporting evidence have been already checked. In case any lack or incorrectness is stated, a non-conformance is identified – depending on how severe it is, it can be classified as either a minor or major non-conformance. As a result, QA audit is not passed. According to BREEAM Operations Manual, a non conformance is typically issued where there is:
- An error/omission resulting in incorrect/invalid assessment,
- No or incorrect documentation/evidence provided,
- Inappropriate/inaccurate referencing of documentation/evidence.

In such a case, appropriate improvements need to be made (as the algorithm shows, one should refer to the feedback and get back to the point of gathering appropriate evidence) and the corrected assessment should be re-submitted for further checks. To prevent from common errors, BRE Global has prepared the list of standard examples of minor and major non-conformances. Described procedure should be repeated until the assessment passes the appropriate level of Quality Assurance. At this stage, the assessment can be finally certified.

## References

1. Building Research Establishment Global materials, www.breglobal.com, (2016)
2. SD 5066A: Issue 1.1 BREEAM Europe Commercial 2009 Assessor Manual (2012)
3. SD 5070 v5.1. BREEAM Operations Manual (2016)
4. The Green Guide Explained, BRE Global (March 2016)
5. N. Ibadov, J. Rosłon,. Archives of Civil Engineering, **61(3)**, 105-118 (2015)
6. M. Krzemiński, M. Książek, Warsztaty Inżynierów Budownictwa,,,Problemy przygotowania i realizacji inwestycji budowlanych", 249-254 (Puławy 2008),
7. M. Książek, P. Nowak, J. Rosłon, T. Wieczorek, Procedia Engineering, **91**, 406-411 (2014)
8. M.V. Książek, P.O. Nowak, S. Kivrak, J.H. Rosłon, L. Ustinovichius, Journal of civil engineering and management, **21(2)**, 248-259 (2015)
9. M. Książek, P. Nowak, Logistyka, **6** (2009)
10. L. Ustinovichius, E. Zavadskas, D. Migilinskas, A. Malewska, P. Nowak, A. Minasowicz, (2006, September). In International Conference on Cooperative Design, Visualization and Engineering, 295-302 (Springer Berlin Heidelberg, 2006),
11. M. Kaftanowicz, M. Krzemiński, Procedia Engineering, **111**, 364-370 (2015)
12. B. Jankowski, J. Prokocki, M. Krzemiński, Procedia Engineering **111**, 351-355 (2015)
13. E. Radziszewska-Zielina, ACE **62(2)**, 167 (2016)
14. E. Radziszewska-Zielina, G. Śladowski, Journal of Cultural Heritage (2017)

# Last planner system in construction

*Katarzyna* Ćwik[1], *Jerzy* Rosłon[1*]

[1] Warsaw University of Technology, Civil Engineering Faculty, Armii Ludowej 16, 00-637 Warsaw, Poland

**Abstract.** The paper presents the set of information on the particular measures of the Last Planner System (LPS) of Production Control. Then, the system as a whole is described and the profits for all relevant parties of investment process arising from adopting the Last Planner System are shown. Paper include comparison between LPS and traditional project control model.

## 1 Introduction

According to data presented by Lean Construction Institute, at the moment even 70% of the projects are over budget and delivered late [10]. The situation in building industry is getting worse every year – the results achieved in construction section are far less efficient than the ones obtained by remaining branches like Education, Health and Social Care, Transportation and Information.

Numerous attempts have been made in order to find a solution for the current inefficiency of building industry [5, 6, 7]. Here comes the solution – Last Planner System of Production Control as a tool developed by Hermann Glenn Ballard since 1992.

In the beginning, the specific attention has been paid to achieving better results in weekly work plans. Then, the lookahead process has been added to control the work flow better. Eventually, the scope of Last Planner System has been extended from construction to design [2]. This meant the shift from improving the productivity to improving the reliability of work flow [3]. There was a substantial influence of Toyota Production System and production theory of Lauri Koskela on this fact.

Since this time, the Last Planner System has been continuously evolving to finally achieve its current form. In the meantime, the Lean Construction Institute (LCI) has been founded in 1997 as a partnership between Gregory A. Howell and Glenn Ballard. The production theory of Koskela has been implemented and the work flow reliability improvement emphasized.

This Last Planner System tool enables for abandoning the concept of traditional approach towards realization of building projects and introduces the new type of thinking. However, to achieve a satisfactory result, one should implement the measures of the Last Planner System in an appropriate way.

* Corresponding author: j.roslon@il.pw.edu.pl

On the website of Lean Construction Institute, the following definition of the Last Planner System can be found [10]:

*"Last Planner System - system for project production planning and control, aimed at creating a workflow that achieves reliable execution, developed by Glenn Ballard and Greg Howell, with documentation by Ballard in 2000. LPS is the collaborative, commitment-based planning system that integrates should-can-will-did planning: pull planning, make-ready look-ahead planning with constraint analysis, weekly work planning based upon reliable promises, and learning based upon analysis of PPC (Project Partnering Contracts) and Reasons for Variance".*

## 2 Last Planner System and traditional project control model

Even though the Last Planner System has been developed for over twenty years already, it is still not well-known and properly understood in some parts of the engineering world. With its simple tools and new adaptations, it offers a perfect solution for the construction and design processes and enables project managers to improve the quality of their projects as well as satisfaction of the clients and owners.

Alan Mossman in his paper Last Planner: 5+1 crucial & collaborative conversations for predictable design & construction delivery mentions eight exemplary reasons for adopting the Last Planner System in construction. These are [11]: to deliver projects more safely; to create a more predictable production program; to reduce project durations; to better manage costs; to reduce stress on project management staff; to help to improve the overall production process; to help to make projects a reliable customer for just-in-time deliveries; it works in a way that traditional Critical Path Methods (CPM) do not.

Traditional way of thinking in project management is based on establishing the sequence of tasks that should be done. In his paper, Ballard sharply criticized this kind of project control model. As he described, the traditional attitude is to control the objects such as time and resources. Controlling of resources consists in their efficient use [8, 9, 12]. This means, that appropriate productivity is set based on the budget that had been prepared for each resource. Taking into consideration the current state of the project, the forecast concerning the resources is made. Time controlling demands planning, scheduling and monitoring. The sequence and duration of activities is determined. Again, based on the progress of works and forecasts, the date of works completion can be estimated. As Ballard emphasized, the objective of time control is production or progress, not productivity [3]. In this philosophy, the focus is on taking actions in case any deviations from the targeted values (time and resources) occur. Quoting Diekmann and Thrush [4], "*this (project control) system must provide the information needed for the project team and project participants to identify and correct problem areas and, ultimately, to keep project costs and schedule 'under control'*".

The assumption of the pull strategy implemented in the Last Planner System is completely opposite. Pulling is a method of introducing materials or information into a production process. This way, from the list of the tasks that should be done, only those that can be done are chosen to be performed. Ballard mentions a very clear example of the activity that cannot be pushed - the concrete delivery on site. The material cannot be ordered long time before it is needed as it is not possible to estimate its amount and time of delivery of concrete in advance.

A traditional Push Planning System is presented in the Fig. 1. A Pull System of the Last Planner System is presented in the Fig. 2. The table 1 presents the comparison of CPM and the LPS.

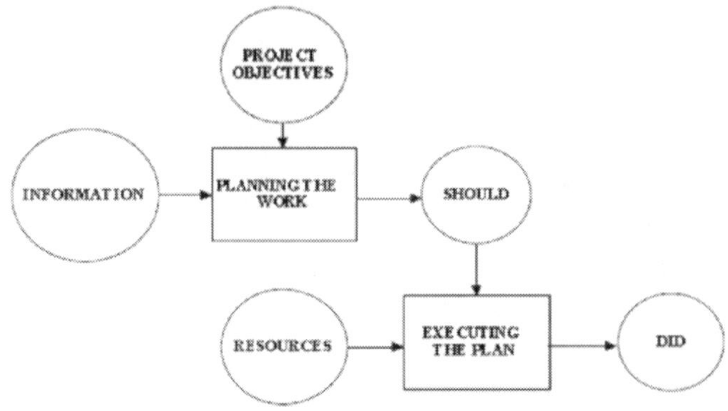


**Fig.1.** A traditional (Push) Planning System [3]

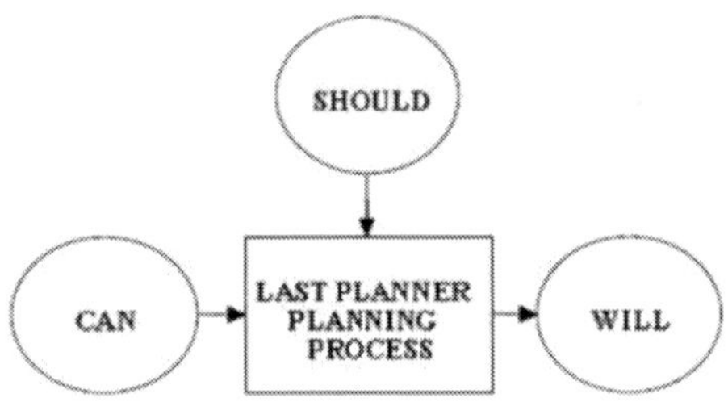


**Fig. 2.** Last Planner: A Pull System [3]

**Table 1.** Comparison of Critical Path Method and the Last Planner System, source: [1]

| Critical Path Method | Last Planner System |
|---|---|
| CPM logic embedded in software | Applied common sense |
| High maintenance | Low maintenance |
| Managing Critical Path | Managing variability |
| Focus on managing work dates | Focus on managing work flow |
| Planning based on contracts | Planning based on interdependencies |

The Critical Path Method has been developed in the late 1950 by the U.S. Navy as a mathematically based algorithm for scheduling a set of activities. It has been evolving to finally reach its current state to be widely implemented in scheduling of works in construction industry. Within CPM, the list of activities is prepared (so called Work Breakdown Structure), the duration of each task is determined and the dependencies between the activities established. In this method, the earliest possible start and finish dates as well as the latest ones are determined. The critical path is set – meaning the sequence of activities that in total gives the longest project duration. In case of any delay of any activity, the finish date of the entire project is shifted (the project is completed later). As Mossman notices [11], the Critical Path Method is a great tool to measure the feasibility of the project in terms of finishing the project according to a given schedule. However, the number of disadvantages of this method exists.

## Summary

The biggest disadvantage of CPM, comparing to the LPS is that in the first method, there is no way to determine which activities indeed can be done. As table 1 shows, the CPM logic is embedded in software and focuses on managing dates. The survey of Alan Mossman [11] showed that as many as 66% UK projects in 2013 and 70% US projects in 2012 based on CPM were delivered late. According to the research of LPS developers, the following weak points of the Critical Path Method have been observed: all plans are forecasts; all forecasts are wrong, the longer the forecast, the more wrong it becomes, the more detailed the forecast, the more wrong it is. The results of aforementioned analysis show that the traditional system of project control is not valid and useful anymore. On the contrary, the Last Planner System applies the common sense in planning and focuses on the smooth work flow as well as takes into account the independencies between activities. This way, it helps to move from push to pull system and supports logistics planning involving much more just than the due dates and the sequence of prior tasks.

## References

1. R.F. Aziz, & S. M. Hafez, Alexandria Engineering Journal, **52(4)**, 679-695 (2013)
2. G. Ballard, G. Howell, Implementing lean construction: improving downstream performance. Lean construction, 111-125 (1997)
3. H. G. Ballard, The last planner system of production control (Doctoral dissertation, The University of Birmingham 2000).
4. J.E. Diekmann, K.B. Thrush, K. B. Project control in design engineering. Construction Industry Institute, University of Texas at Austin (1986)
5. N. Ibadov, J. Rosłon,. Archives of Civil Engineering, **61(3),** 105-118 (2015)
6. M. Krzemiński, M. Książek, Warsztaty Inżynierów Budownictwa,„Problemy przygotowania i realizacji inwestycji budowlanych", 249-254 (Puławy 2008),
7. M. Książek, P. Nowak, J. Rosłon, T. Wieczorek, Procedia Engineering, **91**, 406-411 (2014)
8. M.V. Książek, P.O. Nowak, S. Kivrak, J.H. Rosłon, L. Ustinovichius, Journal of civil engineering and management, **21(2)**, 248-259 (2015)
9. M. Książek, P. Nowak, Logistyka, **6** (2009)
10. Materials presented on the website of Lean Construction Institute: www.leanconstruction.org
11. A. Mossman, Last Planner®: 5+ 1 crucial & collaborative conversations for predictable design & construction delivery. The Change Business Ltd., UK, 26 (2013).
12. L. Ustinovichius, E. Zavadskas, D. Migilinskas, A. Malewska, P. Nowak, A. Minasowicz, (2006, September). In International Conference on Cooperative Design, Visualization and Engineering, 295-302 (Springer Berlin Heidelberg, 2006),
13. M. Kaftanowicz, M. Krzemiński, Procedia Engineering, **111**, 364-370 (2015)
14. B. Jankowski, J. Prokocki, M. Krzemiński, Procedia Engineering **111**, 351-355 (2015)
15. E. Radziszewska-Zielina, ACE **62(2)**, 167 (2016)
16. E. Radziszewska-Zielina, G. Śladowski, Journal of Cultural Heritage (2017)

MATEC Web of Conferences **117**, 00033 (2017) DOI: 10.1051/matecconf/201711700033
*XXVI R-S-P Seminar 2017, Theoretical Foundation of Civil Engineering*

# CFRP reinforcement adhesive joint performance on the transversely damaged steel elements under axial tension

*Alexander* Danilov[1*]

[1] Moscow state university of civil engineering, Yaroslavskoye shosse, 26, Moscow, Russia, 129337

**Abstract.** The analytic solution for analysis and design of bonded CFRP-reinforcement of tension elements with discontinuity type damage are presented. Analytical expressions are obtained from integrating a homogeneous second-order differential equation describing the equilibrium in the adhesive layer. A simplified equilibrium model is considered with only shear stresses in the adhesive layer taken into account. Comparison of numerical results obtained on the basis of analytical expressions and on the basis of FEM in the MSC Nastran software is performed. The graphical and tabular dependences for the joint adhesive layer parameters are presented. It is shown that the bearing capacity of the adhesive joint is significantly influenced not only by the strength of the glue, but also, and to a great extent, by its shear modulus. The corresponding effective length of the adhesive layer is calculated which increase does not result in the bearing capacity of the joint increase.

## 1 Introduction

The traditional approach in repair of corroded or mechanically damaged steel structures is to cut out the damaged areas and replace them with new steel elements using welded and bolted connections, which requires considerable labor, takes a lot of time, and may sometimes require the use of heavy Lifting equipment. Steel reinforcement elements increase the structural weight and are susceptible to corrosion, and there also additional stress concentration zones appear in base metal resulting of welding or drilling [1]. Welding involves additional problems of quality control of welded joints, welding in hard-to-reach places, welding residual stresses, cracking in welded zones subject to heating, and a significant decrease in fatigue strength [2]. The welding application in a highly explosive environment is possible only if strict safety precautions are taken with prolonged interruptions of the object normal operation resulting in a dramatic uplift in the repair cost [3].

Along with the traditional methods of reinforcing steel structural elements using the welded and bolted joints, the application of carbon fiber based polymer materials (CFRP) may prove effective providing just slight increase in weight of the structure, no loss in cross sections, no stress concentration or additional potential corrosion sites, and less labor-intensive reinforcement process compared to that with steel elements. CFRPs are of high

* Corresponding author: alenk904@mail.ru

strength and wear resistance and the resistance to aggressive environment as well. The weight of carbon fiber material is multiple times lower than that of steel plates of similar amplification [4]. At that, the resistance of the carbon fiber to stretching is many times higher than that of steel, and the Young's modulus may be close to or even much higher than that of steel [5]. The application of CFRPs in building structures strengthening is also considered in [6-15].

This paper concerns the axially stretched steel elements of truss or brace systems with significant damages such as transverse cracks. In the numerical examples the ultimate case of cross-section completely disabled is considered for a steel strip. The calculation methods and conclusions are applicable for the symmetrically strengthened elements of any cross section under axial tension.

## 2 Analytical expressions

The equilibrium conditions of the element of adhesion layer of the glued joint lead to the differential equation in the form (1)

$$\tau'' - \beta^2 \tau = 0, \tag{1}$$

where τ - shear stress in the adhesive layer; $G_a$ - adhesive shear modulus; $t_a$ - adhesive layer thickness; $b_p$ - width of the adhesive layer and CFRP tape; $d$ - length of the adhesive layer on one side of the joint; $E_s$ - elasticity modulus of the tension element material; $A_s$ - cross-section area of tension element; $E_p$ - elasticity modulus of the CFRP material; $A_p$ - cross-section area of CFRP material.

$$\beta^2 = \frac{G_a}{t_a} b_p \cdot \left( \frac{1}{E_p A_p} + \frac{1}{E_s A_s} \right) \tag{2}$$

The solution of this equation follows

$$\tau(x) = P\left(B_{11} sh(\beta x) + B_{21} ch(\beta x)\right) \tag{3}$$

$$B_{11} = -\frac{G_a}{\beta \cdot t_a} \cdot \frac{1}{E_s A_s} \qquad B_{21} = \frac{\beta}{b_p} \cdot \frac{1}{sh(\beta d)} + B_{11} \cdot \frac{1 - ch(\beta d)}{sh(\beta d)} \tag{4}$$

The tension force in the CFRP reinforcement tape at the gluing length may be defined from

$$N_p(x) = P \cdot \frac{b_p}{\beta} \cdot \left( \left( B_{11} ch(\beta x) - 1 \right) + B_{21} sh(\beta x) \right) \tag{5}$$

The $x$ coordinate is measured from the left end of the gluing length on the left side of connection.

From the condition (5) of limiting shear stresses in the adhesive, it is easy to obtain a constraint on the applied force $P$.

$$\tau(x) = P \cdot \left[ B_{11} sh(\beta x) + B_{21} ch(\beta x) \right] \le R_{as} \tag{6}$$

## 3 Numerical studies

To analyze the behavior of such connections, an example of the connection of two steel plates is considered, the gap between them is overlapped on both sides by high-strength tapes (based on carbon fiber and similar materials) pasted on them (Fig. 1). A tensile force of 40000 N (≈ 4 t) is applied to the joint. All other parameters of the model are given below. The results are presented in Fig. 2-9.

Connection parameters:

- Combined plates are each 195 mm long, 10 mm gap, steel C255, cross section 50x5. The yield stress tension resistance the of steel is 50x5x250 = 62500 N ≈ 6.25 t, and ultimate stress resistance is 50x5x370 = 92500 N ≈ 9.25 t
- The top and bottom CFRP tapes 200 mm length are taken each of 50x1 mm cross-section ($E_p$ = 300 GPa, $R_{pu}$ = 2000 MPa). Then 2000 MPa x 50 mm2 = 100000 N ≈ 10 t is the ultimate tension force for one tape
- The adhesive layer length on one side of the joint is 95 mm, the total cross-section is 2x50x1 mm$^2$ (the shear module $G_a$ was taken successively at 50, 100, 200, 400, 700, 1000, 2000 and 3000 MPa, the shear strength $R_{as}$ in all variants is 15 MPa)
- The tensile force applied to the steel strip is 40000 N
- The tensile stresses in the carbon fiber tape in the gap are 40000 / (2x50x1) = 400 Mpa.

**Fig. 1**. FE Model of joint

Below (Fig. 2-9) are the results of calculating the shear stresses in the adhesive (the top glue layer to the left of the gap) for different values of the shear modulus obtained with the MSC Nastran software (left) and using the MS EXCEL program (right) with the formulas (2-6) obtained by integration the differential equation (1).

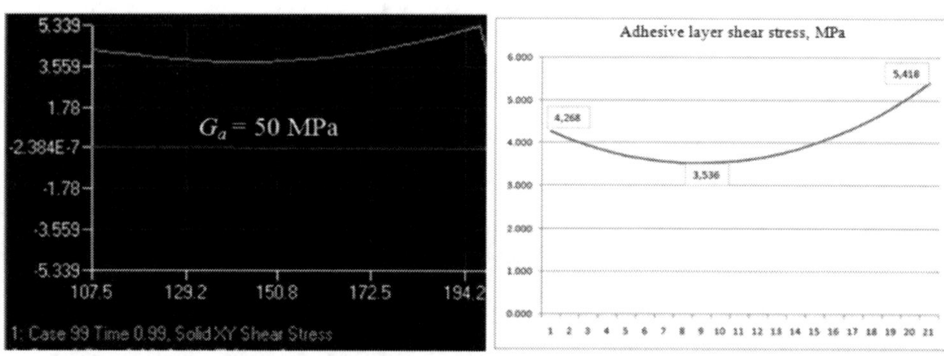


**Fig. 2**. τ in adhesive versus its length for *Ga* = 50 Mpa (on the left - Nastran, right - analytical)

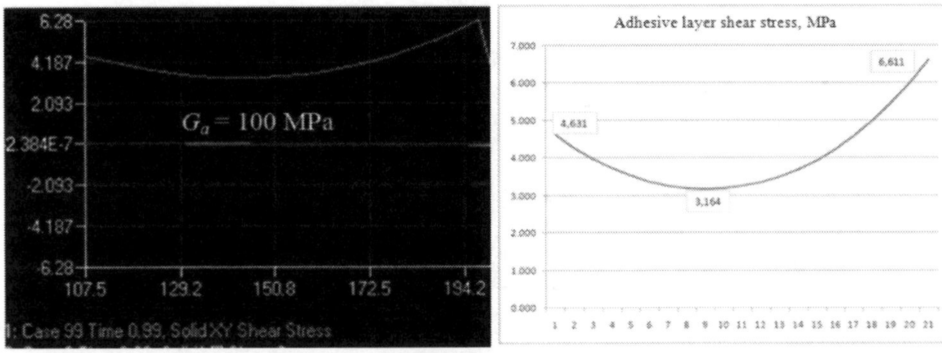


**Fig. 3**. τ in adhesive versus its length for *Ga* = 100 Mpa (on the left - Nastran, right - analytical)

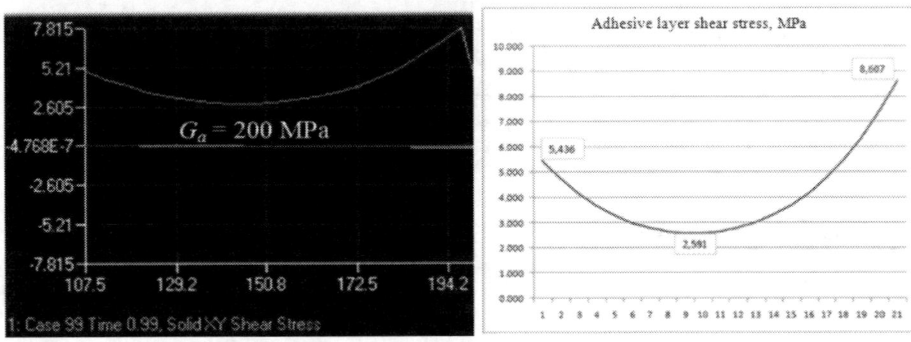


**Fig. 4**. τ in adhesive versus its length for *Ga* = 200 Mpa (on the left - Nastran, right - analytical)

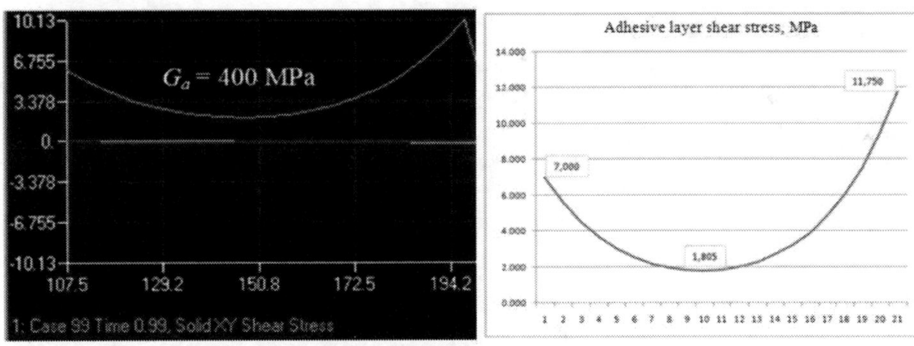


**Fig. 5**. τ in adhesive versus its length for *Ga* = 400 Mpa (on the left - Nastran, right - analytical)

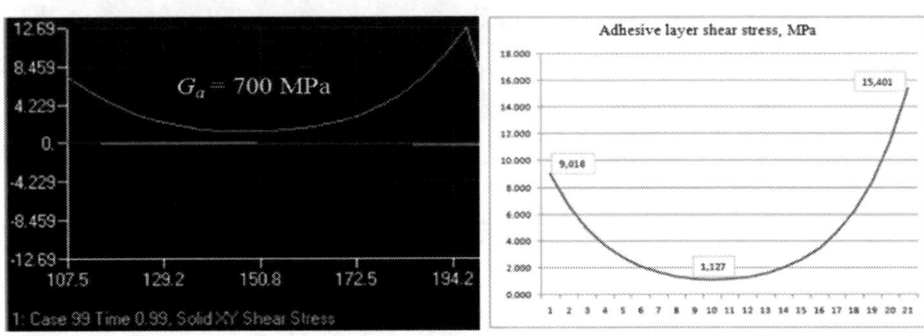


**Fig. 6**. τ in adhesive versus its length for *Ga* = 700 Mpa (on the left - Nastran, right - analytical)

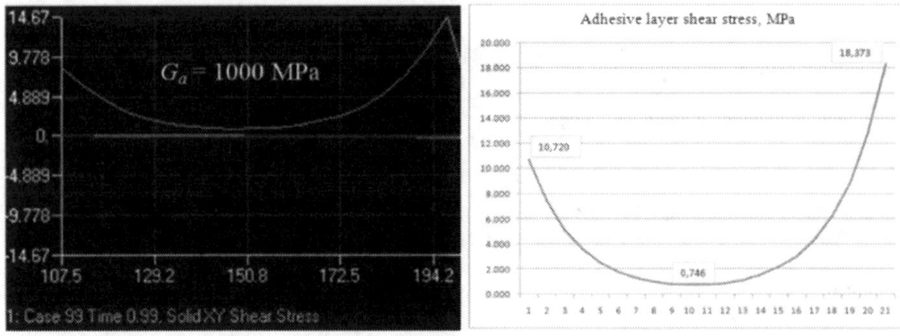


**Fig. 7**. τ in adhesive versus its length for *Ga* = 1000 Mpa (on the left - Nastran, right - analytical)

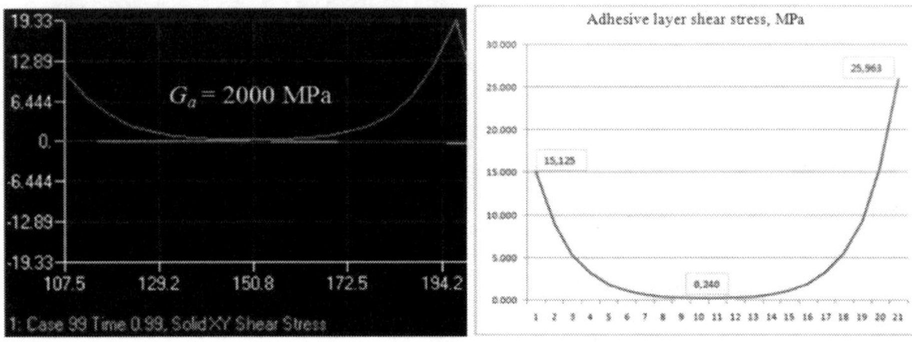


**Fig. 8**. τ in adhesive versus its length for *Ga* = 2000 Mpa (on the left - Nastran, right - analytical)

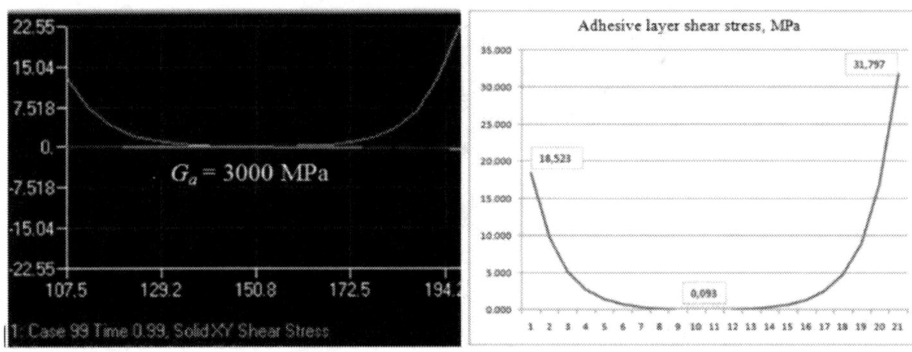


**Fig. 9**. τ in adhesive versus its length for *Ga* = 3000 Mpa (on the left - Nastran, right - analytical)

Based on the analytical solution, it is quite easy to obtain numerically a number of useful dependencies: ultimate value of the tensile force $P_{ult}$ versus shear modulus $G_a$ for a fixed values of the shear strength $R_s$ and the length of gluing (Table 1), and the shear design resistance of the adhesive joint $P_{ult}$ and the corresponding minimum allowable length of gluing $d_{min}$ versus shear modulus $G_a$ (Table 2). The minimum allowable length of gluing $d_{min}$ should provide the strength of glue connection and may be easily obtained for any allowable (for all joint elements) value of $P \leq P_{ult}$. For example, with the shear strength of the adhesive $R_s$ = 15 MPa and the gluing length of 100 mm the dependence is:

Table 1. Shear resistance of the adhesive joint $P_{ult}$ versus shear modulus $G_a$

| $G_a$, MPa | 50 | 100 | 200 | 400 | 700 | 1000 | 2000 | 3000 |
|---|---|---|---|---|---|---|---|---|
| $P_{ult}$, N | 109752 | 90755 | 69712 | 51065 | 38960 | 32657 | 23110 | 18870 |

The results of calculations on the basis of analytical solution of minimum gluing length $d_{min}$, providing the maximum joint shear resistance for the shear strength of the adhesive layer of 15 MPa, is shown in Table 2:

Table 2. Minimum gluing length $d_{min}$ and $P_{ult}$ versus shear modulus $G_a$

| $G_a$, MPa | 50 | 100 | 200 | 400 | 700 | 1000 | 2000 | 3000 |
|---|---|---|---|---|---|---|---|---|
| $d_{min}$, mm | 656 | 545 | 358 | 256 | 169 | 147 | 119 | 91 |
| $P_{ult}$, N | 146161 | 103354 | 73082 | 51677 | 39064 | 32683 | 23111 | 18870 |

Further increase in the length of the gluing does not lead to increase in $P_{ult}$. It follows from Table 2 that with an increase in the shear modulus, the rational length of the gluing decreases sharply, that is, as show the shear stress distribution diagrams on fig. 2-9, the effective zone of the adhesive layer concentrates at its ends. In Table 3 the $P_{ult}$ and the corresponding effective length of the adhesion layer $d_{eff}$ versus glue shear strength $R_{as}$ are presented for the shear modulus value of $G_a$ = 100 MPa.

Table 3. $P_{ult}$ and effective length of adhesion $d_{eff}$ versus $R_{as}$

| $R_s$ клея, MPa | 2 | 5 | 7 | 10 | 13 | 15 | 17 | 20 |
|---|---|---|---|---|---|---|---|---|
| $d_{eff}$, mm | 580 | 471 | 464 | 459 | 457 | 456 | 455 | 454 |
| $P_{ult}$, N | 13781 | 34451 | 48231 | 68901 | 89571 | 103351 | 117131 | 137801 |

For the shear modulus of $G_a$ = 100 MPa and shear strength of the adhesive $R_s$ = 15 MPa, the maximum allowed applied force $P_{ult}$ versus gluing length $d$ is shown in Table 4:

Table 4. Maximum allowed applied force $P_{ult}$ versus gluing length $d$

| $d$, mm | 456 | 400 | 300 | 200 | 150 | 100 | 50 | 10 |
|---|---|---|---|---|---|---|---|---|
| $P_{ult}$, N | 103361 | 103342 | 103231 | 102129 | 99451 | 90755 | 63233 | 14883 |

It is easy to see that here, even with a more than twofold decrease in the length of gluing, the bearing capacity of the joint reduces insignificantly.

These results are of practical importance, since they allow to significantly reduce the surface area of steel, which requires special preparation for gluing the reinforcement elements, which can provide a significant economic effect when repairing and strengthening the stretched elements due to a reduction in material consumption, labor and time.

## 4 Conclusions

1. There is a correlation of numerical data based on the analytical solution and using FEM
2. Calculation by formulas gives higher values of the maximum shear stresses in the glue than the FEM, since the closed solution is based on the simplified approach
3. Reduction of the adhesive shear modulus leads to an increase of bearing capacity of the adhesive joint at the same shear strength of the adhesive
4. This is due to a more uniform distribution of shear stresses along the length of connection with the decrease of the shear modulus value
5. In different material specifications not only the strength of the glue should be indicated, but also the corresponding shear modulus and Poisson ratio, since the adhesive strength alone does not determine the bearing capacity of the adhesive bond completely
6. The analytical solution makes it possible to select the parameters of glued joints for the subsequent refined calculations in specific application cases with minimal effort. For

example, the most effective combination of glue parameters such as shear resistance and shear modulus.

## References

1. M. Tavakkolizadeh, H. Saadatmanesh, Fatigue strength of steel girders strengthened with carbon fiber reinforced polymer patch. *Journal of Structural Engineering, ASCE,* **129,** 186-196 (2003)
2. S. El-Tawil, E. Ekiz, S. Goel, S.-H. Chao, Retraining local and global buckling behavior of steel plastic hinges using CFRP. *Journal of Constructional Steel Research,* **67,** 261-269 (2011)
3. N. G. Tsouvalis, L. S. Mirisiotis, D. N. Dimou, Experimental and numerical study of the fatigue behaviour of composite patch reinforced cracked steel plates. *International Journal of Fatigue,* **31,** 1613-1627 (2009)
4. A. Shaat, D. Schnerch, A. Fam, S. Rizkalla, Retrofit of Steel Structures Using Fiber Reinforced Polymers (FRP): State-of-the-Art. *Centre for Integration of Composites into Infrastructure* (2003)
5. T.-C. Nguyen, Y. Bai, X.-L. Zhao, R. Al-Mahaidi, Mechanical characterization of steel/CFRP double strap joints at elevated temperatures. *Composite Structures,* **93,** 1604-1612 (2011)
6. I.I. Ovchinnikov, I.G. Ovchinnikov, G.V. Chesnokov, D.A. Tatiyev, D.V. Pokulayev, Usilenie metallicheskih konstrukziy fibroarmirovannymi plastikami, Internet-jurnal *Naukovedeniye*, Vypusk **3**, may – june (2014)
7. M. Bocciarelli, P. Colombi, G. Fava, and C. Poggi, Fatigue performance of tensile steel members strengthened with CFRP plates, *Composite Structures* **87** 334-343 (2009)
8. H. Liu, R. Al-Mahaidi, and X. Zhao, Experimental study of fatigue crack growth behaviour in adhesively reinforced steel structures, Compos. Struct. Vol. 90, pp. 12-20 (2009)
9. K. A. Harries, A. J. Peck, and E. J. Abraham, Enhancing stability of structural steel sections using FRP, *Thin-Walled Structure*. Vol. **47**, pp. 1092-1101 (2009)
10. A.K. Patnaik, and C. L. Bauer, Strengthening of steel beams with carbon FRP laminates, *Proceeding of the 4th Advanced Composites for Bridges and* structures *conference,* Calgary, Canada (2004)
11. P. Colombi, A. Bassetti, A. Nussbaumer, Analysis of cracked steel members reinforced by prestressed composite patch, *Fatigue Fract Eng Mater Struct*, Vol. **26**, No. 1, pp. 59-67 (2003)
12. B. Täljsten, C.S. Hansen, J.W. Schmidt, Strengthening of old metallic structures in fatigue with prestressed and non-prestressed CFRP laminates. *Construction and Building Materials*, **23(4)**, pp. 1665-1677 ( 2009)
13. E. Ghafoori, M. Motavalli, J. Botsis, A. Herwig, M. Galli, Fatigue strengthening of damaged metallic beams using prestressed unbonded and bonded CFRP plates. *International Journal of Fatigue*, **44**, pp. 303-315 (2012)
14. E. Ghafoori, A. Schumacher, M. Motavalli, Fatigue behavior of notched steel beams reinforced with bonded CFRP plates: Determination of prestressing level for crack arrest, *Engineering Structures*, **45**, pp. 270-283 (2012)
15. E. Ghafoori, M. Motavalli, Flexural and interfacial behavior of metallic beams strengthened by prestressed bonded plates, *Composite Structures*, **101**, pp. 22-34 (2013)

# Models of formation and spread of fire to increase safety of road tunnels

*Peter* Danisovic[1,*], *Jan* Glasa[2] , *Peter* Weisenpacher[2] , and *Lukas* Valasek[2]

[1]University of Zilina, Faculty of Civil Engineering, Univerzitna 8215/1, 01026 Zilina, Slovakia
[2]Institute of Informatics, Slovak Academy of Sciences, Dubravska cesta 9, 84507 Bratislava, Slovakia

**Abstract.** Fire safety of tunnels is an important issue which must be solved carefully during all phases of preparation, building and operation of each road tunnel. In this paper some particular results of a joint research project of the Institute of Informatics of Slovak Academy of Sciences and the University of Zilina are summarized. Some specific phenomena caused by fire in a real highway tunnel in Slovakia endangering safe evacuation of people from the tunnel are illustrated.

## 1 Introduction

Recent large-scale catastrophic events in underground structures such as large car parks and tunnels have caused enormous material damages, led to tragic losses of human lives and endangered important international traffic structures and systems. To prevent devastating consequences of such events a big effort is paid for providing best practices and adopting efficient fire safety procedures in currently prepared, built as well as operated road tunnels.

Tunnel is generally a very complex and robust structure requiring enormous investment costs. Therefore, a high degree of protection is required for each tunnel in operation. However, the increasing number of cars on roads and in transport structures as well as the number of road tunnels in operation cause a growth of car incidents. One of the most destructive events in tunnel is fire of a car or group of cars. Therefore, the interest to investigate possible course of fire and development of smoke produced by fire in each tunnel is increasing. There are different possibilities how to study the course of fire and its consequences. An important form of fire study is to prepare a fire as a full-scale or small-scale experiment with extensive measuring instrumentations. Full-scale experiment is a copy of a fire in real or very similar geometric dimensions. Small-scale experiment is a copy of a fire received by creating reduced geometric dimensions and/or by including some other input variables. Consistent analysis of obtained experimental data often helps to formulate laws and description of combustion products, their range and other fire properties. However, fire experiment in a tunnel in operation is very expensive because it would cause material damages and even lead to the tunnel failure. Moreover, tunnels differ from each other by their shape, spatial dimensions, slope, ventilation, and so on. For this reason, the results of a full-scale fire experiment in one tunnel cannot be automatically applied to another tunnel. The next possibility is to use theoretical laws and the knowledge

* Corresponding author: peter.danisovic@fstav.uniza.sk

about formulation of physical modelling of fire which can be used as a test or demonstration of fire in given tunnel utilizing the main advantage of this approach, the flexible use of different distributions of flammable materials in the tunnel and different fire scenarios [1].

To ensure the road tunnel safety in the case of fire, various safety systems are used among which ventilation system is especially important. There are two main computer modelling approaches for the design of a tunnel ventilation system: 1D and CFD (Computational Fluid Dynamics) approaches. The 1D approach for ventilation simulation in tunnels [2] is used to simulate steady-state calculations in a tunnel system with low computational requirements. It can be used to simulate a whole tunnel network including chimneys, main tunnels and branches. However, the key disadvantage is that it is useful for bulk flows, but does not adequately represent flows in complex regions or the couplings between multiple variables, e.g., jet fans, tunnel geometry, traffic, environmental influence, heat, etc. With the increasing size and scale of the complexity of modern tunnels, the 1D method alone is increasingly inadequate to model these complex interactions and couplings [3]. The CFD is a more modern technique [4-7] which has been used over the last two decades focused mainly on refining the numerical methods, and verification and validation of the fire simulation systems. CFD allows better representation of complex regions, however, for a larger variety of quantities to be coupled and modelled it is computationally expensive. Therefore for practicality CFD is currently used to only model a select region of a tunnel ventilation section. In last years, a big effort is paid for analysis of applicability of current high-performance computing systems to allow the modelling of longer tunnel sections or the whole tunnel structure to study the coupling or interaction (e.g., between a tunnel section and a station) [3].

Institute of Informatics of Slovak Academy of Sciences and Centrum of Transport Research of the University of Zilina solve a joint research project focused on the study of models of formation and spread of fire to increase safety of road tunnels to increase their safety. The project research results will be applied by National Highway Company, a. s. (NDS) and PPA Engineering, s.r.o. (PPA). The main aims of the project are to develop:

- a series of computer simulations of selected types of fires in selected road tunnels managed by NDS according to negotiated scenarios using the FDS fire simulation program system
- thorough analysis and visualization of the simulation results
- series of visualizations of results of computer simulation of fires in a virtual tunnel in the Tunnel Traffic & Operation Simulator (TTOS) at Centrum of Transport Research of the University of Zilina using the FDS system
- database of evidence of incidents in road tunnels in operation in Slovakia with analysis and statistical processing of the data
- actual risk analysis of tunnels in operation and verification of the impact of transport of emergent goods on final value of risk specified by current legislation

and conduct a series of full-scale fire experiments in a single- and two-tube highway tunnel managed by NDS. The project is specifically focused on the smoke stratification investigation and the impact of ventilation on maintaining and disruption of stratification.

This paper summarizes some particular results of the joint research project on computer modelling of flows in specified highway tunnel in Slovakia and illustrates dangerous phenomena caused by fire which endanger safe evacuation of people from the tunnel.

## 2 Methodology

The aim of this paper is to show the results of computer simulation of some specific phenomena arising during tunnel fire using the FDS simulation system. Preparation and realization of the simulations in a parallel manner as well as the simulation results processing are not trivial tasks requiring a thorough application of various types of methodologies. The project methodology includes:

- gathering data about geometry of the tunnel where fire will be simulated, fire properties of flammable materials, initial and boundary conditions and other input parameters for simulation of fire according to given scenario
- modelling 3D structures in which fire will be simulated and preparation of input data and simulation parameters
- representation of the place and way of the fire origin
- design of calculation parallelization
- realization of parallel calculation on a high-performance computing system
- simulation results processing and analysis
- creation of visualisations
- validation of fire simulation results by full-scale fire experiments.

Within the project, the related standard methodologies commonly used in the fire modelling community are thoroughly applied.

### 2.1 Fire Dynamics Simulator

Fire Dynamics Simulator (FDS) is a CFD-based simulation system intended for modelling of fire-driven fluid flows developed by the National Institute of Standards and Technology, USA with cooperation with VTT Technical Research Centre of Finland [8, 9]. FDS numerically solves a form of Navier-Stokes equations appropriate for the low-speed, thermally-driven flow with an emphasis on the smoke and heat transport from fires. It also includes models of various physical and chemical processes related to fire such as turbulence, thermal radiation, pyrolysis, combustion of the pyrolysis products, conductive heat transfer, fire suppression, etc. FDS enables various kinds of parallelization of calculation utilizing advantages of available computer platforms.

The input data for fire simulation using FDS can be divided into three groups: geometrical data including the size and shape of the tunnel, location, size and shape of all objects in the tunnel; representation of jet fans, data characterizing properties of burning materials (ignition temperature, thermal conductivity, specific heat, density, emissivity, heat of evaporation of unit quantity of material, etc.) and heat release rate (HRR) of fire. The most important output of the gaseous component are usually temperature, velocity, pressure and density of the gas, the concentration of smoke and visibility as well as an estimate of the total HRR [10-13].

### 2.2 HPC cluster

The used high-performance computing cluster (HPC cluster) is located at Institute of Informatics of Slovak Academy of Sciences in Bratislava (Slovakia) [1, 14]. It is a HP blade cluster consisting of 54 compute nodes, each comprising of two 6-core 2.4 GHz processor Intel Xeon X5645 and 48GB of RAM. All nodes are connected by the Infiniband interconnection network with the bandwidth of 40 Gbit/s per link and direction (see Fig. 1).

### 2.3 Tunnel Traffic & Operation Simulator

Tunnel Traffic & Operation Simulator (TTOS) is located at Centrum of Transport Research of the University of Zilina (see Fig. 1) [15-17]. This technology was developed by PPA Engineering s.r.o. in terms of functionality based on real control algorithms used for the management of existing tunnels. It allows to simulate incidents which are rare in the real tunnel traffic with the aim of verification of correctness and philosophy of the management. Visualizations of the tunnel management are the same as on the real operator workplace of a two-tube tunnel. In contrast to real traffic, simulation of video surveillance shows a virtual traffic in the tunnel tubes and in front of them. As most of operator workstations also the operator workstation of the simulator is implemented to operate the tunnel by two operators. The first operator is responsible for management of the tunnel traffic system; the second operator is responsible for the management of the tunnel technology equipment. There are changeable; a single operator is able to control the entire tunnel. This simulation workstation for the road tunnel control and operation is realized as a real tunnel control centre of a virtual 1000 m long two-tube highway tunnel with unidirectional traffic. The technological equipment of the virtual tunnel is in accordance with Slovak legislative regulation. TTOS consists of the central control system (part of automatic tunnel equipment control), manual control module (part of manual control - separated module) and software for prediction of phenomena (EMUT – evidence of tunnel incidents). It can be used for training and education of tunnel specialists as well.

**Fig. 1.** HPC cluster at Institute and Informatics, Slovak Academy of Sciences (left) and Tunnel Traffic & Operation Simulator at Centrum of Transport Research, University of Zilina (middle and right).

## 3 Road tunnel fire modelling

Proper jet fan ventilation strategy is a crucial precondition of efficient safety measures in road tunnels. In unidirectional tunnels, the procedure is straightforward: jet fans create an air flow to push the smoke away from the tunnel and to prevent backlayering in upstream direction (so called critical flow velocity). Conditions downstream of the fire are usually untenable for people as critical velocity does not maintain the smoke stratification assuming that people are able to escape before conditions become untenable. In bidirectional tunnels, people are endangered by smoke on both sides of the fire, therefore the tenability conditions [18, 19] should be maintained long enough to ensure safe people evacuation. The aim of ventilation is to achieve a prescribed target flow velocity maintaining the smoke stratification (the Slovak and Czech regulations recommend the velocity value of 1.5 m/s [20] and 1.2 m/s [21], respectively). However, the impact of buoyancy forces created by fire may lead to significantly higher flow velocities in some regions of the tunnel, therefore the smoke stratification may not be maintained there. Therefore, careful evaluation is needed to predict the smoke movement. This problem is of special importance in Slovakia because of increasing number of road tunnels with longitudinal ventilation [22, 23]. To describe the descent of the smoke layer, entrainment between hot and cold layers [24] is modelled by the entrainment equations [25, 26]. However, there is a lack of knowledge about their general applicability and they must be validated for each tunnel. Therefore, computer simulation of typical fire and ventilation scenarios can be considered as useful guide for tunnel management.

In this paper, the impact of ventilation on smoke stratification in the Polana highway tunnel (Slovakia) is illustrated. The Polana tunnel is a 900 m long single-tube bidirectional highway tunnel with a longitudinal ventilation consisting of four couples of jet fans. We conducted computer simulation of automobile fire in the middle of the tunnel between two emergency lay-bys. The ventilation was set in accordance with Slovak legislation regulations. In Fig. 2, the development of smoke is illustrated in four phases of the fire spread. The figures show the smoke stratification during the first 160 seconds of the fire.

**Fig. 2.** Development of smoke at the 88$^{th}$, 100$^{th}$, 127$^{th}$ and 155$^{th}$ second of fire.

**Fig. 3.** Flow velocity at the 88$^{th}$, 100$^{th}$ , 127$^{th}$ and 155$^{th}$ second of fire (red and blue colour represents the biggest and smallest values, respectively).

At the 88$^{th}$ second of the fire, the tenable conditions are maintained at the left as well as at the right from the fire source. The corresponding vertical section of the flow velocity in the middle of the tunnel in Fig. 3 shows different tendency of air flow at the left and at the right from the fire source. At the right from the fire source, a marked counter-flow is formed. At the 100$^{th}$ second of the fire, the smoke comes to the emergency lay-by located at the right from the fire source. The conditions are tenable both at the left and at the right from the fire source. At the right from the fire source the smoke spread in upper part of the tunnel tube faster to the right tunnel portal because of the impact of buoyancy force. At the same time creation of a lower counter-flow to the fire source can be observed which cools down the smoke layer (see Fig. 3).

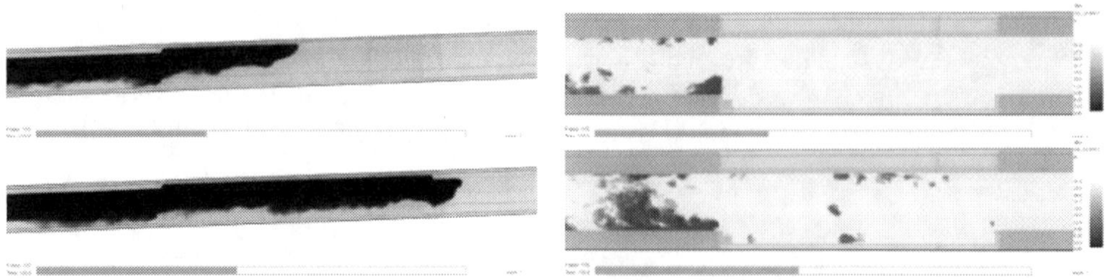

**Fig. 4.** Visibility decrease at the beginning of lay-by between 115th and 135th s of the fire caused by the spread of smoke layer through the tunnel tube with enlarged cross section into lay-by

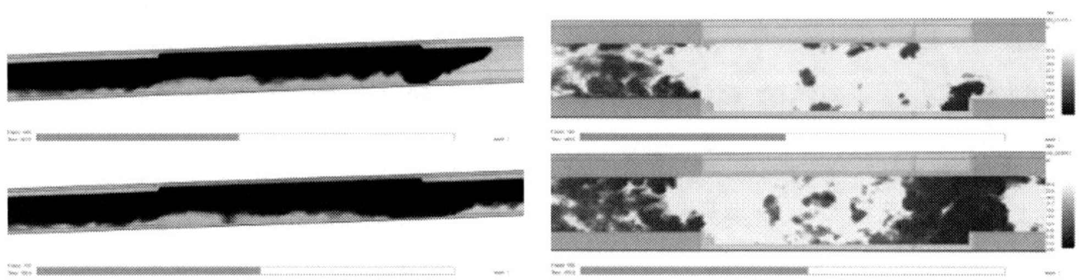

**Fig. 5.** Visibility decrease at the end of lay-by between 140th and 155th s of the fire caused by the spread of smoke layer from lay-by to tunnel tube with narrower cross section.

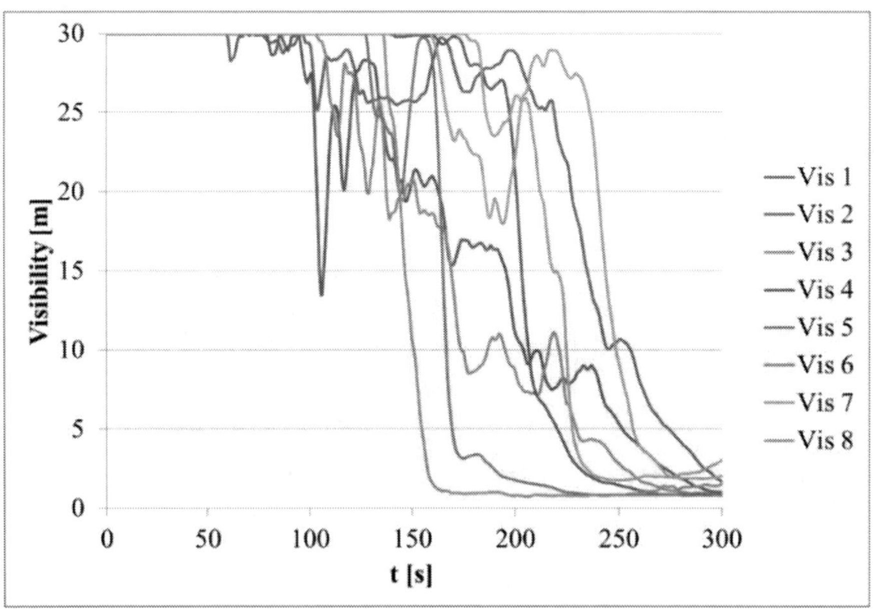


**Fig. 6.** Visibility at head level in eight selected points in the tunnel during the fire (the point are about 25 m far from each other at the right side from the fire source).

At the 127th second of the fire, the smoke came into the emergency lay-by at the right from the fire source and started to fall down creating safety risk. In the velocity section turbulences can be observed in the emergency lay-by. At the left from the fire source, the smoke stratification is maintained. At the 155th second of the fire, the tenable conditions are at the left of the fire source. The conditions in front of the emergency lay-by at the right from the fire source are failing and the safety risk is formed at the end as well as behind the emergency lay-by because of cooling, air blast onto vertical obstruction and marked turbulence at the end of the emergency lay-by (see Fig. 3). As since the 182nd second the air flow velocity to the fire source is increased, untenable conditions will emerge gradually people at the right and then at the left from the fire source. In the considered scenario two emergent places were originated at the right from the fire source (in front of the emergency lay-by and behind the emergency lay-by). In Figures 4 and 5, the breach of the smoke stratification at some places in the tunnel are illustrated. In Fig. 6, the decrease of visibility at selected places in the tunnel is shown.

## 4 Conclusion

In this paper some specific phenomena related to disruption of smoke stratification in the Polana highway tunnel are illustrated. The phenomena endanger the safe evacuation of people from the tunnel are illustrated. The simulation results indicate that the critical places of the tunnel are located in front of and behind the emergency lay-bys.

The authors would like to thank to P. Schmidt (National Highway Company, a.s.) and P. Pospisil (IP Engineering GmbH) for providing us the information about technical specifications of the Polana highway tunnel. This work was supported by the Slovak Research and Development Agency APVV (project No. APVV-15-0340).

## References

1. P. Weisenpacher, J. Glasa, L. Halada, L. Valasek, V. Sipkova, Parallel computer simulation of fire in road tunnel and people evacuation, Comput. Inform. 33(6): 1237-1268 (2014)
2. R.E. Greuer, Modeling the movement of smoke and the effect of ventilation systems in mine shaft fires. Fire Safety J. 9: 81–87 (1985).
3. Ch.D. Ang, G. Rein, J. Peiro, R. Harrison, Simulating longitudinal ventilation flows in long tunnels: Comparison of full CFD and multi-scale modelling approaches in FDS6, Tunnelling and Underground Space Technology. 52: 119-126 (2016)
4. D.F. Fletcher, J.H. Kent, Numerical simulations of smoke movement from a pool fire in a ventilated tunnel. Fire Safety J. 23: 305–325 (1994)
5. P.J. Woodburn, R.E. Britter, CFD Simulation of tunnel fire – part 2. Fire Safety J. 26: 63–90 (1996)
6. P.Z. Gao, S.L. Liu, W.K. Chow, N.K. Fong, Large eddy simulation for studying tunnel smoke ventilation. Tunnel. Undergr. Space Technol. 19: 577–586 (2004).
7. K. McGrattan, S. Hostikka, R. McDermott, J. Floyd, C. Weinschenk, K. Overholt, Fire Dynamics Simulator Technical Reference Guide, Validation, vol. 3, sixth ed. NIST and VTT Technical Research Centre of Finland (2014).

8. K. McGrattan, S. Hostikka, R. McDermott, J. Floyd, C. Weinschenk, K. Overholt, Fire Dynamics Simulator, Technical Reference Guide (National Institute of Standards and Technology, Gaithersburg, Maryland, USA, and VTT Technical Research Centre of Finland, Espoo, Finland, sixth edition, 2017)
9. K. McGrattan, S. Hostikka, R. McDermott, J. Floyd, C. Weinschenk, K. Overholt, Fire Dynamics Simulator, User's Guide (National Institute of Standards and Technology, Gaithersburg, Maryland, USA, and VTT Technical Research Centre of Finland, Espoo, Finland, sixth edition, 2017
10. P. Weisenpacher, J. Glasa, L. Halada, Automobile interior fire and its spread to an adjacent vehicle: parallel simulation, Journal of Fire Sciences 34(4): 305–322 (2016)
11. P. Weisenpacher, J. Glasa, L. Halada, Parallel computation of smoke movement during a car park fire, Comput. Inform. 35(6): 1416-1437 (2016)
12. L. Halada, P. Weisenpacher, J. Glasa, Computer Modelling of Automobile Fires. Chapter 9. In: Advances in Modeling of Fluid Dynamics (LIU, Chaoqun, ed.). InTech Publisher, Rijeka, 203-228 (2012)
13. L. Valasek, J. Glasa, P. Weisenpacher, L. Halada, Impact of vehicles on smoke spread dynamics in the case of fire in road tunnel, J. of Physics: Conf. Ser. 574 (2015)
14. SIVVP – Slovak Infrastructure for High Performance Computing. Available from: http://www.sivvp.sk/ , http://hpc.ui.savba.sk/
15. P. Danisovic, J. Sramek, E. Jancarikova, Special vocational training and education on tunnel traffic & operation simulator, Proc 9th annual International Conference of Education, Research and Innovation, Seville, Spain, pp 997-1003.
16. E. Jancarikova, P. Danisovic, J. Sramek, Increase of road tunnels safety using Tunnel Traffic & Operation Simulator, Proc 5th International Conference "Integration, Partnership and Innovation in Construction Science and Education", Moscow, Russia.
17. P. Danisovic, M. Razga, S. Sedivy, Road Tunnel Operation and Simulation, Procedia Engineering 111: 134-141 (2015)
18. Practice Note for Tenability Criteria in Building Fires (Society of Fire Safety, Engineers Australia, 2014)
19. D.A. Purser, Assessment of Hazards to Occupants from Smoke, Toxic Gases and Heat. In: DiNenno, P. J. (Ed.): The SFPE Handbook of Fire Protection Engineering. 4th Edition, National Fire Protection Association, Quincy, MA 02269, 2009, pp. 2-96 - 2-193.
20. TP 12/2011 Road tunnels ventilation (in Slovak), Ministry of Transport and Construction of the Slovak Republic, 2011.
21. Guideline - Road tunnels ventilation (in Czech), Ministry of Transport of the Czech Republic, 2013.
22. http://ndsas12.cmsromboid.sk/prehlad-tunelov/44346s
23. http://www.sta-ita-aites.sk/index.php?page=tunely_vo_vystavbe
24. H. Ingason, Y.Z. Li, A. Lönnermark, Tunnel fire dynamics (Springer-Verlag, New York, 2015)
25. R.L. Alpert, Fire induced turbulent ceiling-jet. (Report no. 19722-2. Factory Mutual Research Corp., Norwood, 1971)
26. H.Z. You, G.M. Faeth, An investigation of fire plume impingement on a horizontal ceiling 2 – Impingement and ceiling-jet regions, Fire and Materials. 9 (1): 46–56 (1985)

# Testing measurements of airflow velocity in road tunnels

*Peter* Danišovič[1,*], *Juraj* Šrámek[1], *Michal* Hodoň[2] and *Martin* Húdik[2]

[1]University of Žilina, Faculty of Civil Engineering, Department of Construction Management, Univerzitná 8215/1, 01026 Žilina, Slovakia
[2]University of Žilina, Faculty of Management Science and Informatics, Department of Technical Cybernetics, Univerzitná 8215/1, 01026 Žilina, Slovakia

**Abstract.** Within the project entitled "Models of formation and spread of fire to increase safety of road tunnels", it was necessary to devise a method how to record airflow velocity during the fire in situ tests in road tunnels. Project is in first year of its solution so one testing measurement was performed to check the functionality of anemometers selected for this project and the first in situ measurement was also performed just a few days ago.

## 1 Introduction

Fires in road tunnels are the most unfavourable emergency events with catastrophic consequences in enormous material damages and sometimes in losses of human lives. Several recent fires in the last 20 years have shown warning message for humanity and the need to deal with smoke behaviour during the fire in road tunnel. The lack of information about an active fire is one of the most important factors leading to fatal tragic incidents which can be partially eliminated by computer simulation. The aim of applied research project mentioned in the abstract is, among other things, to perform two in situ fire simulations. For these reasons we had to devise a method of airflow velocity recording during the fire simulations because this is extremely important to know the progress of airflow velocity, the change of airflow direction and time of smoke stratification disturbing. Before first in situ fire simulation we decided to check our anemometers in the longest Slovak road tunnel. The first in situ fire simulation was performed a few days ago. The methods of measurements are described in this paper.

## 2 Testing and calibration

Before the first measurement we create own wind tunnel (Fig. 1) to calibrate anemometers, test connection and way of power supply and data recording to flash memory (microSD card).

* Corresponding author: peter.danisovic@fstav.uniza.sk

**Fig. 1.** Wind tunnel for testing and calibration of anemometers.

## 2.1 Testing measurement

First testing measurement was performed in the longest road tunnel of Slovakia (up to 5 km long). It is one tube tunnel with bi-directional traffic and semi-transversal ventilation system (Fig. 2). This system uses air ducts to introduce inside the tunnel set flow rates of fresh air so to accomplish the dilution of the pollutants [1]. During normal operation pollutants can be also extracted by ventilation chimney located in the heart of the tunnel. In case of fire, the fans need to be able to reverse their running direction so to extract the smoke flow rate and grant safe environment to the tunnel users [1]. The contaminated air or fire smoke is extracted through the open air flaps in the tunnel intercooling. With the correct fire scenario, 5 ventilation flaps open in the vicinity of the fire, then the contaminated air spreads in ventilation duct separated from traffic environment.

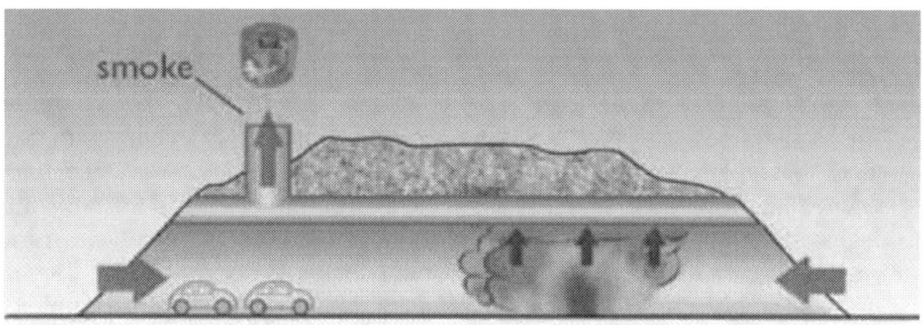


**Fig. 2.** Principle of semi-transversal ventilation system [1].

Our testing measurement was positioned to the ventilation duct to verify the airflow velocity and the required volume rate of airflow produced by fans located in the chimney. There may be some losses from leakage of ventilation system (ventilation flaps which should be closed). We placed 5 anemometers (Fig. 3) in the vicinity of the expected opening of the 5 ventilation flaps based on the location of the simulated fire source (Fig. 4).

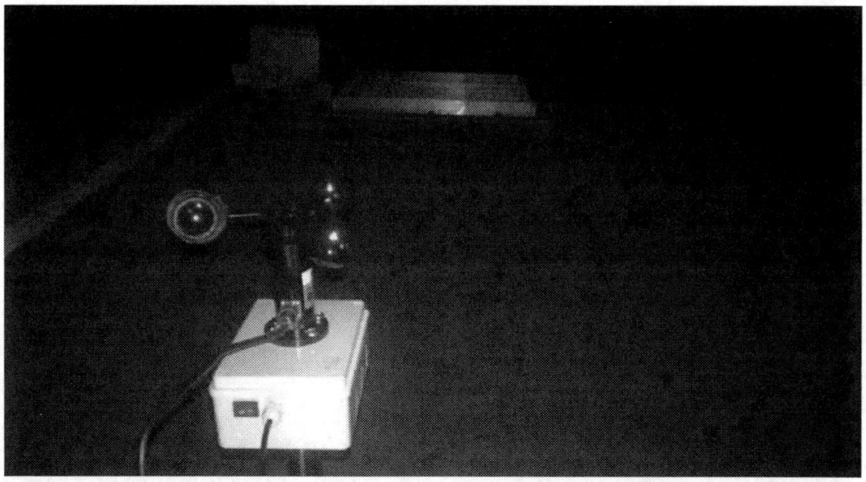

**Fig. 3.** Anemometer in ventilation duct (ventilation flap in the background).

**Fig. 4.** Simulated fire (an action member of fire scenario).

An expected scenario did not occur because fire alarm was activated in the next zone as expected. Six ventilation flaps were opened and only two of our predicted flaps. A maximum speed of about 17 meters per second was measured. Ventilation fans placed in chimney have output volume totally about 188 $m^3.s^{-1}$ so in the cross section of ventilation duct about 14 $m^2$ it is very good result of airflow velocity extraction. Airflow velocity was measured in the middle of ventilation duct and 1,1 m above the ground.

## 3 PROJECT EXPERIMENT IN SITU

Research of proper method of parallelisation of simulation calculation with the aim to achieve substantial speedup of calculation without negative effect on the calculation accuracy, research on the impact of ventilation on stratification of smoke in tunnel tube and realization of experiments in two road tunnels in situ are the main parts of the project [2]. Two full-scale fire experiments in an one tube and two tube tunnel are planned during the project solution. First experiment was performed in a tunnel Poľana (bidirectional traffic) just only one week ago. Tunnel is about 900 m long with a longitudinal ventilation system consisting of four couples of jet fans. In cooperation with company KBK fire we prepared experiment based on "cold smoke" (aerosol, Fig. 4) focused on the smoke stratification investigation and the impact of ventilation on maintaining and disruption of stratification. For this reason, 2 mobile anemometric networks (one consisting of 9 anemometers and 2 potentiometric wind-vanes, Fig. 5), 2 separate anemometers along with potentiometric wind-vanes (Fig. 6) and 3 optical gates (Fig. 7) were efficiently located in the tunnel.

**Fig. 5.** Mobile anemometric network for measuring air flow velocity and direction in tunnels.

**Fig. 6.** Two separate anemometers located behind the end of lay-by in direction of presupposed airflow.

**Fig. 7.** Optical gate installed at the beginning of lay-by.

All measuring equipment were efficiently located (Fig. 8), it means that the anemometric mobile networks were located before and behind the source of aerosol just under the real tunnel anemometers (blocks n. 39 and 47) to compare the measured values. The optical gates and separate anemometers were arranged before and behind lay-by (blocks n. 36, 54 and 58) because this is the place of smoke stratification problems as shown initial computer simulations of automobile fire in the middle of the tunnel Poľana between two emergency lay-bys presented in [3].

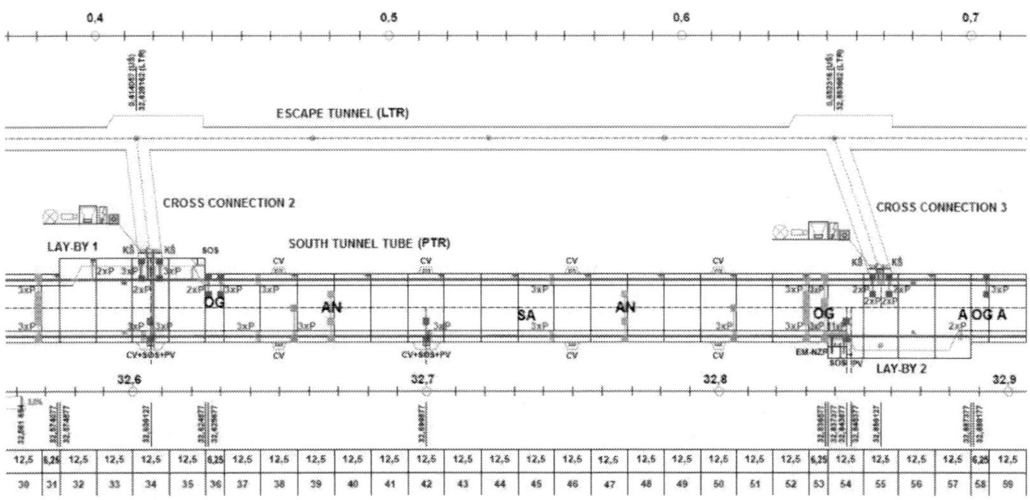


**Fig. 8.** Location of measuring equipment. Legend: OG – optical gate, AN – anemometric network, SA – source of aerosol, A – separate anemometer

## 4 Conclusion

In this paper specific measurements of airflow velocity in different road tunnels are illustrated. Measured values in tunnel Poľana will be used for verification of smoke spread and stratification computer simulations. The aim of the project is to develop a series of computer simulations of selected types of fires in road tunnels and in the virtual tunnel of

Tunnel Traffic & Operation Simulator which can be used as a didactic tool for visualisation of the smoke stratification.

This work was supported by the Slovak Research and Development Agency APVV (project No. APVV-15-0340).

## References

1. *Tunnel Ventilation. Business Area Tunnel Fans*. Catalogue of Systemair S.p.A (2012)
2. *Models of formation and spread of fire to increase safety of road tunnels.* Basic information on the project APVV-15-0340 (2015)
3. P. Danisovic, J. Glasa, P. Weisenpacher, L. Valasek, Models of formation and spread of fire to increase safety of road tunnels. *XXVI Polish – Russian – Slovak Seminar TFoCE* (to be published)

# Improving Fire Protection of Pontoon Tanks or Floating Roof Tanks

*Evgeniy* Degaev[1*], *Dmitry* Korolchenko[1]
[1] Moscow State University of Civil Engineering, Yaroslavskoye shosse 26, Moscow, 129337 Russian Federation

**Abstract:** The paper presents research of new solutions to improve fire protection of pontoon tanks or floating roof tanks. Extinguishing efficiency of the subsurface supply system is boosted by a different configuration of the pontoon or floating roof, and also with a cooling system used for oil product evacuation. Efficiency of the proposed methods is more than doubled compared to the existing subsurface extinguishing method.

## 1 Introduction

Fire protection of tanks holding oil and its products relies on a combination system where the foam is served at the same time to the tank's base and to the annular space between the tank's roof and wall [1-4].

A serious disadvantage of all existing methods of fire extinguishing by serving foam to the tank's base, is its low efficiency when fighting fires inside tank pontoons or floating roofs. As the pontoon is partly flooded, the foam served from below tends to accumulate in the tank's one half but not to flow to the other half that is open on top. If a burst of the vapor-air mix blows the mounted foam generators off, the fire becomes practically unstoppable. Such cases have to use extra quantities of foaming agent [5-7].

Because estimated intensity of foam injection from the top can extinguish the fire only in the narrow gap between the floating roof and the wall, then if the pontoon or floating roof is partly flooded, it becomes impossible to extinguish fire over the entire exposed surface. Quantity of foam served to the tank's base is sufficient to extinguish fire on the entire area, but the tilted roof or pontoon blocks entry of foam into the partitioned area of the oil product surface. This creates a problem with using foam in the subsurface extinguishing system.

Extinguishing devices exist for oil and products burning inside tanks, where check valve are installed on pontoons and floating roofs near the tank walls, on the pontoon side some 1-3 m from the edge. Such check valves prevent propagation of foam supplied from above by foam generators installed in the tank's lower ranges. Such devices are designed to extinguish flame in the ring clearing, but foam jets deflect from the tank walls and fade within on the pontoon surface.

Special focus on fire extinguishing by supplying foam into the tank's base is explained by low reliability of foam supply systems with supply from above, from foam generators

* Corresponding author: degaev@inbox.ru

installed on the upper edge of the tank wall, because when the vapor-air mix explodes inside the tank, it compromises integrity of suspended structures, and no foam can get into the tank [8-11].

The existing systems of fire extinguishing by supplying foam into the tank's base reliably protect tanks that have no pontoons and floating roof. Such tanks create no obstacles preventing foam from rising to the burning surface, and foam can extinguish foam in areas covered from above, for example by a collapsed permanent roof [12-14].

Another problem is organizing cooling for the tank's upper girth as oil products are evacuated from other tanks nearby that stand directly close to the tank in emergency. Water consumption rate to cool the tank's upper girth is very high, and it practically cannot be supplied for the entire duration of fire extinguishing [15, 16].

***The purpose of this research*** is to improve fire protection of tanks with a pontoon or floating roof; this can be achieved by solving the following ***problems:***

- reconfiguring the pontoon or floating roof to redistribute foam over the entire combustion area;
- designing a cooling system for the tank's upper girth for other tanks nearby that stand directly close to the tank in emergency until all oil products can be evacuated.

## 2 Materials and Methods

The proposed device is schematically illustrated in Fig. 1 and 2. We have tested devices with different diagrams and different number of wholes.

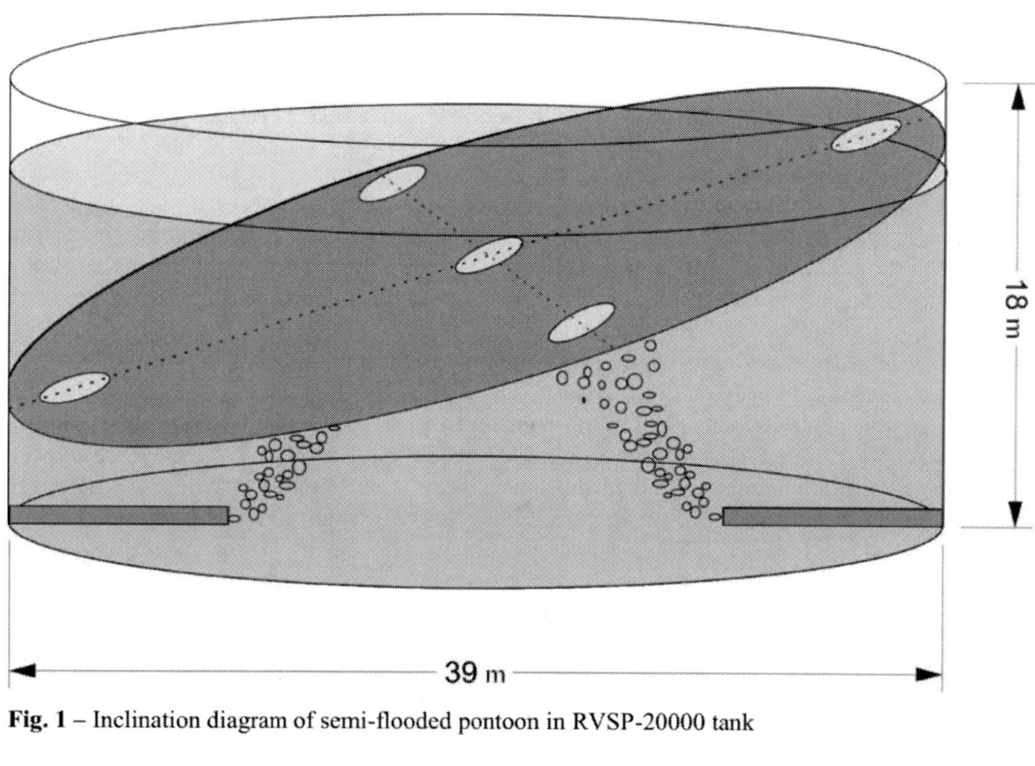


**Fig. 1** – Inclination diagram of semi-flooded pontoon in RVSP-20000 tank

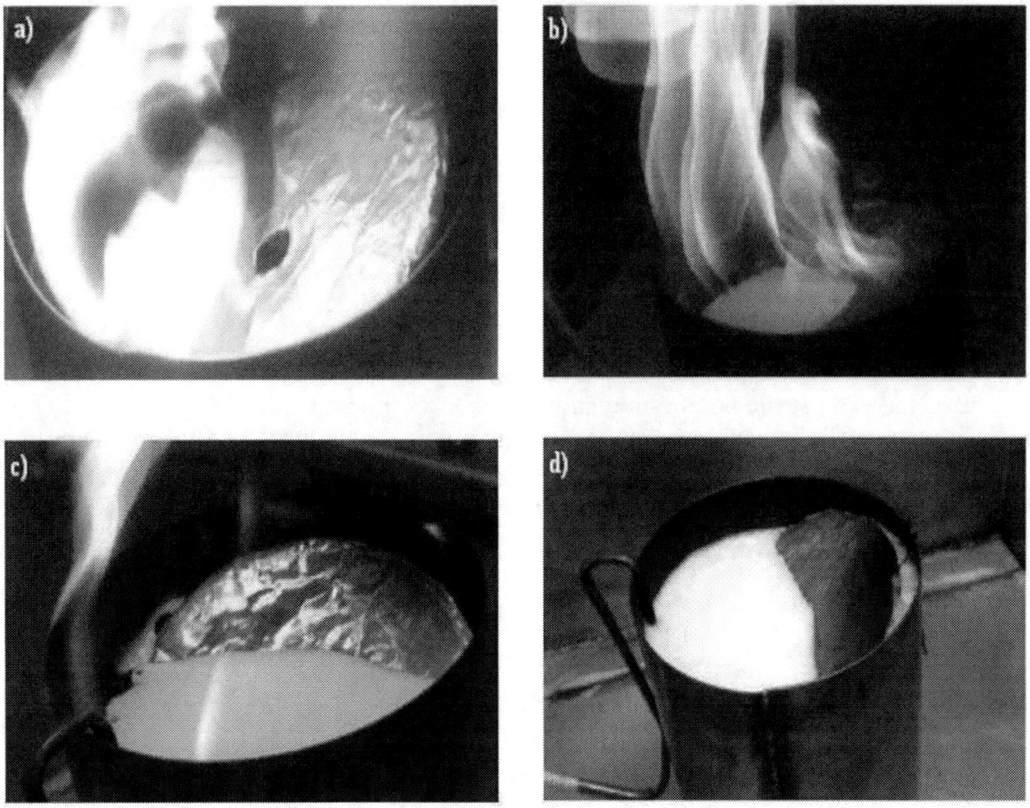


**Fig.2**. Images of heptane flames being extinguished in a pontoon tank model, with a central passage for foam to the open hydrocarbon area: a) Foam is supplied from the tank's bottom; b) Foam flows over to the second half; c) Flame is localized by the protective layer of foam; d) Flame is extinguished in both halves of the tank

We measured the time it took to put out heptane flames with the pontoon flooded to different degrees:

- the centrally located embrasure above the heptane layer;
- the centrally located embrasure semi-submerged in heptane layer;
- the centrally located embrasure fully submerged in heptane layer

In addition, we measured the time it took to extinguish the flames with embrasures of different sizes. The test used the bench design described in GOST R 53280.2-2010 for foaming agents made to extinguish flames of petroleum products burning inside tanks [9].

## 3 Results and Discussion

Extinguishing efficiency in the subsurface supply system in the presence of a pontoon can be boosted by changing the configuration of the pontoon or floating roof. Foam supplied from the tank's base is redistributed through embrasures provided regularly over the pontoon in a tank with diameter more than 25 m (Fig. 1). If the surface area is less, an embrasure is organized in the middle of the pontoon [17, 18].

The results of comparative extinguishing efficiency tests on the proposed design and prototype structure are shown in Table 1. Extinguishing efficiency is measured in units of time it takes to put out the flame at the constant rate of foam supply.

**Table 1.** Results of comparative extinguishing efficiency tests on proposed design and prototype structure for semi-flooded pontoon

| Foam supply rate, kg/(m$^2$·sec) | Flooding of central pontoon embrasure | Extinguishing time using method/ prototype, sec | Extinguishing time using method, sec | Relative efficiency |
|---|---|---|---|---|
| 0.08<br>0.10 | Embrasure above heptane level | > 150<br>>150 | 55<br>45 | 2.7<br>3.3 |
| 0.08<br>0.10 | Embrasure half flooded | >150<br>>150 | 65<br>55 | 2.3<br>2.7 |
| 0.08<br>0.10 | Embrasure fully flooded | >150<br>>150 | 75<br>65 | 2.0<br>2.3 |

Experiments done on the model tank (Fig. 2) have demonstrated that foam cannot migrate over the pontoon's flooded part to the open surface of fuel. Foam is accumulated in half of the tank, while combustion in the tank's other half continues. The layer of foam from under a tilted pontoon reached the tank's upper edge but failed to spill over to the open surface. Therefore, no localization occurred, let alone total extinguishing of the flames. With embrasures, the flame can be extinguished by foam flowing through the open embrasure from the closed area to the tank's other half.

To solve the second problem, we propose using a water cooling system with water circulated through the girth that contains the filler. Surplus water spills down over drain tubes in the upper section of the girth. See Fig. 3 for a diagram of the circulating cooling system.

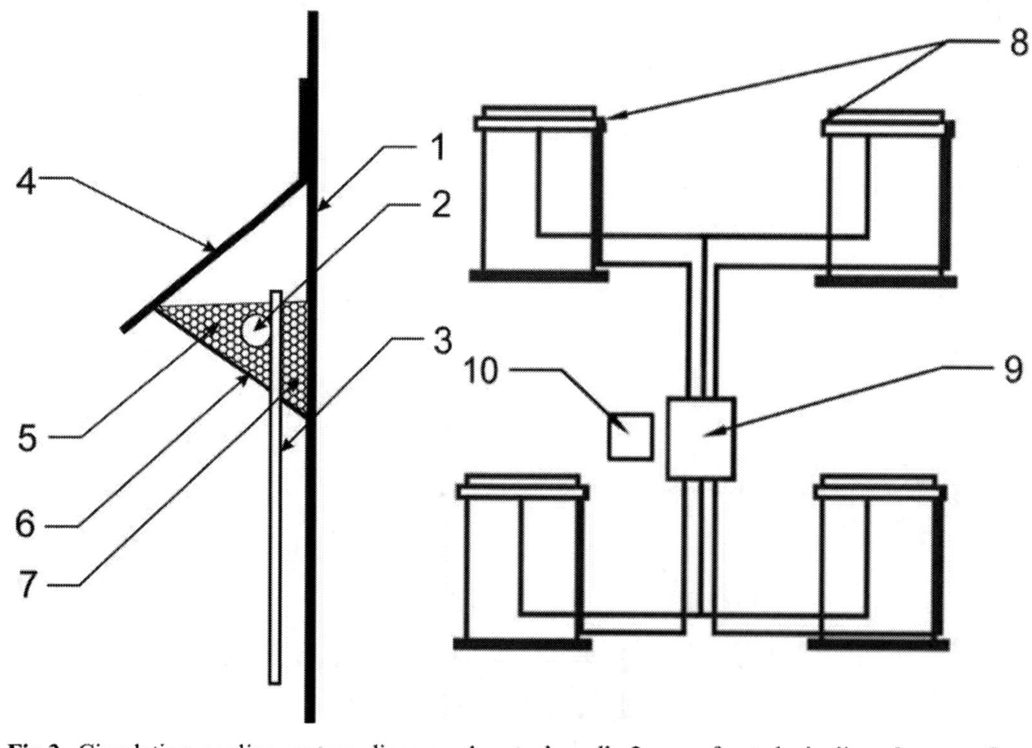


**Fig.3.** Circulation cooling system diagram: 1 – tank wall; 2 – perforated pipeline; 3 – overflow branch; 4 – protection plate; 5 – non-flammable porous filling agent; 6, 8 – annular recess; 7 – porous system's surface layer adjoining the tank wall; 9 - pump; 10 – external expansion tank

The tank is cooled with water solution of wetting agent, which is circulated by a pump between the annual recess on the tank wall and the external expansion tank. From the perforated pipeline, the water solution of wetting agent is injected into the porous non-combustible filling agent directly adjoining the tank wall, while surplus of the water solution in the annular recess returns into the circulation system by the overflow branches. To ensure sufficient efficiency, the level of solution in the annular recess should be above that of the porous filling agent.

Porous filling agent is a product of fiberglass or combustible heat insulation with through pores that prevent evaporation of water and keep feeding the surface layer of the porous system contacting with the tank wall, which cools the metal surface by absorbing heat as the wetting agent solution evaporates. As a result, the metal surface is not heated above the water boiling point; this prevents dangerous overheating of the combustible liquid inside the tank.

In term so the time needed to extinguish heptane flames, efficiency of the proposed cooling system is twice that of the existing one. If the device is used as proposed, the quantity of heat generated by a torch flame is halved compared to the prototype method. Direct records of the heat flow also demonstrated high efficiency of the proposed method that proved 1.5 times more efficient (Table 2).

**Table 2.** Comparative efficiency of methods by quantity of heat

| Burning time, sec | Free board, cm | Quantity of heat, KJ | | Efficiency |
|---|---|---|---|---|
| | | Prototype method | Proposed method | |
| 30 | 5 | 3.7 | 2.1 | 1.9 |
| 60 | 5 | 5.9 | 4.1 | 1.4 |
| 120 | 5 | 11.2 | 7.1 | 1.6 |

## 4 Conclusions

According to the research and resulting experimental data, the authors have proposed methods to improve fire protection of tanks with a pontoon or floating roof with subsurface supply of foam into the tank; the problem is solved by changing the configuration of said pontoons or floating roofs, and by using a cooling system for the nearby tanks. The results of science research have been practically implemented through patent coverage [11, 12].

## References

1. V.A. Markeev, S.S. Voevoda, D.A. Korolchenko, Oil Industry **9**, 83-85 (2006)
2. D.Korolchenko, S. Voevoda, MATEC Web of Conferences, **86**, (2016)
3. A.F. Sharovarnikov, D.A. Korol'chenko D.A., Applied Mechanics and Materials, **475-476**, 1344-1350 (2014)
4. D.A. Korolchenko,A.F. Sharovarnikov, Fire and Explosion Safety, **23**, **12**, 59-68 (2014)
5. A.F. Sharovarnikov, G.I. Punchik, Colloid Journal, **44**, **1**, 180-182 (1982)
6. J.J. Bikerman, Foams. — New York: Springer (1973)
7. Oil Tank Fire Extinguishing // Fire. — October, (1960)
8. C. Jho, Journal of Colloid and Interface Science, **117**, **1**, 139–148 (1987)
9. D.A. Korolchenko, E.N. Degaev, A.F. Sharovarnikov, Determination of the Effectiveness of Extinguishing Foaming Agents in the Laboratory 2015 2nd

International conference on material engineering and application,C 17-22 (ICMEA 2015)

10. D.A. Korolchenko, E.N. Degaev, A.F. Sharovarnikov, Dependence of Fire Extinguishing Efficacy of Low Expansion Foams Solutions Homology Sodium Sulfate on the Molecular Weight of the Surface-active Substances 2015 2nd International conference on material engineering and application (ICMEA 2015)
11. Yu. N, Shebeko, S.G. Tsariichenko, A. Ya. Korolchenko, A.N.Erofeev, Characteristics of combustion of hydrogen - methane - air mixtures in a closed vessel. Fizika Goreniya i Vzryva, **6**, 52-56 (1991)
12. Yu. N, Shebeko, S.G. Tsariichenko, A. Ya. Korolchenko, A.N.Erofeev, Characteristics of combustion of hydrogen - methane - air mixtures in a closed vessel. Fizika Goreniya i Vzryva, **6**, 52-56 (1991)
13. Zu. N. Shebeko, A. Ya. Korolchenko, E.D. Zamishevski, A.V. Trunev, Combustion and Flame, **117 (1-2)**, 438-440 (1999)
14. Xu. Lang - Procedia Engineering, **11**, 189-195 (2011)
15. Q. Zhang, L.Wang, Y. Bi, (...), H. Zhi, P. Qiu, Journal of Hazardous Materials, **287**, 7–92 (2015)
16. H. Ranjbar, B.H. Shahraki, Chemical Engineering and Technology, **36,2,** 295-299 (2013)
17. Patent 2579730. Russian Federation, A62C3/06. Method to Extinguish Fire Emergencies of Petroleum and Products in Tanks with Pontoons or Floating Roof, by Supplying Foam to Tank Base / D.A. Korolchenko, A.F. Sharovarnikov, E.N. Degayev; Applicant and Patent Holder FGBOU VPO MGSU – № 2015101161/12; filed 16.01.2015; published 10.04.2016, Bulletin No. 10 – 9 pages.
18. Patent 2589613. Russian Federation, A62C3/00. Method to Extinguish Fire Emergencies of Petroleum and Products by Preventing Tank Overheating by Torch Flame / D.A. Korolchenko, A.F. Sharovarnikov, E.N. Degayev; Applicant and Patent Holder FGBOU VPO MGSU – № 2015101162/12; filed 16.01.2015; published 10.07.2016, Bulletin No. 9 – 7 pages.

# Evaluation of exploitation suitability of enclosing structures of filling stations

*Marina* Dement'eva[1*]

[1] Moscow state university of civil engineering, Yaroslavskoye shosse, 26, Moscow, Russia, 129337

**Abstract.** The article presents the results of calculating the reliability of enclosing structures of filling station buildings. The aim of the work was to determine the optimal number of inspections of technical condition and to estimate the cost of the operation of the sandwich panel. The problems of searching for hidden failures were analyzed. An overview of the causes and mechanism of wetting the structure was made. An example of a logical-probabilistic structural model for computing the operability for increasing the operational usability of multilayer structures is given. A mathematical model was proposed for determining losses in the selection of various maintenance strategies.

## 1 Introduction

The existing automobile fleet of Russia is constantly increasing [1]. To meet the road transport of petroleum products, has been created an extensive network of filling stations and complexes. According to the data of the largest oil companies [2–5], the number of filling stations exceeds 25,000. Construction and reconstruction of filling stations are carried out taking into account the strict requirements of safety, ecology, energy saving and rational equipment placement in the station territory [6-8]. This approach allows increasing the capacity of filling stations, improve the quality and safety of customer service. At present, the largest number of filling stations are built by standard projects with different space-planning characteristics and standard design solutions [9–10].

Climatic features of our country must be taken into account, therefore, regardless of the volume-planning solutions, a heating system and energy-efficient external enclosing structures are required. As a rule, modern service buildings of the gas station have a bearing metal frame made of steel columns on a monolithic plate with self-supporting outer walls, with bearing elements of a coating made of steel I-beams, supported by bearing columns and steel beams from channels. Between the columns along the contour of the building there is a frame for fastening the sandwich panels.

The sandwich panel consists of two sheets of hard material and a layer of insulation between them. All parts of sandwich panels are glued together by hot or cold pressing. Outside, the facade of the building is made of aluminum facing panels. They are painted with powder paint in accordance with the brand and fixed on a metal frame made of galvanized profile. Inside the walls are covered with plasterboard panels also on a metal

* Corresponding author: 7dem@mail.ru

frame made of galvanized profile. As a thermal insulation and soundproofing mineral incombustible insulation is used (Fig. 1). Walls from sandwich panels are also used in the reconstruction of old filling stations [11].

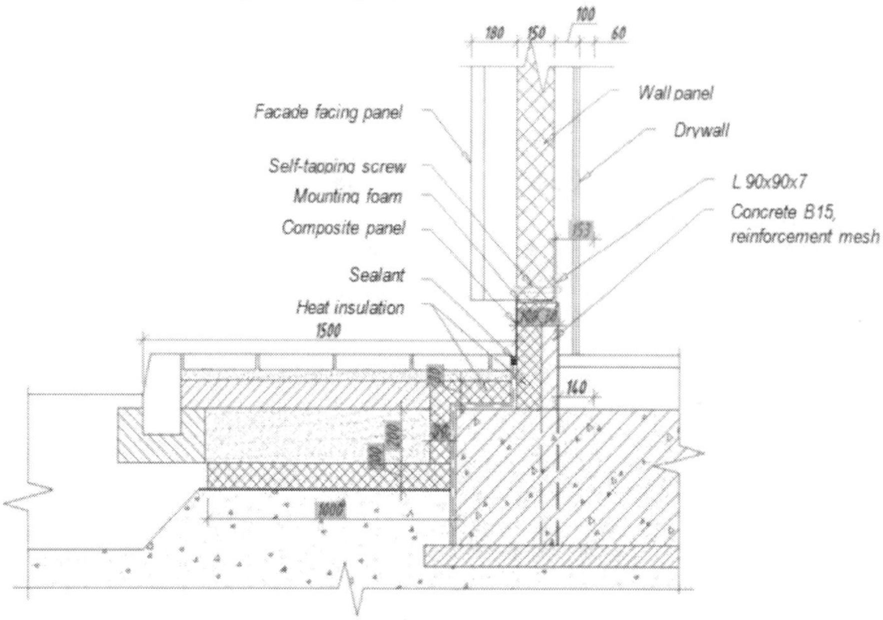


**Fig. 1.** Knot of interface of an external wall to the base and blind area.

It should be noted that violations in the functioning of sandwich panels are almost impossible to detect without specialized equipment, since they are closed from all sides. Therefore, there is a danger of hidden failures. Such failures include: leakage of roofs at the junction with wall sandwich panels, poor waterproofing of the foundation plate, corrosion of the bearing beams. Failures of this type are dependent, since they lead to the failure of the accompanying building elements and additional energy losses. The detection of hidden failures in buildings of this type has not been investigated previously, and their elimination has not been given due attention in connection with the visual absence of problems. Therefore, the object of research of this work is the process of troubleshooting sandwich panels, as well as the calculation of their trouble-free operation.

The subject of the study is the study of logic-probabilistic modeling techniques that contribute to improving the operational fitness of multi-layer structures. This method of calculating reliability can be applied to any distribution of the time between failures, with any logical structure of the system [12].

The choice for the study of the sandwich panel is not accidental, since the wetting of the insulation leads to a loss of thermal conductivity of the enclosing structures. As a result, the operating costs for heating increase, as well as the discomfort of staff and customers. This means that increases the probability of refusal to provide any services at filling stations and financial losses for companies. In this regard, the problem of assessing the operability of buildings of gas stations and determining the frequency of maintenance and repairs is acute.

The planning of the overhaul of the building, the frequency of inspections and maintenance is based on the planned system developed in the state of centralized planning [13] and does not take into account the actual technical condition of the facility. In addition, the assignment to one or another category of technical condition of building and an assessment of its operational availability, according to the existing normative and technical documentation [14], is a qualitative assessment which depending on the qualification level of the appraiser. This approach increases unreasonable operating costs, does not ensure proper reliability and quality of service provision [15].

In addition, the exploitation service should provide timely warning failures, and since such failures are difficult to diagnose, the problem studied in this paper is actual. Thus, the ultimate goal of the study in this work is to justify the periodicity or conditions for assigning buildings of filling stations for maintenance and repair, for example, assessing the operational fitness of sandwich panels.

## 2 Experimental section

One of the objectives of the study was to investigate possible causes of damage to sandwich panels buildings of filling stations. Analysis of information on the problem under consideration [16, 17] indicates that one of the main problems in the operation is water. It through the gaps and cracks gets inside the sandwich panels, as a result of which the wet insulation gradually loses its mechanical strength and its heat-insulating properties. The causes of water ingress into the sandwich panel were analyzed. The results are systematized and presented in Table 1.

**Table 2.** Analysis of damage to the sandwich panel due to moisture at various stages of the life cycle.

| The reasons for wetting the insulation of the sandwich panel | Damage to the sandwich panel and wettings mechanism |
|---|---|
| The period before the start of exploitation | |
| Marriage of sandwich panels | Poor bonding of layers of sandwich panels due to deterioration of equipment at the plant leads to water entering the insulation during rain or snow |
| Mounting of the supporting structure of the building was of poor quality | Inconsistency of the supporting structure of the project (horizontality, verticality, flatness, parallelism), as a result of which it is impossible to mount a sandwich panel without gaps, through which moisture gets to the heat insulation |
| Mounting of the sandwich panel was of poor quality or was not completed | The lack of protection of the upper structural element from external influences during unfinished mounting leads to the humidification of the insulation through the joints |
| Mounting the ebb of the roof was of poor quality | The leaky seam between the ebb leads to the flow of water through on the walls surface, which gets inside the sandwich panel to the heat insulation |
| The period of exploitation | |
| Flooding of the territory | Accidents of external engineering networks or significant precipitation (in the absence of drainage systems or in poor planning of drainage from the territory) lead to flooding of the territory with moisture ingress into the sandwich panel |
| Breakthrough internal pipes of water supply and heating | Poor mounting or operation of internal engineering systems leads to water leaks, which floods the interior of the building and leads to moistening of heat insulation inside the sandwich panel |
| Sewerage clogging | Violation of the rules of use and poor performance lead to the clogging of pipes, treatment facilities, as a result of water floods the interior of the building and leads to moistening of heat insulation inside the sandwich panel |
| Traffic Accident | Hitting a vehicle onto the wall of the building leads to damage to the facing layer of the sandwich panel, as a result of which moisture enters the heat insulation |

Identify during the operation of the listed damage is impossible, without dismantling the facing panels from the outside or plasterboard walls inside the building. Elimination of these faults is only possible when replacing the sandwich panel. Therefore, the next task of the study was to develop a methodology for calculating the optimal period of maintenance and restoration of the operational fitness of sandwich panels based on the application of logic-probabilistic methods for assessing reliability. To do this, the sandwich panel was presented in the form of a complex model consisting of separate elements, information about the reliability of which makes it possible to assess the reliability of the object as a whole.

The logical and probabilistic model of the operability of an object determines its ability to perform its functions in the case of occasional failures of individual elements or groups of elements and their recovery at a certain time interval of operation. For buildings, engineering systems and their elements, this means the realization of their technical purpose, and for organizational objects - the ability to perform the task. In this case, the concept of working capacity assumes the functioning of the object under normal planned operating conditions. Such a model is sufficient to assess the main property of the reliability of the object - its non-failure. At the same time, operational availability implies safe operation of the facility. Therefore, when investigating the safety of an object, along with non-failure, it is necessary to perform analysis of such properties as persistence, survivability, stability, technical risk, expected damage, real efficiency.

The general logical-probabilistic method of system analysis of the listed properties of safe functioning of an object was carried out in the following sequence:

– development of a structural model of the investigated property of an object, the elements of which are represented by binary events;

– determination of probabilistic parameters of elements (for example, probability of failure-free operation or failure, mean time to failure, average recovery time, occurrence and resistance to exorbitant effects, etc.) and other parameters necessary for calculating required system characteristics;

– setting of logical criteria that determine the conditions for the implementation of the properties of the object under consideration, for example, failure-free operation or failure, ensuring or not ensuring a given level of efficiency, the occurrence or non-occurrence of a certain level of damage in an emergency situation, etc.;

– the construction of the logical function of efficiency in the form of a system of equations that describe the conditions for the implementation of the object of a given logical criterion, depending on the state of all its elements;

– construction of the calculated mathematical model of the investigated property of the system in the form of a polynomial of the probability function;

– calculation of probabilistic indicators of the system properties of reliability and safety of the object under investigation and analysis of their compliance with current state standards and guidelines.

Initially, the process of functioning of all elements of the object is subject to formalization, for each of which data is displayed, under what conditions it implements (or does not realize) its output functions. An element is represented by a set of simple binary events (extreme technical states), consisting in the fulfillment or non-fulfillment of the i–th element of its functions:

$$X_i'' = \{x_i, \overline{x_i}\} \tag{1}$$

Where $x_i$ – the i–th element performs its functions;

$\overline{x}_i$ – i–th element does not fulfill its functions.

Further, the element is represented by a set of complex functional events that determine the "success" of the functioning of the i–th element under the influence of the results of the functioning of other elements logically associated with it:

$$Y_i'' = \{y_i, \overline{y_i}\} \tag{2}$$

Where $y_i$ – are the conditions for the realization of functions by the i–th element;

$\bar{y}_i$ – conditions for not implementing functions by the i–th element.

Graphically, the model is the set of vertices $X$ (binary events $X_i''$) and the set of connecting edges $Y$ (output and providing functions $Y_i''$). Such a representation is a scheme of functional integrity of $F\ (X,\ Y)$ of the object under study.

The second step of formalization is the assignment of logical criteria that determine on the scheme of functional integrity the conditions for the realization of the properties of the object under study. Combining all these results, we obtain the desired logical function of the object's operability:

$$Y_F = \{y_i''\} = \{x_i''\} \tag{3}$$

The final step of the formalization is to determine the probabilistic parameters of the results of the functioning of the elements and to construct a polynomial of the calculated probability function that makes it possible to analytically determine the distribution of the time of trouble-free operation:

$$P_F = \{p_i(t), q_i(t)\} \tag{4}$$

Where $p_i(t)$ – is the probability of the realization of functions by the i–th element;
$q_i(t)$ – is the probability of not realizing the functions by the i–th element.

The time to reach the limit state is determined depending on the probability of failure-free operation:

$$T = \int_0^{\infty} \frac{P_F(t)}{dt} \tag{5}$$

Based on this indicator, the maintenance schedule is drawn up. Repair is appointed not according to the normative terms, but taking into account the actual operating time, which takes into account the quality of manufacturing, installation, exploitation and observance of the exploitation conditions of the facility.

In conclusion, an optimal operation plan is drawn up, which will ensure a minimum of losses associated with the cost of inspections and the stay of the system in a state of failure. The object is considered as a system [18], working on the interval $[0,\ T]$. The cost of verification will be denoted by $c$ and we will assume that for t units of the system time in the failure state, we carry the loss $v(t)$. The failure of the system after time $T$ no longer causes losses. Let's designate the maintenance strategy:

$$S_n = \{t_k, k = 1,2,\ldots n\}, 0 < t_1 < \ldots < t_n < T \tag{6}$$

Where $S_n$ – is the maintenance strategy consisting of $n$ checks on the interval $[0,\ T]$;
$t_k$ – is the moment of time to which $k$ checks took place, h.

Then the mathematical expectation of losses with the chosen maintenance strategy:

$$\mathrm{E}[G(S_n)] = \sum_{i=1}^{n} \int_0^T [k \cdot c + v \cdot (t_k - t)] dF(t) \tag{7}$$

Where $G(S_n)$ – losses due to the use of the strategy $S_n$, rub.
$k$ – is the actual number of checks;
$c$ – cost of control of the system's operability, rub;
$v$ – cost of damage from the system staying in a state of failure, rub;
$T$ – time of the system stay in the state of failure, h;
$F(t)$ –function of system health;
$n$ – is the number of scheduled checks.

## 3 Results section

For the presented in Table. 1 reasons for wetting the sandwich panel were composed of structural and logical models of failure-free operation. The failures of the elements that lead to the object's failure as a whole were analyzed. The formalized model of statement of the problem consists in definition of such composition and form of representation of the initial data which allow to carry out logic-probabilistic modeling both manually, and with use of program complexes.

An example of calculating the operability of a sandwich panel for the case of poor-quality installation or repair the ebb of the roof is shown in Fig. 3. Damage to the sandwich panel during a technical inspection can be found by visual inspection or instrumental inspection. If there are no damages, then the panel is considered operable. If there is damage, the degree of damage is assessed and a decision is made on the need for repair. If the panel is not repaired, additional measures are assigned.

A complex functional event $Y_i''$ is defined as the execution of an element of its functions, provided that the "input" of the element receives a signal about the performance of its functions from the elements logically associated with it. If an element is logically connected with several, then a rule is defined, according to which the output function of the element is described for various combinations of signals input to it. For example, the output signal "operability" the panel (element "7" in Fig. 2) will be if two conditions are met:

– the element is operational (in the state $x_7$ or $x_8$);

– the input receives a signal about element's operability for different types of inspections (states $x_3''$, $x_4''$ and $x_9''$).

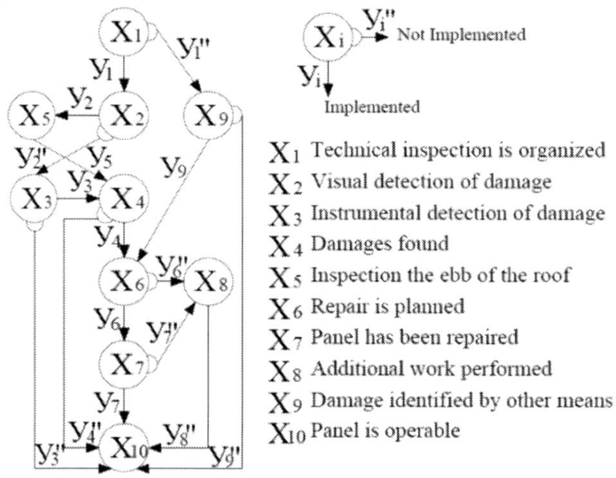


**Fig. 2.** An example of a logically-probabilistic structural model with non-qualitative mounting or repair the ebb of the roof.

The system of logical equations:

$$\begin{cases} Y_1 = X_1 \\ Y_2 = X_2 \cdot Y_1 \\ Y_2 = X_6 \cdot (Y_4 \vee Y_9) = X_2 \cdot X_1 \\ \ldots \\ Y_3'' = X_3'' \vee X_2'' \vee X_1'' \\ Y_4'' = X_4'' \vee Y_3'' \vee Y_5'' = X_5'' \vee X_4'' \vee X_3'' \vee X_2'' \vee X_1'' \\ \ldots \end{cases} \tag{8}$$

The desired function of the object's operability:

$$\begin{aligned} Y_{10} &= Y_3'' \vee Y_4'' \vee Y_7 \vee Y_8 \vee Y_9'' = X_9'' \vee X_5'' \vee X_4'' \vee X_3'' \vee X_2'' \vee X_1'' \\ &\vee (X_9 \cdot X_7 \cdot X_6 \cdot X_5 \cdot X_4 \cdot X_3 \cdot X_2 \cdot X_1) \vee (X_9 \cdot X_8 \cdot X_7 \cdot X_6 \cdot X_5 \cdot X_4 \cdot X_3 \cdot X_2 \cdot X_1) \end{aligned} \tag{9}$$

Using the formulas of Boolean algebra we obtain:

$$Y_F = (X_9'') \vee (X_9 \cdot X_5'') \vee (X_9 \cdot X_5 \cdot X_4'') \vee (X_9 \cdot X_5 \cdot X_4 \cdot X_3'') \vee (X_9 \cdot X_5 \cdot X_4 \cdot X_3 \cdot X_2'') \vee (X_9 \cdot X_5 \cdot X_4 \cdot X_3 \cdot X_2 \cdot X_1'') \vee (X_9 \cdot X_8 \cdot X_7 \cdot X_6 \cdot X_5 \cdot X_4 \cdot X_3 \cdot X_2 \cdot X_1) \vee (X_9 \cdot X_7 \cdot X_6 \cdot X_5 \cdot X_4 \cdot X_3 \cdot X_2 \cdot X_1 \cdot X_4'') \quad (10)$$

We perform the substitution of Boolean variables and logical operations:

$$P_F = q_9 + p_9 \cdot q_5 + p_9 \cdot p_5 \cdot q_4 + p_9 \cdot p_5 \cdot p_4 \cdot q_3 + p_9 \cdot p_5 \cdot p_4 \cdot p_3 \cdot q_2 + p_9 \cdot p_5 \cdot p_4 \cdot p_3 \cdot p_2 \cdot q_1 + p_9 \cdot p_8 \cdot p_7 \cdot p_6 \cdot p_5 \cdot p_4 \cdot p_3 \cdot p_2 \cdot p_1 + p_9 \cdot p_7 \cdot p_6 \cdot p_5 \cdot p_4 \cdot p_3 \cdot p_2 \cdot p_1 \cdot q_8 \quad (11)$$

The spread in the values of $P_F$ for various values of pi and $q_i$ is within the error limits (Fig. 3), except when, with a high probability of damage $p_4$, the probability of repair $p_7$ is small, that is, when the rules of exploitation are violated.

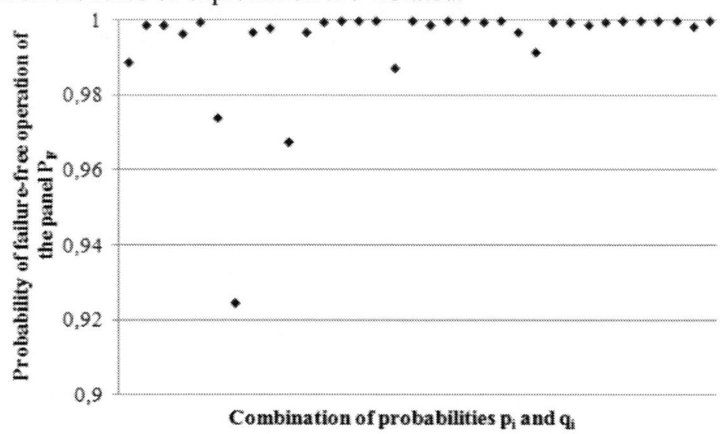


**Fig. 3.** The dispersion graph of $P_F$.

Substituting the upper and lower limits of the probabilities in the formula (7) at a given cost of monitoring and the cost of damage from the electricity consumption for additional heating, due to the loss of sandwich panels associated with damage, we determine the optimal period of maintenance (Fig. 4).

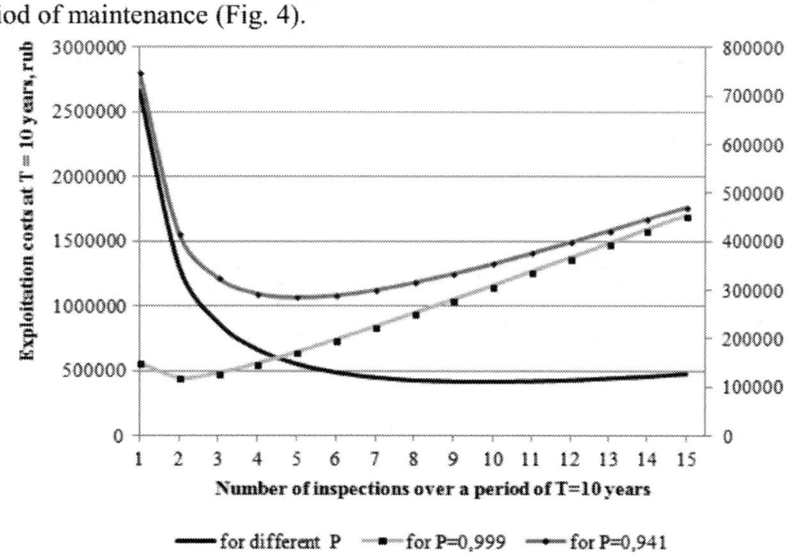


**Fig. 4.** The schedule for determining the optimum number of inspections $k$ with the minimum operating costs $G(S_n)$: $c$=30 000 rub/inspection; $v$=240 000 rub/year.

## 4 Conclusions

In order to investigate the full range of factors affecting the operational suitability of the filling station buildings and to achieve the set goal in this work, the following was accomplished:

–the existing system of technical operation of the petrol station buildings was analyzed, based on the planned designation of repairs and maintenance;

– the possibility of using logical-probabilistic methods for calculating the reliability of sandwich panels of exterior walls of buildings of filling stations is substantiated, to assess their operational suitability and expediency of performing repair works;

– equation of working capacity and probabilistic function of trouble-free operation of the sandwich panel is made.

To identify damage to sandwich panels, it is required to include measures for troubleshooting into the complex of works on the technical operation of buildings. It is also necessary to establish the frequency of detecting damage to sandwich panels so that take measures to eliminate them.

## References

1. URL: http://www.gks.ru/bgd/regl/b07_13/IssWWW.exe/Stg/d04/17-22.htm (2017)
2. URL: http://www.lukoil.ru/Business/Downstream/PetroleumProductSales/Petroleum ProductRetail (2017)
3. URL: https://www.rosneft.ru/about/Glance/ (2017)
4. URL: http://www.gazprom-neft.ru/products-and-services/for-consumers/filling-stations/ (2017)
5. URL: http://www.benzin-price.ru/zapravka.php?region_id=777 (2017)
6. Russian Standard TR "On the safety of buildings and structures" (2009)
7. Russian Standard TR "About fire safety requirements" (2008)
8. Russian Standard FL "On energy saving and on improving energy efficiency" (2009)
9. Russian Standard SP 118.13330.2012*
10. Russian Standard SP 2.2.1.1312-03
11. URL: http://www.ts-panel.ru/inform/practical_application/stroitelstvo_azs/ (2017)
12. N.N. Gordienko, A.V. Oblienko, E.A. Sushko, Scientific Herald of the Voronezh State University of Architecture and Civil Engineering. Construction and Architecture, **3**, 148-153 (2010)
13. Russian Standard RD 153-39.2-080-01
14. Russian Standard GOST 31937-2011
15. M.E. Dement'eva, Housing Construction **8**, 1-4 (2007)
16. I. Lepeshkin, Izvestiya MGTU MAMI, **1**, 136-147 (2010)
17. N.A. Stashevskaya, A.N. Malov, Y.V. Nikolenko, Bulletin of PFUR, *Series: Engineering Research*, **2**, 18-22 (2011)
18. A.N. Bochkarev, Yu.V. Lonchakov, A.V. Osipov, V.A. Ostashkevich, I.A. Fainburg, Scientific Bulletin of the Moscow State Technical University of Civil Aviation, **121**, 116-130 (2007)

# Formation of new methodological approaches to determining the cost of equipment for capital construction projects

*Olga* Didkovskaya[1], *Alexey* Bocharov[1,*], *Olga A.* Mamayeva[1]

[1]Samara State Technical University, Institute of Architecture and Civil Engineering, 194, Molodogvardeyskaya St., 443001, Samara, Samara

**Abstract.** The article presents author's ideas on the transformation of methodological approaches to determining the cost of technological equipment associated with capital construction objects. Diagnostics showed that the estimated norms for the installation of equipment in the current estimate and regulatory database contain hundreds of obsolete items. At the same time monitoring of the equipment market and comparison of its result with the norms show the absence of even more positions necessary for calculating the cost of modern equipment for capital construction projects. The proposed changes in the existing practice of standardizing the costs of installing equipment using obsolete estimates are based on the results of the author's analysis. The analysis is carried out in the form of continuous diagnostics of existing standards and their comparison with modern market indicators and methods of determining the cost of technological equipment, its installation and commissioning. Authors propose to revise the methodological approaches to determining the estimated cost of industrial equipment, bringing it closer to the real market pricing conditions. The formation of current standards should be made on the basis of analysis and processing of information from manufacturers/suppliers and collective peer reviews, i.e. through analytical, statistical and expert methods.

## 1 Introduction

The task of determining the cost of technological equipment for capital construction projects is now classified as complex. Its complexity is due to several reasons. Technological equipment, on the one hand, is functionally connected with objects of capital construction and is a part of them, while on the other hand it has specific features. The life cycles of the capital construction object and equipment differ by several times. The dynamics of changes in the functional, technical and cost characteristics of equipment significantly outstrips similar indicators of construction products. In this connection, the normative approach applied in determining the cost of construction in terms of technological equipment does not give a reliable result [4].

* Corresponding author: kafedra_cen@mail.ru

## 2 Materials and Methods

### 2.1 Budget-normative base as a basis for determining the cost of technological equipment

According to the technological structure of capital investments, the cost of construction includes three basic elements (1):

$$C_{cons} = C_w + C_{eq} + C_{oth}, \quad (1)$$

$C_{cons}$ - cost of construction, roubles;
$C_w$ - cost of work, roubles;
$C_{eq}$ - cost of equipment, roubles;
$C_{oth}$ - other expenses, roubles.

Equipment of capital construction projects is represented by two categories - engineering and technological. Engineering equipment includes equipment that carries a functional load in a building or structure and provides specified parameters for the operation of a building or structure. Technological equipment is the equipment of an enterprise for which construction (reconstruction, overhaul) of capital construction facilities is being carried out, used in technological processes of processing, output of products (rendering services) [3].

The cost of engineering equipment (electrical equipment, sanitary fixtures, etc.) is included in the cost of the relevant work (installation of a power supply system, etc.).

In formula (1), the cost of equipment allocated to an independent component of the cost of construction refers to the cost of technological equipment of the capital construction object.

Currently, the formation of the cost of construction at the stage of preparation of project documentation is carried out using the estimate norms. The current state system of estimate norms includes more than 230 collections of elemental norms, unit prices, estimated prices for building resources, etc.

An estimate and regulatory framework is also used to determine the estimated cost of equipment (compilations of State itemized estimates – CSIEe-2001, and Federal unit rates for equipment installation – FURe-2001). The problem is that the estimate norms for installation of equipment have not been updated for more than 30 years and have long been inadequate for modern developments in the equipment market.

### 2.2 Diagnostics and analysis of estimate norms for equipment installation

Diagnostics of estimate norms for installation of equipment was carried out by the authors in the form of a continuous expert examination of the estimate norms contained in the collections of CSIEe-2001, in accordance with the procedure established for these purposes.

Analysis of information resources of manufacturing companies, suppliers, service companies was conducted to analyze information on the relevance of equipment included in the compilation of estimates (Table 1).

**Table 1.** Primary diagnostics of estimate norms.

| Number and title of the compilation of estimates (CSIEe-2001) | Groups of equipment presented in the collection | Number of enterprises in the industry that produce equipment |
|---|---|---|
| 1 | 2 | 3 |
| 81-02-04-2001 Crushing-grinding, concentrating and agglomeration equipment | Crushing-grinding, concentrating and agglomeration equipment | 65 |
| 81-03-07-2001 Compressor units, pumps and fans | Compressor; compressor, injection and refrigeration; ventilating; pumping | 30 |
| 81-02-09-2001 Electric furnaces | Electric furnaces for heat treatment and drying of industrial purposes; electric furnaces for steel-making | 18 |
| 81-02-15-2001 Gas treating equipment | Purification of gases by filtration; mechanical cleaning of gases; electrical cleaning of gases | 14 |
| 81-03-18-2001 Equipment for chemical and petrochemical industry | Chemical industry; petrochemical industry | 26 |
| 81-03-19-2001 Equipment for coal and peat industry | Coal industry - mine equipment; for open mining; peat industry | 7 |
| 81-03-20-2001 Alarm, centralization, interlocking and contact network equipment in railway transport | Alarms; centralization; blocking; contact network | 38 |
| 81-03-23-2001 Equipment for electrical industry | Manufacture of cable products; electric machine building; electrical porcelain industry; production of electrical insulating materials; production of light sources | 16 |
| 81-02-25-2001 Equipment for the pulp and paper industry | Preparation of wood raw materials for processing; production of cellulose and semi-cellulose; production of thermomechanical, waste paper and wood pulp; dissolution and grinding of fibrous materials; thickening; storage, mixing and transportation of the mass; cleaning of paper and cardboard mass; production of paper, cardboard, commodity pulp and commercial wood pulp; packing and transportation of paper, cardboard and cellulose; application of water-dispersed and film coatings to paper and cardboard; processing of paper and cardboard; preparation of liquors and chemical reagents; fiber collection and effluent treatment | 10 |
| 81-02-27-2001 Equipment for printing industry | Forming equipment; printing equipment; bookbinding and binding equipment; equipment for manufacturing and finishing of cardboard and stationery | 21 |
| 81-02-29-2001 Equipment for theater and entertainment industry | Winches and drives; suspension equipment and supports; ropes; stage floor equipment; fences | 6 |
| 81-02-31-2001 Equipment for cinema industry | Film technology; film copying; film laboratory | 6 |
| 81-03-32-2001 Equipment for electronics and communication industry | Equipment for electronics and communication industry | 7 |
| 83-03-36-2001 Equipment for consumer services and utilities | Laundry; dry cleaning and dyeing; repair of household appliances | 4 |
| **Total** | | 268 |

# 3 Results

## 3.1 Main results of diagnostics and analysis of estimate norms for equipment installation

Diagnostics of estimate norms was made according to the full nomenclature of estimate norms presented in the collections of CSIEe-2001 (2664 items) [2].

Graphical interpretation of the main diagnostic results is shown in Figures 1 and 2.

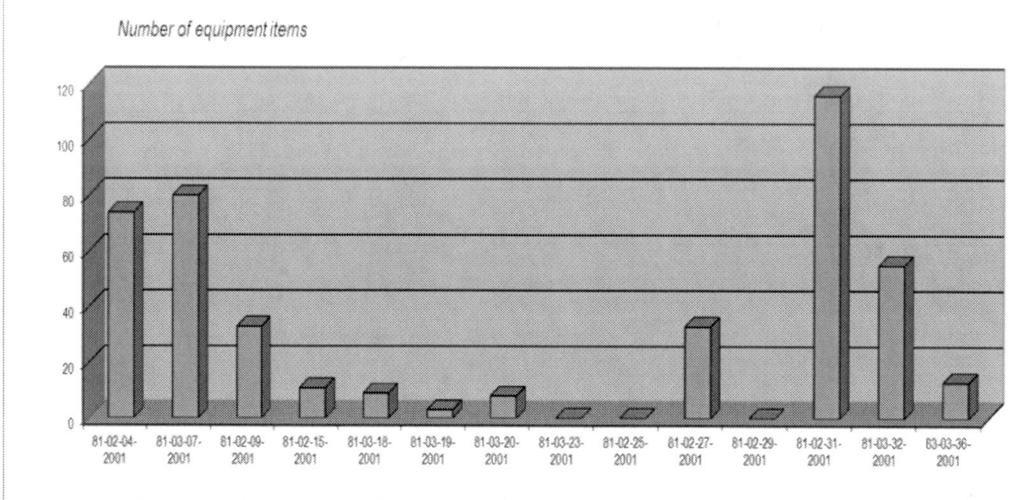


**Fig. 1.** Inactive items in State itemized estimates for equipment installation 2001, based on the diagnostic results in the context of analyzed collections.

Analysis of the current estimate norms for the installation of equipment showed that the collections of CSIEe-2001 contain 435 items of irrelevant equipment, which are present in the estimates but not produced by modern industry (Figure 1).

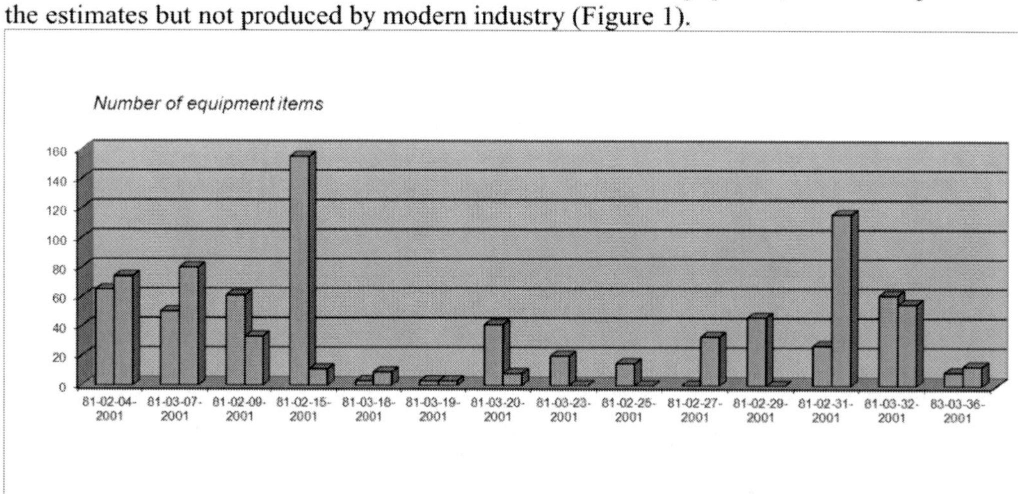


**Fig. 2.** Missing items in State itemized estimates for equipment installation 2001, based on the diagnostic results in the context of analyzed collections.

At the same time, the number of missing items of modern equipment, which is actually used at the capital construction facilities today, is more than 550 (Figure 2).

## 3.2 Features of the study subject, revealed as a result of diagnostics and analysis of estimate norms for installation of equipment

As a result of the analysis of the collections for equipment installation the following features were revealed:

– the need to take into account the industry specificity of collections;

– the need for analysis of modern production and technological processes in various sectors of the economy;

– the need for a comparative analysis of technical and technological characteristics of equipment provided for by estimate norms and equipment used in modern production in the sectoral context;

– market research of technological equipment (in the context of a competitive market) and/or information from producers and suppliers in the sectoral context;

– study of current experience and established practice in terms of equipment installation technologies and cost formation.

The collections of State itemized estimates for equipment installation (CSIEe-2001) and Federal unit rates for equipment installation (FURe-2001) are conventionally classified into the following groups according to the nature of their use:

– estimate norms for the installation of general engineering equipment used to determine the estimated cost of various production and non-production facilities;

– estimate norms for the installation of general-purpose technological equipment used to determine the estimated cost of various production and non-production facilities;

– estimate norms for the installation of industry-specific technological equipment used to determine the estimated cost of highly specialized production facilities.

The proposed classification meets the objectives of further diagnostics and updating of methodological approaches to the formation of the estimated cost of equipment.

We propose to update the collections of CSIEe-2001assigned to the first (8, 10-12) and the second (3, 5-7, 37-40) classification groups in the same general order as the compilation of estimates for construction and special construction work CSIE-2001.

We also propose to revise the methodological approaches to determining the estimated cost of industrial technological equipment (the third group – 1, 2, 4, 9, 13-36).

## 3.3. Trends in the development of pricing in terms of determining the cost of equipment

As a result of the analysis of modern industrial production processes, manufacturers and suppliers of technological equipment the following trends have emerged:

– CSIEe-2001 contains a huge amount of obsolete equipment that has not been produced (since the 1970s) and equipment that is not used at present. A small part of such equipment is replaced by its modernized versions (for example, in the collection of CSIEe 81-02-15-2001 "Gas treating equipment" only three items have modernized and used analogues from the entire range of estimate norms, the remaining items are completely outdated – mechanical and filtration cleaning is done today in other ways with the use of different equipment);

– many modern production and technological processes use equipment that is not listed in CSIEe-2001 (e.g., CSIEe 81-02-04-2001 "Crushing-grinding, concentrating and agglomeration equipment" Section 1 of Division 2 is represented by one type of equipment - disc screen, while in fact currently four types of such equipment are used: disc, inertial, self-balancing, and electric screens);

– for some industry-specific technological equipment there is a functioning competitive producer/supplier market, represented by domestic and import manufacturers (e.g., equipment of textile, printing, food processing, light industry, consumer services and municipal services, etc.); for other types of equipment manufacturers are represented by only a few enterprises (for example, equipment for cleaning gases, etc.);

– a significant share of specialized industrial equipment belongs to the individual delivery category;

– reduction in the mass of modern equipment similar/superior to outdated and unused one in performance and/or other characteristics, as a parameter that determines the cost of its installation;

– changes in the basic equipment parameters that determine the choice/selection of the estimate norm (e.g., "productivity" instead of "mass", etc.);

– a significant share of highly specialized industrial technological equipment is offered by manufacturers/suppliers "in business" (i.e. installed and adjusted);

– collection of information on cost indicators of industry-specific technological equipment. In response to requests, suppliers/producers generally propose to perform a cost estimate for the process parameters specified by the customer, taking into account the installation and commissioning "in the kit".

# 4 Discussion

## Proposals on the formation of actual methodological approaches to determining the cost of equipment for capital construction projects

Formation of norms should be carried out on the basis of analysis and processing of information from producers/suppliers and collective peer reviews, i.e. analytical, statistical and expert methods.

Such approach to determining the estimated cost was established for laboratory equipment [5]: "The cost of installing laboratory equipment and furniture supplied assembled and not requiring preliminary disassembly and subsequent assembly during installation, as well as connection to engineering networks and communications (electricity, sewerage , gas supply, ventilation, etc.) are determined at a rate of 1% of their cost, taking into account the costs of unloading from vehicles, lifting to floors and bringing up to the installation site.

The costs for the installation of laboratory equipment and furniture supplied in disassembled form, or requiring disassembly and subsequent assembly for installation with connection to engineering networks and communications (electricity, water, gas, sewerage, ventilation, etc.) are determined in the amount of 6% of their cost, taking into account the costs of unloading from vehicles, lifting to floors and bringing up to the installation site.

The cost of the same works without connection to engineering networks and communications, is determined at a rate of 4% of the equipment cost."

Similar approaches are also used in appraisal practice with a view to reliable valuation of machinery and equipment [1,6,7,8].

The proposed new approach to the formation of the cost of industry-specific technological equipment can be expressed through the following formula (2):

$$C_{eq}^{est} = P_{eq} \times \frac{N_c}{100}, \tag{2}$$

$C_{eq}^{est}$ - estimated cost of technological equipment, roubles;

$P_{eq}$ - equipment purchase price, roubles;

$N_c$ - comprehensive average norm that determines the ratio of the cost of equipment "in business" (installed and adjusted) to the purchase price of equipment,%.

The purchase price of equipment ( $P_{eq}$ ) can be determined in the following way:

– by types of technological equipment for which there is a functioning competitive market, price indicators can be weighted averages determined on the basis of monitoring carried out in the order established for construction resources;

– by types of technological equipment of the so-called individual delivery price indicators can be determined by calculation in the order and forms established for this purpose.

The main normative and methodological task in establishing such an approach is the formation of complex norms:

– in the sectoral context (by types of industries);

– by kinds of the industry-specific equipment (by homogeneous groups of the equipment in the corresponding sector);

– with the differentiation of complex norms for individual types of costs (parts, installation works, supervised installation, commissioning).

Formation of norms should be carried out on the basis of analysis and processing of information from producers/suppliers and collective peer reviews, i.e. through analytical, statistical and expert methods.

Such an approach is proposed to be adopted, among other things, to determine the costs of commissioning highly specialized technological equipment.

## Conclusions

The implementation of the proposed approach will allow to:

1. Eliminate the need to develop a huge number of new estimate norms for the installation and commissioning of modern industry-specific technological equipment the need for which is rare, thus optimizing budget costs for these processes;
2. Determine the estimated cost of the technological equipment by a separate estimated calculation of the established form, taking into account the full set of costs for equipment "in business" (equipment cost, installation of equipment foundations, installation and commissioning works). The form and procedure of the estimate calculation should be established in the updated procedure for determining the estimated cost of construction (MDS 81-35.2004);
3. Form and allocate the estimated cost of technological equipment in the consolidated estimate of the cost of construction (a separate column) in order to further harmonize with financial statements requirements;
4. Distribute competencies and responsibility of state customers: reliability of the estimated cost of construction of capital construction facilities is the responsibility of the Russian Ministry of Construction, estimated cost of technological equipment is the responsibility of the relevant ministries (Ministry of Health, Ministry of Transport, etc.).

The procedure for determining the value of industrial technological equipment is proposed to be fixed in the updated procedure for determining the estimated cost of construction.

## References

1. O.V. Butkova, R.I. Butkov, *On the types of value of machinery, equipment and vehicles in appraisal practice* V.1, **2**, 30-32 (2017)
2. *State itemized estimates for equipment installation (CSIEe-2001)* (2001)
3. O.V. Didkovskaya, *Economic aspects of management of the building complex in modern conditions: Materials of the 3rd International scientific and practical conference*, SSUACE, Samara, 201-208 (2015)
4. O.A. Mamayeva, M.V. Ilyina, Science Review **9**, 308-311 (2015)
5. *Guidance documents in construction 81-37.2004. Instructions on the application of Federal unit rates for equipment installation (FURe-2001)* (2004)
6. *Price Indices in Construction, Inter-regional information and analytical bulletin* **91**, April (2015)
7. V.M. Sidorov, Economy. Innovation. Quality control **3(16)**, 22-23 (2016)
8. *Directory of the appraiser of machines and equipment. Correction coefficients and characteristics of the machinery and equipment market* (Ed. by Leyfer L.A., Nizhny Novgorod, 2015)

# The determination of values of the specific heat capacity of the selected thermal insulation materials used in track bed structure

*Peter* Dobeš[1,*], *Libor* Ižvolt[1] , *Martin* Mečár[1], and *Jana* Malachová[2]

[1]University of Zilina, Faculty of Civil Engineering, Department of Railway Engineering and Track Management (DRETM), Univerzitná 8215/1, 01026 Žilina, Slovakia
[2]University of Žilina, Institute of Continuing Education, Univerzitná 8215/1, 01026 Žilina, Slovakia

**Abstract.** The report concentrates on the determination of the specific heat capacity of the selected thermal insulation materials (liapor, styrodur, foam concrete). The aim of the report is to gain the necessary input parameters for the numerical modelling of the temperature changes of various track bed structures of the railway line where a part or, if appropriate, the whole protective layer is replaced by the material with better thermal insulation properties. There are described the methods for stating the specific heat capacity as well as the calibration of the calorimeter for stating of the calorimetric constant in the introduction of the report. The parameters needed for calculation of the specific heat capacity of the selected thermal insulation materials are characterized in the second part of the report. There are also introduced the laboratory stated values of the parameter in question. The comparison of the values stated by the laboratory measurements with the values introduced in the technical data sheets from their producers (or if appropriate gained from the foreign sources) is made in the conclusion of the report.

## 1 Introduction

The design and examination of the trackbed structure on the non-traffic load, related to numerical modelling of the frost penetration of the railway line structure (of the track bed), cannot be made without the basic input thermo-technical parameters of the built- in materials. Therefore, the most important characteristics are the heat conductivity coefficient $\lambda$ and the specific heat capacity $c$ of the building materials. The accepted EN STN do not introduce all the values of the thermal conductivity coefficients $\lambda$ and the specific heat capacity $c$ of the materials which are used in trackbed structure nowadays.

The design and examination of the trackbed structure of the railway lines is made according to TNŽ 73 6312 [1]. There are introduced the thermal conductivity coefficients $\lambda$ of various building materials used in the trackbed structure, in this legislative document. These coefficients are presented in the part concentrating on the design and examination of

* Corresponding author: dobes@fstav.uniza.sk

Railways of Slovak Republic on the non-traffic load. With regard to the necessity of amendment to the legislation document [1] and the updating of chart with the proposed values of the thermal conductivity coefficients $\lambda$ of the used building materials it is necessary to verify some of them and for some to determine new materials.

Since 1993 the Department of Railway Engineering and Track Management (DRETM) has been specialising in the problematics of the climatic parameters and the problematics of the frost penetration of the railway line structure. A mutual comparison of the results of frost penetration into the real models of the railway line structure with different material composition and the results obtained by numerical modelling of the models in question has been realised since 2009. The DRETM has been focusing on the problematics of the application of thermal insulation materials in the structure composition of the trackbed (where the thermal insulation material in question (structural element) replaces a part of protective layer) since 2016. There was proceeded an extension of the original experimental stand for monitoring the thermo-technical properties and penetration of the zero isotherm of the railway line structure (of the trackbed) in 2016. There was completed a measuring profile with the application of the thermal insulation layer to the measuring profiles which constituted the standard track bed structures. This experimental workplace, in a work-related context referred to as "Experimental stand DRETM", consists of more measuring profiles nowadays and it will be completed within the year 2017.The first three measuring profiles represent the most frequently used railway line structures for modernisation of the railway infrastructure of Slovak Republic nowadays. The first profile is created by an embankment from the crushed aggregate of size fraction 0/63 mm without any protective layer. The second measuring profile forms an embankment with a protective layer from the crushed aggregate of size fraction 0/31.5 mm and the structural thickness 0.45 m. The third measuring profile is a cutting with the protective layer from crushed aggregate of the same size fraction and structural thickness as in the second measuring profile. In the fourth profile was a part of the protective layer, the crushed aggregate of size fraction 0/31.5 mm and of construction thickness 0.30 m, replaced by a layer of liapor of size fraction 0/16 mm and of construction thickness 0.10 m.

The application of liapor in the structure composition of the track bed resulted from the fact, that the material in question has been used in embankments of traffic structures for several years. The use of liapor is relevant because of its small bulk density (the lightenance of the embankment in case of a poor supporting soil) and its excellent thermal insulation properties. However, liapor is not the only thermal insulation material which seems to be an appropriate structural element for the protection against the frost-susceptible subgrade in the track bed against the adverse frost effects. Therefore, there has been planned an extension of already existing Experimental stand DRETM to other two measuring profiles ($5^{th}$ and $6^{th}$ measuring profile) this year. It is expected that there will be built-in and tested other thermal insulation materials in these measuring profiles. Specifically, there is expected to built-in slabs from styrodur and liapor bound by cement.

Unfortunately, the thermo-technical parameters (especially the specific heat capacity $c$) of the building materials in question are missing on the websites of producers or they are insufficiently introduced. Since these parameters are inevitable for the numerical modelling of the temperature development of railway line in the software *Soilvision*, it was necessary to determine them by laboratory tests. The modification of the procedure for measuring, calibration of the apparatus, as well as the measured parameters themselves, which are inevitable for determination of the specific heat capacity, are introduced in the following text.

# 2 Methodology for the determination of the specific heat capacity *c* of the selected thermal insulation materials of the trackbed

The application of the thermal insulation materials in the construction of the trackbed may seem irrelevant according to the predictions of the models concerning the climatic conditions. As according to the climatologists there is the assumption of the warming to continue, the annual average temperatures to grow and the snow cover to decrease. However, these climatic models do not exclude extreme winter periods - with longer and intense periods of frost and a reached high frost index-which can occur from time to time. The use of the thermal insulation materials is also supported by the fact, that the modernisation of main corridors of the railway lines in Slovakia approaches slowly the coldest areas of Slovakia (the regions of Low and High Tatras). Nowadays there have been reached the values of frost index close to the proposed values of the frost index which were defined for the territory of the Slovak Republic. The application of the heat insulation building materials into the structure of the railway line or the track bed can bring many advantages with itself. To these belong in particular the total heat protections of the frost-active soil against the adverse effects of frost while saving natural materials (crushed aggregate) because of the design with the lesser structural thickness of the frost protective layer.

The experimental measurements of the specific heat capacity $c$ and other parameters, which are necessary for the final determination of the thermal conductivity coefficient $\lambda$, have been realised in the laboratory of DRETM since 2008. The initial measurements of the parameters in question served especially to verify the parameters introduced in [1] and [2]. As mentioned above, nowadays the experimental measurements are concentrated on new building materials which could be applied in the structural composition of the railway line trackbed (e.g. liapor, styrodur, foam concrete, etc.). As the most important thermo-technical parameter – the thermal conductivity coefficient $\lambda$ – is well identified for the building materials in question, the experimental measurements were concentrated only on the determination of the specific heat capacity $c$ of the thermal insulation materials in question. This parameter is either unknown or insufficiently introduced but it is needed for the numerical modelling of the frost penetration of railway line structure (of the track bed).

## 2.1 The adaptation of the equipment for determination of the specific heat capacity *c*

The specific heat capacity $c$ [$J.kg^{-1}.K^{-1}$)] of the monitored building materials is being determined by a calorimetric test. The precise determination of the specific heat capacity $c$ is represented by a complex laboratory process of defining the partial parameters which influence the overall results achieved. As the water, which surrounds the sample of the measured material, is the main medium for the calorimetric test and the thermal insulation materials have a smaller bulk density than water, it was necessary to adapt the structure of the calorimeter.

In contrast to the standard structure of the calorimeter it was necessary to fix a nylon thread to the inner walls of the calorimeter. The nylon thread not only keeps the sample under water during the entire process of measurement but it also prevents the test sample to be sucked into the stirrer. There were used plastic holders (plastic materials consume minimum heat) for fixing the nylon tread to the inner walls of the calorimeter. The further procedure remained unchanged where a polyethylene packaging was used to protect the sample against soaking. The sample was tempered to the temperature of the room (less than 20°C) and the calorimetric test started at the temperature of the + 50 °C ± 2 °C water.

The specific heat capacity of a dry sample $c_0$ was calculated from the following relation on the basis of the experimentally stated parameters:

$$c_0 = \frac{m_v . c_v + K}{m} . \frac{\theta_P - \theta_K}{\theta_K - \theta} \tag{1}$$

Where is: $c_0$ - specific heat capacity of a dry sample , [J/(kg. K)],
$m_v$ - water mass, (kg),
$c_v$ - specific heat capacity of the water, [J/(kg. K)],
$m$ - test sample mass, (kg),
$\theta_P$ - initial temperature of the water, (°C),
$\theta_K$ - final temperature of the system, (°C),
$\theta$ - temperature of the sample before measurement, (°C),
$K$ - heat capacity of the calorimeter, ($J.kg^{-1}$),

The recalculation of the specific heat capacity of the sample from the dry to the wet state was subsequently realised by means of the relation:

$$c = c_0 + (c_v - c_0).\frac{w_m}{100 + w_m} \tag{2}$$

Where is : $c$ - specific heat capacity of a wet material , [J/(kg. K)],
$w_m$ - moisture content in the sample, (%) [3].

Since the equipment for determination of the specific heat capacity $c$ was supplemented by other components, there was necessary a calibration of calorimeter what is introduced in the following part of the text.

### 2.2 Calibration of calorimeter

The experimental monitoring of the temperature decrease of the thermal insulation material (medium) within the calibration was carried out in the identical working procedure as the for the standard building materials of exceeding specific mass with the difference that there was used water as a sample (it has the biggest specific heat capacity from the common substances) with lower temperature than the medium has (less than 20 °C). It should be emphasised that the sample of water was not inserted into the calorimeter in the polyethylene packaging but it was poured freely and the tempering of it to the temperature of the room was realised by placing it into the container with water. The resulting value of the heat capacity of the calorimeter was defined by deriving the parameter K from the relation (1).

Overall, there were realised 5 calibration measurements for calibration of the calorimeter where the water volume was about 10 litres in the calorimeter, its temperature was 50 °C ± 2 °C and the mass of the test sample was from 500 to 700 g. The determined values of the heat capacity *K* are presented in Table 1.

**Table 1.** Determined values of heat capacity *K* of the calorimeter.

| Number of measurement | Sample temperature $\theta$ (°C) | Water temperature $\theta_p$ (°C) | Water temperature $\theta_k$ (°C) | Water mass $m_v$ (kg) | Sample mass $m$ (kg) | Heat capacity $K$ ($J.kg^{-1}$) |
|---|---|---|---|---|---|---|
| 1 | 13.48 | 48.27 | 46.37 | 10.221 | 0.595 | 330 |
| 2 | 14.55 | 48.81 | 46.96 | 10.059 | 0.577 | 207 |
| 3 | 15.15 | 49.15 | 47.11 | 10.635 | 0.683 | 274 |
| 4 | 14.40 | 48.95 | 46.76 | 9.992 | 0.681 | 296 |
| 5 | 13.95 | 47.15 | 45.46 | 10.502 | 0.568 | 331 |

Subsequently, there was calculated, as an arithmetic average, the resulting value of the heat capacity K = 288 $J.kg^{-1}$from the 5 determined values of the heat capacity of the calorimeter. This heat capacity was further used for the calculation of specific heat capacity $c_0$ of other tested building materials.

### 2.3 Determination of specific heat capacity of the selected materials

The specific heat capacity is a quantity characteristic for the substance and it has for various substances and various forms various values. In general, this quantity specifies what amount of heat is to be added to a substance with the mass of one kilogram to increase its temperature in 1 °C or 1 K. For all the substances remains the fact, that the specific heat capacity is being reduced by the decrease of the temperature, however, the reduction is not very big so it is possible to consider the values of specific heat capacity as constant for a certain temperature interval [4].

As mentioned in the introduction of this article, there were tested 3 thermal insulated building materials altogether, namely liapor of size fraction 0/16 mm, styrodur and foam concrete, which are assumed to be built-in to other measuring profiles of the railway line model – of the Experimental stand DRETM.

#### *2.3.1 Determination of specific heat capacity of liapor of size fraction 0/16 mm*

Liapor is lightweight clay aggregate which is produced by burning and expanding clay in rotary kilns at temperatures about 1200 °C. There are created granules of pure natural material with excellent performance that is transferred also to other materials produced of it. Liapor, in the form of ceramic granules, is used in various grain sizes especially for different backfills and subbase where it performs a thermal insulation function. This material is used, because of its low specific weight, into the lightened embankments of line constructions abroad. The value of the thermal conductivity coefficient λ of liapor is from 0.09 to 0.15 $W.m^{-1}.K^{-1}$ depending on its specific weight.

The basic properties of liapor are: low specific weight 350-500 $kg.m^{-3}$, heat conductivity from 0.10 $W.m^{-1}.K^{-1}$, high compressive strength, mechanical resistance, chemical stability, fire-resistance, low absorptivity, high soundness, without organic particles and foreign bodies, wholesomeness [5].

The value of specific heat capacity $c_0$ of liapor was defined from the measured parameters presented in Table 2, where are presented also its specific values for the individual measurements. There was created a sample of mass about 500 g for determination of the specific heat capacity $c_0$ of liapor. The sample was produced by mixing of size fraction 8/16 and 0/4 mm in a ratio 1:1 (the same ratio was used also for the construction of the $4^{th}$ measuring profile of the Experimental stand DRETM.

**Table 2.** Determined values of the specific heat capacity $c_0$ of liapor of size fraction 0/16 mm.

| Number of measure-ment | Sample temperature $\theta$ (°C) | Water temperature $\theta_p$ (°C) | Water temperature $\theta_k$ (°C) | Water mass $m_v$ (kg) | Sample mass $m$ (kg) | Spec. heat capacity $c_0$ ($J.kg^{-1}.K^{-1}$) |
|---|---|---|---|---|---|---|
| 1 | 13.00 | 47.30 | 46.98 | 10.882 | 0.500 | 864 |
| 2 | 15.48 | 46.52 | 46.22 | 10.730 | 0.518 | 852 |
| 3 | 15.20 | 48.70 | 48.36 | 10.403 | 0.519 | 866 |
| 4 | 16.30 | 48.33 | 48.01 | 10.474 | 0.522 | 853 |
| 5 | 16.48 | 46.37 | 46.07 | 10.469 | 0.519 | 862 |

Subsequently, there was defined (as an arithmetic average) the resulting value of the specific heat capacity $c_0$ of liapor of size fraction 0/16 mm from the five measured values of the specific heat capacity. The resulting value is $c_0 = 860$ $J.kg^{-1}.K^{-1}$.

### *2.3.2 Determination of specific heat capacity of styrodur*

Excellent thermal insulation properties, variability of types, various compressive strength and edge performance make styrodur a very necessary and useable material in structural, underground and traffic building industry. Besides of the application for insulation of circumferential walls, roofs and ceilings it is possible to use it also as a frost protection in the road and railway construction. Styrodur is extremely hard extruded polystyrene produced from polystyrene granules by a physical change of form called extrusion. It is characterised by its high compressive strength – up to 700 kPa according to the type by 10 % compression. The heat conductivity coefficient of styrodur is $\lambda = 0.03 - 0.04\ W.m^{-1}.K^{-1}$ [6].

The value of specific heat capacity of styrodur was defined from the measured parameters presented in Table 3, where are presented also its specific values for the individual measurements. There was created a sample of mass about 80 g and with dimensions about 150x150x100 mm for determination of the specific heat capacity of styrodur. The sample was cut out of a styrodur slab (STYRODUR 2800 C).

Subsequently, there was defined (as an arithmetic average) the resulting value of the specific heat capacity $c_0$ of styrodur from the five measured values of the specific heat capacity. The resulting value is $c_0 = 2051\ J.kg^{-1}.K^{-1}$.

**Table 3.** Determined values of the specific heat capacity $c_0$ of styrodur.

| Number of measurement | Sample temperature $\theta$ (°C) | Water temperature $\theta_p$ (°C) | Water temperature $\theta_k$ (°C) | Water mass $m_v$ (kg) | Sample mass $m$ (kg) | Spec. heat capacity $c_0$ ($J.kg^{-1}.K^{-1}$) |
|---|---|---|---|---|---|---|
| 1 | 16.48 | 45.83 | 45.72 | 10.404 | 0.079 | 2090 |
| 2 | 14.76 | 47.59 | 47.47 | 10.578 | 0.078 | 2087 |
| 3 | 14.27 | 47.70 | 47.58 | 10.569 | 0.078 | 2051 |
| 4 | 13.60 | 50.04 | 49.96 | 10.382 | 0.048 | 2007 |
| 5 | 13.80 | 49.91 | 49.83 | 10.349 | 0.048 | 2019 |

### *2.3.3 Determination of specific heat capacity of foam concrete*

Foam concrete is a building material with a good mechanical strength, low thermal conductivity and simple but highly technological processing. It is produced by mixing of binder, water, ingredients and technical foam. Foam concrete contains closed air gaps by which is reached its low specific weight and saving of material inputs. The variability of properties and wet process of its production offer a wide range of uses of foam concrete for compensation layers, fillings of various "deaf" spaces and production of prefabricated panels and blocks. This material can have its application on the places with a low deformation resistance of subgrade as well as in the areas with great frost-penetration depth. There is used foam concrete of low bulk density in practice, however, generally it is possible to produce foam concrete with bulk density from 350 to 1200 $kg.m^{-3}$. On the bulk density and also moisture of the foam concrete depends its value of the thermal conductivity coefficient $\lambda$ which reaches the values from 0.1 to 0.4 $W.m^{-1}.K^{-1}$ [7].

The value of specific heat capacity of foam concrete was defined from the measured parameters presented in Table 4, where are presented also its specific values for the individual measurements. There was created a sample of mass about 500 g for determination of the specific heat capacity $c_0$ of foam concrete. The sample had dimensions 100x100x100 mm and bulk density 500 $kg.m^{-3}$.

Subsequently, there was defined (as an arithmetic average) the resulting value of the specific heat capacity of foam concrete from the five measured values of the specific heat capacity. The resulting value is $c_0 = 1118\ J.kg^{-1}.K^{-1}$.

**Table 4.** Determined values of the specific heat capacity $c_0$ of foam concrete.

| Number of measure-ment | Sample temperature $\theta$ (°C) | Water temperature $\theta_p$ (°C) | Water temperature $\theta_k$ (°C) | Water mass $m_v$ (kg) | Sample mass $m$ (kg) | Spec. heat capacity $c_0$ (J.kg$^{-1}$.K$^{-1}$) |
|---|---|---|---|---|---|---|
| 1 | 13.62 | 49.41 | 48.90 | 10.480 | 0.567 | 1127 |
| 2 | 14.00 | 49.76 | 49.24 | 10.363 | 0.562 | 1148 |
| 3 | 16.84 | 47.26 | 46.83 | 10.415 | 0.565 | 1115 |
| 4 | 13.72 | 46.79 | 46.33 | 10.468 | 0.563 | 1106 |
| 5 | 15.30 | 49.21 | 48.74 | 10.425 | 0.564 | 1095 |

## 3 Conclusion

Thermal insulation building materials find their usage not only in the building industry but also in other various parts of industry. Thermal insulation materials are often used also in the structure composition of layers of linear traffic structures abroad. Also our legislative documents, involved in the design of structure composition of traffic structures, consider these materials. The aim is to increase the heat resistance of pavement structure or track bed. Nowadays an amendment to the legislative document shall be taken [1] which has been in force since 1988. There are not presented the thermal conductivity coefficients of all building materials which are (or can) be used in the trackbed layers in the part of dimensioning of the track bed structure on the non-traffic load. It is desirable to verify in this connection not only the thermal conductivity coefficients λ of standard materials, but also of the ones which have a perspective to be used. The Experimental stand DRETM, which is represented by a model of a railway line on a scale 1:1, serves to verify this parameter within our research. There already is (4$^{th}$ measuring profile) or is intended to be built-in (5$^{th}$ and 6$^{th}$ measuring profile) a thermal insulation layer in some parts of the track bed. In particular, there is assumed to use styrodur and liapor bound by cement. The verification of the temperature development and penetration of zero isotherm into the railway line structure (of the track bed) happens in two ways – by an experimental measurement of the relevant thermo-technical parameters on the real railway line structure, but also by a numerical modelling of the temperature development of the railway line structure in software *SoilVision*. It was essential to obtain the necessary input parameters, to which belongs also the special heat capacity of the individual materials, for the realisation of the experimental monitoring and also the numerical modelling.

The value of the specific heat capacity $c$ of the thermal insulation building materials is not as important to the final customers as the thermal conductivity coefficient λ. Because of this reason, it is often much more difficult or absolutely impossible to obtain these values. At DRETM (in response to the necessary amendment to the legislation document [1]) is paid attention to the verification of the thermal conductivity coefficients λ of the materials used in the track bed layers of the railway line. For this verification and their calculation is necessary a partial parameter what is a specific heat capacity. The values of the specific heat capacity $c_0$ of the individual thermal insulation building materials and their comparison with the values from their producers or from foreign sources are presented in Table 5.

As is evident from Table 5, the difference between the laboratory defined specific heat capacity of the tested thermal insulation building materials and the specific heat capacity declared by producer or gained from foreign sources (foam concrete) is minimal. The biggest difference was identified in case of liapor. However, it is necessary to emphasise that the value defined by producer is allocated to the size fraction 8/16mm but the tested sample in laboratory DRETM was of size fraction 0/16 mm (it was created by mixing of two fractions).

**Table 5.** Measured values of the specific heat capacity *c* and their comparison with the values from producer.

| Building material | Specific heat capacity $c_0$ (determined in the laboratory KŽSTH) ($J.kg^{-1}.K^{-1}$) | Specific heat capacity $c_0$ (technical data sheet of the producer or foreign source) ($J.kg^{-1}.K^{-1}$) |
|---|---|---|
| Liapor fr. 0/16 mm | 860 | 900 [8] |
| Styrodur 2800 C | 2051 | 2060 [9] |
| Foam concrete | 1118 | 900-1100 [10] |

The application of the thermal insulation materials in the structure composition of the railway line appears as undesirable based on the climatic models forecasting constant warming of climate and increase of the average annual temperatures. But at the same time the climatologists do not exclude winter seasons with intense and longer frost periods. In this connection, we have to respect the fact that we have regions in Slovakia where were and are likely to be reached high values of frost index (Podtatranský region, Orava, Kysuce). In response to the necessary modernisation of the railway infrastructure of Railways of the Slovak Republic, the building materials with better thermal insulation properties can find a real application in the track bed structure related to the modernisation of railway lines.

The presented results are the results of solving the VEGA grant project 1/0275/16 *Optimization design of sleeper subgrade due to non-traffic load aspect.*

## References

1. TNŽ 73 6312, Design of subgrade construction layers, Standard of Slovak Railway (2005)
2. STN EN ISO 10 456, Construction materials and products. Hygrothermal properties. Tabulated design (calculated) values and procedures for the determination of declared and design values of thermal quantities (2007)
3. P. Dobeš, Optimization of the subgrade design for non-traffic load (Dissertation thesis), ). University of Žilina, Faculty of Civil Engineering, Department of Railway Engineering and Track Management (2015)
4. http://fyzika.jreichl.com/main.article/view/578-merna-tepelna-kapacita
5. http://finistav.sk/katalog/vyrobok/liapor
6. http://www.prespor.sk/content/view/id/55/tepelne-izolacie-xps-a-eps.html
7. http://www.sircontec.sk/penobetony-pbg
8. http://registrace.novazelenausporam.cz/vyhledavani/vyrobek/SVT3013/lehke-keramicke-kamenivo-liapor/
9. http://www.isover.sk/sites/isover.sk/files/assets/documents/styrodur_2800_c_technicky_list.pdf
10. http://www.jurnalteknologi.utm.my/index.php/jurnalteknologi/article/view/8374/5067

# Rationale for statistical characteristics of road safety parameters

*Tatiana* Dormidontova[1], and *Sergey* Evdokimov [1,*]

[1] Samara State Technical University, Institute of Architecture and Civil Engineering, 194, Molodogvardeyskaya St., 443001, Samara, Russia

**Abstract.** When making engineering decisions at the stage of designing auto-roads and man-made structures it is necessary to take into account the statistical variability of physical and mechanical characteristics of the used materials as well as the different effects on the structures. Thus the rationale for the statistical characteristics of the parameters that determine the reliability of roads and man-made engineering facilities is of particular importance.There are many factors to be considered while designing roads, such as natural climatic factors, the accidental effects of the operating loads, the strength and deformation characteristics of the materials, the geometric parameters of the structure, etc. which affect the strength characteristics of roads and man-made structures. The rationale for statistical characteristics of the parameters can help an engineer assess the reliability of the decision and the economic risk, as well as avoid making mistakes in the design of roads and man-made structures.However, some statistical characteristics of the parameters that define the reliability of a road and man-made structures play a key role in the design. These are the visibility distance in daytime for the peak curve, variation coefficient of radial acceleration, the reliability of visibility distance and other parameters.

## 1 Introduction

Nowadays there exist a lot of problems in road and highway traffic management. This is due to the increase in the number of automobiles, historically established street-road network of cities as well as the parking lots on the driveway. All these factors decrease the traffic safety and the highway traffic capacity causing traffic jams [1-3].

Great improvements in city traffic can be achieved by taking urban planning measures such as bridge, tunnel and highway construction [4-5]. The development of such projects is expensive and time consuming.

The relevance of the technical solutions is based on a comparison of different techniques of traffic management and the ways of their technical implementation. It takes into account the specificities and characteristics of the designed sections of roads and the current domestic and foreign experience of operating similar systems.

* Corresponding author: adisk63@yandex.ru

## 2 Materials and Methods

The paper considers statistical characteristics of parameters determining the reliability of roads and man-made structures in details.

Variation coefficient of curvature radius of vertical curve is so small that it is meaningless to evaluate it. For example, the average radius is 25 km. The maximum possible deviation caused by the inaccuracy of pegging-out is 10m. Variation coefficient is $V = 10/25000 = 1/2500 = 0.04\%$. Its value is too small to be taken into account in calculations.

The visibility distance in daytime for the peak curve is obtained by the following:

$$L = \left(\sqrt{h_1} + \sqrt{h_2}\right) \times \sqrt{2R}\,, \tag{1}$$

where $h_1$ and $h_2$ are, respectively, the elevation of the driver above the road and the rise of the obstacle to be visible; $R$ - radius of the vertical curve.

The average visibility distance $\overline{L}$ and standard visibility distance $\sigma_L$ standard are calculated by the following formulas:

$$\overline{L} = \left(\sqrt{\overline{h_1}} + \sqrt{\overline{h_2}}\right) \times \sqrt{2\overline{R}}\,, \tag{2}$$

$$\sigma_L = \frac{1}{2} \times \sqrt{2R \times \left(\frac{\sigma_{h_1}^2}{h_1} + \frac{\sigma_{h_2}^2}{h_2}\right)}\,. \tag{3}$$

Calculated value of the visibility distance:

$$L = \overline{L} + n \times \sigma_L\,. \tag{4}$$

The average distance from the driver's eyes to the surface of the road is 120 cm. The range of the driver's eyes distance from the road surface is 30 cm. For this indicator the normal distribution is accepted [6,7]. Standard value for a driver's eye elevation above the road surface is $h_1 = 5$ cm. Variation coefficient is $V = 0.166$.

The height of the road bump is taken as the obstruction on the road surface, $h_2 = 20$ cm. The height of the road bump is $15 \div 25$ cm. The default value, $h_2 = 1.6$ cm $V = 0.16$.

Thus, when $R = 70000$ m, $h_1 = 120$ cm and $h_2 = 10$ cm

$$L = \left(\sqrt{120} + \sqrt{10}\right) \times \sqrt{2 \times 7000000} = 527.7\,\text{m}, \tag{5}$$

$$\sigma_L = \frac{1}{2} \times \sqrt{2 \times 7000000 \times \left(\frac{25}{120} + \frac{2.56}{10}\right)} = 12.7\,\text{m}. \tag{6}$$

Variation coefficient $V = 0.024$.

For a concave curve, the car speed $v$ and the speed standard $\sigma$ under the conditions of overloading of the bumpy and the passenger's overall health are calculated by the formulas:

$$V = \sqrt{R \times b}\,, \tag{7}$$

$$\sigma = \frac{\partial v}{\partial b} \times \sigma_b = \frac{\sqrt{R}}{2 \times \sqrt{b}} \times \sigma_b = 0.5\sqrt{\frac{R}{b}} \times \sigma_b, \tag{8}$$

where, $b$ – normative radial acceleration, that is $0.5 \div 0.7$ m/s$^2$.

Variation coefficient is calculated by the formula:

$$V = \frac{0.5 \times \sigma_b \times \sqrt{R}}{\sqrt{b} \times \sqrt{R} \times \sqrt{b}} = 0.5\frac{\sigma_b}{b} = 0.5 v_b. \tag{9}$$

Variation coefficient of radial acceleration can be determined by the following factors, i.e. the range of radial acceleration changes has 6 root-mean-square deviations [8-10]. Consequently, the standard is $0.2/6 = 0.03$ m/s$^2$, while variation coefficient is $V_b = 0.03/0.6 = 0.05$ or 5 %.

Thus, the average car speed value on the concave curve when $R$ = 8000 m and the average value of $b = 0.6$ m/s$^2$ is $\bar{v} = \sqrt{8000 \times 0.6} = 69.3$ m/s = 41,6 km/h; root-mean-square deviations $\sigma = 0.5 \times \sqrt{8000/0.6} \times 0.03 = 1.73$ m/s = 0.104 km/h; variation coefficient of the traffic speed is 0.104/41.6 = 0.0025.

The calculated visibility distance in darkness hours with headlamp lightning is calculated by the formula:

$$L = \sqrt{2R \times h_F}, \tag{10}$$

where, $h_F$ – elevation of the center of the headlamp above the road surface.

Average value, standard and the variation coefficient are calculated by the formulas:

$$\bar{L} = \sqrt{2\bar{R} \times \bar{h}_F}, \tag{11}$$

$$\sigma_L = 0.5 \times \sqrt{\frac{R}{h_F}} \times \sigma_{h_F}, \tag{12}$$

$$V = 0.5 \times V_{h_F}. \tag{13}$$

The following values were analyzed as the parameters of the study: $R = 8000$ m, $h_F = 0.6$ m, $V_{h_F} = 0.01/0.6$.

Then the calculated values are: $\bar{L} = \sqrt{2 \times 8000 \times 0.6} = 98$ m, $\sigma_L = 0.5 \times \sqrt{8000/0.6} \times 0.01 = 1.15$ m and $V = 0.5 \times (0.01/0.6) = 0.017$.

Thus, variation coefficient values of vertical curve parameters are negligible:

- for the visibility distance in daytime for the peak curve $V = 0.024$,
- for car speed on a concave curve $V = 0.0025$,
- for visibility distance in darkness hours with headlamp lightning on a concave curve $V = 0.017$.

When determining the reliability of visibility distance normative values for visibility distance were used.

For Category II road with the traffic speed of 120 km/h the visibility distance is 250m, $V$ is 20 %, thus $\sigma_V = V \times \bar{L} = 250 \times 0.2 = 50$ km/h.

Then the reliability is calculated in the following way:

$$P = f(Z) = f\left(\frac{\bar{L} - L_n}{\sqrt{\sigma_L^2 + \sigma_n^2}}\right) = f\left(\frac{250 - 98}{\sqrt{1.15^2 + 50^2}}\right) = f(3.03) = 0.998 \quad (14)$$

The average value of a road visibility distance is evaluated like this:

$$\bar{S} = \frac{\bar{v}}{3.6} + \frac{\bar{k} \times \bar{v}^2}{254(\varphi \pm i + \bar{f})} + L_b \,. \quad (15)$$

The standard is calculated by the formula:

$$\sigma = \sqrt{\begin{aligned}&\left(\left(\frac{1}{3{,}6} + \frac{2 \times v \times k}{254(\varphi \pm i + f)}\right) \times \sigma_v\right)^2 + \left(\frac{v^2}{254(\varphi \pm i + f)} \times \sigma_k\right)^2 + \\ &+ \left(\frac{k \times v^2}{254(\varphi \pm i + f)^2} \times \sigma_\varphi\right)^2 + \left(\frac{k \times v^2}{254(\varphi \pm i + f)^2} \times \sigma_f\right)^2\end{aligned}} \,, \quad (16)$$

where, $v$ – car speed at the start of the brake; $k$ – braking performance coefficient (minimum value 1.2, maximum 2.3, average value 1.75, standard 0.18, variation coefficient 0.1); $\varphi$ – coefficient of adhesion ($\varphi = 0.2$); $f$ – tractive resistance ($f = 0.35$); $i$ – longitudinal slope of the road; $L_b$ – security clearance.

## 3 Results

On the basis of the obtained data the section of the road can be calculated by the formulas given above and the following results can be got, when $v = 60$ km/h, $\bar{S} = 152.4$ m, $\sigma = 8.04$ m, $V = 0.05$.

The formula for calculating the storm water can be written as follows:

$$Q_L = 16.7 \times a_r \times F \times \alpha \times \varphi \,, \quad (17)$$

Where, $a_r$ – the calculated intensity of the downpour of the same probability of exceedance as the amount of the desired mm/min; $F$ – drainage area; $\alpha$ – run-off coefficient; $\varphi$ – reduction coefficient (imperfection of run-off), thus:

$$\varphi = \frac{1}{\sqrt[4]{10F}} \,, \quad (18)$$

$$a_r = \left(\frac{60 \times v_d}{L}\right)^{2/3} \times a_t = K_T \times a_t \,. \quad (19)$$

Where, $a_t$ – the average intensity of one-hour heavy rain, mm/min; $L$ – basin length, km; $v_d$ – water velocity, km/min; $K_T$ – transition code.

Then, the formula (17) looks like this:

$$Q_L = 16.7 \times a_t \times K_T \times F \times \varphi \,. \quad (20)$$

Taking into account the road area and the natural slope area the following formula can be derived:

$$Q_L = 16.7a_t\left(5.24 \times F_d + K_T \frac{F_o}{\sqrt[4]{10F_o}}\right), \tag{21}$$

Where, $F_d$ – road area; $F_o$ – natural slope area.

On the basis of the got results a graph can be made showing the dependency of reliability $P$ on the average intensity of one-hour heavy rain $a_t$, Fig.1.

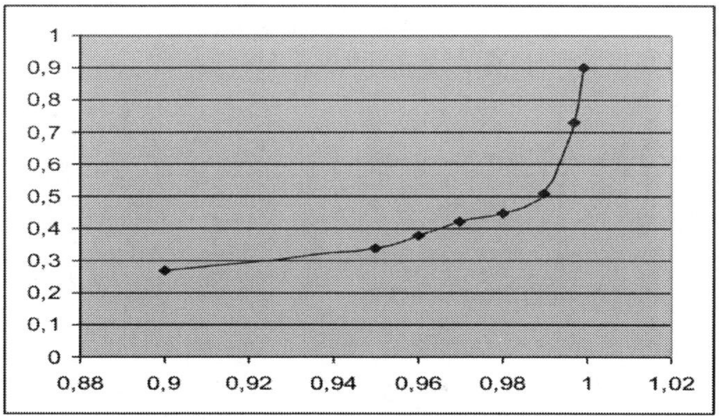


**Fig.1.** The dependency of reliability on the average intensity of one- hour heavy rain.

The values of $\overline{a_t}$ with the probability of 1, 2, 3, and 4% (these values are accepted for calculation depending on the road category) are of interest [11].

$\sigma_{a_t}$ and $V_{a_t}$ for all of four probability values have been calculated and the results are given in Table1.

**Table 1.** Calculated values $\sigma_{a_t}$ and $V_{a_t}$ depending on probability.

| 1 % ($a_t = 0.59$) | | 2 % ($a_t = 0.465$) | | 3 % ($a_t = 0.42$) | | 4 % ($a_t = 0.38$) | |
|---|---|---|---|---|---|---|---|
| $\sigma_{a_t}$ | $V_{a_t}$ | $\sigma_{a_t}$ | $V_{a_t}$ | $\sigma_{a_t}$ | $V_{a_t}$ | $\sigma_{a_t}$ | $V_{a_t}$ |
| 0.093 mm/min | 16 % | 0.03 mm/min | 6.4 % | 0.023 mm/min | 5.5 % | 0.013 mm/min | 5.2 % |

## 4 Discussion

Therefore, the variation coefficient for road area tends to zero, because the error in its measurements is negligible. In order to assess the reliability of the road section, it is necessary to have statistical characteristics of all parameters and determine the reliability of an auto-road in accordance with the known dependences of the reliability theory [12-15]. The algorithms for calculating the statistical characteristics of the parameters that determine the reliability of the road elements are reduced to the use of linearization and brute force methods taking into account the Monte Carlo method [16,17].

The research results are proved by checking all the results for homogeneity and determining the variation coefficient. It has been observed that in the vast majority of cases, individual structures and roads are operated without meeting the requirements. It results in the increased number of defects and rapid decrease of reliability to normalized values. The operation of items with the reliability of below normalized values does not guarantee their safety and efficiency.

## 5 Conclusions

Nowadays the problem of estimating the statistical characteristics of road parameters is quite evident. The road parameters determined by analytical dependences are highly volatile because of the inaccuracy of the defining factors. However, every parameter of physical origin is statistically unstable. It is shown by the variation coefficient of their values. This causes the variability in the parameters that determine the reliability of a road. Their study may result in the changes of general and particular road design rules in order to increase their reliability. The methodology for estimating the statistical parameters that define the operating characteristics of a road has been determined. The statistical characteristics of road parameters have been estimated by the obtained statistical distributions. Statistical information on the material properties, geometric dimensions of the road elements and their statistical distributions acted as the initial data. On their basis the distribution of the analyzed parameter has been made by using normalized dependences.

## References

1. Yu.M. Galitskova, *Tradition and Innovation in Building and Architecture,* 222-223 (2000)
2. L.G. Goverdovskaya, Urban Construction and Architecture **1**, 72-80 (2015) (doi:10.17673/Vestnik.2015.01.12)
3. A.N. Davydov, *Tradition and Innovation in Building and Architecture,* 162-165 (2015)
4. T.V Dormidontova, *Monograph*, Sibak, 152 (2015)
5. T.V. Dormidontova, *Patent to Useful Model* 160415 (2016)
6. S.V. Evdokimov, *Scientific Survey* **22**, 102-107 (2015)
7. B.G. Ivanov, Again on the roads of Russia, 112-117 (2015)
8. B.G. Ivanov, *Tradition and Innovation in Building and Architecture,* 36-37 (2013)
9. B.G. Ivanov, *Tradition and Innovation in Building and Architecture,* 47 (2013)
10. E.I. Marinin, Urban Construction and Architecture **4**, 80-84 (2014) (doi:10.17673/Vestnik.2014.04.12)
11. A.A. Mikhasek, *Prirodoohranyie I gidrotehnicheskie sooruzheniya*, 398-401 (2014)
12. V.A. Pavlova, *Ways of Improving the Quality of Roads,* 158-168 (2015)
13. L.V. Pavlova, Urban Construction and Architecture **2**, 96-103 (2016) (doi:10.17673/Vestnik.2016.02.18)
14. L.V. Pavlova, *Ways of Improving the Quality of Roads,* 168-176 (2015)
15. S.A. Piyavskiy, *Vestnik MGSU*, 54-61 (2012)
16. M.V. Rodionov, Geotec Hanoi, 399-402 (2013)
17. A.V. Filatova *Ways of Improving the Quality of Roads,* 206-211 (2015)

# Assessment of possibility of metal waterjet cutting wastes use in building materials production

*Nataliya* Skanavi[1], *and Timofey* Dovydenko[1,*]

[1]Moscow state university of civil engineering, Yaroslavskoye shosse, 26, Moscow, Russia, 129337

**Abstract.** The paper is aimed at studying the composition and properties of the wastes generated during metal waterjet cutting and assessing the possibility of their use in building materials production. The essence of waterjet cutting process, waste generation mechanism, waste accumulation volumes at enterprises are described. The composition and properties of the used abradant – garnet sand is given and the features of its destruction during cutting are revealed. Waterjet cutting wastes are comprehensively studies: average and bulk density, granulometric composition, chemical composition are determined, various fractions are studied with an electronic microscope. It is revealed that during cutting abradant particles are destroyed, a large amount of dust fraction emerges with the particles of cut metal mixed into it. The metal waterjet cutting wastes are found to be very small, practically monofractive, heavy sands with a high content of dust fraction, which chemical composition is dominated by oxides of iron, silicon and aluminum. This characteristic of the wastes has allowed us to outline possible ways of how to use them: in ceramic items production as thinning agents and fluxing agents, in Portland cement production as a correcting iron-containing agent, as a mortar filler, including special mortars, as a building materials volumetric staining pigment, etc.

## 1 Introduction

### 1.1 Study goal

The use of industrial and consumption wastes is one of the most important areas in environmental protection directly related to natural resources preservation and biosphere contamination prevention. The total amount of accumulated and recorded wastes in Russia as a whole is approximately 31.5 bln tons [1]. The vast majority of accumulated and identified wastes is classified under Hazard Class 5, that is, as non-hazardous – 31.1 bln tons, or almost 99% of their total volume. In 2015, industrial and consumption wastes in Russia were generated in the amount of 5,060.2 mln tons, 4,950.2 mln tons of which were of Hazard Class 5. These figures are close to average annual values. To store such amount of wastes, hundreds of thousands of hectares of land are taken from the economic turnover.

* Corresponding e-mail: dovydenkotimofei@gmail.com

Necessary storage of wastes, arrangement and maintenance of waste dumps require huge expenditures, while the wastes concentrated in dumps, storages, junkyards pollute soil, atmospheric air, surface and ground waters, suppress flora and fauna. All the business costs associated with operation of relevant facilities, plants, machinery and equipment are included in products prime cost.

In 2015, the level of non-hazardous wastes use was approximately 52%, which was extremely inadequate. Therefore, study and involvement of new, previously unused wastes in production is an urgent task. Even the wastes generated in small amounts can fill the gap in resource deficit for production of a particular material. The paper is aimed at studying the composition and properties of the wastes generated during metal waterjet cutting and assessing the possibility of their use in building materials production.

### 1.2 Waterjet cutting principle and waste generation

Waterjet cutting is a process when a material is exposed to a high-speed jet of water mixed with an abradant [2, 3, 4, 5]. The functional concept of this type of cutting is that when passing through a hole of 0.2...0.8 mm in diameter a water flow accelerates up to about 300...900 m/s, mixes with an abradant and is applied to the surface to be cut under a high pressure (200...600 MPa). At the instant of collision with the material, jet kinetic energy transforms into mechanical energy of micro-destruction of the processed material, thus cutting it.

Various materials (ferrous and non-ferrous metals, alloys, ceramics [6], glass, natural stone, plastics, etc.) are cut with the use of this method, but metal waterjet cutting is of greater interest.

Waterjet cutting advantages as compared to other types of cutting (gas-oxygen, plasma-arc, laser cutting) are as follows: low operating temperature in the cutting zone (60...90°C), no thermal impact on the material, low material consumption factor (cutting width of 0.2...3 mm), large thickness of cut work pieces (up to 300 mm and above), no impact on alloyed steels chemical composition, possibility of cutting thin sheet materials stacked in several layers, high process safety due to absence of flammable and explosive substances, environmental friendliness of the process, high quality of treated surface which often requires no additional machining, etc.

Garnet sand is used as an abradant [7]. Garnets are a vast group of minerals which are chemically related to the so-called orthosilicates. Various mineral types are marked out of garnets group, of which ferrous-aluminum garnet – almandine $Fe_3Al_2(SiO_4)_3$ is used in waterjet cutting. This mineral is of a cherry-red color with a high hardness of 7...8 on the Mohs scale, high density of 4.1...4.3 g/cm$^3$, very high destruction resistance, and no toxicity to humans and environment. The garnet sand is mainly produced in India, Australia, South Africa, and Czech Republic.

The waterjet cutting process generates slurry wastes containing garnet sand, particles of the material being cut and water. The slurry is dehydrated, the dry residue is stored within the territory of an enterprise, and is transported to solid domestic waste landfills when accumulated in sufficient amounts.

One medium-powered waterjet cutting machine (Water Jet Sweden NC3015) produces approximately 200 grams of dry waste per minute. The machine is operated round-the-clock and its useful working time amounts to 70...80% (the rest of time is readjustment, cleaning, etc.). Thus, even with one waterjet cutting machine operated, an enterprise produces more than 200 kg of wastes per day. In consideration of decisive advantages of waterjet cutting as compared to other types of cutting, its share in metal cutting area will increase, so will the amount of wastes generated. This challenges enterprises to store and recycle the wastes.

## 2 Study objects and methods

### 2.1 Garnet abradant

We examined the garnet sand of one of the main Russian suppliers R-Garnet CJSC, which had developed garnet abradant specifications [8]. Sand moisture content was 1.5%, bulk density – 2.31 g/cm$^3$, grains density – 3.805 g/cm$^3$, intergrain porosity – 39.29%. The sand was pure, with no organic impurities.

The granulometric composition determined with the use of standard sand sieves (listed in the table) showed no particles larger than 0.63 mm. The fineness modulus was 1.752.

Particles of different sizes with sharp edges were made by crushing the rock. Sand with particle sizes of 180...300 μm is used for waterjet cutting. With the use of one cutting head abradant consumption, depending on the material to be cut, ranged from 50 g/min (for plastics) and 160...200 g/min (for glass) to 300...450 g/min (for steel). Approximate chemical composition of the garnet sand (produced by Garnet "MICROBLAST"), wt %: $SiO_2$ – 31.00%, $Al_2O_3$ – 21.60%, $Fe_2O_3$ – 37.00%, MgO – 7.40%, with 1.84%, $TiO_2$ – 0.55%, as well as chlorides, $Na_2O$, $K_2O$, $MnO_2$, $P_2O_5$ etc. In terms of mineral composition, this sand is composed of 96.9% of garnet, 2.4% of ilmenite, 0.4% of quartz.

### 2.1 Metal waterjet cutting wastes

The ferrous metals waterjet cutting wastes of SK Modul LLC (Khimki city) have been examined. The results are as follows: true density is 3.7 g/cm$^3$, bulk density is 2.2 g/cm$^3$. The granulometric composition has been determined with the use of standard sand sieves, and a fraction of less than 0.16 mm has been sieved through the 008 sieve.

**Table 1.** Comparison of the granulometric composition of the original abradant and waterjet cutting waste s

| Sieve hole dimensions, (mm) | Garnet abradant | | Waterjet cutting wastes | |
|---|---|---|---|---|
| | Individual residues, % | Full residues, % | Individual residues, % | Full residues, % |
| 0.315 | 75.6 | 75.6 | 1.6 | 1.6 |
| 0.16 | 24 | 99.6 | 70.5 | 72.1 |
| Less than 0.16 | 0.3 | 99.9 | 27.1 | 99.2 |
| Total | 99.9 | - | 99.2 | - |

To identify the processes occurring during waterjet cutting with abradant and cut metal, and accordingly, the process of cutting wastes composition generation, a microscopic study of individual waste fractions and determination of their chemical (elemental) composition was performed using spectral analysis method.

## 3 Results

### 3.1. Granulometric composition

The table shows that large particles (larger than 0.315 mm) of the original abradant are destroyed during metal cutting. A large number of small (0.16...0.315 mm) and minute (less than 0.16 mm) particles emerge. However, such a change in granulometric composition is also due to the fact that the particles of the metal being cut join the minute dust fraction.

The dust fraction has been sieved through the 008 sieve, and the residue on it is 31.3%. That is, 69% of the dust fraction particles are less than 80 μm, which is comparable with the Portland cement particles size.

### 3.2 Microscopic examination

The assumption that the cut metal particles enter exactly into the dust fraction has been confirmed by the wastes microscopic examination and chemical element analysis results.

The microscopic examination has shown that the particles ranging in size from 0.16 to 0.315 mm are of "crushed stone", irregular shape, of different sizes, and are destroyed garnet sand. During destruction new abrasive sides emerge and sometimes the garnet abradant is used repeatedly by returning it to production again [9].

Fig. 1 shows a micrograph of 0.08...0.16 mm fraction waste particles. Larger particles of destroyed abradant grains, covered with minute particles – products of cut metal destruction, are seen. The fact that the minute fraction is mainly metal is confirmed by particles shape and structure (Fig.2). The particles are of leaf-shaped, layered structure with sharp edges. The particles are relatively uniform in size and structure, their size is about 90 μm.

| мкм | µm |
|---|---|

**Fig.1.** Micrograph of metal waterjet cutting wastes, fraction 0.08...0.16 mm, x800

| | |
|---|---|
| мкм | µm |

**Fig.2.** Micrograph of fraction less than 0.08 mm, x200

### 3.3 Chemical composition

The chemical composition determined by spectral analysis confirms a high content of iron oxide $Fe_2O_3$ in the wastes – over 35% (Fig.3). It can be higher than in the original abradant due to the cut metal destruction products. In terms of iron content, these wastes meet or exceed pyrite cinders and other iron-containing industrial wastes [10, 11]. The following substances are also present in significant quantities: $SiO_2$ – 31.02%, $Al_2O_3$ – 17.22%, $TiO_2$ – 7.62%, MgO – 6.74%, etc. Titanium dioxide appears in cut steel wastes; in the original abradant its content is only 0.55%. Titanium dioxide in wastes is worth of special study.

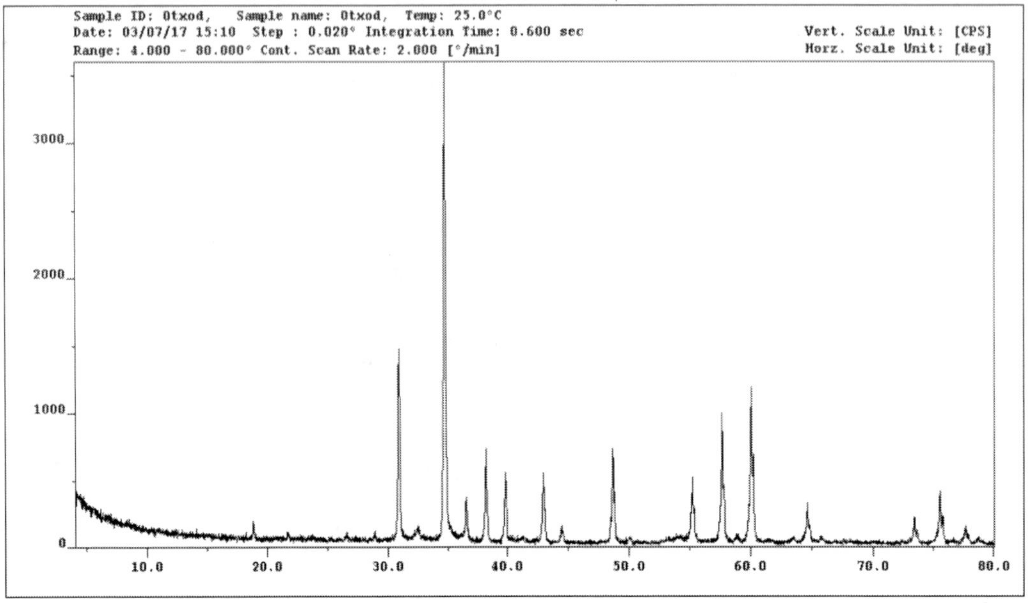


**Fig.3.** Diffractogram of waste

Based on the metal waterjet cutting wastes composition and properties comprehensive study we can describe them as follows.

These are very small, practically monofractive, heavy sands with a high content of dust fraction, which chemical composition is dominated by oxides of iron, silicon and aluminum.

### 3.4 Pilot study of waterjet cutting wastes behavior during firing

To study the waterjet cutting wastes (mainly metal component) behavior during firing in ceramic mixture composition, sample cubes with an edge of 5 cm have been made. A lean montmorillonite clay with the following mineral composition has been used: quartz – 49%, montmorillonite – 25%, hydromica – 7%, microcline – 5%, kaolinite – 3%, hematite – 2%, anorthite – 1%, amorphous phase – 8%. [12]. Test composition of ceramic mixture: 50% of clay and 50% of wastes. Water-solid ratio was......%, which corresponded to the water content during production of plastic molded items. The mixture was laid into a mold without compaction. Samples were dried at a temperature of 105°C, and then exposed to firing at 900°C.

No deformation of the samples, cracking, color anomalies were observed during firing. The average sample density after firing was... 2.4 g/cm$^3$, water absorption – 11.6%, compressive strength – 11.46 MPa, which corresponded to the brand 125. The samples' color was red.

Preliminary study allows to conclude about possible use of metal waterjet cutting wastes for ceramic items production: as a thinning agent for ductile clays, for staining of a ceramic piece (due to large content of chromophore – iron oxide and cherry-red mineral – almandine, which is included in the wastes). High average density is not a negative factor in floor tiles, clinker ceramics production. High content of iron compounds suggests possible use of these wastes as a fluxing agent.

## 4 Conclusion

Our study and the data from other sources make it possible to classify the metal waterjet cutting wastes under Hazard Class 5, that is, as non-hazardous [13]. We have started exploring possible use of these wastes in the following areas:
- in building ceramics production – as a thinning agent and fluxing agent, staining component;
- in cements production – as a correcting iron-containing agent for production of Portland cement clinker as a component of the raw mixtures used to produce high-iron cements;
- as a filler in general mortars and special solutions;
- as a pigment for building materials staining; [14]
- other ways of disposal are also considered.

Looking through the publications on the use of wastes does not result in revealing any information on the use of waterjet cutting wastes, which may be due to the lack of interest in the wastes generated in small amounts. We consider it necessary to seek possibilities of using not only the so-called large-tonnage industrial wastes, but also the wastes generated in moderate amounts, which also contributes to solving the problems of resource saving and environmental protection.

The study was carried out with the support of the Principal Regional Center of Collective Use of Moscow State University of Civil Engineering (National Research University).

## References

1. N. G. Rybalsky, E. V. Muravyeva, D. A. Boriskin, About a state and about environmental protection in the Russian Federation in 2015 (Moscow, 2016)
2. M. Hashish, D.E. Steele, D.H. Bothell, Int'l. J. of M/C Tools and Mfg. **37,** 465-479 (1997)
3. G. V. Borovsky, S. N. Grigoriev, A. R. Maslov, Modern technologies of processing of materials (Moscow, 2015)
4. I. I. Sazanov, A. M. Lyadnik, S. V. Lyadnik, Tech. ME. **3**, 11-14 (2013)
5. D. Krajcarz, Procedia Engineering, **69**, 838-843 (2014)
6. P. Gudimetla, J. Wang, W. Wong, J. of Mat. Proc. Tech. **128**, 123-129, (2002)
7. D. Begic-Hajdarevic, A. Cekic, M. Mehmedovic, A. Djelmic, Procedia Engineering, **100**, 394-399 (2015)
8. TU 3988-002-76245879-2011 Garnet abrasive. Specifications.
9. M. Kantha Babu, O.V. Krishnaiah Chetty, Wear, **254**, 763-773 (2003)
10. I. Netinger Grubeša, I. Barišić, A. Fucic, S.S. Bansode, *Characteristics and Uses of Steel Slag in Building Construction* (2016)
11. M.L.S. Oliveira, C.R. Ward, M. Izquierdo, C.H. Sampaio, I.A.S. de Brum, R.M. Kautzmann, S. Sabedot, X. Querol, L.F.O. Silva, Sci. of The Tot. Env. **430**, 34-47 (2002)
12. D.Yu. Zemlyanushnov, V. N. Sokov, D. V. Oreshkin, N. A. Skanavi, Messenger of INRTU, **92**, 122-126 (2014)
13. L. I. Dvorkin, O. L. Dvorkin, Construction materials from industry waste (Rostov-na-Donu, 2007)
14. A. Yu. Stolboushkin, S. V. Druzhinin, T. I. Storozhenko, V. F. Zavadsky, Consrt. Mat., 5, 95-97 (Moscow, 2008)

# Determining the optimum length of a bridge opening with a specified reliability level of water runoff

*Sergey* Evdokimov [1], and *Tatiana* Dormidontova [1,*]

[1] Samara State Technical University, Institute of Architecture and Civil Engineering, 194, Molodogvardeyskaya St., 443001, Samara, Russia

**Abstract**. Current trends in construction are aimed at providing reliability and safety of engineering facilities. According to the latest government regulations for construction, the scientific approach to engineering research, design, construction and operation of construction projects is a key priority. The reliability of a road depends on a great number of factors and characteristics of their statistical compounds (sequential and parallel). A part of a road with such man-made structures as a bridge or a pipe is considered as a system with a sequential element connection. The overall reliability is the multiplication of the reliability of these elements. The parameters of engineering structures defined by analytical dependences are highly volatile because of the inaccuracy of the defining factors. However each physical parameter is statistically unstable that is evaluated by variable coefficient of their values. It causes the fluctuation in the parameters of engineering structures. Their study may result in the changes in general and particular design rules in order to increase the reliability. The paper gives the grounds for these changes by the example of a bridge. It allows calculating its optimum length with a specified reliability level of water runoff under the bridge.

## 1 Introduction

At present, safety and reliability of construction engineering facilities on roads is a matter of great concern. The paper reports on the reliability of a road that includes the reliability of its elements- road surfacing, geometric elements, man-made structures [1-7].

The reliability is closely related to the probability of failures or risks. The reliability of an auto-road and man-made structures on it is much less important, less durable and has absolute serviceability.

The low reliability of a road depends on a large number of factors and characteristics of their statistical sequential and parallel compounds. A part of a road with such man-made structures as a bridge or a pipe is considered as a system with a sequential connection of elements. The overall reliability is the multiplication of the reliability of these elements. In this regard, the methodology proposed in this research work aimed at determining the

* Corresponding author: adisk63@yandex.ru

optimum length of the bridge opening with a specified reliability level of water runoff seems to be relevant.

## 2 Materials and Methods

In order to provide the reliability of slopes stabilized with soil, its gradient $m$ should correspond to the calculated shift angle $\psi$, Formula (1):

$$tg\,\psi = \frac{1}{K_3}\left(tg\,\varphi + \frac{c}{p_h}\right), \tag{1}$$

where, $K_3$ – assurance coefficient ($K_3 = 1$, as it corresponds to the limit state); $\varphi$ – angle of internal friction; $c$ – adhesion;

$$m = \frac{1}{tg\,\varphi + \dfrac{c}{p_h}}$$

$p_h$ – pressure on ground: $p_h = \cos^2\varphi \cdot h \cdot \gamma \cdot 1rm$; $cos^2\varphi \cdot h$ – ground area on the cross-section of a slope; $\gamma$ – soil density; $1rm$ – running meter.

Let's find the derivatives of the function **m** from $\gamma$, $\varphi$, and $c$ calculated by the formula (2)

$$\frac{\partial m}{\partial \varphi} = \frac{-1}{tg\,\varphi^2 + \dfrac{c}{p_h}} \tag{2}$$

$$\frac{\partial m}{\partial c} = \frac{-1}{tg\,\varphi + \dfrac{c^2}{p_h}}$$

$$\frac{\partial m}{\partial \gamma} = \frac{-1}{tg\,\varphi + \dfrac{c}{F \cdot \gamma^2}}$$

To determine the reliability of a slope the authors use the existing normative values for the gradient.

It should be noted, the reliability is calculated as follows, Formula (3):

$$P = f\,(Z) = f\left(\frac{\overline{m} - m_{norm}}{\sqrt{\sigma_m^{\;2} + \sigma_{norm}^{\;2}}}\right) \tag{3}$$

To determine the reliability of superelevation Formula (4) should be used.

$$i_{superel} = V^2/gR - M = v^2/gR - C_{II} \tag{4}$$

Where: $V$ – car speed in superelevation; g – acceleration of gravity (g = 9,81 m/s$^2$); $R$ – turning radius ($R_{min}$=800 m for a road of Category II); $M$ – sideway force coefficient; $C_{II}$ – coefficient of adhesion of car wheels with the ground.

Then standard for the superelevation gradient should be calculated by (5).

$$\sigma_i = \sqrt{\left(2 \cdot \frac{\overline{V}}{g \cdot R} \cdot \sigma_V\right)^2 + 1 \cdot \sigma_\mu{}^2} \tag{5}$$

Then a bridge capability is calculated. To determine the probability of exceeding the flow through the specified bridge section the following formula should be used (6):

$$Q_C = 1.35 H^{2/3} \cdot b \tag{6}$$

Where, $H = 1{,}45 V_C/g$ – head of water; $V_C$ – flow velocity under the bridge; $g$ – acceleration of gravity ($g = 9{,}81$ m/s$^2$); $b$ – width of the bridge.

Standard for water flow should be calculated by the derivative $V_{C,}$ Formula (7).

$$\sigma_{Q_C} = \sqrt{\left(1.15 \cdot \frac{\overline{V_C}^{1/3}}{g^{2/3}} \cdot b \cdot \sigma_{V_C}\right)^2} \tag{7}$$

It should be noted that the reliability of a road and its engineering facilities depends on a number of parameters described above. The authors have assessed the statistical characteristics and parameters of a road and engineering facilities derived from the statistical distributions [8-10]. The statistical data on the material properties, geometric dimensions of road elements and engineering structures, and their statistical distribution acted as the initial data.
On their basics the distribution of the analyzed parameters is made by using standardized force deformation relationship [11]. The authors prove this statement by the example providing the following results.

## 3 Results

According to statistical research, variation coefficient for soil density ($\gamma$), angle of internal friction (φ) and adhesion (c) is $V$ = 15%. The average values for $\gamma$, $\varphi$ and $c$ are: $\gamma$ = 2000 kg/m$^3$; $\varphi$ =7,5є and $c$ = 0,006.

Then the standards are found for $\gamma$, $\varphi$, $c$

$$\sigma_\gamma = V_\gamma \cdot \overline{\gamma} = 0.15 \cdot 2000 = 300\ kg/m^3$$

$$\sigma_\varphi = V_\varphi \cdot tg\overline{\varphi} = 0.15 \cdot tg7.5 = 0{,}02$$

$$\sigma_c = V_c \cdot \overline{c} = 0.15 \cdot 0.006 = 0.0009$$

Now the standard gradient m, average value and variation coefficient should be found.

$$\sigma_m = \sqrt{\left(\frac{1}{tg\bar{\varphi}^2 + \frac{\bar{c}}{\bar{p}_h}} \cdot \sigma_\psi\right)^2 + \left(\frac{1}{tg\bar{\varphi} + \frac{\bar{c}^2}{\bar{p}_h}} \cdot \sigma_c\right)^2 + \left(\frac{1}{tg\bar{\varphi} + \frac{\bar{c}}{\cos^2\varphi \cdot \bar{\gamma}^2}} \cdot \sigma_\gamma\right)^2} =$$

$$= \sqrt{\left(\frac{1}{(tg\,7.5^{O})^2 + \frac{0.006}{\cos^2 7.5^{O} \cdot 6 \cdot 2000}} \cdot 300\right)^2 + \left(\frac{1}{0.13 + \frac{0.006^2}{0.98 \cdot 6 \cdot 2000}} \cdot 0.0009\right)^2 + \left(\frac{1}{0.13 + \frac{0.006}{0.98 \cdot 6 \cdot 2000}^2} \cdot 0.02\right)^2} = \sqrt{16.66+0+0} = 4.08$$

$$\bar{m} = \frac{1}{tg\bar{\varphi} + \frac{\bar{c}}{\cos^2\varphi \cdot h \cdot \bar{\gamma}}} = \frac{1}{0.13 + \frac{0.006}{0.98 \cdot 6 \cdot 2000}} = 7.69$$

$$v_m = \frac{\sigma_m}{\bar{m}} = \frac{4.08}{7.69} = 0.53$$

To determine the reliability of the slope the normative values for the slope should be used.

For loam the slightest slope is 1:1.5

$V$ is 10%, thus:

$$\sigma_V = V \cdot \bar{m} = 1.5 \cdot 0.1 = 0.15$$

The reliability is calculated by the Formula (3):

$$P = f\left(\frac{7.69 - 1.5}{\sqrt{4.08^2 + 0.15^2}}\right) = f(1.51) = 0.93$$

The reliability of the superelevation can be calculated by the Formula (4) following the statement of the research that provides the best conditions for car passage along the curve when $\mu$=0.05$h$0.1. Then the value of μ and the average values of $\sigma_\mu$ and $v_\mu$ are calculated.

$$\bar{\mu} = \frac{0.1 + 0.05}{2} = 0.075$$

$$\sigma_\mu = \sqrt{\frac{(0.075 - 0.05)^2 + (0.075 - 0.1)^2}{2 - 1}} = 0.035$$

$$v_\mu = \frac{0.035}{0.075} = 0.47$$

For speed V we take variation coefficient $v_V \cong 0.2$, while the average value is $\bar{V}$= 120 km/h.

We get:

$$\sigma_V = \overline{V} \cdot v_V = 120 \cdot 0.2 = 24 km/h$$

Then the standard for the superelevation gradient is calculated by the Formula (5).

$$\sigma_i = \sqrt{\left(2 \cdot \frac{33.33}{9.81 \cdot 800} \cdot 6.67\right)^2 + 0.035^2} = 0.036 = 36\%$$

$$\bar{i} = \frac{\overline{V}^2}{g \cdot R} - \bar{\mu} = \frac{(120/3.6)^2}{9.81 \cdot 800} - 0.075 = 0.067 = 67\,\%_0$$

$$v_i = \frac{\sigma_i}{\bar{i}} = \frac{0.036}{0.067} = 0.53$$

For Category II road with the turning radius from 2000m to 1000m the superelevation gradient is 200 $^0/_{00}$.

$V$ take 10 %, thus:

$$\sigma_V = V \cdot \bar{i} = 20 \cdot 0.1 = 2\ ^0/_0$$

Then the reliability is determined by the following:

$$P = f\,(Z) = f\left(\frac{\overline{m} - m_{norm}}{\sqrt{\sigma_m{}^2 + \sigma_{norm}{}^2}}\right) = f\left(\frac{67 - 20}{\sqrt{36^2 + 2^2}}\right) = f(1{,}3) = 0.903$$

Then the reliability of a bridge capability is calculated, Formula (6).

$$Q_C = 1.35\left(\frac{1.45 \cdot V_C}{g}\right)^{4/3} \cdot b = 1.73 \cdot \frac{V_C{}^{4/3}}{g^{2/3}} \cdot b$$

In the given formula only the flow velocity under the bridge is variable.

The average value of flow velocity, standard and its variation coefficient are the following.

$$\overline{V}_C = \frac{5+7}{2} = 6\,m/s$$

$$\sigma_{\overline{V}_C} = \sqrt{\frac{(6-5)^2 + (6-7)^2}{2-1}} = 1.41\,m/s$$

$$v_{V_C} = \frac{1.41}{6} = 0.23$$

Standard for water flow is determined by the Formula (7).

$$\sigma_{Q_C} = \sqrt{\left(1.15 \cdot \frac{V_C^{1/3}}{g^{2/3}} \cdot b \cdot \sigma_{V_C}\right)^2} = \sqrt{\left(1.15 \cdot \sqrt[3]{\frac{6}{9.81^2}} \cdot 10 \cdot 1.41\right)^2} = 6.42\,m^3/s$$

$$\overline{Q}_C = 1.73 \cdot \left(\frac{\overline{V}_C}{g}\right)^{4/3} \cdot b = 1.73 \cdot \left(\frac{6}{9.81}\right)^{4/3} \cdot 10 = 12.51\,m^3/s$$

$$v_{V_c} = \frac{\sigma_{Vc}}{\overline{Q}_C} = \frac{6.42}{12.51} = 0.51$$

The distribution of statistical characteristics [12-15] should be used to determine the reliability of engineering facilities.

## 4 Discussion

To determine the failure of a bridge, you must subtract one distribution from another, that is $Q_C - Q_L$, Fig.1.

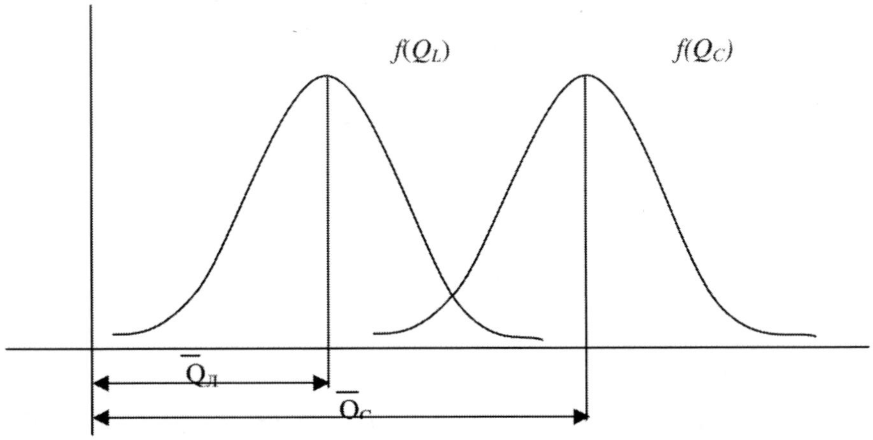


**Fig. 1**. Distribution of bridge failures.

The difference in these distributions will give the characteristic presence of zero on the abscissa axis, which divides the area below the distribution line into two parts, the risk is on the left while the reliability is on the right [16-17].

To determine zero it is necessary to find the value of $Z$.

At that we find $Z$ for $Q_L$ average and $\sigma_Л$ with probability value $\overline{a}_{\text{hour}}$ 1,2,3 и 4%, (8), Fig. 2.

$$Z = \frac{\overline{Q}_C - \overline{Q}_L}{\sqrt{\sigma_C{}^2 + \sigma_L{}^2}} \quad (8)$$

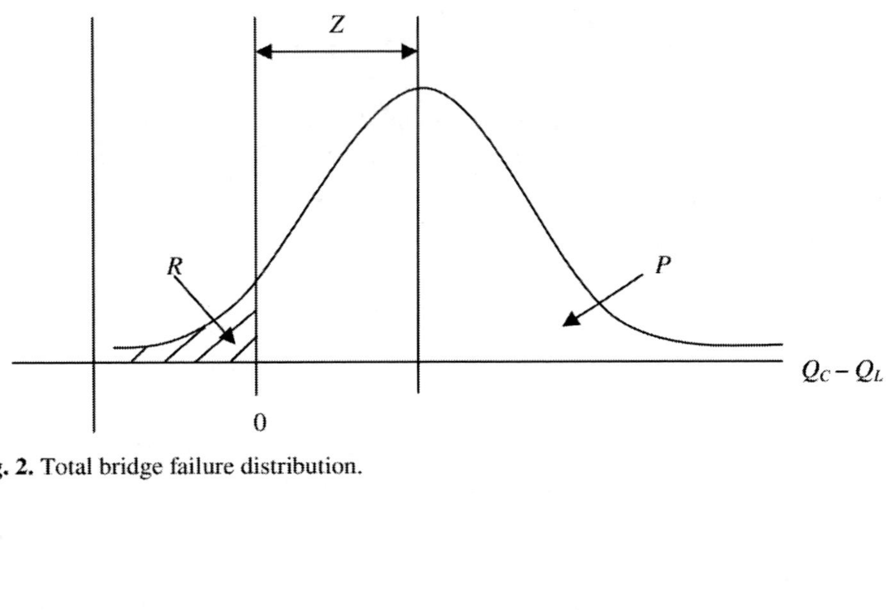


**Fig. 2.** Total bridge failure distribution.

**4%**

$$R = 1 - P = 1 - f(Z) = 1 - f\left(\frac{\overline{Q}_C - \overline{Q}_L}{\sqrt{\sigma_C{}^2 + \sigma_L{}^2}}\right)$$

$$R = 1 - f\left(\frac{12.51 - 7.87}{\sqrt{6.42^2 + 1{,}7^2}}\right) = 1 - f(0.697) = 1 - 0.758 = 0.242$$

**3%**

$$R = 1 - f\left(\frac{12.51 - 8.64}{\sqrt{6.42^2 + 1.78^2}}\right) = 1 - f(0.581) = 1 - 0.7257 = 0.2743$$

**2%**

$$R = 1 - f\left(\frac{12.51 - 9.57}{\sqrt{6.42^2 + 1{,}.9^2}}\right) = 1 - f(0.437) = 1 - 0.6736 = 0.3264$$

**1%**

$$R = 1 - f\left(\frac{12.51 - 12.4}{\sqrt{6.42^2 + 3.07^2}}\right) = 1 - f(0.0155) = 1 - 0.5 = 0.5$$

The probability of exceeding $\overline{Q}_L$ и $\overline{a}_{hour}$ under the bridge for the Category II road is 1%. The bridge failure should be 0.01 or 1%, thus the reliability is 0.99.

$$\sigma_{Q_C} = \sqrt{\left(1.15 \cdot \frac{\overline{V}_C^{1/3}}{g^{2/3}} \cdot b \cdot \sigma_{V_C}\right)^2} = \sqrt{\left(1.15 \cdot \sqrt[3]{\frac{6}{9.81^2}} \cdot 16 \cdot 1.41\right)^2} = 7.51 m^3/s$$

$$\overline{Q}_C = 1.73 \cdot \left(\frac{\overline{V}_C}{g}\right)^{2/3} \cdot b = 1.73 \cdot \left(\frac{6}{9.81}\right)^{2/3} \cdot 16 = 18.32\, m^3/s$$

$$R = 1 - P = 1 - f(Z) = 1 - f\left(\frac{\overline{Q}_C - \overline{Q}_L}{\sqrt{\sigma_C{}^2 + \sigma_L{}^2}}\right)$$

$$R = 1 - f\left(\frac{18.32 - 12.4}{\sqrt{7.51^2 + 3.07^2}}\right) = 1 - f(2.35) = 1 - 0.99 = 0.01$$

It is evident that the specified bridge opening b=10m is not able to provide the passage of the storm drain $\overline{Q}_L$ with the necessary probability degree of failure even if the probability value $\overline{a}_{hour}$ is 4%. Therefore for the given exceedance interval $\overline{a}_{hour}$ and $\overline{Q}_L$ it is necessary to calculate the required bridge length by using a pick method. It has been confirmed by the latest calculation studies.

## 5 Conclusions

The reliability of each of the components of roads and engineering structures on them is a great task. The reliability of the road is different depending upon the season. Thus it should be determined for spring-summer and autumn-winter periods. These estimates are independent. For solving it you should:

- know the factors that make up the parameters of roads and engineering facilities;
- make sequential and parallel connections;
- determine the statistical factors;
- evaluate the reliability of each factor.

The calculations got in the research make it possible to calculate the optimum length of a bridge opening with a specified reliability level of water runoff under the bridge.

## References

1. L.G. Gordovskaya, Urban Construction and Architecture **1**, 72-80 (2015) doi:10.17673/Vestnik.2015.01.12
2. A.N. Davydov, *Tradition and Innovation in Building and Architecture,* 162-165 (2015)
3. A.N. Davydov, *Tradition and Innovation in Building and Architecture,* 116-120 (2016).
4. T.V Dormidontova, *Monograph*, Sibak, 152 (2015)
5. T.V. Dormidontova, Internet journal Naukovedenie **2**, 64-70 (2015)
6. T.V. Dormidontova, Patent to Useful Model 160415 (2016)
7. S.V. Evdokimov, Scientific Survey **22**, 102-107 (2015)
8. B.G. Ivanov, *Ways of Improving the Quality of Roads,* 107-111 (2015)
9. B.G. Ivanov, *Tradition and Innovation in Building and Architecture,* 36-37 (2013)
10. B.G. Ivanov, *Tradition and Innovation in Building and Architecture,* 47 (2013)
11. E.I. Marinin, Urban Construction and Architecture **4**, 80-84 (2014) doi:10.17673/Vestnik.2014.04.12
12. V.A. Pavlova, *Ways of Improving the Quality of Roads,* 158-168 (2015)
13. V.A. Pavlova, *Tradition and Innovation in Building and Architecture,* 156-161 (2015)
14. L.V. Pavlova, Urban Construction and Architecture **2**, 96-103 (2016) doi:10.17673/Vestnik.2016.02.18
15. L.V. Pavlova, *Ways of Improving the Quality of Roads,* 168-176 (2015)
16. A.V. Filatova *Ways of Improving the Quality of Roads,* 206-211 (2015)
17. A.V. Filatova, Scientific Survey **7**, 152-156 (2015)

# Manufacturing processes of cellular concrete products for the construction

*Muhammet* Fakhratov[1], *Ramazan* Erkenov[1], *Akhmat* Kulchaev[1,*], *and Anton* Zavgorodniy[1]

[1] Moscow State University of Civil Engineering, 129337 Moscow, Russia

**Abstract**. Cellular concrete takes the lead in the world of construction as a structural insulation material used in the construction and reconstruction of buildings and constructions of various purposes. In this artificial stone building material, pores are distributed relatively evenly and occupy from 20 to 90% of the concrete volume, ensuring good thermal qualities, which allows cellular concrete houses to keep warmth well. For production of cellular concrete, Portland cement, "burnt lime", and fine-pulverized blast furnace slags, with a hardening activator are used as binders. As silica components, quartz sand or "fly ash" obtained by combustion of pulverized fuel in power plants as well as secondary products of different ore dressing treatments are used. The low density and high thermal insulation properties of cellular concrete enables 3 times lighter wall weight than the weight of brick walls and 1.7 times lighter than the walls of ceramsite concrete. Thermal insulation and mechanical properties of cellular concrete make possible to construct of it single-layer protecting structures with the desired thermal resistance. Cellular concrete is divided into aerated concretes and foam concretes, whose physical/mechanical and operational performance is, ceteris paribus, almost identical. By the method of hydrothermal treatment cellular concretes are divided into two groups: concrete of autoclave and non-autoclave curing.

## 1. Introduction

In 1919, for the first time, metal powder was proposed for gas forming in production of cellular concrete, and currently for this purpose zinc, magnesium, and aluminum powder is used [1]. A reaction of aluminum with calcium oxide hydrate occurs, resulting in a large amount of evolved gas and heat. The reaction binds water, and that accelerates thickening and setting of porous mixture. Aluminium powder forms gas pores of a uniform structure, and that is extremely important for increasing the quality of products and increasing their durability in upkeeping of buildings. Sand, binders and water are uniformly mixed with aluminum powder. The prepared solution not containing gas, so far, is poured into molds, and thereafter a chemical reaction occurs, with isolation of hydrogen. The resultant gas bubbles bloat the solution which is distributed around the bubbles and forms a uniform cellular structure of concrete. One option of production of foam concretes involves the introduction, with stirring, of a freshly prepared foam or foaming agent into the mixture.

By the close of the XX century, the world annual production of cellular concrete products ranged 43-45mln $m^3$. The majority of firms have annual output of 160-200 thousand $m^3$, and the largest enterprise built in 1987 in Hungary under a licence of the

[*]Corresponding author: kulchaev94@gmail.com

Hebel company, has reached a performance 560 thousand $m^3$ of cellular concrete/year. A number of CIS countries, Poland, China, Czech Republic, Slovakia, Denmark, Japan, and others have their own domestic developments and technology they use, along with the licensed. Technologies differ in ways of preparation, molding of porous mixture and cutting concrete blocks into products of specified sizes [2,3].

On base of cutting technology a complete set of cellular concrete products for low-rise construction houses is produced. The precision of goods manufactured is due to different technical solutions applied in cutting equipment and depends on the structural and mechanical properties of a concrete block to be cut, and the sizes of cellular concrete.

By the technology of the Ytong and the Masa-Henke companies, prior cutting the product in specified sizes, a raw cellular concrete block is turned 90° with a mold on its side, and the technology of the Wehrhahn company foresees turning over on "alien" pallet specially stitched up under the lateral surface, on which unbuttoned block is fed under cutting machines, and then to an autoclave and to a finished product warehouse.

In accordance with the technology of the companies Hebel, Durox, Aerok, and Silbet concrete block is unbuttoned and transported with a special holder from the pallet plate onto the table of a cutting machine, cut, then fed on grates into the autoclave and the finished product warehouse.

The technology of the Siporex company foresees lifting the concrete blocks cutting on its slot pallet, afterwards the mold sides lowered into its former place and the mod with a concrete block is fed into the autoclave [4]. Steam treatment is carried out in dead-end and pass-through autoclaves having 2.4-2.8 m diameter and 50 m length, at a pressure min. 1.0 MPa.

The technology of the listed and other foreign companies finished products are sent to the customer in a packaged form. Typically, the density of products manufactured is 400-700kg/$m^3$ and compressive strength of concrete therein is not less than 2.0-5.0MPa respectively.

When delivered to the consumer the humidity of cellular concrete products is about 20% by volume or 30-35% by weight. During the construction and maintenance of cellular concrete walling the humidity drops to equilibrium operation values and is 2-3% by volume, at an average concrete density of 600 kg/$m^3$.

Reinforced products are produced up to 7.2m long, up to 0.75 m wide, and up to 0.375m thick. Pitch of products in length is 5-25 mm and in thickness 25-100 mm and in width of products is usually equal to the height of the molded concrete block.

The length of reinforced products depends on their thickness and design loads. The maximum length of articles is usually equal to the length of the molded or cut concrete block [5]. Non-reinforced products are made of any size, and the block length or the width of the partitions is equal to the height of the concrete block. Dimensions of small blocks correspond to the requirements of STB 1117-98 "Wall blocks of aerated concrete", GOST 25485, DIN 4165, and others.

In the 50s of the last century reinforced products of heat insulating concrete had density 800-1,000 kg/$m^3$, and by 1996 for mass production it dropped to 650-750kg/$m^3$.

For a comprehensive test technology of production of small cellular concrete blocks with a 2 m diameter autoclave, of forming 1.2 m high concrete blocks, and cutting them into products of specified sizes on mold pallets, without moving the block (as it is carried out on the "Universal-60" line), the construction of four plants of 160,000 $m^3$ per year was planned in Belarus (Bobruisk, Luban, Orsha, Petrikov), by two plants in Ukraine and Kazakhstan, as well as seven plants for 40- 400 thousand $m^3$ a year in the Russian Federation [6].

The mentioned program was based on the use of domestic experimental and the newly created equipment for the production of small cellular concrete blocks on shock cutting and vibro-cutting technologies. For the production of reinforced products and small cellular concrete blocks the equipment of the Ytong company was purchased with the option of its reproduction. On base of the domestic produced and reproduced equipment of the Ytong company, 400-450 plants for the production of industrial products from cellular concrete, with production capacity 50-200,000 $m^3$/year, were planned to be built in the USSR in 1989-1995, including 50 complex technological lines of new generation with the production capacity 100-200 thousand $m^3$/year, in 1990.

It was planned to launch in cement plants the manufacture of low water-need binders (VNV) and process equipment for the production of non-autoclave wall blocks of cellular concrete on base of VNV, with output 50,000 $m^3$/year, including - by 40 production lines in 1989-1990.

To coordinate these plans, a program was designed to 1989-1995 "Creating highly effective types of cellular concrete with desired properties based on new binders, the development of a new generation of building structures, technology, equipment, the creation of highly automated manufacturing equipment."

The program not only included the creation of highly automated manufacturing equipment, but, since 1995, intended the development of highly automated pilot production line for the manufacture of cellular concrete products, with the density of 500-550kg/$m^3$ and 600-650kg/$m^3$.

After the collapse of the Soviet Union, in which the USSR Gosstroi (State Committee for Construction) was the sole head of all the works on the development of the production of cellular concrete products, all CIS-wide programs were practically stopped, and the USSR Gosstroi was dissolved.

In some cities of the CIS and Baltic countries (Ekaterinburg, St. Petersburg, Saratov, Tallinn, and others), cellular concrete buildings up to 25 floors were built.

The plants in the CIS country usually release the whole range of reinforced products providing the use of a large number of individual shapes of different sizes. The number of standard sizes of panels for the construction of a series of a residential or public building is up to 100, and the outfit weight is 750-800 t of metal. The filling factor of the autoclave in the manufacture in individual forms of reinforced products equals 0.25-0.30, and by cutting technology it is 0.40-0.45. Steam consumption for the manufacture of concrete with the density 500-600kg/$m^3$, in individual forms, equals to 250-280kg/$m^3$ and by cutting technology - 165-190kg/$m^3$.

The production-line technology of manufacturing reinforced cellular concrete products in individual forms was borrowed from the experience of the manufacture of concrete products in the early days of the production of autoclave materials and plants and it has been used in the CIS by now so far. From volume production of 1.6 mln.$m^3$ reinforced products, in the CIS, only 130,000 $m^3$/year are produced by cutting technology.

The main drawback of the manufacture of reinforced products in individual forms consists in dependence of product manufacture quality upon the outfit quality. High deformability and low strength of cellular raw concrete require the use of hard tooling technology to prevent cracks in the steps of pre-autoclave curing and hydrothermal treatment. It increases the material consumption of production, reduces the volume of the output of finished products from the autoclave. The outfit, under constant mechanical, cyclic and thermal effect, requires frequent preventive and major repairs. Otherwise, the quality of products has steadily decreased over time. The disadvantage of the considered technologies is the need of complete replacement of the forms in the transition to a new product. Horizontal position of forms when the preparatory operations, molding and cropping or stitching "crusts" require large production areas, additional amounts of manual labor and a significant increase in the number of means of mechanization and transportation.

These drawbacks are not evident at the cutting manufacturing technology when in the form with defined and fixed reinforced frames the block is shaped with dimensions 6.0 × (1.2-1.5) × (0.6-0.9)m or 6.0 × 0.9 × 0.9m, and then its cutting is carried out by means of special machines, for products of specified sizes, and that allows to obtain different forms of the same type in the length, thickness and width of the product with profiles, grooves, chamfers and keys. After autoclaving additional mechanical processing can be made: further calibration of products, "groove - comb" cutting and "pockets" to capture with masonry and others.

Technical and economic analysis of the production of reinforced cellular concrete products by cutting technology shows that in comparison with the production in individual forms it is possible to:

— *lower:* metal consumption of forms per 1m$^3$ of finished product, 3 times; specific heat consumption for autoclaving, by 18%; relative capital investments for the production of products, by 16%;

— *improve performance of:* autoclave unit, by 1.5; molding span beam, by 4.7 times; labor per employee of main production, by 21%; labor across the whole enterprise, by 14%.

Unlike some other foreign countries, where finishing of cellular concrete products is carried out on the construction site, in the CIS countries, as a rule, the reinforced wall panels are finished at the factory. With mass standardized public construction, such an approach met the requirements of time, however, the quality and variety of the raw materials used for protective and decorative coatings and coating quality do not allow providing the necessary appearance and durability of building facades.

In recent years, Russia and Belarus began intensively re-use cellular concrete. At present, in Russia, a plant in the village of Sertolovo, Leningrad region, and two plants in Lipetsk, licensed by the Hebel company, produce cellular concrete. The plant "Sibit", in Novosibirsk, the Glavnovosibirsk, JSC, and the Plant on Manufacture of Products of Cellular Concrete, JSC, and the Cottage, JSC, Vodino town, Samara region, mastered the production of cellular concrete products by the Ytong technology. Plants with capacity of 120 thousand m$^3$/year produce wall blocks with density 400-600kg/m$^3$ and reinforced products with density 700 kg/m$^3$ (floor slabs, roofs and intel blocks). In the city of Naberezhnye Chelny the Plant of Cellular Concrete, JSC, has developed, produced and mastered the production line for the production of cellular concrete products, which is similar to the cutting technology of the Ytong or the Masa-Henke companies but without turning over the concrete block.

By injection-molding technology, the block is molded in a vertical snap (6.0 × 0.626 × 1.3 m) consisting of a stationary opening sides and removable tray. After getting the required plastic strength by fresh aerated raw concrete, molds and tray with a vertically standing block are opened, using a special traverse, suspended on the crane, moved under the cutting machine for cutting "crusts", the lateral vertical slitting of the block, horizontal and vertical slitting cut to length.

The Corporation of Building Materials, ZAO, (Moscow), Volgotsemmash, JSC (Togliatti) and P.P. Budnikov VNIISTROM Institute (settlement Kraskovo, Moscow region) developed a conveyor line without a crane of the Vibroblok system, production capacity 30-120 thousand m$^3$/year. Aerated blocks, dimensions 3,000×1,300×1,270 mm, made by vibro-cutting technology. Head sample line with an annual output 100,000 m$^3$ (BKA-100) is manufactured, assembled and mastered by the Volgotsemmash, JSC.

The Silbetindustriya, ZAO, (Moscow) has developed processing equipment for the construction of new and modernization of existing cellular concrete plants using domestic shock technology and different versions of cutting the concrete block into the products of specified sizes.

In the project, the Hebel company has laid claim to the original raw materials, especially to cement and lime that exceed the requirements of GOST and STB, i.e., in the Republic of Belarus and the CIS countries such cement and lime are not made. For example, raw materials of the deposit "Kolyadichi", used for the production of cement by Krasnoselskcement JSC, and the existing clinker production technology with short rotary kilns do not provide clinker with saturation ratio above 0.9. Building industry enterprises do not produce lime containing over 80% of calcium oxide, and lime hydration kinetics does not meet the requirements of DIN 1060.

Specialists of the technical/engineering centre of Zabudova, JSC and UPP Building Structures Plant developed a number of cellular-concrete mixture formulations for 350-700kg/m$^3$ concrete density on the raw materials base of the Republic of Belarus. More than 30 formulations have been introduced in the production, allowing producing aerated concrete products and structures of different density and strength. The plant produces a complete set of home materials of cellular concrete, according to the standards of the Republic of Belarus (STB): non-reinforced blocks (STB 1117-98), roof slabs and floor (STB 1034-96), trough and arched intels (STB 1332-2002), wall panels (STB 1185-99), elements of stairs (STB 1330-2002). The products obtained certificates of conformity of the Republic of Belarus, Russia, Lithuania, Latvia, and others. Production of aerated concrete is certified by the International Quality System ISO 9001. In 2002, the plant was awarded a Prize of the Government of the Republic of Belarus for achievements in the field of quality.

Cellular concrete produced by Zabudova, JSC and UPP (Unitary Production Enterprise) Building Construction Factory is used for the erection of residential, public and social buildings. For example, in Minsk, the cottages are built of aerated concrete in two microdistricts "Bolshaya Slepyanka" and in Gazeta Izvestia avenue. In Moscow - experimental districts "Kurkino", "Mitino", "Edem", etc. Cellular concrete is widely used in enclosing structures of multi-storey buildings. In Minsk, Moscow and in other regions of Russia, as well as in the Baltic countries a number of high-rise buildings were built, including in Moscow - a complex of residential houses in Mosphilmovskaya street, and the building of the British Embassy. The study and critical analysis of the global and domestic experience in the production of cellular autoclaved concrete, as well as the use of shock technology make possible the modernization of plants and increasing the cellular concrete production.

The Republic of Belarus uses modern technology equipment sets, combining domestic and foreign technology shock cutting technologies of the leading German companies - Masa-Henke, Wehrhahn, Hebel, and others.

For the manufacture of aerated concrete binders, siliceous components and gas forming agents *binders:* portland cement; calcium burnt lime; granulated blast furnace slag; ash from the burning of oil shale and some types of brown coal; *silicon components*: quartz sand or silica-containing waste mining and processing industry; ash from the burning of lignite and coal; *gas forming agents*: aluminum powder or paste, which is applied with the addition of surfactants; plasticizing additives, regulating processes of gas forming and thickening are used.

Raw materials used for the manufacture of cellular concrete products, must meet certain requirements. The suitability of the raw materials and properties of porous concrete is determined on base of technological tests. Test scope to be specified in each case [7].

*Binders:*

a) Portland cement (GOST 10178). Brand M400-500 without active mineral additives. The content of tricalcium silicate - at least 50%, tricalcium aluminium - less than 6%. The specific surface of cement should be at least 3000 sm$^2$/g. Initial setting - not later than in 2 hours, the end setting - not later than 4 h after mixing;

b) slaked lime calcium (GOST 9179; SN 277-80). The content of active CaO and MgO - at least 70%, including MgO - less than 5%. Overburning number - less than 2%. blanking speed 5-15min;

c) granulated blast-furnace slag (basic and neutral) (GOST 3476). The content of manganese oxide is not more than 1.5%, sulphide sulfur is not more than 0.1%. Activity module of not less than 0.4, the main modules of at least 0.9;

d) ash from the incineration of oil shale and some types of brown coal. The CaO content is not less than 30% (including free CaO not less than 15%). SiO2 20 to 30%. CO3 not more than 6%. K2O + Na2O - not more than 3.5%. The specific surface of the ash 3000-3500 $sm^2$ /g.

*Silica components* [8]:

a) quartz sand (GOST 8736; SN 277-80). Silica content (unbound SiO2) of not less than 70%; alkali (calculated as Na2O) is not more than 2.7%; sulfuric acid and sulfur compounds (calculated as $SO_3$) not more than 0.5%; mica is not more than 0.5%; dust, silt and clay particles are not more than 5%. The recommended maximum size of sand grains - not more than 3 mm content of grains of 0-1mm in the range of 60-80%;

b) acidic ash from the burning of lignite and coal, captured in the electrostatic precipitators (OST 21-60; SN 277-80). Quartz SiO2 content of at least 45%; CaO is not more than 10%; sulfuric acid and sulfur compounds (calculated as $SO_3$) not more than 2%; the content of unburned fuel residues: not more than 7% for fossil fuels; for brown coal is not more than 5%. The specific surface area of not less than 2,500 $cm^2$/g.

*Gas-forming agent:* Aluminium powder or paste based on it (GOST 5494-71).

*Industrial water:* (GOST 23732; STB 1114) [9].

For the preparation of cellular concrete can be used water suitable for the preparation of conventional concrete, can be used. Maximum content in mg/l: soluble salts – 10,000; SO4 ions -2,700; C1-1 ions – 3,500; suspended solids - 300.

*Chemical additives:* These surfactants are specified in dependence on the type of binder and specified empirically. To control the process of structure formation, increase the plastic strength and accelerate hardening of porous mixture and its plasticizing the following is used: gypsum dihydrate; soda ash; triethanolamine; trisodium phosphate; water glass; sulfanol; superplasticizers, and other substances.

*Bar steel for frames:* for reinforced structures (frames and grids) bar steel is used, which meets the requirements of GOST 10922. Control tests on the strength of steel are carried out in accordance with GOST 12004. As an anticorrosion coating for the reinforcement the following is used: cold cement and bitumen mastic; polystyrene cement paste; latex-cement paste; latex-mineral paste; inhibited shale-bitumen-cement paste; metallosilicate mastic; anti-corrosion coatings for steel [10].

*Anticorrosion coatings* must meet the following requirements: impact strength of not less than 10 kgs/$cm^2$ according to GOST 4765; elasticity of not more than 20 mm in accordance with GOST 10086; layer thickness reinforcement coating after drying 0.4-0.7mm; continuous coating without cracks and through pores; coating must withstand transportation stacking and piling of rebar in a mold.

*Lubricating materials*: to apply the lubricant on material forms the following compositions (in weight ratio) are applied: petrolatum-kerosene mixture 1:2.5; grease or avtol with kerosene 1: 1; mixture of rocker and engine oil 1:3; different separating liquid or film coatings that prevent adhesion of raw cellular concrete with steel.

## References

1. V. Chulkov, *Production and use of construction materials, products and systems* (SvR-ARGUS, Moscow, 2011)

2. I. Rybiev, *Construction materials science* (Moscow, Visshy school, 2002)
3. I. Nanazashvili, I. Bunkin, V. Nanazashvili, *Construction materials and product* (Moscow, Adelant, 2005)
4. L. Suleymanova, V. Lesovik, *Gazobetony non-autoclaved hardening on composite astringents: monograph* (Belgorod, BSTU, 2013)
5. M. Kaftaeva, Sh. Rakhimbayev, *Peculiarities of the technology of production of energy-efficient autoclave silicate aerated concrete* (Belgorod, BGTU, 2015)
6. L. Sychova, *Technology of gypsum binders* (Moscow, RHTU them. D. I. Mendeleyev, 2016)
7. L. Kramar, A. Orlov, *Methods of research of building materials* (Chelyabinsk, Center of SUSU, 2015)
8. A. Suvorova, G. Sycheva, M. Tachaev, *Chemistry of knitting materials* / (Moscow, RGAU-MAHA, 2016)
9. M. Kaftaeva, G. Malichenko, O. Skorokhodova, *Theory and practice of cellular concrete of autoclave hardening* (Belgorod, BSTU Publishing, 2012)
10. V. Belov, Yu. Kuryatnikov, *Dry mixtures for the production of non-autoclaved aerated concrete* (Tver, TSTU, 2010)

# Computational and experimental study of probability of galloping and torsional divergence at span structure

*Anastasia* Fedosova[1,*], *Pavel* Churin[1] and *Nikita* Ageev[1]

[1] Moscow state university of civil engineering, Yaroslavskoye shosse, 26, Moscow, Russia, 129337

**Abstract.** The paper considers study of aeroeslatic stability of span structure using computational and experimental modelling of aerodynamics by wind-tunnel tests and numerical simulation at software package ANSYS CFX. Wind-tunnel test were conducted to to determine steady state wind load coefficients for zero wind attack angle. Using numerical simulation aerodynamic coefficients were obtained for different wind attack angles from -5 to 5 degree with verification to experimental data at zero wind attack angle. Using this approach of combination the accuracy of experimental studies and the convenience of numerical simulation the problems of forecasting phenomena such as galloping and torsion torsional divergence at span structures were solved.

## 1 Introduction

Originally, all the phenomena of aerodynamic instability first of all were associated with the aircraft wings, subsequently it turned out that the same phenomena would be characteristic not only of aircraft structures, but also of bridges. The problem of aerodynamic stability of bridge structures came to the fore in 1940, when at a relatively low wind speed (v=19 m/s) for 1 hour the suspension bridge across the river Tacoma with span 855 m collapsed. The Tacoma catastrophe revealed inadequate structural checks for resistance to wind loads. Destruction of the bridge contributed to research in the field of aerodynamics and aeroelasticity of structures and changes in approaches to the design of all large-span bridges in the world since the 1940s.

Since the Tacoma catastrophe, at designing bridge structures a particular place is given to wind loads, as the span structures have an increased sensitivity to wind forces and the interaction of the span structure with the wind flow can lead to a number of aeroelastic phenomena, the most dangerous are: flutter, buffeting, eddy shedding resonance, parametric resonance and torsional divergence.

In this paper it will be considered two types of aeroelastic phenomena appearing in span structures: galloping and torsional divergence. The acting regulatory documents indicate the need for span structures of complex structures during testing for aerodynamic stability to use the results of testing models of structures in wind tunnels, but such tests are

* Corresponding author: mgsu@broll.ru

laborious and expensive. With the modern development of the mathematical apparatus, numerical methods and bases of commercial licensed software an alternative to physical modelling in a wind tunnel can become the transfer of the wind tunnel to a virtual environment with the replacement of the real purging of the model with a virtual one. In this paper, a computational and experimental approach will be applied to simulate aerodynamics, that approach could combine the accuracy of experimental studies and the convenience of numerical simulation.

## 2 The object of investigation

In the paper, a six-lane bridge with a length of 600 m with a three-meter pedestrian sidewalk provided on it for public transport, located in the Moscow city (Figure 1) is being investigated.

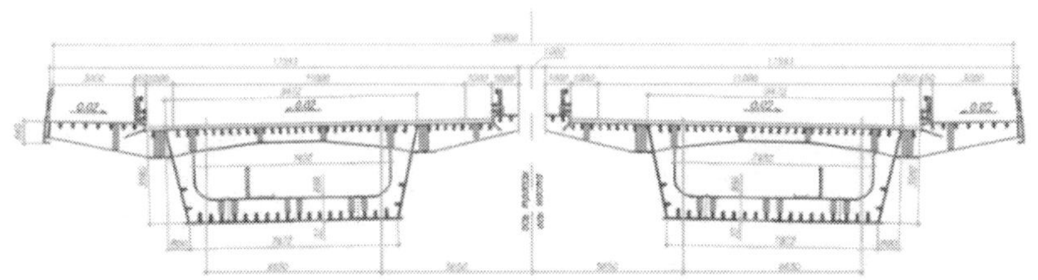

**Fig. 1.** Typical cross-section of the bridge.

To investigate the possibility of galloping and torsional divergence effects, it is required to determine steady state wind load coefficients: coefficient of drag (Cdrag), coefficient of lift (Clift) and coefficient of pitching moment (Cmz) for different wind attack angles from -5° to 5°. According to the proposed method, aerodynamic coefficients will be determined by experiment for zero wind attack angle (corresponding to direction of flow perpendicular to the span structure) and numerical simulation will be performed for following wind angles: -5°, -3°,0°,3° and 5° with verification to experimental data at zero wind attack angle.

## 3 Physical modelling at wind-tunnel

During the experimental studies, static tests will be carried out to determine the coefficient value of lift (Clift), drag (Cdrag) and pitching moment (Cmz).

Taking into account the dimensions of the wind tunnel working part (4 m x 2.5 m), the optimum scale layout was chosen as 1:70 on the assumption of flow blockage effect.

If the section model has the required mass and inertia characteristics, which are determined by similarity laws, then the actual operation of the span structure will be similar and can be directly deduced from the model behaviour in the wind tunnel.

The design of the model in conjunction with the matched material provides not only a geometric similarity, but also the correct distribution of masses, corresponding to the natural object. Thus, the geometric places of the mass centres of the model and the natural object are also corresponded. The weight of the model was calculated before its manufacture, and then it was measured and finally adjusted on the stand in the wind tunnel. Addition of extra weight was not required because of the small deviation of the measured value from the calculated one.

Thus, the projected model (Figure 2) corresponds to the natural object according to the following parameters: geometric similarity; the correspondence of the mass distribution; the correspondence between geometric centres of mass and moments of inertia, which fully meets the requirements of experimental studies [1],[2].

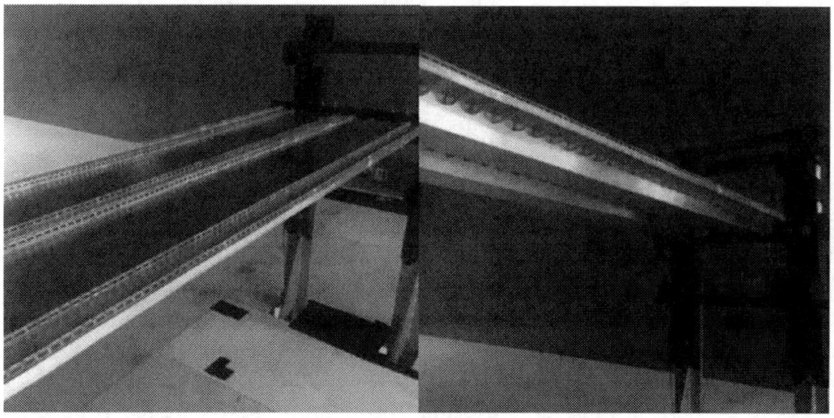

**Fig. 2.** Model of span structure.

The model was firmly fixed with screws to the 6-component tensors and the lifting force, the drag force and the aerodynamic moment were measured.

The model is exposed to wind, the signals about the oscillations of effort are recorded for 1 min, and then their mean values are calculated. Efforts are expressed in a dimensionless form, called the "aerodynamic coefficient of stability", as it follows:

– coefficient of drag

$$Cdrag = \frac{F_x}{\frac{1}{2}\rho U^2 HL}; \quad (1)$$

– coefficient of lift

$$Clift = \frac{F_y}{\frac{1}{2}\rho U^2 HL}; \quad (2)$$

– coefficient of pitching moment

$$Cmz = \frac{Mz}{\frac{1}{2}\rho U^2 BHL}, \quad (4)$$

where $\frac{1}{2}\rho U^2$ is wind pressure measured in a wind tunnel; H=0.053 m (corresponds to 3.71 m in full scale), L = 1,714 m and B = 0.254 m (corresponding to 17.8 m in full scale) are respectively height, length and width of model.

Aerodynamic coefficients (1), (2) and (3) were measured at different wind speeds to check the dependence of the measured value on the velocity in the wind tunnel. The following values of the aerodynamic coefficients for zero wind attack angle are obtained: Cdrag = 1.2275; Clift = 0.585; Cmz = 0.86237.

## 3 Numerical simulation technology

Reynolds number, which is characteristic for the problem of the span flow simulation is about 320 thousand, which corresponds to the turbulent motion of gas. The turbulent motion of the air medium near the body is described by the RANS closed with the additional differential relations of the two-parameter dissipative turbulence model.

The instantaneous values of the flow parameters (such as velocity and pressure) are represented as the sum of the averaged quantity and its pulsation component. Thus, in addition to the usual flow parameters (averaged velocity, pressure) additional parameters appear associated with the characteristics of turbulent pulsations. As a result, the system of basic equations of air flow motion is not closed. Closure of the space is accomplished by using additional equations obtained with empirical data. In this case, closure of the space was accomplished by the two-parameter differential model of Menter's turbulence k-ω SST [3]. The boundary layer is modeled completely turbulent. For the solution a grid method with an implicit finite-difference scheme of conditionally second order (in the region of extrema the first order) approximations for all equations including the equations of the turbulence model was used. The solution was performed using the licensed commercial software package ANSYS CFX, which implements the finite volume method. The established total root-mean-square discrepancies along the entire flow field were about the order of 10 ^ (- 4). To solve the problem, it took 120 iterations of the solver. The time scale depended on the iteration number according to the sigmoidal law, the final time scale was 0.001 seconds, the characteristic number of iterations to the middle of the height of the sigmoid was 100, the initial time scale was 10^(-6) seconds. For numerical simulation, a computational grid was constructed around the model of the bridge used in the experimental studies. The grid is constructed in the computational domain with a width and length of 6, a height of 3 and a width of 4.5 meters (Figure 3).

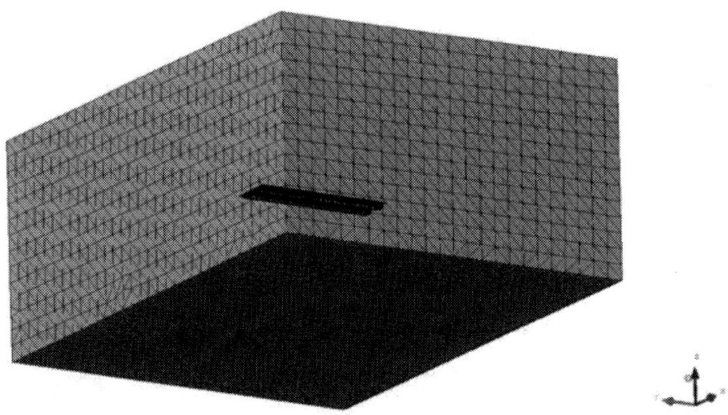

**Fig. 3.** Computational domain.

In the construction of the computational domain, in the first stage, using the Autoblock method, a surface grid was constructed, consisting mainly of quadrangular elements. Then each quadrangular element was divided into two triangular elements and the grid was checked and corrected. On the basis of the surface mesh, a three-dimensional grid was constructed according to the Delaunay triangulation. At the last stage, the prismatic layers were constructed along the bridge surfaces. In the computational domain at a great distance from the model the maximum size of the final grid element is equal to 0.2 m. On the surfaces of the span structure the characteristic element dimension o is equal to 5 mm. The increase ratio of the elements dimensions is equal to 1.2. The number of prismatic layers is 30. The total number of grid elements is 14 million and 311 thousand. The view of the grid on the span structures surfaces is shown at Figure 4.

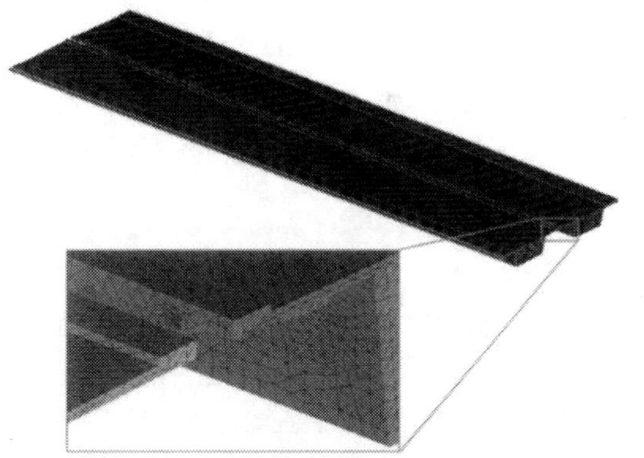

**Fig. 4.** The calculated mesh for span structure surfaces.

The cross-section of the calculated grid by the plane of symmetry is shown at Figure 5. One can clearly see the sections of the prismatic layers near the surfaces of the bridge. When modelling the flow, the size of the wall cell in the direction perpendicular to the surface is important. This size is characterized by the parameter Y +, which is essentially the Reynolds number calculated from the cell height. If this parameter is much larger than 1 in order of magnitude, the modelling of the boundary layer is performed using so-called wall functions obtained semi-empirically. In the case of Y +values is the order of 1, boundary layer modelling is performed using the Reynolds equations directly with the closure of the SST turbulence model.

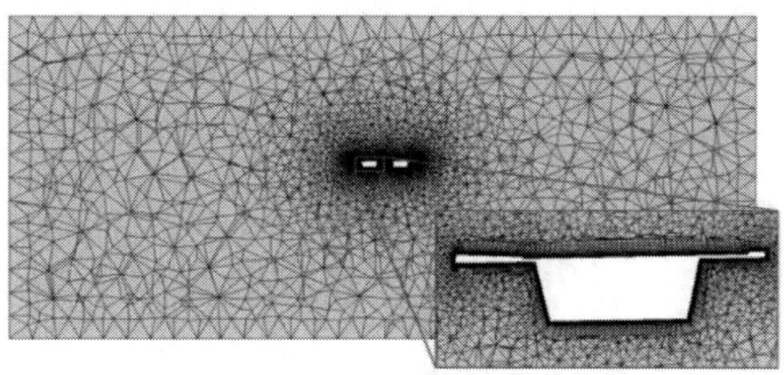

**Fig. 5.** The cross section of the computational grid by the plane of symmetry.

During the modelling, it is necessary to strive to fulfill the condition Y + ~ 1. The distribution of this parameter over the bridge surfaces is shown at Figure 6. It can be seen that this parameter practically does not exceed 1 on the entire surface of the bridge. The average value is about 0.4. The maximum value of the parameter Y + over the entire design area is equal to -2.

At the input boundary of the computational domain, the flow velocity was set at 20.52 m / s. The level of turbulent pulsations at the input boundary was set formally because of their rapid repayment as they departed from the boundary and amounted to 1%. On the model surfaces of span structures no-slip condition was set, the flow velocity was zero. On the output boundary (one or two surfaces, as well as for the input boundary) the relative static pressure on average is equal to zero with the possibility of small deviations from the average (Pressure Profile Blend = 0.05).

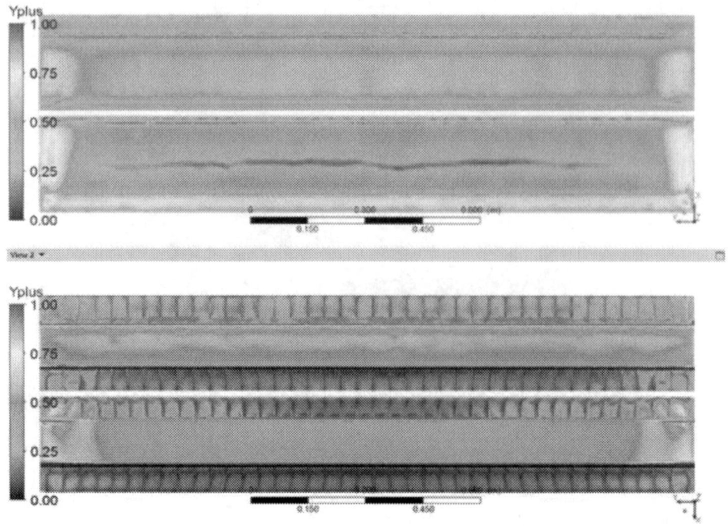

Fig. 6. Distribution of the parameter Y + over the bridge surfaces.

On the side surfaces are open-type conditions (Opening) with a relative static pressure of 0 Pa and zero gradient of the turbulence parameters, which allows the flow both to enter the design area and to leave it. The initial conditions specified in the calculation are the flow in the computational domain with zero relative static pressure and the velocity corresponding to the input boundary. The obtained values of steady state wind load coefficients for different wind angles are given at Table 1, the value for zero wind attack angle was verified with the experimental results, the error between experimental and numerical simulation was no more than 15%.

**Table 1.** Wind load coefficients.

| attack angle | Cdrag | Clift | Cmz | dCmz/dα | Urk(α),m/s |
|---|---|---|---|---|---|
| 5° | 1,26 | 2,17 | 0,69 | -1,30571 | 177,2 |
| 3° | 1,32 | 1,84 | 0,78 | -0,56843 | 268,6 |
| 0° | 1,42 | 0,52 | 0,75 | 1,6476 | 1000 |
| -3° | 1,41 | -0,92 | 0,52 | 5,195764 | 1000 |
| -5° | 1,33 | -1,83 | 0,34 | 8,30128 | 1000 |

## 4 The study of the bridge to the possibility of galloping

Galloping is one of the types of aeroelastic instability, most often poorly streamlined flexible elements with aerodynamically unstable cross sections (square or rectangular) are subject to galloping. Galloping is more dangerous than resonance, since it increases with an increase in the flow velocity even above the critical one, and resonance occurs only at a critical wind speed. Under certain flow conditions it is possible that a vertical oscillation of the cross-section with horizontal wind leads to a further self-excitation. In this case, the total damping $d$ for the vertical oscillation is given by

$$d = 2m\xi\omega_1 + \frac{1}{2}\rho Vl\left[C_L^{'}(0) + C_D(0)\right], \quad (5)$$

where $l$ denotes the normalisation length. The first term on the right side, the mechanical damping, is usually positive. Therefore, the criterion for galloping (the so called Glauert-Den Hartog) criterion is given by

$$\left|C_L^{'}(0) + C_D(0)\right| < 0\,. \quad (6)$$

If the wind direction is initially not horizontal, the above criterion might be generalized for arbitrary angles α.

The figure 7 table holds steady state coefficients for drag, lift and moment dependant on wind attack angle. As can be seen from Figure 8, since the coefficient of lift in the interval increases, the Glauert-Den Hartog criterion stays positive and galloping is impossible. It should be emphasized that this estimate does not allow to determine the actual behaviour of the structure (frequency, amplitude of oscillations), but only establishes the possibility of galloping effect appearance.

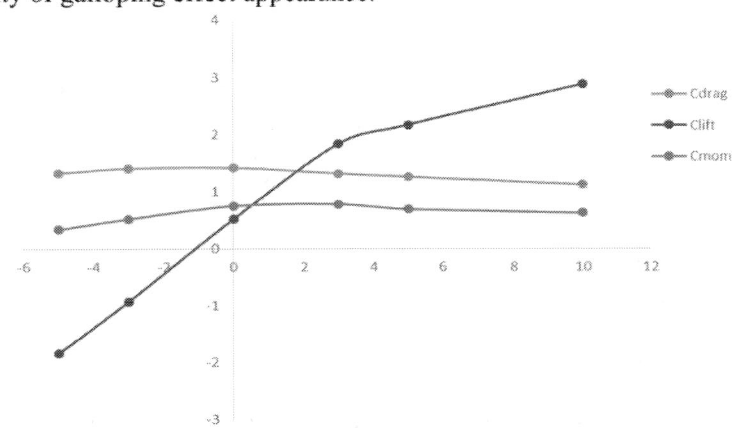


**Fig. 7.** Drag, lift and moment coefficient dependant on wind attack angle.

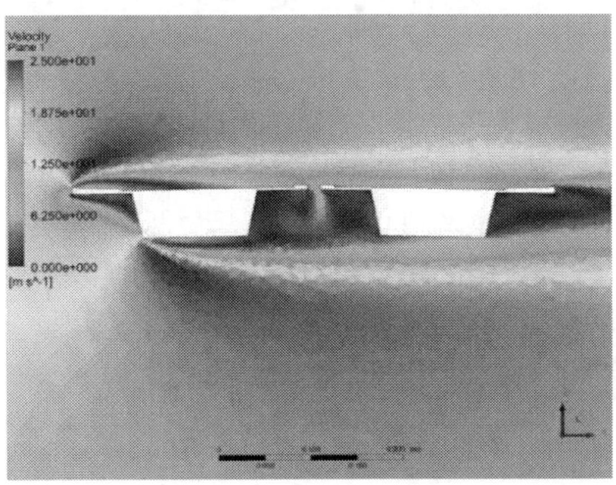

**Fig. 8.** Bridge in the wind flow.

## 5 The study of the bridge to the possibility of torsional divergence

Torsional divergence is an aperiodic phenomenon of twisting of a span structure up to destruction caused by an increase in the aerodynamic moment, (the moment of elastic forces can not balance the moment of aerodynamic forces). The wind speed at which the moment of aerodynamic forces *M(α)* is equal to the moment of elastic forces *Me* is called the critical divergence velocity. When this speed is exceeded, the span structure would be destroyed. The critical torsional divergence velocity for small wind attack angles can be calculated by

$$U_{cr}(\alpha) = \sqrt{\frac{2k_\alpha}{\rho ALC'mz}}, \tag{8}$$

where $k_\alpha$ denotes the spring constant.

The determination of the critical divergence velocity for different wind attack angles will be carried out by equation 8, where $k_\alpha = 1{,}99 \cdot 10^5$ kNm /rad, ρ=1,225 g/m^3 and the characteristics of the span structure in full scale are given at paragraph 3. To determinate the derivatives $dC_{mz}(\alpha)/d\alpha$ following [4] the obtained Cmz dependence on the wind attack angle is approximated by a polynomial, differentiating which we can find the derivative and use formula (8). The obtained values $dC_{mz}(\alpha)/d\alpha$ and torsional divergence speed in dependency of wind attack angle are given at Table 1, where speed of 1000m/s represents no torsional divergence; since for these angles of attack Cmz increases with increasing α. The critical wind speed for torsional divergence is about 177 m/s. To test aerodynamic stability according building code SP 35.13330.2011 "Bridges and pipes" it is necessary to compare the critical speed of torsional divergence with the maximum wind speed possible in the region of construction. According to the meteorological station located in the construction region, for the ten-year period from 2005 to 2015 the maximum recorded wind speed was 17 m / s. Taking into account the margin factor for wind load - 23.8 m/s. Thus, since the critical divergence rate (177 m / s) is greater than the maximum possible wind speed (23.8 m / s) taken with a safety factor of more than 1.5 times, the aerodynamic stability of the span structure in a case of torsional divergence is ensured.

## 7 Acknowledgement

The work was financial supported by the Ministry of Education and Science of the Russian Federation within the framework of the state #7.6075.2017/BCh, Project «Investigation of the phenomena of aerodynamic instability of building structures in aeroelastic statement, including the development of an innovative methodology for analysing meteorologica data to refine the parameters of the wind load».

All tests were carried out using research equipment of Head Regional Collective Research Centre of Moscow State University of Civil Engineering.

## References

1. P. Churin, O.I. Poddaeva, Aerodynamic testing of bridge structures Applied Mechanics and Materials **477-478**, 817-821 (2014)
2. O. Poddaeva, P. Churin, I. Dunichkin, MATEC Web of Conferences **86**, 02012 (2016)
3. F.R. Menter, AIAA journal **32.8**, 1598-1605 (1994)
4. Yu. L. Rutman, VA Meleshko. Engineering and Construction Journal, **6**, 5 – 11 (2011)
5. A.N. Fedosova, International Journal of Applied Engineering Research **10 (20)**, 40844-40845 (2015)

# Experimental investigation of stress-strain conditions in punching zone of flat slabs

*Valery* Filatov[1,*]

[1]Samara State Technical University, Institute of Architecture and Civil Engineering, 194, Molodogvardeyskaya St., 443001, Samara, Russia

**Abstract.** The technique and results of experimental study on the test specimens of the stress-strain condition of flat reinforced concrete slabs of a girderless framework at punching by columns of square and rectangular section are presented. The punching shear strength of slabs at rectangular columns can be lower than at square columns with a similar length of the control perimeter. The research of influence of a form of a column cross-section on the stress-strain condition of a reinforced concrete slab in a punching zone is conducted. The comparative analysis of distribution of concrete and reinforcement deformations of reinforced concrete slabs on column perimeter depending on a ratio of the sides of cross section is made. The results of the research of cracking in a zone of punching of reinforced concrete slabs are presented here. Results of the conducted research allow expanding the base of experimental data of the considered subject and promoting improvement of modern positions in the theory of power resistance of reinforced concrete. New experimental data received in the research corresponded with the earlier offered hypothesis that in a slab there is a zone of the constrained deformation located along the perimeter around the rectangular column.

## 1 Introduction

Columns of rectangular section with the height several times surpassing its width have wide distribution at construction of buildings with monolithic girderless floors [1–3]. The application of rectangular columns (pylons) allows increasing lateral stiffness of the building. The stress-strain condition in a punching zone of flat reinforced concrete slab supported on pylons (in comparison with a square or round columns) is characterized by considerable non-uniformity distribution of strains along the control perimeter around the column. This peculiarity was observed experimentally by several researches such as Moe [4], Hawkins et al. [5], Muttoni, Sagaseta et al. [6, 7] and any others. In experiences of the researchers mentioned above there was a concentration of concrete strains of the slab in zones of the columns corners. At tests of samples with round section columns that concentration of concrete strains wasn't observed.

This deformation peculiarity of flat reinforced concrete slabs, as test results show, can reduce their bearing capacity at punching by rectangular columns (pylons) [8–10]. The

* Corresponding author: vb_filatov@mail.ru

punching shear strength of RC flat slabs at rectangular columns can be overestimated if the non-uniform distribution of strains along the control perimeter around the column is neglected in the calculations. In design codes various approaches are applied to account for this feature. In USA code ACI 318-11 a correction factor is applied which depends on the side ratio of column. In Model Code MC2010 and Swiss design code SIA 262 it is recommended to reduce the size of control perimeter. There are also other suggestions for improvement of calculated techniques [11–15].

The research of power resistance of reinforced concrete constructions in the combined stress (the joint action of normal and cross forces, bending moments) has an empirical basis, owing to concrete work as a composite elastic-plastic material with violation of discontinuity (cracks formation) [16–19]. The structural behavior is examined in the experimental path to check put-forward theoretical prerequisites and suggest subsequent development of computational models and calculation procedures [20–24].

As the calculation of punching shear forces in slabs has an empirical basis, the actual task is the expansion of test data base which allows to improve calculation procedures by taking into account various factors (constructive, force et al) influencing the slabs power resistance to punching.

## 2 Materials and Methods

The test specimens represented a fragment of a reinforced concrete monolithic unit of a column pair of the round or rectangular section and a flat slab. The varied parameter was the aspect ratio of a column section. The four specimens have been tested:
- CMR-D – a column of round section with a diameter of 210 mm.
- CMR-1 – a column of square section with a side size of 200 mm.
- CMR-2.5 – a column of rectangular section of the size 200×500 mm.
- CMR-4 – a column of rectangular section of the size 200×800mm.

The slab thickness of all specimens made 140 mm. Reinforcement at the tensioned side of a slab was carried out by reinforcing grids made of hot-rolled reinforcement bars Class A500 with a 14 mm diameter and 100 mm step. The reinforcement ratio was the same for all the specimens (1.41%). The measured values of effective depth in the specimens ranged from 107 mm to 110 mm as shown in Table 1. Crushed dolomites' limestone was used for the concrete with a maximum aggregate size of 20 mm. The measured concrete cube strength $f_{cu}$ at time of testing ranged from 29 MPa to 40 MPa as shown in Table 1. The detailed description of the construction, the geometrical sizes of specimens and their reinforcing are presented in [10].

**Table 1.** Summary of test specimens and experimental results.

| Test | $d_{avg}$, mm | $f_{cu}$, MPa | $V_{test}$, kN |
|---|---|---|---|
| CMR-D | 110 | 40.0 | 284.9 |
| CMR-1 | 108 | 34.7 | 272.0 |
| CMR-2.5 | 107 | 29.0 | 281.3 |
| CMR-4 | 109 | 35.1 | 328.4 |

Loading of test specimens was carried out in test frame placed underneath the reaction floor. The loading force was created by a hydraulic jack and applied to the lower side of a column. The loading of specimen was made gradually by 5% of a theoretical failure load. The hold time of the specimen under loading at each step was 15 minutes. At each step of loading the indications of strain and dial gages, schemes of crack formation and development were registered. The test frame construction and the technique of specimen loading are presented in [10]. The specimen CMR-D in the process of testing is shown in

**Fig. 1.** Testing of the specimen CMR-D.

The strains of concrete of the slab compressed side and the tensioned reinforcement bars was measured by means of strain gages placed in radial and tangential directions (fig. 2).

a) b)

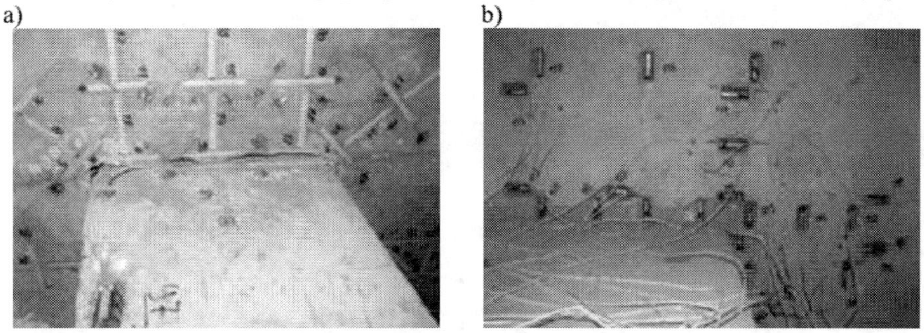

**Fig. 2.** Position of strain gages on concrete (a) and reinforcement (b).

Deflections of slab were measured by means of dial gauges as the difference in slab displacements at a given point and near the surface of the column. Width of cracks opening on a slab surface was measured by means of microscope.

## 3 Results

The destruction of all specimens was fragile, due to punching slabs by the column. The values of failure loads are specified in Table 1.

The crack formation in slabs of all specimens had similar image. The first radial cracks with opening width 0.05 mm were formed at the level of loading 0.25–0.3 from the ultimate load ($P_{ult}$) on the top surface of the slab from the sides of the column and developed to the external contour of the slab. Formation of tangential cracks was noted at the level of loading (0.45–0.5) $P_{ult}$. When the crack formation began the cracks openings was insignificant, the formation of new cracks was more essential. At the level loading 0.85 $P_{ult}$ the intensity of crack formation considerably decreased, the process of their opening was prevailing. On the limit stage before destructing the width of radial cracks reached 0.3 mm, and of tangential cracks (at column sides) 0.7 – 0.8 mm. The greatest intensity of crack formation in the slab was observed on the sites near columns corners.

Based on the results of measuring the strains of concrete and reinforcement of slabs, graphs of the distribution of concrete tangential strains on compressed surface of the slab along the sides of the column are carried out (fig. 3) and the graphs of distribution of strains of the reinforcing bars located along the sides of the column at distance of 50 mm from her faces (fig. 4). The graphs were carried out for a quarter of column section at the loading close to a failure load.

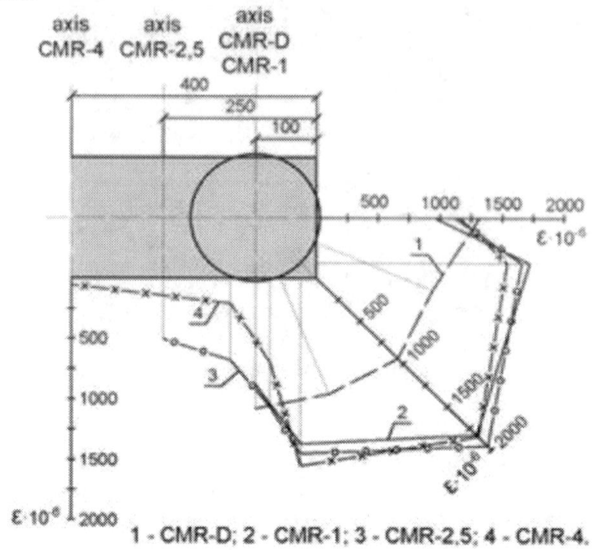


**Fig. 3.** Distribution of tangential strains of concrete on the compressed surface of slab.

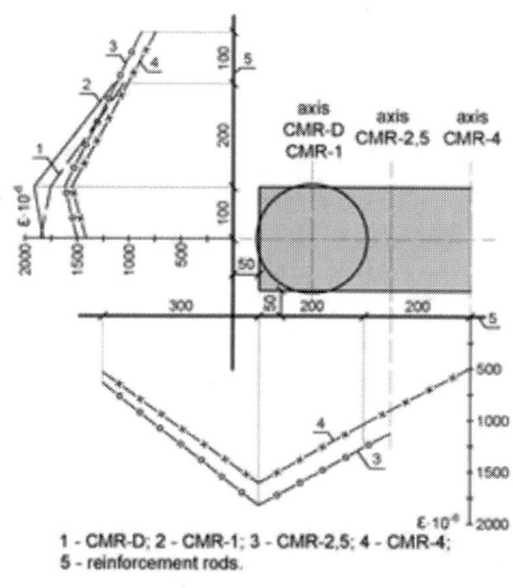


**Fig. 4.** Strain distributions of reinforcement bars located along the column side.

Distributions of radial strains of concrete on the compressed surface of slab and strain distributions of reinforcement bars, perpendicular to the sides of the column have shown a smaller variability at changing the form and sides ratio of the column section, therefore they are not considered in this article.

## 4 Discussion

The analysis of distribution of tangential strains of concrete on the compressed surface of slab (fig. 3) for specimen with round section of column (CMR-D) shows that strains along the perimeter of the column change slightly – the scatter values don't exceed 20%. For specimen with square section of column (CMR-1) the maximum values of tangential strains of the slab concrete are observed near the column corners. In the direction from column corner to its mid-section a considerable strains decrease is observed (from 24% to 45%). For specimens with rectangular section of columns (CMR-2.5 and CMR-4) maximum values of concrete strains of the slab are also observed near column corners. The distribution of tangential concrete strains of the slab along the short side of the column has the same character, as well as for CMR-1 specimen. Along the long side of the column section it is observed that tangential concrete strains of the slab in the direction from the column corner to its mid-section decrease and they reach the minimum there. In comparison with the slab concrete strains at the corner of the column, tangential strains in the middle of the long side of the column section make 26% and 3% for the specimens of CMR-2.5 and CMR-4 respectively.

The peculiarities of tangential strains distribution of concrete on the compressed slab side noted in the experimental study confirm the previously proposed hypothesis about constrained deformation zone that located in the slab along the perimeter around the rectangular column and having the form of an ellipse [14]. The constrained deformation zone is evidenced by the nature of crack on the compressed side of slab at punching by a rectangular column. On the slab compressed side of the CMR-1 specimen no cracks were observed, and in the CMR-4 specimen, at punching, on the compressed side of slab, the characteristic cracks were formed along the long sides of the column section (fig. 5). The trajectory of cracks had an elliptic configuration with the maximum removal from the column face in the middle of the section long side.

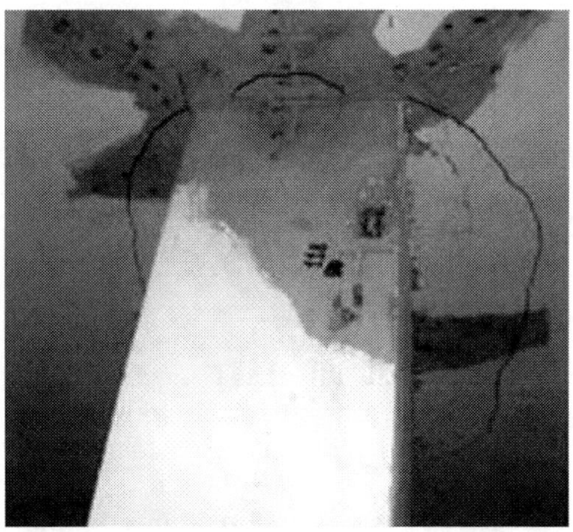

**Fig. 5.** Cracks on the compressed slab side CMR-4 specimen.

The analysis of strains distribution in the reinforcement bars of the slab located along the face of the column section (fig. 4) shows that the maximum value of strains is observed in the section, coinciding with the column side (a peak value). The exception makes a specimen CMR-D for which the maximum value of strains is noted on an axis of symmetry of the specimen. For specimen CMR-1 the strains of slab reinforcement bars on an axis of symmetry of the specimen are slightly (3%) less than the peak value. At the distance of 200 mm from the column face the strains of reinforcement bar make 60% of the peak value for CMR-D and CMR-1 specimens.

For CMR-2.5 and CMR-4 specimens strains distribution in the reinforcement bar of the slab along the short side of the column is similar to the CMR-1 specimen. The strains of reinforcement bar on the axis of symmetry of specimen is 7.4% and 8.8% less than the peak value for the specimens CMR-2.5 and CMR-4 respectively. At the distance of 300 mm from the column face the strains of reinforcement bar make 54% and 48% of the peak value for the specimens of CMR-2.5 and CMR-4 respectively. It should be noted that the peak values of the strains of reinforcement bar along the short side of the column for the specimens of CMR-2.5 and CMR-4 15-19% are lower than for the CMR-1 specimen.

The strains distribution at reinforcement bar of the slab along the long side of the column for the specimens of CMR-2.5 and CMR-4 are characterized by a considerable fall in strains to the middle of the column section compared with the peak value. The value of strains of reinforcement bar of the slab on the axis of symmetry of the specimen makes 60% and 31% of the peak value for specimens of CMR-2.5 and CMR-4 respectively. At the distance of 300 mm from the column face the strains of reinforcement bar make 35% and 34% of the peak value for specimens of CMR-2.5 and CMR-4 respectively.

The analysis results of the slab reinforcement strains along the column sides of the rectangular section show that the strains of reinforcement bar of the slab decrease from the column corner to its mid-section. The decrease value of reinforcement strains of the slab depends on the ratio of the column section sides. Thus, in the constrained deformation zone along the perimeter around rectangular column, not only tangential strains of the compressed slab concrete, but also strains of the tensioned reinforcement bars located parallel to the column face decrease.

## 5 Conclusions

In the research presented here new experimental data about the stress-strain condition of flat reinforced concrete slabs in a punching zone are obtained. The analysis is made about the influence of the form and ratio of the column section sides on the value and nature of distribution of the compressed concrete strains and those of tensioned reinforcement bars in the punching zone. The results analysis of the research presented here allow drawing the following conclusions.

1. Strength reduction of flat reinforced concrete slabs at punching by columns of a rectangular section happens owing to a less intense part of control perimeter of the slab located in the middle of the long side of the column section. With the increase in the aspect ratio of the column section the length of a less intense part of the control perimeter increases as well as the strength reduction.

2. In the research the tangential strains of the compressed slab concrete were considered though shear stresses are a cause of destruction at punching. Comparing the distributions of tangential strains of the compressed concrete of slabs received in the very research with the distributions of shear fields of slabs received by calculation in [6] it is possible to see a considerable similarity of their configuration. It allows to draw a conclusion about the prevailing influence of concrete tangential deformations on the value of shear stresses and, respectively, on the punching strength.

3. The peculiarities of distribution of concrete tangential strains of the slab compressed side and those of reinforcement bars along the column sides mentioned in the experimental study confirm the earlier offered hypothesis that in a slab there is a zone of the constrained deformation located along the perimeter around rectangular column and it has an ellipse form. The concrete and reinforcement bars strains in the zone of the constrained deformation decrease that leads to the insufficient use of strength characteristics of materials and decrease in the punching shear strength. For CMR-2.5 and CMR-4 specimens the decrease in concrete tangential strains of the slab compressed side makes 74% and 97%, the decrease in strains of reinforcement bars along the long side of the column section makes 40% and 69% respectively.

## References

1. V.P. Popov, D.V. Popov, A.Yu. Davidenko, MATEC Web of Conferences **86**, 01014 (2016)
2. A.O. Mateyko, Urban Construction and Architeture **3(24),** 74–78 (2016) DOI: 10.17673/Vestnik.2016.03.12
3. V.P. Generalov, E.M. Generalova, Urban Construction and Architeture **2(10),** 12–16 (2013) DOI: 10.17673/Vestnik.2013.02.2
4. J. Moe, Dev. dep. bul. D47, PCA (1961)
5. N.M. Hawkins, H.B. Falssen, R.C. Hinojosa, ACI Spec. Publ. **SP-30**, 127–146 (1971)
6. J. Sagaseta, L. Tassinari, M. Fernandez Ruiz, A. Muttoni, Eng. Struct., **77**, 17–33 (2014)
7. J. Sagaseta, A. Muttoni, M. Fernandez Ruiz, L. Tassinari, Mag. Conc. Res., **63**(6), 441–457 (2011)
8. T. Susanto, H.K. Cheong, K.L. Kuang, J.Z. Geng, ACI Struct. J., **101**(5), 678–687 (2004)
9. V.B. Filatov, E.P. Bubnov, MATEC Web of Conferences **86**, 01007 (2016)
10. V.B. Filatov, E.P. Bubnov, Industrial and civil engineering **2**, 86–91 (2017)
11. A. Muttoni, ACI Struct. J., **105**(4), 440–450 (2008)
12. C.E. Broms, ACI Struct. J., **87**(3), 292–304 (1990)
13. N.I. Karpenko, B.S. Sokolov, O.V. Radaykin, Industrial and civil engineering **1**, 28–30 (2013)
14. V.B. Filatov, Bulletin of civil engineers **5(40)**, 80–84 (2013)
15. V.B. Filatov, *Proceedings of the Samara scientific center of the Russian Academy of Sciences*, **16**, no. 4-3, 646–649 (2014)
16. E.N. Kodysh, N.N. Trekin, D.N. Trekin, *Concrete and reinforced concrete – glance at future : III All-Russian (II International) Conference on concrete and reinforced concrete*, **I**, 69–75 (2014)
17. V.I. Travush, V.I. Kolchunov, N.V. Klyueva, Industrial and civil engineering **3**, 4–11 (2015)
18. V.I. Travush, S.G. Emel'yanov, V.I. Kolchunov, Industrial and civil engineering **7**, 20–27 (2015)
19. A.G. Tamrazyan, Industrial and civil engineering **7**, 26–27 (2012)
20. N.N. Trekin, D.A. Pekin, Industrial and civil engineering **7**, 17–20 (2014)
21. A.N. Bolgov, A.Z. Sokurov, *Concrete and reinforced concrete – glance at future : III All-Russian (II International) Conference on concrete and reinforced concrete*, **IV**, 139–149 (2014)
22. A.N. Bolgov, D.M. Yakimovich, *Concrete and reinforced concrete – glance at future : III All-Russian (II International) Conference on concrete and reinforced concrete*, **I**, 224–233 (2014)
23. V.B. Filatov, A.A. Suvorov, Procedia Engineering **153**, 144–150 (2016)
24. A.G. Tamrazyan, Yu.N. Zvonov, Industrial and civil engineering **7**, 24–28 (2016)

# Analysis of results of the aggregate microtexture evaluation by volumetric characteristics

*Zuzana* Florková[1,*], *Štefan* Šedivý[2] , and *Ľubomír* Pepucha[1]

[1]University of Zilina, Research Centre, Univerzitna 8215/1, Zilina SK- 010 26, Slovak Republic
[2]University of Zilina, Department of Construction Management, Univerzitna 8215/1, Zilina SK- 010 26, Slovak Republic

**Abstract.** The article is concerned about the evaluation of aggregate microtexture. It is very important parameter from the traffic safety point of view and is described as a configuration of small irregularities on the aggregate surface. The new evaluating approach of aggregate microtexture based on the 3-D method is described in this article. 3-D model of surface of aggregate particle was generated from the measurements by microscope method. New program was developed for quantification of aggregate microtexture by a volumetric characteristics. Usability of this method was tested on samples of two coarse aggregate fraction taken from the eleven different quarries in Slovakia. The results of this evaluation have shown a high variability of values of volumetric characteristics between the fractions of aggregate and also between quarries.

## 1 Introduction

Microtexture of asphalt pavement is significant surface characteristics in term of skid resistance. It is defined by wavelengths in range 1μm to 0.5 mm and is characterized by microtexture of particular aggregate used in pavement. Its importance consist in providing basic level of friction between tire rubber and the pavement surface (significant on dry pavement surface by low speed up to 40 kmph). The effect of microtexture on the pavement skid resistance is in detail described in [1]. From the geometric point of view the microtexture is defined as a configuration of particular peaks on the aggregate surface [2]. It is clear, microtexture depends on the geometric properties of the aggregate grain (shape and size) and also on petrological and physical properties of aggregate grain [3]. Consequently, the type of used aggregate by its properties predetermine the microtexture level of pavement surface.

A lot of different approaches of aggregate microtexture evaluation methods have been developed and a wide range of measurement methods have been developed for its quantifying. However, taking into account their wavelength in micron scale, the possibility to direct accurate quantification is still problematic and the indirect quantification

* Corresponding author: zuzana.florkova@rc.uniza.sk

techniques with the pendulum tester [4] or dynamic friction tester [5] are used as an alternate.

At the present time, laboratory digital image analysis methods (various complex systems, laser scanner, photogrammetry method or microscopes) are extended for direct quantifying of aggregate microtexture. These methods allows to obtain a digital image of aggregate grain that is obviously transferred into 3-D. A video or camera is used to take photos for 3-D image generation in the complex systems developed for evaluation of aggregate microtexture [6 – 10]. The use of a laser scanner for 3-D image of aggregate grain and shape properties evaluation is presented in [11].

The main disadvantage of these above mentioned methods is, although 3-D surface models of aggregate grain are available they allow microtexture evaluation of grains only by 2-D characteristics which are mostly based on the shape of aggregate grain. These characteristics are closely presented in [12-15]. In general, disadvantage of 2-D characteristics based on digital image analysis is that these parameters allow to capture only larger peaks on the surface of aggregate grains. As the result of this fact, only the upper part of aggregate microtexture range is included into the evaluation [16]. Information including the evaluation of aggregate surface microtexture using volumetric characteristics derived from the 3-D model of aggregate surface could be found rarely. The example is the evaluation presented in [17]. In this study, the 3-D model of aggregate grain resulting from the photogrammetry method was used. The peak material volume characteristics (Vmp) was determined from 3-D model surface based on the Abbott-Firestone curve and only top parts of aggregate grains were included into the evaluation of the microtexture.

Lack of knowledge has been a motive for the research in this field. The main purpose was to use the 3-D model of aggregate surface obtained by optical microscopy as a basic evaluation input and find an adequate volumetric characteristics for following evaluation.

## 2 Principle of measurement by microscope method

Optical microscope NIKON AZ 100 (Fig.1) was used to obtain 3-D model of aggregate grains with different surface microtexture. The principle of this measurement rest in scanning of investigated particle by cameras at different Z-axis planes.

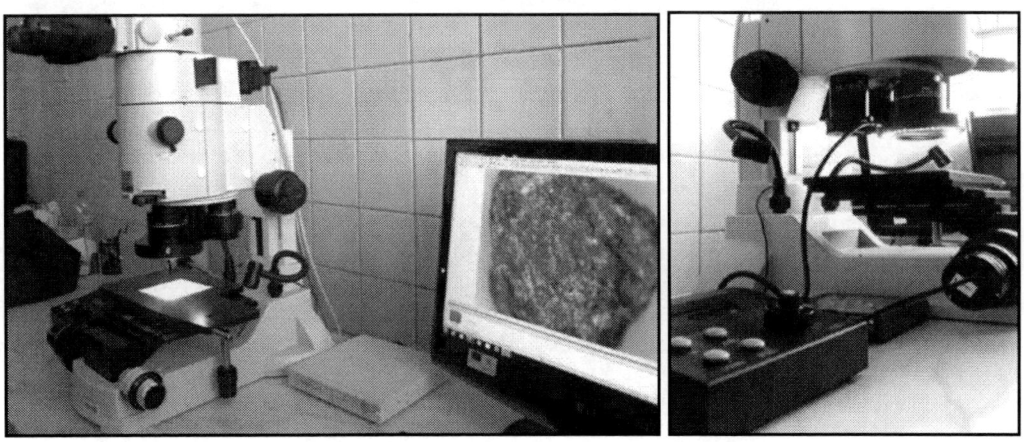

**Fig. 1.** Optical microscope NIKON AZ 100.

The investigated particle is set on the microscope pad. This pad is during scanning vertically moved in Z-axis direction (in micro steps). After scanning process, the microscope imaging software NIS Elements Documentation enables tag together images that have been scanned in a different Z-axis planes and then creating a resultant 3-D model.

This 3-D model of investigated particle can be exported into the Virtual Reality Modelling Language (Wrml) format (matrix form of regular placed height points of investigated grain). It allows to display and process the obtained 3-D model also apart from the appropriate microscope software in the working background of different computing programs. Then, a choices computing program can be used for following evaluation.

# 3 Evaluation of aggregate microtexture

## 3.1 Determination of volumetric characteristics

The new evaluation method was developed in the Matlab computing program with the aim of quantification microtexture of aggregate grain. This method uses as a basic input 3-D model in Wrml format, determined by microscope measurement. The main principle of this method is in definition of volume between particular peaks on the aggregate surface. Basically, it is a volume between "real" and convex surface of aggregate grain (Fig. 2). Then, this volume presented microtexture of aggregate grain and is determined by characteristics Zvd (Z - plane volume difference). It is a new volumetric microtexture evaluating characteristics which determined aggregate microtexture value by percentage difference of two volumes.

**Fig. 2**. 3-D model of aggregate particle.

The particular volumes are calculated from the defined comparative plane. First volume is determined as the volume of the aggregate grain and the second volume is determined as the volume of wrap (volume under the wrapping plane). Then, smaller value of Zvd characteristics defines lower suitable microtexture and vice versa. That higher value of Zvd characteristics defines angular aggregate with more peaks on the aggregate surface in comparison to polished or rounded aggregate. For more simply and practically calculation of Zvd characteristics was developed the program MicoSYS. This program allows evaluation of aggregate microtexture of only a selected part of aggregate particle by options of height adjustment of comparative (cutting) plane in the Z-axis direction. The principle and determination of Zvd characteristics and program MicroSyS are in detail described in [18].

## 3.2 Measurement and evaluation conditions

Usability of this evaluation method was tested on aggregate samples taken from eleven different quarries in Slovakia. The aggregates produced from these quarries fulfill all requirements from aspect of road construction quality. The aggregates differed from petrography point of view and ranged from soft (dolomite) to hard (andesite). Two fractions of aggregate particles (4/8 and 8/11) taken from each quarry were used in investigation. These fractions are the most important to ensure the adequate level of microtexture of pavement wearing course. Each sample of the quarry consists of ten aggregate grains (five grains from each fraction). Particular grains were scanned by the stereomicroscope. The measurement process was the same in the case of all scanning aggregates and was performed with the total magnification of 12.5 for the purpose of capturing the largest surface area of scanned aggregate. This magnification represents accuracy 57 μm in X-axis and Y-axis direction (in relation to dimensions of scanned area). The scanning range was divide into 400 steps. It represents 10-30 μm accuracy in Z-axis direction (in relation to aggregate height). Thereafter, the MicroSYS program was used to calculate value of microtexture volumetric characteristics Zvd for each tested aggregate grain. The position of the comparative plane was chosen in the same position for all aggregate grains to have comparable results.

## 4 Analysis of results

The representative value of aggregate microtexture for each quarry and fraction was obtained as a mean value from five values of volumetric characteristics Zvd determined on particular aggregate grains. The results of evaluation are presented in Figure 3.

It was supposed the different nature of aggregate in the individual quarries (and various portion of soft and hard minerals in these rocks) in combination with equipment used for aggregate production could influence number, shape and evenness of fracture planes and thus differences of microtexture values among the particular kinds of aggregates will be observed. However, when two fractions are compared, a production equipment should not play any role (obviously the same is used for production both fractions) and a microtexture of aggregate surface should only depend on properties of rock.

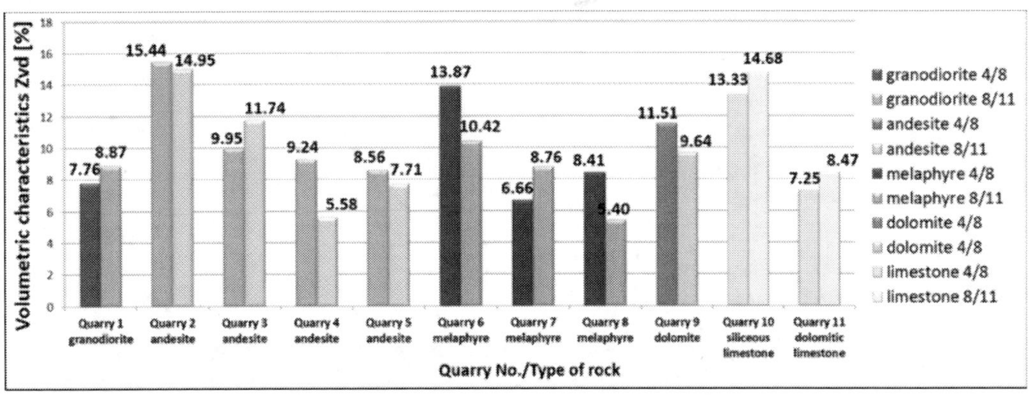


**Fig. 3.** Resultant values of volumetric characteristics Zvd – fraction 4/8 and 8/11.

It is obviously in the Figure 3, that obtained results have shown the considerable variability. There are the differences between fractions of aggregates and also between the quarries. Therefore, it is not possible expressly to type of particulars types of rocks by the values of determined volumetric characteristics Zvd from the microtexture point of view.

Table 1, summarizes a values variability of volumetric charasteristics Zvd in case of all investigated samples of aggregates form particular quarries and fractions. Data in Table 1 are presented by maximum, minimum and mean values of Zvd charcteristics. For better interpretation, these values are also presented in Figure 4. On the basis of results presented in Table. 1 and Figure 4, it can be argue, that the variability of results obtained by this method is very high. In expression as a percentage difference between maximally and minimally obtained values it can achieve for particular sample even 296 %.

**Table 1.** Variability summarization of volumetric characteristics Zvd determined on particular samples of aggregates.

| Zvd [%] | Fraction 4/8 | | | | Fraction 8/11 | | | |
|---|---|---|---|---|---|---|---|---|
| Quarry No. | Mean Zvd | Min. Zvd | Max. Zvd | Percentage difference [%] | Mean Zvd | Min. Zvd | Max. Zvd | Percentage difference [%] |
| 1 | 7.76 | 5.53 | 8.62 | 55.88 | 8.87 | 5.41 | 12.63 | 133.46 |
| 2 | 15.44 | 10.00 | 25.49 | 154.90 | 14.95 | 7.02 | 22.50 | 220.51 |
| 3 | 9.95 | 6.02 | 15.42 | 156.15 | 11.74 | 6.06 | 22.37 | 269.14 |
| 4 | 9.24 | 6.64 | 13.29 | 100.15 | 5.58 | 4.19 | 7.59 | 81.15 |
| 5 | 8.56 | 6.41 | 11.18 | 74.41 | 7.71 | 5.08 | 11.07 | 117.91 |
| 6 | 13.87 | 7.80 | 23.80 | 205.13 | 10.42 | 5.97 | 23.64 | 295.98 |
| 7 | 6.66 | 4.52 | 8.43 | 86.50 | 8.76 | 6.15 | 10.32 | 67.80 |
| 8 | 8.41 | 5.61 | 15.69 | 179.68 | 5.40 | 2.98 | 8.37 | 180.87 |
| 9 | 11.51 | 6.01 | 21.11 | 251.25 | 9.64 | 5.96 | 11.88 | 99.33 |
| 10 | 13.33 | 9.57 | 22.16 | 131.56 | 14.68 | 10.0 | 19.82 | 96.43 |
| 11 | 7.25 | 5.30 | 10.31 | 94.53 | 8.47 | 6.19 | 11.52 | 86.11 |

Possible causes can be several. Firstly, it can relates with a low number of grains representing particular aggregate samples. It confirm also differences among values of characteristics Zvd determined on the particular grains of aggregates within investigated samples (Figure 5). By reason of lucidity, the Figure 5 shows only some of quarry samples but it is sufficient for illustration of values difference within investigated grains.

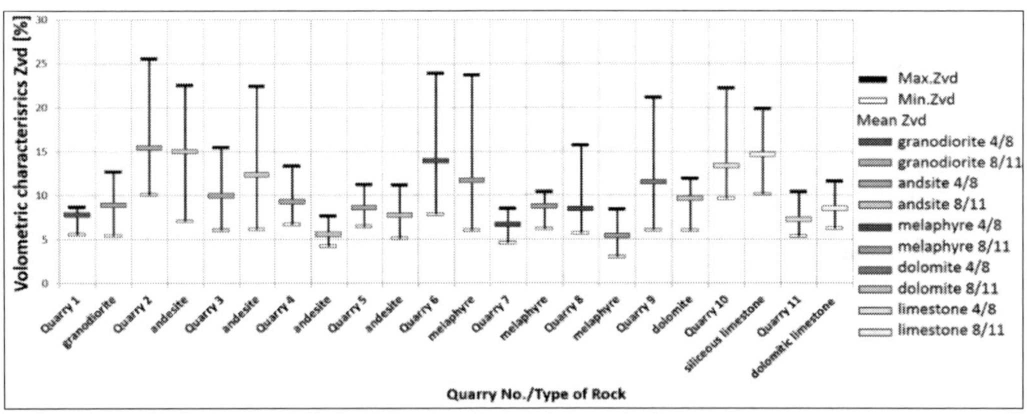


**Fig. 4.** Values variability of volumetric characteristics Zvd determined on particular samples of aggregates.

Another reason can be probably related with grain size of investigated aggregates. Inspite of equal measurement and evaluation conditions by measurements with the total magnification of 12.5 each aggregate grain has different size and thus always different surface area (below which is calculated volume) enters in the resultant evaluation. It causes that basic input entering in the resultant evaluation is always different especially when fractions 4/8 and 8/11 are compared.

It could be influenced also by measurement errors (different reflection or variance light conditions) because of optical microscope method. The insufficient magnification mainly influence the resultant values of characteristics Zvd. It causes that the scanning area of aggregate surface is always different and it affects also the accuracy of obtained microscope output in Wrml format. This format is a standard file format for representing 3-D information and in this case is represented by matrix form of regular placed height points of scanned particle. The mathematic matrix dimensions are 200 points in X-axis direction and 150 points in Y-axis direction. It leads to the lack of microtexture informations by usage a low microscope scanning magnification. On the basis of results obtained so far it can be concluded that the technique of microscope scanning was choose incorrect.

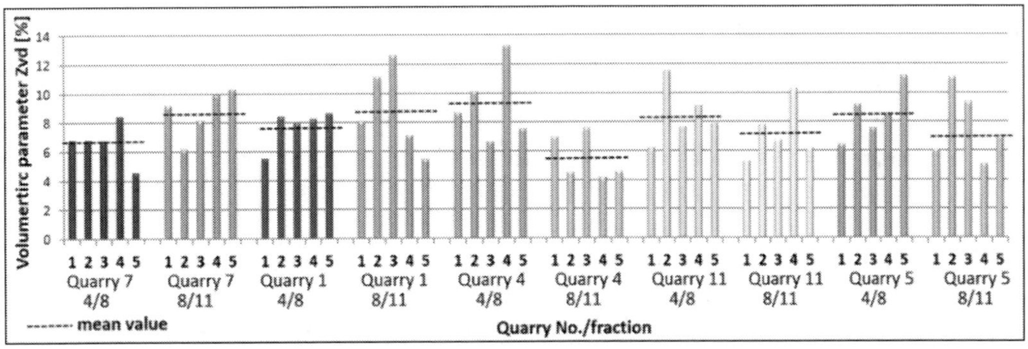


**Fig. 5.** Values variability of characteristics Zvd determined on the particular grains of aggregates within investigated samples.

## 5 Conclusions

The new approach for microtexture quantification and subsequent evaluation on the basis of 3-D outputs obtained by optical microscope was presented in this paper. Presented method uses new program MicroSYS for calculation of characteristics Zvd (Z – plane volume difference) as a volumetric characterization of aggregate surface microtexture. Zvd characteristics determined aggregate microtexture value by percentage difference of two volumes - the volume of the aggregate grain and the volume of wrap (volume under the wrapping plane). Applicability of this evaluation method was tested on aggregate samples taken from eleven different quarries in Slovakia. The aggregates differed from petrography point of view and ranged from soft to hard. Two fractions of aggregate particles (4/8 and 8/11) taken from each quarry were used in investigation.

The presented results have shown, that evaluation by characteristics Zvd incommensurate expected results because of considerable variability. A wide differences between fractions of aggregates and between the quarries was found. Therefore, it is possible to argue that it is not possible expressly to type of particulars types of rocks by the values of determined volumetric characteristics Zvd from the microtexture point of view. The results analysis have shown that the main reason of uncertain results relates with incorrect choosing of technique of microscope scanning. For the purpose of confirmation of usability this characteristics Zvd in microtexture evaluation process next test have to be carried out. It is necessary take measurements with higher magnification because the accuracy of the obtained output depends on the size of scanning area thus on the required magnification. With increasing of magnification is decreasing the size of scanning area and also is decreasing the depth of focus. It leads to higher accuracy of scanning resolution and also to the constant scanning area of aggregate surface for the all investigates grains. In this case is needed to practise a new measurement.

It is also necessary to carry out measurements on a numerous aggregate grains, which have a different level of microtexture (e.g. granodiorite versus limestone) in order to obtain sufficiently representative microtexture value of investigated aggregate by this method and after confirmation that this method makes it possible to distinguish the aggregate surface to carry out additional measurements.

This research was supported by European project Research Centre of the University of Žilina – Second Phase: ITMS 313011D011.

## References

1. J.W. Hall et al, *Guide for Pavement Friction*. NCHRP Web-Only Document 108, February (2009)
2. CEN. EN ISO 13473-5, Characterization of pavement texture by use of surface profiles, Part 5: Determination of megatexture (2009)
3. K. Scharnigg, G. Schwalbe, TYROSAFE - *Report on knowledge gaps and proposals for further research concerning optimisation for road surfaces and tyres for skid resistance rolling resistance and noise emissions* (2010)
4. STN EN 13036-4, Road and airfield surface characteristics, Test methods, Part 4: Method for measurement of slip/skid resistance of a surface: The pendulum test (2012)
5. Q. Lu, B. Steven, Friction Testing of Pavement Preservation Treatments: Literature Review (2006)
6. N. H. Maerz, W. Zhou. Flat and Elongated, Advances Using Digital Image Analysis, In: *Proc. of the 9th Annual Symposium of the International Centre for Aggregates Research (ICAR)* (2001)
7. E. Tutumluer, C. Rao, J. Stefanski, Video Image Analysis of Aggregates, In: *Transportation Engineering Series, **111**, Illinois Cooperative Highway and Transportation Series* (2000)
8. C. Rao, Development of 3-D Image Analysis Techniques to Determine Shape and Size Properties of Coarse Aggregate, *PhD. Thesis*, Department of Civil Engineering, University of Illinois at Urbana-Champaign (2001)
9. E. Masad, *The Development of a Computer Controlled Image Analysis System for Measuring Aggregate Shape Properties*, National Cooperative Highway Research Program NCHRP-IDEA Project 77, Final report, (2003)
10. H. Kim et al, A Prototype Laser Scanner for Characterizing Size and Shape Properties in Aggregates, In: *Proc. of the 9th Annual Symposium of the International Centre for Aggregate Research (ICAR)* (2001)
11. J.K. Anochie-Boateng et al, Evaluation of 3D laser device for characterizing shape and surface properties of aggregates used in pavements, In: *29th Southern Africa Transportation Conference* (2010)
12. E. Masad et al, *Test Methods for Characterizing Aggregate Shape Texture and Angularity*, Final report – appendixes, National Cooperative Highway Research Program, Appendixes to NCHRP Report 555 (2005)
13. Z. Florková, Microtexture of coarse aggregate for asphalt pavements, *Diploma thesis*, University of Žilina, Faculty of Civil Engineering, Department of Highway Engineering (In Slovak) (2012)

14. S. Tafesse et al, Sedimentology, **60, 4**, 1100-1110 (2013)
15. T. Al-Rousan et al., Cons. and Build. Mat, **21, 5,** 978-990 (2007)
16. Z. Florková, Communications, **17**, 4 (2015) p. 70-74.
17. G. Mcquaid et al., Intern. J. of Pavements Conference, Sao Paulo, (2013)
18. Z. Florková, Microtexture evaluation of asphalt pavement surface, *PhD. Thesis*, University of Žilina, Faculty of Civil Engineering, Department of Highway Engineering, (In Slovak) (2016)

# Evaluations of the solution to the homogeneous plane problem of the theory of elasticity in the neighborhood of an irregular boundary point

*Ludmila Frishter*[1,*]

[1]Moscow state university of civil engineering, Yaroslavskoye shosse, 26, Moscow, Russia, 129337

**Abstract.** The most complex stress-strain state (SSS) occurs in the domain of stress concentration due to the shape of the boundary ("the geometric factor") and the finite discontinuities of the specified forced deformations, the mechanical properties, emerging at the irregular point of the boundary of the domain. In this paper, there is a review of the methods for analyzing the particular qualities of the solution to the problem of the theory of elasticity, due to the shape of the boundary or "the geometric factor". The features of the stress-strain state of constructions and structures, possessing "constructive heterogeneity" under the action of discontinuous forced deformations, are stress concentrators, determined on the polymer models of the photoelastic method. For interpreting and decoding of experimentally obtained local SSS in stress concentration domains of structures there are given estimations of the solution to homogeneous plane problem of the elasticity theory in neighborhood of an irregular boundary point.

## 1 Introduction

The stress-strain state (SSS) of complex structures is characterized by a significant concentration of stresses in the domains of conjugation of elements with different options for the constructive design of the boundary: singular lines, incoming angle, etc.

The research of the stress state of complex structures in the domains of conjugation of elements made from materials with different mechanical properties under the action of forced deformations, that are discontinuous along the contact line (surface) of the elements, is an actual problem in the practice of engineering design. The research of the stress-strain state of complex structures in zones of constructive heterogeneity is focused on the consideration of the problem of elasticity theory in the neighborhood of the irregular point of the domain boundary, which includes the discontinuity (jump) of forced deformations.

* Corresponding author: lfrishter@mail.ru

## 2 Methods for analyzing the solution to elastic problems in the neighborhood of an irregular boundary

**2.1.** Questions of the behavior of solutions to the Laplace, Poisson, and elliptic equations for domains with non-smooth boundaries are considered by V.A. Kondratyev, V.V. Fufaev, M.L. Williams, S.Ja. Ufljand, A.I. Kalandii, G.P. Cherepanov, D.B. Bodzhy, O.K. Aksentjan, A.Ja. Aleksandrov, K.S. Chobanjan, I.T. Denisjuk, V.D. Kuliev, etc.

The fundamental work of V.A. Kondratiev [1] proves that the solution to the general elliptic boundary value problem in the neighborhood of irregular points of the domain boundary is represented in the form of an asymptotic expansion and an infinitely differentiable function. The terms of this expansion contain solutions to homogeneous boundary value problems for model domains: a wedge or a cone. They depend on the local characteristics - the magnitude of the solid or flat angle and the type of boundary conditions, the mechanical characteristics for piecewise homogeneous bodies. The magnitudes of the coefficients to the solution expansion in the neighborhood of a singular point are unknown and depend on the problem. The methods for determining these coefficients of decomposition are complex and difficult to be realized in the practical determination of stresses of complex structures having a complex shape of the boundary.

In the scientific papers of I.T. Denisyuk is given the asymptotics of the elastic solution to a plane complex domain with angular points on the separation lines [2]. The work of V.D. Kuliev [3] shows the advantage of a fundamental research of the singular solution to the boundary value problem in the possibility of applying its results to the investigation of problems on residual stresses and crack propagation across the connection boundary of dissimilar materials. It is shown that the order of the stress singularity depends on the approach degree of the crack tip to the connection domain of materials with different Young's modulus.

Mostly referred works of M.L. Williams show that stresses, deformations, the Erie stress function near the vertex of the sector with rectilinear sides have got a polynomial form [4]. The works of A.I. Kalandii, K.S. Chobanjan, L.A. Bagirov, O.K. Aksentjan, G.P. Cherepanov, V.P. Netrebko and many others represent the solution to a homogeneous boundary value problem in a neighborhood of an irregular boundary point in a polynomial form.

**2.2.** The scientific work [5] introduces a local curvilinear coordinate system to research the SSS in the neighborhood of an irregular point on the singular boundary line of an elastic body in which the Lame equations are written in. When approaching to an irregular point of the boundary from the inside the solution to the elastic problem comes down to solving two homogeneous plane problems: plane deformation and antiplane deformation or a transverse shift. It is possible to show that the representation of the solution to an elastic problem in the neighborhood of an irregular point on a singular boundary line in the form of two homogeneous plane problems is valid in the case when: a) given forced deformations and body forces are continuous along the domain of the elastic body; b) given forced deformations and body forces are piecewise-continuous functions, and moreover, the jump in the values of the forced deformations and body forces along the inner surface of the contact of the domains goes to the specific boundary line of the body; c) the contact surface of domains $V_1$ and $V_2$ of the elastic composite body $V$, having different mechanical characteristics $E_i, \nu_i$, $i = 1, 2$, goes to the specific boundary line of the body. In case of c), the solution to the elastic problem in the neighborhood of an irregular point of the boundary comes down to solving plane problems for a composite body.

**2.3.** In [6, 7, 8], similarity theory is used to research the solution to an elastic problem in displacements in the neighborhood of an irregular point on a singular boundary line of the domain.

We consider a small neighborhood of an irregular point on a singular line - the line of discontinuity, for example, boundary conditions or the first derivatives of the surface function of the domain. In a small neighborhood of the irregular point of the surface boundary, there is applied the following similarity group: $x_1 = tx;\quad y_1 = ty;\quad z_1 = z;\quad t > 0,$ when $t$ is the group parameter. Writing the Lame equation in a small neighborhood of a point on a singular line and passing to the limit for $t \to +\infty$, the solution to the elastic problem comes to solving two homogeneous plane problems, as in case 2.2: a plane deformation and an antiplane deformation.

In [6], the concept of a canonical singular problem characterizing the singularity of the SSS in the neighborhood of an irregular boundary is defined, for which the following two theorems are valid:

1) Any canonical singular problem corresponds to a transcendental equation, to each root of which there is corresponded a homogeneous solution, the number of arbitrary real constants in this solution is equal to the multiplicity of the root.

2) In an infinitesimal neighborhood of a singular point the solution to the correct boundary value problem of the theory of elasticity behaves as an asymptotically largest in absolute value eigenfunction of the corresponding canonical singular problem.

## 3 Statement of the problem and estimations of the solution

We consider the solution to the problem of elasticity theory for a homogeneous or piecewise homogeneous body in the neighborhood of an irregular point of the boundary of a plane domain into which a finite discontinuity (jump) of the forced deformations is included.

According to the theoretical analysis [8], the solution to the problem of elasticity theory in the neighborhood of an irregular point of the boundary of a plane domain can be written in the following form:

$$\sigma_{ij} = \sigma_{ij}^{E} + \sigma_{ij}^{S}, \qquad \varepsilon_{ij} = \varepsilon_{ij}^{E} + \varepsilon_{ij}^{S}, \qquad U_i = U_i^{E} + U_i^{S},$$

or

$$\eta = \eta^{E} + \eta^{S}, \tag{1}$$

when $\eta^{E} = (\sigma_{ij}^{E}, \varepsilon_{ij}^{E}, U_{ij}^{E})$ – "eigen" solution to a homogeneous boundary value problem in a neighborhood of an irregular point of the domain boundary characterizes the singularity for the solution;

$\eta^{S} = (\sigma_{ij}^{S}, \varepsilon_{ij}^{S}, U_{ij}^{S})$ – the stress-strain state, determined by the action of given loads, depends on the geometric parameter - the "approximation degree" to the singular point. The representation (1) is also valid in the spatial case for the points on the singular boundary line of the domain.

We consider a small neighborhood of the irregular point $O$ on the singular boundary line of $V$ - body, as a self-balanced part of body $V$ in the following form:

$x^2 + y^2 < \varepsilon_1^2;\ z^2 < \varepsilon_2^2;\ \varepsilon_1, \varepsilon_2$ - are small positive numbers.

The conditions on boundary $L$ of the domain are homogeneous.

The original elastic problem in the neighborhood of an irregular boundary point admits a similarity group:

$$x_1 = tx;\quad y_1 = ty;\quad z_1 = z;$$

$$\sigma_{ij} = t\overline{\sigma_{ij}}; \quad \varepsilon_{ij} = t\overline{\varepsilon_{ij}}; \quad U_i = \overline{U_i},$$

when $i, j = x, y, z;\ i1, j1 = x_1, y_1, z_1;\ t$ - is a group parameter, $t > 0$.

In an infinitesimal neighborhood of a singular point of the boundary, the solution to the correct boundary value problem of elasticity theory behaves as the asymptotically largest one in absolute value eigenfunction of the corresponding canonical singular problem. The singularity of stresses and strains has order $r^{\lambda_0 - 1}$, when $\lambda_0 = \min \operatorname{Re} \lambda$ - the minimal value of the real part of the complex root of the characteristic equation of a homogeneous plane boundary value problem of elasticity theory is determined numerically [9].

We choose a neighborhood $c < r < (1+\alpha)c,\ \ \alpha \in (0,1),\ \ \theta \in (\theta_0, \theta_1)$ in which the nonsingular plane homogeneous problem with homogeneous boundary conditions is valid. The parameter $t = \dfrac{1}{c}$ is great enough and parameter $c$ is small enough so as not to take into account the stresses of the common field $\sigma_{ij}^S$ determined by the given forced deformations.

The stress function of a homogeneous plane elastic problem in a polar coordinate system is written in the following form:

$$\varphi = r^{\lambda_0 + 1} f(\theta), \tag{2}$$

when $\lambda_0 = \min \operatorname{Re} \lambda$ – the minimum value of the real part of the complex root $\lambda$ of the characteristic equation of a homogeneous boundary (singular) problem in the domain $r < \varepsilon_1 < c;\ \varepsilon_1 > 0,\ f(\theta)$ – function of the variable $\theta,\ \theta \in (\theta_0, \theta_1)$.

We introduce a new variable $r_1 = r - c,\ \ r \in [c,\ (1+\alpha)c],\ \ r_1 \in (0, \alpha c)$. The stress function (2) is taken in the following form:

$$\varphi = (r_1 + c)^{\lambda_0 + 1} f(\theta) = c^{\lambda_0 + 1}\left(1 + \frac{r_1}{c}\right)^{\lambda_0 + 1} f(\theta), \tag{3}$$

when $0 < \dfrac{r_1}{c} < \alpha < 1$. We apply the binomial expansion in the following form:

$$\left(1 + \frac{r_1}{c}\right)^{\lambda_0 + 1} = 1 + (\lambda_0 + 1)\left(\frac{r_1}{c}\right) + \frac{(\lambda_0 + 1)\lambda_0}{2!}\left(\frac{r_1}{c}\right)^2 + \\ + \frac{(\lambda_0 + 1)\lambda_0(\lambda_0 - 1)}{3!}\left(\frac{r_1}{c}\right)^3 + \ldots + \frac{(\lambda_0 + 1)\lambda_0(\lambda_0 - 1)\cdots(\lambda_0 - n + 2)}{n!}\left(\frac{r_1}{c}\right)^n \ldots \tag{4}$$

The stress function (3) takes into account the expansion (4) and is rewritten in the following way:

$$\varphi = c^{\lambda_0 + 1}\left[1 + (\lambda_0 + 1)\left(\frac{r_1}{c}\right) + \sum_{n=2}^{\infty} \frac{(\lambda_0 + 1)\lambda_0(\lambda_0 - 1)\cdots(\lambda_0 - n + 2)}{n!}\left(\frac{r_1}{c}\right)^n\right] f(\theta). \tag{5}$$

According to the stress function (5), we write the stresses in the neighborhood of the irregular point of the boundary as follows:

$$\sigma_r = \frac{1}{r}\frac{\partial\varphi}{\partial r} + \frac{1}{r^2}\frac{\partial^2\varphi}{\partial\theta^2} = \frac{c^{\lambda_0+1}}{r} f(\theta)\left[\frac{\lambda_0+1}{c} + \frac{(\lambda_0+1)\lambda_0}{c^2}(r-c) + \right.$$
$$\left. + \sum_{n=3}^{\infty}\frac{(\lambda_0+1)\lambda_0(\lambda_0-1)\cdots(\lambda_0-n+2)}{(n-1)!c^n}(r-c)^{n-1}\right] + \frac{c^{\lambda_0+1}}{r^2} f''(\theta)\left[1 + \frac{(\lambda_0+1)}{c}(r-c) + \right.$$
$$\left. + \sum_{n=2}^{\infty}\frac{(\lambda_0+1)\lambda_0(\lambda_0-1)\cdots(\lambda_0-n+2)}{n!c^n}(r-c)^n\right],$$
$$\sigma_\theta = \frac{\partial^2\varphi}{\partial\theta^2} = c^{\lambda_0+1} f(\theta)\left[\sum_{n=2}^{\infty}\frac{(\lambda_0+1)\lambda_0(\lambda_0-1)\cdots(\lambda_0-n+2)}{(n-2)!c^n}(r-c)^{n-2}\right], \tag{6}$$
$$\tau_{r\theta} = -\frac{\partial}{\partial r}\left(\frac{1}{r}\frac{\partial\varphi}{\partial\theta}\right) = -c^{\lambda_0+1} f'(\theta)\left[\frac{\lambda_0}{r} + \right.$$
$$\left. + \sum_{n=2}^{\infty}\frac{(\lambda_0+1)\lambda_0(\lambda_0-1)\cdots(\lambda_0-n+2)}{n!c^n}\left(n-1+\frac{c}{r}\right)\frac{(r-c)^{n-1}}{r}\right].$$

We estimate the remainder term of the series (5) in the following form:

$$\left|R_n\left(\frac{r_1}{c}\right)\right| < \left|\frac{(\lambda_0+1)\lambda_0(\lambda_0-1)\cdots(\lambda_0+1-n)}{(n+1)!}\left(\frac{r_1}{c}\right)^{n+1}\right| < \delta,$$

when $\frac{r_1}{c} = \frac{r-c}{c} < \alpha \ll 1$, $\lambda_0 \in (0,1)$; $\alpha \in (0,1)$,

or $$\left|R_n\left(\frac{r_1}{c}\right)\right| < \left|\frac{(\lambda_0+1)\lambda_0(\lambda_0-1)\cdots(\lambda_0+1-n)}{(n+1)!}(\alpha)^{n+1}\right| < \delta .$$

With an increase in the angle of the solution between the tangents drawn to the boundary of the domain at the singular point: $\lambda_0 \to 0,5$. To assess the remainder $R_n\left(\frac{r}{c}\right)$ we consider, as an example, $\lambda_0 = 0,5$:

For $n=0$ $|R_n| < \frac{3}{2}(\alpha) < \frac{3}{2}$.

For $n=1$ $|R_n| < \frac{3}{8}(\alpha)^2 < \frac{3}{8} = 0,375$.

For $n=2$ $|R_n| < \frac{9}{48}(\alpha)^3 < 0,09375 < 0,1$.

Therefore, with the necessary accuracy for the stress function, it suffices to take the first two (three) terms of the expansion series for $n$ = 0, 1 (2), i.e.

$$\varphi(r_1,\theta) \cong c^{\lambda_0+1} f(\theta)\left[1 + (\lambda_0+1)\left(\frac{r_1}{c}\right)\right], \quad n=0,1,$$

or

$$\varphi(r,c) \cong c^{\lambda_0+1} f(\theta)\left[-\lambda_0 + (\lambda_0+1)\frac{r}{c}\right]. \tag{7}$$

Taking into account the stress function (7) in the neighborhood $c < r < (1+\alpha)c$, $c \in (0,1)$, $\alpha \in (0,1)$, the stresses (6) are written as following:

$$\sigma_r \cong \frac{c^{\lambda_0}}{r}(\lambda_0+1)c_1\left[f(\theta)+f''(\theta)\right], \qquad \bar{\sigma}_\theta^c \cong 0; \quad \bar{\tau}_{r\theta}^c \cong 0.$$

The terms causing a significant increase in stresses and deformations (an increase in the deformation energy in the neighborhood: $r \in (0, \alpha c)$; $\alpha c \ll 1$; $\frac{1}{r^2} \gg \frac{1}{r}$;) are not considered.

For example, for the stress function $f(\theta) = \left[c_0 \cos(\lambda_0 - 1)\theta - c_1 \cos(\lambda_0 + 1)\theta\right]$ at the first approximation the stresses will be written down in the following way:

$$\sigma_r \cong \frac{c^{\lambda_0} c_1}{r}(\lambda_0+1)\lambda_0[c_2(2-\lambda_0)\cos(\lambda_0-1)\theta + (2+\lambda_0)\cos(\lambda_0+1)\theta],$$

when $c_1, c_2 \in \mathbb{R}$ - the arbitrary constants,

or $\sigma_r \cong \frac{(\lambda_0+1)\lambda_0}{r} c\,[c_2(2-\lambda_0)\cos(\lambda_0-1)\theta + (2+\lambda_0)\cos(\lambda_0+1)\theta], \quad \sigma_\theta \cong 0; \quad \tau_{r\theta} \cong 0.$

Choosing the required number of terms in the expansion series of the stress function (5), it is possible to obtain stresses and strains in the domain of the nonsingular solution to the problem with an accuracy corresponding to the accuracy of the data measurement by the photoelasticity method [10, 11]. The neighborhood of the irregular point of the boundary of a plane domain, where the estimations of the nonsingular solution to the homogeneous problem (5), (7), (6) are valid, are chosen from the experimental data.

In accordance with the accuracy of the photoelasticity method [10], it is sufficient to take the first two (three) terms of the series (5) for the expansion of the stress function for n = 0,1 (2).

## 4 Conclusions

In the solution domain to a nonsingular homogeneous plane elastic problem, there is a possible estimation to the solution, using which we can extrapolate the solution to sections close to an irregular boundary point, taking into account the experimental data and practical accuracy of the measurement by the photoelasticity method.

## References

1. V.A. Kondratyev. *Proceedings of the Moscow Mathematical Society*. Vol.**16**. pp. 209–292 (1967)
2. I. T. Denisjuk. *Proceedings of universities, Mathematics*, **6**, pp.70–74 (2000)
3. V.D. Kuliev. *M.:Science, Singular boundary value problems*. p.719 (2005)
4. M. L. Williams. *J. Appl. Mech.*, v. **19**, **4**, p. 526 (1952)
5. O.K. Aksentian. *Applied Mathematics and Mechanics.*, 1967, Vol.**31**, Issue 1, pp.178-186 (1967)
6. G.P. Cherepanov. *M.:Science, Mechanics of brittle fracture*. p.640 (1974)
7. G.S.Vardanjan, L.Ju.Frishter. *International journal for computational civil and structural engineering*. Volume **3**, Issue 2, pp.75-81 (2007)
8. L.Ju.Frishter. *Proceedings of MGSU*, **1**, pp.169-174 (2008)
9. G.S.Vardanjan, M.L. Mozgaleva, V.N. Savostyanov, L.Ju.Frishter. *Proceedings of universities.Civil Engineering*. **10**, pp.28-31 (2003)

10. G.L. Hesina and others. *M.: Stroyizdat, The photoelasticity method. In three volumes.* Vol.**3,** p.311 (1975)
11. G.S. Vardanjan, V.N. Savostyanov, L.Ju. Frishter. *The development of methods of experimental mechanics. Under. ed. O.N. Makhutova and others.- M.: IMASH RAS*, pp.60-68 (2003)

# Particulate matter mass concentrations produced from pavement surface abrasion

*Dasa* Fullova[1,*], *Daniela* Durcanska[1] , and *Jitka* Hegrova[2]

[1] University of Žilina, Faculty of Civil Engineering, Department of Highway Engineering, Univerzitná 1, 010 26 Žilina, Slovakia
[2] Transport Research Centre, Division of transport infrastructure and the environment, Líšeňská 33a, 636 00 Brno, Czech Republic

**Abstract.** According to the latest findings particulate matter belong to the most significant pollutants in Europe together with ground-level ozone O3 and nitrogen dioxide NO2. Road traffic is one of the main sources of particulate matter. Traffic volume has unpleasant impact on longevity of the pavements and also on the environment. Vehicle motions cause mechanical wearing of the asphalt pavement surface - wearing course by vehicle tyres. The paper deals with abrasion of bituminous wearing courses of pavements. The asphalt mixtures are compared in terms of mechanically separated particulate matter. The samples of asphalt mixtures were rutted in wheel tracking machine. The particulate matter measurements were performed in laboratory conditions. The experimental laboratory measurements make it possible to sample particulates without contamination from exhaust emissions, abraded particles from vehicles, resuspension of road dust and climate affects. The paper offers partial results of measurements on six trial samples of asphalt mixtures with different composition. It presents particulate matter morphology and the comparison of rutted asphalt samples in terms of PM mass concentrations and chemical composition.

## 1 Introduction

Non-exhaust emissions of particulate matter from road traffic may originate from abrasion of vehicle parts (tyres, brake linings, clutch and bodywork) or from pavement surface abrasion and also from resuspension of dust 1, 1. The aim of this research is to monitor the abraded particulate matter from non-combustion processes – from abrasion of pavement surface. A differentiation between PM emissions from road pavement abrasion and resuspension is very difficult due to their similar elemental composition 2. For this reason, the measurements are performed in the laboratory at the Department of Highway Engineering. The issue of abraded particulate matter from pavement surfaces and the contribution of those emissions to air pollution is mostly solved in the Nordic countries 3, 4, 5, 6, 7.

* Corresponding author: dasa.fullova@fstav.uniza.sk

Laboratory particle sampling makes it possible to sample particulates with very low contamination from surrounding sources and meteorological factors. An important phase of the research is to determine chemical elements in the collected particulate matter and to confront it with the representation of these elements in the chemical composition of the asphalt mixture. The next step is to attempt to determine the share of the mechanical separation of PM from the pavement by evaluating the chemical composition of the materials used for studied asphalt mixtures of wearing courses and subsequently collected particulate matter PM. The research is aimed at finding dependence between the composition of asphalt mixture and collected PM.

## 2 Methodology of laboratory tests

The asphalt mixture samples of wearing courses are used for laboratory measurements. The used asphalt mixture samples (plates of size 320 x 260 mm and thicknesses of 40 mm) are rutted in the wheel tracking machine DYNA-TRACK which is used to assess the resistance to rutting of asphalt materials under conditions, which simulate the effect of traffic (Fig. 1) 8.

Air sampling from inside of the machine DYNA-TRACK during the rutting of samples is performed by equipment APS 3321 (Aerodynamic Particle Sizer) and SMPS 3080 (Scanning Mobility Particle Sizer) which intercept and distribute particle range from 0.012 to 20 μm and 2 pieces of Leckel LVS3 which intercept particulate matter $PM_{2.5}$ and $PM_1$ on the nitrocellulose filters for the subsequent evaluation of mass concentrations and chemical analysis. The mixtures are rutted by typified wheel for 12 hours through the steady working conditions. It performs 20 000 cycles during the 12 hours (i.e. 40 000 running gears).

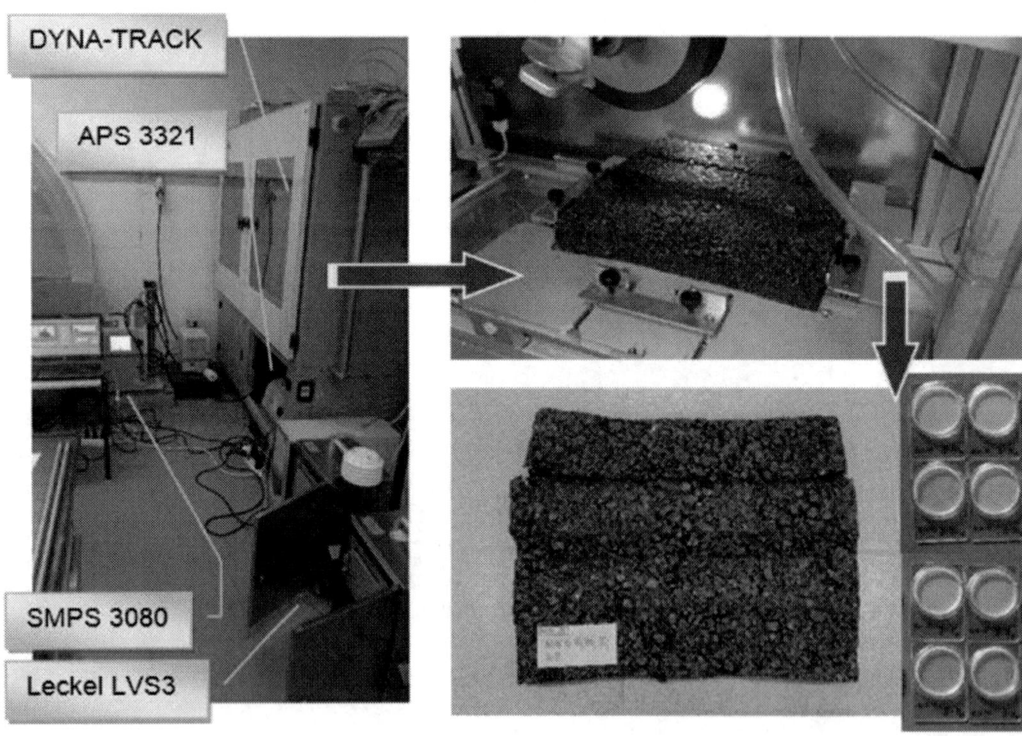


**Fig. 1.** The used machine technology and rutted asphalt mixture sample

The tests of chemical composition of materials (aggregates, asphalt) of samples were carried out before the rutting of these asphalt mixture samples by XRF analyzer (X-ray fluorescence spectroscopy). For the experiments was used the machine SPEKTRO iQ II (AMATEK, Germany) in the laboratory of the Faculty of Civil Engineering of the Technical University in Košice. Chemical analysis of particulate matter intercepted on filters was carried out by ICP - MS - Inductively coupled plasma mass spectrometry in the laboratory at the Transport Research Centre in Brno. Imaging of PM was done by using the scanning electron microscope VEGA TS 5136 LSU (Tescan s.r.o., Czech Republic) that permits both low vacuum and high vacuum mode of measurements with resolution up to 3.5 nm.

# 3 Results of current laboratory tests

Till now, the measurements of particulate matter were performed on six trial samples of asphalt mixtures with different composition. Two plates of each mixture were rutted in the laboratory and each plate was rutted two times. All six samples had the same conditions in the space of rutting during abrasion of asphalt mixtures. The total length of measurement was 12 hours. Marks of six rutted samples:

1. *AC 11 O 50/70, II, D – 5*
- aggregate: quarry Tunezice (siliceous limestone), asphalt: CA 50/70
2. *AC 11 O 50/70, II, R – 9*
- aggregate: quarry Biely Potok (dolomite), asphalt: CA 50/70
3. *AC 11 O PMB 45/80 – 75, I, R – 8*
- aggregate: quarry Maluzina (melaphyre), quarry Biely Potok (dolomite),
- asphalt: PMB 45/80-75
4. *AC 11 O 50/70, II, K – 4*
- aggregate: quarry Biely Potok (dolomite), quarry Badin (andesite),
- asphalt: CA 50/70
5. *AC 11 O 50/70, II, S – 9*
- aggregate: quarry Solosnica (melaphyre), quarry Trstin (dolomite),
- asphalt: CA 50/70
6. *SMA 11 PMB 45/80 – 75, S – 4*
- aggregate: quarry Solosnica (melaphyre), quarry Trstin (dolomite),
- asphalt: PMB 45/80-75

## 3.1 PM mass concentrations during the rutting of asphalt mixture samples

The results of PM mass concentrations are shown in the following figure (Fig. 4). The highest concentrations were measured for the mixture with the mark S – 9 (AC 11 with classic bituminous binder) and the lowest concentrations of particulate matter were measured for the sample D – 5 (AC 11 with classic bituminous binder). The mixtures differ in the use of aggregate.
The concentrations were determined by two methods – spectrometers SMPS, APS and low volume sampler Leckel. The results correlate with each other – Leckel $PM_{2.5}$ and SMPS+APS (r = 0.98) (Fig. 4). In the mixture with the highest concentrations the aggregates melaphyre and dolomite were used and in the mixture with the lowest concentrations the siliceous limestone was used. The average mass concentration of the

sample S - 9 is by 6.30 μg/m$^3$ higher than the average mass concentration of the sample D – 5. These concentrations were performed by means of SMPS and APS. By using gravimetric method the difference is determined at 5.36 μg/m$^3$ for particulates of fraction $PM_{2.5}$ and 4.77 μg/m$^3$ for particulates of fraction $PM_1$.

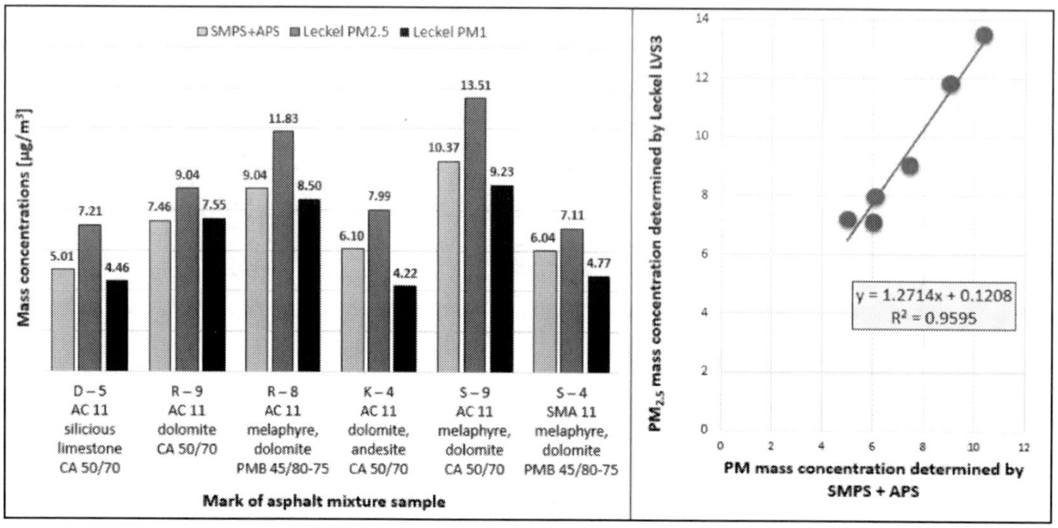


**Fig. 2.** The average measured PM mass concentrations and the correlation between PM mass concentrations determined by spectrometers (SMPS+APS) and low volume sampler (Leckel LVS3)

### 3.2 Chemical composition of particulate matter

For identification of particulate matter originating from the asphalt mixture abrasion the chemical composition of the basic materials of mixtures were carried out. The chemical elements (markers) were determined by analysis of the aggregate as a filler and the asphalt as a binder. Identified elements are subsequently confronted with the elements of the particulate matter collected on the filters.

It has been found that the bituminous binder contains approximately 4.8 % of inorganic elements and organic elements are the rest. The sulphur S has the largest representation, it represents 4.19 %. The silicon Si (0.21%) and the chlorine Cl (0.17%) represent a higher share of identified inorganic elements.

The tests of chemical composition were also carried out for all kinds aggregates used in tested samples. In the samples the aggregates from six quarries were used (Maluzina, Badin, Solosnica, Biely Potok, Tunezice, Zirany). For XRF spectroscopy the powder samples (crushed aggregate) were used. From the analysis of materials the percentage of each selected element has been found. The results have shown that elements silicon Si and calcium Ca have the largest percentage in aggregates. The silicon Si predominates in aggregate from quarries Maluzina (melaphyre), Badin (andesite) and Solosnica (melaphyre) and the calcium Ca predominates in aggregate from quarries Biely Potok (dolomite), Tunezice (siliceous limestone), Zirany (limestone). The elements such as Mg, Al, P, Cl, and K represent a higher share of identified elements.

Chemical analysis of the exposed filters of particulate matter $PM_1$ and $PM_{2.5}$ was carried out by mass spectrometry with inductively coupled plasma ICP-MS. The principle of the device is that the samples are decomposed by acids by the action of microwaves. The prepared (decomposed) samples (51 filters) were analyzed for the presence of selected 13

elements which have been determined by previous analyses of used aggregates and asphalt - *Ca, Si, Mg, Al, Fe, P, S, Cl, K, V, Cr , Mn, Na*.
The following figure (Fig. 5) presents results of analyses of selected elements. Till now, 51 filters were analyzed. Two fraction of particulate matter were intercepted on these filters – fraction $PM_1$ and fraction $PM_{2.5}$. For each mixture (2 plates, 4 ruttings) 4 filters of fraction $PM_1$ and 4 filters of fraction $PM_{2.5}$ were analyzed.

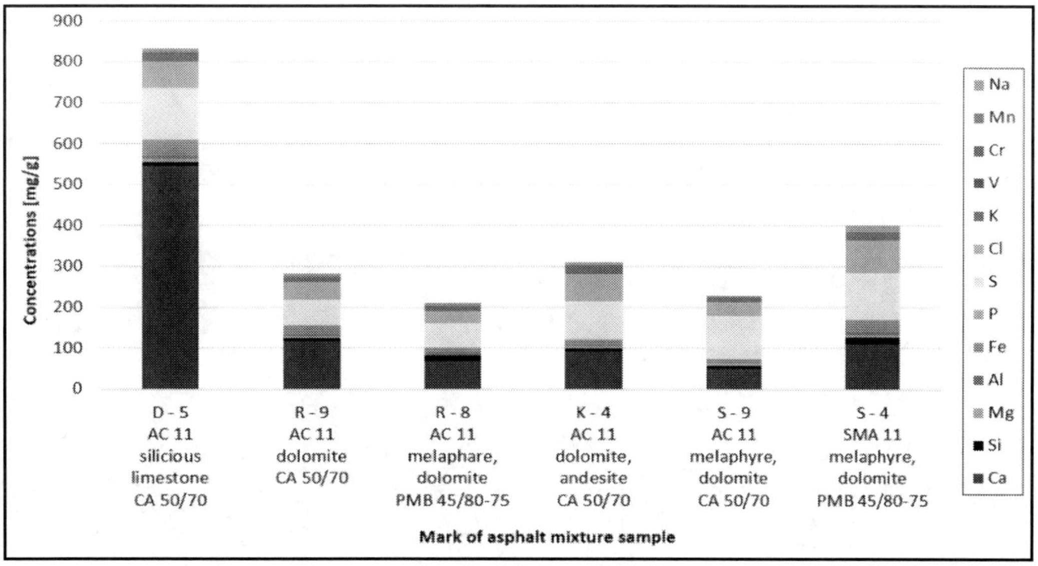


**Fig. 3.** The average mass concentrations of selected chemical elements in $PM_{2.5}$ fraction

The elements Ca, S, Cl have the highest representation in the analysed filters. Calcium Ca has the highest representation in the sample with mark D – 5 and its concentration is 546.22 mg/g $PM_{2.5}$. In this sample of asphalt mixture aggregate siliceous limestone is sued. Sulphur with mass concentration 126.15 mg/g $PM_{2.5}$ has also the highest concentration in the sample D – 5. With regard to the chlorine the highest concentration was measured in the sample S – 4 (81.47 mg/g $PM_{2.5}$). On the other hand, the lowest concentrations were measured for the element vanadium V (in range from 0.01 to 0.04 mg/g $PM_{2.5}$) and for elements such as Cr, Mn and P. The highest concentrations of selected elements were analysed in the sample D – 5. Despite of the fact that the sample with mark D – 5 has the lowest mass concentrations of particulate matter. Sample R - 9 presents the lowest mass concentrations of these elements and it is together 211.77 mg/g PM2.5. The aggregates rich in limestones are prone to polishing of its surface, the aggregates rich in feldspars have an average polishing and the aggregates rich in quartz are quite resistant to polishing. To exemplify, the highest mass concentration of calcium was measured in the sample in which the siliceous limestone is used.
As seen from the figure all selected elements occur in the intercepted particulates. It is obvious that the released particulates from surface abrasion have similar composition as the basic materials of asphalt mixtures (aggregate, bituminous binder).

### 3.3 Morphology of abraded particulate matter

Particulate matter (PM) is considered a mixture of chemical components originated from a variety of sources with specific physical parameters (such as shape, diameter, electric

charge, solubility, surface of the particles) and chemical composition (organic and inorganic compounds). Sources, mechanisms of formation, meteorological conditions, distance from the source and other conditions determine these basic parameters 9. To gain a better understanding of PM origin and behaviour, it is necessary to describe their chemical composition and size distribution.
Imaging of PM was done by using the scanning electron microscope VEGA TS 5136 LSU (Tescan s.r.o., Czech Republic) in Transport Research Centre in Brno. Several types of filters are commonly used in sampling aerosols, but not all are compatible with SEM/EDX analysis 10. Particulate matter was intercepted on special filters with polycarbonate membrane.
Sharp-edged particles were observed on contaminated filters in a numerous representation. These particles can be characterized as fragments of geological materials (aggregates used in the mixtures). As a result, we can say that the particles are abraded from the mixtures during the rutting and the abrasion of aggregates used for asphalt mixtures could be the source of these particles.

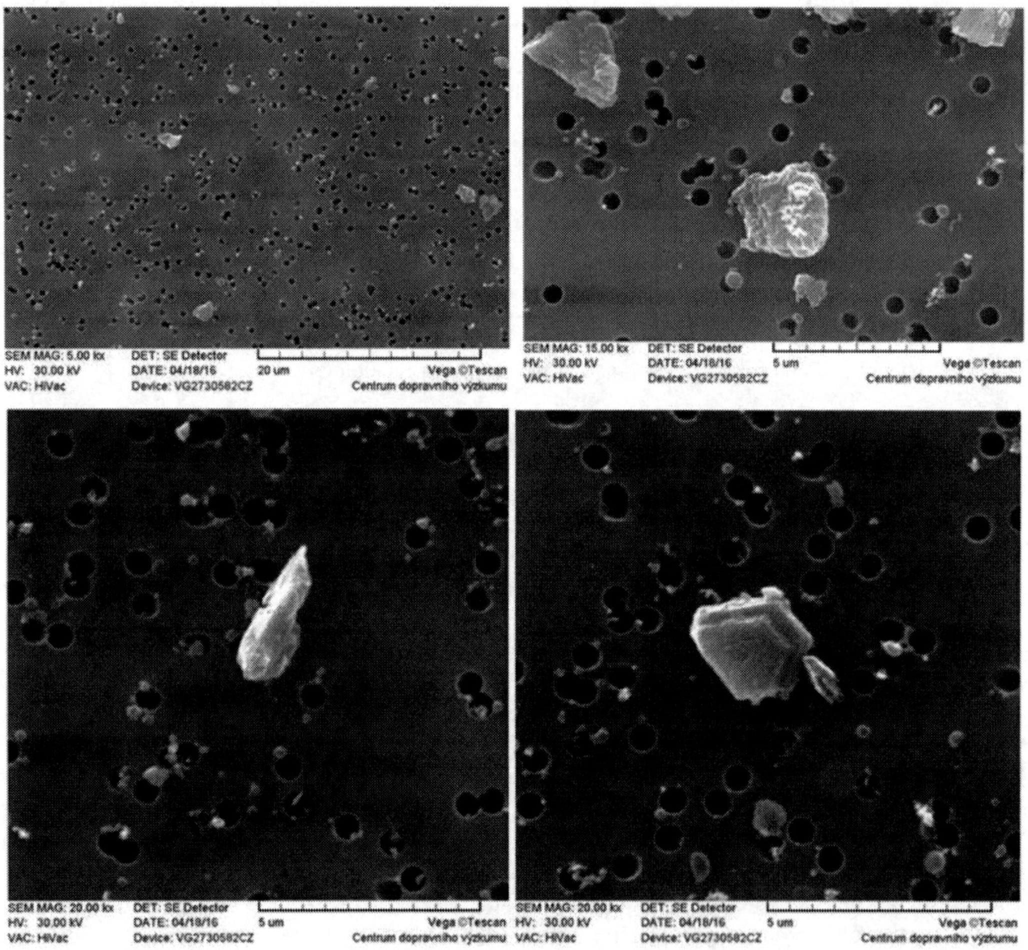


**Fig. 4.** Sharp-edged particles probably geological origin

Scanning electron microscope (SEM) may be used to determine the possible sources based on the size and shape of the particles. Nevertheless, for more proper PM sources identification it is suitable to combine the knowledge on separate PM morphology with the chemical composition of PM. PM origin interpretation only on the basis of morphology characterization without the chemical composition knowledge could be in some cases inaccurate or misleading.

Abrasion of individual components of the mixtures must be confirmed by means of chemical analysis of the particulates collected on nitrocellulose filters and its subsequent comparison with the composition of each tested mixture. Thus, obtained results presented in this paper were compared with data published in scientific literature 9, 10 in which the same PM shapes were observed and assigned to specific source.

## 4 Discussion to the current results

Primordial laboratory measurements of abrasion of asphalt mixture samples were carried out. Number and mass concentrations of particulate matter released during rutting of selected samples were measured. The tests of chemical composition of sample materials (aggregates, asphalt) were carried out before the rutting of these asphalt mixture samples by XRF analyzer (X-ray fluorescence spectroscopy). Also, the chemical analyses of particulate matter intercepted on filters were carried out by ICP - MS (Inductively coupled plasma mass spectrometry).

So far, 24 tests were performed for 6 samples of asphalt mixtures. It follows that each asphalt mixture had 4 available measurements. The mixtures differ in the use of aggregate and bituminous binder. The highest concentrations were measured for the mixture with mark S – 9 (AC 11 with classic bituminous binder) and the lowest concentrations of particulate matter were measured for the sample D – 5 (AC 11 with classic bituminous binder). In the mixture with the highest concentrations the aggregates melaphyre and dolomite were used and in the mixture with the lowest concentrations the siliceous limestone was used.

Imaging of PM was done by using the scanning electron microscope. Scanning electron microscopy (SEM) is an effective tool for the identification of PM sources. However, for more proper PM sources identification it is suitable to combine the knowledge on separate PM morphology with the chemical composition of PM. For examination the particulate matter shapes and sizes from abrasion of asphalt mixture surfaces the electron microscope was used. In numerous representation sharp-edged particles were observed on the filters, probably fragments of geological origin. Therefore we can say that the particulate matter are abraded from asphalt mixture surfaces and it may originate from abrasion of aggregates used in the mixtures.

From chemical analyses a number of facts about content of individual elements in the investigated materials has been revealed. The elements Si and Ca have majority representation in the aggregates. The elements of the under-representation in the aggregates are Mg, Al, and Fe. The bituminous binder contains primarily elements of organic chemistry but sulphur S is majority element from inorganic elements.

From PM chemical analyses the average mass concentrations of selected chemical elements were found for each tested sample. The elements Ca, S, Cl have the highest representation in the analysed filters. Calcium Ca has the highest representation in the sample with mark D – 5 and in this sample the aggregate siliceous limestone is sued. On the other hand, the lowest concentrations were measured for the element vanadium V, Cr, Mn and P. As resulting, the released particulates from surface abrasion have similar composition as the basic materials of asphalt mixtures (aggregate, bituminous binder).

The aim of experimental measurements is to confirm the abrasion of pavement surface as a source of particulate matter and to verify the impact of the composition of asphalt mixture for the PM production during the rutting of samples. The findings from chemical analyses will be further applied in the next phase of research and used for verification of this impact.

The measurements are still being continued and next assessment of PM concentrations and contents of selected elements in PM are expected in order to compare the asphalt mixtures with each other in terms of production PM.

## Acknowledgements

The paper is a result of research supported by research project *KEGA 045ŽU-4/2017 Environmental education of engineers for application in public administration.*

## References

1. Ďurčanská, D., et al., *Analysis of methods of evaluation of the air pollution from road traffic: analysis task.* Žilina: Faculty of Civil Engineering, Department of Highway Engineering (2014).
2. Amato, F., et al., *Urban air quality: The challenge of traffic non-exhaust emissions.* Journal of hazardous materials, 275, 31-36. (2014). Retrieved from www.scopus.com
3. Jandacka, D. : *Contributory assessment of creation of PM10 as impacted by vehicular traffic based on the presence of heavy metals* In: Communications : scientific letters of the University of Žilina. - ISSN 1335-4205. - Vol. 15, no. 3 p. 96-101, (2013).
4. Folkeson, et al., Spens: Sustainable Pavements for European New Member States, Guidelines for the environmental assessment of various pavement types including recommendations to road authorities in NMS, European commission DG research, Sixth framework programme, (2009).
5. Gehrig, R., Zeyer, K., Bukowiecki, N., Lienemann, P., Poulikakos, L. D., Furger, M., & Buchmann, B., *Mobile load simulators - A tool to distinguish between the emissions due to abrasion and resuspension of PM10 from road surfaces.* Atmospheric Environment, 44(38), 4937-4943, (2010).
6. Gustafsson, M., Blomqvist, G., Gudmundsson, A., Dahl, A., Jonsson, P., & Swietlicki, E., *Factors influencing PM10 emissions from road pavement wear.* Atmospheric Environment, 43(31), 4699-4702, (2009).
7. Pirjola, L., Kupiainen, K. J., Perhoniemi, P., Tervahattu, H., & Vesala, H. *Non-exhaust emission measurement system of the mobile laboratory SNIFFER.* Atmospheric Environment, 43(31), 4703-4713, (2009).
8. Tervahattu, H., Kupiainen, K. J., Räisänen, M., Mäkelä, T., & Hillamo, R., *Generation of urban road dust from anti-skid and asphalt concrete aggregates.* Journal of Hazardous Materials, 132(1 SPEC. ISS.), 39-46, (2006).
9. Wheel tracking machine DYNA-TRACK. Cernusco, Italy: Controls S.R.L., (2000).
10. Ličbinský, R., Frýbort, A., Huzlík, J., Adamec, V., Effenberger, K., Mikuška, P., Vojtěšek, M., Křůmal, K. *Usage of Scanning Electron Microscopy for Particulate Matter Sources Identification.* Transactions on transport sciences, Number 3/2010. Ministry of Transport, p. 137-144. ISSN 1802-9876. (2010).
11. Willis, Robert D., Blanchard, Fredrick T., Conner, Teri L. *Guidelines for the Application of SEM/EDX Analytical Techniques to Particulate Matter Samples.* EPA # 600/R-02/070. (2002).

MATEC Web of Conferences 117, 00049 (2017) DOI: 10.1051/matecconf/201711700049
*XXVI R-S-P Seminar 2017, Theoretical Foundation of Civil Engineering*

# Constitutive modelling of fibre reinforced nonhomogenous hyperelastic materials

*Stanisław* Jemioło[1], *Marcin* Gajewski[1,*]

[1]WUT, Civil Engineering Department, Armii Ludowej 16, 00-637 Warsaw, Poland

**Abstract.** In the paper two classes of poly-convex stored energy functions for transversally isotropic hyper-elastic materials have been proposed. Within these classes, a simplified model of hyper-elastic materials with an isotropic matrix reinforced with fibre family was derived and implemented in finite element software ABAQUS/Standard. The two numerical examples proving correctness of the implementation and showing its positive physical features are presented.

## 1 Introduction

Large elastic deformations are typical among others in mechanics of composite materials in which polymeric matrix is reinforced with different types of fibres and in biomechanics of soft tissues [1-3]. Therefore, in this work, the constitutive relations of these materials within the theory of hyper-elasticity are considered. In case of materials with isotropic matrix reinforced with one fibre family the considered relationships belong to the group of transversally isotropic materials. In the paper, the results of the theory of scalar anisotropic tensor functions representation are used [4]. The material heterogeneity result from the fibre distribution in considered region. Authors aim to indicate such constitutive relationships for which the stored/strain energy (SE) function is a poly-convex function [5,6]. The simplest possible examples of poly-convex SE functions and related constitutive relationships are given altogether with presentation of the basic equations needed for implementation in FEM software ABAQUS/Standard [7-9]. The issues presented here may be treated as a continuation and generalization of concepts presented by authors in papers [10-13] related to modelling of fibre reinforced materials.

## 2 Constitutive relationships for transversal isotropy

The SE function $W$ is considered as a function of deformation tensor $\mathbf{C}=\mathbf{F}^T\mathbf{F}\in Sym^+$ ($\mathbf{F}\in Lin^+$ is a deformation tensor gradient and $J=\det\mathbf{F}>0$, while „det" stands for the tensor determinant) with additional tensor parameter $\mathbf{M}=\mathbf{m}\otimes\mathbf{m}$. Unit vector $\mathbf{m}(\mathbf{X})$ is tangent to specific chosen fibre direction, and $\mathbf{X}$ is a vector determining the location of

* Corresponding author: m.gajewski@il.pw.edu.pl

material point in reference configuration. Analysed material in reference configuration is a transversally isotropic one [4].
When the SE function will be postulated in the following form

$$W(\mathbf{X},\mathbf{F}) = W(\mathbf{C},\mathbf{M}) = \hat{W}(I_i) = \hat{W}\left(\mathrm{tr}\,\mathbf{C}, \mathrm{tr}\,\mathrm{Cof}\,\mathbf{C}, \det\mathbf{C}, \mathrm{tr}(\mathbf{MC}), \mathrm{tr}(\mathbf{M}\mathrm{Cof}\mathbf{C})\right), \quad (1)$$

then from basic relationships of continuum mechanics and hyper-elasticity the following constitutive relationship in Lagrange description for the second Piola-Kirchhoff stress tensor may be determined:

$$\mathbf{T} = 2\left.\frac{\partial \hat{W}}{\partial \mathbf{C}}\right|_{\mathbf{C}=\mathbf{C}^T} = 2\beta_1\mathbf{I} + 2\beta_2\left[I_1\mathbf{I} - \mathrm{Cof}\,\mathbf{C}\right]\mathbf{C}^{-1} + 2\left(\beta_3 I_3 + \beta_5 I_5\right)\mathbf{C}^{-1} + \\ 2\beta_4\mathbf{M} - \beta_5 I_3\left(\mathbf{MC}^{-1} + \mathbf{C}^{-1}\mathbf{M}\right), \qquad \beta_i = \frac{\partial \hat{W}}{\partial I_i}. \quad (2)$$

In (2) the following results on derivatives were used:

$$\left.\frac{\partial I_1}{\partial \mathbf{C}}\right|_{\mathbf{C}=\mathbf{C}^T} = \mathbf{I}, \quad \left.\frac{\partial I_2}{\partial \mathbf{C}}\right|_{\mathbf{C}=\mathbf{C}^T} = \left[(\mathrm{tr}\,\mathrm{Cof}\,\mathbf{C})\mathbf{I} - \mathrm{Cof}\,\mathbf{C}\right]\mathbf{C}^{-1}, \quad \left.\frac{\partial I_3}{\partial \mathbf{C}}\right|_{\mathbf{C}=\mathbf{C}^T} = (\det\mathbf{C})\mathbf{C}^{-1}, \\ \left.\frac{\partial I_4}{\partial \mathbf{C}}\right|_{\mathbf{C}=\mathbf{C}^T} = \mathbf{M}, \quad \left.\frac{\partial I_5}{\partial \mathbf{C}}\right|_{\mathbf{C}=\mathbf{C}^T} = \mathrm{tr}(\mathbf{M}\mathrm{Cof}\mathbf{C})\mathbf{C}^{-1} - \frac{1}{2}(\det\mathbf{C})\left(\mathbf{MC}^{-1} + \mathbf{C}^{-1}\mathbf{M}\right). \quad (3)$$

In (1)-(3) symbols „tr" and „Cof" stands respectively for tensor trace and tensor cofactor (transposition of adjugate tensor) [6]. From natural state condition results that:

$$\hat{W}(3,3,1,1,1) = 0 \ \text{ and } \ \beta_i(3,3,1,1,1) = 0, \quad i = 1,\dots,5. \quad (4)$$

After substituting (2) in relationship $\boldsymbol{\sigma} = \frac{1}{J}\mathbf{FTF}^T$ the following constitutive realtionship of transversally isotropic material in Euler description is obtained,

$$\boldsymbol{\sigma} = 2\left(\frac{I_2}{J}\beta_2 + J\beta_3 + \frac{I_5}{J}\beta_5\right)\mathbf{I} + \frac{2\beta_1}{J}\mathbf{B} - 2\beta_2 J\mathbf{B}^{-1} + 2\beta_4\tilde{\mathbf{M}} - J\beta_5\left(\tilde{\mathbf{M}}\mathbf{B}^{-1} + \mathbf{B}^{-1}\tilde{\mathbf{M}}\right), \quad (5)$$

where

$$\mathbf{B} = \mathbf{FF}^T, \quad \tilde{\mathbf{M}} = \mathbf{FMF}^T. \quad (6)$$

Through introduction of the notation $\tilde{\mathbf{M}}$ it was emphasized that the fibres of the material, which were defined in the initial configuration in accordance with preferred specific direction, undergo elongation and rotation according to formula $(6)_2$. It is also worth noting that the following relationships:

$$I_1 = \mathrm{tr}\,\mathbf{C} = \mathrm{tr}\,\mathbf{B}, \quad I_2 = \mathrm{tr}\,\mathrm{Cof}\,\mathbf{C} = \mathrm{tr}\,\mathrm{Cof}\,\mathbf{B}, \quad J = \sqrt{I_3} = \sqrt{\det\mathbf{C}} = \sqrt{\det\mathbf{B}}, \\ I_4 = \mathrm{tr}(\mathbf{MC}) = \mathrm{tr}\,\tilde{\mathbf{M}}, \quad I_5 = \mathrm{tr}(\mathbf{M}\mathrm{Cof}\mathbf{C}) = \mathrm{tr}\,\mathrm{Cof}\,\tilde{\mathbf{M}}, \quad (7)$$

hold.

## 3 Poly-convex stored energy functions

In the literature on transversally isotropic hyper-elastic materials the constitutive relationships in the form resulting from application of the results presented in Spencer's monograph [1] are used. In that case, a slightly different than in (1) invariant basis of tensor C are applied. It is worth underliynig that invariants used in SE function (1) have clear geometrical interpretation (connected with deformation of material line, surfece and volume elements) and are very convinient for polyconvexity testing of SE function.

The SE function $W(\mathbf{X},\mathbf{F})$ is a poly-convex one if for every $\mathbf{X}$ the extended function

$$W(\mathbf{X},\mathbf{F}) = \mathcal{W}(\mathbf{X},\mathbf{F},\mathrm{Cof}\mathbf{F},\det\mathbf{F}) \tag{8}$$

is a convex one. The domain of the function $\mathcal{W}(\cdot,\mathbf{F},\mathrm{Cof}\mathbf{F},\det\mathbf{F})$ is a nineteenth dimension set $Lin \times Lin \times (0,\infty)$. The extension of the elasticity potential is necessary, because for the function $W(\cdot,\mathbf{F})$ tensor $\mathbf{F} \in Lin^+$ is an object from non-convex set $Lin^+$. The convex function $\mathcal{W}(\mathbf{F},\mathrm{Cof}\mathbf{F},\det\mathbf{F}) = W(\mathbf{C})$ have to satisfy also constraints resulting from the requirement for adequate growth of the function $W(\mathbf{C})$ for $\|\mathbf{C}\| \to 0^+$, $\|\mathrm{Cof}\mathbf{C}\| \to 0^+$ and $\det\mathbf{C} \to 0^+$ (and analogically in case when the above quantities are tending to infinity) [6]. Invariants $I_1$ and $I_4$ are convex functions with respect to $\mathbf{F} \in Lin$, while invariants $I_2$ and $I_5$ are convex functions with respect to $\mathrm{Cof}\,\mathbf{F} \in Lin$, what can be easily verified straight from the definition of convex functions. Therefore, it is convenient to use function of the form

$$\hat{W}(I_i) = W_1(I_1) + W_2(I_2) + W_3(I_3) + W_4(I_4) + W_5(I_5), \tag{9}$$

where functions $W_1(I_1)$ and $W_4(I_4)$ are convex functions with respect to tensor $\mathbf{F} \in Lin$, functions $W_2(I_2)$ and $W_5(I_5)$ are convex functions with respect to $\mathrm{Cof}\,\mathbf{F} \in Lin$, while function $W_3(I_3)$ is convex function with respect to $\det\mathbf{F} > 0$. It is worth noting that invariant $\mathrm{tr}\,\mathbf{C}^2\mathbf{M}$ and product of invariants $I_1 I_4 = \mathrm{tr}\,\mathbf{C}\,\mathrm{tr}\,\mathbf{C}\mathbf{M}$ are not convex with respect to $\mathbf{F} \in Lin$. Presented below SE functions are special cases of those proposed in [2]:

$$\hat{W}(I_i) = \sum_{i=1}^{N_1} a_i (I_1 - 3)^{\alpha_i} + \sum_{j=1}^{N_2} b_j (I_2 - 3)^{\beta_j} + \sum_{l=1}^{N_3} c_l (I_4 - 1)^{\gamma_l} + \sum_{m=1}^{N_4} d_m (I_5 - 1)^{\delta_m} + \Gamma(I_3), \tag{10}$$

$$\hat{W}(I_i) = \sum_{i=1}^{N_1} a_i \left(I_1^{\alpha_i} - 3^{\alpha_i}\right) + \sum_{j=1}^{N_2} b_j \left(I_2^{\beta_j} - 3^{\beta_j}\right) + \sum_{l=1}^{N_3} c_l \left(I_4^{\gamma_l} - 1\right) + \sum_{m=1}^{N_4} d_m \left(I_5^{\delta_m} - 1\right) + \Gamma(I_3), \tag{11}$$

are poly-convex when $a_i > 0$ and $\alpha_j \geq 1$, etc. The function $\Gamma(I_3)$ ($\Gamma(1) = 0$) should be assumed as convex function tending to infinity for $J$ tending to zero and infinity as well. A poly-convex SE functions other than (10) and (11) may be found in [14-19].

## 4 Simplified constitutive relationships

According to simplifying assumptions used in [10,12,13], the SE function of hyper-elastic material reinforced with several fibre families was postulated in the following form:

$$\Psi = \left(1 - \sum_{n=1}^{N} p_n\right)\Psi_M + \sum_{n=1}^{N} p_n \Psi_{Zn}, \tag{11}$$

where $\Psi_M(I_1, I_2, J)$ is SE function of matrix (isotropic function), $\Psi_{Zn}$ are SE functions of fiber families, while $p_n$ stands for the volume fraction of the fiber family in unit of material volume. Functions of SE accumulated in n-th fiber family that is working axially in direction determined by $\mathbf{m}_n(\mathbf{X})$ is approximated as:

$$\Psi_{Zn} = \frac{E_{Zn}}{4}(I_{4n} - 1)^2, \tag{12}$$

where

$$I_{4n} = \mathrm{tr}\tilde{\mathbf{M}}_n, \quad \tilde{\mathbf{M}}_n = \mathbf{F}\mathbf{M}_n\mathbf{F}^T, \tag{13}$$

and parametric tensors $\mathbf{M}_n = \mathbf{m}_n \otimes \mathbf{m}_n$ are defined in reference configuration. Parameter $E_{zn}$ may be interpreted as the initial Young's modulus of n-th fiber family.

Among others in works [10,12,13] the special case of SE function, which corresponds to the Ciarlet model for incompressible materials discussed in [6]:

$$\Psi_M = \frac{\mu_o}{2}\left[f(I_1 - 3) + (1 - f)(I_2 - 3)\right] + \frac{1}{4}\left[\lambda_o - 2\mu_o(1 - f)\right]J^2 + \\ -\left[\frac{1}{2}\lambda_o + \mu_o\right]\ln J - \frac{1}{4}\left[\lambda_o - 2\mu_o(1 - f)\right]. \tag{14}$$

was considered. For $f = 1$, from (14) the Ogden model is obtained [10]. Parameters: $\mu_o$ and $\lambda_o$ are Lame's constants (identical like for small deformation theory). Function (14) is poly-convex one and satisfy the proper function growth conditions if and only if: $\mu_o > 0$, $f \in (0,1)$ and $\lambda_o > 2\mu_o(1 - f)$, cf. [6]. The other examples of poly-convex functions $\Psi_M(I_1, I_2, J)$ are given in [3,5,20].

# 5 Numerical implementation in ABAQUS

## 5.1 UMAT procedure

Here we are proceeding analogically as shown in [10-13]. The model presented in section 4 was programmed as the UMAT procedure, which requires the formulation of constitutive relations (5) in incremental form. After counting Kirchhoff's tensor increment (as defined by the objective Jaumann derivative),

$$\overset{\circ}{\boldsymbol{\tau}} = \mathrm{c}.\mathbf{D} = \left\{\left(1 - \sum_{n=1}^{N} p_n\right)\mathrm{c}_M + \sum_{n=1}^{N} p_n E_{Z_n}\left[2\tilde{\mathbf{M}}_n \otimes \tilde{\mathbf{M}}_n + (I_4 - 1)\left(\tilde{\mathbf{M}}_n \Diamond \mathbf{I} + \mathbf{I} \Diamond \tilde{\mathbf{M}}_n\right)\right]\right\}.\mathbf{D}. \tag{15}$$

In case of function (14) the forth order tangent tensor operator $\mathrm{c}_M$ have the following form:

$$\mathrm{C}_M = \mu_0\left[\left(f + I_1(1-f)\right)\left(\mathbf{I}\Diamond\mathbf{B} + \mathbf{B}\Diamond\mathbf{I}\right) + 2\mathbf{B}\otimes\mathbf{B} - \left(\mathbf{I}\Diamond\mathbf{B}^2 + \mathbf{B}^2\Diamond\mathbf{I}\right) - 2\mathbf{B}\Diamond\mathbf{B}\right] + \\ + \left(\frac{\lambda_0}{2} - \mu_0(1-f)\right)J^2\mathbf{I}\otimes\mathbf{I}\,. \tag{16}$$

The UMAT material procedure was tested among others on boundary value problems with homogenous stress and strain fields and known analytical solutions. The main purpose of this type of tests was to demonstrate the correctness of the implementation and the equivalence of the relationship (5) (with potential (11)) and the incremental relationship (15) with the operator (16).

## 5.2 Numerical test – tension test of a plane strain quarter of circular disk

In this section the problem of hydrostatic tension of the plane strain circular disc (with radius a) is studied. The disc is reinforced with fibres symmetrically placed in relation to two planes. Due to the symmetry of the problem, the modelled region is quarter-circle shape. Heterogeneity of the disc is due to different orientation of the preferred specific fibre direction in material. There is seven variants of placement of reinforcement fibre families for which the problem is solved: i) fibers constantly inclined to $x_1$ axis with an angle $\pi/4$, $\mathbf{m}(\mathbf{x}) = 1/\sqrt{2}\,\mathbf{e}_1 + 1/\sqrt{2}\,\mathbf{e}_2$; ii) fibers constantly inclined to $x_1$ axis with an angle $3\pi/4$, $\mathbf{m}(\mathbf{x}) = 1/\sqrt{2}\,\mathbf{e}_1 - 1/\sqrt{2}\,\mathbf{e}_2$; iii) fibers placed in parallel to $x_1$ axis, $\mathbf{m}(\mathbf{x}) = \mathbf{e}_1$; iv) fibers placed in parallel to $x_2$ axis, $\mathbf{m}(\mathbf{x}) = \mathbf{e}_2$; v) fibers placed circumferentially, $\mathbf{m}(\mathbf{x}) = -\sin\varphi\mathbf{e}_1 + \cos\varphi\mathbf{e}_2$, where $\varphi = \arctan(x_2/x_1)$; vi) fibers placed radially, $\mathbf{m}(\mathbf{x}) = \cos\varphi\mathbf{e}_1 + \sin\varphi\mathbf{e}_2$, where $\varphi = \arctan(x_2/x_1)$; vii) without reinforcement.

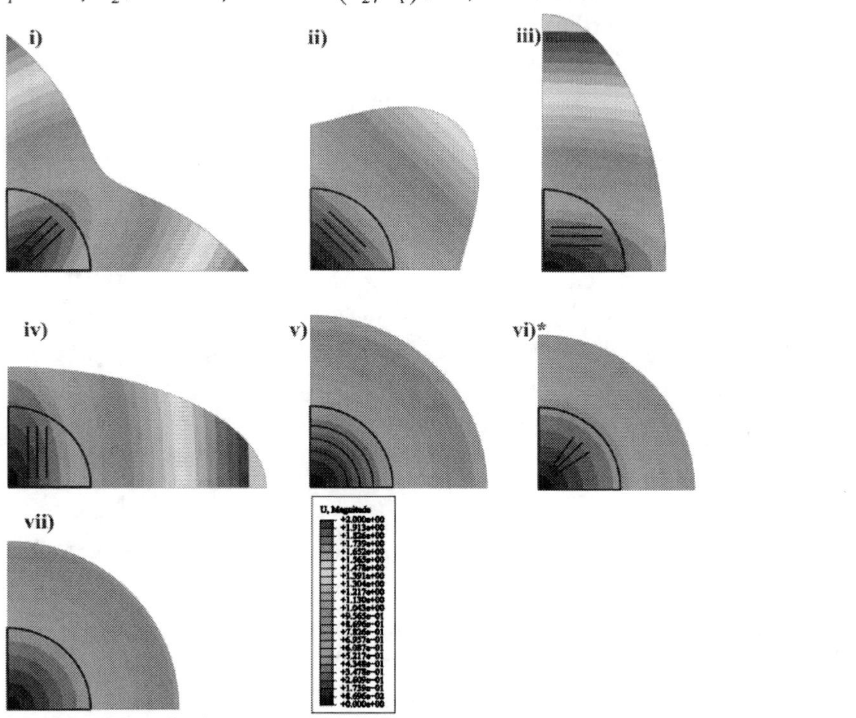


**Fig. 1.** Contour graphs of the displacement vector norm for different variants of fibres placement (*explanations are given in the text below, multiplayer for displacements is radius a).

The region of the quarter of circular disk was modelled with 9516 CPE4 elements (plane strain, four nodes with linear shape functions). The constitutive model for material is described with (15) and (16) where the following parameters are substituted: $p = 0.1$, $\mu_0$=1.0, $\lambda_0$=49.0 $\mu_0$, $E_Z$=26 $\mu_0$ and $f$ =0.9. The hydrostatic loading was assumed as multiple value of the initial shearing modulus equal to $q$ =100.0 $\mu_0$. On the basis of the analysis of the examples presented in Figure 1 it can be stated that only three qualitatively different solutions were obtained. The solutions for variants iii) and iv) are analogous: the circle has deformed into an ellipse (there is a homogeneous orthotropic material in the plane), except that the extreme deformation values in the first case are observed in the direction of the $x_2$ axis while in the second case in $x_1$ direction. With respect to variants v) and vi) the solutions are qualitatively identical and very similar to the solution obtained for non-reinforced material. In these three cases the values of stresses, strains and displacements in some particular locations are not the same. In the case of solution for a radially-reinforced disk (i.e. variant vi) the deformation shown in Figure 1 corresponds to a load value of $0.748\,q$. This is the last value at which the integration algorithm with the same steering paremeters for all variants was able to find a solution with a certain assumed accuracy. In tasks i) and ii) there is solution for the orthotropic circular non-homogeneous disk. In the paper [21], the boundary problem of hydrostatic tension test for sphere made of transversally isotropic hyper-elastic material was considered. For this type of material available in ABAQUS [9], non-physical results were obtained. The sphere deformed into a sphere. The reason for this is the application of too simple constitutive relationships for material components (matrix and fibres) in case of hyper-elasticity of fiber reinforced materials, in which additive decomposition of SE function is assumed.

### 5.3 Numerical test – tension test of a flat bar with heterogeneity

Next numerical example refers to the tension of the thin flat bar with reinforcement placed in plane $OX_1X_2$ according to sinus function, as a one fibre family. The flat bar is modelled as 3D problem, with thickness equal to g, length equal to l=50g and height h=12.5g, see Figure 2. In the analysed problem on one boundary surface (ABCD) all three degrees of freedom are blocked, i.e.: it is assumed that $u_i = 0$, $i = 1,2,3$. On the opposite boundary surface (A'B'C'D') it was assumed that $u_2 = u_3 = 0$. Tensioning of the bar is realized applying displacement $u_1$=25g on that boundary. On all other surfaces, zero stress boundary conditions are assumed. The flat bar region is modelled with 5000 C3D8 elements (eight-node elements with linear shape functions) 100x50x1 (length x height x thickness).The constitutive model of the material is the same like in previous example and the following material parameters are taken: $p = 0.1$, $\mu_0$=1.0, $\lambda_0$=49.0 $\mu_0$, $E_Z$=26 $\mu_0$ and $f$ =0.9. The fibers are placed in $OX_1X_2$ plane according to the function

$$\mathbf{m}(\mathbf{X}) = \cos\varphi\mathbf{e}_1 + \sin\varphi\mathbf{e}_2, \qquad \varphi = \arctan\left(\frac{\partial f}{\partial X_1}\right) \tag{17}$$

where

$$f\left(X_1, m, X_{2p}\right) = X_{2p} + \frac{h}{2}\sin\left(\frac{m\pi}{l}X_1\right). \tag{18}$$

In (18) symbol $X_2^p$ stands for the offset of the function in the direction of the $X_2$ variable. It is worth noting that parameter $X_2^p$ has no influence on components of parametric tensor

$\mathbf{M} = \mathbf{m} \otimes \mathbf{m}$. In initial configuration $X_1 = x_1$ and $X_2 = x_2$. The examplary results obtained for the analysed problem in the form of a contour graphs are presented in Figures 3-4.

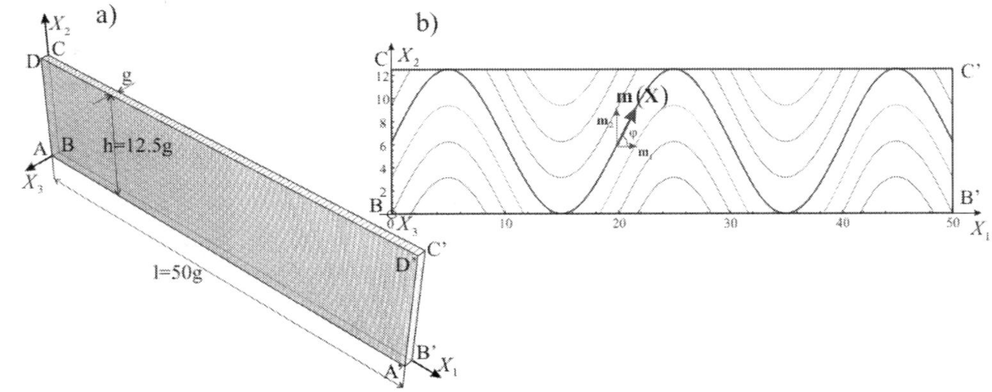


**Fig. 2.** a) Geometry, FEM mesh, dimensions and designation of characteristic points, b) the idea of the fibre placement in tensioned flat bar for n=5 (in case of blue line $X_2^p$=h/2).

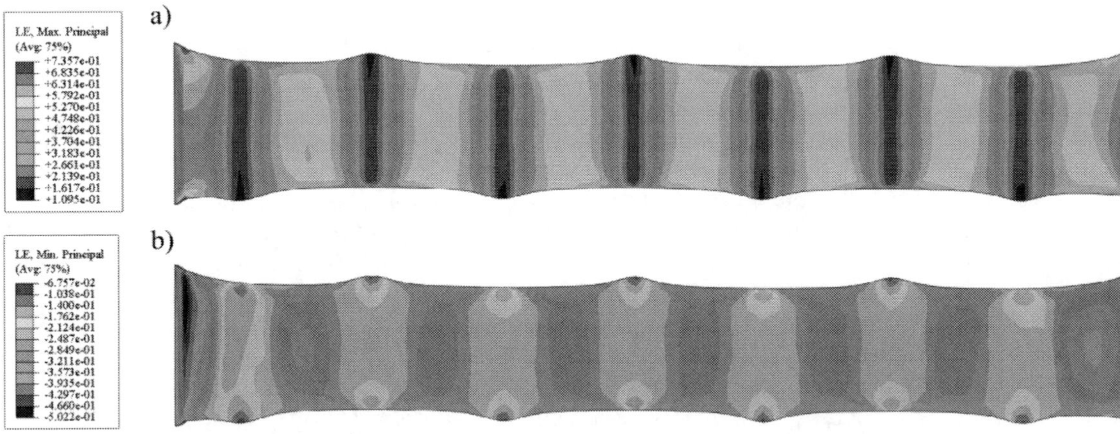


**Fig. 3.** Contour graphs of: a) maximum, b) minimum principal strain (interpreted as $\ln \mathbf{V}$ ).

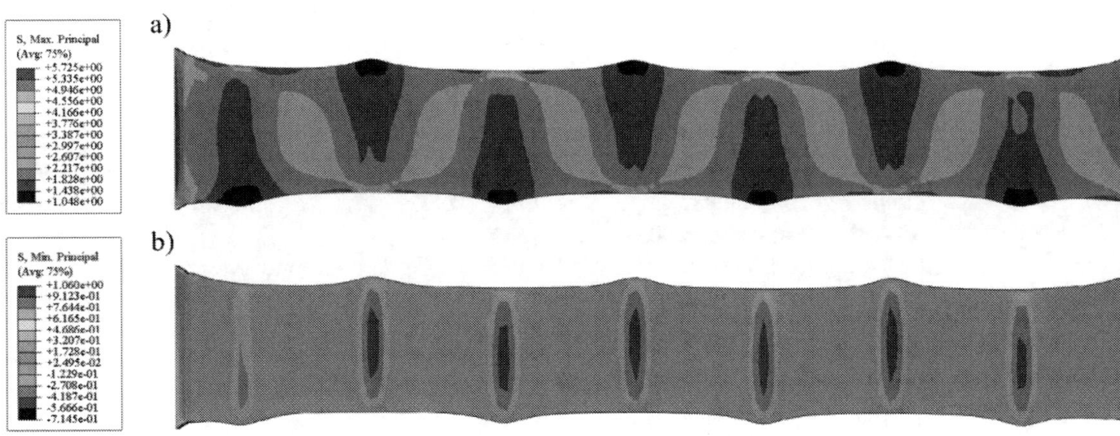


**Fig. 4.** Contour graphs of: a) maximum, b) minimum principal Cauchy's stress (in $\mu_o$ units).

## 6 Final remarks

Two classes of poly-convex SE functions for transversally isotropic materials have been proposed. Within these classes, a simplified model of hyper-elastic materials with isotropic matrix reinforced with fibre families was proposed. In particular, the proposed SE functions are simplifications of the SE functions applied in our earlier works [10,12,13]. These models do not lead to non-realistic results, as shown in [21], where the transversally isotropic material models available in the ABAQUS program [9] were used. The heterogeneity of the material result from the different arrangement of the fibres in the matrix material. The fibres orientation were defined by the parametric tensors, which are used as arguments in the constitutive relation. This facilitates the numerical implementation in the finite element method programs. For example, in the ABAQUS/Standard software, for programming the material model the UMAT user procedure was used.

## References

1. A.J.M. Spencer, *Deformations of fibre-reinforced materials* (Oxf. Univ. Press, 1972)
2. S. Jemioło, J.J. Telega, Eng. Trans. **49(2-3)**, 213 (2001)
3. S. Jemioło, *Constitutive relationships of hyperelasticity* [in Polish], (PAN, KILiW, Warszawa, 2016)
4. J.P. Boehler [ed], *Applications of tensor functions in solid mechanics* (Springer-Verlag, Wien-New York, 1987)
5. J.M. Ball, Arch. Rat. Mech. Anal. **66**, 337 (1977)
6. P.G. Ciarlet, *Mathematical elasticity* (North-Holland, Amsterdam-Tokyo, 1988)
7. ABAQUS v. 6.1, *Theory manual* (Hibbitt, Karlsson and Sorensen, Pawtucket, 2000)
8. ABAQUS/*Standard User's manual, Version 6.11* (Dassault Systèmes, 2011).
9. ABAQUS v. 6.12, *Theory Manual* (Dassault Systèmes, 2012)
10. M. Gajewski, S. Jemioło, *Proc. XVI Slovak-Polish-Russian Seminar Theoretical Foundation of Civil Engineering,* 119[in Polish] (PH MSUCE Moscow 2007)
11. S. Jemioło, M. Gajewski, *Theoretical Foundations of Civil Engineering, XV Polish-Ukrainian-Lithuanian Transactions*, 271 [in Polish] (OW PW Warsaw 2007)
12. S. Jemioło, M. Gajewski, *Theoretical Foundations of Civil Engineering, XVII Polish-Ukrainian-Lithuanian Transactions,* 133 [in Polish] (OW PW Warsaw 2009)
13. S. Jemioło, M. Gajewski, Logistyka **6**, 1203 [in Polish] (2010)
14. D. Balzani, P. Neff, J. Schröder, G.A. Holzapfel, Int. J. Sol. Struct. **43**, 6052 (2006)
15. A. E. Ehret, M. Itskov, PAMM Proc. Appl. Math. Mech. **6**, 367 (2006)
16. M. Itskov, N. Aksel, Int. J. Sol. Struct. **41**, 3833 (2004)
17. J. Schröder, P. Neff, Int. J. Sol. Struct. **40**, 401 (2003)
18. J. Schröder, P. Neff, *PAMM Proc. Appl. Math. Mech*. **2**, 172 (2003)
19. J. Schröder, P. Neff, D. Balzani, Int. J. Sol. Struct. **42**, 4352 (2005)
20. S. Jemioło, *Study of hyperelastic properties of isotropic materials. Modeling and numerical implementation. Scientific Works. Civil Engineering* **140** [in Polish] (OWPW, Warszawa, 2002)
21. L. Vergoni, M. Destrade, P. McGarry, R.W. Ogden, Comput Mech **52**, 1185 (2013)

MATEC Web of Conferences **117**, 00050 (2017) DOI: 10.1051/matecconf/201711700050
*XXVI R-S-P Seminar 2017, Theoretical Foundation of Civil Engineering*

# Spreadsheet solutions of the one-dimensional cutting stock problem

*Robert* Gajewski[1,*]

[1] Warsaw University of Technology, Faculty of Civil Engineering, Al. Armii Ludowej 16, Warsaw 00-637, Poland

**Abstract.** The cutting stock problem encompasses cutting parts available in stock, which are called objects, to produce in specified quantities smaller pieces which are called items and optimizing an objective function. The cutting stock problem is common for many industrial processes – steel, paper tube and construction industries. One of the most important phases in solution of the one-dimensional cutting stock problem is a pattern generating procedure which can be performed in a spreadsheet.

## 1 Introduction and literature review

The cutting stock problem encompasses cutting parts available in stock, which are called objects, to produce in specified quantities smaller pieces which are called items and optimizing an objective function. Objective functions comprise minimizing the total waste, minimizing the cost of cutting the objects, minimizing the total number of objects cut, maximizing profit and minimizing production costs. Sample solution of this problem is called cutting plan. Such plan is provided by a set of cutting patterns and corresponding frequencies which describe how many times each pattern should be used to produce the items, see Fig. 1. Each cutting pattern defines a subset of items which will be cut from an object. More information about this can be found in surveys [1] and [2].

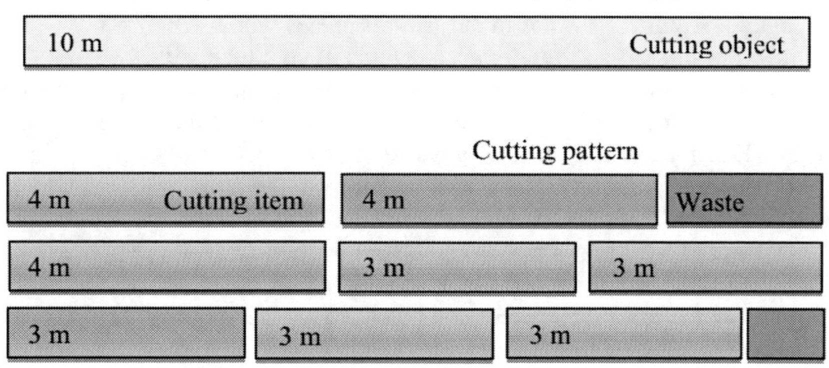


**Fig. 1.** Cutting plan: cutting object, items, patterns and wastes.

* Corresponding author: r.gajewski@il.pw.edu.pl

The cutting stock problem is common for many industrial processes. Armbruster described in [3] a solution procedure for a pattern sequencing problem as part of a one-dimensional cutting stock problem in the steel industry. In [4] Belov presented a cutting plane algorithm for the one-dimensional cutting stock problem with multiple stock lengths. Cherri [5] introduced a heuristic approach to the one-dimensional cutting stock problem with usable leftover. Gau [6] created a problem generator for the standard one-dimensional cutting stock problem. Matsumoto described in [7] the one-dimensional stock cutting problem in the paper tube industry. Ogunranti discussed in [8] how to minimize waste (off-cuts) using cutting stock model for the case of one dimensional cutting stock problem in wood working industry. Poldi presented in [9] heuristics for the one-dimensional cutting stock problem with limited multiple stock lengths. In [10] Shahin used genetic algorithms to solve the one-dimensional cutting stock problem in the construction industry. In [11] Stadtler described the one-dimensional cutting stock optimization with usable leftover - a case of low stock-to-order ratio. In thesis [12] Tomat discussed one-dimensional cutting stock optimization with usable leftover - a case of low stock-to-order ratio. In [13] Umetani solved one-dimensional cutting stock problem to minimize the number of different patterns. In [14] Vance used branch-and-price algorithms for the one-dimensional cutting stock problem.

One of the most important phases in solution of the one-dimensional cutting stock problem is a pattern generating procedure presented by Suliman in [15]. Similar approach was presented by Goulimis [16]. This work was based on enumerating the possible cutting patterns, and solving the associated integer program by a combination of cutting planes and branch and bound.

## 2 Search tree

In order to generate feasible cutting patterns a search tree can be used [15]. This procedure is described by Suliman as follows: "The search tree is constructed by traversing first from bottom root to top and then from left to right. The construction starts at the root (first level of the tree), and continues to move upward in the tree by adding additional sizes to the combinations already specified by the previous branches. While traversing along a specific path on the search tree, the feasibility of that path is maintained by ensuring a non-negative end terminal node which implies adequate material to satisfy the requirements of the path. Each node at any level has as many branches as the maximum number of units that can be cut from the remaining width."

All calculations for a simple case can be performed by hand. Let us consider element which length is 9 units. We want to cut from it pieces which lengths are 5, 4, 3 and 2 units respectively. For this case search tree can be constructed manually.

- How many 5 unit's pieces can be cut? Only one. The remaining part is 4 units.
- How many 4 unit's pieces can be cut from the remaining part which has 4 units? Exactly one and remaining part will be 0.
- How many 3 unit's pieces can be cut from the remaining part which has 4 units? Also one and remaining part will be 1.
- How many 2 unit's pieces can be cut from the remaining part which has 4 units? Exactly two and remaining part will be 0.
- How many 4 unit's pieces can be cut? Two. Remaining part is 1 unit.
- When we start from cutting one 4 unit's piece we have remaining part which has 5 units.

We can continue this and all results can be summarized in a table (see Fig. 2).

| | | | | | | | | | | |
|---|---|---|---|---|---|---|---|---|---|---|
| 5 | 1 | 1 | 1 | 0 | 0 | 0 | 0 | 0 | 0 | 0 |
| 4 | 1 | 0 | 0 | 2 | 1 | 1 | 0 | 0 | 0 | 0 |
| 3 | 0 | 1 | 0 | 0 | 1 | 0 | 3 | 2 | 1 | 0 |
| 2 | 0 | 0 | 2 | 0 | 1 | 2 | 0 | 1 | 3 | 4 |
| | 0 | 1 | 0 | 1 | 0 | 1 | 0 | 1 | 0 | 1 |

**Fig. 2.** Cutting object, items, pattern, plan and wastes.

This solution can also be presented as a tree (see Fig. 3).

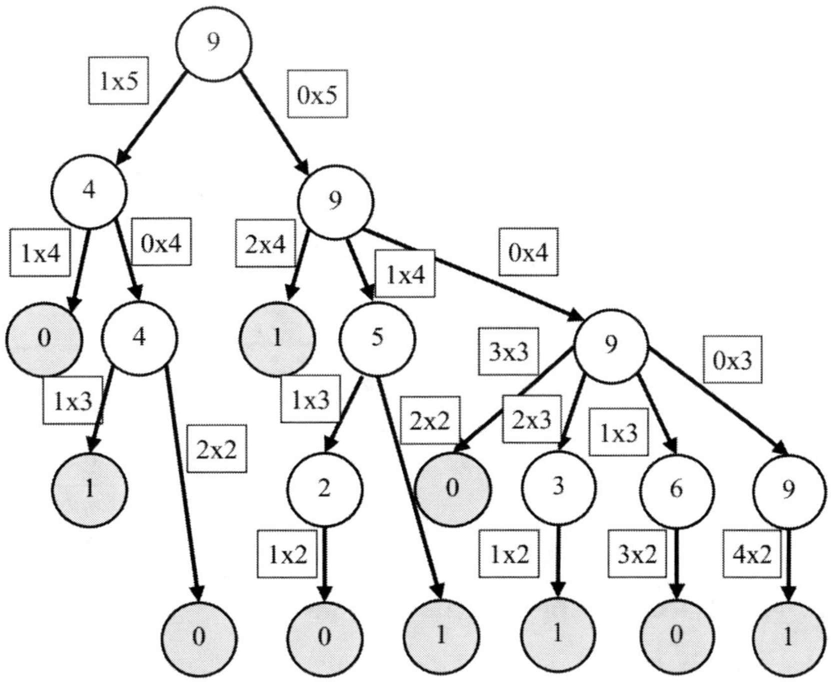


**Fig. 3.** Sample search **tree**.

It is hardly possible to perform all calculations even for slightly more complicated example. When we want to cut pieces which lengths are 5, 4, 3 and 2 units respectively from element which length is 18 units number of different cutting patterns is 40 and algorithmic approach is necessary.

## 3 Pattern generation algorithm

All details of the pattern generation algorithm can be found in [15]. Let us denote by $w$ the length of the element and by $w_1$, $w_2$,...$w_n$ lengths of pieces which should be cut from it. We assume that $w_1>w_2>\ldots>w_n$. For each problem matrix **A** with number of elements $n$x$m$ should be calculated, where $m$ denotes number of different cutting patterns. The first element is given by:

$$a_{11} = \operatorname{div}\left(\frac{w}{w_1}\right) \tag{1}$$

The second element in the first column is:

$$a_{21} = \operatorname{div}\left(\frac{w - a_{11} \cdot w_1}{w_2}\right) \tag{2}$$

The $i$-th element in the first column is:

$$a_{i1} = \operatorname{div} \frac{w - \sum_{k=1}^{i-1} a_{k1} \cdot w_k}{w_i} \tag{3}$$

Each element in the matrix is given by:

$$a_{ij} = \operatorname{div} \frac{w - \sum_{k=1}^{i-1} a_{kj} \cdot w_k}{w_i} \tag{4}$$

We can check these formulas for the given example and compare them with Fig. 2.

$$\begin{gathered}
a_{11} = \operatorname{div}\left(\frac{\mathrm{w}}{\mathrm{w}_1}\right) = \operatorname{div}\left(\frac{9}{5}\right) = 1 \\
a_{21} = \operatorname{div}\left(\frac{w - a_{11} \cdot w_1}{w_2}\right) = \operatorname{div}\left(\frac{9 - 5 \cdot 1}{4}\right) = 1 \\
a_{31} = \operatorname{div} \frac{w - \sum_{k=1}^{2} a_{k1} \cdot w_k}{w_i} = \operatorname{div} \frac{9 - 5 \cdot 1 - 4 \cdot 1}{3} = 0 \\
a_{41} = \operatorname{div} \frac{w - \sum_{k=1}^{3} a_{k1} \cdot w_k}{w_i} = \operatorname{div} \frac{9 - 5 \cdot 1 - 4 \cdot 1 - 3 \cdot 0}{2} = 0
\end{gathered} \tag{5}$$

More information about algorithm and its flowchart can be found in [15].

## 4 From algorithm to program

In [15] a special computer routine for generating cutting patterns was develop which requires knowledge of programming language. This problem can be solved directly inside spreadsheet without programming in Visual Basic for Application. Such approach is not known in literature. It enables fast and robust solution of the whole problem in a spreadsheet. Spreadsheet formulas were tested on many examples. Solution of one of them is presented in Fig. 4. Length of element is 18 units and it is cut into pieces which length is 5, 4, 3 and 2 units respectively. In this case, there are 40 different cutting patterns. This table can also be calculated by hand but this process is time consuming and can be a source of mistakes.

| | | | | | | | | | | | | | | | | | | | | |
|---|---|---|---|---|---|---|---|---|---|---|---|---|---|---|---|---|---|---|---|---|
| 5 | 3 | 3 | 2 | 2 | 2 | 2 | 2 | 2 | 1 | 1 | 1 | 1 | 1 | 1 | 1 | 1 | 1 | 1 | 1 | 1 |
| 4 | 0 | 0 | 2 | 1 | 1 | 0 | 0 | 0 | 3 | 2 | 2 | 1 | 1 | 1 | 1 | 0 | 0 | 0 | 0 | 0 |
| 3 | 1 | 0 | 0 | 1 | 0 | 2 | 1 | 0 | 0 | 1 | 0 | 3 | 2 | 1 | 0 | 4 | 3 | 2 | 1 | 0 |
| 2 | 0 | 1 | 0 | 0 | 2 | 1 | 2 | 4 | 0 | 1 | 2 | 0 | 1 | 3 | 4 | 0 | 2 | 3 | 5 | 6 |

| | | | | | | | | | | | | | | | | | | | |
|---|---|---|---|---|---|---|---|---|---|---|---|---|---|---|---|---|---|---|---|
| 0 | 0 | 0 | 0 | 0 | 0 | 0 | 0 | 0 | 0 | 0 | 0 | 0 | 0 | 0 | 0 | 0 | 0 | 0 | 0 |
| 4 | 3 | 3 | 3 | 2 | 2 | 2 | 2 | 1 | 1 | 1 | 1 | 1 | 0 | 0 | 0 | 0 | 0 | 0 | 0 |
| 0 | 2 | 1 | 0 | 3 | 2 | 1 | 0 | 4 | 3 | 2 | 1 | 0 | 6 | 5 | 4 | 3 | 2 | 1 | 0 |
| 1 | 0 | 1 | 3 | 0 | 2 | 3 | 5 | 1 | 2 | 4 | 5 | 7 | 0 | 1 | 3 | 4 | 6 | 7 | 9 |

**Fig. 4.** Cutting plan.

## 5 Solver solution

In the next step Linear Programming problem is formulated and solved by spreadsheet solver. Two models can be built for such problem. We will denote by $x_i$ number of elements cut according to $i$-th pattern.

In the first model, we minimize only wastes – we do not take into account that number of cut pieces can be greater than required so in all constraints we will have condition ≥. In the second model, we will use additional decision variables $y_j$ denoting number of elements having length $w_j$ cut above the requirements. In this case, we will subtract value of $y_j$ in each constraint and use equality condition. Moreover, in the goal function we will add the total length of these additional elements.

Example which is solved has element which length is 13 units cut into pieces which have length 5, 4, 3 and 2 units. Data from cutting plan can be directly used to build spreadsheet model of the problem.

| | _1 | _2 | _3 | _4 | _5 | _6 | _7 | _8 | _9 | 10 | 11 | 12 | 13 | 14 | 15 | 16 | 17 | 18 | 19 | 20 | Su | No |
|---|---|---|---|---|---|---|---|---|---|---|---|---|---|---|---|---|---|---|---|---|---|---|
| 5 | 2 | 2 | 1 | 1 | 1 | 1 | 1 | 1 | 0 | 0 | 0 | 0 | 0 | 0 | 0 | 0 | 0 | 0 | 0 | 0 | 12 | 12 |
| 4 | 0 | 0 | 2 | 1 | 1 | 0 | 0 | 0 | 3 | 2 | 2 | 1 | 1 | 1 | 1 | 0 | 0 | 0 | 0 | 0 | 19 | 13 |
| 3 | 1 | 0 | 0 | 1 | 0 | 2 | 1 | 0 | 0 | 1 | 0 | 3 | 2 | 1 | 0 | 4 | 3 | 2 | 1 | 0 | 14 | 14 |
| 2 | 0 | 1 | 0 | 0 | 2 | 1 | 2 | 4 | 0 | 1 | 2 | 0 | 1 | 3 | 4 | 0 | 2 | 3 | 5 | 6 | 15 | 15 |
| W | 0 | 1 | 0 | 1 | 0 | 0 | 1 | 0 | 1 | 0 | 1 | 0 | 1 | 0 | 1 | 1 | 0 | 1 | 0 | 1 | 0 | |
| R | 2 | 0 | 8 | 0 | 0 | 0 | 0 | 0 | 0 | 0 | 0 | 3 | 0 | 0 | 0 | 0 | 0 | 0 | 3 | 0 | 24 | |

**Fig. 5.** Solution of the first model.

| | _1 | _2 | _3 | _4 | _5 | _6 | _7 | _8 | _9 | 10 | 11 | 12 | 13 | 14 | 15 | 16 | 17 | 18 | 19 | 20 | y1 | y2 | y3 | y4 | Su | No |
|---|---|---|---|---|---|---|---|---|---|---|---|---|---|---|---|---|---|---|---|---|---|---|---|---|---|---|
| 5 | 2 | 2 | 1 | 1 | 1 | 1 | 1 | 1 | 0 | 0 | 0 | 0 | 0 | 0 | 0 | 0 | 0 | 0 | 0 | 0 | -1 | 0 | 0 | 0 | 12 | 12 |
| 4 | 0 | 0 | 2 | 1 | 1 | 0 | 0 | 0 | 3 | 2 | 2 | 1 | 1 | 1 | 1 | 0 | 0 | 0 | 0 | 0 | 0 | -1 | 0 | 0 | 13 | 13 |
| 3 | 1 | 0 | 0 | 1 | 0 | 2 | 1 | 0 | 0 | 1 | 0 | 3 | 2 | 1 | 0 | 4 | 3 | 2 | 1 | 0 | 0 | 0 | -1 | 0 | 14 | 14 |
| 2 | 0 | 1 | 0 | 0 | 2 | 1 | 2 | 4 | 0 | 1 | 2 | 0 | 1 | 3 | 4 | 0 | 2 | 3 | 5 | 6 | 0 | 0 | 0 | -1 | 15 | 15 |
| W | 0 | 1 | 0 | 1 | 0 | 0 | 1 | 0 | 1 | 0 | 1 | 0 | 1 | 0 | 1 | 1 | 0 | 1 | 0 | 1 | 5 | 4 | 3 | 2 | 11 | |
| R | 4 | 1 | 1 | 1 | 1 | 0 | 0 | 0 | 0 | 3 | 0 | 1 | 0 | 0 | 2 | 0 | 1 | 0 | 0 | 0 | 1 | 0 | 0 | 1 | | |

**Fig. 6.** Solution of the second model.

Results for the first and second model are different. In the first model value of goal function is 11 but total length of not necessary elements is 24. In the second model toal number of unnecessary elements is 11.

## 6 Final conclusions

This paper presents spreadsheet implementation of pattern generation algorithm not known in the literature of the subject. This approach does not require knowledge of programming languages. All operations on matrix elements are performed directly in a spreadsheet.

Excel solver has well known limitations. Number of decision variables cannot exceed 200 for both linear and nonlinear problems. Moreover, using basic Excel Solver one can place constraints on up to 100 cells which are not decision variables. In such a case one can buy Frontline's Premium Solver or use Open Solver, the Open Source linear, integer and nonlinear optimizer for Microsoft Excel.

This research was not sponsored by any grant or authorities. This research was done just for pleasure and satisfaction.

## References

1. A. C. Cherri, M. N. Arenales, H. H. Yanasse, K. C. Poldi, and A. C. Gonçalves Vianna, Eur. J. Oper. Res. **236**, 395 (2014)
2. C. H. Cheng, B. R. Feiring, and T. C. E. Cheng, Int. J. Prod. Econ. **36**, 291 (1994)
3. M. Armbruster, Eur. J. Oper. Res. **141**, 328 (2002)
4. G. Belov and G. Scheithauer, Eur. J. Oper. Res. **141**, 274 (2002)
5. A. C. Cherri, M. N. Arenales, and H. H. Yanasse, Eur. J. Oper. Res. **196**, 897 (2009)
6. T. Gau and G. Wäscher, Eur. J. Oper. Res. **84**, 572 (1995)
7. K. Matsumoto, S. Umetani, and H. Nagamochi, J. Sched. **14**, 281 (2011)
8. G. A. Ogunranti and A. E. Oluleye, J. Ind. Eng. Manag. **9**, 834 (2016)
9. K. C. Poldi and M. N. Arenales, Comput. Oper. Res. **36**, 2074 (2009)
10. A. A. Shahin and O. M. Salem, Can. J. Civ. Eng. **31**, 321 (2004)
11. H. Stadtler, Eur. J. Oper. Res. **44**, 209 (1990)
12. L. Tomat, *One-Dimensional Cutting Stock Optimization with Usable Leftover: A Case of Low Stock-to-Order Ratio* (Doctoral Dissertaton, University of Lublijana, Faculty of Economics, 2013)
13. S. Umetani, M. Yagiura, and T. Ibaraki, Eur. J. Oper. Res. **146**, 388 (2003)
14. P. H. Vance, Comput. Optim. Appl. **9**, 211 (1998)
15. S. M. A. Suliman, Int. J. Prod. Econ. **74**, 293 (2001)
16. C. Goulimis, Eur. J. Oper. Res. **44**, 197 (1990)

MATEC Web of Conferences 117, 00051 (2017) DOI: 10.1051/matecconf/201711700051
*XXVI R-S-P Seminar 2017, Theoretical Foundation of Civil Engineering*

# Building energy modelling and simulations: qualitative and quantitative analysis

*Robert* Gajewski[1,*], *Paweł* Pieniążek[1]

[1] Warsaw University of Technology, Faculty of Civil Engineering, Al. Armii Ludowej 16, Warsaw 00-637, Poland

**Abstract.** Buildings consume half of all energy use and are also responsible for a similar proportion of carbon dioxide emission. The heat transfer across the building envelope - the shell of a house that separates the inside and outside, should generally be minimized. In the paper validation and verification based on Building Energy Simulation Test (BESTEST) of Energy3D computer code is presented. Program proved to be an excellent tool for qualitative and quantitative analysis of buildings.

## 1 Introduction and literature review

Buildings consume between 40% and 60% of all energy use. They are also responsible for a similar proportion of humankind's carbon dioxide emissions. In the first part of twentieth century, calculations of the thermal response of buildings have made an assumption that the boundary conditions and the whole problem are static. Thus a fictitious steady-state problem was defined in order to perform energy calculations. Now building energy modelling is mainly based on CFD (Computational Fluid Dynamics) methods. An excellent review of different modelling methods for energy in building is given by Underwood and Yik in their book [1]. Energy simulation in building design is presented by Clarke in [2]. Building performance simulation for design and operation are described in the monograph [3] edited by Hensen and Lambers. Review of different building simulation applications can be found in [4] and [5]. Practical approach to computational fluid dynamics, which is suitable for civil engineers, can be found in [6]. A practical guide of finite element methods for computational fluid dynamics can be found in [7].

All computer codes used for building energy simulations especially these based on computational fluid dynamics are very complicated. Their proper usage requires deep theoretical and practical knowledge. There are two computer codes – Energy2D and Energy3D created by Doctor Cherles Xie which are very simple and user friendly.

## 2 Building energy – theoretical background

The thermal performance of a building is represented by the efficiency with which it uses energy to maintain the thermal comfort for its occupants. In order to calculate this energy

* Corresponding author: r.gajewski@il.pw.edu.pl

all heat transfer methods are taken into account: thermal conduction, convection and radiation. In thermal conduction the rate of heat flow is governed by the following relationship:

$$\frac{\Delta Q}{\Delta t} = -\lambda A \frac{\Delta T}{\Delta x} \tag{1}$$

where $\Delta Q$ is the thermal energy change of the object within time period of $\Delta t$, $\Delta T$ is the temperature difference across a distance $\Delta x$, $\lambda$ is the thermal conductivity of the material and $A$ is the area. This relation is also known as Fourier's law of thermal conduction in which we use the concept of heat flux $\Phi$.

$$\Phi = -\lambda \frac{\Delta T}{\Delta x} \tag{2}$$

In the process of convection thermal energy passes through a flow of fluid. Convective heat flux is estimated by simple formula

$$\Phi = h(T - T_{\infty}) \tag{3}$$

where $h$ is the convective heat transfer coefficient of a fluid, $T$ is the temperature of the surface and $T_{\infty}$ is the temperature of the fluid. This equation is known as Newton's law of cooling. In the radiation process the energy radiated by an object per unit time is proportional to the fourth power of the absolute temperature

$$\Phi = \varepsilon \sigma T^4 \tag{4}$$

where $\sigma$ is the Stephan-Bolzman constant, $\varepsilon$ is the emissivity of the object's material. This formula is known as Stephan-Bolzman law.

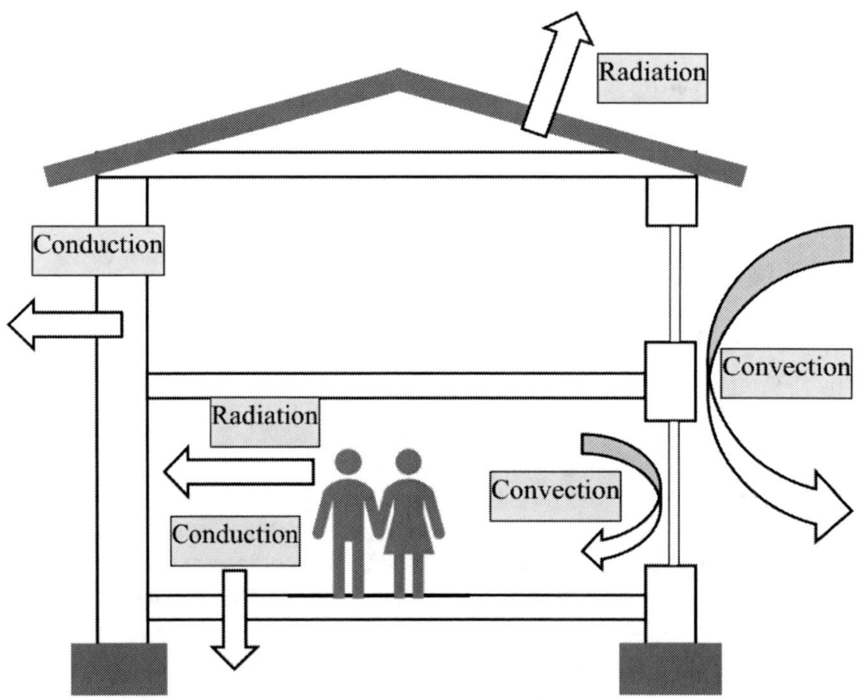


**Fig. 1.** Heat transfer across the building envelope.

The heat transfer across the building envelope - the shell of a house that separates the inside and outside, should generally be minimized (see Fig. 1). The measure of insulation is named $R$-value. It represents the ability to resist heat transfer through the building envelope under given temperature difference. $U$-factor on the other hand represents material's ability to transfer heat.

$$\Phi = \frac{T_2 - T_1}{R} = U(T_2 - T_1) \tag{5}$$

The thermal energy $p_i$ gained or lost per unit time through a surface area $A_i$ of a building component $i$ at time $t$ is:

$$p_i(t) = A_i U_i [T_{in} - T_{out}(t)] \tag{6}$$

The total heat flow across the building envelope is the sum of heat flow through all of its components:

$$p(t) = \sum_i A_i U_i [T_{in} - T_{out}(t)] \tag{7}$$

The total quantity of heat that flows across the entire building envelope in a year is the sum of heat flow through all of its components over 365 days:

$$Q = \sum_{d=1}^{365} \sum_{k=1}^{N} \left| \sum_i A_i U_i [T_{in} - T_{out}(t_d + k\Delta)] \Delta, \quad \Delta = \frac{24 \cdot 60 \cdot 60}{N} \right| \tag{8}$$

where $N$ is the number of time intervals in which a day is divided for this calculation.

## 3 Verification and validation

In addition to all errors that can arise while performing numerical simulation there are also uncertainties due to improper modelling of physics or incorrect computational design. Verification and validation procedures are used in order to properly assess solution. Both terms have distinctive definitions. As stated in [6] "verification can be defined as a process for assessing the numerical simulation uncertainty and when conditions permit, estimating the sign and magnitude of the numerical simulation error and the uncertainty in that estimated error. This procedure concerns primarily the input parameters used for geometry, initial conditions, and boundary conditions." On the other hand as stated in [6] "validation can be defined as a process for assessing simulation model uncertainty by using benchmark experimental data and when, conditions permit, estimating the sign and magnitude of the simulation modelling error itself." This procedure simply means validating the calculations by establishing a range of physical conditions obtained from the calculations and by performing comparisons of the results from the CFD code with experiments that span the range of conditions.

Both processes were performed for Energy3D code in order to check the quality of models and results of simulations. From a validation perspective, comparative tests will show that Energy3D gives solutions that are reasonable compared to other energy simulation programs. Comparative testing is also useful for input debugging. Energy simulation programs have so many inputs and outputs that the results are often difficult to interpret. In order to compare results with known from literature Building Energy Simulation Test (BESTEST) was used. It was described in many publications [8], [9], [10], [11] [12] and [13]. Sample BESTEST building is shown in Fig. 2. The tests described in ANSI/ASHRAE Standard 140-2001, Standard Method of Test for the Evaluation of Building Energy Analysis Computer Programs (ANSI/ASHRAE 2001) were performed.

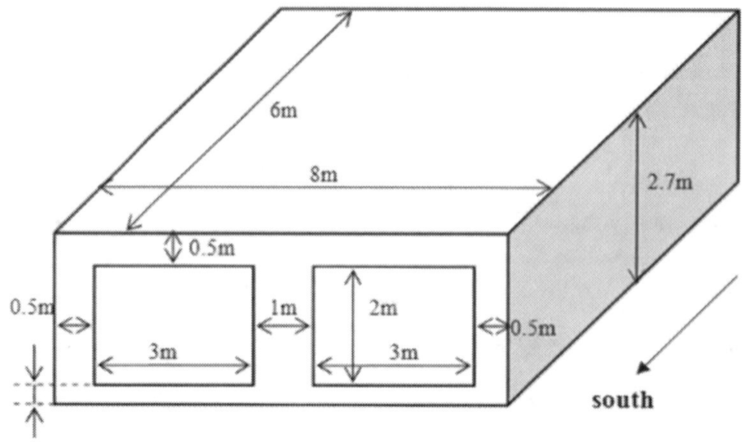


**Fig. 2.** BESTEST case building 600.

The basic test building is a rectangular single zone (8 m wide x 6 m long x 2.7 m high) with no interior partitions and 12 $m^2$ of windows on the south exposure. The building is of lightweight construction with characteristics as described below in Table 1.

**Table 1.** Wall Construction (light weight mass).

| Element | k [W/mK] | Thickness [m] | U [W/m²K] | R [m²K/W] | Density [kg/m3] | $c_p$ [J/kgK] |
|---|---|---|---|---|---|---|
| Int. Surface Coeff. | | | 8.290 | 0.121 | | |
| Plasterboard | 0.160 | 0.012 | 13.333 | 0.075 | 950 | 840 |
| Fiberglass Quilt | 0.040 | 0.066 | 0.606 | 1.650 | 12 | 840 |
| Wood Siding | 0.140 | 0.009 | 15.556 | 0.064 | 530 | 900 |
| Ext. Surface Coeff. | | | 29.300 | 0.034 | | |
| Overall, air-to-air | | | 0.514 | 1.944 | | |

Charts comparing Energy3D results with other whole building energy simulation programs are in Fig. 3. More results can be found in B.Sc. thesis prepared by Pieniążek [14].

## 4 Qualitative and quantitative analysis

Energy3D can be used for both qualitative and quantitative analysis. One can measure and compare such values as:

- Effects of the house size on the energy use;
- Effects of the house shape on the energy use;
- Effects of roof insulation on the energy use of a house;
- Effects of roof colour on the energy use of a house;
- Effects of solar heat gain coefficients of windows on the energy use of a house;
- Effects of orientation on the energy use of a house;

- Effects of the thermostat setting on the energy use of a house;
- Energy use of a house at different locations;
- Effects of environment albedo on the energy use of a house;
- Effects of sun-facing windows on the energy use of a house;
- Effects of trees on the air conditioning of a house;

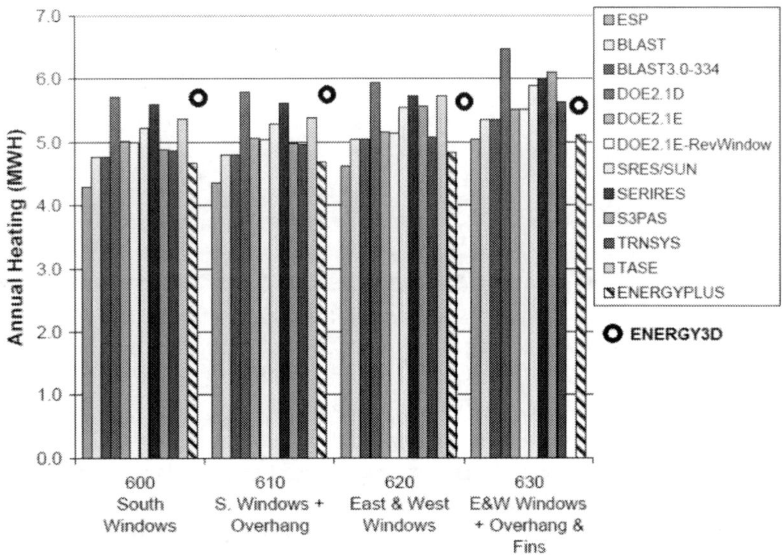


**Fig. 3.** Comparison of Energy3D results with other whole building energy simulation programs.

Results presenting energy use of a house at different locations are in Fig. 4. Differences in climate in Poland are relatively small but annual energy for Białystok (solid line without markers) and Wrocław differ significantly (qualitative analysis).

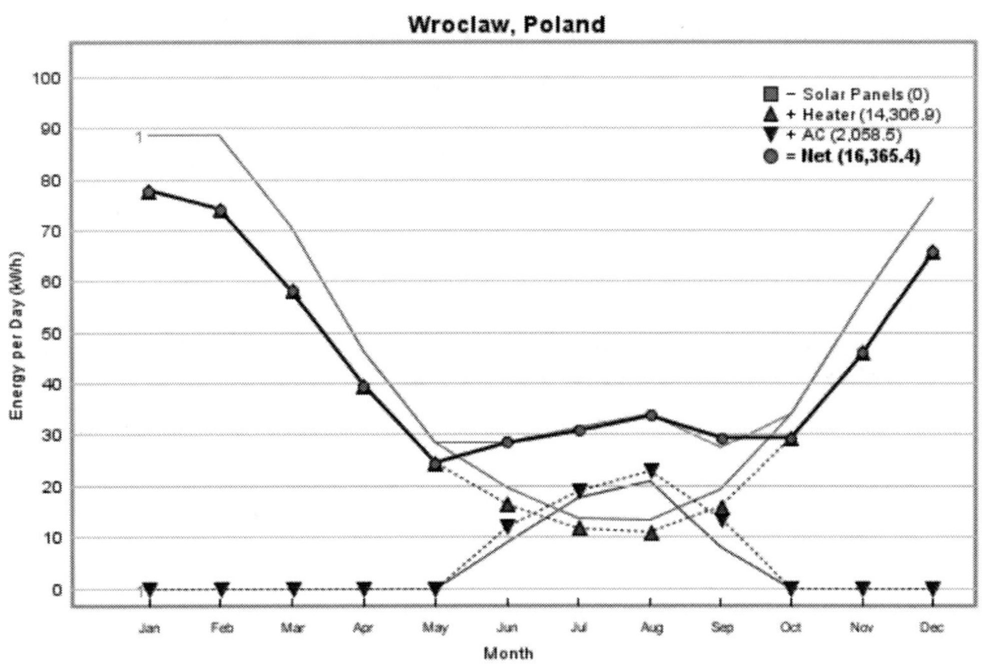


**Fig. 4.** Annual energy for sample building located in Białystok and Wrocław.

## 6 Final conclusions

This paper presents validation and verification of Energy3D code used for building energy modelling and simulations. This simple computer code proved to be an excellent tool for qualitative and quantitative analysis of buildings. Such a program can be an excellent part of a computer supported design environment [2] which takes into account also energy considerations.

This research was not sponsored by any grant or authorities. This research was done just for pleasure and satisfaction.

## References

1. C. Underwood, F. Yik, *Modelling Methods for Energy in Buildings*, 1 edition (Wiley-Blackwell, Oxford Malden, MA, 2004)
2. J. Clarke, *Energy Simulation in Building Design*, 2 edition (Routledge, Oxford, 2001).
3. J. L. M. Hensen and R. Lamberts, editors, *Building Performance Simulation for Design and Operation* (Routledge, London; New York, 2011)
4. M. Baratieri, V. Corrado, A. Gasparella, and F. Patuzzi, editors, *Building Simulation Applications BSA 2013* (Bozen-Bolzano University Press, Bozen-Bolzano, 2013)
5. M. Baratieri, V. Corrado, A. Gasparella, and F. Patuzzi, editors, *Building Simulation Applications BSA 2015* (Bozen-Bolzano University Press, Bozen-Bolzano, 2015)
6. J. Tu, G. Heng, C. Liu, *Computational Fluid Dynamics, Second Edition: A Practical Approach*, 2 edition (Butterworth-Heinemann, Amsterdam, 2012)
7. D. Kuzmin, J. Hamalainen, *Finite Element Methods for Computational Fluid Dynamics: A Practical Guide* (SIAM, 2014)
8. R. H. Henninger, M. J. Witte, *EnergyPlus Testing with ANSI/ASHRAE Standard 140-2001 (BESTEST)* (Lawrence Berkeley National Laboratory, Berkeley, California, 2004)
9. R. Judkoff , J. Neymark, *International Energy Agency Building Energy Simulation Test (BESTEST) and Diagnostic Method* (National Renewable Energy Lab., Golden, CO (US), 1995)
10. R. Judkoff, J. Neymark, *Standard Method of Test for the Evaluation of Building Energy Analysis Computer Programs* (American Society of Heating, Refrigerating and Air-Conditioning Engineers, Inc., 2004)
11. R. Judkoff and J. Neymark, in (Chambery, France August 25-28, 2013, 2013).
12. M. Woloszyn, C. Rode, *IEA Annex 41, MOIST-ENG Subtask 1 – Modelling Principles and Common Exercises* (2007)
13. T. Soubdhan, T. A. Mara, H. Boyer, A. Younes, *World Renew. Energy Congr. VI* (Pergamon, Oxford, 2000)
14. P. Pieniążek, *Computer Simulation of Building Energy* (B.Sc. Thesis, Warsaw University of Technology, Faculty of Civil Engineering, 2016)

# Calculation of the filtration in a heterogeneous porous medium

*Yuri* Galaguz[1], *Galina* Safina[1,*]

[1] Moscow state university of civil engineering, Yaroslavskoye shosse, 26, Moscow, Russia, 129337

**Abstract.** The influence of groundwater filtration on soil strength should be taken into account in the design and construction of bases and foundations. The mathematical model of the filtration of a monodisperse suspension in a heterogeneous porous medium is considered, which takes into account the change in porosity and permeability in the formation of a deposit. A numerical solution of the problem is obtained by finite differences method.

## 1. Introduction

In the design and construction of bases and foundations it is necessary to take into account the influence of groundwater filtration on soil strength. The filtration problem of a suspension in a porous medium describes the change of characteristics of a porous rock during retention of solid particles in the pores [1-3].

A porous medium is a solid body which containing thin hollow channels of various lengths and cross sections (pores). The pores connect mixing chambers where the particles mix and then pass through other pores. When a flow of suspension (carrier fluid with suspended particles) passes through a porous medium, some particles get stuck in the pores and form a deposit. The geometric model of particle retention assumes that the particles get stuck at the entrance of small pores, and freely pass through the pores of a large cross section. The scheme of particle transport and retention is given in Fig. 1.

* Corresponding author: minkinag@mail.ru

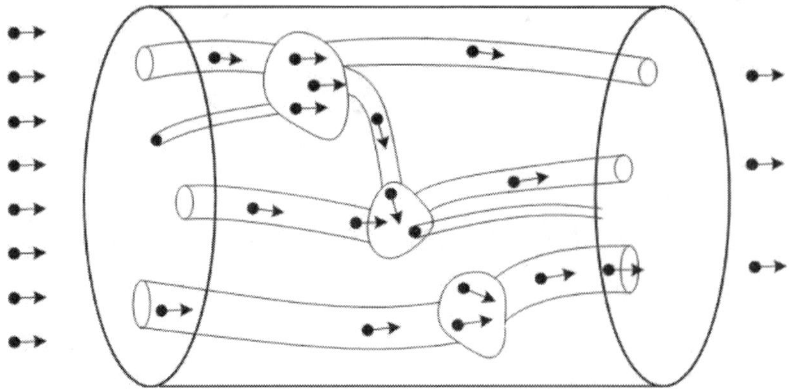

Fig. 1. Scheme of transport and retention of particles in a porous medium.

The basic model of filtration of a monodisperse suspension assumes that the velocity of particles in a porous medium is constant, and the deposit concentration does not affect the porosity and permeability of a porous medium [4-6]. More complex filtration models in a homogeneous porous medium take into account the change in porosity and permeability during the deposit formation and the inconstancy of the suspended particles velocity [7].

In a number of filtration models an exact solution is obtained [7-9], in others an asymptotic is constructed [10-14]. If analytical solution is absent, numerical methods are used [15-17].

In this paper the mathematical model of filtration of a monodisperse suspension in a heterogeneous porous medium is considered, taking into account the change in porosity and permeability during the deposit formation. The suspension of constant concentration is injected into a porous medium that does not contain suspended and retained particles. In the system of filtration equations, the unknowns are the concentrations of suspended $C(x,t)$ and retained $S(x,t)$ particles in a porous medium. A numerical solution of the problem is obtained by finite differences method [18].

## 2. Formulation of the problem

One-dimensional filtration model of suspension in a heterogeneous porous medium with changing porosity and permeability consists of two partial differential equations

$$\frac{\partial(g(x,S)C)}{\partial t}+\frac{\partial(f(x,S)C)}{\partial x}+\frac{\partial S}{\partial t}=0; \qquad (1)$$

$$\frac{\partial S}{\partial t}=\Lambda(x,S)C. \qquad (2)$$

Here the filtration coefficient $\Lambda(x,S)$, the porosity $g(x,S)$ and permeability $f(x,S)$ are continuous functions, $\Lambda(x,S)$ is non-negative, $g(x,S)$ and $f(x,S)$ are strictly positive at $0 \le x \le 1,\ S \ge 0$.

The system of equations (1), (2) are considered in the domain

$$\Omega=\{(x,t): 0<x<1,\ t>0\}.$$

The boundary conditions for the system (1), (2) are given at the filter inlet $x=0$ and at the initial time $t=0$:

$$C(x,t)\big|_{x=0} = 1, \tag{3}$$

$$C(x,t)\big|_{t=0} = 0; \quad S(x,t)\big|_{t=0} = 0. \tag{4}$$

The conditions (3), (4) determine a unique solution of the problem in the domain $\Omega$.

The mobile boundary of the domain filled with particles and the empty part of a porous medium is called the concentration front of the suspended and retained particles. The concentration front is the characteristic line of the equation (1) emerging from the origin.

Since the conditions (3) and (4) do not match at the origin, according to the theory of characteristics the solution $C$ has a strong discontinuity on the concentration front; and the solution $S$ has a weak discontinuity (discontinuity of first-order derivatives) on the concentration front. Behind the concentration front in the domain $\{(x,t): 0<x<1,\, t>x\}$ the solution is positive $C>0,\, S>0$; before the front $\{(x,t): 0<x<1,\, 0<t<x\}$ the problem (1)-(4) has a zero solution $C=0,\, S=0$.

The concentration front of suspended and retained particles propagates in a porous medium with variable velocity

$$v = \frac{f(x,0)}{g(x,0)}. \tag{5}$$

## 3. The exact solution at the filter inlet

The equation (2) at filter inlet $x=0$ has the form

$$\frac{\partial S}{\partial t} = \Lambda(0,S). \tag{6}$$

Division of both sides of the equation (6) by $\Lambda(0,S)$

$$\frac{\partial S/\partial t}{\Lambda(0,S)} = 1 \tag{7}$$

and integration of (7) with respect to the variable $t$ gives

$$\int_0^t \frac{\partial S/\partial t}{\Lambda(0,S)}\, dt = t. \tag{8}$$

Using the condition (4), we transform the integral on the left-hand side of (8)

$$\int_0^{S(0,t)} \frac{dS}{\Lambda(0,S)} = t. \tag{9}$$

The equation (9) determines the concentration of retained particles at the filter inlet.

Consider the most important examples of filtration coefficients. The filter coefficient $\Lambda(x,S)$ is called blocking, if it is positive at $0 \le x \le 1,\ 0 \le S < S_M(x)$, and is zero when $S \ge S_M(x)$. In this case, the retained particles concentration is limited by the value $S_M(x)$.

A) For a linear blocking filtration coefficient

$$\Lambda(x,S)=\lambda(x)\left(S_M(x)-S\right), \tag{10}$$

where $\lambda(x)>0$, $S_M(x)>0$, the integral on the left-hand side of (9) is computed explicitly:

$$\int_0^{S(0,t)} \frac{dS}{\lambda(0)\left(S_M(0)-S\right)}=\frac{1}{\lambda(0)}\ln\left|\frac{S_M(0)}{S_M(0)-S(0,t)}\right|. \tag{11}$$

The concentration of retained particles at the filter inlet for the filtration coefficient (10) has the form

$$S_1(t)=S_M(0)\left(1-e^{-\lambda(0)t}\right). \tag{12}$$

B) For the quadratic blocking filtration coefficient

$$\Lambda(x,S)=\lambda(x)\left(S_M(x)-S\right)^2 \tag{13}$$

the integral (9) is equal to

$$\int_0^{S(0,t)} \frac{dS}{\lambda(0)\left(S_M(0)-S\right)^2}=\frac{S}{\lambda(0)S_M(0)\left(S_M(0)-S(0,t)\right)}.$$

The dependence of the concentration of deposited particles on time at the filter inlet is given by formula

$$S_2(t)=\frac{S_M^2(0)\lambda(0)t}{1+S_M(0)\lambda(0)t}. \tag{14}$$

Formulas (12), (14) show that the function $S(0,t)$ increases monotonically and tends to the limiting value $S_M(0)$ for large values of time $t$.

In Fig. 2 a), b) the graphs of the retained particles concentration at the filter inlet for the blocking filtration coefficients (12), (14) with the parameters $\lambda(0)=1$, $S_M(0)=1$ are shown.

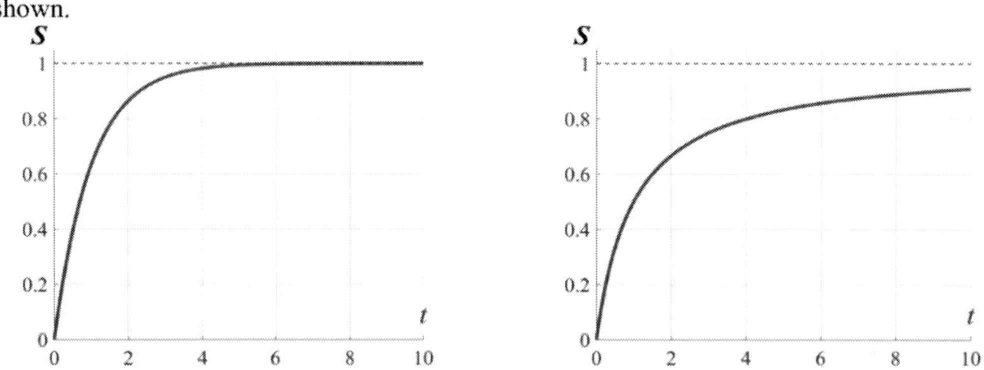


Fig. 2 a) Retained particles concentration $S_1(t)$. b) Retained particles concentration $S_2(t)$.

## 4. Numerical calculation

The calculation is carried out by the finite difference method. For equation (1), the TVD-version of the Lax-Wendroff scheme is applied. For equation (2) the second-order Runge-Kutta method is used. The solution of system (1)-(4) is obtained in the domain $\{(x,t): 0<x<1,\ 0<t<10\}$. Integration step in $x$: $h=0.001$, step in $t$: $\tau=0.0005$. The scheme satisfies the Courant-Friedrichs-Lewy condition

$$\tau \le \min_{x\in[0,1],\ t\in[0,10]} \left( \frac{g(S(x,t))}{f(S(x,t))} \right) \cdot h .$$

Numerical calculation of the problem is carried out for coefficients of equations (1), (2)

$$g(x,S)=(1-0{,}1x)(1-0{,}3S)\,,\quad f(x,S)=(1+0{,}1x)(1-0{,}5S)\,,\quad \Lambda(x,S)=(1+0{,}2x)(1-S)\,.$$

At Fig. 3 a), b) 3-D graphs of suspended and retained particles concentrations are shown.

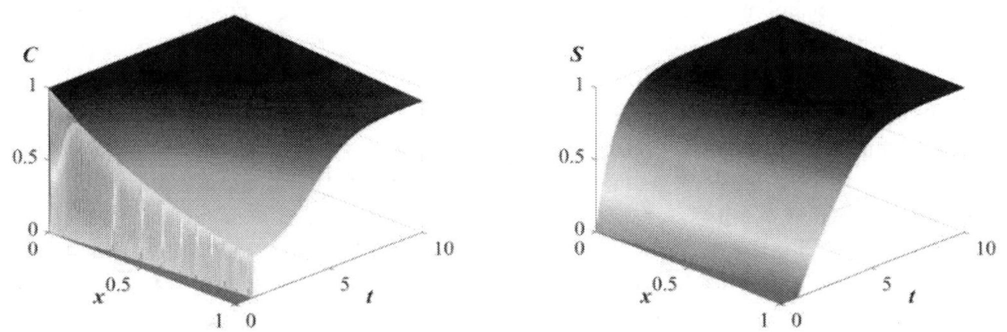


Fig. 3. a) Suspended concentration $C(x,t)$. b) Retained concentration $S(x,t)$.

Graphs of suspended particles concentration at fixed time $t=0.5$ and at fixed distance $x=0.5$ are shown in Fig. 4 a), b).

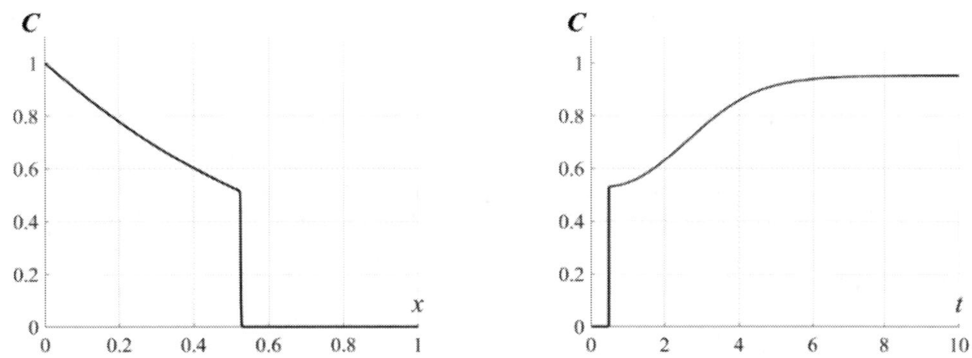


Fig. 4. a) Suspended concentration $C(x,t)\big|_{t=0.5}$. b) Suspended concentration $C(x,t)\big|_{x=0.5}$.

Graphs of retained particles concentration at fixed time $t=0.5$ and at fixed distance $x=0.5$ are shown in Fig. 5 a), b).

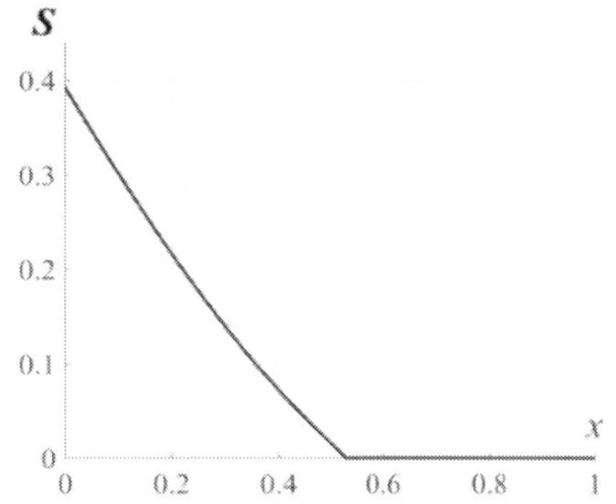


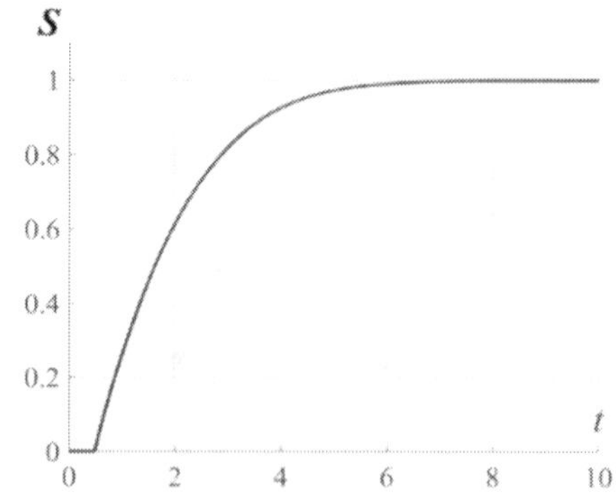


Fig. 5. a) Retained concentration $S(x,t)|_{t=0.5}$. b) Retained concentration $S(x,t)|_{x=0.5}$.

## 5. Conclusion

A numerical solution of the one-dimensional filtration problem for a monodisperse suspension in a heterogeneous porous medium is obtained. Unlike standard models that consider a homogeneous porous medium, in this paper the porosity, the permeability and the filtration coefficient depend not only on the concentration of retained particles $S(x,t)$, but also on the distance $x$ up to the filter inlet. Fig. 3 shows how the suspended and retained particles gradually fill the porous medium, moving from the inlet $x=0$ to the outlet $x=1$. The front of concentrations of suspended and retained particles is the boundary between two parts of a porous medium - empty and filled with particles. It moves with a velocity $v=1$ and reaches a filter outlet at time $t=1$. The larger the value of $x$, the later the concentration front reaches this point and retention begins (Fig. 3). With increasing deposit, the growth rate of retained particles decreases. With increasing time, the concentration of retained particles tends to the limiting value $S_M=1$. Since accumulation of deposit decreases for long times, the concentration of suspended particles tends to the maximum value.

## References

1 K. Khilar, S. Fogler, *Migration of fines in porous media* (Kluwer, 1998)
2 A. Santos, P. Bedrikovetsky, S. Fontoura, J. Memb. Sci. **308** (2008)
3 F. Civan, *Reservoir formation damage: fundamentals, modeling, assessment, and mitigation*, (Gulf Publ. Com., 2007)
4 J. P. Herzig, D. M. Leclerc, P. Legoff, Ind. Eng. Chem. **62** (1970)
5 E.A. Vyazmina, P.G. Bedrikovetskii, A.D. Polyanin, Theor. Found. Chem. Eng. **41**, 5 (2007)
6 L.I. Kuzmina, Yu.V. Osipov, Int. J. Comp. Civ. Str. Eng. **10**, 3 2014
7 Z. You, P. Bedrikovetsky, L. Kuzmina, Abstr. Appl. Anal. **2013**, ID 680693 (2013)
8 L.I. Kuzmina, Yu.V. Osipov, Struc. Mech. Anal. Const. **1** 2017 [in Russian].
9 L.I. Kuzmina, Yu.V. Osipov, Matec Web Conf. **86**, 01005 (2016)
10 Z. You, Y. Osipov, P. Bedrikovetsky, L. Kuzmina, Chem. Eng. J. **258** (2014)
11 L.I. Kuzmina, Yu.V. Osipov, Procedia Eng. **111** (2015)
12 L.I. Kuzmina, Yu.V. Osipov, Procedia Eng. **153** (2016)
13 L.I. Kuzmina, Yu.V. Osipov, Int. J. Comp. Civ. Str. Eng. **12**, 1 2016

14 L.I. Kuzmina, Yu.V. Osipov, Vestnik MGSU, **1** (2015)
15 V.I. Golubev, D.N. Mikhailov, Proc. MIPT **3**, 2 (2011) [in Russian].
16 Yu.P. Galaguz, G.L. Safina, Procedia Eng. **153** (2016)
17 Y. Galaguz, G. Safina, Matec Web Conf. **86**, 03003 (2016)
18 E.F. Toro, *Riemann solvers and numerical methods for fluid dynamics*, (Springer, Dordrecht, 2009)

# Modeling of filtration of 2-types particles suspension in a porous medium

*Yuri* Galaguz[1], *Galina* Safina[1,*]

[1] Moscow state university of civil engineering, Yaroslavskoye shosse, 26, Moscow, Russia, 129337

**Abstract.** The filtration problem describes the process of concreting loose soil with a liquefied concrete solution. The filtration of 2-types particles suspension in a homogeneous porous medium with a size-exclusion particles retention mechanism is considered. The difference in the filtration coefficients of 2-types particles leads to the separation of the filtration domain into two zones, in one of which two types of particles are deposited and in another – only particles of one type are deposited. In this paper, the mobile boundary of two zones is calculated, and numerical solution of the problem is calculated.

## 1. Introduction

The calculation of filtration problem is important for concreting loose soil. Liquefied concrete penetrates into the porous soil and, solidifying, strengthens it. Due to the filtration of solid suspended particles in porous filters, primary purification of liquid industrial wastes and municipal wastewater occurs [1-3].

The classical model of filtration of a suspension in a porous medium considers the process of transport and deposition of identical particles moving with a constant velocity [4, 5]. The mathematical model includes the equation of mass transfer and the kinetic equation that determines the growth of the deposit in proportion to the concentration of suspended particles. The coefficient of proportionality in the kinetic equation is called the filtration coefficient. Modern models take into account the dependence of the porosity and permeability of a porous medium on the deposit size and various mechanisms of particle capture [6-9].

A mechanical model of particles retention, based on a comparison of particle and pore sizes is considered in this paper. It is assumed that the particles freely pass through large pores and get stuck at the entrance of small pores, the dimensions of which are smaller than the particle diameter [10]. In contrast to the classical model, a suspension containing particles of two different types is considered. Particles that differ in size, shape, or density move in a porous medium with the same velocities. The concentrations of suspended and retained particles satisfy kinetic equations with different filtration coefficients. The condition of non-negativity of the filtration coefficients leads to the separation of the filtration domain into two zones. At small times, in the first zone both types of particles are retained. When one of the filter coefficients turns to zero, the corresponding suspended

* Corresponding author: minkinag@mail.ru

particles begin to flow freely through the porous medium. In this zone only the particles of the second type are retained.

For some filtration models, the exact solutions are obtained or the asymptotic formulas are constructed [11-17]. In the general case, the analytical solution is absence and numerical calculations are carried out [18, 19]. In this paper, the mobile boundary of two zones is computed numerically and numerical solution of the problem is obtained.

## 2. The problem formulation

In the domain $\Omega = \{0 < x < 1, t > 0\}$ consider a first-order quasilinear hyperbolic system with respect to the concentrations of suspended and retained particles of two types $C_i(x,t);\ S_i(x,t);\ i = 1,2$

$$\frac{\partial C_i}{\partial t} + \frac{\partial C_i}{\partial x} + \frac{\partial S_i}{\partial t} = 0; \tag{1}$$

$$\frac{\partial S_i}{\partial t} = \Lambda_i(S)C_i;\ i = 1,2 \tag{2}$$

with the boundary condition

$$x = 0 : C_i = p_i \tag{3}$$

and the initial conditions

$$t = 0 : C_i = 0,\ S_i = 0\,. \tag{4}$$

Here $S = S_1 + S_2$ is the total deposit concentration; the filtration coefficients are $\Lambda_i(S) \geq 0,\ i = 1,2$.

At the initial time the porous medium does not contain any particles (condition 3). At $t = 0$ a suspension with constant suspended particles concentrations $p_i$ is injected at the filter inlet (condition 4). The discontinuity of the boundary conditions (3) and (4) at the origin extends over the characteristic $t = x$ of the hyperbolic system (1)-(4). This characteristic is the concentrations front of the suspended and retained particles, moving with a constant velocity $v = 1$. The concentration front divides the domain $\Omega$ into two subdomains $\Omega_0 = \{0 < x < 1, 0 < t < x\}$ and $\Omega_S = \{0 < x < 1, t > x\}$. In the subdomain $\Omega_0$ the system (1)-(4) has a zero solution, in the subdomain $\Omega_S$ the solution is positive.

Assume that the filtration coefficient $\Lambda_1(S)$ turns to zero at a point $S = S_0$, and the coefficient $\Lambda_2(S)$ is positive at this point:

$$\Lambda_1(S) > 0,\ 0 \leq S < S_0;\ \ \Lambda_1(S) = 0,\ S \geq S_0;\ \ \Lambda_2(S) > 0,\ 0 \leq S \leq S_0.$$

When $S < S_0$ particles of two sizes are retained in a porous medium. When $S > S_0$ the filtration coefficient $\Lambda_1(S) = 0$ and particles of the first type pass freely through the porous medium. In this zone only particles of the second type are retained. The mobile boundary of the two filtration zones is given by the equation

$$\Lambda_1\left(S(x,t)\right) = 0\,. \tag{5}$$

## 3. Numerical modeling

A numerical solution of the problem is obtained by the method of finite differences using an explicit difference scheme. To fulfill the convergence conditions and the adequacy of the approximate solution on the discontinuity line $t = x$ steps in time and coordinate are chosen the same $\tau = h = 0,001$ [20].

A numerical calculation is performed for a 2-types particles suspension in the case of linear filtration coefficients

$$\Lambda_i(S) = \lambda_i - a_i S;\ i = 1,2 \tag{6}$$

for initial conditions $p_1 = 1,\ p_2 = 1$.

The parameters of the functions (6) are chosen as follows $\lambda_1 = 1,\ \lambda_2 = 2,\ a_1 = 1,\ a_2 = 1$.

The concentrations front of suspended and retained particles and the mobile boundary of the two filtration zones are shown in Fig. 1. In the case of linear filtration coefficients (6) the boundary line is close to a straight line.

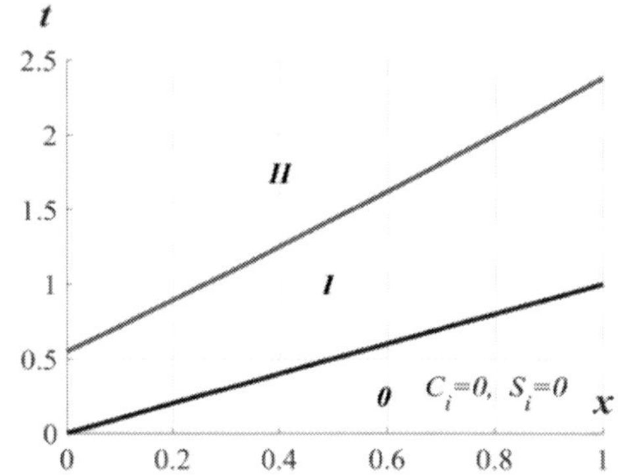


Fig. 1. The concentrations front and the mobile boundary of filtration zones.

In zone 0 (the domain $\Omega_0$) there are no particles and the solution is zero. In zone I the particles of two sizes are retained. In zone II the particles of type 1 pass only through large pores and do not get stuck, only particles of type 2 are retained.

Below 3-dimensional graphs of the suspended particles concentrations $C_1$, $C_2$ (Fig. 2, a), b)), the retained particles concentrations $S_1$, $S_2$ (Fig.3 a), b)) and the total deposit concentration $S$ (Fig. 4) are shown.

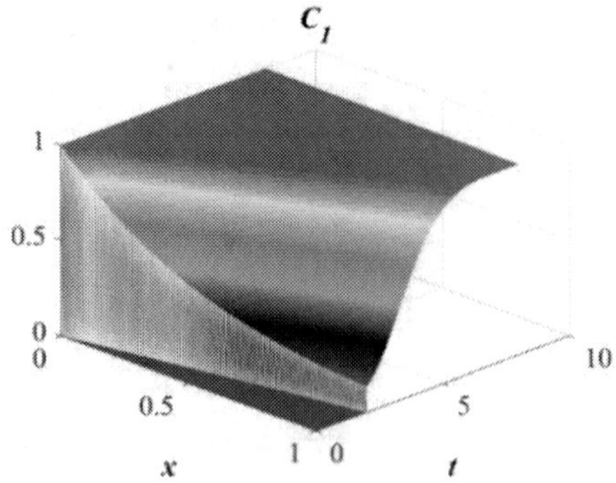


Fig. 2. a) Suspended particles $C_1(x,t)$.

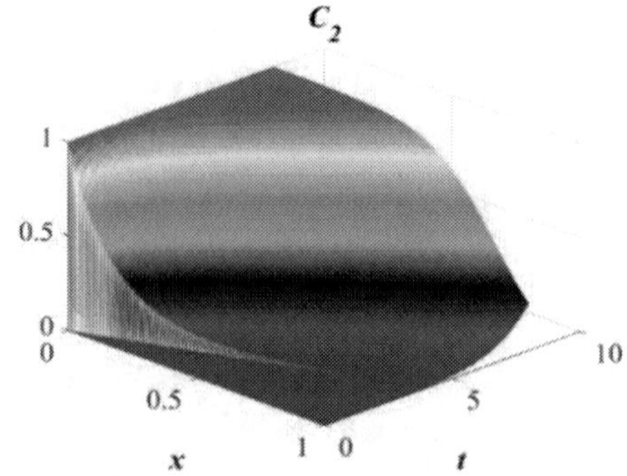


b) Suspended particles $C_2(x,t)$.

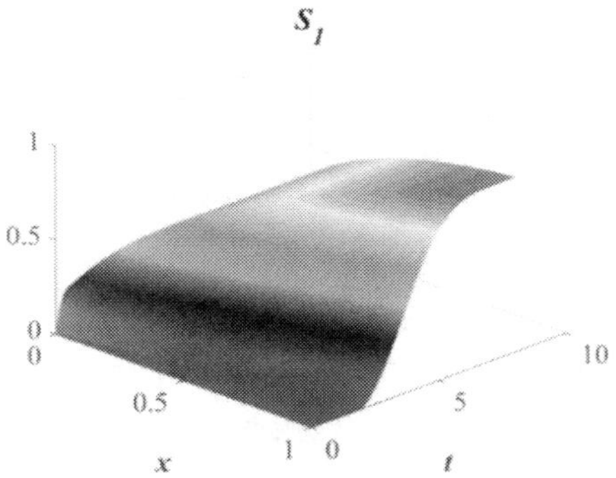


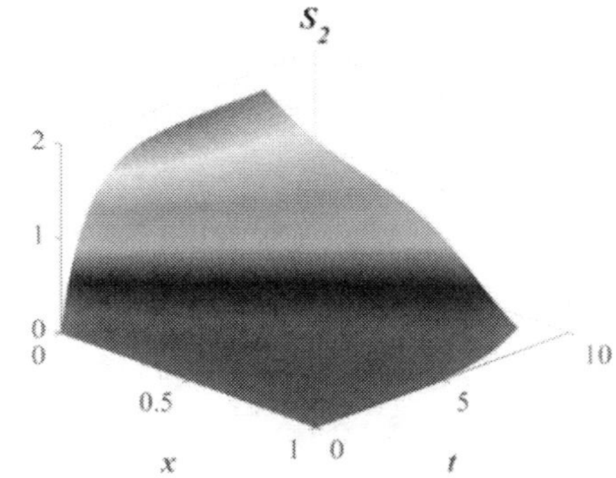


Fig. 3. a) Retained particles $S_1(x,t)$. b) Retained particles $S_2(x,t)$.

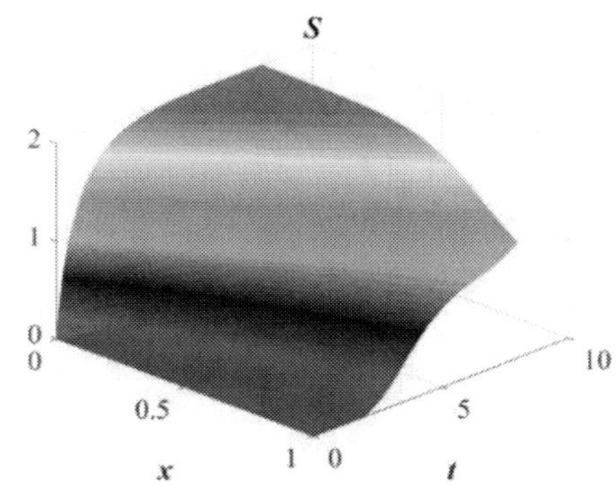


Fig. 4. Total retained particles concentration $S(x,t)$.

Fig. 2 shows that the solution $C_i(x,t),\ i=1,2$ experiences a jump at the concentration front $t=x$; on this line the solution $S_i(x,t),\ i=1,2$ and the total retained particles concentration $S(x,t)$ are continuous and have a break (Fig. 3, 4).

On the boundary (5) of two filtration zones the solution is smooth (Fig. 2-4). On this line the solution undergoes a weak discontinuity: first-order derivatives $\partial S_i/\partial t$, $i=1,2$ have a break and second-order derivatives are discontinuous (Fig. 5).

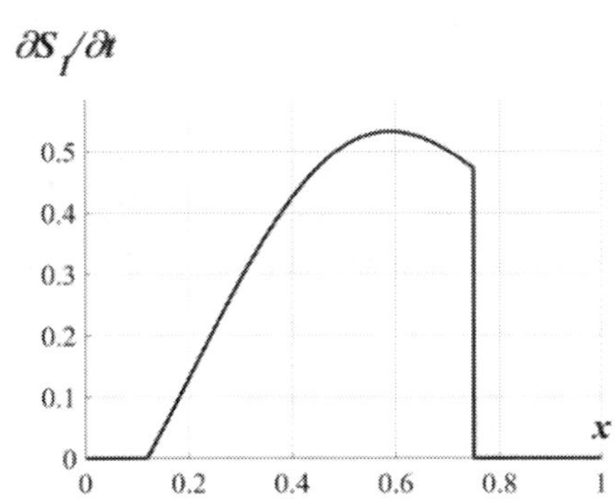


Fig. 5. The derivative $\partial S_1/\partial t|_{t=0.75}$.

The left zero part of the graph corresponds to zone II, in which the particles of the first type do not deposited. The part of the positive derivative corresponds to zone I, in which all types of particles are filtered. At the point $x = 0.75$ the graph reaches the front of the concentrations, there are no particles before of the concentration front on the segment $0.75 \le x \le 1$ and the derivative is zero there.

In Fig. 6, 7 the graphs of suspended and retained particles concentrations at fixed time $t = 1.5$ are presented.

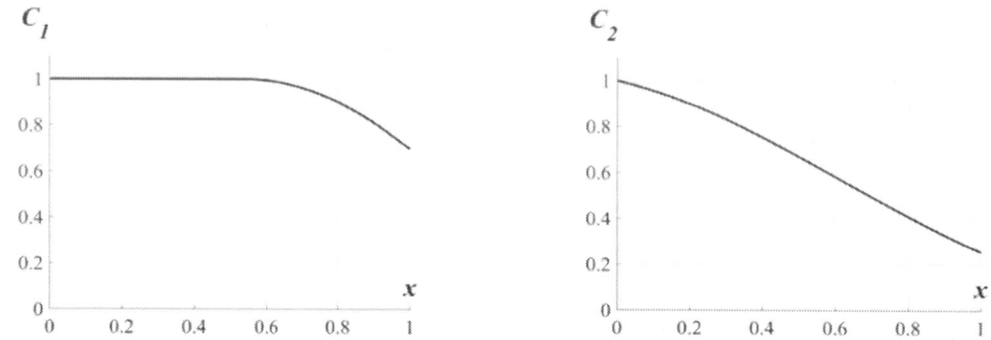


Fig. 6. a) $C_1(x,t)|_{t=1.5}$. b) $C_2(x,t)|_{t=1.5}$.

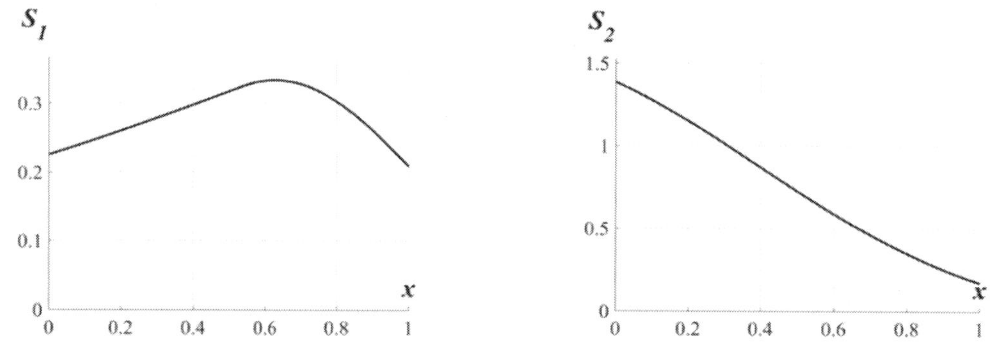


Fig. 7. a) $S_1(x,t)|_{t=1.5}$. b) $S_2(x,t)|_{t=1.5}$.

The concentrations of suspended particles $C_i(x,t)|_{t=1.5}$, $i = 1,2$ decrease as the distance $x$ increases to the filter inlet (Fig. 4). At the filter inlet $x = 0$ the filtration of particles of the first type terminates at the time $t = t_0$ specified by the equation $\Lambda_1(S(0,t_0)) = 0$, and at the filter outlet at the time $t = t_1$, where $\Lambda_1(S(1,t_1)) = 0$. For a fixed $t < t_0$ the concentration $C_1(x,t) < 1$ at $0 < x < 1$. At $t_0 < t < t_1$ the concentration $C_1(x,t) = 1$ at $S(x,t) < S_0$ and $C_1(x,t) < 1$ at $S(x,t) > S_0$. At $t > t_1$ all over the porous medium only particles of the second type are retained and $C_1(x,t) = 1;\ 0 < x < 1$.

The retained particles concentration $S_2(x,t)|_{t=1.5}$ decreases as the distance $x$ increases to the filter inlet (Fig. 7 b)). At a fixed $t < t_0$ the concentrations $S_i(x,t);\ i = 1,2$ and $S(x,t)$ decrease for all $0 < x < 1$. When $t_0 < t < t_1$ the concentration $S_1(x,t)$ reaches the limiting constant value $S_1^0 < S_0$ on a interval $0 \le x \le x_0(t);\ x_0(t) < 1$, where $S(x_0(t),t) = S_0$ (Fig. 7 a)). When $x_0(t) < x < 1$ we have $S(x,t) < S_0$ and the function $S_1(x,t)$ decreases. At $t > t_1$ all over the filter the retained particles concentration of the type 1 is constant: $S_1(x,t) = S_1^0$.

In Fig. 8, 9 the graphs of concentrations of suspended and retained particles at a fixed distance from the filter input $x = 0.5$ are shown.

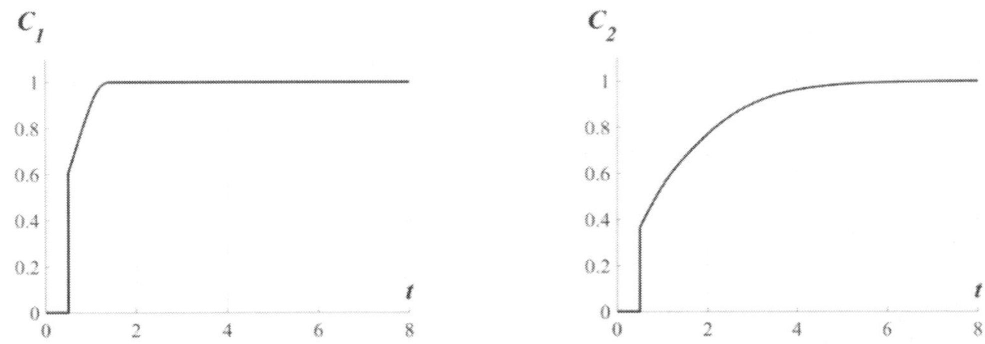


Fig. 8. a) $C_1(x,t)|_{x=0.5}$. b) $C_2(x,t)|_{x=0.5}$.

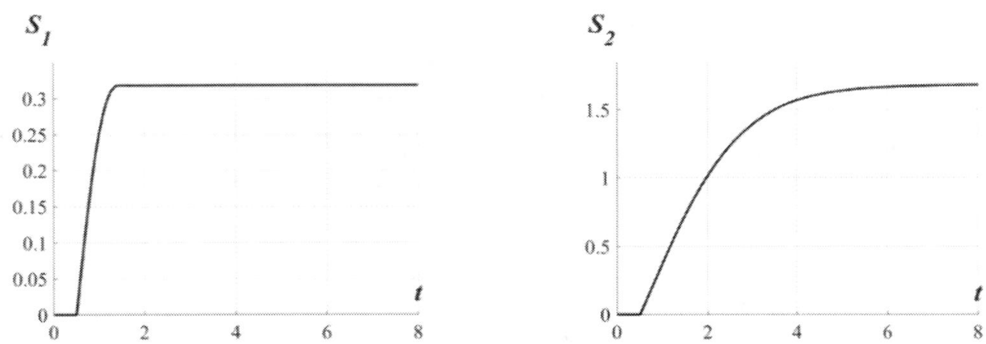


Fig. 9. a) $S_1(x,t)|_{x=0.5}$. b) $S_2(x,t)|_{x=0.5}$.

The concentrations of suspended and retained particles are zero before the concentration front (Fig. 8, 9). At the front of concentration $t = x$, the concentrations of suspended particles $C_i(x,t)$; $i = 1,2$ are discontinuous, and the concentrations of retained particles $S_i(x,t)$; $i = 1,2$ are continuous and fractured. On the boundary line of the two filtration zones the solution does not lose its smoothness.

For a fixed $x$ all concentrations are non-decreasing. When the total deposit reaches the value $S_0$ the concentrations of the suspended and retained particles of the first type become constant: $S_1(x,t) = S_1^0$, $C_1(x,t) = 1$. The concentrations of suspended and retained particles of the second type $S_2(x,t)$, $C_2(x,t)$ constantly increase and tend to limit values.

## 4. Conclusion

A numerical solution of the problem of filtration of a suspension with solid particles of two types in a porous medium for linear and linearly-constant filtration coefficients is constructed. The filtering domain is divided into two zones. In one zone particles of two types are deposited. And in the other - only one type. A mobile border of two zones is constructed; the graphs of concentrations of suspended and retained particles of two types are presented. It is shown that the solution at the boundary of two filtration zones is smooth.

## References

1 P.G. Bedrikovetsky, *Mathematical Theory of Oil and Gas Recovery* (Kluwer 1994)
2 S.A. Bradford, S. Torkzaban, Vad. Zone J. **7** (2008)
3 C. Noubactep, S. Care, Chem. Eng. J. **163** (2010)
4 J. P. Herzig, D. M. Leclerc, P. Legoff, Ind. Eng. Chem. **62** (1970)
5 E.A. Vyazmina, P.G. Bedrikovetskii, A.D. Polyanin, Theor. Found. Chem. Eng. **41**, 5 (2007)
6 Z. You, P. Bedrikovetsky, L. Kuzmina, Abstr. Appl. Anal. **2013**, ID 680693 (2013)
7 H. Yuan, G. Sin, Chem. Eng. J. **168** (2011)
8 A. Santos, P. Bedrikovetsky, Transp. Porous Media **62** (2006)
9 H. Yuan, Z. You, A.A. Shapiro, P. Bedrikovetsky, Chem. Eng. J. **226** (2013)
10 A. Badalyan, Z. You, K. Aji, P. Bedrikovetsky, T. Carageorgos, A. Zeinijahromi, Rev. Sci. Instr. **85** (2014) art. No.: 015111
11 L.I. Kuzmina, Yu.V. Osipov, Int. J. Comp. Civ. Str. Eng. **12**, 3 2016
12 L.I. Kuzmina, Yu.V. Osipov, Matec Web Conf. **86**, 01005 (2016)
13 S. Borazjani, A. J. Roberts, P. Bedrikovetsky, Appl. Math. Lett. **53** (2015)
14 M. Hayek, G. Kozakovski, S. Churakov, Wat. Res. Res. **47** (2011)
15 Z. You, Y. Osipov, P. Bedrikovetsky, L. Kuzmina, Chem. Eng. J. **258** (2014)
16 L.I. Kuzmina, Yu.V. Osipov, Procedia Eng. **111** (2015)
17 L.I. Kuzmina, Yu.V. Osipov, Vestnik MGSU, **2** (2016)
18 Yu.P. Galaguz, G.L. Safina, Procedia Eng. **153** (2016)
19 Y. Galaguz, G. Safina, Matec Web Conf. **86**, 03003 (2016)
20 E.F. Toro, *Riemann solvers and numerical methods for fluid dynamics*, (Springer, Dordrecht, 2009)

# Structural synthesis of electrical engineering complex' control system of a plant for plastic shaping of the ceramic mixture

*Stanislav* Galitskov[1], *Maxim* Nazarov[1*], and *Konstantin* Galitskov[1]

[1]Samara State Technical University, Institute of Architecture and Civil Engineering, 194, Molodogvardeyskaya St., 443001, Samara, Russia

**Abstract.** Production of ceramic bricks with the required strength imposes significant restrictions on the process control of plastic shaping of the ceramic mixture in the auger extruder. It is due to several factors. Firstly, the certain nonstationarity of rheological properties of the source raw materials necessitates the automatic task-oriented changes in combinations of such values as shear rate, ceramic mixture moisture and vacuum pressure in the vacuum chamber of the extruder. To solve this problem it is necessary to maintain a coordinated control of the relevant automatic control systems of the electrical engineering complex. The second problem is the lack of technical tools to measure the values of shear rate in the pressure head of the extruder. And finally, the third factor is a necessity for monitoring and modeling of operating steps in brick production – from shaping to finished product output, that is a necessity to assess the impact of drying and firing processes on the possibility to make bricks of specified strength. The paper considers structural synthesis of the electrical engineering complex' control system for plastic shaping of the ceramic mixture, including the problem of coordinated control: of the vacuum pump' electrical drive, of the solenoid valve for water dosing, of the belt feeder' electrical drives, of the mixer and the auger, as well as the use of digital observers of technological controlled coordinates and models in further phases of brick production.

## 1 Introduction

When manufacturing bricks, plastic shaping of the ceramic mixture in the auger vacuum extruder has gained widespread [1-5], which is considered, with regard to automation of this process, to be the electrical engineering complex, including electromechanical drives of the auger vacuum extruder (the clay mixer, the vacuum pump and the extruder itself) and the electromagnetic dosing device for regulating moister and temperature of the clay mixture in the mixer [6-8]. The practice of ceramic brick production has proved that strength, structure homogeneity and geometric shape defect strongly depend on

* Corresponding author: nazarovm86@yandex.ru

technological parameters of adobe brick shaping, which, in turn, are determined by operational modes of actuating devices [6-9].

Existing control methods of shaping are mainly limited to automatic stabilization of either ceramic mixture pressure in the pressure head, or moister in the clay mixer [10]. In these automated systems the actuating devices are solenoid valves for water and steam dosing. Such limited use of the complex' electrical drives control does not allow achieving stable production of bricks of the required brand in circumstances where one should meet the modern requirements for energy efficiency and plant productivity when original properties of raw materials and technological conditions of production can differ.

The paper deals with modernization of methods and techniques of coordinated control of the complex' electrical drives, which is, in particular, aimed at providing the required strength of manufactured bricks at minimum energy consumption and maximum achievable productivity.

## 2 Materials and methods

The results of the research conducted by Barabanshchikov Yu.G., Tretyakov I.M., Golubovich S.R. [6,11,12] on the influence of the main characteristics of plastic shaping on the strength of the finished brick and occurrence of structure homogeneity defects and geometrical errors assuming the stability of the drying and firing technology, allowed obtaining the three-dimensional space, coordinates of which are technological parameters of shaping – moisture content $w$, vacuum pressure $P_v$ in the vacuum chamber and shear rate $\dot{\gamma}$ of the ceramic mixture in the exit section of the pressure head. Part of this space is allocated for the area controlling the electrical engineering complex of the plant for plastic shaping of the ceramic mixture, the analysis of which allows making a conclusion that to ensure stable production of bricks of the required brand at minimum energy consumption and maximum possible productivity under a number of disturbances it is necessary to have coordinated control of electrical engineering complex' drives as a function of shear rate of the ceramic mixture in the pressure head of the auger extruder, which is the main parameter defining the finished brick strength.

## 3 Results

The structural synthesis of the control system of such a compound object as the complex under consideration highlights two problems [6,13-15]: definition of reference signal values, which ensure the production of ceramic bricks of specified strength and robust stability of the system that controls the auger' electrical drive under the conditions of changing parameters of the clay raw materials.

The first problem is solved by using the generator of reference signals vector. These signals are applied at the stage of forming local control systems. The second problem is solved by applying the ACS of the auger' electrical drive in the form of the dual circuit system with one measured coordinate.

The mixer and the belt feeder of existing models of the auger extruders have unregulated drives, so the change in $\dot{\gamma}$ results in discrepancy between productivities of the belt feeder $G_{bf}$, the mixer $G_m$, the extruder, and consequently, in changes of the filling levels $h_{vc}$ and $h_m$ of the vacuum chamber and the mixer, respectively. This technological unbalance of complex' devices operation causes significant deviation of the shaped brick properties from the required values. Therefore, it is necessary to ensure automatic coordination of the extruder, the mixer and the belt feeder modes. Thus, in addition to the automatic control system of the auger electric drive, the vacuum pump and the solenoid valve for water dosing, that determine shear rate $\dot{\gamma}$, vacuum pressure $P_v$ and moister $w$, respectively, the system of automatic control of this electrical engineering complex (fig. 1) includes the control system of the belt feeder and clay mixer drives that support filling levels $h_m$ of the mixer and $h_{vc}$ of the vacuum chamber.

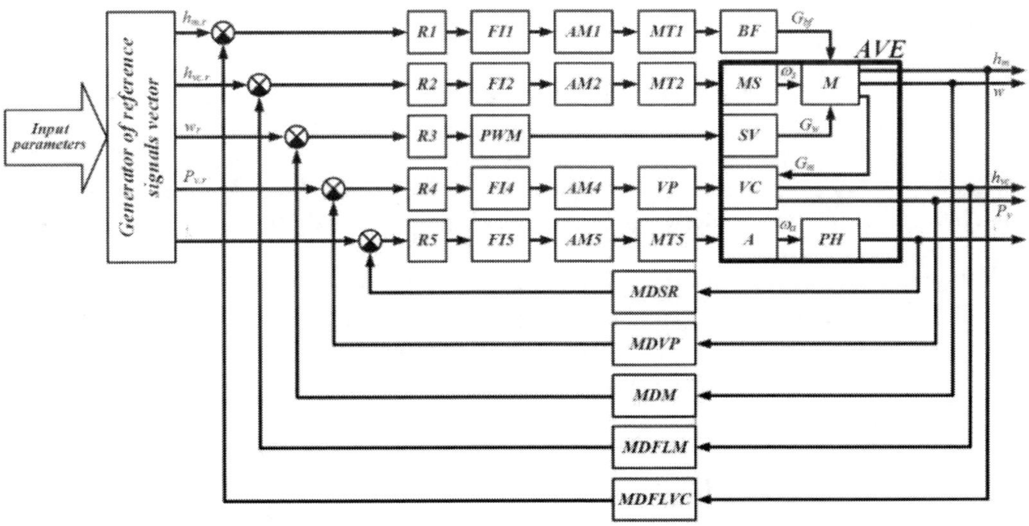


**Fig. 1.** The structure of the electrical engineering complex' control system of the plant for plastic shaping of the ceramic mixture: R1, R2, R3, R4, R5 – the controllers of the local control systems; FI1, FI2, FI4, FI5 – the variable speed drivers; PWM – the pulse-width modulator; AM1, AM2, AM4, AM5 – the asynchronous motors; MT1, MT2, MT5 – the mechanical transmissions; VP – the vacuum pump; VC – the vacuum chamber; BF – the belt feeder; MS – the mixer shafts; M – the mixer; SV – the solenoid valve; A – the auger; PH – the pressure head; AVE – the auger vacuum extruder; MDSR, MDM, MDVP, MDFLM, MDFLVC the measuring devices of shear rate and moister of the ceramic mixture, of vacuum pressure in the vacuum chamber, of filling level in the mixer and vacuum chambers, respectively; $\omega_s$, $\omega_a$ – angular velocity of shafts rotation of the mixer and the auger, respectively; $G_w$ – water flow rate; parameters with the index "$r$" – reference signals.

Practical implementation of the control system is constrained by inability to measure shear rate $\dot{\gamma}$ in modern conditions. So it is important to develop the device, allowing not only to indirectly calculate the value $\dot{\gamma}$, but also to determine (based on the target strength values $R_r$ of brick, constraints on the mixture technological characteristics, constraints conditioned by the technical characteristics of the extruder) the set value $\dot{\gamma}_r$ of the control system of the ceramic mixture shear rate. These problems can be solved by applying the developed algorithm of the setting device, taking into account the subsequent stages of bricks production, and the observer of the ceramic mixture shear rate in the pressure head of the auger vacuum extruder [16].

## 4 Discussion

The proposed structural configuration of the control system is of great practical importance as it allows improving significantly the efficiency of the electrical engineering complex for shaping the ceramic mixture without substantial engineering changes in the structure of the auger vacuum extruder. However, the creation of the automatic control system of the electrical engineering complex for shaping the ceramic mixture should be preceded by solving a number of theoretical problems related to structural synthesis of the complex, which includes the electrical drives of the mixer, the auger, the vacuum pump, the belt feeder and the solenoid valve for water dosing, the algorithm design for effective control of these drives, creation of the new mathematical model of the ceramic mixture shaping in the auger vacuum extruder, synthesis of the generator of reference signals vector' algorithm of the ACS complex, adaptive to nonstationarity of making the ceramic mixture as well as to changing conditions of producing bricks of the limited number of brands.

## 5 Conclusions

A new control system configuration of the electrical engineering complex for shaping the ceramic mixture with the generator of reference signals vector has been developed. In order to solve the complicated problem of stabilizing the bricks strength under the conditions of reducing energy consumption in this system it has been offered to coordinate the control of the belt feeder, the mixer, the vacuum pump and the extruder drives, as well as the solenoid valve for water dosing.

The necessity of controlling the complex as a function of its shear rate in the exit section of the pressure head has been shown, as there exists no measuring device for its determination. Therefore, it is necessary to use the digital observer.

## References

1. F. Handle, *Extrusion in Ceramics* (Springer, Berlin, 2007)
2. N.N. Evstratova, A.V. Rud, *Design of auger extruders for shaping clay bricks* (TNT, Stary Oskol, 2013)
3. V.S. Bogdanov, S.B. Bulgakov, A.S. Ilyin, *Technological complexes and mechanical equipment in construction industry* (Prospekt nauki, Saint Petersburg, 2010)
4. T. Kloft, Building material **4**, 64-65 (2010)
5. V.Ya. Tolkachev, *Technology of high-quality extrusion of products made of clay* (Computer technologies, Krasnoyarsk, 2009)
6. S.Ya. Galitskov, M.A. Nazarov, K.S. Galitskov, International Journal of Applied Engineering **10-20**, 40846-40852 (2015)
7. S.Ya. Galitskov, M.A. Nazarov, K.S. Galitskov, MATEC Web of Conferences **86**, 1-6 (2016) doi: 10.1051/matecconf/20168604010
8. S.V. Zubanov, E.V. Tkachev, Urban Construction and Architecture **3(11)**, 90-96 (2013) doi: 10.17673/Vestnik.2013.03.18
9. O.A. Luneva, E.V. Averyanova, Urban Construction and Architecture **3(11)**, 70-72 (2013) doi: 10.17673/Vestnik.2013.03.14
10. W. Leisenberg, Zigelindustrie International **8**, 25-32 (2000)
11. Yu.G. Barabanshchikov, PhD Thesis (St. Petersburg, 2006)
12. Yu.G. Barabanshchikov, Scientific and technical journal of SPBSPU **6-1**, 170-175 (2006)
13. V.N. Mitroshin, *Problems of control and simulation in complex systems: $18^{th}$ International Conference proceedings*, 333-339 (Ofort, Samara, 2016)
14. E.Ya. Rapoport, Yu.E. Pleshivtseva, A.N. Diligenskaya, *Problems of control and simulation in complex systems: 18th International Conference proceedings*, 317-324 (Ofort, Samara, 2016)
15. J.E. Normey-Rico, E.F. Camacho. *Control of dead-time processes* (London, Springer, 2007).
16. S.Ya. Galitskov, K.S. Galitskov, M.A. Nazarov, Industrial and civil engineering, **9**, 27-30 (2015).

# Efficiency of construction waste recycling

*Yulia* Galitskova[1,*], and *Andrey* Mikhasek[1]

[1]Samara State Technical University, Institute of Architecture and Civil Engineering, 194, Molodogvardeyskaya St., 443001, Samara, Russia.

**Abstract.** Recycling is widely used in practice in various fields of activity. However, the effect of such use does not always cover the costs of the processing. The article considers the problem of recycling waste generated while constructing residential buildings and structures. We present the results of full-scale studies of construction waste generated at construction sites in the city of Samara. We also show the qualitative and quantitative composition of the elements and analyze possible ways of their reuse. In addition, we have calculated economic feasibility of reusing building materials recycled from construction waste.

## 1 Introduction

Currently, there is an increase in the activity of urban development. Adjoining territories are built up and already existing city buildings are renovated as well. The expansion of territories is most noticeable and occurs more actively in large cities. For example, over the past decade, dozens of residential complexes have been built and continue to be built in the territory of the city of Samara: Koshelev project has increased the built-up area of the city territory by 130 hectares, Novaya Samara residential area in Krasnoglinsky district of the city occupies about 58 hectares, Southern City project has expanded the urban development in the southern part of the city by more than 1000 hectares. Such area expansion results from the increase in the number of city residents and their need to obtain comfortable housing with necessary utility systems and social infrastructure [1-4].

It is especially difficult to carry out construction work on infill areas, when worn-out buildings and structures are replaced with more comfortable and new ones (Fig. 1). In this case, all works on preparing the territory, cleaning, planning, erection and improvement are carried out in cramped conditions. During construction, adjacent residential areas suffer from a negative impact, including a large number of various construction wastes. At first, these wastes are stored on the construction site, then they are to be transported to a landfill. It should be noted that at present more and more construction wastes are sorted at the place where they are formed and subsequently transported to specialized enterprises for processing into secondary raw materials [5-9]. This practice takes place in the cities of Moscow, Novokuibyshevsk, Togliatti and others. Despite this, unfortunately, construction waste recycling is not widely spread and is currently developing slowly. The main reason

* Corresponding author: galickova@yandex.ru

for this is lack of economic leverage to influence construction firms or waste management companies.

**Fig. 1**. A worn out building is being demolished.

To determine the efficiency of construction waste recycling and reuse, studies were conducted to determine the component composition of construction waste at city construction sites.

## 2 Materials and Methods

To determine the component composition of construction waste generated during the construction process, a method of field observation was used, followed by statistical analysis of the results obtained. Field observation implied surveying urban construction sites with measuring equipment and photographic fixation, and the results were automatically processed by graphic editors.

The first stage of field inspection was fixating geometric indices of a construction waste dump location. The inspection report noted its location relative to the construction site boundary, settlements, the approximate shape and dimensions of the dump (length, width, average and maximum height of waste storage).

The second stage involved sampling, fractionation and chemical analysis of the samples in the laboratory to determine the composition of the wastes as a percentage of the volume of each type of fraction to the total volume.

The types of fractions were taken in large scale, taking into account the Federal Classification Catalog of Waste acting in the territory of the Russian Federation, namely: paper and cardboard, broken glass, plastic and polyethylene, wood waste, metal waste, broken concrete and reinforced concrete, broken bricks, household waste.

The third stage of the study was determining the criteria for reusing recycled waste products. For this purpose, physical and chemical properties of the recycled semi-finished products were analyzed in the laboratory, feasibility and appropriateness of their further use was settled, and their applicability criteria were established.

## 3 Results

More than 30 construction sites located in different parts of the city of Samara were surveyed. The main criterion for choosing the surveyed sites was the proximity to the existing residential buildings. We selected construction sites located not farther than 100 meters from a residential building.

The surveyed construction sites satisfying the above conditions were concentrated in the central part of the city, which is densely populated; there are both private houses and new multistoried comfortable residential complexes.

The preliminary survey of construction sites showed that in 85% of cases there were no specially designated places for temporary storage of construction waste at the construction site. As a rule, the waste was stored without prior division into components in bulk either in the territory of the construction site, or outside its boundaries. More often the mass of waste was lying along the protective fence and had the following dimensions: 1.5 × 5 m, average height of waste storage was 1.5 m (Fig. 2). The dimensions are shown in Table 1.

**Fig. 2**. Type of waste storage area.

**Table 1.** Typical dimensions of construction waste dumps.

| construction site № | Dimensions of the dump | | | Shape of the dump |
|---|---|---|---|---|
| | Length, m | Width, m | Height, m | |
| 1 | 3 | 2 | 1 | Stretched along the fence |
| 2 | 20 | 3 | 1 | |
| 4 | 2 | 2 | 1,5 | Round |
| 5 | 5 | 3 | 1 | Rectangular |
| 8 | 5 | 2 | 2 | Rectangular |
| 12 | 10 | 3 | 2 | Stretched along the fence |
| 17 | 10 | 6 | 3 | Rectangular |
| 18 | 5 | 4 | 2 | Rectangular |
| 23 | 5 | 4 | 2 | Rectangular |
| 24 | 2 | 2 | 1 | Round |

| | | | | |
|---|---|---|---|---|
| 27 | 5 | 2 | 2 | Rectangular |
| 30 | 6 | 4 | 3 | Stretched along the fence |

Further, the component composition of the stored waste was determined for each construction site. The results of the studies are presented in Table 2.

**Table 2.** Results of component composition analysis of construction waste.

| construction site № | Component composition of construction waste, % | | | | | | |
|---|---|---|---|---|---|---|---|
| | Broken bricks | Broken concrete | Plastic and polyethylene | Wood waste | Iron waste | Paper and cardboard | Other waste |
| 1 | 40 | 30 | 5 | 10 | - | - | 15 |
| 2 | 10 | - | 20 | 50 | - | 5 | 15 |
| 5 | - | - | 30 | 50 | - | 5 | - |
| 7 | - | 30 | 5 | 10 | 50 | 5 | |
| 10 | 70 | - | 5 | 3 | 17 | 5 | |
| 12 | - | 40 | 5 | 10 | 25 | 10 | 10 |
| 16 | - | 50 | - | 15 | 35 | - | - |
| 21 | - | 30 | - | 10 | 20 | 10 | 30 |
| 24 | - | 40 | 5 | 30 | 25 | | |
| 27 | - | - | 80 | - | 5 | - | 15 |
| 29 | 50 | 35 | 5 | 10 | - | - | - |

Based on the results presented in Tables 1 and 2, it is possible to determine the type and volume of construction works, as well as how often the dumped waste is transported away from the site.

For example, the wastes at construction sites № 1, 10 and 29 were formed as a result of dismantling old buildings and clearing the territory for new construction. Wastes at sites № 12, 16 and 24, where there is a large proportion of broken concrete, were most likely formed during the erection of a panel building. This is indicated not only by the highest number of concrete waste, but also by lumpy iron waste.

Since the waste at construction site №27 is mainly represented by plastic and polyethylene, it can be assumed that the construction works at this site are at the final stage: installation of utility systems, windows, and so on.

At construction site № 2, a large amount of waste is stored, which indicates lack of timely waste disposal and non-compliance with sanitary requirements.

Based on all survey results, we propose to divide the construction sites into the following types of waste components: demolition-construction and finishing. Their averaged component compositions are presented in the diagrams (Fig. 3).

a) b)

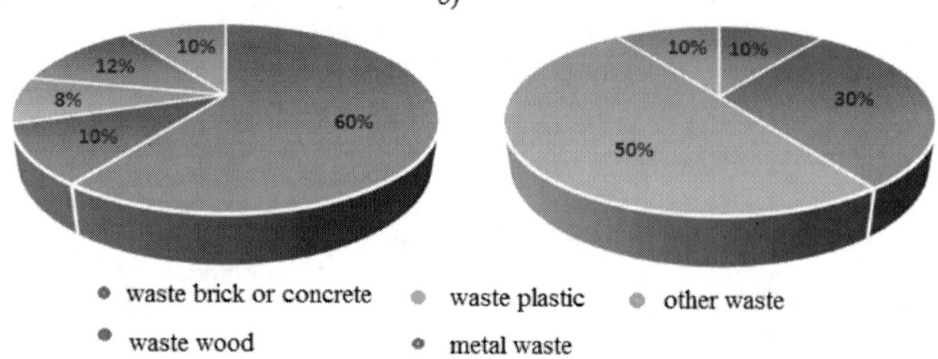


**Fig. 3**. Diagrams of the component composition for: a) demolition-construction, b) finishing.

## 4 Discussion

Analysis of construction sites research data showed that the most common wastes generated both during the dismantling of buildings and in the main construction period are broken bricks, broken concrete and reinforced concrete. It should be noted that in some cases these wastes can reach 70-80% of the volume of stored wastes. Thus, the approximate amount of concrete transported to the landfill ranges from 9 to 120 tons (excluding waste that was disposed of before the survey of construction sites). Such volumes are significant, and waste disposal to the landfill leads to losing valuable materials [5-9]. In the authors' opinion, the use of concrete waste as a material for building temporary roads, for secondary backfilling and in terrain planning [10-12], also leads to losing valuable recyclable materials.

The conducted studies of the resulting broken concrete quality showed sufficient strength of the resulting building products [13] for their use in the construction of new buildings and structures or as a concrete filler for reconstruction of hydraulic structures. Such use of concrete waste will reduce the volume of extracted natural resources, decrease the load on landfills, and also minimize the logistics costs for the transportation of waste and natural resources.

To determine the efficiency of using recycled scree, a comparative cost analysis was carried out, which showed the following. The cost of natural scree, which is used for the most common concrete of class B22.5 and mined in the quarries of the Samara region, is about 500-700 rubles per ton. The cost of processing recycled scree, including preliminary sorting, transportation or cost of receiving waste (50-70 rubles), is about 300-400 rubles per ton. Thus, recycling concrete and reinforced concrete waste is financially beneficial. Even taking into account the purchase of equipment (the cost of installation is about 60 million rubles), the costs will be recouped after processing 300,000 tons of waste, which can happen in 1.5-2 years of good load and development.

## 5 Conclusions

The research showed the following:

1. Large amounts of various wastes are formed on construction sites; their component composition makes it possible to determine the stage and specifics of the construction work.

2. The resulting waste is reused in small volumes, which leads to losing large quantities of valuable materials.

3. Using recycled scree in the production of building materials will minimize the cost of extracting natural resources, reduce the burden on the environment, and also make profit from processing waste concrete.

## References

1. E.O. Smolenskaya, Urban Construction and Architecture **1,** 16-20 (2011)
2. V.M. Melnikova, N.A. Mastalerzh, Urban Construction and Architecture **1,** 31-37 (2011)
3. V.A. Samogorov, O.S. Rybachova, Urban Construction and Architecture **3**, 71-72 (2011)
4. IA. Solomin, Industrial and Civil Engineering **1**, 39-41 (2011)

5. I.I. Romanenko, M.I. Romanenko, I.N. Petrovnina, E.M. Pint, K.A. Elichev, Internet-journal Naukovedenie **1,** 86 (2015)
6. A.V. Seliverstova, Research and Developments of Young Scientists **3**, 107-111 (2013).
7. P.N. Kurochka, R.R. Mirzaliev, Bulletin of the Rostov State University of Railways **3** (47), 140-147 (2012)
8. A.Z. Efimenko, Concrete Technologies **2** (91), 17-21 (2014)
9. G.G. Lunev, Ecological Systems and Devices **2**, 49-55 (2015)
10. V.V. Gasilov, I.V. Kryuchkova, FES: Finance. Economics **11**, 5-8 (2012)
11. E.V. Kashevskaya, K.K. Kostenko, A.K. Kostenko, Science and Technology in the Road Industry **2** (29), 30-33 (2004)
12. A.Yu. Kanayev, Science and the Present **11**, 254-257 (2011)
13. Yu.M. Galitskova, A.A. Mikhasek, Industrial and Civil Engineering **6**, 51-54 (2015)

# Investigation of the ultrasonic sound signals during the cementing materials induration

*Vladimir* Gaponov [1], *Dmitry* Kuznetsov [1], *Vitaly* Dudnik [1], and *Nadiia* Afanasieva[1*]

[1]Don State Technical University, Department of Industrial Safety, 344000 Rostov-on-Don, Russian Federation

**Abstract.** The paper is devoted to investigation mechanism of cement mixture setting and stiffening. Method for registration of the acoustic oscillations induced during the heterophase transitions executing in liquid medium was proposed (method of acoustic emission (AE)). Using of conical vessel serving as a resonator allows the amplification of registered acoustic signal and provides high sensitivity of the registration. The experimental results show that the using of AE-method allows remote monitoring of the dynamics of concrete mixtures condition during setting. Therefore, this method can be successfully applied in building as a non-destructive method for determining the start and end of the solidification of the cement solution.

## 1 Introduction

There is a set of complex physical and chemical processes occur during the cementing substances interaction, especially during the cement clinker and water interaction. Mechanism of the cementing substances setting and stiffening is based on exothermal processes of interaction with water. As a result the water-cement paste transform into a dense stony material [1-2]. Method for registration of the acoustic oscillations induced during the heterophase transitions holding in liquid medium was proposed for the considered processes investigation (method of acoustic emission (AE)). Various authors [2-5] pay attention to the efficiency of AE-method coupled with optical method and computer simulation of crystals forming during the gypsum $CaSO_4 \cdot 0,5H_2O$ and water interaction.

The cement hardening has a complex mechanism. Chemical reactions between the cement components start immediately after the cement mix and water are blending. The cement particles begin to dissolve and simultaneously the hydration and hydrolysis of resolving products starts. The known research methods mostly used for the considered physical and chemical processes such as static method of the structural and mechanical properties investigation, roentgen phase analysis, local microanalysis etc. provide only integral data and do not provide any time-dependent information. Compare to these methods the AE-method provides information on the cement mixture condition every second. The extended theoretical and experimental analysis of the acoustic oscillations

* Corresponding author: nadezda.a.afanasyeva@gmail.com

registration methods during the heterophase transitions holding in liquid medium presented in [1-3].

## 2 Methodology

In the present paper, the processes of mixing the building mixture based on cement were studied using the AE-method coupled with heat monitoring. AE-method is a passive method of monitoring. The ultrasonic oscillations provide information on microscopic motions of elastic waves produced by an abrupt deformation of stressed material. Thus the AE-method allows to record timing dynamics of processes that take place in building mixture during its mixing. Additionally, the heat monitoring allows to obtain a thermogram of the building mixture mixing.

As an experimental was considered a concrete mix with grade of 400 and with next proportion of cement, sand and macadam 1:1,6:3,2. The acoustic emission monitoring was conducted by using the A-line 32D electronic device. The thermogram was obtained by using infrared imager Sat Hotfind D. Compare to the standard methodology of AE-signals registration the methodology presented in current paper supposes to use a conical vessel that serves as a resonator and provide amplification of registered acoustic signal. Capacity was installed on piezosensor that works at frequency range 30-500 kHz. The vessel was made by material with low factor of sound attenuation (quartz). The following AE-parameters was estimated within the study (Figure 1):

Amplitude of emission – the maximum magnitude of emission signal during the given time interval.

The electrical signal duration $T_0$ [s] – the time during which the AE electrical impulse envelope exceeds limiting threshold. The signal time variation range: $10^{-4}...10^{-8}$ s.

Rise time $D\tau_r$ [s] – time interval betwing the moment when AE-impulse envelope appears over limiting threshold and the moment when it achieves its maximum amplitude.

Total count $N_\Sigma$ [pulse] – the number of recorded AE-impulse exceeding the discrimination level (limiting threshold).

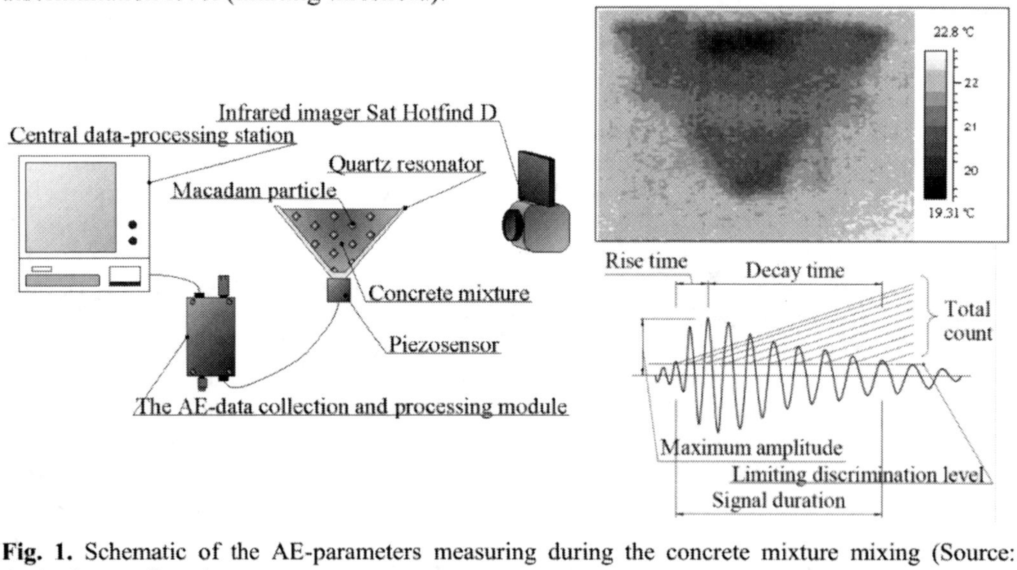


**Fig. 1.** Schematic of the AE-parameters measuring during the concrete mixture mixing (Source: Author's own figure).

Count rate N [pulse/s] – time derivative of the AE total count. Variation range: $0...10^{15}$ [pulse/s].

Electric signal energy Ec [J] – the measured area under the AE electrical signal envelope.

## 3 Results

The obtained results showed that infrared imager Sat Hotfinds do not recorded any temperature decrease. The observation is caused by the heat exchange with environment and by relatively low exothermicity of the reactions. Therefore only acoustic parameters were studying. The plot presented on figure 2 shows the change in integrated index – total count $N_\Sigma$ within 16 hours from the moment of concrete mixture mixing at the temperature of 30 °C. Another integrated index presented here is a compressive strength of concrete sample.

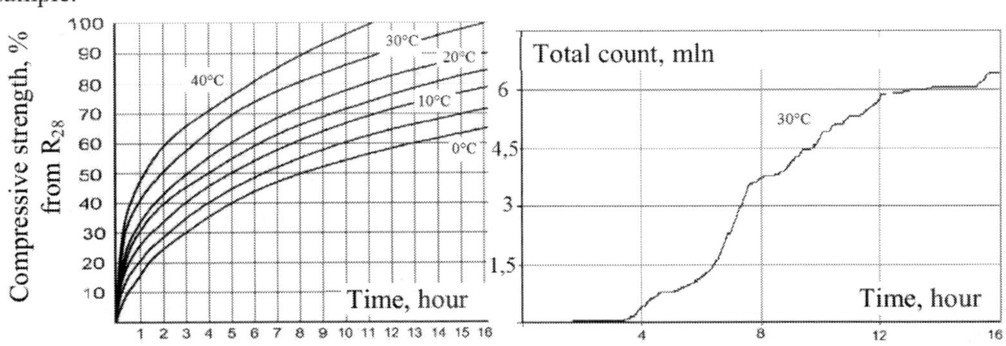


**Fig. 2.** Change in total count and compressive strength within 16 hours from the moment of concrete mixture mixing [1].

The first period (0-3 hours) can be called a period of dissolution or a preparatory period. When the cement interacts with water, a chemical reaction occurs on the surface of the granules. The observed monotectic transformation is accompanied by the emission of acoustic signals in the ultrasonic frequency range. Reaction products pass into the solution until the liquid surrounding the cement granules becomes a saturated solution of the reaction products. During this period, the parameters of the AE change insignificantly, the growth of AE activity is observed only after the 3rd hour. Then begins setting, which ends after 5-10 hours after mixing. During the second period (colloidation), the water-cement paste thickens, loses mobility, but strength is not great yet. During the experiments, it was found that approximately 3 hours after the start of the second stage, it can be saying that the concrete/cement was stiffened. Acoustic data indicate that in the period after the 3rd hour from the moment of mixing, a pronounced peak of AE activity is observed. This process executes a little more than 60 minutes. In addition, this process is clearly visible on the plot of change in AE pulses energy (see Fig. 3). At present, the water-cement paste setting is a well-studied process [5].

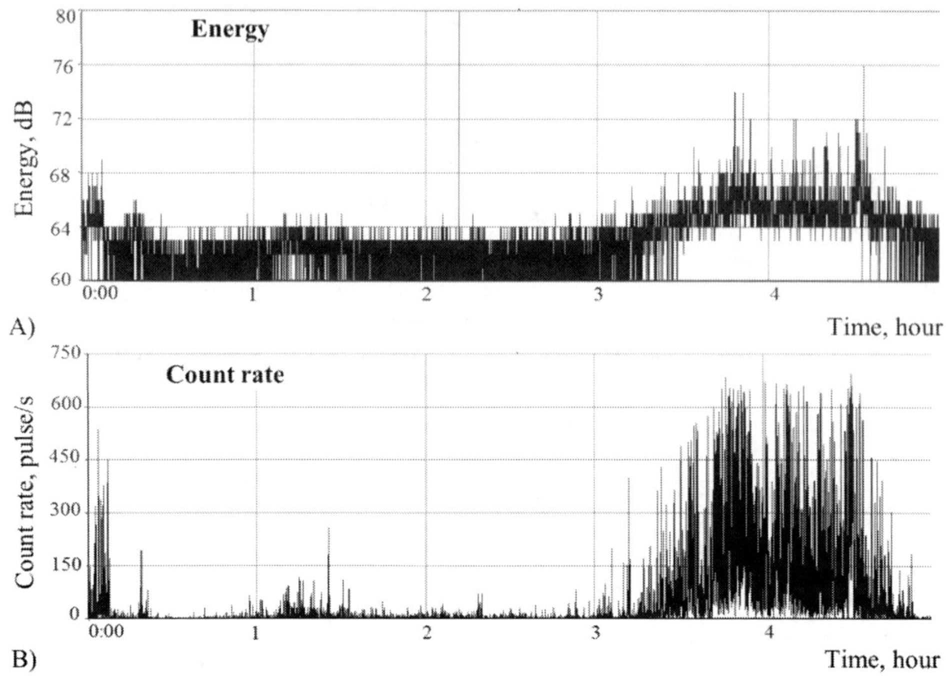


**Fig. 3.** Change in the AE parameters during the initial time period (0-5 hours) after the concrete mixture mixing. A – the change in AE pulses energy $E_c$; B – the change in AE count rate N (Source: Author's own figure).

It should be noted that in process of further hydration, the amount of free water in the water-cement paste decreases, the stickiness of the gel increases. The cement paste is thickening and setting. Then the new formations start to crystallize. AE-method allows observing the peritectic transformation i.e. the process of solid phase formation as a result of interaction between the liquid phase and initial solid phase, which differs from the new one by its composition and structure. This peritectic transformation can be characterized as follows. The crystals formed are splicing together, overgrowing with long needle crystals and finally a crystalline splice is formed, i.e., the setting of cement becomes finished. Hereafter, the cement stone is compacted due to the ongoing reactions of interaction between cement and water, partial dehydration and further crystallization. A detailed description of the process is available in [1-3].

The main ongoing chemical reactions can be described as follows. Tricalcium silicate $3CaO*SiO_2$ (abbreviated as $C_3S$), interacting with water, forms hydrosilicate and calcium hydroxide $2(3CaO*SiO_2)+6H_2O=3CaO*2SiO_2*3H_2O+3Ca(OH)_2$. Then the dicalcium silicate is hydrated $2CaO*SiO_2$ ($C_2S$): $2(2CaO*SiO_2)+4H_2O=3CaO*2SiO_2*3H_2O+Ca(OH)_2$. The tricalcium aluminate, interacting whith water, forms a calcium hydroaluminate: $3CaO*Al_2O_3+6H_2O=3CaO*Al_2O_3*6H_2O$. Four-calcium alumoferrite, interacting with water, break up into hydroaluminate and hydroferrite: $4CaO*Al_2O_3*Fe_2O_3+mH_2O=3CaO*Al_2O_3*6H_2O+CaO*Fe_2O_3*nH_2O$. The hydroaluminate is bonding by the addition of natural gypsum, and the hydroferrite becomes a part of the cement gel. The registered changes in AE-parameters for time intervals from 5 to 10 hours from the moment of cement mixing are presented in the Fig. 4. The data obtained indicate that the described process has a large number of intermediate stages. The observation was expectable due to molecular nature of chemical reactions that take place. Since reactions with a molecular structure of more than three are impossible, the registration of AE-parameters makes it possible to differentiate the complex chemical process in time.

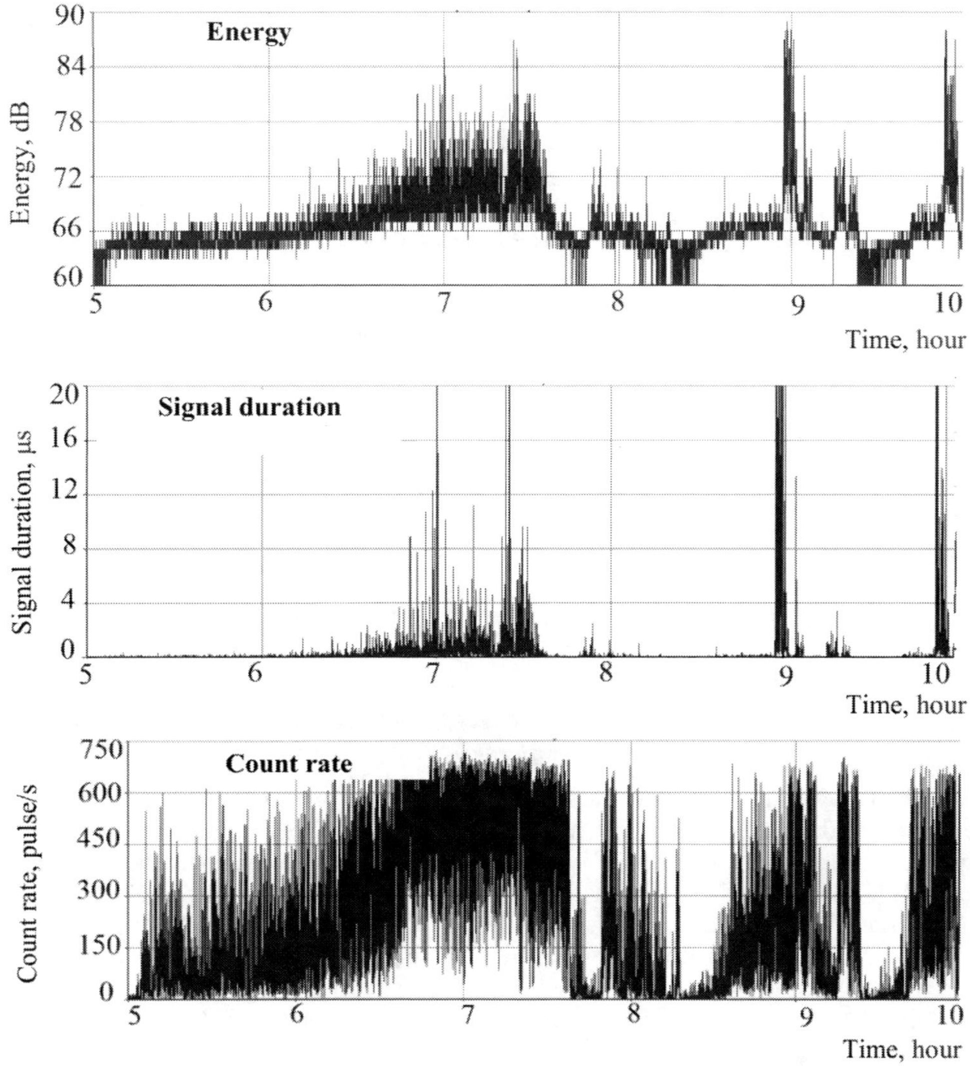


**Fig. 4.** Change in AE-parameters during setting and hardening of concrete mixture during the time intervals from 5 to 10 hours after mixing.

Further observations on the hardening concrete mixture conditions are reflected at Fig. 5. The high sensitivity of the AE-method made it possible to record the acoustic signals induced by the solidifying concrete mixture during the next 10 hours after its mixing. When the concrete mixture is hardened in the open air, the cement stone is additionally hardened as a result of carbonization of calcium hydroxide. The hardened cement stone is a very strong crystalline framework filled with a gel, inside which are the internal layers of cement granules that are not affected by reaction. The pores in the cement stone are filled with air and capillary water. If the cement hydration passes under normal conditions, the resulting calcium hydrosilicates have an alternating composition, that is largely close to amorphous substances and do not form a complete crystalline structure. The resulting material is characterized by good acoustic permeability. Therefore, during the further setting and hardening of cement, that accompanied by insignificant internal volume changes, the AE-signals of shear deformations reach the outer surface of the concrete sample. These deformations are detected by a piezosensor in the frequency range of 30-500 kHz.

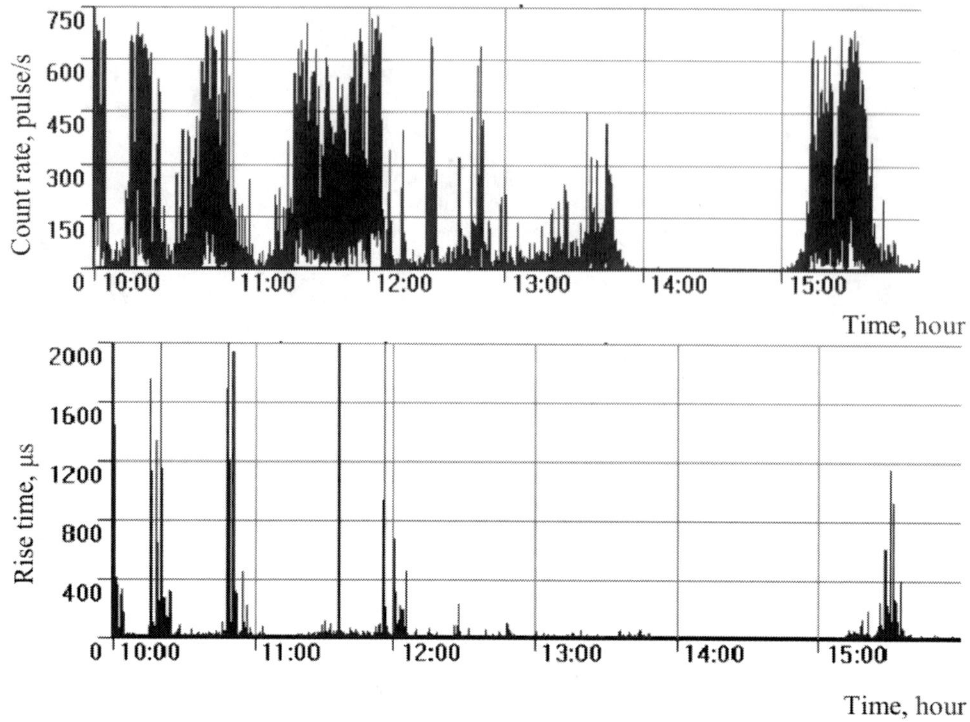


**Fig. 5.** Change in AE-parameters during setting and hardening of concrete mixture over the time intervals from 10 to 16 hours after its mixing (Source: Author's own figure)

The obtained results can be interpreted as follows. As time pass, the solution becomes supersaturated relative to the hydroferrites. Therefore, the conditions for formation of nuclei of the crystals of cement clinker alumina-ferrite phase and their precipitation from the solution appear in the liquid phase. This causes a decrease of hemihydrate concentration in the liquid phase and creates conditions for the dissolution of new portions of this substance and the formation of a supersaturated solution of the alumina-ferrite phase. The crystals grow, intertwine and grow together, causing setting and hardening of cement mixture. Whereupon there is induction of acoustic signals in the frequency range 100-400 kHz.

# 4 Conclusions

The experimental results show a high informativeness of acoustic method for the concrete mixture condition monitoring. Using of AE-method allows remote monitoring of the dynamics of concrete mixtures condition and the eutectoid transformation of the solid phases. In addition, it has been established that it allows also remote monitoring of the shrinkage processes dynamics in finished product, because each change in the equilibrium condition of thermodynamic system is accompanied by the emission of acoustic signals. Therefore, this method can be successfully applied in building as a non-destructive method for determining the start and end of the solidification of the cement solution. This is especially importance for continuous concrete of civil and industrial buildings and structures, precast structure, road-building, and for the hydraulic structures building.

## References

1. Guide on the electro-thermal processing of concrete. M. Stroyizdat, Scientific and research institute of concrete and reinforced concrete Gosstroy USSR, (1974)
2. A. Lemarchand, F. Boudoire, E. Boucard, Th. Chotard, A. Smith, J. Phys. Chem., **116**, (2012)
3. S.I. Builo, D.M. Kuznetsov, V.L. Gaponov, V.V. Trepachev, *Acoustic-Emission Testing and Diagnostics of the Dissolution Kinetics of Crystalline Componet*, Rus. J. of Nondestructiv Testing, **48**, 10 (2012).
4. S. I. Builo, D. M. Kuznetsov, V.L. Gaponov, *Acoustic Emission Testing of Capillary Liquid Flows in Porous Media*, Rus. J. of Nondestructiv Testing, **50**, 7, (2014).
5. S.I. Builo, D.M. Kuznetsov, V.L. Gaponov, *Chapter 13 Acoustic Emission Diagnostics of the Kinetics of Physicochemical Processes in Liquid and Solid Media*, In: Advanced Materials. Studies and Applications. Ivan A. Parinov, Shun-Hsyung Chang and Somnik Theerakulpisut (Eds.), New York: Nova Science Publishers, (2015)

# Cantilever structure in modern construction

*Elena* Generalova[1,*] , *Viktor* Generalov[1], and *Anna* Kuznetsova[1]

[1] Samara State Technical University, Institute of Architecture and Civil Engineering, 194, Molodogvardeyskaya St, 443001, Samara, Russia

**Abstract.** This article discusses innovative ways of using cantilever structures in modern construction. It highlights that the unique design of cantilever structures pushes forward the physical possibilities of architecture and construction, challenges the attractive force and breaks stereotypes. The authors provide an analysis of the best worldwide experience of introducing the cantilever elements in the objects of different functional purposes. Unique engineering solutions with a cantilever span of over 20m are paid special attention to. The paper considers the basic technologies, design systems, materials and testing techniques that make it possible to create "flying" structures. Peculiar features of using the cantilever structures in high-rise buildings and high-technology development of high-rise erection are described. Special attention is paid to the economic, safe and environmentally sound technology for the use of lightweight concrete slabs with void formers reducing the amount of concrete and the mass of slabs without losing strength. As a result, the paper describes the prospects for introducing the cantilever elements in modern construction and their functional use.

## 1 Introduction

A study of the objects of modern architecture and construction technologies used in the engineering and design systems has shown that at the conceptual and architectural stage the creative work of architects is increasingly related to the desire to use innovative design solutions. These solutions make such objects unique. In accordance with article 48, paragraph 2, of the Russian Urban Code, unique buildings and facilities can be referred to capital construction facilities if they have at least one of the following characteristics in their project documentation: 1 - height of over 100m; 2- span of over 100m; 3-cantilever of over 20m; 4- the substructure of over 15m below ground level. The purpose of this research is to study, analyze and identify the specific use of unique cantilever structures in modern buildings.

The first step is to deal with key concepts and terminology. A number of definitions of cantilever elements in architecture can be found in literature. Summarizing these definitions it can be said that a cantilever is a rigid structural element anchored at only one end to a (usually vertical) support from which it protrudes.

* Corresponding author: generalova-a@yandex.ru

The larger the cantilever span the more complex calculations are required for calculating loads, selecting the design and the properties of materials. All of this, of course, increases the cost of such buildings and facilities. Despite of this fact, the cantilevers are actively used in modern facilities of different functional purpose [1-10]. Building cantilever structures helps to find solutions for various tasks, such as urban planning (the problem of the lack of space on the site with difficult terrain), aesthetics (designing an interesting and unusual façade), functional design (creating additional space with unique view characteristics), symbolic (forming a mass image that works as a landmark and an architectural symbol of a city), etc.

## 2 Materials and Methods

It should be noted that nowadays there is a great variety of cantilever architectural and structural elements of buildings that perform certain functional and aesthetic tasks. They can be classified as follows: type 1- a roof that is an element of the building that has enclosure and structural functions (the roof can be walkable); type 2-part of a building's volume of the height of one or more floors (from small oriel windows to great volumes of a building); type 3-open sites that protrude from a facade's flat area (balconies, terraces) and have different functions and equipment (see Fig.1).

**Fig. 1.** (a) Lucerne Culture and Congress Centre (KKL), architect – Jean Nouvel, 2000, (b) BMW Welt, Munich, Germany, architects – Coop Himmelb(l)au, 2007, (c) 28 Social Housing, Paris, France, KOZ Architectes, 2010, (d) UNASUR General Secretary Headquarters in Quito, Ecuador, architect – Diego Guayasamin, 2014, (e) Práter Street 30-32, architects – PLANT Atelier Peter Kis, Budapest, Hungary, 2007, (f) Bosco Verticale Torre, architects – Stefano Boeri Architetti, 2014.

The use of innovative technologies, materials and design systems enables to increase the application of cantilever structures. It allows choosing the best solutions for them and overcoming different risks. In this study the authors would like to highlight the modern designed capabilities of steel (metallic) and reinforced concrete cantilever structures in unique buildings that have won a visual victory over gravity.

## 3 Results

The paper considers in details the examples of the above mentioned classification. The Busan Cinema Center, located in the city of Busan (South Korea) is a unique architectural object completed in 2012. The building was designed by a cooperative architectural design firm Coop Himmelb(l)au, Austria. It is registered in the Guinness World Records for the «longest cantilever roof» (see Fig.2). The roof of the Busan Cinema Center features a cantilever that extends 85 m [11]. "Flying Roof" is 21 m above the ground. It relies on single cone-shaped column support –The Double Cone where the main entrance and a café are located. The structure of this cone is a spatial metallic structure consisting of horizontal rings connected by diagonal links.

**Fig. 2.** Busan Cinema Center, Busan, South Korea, architects – Coop Himmelb(l)au, 2012 [12].

The giant roof is a metal space structure. Its upper surface is flat while the inner surface is of undulated shape. LED saturated outdoor roof elements act as a screen for light shows. The greater part of the metal framework was produced at factories and assembled at the construction site. The roof is anchored to The Double Cone by a structural element named «crown». This design project shows the reinterpretations of the roof as an integral architectural element that forms the system of interrelated open-architecture (the Urban Valley, the Red Carpet Zone, the Walk of Fame, the Memorial Court, the BIFF Canal Park, open cinema for 4000 people) and close-architecture (cinemas, theatres, studious, offices, restaurants, etc.) public spaces. The giant cantilever of the Busan Cinema Center shows not only contemporary capabilities of metal structures but also the new concept of forming the urban environment with hybrid functions.

Another example is «Marina Bay Sands» (Safdie Architects) that realizes the most courageous engineering, architectural and aesthetic fantasies and innovations (see Fig. 3). A distinctive feature of this complex is the unique engineering design of the public platform «Sands SkyPark» located at the height of 200 m (the 57th floor). It can house 3900 people. The platform of 1.2 hectares is 340m length and maximum 40m width. It combines three high-rise towers and is covered with a cantilever of 65m which is one of the largest in the world. «Sands SkyPark» is a real tropical oasis with a large garden (250 trees and 650 plants), recreation areas, treadmills, restaurants and «Infinity Pool» that is the largest outdoor pool in the world, located at such a height (area-1396 m$^2$, length-150 m). One of the main attractions of this platform is the Sands SkyPark public observation deck located just within a 65-metre-long cantilever on the 57th floor. This is an example of creating vertical urban environment with high density. Urban functions that traditionally tend to land (areas, parks, basins, etc.) are actively developed in the high-rise on its walkable roof as a new type of public space.

**Fig. 3.** Marina Bay Sands, Singapore, Safdie Architects, 2010.

To make this concept real a lot of engineering innovations had to be introduced: 1- wind load-the unique spatial metal structure of the platform consisting of mega trusses was designed taking into account the constant movement of the three towers because of wind, for this purpose special clearances and sliding elements with a range of up to 0.5 m were provided; 2- load of the pool. The problem of introducing a large continuous infinity edge pool was solved by making three separate 50-metre stainless steel shells united in a flexible singular whole; 3-assambly problem-the metal structure of the platform consists of 14 separate sections which were produced at a factory and raised one by one to the top of the tower with hydraulic jacks and mounted into a single structure there (the largest sections are two box girders of 80 m long that are the base of the cantilever. It took 16 hours to lift them to the assembling position at 14 m/h; 4-the problem of resonance in the cantilever. One of the solutions to this problem is to use of a 5-ton dampener hidden inside the structure and suspended to the transverse beams at the end of the cantilever [13].

Advanced technologies, materials and constructive systems are tools that can lead not only to economic and efficient use of different resources, but also to a higher and innovative level of functional and architectural planning of a building. The analysis of the following object, Columbia University Medical and Graduate Education Building (CUMGEB) shows the peculiar features of using unusual public space «The Study Cascade» (see Fig. 4). «The Study Cascade» is developed vertically within 14 floors and consists of cantilevered volumes flowing into each other and creating spaces of different volume and function. The spaces are connected by stairs and ramps. A «porous» structure was created to facilitate the communication of students, teachers and the urban environment.

**Fig. 4.** Columbia University Medical and Graduate Education Building (CUMGEB), New York, United States, architects – Diller Scofidio + Renfro, 2016 [14].

The emphasis was made on vertical spatial links, which is against the traditional organization of horizontal communication institutions. It is to be noted that neither new materials nor new technologies were developed for this building. However, an integrated approach to architectural and design solutions, integration of forms and functions made it possible to optimize the existing developments. A post-tensioned concrete structure of high strength concrete, structural steel, high strength rebar and self-consolidating concrete was chosen for «The Study Cascade». Void formers were extensively used in the construction of concrete slabs.

**Fig. 5.** Cobiax Technologies AG (a void former made from plastic, which takes the place of the solid concrete within the reinforced slab).

The technology of reducing the use of concrete in slabs should be considered in details. The examples are the developments of the Canadian company «BubbleDeck» or the Swiss firm Cobiax Technologies AG (its products are manufactured in Russia by franchise company «Ecotechplaza», Moscow). The analogy of this product is the unremovable deck adapted to the Russian rules and regulations produced by OOO Sibforma (Novosibirsk). The common thing to these technologies is the use of plastic hollow void formers to fill reinforced slabs (see Fig5). As a result slabs lose up to 35% of their mass but retain strength properties. The described technology is the ideal solution for floorings with spans and cantilevered concrete structures where large free space is needed when using a minimum number of columns.

## 4 Discussion

Having analyzed the unique buildings with introduced cantilever architectural and structural elements it can be concluded that their aesthetic and symbolic characteristics are subordinated to urban and functional planning tasks and are generally innovative solutions. Surely thanks to impressive appearance they have become visible and meaningful urban landmarks but it is much more important that they form a fresh approach to public urban space. Its main features are multifunctionality; protection from negative exposure to the weather; different functional use; the link to the interior space of a building; the ability to reproduce vertically; the creation of urban environment with high density. This is essential under the conditions of intense global urbanization when the search for new approaches to creating the urban living environment is becoming a vital issue.

## 5 Conclusions

In conclusion it should be pointed out that the considered architectural and engineering elements of the buildings are at the cutting-edge of scientific and technological progress and are a site for introducing not only advanced but experimental constructive systems, materials and technologies as well. All these things make these buildings unique. At the same time when considering the prospects of introducing cantilever elements in modern construction and taking into account extensive functional capabilities it is necessary to reflect on their mass use in urban environment at the level of standardized and harmonized typological and constructive structures.

## References

1. V.P. Generalov, E.M. Generalova, Urban Construction and Architecture **4(21)**, 23-29 (2015) DOI 10.17673/Vestnik.2015.04.3
2. E.M. Generalova, V.P. Generalov, A.A. Kuznetsova, Procedia Engineering **153**, 167-172 (2016)
3. T.Y. Vavilova, N.D. Potienko, I.V. Zhdanova, Procedia Engineering **153**, 938-943 (2016)
4. N.A. Ilyin, D.A. Panfilov, A.P. Shepelev, Panfilov, A.P. Shepelev, Urban Construction and Architecture **4(17)**, 98-106 (2014) DOI 10.17673/Vestnik.2014.02.14
5. A.Y. Zhigulina, N.G. Chumachenko, Urban Construction and Architecture **4(21)**, 94-99 (2015) DOI 10.17673/Vestnik.2015.04.12
6. V.P. Generalov, E.M. Generalova, Urban Construction and Architecture **2(23)**, 85-90 (2016) DOI 10.17673/Vestnik.2016.02.16

7. H.Y. Lee, Z.X. Hou, Procedia Engineering **14**, 1089-1097 (2011)
8. N.X. Liu, X. Zhao, H.H. Sun, Y.M. Zheng, J.M. Ding, Procedia Engineering **14**, 2503-2510 (2011)
9. M. Balzannikov, V. Alpatov, I. Kholopov, A. Saharov, A. Lukin, MATEC Web of Conferences **73**, 01012 (2016)
10. S.A. Mizuryaev, A.G. Chiknovorian, G.S. Solopova, R.V. Demidov, Procedia Engineering **153**, 599-603 (2016)
11. E.M. Generalova, V.P. Generalov, Scientific Survey **11**, 46-51 (2015)
12. Busan Cinema Center / Busan International Film Festival. [Online]. Available: http://www.coop-himmelblau.at/architecture/projects/busan-cinema-center [Accessed: Apr. 3, 2017]
13. M. Safdie, CTBUH Journal **I**, 12-17 (2011)
14. Roy and Diana Vagelos Education Center / Diller Scofidio + Renfro. [Online]. Available: http://www.archdaily.com/793971/roy-and-diana-vagelos-education-center-diller-scofidio-plus-renfro [Accessed: Apr. 3, 2017]

MATEC Web of Conferences **117**, 00058 (2017) DOI: 10.1051/matecconf/201711700058
*XXVI R-S-P Seminar 2017, Theoretical Foundation of Civil Engineering*

# Some considerations on perpendicular to grain stress in double-tapered glulam beams

*Anna* Al Sabouni-Zawadzka, *Wojciech* Gilewski*, and *Jan* Pełczyński

Warsaw University of Technology, Faculty of Civil Engineering, Warsaw, Poland

**Abstract.** The present paper focuses on the analysis of tensile stresses perpendicular to the grain in simply supported double-tapered beams made of glued laminated timber. Two approaches are compared: a code design verification method and a finite element method (FEM) in terms of the linear theory of elasticity with plane stress assumption. The values and distributions of the analysed stresses are determined, depending on the apex height and orthotropic properties of wood.

## 1 Introduction

Simply supported beams are one of the most commonly used structural members in timber engineering. They can be tapered or have constant cross-section. Tapered beams are often used as roof elements, as their geometry naturally provides a desired roof inclination. Furthermore, the material of such beams is used with better efficiency, because the geometry tends to reflect the bending stress distribution under most load cases. In the present paper double-tapered glulam beams are considered, with a standard geometry described in Eurocode 5 [1] (Fig. 1). In the design and analysis of such beams, apart from shear and bending stresses, tensile stress perpendicular to the wood fibres should be taken into account and reinforcement against this stress should be considered if needed [2-4].

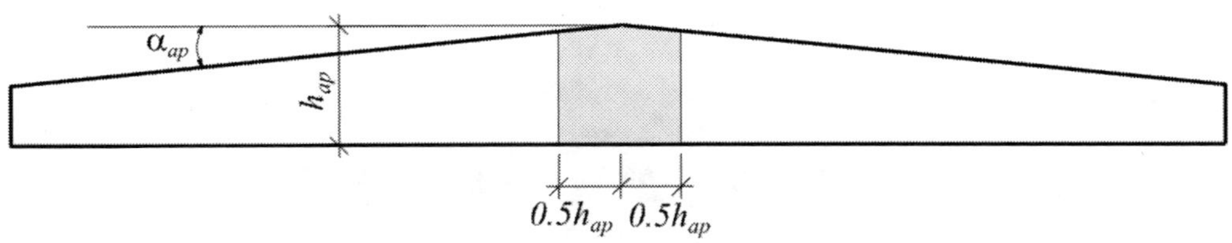


**Fig. 1.** Double-tapered beam considered in Eurocode 5 with the apex zone.

According to the design standard [1] normal stress perpendicular to grain should be evaluated in the apex zone, which is marked in Figure 1. The width of this zone is equal to the apex height. Eurocode 5 provides two alternative formulae for calculation of the stress perpendicular to grain, depending on the load type:

* Corresponding author: w.gilewski@il.pw.edu.pl

$$\sigma_{t,90,d} = k_p \frac{6M_{ap,d}}{bh_{ap}^2}, \tag{1}$$

$$\sigma_{t,90,d} = k_p \frac{6M_{ap,d}}{bh_{ap}^2} - 0,6\frac{p_d}{b}, \tag{2}$$

where: $M_{ap,d}$ is the design moment at the apex, $b$ is the width of the beam, $h_{ap}$ is the height of the beam at the apex, $p_d$ is the uniformly distributed load acting on the top of the beam over the apex area and $k_p = 0.2\tan(\alpha_{ap})$. Whereas the first expression (1) is accepted and recommended (unless the National annex states otherwise) for all load cases, the second one (2) may be used only when the beam is loaded with a uniform load applied to the top of the apex zone.

According to Eurocode 5, the biggest tensile stress perpendicular to the grain should satisfy the expression

$$\sigma_{t,90,d} \le k_{dis} k_{vol} f_{t,90,d}, \tag{3}$$

where: $f_{t,90,d}$ is the design tensile strength perpendicular to grain, $k_{dis}$ is a factor which takes into account the effect of the stress distribution in the apex zone and for double-tapered beams equals 1.4, $k_{vol}$ is a volume factor given by the formula

$$k_{vol} = \left(\frac{V_0}{V}\right)^{0,2}, \tag{4}$$

where: $V_0$ is the reference volume of $0.01m^3$, $V$ is the volume of the apex zone (see Fig. 1) and should not be taken greater than $2V_b/3$, where $V_b$ is the total volume of the beam.

In this paper a simply supported double-tapered glulam beam with varying apex height is considered. Geometry of the beam, support scheme and analysed load cases are presented in Figure 2.

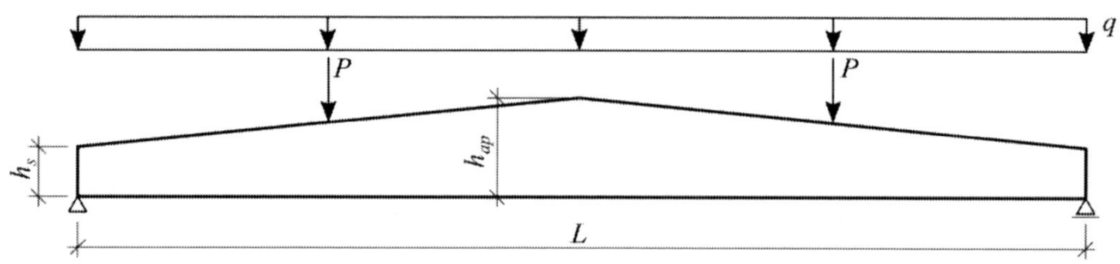


**Fig. 2.** Double-tapered beam with geometry, support scheme and loads.

The purpose of this study is to determine the level and distribution of normal stresses perpendicular to the grain in the analysed example of a glulam beam depending on the apex height and to compare two approaches to the structural analysis of such a system: the design code verification [1] and the finite element method (FEM) [5].

## 2 Analysis

### 2.1 Design code verification

The analysed beam (Fig. 2) is made of glued laminated timber GL28h. Material properties were taken from the currently valid design standard EN 14080 [6] and compared to the ones from the withdrawn standard EN 1194 [7]: characteristic density $\rho_k = 425kg/m^3$ [6] and $\rho_k = 410kg/m^3$ [7], characteristic tensile strength perpendicular to grain

$f_{t,90,k} = 0.5MPa$ [6] and $f'_{t,90,k} = 0.45MPa$ [7]. The design values of strength $f_{t,90,d}$ were calculated using the following expression from Eurocode 5:

$$f_{t,90,d} = k_{\mathrm{mod}} \frac{f_{t,90,k}}{\gamma_M},$$

where: $k_{mod}$ is a modification factor taking into account the effect of load duration and moisture content and was taken as 0.8 (for service class 2 and medium-term action), $\gamma_M$ is the partial factor for material properties and for glulam equals 1.25. The design tensile strengths obtained from the above expression are: $f_{t,90,d} = 0.32MPa$ [6] and $f'_{t,90,d} = 0.29MPa$ [7].

In the analysis the following geometry was adopted: $L = 20m$, $h_s = 1.0m$, $h_{ap} = \{1.25;\ 1.5;\ 1.75;\ 2.00;\ 2.25;\ 2.50;\ 2.75\}m$ (see. Fig. 2) and width $b = 0.2m$. Two load cases were considered: a uniformly distributed load $q = 25kN/m$ acting on the top of the beam and two concentrated forces $P = 25kN$ applied in $1/4$ and $3/4$ of the beam span $L$. The loads were chosen in such a way so that the maximum bending moment had the same value of $1250kNm$ in both load cases. Moreover, deadweight of the beam was considered in some of the performed analyses.

Figure 3 shows how the apex height of the beam influences the maximum value of tensile stress perpendicular to the grain. The calculations were performed twice: in case when concentrated forces expression (1) was used and for the distributed load formula (2) was applied.

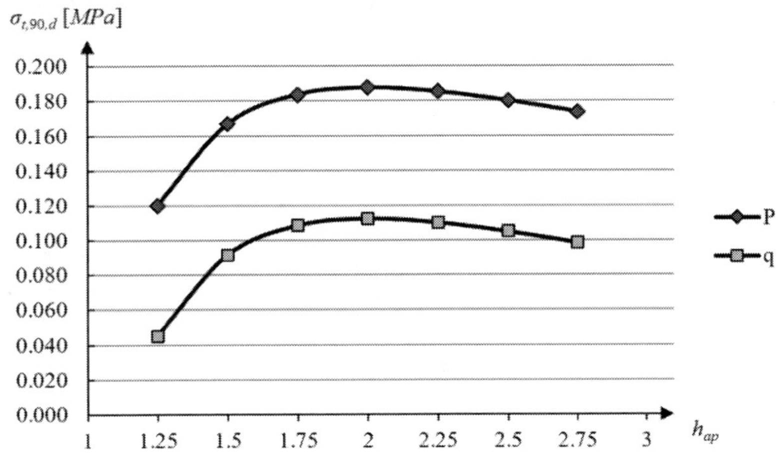


**Fig. 3.** Maximum tensile stress perpendicular to grain for two load cases.

It should be noticed that application of the expression (2) results in obtaining considerably lower values of stresses than in the case of the recommended formula (1). In both cases an extremum can be observed for the apex height of about 2,0 *m*.

In order to check the condition (3) a ratio $\sigma_{t,90,d} / k_{dis} k_{vol}$ was considered (Fig. 4). In the analysed example the condition (3) is fulfilled for all the considered apex heights only in case of the distributed load and application of the expression (2). In case of concentrated forces the condition (3) is satisfied only for lower apex heights – in taller beams additional reinforcement against delamination would have to be applied. Moreover, it can be observed that the maximum value of the beam height for which the reinforcement is needed depends on the design standard that is applied. The new standard [6] gives greater values of strengths perpendicular to grain and thereby allows to design taller beams without reinforcement.

Analysis of the diagram in Figure 4 imposes a question, why the stresses calculated using the expression (2) are so much lower than the ones obtained from the formula (1). For example, the beam with $h_{ap} = 2.25m$ and a distributed load applied on the top does not have to be reinforced if the expression (2) is used but if the formula (1) is applied (which is recommended by Eurocode 5 for all load cases) the reinforcement is required. Such a discrepancy may result from the particular character of a uniformly distributed load acting on the top of the beam in the apex zone, which presses down the upper layers of glulam causing a reduction of tensile stresses.

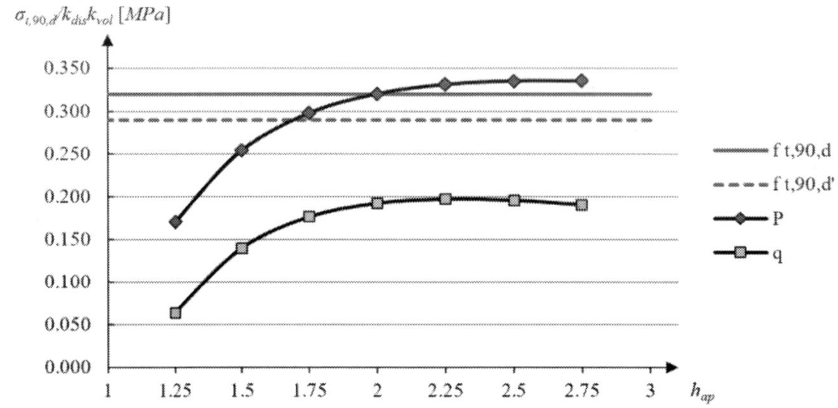


**Fig. 4.** Maximum tensile stress perpendicular to grain for two load cases, including factors $k_{dis}$ and $k_{vol}$.

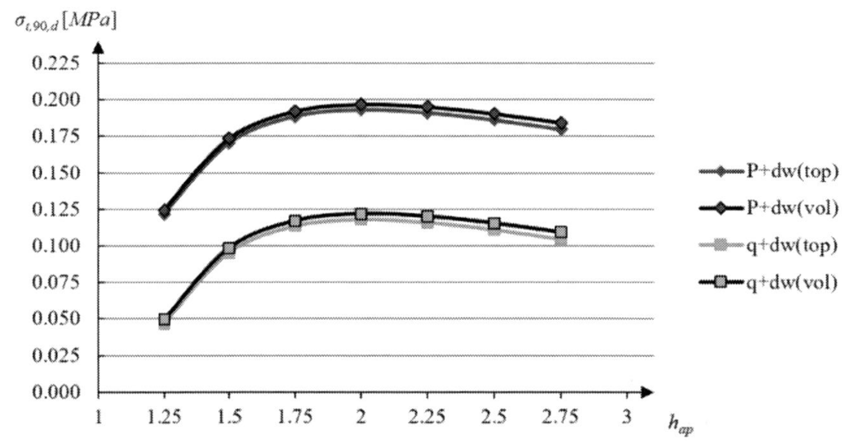


**Fig. 5.** Maximum tensile stress perpendicular to grain for two load cases, including deadweight.

In Figure 5 results of the analysis that included deadweight are presented. Two load cases were considered: concentrated forces with deadweight and distributed load with deadweight. For each case the calculations were performed twice – deadweight was applied either to the whole volume or to the top.

In the analysed example the deadweight of the structure is 3,6% of the total load so its influence on the considered stresses is rather little. Moreover, it practically does not matter if it is applied as volume forces or to the upper surface of the beam – such a hypothetical application of deadweight on the beam top causes insignificant reduction of stresses perpendicular to the grain, so it might be used if needed for simplification of the analysis.

Figure 6 shows how the volume factor $k_{vol}$ depends on the apex height. It might be observed that the value of this factor decreases with an increase of the apex heights, so the taller the beam is, the more rigorous the condition (3) becomes.

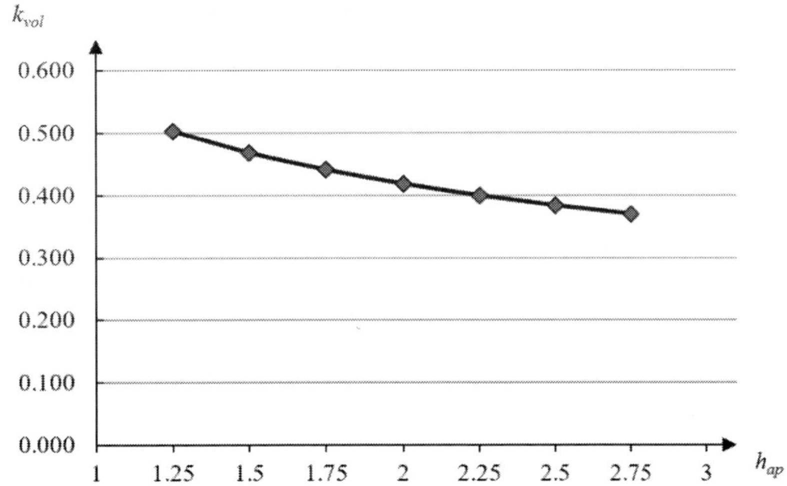


**Fig. 6.** Volume factor.

## 2.2 Finite element analysis

In order to compare two approaches to the analysis of a double-tapered glulam beam, the structure described in the previous section (Fig. 2) was modeled using a 2D linear theory of elasticity in plane stress condition. Four cases of material properties were considered (Tab. 1): two types of orthotropy with the parameters taken from the withdrawn [7] and valid [6] design standard for timber class GL28h, a non-standardized orthotropy and an isotropy as a borderline case. The same geometry, support scheme and loads were applied as in the case of the design code verification. The calculations were performed using the Abaqus FEM software [8]. Two dimensional 8-node biquadratic solid plane stress elements (CPS8R) with reduced integration and seed size of $0.1m$ were used [8].

**Table 1.** Material parameters

| Material | $E_1$ [MPa] | $E_2$ [MPa] | $\nu_{12}$ | $G_{12}$ [MPa] |
|---|---|---|---|---|
| Ort-1 (EN 14080) | 12600 | 300 | 0,35 | 650 |
| Ort-1' (EN 1194) | 12600 | 420 | 0,35 | 780 |
| Ort-2 | 12600 | 2450 | 0,35 | 1428 |
| Iso (Isotropy) | 12600 | 12600 | 0,35 | 4667 |

The FE analysis was performed in order to determine maximum values and distributions of tensile stresses perpendicular to grain and to compare these results with the ones obtained from the calculations according to the design standard.

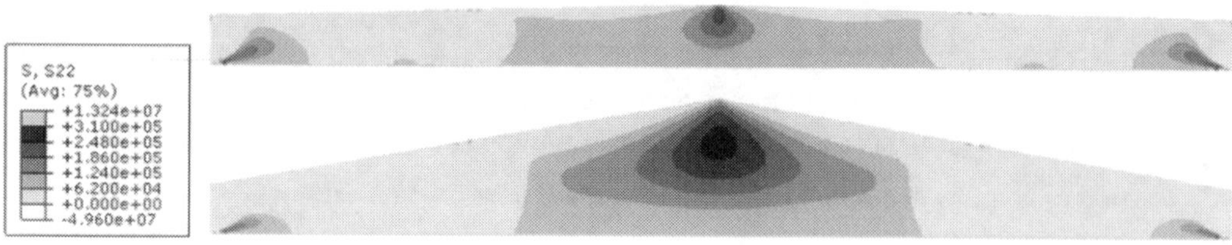


**Fig. 7.** Distribution of positive stress perpendicular to the grain under concentrated forces for the lowest and highest apex.

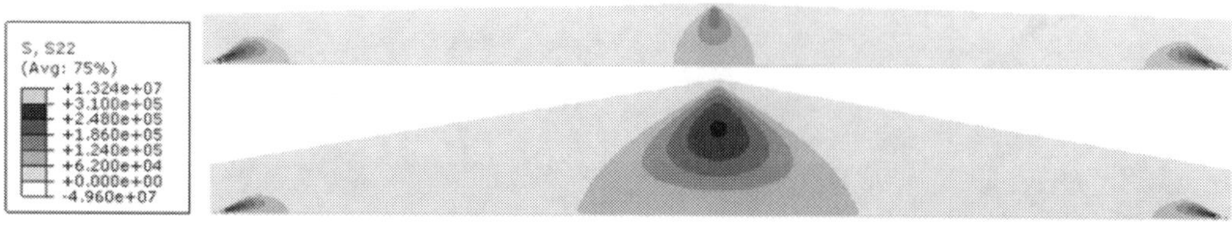


**Fig. 8.** Distribution of positive stress perpendicular to the grain under distributed load for the lowest and highest apex.

Figures 7÷8 present examples of distributions of positive tensile stresses perpendicular to the fibres for the lowest and highest apex considered in the analysis and for two load cases: concentrated forces and distributed load.

The area of positive tensile stress in case of the beam loaded with two concentrated forces (Fig. 7) is much wider than the apex area considered in Eurocode 5 (Fig. 1). This is particularly noticeable for low apex height. However, taking into account the level of stresses, the area of the zone in taller beam is wider. The zone of positive tensile stresses under uniformly distributed load is considerably smaller than in the case of concentrated forces. For distributed load the width of the zone in both taller and lower beams is close to the double apex height, which is also more than the width suggested in the design standard.

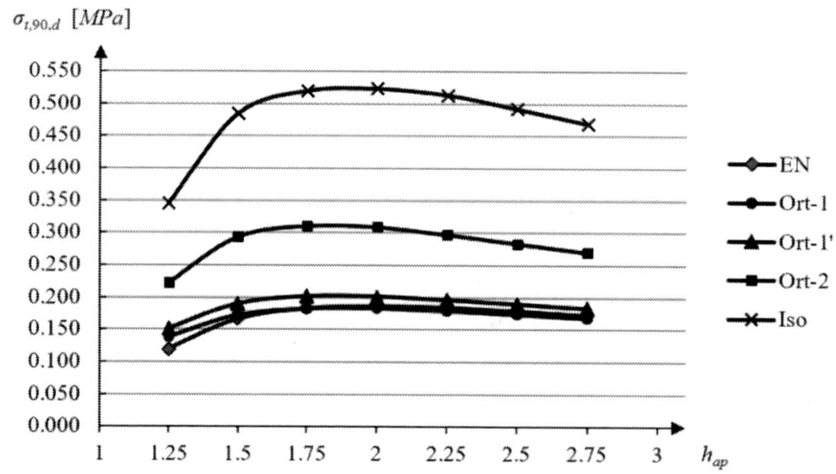


**Fig. 9.** Influence of material properties on tensile stress perpendicular to the grain – concentrated forces.

The maximum values of normal stresses perpendicular to the grain depend on the apex height and adopted material properties (see Fig. 9÷10). Higher stress values are obtained for materials close to isotropy. Assumption of strong orthotropic properties of the material leads to smaller stress variation with regard to the apex height. It is worth mentioning that the greatest value for isotropy occurs when the apex height is two times bigger than the height of the beam in the support zone – the extremum moves in the direction of $h_{ap}$=1,75 *m* for beams with different orthotropic properties, regardless of the type of load (see also [9]).

It should be noticed that the maximum values of stresses from the design code verification (marked "EN" in Fig. 9÷10) are closest to the FE solution obtained for the material parameters taken from the valid design standard [6].

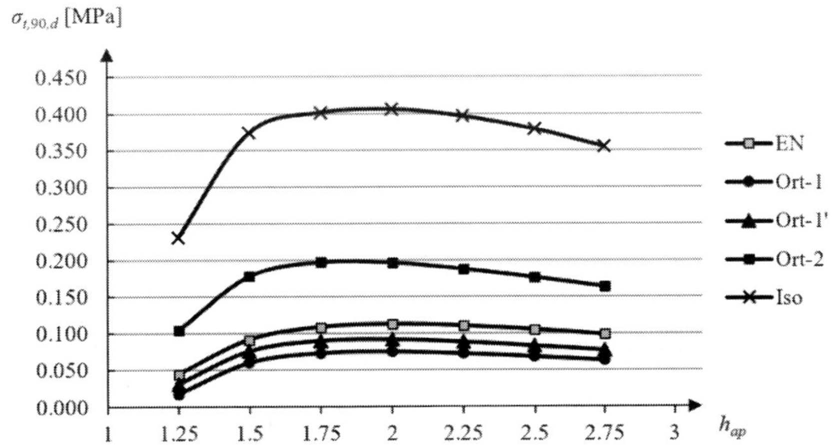


**Fig. 10.** Influence of material properties on tensile stress perpendicular to the grain – distributed load.

Diagrams presented in Figures 11÷12 show how the adopted material properties influence the volume of the zone of positive tensile stresses perpendicular to the grain. Curves marked with "EN" correspond to the apex volume *V* calculated according to Eurocode 5. It can be noticed that the values obtained from the FE analysis are closer to the ones calculated using the design standard in the case of distributed load. When concentrated forces are applied the differences between two considered approaches are more significant.

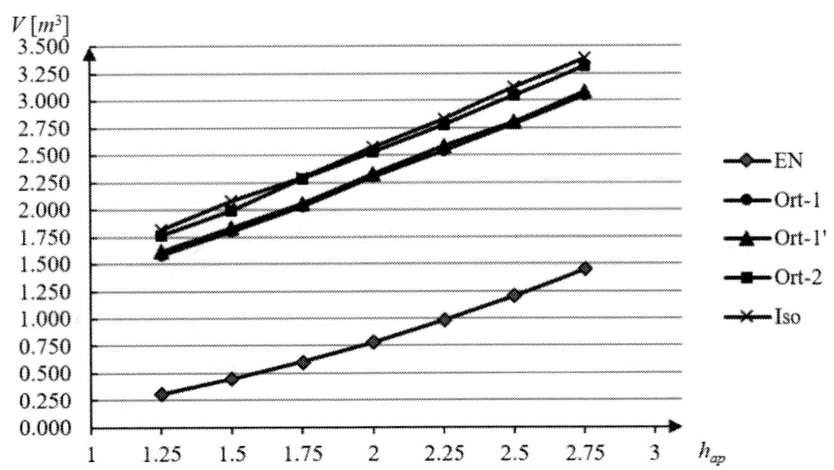


**Fig. 11.** Volume of the zone of positive stresses perpendicular to the grain – concentrated forces.

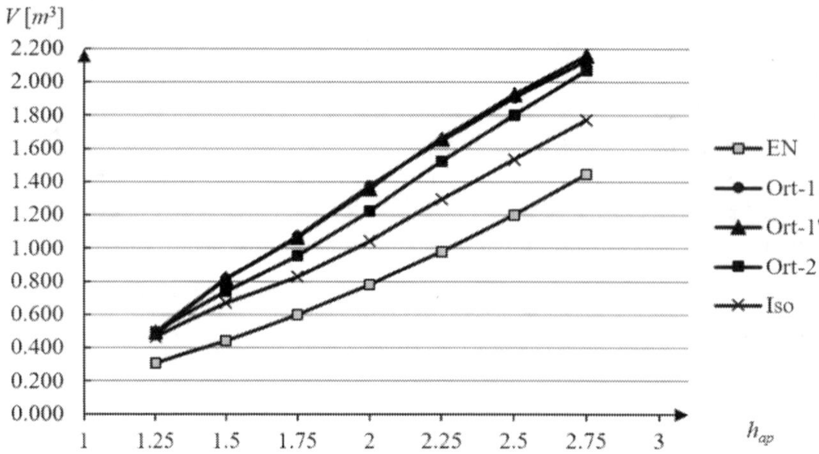


**Fig. 12.** Volume of the zone of positive stresses perpendicular to the grain – distributed load.

## 3 Concluding remarks

The design and structural analysis of double-tapered beams, made of glued laminated timber, require taking into account tensile stress perpendicular to the wood fibres. According to the design standard [1] such stresses should be evaluated in the apex zone, which has a width equal to the apex height. Examples presented in this paper indicate some important differences between two approaches to the structural analysis of such beams: analysis according to the design standard [1] and the finite element method (FEM) [5]. The levels and distributions of tensile stresses perpendicular to the grain are determined depending on the apex height and material properties and differences between Eurocode 5 and FEM are highlighted.

## References

1. EN 1995-1-1:2004 (2004)
2. S. Franke, B. Franke, A.M. Harte, Constr. and Build. Mater. **97**, 2 (2015)
3. S. Thelandersson, H.J. Larsen *Timber Engineering* (John Wiley & Sons, Chichester, England, 2003)
4. S. Vratusa, M. Kitek, V. Kilar, Drewno **54**, 185 (2011)
5. O.C. Zienkiewicz, R.L. Taylor, *The finite element method. Vol. 1. The basis* (Butterworth-Heinemann, New Jersey, 2000)
6. EN 14080:2013 (2013)
7. EN 11994:1999 (1999)
8. ABAQUS Documentation 2017
9. H. Danielsson, *Design and perpendicular to grain tensile stress in double-tapered glulam beams*, [publisher information missing] (2010)

# Concrete Deformations under Simultaneous Impact of Loading and Alternate Freezing and Thawing

*Anatoly* Goncharov[1*]

[1] Moscow state university of civil engineering, Yaroslavskoye shosse, 26, Moscow, Russia, 129337

**Abstract.** To the best knowledge available today, the creep of concrete is understood as slow plastic deformation caused by continuous external factors. Changes occurring in the structure of concrete are however not discussed. Creep as a phenomenon is typically perceived "in purity", in an invariable and favorable environment. Evolution of creep deformations thus has a fading character and usually goes along with consolidation of concrete. It is only under loads near the threshold of concrete's long-term strength that micro-destructions tend to grow during the first phase following application of the load. Creep phenomenon is researched «in pure form», i.e. in constant favorable external conditions. Developing of creep deformations has fading character and accompanied by seal of concrete. In real exploitation conditions of constructions appears changes in temperature and humidity, influencing on development of creep deformation. Alternate freezing and thawing of concrete, which most affects durability, causes in unloaded concrete significant extension deformations. Deformation development under the joint action of loads and alternate freezing and thawingalmost was not researched. In the article presented results of concrete deformation research, differing by frost resistance, under loads in the range from 0,2 to 0,7 of prismatic strength.

## 1. Introduction

The most detailed research of concrete creep was performed in the period from 1960 to 1975 and led by V.M. Moskvin, A.A. Gvozdev, O.Ya. Berg. Recently, significant attention has been paid to structural changes of concrete under load. Rheological models have been offered to predict and calculate creep deformations [1,2,3] in a sufficiently reliable manner and reflect them in design standards [4].

The most detailed knowledge was obtained for concrete creep in the ambient air environment under normal positive temperatures [5;6;7;8]. Humidity impact on creep deformation evolution was reflected in papers [9;10] to allow these factors to be accounted for in design standards. The impact of positive temperature variations was investigated in papers [11,12]. Creep deformation slowdown under negative temperatures was noted in papers [13;14; 15] which is caused by improved strength of frozen concrete [15]. However, in paper [14] investigating the concrete sample creep within the temperature range from

* Corresponding author: ttg38@mail.ru

+20°C to -20°C it was noted that the deformation increased as ice was formed in concrete. The creep stops if the temperature continues decreasing. The creep increase of frozen concrete samples with high water saturation was also noted in paper [16]. This type of creep deformations indicates that they are associated with changes taking place in concrete during freezing.

Destruction of unloaded concrete under alternate freezing and thawing is accompanied by significant expansion deformations [17]. Sufficiently detailed data on deformation evolution caused by the simultaneous impact of tensile stresses and alternate freezing of concrete can be found in paper [18]. These data allow deflection design values of bending reinforced concrete structures to be adjusted.

Deformation development caused by the simultaneous impact of compressive stresses and alternate freezing and thawing of water-saturated concrete has been insufficiently investigated. According to paper [19], expansion deformations along the load axis are not observed under simultaneous compressive stresses (40% of concrete prism strength) and alternate freezing and thawing. Instead, fading compressive deformations 6 times higher than creep deformations under the same load and normal conditions take place. The paper also notes that the difference in deformations flattens under lower load values.

The above data indicate that deformations of reinforced concrete structures subjected to multiple freezing and thawing cycles may develop which are not taken into account in present-day creep calculations. Such deformations in pre-stressed structures may result in the loss of pre-stress exceeding design values.

Loaded concrete deformation under alternate freezing and thawing may be considered as creep deformations. However, this creep has a specific nature as it is related to water-saturated concrete destruction under negative temperatures. Build-up of deformations is accompanied by concrete softening and strength degradation.

## 2. Basic Part

In the light of foregoing, a research was carried out to study deformation data and take them into account when designing structures operating under alternate freezing and thawing.
Techniques. 10 x 10x 40 cm concrete samples mounted in a levered disk spring loading device and 2.5 x 2.5 x 25 cm samples loaded with high-strength rods running through the sample center were subjected to freezing and subsequent thawing in sea water.

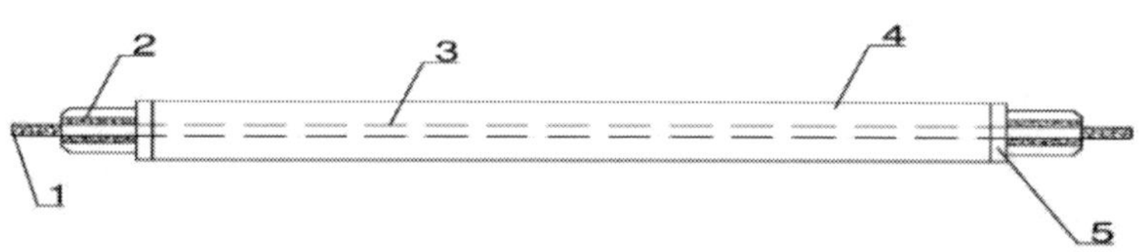


**Fig. 1**. Loading diagram for sample sized 2.5 x 2.5 x 25 cm:
1- taper socket for mounting on the deformation gage;
2 – screw nut; 3- 6 mm rod; 4 - concrete sample; 5 - 6 mm washer.

Before concreting, the rods had been treated with a special substance to prevent their bonding with concrete. The sample was evenly loaded across the complete cross section using 6 mm thick steel butt plates. The load value was determined using three techniques: by reading the gauge indication of the torque wrench used to strain the rod; checking the rod and concrete sample deformations. Prior to this, the torque wrench had been calibrated using a special device equipped with a reference load indicator. The rod elasticity modulus was checked at the same time.

10x10x40 prism deformation was measured using a portable dial gauge strainmeter mounted between prism buttresses (30 cm base). Measurements were performed every 10-15 alternate freezing and thawing cycles. Before commencing the measurement, the strainmeter and reference steel sample were kept exposed to build up the ambient temperature at which the measurements were conducted. Temperature correction for the device deformation was determined using the following formula:

$$\Delta L = aL(t_1 - t_1) + (A_0 - A_1), \text{ where:}$$

a – steel line expansion coefficient;
L – reference standard length;
t - ambient temperature of deformation measurements;
t - constant temperature the measurements were reduced to;
A - reading for steel reference standard at t = t;
A - reading for steel reference standard at t = t.

2.5 x 2.5 x 25 cm prism deformation was also measured every 10-15 alternate freezing and thawing cycles. A desktop dial gage strainmeter was used. Sample contraction deformations over the period between the measurements did not exceed $2 \cdot 10^{-4}$. At the same time concrete stress decreased by $\Delta\sigma < E_a \cdot \varepsilon \cdot F_a / F_б = 1{,}8 \cdot 10^6 \cdot 2 \cdot 10^4 \cdot 0.196/6 = 11.7$ kg/cm$^2$, which amounts to no more than 10% of the original value. After each deformation measurement, the original concrete stress level of the sample was restored followed by the dynamic elasticity modulus measurement.

Results. The deformation measurement results for the 10 x 10 x 40 cm prisms are shown in Fig. 2. The broken line indicates creep deformation of concrete of the given mix which is not subjected to alternate freezing and thawing. As it is seen, there is a significant difference in the development nature of the regular creep and that caused by alternate freezing and thawing.

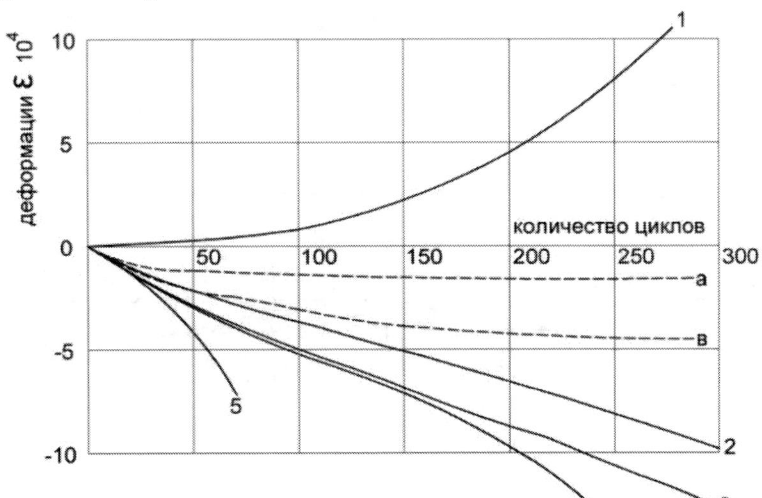


**Fig. 2** - Deformation growth in prisms sized 10 x 10 x 40 cm
1 – σ=0; 2- σ=0.2 $R_{пр}$; 3-σ=0.4 $R_{пр}$; 4- σ=0.6 $R_{пр}$; 5-σ=0.7-0.8 $R_{пр}$;
a – σ=0.2 $R_{пр}$; в- σ=0.6 $R_{пр}$ (under normal conditions)

At σ = 0.2 and 0.4 Rпр, the deformation increase is proportional to the number of cycles. At σ = 0.6 and especially at 0.8 Rпр, deformation development attains a progressive nature which is caused by the formation of micro-cracks and degradation of concrete strength at the given load values. This is confirmed by ultrasonic velocity measurements in the samples.

2.5 x 2.5 x 25 cm prism deformation measurement results are shown in Fig. (additive-free mix) and Fig.4 (mix with modified air entraining resin added). The dotted curves in Fig. 3 reflect the creep development in samples of the identical mix (at σ = 0.5 и 0.7 Rпр), not subjected to alternate freezing and thawing cycles.

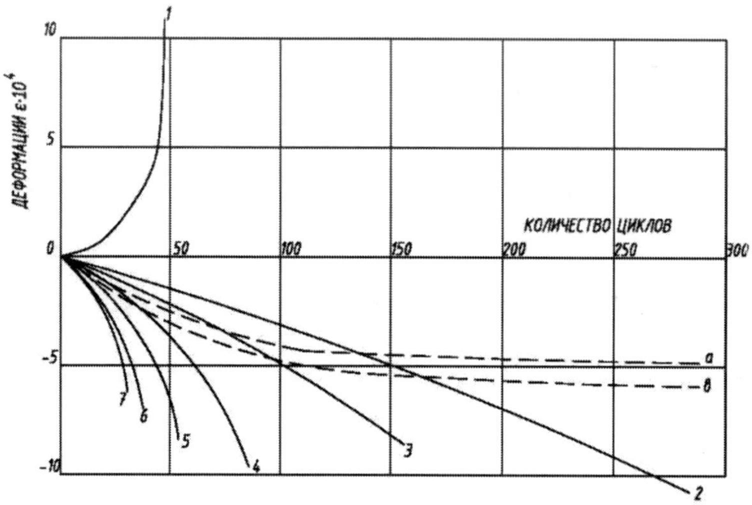


**Fig. 3** – Deformation growth in prisms 2.5 x 2.5 x 25 cm (no additives, $R_{пр}$ = 39 MPa)

1 – σ = 0; 2 – σ = 0.2 $R_{пр}$; 3 – σ = 0.3 $R_{пр}$; 4 – σ = 0.4 $R_{пр}$; 5 – σ = 0.5 $R_{пр}$;

6 – σ = 0.6 $R_{пр}$; 7 – σ = 0.7 $R_{пр}$; a – σ=0.5 $R_{пр}$; в- σ=0.7 $R_{пр}$ (under normal conditions)

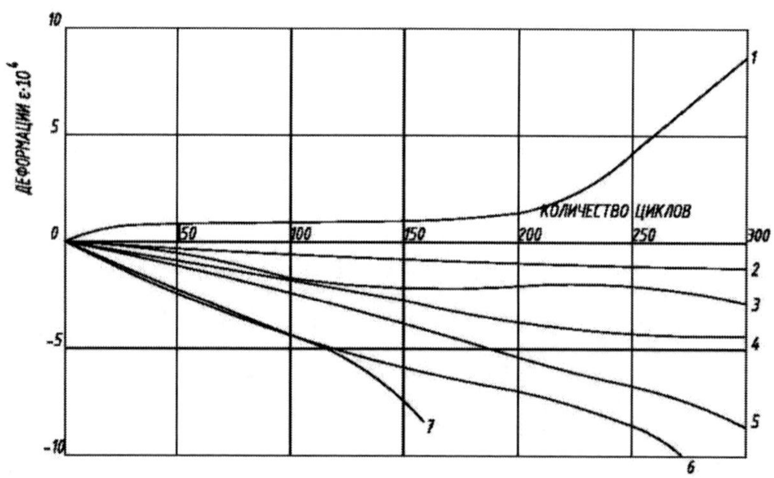


**Fig. 4** Deformation growth in prisms 2.5 x 2.5 x 25 cm (mix with modified air entraining resin added, $R_{пр}$ = 37 MPa) 1 – σ=0; 2- σ=0.2 $R_{пр}$; 3-σ=0.3 $R_{пр}$; 4- σ=0.4 $R_{пр}$; 5-σ=0.5 $R_{пр}$;

6-σ=0.6 $R_{пр}$;7-σ=0.7 $R_{пр}$

As seen from the diagrams, in all the cases the additive-free samples display residual deformations significantly higher than regular creep deformations at the same load value. Deformations develop in a progressive manner at stresses exceeding 0.3-0.5 Rпр for all additive-free sample batches. At the same time, the highest deformations by the moment of destruction were displayed by the samples with the lowest loads which had survived the biggest number of cycles before destruction.

Test results confirm the data found in paper [19]. In the diagram (Fig. 2), one can see, that at 0.4 Rпр the contraction deformations are 5 times higher than creep deformations occurring at the same force level under normal conditions. However, such a variance is typical for the concrete with the strength below 30 MPa. At a higher strength, the difference decreases. At 45-50 MPa, deformations (at $\sigma$=0.4 R) vary by 3-4 times. In concrete types with added modified air entraining resin the difference in deformations flattens. (Fig. 4).

Concrete structure modification by adding modified air entraining resin resulted in a significantly improved freeze-thaw resistance and changes in the nature of deformation development due the fact that the additive created reserve "relatively closed" porosity not filled with water under alternate freezing and thawing (Fig. 4). Absolute deformation values in samples with added modified air entraining resin are 3-5 times lower than in additive-free samples (at the same strength). At $\sigma$ = 0.2-0.4 Rпр, not exceeding the R boundary, the nature of sample deformation development differs insignificantly from the creep under regular conditions. However, as the strain increases, the deformations also grow. Ultrasonic measurements in 10 x 10 x 40 cm samples and elasticity modulus measurements in all tested samples indicate that the creep deformation rate and concrete destruction rate under alternate freezing and thawing are correlated. Creep deformations are the highest in the samples where strength degradation and concrete structure softening occur faster, while expansion deformations in unloaded twin samples are the highest too.

## 3. Conclusion

Based on performed tests, the following key factors affecting the creep nature under compression and alternate freezing and thawing can be identified:

1. During ice formation, moisture migration in the direction of growing ice crystals in the largest pores takes place [20; 21]. At the same time, dehydration of smaller "gel" pores occur which facilitates creep deformations. Therefore, increased creep may take place during concrete freezing without structural damage.
2. It is known that at stresses exceeding the micro-crack formation boundary R, creep rate increase is related to the growth of micro destructions in time. At the same time, if the stresses do not exceed durability limits, micro destructions fade in time and the creep becomes linear. If water–saturated concrete freezes under load, ice formation is accompanied by the development of micro-cracks and new micro-defects which result in persistent nature of creep deformations.
3. Concrete structure softening and concrete strength degradation take place under multiple freezing and thawing. Therefore, the level of relative stress increases from cycle to cycle and results in increased creep.
4. Moisture migration under alternate freezing and thawing is restricted in freeze-thaw resistant concrete (with added modified air entraining resin) and micro-defect either develop slowly or do not develop at all. In this case, under stresses below the R boundary, deformations do not differ significantly from creep under normal conditions.
5. The presented data allow a well-informed selection of concrete specifications during the design process to be performed and potential increase in operational structure

deformations resulting in increased bending and decreased reinforcement pre-stress value to be taken into account.

## References

1. E.E. Goncharov, Modelirovaniye polzuchesti betona v differencialnoy forme s ispolzovaniem reologicheskikh modeley. // Trudi III Vserossiyskoy (II Mezhdunarodnoy) konferencii po betonu i zhelezobetonu, tom1, M. 21-26 (2014)
2. K.Z.Galustov Nelineynaya teoriya polzuchesti betona i raschet zhelezobetonnikh konstrukciy.//Fizmatlit, M., (2006)
3. G.K. Fanouraksis Otnositelnaya rol' koefficienta polzuchesti i modul'a uprugosti v modelakh prognozirovania polzuchesti betona. // Trudi III Vserossiyskoy (II Mezhdunarodnoy) konferencii po betonu i zhelezobetonu, tom1, M. 153-163 (2014)
4. AS 3600-2009 Concrete Structures. //Standards Association of Australia, Sydney.
5. A.A. Gvozdev, Polzuchest' betona. //Trudi 2-go vsezouznogo s'ezda po teoreticheskoy i prikladnoy mekhanike, izd. Nauka, 74-81. (1966)
6. R.S. Sanzharovskiy, Nelineynaya nasledstvennaya teoriya polzuchesti. //J-l Stroitelnaya mekhanika inzhenernikh konstrukciy i sooruzheniy, **1**, 63-68 (2014)
7. R.S. Sanzharovskiy M.M. Manchenko, Polzuchest' betona i ego mgnovennaya nelineinost' deformirovaniya v raschetakh konstrukciy. //J-l Stroitel'naya mekhanika inzhenernikh konstrukciy i sooruzheniy, **2**, 33-40 (2015)
8. R.S. Sanzharovskiy, M.M. Manchenko, Oshibki teorii polzuchesti zhelezobetona i sovremenniye normi. //J-l Stroitelnaya mekhanika inzhenernikh konstrukciy i sooruzheniy, **3**, 25-32 92016)
9. Ukazaniya po proektirovaniu zhelezobetonnikh i betonnikh konstrukciy zheleznodorozhnikh, avtodorozhnikh i gorodskih mostov i trub. SN 365-67., //M., (1967)
10. A.G. Tamrazyan S.G. Esayan, Mekhanika polzuchesti betona. //FGBOU VPO MGSU (2012)
11. S.F. Klovanich, Polzuchest' betona pri slozhnikh rezhimakh nagruzheniya i nagreva. // Trudi III Vserossiyskoy (II Mezhdunarodnoy) konferencii po betonu i zhelezobetonu, tom1, M. 59-68 (2014)
12. O.R. Barani, D. Mostofinejaad, M.M. Saadatpour, M. Shekarchi, The Arabian Journal for Science and Engineering, **35, 28**, 106-121 (2010)
13. I.N. Seregin I.N. Polzuchest' betona v dorozhno-mostovikh sooruzheniyakh. Izd. «Transport», (1965)
14. R. Johansen, C.N. Best, Creep of Concrete mith and mithouf ice in the system. Bulletin Rilem, **16** (1962)
15. A.D. Istomin Deformatsiya polzuchesti betona pri szhatii v usloviyakh malociklovogo silovogo i temperaturnogo nagruzheniya. //Vestnic MGSU, **2-1**, 28-32 (2011)
16. L.N Antonov, Issledovanie vliyaniya nizkih temperature na prochnost' i na deformativnost' betona i zhelezobetona. //Diss., M., (1968)
17. P. Klieger Journal of the American Concrete Institute, **27, 2**, 7-13 (2012)
18. A.M. Podvalniy Polzuchest' pri izgibe poperemenno-zamorazhivaemogo betona. //Trudi koordunatcionnikh soveshaniy po gidrotekhnike, vip.13, (1964)
19. A.M. Podvalniy, O polzuchesti zamorazhivaemogo betona. Dokladi AN SSSR,148, No. 5, (1963)
20. V.M. Moskvin, M.M. Kapkin, B.M. Mazur, A.M. Podvalniy, Stoikost' betona i zhelezobetona pri otricatelnoy temperature. Stroizdat, M, (1967)

# Geometrical analysis of Palesch family chapel in Kľačno, former Gaidel in Western Slovakia.

*Zuzana* Grúňová[1,*], *Michaela* Holešová[2]

[1]Department of Building Engineering and Urban Planning, Faculty of Civil Engineering, University of Žilina, Univerzitná 8215/1, 010 26 Žilina, Slovakia
[2]Department of Structural Mechanics and Applied Mathematics, Faculty of Civil Engineering, University of Žilina, Univerzitná 8215/1, 010 26 Žilina, Slovakia

**Abstract.** Oval and elliptic spaces are one group of central plans, used mainly in sacral architecture and palaces. Paper deals with geometrical analysis of one of the smaller representatives of the neoclassical sacral architecture with oval plan - Palesch family chapel of Virgin Mary in village Kľačno in Western Slovakia. Oval and elliptic forms are not so often used in Slovak historical architecture and they are almost always connected with foreign influence and knowledge brought form Vienna, Paris, Pest, Eger and other education and praxis localities of the builder or architect. This uncommon oval form used in the small chapel is therefore certainly of interest from the point of view of architecture and geometry.

## 1 Introduction

Oval and elliptic spaces in architecture start to appear in the 15th - 16th cent. and evolution of forms continued with refinement of geometric constructions and broadening of mathematical knowledge. Baroque central sacral and hall spaces were the most inventive demonstration of this progress. Oval, circular and elliptic central spaces continued as a part of architectonical vocabulary in Slovakia also in the following periods of neoclassicism and historicist styles, but only few examples of fully developed large spaces with curvilinear forms could be found here and it is therefore a surprise, that oval was chosen for a small chapel. Fine form and well thought proportions are the most certainly connected with person of highly educated Georg Palesch. Further research in the archive of Spiš Diocese in Spišská Kapitula will be able to confirm hypotheses about construction and bring maybe even original drawings and name of author, who is unknown at the present.

## 2 Village Kľačno, its history and architecture of the chapel

Village Kľačno with population of 1100 inhabitants situated in Prievidza district, Trenčín region in the western Slovakia, surrounded by Strážovské mountains, most of the dwellings

* Corresponding author: grunova@fstav.uniza.sk

lie in the narrow valley of Nitra river. Village was founded in the 13th cent. (the first mention in 1413 as Gaydel). It was mostly occupied by German catholic inhabitants, only after the WWII Slovaks become a majority. Name Gaydel (Gaÿdiell ad Baÿmocz[1], Gaydelehota) remained together with Hungarian name Nyitrafő (1913) up to 1948, when village got name Kľačno. Village is rich in term of historical buildings comparing to the size of the settlement - Roman catholic church with preserved renaissance constructions (built 1464) and two chapels (baroque from 1777 and neoclassical from 1824). Patron of the analysed monument - Chapel of Virgin Mary - was Juraj Palesch - Georg Palesch[2], born 1753 in Gaidel (Kľačno) died 1833 in Zipser Kapitel (Spišská Kapitula) [3,4]. He was a roman catholic priest, teacher and pious writer, known as a deeply educated and enlightened personality who founded schools and foundations for education promotion[5]. Palesch founded school also in Kľačno (built 1829) and decided to built here chapel of Virgin Mary, dedicated to memory of his deceased parents (both born and lived in Kľačno). About the period the construction of Palesch family Chapel in Kľačno, that was in the year of 1824 according to[6], village had 254 houses with 1774 inhabitants.

**Fig. 1.** Exterior of the Palesch Family chapel of Virgin Mary in Kľačno. Photo R. Kollárová, 2015.

Chapel was built in the neoclassical style in 1824. It is oval in the plan, rather small building decorated with pilasters and portico with wooden bell tower in the exterior and disproportionally large built-in organ gallery (added in 1827 according to inscribed cartouche "ANNO 1827 JT " on organ encasement) in the interior. Building was not profoundly altered up to date - geometry analysis could be conducted at the original proportions (original masonry shape, volume). Roofing replacement and minor renovations were realized in 1905, 1927. Renovation together with waterproofing and humidity control works were realized in 1995 (project of Ing. M. Pichová). [7]

## 3 Geometrical analysis of chapel

### 3.1 Module consideration and measurement units

Compact mass of the chapel shows clearly well thought proportion ratios indicating primary geometric construction. The very first step in geometric analysis was a search for modularity in plan of structure. Square raster was used with various modules. The most striking was coordination of the most important structural points of chapel with module 600 by 600 mm (fig.2.). Intersections coincide with regulation, main points of circle sections describing curvilinear complex plan. Rectangle circumscribing the main oval/elliptic space is constructed in the ratio 6:9 (approx. 3600 by 5400 mm). Also other main axes and lines lie very closely to or at the modular net - e.g. centres of pillars on the main facade, centres of the concave chamfering between main mass and portico entrance etc.

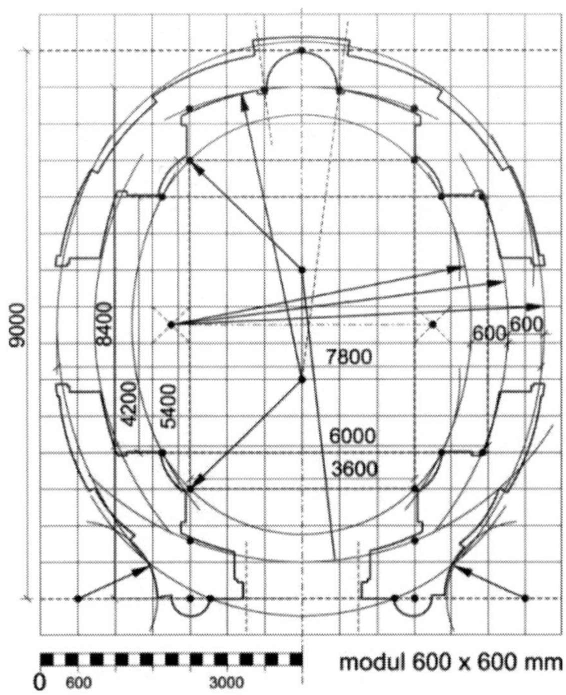


**Fig. 2.** Analysis of possible modular net applied to plan of the chapel.

Modularity close to 300 or 600 mm would be logical, because of the most probably used measurement units of the Austrian Empire - the Vienna fathom ((Kl) 6 ft = 1896 mm ) and its fraction - foot, Fuß Another system of units that could be used is Roman cubit (römische Elle), Roman foot (pes) , Vienna cubit (Wiener Elle) or German elbow (elle). Any system used would be approximation of human proportions. Further analysis of historical measurement units depends on if original plans could be found in the future. Also overall mass of the chapel structure is so small comparing to the possible observational error and deviations caused by building process, that it is impossible to relevantly proof difference between possible historical measurement units.

**Table 1.** Deviations of the multiples of historical units comparing to measurements in mm.

| | mm | multiple | width | deviation | multiple | length | deviation |
|---|---|---|---|---|---|---|---|
| 600 mm | | **6** | 3600 | **0,00 %** | **9** | 5400 | **0,00 %** |
| Vienna ft | 316.080 | **11,5** | 3634.92 | **0,97 %** | **17** | 5373.36 | **0,49 %** |
| Roman cubit | 444.000 | **8** | 3552 | **1,33 %** | **12** | 5328 | **1,33 %** |
| Roman foot | 296.352 | **12** | 3556.224 | **1,22 %** | **18** | 5334.34 | **1,22 %** |
| Vienna cubit | 779.300 | **4,5** | 3506.85 | **2,59 %** | **7** | 5455 | **1,02 %** |
| German elle | 600.000 | **6** | 3600 | **0,00 %** | **9** | 5400 | **0,00 %** |

Dimensions like width of the masonry could be attributed to the dimension of the brick used. In this case it is brick 55 by 285 by 135 mm, with plastic inscription CFP - signature means COMES FRANCISCUS PALFFY. Bricks were produced in the kiln of Palffy Bojnice estate in the first third of the 19th century. This kind of bricks can produce wall thickness as a multiple of approx. 150 mm.

### 3.2 Hypothesis A - Ellipse was used in construction drawing and on site

Elliptic constructions are rather common in technical praxis, mainly because of its interesting properties - sum of distances between any point on the ellipse to each of its focuses is constant. But it is not so easy to draw. Various drawing aids were invented to help the process of drawing an elliptic forms in technical design. One of the oldest so called "ellipsographs" originated from the idea of ancient Greek trammel of Archimedes (maybe even from Proclus time). Later on this idea was transformed into many simpler or more sophisticated variations. One of them, based on ellipsograph of Guidobaldo del Monte is described in [1]. Description only deals with mechanism and author took as already clear, that final drawing is really ellipse in the geometrical sense. Proof is in fact already implicated by the means of construction of the mechanism.

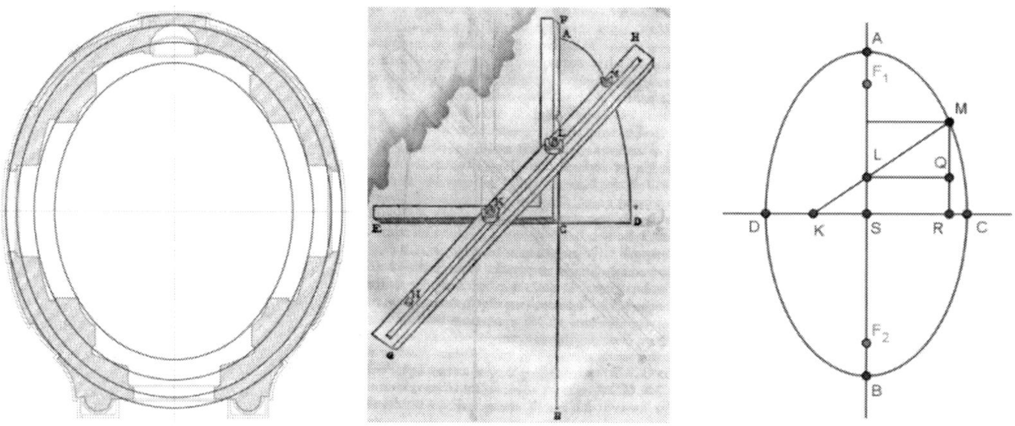


**Fig. 3.** Left are ellipses superimposed on the chapel´s plan, drawn by Grúňová, in the middle is L'ellissografo, Planisphaeriorum universalium theorica, Pesaro 1579 presented in [8]. Right is geometric reconstruction of L'ellissografo created by Holešová.

Ellipsograph consists of two long flat boards, perpendicular to each other with length not shorter than length of minor axis of the ellipse. Third part is movable flat board with two sharp points - two screws, marked as K, L see fig.3 that could be moved on it and fixed in the distances $a$ and $a$-$b$ ($a$ is equal to the major axis and $b$ is equal to the minor axis of the ellipse) from the drawing end M, which will inscribe ellipse. It is clear that distance $|LM| = b$. Points K, L are fixed on the third board, but they have to be fluently movable along two basic perpendicular boards, so K is movable along board with points C,D and L is movable along board marked by A, B. End point M will describe the ellipse during motion, what can be easily proved. From the similarity of the triangles ΔKRM and ΔLQM and at the designation $|KS|= v$, $|SR|= x$, $|RM|= y$ is

$$\frac{v+x}{a} = \frac{x}{b} \tag{1}$$

We have for the rectangular triangle ΔKRM and the lengths of its sides that

$$(v + x)^2 + y^2 = a^2 \quad (2)$$

Equations (1) and (2) show that $\frac{x^2}{b^2} + \frac{y^2}{a^2} = 1$ and point M is already located on an ellipse whose centre is at point S and the length of the major and minor axis are $a$ and $b$, because $x$ and $y$ are the coordinates of M.

### 3.2 Hypothesis B - Mandorla and pole-rope method on site.

Mandorla (also called Vesica Piscis) is geometrical shape also often with symbolical meaning, used in the medieval European art and architecture, but with an ancient origin. Simply constructed as intersection of the two same circles with centres on their radiuses. Intersection is almond shape with symbolical meaning of origin of life. Combination of two mandorlas with their axis oriented perpendicularly is called "Eye of the God". Mandorla is simple construction, easily drawn using poles and rope, therefore practical for building site construction.

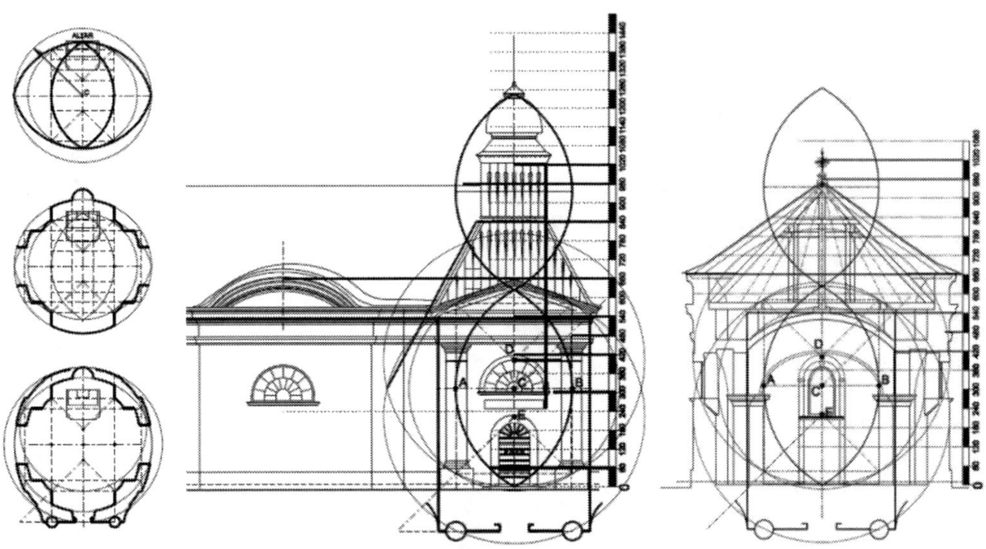

**Fig.4.** Oval constructions by Serlio. Source [3] redrawn by Grúňová

All of the points of the mandorla based geometrical construction are derived from two points of origin A and B with distance of approximately 3600 mm, that create a hypotenuse of the first smaller mandorla. All derived points could be drawn using circles or their section and divisions of segments to 2 or 3 parts. Construction of the "God's eye" is modified, more "rounded" - bigger mandorla has centres of its circles at the points with the distance of 1/6 from endpoints of smaller mandorla hypotenuse. Length of basic segment A, B close to 3600 mm also means transfer of hypothetic module close to 600 mm to all derived distances. The same mandorlas could be attributed to position of important points of exterior architecture of the Chapel.

### 3.3 Hypothesis C - Oval based construction and pole-rope method on site.

Commonly used approximation of the ellipse are various oval shapes consisting of combination of the arcs - more easy to construct basic oval and parallel smaller and bigger oval arcs. The most famous geometrical constructions of oval (used also in baroque and later) are Serlio´s two-centre ovals. (fig.5).

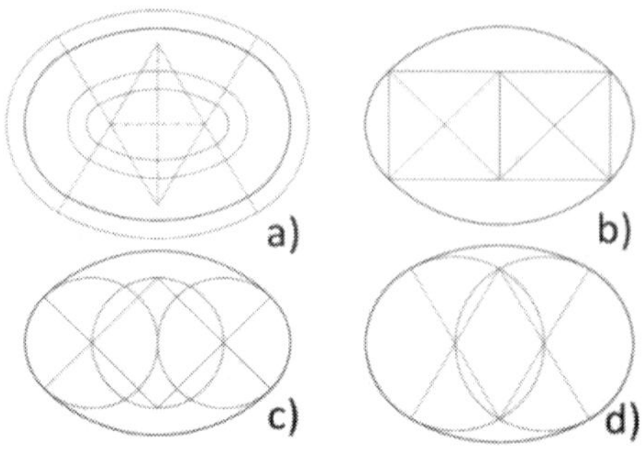


**Fig.5.** Oval constructions by Serlio. Source [9] redrawn by Grúňová

All these constructions retain the length of the major axis for the given ellipse. Ellipse is approximated by these constructs and for constructions (b) to (d) the length of the major axis gives the length of the minor axis. If we denote the height of the triangle ΔUHK as $k$ and half of the side of the triangle ΔUHK as $h$ (see Figure 6), we can express these lengths for each type of Serlio's constructions. We can also express the length of the minor axis

$$b = a + \sqrt{h^2 + k^2} - (h + k). \tag{4}$$

In the construction in Figure 5a), we can approach two ways. The first is determining the length of the minor axis $b$, thus finding the length of the triangle side $2h$.
The second is determining the length of the side $2h$ of the triangle ΔUHK and calculating the length of the minor axis $b = a + h(1 - \sqrt{3})$.

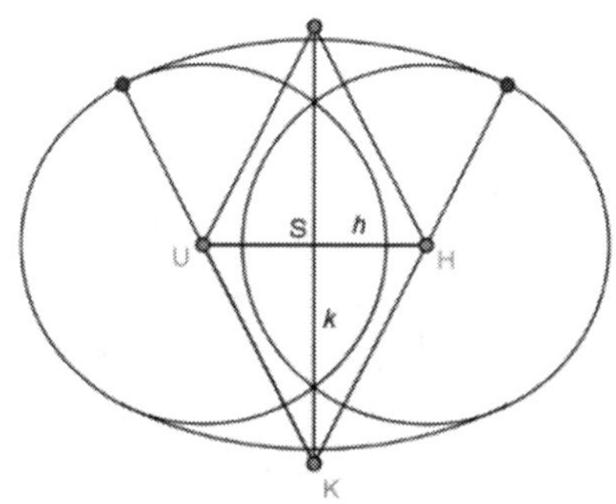


**Fig. 6.** Construction of two-centre oval, based upon Serlio's treatise redrawn by Holešová.

ΔUHK is the equilateral triangle for the construction of the Fig. 5 a) and when we know $a$, $b$ is $h = \frac{a-b}{\sqrt{3}-1}$, $k = \sqrt{3}h$. When we have two squares as in Fig. 3b) is $h$=$k = \frac{a}{\sqrt{2}+1}$, $b = a(5 - 3\sqrt{2})$. For Fig. 3c) is $h = k = a/2$, $b = \frac{\sqrt{2}}{2}a$ and in case Fig. 5 d) is $h = a/3$, $k = \sqrt{3}h$, $b = \frac{4-\sqrt{3}}{3}a$.

Length (longer distance) of the chapel was chosen as the constant 2a to be applied as a basis for construction of all types of ovals, constructed after Serlio. Resulting ovals were positioned on the plan of the chapel, see fig.7.

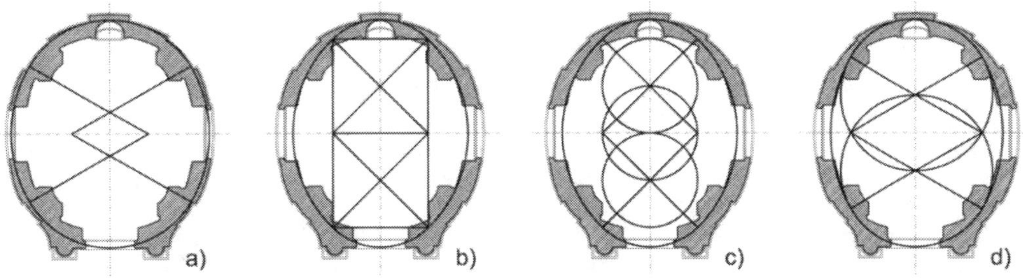


**Fig. 7.** Comparison of resulting oval construction after Serlio (fig.5) to the actual plan of the Chapel in Kľačno, using length of chapel as 2a (double major axis) as a defining base. Drawing by authors.

## 4 Conclusions

**Hypothesis A Ellipse** - construction is simple, but not easily applicable on building site - maybe on small scale, on larger scale is not applicable at all. If similar drawing method would be used on building site of the Palesch family Chapel in Kľačno movable board with drawing point M would had be almost 6 m long.

**Hypothesis B Mandorla based construction** - easily executable, but highly complicated one, therefore it is much less probable that it would be the first choice for a small scale design of the Chapel (and no visible mention in the form of inscription of symbolism on the Chapel itself or in the chronicle is found by research so far).

**Hypothesis C Oval based construction** - comparison on the fig.7. clearly shows that three of Serlio's methods result in too narrow ovals to be applicable on the chapel's plan. The first construction is much more flexible, so it is possible to use reverse-engineering to find the most suitable pair of equilateral triangles.

Three main ways of possible construction method were chosen and those divided to method of construction "on a paper" and on the building site. Chapel could be designed using "true" ellipse construction, mandorla based, heavily symbolic construction and by oval based construction method. Comparison of hypothesis A, B and C shows the logical conclusion - two of the methods A and B are too difficult, rather complex, therefore not easily repeatable, and un-practical to construct on the building site. Serlio's constructions method were certainly known to builder and designer of the chapel as a common practice. They are easily executed and method fig. 5 a) and fig. 7a) oval is the most probably used.

Further analysis of the chapel in Kľačno will concentrate on finding more connections of Georg Palesch to possible architects - probably foreign or builders in the region with aim to find if possible original documentation. The second aim will be to make comparison as broad as possible to similar historical buildings in Slovakia with the goal to find possible used methods of construction known in the first half of 19th century.

## References

1. List of the houses in Nitra region, in latin, 1598. Archive of Magyar Nemzeti Levéltár Budapest, Magyar Kamara Archívuma. Sectio E 158, nr. 37. (slovak, latin) Published in: Štúdie o dejinách/Historia Nova 6, STIMUL 2013. ISBN 978-80-8127-082-6.
2. Mentioned as Paul Henrik Pales (1756-1835) in: Szinnyei József: Magyar írók élete és munkái XIII. (Steiner–Télfy). Budapest: Hornyánszky. 1909. (hungarian) [online] Available at: http://mek.oszk.hu/03600/03630/html/index.htm
3. Palles, Georg. Österreichisches Biographisches Lexikon und Dokumentation 1815-1950, Bd. 7 (Lfg. 34, 1977), S. 304. ISBN: 978-3-7001-3213-4 [online] Available at: http://www.biographien.ac.at/oebl/oebl_P/Palles_Georg_1753_1833.xml
4. Spišská Kapitula: Matrica Defunctorum 1819-1843, page 62. (latin)
5. Lacko, R. - Strečanská, Z. - Struhár, M.: Kľačno : 600 rokov prvej písomnej zmienky o obci : monografia obce, Martin, FOMI, 2013. ISBN: 9788097022679, p. 152. (slovak)
6. Archive of the Banská Bystrica dioceze, book of official documents facsimiles, 1824, page 242.
7. Kollárová, R. - Grúňová, Z. - Žiak, J.: Architectonic-historical and art-historical research of the Slovak national monument, Chapel of Virgin Mary in Kľačno, 2016. (slovak). Unpublished research.
8. A. Becchi, D.B. Meli, E. Gamba, Guidobaldo del Monte (1545-1607): Theory and Practice of the Mathematical Disciplines from Urbino to Europe, 380 (epubli Berlin, 2013)
9. S. Serlio, Il primo [-secondo] libro d'architettura, (Venice, 1545) (italian). [online] Available at: https://archive.org/details/gri_33125008448702

# Technogenic resources for nanotechnologies in construction

*Alexander* Guryanov [1,*], *Sofia* Korenkova[1] , and *Lyubov* Bezgina[1]

[1]Samara State Technical University, Institute of Architecture and Civil Engineering, 194, Molodogvardeyskaya St., 443001, Samara, Russia

**Abstract.** A large group of industrial wastes from water purification, treatment and softening has been considered. The reference is made to the fact that slurries are a unique man-made product technologically ready for use and constantly restocking. Slurries have been classified by conditions of the formation, composition, chemical and physical properties. The nanoscale structural parameters of the slurry particles (the size distribution and fractal dimension) have been determined by the neutron small-angle scattering method. The conducted researches allow slurries to be considered a nanotechnogenic raw material. The use of slurries as additives in the building material production process reduces the harmful influence of industrial wastes on the environment and creates a significant reserve of material and energy resources in the construction industry.

## 1. Introduction

Recycling of the industrial waste is an important environmental and economic problem [1]. Among a variety of industrial wastes a special mention should go to a large group of them, i.e. slurries resulting from the chemical treatment of the industrial waste water in machine-building, tool engineering, metal working, ferrous and non-ferrous metallurgy, oil production and refining [2]. The formation of the slurry occurs as a result of processes of chemical reactions, dispersion and adsorption, as well as mixing, setting and coagulation of solid particles from supersaturated solutions of the waste water at its cleaning or softening. Conditions for the formation of slurries are sol-gel-assisted. The slurry is a suspension which is characterized by the stable structure and composition.

Most slurry wastes find no practical application. To store them in the dumps is one of the sources of environmental pollution and disruption of the ecological balance [3]. The consequences of the slurry waste negative impact on nature and people can be eliminated by the creation of the developed system of the slurry recycling as one of the components in the production of materials for various purposes [4]. Slurries are a unique man-made product. The use of them as additives increases the strength, moisture resistance and durability of building materials with cement [2, 5].

This paper summarizes the results of the classification of the slurry waste. The neutron small-angle scattering method (SANS) [6] was used to determine structural parameters of

* Corresponding author: gurjanovam@mail.ru

basic types of nanoscale slurries. This definitely makes it possible to consider slurry wastes to be nanotechnogenic raw materials. The qualitative improvement of properties of building materials with cement thanks to slurry additives is due to the fact that they affect the formation of structural parameters of the final product at the micro and nanoscale [7, 9].

## 2 Materials and methods

The slurry wastes can be divided into groups according to the following main factors: the condition of the formation, the chemical and mineralogical composition, the action tendency when used as additives to building materials with cement.

The industrial discharged waters are very diverse. At the enterprises of mechanical engineering and energy the waste water formation is associated with washing operations, application of technological mixtures, emulsions and electrolytes. The slurry is the end product of the waste water which has gone through crystallization, settling and water removal. The slurries differ mainly in their chemical and mineralogical composition. We may distinguish mineral and organic- mineral slurries.

The basis of mineral slurries is oxides, carbonates and hydroxides of aluminum, magnesium, iron and silicon. They represent a heterogeneous system in which water is the dispersion medium and solid particles are the dispersed phase.

A distinctive feature of the organic-mineral sludge (oil slurries) is the presence of the residual oil, petrochemical products and emulsified particles of mineral oils in their composition. The composition of some mineral slurries is given in Table 1.

**Table 1.** Composition of mineral slurries.

| Slurry sample | $Al_2O_3$ [%] | CaO [%] | MgO [%] | $Fe_2O_3$ [%] | $SiO_2$ [%] | $SO_3$ [%] | Losses during steaming [%] |
|---|---|---|---|---|---|---|---|
| Carbonate | 0-14 | 30-48 | 2-12 | 2-12 | 2-5 | 0-7 | 32-42 |
| Aluminum- alkaline | 42-58 | 0-10 | 0-5 | 0-6 | 0-2 | 0-7 | 30-58 |
| Aluminum-calcium | 12-36 | 10-21 | 2-7 | 1-7 | 2-8 | 1-6 | 34-52 |

Despite the variation in the percentage of the individual components of the slurry, the high technological readiness for use and the slurry constant restocking enable to speak about their efficiency in the production of building materials.

In regards to the action tendency slurries can be attributed to filling (e.g. carbonate slurries) or structure-forming additives (e.g. aluminum-alkaline and aluminum-calcium slurries) [7-9].

The hydration of the cement clinker is associated with the formation of the skeleton of the hydrated calcium silicate particles [10, 11]. The parameters characterizing the structure of the hydrated calcium silicate particles, the ratio of the free and bound water the presence of pores and their shape and size distribution, i.e. the structural parameters of the material at micro and nanoscale [7-9], affect the properties of the resulting cement brick [12, 13]. It is possible to obtain certain specified properties of building materials with cement by using sludge as additives to be fillers or modifiers. It shows up most vividly if there is a nano-dispersed component in additives introduced into the cement compositions.

To determine the material structural parameters within the limits from 1 nm to 100 nm, it is possible to use the neutron small-angle scattering method (SANS) [6]. This method is appropriate to investigate nanoparticles of different physical nature. It is also successfully used to study at the nanoscale a variety of nanodispersed materials and to determine structural parameters of nanomaterials, as well as bulk ones containing nanodispersed components, e.g. hydrated cement compositions [14-16].

The SANS intensity *I(q)* depends on the scattering power of the compositional inhomogeneity of the matter and the contrast at the phase boundary. It is the function of the neutron pulse sent at scattering

$$q = \frac{4\pi}{\lambda}\sin\frac{\theta}{2}, \tag{1}$$

where $\theta$ is the angle of neutron scattering.

The SANS intensity *I(q)* carries information about the structure of scattering particles (scattering centres) over the atom. Provided that $qR_c < 1$ (Guinier regime), the intensity of the small-angle scattering *I(q)* is determined by the average (characteristic) size and shape of scattering particles:

$$I(q) = I(0)\exp\left(-\frac{R_g^2 q^2}{3}\right), \tag{2}$$

where $R_g$ is the particle radius of gyration. From the experimental data on small-angle scattering in the Gunier regime it is possible to assess the particle radius of gyration $R_g$ and their typical size $R_c$.

If the condition $qR_c > 1$ (Porod regime) for the intensity of the small-angle scattering *I(q)* on fractal objects has been fulfilled, there happens to be a power characteristic from the transmitted neutron pulse:

$$I(q) \propto q^{-n}. \tag{3}$$

The deviation of the exponent $n$ from the Porod asymptotic behavior ($n = 4$) indicates fractal properties of scattering objects and makes it possible to determine their fractal dimension. In the case of volumetric or mass fractals the exponent coincides with the fractal dimension $n=D_V$ and takes on the value from the interval $1 < D_V < 3$. In the case of scattering by objects with a fractal surface, the fractal dimension is $D_S = 6 - n$ it takes values from the interval $2 < D_S < 3$. The exponent is respectively in the interval $3 < n < 4$.

Thus, by the slope of log-log linear segments of the SANS dependencies it is possible to determine the scattering objects of a particular fractal type (volume or surface) and the fractal dimension [17-19].

The wavelength $\lambda$ of slow neutrons in the neutron small-angle scattering method is in the interval which allows determining structural parameters of scattering particles in the nanometer interval and to clearly relate test materials to nanomaterials.

For polydisperse systems according to small-angle scattering we can recover the distribution function of the scattering objects in terms of sizes $G(R)$ on the basis of the spectra of pair correlations $\gamma(R)$ obtained by the Fourier transformation of the data for cross sections of neutron scattering $\sigma(R)$. This function is related to the shape of the particles and describes quantitatively the set of segments connecting elements of the particle volume and depends on the distribution of inhomogeneities inside the particles. The size distribution function *G(R)* has the following form:

$$G(R) = R^2\gamma(R) = \frac{1}{2\pi^2}R^2\int\frac{d\sigma(q)}{d\Omega}\frac{\sin(qR)}{qR}q^2\,dq. \tag{4}$$

The function $G(R)$ shows if there are particles of the given size (radius) in the test material. All SANS experimental data is implemented using the software ATSAS 2.8.0 [20, 21].

## 3 Results

The SANS intensity spectra from the sludge samples were measured using the diffractometer "Membrana-2" installed at the reactor WWR-M in the Petersburg Nuclear Physics Institute (Gatchina) [22]. The wavelength of slow neutrons was equal to $\lambda$ = 0,3 nm. The range of the recorded transmitted neutron pulses $q$ varied from 0, 03 $nm^{-1}$ to 0, 8 $nm^{-1}$. The samples of scattered neutrons were recorded in the range of angles $\theta = \pm 0{,}017$ rad. The detector consisted of forty-one $^3$He-counters. The intensity of the beam on the sample was measured by two counters (monitors) installed in front of the sample above and below the axis of the neutron beam.

The samples of the test sludge were placed in the target device that is a cavity in a cadmium plate. The volume of the cavity equaled 0,52 $sm^3$. The sample layer thickness was equal to the thickness of the cadmium plate $d_S$ = 2 mm. Loose materials were placed in the cavity of the target device and covered with an aluminum foil of 6 mkm. The transmission coefficient of the neutron beam was $T$ = 0,45 – 0,85.

The SANS data resulting from the experiment were normalized in accordance with the monitor readings and the transmission intensity. We also conducted the subtraction of background and contribution of neutrons that had passed through the sample without scattering. The obtained distributions of scattering intensities $I_S(q)$ were normalized to those obtained under the similar conditions of the intensity distribution $I_{ST}(q)$ of the standard sample which was the light water layer $d_{ST}$ = 1 mm. Differential cross sections of the sample scattering based on the 1 $sm^3$ of the sample volume in absolute units were calculated according to the following formula

$$\frac{d\sigma(q)}{d\Omega} = \frac{I_S(q)}{I_{ST}(q)} \frac{d_{ST}}{d_S} \frac{d\sigma_{ST}}{d\Omega}. \tag{5}$$

From differential cross sections of neutron scattering the distribution function of the scattering center (particles) sizes $G(R)$ in the approximation of homogeneous spheres was recovered using the Fourier transform.

Analyzing the experimental data we selected the maximum radius of the scattering objects. When its value was 50 nm, the points corresponding to minimum neutron pulses transferred during the experiment collapsed (were not described theoretically).The points corresponding to maximum pulses transferred collapsed when the radius value was 100 nm. The description of the SANS experimental data was correct selecting the maximum size of scattering objects equal to about 80 nm. The average radius of scattering objects for test samples was in the range from 30 nm to 65 nm.

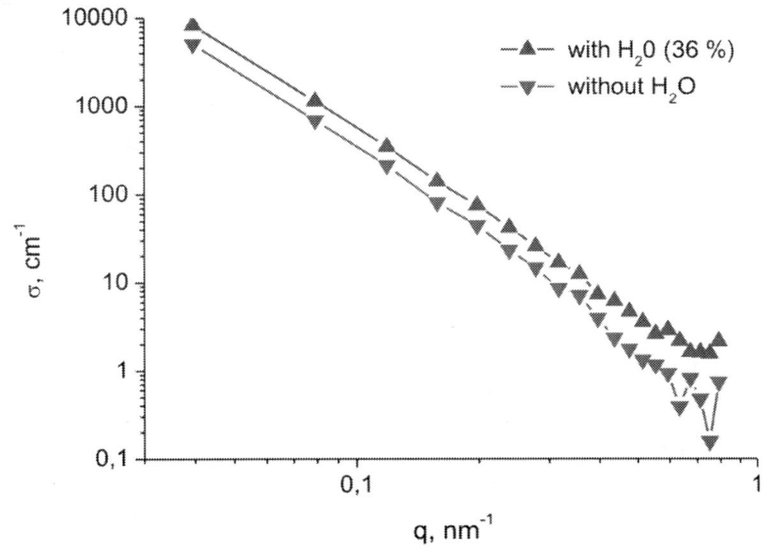


**Fig. 1.** Experimental spectra of the SANS differential cross sections for the carbonate sludge in two states (wet and dry).

Figure 1 shows spectral plots of the differential cross sections of the neutron scattering of the carbonate sludge sample. Figure 2 shows graphs of the calculated distribution functions of the scattering objects by sizes for the carbonate sludge sample.

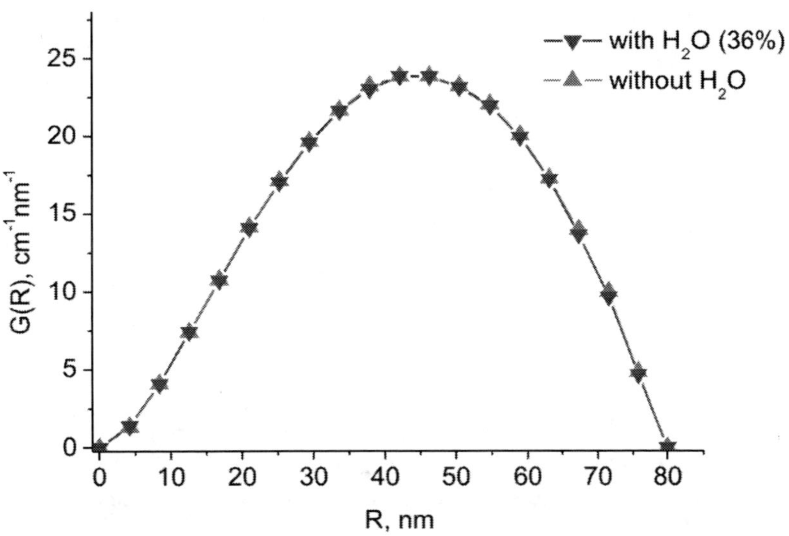


**Fig. 2** The distribution function of scattering objects by sizes for the carbonate sludge sample in two states (wet and dry).

According to experimental data of neutron scattering intensities the following calculations were performed: the average radii were found using formula (2); formula (3) was used for fractal dimensions; formula (4) for differential scattering cross sections and formula (5) for scattering object distribution functions by sizes.

For all sludge samples it was possible to observe the power dependence of the neutron scattering intensity from the transmitted neutron pulse. Table 2 shows the description of test sludge samples, the average size of particles $R_c$, and values of the fractal dimension $D_S$.

**Table 2.** Structural parameters of sludge samples.

| Sample number | Slurry type | $R_c$, nm | $D_S$ |
|---|---|---|---|
| 1 | Carbonate (wet, 36 % $H_2O$) | 45 | 2,61 |
| 2 | Carbonate(dry) | 47 | 2,65 |
| 3 | Aluminum-alkaline(wet, 38 % $H_2O$) | 38 | 2,54 |
| 4 | Aluminum- alkaline(dry) | 40 | 2,58 |
| 5 | Aluminum-calcium(wet, 35 % $H_2O$) | 44 | 2,56 |
| 6 | Aluminum-calcium(dry) | 46 | 2,60 |

## 4 Discussion

The SANS method was used to determine at the nanoscale structural parameters of some basic types of slurries. The obtained results show that slurries are composed of the nanodispersed component and it is quite possible to refer them to nanotechnogenic raw materials. A slight change in the structural parameters of slurries at various stages of completion (wet or dry) demonstrates their stability.

## 5 Conclusions

Slurry waste resulting from the industrial enterprise sewage treatment are nanotechnogenic raw materials. The slurries affect the formation of structural parameters of building materials with cement at the micro and nanoscale acting as fillers or modifying additives, efficiently and purposefully improving their properties

Furthermore, the use of the slurry waste for the production of building materials makes it possible to solve the important environmental and economic problem of the industrial waste utilization, to avoid environmental pollution and disruption of the ecological balance.

## References

1. K.L. Chertes, A.A. Savelyev, E.G. Martynenko, O.V. Tupitsyna, A.A. Mikhasek, Urban Constrution and Architecture **1**, 49-57 (2016) DOI: 10.17673/Vestnik.2016.01.8
2. S.F. Korenkova, T.V. Sheina, *The basics and the concept of chemical precipitation of the industrial waste in the construction industry* (SGASU, Samara, 2004)
3. K. Chertes, O. Tupitsyna, V. Pystin, A. Savelyev, E. Martynenko, MATEC Web of Conferences **86**, 06003 (2016)
4. K.L. Chertes, O.V. Tupitsyna, V.N. Pystin, A.M. Shterenberg, Urban Construction and Architecture **4**, 78-84 (2015) DOI: 10.17673/Vestnik.2015.04.10
5. A.M. Guryanov, Industrial and civil engineering **6**, 55-58 (2015)
6. M.V. Avdeev, V.L. Aksenov, Advances in physical sciences **180-10**, 1009-1034 (2010)
7. A.M. Guryanov, Procedia Engineering **111**, 283-289 (2015)
8. A.M. Guryanov, Procedia Engineering **153**, 217-222 (2016)
9. A. Guryanov, S. Korenkova, Yu. Sidorenko, MATEC Web of Conference **86,** 04011 (2016) DOI: 10.1051/matecconf/20168604011
10. J.W. Bullard, H.M. Jennings, R.A. Livingston, A. Nonat, G.W. Scherer, J.S. Schweitzer, K.L. Scrivener, J.J. Thomas, Cement and Concrete Research **41**, 1208-1223 (2011)
11. A. Nonat, Cement and Concrete Research **34**, 1521-1528 (2004)

12. L.B. Skinner, S.R.Chae, C.J. Benmore, H.R. Wenk, P.J.M. Monteiro, Physical Review Letters **104**, 195502 (2010)
13. L. Raki, J. Beaudoin, R. Alizadeh, J. Makar, T. Sato, Materials **3**, 918-942 (2010)
14. J.J. Thomas, J.J. Chen, A.J. Allen, H.M. Jennings, Cement and Concrete Research **34**, 2297-2307 (2004)
15. A.J. Allen, J.J. Thomas, H.M. Jennings, Nature Materials **6**, 311-316 (2007)
16. A.J. Allen, J.J. Thomas, Cement and Concrete Research **37**, 319-324 (2007)
17. T. Ficker, Theoretical and Applied Fracture Mechanics **50**,167-171 (2008)
18. Q. Zeng, M. Luo, X. Pang, L. Li, K. Li. Applied Surface Science **282**, 302-307 (2013)
19. Y. Gao, J.Jiang, G.D. Schutter, G. Ye, W. Sun, Construction and Building Materials **69**, 253-261 (2014)
20. P.V. Konarev, M.V. Petoukhov, V.V. Volkov, D.I. Svergun, J. Appl. Cryst. **39**, 277-286 (2006)
21. M.V. Petoukhov, D. Franke, A.V. Shkumatov, G. Tria, A.G. Kikhney, M. Gajda, C. Gorba, H.D.T. Mertens, P.V. Konarev and D.I. Svergun, J. Appl. Cryst. **45**, 342-350 (2012)
22. V.M. Lebedev, B.T. Lebedev, I.N. Ivanova, M.R. Kolhidashvili, D.N. Orlova, The analytical measuring system for the small-angle diffractometer "Membrane-2", Preprint PNPI-2785 (2008)

# Lightweight building envelopes in prefabricated buildings in terms of fire resistance

*Agnes* Iringová[1*]

[1]University of Žilina, Faculty of Civil Engineering, Univerzitná 1, 010 26 Žilina, Slovakia

**Abstract.** Dimensioning the lightweight, building envelopes in prefabricated buildings in terms of fire protection within the passive standard. Legislation. Optimizing the protection of a supporting system in prefabricated wood-based walls in terms of thermal and fire protection according to Eurocodes.

## 1 Introduction

The design and material solution has a major impact on the building's microclimate and its fire safety. In lightweight prefabricated building envelopes it is important to choose the thermal insulation and its location in relation to the supporting structure, both in terms of heat and moisture transfer, as well as in terms of fire protection. Due to the increasing requirements for thermal protection of buildings, the composition and thickness of the envelopes in prefabricated buildings changes. Depending on the type of thermal insulation used, the building´s fire safety is better or worse. The paper is to evaluate the influence of the choice and location of the thermal insulation in lightweight envelopes according to the European legislation relating to the fire resistance of supporting elements using calculation in Eurocodes.

## 2 Design and material solution of prefabricated supporting systems in terms of fire safety considering the building´s function and fire height

The supporting structure of lightweight sandwich envelopes usually consists of timber or steel framework. Both types of supporting structure must be fire protected. At a minimal fire height, the required fire resistance is 30 minutes. The increase in fire height typically results in higher fire resistance requirements. The paper analyses the possibilities of using different types of thermal insulation and optimal lightweight wood-based envelopes in terms of fire resistance.

The use of wood in building constructions is wide - from supporting systems including partition walls and ceilings to the final furnishing. Timber buildings are the buildings with

* Corresponding author: agnes.iringova@fstav.uniza.sk

the wood-based supporting system and may be divided into log buildings, half-timbered buildings, timber-framed buildings (Balloon-Frame, Platform-Frame), skeletal structures, and panel structures. Usability of the tested compositions is in the last three above-mentioned systems. [6]

Nowadays, the timber framework structures are the most used ones in constructing low-storey houses. The timber framework may be interrupted at the ceiling, or continuous along its entire height. If the wall load-bearing framework is interrupted in the ceiling, i.e. the wall of the next floor is fastened directly on the ceiling structure - it is a platform-frame structure. In the balloon-frame structure, the pillars go continuously along its entire height. Pre-engineered timber buildings are similar to this system. [4]

Timber skeleton frame structures are the counterpart of the buildings made of solid timber and framework. The buildings of solid timber and framework have the linear load-bearing construction and the load from the roof and ceiling is transferred by supporting peripheral and internal walls. The load from ceilings and roofs in skeletal systems is transferred by pillars. These buildings have a more variable layout. The timber load-bearing members are typically made of glued laminated wood with the steel joint elements embedded. The building envelope, anchored to them, is generally designed as a lightweight sandwich self-supporting panel structure, e.g. the timber framework with thermal insulation. It may be placed in front of the load-bearing system, or embedded in the pillars - the ideal solution in terms of thermal protection. As for fire protection, the building envelope is generally designed as a fire-compact structure with the required fire resistance EI, or EW for wood-based ones.

## 3 Legislative limitations in using construction systems for wood-based buildings in terms of fire regulations

All of the above wood-based construction systems are in terms of fire safety classified as combustible structural units. [7] The fire height of the buildings with the timber supporting system according to legislation valid in Slovakia is limited to maximal 9 m, excepting the buildings intended for housing and accommodation (typically four-storeyed). This goes provided that the fire load calculated in the most critical fire section of such buildings is up to 60 kg/m$^2$. If the fire load is higher than 60 kg/m$^2$, the maximal allowable fire height is 4 m. The requirement for maximal fire resistance of the load-bearing members is REI 90. According to the currently valid standards, residential buildings and houses may contain wood-based load-bearing systems up to the second floor (see Table1, column 3). At the present time, the modification of STN 920201-2 is being approved; it will allow using wood-based construction systems for upper floors (see Table1, column 4).

**Table 1.** Possibility to use wooden sandwich panels in residential and apartment buildings made up of combustible and combined construction system within the current and future legislation.

| **Construction system** | **Maximal number of above ground storeys** | **The lowest degree of fire safety, fire zones** | |
|---|---|---|---|
| | | **Current state within standard STN 920201-2 valid for above ground storeys + 1st basement** | **State after modification of standard STN 920201-2 valid for above ground storeys + 1st basement** |
| combined | 2 | I | I |
| | 3 | II | II |
| | **5** | unacceptable | **III** |

| | | | |
|---|---|---|---|
| combustible | 1<br>2<br>**3** | I<br>II<br>unacceptable | I<br>II<br>**III** |

## 4 Requirements for the design of wood-based lightweight envelope in terms of thermal protection

Current legislative conditions to optimize the energy performance of buildings are set in STN 735040-2: 2012 Z1 - 2016 so that the building would be supposed to meet the global indicator required - primary energy consumption in kWh/(m$^2$.a) in the energy performance class A1. The limit values are defined depending on the building´s category - its main functional use. As of 2020, all the buildings designed, including those for housing (family houses and apartment buildings), will have to meet the national requirements for energy consumption in the category of nearly-zero-energy buildings within the limit of the global indicator - primary energy consumption in kWh/(m$^2$.a) in the energy performance class A0. In terms of the heat loss reduction, the current standard requirement for heat transfer through the building envelope is 0.22 W/m$^2$K.

Nowadays, the construction of wood-based buildings often uses the load-bearing systems of timber framework. [3] The framework in low-storey buildings serves as a building´s supporting system and at the same time as a load-bearing structure for the envelope. The load-bearing structure for the envelope consists of vertical wooden pillars with an air gap filled with thermal insulation. Its exterior side may be covered with vapour permeable formwork of thermal insulation (Fig. 1), or it may be covered with OSB boards, subsequently thermo- insulated and finished with a thin-layer plaster. Its interior side may be covered with OSB boards having vapour barrier with a grid and a pre-wall of plasterboard fastened to it. Technical wiring may be situated in the air gap between the pre-wall and plasterboard. The plasterboard thickness is designed according to the fire resistance requirements for the supporting wall and the envelope itself. The pre-wall also makes the acoustic insulation better. [8]

Choosing the thermal insulation for sandwich timber structures is conditioned by its thermo-insulating efficiency, combustibility class, durability, and recyclability. The building envelope and timber load-bearing framework must be designed so that the water condensation would not occur on joints of timber elements. As for hygienic demands, the layers of thermal insulation and its physical parameters must be designed to keep the minimal surface temperature of the interior wall above the critical surface temperature for mould growing, depending on the boundary environmental conditions and the type of heating. The type and thickness of thermal insulation is essential for the fire resistance of a timber panel. Its melting temperature should be above 1000 ° C and a minimal mass weight should be above 30 kg/ m$^3$. Thermal insulations m based on mineral-wool or glass-fibre meet these physical parameters. [5]

**Table 2.** Optimization of a timber panel - thermal insulation in terms of heat loss for structures built by 2020.

| Thermo-insulating material | λ W/mK | ρ [kg/m3] | Combustibility class | Hr. TI [m] | Uv ≤0,22 W/(m2.K) |
|---|---|---|---|---|---|
| Gypsum plaster board, at plant | 0.22 | 750 | A1 | 0.015 | |
| Isover Woodsil<br>+ vapour barrier<br>in the grid of an installation pre-wall | 0.038 | 37 | A1 | 0.06 | |
| Oriented strand board, at plant | 0.130 | 650 | E | 0.015 | |

| | | | | | |
|---|---|---|---|---|---|
| Variant 1 Isover Woodsil<br>Variant 2 fibreboard soft, at plant<br>Variant 3 blown cellulos | 0.038<br>0.040<br>0,044 | 37<br>55<br>50 | A1<br>E<br>E | 0.12<br>0,13<br>0,14 | cca 0.17 |
| Fibreboard soft, at plant | 0.043 | 55 | E | 0.06 | |
| Thin-layer plaster with reinforcement fabric | 0.75 | 1600 | A1 | 0.001 | |
| Sawn timber, softwood, raw, kiln dried, u=10% | 0,018 | 300 | E | | |

In the lightweight sandwich structures - timber panels, the currently convenient thickness of thermal insulation based on mineral wool, fibreboards, blown cellulose, or sheep wool is about 220 mm, depending on the mass weight and heat conductivity coefficient. As of 2020, the requirement for the thickness of thermal insulation of building envelopes in the passive houses will be a quarter higher. [9]

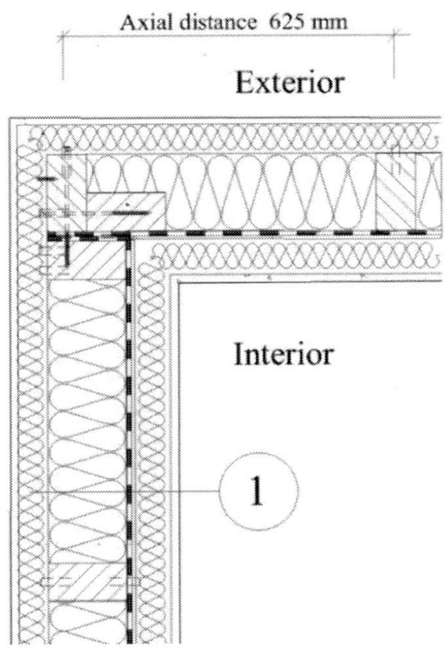


**Composition 1** -from the interior side
- Gypsum fiber board thickness of 12,5 mm
- Mineral wool thickness of 60 mm in the installation cavity
- Vapor barrier
- OSB boards thicknesss of 15 mm
- Supporting Wooden columnar structure 60/120 mm inserted in the air gap insulation thickness of 120 mm
- Fibreboards insulation thickness of 60 mm
- Exterior plaster thickness of 10 mm

**Fig. 1.** Composition of timber panel optimized in terms of higher fire resistance

## 5. Fire resistance of timber sandwich panels according to Eurocode

The required time of fire resistance and the criteria for limit states of timber load-bearing elements of external walls are directly related to the static load. In case of frameworks, the timber vertical elements are part of the supporting walls and are linked up by the horizontal ceiling or roof supporting elements. These structures, including all those of bracing and anchor elements, must meet the REI criterion.
The composite supporting elements of the fire-separating structures, as well as all the elements related to the supporting ones in terms of load, are required to meet the criteria for the fire resistance time. The interaction of all the elements provides the fire resistance of the

whole structure. In case of timber supporting elements, the value of fire resistance can be determined by calculation methods depending on the static load using STN EN 1995-1-1 (Eurocode 5). The fire resistance of timber or wood-based load-bearing and construction elements depends on the burning and charring speed.

Methods to calculate fire resistance of structures in terms of Eurocodes employ the analytical approach to design buildings considering the fire effect. It takes into account the function of a structural system at higher temperatures in terms of static load of fire separating structures as well as increases in thermal protection by active or passive elements of fire protection systems. When designing or assessing load-bearing fire barrier according to EN 1991-1-2, the appropriate model of fire situation is chosen. Subsequently, the corresponding fire scenario is determined. According to standard temperatures in the affected fire zone, it is calculated temperature course in load-bearing structural elements which are stressed both statically and thermally. The resulting static load of a structure in a critical fire situation is calculated and the cross-section is evaluated - whether it meets supposed thermal stress in the required time.

The mechanical behaviour of statically stressed structures during fire depends on thermal load defined by fire risk, thermal effects on material properties - flammability, indirect mechanical load (release or fall of facing etc.), and direct effects of static load. Thermal analysis is carried out according to STN EN 1991-1-1-2. Static analysis is made according to STN EN 1991-1-1-2 and other STN, depending on the material type.

In accordance with the procedure, specified in STN EN 1991-1-1-2, the model fire situation was determined on the basis of an estimation or calculation of fire risk. The model (normative) fire in a fire sector (FS) is determined for the model fire situation. It is always considered only for one FS in given time; in this case the entire object is one fire sector. When assessed fire resistance of a structure, calculations considered the model (normative) fire with nominal fire curves. [1]

The charring depth and the subsequent static weakening of the cross-section must consider all the surface elements directly exposed to fire. Surface charring need not consider the surface of the timber load-bearing elements protected during the calculated fire load by other elements. Likewise, it need not be considered at the elements protected by the fire facing provided that: $t_{pr} \geq t_{fi,req;}$ where $t_{pr}$ is the time to break the protective plate or other protective material and $t_{fi,req}$ is the time required for fire protection at standard fire load.

The timber load-bearing elements of frameworks have typically double-sided covering with boards and the air gap is filled with a thermal insulator. The OSB or other wood-based boards are used for covering. Since their surface is burning more slowly than that of solid timber, they increase the fire resistance of the load-bearing elements.

Most board materials that are construction elements of the supporting system of the envelope may be also used as an additional fire protection of timber load-bearing members. They also serve as reinforcing wall elements and a foundation for the final finishing. If they are considered as a protective fire facing, the time when the breakage occurs must be determined. In such cases, the time to break is the time within the temperature on the non-radiated side of the fire barrier will increase by more than 180 K in combustible materials and by more than 500K in non-combustible materials. These time limits are being determined by tests.

## 5.1 Fire resistance of the tested timber sandwich envelopes

The testing of envelope compositions considers covering using OSB boards on the interior side and fibreboards with a thin-layer plaster on the exterior side. Before completing the structure, the plasterboard pre-wall with an installation gap will be fastened on a wooden grid on the interior side. After wiring, the gap will be filled with mineral wool.

This adjustment will also serve as a passive fire protection of the timber elements in the load-bearing system.

The time to break a fire protection by these structures up to the surface of timber supporting elements will be the sum of the partial values of the time when the fire damage occur. [3]

The time to break the timber and wood-based fire-protecting boards may be determined using the tests according to the relation:

$$t_{pr} = t_p/\beta_0 - t_r \ (min) \tag{1}$$

$\beta_0$ - charring speed

$t_p$ - thickness of the facing

$t_r$ = 4 mm; it is the reserve to prevent boards from falling down or to prevent fire from spreading into cavities (e.g. installation gaps)

The multi-layer fire facing must always be anchored to the protected supporting element.

The time to break OSB boards in the composition tested (see Fig. 2) is 13.4 min.

The time to break the non-flammable facing and boards is the time when the temperature of the surface that is not exposed to fire rises by more than 500K.

For F-type plasterboards, 15 mm thick, with increased cohesion of the core at the high temperature, it can be determined according to the relation:

$$t_{pr}=1{,}9 .\ \xi\ .\ t_p \ (min) \tag{2}$$

The time to break F-type plasterboards is 28.5 min

The time to break non-flammable insulation materials, more than 20 mm thick with mass density above 30 kg/m$^3$ that remain compact to 1000 ° C, can be calculated according to the relation:

$$t_{pr} = 0{,}07(t_{ins} - 20)\ .\ \sqrt{\rho_{ins}} \ (min) \tag{3}$$

$t_{ins}$ - thickness of the insulation material in mm

$\rho_{ins}$ - mass weight of the insulation material in kg/m$^3$

In this case, mineral wool with a specific weight of 37 kg/m$^3$ is used in both the pre-wall cavity.

Supporting fire barrier wall - **option 1a.** $t_{pr}$ = 17 min

The resultant value of $\Sigma t_{prj}$ for verified composition is 55 minutes. The condition corresponds to the fire resistance of a filling wall REI 30.

Curtain framework - **option 1b.** $t_{pr}$ =68 min

The resultant value of $\Sigma t_{prj}$ for verified composition is 109 minutes. The condition corresponds to the fire resistance of a filling wall EI 60.

The resultant value of $\Sigma t_{prj}$ for verified composition is 109 minutes. The condition corresponds to the fire resistance of a filling wall REI 30.

If the air gap in the framework is filled with a combustible thermal insulation with the melting temperature of 1000 °C - **option 1b, 2, and 3.** $t_{pr}$ = 0.07 (60 -20). $\sqrt{37}$ = 17 min. The solution corresponds to the fire resistance of a supporting wall REI 30.

**Criterion I.** - verification of the temperature rise on the side that is not exposed to fire, up to 140K:

$\Sigma t_{prj} \geq t_{fi,req} + 5$ = 109 > 60 +15 .... EI 60 ....... option 1b, 2,3.

$\Sigma t_{prj} \geq t_{fi,req} + 5 = 55 > 30 + 15$ .... REI 30 ...... option 1a.

**Criterion II.** - verification of the maximal temperature rise in any spot is limited up to 180°C, the condition is met if the inequation is true:
$\Sigma t_{prj} \geq t_{fi,req} + 5 = 80.8 > 60 + 5$ EI ...... option 1b, 2,3.
$\Sigma t_{prj} \geq t_{fi,req} + 5 = 45.5 > 30 + 5$ REI ...... option 1a.

**Table 3.**Fire protection of wood-based sandwich panels according to the thermal insulation used.

| Thermo-insulating material | ρ [kg/m$^3$] | Combustibility class | Melting temperature °C | Hr. TI [m] | Δ Fire resistance min. |
|---|---|---|---|---|---|
| Gypsum plaster board, at plant | 750 | A1 | | 0.015 | 28,5 |
| Isover Woodsil + vapour barrier in the grid of an installation pre-wall | 37 | A1 | >1000 | 0.06 | 17 |
| Oriented strand board, at plant | 650 | E | | 0.015 | 13,4 |
| Variant 1 Isover Woodsil<br>Variant 2 fibreboard soft, at plant<br>Variant 3 blown cellulos | 37<br>55 | A1<br>E<br>E | >1000<br><1000<br><1000 | 0.12 | 25<br>8<br>0 |
| Fibreboard soft, at plant | 55 | E | <1000 | 0.06 | - |
| Thin-layer plaster with reinforcement fabric | 1600 | A1 | | 0.001 | - |
| Sawn timber, softwood, raw, kiln dried, u=10% | 300 | E | | | - |

## 6 Conclusion

The increasing thickness of a thermal insulation, in terms of energy performance optimization, causes increasing fire resistance of the building envelopes where the non-combustible thermal insulation with the melting temperature of 1000 °C is used. Low-storey panel buildings, as well as kit homes with framework systems, are provided with the adequate fire-resistant load-bearing members of a building envelope when choosing interior thermal insulation optimally. In timber skeleton frame structures, where the timber load-bearing members are not fire-protected, it is possible to achieve the fire resistance by over-dimensioning the cross-section of load-bearing glued timber elements according to EC:5 methodology. The conditions valid for the envelope in the panel buildings are similar to those valid for wood-based filling masonry.

### Acknowledgements
The paper presents results of the project VEGA 1/0945/16.

## References

1. STN EN 1995-1-2, Eurocode 5, Design of wooden structures - Part 1 - 2, General rules – Design of structures for fire effect. SUTN, Bratislava (2008)
2. P. Ďurica: Defects in buildings - diagnostics and redevelopment (in Slovak), 1st edition. Žilina : Žilinská univerzita, (2012)

3. Timber structures according to Eurocode 5 - step 2, Designing details and supporting systems (in Czech). ISBN 80-86 769-13-5, Praha (2004)
4. P. Durica, P. Juras, D. Staffenova, J. Rybarik, Proceedings of the Conference enviBUILD, Brno (2016)
5. L. Makovická -Osvaldová, L. – S. Gašpercová, Civil and Environmental Engineering: scientific technical journal, **11/2**, 142 – 146 (2015)
6. K. Ďurian, P. Krušinský, R. Korenková, D.Zacharová, Fireco : X. international conference : fire protection (2013)
7. Trenčín, Slovakia, 2nd - 3rd May 2013 : conference proceedings. Bratislava: Požiarnotechnický a expertízny ústav MV SR, (2013)
8. V. Mózer, Proceedings of the 2nd European symposium on Fire safety science: Nicosia, June 16th-18th 2015. Nicosia : CERISE, European University Cyprus, 289-293 (2015)
9. D. Katunský, M. Lopušniak, M. Bagoňa, E. Dolníková, J. Katunská, M. Vertaľ, Journal of Theoretical and Applied Information Technology. **44/1,** 40-50 (2012)
10. P. Durica, P. Juras, V. Gaspierik, J. Rybarik, Procedia Engineering **111**, 176-182 (2015)

# The railway superstructure monitoring in bratislava tunnel no. 1 – section of ballastless track and its transition areas

*Libor* Izvolt[1], *Janka* Sestakova[1*], and *Michal* Smalo[1]

[1]University of Zilina, Faculty of Civil Engineering, Department of Railway Engineering and Track Management, Univerzitna 8215/1, 010 26 Zilina, Slovak Republic

**Abstract.** The requirements for the quality of the railway track with emphasis on its long service life, low maintenance costs and the required parameters of the track geometry have led railway infrastructure manager on the territory of the Slovak Republic – Zeleznice Slovenskej republiky (ZSR) – to modernize or to realize reconstructions of railway lines on the appropriate track sections with the application of the unconventional construction of the railway superstructure (ballastless track). The Department of Railway Engineering and Track Management currently performs the monitoring of two sections with the application of ballastless track construction. This paper is aimed to the monitoring of the ballastless track section and presents the results obtained so far in one of the monitored sections in the Bratislava tunnel No. 1.

## 1 Introduction

The ballastless track has become a standard construction for a railway superstructure on some sections of the modernized railway infrastructure of Railways of Slovak Republic (Zeleznice Slovenskej republiky – ZSR). The ZSR network has its application in some new constructions of bridges and tunnels. Supposed service life, easier and less costly maintenance, and thinner structural thickness in comparison with ballasted track construction, define the progressive use of a ballastless track even for more extensive reconstruction of these objects. The design and operating principles and advantages of a ballastless track compared to ballasted track have already been characterized in [1] and [2].

Within the framework of the cooperation of the Department of Railway Engineering and Track Management (KZSTH) of the Faculty of Civil Engineering of the University of Zilina with the ZSR, monitoring of two experimental sections is currently being carried out, in the tunnel Turecky vrch and around the tunnel portals (in four subsections of a total length 1.640 meters of track is monitored) and in the Bratislava tunnel No. 1 and around its portals (647 m long section).

Diagnostics by continuous monitoring and evaluation of its results, which were obtained by measurements in Bratislava tunnel No. 1 between 2015 and 2017 will be further characterized.

* Corresponding author: janka.sestakova@fstav.uniza.sk

## 2 Experimental Section in Bratislava Tunnel No. 1

The track section, which is the subject of track diagnostics, is located in track No. 1 from km 53.063 to km 53.710 of double-track railway line Devinska Nova Ves – Sturovo. The construction of the railway superstructure was built during the reconstruction of the single-track tunnel in 2014 and consists of ballasted track, ballastless track of the prefabricated *ÖBB-PORR* slab system and the transition area between these structures. The construction of the railway superstructure of the section is described in detail in [3].

## 3 Monitoring of the Experimental Section

Experimental section is monitored since 2015 with aim of track geometry quality diagnostics. Monitoring is carried out by the measuring trolley KRAB™–Light with continuous data recording system:

- gauge tolerance RK (after calculating the change of gauge ZR is also recorded),
- alignment of right rail SP (after calculating the alignment of left rail SL is also recorded),
- rail top level of right rail VP (after calculating the rail top level of left rail VL is also recorded),
- cant PK,
- quasi-twist on a short base (calculated to a quasi-twist on a base of 3.0 m long – $ZK_{3.0}$, 6.0 m long – $ZK_{6.0}$ and 12.0 m long – $ZK_{12.0}$),
- distance,
- measuring device velocity.

The diagnostics of the track geometry was carried out on semi-annual basis as:

- measurements before putting sections into operation (MSO) 29. 4. 2015,
- first operational measurement (PO1) 22. 10. 2015,
- second operational measurement (PO2) 11. 4. 2016,
- third operational measurement (PO3) 11. 10. 2016,
- fourth operational measurement (PO4) 6. 4. 2017.

## 4 Evaluation of Measurements

In the track, the measured data (their primary values) are recorded on the on-board computer PDA-TDS NOMAD. The onboard computer program allows reading and storing values in a data file with a step of 0.25 m. Subsequent processing of values of geometric quantities obtained by measurement is carried out in the KRAB 8.1 program, from a set of primary data (.krx). After editing the evaluation parameters (description of the evaluated section, assignment of the track geometry parameters values, speed range and evaluation limits), the track geometry parameters are calculated by transforming the primary values. These resulting values, taking into account the dynamic component of measurement, are the basis for assessing the track geometry quality of the section in the form of:

- local errors table with sections where the sections with exceedances of the allowed variations of the individual variables are highlighted,
- section evaluation table containing standard tolerances of variables (SDV), quality marks (QM) and quality number of evaluated section (QN),
- graph of track geometry parameters and values of permissible deviations.

Experimental section is assigned to speed zone 1 (RP1) according to line speed (V = 60 kmph) [4]. Section is defined as straight line with a short curve of large radius R = 2000 m in area of Lamac portal. Section was evaluated against allowed deviations, resp. against allowed values (*table 1*) for speed zone 1 for straight line (limited allowed deviations for parameter RK evaluated according to operational diagnostics).

**Table 1.** The deviations of the relative geometric parameters of the track for RP1. [5, 6]

| Measured parameter | Limit input deviations | | Operational deviations (AL) | | Operational deviations (IL) | | Limit operational deviations (IAL) | |
|---|---|---|---|---|---|---|---|---|
| RK (in straight track) (mm) | -3 | 3 | -5 | 17 | -7 | 20 | -9 | 25 |
| | *-2* | *3* | *-4* | *10* | *not evaluated* | | *-5* | *20* |
| ZR (mm/m) | -3 | 3 | not evaluated | | -5 | 5 | -6 | 6 |
| | *-3* | *3* | *-4* | *4* | *not evaluated* | | *-5* | *5* |
| PK (mm) | -3 | 3 | -13 | 13 | -15 | 15 | -20 | 20 |
| | *-3* | *3* | *-15* | *15* | *not evaluated* | | *-20* | *20* |
| VP, VL (mm) | -5 | 5 | -15 | 15 | -20 | 20 | -28 | 28 |
| | *-4* | *4* | *-15* | *15* | *not evaluated* | | *-20* | *20* |
| SP, SL (mm) | -5 | 5 | -15 | 15 | -18 | 18 | -20 | 20 |
| | *-5* | *5* | *-15* | *15* | *not evaluated* | | *-20* | *20* |
| Measured parameter | Limit input value | | Operational value (AL) | | Operational value (IL) | | Limit operational value (IAL) | |
| ZK $_{3.0}$ (mm / 3.0 m; 1:n) | 4.5 | 1:667 | 12 | 1:250 | 15 | 1:200 | 18 | 1:167 |
| | *7.2** | *1:250** | *7.2** | *1:250** | *not evaluated* | | *10.8** | *1:167** |
| ZK $_{6.0}$ (mm / 6.0 m; 1:n) | not evaluated | | 16 | 1:375 | 20 | 1:300 | 24 | 1:250 |
| | *18.8* | *1:320* | *not evaluated* | | *not evaluated* | | *24* | *1:250* |
| ZK $_{12.0}$ (mm / 12.0 m; 1:n) | not evaluated | | 31 | 1:387 | 34 | 1:353 | 36 | 1:333 |
| | *not evaluated* | | *not evaluated* | | *not evaluated* | | *36* | *1:333* |
| NOTE: Values written in red and cursive are by [5]. * at measuring plant 1.8 m. | | | | | | | | |

Because of the ZSR requirements were measurements carried out until 31. 12. 2016 (MSO, PO1, PO2 a PO3) evaluated according to [5], [7]. Measurements carried out since 1. 1. 2017 (PO4) are evaluated according to [6], [8]. For the comparison of values, measurements of MSO, PO1, PO2 and PO3 were also evaluated according to [6], [8]. The limit values for standard deviations and quality numbers for section evaluation of the measured quantities are in *table 2*.

**Table 2.** Limit value of standard tolerances of variables and quality number for RP1. [7, 8]

| Limit input deviations | | | | | Limit operational deviations (IAL) | | | | |
|---|---|---|---|---|---|---|---|---|---|
| $SDV_{RK}$ | $SDV_{PK}$ | $SDV_{VK}$ | $SDV_{SK}$ | QN | $SDV_{RK}$ | $SDV_{PK}$ | $SDV_{VK}$ | $SDV_{SK}$ | QN |
| 1.5 | 1.8 | 2.0 | 1.5 | 4.1 | 2.7 | 2.9 | 3.4 | 3.1 | 7.2 |
| *1.5* | *1.8* | *2.0* | *1.5* | *4.1* | *2.9* | *2.9* | *3.4* | *2.9* | *7.2* |
| NOTE: Values written in red and cursive are by [7]. | | | | | | | | | |

The overview of quality marks RK, PK, VK and SK (*figure 1*) and quality numbers (*figure 2*) of the monitored section displays stabilized parameter values. Among the MSO and PO1 and PO1 and PO2 measurements, the ZSR staff performed a set of corrective interventions that affected these values in the section and therefore it is not yet possible to fully determine the assumptions for further development of the quality of the section. Maintenance performed after MSO removed track geometry parameters errors before putting the track into operation. Post-PO1 corrective interventions were taken to eliminate mistakes due to an accident.

Within the MSO measurement, local errors (3x SL, 3x SP, 3x RK, 1x ZR) were detected, but after the next corrective intervention they were no longer detected in the operational evaluation. From the parameters whose quality mark enters the quality assessment of the section, the lowest quality has quality mark of track alignment $QM_{SK}$.

The $QM_{SK}$ reaches slightly above the upper limit of the quality mark (2.00), which means the section quality in terms of track geometry is satisfactory. The PO3 rating was slightly higher than 2.00, but PO4 detected a decrease in $QM_{SK}$ below this value (according to [5] - - [8]). Quality marks of other parameters achieve satisfactory values up to 0.70.

The quality number of the monitored section is stabilized between to 1.97 to 2.11 (evaluated according to [5], [7]), respectively 1.96 to 2.02 (evaluated according to [6], [8]), which represents the satisfactory quality of the section, without the need to plan and carry out maintenance in the near term.

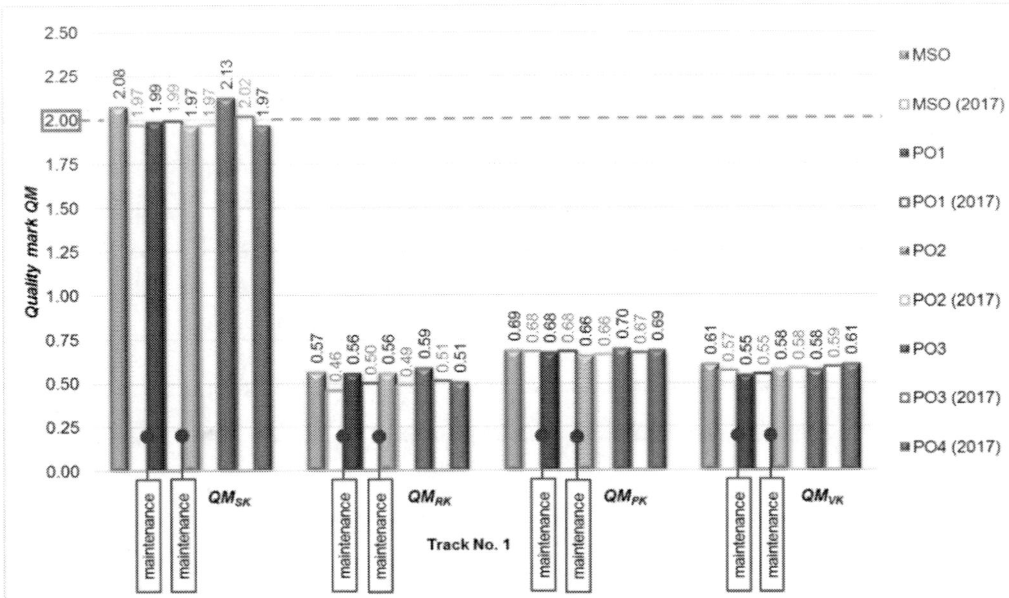


**Fig. 1.** Quality marks of the parameters of the track geometry – measurement carried out by measuring trolley KRAB$^{TM}$–Light.

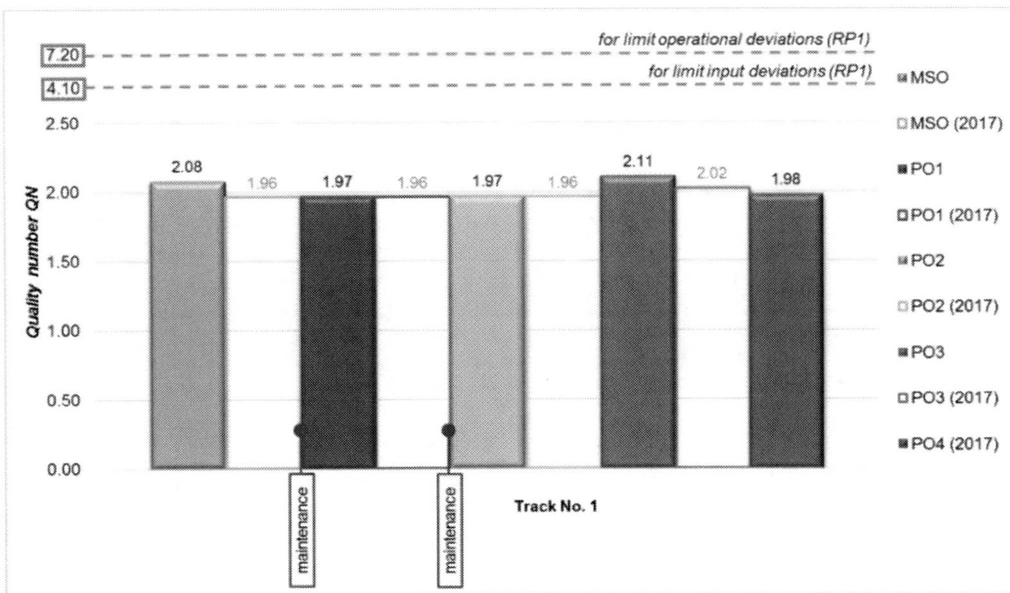


**Fig. 2.** Quality number of the experimental sections.

In relation to traffic load of the structure, the transition areas between ballasted track and ballastless track (eg [9]) appear to be risky. For a detailed assessment of construction behaviour in the transition areas of the Lamac portal of length 18.020 m is processed deviation course overview of key parameters of track geometry involved to section assessment: RK, which is in KRAB 8.1 data represented by values RK_celk, PK (PK_celk), VP /VL/ (VP_dyn_P /VL_dyn_L/), SP /SL/ (SP_dyn_P / SL_dyn_L/) – *figures 3* to *8*.

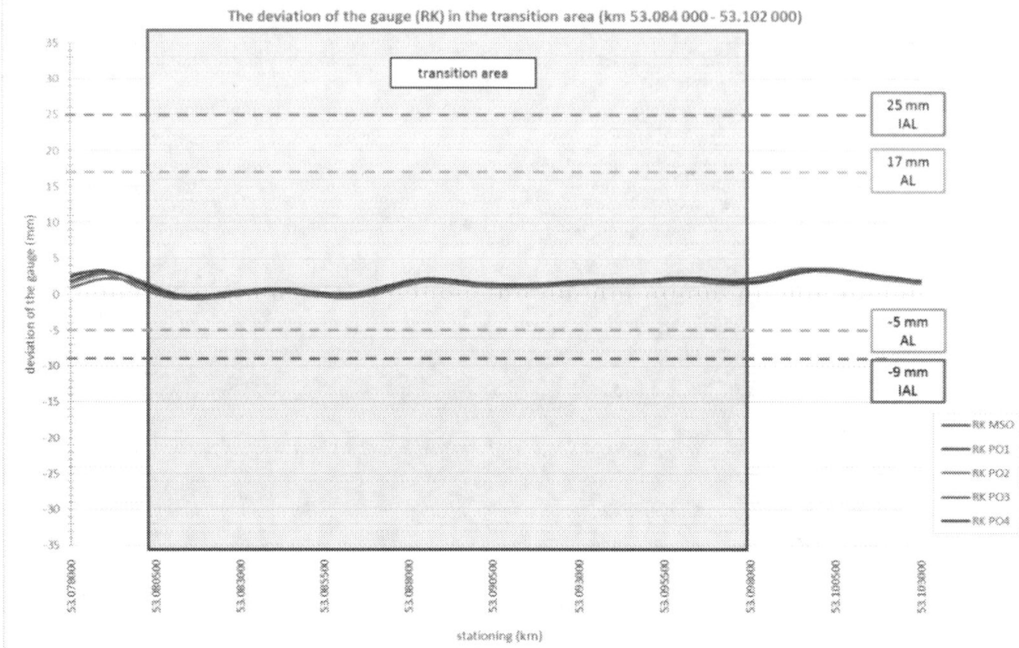


**Fig. 3.** The deviation of the gauge in the transition area.

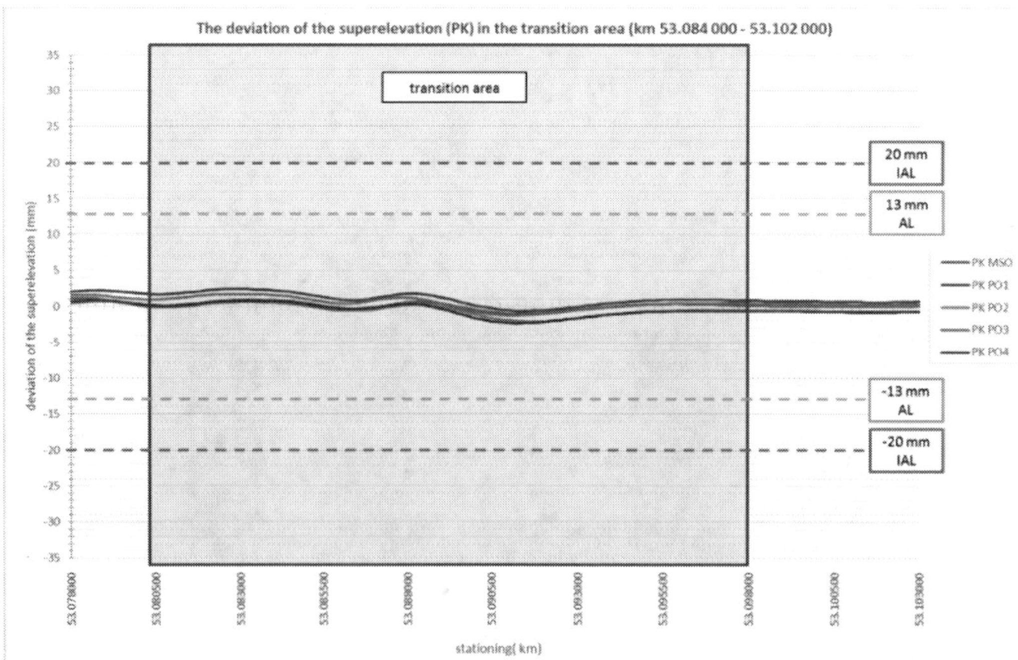


**Fig. 4.** The deviation of the superelevation in the transition area.

Deviations of PK parameter measured in PO4 have a value of less than about 2 mm compared to deviations in PO1, but given the almost parallel curve behaviour, it can be assumed that the change is a combination of the impact of track consolidation after corrective intervention and measurement uncertainty of the measuring device calibrated before measurement MSO and before measurement PO4.

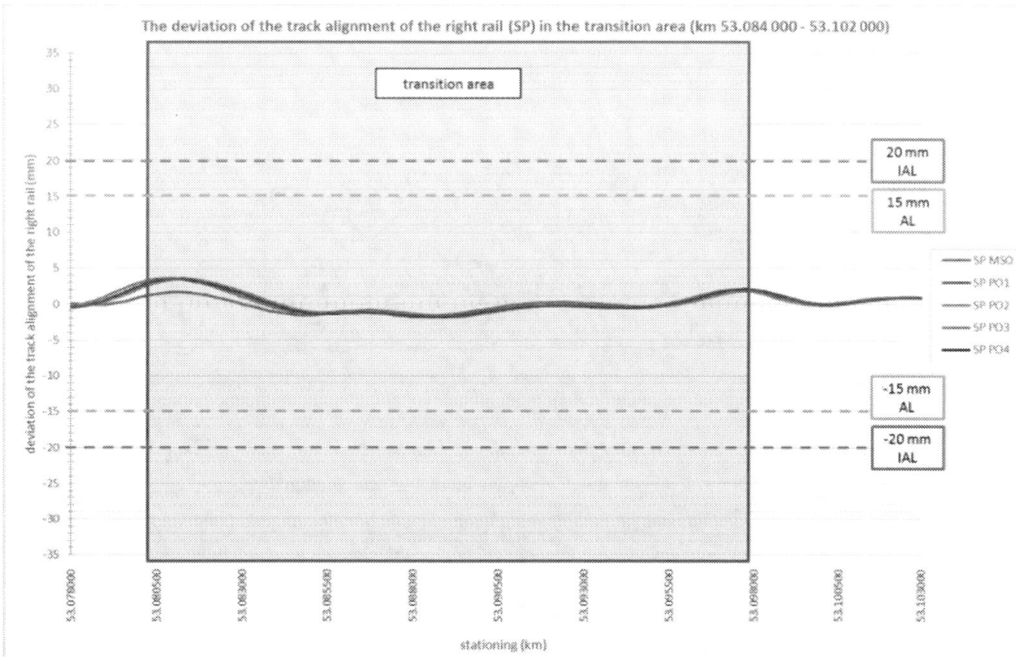


**Fig. 5.** The deviation of the top level of the right rail in the transition area.

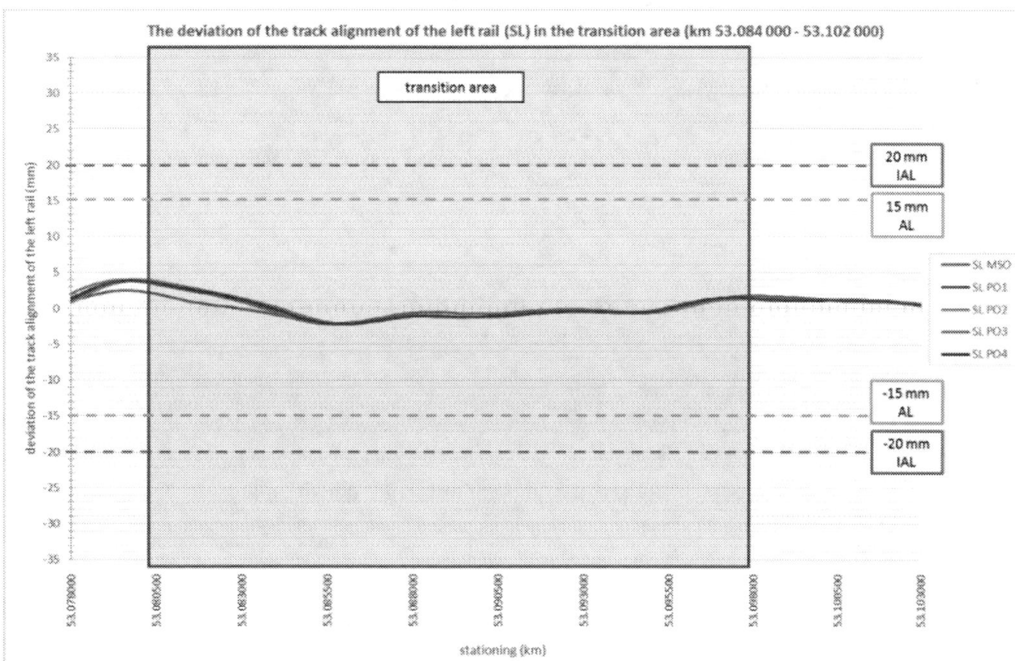


**Fig. 6.** The deviation of the top level of the left rail in the transition area.

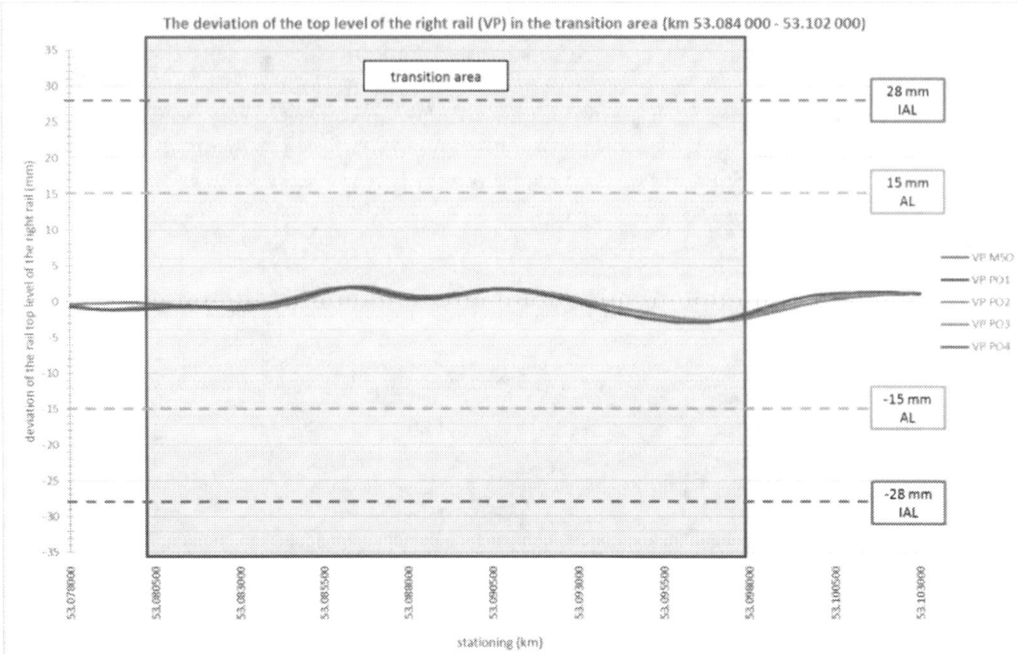


**Fig. 7.** The deviation of the track alignment of the right rail in the transition area.

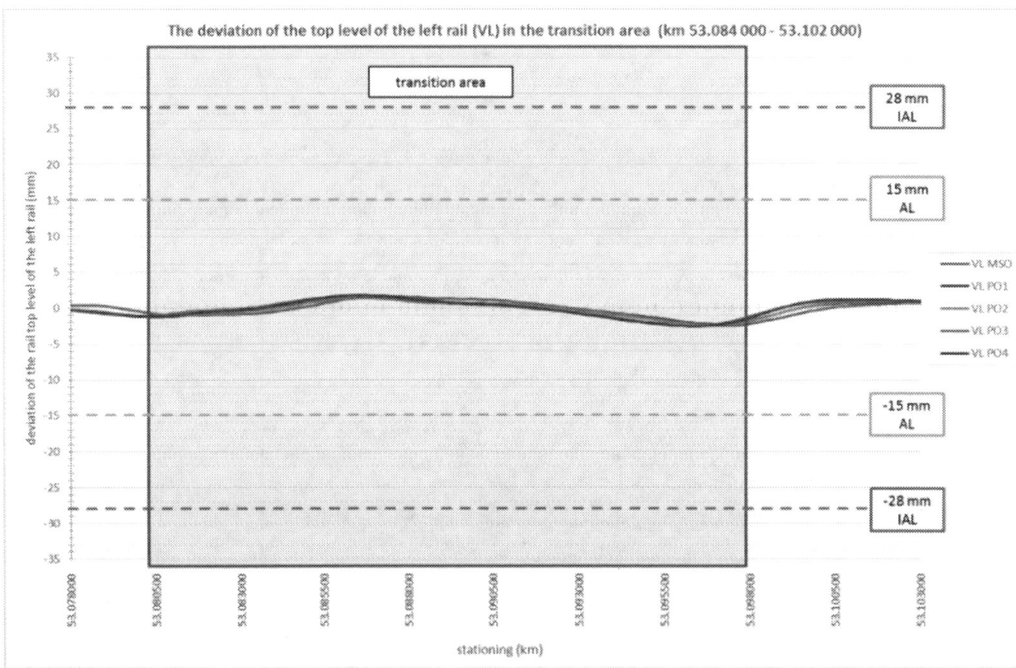


**Fig. 8.** The deviation of the track alignment of the left rail in the transition area.

## 5 Conclusions

Monitoring of the track geometry parameters of the track on the KZSTH experimental section in the Bratislava tunnel No. 1 confirms the sustained state at the required level of quality. Three quality marks (RK, PK, VK) of the four evaluated variables affecting the section quality number reach 35 % of the maximum value of the quality mark (2.00) in the interval of satisfactory evaluation of the track geometry of the section. Marks of the quality of the remaining quantity (SK), within the measurements MSO, PO1, PO2 and PO3, approached or exceeded the value of 2.00 (in the interval 98.5 % to 106.5 % of this value), but in last realized measurement (PO4) again fell to the range of satisfactory evaluation (value 1.97). The section quality number is 27.2 % to 29.3 % of the maximum permissible value of the quality number (7.2) for the sectional evaluation of the measured quantities according to the immediate action limit (IAL). The satisfactory quality of the whole section is also achieved by the high length of the ballastless track construction (approx. 94 %), where track geometry quantities almost do not change.

From the evaluation of the exact deviation course of the values RK, PK, SP (SL) and VP (VL) from the projected values, we can state that there are no changes in the track geometry parameters in the transition area of the Lamac portal. The graphical representation of the deviations in the transition area (km 53.080 to km 53.098) shows a stabilized deviation course (maximum values are within ± 5 mm for all variables) over the entire length of the transition area, without the tendency to quality degradation. These positions are retained since the construction was put into operation and also after maintenance carried out after measurement MSO for all evaluated variables. The railway track superstructure in the transition area, according to the results obtained so far, appears to be suitable for a track with a track speed (60 kmph) assigned to RP1.

The paper contains results of the grant VEGA 1/0275/17 "Application of numerical methods to define the changes of geometrical track position".

## References

1. E. Darr, W. Fiebig, W. *Feste Fahrbahn. Konstruktion und Bauarten für Eisenbahn und Straßenbahn* (Eurailpress Tetzlaff-Hestra GmbH&Co. KG, Hamburg, 2006)
2. O. Plasek, P. Zverina, R. Svoboda, M. Mockovciak *Railway Constructions. Railway Subgrade and Superstructure* (Akademicke nakladatelstvi CERM, Brno, 2014) (in Czech)
3. L. Izvolt, M. Smalo, J. Malchova *The Quality Evaluation of the Ballastless Track Construction in the Area of Bratislava Tunnel No. 1*, MATEC Web of Conferences **86**, 05001 (2016)
4. Technical Standard STN 73 6360-1 *Railway Applications. Track. Part 1: Geometrical Position Arrangement of 1 435 mm Gauge Railways*, (2015) (in Slovak)
5. Technical Standard STN 73 6360 (1999) *Track Alignment Design and Track Geometry of Normal-gauge Tracks* (1999) *and Amendment 1* (2002) (in Slovak)
6. Technical Standard STN 73 6360-2 *Railway Applications. Track. Part 2: Acceptance of Construction Works, Maintenance Works and Assessment of Service Condition Track Gauges 1 435 mm* (2015) (in Slovak)
7. Regulation of ZSR SR 103-7 (S) *Measurement and Evaluation of Track Geometry by Measuring Trolley KRAB* (2008) (in Slovak)
8. Regulation of ZSR SR 103-7 (TS) *Measurement and Evaluation of Track Geometry of Gauge 1 435 mm and 1 520 mm by Measuring Trolley KRAB* (2016) (in Slovak)
9. M. Smalo *The Use of Ballastless Track Monitoring in its Maintenance* (Dissertation) (University of Zilina, Faculty of Civil Engineering, 2017) (in Slovak)

# Deformation Measurements of Ballast-less Railway Line

*Jana* Ižvoltová*, *Tomáš* Cesnek
Department of Geodesy, University of Žilina, Univerzitná 8215/1, 010 26 Žilina

**Abstract:** New technologies of absolute and relative measurements of a railway line based on continual observation techniques bring a new approach to data processing. The paper is devoted to the process of data manipulation, evaluation and analysis by using methods of numerical mathematics with aim to obtain precise, reliable and objective results, which define the track geometry in conformity with the appropriate technical standards and instructions.

## 1. Introduction

Present monitoring the railway line involves absolute and relative techniques, which are based on discrete and continual technologies of data acquisition. While the discrete measurements belong to the conventional methods, the modern one exchange manual measuring devices for automatized measuring vehicles with the continual data recording. However, the observation process has become more time efficient, the evaluation procedures have changed due to clouds of data recording. Besides, the new measuring technologies bring also changes into appropriate technical standards, which modify their requirements on diagnostics of a railway line. Specifically, the Slovak Technical Standard STN 73 6360 [1] primary recommends the continual measurements of track geometry, though the discrete observations are accepted in rare cases. The complex diagnostics of the

*Corresponding author: jana.izvoltova@fstav.uniza.sk

actual railway condition anticipate the combination both of relative and absolute observation techniques, which lead to determining the relative track parameters as track gauge, cross-level and track gradient and absolute track characteristics represented by the longitudinal, directional and height track position. defined in the global coordination and height systems. The paper is devoted to the process of evaluation of clouds of data acquired by the absolute geodetic techniques, which involves the numerical solutions need to control the stability of benchmarks, to construct the mathematical model of railway line by the regression method and to qualify the actual track condition by the method of deformation analysis.

## 2. Methods of Railway Line Diagnostics

Absolute conventional methods of railway line utilized geodetic discrete measurements of polar coordinates (angle, distance). Three basic methods used to be utilized in the railway line measurements, in the past:

- *Method of arc coordinates*, which utilized mathematical relations between arc chord, chamber and radius.
- *Polygonal method* based on measurement of polar coordinates in polygon stabilized right in the railway line axes.
- *Long chord method* based on the relationship between height of arc and it´s long chord.

In case of deformation analysis of track geometry, the evaluation of discrete measurements process was based on mutual comparison of the appropriate discrete points obtained by the periodical measurements. So, the observations process assumed to minimize a pointing error by adding precise signalling devices fixed at the rail body (see fig 1.) or placed on the gauge bar right in the track axis.

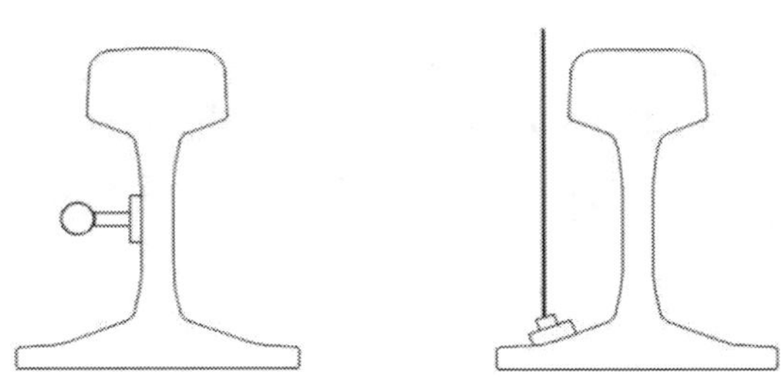

**Fig. 1** An example of additive devices for discrete measurement of track geometry

The present relative track diagnostics is realized by a gauge vehicle to measure the main relative track characteristics as gauge, vertical and horizontal alignment, cant and twist. Sophisticated software recalculates measured data to real track geometry, and displays measurement history, events, local defects and quality indices. The actual absolute track diagnostics represent precise levelling, electronical surveying, 3D laser scanning, satellite observations, etc.. The advanced track diagnostics technology is based on multisensory, which combine both of relative and absolute techniques into automated track vehicle for continual recording and surveying the railway body and it´s environment. Such a track "trolley" is characterized by an amount of recording data, which require a special hardware and software to manipulate and evaluate with them by using methods of numerical mathematics. The actual diagnostics method of track geometry is realized in conformity with the appropriate technical standards [4], [5] and [6].

## 3. Numerical Analysis of Reference System

Geodetic measurements of railway line are connected to the National Spatial Network via railway benchmarks formed into spatial reference system. The positional and height stability of the reference system belongs to the first requisite to ensure the reliable results. Numerical mathematics provides several methods to control the stability of the spatial reference system. Very useful seems to be *seven parameters similarity transformation method*, which can be expressed as:

$$\mathbf{X}' = q.\mathbf{R}\,\mathbf{X} + \mathbf{X_0}, \tag{1}$$

where $\mathbf{X}'$ is the transformed vector, $\mathbf{X}$ is the initial vector, $\mathbf{X_0}$ is translation vector, $\mathbf{R}$ is rotation matrix and $q$ is a scale factor. An application of this method requires an adequate number of identical points. In case of redundancy the least square method is recommended to find an unbiased estimation. The estimated vector of residuals give the view on a possible spatial discrepancy of a benchmark, which should be analysed by a method of outlier diagnostics.

Marcak and Kubacek published in [2] sophisticated method used to control vertical reference network by the method of regression analysis. The method involves parameter estimation of a regression line constructed from a cumulative function of elevation differences (superelevations) between all pairs of the reference points:

$$\Delta^{j}_{i,i+1} = \sum_{k=i}^{n-1} \Delta h^{j}_{k,k+1}, \tag{2}$$

where the upper index $j$ is the series of measurement, lower index $i,i+1$ represents the pairs of points, in which the superelevation $\Delta$ is calculated and $n$ is the number of all reference points. The elevation $\Delta h$ represents the levelling difference between a point of an actual measurement and the appropriate one comes from the first measurement as follows:

$$\Delta h^{j}_{k,k+1} = h^{j}_{k,k+1} - h^{1}_{k,k+1}, \tag{3}$$

Providing, the superelevations are normally distributed, the regression line can be estimated by least square method, in other case a robust regression is recommended (Cebecauer [6], Rousseeuw [7]).

The final advisement of the reference network stability is realized by the statistical hypothesis testing. A null hypothesis is proposed for the statistical relationship between an extreme value and the critical one, both from the same probability distribution:

$$H_0 : x_{ex} > t_\alpha \sigma , \tag{4}$$

where $t_\alpha$ is a quantile of Student's $t$-distribution and $\sigma$ is standard deviation, which defines an internal accuracy of the reference system.

## 4. Regression analysis

A regression model is used to substitute the real continual measurement with a mathematical function with aim to study the track directional and longitudinal parameters to predict the railway line behaviour. Developing of the mathematical function depends on the directional and height design parameters of the railway line.

In case of curve section of the railway line, the quadratic nonlinear regression model used to be applied to reconstruct the curve parameters of the railway line:

$$R^2 = (x_i - X_S)^2 + (y_i - Y_S)^2, \tag{5}$$

where R is an unknown radius of a curve, $X_s$, $Y_s$ are also unknown coordinates of a curve centre and $x_i$, $y_i$ are known coordinates of a $i$-th measured point.

In case of spiral section of railway line, the cubic nonlinear model is developed to estimate the unknown parameters $a$, $b$, $c$, $d$ of cubic parabola, as follows:

$$y_i = ax_i^3 + bx_i^2 + cx_i + d . \tag{6}$$

The linear section of railway line can be estimated by linear function:

$$y_i = a + bx_i , \tag{7}$$

where $a$, $b$ represents intercept and slope of line.

Evaluation of the above mentioned mathematical functions anticipates an application of a method of estimation theory to find the unknown parameters. An unbiased estimation is possible to achieve by the application of the collocation method, least square method, robust method, maximum like hood estimation, unbiased quadratic estimation, etc.

## 5. Deformation Analysis

The analysis of track changes belongs to the statistical analysis testing based on postulate of null hypothesis:

$$H_0 : \Delta Z = 0, \tag{8}$$

where $\Delta Z$ is real track displacement, which depends on real observation error $\varepsilon_{\Delta z}$ and the measured displacement $\Delta z$

$$\Delta Z = \Delta z + \varepsilon_{\Delta z}. \tag{9}$$

Suppose, the measured values are normal distributed the confidential interval for identification the measured displacement can be defined as follows:

$$P\left(\left|\varepsilon_{\Delta z}\right| > t_\alpha \sigma_{\Delta z}\right) = P\left(\left|\Delta z\right| > t_\alpha \sigma_{\Delta z}\right) = \alpha, \tag{10}$$

where $\sigma_{\Delta z}$ is standard deviation depended on random variances of both of series of observations:

$$\sigma_{\Delta z} = \sqrt{\sigma_{z1}^2 + \sigma_{zi}^2} \tag{11}$$

Deformation analysis provides three possibilities to identify the track displacement [7]:

- If $|\Delta z| < \sigma_{\Delta z}$ null hypothesis is accepted and a point displacement is not evident
- If $\sigma_{\Delta z} \leq |\Delta z| < 2\sigma_{\Delta z}$ a point, displacement is regarded to be possible
- If $|\Delta z| > 2\sigma_{\Delta z}$ null hypothesis is not accepted and a point displacement is evident

## 6. Mathematical Modelling

In Slovakia, the railway modernization brings except of social-economic and environmental advantages also the technological innovations of construction layers of railway body. Such a new technology consist in replacement of the traditional railway body by the ballast-less track construction. The specialists of the Department of Geodesy and Department of Railway Construction of the University of Zilina have been asked to monitor such a new railway construction with aim to diagnose its behaviour in time of traffic loading. Up to time, five periodical measurements of absolute and relative track measurements were realized to define track geometry of about 2 km long ballast-less railway line, see fig. 1, 2.

**Fig. 2** Bridge part of ballast-less track construction

**Fig. 3.** Tunnel part of ballast-less track construction

Evaluation process of continual observations realized by the automatized railway vehicle starts with mathematical modelling by applying mixed Gauss-Markov model [8], [9], which deterministic formula can be written as:

$$\mathbf{B}^T\mathbf{v} + \mathbf{Adx} + \mathbf{u} = 0\,, \tag{12}$$

where **A**, **B** and **u** represent matrices of known parameters, **dx** is a vector of unknown values and **v** is vector of residuals. Least square method was applied to estimate unknown parameters of all railway line sections (curve, transit curve, line) and accuracy parameters:

$$\Omega = \mathbf{v}^T\mathbf{P}\mathbf{v} + 2\mathbf{k}^T\left(\mathbf{B}^T\mathbf{v} + \mathbf{A}d\mathbf{x} + \mathbf{u}\right) = \min\,. \tag{13}$$

The solution of equation (13) leads to multiplying block matrices:

$$\begin{pmatrix} \mathbf{k} \\ -\mathbf{dx} \end{pmatrix} = \begin{pmatrix} \mathbf{B}^T\mathbf{QB} & \mathbf{A} \\ \mathbf{A}^T & 0 \end{pmatrix}^{-1} \begin{pmatrix} \mathbf{u} \\ 0 \end{pmatrix}, \tag{14}$$

where **k** is unknown vector of Lagrange coefficients. Accuracy parameters are represented by the covariance matrix, which is estimated according to the equation:

$$\mathbf{D(x)} = \sigma^2\left(-\left(\mathbf{A}^T\left(\mathbf{B}^T\mathbf{QB}\right)^{-1}\mathbf{A}\right)^{-1}\right). \tag{15}$$

The estimation of unit variance depends inter alia on the number of measured values $n$ and the number of unknown parameters $k$ as follows:

$$\sigma^2 = \frac{\mathbf{v}^T \mathbf{P} \mathbf{v}}{n-k}, \qquad (16)$$

## Conclusion

Standard deviation subjected to the precision analysis in equation (21) originates from the diagonal elements of covariance matrix in (13) and represents the a-posteriori accuracy of transversal track changes. It involves the error of multitrack target positioning in regard of track axis, which is ±2 mm and accuracy of estimation process, which is about ± 2-3 mm. As we can see in fig. 1 the transversal displacement of ballast-less track was not confirmed and the estimated values of track changes are in the frame of measuring errors.


## Acknowledgements

*This article is the result of the implementation of the project VEGA 1/0275/17 "Application of numerical methods to define the changes of geometrical track position" supported by the Scientific Grant Agency of the Ministry of Education, science, research and sport of the Slovak Republic and the Slovak Academy of Sciences.*

*This article is the result of the implementation of the project ITMS 26220220156 "Broker centre of air transport for transfer of technology and knowledge into transport and transport infrastrukture" supported by the Research & Development Operational Programme funded by the ERDF.*


## References


1. STN 73 6360 Railway applications. Track. Part 1: Geometrical position and arrangement of 1 435 mm gauge Railways, Slovak Office of Standards, Metrology and Testing, Bratislava (2015)
2. P Marcak, L. Kubacek. L.: Studia Geophysica et Geodaetica, **18**, 33-46, 1974
3. L. Ižvolt, S. Hodás, J. Šestáková, Railway buildings 1. Design, building, construction of railway lines and stations. Handbook, EDIS (2015)
4. STN 73 0275 Control measurements of line buildings. Precision of geometrical parameters in construction. SÚTN (1985)
5. STN 73 0405 Measurement of building deformations. SÚTN (1985)
6. Direction SR 103- (S) Common demands on design, building, repair, maintenance and acceptance of work realized on ballast-less track construction. Bratislava
7. J. Gašinec, S. Gašincová, J. Sabová, G., 14. conference of Mining Surveyors and Geologists, Ostrava VŠB TU, (2007)

8. W. Niemeier, Ausgleichungsrechnung. Walter de Gruyter GmbH & Co. Berlin (2001)
9. A. Villim, J. Mužík, Analysis of satellite and terrestrial measurements in 3D geodetic network for transport infrastructure, Juniorstav 15, Brno (2013)

MATEC Web of Conferences 117, 00065 (2017) DOI: 10.1051/matecconf/201711700065
*XXVI R-S-P Seminar 2017, Theoretical Foundation of Civil Engineering*

# Probabilistic analysis of the safety margin assured by shear strength models of stirrup reinforced concrete beams

*Roman* Jaskulski[1,*], and *Piotr* Wiliński[1]

[1]Warsaw University of Technology, Faculty of Building, Mechanics and Petrochemistry, 17 Łukasiewicza Street, 09-402 Płock, Poland

**Abstract.** The aim of this study was to assess the safety margin assured by stirrups of reinforced concrete elements subjected to shear. The safety margin was assessed by means of Monte Carlo simulation. The impact of the strength of steel and spacing of stirrups on the results was analysed as well. The used models of shear resistance were taken from the Polish standards PN-84/B-03264 and PN-EN 1992-1-1:2008. The safety margin, expressed as the logarithm of the probability of not achieving the design value of the design ultimate shear capacity was analysed as well as the impact of different assumptions on the obtained results. An attempt was made to assess the "sensitivity" of models to changes of the basic parameters of the probability distribution function of their selected variables.

## 1 Introduction

The methods of probabilistic structural analysis enable assessment of the safety margin of existing structures and the design of structures with a specific safety margin. Inter alia the work [1] is devoted to the second goal of probabilistic analysis. The methods can also be used to determine the safety margin of the shear strength models included in the standards. Full model analysis combined with their comparison is a complex and laborious task due to significant differences in the assumptions of individual models, as well as variations in their parameters. Therefore, the authors of this paper have set themselves a slightly less complex task, which is reduced to the parallel analysis of two significantly differing normative models. This analysis was not intended to directly compare selected models and the safety margin assured by them, but to compute safety margins of the same structure element using selected models with possibly similar assumptions, and to determine the impact of changing the assumptions on the level of its safety margin. Having independently prepared software with built-in models of shear strength of stirrup reinforced concrete element, an analysis of selected models was conducted in terms of their "sensitivity" to change the parameters of distribution of selected variables.

* Corresponding author: Roman.Jaskulski@pw.edu.pl

This topic has not been widely taken up so far. Results of probabilistic analysis of the shear strength of elements cross-sections was presented i.a. in [2-4]. Some attempts of this kind of probabilistic were also undertaken in [5]. Due to the complexity of the modelling of shear strength in reinforced concrete elements, the paper is limited to the models of shear strength of rectangular cross-section beam reinforced with single stirrups perpendicular to the axis of the element. Two models derived from completely different assumptions were selected for analysis. One of them was included in PN-84/B-03264 (denoted further as PN-84) [6], which was a standard used for a long time in Poland, and the second one is included in the currently used Polish standard PN-EN 1992-1-1:2008 (denoted further as EC-2) [7].

## 2 Analyzed shear strength models

In the literature a large number of models of shear resistance of reinforced concrete elements are described (i.a. in [8, 9]). In this work authors limited their choice to two models based on different concepts and assumptions. Model in the PN-84 standard is based on the Boriszanski method [10], developed in the former USSR. In this model the destruction of the element is due to the rotation of the part of a beam cut-off by a diagonal crack around the point located over the end of the crack. More detailed description of this method can be found in [8, 9, 11]. The model included in EC-2 standard is based on the truss analogy and was originally developed by Mörsch [12]. The Kupfer concept [13] was finally adopted in standards which is the development of the classic truss method.

### 2.1 The PN-84 model

The following formulas were given in [6] and explained in [11, 14]. The ultimate shear force in the case of elements reinforced only with stirrups (Qsb) is described by the formula:

$$Q_{sb} = \sqrt{4 \cdot \beta_s \cdot R_b \cdot q_s \cdot b \cdot h_o^2} - R_{as} \cdot F_s \tag{1}$$

where: $Q_{sb}$ is the ultimate shear force carried by concrete and stirrups, $\beta_s$ is a coefficient dependent on how the element is supported and loaded, $R_b$ is the design compressive strength of concrete, $q_s$ is the ultimate force carried by stirrups (per unit length), $b$ is the cross-section width, $h_o$ is the effective height of the cross-section, $R_{as}$ is the design tensile strength of steel of the stirrups equal 80% of design yield strength of steel and $F_s$ is the cross-section area of one stirrup.

The ultimate force $q_s$ is calculated using formula:

$$q_s = \frac{R_{as} \cdot F_s}{s} \tag{2}$$

where $s$ is the spacing of stirrups.

### 2.2 The EC-2 model

The design shear force carried by the stirrups ($V_{Rd,s}$) in the elements with vertical shear reinforcement shall be calculated using the formula (3) which is given in [7]. And the sole model is described in [15].

$$V_{Rd,s} = \frac{A_{sw}}{s} \cdot z \cdot f_{ywd} \cdot \cot\theta \qquad (3)$$

In the formula (3) $V_{Rd,s}$ is design shear force carried by the shear reinforcement, $A_{sw}$ is the cross-section area of the stirrups which are located in one plane perpendicular to the axis of the element, $z$ is the shear zone height (can be assumed as $0{,}9 \cdot d$), $d$ is effective height of the cross-section, $f_{ywd}$ is the design yield strength of steel of stirrups, $\cot\theta$ is cotangent of angle $\theta$ of inclination of compressed concrete struts (arbitrarily taken from interval <1,2>), $s$ is spacing of stirrups perpendicular to the axis of the element.

## 3 Assumptions adopted in calculations

The variables in the formulas (1), (2) and (3) were divided into two groups. The first included: the spacing of the stirrups, compressive strength of concrete, yield strength of steel and the cotangent θ. For these variables, at least two different variants of their distribution parameters were used in the simulations. The second group of variables included: the height and width of the element cross-section, thickness of the reinforcement cover and the cross-section area of the single stirrup. In this group, one variant of distribution parameters was used in all simulations.

**Table 1.** Distribution parameters of the first group of variables.

| Variant | Parameter | $x_{nom}$ | $x_m$ | $\sigma_X$ | $v_X$ | Type of distibution |
|---|---|---|---|---|---|---|
| SM | stirrups spacing $s_1$ [mm] | 70 | 70 | 5 | 7,14% | normal |
| SD | | 150 | 150 | 5 | 3,57% | |
| FM | yield strength of steel $f_{yw}$ [MPa] | 210 | 341,6 | 10,25 | 3,0% | normal |
| FD | | 210 | 341,6 | 44,38 | 13,0% | normal |
| C16/20NM | compressive strength of concrete $f_c$ [MPa] | 10,8 | 21,30 | 0,85 | 4.0% | normal |
| C40/50NM | | 26,7 | 53,20 | 2,13 | 4,0% | |
| C16/20ND | | 10,8 | 26,20 | 4,19 | 16,0% | |
| C40/50ND | | 26,7 | 65,50 | 10,05 | 15,3% | |
| C16/20LM | | 10,8 | 21,30 | 0,85 | 4,0% | lognormal |
| C40/50LM | | 26,7 | 53,20 | 2,13 | 4,0% | |
| C16/20LD | | 10,8 | 26,20 | 4,19 | 16,0% | |
| C40/50LD | | 26,7 | 65,50 | 10,05 | 15,3% | |
| T1 | $\cot\theta$ [-] | 1 | – | – | – | --- |
| T2 | | 2 | – | – | – | --- |

A rectangular cross-section of nominal dimensions: $b$ = 0.30 m and $h$ = 0.60 m was adopted in simulations. The thickness of the reinforcement cover $a_l$ = 0.05 m was assumed. The distribution parameters of the above variables were assumed according to the recommendations in [16, 17]. Values of steel yield strength distribution parameters were taken from [11]. The values given therein have been verified in own tests and the study description with results is given in [18]. After analysing the results of the own research it was decided to assume two values of coefficient of variation of yield strength of stirrup steel: 3% and 13%. It was assumed that the cross-section would be reinforced with stirrups with a diameter of 8 mm, so the mean and nominal value of the cross-sectional area of the single stirrup was assumed to be 1.0 $cm^2$. Cross-sectional variation coefficient was assumed to be 3% according to [11]. Two nominal values of stirrup spacing were assumed: 0.07 m and 0.15 m. The mean values of stirrup spacing were assumed to be equal to the nominal values and the standard deviation was assumed in both cases at 0.005 m.

Four sets of compressive strength distribution parameters were adopted. Two of them corresponded to the strength class C16/20 and two class C40/50. Within one class of concrete, the values of variation coefficients were additionally varied. The cot$\theta$ values were assumed a deterministic variables as cot$\theta$ = 1 and cot$\theta$ = 2.

Normal distribution was used to for all random variables as a base one. Some additional simulations were made using lognormal distribution of compressive strength of concrete. The variable distribution parameters are given in Tables 1 and 2.

**Table 2.** Distribution parameters of the second group of variables.

| Parameter | $x_{nom}$ | $x_m$ | $\sigma_X$ | $v_X$ | Type of distibution |
|---|---|---|---|---|---|
| width $b$ [mm] | 300 | 300,9 | 5,8 | 1,93% | normal |
| height $h$ [mm] | 600 | 601,8 | 7,6 | 1,26% | normal |
| concrete cover $a_l$ [mm] | 50 | 50 | 3 | 6,0% | normal |
| stirrups cross-section area $A_{sw}$ [$cm^2$] | 1,00 | 1,00 | 0,03 | 3,0% | normal |

## 4 Monte Carlo simulations

The simulation program developed by Jaskulski [1] was used. The program generates random variable values based on a pseudorandom number generator with a uniform distribution on the interval (0,1), which is an integral part of the Pascal language. The program allows to calculate the shear strength as described above, and also gives the value of the reliability index $\beta_R$ defined by the formula (4)

$$\beta_R = \frac{R_m - R_d}{\sigma_R} \tag{4}$$

where $R_m$ is the average shear strength (based on the simulation), $R_d$ is design shear strength, $\sigma_R$ is standard deviation of shear strength (based on the simulation). The design strength $R_d$ is entered as a parameter when starting the software. The value $\beta_R$ serves as the boundary of the integral (5) calculating the probability $p_f$ obtaining the value of the strength lower than the calculated value $R_d$. Integrated function is the NG probability distribution described i.a. in [19]

$$p_f = \int_{-\infty}^{-\beta_R} f_{NG}(t)dt \tag{5}$$

In the formula (5) $f_{NG}$ is the function of NG distribution of the random variable, $t$ is normalized random variable.

The simulations were divided into two groups: basic simulations and so called "zero" simulations. In the basic simulations, all random variables were generated according to the probability distributions assigned to them. The "zero" simulations determined the effect of the random variation of each variable individually on the safety margin of the element cross section. For this purpose, only one selected variable was generated randomly and the remaining values remained constant during simulation (average values were assumed). The exception to this rule was the effective height $d$ where two variables were assumed to be randomly variable: the section height $h$ and the thickness of the concrete cover $a_l$. In the "zero" simulations, only one variant of distribution parameters was used for each variable. The values of the parameters accepted in the null simulations are summarized in Table 3.

**Table 3.** Random variables distribution parameters in the "zero" simulations.

| **Parameter** | $x_{nom}$ | $x_m$ | $\sigma_X$ | $v_X$ | **Type of distribution** |
|---|---|---|---|---|---|
| width $b$ [mm] | 300 | 300,9 | 5,8 | 1,93% | normal |
| height $h$ [mm] | 600 | 601,8 | 7,6 | 1,26% | normal |
| concrete cover $a_l$ [mm] | 50 | 50 | 3 | 6,0% | normal |
| yield strength of steel $f_{yw}$ [MPa] | 210 | 341,6 | 44,38 | 13,0% | normal |
| stirrups cross-section area $A_{sw}$ [cm$^2$] | 1,00 | 1,00 | 0,03 | 3,0% | normal |
| stirrups spacing $s_l$ [mm] | 70 | 70 | 5 | 7,14% | normal |
| compressive strength of concrete $f_c$ [MPa] | 10,8 | 21,30 | 0,85 | 4.0% | normal |
| cot $\theta$ [-] | 1 | – | – | – | --- |

For each data set, both in the basic and the "zero" simulations, five independent simulations of 2,000,000 randomly generated values for each random variable were performed. The results of these five simulations were then averaged. The detailed results are included in [S].

## 5 Results and discussion

Some difficulties in analysing the obtained results are due to the fact that both used models have only two common parameters that were used in more than one variant: the spacing of the stirrups and the yield strength of steel. Additionally, for each of these models there is a parameter that is not present in the second model. These are the compressive strength of concrete in the PN-84 model and the angle of inclination of compressed concrete struts in the EC-2 model. This makes it difficult to interpret the identified differences. Therefore, the results obtained, with the exception of mean ultimate shear strength, were analyzed in parallel without direct comparing them.

Mean values of shear strength obtained from the simulation is given in Table 4. The results obtained using the lognormal distribution were omitted because the type of distribution did not influence the obtained mean values. Analysis of mean values of shear strength leads to the conclusion that with the stirrup spacing of 0.07 m, the least load capacity (calculated using the model EC-2 at $\cot\theta = 1$) is almost five times lower than the maximum value (calculated according to PN-84 model assuming C40/50 concrete and greater variability of its strength). When the spacing increases to 0.15 m, the difference between models is up to almost seven times. Such great differences result primarily from the fact that the PN-84 model, in contrast to the EC-2, assumes the cooperation of concrete in shear force carrying. The adoption of the C16/20 concrete and the more favourable value $\cot\theta = 2$ in the EC-2 model reduces the advantage of the first of these models to 33%.

**Table 4.** Mean values of shear strength.

| Spacing of stirrups [m] | Steel yield strength variation variant | $V_{R,s}$ (the EC-2 model) | | $Q_{sb}$ (the PN-84 model) | | | |
|---|---|---|---|---|---|---|---|
| | | T1 | T2 | C16/20 | | C40/50 | |
| | | | | NM | ND | NM | ND |
| 0,07 | FM | 244 | 487 | 650 | 721 | 1043 | 1157 |
| | FD | 244 | 487 | 648 | 720 | 1041 | 1154 |
| 0,15 | FM | 113 | 226 | 435 | 483 | 703 | 780 |
| | FD | 113 | 226 | 434 | 482 | 702 | 779 |

The level of the safety margin of the concrete element subjected to shear forces obtained in the simulation was also analysed. The partial reliability exponent $\Delta_R$ defined by the formula (6) was used for this purpose.

$$\Delta_R = -\log\left[p_f\left(R \le R_d\right)\right] \tag{6}$$

With such a useful measure of safety margin, the analysis is convenient. An alternative is to use the probability values pf or the reliability index $\beta_R$. Both of these measures have their limitations. The reliability index $\beta_R$ does not allow for the impact of shear strength distribution skewness to be taken into account, which may sometimes significantly influence the level of safety margin. On the other hand, using of the probability values $p_f$ is impracticable because of the differences which often reach more than a dozen orders of magnitude. This makes it difficult for direct comparisons and reception of results. The values of the partial reliability exponent $\Delta_R$ are presented in Table 5.

**Table 5.** Values of partial reliability index $\Delta_R$.

| Spacing of stirrups [m] | Steel var. | $V_{R,s}$ | | $Q_{sb}$ (Normal) | | | | $Q_{sb}$ (Lognormal) | | | |
|---|---|---|---|---|---|---|---|---|---|---|---|
| | | T1 | T2 | C16/20 | | C40/50 | | C16/20 | | C40/50 | |
| | | | | NM | ND | NM | ND | LM | LD | LM | LD |
| **0,07** | **FM** | 5,78 | 11,7 | 18,2 | 4,66 | 18,7 | 5,09 | 18,2 | 7,35 | 18,7 | 7,95 |
| | **FD** | 2,38 | 2,38 | 7,99 | 5,26 | 7,97 | 5,53 | 7,99 | 5,44 | 7,97 | 5,69 |
| **0,15** | **FM** | 5,79 | 11,7 | 30,4 | 4,24 | 31,1 | 4,48 | 30,4 | 8,03 | 31,1 | 8,82 |
| | **FD** | 2,63 | 2,63 | 5,17 | 5,62 | 5,11 | 5,93 | 5,20 | 5,83 | 5,13 | 6,12 |

Analysis of partial reliability exponent values shows that the level of safety margin achieved for different data variants in the EC-2 model is less varied than that of the PN-02 model. Given the high variability of the average yield strength of steel, the safety margin determined using the EC-2 model is lower than that of the PN-02 model. On the other hand, with the low variability of the yield strength of steel, the obtained safety margin according to the EC-2 model is within the range obtained using the PN-02 model.

For the EC-2 model, the calculated value of the partial reliability exponent depends on all analysed variables occurring in more than one variant. In the case of low variability of yield strength of steel (the FM variant), the obtained exponent value was 2 to almost 5 times higher than the value obtained for high variability of the steel yield strength (the FD variant). The stirrup spacing did not affect the results, and the higher cotθ resulted in than double increase of the partial reliability exponent. This may be due to the more favourable positioning of the concrete compressive struts relative to the shear force direction at the higher cotθ value. In the case of high variability of higher variability of yield strength of steel (the FD variant), the greater stirrup spacing resulted in an increase in the partial reliability exponent, and this time the cotθ value did not affect the results. More dense stirrups spacing in the element causes the increase in steel shear in the assuring of the element's safety margin, thus the high variability of its yield strength is more likely to translate into variability in shear strength and as a result, the safety margin decrease.

Analysis of the partial reliability exponent values obtained with the PN-02 model shows that the combination of the low variability of steel strength variant (FM) with low variability in compressive strength of concrete results in a significant increase in the partial reliability exponent (from more than 2 to almost 6 times) in comparison to the FD variant. This effect is not dependent on the type of concrete strength distribution. With high variability of compressive strength of concrete, two tendencies can be distinguished, depending on the type of the distribution adopted. In the case of normal distribution, the partial reliability exponent decreases by a maximum of 30%, whereas in the case of lognormal distribution it increases by approximately 40%.

Using two different types of concrete strength distribution is reflected in the differences in the value of the partial reliability factor in the PN-02 model at a moderate level, if analysed together with the FD variant. In the case of low variability in concrete strength differences are practically negligible. In the case of low variability of the steel yield strength (FM), combined with the high variability of concrete strength, the effect of the type of concrete strength distribution is revealed. The partial reliability exponent is almost doubling, which may indicate a specific case of synergy.

Assuming of different values of the coefficient of variation of concrete strength, coupled with the high variability of the yield strength of steel, leads to distinct differences in the estimated safety level. There are two different tendencies here. Assuming the smaller spacing of stirrups and increasing the coefficient of variation in concrete strength fourfold decreases the $\Delta_R$ value by approximately 35%. On the other hand, assuming the higher spacing of stirrups and increasing the coefficient of variation of concrete strength fourfold leads to an increase of $\Delta_R$ by a maximum of about 30%.

In addition to the mean values also the coefficient of skewness $a_R$ was calculated in simulations. It is important in probabilistic estimating of safety margin by integrating the ultimate strength distribution. If the strength distributions has skewness which significantly differs from zero, then omission of this parameter in calculating of safety margins results in incorrect estimation. The values of the coefficient of skewness $\Delta_R$ are presented in Table 6.

Analysis of values of the skewness coefficient shows that in the case of the PN-02 model, assuming a high steel yield strength variation (FD), in most cases the skewness coefficient is close to zero. Derogation from this rule only occurred when assuming C40/50 concrete and lognormal distribution of its strength. Under such assumptions, the skewness coefficient ranged from 0.23 to 0.27. There was also a tendency observed that the value of the coefficient of skewness decreases as the spacing of the stirrup increases.

**Table 6.** Values of coefficient of skewness $a_R$.

| Spacing of stirrups [m] | Steel var. | $V_{R,s}$ | | $Q_{sb}$ (Normal) | | | | $Q_{sb}$ (Lognormal) | | | |
|---|---|---|---|---|---|---|---|---|---|---|---|
| | | T1 | T2 | C16/20 | | C40/50 | | C16/20 | | C40/50 | |
| | | | | NM | ND | NM | ND | LM | LD | LM | LD |
| 0,07 | FM | 0,37 | 0,37 | 0,22 | 0,22 | -0,04 | - 0,02 | 0,23 | 0,23 | 0,28 | 0,28 |
| | FD | 0,24 | 0,24 | 0,07 | 0,06 | 0,09 | 0,10 | 0,07 | 0,06 | 0,27 | 0,26 |
| 0,15 | FM | 0,16 | 0,16 | 0,09 | 0,09 | -0,13 | - 0,12 | 0,11 | 0,11 | 0,25 | 0,24 |
| | FD | 0,09 | 0,09 | -0,04 | -0,04 | 0,03 | 0,03 | -0,04 | - 0,04 | 0,25 | 0,23 |

Assuming the low variation of the yield strength of steel (FM), the value of this factor increases significantly (almost fourfold) compared to the high variation (FD) variant. Derogation from this rule only occurred when concrete class C40/50 was assumed. In the case of assumption of the normal distribution of concrete strength in the simulations, the values of the coefficient of change the sign and grow or decrease almost fourfold. In the case of assumption of the lognormal distribution of concrete strength, the skewness coefficients slightly increase. Also the tendency of decrease of the coefficient of skewness with the increasing stirrup spacing was observed (except for the assumption of C40/50 concrete with normal distribution of its strength).

Also in the case of the EC-2 model, the coefficient of skewness obtained by assuming the larger stirrup spacing is smaller, with the differences here being very clear. With the small stirrup spacing, the coefficient of variation was approximately 2.5 times greater than that for the larger span, and this ratio is independent of the variability of the steel yield strength. The value of the coefficient of skewness also does not depend on the assumed value of the $\cot\theta$.

The last analyzed dependency was the influence of the variability of the individual parameters of the shear strength models on the coefficient of variation of the shear strength distribution obtained by the simulation. For this purpose, the previously mentioned "zero" simulations were performed. For the purpose of the analysis, the values of three coefficients of variation were calculated (and are presented in Table 7):

**Table 7.** Results of the "zero" simulations.

| Model | Variable | $v_R$ | $v_{Xi}$ | $v_{R,Xi}$ | $\frac{v_{R,Xi}}{v_{Xi}}$ | $\frac{v_{R,Xi}}{v_R}$ |
|---|---|---|---|---|---|---|
| PN-02 | b | 0,0797 | 0,0193 | 0,0100 | 52,0% | 12,6% |
| | d | | 0,0148 | 0,0154 | 104,3% | 19,4% |
| | s | | 0,0714 | 0,0376 | 52,7% | 47,2% |
| | $f_{yw}$ | | 0,1300 | 0,0629 | 48,3% | 78,8% |
| | $f_c$ | | 0,0400 | 0,0208 | 52,0% | 26,1% |
| | $A_{sw}$ | | 0,0300 | 0,0144 | 47,9% | 18,0% |
| EC-2 | d | 0,1528 | 0,0148 | 0,0148 | 100,0% | 9,7% |
| | s | | 0,0714 | 0,0725 | 101,5% | 47,4% |
| | $f_{yw}$ | | 0,1300 | 0,1299 | 99,9% | 85,0% |
| | $A_{sw}$ | | 0,0300 | 0,0300 | 100,0% | 19,6% |

- $v_{Xi}$ – coefficient of variation of the single variable,
- $v_R$ – coefficient of variation of the strength distribution (all variables as random),
- $v_{R,Xi}$ – coefficient of variation of the strength distribution (single random variable).

Analiza danych zawartych w tablicy 7 prowadzi do wniosku, że w obu modelach dominującym parametrem jest granica plastyczności stali zbrojenia poprzecznego, choć znaczący wpływ ma również rozstaw strzemion.

The analysis of the data in Table 7 leads to the conclusion that in both models the yield strength of stirrup steel is the dominant parameter, although the stirrup spacing is also a significant factor.

## 6 Conclusions

The performed simulations and the analysis of the results allowed us to formulate the following conclusions:

The mean values of the shear strength simulated according to the analysed models show great variation. The obtained minimum values (the EC-2 model) were 5 to 7 times lower than the maximum values (the PN-02 model).

The values of the partial reliability exponent $\Delta_R$ indicate a considerable variation in the level of safety margin. The values calculated for the EC-2 model are in the range of 2.38 to 11.73, with the smallest for the lower stirrup spacing and greater variability of the steel yield strength and the highest at lower steel yield strength variability and $\cot\theta = 2$. For the PN-02 model, extreme values of 4.24 and 31.11 were obtained. Low values are conducive to high variability of the strengths of materials, although trends are so complex that there are exceptions from this general rule.

Analysis of skewness coefficients leads to the conclusion that regardless of the assumed distribution of the compressive strength of compressive concrete, the high variability of the yield strength of steel leads to a low skewness of the shear strength distribution (except for C40/50 concrete). And low variability of the yield strength of steel may lead to a significant increase in the coefficient of skewness and to a simultaneous change of its sign (except for C40/50 concrete and lognormal distribution of its strength, when the values of the coefficient of skewness changes slightly).

In both the models, the dominant parameter is the yield strength of steel and the stirrup spacing is also significant.

The performed analysis does not allow to estimate the total safety margin and the probability of the failure of the construction element as a result of exceeding of the shear strength. This issue may be the subject of further research and analysis covering the parameters of the reliability function Z = R – S distribution.

## References

1. R. Jaskulski, *Wpływ wybranych założeń obliczeniowych na zapas bezpieczeństwa zginanych i ściskanych żelbetowych elementów prętowych, wyznaczany metodą Monte Carlo*, PhD thesis (Warsaw University of Technology, Płock, 2005)
2. R. Jaskulski, P. Wiliński, Proc. Eng., **153**, 232 (2016)
3. R. Jaskulski, P. Wiliński, Proc. Eng., **172**, 449 (2017)
4. R. Jaskulski, P. Wiliński, App. Mech. and Mat., **797**, 35 (2016)
5. K. Wróbel, *XLIX Scientific Conference „Krynica 2003"*, **III**, (2003)
6. PN-84/B-03264, *Concrete, reinforced concrete and prestressed concrete. Design rules*

7. PN-EN 1992-1-1:2008, *Eurocode 2: Design of concrete structures - Part 1-1: General rules and rules for buildings*
8. T. Godycki-Ćwirko, *Mechanika betonu* (Arkady, Warsaw, 1982)
9. T. Godycki-Ćwirko, *Ścinanie w żelbecie* (Arkady, Warsaw, 1968)
10. M.S. Borishanskiy, *Расчет отогнутых стержней и хомутов в изгибаемых железобетонных элементах по стадии разрушения* (Stroyizdat, Moscow, 1946)
11. W. Kukulski (Ed.), *Obliczanie konstrukcji żelbetowych i sprężonych metodą stanów granicznych wg PN-76/B-03264* (Arkady, Warsaw, 1976)
12. E. Mörsch, *Concrete-steel Construction*, Authorized translation from the 3rd German ed. (The Engineering News Pub. Co., New York, 1909)
13. H. Kupfer, CEB Bull. d'information, **40**, 44 (1964)
14. J. Kobiak, W. Stachurski, *Konstrukcje żelbetowe* (Arkady, Warsaw, 1995)
15. M. Knauf (Ed.), *Podstawy projektowania konstrukcji żelbetowych i sprężonych według Eurokodu 2* (DWE, Wrocław, 2006)
16. F. Casciati, I. Negri, G. Rackwitz, *Geometrical Variability in Structural Members and Systems*, JCSS Working Document (1991)
17. *General Principles on the Specification on Safety Requirements for Structures* (German Institute for Standardisation DIN, Beuth, 1981)
18. P. Wiliński, R. Jaskulski, J. Pawlikowski, *Probabilistyczna analiza zapasów bezpieczeństwa przekrojów ukośnych w elementach żelbetowych*, report from statutory research work nr 504G/7701/0250/001 (not published, Warsaw University of Technology, Faculty of Building, Mechanics and Petrochemistry, Płock, 2007)
19. J. Pawlikowski, P. Wiliński, R. Jaskulski, Inż. i Bud., **9**, 496 (2005)

MATEC Web of Conferences 117, 00066 (2017) DOI: 10.1051/matecconf/201711700066
*XXVI R-S-P Seminar 2017, Theoretical Foundation of Civil Engineering*

# Predicting of the compressive strength of RCA concrete

*Roman* Jaskulski[1,*], *Wojciech* Kubissa[1] , *Peter* Koteš[2], and Miroslav Brodňan[2]

[1]Warsaw University of Technology, Faculty of Building, Mechanics and Petrochemistry, Łukasiewicza 17, 09-402 Płock, Poland
[2]University of Žilina, Univerzitná 1, 010 26 Žilina, Slovakia

**Abstract.** The paper presents the results of predicting the strength of 61 concretes made with the use of recycled concrete aggregate (RCA). Five models in the form of first-order polynomials containing two to six variables characterizing the composition of the mixture were formulated for this purpose. Factors for unknowns were selected using linear regression in two variants: with and without additional coefficient. For each model, the average absolute error of the concrete strength estimation was determined. Because of the various consequences of underestimation and overestimation of the results, the analysis of models quality was carried out with the distinction of the two cases. The results indicate that the key to improving the quality of models is to take into account the quality of the aggregate expressed by the ACV parameter. Better match results were also obtained for models with more variables and the additional coefficient.

## 1 Introduction

Sustainable development is one of the most important contemporary ideas. This term implies a way of managing natural resources that meets current needs without compromising the ability to meet future needs. It assumes a reduction in the exploitation of natural resources for secondary raw materials that have not been re-processed and would have to be stored negatively impacting on the environment. The need to reduce the use of natural resources and energy, reduce greenhouse gas emissions while increasing the capacity to meet needs, affects most areas of life and industry.

The cement industry consumes 20 billion tons of aggregate annually, 1.5 billion tons of cement and 800 million tons of water. It generates 5-7% of the world's $CO_2$ emissions [1, 2]. It should be assumed that cement and concrete production will continue to increase, so that emissions will also increase. One of the possible ways to reduce the use of natural resources is to use RCA (Recycled Concrete Aggregate) in concrete production. The use of recycled aggregates for the production of concrete can also be a response to the shrinking resources of natural aggregates in many countries and to the rising cost of obtaining them. It should be assumed that RCA will be increasingly used.

* Corresponding author: Roman.Jaskulski@pw.edu.pl

The use of concrete waste as aggregate is associated with quality requirements. On the other hand, the quality of the aggregate depends primarily on the origin and quality of the concrete from which it was obtained. Numerous researches preceding the introduction of RCA for concrete production have allowed to classify it in terms of quality and to formulate the requirements for their properties to ensure the quality of good concrete [3, 4, 5, 6]. The formulation of qualitative requirements for aggregates is to ensure that the RCA used to produce the concrete will not significantly reduce its parameters.

RCA concrete is a relatively new material and, therefore, there are no recognized, proven methods of designing its composition. Taking into account the greater variability of the RCA properties, the higher water demand (due to high absorbability) than in case of natural aggregate and lower strength, the use of traditional concrete mix design methods has limits. With the use of a small amount of RCA of a suitable quality and taking into account in the calculations of additional water absorbed by the aggregate, it is possible to use in the design of mixtures methods which are used in the case of concretes with natural aggregates. While some researchers [7] are of the opinion that even with the replacement of 80% of natural aggregate with RCA and the associated compressive strength decrease, there is no need to use different mixture design methods as the strength of the concrete can still achieves its assumed values. However, it is still advisable to develop methods to better predict the properties of RCA concrete and to take into account the specificity of this aggregate in designing the composition of the mix.

The research allowed to develop methods of designing RCA concrete considering properties of the aggregate. Some of the methods are based on the aggregate properties determined by the Aggregate Crushing Value (ACV) [4] or the toughness of the aggregate determined in the Los Angeles abrasion test. Also used in such methods is the aggregate resistance to abrasion, denoted as Micro-Deval [8]. The method presented in [4] also includes the W/C ratio and the increased water demand of RCA. The authors propose that the concrete mix be designed using the "Three Equations Method" (Bolomey Method) with modified coefficients A. A slightly different approach is presented in [8], where a correction factor is proposed depending on the required concrete strength class and the RCA aggregate quality in relation to the strength of the concrete of the same composition containing only natural aggregate. It was also proposed to use up to 100% RCA as a coarse aggregate in concrete of C35/45 strength class or lower when using good quality RCA. The EMV method (Equivalent Mortar Volume) described in [9, 10, 11] is based on the assumption that the mortar contained in the RCA should be included in the total amount of mortar in the mix. This method gives good results, however, it requires precise determination of the mortar content in the RCA. In addition, its use may be problematic if various RCA with different mortar content are used.

The neural networks [12, 13] are also used for predicting the compressive strength and other properties of RCA concrete. However, they have the disadvantage that a large number of sets of differentiated data are required to properly train the network. Same as later to verify its correctness. This may limit the usefulness of using neural networks in practical applications. In addition, neural networks are much better at classifying data than predicting results based on the data provided [14]. Nevertheless, this does not negate the idea of using this tool in predicting the properties of concrete with RCA or identifying parameters that have a significant effect on these properties.

An interesting approach is to use the Model Tree (MT) algorithm to create decision trees and models for predicting the properties of RCA [15, 16] as well as natural aggregates concretes [17]. These algorithms do not require too much detailed data on materials and composition of concrete to create a decision tree to estimate the properties of concrete. Such data can be obtained from the literature and a properly working model can be developed using just over 200 results sets [15]. This is a great advantage of these methods due to the generally limited amount of data coming either from own research or from the literature on the composition and properties of both materials and concrete. The assembly of the required number of results in this case is easier than in the case of neural networks.

Another approach to designing or predicting the properties of RCA concrete with most of the advantages of the MT algorithm, but which is even more simplicity, is to build single models based on mix and RCA parameters using linear regression. This is the approach that is the subject of this paper.

The paper presents the results of applying models obtained by linear regression method to predict RCA concrete strength. Models with increasing numbers of two to six parameters are considered. Each model is additionally formulated in two versions. In the first an additional coefficient is used that took into account the influence of parameters not explicitly included in the model. In the second only the parameters were used which values were assumed during the design of the mixture or measured.

## 2 Properties of analysed concrete mixtures

Parameters of sixty concrete mixtures were used to formulate and test the models. In forty of them Portland cement CEM I 32.5R was used and in the rest twenty Portland-fly ash cement CEM II/B-V 32.5N. Both cements were from Ożarów Cement Plant. As a natural aggregate, river sand 0 – 2 mm and pebble gravel 2 – 16 mm were used.

RCA in the first forty series was prepared from the remaining specimens after the compressive strength tests made of 11 concretes with varying strengths ranging from 10 to 50 MPa. Concrete at the time of crushing had a minimum of 90 days. Aggregate obtained from these concretes had an average ACV ranging from 16.5 to 25.1.

The RCA in the remaining twenty one series was obtained from similar specimens of concrete of strength classes from C25/30 to C50/60, which were crushed after 1 year to 4 years after preparing. The aggregate during the preparation was mixed several times, thus its properties have averaged. The average value of ACV of this aggregate was 17.3. Concrete specimens were crushed in a laboratory jaw crusher and were sieved to obtain a fraction of 4 – 16 mm. The fraction below 4 mm, due to the large amount of dust and cement paste and associated with it very high water demand, was not used.

The mixes were prepared in a 250-liter free-fall concrete mixer. Tap water was used as the mixing water. The ingredients were dosed with the accuracy required by EN 206. Table 1 gives the limit values of the compressive strength of the concrete obtained in the tests and other parameters characterizing the blends taken into account in the presented models.

**Table 1.** Limit values of selected parameters of analysed concrete mixtures

| Parameter | $f_{cm}$ [MPa] | W/C [-] | C | W | FA/TA [-] | ACV [%] | CA/C [-] | r [-] |
|---|---|---|---|---|---|---|---|---|
| Maximum value | 62.6 | 1.0 | 540 | 261 | 0.43 | 25.1 | 6.47 | 1.0 |
| Minimum value | 14.6 | 0.4 | 200 | 111 | 0.30 | 12.1 | 1.92 | 0.5 |

## 3 Models

The analysed models were based on selected parameters such as: content of cement C, the effective water content $W_{eff}$, taking into account the amount of water absorbed by RCA, the proportion of fine aggregate to total aggregate mass FA/TA, the proportion of coarse aggregate to cement CA/C, the ACV value and volumetric share of RCA in total aggregate volume r calculated according to formula (1) where *M* is the weight of natural (index NA) and recycling (index RA) aggregate in 1 $m^3$ of concrete, and *SSD* is the density of aggregate in dry state.

$$r = \frac{\frac{M_{RA}}{SSD_{RA}}}{\frac{M_{RA}}{SSD_{RA}} + \frac{M_{NA}}{SSD_{NA}}} \cdot 100\% \quad (1)$$

Each model has the form of a polynomial, the sum of the aforementioned parameters multiplied by the coefficients expressing their influence on the final result. Models differ in the number of parameters included (from 2 to 6) and the presence of a additional coefficient that expresses the influence of unknown or unidentifiable parameters. All possible combinations of parameters used in the model were not analyzed, but the output model containing only two of them (water and cement content) was expanded sequentially by the selected additional parameters. The general form of the models is presented below in the form of the general formula (2) in the case of models without additional coefficient and in formula (3) in the case of models with additional coefficient.

$$f_{cm,1} = m_1 C + m_2 W_{eff} + m_3 \frac{FA}{TA} + m_4 \frac{CA}{C} + m_5 ACV + m_6 r \quad (2)$$

$$f_{cm,2} = m_1 C + m_2 W_{eff} + m_3 \frac{FA}{TA} + m_4 \frac{CA}{C} + m_5 ACV + m_6 r + \mathrm{m} \quad (3)$$

## 4 Results and discussion

The coefficients in the models were calculated by linear regression based on parameters of 46 concrete mixtures with RCA blends. Table 2 summarizes the obtained parameters.

**Table 2.** Values of the models coefficients

| **Number of variables** | **Parameter values** | | | | | | |
|---|---|---|---|---|---|---|---|
| | $m_1$ | $m_2$ | $m_3$ | $m_4$ | $m_5$ | $m_6$ | m |
| models without the additional coefficient | | | | | | | |
| 2 | 0.137 | -0.071 | 0 | 0 | 0 | 0 | --- |
| 3 | 0.149 | -0.291 | 98.8 | 0 | 0 | 0 | --- |
| 4 | 0.161 | -0.311 | -118.5 | -0.416 | 0 | 0 | --- |
| 5 | 0.201 | -0.278 | 33.4 | -0.69 | 4.13 | 0 | --- |
| 6 | 0.175 | -0.214 | 30.5 | -1.66 | 3.16 | 23.6 | --- |
| models with the additional coefficient | | | | | | | |
| 2 | 0.134 | -0.259 | 0 | 0 | 0 | 0 | 34.5 |
| 3 | 0.135 | -0.263 | 5.54 | 0 | 0 | 0 | 32.9 |
| 4 | 0.147 | -0.280 | -42.9 | -0.935 | 0 | 0 | 65.7 |
| 5 | 0.114 | -0.286 | -68.2 | -1.02 | -2.64 | 0 | 98.0 |
| 6 | 0.104 | -0.237 | -60.4 | -1.72 | -2.70 | 17.8 | 88.4 |

The models thus obtained were then tested with use of the parameters of the following 15 mixtures. Several parameters were calculated to determine the quality of the models.

The basis for the model quality assessment was the difference between the compressive strength values obtained from the test and predicted on the basis of models. As the average error, the mean of these differences was calculated regarding their sign.

Due to the fact that the underestimation of the compressive strength of the concrete is an error in favor for the safety of the structure made of such a material, as opposed to the overestimation of this value, further evaluation of the two cases was made. A positive average error was calculated on the basis of differences greater than zero (cases of model underestimation) and the negative average error on the basis of the values of differences smaller than zero (cases of overestimation). The determination coefficient $R^2$ that determines the quality of the model's fit to the results was also calculated. The values obtained are given in Table 3 (models without the additional coefficient) and in Table 4 (models with the additional coefficient). In the above mentioned tables also values of the maximum positive and negative error for each model were included.

**Table 3.** Values of the parameters in the models without the additional coefficient

| **Model quality assesment parameter** | | **Number of the variables in the model** | | | | |
|---|---|---|---|---|---|---|
| | | **2** | **3** | **4** | **5** | **6** |
| average error [MPa] | general | -0.22 | -0.40 | -0.07 | -2.02 | -1.56 |
| | positive | 4.88 | 7.86 | 4.26 | 2.78 | 4.49 |
| | negative | -6.05 | -7.63 | -6.58 | -5.22 | -3.75 |
| maximum error [MPa] | positive | 8.00 | 16.49 | 7.35 | 5.41 | 7.70 |
| | negative | -14.69 | -20.95 | -12.78 | -11.95 | -11.93 |
| coefficient of determination $R^2$ [--] | | 0.45 | 0.64 | 0.67 | 0.81 | 0.89 |

**Table 4.** Values of the parameters in the models with the additional coefficient

| **Model quality assesment parameter** | | **Number of the variables in the model** | | | | |
|---|---|---|---|---|---|---|
| | | **2** | **3** | **4** | **5** | **6** |
| average error [MPa] | general | -1.59 | -1.66 | -2.63 | -2.65 | -0.81 |
| | positive | 4.30 | 4.24 | 2.50 | 1.55 | 3.36 |
| | negative | -5.52 | -5.59 | -3.91 | -4.17 | -3.60 |
| maximum error [MPa] | positive | 6.31 | 6.17 | 3.07 | 2.56 | 8.77 |
| | negative | -13.70 | -13.72 | -8.40 | -6.77 | -1.25 |
| coefficient of determination $R^2$ [--] | | 0.74 | 0.74 | 0.92 | 0.93 | 0.97 |

As it can be seen, in case of models without the additional coefficient the average error values are not forming any clear trends. The lowest general average error was obtained for the model with four parameters, the lowest average positive error for the model with five variables and the lowest average negative error for the model with six parameters. The best fit for these models expressed by the $R^2$ value was obtained for the model with the maximum number of parameters.

If results for models with the additional coefficient are analyzed, the situation is a little more clear. The lowest values of the average general and negative errors and the lowest maximum value of the negative error were obtained for the model with six parameters. The lowest value of the average positive error and the lowest maximum value of the error were obtained for the model with five parameters. The quality of the models' fit is growing with the increasing number of parameters included in the model formula. And the values of the $R^2$ determination coefficient are clearly higher for these models than for models without the additional coefficient.

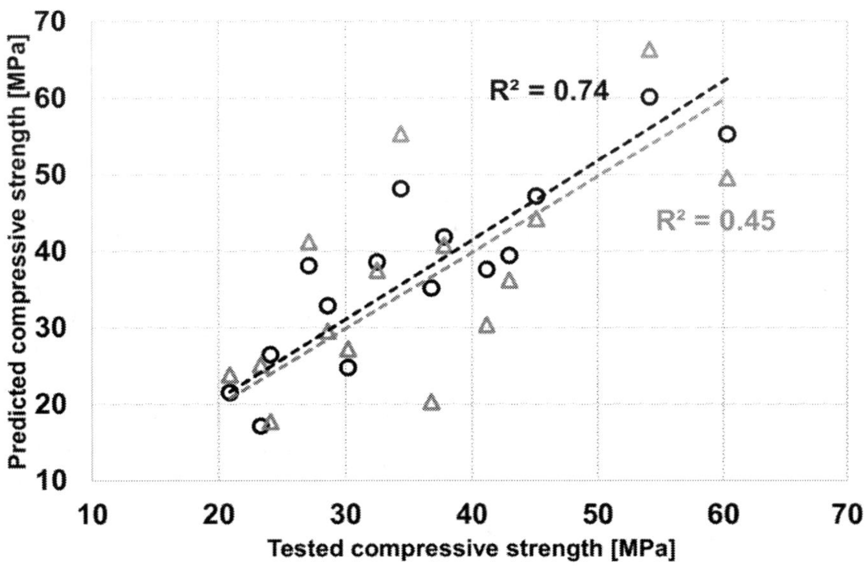


**Fig. 1.** Comparison of the concrete compressive strength values obtained from tests and predicted by models with two parameters: black – model with the additional coefficient, grey – model without the additional parameter.

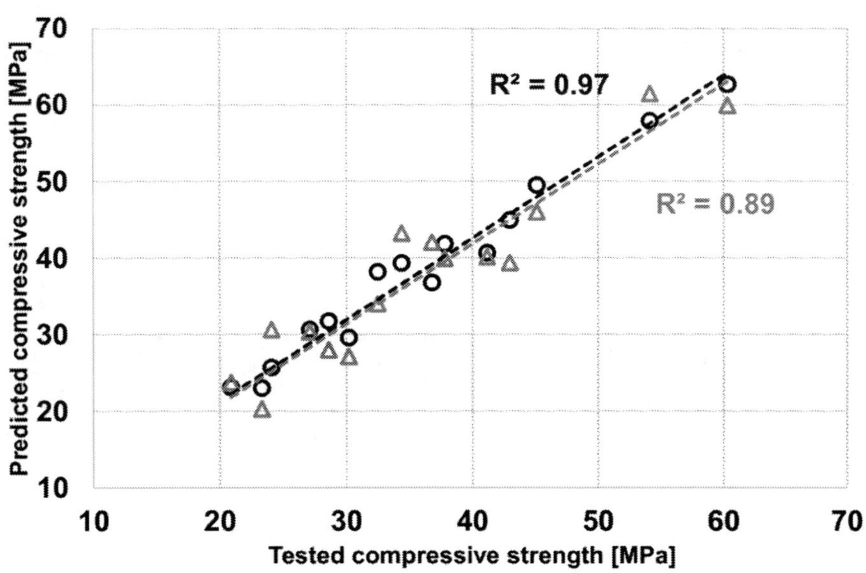


**Fig. 2.** Comparison of the concrete compressive strength values obtained from tests and predicted by models with six parameters: black – model with the additional coefficient, grey – model without the additional parameter.

The differences between two regarded kinds of models are also presented graphically in the Figures 1 and 2. In both the figures regression lines are presented, the black line for model with the additional coefficient and the grey one for the model without it. In the Figure 1 models with two parameters are compared and the value of determination coefficient for both cases are presented. In the Figure 2 in the same way two models with the maximum number of parameters are compared. It can be clearly seen, that the models with six parameters better predict the values of the compressive strength of concretes with RCA than models with only two parameters. In the first case the results are less scattered around the regression line.

## 5 Summary and conclusions

The calculations and analyses of the formulated predictive models allowed to formulate the following conclusions.

It seems possible to formulate a model that allows to estimate the compressive strength of RCA concretes using the linear regression method with satisfactory accuracy. The models obtained by this method allow for proper adjustment measured by the determination coefficient $R^2$, of the RCA concrete compressive strength calculated values to the values obtained from the strength test,. The quality of the fit increases with increasing number of parameters taken into account. Similarly, it is preferable to include an additional parameter that takes into account the effects of unknown or unmeasured variables. It seems that it can express, for example, the effect of the type of cement used in the concrete from which the RCA was obtained or the coefficient w/c of the concrete. Such parameters are usually difficult or even impossible to determine, but it cannot be assumed that these factors do not affect the strength of RCA concrete.

The question that remains to be clarified is the value of this additional parameter. Should this be a value taken straight from the calculations by means of linear regression, which maximizes the quality of matching results to the model? Or should it be additionally adjusted arbitrarily to minimize the risk of overestimate of the strength value? This may be one of the areas of further consideration and development of the topic.

But that is not the only question that remains open. Analysis of estimation errors calculated for each model indicates, unlike fit quality analysis, that adding additional variables does not necessarily lead to error reduction. It is worth to find in this situation the answer to the question, which of the parameters included allow to minimize errors, especially in cases where the model overestimates the value of strength. For this purpose, it is worthwhile to carry out an analysis of models containing different combinations of variables already taken into account, but also those that have not been included (e.g. aggregate water absorbability, its porosity, etc.).

The authors also see the need to formulate models at a higher level of generality based on dimensionless variables. It would be useful to adopt some technologically justified limit values for the variables involved, then to standardize (in a statistical sense) the data to the models using those limits, and then to formulate the models in new form based on such data. In summary, the problem presented in the paper can certainly not be considered solved, but the results presented are quite a big step towards such a solution.

## References

1. L. Czarnecki, M. Kapron, Int. Jour. of the Soc. of Mat. Engin. for Res. **17**, 99 (2010)
2. L. Czarnecki, W. Kurdowski, Budown., Technol., Archit., **37**, 50 (2007)

3. M.N. Fardis (Ed.), *Innovative materials and techniques in concrete construction. ACES Workshop*, (Springer Netherlands, Dordrecht, 2012)
4. A. Koper, M. Koper, W. Kubissa. Key Engin. Mater. **677**, 266 (2016)
5. W. Kubissa, R. Jaskulski, A. Koper, M. Supera, Key Engin. Mater. **677**, 233 (2016)
6. RILEM TC 121-DRG, Mater. and Struct., **173**, 557 (1994)
7. W.H. Kwan, M. Ramli, K.J. Kam, M.Z. Sulieman, Constr. and Build. Mater. **26**, 565 (2012)
8. T. Kausay, T. Simon, G. Balázs, Concr. Struct, **9**, 45 (2008)
9. A. Abbas, G. Fathifazl, O.B. Isgor, A.G. Razaqpur, B. Fournier, S. Foo, Cem. and Concr. Comp. **31**, 555 (2009)
10. G. Fathifazl, A. Abbas, A.G. Razaqpur, O.B. Isgor, B. Fournier, F. Simon, J. of Mater. in C. Engin. **21**, 601 (2009)
11. C. Jiménez, M. Barra, S. Valls, D.F. Aponte, E. Vázquez, Mater. de Constr. **313**, e006 (2014)
12. A. Dantas Trocoli Abdon, M. Batista Leite, K. de Jesus Nagahama, Construction and Building Materials **38**, 717 (2013)
13. N. Deshpande, S. Londhe, S. Kulkarni, Intern. J. of Sust. Built Env. **3**, 187 (2014)
14. R. Tadeusiewicz, *Elementarne wprowadzenie do techniki sieci neuronowych z przykładowymi programami* (Akademicka Oficyna Wydawnicza PLJ, Warszawa,1998)
15. A. Behnood, J. Olek, M.A. Glinicki, Proceedings of *11th International Symposium on Brittle Matrix Composites 2015*, 381 (2015)
16. A. Behnood, J. Olek, M.A. Glinicki, Constr. and Build. Mater. **94**, 137 (2015)
17. C. Deepa C., K. SathiyaKumari, V. Pream Sudha, Int. J. of Comp. App. **6**, 18 (2010)

# Limitation of Stresses in Concrete According to Eurocode 2

*Marek* Jędrzejczak[1,*], and *Krzysztof* Klempka[1]

[1]The Faculty of Geodesy, Geospatial and Civil Engineering University of Warmia and Mazury Olsztyn, Poland

**Abstract.** When stresses in concrete in reinforced concrete beams with the reinforcement ratio being high due to the ultimate limit state are calculated – according to the linear elasticity theory – from strains of a characteristic combination or even a quasi-permanent one, in some cases compressive stresses in extreme concrete fibres can be even higher than the characteristic compressive strength of concrete. In the following report, it is postulated to calculate these stresses with the help of the non-linear stress-strain relationship for concrete. It has also been demonstrated that the requirements specified in points 7.2(1)P and 7.2(2) of the standard *PN-EN 1992-1-1:2008/NA:2010 Eurocode 2:Design of concrete structures. Part 1-1: General rules and rules for buildings,* are useless and do not suit the limit state analysis, on which Eurocodes are based.

## 1 Introduction

Eurocode 2 [1] comprises regulations – not completely unambiguous ones – regarding stresses in concrete. Point 7.2(1)P of the norm [1] states 'the compressive stress in the concrete shall be limited in order to avoid longitudinal cracks, micro-cracks or high levels of creep, where they could result in unacceptable effects on the function of the structure'. Point 7.2(2) further specifies that 'longitudinal cracks can appear if the stress level under a characteristic combination of loads exceeds a critical value'. Such cracks can deteriorate the durability of a structure. If no other measures are taken, e.g. an increase in the cover of the reinforcement in the compressive zone or confinement of the concrete by transverse reinforcement, then it may appropriate to limit compressive stress to a value k1fck in areas exposed to the influence of environments of exposure classes XD, XF and XS. The recommended value is k1 = 0.6, but the norm [1] allows the specification of another value of this factor in a National Annex. The Polish National Annex NA.5 assumes that k1 = 1.0.

As noted by M. Knauff in [3], the above regulation is made ambiguous by the phrase: 'it may be appropriate (does it mean mandatory?) to limit the compressive stresses to a value $k_1 f_{ck}$'. It is also uncertain whether an increase in the thickness of the cover up to the one required because of more unfavourable environments, such as XD, XF and XS, can be equated with the increase which abolishes the obligation to monitor compressive stresses in concrete.

* Corresponding author: mjed@uwm.edu.pl

Before Eurocode 2 [1] was adopted in Poland, when designing reinforced concrete structures according to the Polish standard PN-B-03264:1984 [2], it was necessary to check the ultimate limit state (ULT) as well as bendings and widths of cracks caused by quasi-permanent loadings. Meanwhile, compressive stresses in concrete caused by a characteristic combination of loads were not monitored. This approach did not lead to any damage to a designed structure.

## 2 Calculation of Compressive Stresses in Compressed Concrete

As in [4], stresses in concrete for non-linear structural analysis were calculated as recommended by Eurocode 2 [1]. By employing the stress-strain relation in concrete (3a – 3c), given in point 3.1.5 of Eurocode 2 and shown in fig. 1, as well as the values illustrated in fig.2, two equilibrium equations in a cross-section being in the serviceability state can be written, i.e. the sum of projection forces onto the axis of a beam (1) and the sum of bending moments of the cross-section relative to the resultant force of the concrete's compression zone (2).

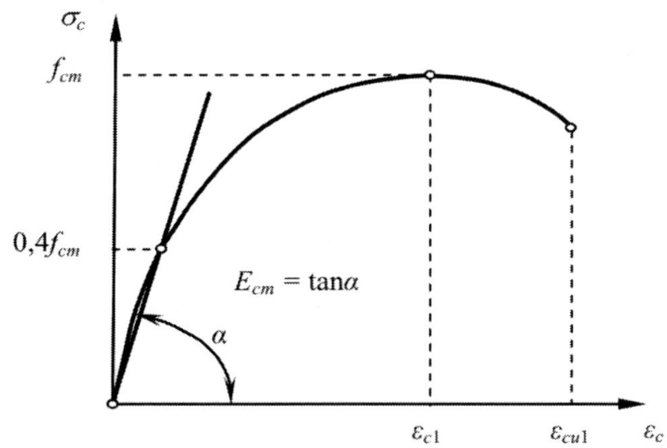


**Fig. 1.** Schematic representation of the stress-strain relation in concrete for structural analysis

$$A_{s1}\sigma_{s1} = P_c = \int_0^x \sigma_c(\varepsilon_c) b dr \tag{1}$$

$$A_{s1}\sigma_{s1}(d - x + r_0) = M \, , \tag{2}$$

where:

$$\sigma_c(\varepsilon_c) = f_{cm}\left[\frac{k\eta - \eta^2}{1 + (k-2)\eta}\right] \tag{3a}$$

$$k = 1{,}05 E_{cm} \frac{\varepsilon_{c1}}{f_{cm}} \tag{3b}$$

$$\eta = \frac{\varepsilon_c}{\varepsilon_{c1}}, \tag{3c}$$

$$r_0 = \frac{\int_0^x \sigma_c(\varepsilon_c) b r dr}{\int_0^x \sigma_c(\varepsilon_c) b dr}, \tag{4}$$

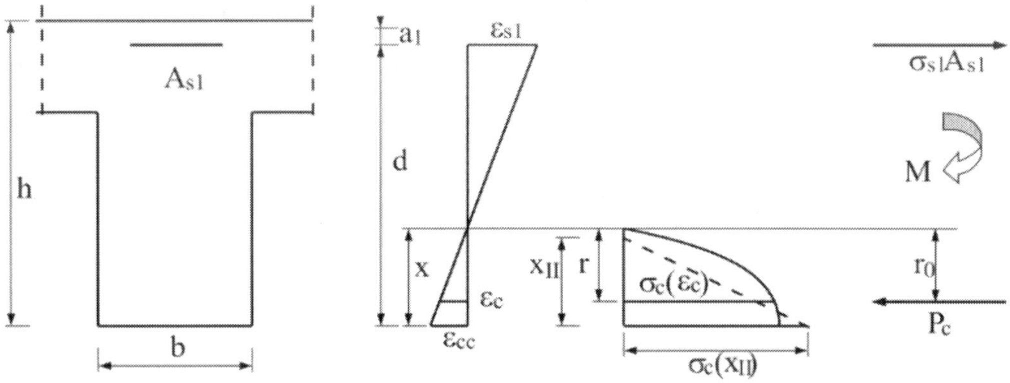


**Fig. 2.** Stress-strain diagram in a bending cross-section in the serviceability state. For comparison, the picture shows the location of the cross-section centre of gravity in phase II - $x_{II}$ and the diagram of strains in concrete determined according to the linear theory - $\sigma_c(x_{II})$.

In line with the Bernoulli's hypothesis regarding flat cross-sections, the following are calculated from the relationships between strains:

$$\varepsilon_c = \varepsilon_{cc}\frac{r}{x}, \quad \varepsilon_{s1} = \varepsilon_{cc}\frac{d-x}{x}, \tag{5}$$

And afterwards, stresses $\sigma_{s1}$ in reinforcement $A_{s1}$ are derived:

$$\sigma_{s1} = E_s \varepsilon_{s1}. \tag{6}$$

By substituting expressions (3) ÷ (6) to (1) and (2), we obtain a set of two equations with two unknowns $x$ and $\varepsilon_{cc}$. The problem can be solved numerically (integration leads to the formulation of implicit equations). In a programme written in the Fortran language, for a given $\varepsilon_{cc} \in (0, \varepsilon_{c1})$ that changes by $0.001\varepsilon_{c1}$ at a step, the subsequent values $x \in (0, 0.5d)$ changing by $0.5d/1000$ at a step are assumed. Then, it is checked whether the left side of equation (1) is equal to the right side, and when this agreement is sufficient, the appropriateness of the achieved values $x$ and $\varepsilon_{cc}$ is verified with equation (2). If the verification fails, the programme resumes the search (at the values of $\varepsilon_{cc}$ and $x$ at which it stopped) for a new pair of $x$ and $\varepsilon_{cc}$, again solving equation (1), then verifying whether a new pair of values satisfies equation (2). Integers present in expression (4) are calculated by the programme with the Simpson method. Having derived the value of $\varepsilon_{cc}$ from equations (1) and (2), equations (3a – 3c) are used to determine the stress in the extreme compressive fibre of the cross-section $\sigma_c(\varepsilon_c = \varepsilon_{cc})$.

## 3 Examples

The solid reinforced concrete flooring shown in fig. 3, of the thickness 0.2 m and the span 5.1 m, rests on the main T cross-section beam, which is a two-span continuous beam of the span length equal 8.1 m. The cross-section of the main beam: height h=0.70 m, width of the web of the T beam b=0.35 m, distance from the reinforcement centre of gravity to the tensile edge a1=0.05 m; material: concrete C16/20, the secant modulus of elasticity of concrete Ecm = 29 GPa, design yield strength of reinforcement fyd = 350 MPa, the modulus of elasticity of reinforcing steel Es = 200 GPa. The live action of the warehouse equals 4.0 kN/m2, the value of the quasi-permanent coefficient of variable action is ψ2 = 0.8. The following were determined in the main beam above the support: design value of the applied bending moment MEd = 591.6 kNm and the cross-sectional area of reinforcement As1 = 34.00 cm2. The planned reinforcement consisted of 7ϕ25 with As1 = 34.96 cm2, which yields the reinforcement ratio ρL = As1/(bd) = 0.0154.

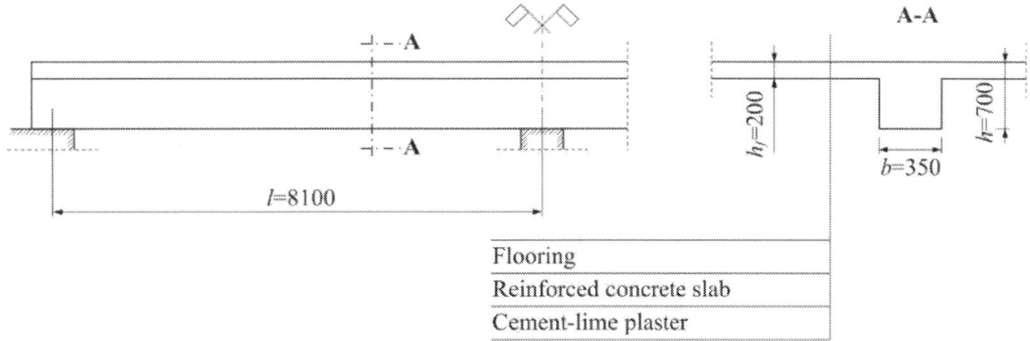


**Fig. 3.** Cross-sections of the floor

The moment from quasi-permanent actions is $M = 430.65$ kNm. Having solved a set of equations (1) and (2), the following were determined: $x = 0.2535$ m and $\varepsilon_{cc} = 0{,}0007011$, and when $\varepsilon_{cc}$ was substituted to equations (3a – 3c) the value $\sigma_c = 15.70$ MPa was obtained.

It is worth underlining that calculations of widths of cracks of strains in compressed concrete and tensile steel at $A_{s1}$ – in the cracked cross-section (being in phase II) - are traditionally made having assumed the linear elasticity theory (e.g. [5]). However, in cross-sections exposed to very heavy actions, stresses in concrete are impossible to be calculated based on the assumption of their linear distribution, as this would generate absurd results, i.e. not only higher than the ones derived from the set of equations (1) and (2), but also, and not infrequently, even exceeding the characteristic compressive strength of concrete, as it is illustrated underneath – treating actions as short-term ones (immediately after the application of the moment $M = 430.65$ kNm), for which $\alpha_e = E_s/E_{c,eff} = 200/29 = 6.90$.

$$x_{II} = \frac{\alpha_e}{b}[-A_{s1} + \sqrt{(A_{s1})^2 + \frac{2b}{\alpha_e} A_{s1} d}\,] = 0.238m \tag{7}$$

$$I_{II} = \frac{b x_{II}^3}{3} + \alpha_e A_{s1}(d - x_{II})^2 = 0.00567m^4 \tag{8}$$

$$\sigma_c = \frac{M}{I_{II}} x_{II} = 18.10MPa > f_{ck} = 16MPa \tag{9}$$

An example of the difference in the distribution of stresses in the compression zone of concrete $\sigma_c$ calculated from the set of equations (1) and (2) and in agreement with the linear theory is shown in figure 2.

While proceeding with further calculations, it was assumed that RH = 50% and $h_0$ = 233 mm from [3] (as it is simpler and more accurate than according to [1]), after which the creep coefficient was calculated: $\varphi(\infty, 28) = 3.04$, and next, the following were calculated at $\sigma_c$ = 15.70 MPa: $\varphi_{nl}(\infty, 28) = 3.04 \cdot 2{,}218 = 6.74$ (from equation (3.7) from [1]), $E_{c,eff}$ = 3.74 GPa (equation (7.20) from [1]), $\alpha_{e,t} = E_s/E_{c,eff} = 200/3.74 = 53.41$, as well as:

$$x_{II} = \frac{\alpha_{e,t}}{b}[-A_{s1} + \sqrt{(A_{s1})^2 + \frac{2b}{\alpha_{e,t}} A_{s1} d}] = 0.4556m \tag{10}$$

$$\sigma_{s1} = \frac{M}{(d - \frac{x_{II}}{3})A_{s1}} = 247.3MPa \tag{11}$$

The above calculations – from equations (10) and (11) – of stresses in tensile reinforcement $\sigma_{s1}$ in the elasticity range are sufficiently accurate because – having accounted for the creep - the depth of the compression zone $x_{II}$ increased significantly, i.e. from 0.238 m to 0.4556 m.

Next, according to [1] in fig. 7.1 and from equations (7.10), (7.9), (7.11) and (7.8), the following were calculated: $h_{c,ef}$ = 0.0814 m, $\rho_{p,eff}$ = 0.1226, $(\varepsilon_{sm} - \varepsilon_{cm}) = 0.00118$, $s_{r,max}$ = 162 mm and $w_k$ = 0.191 mm.

In the analysed cross-section, the moment of the characteristic combination was $M$ = 464.11 kNm. The stress calculated in line with the procedure described in point 2 was $\sigma_c$ = 16.58 MPa, which is higher than $f_{ck}$ = 16.0 MPa! Should we calculate these stresses in the elasticity range, as soon as the actions are applied ($\alpha_e$ = 6.90), an even higher excess of the characteristic compressive strength of concrete would be achieved, which is shown below ($x_{II}$ = 0.238 m, $I_{II}$ = 0.00567 m$^4$):

$$\sigma_c = \frac{M}{I_{II}} x_{II} = 19.48MPa > f_{ck} = 16MPa \tag{12}$$

If the concrete in the above example is C30/37 concrete, for which $E_{cm}$ = 32 GPa, then the required reinforcement with the same steel will be $A_{s1}$ = 29.39 cm$^2$. Assuming that it is 6$\phi$25 with $A_{s1}$ = 29.45 cm$^2$, the lower reinforcement ratio is achieved: $\rho_L$ = 0.0129.

For the moment from quasi-permanent actions equal $M$ = 430.65 kNm, the following were determined from equations (1) and (2) as well as (3),: $x$ = 0.21905 m, $\varepsilon_{cc}$ = 0.000649 and $\sigma_c$ = 18.80 MPa.

When calculating the stresses from the above moment that appear immediately after its application (and considering them to be short-lived ones, for which $\alpha_e$ = 200/32 = 6.25), the following values were obtained from equations (7), (8) and (9): $x_{II}$ = 0.2141 m, $I_{II}$ = 0.00464 m$^4$ and $\sigma_c$ = 19.86 MPa, which is higher than $\sigma_c$ = 18.80 MPa.

In the subsequent calculations, it was assumed as before that RH = 50% and at $h_0$ = 233 mm, hence the creep coefficient was calculated from [3] as equal $\varphi(\infty, 28) = 2.33$, and after that the following were calculated for $\sigma_c$ = 18.80 MPa: $\varphi_{nl}(\infty, 28) = 2.33 \cdot 1.304 = 3.04$, $E_{c,eff}$ = 7.92 GPa and $\alpha_{e,t} = E_s/E_{c,eff} = 25.25$. Equations (10) and (11) yielded: $x_{II}$ = 0.3544 m and $\sigma_{s1}$ = 274.9 MPa.

Next, according to [1] in fig. 7.1 and from equations (7.10), (7.9), (7.11) and (7.8), the following were determined: $h_{c,ef}$ = 0.1152 m, $\rho_{p,eff}$ = 0.073, $(\varepsilon_{sm} - \varepsilon_{cm}) = 0.00126$, $s_{r,max}$ = 186 mm and $w_k$ = 0.234 mm.

For the moment from the characteristic combination $M$ = 464.11 kNm, using the procedure described in point 2, the calculated stress was $\sigma_c$ = 19,91 MPa, which in this case does not exceed $f_{ck}$ = 32.0 MPa. If these stresses are calculated in the elasticity range, for $\alpha_e$ = 6.25, $x_{II}$ = 0.2141 m, $I_{II}$ = 0.00464 m$^4$, then, naturally, higher values will be obtained from equation (12), i.e. $\sigma_c$ = 21.42 MPa.

## 4 Conclusions

The above examples suggest that sometimes in a cross-section that fulfils the ultimate limit state and has cracks not exceeding the limit value compressive stresses in concrete caused by a characteristic combination exceed the characteristic compressive strength of concrete $f_{ck}$. The ambiguously worded requirement in point 7.2(2) [1] – 'it may be appropriate to limit the compressive stresses to a value $k_1 f_{ck}$' – does not resolve the dilemma whether a given cross-section satisfies the requirements set in Eurocode 2 [1]. The above statement seems to be confounding and it can be claimed to be useless. Limiting stresses down to a level below the limit stresses would be more suitable for the method of designing allowable stresses rather than for the ultimate limit analysis, on which Eurocodes are based.

The excess stresses demonstrated in the discussed example arise solely from certain imperfections of the calculation models, which we have at our disposal. Limiting these stresses to the value in agreement with the original version of Eurocode 2 would be even more absurd, because than the limit stress in the first example would equal merely:

$$\sigma_c = k_1 f_{ck} = 0.6 \cdot 16.0 = 9.6MPa\,,$$

and the width of cracks – in line with Eurocode 2 – is calculated for stresses even equal to $f_{ck}$ (in our case $f_{ck}$ = 16 MPa), and these are stresses triggered by a combination from quasi-permanent loads, which causes smaller generalised internal strengths and stresses in a cross-section than a characteristic combination. Also, in the second case – if the original version of Eurocode 2 was binding in Poland – stresses in concrete $\sigma_c$ = 19.91 MPa would be above the recommended value:

$$\sigma_c = k_1 f_{ck} = 0.6 \cdot 30.0 = 18.0MPa\,.$$

## References

1. PN-EN 1992-1-1:2008/NA:2010 *Eurokod 2: Projektowanie konstrukcji z betonu. Część 1-1: Reguły ogólne i reguły dla budynków*
2. PN-B-03264:1984 *Konstrukcje betonowe, żelbetowe i sprężone – Obliczenia statyczne i projektowanie*
3. M. Knauff., *Obliczanie konstrukcji żelbetowych według Eurokodu 2* (Wydawnictwo Naukowe PWN, Warszawa, 2012)
4. P. Simao, H. Barros, C. Ferreira, T. Marques, Eng. Struc., **129**, 67 (2016)
5. H. Barros, C. Ferreira, J. Figueiras, Eng. Struc., **49**, 600 (2013).

# A Constitutive Model of Pseudo-Hyperelasticity for Description of Rubber-Like Materials

*Stanisław* Jemioło[1,*]

[1]WUT The Faculty of Civil Engineering, Al. Armii Ludowej 16, 00-637 Warszawa, Poland

**Abstract.** A consistent thermomechanical model applicable to pseudo-elastic behaviour of rubber-like materials has been proposed. The model is based on the Naumann and Ihlemann's approach [1] and the concept of material stiffening given by Ogden and Roxburg [2,3]. The constitutive relationships of psedo-hyperelasticity are derived for incompressible isotropic materials. Futhermore, a specific model describing the Mullins effect was proposed. The homogeneous deformations are discussed and the obtained results are compared with Mullins' and Tobin's [4] experimental data. Good correlation between theoretical and experimental data has been observed.

## 1 Introduction

Rubber-like materials exhibit a highly nonlinear elastic behaviour but in cyclic loading the stress-softening phenomenon is observed [4]. This effect is now widely known as the Mullins effect [5]. Therefore, the rubber-like materials cannot have a strain-energy function in the hyperelastic sense [6]. However, in a cyclic loading and unloading the stress-strain relationship does not vary much with the strain rate. If the strain-rate effect is ignored altogether; then the loading curve and the unloading curve (they are unequal) can be separately treated as a uniquely defined stress-strain relationship, which is associated with a strain-energy function. Each of these curves is a pseudo-elastic curve. During successive reloading the material responds softer. This is achieved by scaling the stress with an appropriate softening function whose material parameters can be directly calculated from experimental data [2,3,5,7,8]. These pseudo-elastic models were also widely used in biomechanics [9-12] for constitutive modelling of isotropic and anisotropic soft tissue.

Recently, it was shown in the paper [1] that the softening function proposed in [2,3] and [8] leads to thermomechanically consistent material models. The aim of this paper is to introduce a new phenomenological model for the Mullins effect based on the strain-energy function of isotropic, incompressible rubber-like materials [13,14] and the free energy potential given by Naumann and Ihlemann [1]. It is an alternative approach to Ogden and Roxburg [2,3] which enables a consistent deduction of pseudo-elasticity models by introducing a suitable free energy potential. The insertion into the Clausius-Duhem inequality directly yields the constitutive relationships for the stress tensor and the dissipation.

* Corresponding author: s.jemiolo@il.pw.edu.pl

## 2 Basic assumptions and constitutive relationships of pseudo-hyperelasticity

The Clausius-Duhem [6] inequality for purely mechanical material models can be simplified to

$$D = \frac{1}{2}\mathbf{T}\cdot\dot{\mathbf{C}} - \dot{\Psi} \geq 0. \tag{1}$$

In the inequality (1) $D$ denotes the mechanical dissipation, $\mathbf{T}$ the second Piola-Kirchhoff stress tensor, $\mathbf{C} = \mathbf{F}^T\mathbf{F}$ the right Cauchy-Green tensor, and $\Psi$ the specific free energy. Here $\mathbf{F}$ denotes the deformation gradient with $J = \det\mathbf{F} > 0$, and $\mathbf{F}^T$ stands for the transpose of $\mathbf{F}$. Following the approach of Naumann and Ihlemann [1], the key idea is to define the free energy for pseudo-hyperelastic material

$$\Psi(W_0, W_{max}) \equiv \int_0^{W_0} \eta(\xi, W_{max})\,\mathrm{d}\xi, \tag{2}$$

where $W_0(\mathbf{C})$ denotes the strain-energy function of the virgin material model (it is assumed that in the undeformed configuration $W_0(\mathbf{I}) = 0$), and

$$W_{max}(t) \equiv \max\{W_0(\tau), \tau \leq t\}. \tag{3}$$

The definition directly leads to $\Psi(W_0(\mathbf{I}) = 0, W_{max}) = 0$, and (after differentiation of (2)) to the interpretation of the softening function

$$\eta(W_0, W_{max}) = \frac{\partial \Psi(W_0, W_{max})}{\partial W_0}. \tag{4}$$

Inserting the time derivative of $\Psi$:

$$\dot{\Psi} = \frac{\partial \Psi(W_0, W_{max})}{\partial W_0}\dot{W}_0 + \frac{\partial \Psi(W_0, W_{max})}{\partial W_{max}}\dot{W}_{max}, \tag{5}$$

and (3) into the inequality (1) we obtain

$$D = \left(\frac{1}{2}\mathbf{T} - \eta\frac{\partial W_0}{\partial \mathbf{C}}\bigg|_{\mathbf{C}=\mathbf{C}^T}\right)\cdot\dot{\mathbf{C}} - \frac{\partial \Psi(W_0, W_{max})}{\partial W_{max}}\dot{W}_{max} \geq 0. \tag{6}$$

This inequality is fulfilled for arbitrary $\dot{\mathbf{C}}$, if the following conditions hold:

$$\mathbf{T} = 2\eta\frac{\partial W_0(\mathbf{C})}{\partial \mathbf{C}}\bigg|_{\mathbf{C}=\mathbf{C}^T}, \qquad \frac{\partial \Psi(W_0, W_{max})}{\partial W_{max}}\dot{W}_{max} \leq 0. \tag{7}$$

Due to the definition (3), $\dot{W}_{max}$ is always non-negative. Then, a sufficient condition for thermomechanical consistency $(7)_2$ is

$$\frac{\partial\eta(W_0,W_{max})}{\partial W_{max}}\leq 0. \tag{8}$$

Every function (4) which is a monotonically decreasing function of $W_{max}$ leads to a possitive dissipation.

Since the Kirchhoff stress tensor is defined as follows

$$\boldsymbol{\tau}\equiv\boldsymbol{\sigma}J=\mathbf{F}\mathbf{T}\mathbf{F}^T, \tag{9}$$

therefore in the deformed configuration we have

$$\boldsymbol{\tau}=2\eta\,\mathbf{F}\left.\frac{\partial W_0(\mathbf{C})}{\partial\mathbf{C}}\right|_{\mathbf{C}=\mathbf{C}^T}\mathbf{F}^T. \tag{10}$$

For the softening functions used in [2,3,7,8] :

$$\eta=1-\frac{1}{c_2}\mathrm{Erf}\left[\frac{1}{c_1}(W_0-W_{max})\right]\quad\text{or}\quad\eta=1-\frac{1}{c_2}\mathrm{Tanh}\left[\frac{1}{c_1}(W_0-W_{max})\right], \tag{11}$$

where $c_1$ and $c_2$ are positive constants, the related free energy $\Psi$ (2) of the pseudo-hyperelastic model can be computed by the integration of (11). These functions lead to a positive dissipation. The free energies related to the softening functions (11) can be given by

$$\Psi(W_0,W_{max})=\left.\frac{c_1}{c_2\sqrt{\pi}}\mathrm{Exp}\left[-\frac{1}{c_1^2}(\xi-W_{max})^2\right]+\xi-\frac{1}{c_2}(\xi-W_{max})\mathrm{Erf}\left[\frac{1}{c_1}(\xi-W_{max})\right]\right|_0^{W_0} \tag{12}$$

$$\Psi(W_0,W_{max})=\left.\xi+\frac{c_1}{c_2}\mathrm{Log}\left[\mathrm{Cosh}\left[\frac{1}{c_1}(W_{max}-\xi)\right]\right]\right|_0^{W_0}. \tag{13}$$

## 3 Pseudo-hyperelastic models of isotropic rubberlike materials

Consider the simplest class of constitutive isotropic models of rubberlike materials [13,14]. We assume the rubberlike materials are incompressible, i.e.

$$J-1=0. \tag{14}$$

Consequently, the potential (2) has to incorporate the constraint (14) with the Lagrange multiplier $p$, being the hydrostatic pressure or the spherical part of the Cauchy stress $\boldsymbol{\sigma}$. In the case of incompressible isotropic materials the stored energy function $\Psi_0$ is a function of only two independent invariants of isochoric deformations. The tensor $\mathbf{F}$ is decomposed into the isochoric and volumetric part as follows

$$\mathbf{F}=J^{1/3}\bar{\mathbf{F}}=J^{1/3}\mathbf{R}\bar{\mathbf{U}}=J^{1/3}\bar{\mathbf{V}}\mathbf{R},\quad\det\bar{\mathbf{F}}=\det\bar{\mathbf{U}}=\det\bar{\mathbf{V}}=1. \tag{15}$$

Further, we introduce

$$\mathbf{B}=\mathbf{F}\mathbf{F}^T,\quad \overline{\mathbf{B}}=\overline{\mathbf{F}}\overline{\mathbf{F}}^T,\quad \overline{\mathbf{C}}=\overline{\mathbf{F}}^T\overline{\mathbf{F}}. \tag{16}$$

where $\mathbf{B}$ denotes the left Cauchy-Green strain tensor, cf. [13,14] and the references therein. For incompressible isotropic materials the stored energy function has the form,

$$W_0=\overline{W}_0\left(\overline{\mathbf{C}}\right)=\overline{W}_0\left(\overline{\mathbf{B}}\right)=\overline{W}_0\left(\overline{I}_1,\overline{I}_2\right), \tag{17}$$

where

$$\overline{I}_1=\mathrm{tr}\overline{\mathbf{B}}=\mathrm{tr}\overline{\mathbf{C}},\quad \overline{I}_2=\mathrm{tr}\overline{\mathbf{B}}^{-1}=\mathrm{tr}\overline{\mathbf{C}}^{-1}. \tag{18}$$

From (10), (14) and (17) we obtain the following constitutive relationship of the pseudo-hyperelasticity in the reference configuration

$$\boldsymbol{\sigma}=-p\mathbf{I}+\eta\left(\overline{\beta}_1\overline{\mathbf{B}}+\overline{\beta}_{-1}\overline{\mathbf{B}}^{-1}\right), \tag{19}$$

where

$$\overline{\beta}_1=2\frac{\partial\overline{W}_0\left(\overline{I}_1,\overline{I}_2\right)}{\partial\overline{I}_1},\quad \overline{\beta}_{-1}=-2\frac{\partial\overline{W}_0\left(\overline{I}_1,\overline{I}_2\right)}{\partial\overline{I}_2}. \tag{20}$$

We observe that equations (19) and (20) can be applied to arbitrary stored energy functions describing an isotropic incompressible material. For the details on such functions the reader is referred to [1-3,6-9,13,14].

In the paper [6], following the approach of Ogden and Roxburgh [2,3], we proposed a phenomenological pseudo-hyperelastic model to characterize the Mullins effect in filled rubber. We assumed: the stored energy function in the form

$$\overline{W}_0\left(\overline{I}_1,\overline{I}_2\right)=\frac{1}{2}\left[a_1\left(\overline{I}_1-3\right)+\frac{1}{2}a_2\left(\overline{I}_1^2-9\right)+\frac{1}{3}a_3\left(\overline{I}_1^3-27\right)+a_4\left(\overline{I}_2-3\right)+a_5\left(\overline{I}_1\overline{I}_2-3\right)\right], \tag{21}$$

where $a_i$ are constants, the constitutive relationship is given by Eq. (19) and the softening function is defined by Eq. $(11)_1$. It can be proved [13] that the function (21) is a particular case of the polynomial Rivlin form of the stored energy function available in Abaqus [15]. Somone can obtained the pseudo-hyperelastic model with the stored energy function (21) and the softening function $(11)_1$ by using two options : HYPERELASTIC and MULLINS EFECT.

## 4 An example

To illustrate a simple pseudo-elastic model of rubberlike materials we used the stored energy function proposed in our monograph [13]:

$$\overline{W}_0\left(\overline{I}_1\right)=\frac{\overline{\mu}_0}{2}\left(\overline{I}_1-3\right)+\sum_{i=1}^{I}\frac{\mu_i}{2\alpha_i}\left(\overline{I}_1^{\alpha_i}-3^{\alpha_i}\right), \tag{22}$$

where $\overline{\mu}_0, \mu_i$ and $\alpha_i\neq 1\left(i=1,\ldots,I\right)$ are positive material parameters. The shear modulus in the undeformed, natural state of the virgin material depends on all material parameters

$$\mu_0=\overline{\mu}_0+\sum_{i=1}^{I}\mu_i 3^{\alpha_i-1}. \tag{23}$$

For $\mu_i=0$ we obtained the neo-Hookean material.

In view of (19) and (22), the stress-stretch relations are given by:

$$\boldsymbol{\sigma} = -p\mathbf{I} + \eta\bar{\beta}_1\left(\bar{I}_1\right)\bar{\mathbf{B}}, \qquad \bar{\beta}_1\left(\bar{I}_1\right) = \bar{\mu}_0 + \sum_{i=1}^{I}\mu_i \bar{I}_1^{\alpha_i - 1} . \tag{24}$$

From the plane stress state, Eq. (24) and $\sigma_3 = 0$, we get

$$p = \eta\left[\bar{\mu}_0 + \sum_{i=1}^{I}\mu_i\left(\tilde{I}_1 + \tilde{J}^{-2}\right)^{\alpha_i - 1}\right]\tilde{J}^{-2} . \tag{25}$$

Moreover, the constitutive relationship takes the form

$$\tilde{\boldsymbol{\sigma}} = \eta\left[\bar{\mu}_0 + \sum_{i=1}^{I}\mu_i\left(\tilde{I}_1 + \tilde{J}^{-2}\right)^{\alpha_i - 1}\right]\left(-\tilde{J}^{-2}\tilde{\mathbf{I}} + \tilde{\mathbf{B}}\right) , \tag{26}$$

where $\tilde{J} = \det\tilde{\mathbf{F}}$, $\tilde{\mathbf{F}}$ denotes the plane deformation gradient and $\tilde{I} = \operatorname{tr}\tilde{\mathbf{B}} = \operatorname{tr}\tilde{\mathbf{C}}$. We observe that eq. (26) can also be used for membranes and shells.

Both the uniaxial tension test and biaxial tension test are plane stress tests. For example, we get

$$\sigma_1 = \eta\left[\bar{\mu}_0 + \sum_{i=1}^{I}\mu_i\left(\bar{\lambda}_1^2 + \frac{2}{\bar{\lambda}_1}\right)^{\alpha_i - 1}\right]\left(\bar{\lambda}_1^2 - \frac{1}{\bar{\lambda}_1}\right), \tag{27}$$

in the case of a simple uniaxial extension.

The comparison between experimental data [4] and model predictions is depicted in Fig.1. Best-fit parameters are summarized in Table 1. They are calculated by an error minimization procedure which simultaneously utilizes both loading and unloading experimental data. The model with a simplified stored energy function (22) and with two alternative stress softening functions has been used. For both cases good correlation between the theoretical and the experimental data has been observed.

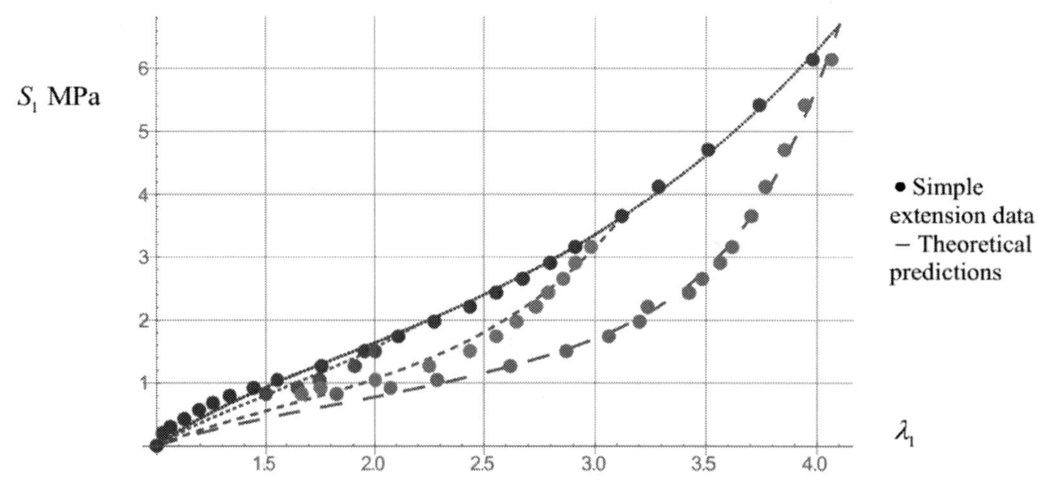


**Fig. 1.** Nominal stress $S_1 = \sigma_1 / \lambda_1$ versus stretch $\lambda_1$ for loading and unloading uniaxial tension tests. Comparison of theoretical predictions (27) with Mullins and Tobin [4] uniaxial extension data.

**Table 1.** Model functions and parameter values

| Energy and softening functions | Value of material parameters |
|---|---|
| $\bar{W}_0(\bar{I}_1)=\frac{\bar{\mu}_0}{2}(\bar{I}_1-3)+\frac{\bar{\mu}_1}{2\alpha_1}(\bar{I}_1^{\alpha_1}-3^{\alpha})$ | $\bar{\mu}_0$=0.79976 MPa, $\mu_1=0.01246$ MPa, $\alpha_1=2.48471$ |
| $\eta=1-\frac{1}{c_2}\mathrm{Erf}\left[\frac{1}{c_1}(W_0-W_{\max})\right]$ | $c_1=4.18084$, $c_2=1.88581$ |
| $\eta=1-\frac{1}{c_2}\mathrm{Tanh}\left[\frac{1}{c_1}(W_0-W_{\max})\right]$ | $c_1=3.83054$, $c_2=1.79701$ |

## 5 Final remarks

The constitutive modelling of isotropic incompressible pseudo-hyperelastic materials can be extended to compressible and anisotropic materials, for instance orthotropic or transversally isotropic. Then in the pseudo-potentials (12) or (13) one has to substitute an appropriate stored energy function similarly, cf. papers [10-12,14].

## References

1. Ch. Naumann, J. Ihlemann, Int. J. Solids Struct. **69-70**, 360 (2015)
2. R.W. Ogden, D.G. Roxburg, Proc. R. Soc. London, Ser. A : Math. Phys. Eng. Sci. **455**, 2861 (1999)
3. R.W. Ogden, D.G. Roxburgh, *Constitutive models for rubber, Dorfman A., Muhr A. (eds), Proc. The First European Conference on Constitutive Models for Rubber*, 23 (A.A. Balkema, Roterdam-Brookfield, 1999)
4. L. Mullins, N.R. Tobin, Rubber Chem. Technol. **30**, 551 (1957)
5. J. Diani, B. Fayolle, P. Gilormini, Eur. Polymer J. **45**, 601 (2009)
6. S. Jemioło, *Theoretical Foundations of Civil Enginierings, Rymsza B. [ed], Proc. The X Russian-Polish Semminar*, 89 [in Polish] (OW PW, Warszawa 2001)
7. M. Šilhavý, *The mechanics and thermodynamics of continuous media* (Springer, Berlin – Tokyo 1997)
8. A. Dorfman, R.W. Ogden, Int. J. Solids Struct., **41**, 1855 (2004)
9. S. Jemioło, J.J. Telega, Acta Bioeng. Biomech. **1 (Suppl. 1)**, 207 (1999)
10. S. Jemioło, J.J. Telega, Eng. Trans., **49(2-3)**, 213 (2001)
11. S. Jemioło, J.J. Telega, Acta Bioeng. Biomech. **3 (Suppl. 2)**, 209 (2001)
12. S. Jemioło, J.J. Telega, C. Michalak, Acta Bioeng. Biomech. **4 (Supp.1)**, 470 (2002)
13. S. Jemioło, *Study of hyperelastic properties of isotropic materials. Modeling and numerical implementation. Scientific Works. Civil Engineering* **140** [in Polish] (OWPW, Warszawa, 2002)
14. S. Jemioło, *Constitutive relationships of hyperelasticity* [in Polish], (PAN, KILiW, Warszawa, 2016)
15. Abaqus v. 6.12, *Theory Manual* (Simulia, 2012)

# Secondary consolidation modelling by using rheological schemes

*Kazimierz* Józefiak[1,*], *Artur* Zbiciak[1]

[1]Warsaw University of Technology, Faculty of Civil Engineering, Institute of Roads and Bridges, 16 Armii Ludowej Ave., 00-637 Warsaw, Poland

**Abstract.** The paper presents the possibility of the secondary consolidation modelling in soil mechanics by using rheological schemes. Non-elastic rheological models represent time-dependent behaviour of the soil skeleton and should be expressed by appropriate constitutive relationships in the form of ordinary differential equations. On the other hand, the primary consolidation phenomenon caused by dissipation of the excess pore pressure is modelled by applying the well-known formulation proposed by Terzaghi and expressed by the partial differential equation. The resulting initial boundary value problem was solved numerically for three selected rheological schemes. Algorithms implemented in Mathematica were used for this purpose.

## 1 Introduction

In order to provide some physical insight into non-elastic and time-dependent behaviour of engineering materials, rheological schemes consisting of springs, dashpots and sliders are used very often [1]. In case of many viscoelastic materials an accurate scheme is the Standard Model, which consists of a spring in series with a parallel combination of a spring and dashpot. Viscoelastic properties of asphaltic materials can be modelled by using the Burgers scheme (Maxwell and Kelvin-Voigt in series) [2]. The application of rheological schemes consisting of additional sliders, gives the possibility of modelling plasticity phenomenon making a material capable to exhibit additional permanent deformations.

Rheological schemes can also be applied to model a consolidation problem in soil mechanics. This time-dependent process depends on permeability and compressibility of the soil as well as the length of the drainage path. Generally, consolidation process consists of two parts: (i) primary consolidation settlement, caused by dissipation of the excess pore pressure generated by load application and (ii) secondary compression, caused by the time-dependent deformation behaviour of soil particles [3].

The theory of one-dimensional consolidation was first proposed by Terzaghi (1925) basing on the linear elastic relationship between effective stress and strain [4]. Such an approach describes only the primary consolidation process. By using non-elastic and time-dependent rheological schemes, modelling the behaviour of the soil skeleton, it is possible to describe secondary consolidation [5-8].

* Corresponding author: k.jozefiak@il.pw.edu.pl

The main goal of this paper is to demonstrate various rheological schemes suited for the secondary consolidation modelling. The systems of differential equations, defining the initial boundary value problem (IBVP) in case of one-dimensional consolidation is formulated. Numerical solution of the IBVP was obtained by using algorithms implemented in Mathematica [9].

## 2 One-dimensional consolidation problem

In soil mechanics, one-dimensional consolidation problem is considered as a consolidation of an infinite saturated homogeneous clay layer, loaded by an infinite and uniform pressure. The soil layer exhibiting consolidation process can be drained on both sides (Fig. 1a), when it is located between two layers of granular soil, or on one side (Fig. 1b). In either cases volume change of the granular soil layers is neglected and the whole half-space rests on a non-deformable rock formation. Aforementioned simplification leads to, so called, oedometer conditions in which horizontal strains are zero and the change in volume of the clay layer is related only to its vertical strain. For a two-side drainage, the flow distance $H_{dr}=H/2$.

For simplification, we consider only excess pore water pressure which arises after application of the external load (the uniform pressure), neglecting the overburden pressure state.

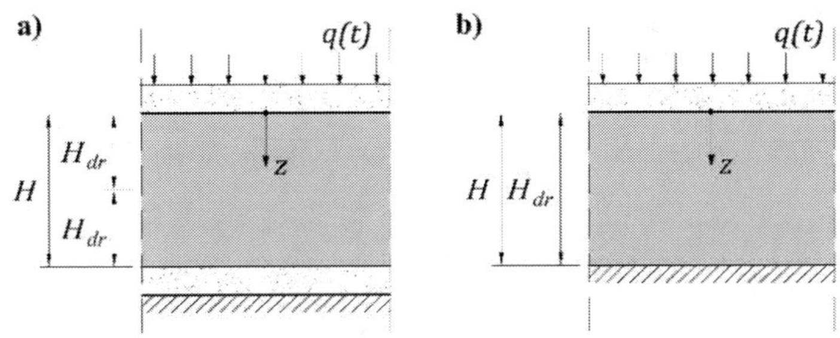


**Fig. 1.** One-dimensional consolidation problem: a) two-side drainage, b) one-side drainage.

Continuity of flow (filtration) in the cohesive soil layer along with the Darcy's law gives the well-known differential equation of one-dimensional consolidation

$$-\frac{k}{\gamma_w}\frac{\partial^2 u(z,t)}{\partial z^2}=\frac{1}{1+e_0}\frac{\partial e(z,t)}{\partial t} \tag{1}$$

where: $k$ – coefficient of permeability [m/s], $\gamma_w$ – unit weight of water [N/m$^3$], $e_0$ – initial void ratio, $u(z,t)$ and $e(z,t)$ – functions of pore water pressure and void ratio respectively.

Applying a relationship between the vertical strain and the void ratio $e=\varepsilon(1+e_0)+e_0$, Eq. (1) can be rewritten as follows

$$\frac{k}{\gamma_w}\frac{\partial^2 u(z,t)}{\partial z^2}=-\frac{\partial \varepsilon(z,t)}{\partial t} \tag{2}$$

In order to consider different loading programmes, Eq. (2) has to be formulated in effective stress $\sigma'$. Assuming that for 1D-consolidation, the applied total stress $\sigma(z,t)$ can vary only linearly with depth, substituting pore pressure $u=\sigma-\sigma'$ gives

$$\frac{k}{\gamma_w}\frac{\partial^2\sigma'(z,t)}{\partial z^2}=\frac{\partial\varepsilon(z,t)}{\partial t} \tag{3}$$

Setting a linear constitutive relationship for the soil skeleton, for example $\varepsilon=\sigma'/M_0$, where $M_0$ is soil's oedometric modulus, we can model primary consolidation phenomenon which is related only to dissipation of pore water pressure and drainage. The modulus $M_0$ can be determined from a simple oedometer consolidation laboratory test. For a one-side drainage, Eq. (3) has to be solved with the following boundary conditions (see Fig. 1b)

$$\left.\frac{\partial\sigma'}{\partial z}\right|_{z=H}=0 \qquad \sigma'\,|_{z=0}=q(t) \qquad 0<t\leq\infty \tag{4}$$

where $q(t)$ is a function of the applied external pressure.

Classically, the above problem was solved for a constant pressure, without unloading. The solution for excess pore water pressure, assuming the boundary condition $\sigma'\,|_{z=0}=q_0=const.$ is shown in Fig. 2.

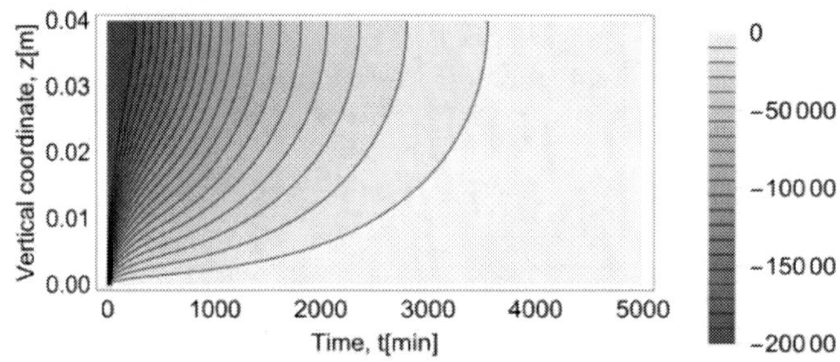


**Fig. 2.** Contours of excess pore water pressure for primary consolidation (linear constitutive relationship for soil skeleton).

## 3 Secondary consolidation modelling

In order to take into account secondary consolidation effects, related to time-dependent behaviour of the soil skeleton, the constitutive relationship in the space of the effective stress can be formulated by using rheological schemes (see Fig. 3). Non-elastic constitutive relationships of the soil skeleton should be expressed in the form of ordinary differential equations with respect to time.

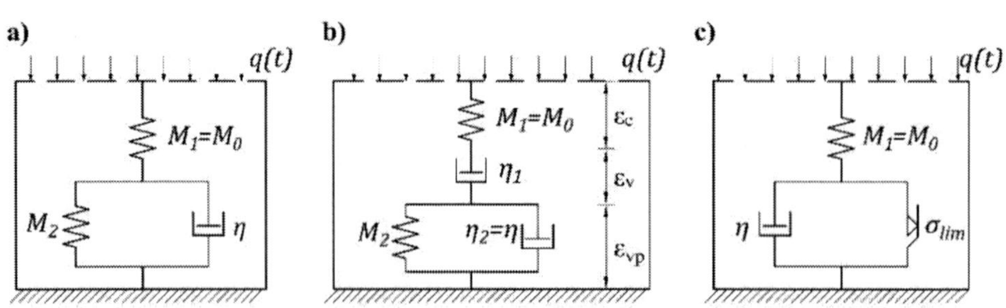


**Fig. 3.** Rheological schemes applied for soil skeleton: a) Standard Model, b) Burgers Model, c) Bingham Model.

Gibson and Lo [6] presented an analytical solution of one-dimensional consolidation in which the relationship between the effective stress and vertical strain was formulated according to the Standard Model shown in Fig. 3a. Viscoelastic constitutive relationship of this model can be expressed as follows

$$\eta M_0 \frac{\partial \varepsilon}{\partial t} + M_0 M_2 \varepsilon = \eta \frac{\partial \sigma'}{\partial t} + \left(M_0 + M_2\right)\sigma' \tag{5}$$

The upper spring shown in Fig. 3a accounts for the primary consolidation and the bottom viscoelastic part represents the secondary consolidation. In order to find the function of vertical strain $\varepsilon(z,t)$, the system of differential equations (3) and (5) has to be solved with the set of boundary conditions (4) and the following initial condition

$$\varepsilon(z,t)\,|_{t=0} = 0 \tag{6}$$

Solving the IBVP allows the function of displacement (settlement) $s(z,t)$ to be found by integrating the strain function

$$\frac{\partial}{\partial z} s(z,t) = \varepsilon(z,t) \qquad s(z,t)\,|_{z=H} = 0 \tag{7}$$

**Tab. 1.** Parameters assumed for the comparison of different models.

| $k$ | $M_1 = M_0$ | $M_2$ | $\eta = \eta_2$ | $\eta_1$ | $H$ | $q_0$ | $\sigma_{lim}$ |
|---|---|---|---|---|---|---|---|
| [cm/s] | [kPa] | [kPa] | [Ns/m$^2$] | [Ns/m$^2$] | [cm] | [kPa] | [kPa] |
| $10^{-8}$ | 1000 | 5000 | $5\cdot10^{10}$ | $1\cdot10^{11}$ | 40.0 | 200.0 | 50.0 |

The analytical solution presented in [6] was used to test a numerical solution. The system of partial differential equations was solved using the numerical method of lines implemented in Mathematica software [9]. The resulting system of ordinary differential equations was solved using LSODA method in Mathematica [9]. For the numerical solution, the following loading programme was assumed (Fig. 4):

$$q(t) = \begin{cases} q_0 & 0 \le t \le t_0 \\ 0 & t < 0 \wedge t > t_0 \end{cases} \tag{8}$$

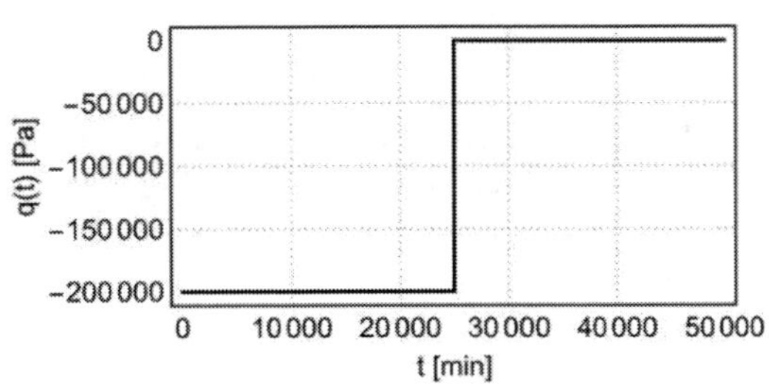


**Fig. 4.** Loading programme.

Fig. 5 presents comparison between numerical and analytical solutions. Assumed material parameters are shown in Tab. 1. It should be emphasised, however, that these parameters were chosen to present only the capabilities of the approach. In the future work curve-fitting of more complicated models to actual laboratory test data is planned.

In case of the Burgers model (Fig. 3b), the constitutive relationship can be formulated as follows

$$\eta M_0 \frac{\partial}{\partial t}\varepsilon + M_0 M_2 \varepsilon = \eta \frac{\partial}{\partial t}\sigma' + M_2 \sigma' + M_0 M_2 \varepsilon_v \tag{9}$$

where the additional internal variable $\varepsilon_v$ should be evaluated by the following differential equation

$$\frac{\partial}{\partial t}\varepsilon_v = \frac{1}{\eta_1}\sigma', \quad \varepsilon_v(z,t)\,|_{t=0} = 0 \tag{10}$$

The system of differential equations (3), (9) and (10) was solved with the boundary conditions (4) and zero initial conditions for $\varepsilon$ and $\varepsilon_v$ (see Fig. 5).

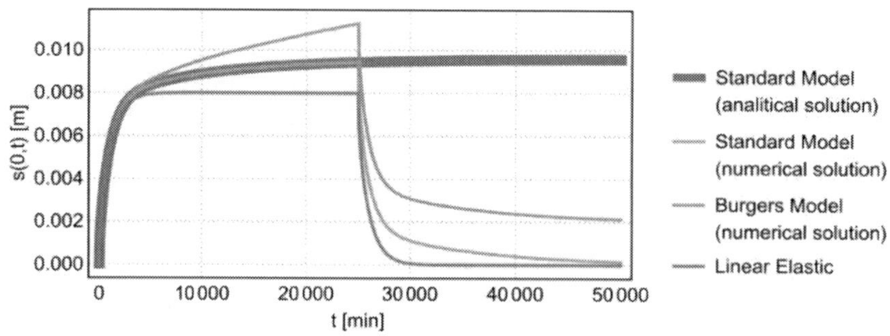


**Fig. 5.** Comparison of settlement curves for different solutions and viscoelastic rheological models.

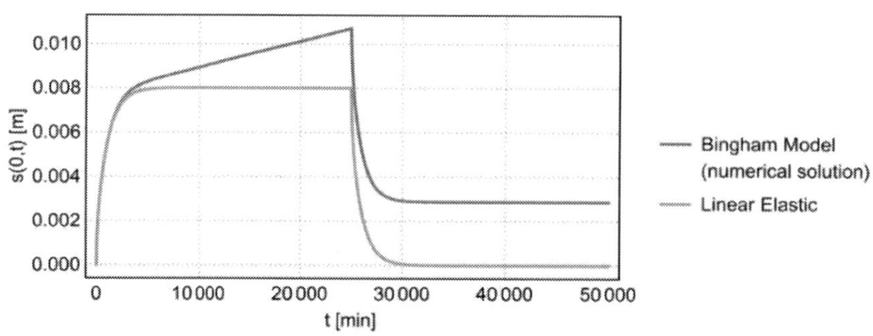


**Fig. 6.** Comparison of settlement curves between linear elastic solution and Bingham Model.

Let us analyse another rheological scheme, called the Bingham model, representing elastic-visco-plastic behaviour of the soil skeleton (Fig. 3c). Constitutive relationship defining this model can be expressed by the following non-linear equation

$$\frac{\partial}{\partial t}\varepsilon = \begin{cases} \dfrac{1}{M_0}\dfrac{\partial}{\partial t}\sigma' & \text{if} \quad |\sigma'| < \sigma_{lim} \\ \dfrac{1}{M_0}\dfrac{\partial}{\partial t}\sigma' + \dfrac{|\sigma'| - \sigma_{lim}}{\eta}\,\mathrm{sgn}(\sigma') & \text{if} \quad |\sigma'| \geq \sigma_{lim} \end{cases} \tag{11}$$

The solution of the IBVP for the Bingham Model is visualized in Fig. 6, showing a difference between non-elastic behaviour of the soil skeleton and elastic one.

Finally, the comparison of viscoelastic rheological schemes analysed in our paper is shown in Fig. 7. For this calculation, $\eta_1 = 5 \cdot 10^{11}\,\text{Ns/m}^2$ and $q(t) = q_0 = const$ were assumed in order to present the solution in a way similar to oedometric laboratory test results.

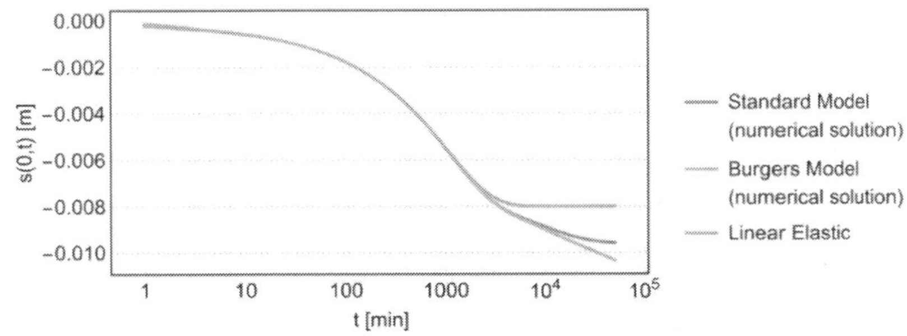


**Fig. 7.** Comparison of settlement curves for viscoelastic rheological models shown on semi-logarithmic graph.

## 4 Conclusions

The mathematical formulation of the soil mechanics consolidation problem was presented in the paper. Viscoelastic and elastic-visco-plastic rheological schemes can be used in order to represent secondary consolidation of the soil skeleton. The results of numerical solutions demonstrate the validity of the proposed method.

The realistic modelling of engineering materials needs a three-dimensional description to be applied which leads to the strain-stress relationships involving the notion of tensors [10, 11]. The authors are currently underway the work in the area of 3D constitutive modelling of consolidation problems combining Finite Element Method simulations along with laboratory experiments.

## References

1. W. Nowacki, *Creep Theory*, in Polish (Arkady, Warsaw, 1963)
2. A. Zbiciak, Bulletin of the Polish Academy of Sciences: Technical Sciences **61(1)**, 65 (2013)
3. B. M. Das, *Advanced Soil Mechanics* (CRC Press, 2014)
4. K.Terzaghi, *Erdbaumechanik auf bodenphysikalischer Grundlage* (F. Deuticke, Leipzig-Wien, 1925)
5. L. Barden, Géotechnique **18**, 3 (1968)
6. R. E. Gibson, K. Y. Lo, *Acta Polytechnica Scandinavica: Civil engineering and building construction series* **296** (Norges Tekniske Vitenskapsakad, Trondheim, 1961)
7. R. L. Shiffman, C. C. Ladd, A. T. Chen, *I.U.T.A.M. Symposium on Rheological Soil Mechanics* (Grenoble, France, 1964)
8. D. W. Taylor, W. Merchant, J. Math. Phys. **19** (1940)
9. *Wolfram Research, Inc., Mathematica, Version 10.2, Champaign, IL (2015)*
10. M. Superczyńska, K. Józefiak, A. Zbiciak, ACE **LXII (3)**, 207 (2016)
11. M. Mitew-Czajewska, ACE **LXII (4)**, 73 (2016)

MATEC Web of Conferences 117, 00070 (2017) DOI: 10.1051/matecconf/201711700070
*XXVI R-S-P Seminar 2017, Theoretical Foundation of Civil Engineering*

# Comparison of different windows for low-energy houses

*Peter* Juras[1,*], *Daniela* Staffenova[1] and *Pavol* Durica[1]

[1]Department of Building Engineering and Urban Planning, Faculty of Civil Engineering, University of Zilina, Univerzitna 8215/1, 010 26 Zilina, Slovakia

**Abstract.** This paper deals with an outcome of long-term experimental measurement of windows suitable for low-energy or zero houses. Three different windows are evaluated since 2011 in the Laboratory of the Department of Building Engineering and Urban Planning, Faculty of Civil Engineering, University of Zilina. All windows have implemented triple glazing, but with various frame type. Two frames are made from plastic and one from wood. The plastic frames differ from each other with use of thermal insulation inside one of mentioned frames. The test room has adjustable indoor boundary conditions and is exposed to real outdoor climate conditions measured through the own detached experimental weather station. Temperatures and heat fluxes are continuously measured on glazings and frames. This paper compares the windows by the requirements of Slovak standard on the inner surface temperatures and differences of the thermal transmittance for glazing and sash.

## 1 Introduction

With the increased thermal insulation and protection of buildings it is also necessary to minimize the loss through the openings (windows, doors, etc.). Windows are complicated construction and during the research through the centuries made big advance in the technology basis, material and construction. On contrary to the non-transparent parts of the building envelope (roof, wall, etc.), window has more function than to isolate the harsh climate conditions and to protect the indoor climate and people. Primary function of windows is solar insulation of the space. Another crucial is air ventilation, contact with outdoor etc. With the more strict requirements of thermal properties it is necessary to increase the thermal properties of used materials, both for glazings (triple, gas fillings, heat mirrors and low emissivity surfaces) and frames (wooden, aluminium, plastic with thermo modules, more chambers etc.) Nonetheless is the advance with incorporating the photovoltaics into the glazing or usage of transparent thermal insulations [1, 2].

The Slovak standard [3] dealing with thermal protection (STN 73 0540:2012) and EC directives regarding to the thermal protection outlines the requirements for windows and low energy houses. This article deals with testing of such windows, which meet the required properties for windows (valid from January 2016 - $U_w < 1.0$ W/(m$^2$.K]). Windows

* Corresponding author: peter.juras@fstav.uniza.sk

in this experimental measurement were commercially available back than in 2011, where were built-in into the experimental laboratory of Department of Building Engineering and Urban Planning.

### 1.1 Description of laboratory and measuring equipment

The window constructions evaluated in this article are built into the testing laboratory of the Department. This is so-called laboratory of the pavilion type, which nowadays consists of three different rooms. Two of them contain lightweight timber framed wall [4]. In the third experimental room, there are three different windows built into the wall (Fig. 1). The windows are oriented to the south with slight declination to the west (15°). From outside, windows are exposed to the real climate boundary conditions. The outdoor climate is monitored and recorded by the detached experimental weather station placed on the nearby building`s roof [5]. Exact monitoring of outdoor boundary conditions in-situ enables the possibility for more precise measurement and simulations. The indoor climate is controlled by the AC unit, which is set to maintain the Slovak standard indoor boundary conditions: 20 °C, 50 %.

Set of sensors used for measurements itself contains of thermocouples [6] and heat flux plates [7]. Totally 36 thermocouples, uniformly placed on three windows on the critical places (contact between glazing and window sash, corners, etc.). That is 6 positions on each window and also each position consists of two sensors (one inside, one outside). There are two types of heat flux plates [7]. There are some standard ones (120 x 120 mm) and half sized for window sashes and frames. Positions of thermocouples and heat flux plates are shown in Fig. 2.

**Fig. 1.** Outside view on the three tested windows in the laboratory belonging to the Department of Building Engineering and Urban Planning. Southward façade, second floor

The datalogger and both sensor types are from Ahlborn. All the windows were built into the wall in 2011, equipped with thermocouples for monitoring the temperatures [5]. Since 2013 the outdoor climate has been monitored more precise by the experimental weather station [5]. Finally, from autumn 2015 there are also the heat flux plates (HFP) fixed on the inner surfaces, which creates the possibility to measure the heat flux and thermal conductance coefficient of the glass and frame, depending on the position of the plates. For needs of this article, seven HFP were used. Three of them have a standard dimensions, placed in the centre of glazing, three small ones placed on the windows sashes and another one on the frame (wooden window C). Positions are marked in Fig. 2. Each HFP measures the thermal flow and also the surface temperature.

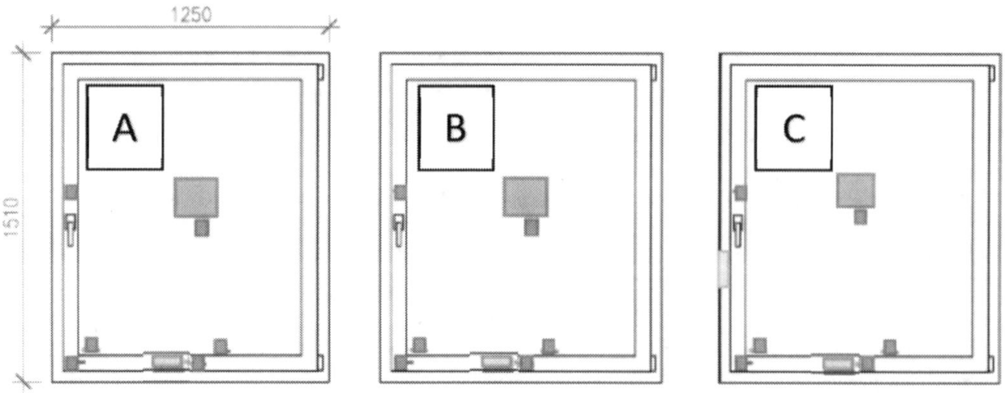


**Fig. 2.** Measuring sensors placement scheme (view from the interior) on the windows. Windows from left: A – plastic, B – plastic with thermal insulation (thermo modules), C – wooden. Small squares are thermocouples from both sides, HFP are placed from inside – centre of glazing and window sash. The last one is placed on the frame of window C

## 1.2 Description of measured windows

As it was mentioned above, three different windows were long-time tested. Outside view is shown in Fig. 1 and a schematic view from inside is in Fig. 2. Windows are marked from left (from inside): window A: plastic, 6 chambers in profile; window B: plastic, 6 chambers in profile with integrated thermo modules (insulation in chambers). The last window is a wooden one with thickness of profile 78 mm. It has declared best thermal properties of frame and slightly worse for glazing. Other parameters, such as gas filling, heat gain or calculated $U_w$- values are summarized in Tab. 1.

**Table 1**. Properties of the measured experimental windows

| Window | Material | No. of chambers | Glazing | Gas | Heat gain | $U_w$ | $U_f$ | $U_g$ |
|---|---|---|---|---|---|---|---|---|
| | | | | | [%] | [W/(m$^2$.K)] | | |
| A | plastic | 6 | triple | Ar | 36 | 0.8 | 1.0 | 0.5 |
| B | plastic with thermo modules | 6 | triple | Kr | 47 | 0.78 | 0.85 | 0.5 |
| C | wood | - | triple | Ar | 48 | 0.79 | 0.8 | 0.6 |

## 1.3 Measuring principle with use of heat flux plates

The thermal transmittance coefficient ($U$-value) of a structural element describes the quantity of thermal energy that passes through it from one side to the other per second and per square meter surface at a constant difference in ambient temperature of 1°K. A constructions $U$-value (unlike its thermal conductance $\Lambda$) includes the thermal transfer characteristics on its surfaces ($h_i$ and $h_e$). The thermal transmittance coefficient of the structure is the reciprocal of its total thermal transmittance resistance $R_o$. $R_o$ is the sum of the thermal conductance resistances of the structure`s various contiguous layers plus the thermal surface transfer resistances between the structure and the ambient media on either side, in our case air on both sides. The relationship is shown in Eq.1:

$$U = \frac{1}{R_o} = \frac{1}{(R_{si}+R+R_{se})} = \frac{1}{\frac{1}{h_i}+\frac{1}{\Lambda}+\frac{1}{h_e}} = \frac{q}{\theta_{ai}-\theta_{ae}} \quad (1)$$

$U$ – thermal transmittance coefficient [W/(m$^2$.K)]
$R_o$ – total thermal transmittance resistance [(m$^2$.K)/W]
$R_{si}$ – thermal surface transfer resistance on the inner surface [(m$^2$.K)/W]
$R_{se}$ – thermal surface transfer resistance on the outer surface [(m$^2$.K)/W]
$R$ – thermal conductance resistance of all layers [(m$^2$.K)/W]
$h_i$ – surface coefficient of heat transfer, inside [W/(m$^2$.K)]
$h_e$ – surface coefficient of heat transfer, outside [W/(m$^2$.K)]
$\Lambda$ – thermal conductance coefficient [W/(m$^2$.K)]

The heat flux plate`s measuring principle is described in the Eq. 2:

$$q = C.U_{th} \tag{2}$$

$q$ – heat flow density [W/(m$^2$.K)]
$C$ – calibration constant of the heat flux plate [W/(m$^2$.mV)]
$U_{th}$ – voltage measured by heat flux plate [mV]

Thermal surface transfer coefficient can be calculated according to Eq. 3 and 4:

$$h_i = \frac{q}{\theta_{si}-\theta_{ai}} \tag{3}$$

$$h_e = \frac{q}{\theta_{se}-\theta_{ae}} \tag{4}$$

$\theta_{si}$ – temperature of inner surface [°C]
$\theta_{ai}$ – temperature of indoor air [°C]
$\theta_{se}$ – temperature of outer surface [°C]
$\theta_{ae}$ – temperature of outdoor air [°C]

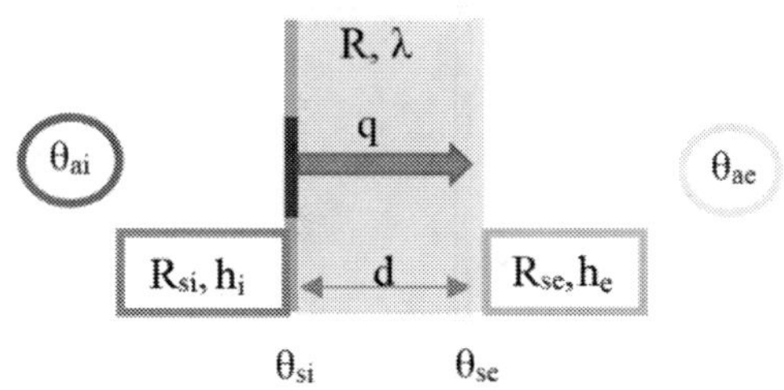


**Fig. 3.** Scheme of factors, which influence the heat flux through the structure

## 2 Measurement results

First result is overview of the surface temperatures – if they meet the Slovak standard for inner surface temperature for windows (9.6 °C) and the second one is a comparison of design or manufacturer values for frame and glass to the measured ones. Also cross comparison between the HFP itself is possible and among the measured surface temperatures from thermocouples and HFPs, too. Last comparisons are temperatures during the lifetime (2012-2017) of the windows.

### 2.1 Surface temperatures

For temperature measurement 28th January 2017 (0:00-24:00) was chosen and graphs representing the courses of surface temperatures are shown in Fig. 4. The outdoor temperature during the night was almost steady for approximately -12 °C and the day was a typically winter one with bright sky and a lot of sunshine. This can be seen on the surface temperatures high above 25 °C. Although it was freezing also during the whole day.

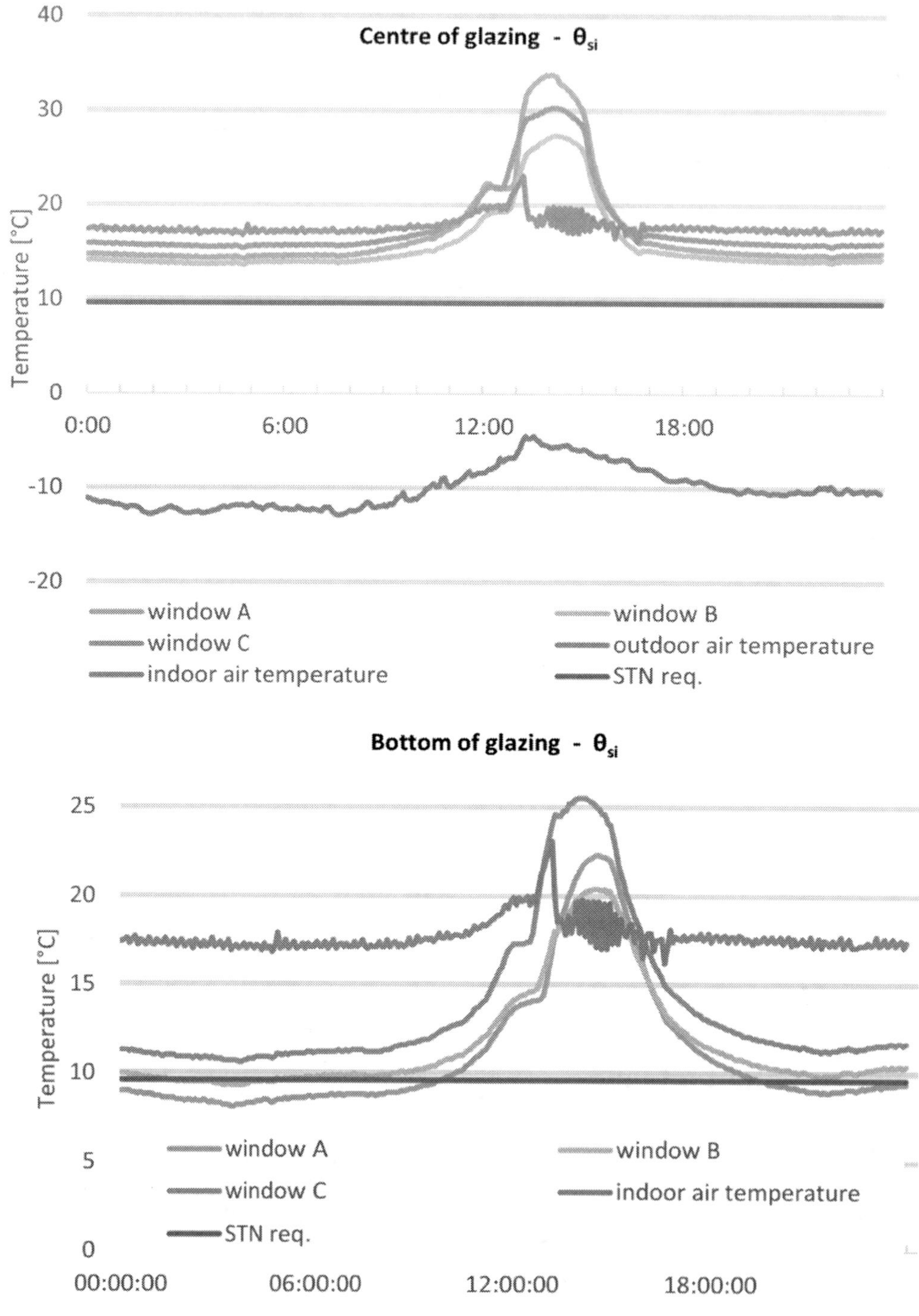


**Fig. 3.** Results of surface temperature measurement on 28[th] January 2017 (0:00-24:00). Graph above represents a daily course for position in the centre of glazing and bottom's graph the worst position – bottom of the glazing / contact with window sash.

## 2.2 Heat fluxes

Heat flux measurement required following criteria: difference of temperatures as high as possible (up to 30 °C in this case), small fluctuations over the time period, no solar radiation (during night) and short time interval (5-minute time period used). For calculation of heat fluxes were selected two time periods, both on the 30th January 2017. The indoor and outdoor temperatures for these two periods are shown in Fig. 4.

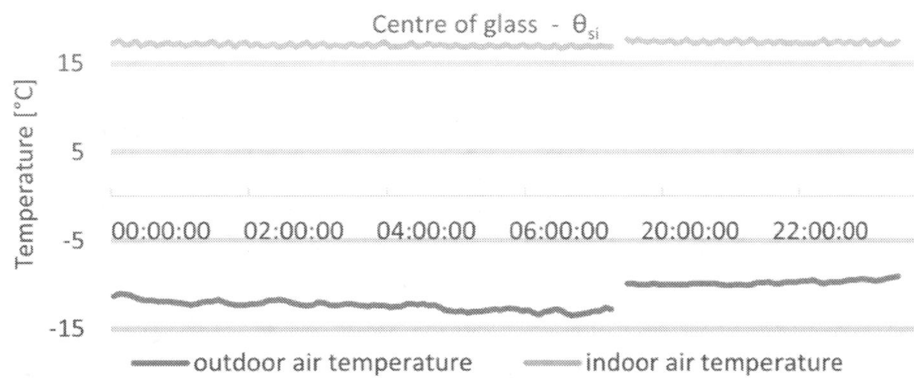


**Fig. 4.** Boundary conditions used for calculation of thermal surface transfer coefficient and thermal transmittance coefficient – night of 30th January 2017 – almost steady outside temperature -11 °C and inside 18 °C

For calculation of the surface coefficient of heat transfer with use of HFPs, the Eq. 3 and 4 were used and the results for all HFP are summarized in the Tab. 2. The results show a big difference between measured and standard values in some cases (exterior mostly), which could be influenced by the measuring method itself, unsteady state, lower accuracy of the thermocouples or imperfect contact between the sensor and surface. For exterior coefficient has also high impact the unsteady condition of outdoor: wind and long wave radiation of night sky. It seems that a considerable uncertainty is also caused by the AC unit, because values differ with distance from AC unit outlet. In case of the surface temperature being measured by HFP and thermocouples ($h_i$), both values are stated in Tab. 2. Calculated heat fluxes in form of thermal conductance coefficient based on the measurement are shown in Tab. 3.

**Table 2.** Surface coefficient of heat transfer defined in the standard STN 73 0540 and the calculated one using Eq. 3 and 4.

| surface coefficient of heat transfer [W/(m$^2$.K)] | STN 73 0540 | A glazing | B glazing | C glazing | A sash | B sash | C sash |
|---|---|---|---|---|---|---|---|
| $h_i$ (thermocouples) | 7.62 | 7.95 | 6.63 | 15.1 | 4.0 | 3.9 | 4.9 |
| $h_i$ (HFP) | | 10.3 | 8.1 | 8.1 | 4.7 | 4.9 | 6.1 |
| $h_e$ | 25 | 13.7 | 5.96 | 8.2 | 6.8 | 6.65 | 7.5 |

**Table 3.** Measured heat fluxes in form of thermal conductance coefficient [W/(m$^2$.K)]

| Window | Frame [W/(m$^2$.K)] | Glazing [W/(m$^2$.K)] |
|---|---|---|
| A | 0.76 | 0.9 |
| B | 0.73 | 1.05 |
| C | 0.65 | 0.67 |

### 2.3 Temperature comparison through the years of measurement

The windows are constantly measured since 2011. In the Tab. 4 there are compared results of average surface temperatures (both positions: centre of the glazing and the worst position based on the research – bottom of the glazing). Through the years there were selected specific hours during the year with similar outdoor air temperature (minimal 10 values, that are 10 x 15 minutes interval, during night). One of the comparative temperature is -14.2 °C (values from years 2012, 2014, 2016 and 2017) and second one is -21,2 °C (2012 and 2017).

**Table 4.** Measured values of surface temperatures during years with similar outdoor and indoor air temperatures

| Outdoor air temperature | Year | Windows Glazing | | | Windows Sash | | | Indoor air temperature | Measured values |
|---|---|---|---|---|---|---|---|---|---|
| [°C] | | A | B | C | A | B | C | [°C] | [n] |
| -14.3 | 2012 | 14.8 | 14.6 | 16.4 | 7.7 | 9.5 | 10.3 | 18.8 | 10 |
| -14.2 | 2014 | 15.1 | 14.6 | 16.4 | 8.5 | 10.0 | 10.8 | 19.1 | 11 |
| -14.3 | 2016 | 14.4 | 14.4 | 16.2 | 7.7 | 9.5 | 10.4 | 18.9 | 14 |
| -14.2 | 2017 | 13.1 | 13.0 | 14.9 | 7.1 | 8.6 | 9.1 | 16.7 | 30 |
| -21.2 | 2012 | 13.8 | 13.4 | 15.4 | 5.0 | 6.7 | 8.4 | 18.7 | 10 |
| -21.2 | 2017 | 11.6 | 11.6 | 13.7 | 3.7 | 5.0 | 6.6 | 15.8 | 22 |

## Results and discussion

Results from surface temperatures measurement are shown in Fig. 3. All points on the windows have to meet the surface temperature higher than 9.6 °C, which is required by STN 73 0540:2012. In this paper, there are presented courses for two positions from six measured. Centre of glazing, where the value should not be influenced by the sash or frame. Based on former research, the worst position is the contact between glazing and sash, but not in the corner but at the bottom.

From the courses it is clear, that for the first case all windows meets the required value and overestimate it about 5 °C. Another case is the second position. The temperature for the window A is low during the whole night (outdoor air temperature around -12 °C), for window B the outdoor air temperature around -13 °C would be critical. The wooden window C stays above the limit with reserve. If the temperature fell down more (lowest than -16 ~ -17 °C) there is a high possibility to underpass the required value. On the other hand, the indoor air temperature oscillated during the night at around 18.5 °C instead of 20 °C, which is defined in the standard. Defined outdoor air temperature for Zilina in the standard is -15 °C [3].

Heat fluxes calculated as a ratio between heat flux and differences between air temperatures (Eq. 1) in the form of $U$-values [W/(m$^2$.K)] are summarized in Tab. 3. These values cannot be strictly compared to the design values ($U_{f}$,) because of the unsteady thermal state. The results of sash itself take into account the glazing, so the heat flux is deformed and influenced by the edge of glazing, spacer etc. On the other hand, the design values are measured according to the STN EN 12412-2 in the climate chamber equipped with hotbox in steady state. By the window A, the heat flux through the glazing is far worse than by the window B or C. $U_g$ values from manufacturers (Tab. 1) should be almost the same. Only the window C is close to the design value and with best results, which is also shown in the surface temperature results. According to the window sashes, best is the wooden window followed by the plastic window with thermo modules. These modules slightly increased the $U_f$, but it cannot be calculated exactly because of the uncertainty during measurement. As mentioned above, the correct comparison of $U$-values cannot be made by the unsteady thermal state and the difference between measurement in situ and in the laboratory.

Long-time measurement results comparison in the Tab. 4 shows very similar results through the years of measurements. Windows A and B had problem to meet the requirements for the critical position – bottom of the glazing. It should be noted, that the indoor air temperature was lower, typically 1 - 2 °C than the required one 20 °C. For example, critical temperature for the air temperature 18 °C is only 7.8 °C [3]. With the outdoor air temperature around -21 °C is the sash of window A really cold, but again, the indoor temperature was lower than 20 °C.

The results of temperature measurement presented in Fig. 3 and Tab.4 show that in the centre of glazing the surface temperature is above the required standard even though the outside temperature was lower. In the lower corner, the temperatures are above the standard as well, but not with such limit than before. This is not surprising. Surprising are the results on the bottom of glazing, because both plastic windows, which should be suitable for passive buildings, did not reach the required temperature, so there is a possibility of dew occurrence or condensation of water vapour. The problem with measured temperatures could be caused by the $U_g$ value, according to the results in Tab. 3, where it could be clearly seen that the measured value is 70 % lower than the designed one (without taking into account the unsteady temperature of exterior air creating unsteady thermal state). This finding supports the idea that the window probably lost the gas filling in the glazing.

## Conclusions

Presented results could be confusing. All windows should meet the requirements for use in low-energy houses, but results of long-time measurement show big differences between the design values of the plastic windows, especially in the glazing, which caused that the surface temperature at the bottom of the glazing did not meet the Slovak standard requirement. It should be noted again, that this type of measurement causes mistakes itself and the measured heat fluxes cannot be compared to the declared ones exactly, but the practical outcome and comparison between windows is significant. Results from the measurement show that both plastic windows did not meet the requirement of the Slovak national standard STN 73 0540:2012 in terms of minimal surface temperature. With such low temperatures at bottom of the glazing is very high probability of vapour condensation. On the other hand, the wooden window C reached good results for the glazing $U$-value and also for surface temperatures on the frame/ sash and glazing, so it proves its design values in the experimental measurement.

The research is supported by the research project VEGA No. 1/0729/13 and VEGA No. 1/0945/16

## References

1. S. Kumar Pal et al., Appl. Energy **162**, 11-20 (2016)
2. A. Hepbasli et al.: J. Estech. (to be published)
3. Slovak standard STN 730540:2012, Parts 1, 2, 3, SUTN Bratislava (2012)
4. P. Durica et. al. Procedia Eng. **111**, 176-182 (2015)
5. D. Staffenova et. al., Adv. Mat. R. **1041** 129-134, (2014)
6. NiCr-Ni thermocouple T190-0, website < http://ahlborn.com/getfile.php?2052.pdf>

7. Thermal Flux Plates, website <http://www.ahlborn.com/download/pdfs/kap13/WflPlatten.pdf>

# Modelling post-critical deformation processes of flat reinforced concrete elements under biaxial stresses

*Oleg* Kabantsev[1,*], *Bozidar* Mitrovic[1,*]

[1] Moscow State University Of Civil Engineering 129337, 26, Yaroslavskoe shosse, Moscow, Russian Federation

**Abstract.** This paper examines the issue of mathematical modelling of local destruction processes of flat reinforced concrete structures beyond design basis impacts. Analysis of the peculiarities of the local failure mechanisms, which are formed in wall concrete structures, was conducted. The paper shows that the processes of post-critical deformation of flat concrete structures occur through several mechanisms with the relevant strength criteria. The method of post-critical deformation modelling of reinforced concrete structures under biaxial stresses is proposed and justified. The modelling is conducted by the gradual change of the analysis model in the areas of failure using finite elements with stiffness parameters, which corresponds to the conditions of interaction between structure parts separated by a crack.

**Keywords:** modelling, load-carrying structures, stress-and-strain state, design prediction, calculation technique, calculation model, model of external constraints.

## 1 Introduction

The prediction of load-carrying building systems stress-strain state within the framework of project impacts is a standard procedure of engineering analysis and has a set of well-developed techniques based on the numerical methods. It is assumed that the load-carrying structure elements have such a level of load-carrying ability that provides the possibility of calculation analysis within straight-line theory. However, in some cases it is necessary to predict the behaviour of building structures and constructions under beyond design basis impacts. For example, in the destruction (failure) of the element of the load-carrying system. Such failure leads to the fact that in other structural elements of the load-carrying system forces (stresses) arise with values exceeding their load-carrying ability, which determines the deformation transition of such structures to the nonlinear stage. Nonlinear deformation occurs with the formation of local damages, the accumulation of their volume

* Corresponding authors: ovk531@gmail.com, bozidarm@rambler.ru

and the formation of main structural damage zones. Such disruptions can lead to two cases: to an increase in the volume of destruction in the building load-carrying system with the transition to global destruction of the system or to the localization of damages in the destruction zone without the spreading of destruction process in the load-carrying structure system. This kind of problems are considered within the framework of the assessment of the structural survivability and load-carrying system beyond design basis impacts (see papers [1, 2]), and also for the assessment of the load-carrying constructive system ability in the presence of local individual element defects (see, e.g., [3]).

For a reasonable stress-and-strain state prediction of the load-carrying system beyond design basis impacts, a method of modeling inelastic deformation of load-carrying structures is necessary. The correctness of the method of modeling inelastic deformation of load-carrying structures is provided by the reflection of the most important proces regularities that arise when one leaves the linear phase of the structure work.

Thus, the research of the regularities of inelastic deformation of reinforced concrete structures (as the main type of materials of load-carrying building systems), including the phase formed after the occurrence of over-standard crack opening, is an important and urgent problem. Inelastic deformation of reinforced concrete structures in the phase of increasing the volume of cracks and increasing the width of their opening can be defined as the phase of post-critical deformation of reinforced concrete structures. In the phase of post-critical deformation, the conditions for the transition to macro-destruction both separate constructive elements, and the constructive system as a whole are formed. The account of post-critical deformation of the structure allows to consider the decrease in the volume of elastic energy of the building load-carrying system in the formation process of inelastic deformations (local and main destructions). Using of post-critical structure deformation account allows to form a work model of a «softening» constructive system, i.e. a system with a rigidity degrading on certain mechanisms.

The importance of developing of the methods of the reinforced concrete walls failure is determined by the fact that such structures allow the creation of the increased rigidity floors that make it possible to ensure the buildings resistance to a progressive collapse (the general principle is proposed and substantiated in [4], used in a number of other papers [5, 6] ]).

## 2 Analysis of peculiarities of post-critical deformation processes in reinforced concrete elements

For certain types of reinforced concrete structures (for example, for bent beam reinforced concrete elements), the phase of inelastic deformation is well studied. Based on the research results, reasonable approaches to assessing the possibilities of bar reinforced concrete elements to take a certain volume of plastic deformations are formed (see, for example, [7, 8]). For bar reinforced concrete elements, the post-critical deformation modeling methods are proposed and substantiated. For example, the method of modelling local destruction (failure) in bar reinforced concrete elements mainly by reducing the total rigidity of the final element - without detailing by the mechanisms of destruction is used for a finite-element approach (see [9, 10]). This method of modeling can be considered quite acceptable (with an amount of essential refinements - see [6]) to account the failures of bar elements.

However, flat (wall) structures under biaxial stress state have a number of significant differences (in relation to the bar elements) in the realization of post-critical deformation processes: in the forms of failures manifestation, one or another kind of failure criteria formation, in the volumes of local and main destruction possible realization.

For stone structures under biaxial stressed state, the processes of post-critical deformation were investigated in [11, 12].According to the research results, it is established that several mechanisms of local destruction are formed, which have a mutual influence on the development of any of them. The mechanisms are formed according to various strength criteria, which determines the impossibility of using simplified one-parameter methods of the plane-stressed stone structures supercritical deformation processes modeling. Research has shown that the implementation of local destruction mechanisms occurs simultaneously and in different volumes by separate mechanisms, that ensures the stone wall structure workability even when a large volume of local fractures is formed by one of the mechanisms (criteria).

Wall reinforced concrete structures under biaxial stress state also have a number of post-critical deformation features, which, if not taken into account, can lead to a significant decrease in the quality of the stress-and-strain state prediction of such structures in the post-critical deformation phase. The largest volume of deformation and destruction of flat-stressed reinforced concrete walls research was carried out within the framework of seismic impacts on wall structures (the results of the research are presented widely and in detail in the monograph [13]). Within the framework of the research it is established that the destruction of reinforced concrete walls under biaxial stressed state occurs in several stages. At the first stage, a crack is formed in the concrete wall, that can be determined based on general criteria of the plane stress state strength. At the following stages, the local crack development is determined by the work of the reinforcement that crosses the local concrete destruction. Mechanisms of interaction of wall parts, separated by a crack, are schematically represented in Fig. 1. When an expansion occurs in the crack zone, an "open" crack is formed (see Fig. 2), the expansion of which is prevented by the reinforcing bars that crosses the crack (the interaction of the crack sides can be neglected).

In the zone of the "open" crack, due to the shift of the crack sides and the inclusion of reinforcing bars into the work, a dowel action is formed that provides the plane-stressed structural element load-carrying ability level (in some cases - quite large), which is observed right up to the destruction of the reinforcement or crushing of concrete in the zone of its anchorage. The dowel action of the reinforcement work in the crack zone can be taken into account by various methods, for example, proposed in [14].

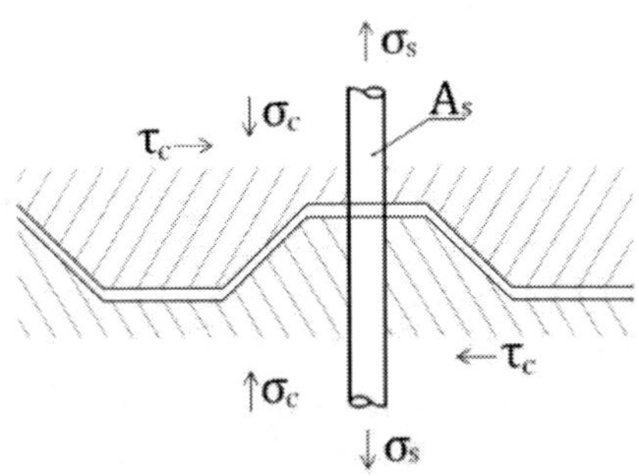


**Fig. 1.** General scheme of interaction of wall parts, divided by a crack.

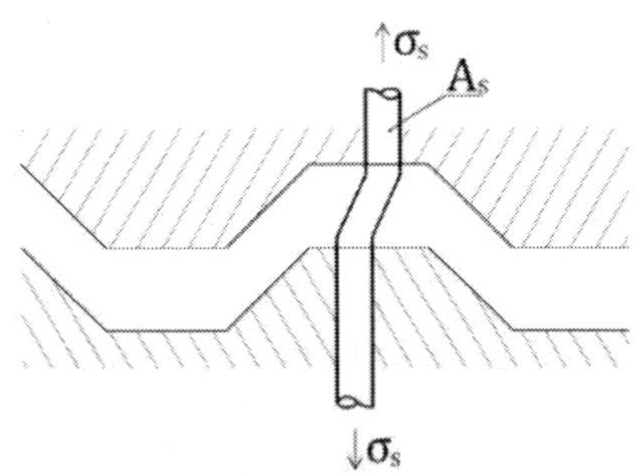


**Fig. 2.** Scheme of interaction of wall parts, separated by an "open" crack.

The development of reinforced concrete wall destruction is not an irreversible process - with the growth of the crack width in one local zone in other local zones a crack "closure" can occur, which forms the conditions for crack sides interaction in the dry friction regime.

Thus, the physical realization of processes at the post-critical deformation stage of reinforced concrete walls under biaxial stress state has a multi-regime nature, which requires appropriate detailing in the modeling. Obviously, the method of modeling the post-critical flat-stressed walls deformation by reducing the generalized rigidity in some local model region cannot reflect such a complex and multifactor process correctly.

# 3 Methods of modelling of the post-critical deformation of flat-stressed reinforced concrete elements

The flat-concrete reinforced concrete structures post-critical deformation processes occur within the framework of several local failure mechanisms types, which requires the use of several strength criteria. The using of strength criteria set (system) makes it possible to take into account the various failure mechanisms in evaluating the state of a structure. The critical condition assessment concept of a material based on the durability criteria set is developed and proved in works of A.A. Ilyushin [15] and Ya.B. Friedman [16] and is widely used for a long time.

Based on the research results of A.A. Ilyushin [17] a general approach to the strength criteria set formation is proposed, the essence of which is as follows: let the measures of the damage tensor Mn (Ω), which are the components Ω functions, can be used to formulate the destruction criteria corre-sponding to different failure mechanisms. We define that there exist critical damage constants for the material Ωcrn such that if for any **n**

$$M_n(\Omega) < \Omega_n^{cr} \quad (1)$$

the state of the particle (element) will be durable, and if for some **n=k**

$$M_k(\Omega) < \Omega_k^{cr} \quad (2)$$

**k** type of destruction will occur.

For isotropic materials, the functions **k** and **q** are invariant measures of the tensor Ω, which in turn express the change of the deformation properties that determine the behavior of materials (elements) under hydrostatic pressure and pure shear, respectively.

It is possible that the element destruction (the local zone of the element) according to one of the criteria adopted within the framework of the criteria system will determine the impossibility of resisting only a certain type of impact, whereas for other types of impacts the particle can work as a part of a general particles ensemble (load-carrying system elements). As shown in [17], formally this can be expressed in an abrupt increase to unity of some damage tensor Mn(Ω) components. The particles work analysis (elements of the calculation model) based of the strength criteria set (system) makes it possible to distinguish the destruction mechanisms and to consider correctly the elements state in the global structure system.

So, for example, when a particle/element is broken according to the shift criteria, this particle can turn out in two state types: in the case of a compressive stress in the zone, the particle will perceive such stresses and participate in the particles ensemble work by the shift criteria within the framework of the friction mechanism; in the case of tensile stresses in the particle zone, this particle disappears from the particles ensemble of the investigated body.

It seems possible to apply the structural local fractures modeling principles, proposed and substantiated in [18], which allow to take into account the performance of the damaged structure zone according to the separate strength criteria types as a general approach to the flat-stressed reinforced concrete constructive element failures.

The flat-stressed rainforced concrete wall destruction multi-mode mechanism modeling can be performed by means of a multi-stage finite element analysis, which takes into account the changing of the calculation model [19]. Numerical investigation of post-critical deformation processes is performed within the framework of the linear problems successive solution process. At each stage, the state of each final calculation model element is evaluated using the strength criteria system.

The finite element model state change in the post-critical deformation phase is modeled by the method of parallel installation of a set of different rigidity plate finite elements in the mutual nodes. At the initial stages of the calculation, the finite element with concrete stiffness parameters. If it is found that the strength criteria for concrete are exceeded in in any finite element of the model, that finite element must be excluded from the model (deactivated). On the deactivated finite element place in the ensemble of the calculation model, one of the finite element from the parallel layer is installed, the characteristics of which correspond to the conditions of operation of concrete in the zone of local destruction, is included.

When acting in the zone of local destruction of tensile stresses, the characteristics of the "new" active finite element must reflect the possibilities of the operation of the reinforcement both in tension and dowel regime, which prevents the shift of the crack sides. When acting in the zone of local destruction of compressive stresses, the characteristics of the active finite element must additionally take into account the interaction of the crack sides in the dry friction regime. In a simplified form, such a scheme (in the form of two parallel finite elements instaled in the mutual nodes) is shown in Fig. 3.

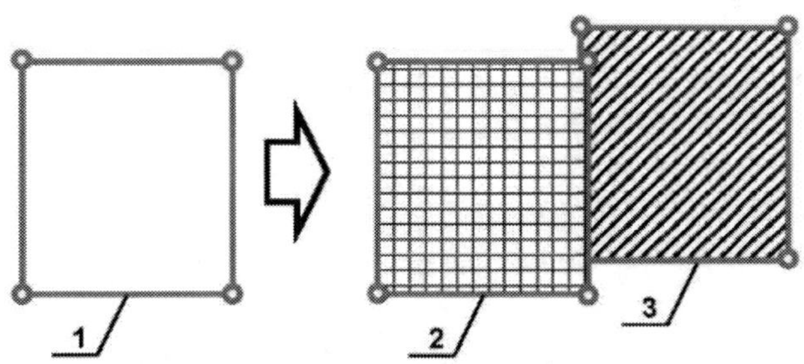


**Fig. 3.** Scheme of modelling of the post-critical wall deformation process by the method of parallel installed finite elements with different rigidity characteristics.
1 – finite element, approximating the concrete; 2 - finite element, approximating the work of the damaged concrete under the action of tensile stresses; 3 - finite element, approximating the work of damaged concrete under the action of compressive stresses

At the subsequent stages of the calculation, the analysis of the state of the finite elements that are part of the ensemble of the model is performed. If the stress sign changes in the zone of the local crack, it is necessary to replace the finite element with the type corresponding to the new stress sign. It is also necessary to monitor the level of stresses in the finite elements, approximating the work of reinforcing bars, which will allow to determine the moment of the transition of the reinforcement to plastic deformation or destruction. The dowel mode of reinforcing bars also has a definite limit. To track it, it is necessary to control the strength of concrete in the crushing zone, which is a technologically difficult design problem.

## Conclusion

The proposed modelling method allows taking into account the key features of the process of post-critical deformation of wall reinforced concrete structures under biaxial stress state. The application of the proposed modeling technology creates conditions for assessment of the evolution of the destruction process of the reinforced concrete wall structure, which is the basis for the prediction of the stability of the load-carrying system beyond design basis impacts.

## References

1. V. I. Kolchunov, N. V. Klyuyeva, N. B. Androsova, A. S. Bukhtiyarova, *Zhivuchest' zdaniy i sooruzheniy pri zaproyektnykh vozdeystviyakh* (Izdatel'stvo ASV, Moskva, 2014)
2. V. O. Almazov, Kkhoy Kao Zuy, *Dinamika progressiruyushchego razrusheniya mnogoetazhnykh karkasov* (Izdatel'stvo ASV, Moskva, 2013)
3. O. V. Kabantsev, S. V. Gorbatov, K. O. Pesin, *Otsenka vliyaniya lokal'nykh defektov perekrytiya na osnove ucheta poetapnogo izmeneniya raschetnoy skhemy pod nagruzkoy*, Vestnik Tomskogo gosudarstvennogo arkhitekturno-stroitel'nogo universiteta **2,** 89 (2015)
4. O.V. Kabantsev, *Nekotoryye voprosy metodiki normirovaniya raschetov zhelezobetonnykh konstruktsiy vysotnykh zdaniy,* 36 (Mezhregional'naya obshchestvennaya organizatsiya «Sodeystviye razvitiyu i primeneniyu prostranstvennykh konstruktsiy v stroitel'stve», Tez. dokladov naunoy sessii, Moskva, 2009)
5. D. V. Rudenko, V. V. Rudenko, *Zashchita karkasnykh zdaniy ot progressiruyushchego obrusheniya*, Inzhenerno-stroitel'nyy zhurnal **3**, 38 (2009)
6. Ye. V. Domarova *Raschetno-konstruktivnyye metody zashchity ot progressiruyushchego obrusheniya zhelezobetonnykh monolitnykh karkasnykh zdaniy*, Vestnik IrGTU **10,**123 (2015)
7. N. N. Popov, *Dinamicheskiy raschet zhelezobetonnykh konstruktsiy*, 207 ( SI, Moskva, 1974)
8. N. N. Popov, B. S. Rastorguyev, A. V. Zabegayev, *Raschet konstruktsiy na dinamicheskiye i spetsial'nyye nagruzki,* 319 (Vysshaya shkola, Moskva, 1992)
9. V. A. Lyublinskiy, YU. V. Guseynova, *Progressiruyushcheye obrushenii nesushchikh sistem mnogoetazhnykh zdaniy,* Yestestvennyye i inzhenernyye nauki – razvitiyu regionov Sibiri **2**, 164 (Trudy Bratskogo gosudarstvennogo universiteta, Bratsk)
10. L. A. Tsyganenko, *Osoblivosti rozrukhanku visotnikh budinkiv ne nesuchu zdatnist', stiykist' ta opir progresuyuchomu obvaleniyu*, Seria «Stroitel'stvo» **5(16)**, 143 (Vestnik Sumskogo natsional'nogo agrarnogo universiteta, Ukraina, 2012)
11. O. Kabantsev, *Modeling Nonlinear Deformation and Destruction Masonry under Biaxial Stresses. Part 1. Masonry as Simulation Object,* 681, 725 (Applied Mechanics and Materials, 2015)
12. O. Kabantsev, *Modeling Nonlinear Deformation and Destruction Masonry under Biaxial Stresses. Part 2. Strength Criteria and Numerical Experiment*, 725, 808 (Applied Mechanics and Materials, 2015)
13. G. N. Ashkenadze, M. Ye. Sokolov, L. D. Martynova i dr., *Zhelezobetonnyye steny seysmostoykikh zdaniy. Issledovaniya i osnovy proyektirovaniya. Pod red. G. N. Ashkenadze i M. Ye. Sokolova,* 504 (Stroyizdat, Moskva, 1988)
14. O. V. Kabantsev, *The design methods of masonry structures seismic strengthening by means of the cast in-situ reinforced concrete applications*, Nauchno-tekhnicheskiy vestnik MGSU **2**, 51 (2011)

15. A. A. Il'yushin, *Mekhanika sploshnoy sredy,* 310 ( Izdatel'stvo Moskovskogo universiteta 3-ye izd., Moskva, 1990)
16. YA. B. Fridman, *Mekhanicheskiye svoystva metallov. CH. 1. Deformatsiya i razrusheniye*, 472 (Mashinostroyeniye, Moskva, 1974)
17. V. E. Vil'deman, YU. V. Sokolkin, A. A. Tashkinov, *Mekhanika neuprugogo deformirovaniya i razrusheniya kompozitsionnykh materialov. Pod red. YU. V. Sokolkina,* 228 (Nauka. Fizmalit, Moskva, 1997)
18. O. V. Kabantsev, *Diskretnaya model' kamennoy kladki v usloviyakh dvukhosnogo napryazhennogo sostoyaniya*, Vestnik Tomskogo gosudarstvennogo arkhitekturno-stroitel'nogo univesiteta **4,** 113 (2015)
19. A. V. Perel'muter, O. V. Kabantsev, *Analiz konstruktsiy s izmenyayushcheysya raschetnoy skhemoy,* 148 (Izdatel'stvo SKAD SOFT, Izdatel'skiy dom ASV, Moskva, 2015)

MATEC Web of Conferences 117, 00072 (2017) DOI: 10.1051/matecconf/201711700072
*XXVI R-S-P Seminar 2017, Theoretical Foundation of Civil Engineering*

# Non-linear shape functions over time in the space-time finite element method

*Zbigniew* Kacprzyk[1,*]

[1]Warsaw University of Technology, Department of Civil Engineering, al. Armii Ludowej 16, 00-637 Warsaw, Poland

**Abstract.** This work presents a generalisation of the space-time finite element method proposed by Kączkowski in his seminal of 1970's and early 1980's works. Kączkowski used linear shape functions in time. The recurrence formula obtained by Kączkowski was conditionally stable. In this paper, non-linear shape functions in time are proposed.

## 1 Introduction

Kączkowski presented the basis of the space-time finite element method in works [1-4]. The recurrence formula obtained by Kączkowski was conditionally stable. This means that the maximal value of the step of integration in time had to be selected adequately and it had to meet the conditions of method stability, cf.[5]. A very small integration step should be selected in practical dynamical analyses of engineering structures - which is very costly (in terms of the number of numerical operations) and counter-competitive compared to other methods of numerical integration of motion equations, e.g. compared to the Newmark method [6]. Kacprzyk presented a particular method of integration step incrementation [7]. This method requires additional operations to be performed and applies to linear dynamics topics only. Application of a space-time superelement removes the limitation in acceptance of the step of integration in time and allows any integration step to be accepted without loss of the stability of the recurrence formula.

In his work [8], and later, in works [9,10] Kacprzyk presented an unconditionally stable formulation of the space-time finite element method. These works provide a generalisation of the derivation presented by Kączkowski and Langer [4], which is, however, quite different from the basic idea of the method presented in works of Kączkowski [1-3]. In this work, we will use the derivation of Kączkowski presented in work [3]. Kączkowski assumes linear shape functions in time in this fundamental work. An attempt at introduction of a non-linear shape function was presented by Pelc in [11].The non-linear shape functions assumed in the mentioned work, however, do not lead to a stable recurrence formulae in general, cf. [12].

In this work, we will repeat the derivation presented by Kączkowski in [2] but assuming non-linear shape functions in time for unknown displacements and linear functions for virtual displacements.

* Corresponding author: z.kacprzyk@il.pw.edu.pl

It should be noted that the presented work is not a full generalisation of the space-time finite element method formulated by Kączkowski. This work generalises this method in the so called stationary formulation. The stationary formulation assumes that the division of a structure into finite elements does not change during analysis of the structure. Kączkowski presented in [13] the space-time finite element method, assuming that the division of a structure into elements changes in time. This is the so called non-stationary formulation of the space-time finite element method. This approach was developed by Bajer in [14,15].

The proposed method may be applied, with slight changes, to the non-stationary formulation of the space-time finite element method. A generalisation of this formulation shall provide the basis for our next work.

## 2 Displacement

Any finite element can be analysed at two time instants: $e$ (exordium) and $f$ (finis). Combination of appropriate nodes for moments $e$ and $f$ yields a space-time element (cf. Fig. 1.), cf. Kączkowski [2].

The following notation is adopted: $x_i, i = 1,2,3$ - spatial coordinates, $t$ - time coordinate, $\tau$ - dimensionless coordinate in time $(\tau = t / h)$ with values in the range between $-1$ and $1$.

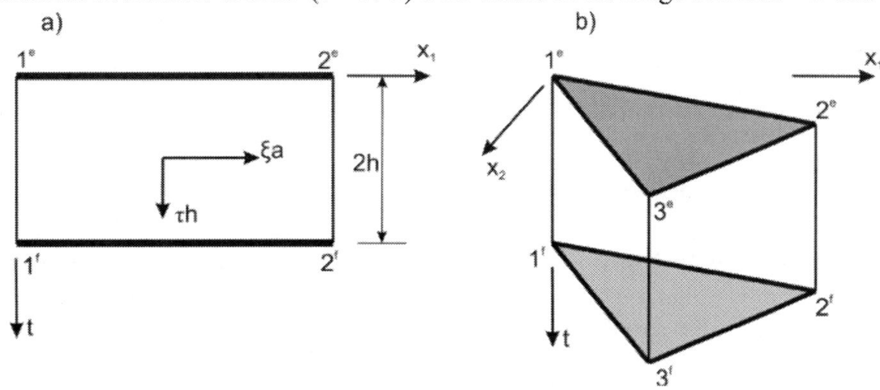


**Fig. 1.** Space-time finite elements.

Fig. 1 presents two- and three-dimensional finite elements. A four-dimensional element cannot be drawn, but all the considerations presented below also apply to four-dimensional elements.

Each finite element in space-time has twice the number of nodes it has in a one-, two- or three-dimensional space.

Displacements of nodes in a finite space-time element form a vector, which can be decomposed as below:

$$\mathbf{q}_e = \begin{Bmatrix} \mathbf{q}_e^e \\ \mathbf{q}_e^f \end{Bmatrix} \tag{1}$$

Displacements in space-time can be expressed conventionally:

$$\mathbf{f}(\mathbf{x},t) = \mathbf{N}(\mathbf{x},t)\mathbf{q}_e \tag{2}$$

where

$$\mathbf{x} = \begin{Bmatrix} x_1 \\ x_2 \\ x_3 \end{Bmatrix} \tag{3}$$

The shape matrix $\mathbf{N}$ will be written as a tensor product of two matrices

$$\mathbf{N}(\mathbf{x},t)=\mathbf{N}_t(t)\otimes\mathbf{N}_x(\mathbf{x}) \tag{4}$$

In this product, the $\mathbf{N}_t(t)$ matrix has a dimensions 1x2 and depends on the time coordinate $t$ only, whilst the $\mathbf{N}_x(\mathbf{x})$, matrix, dependent on spatial coordinates, is a typical shape matrix of an element applied to static problems.

## 3 Strains and stresses

The vector of strain components $\boldsymbol{\varepsilon}_x$ is obtained by appropriate differentiations over spatial coordinates of displacements vector $\mathbf{f}$ .

$$\boldsymbol{\varepsilon}_x=\partial_x\mathbf{f}=\partial_x(\mathbf{N}_t\otimes\mathbf{N}_x)\mathbf{q}_e=(\mathbf{N}_t\otimes\partial_x\mathbf{N}_x)\mathbf{q}_e \tag{5}$$

denoting

$$\mathbf{B}_x=\partial_x\mathbf{N}_x \tag{6}$$

we eventually obtain

$$\boldsymbol{\varepsilon}_x=(\mathbf{N}_t\otimes\mathbf{B}_x)\mathbf{q}_e \tag{7}$$

A finite space-time element is also subjected to "space-time strains" $\boldsymbol{\varepsilon}_t$, namely to motion velocities. These are obtained by differentiating the displacements in time.

$$\boldsymbol{\varepsilon}_t=\frac{\partial}{\partial t}\mathbf{f}=\frac{\partial}{\partial t}(\mathbf{N}_t\otimes\mathbf{N}_x)\mathbf{q}_e=(\frac{d}{dt}\mathbf{N}_t\otimes\mathbf{N}_x)\mathbf{q}_e \tag{8}$$

By assuming the following notation

$$\dot{\mathbf{N}}_t=\frac{d}{dt}\mathbf{N}_t \tag{9}$$

we obtain

$$\boldsymbol{\varepsilon}_t=(\dot{\mathbf{N}}_t\otimes\mathbf{N}_x)\mathbf{q}_e \tag{10}$$

Generalised internal forces comprising a single column matrix $\boldsymbol{\sigma}_x$ depend on element strains. Assuming that the mechanical properties of a viscous-elastic material can be modeled using the Kelvin-Voigt approach, we can write

$$\boldsymbol{\sigma}_x=(\mathbf{E}+\boldsymbol{\eta}_w\frac{\partial}{\partial t})\boldsymbol{\varepsilon}_x, \tag{11}$$

where $\mathbf{E}$ - elastic constants matrix, $\boldsymbol{\eta}_w$ - square matrix containing viscosity coefficients. Substituting relationships (7) in to the formula (11) we obtain

$$\boldsymbol{\sigma}_x=[\mathbf{N}_t\otimes(\mathbf{E}\mathbf{B}_x)]\mathbf{q}_e+[\dot{\mathbf{N}}_t\otimes(\boldsymbol{\eta}_w\mathbf{B}_x)]\mathbf{q}_e. \tag{12}$$

In addition to these conventional, internal forces, some additional quantities are present in space-time, namely momentums (cf. {1,2]). These are obtained from the relationship:

$$\boldsymbol{\sigma}_t=-\mathbf{m}\boldsymbol{\varepsilon}_t=-[\dot{\mathbf{N}}_t\otimes(\mathbf{m}\mathbf{N}_x)]\mathbf{q}_e, \tag{13}$$

where $\mathbf{m}$ is a square matrix of masses.

## 4 Space-time stiffness matrix

According to the virtual four-work theorem, cf. [1], and taking into account linear, viscous external damping, we can obtain the following equation:

$$\bar{\mathbf{q}}_e^T\mathbf{Q}_e+\int_{-1}^{1}\int_V\bar{\mathbf{f}}^T(\mathbf{p}-\boldsymbol{\eta}_z\frac{\partial}{\partial t}\mathbf{f})dVhd\tau=\int_{-1}^{1}\int_V(\bar{\boldsymbol{\varepsilon}}_x^T\boldsymbol{\sigma}_x+\bar{\boldsymbol{\varepsilon}}_t^T\boldsymbol{\sigma}_t)dVhd\tau, \tag{14}$$

where $V$ is a volume of the element in a three-dimensional space.

The virtual state of a translation may be defined as follows, analogously to (2):

$$\bar{\mathbf{f}}(\mathbf{x},t) = [\mathbf{W}_t(t) \otimes \mathbf{N}_x(x)]\bar{\mathbf{q}}_e \tag{15}$$

The shape matrix $\mathbf{W}_t(t)$ is a function of the time variable. Thus, and per analogy to (7) and (10) we obtain

$$\bar{\boldsymbol{\varepsilon}}_x = (\mathbf{W}_t \otimes \mathbf{B}_x)\bar{\mathbf{q}}_e, \tag{16}$$

$$\bar{\boldsymbol{\varepsilon}}_t = (\dot{\mathbf{W}}_t \otimes \mathbf{N}_x)\bar{\mathbf{q}}_e, \tag{17}$$

where

$$\dot{\mathbf{W}}_t = \frac{d}{dt}\mathbf{W}_t \tag{18}$$

By executing the assigned substitution and using the formula (14) we obtain

$$\mathbf{Q}_e = \mathbf{K}_e \mathbf{q}_e - \mathbf{Q}_{ep} \tag{19}$$

where

$$\mathbf{K}_e = \int_{-1}^{1} \mathbf{W}_t^T \mathbf{N}_t h d\tau \otimes \int_V \mathbf{B}_x^T \mathbf{E} \mathbf{B}_x dV - \int_{-1}^{1} \dot{\mathbf{W}}_t^T \dot{\mathbf{N}}_t h d\tau \otimes \int_V \mathbf{N}_x^T \mathbf{m} \mathbf{N}_x dV + \int_{-1}^{1} \mathbf{W}_t^T \dot{\mathbf{N}}_t h d\tau \otimes \left(\int_V \mathbf{B}_x^T \boldsymbol{\eta}_w \mathbf{B}_x dV + \int_V \mathbf{N}_x^T \boldsymbol{\eta}_z \mathbf{N}_x dV\right) \tag{20}$$

and

$$\mathbf{Q}_{ep} = \int_{-1}^{1}\int_V (\mathbf{W}_t^T \otimes \mathbf{N}_x^T) p \, dV h \, d\tau \tag{21}$$

Let us introduce the following notation:

$$\mathbf{S}_e = \int_V \mathbf{B}_x^T \mathbf{E} \mathbf{B}_x dV, \quad \mathbf{M}_e = \int_V \mathbf{N}_x^T \mathbf{m} \mathbf{N}_x dV$$
$$\mathbf{C}_e^w = \int_V \mathbf{B}_x^T \boldsymbol{\eta}_w \mathbf{B}_x dV, \quad \mathbf{C}_e^z = \int_V \mathbf{N}_x^T \boldsymbol{\eta}_z \mathbf{N}_x dV \tag{22}$$

The $\mathbf{S}_e$ is a static stiffness matrix of the element $\mathbf{M}_e$ - a matrix of element mass, $\mathbf{C}_e^w$ and $\mathbf{C}_e^z$ are matrices of internal and external damping, respectively.

Taking into account the relationships (22) formula (20) takes the following form:

$$\mathbf{K}_e = \int_{-1}^{1} \mathbf{W}_t^T \mathbf{N}_t h d\tau \otimes \mathbf{S}_e - \int_{-1}^{1} \dot{\mathbf{W}}_t^T \dot{\mathbf{N}}_t h d\tau \otimes \mathbf{M}_e + \int_{-1}^{1} \mathbf{W}_t^T \dot{\mathbf{N}}_t h d\tau \otimes (\mathbf{C}_e^w + \mathbf{C}_e^z) \tag{23}$$

We assume non-linear shape functions in time $\mathbf{N}_t$, cf. Fig. 2.

$$\mathbf{N}_t = \left[\frac{1}{2}(1-\tau) + \varsigma(\tau - \tau^3), \ \frac{1}{2}(1+\tau) - \varsigma(\tau - \tau^3)\right] \tag{24}$$

where $\varsigma$ is a number (parameter), whilst $\tau = -1$ at time e and $\tau = 1$ at time f .

On the other hand, in the case of virtual displacements $\mathbf{W}_t$, we will assume linear shape functions, cf. Fig. 3.

$$\mathbf{W}_t = \frac{1}{2}[1-\tau, \ 1+\tau] \tag{25}$$

Derivatives of the shape function in time are as follows:

$$\dot{\mathbf{N}}_t = \frac{1}{h}\left[-\frac{1}{2} + \varsigma(\tau - 3\tau^2), \ \frac{1}{2} - \varsigma(\tau - 3\tau^2)\right]. \tag{26}$$

and

$$\dot{\mathbf{W}}_t = \frac{1}{2h}[-1,\ 1] \tag{27}$$

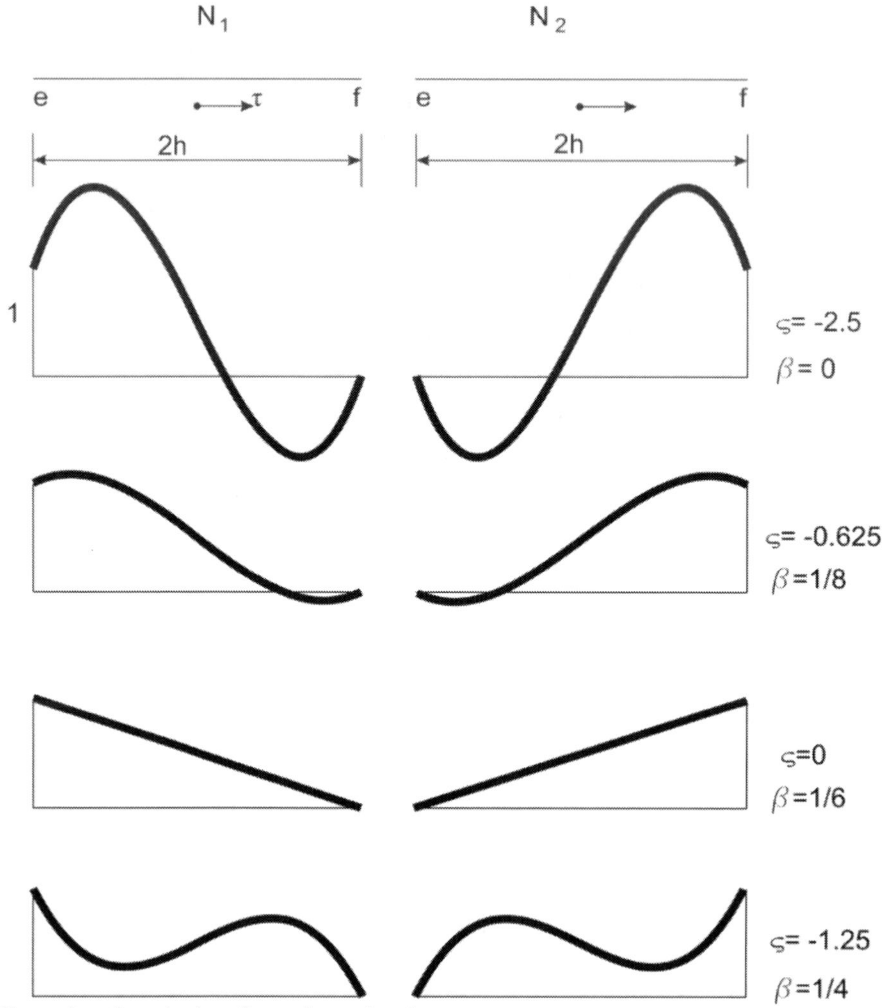


**Fig.2.** Shape functions in time for various $\varsigma$ values

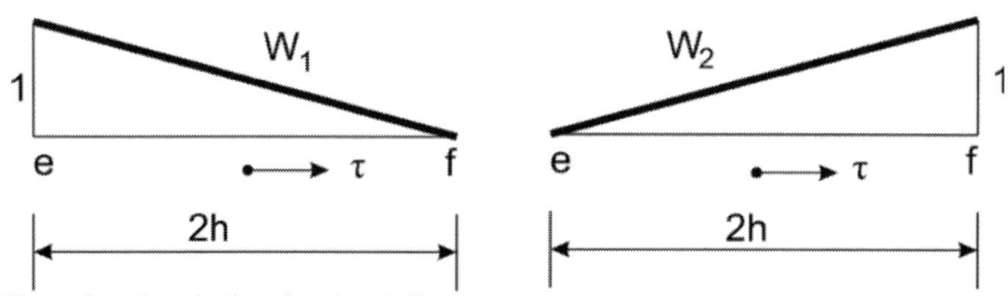


**Fig. 3.** Shape functions in time for virtual displacements

Integrals present in formula (23) can be easily calculated. Thus

$$\int_{-1}^{1} \mathbf{W}_t^T \mathbf{N}_t h\, d\tau = 2h \begin{bmatrix} \frac{1}{3} - \frac{1}{15}\varsigma & \frac{1}{6} + \frac{1}{15}\varsigma \\ \frac{1}{6} + \frac{1}{15}\varsigma & \frac{1}{3} - \frac{1}{15}\varsigma \end{bmatrix}, \tag{28}$$

$$\int_{-1}^{1} \dot{\mathbf{W}}_t^T \dot{\mathbf{N}}_t h\, d\tau = \frac{1}{2h}\begin{bmatrix} 1 & -1 \\ -1 & 1 \end{bmatrix}, \qquad \int_{-1}^{1} \mathbf{W}_t^T \dot{\mathbf{N}}_t h\, d\tau = \frac{1}{2h}\begin{bmatrix} -1 & 1 \\ -1 & 1 \end{bmatrix},$$

It should be noted that the reverse substitution, that of a non-linear shape function for virtual displacements and of a linear shape function for real displacements leads to the same integration results.

By using the substitution (such substitutions facilitate comparisons of the resulting formulae with other methods)

$$\varsigma = 15(\beta - \frac{1}{6}) \tag{29}$$

we obtain

$$\mathbf{K}_e = 2h \begin{bmatrix} \frac{1}{2} - \beta & \beta \\ \beta & \frac{1}{2} - \beta \end{bmatrix} \otimes \mathbf{S}_e - -\frac{1}{2h}\begin{bmatrix} 1 & -1 \\ -1 & 1 \end{bmatrix} \otimes \mathbf{M}_e + +\frac{1}{2}\begin{bmatrix} -1 & 1 \\ -1 & 1 \end{bmatrix} \otimes (\mathbf{C}_e^w + \mathbf{C}_e^z). \tag{30}$$

Let

$$\mathbf{K}_e = \begin{bmatrix} \mathbf{A}_e & \mathbf{B}_e \\ \mathbf{B}_e^{\mathrm{i}} & \mathbf{D}_e \end{bmatrix}. \tag{31}$$

Thus

$$\begin{aligned} \mathbf{A}_e &= 2h(\frac{1}{2} - \beta)\mathbf{S}_e - \frac{1}{2h}\mathbf{M}_e - \frac{1}{2}(\mathbf{C}_e^w + \mathbf{C}_e^z), \qquad & \mathbf{B}_e &= 2h\beta\mathbf{S}_e + \frac{1}{2h}\mathbf{M}_e + \frac{1}{2}(\mathbf{C}_e^w + \mathbf{C}_e^z), \\ \mathbf{B}_e^{\mathrm{i}} &= 2h\beta\mathbf{S}_e + \frac{1}{2h}\mathbf{M}_e - \frac{1}{2}(\mathbf{C}_e^w + \mathbf{C}_e^z), \qquad & \mathbf{D}_e &= 2h(\frac{1}{2} - \beta)\mathbf{S}_e - \frac{1}{2h}\mathbf{M}_e + \frac{1}{2}(\mathbf{C}_e^w + \mathbf{C}_e^z) \end{aligned} \tag{32}$$

Let us also introduce the following notation:

$$\mathbf{A}_e + \mathbf{D}_e = \mathbf{H}_e = 4h(\frac{1}{2} - \beta)\mathbf{S}_e - \frac{1}{h}\mathbf{M}_e \tag{33}$$

## 5 System solution and recurrent formulae

Space-time elements contained between instants e and f can be combined into a structure observed within the same time range. We thus obtain a space-time element of a single time layer. Matrices $\mathbf{S}$, $\mathbf{M}$, $\mathbf{C}^w$, $\mathbf{C}^z$ for the entire structure must be formed in order to obtain the stiffness matrix for such a space-time element. Space-time elements for individual time layers are then combined and the stiffness matrix for the space-time structure over the entire observation time is observed.

$$\mathbf{K} = \begin{bmatrix} \mathbf{A} & \mathbf{B} & & \\ \mathbf{B}^{\mathrm{i}} & \mathbf{H} & \mathbf{B} & \\ & \mathbf{B}^{\mathrm{i}} & \mathbf{H} & \mathbf{B} \\ & & \dots & \dots & \dots \end{bmatrix} \tag{34}$$

The number of rows in matrix $\mathbf{K}$ is equal to the number of time layers. Finally, the system of equations of the space-time element method has the following form.

$$\begin{bmatrix} \mathbf{A} & \mathbf{B} & & & \\ \mathbf{B}^{\mathrm{i}} & \mathbf{H} & \mathbf{B} & & \\ & \mathbf{B}^{\mathrm{i}} & \mathbf{H} & \mathbf{B} & \\ & & \dots & \dots & \dots \end{bmatrix} \begin{bmatrix} \mathbf{q}^0 \\ \mathbf{q}^1 \\ \mathbf{q}^2 \\ \dots \end{bmatrix} = \begin{bmatrix} \mathbf{Q}^0 \\ \mathbf{Q}^1 \\ \mathbf{Q}^2 \\ \dots \end{bmatrix}, \tag{35}$$

where $\mathbf{q}^i$, $\mathbf{Q}^i$ denote vectors of displacements and of impulses, respectively. The dimension of these vectors is equal to the number of degrees of freedom of the structure.

While

$$\begin{aligned} \mathbf{A} &= h(1-2\beta)\mathbf{S} - \frac{1}{2h}\mathbf{M} - \frac{1}{2}(\mathbf{C}^w + \mathbf{C}^z), \\ \mathbf{B}, \mathbf{B}^{\mathrm{i}} &= 2h\beta\mathbf{S} + \frac{1}{2h}\mathbf{M} \pm \frac{1}{2}(\mathbf{C}^w + \mathbf{C}^z), \\ \mathbf{H} &= 2h(1-2\beta)\mathbf{S} - \frac{1}{h}\mathbf{M}. \end{aligned} \tag{36}$$

System (35) is solved using the recurrent formula:

$$\begin{aligned} \mathbf{B}\mathbf{q}^1 &= \mathbf{Q}^0 - \mathbf{A}\mathbf{q}^0, \\ \mathbf{B}\mathbf{q}^{i+1} &= \mathbf{Q}^i - \mathbf{H}\mathbf{q}^i - \mathbf{B}^{\mathrm{i}}\mathbf{q}^{i-1}, \qquad i = 1,2,3,\dots \end{aligned} \tag{37}$$

It should be noted that for $\beta = 0$ the stiffness matrix $\mathbf{K}$ is multiplied by zero (cf. relationship (36)) and does not affect the values of components the $\mathbf{B}, \mathbf{B}^{\mathrm{i}}$ matrices. In the case of undamped vibrations and assuming the diagonality of the mass matrix $\mathbf{M}$ an explicit recurrent diagram is obtained - the matrix of $\mathbf{B}$ indices is diagonal.

Displacements are the unknown variables in the recurrent scheme described according to equation (37). Velocities and accelerations are obtained from the respective differential equations, cf. [8,10].

Assuming $\varsigma = 0$ we obtain $\mathbf{W}_t = \mathbf{N}_t$ and $\beta = 1/6$ . Formulae (36) and (37) are then identical to the formulae obtained in the work of Kączkowski [2].

## 6 Summary

The recurrent formulae obtained in this work have been studied by numerous authors, [16]. Numerical verifications show that no single best scheme can be indicated. The unconditionally stable scheme is the most popular one. However, errors in the solution resulting from assumption of an excessive integration step should be taken into account when using this algorithm, cf. [9,10].

Work [10] included numerical testing of the aforementioned formulae. Each of the proposed recurrent formulae leads to correct results. The recurrent formulae differ mostly in the permitted integration step in time (a $2h$ time step) and in costs of the solution (understood as the number of numerical operations). There are also some slight differences in solutions caused by numerical dumping. However, these differences are insignificant in technical applications.

Work [5] presents a wide description of various numerical integration methods applied to the motion equation. This work provided an analytical proof of equivalence of numerous recurrent formulae. According to this work, it may be shown that for $\beta = 1/4$ the generalised space-time finite element method is identical to the Newmark method. On the other hand, for $\beta = 0$ and assuming the mass matrix $\mathbf{M}$ and taking damping into account, as well as diagonal damping matrices $\mathbf{C}^w$ and $\mathbf{C}^z$ an explicit recurrent diagram is obtained. This means that we no longer have to solve a system of equations during the recurrence.

## References

1. Z. Kączkowski, J. Tech. Phys. **16 (1)**, 69 (1975)
2. Z. Kączkowski, Arch. Inż. Ląd. **22 (3)**, 365 (1976) (in Polish)
3. Z. Kączkowski, Arch. Inż. Ląd., **25 (3)**, 351 (1979)
4. Z. Kączkowski, J. Langer, Arch. Inż. Ląd. **26(1)**, 11 (1980)
5. Z. Kacprzyk, T. Lewiński, Engineering Transactions **31(2)**, 213 (1983)
6. N. M. Newmark, J. Eng. Div., ASCE **85 (EM3)**, 67 (1959)
7. Z. Kacprzyk, Arch. Inż. Ląd. **28 (1)**, 47 (1982)
8. Z. Kacprzyk, Zeszyty Naukowe Politechniki Warszawskiej, s. Budownictwo **85**, 83 (in Polish) (1984)
9. Z. Kacprzyk, Proc. Engineering **153**, 248 (2016)
10. Z. Kacprzyk, Structural Mechanics **77** (2016)
11. J. Pelc, Arch. Inż. Ląd. **30 (1)**, 53 (in Polish) (1984)
12. J. Pelc, Comput. Methods Appl. Mech. Engrg **190**, 3915 (2001)
13. Z. Kączkowski, Journal of Theoretical and Applied Mechanics **21 (4)**, 531 (in Polish), (1983)
14. C. Bajer, Int. Jou. for Num. Methods in Eng. **23 (11)**, 2031 (1986)
15. C. Bajer, *The space-time finite elements method in dynamic structures calculation* (in Polish), (IPPT PAN, 2009)
16. G. Rakowski, Z. Kacprzyk, *Finite elements method in structure mechanics*, third ed. (in Polish), (Oficyna Wydawnicza Politechniki Warszawskiej, 2016)

# The problem of choosing risk management methodology at the example of railway construction.

*Krzysztof* Kaczorek[1,*], *Michał* Krzemiński[1], and *Nabi* Ibadov[1]

[1] WUT, The Faculty of Civil Engineering, Institute of Building Engineering, Division of Production Engineering and Construction Management, al. Armii Ludowej 16, 00-637 Warsaw, Poland

**Abstract.** One of the most expensive projects in the construction industry are railway projects. High costs and specificity of implementation involve high risks throughout the investment and construction process. Over the years, project management methodologies have been developed that facilitate, among other things, the management of risk factors This article reviews the recommendations for risk management in each methodology. In addition, their usefulness have been evaluated in railway construction.

## 1 Risk issue in railway projects

One of the most expensive projects in the construction industry are railway projects [1, 2, 3]. The reason for this is primarily from the very wide range of works being carried out [4]. It also causes a long investment period [5, 6, 7]. These values, i.e. cost, range and time [8, 9] are, according to many recognized methodologies(for example: PRINCE2, PMBOK, AGILE, IPMA, ITIL, SCRUM or JASPERS) three basic parameters of the project. Moreover, they are closely related, so that modifying the value of one parameter results in at least one of the other two. These three methodologies supplement the basic parameters with additional parameters. The parameters differ depending on the methodology, but each of them takes into account the issues related to the management of risk factors. In the article, the authors attempt to compare the ways of managing the risk factors represented by the individual methodologies, and also assess which of them would be most appropriate for the implementation into railway contracts.

## 2 Methodologies used in project management

The year of the first project management methodology is considered to be 1942, when relevant studies were prepared for the Manhattan Engineering District Project. With the passing of time, new methodologies have appeared while the existing ones have been developing.. As a result of this process, there are a whole array of different methodologies today. What is the reason for such a large development of the area of project management knowledge? The reasons are mainly three. The first one is the geographic cause. In different

* Corresponding author: k.kaczorek@il.pw.edu.pl

parts of the globe, in parallel with the needs of individual societies, methodologies were developed to meet their expectations (for examaple American PMBOK, British PRINCE2 or Swiss IPMA). The second reason was the adaptation of methodologies to the needs of individual industries. Most of this type of procedures were done for the IT industry (for example AGILE, ITIL or SCRUM). The third reason for generating new work is the fact that special programs are being developed within international organizations that require appropriate standards (for example materials issued by JASPERS to improve the preparation and implementation of projects applying for EU funding). In the remainder of the paper, the authors focused on three methods: PMBOK, PRINCE2 and JASPERS (Blue Book), as applicable to railway contracts.

## 3 Recommendations of individual methodologies for subsequent stages of risk management

In the PMBOK, risk management is described in the "Risk Management" knowledge area, PRINCE2 under the "Risk" topic, and JASPERS in the "Risk and Sensitivity Analysis". The next steps in the risk management process are presented at Fig. 5.

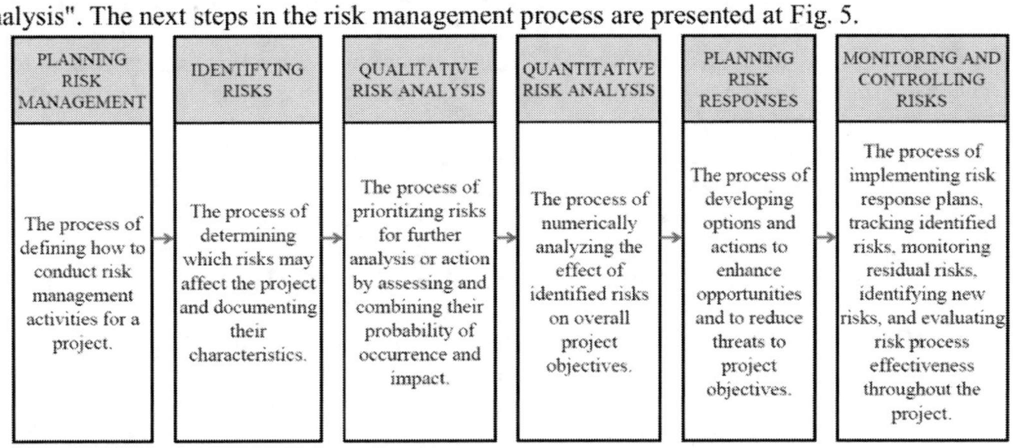


**Fig. 1.** The next steps in the risk management according to the PMBOK. [10].

### 3.1 Planning risk management

PMBOK defines planning risk management as „the process of defining how to conduct risk management activities for a project" [11]. The tools that are recommended at this stage are planning meetings and analysis. Participants who should attend the meetings are: the project manager, selected project team members and stakeholders, anyone in the organization with responsibility to manage the risk planning and execution activities, and others, as needed. The result of working at this stage should be a risk management plan that contains: methodology, roles and responsibilities, budgeting, timing and risk categories [11].

PRINCE2 recommends that you thoroughly review the documentation at the organization and program level and then develop a Risk Management Strategy for the project. The main purpose of developing this strategy is to show how the risk management will be integrated into the project management activities. The Risk Management Strategy consists of the following elements: introduction, risk management procedure, tools and techniques, required records, reporting, deadlines for risk management activities, roles and responsibilities, rating scales, proximity and risk categories, risk response categories, early indicators warning, risk tolerance and risk budget [12, 13, 14].

JASPERS does not undertake risk management planning.

In conclusion, the first two methodologies provide a solid foundation (in the form of a PMBOK risk management plan or PRINCE2 Risk Management Strategy) for subsequent risk management processes. JASPERS does not undertake risk management planning because it is a dedicated field-specific methodology (rail infrastructure), so the adaptive needs of the methodology for the project are significantly smaller..

### 3.2 Identifying risks

PMBOK defines identifying risks as „the process of determining which risks may affect the project and documenting their characteristics" [11]. Methodology proposes a whole range of tools: documentation reviews, information gathering techniques (brainstorming, Delphi technique, interviewing, root cause analysis), checklist analysis, assumptions analysis, diagramming techniques (cause and effect diagrams, system of process flow charts, influence diagrams), SWOT analysis and expert judgement. As a result of the tools are the following products: list of identified risks, list of potential responses, root causes of risks, updated risk categories [11, 15, 16].

PRINCE2 recommends the so-called "Risk workshop" – a group session that is organized to identify hazards and opportunities. The risk identification techniques proposed by the methodology include: experience review, risk checklists, risk categorization lists, brainstorming and risk sharing structure diagram. The result of the activities should be the Risk Register [12, 18].

JASPERS proposes a ready-made list of risk factors that, based on the experience of the members of the association, have been aggregated and allocated to specific groups of risk factors.. These groups are: demand risks, design risks, administrative risks, land acquisition risks, procurement risks, performance risks, operational risks, regulatory risks, financial risks, management risks, political risks and other risks. It is important that the list remains open, so despite the extensive list, further proposals for risk factors [17].

In conclusion, the first two methodologies offer a wide range of tools for identifying risk factors. However, they do not offer any sample list. This is not the case for JASPERS, which offers a prepared list of factors. Unfortunately, beyond the list, there is no information on tools to identify potential factors that are not included in the enclosed list. It is obvious that the most rational approach to the topic would be to supplement the JASPERS list with the tools proposed by PMBOK or PRINCE2.

### 3.3 Qualitative risk analysis

PMBOK defines qualitative risk analysis as „the process of prioritizing risks for further analysis or action by assessing and combining their probability of occurrence and impact" [11]. Methodology at this stage proposes the following tools: risk probability and impact assessment, probability and impact matrix, risk data quality assessment, risk categorization, risk urgency assessment and expert judgement. The purpose of this stage is to update the risk register, which consists in complementing it with the following elements: relative ranking - list of project risks sorted by relevance, grouping risks by category, list of risks requiring in the near term, list of risks requiring additional analysis and response, list of observed risks of low significance and trends characterizing the results of qualitative risk analysis [11, 19, 20].

PRINCE2 recommends considering: the probability of threats and opportunities in terms of how likely they are to occur; the impact of each threat and each opportunity with respect to the objectives of the project; the proximity of these threats and opportunities; scope of changes in the impact of threats and opportunities during the life of the project. Methodology proposes two techniques to achieve the stated objectives: the Pareto principle and the probability / impact matrix. The result of the work is a completed and updated Risk Register [12, 21].

During qualitative analysis, JASPERS requires that the following characteristics be described for each risk factor: cause, effect, risk manager, project phase at risk, probability, impact strength, and risk level. The methodology includes tables that provide ready-to-use compartments to qualify individual risk factors. The effect of the operation is to locate the individual factors on the matrix level of risk [17].

The first two methodologies are very broadly concerned with the qualitative analysis of risk factors. They propose a whole range of tools to help in making the right analysis. The last method is limited to the most common tool – the risk matrix.

### 3.4 Quantitative risk analysis

PMBOK defines quantitative risk analysis as „the process of numerically analyzing the effect of identified risks on overall project objectives" [11]. Methodology recommends the implementation of tools from one of two groups of techniques: Data gathering and representation techniques (surveys, probability distributions, expert opinions), quantitative risk analysis and modeling techniques (sensitivity analysis, expected monetary value analysis, decision tree analysis, modeling and simulation) and expert judgement. The result of this stage of risk management process is the next update of the risk register. The register is supplemented with a probabilistic analysis of the project, the probability of achieving cost and time goals, a list of measured quantitative risks classified by relevance, and trends characterizing quantitative risk analysis results [12, 22].

PRINCE2, as in qualitative analysis, offers only two tools for quantitative analysis. These are the probability trees and the monetary value worked out. In addition, risk models, such as Monte Carlo analysis, are recommended during the evaluation. Similarly to qualitative analysis, quantitative analysis is also used to supplement and update the Risks Register [12].

JASPERS recommends quantitative analysis of risk factors based on probability distributions. These can be compiled according to Annex "R" to the Blue Book, which provides a list of historical data that will be needed to properly analyze. After obtaining probability distributions, the methodology recommends using the Monte Carlo method or other simulation method [17].

As with the qualitative analysis of risk factors, the first two methodologies propose a similar, very broad approach in quantitative risk analysis. Moreover, the choice of parameters and characteristics to be taken into account in the analysis leaves the decisions of the employees involved in project management. JASPERS, however, makes it very clear what inputs are needed to perform a meaningful quantitative evaluation using the selected simulation method.

### 3.5 Planning risk responses

PMBOK defines qualitative risk analysis as „the process of developing options and actions to enhance opportunities and to reduce threats to project objectives" [11]. Methodology proposes four tools that can be applied at this stage: strategies for threats (avoidance, transfer, mitigation), strategies for opportunities (use, access, reinforcement), threats and opportunities strategies (acceptance), contingent response strategies and expert judgement. Thanks to the strategies used, an updated risk register and a project management plan are developed as well as contractual risk agreements [11].

PRINCE2 presents potential responses to risk factors in tabular form (for threats: avoidance, reduction, reserve plan, transfer, acceptance; for threats or opportunities: sharing; for opportunities: use, reinforcement, rejection. In addition, it requires that the owner and the contractor assign risk responses to the individual factors [12].

JASPERS also includes the proposed responses to project risk. Unlike the previous two methodologies, the Blue Book sets out only countermeasures for threats (prevention, limitation, transfer and tolerance). There is no information about responding to opportunities [17].

In conclusion, the first two methodologies describe the topic of responding to risk factors very broadly. They anticipate reactions both to threats and to opportunities. JASPERS methodology is limited to responding to threats only.

### 3.6 Monitoring and controlling risks

PMBOK defines monitoring and controlling risks as „the process of implementing risk response plans, tracking identified risks, monitoring residual risks, identifying new risks, and evaluating risk process effectiveness throughout the project" [11]. Methodology proposes the following tools and techniques for monitoring and controlling risk factors: risk reassessment, risk audits, variance and trend analysis, technical performance measurement, reserve analysis and status meetings. Thanks to used tools and techniques there are developed: risk register updates, organizational process assets updates, change requests, project management plan updates and project document updates [11, 22].

PRINCE2 recommends, first and foremost, extensive communication and reliable reporting, based on which decisions are made to correct previous assumptions. The reports listed in the methodology are: Control Point Reports, Periodic Reports, Stage Final Reports, Project Final Reports, and Experience Reports [12].

JASPERS does not impose or recommend any standards for the monitoring and control of risk factors. It is limited to suggesting the development of brief descriptions of monitoring procedures and templates of protocols [17].

The broader issue of monitoring and control of risk factors has been developed in the PMBOK methodology. PRINCE2 is limited to the list of documents (including sample examples) to be prepared during the implementation of the project. The least recommendations on the topic include the JASPERS method, which is limited to just a few sentences suggesting the development of appropriate procedures.

## 4 Tabular comparison of methodologies

In order to increase the transparency of the analysis from third point, the authors decided in a synthetic way to present the results of their work in tabular form. With this approach, the decision-making process associated with the choice of methodology for the various stages of risk management is facilitated.

**Table 1.** Tabular comparison of methodologies for stage "Planning risk management" (where: „M" – methodology; "PM" – PMBOK; "P2" – PRINCE2; "J" – JASPERS) [own elaboration]

| M | Tools and techniques | Outputs | Advantages | Disadvantages |
|---|---|---|---|---|
| PM | Planning meetings and analysis. | Risk management plan. | The plan provides a solid foundation for further risk management and allows for the adaptation of the methodology to the project. | Time-consuming. |
| P2 | Analysis of documentation. | Risk Management | The Strategy provides a solid foundation for further risk management and allows for the | Time-consuming. |

| | | | | |
|---|---|---|---|---|
| | | Strategy. | adaptation of the methodology to the project. | |
| J | None. | None. | Saving time. | Lack of plan or strategy limits the field of methodology application only to railway projects. |

**Table 2.** Tabular comparison of methodologies for stage "Identifying risks" (where: „M" – methodology; "PM" – PMBOK; "P2" – PRINCE2; "J" – JASPERS) [own elaboration]

| M | Tools and techniques | Outputs | Advantages | Disadvantages |
|---|---|---|---|---|
| PM | Documentation reviews, information gathering techniques, checklist analysis, assumptions analysis, diagramming techniques, SWOT analysis, expert judgement. | list of identified risks, list of potential responses, root causes of risks, updated risk categories. | The described tools allow for a very thorough development of the problem. | All factors should be identified from the beginning. There is no initial list with risk factors. |
| P2 | Risk workshop. | Risk register. | The described tools allow for a very thorough development of the problem | All factors should be identified from the beginning. There is no initial list with risk factors. |
| J | None. | A ready-made list of risk factors is provided. | Saving time. | It may be necessary to use tools from another methodology to supplement the list. |

**Table 3.** Tabular comparison of methodologies for stage "Qualitative risk analysis" (where: „M" – methodology; "PM" – PMBOK; "P2" – PRINCE2; "J" – JASPERS) [own elaboration]

| M | Tools and techniques | Outputs | Advantages | Disadvantages |
|---|---|---|---|---|
| PM | Risk probability and impact assessment, probability and impact matrix, risk data quality assessment, risk categorization, risk urgency assessment and expert judgement. | Risk register updates. | The described tools allow for a very thorough development of the problem. | Time-consuming. |
| P2 | The Pareto principle and the probability / impact matrix. | Completed and updated Risk Register. | The described tools allow for a very thorough development of the problem. | Time-consuming. |
| J | Tables that provide ready-to-use compartments to qualify individual risk factors. | Feedback on the potential impact of each factor. | Saving time. | It may be necessary to use tools from another methodology for further analysis. |

**Table 4.** Tabular comparison of methodologies for stage "Quantitative risk analysis" (where: „M" – methodology; "PM" – PMBOK; "P2" – PRINCE2; "J" – JASPERS) [own elaboration]

| M | Tools and techniques | Outputs | Advantages | Disadvantages |
|---|---|---|---|---|
| PM | Data gathering and representation techniques, quantitative risk analysis and modeling techniques, expert judgement. | Risk register updates. | The described tools allow for a very thorough development of the problem. | Time-consuming. |
| P2 | Probability trees and the monetary value worked out. | Completed and updated Risk Register. | The described tools allow for a very thorough development of the problem. | Time-consuming. |
| J | Quantitative analysis of risk factors based on probability distributions. | Feedback on the potential impact of each factor. | Ready-made materials make it easy to get input to develop probability distributions. | Requires historical data. It may be necessary to use tools from another methodology for further analysis. |

**Table 5.** Tabular comparison of methodologies for stage "Planning risk responses" (where: „M" – methodology; "PM" – PMBOK; "P2" – PRINCE2; "J" – JASPERS) [own elaboration]

| M | Tools and techniques | Outputs | Advantages | Disadvantages |
|---|---|---|---|---|
| PM | Strategies for each types of risk factors. | Updated risk register, updated project management plan, contractual risk agreements. | The recommendations concern both opportunities and threats. | Time-consuming. |
| P2 | Tabular list of potential reactions for risk factors. | Assignment to the each factors: reaction, owner and contractor reaction. | The recommendations concern both opportunities and threats. | Time-consuming. |
| J | Tabular list of potential reactions only for risk threats. | Assignment reactions to the each factors. | Saving time. | Lack of reaction to opportunities. |

**Table 6.** Tabular comparison of methodologies for stage "Monitoring and controlling risks" (where: „M" – methodology; "PM" – PMBOK; "P2" – PRINCE2; "J" – JASPERS) [own elaboration]

| M | Tools and techniques | Outputs | Advantages | Disadvantages |
|---|---|---|---|---|
| PM | Risk reassessment, risk audits, variance and trend analysis, technical performance measurement, reserve analysis and status meetings. | risk register updates, organizational process assets updates, change requests, project management plan updates and project document updates. | The described tools allow for a very thorough development of the problem. | Time-consuming. |
| P2 | Extensive communication and reliable reporting. | Control Point Reports, Periodic Reports, Stage Final Reports, Project Final Reports, and Experience Reports. | The described tools allow for a very thorough development of the problem. | Time-consuming. |
| J | None. Only general recommendations are included. | Descriptions of monitoring procedures and templates of protocols used. | Saving time. | It is necessary to use tools from another methodology for further analysis. |

## 5 Summary and conclusions

Railway investments are large in scope. This causes high investment costs. The high costs generate the need for reliable management of risk factors. There are many methodologies that allow the management of rail projects and, therefore, risk factors. Some methodologies are universal, and their approach to project management is very broad. Thanks to that, they have very great adaptability to various projects. Unfortunately, this involves additional time, money and human resources. Other methodologies are dedicated to specific issues. Their scope is much narrower, but they often have ready-made solutions to streamline the management process.

In the analyzed case of railway constructions, it is best to base on the methodology of the JASPERS Blue Book and complement it with tools offered by the other two methodologies. Proposed solution, will allow for significant time savings, and what is associated with it, also costs savings. In addition, the areas described in the JASPERS Blue Book insufficiently can be supplemented by tools from PMBOK or PRINCE2. This creates a complete risk management project for the railway investments while minimizing the time, cost and human effort required.

## References

1. H. Anysz, A. Zbiciak, Autobusy: Technika, Eksploatacja, Systemy Transportowe, **3**, 963 (2013)
2. E. Plebankiewicz, K. Zima, D. Wieczorek, Archives of Civil Engineering, **62(2)**, 149, (2016).
3. J. Korytarova, T. Hanak, R. Kozik, E. Radziszewska-Zielina E., Elsevier, *Procedia Engineering*, **123**, 276 (2015)
4. B. Bogdaniuk, K. Towpik, *Budowa, modernizacja i naprawy dróg kolejowych* (2010)
5. N. Ibadov , Archives of Civil Engineering, **61(2)**, 23 (2015)
6. N. Ibadov , Technical Transactions. Civil Engineering, **1-B(2)**, 139 (2010)
7. N. Ibadov, J. Kulejewski, Technical Transactions. Civil Engineering, **1-B(2)**, 231 (2010)
8. M. Krzemiński, M. ACTA PHYSICA POLONICA A, **130(6)**, 1439 (2016)
9. M. Kaftanowicz, M. Krzemiński, Procedia Engineering, XXIV R-S-P Seminar, Theoretical Foundation Of Civil Engineering (24RSP) (TFOCE 2015), **111**, 364 (2015)
10. 10. K. Kaczorek, Technical Transactions. Civil Engineering (to be published)
11. *A Guide to the Project Management Body of Knowledge (PMBOK® Guide)* (PMI, 2004)
12. *Managing Successful Projects with PRINCE2®* (OGC, 2010)
13. M. Górski, A. Dziadosz, D. Skorupka, Zeszyty Naukowe Wyższej Szkoły Oficerskiej Wojsk Lądowych, **157(3)**, 41 (2010)
14. J. Kulejewski, Przegląd Budowlany, **80(6)**, 43 (2009)
15. M. Górski, D. Skorupka, Zeszyty Naukowe Wyższej Szkoły Oficerskiej Wojsk Lądowych, **162(4)**, 307 (2011)
16. N. Ibadov, Acta Physica Polonica, **130(1)**, 107 (2016)
17. *Blue Book. New edition 2014-2020. Railway sector. Railway infrastructure.* (JASPERS, 2015)
18. N. Ibadov, ICCESEN. 2nd International conference on computational and experimental science and engineering. Abstract book, 239 (2015)
19. M. Krzemiński, IV Konferencja Naukowa ARCHBUD (2011)
20. M. Krzemiński, W. Karczmarczyk, Oficyna Wydawnicza Politechniki Warszawskiej, Theoretical foundations of civil engineering: Polish-Ukraïnian transactions, 413 (2013)
21. M. Krzemiński, P. Nowak, Oficyna Wydawnicza Politechniki Warszawskiej, .XXI Russian-Slovak-Polish Seminar "Theoretical foundation of civil engineering". Proceedings, 545 (2012).
22. J. Kulejewski, Przegląd Budowlany, **80(4)**, 47 (2009)

# Analysis of the causes and recommendations on elimination of biological damage of structures during the repair and reconstruction of the State Biological Museum in Moscow.

*Victor* Kamskov[1,*]

[1] Moscow state university of civil engineering, Yaroslavskoye shosse, 26, Moscow, Russia, 129337

**Abstract.** The article presents the results of mycological research on buildings of the State Biological Museum located in Moscow. The problems have been considered as for a complex of buildings of the State Biological Museum built in the late nineteenth century which, to the present time, has been operated almost in its original form. Over time, the building maintenance conditions have worsened, in particular because of construction of high-rise buildings in the immediate vicinity of the museum, as well as construction of a greenhouse above the underground passage tunnel between buildings 1 and 2. Over the years, the temperature gradients, high humidity, wear and damage of wall waterproofing and foundations have caused leaks in the underpass tunnel and the biological corrosion of stone, wood and metal structures in indoor exhibition halls. In this connection, part of the survey was to determine the types and size of biological lesions in structures, determination of the causes of biological damage, and the development of measures to eliminate the mycological problems during repair and reconstruction works in the museum.

## 1 Introduction.

The main exhibition halls of the museum located in the building (building 1) designed by the architect A.E.Erichson in 1898. Until the present time, basements and premises of the 1st and 2nd floors have been used. In the building (Building 2), designed by the architect B.V.Freudenberg, museum administration, research laboratories, restoration workshops, scientific library, shopping service are located. To connect the buildings 1 and 2 the architect A.E.Erichson designed and built an underground passage shown in Fig. 1. Later on, above the tunnel a greenhouse was built for year-round operation with electric lighting, hot and cold water supply, sewerage. The tunnel is equipped with manholes, which are destined for lighting the tunnel interior space with natural light. Over the years, the temperature gradients, high humidity, worn waterproofing caused leakage through the tunnel body. Mould colonies appeared (Figure 2), which caused the biological corrosion of stone materials in the tunnel. (1,2,8)

* Corresponding author: kamskov@mail.bg

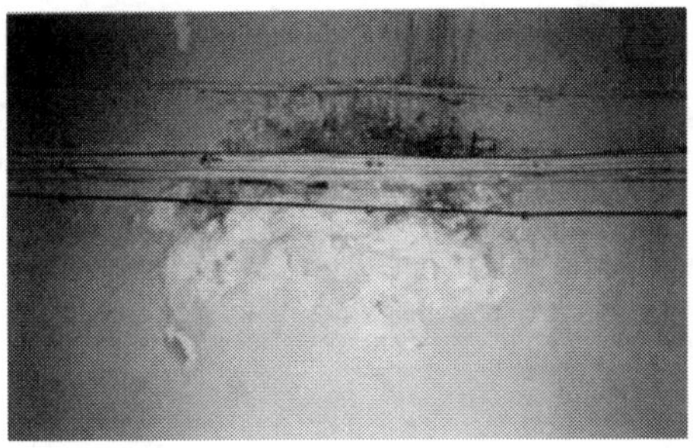

**Fig. 1.** Mould colony in the underground passage.

The biological corrosion mostly damaged internal and external walls of buildings 1 and 2, as well as wooden structures of attic rooms, and metal structural components of cellars. The aim of research was determining type and size of biological damage and making recommendations for its elimination in anticipation of the reconstruction works in the museum. (7)

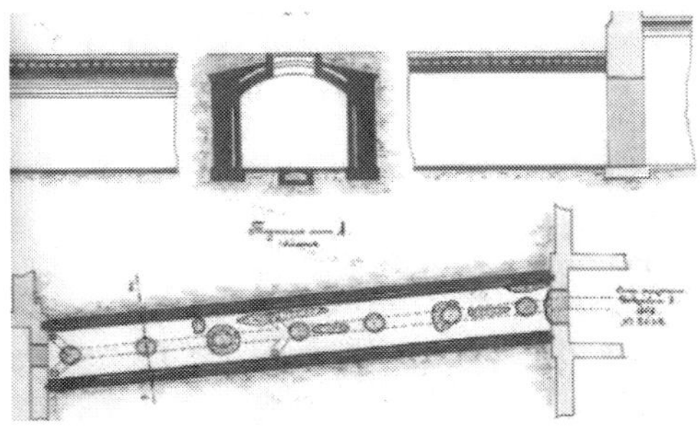

**Fig. 2.** Location of mould colonies in the underground passage between the buildings 1 and 2.

## 2 Problems.

The aim of this research was to determine the real technical condition of the buildings of the State Biological Museum in terms of biological and chemical corrosion and to define "treatment" methods of building structures for restoration and reconstruction works. In this connection, for research the most affected buildings 1 (1898) and 2 (1892-93) and the underpass between the buildings were selected. (3,4,5,6).

Building 1 has a basement floor and two underground floors and an attic room located in the space. A general view of the structure 1 and a greenhouse on an underground passage between buildings 1 and 2 is presented in Figure 3.

a) 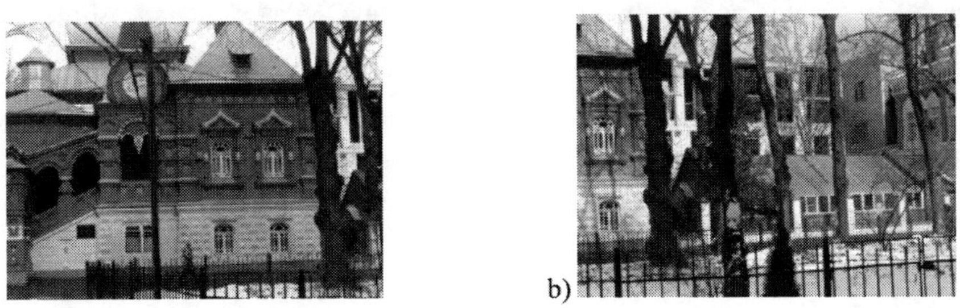 b)

**Fig. 3.** General view of the building 1 and the greenhouse above the underground passage between buildings 1 and 2 { a) and b) }.

A mycological research showed that moulds destroy plasters and concrete surface, on the one hand, and, on the other, adversely affect the health of the personnel permanently working in these areas. Various kind of corrosive media (natural and industrial) damage 15% to 75% of building structures of buildings. In addition to inorganic corrosive environments, significant devastating impact on building materials and construction of buildings and structures is exercised by animals and plant macro- and microorganisms and their derivatives, i.e. by biogenic aggressive media.

The bio corrosion can be divided in:

- impact of bacteria;
- impact of moulds;
- impact of moss, lichen, and lower algae.

Chemical destruction of stone is the result of the action of acid products on it, produced during the growth of microorganisms of inorganic and organic acids and carbon dioxide. The most severe destruction of stone materials is produced by nitric and sulfuric acid, microbial oxidation products of gases, present in the atmosphere.

Nitrate bacteria oxidizing ammonia to nitrate, convert limestone chemical chalk into soluble nitrate. As a result, in the chemical process, limestone fully loses adhesion and gets converted to powder. This powder is blown by the wind and washed away by water. Silicate rocks are more resistant to the action of nitric acid. However, their destruction is greatly accelerated in the presence of nitrate bacteria. Not less destruction is produced by sulfuric acid from sulfur-oxidizing bacteria. For example, it reacts with calcium hydroxide $Ca(OH)_2$ formed during cement hardening. This gives rise to new compounds - gypsum $CaSO_4 \bullet 2H_2O$ - which occupies a volume 2-fold greater than the $Ca(OH)_2$ or calcium hydrosulfoaluminate $AL_2O_3{\cdot}3CaO{\cdot}3CaSO_4{\cdot}32\ H_2O$. The formation of such salts creates internal tensions in concrete and leads to cracking. In the presence of organic compounds on stone materials mainly fungi and heterotrophic bacteria develop. Most of the organic matter is decomposed by them into carbon dioxide and water.

Carbon dioxide plays a big role in the destruction of stone. This process is described by a simple chemical reaction. Calcium carbonate, which is part of limestone and concrete, is converted in presence of carbon dioxide and water to calcium bicarbonate: $CaCO_3 + CO_2 + H_2O \rightarrow Ca(HCO_3)_2$. Assuming that the solubility of $CaCO_3$ equals to 0.031 g/1dm$^3$ of water, solubility of the calcium bicarbonate $Ca(HCO_3)_2$ is 1.89 g/1 dm$^3$ of water and easily washed away with water from a stone. Similarly, although somewhat slower, silicate rocks get destructed. Feldspar and mica, a part of granite, also form water-soluble compounds under the action of carbon dioxide and water.

Organic substances present on the surface and in the structure of stone materials, may be not fully degraded by micro-organisms, so along with the carbon dioxide organic acids are formed. They penetrate the substrate and reduce the pH in the growth zone. However, in terms of biodegradation of stone, chelating ability of organic acids is more important than acidifying ability since the release of cations from rock minerals by forming water soluble complexes leads to more rapid disruption of the integrity of sandstone, limestone and concrete.

The mechanism of destruction of cement materials by fungi is the following. Fungi throughout their life secrete various mineral and organic acids (acetic, citric, lactic, butyric, formic, malic, etc.) which interact with basic (alkaline) cement stone compounds and destroy it, with turning hydrosilicates and hydroaluminates to non-astringent salts. Cement stone degenerates and loses its construction and technical properties. Fungi, algae, lichens insert destroy stone materials not only chemically, by the action of metabolism products, but also mechanically.

The biomass growth of microorganisms penetrated into pores and micro cracks, contributes to their expansion. Intermittent wetting and drying lichens, accompanied by a significant change in the cell volume, results in cyclical pressure on walls of cracks and in fatigue failure of stone. The ability of the fungal mycelium to condense on its surface an increased amount of moisture increases the distraction effect of cyclic freezing and thawing of water in stone pores and cracks. There are three main groups of methods of protection of building structures from the destructive action of fungi: performance/preventive, design, and construction/technology methods.

The first group includes surface disinfection methods, regulation of heat and humidity environment mode in premises, and ventilation. Constructive measures include imparting to the structure surfaces the forms in which wetting of structures and accumulation of dust on the surface decreases. Construction/ technology activities include the use of materials resistant to fungi.

Fouling of structures with microscopic fungi, breeding of bacteria in porous construction material creates hygienic problems. Bacteria and fungal spores, substances secreted by bacteria and fungi can cause serious disease in humans, living or working in the affected areas, in particular, the development of allergic and respiratory diseases, as well as pneumonia and liver cirrhosis, especially in children. Some of substances secreted by microorganisms are dangerous toxins. In any case, it is believed that the reason for the seemingly inexplicable people's death, who opened tombs of Egyptian pharaohs, is the so-called yellow mould releasing hazardous substance aflatoxin.

In the buildings constructed in the first half of the 19th century and earlier, as a rule, there is no horizontal waterproofing. Also, when building-up this cultural layer waterproofing ceases to perform its function. This leads to dampen the walls due to capillary moisture suction. This reduces the insulating properties of the walls and their longevity because of destruction of freeze-thaw, salt corrosion and biocorrosion.

From construction physics it is known that an increase in humidity of outer brick walls only by 3% increases their heat conductivity by 50%. The temperature on the inner wall surface of premises with normal indoor air parameters, that is, at temperature of 18 °C and humidity of 55...60% - is normally 14...16°C. This temperature corresponds to the standard of thermal acceptability and does not cause the appearance of condensate (at the normal natural ventilation removing excess household moisture). If the wall is waterlogged, the temperature on its surface drops to +9 °C and it inevitably results in household condensate and moisture exhaled by people who are in the room. The condensate precipitated is absorbed into the wall, and its heat conductivity is even greater, and so the process is repeated.

Because of capillary suction of moisture at the masonry seams of foundations and walls the construction material is exposed to salt corrosion, which is manifested in the form of leaching or, conversely, salinization of these materials. At the capillary suction of groundwater followed by crystalhydrate addition of water to anhydrous salts of building materials formed such minerals as $CaSO_4 \cdot 2H_20$ (plaster), $MgCl_2 \cdot 6H_20$ (bischofite), $CaCl_2 \cdot 2H_20 \cdot MgSO_4 \cdot 2H_20$ (kieserite), $MgSO_4 \cdot 6H_20$ (hexahydrite), $Na_2CO_3 \cdot 10H_20$ (soda), $NaCl \cdot 2H_20$ ( hydrohalite), $Na_2SO_4 \cdot 10H_20$ (mirabilite) are formed.

Upon addition of water by anhydrous salts to form crystalhydrates the volume of the solid phase augments and a tensile stress appears in the material pore walls that reaches the values of 0.21 - 0.44 MPa. Apart of that, the pressure of crystalhydrates is a permanent load, and that reduces the durability of material. Thus, the strength of ceramic bricks comprising crystalhydrates of the above salts is reduced by 30 - 40% and the frost resistance of salinized cement concrete is reduced by 2.5 times. Following methods are used for desalting the structures:

- electroosmosis;

- use of sanitizing plasters;
- conversion of salts comprised in materials into insoluble compounds.

Mycological studies of structural elements of the building, for example, of walls, beams, expansion joints, etc. give rise to the development of recommendations for combating molds.

## 3 Results

Visual inspection results as for biocorrosion are demonstrated in the table 1.

**Table 1** Visual inspection results as for biocorrosion

| Building cellar. | Biocorrosion behaviour | Place and area of lesion, m2 |
|---|---|---|
| Work rooms | White mould and green mould | Walls, 12 m2 |
| Work rooms | Green and black mould | Walls , 10 m2 |
| Cellar | Green and black mould | Walls, 20 m2 |

The figure 4 shows colonies of mould fungi in their location places.

a) 

b) 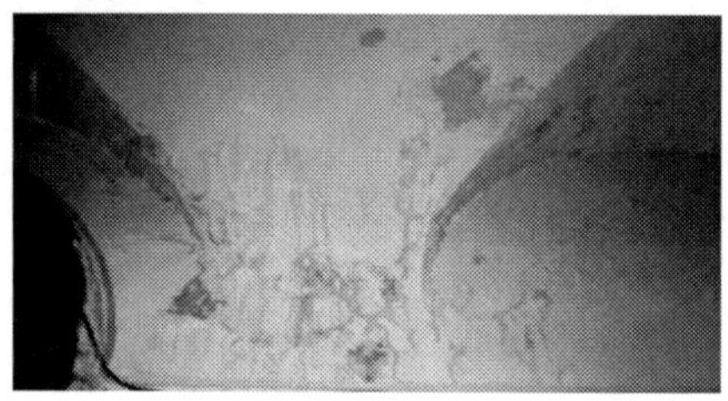

a) on the wall; b) in the vestibule ceiling

**Fig. 4.** Colonies of mould fungi in their location places of the building 2.

It was noted earlier that mould fungi show up on the surface of building materials at following conditions: humidity of wall materials approx. 10%; ambient temperature approx. + 15…18$^0$C.

The development of fungi is facilitated by poor ventilation of premises. These factors shown are observed in the investigated areas.

Analysis of the places of location of fungal lesions (Fig. 4) showed that in the centers of biocorrosion paint and plaster coatings are mostly destroyed. Additional danger of colonies consists in that they are a background for germing of the fungus Serpula Lakrimas, which damages books and other materials.

Microscopic analysis of samples taken in different rooms has shown that in this building there are the following types of fungi: white mould (Penicilium), green mould (Aspergilium), black mould (Cladosporium). Mould colonies develop in the specified order. Note that the plastered surfaces in the building are painted with oil paint, which with time forms shrinkage cracks. Spores colonize the cracks, leading to the formation of white mould colonies, which detach oil paint. Afterwards green mould colonies arise that create a breeding ground for black mould.

Moulds develop at substrate humidity 8...10% and at the temperature above 10C. The humidity material is provided by the absence of cut-off waterproofing and by basement premises conditions as well as by the presence of structural defects, at the inside temperature over 18 °C. Especially intense development of moulds takes place in rooms with high humidity and in unventilated areas. Fig. 5 shows micrographs of plaster solution samples with various types of moulds, made by means of the electronic microscope «Camebax» (Japan), 2000 x zoom.

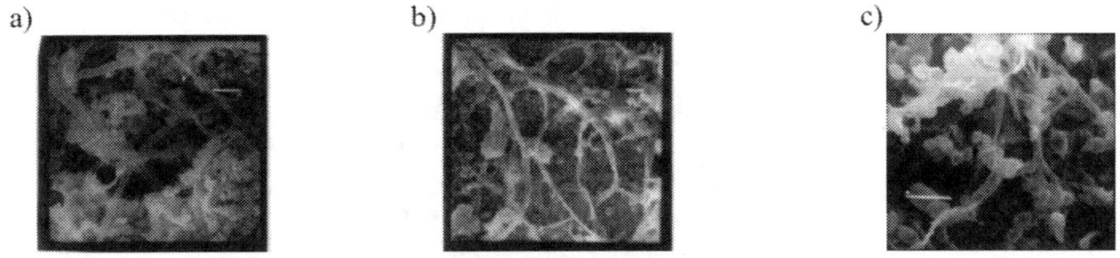


a) white mould; b) green mould, c) black mould

**Fig. 5**. Dissolved plaster samples with different types of mould.

The data (Fig. 5) prove the development activity of fungi lesions (presence of mycelium, punks). In this case, it is necessary to note the increase in density of their biomass, which in itself is a destructive factor. Note that the biocorrosion mainly affects sand-cement compositions (plaster coverings) because a breeding ground for them is calcium hydroxide $Ca(OH)_2$. In addition, the plaster samples taken demonstrated the lowest durability of soluted cement-sand plaster.

Next, we carried out the study on wooden structures of the building tested using an electron probe microanalyzer. The electron-probe microanalyzer provides X-ray radiation of the sample researched, excited with an electron beam. For the identification of elements contained in the sample and determination of their concentration, the wavelength and intensity of the respective X-ray spectral lines were measured for a certain time, for example, for 100 seconds. For determining the material composition of the material it is necessary to register in the X-ray spectrum of the analyzed line elements in the 1 - A 12 range and compare them with X-ray lines emitted by reference samples of a known composition. Characteristic X-ray emission lines are formed, the lengths of which uniquely characterize a chemical element whose atoms emit these lines.

Putrefaction or decomposition of wood is accompanied by influence of "cytase" enzymes secreted by microorganisms or wood destroying fungi. Cytase as a catalyst, converts a water-insoluble cellulose in insoluble substance - glucose, in the following reaction:

$$(C_6H_{10}O_5)n + H_2O + cytase = (C_6H_{12}O_6)n \quad (1)$$

Glucose is oxydized by ambient oxygene, with forming carbon dioxide and water. The reaction:

$$(C_6H_{12}O_6)n + 6O_2 = 6CO_2 + 6H_2O \quad (2)$$

The chemical composition is a determinant of defeat of wood by fungi. At the same time, the sum of oxides fairly accurately characterizes the type of lesion, and iron oxide $Fe_2O_3$ comprised in the chemical composition indicates the presence of a strong oxidant, which is cytase. Mikrofotographing a test sample confirms the chemical analysis.

Threadlike spores with oval heads characterize the "real wood fungus" growing mainly on dead wood, which is the most dangerous of all destroying fungi. The filamentous fungal spores of sap-wood rot found in the wood walls of the building tested are shown in Fig. 6.

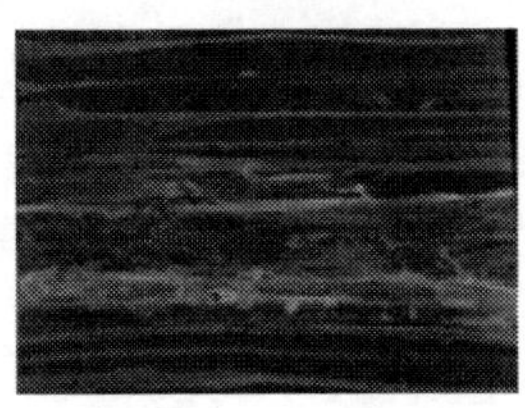

**Fig. 6.** Threadlike fungi spores of sap-wood rot, found in wooden walls of the building tested.

Another timber destroyer - "mine fungus" has spores in the form of filamentary chains, pierced with smooth fibers, binding all in a single unit (Figure 7).

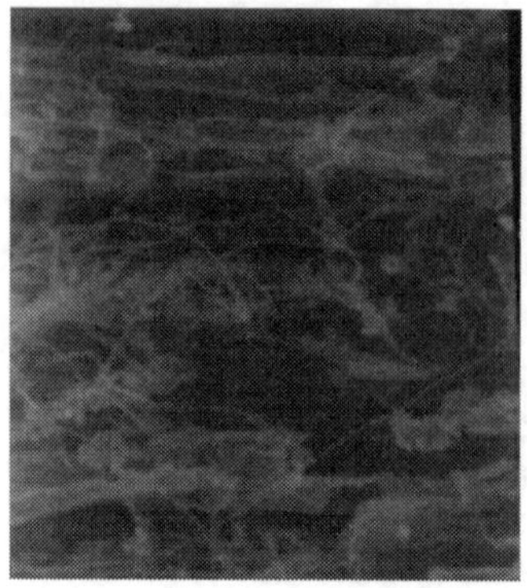

**Fig. 7.** Thread fungi spores found in wooden walls of the building.

Wood failure occurs as a result of mineralization of pore space with gypsum. It is found that the presence of sulfur oxide is 4.51% and of calcium oxide 6.44%. These are exactly the components of the formula of gypsum $CaSO_4$ x $2H_2O$. The concentration of sodium oxide is 0.722% and chlorine oxide is 0.773%, and that indicates the presence of NaCl crystals in the wood pores. The growth of salt crystals is accompanied by an increase in their volume and the appearance of internal stresses that reduce the strength of wood structural elements.

The study of metal structures was conducted using X-ray diffraction method. Figure 8 is an X-rayogram of rust, efflorescence scale and the most damaged portions of steel structures that are complex salts containing NaCl, $Na_2CO_3$, $MgCO_3$, and the silica gel $Si(OH)_4$. These include such as $Fe_2O_3xCO_3$; $Fe_2Si_2O_5\,(OH)_4\,2H_2O$; $Na_3Mg\,(CO_3)_2\,Cl$.

Chlorides of sodium and magnesium metals in combination with the moisture are steel corrosion process activators. They are adsorbed on the surface of structures, displacing oxygen and causing pitting corrosion of steel.

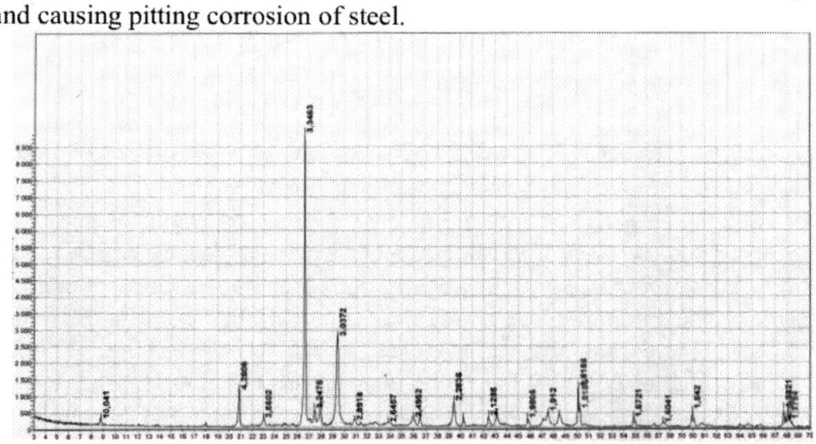

**Fig. 8.** X-rayogram of the microstructure of the most damaged parts of steel structures.

Sodium carbonate is the product of artificial origin and, accumulating in the pore space, it increases its volume, creating internal stresses.

α FeO (OH) and γFeO (OH) - alpha and gamma hydrohematite – the substances having high absorption properties and increasing the initial volume. Thus, neoplasms Fe $(HCO_3)$, α

FeO (OH) and γFeO(OH) provoke internal stresses and destroy steel bearing components of the building.

# 4 Conclusions and recommendations

While developing measures to eliminate mould, you must first remove conditions favorable for their development, i.e. surface humidity of wall materials exceeding 10%. Otherwise, the fungicidal treatment of wall materials will be effective for not longer than 2 years.

For fungicidal treatment of premises the approved preparation "Cartocide-Compound" developed by the "NIOPIK" (SRI of dyes and semi-finished products) is recommended. The preparation comprises the following active ingredients: *Cartocide, yuglog, permethrin and glisol*. Furthermore, all the brand preparations contain acetate buffer and acetic acid to create a certain pH level and surfactants as well.

All *"Cartocide - Compound*" brands are to be shaken before use, and then mixed with water in desired ratios, and applied to the object to be protected by any known method (brushing, spraying, impregnating, dipping, soaking, etc.).

All preparation brands due to the surfactants comprised have high wettability, which increases the total penetrating (impregnating) action. The most popular brands are "Cartocide-compound" B-1 and B-1k, used for processing internal wooden, concrete and brick surfaces from staining and destructive fungal and bacterial diseases.

Brand «B-1» - 5% aqueous solubilizate (solution) of C*artocid* and *Yuglog*, prepared of 98% C*artotsid* dry powder or of its 50% aqueous solution with addition of surfactant and acetate buffer.

Brand «B-1 k» - 30% aqueous solubilizate of *Cartotsid*e and *Yuglog*, prepared of 98% Cartotside dry powder or of its 50% aqueous solution with addition of surfactant and acetate buffer.

# References

1. V.P. Kamskov, V.V. Kozlov, *Hydroisolation materials*. Nautchnoye izdanie. Moscow, (2014)
2. V.P. Kamskov, V.S. Semionov, I.V. Balandina, A.A. Pakhratdinov, Investigation of Reasons for Presence of Gaseous Products in Concrete of Monolith Ferroconcrete Structures of a School Building in Moscow. Nautchnoye obozrenie, **19**, 77-81 (2015)
3. V.P. Kamskov, R.F. Serova, E.A. Stasilovich, Investigation of Reasons of Splitting-off of Concrete from Ferroconcrete Bearing Structures of a Frame Building. Vestnik sovremennoi Nauki, **12-1 (12)**, 39-43 (2015)
4. V.P. Kamskov, I.V. Balandina, D.Yu. Zemlyanushnov, Research of Reasons of Destruction of Cement/Sand Screed for the Floor and Polymer Coating. Methods of their Repair. *Sovremennye tendentsii razvitiya nauki i tekhnologiy*, **9-3**, 95-99 (2015)
5. A. Grishina, E.V. Korolev, Chemical Composition of Biocide Silicate-based Modifier. Vestnik MGSU, **11**, 58-67 (2016)
6. A. Grishina, E.V. Korolev, AIP Conference Proceedings 13. Ser. "Prospects of Fundamental Sciences Development, PFSD-2016: Proceedings of the XIII International Conference of Students and Young Scientists, 20-26 (2016)
7. A. Grishina, E.V. Korolev, Strength of Gypsum Stone comprising a Biocide Modifier// Evraziyskiy coyuz uchyonykh. **30-2,** 21-24 (2016)

# To the question on the factors of formation of the sands density and their potential accounting by cone penetration test

*Pavel* Kashperyuk[1,*], *Kristina* Nikitina[1], and *Aleksandr* Yulin[2]

[1]Moscow state university of civil engineering, Yaroslavskoye shosse, 26, Moscow, Russia, 129337
[2]Chief engineer of LLC NPF "SIVS", 117393, 45A, street of Architect Vlasova, Moscow, Russia

**Abstract.** Undisturbed sampling of sands from boreholes is difficult. It is important to find methods for determining the sands density in-situ, in particular, by cone penetration test. According to the results of cone penetration test can be determined only the categories of relative density of sands depending on their size distribution and probing depth. Based on an extensive literary material and long-term studies of the authors were considered a number of factors, which form the natural density of sands with different particle size distribution and genesis. For improving the accuracy, it is necessary to consider not only the particle size distribution, but also morphological features of grains, primarily depending on the genesis. The most versatile and effective indicator of the morphological features of grains is generalized morphological index λ (by A.D. Potapov). Taking into account that the empirical formula for quantitative assessment of the sands density by cone resistance (by L.G. Mariupolskiy), in our opinion, applicable only for average-size sands without considering of their uniformity, genesis and generalized morphological index, content of silt and clay fraction, water saturation and presence of cohesion-hardening, we offer new correlations between dry soils density and cone resistance for different genetic types of sand.

For many engineering surveys it is a problem to determine the value of sands density by practical consideration. This problem becomes unsolvable using only direct methods of borehole sampling with increasing depth and changes in water saturation of sands.

Current laboratory methods for determining the density of sandy soils due to their complexity and often - the impossibility of undisturbed sampling, with the sufficient reliability do not allow to estimate their main physical parameters, in particular, values of densities.

An analysis of archival materials of engineering surveys in Moscow has shown that more than 90% of laboratory determinations of physical properties of sands contains no direct measurements of density, limited the definition of particle size distribution, water

* Corresponding author: npf-sivs@yandex.ru

content, solid particles density ($\rho_s$), maximum and minimum dry density ($\rho_{dmax}$ и $\rho_{dmin}$) and friction angle.

In such cases, as a rule, the density of sand can be obtained by reverse calculation by setting the value of the void ratio (e) after the provisional qualitative assessment the studied density of sand and calculating its degree of water-saturation. The qualitative assessment (category) of the relative density of sand associating with the cone resistance ($q_s$) can be carried out on the table I.1, SP 47.13330.2012 [1].

However, using the specified table, to set the value of the void ratio for the sample of sand with the value $q_s$, for example (for fine sand of medium density) equal to 8 MPa, with the change of its value $q_s$ in a range from 4 MPa to 12 MPa, and void ratio from 0.75 to 0.60, the direct interpolation method is not possible, as for the ranges of variations the above values of e and $q_s$, there is no direct correlation, or at least its presence in each case requires proof. If the medium density sand with interpolation similarity linear dependence between e and $q_s$ can be assumed, for loose and dense sand such a device is impossible because of lack in the table and in practice (in-situ) determinations of void ratio (e) extreme (max and min) values of these indicators.

According to numerous publications, current research of properties of sandy soils shows that in addition to the particle size distribution of a significant impact on the density have age, genesis, mineral composition, natural water content, depth, degree of rounding of the particles, the surface shape of the particles, the presence of cover ("shirts"), the roughness of the grains, and the spatial arrangement, etc. [2, 3, 4, 5, 6, and 7]. It is known that in previously examined table I.1, SP 47.13330.2012 [1], none of the above factors affecting the sands density, except the particle size distribution and depth of probing, is not considered.

During penetration testing some researchers have noted a number of factors influence the value of cone resistance ($q_s$) and the sleeve friction ($f_s$). So, Yu.G. Trofimenkov and L.N. Vorobkov [8] noted the impact of the roughness of grains surface (R), the uniformity coefficient ($C_u$) and even the presence of small amount of dust particles ($\Sigma d < 0.05$ mm), which can cause substantial deviations of cone resistance from the mean values, listed in the table I.1, SP 47.13330.2012 [1].

Research of sands by V.I. Kashirskiy [9] made it possible to establish a connection between the varieties of sandy soils and their performance of cone resistance and friction ratio ($R_f$). Thus, for each studied variety of sands (from gravel to silty) change in the value of cone resistance ($q_s$) from 0.5 MPa to 40.0 MPa and friction ratio ($R_f$) from 0.3% to 4.0% he explains by changing of their density and quantitative content in the sands of silty clay fraction. Indeed, porosity decreases with increasing of grain-size and density, what leads to increasing of cone resistance. With decreasing the grain-size of sand and increasing the density increases the number of contacts of the sand grains with a friction-coupling surface, leading to an increase in friction ratio ($R_f$).

P.L. Ivanov and L.V. Schultz [4] studied the influence of the morphology of the sand grains on the value of the min and max density based on an evaluation of roundness ($\alpha$) and sphericity ($\beta$) of grains. They found (Fig.1) that with an increase in the treatment of the grains surface extreme values of the void ratio is reduced, so sands fit more tightly, which complicates the penetration of cone into the soil and leads to an increasing values of $q_s$ and $R_f$.

The influence of the morphological characteristics of the sands on their properties were particularly studied by I.V. Dudler and A.D. Potapov [3]. A.D. Potapov suggested using the generalized morphological index ($\lambda$), which takes into account not only the roundness and sphericity of grains, but also the roughness of their surfaces. He identified six categories of the treatment of sands grains, and the study of sands of various origins allowed establishing its influence on morphological characteristics of the sands. According to A.D. Potapov [5,

10], the treatment of grains increases in the following sequence: volcanic → aeolian → fluvioglacial → alluvial → limnoglacial → marine.

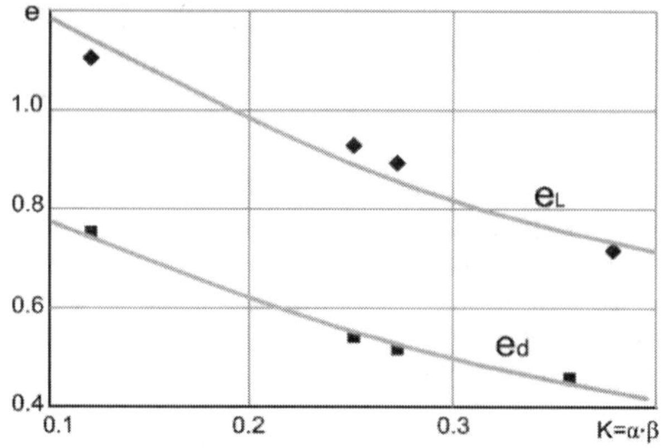


**Fig.1.** Influence of the particles (grains) shape on values of void ratio (according to P.L. Ivanov, L.V. Shulc, [4]). K - form coefficient (generalized index of roundness and sphericity of grains).

Our study (according to the materials of long-term surveys LLC NPF "SIVS", LLC "Investroy", LLC "IGC" in Moscow and the near suburbs) of the influence of grains morphology in alluvial sands (fine and medium) on their density showed that with increasing of roundness, sphericity, and the generalized morphological index, increases the density and decreases porosity (Fig. 2). At the same time, with angular and very non-spherical grains the average value of the void ratio in a very loose state is 0.95, and in a very dense - 0.55. For fine sands with poorly rounded, a little spherical grains the void ratio decreasing to 0.75 and 0.45 accordingly.

Similar dependences are for medium sands with the changing of their grains shape from slightly angular, anisometric to poorly rounded, a little spherical. Influence of roundness and sphericity of sands grains taking into account changes in roughness of their surface from the rough to very weakly rough, on the range of variation of the void ratio in a very loose and dense state was estimated by the generalized morphological index (λ) and is shown in Figure 2.

Analyzing the experimental results, it can be noted that the presence of untreated grains of sand (II category on λ) increase the range of the void ratio variation from $e_L$ to $e_d$, equal to 0.4. With the increasing of treated of grains (to a category IV on λ) the range between $e_L$ and $e_d$ reduces to 0.3 for medium sand and to 0.35 for fine sand. Installed in GOST 25100-2011 [11] the boundary values of the void ratio (0.55 and 0.70 for a medium sand, and 0.60 and 0.75 for a fine sand), as mentioned above, does not take into account the complex of significant factors (grain-size uniformity, the content of fine-grained fractions, mineral composition, hardening of sand, etc.), including morphological features of grains. As shown in Figure 2, this could lead to an incorrect assessment of the relative density of sand. For assessing the relative density of sands in a view of the morphological features of their grains curves of the boundary values of the void ratio, in our opinion, should be parallel in loose and dense state.

The study of the influence of morphology of alluvial sands (fine and medium ) on their density showed that with increasing rates of roundness (α), sphericity (β) and generalized morphological index (λ), which takes into account the surface roughness of the grains, the density of sand increases and the void ratio decreases.

The content of fraction less than 0.1 mm in the sands effects the particle size distribution. Thus, the change in a content of this fraction in a fine sand from 1% to 10.5% causes a change of the uniformity coefficient in range from 3.1 to 1.8, and in medium sand with less range of the change of a content of fraction < 0.1mm from 0.5% to 7.5% a change of the uniformity coefficient values from 3.2 to 1.8. Thus, the influence of the changes in a content of fine fraction of sand on the uniformity coefficient more significantly manifested in medium sands. Non-uniform sands are able to condense more than uniform. Thus (if not take into account the age-hardening of sands), it can be argued that the sands density and the corresponding value of penetration parameters depends on the grain-size of sands, content of the fraction less than 0.1 mm and morphological features typical for the relevant genetic types of sand.

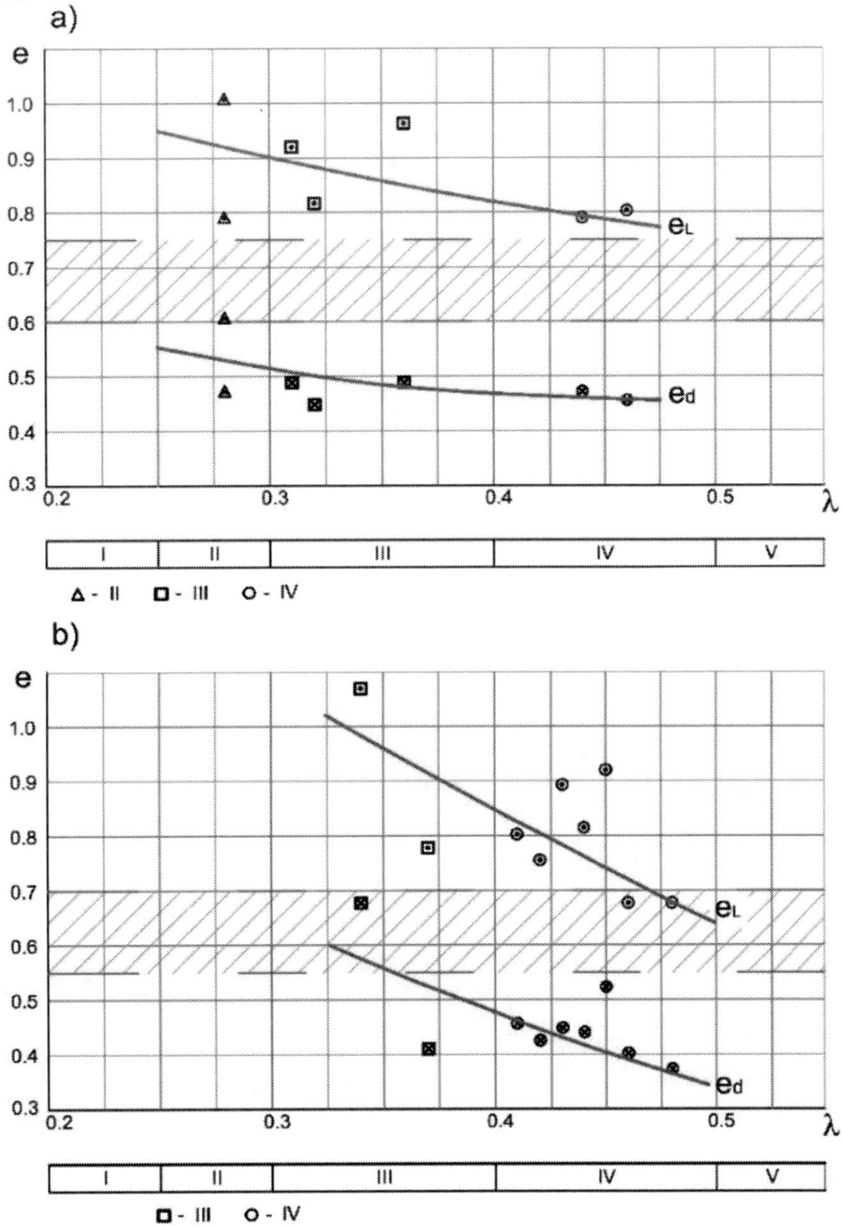


**Fig.2.** The influence of the generalized morphological index on density of fine (a) and medium (b) sands. Categories of the treatment of grains: I-very rough, II-raw, III-almost raw, IV-soft raw, V-treated (the values of correlation coefficients vary from 0.74 to 0.94). The shaded area on the graphs – the values of medium density sands (according to GOST).

To quantify the density of sandy soils according to the data of cone penetration test was proposed a number of empirical relationships [8, 12, 13, and 14]. In particular, L.G. Mariupolskiy [12] recommends to determine the density of sand from the values of the solid particles density ($\rho_s$), moisture weight (W) and a cone resistance ($q_s$) using the following equation:

$$\rho_d = \frac{\rho_s}{1.765-0.185\lg(\frac{q_s}{q_s^0})}. \quad (1)$$

where $q_s^0$=1MPa.

In our view, recommended equation (1) can be used only for variety of sands averaged by size, without taking into account their uniformity coefficient, genetic affiliation, water saturation, hardening and generalized morphological index.

To verify the correctness of our arguments, results of the cone penetration test (according to the materials of long-term surveys LLC NPF "SIVS", LLC "Investroy", LLC "IGC" in Moscow and the near suburbs) in water-saturated sands (silt, fine and medium-size) of two genetic types: alluvial (a) and fluvioglacial (f), were analyzed. A comparison of the values of cone resistance ($q_s$) with the values of dry soil density ($\rho_d$) in a view of their origin and size shown in Figure 3 (a, b, c).

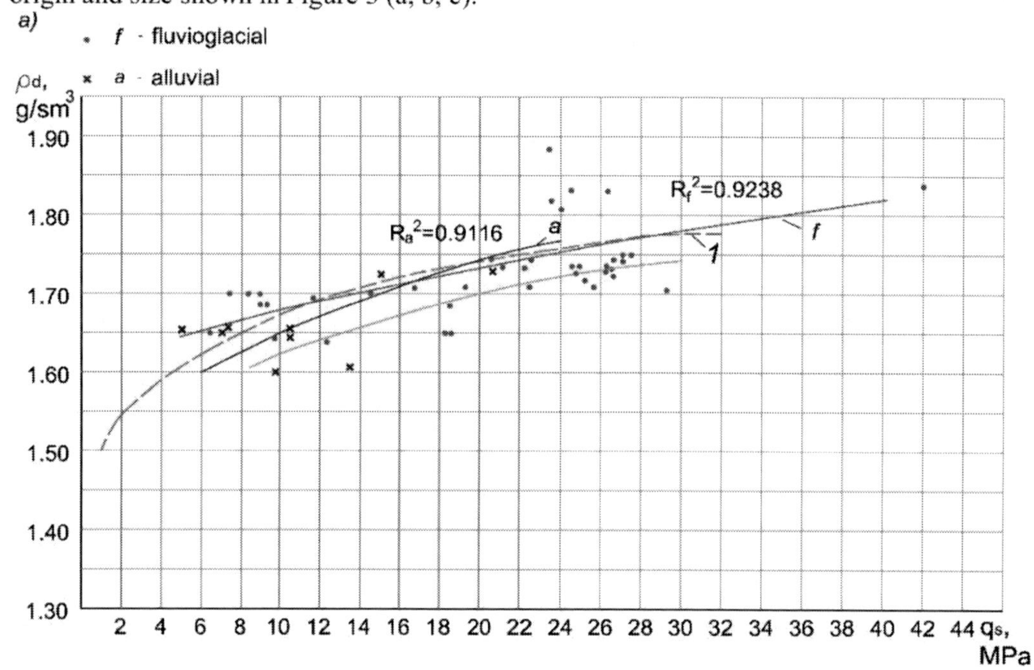

b)

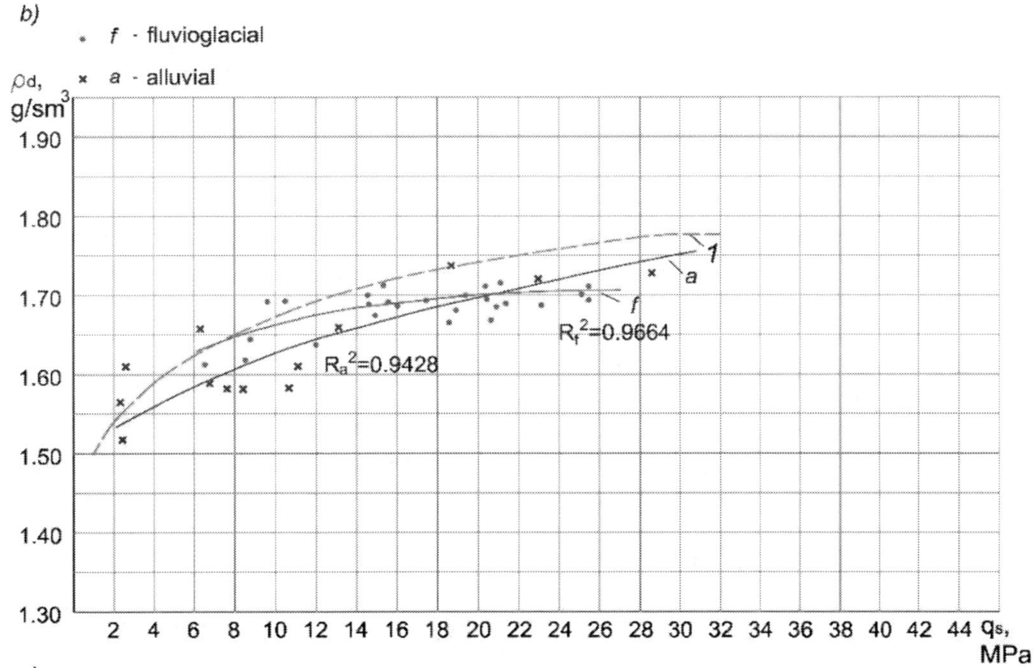


c)

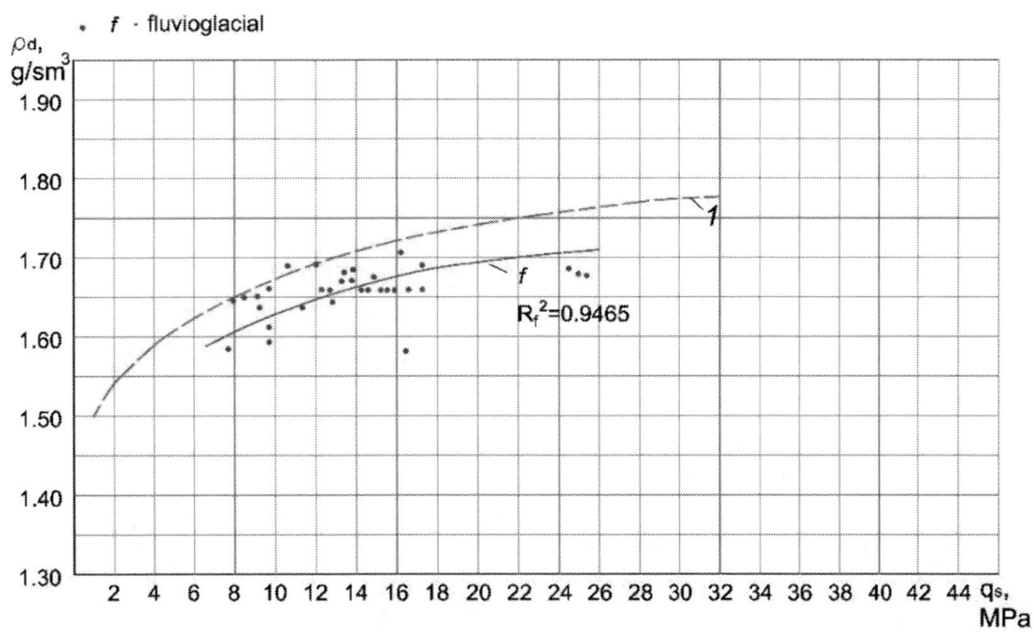


**Fig.3.** Relationship between dry soil density and cone resistance for water-saturated medium (a), fine (b) and silt (c) sands: alluvial (a) and fluvioglacial (f). 1 - theoretical curve by L.G. Mariupolskiy [12].

Figure 3 (a, b, c) shows, that the same values of cone resistance of different size sands correspond to different values of the dry soil density. Thus, the finer sand, the lower the absolute value of dry soil density and more differences with their theoretical curve. For each studied species of sands grain-size was evaluated the effect of the morphology of the sand grains on their density. According to the data performed above (Fig. 2) the more treated grains fit tighter, that confirmed by a graph of the relationship between dry soil density and cone resistance for the studied genetic types and varieties of grain-size sands (Fig. 3). The scatter of the experimental points explains with the difficulty of the undisturbed sampling of sand in water-saturated state, as well as changes in their morphological parameters that affect the sands density, although not in a large range, even for the same genetic type. In addition, we did not take into account the uniformity coefficient of sand. Thus, it becomes evidence that for the determining the density of quaternary sands necessary to consider their grain-size, uniformity, structural-textural features (hardening), morphological parameters.

Established correlations between the dry soil density of the studied sands and the cone resistance describes by equations shown in the Table 1. Data are given according to the range of change of cone resistance from 6.0 MPa to 26.0 MPa.

**Table 1.** Correlations between dry sands density and cone resistance, obtained on the basis of the empirical formula L. G. Mariupolskiy (for the quaternary sands of Moscow).

| Grain size | Genetic type of sand | |
|---|---|---|
| | alluvial (a) | fluvioglacial (f) |
| Medium | $\rho_d = \frac{\rho_s}{1.765 - 0.170\lg(\frac{q_s}{q_s^0})}$ | $\rho_d = \frac{\rho_s}{1.765 - 0.190\lg(\frac{q_s}{q_s^0})}$ |
| Fine | $\rho_d = \frac{\rho_s}{1.765 - 0.142\lg(\frac{q_s}{q_s^0})}$ | $\rho_d = \frac{\rho_s}{1.765 - 0.168\lg(\frac{q_s}{q_s^0})}$ |
| Silt | - | $\rho_d = \frac{\rho_s}{1.765 - 0.140\lg(\frac{q_s}{q_s^0})}$ |

## Summary

1. Establishing the value of the sands density is an important practical task during the engineering surveys, but very difficult to solve because of the difficulty of undisturbed sampling, especially in water-saturated sands. The established categories of sands density by cone resistance ($q_s$) provide only a qualitative picture on sands density. Proposed earlier dependence (by L.G. Mariupolskiy) to quantifying the density of sand in terms of the cone resistance does not take into account a number of factors that influence the formation of the sands density, which leads in some cases to large differences in the density values identified by direct and indirect methods.
2. To improve the accuracy of determining the sands density is important to consider not only their particle size distribution (the size and uniformity), hardening of the soil due to the formation of structural links, but also the genesis of the sands, which determines the morphological features of the sands grains.
3. For each variety of sand (by particle size and uniformity) is necessary to install the morphological characteristics of their grains, for what the generalized morphological index $\lambda$ (proposed by A.D. Potapov) can be used. These works will allow to make a series of correlations at sites, to describe them with the equations and use in the engineering surveys for quantifying the sands density.
4. It should be noted that such an approach to the determination of the sands density by cone penetration test valid for quaternary sediments in which virtually no cementitious hardening of sand, except highly ferruginized differences. In assessing the density of pre-

quaternary sands should be taken into account also the presence of clay and silt layers, sections, cemented siliceous phosphate cement, which will have an additional impact to cone penetration, increasing the value of cone resistance. As a result, for the same size sands will be formed their own correlations on their density and cone resistance because of the differences in their structural and textural features inherent for sediments of different ages.

**References**

1. *SP 47. 13330. 2012. Engineering surveys for construction*, Moscow (2013)
2. *Genesis and model of the formation of soil properties*/ Edited by V. T. Trofimov and V. A. Korolev, M.: Publisher MSU, **165** (1998)
3. I. V. Dudler, A. D. Potapov, A. N. Yulin. Some features of the grains of sand of different origins and methods of their study. Collection of Engineering Geology, №111, Moscow, MISI, **109-123** (1974)
4. P. L. Ivanov, L. V. Shulc. Influence of particle shape on the properties of alluvial sands facilities. Hydraulic Engineering, №11, Moscow, **47-49** (1972)
5. A. D. Potapov. Morphological study of sands in geotechnical purposes. Autoabstract, Moscow, **24** (1982)
6. *Recommendations for the integrated study and evaluation of building properties of sandy soils*. M.: Stroyizdat, **212** (1984)
7. V. T. Trofimov, V. A. Korolev, E. A. Voznesenskiy, G. A. Golodkovskaya, Yu. K. Vasilchuk, R. S. Ziangirov. *Soil science*/ Edited by V. T. Trofimov. M.: Publisher MSU, **1024** (2005)
8. Yu. G. Trofimenkov, L. N. Vorobkov. *Field research methods of construction properties of soils*. M.: Stroyizdat, **215** (1981)
9. V. I. Kashirskiy. Methods of studying the composition and properties of fine-grained soils by field methods in a metropolis (on the example of Moscow). Autoabstract. Moscow (2005)
10. A. D. Potapov Method of rapid determination of the morphology of the grains of sand with different genesis. Deposited manuscript VINITI, №10-76, Moscow (1975)
11. *GOST 25100-2011. Soils. Classification*. Moscow (2013)
12. L. G. Mariupolskiy. *Soil analysis for the design and construction of pile foundations*. M, Стройиздат, **199** (1989)
13. BS 5930-1999. Code of practice for site investigations.
14. T. Lunne, P. K. Robertson, J. J. M. Powel. Cone Penetration Testing in geotechnical practice, **312** (1997)

# On Preparing Offers for Targeted Development Programs in Housing and Communal Services

*Rustam* Khayrullin[1,*]

Moscow state university of civil engineering, Yaroslavskoye shosse, 26, Moscow, Russia, 129337

**Abstract.** The article offers a mathematic development model for economically sound and close to optimum park of technical facilities used in housing and communal services, that implies purchasing new samples of these facilities and upgrade items already existing that have quite low level of technical perfection. A level of technical perfection of the item is a complex (integral) characteristic that relies on the system of indexes that describe its main properties, based on its purpose, and is reflective of its technical progressiveness. Level of technical perfection of the park is a mean weighted sum of technical perfection parameters for all the items in the park. The model is based on algorithm for presentation of technical facilities development park and is made as a standard task of linear programming, solution methods with limitations in the form of equalities and inequalities, including methods of integer linear programming. The criterion used is an indicator that characterize reduction of degradation, a value inverse to the technical perfection level. The author analysed economic expediency of combined purchases and upgrades for various types of technical facilities under various amounts of financing. The article presents results of mathematic model approbation.

## 1 Introduction

Housing and communal services (HCS) present one of the most promising sector of Russian economy in terms of increasing power efficiency and energy saving. The data of Russian Federation Ministry of Energy, HSS consumes 16% of fuel total amount of fuel used in the country.

One of the most promising trends for solving the energy efficiency problem is the one that supported from the ‘smart house-smart city-smart country’ concept, among other things, in the frames of long-term target programs [1], like «Development of housing and communal services for years 2014-2018», «Development strategy for housing and communal services of Russian Federation by year 2020» and other territorial programs of HCS development.

Implementation of these programs requires introducing and operation of instruments, tools and equipment (ITE) of sufficiently high technical perfection, targeted management of instruments, tools and equipment park (ITEP) used in HCS sector.

Topical is a task of developing a method that can obtain rational structure of ITEP, develop versions of ITEP re-equipment in accordance with modern requirements, manage

* Corresponding author: KhairullinRZ@mgsu.ru

purchasing and upgrade of facilities samples both for making offers for targeted development programs and for control of implementation of these programs.

At the same time, the method should be quite easy to use, be applicable for tasks of various dimensionality, fit for optimum management of the whole ITEP and/or for its part.

The article presents the development results of methodical approaches [2-4] to ITE used by HCS sector. The article offers the mathematic model for optimum strategies of purchasing and upgrade that assure maximum level of technical perfection (minimum level of degradation) for the planned date, at the required level of ITE, with regard to limitations for production capabilities and total cost of purchases and upgrades.

[2] implements the approach to finding a technical perfection of the ITE sample based on comparison of main technical and operation parameters the group of experts made between the sample researched and some base sample. To assess technical perfection of the ITE sample, the article uses compression of indicators like reliability, accuracy, automation level, autonomy etc. Technical perfection level of ITE park is generally understood as a mean weighted sum of technical perfection indicators of all items that make ITE park.

In [4], classic theory of optimal control and approximate methods for solution of optimization tasks [5-7] contribute to posing and solving the task of targeted management of efficiency indicators for metrological equipment park.

Direct use of [4] methods and results that would model close to optimum HCS-used HITEP renewal processes is hindered by absence of models for joint records of purchase and upgrade, with regard to their time and limited production capacities of instrument-making industry, as well as of specific character of linear programming problem (LPP), particularly, in the fact that LPP matrix is inherently heterogeneous and highly dimensional [8].

The article assumes the modernization takes several years and receives certain amounts of financing throughout its time. Upgraded technical facility takes a year to be put in operation. Unlike upgrade, purchase of new ITE sample is instant. Purchased sample is put in operation in the year of when it was bought.

# 2 Posing the task for joint management of purchases and upgrades

Let's denote:

$TZ$ - number of ITE types that can be purchased, $tz = 1,2,...,TZ$,

$TM$ - number of ITE types that can be upgraded, $tm = 1,2,...,TM$,

$kz^{tz}$ - number of park ITE samples of type ($tz$, $tz = 1,2,...,TZ$), having degradation levels of $nz^{tz}$, instead of which new ITE samples may be purchased,

$km^{tm}$ - number of park ITE samples of type ($tm, tm = 1,2,...,TM$) having degradation levels of $nm^{tm}$, that can be upgraded,

$p$ - duration of planning period,

$pm$ - duration of upgrade,

$CZ_i^{tz}$ - costs of $tz$ - type ITE samples purchased in *i* year,

$CM_i^{tm}$ - costs of $tm$ -type ITE samples upgraded in *i* year,

$S_{tot} = C_1 + C_2 + ... + C_p$ - limitation of total cost for all purchased and upgraded ITE samples for the whole period of planning.

$(P_i^{tm},\ tm = 1,2,...,TM\ ,\ i = 1,2,...,p)$ - limitation for production capacities of instrument-making industry;

$C_{\deg}^{tz}$ - upper limitation for $tz$ - type ITE level of degradation,

$C_{\deg}^{tm}$ - upper limitation for $tm$ - type ITE level of degradation,

$\alpha^{tz}$ , $\beta^{tm}$ - weight factors with regard to importance for upgrade of various ITE types.

Now, let's introduce variables:

$x_i^{tk}$ - number of $tk$ -type ITE samples purchased in $i$ year,

$z_i^{tm}$ - number of $tm$ -type ITE samples upgraded in $i$ year.

Formulate the task of ITEP renewal as follows: we are to obtain $x_i^{tz}$, $z_i^{tm}$ so that the general level of ITEP degradation would take on minimum values:

$$F = \sum_{tz=1}^{TZ} \alpha^{tz} \cdot kz^{tz} \cdot nz^{tz} + \sum_{tm=1}^{TM} \beta^{tm} \cdot km^{tm} \cdot nm^{tm} -$$
$$- \sum_{tz=1}^{TZ} \sum_{i=1}^{p} \alpha^{tz} \cdot x_i^{tz} \cdot nz^{tz} - \sum_{tm=1}^{TM} \sum_{i=1}^{p} \beta^{tm} \cdot z_i^{tm} \cdot nm^{tm} \to \min \ , \tag{1}$$

with limitations for the number of various types ITE to be purchased (quantity of various types ITEs:

$$\sum_{i=1}^{p} x_i^{tz} \le kz^{tz},\ tz = 1,2,...,TZ\ , \tag{2}$$

and quantity of ITE samples to be upgraded:

$$\sum_{i=1}^{p} z_i^{tm} \le km^{tm},\ tm = 1,2,...,TM\ ; \tag{3}$$

It is presumed that in every year of planning $i = (1,..., p)$ total cost of purchases and upgrades is limited by the amount of funds $C_1, C_2, ..., C_p$ allocated for the year planned:

$$\sum_{tz=1}^{TZ} CZ_1^{tz} x_1^{tz} + \sum_{tm=1}^{TM} CM_1^{tm} z_{pm+1}^{tm} \le C_1\ , \tag{4}$$

$$\sum_{tz=1}^{TZ} CZ_2^{tz} x_2^{tz} + \sum_{tm=1}^{TM} CM_1^{tm} z_{pm+1}^{tm} + \sum_{tm=1}^{TM} CM_2^{tm} z_{pm+2}^{tm} \le C_2\ ,$$

....

$$\sum_{tz=1}^{TZ} CZ_i^{tz} x_i^{tz} + \sum_{j=1}^{i} \sum_{tm=1}^{TM} CM_j^{tm} z_{pm+j}^{tm} \le C_i\ ,\ i = 1,2,..., pm;$$

....

$$\sum_{tz=1}^{TZ} CZ_i^{tz} x_i^{tz} + \sum_{j=1}^{pm} \sum_{tm=1}^{TM} CM_{i+j-pm}^{tm} z_{i+j}^{tm} \le C_i\ ,\quad i = pm+1,...,p-pm;$$

...

$$\sum_{tz=1}^{TZ} CZ_i^{tz} x_i^{tz} + \sum_{j=i}^{p-1} \sum_{tm=1}^{TM} CM_{j-pm+1}^{tm} z_{j+1}^{tm} \le C_i\ ,\quad i = p-pm+1, p-pm+2,...,p-1;$$

.....

$$\sum_{tz=1}^{TZ} CZ_{p-2}^{tz} x_{p-2}^{tz} + \sum_{tm=1}^{TM} CM_{p-pm-1}^{tm} z_{p-1}^{tm} + \sum_{tm=1}^{TM} CM_{p-pm}^{tm} z_p^{tm} \le C_{p-2},$$

$$\sum_{tz=1}^{TZ} CZ_{p-1}^{tz} x_{p-1}^{tz} + \sum_{tm=1}^{TM} SM_{p-pm}^{tm} z_p^{tm} \le C_{p-1},$$

$$\sum_{tz=1}^{TZ} CZ_p^{tz} x_p^{tz} \le C_p\ ;$$

with limitations to industry production capacities in upgrade of ITE samples:

$$z_i^{tm} \le P_i^{tm},\ i = pm+1, pm+2,...,p\ ; \tag{5}$$

and limitations for degradation level of each ITE type:

$$\alpha^{tz} \cdot kz^{tz} \cdot nz^{tz} - \sum_{i=1}^{p} \alpha^{tz} \cdot x_i^{tz} \cdot nz^{tz} \le C_{\deg}^{tz}\ ,\qquad tz = 1,2,...,TZ\ ; \tag{6}$$

$$\beta^{tm} \cdot km^{tm} \cdot nm^{tm} - \sum_{i=1}^{p} \beta^{tm} \cdot z_i^{tm} \cdot nm^{tm} \le C_{\deg}^{tm}\ ,\quad tm = 1,2,...,TM\ . \tag{7}$$

Note that (1-7) is a task of integer linear programming.

Moreover, inequalities system (4) describes financial limitations with the assumption that various types of ITE have the same duration. Typical features of forming the LPP matrix for different duration of upgrade for various ITE types are described below.

# 3 Approbation results

Initial data: ITE assortment, averaged costs of purchases and upgrades, planning period, upgrade duration, financing distribution broken down by years of planning, production limitations, limitations of allowable level of degradation for each ITE type, $TZ = 2$, $TM = 1$, $\alpha^1 = 1, \alpha^2 = 1$, $\beta^1 = 1$ , $k_1^1 = 5$, $n_1^1 = 1{,}1$, $k_1^2 = 15$, $n_1^2 = 1{,}3$, $k_1^3 = 10$, $n_1^3 = 1{,}1$, $CZ^1 = 10$, $CZ^2 = 15$, $CM^1 = 3$, $pm = 4$, $p = 20$.

Calculations were made for different amounts of financing $S_{tot}$, with $C_1 = C_2 = ... = C_p = S_{tot} / p$. The article presents results of dimensionless calculations for limited ITEP used for HCS service automation: detectors, controllers and automated work places. The predictions proved the limitations (5)-(7) fulfil as a strict inequality, as the model was approbated for the ITEP part with the highest level of degradation. Purchase cost was assumed to be not related to the year of planning.

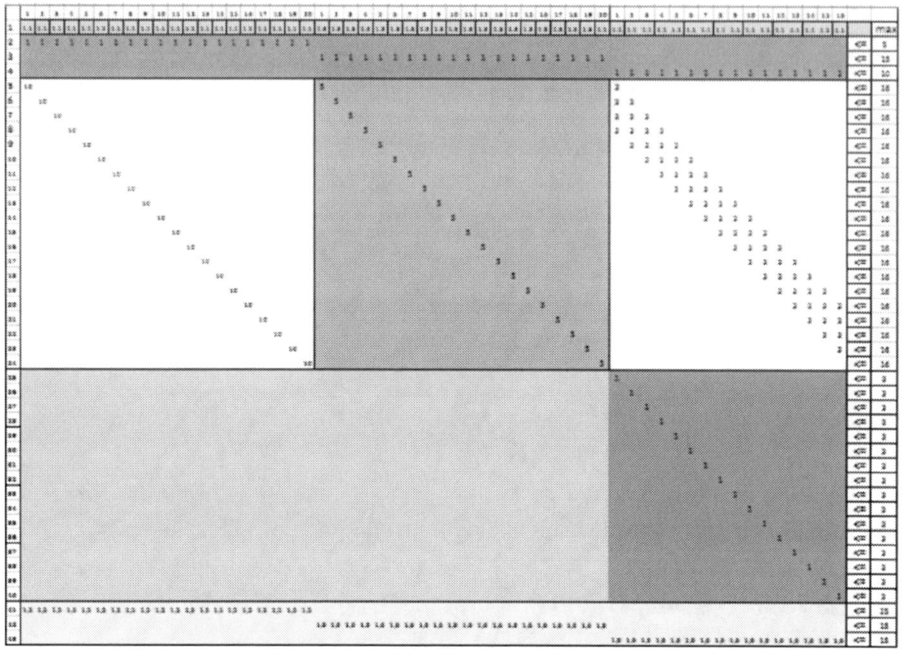

**Figure 1**. Structure of matrix of line program problem

The view of LPP matrix [4, 7] is shown at Fig. 1. Note that the matrix has a block structure and most of the matrix elements are equal to zero. The first line of the matrix carries objective functions coefficients. Next three lines represent inequalities limited by three types of ITE samples. Below, there are five blocks, with three of them being diagonal. First two blocks represent the purchases of two ITE types. The third block that represents includes two column vectors with nonzero elements arranged in a row. The fourth block consists of zeroes. The fifth, diagonal-shaped block, represents limitations for industry production capacities. Last three rows of the matrix represent limitations for degradation level allowable for each ITE type.

Note that the amount of nonzero elements in the columns of the blocks that represent upgrade is the same as the upgrade period in years. In case various ITE types are upgraded in the same year, and/or various ITE samples having various duration of upgrade, LPP matrix contains several blocks of type three with various amount of nonzero column and various amount of nonzero elements in appropriate columns.

The solution of linear programing employed both Homory method and standard simplex method. The calculations proved that higher amount of financing reduces ITEP degradation level from 31.6 to zero by almost linear law. The explanation is that the degradation indicators for various ITE types are very close to each other.

Amount of financing $S_{tot} = 395$ is sufficient for complete renewal of ITEP. With lower financing, complete renewal of ITEP is impossible.

Fig. 2 offers optimum purchase-upgrade combinations by years of planning for various types of ETI.

DOI: 10.1051/matecconf/201711700076


If financing is low, it is wise to purchase of first type ITEPs first: ITE-1, and funds remained should be used to upgrade ITE-U. Medium-size financing first pays the purchase of ITE-1 until complete renewal of its park, then pays upgrade of ITE-U until complete renewal of its park, and, at last, ITE-2 is purchased. When the financing is large enough, purchase of ITE-1 is financed first, and, before renewal of ITE-1 is completed, some funds are allocated for ITE-U renewal. At last, after complete renewal of ITE-1 park and partial renewal of ITE-U, purchase of ITE-2 is financed.

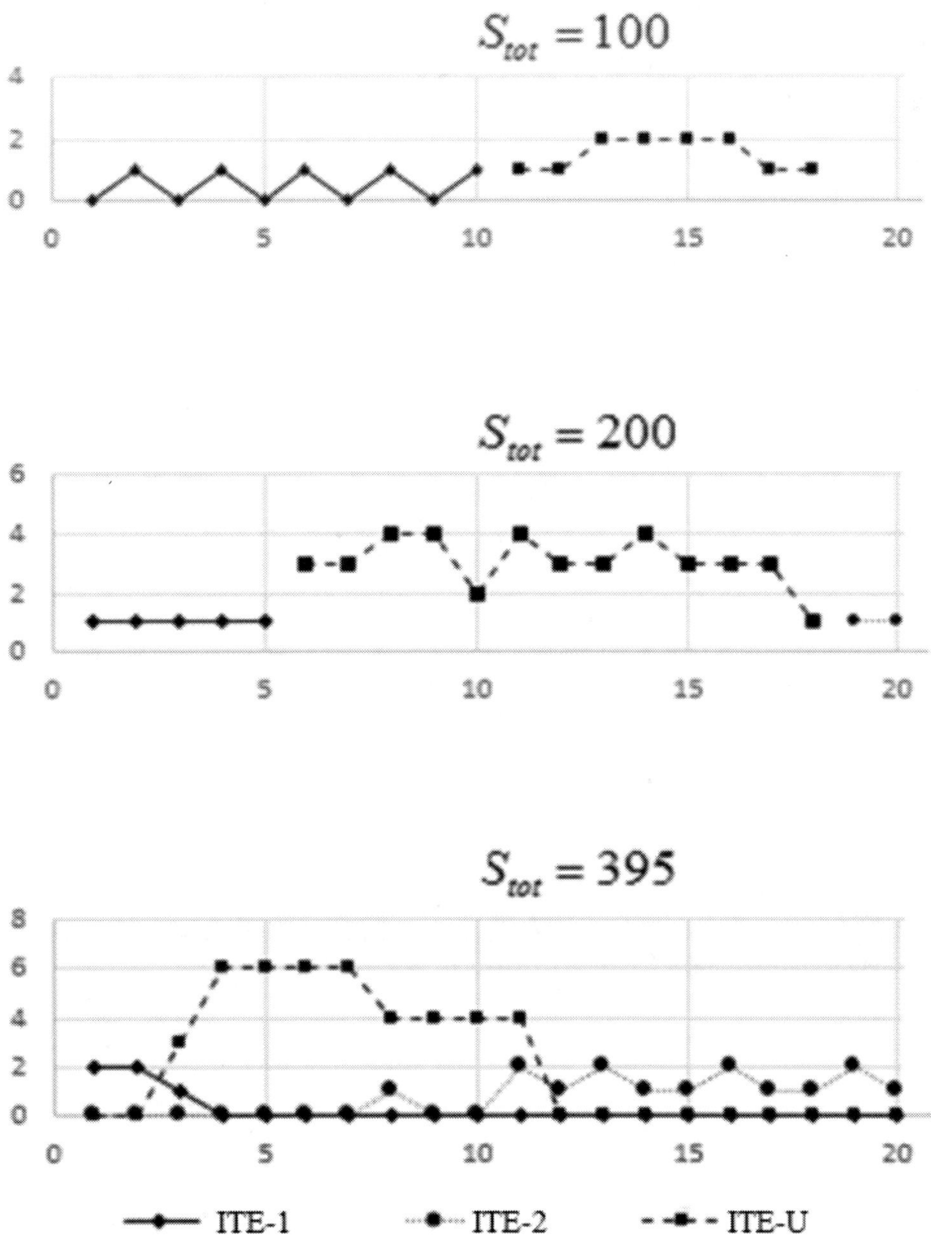


**Figure 2**. Procurements and modernization of different types of ITE depends on planning period

## 4 Conclusions

1. Software mathematic support described in the article allows posing and solving linear programing tasks both for general purposes and for integer linear programing. Solution of the latter, especially the tasks of dimensionality, usually requires considerable calculating resources and use of specialized software [8].
2. The experience in making offers for targeted industrial programs and operating control of their implementation has proven the expediency of the model offered in the article. The model based on the solution of integer LPP fits for operating control of purchase and upgrade, as well as for correction of annual plans, whereas long-term (strategic) planning requires standard LPP solutions , with amounts of purchase and upgrade of various ITE types taken as 'managements'.
3. The model developed is a system of linear limitations expressed as equalities (inequalities) and optimum criteria, affords, when required, inclusion of additional qualifying information by changing the vertical dimension of LPP to introduce additional limitations expressed as equalities (inequalities), or be changing LPP horizontal dimension to account for additional ITE types (subtypes).

## References

1. N. Yu. Yas'kova, D. M. Karasik. Programme-oriented methods of construction development, modern format city target programs // the Bulletin]. No. **1**. P. 182-186 (2014)
2. V. M. Burenok, R. N. Pogrebnyak, A. P. Skotnikov. Methodology of substantiation of prospects of development of means of armed struggle, General purpose]. Mechanical engineering (2010)
3. A. I. Buravlev, A. A. Nesterov. Methodology for military – economic analysis of the feasibility of procurement of weapons and military equipment // Weapons and the economy. No. **2** (35). P. 83-89 (2016)
4. R. Z. Khayrullin. Targeted management performance indicators Park metrological equipment // Scientific review. No. **10**. P. 300-303 (2016)
5. L. S. Pontryagin, V. G. Boltyansky, R. V. Gamkrelidze, E. F. Mishchenko. The mathematical theory of optimal processes]. Moscow. Science (1961)
6. Bryson A., Ho, Y. *Applied Optimal Control.* Blaisdell Publishing, Walthman, MA (1969)
7. R. P. Fedorenko. Approximate solution of optimal control problems // Moscow. Science (1978)
8. R. Z. Khayrullin. To the optimization of high dimensionality systems using ILOG software components // Vestnik MGSU. No. **8**. P. 157-163 (2013)

# Ways to improve physical and thermal performance of refractory lining materials

*A.I.* Khlystov[1]*, *E.M.* Vlasova[1], *D.I.* Isaev[1]

[1]Samara State Technical University, Institute of Architecture and Civil Engineering, 194, Molodogvardeyskaya St., 443001, Samara, Russia

**Abstract.** Refractory lining materials, which include ceramic refractories and nonfired heat-resistant concretes, have a very short lifespan during the turnaround time measured in years and sometimes months. Therefore, increasing the service life of thermal generating units by 1.5-2 times will bring significant economic benefits. The main factor that determines the durability of refractory lining materials is the thermal resistance. It is possible to increase the thermal resistance by improving such physical and mechanical properties as strength and density. As for the ceramic refractory performance improvement, such technological methods as their structural and chemical modification by phosphate binder impregnation, as well as introduction of phosphate components into the ceramic batches during the molding process increase, in particular, their thermal stability. The use of aluminous and high-alumina cements contributes to a significant increase of not only strength, but also physical and thermal performance of heat-resistant concretes with different fillers. Switching to the use of chemical binders in the compositions of heat-resistant concretes (liquid glass with effective hardeners; silicate-block and phosphate binders) enables to develop high-heat resistant materials which do not soften in a wide range of heating temperatures from 400 °C to 1600 °C. The positive results on increasing the thermal resistance of heat-resistant composites can be obtained by reinforcing them with high temperature fibers.

## 1 Introduction

Traditional ceramic refractories including fired ceramic concretes, as well as nonfired heat-resistant concretes with different binders belong to refractory lining materials. Piece ceramic refractories made by the traditional ceramic technology (molding → drying → firing) have various chemical and mineralogical compositions affecting ultimately their physical and thermal operation factors (strength properties, deformation temperature under the load, heat and chemical resistance, etc. For example, the thermal resistance of piece refractories, i.e. the main property that determines their durability, varies from 10 water thermal cycles for fireclay refractories to 100 for corundum ceramics. And the limit of the compressive strength ranges from 15 to 100 MPa, respectively.

* Corresponding author: alex-x1950@yandex.ru

However, lining constructions made of piece ceramic refractories have a large number of joints which are the "neck stage" between piece refractories in the brick masonry, where the entire lining failure begins.

Since the amount of refractory lining works increases annually, heat-resistant concretes have been used recently to reduce the number of these joints.

Heat-resistant concretes are modern refractory lining materials made with the use of the known hydraulic binders (Portland cement, alumina cements) and special chemical binders (liquid glass, silicate – block, phosphates). Crushed heavy and light piece refractories, as well as artificial fired porous aggregates (expanded clay gravel, haydite, perlite, vermiculite, etc.) serve as aggregates for heat-resistant concretes. On the basis of binders listed above it is also possible to produce heat-resistant cellular concretes.

Heat – resistant concretes, as well as conventional ones can be used in the manufacture of thermal unit linings in precast and monolithic variants. These variants of heat-resistant concrete use during refractory lining works enable to obtain thermal unit linings with minimum joints. Reducing the number of joints in lining constructions leads not only to the improvement of their properties and durability, but also to the saving of material resources, because it reduces the number of routine, average and general repairs required in practice of works with refractories. According to the information provided above, the development of ways to improve physical and thermal performance of refractory lining materials, including the thermal stability, is regarded as a priority [1÷3].

## 2 Materials and methods

The Portland cement is one of the widely used kinds of binders for chemically bonded refractory composites. The Portland cement as a hydraulic binder contains 75-80% of the highly basic calcium silicate, 5-15% of the highly basic calcium aluminate and 10-20% of the tetracalcium aluminoferrite. The major feature of the Portland cement as a binder for making heat-resistant composites is the formation of significant quantities of the calcium hydroxide $Ca(OH)_2$ (up to 15% by weight) in the cement stone as a result of hardening. During drying and the first heating at a temperature of about 500°C calcium hydroxide decomposes into calcium oxide and water. Re–slaking of the newly formed calcium oxide with the water vapor contained in the air, leads to the destruction of the cement stone and concrete respectively, due to the formation of $Ca(OH)_2$ from CaO. To prevent this process, agents capable of binding the calcium oxide to more heat stable compounds are added to the Portland cement. For this purpose traditional fine-ground refractory additives are used: granular blast–furnace slag, alumina-chromium waste in the form of the dead petrochemical catalyst, chamotte, cordierite, volcanic ash, fly ash, alumina cement and other substances capable of reacting with the calcium oxide CaO. Positive results, in particular the strength increase, are provided by the addition of the sodium tripolyphosphate after its introduction into the Portland cement binders, as well as classical plasticizers of C-3 type or nanotehnogenic high–alumina sludge–waste of non-ferrous metallurgy [3-5].

The maximum operating temperature for the widely used fireclay refractory concrete on the Portland cement does not exceed 1200°C. When using a heat-resistant concrete on the Portland cement as a fine-ground refractory additive and fillers of more highly refractory materials, such as chromomagnesite or periclase products, its operation temperature rises to 1600°C [3].

Aluminate binders, including alumina and high-alumina cements, consist of low-basic calcium aluminates. The total content of the aluminum oxide in the aluminous cement is not more than 50%, and in the high-alumina cement is not less than 17%. These cements are quick-hardening and very active, i.e. high-strength binders. During mixing with water alumina cements do not form the calcium hydroxide $Ca(OH)_2$ and other compounds that reduce the performance of heat-resistant composites. The cement stone based on the alumina cement is highly refractory in comparison with the Portland cement due to the increased content of the aluminum oxide $Al_2O_3$ [6]. The heat-resistant concrete on aluminous cement has higher performance characteristics than concrete on the Portland cement. This type of the heat-resistant concrete was tested at temperatures of 1300-1400°C in neutral and weakly reducing environments. However, when working with this type of the heat-resistant concrete on the installation site, the ambient temperature must be taken into account.

The heat-resistant concrete on the high-alumina cement and high-alumina aggregate (e.g., a white electrofused corundum) can be used not only in oxidizing, but also reducing environments at an ambient temperature of not more than 1700°C. After heating to a temperature above 800°C, the strength of concretes on the high-alumina cement decreases. However, their residual strength is higher than that of concretes on the Portland and alumina cements and averages 40-45% of the grade value [2; 6].

The dense concrete on the high-alumina cement and corundum can have a grade of 600 and 650 and a residual strength of 25-30 MPa when the average density of the normally compacted concrete mix is 2900-3100 $kg/m^3$.

This type of concrete is sufficiently stable both in oxidizing and reducing environments and it is resistant to abrasive particles.

It is also possible to increase physical and thermal characteristics of heat-resistant composites (concretes, solutions, shotcrete masses) by using liquid-glass and phosphate binders.

## 3 Experiments and results

The liquid glass used in heat-resistant concrete compositions is an aqueous solution of the sodium silicate of 1,3-1,5 $g/sm^3$ density. The ratio of silica to the sodium oxide is usually 2-3.5. The hardening of the liquid glass and concretes on its basis is the result of their dehydration or action of hardening initiators. The solidification of materials by drying is possible only if they contain a large number of pores and the exposed surface of the concrete. In dense concretes one can observe a rapid formation of a vapor- tight surface film of the liquid glass which prevents drying and hardening of deep sections of the concrete. In this regard, the solidification of liquid-glass materials by introducing chemical hardeners which ensure the volume curing is more promising. The sodium silicofluoride, self-separating slag of ferrochromium production, nepheline sludge, Portland cement, aluminous cement and nepheline flame retardant can be used as hardeners [7].

The melting point of sodium and potassium silicates is relatively low (about $900^0$ C) and the liquid glass in the composition of the lining material is a fluxing agent, but in the combination with aluminosilicates, corundum, magnesite and other compounds with high refractory properties, it allows obtaining concretes and protective coatings that withstand the temperature of about 1600°C [8; 9].

Concretes and solutions based on the liquid glass retain for a long time the ductility under the load which is much less than their strength limit. The reduction and complete elimination of this defect is achieved by using high-temperature drying (at the temperature above 120°C).

The heat resistant concrete on the liquid glass and fireclay filler is used at the temperature of not more than 1300°C. This type of concrete properly resists the action of acidic media, provided that the sodium silicofluoride is used as a hardener for the liquid glass. Therefore, concretes on the liquid glass with the sodium silicofluoride can be used for linings of thermal units in acidic corrosive media.

The concrete on the liquid glass can be placed both by vibrating and by shotcreting. When replacing the sodium silicofluoride with the self-separating ferrochrome slag with the fine-ground aluminosilicate powder, physical and mechanical properties of the concrete are improved. And along with this its residual strength increases by about 10% and before the first heating at the operating temperature it becomes less sensitive to moisture.

One of the main advantages of the heat-resistant concrete on the liquid glass is that it unlike concrete on hydraulic binders does not practically reduce the strength when heated. In certain cases, for example, when the ferrochrome slag or aluminous cement are used as liquid glass hardeners and when a finely ground aluminosilicate additive is introduced into the concrete mix; the heat-resistant concrete after heating it to 800◦C increases its mechanical strength. The result is that its residual strength becomes higher than the grade one.

It is possible to consider the heat-resistant concrete on the silicate-block, i.e. on the same sodium metasilicate, but in the form of a dry fine powder, to be an analogue to the heat-resistant concrete on the liquid glass. The approximate composition of the heat – resistant concrete on the silicate –block is as follows, % by weight: silicate- block 6.0; technical barium chloride 0.6; finely ground fireclay additive 34.4; fireclay sand 30; crushed fireclay 30; water of mixing 300 liters per $1m^3$ of the concrete mix. The concrete mix is prepared in fixed-drum concrete mixers by mixing dry components during 1,0-1,5 min and placed by vibration. The concrete takes 4-6 hours to harden at 100-150°C [8; 9].

The basic physical and technical properties of the heat-resistant concrete on the silicate –block are: the usage temperature limit- 1200°C-1600°C; grade strength- 20-30 MPa; residual strength- 100%; average density- 1900 $kg/m^3$; high-temperature strength-23 water-based thermal cycles (water is 800°C). The linear shrinkage after heating to 1200°C does not exceed 0.5%. The characteristic feature of this type of concrete is the increased high-temperature strength under acidic gases.

The heat-resistant concrete on the silicate-block differs from other types of heat-resistant concretes by the increased strength and thermal stability. In addition, the silicate-block is cheaper than the liquid glass and its consumption per $1m^3$ of the concrete mix calculated on the dried basis is approximately 20% less. However, the heat-resistant concrete on the silicate-block can only be effectively used for the manufacture of blocks and individual elements.

Heat-resistant compositions on phosphate cements and binders have very high physical and thermal parameters. Phosphate cements are formed during the reaction of the orthophosphoric acid with metal oxides or metallic mineral materials (including ashes, slags, natural and artificial silicates, etc.). Depending on the activity of the material used against the phosphoric acid, the binding properties of the resulting cements prove themselves at normal, elevated or high temperatures. At the temperature of about $20^0$C the hardening phosphate compositions can be prepared on the basis of magnesium and zinc oxides, titanium, aluminum and zirconium hydroxides previously passivated at 1100-$1200^0$C. Binding properties of phosphate cements based on oxides of titanium, aluminum, zirconium, chromium and a number of other metals appear only when heated to 200-$500^0$ C [10;11].

The settling time of the cold hardening phosphate compositions can be controlled by changing the specific surface of the solid component (e.g. to change its grinding fineness and the reaction activity of grain surface either by thermal or chemical treatment, or pre-neutralizing of the liquid component.

Industrial water solutions of aluminum phosphate or aluminophosphate are the most common liquid phosphate binders. The density of the aluminophosphate binder is about $1,5r/sm^3$. Aluminum chromophosphates have approximately the same density and are a sticky liquid of green color. In addition, partially or completely dehydrated powders of aluminophosphate and aluminochromophosphate are sometimes used. Binding properties of partially dehydrated powders and water solutions appear at the same temperatures and those of completely dehydrated ones at elevated and high temperatures.

Solidification of liquid phosphate binders occurs due to their drying and subsequent polymerization and condensation or introduction of hardening initiators reacting with the acid group present in the composition of binders. Here hardly soluble compounds are formed.

The strength increase of phosphate binders together with that of the heating temperature is their characteristic feature. In the production of wear-resistant, highly plastic and corrosion-resistant coatings (slag- resistant included), the mechanical characteristics of materials based on them are taken into account.

The heat- resistant concrete with an alumochromophosphate binder (AChPhB) and zircon aggregate can be used at temperature of not more than 1500$^0$C. The characteristic feature of this type of concrete is its basic and weakly acidic slag repellency. Therefore, it can be effectively used to line the bottom of boiler furnaces with the liquid ash removal, non-ferrous melting furnaces and other thermal units.

Unlike heat- resistant concretes on orthophosphoric acid, concretes on the alumochromophosphoric binder are more technologically advanced, since it is unnecessary to perform heat-treatment to provide the structural strength to constructions manufactured from them. They can harden at a normal temperature above 15$^0$C if certain chemical additives are introduced into the concrete mix as a hardener initiator and they do not impair the operation properties of the concrete (i.e. they do not lead to corrosion of the reinforcement, the concrete strength decrease when heated, etc.) One of such additives is the waste of the dead chromia-alumina catalyst IM-2201, which initiates hardening of AChPhS and simultaneously acts as a fine-ground additive. The concrete on the alumochromophosphate binder finally gains strength during the process of drying and the first heating of the structure made of it, thus it can be used to form larger blocks and shields, as well as during monolithic concreting and semi-dry shotcreting of structure lining where the boiler is installed [10; 11].

## 4 Discussion

The use of alumina-containing components in concretes and concrete grouts promotes the increase of the chemical resistance of refractory composites.

It is also possible to increase the strength characteristics of heat-resistant composites by producing individual products from refractory ramming mixtures. Earlier, ramming mixtures were used only for relining in thermal units.

The ramming mixtures in contrast to the developed heat-resistant concretes for manufacturing articles and relining have a number of specific differences due to the technological features of their compaction during the molding of individual blocks and maintenance work in furnaces and other thermal units. The consolidation of ramming mixtures (monolithic concretes) during these works is carried out only by the method of tamping which results in the reduction of the amount of the grouting fluid necessary for the high-quality placing of the concrete mix.

To improve the technological parameters necessary for the high-quality compaction of refractory ramming mixtures, refractory clay or kaolin are introduced into-their composition. Considering the fact that these components are scarce and costly (practically most of the deposits of refractory clays and kaolins are concentrated on the territory of Ukraine). We have made an attempt to replace plastic components with some wastes from the ceramsite industry.

At a number of sintering plants both in the Samara region and the Russian Federation, where double-drum rotary kilns are used for the production of the claydite, a coproduct in the form of the ceramsite dust falls out from the firing unit at the junction of two drums. It was found that this coproduct is about 20-30% clay bond along with fired components.

However, the use of phosphate binders (orthophosphoric acid or liquid alumochromophosphate binder) as grouting fluids of refractory ramming mixtures with the ceramsite dust, allowed, as a result of chemical reactions between active liquid phosphates and components (oxides) of mineral fillers (including dust), obtaining high –temperature compounds in the form of the of ferric phosphate $FePO_4$; aluminum phosphate $AlPO_4$; calcium phosphate $Ca_3(PO_4)_2$ and others.

The optimum compositions of monolithic heat-resistant concretes (repair ramming mixtures) are as follows: Composition 1: clay-containing ceramsite dust -10%; aluminum chromium waste. IM 2201-30%; corundum fine (fr<1.25) aggregate -20%; high-alumina sand -40%. The grouting fluid is the orthophosphoric acid or an acidic alumochromophosphate binder (Liquid/Solid ratio= 0.15-0.2).

Composition 2: clay-containing ceramsite dust -10%; zircon concentrate CZ - 1 -35%; corundum fine (fr< 1.25) aggregate -20% ; high-alumina or fireclay sand -40%.

Grouting fluids and ramming mixture consistency are as in composition 1.

Such compositions of ramming mixtures are recommended to be used for the manufacture of both large-sized blocks and small piece critical products required for the installation and repair of thermal unit linings. In the case of manufacturing individual blocks, parts and products with the use of ramming mixtures the method of the immediate stripping is quite acceptable.

These heat-resistant composites obtain the grade strength directly in the thermal unit after the first lining heating to the operating temperature.

For the further growth of strength and other physical and thermal parameters of heat-resistant phosphate hardening composites it is recommended to treat products or carry out the structural and chemical modification by impregnation with the solution of the liquid alumophosphate binder.

The same technology can be used to improve physical and thermal parameters of fireclay and high-alumina refractories [3].

The major physical and thermal characteristics of the developed refractory ramming mixtures suitable for the production of individual products and blocks, as well as for thermal unit relining are given in Table 1.

**Table1.** Physical and thermal characteristics of ramming mixtures on phosphate binders

| **Properties** | **Unit of measure** | **Characteristics** | |
|---|---|---|---|
| | | **Composition 1** | **Composition 2** |
| Average density after thermal treatment(200 °C) | $kg/m^3$ | $\frac{2550-2600}{2600-2650}$ | $\frac{2650-2700}{2700-2750}$ |
| Compressive strength after drying at 200 °C | MPa | $\frac{41-47}{47-53}$ | $\frac{48-50}{54-57}$ |
| Compressive strength after burning at 1500 °C | MPa | $\frac{54-59}{60-65}$ | $\frac{58-62}{66-72}$ |
| Thermal resistance | A number of water thermal cycles | 44-48 | 49-51 |
| Fire shrinkage | % | 0,25 | 0,19 |

| | | | |
|---|---|---|---|
| | | 0,21 | 0,16 |
| Thermal coefficient of linear expansion | $10^{-6}\ deg^{-1}$ | 9,81<br>9,64 | 8,75<br>8,57 |
| Temperature of initial deformation under the load of 0,2 MPa | °C | 1500<br>1540 | 1550<br>1590 |

Note: Above the bar are ramming mixture properties; below the bar are properties of ramming mixtures modified with acidic aluminophosphates after the heat treatment at t=200 °C.

## 5 Conclusion

Thus, the use of the industrial waste as fine-grained additives and fillers significantly reduces the cost of heat-resistant concretes and extends their use.

Increasing strength and correspondently thermal resistance of small-piece heat-resistant parts and products based on phosphate compositions can also be carried out using at the pressure of 5.0-15.0 MPa. The composition of mixtures in addition to the refractory clay can include both finely ground high-alumina fillers and small aggregates of aluminosilicate or high-alumina compositions. Almost all phosphate binders serve as grouting fluids, but the advantage is given to water-soluble aluminophosphate compounds [3; 5; 10].

The method of increasing the thermal stability of heat-resistant concretes practically on all types of binders considered, which consists of the introduction of refractory high-temperature fibers into the composition of concrete mixes, is quite simple. As a result of this technology heat-resistant concretes reinforced with fibers are formed. Kaolin wool, basalt fibers and also metallic fibers prepared from high-temperature alloys, e.g. nickrochrome, are possible to be used as fibrous refractory materials [12].

It has been established that the introduction of above-mentioned refractory inorganic fibers and metallic fibers in the amount of 3-4% of the total weight of the concrete mix increases its thermal stability by 1.2÷1.5 times.

## References

1. V.I. Krayuhin, *Heat-resistant concretes, Compositions, Operation, Restoration of main properties after a long-time operation* (2014)
2. N.G. Chumachenko, V.V. Tyurnikov, D. V. Kirillov, News of Higher Educational Institutions, Construction **7**, 38-42 (2004)
3. A.I. Khlystov, *Efficiency upgrading and quality improvement of refractory lining materials* (Samara, 2004)
4. Yu.M Bazhenov, V.S. Demyanova, V. I. Kalashnikov, *Modified high quality concretes* (Publisher Association of civil engineerig universities and colleges, Moscow, 2006)
5. A.I. Khlystov, M.V. Konnov, A.V. Vlasov, E.A. Chernova, Urban Construction and Architecture **4**, 87-92 (2011)
6. N.G. Chumachenko, V.V. Tyurnikov, D. V. Kirillov, News of Higher Educational Institutions, Construction **6**, 25-31 (2004)
7. A.A. Ovchinnikov, *Development of compositions of the heat-resistant concrete based on the liquid glass with the superplastizer*: Ph D. thesis in Engineering Science, 05.23.05 (Ivanovo, 2003)
8. A. B. Toturbiyev, Concrete and reinforced concrete **3**, 5-8 (2010)

9. Z.A. Manturov, Engineer-construction journal **3** (2009)
10. L.G. Sudakas, *Phosphate binding systems* (Ria "Quintet", Saint-Petersburg, 2008)
11. B.L Krasny, *Fire resistant building materials based on phosphate binders* (2003)
12. A. I. Khlystov, V.V. Suldin, Association of Scientists (NAS), **3**(8), P. 3, 131-136 (2015)

# Some features of calculation of the reinforced concrete bent elements and economic assessment of design

*Asmik* Klochko[1*], *Aleksej* Klochko[2]

[1] Moscow state university of civil engineering, Yaroslavskoye shosse, 26, Moscow, Russia, 129337
[2] Moscow state university of civil engineering, Yaroslavskoye shosse, 26, Moscow, Russia, 129337

**Abstract.** At the design stage of the reinforced concrete bent elements, and also at their strengthening by method of accumulation of section, it is necessary to aspire to that the settlement strain-stress distribution of an element corresponded to actual. Importance of this circumstance is that their discrepancy can lead to increase in both economic, and not economic risks at exploitation of construction. In this article some features of calculation which are directly influencing profitability during creation new or strengthening of the rod-shaped bending structure are considered.

## 1. About the provision of the neutral axis of section of the bent element

According to the provisions formulated in [1], the bending moment perceived by the bending element of rectangular section $M_{ult}$, at observance of a condition:

$$\xi = \frac{x}{h_0} \leq \xi_R \tag{1}$$

is defined by a formula:

$$M_{ult} = R_b \cdot b \cdot x \cdot (h_0 - 0.5x) + R_{sc} \cdot A_s' \cdot (h_0 - a'), \tag{2}$$

where height of the compressed zone (h) is defined from a balance condition:

$$R_b \cdot b \cdot x + R_{sc} \cdot A_s' = R_s \cdot A_s \tag{3}$$

It is important that in the equation (3) the area of compression reinforcement $A_s'$ fulfill conditions:

* Corresponding author: asmik1985@mail.ru

$$A'_s < \frac{R_s}{R_{sc}} \cdot A_s. \quad (4)$$

Otherwise, if:

$$A'_s \geq \frac{R_s}{R_{sc}} \cdot A_s \quad (5)$$

that height of the compressed zone – x, calculated from the equation (3), will turn out zero or negative which physical sense can't be understood. It cannot be assumed that the compression pressure in fittings can reach the extreme $R_{sc}$, value if it is in concrete in which tension steadily remains zero. It is obvious that at the overestimated values of the area of the compressed fittings (5) the settlement strain-stress distribution of a design described by a formula (3) doesn't correspond to actual state and therefore, it will be correct to present it in the form:

$$\sigma_b \cdot b \cdot x + \sigma_{sc} \cdot A'_s = R_s \cdot A_s \,, \quad (6)$$

meaning that at the moment when tension in the stretched fittings reaches the limit state, tension in materials of the compressed zone doesn't manage to reach extreme values, and height of the compressed zone (x) surely covers the compressed fittings:

$$\sigma_b < R_b; \quad \sigma_{sc} < R_{sc}; \quad x > a', \quad (7)$$

where $a'$ - distance from the compressed element side to the center of gravity of the compressed fittings.

This circumstance isn't threat for reliable operation of a construction. It testifies only to excessive margin of safety of materials of the compressed zone and, so, to inefficient designing. Therefore, from the economic point of view, it is expedient that the area of the compressed fittings has been chosen within (4).

However researches show that similar (7) is observed as well at some values from a condition (4) at which the neutral axis of section gets to an interval $0 \div a'$.

We will show it on the example of 1.

Example 1.

It is required to define the limit bending $M_{ult}$ moment for a beam with b=40sm section sizes; h=45sm; made of class B20 concrete. The stretched fittings are accepted by $A_s$=15.2 cm2 (4Ø22 A400 s). For the compressed fittings we will consider two options:

option-1: $A'_s$ =12.32 cm2 (2Ø28 A400 s)
option-2: $A'_s$ = 5.09 cm2 (2Ø18 A400 s).

Accounting results are given in table 1.

**Table 1.** Accounting results

| Example 1 | x | ξ | $\xi_R$ | $A'_s$ | $A_s$ | Expense fittings | $M_{ult}$ |
|---|---|---|---|---|---|---|---|
| | cm | | | cm 2 | cm 2 | kg / m3 | kNm |
| Option-1 | 2.2 | 0.056 | 0.589 | 12.32 | 15.2 | 120.0 | 198.3 |
| Option-2 | 7.9 | 0.197 | 0.589 | 5.09 | 15.2 | 88.5 | 198.1 |

We will notice that in both options the condition (1) is satisfied. However in the first option the compressed zone of section doesn't cover the compressed fittings: x=2.2sm<$a'$=5.0sm. It means that the condition of balance of section is more precisely described by the equation (6), but not the equation (3) because the compressed fittings can't be out of a zone of the compressed concrete. So the case described in (7) when in a limit condition of a design of tension in materials of the compressed zone don't reach the extreme values takes place. The fact that such designing isn't economic becomes obvious of comparison of settlement these both options.

So, in the second option the compressed section zone covers the compressed fittings now: x=7.9 cm <a′=5.0 cm. In other words in a extreme limit state of an element the tension in materials of the compressed zone reach the extreme values, so, the condition of balance (3) is fair. It must be noted that the reason for that was that the area of the compressed fittings $A'_s$ has been lowered with 12.32sm2 on 5.09sm2. As a result, apparently from the table−1, the limit bending moment $M_{ult}$ remained almost invariable, and the specific expense of fittings has decreased by 27% - from 120.0 kg/m3 on 88.5 kg/m3, hat testifies to an economic benefit.

Thus, it is possible to conclude that, from the point of view of profitability, when calculating of the bent rod elements:

• it is necessary to aspire that the area of the compressed fittings was less than the area of the stretched fittings;

• control over the provision of a neutral axis by means of only a condition (1) is insufficient. It is expedient to demand also that in a limit condition of a design the compressed zone of cross section has captured the compressed fittings.

Importance of this conclusion is especially shown in problems of strengthening of the rod bent designs by method of accumulation of section.

## 2. About the location of the section neutral axis when strengthening of the bent element by method of section growth

The reinforced concrete rod bent elements strengthened by cross section growth, should be calculated as monolithic and are carried out according to instructions [2]. However researches show that except observance of the recommendations stated in [2] it is necessary to pay attention as well to some other features of calculation which outright influence profitability of strengthening. One of such feature is the additional control of neutral axis location of the strengthened design section which is carried out from a position of profitability of strengthening [3,4].

Usually when strengthening of the bent elements the neutral axis location of cross section of the strengthened construction is controlled by a condition (1). However it is not infrequent that at observance of this condition, the neutral axis appears above the compressed fittings of the strengthened element. From the point of view of reliability such circumstance is admissible. However researches show that in such cases the cumulative durability of the compressed zone materials of the strengthened construction remains with an excessive stock and, therefore, strengthening become inefficient.

We will show it on the example of 2.

Example 2.

Parameters of section of the strengthened construction are given: b=30sm width; h=50sm height; class B20 concrete; the stretched$A_s$=18.47 sm2 (3Ø28) fittings: the compressed $A'_s$=6.28 sm2 (2Ø20); A400s fittings class. It is required to strengthen a beam a reinforced concrete holder with finishing her bearing ability on the bending moment to $M_{ult}$=880 кНм.

We will consider option of strengthening at which section is increased from below on $y_1$=17.0sm; from above on $y_2$=8.0sm and on each side on $y_3$=5.0sm. Total, the sizes of section of the strengthened element make $b_{ad} \times h_{ad}$ = 40×75sm. Concrete of the strengthening element is accepted the class B25, class A400s fittings. For further comparison reinforcing of the strengthening element it is feasible in two options:

option-1: $A_{s,ad}$=24.55sm2 (5Ø25) и $A'_{s,ad}$ =18.60sm2 (2Ø20+2Ø28),
option-2: $A_{s,ad}$=26.08sm2 (5Ø25+1Ø14) и $A'_{s,ad}$ =10.31sm2 (2Ø20+2Ø16).

Calculations are executed according to the settlement model given in [2] according to which the bearing ability of the strengthened design is determined by the bending moment by a formula:

$$M_{ult} = R_{b,red} \cdot b_{ad} \cdot x \cdot (h_{0,red} - 0.5x) + R_{sc,ad} \cdot A'_{s,red} \cdot (h_{0,red} - a'_{red}), \quad (8)$$

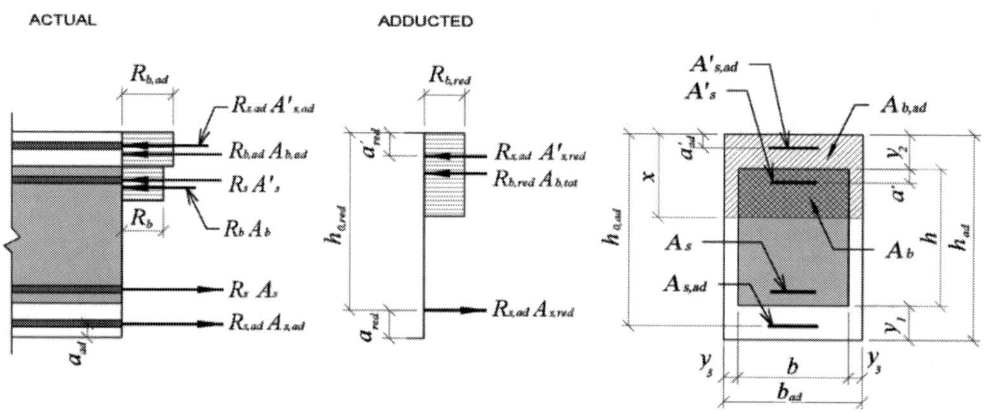


**Fig. 1.** The scheme of efforts in the cross section of the strengthened bent element.

where height of the compressed zone (x) is defined from a balance condition:

$$R_{b,red} \cdot b_{ad} \cdot x + R_{sc,ad} \cdot A'_{s,red} = R_{s,ad} \cdot A_{s,red}. \quad (9)$$

Results of calculations are given in the table-2.

From the table-2 it is visible that in both options the condition (1) is satisfied. However the neutral axis of cross section is in the first option above the compressed fittings of the strengthened element: x=11.8sm<$a' + y_2$=13.0sm (Fig. 2). Means, in a limit condition of the strengthened tension design in materials of the compressed zone of the strengthened element don't reach the extreme values. It contradicts the value accepted in a condition of balance of efforts (9) where are accepted $\sigma_{b,red} = R_{b,red}$ и $\sigma_{sc} = R_{sc}$. That is the settlement intense deformed condition of the strengthened design doesn't correspond to actual state and therefore, also R _ will be correct in a formula (9) parameters $R_{b,red}$ и $R_{sc,ad}$ o replace with values$\sigma_{b,red}$ и $\sigma_{sc,ad}$.

In the second option, apparently from the table-2, the neutral axis is below the compressed fittings of the strengthened element: $x$=19.0sm< $a' + y_2$=13.0sm (Fig. 2). Therefore, durabilities of materials of the compressed zone of section of the strengthened element in the general work of the strengthened design are used fully that is, the condition of balance (9) is fair.

Thus, the task of increase in the bearing ability of a design of the bending moment to $M_{ult}$=880 kNm in both options is carried out. However, apparently from comparison of results of calculation, on a specific expense of fittings the second option in relation to the first is effective for 16% thanks to a right choice of provision of a neutral axis.

One more feature of strengthening of the bent designs, noteworthy, the choice of the direction of building is. We will show it on the example reviewed above where the section of the strengthened design it was increased mainly from the stretched zone: $y_1$=8.0sm, $y_2$=17.0sm. Now we will consider option at which building is made mainly from the compressed section zone: $y_1$=8.0sm, $y_2$=17.0sm (Fig. 2). For descriptive reasons we will accept all other data accepted in the second option without change, except for the cross-sectional area of the additional compressed fittings $A'_{s,ad}$, which is defined by calculation.

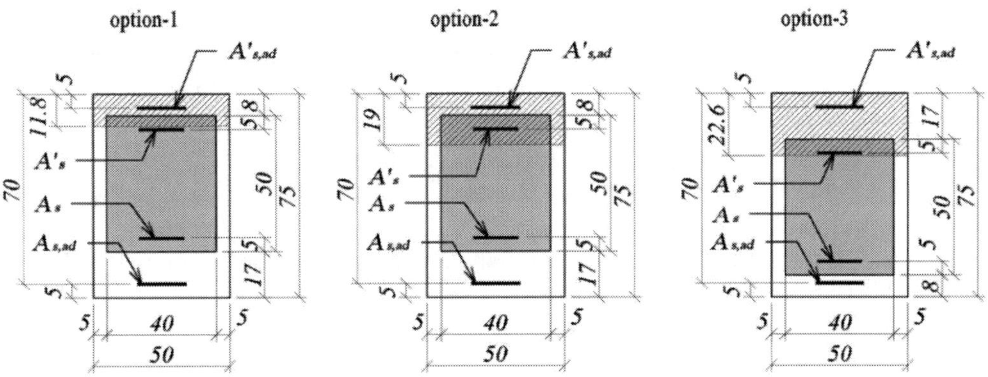


**Fig. 2** Illustrations for example 2

From the table-2 it is visible that in the third option the condition on control of provision of a neutral axis offered above is satisfied again: $x$=22.6sm< $a' + y_2$=22.0sm, therefore, durabilities of materials of the compressed zone of section of the strengthened element in the general work of the strengthened design are fully used by a. Besides, from comparison with the second option it becomes obvious that the third option of strengthening is more effective as at an identical consumption of concrete the identical bearing ability is reached by a smaller expense of fittings 3.08sm2 against 10.31sm2. That is the specific consumption of fittings was cut only by one change of the direction of accumulation of section for 20% (152.6 kg/m3 against 190.4 kg/m3).

**Table 2.** Accounting results

| Ex. 2 | $R_{b,red}$ | $A'_{s,red}$ | $A_{s,red}$ | $a'_{red}$ | $a_{red}$ | $a' + y_2$ | x | $\xi$ | $\xi_R$ | $A'_{s,ad}$ | $A_{s,ad}$ | Elementary discharge of fittings | $M_{ult}$ |
|---|---|---|---|---|---|---|---|---|---|---|---|---|---|
| | Kg / $sm^2$ | $sm^2$ | $sm^2$ | sm | sm | sm | sm | | | $sm^2$ | $sm^2$ | kg/$m^3$ | kNm |
| Opt. 1 | 141.0 | 24.88 | 43.02 | 7.0 | 12.3 | 13.0 | 11.8 | 0.188 | 0.589 | 18.60 | 24.55 | 225.8 | 884.0 |

| | | | | | | | | | | | | | |
|---|---|---|---|---|---|---|---|---|---|---|---|---|---|
| Opt. 2 | 135.0 | 16.59 | 44.55 | 8.0 | 12.0 | 13.0 | 19.0 | 0.302 | 0.589 | 10.31 | 26.08 | 190.4 | 882.0 |
| Opt. 3 | 142.0 | 9.36 | 44.55 | 16.4 | 8.3 | 22.0 | 22.6 | 0.339 | 0.589 | 3.08 | 26.08 | 152.6 | 883.0 |

It is immediately apparent from the foregoing that when calculating strengthening of the bent rod elements by method of accumulation of section:

• it is expedient to demand that in a limit condition of the strengthened design the compressed zone of cross section has captured the compressed fittings of the strengthened construction.

• important economic value has a right choice of the direction of building of the strengthened construction.

## References

1. Rules and regulations 63.13330.2012 Concrete and reinforced concrete structures. Basic provisions. 152 pages (2012)
2. Recommendations about design of buildings and constructions concrete structures strengthening of the reconstructed enterprises. Elevated designs and constructions (Kharkiv Promstroyniiproyekt, NIIZhB. - M.: Sroyizdat. - 1992)
3. R. G. Arutyunyan Strengthening of reinforced concrete designs by method of search optimization (Housing construction, **11**, pp.12-13, 2000).
4. A.G.Tamrazyan, M.A. Orlova. Stress-strain state of concrete bending elements with cracks (Bulletin of the Tomsk state architectural and construction university, **6 (53)**, pp. 98-105, 2015)

# The influence of self-stress on the behavior of tensegrity-like real structure

*Wojciech* Gilewski[1], *Joanna* Kłosowska[2,*] and *Paulina* Obara[2]

[1]Warsaw University of Technology, Faculty of Civil Engineering, Al. Armii Ludowej 16, Warsaw 00-637, Poland
[2]Kielce University of Technology, Faculty of Civil Engineering and Architecture, Al. Tysiąclecia PP 7, Kielce 25-314, Poland

**Abstract.** Full-scale tensegrity skeleton of White Rhino is considered in this paper. The influence of self-stress state on structure behaviour is studied. Few models that consist of 15, 16, 17 or 18 elements are analysed. For the models self-stress states and infinitesimal mechanisms are considered. The impact of the level of self-stress state on displacement is investigated. Moreover a failure of White Rhino is considered. The influence of a damage of one, two or three cables on displacements is examined. Analyses are performed using the second order theory in the *Mathematica* environment and the third order theory using the *Sofistic* program.

## 1 Introduction

The term tensegrity concerns complex systems which consist of elements which are only in compression (struts) or tension (cables) [1]. Cables are foldable, light and strong. Thanks to these properties tensegrity structures gain the potential to be light but as well as deployable. There are many benefits of tensegrities for example tension stabilization, efficient structures development, deployability and easily tunable properties [2]. That are the main reasons of wide applications of tensegrity structures is appropriate in various areas of civil engineering [3]. Due to construction tensegrities can be clasiffied as linear systems (1D) like towers, sufrace systems (2D) like plates or shalls and single modules (3D). The example of the last one is building which is covered with membrane roofs supported by two tensegrity skeletons called White Rhino [4, 5].

White Rhino has been constructed at Chiba in Japan in June, 2001. The name of a structure, comes from the exterior appearance of the roofs. The membrane roof is white colour and it is supported by two tensegrity trusses varying in height which looks like two "horns" (Fig. 1a). The height of the bigger tensegrity structure is about ten meters, the smaller is seven meters in high. Both skeletons are modified simplex trusses (Fig. 1b). The simplex is three dimensional truss consists of the three struts and nine cables. This truss is the simplest example of tensegrity. It has one infinitesimal mechanism which is stabilized by self-stress state. The constructors of White Rhino added seven additional members – six

* Corresponding author: jklosowska@tu.kielce.pl

cables and one strut. Three additional strings there are between unconnected six points of Simplex. These members do not change the appearance of the tensegrity but they improve much the rigidity of the structure. An isolated strut which is connected with truss by three another cables are a kind of support for the roof. These elements absorb large deformation of membrane roof and transmits the force from membrane roof to the tensegrity skeleton. Moreover White Rhino's general shape differs from Simplex model. It is rather trapezoidal than prismatic. The upper triangle is smaller than the bottom triangle.

a) b)

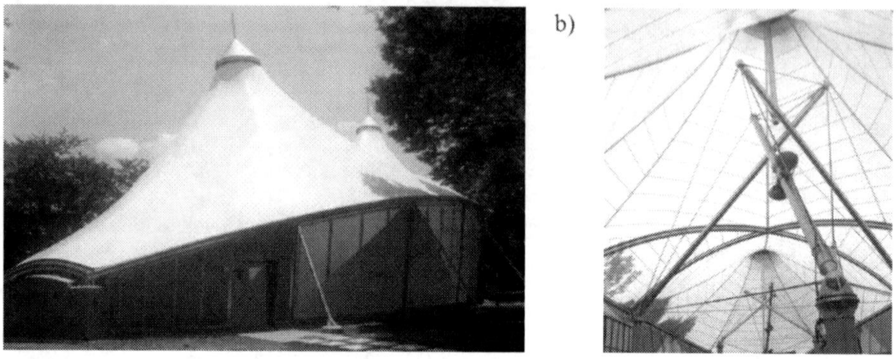

**Fig. 1.** White Rhino: a) exterior appearance, b) view inside [4].

Extended Simplex tensegrity skeleton of White Rhino is considered in this paper. According to the paper [4] real geometry and material characteristics are taking into account. The influence of damage of some cables on behaviour of the truss is discussed. Additionally, the impact of the level of self-stress on displacements in the structure are analysed.

Analyses were performed using the second order theory and the third order theory. In the first case calculations were made in the *Mathematica* environment – the computational program based on the finite element method was written. In the second case calculations were made using the program *Sofistik* that served to accomplish fully geometrical non-linear analysis.

## 2 Formulation of the problem

The aim of this paper is study behaviour of structure stabilized by cables tension. To do this infinitesimal mechanisms and self-stress states should be determined and next the influence of the level of self-stress on displacements should be analysed. Identification infinitesimal mechanisms and a self-stresses in structures is possible by using the second order theory. In order to take account the effect of additional compression the third order theory should be used.

### 2.1 Mathematical formulation

The equilibrium equation for geometrically nonlinear model (third order theory) is presented in the form:

$$\left(\mathbf{K}_L + \mathbf{K}_{NL}^{I} + \mathbf{K}_{NL}^{II}\right)\mathbf{q} = \mathbf{P} \tag{1}$$

where **q** is displacement vector and **P** is the load vector, $\mathbf{K}_L$ is the linear stiffness matrix and $\mathbf{K}^{I}_{NL}$ is the pre-stress stiffness matrix called the geometric stiffness matrix and $\mathbf{K}^{II}_{NL}$ is the initial strain matrix. For trusses the linear stiffness matrix can be write as:

$$\mathbf{K}_L = \mathbf{B}^T \mathbf{E} \mathbf{B} \tag{2}$$

where **B** is the compatibility matrix and **E** is the elasticity matrix.

Based on singular value decomposition of compatibility matrix **B** infinitesimal mechanisms and self-stress states are identified [6, 7]. Next the pre-stress stiffness matrix depended on the self-stress state is calculated and spectral analysis taking into account this matrix is used. Geometric non-linear analysis, including the initial strain matrix, is made using the program *Sofistic*.

## 2.2 Model of the White Rhino

A modified model of the full-scale extended Simplex tensegrity skeleton of the White Rhino is considered. This model consists of 18 elements. An isolated strut which transmits the force from membrane roof to the tensegrity skeleton is omitted. This rod is modelled as a load. Several models of the structure are analysed.

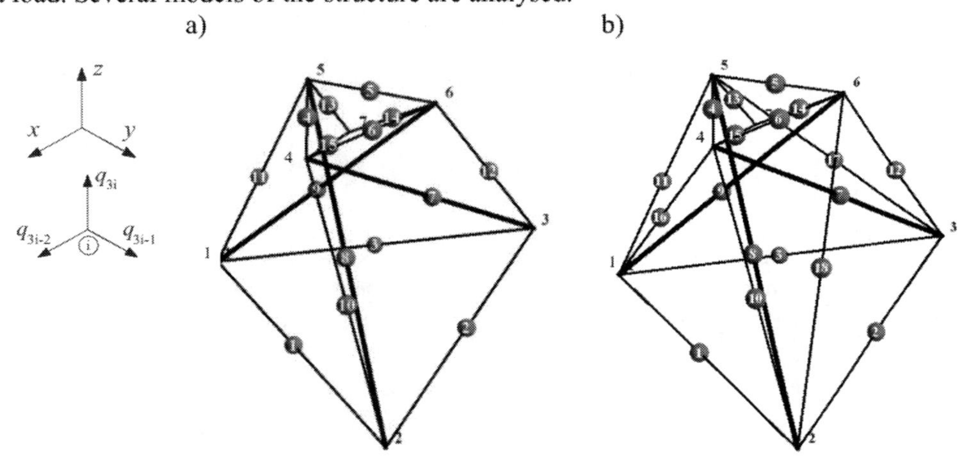


**Fig. 2.** Scheme of the White Rhino: a) model I, b) model IV.

**Table 1.** Node coordinate of the White Rhino model.

| No. of node | *x* [m] | *y* [m] | *z* [m] |
|---|---|---|---|
| 1 | -15.283 | 0.000 | 3.000 |
| 2 | -8.868 | -3.748 | 4.000 |
| 3 | -8.868 | 3.748 | 4.000 |
| 4 | -11.805 | - 1.709 | 8.796 |
| 5 | -13.267 | 0.854 | 8.568 |
| 6 | -10.343 | 0.854 | 9.020 |
| 7 | -11.805 | 0.000 | 8.300 |

The first examined model (model I) consists of the 15 elements (Fig. 2a). It is a modified Simplex truss by added three cables (13, 14 and 15), which carry load from membrane roof. The next models are built by adding additional cables. Model II consists of the 16 elements (16th cable was added). Model III consists of the 17 elements (16th and 17th cables were added) and last – model IV consists of the 18 elements (16th, 17th and 18th cables were added) (Fig. 2b). The structure is supported in nodes 1, 2, 3 (six degrees of freedom are locked, in the direction of displacements: $q_1, q_2, q_4, q_6, q_8, q_9$). For all models self-stress states and infinitesimal mechanisms are defined. For the model I and the model IV the impact of the level of self-stress state on displacements is investigated.

**Table 2.** Member properties of steal (Young's modulus $E = 210$GPa).

| No. of elements | Kind of elements | Diameter [m] | Sectional area [m$^2$] | Length [m] | Yielding stress [MPa] | Load capacity [kN] |
|---|---|---|---|---|---|---|
| 1, 2, 3 | cables | 0.032 | $8.04 \cdot 10^{-4}$ | 7.497 | 460 | 369.8 |
| 4, 5, 6 | cables | 0.028 | $6.16 \cdot 10^{-4}$ | 2.959 | 460 | 283.4 |
| 7, 8, 9 | struts | 0.216 | $5.36 \cdot 10^{-3}$ | 7.836 | 320 | -938.0 |
| 10, 11, 12 | cables | 0.032 | $8.04 \cdot 10^{-4}$ | 5.982 | 460 | 369.8 |
| 13, 14, 15 | cables | 0.025 | $4.91 \cdot 10^{-4}$ | 1.840 | 460 | 225.9 |
| 16, 17, 18 | cables | 0.025 | $4.91 \cdot 10^{-4}$ | 6.972 | 460 | 225.9 |

In the next step of the analysis the failure of the White Rhino is considered. The impact of a damage of one, two or three cables of (1, 2, 3, 4, 5, 6, 10, 11, 12, 16, 17, 18) on displacement $q_{10}$ is examined (Table 4,5).

Real geometry (Table 1) and material characteristics (Table 2) are taking into account according to the paper [1]. Concentrated force $P_{21}$ at the node 7, which is a simulation of the load from the roof, is considered. Since the model is nonlinear three values of this force are taking into account: $P_{21}$=-20 kN, $P_{21}$=-60 kN and $P_{21}$=-100 kN.

## 3 Results of the analysis

### 3.1 Identification of self-stress states and infinitesimal mechanisms

For all models self-stress states (Tab. 3) and infinitesimal mechanisms are defined. For the model I and the model IV the impact of the level of self-stress state on displacements for different values of external load is presented in Fig. 3-5.

Analysing the first model one infinitesimal mechanism which is stabilised by one identified self-stress state (SS) is determined. One additional cable (model II) causes that the structure loses some of features which are typical for tensegrities. There is no mechanism identified but there is still one self-stresses state (SS) the same like in the model I. In the case of the model III no mechanisms are identified but there are two self-stress states (1th SS, 2th SS – in the Tab. 3). These states need to be considered together. Superposition and normalisation of self-stress states are performed and the self-stress state (SS – in the Tab. 3) is obtained which is equal to self-stresses identified in previous models. The analysis of the model IV shows that there are no mechanisms, like previously, but there are three self-stress states (1th SS, 2th SS, 3th SS). In this case a superposition and a normalisation of three states allow to gain the one self-stress state (SS). This self-stress state is the same like in previous models. Further analyses are performed for this self-stress state.

Dependence of self-stress factor $S$ on displacement $q_{10}$ for the model I and model IV is shown in Fig. 3, 4 and 5. Three values of the external force are analysed: $P_{21}$=-20 kN (Fig. 3) $P_{21}$=-60 kN (Fig. 4) and $P_{21}$=-100 kN (Fig. 5). Values of the self-stress factor $S$ are limited by load capacity of struts.

**Table 3.** Self-stress states [*S*].

| No. of element | Model I | Model II | Model III | | | Model IV | | | |
|---|---|---|---|---|---|---|---|---|---|
| | SS | SS | 1 th SS | 2 th SS | SS | 1 th SS | 2 th SS | 3 th SS | SS |
| 1 | 0.22 | 0.22 | -0,09 | 0,08 | 0.22 | -0,17 | 0,03 | 0,08 | 0.22 |
| 2 | 0.22 | 0.22 | -0,09 | 0,08 | 0.22 | 0,10 | -0,02 | 0,16 | 0.22 |
| 3 | 0.22 | 0.22 | -0,09 | -0,17 | 0.22 | -0,02 | -0,18 | -0,05 | 0.22 |
| 4 | 0.55 | 0.55 | -0,23 | 0,42 | 0.55 | -0,10 | 0,16 | 0,44 | 0.55 |
| 5 | 0.55 | 0.55 | -0,23 | -0,21 | 0.55 | -0,41 | -0,24 | -0,09 | 0.55 |
| 6 | 0.55 | 0.55 | -0,23 | -0,21 | 0.55 | 0,28 | -0,37 | 0,12 | 0.55 |
| 7 | -1.00 | -1.00 | 0,42 | 0,01 | -1.00 | -0,27 | 0,35 | -0,41 | -1.00 |
| 8 | -1.00 | -1.00 | 0,42 | -0,38 | -1.00 | 0,37 | -0,06 | -0,48 | -1.00 |
| 9 | -1.00 | -1.00 | 0,41 | 0,39 | -1.00 | 0,32 | 0,51 | 0,04 | -1.00 |
| 10 | 0.76 | 0.76 | -0,32 | 0,29 | 0.76 | 0,03 | -0,02 | 0,46 | 0.76 |
| 11 | 0.76 | 0.76 | -0,32 | 0,00 | 0.76 | -0,42 | -0,14 | 0,12 | 0.76 |
| 12 | 0.76 | 0.76 | -0,31 | -0,30 | 0.76 | 0,07 | -0,45 | 0,07 | 0.76 |
| 13 | 0.00 | 0.00 | 0.152 | 0.152 | 0.00 | 0.00 | 0.00 | 0.00 | 0.00 |
| 14 | 0.00 | 0.00 | 0.00 | 0.00 | 0.00 | 0.00 | 0.00 | 0.00 | 0.00 |
| 15 | 0.00 | 0.00 | 0.00 | 0.00 | 0.00 | 0.00 | 0.00 | 0.00 | 0.00 |
| 16 | - | 0.00 | 0,00 | -0,34 | 0.00 | 0,20 | -0,29 | -0,17 | 0.00 |
| 17 | - | - | 0,00 | 0,34 | 0.00 | 0,16 | 0,22 | 0,28 | 0.00 |
| 18 | - | - | - | - | - | -0,37 | 0,07 | -0,11 | 0.00 |

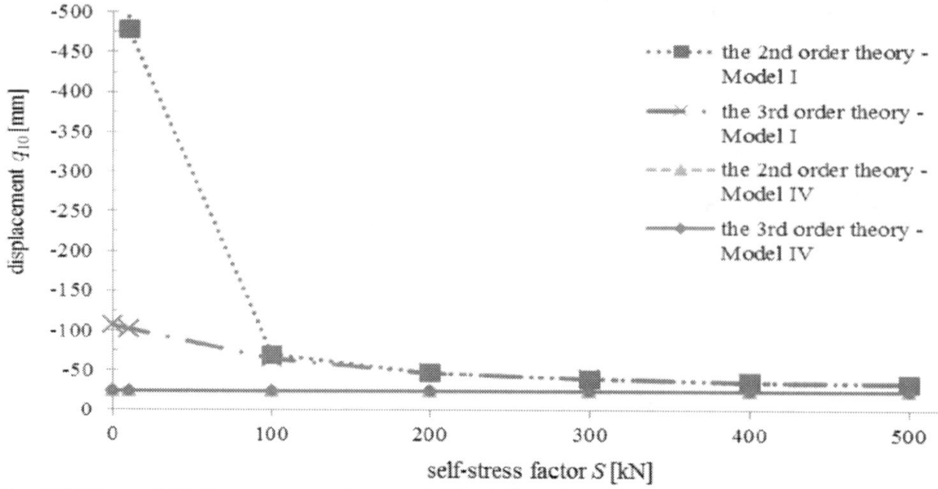


**Fig. 3.** Values of displacement $q_{10}$ for a nodal load $P_{21}$=-20kN.

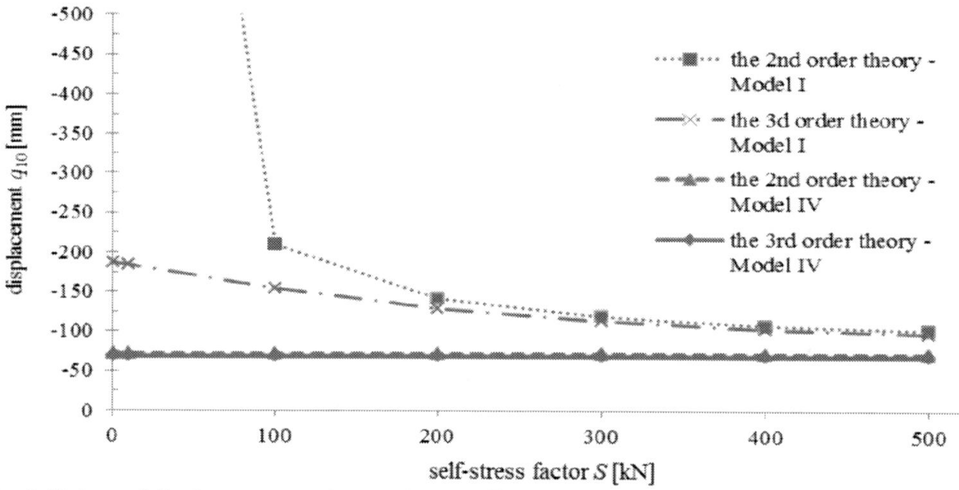


**Fig. 4.** Values of displacement $q_{10}$ for a nodal load $P_{21}$=-60kN.

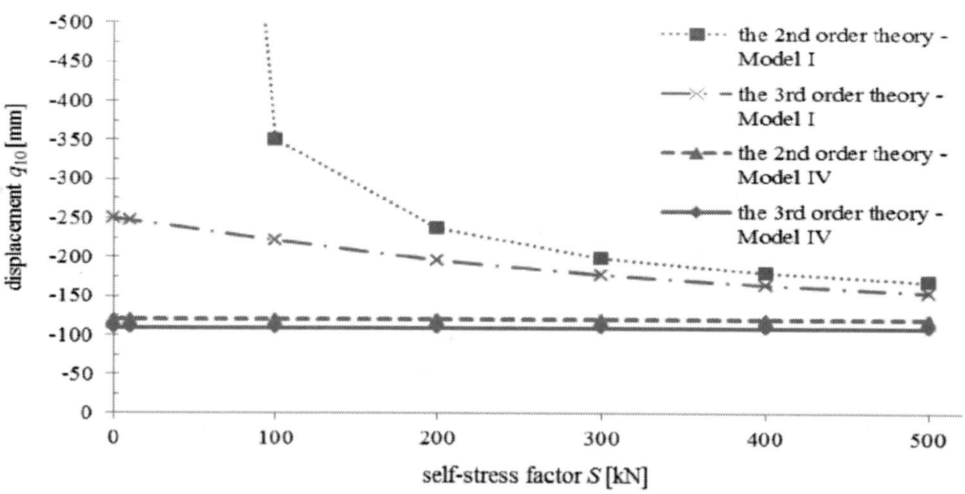


**Fig. 5.** Values of displacement $q_{10}$ for a nodal load $P_{21}$=-100kN.

The analysis of selected displacement in case of damage of some elements are presented in the Tab. 4 for the structure supported with six d.o.f. locked on the support. The displacement is relatively large, then the analysis with nine d.o.f. locked is performed and presented in the Tab. 5. The displacement is significantly smaller in the second model of damage.

**Table 4.** Values of displacement $q_{10}$ [mm] in dependence on the self-stress factor $S$ for a nodal load $P_{21}$=-60kN (six degrees of freedom are locked).

| The number of damaged elements | No. of damaged elements | The self-stress factor $S$ | |
|---|---|---|---|
| | | $S$=0 kN | $S$=500 kN |
| 1 | 1 or 2 or 3 | -586.273 | -835.759 |
| | 4 or 5 or 6 | -72.494 | -61.959 |
| | 10 or 11 or 12 | -69.263 | -83.482 |
| | 16 or 17 or 18 | -68.996 | -68.933 |
| 2 | 1,2 or 2,3 or 1,3 | -665.817 | -762.567 |
| | 4,5 or 5,6 or 4,6 | -76.850 | -42.715 |
| | 10,11 or 11,12 or 10,12 | -73.081 | -52.496 |
| | 16,17 or 17,18 or 16,18 | -68.946 | -68.897 |
| 3 | 1, 2, 3 | -663.711 | -741.911 |
| | 4, 5, 6 | -75.974 | -66.571 |
| | 10, 11, 12 | -75.580 | -71.866 |
| | 16, 17, 18 | -187.673 | -97.552 |

**Table 5.** Values of displacement $q_{10}$ [mm] in dependence on the self-stress factor $S$ for a nodal load $P_{21}$=-60kN (nine degrees of freedom are locked).

| The number of damaged elements | No. of damaged elements | The factor of self-stress $S$ | |
|---|---|---|---|
| | | $S$=0 kN | $S$=500 kN |
| 3 | 1, 2, 3 | -0.772 | -0.768 |
| | 4, 5, 6 | -1.400 | 4.057 |
| | 10, 11, 12 | -1.151 | -8.731 |
| | 16, 17, 18 | -109.210 | -26.341 |

## 4 Conclusions

The influence of self-stress on the behaviour of the tensegrity-like structure is studied in this paper. The full-scale tensegrity skeleton of White Rhino is considered. Four models which consist of 15, 16, 17 or 18 elements are analysed.

For the model I, the influence of the self-stress factor on nodal displacements is very noticeable. Nodal displacements decrease significantly with increase of internal forces of self-stress as well as of the influence of geometrical nonlinearity is clearly seen. If the value of self-stress equals zero, displacements tend to the infinity. An adjusting of prestressing forces allows to control displacements of nodes.

Additionally the analysis shows that the effect of nonlinearity is the most significant at low values of self-stress forces. With increasing of self-stress the differences in the values of displacements calculated according to the second and third order theory are getting smaller.

In the case of the model IV three extra cables cause that the structure lost some of features which are typical for tensegrities. There are no mechanisms identified but there are three self-stresses states which is just one both in the case of the original Simplex and in the model I. Elimination of the mechanism makes the skeleton very rigid. The model IV is much less sensitive for changes of the self-stress factor.

Moreover, comparing obtained graphs (Fig. 3-5) it can be noticed that nodal displacements increase nonlinearly with the increase of external load.

In the analysis of damage of White Rhino, the failure of one, two and three cables is considered. Lower cables (1,2,3), upper cables (4,5,6), diagonal cables (10,11,12) and additional cables (16,17,18) are analysed (Tab. 4). The analyses performed reveal that the destruction of one, two or even three cables from (4,5,6,10,11,12,16,17,18) does not damage the whole structure. In such cases, the higher factor of self-stress improves a reliability slightly.

However, the failure of one of the lower cables (1,2,3) would be equivalent to the damage of the truss. In case of such the damage, the pre-stress of the elements additionally increases nodal displacements. The values of displacements are relatively large in case of the failure lower tendons what was the reason that the model supported by nine ties was analysed (Tab. 5). Safety of the structure increases significantly in this case. That is why it is a recommended way to support the structure.

## References

1. R. Motro, *Tensegrity. Structural Systems for the Future* (Kogan Page, London-Sterling, 2003)
2. Skelton R.E., de Oliveira M.C., *Tensegrity systems* (Springer, London, 2009)
3. V. Gómez-Jáuregui, *Servicio de Publicaciones de la Universidad de Cantabria*, pp. 296 (2010)
4. K. Kawaguchi, Z-Y. Lu, *Proceedings of 5th International Conference on Space Structures, edit. by G. Parke & p. Disney* **1**, 1 (2002)
5. K. Kawaguchi, S. Ohya, S. *Vormus, 5th Annual Symposium of IABSE/52nd Annual Symposium of IASS / 6th International Conference on Space Structures*. London, (2011)
6. W. Gilewski, J. Kłosowska, P. Obara, *Procedia Engineering* **111**, 242 (2015)
7. W. Gilewski, J. Kłosowska, P. Obara, *Advances in Mechanics: Theoretical, Computational And Interdisciplinary Issues*, 191 (2016)

# Risks of the Durability of Large-Panel Buildings Elevations in Reference to the Conclusions from Technical Conditions Audits

*Piotr* Knyziak [1,*], *Janusz R.* Krentowski [2] and *Piotr* Bieranowski[3]

[1]Warsaw University of Technology, Al. Armii Ludowej 16, 00-637 Warsaw, Poland
[2]Bialystok University of Technology, Wiejska str. 45, 15-351 Bialystok, Poland
[3]University of Warmia and Mazury in Olsztyn, Oczapowskiego str. 11, 10-719 Olsztyn, Poland

**Abstract.** In article the problem of durability of elevations of large-panel technology buildings is analyzed. In Poland this is a serious technical issue because it has an impact on the whole country. Heat loss through infiltration, the possibility of mold growth and general breakdowns are also reasons for increasing user discomfort. In article the results of the diagnostics presented in the literature and the authors' analysis and diagnostic results are presented. Article also presents conclusions regarding to the durability of applied insulation systems ETICS and the need to strengthening the layers connections of the facade panels.

## 1 Introduction

Building prefabrication in Poland has a bad reputation, especially in the area of residential buildings. From one side it is result of the real mistakes in designing and manufacturing process, on the other side such point of view was formed in newspapers articles where the problem of defects has been highlighted and maximizing.

In buildings made of precast concrete plates the influence of unfavourable factors can not disclose for a very long time. Prefabricated buildings build in Poland in XX century are especially sensitive in regard to the layered structure of cladding panels of which were made external walls. They consist of a structural layer, an insulating layer made of polystyrene or mineral wool with varying thicknesses and a finishing layer. Thickness of the designed layers were different in different systems and they took into account the climatic zones, the number of storeys and the type of connecting elements. The technical condition of connections of these layers is the main problem, because often to achieve savings ordinary steel were used in them instead of stainless steel.

Both insulation and textured finishing layers were fixed to the structural panel by means of fasteners specified as "hangers" (Figure 1), made of steel bars. The effectiveness of the connection of the two concrete layers determines the durability of such external wall.

In the 1980s, as a result of research carried out on the Building Research Institute in Warsaw, hangers made of H13N4G9 steel were recommended. It was found that connectors made of St3SX steel had actually been used. Moreover, the presence of galvanized bars of StOS steel was confirmed [1]. There were also errors in the layout, number and diameters

* Corresponding author: p.knyziak@il.pw.edu.pl

of the hangers used. In addition, it was found that the thickness of concrete in the façade layers often exceeded the designed values, resulting in both an increase in the weight and a reduction in the thickness of the thermal insulation layer. Consequently, it resulted in a reduction in the thermal insulation of the shielding wall and an increase in the stresses in the hangers due to both the heavier load of the façade layer and the thermal deformation of the façade panel exposed to direct weather and temperature stress. In an aggressive environment, in case of leakage of the façade layer, rain water, containing dissolved substances accelerating the corrosion process of the steel hangers, penetrates into the insulation space.

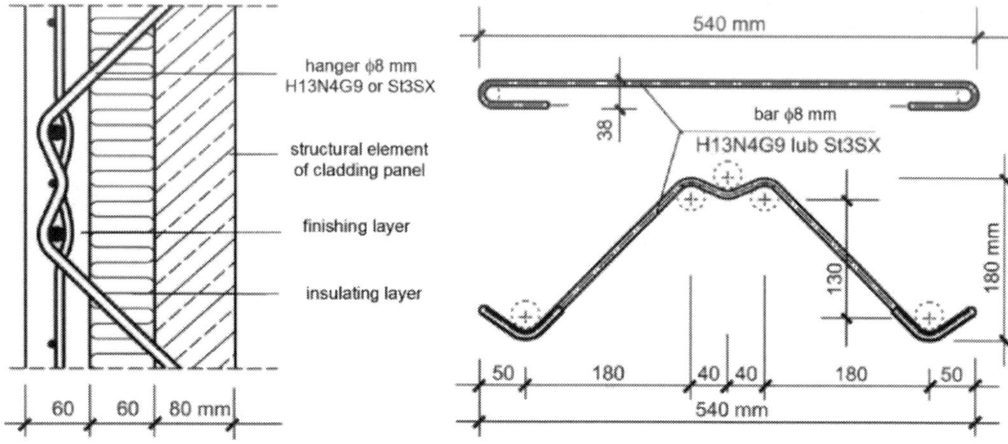


**Fig 1.** Construction of a connection of prefabricated wall elements in a Wk-70 system

Nowadays the high thermal conductivity of panels and connections leak causing moisture of precast elements are very important problems (Figure 2). Often wetting of the structure resulting from the leakage of connection is a stimulator of many degradation processes. They have a significant impact on the condition of not only individual panels or components, but the entire building. The negative effects of moisture have an influence on the health of residents by supporting the development of fungi and molds.

**Fig. 2.** Building façades in Wk-70 technology, Warsaw (photos by Janusz Krentowski)

It can be noted that after the executed thermorenovations of large-panel buildings [2], increased heat loss through contact zones of windows' frames and walls, prefabricated facade panels connections/joints, loggias' and balconies' bonding zones and plinth zones may be observed. Such repairs can be regarded as barely sufficient, but the results of their execution do not comply with the economic requirements of thermal protection. In the further use of buildings, the elimination of thermal bridges still remains a major problem.

Housing co-operatives in Poland, using state financial aid launched in 1982, have warmed/insulated many buildings or at least their gable walls. Up to 1997, it was financed from subsidies on housing cooperatives for the removal of technological defects. Unfortunately, especially in the first years, the insulation thickness was 2÷3-fold smaller than the current requirements of buildings thermal protection. In addition, in some cases a heavy dry method or insulation of fiber-cement boards was used, which did not allow for the addition of another insulation layer. As a consequence of a low thickness of styrofoam or mineral wool used in the first layers of insulation, many of the evaluated buildings had to be insulated repeatedly. At present, the typical insulation thickness is 16 cm of styrofoam.

## 2 Risks of the elevation durability indicated in the literature

Problems with the elements forming the facade of the large-panel buildings have been analyzed by many authors.

The research on panel layer fasteners were carried out on a wide scale around 1985 at the Silesian Technical University in Gliwice [3]. Hangers and pins from the current production of prefabrication factories made of St3SX steel were used for the tests. As a result of the study, images of the deformations of the hangers and displacements of the façade layer were obtained. What was important, it was found that during subsequent phases of work, the system connecting the wall-supporting plate with the façade layer indicates the failure by large displacements resulting from increasing loads.

Conducted studies and observations of existing buildings [4] have shown that sandwich panels have been made and assembled poorly. However, the assessed precast concrete was generally airtight, excluding sporadic cases, and the depth of the carbonation of concrete in façade panels did not exceed 5 mm. In terms of sustainability, the working conditions of the plates are good, as the structural layer of the boards is not directly affected by atmospheric factors. From the outside of the building, it contacts with the insulating material, which is usually dry. In the examined buildings, it was found that mineral wool and styrofoam were dry or only slightly damp. After the winter-spring season, the maximum moisture content of mineral wool did not exceed 1.5-2.0%, which indicated that the working conditions of plain steel hangers were more favorable than initially estimated. Occasional pitting corrosion, e.g. after 8 years of operation, pitting was 0.2 mm deep and did not significantly reduce the cross-section of the hangers.

The durability of a well-executed construction layer is expected to be over 100 years. The warming layer is generally thinner than required, which is the result of the insufficient density of the products used for warming and their compression during plate production. As a result, the walls are frosted and dampened. The elements that are important for the safety of the whole building, and therefore its life expectancy, are all kinds of joints, especially when they are not made of suitable materials.

The assessment of the technical condition and identification of damaged components [5] of panel buildings have revealed many attempts to repair the defects in joints. However, many repairs have been made incorrectly. In many cases lack of knowledge regarding the issues related to the construction of the discussed panel structures and failure to carry out more accurate analyzes caused fundamental technical errors. Movements in the joints of panel elements that constitute the façade of a building often negatively affect thermal insulation (moisture). As a result of the neglect, there were economic losses due to the need for further repairs or increased expenditures on current repairs.

Strengthening and repair works in buildings erected using the industrialized method, in most cases have been executed without prior investigations, morphology of existing scratches and defects [6]. ETICS insulation layers are usually made in the standard way, without any additional work to remove the cause of the cracks and damage. Covered façade of the walls does not allow for the control of the state of important structural façade wall elements.

Investigations conducted in buildings with visible textured façade layer [7] in many Polish cities showed: the overvalued concrete thickness in the façade layer, the incorrect position of the hangers, and incorrect anchoring of the hangers. The steel of the built-in hangers (except H13N4G9 steel) did not meet the technical requirements, contained in the COBPBO Manual (technical guidelines for this type of buildings). In the analyzed cases, there were no corrosion defects of the hangers affecting the safety of the structural part of the multi-layer walls.

## 3 Results of technical inspections in years 2005-06 and 2016-17 in Warsaw

In 2005-06 95 buildings in Warsaw were inspected [8, 9], and in 2016-17 110 buildings were examined. Data regarding large-block and large-panel dwellings and mixed technology (using prefabricated small-size elements and large-size prefabricated roof panels) was collected in representative quantities and quality. Buildings were chosen in terms of representativeness, so they are good examples of all the buildings in the settlements from which they are derived. Conclusions from the investigated buildings may be extrapolated to at least a three-fold higher number of buildings.

Few buildings in Warsaw made of prefabricated elements remain non-insulated. Out of the 110 buildings evaluated, 15 buildings have already been insulated twice (in the 2005-06 inspections, two buildings were already insulated twice, and one had double-insulated gable walls). Further layers have been added to the initial thermal insulation to achieve greater thermal energy savings. Some of the buildings were initially insulated only on the gable walls (without windows), and later, during the addition of insulation on the remaining walls, an additional layer was made on the first.

Another 21 buildings have been insulated twice, but the insulation works concerned different parts of the façade. In the first stage, only the gable walls (without windows) were insulated, because the unit cost was the lowest, and, as a result, the thermal conditions of the walls with the greatest frost penetration problems were corrected. Other walls, with windows, were warmed with radiators typically placed under the windows. In the case of 59 buildings, works covering the entire building or only gable walls were made once. The remaining 15 buildings have not had any insulation work done so far. The majority of them are buildings that were put into use in the 1990s. As they were built from upgraded system prefabricated materials, they are characterized by better insulation of exterior walls and roofs.

In addition, it has been found that not-isulated buildings are characterized by a relatively good quality of prefabricated materials, and the cases of leakage occur sporadically, between adjacent plates rather than in many plates. The frost's penetrations were present in few corners of examined buildings. The buildings were relatively new, and the façades are not technically problematic (Figure 3), so up to now, the co-operatives managing these buildings have not considered the insulation works economically and technically feasible.

**Fig. 3.** Elevation of a non-insulated large-panel building (photo by Piotr Knyziak, 2016)

In 104 out of the 110 assessed buildings, the reinforcement of joints in laminated cladding boards was not performed. In some of the buildings in these settlements, the earlier versions of thermal insulation (heavy-dry or light-dry) were dismantled and the buildings were re-insulated using ETICS method. In three buildings located in the analyzed areas (not only the evaluated buildings), there were cases of wind rupture on the large façade surfaces (Figure 4). None of the failures observed in the ETICS insulation systems caused the failure of the structures of the sandwich panels. In cases of gas explosions in flats in paneled buildings, there were plate damages only in the immediate vicinity of the explosion site. In the distant parts of the building, rupture of the interlayer hangers did not occur.

**Fig. 4.** ETICS system failures (photos by Piotr Knyziak, 2016)

Thermal insulation works, replacement of insulation, re-insulation, drilling of anchor holes to reinforce bonded joints in any of the buildings in the surveyed estates did not lead to the damage of the external façade panels. It should be emphasized that the process of drilling the holes designed for reinforcing anchors could have caused the vibrations and finally damages to the boards if they had already been weakened.

The fact that reinforcing the joints of multipanel layers during buildings insulation is not commonly used is not the result of lack of solutions to apply. There are many options from different manufacturers. One of the co-authors of the present article has proposed his own patented solution to the problem of reinforcing the bonding of prefabricated slab layers (Figure 5).

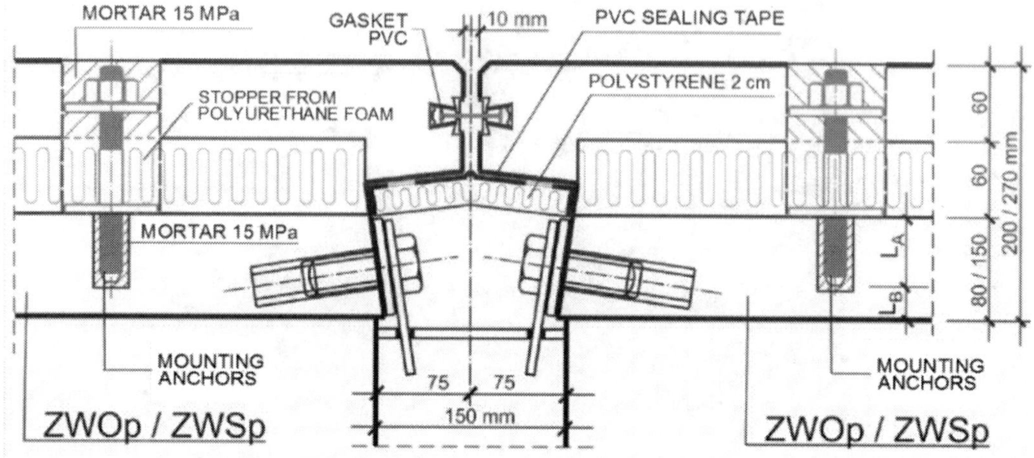


**Fig. 5.** The concept of connection reinforcing of prefabricated slabs (Janusz Krentowski)

Avoidance of reinforcing the layer connections is most often the result of significant costs (more than 10% of the value of the works) and executive issues that arise during the work. In one of the areas, after the initial experiences with the first reinforced building, the reinforcement of the joints was discontinued due to brittle concrete problems in the prefabricated buildings. During the drilling, there were extremely fragile breakings of concrete cones inside the apartments. A similar phenomenon can be sometimes observed in stairwells where residents made holes for dowels (Figure 6).

In addition, the limited number of tests regarding the condition of the connecting elements of the façade have brought positive results. The corrosion level of the fasteners was low, and the arrangement and number of fasteners allowed for the proper use of façade panels. The above conclusions were drawn on the basis of the openings made in the multilayer panels designed both for insulation projects and for the administrative decision of housing estates prior to the execution of thermal insulation. Due to the limited scope of research, its results cannot be the basis for general conclusions.

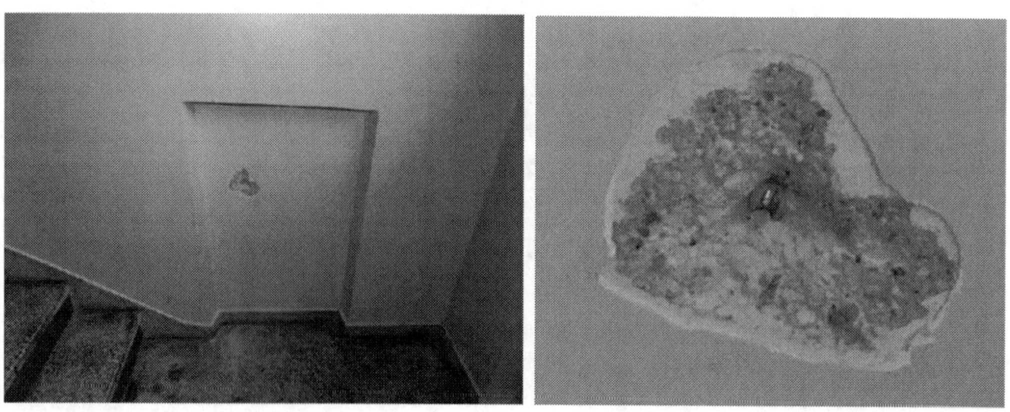

**Fig. 6.** Brittle breaking of concrete cones during drilling (photos by Piotr Knyziak, 2016)

Important problems are connected with the surface damages of the ETICS insulations. They often embraced only a piece of façade surface, but are difficult to remove, because they are found on higher floors of buildings. The majority of the problems with ETICS systems are local, most often concern the oldest investments and are limited to painted surfaces. In some cases, defects are caused by poor execution process, non-compliance with the manufacturers' recommendations of ETICS system. Individual holes have been found, resulting in the insects and birds nesting.

The basic conclusion of the investigations, referring to the aforementioned research conducted in Gliwice [3], is the observation that in both technical inspections no visible deformation of the textured façade panels was noted. During the 2005-06 survey, many of the buildings were still not insulated and the state of the façade layer could be seen directly.

In 2016-17 surveys, the symptoms of incorrect operation of the ETICS system included deformation of the surface layer, scorching of the plaster in the outlines of styrofoam boards (excessive moisture transfer in the partition), unusual fractures (many plaster cracks due to incorrect execution or passes of time). The remark concerns not only the investigated 110 buildings, but also many neighboring buildings.

A well-made and preserved insulation layer on the façade wall significantly improves the operating conditions of façade boards and reduces the corrosion rate of interlayer fasteners. This is important for the aforementioned results [1], which speak of the layout mismatches and the type of steel grade. A significant improvement in the façade's technical condition is observed (Figure 7). The vast majority of buildings received a good rating indicating that the façade layers did not show any significant damage nor need repair.

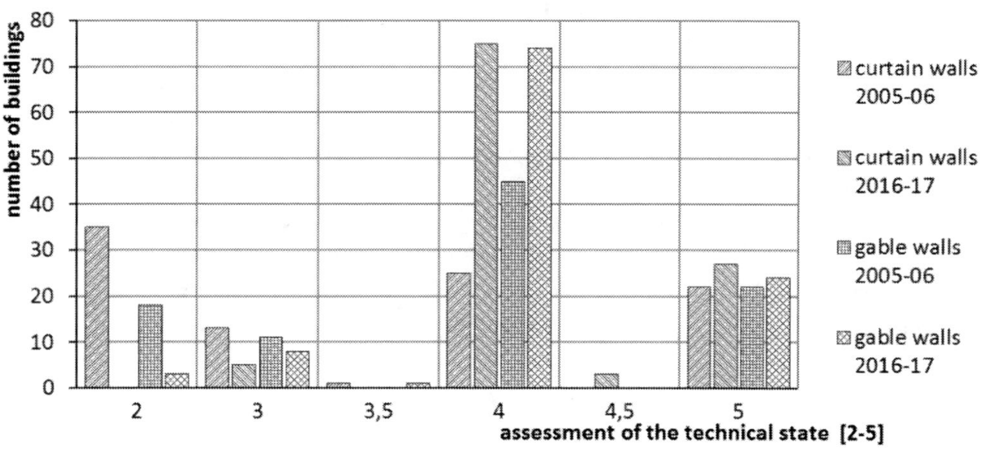


**Fig. 7.** Evaluation of the state of curtain and gable walls (data collected by Piotr Knyziak)

## Summary

Inspections have shown good overall condition of the analysed buildings. Hasn't been noticed visible signs which would be point of any security risk. In buildings, however, it can be distinguishing installations or elements in visible unsatisfactory condition, waiting for repair or replacement.

Carried out authors studies confirm that the quality of the precast concrete residential buildings were built often did not as good as it should be. Often it was the result of imperfect technical knowledge but also often a hurry and not paying attention to the quality of the work. These studies also confirm the fact that the reliability of these residential buildings is high. There are no visible signs of progressive deterioration of the main structure. Elevations of buildings are properly maintained and the number of fully insulated and even twice insulated buildings increases.

Previous authors studies [1, 8-11] and the wide current work allow concluding that it is possible long-term safe use of prefabricated large-panel buildings. Durability of buildings properly exploited, who are under constant supervision is estimated as comparable to buildings made of monolithic technology.

Prior to the implementation phase of the extension up, enlargement or thermorenovation, it is important to define the method of elimination of the identified structural defects of the object. The fixing technology for additional thermal insulation and façade layers should take into account, in addition to the verification of the state of system connectors, the concept of reinforcement of the bonding and prefabrication zone of the prefabricated curtain wall and the preparation of the structural element for transferring additional load resulting from thermo-modernization or expansion.

## References

1. J. Krentowski, R. Tribiłło, Przeg. Bud. **2**, 34 (2008)
2. A. Ostańska, Przeg. Bud. **9**, 68 (2011)
3. Z. Dzierżewicz, W. Starosolski, *Systemy budownictwa wielkopłytowego w Polsce w latach 1970-1985* (Oficyna Wolters Kluwer Business, Warszawa, 2010)
4. Z. Ściślewski, Mat. Bud. **11,** 3 (2004)
5. W. Ligęza, J. Dębowski, *Konferencja Naukowo-Techniczna: Awarie Budowlane 2007*, 741 (2007)
6. W. Ligęza, *Budownictwo wielkopłytowe po latach. Wybrane problemy remontowe.* Bud. i Arch. **3**, 15 (2014)
7. M. Wójtowicz, *Konferencja Naukowo-Techniczna: Awarie Budowlane 2011*, 533 (2001)
8. P. Knyziak, *Analiza stanu technicznego prefabrykowanych budynków mieszkalnych za pomocą sztucznych sieci neuronowych* (PhD thesis. Warsaw University of Technology, Warsaw, 2007)
9. P. Knyziak, M. Witkowski, Inż. i Bud. **12,** 639 (2007)
10. P. Knyziak, Procedia Engineering **91**, 394 (2014)
11. P. Knyziak, Key Eng. Mat. **691**, 420 (2016).

# Impact of corrosion processes in the basement level on the durability of the construction of large-panel buildings

*Piotr* Knyziak [1,*], *Piotr* Bieranowski [2] and *Janusz R.* Krentowski [3]

[1]Warsaw University of Technology, Al. Armii Ludowej 16, 00-637 Warsaw, Poland
[2]University of Warmia and Mazury in Olsztyn, Oczapowskiego str. 11, 10-719 Olsztyn, Poland
[3]Bialystok University of Technology, Wiejska str. 45, 15-351 Bialystok, Poland

**Abstract.** The paper presents results of research on the assessment of the corrosion progress level in structural elements of buildings made of large-panel technology. The symptoms of corrosion processes and the results of other authors research are presented.

## 1 Background of the study

The importance of proper maintenance and modernization of prefabricated buildings has been appreciated a long time ago. Published in 1978 the book [1] involved in all issues of maintenance of the building envelope. Already then noted that, despite the passage of about 20 years from the erection of the first buildings of the large-panel buildings, these buildings were not subjected to maintenance or repair. Similar conclusions were derived from the 2005-06 technical audits for buildings used for about 20 years [2].

Steel corrosion is the basic process of destroying reinforced concrete. The rate of this process depends on the possibility of penetration into the concrete of aggressive substances [3]. Correctly designed cover of the rods and appropriate tightness of the concrete ensure long-term use of the structure before the reinforcement corrosion [4]. The ability to protect reinforcing steel through a concrete cover depends on the pH of the concrete and the potential for corrosion-induced electrical charges (Figure 1). As a result of corrosion progress, the volume of corrosion products increases (Figure 2). On the surface of reinforced concrete elements, this process in the initial phase is manifested as cracks along the corroding reinforcement bars, and in the final phase as the concrete peeling and the direct exposure of the pretax.

The progress of corrosion processes in the construction of large-panel buildings is most noticeable in the cellars of buildings, because in most cases the surface of prefabricated and monolithic elements, both wall and ceiling remains unfinished with plaster. The condition of the joints of prefabricated panels and monolithic wreaths in the construction of large-panel buildings can be estimated only by point outcrops or signs of work (cracks, angles of rotation of the joints).

* Corresponding author: p.knyziak@il.pw.edu.pl

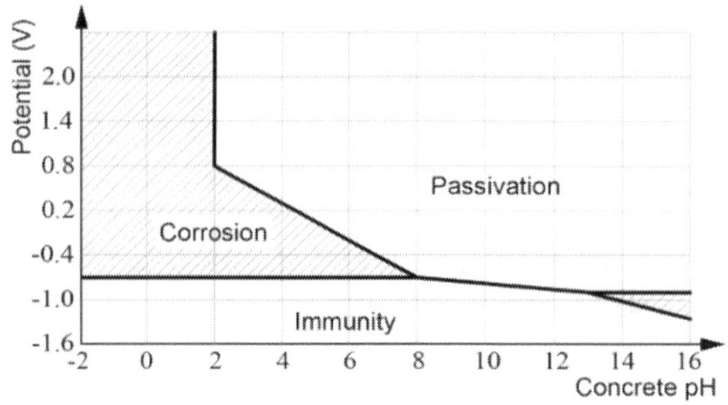


**Fig. 1.** Steel corrosion vs concrete pH value and potential

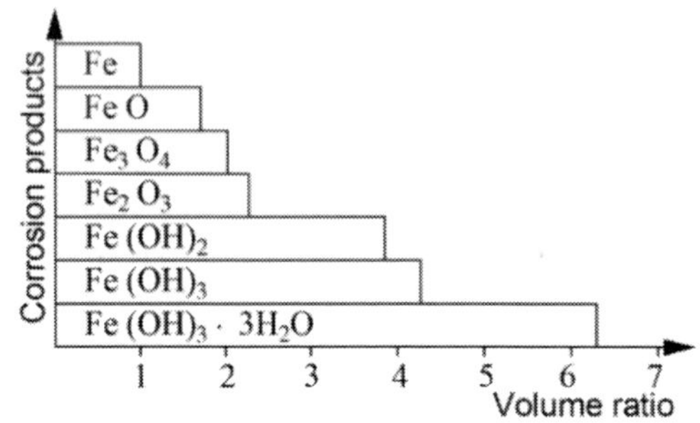


**Fig. 2.** Relative volume of steel and corrosion products [5]

## 2 Risks of the corrosion indicated in the literature

The results of conducted technical inspections and studies confirm lower level of manufacturing and embedding of sandwich panels [6, 7]. The depth of the carbonation of concrete in facade plates found in outcrops does not exceed 5 mm [6]. The report of these studies indicates the moisture of mineral wool in winter-spring period. In this case increased humidity led to pitting corrosion of hangers (members connecting layers of cladding panels). After 8 years of operation, the depth of corrosion pits reached 0.2 mm. High humidity is also due to the possibility of an incomplete seasonal evaporates of moisture.

A major problem [8] is the quality of prefabricated joints. The disadvantages are both the partial absence of reinforcement and the cavities in cross section of the elements, and a danger of corrosion of the reinforcing bars in RC sections. Chemical analysis of the amount of aggressive ions in the aspect of impact on concrete and steel, according to chemical quantitative analysis, showed that concrete from samples taken in joints and prefabricated panels contained an average of 36% of soluble in acid parts. Samples exposed to more intense external atmospheric agents exhibited 30% soluble fraction. Chloride and sulfate ions in the samples were not present in increased quantity. The pH of the samples was in the range 11.6 ÷ 12.1.

Characteristic damage in the context of corrosion, also include damage and cracks of the finishing layer. They occur on the surface of the element, rarely in the area of the joints. At work [9] the measured crack width was in the range of 0.5 ÷ 3.0 mm. There have been also cases of improper plate assembly where the displacement of the joined elements was within 2 to 4 cm. The occurrence of damage of the cladding panels has an explicit effect on corrosion of hangers joining finishing and structural parts of precast elements.

Non-invasive testing methods are widely used in diagnostics [10]. They should be conducted in stages: stage I - global diagnostics - localization of building structures potentially threatened with destruction (e.g. thermovision); stage II - conducting research in specific locations in the context of the occurrence of destruction (e.g. radiological, electrochemical methods); - stage III - sublocal diagnosis - identification of damage: type, intensity and extent of impact). Especially useful is the thermovision, allow on identification of places of cooling on the building's elevation, thus the critical points in the context of excessive moisture, which together with the chemical composition of building materials is an environment and a medium suitable for the development of molds. The greatest moisture can be found in the cellar walls. This is often caused by badly profiled bands and cracks in the plinth zones, as well as the destruction of insulation in the cellar floor. Conduct inspection, allows to identify this type of damage. It can also often notice incorrectly performed installation repair work, such as the careless execution of installation passes between and through precast elements in the horizontal and vertical plane.

The non-destructive electrochemical method used in [10] allowed to determine the probability of corrosion of reinforcement in reinforced concrete elements. It has been found that the probability of corrosion of reinforcement reaches 95% in the exterior walls to a height of about 0.5 m above the floor. Up to 1.30m this value reached 50%. In the basement's internal walls up to 1.30 m above the floor, the probability of corrosion was estimated at 5%.

## 3 Results of technical inspections in years 2005-06 and 2016-17

In the years 2005-06, technical inspections were carried out in 95 buildings [11, 12] and in 2016-17 in 110 buildings. Their results are reported below.

One of the basic processes to be checked during technical inspections is corrosion of steel. The main task of technical inspections is to check the progress of corrosion in the elements of the building, because most often they are components or installations that have been subject to this process from the beginning (e.g. tinplate). Regarding to reinforced concrete elements, cracks control and corrosion symptoms as discoloration and cracking of the reinforcement cover should be observed. Damage points should also be indicated to protect them. Despite the many annual and five-year inspections in most buildings, many of the reinforced concrete elements in cellar level in multi-family buildings remain not protected since construction moment (Figure 3).

**Fig. 3.** Reinforcement not protected since construction moment (photos by Piotr Knyziak, 2016)

There are no symptoms of poor operation of these components, although the reinforcement steel has no corrosion protection on site and the requirements for the transmission of stresses between steel and concrete are not met. In most of the cases they were small fragments ceilings made of monolithic construction or parts of basement walls (elements using a low bearing capacity). In a few cases it was possible to find a precast elements with improperly made reinforcement cover - visible abnormal thickening of concrete around the bar located just off the surface of the element, or even less frequently, bars visible on the surface of the precast elements.

**Fig. 4.** Reinforcement exposed during installation work and not secured (Piotr Knyziak, 2016)

In many cases, installers performing upgrades made breakthrough of reinforced concrete elements, including exposing the reinforcing bars and the cutting of the bars (Figure 4). Mostly damaged are the external parts of the walls in the basement corridors and the lintel sections of the prefabricated sections with a communication hole. In some cases, the damage is covered with new mortar or putty, probably lacking the protective properties in relation to steel. In four of the tested buildings the basement ceilings were insulated (Figure 5), which prevented the cracks and corrosion control of the slabs. They have also been recorded cases of white paint on the basement, but paint with the addition of gypsum.

**Fig. 5.** Insulation of ceiling and ceiling with exposed reinforcement painted with the addition of gypsum (photos by Piotr Knyziak, 2016)

During the inspection of buildings were observed a low level of corrosion of the most parts of the exposed reinforcement and other steel elements. Bars not concreted during construction found in the basement level in concrete elements made on site, in almost all cases, were covered only with a thin corrosion coating, and often on a part of the surface of the rod there was no corrosive attack at all (Figure 6).

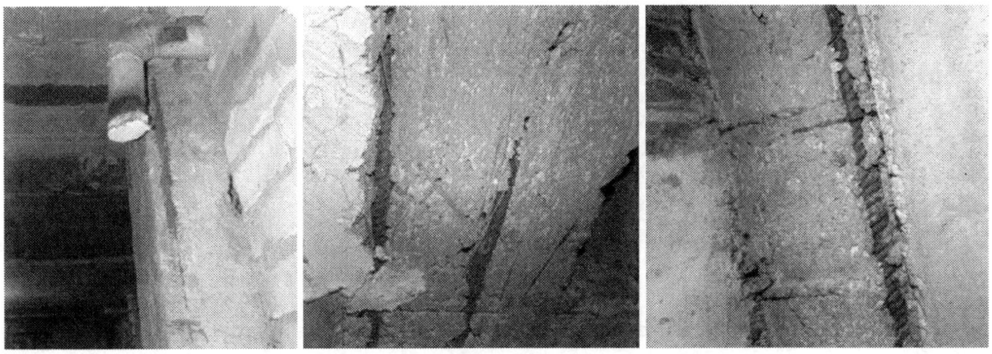

**Fig. 6.** No corrosion on reinforcement surface (photos by Piotr Knyziak, 2016)

Basement conditions can be considered as unfavorable to the development of corrosion, due to the low humidity of the air, which is heated practically throughout the year by hot water circuits and in winter also by central heating. In many buildings there are incidental flooding of cellars from the rain water installation, after the downpour, but not little impact on the progress of corrosion in the current period of use.

During the technical inspections of 110 prefabricated buildings particular attention was paid to the correctness of prefabrication, the occurrence of symptoms of too thin cover of reinforcement bars, the level of corrosion of visible sections of reinforcing bars. The results of the technical inspections are summarized (Figure 7) by the following indications: 0 - elements without symptoms of poorly made cover and without symptoms of low thickness of the cover, 1 - symptoms of low thickness of reinforcement cover in the elements, 2 - symptoms of corrosion of reinforcement under the cover, 3 - cover partial deficiencies, bars partially corroded, 4 - cover partial deficiencies, rods coated by low corrosion, 4 - cover partial deficiencies, rods coated by high corrosion, 6 - insulation of elements that do not determine the level of corrosion.

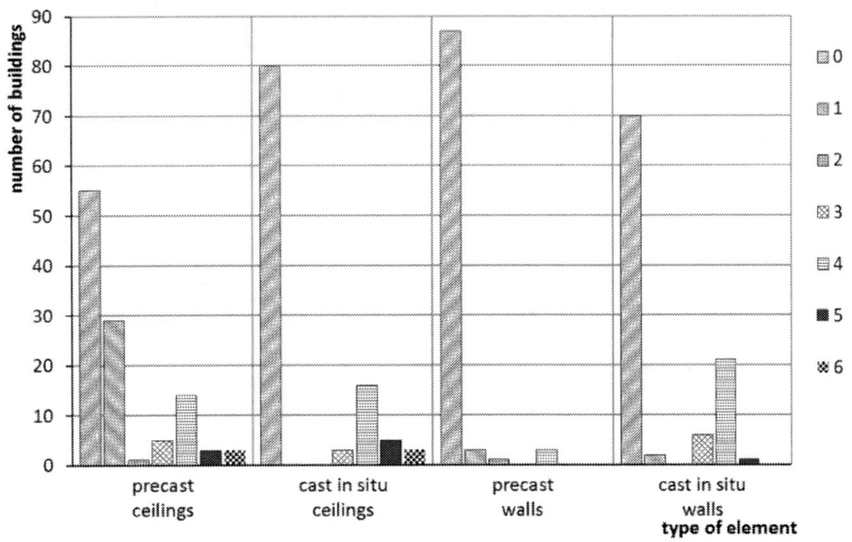


**Fig. 7.** Shortcomings and corrosion ratio in elements (data collected by Piotr Knyziak)

Results of technical inspections indicate that in most of the reviewed buildings, no cases of improperly constructed prefabricated or monolithic elements were found. In most cases, there are also no signs of corrosion under the concrete cover.

## 5 Summary

Evaluation of the technical condition based on the indications visible on the surface of the elements makes the most sense. It allows quickly in many buildings to check whether corrosion has crossed the initial stage. Taking into account the results of the technical inspections and the research on carbonation of reinforcement cover, it can be concluded that under very favorable conditions in prefabricated buildings, so made of higher classes of concrete, very low corrosion progress and technical wear can occur. Should be the rule to repair damaged elements without delay and periodically control the condition of buildings elements.

Previous author studies [11, 12] and the wide current work allow concluding that it is possible long-term safe use of prefabricated buildings. Durability of buildings properly exploited, who are under constant supervision is estimated as comparable to buildings made of monolithic technology.

## References

1. W. Konecki, J. Sitkowski, A. Ulatowski, *Remonty budynków mieszkalnych wznoszonych metodami uprzemysłowionymi. Przegrody zewnętrzne* (Arkady, Warszawa, 1978)
2. P. Knyziak, *Trwałe metody naprawcze w obiektach budowlanych,* 206 (Dolnośląskie Wydawnictwo Edukacyjne, 2015)
3. A. M. Neville, *Właściwości betonu* (Polski Cement, Kraków, 2000)
4. M. Jaśniok, A. Zybura, Przegląd Budowlany **1** (2007)
5. S.J. Han et al., Constr. and Build. Mat. **54,** 572 (2014)
6. Z. Ściślewski, Mat. Bud. **11,** 3 (2004)
7. J. Krentowski, R. Tribiłło, Przeg. Bud. **2**, 34 (2008)
8. M. Wójtowicz, T. Możaryn, *Konferencja Naukowo-Techniczna: Awarie Budowlane 2013*, 889 (2013)
9. W. Ligęza, *Konferencja Naukowo-Techniczna: Awarie Budowlane 2007*, 741 (2007)
10. J. Sobczak-Piąstka, *Konferencja Naukowo-Techniczna: Awarie Budowlane 2015*, 695 (2015)
11. P. Knyziak, *Analiza stanu technicznego prefabrykowanych budynków mieszkalnych za pomocą sztucznych sieci neuronowych* (PhD thesis. Warsaw University of Technology, Warsaw, 2007)
12. P. Knyziak, M. Witkowski, Inż. i Bud. **12,** 639 (2007)

# Determining the effectiveness of protective fences from explosions of terroristic orientation

*Alexander* Komarov[1], *Phan* Tuan Anh[1, *]

[1]Moscow State University of Civil Engineering, 26, Yaroslavskoe Shosse, Moscow, 129337, Russia

**Abstract.** The article provides a general methodology for determining the effectiveness of protective fence against explosive devices of a terroristic nature. It also presents an example of using the developed methodology to create a more optimal solution for protective fence for a particular enclosure.

***Keywords***: protective fences, explosion, pressure, blast shock wave, probability of human casualty

## 1 Introduction

Protective fences (fences) are in places with massive congestions to reduce the probability of human casualty to explosive devices (ED) which are used for terroristic purposes. To evaluate the effectiveness of the protective screens (PS), it is necessary to determine in advance the following initial data: equivalent weight in TNT to the explosive substance (ES) in the explosive device; area of the protected territory and the permissible probability of a person being killed in the protected area.

The main damaging factors from explosive devices include an blast shock wave (BSW) accompanying the explosion; the fragmentation of the debris and fragments from the ED; significant thermal emission near the explosion; secondary damaging factors, which are associated with the destruction of the structures near the site of the explosion. This article examines the optimal geometric parameters for protective screens in terms of ensuring people's protection from BSW. Durability characteristics and the screen's efficiency for protection from debris in this article are not examined. Therefore, when carrying out calculations it is assumed that the screens are considered rigid and resistant to explosion.

## 2 Method for determining the effectiveness of protective fences from explosions

Let's take a look at the specifics of the calculation of the effectiveness of protective screens (PS) using the example of a protected territory located outside the parking lot, away from blast shock waves that form during explosions which can be located in a car in the parking lot.

* Corresponding author: ptuananh26789@gmail.com

It is established that the source of threat, when the explosive device (ED) is delivered by car, the maximum mass of the explosive substance (ES) in TNT equivalent weight cannot exceed 50 kg. In cases where there are visual inspections of a car, delivery of charges with greater mass may be excluded. The given mass of ED is compactly packed and during the explosion, it does not use any energy to open the caches (for example, opening the hidden cavities of the car, its trunk, etc.). I.e. in the parking lot, the biggest blast shock wave that can be formed is created in an above ground explosion of 50 kg of compactly packed TNT located in an open space.

To determine the protective screen's effectiveness, it is necessary to determine the parameters of the blast shock wave (BSW). For free spaces, the BSW parameters can be determined from empirical relations [1-4], and for cluttered spaces (there are protective screens on the BSW's path), it is necessary to use numerical calculation methods. To determine BSW's parameters, we use a numeric scheme from Godunov's scheme based on the breakdown of an arbitrary discontinuity [5-9].

Knowing temporal dependencies of the BSW into all points of space, one can determine the probability of an event: the complete destruction of buildings; average destruction of buildings; human casualties; eardrum rupture; human spatial movement due to BSW, etc.

The probability of human casualty was determined from known ratios using the probit function. The value of the probit function, which determines the probability of dealing damage (in this case, the probability of human casualty) is:

$$Pr_1 = 5 - 0.26 \cdot lnV_1, \tag{1}$$

Where factor V1 equals:

$$V_1 = \left(\frac{17500}{P(Pa)}\right)^{8.4} + \left(\frac{290}{I(Pa\cdot s)}\right)^{9.3}, \tag{2}$$

Where P (Pa) is the maximum pressure in the BSW, I(Pa*s) – the momentum of the compression phase in the BSW [10,11].

## 3 Calculation result

Let us look at the results of the calculation for the first schematics of the location of the EDand the fence dimensions (Option 1).

It is established that the ED is located at a point with a coordinate y=1.8 m. The first fence with a height of 1.5 m is 1.2 m away from the charge (y=3 m). This fence should destroy the BSW at the initial stage and perform its function of intercepting the fragments that arise during the explosion. The second fence height 4.5 m is located at a distance of 7.2m from the charge (y=9m). The second fence should provide a reduction in the intensity of the BSW to an acceptable (safe) level on the outside of the enclosure, i.e. outside the parking lot. In the first arrangement option of fences, it is assumed that the space between the first and second fencing will be used only for driving cars in the parking lot. With this arrangement option of the protective screens parking will be carried out only after the first (H = 1.5 m) fence.

Fig.1 shows isolines of equal explosion pressure in the examined region with the first option of the fence arrangement. Additionally, Fig. 1 shows the location of the ES, the fences are shown and the points of the calculated temporal dependencies of the surplus pressure P(t) are marked.

From the figure below, it can be seen that the fence significantly reduces the levels of BSW outside the parking lot. The quantitative attenuation of the BSW is illustrated in Fig. 2, where the ratio is given of the maximum levels of the BSW in the presence of a fence to the maximum levels of BSW in free space, i.e. without fencing.

It can be easily seen that directly behind the fence, the parameters of the BSW is reduced by 5 times, because the ratio of the maximum levels of BSW in the presence of a guard to the maximum levels of BSW in free space is less than 20%.

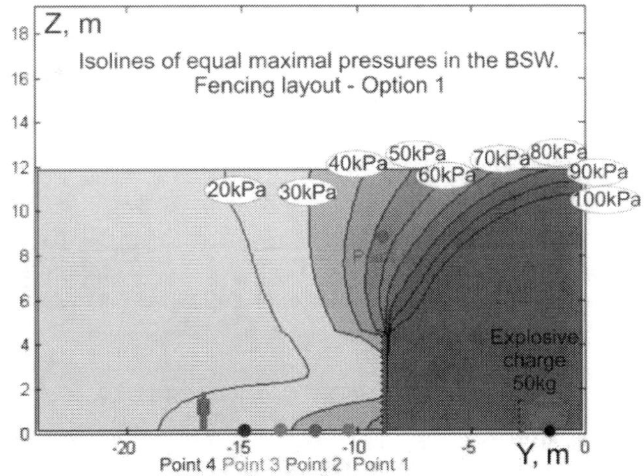


**Fig.1.** Isolines of equal pressurse in BSW. Option 1.

Here we have the levels of maximum pressures in the BSW at points 1-5: 37.2; 32.6; 29.0; 25.9; 54.3 kPa.

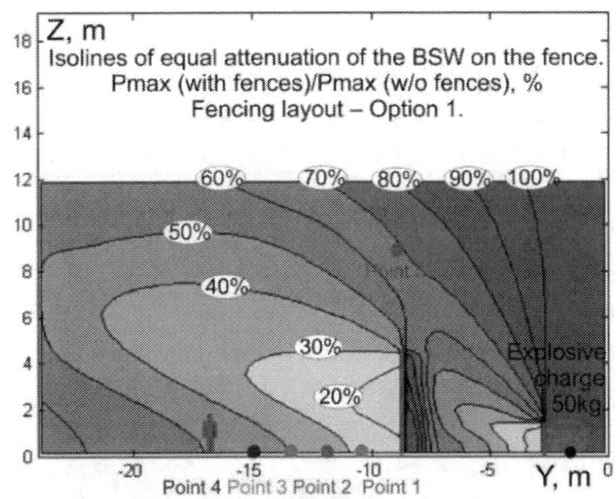


**Fig.2.**Isolines of equal attenuation of the BSW on the fence. Option 1

Levels of maximum pressures in the shock wave for the same points 1-5, but in the absence of protective guards, respectively are equaled to: 134.6; 96.7; 73.7; 58.6; 78.9 kPa. The probability of human casualty at points 1-5, determined by the relationships between (1) and (2), has the following values: 69.5; 52.7; 36.8; 24.8; 40.7%.

Fig. 3 shows isolines of equal probabilities of human casualty in the presence of a guard in the first option. The probabilities of human casualty at points 1-5, determined by the ratios between (1) and (2) and have the following values: 0; 0; 0; 0; 12.6%.

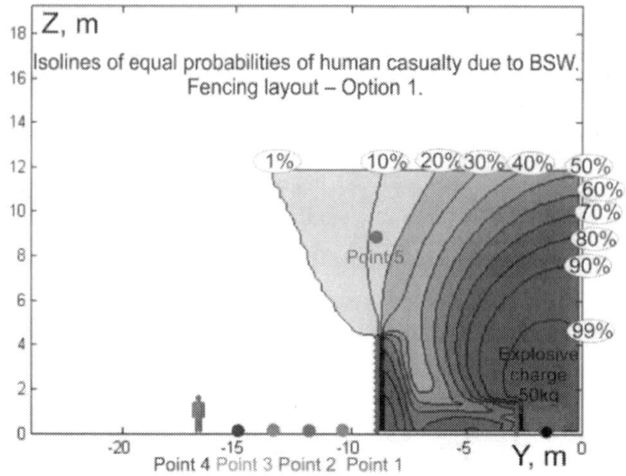


**Fig.3.** Isolines of equal probabilities of human casualty due to BSW. Option 1.

Thus, the calculations showed that the probability of human casualty outside the parking lot using the protective screens by option 1 would be close to zero. This type of fence can be used in places that require a wide (6m or more) transport passage, i.e. this fencing design limits the use of parking space.

Now let us look at the effectiveness of protective fences for a situation where parking should be located closer to the fence. The second option of fencing design was examined, which assumes that the distance between the explosive charge and the fence is 4.2m, i.e. between a parked car and a fence, the journey is only 3 m. There is only one fence and its height is 4m.

Let's take a look at the results of the calculations for the second schematic of the layout of the ED and the dimensions of the enclosure (Option 2). Fig. 4 shows isolines of equal explosion pressure in the examined region with the second option in the fence arrangement of the enclosure.

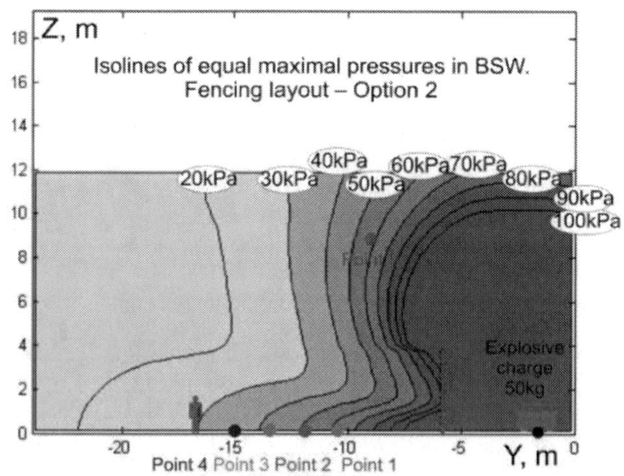


**Fig. 4.** Isolines of equal explosion pressures in BSW. Option 2.

It can be seen from the figure that although the screen reduces the levels of the BSW outside the parking lot, but their numerical values are still quite large. The quantitative attenuation of the BSW is illustrated in Fig. 5.

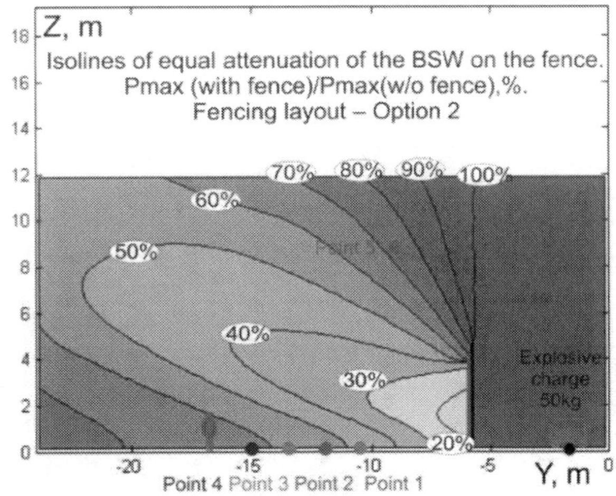


**Fig. 5.** Isolines of equal attenuation of the BSW on the fence. Option 2

Figure 5 shows the ratio of the maximum levels of BSW in the presence of a guard to the maximum levels of BSW in free space, i.e. without fencing.

Significant high levels of pressure outside the parking lot leads to an increase in the likelihood of deaths in the explosion of the ED in the parking lot.

Figure 6 shows the isolines of equal probabilities of human casualty in the presence of a guard in the second option.

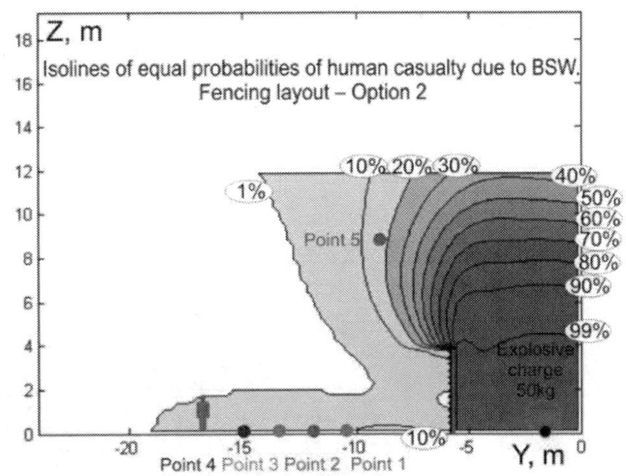


**Fig. 6.** Isolines of equal probabilities of human casualty due to BSW. Option 2.

From the figure above, it can be concluded that the probability of hitting people outside the parking lot at distances of 15-20m from the fence will be more than 1%, and in some places will reach 10%, which is at an unacceptable level from the point of view of explosion safety.

Levels of maximum pressures in the BSW when using the enclosure of option 2 for points 1-5 are respectively: 63.7; 51.6; 42.7; 36.0; 59.1 kPa. The probability of human casualty at the same points 1-5 (with the second layout option of the fence) has the following values: 9.6; 7.2; 4.8; 3.2; 17.4%.

To reduce the explosive pressure significantly outside the parking lot, it is possible, with the same dimensions, but it is necessary to place the fence at an angle to the horizon (making it inclined). The outing of the upper point should be 1.5 m (option 3).

Figure 7 shows the isolines of equal explosion pressure in the examined region using the third fence layout option in the enclosure.

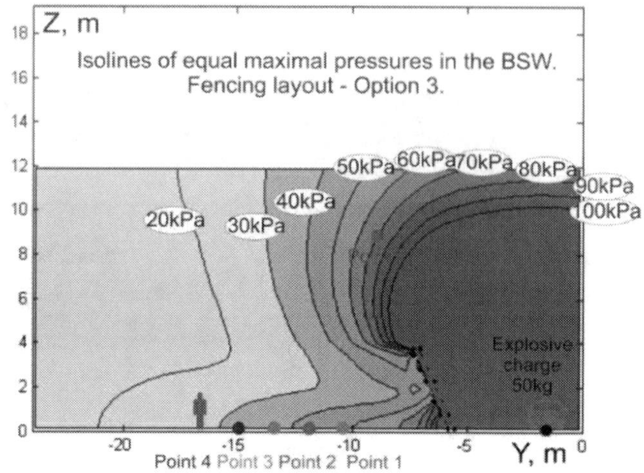


**Fig. 7.** Isolines of equal pressures in the BSW. Option 3.

From the figure above, it is seen that a slight slope of the fence leads to some reflection of the BSW, which reduces the levels of the BSW outside the parking lot.

In addition, with an inclined arrangement of the fence, the impact load is reduced, and the support structure is more resistant to impact loads.

Fig. 8 shows the isolines of equal probabilities of human casualty in the presence of a guard in the third option. From the data shown in Fig. 8, it can be interpreted that the probability of hitting people outside the parking lot at distances of 10-12 m from the fence is reduced to 1%.

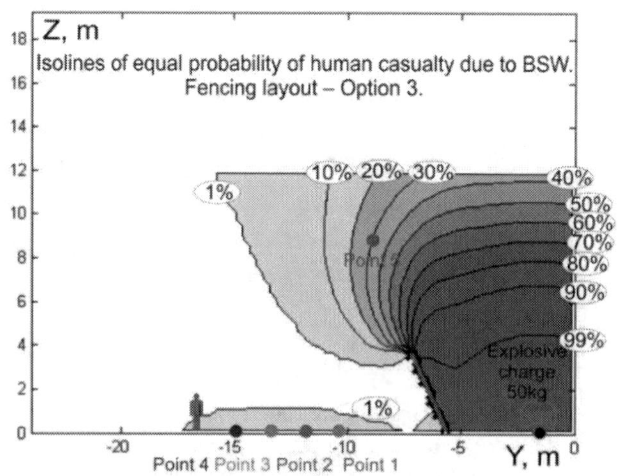


**Fig. 8.** Isolines of equal probability of human casualty due to BSW. Option 3.

Levels of maximum pressures in the BSW when using the fence layout of option 3 for points 1-5 are respectively equal to: 48.9; 43.0; 37.5; 32.6; 70.6kPa. The probability of human casualty at the same points 1-5 (with the third version of the fence) has the following values: 2.8; 3.0; 2.5; 1.9; 30.7%.

Let us end on the procedure for selecting the necessary schematic of protective structures for the situation in question.

Following from the above calculations, reducing the probability of human casualty behind the fence to zero leads to certain difficulties when using parking spaces and certain costs for the construction of fences (Option 1). In this case, the parameters of the explosive device are unknown in advance and the probability of its appearance in the parking lot. Therefore, to use this or that option of the fencing layout follows from the considerations of presence of people in the possible zones of casualty. For example, if a human flow enters the affected areas, then it makes sense to use a more expensive, but correspondingly much more effective, first option for fencing layout. Otherwise, you should use the fence in the third option.

## 4 Conclusions

The article provided a general methodology for determining the effectiveness of the protective fences against explosive devices of a terroristic orientation.

An example is also provided about using the developed methodology for developing the most optimal solution for protective fences for a particular facility.

## References

1. Sadovsky M.A. The mechanical effect of blast shock waves from explosions according to experimental studies – in book Blast physics., №1, M., pub. AN USSR, 1952.
2. Polandov Y., Korolchenko A., Dobrikov S. Gas explosion in a room with a window and passage to an adjacent room. (2016) MATEC Web of Conferences, 86, art. no. 04031. DOI: 10.1051/matecconf/20168604031.
3. Shebeko Yu.N., Smolin I.M., Korolchenko A.Ya., Shevchuk A.P., Borodkin A.N., Malkin V.L., Simonov O.A., Gurinovich L.V., Popov S.A., Kolosov V.A., Smirnov E.V. Some aspects of fire and explosion hazards of large LPG storage vessels. (1995) Journal of Loss Prevention in the Process Industries, 8 (3), pp. 163-168. DOI: 10.1016/0950-4230(95)00011-O
4. Shebeko Yu.N., Korolchenko A.Ya., Tsarichenko S.G., Navtsenya V.Yu., Malkin V.L. Effects of the initial prssure and temperature on the combustion characteristics of hydrogen-containing mixtures. (1989) Combustion, Explosion, and Shock Waves, 25 (3), pp. 289-292. DOI: 10.1007/BF00788799 .
5. Numerical solution of multidimensional problems of gas dynamics. Edited by Godunova S.K. M. Science. 1976. P.400.
6. Komarov A.A. Forecasting impact loads from disastrous deflagrating explosions and assessing the consequences of their impact on buildings and structures. Thesis for the degree of Doctor of Technical Sciences. M. MGSU. 2001. -460p.
7. Lyapin A., Korolchenko A., Meshalkin E. Expediency of application of explosion-relief constructions to ensure explosion resistance of production buildings. (2016) MATEC Web of Conferences, 86, art. no. 04029. DOI: 10.1051/matecconf/20168604029"

8. Korolchenko D., Voevoda S. Influence of spreading structure in an aqueous solution-hydrocarbon system on extinguishing of the flame of oil products, MATEC Web of Conferences, Volume 86, 2016, DOI: 10.1051/matecconf/20168604038.
9. Markeev V.A., Voevoda S.S., Korolchenko D.A. Fire protection of tank farm objects of Rosneft NK OAO, Neftyanoe khozyaystvo - Oil Industry Issue 9, 2006, Pages 83-85.
10. RD 03-409-01 Methodology for assessing the consequences of disastrous explosions of fuel-air mixtures. Collection of documents from ROSTECHNADZOR of Russia, SEC "Industrial Safety", series 27, issue 2. - M .: 2001. - 224 p.
11. GOST R 12.3.047-2012 SSBT "Fire safety of technological processes". -M .: GOSTSTANDART of Russia. – 85p.

# Structural behavior of suspension bridge with a stabilizing cable

*Anatoliy* Konovalov[1], *Marina* Pustovalova[2*]

[1]Structural Mechanics and Strength of Materials Department, Head of Department
[2]Structural Mechanics and Strength of Materials Department, Northern (Arctic) Federal University named after M.V. Lomonosov, Arhangelsk, Russia

**Abstract.** Suspended structures are commonly used in construction of motorway and pedestrian bridges. These structures allow wide spans without the need for intermediate supports. Suspension bridges are noted for lower structural stiffness as compared to beam bridges and arc bridges. The stiffness control depending on the environment and external effects (moving loads, wind, seismic forces, etc.) is a real-life challenge [1, 2]. The authors of this paper have evaluated the use of stabilizing cable installed in the central span under the stiffening girder as the means of stiffness control. A plane 3D model of a suspension bridge was developed using the ANSYS software. The study compared the stress deformed state and dynamic properties of the models with and without a stabilizing cable. The displacement in the model equipped with a stabilizing cable, as compared to the one without, was noted to be lower in all relevant sections: 2.6 times in the middle of the central span of the stiffening girder; 15 times in the middle of the end span; and displacement of the tower top was 3.5 times lower.

## 1 Structural design

This paper presents the structural design of a suspension bridge as referenced to the actual location of proposed bridge across the Northern Dvina River in Arkhangelsk [3]. The 3D plane finite element model was built using ANSYS software. It is a full-scale model with the end spans of 200 m and the central span of 450 m. The tower height was 90 meters including 30 meters of the bridge height. The height of the tower from the stiffening girder to the tower top was 60 m.

Each tower comprised a box section with the stiffness expressed as $EA_t$= 1.907E+5 MN, $EI_{maxt}$=1.7E+6MN*m$^2$, $EI_{mint}$=5.8E+5MN*m$^2$. The tower was rigidly bonded to the ground and was a continuous structure.

The steel stiffness girder was a closed box-type structure [1] with the stiffness expressed as $EA_g$= 5.4E+5 MN, $EI_{maxg}$=5E+7MN*m$^2$, $EI_{ming}$=2E+6MN*m$^2$. The stiffness girder rested on the ground though a hinged immovable support on one side, and through a hinged movable support on the other side.

The main cable was assumed with a stiffness of $EA_c$= 2.1E+5 MN, and was fixed to independent anchors independently from the girder. The main cable was a continuous structure having no rigid ties with the towers. Stay cables were assumed to have the

* Corresponding author: m.pustovalova@narfu.ru

stiffness expressed as $EA_s$= 5.3E+4 MN, and were pin-jointed to the main cable and to the girder. In the middle third of the central span, stay cables had crosswise configuration with 10 m spacing; and vertical configuration with the same spacing in the rest of the model. The stabilizing cable with the stiffness equal to that of the main cable was fixed by means of parralel vertical suspension cables (hangers) in the central span under the stiffening girder (fig. 1).

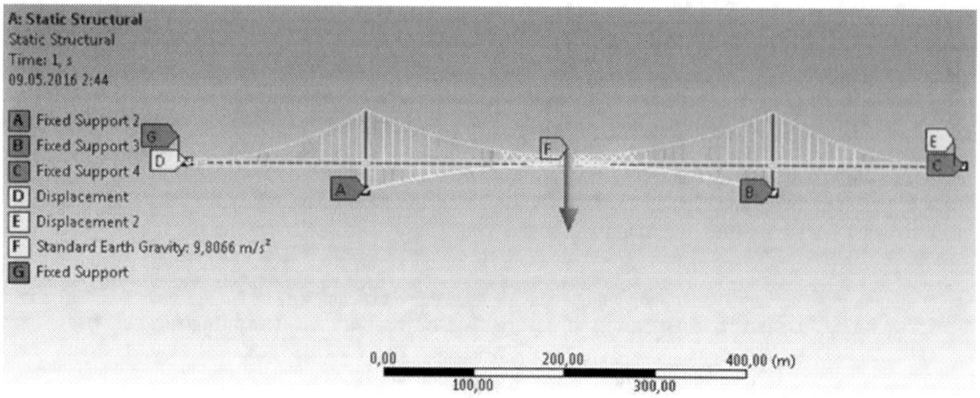


**Fig. 1.** The schematic structural design of 3D plane FE model of suspension bridge with a stabilizing cable.

## 2 Comparative analysis of stress deformed state

The comparative analysis of the stress deformed state between the model of the bridge equipped with a stabilizing cable and the model of the bridge without a stabilizing cable (initial model) caused by the dead-load is presented below in Table 1.

As shown in Table 1, tensile axial forces in the main cable of the model equipped with a stabilizing cable are 36% lower, than in the model which lacks a stabilizing cable, all other conditions being equal. In the stiffening girder, the axial forces are compressive throughout the entire length of the girder. In the middle of the central span, these forces are of the same order of magnitude as in the model discussed above, but their direction has changed. In the model with a stabilizing cable, the axial forces have appeared to be 5.4 times less than in the model without the stabilizer.

**Table 1.** The axial forces and displacements in compared models.

| Parameter | The model without a stabilizing cable (initial) | The model with a stabilizing cable |
|---|---|---|
| The axial forces in the main cable, $\times 10^8$N | 1,741 | 1,11 |
| The axial forces in the stiffening girder in the middle of the span, $\times 10^8$N | 0,430 | 0,697 |
| The axial forces in the stiffening girder near the tower, $\times 10^8$N | -0,291 | -0,054 |
| Deflection in the middle of the central span, m | 2,030 | 0,772 |

| | | |
|---|---|---|
| Deflection in the middle of the end span, m | 0,724 | 0,046 |
| Displacement of the tower top, m | 0,457 | 0,130 |

## 2.1 The Main Cable and Stabilizing Cable Pre-Stressing

A change in simultaneous tension of the main cable and the stabilizing cable from zero to the final value of 1.3·10^8 N results in a 42% increase of the axial tensile force in the main cable. There is almost no change of the axial forces in the stiffening girder near the tower. In the middle of the central span of the girder pre-stressed to 1.04·10^8 N, the internal axial forces are compressive, and show only a slight change in response to increasing force. Yet, once pre-stressed force of 1.3·10^8 N is achieved, the tensile force of the same order of magnitude occurs in the middle section (fig. 2).

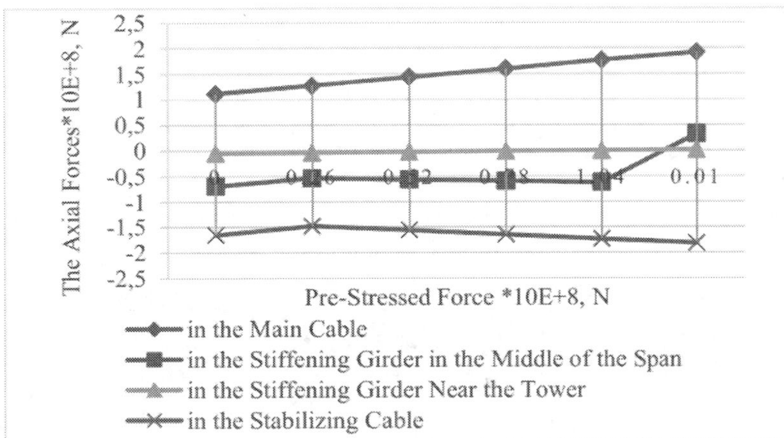


**Fig. 2.** The influence of the main cable and stabilizing cable pre-stressing on the axial forces.

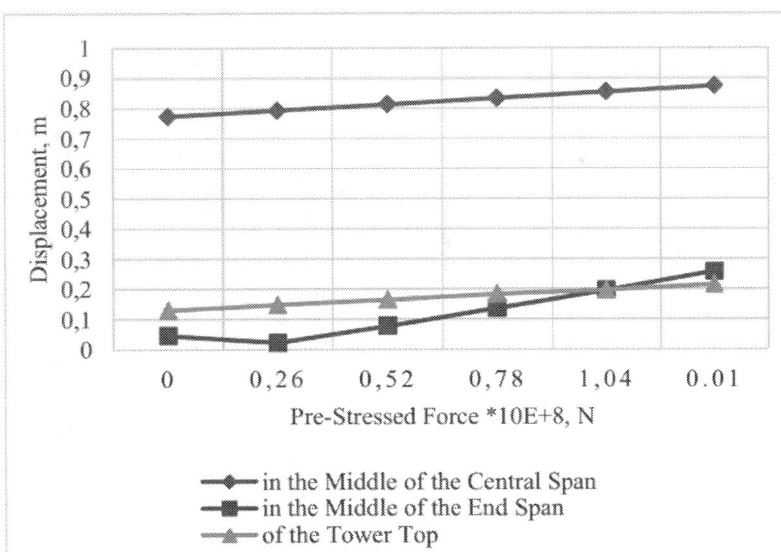


**Fig. 3.** The influence the main cable and stabilizing cable pre-stressing on displacement.

## 2 Modal Analysis

The use of ANSYS system allows displaying a wide range of natural vibrations and identifying the influence of the structural design on the model's behavior. Table 2 contains the natural vibration frequencies of the fundamental period output by the initial model and the model with a stabilizing cable.

**Table 2.** The natural vibration frequencies of the compared models.

| Parameter | The model without a stabilizing cable (initial) | The model with a stabilizing cable |
|---|---|---|
| Transverse vibrations frequency, Hz | 0,251 | 0,156 |
| Vertical vibrations frequency, Hz | 0,254 | 0,497 |
| Torsional vibrations frequency, Hz | 0, 251 | 0,282 |

The modal analysis has shown the model with a stabilizing cable to have the frequency of transverse vibrations (see fig. 4, mode 3) 1.6 times lower than a similar model without the stabilizer. At the same time, the frequency of vertical vibrations has increased two fold (see fig. 4, mode 10), while torsional vibrations showed a 6% increase (fig. 4, mode 4). The natural vibration frequencies of the fundamental period have changed in magnitude and direction as compared to the initial model without a stabilizing cable.

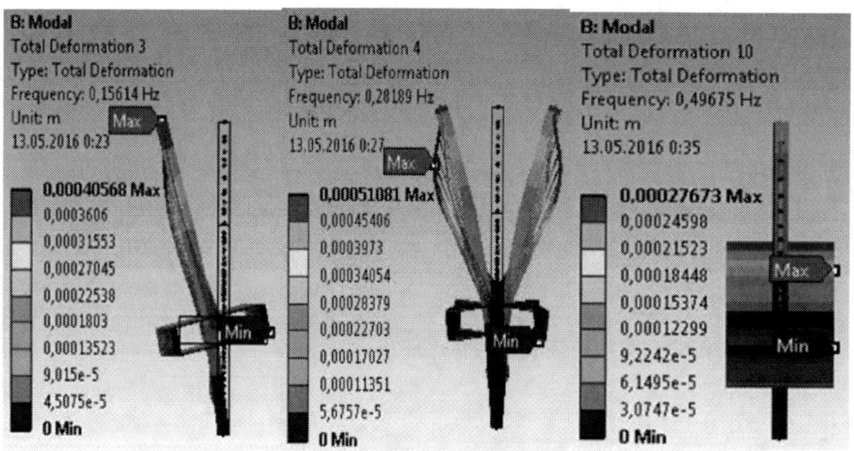


**Fig. 4.** The modal analysis of suspension bridge with a stabilizing cable – modes 3, 4, and 10.

The modal analysis of a suspension bridge with simultaneous stressing of the main and stabilizing cables results in the change of the natural vibration frequency of the fundamental period both in values (fig. 5) and directions of vibrations. As pre-stressing force increases to 0.52·〖10〗^8 N, the first shear mode of vibrations disappears to be replaced by other modes, while the primary mode of vertical vibrations is shifted in the spectrum and corresponds to mode 10. An increase of pre-stressing force eliminates torsional vibrations (fig. 5), but new combined modes of vibrations emerge instead.

As pre-stressing force continues to increase, vertical vibrations move to position 12 with almost no change to the frequency. Figure 6 shows the spectrum of vibration modes of the stabilized suspension bridge model after simultaneous pre-stressing of the main cable and stabilizing cable to force 1.3· 〚10〛 ^8 N.

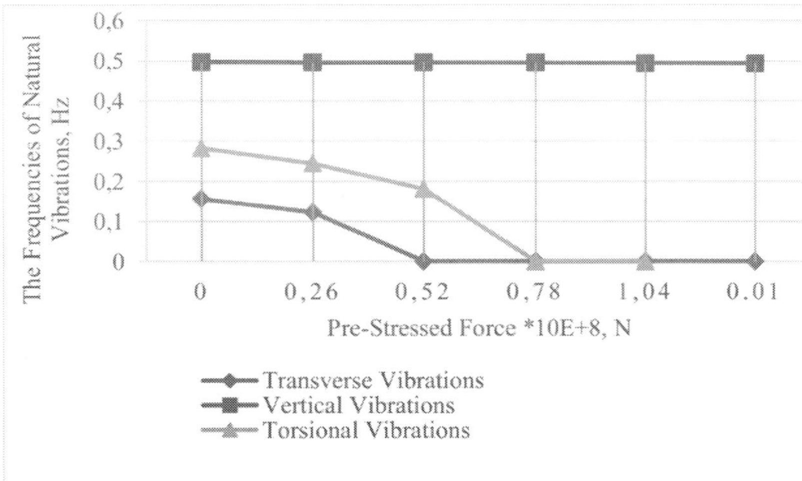


**Fig. 5.** The influence of the main cable and stabilizing cable pre-stressing on the frequencies of natural vibrations.

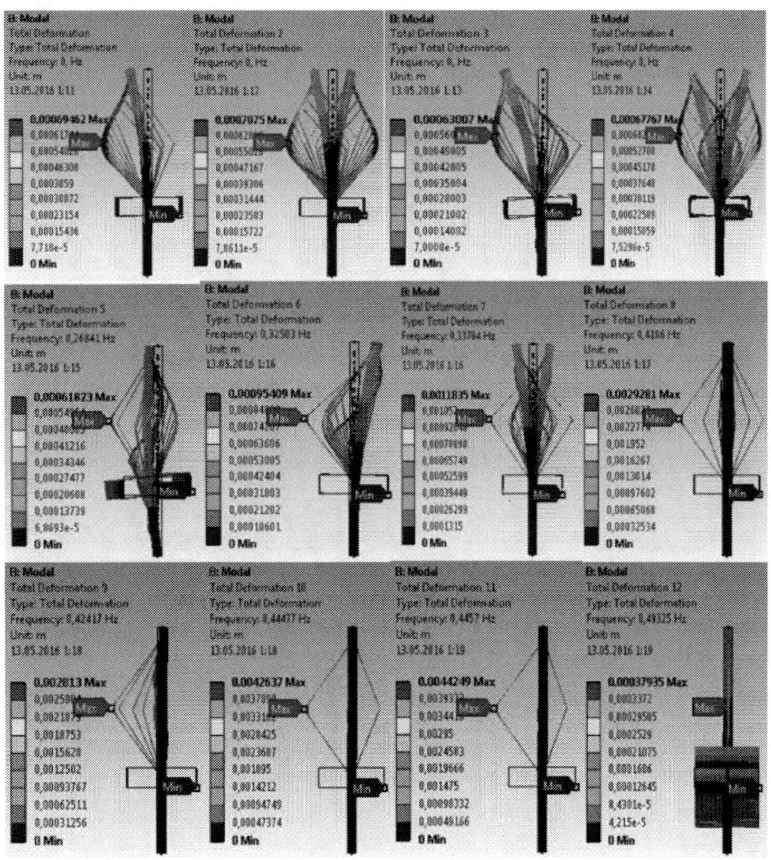

**Fig. 6.** Vibration modes 1-12, pre-stressing of the main cable and stabilizing cable to 1.3·〚10〛^8 N

## 3 Summary

The use of a stabilizing cable in a combined suspension bridge reduced the displacement in all relevant sections: 2.6 times in the middle of the central span of the stiffening girder; 15 times in the middle of the end span, 3.5 times in the tower top; and also changes the behavior of natural vibrations. The main frequencies of natural vibrations having effect on the design analysis of suspension bridges move in the spectrum. While the main modes of vibrations are dampened, new combined vibration modes emerge having no effect on the structural behavior of the bridge. From the engineering standpoint, the use of a stabilizing cable appeares to be an advantageous tool for controlling vibrations by pre-tensioning of the main cable and the stabilizing cable.

## References

1. N.P. Abovsky, Upravlyaemye konstruktsii/ N.P. Abovskiy – SAU NDS. KISI, Krasnoyarsk,. - p.125 (1995)
2. A.S. Dorogan, Raspornye visyachie mosty povyshennoi zhyostkosti // Stroitelnaya mekhanika i raschyot sooruzheniy, 4, p. 4-10 (2011)
3. A. Yu. Konovalov, M.A. Pustovalova, Vliyanie usloviy zakrepleniya nesuschego kabelya i orientatsii vant na napryazhionno deformirovannoye sostoyanie visyachego mosta // NAUKA, OBRAZOVANIE I INNOVATSII: colection of articles of the International Conference and Workshop (October 15, 2016. Ekaterinburg). 3 volumes, vol. 2/ - Ekaterinburg: AETERNA, p. 31-34 (2016)

# Simulating operational control of production in lumber house building businesses

*Dmitriy* Korolchenko[1], and *Andrey* Pizhurin[1,*]

[1]Moscow state university of civil engineering, Yaroslavskoye shosse, 26, Moscow, Russia, 129337

**Abstract.** Substantial reserves for higher operation efficiency in wooden house building businesses exist in improvements of production control system through methods of simulation and optimization. This report presents analysis of key factors of lumber house building processes that influence approaches to production control simulation. The need to factor in interconnection and dependence of lumber and house part production processes is highly difficult for operational control of the production system in question. The optimization mathematical models that we developed will help to solve the tasks of operational scheduling at all stages of wooden house manufacturing, from log cutting to production of finished assembly sets.

## 1 Introduction

The current economic situation in the wooden housebuilding sector in Russia is experiencing a shortage of funds for radical upgrade of production equipment, as well as persistent deficit of working capital. The market forces businesses to expand their product ranges and move on to small-scale production and to customer-built houses. All this calls for improved control methods, primarily ensuring flexible and optimal production processes in real time [1].

Under the circumstances, businesses are confronted with difficult problems of ensuring timely deliveries to users, well-paced manufacturing, optimized equipment load distribution, creating regulated reserves, balanced production plans and resources, ensuring strict synchronization of warehouses, motor vehicles and equipment, production facilities and operation sites [2].

Solution of the above tasks should be addressed out by a production control system built with optimization mathematical models.

There have been numerous studies to develop efficient operation control systems, and such research goes well back in time [3-9]. However, studies of operation control in woodworking business are rather few, and those available fail to exhaustively cover the capability of today's information technologies and production control system simulation methods. Besides, most of such studies mainly focus on operation management of sawmills [10-14], pulp-and-paper mills [15, 16] or furniture manufacturing businesses [17].

* Corresponding author: ananpi@yandex.ru

The main purpose of this paper is to design mathematical optimization models addressing issues of time operation planning for multi-stage processes of wooden house building.

## 2 Problem description

As its operation control object, this study examines production of wooden houses that uses processes to transform lumber and other materials and power to semi-finished and finished products within a complex business. As the key feature, each such business must include sawmills and woodworking units to manufacture joinery products, wooden structures, and wooden houses assembly sets. Such production system is a complex one because both a sawmill and a woodworking shop use highly complicated processes, and when they are integrated into a united interrelated production process the system's complexity rises to a whole new level. Wooden house manufacturing is a multi-stage process that includes lumber shaping to finished-product set delivery or on-site assembly. The stages of processing are handled by a central logistical system with many warehouses to accumulate stock between steps (saw logs, sawn lumber, blanks, parts, and assembly units).

Unlike in other industries, material is a major extra randomization factor for the production environment of wooden house building, because as a product of living nature, wood is highly diverse and unpredictable in terms of dimensional and qualitative characteristics.

One of the main objectives for production control systems is to achieve coordinated control of sawmilling and woodworking in the context of supply of lumber, blanks and parts within the premises, and thus organizing a comprehensive production process.

## 3 Modelling

To work out some rational managerial solutions, a wooden house building business needs to prepare mathematical models addressing the problems of optimized time scheduling for each stage of house manufacturing, from lumber cutting to specific finished parts.

The task of operation time scheduling for log sawing is stated as follows.

The sawmill uses $L$ parallel production flows to process logs. Let $l = 1......L$ be the numbers of the process flows.

Timber logs are sorted to form $M$ dimension groups. Let $m = 1, \ldots, M$ – be the numbers of the log size groups.

A variety of cutting patterns can be used to mill the timber. Total number of usable rational cutting patterns is $P$. Let then $p = 1,\ldots, P$ be the serial number of any pattern in the list of cutting patterns.

Specification for lumber to be produced during the planned time period contains $K$ of various type-sizes of boards. Let $k = 1, \ldots, K$ be the number of lumber type-sizes in the specification.

The entire scheduled operation time period is divided into $T$ time intervals of equal length, such as half-shift. Let $t = 1, \ldots, T$ is the serial number of the respective planning interval (control step).

For the planning period, we set the deadlines $\tau_k$ and volumes $V_{k,}$ of lumber of $k$-th type-size, to be produced by the deadlines. We also assume permissible above target $V_k^+$ and below target $V_k^-$ for production of $k$-th type-size lumber during the planning period.

We know number $Q_m$ of logs in each $m$-th group, stored at the stock warehouse during the planning period.

During the same interval of the planning period, each flow may only use one mill (cutting shape) to process material.

Total quantity of type-size lumber produced during each interval of the planning period is limited by number $N$ of pallets in the sorting bay (or recesses in the sorting line). In addition, some pairs of lumber type-sizes exist that are illegal for simultaneous production, because they are hard to tell apart during sorting. We assume that we have a total $S$ of such pairs, while $s = 1, \dots, S$ is the serial number of the pair in the list. Let us take that $\Omega_s$ is a set of numbers of lumber type-sizes, which stands for a pair illegal for simultaneous production.

Now we need to generate a calendar time schedule for stock milling: we therefore need to know which group of logs will be milled during each interval of the planning period, using which cutting pattern and as part of which flow. The following constraints of technology and planning apply.

The quality of any time schedule is decided by two criteria. First, economic criterion $F_1$ requires maximized total cost of all production output milled during the planning period. Second, technological criterion $F_2$ represents the requirement of least changes in the product range during the planning period, helping to rapidly accumulate batches of boards in the same type-size and raise equipment load efficiency during lumber drying and finishing.

Let us adopt the following notation:

$x^t_{lmp}$ –the problem's variable that is 1, if during the $t$-th planning period in $l$-th flow to mill the material of $m$-th group using $p$-th cutting pattern; else the value is 0.

$a_{kmp}$ – quantity of $k$-th type-size lumber, made of logs of $m$-th group using $p$-th cutting pattern.

$b_{lmp}$ – output capacity of $l$-th flow while milling logs of $m$-th group using $p$-th cutting pattern during the planning interval.

$C_k$ – cost of a unit of lumber of $k$-th type-size;

sgn ($d$) – sign $d$ – function defined as:

$$\text{sgn}(d) = \begin{cases} 1, \text{ if } d > 0 \\ 0, \text{ if } d \le 0. \end{cases}$$

Now the mathematical model for the task of generating an optimal time schedule for material milling will appear as follows.

Criteria of optimization:

$$F_1 = \max \sum_{t=1}^{T} \sum_{k=1}^{K} \sum_{l=1}^{L} \sum_{m=1}^{M} \sum_{p=1}^{P} C_k a_{kmp} b_{lmp} x^t_{lmp} ; \tag{1}$$

$$F_2 = \min \left\{ \sum_{t=1}^{T-1} \sum_{k=1}^{K} \left| \text{sgn}\left( \sum_{l=1}^{L} \sum_{m=1}^{M} \sum_{p=1}^{P} a_{kmp} b_{lmp} x^t_{lmp} \right) - \text{sgn}\left( \sum_{l=1}^{L} \sum_{m=1}^{M} \sum_{p=1}^{P} a_{kmp} b_{lmp} x^t_{lmp} \right) \right| \right\} . \tag{2}$$

Limitations:

1. volume of material to mill

$$\sum_{t=1}^{T} \sum_{l=1}^{l} \sum_{p=1}^{P} b_{lmp} x^t_{lmp} \le Q_m , \; m = 1,\dots,M; \tag{3}$$

2. volume of lumber to produce

$$V_k - V_k^- \le \sum_{t=1}^{\tau_k} \sum_{l=1}^{L} \sum_{m=1}^{M} \sum_{p=1}^{P} a_{kmp} b_{lmp} x^t_{lmp} \le V_k + V_k^+ , \; k = 1,\dots,K; \tag{4}$$

3. number of lumber type-sizes produced simultaneously

$$\sum_{k=1}^{K} \operatorname{sgn}\left( \sum_{l=1}^{L} \sum_{m=1}^{M} \sum_{p=1}^{P} a_{kmp} b_{lmp} x_{lmp}^{t} \right) \leq N, \qquad t = 1,\ldots,T; \tag{5}$$

4. individual lumber type-sizes produced simultaneously

$$\sum_{k \in \Omega_s} \operatorname{sgn}\left( \sum_{l=1}^{L} \sum_{m=1}^{M} \sum_{p=1}^{P} a_{kmp} b_{lmp} x_{lmp}^{t} \right) \leq 1, \ s = 1,\ldots,S; \ t = 1,\ldots,T; \tag{6}$$

5. use of only one milling pattern for each flow during any single planning interval

$$\sum_{m=1}^{M} \sum_{p=1}^{P} x_{lmp}^{t} = 1, \ l = 1,\ldots,L; \ t = 1,\ldots,T; \tag{7}$$

6. variables of the mathematical model

$$x_{lmp}^{t} = 1 \vee 0, \ p = 1,\ldots P; \ m = 1,\ldots,M; \ l = 1,\ldots,L; \ t = 1,\ldots,T. \tag{8}$$

The production process to make assembly sets of wooden houses from lumber is treated as a multi-stage line production system. The system includes the stages of lumber patterns cutting to make blanks, blank processing to make parts, and preparing sets of parts as per respective specification. Certain cases, depending on the wooden house design, may include the stage that assembles wooden panels. Each such stage can include multiple steps of blank and part processing. The number of steps and the sequence for the blanks and parts to follow through depend on engineering and art design of the housing product, and the technology adopted by the business. Manipulations of the production process can be performed on several machines working in parallel. To coordinate operation of various stages and steps of the production process, the business must establish in-process stocks to store blanks. Such stock enables control of all blanks passing each processing stage, because blanks can enter a specific stage from both the previous processing stage and from the warehouse. The blank/part processing stage is simulated as an in-process stock control system. For ongoing control of the production process, the business must draw a time schedule of operations to plan for changes in sizes of in-process stock with time.

Let us examine the problem of operation time schedule planning for the woodworking stages.

We adopt the following notation:

$T$ – planning horizon – planned time interval expressed in work shifts;

$t$ – number of planned work shift, $t = 1, \ldots, T$;

$m$ – number of work shift in the planned period, $m = 1, \ldots, T$;

$L$ – total number of blank/part processing stages at the business;

$l$ – serial number of processing stage, $l = 1, \ldots, L$;

$R^{(l)}$ – number of machines/lines in $l$-th processing stage, used in parallel;

$r$ – serial number of machine/line in $l$-th processing stage, $r = 1, \ldots, R^{(l)}$;

$T_r^{(l)}$ – effective run time of each $r$-th machine/line in $l$-th stage;

$N$ – total number of wooden house projects /products made by the business;

$j$ – number of project/product in the business's product range, $j = 1, \ldots, N$;

$M$ – total number of type-sizes of parts/blanks made by the business;

$i$ – number of part/blank as per the manufacturer's general specification, $i = 1, \ldots, M$;

$a_{ij}$ – number of parts/blanks of $i$-th type-size in $j$-th product;

$W_{1i}^{(l)}$ – number of $i$-th type-size blanks existing before $l$-th processing stage at the beginning of the planning period;

$W_{2i}^{(l)}$ – number of $i$-th type-size blanks existing after $l$-th processing stage at the beginning of the planning period;

$Q_{1i}^{(l)-}, Q_{1i}^{(l)+}$ – least/greatest number of $i$-th type-size part/blank that may be stored before processing at $l$-th stage;

$Q_{2i}^{(l)-}, Q_{2i}^{(l)+}$ – least/greatest number of $i$-th type-size part/blank that may be stored after processing at $l$-th stage;

$S_1^{(l)}, S_2^{(l)}$ –warehouse space before and after $l$-th processing stage;

$H_1^{(l)}, H_2^{(l)}$ – warehouse storage capacity coefficients before and after $l$-th processing stage;

$s_i$ – area of i-th type-size part/blank;

$h_i$ – number of i-th type-size part/blank in the stack;

$\tau_{ir}^{(l)}$ – time to process i-th type-size part/blank by $r$-th machine/line of $l$-th stage;

$\tau_r^{(l)}$ – time to readjust $r$-th machine/line of $l$-th stage;

$N_{\max r}^{(l)}$ – possible greatest number of readjustments of $r$-th machine/line of $l$-th stage per work shift;

$З_{ir}^{(l)}$ – cost to process one $i$-th type-size blank by $r$-th machine/line of $l$-th stage;

$З_{1i}^{(l)}$, $З_{2i}^{(l)}$ – storage cost per $i$-th type-size blank warehouses respectively before and after $l$-th stage during the work shift.

Let us introduce the following variables into our mathematical model:

$$\lambda_{irm}^{(l)} = \begin{cases} 1, \text{ if } i\text{-th type-size blanks are processed by } r\text{-th machine of } l\text{-th stage} \\ \quad \text{during } m\text{-th shift} \\ 0, \text{ otherwise;} \end{cases}$$

$x_{irm}^{(l)}$ – number of $i$-th type-size blanks processed during $m$-th shift by $r$-th machine of $l$-th stage;

$V_{im}^{(l)}$ – number of $i$-th type-size blanks, entering $l$-th stage from previous stages during $m$-th shift;

$U_{im}^{(l)}$ – number of $i$-th type-size blanks, exported from $l$-th stage to subsequent processing stages during $m$-th shift.

The resulting mathematical model for the problem of operation time schedule planning to manufacture assembly part sets of wooden houses for $l$-th processing stage appears as follows:

The target function of the problem is to minimize total costs of blank processing and storage at warehouses in between operations:

$$F_3 = \left\{ \sum_{m=1}^{T} \sum_{r=1}^{R^{(l)}} \sum_{i=1}^{M} З_{ir}^{(l)} \lambda_{irm}^{(l)} x_{irm}^{(l)} + \sum_{t=1}^{T} \sum_{i=1}^{M} З_{1i}^{(l)} \cdot \left( W_{1i}^{(l)} + \sum_{m=1}^{t} \left( V_{im}^{(l)} - \sum_{r=1}^{R^{(l)}} \lambda_{irm}^{(l)} x_{irm}^{(l)} \right) \right) + \right.$$

$$\left. + \sum_{t=1}^{T} \sum_{i=1}^{M} З_{2i}^{(l)} \cdot \left( W_{2i}^{(l)} + \sum_{m=1}^{t} \left( \sum_{r=1}^{R^{(l)}} \lambda_{irm}^{(l)} x_{irm}^{(l)} - U_{im}^{(l)} \right) \right) \right\} \to \min. \quad (9)$$

Limitations:

1. number of blanks warehoused before processing of $l$-th stage

$$Q_{1i}^{(l)-} \le W_{1i}^{(l)} + \sum_{m=1}^{t} \left( V_{im}^{(l)} - \sum_{r=1}^{R^{(l)}} \lambda_{irm}^{(l)} x_{irm}^{(l)} \right) \le Q_{1i}^{(l)+}, \; i = 1,\ldots,M \,;\; t = 1,\ldots,T \,; \quad (10)$$

2. number of blanks warehoused after processing of $l$-th stage

$$Q_{2i}^{(l)-} \le W_{2i}^{(l)} + \sum_{m=1}^{t}\left(\sum_{r=1}^{R^{(l)}} \lambda_{irm}^{(l)} x_{irm}^{(l)} - U_{im}^{(l)}\right) \le Q_{2i}^{(l)+}, i = 1,\dots,M;\ t = 1,\dots,T; \tag{11}$$

3. capacity of stock warehouse before processing of $l$-th stage

$$\sum_{i=1}^{M} \frac{s_i}{h_i}\left(W_{1i}^{(l)} + \sum_{m=1}^{t}\left(V_{im}^{(l)} - \sum_{r=1}^{R^{(l)}} \lambda_{irm}^{(l)} x_{irm}^{(l)}\right)\right) \le H_1^{(l)} S_1^{(l)},\ t = 1,\dots,T; \tag{12}$$

4. capacity of stock warehouse after processing of $l$-th stage

$$\sum_{i=1}^{M} \frac{s_i}{h_i}\left(W_{2i}^{(l)} + \sum_{m=1}^{t}\left(\sum_{r=1}^{R^{(l)}} \lambda_{irm}^{(l)} x_{irm}^{(l)} - U_{im}^{(l)}\right)\right) \le H_2^{(l)} S_2^{(l)},\ t = 1,\dots,T; \tag{13}$$

5. output of machines used to process blanks during $l$-th stage per shift

$$\sum_{i=1}^{M}\left(\tau_{ir}^{(l)} x_{irm}^{(l)} + \tau_r^{(l)}\right)\lambda_{irm}^{(l)} \le T_r^{(l)},\ r = 1,\dots,R^{(l)};\ m = 1,\dots,T; \tag{14}$$

6. readjustment cycles for machines used to process blanks during $l$-th stage per shift

$$\sum_{i=1}^{M} \lambda_{irm}^{(l)} \le N_{\max r}^{(l)},\quad r = 1,\dots,R^{(l)};\ m = 1,\dots,T; \tag{15}$$

7. correlation of variables

$$x_{irm}^{(l)} \in Z_+;\ V_{im}^{(l)} \in Z_+;\ U_{im}^{(l)} \in Z_+;\ \lambda_{irm} \in \{0;1\} : (1 - \lambda_{irm})x_{irm} = 0 \wedge (x_{irm} - 1)\lambda_{irm} \ge 0;$$
$$i = 1,\dots,M;\ r = 1,\dots,R^{(l)};\ m = 1,\dots,T. \tag{16}$$

## 4 Conclusions

Mathematical analysis of the resulting models demonstrates that they can be reduced to the problem of unconditional pseudo-Boolean optimization with target functions preset algorithmically as a problem of conditional optimization with continuous or discrete variables. Coefficients for Boolean variables are not constant values, but they represent functions of the same Boolean variables. To solve this problem, we use a hybrid adaptive heuristic search algorithm.

The proposed set of interrelated mathematical models is basis for the designed system to support operational control of wooden house manufacturing.

Future studies are expected to consider the uncertainty of the production environment of the businesses of wooden house building using the methods of fuzzy mathematical programming.

## References

1. A. S. Ermakov, O. G. Mukhamedzhanova, To the enlightened development of the construction industry, *Integration, partnership and innovation in building science and education.The collection of materials of the International Scientific Conference*, 329–333, (2017)
2. B. Skołud, Market oriented approach to the production management on the operational level, Adv. Mater. Res., **837**, 663-668, (2014)
3. K. N. McKay, G. W. Black, The evolution of a production planning system: A 10-year case study, Comput. Ind., **58**(8–9), 756–771, (2007)

4. C. T. Maravelias, C. Sung, Integration of production planning and scheduling: Overview, challenges and opportunities, Comput. Chem. Eng., **33**(12), 1919–1930, (2009)
5. I. C. Dima, M. Man, Managerial Systems, Methods, and Techniques Used in Scheduling Industrial Production, *Industrial Production Management in Flexible Manufacturing Systems* (IGI Global, 345–369, 2013)
6. I. Harjunkoski, C. T. Maravelias, P. Bongers, P. M. Castro, S. Engell, I. E. Grossmann, J. Hooker, C. Méndez, G. Sand, J. Wassick, Scope for industrial applications of production scheduling models and solution methods, Comput. Chem. Eng., **62**, 161–193, (2014)
7. M. Baldea, I. Harjunkoski, Integrated production scheduling and process control: A systematic review, Comput. Chem. Eng., **71**, 377–390, (2014)
8. D. A. Korolchenko, E. N. Degaev, A. F. Sharovarnikov, Dependence of Fire Extinguishing Efficacy of Low Expansion Foams Solutions Homology Sodium Sulfate on the Molecular Weight of the Surface-active Substances, *2nd International conference on material engineering and application* (ICMEA 2015)
9. D. Korolchenko, S. Voevoda, Influence of spreading structure in an aqueous solution-hydrocarbon system on extinguishing of the flame of oil products, MATEC Web of Conferences, **86**, (2016)
10. S. Maturana, E. Pizani, J. Vera, Scheduling production for a sawmill: A comparison of a mathematical model versus a heuristic, Comput. Ind. Eng., **59**(4), 667–674, (2010)
11. M. K. Zanjani, M. Nourelfath, D. Ait-Kadi, Sawmill Production Planning Under Uncertainty: Modelling and Solution Approaches, Stochastic Programming, 347–395, (2012)
12. M. K. Zanjani, D. A. Kadi, M. Nourelfath, A stochastic programming approach for sawmill production planning, IJMOR, **5**(1), 1, (2013)
13. M. Varas, S. Maturana, R. Pascual, I. Vargas, J. Vera, Scheduling production for a sawmill: A robust optimization approach, Int. J. Prod. Econ., **150**, 37–51, (2014)
14. A. Lobos, J. R. Vera, Intertemporal stochastic sawmill planning: Modeling and managerial insights, Comput. Ind. Eng., **95**, 53–63, (2016)
15. D. Bredstrom, M. Ronnqvist, Integrated production planning and route scheduling in pulp mill industry, *Proceedings of the 35th Annual Hawaii International Conference on System Sciences*, (2002)
16. G. Figueira, P. Amorim, L. Guimarães, M. Amorim-Lopes, F. Neves-Moreira, B. Almada-Lobo, A decision support system for the operational production planning and scheduling of an integrated pulp and paper mill, Comput. Chem. Eng., **77**, 85–104, (2015)
17. J. C. Vidal, M. Mucientes, A. Bugarín, M. Lama, Machine scheduling in custom furniture industry through neuro-evolutionary hybridization, Appl. Soft Comput., **11**(2), 1600–1613, (2011)

# Investigation of structure formation processes at the hydration of modified gypsum binders and mortars

*Vasili* Korovyakov[1], *Anna* Zaikina[1] , and *Vadim* Solovyov[2,*]

[1]Research Institute Mosstroy, Moscow, Russian Federation
[2] Moscow state university of civil engineering, Yaroslavskoye shosse, 26, Moscow, Russia, 129337

**Abstract.** The results of investigations of structure formation processes of a mortar while hardening have been presented in the article. The results obtained indicate that during the period of structure formation and the heat release of solidifying composite gypsum binders and solutions, the heat release in hardening of composite gypsum binders (CGB) and modified gypsum mortars (MGM) proceeds at different rates, and is characterized, in some cases, by the rapid release of hydration heat. As studies have shown, modified gypsum binders and mortars, in order to avoid the appearance of structure defects, should contain setting-time additives, promoting to a decrease in the intensity of heat release and, as consequence, reducing risks of destructive processes. With a large amount of a setting-time retarder, hydration process may be disturbed, and that can be accompanied by following abnormalities (abundant bleeding, slow gain in strength, etc.). Significant heat release during hardening allows using the developed mortars even at negative temperatures, as also evidenced by the results of other researchers.

## 1 Introduction

The aim of the studies performed: the definition of regularities in the process of structure formation at hardening of multicomponent binders and mortars, depending on compositions, type of components and their nature.

Currently, the assessment of quality indices of construction mortars is made in accordance with Russian Standard GOST 28013-98 "Mortars. General specifications" and Russian Standard GOST 31377-2008 "Dry building levelling plaster mixtures based on gypsum binder. Specifications”. The main parameters of the quality of a mortar mixture according to these standards are: flowability, water retention ability, sedimentation, compressive strength, density and frost resistance [1, 2].

Of great importance for the development of technological properties of plaster mortars is the knowledge of structure formation processes in a hardening mortar, especially a multicomponent one, which is composite gypsum binders and mortars based on them.

* Corresponding author: s_vadim_g@mail.ru

One of the most effective finishing mortars is gypsum modified mortars [3, 4, 5].

## 2 Materials and mixture composition

Studies have been conducted to determine the period of structure formation of gypsum systems and, accordingly, the patterns of structure formation with heat losses to the environment. For the studies, compositions consisting of a hemihydrate gypsum binder (GB) and an organomineral modifier were prepared [6 - 9].

As a gypsum binder the binder of grade G-7 BI, manufactured by the Peshelansky plant, in accordance with Russian Standard GOST 125-79, was used. The organomineral modifier was prepared from Portland cement of CEM I 42.5 N by Russian Standard GOST 31108-2003 of Volsky cement plant, microsilica (Russian Technical conditions 5743-048-02495332-96 "Microsilica condensed", and C-3 superplasticizer based on naphthalene sulfonic acids according to Russian Technical conditions 6-36-0204229-625-90 "Plasticizer C-3".

## 3 Results

The process of structure formation while hardening of modified compositions consisting of a gypsum binder, a hydraulic component, various additives and modifiers, is of interest [10-12]. In the present studies, the structure formation process was studied in terms of the kinetics of heat release while hardening of binders [13, 14, 15, 16].

The reactions of the interaction of hemihydrate calcium sulfate with water are exothermic, i.e. proceed with the release of heat. From a production point of view, it is important to know both the absolute amount of heat that can be released during the binder hydration, and the kinetics of heat release.

Figures 1 - 4 show the kinetics of temperature change of solidifying systems. The analysis of curves showing changes in the temperature of mortars with a content of aggregate of 5%, 50%, optimal composition, and without the addition of setting-time retarders in the systems showed the following:

- the composition with content of 5% aggregate has the most intensive character of heat release. In the composition, the maximum temperature of 44.4 °C is reached after 24.5 minutes, with a treatment period of 9 - 15 min;
- composition with content of 50% aggregate (by weight) has a slightly less pronounced heat release (43 °C after 32.5 minutes of observation) with a mortar setting time of 8 ... 14 minutes;
- a mortar of an optimal composition of 1:1 based on a modified gypsum binder and sand with Mk = 2.0 falls in between above compositions (the temperature is only 40.2 °C after 23 minutes of the initial formation of the structure of stone with its time terms only 4 - 7 minutes).

The kinetics of temperature changes in MGM compositions with a setting-time retarder in amount of 0.2% of the weight of MGB and the content of aggregate of various sizes is almost similar.

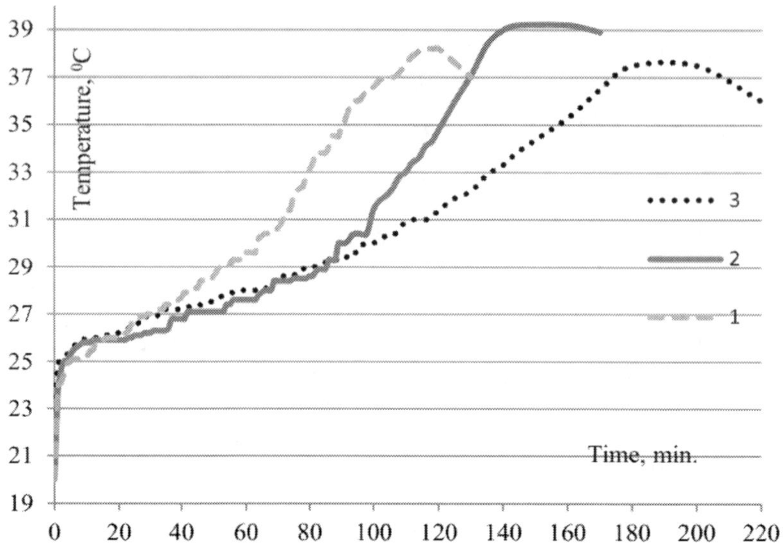


**Fig. 1.** Hardening behaviour of modified gypsum binder (MGB) with 0.2% retarder: 1 - gypsum binder with 20% organomineral modifier; 2 - gypsum binder with 25% organomineral modifier; 3 - gypsum binder with 25% organomineral modifier.

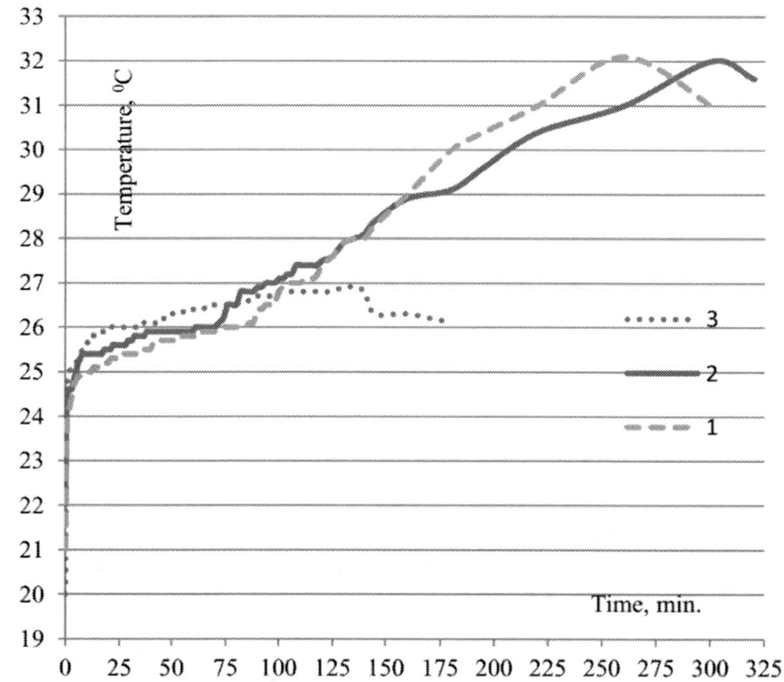


**Fig.2.** Hardening behaviour of modified gypsum binder (MGB) with 0.3% retarder: 1 - gypsum binder with 20% organomineral modifier; 2 - gypsum binder with 25% organomineral modifier; 3 - gypsum binder with 25% organomineral modifier.

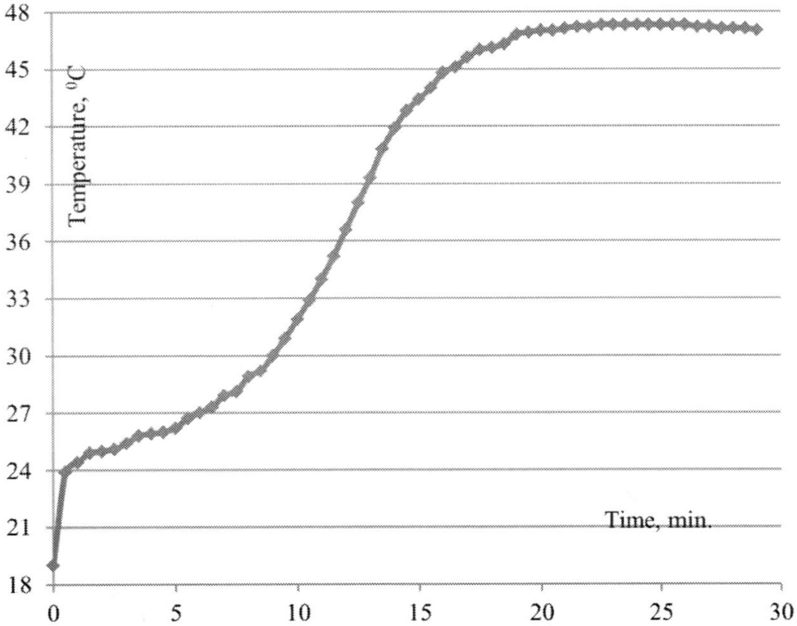


**Fig. 3.** Hardening behaviour of the optimal composition of modified gypsum binder (MGB).

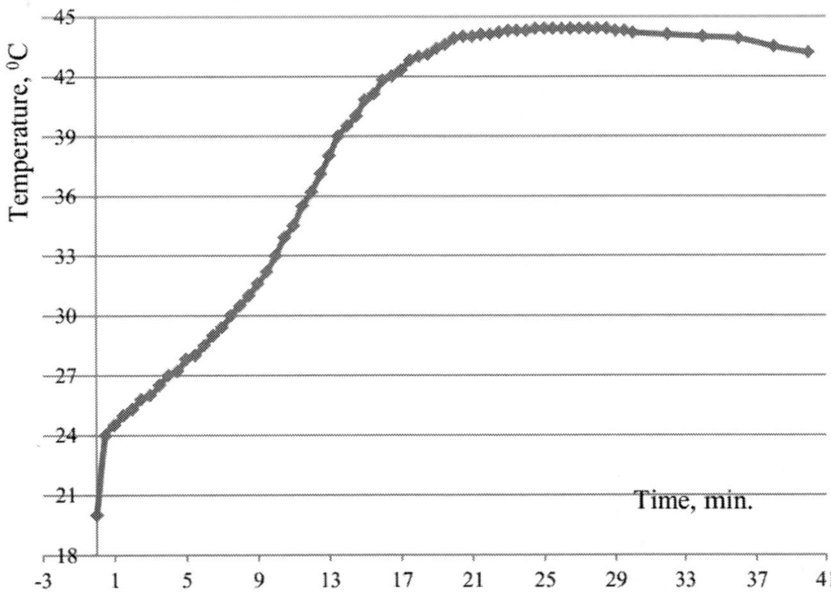


**Fig. 4.** Hardening behaviour of the optimal composition of modified gypsum mortar: binder + fine aggregate (95+5%).

Hardening and structuring of the mortar is the most important stage determining the further quality of the solidified mortar. The control of this process is necessary both at the preparation stage and at the stage of application of the mortar mixture. As can be seen from the results obtained, the complex of technological factors (amount of additives introduced, ratio of components, size of the aggregates, etc.) affects the period of the structure formation and the heat release temperature of the solidifying composite gypsum binders and mortars.

## 4 Conclusion

Thus, analyzing the obtained data, the heat release process can develop at different rates and is characterized by the rapidly released hydration heat. In systems without regulation of the setting (hardening) time, a drastic increase in temperature is noted, therefore internal pressure and an emerging temperature gradient may arise (especially on massive surfaces treated). In addition, exothermic processes take place at the end of workability, and the structure may not reach the initial critical strength capable of withstanding internal tensile stresses when heated, the same effect can be observed in mortar systems with a filler, but without additives regulating the setting time. The main principle of the processes of normal structure formation (especially in the technological application) is provided by the expediency of using additives regulating the kinetics of hardening due to the formation (in given ranges) of hardly soluble and slightly dissociable compounds. As the studies have shown, the modified gypsum binders and mortars, in order to avoid the appearance of structure defects, should contain setting-time regulators that help to reduce the intensity of heat release and, as a consequence, reduce the risks of destructive processes.

Summarizing the above, it can be noted that the choice of the optimal composition of the mortar mixture is determined not only by solving the problems of obtaining a mortar with predetermined physical and mechanical properties, but also by effectively controlled processes of structure formation, which are the most important factor determining the further quality of the solidified mortar, as well as its workability. The control of these processes is necessary to define the viability of the mortar mixture, as well as the timing (age) of testing the solution and determining its moisture content. A significant rise in temperature makes it possible to apply the mortars developed at negative temperatures, too, as evidenced by the results of other researchers.

## References

1 G. Severinova, Walls and facades (in Russian - Steny i fasady), **6,** 45 (2006)
2 V. Korovyakov, Dry mixtures (in Russian - Suhie stroitelnye smesi), 1 (2007)
3 M. Muller, H. Fischer, Conductometric description of processes of interaction of calcium sulfate with water (Konduktometrische Beschreibung der Prozesse beim Kontakt mit Wasser von Calciumsulfaten). //SB. Gypsum, its study and application. - Kraskovo, 2005.
4 A. Adamtsevich, S. Pashkevich, A. Pustovgar, Magazine of Civil Engineering (in Russian - Inzhenerno-stroitelnyj zhurnal), 3 (2016)
5 V. Korovyakov, Russian chemical journal (in Russian - Rossijskij himicheskij zhurnal), 4 (2003)
6 Q. Yu , H. Brouwers, Cem. and Conc. Comp., **34**, 9 (2012)
7 R. Magallanes-Rivera, C. Juarez-Alvarado, P. Valdez, J. Mendoza-Rangel, Cons. and Build. Mat, 37 (2012)
8 M. Singh, M. Garg, Cem. and Conc. Res., **23**, 1 (1993).
9 A. Colak, Cem. and Conc. Comp., 22, 3 (2000)
10 V Soloviev, A. Eremin, D. Eliseev, A. Buryanov, Building materials (in Russian - Stroitelnye materialy), 1-2 (2017)
11 E. Boldygeva, A. Buryanov, N. Galtseva, V. Soloviev, Building materials (in Russian - Stroitelnye materialy), 6 (2015)

12 E. Boldygeva, Galtseva N., A. Buryanov, V. Petropavlovskaya, Journal of Modern problems of science and education (in Russian - Aktualnye problemy nauki i obrazovanija), **2**, 71 (2013)
13 G. Vasconcelosa, P. Lourencoa, A. Camoesb, A. Martinsa, S. Cunha, Cem. and Conc. Comp., **58** (2015)
14 J. Escalante-García, M. Rios-Escobar, A. Gorokhovsky, A. Fuentes, Cem. and Conc. Comp., **30,** 2 (2008)
15 M. Garg, A. Pundir, Cem. and Conc. Comp., **45** (2014)
16 V. Nilles, J. Plank, Cem. and Conc. Res., **42**, 5 (2012)

# Dynamic stability of a viscoelastic plate

*Elena* Kosheleva[1,*]

[1] Moscow State University of Civil Engineering (National Research University)
Yaroslavskoe Shosse 26, Moscow, Russia

**Abstract.** The paper considers the problem of stability of a viscoelastic plate subjected to intermittent load. A differential equation was designed, which describes the dynamic stability of the plate stressed by permanent and alternating loads across the plate's plane. A solution in the form of a series with separated variables has been examined. Linear differential equations of the third order were obtained for the time function. A critical frequency equation was obtained as well. The major domains of dynamic instability were investigated. Another analysis was dedicated to the way the position of the prime instability domain is affected by relaxation time as well as correlation between long-term and instantaneous moduli of elasticity.

## 1 Problem setting

Let us consider a rectangular plate exposed to intermittent compressing forces across its plane, with these forces evenly distributed along the edges:

$$N_x = N_{x0} + N_{xt}\cos\theta t, \qquad N_y = N_{y0} + N_{yt}\cos\theta t \tag{1}$$

The plate material is subject to a linear creep law, which is given by [1] for a flat strained state:

$$\begin{aligned}
\sigma_x + n\dot{\sigma}_x &= \frac{H}{1-\nu^2}\left(\varepsilon_x + \nu\varepsilon_y\right) + \frac{En}{1-\nu^2}\left(\dot{\varepsilon}_x + \nu\dot{\varepsilon}_y\right),\\
\sigma_y + n\dot{\sigma}_y &= \frac{H}{1-\nu^2}\left(\varepsilon_y + \nu\varepsilon_x\right) + \frac{En}{1-\nu^2}\left(\dot{\varepsilon}_y + \nu\dot{\varepsilon}_x\right),\\
\tau_{xy} + n\dot{\tau}_{xy} &= \frac{H}{2(1+\nu)}\gamma_{xy} + \frac{En}{2(1+\nu)}\dot{\gamma}_{xy},
\end{aligned} \tag{2}$$

where $n$ - relaxation time, $E$ - instantaneous elasticity modulus, $H$ - long-term elasticitymodulus. Plate bending equation with (2) taken account of takes the form

$$\frac{H}{E}D\nabla^2\nabla^2 w + Dn\frac{\partial}{\partial t}\nabla^2\nabla^2 w = q + n\frac{\partial q}{\partial t}, \tag{3}$$

* Corresponding author: elleok@mail.ru

where $D = \frac{Eh^3}{(1-\nu^2)}$ - instantaneous bending rigidity. Load and inertial forces inserted into (3) produce a differential equation of the third order:

$$\frac{H}{E} D\nabla^2\nabla^2 w + Dn\frac{\partial}{\partial t}\nabla^2\nabla^2 w + m\frac{\partial^2 w}{\partial t^2} + nm\frac{\partial^3 w}{\partial t^3} + N_x\frac{\partial^2 w}{\partial x^2} + \\ N_y\frac{\partial^2 w}{\partial y^2} + n\left(\frac{\partial N_x}{\partial t}\frac{\partial^2 w}{\partial x^2} + \frac{\partial^3 w}{\partial x^2 \partial t}N_x + \frac{\partial N_y}{\partial t}\frac{\partial^2 w}{\partial y^2} + \frac{\partial^3 w}{\partial y^2 \partial t}N_y\right) = 0. \tag{4}$$

## 2 Solution technique

The solution to this equation can be found by separating the variables:

$$w = \sum f_i(t) X_i(x) Y_i(y) \tag{5}$$

where $f_i(t)$ - the unknown time functions, $X_i(x)$ and $Y_i(y)$ - functions fulfilling the boundary conditions along the edges of the rectangular outline relative to $w$. By inserting the solution (5) to (4) and by utilizing the Bubnov-Galerkin method, an equation with alternating coefficients is obtained:

$$f''' + \frac{1}{n\theta}f'' + \frac{\Omega^2}{\theta^2}(1 - 2\mu\cos\tau)f' + \frac{\Omega^2}{n\theta^2}(\xi - 2\mu\cos\tau + 2\mu n\theta\sin\tau)f = 0, \tag{6}$$

where the prime denotes a nondimensional time derivative $\tau = \theta t$, $\Omega$ - proper vibration frequency of the plate exposed to the constant components of longitudinal forces, $\mu$ - excitation coefficient, $\xi$ - nondimensional parameter determined by the correlation between the instantaneous and long-haul elasticity modulus;

$$\Omega^2 = \omega^2\left(1 - \frac{N_{x0}N_{2*} + N_{y0}N_{1*}}{N_{1*}N_{2*}}\right),$$

$$\mu = \frac{1}{2}\frac{N_{xt}N_{2*} + N_{yt}N_{1*}}{N_{1*}N_{2*} - N_{x0}N_{2*} - N_{y0}N_{1*}}, \quad \xi = \frac{\frac{H}{E}N_{1*}N_{2*} - N_{x0}N_{2*} - N_{y0}N_{1*}}{N_{1*}N_{2*} - N_{x0}N_{2*} - N_{y0}N_{1*}}, \tag{7}$$

$N_{1*}$ и $N_{2*}$ - critical values of strains $N_{x0}$ and $N_{y0}$ provided that their statical action is independent, $\omega$-frequency of natural vibration of an unloaded plate,

$$\omega^2 = \frac{DI_1}{mI_2},\ N_{1*} = \frac{I_1}{I_3}D,\ N_{2*} = \frac{I_1}{I_4}D,\ I_1 = \int_0^a\int_0^b\left(\frac{d^4X}{dx^4}Y + 2\frac{d^2X}{dx^2}\frac{dY^2}{dy^2} + \frac{dY^4}{dy^4}X\right)XYdxdy,$$

$$I_2 = \int_0^a\int_0^b X^2Y^2dxdy,\ I_3 = -\int_0^a\int_0^b X\frac{d^2X}{dx^2}Y^2dxdy,\ I_4 = -\int_0^a\int_0^b X^2\frac{d^2Y}{dy^2}Ydxdy. \tag{8}$$

The equation features four nondimensional coefficients $\mu$, $\xi$, $\alpha = \frac{1}{n\theta}$, $\varphi = \frac{\Omega^2}{\theta^2}$. Changes in these coefficients impact the solution to the equation (6). Provided certain correlations between the coefficients are in place, this equation may have infinitely increasing solutions, which take up entire domains on the parameter plane $\mu, \frac{\theta}{2\Omega}$. Defining locations of dynamic instability domains represents one of the major tasks of the dynamic stability theory. The domains of boundlessly increasing solutions are separated from the domains of stability by intermittent solutions at intervals $T$ and $2T$. Two solutions of the same period limit the domain of instability, two solutions of different periods do same for the domain of stability. The search for instability domains comes down to determination of conditions under which the equation (6) has periodic solutions with the period being $T$ and $2T$. The solution to the equation (6) with the period $T$ is searched for in the form of a seriesse

$$f(t) = \sum_{k=1,3,5}^{\infty} \left( a_k \sin \frac{k\tau}{2} + b_k \cos \frac{k\tau}{2} \right), \tag{9}$$

while the solution with the $2T$ period is searched for in the form of a series

$$f(t) = \sum_{k=2,4,6}^{\infty} \left( a_k \sin \frac{k\tau}{2} + b_k \cos \frac{k\tau}{2} \right) + b_0 . \tag{10}$$

The first region of instability is of paramount importance, hence we insert the solution (9) into the equation (6) :

$$\begin{gathered}\sum_{k=1,3}^{\infty}\left(-a_k \frac{k^8}{8}\sin\frac{k\tau}{2} + b_k \frac{k^3}{8}\cos\frac{k\tau}{2}\right) + \alpha \sum_{k=1,3}^{\infty}\left(-a_k\frac{k^2}{4}\sin\frac{k\tau}{2} - b_k\frac{k^2}{4}\cos\frac{k\tau}{2}\right) + \\ \varphi\left(1 - 2\mu\cos\tau + \frac{2\mu}{\alpha}\sin\tau\right)\sum_{1,3}^{\infty}\left(a_k \sin\frac{k\tau}{2} + b_k\cos\frac{k\tau}{2}\right) = 0.\end{gathered} \tag{11}$$

After transforming the equation (11) and equalizing the coefficiens, $\sin\frac{k\tau}{2}$ and $\cos\frac{k\tau}{2}$ being identical, we obtain a system of linear equations in respect of $a_k$ and $b_k$ :

$$\begin{gathered}a_1\alpha\left(-\frac{1}{4} + \varphi\xi + \varphi\mu\right) - \alpha\varphi\mu a_3 - \frac{1}{2}b_1\left(-\frac{1}{4} + \varphi - \varphi\mu\right) + \frac{1}{2}\varphi\mu b_3 = 0, \\ \frac{1}{2}a_1\left(-\frac{1}{4} + \varphi + \varphi\mu\right) + \alpha b_1\left(-\frac{1}{4} + \varphi\xi - \varphi\mu\right) - \frac{1}{2}\varphi\mu a_3 - \alpha\varphi\mu b_3 = 0, \\ \alpha\left(-\frac{k^2}{4} + \varphi\xi\right)a_k - \frac{k}{2}b_k\left(-\frac{k^2}{4} + \varphi\right) - \alpha\varphi\mu(a_{k+2} + a_{k-2}) + \varphi\mu\frac{k}{2}(b_{k+2} - b_{k-2}) = 0, \\ \frac{k}{2}a_k\left(-\frac{k^2}{4} + \varphi\right) + \alpha b_k\left(-\frac{k^2}{4} + \varphi\xi\right) - \varphi\mu\frac{k}{2}(a_{k+2} + a_{k-2}) - \alpha\varphi\mu(b_{k+2} + b_{k-2}) = 0, \\ (k = 1,3,5\ldots)\end{gathered} \tag{12}$$

An equation of critical frequencies is obtained by setting the equation determinant to zero. By critical frequencies we understand the frequencies of the external load $\theta_*$, corresponding to the boundaries of the instability domains. By substituting the coefficient

$\varphi$ in (12) by $p^2 = \dfrac{1}{4\varphi}$, we transform the coefficient $\alpha = \dfrac{1}{n\theta} = \dfrac{r}{p}$, where $r = \dfrac{1}{2\Omega n}$.

Consequently, we obtain

$$\begin{vmatrix} \frac{r}{p}\left(-p^2+\xi+\mu\right) & -\frac{1}{2}\left(-p^2+1-\mu\right) & -\frac{r}{p}\mu & \frac{1}{2}\mu & \ldots \\ \frac{1}{2}\left(-p^2+1+\mu\right) & \frac{r}{p}\left(-p^2+\xi-\mu\right) & -\frac{1}{2}\mu & -\frac{r}{p}\mu & \ldots \\ -\frac{r}{p}\mu & \frac{3}{2}\mu & \frac{r}{p}\left(-9p^2+\xi\right) & -\frac{3}{2}\left(-9p^2+1\right) & \ldots \\ -\frac{3}{2}\mu & -\frac{r}{p}\mu & \frac{3}{2}\left(-9p^2+1\right) & \frac{r}{p}\left(-9p^2+\xi\right) & \ldots \\ \ldots & \ldots & \ldots & \ldots & \ldots \end{vmatrix} = 0. \qquad (13)$$

This equation facilitates finding the domains of dynamic instability which lie in the proximity of the frequencies $\theta_* = \dfrac{2\Omega}{k}, k = 1,3,5\ldots$ Next we are going to consider the principal region of instability in the proximity of $\theta_* = 2\Omega$, which is the most dangerous and for this reason has the most significant practical implications.

## 3 The findings

Three iterations were made to build the main instability domain, i.e. determinants of the second, fourth and sixth order were considered. The second and third iterations yielded close results, and for this reason the determinant of the fourth order was used to investigate the instability domain. The equation (13) encompasses coefficients $r$ and $\xi$, one of which depends on the relaxation time, while the other hinges on the correlation between the instantaneous and long-term moduli of elasticity. Changes in these coeffictions impact the position of the instability domains. To analyze the impact of the coefficients we are going to build the domains of dynamic instability where $\xi$ is constant and specific while $r$ is variable. Figure 1 displays the domains for the value $\xi = 0.5$, while $r$ assumes values from zero to infinitely high values. The figure makes it clear that with $r$ increasing, the boundaries of the domain initially depart from the $p$ axis and move to the right, and at the same time displace vertically towards the reduction of $p$, and then approach the axis again.

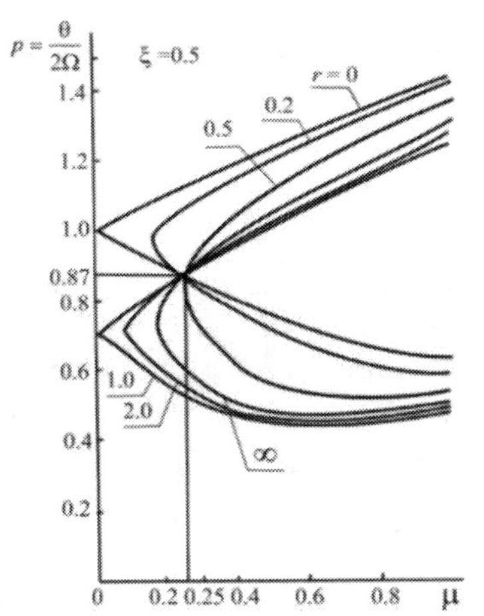

Figure 1 Major areas of sustainability

The instability domain with $r = 0$ begins from the point, the coordinate of which is $p = 1$, while the domain stems from the point with coordinates $p = \sqrt{\xi} = 0.707$ given $r = \infty$. With $r = 0$, which implies infinitely long relaxation time, the material showcases elasticity with an instantaneous elasticity modulus $E$. In constrast, $r = \infty$, which corresponds to $n = 0$, makes the material perform with elasticity, $H$ being the respective long-haul elasticity modulus. Qualitative similarity may be noticed if a comparison is drawn between the outcome displayed in Figure 1 and the borders of the domains built for a beam [3]. In what follows we consider the dependency between $\xi$ and $r$. Figure 2 presents the results of such calculations. The impact of $\xi$ on the position of the domain depends, as the graphs indicate, on the magnitude of $r$.

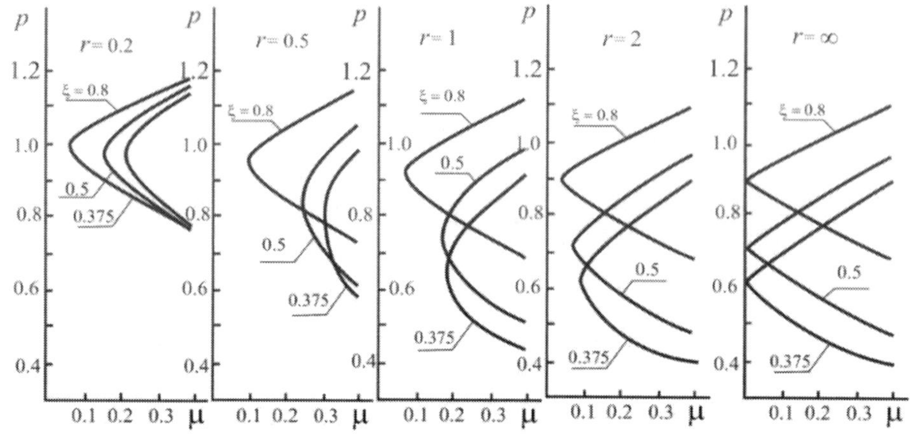


Figure 2 Influence of $\xi$ on the principal domains of dynamic stability, depending on the value of $r$

Decrease in $\xi$ with $r$ being constant, causes the boundary to displace towards the increasing $\mu$. With $r$ increasing, such displacement reduces, while $r = \infty$ means that all the curves border upon the $p$ axis. At the same time, increasing $r$ combined with decreasing $\xi$ displaces the curves towards the $\mu$ axis. The parameter $\xi$ depends on the correlation of the elasticity moduli; if this parameter goes down against the backdrop of the constant relaxation rate (unless it is equal to zero or is infinitely high), the minimum value of the excitation coefficient $\mu$ increases, and thus the minimum amplitudes of periodic forces $N_{xt}$ and $N_{yt}$ rise as well, while the frequency $\theta$ goes down. If the relaxation rate is equal to zero, the material displays elastic behavior, $E$ representing the instantaneous elastic modulus. If the relaxation rate assumes an infinitely high value, decreasing $\xi$ is accompanied by the lowering load frequency, which leads to stability being lost.

We thus have investigated the way the parameter $\xi$ determines the position of the boundaries of the main instability domain. In case the correlation between the elasticity moduli narrows down, and hence $\xi$ decreases when the relaxation rate is constant, unless it is zero or infinitely high, the minimum value of the excitation coefficient as well as the minimum amplitudes of periodic forces increase, while the frequencies of these forces contract. Infinitely high speed of relaxation along with reduction of the correlation in question causes a fall in load frequency, which accounts for a loss of dynamic stability.

## 4 Conclusion

The paper has explored the way relaxation time and correlation between elasticity moduli impact the position of the principal dynamic instability domain. It has been demonstrated that the lower the correlation between the elasticity moduli with relaxation speed being steady (except for its zero and infinitely high values), the higher the minimum value of the excitation coefficient, the higher the minimum amplitudes of periodic forces, while the frequency of these forces goes down. Viscoelastic behavior of structural elements is most relevant for polymers and composite materials 0[4]. Recent time has seen multiple papers exhibiting new results of research dedicated to viscoelastic characteristics of materials and development of computing techniques for respective structures [5 – 9 etc.]

## References

1. V.V. Bolotin, *Dynamic Stability of Elastic Systems* (State publishing house of theoretic technical literature, 1956)
2. A.R. Rzhanitsyn, *Creep theory*. (Stroyizdat M., 1968)
3. V.I. Matyash, Mechanics of polymers **1**, p.116-121(1974)
4. R.A. Turusov, V.I. Andreyev, B.M. Yaziyev, *Selected Problems and Techniques of Mechanics of Macroheterogenous Viscoelastic Environment*. (RGSU Publishing, Rostov-on-Don ,2009)
5. N.Yu.Tsybin, R.A.Turusov, V.I.Andreev, Comparison of creep in free polymer rod and creep in polymer layer of the layered composite. Procedia engineering **153** ( 2016 ) 51 – 58
6. V.I. Andreyev, B.M. Yaziyev, Buckling of Lengthwise Clamped Bars of Alternating Rigidity under Creep. Inzhenerny Vestnik Dona (Engineering Reporter of Rostov-on-Don) 4 (2012) 142-148.
7. V.I. Andreev, A.S. Chepurnenko, B.M. Yazyev, On the Bending of a Thin Plate at Nonlinear Creep. Advanced Materials Research, **900** (2014) 707-710
8. V.I. Andreev, A.S. Chepurnenko, B.M. Yazyev, Energy Method in the Calculation Stability of Compressed Polymer Rods Considering Creep. Advanced Materials Research 1004-1005 (2014) 257-260
9. V.I. Andreev, A.S. Chepurnenko, B.M. Yazyev, A.N. Beskopylny Determination of Rheological Parameters of Polyvinylchloride at Different Temperatures. MATEC Web of Conferences **67** (2016) No 06059

# On analysis of operating efficiency of autonomous ventilation systems

*Arman* Kostuganov [1*], and *Yuri* Vytchikov[1]

[1]Samara State Technical University, Institute of Architecture and Civil Engineering, 194, Molodogvardeyskaya St., 443001, Samara, Samara

**Abstract.** The paper deals with the causes and consequences of malfunctioning of natural and mechanical ventilation systems in civil buildings of Russia. Furthermore it gives their classification and analysis based on the literature review. On the basis of the analysis technical solutions for improving the efficiency of ventilation systems in civil buildings are summarized and the field of their application is specified. Among the offered technical solutions the use of autonomous ventilation systems with heat recovery is highlighted as one of the most promising and understudied. Besides it has a wide range of applications. The paper reviews and analyzes the main Russian and foreign designs of ventilation systems with heat recovery that are mostly used in practice. Three types of such systems: UVRK-50, Prana-150, TeFo are chosen for consideration. The sequence of field tests of selected autonomous ventilation systems have been carried out in order to determine the actual air exchange and efficiency of heat recovery. The paper presents the processed results of the research on the basis of which advantages and disadvantages of the tested ventilation systems are identified and recommendations for engineering and manufacturing of new design models of autonomous ventilation systems with heat recovery are formulated.

## 1 Introduction

Today, we can come across the term "sick building syndrome" not only in the specialized literature. Unfortunately, this is a widespread phenomenon in most civil buildings of Russia [1-6]. The main causes of the ventilation systems malfunctioning in premises of civil buildings are ambiguous and depend on lots of factors. The authors of this paper systematize the causes of the ventilation systems malfunctioning and classify them in table 1. Climatic causes arise in the case of using ventilation systems with gravity (natural) impulse of air movement. The papers [7-9] present quite reliable and detailed facts that for the largest and most densely-populated part of Russia the average insufficiency of the systems' natural ventilation is 5-6 months in a year. Regulatory causes – the papers [10 - 14] provide facts that as the requirements for thermal protection of buildings were reviewed, since 1995, the degree of air-tightness has been considerably increased. Codes of structural design were also changed, however, a number of factors affecting the estimated

* Corresponding author: kostuganow@mail.ru

and actual values of the air exchange remain unconsidered as far as contemporary standards are concerned [16, 18, 19].

**Table 1**. Classification of the causes of the ventilation systems malfunctioning in civil buildings.

| Type of causes | Causes' subject-matter |
|---|---|
| Climatic | Insufficiency of year-round sustainable difference of gravitational pressures of inside and outside air due to natural-climatic conditions |
| Regulatory | a) Strict requirements for thermal protection of building enclosing structures, which result in their significant air-tightness<br>b) Regulations contain no clear procedural guidelines for ventilation systems installation<br>c) There is no standardization of gas and ion air composition indicators<br>d) There are no data on the impact of numerous chemical air pollutants in modern residential buildings on the human body<br>e) Standardization of air environment parameters is carried out on the bases of the sanitary-hygienic principle, without taking into account the complex physiological impact of the air environment on the human body |
| Design | Errors in the system design due to the lack of basic education, theoretical knowledge and design experience, as well as the lack of clear procedural guidelines for the design. |
| Construction and installation | a) Poor quality of construction and installation works<br>b) Nonconformance of the completed construction, installation works to the system design<br>c) Filling of ventilation shafts with construction waste |
| Town-planning | Planning the development of settlements without taking into account the aerodynamic influence of air flows on buildings. |
| Operating | a) Replanning of premises including the reconstruction of ventilation ducts<br>b) Installation of air coolers for working indoors in the warm season<br>c) Installation of sealed windows and doors during the premises renovation<br>d) Connection of hoods' fans and axial ventilators' fans in bathrooms to the communal exhaust ventilation system<br>e) Incorrect connection of fan motor phases |

Design causes come up quite frequently in practice not only in Russia but also in foreign countries [6, 13, 15]. The same factors are mainly behind town-planning causes. Construction and installation causes are also wide-spread in practice due to lack of skilled workers, low level of work organization, deviations from the design. Operating causes are, perhaps, the most common for existing buildings both in Russia and foreign countries [13]. It is the result of technical illiteracy of premises' owners or specialists of management companies. Climatic, regulatory, design and town-planning causes can and should be eliminated at the stage of systems design by relevant research and design organizations.

Today one of the most common methods to ventilate civil buildings (especially residential buildings) is installation of the low-effective system of natural exhaust ventilation with unorganized air intake due to its relative cheapness and minimum operating costs. To eliminate existing defects in the system of natural exhaust ventilation, combined or hybrid ventilation is used. It includes elements of balanced natural and mechanical ventilation systems. One of the most promising and the least investigated methods of providing air exchange in premises of civil buildings is the installation of autonomous (decentralized) ventilation systems for separate rooms. This method has both advantages and disadvantages. The disadvantages include relatively high cost of systems' installation and operation, the occurrence of additional noise sources, some disruption of the interior space. On the other hand, the advantages include the possibility of organizing a sustainable air exchange in the room as well as the air heat recovery. If we take into consideration that according to various estimates the heat loss with air exchange in modern buildings can reach 25% of the total heat loss [17], the latter advantage becomes even more significant. It is also worth mentioning that this method of air exchange is applicable for all civil buildings regardless of their number of stories. Because of the mentioned above prospects, insufficient awareness and a wide range of applications for civil buildings, the authors of this work examine the effectiveness of using some models of autonomous ventilation systems according to the criteria of the created air exchange and the effectiveness of heat transmission defined in the course of field tests.

The purposes of this study are:

1. To determine the actual air exchange and the effectiveness of heat transmission in some models of autonomous ventilation systems in the course of field tests;
2. To compare the received values with the regulatory requirements;
3. To identify additional evaluation criteria for these systems on the basis of all data obtained during the tests;
4. To formulate recommendations for the design and manufacturing of new models of autonomous ventilation systems based on the results of the study.

## 2 Materials and methods

At present there are quite a lot of manufacturers of ventilation equipment who offer similar systems, for example, the models "UVRK", "Prana", "Tefo", "Reventa", "Aeroeco", "Meltem", "AirLaska", "Frivent" and others. Based on the criteria that these systems have various heat exchangers for heat recovery, they are adapted to the climatic conditions of Russia and represented in the Russian market of goods at affordable prices we have chosen the following systems for testing: UVRK-50 (manufacturer company – LLC Research and Production Company "Ekoterm", the city of Omsk, Russia), Prana-150 (manufacturer company - "Prana"; the city of Lvov, Ukraine), Tefo (manufacturer company – LLC "Teploobmen"; the city of Sevastopol, Russia). The selected system models are shown in figures 1 - 3.

**Fig. 1.** UVRK-50. System.

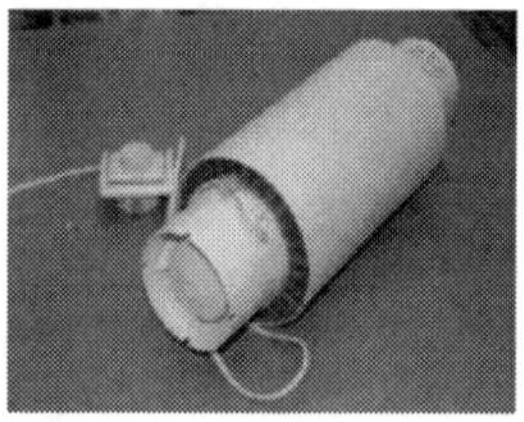

**Fig. 2.** Prana-150 system.

**Fig. 3.** Tefo system.

Field tests of the systems have been carried out by the authors of this research in the period of 2014 - 2016 in the laboratory of the Heat, Gas Supply, Ventilation and Hydro-Mechanics Department of the Orenburg State University. For purposes of testing a hole was made in the outer wall of the laboratory with a diameter of 250 mm (figures 4-6), in which the tested system models were installed (figures 7-8).

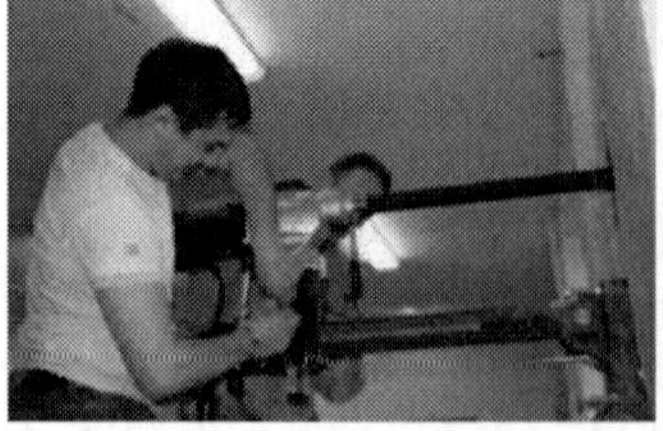

**Fig. 4.** Installing the equipment.

**Fig. 5.** Drilling a hole.

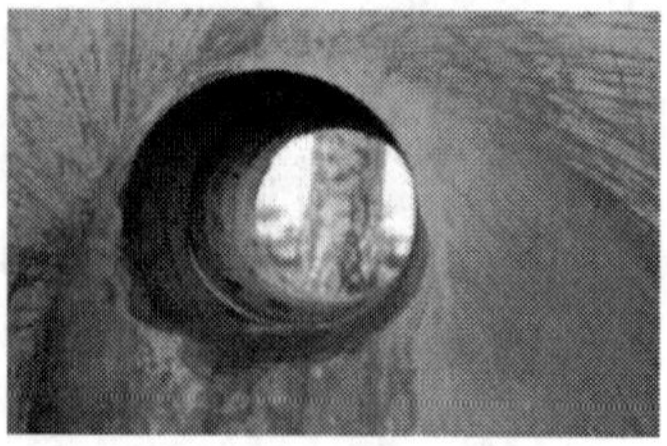

**Fig. 6.** Resulting hole.

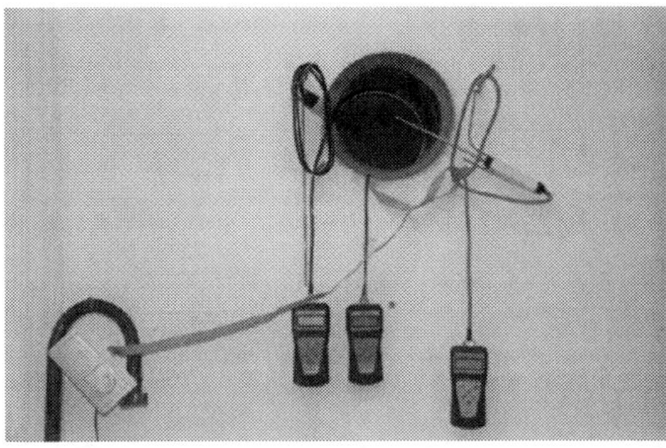

**Fig. 7.** UVRK-50 system (view from inside).

**Fig. 8.** UVRK-50 system (view from outside).

Thus, when conducting field tests all systems were installed in the hole made in the outer wall. The locations of measuring instruments are listed in table 2.

**Table 2.** The locations of measuring instruments for systems' field tests.

| Name of the system | Name of the measuring instrument | Location of the sensor | Location of the module |
|---|---|---|---|
| "UVRK-50" | Anemometer ISP MG-4 | Inside | Inside |
| | Thermohygrometer THZ MG-4.01 | Outside | Inside |
| | Thermohygrometer THZ MG-4.01 | Inside | Inside |
| "Prana-150" | Anemometer ISP MG-4 | Inside | Inside |
| | Anemometer ISP MG-4 | Outside | Inside |
| | Thermohygrometer THZ MG-4.01 | Outside | Inside |
| | Thermohygrometer THZ MG-4.01 | Inside | Inside |

Field tests of systems were carried out in winter 2014, 2015 when the average daily temperatures of outdoor air reached stable negative values of 6.1 ° C below zero (which is the average temperature during the heating period for the city of Orenburg). Test methods included recording of the results by operators who observed temperature, relative humidity, air flow rate during the system testing over the interval 3...5 seconds. The duration of experimental runs was 20 ... 25 minutes. The beginning of the tests was preceded by the systems' trial run (at least for an hour) to start the operating mode close to the stationary mode. The tests were performed in the modes corresponding to average and maximum air flow rate formed at the outlets of the air distribution holes of the systems. The measurements were carried out in sections defined beforehand where the air parameters are stable and correspond to average values for both modes of testing. Tests with minimum capacity were not carried out, since during the prearranged experiments it was proved that air exchange in systems at this mode practically stops.

As a result of several field tests the test protocols were formed. Their data were used when calculating the values of actual air exchange and effectiveness of heat transmission in the tested models of autonomous ventilation systems. The actual air exchange values were calculated according to the formula:

$$L=3600\cdot F\cdot v_{s} \quad (1)$$

$L$ - the actual volumetric air consumption, $m^3/h$; $F$ – the cross-sectional area, in which the rate was measured, $m^2$; $v_s$ - the average value of air flow rate, m/s;

The average value of air flow rate was calculated as the average value of the following mathematical estimates of the distribution of variables: the arithmetic average, the arithmetic average 90% of the sample, the median, the median amplitude, the amplitude center.

The efficiency of heat transmission in the heat exchanger - heat recovery unit was determined by the expression:

$$\eta = \frac{c_p \cdot G_{ud} \cdot (t_{ud1} - t_{ud2})}{c_p \cdot G_{min} \cdot (t_{ud1} - t_{pr1})} \quad (2)$$

$$\eta = \frac{c_p \cdot G_{pr} \cdot (t_{pr2} - t_{pr1})}{c_p \cdot G_{min} \cdot (t_{ud1} - t_{pr1})} \quad (3)$$

$\eta$ - the efficiency of heat transmission in the heat exchanger - heat recovery unit; $G_{ud}, G_{pr}$ - mass consumption of intake and removed air respectively, kg/s; $G_{min}$ - the lower value of mass air consumption, kg/s; $c_p$ - mass specific heat of air at constant pressure, J/(kg· ° C); $t_{pr1}, t_{pr2}$ - intake air temperature respectively at the inlet and at the outlet of the heat exchanger - heat recovery unit, ° C; $t_{ud1}, t_{ud2}$ - temperatures of the removed air at the inlet and at the outlet of the heat exchanger-heat recovery unit respectively, ° C;

The results of the conducted research are presented in the next section of this paper.

## 3 Results

In accordance with the purposes and objectives of the research the values of actual air exchange and efficiency of heat transmission in the tested system models were identified. The average intervals of the obtained results are presented in tables 3 and 4.

**Table 3**. The results of calculating the actual values of air exchange and efficiency of heat transmission in the tested system models at average capacity mode.

| Name of the system | Average temperature of intake air at the inlet of the heat exchanger, °C | Average temperature of intake air at the outlet of the heat exchanger, °C | Average temperature of removed air at the inlet of the heat exchanger, °C | Average temperature of removed air at the outlet of the heat exchanger, °C | Average intake air consumption, m³/h | Average removed air consumption, m³/h | Efficiency of heat transmission |
|---|---|---|---|---|---|---|---|
| UVRK-50 | -1,0 ... -13,0 | +17,0 ... +18,0 | +18,0 ... +21,0 | 0,0 .... -12,0 | 30 ...35 | 30 .... 35 | 0,7 ... 0,8 |
| TeFo | -25,0 ... -25,5 | -4,9 ... -10,7 | +23,7 ... +26,4 | -23,7 ... -24,3 | 11,4 ... 12,2 | 18,5 ...21,0 | 0,3 ... 0,4 |
| Prana 150 | -17,9 ... -21,7 | -6,9 ... -9,7 | +15,9 ... +16,4 | -4,6 ... -14,7 | 48,1 ...51,2 | 18,2 ... 30,0 | 0,3 ... 0,5 |

**Table 4**. The results of calculating the actual values of air exchange and efficiency of heat transmission in the tested system models at maximum capacity mode.

| Name of the system | Average temperature of intake air at the inlet of the heat exchanger, °C | Average temperature of intake air at the outlet of the heat exchanger, °C | Average temperature of removed air at the inlet of the heat exchanger, °C | Average temperature of removed air at the outlet of the heat exchanger, °C | Average intake air consumption, m³/h | Average removed air consumption, m³/h | Efficiency of heat transmission |
|---|---|---|---|---|---|---|---|
| UVRK -50 | -1,0 … -13,0 | +17,0 … +18,0 | +18,0 … +21,0 | 0,0 …. -12,0 | 50 … 55 | 50 … 55 | 0,7 … 0,8 |
| TeFo | -26,0 … -26,2 | -2,0 … -3,6 | +27,0 | -21 … -22 | 14,8 … 33,3 | 25,3 … 27,0 | 0,4 … 0,5 |
| Prana 150 | -19,8… -20,8 | -5,2 …. -9,2 | +16,8 …. +23,6 | -0,3 … -4,7 | 24,7 … 30,4 | 28,2 …. 48,6 | 0,3 … 0,5 |

When processing the test results the values of variables' measurement errors were also determined.

**Table 5.** Maximum absolute error values recorded by measuring instruments, when determining the values of air rate and temperature

| Name of the system | Name of the instrument | Measured parameter | Value of absolute error |
|---|---|---|---|
| UVRK-50 | ISP 4 MG | Air temperature | ±0,5°C |
| | | Air rate | ±0,2 … 0,3 м/с m/s |
| | THZ MG 4.01 | Air temperature | ±0,5°C |
| Prana-150 | ISP 4 MG | Air temperature | ±0,5°C |
| | | Air rate | ±0,15 …0,3 м/с m/s |
| | THZ MG 4.01 | Air temperature | ±0,5°C |
| TeFo | ISP 4 MG | Air temperature | ±0,5°C |
| | | Air rate | ±0,15 …0,3 м/с m/s |
| | THZ MG 4.01 | Air temperature | ±0,5°C |

As table 5 shows the temperature measurements were carried out with the maximum indicated error not exceeding the value of ±0,5 ° C. That does not exceed 10% of the relative measurement error. It should be noted that when calculating the rate measurement errors for the ISP MG4 instruments the errors reach maximum values (from 5% ... to 20%). That is why the control measurement of rate was carried out with the multi-function instrument Testo 480. In general, the maximum value of the final results' relative error does not exceed 15%.

The tables above show that the temperature values at the air inlet to the room for some system models are quite low (from -2.0 ° C and to -10.7 °C). That will create thermal discomfort during continuous operation of the system. The average of actual air consumption is also less than the standard values and the values declared by the manufacturers of systems. If we take the interval of minimum standard values of the intake air consumption per one person from 20 to 60 m$^3$/h, as it is clear from tables 4 and 5, some tested system fall within this interval only at maximum capacity mode.

It should be mentioned that when preforming tests the fluctuation of room air temperature ranged from 0.5 ° C to 3 ° C, fluctuations in relative humidity ranged from 5%

to 15%, air mobility varied from 0.12 m/s to 0.35 m/s, and noise level was from 38.6 dBA to 50.1 dBA.

## 4 Discussion

As the result of the analysis of indicators' values obtained during the study, we can formulate the following recommendations for design and manufacture of new models of such systems:

1 In the design of new system models as the driving force of air movement the centrifugal fan must be used not the axial one. It is this particular design feature of the system, according to the authors, that causes low values of the actual air exchange and the efficiency of heat transmission of the tested models.

2 New system models must optimally combine two characteristics – the surface area of heat exchange should be as large as possible while the wall surface in the room should be as small as possible.

3 The design of system heat exchanger must be easily manufactured and effective when cleaning the surface of heat exchanger (the example of heat exchanger system design "TeFo").

4. The design of new system models must include filters for cleaning of the air coming from the street and the air removed from the premises.

5. To reduce noise levels up to standard values the installation of fans should be made outside, or as far as possible from the room air outlet.

6 New models of systems must be suitable for repairing to the maximum extent.

7 System user control should be as simple as possible.

8 The warranty period of new system models should be within minimum one year.

9 For easy installation the degree of readiness of new system models should be maximized.

It should be mentioned that around the same time the Kazan State University of Architecture and Civil Engineering was conducting similar field tests of some other types of devices for organizing the outside air intake including UVRK-50 [20]. In many aspects the similar results were obtained.

## 5 Conclusions

As a result of the conducted research the main quantitative parameters that must be met in new models of autonomous ventilation systems have been set. These include:

1 Guaranteed air exchange at the level of sanitary norms (minimum 30 $m^3$/hour).

2 Share of heat return from 50% to 75% depending on the mode of equipment operation, as well as the external and internal environment;

3 Energy consumption at the level of modern household appliances;

4 Compactness - the system should fit in the average size window;

5 The systems shall be available to users for maintenance and minor repairs.

Based on the results of the research and considering the above recommendations a new model of autonomous ventilation systems for wall mounting was developed at the premises of LLC " SPE "Pneumax". As part of further research of this topic it is planned to carry out tests (field and in climatic chamber) of the model of autonomous ventilation systems, as well as to develop a type of such system with heat recovery suitable for installation under the window sill.

## Acknowledgement

The authors express their deep gratitude for providing invaluable assistance in conducting the study to Candidate of Technical Sciences, Professor, Head of the Department "Heat and Gas Supply and Ventilation" at the Novosibirsk State University of Architecture and Civil Engineering (Sibstrin) Rustam Shamilyevich Mansurov.

The authors also express their gratitude to the Foundation for Assistance to Small Innovative Enterprises in Science and Technology for the support and grant-making to conduct further research on the topic within a framework of the all-Russia innovation competition "U. M. N. I. K.".

## References

1. O. Fanger, AVOK **2**, 12-19 (2006)
2. Yu.A. Tabunshchikov, AVOK **5**, 4-11 (2008)
3. Yu.A. Tabunshchikov, AVOK **4**, 4-7 (2007)
4. A.B. Kostuganov, Yu.S. Vytchikov, *Traditions and innovations in construction and architecture. Building technologies. Collection of articles,* 309-313 (Samara State University of Architecture and Civil Engineering, 2016)
5. B.M. Romeyko, D.V. Zelentsov, *Traditions and innovations in construction and architecture. Building technologies. Collection of articles,* 783-784 (Samara State University of Architecture and Civil Engineering, 2014)
6. N.P. Tyurin, E.P. Lapin, G.I. Titov, Yu.E. Zakharova, *Traditions and innovations in construction and architecture. Building technologies. Collection of articles,* 267-271 (Samara State University of Architecture and Civil Engineering, 2013)
7. M.V. Bodrov, V.Yu. Kuzin, M.S. Morozov, S. O. K. **1**, 74-77 (2016)
8. M.V. Bodrov, V.Yu. Kuzin, D.Yu. Kuzin, Privolzhsky scientific journal **4,** 26-32 (2016)
9. M.V. Bodrov, V.P. Boldin, V.Yu. Kuzin, M.N. Kucherenko, Privolzhsky scientific journal **1**, 54-59 (2015)
10. Yu.S. Vytchikov, A.V. Sidorova, *Traditions and innovations in construction and architecture. Building technologies. Collection of articles,* 284-286 (Samara State University of Architecture and Civil Engineering, 2013)
11. Yu.S. Vytchikov, A.V. Sidorova, Urban Construction and Architecture **4**, 87-94 (2013) DOI 10.17673/Vestnik.2013.04.15
12. V.I. Vesnin, Urban Construction and Architecture **3**, 10-16 (2016) DOI 10.17673/Vestnik.2016.03.2
13. R. Schulte, B. Bridges, D. Grimsrud, AVOK **8**, 36-47 (2005)
14. A.M. Saifutdinova, V.N. Kupriyanov, Proceedings of the Kazan State University of Architecture and Civil Engineering **1**, 113-118 (2014)
15. N.P. Tyurin, E.P. Lapin, Traditions and innovations in construction and architecture. Proceedings of the 70th anniversary all-Russian scientific and technical conference. Samara State University of Architecture and Civil Engineering, pp. 265-266 (2013)
16. A.G. Rymarov, S. O. K. **11**, 75-77 (2012)
17. V.G. Gagarin, V.V. Kozlov, Vestnik MGSU **7**, 59-66 (2011)
18 A.G. Rymarov, V.V. Savichev, News of higher educational institutions. Civil Engineering **4**, p. 99-102 (2014)

19. V.N. Kupriyanov, A.M. Saifutdinova, Proceedings of the Kazan State University of Architecture and Civil Engineering **1**, 109-119 (2013)
20. V.N. Kupriyanov, A.M. Saifutdinova, A.M. Ziganshin, I.Sh. Safin, Bulletin of the Volga regional branch of the Russian Academy of architecture and construction sciences **16**, 245-254 (2013)

MATEC Web of Conferences 117, 00088 (2017) DOI: 10.1051/matecconf/201711700088
*XXVI R-S-P Seminar 2017, Theoretical Foundation of Civil Engineering*

# Calibration of partial safety factors according to Eurocodes

*Peter* Koteš[1,*], *Jozef* Prokop[1] , *Miroslav* Strieška[1] , and *Josef* Vičan[1]

[1]University of Žilina, Faculty of civil engineering, Univerzitná 8215/1, 010 26 Žilina, Slovakia

**Abstract.** Recommended reliability levels and their adequate partial safety factors can be found in Eurocodes only for design of new structures and bridges. Modified reliability levels and partial safety factors for verifying bridge structures for their remaining lifetime were determined by theory of reliability at Department of Structures and Bridges at Civil Engineering Faculty, University of Žilina. This paper is focused on verification of modified partial safety factors of materials and load effects using calibration.

## 1 Introduction

Each building and civil engineer should ensure that the design of the building and civil structures is reliable, in this case the reliability of the bridge structures. The reliability is the ability of the construction to perform the required functions to which it was designed under normal conditions and maintenance without any necessary repairs and reconstructions throughout its design lifetime $T_d$.

For newly designed structures, the decisive role of the properties, as load-bearing capacity, safety, serviceability, durability and robustness, which are closely related to each other, are crucial to ensuring the required reliability. We consider the probability of failure-free operation during the operation, which is also called as the reliability level and is denoted as $P_r$. The opposite value of the reliability level is the failure probability $P_f$ and it expresses the probability of occurrence of a failure over the lifetime of the structure and the following should valid

$$P_r = 1 - P_f \,. \tag{1}$$

Any building construction is not completely safe and reliable, because its resistance and load are affected by uncertainties that can be of different character.

## 2 Existing bridge structures – calibration of partial safety factors

The partial safety factors for the design of new bridge structures are given in codes [1-5]. The standards are an important part in designing new bridge structures, but also in

* Corresponding author: kotes@fstav.uniza.sk

evaluating existing bridges. The actions of bridges according to the code [1] are defined by a representative (e.g. characteristic) value and the partial safety factors for loads. This standard also defines load combinations. However, these partial safety factors can be used only for the design of new bridge structure for the whole design lifetime $T_d = 100$ years, but not for the evaluation of existing bridge structure for their remaining lifetime $t_r$. The modified reliability levels given by the reliability index $\beta_t$ or by the probability of failure $P_{ft}$ and the corresponding partial safety factors of material properties and load effects were determined in the frame of the research activities of the Department of Structures and Bridges of University of Žilina [6, 7, 8]. The recommended modified reliability levels for the planned remaining lifetime of the bridge structures in the simplified form are shown in Tab. 1.

**Table 1.** Modified values of reliability index $\beta_t$ and the probability of failure $P_{ft}$ for remaining lifetime.

| Planned remaining lifetime $t_r$ [years] | Inspection in time – age of the bridge structure/member | | | | | |
|---|---|---|---|---|---|---|
| | 40 years | | 60 years | | 70 years | |
| | $\beta_t$ | $P_{ft}$ | $\beta_t$ | $P_{ft}$ | $\beta_t$ | $P_{ft}$ |
| < 2 | 2.767 | $2.830 \cdot 10^{-3}$ | 2.632 | $4.248 \cdot 10^{-3}$ | 2.578 | $4.975 \cdot 10^{-3}$ |
| 2 - 10 | 3.209 | $6.666 \cdot 10^{-4}$ | 3.108 | $9.439 \cdot 10^{-4}$ | 3.066 | $1.085 \cdot 10^{-3}$ |
| 10 - 20 | 3.354 | $3.984 \cdot 10^{-4}$ | 3.271 | $5.362 \cdot 10^{-4}$ | 3.236 | $6.061 \cdot 10^{-4}$ |
| > 20 | 3.424 | $3.093 \cdot 10^{-4}$ | 3.352 | $4.012 \cdot 10^{-4}$ | 3.322 | $4.477 \cdot 10^{-4}$ |

## 2.1 Partial safety factors of loads effects

Individual loads are considered as random variables and they are described by the coefficient of variation v and the bias factor $\lambda$, which expresses the ratio between the mean value and the characteristic value [9, 10]. The modified reliability levels for the evaluation of existing bridge structures affect the modified values of the partial safety factors of the load effects and material properties. For this calculation, it is necessary to know the characteristics of the statistical distributions of individual loads, as the distribution functions, density functions, mean values, standard deviations, etc.

### *2.1.1 Permanent load effects*

The mean value $m_G$ is the characteristic value $G_k$ of permanent load if the coefficient of variation is $v_G \leq 0.10$ (normally ranging in $v_G = 0.02 \sim 0.05$). Permanent loads of bridge structures include the self-weight of structural and non-structural members of the structures. According to Amendment A1 of Annex A2 [11] of Standard [3], two values for the partial safety factors for permanent loads of the structural and non-structural members of the bridges are recommended, depending on whether they are factory-made products ($\gamma_G = 1.25$), or they are cast-in-place products ($\gamma_G = 1.35$) due to possible different variations in the weights of the structural elements.

When considering the normal distribution of random variables of constant loads, then the partial safety factor is given by the equation

$$\gamma_{G,i} = \gamma_{Sd} \cdot \frac{G_d}{G_k} = \gamma_{Sd} \cdot \frac{m_{G,i} \cdot \left(1 + \alpha_E . \beta_t . v_{G,i}\right)}{m_{G,i}}, \tag{2}$$

where $G_d$ is the design value of permanent load,
$G_k$ is the characteristic value of permanent load,
$m_{G,i}$ is the mean value of each permanent loads,
$v_{G,i}$ is the coefficient of variation of each permanent loads,
$\gamma_{Sd} = 1.05$ is the partial safety factor for model uncertainties,
$\alpha = 0.70$ is the sensitivity coefficient of the load,
$\beta_t$ is the modified value of the reliability index for evaluation (Tab. 1).

The recommended modified partial safety factors of the factory-made products are given in Tab. 2. These include, for example, bridge rails, guard-rails, vehicle parapets, waterproofing (asphalt) and prefabricated parts of the superstructure.

**Table 2.** Partial safety factors of permanent loads – factory-made products (basic value $\gamma_G = 1.25$)

| Planned remaining lifetime $t_r$ [years] | $\gamma_{G,i}$ for age of the bridge structure/member | |
|---|---|---|
| | ≤ 60 years | > 60 years |
| ≤ 2 | 1.20 | 1.20 |
| 2 – 10 | 1.25 | 1.20 |
| > 10 | 1.25 | 1.20 |

The recommended modified partial safety factors of the permanent loads of the elements cast-in-place in relation to the planned remaining lifetime of the bridge are given in Tab. 3. These loads include the actual weight of the monolithic parts of the superstructures, the pavement, the ballast bed, and the like.

**Table 3.** Partial safety factors of permanent loads – cast-in-place products (basic value $\gamma_G = 1.35$)

| Planned remaining lifetime $t_r$ [years] | $\gamma_{G,i}$ for age of the bridge structure/member | |
|---|---|---|
| | ≤ 60 years | > 60 years |
| ≤ 2 | 1.30 | 1.25 |
| 2 – 10 | 1.30 | 1.30 |
| > 10 | 1.35 | 1.30 |

### *2.1.2 Variable load effects*

Variable loads (traffic loads) on road bridges are given in code [1] by load models LM1 to LM4. For the calibration of bridge constructions, two decisive load models were considered:

- LM1, given by three-lane concentric axle loads and a continuous equivalent load, which represent most of the effects of normal traffic,
- LM3, which is given by a grouping of axle loads representing the special vehicles that can travel along the routes.

The decisive load model was considered from two of those that had greater unfair effects on the structure (LM1 or LM3). The variable load effects (traffic loads on road bridges) are random variables with a Gumbel's distribution [3]. For this reasoning, the partial safety factors of the variable load effects is defined by the relation

$$\gamma_Q = \gamma_{Sd} \cdot \frac{Q_d}{Q_k} = \gamma_{Sd} \cdot \frac{m_Q \cdot \left(1 - v_Q\left[0{,}449 + 0{,}778.\ln\left(-\ln\Phi\left(\alpha_E . \beta_t\right)\right)\right]\right)}{m_Q \cdot \left(1 - v_Q\left[0{,}449 + 0{,}778.\ln\left(-\ln\left(0{,}95\right)\right)\right]\right)}, \tag{3}$$

where $Q_d$ is the design value of variable load effects,
$Q_k$ is the characteristic value of variable load effects,
$m_Q$ is the mean value of variable load effects,

$v_Q$ is the coefficient of variation of variable load effects,
$\Phi$ is the distribution function of random variable.

The recommended modified partial safety factors of the variable load effects taken into account the planned remaining lifetime of the bridge are given in Tab. 4.

**Table 4.** Partial safety factors of variable load effects (basic value $\gamma_Q = 1.35$)

| Planned remaining lifetime $t_r$ [years] | $\gamma_{G,i}$ for age of the bridge structure/member | |
|---|---|---|
| | ≤ 60 years | > 60 years |
| ≤ 2 | 1.15 | 1.10 |
| 2 – 10 | 1.25 | 1.20 |
| > 10 | 1.30 | 1.20 |

## 2.2 Calibration – theoretical approach

Verification of reliability is based on limit states that represent the boundary between reliability and failure of structure/member. The paper focuses on the ultimate limit state - bending resistance. In principle, we always try to find a design point that we consider to be the limit of failure. In this case, the Rackwitz-Fiessler iterative method was applied, which was also used to calibrate the partial safety factors for loads and material resistance for the American and Canadian standards [9,10]. The design point lies at the failure boundary and shows the greatest probability of failure, the failure boundary being given by equation

$$M = R - E, \tag{4}$$

for the design point D ($R_D$,$E_D$) is valid $R_D = E_D$.

$F_R$ is the cumulative distribution function and $f_R$ is the probability density function for R. Similarly, $F_E$ and $f_E$ are valid for load effects. Firstly, an initial value of design point $R_D$ is guessed. The method assumes approximation of cumulative distribution function $F_R$ and $F_E$ on normal distribution function $\overline{F_R}$ , $\overline{F_E}$ . Then, the standard deviation and the mean value of $R_D$ are

$$\overline{\sigma_R} = \varphi_n\left\{\phi^{-1}\left[F_R(R_D)\right]\right\}/f_R(R_D), \tag{5}$$

$$\overline{\mu_R} = \overline{R_D} - \overline{\sigma_R}.\phi^{-1}\left[F_R(R_D)\right], \tag{6}$$

where $\phi$ is distribution function of the standard normal distribution,
$\varphi_n$ is density function of standard normal random variable.

Similarly, the same formulas can be derived for $\overline{\mu_E}$ and $\overline{\sigma_E}$ of the load effects

$$\overline{\sigma_E} = \varphi_n\left\{\phi^{-1}\left[F_E(R_E)\right]\right\}/f_E(R_E), \tag{7}$$

$$\overline{\mu_E} = \overline{R_E} - \overline{\sigma_E}.\phi^{-1}\left[F_E(R_E)\right]. \tag{8}$$

It is now possible to determine the reliability index β, which defines the distance of the design point D from the beginning in shape

$$\beta = \left(\overline{\mu_R} - \overline{\mu_E}\right)/\left(\left(\overline{\sigma_R}\right)^2 + \left(\overline{\sigma_E}\right)^2\right)^{0,5}. \tag{9}$$

Next, a new design point can be calculated from the following equations

$$R_D = \overline{\mu_R} - \beta.\left(\overline{\sigma_R}\right)^2 / \left(\left(\overline{\sigma_R}\right)^2 + \left(\overline{\sigma_E}\right)^2\right)^{0,5}, \tag{10}$$

$$E_D = \overline{\mu_E} - \beta.\left(\overline{\sigma_E}\right)^2 / \left(\left(\overline{\sigma_R}\right)^2 + \left(\overline{\sigma_E}\right)^2\right)^{0,5}. \tag{11}$$

So, the second iteration can begin. The approximating normal distributions are found for $F_R$ and $F_E$ at the new design point. The reliability index was calculated using formula (9) and the next design point is found from (10) and (11). Approximation is continued until $R_D$ and $E_D$ do not change in consecutive iterations. In our case of calibration, the resistance of reinforced concrete girders was assumed as lognormal random variable and the load effects were assumed as normal random variables. From this reason, it is not needed to do their approximation.

The design reliability levels, which are projected into partial safety factors, were calibrated on reinforced concrete bridges, three slabs and one frame bridge. These are already existing bridge structures near town Nitra.

The calibration was performed on the basis of the actual traffic load, which was determined by the traffic count in 2010 by the Slovak Road Administration (SRA). The maximal positive moments in the middle of the span, respectively, at the point of maximum positive moments, if they are not in the middle of the span, were considered as the load effects for calibration.

Calculation of load effects from individual load models was performed in programme SCIA Engineer on slab models. The real traffic loads on the bridge was simulated by passage cars for 5 standard categories of vehicles N1, N2, N3, PN3 and NS.

Since the actual age of the selected bridge constructions was not known exactly, the calibration was performed on the assumption that they were less than 60 years of age and equal to or greater than 60 years and compared the residual life for both cases.

A frame bridge was also included in the set of bridges. Specifically, these are objects no. 51-124 at the village of Babindol, no. 65-012 near the village Tesare, no. 51-134 in the village of Čifáre and no. 51-125 at the village Klasov. The brief characteristics of the individual bridges are:

- Object no. 51-124 at the village of Babindol - it is a single span slab bridge with a span of 2.30 m. The width of the road on the bridge is 8.24 m.
- Object no. 65-012 near the village of Tesare - it is a single span slab structure with a span of 13.60 m. The width of the road on the bridge is 10.30 m.
- Object no. 51-134 in the village of Čifáre - it is a single span slab structure with a span of 4.30 m (total length is 4.60 m). The width of the road to the bridge is 7.80 m.
- Object no. 51-125 at the village of Klasov - it is a single span frame structure with a span of 4.45 m and a height of 2.80 m. The width of the road on the bridge is 7.90 m.

This approach should take into account the actual parameters of the bridge objects as well as the possible degradation of the materials [14-18], eventually the possible fatigue failure [19,20]. For this reason, diagnostics were first performed on the bridge objects. Diagnostic results were included in bridged objects models - measured geometric and material properties with regard to failures. As the observed support element was considered the most stressed part of the slab width of 1.0 m. The bridge slabs were modelled as slabs.

Due to the character of the reinforced concrete, the cross-section resistances were considered as random variables with normal distribution. In the partial safety factor method, for the basic load combination with one variable short-term load, the equation for the element subjected to bending is

$$\gamma_{G,1}.M_{G1,k} + \gamma_{G,2}.M_{G2,k} + \gamma_{F,t}.\delta.M_{t,k} \leq M_k / \gamma_{M,c}, \tag{12}$$

kde $\gamma_{G,1}$ is the partial safety factor for cast-in-place products - permanent load,
$\gamma_{G,2}$ is the partial safety factor for factory-made products - permanent load,
$\gamma_{F,t}$ is the partial safety factor for variable loads,
$\gamma_{M,c}$ is the partial safety factor for reinforced concrete material,
$M_{Gi,k}$ are the characteristic values of bending moments due to permanent loads,
$M_{t,k}$ is the characteristic values of bending moments due to variable loads,
$M_k$ is the characteristic values of bending resistance,
$\delta$ is the dynamic coefficient, in the case of road bridges is equal to 1.0.

The partial safety factor for material for reinforced concrete element subjected to bending $\gamma_{M,c}$ as a whole cross-section includes the individual partial safety factors of the concrete for compression $\gamma_{c,c}$ and the reinforcement $\gamma_s$. For this reason, the partial safety factor was obtained from the ration between the characteristic moment of resistance of the reinforced concrete cross-section $M_k$ and the design moment of resistance of the same cross-section $M_d$ calculated according to the standard [5]

$$\gamma_{M,c} \leq M_k / M_d. \tag{13}$$

Since the reliability index has been calculated for several possible remaining lifetimes, it was also necessary to calculate the material reliability factor $\gamma_{M,c}$ for each of them. The resulting material reliability factors for slab bridges can be found in Tab. 5. In order to calculate the reliability indices, it was also necessary to determine the statistical characteristics of the load effects for each bridge object separately, namely the mean value $\mu_E$ and the standard deviation $\sigma_E$. Actual traffic data on selected bridges has been provided to us by the Slovak Road Administration. The resulting reliability index values for slab bridges are shown in Tab. 6.

When comparing calculated reliability indices with modified reliability indices according to [6,7,8], we find that they are much higher. This means that the selected slab bridges are suitable for a remaining lifetime of at least 20 years.

**Table 5.** Partial safety factors of materials for slab bridges

| **Bridge structure / Planned remaining lifetime $t_r$ [years]** | | $\gamma_{M,c}$ | |
|---|---|---|---|
| | | **age of the bridge ≤ 60 years** | **age of the bridge > 60 years** |
| n. 51-124 Babindol | > 20 | 1.167 | 1.117 |
| | $10 < t_r \leq 20$ | 1.164 | 1.114 |
| | $2 < t_r \leq 10$ | 1.114 | 1.111 |
| | ≤ 2 | 1.106 | 1.056 |
| n. 51-125 Klasov | > 20 | 1.155 | 1.105 |
| | $10 < t_r \leq 20$ | 1.154 | 1.104 |
| | $2 < t_r \leq 10$ | 1.104 | 1.103 |
| | ≤ 2 | 1.102 | 1.052 |
| n. 51-134 Čifáre | > 20 | 1.176 | 1.126 |
| | $10 < t_r \leq 20$ | 1.172 | 1.122 |
| | $2 < t_r \leq 10$ | 1.122 | 1.117 |
| | ≤ 2 | 1.109 | 1.059 |
| n. 65-012 Tesáre | > 20 | 1.202 | 1.152 |
| | $10 < t_r \leq 20$ | 1.193 | 1.143 |
| | $2 < t_r \leq 10$ | 1.143 | 1.134 |
| | ≤ 2 | 1.117 | 1.067 |

**Table 6.** Values of reliability indices β for remaining lifetime on slab bridges

| Bridge structure | Age of the bridge ≤ 60 years | | | | Age of the bridge > 60 years | | | |
|---|---|---|---|---|---|---|---|---|
| | Planned remaining lifetime $t_r$ [years] | | | | Planned remaining lifetime $t_r$ [years] | | | |
| | ≤ 2 | $2 < t_r \leq 10$ | $10 < t_r \leq 20$ | > 20 | ≤ 2 | $2 < t_r \leq 10$ | $10 < t_r \leq 20$ | > 20 |
| n. 51-124 Babindol | 8.166 | 8.364 | 8.566 | 8.571 | 7.897 | 8.272 | 8.279 | 8.086 |
| n. 51-125 Klasov | 6.770 | 6.888 | 7.024 | 7.025 | 6.590 | 6.832 | 6.833 | 6.834 |
| n. 51-134 Čifáre | 8.717 | 8.960 | 9.203 | 9.213 | 8.403 | 8.849 | 8.862 | 8.874 |
| n. 65-012 Tesáre | 7.579 | 7.962 | 8.437 | 8.480 | 7.053 | 7.794 | 7.840 | 7.887 |
| **Recomm. value** | **2.632** | **3.108** | **3.271** | **3.352** | **2.578** | **3.066** | **3.236** | **3.322** |

## 3 Conclusions

In the paper, modified reliability levels and partial safety factors of load effects and reliability of materials on existing reinforced concrete slab bridges (one bridge object was frame structure) were verified by means of calibration. From the calculated reliability indices, we can see that all observed slab bridges have a significantly higher level of reliability than the recommended values. This means that even if they do not know their true age, they are still suitable for the remaining lifetime (up to and over 20 years).

Furthermore, it is possible to observe from this statistical set that the reliability level is much higher than the recommended values for most bridge objects. This means that the load given in codes is far greater than the actual load that occurs on road communications. If the same assumption were to be found with a larger number of bridge structures, it would be worthwhile to reduce the given load in the design of new constructions by the categorization coefficients $\alpha_q$ and $\alpha_Q$, which could result in economically more efficient design of bridge structures on road communications. This would, however, need to make a recalculation on a substantially larger set of different types of bridge objects.

The paper presents results of the research project APVV-14-0772 supported by the Slovak Research and Development Agency, the project 012ŽU-4/2016 supported by the Cultural and Education Agency of Slovak Republic, by the European Regional Development Fund and the Slovak state budget for the project "Research Centre of University of Žilina" ITMS 26220220183 and by Research Project No. 1/0566/15 of Slovak Grant Agency

## References

1. STN EN 1991-2 (73 6203): Eurocode 1. Actions on structures. Part 2: Traffic loads on bridges. SÚTN (2006) (in Slovak)
2. STN EN 1991-1-1 (73 0035): Eurocode 1. Actions on structures. Part 1-1: General actions – Densities, self-weight, imposed loads for buildings. SÚTN (2007) (in Slovak)
3. STN EN 1990 (73 0031): Eurocode. Basis of structural design. SÚTN (2009) (in Slovak)
4. STN EN 1992-1-1 (73 1201): Eurocode 2. Design of concrete structures. Part 1-1: General rules and rules for buildings. SÚTN (2007) (in Slovak)

5. STN EN 1992-2 (73 6206): Eurocode 2. Design of concrete structures. Part 2: Concrete bridges – Design and detailing rules. SÚTN (2007) (in Slovak)
6. Koteš, P., Reliability of existing bridge structures and possibilities of its increasing. Habilitation thesis, Žilina, 143 p. (2012) (in Slovak)
7. Koteš, P., Vičan, J., Reliability-based evaluation of existing concrete bridges in Slovakia according to Eurocodes. In: The Fourth International fib Congress 2014, Mumbai, „Improving Performance of Concrete Structures", Proceedings, Mumbai, India, 2014, © IMC-FIB, pp. 227-229 (2014)
8. Koteš, P. – Vičan, J., Recommended reliability levels for the evaluation of existing bridges according to eurocodes. In: Structural Engineering International – International Association for Bridge and Structural Engineering (IABSE), SEI Volume 23, **4**, p. 411-417 (2013)
9. Nowak, A.S., Calibration of LFRD Bridge Code. In: Journal of Structural Enfineering, pp. 1245-1251 (1995)
10. Nowak,A.S., Grouni, H.N., Calibration of the Ontario Highway Bridge Code 1991 edition. In: Canadian Journal of Ci. Engineering **21**, pp. 25-35 (1994)
11. STN EN 1990 (73 0031): (in Slovak). Basis of structural design. Amendment A1. Annex A2: Use for bridges. SÚTN (2006) (in Slovak)
12. Benko, V.: Reliability of structures (according to Eurocodes). SKSI, 86. P, (2010)
13. Bilčík, J., Fillo, Ľ., Benko, V., Halvoník, J., Concrete structures. Design according to STN EN 1992-1-1. Editorship STU in Bratislava, 374 p. (2008)
14. Macho, M., Ryjaček, P., The impact of the severe corrosion on the structural behavior of steel bridge members. Advances and Trends in Engineering Sciences and Technologies - Proceedings of the International Conference on Engineering Sciences and Technologies, ESaT 2015, pp. 123-128 (2015)
15. Ryjaček, P., Macho, M., Stančík, V., Polák, M., The Deterioration and assessment of steel bridges. Maintenance, Monitoring, Safety, Risk and Resilience of Bridges and Bridge Networks - Proceedings of the 8th International Conference on Bridge Maintenance, Safety and Management, IABMAS 2016, pp. 1188-1195 (2016)
16. Hollý, I., Bilčík, J., Gajdošová, K., Numerical modeling of reinforcement corrosion on bond behaviour. International Multidisciplinary Scientific GeoConference Surveying Geology and Mining Ecology Management, SGEM, **249**, pp. 191-196 (2016)
17. Paulík, P., Bačuvčík, M., Ševčík, P., Janotka, I., Gajdošová, K., Experimental evaluation of properties of 120 years old concretes at two concrete bridges in Slovakia. International Multidisciplinary Scientific GeoConference Surveying Geology and Mining Ecology Management, SGEM, **249**, pp. 227-234 (2016)
18. Kala, V., Valeš, J., Stochastic analysis of the lateral beam buckling of beams with initial imperfections. Safety and Reliability of Complex Engineered Systems - Proceedings of the 25th European Safety and Reliability Conference, ESREL 2015, pp. 2547-2552 (2015)
19. Krejsa, M., Koubova, L., Flodr, J., Protivinsky, J., Nguyen, Q.T., Probabilistic prediction of fatigue damage based on linear fracture mechanics. Frattura ed Integrita Strutturale, **11**, Issue **39**, Volume **249**, pp. 143-159 (2017)
20. Krejsa, M., Probabilistic reliability assessment of steel structures exposed to fatigue. Safety, Reliability and Risk Analysis: Beyond the Horizon - Proceedings of the European Safety and Reliability Conference, ESREL 2013, pp. 2671-2679 (2013)

# Modal approach in structural-dynamic equations

*Oleg* Kovalchuk[1,*]

[1]Moscow state university of civil engineering, Yaroslavskoye shosse, 26, Moscow, Russia, 129337

**Abstract.** Some aspects of the solution of inhomogeneous problems of rods and plates are considered in the article. Dynamic effects from the effects of short-term loads are revealed with the help of modal decomposition under the assumption of the constancy of the elastic properties of concrete corresponding to its age during the application of the perturbing dynamic effect. Superposition of reactions to long-and-short-term effects are possible due to the inadmissibility of finite deformations in building structures.

## 1 Introduction

For the solution of the dynamic problems known averaging method [1, 2, 3], where there is a limitation: for the implementation of the method requires that the material have low relaxation or the step of falling stress is a "small parameter" in the experience of relaxation. The method of expansion in eigenfunctions of a linear homogeneous operator with constant coefficients has no such restrictions, which form the system of orthogonal functions [4], satisfying the conditions of theorem Steklov [4, 5, 6], i.e. it is possible to build an absolutely convergent generalized Fourier series [7]. As such a system, one can effectively use the solution of the problem of free vibrations of linearly elastic bodies with homogeneous kinematic conditions [8, 9]. In the literature the forms of free oscillations is often called fundamental mode oscillation, then this method is called modal approach. In structural mechanics there are many analytical solutions of problems of free oscillations of rods and plates. The based on them solutions of the inhomogeneous problems (forced oscillations, vibrations, impacts with different pulse shape) you can also get analytical. Further development of the proposed theory envisages the construction of analytical expressions for reinforced concrete structural elements.

## 2 Evolutionary model of hereditary-viscoelastic material

The most suitable for describing the rheological properties of the material should be considered viscoelastic the models hereditary ageing [10]. For such models, an analytical

* Corresponding author: oko44@mail.ru

expression has the form:

$$\sigma(t)=E(t)\varepsilon(t)-\int_{\tau_0}^{t}\sigma(\tau)K(t,\tau)d\tau \tag{1}$$

where $K(t,\tau)$ – creep memory function, $\sigma(\tau)$ – development of stress, $\varepsilon(t)$ – deformation, $E(t)$ – modulus of elasticity of instantaneous deformation.

Thus, in order to find the stress, it is necessary to solve the integral equation (1).

Using the results of [1], one can take the relations for relaxation in the form:

$$\sigma(t)=E(t)\varepsilon(t)-\int_{0}^{t}\Gamma(t,\tau)\varepsilon(\tau)d\tau=E(t)\left[\varepsilon(t)-\int_{0}^{t}T(t-\tau)\varepsilon(\tau)d\tau\right] \tag{2}$$

where $t$ – moment of observation; $E(t)$ – modulus of elasticity, which depends on time; $\varepsilon(t)$ – deformation, small, i.e. $\varepsilon >> \varepsilon^2$; $\sigma(t)$ – Conditional stress; $\Gamma(t,\tau)$ – gamma function or dimensional kernel of relaxation; $T(t-\tau)$ – a positive function belonging to the space of L1 functions [6] or space of integrable functions on the interval $(0, \infty)$, i.e.

$$\int_{0}^{\infty}T(\tau)d\tau=C<\infty \tag{3}$$

The latter property can be enhanced, bearing in mind of equilibrium stress that in the experiment on the relaxation is not zero:

$$\int_{0}^{\infty}T(\tau)d\tau=C<1 \tag{4}$$

In contrast to [1] modulus elasticity in (2) is assumed to be independent of the moment of observation. Dependence of modulus elasticity on time, following [9], written in the form:

$$E(t)=E_0\left(1-\beta e^{\alpha t}\right) \tag{5}$$

For the parameters of this ratio, there is a clear physical meaning: $E_0$ – "equilibrium" module that corresponds to the ending of the aging process (при $t\to\infty$); $\beta$ – is the relative change of modulus with age t, where the parameter $\alpha$ determines the speed of aging:

$$\beta=\frac{E_0-E_\infty}{E_0}$$

In [11, table 4] given the data to determine the parameters of the formula (5).

## 3 Modal approach

Consider a viscoelastic body whose properties are described by expression $\sigma(t)=E(t)\varepsilon(t)-\int_{0}^{t}\Gamma(t,\tau)\varepsilon(\tau)d\tau=E(t)\left[\varepsilon(t)-\int_{0}^{t}T(t-\tau)\varepsilon(\tau)d\tau\right]$ (2), written in the form:

$$\hat{\sigma}(t)=\hat{\hat{\mathrm{E}}}\cdot\cdot\varepsilon(t)-\int_{0}^{t}\hat{\hat{\mathrm{T}}}_{dyn}(t-\tau)\cdot\cdot\hat{\varepsilon}(\tau)d\tau \tag{6}$$

Here $\hat{\sigma}$, $\hat{\varepsilon}$ – tensors of rank II of the stresses and small deformations; $\hat{\hat{\mathrm{E}}}$, $\hat{\hat{\mathrm{T}}}$ – tensors of rank IV instantaneous modulus of elasticity and of relaxation memory function.

In the General case the tensor of elastic properties $\hat{\mathrm{E}}$ is considered constant by in the equation (5), moreover, the start timing is the moment occurrence of dynamic programming process, a but age of concrete has its start timing; the tensor of kernels of relaxation is calculated at a certain age of the concrete from wave or other dynamic experiments. To determine the dynamic states of an object, we use the Lagrange-d'Alembert variational equation [12]:

$$\begin{gathered}\int_V \delta\hat{\varepsilon}(t)\cdot\cdot\left[\hat{\hat{\mathrm{E}}}\cdot\cdot\hat{\varepsilon}(t)-\int_0^t \hat{\hat{\mathrm{T}}}_{dyn}(t-\tau)\cdot\cdot\hat{\varepsilon}(\tau)d\tau\right]dV+\int_V \delta\mathbf{u}(t)\cdot\rho_0\ddot{\mathbf{u}}(t)dV- \\ -\int_V \delta\mathbf{u}\cdot\rho_0\mathbf{k}dV-\int_S \delta\mathbf{u}\cdot\mathbf{f}_0 dS=0\end{gathered} \tag{7}$$

Here $\rho_0$ – density; **u** – вектор перемещений; **k, f**$_0$ – the vectors of mass and surface loads, in the adopted configuration; δ – denotes kinematically admissible variation; character ( · ) denotes tensor convolution; the integrals are computed on the undeformed volume and surface of the body. Acceleration of the material point is identified with the second partial time derivative. It is assumed that the properties of the material and the kinematic boundary conditions are unchanged during the deformation process.

Suppose that an analytic solution of the problem of free vibrations of a linearly elastic body is known (so in (7) we must eliminate the summand, which responsible for the viscoelasticity), i.e. known countable set the frequencies of free vibration $\omega_k$ and countable set of modes of vibration $\mathbf{Y}_k(\mathbf{R})$ ($k$=1, 2, 3…$N$), $N<\infty$; $\boldsymbol{R}$ – the radius-vector of the current point. These forms have properties:

$$\begin{gathered}\int_V \rho_0\mathbf{Y}_k(\mathbf{R})\cdot\mathbf{Y}_m(\mathbf{R})dV=\begin{cases}1\forall k=m\\ 0\forall\neq m\end{cases} \\ \int_V def(\mathbf{Y}_k(\mathbf{R}))\cdot\cdot\hat{\hat{\mathrm{E}}}\cdot def(\mathbf{Y}_m(\mathbf{R}))dV=\begin{cases}\omega_{0n}^2\forall k=m\\ 0\forall\neq m\end{cases}\end{gathered} \tag{8}$$

Here $def(\boldsymbol{u})$ denotes a tensor of the II rank calculated by the Cauchy formula [12] for small deformation.

We will record the displacement vector as the sum of product their own forms $\boldsymbol{Y}_k$ on of time functions $\Phi_k(t)$ (*modal approach*):

$$\begin{gathered}\mathbf{u}(\mathbf{R},t)=\sum_{k=1}^{N}\mathbf{Y}_k(\mathbf{R})\Phi_k(t)=\Psi(\mathbf{R})\mathbf{\Phi}(t) \\ \Psi(\mathbf{R})^T\Psi(\mathbf{R})=\mathrm{I},\ def(\Psi(\mathbf{R}))^T def(\Psi(\mathbf{R}))=diag(\omega^2)\end{gathered} \tag{9}$$

Here I – identity matrix; $def(\Psi(\mathbf{R})$ – vector (column), made up of operators $def(\boldsymbol{Y}_k)$, i.e. each component of this column is the tensor $def(\mathbf{Y}_k)$; $\Psi(\mathbf{R})$ – rectangular matrix which has as many rows as components of the vector **u** and $N$ columns; $diag(\omega^2)$ – diagonal matrix which is composed of squares of natural frequencies.

Substituting in (7) modal decomposition (9) and taking into account properties (8), we obtain integro-differential equations for functions of time $\Phi_k(t)$, which combine into the vector $\mathbf{\Phi}(t)$:

$$\ddot{\mathbf{\Phi}} + diag\left(\omega^2\right)\mathbf{\Phi} - \int_0^t \mathrm{T}_{\mathrm{mod}}(t-\tau)\mathbf{\Phi}dt = \mathbf{P}(t)$$

$$\mathrm{T}_{\mathrm{mod}}(t) = \int_V def\left[\Psi(\mathbf{R})\right]^T \hat{\hat{\mathrm{T}}}_{dyn}(t) def\left[\Psi(\mathbf{R})\right] dV \tag{10}$$

$$\mathbf{P}(t) = \int_V \rho_0 \Psi^T(\mathbf{R})\mathbf{k}(\mathbf{R},t)dV + \int_S \Psi^T(\mathbf{R})\mathbf{f}(\mathbf{R},t)dS$$

In (10) used the matrix form of record where direct font denotes a rectangular matrix, bold italics – column matrix, the multiplication operations are performed according to the rules of matrix multiplication, the upper symbol $T$ denotes the operation of matrix transposition. When performing matrix operations multiplication the of individual actions above the components of matrices are performed according to the rules prescribed by the meaning component. For example, when calculating the modal matrix relaxation $\mathrm{T}_{\mathrm{mod}}$ its size is determined by the rules of matrix multiplication: a tensor of rank IV $\hat{\hat{\mathrm{T}}}_{dyn}$ is considered as a scalar. Therefore, the dimension of the result is determined by the dimension of the vectors-columns $def\left[\Psi(\mathbf{R})\right]$: i.e. $\mathrm{T}_{\mathrm{mod}}$ there is a square matrix, where the number of rows and columns equal to the number modes, which are retained in the expansion (9) (N×N); each element is calculated according to the rules of tensor multiplication:

$$\left(\mathrm{T}_{\mathrm{mod}}\right)_{k,n} = \int_V def\left(\mathbf{\Psi}_k\right) \cdot\cdot \hat{\hat{\mathrm{T}}}_{dyn} \cdot\cdot def\left(\mathbf{\Psi}_n\right) dV, \ k,n = 1,2,3..N$$

Similarly, when computing the vector of modal forces (the of volume and surface forces **k** and **f**, which are specified in the formula (7) and below) are considered as scalars; here the dimension of the vector of modal forces $P$ corresponds to the number of modes specified in (9), i.e. $N$. Components of the modal forces are calculated according to the formula:

$$P_k = \int_V \mathbf{\Psi}_k \cdot \mathbf{k}\rho_0 dV + \int_S \mathbf{\Psi}_k \cdot \mathbf{f} dS$$

The solution of equation (10) is easily obtained by using the integral Laplace transform with respect to time:

$$\mathbf{\Phi}^*(s) = \mathrm{W}^*(s)\left[\mathbf{P}^*(s) - \mathbf{C}(s)\right]$$

$$\mathrm{W}^*(s) = \left[s^2\mathrm{I} + diag\left(\omega^2\right) - \mathrm{T}^*_{\mathrm{mod}}(s)\right]^{-1} \tag{11}$$

$$\mathbf{C}(s) = \int_V \Psi(\mathbf{R})\left[\dot{\mathbf{u}}(\mathbf{R},0) + s\mathbf{u}(\mathbf{R},0)\right]dV = \mathbf{C}_v(0) + s\mathbf{C}_u(0)$$

Here I – identity matrix, $s$ – conversion parameter, $\mathrm{W}^*(s)$ – transfer functions matrix (a term from theory of automatic control), $\mathbf{C}(s)$ – vector, defined as expansion mode Y of coefficients of the initial conditions for displacement $\mathbf{u}(\mathbf{R},0)$ and the speed $\dot{\mathbf{u}}(\mathbf{R},0)$ (when calculating the decomposition of the initial conditions used the same rule as when calculating modal forces, see above).

The solution to (11) is obtained using the theorems of convolution and differentiation of the original [13, 14]:

$$\mathbf{\Phi}(t) = \mathrm{W}(t)\mathbf{C}_v(0) + \dot{\mathrm{W}}(t)\mathbf{C}_u(0) + \int_0^t \mathrm{W}(t-\tau)\mathbf{P}(\tau)d\tau \tag{12}$$

In the last formula the first two terms are the General solution of (10) defined by the initial conditions, the last term (integral Duhamel) is a particular solution corresponding to given loads. Then the function $\mathrm{W}(t)$ is the normalized matrix of fundamental solutions having the obvious properties of (12):

$$\mathrm{W}(0) = 0; \quad \dot{\mathrm{W}}(t) = \mathrm{I} \tag{10}$$

Components of tensor of small deformations and the stress tensor can be obtained from expansion mode (9) and physical law (7):

$$\hat{\varepsilon}(\mathbf{R},t) = def\left[\mathbf{Y}(\mathbf{R})\right]\mathbf{\Phi}(t)$$

$$\hat{\sigma}(\mathbf{R},t) = \hat{\hat{\mathrm{E}}} \cdot\cdot def\left[\mathbf{Y}(\mathbf{R})\right]\left[\mathbf{\Phi}(t) - \int_0^t \hat{\hat{\mathrm{T}}}_{dyn}(t-\nu)\mathbf{\Phi}(\tau)d\tau\right] \tag{11}$$

Thus, if the known oscillation modes Y and normalized matrix of fundamental solutions $\mathrm{W}(t)$, the components of the stress-strain state (SSS) are determined uniquely. If of oscillation modes and the matrix of fundamental solutions is determined analytically, then the components of the SSS are also determined analytically.

## 4 Conclusion

Dynamic effects of action of short-term loads can be identified by using of modal approach [15, 16] with the assumption of the constancy of the elastic properties of concrete appropriate to the age by the time of application of the disturbing. Superposition of reactions to long - and short-term impacts are possible due to the inadmissibility of finite deformations in building structures.

## References

1. A. Ilyushin, B. Pobedrya, *Foundations of the mathematical theory of thermoviscoplastic.* (Science, Moscow, 1970)
2. A. Ilyushin, G. Larionov, A. Filatov, *To averaging in systems of integro-differential equations.* Report Akad. Sciences USSR **188** (1), 49–52 (1969)
3. Yu. Mitropolsky, A. Filatov, Ukr. Mat. Journal **24** (1), 30–48 (1972)
4. V. Vladimirov, *Equations of mathematical physics.* (Science, Moscow, 1981)
5. S. Algazin, Mathematical modeling **24** (3), 65-69 (2012)
6. A. Kolmogorov, S. Fomin, *Elements of the theory of functions and functional analysis.* (Science, Moscow, 1976)
7. R. Trigub, *Absolute convergence of Fourier integrals, summability of Fourier series, and polynomial approximation of functions on the torus.* Izv. Akad. Sciences SSSR **44** (6), 1378–1409 (1980)
8. V.Yeltkov, D. Komolov, N. Khromova, *Some of the features of automation of calculations of dynamics of viscoelastic systems.* Izv. Tula State University **1** (2), 58–69 (1995)
9. V.Yeltkov, L. Tolokonnikov, N. Khromova, *Transition functions in the dynamics of viscoelastic bodies.* Report Russia Akad. Sciences **329** (6),718 - 719 (1993)

10. A. Tamrazyan, N. Yesayan, *Mechanics of creep of concrete.* (National Research University, Moscow, 2012)
11. N. Arutyunyan, *Some questions of creep theory.* (GITTL, Moscow- Ленинград, 1952)
12. A. Lurie, *Theory of Elasticity.* (Science, Moscow, 1970)
13. V. Ditkin, A. Prudnikov, *Integral transformations and operational calculus.* (Fizmatgiz, Moscow, 1961)
14. M. Lavrent'ev, B. Shabat, *Methods of the theory of functions of a complex variable.* (Science, Moscow, 1987)
15. R. Brincker, L. Zhang, P. Andersen, Smart Materials andStructures 10, 441–444 (2001)
16. O. Balakshin, Problems of mechanical eng. and reliability of machines 6, 16-23 (2011)

# Mechanical modelling and application of vibroacoustic isolators in railway tracks

*Artur* Zbiciak[1], *Cezary* Kraśkiewicz [1,*], *Wojciech* Oleksiewicz[1], *Monika* Płudowska-Zagrajek[1] and *Cezary* Lipko[2]

[1]Warsaw University of Technology, Faculty of Civil Engineering, Institute of Roads and Bridges, 16 Armii Ludowej Ave., 00-637 Warsaw, Poland
[2]TINES Capital Group SA, TINES Railway, 23 Ciepłownicza St., 31-574 Cracow, Poland

**Abstract.** The paper presents systematization and description of vibroacoustic isolators used in railway tracks (due to track structure type), with special attention paid to resilient mats. As in the second part of the paper the state-space mechanical model of a system with Under-Ballast Mat is formulated. Also some numerical problems arising from the mass matrix singularity are discussed. The poles of the system were calculated by using Matlab. Moreover, the influence of various parameters on the system's insertion loss and its transmissibility was visualized in figures.

## 1 Introduction

Rail track structure consists of many components, with different functions and material characteristics. As rail traffic generates dynamic effects, such as vibrations and structure-borne noise, during rail vehicle movement, it is appropriate to use additional elements to reduce those negative effects called vibroacoustic isolators. Due to track structure type (ballasted or ballastless track system), elements location in the track structure or material characteristics, we can identify different types of isolators (i.e. rail pads, under sleeper pads, under-ballast mats).

## 2 Application areas of vibroacoustic isolators in railway tracks

Classifying kinds of rail track superstructures it is appropriate to assume as the main criterion occurrence in a given structure of the ballast, as the main superstructure layer and its function of shaping the geometry of the track. If this feature is fulfilled, track structure is qualified as ballasted track system (Fig. 1). Otherwise, it is qualified as ballastless track system (Fig. 2). It is important for ballast layer to meet both of the above conditions, because ballast layer can occur in ballastless track system - in that case it has different functions i.e.: drainage function, thermal insulation function or noise reduction function.

[*] Corresponding author: c.kraskiewicz@il.pw.edu.pl

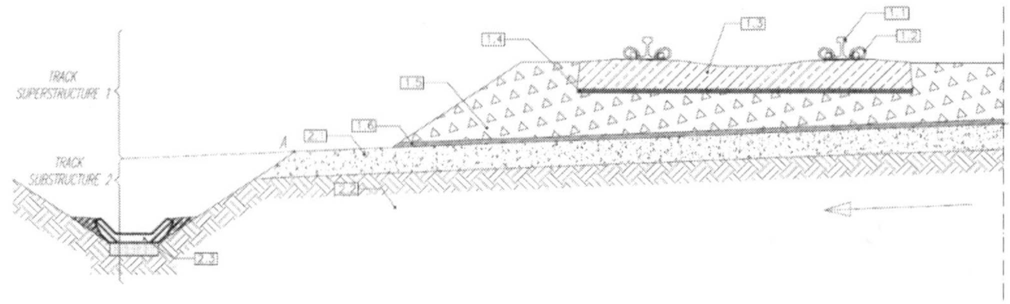


**Fig. 1.** Components of ballasted track system. Key: Track superstructure (1): 1.1 – Vignole rail profile, 1.2 – rail fastening system (type SB or W14), 1.3 – rail sleeper, 1.4 – (option) under sleeper pad, 1.5 - ballast, 1.6 – (option) under-ballast mat. Track substructure (2): 2.1 – blanket layer, 2.2 – subgrade, 2.3 – surface drainage. Note: In case of use of under-ballast mat (UBM), no under sleeper pads (USP) is applied.

The following figure (Fig. 2) shows a use of a ballastless track system in a tunnel on example of the Warsaw Metro Line II.

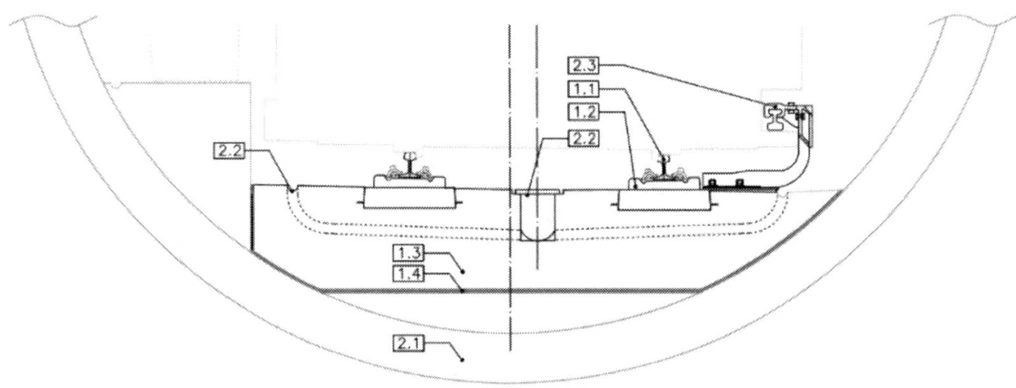


**Fig. 2.** Components of ballastless track system in a tunnel on example of the Warsaw Metro line II. Key: 1.1 – Vignole rail profile type 49E1; 1.2 – rail fastening system – Embedded Blocks System (EBS); 1.3 – concrete track slab; 1.4 – slab-track mat; 2.1 – concrete inverted slab; 2.2 – surface drainage; 2.3 – conductor-rail.

Application areas of vibroacoustic isolators depend on type of rail track structure, rail fastening system or theirs functions. At Fig. 3 are shown components of vibroacoustic isolators system and their main application areas:

1. resilient fastening (fixing) of rail foot: resilient clamps or grouting;
2. resilient fastening (supporting) of rail foot: rail pad, continuous rail pad, discrete or continuous grouting;
3. resilient fixing in indirect rail fastening system: tension clamps, anchor bolts with compression spring;
4. resilient support of rail supporting structure in indirect rail fastening system: base plate pad, discrete grouting;
5. resilient support or fixing of rail supporting structure (sleeper or block): under sleeper pad, under block pad, grouting;
6. resilient fixing and support of railway track: under-ballast mats or slab-track mats – which with rail track structure compose floating slab track system;

7. filling gaps between fishplate and rail web (noise reduction): grouting or rail web filler block.

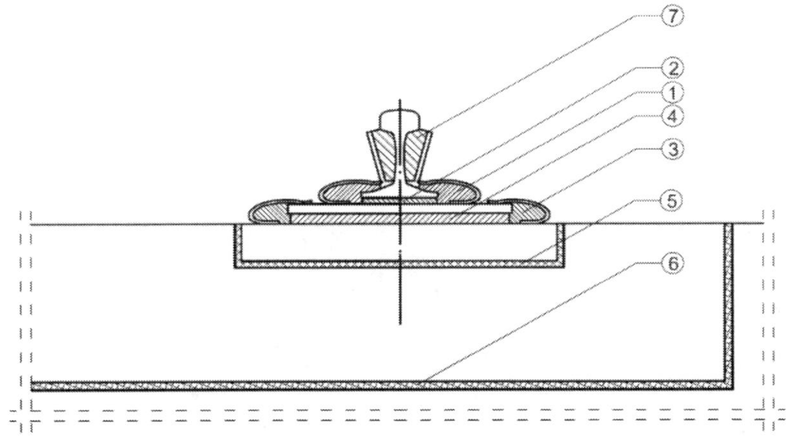


**Fig. 3.** Application areas of vibroacoustic isolators in railway tracks.

Resilient mats can be grouped into two categories, because of the range of applications in various types of railways track construction:
a) Under-Ballast Mats (UBM)/ Sub-Ballast Mats (SBM) or ger. Unterschot-termatten (USM) – used in ballast railway track (Fig. 1); among which two varieties can be classified due to the main purpose of their use:
- used primarily for isolation from vibration,
- used primarily for stress reduction in ballast.
b) Slab-Track Mats (STM) – used in ballastless railway track (applied under or from the sides of concrete track base plate – Fig. 2), with three varieties of slab support system according to standard [4]:
- discrete support (steel springs or elastomeric pads),
- linear support (strip mats),
- continuous support (elastomeric mats) – Fig. 2.

Characteristics of resilient mats important for vibration isolation and their test methods were given in standard [3, 4] and [8, 9, 10].

The further part of the paper will address mechanical model of ballasted track system with Under-Ballast Mat.

## 3 Mechanical model

Mechanical model of the system with Under-Ballast Mat (UBM) is shown in Fig. 4. Such a model was partially analysed in [15] and is also presented in Standard [2] for calculation of insertion loss. The objective of this analysis is to present the state-space formulation of differential equations representing this model.

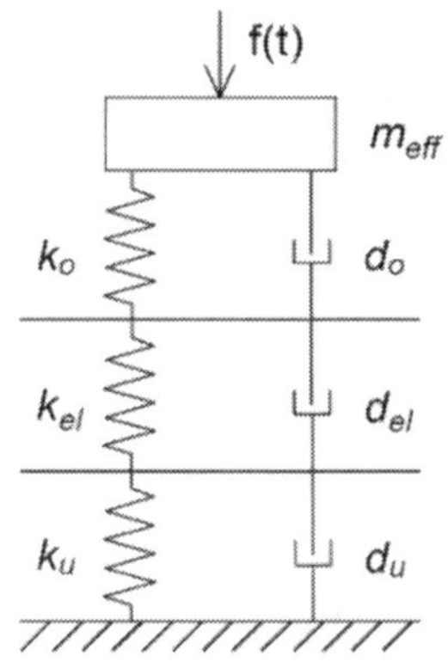


**Fig. 4.** Mechanical model of a system with Under-Ballast Mat.

Material parameters of mechanical model shown in Fig. 4 are presented in Table 1. Let us note that instead of damping coefficients $d_o$ and $d_{el}$, Table 1 gives loss factors $\eta_o$ and $\eta_{el}$. Further analysis will be conducted assuming $\eta = 2\xi$ (see Standards [1-4]), where $\xi = \dfrac{d}{2\sqrt{km}}$ denotes damping ratio [5]. Thus, damping coefficients can be calculated as follows $d = \eta\sqrt{km}$.

**Table 1.** Parameters of mechanical model shown in Fig. 4.

| $m_{eff}$ [kg] | $k_o$ [MN/m] | $k_{el}$ [MN/m] | $k_u$ [MN/m] | $\eta_o$ [–] | $\eta_{el}$ [–] | $d_u$ [MNs/m] |
|---|---|---|---|---|---|---|
| 2600 | 220 | 140 | 1500 | 0,35 | 0,05 | 1,2 |

Mechanical model representing a rail track system with vibroacoustic elements can be described by inertia $\mathbf{M}$, damping $\mathbf{C}$ and stiffness $\mathbf{K}$ matrices, as follows

$$\mathbf{M}\ddot{\mathbf{x}} + \mathbf{C}\dot{\mathbf{x}} + \mathbf{K}\mathbf{x} = \mathbf{f} \tag{1}$$

where $\mathbf{f}$ represents loading vector. In case of analysed system, we obtain

$$\mathbf{M} = \begin{bmatrix} m_{eff} & 0 & 0 \\ 0 & 0 & 0 \\ 0 & 0 & 0 \end{bmatrix}, \quad \mathbf{K} = \begin{bmatrix} k_o & -k_o & 0 \\ -k_o & k_o + k_{el} & -k_{el} \\ 0 & -k_{el} & k_{el} + k_u \end{bmatrix}, \quad \mathbf{C} = \begin{bmatrix} d_o & -d_o & 0 \\ -d_o & d_o + d_{el} & -d_{el} \\ 0 & -d_{el} & d_{el} + d_u \end{bmatrix} \tag{2}$$

In many cases the system of 2$^{nd}$ order differential equations (1) should be represented in the following state-space form

$$\begin{aligned} \dot{\mathbf{y}} &= \mathbf{A}^s\mathbf{y} + \mathbf{B}^s\mathbf{u} \\ \mathbf{z} &= \mathbf{C}^s\mathbf{y} + \mathbf{D}^s\mathbf{u} \end{aligned} \tag{3}$$

where $\mathbf{A}^s$ denotes the system matrix, $\mathbf{B}^s$ is the input matrix, $\mathbf{C}^s$ is the output matrix and $\mathbf{D}^s$ is the feed through matrix. Comparing Eq. (1) and Eq. (3) gives

$$\mathbf{y}=\begin{bmatrix}\mathbf{x}\\ \dot{\mathbf{x}}\end{bmatrix},\quad \mathbf{A}^s=\begin{bmatrix}\mathbf{0} & \mathbf{I}\\ -\mathbf{M}^{-1}\mathbf{K} & -\mathbf{M}^{-1}\mathbf{C}\end{bmatrix},\quad \mathbf{B}^s=\begin{bmatrix}\mathbf{0}\\ \mathbf{M}^{-1}\end{bmatrix},\quad \mathbf{C}^s=\begin{bmatrix}\mathbf{C}^{ox} & \mathbf{C}^{ov}\end{bmatrix},\quad \mathbf{D}^s=\mathbf{0} \tag{4}$$

The inertia matrix shown in Eq. (2) is singular. On the other hand, in order to obtain the system matrix $\mathbf{A}^s$ applying Eq. (4), needs the mass matrix to be inverted. The reason of such a problem is the presence of degrees of freedom without inertia, so called half degrees of freedom. This problem can be solved by using the method presented in [11], where a very useful and easy to be implemented in a computer routine algorithm was invented. This algorithm was programmed in Matlab, giving the following results for the system matrix and the input matrix

$$\mathbf{A}^s=\begin{bmatrix}0 & 0 & 0 & 1\\ 747.96 & -1212.26 & 339.25 & 0.90\\ 18.34 & 84.08 & -1324.83 & 0.02\\ -8465.30 & -38805.52 & 34538.85 & -10.19\end{bmatrix},\quad \mathbf{B}^s=\begin{bmatrix}0 & 0 & 0\\ 0 & 3.40 & 0.08\\ 0 & 0.08 & 0.82\\ 384.62 & 346.14 & 8.49\end{bmatrix} \tag{5}$$

Using the functions implemented in Control System Toolbox of Matlab software, the poles of analysed system were calculated (see Fig. 5). Poles show the frequencies where the system will amplify inputs. The poles depend only on the distribution of mass, stiffness and damping throughout the system. They do not depend on where the forces are applied or where displacements are measured.

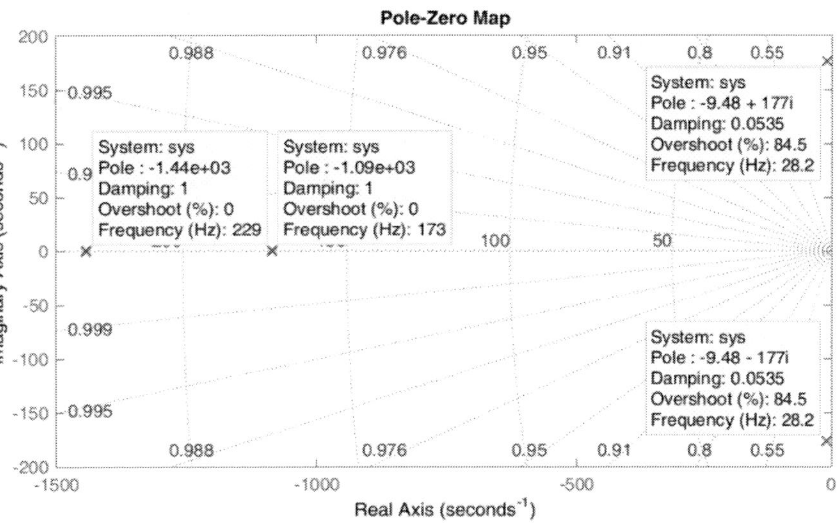


**Fig. 5.** Pole-Zero Map of analysed dynamic system.

Let us analyse additional properties of mechanical system shown in Fig. 4. Paper [15] presents formulas to calculate transmissibility and insertion loss of such system. Using these formulas we will show the influence of certain parameters on its dynamic properties.

Fig. 6 visualizes insertion loss diagrams for various mass parameters $m_{eff}$. Horizontal axis used in Fig. 6 (also in Fig. 7) represents centre frequencies of 1/3 octave bands. The plots depicted in Fig. 6 indicate regions of vibration effectiveness where $IL > 0$ dB. Increasing the effective mass results in increase of vibration effectiveness region.

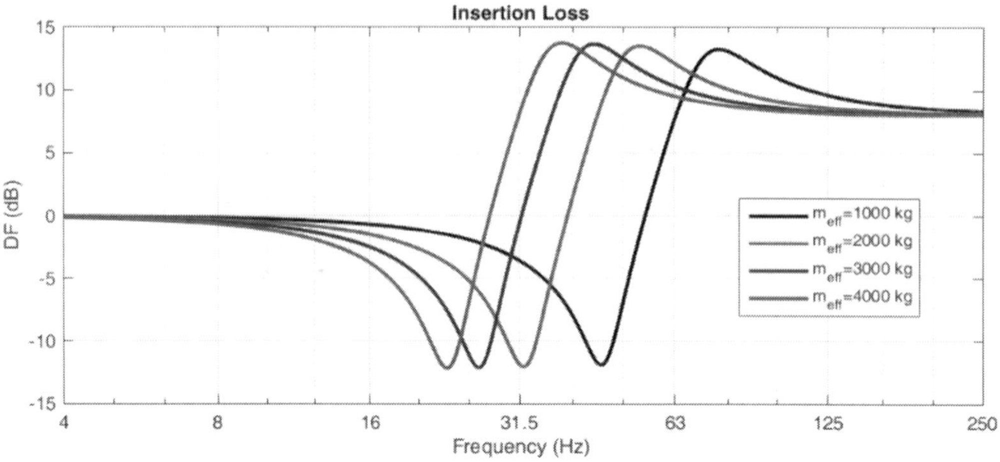


**Fig. 6.** Insertion loss diagrams for various effective mass parameters.

Insertion loss diagrams for various mat stiffness parameters $k_{el}$ are given in Fig. 7. Increase of mat's stiffness makes the effectiveness region smaller and decreases the amplitudes of insertion loss.

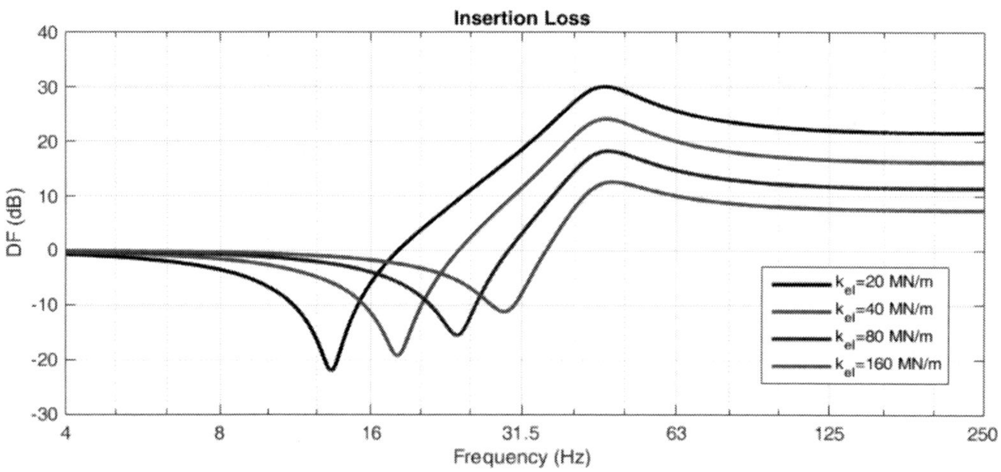


**Fig. 7.** Insertion loss diagrams for various mat stiffness parameters.

Finally, Fig. 8 visualizes transmissibility of the system for various mat loss factors $\eta_{el}$. Such diagrams are presented in many papers [12-14] for various mechanical models representing railway track systems with vibroacoustic isolators.

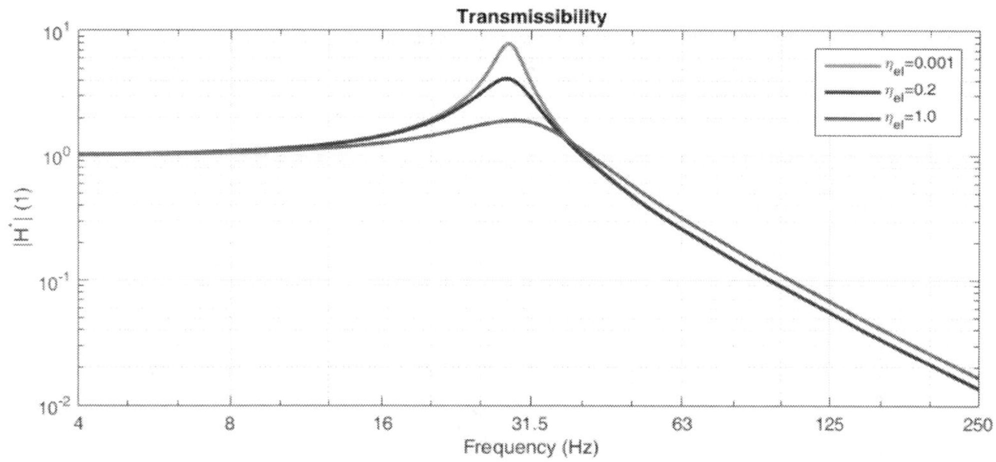


**Fig. 8.** Transmissibility diagram for various loss factors.

## 4 Conclusions

Vibroacoustic isolators are components used to limit the negative effects (vibration and noise) generated by rail traffic on the track, and vary depending on material characteristics, application area or due to track structure type (ballasted or ballastless track system). One of goals of this paper was to systematize vibroacoustic isolators used as close as possible to the place of vibration generation during the movement of rail vehicles. As there are reasonable grounds to conclude that those elements will find increasingly widespread use in rail track structure along with further advances in engineering and material engineering, and taking into consideration increased requirements for protecting surroundings of transport routes from negative transport effects.

The results of calculations presented in this paper were obtained by applying a mechanical system containing half degrees of freedom. It results in singularity of inertia matrix. The problem was solved using a method proposed in literature [11]. It would be possible to use the same methodology in case of multi degree of freedom linear models representing rail track system with resilient elements. In such systems, half degrees of freedom can be associated with more general models of vibroacoustic elements, described by fractional-order derivatives [6], as it was suggested in [15]. Such an approach would make possible to fit better experimental results. More advanced models of railway tracks systems with vibroacoustic isolators can be formulated by using Finite Element Method [7].

## References

1. *DIN 45673-1:2010-08 Mechanical vibration. Resilient elements used in railway tracks. Part 1: Terms and definitions, classification, test procedures*
2. *DIN 45673-4:2008-07 Mechanical vibration. Resilient elements used in railway tracks. Part 4: Analytical evaluation of insertion loss of mounted track systems*
3. *DIN 45673-5:2010-08 Mechanical vibration. Resilient elements used in railway tracks. Part 5: Laboratory test procedures for under-ballast mats*
4. *DIN 45673-7:2010-08 Mechanical vibration - Resilient elements used in railway tracks - Part 7: Laboratory test procedures for resilient elements of floating slab track systems*

5. R. W. Clough, J. Penzien, *Dynamics of Structures* (McGraw-Hill, 1975)
6. W. Grzesikiewicz, A. Wakulicz, A. Zbiciak, IJMS **70**, 90 (2013)
7. C. Kraśkiewicz, R. Michalczyk, K. Brzeziński, M. Płudowska, Proc. Eng. **111**, 462 (2015)
8. C. Kraśkiewicz, C. Lipko, W. Oleksiewicz, A. Zbiciak, *Materiały Konferencyjne, Stowarzyszenie Inżynierów i Techników Komunikacji Rzeczpospolitej Polskiej. Oddział w Krakowie* **2(106)**, 89 (2015)
9. C. Kraśkiewicz, C. Lipko, W. Oleksiewicz, A. Zbiciak, Przegląd Komunikacyjny **9**, 76 (2015)
10. C. Kraśkiewicz, C. Lipko, M. Płudowska, W. Oleksiewicz, A. Zbiciak, Proc. Eng. **153**, 317 (2016)
11. V. G. S. Simionatto, H. H. Miyasato, F. Melo, M. Junior, *Proceedings 21$^{st}$ Int. Congr. Mech. Engin.* (2011)
12. R. G. Wettschureck, U. J. Kurze, ACUSTICA **58**, 177 (1985)
13. R. G. Wettschureck, DAGA **87**, 217 (1987)
14. R. G. Wettschureck, M. Heim, S. Mühlbachler, *Proceedings Inter-noise* **97**, 577 (1997)
15. A. Zbiciak, C. Kraśkiewicz, C. Lipko, W. Oleksiewicz, *IPICSE-2016, MATEC Web of Conferences* **86**, 01015 (2016)

# Monitoring of temperature gradient development of highway concrete bridge

*Lukas* Krkoska[1,*], and *Martin* Moravcik[1]

[1] University of Zilina, Faculty of Civil Engineering, Department of Structures and Bridges, Univerzitna 2, 01026 Zilina, Slovakia

**Abstract.** Thermal effect is one of very important from the large bridge design procedure point of view. Especially vertical temperature gradient is being crucial. There were realized some research works of the thermal effects monitoring on the concrete bridges in the world. More of them were performed in the USA but only a few at European bridges. The short overview of our long-term monitoring of the temperature load on chosen concrete bridge is presented in this paper. We decided to analyse one concrete box girder bridge that was built by incremental launching method on highway D1 at Slovakia near Zilina city. Recorded temperature gradient was compared with thermal gradients for the concrete box girder bridge recommended by EC 1991-1-5 design specifications.

## 1 Introduction

Bridges are designed for the 100 years of the service life but they are under traffic even longer. They are exposed to thermal effects daily, seasonal and yearly dependent on the climatic conditions. Thermal effect has been primary characterized as a non-stationary action in time with strong variations caused by solar radiation, ambient air temperature, humidity and wind. The next significant influence that is affecting secondary thermal condition of the bridge being is the bridge dimension itself. Mainly it is structure orientation exposed to the sunshine, material of the deck surface, structure dimensions and cross-section geometry.

Thermal actions have significant effect on bridge structures and can be reason for more than 30% of the reinforcement used from that point of view. That is why it cannot be neglected in the bridge design process. Thermal actions differ from other load types through being a constraining load. If these deformations are restrained, temperature variations result in thermal stresses, what may lead to stress field changing along the length of structures and cross-section too. A temperature profile can be divided into a constant part which affects the bridge with a linear expansion and a non-uniform part that will induce vertical or horizontal curve shape of the bridge superstructure. Uniform temperature changes cause mostly length changes, expansion or contraction. It is important for the expansion joints but for example it can change the internal forces too of the frame structures. Vertical and horizontal temperature gradient results in additional bending

* Corresponding author: lukas.krkoska@fstav.uniza.sk

moments in vertical and horizontal plane, respectively. This makes thermal actions in bridge design interesting for further investigations.

## 2 Thermal effects on concrete structures

The temperature distribution within the concrete bridge deck may be generally split into the four essential constituent components, which are uniform temperature component (effective bridge temperature), $\Delta T_{unif}$, linearly temperature difference component about the z-axis, $\Delta T_{My}$, and about the y-axis, $\Delta T_{Mz}$ and non-linear temperature difference component, $\Delta T_E$. It results in a system of self-equilibrated stresses.

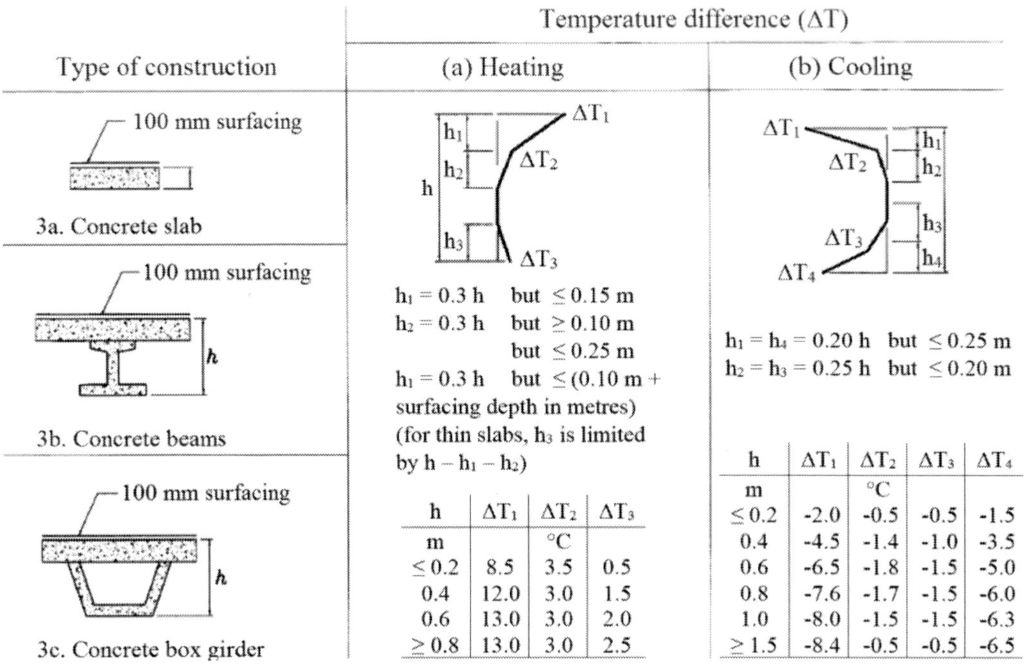


(a) Heating

| h | $\Delta T_1$ | $\Delta T_2$ | $\Delta T_3$ |
|---|---|---|---|
| m | | °C | |
| ≤ 0.2 | 8.5 | 3.5 | 0.5 |
| 0.4 | 12.0 | 3.0 | 1.5 |
| 0.6 | 13.0 | 3.0 | 2.0 |
| ≥ 0.8 | 13.0 | 3.0 | 2.5 |

(b) Cooling

| h | $\Delta T_1$ | $\Delta T_2$ | $\Delta T_3$ | $\Delta T_4$ |
|---|---|---|---|---|
| m | | °C | | |
| ≤ 0.2 | -2.0 | -0.5 | -0.5 | -1.5 |
| 0.4 | -4.5 | -1.4 | -1.0 | -3.5 |
| 0.6 | -6.5 | -1.8 | -1.5 | -5.0 |
| 0.8 | -7.6 | -1.7 | -1.5 | -6.0 |
| 1.0 | -8.0 | -1.5 | -1.5 | -6.3 |
| ≥ 1.5 | -8.4 | -0.5 | -0.5 | -6.5 |

**Fig. 1.** Temperature differences for bridge deck, [7] – case 3c)

The uniform temperature component depends on the minimum and maximum temperature which a bridge will reach. This results in changes of uniform temperature which in an unrestrained structure would result in a change in element length. The values of minimum and maximum uniform bridge temperature components □TN for restraining forces shall be derived from the minimum ($T_{min}$) and maximum ($T_{max}$) shade air temperatures, that can be gain from the isotherms in accordance to the bridge exposition.

Over a time period heating and cooling of a bridge deck upper surface will result in a maximum heating (top surface warmer) and a maximum cooling (bottom surface warmer) temperature variation, ($\Delta T_{M,heat}$ and $\Delta T_{M,cool}$). The effect of the vertical temperature differences should be considered by including a non-linear temperature difference component, $\Delta T_E$. Fig. 1 represents the temperature differences along the cross section on bridge decks. Our case is the case 3c). These gradient incorporates parts of $\Delta T_M$ and $\Delta T_E$ and the small part of the component $\Delta T_N$.

# 3 Bridge Geometry and Measuring Equipment Description

In this period we are monitoring a concrete box girder bridge SO 201-10 which is situated on the highway near the Zilina airport. We had started to observe it June 2015. Bridge has nine spans (48 m + 3 x 58 m + 60 m + 2 x 68 m + 54 m + 37.1 m) and it is built by using incremental launching method with 39 m long steel launching nose, see Fig. 2. The overall span of the bridge is 835.1 m and with his main span it is crossing the highway and railway. The bridge superstructure is made with a continuous box girder simple supported by the piers and abutments. Hight of the cross-section varies from 3.85 m to 4.5 m.

Concrete strength class has been designed C45/55 and prestressing tendons with characteristic strength of 1860 MPa. The tendons consist of 19 strands (bonded tendons) and 27 strands (free tendons) of diameter 15.7 mm.

**Fig. 2.** Bridge 201-10 – construction stage with two spans completed

In the construction stage of the bridge was installed measuring equipment with 19 strain gauges type TES/5.5/T embedded in concrete to measure strain and temperature, 8 thermocouples to measure surface temperature and 2 thermometers for measuring air temperature. These sensors are divided into two measured cross sections of the first span, see Figs. 6 ~ 10. Measured cross-section (MP1) and measured cross-section (MP2). Cross-section MP1 is placed at 3/5 of the first span and includes strain gauges (S1 ~ S7). Cross-section MP2 is placed very next to the crossbeam and includes strain gauges (S1 ~ S11), thermocouples (TP1 ~ TP7), interior and exterior air thermometers (TE1, TE2) and one thermometer embedded in depth 400 mm (TC400). All of these sensors have been continually monitored and recorded.

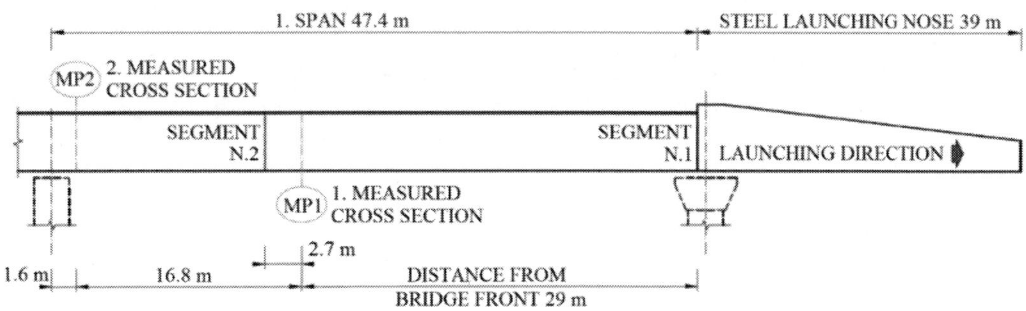


**Fig. 3.** Placing of measured cross sections in the first span

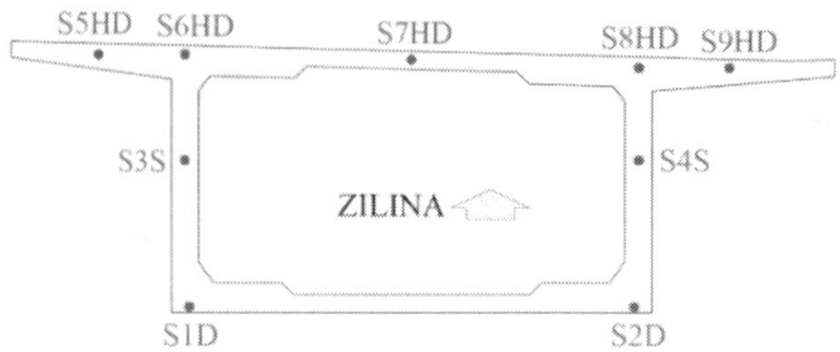


**Fig. 4.** Strain gauges layout at MP1

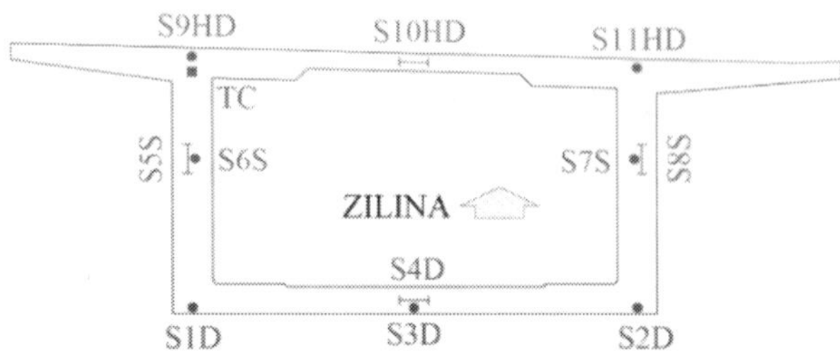


**Fig. 5.** Strain gauges layout at MP2

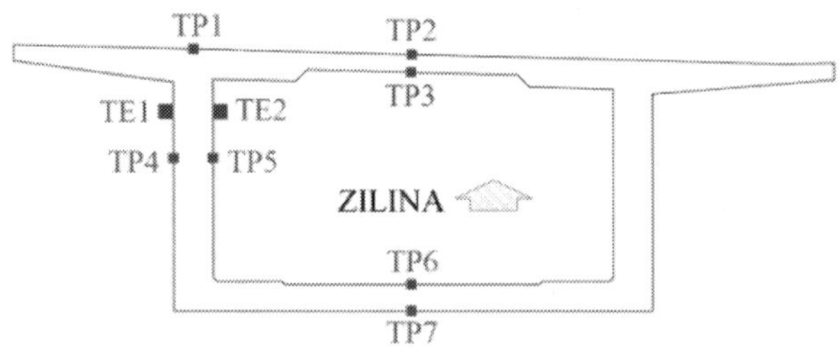


**Fig. 6.** Thermocouples layout at MP2

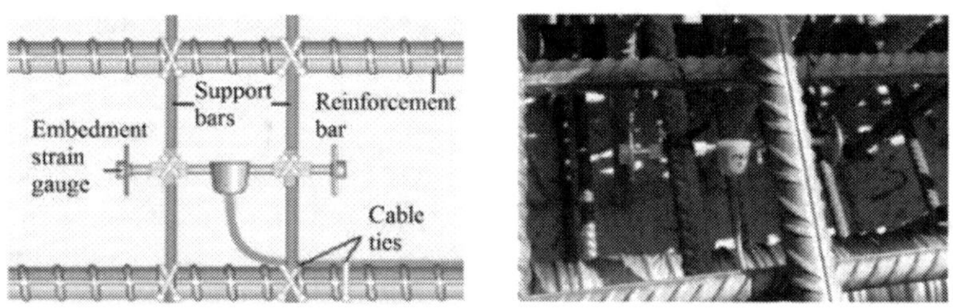


**Fig. 7.** Strain gauges fixing to the reinforcement bars.

## 4 Temperature changes due to temperature of the air

Last year we described temperature field in concrete box cross section during the summer days with maximum temperature. In this article we will describe measured temperatures during the winter days. Fig. 8 shows diagrams of temperature changes measured during the days from 5th January 2016 to 7th January 2016 for parts of the cross section separately. These diagrams represent period when the temperature of the air reached the minimum value from the date of applying surface thermocouples (TP1 – TP7) to surface of the cross section parts. All the measure temperature have linear trend. The temperature of top slab inner surface (TP3) is lower (up to 3 °C) than the temperature inside the box (TE2). Temperatures measured by strain gauge inside the concrete (S10HDt) also copies the temperature inside box with the deviation up to 5 °C. Temperature of top slab outer surface is almost identical with the temperature of outside air (TE1).

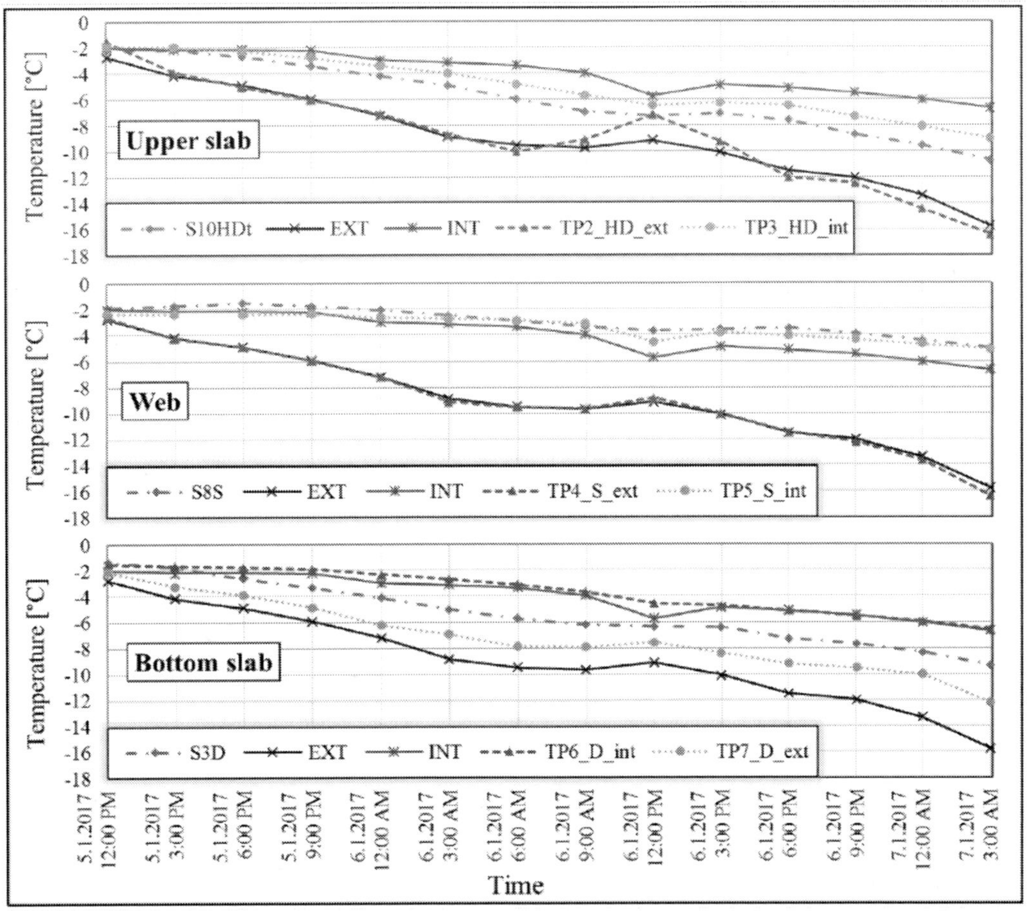


**Fig. 8.** Temperature variations of concrete over the cross section.

Temperature field of the web is similar to the top slab, but the values measured by strain gauge inside the concrete (S8S) and by surface thermocouple at the inner surface (TP5) are slightly higher than the temperature of the air inside the box (TE2). Outer surface temperature of the web is identical with the temperature of outside air (TE1).

The main difference in temperature field of the bottom slab is, that the temperature of outer surface (TP7) is higher (up to 4 °C) than outside air temperature. This leads to development of negative temperature gradient in the cross section. Temperature measured by strain gauge inside the concrete (S3D) is by 2 °C higher than temperature of outer surface (TP7) and inner surface temperature is identical to the temperature inside the box (TE2).

## 5 Temperature gradient

After the monitoring of temperature changes described in previous paragraphs we tried to create the negative vertical temperature gradient along the cross section of recorded values of temperature in characteristic points. We can create negative vertical temperature gradient by dividing a minimum obtained value of temperature (absolute value) from measured temperatures along the cross section. Furthermore we have compared obtained temperature gradient with negative temperature gradient recommended in Slovak annex of the Eurocode STN EN 1991-1-5 for our type of concrete cross section. Comparison presented in Fig. 9 shows that negative temperature gradient obtained from the measurement reach 11.6 °C and did not exceed the recommended value in Eurocode. It is necessary to say that the temperature of the air at night of 7th January 2017 did reach value -23 °C but we were not able to measure these values because of measuring system power source fail. We can suppose that negative vertical temperature gradient could extend the recommended value in Eurocode.

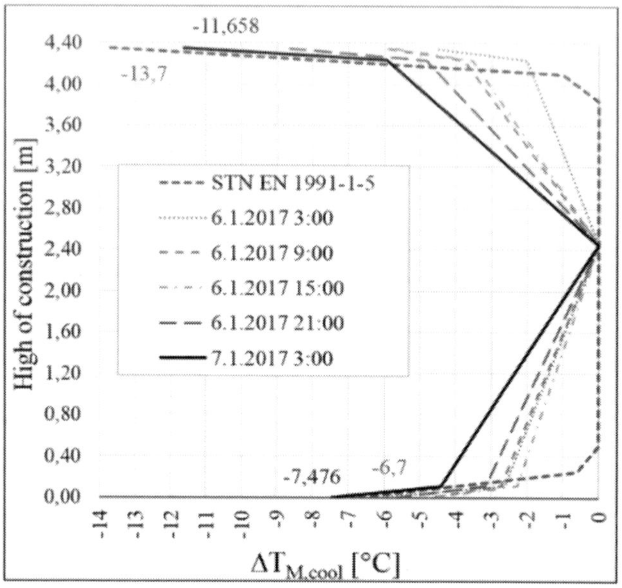


**Fig. 9.** Comparison of measured temperature gradients with Eurocode recommendation.

## Conclusion

Variations of air temperature affect overall temperature distribution of cross section and large temperature gradient can occur. This temperature gradient, especially vertical temperature gradient can exceed the value in bridge design standard. Underestimation of this problem can affect significantly stresses distribution of the bridge and tensile cracks development. Therefore it is necessary to collect specific temperature gradient data during longer time period continually on various type of bridges. It can allow to do more precise structural analysis of bridge structures in the future.

## Acknowledgements

The research is supported by the Slovak Research and Development Agency under contract No. APVV-0736-12 and by Research Project No. 1/0336/15 and 2/0033/15 of Slovak Grant Agency VEGA.

## References

1. M. J. N. Priestley: Design of concrete bridges for temperature gradients, American Concrete Insititute Journal, May, pp. 209-217, (1978).
2. J. H. Emanuel, J. L. Hulsey: Temperature distribution in composite bridges, Journal of the Structural Division, ASCE, vol.104.ST1, pp.65-78, (1987).
3. M. M. Elbadry, A. Ghali: Temperature variations in concrete bridges, Journal of the Structural Engineering, ASCE, vol. 109(10), pp.2355-2374, (1983).
4. M. Emerson: Thermal movements of concrete bridges: field measurements and methods of prediction, Transportation Road Research Laboratory Supplementary Report 747, pp.5-10, (1982).
5. J. Rimal: Comparison of temperature loadings of bridge girders, Czech Technical University in Prague, vol. 48(05), pp.22-8, (2008).
6. Guang-Dong Zhou, Ting-Hua Yi, Thermal Load in Large-Scale Bridges: A State-of-the-Art Review. International Journal of Distributed Sensor Networks, (2014).
7. STN EN 1991-1-5: Actions on Structures. Part 1-5: General Actions – Thermal actions.
8. STN EN 1990: Eurocode. Basis of structural design.
9. Moravcik, M. – Bahleda, F. – Jost, J. – Dolinajova, K.: Monitoring of Long-Term Effects on Concrete Bridge Realised by Balanced Cantilever Method. Proceedings of the International Conference on Civil, Structural and Transportation Engineering (ICCSTE' **15**), http://avestia.com/ICCSTE2015 Proceedings, Ottawa, Ontario, Canada, May 4-5, (2015).
10. CSN 73 6203: Actions on Bridges.
11. BS 5400-2: Steel, Concrete and Composite Bridges. Part 2: Specification for Loads.
12. DIN 1072: Road and Food Bridges: Design Loads.
13. ASSHTO 2012: LRFD Bridge Design Specification.

# Analysis of proportional conception of the historical truss Church of St. Martin in Dolní Újezd

*Peter* Krušinský[1,*], *Michaela* Holešová[2] , *Eliška* Racková[3]

[1] University of Žilina, Faculty of Civil Engineering, Žilina, Univerzitna 8215/1, Slovak Republic
[2] National heritage institute CR, Pardubice office, Zámek 5, Pardubice, Czech Republic

**Abstract.** The late-Romanesque St. Martin´s Church in Dolní Újezd was probably built in the mid-13th century. The nave has a wooden joist ceiling whose beams serve as tie beams too. The Gothic vaults above the presbytery date from ′30s of the 15th century. The truss of the church is dendro-dated to 1420 - 1421. Some members of a truss structure are dendro-dated to 1823 - 1824, when the repair was undertaken. The examined truss is a rafter-collar tie structure with collar beams and pedestal struts. It has a characteristic shape and may be typologically classified as a structure of the earlier period. This type of a truss structure was typically built in this region and represents one of the features occurring in the particular social and culture circles at the time.

## 1 Introduction

Analysis of proportional conception is based on historical-structural, archival and dendrochronological research of the object as a whole. The method of geometrical analysis of the concept of the roof is based of acquiring information from period literature and from previous analyses. The Proportional Relationship of historic trusses was analysed mainly from Slovakia and Czechia. The results show further comparisons. Object of research is analysing the proportion of historical structure and relation into floor plan of building and layout of the building solution. Building space solution is also built on the concept of proportional relationships. Significant part of method is determining the roof height. The roof height is proportionally and geometrically derived from the ground plan of the building. This principle has also been identified in other historical roof trusses of Slovakia and Czechia, which we were analysing (eg. roof trusses on a churches in the village of Bela Dulice, Abramova, Okolicne, Klimkovice, Vlčovice and others [3,4,6]).

## 2 History of St. Martin´s Church in Dolní Újezd

The church was probably built in the mid-13th century. It has a late-Romanesque core dated from this century - only the tower and west nave survived to this day. In the late-14th

* Corresponding author: krusop@wp.pl

century, a new presbytery arose on the church's eastern elevation; it is built in high-level architecture (shafts and the lower part of arch ribs survived). The old presbytery was demolished and the new one was interconnected with a late-Romanesque nave, which can be seen in the irregular floorplan. At the time of the Hussite wars, the church was destroyed by the running armies, like many other churches in this region. Soon after, today's truss, dated from 1420 - 1421d, was built. Sometimes later, probably in the '30s of the 15th century when the situation became stable, the presbytery was newly vaulted with a brick vault and stone ribs. Further modifications were made after 1533 - an addition of a northern antechapel, knocking out of windows and southern entrance. Several small modifications were made later, but they did not affect the overall appearance of the church. [1,2] (Fig. 1.)

**Fig. 1.** The researched object – The St. Martin's Church in Dolní Újezd.

## 3 Truss construction

The church has a gable roof with a single-level ridge above the nave and sanctuary. The roof has polygonal hipped end above the polygonal end of the sanctuary. Above the sanctuary, there is a polygonal turret. The truss above the nave is a rafter-collar structure with two levels of collar beams and heel struts. The truss is longitudinally tied by a central frame that is divided into two parts by a turret. The truss as a whole contains three types of ties. The first type consists of rafters, two levels of collar beams, and a central post. The central post of the first collar beam, into which it is halved, is jointed with the second collar beam using the straight halving; the top of a post is also halved into the rafters. The post is cross-braced by the heel struts into the first level of the collar beams. The heel struts are halved into the tie beam and rafters. The collar beams are dovetailed into rafters. The second type, unlike the first one, does not contain a central post with heel struts. The cross-bracing is made by a saltire cross dovetailed into the tie beam and rafters below the level of the second collar beam. The third type is similar to the second one; except to saltire crosses at the level of the first collar beam which are replaced by heel struts dovetailed into the collar beam and rafters. The truss is longitudinally braced using span pieces located on the upper and lower collar beams. The structure is longitudinally braced by saltire crosses halved and dovetailed into the posts and span pieces. (Figure 2)

**Fig. 2.** The inner view of the truss

# 4 Geometric concepts and proportional relationships within the truss

Previous examination of historic trusses, mainly those of sacral buildings, suggests that one of the most commonly method used to determine the basic truss dimensions is the principle that may be named as „n plus k", where k = 1, 2, 3. The principle is used to determine the truss height on the basis of its width by the ratio method. The truss width is divided into 2n of x units and the truss height is then (n+k) of x units. It is this principle where k = 3 that was used in the St Martin´s Church in the village of Dolní Újezd. The basic concept of the truss above the nave in the cross-section is derived from the basic ABCD square (Fig x). The AB side of the square stands for nx (within the principle n plus 3). If we divide it into n-parts, we get the measure unit x (n = 4 in this case). We can determine the truss height using the principle (n+3)x, on the basis of which the truss height is 7/4 of the AB side of the ABCD square. The truss width/height ratio is, using this method, 8:7. The truss above the nave contains three types of ties - A, B, and C. In all the ties (Figure 3), the position of a lower collar beam is defined as 5/8 of the length of the BC side of the basic ABCD square; the position of an upper collar beam is 9/8 of the BC side of the basic ABCD square. It follows that the ratio of the distance of a tie beam from a lower collar beam to the distance of both collar beams is 5:4. The heel strut is located on a tie beam at a distance of x from the position of a rafter heel (i.e. 1/8 of the truss width) and positioned on a rafter in a height of 3/8 of the length of the BC side of the ABCD square (Figure 4). The joints of a heel strut of a post in the A-tie are located on a post at a distance of 11/12 of a length of the BC side of the ABCD square from the tie beam end. The heel struts of a post are horizontally positioned on a lower collar beam at a distance of 1/6 of a length of the AB side.

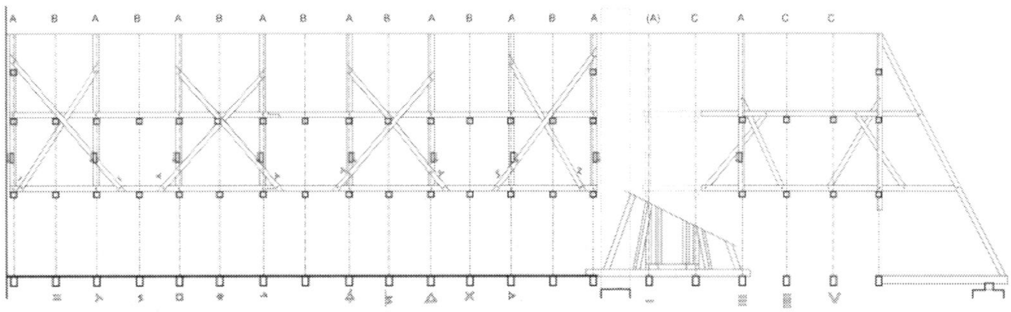

**Fig. 3.** Longitudal truss section.

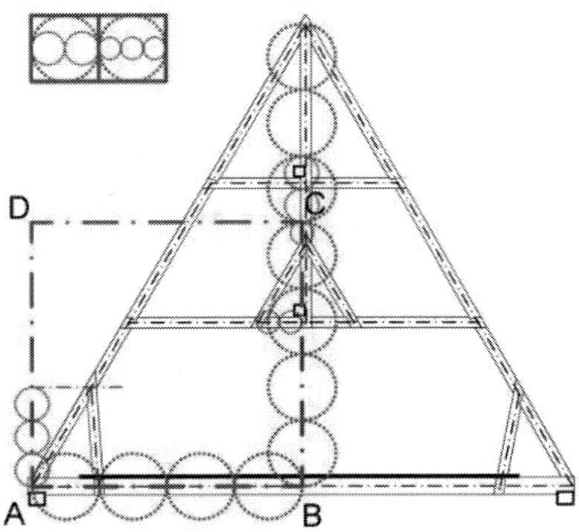


**Fig. 4.** Longitudal section of the truss.

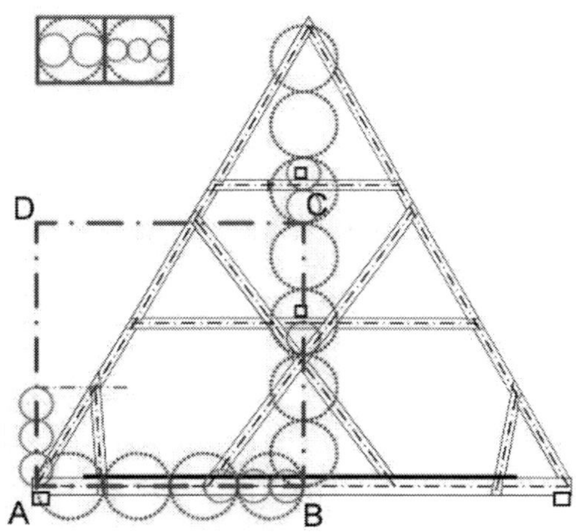


**Fig. 5.** B cross-section of the truss.

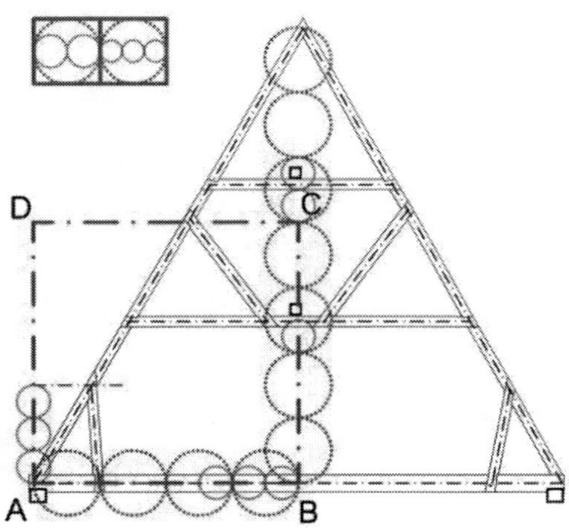


**Fig. 6.** C cross-section of the truss.

The transversal B-tie is braced by a saltire cross. The strut joint is located on a tie beam at a distance of 5/8 of the AB side of the ABCD square from a rafter heel. The rafter heel joint is determined by the basic ABCD square. (Figure 5) The transversal C-tie of a truss is derived from the B-tie; the heel struts at a level of a lower collar beam respect the slope and position of the struts in a saltire cross of the B-tie (Figure 6).

## Conclusion

The method of a ratio conception to design a truss in relation to the floorplan is visible and identified in the examined building. The foundation is to use a period system to determine the height and position of individual truss members, as known from later literature. The ratio analysis of a truss structure and building itself may help study its historic development. [7] Identification of historic measures (on the basis of a division), ratios, and construction used may reveal how the cultural circles and socio-economic ties influenced regional architecture. Information on proportional and geometric conception may be a guideline for a recovery methodology in relation to the conservation of historic buildings. [8,9] The similar analysis of the truss and its proportional relations in the All Saints Church in Vlčovice in Silesia revealed the similar method "n plus 2" was used to determine the truss height (Figure 7), in the rural house in Vápenná [4,5] as well as in the Trinity Church in Klimkovice (also situated in Silesia) is the use of the similar principle to determine a truss height - the principle of "n plus one". The main difference is in the number of x units within a side of the basic square defining the truss structure. Individual divisions of the basic square refer to the potential use of a linear measure system, which should be subjected to further detailed examination. To understand time and spatial relationships better, it is necessary to carry out more analyses focused on the truss designing and conceptions in order to create an extensive statistical database for numerical analyses.

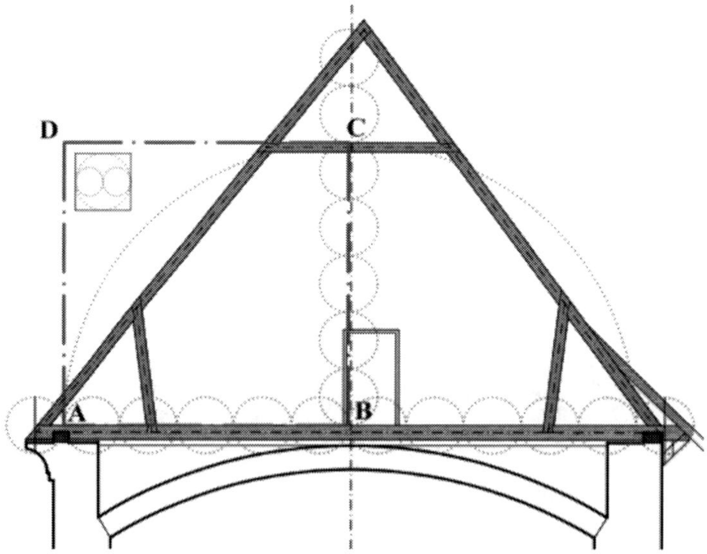


**Fig. 7.** Cross-section of the truss The All Saints Church in Vlčovice.

## References

1. D. Líbal: Katalog gotické architektury v České republice do husitských válek. 1. vyd. Praha: Unicornis (2001)
2. F. Václavík, Nové poznatky ze stavební historie kostela sv. Martina v Dolním Újezdě okr. Svitavy, in: Výroční zpráva za rok 2000. V Pardubicích: Památkový ústav v Pardubicích, 201-203 (2000)
3. P. Krušinský, E. Capková, J. Gocál, Advanced Materials Research. **1122**, 243-248 (2015)
4. P. Krušinský, E. Capková, J. Gocál, M. Holešová, Civil and environmental engineering: Scientific technical journal **11/2,** 136-141 (2015)
5. P. Krušinský, E. Capková, L. Augustinková, R. Korenková, Civil and Environmental Engineering. **12/2**, 111–115 (2016)
6. J. Gocál, P. Krušinský, E. Capková, M. Kekeliak, Advanced materials research. **969**, 199-207 (2014)
7. A. Struhár, "Geometric harmony of historical architecture in Slovakia." Bratislava. Pallas, (1977)
8. O. Makýš, "Včasnohistorické a rannostredoveké drevené a drevozemné pamiatky : hypotetické rekonštrukcie a konzervácia." 1. vyd. Pezinok : Renesans (2014)
9. A. Iringová, R. Idunk, Civil and environmental engineering: Scientific technical journal. **12/2**, 84-93 (2016)

# AUTHOR INDEX

# AUTHOR INDEX

# AUTHOR INDEX

# AUTHOR INDEX

**EDP Sciences – Web of Conferences**
17, Avenue du Hoggar
Parc d'Activité de Courtabouf
BP 112
F-91944 Les Ulis Cedex A
France

ISBN 978-1-5108-4515-2

CURRAN ASSOCIATES INC.
proceedings
.com